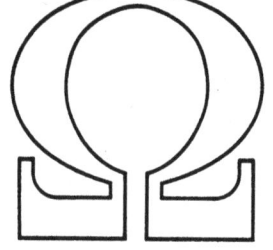
Perspectives in Mathematical Logic

Ω-Group:

R. O. Gandy H. Hermes A. Levy G. H. Müller
G. E. Sacks D. S. Scott

Ω-Bibliography of Mathematical Logic

Edited by Gert H. Müller

In Collaboration with Wolfgang Lenski

Volume III

Model Theory

Heinz-Dieter Ebbinghaus (Editor)

Springer-Verlag Berlin Heidelberg GmbH

Gert H. Müller
Wolfgang Lenski
Mathematisches Institut, Universität Heidelberg
Im Neuenheimer Feld 288, D-6900 Heidelberg

Heinz-Dieter Ebbinghaus
Mathematisches Institut, Abteilung für mathematische Logik
Universität Freiburg, Albertstr. 23 b, D-7800 Freiburg

The series *Perspectives in Mathematical Logic* is edited by the Ω-group of the Heidelberger Akademie der Wissenschaften. The Group initially received a generous grant (1970–1973) from the Stiftung Volkswagenwerk and since 1974 its work has been incorporated into the general scientific program of the Heidelberger Akademie der Wissenschaften (Math. Naturwiss. Klasse).

ISBN 978-3-662-09060-2   ISBN 978-3-662-09058-9 (eBook)
DOI 10.1007/978-3-662-09058-9

Library of Congress Cataloging in Publication Data
[Omega]-bibliography of mathematical logic.
(Perspectives in mathematical logic)
Includes indexes.
Contents: v. 1. Classical logic / Wolfgang Rautenberg, ed. – v. 2. Non-classical logics / Wolfgang Rautenberg, ed. – v. 3. Model theory / Heinz-Dieter Ebbinghaus, ed. [etc.]
1. Logic, Symbolic and mathematical–Bibliography. I. Müller, G. H. (Gert Heinz), 1923 –. II. Lenski, Wolfgang, 1952 –. III. Title: Bibliography of mathematical logic. IV. Series.
Z6654.M26047 1987 [QA9] 016.5113 86-31426

This work is subject to copyright. All rights are reserved, whether the whole or part of the material is concerned, specifically those of translation, reprinting, re-use of illustrations, broadcasting, reproduction by photocopying machine or similar means, and storage in data banks. Under § 54 of the German Copyright Law where copies are made for other than private use, a fee is payable to "Verwertungsgesellschaft Wort", Munich.

© Springer-Verlag Berlin Heidelberg 1987
Originally published by Springer-Verlag Berlin Heidelberg New York in 1987.
Softcover reprint of the hardcover 1st edition 1987
2141/3140-543210

*Dedicated
to
Alonzo Church*

whose bibliographic work for the
Journal of Symbolic Logic
was a milestone in the
development of modern logic.

Table of Contents

Preface IX

Introduction XV

User's Guide XXV

Ω-Classification Scheme XXXIII

Subject Index 1

    1. The classical part 3

1. Basic properties of first-order languages and structures C07 3
2. Denumerable structures C15 14
3. Ultraproducts and related constructions C20 18
4. Model-theoretic forcing C25 29
5. Other model constructions C30 36
6. Categoricity and completeness of theories C35 45
7. Interpolation, preservation, definability C40 57
8. Stability and related concepts C45 69
9. Models with special properties C50 75
10. Properties of classes of models C52 85

    2. Aspects of effectiveness 94

11. Elimination of quantifiers and decidability . . . . C10, B25 . . . 94
12. Finite structures C13 116
13. Recursion-theoretic model theory C57 121

    3. Algebraic aspects and special theories 129

14. Equational classes, universal algebra (partial) C05 129
15. Model-theoretic algebra C60 142
16. Models of arithmetic and set theory . . . . C62, H15 . . 169
17. Models of other mathematical theories C65 184

    4. Further aspects and extended model theory 195

18. Set-theoretic model theory C55 195
19. Admissible sets and infinitary languages . . . . C70, C75 . . 204
20. Other extended logics and generalized model theory . . C80, C85, C95 217
21. Nonclassical models C90 231

    5. General literature 242

22. Proceedings C97 242
23. Textbooks, surveys C98 244

Author Index 251

Source Index 529

- Journals 531
- Series 548
- Proceedings 552
- Collection volumes 570
- Publishers 577

Miscellaneous Indexes 583

- External classifications 585
- Alphabetization and alternative spellings of author names 613
- International vehicle codes 615
- Transliteration scheme for Cyrillic 617

Preface

Gert H. Müller

The growth of the number of publications in almost all scientific areas, as in the area of (mathematical) logic, is taken as a sign of our scientifically minded culture, but it also has a terrifying aspect. In addition, given the rapidly growing sophistication, specialization and hence subdivision of logic, researchers, students and teachers may have a hard time getting an overview of the existing literature, particularly if they do not have an extensive library available in their neighbourhood: they simply do not even know what to ask for! More specifically, if someone vaguely knows that something vaguely connected with his interests exists somewhere in the literature, he may not be able to find it even by searching through the publications scattered in the review journals. Answering this challenge was and is the central motivation for compiling this Bibliography.

The Bibliography comprises (presently) the following six volumes (listed with the corresponding Editors):

    I.   Classical Logic   W. Rautenberg
    II.   Non-classical Logics   W. Rautenberg
    III. Model Theory   H.-D. Ebbinghaus
    IV. Recursion Theory   P. G. Hinman
    V.   Set Theory   A. R. Blass
    VI. Proof Theory; Constructive Mathematics
                           J. E. Kister; D. van Dalen & A. S. Troelstra.

Each volume is divided into four main parts:

1) The *Subject Index* is arranged in sections by topics, usually corresponding to sections in the classification scheme; each section is ordered chronologically by year, and within a given year the items are listed alphabetically by author with the titles of the publications and their full classifications added.

2) The *Author Index* is ordered alphabetically by author, and contains the full bibliographical data of each publication together with its review numbers in Mathematical Reviews (MR), Zentralblatt für Mathematik und ihre Grenzgebiete (Zbl), Journal of Symbolic Logic (JSL), and Jahrbuch über die Fortschritte der Mathematik (FdM). We much regret that we were not able to include reviews from Referativnyj Zhurnal Matematika in this edition.

3) The *Source Index* gives the full bibliographical data of each source (journals and books) for which only abbreviated forms are used in the Author Index.

4) The *Miscellaneous Indexes* contain various further indexes and tables to aid the reader in using the Bibliography.

For a more detailed technical description of the Bibliography see the *Table of Contents* and the *User's Guide*.

The uniform classification of all entries is a central feature of the Bibliography. The basic framework is the 03 section of the (1985 version of the) 1980 classification scheme of Mathematical Reviews and Zentralblatt für Mathematik und ihre Grenzgebiete. However, this has been modified in a number of ways. Indeed, the 1980 scheme was designed for the classification of works written after 1980, whereas the majority of entries in the Bibliography come before this date. In some areas this has made the classification of older works difficult, and we have tried to cope with this by adding a few new sections and altering slightly the interpretation of others. We have not designated the classifications assigned to a work as primary and secondary, because of the difficulty in doing so in many cases. Each volume contains the full annotated classification scheme together with a description of its general features. In their *introductions* the Editors discuss specifically their interpretations of the classification sections falling in their respective volumes.

The Subject Index is another central feature of the Bibliography. Reading through this Index gives a *historical perspective* for each classification section and provides a rather quick overview of the literature in it. By browsing through the entries of a specific area the reader may be rewarded by finding things (literature, subjects, questions) he was not aware of or had forgotten.

An obvious question now is the extent to which one can rely on the *completeness* and *correctness* of the Bibliography and on the *accuracy of the classifications*. We comment on each of these aspects separately.

In an effort to be as complete as possible, we consulted all sources available to us and decided in favour of inclusion in doubtful cases (so that certainly some papers with little bearing on mathematical logic are listed here and there). As the historical starting point for the Bibliography we chose the appearance of Frege's *Begriffsschrift* (1879). A certain restriction on scope stems from our decision to concentrate on mathematical logic and in particular on those areas defined by the titles of the six volumes. A major source of material was provided by the review journals mentioned above; we used them both to identify publications in the less known journals and to find review numbers and other bibliographical data of items found in other sources. We also made use of various lists of literature contained in books, survey articles, mimeographed notes, etc. Some especially valuable newer sources were:

W. Hodges: A Thousand Papers in Model Theory and Algebra

M. A. McRobbie, A. Barcan and P. B. Thistlewaite: Interpolation Theorems: A Bibliography

D. S. Scott and J. M. B. Moss †: A Bibliography of Books on Symbolic Logic, Foundations of Mathematics and Related Subjects

C. A. B. Peacocke and D. S. Scott: A Selective Bibliography of Philosophical Logic.

Various strategies and crosschecks were used to ensure the completeness of the bibliographical data and in particular of the reviews mentioned above. For each item listed in the Bibliography we tried to include any translations, reprintings in alternative sources and errata, and to give cross references for a work appearing in several parts.

On the whole this Bibliography was compiled and organized for use by the practising mathematician; there is no claim that the most rigorous standards of librarianship are met.

It is hard to say how successful our striving for completeness was. This is especially true for the most recent literature. No 1986 items were included. We checked all the main journals in logic, the reviews in MR, Zbl and JSL and Current Mathematical Publications for literature published up to the end of 1985, but undoubtedly some gaps remain.

As for correctness, in any ordinary book we can tolerate a number of printing errors because of our knowledge of the language and the context, but, when one organizes data connected by (abstract) pointers in a computer program, almost every typing error has far-reaching consequences. Various consistency tests were used to check the program and the input data. There are, however, many other sources for mistakes and errors.

For some items our references contained incomplete or ambiguous information. Although we tried to complete the bibliographical data, this was often difficult, particularly in cases where, for example, the source was obscure or the publisher was given only by location. Another source of errors lies in the identification of author names. An author may publish using abbreviations of his first, his second or both of his given names. This is generally not a problem for authors with uncommon surnames, but if the surname is, e.g., Smith or Brown the possibility of misidentification arises. We may have identified two different authors or failed to identify two or more different forms of an author name.

It is unavoidable in a project of this scope that there will be errors, particularly in the classification, so perhaps it is worthwhile to explain briefly the process by which the classification was done. Items entered before 1981 were originally classified according to a scheme unrelated to the current one. To begin the conversion to the 1980 scheme we used the computer to change old categories to their new versions wherever there was a well-defined correspondence. Then every entry was checked and if necessary reclassified by hand. From 1981 each new entry was classified shortly after being entered in the database. For the most part this was done on the basis of titles, reviews, and other information, but without consulting the works themselves. This was necessary to preserve the finiteness of the enterprise, but it has inevitably led to errors, certainly in some cases egregious ones. These were constantly being corrected during the final editing process, but many will remain.

Although the Editors have to some extent used different strategies in classifying the entries falling into their respective volumes, finally a reasonable degree of uniformity has arisen. The user is referred to the Editors' introductions for further details on the classifying procedure.

A special apology goes to the native speakers of languages with diacritical marks. Our central difficulty was to get the right spelling of names used in different forms in such a variety of sources. In addition, entering diacritical marks in a computer introduces yet another source of errors; so they have almost all been ignored (the User's Guide and the Miscellaneous Indexes contain details of those that have been transliterated). We appreciate that, although the absence of, for example, accents in the text of a French title may not create undue problems, the lack of diacritical marks in author names is particularly unfortunate. We hope that this omission will not be too misleading.

The future

By its nature a bibliography has lasting value to the extent it succeeds in "completing the past". But it should also serve for some years as an aid to current research. We have various plans to extend the scope of the present Bibliography by including new areas such as universal algebra, sheaves, philosophical logic (subdividing the present volumes I and II appropriately), and philosophy of mathematics. The present six volumes cover only approximately 80% of the data on our computer files.

The possibility of extending the classification scheme by developing a so-called thesaurus system was discussed on several occasions. Certainly this would be desirable; to some extent *Alonzo Church* tried to create such a system in connection with his bibliography in the Journal of Symbolic Logic. However there are difficult scientific problems connected with the creation of such a system and their solution requires much time and expertise.

Another way to extend the Bibliography which would perhaps better serve the purpose of providing an overview of certain special areas would be to commission a series of survey papers to appear from time to time as, say, an additional issue of the Journal of Symbolic Logic; each paper would include an annotated listing of the literature taken from the Bibliography.

There are plans to establish a bibliographical centre for Mathematical Logic and adjacent areas. A central function of such a centre would be to collect information on all new publications (including mimeographed notes, theses, etc.) as well as to correct errors and omissions in the current data. It is hoped that all logicians would provide information concerning their own publications as they appear. A continuation of the Bibliography together with supplements (to appear periodically) would be prepared at the centre. We also hope to make available an on-line system. From these activities and the flow of information from the individual logician to the centre and vice versa a "living Bibliography" would emerge. This would provide a way to determine the main trends in the progress (or decline) of specified directions of work. So a centre would exist at which it would be possible to gain some oversight of the rapidly developing field of mathematical logic.

Acknowledgements

Work on the Bibliography started at the same time as the Ω-Group came into being, early in 1969. To begin with, index cards were used for storing bibliographical data; it was *Horst Zeitler* and *Diana Schmidt* who convinced me that we are living in the 20th century and that the data should be computerized. They, together with *Ann Singleterry-Ferebee,* first brought the Bibliography to a workable computerized form at the end of the seventies. In this period I also had the help of *Ulrich Felgner* and *Klaus Gloede,* in particular in classifying the literature. At about this time, others contributed in many useful ways. In particular, important problems of principle were highlighted by a long list of intriguing questions from *Dana Scott:* "How do you classify this or that item...?" *Robert Harrison* worked faithfully collecting data for the Source Index.

The second period, beginning in the early eighties, was characterized by the programming necessary to manage the data. This was carried out by *Ulrich Burkhardt* (†) and *Werner Wolf* and finally by the *outstanding work* of *Rolf Bogus.* In this period we also changed the classification system and for this I had the continuous and intensive help of *Andreas Blass* and *Peter Hinman.* In addition, both of them, together with *Heinz-Dieter Ebbinghaus,* gave me much advice about organization and technical arrangements. Over the last four years the work of large groups of students has been essential for collecting reviews, entering corrections and new items into the computer, etc. Again and again I have been overwhelmed by their idealism and energy. Among them I wish to mention particularly the continuous help of *Elisabeth Wette* and *Ulrike Wieland.*

The Bibliography would not have reached publishable form without the work of my collaborator *Wolfgang Lenski* (in the second period). It would have been unthinkable for me to interfere anywhere in the process of the growth of the Bibliography without discussing the matter with him beforehand. He has accumulated a detailed knowledge of every aspect of the project and has devoted his talents for many years to the common enterprise.

My secretary, *Elfriede Ihrig,* has willingly assisted in the work of the Ω-Group and the Bibliography from the beginning, over many years, filled with ups and downs and with all kinds of tasks. She has always maintained her warmhearted balance.

To all I express my personal warm thanks!

The *Journal of Symbolic Logic* sent information concerning papers for which reviews were never published. We also acknowledge permission to use computer tapes with lists of literature covering certain periods of time from *Mathematical Reviews* and from *Zentralblatt für Mathematik und ihre Grenzgebiete.*

Yuzuru Kakuda and *Tosiyuki Tugue* collected and prepared the Japanese literature for us. *Petr Hajek* and *Gerd Wechsung* helped us with updating the bibliographical references of so many sources not available to us. *Mo Shaokui* corrected data on the Chinese literature and added items of which we were not aware.

The Editors filled many gaps, corrected all mistakes which came to their attention and undertook the burden of checking – and changing if necessary – the classification of the entries in their special areas. Here again I would like to mention *Rolf Bogus* and *Wolfgang Lenski* who organized the enormous exchange service for the transfer of literature among the Editors and the inputting of the many changes and corrections. *Andreas Blass* and *Peter Hinman* were also instrumental in this exchange; their preliminary classification of each item added to the Bibliography during the final editing process meant that the Editors had mainly to look at items inside their own areas. *Jane Kister* read through the whole Source Index correcting mistakes and suggesting valuable changes in it.

In collecting and organizing the data for the Bibliography we have received much help from various sources, and especially in letters from colleagues all over the world, containing information and suggestions. I apologize for being unable to answer them all individually, but all were read carefully.

We thank all those concerned.

As everybody can guess, the whole enterprise was indeed expensive. *Financial support was provided by the Heidelberger Akademie der Wissenschaften in the framework of the Ω-Group project.*

Special thanks go to the firm APPL, who transformed our computer tapes to the present printed form, and to the editorial and production staff of SPRINGER-Verlag for their continuous help, notably in the traditionally fine realization of the six volumes.

Finally, through working so many years on this project I have come to understand and appreciate more and more the immense work of Alonzo Church in building his Bibliography of Logic and its adjacent areas, together with a detailed classification, that is contained in so many volumes of the Journal of Symbolic Logic. Understanding comes from doing.

Introduction

Heinz-Dieter Ebbinghaus

One of the main features of the present series of bibliographies in mathematical logic is the detailed classification of each paper together with a separate listing for each classification number. The value of this system depends essentially on the quality of the classifications. However, there are difficulties in this respect. Apart from those described in the general preface there is another, intrinsic, one due to the fact that classifying has been an individual process performed by the editor, whereas the reader, at least in principle, is right to expect objective information. The present preface is intended to describe the guidelines and principles which I followed when I was classifying the papers listed; let us hope it will build bridges between author and user.

A first section is devoted to remarks on the scope of this volume and its subdivisions. The second section will describe and motivate the general lines I adopted during the classification process. Section three contains detailed comments on each classification number besides those laid down in the Blass-Hinman classification scheme; this section is not intended to be read over as a whole, but can be consulted on occasion for specific information.

1. The Contents

Model theory or the *theory of models,* as it was first named by Tarski (1954), may be considered as that part of the semantics of formalized languages which is concerned with the interplay between syntactic properties of axioms and, for instance, algebraic or set-theoretic properties of their models. Although its roots go back to the first half of the century, as is illustrated by the Löwenheim-Skolem theorem (Löwenheim 1915, Skolem 1920) or the compactness theorem for first-order languages (Gödel 1930 for the countable case, Mal'tsev 1936 for the general case), the theory evolved as a discipline in its own right only in the early fifties. The pioneering work by Tarski on one hand and A. Robinson on the other hand reflects what may be considered the two sides of the model-theoretic coin: the pure theory, sometimes – e.g. by Chang and Keisler (1973) – characterized as a marriage of universal algebra and logic, and the algebraically influenced theory, created by the insight that model theory may lead to new methods and new views in investigations of specific mathematical structures. Of course, both aspects are intimately interwoven. During the last decade the mathematical orientation has gained an increasing influence which leads, for instance, to the extraordinary size of the C60 section. However, by also taking account of the rapid development of "pure" areas such as stability theory, the fruitful investigations on extended logical systems and the beginning rise of finite model theory, model theory as a whole appears as a lively and flourishing area.

Roughly, the topics covered in the present volume coincide with section 03C of the AMS classification scheme. The detailed classification follows the Blass-Hinman variant for mathematical logic with three exceptions:

- Universal algebra, C05, has been incorporated only so far as it has model-theoretic aspects. A sequel volume is planned for this series that will be concerned with universal algebra in general and with algebraic logic.

- Section B25 (Decidability of theories and of sets of sentences) appears in this volume instead of volume I (Classical logic) because of its relationship to section C10 (Quantifier elimination and related topics).
- Similarly, section H15 (Non-standard models of arithmetic) has been incorporated in this volume; it is combined with section C62 (Models of arithmetic and set theory) because there is a continuous transition from a model-theoretic treatment of arithmetic – as emphasized in C62 – to the viewpoint of non-standardness – as emphasized in H15. Sections H05 and H10 (Non-standard analysis and its applications), however, appear in volume I.

Even outside sections such as B25, H15 or C75 (with respect to infinitary terms or syntax) the bibliography contains papers that are not concerned with model theory in the strict sense. The motivation for keeping them may be seen in the fact that they are interesting for methodological or historical reasons or give some indication of how model-theoretic notions occur in a non-technical context.

For the Subject Index the different sections have been grouped into twenty three chapters which follow the Blass-Hinman classification scheme essentially in an increasing order, sometimes comprising several classification items. To underline the internal structure, they are grouped in five blocks which I briefly describe in the following paragraphs.

Block 1: The classical part

This block contains the bulk of first-order model theory, but of course also the analogues in extended model theory. Its areas have grown from the basic aims of the theory that may be formulated as follows:

(i) Methodological: Provide methods for constructing models with specific properties.

(ii) In substance: Given first-order logic or another formalized language, answer questions such as: Which classes of structures are axiomatizable? What are the crucial properties of axiomatizable classes and how are such properties linked to the axiomatization? Is a given axiomatization complete? Is a given theory categorical in some sense? In full generality: What are the models of a given theory?

C07 collects the pioneering work, but also later papers, that are concerned with the fundamental notions of model theory. Obviously, C20, C25, and C30 refer to (i), C35, C45, and C52 to (ii), where, e.g., stability theory as a way to approach the program of Shelah's classification theory, is concerned with the last question under (ii).

When working on these aims, structures with special properties, such as those that are homogeneous, universal, or saturated, turn out to be methodologically fruitful, but have also been studied for their own sake. They are classified in C50. The fact that, on one hand, many model-theoretic methods are most effective in the countable case and that, on the other hand, many problems can be reduced to this case results in the methodological importance of countable structures; this is reflected in C15.

The definability theorem (Beth 1953) and the related interpolation theorem (Craig 1957) are also accessible via model theory and, moreover, have gained methodological influence themselves, for instance in proofs of preservation results. The related papers are listed under C40.

Of course, there are various connections and overlaps among the chapters described above or following. These interrelations among the different classification numbers have a strong bearing on the classification as will be discussed later. Let me give some examples that may also serve as illustrations for the blocks that follow.

- C15 is, of course, linked to C57 (as effectively given structures have countable domains), but also to C75 (think, for instance, of $\mathfrak{L}_{\omega_1\omega}$ as an appropriate framework for Vaught's conjecture).
- Morley's notions of rank and total transcendence, developed in proving the Łos conjecture (1965), deeply influenced investigations on categoricity in power and initiated stability theory, thus building a bridge between C35 and C45. Moreover, C35 is related to B25 as the completeness of an enumerable theory yields its decidability, and to C10 via model completeness.
- Usually, theories of classes of algebraic structures are far from being complete (e.g., the theory of fields); there may exist, however, algebraically natural and valuable "companion" theories (e.g., the theory of algebraically closed fields) that are complete in the strict or in some usefully weaker sense. By adapting the forcing technique of set theory to model theory, A. Robinson initiated a powerful method for obtaining such completions, thereby linking C25, C35, and C60.

Block 2: Aspects of effectiveness

There are some natural bridges between model theory and recursion theory. One of them is based on the recursion-theoretic consequences of Gödel's completeness theorem: model-theoretic proofs of the completeness of an axiomatizable theory yield decidability. Decidability may also be obtained by effective quantifier elimination. Results of this kind form a major part of B25 and C10.

A second way to recursion theory starts from the effective content of Henkin's completeness proof for first-order logic. It leads to the theory of effectively given structures, C57. Finally, finite structures are manageable objects themselves, and many model-theoretic questions concerning them are connected with automata theory as is illustrated by the spectrum problem. This is one reason to include C13 in block 2.

Block 3: Algebraic aspects and special theories

As already mentioned in the beginning, the pioneering work of A. Robinson was strongly influenced by motivations from and applicability in algebra. Model theory of this kind is also a reasonable enterprise from the algebraic viewpoint: As the ultimate aim of algebra, the classification of algebraic structures up to isomorphism, often faces unsurmountable problems, a more modest and nevertheless useful aim can be seen in the classification of structures up to elementary equivalence or some related kind of indiscernibility. Hence the model-theoretic treatment may add to the supply of algebraic methods and may lead to new insights of both a logical and an algebraic character. Convincing examples are Ax and Kochen's proof of Artin's conjecture (1965) or the generalization of Ulm's theorem by Barwise and Eklof (1970). Besides model-theoretic work in algebra (C60), block 3 represents model-theoretic developments in various areas including universal algebra (C05), number theory (C62, H15), set theory (C62), and other mathematical theories (C65).

Block 4: Further aspects and extended model theory

The scope of first-order model theory is limited because of the limited expressive power of first-order language. Model theorists have tried various ways to alter this situation. One possibility is the incorporation of stronger concepts in the first-order framework. An example is given by the omitting types technique which makes parts of $\mathfrak{L}_{\omega_1\omega}$ accessible to first-order methods; this technique falls under C75 or – for the basic part – under C07. More directly, one can try to build up a model theory for stronger languages. In large part, block 4 is concerned with this direction. Languages considered include the infinitary languages (C75); they have led to a convincing theory especially in the case of countable admissible fragments (C70), as demonstrated by the work of Barwise (1969). Other examples are higher order languages (C85) and languages (such as first-order or infinitary languages) with additional quantifiers (C80).

The large variety of quantifiers investigated calls for a unified treatment, and the large variety of model-theoretic languages for a general investigation of model-theoretic possibilities. Initiated by Mostowski's work on generalized quantifiers (1957) and with a first highlight in the form of Lindström's theorems (1969), the pursuance of these ideas led to abstract model theory, C95.

All the developments sketched above are based on the notion of classical truth in the usual structures that consist of universal domains, relations, total functions, and constants. This framework has been departed from in various directions, e.g. by using non-two-valued semantics or by considering unusual structures such as topological ones. The papers concerned are collected under C90.

As mentioned above, model theory is also concerned with set-theoretic properties of structures. Moreover, many of the additional quantifiers - such as cardinality quantifiers - are of a set-theoretic character. For reasons such as these, set-theoretic methods play a major role in numerous model-theoretic investigations. This connection with set theory is taken into account by C55.

Block 5: General literature
This block contains the literature of C97 and C98.

To conclude, some critical remarks might be in order. As the attentive reader will have noticed, the grouping proposed above should be taken only with a grain of salt. Certainly, alternatives (and maybe even better ones) are possible. Instead of discussing such possibilities, however, we close by describing some of the difficulties that might be met with any alternative.

- First, the borderlines of a careful grouping may divide some natural classification areas. To give an illustration, section C55 might be a candidate for such a division: Incorporated in block 4 it contains numerous papers on ultrafilters and ultraproducts (thus touching C20 in block 1) and numerous papers on models of set theory (falling under C62 in block 3), and of course also many papers on set-theoretically defined quantifiers (belonging to C80 in block 4).
- Second, as an effect of the dimension of time there is a more fundamental problem: The bibliography comprises several decades of intensive work in model theory that have seen the maturity of old areas, the rise of new ones, the shifting of main interests, and the continuing development and reshaping of methods and ideas. Therefore, in principal, each period is likely to demand its own system of classification. From this viewpoint, for instance, the title of block 1 may be said to pay its tribute to the past. Taking this aspects seriously, the user is invited to develop his own understanding independent from the organization of the contents, simultaneously creating his own feel for historical lines by reading the chronological lists of the Subject Index.

2. Classifying

The present section contains some general remarks on the principles of classification followed in this volume. Detailed comments on specific sections will be given in 3.

Usually a paper has been assigned several classification numbers according to the different areas with which it is concerned. For instance, a paper that proves the decidability of some algebraic theory by using ultraproducts has, at least, been classified C60, C20 and B25. Or a paper on models of set theory that contains an application to some logic with a generalized quantifier (which is more than only a side remark) has also been classified C80 besides C62 and some appropriate En. As an important consequence, the heading of a chapter or a classification number does not necessarily reflect (all) the main features of the papers listed under it.

When following this principle of multiple classification, I have tried to go some way between the two extremal possibilities of attributing only very few numbers (thus maybe losing information) and of attributing many numbers (thus taking into account many aspects, but paying for this universality with decreasing clarity). In questionable cases, however, I have tended to more information. Hence the user should keep in mind that the classification numbers of a paper may reflect topics and aspects of rather different weight. In many cases, the combination of classification numbers attributed to a paper, together with the title, may give more information than the sum of the isolated contributions of the individual numbers. For instance, C80 together with C55 may indicate (and really does so in most cases), that the quantifiers treated are of a set-theoretic character; C80 (or C75) together with A05 may indicate that the quantifiers (or infinitary languages) appear only in a non-technical sense.

Some critical remarks are in order about the exactness of classification. In some sense each classification has been an individual act influenced by many circumstances such as

- the momentary picture I had of a classification section;
- the classifications that had been made before by others;
- the opinion a reviewer has incorporated in his review.

To illustrate this situation by an extreme example, let me mention a paper that had been given four classifications (following maybe a review of Mathematical Reviews), whereas one of its translations had been given five different classifications (following maybe a review of Zentralblatt). Having looked through the paper, I added a further classification and deleted six old ones. Due to neverending questionable cases, the picture I had of the various classification sections changed during the process of classifying: further aspects appeared, others changed their character. Together with typographical errors these circumstances will have led to mistakes and inconsistencies. Nevertheless I hope that the user will be able to develop his own picture by experience.

3. Comments on Specific Sections

The various classification sections are of different character. Some of them have a rather specific and definite meaning such as C20, C40, or C80. Others are of a more "collective" character. Probably the most characteristic example of the latter kind is C52. Its range covers typical properties of classes (such as the amalgamation property), (non-)axiomatizability results, results about many non-isomorphic models, etc., sometimes in addition with a continuous transition to other sections, for example to C40 (with respect to preservation theorems). Moreover, there are sections of a somewhat "else"-character such as C90 that forms a collection of different nonclassical topics.

In the following the sections are listed in the Subject Index order, together with some detailed comments beyond those given in the Blass-Hinman classification scheme. The comments are to serve the double purpose of exhibiting the interpretation of the classification and of giving some concrete guidelines of where to search for specific information.

Basic properties of first-order languages and structures (C07)

The fact that a paper is placed here does not necessarily mean that it is of a basic character. However, papers treating omitting types phenomena beyond the ordinary omitting types theorems, such as Morley numbers or omitting types in stronger languages, should be looked for under C75. General properties of first-order theories include, for instance, elementary equivalence and partial isomorphisms; generalizations concerning similar notions in the context of stronger logics will be found under the corresponding sections (such as C75 or C80). Papers dealing with the automorphism groups of first-order structures are listed primarily here; however, they can also be found in sections such as C30 (if indiscernibles are involved) or in C60 (if special algebraic structures are treated).

Denumerable structures (C15)

This section comprises a spectrum going from a purely model-theoretic treatment of countable structures to countable models of specific theories, e.g. groups, rings, graphs, etc.. It also includes papers on denumerable models with special properties, thereby overlapping with C50. Papers that emphasize aspects of effectiveness (for instance enumerable models) will be found under C57.

Ultraproducts and related constructions (C20)

C20 contains papers on ultraproducts and related constructions together with their applications in various theories of algebraic, analytic or other character, even if the underlying language is not first-order. In particular it also covers ultraproduct constructions for non-classical logics. With respect to set theory there is a connection with C55 (which contains set-theoretic papers about ultrafilters if they have a model-theoretic component) and with E05.

Model-theoretic forcing (C25)

This section also contains papers on algebraically closed structures except for algebraically closed fields (these are treated in C60). There is access to papers on model companions and model completions via C35, too.

Other model constructions (C30)

This section covers all model constructions different from those falling under C20 or C25, ranging from the elementary method of diagrams to various kinds of products (including Boolean ones) and very specific constructions. Papers, however, that treat products in the sense of universal algebra, are not found here (nor under C05), unless they have model-theoretic content. Constructions (such as end extensions) of models of specific theories fall under the section treating the theory concerned; often this is C62 (set theory and arithmetic). Besides constructions involving indiscernibles, C30 contains papers that are concerned with indiscernibility itself. However, for the role of indiscernibles in connection with questions of stability one should consult C45.

Categoricity and completeness of theories (C35)

As indicated in the classification scheme annotations, C35 does not cover completeness of axiomatizations of logics; in this respect, the user should pass to the appropriate Bn or Fn. Papers, however, treating the completeness of logics from a model-theoretic viewpoint, should be found under C07 (if first-order logic is concerned) or under the appropriate sections C80, C90, etc. (if other logics are concerned). Non-categoricity results are, in general, listed elsewhere, for example under C45, C52, C55, etc., according to their main character.

Interpolation, preservation, definability (C40)

C40 covers interpolation and definability results in all kinds of logics, including non-classical ones. The definability results listed range from Beth's definability theorem in various logics and abstract model theory to questions of definability in specific structures, but only if they are sufficiently general. Hence, for instance, papers that show the undecidability of a theory by defining the natural numbers in its models, are to be looked for under D35; and papers proving results on the (mutual) definability of geometrical notions are to be found in section B30. Papers that treat interpretations, for example, interpretations of one theory in another, have primarily been listed in F25. Results such as those on definable types, typical for stability theory, can be found in section C45. The treatment of preservation theorems is (necessarily) somewhat vague; the papers concerned have been listed here only reluctantly. Related sections in this respect are C52 or, for example, C30 (for preservation under products).

Stability and related concepts (C45)

C45 contains papers on stability theory and its applications in other areas, such as algebra, and also papers that employ concepts of stability theory as a methodological means. Moreover, this section covers the theory of types and of indiscernibles as far as it is concerned with questions of stability. For results on categoricity in power, C35 is the more appropriate section.

Models with special properties (C50)

Besides the properties mentioned in the annotated classification scheme and others such as homogeneity, minimality, universality or resplendency, section C50 also lists papers that are concerned with even more specific properties, if these are not introduced only for technical reasons and form an important part of the paper. Examples are maximal models of a theory and models without end extensions. However, for existential completeness and related properties, C25 is the appropriate section.

Properties of classes of models (C52)

As already mentioned above, C52 has a broad scope without sharp borderlines. Topics falling under this section are, for instance: the amalgamation and the joint embedding property, compactness of classes, questions of axiomatizability in certain languages or fragments, first-order properties of classes being themselves not first-order (e.g. defined algebraically), classes (or their theories, respectively) having many models of some fixed cardinality (if not falling under C45 or C55).

Elimination of quantifiers and decidability (C10, B25)

A paper has been listed here if it proves decidability or quantifier elimination for a set of sentences in some classical or non-classical logic, or if it makes use of such a result in an essential manner. Model-theoretic papers, however, that treat decidable theories from a recursion-theoretic viewpoint, are listed under C57. Papers that are concerned with the reduction of the decision problem of first-order logic – even if they were written before undecidability was known – are not primarily listed here but in B20 or D35. Concerning decidability results for specific non-classical logics, the user may be well advised to look also through the appropriate Bn section in volume II. Concerning computer-oriented logics such as dynamic logic, he is urged to consult section B75. For decidability results about small fragments such as term equations, D40 (word problems) may be the more appropriate section.

Finite structures (C13)

C13 contains papers in which finite models form an essential part. Entries on spectra also cover generalized spectra in higher order logic. Papers on spectra in the context of universal algebra, however, have been listed only reluctantly. Papers concerning the finite model property in classical or non-classical logics have to be looked for in the appropriate Bn sections treating these logics or in section B25 (if decidability results are the main aim). Analogous remarks apply to computational logics.

Recursion-theoretic model theory (C57)

This section covers all aspects of recursive model theory, e.g. models with enumerations that allow a recursion-theoretic treatment of model-theoretic questions, the effective treatment of algebraic or topological structures, and recursively saturated models (accessible also via C50). The section is closely related to D45 which emphasizes more strongly the recursion-theoretic aspects. A combined list of D45 and C57 can be found in volume IV.

Equational classes, universal algebra (partial) (C05)

As already stated at the beginning, section C05 in this volume contains only papers that have a model-theoretic flavour or treat some other aspect covered here, such as questions of decidability or infinitary syntax. In particular, papers dealing with products or similar constructions solely from the viewpoint of universal algebra have been listed neither here nor in C30. Likewise, papers limited to a purely algebraic discussion of quasi-varieties have been left out.

Model-theoretic algebra (C60)

This section contains papers that deal with algebraic structures or theories, aiming at model-theoretic results or using model-theoretic means. The structures that are covered include also vector spaces, differential fields, and topological groups and fields (if the algebraic viewpoint dominates; see also the comments on section C65). I have been rather liberal when weighting the model-theoretic content of an essentially algebraic paper.

Models of arithmetic and set theory (C62, H15)

As to models of arithmetic, papers that treat reducts or successor structures are classified C62 or C65 according to their arithmetical content. If models of set theory are merely used as a model-theoretic tool (for instance in the sense of model theory via set theory), C55 should be taken into account, too.

Models of other mathematical theories (C65)

Other mathematical theories in the sense of this section include, for instance, classical theories of analytical nature, geometry, the theory of graphs, of orderings, and of topological or normed spaces. In general, the ordering of models is classified in C55. A paper on topological structures is also listed in section C80 (if the topology is condensed into a quantifier) or in C90 (if topological model theory dominates). Papers on non-standard analysis can be found in H05 and H10, that is in volume I.

Set-theoretic model theory (C55)

Papers are listed here if they use set-theoretic methods to solve model-theoretic questions (and no other Cn, such as C20, predominates) or if they treat model-theoretic questions of set-theoretic character. Typical examples of the first kind include the extensive use of cardinal arithmetic or the use of models of set theory; typical examples of the latter kind are Löwenheim-Skolem phenomena or questions in connection with set-theoretically defined quantifiers. Related sections include C20 (concerning ultrafilters), C65 (concerning orderings), C80 (concerning set-theoretically defined quantifiers), C95 (concerning a set-theoretical treatment of logical systems) and the appropriate En sections such as E75 or E05.

Admissible sets and infinitary languages (C70, C75)

The recursion-theoretic counterpart of C70 is D60. A paper is listed in C75 if it contains a reasonable amount of information about an infinitary version of some classical or non-classical logical system in a model-theoretic, algebraic or proof-theoretic context. In particular, C75 collects papers on omitting types (if not elementary and hence in C07) and papers treating any kind of infinitary quantifiers, terms or syntax. Concerning the role of infinitary languages in computational logics, the user is also referred to B75 in volume II.

Other extended logics and generalized model theory (C80, C85, C95)

C80 takes into account all kinds of quantifiers, such as cardinality quantifiers, the Henkin quantifier and other Lindström quantifiers, or higher order quantifiers different from $\forall$, $\exists$. Quantifiers of the latter kind include, for instance, topological quantifiers or quantifiers over monotone systems. The area covered extends to quantifiers of a linguistic character. Infinitary quantifiers, however, occur in section C75.

A paper on second or higher order logic has been listed under C85 if it has a model-theoretic aspect; otherwise section B15 should be consulted. In questionable cases, however, I have been liberal.

Section C95 contains papers on model-theoretic languages and their properties, on characterizations of logical systems, general notions of quantifiers and related topics. Papers on abstract deductive systems or consequence relations, typical for the Polish school, are, however, listed in section B22.

Nonclassical models (C90)

As already mentioned in the introduction, this section collects papers on different topics. Besides the subjects mentioned in the classification scheme, C90 also covers papers on Borel logic, on many-valued logics (if model-theoretic questions are treated), and on logics of partial models (if not exclusively in the range of universal algebra or general logic). Logics of monotone systems are to be found under C80. For Kripke models the user is urged to consult also sections B45 and F50. For the borderlines with sheaf theory or category or topos theory, the classifications are not very restrictive; however, for papers on the latter subject G30 in volume V is the more appropriate section.

Proceedings (C97)

Proceedings are listed only in case they contain a substantial part (roughly at least one third) falling under the scope of this volume. Otherwise they are accessible via their source number in the Source Index. Collected works are not listed in the Subject Index; they can be found in the Author Index.

Textbooks, surveys (C98)

Besides text books and pure survey papers, C98 contains also papers, that are only partially surveys, and papers of expository character.

Acknowledgements

First of all I would like to thank the co-editors for their help and constructive criticisms, and for the good atmosphere they have maintained during the whole cooperation. Gert H. Müller und Wolfgang Lenski, moreover, were always ready to provide editorial assistance, data base material, and any other support. Of course, thanks have to go also to all those in Heidelberg, who worked so carefully on the immense amount of data that had to be handled. I am indepted to Otto H. Kegel for kindly informing me about numerous papers on the borderline of algebra and model theory. Furthermore, I thank Monika Bayer for her intensive work on the identification of authors, Erika Wagner for all her support in providing review material, and Susanne who sacrificed her precious Saturdays free from school. In particular, I would like to thank Thomas Geiß: with carefulness, enthusiasm, and patience, he contributed substantially to corrections and improvements. It is deep gratitude that I express to my wife Heidi; through all the work she has accompanied me with warm understanding.

User's Guide

Wolfgang Lenski & Gert H. Müller

§1. Introduction

After some opening remarks, the organization of this Guide follows the main division of the volume: *Subject Index, Author Index, Source Index, Miscellaneous Indexes*. For each part we give first a general explanation followed by a more detailed description of typical entries in the index in question. The reader will probably find the User's Guide most helpful when he comes across an unclear entry in the Bibliography: he can then turn directly to the corresponding section in this Guide for an explanation of the abbreviations and conventions used.

§2. General remarks

The main languages of the Bibliography are English, French, German, Italian, and Spanish. For other languages translations (of titles, names of sources, etc.) are used – with some few exceptions in cases for which we had no translation. These translations were taken from various available sources or made by the Editors.

For practical reasons, all entries are in the Roman alphabet and diacritical marks have not been used. Thus, for languages other than English certain conventions have been adopted.

The *transliteration of Cyrillic* names and titles, the treatment of *diacritical marks* and the *alphabetization and alternative spelling* of author names are explained in detail in the Miscellaneous Indexes.

The *abbreviations* of *sources* were either taken from one of the various reviewing journals or invented by us. Although we had to abbreviate long titles, we hope that in most cases the abbreviation will suggest the full title in a sufficiently understandable way. How successful we were is left to the user to decide.

The *review numbers* given with the entries in the Author Index are from *Mathematical Reviews* (MR), *Zentralblatt für Mathematik und ihre Grenzgebiete* (Zbl), *Journal of Symbolic Logic* (JSL), and *Jahrbuch über die Fortschritte der Mathematik* (FdM). We made a serious attempt to include all reviews of any given item but we have doubts concerning our success. We also tried to avoid listing two reviews for a given item in those cases in which the second "review" simply points to the original review and does not give any additional information.

In case of *multipart publications* pointers are given to the other parts, as far as they are known, in the *Remarks* to the publication in question. It is not always the case that the different parts of a publication all have the same classifications. Thus it may happen that, for example, part I has a classification in this volume and part II does not. In this case the Remarks for part I indicate the author(s) and year of publication of part II. The user will need to consult the other volumes for further bibliographic information on part II.

The general way to search through the Bibliography is to use certain *pointers*: From the Subject Index to the Author Index the pointer is [Author, Year, Title]; from the Author Index to the Source Index the pointers are 5-digit codes; e.g. (**J** 1234) is a code which sends the user to the J-section of the Source Index.

A *word* of *caution*: In order to use the Bibliography for quotations in future publications it is necessary to use both the Author Index and the Source Index; it is not generally sufficient to quote just from the Author Index. For example, for the paper "AANDERAA, S.O. [1974] *On k-tape versus (k−1)-tape real time computation*" a quotation of the source of this paper as given in the Author Index listing of this item, "Complexity of Computation; 1973 New York 75-96", would with high probability be misleading: one might try to find a book of this title published in New York in 1973 whereas in fact "1973 New York" denotes the date and place of the conference in the proceedings of which Aanderaa's paper appears. The volume was actually published in 1974. The source code (**P** 0761) for "Complexity of Computation; 1973 New York" should be used to find the full details of the source in the Source Index. The abbreviations of the sources may themselves be misleading without the corresponding additional details (e.g. country codes) given in the Source Index. For example, many abbreviations for conference proceedings do not include an abbreviation for "Proceedings of ...". Thus "Proceedings of the Third Brazilian Conference on Mathematical Logic" is abbreviated by "Brazil Conf Math Logic (3); 1979 Recife"; a reader without the Bibliography at hand might search in vain for the volume under "Brazil" in his library whereas in fact it might be found alphabetically under "Proceedings".

The Source Index includes, as far as they are known to us, *International Standard Serial Numbers* (ISSN) or *Book Numbers* (ISBN) and *Library of Congress* (LC) numbers. They may help in finding the source in question in libraries or bookstores.

To facilitate searches for works spanning two or more of the major subfields of logic, the first of the Miscellaneous Indexes lists the entries in the present volume that also occur in other volumes of the Bibliography.

Accidental occurrences of features not explained in the User's Guide are left as exercises to the user. HINT: Write to us (in any case), please.

§3. Subject Index

This is a listing of publication items ordered

> first by the (special) *classification sections,*
> then by the *year of appearance,*
> and finally *alphabetically* by *author,*

showing the author, title and the codes of all classification sections which apply to the given publication.

• The *titles* are given in the main languages of the Bibliography; if the original title is in another language, this is indicated in parentheses, e.g. (Russian), but only a translation of the title is given. Information on summaries in languages other than the original is included.
• If a publication is by *multiple authors,* it occurs only *once,* under the alphabetically first name. (But see also the Author Index.)
• In order to get the full bibliographical data of a publication, use the author, year and title to find the item in the Author Index.
• The classification sections listed in each volume have been selected by the individual editors. Sections B96-F96 have been systematically omitted; for the collected works of an author refer to the Author Index.

User's Guide

§4. Author Index

This is a listing of publication items ordered

> alphabetically by *author*, and for a given author
> chronologically by the *year of appearance*, and therein
> alphabetically by *title* of the item.

• The titles are given in the main languages as in the Subject Index.
• The *names* of the *authors* are written in the Roman alphabet using the Transliteration Table (see Miscellaneous Indexes) if necessary. There may be many versions of the name in use for a given author; (e.g. different combinations of the given name(s) or initials; different names used before and after marriage; different transliterations). The Miscellaneous Indexes include a table of different versions (known to us) and the corresponding form used in this Bibliography.
• Here publications with *multiple authors* are listed under *each* author but in the alphabetically later cases only the year is given and there is a pointer to the full entry given under the first author.
• The last entries for an author may contain a *reference to other name(s)* under which he/she also has publications in the Bibliography or *to other volumes* of the Bibliography where he/she has publications not mentioned in the present volume. A complete list of the author's papers contained in the six volumes is obtained by consulting the other volumes.
• In the following we explain the *individual entries* in more detail by giving an *idealized* example using fictitious names and sources showing all features that might occur; in a given case some features may not appear either because they do not apply or because our information is incomplete. The typefaces of the example and the order of its fields are as in the Author Index but, for *expository reasons only, here* we list all features on separate lines numbered by (1), (2), ...; we list explicitly those fields that begin a new line in the Author Index itself. (The foregoing description applies not only to the explanation of the Author Index treated here but also to the explanation of the Source Index later on.)

Example

(1) AUTHOR, K.J. & COMPANION, CECIL X. [1972]
(2) *On coding and decoding (Russian) (English and French summaries)*
(3) (**J** 9999) or (**S** 9998) or (**P** 9997) or (**C** 9996) or (**X** 9995)
(4) J Math 1∗1-10
    or Math Logic Series 1
    or Logic Conf; 1999 London 3-10
    or Math Publ xxv+200pp
(5) • ERR/ADD ibid 2∗3-4 or (**J** 8888) Arch of Logic 2∗3-4
(A new line begins here.)
(6) • LAST ED [1983] (**X** 9900) Logic Publ xx+100pp
(7) • REPR [1981] (**J** 9901) Math Logic J 2∗3-8
(8) • TRANSL [1979] (**J** 9902) Math Transl 1∗4-8
(A new line begins here.)
(9) ◇ B05 B20 C12 ◇
(10) • REV MR 99a:03001 Zbl 999#03001 JSL 99.321 FdM 99.123
(11) • REM This is an illustrative example
(12) • ID 12345

Explanations

(1) lists the authors followed by the year (in brackets) of publication of the item. Exceptionally a full given name (e.g. CECIL) is used to distinguish several authors with the same surname and initials.

(2) gives the title of the item followed, if the original is not in an official language of the bibliography, by the original language in parentheses and an indication of summaries in languages other than the original.

(3) is a pointer (or "source code") to the *Source Index;* there are five types: *Journal* (**J**), *Series* (**S**), *Proceedings Volume* (**P**), *Collection Volume* (**C**), and *Publisher* (**X**); one such code appears in (3). In order to find the full bibliographical data of the source use the pointer to locate the source in the Source Index; e.g. (**J** 9999) is given in the J-section of the Source Index.

Note: For a small number of items the source code is 0000, 1111, 2222 or 3333 (*not* preceded by J,S,P,C, or X). The code 0000, respectively 1111, indicates that the item is a thesis, respectively technical report. The code 2222, respectively 3333, is used for those cases in which the source, respectively publisher, is unknown. In each such case any further source information available is given in the Remarks (see line (11)).

(4) contains the abbreviation of the source indicated by the code in (3) followed by the paging as appropriate. Certain uniform features of the form of abbreviation used for proceedings and collection volumes should help the reader to recognise the volume. Abbreviations for proceedings (**P**) volumes end with an indication of the year and place of the corresponding conference, e.g. 1973 New York. Likewise, a name in parentheses, e.g. (Goedel), in an abbreviation of a collection (**C**) volume indicates the honorand to whom the volume is dedicated. A name followed by a colon, e.g. "Wang:", at the beginning of a collection volume abbreviation, indicates the author of all papers in the collection. The paging takes one of the following forms:

1*1-10   : Volume 1, pages 1-10 (for journals or series)
1/2*1-10 : Volume 1, Issue 2, pages 1-10 (for journals)
3-10   : pages 3-10 (for proceedings or collection volumes)
xx+200pp : initial paging + paging of a book (following a publisher or series)

(5) The • here and later is intended to make the entries easier to read. It is used to separate different types of information. After the • is the bibliographical information for published errata or addenda to the item. The two ways ERR/ADD can be given correspond to the cases in which its source is the same as in (4) (indicated by "ibid") and that in which it is in a different source; in the latter case the entry is of the same form as in (3) and (4).

The remaining information is not strictly part of the bibliographical data but contains useful additions.

(6), (7), (8) list the most recent edition, reprintings and translations, respectively, given by source as in (3) and (4); note that (7) and/or (8) may contain several entries for one publication.

(9) The classification codes enclosed in ◇ always begin a new line. Note that the codes are given in alphabetical/numerical order; no distinction of primary and secondary classification is made. (The classifications often differ from those assigned to the item in MR or Zbl.)

(10) lists the reviews. Sometimes two reviews are given from one reviewing journal. This may happen, e.g., when an item and its erratum/addendum are reviewed separately or when two different editions of a book have independent reviews.

(11) contains additional information not appropriate for coding in one of the standard fields.

(12) Each entry ends with its *identification number*. It is not used elsewhere in the main body of this volume except occasionally in the Introduction and the Remarks of another item where it may be used to pinpoint an item not uniquely identified by author(s) and year. The identification number is used (together with author(s) and year) as a pointer in the External Classification Code Index. We ask that the identification number be used in any correspondence with the Editors concerning this publication, as the bibliographical data base is indexed by these numbers.

§5. Source Index

This index contains the bibliographical data of the sources of the publications listed in this volume. It is subdivided into the following parts.

J (*Journals*), **S** (*Series*), **C** (*Collection volumes*), **P** (*Proceedings*), **X** (*Publishers*).

- Each part is ordered by the 4-digit source code numbers. (There is no significance to the particular 4-digit number assigned to a given source other than as a way to find the entry in the source index. Numbers were assigned as the sources were entered into the data base and so the numbering does not correspond to alphabetical order or order of publication.) Each 4-digit number is used *only once* as a source code so that, e.g., 0007 is a source code for a journal and the number 0007 is not used as a code for a series, proceedings, collection volume or publisher.
- Titles are given in the original language, using the transliteration system (see Miscellaneous Indexes) where necessary, followed, if necessary, by a translation into one of the main languages in parentheses. Sometimes if the original title is unknown to us, we give only a translated title in parentheses. Sometimes a source, e.g., a journal, has more than one title (English, French, German); in this case all titles are given, separated by *. These measures were taken to ease the search in libraries. In order to explain the entries in the Source Index we again use idealized examples and apply the conventions described in §4 above.

Journals

Example of a journal entry:

    (1)   **J** 8888 Math Div • F

(A new line begins here)

    (2)   *Mathematica Diversa * Mathematiques Diverses*
    (3)   [1900ff] or [1905-1935] ISSN 0007-0882

(A new line begins here.)

    (4)   • CONT OF (**J** 8885) J Math Ser A
    (5)   • CONT AS (**J** 8887) J Math Ser C
    (6)   • TRANSL IN (**J** 9904) Math Transl
    (7)   • TRANSL OF (**J** 9905) Matemat

((4) - (7) may contain more than one entry)

(A new line begins here.)

    (8)   • REL PUBL (**J** 9903) Mathematica (Subseria)
    (9)   • REM This journal is a fiction

Explanations

(1) Source code and abbreviation of the journal as used in the Author Index followed by the *international vehicle code* of the country in which the journal is published. A list of these codes is included in the Miscellaneous Indexes.

(2) The form of title(s) (and translations) are explained above.

(3) [1900ff] indicates that this journal has appeared continuously since 1900; [1905-1935] indicates that the journal appeared from 1905 to 1935. The International Standard Serial Number (ISSN) is given whenever possible.

(4), (5) give the predecessors (continuation of) and successors (continued as) of the journal in (2). In some cases in (4) or (5) the source code may be missing; this means that there are no entries in the Author Index which refer to the continued source. (It is mentioned, however, for the convenience of the user.)

(6) lists the translation journals of (2) and (7) gives the journal of which (2) is a translation; the source code is shown only if the translation in question is used as a source in this Bibliography. (6) and (7) do not both occur in a single journal entry.

(8) lists further entries in the Bibliography related to this journal, e.g. a subseries of the journal.

(9) is intended for additional information of various kinds.

Series

It is often hard to determine what should and what should not be characterised as a series. Some serials that we have chosen to treat as series may elsewhere be considered to be journals. In other cases, in particular certain publication series of university mathematics departments, the series includes all publications of its publisher and so might reasonably be identified with the publisher. Despite these considerations, we have chosen to list series separately to accord with the form of quotation often used in the modern literature.

Example of a series entry:

 (1) S 8999 Notae Log • NL
(A new line begins here)
 (2) *Notae Logicae* * *Notas Logicas*
 (3) [1900ff] or [1905-1935]
 (4) • ED: EDITOR, A.A. & COEDITOR, B.B.
 (5) • SER (S 8998) Notes in Phil
 (6) • PUBL (X 9950) Logic Publ Co: Heidelberg
 (7) • ALT PUBL (X 9951) Math Publ Inc: London
(A new line begins here.)
 (8) • CONT OF (S 9975) Notes in Logic A
 (9) • CONT AS (S 9901) Notes in Logic B
 (10) • TRANSL IN (S 9902) Notes de Logique
 (11) • TRANSL OF (S 9903) Logical Notes
(A new line begins here.)
 (12) • ISSN 0011-11122 (or ISBN 0011-11123) LC-No 73-10000
 (13) • REL PUBL (S 9900) Notae Logicae (Subseria)
 (14) • REM The origins of this series are somewhat obscure

Explanations

Entries (1), (2), (3), (8)-(11), (13), and (14) correspond to (1), (2), (3), (4)-(7), (8), (9), respectively, of the *journal entry* described above.

(4) lists the editors of the series (given in the same form as in line (1) of the Author Index example).

(5) Occasionally a series is itself a subseries of another series or journal. This is indicated in (5) (with an S or J as appropriate).

(6) gives the publisher of (2). For those publishers not listed in the publisher section of the Source Index, an abbreviation is sometimes used if either the abbreviation is readily understandable or the full name is not known.

(7) Some sources are published by two or more publishers; ALT PUBL lists the alternative publisher(s).

(12) lists the ISSN (or ISBN) and the Library of Congress number.

User's Guide   XXXI

Proceedings and Collection Volumes

Example of a proceedings or collection volume:

 (1) P 9920 Atti Congr Mat; 1971 London, ON • CDN
   or
   C 9921 Atti Congr Mat • D

(A new line begins here.)
 (2) [1972]
 (3) *Atti del Congresso di Matematica * Actes du Congres de Mathematique*
 (4) • ED: EDITOR, A.A. & COEDITOR, B.B.
 (5) • SER (S 8999) Notes in Logic
 (6) • PUBL (X 9950) Logic Publ Co: Heidelberg
 (7) • ALT PUBL (X 9951) Math Publ Co: London

(A new line begins here.)
 (8) • DAT&PL 1971 Aug;London, ON, CDN
 (9) • ISBN 0-012-34567-X, LC-No 84-98765
 (10) • REL PUBL (P 9947) Atti Congr Mat Vol Spez

(A new line begins here.)
 (11) • TRANSL IN [1973] Conf de Logique Math (3); London, ON, CDN
   • PUBL (X 9949) Livres: Paris
 (12) • TRANSL OF [1971] Konf Math Logik (3); London, ON, CDN
   • PUBL (X 9948) Buchverlag: Stuttgart

(A new line begins here.)
 (13) • REM Not all the articles appear in the translation

Explanations

(1), (3), (4) - (7), (9), (11), (12), (13) correspond to (1), (2), (4)-(7), (12), (10), (11) and (14), respectively, of a *series entry*. In (11), (12) PUBL denotes the publisher of the translation or original, respectively.

(2) denotes the year of publication of the volume (and not, in the case of a proceedings, the year of the conference).

(8) is used for proceedings volumes to indicate the date (year and month) and place of the conference, given by the city, the state (for the USA and elsewhere) and the country using its code as defined above. Note in case of *Proceedings* (**P**) volumes in (1) the country code of the place of the conference is repeated for conformity reasons, whereas for *Collection* (**C**) volumes the country code in (1) refers to the location of the publisher as in the case of *Journals* and *Series*.

(10) lists further entries in the Bibliography related to this volume, e.g. another proceedings volume of the same conference or a journal of which the volume is a special issue.

Publisher

Example of a publisher entry:

 (1) X 9950 *Logic Publishing Company* (Heidelberg, D & London, GB) ISBN 0-01
 (2) • REL PUBL (X 9930) Editions Logiques: Paris, F
 (3) • REM In London called Logic Publishing Corporation

Explanations

(1) lists the source code and full name of the publisher followed, in parenthesis, by the cities from which the publisher publishes and the ISBN. As in (8) of a **P** or **C** entry, codes are used for countries (see Miscellaneous Indexes).

(2) lists those publishers who have connections with the publisher listed in (1).

§6. Miscellaneous Indexes

This part contains the following indexes:

1. External classifications
2. Alphabetization and alternative spellings of author names
3. International vehicle codes
4. Transliteration scheme for Cyrillic

In each case a description of the contents and use are given in the corresponding introductory texts.

Ω-Classification Scheme

Andreas R. Blass
Peter G. Hinman

The classification scheme used for the Ω-Bibliography is a modified version of the section "03: Mathematical Logic and Foundations" of the 1985 Mathematics Subject Classification of *Mathematical Reviews* and *Zentralblatt für Mathematik und ihre Grenzgebiete*. For the sake of uniformity we have labeled all sections with a letter followed by a two-digit number; the prefix 03 is superfluous and therefore omitted. This decision has led to the creation of new sections to replace 03-01 through 03-06 (cf. X96-X98 and A10) and several sections with prefix other than 03 which have substantial logical content. Examples of the latter sort are B70 (to replace 94C10) and B75 (to replace the "logical part" of 68B10) (68Q55 and 68Q60 since 1985).

An important category of differences between the two schemes arises from the fact that whereas the MR/Zbl system is intended to classify works written after 1980, the majority of entries in the Ω-Bibliography were written before 1980. The subject matter of Mathematical Logic has, of course, changed immensely over the years, and today's categories are not always sufficient to distinguish properly important lines of earlier research. To deal with this problem we have added a few new sections (e.g. B22, B28, B65, C07, E07, and E47), renamed others (e.g. B35, C35, and E10), and altered slightly the interpretation of others (e.g. B25 and D65). To aid the reader in learning our conventions we have added descriptors to the section names. Topics preceded by a + (−) sign are specifically included (excluded) from a section. When this is in conflict with current MR/Zbl practice, this fact is also noted.

A

A05   Philosophical and critical

A10   History, Biography, Bibliography
   MR uses 03-03 and 01A for history and biography
   MR puts bibliography under specific fields.

B   GENERAL LOGIC

B03   Syntax of logical languages

B05   Classical propositional logic and boolean functions
   + Axiomatizations of classical propositional logic
   + Boolean functions (machine manipulation is also in B35); MR puts these in G05 and in 06E30 and 94C10.
   − Fragments of propositional logic: see B20
   − Switching circuits: see B70; MR also uses 94C10

B10   Classical first-order logic
   + Many-sorted logic
   + Syntax and semantics up to the Completeness Theorem
   − Model theory: see Cn, particularly C07
   − Proof theory: see Fn

B15 Higher-order logic and type theory
   + Higher-order algebraic and other theories
   − Higher-order model theory: see C85
   − Set theory with classes: see E30 and E70
   − Intuitionistic theory of types: see F35

B20 Fragments of classical logic
   + Fragments of propositional and of first-order logic
   + Fragments used in model theory, set theory, etc.
   + Syllogistic
   − Classical propositional logic: see B05
   − Weak axiomatizations without restrictions on formulas: see B55, B60, F50 ("Fragment" refers to reduced expressive power, not reduced deductive power; MR heading "Subsystems of classical logic" includes both)

B22 Abstract deductive systems
   + Consequence relations
   MR uses B99

B25 Decidability of theories and sets of sentences
   + Decidability of satisfiability
   + Decidable Diophantine problems
   − Decidable word problems: see D40
   − Other decidability results: see subject of problem, e.g. D05, or D80; MR includes these results here.
   − Undecidability results: see D35, D40, D80, etc.

B28 Classical foundations of number systems
   + Natural numbers, real numbers, ordinal numbers
   + Axiomatic foundations and set-theoretic foundations
   MR uses B30

B30 Logical foundations of other classical theories; axiomatics
   + Axiomatic method
   + Geometry, probability, physics, etc.
   + Models for non-mathematical theories
   − Foundations of parts of logic: see that part.
   MR heading: "Foundations and axiomatics of classical theories" includes also B28

B35 Mechanization of proofs and logical operations
   + Theorem proving, proof checking by machine
   + Minimization algorithms for Boolean functions
   + Optimization of logical operations
   MR sometimes uses 03-04 or 68G15 (68T15 since 1985)

B40 Combinatory logic and lambda-calculus
   + Models of lambda-calculus

B45 Modal and tense logic
   + Intensional logic; see also A05
   + Normative and deontic logic
   + Other non-truth-functional systems

B46 Relevance and entailment
   + Fragments
   − Primarily modal logic
   MR uses B45

B48 Probability and inductive logic
See also A05 and C90
+ Confirmation theory
− Foundations of probability: see B30; MR uses B48

B50 Many-valued logic
+ Matrix interpretations of propositional connectives unless used only as a tool for investigating classical propositional logic.
− Boolean valued set theory: see E40
− Probability logic: see B48 or C90

B51 Quantum logic
− Algebraic study of Quantum logic: see G12
MR uses only G12

B52 Fuzzy logic
+ Vagueness logic
− Papers demonstrating the fuzziness of the author's thought processes

B53 Paraconsistent logic
+ Discussive and dialectical logic
MR uses B60

B55 Intermediate and related logics
+ (Fragments of) propositional and predicate logics between intuitionistic or minimal and classical

B60 Other logics
− Intuitionistic logic: see F50 (MR uses B20)

B65 Logic of natural languages
− Computer languages: see B75
− Formal grammars unless applied to natural languages: see D05
− Natural language as a tool for the study of thought, reality, etc.: see A05
MR uses B65, B99, and 68Fn (68Sn since 1985)

B70 Logic in computer design; switching circuits
+ Hardware related to logic
MR uses 94Cn

B75 Logic of algorithmic and programming languages
+ Algorithmic and dynamic logic; MR uses B70 (formerly B45)
+ Logical analysis of programs
+ Logical aspects of database query languages and information retrieval
+ Semantics of programming languages related to logic
+ Software related to logic
− Specific algorithms: see subject of algorithm
MR uses B60, B70, 68Bn, 68Fn, and 68H05 (68Pn, 68Qn, and 68Tn since 1985)

B80 Other applications of logic
MR uses B99

B96 Collected works
+ Selected works
− Collections (almost) entirely in one subfield: see that subfield
MR uses 01A75, 03-03, and 03-06

B97 Proceedings
- \+ Collections of papers by various authors, even if they do not derive from any actual conference
- − Proceedings (almost) entirely in one subfield: see that subfield
- − Proceedings not concentrated in this field: see Source Index

MR uses 03–06

B98 Textbooks, surveys

MR uses 03–01 and 03–02

B99 None of the above or uncertain, but in this section

C MODEL THEORY

C05 Equational classes, universal algebra
- \+ Quasi-varieties, if the emphasis is algebraic
- − Word problems: see D40

C07 Basic properties of first-order languages and structures
- \+ Completeness, compactness, Löwenheim-Skolem, and omitting types theorems for ordinary first-order logic; MR uses C50 for omitting types
- \+ General properties of first-order theories
- \+ Homomorphisms, automorphisms, and isomorphisms of first-order structures
- − Analogues of these for stronger languages: see C55, C70, C75, etc.

C10 Quantifier elimination and related topics

C13 Finite structures
- \+ The spectrum problem
- \+ Probabilities of sentences being true in finite structures

C15 Denumerable structures

C20 Ultraproducts and related constructions
- \+ Applications of ultraproducts
- \+ Reduced products, limit ultrapowers, etc.
- − General products: see C30

C25 Model-theoretic forcing
- \+ Existentially closed structures, model companions, etc.
- − Model complete theories: see C35
- − Set-theoretic forcing: see E35, E40

C30 Other model constructions
- \+ Contructions involving indiscernibles
- \+ Products, diagrams

C35 Categoricity and completeness of theories
- \+ Model completeness
- − Gödel's completeness theorem: see C07
- − Completeness of axiomatizations of other logics: see those logics, e.g., B45

C40 Interpolation, preservation, definability
- \+ Definability in classes of structures
- − Definability in recursion theory: see appropriate Dn.
- − Definability in set theory: see E15, E45, and E47

C45 Stability and related concepts
+ Rank, total transcendence (even before stability was defined)

C50 Models with special properties
+ Saturated, rigid, etc.

C52 Properties of classes of models

C55 Set-theoretic model theory
+ Cardinality and ordering of models
+ Generalized Löwenheim-Skolem results
− Applications of set theory to some part of model theory: see that part
− Models of set theory: see C62
− Original Löwenheim-Skolem theorem: see C07

C57 Recursion-theoretic model theory
+ Model theory of recursive, arithmetical, etc. structures, types, etc.
− Recursion theory without substantial model-theoretic content: see D45
MR uses D45

C60 Model-theoretic algebra
+ Applications of model theory to specific algebraic theories
− Applications of set theory to algebra: see E75
− Decidability questions for algebraic theories: see B25, D35, and D40
− Model theory of orderings: see C65
− Universal algebra: see C05

C62 Models of arithmetic and set theory
+ Admissible sets as models: see also C70 and D60
+ Nonstandard models of arithmetic, when model theory is emphasized
+ Omega-models of higher-order arithmetic
− Models introduced only for consistency results: see F25 and E35
− Nonstandard models of arithmetic, when non-standardness is emphasized: see H15 or H20
MR uses C62, C65, F30, or H15

C65 Models of other mathematical theories
+ Other applications of model theory outside logic
+ Theories of orderings
− Uses of models for purely foundational studies: see B30

C70 Logic on admissible sets
+ All sorts of "effective" infinitary logic

C75 Other infinitary logic
+ Infinitary logic even if not model theory, e.g., infinite terms in proof theory and infinitary definability in set theory

C80 Logic with extra quantifiers and operators
− Hilbert epsilon-theorems: see B10
− Modal or many-valued operators: see B45 or B50

C85 Second- and higher-order model theory
+ Weak second-order theories (quantification over finite sets)

C90 Nonclassical models
- + Boolean-valued models
- + Sheaf models
- + Kripke models (also in B45 or F50)
- + Probability models (often also in B48)
- + Topological models (unless the topological structure is condensed into a quantifier: see C80); MR uses C85
- − Models of lambda calculus: see B40

C95 Abstract model theory
- + Lindström's theorem, delta-logics, etc.

C96 Collected works
- + Selected works
- − Collections (almost) entirely in one subfield: see that subfield

MR uses 01A75, 03-03, and 03-06

C97 Proceedings
- + Collections of papers by various authors, even if they do not derive from any actual conference
- − Proceedings (almost) entirely in one subfield: see that subfield
- − Proceedings not concentrated in this field: see Source Index

MR uses 03-06

C98 Textbooks, surveys

MR uses 03-01 and 03-02

C99 None of the above or uncertain, but in this section

D RECURSION THEORY

D03 Thue and Post systems, etc.
- + Markov's normal algorithms

D05 Automata and formal grammars in connection with logical questions
- + Cellular automata
- + Finite automata
- + Generalized automata
- + Regular events
- − Grammar of natural languages: see B65

MR uses 68 for most of these topics

D10 Turing machines and related notions
- + Potentially infinite automata
- + Probabilistic Turing machines

D15 Complexity of computation
- + Chaitin-Kolmogorov-Solomonoff complexity
- + Finer classification of decidable problems
- + Generalized complexity
- + Resource-bounded computability and reducibility
- + Speed-up theorems
- − Complexity of derivations and proofs: see F20
- − Complexity of specific non-logical problems (excluded from the Ω-Bibliography)
- − Syntactic complexity, complexity of Boolean functions, etc.

MR uses also 68Q15

D20 Recursive functions and relations, subrecursive hierarchies
+ Computable functions of real numbers; MR uses D65 and F60
+ General theory of algorithms
+ Partial recursive functions
+ Primitive recursion

D25 Recursively enumerable sets and degrees
+ Finer classification of undecidable r.e. problems
+ Many-one, truth table, etc., degrees of r.e. sets
+ Sets whose theory is closely related to that of r.e. sets, e.g., productive sets: see also D50
− Generalizations of recursive enumerability: see D60 and D65
− Partial functions with r.e. graphs: see D20

D30 Other degrees; reducibilities
+ Degrees in generalized recursion and constructibility: see also D55, D60, D65, and E45
+ Jump operators
− Subrecursive reducibilities: see D15 and D20

D35 Undecidability and degrees of sets of sentences
+ Hilbert's tenth problem and extensions
+ Reduction classes of the predicate calculus (also in B20)
− Decidability results: see B25
− Halting problems, word problems, etc.: see D03, D05, D10, D30, D40, or D80

D40 Word problems, etc.
+ Conjugacy, isomorphism, and other algorithmic problems in algebra
+ Decidability and undecidability
+ Other algorithmic questions in classical algebra
− Problems concerning production systems or formal grammar: see D03 and D05
− Recursive functions on words: see D20

D45 Theory of numerations, effectively presented structures
+ Numberings of (partial) recursive functions
+ Numerations in the sense of Ershov
+ Recursive algebra, except when it is about recursive equivalence types: see D50
+ Recursive order types
− Classical recursive analysis: see F60
− Model theory of recursive structures: see C57
− Recursive arithmetic: see F30

D50 Recursive equivalence types of sets and structures, isols
+ Concepts traditionally associated with isols, e.g., regressiveness and immuneness

D55 Hierarchies
+ Arithmetical, Borel, analytical, projective, etc. hierarchies
− Descriptive Set Theory in which hierarchical questions are not central: see E15
− Hierarchies of definability in set theory: see E47
− Incidental use of hierarchies outside recursion theory
− Subrecursive hierarchies: see D15 and D20

D60 Recursion theory on ordinals, admissible sets, etc.
+ Beta-recursion on inadmissible ordinals
− Classification of ordinary recursive functions using ordinals: see D20
− Ordinal notations: see D45 and F15
− Other aspects of admissibility: see C62, C70, or E45

D65 Higher-type and set recursion
+ Primitive recursive set functions
− Functionals in Proof Theory: see F10
− Recursion on the hereditarily finite sets: see D20
− Recursion with all arguments and parameters of type ≤ 1: see D20; MR includes this in D65 as long as there are type 1 arguments

D70 Inductive definability
+ Constructions equivalent to inductive definitions, e.g. set derivatives, game sentences, etc.
+ Recursion theory of inductive definitions and their duals
− Inductive definitions in proof theory: see F35 and F50
− Mechanics of inductive definitions: see B28, E20, or E30

D75 Abstract and axiomatic recursion theory
+ Algebras of (partial) recursive functions; MR uses D20
+ Recursion over general structures

D80 Applications
+ Decidability or undecidability results in areas outside logic and algebra
+ Effective versions of problems outside logic and algebra

D96 Collected works
+ Selected works
− Collections (almost) entirely in one subfield: see that subfield

MR uses 01A75, 03-03, and 03-06

D97 Proceedings
+ Collections of papers by various authors, even if they do not derive from any actual conference
− Proceedings (almost) entirely in one subfield: see that subfield
− Proceedings not concentrated in this field: see Source Index

MR uses 03-06

D98 Textbooks, surveys

MR uses 03-01 and 03-02

D99 None of the above or uncertain, but in this section

E SET THEORY

E05 Combinatorial set theory
+ Partition relations, ideals, ultrafilters, trees named after people; MR uses also 04A20
− Finite combinatorics (excluded from the Ω-Bibliography); MR uses 05Xn

E07 Relations and orderings
+ Relation algebras: see also G15; MR uses G15
− Theories about ordering: see C65

MR uses E20, 04A05, 04A20, or 06An

E10 Ordinal and cardinal numbers
  + Cardinal algebras, ordinal algebras
  + Dedekind finite cardinals
  − Cardinal exponentiation and the (generalized) continuum hypothesis: see E50; MR sometimes uses 04A10
  − Combinatorial aspects of cardinals and ordinals: see E05
  − Large cardinals: see E55

E15 Descriptive set theory
  + Definability properties of sets (in the real line or similar spaces)
  + Effective descriptive set theory
  − General topology, measure theory, etc.: see E75
  MR sometimes uses 04A15
  See also D55

E20 Other classical set theory
  + Set algebra

E25 Axiom of choice and related propositions
  + Weak axioms of choice and their negations
  MR sometimes uses 04A25

E30 Axiomatics of classical set theory and its fragments
  + Zermelo-Fraenkel set theory and minor variants
  + Gödel-Bernays set theory (also in E70)
  − Morse-Kelley set theory (a second order theory: see E70)
  − New Foundations, etc.: see E70

E35 Consistency and independence results
  + Forcing used to prove consistency

E40 Other aspects of forcing and Boolean-valued models
  + Forcing in generalized recursion theory: see also D60 and D65
  − Model theoretic forcing: see C25

E45 Constructibility, ordinal definability and related notions
  + Other inner models, e.g. the core model

E47 Other notions of set-theoretic definability
  + Lévy hierarchy, indescribability
  − Formalization of branches of mathematics within set theory

E50 Continuum hypothesis and Martin's axiom
  + Cardinal exponentiation
  + Variants of Martin's axiom
  MR sometimes uses 04A30

E55 Large cardinals
  + Effective (denumerable) analogues of large cardinals
  + Weakly inaccessible and larger cardinals
  − Axioms of infinity provable in ZFC
  − Large proof-theoretic ordinals: see F15

E60 Determinacy and related principles which contradict the axiom of choice
  + Infinite exponent partition relations
  + Projective determinacy, definable determinacy
  + Other uses of infinite games in set theory and logic
  − Applications of games outside set theory and logic
  − Weak axioms that merely contradict choice

E65 Other hypotheses and axioms
+ Reflection principles
+ Combinatorial principles

E70 Nonclassical and second-order set theories
+ Leśniewski's Ontology and Mereology; MR uses B60
+ Nonstandard theories, e.g. New Foundations, Ackermann
+ Set theories formulated in non-classical logic
+ Theory of real classes (Morse-Kelley, and Gödel-Bernays set theory); MR uses E30

E72 Fuzzy sets

E75 Applications
+ Independence from set theory of mathematical propositions (also in E35)
+ Results in other branches of mathematics obtained by set theoretic methods
− Set-theoretical foundations of mathematics: see B28 and B30

E96 Collected works
+ Selected works
− Collections (almost) entirely in one subfield: see that subfield
MR uses 01A75, 03-03, and 03-06

E97 Proceedings
+ Collections of papers by various authors, even if they do not derive from any actual conference
− Proceedings (almost) entirely in one subfield: see that subfield
− Proceedings not concentrated in this field: see Source Index
MR uses 03-06

E98 Textbooks, surveys
MR uses 03-01 and 03-02

E99 None of the above or uncertain, but in this section

F   PROOF THEORY AND CONSTRUCTIVE MATHEMATICS

F05 Cut elimination and normal form theorems
+ Hilbert's epsilon symbol
− Cut elimination and normal form theorems for modal systems: see B45

F07 Structure of proofs
− Proof schemas used rather than studied: see B10, C07, etc.

F10 Functionals in proof theory
− Typed lambda-calculus: see B40

F15 Recursive ordinals and ordinal notations
+ Ordinal notations even if not proof theory
+ Transfinite progressions of theories (Turing, Feferman; also in F30)

F20 Complexity of proofs
− Complexity of non-proof-theoretic procedures: see D15
− Purely qualitative (rather than quantitative) properties of proofs: see F07

F25 Relative consistency and interpretations
- Consistency of systems of arithmetic: see F30 and F35
- Set theoretic consistency results: see E35

F30 First-order arithmetic and fragments
+ Gödel incompleteness theorems
+ Metamathematics of intuitionistic arithmetic
+ Provability logic; MR uses also B45 and F40
+ Provably recursive functions; MR uses also D20
+ Recursive arithmetic
- Model theory of arithmetic: see C62 and H15

F35 Second- and higher-order arithmetic and fragments
+ Metamathematics of intuitionistic analysis
+ Proof theory of systems of type theory
+ Proof theory of generalized inductive definitions
- Model theory : see C62

F40 Gödel numberings in proof theory
+ Any use of Gödel numbering of syntax
- Gödel numberings in recursion theory: see D20 and D45

F50 Metamathematics of constructive systems
+ Intuitionistic logic and subsystems; MR uses also B20
+ Model theoretic methods applied to constructive systems
+ Realizability
- Metamathematics of predicative systems: see F65

F55 Constructive and intuitionistic mathematics
+ Bishop school of constructivism
- Metamathematics: see F50

F60 Constructive recursive analysis
+ Classical recursive analysis
+ Soviet school of constructivism
- Metamathematics: see F50

F65 Other constructive mathematics
+ Constructive trends not covered by F55 or F60
+ Predicative mathematics
+ Metamathematics of predicative systems
- Other metamathematics: see F50

F96 Collected works
+ Selected works
- Collections (almost) entirely in one subfield: see that subfield
MR uses 01A75, 03-03, and 03-06

F97 Proceedings
+ Collections of papers by various authors, even if they do not derive from any actual conference
- Proceedings (almost) entirely in one subfield: see that subfield
- Proceedings not concentrated in this field: see Source Index
MR uses 03-06

F98 Textbooks, surveys
MR uses 03-01 and 03-02

F99 None of the above or uncertain, but in this section

G ALGEBRAIC LOGIC

G05 Boolean algebras
+ Boolean rings, etc.
− Boolean functions : see B05; MR puts Boolean functions in G05, 06E30, and sometimes 94C10
− Pseudo-Boolean algebras : see G10

G10 Lattices and related structures
+ Heyting algebras; MR uses also 06D20
+ Semilattices, continuous lattices; MR uses 06B35
− Studies of "The lattice of..." where the lattice structure is not the main point

G12 Quantum logic
See also B51

G15 Cylindric and polyadic algebras, relation algebras

G20 Łukasiewicz and Post algebras
+ Lattices (or weaker structures) corresponding to many-valued logic

G25 Other algebras related to logic
+ Boolean algebras with provability and other operators
+ Implicative algebras, BCK algebras, etc.

G30 Categorical logic, topoi
+ Almost any connection between categories and logic, e.g. categories of models, logical foundations of category theory
− Pure category theory (Excluded from the Ω-Bibliography); MR uses 18Xn

G96 Collected works
+ Selected works
− Collections (almost) entirely in one subfield: see that subfield
MR uses 01A75, 03-03, and 03-06

G97 Proceedings
+ Collections of papers by various authors, even if they do not derive from any actual conference
− Proceedings (almost) entirely in one subfield: see that subfield
− Proceedings not concentrated in this field: see Source Index
MR uses 03-06

G98 Textbooks, surveys
MR uses 03-01 and 03-02

G99 None of the above or uncertain, but in this section

H NONSTANDARD MODELS

H05 Infinitesimal analysis in pure mathematics

H10 Other applications of infinitesimal analysis
+ Economics, physics, etc.

H15 Nonstandard models of arithmetic
+ Work emphasizing nonstandard methods
− Work emphasizing model theory : see C62

H20 Other nonstandard models

H96 Collected works
+ Selected works
− Collections (almost) entirely in one subfield: see that subfield
MR uses 01A75, 03-03, and 03-06

H97 Proceedings
+ Collections of papers by various authors, even if they do not derive from any actual conference
− Proceedings (almost) entirely in one subfield: see that subfield
− Proceedings not concentrated in this field: see Source Index
MR uses 03-06

H98 Textbooks, surveys
MR uses 03-01 and 03-02

H99 None of the above or uncertain, but in this section

Subject Index

C07 Basic properties of first-order languages and structures

1915
LOEWENHEIM, L. *Ueber Moeglichkeiten im Relativkalkuel*
⋄ B25 C07 C10 C85 E07 G15 ⋄

1920
SKOLEM, T.A. *Logisch-kombinatorische Untersuchungen ueber die Erfuellbarkeit oder Beweisbarkeit mathematischer Saetze nebst einem Theorem ueber dichte Mengen* ⋄ B10 B25 C07 C65 C75 ⋄

1923
SKOLEM, T.A. *Einige Bemerkungen zur axiomatischen Begruendung der Mengenlehre*
⋄ A05 B30 C07 C62 E30 ⋄

1928
SKOLEM, T.A. *Ueber die mathematische Logik*
⋄ A05 B25 B98 C07 C10 C35 ⋄

1929
SKOLEM, T.A. *Ueber einige Grundlagenfragen der Mathematik* ⋄ A05 A10 B30 C07 E30 ⋄

1930
GOEDEL, K. *Die Vollstaendigkeit der Axiome des logischen Funktionenkalkuels* ⋄ B10 C07 ⋄

1932
TARSKI, A. *Der Wahrheitsbegriff in den Sprachen der deduktiven Disziplinen*
⋄ A05 B30 C07 C40 F30 ⋄

1933
TARSKI, A. *On the notion of truth in reference to formalized deductive sciences (Polish)*
⋄ A05 B15 B30 C07 C40 F30 ⋄

1934
KALMAR, L. *Ueber einen Loewenheimschen Satz*
⋄ C07 ⋄

1935
BIRKHOFF, GARRETT *On the structure of abstract algebras*
⋄ C05 C07 G25 ⋄
LINDENBAUM, A. & TARSKI, A. *Ueber die Beschraenktheit der Ausdrucksmittel deduktiver Theorien*
⋄ B30 C07 C35 C65 ⋄
TARSKI, A. *Grundzuege des Systemkalkuels. Teil I*
⋄ B22 C07 ⋄

1936
MAL'TSEV, A.I. *Untersuchungen aus dem Gebiete der mathematischen Logik (Russian summary)*
⋄ B10 C07 ⋄
TARSKI, A. *Grundzuege des Systemkalkuels. Teil II*
⋄ A05 B22 C07 C35 G15 ⋄

TARSKI, A. *On the concept of logical consequences (Polish)*
⋄ A05 C07 ⋄
TARSKI, A. *The establishment of scientific semantics (Polish)* ⋄ A05 B10 B30 C07 ⋄

1938
KRASNER, M. *Une generalisation de la notion de corps*
⋄ C07 C60 C75 ⋄

1941
MAL'TSEV, A.I. *A general method for obtaining local theorems in group theory (Russian)* ⋄ C07 C60 ⋄
PETER, R. *The bounds of the axiomatic method (Hungarian) (German summary)* ⋄ A05 C07 ⋄
SKOLEM, T.A. *Sur la portee du theoreme de Loewenheim-Skolem* ⋄ C07 C62 E30 ⋄

1944
TARSKI, A. *The semantic conception of truth and the foundations of semantics* ⋄ A05 B10 B30 C07 ⋄

1945
KRASNER, M. *Generalisation et analogues de la theorie de Galois* ⋄ C07 C60 C75 ⋄

1946
BIRKHOFF, GARRETT *Sobre los grupos de automorfismos*
⋄ C05 C07 G10 ⋄

1949
FRAISSE, R. *Sur une classification des systemes de relations faisant intervenir les ordinaux transfinis*
⋄ C07 E07 ⋄
HENKIN, L. *The completeness of the first-order functional calculus* ⋄ B10 C07 ⋄

1950
FRAISSE, R. *Sur les types de polyrelation et sur une hypothese d'origine logistique*
⋄ B10 C07 E07 E10 ⋄
FRAISSE, R. *Sur une nouvelle classification des systemes de relations* ⋄ C07 E07 ⋄
KRASNER, M. *Generalisation abstraite de la theorie de Galois* ⋄ C07 C60 C75 ⋄
NOVAK, I.L. *A construction for models of consistent systems* ⋄ B15 C07 E70 ⋄
RASIOWA, H. & SIKORSKI, R. *A proof of the completeness theorem of Goedel* ⋄ C07 G05 ⋄

1951
BETH, E.W. *A topological proof of the theorem of Loewenheim-Skolem-Goedel* ⋄ C07 ⋄
RASIOWA, H. & SIKORSKI, R. *A proof of the Skolem-Loewenheim theorem* ⋄ C07 G05 ⋄
RIEGER, L. *On countable generalized σ-algebras, with a new proof of Goedel's completeness theorem (Czech)*
⋄ C07 G05 ⋄

RIEGER, L. *On free $\aleph_\xi$-complete boolean algebras (with an application to logic)* ◇ C07 G05 ◇

ROBINSON, A. *On the metamathematics of algebra* ◇ C07 C60 C98 ◇

1952

DREBEN, B. *On the completeness of quantification theory* ◇ C07 ◇

HASENJAEGER, G. *Topologische Untersuchungen zur Semantik und Syntax eines erweiterten Praedikatenkalkuels* ◇ B10 C07 G05 ◇

RASIOWA, H. *A proof of the compactness theorem for arithmetical classes* ◇ C07 G05 ◇

TARSKI, A. *Some notions and methods on the borderline of algebra and metamathematics* ◇ C07 C35 C52 C60 ◇

1953

BERRY, GEORGE D.W. *Symposium: On the ontological significance of the Loewenheim-Skolem theorem* ◇ A05 C07 ◇

BETH, E.W. *Some consequences of the theorem of Loewenheim-Skolem-Goedel-Malcev* ◇ C07 ◇

BETH, E.W. *Sur la description de certains modeles d'un systeme formel* ◇ C07 ◇

CRAIG, W. *On axiomatizability within a system* ◇ C07 D20 ◇

HASENJAEGER, G. *Eine Bemerkung zu Henkin's Beweis fuer die Vollstaendigkeit des Praedikatenkalkuels der ersten Stufe* ◇ B10 C07 C57 D55 ◇

HENKIN, L. *Some interconnections between modern algebra and mathematical logic* ◇ C07 C60 C85 ◇

HINTIKKA, K.J.J. *Distributive normal forms in the calculus of predicates* ◇ B10 C07 ◇

SCHROETER, K. *Theorie des mathematischen Schliessens* ◇ C07 ◇

WANG, HAO *Quelques notions d'axiomatique* ◇ B25 B30 C07 C35 E30 F30 ◇

1954

FRAISSE, R. & POSSEL DE, R. *Sur certaines suites d'equivalences dans une classe ordonnee, et sur leur application a la definition des parentes entre relations* ◇ C07 E07 ◇

FRAISSE, R. *Sur l'extension aux relations de quelques proprietes des ordres* ◇ C07 C30 C50 E07 ◇

FRAISSE, R. *Sur quelques classifications des systemes de relations (English summary)* ◇ C07 E07 ◇

FUENTES MIRAS, J.R. & JOSE, R. *"Deductive truth" according to Tarski (Spanish)* ◇ C07 C98 ◇

HENKIN, L. *A generalization of the notion of ω-consistency* ◇ B10 C07 F30 ◇

LOS, J. *Sur le theoreme de Goedel pour les theories indenombrables* ◇ C07 E25 G05 ◇

VAUGHT, R.L. *Applications of the Loewenheim-Skolem-Tarski theorem to problems of completeness and decidability* ◇ B25 C07 C35 ◇

VAUGHT, R.L. *Topics in the theory of arithmetical classes and boolean algebras* ◇ C07 G05 ◇

1955

ASSER, G. *Eine semantische Charakterisierung der deduktiv abgeschlossenen Mengen des Praedikatenkalkuels erster Stufe* ◇ C07 ◇

BETH, E.W. *Semantic entailment and formal derivability* ◇ B10 C07 ◇

FRAISSE, R. *La construction des γ-operateurs et leur application au calcul logique du premier ordre* ◇ B10 C07 E07 ◇

FRAISSE, R. *Sur certains operateurs dans les classes de relations* ◇ C07 E07 ◇

FRAISSE, R. *Sur quelques classifications des systemes de relations* ◇ C07 E07 ◇

FRAISSE, R. *Sur quelques classifications des relations basees sur des isomorphismes restreints I: Etude generale. II: Applications aux relations d'ordre, et construction d'exemples montrant que ces classifications sont distinctes* ◇ C07 C20 C65 E07 ◇

HASENJAEGER, G. *On definability and derivability* ◇ B10 B15 C07 D55 ◇

LOS, J. *The algebraic treatment of the methodology of elementary deductive system(Polish and Russian summaries)* ◇ B22 C07 ◇

REICHBACH, J. *Completeness of the functional calculus of first order (Polish, Russian) (English summary)* ◇ B10 C07 ◇

RIEGER, L. *On a fundamental theorem of mathematical logic (Czech) (Russian and English summaries)* ◇ B10 C07 G05 G25 ◇

ROSSER, J.B. *Les modeles des logiques formelles* ◇ C07 ◇

1956

DESUA, F. *Consistency and completeness - a resume* ◇ C07 ◇

EHRENFEUCHT, A. & MOSTOWSKI, ANDRZEJ *Models of axiomatic theories admitting automorphisms* ◇ C07 C30 E75 ◇

FRAISSE, R. *Application des γ-operateurs au calcul logique du premier echelon* ◇ B10 C07 C65 ◇

FRAISSE, R. *Etude de certains operateurs dans les classes de relations, definis a partir d'isomorphismes restreints* ◇ C07 E07 ◇

FRAISSE, R. *Sur quelques classifications des relations, basees sur des isomorphismes restreints. III. Comparaison des parentes introduites dans la premiere partie avec des parentes precedemment etudiees* ◇ C07 E07 ◇

HENKIN, L. *La structure algebrique des theories mathematiques* ◇ C07 F50 G05 G10 G15 ◇

KLEIN-BARMEN, F. *Zur Theorie der Strukturen und Algebren* ◇ C07 ◇

LOS, J. & MOSTOWSKI, ANDRZEJ & RASIOWA, H. *A proof of Herbrand's theorem* ◇ B10 C07 F05 F50 ◇

MAL'TSEV, A.I. *Representations of models (Russian)* ◇ C07 C60 ◇

OREY, S. *On ω-consistency and related properties* ◇ C07 F30 ◇

ZUBIETA RUSSI, G. *Definiciones formales de numerabilidad* ◇ C07 ◇

1957

BERNAYS, P. *Betrachtungen zum Paradoxon von Thoralf Skolem* ⋄ A05 C07 ⋄

COHN, P.M. *Groups of order automorphisms of ordered sets* ⋄ C07 C65 E07 ⋄

CRAIG, W. *Linear reasoning. A new form of the Herbrand-Gentzen theorem*
⋄ B10 C07 C40 F05 ⋄

CRAIG, W. *Three uses of the Herbrand-Gentzen theorem in relating model theory and proof theory*
⋄ B10 C07 C40 F05 ⋄

EHRENFEUCHT, A. *Application of games to some problems of mathematical logic (Russian summary)*
⋄ C07 C35 C65 E60 ⋄

FEFERMAN, S. *Some recent work of Ehrenfeucht and Fraisse* ⋄ B10 C07 C65 E10 ⋄

FRAISSE, R. *Obtention en theorie des relations de certaines classes d'origine logique* ⋄ C07 ⋄

HENKIN, L. *A generalization of the concept of ω-completeness* ⋄ B10 C07 F30 ⋄

LYNDON, R.C. *Sentences preserved under homomorphisms; sentences preserved under subdirect products* ⋄ C07 C30 C40 C52 ⋄

RASIOWA, H. *Sur la methode algebrique dans la methodologie des systemes deductifs elementaires*
⋄ B10 C07 G05 G10 ⋄

TAKEUTI, G. *On Skolem's theorem* ⋄ C07 F25 ⋄

TAKEUTI, G. *Remark on my paper: On Skolem's theorem*
⋄ C07 F25 F30 ⋄

TARSKI, A. & VAUGHT, R.L. *Arithmetical extensions of relational systems* ⋄ C05 C07 C55 C60 ⋄

TARSKI, A. & VAUGHT, R.L. *Elementary (arithmetical) extensions* ⋄ C05 C07 C35 C55 C60 ⋄

ZUBIETA RUSSI, G. *Clases aritmeticas definidas sin igualdad* ⋄ B10 B20 C07 C52 ⋄

1958

ENGELER, E. *Untersuchungen zur Modelltheorie*
⋄ C07 C75 ⋄

FRAISSE, R. *Sur une extension de la polyrelation et des parentes tirant son origine du calcul logique du $k^{\grave{e}me}$-echelon* ⋄ B15 C07 C85 E07 ⋄

HAERTIG, K. *Einbettung elementarer Theorien in endlich axiomatisierbare* ⋄ C07 ⋄

KRASNER, M. *Les algebres cylindriques*
⋄ C07 C60 G15 ⋄

LEVY, A. *Comparison of subtheories*
⋄ C07 C40 E30 F30 ⋄

MAEHARA, S. *Another proof of Takeuti's theorems on Skolem's paradox* ⋄ C07 F25 F30 ⋄

MAL'TSEV, A.I. *Homomorphisms onto finite groups (Russian)* ⋄ C07 C13 C60 ⋄

REICHBACH, J. *On the first-order functional calculus and the truncation of models (Polish and Russian summaries)* ⋄ B25 C07 C90 ⋄

ROBINSON, A. *Outline of an introduction to mathematical logic I,II,III* ⋄ B98 C07 G05 ⋄

SCHROETER, K. *Theorie des logischen Schliessens II*
⋄ B10 C07 F05 F07 ⋄

SPECKER, E. *Dualitaet* ⋄ B30 C07 E70 ⋄

1959

ENGELER, E. *Aequivalenzklassen von n-Tupeln* ⋄ C07 ⋄

LYNDON, R.C. *Properties preserved under homomorphism*
⋄ C07 C40 C52 ⋄

LYNDON, R.C. *Properties preserved under algebraic constructions* ⋄ C07 C30 C40 C52 ⋄

MAL'TSEV, A.I. *Model correspondences (Russian)*
⋄ B15 C05 C07 C52 C60 C85 ⋄

MONTAGUE, R. & VAUGHT, R.L. *A note on theories with selectors* ⋄ C07 ⋄

ROBINSON, A. *Outline of an introduction to mathematical logic IV* ⋄ B10 B98 C07 C98 G05 ⋄

1960

BETH, E.W. *Completeness results for formal systems*
⋄ B10 C07 ⋄

FRAISSE, R. *Les modeles et l'algebre logique* ⋄ C07 ⋄

KEISLER, H.J. *Theory of models with generalized atomic formulas* ⋄ C07 C30 C52 ⋄

LIS, Z. *Logical consequence, semantic and formal*
⋄ B10 C07 ⋄

LOS, J. & SLOMINSKI, J. & SUSZKO, R. *On the extending of models V: Embedding theorems for relational models*
⋄ C07 C30 C52 ⋄

RASIOWA, H. & SIKORSKI, R. *On the Gentzen theorem*
⋄ B10 C07 F05 ⋄

SVENONIUS, L. *On minimal models of first-order systems*
⋄ C07 C50 ⋄

SVENONIUS, L. *Some problems in logical model-theory*
⋄ C07 C52 C95 ⋄

1961

EHRENFEUCHT, A. & MOSTOWSKI, ANDRZEJ *A compact space of models of first order theories* ⋄ C07 C52 ⋄

EHRENFEUCHT, A. *An application of games to the completeness problem for formalized theories*
⋄ C07 C65 C85 E10 E60 ⋄

LEVY, A. *Axiomatization of induced theories* ⋄ C07 ⋄

POGORZELSKI, W.A. & SLUPECKI, J. *A variant of the proof of the completeness of the first order functional calculus (Polish and Russian summaries)* ⋄ B10 C07 ⋄

ROBINSON, A. *On the construction of models*
⋄ C07 C20 C30 E05 ⋄

SIKORSKI, R. *A topological characterization of open theories* ⋄ B20 C07 G05 ⋄

1962

FRASNAY, C. *Relations invariantes dans un ensemble totalement ordonne* ⋄ C07 C30 E07 ⋄

LUXEMBURG, W.A.J. *A remark on a paper by N.G.de Bruijn and P.Erdoes* ⋄ C07 C20 E05 E25 ⋄

MONTAGUE, R. *Theories incomparable with respect to relative interpretability* ⋄ C07 F25 F30 ⋄

MOSTOWSKI, ANDRZEJ *L'espace des modeles d'une theorie formalisee et quelques-unes de ses applications*
⋄ B50 C07 C40 C57 C80 C85 C90 D45 ⋄

RABIN, M.O. *Classes of models and sets of sentences with the intersection property* ⋄ C07 C52 ⋄

ROBINSON, A. *A note on embedding problems*
⋄ C07 C60 ⋄

SIKORSKI, R. *Algebra of formalized languages*
⋄ C05 C07 G15 G25 ⋄

SIKORSKI, R. *On open theories* ⋄ B20 C07 G05 ⋄

1963

CANTY, J.T. *Completeness of Copi's method of deduction* ⋄ B05 C07 ⋄

FLETCHER, T.J. *Models of many valued logics* ⋄ C07 ⋄

FRASNAY, C. *Groupes de permutations finies et familles d'ordres totaux, application a la theorie des relations* ⋄ C07 C30 E07 ⋄

SIKORSKI, R. *Products of generalized algebras and products of realizations* ⋄ C05 C07 G25 ⋄

1964

FRASNAY, C. *Groupes de compatibilite de deux ordres totaux; application aux n-morphismes d'une relation m-aire* ⋄ C07 C30 E07 ⋄

FRASNAY, C. *Partages d'ensembles de parties et de produits d'ensembles, applications a la theorie des relations* ⋄ C07 C30 E07 ⋄

HINTIKKA, K.J.J. *Distributive normal forms and deductive interpolation* ⋄ B10 C07 C40 ⋄

MO, SHAOKUI *Enumeration quantifiers and predicate calculus (Chinese)* ⋄ B10 C07 C80 ⋄

MOREL, A.C. *Ordering relations admitting automorphisms* ⋄ C07 C65 E07 ⋄

RIEGER, L. *Zu den Strukturen der klassischen Praedikatenlogik* ⋄ B10 C07 G05 ⋄

1965

FRAISSE, R. *A hypothesis concerning the extension of finite relations and its verification for certain special cases* ⋄ C07 C13 E07 ⋄

FRASNAY, C. *Quelques problemes combinatoires concernant les ordres totaux et les relations monomorphes* ⋄ C07 C30 E07 ⋄

HANF, W.P. *Model-theoretic methods in the study of elementary logic* ⋄ B25 C07 D20 D25 D35 F25 ⋄

HINTIKKA, K.J.J. *Distributive normal forms in first-order logic* ⋄ B10 C07 ⋄

JENSEN, A. *Some results concerning a general set-theoretical approach to logic* ⋄ C07 ⋄

MONTAGUE, R. *Interpretability in terms of models* ⋄ C07 F25 ⋄

MUELLER, GERT H. *Der Modellbegriff in der Mathematik* ⋄ C07 C98 ⋄

MYCIELSKI, J. *On unions of denumerable models (English)* ⋄ C07 C15 ⋄

RABIN, M.O. *Universal groups of automorphisms of models* ⋄ C07 C30 ⋄

SCHOCK, R. *Another proof of the strong Loewenheim-Skolem theorem* ⋄ C07 ⋄

VAUGHT, R.L. *The Loewenheim-Skolem theorem* ⋄ C07 C55 C80 ⋄

1966

BOURTOT, B. & CUSIN, R. *Une demonstration du theoreme de Loewenheim-Skolem* ⋄ C07 ⋄

CROSSLEY, J.N. *Some theorems in logic* ⋄ B10 C07 ⋄

HAJEK, P. *Generalized interpretability in terms of models. Note to a paper of R.Montague (Czech and Russian summaries)* ⋄ C07 F25 ⋄

LINDSTROEM, P. *On relations between structures* ⋄ C07 C40 C52 ⋄

MAL'TSEV, A.I. *On standard notations and terminology in the theory of algebraic systems (Russian)* ⋄ C05 C07 ⋄

PONASSE, D. *Une demonstration du theoreme de completude de Goedel* ⋄ C07 G05 ⋄

ROEDDING, D. *Anzahlquantoren in der Praedikatenlogik* ⋄ C07 C80 ⋄

1967

CALAIS, J.-P. *Relation et multirelation pseudohomogene* ⋄ C07 C50 ⋄

CHANG, C.C. *Omitting types of prenex formulas* ⋄ C07 C52 ⋄

CHOBANOV, I. *A model-axiomatizing scheme (Bulgarian) (English summary)* ⋄ C07 ⋄

ENGELER, E. *On the structure of elementary maps* ⋄ C07 C20 ⋄

FRAISSE, R. *L'algebre logique et ses rapports avec la theorie des relations* ⋄ C07 C50 E07 ⋄

FRAISSE, R. *La classe universelle: Une notion a la limite de la logique et de la theorie des relations* ⋄ C07 C30 C52 ⋄

FRAISSE, R. *Une hypothese sur l'extension des relations finies et sa verification dans certaines classes particulieres. II* ⋄ C07 C13 E07 ⋄

HAUSCHILD, K. *Verallgemeinerung eines Satzes von Cantor* ⋄ C07 C15 C35 ⋄

HEIDEMA, J. & WALT VAN DER, A.P.J. *Contributions to the metamathematical theory of ideals I. Domains with dense kernel* ⋄ C07 C60 ⋄

JEAN, M. *Relations monomorphes et classes universelles* ⋄ C07 C30 C52 ⋄

KREISEL, G. *Informal rigour and completeness proofs* ⋄ A05 B30 C07 E30 E50 F50 ⋄

LOVASZ, L. *Operations with structures* ⋄ C07 C13 C30 ⋄

PONASSE, D. *Quelques remarques topologiques sur l'espace des interpretations* ⋄ C07 G05 ⋄

RYLL-NARDZEWSKI, C. & WEGLORZ, B. *Compactness and homomorphisms of algebraic structures* ⋄ C07 C50 C75 ⋄

TAJMANOV, A.D. *Automorphisms of an algebra having one unary operation in the signature (Russian)* ⋄ C07 C20 C40 ⋄

TAJMANOV, A.D. *Remarks to the Mostowski-Ehrenfeucht theorem (Russian) (English summary)* ⋄ C07 C20 C30 C52 E50 ⋄

VAUGHT, R.L. *Axiomatizability by a schema* ⋄ C07 C52 ⋄

1968

CHANG, C.C. *Some remarks on the model theory of infinitary languages* ⋄ C07 C30 C55 C75 ⋄

DALLA CHIARA SCABIA, M.L. *Modelli sintattici e semantici delle teorie elementari* ⋄ C07 C98 E35 E45 F25 F98 ⋄

FRAASSEN VAN, B.C. *A topological proof of the Loewenheim-Skolem, compactness, and strong completeness theorems for free logic* ⋄ B60 C07 ⋄

HATCHER, W.S. *Remarques sur l'equivalence des systemes logiques* ◇ C07 G05 ◇

HAUSCHILD, K. & WOLTER, H. *Zur Charakterisierung von Theorien mit syntaktisch beschreibbarer Auswahlfunktion* ◇ C07 C20 ◇

HEIDEMA, J. *Contributions to the metamathematical theory of ideals II. Metamathematical prime ideals and radicals* ◇ C07 C60 ◇

JONSSON, B. *Algebraic structures with prescribed automorphism groups* ◇ C05 C07 ◇

KANGER, S. *Equivalent theories* ◇ C07 G05 ◇

KUEKER, D.W. *Definability, automormophisms and infinitary languages* ◇ C07 C40 C75 ◇

LINDSTROEM, P. *Remarks on some theorems of Keisler* ◇ C07 C20 C30 ◇

POUR-EL, M.B. *Independent axiomatization and its relation to the hypersimple set* ◇ C07 D25 ◇

RAUTENBERG, W. *Unterscheidbarkeit endlicher geordneter Mengen mit gegebener Anzahl von Quantoren* ◇ B20 C07 C13 C65 D35 ◇

SAYEKI, H. *Some consequences of the normality of the space of models* ◇ C07 C40 C52 ◇

SURMA, S.J. *Four studies in metamathematics (Polish) (Russian and English summaries)* ◇ B22 C07 C35 E25 ◇

TAJTSLIN, M.A. *On the isomorphism problem for commutative semigroups (Russian)* ◇ C05 C07 C60 D40 ◇

WEGLORZ, B. & WOJCIECHOWSKA, A. *Summability of pure extensions of relational structures* ◇ C05 C07 C52 ◇

1969

BOUVERE DE, K.L. *Remarks on classification of theories by their complete extensions* ◇ C07 C35 ◇

CALAIS, J.-P. *Isomorphismes locaux generalises* ◇ C07 ◇

COMER, S.D. & TOURNEAU LE, J.J. *Isomorphism types of infinite algebras* ◇ C07 ◇

FELSCHER, W. *On the algebra of quantifiers (Russian summary)* ◇ B10 C05 C07 G15 ◇

GLEBSKIJ, YU.V. & KOGAN, D.I. & LIOGON'KIJ, M.I. & TALANOV, V.A. *Range and degree of realizability of formulas in the restricted predicate calculus (Russian)* ◇ B10 C07 C13 ◇

HENSEL, G. & PUTNAM, H. *Normal models and the field Σ_1^** ◇ C07 C30 C57 D55 ◇

LADRIERE, J. *Le theoreme de Loewenheim-Skolem* ◇ C07 ◇

LEBLANC, H. *Three generalizations of a theorem of Beth's* ◇ B15 C07 ◇

POUR-EL, M.B. *A recursion-theoretic view of axiomatizable theories* ◇ C07 D25 D35 ◇

POUR-EL, M.B. *Independent axiomatization and its relation to the hypersimple set* ◇ C07 D25 D35 ◇

REICHBACH, J. *Propositional calculi and completeness theorem* ◇ B05 C07 ◇

RIBEIRO, H. *Elementary languages and mathematical structures (Portuguese)* ◇ B10 C07 ◇

SURMA, S.J. *On the axiomatic treatment of the theory of models I: Theory of models as an extension of Tarski's consequence theory (Polish summary)* ◇ B22 C07 ◇

TAKHTADZHYAN, K.A. *Logical formulae that are preserved under a strong homomorphism (Armenian) (Russian and English summaries)* ◇ C07 C40 ◇

1970

BENEJAM, J.-P. *Quelques proprietes algebriques des isomorphismes locaux* ◇ C07 C52 E07 G15 ◇

BENEJAM, J.-P. *Une classification des isomorphismes locaux et son application a la logique* ◇ C07 G15 ◇

BOOLOS, G. *A proof of the Loewenheim-Skolem theorem* ◇ C07 ◇

FRAISSE, R. *Aspects du theoreme de completude selon Herbrand* ◇ B10 C07 ◇

KOSSOWSKI, P. *On the axiomatic treatment of the theory of models IV: Independence of some axiom system of model theory (Polish summary)* ◇ B22 C07 ◇

KRASNER, M. *Endotheorie de Galois abstraite* ◇ C07 C60 C75 ◇

MAEHARA, S. *A general theory of completeness proofs* ◇ C07 C90 F50 ◇

PABION, J.F. *Theorie des modeles sans identite* ◇ C07 ◇

POUR-EL, M.B. *A recursion-theoretic view of axiomatizable theories* ◇ C07 D25 D35 ◇

REYES, G.E. *Local definability theory* ◇ C07 C15 C40 C50 C55 C75 ◇

SENFT, J.R. *On weak automorphisms of universal algebras* ◇ C05 C07 ◇

SMULLYAN, R.M. *Abstract quantification theory* ◇ B10 C07 F07 ◇

SURMA, S.J. *On the axiomatic treatment of the theory of models II: Syntactical characterization of a fragment of the theory of models (Polish summary)* ◇ B22 C07 C75 ◇

SURMA, S.J. *On the axiomatic treatment of the theory of models III: The function of model content (Polish summary)* ◇ B22 C07 ◇

1971

BENEJAM, J.-P. *La methode des s-ludions et les classes logiques* ◇ C07 C52 ◇

DOETS, H.C. *A theorem on the existence of expansions (Russian summary)* ◇ C07 C20 E25 ◇

DRABBE, J. *Quelques notions de la theorie des modeles* ◇ C07 C98 ◇

EHRENFEUCHT, A. & FUHRKEN, G. *A finitely axiomatizable complete theory with atomless $F_1(T)$* ◇ C07 ◇

FELSCHER, W. *An algebraic approach to first order logic* ◇ B10 C05 C07 C20 C90 G15 ◇

FRAISSE, R. & POUZET, M. *Interpretabilite d'une relation par une chaine* ◇ C07 C30 C65 E07 ◇

FRAISSE, R. & POUZET, M. *Sur une classe de relations n'ayant qu'un nombre fini de bornes* ◇ C07 C30 C65 E07 ◇

FRASNAY, C. *Relations interpretables par une chaine, relations monomorphes* ◇ C07 E07 ◇

LUSTIG, M. *On isotone and homomorphic maps of ordered unary algebras* ◇ C05 C07 ◇

NADIU, G.S. *Elementary logic and models* ◇ C07 C20 C30 ◇

NAKANO, Y. *An application of A.Robinson's proof of the completeness theorem* ⋄ C07 ⋄

PABION, J.F. *Problemes d'immersions en algebre* ⋄ C07 C60 ⋄

PAILLET, J.L. *Quelques rapports entre l'interpretabilite et l'extension logique* ⋄ C07 C40 ⋄

POIZAT, B. *Theorie de Galois des relations* ⋄ C07 C60 C75 E07 ⋄

PREVIALE, F. *Applicazioni del concetto di diagramma nella teoria dei modelli di linguaggi con piu specie di variabili* ⋄ C07 C40 ⋄

PRZELECKI, M. & WOJCICKI, R. *Inessential parts of extensions of first-order theories (Polish and Russian summaries)* ⋄ A05 C07 ⋄

SHELAH, S. *Every two elementarily equivalent models have isomorphic ultrapowers* ⋄ C07 C20 E05 ⋄

SUSZKO, R. *Quasi-completeness in non-Fregean logic (Polish and Russian summaries)* ⋄ B60 C07 ⋄

ZYGMUNT, J. *A survey of the methods of proof of the Goedel-Malcev's completeness theorem* ⋄ A10 C07 C98 ⋄

1972

ADLER, A. *Representation of models of full theories* ⋄ C07 C20 ⋄

BOWEN, K.A. *A note on cut elimination and completeness in first order theories* ⋄ B10 C07 F05 ⋄

BULLOCK, A.M. & SCHNEIDER, H.H. *A calculus for finitely satisfiable formulas with identity* ⋄ B10 C07 C13 ⋄

CALAIS, J.-P. *Partial isomorphisms and infinitary languages* ⋄ C07 C65 C75 ⋄

CALL, R.L. *The Goedel-Herbrand theorems* ⋄ B10 C07 ⋄

DENISOV, S.D. *Models of a noncontradictory formulas and the Ershov hierarchy (Russian)* ⋄ C07 C57 D55 ⋄

FELSCHER, W. *Modelltheorie als universelle Algebra* ⋄ C05 C07 ⋄

FINE, K. *For so many individuals* ⋄ C07 C80 ⋄

FRAISSE, R. *Chaines compatibles pour un groupe de permutations. Application aux relations enchainables et monomorphes* ⋄ C07 C30 C52 C65 E07 ⋄

FRAISSE, R. *Reflexions sur la completude selon Herbrand* ⋄ B10 C07 ⋄

GOODSTEIN, R.L. *A new proof of completeness* ⋄ B05 C07 ⋄

GRANT, J. *Recognizable algebras of formulas* ⋄ C07 C40 G05 G15 G25 ⋄

GRILLIOT, T.J. *Omitting types; applications to recursion theory* ⋄ C07 C62 D30 D55 D60 D75 ⋄

HAUCK, J. & HERRE, H. & POSEGGA, M. *Zur Metatheorie formaler Systeme* ⋄ B10 C07 F30 ⋄

KRAUSS, P.H. *Extending congruence relations* ⋄ C05 C07 C50 ⋄

PABION, J.F. *Relations prehomogenes* ⋄ C07 C35 C50 ⋄

PAILLET, J.L. *La methode des "systemes logiquement inductifs"* ⋄ C07 C30 ⋄

SELMAN, A.L. *Completeness of calculi for axiomatically defined classes of algebras* ⋄ B55 C05 C07 ⋄

SMIRNOV, V.A. *Formal derivation and logical calculi (Russian)* ⋄ B98 C07 C40 C98 F05 F07 F98 ⋄

SUSZKO, R. *Equational logic and theories in sentential language* ⋄ B05 C05 C07 ⋄

VINNER, S. *A generalization of Ehrenfeucht's game and some applications* ⋄ B25 C07 C55 C80 E60 ⋄

1973

ANDREKA, H. & NEMETI, I. *On the equivalence of sets definable by satisfaction and ultrafilters* ⋄ C07 C20 ⋄

ANDREKA, H. & GERGELY, T. & NEMETI, I. *Some questions on languages of order n (Hungarian)* ⋄ B15 C07 C85 ⋄

BALDWIN, JOHN T. *The number of automorphisms of models of $\aleph_1$-categorical theories* ⋄ C07 C35 C50 ⋄

BENNETT, D.W. *An elementary completeness proof for a system of natural deduction* ⋄ B10 C07 F07 ⋄

BLOOM, S.L. *Extensions of Goedel's completeness theorem and the Loewenheim-Skolem theorem* ⋄ C07 ⋄

BUECHI, J.R. & DANHOF, K.J. *Variations on a theme of Cantor in the theory of relational structures* ⋄ C07 C35 C50 C75 ⋄

BULLOCK, A.M. & SCHNEIDER, H.H. *On generating the finitely satisfiable formulas* ⋄ B10 C07 C13 D25 ⋄

FRASNAY, C. *Interpretation relationniste du seuil de recollement d'un groupe naturel* ⋄ C07 C30 E07 ⋄

GRANT, J. *Automorphisms definable by formulas* ⋄ C07 C40 C75 ⋄

HAIMO, F. *The index of an algebra automorphism group* ⋄ C05 C07 ⋄

HINTIKKA, K.J.J. *Logic, language-games and information. Kantian themes in the philosophy of logic* ⋄ A05 C07 ⋄

HINTIKKA, K.J.J. & NIINILUOTO, I. *On the surface semantics of quantificational proof procedures* ⋄ B35 C07 F07 ⋄

JERVELL, H.R. *Skolem and Herbrand theorems in first order logic* ⋄ B10 C07 F05 ⋄

KEISLER, H.J. & WALKOE JR., W.J. *The diversity of quantifier prefixes* ⋄ B10 C07 C13 D55 ⋄

KNIGHT, J.F. *Complete types and the natural numbers* ⋄ C07 C75 ⋄

KOTLARSKI, H. *Some simple results on automorphisms of models (Russian summary)* ⋄ C07 E25 ⋄

PAILLET, J.L. *La methode des systemes logiquement inductifs* ⋄ B10 C07 C30 ⋄

POUZET, M. *Une remarque sur la libre interpretabilite d'une relation par une autre* ⋄ C07 E07 ⋄

RANTALA, V. *On the theory of definability in first order logic* ⋄ C07 C40 ⋄

RUBIN, J.E. *The compactness theorem in mathematical logic* ⋄ C07 ⋄

SIMMONS, H. *An omitting types theorem with an application to the construction of generic structures* ⋄ C07 C25 C30 ⋄

SZCZERBA, L.W. *The notion of an elementary subsystem for a boolean-valued relational system* ⋄ C07 C90 ⋄

TABATA, H. *An application of a certain argument about isomorphisms of α-saturated structures* ⋄ C07 C50 ⋄

1974

ADAMEK, J. & REITERMAN, J. *Fixed-point property of unary algebras* ⋄ C05 C07 ⋄

ANDREKA, H. & GERGELY, T. & NEMETI, I. *Some questions on languages of order n I,II (Russian) (English summaries)* ⋄ B15 C07 C85 ⋄

ANDREKA, H. & GERGELY, T. & NEMETI, I. *Sufficient and necessary condition for the completeness of a calculus* ⋄ B10 C07 G05 ⋄

BACSICH, P.D. *Model theory of epimorphisms* ⋄ C07 G30 ⋄

BACSICH, P.D. & ROWLANDS-HUGHES, D. *Syntactic characterizations of amalgamation, convexity and related properties* ⋄ C05 C07 C60 C75 ⋄

BENTHEM VAN, J.F.A.K. *Semantic tableaus* ⋄ B10 C07 F07 ⋄

COLLET, J.-C. *Multirelations (k,p)-homogenes et (k,p)-ages* ⋄ C07 C50 ⋄

CROSSLEY, J.N. *Satisfaction (a brief survey of model theory)* ⋄ C07 C98 ⋄

DANHOF, K.J. *A proof of the compactness theorem* ⋄ C07 ⋄

DOKAS, L. *The notion of mathematical structure (Greek) (French summary)* ⋄ C07 ⋄

EBBINGHAUS, H.-D. & FLUM, J. *Algebraische Charakterisierung der elementaren Aequivalenz* ⋄ C07 C35 ⋄

EHDEL'MAN, G.S. *Certain radicals of metaideals (Russian)* ⋄ B10 C07 C60 ⋄

FRAISSE, R. *Isomorphisme local et equivalence associes a un ordinal; utilite en calcul des formules infinies a quanteurs finis* ⋄ B15 C07 C75 C85 ⋄

FRAISSE, R. *Multirelation et age 1-extensifs* ⋄ C07 E07 ⋄

FRAISSE, R. *Problematique apportee par les relations en theorie des permutations* ⋄ C07 C30 E07 ⋄

FRASNAY, C. *Theoreme de G-recollement (d'une famille d'ordres totaux). Application aux relations monomorphes, extension aux multirelations* ⋄ C07 C30 E07 ⋄

FRODA-SCHECHTER, M. *Sur les homomorphismes des structures relationnelles I* ⋄ C07 ⋄

GOODSTEIN, R.L. *Satisfiable in a larger domain* ⋄ B10 C07 ⋄

MARCJA, A. & TULIPANI, S. *Questioni di teoria dei modelli per linguaggi universali positivi (English summary)* ⋄ B20 C07 C10 ⋄

MARKOV, A.A. *The completeness of the classical predicate calculus in constructive logic (Russian)* ⋄ C07 F50 ⋄

MYERS, D.L. *The back-and-forth isomorphism construction* ⋄ C07 C15 G15 G30 ⋄

PAILLET, J.L. *Quelques rapports entre la fini-model-completabilite et des proprietes approchantes* ⋄ C07 C35 ⋄

PAILLET, J.L. *Une etude sur les theories "finiment model-completables"* ⋄ C07 C35 ⋄

POWELL, W.C. *Variations of Keisler's theorem for complete embeddings* ⋄ C07 C62 ⋄

REINHARDT, W.N. *Remarks on reflection principles, large cardinals, and elementary embeddings* ⋄ C07 E47 E55 E65 ⋄

SUSZKO, R. *Equational logic and theories in sentential languages* ⋄ B05 C05 C07 ⋄

THARP, L.H. *Continuity and elementary logic* ⋄ C07 C80 C95 D30 D55 ⋄

WEAVER, G.E. *Finite partitions and their generators* ⋄ B20 B25 C07 C40 ⋄

1975

ANDREKA, H. & NEMETI, I. *A simple, purely algebraic proof of the completeness of some first order logics* ⋄ C07 G15 ⋄

BEAUCHEMIN, P. *Un theoreme de Loewenheim-Skolem-Tarski pour les structures homogenes (English summary)* ⋄ C07 C50 E50 ⋄

COSTA DA, N.C.A. & DRUCK, I.F. *Sur les "VBTOs" selon M.Hatcher (English summary)* ⋄ C07 C40 C80 ⋄

CROSSLEY, J.N. *A brief survey of model theory* ⋄ C07 C98 ⋄

ELLENTUCK, E. & HALPERN, F. *Theories having many extensions* ⋄ C07 C75 C80 E75 ⋄

FINE, K. *Some connections between elementary and modal logic* ⋄ B45 C07 C90 ⋄

GOULD, M. *Endomorphism and automorphism structure of direct squares of universal algebras* ⋄ C05 C07 ⋄

HALKOWSKA, K. & PIROG-RZEPECKA, K. *On proofs in the theories containig conditional definition (Polish) (English summary)* ⋄ B20 C07 ⋄

HAUSCHILD, K. *Ueber zwei Spiele und ihre Anwendung in der Modelltheorie (Russian, English and French summaries)* ⋄ B25 C07 C30 C35 C85 ⋄

KNIGHT, J.F. *Types omitted in uncountable models of arithmetic* ⋄ C07 C62 ⋄

KREISEL, G. *Observations on a recent generalization of completeness theorems due to Schuette* ⋄ A05 C07 C57 F05 F20 F35 F50 ⋄

KREISEL, G. & MINTS, G.E. & SIMPSON, S.G. *The use of abstract language in elementary metamathematics: Some pedagogic examples* ⋄ A05 C07 C57 C75 F05 F07 F20 F50 ⋄

MANASTER, A.B. *Completeness, compactness, and undecidability: an introduction to mathematical logic* ⋄ B98 C07 D35 D98 ⋄

RANTALA, V. *Urn models: A new kind of non-standard model for first-order logic* ⋄ A05 B10 C07 C90 ⋄

ROCKINGHAM GILL, R.R. *A note on the compactness theorem* ⋄ C07 C20 E25 ⋄

SIMMONS, H. *The complexity of T^f and omitting types in F_T* ⋄ C07 C25 ⋄

SMOLIN, V.P. *Monadic algebras and the one-place predicate calculus* ⋄ B20 C07 G15 ⋄

TULIPANI, S. *Questioni di teoria dei modelli per linguaggi universali positivi II: Metodi di "back and forth" (English summary)* ⋄ B20 C07 C40 C50 ⋄

VENNERI, B.M. *Semantic implications of Herbrand's theory of fields* ⋄ A05 B10 C07 ⋄

ZIEGLER, M. *A counterexample in the theory of definable automorphisms* ⋄ C07 C60 ⋄

1976

ANDREKA, H. & DAHN, B.I. & NEMETI, I. *On a proof of Shelah (Russian summary)* ⋄ C07 C20 C90 ⋄

BLANC, G. & RAMBAUD, C. *Theories formelles sur graphe (English summary)* ⋄ C07 C65 G30 ⋄

BLOOM, S.L. *Varieties of ordered algebras* ⋄ C05 C07 ⋄

CUTLAND, N.J. *Compactness without languages* ⋄ C07 C20 ⋄

GILLAM, D.W.H. *Saturation and omitting types* ⋄ C07 C30 C50 ⋄

HALKOWSKA, K. & PIROG-RZEPECKA, K. & SLUPECKI, J. *Mathematical logic (Polish)* ⋄ B98 C07 C98 E98 ⋄

KLENK, V. *Intended models and the Loewenheim-Skolem theorem* ⋄ A05 C07 ⋄

KNIGHT, J.F. *Hanf numbers for omitting types over particular theories* ⋄ C07 C55 C62 ⋄

KNIGHT, J.F. *Omitting types in set theory and arithmetic* ⋄ C07 C62 E40 ⋄

KRASNER, M. *Endotheorie de Galois abstraite et son theoreme d'homomorphie (English summary)* ⋄ C07 C60 C75 ⋄

KRASNER, M. *Polytheorie de Galois abstraite dans le cas infini general* ⋄ C05 C07 C60 ⋄

MANSOUX, A. *Logique sans egalite et (k,p)-quasivalence (English summary)* ⋄ C07 C40 C75 ⋄

MANSOUX, A. *Theories de Galois locales (English summary)* ⋄ C07 C50 E07 ⋄

PAILLET, J.L. *Une etude sur des structures ayant une certaine propriete de seuil pour les automorphismes elementaires (English summary)* ⋄ C07 C35 C50 ⋄

PAULOS, J.A. *A model-theoretic semantics for modal logic* ⋄ B45 C07 C90 ⋄

1977

BARWISE, J. *An introduction to first-order logic* ⋄ B10 C07 C98 ⋄

BAUDISCH, A. & TUSCHIK, H.-P. *Eine Bemerkung ueber Vermeidung von Typen in ueberabzaehlbaren Modellen* ⋄ C07 ⋄

BENNETT, D.W. *A note on the completeness proof for natural deduction* ⋄ B10 C07 F07 ⋄

BRADY, R.T. *Unspecified constants in predicate calculus and first-order theories* ⋄ B10 C07 ⋄

BRIDGE, J. *Beginning model theory. The completeness theorem and some consequences* ⋄ C07 C98 ⋄

CASCANTE DAVILA, J.M. & PLA I CARRERA, J. *A precise proof of criterion CST.6 in Chapter IV of Nicolas Bourbaki's Theorie des ensembles (Spanish)* ⋄ C07 E30 ⋄

DAWES, A.M. *End extensions which are models of a given theory* ⋄ C07 C15 C30 C62 ⋄

DONNADIEU, M.-R. & RAMBAUD, C. *Theoremes de completude dans les theories sur graphes orientes (English summary)* ⋄ C07 C65 G30 ⋄

FAGIN, R. *The number of finite relational structures* ⋄ C07 C13 ⋄

FLUM, J. *Distributive normal forms* ⋄ B25 C07 ⋄

FRAISSE, R. *Present problems about intervals in relation-theory and logic* ⋄ C07 E07 ⋄

FRANCI, R. *Un'osservazione sulle connessioni di Galois (English summary)* ⋄ C07 E20 ⋄

FRIED, E. *Automorphism group of integral domains fixing a given subring* ⋄ C07 C60 ⋄

HIGMAN, G. *Homogeneous relations* ⋄ C07 C30 C65 E05 ⋄

LADNER, R.E. *Application of model theoretic games to discrete linear orders and finite automata* ⋄ B15 B25 C07 C13 C65 C85 D05 E07 E60 ⋄

MANIN, YU.I. *A course in mathematical logic* ⋄ B98 C07 D98 E35 E50 F30 G12 ⋄

PINTER, C. *Some theorems on omitting types, with applications to model completeness, amalgamation, and related properties* ⋄ C07 C25 C35 C52 C75 ⋄

RANTALA, V. *Aspects of definability* ⋄ A05 C07 C40 C75 ⋄

RAO, A.P. *A more natural alternative to Mostowski's (MFL)* ⋄ B60 C07 ⋄

ROISIN, J.-R. *Introduction a une approche categorique en theorie des modeles* ⋄ C07 G30 ⋄

SCHOENFELD, W. *Eine algebraische Konstruktion abzaehlbarer Modelle* ⋄ C07 C15 ⋄

SNAPPER, E. *Omitting models* ⋄ C07 C15 C50 ⋄

TWER VON DER, T. *Pseudoelementare Relationen und Aussagen vom Typ des Bernstein'schen Aequivalenzsatzes* ⋄ C07 C52 ⋄

1978

ANAPOLITANOS, D.A. *A new kind of indiscernibles* ⋄ C07 C30 C50 ⋄

FERRO, A. & MICALE, B. *Sui modelli di una teoria del 1º ordine nei quali certe formule definiscono un modello di un'altra teoria* ⋄ C07 ⋄

GERGELY, T. & VERSHININ, K.P. *Model theoretical investigation of theorem proving methods* ⋄ B35 C07 ⋄

GRANT, J. *Classifications for inconsistent theories* ⋄ B53 C07 ⋄

KRABBE, E.C.W. *The adequacy of material dialogue-games* ⋄ A05 B60 C07 ⋄

KRASNER, M. *Abstract Galois theory* ⋄ C07 C10 C60 C75 ⋄

KUEHNRICH, M. *On the Hermes term logic* ⋄ C07 C20 ⋄

LABLANQUIE, J.-C. *Forcing infini et theoreme d'omission des types de Chang* ⋄ C07 C25 ⋄

LOPEZ, G. *L'indeformabilite des relations et multirelations binaires* ⋄ C07 C13 E07 ⋄

MONK, J.D. *Omitting types algebraically* ⋄ C07 G15 ⋄

PAILLET, J.L. *Une etude sur des structures ayant une certaine propriete de seuil pour les automorphismes elementaires* ⋄ C07 C35 C50 ⋄

PINTER, C. *A note on the decomposition of theories with respect to amalgamation, convexity, and related properties* ⋄ C07 C52 ⋄

PRAZMOWSKI, K. *Conciseness and hierarchy of theories (Russian summary)* ⋄ C07 F25 ⋄

PRIJATELJ, N. *Formal construction of a logical language (Slovenian) (English summary)* ⋄ C07 ⋄

RANTALA, V. *The old and the new logic of metascience*
  ◇ A05 C07 ◇
SHELAH, S. *Models with second order properties. I: Boolean algebras with no definable automorphisms*
  ◇ C07 C30 C50 C55 E50 G05 ◇
TAMTHAI, M. *A note on conservative extensions* ◇ C07 ◇

1979

ANAPOLITANOS, D.A. *Automorphisms of finite order*
  ◇ C07 C62 E25 E35 ◇
CZELAKOWSKI, J. *A purely algebraic proof of the omitting types theorem* ◇ C07 C15 C30 G99 ◇
DAHN, B.I. *Constructions of classical models by means of Kripke models (survey)*
  ◇ C07 C20 C90 E35 E45 ◇
DANILOVA, S.A. *The group of automorphisms of a model (Russian)* ◇ C07 C30 ◇
FOO, N.Y. *Homomorphisms in the theory of modelling*
  ◇ C07 C40 ◇
FRASNAY, C. *Sur quelques classes universelles de relations m-aires* ◇ C07 C30 C52 E07 ◇
HANAZAWA, M. *An interpretation of Skolem's paradox in the predicate calculus with ε-symbol*
  ◇ B10 C07 F05 F25 ◇
MARKOVIC, Z. *An intuitionistic omitting types theorem*
  ◇ C07 C90 F50 ◇
OIKKONEN, J. *Mathematical applications of model theory (Finnish) (English summary)* ◇ C07 C60 ◇
PLONKA, J. *On automorphism groups of relational systems and universal algebras* ◇ C05 C07 E20 ◇
RAMBAUD, C. *Completude en theorie sur graphes orientes*
  ◇ C07 C65 ◇
ROISIN, J.-R. *On functorializing usual first-order model theory* ◇ C07 G30 ◇
RUBIN, M. *On the automorphism groups of homogeneous and saturated boolean algebras*
  ◇ C07 C50 C85 G05 ◇
SCOTT, D.S. *A note on distributive normal forms*
  ◇ B10 C07 ◇
THOMAS, WOLFGANG *Star-free regular sets of ω-sequences* ◇ C07 D05 ◇

1980

ADAMSON, A. *A note on the realization of types*
  ◇ C07 C50 ◇
ALCANTARA DE, L.P. & CORRADA, M. *Notes on many-sorted systems* ◇ B10 C07 ◇
ANAPOLITANOS, D.A. & VAEAENAENEN, J. *On the axiomatizability of the notion of an automorphism of a finite order* ◇ C07 C65 ◇
COSTA DA, N.C.A. *A model-theoretical approach to variable binding term operators* ◇ C07 C80 ◇
FOELDES, S. *On intervals in relational structures*
  ◇ C07 E07 ◇
HEIDEMA, J. & MALAN, S.W. *Congruences on relational structures* ◇ C05 C07 ◇
HENNESSY, M. *A proof system for the first-order relational calculus* ◇ B35 C07 ◇
HOLLAND, W.C. *Trying to recognize the real line*
  ◇ C07 C60 C65 ◇
JAMBU-GIRAUDET, M. *Theorie des modeles de groupes d'automorphismes d'ensembles totalement ordonnes 2-homogenes (English summary)*
  ◇ C07 C60 C62 C65 D35 E07 F25 F35 ◇
MYERS, R.W. *Complexity of model-theoretic notions*
  ◇ C07 C35 C57 D55 ◇
PINTER, C. *Topological duality theory in algebraic logic*
  ◇ C07 G25 ◇
RUBIN, M. *On the automorphism groups of countable boolean algebras* ◇ C07 C15 C85 E45 G05 ◇
RUBIN, M. & SHELAH, S. *On the elementary equivalence of automorphism groups of boolean algebras; downward Skolem Loewenheim theorems and compactness of related quantifiers*
  ◇ C07 C30 C75 C80 C95 G05 ◇
SZCZERBA, L.W. *Interpretations with parameters*
  ◇ B25 C07 C35 C45 F25 ◇
WEAVER, G.E. *A note on the compactness theorem in first order logic* ◇ B10 C07 ◇
WEINBERG, E.C. *Automorphism groups of minimal η_α-sets* ◇ C07 C65 E07 ◇

1981

ANAPOLITANOS, D.A. *Cyclic indiscernibles and Skolem functions* ◇ C07 C30 ◇
BABAI, L. *On the abstract group of automorphisms*
  ◇ C07 C65 ◇
CASARI, E. *Positively omitting types* ◇ B10 C07 ◇
EHDEL'MAN, G.S. *π-decomposition of metaideals (Russian)* ◇ C07 C60 ◇
GLASS, A.M.W. & GUREVICH, Y. & HOLLAND, W.C. & JAMBU-GIRAUDET, M. *Elementary theory of automorphism groups of doubly homogeneous chains*
  ◇ C07 C50 C65 E07 ◇
GLASS, A.M.W. *Elementary types of automorphisms of linearly ordered sets -- a survey* ◇ C07 C65 C98 ◇
GLASS, A.M.W. *Ordered permutation groups*
  ◇ C07 C60 C65 C98 ◇
GLASS, A.M.W. & GUREVICH, Y. & HOLLAND, W.C. & SHELAH, S. *Rigid homogeneous chains*
  ◇ C07 C50 C65 E07 E35 E50 ◇
GUREVICH, Y. & HOLLAND, W.C. *Recognizing the real line* ◇ C07 C65 ◇
IMMERMAN, N. *Number of quantifiers is better than number of tape cells* ◇ C07 D15 ◇
JAMBU-GIRAUDET, M. *Interpretations d'arithmetiques dans des groupes et des treillis*
  ◇ C07 C62 C65 D35 E07 F25 F30 G10 ◇
MUNDICI, D. *A group-theoretical invariant for elementary equivalence and its role in representations of elementary classes* ◇ B10 C07 C52 ◇
POUZET, M. *Application de la notion de relation presque-enchainable au denombrement des restrictions finies d'une relation* ◇ C07 C30 E07 ◇
RADU, A.G. & RADU, G.G. *Existential and universal quantifiers in an elementary topos* ◇ C07 G30 ◇
RODRIGUEZ ARTALEJO, M. *A method of constructing structures based on the distributive normal form (Spanish)* ◇ C07 C30 ◇
RODRIGUEZ ARTALEJO, M. *Eine syntaktisch-algebraische Methode zur Konstruktion von Modellen*
  ◇ B10 C07 C30 C57 ◇

Rothmaler, P. *A note on I-types* ◇ C07 C45 C75 ◇

1982

Bloom, S.L. *A note on the logic of signed equations* ◇ B60 C05 C07 ◇

Comer, S.D. *Boolean combinations of monadic formulas* ◇ C07 C90 G15 ◇

Fraisse, R. *La zerologie, une recherche aux frontieres de la logique et de l'art; application a la logique des relations de base vide* ◇ B60 C07 ◇

Furmanowski, T. *The logic of algebraic rules as a generalization of equational logic* ◇ C05 C07 ◇

Gaifman, H. *On local and non-local properties* ◇ C07 C13 C65 E47 ◇

Glubrecht, J.-M. *Ein Vollstaendigkeitsbeweis fuer schnittfreie Kalkuele mit der Maximalisierungsmethode von Henkin* ◇ B10 C07 F05 ◇

Hickman, J.L. *Automorphisms of medial fields* ◇ C07 C60 E25 E75 ◇

Hugly, P. & Sayward, C.W. *Indenumerability and substitutional quantification* ◇ A05 B10 C07 ◇

Makkai, M. *Stone duality for first order logic* ◇ C07 G30 ◇

Motohashi, N. *An axiomatization theorem* ◇ B10 C07 C40 C75 ◇

Mycielski, J. *An essay about old model theory* ◇ C07 C52 ◇

Nakano, Y. *On the theories with countably many unary predicate symbols* ◇ C07 ◇

Pinus, A.G. *Complete embeddings of categories of algebraic systems and definability of a model for its semigroup of endomorphisms (Russian)* ◇ C07 F25 G30 ◇

Poizat, B. *Deux ou trois choses que je sais de L_n* ◇ B20 C07 C13 C50 ◇

Shtejnbuk, V.B. *Semigroups of endomorphisms of filters (Russian)* ◇ C07 E05 ◇

Verdu i Solans, V. *The ultrafilter theorem and the adequacy theorem (Spanish) (English summary)* ◇ C07 C20 E25 ◇

Wang, Shiqiang *An omitting types theorem in lattice-valued model theory (Chinese)* ◇ C07 C75 C90 ◇

1983

Baranskij, V.A. *Algebraic systems whose elementary theory is compatible with an arbitrary group (Russian)* ◇ C07 C65 G10 ◇

Furmanowski, T. *The logic of algebraic rules as a generalization of equational logic* ◇ C05 C07 C13 ◇

Jambu-Giraudet, M. *Bi-interpretable groups and lattices* ◇ C07 C65 G10 ◇

Koppelberg, S. *Groups of permutations with few fixed points* ◇ C07 C50 E05 G05 ◇

Kuehnrich, M. *On the Hermes term logic* ◇ B10 C07 C20 ◇

Lopez, G. *Reconstruction d'une S-expansion* ◇ C07 C13 E07 ◇

Markusz, Z. *On first order many-sorted logic* ◇ B10 C07 ◇

Millar, T.S. *Omitting types, type spectrums, and decidability* ◇ C07 C57 ◇

Morozov, A.S. *Groups of recursive automorphisms of constructive boolean algebras (Russian)* ◇ C07 C57 D45 F60 G05 ◇

Rudak, W. *A completeness theorem for weak equational logic* ◇ C05 C07 ◇

Steinhorn, C.I. *A new omitting types theorem* ◇ C07 C15 C45 C50 C57 ◇

Szczerba, L.W. *Interpretations* ◇ B30 C07 F25 ◇

1984

Blass, A.R. *There are not exactly five objects* ◇ B20 C07 ◇

Cameron, P.J. *Aspects of random graph* ◇ C07 C60 C65 ◇

Fraisse, R. *L'intervalle en theorie des relations; ses generalisations, filtre intervallaire et cloture d'une relation (English summary)* ◇ C07 E07 ◇

Frasnay, C. *Relations enchaineables, rangements et pseudo-rangements (English summary)* ◇ C07 C13 C30 E07 ◇

Goldblatt, R.I. *An abstract setting for Henkin proofs* ◇ C07 C75 ◇

Hebert, M. *Un theoreme de preservation pour les factorisations* ◇ C07 C40 C52 ◇

Jurie, P.-F. & Touraille, A. *Ideaux elementairement equivalents dans une algebre Booleenne* ◇ C07 G05 ◇

Lascar, D. *Sous-groupes d'automorphismes d'une structure saturee* ◇ C07 C35 C45 C50 ◇

Manca, V. & Salibra, A. *First-order theories as many-sorted algebras* ◇ C05 C07 ◇

Morozov, A.S. *Group $Aut_r(Q, \leq)$ is not constructivizable (Russian)* ◇ C07 C57 F60 ◇

Motohashi, N. *A normal form theorem for first order formulas and its application to Gaifman's splitting theorem* ◇ C07 C62 ◇

Motohashi, N. *A normal form theorem for first order formulas and its application to Gaifman's splitting theorem* ◇ B10 C07 C62 ◇

Seremet, Z. *On automorphisms of resplendent models of arithmetic* ◇ C07 C50 C62 ◇

Sette, A.-M. *Partial isomorphism extension method and a representation theorem for Post-language* ◇ C07 G20 ◇

Thomas, Wolfgang *An application of the Ehrenfeucht-Fraisse game in formal language theory* ◇ C07 D05 ◇

Tokarenko, A.I. *Two variants of the compactness theorem (Russian)* ◇ C07 C20 ◇

Weese, M. *The theory of Boolean algebras extended by a group of automorphisms* ◇ C07 D35 G05 ◇

1985

Baranskij, V.A. *Independence of equational theories and automorphism groups of lattices* ◇ C05 C07 G10 ◇

D'Ottaviano, I.M.L. *The model extension theorems for Π_3-theories* ◇ B50 C07 C52 ◇

Droste, M. & Shelah, S. *A construction of all normal subgroup lattices of 2-transitive automorphism groups of linearly ordered sets* ◇ C07 C60 C65 E07 ◇

DROSTE, M. *The normal subgroup lattice of 2-transitive automorphism groups of linearly ordered sets*
⋄ C07   C60   C65 ⋄

GEORGE, A. *Skolem and the Loewenheim-Skolem theorem: a case study of the philosophical significance of mathematical results* ⋄ A05   A10   C07 ⋄

GOGUEN, J.A. & MESEGUER, J. *Completeness of many-sorted equational logic* ⋄ B20   C05   C07 ⋄

GOLDBLATT, R.I. *On the role of the Baire category theorem and dependent choice in the foundations of logic*
⋄ C07   E25   E40   E75   G30 ⋄

GRILLIOT, T.J. *Disturbing arithmetic*
⋄ C07   C30   C40   C62 ⋄

HOOK, J.L. *A note on interpretations of many-sorted theories* ⋄ B10   C07   F25 ⋄

JAMBU-GIRAUDET, M. *Quelques remarques sur l'equivalence elementaire entre groupes ou treillis d'automorphismes de chaines 2-homogenes*
⋄ C07   C65 ⋄

JEZEK, J. *Elementarily nonequivalent infinite partition lattices* ⋄ C07   E20   G10 ⋄

KOPPELBERG, S. *Homogeneous boolean algebras may have non-simple automorphism groups*
⋄ C07   C50   E50   E75   G05 ⋄

MLCEK, J. *Some automorphisms of natural numbers in the alternative set theory* ⋄ C07   C30   E70   H15 ⋄

MOORE, A.W. *Set theory, Skolem's paradox and the "Tractatus"* ⋄ A05   C07   E30 ⋄

MOROZOV, A.S. *Constructive Boolean algebras with almost identical automorphisms (Russian)*
⋄ C07   C57   G05 ⋄

OLIVEIRA DE, A.J.F. & SEBASTIAO E SILVA, J. *On automorphisms of arbitrary mathematical systems*
⋄ C07 ⋄

PAUL, E. *Equational methods in first order predicate calculus* ⋄ B75   C05   C07 ⋄

PRELLER, A. *A language for category theory, in which natural equivalence implies elementary equivalence of models* ⋄ C07   G30 ⋄

SETTE, A.-M. & SZCZERBA, L.W. *Characterizations of elementary interpretations in category theory*
⋄ C07   F25   G30 ⋄

THOMAS, S. *The automorphism tower problem*
⋄ C07   C60   E75 ⋄

TRUSS, J.K. *The group of the countable universal graph*
⋄ C07   C50   C65 ⋄

C15 Denumerable structures

1937
MOSTOWSKI, ANDRZEJ *Abzaehlbare Boolesche Koerper und ihre Anwendung auf die allgemeine Metamathematik* ⋄ C15 C50 G05 ⋄

1948
TARSKI, A. *Axiomatic and algebraic aspects of two theorems on sums of cardinals*
⋄ B28 C15 E10 E25 G05 ⋄

1957
HANF, W.P. *On some fundamental problems concerning isomorphism of boolean algebras*
⋄ C05 C15 C30 G05 ⋄

1961
ENGELER, E. *Unendliche Formeln in der Modell-Theorie* ⋄ C15 C35 C40 C75 ⋄
VAUGHT, R.L. *Denumerable models of complete theories (Polish)* ⋄ C15 C35 C50 ⋄

1962
FUHRKEN, G. *Bemerkung zu einer Arbeit E.Engelers* ⋄ C15 C40 C75 ⋄
MOSTOWSKI, ANDRZEJ *A problem in the theory of models* ⋄ C15 C52 ⋄

1964
ONYSZKIEWICZ, J. *On the different topologies in the space of models* ⋄ C15 C52 C85 ⋄

1965
MYCIELSKI, J. *On unions of denumerable models (English)* ⋄ C07 C15 ⋄
SCOTT, D.S. *Logic with denumerably long formulas and finite strings of quantifiers* ⋄ C15 C40 C75 C98 ⋄
SVENONIUS, L. *On the denumerable models of theories with extra predicates* ⋄ C15 C40 C52 ⋄

1967
HAUSCHILD, K. *Verallgemeinerung eines Satzes von Cantor* ⋄ C07 C15 C35 ⋄
MORLEY, M.D. *Countable models of $\aleph_1$-categorical theories* ⋄ C15 C35 ⋄

1968
HAUSCHILD, K. *Metatheoretische Eigenschaften gewisser Klassen von elementaren Theorien*
⋄ B28 C15 C35 E30 E70 ⋄

1969
BARWISE, J. *Remarks on universal sentences of $L_{\omega_1,\omega}$*
⋄ C15 C40 C75 ⋄
BONNET, R. *Stratifications et extension des genres de chaines denombrables* ⋄ C15 C65 E07 ⋄

1970
MORLEY, M.D. *The number of countable models*
⋄ C15 C75 ⋄
MOSCHOVAKIS, Y.N. *The Suslin-Kleene theorem for countable structures* ⋄ C15 C40 D55 D70 D75 ⋄
MUSTAFIN, T.G. & TAJMANOV, A.D. *Countable models of theories which are categorical in power $\aleph_1$ but not in power $\aleph_0$ (Russian)* ⋄ C15 C35 ⋄
REYES, G.E. *Local definability theory*
⋄ C07 C15 C40 C50 C55 C75 ⋄
SUZUKI, Y. *Orbits of denumerable models of complete theories* ⋄ C15 C35 C52 C62 ⋄

1971
CARPINTERO ORGANERO, P. *Numero de tipos diferentes de algebras de Boole de cardinal m infinito (English summary)* ⋄ C15 G05 ⋄
HENSON, C.W. *A family of countable homogeneous graphs* ⋄ C13 C15 C50 C65 ⋄
PABION, J.F. *Etude des bases dans les modeles des theories $\aleph_1$-categoriques* ⋄ C15 C35 C50 ⋄
ROSENSTEIN, J.G. *A note on a theorem of Vaught*
⋄ C15 C35 C50 ⋄

1972
BLASS, A.R. *Theories without countable models*
⋄ C15 E50 ⋄
BOFFA, M. & PRAAG VAN, P. *Sur les sous-champs maximaux des corps generiques denombrables*
⋄ C15 C25 C60 ⋄
HENSON, C.W. *Countable homogeneous relational structures and $\aleph_0$-categorical theories*
⋄ C10 C15 C35 C50 D35 ⋄
MOTOHASHI, N. *Countable structures for uncountable infinitary languages* ⋄ C15 C40 C75 ⋄
POUZET, M. *Modele universel d'une theorie n-complete*
⋄ C15 C35 C50 ⋄
POUZET, M. *Modele universel d'une theorie n-complete: modele uniformement prehomogene*
⋄ C15 C35 C50 ⋄
POUZET, M. *Modele universel d'une theorie n-complete: modele prehomogene* ⋄ C15 C35 C50 ⋄

1973
BALDWIN, JOHN T. & BLASS, A.R. & GLASS, A.M.W. & KUEKER, D.W. *A "natural" theory without a prime model* ⋄ C15 C50 C62 C65 ⋄
BARWISE, J. *Back and forth through infinitary logics*
⋄ C15 C75 ⋄
BURRIS, S. *Scott sentences and a problem of Vaught for mono-unary algebras* ⋄ C05 C15 C75 ⋄

FRIEDMAN, H.M. *Countable models of set theories*
◇ C15 C62 E45 F35 H20 ◇

LACHLAN, A.H. *On the number of countable models of a countable superstable theory* ◇ C15 C45 ◇

MORLEY, M.D. *Countable models with standard part*
◇ C15 C50 C52 ◇

PODEWSKI, K.-P. *Zur Vergroesserung von Strukturen*
◇ C15 C30 ◇

VAUGHT, R.L. *Descriptive set theory in* $L_{\omega_1,\omega}$
◇ C15 C70 C75 D55 D70 E15 ◇

1974

BALDWIN, JOHN T. & PLOTKIN, J.M. *A topology for the space of countable models of a first order theory*
◇ C15 ◇

BARWISE, J. *Mostowski's collapsing function and the closed unbounded filter* ◇ C15 C55 C75 E47 ◇

BENDA, M. *Remarks on countable models*
◇ C15 C50 ◇

CUSIN, R. *The number of countable generic models for finite forcing* ◇ C15 C25 ◇

KHISAMIEV, N.G. *Strongly constructive models of a decidable theory (Russian)* ◇ C15 C57 F60 ◇

LASCAR, D. *Sur les theories convexes "modeles completes"*
◇ C15 C35 C52 ◇

MYERS, D.L. *The back-and-forth isomorphism construction* ◇ C07 C15 G15 G30 ◇

NADEL, M.E. *Scott sentences and admissible sets*
◇ C15 C65 C70 C75 ◇

ONYSZKIEWICZ, J. *Dimensions of PC space of models (Russian summary)* ◇ C15 C52 C85 ◇

RUBIN, M. *Theories of linear order*
◇ C15 C35 C50 C65 E07 ◇

SIMPSON, S.G. *Forcing and models of arithmetic*
◇ C15 C25 C62 E40 E47 ◇

1975

CHIKHACHEV, S.A. *Generic models of countable theories (Russian)* ◇ C15 C25 ◇

FRIEDMAN, H.M. *Large models of countable height*
◇ C15 C55 C62 E25 E40 ◇

LAKE, J. *Characterising the largest, countable partial ordering* ◇ C15 C50 C65 E07 ◇

MALITZ, J. *Complete theories with countably many rigid nonisomorphic models* ◇ C15 C35 C50 ◇

METAKIDES, G. & PLOTKIN, J.M. *An algebraic characterization of power set in countable standard models of ZF* ◇ C15 C62 E35 G05 ◇

SIMMONS, H. *Counting countable e.c. structures*
◇ C15 C25 C50 C98 ◇

1976

BARWISE, J. & SCHLIPF, J.S. *An introduction to recursively saturated and resplendent models*
◇ C15 C40 C50 C57 ◇

HARNIK, V. & MAKKAI, M. *Applications of Vaught sentences and the covering theorem*
◇ C15 C40 C45 C50 C52 C75 D70 E15 ◇

HUTCHINSON, J.E. *Elementary extensions of countable models of set theory* ◇ C15 C62 ◇

LABLANQUIE, J.-C. *Le nombre des structures generiques d'une propriete de "forcing" (English summary)*
◇ C15 C25 ◇

LASCAR, D. *Ranks and definability in superstable theories*
◇ C15 C45 ◇

MARCUS, L. *Elementary separation by elementary expansion* ◇ C15 C50 ◇

NADEL, M.E. *On models* $\equiv_{\infty\omega}$ *to an uncountable model*
◇ C15 C70 C75 ◇

SOPRUNOV, S.F. *Denumerable non-standard models of arithmetic (Russian)* ◇ C15 C62 ◇

WOODROW, R.E. *A note on countable complete theories having three isomorphism types of countable models*
◇ C15 C35 ◇

1977

CUTLAND, N.J. *Some theories having countably many countable models* ◇ C15 C35 ◇

DAWES, A.M. *End extensions which are models of a given theory* ◇ C07 C15 C30 C62 ◇

DEISSLER, R. *Minimal models*
◇ C15 C40 C50 C75 ◇

DROBOTUN, B.N. *On countable models of decidable almost categorical theories (Russian)*
◇ B25 C15 C35 C57 ◇

DUBIEL, M. *Generalized quantifiers and elementary extensions of countable models* ◇ C15 C65 C80 ◇

FRAISSE, R. *Deux relations denombrables, logiquement equivalentes pour le second ordre, sont isomorphes*
◇ B15 C15 C85 E07 E45 ◇

GRANT, P.W. *Strict* Π_1^1*-predicates on countable and cofinality* ω *transitive sets*
◇ C15 C40 C70 D55 D60 D70 E47 E60 ◇

HARNIK, V. & MAKKAI, M. *A tree argument in infinitary model theory* ◇ C15 C75 ◇

KNIGHT, J.F. *Skolem functions and elementary embeddings* ◇ C15 C30 ◇

KUEKER, D.W. *Countable approximations and Loewenheim-Skolem theorems*
◇ C15 C55 C75 E05 ◇

LABLANQUIE, J.-C. *Le nombre de classes d'equivalence elementaire des modeles generiques d'une propriete de "forcing" (English summary)* ◇ C15 C25 C35 ◇

LABLANQUIE, J.-C. *Proprietes de forcing et modeles generiques* ◇ C15 C25 C70 ◇

LANDRAITIS, C.K. *Definability in well quasi-ordered sets of structures* ◇ C15 C65 C75 ◇

LASCAR, D. *Generalisation de l'ordre de Rudin-Keisler aux types d'une theorie* ◇ C15 C35 C45 E05 ◇

LASCAR, D. *Sur la hauteur des structures denombrables (English summary)* ◇ C15 C50 C75 ◇

LITMAN, A. & SHELAH, S. *Models with few isomorphic expansions* ◇ C15 C35 C50 C55 ◇

MAKKAI, M. *An "admissible" generalization of a theorem on countable* Σ_1^1 *sets of reals with applications*
◇ C15 C40 C50 C70 D55 E15 ◇

SCHLIPF, J.S. *A guide to the identification of admissible sets above structures* ◇ C15 C50 C70 ◇

SCHOENFELD, W. *Eine algebraische Konstruktion abzaehlbarer Modelle* ◇ C07 C15 ◇

SHELAH, S. *Existentially-closed groups in* $\aleph_1$ *with special properties* ◇ C15 C25 C60 C75 ◇

SNAPPER, E. *Omitting models* ⋄ C07 C15 C50 ⋄

ZARACH, A. *Extension of ZF-models to models with the scheme of choice* ⋄ C15 C62 E25 E30 E35 ⋄

1978

KETONEN, J. *The structure of countable boolean algebras* ⋄ C15 G05 ⋄

KNIGHT, J.F. *An inelastic model with indiscernibles* ⋄ C15 C30 C50 ⋄

LIPSHITZ, L. & NADEL, M.E. *The additive structure of models of arithmetic* ⋄ C15 C50 C57 C62 ⋄

MILLER, DOUGLAS E. *The invariant Π^0_α separation principle* ⋄ C15 C70 C75 D55 E15 ⋄

MLCEK, J. *End-extensions of countable structures and the induction schema* ⋄ C15 C62 F30 ⋄

PILLAY, A. *Number of countable models* ⋄ C15 C50 ⋄

SCHLIPF, J.S. *Toward model theory through recursive saturation* ⋄ C15 C40 C50 C57 C70 ⋄

SCHMERL, J.H. *$\aleph_0$-categoricity and comparability graphs* ⋄ C15 C35 C65 ⋄

SHELAH, S. *End extensions and numbers of countable models* ⋄ C15 C30 C55 C62 E50 ⋄

SHELAH, S. *On the number of minimal models* ⋄ C15 C50 ⋄

STEEL, J.R. *On Vaught's conjecture* ⋄ C15 C65 C70 C75 ⋄

TAJMANOV, A.D. *On topologizing countable algebras (Russian)* ⋄ C05 C15 C65 ⋄

WOODROW, R.E. *Theories with a finite number of countable models* ⋄ C15 ⋄

1979

AJTAI, M. *Isomorphism and higher order equivalence* ⋄ C15 C55 E35 E45 ⋄

CZELAKOWSKI, J. *A purely algebraic proof of the omitting types theorem* ⋄ C07 C15 C30 G99 ⋄

CZELAKOWSKI, J. *A remark on countable algebraic models* ⋄ C15 C20 G05 G99 ⋄

LASCAR, D. *La conjecture de Vaught* ⋄ C15 C50 C75 ⋄

LASCAR, D. *Les modeles denombrables d'une theorie superstable ayant des fonctions de Skolem (English summary)* ⋄ C15 C45 C52 ⋄

LIPSHITZ, L. *Diophantine correct models of arithmetic* ⋄ C15 C62 H15 ⋄

MAKKAI, M. *There are atomic Nadel structures* ⋄ C15 C70 ⋄

SACKS, G.E. *Effective bounds on Morley rank* ⋄ C15 C35 C45 C70 ⋄

SCHMERL, J.H. *Countable homogeneous partially ordered sets* ⋄ C15 C50 C65 E07 ⋄

1980

BAJZHANOV, B.S. & OMAROV, B. *On omitting models (Russian) (Kazakh summary)* ⋄ C15 C50 ⋄

HAJEK, P. & PUDLAK, P. *Two orderings of the class of all countable models of Peano arithmetic* ⋄ C15 C62 ⋄

HERRMANN, E. & WOLTER, H. *Untersuchungen zu schwachen Logiken der zweiten Stufe* ⋄ B15 C15 C85 ⋄

KIERSTEAD, H.A. *Countable models of ω_1-categorical theories in admissible languages* ⋄ C15 C35 C70 ⋄

KUDAJBERGENOV, K.ZH. *On constructive models of undecidable theories (Russian)* ⋄ C15 C35 C57 D35 ⋄

KULUNKOV, P.A. *Questions of definability in ordered sets (Russian) (English summary)* ⋄ C15 C65 E07 ⋄

LACHLAN, A.H. & WOODROW, R.E. *Countable ultrahomogeneous undirected graphs* ⋄ C15 C50 C65 ⋄

LUO, LIBO *The union and product of models, and homogeneous models (Chinese) (English summary)* ⋄ C15 C30 C50 ⋄

LYNCH, J.F. *Almost sure theories* ⋄ C13 C15 C30 ⋄

MARCUS, L. *The number of countable models of a theory of one unary function* ⋄ C15 ⋄

PARIS, J.B. *A hierarchy of cuts in models of arithmetic* ⋄ C15 C62 F30 ⋄

PERETYAT'KIN, M.G. *Theories with three countable models (Russian)* ⋄ C15 ⋄

PILLAY, A. *Instability and theories with few models* ⋄ C15 C45 C50 ⋄

PILLAY, A. *Theories with exactly three countable models and theories with algebraic prime models* ⋄ C15 C50 ⋄

RUBIN, M. *On the automorphism groups of countable boolean algebras* ⋄ C07 C15 C85 E45 G05 ⋄

SCHMERL, J.H. *Decidability and $\aleph_0$-categoricity of theories of partially ordered sets* ⋄ B25 C15 C35 C65 D35 ⋄

1981

CLOTE, P. *A note on decidable model theory* ⋄ B25 C15 C50 C57 D30 ⋄

GIVANT, S. *The number of nonisomorphic denumerable models of certain universal Horn classes* ⋄ C05 C15 C35 ⋄

KOGALOVSKIJ, S.R. & SOLDATOVA, V.V. *Remarks on countably local classes (Russian)* ⋄ C15 C52 ⋄

LASCAR, D. *Les modeles denombrables d'une theorie ayant des fonctions de Skolem (English summary)* ⋄ C15 C45 C50 ⋄

MILLAR, T.S. *Stability, complete extensions, and the number of countable models* ⋄ C15 C35 C45 ⋄

MILLAR, T.S. *Vaught's theorem recursively revisited* ⋄ C15 C57 D45 ⋄

MILLER, A.W. *Vaught's conjecture for theories of one unary operation* ⋄ C15 C75 ⋄

MILLER, DOUGLAS E. *The metamathematics of model theory: Discovering language in action* ⋄ C15 E15 ⋄

MUNDICI, D. *An algebraic result about soft model theoretical equivalence relations with an application to H. Friedman's fourth problem* ⋄ C15 C40 C95 ⋄

MUSTAFIN, T.G. *On the number of countable models of a countable complete theory (Russian)* ⋄ C15 C45 ⋄

RABINOVICH, E.B. *Inductive limits of symmetric groups and universal groups (Russian) (English summary)* ⋄ C15 C50 C60 ⋄

SAFFE, J. *Einige Ergebnisse ueber die Anzahl abzaehlbarer Modelle superstabiler Theorien (Dissertation)* ⋄ C15 C45 ⋄

1982

DOBBERTIN, H. *On Vaught's criterion for isomorphisms of countable boolean algebras* ◇ C05 C15 G05 ◇

KUDAJBERGENOV, K.ZH. *A remark on omitting models (Russian)* ◇ C15 C50 ◇

MOROZOV, A.S. *Countable homogeneous boolean algebras (Russian)* ◇ C15 C50 C57 G05 ◇

PILLAY, A. *Dimension theory and homogeneity for elementary extensions of a model* ◇ C15 C45 C50 ◇

PILLAY, A. *Weakly homogeneous models*
◇ C15 C45 C50 ◇

SAMI, R.L. *Sur la conjecture de Vaught en theorie descriptive (English summary)* ◇ C15 E15 ◇

SARACINO, D.H. & WOOD, C. *QE nil-2 groups of exponent 4* ◇ C10 C15 C60 ◇

SMORYNSKI, C.A. *Back-and-forth inside a recursively saturated model of arithmetic*
◇ C15 C50 C57 C62 ◇

WAGNER, C.M. *On Martin's conjecture*
◇ C15 C35 C45 C65 C75 ◇

WILSON, J.S. *The algebraic structure of $\aleph_0$-categorical groups* ◇ C15 C35 C60 ◇

1983

BOUSCAREN, E. & LASCAR, D. *Countable models of nonmultidimensional $\aleph_0$-stable theories*
◇ C15 C45 C50 ◇

BOUSCAREN, E. *Countable models of multidimensional $\aleph_0$-stable theories* ◇ C15 C45 C50 ◇

EISENMENGER, H. *Some local definability results on countable topological structures* ◇ C15 C40 C90 ◇

KNIGHT, J.F. & WOODROW, R.E. *A complete theory with arbitrarily large minimality ranks*
◇ C15 C40 C50 ◇

LASCAR, D. *Les corps differentiellement clos denombrables* ◇ C15 C45 C60 ◇

LUO, LIBO *On the number of countable homogeneous models* ◇ C15 C50 ◇

MIJAJLOVIC, Z. *On Σ_1^0-extensions of ω*
◇ C15 C30 C62 ◇

MILLAR, T.S. *Persistently finite theories with hyperarithmetic models* ◇ C15 C50 C57 D30 ◇

MILLER, A.W. *On the Borel classification of the isomorphism class of a countable model*
◇ C15 D55 E15 ◇

OMAROV, B. *A question of V.Harnik (Russian)*
◇ C15 C35 C50 ◇

OMAROV, B. *Nonessential extensions of complete theories (Russian)* ◇ C15 C35 ◇

PILLAY, A. *A note on tight stable theories*
◇ C15 C35 C45 ◇

PILLAY, A. *Countable models of stable theories*
◇ C15 C45 ◇

SACKS, G.E. *On the number of countable models*
◇ C15 C70 C75 ◇

STEINHORN, C.I. *A new omitting types theorem*
◇ C07 C15 C45 C50 C57 ◇

1984

BOUSCAREN, E. *Martin's conjecture for ω-stable theories*
◇ C15 C45 ◇

GIORGETTA, D. & SHELAH, S. *Existentially closed structures in the power of the continuum*
◇ C15 C25 C30 C52 C55 C60 C75 ◇

HARRINGTON, L.A. & MAKKAI, M. & SHELAH, S. *A proof of Vaught's conjecture for ω-stable theories*
◇ C15 C35 C45 ◇

KAUFMANN, M. *On expandability of models of arithmetic and set theory to models of weak second-order theories*
◇ C15 C30 C62 C70 C75 E30 E70 ◇

LACHLAN, A.H. *Binary homogeneous structures. I*
◇ C10 C13 C15 C45 C50 C57 ◇

LACHLAN, A.H. *Countable homogeneous tournaments*
◇ C15 C50 C65 ◇

LACHLAN, A.H. *On countable stable structures which are homogeneous for a finite relational language*
◇ C10 C13 C15 C45 C50 ◇

LACHLAN, A.H. & SHELAH, S. *Stable structures homogeneous for a binary language*
◇ C10 C13 C15 C45 C50 ◇

MILLAR, T.S. *Decidability and the number of countable models* ◇ C15 C57 ◇

PILLAY, A. *Countable modules*
◇ C15 C35 C45 C52 C60 ◇

SCHMERL, J.H. *$\aleph_0$-categorical partially ordered sets (French summary)*
◇ B25 C15 C35 C65 C98 E07 ◇

1985

CHIKHACHEV, S.A. *An example of an atomic model which is not algebraically prime (Russian)* ◇ C15 C50 ◇

FRAISSE, R. *Deux relations denombrables, logiquement equivalentes pour le second ordre, sont isomorphes (modulo un axiome de constructibilite)*
◇ B15 C15 C85 E45 ◇

HUYNH, D.T. *The complexity of equivalence problems for commutative grammars* ◇ C05 C15 ◇

KOTLARSKI, H. *Bounded induction and satisfaction classes* ◇ C15 C50 C57 C62 ◇

LASCAR, D. *Why some people are excited by Vaught's conjecture* ◇ C15 C70 C75 ◇

MILLAR, T.S. *Decidable Ehrenfeucht theories*
◇ B25 C15 C57 C98 D30 D35 D55 ◇

PILLAY, A. & STEINHORN, C.I. *A note on nonmultidimensional superstable theories*
◇ C15 C45 ◇

TSUBOI, A. *On theories having a finite number of nonisomorphic countable models* ◇ C15 C35 C45 ◇

TUSCHIK, H.-P. *Algebraic connections between definable predicates* ◇ C15 C30 C40 ◇

C20 Ultraproducts and related constructions

1933
SKOLEM, T.A. *Ueber die Unmoeglichkeit einer Charakterisierung der Zahlenreihe mittels eines endlichen Axiomensystems*
⋄ B28 C20 C62 F30 H15 ⋄

1934
SKOLEM, T.A. *Ueber die Nichtcharakterisierbarkeit der Zahlenreihe mittels endlich oder abzaehlbar unendlich vieler Aussagen mit ausschliesslich Zahlenvariablen*
⋄ B28 C20 C62 F30 H15 ⋄

1948
HEWITT, E. *Rings of real-valued continuous functions*
⋄ C20 C50 C60 C65 E75 ⋄

1955
FRAISSE, R. *Sur quelques classifications des relations basees sur des isomorphismes restreints I: Etude generale. II: Applications aux relations d'ordre, et construction d'exemples montrant que ces classifications sont distinctes* ⋄ C07 C20 C65 E07 ⋄
LOS, J. *Quelques remarques, theoremes et problemes sur les classes definissables d'algebres*
⋄ C05 C20 C30 C52 ⋄
SKOLEM, T.A. *Peano's axioms and models of arithmetic*
⋄ B28 C20 C62 F30 H15 ⋄

1959
FEFERMAN, S. & VAUGHT, R.L. *The first order properties of products of algebraic systems*
⋄ B25 C10 C20 C30 G05 ⋄
RABIN, M.O. *Arithmetical extensions with prescribed cardinality* ⋄ C20 C30 C55 ⋄

1961
KEISLER, H.J. *Ultraproducts and elementary classes*
⋄ C20 C40 C50 C52 C55 E05 E50 ⋄
KOCHEN, S. *Ultraproducts in the theory of models*
⋄ C20 C35 C52 C60 ⋄
MACDOWELL, R. & SPECKER, E. *Modelle der Arithmetik*
⋄ B28 C20 C55 C62 ⋄
ROBINSON, A. *On the construction of models*
⋄ C07 C20 C30 E05 ⋄
SCOTT, D.S. *Measurable cardinals and constructible sets*
⋄ C20 E45 E55 ⋄
SCOTT, D.S. *On constructing models for arithmetic*
⋄ C20 C57 C62 H15 ⋄

1962
CHANG, C.C. & KEISLER, H.J. *Applications of ultraproducts of pairs of cardinals to the theory of models* ⋄ C20 C55 E05 E50 ⋄
CHANG, C.C. & KEISLER, H.J. *Model theories with truth values in a uniform space* ⋄ B50 C20 C50 C90 ⋄

FRAYNE, T.E. & MOREL, A.C. & SCOTT, D.S. *Reduced direct products* ⋄ B20 C20 C55 ⋄
KEISLER, H.J. *Some applications of the theory of models to set theory* ⋄ C20 C55 E55 ⋄
LUXEMBURG, W.A.J. *A remark on a paper by N.G.de Bruijn and P.Erdoes* ⋄ C07 C20 E05 E25 ⋄
LUXEMBURG, W.A.J. *Nonstandard analysis. Lectures on A. Robinson's theory of infinitesimals and infinitely large numbers* ⋄ C20 H05 H98 ⋄
LUXEMBURG, W.A.J. *Two applications of the method of construction by ultrapowers to analysis*
⋄ C20 E05 E75 H05 ⋄
TAJMANOV, A.D. *On a theorem of Beth and Kochen (Russian)* ⋄ C20 C40 C52 ⋄
TAJMANOV, A.D. *The characteristics of axiomatizable classes of models (Russian)* ⋄ C20 C52 ⋄
VOPENKA, P. *A method of constructing a nonstandard model of the Bernays-Goedel axiomatic theory of sets (Russian)* ⋄ C20 C62 E30 E70 H20 ⋄
VOPENKA, P. *Construction of models for set theory by the method of ultra products (Russian) (German summary)*
⋄ C20 C62 E05 E30 E55 E70 ⋄

1963
FLEISCHER, I. *Elementary properties of inductive limits*
⋄ C05 C20 C30 ⋄
KEISLER, H.J. *Limit ultrapowers* ⋄ C20 C55 ⋄
KOGALOVSKIJ, S.R. *Quasiprojective classes of models (Russian)* ⋄ C20 C52 C60 ⋄
VOPENKA, P. *Construction of non-standard, non-regular models in set theory (Russian) (German summary)*
⋄ C20 C62 E30 E70 ⋄

1964
FENSTAD, J.E. *Model theory, ultraproducts and topology*
⋄ C20 C65 E75 ⋄
KEISLER, H.J. *Good ideals in fields of sets*
⋄ C20 C55 E05 ⋄
KEISLER, H.J. *On cardinalities of ultraproducts*
⋄ C20 C55 E05 ⋄
KEISLER, H.J. *Ultraproducts and saturated models*
⋄ C20 C50 C55 ⋄
KOGALOVSKIJ, S.R. *A remark on ultraproducts (Russian)*
⋄ C20 C52 ⋄
MAKKAI, M. *On PC_Δ-classes in the theory of models*
⋄ C20 C40 C52 ⋄
ROBINSON, A. *On generalized limits and linear functionals*
⋄ C20 H05 ⋄
VOPENKA, P. *Submodels of models of set theory (Russian) (German summary)* ⋄ C20 C62 E30 E55 E70 ⋄

1965

ALLING, N.L. *Rings of continuous integer-valued functions and nonstandard arithmetic* ⋄ C20 C60 H15 ⋄

ALMAGAMBETOV, ZH.A. *On classes of axioms closed with respect to the given reduced products and powers (Russian)* ⋄ B25 C20 C52 D35 ⋄

AX, J. & KOCHEN, S. *Diophantine problems over local fields I,II (II: A complete set of axioms for p-adic number theory)* ⋄ B25 C20 C35 C60 ⋄

KEISLER, H.J. *A survey of ultraproducts*
⋄ C20 C55 C98 E05 ⋄

KEISLER, H.J. *Ideals with prescribed degree of goodness*
⋄ C20 C55 E05 ⋄

KEISLER, H.J. *Limit ultraproducts* ⋄ C20 C55 E55 ⋄

KEISLER, H.J. *Reduced products and Horn classes*
⋄ B20 C20 C52 C55 ⋄

KOCHEN, S. *Topics in the theory of definition*
⋄ C20 C40 C52 ⋄

KOGALOVSKIJ, S.R. *Certain remarks on classes axiomatizable in a nonelementary way (Russian)*
⋄ C20 C52 C85 ⋄

KOGALOVSKIJ, S.R. *Some remarks on ultraproducts (Russian)* ⋄ C20 C52 E55 ⋄

MONK, J.D. *Model-theoretic methods and results in the theory of cylindric algebras* ⋄ C05 C20 C52 G15 ⋄

SCHEIN, B.M. *Relation algebras* ⋄ C05 C20 C52 ⋄

STONE, A.L. *Extensive ultraproducts and Haar measures*
⋄ C20 C65 E75 H05 ⋄

VOPENKA, P. *On ∇-model of set theory*
⋄ C20 C62 C90 E30 E35 E40 E70 ⋄

VOPENKA, P. *The limits of sheaves and applications on constructions of models*
⋄ C20 C62 C90 E25 E30 E35 E40 E70 ⋄

1966

AX, J. & KOCHEN, S. *Diophantine problems over local fields, III: Decidable fields*
⋄ B25 C10 C20 C35 C60 C65 ⋄

BUKOVSKY, L. & PRIKRY, K. *Some metamathematical properties of measurable cardinals*
⋄ C20 E47 E55 ⋄

CHANG, C.C. & KEISLER, H.J. *Continuous model theory*
⋄ B50 C05 C20 C50 C90 C98 ⋄

ENGELER, E. *On structures defined by mapping filters*
⋄ C20 E75 G30 ⋄

FRAISSE, R. *Une generalisation de l'ultraproduit*
⋄ C20 ⋄

HRBACEK, K. & VOPENKA, P. *On strongly measurable cardinals* ⋄ C20 E35 E45 E55 ⋄

OHKUMA, T. *Ultrapowers in categories*
⋄ C05 C20 G30 ⋄

WEGLORZ, B. *Equationally compact algebras I*
⋄ C05 C20 C50 ⋄

1967

AX, J. *Solving diophantine problems modulo every prime*
⋄ B25 C13 C20 C60 ⋄

BUNYATOV, M.R. *Dirac spaces (Russian)*
⋄ C20 C60 H05 H10 ⋄

CHANG, C.C. *Ultraproducts and other methods of constructing models* ⋄ C20 C30 ⋄

ENGELER, E. *On the structure of elementary maps*
⋄ C07 C20 ⋄

GAIFMAN, H. *Uniform extension operators for models and their applications* ⋄ C20 C30 C62 ⋄

GALVIN, F. *Reduced products, Horn sentences, and decision problems*
⋄ B20 B25 C20 C30 C40 D35 ⋄

KEISLER, H.J. *Ultraproducts which are not saturated*
⋄ C20 C50 C55 ⋄

KEISLER, H.J. *Ultraproducts of finite sets*
⋄ C13 C20 C55 E05 ⋄

KOGALOVSKIJ, S.R. *On compactness-preserving algebraic constructions (Russian)* ⋄ C20 C30 C52 ⋄

OMAROV, A.I. *Filtered products of models (Russian) (English summary)* ⋄ C20 C30 C40 C50 C52 ⋄

OMAROV, A.I. *On compact classes of models (Russian) (English summary)* ⋄ C20 C52 ⋄

ROBINSON, A. *Non-standard theory of Dedekind rings*
⋄ C20 C60 H15 H20 ⋄

ROBINSON, A. *Nonstandard arithmetic*
⋄ C20 C60 H15 H20 ⋄

TAJMANOV, A.D. *Automorphisms of an algebra having one unary operation in the signature (Russian)*
⋄ C07 C20 C40 ⋄

TAJMANOV, A.D. *Remarks to the Mostowski-Ehrenfeucht theorem (Russian) (English summary)*
⋄ C07 C20 C30 C52 E50 ⋄

TAJMANOV, A.D. *Systems with a solvable universal theory (Russian)* ⋄ B20 B25 C20 ⋄

VENNE, M. *Ultraproduits de structures d'ordres superieurs*
⋄ B15 C20 ⋄

WEGLORZ, B. *Equationally compact algebras III*
⋄ C05 C20 C50 ⋄

1968

AX, J. *The elementary theory of finite fields*
⋄ B25 B30 C13 C20 C60 ⋄

BARTOCCI, U. *Sulla generalizzazione di un teorema di Kochen (English summary)* ⋄ C20 C60 ⋄

BENDA, M. *Ultraproducts and non-standard logics (Russian summary)* ⋄ C20 C50 C75 ⋄

CHOQUET, G. *Construction d'ultrafiltres sur N*
⋄ C20 E05 E50 ⋄

CHOQUET, G. *Deux classes remarquables d'ultrafiltres sur N* ⋄ C20 E05 E50 ⋄

FELSCHER, W. *On criteria of definability* ⋄ C20 C52 ⋄

GEISER, J.R. *A result on limit ultra powers* ⋄ C20 ⋄

GEISER, J.R. *Nonstandard logic* ⋄ C20 H10 H15 ⋄

HAUSCHILD, K. & WOLTER, H. *Zur Charakterisierung von Theorien mit syntaktisch beschreibarer Auswahlfunktion* ⋄ C07 C20 ⋄

JONSSON, B. & OLIN, P. *Almost direct products and saturation* ⋄ C05 C20 C50 G05 ⋄

LINDSTROEM, P. *Remarks on some theorems of Keisler*
⋄ C07 C20 C30 ⋄

MYCIELSKI, J. & RYLL-NARDZEWSKI, C. *Equationally compact algebras II* ⋄ C05 C20 C50 C65 ⋄

OMAROV, A.I. *The $M(\alpha)$-class of models (Russian)*
⋄ C20 C50 ⋄

SLOMSON, A. *The monadic fragment of predicate calculus with the Chang quantifier and equality*
⋄ B25 C20 C55 C80 ⋄

WASZKIEWICZ, J. & WEGLORZ, B. *Some models of theories of reduced powers (Russian summary)* ◊ C20 ◊

WEGLORZ, B. *Limit generalized powers* ◊ C20 ◊

WOLTER, H. *Eine Erweiterung der klassischen Analysis* ◊ C20 H05 ◊

1969

BELL, J.L. & SLOMSON, A. *Models and ultraproducts: An introduction* ◊ C20 C98 ◊

BENDA, M. *Reduced products and nonstandard logics* ◊ C20 C50 C75 ◊

EKLOF, P.C. *Resolutions of singularities in prime characteristic for almost all primes* ◊ C20 C60 ◊

HOMMA, K. *Some results on saturated models* ◊ C20 C50 ◊

LOEWEN, K. *Some examples of ultraproducts (Polish and Russian summaries)* ◊ C20 ◊

LUXEMBURG, W.A.J. *A general theory of monads* ◊ B15 C20 H05 ◊

LUXEMBURG, W.A.J. *Reduced powers of the real number system and equivalents of the Hahn-Banach extension theorem* ◊ C20 E25 E75 H05 ◊

MARCJA, A. & MARONGIU, G. *Estensioni booleane di strutture relazionali (English summary)* ◊ C20 C30 G05 ◊

REILLY, N.R. *Some applications of wreath products and ultraproducts in the theory of lattice ordered groups* ◊ C20 C30 C60 ◊

RIBENBOIM, P. *Boolean powers* ◊ C20 C30 G05 ◊

ROBINSON, A. & ZAKON, E. *A set-theoretical characterization of enlargements* ◊ C20 C62 E75 H05 H20 ◊

SHAFAAT, A. *A remark on a paper of Taimanov* ◊ B25 C20 ◊

WASZKIEWICZ, J. & WEGLORZ, B. *On products of structures for generalized logics (Polish and Russian summaries)* ◊ B50 C20 C30 C90 ◊

WOJCIECHOWSKA, A. *Generalized limit powers (Russian summary)* ◊ C20 ◊

WOJCIECHOWSKA, A. *Limit reduced powers* ◊ C20 ◊

1970

BENDA, M. *Reduced products filters and Boolean ultrapowers* ◊ C20 ◊

BOOTH, D. *Ultrafilters on a countable set* ◊ C20 E05 ◊

CHAUBARD, A. *Ultraproduit de suites generiques et application* ◊ C20 C25 ◊

CONNES, A. *Ultrapuissances et applications dans le cadre de l'analyse non standard* ◊ C20 H05 ◊

GALVIN, F. *Horn sentences* ◊ B20 C20 C30 C40 D35 ◊

GUDJONSSON, H. *Remarks on limitultrapowers* ◊ C20 E05 ◊

KAISER, KLAUS *Zur Einbettungsbedingung im Zusammenhang mit einer Programmiersprache von Engeler* ◊ B75 C20 C35 C52 ◊

KAISER, KLAUS *Zur Theorie algorithmisch abgeschlossener Modellklassen* ◊ C20 C52 D75 ◊

KUNEN, K. & PARIS, J.B. *Boolean extensions and measurable cardinals* ◊ C20 C62 E35 E40 E50 E55 ◊

KUNEN, K. *Some applications of iterated ultrapowers in set theory* ◊ C20 C55 C62 E05 E35 E45 E50 E55 ◊

MANSFIELD, R. *The theory of boolean ultrapowers* ◊ C20 C90 G05 ◊

MARCJA, A. *Alcune questioni intorno alle ultraestensioni di strutture realizionali (English summary)* ◊ C20 G05 ◊

MARCJA, A. *Tipi di ultrafiltri in algebre di Boole complete con alcune applicazioni alle ultraestensioni* ◊ C20 G05 ◊

NAMBA, K. *On arithmetical extension operators* ◊ C20 E47 E55 ◊

NAMBA, K. *On arithmetical extension operators (Japanese)* ◊ C20 E47 E55 ◊

OLIN, P. *Elementary extensions within reduced powers* ◊ C20 ◊

OMAROV, A.I. *A certain property of Frechet filters (Russian)* ◊ C20 C30 C50 C52 ◊

OSSWALD, H. *Modelltheoretische Untersuchungen in der Kripke-Semantik* ◊ C20 C90 F50 ◊

PACHOLSKI, L. & RYLL-NARDZEWSKI, C. *On countably compact reduced products I* ◊ C20 C50 ◊

PACHOLSKI, L. *On countably compact reduced products II (Russian summary)* ◊ C20 C50 ◊

RASIOWA, H. *Ultraproducts of m-valued models and a generalization of the Loewenheim-Skolem-Goedel-Mal'cev theorem for theories based on m-valued logics (Russian summary)* ◊ B50 C20 C90 ◊

ROBINSON, A. *Elementary embeddings of fields of power series* ◊ C20 C60 H20 ◊

SABBAGH, G. *Aspects logiques de la purete dans les modules* ◊ C20 C30 C60 ◊

SHELAH, S. *On the cardinality of ultraproduct of finite sets* ◊ C13 C20 C55 E05 ◊

VOL'VACHEV, R.T. *Linear groups as axiomatizable classes of models (Russian)* ◊ C20 C52 C60 ◊

WASZKIEWICZ, J. *A remark on Engeler's filter-images* ◊ C20 ◊

1971

ADLER, A. *The cardinality of ultrapowers - an example* ◊ C20 E05 E55 ◊

ARMBRUST, M. & KAISER, KLAUS *Remarks on model classes acting on one another* ◊ C05 C20 C30 C52 ◊

AX, J. *A metamathematical approach to some problems in number theory* ◊ C20 C60 C98 ◊

BENDA, M. *On saturated reduced products* ◊ C20 C50 E05 ◊

CHAUBARD, A. *Puissances ultrafiltrees* ◊ C20 ◊

DAIGNEAULT, A. *Boolean powers in algebraic logic* ◊ C20 C30 G05 G15 ◊

DOETS, H.C. *A theorem on the existence of expansions (Russian summary)* ◊ C07 C20 E25 ◊

EKLOF, P.C. & SABBAGH, G. *Definability problems for modules and rings* ◊ C20 C52 C60 C75 ◊

ELLERMAN, D.P. *Sheaves or relation structures and ultraproducts* ◊ C20 C90 ◊

FELSCHER, W. *An algebraic approach to first order logic* ⋄ B10 C05 C07 C20 C90 G15 ⋄

GRZEGORCZYK, A. *An unfinitizability proof by means of restricted reduced power* ⋄ C20 C62 ⋄

HAUSCHILD, K. *Nichtaxiomatisierbarkeit von Satzmengen durch Ausdruecke spezieller Gestalt* ⋄ B28 C20 C40 C52 F30 ⋄

ITHIER, P. *Sur les formules stables par produit reduit suivant certains filtres* ⋄ C20 ⋄

LASSAIGNE, R. *Produits reduits et langages infinis* ⋄ C20 C75 E05 E55 ⋄

MAGIDOR, M. *There are many normal ultrafilters corresponding to a supercompact cardinal* ⋄ C20 C55 E05 E55 ⋄

MARCO DE, G. & RICHTER, M.M. *Rings of continuous functions with values in a non-archimedean ordered field* ⋄ C20 C60 C65 E75 ⋄

NADIU, G.S. *Elementary logic and models* ⋄ C07 C20 C30 ⋄

OMAROV, A.I. *Products with respect to an ω-universal filter (Russian) (Kazakh summary)* ⋄ C20 C50 G05 ⋄

PACHOLSKI, L. *On countably compact reduced products III* ⋄ C10 C20 C50 G05 ⋄

POTTHOFF, K. *Representation of locally finite polyadic algebras and ultrapowers* ⋄ C20 G15 ⋄

PURITZ, C.W. *Ultrafilters and standard functions in non-standard arithmetic* ⋄ C20 E05 H15 ⋄

RICHTER, M.M. *Limites in Kategorien von Relationalsystemen* ⋄ C20 C30 G30 ⋄

SABBAGH, G. *A note on the embedding property* ⋄ C20 C52 ⋄

SABBAGH, G. *On properties of countable character* ⋄ C20 C50 C52 ⋄

SHELAH, S. *Every two elementarily equivalent models have isomorphic ultrapowers* ⋄ C07 C20 E05 ⋄

SILVER, J.H. *Measurable cardinals and Δ_3^1 well-orderings* ⋄ C20 C30 D55 E05 E15 E35 E45 E55 ⋄

1972

ADLER, A. & JORGENSEN, M. *Descendingly incomplete ultrafilters and the cardinality of ultrapowers* ⋄ C20 E05 E50 ⋄

ADLER, A. *Representation of models of full theories* ⋄ C07 C20 ⋄

ARMBRUST, M. & KAISER, KLAUS *On quasi-universal model classes* ⋄ B15 C05 C20 C52 C85 ⋄

BENDA, M. *On reduced products and filters* ⋄ C20 C55 E05 E50 ⋄

BERNAYS, P. *Bemerkung ueber eine Ordnung der zahlentheoretischen Funktionen mittels eines additiven 0-1-Masses fuer die Menge der natuerlichen Zahlen* ⋄ C20 E05 E50 ⋄

BITTER-RUCKER VON, R. & ELLENTUCK, E. *Martin's axiom and saturated models* ⋄ C20 C50 E05 E50 ⋄

BOFFA, M. *Ultraproduits et application a l'algebre* ⋄ C20 C60 ⋄

BUECHI, J.R. & DANHOF, K.J. *Model theoretic approaches to definability* ⋄ C20 C40 ⋄

BUKOVSKY, L. *Models of set theory with axiom of constructibility which contain non-constructible sets* ⋄ C20 C62 E45 ⋄

CHAUBARD, A. *Sur les puissances ultrafiltrees* ⋄ C20 ⋄

CHERLIN, G.L. & HIRSCHFELD, J. *Ultrafilters and ultraproducts in non-standard analysis* ⋄ C20 E05 H05 ⋄

CUTLAND, N.J. *Σ_1-compactness and ultraproducts* ⋄ C20 C70 ⋄

DACUNHA-CASTELLE, D. & KRIVINE, J.-L. *Applications des ultraproduits a l'etude des espaces et des algebres de Banach* ⋄ C20 C65 E75 ⋄

DACUNHA-CASTELLE, D. *Applications des ultraproduits a la theorie des plongements des espaces de Banach* ⋄ C20 C65 E75 ⋄

DACUNHA-CASTELLE, D. *Ultraproduits d'espaces de Banach* ⋄ C20 C65 E75 ⋄

DACUNHA-CASTELLE, D. *Ultraproduits d'espaces L^p et d'espaces d'Orlicz* ⋄ C20 C65 E75 ⋄

FAKIR, S. *Categories localement coherentes et objets α-injectifs* ⋄ C20 G30 ⋄

FAKIR, S. & HADDAD, L. *Objets coherents et ultraproduits dans les categories* ⋄ C20 G30 ⋄

GABBAY, D.M. *Model theory for intuitionistic logic* ⋄ B55 C20 C90 F50 ⋄

GLASS, A.M.W. *An application of ultraproducts to lattice-ordered groups* ⋄ C20 C60 G10 ⋄

HIRSCHELMANN, A. *An application of ultra-products to prime rings with polynomial identities* ⋄ C20 C60 ⋄

JANSSEN, G. *Restricted ultraproducts of finite von Neumann algebras* ⋄ C20 C60 H05 ⋄

JARDEN, M. *Elementary statements over large algebraic fields* ⋄ C20 C60 ⋄

KASHIWAGI, T. *On (m, A)-implicational classes* ⋄ C20 C52 C75 ⋄

KETONEN, J. *On nonregular ultrafilters* ⋄ C20 C55 E05 E55 G05 ⋄

KOPFERMANN, K. *Ultraprodukte endlicher Koerper* ⋄ C13 C20 C60 ⋄

MALCOLM, W.G. *On a compactness theorem of A.Shafaat* ⋄ C20 C75 ⋄

MALCOLM, W.G. *Variations in definition of ultraproducts of a family of first order relational structures* ⋄ C20 ⋄

MANSFIELD, R. *Horn classes and reduced direct products* ⋄ C20 C52 C90 ⋄

MARCJA, A. *Una generalizzazione del concetto di prodotto diretto* ⋄ C20 C30 C40 G05 ⋄

MATHIAS, A.R.D. *Solution of problems of Choquet and Puritz* ⋄ C20 D55 E05 E15 ⋄

NEGREPONTIS, S. *Adequate ultrafilters of special boolean algebras* ⋄ C20 E05 G05 ⋄

OLIN, P. *$\aleph_0$-categoricity of two-sorted structures* ⋄ C20 C30 C35 C60 ⋄

OSDOL VAN, D.H. *Truth with respect to an ultrafilter or how to make intuition rigorous* ⋄ C20 H05 ⋄

PACHOLSKI, L. & WASZKIEWICZ, J. *On compact classes of models* ⋄ C20 C52 ⋄

POTTHOFF, K. *Ordnungseigenschaften von Nichtstandardmodellen* ⋄ C20 C62 H05 H15 ⋄

SCHREIBER, M. *Quelques remarques sur les caracterisations des espaces L^p, $0 \leq p < i$ (English summary)* ⋄ C20 C65 ⋄

SHELAH, S. *For what filters is every reduced product saturated?* ⋄ C20 C50 C55 ⋄

SHELAH, S. *Saturation of ultrapowers and Keisler's order* ⋄ C20 C35 C45 C50 C55 ⋄

WATANABE, T. *Ultraproducts and diophantine problems* ⋄ C20 C60 ⋄

WOLTER, H. *Ueber Mengen von Ausdruecken der Logik hoeherer Stufe, fuer die der Endlichkeitssatz und der Satz von Loewenheim-Skolem gelten* ⋄ B15 C20 C85 ⋄

WOLTER, H. *Untersuchung ueber Algebren und formalisierte Sprachen hoeherer Stufen (Russian, English and French summaries)* ⋄ B15 C20 C30 C40 C85 ⋄

1973

ADLER, A. & HAMILTON, J. *Invariant means via the ultrapower* ⋄ C20 H05 ⋄

ANDREKA, H. & NEMETI, I. *On the equivalence of sets definable by satisfaction and ultrafilters* ⋄ C07 C20 ⋄

EKLOF, P.C. *The structure of ultraproducts of abelian groups* ⋄ C20 C60 ⋄

FITTLER, R. *Categories with ultraproducts* ⋄ C20 G30 ⋄

KAISER, KLAUS *Diagonal-like embeddings* ⋄ C20 C52 ⋄

KAISER, KLAUS *On generalized projective classes* ⋄ C20 C52 ⋄

KAISER, KLAUS *On reductive and projective classes* ⋄ C20 C52 ⋄

KASAHARA, S. *A characterization of nonstandard real fields* ⋄ C20 C60 H05 ⋄

KFOURY, D. *Comparing algebraic structures up to algorithmic equivalence* ⋄ B75 C20 C35 ⋄

KOPFERMANN, K. *Ultraprodukte endlicher Koerper* ⋄ C13 C20 C60 ⋄

KUEKER, D.W. *A note on the elementary theory of finite abelian groups* ⋄ C13 C20 C60 ⋄

MALCOLM, W.G. *Application of higher-order ultraproducts to the theory of local properties in universal algebras and relational systems* ⋄ C05 C20 C85 E40 ⋄

MEISTERS, G.H. & MONK, J.D. *Construction of the reals via ultrapowers* ⋄ C20 H05 ⋄

MOSTOWSKI, ANDRZEJ *A contribution to teratology (Russian)* ⋄ C20 C62 ⋄

OHKUMA, T. *Finitary objects and ultrapowers* ⋄ C20 G30 ⋄

OMAROV, A.I. *On subsystems of reduced powers (Russian)* ⋄ C20 G05 ⋄

PACHOLSKI, L. *On countable universal boolean algebras and compact classes of models* ⋄ C20 C50 C52 G05 ⋄

REICHBACH, J. *Generalized models for intuitionistic and classical predicate calculi with ultraproducts* ⋄ C20 C25 C80 C90 F50 ⋄

ROBINSON, A. *Nonstandard points on algebraic curves* ⋄ C20 C60 H20 ⋄

SILVER, J.H. *The bearing of large cardinals on constructibility* ⋄ C20 C30 E05 E45 E55 ⋄

VASIL'EV, EH.S. *The elementary theories of complete torsion-free abelian groups with p-adic topology (Russian)* ⋄ B25 C20 C60 D35 ⋄

WASZKIEWICZ, J. *ω_0-categoricity of generalized products* ⋄ C20 C30 C35 ⋄

ZYGMUNT, J. *On the sources of the notion of the reduced product* ⋄ A10 C20 ⋄

1974

ABIAN, A. *Nonstandard models for arithmetics and analysis* ⋄ C20 H05 H15 ⋄

ADLER, A. *An application of elementary model theory to topological boolean algebras* ⋄ C20 G05 H20 ⋄

ARMBRUST, M. & KAISER, KLAUS *On some properties a projective model class passes on to the generated axiomatic class* ⋄ C20 C52 ⋄

ASH, C.J. *Reduced powers and boolean extensions* ⋄ B25 C20 C30 G20 ⋄

BENDA, M. *Note on boolean ultrapowers* ⋄ C20 C50 C90 ⋄

BLASS, A.R. *Ultrafilter mappings and their Dedekind cuts* ⋄ C20 E05 E50 ⋄

CHUDNOVSKY, G.V. *Transversals and properties of compactness type (Russian)* ⋄ C20 C75 E05 E55 ⋄

COMFORT, W.W. & NEGREPONTIS, S. *The theory of ultrafilters* ⋄ C20 C55 C98 E05 E55 E75 E98 G05 ⋄

ELLERMAN, D.P. *Sheaves of structures and generalized ultraproducts* ⋄ C20 C25 C90 G30 ⋄

FITTLER, R. *Saturated models of incomplete theories* ⋄ C20 C50 C52 ⋄

FLUM, J. *On Horn theories* ⋄ B20 C20 C80 ⋄

GAIFMAN, H. *Operations on relational structures, functors and classes I* ⋄ C20 C30 C40 G30 ⋄

JARDEN, M. *An injective rational map of an abstract variety into itself* ⋄ C20 C60 ⋄

KEISLER, H.J. & PRIKRY, K. *A result concerning cardinalities of ultraproducts* ⋄ C20 C55 E05 ⋄

KLINGEN, N. *Elementar aequivalente Koerper und ihre absolute Galoisgruppe* ⋄ C20 C60 H20 ⋄

MALCOLM, W.G. *Some results and algebraic applications in the theory of higher-order ultraproducts* ⋄ C20 C60 C85 G05 ⋄

NARENS, L. *Minimal conditions for additive conjoint measurement and qualitative probability* ⋄ C20 ⋄

PETRESCU, I. *A theorem of Loewenheim-Skolem type for topological models (Romanian) (English summary)* ⋄ C20 C90 ⋄

PIETSCH, A. *Ultraprodukte von Operatoren in Banachraeumen* ⋄ C20 C65 E75 H20 ⋄

POTTHOFF, K. *Boolean ultrapowers* ⋄ C20 C90 ⋄

SALIJ, V.N. *Reduced extensions of models (Russian)* ⋄ C20 C30 ⋄

SILVER, J.H. *Indecomposable ultrafilters and $O^{\#}$* ⋄ C20 E05 E45 E55 ⋄

SOLOVAY, R.M. *Strongly compact cardinals and the GCH* ⋄ C20 E05 E50 E55 E65 ⋄

SOPRUNOV, S.F. *On the power of a real-closed field (Russian) (English summary)* ⋄ C20 C60 D35 ⋄

SPIVAKOV, YU.L. *Algebraic extensions of the field of formal power series $F_D((t))$ (Russian) (Uzbek summary)* ⋄ C20 C60 H05 ⋄

STEPANEK, P. *Contributions to the theory of semisets. IV* ⋄ C20 C62 E35 E70 ⋄

STERN, J. *Sur certaines classes d'espaces de Banach caracterisees par des formules* ⋄ C20 C65 ⋄

WIERZEJEWSKI, J. *A note on stability and products (Russian summary)* ⋄ C20 C30 C45 ⋄

ZAKON, E. *A new variant of non-standard analysis* ⋄ C20 H05 ⋄

ZYGMUNT, J. *A note on direct products and ultraproducts of logical matrices* ⋄ B50 C05 C20 ⋄

1975

BELESOV, A.U. *α-atomic compactness of filtered products (Russian)* ⋄ C05 C20 C50 ⋄

BERTHIER, D. *Stability of non-model-complete theories; products, groups* ⋄ C20 C30 C45 C60 ⋄

BLOOM, S.L. *Some theorems on structural consequence operations* ⋄ B22 C05 C20 ⋄

BLOOM, S.L. & SUSZKO, R. *Ultraproducts of SCI models* ⋄ C20 ⋄

BROESTERHUIZEN, G. *A generalized Loś ultraproduct theorem* ⋄ C20 C80 ⋄

DACUNHA-CASTELLE, D. & KRIVINE, J.-L. *Sous-espaces de L^1* ⋄ C20 C65 E75 ⋄

DAGUENET, M. *Rapport entre l'ensemble des ultrafiltres admettant un ultrafiltre donne pour image et l'ensemble des images de cet ultrafiltre* ⋄ C20 E05 ⋄

DEHORNOY, P. *Solution d'une conjecture de Bukovsky (English summary)* ⋄ C20 C62 E40 E55 ⋄

FLEISSNER, W.G. *Lemma on measurable cardinals* ⋄ C20 C62 E45 E55 ⋄

FLUM, J. *L(Q)-preservation theorems* ⋄ C20 C40 C55 C80 E60 ⋄

GABBAY, D.M. *Model theory for tense logics* ⋄ B45 C20 C90 ⋄

GOLDBLATT, R.I. *First order definability in modal logic* ⋄ B45 C20 C90 ⋄

HALPERN, F. *Transfer theorems for topological structures* ⋄ C20 C40 C90 ⋄

HOWARD, P.E. *Los' theorem and the boolean prime ideal theorem imply the axiom of choice* ⋄ C20 E25 G05 ⋄

JARDEN, M. & KIEHNE, U. *The elementary theory of algebraic fields of finite corank* ⋄ B25 C10 C20 C60 ⋄

JORGENSEN, M. *Regular ultrafilters and long ultrapowers* ⋄ C20 E05 ⋄

KLINGEN, N. *Zur Idealstruktur in Nichtstandardmodellen von Dedekindringen* ⋄ C20 C60 H15 H20 ⋄

KOCHEN, S. *The model theory of local fields* ⋄ B25 C20 C35 C60 ⋄

MAGID, A.R. *Ultrafunctors* ⋄ C20 C60 ⋄

ROCKINGHAM GILL, R.R. *A note on the compactness theorem* ⋄ C07 C20 E25 ⋄

SANERIB JR., R.A. *Ultraproducts and elementary types of some groups related to infinite symmetric groups* ⋄ C20 C60 ⋄

SCHWARTZ, DIETRICH *Ultraprodukte in der Theorie der logischen Auswahlfunktionen* ⋄ B10 C20 ⋄

SOCHOR, A. & VOPENKA, P. *Contributions to the theory of semisets. V: On axiom of general collapse* ⋄ C20 C62 E35 E70 ⋄

SOCHOR, A. *Contributions to the theory of semisets. VI: Non-existence of the class of all absolute natural numbers* ⋄ C20 C62 E35 E70 ⋄

SOCHOR, A. *Real classes in the ultrapower of hereditarily finite sets* ⋄ C20 E70 H15 ⋄

STERN, J. *Proprietes locales et ultrapuissances d'espaces de Banach* ⋄ C20 C65 ⋄

VOLGER, H. *Logical categories, semantical categories and topoi* ⋄ C20 C85 G30 ⋄

VOLGER, H. *Ultrafilters, ultrapowers and finiteness in a topos* ⋄ C20 G30 ⋄

WEGLORZ, B. *Substructures of reduced powers* ⋄ C20 G05 ⋄

1976

ANDREKA, H. & DAHN, B.I. & NEMETI, I. *On a proof of Shelah (Russian summary)* ⋄ C07 C20 C90 ⋄

CUTLAND, N.J. *Compactness without languages* ⋄ C07 C20 ⋄

DEHORNOY, P. *Intersections d'ultrapuissances iterees de modeles de la theorie des ensembles (English summary)* ⋄ C20 C62 E25 E55 ⋄

GRANDY, R.E. *On the relation between free description theories and standard quantification theory* ⋄ B60 C20 ⋄

HENSON, C.W. *Ultraproducts of Banach spaces* ⋄ C20 C65 H05 ⋄

JARDEN, M. *Algebraically closed fields with distinguished subfields* ⋄ C20 C60 ⋄

JARDEN, M. *The elementary theory of ω-free Ax fields* ⋄ C20 C60 ⋄

JECH, T.J. & PRIKRY, K. *On ideals of sets and the power set operation* ⋄ C20 E05 E40 E50 E55 ⋄

KAISER, KLAUS *Quasi-axiomatic classes* ⋄ B15 C20 C52 C75 C85 ⋄

KASHIWAGI, T. *A generalizaton of a certain compactness property* ⋄ C20 C52 C75 ⋄

KASHIWAGI, T. *On certain compactness properties for classes of structures* ⋄ C20 C52 ⋄

KASHIWAGI, T. *On locally definable classes closed under reduced products* ⋄ C20 C75 ⋄

KEISLER, H.J. *Foundations of infinitesimal calculus* ⋄ C20 H05 H98 ⋄

KOPPELBERG, B. & KOPPELBERG, S. *A boolean ultrapower which is not an ultrapower* ⋄ C20 E05 ⋄

KRIVINE, J.-L. *Sous-espaces de dimension finie des espaces de Banach reticules* ⋄ C20 C65 ⋄

MANGANI, P. *Alcune questioni di teoria dei modelli per linguaggi predicativi generali (English summary)* ⋄ B50 C20 C90 ⋄

MURZIN, F.A. *Two arbitrary elementarily equivalent models of continuous logic have isomorphic ultra-powers (Russian)* ⋄ C20 C90 ⋄

NITA, A. *Quelques observations sur les ultraproduits d'anneaux* ⋄ C20 C60 ⋄

PACHOLSKI, L. *On limit reduced powers, saturatedness and universality* ⋄ C20 C50 ⋄

POTTHOFF, K. *A simple tree lemma and its application to a counterexample of Philips* ⋄ C20 C62 H15 ⋄

SAITO, M. *Ultraproducts and nonstandard analysis (Japanese)* ⋄ C20 H05 ⋄

SARACINO, D.H. *Existentially complete nilpotent groups* ⋄ C20 C25 C35 C60 ⋄

STERN, J. *Some applications of model theory in Banach space theory* ⋄ C20 C65 ⋄

STERN, J. *The problem of envelopes for Banach spaces* ⋄ C20 C65 ⋄

VOLGER, H. *The Feferman-Vaught theorem revisited* ⋄ C20 C30 C90 ⋄

WIERZEJEWSKI, J. *On stability and products* ⋄ C20 C30 C45 ⋄

1977

BANASCHEWSKI, B. & NELSON, EVELYN *Elementary properties of limit reduced powers with applications to boolean powers* ⋄ C05 C20 C30 C90 ⋄

BANKSTON, P. *Topological reduced products and the GCH* ⋄ C20 C65 E50 E75 ⋄

BANKSTON, P. *Ultraproducts in topology* ⋄ C20 C65 E75 ⋄

BUKOVSKY, L. *Iterated ultrapower and Prikry's forcing* ⋄ C20 C62 E40 E55 ⋄

CHADWICK, J.J.M. & WICKSTEAD, A.W. *A quotient of ultrapowers of Banach spaces and semi-Fredholm operators* ⋄ C20 C65 H05 ⋄

DACUNHA-CASTELLE, D. & KRIVINE, J.-L. *Sous-espaces de L^1* ⋄ C20 C65 E75 ⋄

DIARRA, B. *Ultra-produits de corps munis d'une valeur absolue ultra-metrique (English summary)* ⋄ C20 C60 ⋄

EKLOF, P.C. *Ultraproducts for algebraists* ⋄ C20 C60 C98 ⋄

JEHNE, W. & KLINGEN, N. *Superprimes and a generalized Frobenius symbol* ⋄ C20 C60 E50 E75 ⋄

KAMO, S. *Reduced powers and models for infinitely logic* ⋄ C20 C75 C90 ⋄

KLEINBERG, E.M. *Infinitary combinatorics and the axiom of determinateness* ⋄ C20 E05 E25 E55 E60 E98 ⋄

KUCIA, A. *On saturating ultrafilters on N* ⋄ C20 E05 E50 ⋄

MAKOWSKY, J.A. & TULIPANI, S. *Some model theory for monotone quantifiers* ⋄ C20 C40 C80 ⋄

MARCJA, A. *Prodotti booleani e prodotti booleani ridotti* ⋄ C20 C30 C90 ⋄

MARKOVIC, Z. *Reduced products of saturated intuitionistic theories* ⋄ C20 C90 F50 ⋄

PACHOLSKI, L. *Homogeneity, universality and saturatedness of limit reduced powers III* ⋄ C20 C50 ⋄

PACHOLSKI, L. *Saturated limit reduced powers* ⋄ C20 C50 C55 ⋄

PRESTEL, A. *An introduction to ultraproducts* ⋄ C20 C98 ⋄

RAKOV, S.A. *Ultraproducts and the "three spaces problem" (Russian)* ⋄ C20 E75 ⋄

SGRO, J. *Completeness theorems for topological models* ⋄ C20 C40 C75 C80 C90 ⋄

SGRO, J. *Maximal logics* ⋄ C20 C40 C80 C90 C95 ⋄

STERN, J. *Examples d'application de la theorie des modeles a la theorie des espaces de Banach* ⋄ C20 C65 ⋄

STUETZER, H. *Die elementare Formulierbarkeit von Dimension und Grad algebraischer Mannigfaltigkeiten* ⋄ C20 C60 ⋄

TAKEUCHI, Y. *Construction of nonstandard numbers from the rationals (Spanish)* ⋄ C20 H05 ⋄

VAEAENAENEN, J. *On the compactness theorem* ⋄ C20 C40 C95 ⋄

VETULANI, Z. *Some remarks on ultrapowers of extendable models of ZF* ⋄ C20 C62 E30 E70 H20 ⋄

WEGLORZ, B. *Homogeneity, universality and saturatedness of limit reduced powers II* ⋄ C20 C50 ⋄

WIERZEJEWSKI, J. *Homogeneity, universality and saturatedness of limit reduced powers. I* ⋄ C20 C50 ⋄

WILKIE, A.J. *On models of arithmetic having non-modular substructure lattices* ⋄ C20 C62 ⋄

1978

ANDREKA, H. & NEMETI, I. *Los lemma holds in every category* ⋄ C20 G30 ⋄

ANDREKA, H. & NEMETI, I. *Neat reducts of varieties* ⋄ C05 C20 G15 G25 G30 ⋄

BENDA, M. *Compactness for omitting of types* ⋄ C20 C30 C55 C70 C75 E55 ⋄

BLASS, A.R. *A model-theoretic view of some special ultrafilters* ⋄ C20 E05 H15 ⋄

BRYANT, R.M. & GROVES, J.R.J. *Wreath products and ultraproducts of groups* ⋄ C20 C30 C60 ⋄

BURRIS, S. & JEFFERS, E. *On the simplicity and subdirect irreducibility of boolean ultrapowers* ⋄ C05 C20 C50 ⋄

CHARRETTON, C. *Type d'ordre des ordinaux de modeles non denombrables de la theorie des ensembles* ⋄ C20 C50 C62 C70 E10 E50 H20 ⋄

COHEN, P.E. *Some continuity properties for ultraproducts* ⋄ C20 C75 ⋄

DEHORNOY, P. *Iterated ultrapowers and Prikry forcing* ⋄ C20 C62 E25 E40 E55 ⋄

EHRSAM, S. *L'ordre fondamental, le theoreme de la borne, le theoreme de la relation d'equivalence finie* ⋄ C20 C45 C98 ⋄

GARAVAGLIA, S. *Homology with equationally compact coeffcients* ⋄ C05 C20 ⋄

GARAVAGLIA, S. *Model theory of topological structures* ⋄ C20 C40 C65 C80 C90 ⋄

JOVANOVIC, A. *A note on ultrapower cardinality* ⋄ C20 C55 E05 E10 ⋄

KAISER, KLAUS *On relational selections for complete theories* ⋄ C20 C35 C50 C60 ⋄

KLEINBERG, E.M. *A combinatorial characterization of normal M-ultrafilters* ⋄ C20 C62 E05 E40 E55 ⋄

KOTLARSKI, H. *Some remarks on well-ordered models*
⋄ C20 C50 C55 E07 ⋄
KUEHNRICH, M. *On the Hermes term logic*
⋄ C07 C20 ⋄
LOLLI, G. *Alcune applicazioni della compattezza*
⋄ C20 C60 H05 ⋄
NEMETI, I. *From hereditary classes to varieties in abstract model theory and partial algebra*
⋄ C05 C20 C52 C90 C95 ⋄
POTTHOFF, K. *Orderings of types of countable arithmetic*
⋄ C20 C62 H15 ⋄
SOLOVAY, R.M. & REINHARDT, W.N. & KANAMORI, A. *Strong axioms of infinity and elementary embeddings*
⋄ C20 E35 E45 E55 E65 ⋄
STERN, J. *Ultrapowers and local properties of Banach spaces* ⋄ C20 C65 ⋄
TOKARENKO, A.I. *The group-theoretic properties N_0, N, and $\tilde{N}$ are not axiomatizable (Russian)*
⋄ C20 C52 C60 ⋄
VOJVODIC, G. *Some theorems for model theory of mixed-valued predicate calculi*
⋄ B50 C20 C35 C90 ⋄

1979

ANDREKA, H. & NEMETI, I. *Formulas and ultraproducts in categories* ⋄ C05 C20 G30 ⋄
ANDREKA, H. & NEMETI, I. *Not all representable cylindric algebras are neat reducts* ⋄ C20 G15 ⋄
ANDREKA, H. & MAKAI JR., E. & MARKI, L. & NEMETI, I. *Reduced products in categories* ⋄ C20 G30 ⋄
BANKSTON, P. *Note on "Ultraproducts in topology"*
⋄ C20 C65 E75 ⋄
BANKSTON, P. *Topological reduced products via good ultrafilters* ⋄ C20 C65 E35 E50 E75 ⋄
BECKER, J.A. & DENEF, J. & DRIES VAN DEN, L. & LIPSHITZ, L. *Ultraproducts and approximation in local rings. I* ⋄ C20 C60 ⋄
BERNAL, E. *On classes of groups closed under ultraproducts* ⋄ C20 C60 ⋄
BOYD, M.W. & TRANSUE, W.R. *Properties of ultraproducts* ⋄ C20 C65 ⋄
BUKOVSKY, L. *Any partition into Lebesgue measure zero sets produces a non-measurable set*
⋄ C20 E40 E75 ⋄
BURRIS, S. & WERNER, H. *Sheaf constructions and their elementary properties*
⋄ B25 C05 C20 C25 C30 C60 C90 ⋄
BUTTON, R.W. *When do two topologies have the same monads?* ⋄ C20 H05 ⋄
CZELAKOWSKI, J. *A remark on countable algebraic models*
⋄ C15 C20 G05 G99 ⋄
CZELAKOWSKI, J. *Large matrices which induce finite consequence operations* ⋄ B22 C20 ⋄
DAGUENET-TEISSIER, M. *Ultrafiltres a la facon de Ramsey* ⋄ C20 E05 E50 ⋄
DAHN, B.I. *Constructions of classical models by means of Kripke models (survey)*
⋄ C07 C20 C90 E35 E45 ⋄
ERSHOV, YU.L. *Relatively complemented, distributive lattices (Russian)* ⋄ C20 G05 G10 ⋄

GREITHER, C. *Struktur und Transzendenzgrad von $\eta_{\alpha+1}$-Koerpern und Ultrapotenzen von Koerpern*
⋄ C20 C60 ⋄
JECH, T.J. & PRIKRY, K. *Ideals over uncountable sets: Application of almost disjoint functions and generic ultrapowers* ⋄ C20 E05 E40 E50 E55 ⋄
KIEHNE, U. *Bounded products in the theory of valued fields* ⋄ C20 C60 ⋄
MARTIN, D.A. & MITCHELL, W.J. *On the ultrafilter of closed, unbounded sets*
⋄ C20 C62 E05 E35 E55 E60 ⋄
MIJAJLOVIC, Z. *Saturated boolean algebras with ultrafilters* ⋄ C20 C35 C50 G05 ⋄
NITA, A. *The behavior of ultraproducts of modules under passage to fractions (Romanian) (English summary)*
⋄ C20 C60 ⋄
NITA, A. *Ultraproducts of rings which retain factor properties (Romanian) (French summary)*
⋄ C20 C60 ⋄
OMAROV, A.I. *$\triangleleft_t$-minimality and the finite cover property (Russian)* ⋄ C20 C45 C50 ⋄
PELLER, V. *Applications of ultraproducts in operator theory. A simple proof of E. Bishop's theorem (Russian) (English summary)* ⋄ C20 C65 ⋄
POPESCU, D. *Algebraically pure morphisms*
⋄ C20 C60 ⋄
POTTHOFF, K. *Ultraproduits des groupes finis et applications a la theorie de Galois*
⋄ C13 C20 C60 ⋄
SHELAH, S. *Hanf number of omitting type for simple first-order theories* ⋄ C20 C50 C55 C75 ⋄
TOKARENKO, A.I. *Axiomatizability of the group-theoretic property $\bar{Z}$ (Russian)* ⋄ C20 C52 C60 ⋄
ZLATOS, P. *A characterization of some boolean powers*
⋄ C05 C20 C30 C75 ⋄

1980

BALL, R.N. *Cauchy completions are homomorphic images of submodels of ultrapowers* ⋄ C20 ⋄
BANASCHEWSKI, B. & NELSON, EVELYN *Boolean powers as algebras of continuous functions*
⋄ C05 C20 C30 C50 ⋄
BARONE, E. & GIANNONE, A. & SCOZZAFAVA, R. *On some aspects of the theory and applications of finitely additive probability measures* ⋄ C20 H05 ⋄
BERNARDI, C. *Lo spazio duale di un prodotto di algebre di Boole e le compattificazioni di Stone (English summary)*
⋄ C20 G05 ⋄
BLASS, A.R. *Conservative extensions of models of arithmetic* ⋄ C20 C62 E05 ⋄
COHN, P.M. *On semifir constructions*
⋄ C05 C20 C52 C60 ⋄
CZELAKOWSKI, J. *Reduced products of logical matrices*
⋄ B22 C05 C20 ⋄
DENEF, J. & LIPSHITZ, L. *Ultraproducts and approximation in local rings. II* ⋄ C20 C60 ⋄
FELGNER, U. *Horn-theories of Abelian groups*
⋄ C20 C60 ⋄
HEINRICH, S. *The isomorphic problem of envelopes*
⋄ C20 C65 H05 ⋄

HEINRICH, S. *Ultraproducts in Banach space theory*
  ◇ C20 C65 C98 H05 ◇
HIRSCHFELD, J. *Generalized ultrapowers* ◇ C20 ◇
HODGES, W. *Functorial uniform reducibility*
  ◇ C20 C40 C75 ◇
IMAZEKI, K. *Characterizations of compact operators and semi-Fredholm operators* ◇ C20 C65 ◇
JOVANOVIC, A. *A note on the two-cardinal problem*
  ◇ C20 C55 E10 E55 ◇
JOVANOVIC, A. *Ultraproducts of well orders*
  ◇ C20 C55 E05 E07 ◇
KOPPELBERG, B. *Ultrapowers and boolean ultrapowers of ω and ω_1* ◇ C20 E05 E50 E55 E65 ◇
KOPPELBERG, S. *Cardinalities of ultraproducts of finite sets* ◇ C13 C20 C55 E05 E10 ◇
KOTLARSKI, H. *On Skolem ultrapowers and their non-standard variant* ◇ C20 C62 E45 ◇
NISHIMURA, H. *A preservation theorem for tense logic*
  ◇ B45 C20 C40 C90 ◇
NITA, A. *Sur les ultraproduits des modules*
  ◇ C20 C60 ◇
NITA, A. *Ultraproducts for certain classes of rings (Romanian) (English summary)* ◇ C20 C60 ◇
PARMIGIANI BEDOGNA, D. *Le formule SH nelle potenze generalizzate ridotte (English summary)*
  ◇ C20 C40 G30 ◇
PRZYMUSINSKA, H. *Some results concerning v-valued infinitary predicate calculi (Russian summary)*
  ◇ C20 C75 ◇
RAUSZER, C. *An algebraic and Kripke-style approach to a certain extension of intuitionistic logic*
  ◇ B55 C20 C90 F50 G25 ◇
SGRO, J. *Ultraproduct invariant logics*
  ◇ C20 C40 C80 C95 ◇
SPECTOR, M. *A measurable cardinal with a nonwellfounded ultrapower* ◇ C20 E25 E35 E40 ◇
TAKAHASHI, MAKOTO *Topological powers and reduced powers* ◇ C20 C30 C90 ◇
TOFFALORI, C. *Sugli anelli nilreali (English summary)*
  ◇ C20 C25 C60 ◇
UROSU, C. *On filter products of congruences (Romanian) (French summary)* ◇ C05 C20 ◇

1981

ANTONOVSKIJ, M.YA. & CHUDNOVSKY, D.V. & CHUDNOVSKY, G.V. & HEWITT, E. *Rings of real-valued continuous functions II*
  ◇ C20 C50 C60 C65 E35 E50 E75 ◇
BLASS, A.R. *Some initial segments of the Rudin-Keisler ordering* ◇ C20 E05 E50 ◇
HEINRICH, S. *Ultraproducts of L_1-predual spaces*
  ◇ C20 C65 H05 ◇
HIEN, BUIHUY & NEMETI, I. *Problems with the category theoretic notions of ultraproducts* ◇ C20 G30 ◇
HODGES, W. & SHELAH, S. *Infinite games and reduced products* ◇ C20 C40 C75 E55 E60 ◇
HORIGUCHI, H. *A definition of the category of boolean-valued models* ◇ C20 C30 C90 G30 ◇
KARTTUNEN, M. *Model theoretic results for infinitely deep languages* ◇ C20 C75 ◇

KLEINBERG, E.M. *Producing measurable cardinals beyond κ* ◇ C20 C55 E05 E60 ◇
LAWRENCE, J. *Primitive rings do not form an elementary class* ◇ C20 C52 C60 ◇
LU, JINGBO & WANG, SHIQIANG *On the fundamental theorem of ultraproducts for lattice-valued models (Chinese)* ◇ C20 C90 ◇
MAREK, W. *On cores of iterated ultrapowers*
  ◇ C20 E45 E55 ◇
PACHOLSKI, L. *Homogeneous limit reduced powers*
  ◇ C20 C50 ◇
PELC, A. *Ideals on the real line and Ulam's problem*
  ◇ C20 E05 E40 ◇
RAYNAUD, Y. *Deux nouveaux exemples d'espaces de Banach stables (English summary)* ◇ C20 C65 ◇
RAYNAUD, Y. *Espaces de Banach superstables (English summary)* ◇ C20 C65 ◇
ROMO SANTOS, C. *Two methods for the resolution of singularities in any characteristic: ultraproducts and maximal contact (Spanish)* ◇ C20 C60 ◇
SOCHOR, A. & VOPENKA, P. *The axiom of reflection*
  ◇ C20 E65 E70 H20 ◇
STACHNIAK, Z. *Introduction to model theory for Lesniewski's ontology* ◇ C20 C90 E70 ◇

1982

ANDREKA, H. & NEMETI, I. *A general axiomatizability theorem formulated in terms of cone-injective subcategories* ◇ C05 C20 C90 G30 ◇
BANKSTON, P. *Some obstacles to duality in topological algebra* ◇ C05 C20 C65 G30 ◇
FACENDA AGUIRRE, J.A. *Ultraproducts of locally convex spaces (Spanish) (English summary)* ◇ C20 C65 ◇
GIANNONE, A. *Probabilita non σ-additive e analisi non-standard (English summary)* ◇ C20 H10 ◇
HEINRICH, S. & MANKIEWICZ, P. *Applications of ultrapowers to the uniform and Lipschitz classification of Banach spaces* ◇ C20 C65 H05 ◇
HEINRICH, S. *The isomorphic problem of envelopes*
  ◇ C20 C65 H05 ◇
HIEN, BUIHUY *On a problem in algebraic model theory*
  ◇ C20 G30 ◇
JENSEN, C.U. & LENZING, H. *Homological dimensions and representation type of algebras under base field extension* ◇ C20 C60 ◇
KARTTUNEN, M. *Model theoretic results for infinitely deep languages* ◇ C20 C75 ◇
KOCHEN, S. & KRIPKE, S.A. *Nonstandard models of Peano arithmetic* ◇ C20 C62 ◇
LAVER, R. *Saturated ideals and nonregular ultrafilters*
  ◇ C20 C55 E05 E55 E65 ◇
MALLIAVIN, M.P. *Ultra-produits d'algebre de Lie*
  ◇ C20 C65 ◇
MIRAGLIA, F. *On the preservation of elementary equivalence and embedding by bounded filtered powers and structures of stable continuous functions*
  ◇ C20 C40 ◇
MOSCATELLI, V.B. *Ultraprodotti e super-riflessivita in teoria degli spazi di Banach* ◇ C20 C65 ◇
NELSON, GEORGE C. *The periodic power of $\mathfrak{A}$ and complete Horn theories* ◇ C20 C30 C35 ◇

NEMETI, I. & SAIN, I. *Cone-implicational subcategories and some Birkhoff-type theorems*
◊ C05 C20 C90 G30 ◊

PACHOLSKI, L. & TOMASIK, J. *Reduced products which are not saturated* ◊ C20 C50 ◊

PETRY, A. *Une characterisation algebrique des structures satisfaisant les memes sentences stratifees*
◊ C20 C62 E70 ◊

RICHTER, M.M. *Ideale Punkte, Monaden und Nichtstandard-Methoden*
◊ C20 C60 E70 H05 H98 ◊

ROITMAN, J. *Non-isomorphic hyper-real fields from non-isomorphic ultrapowers* ◊ C20 H05 H20 ◊

ROMO SANTOS, C. *Ultraproducts of nonsingular varietes (Spanish)* ◊ C20 C60 ◊

SAIN, I. *On classes of algebraic systems closed with respect to quotients* ◊ C05 C20 C30 C52 ◊

SOCHOR, A. *Metamathematics of the alternative set theory II* ◊ C20 C62 E35 E70 ◊

TAHA, F. *Algebres simples centrales sur les corps ultra-produits de corps p-adiques* ◊ C20 C60 ◊

TOFFALORI, C. *Strutture esistenzialmente complete per certe classi di anelli (English summary)*
◊ C20 C25 C60 ◊

TULIPANI, S. *On classes of algebras with the definability of congruences* ◊ C05 C20 C52 ◊

VERDU I SOLANS, V. *The ultrafilter theorem and the adequacy theorem (Spanish) (English summary)*
◊ C07 C20 E25 ◊

YANG, SHOULIAN *Ultrafilters with chain bases and their applications (Chinese)* ◊ C20 E05 ◊

ZAYED, M. *Characterisation des algebres de representation finie sur des corps algebriquement clos* ◊ C20 C60 ◊

1983

ANDREKA, H. *Sharpening the characterization of the power of Floyd method* ◊ B75 C20 ◊

BANKSTON, P. *Obstacles to duality between classes of relational structures* ◊ C05 C20 E55 G30 ◊

BANKSTON, P. & FOX, R. *On categories of algebras equivalent to a quasivariety* ◊ C05 C20 G30 ◊

BANKSTON, P. *On productive classes of function rings*
◊ C05 C20 C65 E75 G30 ◊

DEHORNOY, P. *An application of ultrapowers to changing cofinality* ◊ C20 E40 E55 ◊

DEVLIN, K.J. *Reduced powers of $\aleph_2$-trees*
◊ C20 E05 E45 ◊

GUERRE, S. & LEVY, M. *Espaces l^p dans les sous-espaces de L^1* ◊ C20 C65 ◊

HEINRICH, S. & HENSON, C.W. & MOORE JR., L.C. *Elementary equivalence of L_1-preduals* ◊ C20 C65 ◊

HEINRICH, S. & MANKIEWICZ, P. *Some open problems in the nonlinear classification of Banach spaces*
◊ C20 C65 ◊

HEINRICH, S. *Ultrapowers of locally convex spaces and applications I* ◊ C20 C65 H10 ◊

HEINRICH, S. *Ultrapowers of locally convex spaces and applications II* ◊ C20 C65 H10 ◊

HERNANDEZ, R. *Espaces L^p, factorisation et produits tensoriels dans les espaces de Banach* ◊ C20 ◊

HIEN, BUIHUY & SAIN, I. *Category theoretic notions of ultraproducts* ◊ C20 C90 G30 ◊

HIEN, BUIHUY & SAIN, I. *In which categories are first-order axiomatizable hulls characterizable by ultraproducts?* ◊ C20 C90 G30 ◊

JOVANOVIC, A. *On the two-cardinal problem*
◊ C20 C55 E05 ◊

KARTTUNEN, M. *Model theoretic results for infinitely deep languages* ◊ C20 C75 ◊

KUEHNRICH, M. *On the Hermes term logic*
◊ B10 C07 C20 ◊

KUERSTEN, K.-D. *Local duality of ultraproducts of Banach lattices* ◊ C20 C65 ◊

MAGAJNA, B. *Infinitesimals (Slovenian) (English summary)* ◊ C20 H05 ◊

MAKOWSKY, J.A. & SHELAH, S. *Positive results in abstract model theory: a theory of compact logics*
◊ C20 C52 C95 E05 E55 ◊

MARKOVIC, Z. *On reduced products of Kripke models*
◊ C20 C90 ◊

MARONGIU, G. *Definibilita al primo ordine delle congruenze principali e delle congruenze compatte*
◊ C05 C20 C52 ◊

MIRAGLIA, F. *On the preservation of elementary equivalence and embedding by bounded filtered powers and structures of stable continuous functions*
◊ C20 C40 ◊

NELSON, GEORGE C. *Logic of reduced power structures*
◊ C20 ◊

ONO, H. *Model extension theorem and Craig's interpolation theorem for intermediate predicate logics*
◊ B55 C20 C40 C90 ◊

ROSEN, N.I. *A characterization of 2-square ultrafilters*
◊ C20 E05 ◊

SAIN, I. *Applicability of category-theoretic notions of ultraproducts (Hungarian)* ◊ C20 G30 ◊

WALD, B. *On κ-products modulo μ-products*
◊ C20 C60 E45 E75 ◊

XUEH, YUANGCHE *Lattices of width three generate a nonfinitely based variety* ◊ C20 G10 ◊

1984

ANDREKA, H. & NEMETI, I. *Importance of universal algebra for computer science*
◊ C05 C20 C95 G15 ◊

BENTHEM VAN, J.F.A.K. & PEARCE, D.A. *A mathematical characterization of interpretation between theories*
◊ C20 F25 ◊

CONTESSA, M. *Ultraproducts of pm-rings and mp-rings*
◊ C20 C60 G05 ◊

DIARRA, B. *Ultraproduits ultrametriques de corps values*
◊ C20 C60 ◊

GORBUNOV, V.A. *Axiomatizability of replete classes (Russian)* ◊ C05 C20 C52 ◊

HENLE, J.M. & KLEINBERG, E.M. & WATRO, R.J. *On the ultrafilters and ultrapowers of strong partition cardinals*
◊ C20 C55 E05 E60 ◊

HIEN, BUIHUY *Elementary classes in the injective subcategories approach to abstract model theory*
◊ C20 C90 G30 ◊

KANAMORI, A. & TAYLOR, A.D. *Separating ultrafilters on uncountable cardinals* ◇ C20 C55 E05 E55 ◇

KARTTUNEN, M. *Model theory for infinitely deep languages* ◇ C20 C75 ◇

KAUFMANN, M. & KRANAKIS, E. *Definable ultrapowers and ultrafilters over admissible ordinals* ◇ C20 C62 D60 E05 E45 ◇

KIRBY, L.A.S. *Ultrafilters and types on models of arithmetic* ◇ C20 C62 E05 H15 ◇

KRUPA, A. & ZAWISZA, B. *Applications of ultrapowers in analysis of unbounded selfadjoint operators* ◇ C20 C65 H10 ◇

KUERSTEN, K.-D. *Lokale Reflexivitaet und lokale Dualitaet von Ultraprodukten fuer halbgeordnete Banachraeume (English and Russian summaries)* ◇ C20 C65 ◇

LEVY, M. & RAYNAUD, Y. *Ultrapuissances des espaces $L^p(L^q)$ (English summary)* ◇ C20 C65 H10 H20 ◇

MATEESCU, C. & POPESCU, D. *Ultraproducts and big Cohen-Macaulay modules* ◇ C20 C60 ◇

NELSON, GEORGE C. *Boolean powers, recursive models, and the Horn theory of a structure* ◇ C05 C20 C30 C57 G05 ◇

OMAROV, A.I. *Elementary theory of D-degrees (Russian)* ◇ C20 C30 C50 D30 G10 ◇

ROBERTS, I. *Ultrapowers in the Lipschitz and uniform classification of Banach spaces* ◇ C20 C65 ◇

TOKARENKO, A.I. *Two variants of the compactness theorem (Russian)* ◇ C07 C20 ◇

VOLGER, H. *Filtered and stable boolean powers are relativized full boolean powers* ◇ B25 C20 C30 C90 ◇

WANG, CHUSHENG *On ultraproducts in the variety of fields (Chinese)* ◇ C20 C60 ◇

1985

APTER, A.W. *A cardinal structure theorem for an ultrapower* ◇ C20 E35 E55 ◇

BLASS, A.R. *Kleene degrees of ultrafilters* ◇ C20 D65 E05 ◇

CONTESSA, M. *A note on ultraproducts of complete Boolean algebras* ◇ C20 G05 ◇

DUGAS, M. *On reduced products of abelian groups* ◇ C20 C60 E75 ◇

FRANEK, F. *Certain values of completeness and saturatedness of a uniform ideal rule out certain sizes of the underlying index set* ◇ C20 E05 ◇

FUCHS, P.H. & PILZ, G. *Ultraproducts and ultra-limits of near rings* ◇ C20 C60 ◇

GILMAN, R.H. *An application of ultraproducts to finite groups* ◇ C13 C20 C60 ◇

LIN, PEIKEE *Unconditional bases and fixed points of nonexpansive mappings* ◇ C20 C65 H05 ◇

LIPPARINI, P. *Compactness properties of logics generated by monadic quantifiers and cardinalities of limit ultrapowers* ◇ C20 C80 C95 ◇

LIPPARINI, P. *Robinson equivalence relations through limit ultrapowers (Italian summary)* ◇ C20 C95 ◇

MAKKAI, M. *Ultraproducts and categorical logic* ◇ C20 G30 ◇

MIJAJLOVIC, Z. *On ε-bounded ultraproducts* ◇ C20 ◇

NITA, A. & NITA, C. *Sur l'ultraproduit des anneaux de matrices* ◇ C20 C60 ◇

PAPPAS, P. *The model-theoretic structure of abelian group rings* ◇ B25 C20 C60 C90 ◇

PINO, R. & RESSAYRE, J.-P. *Definable ultrafilters and elementary end extensions* ◇ C20 C30 C62 E05 E45 E47 ◇

SPECTOR, M. *Model theory under the axiom of determinateness* ◇ C20 C55 C75 E60 ◇

ZLATOS, P. *On conceptual completeness of syntactic-semantical systems* ◇ C05 C20 C90 G15 G30 ◇

C25 Model-theoretic forcing

1950
SZELE, T. *Ein Analogon der Koerpertheorie fuer abelsche Gruppen* ⋄ C25 C60 ⋄

1951
SCOTT, W.R. *Algebraically closed groups* ⋄ C25 C60 ⋄

1952
NEUMANN, B.H. *A note on algebraically closed groups* ⋄ C25 C60 ⋄

1959
ERDELYI, M. *Systems of equations over non-commutative groups (Ungarisch) (Engl. Zusammenfassung)* ⋄ C25 C60 ⋄

HALL, P. *Some constructions for locally finite groups* ⋄ C25 C60 ⋄

1962
KRUSE, A.H. *Completion of mathematical systems* ⋄ C25 C35 C62 ⋄

1967
SKORDEV, D.G. *Certain algebraic aspects of mathematical logic (Bulgarian) (German summary)* ⋄ C25 C90 G10 ⋄

1969
KAISER, KLAUS *Ueber eine Verallgemeinerung der Robinsonschen Modellvervollstaendigung I,II (II: Abgeschlossene Algebren)* ⋄ C05 C25 C35 ⋄

PABION, J.F. *Sur la methode du "forcing" de P.J.Cohen* ⋄ C25 E40 ⋄

1970
BARWISE, J. & ROBINSON, A. *Completing theories by forcing* ⋄ C25 C35 ⋄

CHAUBARD, A. *Ultraproduit de suites generiques et application* ⋄ C20 C25 ⋄

EKLOF, P.C. & SABBAGH, G. *Model-completions and modules* ⋄ C25 C35 C52 C60 ⋄

1971
BOFFA, M. *Forcing and reflection (Russian summary)* ⋄ C25 E65 ⋄

CHERLIN, G.L. *A new approach to the theory of infinitely generic structures* ⋄ C25 ⋄

COVEN, C. *Forcing in infinitary languages* ⋄ C25 C75 ⋄

CUSIN, R. *Recherche du forcing-compagnon et du modele-compagnon d'une theorie liee a l'existence de modeles $\aleph_\alpha$-universels* ⋄ C25 C35 C50 ⋄

CUSIN, R. & PABION, J.F. *Structures generiques associees a une classe de theories* ⋄ C25 C50 ⋄

CUSIN, R. *Structures prehomogenes et structures generiques* ⋄ C25 C50 ⋄

CUSIN, R. *Sur le "forcing" en theorie des modeles* ⋄ C25 ⋄

HENRARD, P. *Une theorie sans modele generique* ⋄ C25 ⋄

REICHBACH, J. *A note about generalized models and the origin of forcing in the theory of models* ⋄ B25 C25 C90 ⋄

REICHBACH, J. *Generalized models and decidability - weak and strong statistical decidability of theorems* ⋄ B25 C25 C90 ⋄

ROBINSON, A. *Forcing in model theory* ⋄ C25 C35 C52 ⋄

ROBINSON, A. *Forcing in model theory* ⋄ C25 ⋄

ROBINSON, A. *Infinite forcing in model theory* ⋄ C25 C35 C52 ⋄

ROBINSON, A. *On the notion of algebraic closedness for noncommutative groups and fields* ⋄ C25 C60 ⋄

SABBAGH, G. *Embedding problems for modules and rings with application to model-companions* ⋄ C25 C35 C60 ⋄

SABBAGH, G. *Sous-modules purs, existentiellement clos et elementaires* ⋄ C25 C35 C60 ⋄

1972
BACSICH, P.D. *Cofinal simplicity and algebraic closedness* ⋄ C05 C25 C50 C60 C75 G10 ⋄

BOFFA, M. *Modeles universels homogenes et modeles generiques* ⋄ C25 C50 ⋄

BOFFA, M. & PRAAG VAN, P. *Sur les corps generiques* ⋄ C25 C60 ⋄

BOFFA, M. & PRAAG VAN, P. *Sur les sous-champs maximaux des corps generiques denombrables* ⋄ C15 C25 C60 ⋄

CHERLIN, G.L. *The model-companion of a class of structures* ⋄ C25 ⋄

CUSIN, R. *L'algebre des formules d'une theorie complete et "forcing-compagnon"* ⋄ C25 C35 ⋄

FISHER, E.R. & ROBINSON, A. *Inductive theories and their forcing companions* ⋄ C25 G05 G10 ⋄

HENRARD, P. *Forcing with infinite conditions* ⋄ C25 ⋄

MACINTYRE, A. *Omitting quantifier-free types in generic structures* ⋄ C25 C57 C60 C75 D30 D40 ⋄

MACINTYRE, A. *On algebraically closed groups* ⋄ C25 C57 C60 D40 ⋄

REICHBACH, J. *Generalized models, probability of formulas, decisions and statistical decidability of theorems* ⋄ B25 C25 C90 ⋄

ROBINSON, A. *Generic categories* ⋄ C25 C52 C60 G30 ⋄

SHELAH, S. *A note on model complete models and generic models* ◊ C25 C35 ◊

SIMMONS, H. *A possible characterization of generic structures* ◊ C25 ◊

SIMMONS, H. *Existentially closed structures* ◊ C25 C35 C50 ◊

STERN, J. *Theoremes d'interpolation obtenus par la methode du "forcing"* ◊ C25 C40 ◊

WHEELER, W.H. *Algebraically closed division rings, forcing, and the analytic hierarchy* ◊ C25 C60 D55 ◊

WOOD, C. *Forcing for infinitary languages* ◊ C25 C75 ◊

YASUHARA, M. *The generic structures as a variety (Russian summary)* ◊ C25 C52 ◊

1973

BACSICH, P.D. *Primality and model-completion* ◊ C05 C25 C35 ◊

BOFFA, M. *Structures extensionnelles generiques* ◊ C25 C62 E70 ◊

BOWEN, K.A. *Infinite forcing and natural models of set theory (Russian summary)* ◊ C25 C55 C62 C75 C80 E55 G30 ◊

CARSON, A.B. *The model-completion of the theory of commutative regular rings* ◊ C25 C35 C60 ◊

CHERLIN, G.L. *Algebraically closed commutative rings* ◊ C25 C60 ◊

CUSIN, R. *Sur les structures generiques en theorie des modeles* ◊ C10 C25 C35 C50 ◊

FAKIR, S. *Monomorphismes et objects algebriquement et existentiellement clos dans les categories localement presentables. Applications aux monoides, categories et groupoides* ◊ C05 C25 C30 G30 ◊

GOLDREI, D.C. & MACINTYRE, A. & SIMMONS, H. *The forcing companions of number theories* ◊ C25 C62 ◊

HENRARD, P. *Le "forcing-compagnon" sans "forcing"* ◊ C25 ◊

KEISLER, H.J. *Forcing and the omitting types theorem* ◊ C25 C75 ◊

KNIGHT, J.F. *Generic expansions of structures* ◊ C25 C62 C85 ◊

KRIVINE, J.-L. & McALOON, K. *Forcing and generalized quantifiers* ◊ C25 C80 ◊

LIPSHITZ, L. & SARACINO, D.H. *The model companion of the theory of commutative rings without nilpotent elements* ◊ B25 C25 C35 C60 ◊

MACINTYRE, A. *Martin's axiom applied to existentially closed groups* ◊ C25 C55 C60 C75 C80 E50 ◊

NEUMANN, B.H. *The isomorphism problem for algebraically closed groups* ◊ C25 C60 D40 ◊

POUZET, M. *Extensions completes d'une theorie forcing complete* ◊ C25 C52 ◊

REICHBACH, J. *Generalized models for intuitionistic and classical predicate calculi with ultraproducts* ◊ C20 C25 C80 C90 F50 ◊

ROBINSON, A. *Model theory as a framework for algebra* ◊ C25 C35 C60 C98 ◊

ROBINSON, A. *Nonstandard arithmetic and generic arithmetic* ◊ C25 C60 C62 C98 H15 ◊

ROBINSON, A. *Standard and nonstandard number systems* ◊ C25 C60 C98 H05 H15 H20 H98 ◊

SARACINO, D.H. *Model companions for $\aleph_0$-categorical theories* ◊ C25 C35 C60 ◊

SETTE, A.-M. *A propos d'une condition conduisant a une theorie forcee saturee* ◊ C25 C50 ◊

SIMMONS, H. *An omitting types theorem with an application to the construction of generic structures* ◊ C07 C25 C30 ◊

SIMMONS, H. *Le nombre de structures generiques d'une theorie* ◊ C25 ◊

SIMMONS, H. *Une construction du "forcing-compagnon" d'une theorie* ◊ C25 ◊

TAJTSLIN, M.A. *Existentially closed regular commutative semigroups (Russian)* ◊ C25 C35 C52 C60 ◊

WEISPFENNING, V. *Infinitary model-theoretic properties of κ-saturated structures* ◊ C10 C25 C35 C50 C60 C65 C75 ◊

WOOD, C. *The model theory of differential fields of characteristic $p \neq 0$* ◊ C25 C35 C45 C60 ◊

1974

BELEGRADEK, O.V. *Algebraically closed groups (Russian)* ◊ C25 C60 D40 ◊

BELEGRADEK, O.V. *Definability in algebraically closed groups (Russian)* ◊ C25 C40 C60 ◊

BELEGRADEK, O.V. *On algebraically closed groups (Russian)* ◊ C25 C60 ◊

BELEGRADEK, O.V. & ZIL'BER, B.I. *The model companion of an $\aleph_1$-categorical theory (Russian)* ◊ C25 C35 ◊

BELYAEV, V.YA. *Elementary types of algebraically closed semigroups (Russian)* ◊ C25 C60 ◊

BERESTOVOY, S. *Modeles homogenes d'une theorie forcee* ◊ C25 C35 C50 ◊

BOWEN, K.A. *Forcing in a general setting* ◊ C25 C75 C80 C85 E40 ◊

CARSON, A.B. *Algebraically closed regular rings* ◊ C25 C60 ◊

CUSIN, R. *The number of countable generic models for finite forcing* ◊ C15 C25 ◊

DRABBE, J. *Un concept de forcing* ◊ C25 ◊

ELLERMAN, D.P. *Sheaves of structures and generalized ultraproducts* ◊ C20 C25 C90 G30 ◊

GNANVO, C. *Theorie forcee et "forcing-compagnon"* ◊ C25 ◊

LABLANQUIE, J.-C. *Quelques proprietes des enonces universellement forces* ◊ C25 ◊

MACINTYRE, A. *A note on axioms for infinite-generic structures* ◊ C25 C60 C62 C75 E07 ◊

ROBINSON, A. *A decision method for elementary algebra and geometry - revisited* ◊ C25 C60 ◊

SARACINO, D.H. *m-existentially complete structures* ◊ C25 ◊

SARACINO, D.H. *Wreath products and existentially complete solvable groups* ◊ C25 C35 C60 ◊

SIMMONS, H. *Companion theories (Forcing in model theory)* ◊ C25 C98 ◊

SIMMONS, H. *The use of injective-like structures in model theory* ◊ C25 C35 C50 ◊

SIMPSON, S.G. *Forcing and models of arithmetic*
⋄ C15 C25 C62 E40 E47 ⋄
TAJTSLIN, M.A. *Existentially closed commutative rings*
⋄ C25 C60 ⋄
YASUHARA, M. *The amalgamation property, the universal-homogeneous models, and the generic models*
⋄ C25 C50 C52 ⋄

1975

BOFFA, M. *A note on existentially complete division rings*
⋄ C25 C60 ⋄
BURRIS, S. *An existence theorem for model companions*
⋄ C25 C35 ⋄
CHERLIN, G.L. *Second order forcing, algebraically closed structures, and large cardinals* ⋄ C25 C85 E55 ⋄
CHIKHACHEV, S.A. *A question of Simmons (Russian)*
⋄ C25 C50 ⋄
CHIKHACHEV, S.A. *Generic models (Russian)*
⋄ C25 C35 ⋄
CHIKHACHEV, S.A. *Generic models of countable theories (Russian)* ⋄ C15 C25 ⋄
COHN, P.M. *Presentation of skew fields. I. Existentially closed skew fields and the "Nullstellensatz"*
⋄ C25 C60 ⋄
FISHER, E.R. & SIMMONS, H. & WHEELER, W.H. *Elementary equivalence classes of generic structures and existentially complete structures* ⋄ C25 ⋄
FITTLER, R. *Closed models and hulls of theories*
⋄ C25 C52 C60 ⋄
GEORGESCU, G. & PETRESCU, I. *Model-completion in categories* ⋄ C25 C35 G30 ⋄
HIRSCHFELD, J. *Finite forcing and generic filters in arithmetic* ⋄ C25 C60 C62 ⋄
HIRSCHFELD, J. & WHEELER, W.H. *Forcing, arithmetic, division rings* ⋄ C25 C60 C62 C98 ⋄
HIRSCHFELD, J. *The model companion of ZF*
⋄ C25 C62 E30 ⋄
KEISLER, H.J. *Constructions in model theory*
⋄ C25 C30 C50 C75 C95 C98 ⋄
LEE, V. & NADEL, M.E. *On the number of generic models*
⋄ C25 C70 ⋄
MACINTYRE, A. & SIMMONS, H. *Algebraic properties of number theories*
⋄ C25 C40 C52 C62 D25 H15 ⋄
PRAAG VAN, P. *Sur les centralisateurs des corps de type fini dans les corps existentiellement clos (English summary)*
⋄ C25 C60 ⋄
REICHBACH, J. *Infinite proof methods in decidability problems with Ulam's models and classification of predicate calculi* ⋄ B25 C25 C90 ⋄
SABBAGH, G. *Sur les groupes qui ne sont pas reunion d'une suite croissante de sous-groupes propres (English summary)* ⋄ C25 C60 ⋄
SACERDOTE, G.S. *Infinite coforcing in model theory*
⋄ C25 C30 C60 ⋄
SACERDOTE, G.S. *Projective model theory and coforcing*
⋄ C25 C30 C35 C40 C50 C60 ⋄
SARACINO, D.H. *A counterexample in the theory of model companions* ⋄ C25 C35 ⋄
SARACINO, D.H. & WEISPFENNING, V. *Commutative regular rings without prime model extensions*
⋄ C25 C35 C60 ⋄

SARACINO, D.H. & WEISPFENNING, V. *On algebraic curves over commutative regular rings*
⋄ C10 C25 C35 C60 ⋄
SCHMERL, J.H. *The number of equivalence classes of existentially complete structures* ⋄ C25 ⋄
SETTE, A.-M. *Multirelation fortement homogene et ses rapports avec les theories forcees (English summary)*
⋄ C25 C35 C50 ⋄
SHELAH, S. *The lazy model theoretician's guide to stability*
⋄ C25 C35 C45 C50 C60 C98 ⋄
SIMMONS, H. *Counting countable e.c. structures*
⋄ C15 C25 C50 C98 ⋄
SIMMONS, H. *The complexity of T^f and omitting types in F_T* ⋄ C07 C25 ⋄
STERN, J. *A new look at the interpolation problem*
⋄ C25 C40 C70 ⋄
TROFIMOV, M.YU. *Definability in algebraically closed systems (Russian)* ⋄ C25 C40 C60 C85 ⋄
WINKLER, P.M. *Model-completeness and Skolem expansions* ⋄ C25 C30 C35 C60 ⋄

1976

BAUMSLAG, B. & LEVIN, F. *Algebraically closed torsion-free nilpotent groups of class 2* ⋄ C25 C60 ⋄
CHERLIN, G.L. *Model theoretic algebra* ⋄ C25 C60 ⋄
CHERLIN, G.L. *Model theoretic algebra. Selected topics*
⋄ C25 C60 C98 ⋄
COMER, S.D. *Complete and model-complete theories of monadic algebras*
⋄ B25 C05 C25 C35 C60 G15 ⋄
FLUM, J. *El problema de las palabras en la teoria de grupos* ⋄ C25 C60 D40 ⋄
FRIEDRICHSDORF, U. *Einige Bemerkungen zur Peano-Arithmetik* ⋄ C25 C52 C62 F30 ⋄
GEORGESCU, G. *Algebraic structures introduced by forcing (Romanian) (English summary)* ⋄ C25 G25 ⋄
HARNIK, V. *Approximation theorems and model theoretic forcing* ⋄ C25 C75 ⋄
HIRSCHFELD, J. *Another approach to infinite forcing*
⋄ C25 ⋄
KLEJNER, M.L. *On one question of Simmons (Russian)*
⋄ C25 ⋄
LABLANQUIE, J.-C. *Le nombre des structures generiques d'une propriete de "forcing" (English summary)*
⋄ C15 C25 ⋄
MACINTYRE, A. *Existentially closed structures and Jensen's principle* ⋄ ⋄ C25 C60 E45 E65 ⋄
MACINTYRE, A. & SHELAH, S. *Uncountable universal locally finite groups* ⋄ C25 C50 C52 C55 C60 ⋄
MANEVITZ, L.M. *Robinson forcing is not absolute*
⋄ C25 E35 E45 E50 ⋄
POIZAT, B. *"Forcing" et relations generales sur une structure ω-categorique (English summary)*
⋄ C25 C35 ⋄
POIZAT, B. *Remarques sur le "forcing" en theorie des modeles (English summary)* ⋄ C25 ⋄
SARACINO, D.H. *Existentially complete nilpotent groups*
⋄ C20 C25 C35 C60 ⋄
SCHMITT, P.H. *The model-completion of Stone algebras*
⋄ C25 C35 C60 G10 ⋄

SIMMONS, H. *Each regular number structure is biregular*
 ⋄ C25 C62 ⋄
SIMMONS, H. *Large and small existentially closed structures* ⋄ C25 C35 C50 ⋄
TULIPANI, S. *Forcing infinito generalizzato in teoria dei modelli (English summary)* ⋄ C25 ⋄
WEISPFENNING, V. *Negative-existentially complete structures and definability in free extensions*
 ⋄ C25 C40 C60 C75 D40 G05 ⋄
WHEELER, W.H. *Model-companions and definability in existentially complete structures*
 ⋄ C25 C35 C60 C98 ⋄

1977

BELYAEV, V.YA. *Algebraically closed semigroups (Russian)* ⋄ C25 C60 ⋄
DRIES VAN DEN, L. *Artin-Schreier theory for commutative regular rings* ⋄ B25 C10 C25 C60 ⋄
FORREST, W.K. *Model theory for universal classes with the amalgamation property: A study in the foundations of model theory and algebra*
 ⋄ C05 C25 C30 C35 C45 C52 C60 ⋄
GEORGESCU, G. *Forcing infini en theorie des modeles booleens* ⋄ C25 ⋄
GEORGESCU, G. *Structures algebriques introduites par forcing* ⋄ C25 G25 ⋄
GNANVO, C. *Γ-cotheories et elimination des quantificateurs (English summary)*
 ⋄ C10 C25 C35 C52 ⋄
KAISER, KLAUS *On generic stalks of sheaves*
 ⋄ C25 C90 ⋄
LABLANQUIE, J.-C. *Le nombre de classes d'equivalence elementaire des modeles generiques d'une propriete de "forcing" (English summary)* ⋄ C15 C25 C35 ⋄
LABLANQUIE, J.-C. *Proprietes de forcing et modeles generiques* ⋄ C15 C25 C70 ⋄
LEE, V. & NADEL, M.E. *Remarks on generic models*
 ⋄ C25 C70 ⋄
MACINTYRE, A. *Model completeness*
 ⋄ C25 C35 C60 C98 ⋄
PETRY, A. *A propos des centralisateurs de certains sous-corps des corps generiques (English summary)*
 ⋄ C25 C60 ⋄
PINTER, C. *Some theorems on omitting types, with applications to model completeness, amalgamation, and related properties* ⋄ C07 C25 C35 C52 C75 ⋄
SCHMERL, J.H. *Theories with only a finite number of existentially complete models* ⋄ C25 C52 ⋄
SHELAH, S. *Existentially-closed groups in $\aleph_1$ with special properties* ⋄ C15 C25 C60 C75 ⋄
SIMMONS, H. *Existentially closed models of basic number theory* ⋄ C25 C62 ⋄
TAJTSLIN, M.A. *Existentially closed commutative semigroups (Russian) (English summary)*
 ⋄ C25 C52 C60 ⋄
TOFFALORI, C. *Alcune proprieta di teorie di campi con un sottocampo privilegiato (English summary)*
 ⋄ C25 C35 C60 ⋄
WEISPFENNING, V. *Nullstellensaetze - a model theoretic framework* ⋄ C10 C25 C60 ⋄

1978

BASARAB, S.A. *Quelques proprietes modele-theoriques des corps values henseliens (English summary)*
 ⋄ C25 C35 C60 ⋄
BELEGRADEK, O.V. *Elementary properties of algebraically closed groups (Russian) (English summary)*
 ⋄ C25 C60 D40 ⋄
BELYAEV, V.YA. *On algebraic closure in classes of semigroups and groups (Russian)* ⋄ C25 C60 ⋄
BELYAEV, V.YA. *On maximal subgroups of an algebraically closed semigroup (Russian)*
 ⋄ C25 C60 ⋄
BOKUT', L.A. *On algebraically closed and simple Lie algebras (Russian)* ⋄ C25 C60 ⋄
BRUCE, KIM B. *Model-theoretic forcing in logic with a generalized quantifier* ⋄ C25 C70 C75 C80 ⋄
DRIES VAN DEN, L. *Model theory of fields*
 ⋄ B25 C10 C25 C35 C60 H20 ⋄
FUJIWARA, T. *Algebraically closed algebraic extensions in universal classes* ⋄ C05 C25 ⋄
HIRSCHFELD, J. *Examples in the theory of existential completeness* ⋄ C25 C35 C50 ⋄
LABLANQUIE, J.-C. *Forcing infini et theoreme d'omission des types de Chang* ⋄ C07 C25 ⋄
LOULLIS, G. *Infinite forcing for boolean valued models*
 ⋄ C25 C35 C90 ⋄
MCALOON, K. *Completeness theorems, incompleteness theorems and models of arithmetic*
 ⋄ C25 C62 F30 ⋄
MILLS, G. *A model of Peano arithmetic with no elementary end extension* ⋄ C25 C62 H15 ⋄
PETRESCU, I. *Model-companions for logical functors*
 ⋄ C25 C35 G30 ⋄
PINTER, C. *Properties preserved under definitional equivalence and interpretations*
 ⋄ C25 C35 C50 F25 ⋄
POIZAT, B. *Etude d'un forcing en theorie des modeles*
 ⋄ C25 C35 ⋄
PREST, M. *Some model-theoretic aspects of torsion theories* ⋄ C25 C35 C60 ⋄
SARACINO, D.H. *Existentially complete torsion-free nilpotent groups* ⋄ C25 C60 ⋄
SINGER, M.F. *The model theory of ordered differential fields* ⋄ B25 C25 C35 C60 C65 D80 ⋄
VEIT RICCIOLI, B. *Il forcing come principio logico per la costruzione dei fasci. I,II (French summaries)*
 ⋄ C25 C90 E40 G30 ⋄
VELLUTINI, F. *Le "forcing" en theorie des modeles; quelques resultats sur les relations generales (English summary)* ⋄ C25 ⋄
WEISPFENNING, V. *A note on $\aleph_0$-categorical model-companions*
 ⋄ C25 C35 C50 C52 C65 G05 G10 ⋄
WERNER, H. *Discriminator-algebras. Algebraic representation and model theoretic properties*
 ⋄ B25 C05 C25 C35 C90 ⋄
WHEELER, W.H. *A characterization of companionable, universal theories* ⋄ C25 C35 C60 C65 ⋄

1979

BASARAB, S.A. *A model-theoretic transfer theorem for henselian valued fields* ⋄ C25 C35 C60 ⋄

BASARAB, S.A. *Model-theoretic methods in the theory of henselian valued fields. I,II (Romanian) (English summaries)* ⋄ C25 C35 C60 ⋄

BELYAEV, V.YA. & TAJTSLIN, M.A. *On elementary properties of existentially closed systems (Russian)* ⋄ C25 C40 C60 C75 C85 D55 ⋄

BRUCE, KIM B. & KEISLER, H.J. $L_A(\exists)$
⋄ C25 C70 C75 C80 ⋄

BURRIS, S. & WERNER, H. *Sheaf constructions and their elementary properties*
⋄ B25 C05 C20 C25 C30 C60 C90 ⋄

CHERLIN, G.L. *Lindenbaum algebras and model companions* ⋄ C25 C35 ⋄

FRAISSE, R. *Le forcing faible; son utilisation pour caracteriser les relations unaires generales pour la chaine ou la consecutivite des entiers naturels (resultat de R. Solovay, 1976)* ⋄ C25 E40 ⋄

LABLANQUIE, J.-C. *Proprietes de consistance et forcing*
⋄ C25 C40 C75 ⋄

LACAVA, F. *Alcune proprieta delle L-algebre e delle L-algebre esistenzialmente chiuse*
⋄ C25 C35 C60 G20 ⋄

PODEWSKI, K.-P. & REINEKE, J. *Algebraically closed commutative local rings* ⋄ C25 C35 C60 ⋄

POIZAT, B. *Le "forcing" de Fraisse dans un contexte arithmetique (English summary)* ⋄ C25 ⋄

PREST, M. *Model-completions of some theories of modules*
⋄ C25 C35 C50 C60 ⋄

SARACINO, D.H. & WOOD, C. *Periodic existentially closed nilpotent groups* ⋄ C25 C35 C45 C60 ⋄

SCHMID, J. *Algebraically and existentially closed distributive lattices* ⋄ C05 C25 G10 ⋄

SETTE, A.-M. *Fraisse and Robinson's forcing (a comparative study)* ⋄ C25 ⋄

SHELAH, S. & ZIEGLER, M. *Algebraically closed groups of large cardinality* ⋄ C25 C60 C75 ⋄

STEGBAUER, W. *A generalized model companion for a theory of partially ordered fields* ⋄ C25 C35 C60 ⋄

TOFFALORI, C. *Alcune osservazioni sugli anelli commutativi esistenzialmente chiusi (English summary)*
⋄ C25 C60 ⋄

TOFFALORI, C. *Semigruppi archimedei esistenzialmente chiusi (English summary)* ⋄ C25 C60 ⋄

TRANSIER, R. *Verallgemeinerte formal $\mathfrak{p}$-adische Koerper*
⋄ C25 C60 ⋄

WEISPFENNING, V. *Lattice products*
⋄ C10 C25 C30 C35 C90 ⋄

WHEELER, W.H. *Amalgamation and elimination of quantifiers for theories of fields*
⋄ C10 C25 C52 C60 ⋄

WHEELER, W.H. *Model-complete theories of pseudo-algebraically closed fields*
⋄ C25 C35 C60 ⋄

WHEELER, W.H. *The first order theory of N-colorable graphs* ⋄ B30 C10 C25 C35 C65 ⋄

WOOD, C. *Notes on the stability of separably closed fields*
⋄ C25 C35 C45 C60 ⋄

1980

BASARAB, S.A. & NICA, V. & POPESCU, D. *Approximation properties and existential completeness for ring morphisms* ⋄ C25 C60 H15 ⋄

BAUMSLAG, G. & DYER, E. & HELLER, A. *The topology of discrete groups* ⋄ C25 C60 ⋄

BELEGRADEK, O.V. *Decidable fragments of universal theories and existentially closed models (Russian)*
⋄ B25 C25 D40 ⋄

BOUSCAREN, E. *Existentially closed modules: Types and prime models* ⋄ C25 C60 ⋄

BRUCE, KIM B. *Model constructions in stationary logic I. Forcing* ⋄ C25 C55 C70 C75 C80 ⋄

DICKMANN, M.A. *Types remarquables et extensions de modeles dans l'arithmetique de Peano I*
⋄ C25 C30 C62 ⋄

ERSHOV, YU.L. *Multiply valued fields (Russian)*
⋄ C25 C35 C60 ⋄

FUJIWARA, T. *Existentially closed structures and algebraically closed structures* ⋄ C25 C75 ⋄

GLASS, A.M.W. & PIERCE, K.R. *Equations and inequations in lattice-ordered groups*
⋄ C25 C60 G10 ⋄

GLASS, A.M.W. & PIERCE, K.R. *Existentially complete abelian lattice-ordered groups* ⋄ C25 C60 G10 ⋄

GLASS, A.M.W. & PIERCE, K.R. *Existentially complete lattice-ordered groups* ⋄ C25 C35 C60 G10 ⋄

GRULOVIC, M. *A comment on forcing (Serbo-Croatian) (English summary)* ⋄ C25 ⋄

HICKIN, K.K. & MACINTYRE, A. *Algebraically closed groups: embeddings and centralizers*
⋄ C25 C60 D40 ⋄

HIRSCHFELD, J. *Finite forcing, existential types and complete types* ⋄ C25 C35 ⋄

HODGES, W. *Interpreting number theory in nilpotent groups* ⋄ C25 C60 C62 F25 F35 ⋄

JACOB, B. *The model theory of "R-formal" fields*
⋄ B25 C10 C25 C35 C60 ⋄

LABLANQUIE, J.-C. *Modeles premiers et modeles generiques (English summary)* ⋄ C25 C50 ⋄

LACAVA, F. *Osservazioni sulla teoria dei gruppi abeliani reticolari* ⋄ C25 C35 C60 ⋄

MANDERS, K.L. *Theories with the existential substructure property* ⋄ C25 C35 ⋄

MEAD, J. & NELSON, GEORGE C. *Model companions and k-model completeness for the complete theories of boolean algebras* ⋄ C25 C35 G05 ⋄

TOFFALORI, C. *Sheaves of pairs of real closed fields*
⋄ C25 C35 C60 C90 ⋄

TOFFALORI, C. *Sugli anelli nilreali (English summary)*
⋄ C20 C25 C60 ⋄

TOFFALORI, C. *Sul model-completamento di certe teorie di coppie di anelli (English summary)*
⋄ C10 C25 C35 C60 ⋄

TUSCHIK, H.-P. *An application of rank-forcing to ω_1-categoricity* ⋄ C25 C35 ⋄

WHEELER, W.H. *Model theory of strictly upper triangular matrix rings* ⋄ C25 C35 C60 ⋄

ZIEGLER, M. *Algebraisch abgeschlossene Gruppen (English summary)* ⋄ C25 C60 D30 D40 ⋄

1981

BELEGRADEK, O.V. *Classes of algebras with inner mappings (Russian)* ⋄ C05 C25 C50 ⋄

KASHIWAGI, T. *On classes which have the weak amalgamation property* ⋄ C25 C52 ⋄

KEGEL, O.H. *Existentially closed projective planes* ⋄ C25 C65 ⋄

LABLANQUIE, J.-C. *Modeles generiques et compacite (English summary)* ⋄ C25 C52 ⋄

MACINTYRE, A. *A theorem of Rabin in a general setting* ⋄ C25 C62 ⋄

MAIER, B.J. *Existenziell abgeschlossene lokal endliche p-Gruppen* ⋄ C25 C60 ⋄

MANASTER, A.B. & REMMEL, J.B. *Partial orderings of fixed finite dimension: Model companions and density* ⋄ C25 C65 D35 E07 ⋄

PETRESCU, I. *Existential morphisms and existentially closed models of logical categories* ⋄ C25 G30 ⋄

PODEWSKI, K.-P. & REINEKE, J. *On algebraically closed commutative indecomposable rings* ⋄ C25 C35 C60 ⋄

POIZAT, B. *Degres de definissabilite arithmetique des generiques (English summary)* ⋄ C25 C40 D30 D55 ⋄

PRESTEL, A. *Pseudo real closed fields* ⋄ C10 C25 C35 C52 C60 ⋄

SETTE, A.-M. *Fraisse and Robinson's forcing* ⋄ C25 ⋄

TULIPANI, S. *Model-completions of theories of finitely additive measures with values in an ordered field* ⋄ C10 C25 C35 C65 G05 ⋄

WEISPFENNING, V. *The model-theoretic significance of complemented existential formulas* ⋄ C25 ⋄

1982

BELYAEV, V.YA. *On algebraic closure and amalgamation of semigroups* ⋄ C05 C25 C60 ⋄

ERSHOV, YU.L. *Multiply valued fields (Russian)* ⋄ B25 C25 C35 C60 C98 ⋄

ERSHOV, YU.L. *Regularly r-closed fields (Russian)* ⋄ B25 C25 C35 C60 ⋄

GRULOVIC, M. *A note on forcing and weak interpolation theorem for infinitary logics (Serbo-Croatian summary)* ⋄ C25 C40 C75 ⋄

KASHIWAGI, T. *Note on algebraic extensions in universal classes* ⋄ C05 C25 C30 ⋄

LABLANQUIE, J.-C. *Extensions existentielles forcing-completes de theories forcing-completes (English summary)* ⋄ C25 ⋄

LIPPARINI, P. *Existentially complete closure algebras (Italian summary)* ⋄ C25 G25 ⋄

LIPPARINI, P. *Locally finite theories with model companion (Italian summary)* ⋄ C25 C35 C52 ⋄

MEZ, H.-C. *Existentially closed linear groups* ⋄ C25 C35 C60 ⋄

MOENS, J.L. *Forcing et semantique de Kripke-Joyal* ⋄ C25 C90 E40 G30 ⋄

PRESTEL, A. *Decidable theories of preordered fields* ⋄ B25 C25 C35 C60 ⋄

REINEKE, J. *On algebraically closed models of theories of commutative rings* ⋄ C25 C60 ⋄

SCHMID, J. *Algebraically closed distributive p-algebras* ⋄ C05 C25 G05 G10 G20 ⋄

SCHMID, J. *Model companions of distributive p-algebras* ⋄ C05 C25 C35 G10 ⋄

TOFFALORI, C. *Strutture esistenzialmente complete per certe classi di anelli (English summary)* ⋄ C20 C25 C60 ⋄

TOFFALORI, C. *Teoria dei modelli per alcune classi di anelli locali* ⋄ C10 C25 C35 C60 ⋄

WEISPFENNING, V. *Model theory and lattices of formulas* ⋄ C25 C35 C52 C75 G10 ⋄

WERNER, H. *Sheaf constructions in universal algebra and model theory* ⋄ B25 C05 C25 C30 C35 C90 G05 ⋄

1983

ARAGANE, K. & FUJIWARA, T. *Elementary classes whose every model is algebraically closed* ⋄ C25 C52 ⋄

BLASS, A.R. & SCEDROV, A. *Classifying topoi and finite forcing* ⋄ C25 G30 ⋄

DAHN, B.I. & WOLTER, H. *On the theory of exponential fields* ⋄ C25 C60 C65 ⋄

ERSHOV, YU.L. *Regularly r-closed fields (Russian)* ⋄ B25 C25 C35 C60 ⋄

GERLA, G. *Model theory and modal logic (Italian)* ⋄ B45 C25 C75 C90 ⋄

GRULOVIC, M. *On n-finite forcing* ⋄ C25 E40 ⋄

HOOGEWIJS, A. *The non-definedness notion and forcing* ⋄ C25 C90 ⋄

JARDEN, M. *On the model companion of the theory of e-fold ordered fields* ⋄ B25 C25 C35 C60 ⋄

KATZ, M. *Deduction systems and valuation spaces* ⋄ C25 C90 G25 ⋄

MAIER, B.J. *On existentially closed and generic nilpotent groups* ⋄ C25 C60 ⋄

MARKOVIC, Z. *Some preservation results for classical and intuitionistic satisfiability in Kripke models* ⋄ B20 C25 C90 F50 ⋄

MIJAJLOVIC, Z. *On the embedding property of models* ⋄ C25 C35 C52 G05 G10 ⋄

PARIGOT, M. *Le modele compagnon de la theorie des arbres* ⋄ C10 C25 C35 C50 C65 ⋄

PREST, M. *Existentially complete prime rings* ⋄ C25 C60 ⋄

SARACINO, D.H. & WOOD, C. *Finitely generic abelian lattice-ordered groups* ⋄ C25 C60 ⋄

SCHMITT, P.H. *Algebraically complete lattices* ⋄ C25 C35 G10 ⋄

TOFFALORI, C. *Teoria dei modelli per una classe di anelli differenziali* ⋄ C25 C60 ⋄

1984

ALIEV, V.G. *Model companions of theories of semigroup classes (Russian) (English and Azerbaijani summaries)* ⋄ C05 C25 C35 ⋄

BASARAB, S.A. *Axioms for pseudo-real-closed fields* ⋄ C25 C60 ⋄

BASARAB, S.A. *Definite functions on algebraic varieties over ordered fields* ⋄ C25 C60 ⋄

BASARAB, S.A. *On some classes of Hilbertian fields* ⋄ B25 C10 C25 C35 C60 ⋄

DAHN, B.I. & WOLTER, H. *Ordered fields with several exponential functions* ⋄ C25 C35 C65 D35 ⋄

DELON, F. *Corps equivalents a leurs corps de series* ⋄ C25 C60 ⋄

DELON, F. *Espaces ultrametriques* ⋄ C25 C35 C45 C50 C65 ⋄

EKLOF, P.C. & MEZ, H.-C. *Additive groups of existentially closed rings* ⋄ C25 C60 ⋄

GERLA, G. & VACCARO, V. *Modal logic and model theory* ⋄ B45 C25 C75 C90 ⋄

GIORGETTA, D. & SHELAH, S. *Existentially closed structures in the power of the continuum* ⋄ C15 C25 C30 C52 C55 C60 C75 ⋄

HODGES, W. *Groupes nilpotents existentiellement clos de classe fixee* ⋄ C25 C60 ⋄

MAIER, B.J. *Existentially closed torsion-free nilpotent groups of class three* ⋄ C25 C60 ⋄

MANDERS, K.L. *Interpretations and the model theory of the classical geometries* ⋄ C25 C35 C65 ⋄

PHILLIPS, R.E. *Existentially closed locally finite central extensions; multipliers and local systems* ⋄ C25 C60 ⋄

RIBENBOIM, P. *Remarks on existentially closed fields and diophantine equations* ⋄ C25 C60 ⋄

SARACINO, D.H. & WOOD, C. *An example in the model theory of abelian lattice-ordered groups* ⋄ C25 C60 G10 ⋄

SHEN, FUXING *Forcing in lattice-valued model theory (Chinese)* ⋄ C25 C90 ⋄

TOFFALORI, C. *On a class of differential local rings (Italian) (English summary)* ⋄ C25 C60 ⋄

WOLTER, H. *Some remarks on exponential functions in ordered fields* ⋄ C25 C65 D35 ⋄

1985

ARAGANE, K. & FUJIWARA, T. *Elementary classes whose every model is α-existentially closed* ⋄ C25 C52 ⋄

ARAGANE, K. & FUJIWARA, T. *Elementary classes of α-existentially closed structures* ⋄ C25 C52 ⋄

BASARAB, S.A. *The absolute Galois group of pseudoreal closed field with finitely many orders* ⋄ C25 C60 ⋄

BASARAB, S.A. *Transfer principles for PRC fields* ⋄ C25 C60 ⋄

EKLOF, P.C. & MEZ, H.-C. *The ideal structure of existentially closed algebras* ⋄ C25 C60 ⋄

GEORGESCU, G. & VOICULESCU, I. *Eastern model-theory for boolean-valued theories* ⋄ C25 C35 C40 C90 ⋄

HODGES, W. *Building models by games* ⋄ C25 C55 C60 C80 C98 E60 ⋄

LASLANDES, B. *Modele-compagnons de theories de corps munis de n ordres* ⋄ C25 C35 C60 ⋄

LEINEN, F. *Existentially closed groups in locally finite group classes* ⋄ C25 C60 ⋄

POINT, F. *Finitely generic models of T_{UH}, for certain model companionable theories T* ⋄ C25 C60 ⋄

POINT, F. *Transfer properties of discriminator varieties* ⋄ C10 C25 C30 C35 ⋄

SCHMID, J. *Algebraically closed p-semilattices* ⋄ C25 G10 ⋄

SCHMIDT, K. *Nonstandard methods in field theory. Bounds and definability, and specialization properties in prime characteristic* ⋄ C25 C60 H20 ⋄

SHELAH, S. *Uncountable constructions for B.A.,e.c. groups and Banach spaces* ⋄ C25 C60 C65 E05 E65 E75 G05 ⋄

THOMAS, S. *Complete existentially closed locally finite groups* ⋄ C25 C60 ⋄

VARADARAJAN, K. *Pseudo-mitotic groups* ⋄ C25 C60 ⋄

C30 Other model constructions

1946
TARSKI, A. *A remark on functionally free algebras*
⋄ C05 C30 G25 ⋄

1949
JONSSON, B. & TARSKI, A. *Cardinal products of isomorphism types* ⋄ C05 C30 E75 ⋄

1951
HORN, A. *On sentences which are true of direct unions of algebras* ⋄ C05 C30 C40 ⋄

1952
MOSTOWSKI, ANDRZEJ *On direct products of theories*
⋄ B25 C10 C30 ⋄

1953
FOSTER, A.L. *Generalized "boolean" theory of universal algebras, part I: subdirect sums and normal representation theorem* ⋄ C05 C30 ⋄
FOSTER, A.L. *Generalized "boolean" theory of universal algebras, II: identities and subdirect sums of functionally complete algebras* ⋄ C05 C30 ⋄
KRISHNAN, V.S. *Closure operations on c-structures*
⋄ C05 C30 E75 ⋄

1954
EHRENFEUCHT, A. & LOS, J. *Sur les produits cartesiens des groupes cycliques infinis* ⋄ C30 C60 ⋄
FRAISSE, R. *Sur l'extension aux relations de quelques proprietes des ordres* ⋄ C07 C30 C50 E07 ⋄
TARSKI, A. *Contributions to the theory of models I,II*
⋄ C13 C30 C52 ⋄
VAUGHT, R.L. *On sentences holding in direct products of relational systems* ⋄ B20 B25 C30 ⋄

1955
LOS, J. *On the extending of models I* ⋄ C30 C52 ⋄
LOS, J. & SUSZKO, R. *On the extending of models II*
⋄ C05 C30 C52 ⋄
LOS, J. & SUSZKO, R. *On the infinite sums of models*
⋄ C30 C40 C52 ⋄
LOS, J. *Quelques remarques, theoremes et problemes sur les classes definissables d'algebres*
⋄ C05 C20 C30 C52 ⋄
ROBINSON, A. *On ordered fields and definite functions*
⋄ C30 C60 ⋄
ROBINSON, A. *Ordered structures and related concepts*
⋄ C30 C35 C50 C60 ⋄

1956
EHRENFEUCHT, A. & MOSTOWSKI, ANDRZEJ *Models of axiomatic theories admitting automorphisms*
⋄ C07 C30 E75 ⋄
ROBINSON, A. *Completeness and persistence in the theory of models* ⋄ C30 C35 C52 ⋄
ROBINSON, A. *Note on a problem of L.Henkin*
⋄ C30 C52 ⋄
SLOMINSKI, J. *On the extending of models III. Extensions in equationally definable classes of algebras*
⋄ C05 C30 C60 G05 ⋄

1957
CHANG, C.C. *Unions of chains of models and direct products of models* ⋄ C30 C52 ⋄
EHRENFEUCHT, A. *On theories categorical in power*
⋄ C30 C35 ⋄
HANF, W.P. *On some fundamental problems concerning isomorphism of boolean algebras*
⋄ C05 C15 C30 G05 ⋄
JONSSON, B. *On direct decompositions of torsion-free abelian groups* ⋄ C05 C30 C60 ⋄
JONSSON, B. *On isomorphism types of groups and other algebraic systems* ⋄ C05 C30 C60 ⋄
LOS, J. & SUSZKO, R. *On the extending of models IV: Infinite sums of models* ⋄ C30 C52 ⋄
LYNDON, R.C. *Sentences preserved under homomorphisms; sentences preserved under subdirect products* ⋄ C07 C30 C40 C52 ⋄
TARSKI, A. *Remarks on direct products of commutative semigroups* ⋄ C05 C30 G05 ⋄

1958
CHANG, C.C. & MOREL, A.C. *On closure under direct product* ⋄ C30 C40 ⋄
GAL, L.N. *A note on direct products* ⋄ B25 C30 D35 ⋄
MOSTOWSKI, ANDRZEJ *Quelques observations sur l'usage des methodes non finitistes dans la meta-mathematique*
⋄ A05 C30 C62 C80 E30 E35 F99 ⋄
OBERSCHELP, A. *Ueber die Axiome produkt-abgeschlossener arithmetischer Klassen*
⋄ B20 B25 C30 C40 C52 ⋄
TAJMANOV, A.D. *On classes of models, closed under direct product (Russian)* ⋄ C30 C52 D35 ⋄

1959
APPEL, K.I. *Horn sentences in identity theory*
⋄ C13 C30 C52 ⋄
CLARKE, A.B. *On the representation of cardinal algebras by directed sums* ⋄ C05 C30 ⋄
COHN, P.M. *On the free product of associative rings*
⋄ C30 C60 ⋄
ENGELER, E. *Eine Konstruktion von Modellerweiterungen*
⋄ C30 ⋄
FEFERMAN, S. & VAUGHT, R.L. *The first order properties of products of algebraic systems*
⋄ B25 C10 C20 C30 G05 ⋄

LYNDON, R.C. *Existential Horn sentences*
  ⋄ B20 C30 C40 C52 ⋄
LYNDON, R.C. *Properties preserved under algebraic constructions* ⋄ C07 C30 C40 C52 ⋄
LYNDON, R.C. *Properties preserved in subdirect products*
  ⋄ C30 C40 C52 ⋄
MAL'TSEV, A.I. *Regular products of models (Russian)*
  ⋄ C30 ⋄
RABIN, M.O. *Arithmetical extensions with prescribed cardinality* ⋄ C20 C30 C55 ⋄
SVENONIUS, L. *A theorem on permutations in models*
  ⋄ C30 C40 ⋄
TAJMANOV, A.D. *On a class of models closed with respect to direct products (Russian)* ⋄ C30 C52 ⋄

1960

BRYANT, S.J. & MARICA, J.G. *Unary algebras*
  ⋄ C05 C30 ⋄
COHN, P.M. *On the free product of associative rings. II: the case of (skew) fields* ⋄ C30 C60 ⋄
KEISLER, H.J. *Theory of models with generalized atomic formulas* ⋄ C07 C30 C52 ⋄
LOS, J. & SLOMINSKI, J. & SUSZKO, R. *On the extending of models V: Embedding theorems for relational models*
  ⋄ C07 C30 C52 ⋄
MACHOVER, M. *A note on sentences preserved under direct products and powers* ⋄ C30 C40 ⋄
TAJMANOV, A.D. *On a class of models closed with respect to direct products (Russian)* ⋄ C30 C52 ⋄

1961

ROBINSON, A. *On the construction of models*
  ⋄ C07 C20 C30 E05 ⋄

1962

FRASNAY, C. *Relations invariantes dans un ensemble totalement ordonne* ⋄ C07 C30 E07 ⋄
LOS, J. *Common extension in equational classes*
  ⋄ C30 C52 ⋄

1963

FLEISCHER, I. *Elementary properties of inductive limits*
  ⋄ C05 C20 C30 ⋄
FRASNAY, C. *Groupes de permutations finies et familles d'ordres totaux, application a la theorie des relations*
  ⋄ C07 C30 E07 ⋄
GAO, HENGSHAN *Remarks on Los and Suszko's paper "On the extending of models. IV." (Chinese)* ⋄ C30 C52 ⋄

1964

CHANG, C.C. & JONSSON, B. & TARSKI, A. *Refinement properties for relational structures* ⋄ C05 C30 ⋄
CRAWLEY, P. & JONSSON, B. *Refinements for infinite direct decompositions of algebraic systems*
  ⋄ C05 C30 C60 ⋄
ERSHOV, YU.L. *Decidability of certain non-elementary theories (Russian)* ⋄ B25 C30 C85 G05 ⋄
FRASNAY, C. *Groupes de compatibilite de deux ordres totaux; application aux n-morphismes d'une relation m-aire* ⋄ C07 C30 E07 ⋄
FRASNAY, C. *Partages d'ensembles de parties et de produits d'ensembles, applications a la theorie des relations* ⋄ C07 C30 E07 ⋄

GRAETZER, G. *On the class of subdirect powers of a finite algebra* ⋄ C05 C30 ⋄
KEISLER, H.J. *Unions of relational systems*
  ⋄ C30 C52 ⋄

1965

CHANG, C.C. *A simple proof of the Rabin-Keisler theorem*
  ⋄ C30 E05 ⋄
FRASNAY, C. *Quelques problemes combinatoires concernant les ordres totaux et les relations monomorphes* ⋄ C07 C30 E07 ⋄
FREYD, P. *The theories of functors and models*
  ⋄ C30 G30 ⋄
JONSSON, B. *Extensions of relational structures*
  ⋄ C30 C50 C52 ⋄
LOS, J. *Free product in general algebras* ⋄ C05 C30 ⋄
MAKKAI, M. *A compactness result concerning direct products of models* ⋄ C05 C30 C52 ⋄
MORLEY, M.D. *Omitting classes of elements*
  ⋄ C30 C55 C75 ⋄
RABIN, M.O. *Universal groups of automorphisms of models* ⋄ C07 C30 ⋄
WEINSTEIN, J.M. *First order properties preserved by direct product* ⋄ C30 ⋄

1966

KOGALOVSKIJ, S.R. *On the properties preserved under algebraic constructions (Russian)*
  ⋄ B25 C30 C52 C60 D35 D45 ⋄
LOS, J. *Direct sums in general algebras* ⋄ C05 C30 ⋄
MACZYNSKI, M.J. *Generalized free m-products of m-distributive boolean algebras with an m-amalgamated subalgebra* ⋄ C30 G05 ⋄
TANG, C.Y. *An existence theorem for generalized direct products with amalgamated subgroups* ⋄ C30 C60 ⋄

1967

CHANG, C.C. *Cardinal factorization of finite relational structures* ⋄ C05 C13 C30 ⋄
CHANG, C.C. *Ultraproducts and other methods of constructing models* ⋄ C20 C30 ⋄
FRAISSE, R. *La classe universelle: Une notion a la limite de la logique et de la theorie des relations*
  ⋄ C07 C30 C52 ⋄
GAIFMAN, H. *Uniform extension operators for models and their applications* ⋄ C20 C30 C62 ⋄
GALVIN, F. *Reduced products, Horn sentences, and decision problems*
  ⋄ B20 B25 C20 C30 C40 D35 ⋄
JEAN, M. *Relations monomorphes et classes universelles*
  ⋄ C07 C30 C52 ⋄
KOGALOVSKIJ, S.R. *On compactness-preserving algebraic constructions (Russian)* ⋄ C20 C30 C52 ⋄
LOPEZ-ESCOBAR, E.G.K. *On a theorem of J.I.Malitz*
  ⋄ C30 C40 C85 ⋄
LOVASZ, L. *Operations with structures*
  ⋄ C07 C13 C30 ⋄
OMAROV, A.I. *Filtered products of models (Russian) (English summary)* ⋄ C20 C30 C40 C50 C52 ⋄
PLONKA, J. *On a method of construction of abstract algebras* ⋄ C05 C30 ⋄

TAJMANOV, A.D. *Remarks to the Mostowski-Ehrenfeucht theorem (Russian) (English summary)*
⋄ C07 C20 C30 C52 E50 ⋄

1968

BRYCE, R.A. *A note on free products with normal amalgamation* ⋄ C05 C30 ⋄

CHANG, C.C. *Infinitary properties of models generated from indiscernibles* ⋄ C30 C75 ⋄

CHANG, C.C. *Some remarks on the model theory of infinitary languages* ⋄ C07 C30 C55 C75 ⋄

FEFERMAN, S. *Persistent and invariant formulas for outer extensions* ⋄ C30 C40 C70 C75 F35 ⋄

KOGALOVSKIJ, S.R. *On compact classes of algebraic systems (Russian)* ⋄ C30 C52 ⋄

KOGALOVSKIJ, S.R. *Remarks on compact classes of algebraic systems (Russian)* ⋄ C30 C52 ⋄

LINDSTROEM, P. *Remarks on some theorems of Keisler* ⋄ C07 C20 C30 ⋄

MORLEY, M.D. *Partitions and models*
⋄ C30 C50 C55 C62 C75 E05 E45 E55 ⋄

OLIN, P. *An almost everywhere direct power*
⋄ C30 C62 E50 ⋄

SCHMIDT, J. *Universelle Halbgruppe, Kategorien, freies Produkt* ⋄ C05 C30 G30 ⋄

TIMOSHENKO, E.I. *Preservation of elementary and universal equivalence under the wreath product (Russian)* ⋄ C30 C60 ⋄

WEINSTEIN, J.M. *(ω_1, ω) properties of unions of models*
⋄ C30 C40 C52 C75 ⋄

1969

CHEREMISIN, A.I. *Structural characterization of certain classes of models (Russian)* ⋄ C05 C30 C52 ⋄

HENSEL, G. & PUTNAM, H. *Normal models and the field Σ_1^** ⋄ C07 C30 C57 D55 ⋄

MARCJA, A. & MARONGIU, G. *Estensioni booleane di strutture relazionali (English summary)*
⋄ C20 C30 G05 ⋄

MOTOHASHI, N. *On normal operations on models*
⋄ C30 C40 C50 ⋄

PACHOLSKI, L. *Elementary substructures of direct powers (Russian summary)* ⋄ C30 C62 ⋄

PACHOLSKI, L. *On products of first order theories (Russian summary)* ⋄ C30 C50 ⋄

PARIKH, R. *A conservation result* ⋄ C30 H20 ⋄

PAYNE, T.H. *An elementary submodel never preserved by Skolem expansions* ⋄ C30 ⋄

PLONKA, J. *On sums of direct systems of boolean algebras*
⋄ C30 G05 ⋄

REILLY, N.R. *Some applications of wreath products and ultraproducts in the theory of lattice ordered groups*
⋄ C20 C30 C60 ⋄

RESSAYRE, J.-P. *Sur les theories du premier ordre categorique en un cardinal* ⋄ C30 C35 C55 ⋄

RIBENBOIM, P. *Boolean powers* ⋄ C20 C30 G05 ⋄

WASZKIEWICZ, J. & WEGLORZ, B. *On ω_0-categoricity of powers (Russian summary)* ⋄ C30 C35 C50 ⋄

WASZKIEWICZ, J. & WEGLORZ, B. *On products of structures for generalized logics (Polish and Russian summaries)* ⋄ B50 C20 C30 C90 ⋄

WOJCIECHOWSKA, A. *Generalized products for Q_α-languages (Russian summary)*
⋄ C30 C55 C80 ⋄

1970

AOYAMA, K. *On the structure space of a direct product of rings* ⋄ C30 C60 ⋄

DRABBE, J. *Sur une propriete des formules negatives*
⋄ C30 C40 C52 ⋄

EKLOF, P.C. *On the existence of $L_{\infty,\kappa}$ indiscernibles*
⋄ C30 C50 C75 ⋄

FITTLER, R. *Direct limits of models* ⋄ C30 ⋄

FUJIWARA, T. & NAKANO, Y. *On the relation between free structures and direct limits* ⋄ C05 C30 ⋄

GAIFMAN, H. *On local arithmetical functions and their application for constructing types of Peano's arithmetic*
⋄ C30 C62 H15 ⋄

GALVIN, F. *Horn sentences*
⋄ B20 C20 C30 C40 D35 ⋄

KEISLER, H.J. *Logic with the quantifier "there exist uncountably many"* ⋄ C30 C55 C62 C75 C80 ⋄

KIRCH, A.M. *Generalized restricted direct products*
⋄ C05 C30 ⋄

OLIN, P. *Some first order properties of direct sums of modules* ⋄ C30 C40 C60 ⋄

OMAROV, A.I. *A certain property of Frechet filters (Russian)* ⋄ C20 C30 C50 C52 ⋄

SABBAGH, G. *Aspects logiques de la purete dans les modules* ⋄ C20 C30 C60 ⋄

1971

ARMBRUST, M. & KAISER, KLAUS *Remarks on model classes acting on one another*
⋄ C05 C20 C30 C52 ⋄

BALBES, R. & DWINGER, P. *Uniqueness of representations of a distributive lattice as a free product of a boolean algebra and a chain* ⋄ C05 C30 G10 ⋄

DAIGNEAULT, A. *Boolean powers in algebraic logic*
⋄ C20 C30 G05 G15 ⋄

EBBINGHAUS, H.-D. *On models with large automorphism groups* ⋄ C30 C55 C80 ⋄

FRAISSE, R. & POUZET, M. *Interpretabilite d'une relation par une chaine* ⋄ C07 C30 C65 E07 ⋄

FRAISSE, R. & POUZET, M. *Sur une classe de relations n'ayant qu'un nombre fini de bornes*
⋄ C07 C30 C65 E07 ⋄

GOETZ, A. *A generalization of the direct product of universal algebras* ⋄ C05 C30 ⋄

HILL, P. *Classes of abelian groups closed under taking subgroups and direct limits* ⋄ C30 C60 ⋄

KHISAMIEV, N.G. *Strongly constructive models (Russian) (Kazakh summary)* ⋄ C30 C57 ⋄

KRATOCHVIL, P. *Note on a representation of universal algebras as subdirect powers* ⋄ C05 C30 ⋄

KRAUSS, P.H. *Universally complete universal theories*
⋄ C05 C30 C35 ⋄

MAKKAI, M. *Algebraic operations on classes of structures and their connections with logical formulas I,II (Hungarian) (English summary)*
⋄ C30 C40 C52 C75 ⋄

MALITZ, J. *Infinitary analogs of theorems from first order model theory* ⋄ C30 C40 C55 C75 ⋄

MCKENZIE, R. *Cardinal multiplication of structures with a reflexive relation* ⋄ C05 C30 C52 ⋄

MONTALI, T. *Pseudolimiti e formule SH*
⋄ B20 C30 G30 ⋄

MORLEY, M.D. *The Loewenheim-Skolem theorem for models with standard part* ⋄ C30 C55 ⋄

NADIU, G.S. *Elementary logic and models*
⋄ C07 C20 C30 ⋄

OLIN, P. *Direct multiples and powers of modules*
⋄ C30 C40 C60 ⋄

PABION, J.F. *Une notion de transcendance dans les structures relationnelles* ⋄ C30 C45 ⋄

PLATT, C.R. *Iterated limits of universal algebras*
⋄ C05 C30 ⋄

RICHTER, M.M. *Limites in Kategorien von Relationalsystemen* ⋄ C20 C30 G30 ⋄

SHELAH, S. *Two cardinal compactness*
⋄ C30 C52 C55 C80 ⋄

SILVER, J.H. *Measurable cardinals and Δ^1_3 well-orderings*
⋄ C20 C30 D55 E05 E15 E35 E45 E55 ⋄

SILVER, J.H. *Some applications of model theory in set theory* ⋄ C30 C55 D55 E05 E45 E55 ⋄

TRNKOVA, V. *On descriptive classification of set-functors. I,II* ⋄ C30 E55 G30 ⋄

1972

CLARK, D.M. & KRAUSS, P.H. *Monomorphic relational systems* ⋄ C05 C30 C52 ⋄

FEFERMAN, S. *Infinitary properties, local functors, and systems of ordinal functions*
⋄ C30 C40 C75 E10 F15 G30 ⋄

FITTLER, R. *A characteristic direct limit property*
⋄ C30 ⋄

FITTLER, R. *Some categories of models*
⋄ C30 C52 G30 ⋄

FLUM, J. *Die Automorphismenmengen der Modelle einer L_{Q_κ}-Theorie* ⋄ C30 C55 C80 ⋄

FLUM, J. *Hanf numbers and well-ordering numbers*
⋄ C30 C55 C75 C80 ⋄

FRAISSE, R. *Chaines compatibles pour un groupe de permutations. Application aux relations enchainables et monomorphes* ⋄ C07 C30 C52 C65 E07 ⋄

HODGES, W. *On order-types of models* ⋄ C30 C55 ⋄

MACINTYRE, A. *Direct powers with distinguished diagonal*
⋄ C30 ⋄

MARCJA, A. *Una generalizzazione del concetto di prodotto diretto* ⋄ C20 C30 C40 G05 ⋄

MARCUS, L. *A minimal prime model with an infinite set of indiscernibles* ⋄ C30 C50 C75 ⋄

MCKENZIE, R. *A method for obtaining refinement theorems with an application to direct products of semigroups* ⋄ C05 C30 ⋄

MONTEVERDI, D. *SH relative e pseudolimiti (English summary)* ⋄ C30 G30 ⋄

OLIN, P. *$\aleph_0$-categoricity of two-sorted structures*
⋄ C20 C30 C35 C60 ⋄

OLIN, P. *First order preservation theorems in two-sorted languages* ⋄ C30 C40 C60 ⋄

OLIN, P. *Products of two-sorted structures*
⋄ C30 C40 C75 ⋄

PAILLET, J.L. *La methode des "systemes logiquement inductifs"* ⋄ C07 C30 ⋄

SCHMERL, J.H. & SHELAH, S. *On power-like models for hyperinaccessible cardinals* ⋄ C30 C55 E55 ⋄

SHELAH, S. *Uniqueness and characterization of prime models over sets for totally transcendental first-order-theories* ⋄ C30 C45 C50 ⋄

TIMOSHENKO, E.I. *On the question of the elementary equivalence of groups (Russian)* ⋄ C30 C60 ⋄

WOLTER, H. *Untersuchung ueber Algebren und formalisierte Sprachen hoeherer Stufen (Russian, English and French summaries)*
⋄ B15 C20 C30 C40 C85 ⋄

1973

BACSICH, P.D. *An epi-reflector for universal theories*
⋄ C05 C30 C60 G30 ⋄

CHEPURNOV, B.A. *Some remarks on quasiuniversal classes (Russian)* ⋄ C30 C52 ⋄

FAKIR, S. *Monomorphismes et objects algebriquement et existentiellement clos dans les categories localement presentables. Applications aux monoides, categories et groupoides* ⋄ C05 C25 C30 G30 ⋄

FRASNAY, C. *Interpretation relationniste du seuil de recollement d'un groupe naturel* ⋄ C07 C30 E07 ⋄

FUJIWARA, T. *A generalization of Lyndon's theorem characterizing sentences preserved in subdirect products*
⋄ C05 C30 C40 ⋄

GREGORY, J. *Uncountable models and infinitary elementary extensions* ⋄ C30 C70 ⋄

HODGES, W. *Models in which all long indiscernible sequences are indiscernible sets* ⋄ C30 C55 ⋄

KOPPELBERG, S. *Injective hulls of chains* ⋄ C30 G05 ⋄

MONTEVERDI, D. *SH a piu sorte di variabili*
⋄ B20 C30 G30 ⋄

PAILLET, J.L. *La methode des systemes logiquement inductifs* ⋄ B10 C07 C30 ⋄

PLONKA, J. *On the sum of a direct system of relational systems* ⋄ C05 C30 ⋄

PODEWSKI, K.-P. *Zur Vergroesserung von Strukturen*
⋄ C15 C30 ⋄

SHAFAAT, A. *On products of relational structures*
⋄ B15 C30 ⋄

SILVER, J.H. *The bearing of large cardinals on constructibility* ⋄ C20 C30 E05 E45 E55 ⋄

SIMMONS, H. *An omitting types theorem with an application to the construction of generic structures*
⋄ C07 C25 C30 ⋄

WASZKIEWICZ, J. *ω_0-categoricity of generalized products*
⋄ C20 C30 C35 ⋄

1974

ASH, C.J. *Reduced powers and boolean extensions*
⋄ B25 C20 C30 G20 ⋄

COMER, S.D. *Restricted direct products and sectional representations* ⋄ C05 C30 C90 ⋄

FRAISSE, R. *Problematique apportee par les relations en theorie des permutations* ⋄ C07 C30 E07 ⋄

FRASNAY, C. *Theoreme de G-recollement (d'une famille d'ordres totaux). Application aux relations monomorphes, extension aux multirelations* ⋄ C07 C30 E07 ⋄

GAIFMAN, H. *Operations on relational structures, functors and classes I* ⋄ C20 C30 C40 G30 ⋄

GRAETZER, G. & SICHLER, J. *Agassiz sum of algebras* ⋄ C05 C30 C40 ⋄

JAKUBOWICZ, C.A. *On generalized products preserving compactness (Russian)* ⋄ C30 C52 G05 ⋄

KAISER, KLAUS *Direct limits in quasi-universal model classes* ⋄ B15 C30 C52 C85 ⋄

KOPPELBERG, S. & TITS, J. *Une propriete des produits directs infinis de groupes finis isomorphes* ⋄ C30 C60 E75 ⋄

KRAJEWSKI, S. *Predicative expansions of axiomatic theories* ⋄ C30 C62 E30 E70 F25 ⋄

MONTAGNA, F. *Sui limiti di certe teorie (English summary)* ⋄ C30 C35 C50 C62 H15 ⋄

NOVAK, VITEZSLAV & NOVOTNY, M. *Abstrakte Dimension von Strukturen* ⋄ C30 C65 ⋄

PARIS, J.B. *Patterns of indiscernibles* ⋄ C30 E45 E55 ⋄

PASHENKOV, V.V. *Duality of topological models (Russian)* ⋄ C30 C90 ⋄

PLONKA, J. *On connections between the decomposition of an algebra into sums of direct systems of subalgebras* ⋄ C05 C30 ⋄

PLONKA, J. *On the subdirect product of some equational classes of algebras* ⋄ C05 C30 ⋄

RACKOFF, C.W. *On the complexity of the theories of weak direct products: a preliminary report* ⋄ B25 C13 C30 C60 D15 ⋄

SALIJ, V.N. *Reduced extensions of models (Russian)* ⋄ C20 C30 ⋄

WIERZEJEWSKI, J. *A note on stability and products (Russian summary)* ⋄ C20 C30 C45 ⋄

1975

ANDLER, D. *Semi-minimal theories and categoricity* ⋄ C30 C35 C55 ⋄

BENDA, M. *Construction of models from groups of permutations* ⋄ C30 ⋄

BERTHIER, D. *Stability of non-model-complete theories; products, groups* ⋄ C20 C30 C45 C60 ⋄

BURRIS, S. *Boolean powers* ⋄ B25 C05 C30 C90 ⋄

CLARE, F. *Embedding theorems for infinite symmetric groups* ⋄ C30 C60 ⋄

GREEN, J. *Consistency properties for finite quantifier languages* ⋄ C30 C40 C70 C75 ⋄

HALPERN, J.D. *Nonstandard combinatorics* ⋄ C30 E05 G05 ⋄

HAUSCHILD, K. *Ueber zwei Spiele und ihre Anwendung in der Modelltheorie (Russian, English and French summaries)* ⋄ B25 C07 C30 C35 C85 ⋄

JERVELL, H.R. *Conservative endextensions and the quantifier "there exist uncountably many"* ⋄ C30 C80 ⋄

KEISLER, H.J. *Constructions in model theory* ⋄ C25 C30 C50 C75 C95 C98 ⋄

LOPEZ, G. *Le probleme de l'isomorphie des restrictions strictes, pour les extensions a un element de relations enchainables non constantes (English summary)* ⋄ C13 C30 E07 ⋄

PASHENKOV, V.V. *On dualities (Russian)* ⋄ C30 C90 ⋄

PLONKA, J. *On relations between decompositions of an algebra into the sums of direct systems of subalgebras* ⋄ C05 C30 ⋄

PLONKA, J. *Remark on direct products and the sums of direct systems of algebras (Russian summary)* ⋄ C05 C30 ⋄

SACERDOTE, G.S. *Infinite coforcing in model theory* ⋄ C25 C30 C60 ⋄

SACERDOTE, G.S. *Projective model theory and coforcing* ⋄ C25 C30 C35 C40 C50 C60 ⋄

SICHLER, J. *Note on free products of Hopfian lattices* ⋄ C30 G10 ⋄

WEISPFENNING, V. *Model-completeness and elimination of quantifiers for subdirect products of structures* ⋄ C10 C30 C35 C52 C60 C90 ⋄

WINKLER, P.M. *Model-completeness and Skolem expansions* ⋄ C25 C30 C35 C60 ⋄

1976

APPLESON, R.R. *Zero-divisors among finite structures of a fixed type* ⋄ C05 C13 C30 ⋄

GAIFMAN, H. *Models and types of Peano's arithmetic* ⋄ C30 C62 H15 ⋄

GILLAM, D.W.H. *Saturation and omitting types* ⋄ C07 C30 C50 ⋄

GNANVO, C. *Γ-extension logique (English summary)* ⋄ C30 C52 ⋄

HUTCHINSON, J.E. *Order types of ordinals in models of set theory* ⋄ C30 C62 E10 ⋄

JONSSON, B. & OLIN, P. *Elementary equivalence and relatively free products of lattices* ⋄ C05 C30 G10 ⋄

OLIN, P. *Free products and elementary types of boolean algebras* ⋄ C30 C50 G05 ⋄

PARIS, J.B. *Sets satisfying $O^{\#}$* ⋄ C30 E45 E55 E65 ⋄

PINCUS, D. *Two model theoretic ideas in independence proofs* ⋄ C30 E25 E35 ⋄

RACKOFF, C.W. *On the complexity of the theories of weak direct powers* ⋄ B25 C13 C30 C60 D15 ⋄

SCHMERL, J.H. *Remarks on self-extending models* ⋄ C30 C55 E55 ⋄

SCHUMACHER, D. *Absolutely free algebras in a topos containing an infinite object* ⋄ C05 C30 G30 ⋄

TOMASIK, J. *On products of neat structures* ⋄ C30 C50 ⋄

VOLGER, H. *The Feferman-Vaught theorem revisited* ⋄ C20 C30 C90 ⋄

WIERZEJEWSKI, J. *On stability and products* ⋄ C20 C30 C45 ⋄

1977

BANASCHEWSKI, B. & NELSON, EVELYN *Elementary properties of limit reduced powers with applications to boolean powers* ⋄ C05 C20 C30 C90 ⋄

DAWES, A.M. *End extensions which are models of a given theory* ⋄ C07 C15 C30 C62 ⋄

ELLENTUCK, E. *Boolean valued rings*
⋄ C30 C60 C90 D50 ⋄

FORREST, W.K. *Model theory for universal classes with the amalgamation property: A study in the foundations of model theory and algebra*
⋄ C05 C25 C30 C35 C45 C52 C60 ⋄

HARRINGTON, L.A. & PARIS, J.B. *A mathematical incompleteness in Peano arithmetic*
⋄ C30 C62 E05 F30 ⋄

HIGMAN, G. *Homogeneous relations*
⋄ C07 C30 C65 E05 ⋄

HODGES, W. & LACHLAN, A.H. & SHELAH, S. *Possible orderings of an indiscernible sequence* ⋄ C30 E07 ⋄

KNIGHT, J.F. *Skolem functions and elementary embeddings* ⋄ C15 C30 ⋄

LOLLI, G. *Indiscernibili in teoria dei modelli e in teoria degli insiemi (English summary)*
⋄ C30 C98 E05 E45 E55 ⋄

MACNAB, D.S. *Some applications of double-negation sheafification* ⋄ C30 C90 F50 G30 ⋄

MARCJA, A. *Prodotti booleani e prodotti booleani ridotti*
⋄ C20 C30 C90 ⋄

MIJAJLOVIC, Z. *A note on elementary end extension*
⋄ C30 C62 ⋄

MORLEY, M.D. *Homogenous sets*
⋄ C30 C55 C98 E05 E55 ⋄

OLIN, P. *Elementary properties of V-free products of groups* ⋄ C05 C30 C60 ⋄

WIERZEJEWSKI, J. *A note on products and degree of types*
⋄ C30 C45 ⋄

1978

ABRAMSON, F.G. & HARRINGTON, L.A. *Models without indiscernibles*
⋄ C30 C50 C55 C62 C65 C75 E05 E40 H15 ⋄

ANAPOLITANOS, D.A. *A new kind of indiscernibles*
⋄ C07 C30 C50 ⋄

ANAPOLITANOS, D.A. *A theorem on absolute indiscernibles* ⋄ C30 C65 ⋄

ANAPOLITANOS, D.A. *Absolute indiscernibles and standard models of ZFC* ⋄ C30 C62 E40 ⋄

BENDA, M. *Compactness for omitting of types*
⋄ C20 C30 C55 C70 C75 E55 ⋄

BRYANT, R.M. & GROVES, J.R.J. *Wreath products and ultraproducts of groups* ⋄ C20 C30 C60 ⋄

BURRIS, S. *Bounded boolean powers and $\equiv_n$*
⋄ C30 G05 ⋄

BURRIS, S. *Rigid boolean powers*
⋄ C05 C30 C50 G05 ⋄

CAMERON, P.J. *Orbits of permutation groups on unordered sets* ⋄ C30 C60 E05 ⋄

HALKOWSKA, K. & PLONKA, J. *On formulas preserved by the sum of a direct system of relational systems*
⋄ C30 C40 ⋄

KNIGHT, J.F. *An inelastic model with indiscernibles*
⋄ C15 C30 C50 ⋄

LACHLAN, A.H. *Skolem functions and elementary extensions* ⋄ C30 C50 ⋄

LUCAS, F. *Equivalence elementaire et produits (English summary)* ⋄ C30 ⋄

MAKOWSKY, J.A. *Some observations on uniform reduction for properties invariant on the range of definable relations* ⋄ C30 C40 C75 C95 ⋄

MLCEK, J. *A note on cofinal extensions and segments*
⋄ C30 C62 C65 ⋄

POIZAT, B. *Suites d'indiscernables dans les theories stables*
⋄ C30 C45 ⋄

REGGIANI, C. *Le SH nei limiti diretti di strutture e di interpretazioni (English summary)*
⋄ C05 C30 C40 ⋄

SCHMERL, J.H. *On $\aleph_0$-categoricity of filtered boolean extensions* ⋄ C30 C35 ⋄

SHELAH, S. *End extensions and numbers of countable models* ⋄ C15 C30 C55 C62 E50 ⋄

SHELAH, S. *Models with second order properties. I: Boolean algebras with no definable automorphisms*
⋄ C07 C30 C50 C55 E50 G05 ⋄

1979

BAJZHANOV, B.S. & OMAROV, B. *Restriction of a theory to a formula (Russian)* ⋄ C30 C45 ⋄

BROESTERHUIZEN, G. *Extensions of automorphisms*
⋄ C30 C50 ⋄

BURRIS, S. & WERNER, H. *Sheaf constructions and their elementary properties*
⋄ B25 C05 C20 C25 C30 C60 C90 ⋄

CLARK, D.M. & KRAUSS, P.H. *Global subdirect products*
⋄ C05 C30 C90 ⋄

COSTE, M. *Localisation, spectra and sheaf representation*
⋄ C30 C90 ⋄

CZELAKOWSKI, J. *A purely algebraic proof of the omitting types theorem* ⋄ C07 C15 C30 G99 ⋄

DANILOVA, S.A. *The group of automorphisms of a model (Russian)* ⋄ C07 C30 ⋄

DAVEY, B.A. & WERNER, H. *Injectivity and boolean powers* ⋄ C30 G05 ⋄

FRASNAY, C. *Sur quelques classes universelles de relations m-aires* ⋄ C07 C30 C52 E07 ⋄

FRIEDMAN, H.M. *On the naturalness of definable operations* ⋄ C30 C62 E35 E47 ⋄

GARAVAGLIA, S. *Direct product decomposition of theories of modules* ⋄ C30 C45 C60 ⋄

HICKIN, K.K. & PHILLIPS, R.E. *A construction of locally finite simple groups* ⋄ C30 C60 ⋄

REGGIANI, C. *Conservazione delle SH di tipo ∀∃ nei limiti diretti (English summary)* ⋄ C05 C30 C40 ⋄

TIMOSHENKO, E.I. *On elementary theories of wreath products (Russian)* ⋄ C30 C60 ⋄

VOLGER, H. *Sheaf representations of algebras and transfer theorems for model theoretic properties*
⋄ B25 C05 C30 C35 C90 ⋄

WEISPFENNING, V. *Lattice products*
⋄ C10 C25 C30 C35 C90 ⋄

ZLATOS, P. *A characterization of some boolean powers*
⋄ C05 C20 C30 C75 ⋄

1980

ARAGANE, K. & FUJIWARA, T. *Free product decompositions in a regular class* ⋄ C30 ⋄

BAJRAMOV, R.A. & MAMEDOV, O.M. *Σ-subdirect representations in axiomatic classes (Russian) (English summary)* ⋄ C05 C30 C52 ⋄

BANASCHEWSKI, B. & NELSON, EVELYN *Boolean powers as algebras of continuous functions*
⋄ C05 C20 C30 C50 ⋄

BENIAMINOV, E.M. *An algebraic approach to models of data bases of the relational type (Russian)*
⋄ B75 C30 ⋄

BURRIS, S. & WERNER, H. *Remarks on boolean products*
⋄ C05 C30 C52 G05 ⋄

CHARRETTON, C. & POUZET, M. *Les chaines dans les modeles d'Ehrenfeucht-Mostowski (English summary)*
⋄ C30 C55 ⋄

DICKMANN, M.A. *Types remarquables et extensions de modeles dans l'arithmetique de Peano I*
⋄ C25 C30 C62 ⋄

EDA, K. *Limit systems and chain condition*
⋄ C30 E40 G05 ⋄

HINNION, R. *Contraction de structures et application a NFU (les "New Foundations" des Quine avec extensionnalite pour les ensembles non vides). Definition du "degre de nonextensionnalite" d'une relation quelconque (English summary)*
⋄ C30 C62 E35 E70 ⋄

LUO, LIBO *The union and product of models, and homogeneous models (Chinese) (English summary)*
⋄ C15 C30 C50 ⋄

LYNCH, J.F. *Almost sure theories* ⋄ C13 C15 C30 ⋄

MIRAGLIA, F. *Relations between structures of stable continuous functions and filtered powers*
⋄ C30 C35 C90 ⋄

OLIN, P. *Varietal free products of bands*
⋄ C05 C30 G10 ⋄

PINUS, A.G. *Weak commutativity of scattered summation of linear orders (Russian)* ⋄ C30 E07 ⋄

PRESIC, M.D. *Some results on the extending of models*
⋄ C30 ⋄

RESSAYRE, J.-P. *Types remarquables et extensions de modeles dans l'arithmetique de Peano. II (Avec un appendice de M.A.Dickmann)* ⋄ C30 C62 E05 ⋄

RUBIN, M. & SHELAH, S. *On the elementary equivalence of automorphism groups of boolean algebras; downward Skolem Loewenheim theorems and compactness of related quantifiers*
⋄ C07 C30 C75 C80 C95 G05 ⋄

RYCHKOV, S.V. *On direct products of abelian groups (Russian)* ⋄ C30 C60 E55 E75 ⋄

TAKAHASHI, MAKOTO *Topological powers and reduced powers* ⋄ C20 C30 C90 ⋄

1981

ANAPOLITANOS, D.A. *Cyclic indiscernibles and Skolem functions* ⋄ C07 C30 ⋄

BURRIS, S. & MCKENZIE, R. *Decidability and boolean representations* ⋄ B25 C05 C30 C60 ⋄

CAMERON, P.J. *Orbits of permutation groups on unordered sets II* ⋄ C30 C60 E05 ⋄

CHATZIDAKIS, Z. *La representation en termes de faisceaux des modeles de la theorie elementaire de la multiplication des entiers naturels*
⋄ C10 C30 C62 C90 ⋄

FRANZEN, B. *Algebraic compactness of filter quotients*
⋄ C05 C30 C60 ⋄

GUZICKI, W. *The equivalence of definable quantifiers in second order arithmetic*
⋄ C30 C55 C62 C80 E40 E45 F35 ⋄

HINNION, R. *Extensional quotients of structures and applications to the study of the axiom of extensionality*
⋄ C30 C62 E30 E35 E70 ⋄

HORIGUCHI, H. *A definition of the category of boolean-valued models* ⋄ C20 C30 C90 G30 ⋄

MURAWSKI, R. *A simple remark on satisfaction classes, indiscernibles and recursive saturation*
⋄ C30 C50 C57 H15 ⋄

PILLAY, A. *Partition properties and definable types in Peano arithmetic* ⋄ C30 C62 E05 ⋄

PINUS, A.G. *The spectrum of rigid systems of Horn classes (Russian)* ⋄ C05 C30 C50 ⋄

POUZET, M. *Application de la notion de relation presque-enchainable au denombrement des restrictions finies d'une relation* ⋄ C07 C30 E07 ⋄

PRESIC, M.D. & PRESIC, S.B. *Some embedding theorems*
⋄ C30 ⋄

RODRIGUEZ ARTALEJO, M. *A method of constructing structures based on the distributive normal form (Spanish)* ⋄ C07 C30 ⋄

RODRIGUEZ ARTALEJO, M. *Eine syntaktisch-algebraische Methode zur Konstruktion von Modellen*
⋄ B10 C07 C30 C57 ⋄

TAKAHASHI, MAKOTO *Iterated boolean powers (Japanese)*
⋄ C30 E40 ⋄

WANG, SHIQIANG *Two cases of application of the method of construction by constants in lattice-valued model theory (Chinese)* ⋄ C30 C90 ⋄

1982

ANELLIS, I.H. *Boolean groups* ⋄ C30 C60 G05 ⋄

APPS, A.B. *Boolean powers of groups*
⋄ C13 C30 C60 ⋄

BROESTERHUIZEN, G. *Locally orderable structures*
⋄ C30 ⋄

CAMERON, P.J. *Orbits and enumeration* ⋄ C30 C60 ⋄

DZGOEV, V.D. *Constructivizations of direct products of algebraic systems (Russian)*
⋄ C30 C57 D45 G05 G10 ⋄

EDA, K. *A boolean power and a direct product of abelian groups* ⋄ C30 C60 C90 E40 E75 ⋄

KASHIWAGI, T. *Note on algebraic extensions in universal classes* ⋄ C05 C25 C30 ⋄

MCKENZIE, R. *Subdirect powers of non-abelian groups*
⋄ C30 C60 ⋄

MESKHI, S. *Complete boolean powers of quasiprimal algebras with lattice reducts* ⋄ C05 C30 ⋄

NELSON, GEORGE C. *The periodic power of $\mathfrak{A}$ and complete Horn theories* ⋄ C20 C30 C35 ⋄

PUEHRINGER, C. *Ein Vollstaendigkeitsbeweis fuer die Theorie der atomlosen ∩ − ^{c}-partiellen Booleschen Algebren* ⋄ C30 G05 ⋄

SAIN, I. *On classes of algebraic systems closed with respect to quotients* ⋄ C05 C20 C30 C52 ⋄

SAIN, I. *Weak products for universal algebra and model theory* ⋄ C05 C30 ⋄

WEISPFENNING, V. *Valuation rings and Boolean products*
⋄ C10 C30 C60 ⋄

WERNER, H. *Sheaf constructions in universal algebra and model theory*
⋄ B25 C05 C25 C30 C35 C90 G05 ⋄

1983

APPS, A.B. $\aleph_0$-*categorical finite extensions of boolean powers* ⋄ C30 C35 C60 ⋄

BAKER, K.A. *Nondefinability of projectivity in lattice varieties* ⋄ C05 C30 C40 G10 ⋄

BALDWIN, JOHN T. & SHELAH, S. *The structure of saturated free algebras* ⋄ C05 C30 C45 C50 ⋄

BENTHEM VAN, J.F.A.K. & HUMBERSTONE, I.L. *Hallden-completeness by gluing of Kripke frames*
⋄ B45 C30 C90 ⋄

BURRIS, S. *Boolean constructions*
⋄ A10 B25 C05 C30 C98 D35 G05 ⋄

CAMERON, P.J. *Orbits of permutation groups on unordered sets. III: Imprimitive groups. IV: Homogeneity and transitivity* ⋄ C30 C60 E05 ⋄

CHARRETTON, C. & POUZET, M. *Chains in Ehrenfeucht-Mostowski models* ⋄ C30 ⋄

CHARRETTON, C. & POUZET, M. *Comparaison des structures engendrees par des chaines*
⋄ C30 C65 G05 ⋄

COMPTON, K.J. *Some useful preservation theorems*
⋄ C30 C40 C65 ⋄

EDA, K. *Infinite products of algebras and measurable cardinals (Japanese)* ⋄ C30 E55 E75 ⋄

EDA, K. *On a boolean power of a torsion free abelian group* ⋄ C30 C60 C90 E40 E75 G05 ⋄

GUZICKI, W. *Definable quantifiers in second order arithmetic and elementary extensions of ω-models*
⋄ C30 C55 C62 C80 E40 E50 F35 ⋄

KIERSTEAD, H.A. & REMMEL, J.B. *Indiscernibles and decidable models* ⋄ C30 C45 C57 C80 ⋄

KRANAKIS, E. *Definable Ramsey and definable Erdoes ordinals* ⋄ C30 D60 E05 E45 E55 ⋄

MIJAJLOVIC, Z. *On Σ_1^0-extensions of ω*
⋄ C15 C30 C62 ⋄

1984

BAJRAMOV, R.A. & MAMEDOV, O.M. *Nonstandard subdirect representations in axiomatizable classes (Russian)* ⋄ C05 C30 C52 ⋄

BAUDISCH, A. *Tensor products of modules and elementary equivalence* ⋄ C30 C52 C60 ⋄

BUDKIN, A.I. *Quasiidentities and direct wreaths of groups (Russian)* ⋄ C05 C30 C60 ⋄

CLARK, D.M. $\aleph_0$-*categoricity in infra primal varieties*
⋄ C05 C30 C35 E05 ⋄

EDA, K. & HIBINO, K. *On boolean powers of the group Z and (ω, ω)-weak distributivity*
⋄ C30 C60 C90 E35 E50 E75 G05 ⋄

ENAYAT, A. *On certain elementary extensions of models of set theory* ⋄ C30 C62 E55 ⋄

FRASNAY, C. *Relations enchaineables, rangements et pseudo-rangements (English summary)*
⋄ C07 C13 C30 E07 ⋄

GARAVAGLIA, S. & PLOTKIN, J.M. *Separation properties and boolean powers* ⋄ C30 E50 G05 G10 ⋄

GIORGETTA, D. & SHELAH, S. *Existentially closed structures in the power of the continuum*
⋄ C15 C25 C30 C52 C55 C60 C75 ⋄

GRAETZER, G. & KELLY, DAVID *Free m-products of lattices I* ⋄ C30 G10 ⋄

HEINDORF, L. *Regular ideals and boolean pairs*
⋄ C30 C85 D35 G05 ⋄

HODGES, W. *Models built on linear orderings (French summary)* ⋄ C30 C55 C98 E07 ⋄

ISKANDER, A.A. *Extensions of algebraic systems*
⋄ C05 C30 ⋄

JOKISZ-ZABILSKA, I. *On a construction of many-sorted structures* ⋄ C30 ⋄

KAUFMANN, M. *On expandability of models of arithmetic and set theory to models of weak second-order theories*
⋄ C15 C30 C62 C70 C75 E30 E70 ⋄

LEBLANC, H. *A new semantics for first-order logic, multivalent and mostly intensional* ⋄ B10 C30 ⋄

MCKENZIE, R. *A new product of algebras and a type reduction theorem* ⋄ C05 C30 ⋄

NELSON, GEORGE C. *Boolean powers, recursive models, and the Horn theory of a structure*
⋄ C05 C20 C30 C57 G05 ⋄

NOVAK, VITEZSLAV & NOVOTNY, M. *On a power of cyclically ordered sets (Russian and Czech summaries)*
⋄ C30 E07 ⋄

OMAROV, A.I. *Elementary theory of D-degrees (Russian)*
⋄ C20 C30 C50 D30 G10 ⋄

RICHTER, M.M. *Some aspects of nonstandard methods in general algebra* ⋄ C05 C30 H05 ⋄

ROSENTHAL, D. *The order indiscernibles of divisible ordered abelian groups* ⋄ C30 C60 ⋄

TSUBOI, A. *Two theorems on the existence of indiscernible sequences* ⋄ C30 ⋄

VOLGER, H. *Filtered and stable boolean powers are relativized full boolean powers*
⋄ B25 C20 C30 C90 ⋄

1985

BALDWIN, JOHN T. *Definable second-order quantifiers*
⋄ C30 C80 C85 C98 ⋄

ENAYAT, A. *Weakly compact cardinals in models of set theory* ⋄ C30 C62 E50 E55 E65 ⋄

GAVALEC, M. *Local properties of complete Boolean products* ⋄ C30 C62 G05 ⋄

GRILLIOT, T.J. *Disturbing arithmetic*
⋄ C07 C30 C40 C62 ⋄

HORIGUCHI, H. *The category of boolean-valued models and its applications* ⋄ C30 C90 E40 G30 ⋄

KIERSTEAD, H.A. & REMMEL, J.B. *Degrees of indiscernibles in decidable models*
⋄ C30 C35 C57 D30 ⋄

LAVENDHOMME, R. & LUCAS, T. *A non-boolean version of Feferman-Vaught's theorem* ⋄ C30 C90 ⋄

MLCEK, J. *Some automorphisms of natural numbers in the alternative set theory* ⋄ C07 C30 E70 H15 ⋄

NOVAK, VITEZSLAV *On a power of relational stuctures*
⋄ C05 C30 ⋄

PINO, R. & RESSAYRE, J.-P. *Definable ultrafilters and elementary end extensions*
⋄ C20 C30 C62 E05 E45 E47 ⋄

PINUS, A.G. *Applications of boolean powers of algebraic systems (Russian)*
⋄ C05 C30 C55 C57 C80 C85 D35 E50 G05 ⋄

POINT, F. *Transfer properties of discriminator varieties*
⋄ C10 C25 C30 C35 ⋄

SCHMERL, J.H. *Recursively saturated models generated by indiscernibles* ⋄ C30 C50 C57 ⋄

TAKAHASHI, MAKOTO *Iterated boolean powers*
⋄ C30 E40 ⋄

TUSCHIK, H.-P. *Algebraic connections between definable predicates* ⋄ C15 C30 C40 ⋄

C35 Categoricity and completeness of theories

1917

HUNTINGTON, E.V. *Complete existential theory of the postulates for serial order* ⋄ B30 C13 C35 E07 ⋄

HUNTINGTON, E.V. *Complete existential theory of the postulates for well ordered sets*
⋄ B30 C13 C35 E07 ⋄

1926

LANGFORD, C.H. *Analytic completeness of postulate sets*
⋄ B20 C35 C65 E07 ⋄

LANGFORD, C.H. *Some theorems on deducibility*
⋄ B30 C10 C35 C65 ⋄

LANGFORD, C.H. *Theorems on deducibility (second paper)*
⋄ B30 C10 C35 C65 ⋄

1927

LANGFORD, C.H. *On a type of completeness characterizing the general laws for separation of point pairs*
⋄ C35 C65 E07 ⋄

1928

HAERLEN, H. *Ueber die Vollstaendigkeit und Entscheidbarkeit* ⋄ A05 C35 ⋄

SKOLEM, T.A. *Ueber die mathematische Logik*
⋄ A05 B25 B98 C07 C10 C35 ⋄

1930

PRESBURGER, M. *Ueber die Vollstaendigkeit eines gewissen Systems der Arithmetik ganzer Zahlen, in welchem die Addition als einzige Operation hervortritt*
⋄ B25 B28 C10 C35 F30 ⋄

1933

WAJSBERG, M. *Beitrag zur Metamathematik*
⋄ A05 B10 B30 B96 C35 F99 ⋄

1934

TARSKI, A. *Some methodological investigations on the definability of terms (Polish)* ⋄ B10 C35 C40 ⋄

1935

LINDENBAUM, A. & TARSKI, A. *Ueber die Beschraenktheit der Ausdrucksmittel deduktiver Theorien*
⋄ B30 C07 C35 C65 ⋄

1936

TARSKI, A. *Grundzuege des Systemkalkuels. Teil II*
⋄ A05 B22 C07 C35 G15 ⋄

1939

LANGFORD, C.H. *A theorem on deducibility for second-order functions* ⋄ B15 C35 C65 C85 ⋄

1948

TARSKI, A. *A decision method for elementary algebra and geometry* ⋄ B25 B30 C10 C35 C60 ⋄

1949

KUSTAANHEIMA, P. *Ueber die Vollstaendigkeit der Axiomensysteme mit einem endlichen Individuenbereich* ⋄ C13 C35 C85 ⋄

1950

JANICZAK, A. *A remark conerning decidability of complete theories* ⋄ B25 C35 ⋄

1952

TARSKI, A. *Some notions and methods on the borderline of algebra and metamathematics*
⋄ C07 C35 C52 C60 ⋄

1953

WANG, HAO *Quelques notions d'axiomatique*
⋄ B25 B30 C07 C35 E30 F30 ⋄

1954

LOS, J. *On the categoricity in power of elementary deductive systems and some related problems* ⋄ C35 ⋄

VAUGHT, R.L. *Applications of the Loewenheim-Skolem-Tarski theorem to problems of completeness and decidability* ⋄ B25 C07 C35 ⋄

1955

HENKIN, L. *On a theorem of Vaught* ⋄ B25 C35 ⋄

ROBINSON, A. *Ordered structures and related concepts*
⋄ C30 C35 C50 C60 ⋄

1956

RIBEIRO, H. *The notion of universal completeness*
⋄ C05 C35 ⋄

ROBINSON, A. *Complete theories*
⋄ B25 C35 C60 C98 ⋄

ROBINSON, A. *Completeness and persistence in the theory of models* ⋄ C30 C35 C52 ⋄

ROZHANSKAYA, YU.A. *On the equivalence of two definitions of completeness of a system of axioms of A. Kolmogorov and H. Weyl (Russian)* ⋄ C35 ⋄

SCHWABHAEUSER, W. *Ueber die Vollstaendigkeit der elementaren euklidischen Geometrie*
⋄ B30 C35 C65 ⋄

1957

EHRENFEUCHT, A. *Application of games to some problems of mathematical logic (Russian summary)*
⋄ C07 C35 C65 E60 ⋄

EHRENFEUCHT, A. *On theories categorical in power*
⋄ C30 C35 ⋄

EHRENFEUCHT, A. *Two theories with axioms built by means of pleonasmus* ⋄ B25 C35 D35 ⋄

KOCHEN, S. *Completeness of algebraic systems in higher order calculi* ⋄ B15 C35 C60 C85 ⋄

RIBEIRO, H. *Universal completeness* ⋄ C05 C35 ⋄
ROBINSON, A. *Relative model-completeness and the elimination of quantifiers* ⋄ C10 C35 ⋄
ROBINSON, A. *Some problems of definability in the lower predicate calculus* ⋄ C35 C40 C60 ⋄
TARSKI, A. & VAUGHT, R.L. *Elementary (arithmetical) extensions* ⋄ C05 C07 C35 C55 C60 ⋄

1958
ROBINSON, A. *Relative model-completeness and the elimination of quantifiers (English, French, and German summaries)* ⋄ C10 C35 C98 ⋄
SCOTT, D.S. *Convergent sequences of complete theories* ⋄ C35 C52 ⋄

1959
ROBINSON, A. *On the concept of a differentially closed field* ⋄ C35 C60 ⋄
ROBINSON, A. *Solution of a problem of Tarski* ⋄ B25 C35 C60 ⋄
ROZHANSKAYA, YU.A. *On the equivalence of two definitions of the completeness of a system of axioms (Russian)* ⋄ C35 ⋄
RYLL-NARDZEWSKI, C. *On the categoricity in power $\leq \aleph_0$ (Russian summary)* ⋄ C35 ⋄
SCHWABHAEUSER, W. *Entscheidbarkeit und Vollstaendigkeit der elementaren hyperbolischen Geometrie* ⋄ B25 B30 C35 C65 ⋄
SVENONIUS, L. $\aleph_0$-*categoricity in first-order predicate calculus* ⋄ C35 ⋄

1960
ROBINSON, A. & ZAKON, E. *Elementary properties of ordered abelian groups* ⋄ B25 C35 C60 C65 ⋄
SCOTT, D.S. *On a theorem of Rabin* ⋄ C35 C57 C62 D35 ⋄

1961
ENGELER, E. *Unendliche Formeln in der Modell-Theorie* ⋄ C15 C35 C40 C75 ⋄
KOCHEN, S. *Ultraproducts in the theory of models* ⋄ C20 C35 C52 C60 ⋄
RIBEIRO, H. *On the universal completeness of classes of relational systems* ⋄ C05 C35 ⋄
ROBINSON, A. *Model theory and non-standard arithmetic* ⋄ C35 C50 C57 C60 C62 H15 ⋄
VAUGHT, R.L. *Denumerable models of complete theories (Polish)* ⋄ C15 C35 C50 ⋄

1962
GILMORE, P.C. *Some forms of completeness* ⋄ C35 ⋄
GRZEGORCZYK, A. *A kind of categoricity* ⋄ C35 C50 C62 C65 ⋄
GRZEGORCZYK, A. *On the concept of categoricity (Polish and Russian summaries)* ⋄ C35 ⋄
KARGAPOLOV, M.I. *On the elementary theory of abelian groups (Russian)* ⋄ C35 C60 ⋄
KRUSE, A.H. *Completion of mathematical systems* ⋄ C25 C35 C62 ⋄
SCHWABHAEUSER, W. *On completeness and decidability and some non-definable notions of elementary hyperbolic geometry* ⋄ B25 C35 C65 ⋄

1963
MORLEY, M.D. *On theories categorical in uncountable powers* ⋄ C35 C45 ⋄

1964
KEISLER, H.J. *Complete theories of algebraically closed fields with distinguished subfields* ⋄ B25 C35 C60 ⋄
LINDSTROEM, P. *On model-completeness* ⋄ C35 ⋄
ROWBOTTOM, F. *The Los conjecture for uncountable theories* ⋄ C35 C45 ⋄

1965
AX, J. & KOCHEN, S. *Diophantine problems over local fields I,II (II: A complete set of axioms for p-adic number theory)* ⋄ B25 C20 C35 C60 ⋄
BUECHI, J.R. *Relatively categorical and normal theories* ⋄ C35 C40 C60 ⋄
ERSHOV, YU.L. *On elementary theories of local fields (Russian)* ⋄ B25 C35 C60 ⋄
ERSHOV, YU.L. *On the elementary theory of maximal normed fields (Russian)* ⋄ B25 C35 C60 ⋄
MORLEY, M.D. *Categoricity in power* ⋄ C35 C45 ⋄
RIBEIRO, H. & SCHWABAUER, R. *A remark on equational completeness* ⋄ C05 C35 ⋄

1966
APELT, H. *Axiomatische Untersuchungen ueber einige mit der Presburgerschen Arithmetik verwandte Systeme* ⋄ B28 C35 C80 F30 ⋄
AX, J. & KOCHEN, S. *Diophantine problems over local fields, III: Decidable fields* ⋄ B25 C10 C20 C35 C60 C65 ⋄
ERSHOV, YU.L. *On the elementary theory of maximal normed fields II (Russian)* ⋄ C35 C60 ⋄
KEISLER, H.J. *Some model theoretic results for ω-logic* ⋄ C35 C50 C55 C75 ⋄
KHISAMIEV, N.G. *Certain applications of the method of extension of mappings (Russian) (Kazakh summary)* ⋄ C35 C52 ⋄

1967
ERSHOV, YU.L. *Fields with a solvable theory (Russian)* ⋄ B25 C35 C60 ⋄
ERSHOV, YU.L. *On the elementary theory of maximal normed fields III (Russian) (English summary)* ⋄ C35 C60 ⋄
HAUSCHILD, K. *Verallgemeinerung eines Satzes von Cantor* ⋄ C07 C15 C35 ⋄
MORLEY, M.D. *Countable models of $\aleph_1$-categorical theories* ⋄ C15 C35 ⋄
TARSKI, A. *The completeness of elementary algebra and geometry* ⋄ B25 B30 C10 C35 C60 C65 ⋄

1968
GRZEGORCZYK, A. *Logical uniformity by decomposition and categoricity in $\aleph_0$* ⋄ C35 ⋄
HAUSCHILD, K. *Metatheoretische Eigenschaften gewisser Klassen von elementaren Theorien* ⋄ B28 C15 C35 E30 E70 ⋄
ROSENSTEIN, J.G. *Theories which are not $\aleph_0$-categorical* ⋄ C35 C65 ⋄

SURMA, S.J. *Four studies in metamathematics (Polish) (Russian and English summaries)*
◇ B22 C07 C35 E25 ◇

1969

BOUVERE DE, K.L. *Remarks on classification of theories by their complete extensions* ◇ C07 C35 ◇

KAISER, KLAUS *Ueber eine Verallgemeinerung der Robinsonschen Modellvervollstaendigung I,II (II: Abgeschlossene Algebren)* ◇ C05 C25 C35 ◇

KOKORIN, A.I. & KOZLOV, G.T. *Elementary theory of abelian groups without torsion, with a predicate for selecting subgroups (Russian)* ◇ B25 C35 C60 ◇

KOPPERMAN, R.D. *Applications of infinitary languages to analysis* ◇ C35 C60 C65 C75 ◇

RESSAYRE, J.-P. *Sur les theories du premier ordre categorique en un cardinal* ◇ C30 C35 C55 ◇

ROSENSTEIN, J.G. $\aleph_0$-*categoricity of linear orderings*
◇ C35 C65 E07 ◇

SHELAH, S. *Categoricity of classes of models* ◇ C35 ◇

SHELAH, S. *Stable theories* ◇ C35 C45 ◇

WASZKIEWICZ, J. & WEGLORZ, B. *On ω_0-categoricity of powers (Russian summary)* ◇ C30 C35 C50 ◇

1970

BARWISE, J. & ROBINSON, A. *Completing theories by forcing* ◇ C25 C35 ◇

BREDIKHIN, S.V. & ERSHOV, YU.L. & KAL'NEJ, V.E. *Fields with two linear orderings (Russian)* ◇ C35 C60 ◇

CHUDNOVSKY, G.V. *Problems of the theory of models, related to categoricity (Russian)*
◇ C35 C45 C50 C52 C55 C75 ◇

CUSIN, R. *Theories quasi-completes* ◇ C35 G05 ◇

EKLOF, P.C. & SABBAGH, G. *Model-completions and modules* ◇ C25 C35 C52 C60 ◇

GRZEGORCZYK, A. *Decision procedures for theories categorical in $\aleph_0$* ◇ B25 C35 ◇

KAISER, KLAUS *Zur Einbettungsbedingung im Zusammenhang mit einer Programmiersprache von Engeler* ◇ B75 C20 C35 C52 ◇

MUSTAFIN, T.G. & TAJMANOV, A.D. *Countable models of theories which are categorical in power $\aleph_1$ but not in power $\aleph_0$ (Russian)* ◇ C15 C35 ◇

SHELAH, S. *A note on Hanf numbers*
◇ C35 C55 C75 ◇

SHELAH, S. *Finite diagrams stable in power*
◇ C35 C45 C50 C55 ◇

SHELAH, S. *On theories T categorical in |T|*
◇ C35 C45 ◇

SUZUKI, Y. *Orbits of denumerable models of complete theories* ◇ C15 C35 C52 C62 ◇

1971

BALDWIN, JOHN T. & LACHLAN, A.H. *On strongly minimal sets* ◇ C35 C45 C50 ◇

CUSIN, R. *Recherche du forcing-compagnon et du modele-compagnon d'une theorie liee a l'existence de modeles $\aleph_\alpha$-universels* ◇ C25 C35 C50 ◇

GLASSMIRE, W. *There are $2^{\aleph_0}$ countable categorical theories* ◇ C35 ◇

HARNIK, V. & RESSAYRE, J.-P. *Prime extensions and categoricity in power* ◇ C35 C50 ◇

KEISLER, H.J. *On theories categorical in their own power*
◇ C35 C55 ◇

KRAUSS, P.H. *Universally complete universal theories*
◇ C05 C30 C35 ◇

MACINTYRE, A. *On ω_1-categorical theories of abelian groups* ◇ C35 C45 C60 ◇

MACINTYRE, A. *On ω_1-categorical theories of fields*
◇ C35 C60 ◇

PABION, J.F. *Etude des bases dans les modeles des theories $\aleph_1$-categoriques* ◇ C15 C35 C50 ◇

PALYUTIN, E.A. *Boolean algebras with a categorical theory in a weak second order logic (Russian)*
◇ B15 C35 C85 D35 G05 ◇

PALYUTIN, E.A. *Models with countably categorical universal theories (Russian)* ◇ C35 C45 G05 ◇

ROBINSON, A. *Forcing in model theory*
◇ C25 C35 C52 ◇

ROBINSON, A. *Infinite forcing in model theory*
◇ C25 C35 C52 ◇

ROSENSTEIN, J.G. *A note on a theorem of Vaught*
◇ C15 C35 C50 ◇

ROSENTHAL, J.W. *Relations not determining the structure of L* ◇ C35 C62 E45 ◇

SABBAGH, G. *Embedding problems for modules and rings with application to model-companions*
◇ C25 C35 C60 ◇

SABBAGH, G. *Sous-modules purs, existentiellement clos et elementaires* ◇ C25 C35 C60 ◇

SHELAH, S. *Stability, the f.c.p., and superstability: model theoretic properties of formulas in first order theory*
◇ C35 C45 C55 ◇

SZCZERBA, L.W. *Incompleteness degree of elementary Pasch-free geometry* ◇ C35 C65 ◇

1972

ABAKUMOV, A.I. & PALYUTIN, E.A. & SHISHMAREV, YU.E. & TAJTSLIN, M.A. *Categorial quasivarieties (Russian)*
◇ C05 C35 C52 ◇

BACSICH, P.D. *Injectivity in model theory*
◇ C05 C35 C50 G05 G10 G30 ◇

BALDWIN, JOHN T. *Almost strongly minimal theories I,II*
◇ C35 C45 ◇

BELEGRADEK, O.V. *Categoricity in nondenumerable powers and $\aleph_1$-homogeneous models (Russian)*
◇ C35 C50 ◇

BURRIS, S. *Models in equational theories of unary algebras* ◇ B25 C05 C13 C35 ◇

CUSIN, R. *L'algebre des formules d'une theorie complete et "forcing-compagnon"* ◇ C25 C35 ◇

CUSIN, R. *Quasi-complete theories* ◇ C35 ◇

EHRENFEUCHT, A. *There are continuum ω_0-categorical theories* ◇ C35 ◇

EKLOF, P.C. *Some model theory of abelian groups*
◇ C35 C60 ◇

EKLOF, P.C. & FISHER, E.R. *The elementary theory of abelian groups* ◇ B25 C10 C35 C50 C60 ◇

HENSON, C.W. *Countable homogeneous relational*

structures and $\aleph_0$-categorical theories
◇ C10 C15 C35 C50 D35 ◇

LASSAIGNE, R. & RESSAYRE, J.-P. *Algebres de Boole et langages infinis* ◇ C35 C75 G05 ◇

MAKOWSKY, J.A. *Note on almost strongly minimal theories (Russian summary)* ◇ C35 C52 ◇

OLIN, P. $\aleph_0$-*categoricity of two-sorted structures*
◇ C20 C30 C35 C60 ◇

PABION, J.F. *Relations prehomogenes*
◇ C07 C35 C50 ◇

PALYUTIN, E.A. *On complete quasimanifolds (Russian)*
◇ C05 C35 ◇

POUZET, M. *Modele universel d'une theorie n-complete*
◇ C15 C35 C50 ◇

POUZET, M. *Modele universel d'une theorie n-complete: modele uniformement prehomogene*
◇ C15 C35 C50 ◇

POUZET, M. *Modele universel d'une theorie n-complete: modele prehomogene* ◇ C15 C35 C50 ◇

ROSENSTEIN, J.G. *On $GL_2(R)$ where R is a boolean ring*
◇ C35 C60 G05 ◇

ROSENTHAL, J.W. *A new proof of a theorem of Shelah*
◇ C35 C45 C55 E45 E50 ◇

SHELAH, S. *A note on model complete models and generic models* ◇ C25 C35 ◇

SHELAH, S. *Saturation of ultrapowers and Keisler's order*
◇ C20 C35 C45 C50 C55 ◇

SHISHMAREV, YU.E. *Categorical theories of a function (Russian)* ◇ C35 ◇

SIMMONS, H. *Existentially closed structures*
◇ C25 C35 C50 ◇

TULIPANI, S. *Sulla completezza e sulla categoricita della teoria delle W-algebre semplici (English summary)*
◇ C05 C35 C50 ◇

WEESE, M. *Zur Modellvollstaendigkeit und Entscheidbarkeit gewisser topologischer Raeume (Russian, English and French summaries)*
◇ B25 C35 C65 D80 ◇

ZIEGLER, M. *Die elementare Theorie der Henselschen Koerper* ◇ B25 C10 C35 C60 ◇

1973

BACSICH, P.D. *Primality and model-completion*
◇ C05 C25 C35 ◇

BALDWIN, JOHN T. α_T *is finite for $\aleph_1$-categorical T*
◇ C35 C45 ◇

BALDWIN, JOHN T. *A sufficient condition for a variety to have the amalgamation property* ◇ C05 C35 C52 ◇

BALDWIN, JOHN T. & LACHLAN, A.H. *On universal Horn classes categorical in some infinite power*
◇ C05 C10 C35 C45 ◇

BALDWIN, JOHN T. *The number of automorphisms of models of $\aleph_1$-categorical theories*
◇ C07 C35 C50 ◇

BELEGRADEK, O.V. *Almost categorical theories (Russian)*
◇ C35 C45 C55 ◇

BELEGRADEK, O.V. *Certain nonelementary properties of models of categorical theories (Russian)* ◇ C35 ◇

BUECHI, J.R. & DANHOF, K.J. *Variations on a theme of Cantor in the theory of relational structures*
◇ C07 C35 C50 C75 ◇

CARSON, A.B. *The model-completion of the theory of commutative regular rings* ◇ C25 C35 C60 ◇

CUSIN, R. *Sur les structures generiques en theorie des modeles* ◇ C10 C25 C35 C50 ◇

DAHN, B.I. $\aleph_0$-*kategorische zyklenbeschraenkte Graphen*
◇ C13 C35 C65 ◇

DAHN, B.I. $\aleph_0$-*kategorische zyklenbeschraenkte Graphen (English summary)* ◇ C13 C35 C65 ◇

DICKMANN, M.A. *The problem of non-finite axiomatizability of $\aleph_1$-categorical theories*
◇ C35 C52 C98 ◇

KFOURY, D. *Comparing algebraic structures up to algorithmic equivalence* ◇ B75 C20 C35 ◇

LIPSHITZ, L. & SARACINO, D.H. *The model companion of the theory of commutative rings without nilpotent elements* ◇ B25 C25 C35 C60 ◇

PALYUTIN, E.A. *Categorical quasivarieties of arbitrary signature (Russian)* ◇ C05 C35 ◇

PERETYAT'KIN, M.G. *A strongly constructive model without elementary submodels and extensions (Russian)*
◇ C35 C57 D45 ◇

ROBINSON, A. *Model theory as a framework for algebra*
◇ C25 C35 C60 C98 ◇

ROSENSTEIN, J.G. $\aleph_0$-*categoricity of groups*
◇ C35 C60 ◇

SARACINO, D.H. *Model companions for $\aleph_0$-categorial theories* ◇ C25 C35 C60 ◇

TAJTSLIN, M.A. *Existentially closed regular commutative semigroups (Russian)* ◇ C25 C35 C52 C60 ◇

WASZKIEWICZ, J. ω_0-*categoricity of generalized products*
◇ C20 C30 C35 ◇

WASZKIEWICZ, J. *On cardinalities of algebras of formulas for ω_0-categorical theories* ◇ C13 C35 G05 ◇

WEISPFENNING, V. *Infinitary model-theoretic properties of κ-saturated structures*
◇ C10 C25 C35 C50 C60 C65 C75 ◇

WOOD, C. *The model theory of differential fields of characteristic $p \neq 0$* ◇ C25 C35 C45 C60 ◇

1974

BALDWIN, JOHN T. *Atomic compactness in $\aleph_1$-categorical Horn theories* ◇ C35 C50 ◇

BAUMGARTNER, J.E. *The Hanf number for complete $L_{\omega_1,\omega}$-sentences (without GCH)*
◇ C35 C50 C55 C75 ◇

BELEGRADEK, O.V. & ZIL'BER, B.I. *The model companion of an $\aleph_1$-categorical theory (Russian)* ◇ C25 C35 ◇

BERESTOVOY, S. *Modeles homogenes d'une theorie forcee*
◇ C25 C35 C50 ◇

DORACZYNSKA, E. *The complete theory of the midpoint operation (Russian summary)* ◇ C35 C60 ◇

EBBINGHAUS, H.-D. & FLUM, J. *Algebraische Charakterisierung der elementaren Aequivalenz*
◇ C07 C35 ◇

ERIMBETOV, M.M. *Categoricity in power and the invalidity of the two-cardinal theorem for a formula of finite rank (Russian)* ◇ C35 C45 C55 ◇

ERIMBETOV, M.M. *On almost categorical theories (Russian)* ⋄ C35 C45 ⋄

FELGNER, U. *$\aleph_1$-kategorische Theorien nicht-kommutativer Ringe* ⋄ C35 C45 C60 ⋄

HARRINGTON, L.A. *Recursively presentable prime models* ⋄ C35 C50 C57 C60 ⋄

KEIMEL, K. *Representation des algebres universelles par des faisceaux* ⋄ C05 C35 C60 C90 ⋄

LACHLAN, A.H. *Two conjectures regarding the stability of ω-categorical theories* ⋄ C35 C45 ⋄

LASCAR, D. *Sur les theories convexes "modeles completes"* ⋄ C15 C35 C52 ⋄

MACINTYRE, A. *Model-completeness for sheaves of structures* ⋄ C35 C90 ⋄

MAKOWSKY, J.A. *On some conjectures connected with complete sentences* ⋄ C35 C45 C52 C98 ⋄

MONTAGNA, F. *Sui limiti di certe teorie (English summary)* ⋄ C30 C35 C50 C62 H15 ⋄

MORLEY, M.D. *A remark on a paper by Abian* ⋄ C35 G05 ⋄

MURZIN, F.A. *Totally transcendental and $\aleph_1$-categorical theories (Russian)* ⋄ C35 C45 ⋄

NURTAZIN, A.T. *Theories categorical in an uncountable power without principal model-complete extensions (Russian)* ⋄ C35 ⋄

PAILLET, J.L. *Quelques rapports entre la fini-model-completabilite et des proprietes approchantes* ⋄ C07 C35 ⋄

PAILLET, J.L. *Une etude sur les theories "finiment model-completables"* ⋄ C07 C35 ⋄

PALYUTIN, E.A. *The algebras of formulae of countably categorical theories (Russian)* ⋄ B25 C35 ⋄

PODEWSKI, K.-P. & REINEKE, J. *An ω_1-categorical ring which is not almost strongly minimal* ⋄ C35 C50 C60 ⋄

QUINN, C.F. *An analysis of the concept of constructive categoricity* ⋄ C35 ⋄

RUBIN, M. *Theories of linear order* ⋄ C15 C35 C50 C65 E07 ⋄

SACERDOTE, G.S. *Projective model completeness* ⋄ C35 ⋄

SARACINO, D.H. *Wreath products and existentially complete solvable groups* ⋄ C25 C35 C60 ⋄

SHELAH, S. *Categoricity of uncountable theories* ⋄ C35 C45 C50 ⋄

SIMMONS, H. *The use of injective-like structures in model theory* ⋄ C25 C35 C50 ⋄

SMIRNOVA, O.S. *Completeness and decidability of the axiomatic theory SG for spherical geometry (Russian)* ⋄ B25 C35 C65 ⋄

WOOD, C. *Prime model extensions for differential fields of characteristic $p \neq 0$* ⋄ C10 C35 C60 ⋄

ZAKON, E. *Model-completeness and elementary properties of torsion free abelian groups* ⋄ C35 C60 ⋄

ZIL'BER, B.I. *Rings with $\aleph_1$-categorical theories (Russian)* ⋄ C35 C60 ⋄

ZIL'BER, B.I. *The transcendental rank of the formulas of an $\aleph_1$-categorical theory (Russian)* ⋄ C35 C45 ⋄

1975

ANDLER, D. *Finite-dimensional models of categorical semi-minimal theories* ⋄ C35 C50 ⋄

ANDLER, D. *Semi-minimal theories and categoricity* ⋄ C30 C35 C55 ⋄

BAUR, W. *$\aleph_0$-categorical modules* ⋄ C35 C60 ⋄

BELEGRADEK, O.V. & TAJTSLIN, M.A. *Categorical varieties of groupoids (Russian)* ⋄ C05 C35 ⋄

BERLINE, C. *Categoricite en $\aleph_0$ du groupe lineaire d'un anneau de Boole (English summary)* ⋄ C35 C60 ⋄

BURRIS, S. *An existence theorem for model companions* ⋄ C25 C35 ⋄

CHIKHACHEV, S.A. *Categorical universal classes of semigroups (Russian)* ⋄ C05 C35 C52 C60 ⋄

CHIKHACHEV, S.A. *Generic models (Russian)* ⋄ C25 C35 ⋄

ERIMBETOV, M.M. *Complete theories with 1-cardinal formulas (Russian)* ⋄ C35 C45 C55 ⋄

ERIMBETOV, M.M. *On almost categorical theories (Russian)* ⋄ C35 C45 ⋄

FELGNER, U. *On $\aleph_0$-categorical extra-special p-groups* ⋄ C35 C60 ⋄

GEORGESCU, G. & PETRESCU, I. *Model-completion in categories* ⋄ C25 C35 G30 ⋄

HAUSCHILD, K. *Ueber zwei Spiele und ihre Anwendung in der Modelltheorie (Russian, English and French summaries)* ⋄ B25 C07 C30 C35 C85 ⋄

JENSEN, F.V. *On completeness in cardinality logics* ⋄ C10 C35 C55 C60 C80 ⋄

KOCHEN, S. *The model theory of local fields* ⋄ B25 C20 C35 C60 ⋄

KOKORIN, A.I. & KOZLOV, G.T. *Proof of a lemma on model completeness (Russian)* ⋄ C35 C60 ⋄

KRAUSS, P.H. *Quantifier elimination* ⋄ C10 C35 C50 ⋄

LACAVA, F. *Teoria dei campi differenziali ordinati* ⋄ C10 C35 C60 ⋄

LACHLAN, A.H. *Theories with a finite number of models in an uncountable power are categorical* ⋄ C35 C45 ⋄

MACINTYRE, A. *Dense embeddings. I. A theorem of Robinson in a general setting* ⋄ B25 C10 C35 C60 D40 ⋄

MALITZ, J. *Complete theories with countably many rigid nonisomorphic models* ⋄ C15 C35 C50 ⋄

MAZOYER, J. *Sur les theories categoriques finiment axiomatisables (English summary)* ⋄ C10 C35 C45 ⋄

OLIN, P. *$\aleph_0$-categoricity of products of 1-unary algebras* ⋄ C05 C35 ⋄

PALYUTIN, E.A. *Description of categorical quasivarieties (Russian)* ⋄ C05 C35 ⋄

PARIKH, R. *An $\aleph_0$-categorical theory whose language is countably infinite* ⋄ C35 ⋄

ROSENTHAL, J.W. *Partial $\aleph_1$-homogeneity of the countable saturated model of an $\aleph_1$-categorical theory* ⋄ C35 C50 ⋄

SABBAGH, G. *Categoricite en $\aleph_0$ et stabilite: constructions les preservant et conditions de chaine (English summary)* ⋄ C35 C45 C60 ⋄

SABBAGH, G. *Categoricite et stabilite: quelques examples parmi les groupes et anneaux (English summary)*
    ◊ C35 C45 C60 ◊
SACERDOTE, G.S. *Projective model theory and coforcing*
    ◊ C25 C30 C35 C40 C50 C60 ◊
SARACINO, D.H. *A counterexample in the theory of model companions* ◊ C25 C35 ◊
SARACINO, D.H. & WEISPFENNING, V. *Commutative regular rings without prime model extensions*
    ◊ C25 C35 C60 ◊
SARACINO, D.H. & WEISPFENNING, V. *On algebraic curves over commutative regular rings*
    ◊ C10 C25 C35 C60 ◊
SCHMITT, P.H. *Categorical lattices* ◊ C35 C60 G10 ◊
SETTE, A.-M. *Multirelation fortement homogene et ses rapports avec les theories forcees (English summary)*
    ◊ C25 C35 C50 ◊
SHELAH, S. *Categoricity in $\aleph_1$ of sentences in $L_{\omega_1\omega}(Q)$*
    ◊ C35 C55 C75 C80 ◊
SHELAH, S. *The lazy model theoretician's guide to stability*
    ◊ C25 C35 C45 C50 C60 C98 ◊
VINNER, S. *Model-completeness in a first order language with a generalized quantifier* ◊ C35 C55 C80 ◊
WEISPFENNING, V. *Model-completeness and elimination of quantifiers for subdirect products of structures*
    ◊ C10 C30 C35 C52 C60 C90 ◊
WINKLER, P.M. *Model-completeness and Skolem expansions* ◊ C25 C30 C35 C60 ◊

1976

ADLER, A. & KIEFE, C. *Pseudofinite fields, procyclic fields and model-completion* ◊ C35 C60 ◊
COMER, S.D. *Complete and model-complete theories of monadic algebras*
    ◊ B25 C05 C25 C35 C60 G15 ◊
ELLENTUCK, E. *Categoricity regained*
    ◊ C35 C75 C90 E40 ◊
GILLAM, D.W.H. *Relatively complete theories* ◊ C35 ◊
GNANVO, C. *Γ-cotheories, test de Robinson (English summary)* ◊ C35 C52 ◊
GROSS, W.F. *Models with dimension*
    ◊ C35 C50 C60 E25 E35 ◊
HERRING, J.M. *Equivalence of several notions of theory completeness in a free logic* ◊ B20 C35 ◊
HERRMANN, E. *On Lindenbaum functions of $\aleph_0$-categorical theories of finite similarity type*
    ◊ B25 C35 C57 ◊
LACAVA, F. & SAELI, D. *Proprieta e model-completamento di alcune varieta di algebre di Lukasiewicz (English summary)* ◊ C35 G20 ◊
LACHLAN, A.H. *Dimension and totally transcendental theories of rank 2* ◊ C35 C45 ◊
MACINTYRE, A. & ROSENSTEIN, J.G. *$\aleph_0$-categoricity for rings without nilpotent elements and for boolean structures* ◊ C35 C60 ◊
MURZIN, F.A. *$\aleph_1$-categorical theories (Russian)*
    ◊ C35 C45 ◊
PAILLET, J.L. *Une etude sur des structures ayant une certaine propriete de seuil pour les automorphismes elementaires (English summary)* ◊ C07 C35 C50 ◊

PALLADINO, D. *Questioni di categoricita delle teorie matematiche* ◊ C35 ◊
PINUS, A.G. *Theories of boolean algebras in a calculus with the quantifier "infinitely many exist" (Russian)*
    ◊ B25 C10 C35 C57 C80 G05 ◊
POIZAT, B. *"Forcing" et relations generales sur une structure ω-categorique (English summary)*
    ◊ C25 C35 ◊
PRAZMOWSKI, K. & SZCZERBA, L.W. *Interpretability and categoricity (Russian summary)* ◊ C35 F25 ◊
ROSENSTEIN, J.G. *$\aleph_0$-categoricity is not inherited by factor groups* ◊ C35 C60 ◊
SARACINO, D.H. *Existentially complete nilpotent groups*
    ◊ C20 C25 C35 C60 ◊
SCHMERL, J.H. *The decidability of some $\aleph_0$-categorical theories* ◊ B25 C35 ◊
SCHMITT, P.H. *The model-completion of Stone algebras*
    ◊ C25 C35 C60 G10 ◊
SIMMONS, H. *Large and small existentially closed structures* ◊ C25 C35 C50 ◊
VETULANI, Z. *Categoricity relative to ordinals for models of set theory and the nonabsoluteness of L*
    ◊ C35 C62 E45 H20 ◊
WEGLORZ, B. *A note on: "Atomic compactness in $\aleph_1$-categorical Horn theories" by John T. Baldwin*
    ◊ C05 C35 C50 C52 ◊
WEISPFENNING, V. *On the elementary theory of Hensel fields* ◊ B25 C10 C35 C60 ◊
WHEELER, W.H. *Model-companions and definability in existentially complete structures*
    ◊ C25 C35 C60 C98 ◊
WOODROW, R.E. *A note on countable complete theories having three isomorphism types of countable models*
    ◊ C15 C35 ◊

1977

BALDWIN, JOHN T. & ROSE, B.I. *$\aleph_0$-categoricity and stability of rings* ◊ C35 C45 C60 ◊
CHIKHACHEV, S.A. *$\aleph_1$-categorical commutative rings (Russian)* ◊ C35 C60 ◊
CLARK, D.M. & KRAUSS, P.H. *Relatively homogeneous structures* ◊ C10 C35 C50 C60 C65 G05 ◊
CUTLAND, N.J. *Some theories having countably many countable models* ◊ C15 C35 ◊
DROBOTUN, B.N. *On countable models of decidable almost categorical theories (Russian)*
    ◊ B25 C15 C35 C57 ◊
FELGNER, U. *Stability and $\aleph_0$-categoricity of nonabelian groups* ◊ C35 C45 C60 ◊
FORREST, W.K. *Model theory for universal classes with the amalgamation property: A study in the foundations of model theory and algebra*
    ◊ C05 C25 C30 C35 C45 C52 C60 ◊
GNANVO, C. *Γ-cotheories et elimination des quantificateurs (English summary)*
    ◊ C10 C25 C35 C52 ◊
GROSS, W.F. *Dimension and finite closure*
    ◊ C35 C50 C60 E25 ◊

HERRMANN, E. *Ueber Lindenbaumfunktionen von $\aleph_0$-kategorischen Theorien endlicher Signatur*
  ◇ B25 C35 C57 ◇
KRAUSS, P.H. *Homogeneous universal models of universal theories* ◇ C05 C35 C50 C52 ◇
LABLANQUIE, J.-C. *Le nombre de classes d'equivalence elementaire des modeles generiques d'une propriete de "forcing" (English summary)* ◇ C15 C25 C35 ◇
LACAVA, F. & SAELI, D. *Sul model-completamento della teoria delle L-catene* ◇ C35 C60 G20 ◇
LASCAR, D. *Generalisation de l'ordre de Rudin-Keisler aux types d'une theorie* ◇ C15 C35 C45 E05 ◇
LITMAN, A. & SHELAH, S. *Models with few isomorphic expansions* ◇ C15 C35 C50 C55 ◇
MACINTYRE, A. *Model completeness*
  ◇ C25 C35 C60 C98 ◇
MAKKAI, M. & MYCIELSKI, J. *An $L_{\omega_1\omega}$-complete and consistent theory without models* ◇ C35 C75 E75 ◇
MIJAJLOVIC, Z. *Homogeneous-universal models of theories which have model completions* ◇ C35 C50 C52 ◇
MURZIN, F.A. *$\aleph_1$-categorical theories (Russian)*
  ◇ C35 C45 ◇
PALYUTIN, E.A. *Number of models in $L_{\infty\omega_1}$-theories I,II (Russian)* ◇ C35 C52 C55 C75 E45 E65 ◇
PINTER, C. *Some theorems on omitting types, with applications to model completeness, amalgamation, and related properties* ◇ C07 C25 C35 C52 C75 ◇
SCHLIPF, J.S. *Ordinal spectra of first-order theories*
  ◇ C35 C50 C55 C62 C70 D15 D30 D70 ◇
SCHMERL, J.H. *$\aleph_0$-categoricity of partially ordered sets of width 2* ◇ C35 C65 ◇
SCHMERL, J.H. *On $\aleph_0$-categoricity and the theory of trees*
  ◇ B25 C35 C65 ◇
TOFFALORI, C. *Alcune proprieta di teorie di campi con un sottocampo privilegiato (English summary)*
  ◇ C25 C35 C60 ◇
TUSCHIK, H.-P. *Elimination verallgemeinerter Quantoren in ω_1-kategorischen Theorien (Russian, English and French summaries)* ◇ B25 C10 C35 C55 C80 ◇
ZIL'BER, B.I. *Groups and rings, the theory of which is categorical (Russian) (English summary)*
  ◇ C35 C45 C60 ◇
ZIL'BER, B.I. *The structure of a model of a categorical theory and finite-axiomatizability problem (Russian)*
  ◇ C35 C52 ◇

1978

BASARAB, S.A. *Quelques proprietes modele-theoriques des corps values henseliens (English summary)*
  ◇ C25 C35 C60 ◇
BASARAB, S.A. *Some model theory for henselian valued fields* ◇ C35 C60 ◇
CHERLIN, G.L. & ROSENSTEIN, J.G. *On $\aleph_0$-categorical abelian-by-finite groups* ◇ C35 C60 ◇
DRIES VAN DEN, L. *Model theory of fields*
  ◇ B25 C10 C25 C35 C60 H20 ◇
EHRSAM, S. *Theories ω_1-categoriques* ◇ C35 C45 ◇

FELGNER, U. *$\aleph_0$-categorical stable groups*
  ◇ C35 C45 C60 ◇
FRIEDMAN, H.M. *Categoricity with respect to ordinals*
  ◇ C35 C62 E45 ◇
GIVANT, S. *Universal Horn classes categorical or free in power* ◇ B20 C05 C35 ◇
GOGOL, D. *The $\forall_n\exists$-completeness of Zermelo-Fraenkel set theory* ◇ B25 C35 E30 ◇
GONCHAROV, S.S. *Constructive models of $\aleph_1$-categorical theories (Russian)* ◇ C35 C57 ◇
HAUSCHILD, K. *Gruppengraphen und endliche Axiomatisierbarkeit* ◇ C35 C52 C60 ◇
HIRSCHFELD, J. *Examples in the theory of existential completeness* ◇ C25 C35 C50 ◇
KAISER, KLAUS *On relational selections for complete theories* ◇ C20 C35 C50 C60 ◇
LACHLAN, A.H. *Spectra of ω-stable theories*
  ◇ C35 C45 ◇
LOULLIS, G. *Infinite forcing for boolean valued models*
  ◇ C25 C35 C90 ◇
OLIN, P. *Aleph-zero categorical Stone algebras*
  ◇ C35 G10 ◇
OLIN, P. *Urn models and categoricity*
  ◇ A05 C35 C90 ◇
PAILLET, J.L. *Theories completes $\aleph_0$-categoriques et non finiment modele-completables (English summary)*
  ◇ C35 C45 ◇
PAILLET, J.L. *Une etude sur des structures ayant une certaine propriete de seuil pour les automorphismes elementaires* ◇ C07 C35 C50 ◇
PERETYAT'KIN, M.G. *Criterion for strong constructivizability of a homogeneous model (Russian)*
  ◇ C35 C50 C57 D45 ◇
PETRESCU, I. *Model-companions for logical functors*
  ◇ C25 C35 G30 ◇
PINTER, C. *Properties preserved under definitional equivalence and interpretations*
  ◇ C25 C35 C50 F25 ◇
POIZAT, B. *Etude d'un forcing en theorie des modeles*
  ◇ C25 C35 ◇
POIZAT, B. *Une preuve par la theorie de la deviation d'un theoreme de J. Baldwin (English summary)*
  ◇ C35 C45 ◇
POIZAT, B. *Une theorie ω_1-categorique a α_T fini*
  ◇ C35 C45 ◇
PREST, M. *Some model-theoretic aspects of torsion theories* ◇ C25 C35 C60 ◇
ROSE, B.I. *Model theory of alternative rings*
  ◇ C10 C35 C45 C60 ◇
ROSE, B.I. *The $\aleph_1$-categoricity of strictly upper triangular matrix rings over algebraically closed fields*
  ◇ C35 C45 C60 ◇
ROTHMALER, P. *Total transzendente abelsche Gruppen und Morley-Rang (English and Russian summaries)*
  ◇ C35 C45 C60 ◇
SCHMERL, J.H. *$\aleph_0$-categoricity and comparability graphs*
  ◇ C15 C35 C65 ◇

SCHMERL, J.H. *A decidable $\aleph_0$-categorical theory with a non-recursive Ryll-Nardzewski function*
　◇ B25　C35　C57　D30 ◇

SCHMERL, J.H. *On $\aleph_0$-categoricity of filtered boolean extensions*　◇ C30　C35 ◇

SCHREIBER, P. *Kategorizitaetsbeweise in der Geometrie*
　◇ C35 ◇

SINGER, M.F. *The model theory of ordered differential fields*　◇ B25　C25　C35　C60　C65　D80 ◇

VOJVODIC, G. *Some theorems for model theory of mixed-valued predicate calculi*
　◇ B50　C20　C35　C90 ◇

WEISPFENNING, V. *A note on $\aleph_0$-categorical model-companions*
　◇ C25　C35　C50　C52　C65　G05　G10 ◇

WERNER, H. *Discriminator-algebras. Algebraic representation and model theoretic properties*
　◇ B25　C05　C25　C35　C90 ◇

WHEELER, W.H. *A characterization of companionable, universal theories*　◇ C25　C35　C60　C65 ◇

YUTANI, H. *Categoricity in the theory of BCK-algebras*
　◇ C35　G25 ◇

1979

BASARAB, S.A. *A model-theoretic transfer theorem for henselian valued fields*　◇ C25　C35　C60 ◇

BASARAB, S.A. *Model-theoretic methods in the theory of henselian valued fields. I,II (Romanian) (English summaries)*　◇ C25　C35　C60 ◇

BAUDISCH, A. *Uncountable n-cubes in models of $\aleph_0$-categorical theories*　◇ C35　C80 ◇

BAUR, W. & CHERLIN, G.L. & MACINTYRE, A. *Totally categorical groups and rings*　◇ C35　C45　C60 ◇

CHERLIN, G.L. *Lindenbaum algebras and model companions*　◇ C25　C35 ◇

COWLES, J.R. *The theory of Archimedean real closed fields in logics with Ramsey quantifiers*
　◇ C10　C35　C60　C80 ◇

GIVANT, S. *A representation theorem for universal Horn classes categorical in power*　◇ B20　C05　C35 ◇

GOGOL, D. *Sentences with three quantifiers are decidable in set theory*　◇ B25　C35　E30 ◇

LACAVA, F. *Alcune proprieta delle L-algebre e delle L-algebre esistenzialmente chiuse*
　◇ C25　C35　C60　G20 ◇

LERMAN, M. & SCHMERL, J.H. *Theories with recursive models*　◇ C35　C57　C65 ◇

LOULLIS, G. *Sheaves and boolean valued model theory*
　◇ C35　C52　C60　C90 ◇

MARCUS, L. *An expansion of an $\aleph_0$-categorical model*
　◇ C35 ◇

MEKLER, A.H. *Model complete theories with a distinguished substructure*　◇ C35　C45　C60 ◇

MIJAJLOVIC, Z. *Saturated boolean algebras with ultrafilters*　◇ C20　C35　C50　G05 ◇

PALYUTIN, E.A. *Categorical positive Horn theories (Russian)*　◇ C05　C35　C52 ◇

PODEWSKI, K.-P. & REINEKE, J. *Algebraically closed commutative local rings*　◇ C25　C35　C60 ◇

PREST, M. *Model-completions of some theories of modules*
　◇ C25　C35　C50　C60 ◇

ROSENTHAL, J.W. *On the dimension theory of $\aleph_1$-categorical theories with the nontrivial strong elementary intersection property*　◇ C35 ◇

SACKS, G.E. *Effective bounds on Morley rank*
　◇ C15　C35　C45　C70 ◇

SARACINO, D.H. & WOOD, C. *Periodic existentially closed nilpotent groups*　◇ C25　C35　C45　C60 ◇

SCHWABHAEUSER, W. *Modellvollstaendigkeit der Mittelpunktsgeometrie und der Theorie der Vektorgruppen*　◇ B25　C35　C65 ◇

SHISHMAREV, YU.E. *Categorical quasivarieties of quasigroups (Russian)*　◇ C05　C35 ◇

STEGBAUER, W. *A generalized model companion for a theory of partially ordered fields*　◇ C25　C35　C60 ◇

SUKHORUTCHENKO, I.I. *Programmably categorical abelian p-groups I,II (Russian)*　◇ B75　C35　C60 ◇

SZCZERBA, L.W. & TARSKI, A. *Metamathematical discussion of some affine geometries*
　◇ B25　B30　C35　C52　C65　D35 ◇

TAJTSLIN, M.A. *A description of algebraic systems in weak logic of order ω and in program logic (Russian)*
　◇ B15　B75　C35　C85 ◇

TUSCHIK, H.-P. *A reduction of a problem about ω_1-categoricity*　◇ C35　C52 ◇

VOLGER, H. *Sheaf representations of algebras and transfer theorems for model theoretic properties*
　◇ B25　C05　C30　C35　C90 ◇

WEISPFENNING, V. *Lattice products*
　◇ C10　C25　C30　C35　C90 ◇

WHEELER, W.H. *Model-complete theories of pseudo-algebraically closed fields*
　◇ C25　C35　C60 ◇

WHEELER, W.H. *The first order theory of N-colorable graphs*　◇ B30　C10　C25　C35　C65 ◇

WOOD, C. *Notes on the stability of separably closed fields*
　◇ C25　C35　C45　C60 ◇

ZIL'BER, B.I. *Definability of algebraic systems and the theory of categoricity (Russian)*　◇ C35　C45 ◇

1980

BAJZHANOV, B.S. *Questions on the spectra of totally transcendental theories of finite rank (Russian)*
　◇ C35　C45 ◇

BAJZHANOV, B.S. *Some properties of totally transcendental theories (Russian)*　◇ C35　C45 ◇

BALDWIN, JOHN T. & KUEKER, D.W. *Ramsey quantifiers and the finite cover property*
　◇ C10　C35　C45　C80 ◇

BECKER, J.A. & HENSON, C.W. & RUBEL, L.A. *First-order conformal invariants*　◇ C35　C65　D35　E35　E75 ◇

CHERLIN, G.L. & DICKMANN, M.A. *Anneaux reels clos et anneaux de fonctions continues (English summary)*
　◇ C10　C35　C60　C65 ◇

CHERLIN, G.L. *On $\aleph_0$-categorical nilrings*
　◇ C35　C60 ◇

CHERLIN, G.L. *On $\aleph_0$-categorical nilrings. II*
　◇ C35　C60 ◇

CORCORAN, J. *Categoricity* ⋄ A05 B15 C35 C85 ⋄
ERSHOV, YU.L. *Multiply valued fields (Russian)*
  ⋄ C25 C35 C60 ⋄
ERSHOV, YU.L. *Regularly closed fields (Russian)*
  ⋄ B25 C35 C60 D35 ⋄
FELGNER, U. *Kategorizitaet* ⋄ C35 C45 C60 C98 ⋄
FELGNER, U. *The model theory of FC-groups*
  ⋄ C35 C45 C60 D35 ⋄
GARAVAGLIA, S. *Decomposition of totally transcendental modules* ⋄ C10 C35 C45 C60 ⋄
GIVANT, S. *Union decompositions and universal classes categorical in power* ⋄ C05 C35 ⋄
GLASS, A.M.W. & PIERCE, K.R. *Existentially complete lattice-ordered groups* ⋄ C25 C35 C60 G10 ⋄
HERRE, H. *Modelltheoretische Eigenschaften endlichvalenter Graphen* ⋄ C35 C45 C50 C65 ⋄
HIRSCHFELD, J. *Finite forcing, existential types and complete types* ⋄ C25 C35 ⋄
JACOB, B. *The model theory of "R-formal" fields*
  ⋄ B25 C10 C25 C35 C60 ⋄
KIERSTEAD, H.A. *Countable models of ω_1-categorical theories in admissible languages* ⋄ C15 C35 C70 ⋄
KOMORI, Y. *Completeness of two theories on ordered abelian groups and embedding relations*
  ⋄ B25 C10 C35 C60 ⋄
KUDAJBERGENOV, K.ZH. *On constructive models of undecidable theories (Russian)*
  ⋄ C15 C35 C57 D35 ⋄
LACAVA, F. *Osservazioni sulla teoria dei gruppi abeliani reticolari* ⋄ C25 C35 C60 ⋄
LACHLAN, A.H. *Singular properties of Morley rank*
  ⋄ C35 C45 ⋄
MANDERS, K.L. *Theories with the existential substructure property* ⋄ C25 C35 ⋄
MANGANI, P. & MARCJA, A. *Shelah rank for boolean algebras and some application to elementary theories I*
  ⋄ C35 C40 C45 C50 G05 ⋄
MEAD, J. & NELSON, GEORGE C. *Model companions and k-model completeness for the complete theories of boolean algebras* ⋄ C25 C35 G05 ⋄
MEJREMBEKOV, K.A. *Remarks on theories with a bounded spectrum (Russian)* ⋄ C35 C45 C60 ⋄
MIRAGLIA, F. *Relations between structures of stable continuous functions and filtered powers*
  ⋄ C30 C35 C90 ⋄
MUSTAFIN, T.G. *A non-two-cardinal set of stable types (Russian)* ⋄ C35 C45 C55 ⋄
MYERS, R.W. *Complexity of model-theoretic notions*
  ⋄ C07 C35 C57 D55 ⋄
OMAROV, B. *Ranks and Deg of countable ω_0-categorical theories (Russian)* ⋄ C35 C45 ⋄
PALYUTIN, E.A. *Categorical Horn classes I (Russian)*
  ⋄ C05 C35 C45 C52 ⋄
PALYUTIN, E.A. *Description of categorical positive Horn classes (Russian)* ⋄ C05 C35 C52 ⋄
PERETYAT'KIN, M.G. *Example of an ω_1-categorical complete finitely axiomatizable theory (Russian)*
  ⋄ C35 ⋄
PILLAY, A. *$\aleph_0$-categoricity in regular subsets of $L_{\omega_1\omega}$*
  ⋄ C35 C75 ⋄

ROSE, B.I. *On the model theory of finite-dimensional algebras* ⋄ B25 C35 C45 C60 ⋄
SCHMERL, J.H. *Decidability and $\aleph_0$-categoricity of theories of partially ordered sets*
  ⋄ B25 C15 C35 C65 D35 ⋄
SZCZERBA, L.W. *Interpretations with parameters*
  ⋄ B25 C07 C35 C45 F25 ⋄
TOFFALORI, C. *Fasci di coppie di campi algebricamente chiusi* ⋄ B25 C35 C60 C90 ⋄
TOFFALORI, C. *Sheaves of pairs of real closed fields*
  ⋄ C25 C35 C60 C90 ⋄
TOFFALORI, C. *Sul model-completamento di certe teorie di coppie di anelli (English summary)*
  ⋄ C10 C25 C35 C60 ⋄
TUSCHIK, H.-P. *An application of rank-forcing to ω_1-categoricity* ⋄ C25 C35 ⋄
WHEELER, W.H. *Model theory of strictly upper triangular matrix rings* ⋄ C25 C35 C60 ⋄
ZIL'BER, B.I. *Solution of the problem of finite axiomatizability for theories that are categorical in all infinite powers (Russian)* ⋄ C35 C45 C52 ⋄
ZIL'BER, B.I. *Strongly minimal countably categorical theories (Russian)* ⋄ C35 C45 C52 ⋄
ZIL'BER, B.I. *Totally categorical theories: Structural properties and the non-finite axiomatizability*
  ⋄ C35 C45 C52 ⋄

1981

BERLINE, C. *Stabilite et algebre I. Pratique de la stabilite et de la categoricite* ⋄ C35 C45 C98 ⋄
BERLINE, C. *Stabilite et algebre II. Groupes Abeliens et modules* ⋄ C35 C45 C60 C98 ⋄
BUNGE, MARTA C. & REYES, G.E. *Boolean spectra and model completions* ⋄ C35 C90 G30 ⋄
BUNGE, MARTA C. *Sheaves and prime model extensions*
  ⋄ C35 C60 C90 ⋄
CORCORAN, J. *From categoricity to completeness*
  ⋄ A10 C35 ⋄
GIVANT, S. *The number of nonisomorphic denumerable models of certain universal Horn classes*
  ⋄ C05 C15 C35 ⋄
HAUSSLER, D. *Model completeness of an algebra of languages* ⋄ B25 C35 D05 ⋄
HUTCHINSON, G. *A complete logic for n-permutable congruence lattices* ⋄ C05 C35 G10 ⋄
JACOB, B. *The model theory of generalized real closed fields* ⋄ B25 C35 C60 ⋄
LASCAR, D. *Dans quelle mesure la categorie des modeles d'une theorie complete determine-t-elle cette theorie?*
  ⋄ C35 C90 G30 ⋄
LAVENDHOMME, R. & LUCAS, T. *A propos d'un theoreme de Macintyre* ⋄ C35 C90 ⋄
MAKKAI, M. *The topos of types* ⋄ C35 C90 G30 ⋄
MILLAR, T.S. *Counterexamples via model completions*
  ⋄ C10 C35 C50 C57 ⋄
MILLAR, T.S. *Stability, complete extensions, and the number of countable models* ⋄ C15 C35 C45 ⋄
MYCIELSKI, J. & PERLMUTTER, P. *Model completeness of some metric completions of absolutely free algebras*
  ⋄ C05 C35 C50 ⋄

PILLAY, A. *A class of $\aleph_0$-categorical theories*
⋄ C35 C45 ⋄

PODEWSKI, K.-P. & REINEKE, J. *On algebraically closed commutative indecomposable rings*
⋄ C25 C35 C60 ⋄

PRESTEL, A. *Pseudo real closed fields*
⋄ C10 C25 C35 C52 C60 ⋄

REMMEL, J.B. *Recursively categorical linear orderings*
⋄ C35 C57 C65 ⋄

SCHMERL, J.H. *Decidability and finite axiomatizability of theories of $\aleph_0$-categorical partially ordered sets*
⋄ B25 C35 C65 ⋄

SMITH, KENNETH W. *Stability and categoricity of lattices*
⋄ C35 C45 G10 ⋄

TULIPANI, S. *Model-completions of theories of finitely additive measures with values in an ordered field*
⋄ C10 C25 C35 C65 G05 ⋄

ZIL'BER, B.I. *Solution of the problem of finite axiomatizability for theories that are categorical in all infinite powers (Russian)* ⋄ B30 C35 C45 C52 ⋄

ZIL'BER, B.I. *Some problems concerning categorical theories (Russian summary)* ⋄ C35 C45 ⋄

ZIL'BER, B.I. *Totally categorical structures and combinatorial geometries (Russian)* ⋄ C35 C45 ⋄

1982

BEKENOV, M.I. *On the spectrum of quasitranscendental theories (Russian)* ⋄ C35 C45 C52 ⋄

BUNGE, MARTA C. *On the transfer of an abstract Nullstellensatz* ⋄ C35 C60 C90 G30 ⋄

COOPER, GLEN R. *On complexity of complete first-order theories* ⋄ C35 C45 C52 ⋄

ERSHOV, YU.L. *Multiply valued fields (Russian)*
⋄ B25 C25 C35 C60 C98 ⋄

ERSHOV, YU.L. *Regularly r-closed fields (Russian)*
⋄ B25 C25 C35 C60 ⋄

GONDARD, D. *Theorie des modeles et fonctions definies positives sur les varietes algebriques reelles (English summary)* ⋄ C35 C60 ⋄

HENDRY, H.E. *Complete extensions of the calculus of individuals* ⋄ C35 G05 ⋄

LASCAR, D. *On the category of models of a complete theory* ⋄ C35 C90 G30 ⋄

LIPPARINI, P. *Locally finite theories with model companion (Italian summary)* ⋄ C25 C35 C52 ⋄

MAKKAI, M. *Full continuous embeddings of toposes*
⋄ C35 C90 G30 ⋄

MARONGIU, G. *Alcuni risultati sulla $\aleph_0$-categoricita per strutture con un numero finito di relazioni di equivalenza (English summary)* ⋄ C35 E07 ⋄

MEZ, H.-C. *Existentially closed linear groups*
⋄ C25 C35 C60 ⋄

NELSON, GEORGE C. *The periodic power of $\mathfrak{A}$ and complete Horn theories* ⋄ C20 C30 C35 ⋄

PRESTEL, A. *Decidable theories of preordered fields*
⋄ B25 C25 C35 C60 ⋄

SCHMID, J. *Model companions of distributive p-algebras*
⋄ C05 C25 C35 G10 ⋄

SHEN, FUXING *Elementary extensions and elementary chains of lattice-valued models* ⋄ C35 C40 C90 ⋄

TOFFALORI, C. *Stabilita, categoricita ed eliminazione dei quantificatori per una classe di anelli locali*
⋄ C10 C35 C45 C60 ⋄

TOFFALORI, C. *Teoria dei modelli per alcune classi di anelli locali* ⋄ C10 C25 C35 C60 ⋄

TYUKAVKIN, L.V. *On the model completion of certain theories of modules (Russian)* ⋄ C35 C60 ⋄

URZYCZYN, P. *An example of a complete first-order theory with all models algorithmically trivial but without locally finite models* ⋄ C35 C50 ⋄

WAGNER, C.M. *On Martin's conjecture*
⋄ C15 C35 C45 C65 C75 ⋄

WEISPFENNING, V. *Model theory and lattices of formulas*
⋄ C25 C35 C52 C75 G10 ⋄

WERNER, H. *Sheaf constructions in universal algebra and model theory*
⋄ B25 C05 C25 C30 C35 C90 G05 ⋄

WILSON, J.S. *The algebraic structure of $\aleph_0$-categorical groups* ⋄ C15 C35 C60 ⋄

ZIL'BER, B.I. *Uncountable categorical nilpotent groups and Lie algebras (Russian)* ⋄ C35 C60 ⋄

1983

APPS, A.B. *$\aleph_0$-categorical finite extensions of boolean powers* ⋄ C30 C35 C60 ⋄

APPS, A.B. *On $\aleph_0$-categorical class two groups*
⋄ C35 C60 ⋄

APPS, A.B. *On the structure of $\aleph_0$-categorical groups*
⋄ C35 C60 ⋄

BALDWIN, JOHN T. & GIVANT, S. *Model complete universal Horn classes* ⋄ C05 C35 ⋄

BAUDISCH, A. *A remark on $\aleph_0$-categorical stable groups*
⋄ C35 C45 C60 ⋄

BLASS, A.R. & SCEDROV, A. *Boolean classifying topoi*
⋄ C35 G30 ⋄

BUECHI, J.R. & SIEFKES, D. *The complete extensions of the monadic second order theory of countable ordinals*
⋄ B15 C35 C85 D05 E10 ⋄

DAHN, B.I. & GEHNE, J. *A theory of ordered exponential fields which has no model completion*
⋄ C35 C60 C65 ⋄

DANKO, W. *Interpretability of algorithmic theories*
⋄ B25 B75 C35 F25 ⋄

DURET, J.-L. *Stabilite des corps separablement clos*
⋄ C35 C45 C60 ⋄

ERSHOV, YU.L. *Regularly r-closed fields (Russian)*
⋄ B25 C25 C35 C60 ⋄

IYANAGA, S. *Problems in foundation of mathematics: Complete and uncomplete theories (Japanese)*
⋄ C35 ⋄

JARDEN, M. & WHEELER, W.H. *Model-complete theories of e-free Ax fields* ⋄ C35 C60 ⋄

JARDEN, M. *On the model companion of the theory of e-fold ordered fields* ⋄ B25 C25 C35 C60 ⋄

KNIGHT, J.F. *Degrees of types and independent sequences*
⋄ C35 C45 D30 ⋄

LUO, LIBO *The τ-theory for free groups is undecidable*
⋄ C35 C60 D35 ⋄

MARCJA, A. *Categoricita e algebra* ⋄ C35 C45 C98 ⋄

MIJAJLOVIC, Z. *On the embedding property of models*
  ⋄ C25 C35 C52 G05 G10 ⋄
MLCEK, J. *Compactness and homogeneity of saturated structures I,II* ⋄ C35 C50 C60 C62 ⋄
NURTAZIN, A.T. *An elementary classification and some properties of complete orders (Russian)*
  ⋄ B25 C35 C65 C75 E10 ⋄
OMAROV, B. *A question of V.Harnik (Russian)*
  ⋄ C15 C35 C50 ⋄
OMAROV, B. *Nonessential extensions of complete theories (Russian)* ⋄ C15 C35 ⋄
PARIGOT, M. *Le modele compagnon de la theorie des arbres* ⋄ C10 C25 C35 C50 C65 ⋄
PILLAY, A. $\aleph_0$-*categoricity over a predicate* ⋄ C35 ⋄
PILLAY, A. *A note on tight stable theories*
  ⋄ C15 C35 C45 ⋄
POIZAT, B. *Une theorie finiement axiomatisable et superstable* ⋄ C35 C45 ⋄
PRESTEL, A. & ROQUETTE, P. *Lectures on formally p-adic fields* ⋄ B25 C10 C35 C60 C98 ⋄
RICHARDSON, D.B. *Roots of real exponential functions*
  ⋄ C35 C65 ⋄
ROTHMALER, P. *Some model theory of modules. I: On total transcendence of modules. II: On stability and categoricity of flat modules* ⋄ C35 C45 C60 ⋄
ROTHMALER, P. *Stationary types in modules*
  ⋄ C35 C45 C60 ⋄
SCHMERL, J.H. $\aleph_0$-*categorical distributive lattices of finite breadth* ⋄ C35 G10 ⋄
SCHMITT, P.H. *Algebraically complete lattices*
  ⋄ C25 C35 G10 ⋄
TOFFALORI, C. *Questioni di teoria dei modelli per coppie di campi* ⋄ C10 C35 C45 C60 ⋄
TULIPANI, S. *On the size of congruence lattices for models of theories with definability of congruences*
  ⋄ C05 C35 G10 ⋄
TUSCHIK, H.-P. & WEESE, M. *An unstable $\aleph_0$-categorical theory without the strict order property* ⋄ C35 C45 ⋄
WHEELER, W.H. *Model complete theories of formally real fields and formally p-adic fields* ⋄ C35 C60 ⋄

1984

ALIEV, V.G. *Model companions of theories of semigroup classes (Russian) (English and Azerbaijani summaries)*
  ⋄ C05 C25 C35 ⋄
BALDWIN, JOHN T. *First-order theories of abstract dependence relations*
  ⋄ C35 C45 C57 C60 D25 D45 ⋄
BASARAB, S.A. *On some classes of Hilbertian fields*
  ⋄ B25 C10 C25 C35 C60 ⋄
BECKER, E. *Extended Artin-Schreier theory of fields*
  ⋄ B25 C35 C60 ⋄
BUECHLER, S. *Kueker's conjecture for superstable theories*
  ⋄ C35 C45 C50 ⋄
BURRIS, S. *Model companions for finitely generated universal Horn classes*
  ⋄ B25 C05 C10 C35 D20 ⋄
CHERLIN, G.L. *Totally categorial structures*
  ⋄ C35 C45 C98 ⋄

CLARK, D.M. $\aleph_0$-*categoricity in infra primal varieties*
  ⋄ C05 C30 C35 E05 ⋄
DAHN, B.I. & WOLTER, H. *Ordered fields with several exponential functions* ⋄ C25 C35 C65 D35 ⋄
DELON, F. *Espaces ultrametriques*
  ⋄ C25 C35 C45 C50 C65 ⋄
DELON, F. *Theories completes de corps*
  ⋄ C35 C60 C98 ⋄
GILMAN, R.H. *Characteristically simple $\aleph_0$-categorical groups* ⋄ C35 C60 ⋄
HARRINGTON, L.A. & MAKKAI, M. & SHELAH, S. *A proof of Vaught's conjecture for ω-stable theories*
  ⋄ C15 C35 C45 ⋄
HEINDORF, L. *Beitraege zur Modelltheorie der Booleschen Algebren (English and Russian summaries)*
  ⋄ B25 C10 C35 C55 C80 G05 ⋄
HEINEMANN, B. & PRESTEL, A. *Fields regularly closed with respect to finitely many valuations and orderings*
  ⋄ C35 C60 ⋄
IVANOV, A.A. *Complete theories of unars (Russian)*
  ⋄ B25 C35 ⋄
KALFA, C. *Decidable properties of finite sets of equations in trivial languages* ⋄ C05 C35 C52 ⋄
KREMER, E.M. *Modules with an almost categorical theory (Russian)* ⋄ C35 C45 C60 ⋄
KREMER, E.M. *Rings over which all modules of a given type are almost categorical (Russian)*
  ⋄ C35 C45 C60 ⋄
LASCAR, D. *Sous-groupes d'automorphismes d'une structure saturee* ⋄ C07 C35 C45 C50 ⋄
LERMAN, M. & SHORE, R.A. & SOARE, R.I. *The elementary theory of the recursively enumerable degrees is not $\aleph_0$ categorical* ⋄ C35 D25 ⋄
LOVEYS, J. *The uniqueness of envelopes in $\aleph_0$-categorical, $\aleph_0$-stable structures* ⋄ C35 C45 ⋄
MANDERS, K.L. *Interpretations and the model theory of the classical geometries* ⋄ C25 C35 C65 ⋄
MARCJA, A. & TOFFALORI, C. *On pseudo-$\aleph_0$-categorical theories* ⋄ C35 C40 C45 G05 ⋄
MARCJA, A. & TOFFALORI, C. *On Cantor-Bendixson spectra containing (1,1) I* ⋄ C35 C40 C45 G05 ⋄
PILLAY, A. & SROUR, G. *Closed sets and chain conditions in stable theories* ⋄ C35 C45 ⋄
PILLAY, A. *Countable modules*
  ⋄ C15 C35 C45 C52 C60 ⋄
ROQUETTE, P. *Some tendencies in contemporary algebra*
  ⋄ C10 C35 C60 C98 ⋄
ROTHMALER, P. *Some model theory of modules III. On infiniteness of sets definable in modules*
  ⋄ C10 C35 C60 C80 ⋄
SAFFE, J. *Categoricity and ranks* ⋄ C35 C45 ⋄
SCHMERL, J.H. $\aleph_0$-*categorical partially ordered sets (French summary)*
  ⋄ B25 C15 C35 C65 C98 E07 ⋄
SCHMITT, P.H. *Model- and substructure complete theories of ordered abelian groups* ⋄ C10 C35 C60 ⋄
STEPANOVA, A.A. *Categorical quasivarieties of abelian groupoids and quasigroups (Russian)* ⋄ C05 C35 ⋄

STEPANOVA, A.A. *Groupoids of rank 2 generating categorical Horn classes (Russian)* ⋄ C05 C35 ⋄
TOFFALORI, C. *Anelli regolari separabilmente chiusi (English summary)* ⋄ C35 C45 C60 ⋄
WEISPFENNING, V. *Quantifier elimination and decision procedures for valued fields* ⋄ B25 C10 C35 C60 ⋄
ZIL'BER, B.I. *Strongly minimal countably categorical theories II (Russian)* ⋄ C35 C45 ⋄
ZIL'BER, B.I. *Strongly minimal countably categorical theories. III (Russian)* ⋄ C35 C45 ⋄
ZIL'BER, B.I. *The structure of models of uncountably categorical theories* ⋄ C35 C98 ⋄

1985

APPS, A.B. *Two counterexamples in $\aleph_0$-categorical groups* ⋄ C35 C60 ⋄
BUECHLER, S. *Invariantes for ω-categorical, ω-stable theories* ⋄ C35 C45 ⋄
BUECHLER, S. *One theorem of Zil'ber's on strongly minimal sets* ⋄ C35 C45 ⋄
CHERLIN, G.L. & HARRINGTON, L.A. & LACHLAN, A.H. *$\aleph_0$-categorical, $\aleph_0$-stable structures* ⋄ C35 C45 C52 ⋄
GEORGESCU, G. & VOICULESCU, I. *Eastern model-theory for boolean-valued theories* ⋄ C25 C35 C40 C90 ⋄
KIERSTEAD, H.A. & REMMEL, J.B. *Degrees of indiscernibles in decidable models* ⋄ C30 C35 C57 D30 ⋄
LASCAR, D. *Les groupes ω-stables de rang fini* ⋄ C35 C45 C60 ⋄
LASLANDES, B. *Modele-compagnons de theories de corps munis de n ordres* ⋄ C25 C35 C60 ⋄
MARCJA, A. & TOFFALORI, C. *On Cantor-Bendixon spectra containing (1,1) II* ⋄ C35 C40 C45 G05 ⋄
MARONGIU, G. *Alcune osservazioni sulla $\aleph_0$-categoricita degli ordini circolari* ⋄ C35 C65 ⋄
PALYUTIN, E.A. & SAFFE, J. & STARCHENKO, S.S. *Models of superstable Horn theories* ⋄ C05 C35 C45 ⋄
PERYAZEV, N.A. *Positive indistinguishability of algebraic systems and completeness of positive theories (Russian)* ⋄ C35 ⋄
POINT, F. *Transfer properties of discriminator varieties* ⋄ C10 C25 C30 C35 ⋄
SHELAH, S. *Classification of first order theories which have a structure theorem* ⋄ C35 C45 C52 C75 ⋄
SHEN, FUXING *Lattice-valued model complete theory (Chinese) (English summary)* ⋄ C35 C90 ⋄
STEPANOVA, A.A. *Semigroups generating categorical Horn classes (Russian)* ⋄ C05 C35 ⋄
TOFFALORI, C. *Teorie p-$\aleph_0$-categoriche* ⋄ C35 C45 C60 ⋄
TSUBOI, A. *On theories having a finite number of nonisomorphic countable models* ⋄ C15 C35 C45 ⋄
WEISPFENNING, V. *Quantifier elimination for modules* ⋄ C10 C35 C57 C60 ⋄

C40 Interpolation, preservation, definability

1902
PADOA, A. *Essai d'une theorie algebrique des nombres entiers, precede d'une introduction logique a une theorie deductive quelconque* ⋄ A10 B30 C40 ⋄

1927
LINDENBAUM, A. & TARSKI, A. *Sur l'independance des notions primitives dans les systemes mathematiques*
⋄ B30 C40 E75 ⋄

1931
TARSKI, A. *Sur les ensembles definissables de nombres reels I*
⋄ B25 B28 C10 C40 C60 C65 D55 E15 E47 F35 ⋄

1932
TARSKI, A. *Der Wahrheitsbegriff in den Sprachen der deduktiven Disziplinen*
⋄ A05 B30 C07 C40 F30 ⋄

1933
TARSKI, A. *On the notion of truth in reference to formalized deductive sciences (Polish)*
⋄ A05 B15 B30 C07 C40 F30 ⋄

1934
TARSKI, A. *Some methodological investigations on the definability of terms (Polish)* ⋄ B10 C35 C40 ⋄

1948
TARSKI, A. *A problem concerning the notion of definability*
⋄ B10 B15 C40 ⋄

1951
HORN, A. *On sentences which are true of direct unions of algebras* ⋄ C05 C30 C40 ⋄
MARCZEWSKI, E. *Sur les congruences et les proprietes positives d'algebres abstraites* ⋄ C05 C40 ⋄

1952
CRAIG, W. & QUINE, W.V.O. *On reduction to a symmetric relation* ⋄ B10 C40 ⋄

1953
BETH, E.W. *On Padoa's method in the theory of definition*
⋄ B10 C40 ⋄

1955
BING, K. *On arithmetical classes not closed under direct union* ⋄ C05 C40 C52 ⋄
LOS, J. & SUSZKO, R. *On the infinite sums of models*
⋄ C30 C40 C52 ⋄
SVENONIUS, L. *Definability and simplicity* ⋄ C40 ⋄

1956
ROBINSON, A. *A result on consistency and its application to the theory of definition* ⋄ C40 ⋄

1957
CRAIG, W. *Linear reasoning. A new form of the Herbrand-Gentzen theorem*
⋄ B10 C07 C40 F05 ⋄
CRAIG, W. *Three uses of the Herbrand-Gentzen theorem in relating model theory and proof theory*
⋄ B10 C07 C40 F05 ⋄
LYNDON, R.C. *Sentences preserved under homomorphisms; sentences preserved under subdirect products* ⋄ C07 C30 C40 C52 ⋄
ROBINSON, A. *Some problems of definability in the lower predicate calculus* ⋄ C35 C40 C60 ⋄

1958
CHANG, C.C. & MOREL, A.C. *On closure under direct product* ⋄ C30 C40 ⋄
LEVY, A. *Comparison of subtheories*
⋄ C07 C40 E30 F30 ⋄
OBERSCHELP, A. *Ueber die Axiome produkt-abgeschlossener arithmetischer Klassen*
⋄ B20 B25 C30 C40 C52 ⋄

1959
BOUVERE DE, K.L. *A method in proofs of undefinability, with applications to functions in the arithmetic of natural numbers* ⋄ C40 F30 ⋄
CHANG, C.C. *On unions of chains of models*
⋄ C40 C52 ⋄
LYNDON, R.C. *An interpolation theorem in the predicate calculus* ⋄ C40 ⋄
LYNDON, R.C. *Existential Horn sentences*
⋄ B20 C30 C40 C52 ⋄
LYNDON, R.C. *Properties preserved under homomorphism*
⋄ C07 C40 C52 ⋄
LYNDON, R.C. *Properties preserved under algebraic constructions* ⋄ C07 C30 C40 C52 ⋄
LYNDON, R.C. *Properties preserved in subdirect products*
⋄ C30 C40 C52 ⋄
OBERSCHELP, A. *Ueber die Axiome arithmetischer Klassen mit Abgeschlossenheitsbedingungen*
⋄ B25 C40 C52 ⋄
ROBINSON, A. *Obstructions to arithmetical extension and the theorem of Los and Suszko* ⋄ C40 C52 ⋄
SVENONIUS, L. *A theorem on permutations in models*
⋄ C30 C40 ⋄

1960
MACHOVER, M. *A note on sentences preserved under direct products and powers* ⋄ C30 C40 ⋄

MAEHARA, S. *On the interpolation theorem of Craig (Japanese)* ◇ B10 C40 F05 ◇

1961

ENGELER, E. *Unendliche Formeln in der Modell-Theorie* ◇ C15 C35 C40 C75 ◇

GRZEGORCZYK, A. & MOSTOWSKI, ANDRZEJ & RYLL-NARDZEWSKI, C. *Definability of sets in models of axiomatic theories* ◇ C40 D55 ◇

KEISLER, H.J. *Ultraproducts and elementary classes* ◇ C20 C40 C50 C52 C55 E05 E50 ◇

MAEHARA, S. & TAKEUTI, G. *A formal system of first-order predicate calculus with infinitely long expressions* ◇ C40 C75 F05 ◇

VAUGHT, R.L. *The elementary character of two notions from general algebra* ◇ C05 C40 C52 C60 ◇

1962

ADDISON, J.W. *Some problems in hierarchy theory* ◇ C40 D55 E15 ◇

ADDISON, J.W. *The theory of hierarchies* ◇ C40 C52 D55 E15 ◇

BETH, E.W. *Observations concernant la theorie de la definition* ◇ C40 ◇

FUHRKEN, G. *Bemerkung zu einer Arbeit E.Engelers* ◇ C15 C40 C75 ◇

LYNDON, R.C. *Metamathematics and algebras: an example* ◇ C40 C60 ◇

MOSTOWSKI, ANDRZEJ *L'espace des modeles d'une theorie formalisee et quelques-unes de ses applications* ◇ B50 C07 C40 C57 C80 C85 C90 D45 ◇

MOSTOWSKI, ANDRZEJ *On invariant, dual invariant and absolute formulas* ◇ B15 C40 C85 ◇

SCHUETTE, K. *Der Interpolationssatz der intuitionistischen Praedikatenlogik* ◇ C40 F50 ◇

SCOTT, D.S. *Algebras of sets binumerable in complete extensions of arithmetic* ◇ C40 C62 F30 ◇

TAJMANOV, A.D. *On a theorem of Beth and Kochen (Russian)* ◇ C20 C40 C52 ◇

TAKEUTI, G. *On the weak definability in set theory* ◇ C40 E25 E47 ◇

1963

BOUVERE DE, K.L. *A mathematical characterization of explicit definability* ◇ C40 ◇

GRINDLINGER, M.D. *Inductive definition of a function (Russian)* ◇ C40 C62 ◇

HENKIN, L. *An extension of the Craig-Lyndon interpolation theorem* ◇ C40 ◇

KROM, MELVEN R. *Separation principles in the hierarchy theory of pure first-order logic* ◇ B10 C40 C52 D55 ◇

REZNIKOFF, I. *Chaines de formules* ◇ B05 C40 C90 F50 ◇

1964

CHANG, C.C. *Some new results in definability* ◇ C40 ◇

DAIGNEAULT, A. *Freedom in polyadic algebras and two theorems of Beth and Craig* ◇ C05 C40 G15 ◇

HINTIKKA, K.J.J. *Distributive normal forms and deductive interpolation* ◇ B10 C07 C40 ◇

MAKKAI, M. *On PC_Δ-classes in the theory of models* ◇ C20 C40 C52 ◇

MAKKAI, M. *On a generalization of a theorem of E.W.Beth* ◇ C40 C50 ◇

MAKKAI, M. *Remarks on my paper "On PC_Δ-classes in the theory of models"* ◇ C40 C52 ◇

PUTNAM, H. *On families of sets represented in theories* ◇ C40 D25 F30 ◇

1965

ADDISON, J.W. *The method of alternating chains* ◇ C40 D55 D75 E15 ◇

BOUVERE DE, K.L. *Logical synonymity* ◇ A05 C40 C52 ◇

BOUVERE DE, K.L. *Synonymous theories* ◇ B30 C40 ◇

BUECHI, J.R. *Relatively categorical and normal theories* ◇ C35 C40 C60 ◇

CRAIG, W. *Satisfaction for n-th order languages defined in n-th order languages* ◇ B15 C40 ◇

KEISLER, H.J. *Finite approximations of infinitely long formulas* ◇ C40 C50 C75 ◇

KEISLER, H.J. *Some application of infinitely long formulas* ◇ C40 C50 C75 ◇

KOCHEN, S. *Topics in the theory of definition* ◇ C20 C40 C52 ◇

KREISEL, G. *Model-theoretic invariants: applications to recursive and hyperarithmetic operations* ◇ C40 C62 D55 D60 ◇

LOPEZ-ESCOBAR, E.G.K. *An interpolation theorem for denumerably long sentences* ◇ C40 C75 ◇

MALITZ, J. *Problems in the model theory of infinite languages* ◇ C40 C75 ◇

REZNIKOFF, I. *Tout ensemble de formules de la logique classique est equivalent a un ensemble independant* ◇ B05 C40 ◇

SCHOCK, R. *On definitions* ◇ A05 C40 ◇

SCOTT, D.S. *Logic with denumerably long formulas and finite strings of quantifiers* ◇ C15 C40 C75 C98 ◇

SVENONIUS, L. *On the denumerable models of theories with extra predicates* ◇ C15 C40 C52 ◇

1966

CUSIN, R. *Une demonstration concernant le theoreme d'interpolation generalise aux ensembles d'enonces* ◇ C40 ◇

FEFERMAN, S. & KREISEL, G. *Persistent and invariant formulas relative to theories of higher order* ◇ B15 C40 C52 C85 ◇

KINO, A. *On definability of ordinals in logic with infinitely long expressions* ◇ C40 C75 E10 E47 ◇

KOGALOVSKIJ, S.R. *On higher order logic (Russian)* ◇ B15 C40 C85 ◇

KOGALOVSKIJ, S.R. *On the semantics of the theory of types (Russian)* ◇ B15 C40 C85 ◇

LEBLANC, L. *Representabilite et definissabilite dans les algebres transformationnelles et dans les algebres polyadiques* ◇ C40 C62 C90 G15 ◇

LINDSTROEM, P. *On relations between structures* ◇ C07 C40 C52 ◇

NAGASHIMA, T. *An extension of the Craig-Schuette interpolation theorem* ◇ C40 F50 ◇

REZNIKOFF, I. *Sur les ensembles denombrables de formules en logique intuitionniste* ◇ C40 C90 F50 ◇

YASUHARA, M. *Syntactical and semantical properties of generalized quantifiers* ⋄ C40 C55 C80 ⋄

1967

BLACK, M. *Craig's theorem* ⋄ A05 C40 ⋄

BOUVERE DE, K.L. *Some remarks about synonymity and the theorem of Beth* ⋄ C40 G15 ⋄

DREBEN, B. & PUTNAM, H. *The Craig interpolation lemma* ⋄ C40 ⋄

GALVIN, F. *Reduced products, Horn sentences, and decision problems*
⋄ B20 B25 C20 C30 C40 D35 ⋄

HAUSCHILD, K. *Ueber ein Definierbarkeitsproblem in der Koerpertheorie (English and French summaries)*
⋄ C40 C60 ⋄

LOPEZ-ESCOBAR, E.G.K. *On a theorem of J.I.Malitz*
⋄ C30 C40 C85 ⋄

LOPEZ-ESCOBAR, E.G.K. *Remarks on an infinitary language with constructive formulas*
⋄ C40 C70 C75 F50 ⋄

OMAROV, A.I. *Filtered products of models (Russian) (English summary)* ⋄ C20 C30 C40 C50 C52 ⋄

TAJMANOV, A.D. *Automorphisms of an algebra having one unary operation in the signature (Russian)*
⋄ C07 C20 C40 ⋄

WEGLORZ, B. *Some preservation theorems* ⋄ C40 ⋄

1968

BARWISE, J. *Implicit definability and compactness in infinitary languages* ⋄ C40 C70 ⋄

BORKOWSKI, L. *Some remarks about the notion of definition (Polish) (Russian and English summaries)*
⋄ A05 B10 C40 ⋄

BOUVERE DE, K.L. *On formal ambiguity* ⋄ C40 ⋄

CHUDNOVSKY, G.V. *Some results in the theory of infinitely long expressions (Russian)* ⋄ C40 C75 ⋄

FEFERMAN, S. *Lectures on proof theory*
⋄ C40 C70 C75 F05 F15 F98 ⋄

FEFERMAN, S. *Persistent and invariant formulas for outer extensions* ⋄ C30 C40 C70 C75 F35 ⋄

FUHRKEN, G. *A model with exactly one undefinable element* ⋄ C40 C50 ⋄

HODES, L. & SPECKER, E. *Lengths of formulas and elimination of quantifiers I* ⋄ B35 C10 C40 ⋄

KEISLER, H.J. *Formulas with linearly ordered quantifiers*
⋄ C40 C50 C75 C80 ⋄

KOGALOVSKIJ, S.R. *Some remarks on higher order logic (Russian)* ⋄ B15 C40 C85 F35 ⋄

KREISEL, G. *Choice of infinitary languages by means of definability criteria: generalized recursion theory*
⋄ C40 C70 C75 D60 D75 ⋄

KROM, MELVEN R. *Some interpolation theorems for first-order formulas in which all disjunctions are binary*
⋄ B20 C40 ⋄

KUEKER, D.W. *Definability, automorphisms and infinitary languages* ⋄ C07 C40 C75 ⋄

KUNEN, K. *Implicit definability and infinitary languages*
⋄ C40 C70 C75 D70 E47 E55 ⋄

MONTAGUE, R. *Recursion theory as a branch of model theory* ⋄ C40 D75 ⋄

MOSTOWSKI, ANDRZEJ *Craig's interpolation theorem in some extended systems of logic* ⋄ C40 C95 ⋄

OBERSCHELP, A. *On the Craig-Lyndon interpolation theorem* ⋄ C40 ⋄

PACHOLSKI, L. & WEGLORZ, B. *Topologically compact structures and positive formulas* ⋄ B20 C40 C50 ⋄

SAYEKI, H. *Some consequences of the normality of the space of models* ⋄ C07 C40 C52 ⋄

TAKEUTI, G. *A determinate logic* ⋄ C40 C75 E60 ⋄

WEINSTEIN, J.M. *(ω_1, ω) properties of unions of models*
⋄ C30 C40 C52 C75 ⋄

WESTRHENEN VAN, S.C. *Statistical estimation of definability* ⋄ B35 C40 ⋄

1969

BARWISE, J. *Applications of strict Π_1^1 predicates to infinitary logic* ⋄ C40 C70 D60 ⋄

BARWISE, J. *Remarks on universal sentences of $L_{\omega_1,\omega}$*
⋄ C15 C40 C75 ⋄

BUECHI, J.R. & LANDWEBER, L.H. *Definability in the monadic second-order theory of successor*
⋄ B15 B25 C40 C85 D05 F35 ⋄

FITTLER, R. *Categories of models with initial objects*
⋄ C40 G30 ⋄

LOPEZ-ESCOBAR, E.G.K. *A non-interpolation theorem (Russian summary)* ⋄ C40 C75 ⋄

MAKKAI, M. *On the model theory of denumerable long formulas with finite strings of quantifiers*
⋄ C40 C70 C75 ⋄

MALITZ, J. *Universal classes in infinitary languages*
⋄ C40 C52 C75 ⋄

MOSCHOVAKIS, Y.N. *Abstract computability and invariant definability* ⋄ C40 D70 D75 ⋄

MOTOHASHI, N. *On normal operations on models*
⋄ C30 C40 C50 ⋄

NADIU, G.S. *The interpolation theorem in strict positive logic* ⋄ B20 C40 ⋄

PRELLER, A. *Interpolation et amalgamation*
⋄ C40 C52 C75 G15 G25 ⋄

ROBINSON, A. *Compactification of groups and rings and non-standard analysis* ⋄ C40 C60 H05 H20 ⋄

TAKHTADZHYAN, K.A. *Logical formulae that are preserved under a strong homomorphism (Armenian) (Russian and English summaries)* ⋄ C07 C40 ⋄

WEGLORZ, B. *Some preservation theorems. II* ⋄ C40 ⋄

1970

CHAUBARD, A. *Definissabilite explicite* ⋄ C40 ⋄

DRABBE, J. *Sur une propriete des formules negatives*
⋄ C30 C40 C52 ⋄

GALVIN, F. *Horn sentences*
⋄ B20 C20 C30 C40 D35 ⋄

HAJEK, P. *Logische Kategorien*
⋄ B10 C40 E40 F25 G30 ⋄

HERRE, H. & RAUTENBERG, W. *Das Basistheorem und einige Anwendungen in der Modelltheorie*
⋄ C40 C52 G05 ⋄

HINTIKKA, K.J.J. & TUOMELA, R. *Toward a general theory of auxiliary concepts and definability in first-order theories* ⋄ A05 C40 ⋄

HODES, L. *The logical complexity of geometric properties on the plane* ⋄ B35 C40 ⋄

ISSEL, W. *Semantische Untersuchungen ueber Quantoren II,III* ⋄ C40 C55 C80 ⋄

KUEKER, D.W. *Generalized interpolation and definability* ⋄ C40 C50 C75 ⋄

KULIK, V.T. *Sentences preserved under homomorphisms that are one-to-one at zero (Russian)* ⋄ C05 C40 C60 ⋄

LEVY, P. *Un paradoxe relatif a l'ensemble des nombres definissables* ⋄ C40 E47 ⋄

MOSCHOVAKIS, Y.N. *The Suslin-Kleene theorem for countable structures* ⋄ C15 C40 D55 D70 D75 ⋄

MOTOHASHI, N. *A theorem in the theory of definition* ⋄ C40 C50 ⋄

OLIN, P. *Some first order properties of direct sums of modules* ⋄ C30 C40 C60 ⋄

RABIN, M.O. *Weakly definable relations and special automata* ⋄ B15 B25 C40 C85 D05 ⋄

REICHBACH, J. *On statistical tests in generalizations of Herbrand's theorems and asymptotic probabilistic models (main theorems of mathematics)* ⋄ B10 C40 C90 ⋄

REYES, G.E. *Local definability theory* ⋄ C07 C15 C40 C50 C55 C75 ⋄

ROCKINGHAM GILL, R.R. *The Craig-Lyndon interpolation theorem in 3-valued logic* ⋄ B50 C40 ⋄

SCHWARTZ, DIETRICH *Nichtdefinierbarkeit der Stufenbeziehung durch die Elementbeziehung in der allgemeinen Mengenlehre von Klaua* ⋄ C40 E47 E70 ⋄

SHELAH, S. *Remark to "Local definability theory" of Reyes* ⋄ C40 C55 ⋄

SLAGLE, J.R. *Interpolation theorems for resolution in lower predicate calculus* ⋄ B35 C40 ⋄

TAKEUTI, G. *A determinate logic* ⋄ C40 C75 E60 ⋄

1971

BARWISE, J. & GANDY, R.O. & MOSCHOVAKIS, Y.N. *The next admissible set* ⋄ C40 C62 D55 D60 D70 E45 E47 ⋄

CARSTENGERDES, W. *Mehrsortige logische Systeme mit unendlich langen Formeln I, II* ⋄ C40 C75 F05 ⋄

CHANG, C.C. *Two interpolation theorems* ⋄ C40 C50 C75 ⋄

DRABBE, J. *Sur une propriete de preservation* ⋄ C40 ⋄

DRAGALIN, A.G. & FUKSON, V.I. & LYUBETSKIJ, V.A. *Definable sequences of countable ordinals (Russian)* ⋄ C40 C62 E10 E35 E45 E47 ⋄

EHRENFEUCHT, A. & FUHRKEN, G. *On models with undefinable elements* ⋄ C40 C50 ⋄

GABBAY, D.M. *Semantic proof of the Craig interpolation theorem for intuitionistic logic and extensions I,II* ⋄ B55 C40 C90 F50 ⋄

GERMANO, G. *Incompleteness and truth definitions* ⋄ C40 D35 F30 ⋄

HAUSCHILD, K. *Nichtaxiomatisierbarkeit von Satzmengen durch Ausdruecke spezieller Gestalt* ⋄ B28 C20 C40 C52 F30 ⋄

KUEHNRICH, M. *Ein Beweis der wesentlichen Nichtdefinierbarkeit* ⋄ C40 E47 E70 ⋄

LINDNER, C.C. *Identities preserved by singular direct product* ⋄ C05 C13 C40 ⋄

MAEHARA, S. & TAKEUTI, G. *Two interpolation theorems for a Π_1^1 predicate calculus* ⋄ B15 C40 F35 ⋄

MAKKAI, M. *Algebraic operations on classes of structures and their connections with logical formulas I,II (Hungarian) (English summary)* ⋄ C30 C40 C52 C75 ⋄

MALITZ, J. *Infinitary analogs of theorems from first order model theory* ⋄ C30 C40 C55 C75 ⋄

MCKENZIE, R. *Definability in lattices of equational theories* ⋄ C05 C40 G10 ⋄

OLIN, P. *Direct multiples and powers of modules* ⋄ C30 C40 C60 ⋄

PABION, J.F. *Remarques sur la definition d'un ensemble de parametres* ⋄ C40 ⋄

PAILLET, J.L. *Quelques rapports entre l'interpretabilite et l'extension logique* ⋄ C07 C40 ⋄

PLATEK, R.A. *The converse to a metatheorem in Goedel set theory* ⋄ C40 E30 E47 E70 ⋄

PREVIALE, F. *Applicazioni del concetto di diagramma nella teoria dei modelli di linguaggi con piu specie di variabili* ⋄ C07 C40 ⋄

RABIN, M.O. *Decidability and definability in second-order theories* ⋄ B15 B25 C40 C85 D05 ⋄

ROMOV, B.A. *Definability by formulas for predicates on a finite model (Russian) (English summary)* ⋄ C13 C40 C60 ⋄

SERVI, M. *Una questione di teoria dei modelli nelle categorie con prodotti finiti (English summary)* ⋄ C40 G30 ⋄

1972

AFRICK, H. *A proof theoretic proof of Scott's general interpolation theorem* ⋄ C40 F07 ⋄

ARMBRUST, M. & KAISER, KLAUS *Some remarks on projective model classes and the interpolation theorem* ⋄ C40 C52 ⋄

BUECHI, J.R. & DANHOF, K.J. *Model theoretic approaches to definability* ⋄ C20 C40 ⋄

CORNMAN, J.W. *Craig's theorem, Ramsey-sentences, and scientific instrumentalism* ⋄ A05 C40 ⋄

FEFERMAN, S. *Infinitary properties, local functors, and systems of ordinal functions* ⋄ C30 C40 C75 E10 F15 G30 ⋄

GABBAY, D.M. *Craig's interpolation theorem for modal logics* ⋄ B45 C40 C90 ⋄

GARLAND, S.J. *Generalized interpolation theorems* ⋄ C40 C75 C85 D55 E15 ⋄

GRANT, J. *Recognizable algebras of formulas* ⋄ C07 C40 G05 G15 G25 ⋄

HAUSCHILD, K. & MELIS, E. *Zusammenhaenge zwischen Axiomatisierbarkeit und Definierbarkeit in elementaren Theorien mit eingeschraenktem Wohlordnungsschema* ⋄ C40 C52 ⋄

HINTIKKA, K.J.J. *Constituents and finite identifiability* ⋄ C40 ⋄

KOGALOVSKIJ, S.R. *On the degrees of indefinability (Russian)* ⋄ C40 ⋄

KUEKER, D.W. *Loewenheim-Skolem and interpolation theorems in infinitary languages* ⋄ C40 C55 C75 E05 ⋄

LINDNER, C.C. *Identities preserved by the singular direct product II* ◇ C05 C13 C40 ◇

MAKKAI, M. *Svenonius sentences and Lindstroem's theory on preservation theorems* ◇ C40 C52 C75 ◇

MARCJA, A. *Una generalizzazione del concetto di prodotto diretto* ◇ C20 C30 C40 G05 ◇

MOTOHASHI, N. *A new theorem on definability in a positive second order logic with countable conjunctions and disjunctions* ◇ B15 C40 C75 C85 ◇

MOTOHASHI, N. *Countable structures for uncountable infinitary languages* ◇ C15 C40 C75 ◇

MOTOHASHI, N. *Interpolation theorem and characterization theorem* ◇ C40 C75 F05 ◇

NEBRES, B.F. *Herbrand uniformity theorems for infinitary languages* ◇ C40 C70 C75 F05 ◇

NEBRES, B.F. *Infinitary formulas preserved under unions of models* ◇ C40 C70 C75 ◇

OLIN, P. *First order preservation theorems in two-sorted languages* ◇ C30 C40 C60 ◇

OLIN, P. *Products of two-sorted structures* ◇ C30 C40 C75 ◇

OSSWALD, H. *Homomorphie-invariante Formeln in der intuitionistischen Logik* ◇ C40 C90 F50 ◇

PIGOZZI, D. *Amalgamation, congruence-extension, and interpolation properties in algebras* ◇ C05 C40 C52 G15 ◇

RASIOWA, H. *The Craig interpolation theorem for m-valued predicate calculi (Russian summary)* ◇ B50 C40 C90 ◇

SMIRNOV, V.A. *Formal derivation and logical calculi (Russian)* ◇ B98 C07 C40 C98 F05 F07 F98 ◇

SMOLIN, V.P. *On the question of definability in the lower predicate calculus on a finite model without equality (Russian) (English summary)* ◇ C13 C40 ◇

STERN, J. *Theoremes d'interpolation obtenus par la methode du "forcing"* ◇ C25 C40 ◇

WOLTER, H. *Untersuchung ueber Algebren und formalisierte Sprachen hoeherer Stufen (Russian, English and French summaries)* ◇ B15 C20 C30 C40 C85 ◇

1973

BACSICH, P.D. *Defining algebraic elements* ◇ C40 C52 C75 ◇

BARWISE, J. *A preservation theorem for interpretations* ◇ C40 F25 ◇

BOOLOS, G. *A note on Beth's theorem (Russian summary)* ◇ C40 C62 ◇

BUECHI, J.R. & DANHOF, K.J. *Definability in normal theories* ◇ C40 ◇

CHANG, C.C. *What's so special about saturated models?* ◇ C40 C50 C55 ◇

FRIEDMAN, H.M. *Beth's theorem in cardinality logics* ◇ C40 C55 C80 ◇

FUJIWARA, T. *A generalization of Lyndon's theorem characterizing sentences preserved in subdirect products* ◇ C05 C30 C40 ◇

FUJIWARA, T. *Note on generalized atomic sets of formulas* ◇ B20 C05 C40 ◇

GEORGIEVA, N.V. *On the logical signs in Schuette's interpolation formulae (Bulgarian) (Russian and English summaries)* ◇ C40 F50 ◇

GERMANO, G. *Incompleteness theorem via weak definability of truth: a short proof* ◇ C40 F30 ◇

GRANT, J. *Automorphisms definable by formulas* ◇ C07 C40 C75 ◇

GRILLIOT, T.J. *Implicit definability and hyperprojectivity* ◇ C40 C75 D75 ◇

ISBELL, J.R. *Functorial implicit operations* ◇ C05 C40 ◇

KOGALOVSKIJ, S.R. & RORER, M.A. *On the question of the definability of the concept of definability (Russian)* ◇ B15 C40 C85 ◇

MAKKAI, M. *Global definability theory in $L_{\omega_1,\omega}$* ◇ C40 C75 ◇

MAKKAI, M. *Preservation theorems for pseudo-elementary classes (Russian)* ◇ C40 C52 ◇

MAKKAI, M. *Vaught sentences and Lindstroem's regular relations* ◇ C40 C52 C75 D55 D70 ◇

MARCHINI, C. *Funtori che conservano e riflettono le formule di Horn* ◇ C40 C52 G30 ◇

MOTOHASHI, N. *An extended relativization theorem* ◇ C40 C75 ◇

MOTOHASHI, N. *Model theory on a positive second order logic with countable conjunctions and disjunctions* ◇ B15 C40 C75 C85 ◇

MOTOHASHI, N. *Two theorems on mix-relativization* ◇ B10 C40 C75 ◇

RANTALA, V. *On the theory of definability in first order logic* ◇ C07 C40 ◇

RESSAYRE, J.-P. *Boolean models and infinitary first order languages* ◇ C40 C50 C55 C70 C75 C80 C90 ◇

SHELAH, S. *Weak definability in infinitary languages* ◇ C40 C55 C75 ◇

TAYLOR, W. *A heterogeneous interpolant* ◇ C40 C75 ◇

WHITELEY, W. *Homogeneous sets and homogeneous formulas* ◇ C40 C60 ◇

1974

AFRICK, H. *Scott's interpolation theorem fails for $L_{\omega_1,\omega}$* ◇ C40 C75 ◇

BELEGRADEK, O.V. *Definability in algebraically closed groups (Russian)* ◇ C25 C40 C60 ◇

COLLET, J.-C. *Multirelation (k,p)-homogene et elimination de quanteurs* ◇ C10 C40 C50 ◇

DAHN, B.I. *Meta-mathematics of some many-valued calculi* ◇ B50 C40 C90 ◇

FEFERMAN, S. *Applications of many-sorted interpolation theorems* ◇ C40 C95 ◇

FEFERMAN, S. *Two notes on abstract model theory I. Properties invariant on the range of definable relations between structures* ◇ C40 C95 ◇

FRODA-SCHECHTER, M. & KWATINETZ, M. *Some remarks on a paper of R. C. Lyndon: "Properties preserved under homomorphism"* ◇ C05 C40 C52 ◇

GAIFMAN, H. *Finiteness is not a Σ_0-property* ◇ C40 E47 ◇

GAIFMAN, H. *Operations on relational structures, functors and classes I* ◇ C20 C30 C40 G30 ◇

GARLAND, S.J. *Second-order cardinal characterizability* ◇ B15 C40 C55 C85 D55 E10 E47 E55 ◇

GRAETZER, G. & SICHLER, J. *Agassiz sum of algebras*
⋄ C05 C30 C40 ⋄

GREGORY, J. *Beth definability in infinitary languages*
⋄ C40 C75 ⋄

HODGES, W. *A normal form for algebraic constructions*
⋄ C40 C75 E47 ⋄

JENSEN, F.V. *Interpolation and definability in abstract logics* ⋄ C40 C95 ⋄

KARP, C.R. *Infinite-quantifier languages and ω-chains of models* ⋄ C40 C75 ⋄

KOGALOVSKIJ, S.R. *Certain simple consequences of the axiom of constructibility (Russian) (English summary)*
⋄ C40 C62 C85 E45 ⋄

KRIVINE, J.-L. *Langages a valeurs reelles et applications*
⋄ B50 C40 C60 C90 ⋄

MAKKAI, M. *Generalizing Vaught sentences from ω to strong cofinality ω* ⋄ C40 C75 ⋄

MIYAMA, T. *The interpolation theorem and Beth's theorem in many-valued logics* ⋄ B50 C40 ⋄

SERVI, M. *Su alcuni funtori che conservano le SH*
⋄ C40 G30 ⋄

SMIRNOV, V.A. *On the question of the definability of predicates introduced via bilateral reduction sentences (Russian)* ⋄ C40 ⋄

WEAVER, G.E. *Finite partitions and their generators*
⋄ B20 B25 C07 C40 ⋄

1975

BACSICH, P.D. *Amalgamation properties and interpolation theorems for equational theories*
⋄ B20 C05 C40 C52 ⋄

BARWISE, J. *Admissible sets and structures. An approach to definability theory*
⋄ B98 C40 C70 C98 D60 D98 E30 E98 ⋄

BRYLL, G. *Classes of asymmetric and bounded graphs defined by the simplest expressions of the predicate calculus (Polish. English translation)* ⋄ C40 ⋄

COSTA DA, N.C.A. & DRUCK, I.F. *Sur les "VBTOs" selon M.Hatcher (English summary)* ⋄ C07 C40 C80 ⋄

CUNNINGHAM, E. *Chain models: applications of consistency properties and back-and-forth techniques in infinite-quantifier languages* ⋄ C40 C75 ⋄

DAHN, B.I. *Eine Anwendung des Basistheorems in der nichtklassischen Logik* ⋄ B55 C40 C90 ⋄

EBBINGHAUS, H.-D. *Zum L(Q)-Interpolationsproblem*
⋄ C40 C55 C80 ⋄

FEFERMAN, S. *Two notes on abstract model theory II. Languages for which the set of valid sentences is semi-invariantly implicitly definable*
⋄ C40 C70 C95 ⋄

FERRO, R. *Limits to some interpolation theorems*
⋄ C40 C75 ⋄

FLUM, J. *First-order logic and its extensions*
⋄ C40 C55 C75 C95 C98 ⋄

FLUM, J. *L(Q)-preservation theorems*
⋄ C20 C40 C55 C80 E60 ⋄

FUJIWARA, T. *Universal sentences preserved under certain extensions* ⋄ B20 C40 ⋄

GREEN, J. *Consistency properties for finite quantifier languages* ⋄ C30 C40 C70 C75 ⋄

GREGORY, J. *On a finiteness condition for infinitary languages* ⋄ C40 C62 C70 C75 E40 ⋄

HALPERN, F. *Transfer theorems for topological structures*
⋄ C20 C40 C90 ⋄

HINTIKKA, K.J.J. & RANTALA, V. *Systematizing definability theory* ⋄ C40 ⋄

HODGES, W. *A normal form for algebraic constructions II*
⋄ C40 C75 E25 G30 ⋄

INDERMARK, K. *Gleichungsdefinierbarkeit in Relationstrukturen* ⋄ C05 C40 ⋄

KANOVICH, M.I. *A hierachical semantic system with set variables (Russian)* ⋄ B20 B65 C40 D05 F50 ⋄

KINO, A. & MYHILL, J.R. *A hierarchy of languages with infinitely long expressions* ⋄ C40 C75 E10 F15 ⋄

KUEKER, D.W. *Core structures for theories*
⋄ C40 C50 C52 ⋄

MACINTYRE, A. & SIMMONS, H. *Algebraic properties of number theories*
⋄ C25 C40 C52 C62 D25 H15 ⋄

MOTOHASHI, N. *On a theorem of Schoenfield (Japanese)*
⋄ C40 ⋄

MOTOHASHI, N. *Some proof-theoretic properties of dense linear orderings and countable well-orderings*
⋄ C40 C75 ⋄

POIZAT, B. *Theoremes globaux (English summary)*
⋄ B20 C40 ⋄

RICKEY, V.F. *Creative definitions in propositional calculi*
⋄ B05 C40 ⋄

RICKEY, V.F. *On creative definitions in the Principia Mathematica* ⋄ B15 C40 ⋄

ROBINSON, A. *Algorithms in algebra*
⋄ C40 C57 C60 D45 D75 ⋄

ROSENTHAL, J.W. *Truth in all of certain well-founded countable models arising in set theory*
⋄ C40 C62 C70 E45 E47 ⋄

SACERDOTE, G.S. *Projective model theory and coforcing*
⋄ C25 C30 C35 C40 C50 C60 ⋄

SHIRAI, K. *Intuitionistic version of the Los-Tarski-Robinson theorem* ⋄ C40 F05 F50 ⋄

STERN, J. *A new look at the interpolation problem*
⋄ C25 C40 C70 ⋄

TROFIMOV, M.YU. *Definability in algebraically closed systems (Russian)* ⋄ C25 C40 C60 C85 ⋄

TULIPANI, S. *Questioni di teoria dei modelli per linguaggi universali positivi II: Metodi di "back and forth" (English summary)* ⋄ B20 C07 C40 C50 ⋄

VOLGER, H. *Completeness theorem for logical categories*
⋄ C40 G30 ⋄

WEAVER, G.E. *Uniform compactness and interpolation theorems in sentential logic* ⋄ B05 C40 ⋄

1976

BARWISE, J. & SCHLIPF, J.S. *An introduction to recursively saturated and resplendent models*
⋄ C15 C40 C50 C57 ⋄

BARWISE, J. *Some applications of Henkin quantifiers*
⋄ C40 C50 C60 C75 C80 ⋄

CSAKANY, B. *Conditions involving universally quantified function variables* ⋄ C05 C40 C85 ⋄

FELSCHER, W. *On interpolation when function symbols are present* ⋄ C40 C75 F50 ⋄

FREYD, P. *Properties invariant within equivalence types of categories* ◇ C40 G30 ◇

FRIEDMAN, H.M. *The complexity of explicit definitions* ◇ C40 C57 ◇

GOSTANIAN, R. & HRBACEK, K. *On the failure of the weak Beth property* ◇ C40 C75 C80 ◇

HARNIK, V. & MAKKAI, M. *Applications of Vaught sentences and the covering theorem*
◇ C15 C40 C45 C50 C52 C75 D70 E15 ◇

HINTIKKA, K.J.J. & RANTALA, V. *A new approach to infinitary languages* ◇ C40 C75 E60 ◇

HUTCHINSON, J.E. *Model theory via set theory*
◇ C40 C55 C62 C70 C75 C80 C95 E75 ◇

JENSEN, F.V. *Souslin-Kleene does not imply Beth*
◇ C40 C95 ◇

LUCAS, T. *Une utilisation du langage des categories pour la presentation de theoremes de theorie des modeles. I*
◇ C40 G30 ◇

MAKOWSKY, J.A. & SHELAH, S. & STAVI, J. *Δ-logics and generalized quantifiers* ◇ C40 C55 C80 C95 ◇

MANSOUX, A. *Logique sans egalite et (k,p)-quasivalence (English summary)* ◇ C07 C40 C75 ◇

PAULOS, J.A. *Noncharacterizability of the syntax set*
◇ C40 C95 ◇

QUINE, W.V.O. *Grades of discriminability*
◇ A05 C40 ◇

SCHULTE-MOENTING, J. *Interpolation formulae for predicates and terms which carry their own history*
◇ C40 ◇

SGRO, J. *Completeness theorems for continuous functions and product topologies*
◇ C40 C65 C80 C90 E75 ◇

WEISPFENNING, V. *Negative-existentially complete structures and definability in free extensions*
◇ C25 C40 C60 C75 D40 G05 ◇

ZIEGLER, M. *A language for topological structures which satisfies a Lindstroem-theorem*
◇ B25 C40 C50 C65 C80 C90 C95 D35 ◇

1977

BADGER, L.W. *An Ehrenfeucht game for the multivariable quantifiers of Malitz and some applications*
◇ C40 C80 E60 ◇

BURRIS, S. *An example concerning definable principal congruences* ◇ C05 C40 ◇

DEISSLER, R. *Minimal models*
◇ C15 C40 C50 C75 ◇

DICKMANN, M.A. *Structures Σ-saturees*
◇ C40 C50 C70 ◇

GABBAY, D.M. *Craig interpolation theorem for intuitionistic logic and extensions part III*
◇ B55 C40 C90 F50 ◇

GRANT, P.W. *Strict Π_1^1-predicates on countable and cofinality ω transitive sets*
◇ C15 C40 C70 D55 D60 D70 E47 E60 ◇

HAJEK, P. *Generalized quantifiers and finite sets*
◇ B25 B50 C13 C40 C80 ◇

KOGALOVSKIJ, S.R. & SOLDATOVA, V.V. *On definability by intension (Russian)* ◇ C40 ◇

KRZYSTEK, P.S. & ZACHOROWSKI, S. *Lukasiewicz logics have not the interpolation property* ◇ B50 C40 ◇

MAKKAI, M. *An "admissible" generalization of a theorem on countable Σ_1^1 sets of reals with applications*
◇ C15 C40 C50 C70 D55 E15 ◇

MAKOWSKY, J.A. & TULIPANI, S. *Some model theory for monotone quantifiers* ◇ C20 C40 C80 ◇

MAKSIMOVA, L.L. *Craig's interpolation theorem and amalgamable varieties (Russian)*
◇ B22 B55 C05 C40 C90 G10 ◇

MAKSIMOVA, L.L. *Craig's theorem in superintuitionistic logics and amalgamable varieties of pseudo-boolean algebras (Russian)*
◇ B55 C05 C40 C90 G05 G10 ◇

MANSFIELD, R. *Sheaves and normal submodels*
◇ C40 C90 E15 G05 G10 ◇

MIZUTANI, C. *Monadic second order logic with an added quantifier Q* ◇ B15 B25 C40 C55 C80 ◇

MOLDESTAD, J. *On monotone second order relations*
◇ C40 C85 ◇

MOTOHASHI, N. *A remark on Scott's interpolation theorem for $L_{\omega_1\omega}$* ◇ C40 C75 ◇

NABEBIN, A.A. *Expressibility in restricted second-order arithmetic (Russian)* ◇ B15 C40 D05 F35 ◇

PRZELECKI, M. *On identifiability in extended domains*
◇ C40 ◇

RANTALA, V. *Aspects of definability*
◇ A05 C07 C40 C75 ◇

SADOVSKIJ, V.N. & SMIRNOV, V.A. *Definability and identifiability: certain problems and hypotheses*
◇ C40 ◇

SCHIEMANN, I. *Eine Axiomatisierung des monadischen Praedikatenkalkuels mit verallgemeinerten Quantoren (Russian, English and French summaries)*
◇ B25 C40 C55 C80 ◇

SGRO, J. *Completeness theorems for topological models*
◇ C20 C40 C75 C80 C90 ◇

SGRO, J. *Maximal logics*
◇ C20 C40 C80 C90 C95 ◇

VAEAENAENEN, J. *On the compactness theorem*
◇ C20 C40 C95 ◇

VAEAENAENEN, J. *Remarks on generalized quantifiers and second order logics* ◇ B15 C40 C80 C85 C95 ◇

WEESE, M. *Definierbare Praedikate in booleschen Algebren. I* ◇ C40 C80 G05 ◇

ZACHOROWSKI, S. *Intermediate logics without the interpolation property* ◇ B55 C40 ◇

1978

ABRAMSON, F.G. *Interpolation theorems for program schemata* ◇ B70 C40 ◇

AMERBAEV, V.M. & KASIMOV, YU.F. *Formal expressibility of order relations in algebraic systems (Russian) (Kazakh summary)* ◇ C40 C60 ◇

APT, K.R. *Inductive definitions, models of comprehension and invariant definability*
◇ C40 C62 D55 D70 F35 ◇

BARWISE, J. & MOSCHOVAKIS, Y.N. *Global inductive definability* ◇ C40 C70 D70 ◇

BENTHEM VAN, J.F.A.K. *Ramsey eliminability*
⋄ A05   C40 ⋄

BERMAN, J. & GRAETZER, G. *Uniform representations of congruence schemes* ⋄ C05   C40 ⋄

BLANC, G. *Equivalence naturelle et formules logiques en theorie des categories* ⋄ C40   G30 ⋄

BRUCE, KIM B. *Ideal models and some not so ideal problems in the model theory of L(Q)* ⋄ C40   C80 ⋄

FERRO, R. *Interpolation theorem for $L^{2+}_{k,k}$*
⋄ B15   C40   C75   C85 ⋄

FINE, K. *Model theory for modal logic. part I: the de re/de dicto distinction. part II: the elimination of de re modality* ⋄ A05   B45   C40   C90 ⋄

FUJIWARA, T. *A variation of Lyndon-Keisler's homomorphism theorem and its applications to interpolation theorems* ⋄ C40 ⋄

GARAVAGLIA, S. *Model theory of topological structures*
⋄ C20   C40   C65   C80   C90 ⋄

HALKOWSKA, K. & PLONKA, J. *On formulas preserved by the sum of a direct system of relational systems*
⋄ C30   C40 ⋄

KOMORI, Y. *Logics without Craig's interpolation property*
⋄ B55   C40 ⋄

KUEKER, D.W. *Uniform theorems in infinitary logic*
⋄ C40   C55   C75 ⋄

MAKOWSKY, J.A. *Some observations on uniform reduction for properties invariant on the range of definable relations* ⋄ C30   C40   C75   C95 ⋄

MIKENBERG, I. *From total to partial algebras*
⋄ C05   C40   C52 ⋄

MOTOHASHI, N. *An elimination theorem of uniqueness condition* ⋄ B10   C40 ⋄

OIKKONEN, J. *Second-order definability, game quantifiers and related expressions*
⋄ C40   C75   C80   C85   E60 ⋄

REGGIANI, C. *Le SH nei limiti diretti di strutture e di interpretazioni (English summary)*
⋄ C05   C30   C40 ⋄

RICKEY, V.F. *On creative definitions in first order functional calculi* ⋄ B10   C40 ⋄

SCHLIPF, J.S. *Toward model theory through recursive saturation* ⋄ C15   C40   C50   C57   C70 ⋄

SMORYNSKI, C.A. *Beth's theorem and self-referential sentences* ⋄ B45   C40   F30 ⋄

SZABO, L. *Concrete representation of related structures of universal algebras I* ⋄ C05   C40 ⋄

VAEAENAENEN, J. *Two axioms of set theory with applications to logic*
⋄ C40   C55   C80   C85   E35   E47   E65 ⋄

WEESE, M. *Definierbare Praedikate in booleschen Algebren. II* ⋄ B25   C10   C40   C80   G05 ⋄

ZACHOROWSKI, S. *Dummett's LC has the interpolation property* ⋄ B55   C40   C90 ⋄

ZIEGLER, M. *Definable bases of monotone systems*
⋄ C40   C80 ⋄

1979

BELYAEV, V.YA. & TAJTSLIN, M.A. *On elementary properties of existentially closed systems (Russian)*
⋄ C25   C40   C60   C75   C85   D55 ⋄

BURRIS, S. & LAWRENCE, J. *Definable principle congruences in varieties of groups and rings*
⋄ C05   C40 ⋄

CSIRMAZ, L. *On definability in Peano arithmetic*
⋄ C40   C62   F30 ⋄

FOO, N.Y. *Homomorphisms in the theory of modelling*
⋄ C07   C40 ⋄

GUMB, R.D. *An extended joint consistency theorem for free logic with equality* ⋄ B60   C40 ⋄

HARNIK, V. *Refinements of Vaught's normal form theorem* ⋄ C40   C75 ⋄

HEINTZ, JOOS *Definability bounds of first order theories of algebraically closed fields*
⋄ B25   C40   C60   D10   D15 ⋄

HIROSE, K. & TAKAHASHI, S. *Sentences preserved by sheaves of structures (Japanese)* ⋄ C40   C90   G30 ⋄

KRYNICKI, M. & LACHLAN, A.H. *On the semantics of the Henkin quantifier*
⋄ B15   B25   C40   C55   C80   C85 ⋄

LABLANQUIE, J.-C. *Proprietes de consistance et forcing*
⋄ C25   C40   C75 ⋄

LILLO DE, N.J. *Models of an extension of the theory ORD*
⋄ B28   C40   C65   E10   E30 ⋄

MAKOWSKY, J.A. & SHELAH, S. *The theorems of Beth and Craig in abstract model theory I: The abstract setting*
⋄ C40   C75   C80   C95 ⋄

MAKSIMOVA, L.L. *Interpolation theorems in modal logics and amalgamable varieties of topological boolean algebras (Russian)* ⋄ B45   C05   C40   C90   G05 ⋄

MILLER, DOUGLAS E. *An application of invariant sets to global definability* ⋄ C40   C70   C75 ⋄

MOTOHASHI, N. *A remark on Africk's paper on Scott's interpolation theorem for $L_{\omega_1,\omega}$* ⋄ C40   C75 ⋄

MUNDICI, D. *Compactness + Craig interpolation = Robinson consistency in any logic* ⋄ C40   C95 ⋄

MUNDICI, D. *Compattezza = JEP in ogni logica*
⋄ C40   C52   C95 ⋄

MUNDICI, D. *Robinson consistency theorem in soft model theory (Italian summary)* ⋄ C40   C55   C80   C95 ⋄

OIKKONEN, J. *Hierarchies of model-theoretic definability - an approach to second-order logics*
⋄ B15   C40   C70   C75   C80   C85 ⋄

REGGIANI, C. *Conservazione delle SH di tipo ∀∃ nei limiti diretti (English summary)* ⋄ C05   C30   C40 ⋄

ROMAN'KOV, V.A. *On definable subgroups of solvable groups* ⋄ C40   C60 ⋄

ROSENTHAL, J.W. *Truth in all of certain well-founded countable models arising in set theory II*
⋄ C40   C62   C70   E45   E47 ⋄

ROWLANDS-HUGHES, D. *Back and forth methods in abstract model theory* ⋄ C40   C80   C95 ⋄

SERVI, M. *A generalization of the exponential functor and its connections with the SH-formulas (Italian summary)*
⋄ C40   G30 ⋄

VAEAENAENEN, J. *Abstract logic and set theory. I. Definability* ⋄ C40   C55   C95   E75 ⋄

VOLGER, H. *Preservation theorems for limits of structures and global sections of sheaves of structures*
⋄ C40   C60   C90 ⋄

WASILEWSKA, A. *A constructive proof of Craig's interpolation lemma for m-valued logic* ◊ B50 C40 ◊

1980

ADAMSON, A. *Saturated structures, unions of chains, and preservation theorems* ◊ C40 C50 C70 ◊

BADGER, L.W. *Beth's property fails in $L^{<\omega}$* ◊ C40 C80 ◊

BOWEN, K.A. *Interpolation in loop-free logic* ◊ B75 C40 ◊

CANTINI, A. *A note on three-valued logic and Tarski theorem on truth definitions* ◊ B50 C40 F30 ◊

CHLEBUS, B.S. *Decidability and definability results concerning well-orderings and some extensions of first order logic* ◊ B15 B25 C40 C65 C80 D35 E07 ◊

CORCORAN, J. *On definitional equivalence and related topics* ◊ A10 C40 F25 ◊

DYWAN, Z. & STEPIEN, T. *Every two-valued propositional calculus has the interpolation property* ◊ B20 C40 ◊

GUICHARD, D.R. *A many-sorted interpolation theorem for L(Q)* ◊ C40 C75 C80 ◊

HARNIK, V. *Game sentences, recursive saturation and definability* ◊ C40 C50 C57 C70 ◊

HARRINGTON, L.A. *Extensions of countable infinitary logic which preserve most of its nice properties* ◊ C40 C55 C70 C75 C95 ◊

HODGES, W. *Functorial uniform reducibility* ◊ C20 C40 C75 ◊

LUO, LIBO *The interpolation theorem for the $P(\kappa)$ of a strongly inaccessible cardinal (Chinese)* ◊ C40 C75 E50 E55 ◊

MAKKAI, M. *On full embeddings I* ◊ C40 G30 ◊

MANGANI, P. & MARCJA, A. *Shelah rank for boolean algebras and some application to elementary theories I* ◊ C35 C40 C45 C50 G05 ◊

MUNDICI, D. *Natural limitations of algorithmic procedures in logic (Italian summary)* ◊ C40 D05 D80 ◊

NISHIMURA, H. *A preservation theorem for tense logic* ◊ B45 C20 C40 C90 ◊

NISHIMURA, H. *Saturated and special models in modal model theory with applications to the modal and de re hierarchies* ◊ B45 C40 C50 C90 ◊

PARMIGIANI BEDOGNA, D. *Le formule SH nelle potenze generalizzate ridotte (English summary)* ◊ C20 C40 G30 ◊

PRZYMUSINSKA, H. *Craig interpolation theorem and Hanf number for v-valued infinitary predicate calculi (Russian summary)* ◊ B50 C40 C55 C75 G20 ◊

RUMELY, R. *Undecidability and definability for the theory of global fields* ◊ C40 C60 D35 ◊

SGRO, J. *Interpolation fails for the Souslin-Kleene closure of the open set quantifier logic* ◊ C40 C80 C90 C95 ◊

SGRO, J. *The interior operator logic and product topologies* ◊ C40 C55 C80 C90 ◊

SGRO, J. *Ultraproduct invariant logics* ◊ C20 C40 C80 C95 ◊

VAEAENAENEN, J. *A quantifier for isomorphisms* ◊ C40 C80 C85 C95 ◊

VOJVODIC, G. *The Craig interpolation theorem for mixed-valued predicate calculi (Serbo-Croatian summary)* ◊ B50 C40 C90 ◊

1981

BAKER, K.A. *Definable normal closures in locally finite varieties of groups* ◊ C05 C40 C60 ◊

BALDWIN, JOHN T. *Definability and the hierarchy of stable theories* ◊ C40 C45 ◊

BERGSTRA, J.A. & TIURYN, J. *Implicit definability of algebraic structures by means of program properties* ◊ B75 C40 C70 ◊

CAICEDO, X. *Independent sets of axioms in $L_{\kappa\alpha}$* ◊ C40 C75 ◊

FERRO, R. *An analysis of Karp's interpolation theorem and the notion of k-consistency property (Italian summary)* ◊ C40 C75 ◊

HODGES, W. & SHELAH, S. *Infinite games and reduced products* ◊ C20 C40 C75 E55 E60 ◊

JEZEK, J. *The lattice of equational theories I: Modular elements. II: The lattice of full sets of terms* ◊ C05 C40 G10 ◊

KNIGHT, J.F. *Algebraic independence* ◊ C40 C55 C75 ◊

LISOVIK, L.P. *Sets that are expressible in theories of commutative semigroups (Russian) (English Summary)* ◊ B25 C05 C40 ◊

LOPEZ-ESCOBAR, E.G.K. *On the interpolation theorem for the logic of constant domains* ◊ B55 C40 C90 F50 ◊

MAKOWSKY, J.A. & SHELAH, S. *The theorems of Beth and Craig in abstract model theory II: Compact logics* ◊ C40 C75 C80 C95 ◊

MAKOWSKY, J.A. & ZIEGLER, M. *Topological model theory with an interior operator: consistency properties and back-and-forth arguments* ◊ C40 C75 C80 ◊

MALINOWSKI, G. & MICHALCZYK, M. *Interpolation properties for a class of many-valued propositional calculi* ◊ B50 C40 ◊

MOTOHASHI, N. *Homogeneous formulas and definability theorems* ◊ C40 ◊

MUNDICI, D. *An algebraic result about soft model theoretical equivalence relations with an application to H. Friedman's fourth problem* ◊ C15 C40 C95 ◊

MUNDICI, D. *Applications of many-sorted Robinson consistency theorem* ◊ C40 C80 C95 ◊

MUNDICI, D. *Complexity of Craig's interpolation* ◊ C40 D15 ◊

MUNDICI, D. *Craig's interpolation theorem in computation theory* ◊ C40 D15 ◊

MUNDICI, D. *Robinson's consistency theorem in soft model theory* ◊ C40 C80 C95 ◊

MUNDICI, D. *Variations on Friedman's third and fourth problem* ◊ C40 C75 C80 C95 ◊

NISHIMURA, H. *Model theory for tense logic: Saturated and special models with applications to the tense hierarchy* ◊ B45 C40 C50 C90 ◊

NISHIMURA, H. *The semantical characterization of de dicto in continuous modal model theory* ◊ B45 C40 C90 ◊

OLIVEIRA DE, A.J.F. *A note on cyclic boolean algebras (Portuguese) (English summary)* ⋄ C40 G05 ⋄

POIZAT, B. *Degres de definissabilite arithmetique des generiques (English summary)*
⋄ C25 C40 D30 D55 ⋄

RENARDEL DE LAVALETTE, G.R. *The interpolation theorem in fragments of logics* ⋄ B20 B55 C40 F50 ⋄

SHORE, R.A. *The degrees of unsolvability: global results*
⋄ C40 D30 D35 F35 ⋄

STEPIEN, T. *Craig-Goedel-Lindenbaum's property and Sobocinski-Tarski's property in propositional calculi*
⋄ B22 C40 ⋄

VOLGER, H. *A unifying approach to theorems on preservation and interpolation for binary relations between structures* ⋄ C40 ⋄

1982

ARNON, D.S. & MCCALLUM, S. *Cylindrical algebraic decomposition by quantifier elimination*
⋄ C10 C40 ⋄

BALDWIN, JOHN T. & MILLER, DOUGLAS E. *Some contributions to definability theory for languages with generalized quantifiers* ⋄ C40 C80 ⋄

CZELAKOWSKI, J. *Logical matrices and the amalgamation property* ⋄ B22 C40 C52 C90 ⋄

DIMITRACOPOULOS, C. & PARIS, J.B. *Truth definitions for Δ_0 formulae* ⋄ C40 C62 H15 ⋄

EBBINGHAUS, H.-D. & ZIEGLER, M. *Interpolation in Logiken monotoner Systeme* ⋄ C40 C80 ⋄

ECSEDI-TOTH, P. & TURI, L. *On enumerability of interpolants* ⋄ B10 C40 ⋄

GRULOVIC, M. *A note on forcing and weak interpolation theorem for infinitary logics (Serbo-Croatian summary)*
⋄ C25 C40 C75 ⋄

JEZEK, J. *The lattice of equational theories III: definability and automorphisms* ⋄ C05 C40 G10 ⋄

KAWAI, H. *Eine Logik erster Stufe mit einem infinitaeren Zeitoperator* ⋄ B45 C40 C75 C90 ⋄

LYNCH, J.F. *On sets of relations definable by addition*
⋄ C40 C62 ⋄

MAKSIMOVA, L.L. *Absence of the interpolation property in modal companions of a Dummett logic (Russian)*
⋄ B45 C40 C90 ⋄

MAKSIMOVA, L.L. *Interpolation properties of superintuitionistic, positive and modal logics*
⋄ B55 C40 C90 ⋄

MAKSIMOVA, L.L. *Lyndon's interpolation theorem in modal logics (Russian)* ⋄ B45 C40 C90 ⋄

MALINOWSKI, G. & MICHALCZYK, M. *That SCI has the interpolation property* ⋄ B60 C40 ⋄

MANGANI, P. & MARCJA, A. $\aleph_1$-*boolean spectrum and stability (Italian summary)* ⋄ C40 C45 ⋄

MARCJA, A. *An algebraic approach to superstability (Italian summary)* ⋄ C40 C45 G05 ⋄

MIRAGLIA, F. *On the preservation of elementary equivalence and embedding by bounded filtered powers and structures of stable continuous functions*
⋄ C20 C40 ⋄

MOTOHASHI, N. ε-*theorems and elimination theorems of uniqueness conditions* ⋄ B10 C40 F05 ⋄

MOTOHASHI, N. *An axiomatization theorem*
⋄ B10 C07 C40 C75 ⋄

MUNDICI, D. *Compactness, interpolation and Friedman's third problem* ⋄ C40 C75 C95 ⋄

MUNDICI, D. *Complexity of Craig's interpolation*
⋄ C40 D15 ⋄

MUNDICI, D. *Duality between logics and equivalence relations* ⋄ C40 C95 ⋄

MUNDICI, D. *Interpolation, compactness and JEP in soft model theory* ⋄ C40 C52 C95 ⋄

RANTALA, V. *Infinitely deep game sentences and interpolations* ⋄ C40 C75 ⋄

SHEN, FUXING *Elementary extensions and elementary chains of lattice-valued models* ⋄ C35 C40 C90 ⋄

WEAVER, G.E. *A note on the interpolation theorem in first order logic* ⋄ C40 ⋄

1983

ANDREKA, H. & NEMETI, I. *Generalization of the concept of variety and quasivariety to partial algebras through category theory*
⋄ C05 C40 C75 C90 C95 G30 ⋄

BAKER, K.A. *Nondefinability of projectivity in lattice varieties* ⋄ C05 C30 C40 G10 ⋄

BALDO, A. *Complete interpolation theorems for L_{kk}^{2+} (Italian summary)* ⋄ C40 C75 C85 ⋄

COMPTON, K.J. *Some useful preservation theorems*
⋄ C30 C40 C65 ⋄

DANKO, W. *Algorithmic properties of finitely generated structures* ⋄ B75 C40 C52 C75 ⋄

DIMITRACOPOULOS, C. & PARIS, J.B. *A note on the undefinability of cuts* ⋄ C40 C62 F30 H15 ⋄

EISENMENGER, H. *Some local definability results on countable topological structures* ⋄ C15 C40 C90 ⋄

FERRO, R. *Seq-consistency property and interpolation theorems* ⋄ C40 C75 ⋄

GAO, HENGSHAN *Comments on "The interpolation theorem for the propositional calculus $P(\kappa)$ when κ is a strongly inaccessible cardinal" by Luo, Libo (Chinese)*
⋄ B05 C40 C75 E55 ⋄

GUREVICH, Y. & SHELAH, S. *Rabin's uniformization problem* ⋄ B15 C40 C65 C85 E40 ⋄

HEINTZ, JOOS *Definability and fast quantifier elimination in algebraically closed fields*
⋄ B25 C10 C40 C60 D10 D15 ⋄

KNIGHT, J.F. & WOODROW, R.E. *A complete theory with arbitrarily large minimality ranks*
⋄ C15 C40 C50 ⋄

LEISS, H. *Implizit definierte Mengensysteme*
⋄ C40 C65 C90 ⋄

LOPEZ-ESCOBAR, E.G.K. *A second paper on the interpolation theorem for the logic of constant domains*
⋄ B55 C40 C90 F05 F50 ⋄

LUE, QICI *Counterexample to Bowen's theorem and failures of interpolation theorem in modal logic (Chinese)* ⋄ B45 C40 ⋄

MIRAGLIA, F. *On the preservation of elementary equivalence and embedding by bounded filtered powers and structures of stable continuous functions*
⋄ C20 C40 ⋄

MUNDICI, D. *A lower bound for the complexity of Craig's interpolants in sentential logic* ◇ B20 C40 D15 ◇
MUNDICI, D. *Compactness = JEP in any logic* ◇ C40 C52 C95 ◇
MUNDICI, D. *Inverse topological systems and compactness in abstract model theory* ◇ C40 C80 C95 ◇
ONO, H. *Model extension theorem and Craig's interpolation theorem for intermediate predicate logics* ◇ B55 C20 C40 C90 ◇
PINUS, A.G. *Calculus with the quantifier of elementary equivalence (Russian)* ◇ C40 C55 C80 D35 ◇
PITTS, A.M. *Amalgamation and interpolation in the category of Heyting algebras* ◇ C40 C52 G30 ◇
PITTS, A.M. *An application of open maps to categorical logic* ◇ C40 C52 G30 ◇
RAUTENBERG, W. *Modal tableau calculi and interpolation* ◇ B45 C40 ◇
VAEAENAENEN, J. *Δ-extension and Hanf-numbers* ◇ C40 C55 C80 C95 ◇

1984
ALVES, E.H. *Paraconsistent logic and model theory* ◇ B53 C40 C52 C90 ◇
CHERLIN, G.L. *Definability in power series rings of nonzero characteristic* ◇ C40 C60 D35 ◇
DRIES VAN DEN, L. *Algebraic theories with definable Skolem functions* ◇ C10 C40 C60 ◇
ECSEDI-TOTH, P. & TURI, L. *On the number of zero order interpolants* ◇ C40 ◇
FITTING, M. *Linear reasoning in modal logic* ◇ B45 C40 ◇
GUMB, R.D. *"Conservative" Kripke-closures* ◇ C40 C90 ◇
HEBERT, M. *Un theoreme de preservation pour les factorisations* ◇ C07 C40 C52 ◇
KAUFMANN, M. *Filter logics on ω* ◇ C40 C75 C80 C95 ◇
MAGIDOR, M. & SHELAH, S. & STAVI, J. *Countably decomposable admissible sets* ◇ C40 C62 C70 E30 E45 E47 ◇
MARCJA, A. & TOFFALORI, C. *On pseudo-$\aleph_0$-categorical theories* ◇ C35 C40 C45 G05 ◇
MARCJA, A. & TOFFALORI, C. *On Cantor-Bendixson spectra containing (1,1) I* ◇ C35 C40 C45 G05 ◇
MARKER, D. *A model theoretic proof of Feferman's preservation theorem* ◇ C40 ◇
MOTOHASHI, N. *Approximation theory of uniqueness conditions by existence conditions* ◇ C40 C75 C80 F07 F50 ◇
MOTOHASHI, N. *Equality and Lyndon's interpolation theorem* ◇ B20 C40 C75 ◇
MUNDICI, D. *Δ-tautologies, uniform and nonuniform upper bounds in computation theory (Italian summary)* ◇ B75 C40 D15 ◇
MUNDICI, D. *Abstract model theory and nets of C_*-algebras: noncommutative interpolation and preservation properties* ◇ C40 C95 ◇
MUNDICI, D. *NP and Craig's interpolation theorem* ◇ C40 D15 ◇
MUNDICI, D. *Tautologies with a unique Craig interpolant, uniform vs. nonuniform complexity* ◇ C40 D15 ◇
PARIGOT, M. *Omission des types dans les theories $\Pi_2^0(\Delta)$* ◇ C40 C75 ◇
PILLAY, A. & STEINHORN, C.I. *Definable sets in ordered structures* ◇ C40 C60 C65 ◇
RADEV, S.R. *Interpolation in the infinitary propositional normal modal logic KL_{ω_1}* ◇ B45 C40 C75 ◇
RUDNIK, K. *A generalization of the interpolation theorem for the many-sorted calculus* ◇ C40 ◇
TIURYN, J. *Implicit definability of finite binary trees by sets of equations* ◇ C40 ◇
WRONSKI, A. *Interpolation and amalgamation properties of BCK-algebras* ◇ C40 C52 G10 G25 ◇

1985
BALDWIN, JOHN T. & SHELAH, S. *Second-order quantifiers and the complexity of theories* ◇ C40 C55 C65 C75 C85 ◇
CAICEDO, X. *Failure of interpolation for quantifiers of monadic type* ◇ C40 C75 C80 C95 ◇
CZELAKOWSKI, J. *Sentential logics and Maehara interpolation property* ◇ B22 C05 C40 C52 ◇
EBBINGHAUS, H.-D. *Extended logics: The general framework* ◇ C40 C55 C80 C95 C98 ◇
FAJARDO, S. *Probability logic with conditional expectation* ◇ B48 C40 C90 ◇
FLUM, J. & MARTINEZ, JUAN CARLOS *κ-local operations for topological structures* ◇ C40 C75 C80 C90 ◇
GEORGESCU, G. & VOICULESCU, I. *Eastern model-theory for boolean-valued theories* ◇ C25 C35 C40 C90 ◇
GORANKO, V.F. *Propositional logics with strong negation and the Craig interpolation theorem* ◇ B55 C40 ◇
GORANKO, V.F. *The Craig interpolation theorem for propositional logics with strong negation* ◇ B55 C40 ◇
GRILLIOT, T.J. *Disturbing arithmetic* ◇ C07 C30 C40 C62 ◇
HOOVER, D.N. *A probabilistic interpolation theorem* ◇ C40 C70 C90 ◇
KALANTARI, I. & WEITKAMP, G. *Effective topological spaces I: A definability theory. II: A hierarchy* ◇ C40 C57 C65 D45 D75 D80 ◇
KEISLER, H.J. *Probability quantifiers* ◇ C40 C80 C90 C98 ◇
KOLAITIS, P.G. *Game quantification* ◇ C40 C70 C75 C98 D60 D65 D70 E60 ◇
MAKOWSKY, J.A. *Compactness, embeddings and definability* ◇ C40 C95 C98 ◇
MARCJA, A. & TOFFALORI, C. *On Cantor-Bendixon spectra containing (1,1) II* ◇ C35 C40 C45 G05 ◇
MCCARTHY, T. *Abstraction and definability in semantically closed structures* ◇ A05 C40 C62 C90 ◇
MEKLER, A.H. & SHELAH, S. *Stationary logic and its friends I* ◇ C40 C55 C75 C80 E35 E50 ◇
PILLAY, A. & SHELAH, S. *Classification theory over a predicate I* ◇ C40 C45 C50 C55 ◇
POREBSKA, M. *Interpolation for fragments of intermediate logics* ◇ B55 C40 ◇

Rauszer, C. *Formalizations of certain intermediate logics. I* ⋄ B55 C40 C90 ⋄

Shelah, S. *Remarks in abstract model theory* ⋄ C40 C50 C52 C55 C80 C95 ⋄

Tuschik, H.-P. *Algebraic connections between definable predicates* ⋄ C15 C30 C40 ⋄

Vaeaenaenen, J. *Set-theoretical definability of logics* ⋄ C40 C55 C80 C95 C98 E47 ⋄

Vincenzi, A. *Alcune proprieta della logica di Chang* ⋄ B45 C40 ⋄

Wasilewska, A. *Monadic second-order definability as a common characterization of finite automata, certain classes of programs and logics* ⋄ B15 B75 C40 D05 ⋄

Wronski, A. *On a form of equational interpolation property* ⋄ C05 C40 ⋄

C45 Stability and related concepts

1963
MORLEY, M.D. *On theories categorical in uncountable powers* ⋄ C35 C45 ⋄

1964
ROWBOTTOM, F. *The Los conjecture for uncountable theories* ⋄ C35 C45 ⋄

1965
MORLEY, M.D. *Categoricity in power* ⋄ C35 C45 ⋄

1969
SHELAH, S. *Stable theories* ⋄ C35 C45 ⋄

1970
CHUDNOVSKY, G.V. *Problems of the theory of models, related to categoricity (Russian)*
⋄ C35 C45 C50 C52 C55 C75 ⋄

MUSTAFIN, T.G. *Models of totally transcendental theories (Russian) (Kazakh summary)* ⋄ C45 C50 ⋄

SHELAH, S. *Finite diagrams stable in power*
⋄ C35 C45 C50 C55 ⋄

SHELAH, S. *On theories T categorical in |T|*
⋄ C35 C45 ⋄

1971
BALDWIN, JOHN T. & LACHLAN, A.H. *On strongly minimal sets* ⋄ C35 C45 C50 ⋄

LACHLAN, A.H. *The transcendental rank of a theory*
⋄ C45 ⋄

MACINTYRE, A. *On ω_1-categorical theories of abelian groups* ⋄ C35 C45 C60 ⋄

PABION, J.F. *Une notion de transcendance dans les structures relationnelles* ⋄ C30 C45 ⋄

PALYUTIN, E.A. *Models with countably categorical universal theories (Russian)* ⋄ C35 C45 G05 ⋄

SHELAH, S. *Stability, the f.c.p., and superstability: model theoretic properties of formulas in first order theory*
⋄ C35 C45 C55 ⋄

SHELAH, S. *The number of non-isomorphic models of an unstable first-order theory* ⋄ C45 C55 ⋄

1972
BALDWIN, JOHN T. *Almost strongly minimal theories I,II*
⋄ C35 C45 ⋄

BELEGRADEK, O.V. *On unstable theories of groups (Russian)* ⋄ C45 C60 ⋄

LACHLAN, A.H. *A property of stable theories*
⋄ C45 C55 ⋄

ROSENTHAL, J.W. *A new proof of a theorem of Shelah*
⋄ C35 C45 C55 E45 E50 ⋄

SHELAH, S. *A combinatorial problem; stability and order for models and theories in infinitary languages*
⋄ C45 C55 C75 E05 ⋄

SHELAH, S. *Saturation of ultrapowers and Keisler's order*
⋄ C20 C35 C45 C50 C55 ⋄

SHELAH, S. *Uniqueness and characterization of prime models over sets for totally transcendental first-order-theories* ⋄ C30 C45 C50 ⋄

1973
BALDWIN, JOHN T. *α_T is finite for $\aleph_1$-categorical T*
⋄ C35 C45 ⋄

BALDWIN, JOHN T. & LACHLAN, A.H. *On universal Horn classes categorical in some infinite power*
⋄ C05 C10 C35 C45 ⋄

BELEGRADEK, O.V. *Almost categorical theories (Russian)*
⋄ C35 C45 C55 ⋄

CUTLAND, N.J. *Model theory on admissible sets*
⋄ C45 C55 C70 ⋄

GONCHAROV, S.S. & NURTAZIN, A.T. *Constructive models of complete solvable theories (Russian)*
⋄ C45 C50 C57 G05 ⋄

LACHLAN, A.H. *On the number of countable models of a countable superstable theory* ⋄ C15 C45 ⋄

LASCAR, D. *Types definissables et produits de types*
⋄ C45 C55 ⋄

PODEWSKI, K.-P. *Theorien mit relativ konstruktiblen saturierten Modellen* ⋄ C45 C50 E45 ⋄

SHELAH, S. *Differentially closed fields* ⋄ C45 C60 ⋄

WOOD, C. *The model theory of differential fields of characteristic $p \neq 0$* ⋄ C25 C35 C45 C60 ⋄

1974
BALDWIN, JOHN T. & BLASS, A.R. *An axiomatic approach to rank in model theory* ⋄ C45 C50 ⋄

ERIMBETOV, M.M. *Categoricity in power and the invalidity of the two-cardinal theorem for a formula of finite rank (Russian)* ⋄ C35 C45 C55 ⋄

ERIMBETOV, M.M. *On almost categorical theories (Russian)* ⋄ C35 C45 ⋄

FELGNER, U. *$\aleph_1$-kategorische Theorien nicht-kommutativer Ringe* ⋄ C35 C45 C60 ⋄

LACHLAN, A.H. *Two conjectures regarding the stability of ω-categorical theories* ⋄ C35 C45 ⋄

MAKOWSKY, J.A. *On some conjectures connected with complete sentences* ⋄ C35 C45 C52 C98 ⋄

MURZIN, F.A. *Totally transcendental and $\aleph_1$-categorical theories (Russian)* ⋄ C35 C45 ⋄

SHELAH, S. *Categoricity of uncountable theories*
⋄ C35 C45 C50 ⋄

WIERZEJEWSKI, J. *A note on stability and products (Russian summary)* ⋄ C20 C30 C45 ⋄

WIERZEJEWSKI, J. *On κ-universal domains*
⋄ C45 C50 ⋄

ZIL'BER, B.I. *The transcendental rank of the formulas of an $\aleph_1$-categorical theory (Russian)* ⋄ C35 C45 ⋄

1975

BALDWIN, JOHN T. *Conservative extensions and the two cardinal theorems for stable theories* ⋄ C45 C55 ⋄

BERTHIER, D. *Stability of non-model-complete theories; products, groups* ⋄ C20 C30 C45 C60 ⋄

ERIMBETOV, M.M. *Complete theories with 1-cardinal formulas (Russian)* ⋄ C35 C45 C55 ⋄

ERIMBETOV, M.M. *On almost categorical theories (Russian)* ⋄ C35 C45 ⋄

HARNIK, V. *A two cardinal theorem for sets of formulas in a stable theory* ⋄ C45 C50 C55 ⋄

HARNIK, V. *On the existence of saturated models of stable theories* ⋄ C45 C50 ⋄

LACHLAN, A.H. *A remark on the strict order property* ⋄ C45 ⋄

LACHLAN, A.H. *Theories with a finite number of models in an uncountable power are categorical* ⋄ C35 C45 ⋄

LASCAR, D. *Definissabilite dans les theories stables* ⋄ C45 ⋄

LASCAR, D. *Quelques remarques sur la notion de bifurcation (forking) (English summary)* ⋄ C45 ⋄

MAZOYER, J. *Sur les theories categoriques finiment axiomatisables (English summary)* ⋄ C10 C35 C45 ⋄

SABBAGH, G. *Categoricite en $\aleph_0$ et stabilite: constructions les preservant et conditions de chaine (English summary)* ⋄ C35 C45 C60 ⋄

SABBAGH, G. *Categoricite et stabilite: quelques examples parmi les groupes et anneaux (English summary)* ⋄ C35 C45 C60 ⋄

SHELAH, S. *The lazy model theoretician's guide to stability* ⋄ C25 C35 C45 C50 C60 C98 ⋄

SHELAH, S. *Why there are many nonisomorphic models for unsuperstable theories* ⋄ C45 C50 C55 C60 C65 G05 ⋄

1976

BALDWIN, JOHN T. & SAXL, J. *Logical stability in group theory* ⋄ C45 C60 ⋄

CHERLIN, G.L. & REINEKE, J. *Categoricity and stability of commutative rings* ⋄ C45 C60 ⋄

HARNIK, V. & MAKKAI, M. *Applications of Vaught sentences and the covering theorem* ⋄ C15 C40 C45 C50 C52 C75 D70 E15 ⋄

KEISLER, H.J. *Six classes of theories* ⋄ C45 C55 ⋄

KOREC, I. & RAUTENBERG, W. *Model-interpretability into trees and applications* ⋄ B25 C45 C60 C65 F25 ⋄

LACHLAN, A.H. *Dimension and totally transcendental theories of rank 2* ⋄ C35 C45 ⋄

LASCAR, D. *Ranks and definability in superstable theories* ⋄ C15 C45 ⋄

MURZIN, F.A. *$\aleph_1$-categorical theories (Russian)* ⋄ C35 C45 ⋄

WIERZEJEWSKI, J. *On stability and products* ⋄ C20 C30 C45 ⋄

WIERZEJEWSKI, J. *Remarks on stability and saturated models* ⋄ C45 C50 ⋄

1977

BALDWIN, JOHN T. & ROSE, B.I. *$\aleph_0$-categoricity and stability of rings* ⋄ C35 C45 C60 ⋄

DURET, J.-L. *Instabilite des corps formellement reels* ⋄ C45 C60 ⋄

FELGNER, U. *Stability and $\aleph_0$-categoricity of nonabelian groups* ⋄ C35 C45 C60 ⋄

FORREST, W.K. *Model theory for universal classes with the amalgamation property: A study in the foundations of model theory and algebra* ⋄ C05 C25 C30 C35 C45 C52 C60 ⋄

LASCAR, D. *Generalisation de l'ordre de Rudin-Keisler aux types d'une theorie* ⋄ C15 C35 C45 E05 ⋄

MURZIN, F.A. *$\aleph_1$-categorical theories (Russian)* ⋄ C35 C45 ⋄

MUSTAFIN, T.G. *A strong base of the elementary types of theories (Russian)* ⋄ C45 C50 ⋄

WIERZEJEWSKI, J. *A note on products and degree of types* ⋄ C30 C45 ⋄

ZIL'BER, B.I. *Groups and rings, the theory of which is categorical (Russian) (English summary)* ⋄ C35 C45 C60 ⋄

1978

BELEGRADEK, O.V. *On nonstable group theories (Russian)* ⋄ C45 C60 ⋄

BOUSCAREN, E. & CHATZIDAKIS, Z. *Modeles premiers et atomiques: theoreme des deux cardinaux dans les theories totalement transcendantes* ⋄ C45 C50 C55 ⋄

CHERLIN, G.L. *Superstable division rings* ⋄ C45 C60 ⋄

DELON, F. *Types localement isoles et theoreme des deux cardinaux dans les theories stables denombrables* ⋄ C45 C50 C55 ⋄

EHRSAM, S. *Expose elementaire de la theorie de la stabilite* ⋄ C45 C98 ⋄

EHRSAM, S. *L'ordre fondamental, le theoreme de la borne, le theoreme de la relation d'equivalence finie* ⋄ C20 C45 C98 ⋄

EHRSAM, S. *Theories ω_1-categoriques* ⋄ C35 C45 ⋄

FELGNER, U. *$\aleph_0$-categorical stable groups* ⋄ C35 C45 C60 ⋄

KEISLER, H.J. *The stability function of a theory* ⋄ C45 C55 ⋄

LACHLAN, A.H. *Spectra of ω-stable theories* ⋄ C35 C45 ⋄

PAILLET, J.L. *Theories completes $\aleph_0$-categoriques et non finiment modele-completables (English summary)* ⋄ C35 C45 ⋄

PODEWSKI, K.-P. & ZIEGLER, M. *Stable graphs* ⋄ C45 C65 ⋄

POIZAT, B. (ED.) *Groupe d'etude de theories stables. 1re annee: 1977/78. Exposes 1 a 9* ⋄ C45 C97 C98 ⋄

POIZAT, B. *Rangs des types dans les corps differentiels* ⋄ C45 C60 ⋄

POIZAT, B. *Suites d'indiscernables dans les theories stables* ⋄ C30 C45 ⋄

POIZAT, B. *Une preuve par la theorie de la deviation d'un theoreme de J. Baldwin (English summary)* ⋄ C35 C45 ⋄

POIZAT, B. *Une theorie ω_1-categorique a α_T fini*
  ⋄ C35 C45 ⋄
ROSE, B.I. *Model theory of alternative rings*
  ⋄ C10 C35 C45 C60 ⋄
ROSE, B.I. *The $\aleph_1$-categoricity of strictly upper triangular matrix rings over algebraically closed fields*
  ⋄ C35 C45 C60 ⋄
ROTHMALER, P. *Total transzendente abelsche Gruppen und Morley-Rang (English and Russian summaries)*
  ⋄ C35 C45 C60 ⋄
SHELAH, S. *Classification theory and the number of nonisomorphic models* ⋄ C45 C98 ⋄

1979

BAJZHANOV, B.S. & OMAROV, B. *Restriction of a theory to a formula (Russian)* ⋄ C30 C45 ⋄
BALDWIN, JOHN T. *Stability theory and algebra*
  ⋄ C45 C60 ⋄
BAUR, W. & CHERLIN, G.L. & MACINTYRE, A. *Totally categorical groups and rings* ⋄ C35 C45 C60 ⋄
BEKENOV, M.I. & HERRE, H. *A remark concerning a generalized Morley-rank* ⋄ C45 ⋄
BEKENOV, M.I. & MUSTAFIN, T.G. *On rank functions, definable and indecomposable types in stable theories (Russian)* ⋄ C45 ⋄
CHERLIN, G.L. *Groups of small Morley rank*
  ⋄ C45 C60 ⋄
CHERLIN, G.L. *Stable algebraic theories* ⋄ C45 C60 ⋄
FORREST, W.K. *Some basic results in the theory of ω-stable theories* ⋄ C45 ⋄
GARAVAGLIA, S. *Direct product decomposition of theories of modules* ⋄ C30 C45 C60 ⋄
HERRE, H. *Transzendente Theorien* ⋄ C45 C50 E75 ⋄
LASCAR, D. & POIZAT, B. *An introduction to forking*
  ⋄ C45 ⋄
LASCAR, D. *Les modeles denombrables d'une theorie superstable ayant des fonctions de Skolem (English summary)* ⋄ C15 C45 C52 ⋄
MEKLER, A.H. *Model complete theories with a distinguished substructure* ⋄ C35 C45 C60 ⋄
OMAROV, A.I. *⊲$_t$-minimality and the finite cover property (Russian)* ⋄ C20 C45 C50 ⋄
SACKS, G.E. *Effective bounds on Morley rank*
  ⋄ C15 C35 C45 C70 ⋄
SARACINO, D.H. & WOOD, C. *Periodic existentially closed nilpotent groups* ⋄ C25 C35 C45 C60 ⋄
SHELAH, S. *On uniqueness of prime models*
  ⋄ C45 C50 ⋄
WOOD, C. *Notes on the stability of separably closed fields*
  ⋄ C25 C35 C45 C60 ⋄
ZIL'BER, B.I. *Definability of algebraic systems and the theory of categoricity (Russian)* ⋄ C35 C45 ⋄

1980

BAJZHANOV, B.S. *Questions on the spectra of totally transcendental theories of finite rank (Russian)*
  ⋄ C35 C45 ⋄
BAJZHANOV, B.S. *Some properties of totally transcendental theories (Russian)* ⋄ C35 C45 ⋄
BALDWIN, JOHN T. & KUEKER, D.W. *Ramsey quantifiers and the finite cover property*
  ⋄ C10 C35 C45 C80 ⋄
BAUDISCH, A. *Application of the theory of Shirshov-Witt in the theory of groups and Lie algebras*
  ⋄ B25 C45 C60 ⋄
BEKENOV, M.I. *Theories with a basis (Russian)*
  ⋄ C45 C52 ⋄
CHERLIN, G.L. & SHELAH, S. *Superstable fields and groups* ⋄ C45 C60 ⋄
DURET, J.-L. *Les corps pseudo-finis ont la propriete d'independance (English summary)* ⋄ C45 C60 ⋄
DURET, J.-L. *Les corps faiblement algebriquement clos non separablement clos ont la propriete d'independance*
  ⋄ C45 C60 ⋄
FELGNER, U. *Kategorizitaet* ⋄ C35 C45 C60 C98 ⋄
FELGNER, U. *The model theory of FC-groups*
  ⋄ C35 C45 C60 D35 ⋄
GARAVAGLIA, S. *Decomposition of totally transcendental modules* ⋄ C10 C35 C45 C60 ⋄
GONCHAROV, S.S. *A totally transcendental decidable theory without constructivizable homogeneous models (Russian)* ⋄ C45 C50 C57 ⋄
GONCHAROV, S.S. *Totally transcendental theory with non-constructivizable prime model (Russian)*
  ⋄ C45 C50 C57 D45 ⋄
HERRE, H. *Modelltheoretische Eigenschaften endlichvalenter Graphen* ⋄ C35 C45 C50 C65 ⋄
LACHLAN, A.H. *Singular properties of Morley rank*
  ⋄ C35 C45 ⋄
MANGANI, P. & MARCJA, A. *Shelah rank for boolean algebras and some application to elementary theories I*
  ⋄ C35 C40 C45 C50 G05 ⋄
MEJREMBEKOV, K.A. *Remarks on theories with a bounded spectrum (Russian)* ⋄ C35 C45 C60 ⋄
MUSTAFIN, T.G. *A non-two-cardinal set of stable types (Russian)* ⋄ C35 C45 C55 ⋄
MUSTAFIN, T.G. *Rank functions in stable theories (Russian)* ⋄ C45 ⋄
MUSTAFIN, T.G. *Theories with a two-cardinal formula (Russian)* ⋄ C45 C50 C52 C55 ⋄
OMAROV, B. *Ranks and Deg of countable ω_0-categorical theories (Russian)* ⋄ C35 C45 ⋄
PALYUTIN, E.A. *Categorical Horn classes I (Russian)*
  ⋄ C05 C35 C45 C52 ⋄
PILLAY, A. *Instability and theories with few models*
  ⋄ C15 C45 C50 ⋄
POIZAT, B. *Les equations differentielles vues par un logicien* ⋄ C45 C60 ⋄
ROSE, B.I. *On the model theory of finite-dimensional algebras* ⋄ B25 C35 C45 C60 ⋄
SHELAH, S. *Independence results*
  ⋄ C45 C55 E07 E35 E40 ⋄
SHELAH, S. *Simple unstable theories*
  ⋄ C45 C50 C55 ⋄
SZCZERBA, L.W. *Interpretations with parameters*
  ⋄ B25 C07 C35 C45 F25 ⋄
ZIL'BER, B.I. *Solution of the problem of finite axiomatizability for theories that are categorical in all infinite powers (Russian)* ⋄ C35 C45 C52 ⋄
ZIL'BER, B.I. *Strongly minimal countably categorical theories (Russian)* ⋄ C35 C45 C52 ⋄

ZIL'BER, B.I. *Totally categorical theories: Structural properties and the non-finite axiomatizability* ⋄ C35 C45 C52 ⋄

1981

BALDWIN, JOHN T. *Definability and the hierarchy of stable theories* ⋄ C40 C45 ⋄

BAUDISCH, A. *There is no module having the finite cover property* ⋄ C10 C45 C60 C80 ⋄

BEKENOV, M.I. & MUSTAFIN, T.G. *Properties of m-types in stable theories (Russian)* ⋄ C45 ⋄

BERLINE, C. *Stabilite et algebre I. Pratique de la stabilite et de la categoricite* ⋄ C35 C45 C98 ⋄

BERLINE, C. *Stabilite et algebre II. Groupes Abeliens et modules* ⋄ C35 C45 C60 C98 ⋄

BERLINE, C. *Stabilite et algebre III. Anneaux* ⋄ C45 C60 C98 ⋄

BERLINE, C. *Stabilite et algebre IV. Groupes* ⋄ C45 C60 C98 ⋄

CHATZIDAKIS, Z. *Forking et rangs locaux selon Shelah* ⋄ C45 ⋄

DELON, F. *Types sur C((X))* ⋄ C10 C45 C60 ⋄

ERIMBETOV, M.M. *A property of countable stable theories (Russian)* ⋄ C45 ⋄

GARAVAGLIA, S. *Forking in modules* ⋄ C45 C60 ⋄

HODGES, W. *Encoding orders and trees in binary relations* ⋄ C13 C45 ⋄

LASCAR, D. *Les modeles denombrables d'une theorie ayant des fonctions de Skolem (English summary)* ⋄ C15 C45 C50 ⋄

MACINTYRE, A. *The complexity of types in field theory* ⋄ C45 C50 C57 C60 ⋄

MEJREMBEKOV, K.A. *Spectra of superstable abelian groups (Russian)* ⋄ C45 C60 ⋄

MEKLER, A.H. *Stability of nilpotent groups of class 2 and prime exponent* ⋄ C45 C60 ⋄

MILLAR, T.S. *Stability, complete extensions, and the number of countable models* ⋄ C15 C35 C45 ⋄

MUSTAFIN, T.G. *On the number of countable models of a countable complete theory (Russian)* ⋄ C15 C45 ⋄

MUSTAFIN, T.G. *Principles of normalization of formulas (Russian)* ⋄ C45 ⋄

NURMAGAMBETOV, T.A. *On almost ω-stable theories (Russian) (Kazakh summary)* ⋄ C45 ⋄

OMAROV, A.I. *A sufficient condition for instability of theories (Russian)* ⋄ C45 ⋄

OMAROV, B. *Properties of rank and Deg of complete countable stable theories (Russian)* ⋄ C45 ⋄

PILLAY, A. *A class of $\aleph_0$-categorical theories* ⋄ C35 C45 ⋄

PILLAY, A. *Finitely generated models of ω-stable theories* ⋄ C45 C50 ⋄

PILLAY, A. *Number of models of theories with many types* ⋄ C45 ⋄

PILLAY, A. *Prime models over subsets* ⋄ C45 C50 ⋄

POIZAT, B. *Exercices en stabilite* ⋄ C45 ⋄

POIZAT, B. (ED.) *Groupe d'etude de theories stables. 2e annee 1978/1979. Exposes 1 a 11* ⋄ C45 C97 C98 ⋄

POIZAT, B. *Le rang selon Lascar* ⋄ C45 C98 ⋄

POIZAT, B. *Modeles premiers d'une theorie totalement transcendante* ⋄ C45 C50 ⋄

POIZAT, B. *Sous-groupes definissables d'un groupe stable* ⋄ C45 C60 ⋄

POIZAT, B. *Theories instables* ⋄ C45 ⋄

ROGERS, P. *Preservation of saturation and stability in a variety of nilpotent groups* ⋄ C45 C50 C60 ⋄

ROTHMALER, P. *A note on I-types* ⋄ C07 C45 C75 ⋄

SAFFE, J. *Einige Ergebnisse ueber die Anzahl abzaehlbarer Modelle superstabiler Theorien (Dissertation)* ⋄ C15 C45 ⋄

SHELAH, S. *On saturation for a predicate* ⋄ C45 C50 C55 C65 E50 ⋄

SMITH, KENNETH W. *Stability and categoricity of lattices* ⋄ C35 C45 G10 ⋄

VIKENT'EV, A.A. *Classes of theories with Lachlan dimensions (Russian)* ⋄ C45 ⋄

VIKENT'EV, A.A. *Models of restrictions of theories (Russian)* ⋄ C45 ⋄

ZIL'BER, B.I. *Solution of the problem of finite axiomatizability for theories that are categorical in all infinite powers (Russian)* ⋄ B30 C35 C45 C52 ⋄

ZIL'BER, B.I. *Some problems concerning categorical theories (Russian summary)* ⋄ C35 C45 ⋄

ZIL'BER, B.I. *Totally categorical structures and combinatorial geometries (Russian)* ⋄ C35 C45 ⋄

1982

BAUDISCH, A. *Decidability and stability of free nilpotent Lie algebras and free nilpotent p-groups of finite exponent* ⋄ B25 C45 C60 ⋄

BEKENOV, M.I. *On the spectrum of quasitranscendental theories (Russian)* ⋄ C35 C45 C52 ⋄

COOPER, GLEN R. *On complexity of complete first-order theories* ⋄ C35 C45 C52 ⋄

LASCAR, D. *Ordre de Rudin-Keisler et poids dans les theories stables* ⋄ C45 ⋄

MANGANI, P. & MARCJA, A. $\aleph_1$-*boolean spectrum and stability (Italian summary)* ⋄ C40 C45 ⋄

MARCJA, A. *An algebraic approach to superstability (Italian summary)* ⋄ C40 C45 G05 ⋄

MUSTAFIN, T.G. & NURMAGAMBETOV, T.A. *Separable types and rank functions in stable theories (Russian)* ⋄ C45 ⋄

PALYUTIN, E.A. *Number of models in complete varieties* ⋄ C05 C45 C52 ⋄

PARIGOT, M. *Theories d'arbres* ⋄ C45 C65 ⋄

PERETYAT'KIN, M.G. *Finitely axiomatizable totally transcendental theories (Russian)* ⋄ C45 C57 ⋄

PILLAY, A. *Dimension theory and homogeneity for elementary extensions of a model* ⋄ C15 C45 C50 ⋄

PILLAY, A. *Weakly homogeneous models* ⋄ C15 C45 C50 ⋄

ROTHMALER, P. & TUSCHIK, H.-P. *A two cardinal theorem for homogeneous sets and the elimination of Malitz quantifiers* ⋄ C10 C45 C80 ⋄

SAFFE, J. *A superstable theory with the dimensional order property has many models* ⋄ C45 ⋄

SHELAH, S. *The spectrum problem I: $\aleph_\varepsilon$-saturated models, the main gap. II: Totally transcendental and the infinite depth* ⋄ C45 C50 C52 ⋄

TOFFALORI, C. *Stabilita, categoricita ed eliminazione dei quantificatori per una classe di anelli locali*
⋄ C10 C35 C45 C60 ⋄

TUSCHIK, H.-P. *Elimination of cardinality quantifiers*
⋄ C10 C45 C55 C60 C80 ⋄

WAGNER, C.M. *On Martin's conjecture*
⋄ C15 C35 C45 C65 C75 ⋄

1983

BALDWIN, JOHN T. & SHELAH, S. *The structure of saturated free algebras* ⋄ C05 C30 C45 C50 ⋄

BAUDISCH, A. *A remark on $\aleph_0$-categorical stable groups*
⋄ C35 C45 C60 ⋄

BAUDISCH, A. & ROTHMALER, P. *Vorlesungen zur Einfuehrung in die Theorie des Shelah'schen "forking"*
⋄ C45 C98 ⋄

BERLINE, C. *Deviation des types dans les corps algebriquement clos* ⋄ C45 C60 ⋄

BOUSCAREN, E. & LASCAR, D. *Countable models of nonmultidimensional $\aleph_0$-stable theories*
⋄ C15 C45 C50 ⋄

BOUSCAREN, E. *Countable models of multidimensional $\aleph_0$-stable theories* ⋄ C15 C45 C50 ⋄

DURET, J.-L. *Stabilite des corps separablement clos*
⋄ C35 C45 C60 ⋄

HERRE, H. & MEKLER, A.H. & SMITH, KENNETH W. *Superstable graphs*
⋄ C45 C65 ⋄

KIERSTEAD, H.A. & REMMEL, J.B. *Indiscernibles and decidable models* ⋄ C30 C45 C57 C80 ⋄

KNIGHT, J.F. *Degrees of types and independent sequences*
⋄ C35 C45 D30 ⋄

LASCAR, D. *Les corps differentiellement clos denombrables* ⋄ C15 C45 C60 ⋄

MARCJA, A. *Categoricita e algebra* ⋄ C35 C45 C98 ⋄

PILLAY, A. *A note on tight stable theories*
⋄ C15 C35 C45 ⋄

PILLAY, A. *A note on finitely generated models*
⋄ C45 C50 ⋄

PILLAY, A. *An introduction to stability theory*
⋄ C45 C98 ⋄

PILLAY, A. *Countable models of stable theories*
⋄ C15 C45 ⋄

PILLAY, A. & PREST, M. *Forking and pushouts in modules*
⋄ C45 C60 ⋄

PILLAY, A. *The models of a non-multidimensional ω-stable theory* ⋄ C45 ⋄

PILLAY, A. *Well-definable types over subsets* ⋄ C45 ⋄

POIZAT, B. *Beaucoup de modeles a peu de frais*
⋄ C45 C50 C52 C55 ⋄

POIZAT, B. *C'est beau et chaud* ⋄ C45 C60 C98 ⋄

POIZAT, B. (ED.) *Groupe d'etude de theories stables. 3e annee 1980-1982* ⋄ C45 C97 C98 ⋄

POIZAT, B. *Groupes stables, avec types generiques reguliers* ⋄ C45 C60 ⋄

POIZAT, B. *Paires de structures stables* ⋄ C45 ⋄

POIZAT, B. *Post-scriptum a "Theories instables"* ⋄ C45 ⋄

POIZAT, B. *Proprietes "modele-theoriques" d'une theorie du premier ordre* ⋄ C45 C52 C95 ⋄

POIZAT, B. *Travaux publies par les membres du groupe "theories stables"* ⋄ C45 C98 ⋄

POIZAT, B. *Une theorie finiement axiomatisable et superstable* ⋄ C35 C45 ⋄

POIZAT, B. *Une theorie de Galois imaginaire*
⋄ C45 C60 ⋄

ROTHMALER, P. *Another treatment of the foundations of forking theory* ⋄ C45 ⋄

ROTHMALER, P. *Some model theory of modules. I: On total transcendence of modules. II: On stability and categoricity of flat modules* ⋄ C35 C45 C60 ⋄

ROTHMALER, P. *Stationary types in modules*
⋄ C35 C45 C60 ⋄

SAFFE, J. *The number of uncountable models of ω-stable theories* ⋄ C45 C50 ⋄

SHELAH, S. *Classification theory of non-elementary classes, Part I: The number of uncountable models of $\psi \in L_{\omega_1,\omega}$* ⋄ C45 C52 C55 C75 C80 E50 ⋄

STEINHORN, C.I. *A new omitting types theorem*
⋄ C07 C15 C45 C50 C57 ⋄

TOFFALORI, C. *Questioni di teoria dei modelli per coppie di campi* ⋄ C10 C35 C45 C60 ⋄

TULIPANI, S. *On the congruence lattice of stable algebras with definability of compact congruences*
⋄ C05 C45 ⋄

TUSCHIK, H.-P. & WEESE, M. *An unstable $\aleph_0$-categorical theory without the strict order property* ⋄ C35 C45 ⋄

1984

BALDWIN, JOHN T. *First-order theories of abstract dependence relations*
⋄ C35 C45 C57 C60 D25 D45 ⋄

BALDWIN, JOHN T. *Strong saturation and the foundations of stability theory* ⋄ C45 C50 ⋄

BAUDISCH, A. *On elementary properties of free Lie algebras* ⋄ C45 C60 ⋄

BAUDISCH, A. & ROTHMALER, P. *The stratified order in modules* ⋄ C45 C60 ⋄

BAUDISCH, A. & ROTHMALER, P. *Vorlesungen zur Einfuehrung in die Theorie des Shelah'schen "forking" 2. Semester* ⋄ C45 C98 ⋄

BOUSCAREN, E. *Martin's conjecture for ω-stable theories*
⋄ C15 C45 ⋄

BUECHLER, S. *Expansions of models of ω-stable theories*
⋄ C45 C50 ⋄

BUECHLER, S. *Kueker's conjecture for superstable theories*
⋄ C35 C45 C50 ⋄

BUECHLER, S. *Resplendency and recursive definability in ω-stable theories* ⋄ C45 C50 C57 ⋄

CHERLIN, G.L. *Totally categorical structures*
⋄ C35 C45 C98 ⋄

DELON, F. *Espaces ultrametriques*
⋄ C25 C35 C45 C50 C65 ⋄

GUREVICH, Y. & SCHMITT, P.H. *The theory of ordered abelian groups does not have the independence property*
⋄ C45 C60 ⋄

HARNIK, V. & HARRINGTON, L.A. *Fundamentals of forking* ⋄ C45 ⋄

HARRINGTON, L.A. & MAKKAI, M. & SHELAH, S. *A proof of Vaught's conjecture for ω-stable theories*
⋄ C15 C35 C45 ⋄

KREMER, E.M. *Modules with an almost categorical theory (Russian)* ⋄ C35 C45 C60 ⋄

KREMER, E.M. *Rings over which all modules of a given type are almost categorical (Russian)* ⋄ C35 C45 C60 ⋄

LACHLAN, A.H. *Binary homogeneous structures. I* ⋄ C10 C13 C15 C45 C50 C57 ⋄

LACHLAN, A.H. *On countable stable structures which are homogeneous for a finite relational language* ⋄ C10 C13 C15 C45 C50 ⋄

LACHLAN, A.H. & SHELAH, S. *Stable structures homogeneous for a binary language* ⋄ C10 C13 C15 C45 C50 ⋄

LASCAR, D. *Relation entre le rang U et le poids (English summary)* ⋄ C45 ⋄

LASCAR, D. *Sous-groupes d'automorphismes d'une structure saturee* ⋄ C07 C35 C45 C50 ⋄

LASCAR, D. *Theorie de la classification* ⋄ C45 ⋄

LOVEYS, J. *The uniqueness of envelopes in $\aleph_0$-categorical, $\aleph_0$-stable structures* ⋄ C35 C45 ⋄

MAKKAI, M. *A survey of basic stability theory, with particular emphasis on orthogonality and regular types* ⋄ C45 C98 ⋄

MARCJA, A. & TOFFALORI, C. *On pseudo-$\aleph_0$-categorical theories* ⋄ C35 C40 C45 G05 ⋄

MARCJA, A. & TOFFALORI, C. *On Cantor-Bendixson spectra containing (1,1) I* ⋄ C35 C40 C45 G05 ⋄

PILLAY, A. & SROUR, G. *Closed sets and chain conditions in stable theories* ⋄ C35 C45 ⋄

PILLAY, A. *Countable modules* ⋄ C15 C35 C45 C52 C60 ⋄

PILLAY, A. *Regular types in nonmultidimensional ω-stable theories* ⋄ C45 C50 ⋄

POIZAT, B. *Deux remarques a propos de la propriete de recouvrement fini* ⋄ C45 C50 C60 C62 ⋄

POIZAT, B. *La structure geometrique des groupes stables* ⋄ C45 C60 C98 ⋄

PREST, M. *Rings of finite representation type and modules of finite Morley rank* ⋄ C45 C60 ⋄

SAFFE, J. *Categoricity and ranks* ⋄ C35 C45 ⋄

TOFFALORI, C. *Anelli regolari separabilmente chiusi (English summary)* ⋄ C35 C45 C60 ⋄

ZIEGLER, M. *Model theory of modules* ⋄ B25 C10 C45 C60 ⋄

ZIL'BER, B.I. *Strongly minimal countably categorical theories II (Russian)* ⋄ C35 C45 ⋄

ZIL'BER, B.I. *Strongly minimal countably categorical theories. III (Russian)* ⋄ C35 C45 ⋄

1985

BAUDISCH, A. *On Lascar rank in nonmultidimensional ω-stable theories* ⋄ C45 C60 ⋄

BUECHLER, S. *Coordinatization in superstable theories I. Stationary types* ⋄ C45 ⋄

BUECHLER, S. *Invariantes for ω-categorical, ω-stable theories* ⋄ C35 C45 ⋄

BUECHLER, S. *One theorem of Zil'ber's on strongly minimal sets* ⋄ C35 C45 ⋄

BUECHLER, S. *The geometry of weakly minimal types* ⋄ C45 ⋄

CHERLIN, G.L. & HARRINGTON, L.A. & LACHLAN, A.H. *$\aleph_0$-categorical, $\aleph_0$-stable structures* ⋄ C35 C45 C52 ⋄

HARNIK, V. *Stability theory and set existence axioms* ⋄ C45 F35 ⋄

HARRINGTON, L.A. & MAKKAI, M. *An exposition of Shelah's "main gap": counting uncountable models of ω-stable and superstable theories* ⋄ C45 C50 C52 ⋄

LASCAR, D. *Les groupes ω-stables de rang fini* ⋄ C35 C45 C60 ⋄

LASCAR, D. *Quelques precisions sur la d.o.p. et la profondeur d'une theorie (English summary)* ⋄ C45 C75 ⋄

MARCJA, A. & TOFFALORI, C. *On Cantor-Bendixson spectra containing (1,1) II* ⋄ C35 C40 C45 G05 ⋄

MUSTAFIN, T.G. *Classification of superstable theories by rank functions (Russian)* ⋄ C45 ⋄

PALYUTIN, E.A. & SAFFE, J. & STARCHENKO, S.S. *Models of superstable Horn theories* ⋄ C05 C35 C45 ⋄

PILLAY, A. & STEINHORN, C.I. *A note on nonmultidimensional superstable theories* ⋄ C15 C45 ⋄

PILLAY, A. & SHELAH, S. *Classification theory over a predicate I* ⋄ C40 C45 C50 C55 ⋄

PREST, M. *The generalised RK-order, orthogonality and regular types for modules* ⋄ C45 C60 ⋄

RAPP, A. *Elimination of Malitz quantifiers in stable theories* ⋄ C10 C45 C55 C80 ⋄

SHELAH, S. *Classification of first order theories which have a structure theorem* ⋄ C35 C45 C52 C75 ⋄

SHELAH, S. *Monadic logic and Loewenheim numbers* ⋄ C45 C55 C65 C75 C85 E45 ⋄

TOFFALORI, C. *Teorie p-$\aleph_0$-categoriche* ⋄ C35 C45 C60 ⋄

TSUBOI, A. *On the number of independent partitions* ⋄ C45 ⋄

TSUBOI, A. *On theories having a finite number of nonisomorphic countable models* ⋄ C15 C35 C45 ⋄

C50 Models with special properties

1908
HAUSDORFF, F. *Grundzuege einer Theorie der geordneten Mengen* ⋄ C50 E07 E10 ⋄

1925
URYSOHN, P. *Sur un espace metrique universel* ⋄ C50 C65 E75 ⋄

1927
URYSOHN, P. *Sur un espace metrique universel I,II* ⋄ C50 C65 E75 ⋄

1937
MOSTOWSKI, ANDRZEJ *Abzaehlbare Boolesche Koerper und ihre Anwendung auf die allgemeine Metamathematik* ⋄ C15 C50 G05 ⋄

1938
MOSTOWSKI, ANDRZEJ *Ueber gewisse universelle Relationen* ⋄ C50 ⋄

1944
McKINSEY, J.C.C. & TARSKI, A. *The algebra of topology* ⋄ C05 C50 C65 G05 G25 ⋄

1948
HEWITT, E. *Rings of real-valued continuous functions* ⋄ C20 C50 C60 C65 E75 ⋄

1954
FRAISSE, R. *Sur l'extension aux relations de quelques proprietes des ordres* ⋄ C07 C30 C50 E07 ⋄

1955
GILLMAN, L. *Remarque sur les ensembles η_α* ⋄ C50 E07 ⋄
ROBINSON, A. *Ordered structures and related concepts* ⋄ C30 C35 C50 C60 ⋄

1956
BAGEMIHL, F. & GILLMAN, L. *Some cofinality theorems on ordered sets* ⋄ C50 E07 E10 ⋄
GILLMAN, L. *Some remarks on η_α-sets* ⋄ C50 E07 ⋄
JOHNSTON, J.B. *Universal infinite partially ordered sets* ⋄ C50 C65 E07 E50 ⋄
JONSSON, B. *Universal relational systems* ⋄ C50 ⋄

1958
MENDELSON, E. *On a class of universal ordered sets* ⋄ C50 E07 ⋄

1960
JONSSON, B. *Homogeneous universal relational systems* ⋄ C50 ⋄
SVENONIUS, L. *On minimal models of first-order systems* ⋄ C07 C50 ⋄

1961
KEISLER, H.J. *Ultraproducts and elementary classes* ⋄ C20 C40 C50 C52 C55 E05 E50 ⋄
ROBINSON, A. *Model theory and non-standard arithmetic* ⋄ C35 C50 C57 C60 C62 H15 ⋄
VAUGHT, R.L. *Denumerable models of complete theories (Polish)* ⋄ C15 C35 C50 ⋄

1962
CHANG, C.C. & KEISLER, H.J. *Model theories with truth values in a uniform space* ⋄ B50 C20 C50 C90 ⋄
GRZEGORCZYK, A. *A kind of categoricity* ⋄ C35 C50 C62 C65 ⋄
MORLEY, M.D. & VAUGHT, R.L. *Homogeneous universal models* ⋄ C50 ⋄

1963
GRAETZER, G. *A theorem on doubly transitive permutation groups with application to universal algebras* ⋄ C05 C50 ⋄

1964
KEISLER, H.J. *Ultraproducts and saturated models* ⋄ C20 C50 C55 ⋄
MAKKAI, M. *On a generalization of a theorem of E.W.Beth* ⋄ C40 C50 ⋄
MYCIELSKI, J. *Some compactifications of general algebras* ⋄ C05 C50 ⋄

1965
EHRENFEUCHT, A. *Elementary theories with models without automorphisms* ⋄ C50 ⋄
JONSSON, B. *Extensions of relational structures* ⋄ C30 C50 C52 ⋄
KEISLER, H.J. *Finite approximations of infinitely long formulas* ⋄ C40 C50 C75 ⋄
KEISLER, H.J. *Some application of infinitely long formulas* ⋄ C40 C50 C75 ⋄
SHOENFIELD, J.R. *Applications of model theory to degrees of unsolvability* ⋄ C50 D25 D30 ⋄
WEGLORZ, B. *Compactness of algebraic systems (Russian summary)* ⋄ C05 C50 ⋄

1966
AMITSUR, S.A. *Rational identities and applications to algebra and geometry* ⋄ C50 C60 C65 ⋄
CHANG, C.C. & KEISLER, H.J. *Continuous model theory* ⋄ B50 C05 C20 C50 C90 C98 ⋄
FLEISCHER, I. *A note on universal homogenous models* ⋄ C50 ⋄
FUHRKEN, G. *Minimal- und Primmodelle* ⋄ C50 ⋄
ISBELL, J.R. *Epimorphisms and dominions* ⋄ C05 C50 ⋄

KEISLER, H.J. *Some model theoretic results for ω-logic*
  ⋄ C35 C50 C55 C75 ⋄
KEISLER, H.J. *Universal homogeneous boolean algebras*
  ⋄ C50 G05 ⋄
MACZYNSKI, M.J. *A property of retracts of α^+-homogeneous α^+-universal boolean algebras*
  ⋄ C05 C50 G05 ⋄
WEGLORZ, B. *Equationally compact algebras I*
  ⋄ C05 C20 C50 ⋄

1967

CALAIS, J.-P. *Relation et multirelation pseudohomogene*
  ⋄ C07 C50 ⋄
FRAISSE, R. *L'algebre logique et ses rapports avec la theorie des relations* ⋄ C07 C50 E07 ⋄
HOWIE, J.M. & ISBELL, J.R. *Epimorphisms and dominions. II* ⋄ C05 C50 ⋄
KEISLER, H.J. & MORLEY, M.D. *On the number of homogeneous models of a given power*
  ⋄ C50 C55 C75 E50 ⋄
KEISLER, H.J. *Ultraproducts which are not saturated*
  ⋄ C20 C50 C55 ⋄
OMAROV, A.I. *Filtered products of models (Russian) (English summary)* ⋄ C20 C30 C40 C50 C52 ⋄
RYLL-NARDZEWSKI, C. & WEGLORZ, B. *Compactness and homomorphisms of algebraic structures*
  ⋄ C07 C50 C75 ⋄
WEGLORZ, B. *Equationally compact algebras III*
  ⋄ C05 C20 C50 ⋄

1968

BENDA, M. *Ultraproducts and non-standard logics (Russian summary)* ⋄ C20 C50 C75 ⋄
FAJTLOWICZ, S. & HOLSZTYNSKI, W. & MYCIELSKI, J. & WEGLORZ, B. *On powers of bases in some compact algebras* ⋄ C05 C50 ⋄
FUHRKEN, G. *A model with exactly one undefinable element* ⋄ C40 C50 ⋄
JONSSON, B. & OLIN, P. *Almost direct products and saturation* ⋄ C05 C20 C50 G05 ⋄
KEISLER, H.J. *Formulas with linearly ordered quantifiers*
  ⋄ C40 C50 C75 C80 ⋄
MORLEY, M.D. *Partitions and models*
  ⋄ C30 C50 C55 C62 C75 E05 E45 E55 ⋄
MYCIELSKI, J. & RYLL-NARDZEWSKI, C. *Equationally compact algebras II* ⋄ C05 C20 C50 C65 ⋄
OMAROV, A.I. *The $M(\alpha)$-class of models (Russian)*
  ⋄ C20 C50 ⋄
PACHOLSKI, L. & WEGLORZ, B. *Topologically compact structures and positive formulas* ⋄ B20 C40 C50 ⋄

1969

BENDA, M. *Reduced products and nonstandard logics*
  ⋄ C20 C50 C75 ⋄
HEDRLIN, Z. *On universal partly ordered sets and classes*
  ⋄ C50 E07 G30 ⋄
HOMMA, K. *Some results on saturated models*
  ⋄ C20 C50 ⋄
MOTOHASHI, N. *On normal operations on models*
  ⋄ C30 C40 C50 ⋄
NEGREPONTIS, S. *The Stone space of the saturated boolean algebras* ⋄ C50 G05 ⋄

PACHOLSKI, L. *On products of first order theories (Russian summary)* ⋄ C30 C50 ⋄
TAYLOR, W. *Atomic compactness and graph theory*
  ⋄ C05 C50 C65 ⋄
WASZKIEWICZ, J. & WEGLORZ, B. *On ω_0-categoricity of powers (Russian summary)* ⋄ C30 C35 C50 ⋄
WHALEY, T.P. *Algebras satisfying the descending chain condition for subalgebras* ⋄ C05 C50 ⋄

1970

BAZHAEV, YU.V. *Universal groups (Russian)*
  ⋄ C50 C60 ⋄
CHUDNOVSKY, G.V. *Problems of the theory of models, related to categoricity (Russian)*
  ⋄ C35 C45 C50 C52 C55 C75 ⋄
EKLOF, P.C. *On the existence of $L_{\infty,\kappa}$ indiscernibles*
  ⋄ C30 C50 C75 ⋄
KARGAPOLOV, M.I. *Universal groups (Russian)*
  ⋄ C50 C60 ⋄
KUEKER, D.W. *Generalized interpolation and definability*
  ⋄ C40 C50 C75 ⋄
MOTOHASHI, N. *A theorem in the theory of definition*
  ⋄ C40 C50 ⋄
MUSTAFIN, T.G. *Models of totally transcendental theories (Russian) (Kazakh summary)* ⋄ C45 C50 ⋄
OMAROV, A.I. *A certain property of Frechet filters (Russian)* ⋄ C20 C30 C50 C52 ⋄
PACHOLSKI, L. & RYLL-NARDZEWSKI, C. *On countably compact reduced products I* ⋄ C20 C50 ⋄
PACHOLSKI, L. *On countably compact reduced products II (Russian summary)* ⋄ C20 C50 ⋄
REYES, G.E. *Local definability theory*
  ⋄ C07 C15 C40 C50 C55 C75 ⋄
SHELAH, S. *Finite diagrams stable in power*
  ⋄ C35 C45 C50 C55 ⋄
TAYLOR, W. *Compactness and chromatic number*
  ⋄ C05 C50 C65 ⋄
WENZEL, G.H. *Subdirect irreducibility and equational compactness in unary algebras $\langle A;f \rangle$* ⋄ C05 C50 ⋄

1971

BALDWIN, JOHN T. & LACHLAN, A.H. *On strongly minimal sets* ⋄ C35 C45 C50 ⋄
BENDA, M. *On saturated reduced products*
  ⋄ C20 C50 E05 ⋄
CHANG, C.C. *Two interpolation theorems*
  ⋄ C40 C50 C75 ⋄
CUSIN, R. *Recherche du forcing-compagnon et du modele-compagnon d'une theorie liee a l'existence de modeles $\aleph_\alpha$-universels* ⋄ C25 C35 C50 ⋄
CUSIN, R. & PABION, J.F. *Structures generiques associees a une classe de theories* ⋄ C25 C50 ⋄
CUSIN, R. *Structures prehomogenes et structures generiques* ⋄ C25 C50 ⋄
EHRENFEUCHT, A. & FUHRKEN, G. *On models with undefinable elements* ⋄ C40 C50 ⋄
ELLENTUCK, E. *A minimal model for strong analysis*
  ⋄ C50 C62 C85 E35 ⋄
FITTLER, R. *Generalized prime models* ⋄ C50 ⋄
FUHRKEN, G. & TAYLOR, W. *Weakly atomic-compact relational structures* ⋄ C05 C50 ⋄

HARNIK, V. & RESSAYRE, J.-P. *Prime extensions and categoricity in power* ⋄ C35 C50 ⋄

HENSON, C.W. *A family of countable homogeneous graphs* ⋄ C13 C15 C50 C65 ⋄

LOPEZ, G. *Un exemple d'une χ-classe M ne contenant pas de multirelation (M-χ)-universelle de cardinal χ (χ nombre beth)* ⋄ C50 C55 ⋄

NEGREPONTIS, S. *A property of the saturated boolean algebras* ⋄ C50 G05 ⋄

OMAROV, A.I. *Products with respect to an ω-universal filter (Russian) (Kazakh summary)* ⋄ C20 C50 G05 ⋄

PABION, J.F. *Etude des bases dans les modeles des theories $\aleph_1$-categoriques* ⋄ C15 C35 C50 ⋄

PACHOLSKI, L. *On countably compact reduced products III* ⋄ C10 C20 C50 G05 ⋄

PALYUTIN, E.A. *N*-homogeneous models (Russian)* ⋄ C50 ⋄

ROSENSTEIN, J.G. *A note on a theorem of Vaught* ⋄ C15 C35 C50 ⋄

ROTMAN, B. *Remarks on some theorems of Rado on universal graphs* ⋄ C50 C65 ⋄

SABBAGH, G. *On properties of countable character* ⋄ C20 C50 C52 ⋄

SEIFERT, R.L. *On prime binary relational structures* ⋄ C05 C50 ⋄

SMITH JR., D.B. *Universal homogeneous algebras* ⋄ C05 C50 ⋄

TAYLOR, W. *Atomic compactness and elementary equivalence* ⋄ C05 C13 C50 C65 ⋄

TAYLOR, W. *Some constructions of compact algebras* ⋄ C05 C50 C55 C65 ⋄

VILLE, F. *Complexite des structures rigidement contenues dans une theorie du premier ordre* ⋄ C50 C57 ⋄

WENZEL, G.H. *On ($\mathfrak{S},\mathfrak{A},\mathfrak{m}$)-atomic compact relational systems* ⋄ C05 C50 ⋄

1972

BACSICH, P.D. *Cofinal simplicity and algebraic closedness* ⋄ C05 C25 C50 C60 C75 G10 ⋄

BACSICH, P.D. *Diagrammatic construction of homogenous universal models* ⋄ C50 C60 ⋄

BACSICH, P.D. *Injectivity in model theory* ⋄ C05 C35 C50 G05 G10 G30 ⋄

BANASCHEWSKI, B. & NELSON, EVELYN *Equational compactness in equational classes of algebras* ⋄ C05 C50 ⋄

BELEGRADEK, O.V. *Categoricity in nondenumerable powers and $\aleph_1$-homogeneous models (Russian)* ⋄ C35 C50 ⋄

BELEGRADEK, O.V. & TAJTSLIN, M.A. *Two remarks on the varieties $\mathfrak{A}_{m,n}$ (Russian)* ⋄ C05 C50 ⋄

BITTER-RUCKER VON, R. & ELLENTUCK, E. *Martin's axiom and saturated models* ⋄ C20 C50 E05 E50 ⋄

BOFFA, M. *Modeles universels homogenes et modeles generiques* ⋄ C25 C50 ⋄

BOFFA, M. *Sur l'existence des corps universels-homogenes* ⋄ C50 C55 C60 ⋄

BULMAN-FLEMING, S. *On equationally compact semilattices* ⋄ C05 C50 ⋄

COLLET, J.-C. *Multirelation homogene et elimination de quanteurs* ⋄ C10 C50 ⋄

EKLOF, P.C. *Homogeneous universal modules* ⋄ C50 C60 ⋄

EKLOF, P.C. & FISHER, E.R. *The elementary theory of abelian groups* ⋄ B25 C10 C35 C50 C60 ⋄

ERDOES, P. & HAJNAL, A. & MILNER, E.C. *Partition relations for η_α and for $\aleph_\alpha$-saturated models* ⋄ C50 E05 E07 E50 ⋄

FITTLER, R. *Verallgemeinerte Prim-Modelle (Zusammenfassung)* ⋄ C50 ⋄

FLUM, J. *Bemerkungen ueber minimale Modelle* ⋄ C50 ⋄

HENSON, C.W. *Countable homogeneous relational structures and $\aleph_0$-categorical theories* ⋄ C10 C15 C35 C50 D35 ⋄

KELLY, DAVID *A note on equationally compact lattices* ⋄ C05 C50 G10 ⋄

KRAUSS, P.H. *Extending congruence relations* ⋄ C05 C07 C50 ⋄

MALITZ, J. & REINHARDT, W.N. *Maximal models in the language with quantifier "there exist uncountably many"* ⋄ C50 C55 C80 E55 ⋄

MARCUS, L. *A minimal prime model with an infinite set of indiscernibles* ⋄ C30 C50 C75 ⋄

PABION, J.F. *Relations prehomogenes* ⋄ C07 C35 C50 ⋄

PARIKH, R. *A note on rigid substructures* ⋄ C50 C57 ⋄

POUZET, M. *Modele universel d'une theorie n-complete* ⋄ C15 C35 C50 ⋄

POUZET, M. *Modele universel d'une theorie n-complete: modele uniformement prehomogene* ⋄ C15 C35 C50 ⋄

POUZET, M. *Modele universel d'une theorie n-complete: modele prehomogene* ⋄ C15 C35 C50 ⋄

SHELAH, S. *For what filters is every reduced product saturated ?* ⋄ C20 C50 C55 ⋄

SHELAH, S. *Saturation of ultrapowers and Keisler's order* ⋄ C20 C35 C45 C50 C55 ⋄

SHELAH, S. *Uniqueness and characterization of prime models over sets for totally transcendental first-order-theories* ⋄ C30 C45 C50 ⋄

SHILLETO, J.R. *Minimum models of analysis* ⋄ C50 C62 D55 E45 ⋄

SIMMONS, H. *Existentially closed structures* ⋄ C25 C35 C50 ⋄

TAYLOR, W. *Fixed points of endomorphisms* ⋄ C05 C50 ⋄

TAYLOR, W. *On equationally compact semigroups* ⋄ C05 C50 ⋄

TAYLOR, W. *Residually small varieties* ⋄ C05 C50 C55 ⋄

TULIPANI, S. *Sulla completezza e sulla categoricita della teoria delle W-algebre semplici (English summary)* ⋄ C05 C35 C50 ⋄

1973

BALDWIN, JOHN T. & BLASS, A.R. & GLASS, A.M.W. & KUEKER, D.W. *A "natural" theory without a prime model* ⋄ C15 C50 C62 C65 ⋄

BALDWIN, JOHN T. *The number of automorphisms of models of $\aleph_1$-categorical theories*
⋄ C07 C35 C50 ⋄

BANASCHEWSKI, B. & NELSON, EVELYN *Equational compactness in infinitary algebras* ⋄ C05 C50 ⋄

BUECHI, J.R. & DANHOF, K.J. *Variations on a theme of Cantor in the theory of relational structures*
⋄ C07 C35 C50 C75 ⋄

CHANG, C.C. *What's so special about saturated models?*
⋄ C40 C50 C55 ⋄

CUSIN, R. *Sur les structures generiques en theorie des modeles* ⋄ C10 C25 C35 C50 ⋄

EHRENFEUCHT, A. *Discernible elements in models for Peano arithmetic* ⋄ C50 C62 ⋄

EMDE BOAS VAN, P. *Mostowski's universal set-algebra*
⋄ C50 C57 C65 D45 E07 E20 G05 ⋄

GANTER, B. & PLONKA, J. & WERNER, H. *Homogeneous algebras are simple* ⋄ C05 C50 ⋄

GONCHAROV, S.S. & NURTAZIN, A.T. *Constructive models of complete solvable theories (Russian)*
⋄ C45 C50 C57 G05 ⋄

HECHLER, S.H. *Large superuniversal metric spaces*
⋄ C50 C65 E75 ⋄

MORLEY, M.D. *Countable models with standard part*
⋄ C15 C50 C52 ⋄

OSTASZEWSKI, A.J. *Saturated structures and a theorem of Arhangelskij* ⋄ C50 C65 E75 ⋄

PACHOLSKI, L. *On countable universal boolean algebras and compact classes of models*
⋄ C20 C50 C52 G05 ⋄

PARIS, J.B. *Minimal models of ZF* ⋄ C50 C62 E45 ⋄

PODEWSKI, K.-P. *Theorien mit relativ konstruktiblen saturierten Modellen* ⋄ C45 C50 E45 ⋄

POL, R. *There is no universal totally disconnected space*
⋄ C50 C65 ⋄

RESSAYRE, J.-P. *Boolean models and infinitary first order languages*
⋄ C40 C50 C55 C70 C75 C80 C90 ⋄

ROSENBERG, I.G. *Strongly rigid relations*
⋄ B50 C05 C50 ⋄

SETTE, A.-M. *A propos d'une condition conduisant a une theorie forcee saturee* ⋄ C25 C50 ⋄

TABATA, H. *An application of a certain argument about isomorphisms of α-saturated structures*
⋄ C07 C50 ⋄

WEISPFENNING, V. *Infinitary model-theoretic properties of κ-saturated structures*
⋄ C10 C25 C35 C50 C60 C65 C75 ⋄

WENZEL, G.H. *Eine Charakterisierung gleichungskompakter universeller Algebren*
⋄ C05 C50 ⋄

1974

BALDWIN, JOHN T. & BLASS, A.R. *An axiomatic approach to rank in model theory* ⋄ C45 C50 ⋄

BALDWIN, JOHN T. *Atomic compactness in $\aleph_1$-categorical Horn theories* ⋄ C35 C50 ⋄

BANASCHEWSKI, B. *Equational compactness of G-sets*
⋄ C05 C50 C60 ⋄

BANASCHEWSKI, B. *On equationally compact extensions of algebras* ⋄ C05 C50 ⋄

BAUMGARTNER, J.E. *The Hanf number for complete $L_{\omega_1,\omega}$-sentences (without GCH)*
⋄ C35 C50 C55 C75 ⋄

BAUR, W. *Ueber rekursive Strukturen*
⋄ C50 C57 D45 ⋄

BENDA, M. *Note on boolean ultrapowers*
⋄ C20 C50 C90 ⋄

BENDA, M. *Remarks on countable models*
⋄ C15 C50 ⋄

BERESTOVOY, S. *Modeles homogenes d'une theorie forcee*
⋄ C25 C35 C50 ⋄

BULMAN-FLEMING, S. *A note on equationally compact algebras* ⋄ C05 C50 ⋄

BULMAN-FLEMING, S. *Algebraic compactness and its relations to topology* ⋄ C05 C50 ⋄

COLLET, J.-C. *Multirelation (k,p)-homogene et elimination de quanteurs* ⋄ C10 C40 C50 ⋄

COLLET, J.-C. *Multirelations (k,p)-homogenes et (k,p)-ages*
⋄ C07 C50 ⋄

FITTLER, R. *Saturated models of incomplete theories*
⋄ C20 C50 C52 ⋄

HARRINGTON, L.A. *Recursively presentable prime models*
⋄ C35 C50 C57 C60 ⋄

KOTLARSKI, H. *On the existence of well-ordered models*
⋄ C50 C55 ⋄

MARCUS, L. *Minimal models of theories of one function symbol* ⋄ C50 ⋄

McKENZIE, R. & SHELAH, S. *The cardinals of simple models for universal theories*
⋄ C05 C50 C55 E05 ⋄

MONTAGNA, F. *Sui limiti di certe teorie (English summary)*
⋄ C30 C35 C50 C62 H15 ⋄

MOTOHASHI, N. *Atomic structures and Scott sentences (Japanese)* ⋄ C50 C75 ⋄

NELSON, EVELYN *Infinitary equational compactness*
⋄ C05 C50 C75 ⋄

NELSON, EVELYN *Not every equational class of infinitary algebras contains a simple algebra* ⋄ C05 C50 ⋄

PODEWSKI, K.-P. & REINEKE, J. *An ω_1-categorical ring which is not almost strongly minimal*
⋄ C35 C50 C60 ⋄

RUBIN, M. *Theories of linear order*
⋄ C15 C35 C50 C65 E07 ⋄

SHELAH, S. *Categoricity of uncountable theories*
⋄ C35 C45 C50 ⋄

SIMMONS, H. *The use of injective-like structures in model theory* ⋄ C25 C35 C50 ⋄

TAYLOR, W. *Pure-irreducible mono-unary algebras*
⋄ C05 C50 G10 ⋄

WIERZEJEWSKI, J. *On κ-universal domains*
⋄ C45 C50 ⋄

YASUHARA, M. *The amalgamation property, the universal-homogeneous models, and the generic models*
⋄ C25 C50 C52 ⋄

1975

ANDLER, D. *Finite-dimensional models of categorical semi-minimal theories* ⋄ C35 C50 ⋄

BARWISE, J. & SCHLIPF, J.S. *On recursively saturated models of arithmetic* ⋄ C50 C57 C62 D80 H15 ⋄

BEAUCHEMIN, P. *Un theoreme de Loewenheim-Skolem-Tarski pour les structures homogenes (English summary)* ⋄ C07 C50 E50 ⋄

BELESOV, A.U. *α-atomic compactness of filtered products (Russian)* ⋄ C05 C20 C50 ⋄

CHIKHACHEV, S.A. *A question of Simmons (Russian)* ⋄ C25 C50 ⋄

CROSSLEY, J.N. *Saturated models* ⋄ C50 ⋄

GONCHAROV, S.S. *Autostability and computable families of constructivizations (Russian)* ⋄ C50 C57 D45 ⋄

HARNIK, V. *A two cardinal theorem for sets of formulas in a stable theory* ⋄ C45 C50 C55 ⋄

HARNIK, V. *On the existence of saturated models of stable theories* ⋄ C45 C50 ⋄

KEISLER, H.J. *Constructions in model theory* ⋄ C25 C30 C50 C75 C95 C98 ⋄

KRAUSS, P.H. *Quantifier elimination* ⋄ C10 C35 C50 ⋄

KUEKER, D.W. *Core structures for theories* ⋄ C40 C50 C52 ⋄

LAKE, J. *Characterising the largest, countable partial ordering* ⋄ C15 C50 C65 E07 ⋄

MALITZ, J. *Complete theories with countably many rigid nonisomorphic models* ⋄ C15 C35 C50 ⋄

MARCUS, L. *A theory with only non-homogenous minimal models* ⋄ C50 ⋄

MARCUS, L. *A type-open minimal model* ⋄ C50 ⋄

NEBRES, B.F. *Elementary invariants for abelian groups (uses of saturated model theory in algebra)* ⋄ C50 C60 ⋄

NELSON, EVELYN *Injectivity and equational compactness in the class of $\aleph_0$-semilattices* ⋄ C05 C50 G10 ⋄

NELSON, EVELYN *On the adjointness between operations and relations and its impact on atomic compactness* ⋄ C05 C50 G30 ⋄

NELSON, EVELYN *Semilattices do not have equationally compact hulls* ⋄ C05 C50 G10 ⋄

NELSON, EVELYN *Some functorial aspects of atomic compactness* ⋄ C05 C50 E50 G30 ⋄

REINEKE, J. *Minimale Gruppen* ⋄ C50 C60 ⋄

ROSENTHAL, J.W. *Partial $\aleph_1$-homogeneity of the countable saturated model of an $\aleph_1$-categorical theory* ⋄ C35 C50 ⋄

SACERDOTE, G.S. *Projective model theory and coforcing* ⋄ C25 C30 C35 C40 C50 C60 ⋄

SCHLIPF, J.S. *Some hyperelementary aspects of model theory* ⋄ C50 C62 C70 D55 ⋄

SETTE, A.-M. *Multirelation fortement homogene et ses rapports avec les theories forcees (English summary)* ⋄ C25 C35 C50 ⋄

SHELAH, S. *The lazy model theoretician's guide to stability* ⋄ C25 C35 C45 C50 C60 C98 ⋄

SHELAH, S. *Why there are many nonisomorphic models for unsuperstable theories* ⋄ C45 C50 C55 C60 C65 G05 ⋄

SIMMONS, H. *Counting countable e.c. structures* ⋄ C15 C25 C50 C98 ⋄

TULIPANI, S. *Questioni di teoria dei modelli per linguaggi universali positivi II: Metodi di "back and forth" (English summary)* ⋄ B20 C07 C40 C50 ⋄

WILMERS, G.M. *Non-standard models and their application to model theory* ⋄ C50 C62 H20 ⋄

1976

BARWISE, J. & SCHLIPF, J.S. *An introduction to recursively saturated and resplendent models* ⋄ C15 C40 C50 C57 ⋄

BARWISE, J. *Some applications of Henkin quantifiers* ⋄ C40 C50 C60 C75 C80 ⋄

BELESOV, A.U. *Injective and equationally compact modules (Russian)* ⋄ C05 C50 C60 ⋄

ENGELER, E. *Lower bounds by Galois theory* ⋄ B75 C50 ⋄

FLEISCHER, I. *Concerning my note on universal homogeneous models* ⋄ C50 ⋄

GALVIN, F. & PRIKRY, K. *Infinitary Jonsson algebras and partition relations* ⋄ C05 C50 C55 C75 E05 ⋄

GILLAM, D.W.H. *Saturation and omitting types* ⋄ C07 C30 C50 ⋄

GOLD, J.M. *A reflection property for saturated models* ⋄ C50 C55 ⋄

GROSS, W.F. *Models with dimension* ⋄ C35 C50 C60 E25 E35 ⋄

HARNIK, V. & MAKKAI, M. *Applications of Vaught sentences and the covering theorem* ⋄ C15 C40 C45 C50 C52 C75 D70 E15 ⋄

MACINTYRE, A. & SHELAH, S. *Uncountable universal locally finite groups* ⋄ C25 C50 C52 C55 C60 ⋄

MANSOUX, A. *Theories de Galois locales (English summary)* ⋄ C07 C50 E07 ⋄

MARCUS, L. *Elementary separation by elementary expansion* ⋄ C15 C50 ⋄

MARCUS, L. *The ≺-order on submodels* ⋄ C50 C52 ⋄

MORLEY, M.D. *Decidable models* ⋄ C50 C57 ⋄

NADEL, M.E. *A transfer principle for simple properties of theories* ⋄ C50 C70 C75 ⋄

OLIN, P. *Free products and elementary types of boolean algebras* ⋄ C30 C50 G05 ⋄

PACHOLSKI, L. *On limit reduced powers, saturatedness and universality* ⋄ C20 C50 ⋄

PAILLET, J.L. *Une etude sur des structures ayant une certaine propriete de seuil pour les automorphismes elementaires (English summary)* ⋄ C07 C35 C50 ⋄

POIZAT, B. *Une relation particulierement rigide (English summary)* ⋄ C50 E07 E55 ⋄

SHELAH, S. *Refuting Ehrenfeucht conjecture on rigid models* ⋄ C50 C55 E50 ⋄

SIMMONS, H. *Large and small existentially closed structures* ⋄ C25 C35 C50 ⋄

TOMASIK, J. *On products of neat structures* ⋄ C30 C50 ⋄

WEGLORZ, B. *A note on: "Atomic compactness in $\aleph_1$-categorical Horn theories" by John T. Baldwin* ⋄ C05 C35 C50 C52 ⋄

WIERZEJEWSKI, J. *Remarks on stability and saturated models* ⋄ C45 C50 ⋄

ZIEGLER, M. *A language for topological structures which satisfies a Lindstroem-theorem*
⋄ B25 C40 C50 C65 C80 C90 C95 D35 ⋄

1977

BELESOV, A.U. *Equationally compact systems (Russian)*
⋄ C05 C50 C60 ⋄

BOZOVIC, N.B. *On Markov properties of finitely presented groups* ⋄ C05 C50 D45 ⋄

BROESTERHUIZEN, G. & WEGLORZ, B. & WIERZEJEWSKI, J. *Remarks on rigid structures*
⋄ C50 ⋄

CLARK, D.M. & KRAUSS, P.H. *Relatively homogeneous structures* ⋄ C10 C35 C50 C60 C65 G05 ⋄

DEISSLER, R. *Minimal models*
⋄ C15 C40 C50 C75 ⋄

DICKMANN, M.A. *Structures Σ-saturees*
⋄ C40 C50 C70 ⋄

ESTERLE, J. *Solution d'un probleme d'Erdoes, Gillman et Henriksen et application a l'etude des homomorphismes de $\mathscr{C}(K)$* ⋄ C50 C60 E50 E75 ⋄

GONCHAROV, S.S. *The quantity of nonautoequivalent constructivizations (Russian)* ⋄ C50 C57 D45 ⋄

GROSS, W.F. *Dimension and finite closure*
⋄ C35 C50 C60 E25 ⋄

KAUFMANN, M. *A rather classless model*
⋄ C50 C57 C62 E65 ⋄

KRAUSS, P.H. *Homogeneous universal models of universal theories* ⋄ C05 C35 C50 C52 ⋄

LASCAR, D. *Sur la hauteur des structures denombrables (English summary)* ⋄ C15 C50 C75 ⋄

LITMAN, A. & SHELAH, S. *Models with few isomorphic expansions* ⋄ C15 C35 C50 C55 ⋄

MAKKAI, M. *An "admissible" generalization of a theorem on countable Σ_1^1 sets of reals with applications*
⋄ C15 C40 C50 C70 D55 E15 ⋄

MIJAJLOVIC, Z. *Homogeneous-universal models of theories which have model completions* ⋄ C35 C50 C52 ⋄

MIJAJLOVIC, Z. *Some remarks on boolean terms - model theoretic approach* ⋄ B05 C50 G05 ⋄

MUSTAFIN, T.G. *A strong base of the elementary types of theories (Russian)* ⋄ C45 C50 ⋄

PACHOLSKI, L. *Homogeneity, universality and saturatedness of limit reduced powers III*
⋄ C20 C50 ⋄

PACHOLSKI, L. *Saturated limit reduced powers*
⋄ C20 C50 C55 ⋄

PAILLET, J.L. *Resultats relatifs a l'existence de structures n-elementaires (English summary)* ⋄ C50 C60 ⋄

PARIS, J.B. *Models of arithmetic and the 1-3-1 lattice*
⋄ C50 C62 G10 ⋄

RESSAYRE, J.-P. *Models with compactness properties relative to an admissible language*
⋄ C50 C55 C70 C80 D55 D60 E15 ⋄

SCHLIPF, J.S. *A guide to the identification of admissible sets above structures* ⋄ C15 C50 C70 ⋄

SCHLIPF, J.S. *Ordinal spectra of first-order theories*
⋄ C35 C50 C55 C62 C70 D15 D30 D70 ⋄

SNAPPER, E. *Omitting models* ⋄ C07 C15 C50 ⋄

WEGLORZ, B. *Homogeneity, universality and saturatedness of limit reduced powers II*
⋄ C20 C50 ⋄

WIERZEJEWSKI, J. *Homogeneity, universality and saturatedness of limit reduced powers. I*
⋄ C20 C50 ⋄

1978

ABRAMSON, F.G. & HARRINGTON, L.A. *Models without indiscernibles*
⋄ C30 C50 C55 C62 C65 C75 E05 E40 H15 ⋄

ADAMSON, A. *Admissible sets and the saturation of structures* ⋄ C50 C70 C75 E40 ⋄

ANAPOLITANOS, D.A. *A new kind of indiscernibles*
⋄ C07 C30 C50 ⋄

BOUSCAREN, E. & CHATZIDAKIS, Z. *Modeles premiers et atomiques: theoreme des deux cardinaux dans les theories totalement transcendantes*
⋄ C45 C50 C55 ⋄

BURRIS, S. & JEFFERS, E. *On the simplicity and subdirect irreducibility of boolean ultrapowers*
⋄ C05 C20 C50 ⋄

BURRIS, S. *Rigid boolean powers*
⋄ C05 C30 C50 G05 ⋄

CHARRETTON, C. *Type d'ordre des ordinaux de modeles non denombrables de la theorie des ensembles*
⋄ C20 C50 C62 C70 E10 E50 H20 ⋄

DELON, F. *Types localement isoles et theoreme des deux cardinaux dans les theories stables denombrables*
⋄ C45 C50 C55 ⋄

FORREST, W.K. *A note on universal classes with applications to the theory of graphs*
⋄ C05 C50 C52 C65 ⋄

GONCHAROV, S.S. *Strong constructivizability of homogeneous models (Russian)* ⋄ C50 C57 D45 ⋄

HICKIN, K.K. *Complete universal locally finite groups*
⋄ C50 C60 ⋄

HIRSCHFELD, J. *Examples in the theory of existential completeness* ⋄ C25 C35 C50 ⋄

KAISER, KLAUS *On relational selections for complete theories* ⋄ C20 C35 C50 C60 ⋄

KNIGHT, J.F. *An inelastic model with indiscernibles*
⋄ C15 C30 C50 ⋄

KNIGHT, J.F. *Prime and atomic models* ⋄ C50 C55 ⋄

KOTLARSKI, H. *A note on countable well-ordered models*
⋄ C50 C55 E47 ⋄

KOTLARSKI, H. *Some remarks on well-ordered models*
⋄ C20 C50 C55 E07 ⋄

LACHLAN, A.H. *Skolem functions and elementary extensions* ⋄ C30 C50 ⋄

LIPSHITZ, L. & NADEL, M.E. *The additive structure of models of arithmetic* ⋄ C15 C50 C57 C62 ⋄

LOATS, J.T. & RUBIN, M. *Boolean algebras without nontrivial onto endomorphisms exist in every uncountable cardinality* ⋄ C50 G05 ⋄

MILLAR, T.S. *Foundations of recursive model theory*
⋄ C50 C57 ⋄

PAILLET, J.L. *Une etude sur des structures ayant une certaine propriete de seuil pour les automorphismes elementaires* ⋄ C07 C35 C50 ⋄

PERETYAT'KIN, M.G. *Criterion for strong constructivizability of a homogeneous model (Russian)*
⋄ C35 C50 C57 D45 ⋄

PILLAY, A. *Number of countable models* ⋄ C15 C50 ⋄

PINTER, C. *Properties preserved under definitional equivalence and interpretations*
⋄ C25 C35 C50 F25 ⋄

SCHLIPF, J.S. *Toward model theory through recursive saturation* ⋄ C15 C40 C50 C57 C70 ⋄

SCHWARTZ, N. η_α-*Strukturen*
⋄ C50 C55 C65 E07 ⋄

SHELAH, S. *Models with second order properties. I: Boolean algebras with no definable automorphisms*
⋄ C07 C30 C50 C55 E50 G05 ⋄

SHELAH, S. *Models with second order properties. II: Trees with no undefined branches*
⋄ C50 C55 C62 C75 C80 C95 E05 E35 E65 ⋄

SHELAH, S. *On the number of minimal models*
⋄ C15 C50 ⋄

WEISPFENNING, V. *A note on $\aleph_0$-categorical model-companions*
⋄ C25 C35 C50 C52 C65 G05 G10 ⋄

1979

ALKOR, C. *The problem of extendability (Italian summary)* ⋄ C50 C62 E70 ⋄

BROESTERHUIZEN, G. *Extensions of automorphisms*
⋄ C30 C50 ⋄

CZELAKOWSKI, J. ω-*saturated matrices* ⋄ C50 ⋄

DEISSLER, R. *Minimal and prime models of complete theories of torsion free abelian groups* ⋄ C50 C60 ⋄

FEFERMAN, S. *Generalizing set-theoretical model theory and an analogue theory on admissible sets*
⋄ C50 C55 C70 E70 ⋄

HERRE, H. *Transzendente Theorien* ⋄ C45 C50 E75 ⋄

LASCAR, D. *La conjecture de Vaught*
⋄ C15 C50 C75 ⋄

MEAD, J. *Recursive prime models for boolean algebras*
⋄ C50 C57 G05 ⋄

MIJAJLOVIC, Z. *Saturated boolean algebras with ultrafilters* ⋄ C20 C35 C50 G05 ⋄

MONK, J.D. & RASSBACH, W. *The number of rigid boolean algebras* ⋄ C50 C55 E75 G05 ⋄

OMAROV, A.I. ⊲$_t$-*minimality and the finite cover property (Russian)* ⋄ C20 C45 C50 ⋄

PERMINOV, E.A. & VAZHENIN, YU.M. *On rigid lattices and graphs (Russian)* ⋄ C50 C52 C65 G10 ⋄

POUZET, M. *Relation minimale pour son age*
⋄ C50 E07 ⋄

PREST, M. *Model-completions of some theories of modules*
⋄ C25 C35 C50 C60 ⋄

RUBIN, M. *On the automorphism groups of homogeneous and saturated boolean algebras*
⋄ C07 C50 C85 G05 ⋄

SCHMERL, J.H. *Countable homogeneous partially ordered sets* ⋄ C15 C50 C65 E07 ⋄

SHELAH, S. *Hanf number of omitting type for simple first-order theories* ⋄ C20 C50 C55 C75 ⋄

SHELAH, S. *On uniqueness of prime models*
⋄ C45 C50 ⋄

TODORCEVIC, S.B. *Rigid boolean algebras*
⋄ C50 E05 E75 G05 ⋄

WOODROW, R.E. *There are four countable ultrahomogeneous graphs without triangles*
⋄ C10 C50 C65 E05 ⋄

1980

ADAMSON, A. *A note on the realization of types*
⋄ C07 C50 ⋄

ADAMSON, A. *Saturated structures, unions of chains, and preservation theorems* ⋄ C40 C50 C70 ⋄

BAJZHANOV, B.S. & OMAROV, B. *On omitting models (Russian) (Kazakh summary)* ⋄ C15 C50 ⋄

BANASCHEWSKI, B. & NELSON, EVELYN *Boolean powers as algebras of continuous functions*
⋄ C05 C20 C30 C50 ⋄

BONNET, R. *Very strongly rigid boolean algebras, continuum discrete set condition, countable antichain condition I* ⋄ C50 E50 E75 G05 ⋄

CHITI, E. *Modelli saturati per particolari teorie di gruppi abeliani ordinati (English summary)* ⋄ C50 C60 ⋄

DROBOTUN, B.N. & GONCHAROV, S.S. *Numerations of saturated and homogeneous models (Russian)*
⋄ B25 C50 C57 D45 ⋄

GOLD, J.M. *Some results on the structure of the φ-spectrum* ⋄ C50 C55 E55 ⋄

GONCHAROV, S.S. *A totally transcendental decidable theory without constructivizable homogeneous models (Russian)* ⋄ C45 C50 C57 ⋄

GONCHAROV, S.S. *Autostability of models and abelian groups (Russian)* ⋄ C50 C57 C60 D45 ⋄

GONCHAROV, S.S. *Totally transcendental theory with non-constructivizable prime model (Russian)*
⋄ C45 C50 C57 D45 ⋄

HARNIK, V. *Game sentences, recursive saturation and definability* ⋄ C40 C50 C57 C70 ⋄

HERRE, H. *Modelltheoretische Eigenschaften endlichvalenter Graphen* ⋄ C35 C45 C50 C65 ⋄

LABLANQUIE, J.-C. *Modeles premiers et modeles generiques (English summary)* ⋄ C25 C50 ⋄

LACHLAN, A.H. & WOODROW, R.E. *Countable ultrahomogeneous undirected graphs*
⋄ C15 C50 C65 ⋄

LUO, LIBO *The union and product of models, and homogeneous models (Chinese) (English summary)*
⋄ C15 C30 C50 ⋄

MANGANI, P. & MARCJA, A. *Shelah rank for boolean algebras and some application to elementary theories I*
⋄ C35 C40 C45 C50 G05 ⋄

MILLAR, T.S. *Homogeneous models and decidability*
⋄ C50 C57 ⋄

MURAWSKI, R. *Some remarks on the structure of expansions* ⋄ C50 C62 E45 ⋄

MUSTAFIN, T.G. *Theories with a two-cardinal formula (Russian)* ⋄ C45 C50 C52 C55 ⋄

NADEL, M.E. *On a problem of MacDowell and Specker*
⋄ C50 C62 E50 ⋄

NISHIMURA, H. *Saturated and special models in modal model theory with applications to the modal and de re hierarchies* ⋄ B45 C40 C50 C90 ⋄

PILLAY, A. *Instability and theories with few models*
⋄ C15 C45 C50 ⋄

PILLAY, A. *Theories with exactly three countable models and theories with algebraic prime models*
⋄ C15 C50 ⋄

RICHARD, D. *Saturation des modeles de Peano*
⋄ C50 C62 H15 ⋄

SCHLIPF, J.S. *Recursively saturated models of set theory*
⋄ C50 C57 C62 E70 H20 ⋄

SHELAH, S. *Simple unstable theories*
⋄ C45 C50 C55 ⋄

SMORYNSKI, C.A. & STAVI, J. *Cofinal extension preserves recursive saturation* ⋄ C50 C57 C62 ⋄

STURM, T. *Universal coretracts in some categories of chains and boolean algebras* ⋄ C50 E07 G05 ⋄

TODORCEVIC, S.B. *Very strongly rigid boolean algebras*
⋄ C50 E05 G05 ⋄

WILMERS, G.M. *Minimally saturated models* ⋄ C50 ⋄

1981

ANTONOVSKIJ, M.YA. & CHUDNOVSKY, D.V. & CHUDNOVSKY, G.V. & HEWITT, E. *Rings of real-valued continuous functions II*
⋄ C20 C50 C60 C65 E35 E50 E75 ⋄

BALDWIN, JOHN T. & KUEKER, D.W. *Algebraically prime models* ⋄ C50 ⋄

BELEGRADEK, O.V. *Classes of algebras with inner mappings (Russian)* ⋄ C05 C25 C50 ⋄

BERTONI, A. & MAURI, G. & MIGLIOLI, P.A. *Model theoretic aspects of abstract data specification*
⋄ B75 C50 C57 ⋄

CLOTE, P. *A note on decidable model theory*
⋄ B25 C15 C50 C57 D30 ⋄

GLASS, A.M.W. & GUREVICH, Y. & HOLLAND, W.C. & JAMBU-GIRAUDET, M. *Elementary theory of automorphism groups of doubly homogeneous chains*
⋄ C07 C50 C65 E07 ⋄

GLASS, A.M.W. & GUREVICH, Y. & HOLLAND, W.C. & SHELAH, S. *Rigid homogeneous chains*
⋄ C07 C50 C65 E07 E35 E50 ⋄

KIRBY, L.A.S. & MCALOON, K. & MURAWSKI, R. *Indicators, recursive saturation and expandability*
⋄ C50 C57 C62 H15 ⋄

KOPPELBERG, S. *A lattice structure on the isomorphism types of complete Boolean algebras*
⋄ C50 C90 G05 G10 ⋄

KOTLARSKI, H. & KRAJEWSKI, S. & LACHLAN, A.H. *Construction of satisfaction classes for nonstandard models* ⋄ C50 C62 ⋄

KOTLARSKI, H. *On elementary cuts in models of arithmetic* ⋄ C50 C57 C62 ⋄

LACHLAN, A.H. *Full satisfaction classes and recursive saturation* ⋄ C50 C57 C62 ⋄

LASCAR, D. *Les modeles denombrables d'une theorie ayant des fonctions de Skolem (English summary)*
⋄ C15 C45 C50 ⋄

MACINTYRE, A. *The complexity of types in field theory*
⋄ C45 C50 C57 C60 ⋄

MILLAR, T.S. *Counterexamples via model completions*
⋄ C10 C35 C50 C57 ⋄

MURAWSKI, R. *A simple remark on satisfaction classes, indiscernibles and recursive saturation*
⋄ C30 C50 C57 H15 ⋄

MYCIELSKI, J. & PERLMUTTER, P. *Model completeness of some metric completions of absolutely free algebras*
⋄ C05 C35 C50 ⋄

NARENS, L. *On the scales of measurements* ⋄ C50 ⋄

NISHIMURA, H. *Model theory for tense logic: Saturated and special models with applications to the tense hierarchy* ⋄ B45 C40 C50 C90 ⋄

PABION, J.F. *Saturation en arithmetique et en analyse*
⋄ C50 C62 ⋄

PACHOLSKI, L. *Homogeneous limit reduced powers*
⋄ C20 C50 ⋄

PILLAY, A. *Finitely generated models of ω-stable theories*
⋄ C45 C50 ⋄

PILLAY, A. *Prime models over subsets* ⋄ C45 C50 ⋄

PINUS, A.G. *The spectrum of rigid systems of Horn classes (Russian)* ⋄ C05 C30 C50 ⋄

POIZAT, B. *Modeles premiers d'une theorie totalement transcendante* ⋄ C45 C50 ⋄

RABINOVICH, E.B. *Inductive limits of symmetric groups and universal groups (Russian) (English summary)*
⋄ C15 C50 C60 ⋄

RICHARD, D. *De la structure additive a la saturation des modeles de Peano et a une classification des sous-langages de l'arithmetique* ⋄ C50 C62 ⋄

ROGERS, P. *Preservation of saturation and stability in a variety of nilpotent groups* ⋄ C45 C50 C60 ⋄

ROSE, B.I. & WOODROW, R.E. *Ultrahomogeneous structures* ⋄ C10 C50 ⋄

SCHMERL, J.H. *Arborescent structures I: Recursive models*
⋄ C50 C57 C65 ⋄

SCHMERL, J.H. *Arborescent structures II: Interpretability in the theory of trees* ⋄ B25 C50 C65 C75 ⋄

SCHMERL, J.H. *Recursively saturated, rather classless models of Peano arithmetic*
⋄ C50 C57 C62 E45 E55 ⋄

SHELAH, S. *On endo-rigid strongly $\aleph_1$-free abelian groups in $\aleph_1$* ⋄ C50 C55 C60 E05 E50 E75 ⋄

SHELAH, S. *On saturation for a predicate*
⋄ C45 C50 C55 C65 E50 ⋄

SMORYNSKI, C.A. *Cofinal extensions of nonstandard models of arithmetic* ⋄ C50 C57 C62 ⋄

SMORYNSKI, C.A. *Elementary extensions of recursively saturated models of arithmetic* ⋄ C50 C57 C62 ⋄

SMORYNSKI, C.A. *Recursively saturated nonstandard models of arithmetic* ⋄ C50 C57 C62 ⋄

WANG, SHIQIANG *Lattice-valued atomic and countable saturated models (Chinese)* ⋄ C50 C90 ⋄

1982

CEGIELSKI, P. & MCALOON, K. & WILMERS, G.M. *Modeles recursivement satures de l'addition et de la multiplication des entiers naturels (English summary)*
⋄ C50 C57 C62 F30 H15 ⋄

GORETTI, A. *Alcune osservazioni sul linguaggio $L(Q_1)$ (English summary)* ⋄ C50 C55 C80 ⋄

KNIGHT, J.F. & NADEL, M.E. *Expansions of models and Turing degrees* ⋄ C50 C57 D30 ⋄

KNIGHT, J.F. *Theories whose resplendent models are homogeneous* ◇ C50 ◇

KUDAJBERGENOV, K.ZH. *A remark on omitting models (Russian)* ◇ C15 C50 ◇

LACHLAN, A.H. *Finite homogeneous simple digraphs* ◇ C13 C50 C65 ◇

LU, JINGBO & SHEN, CHENGMIN *Saturated models in lattice-valued model theory (Chinese) (English summary)* ◇ C50 C90 ◇

MILLAR, T.S. *Type structure complexity and decidability* ◇ C50 C57 ◇

MOROZOV, A.S. *Countable homogeneous boolean algebras (Russian)* ◇ C15 C50 C57 G05 ◇

MOROZOV, A.S. *Strong constructivizability of countable saturated boolean algebras (Russian)* ◇ C50 C57 G05 ◇

PABION, J.F. Π_2-*theorie des ensembles* ◇ C50 C62 E30 E45 ◇

PABION, J.F. *Saturated models of Peano arithmetic* ◇ C50 C62 ◇

PACHOLSKI, L. & TOMASIK, J. *Reduced products which are not saturated* ◇ C20 C50 ◇

PILLAY, A. *Dimension theory and homogeneity for elementary extensions of a model* ◇ C15 C45 C50 ◇

PILLAY, A. *Weakly homogeneous models* ◇ C15 C45 C50 ◇

POIZAT, B. *Deux ou trois choses que je sais de L_n* ◇ B20 C07 C13 C50 ◇

SHELAH, S. *The spectrum problem I: $\aleph_\varepsilon$-saturated models, the main gap. II: Totally transcendental and the infinite depth* ◇ C45 C50 C52 ◇

SMORYNSKI, C.A. *A note on initial segment constructions in recursively saturated models of arithmetic* ◇ C50 C57 C62 ◇

SMORYNSKI, C.A. *Back-and-forth inside a recursively saturated model of arithmetic* ◇ C15 C50 C57 C62 ◇

TSUBOI, A. *On M-recursively saturated models of arithmetic* ◇ C50 C57 C62 ◇

TULIPANI, S. *Sui reticolo di congruenze di strutture algebriche modelli di teorie universali positive (English summary)* ◇ C05 C50 C55 ◇

URZYCZYN, P. *An example of a complete first-order theory with all models algorithmically trivial but without locally finite models* ◇ C35 C50 ◇

WILKIE, A.J. *On core structures for Peano arithmetic* ◇ C50 C62 F30 H15 ◇

1983

BALDWIN, JOHN T. & SHELAH, S. *The structure of saturated free algebras* ◇ C05 C30 C45 C50 ◇

BERTONI, A. & MAURI, G. & MIGLIOLI, P.A. *On the power of model theory in specifying abstract data types and in capturing their recursiveness* ◇ B75 C50 C57 D20 ◇

BOUSCAREN, E. & LASCAR, D. *Countable models of nonmultidimensional $\aleph_0$-stable theories* ◇ C15 C45 C50 ◇

BOUSCAREN, E. *Countable models of multidimensional $\aleph_0$-stable theories* ◇ C15 C45 C50 ◇

GONCHAROV, S.S. *Universal recursively enumerable boolean algebras (Russian)* ◇ C50 C57 D45 G05 ◇

GRAMAIN, F. *Non-minimalite de la cloture differentielle I: La preuve de S.Shelah* ◇ C50 C60 C65 ◇

GRAMAIN, F. *Non-minimalite de la cloture differentielle II: La preuve de M.Rosenlicht* ◇ C50 C60 C65 ◇

ISBELL, J.R. *Generic algebras* ◇ C05 C50 E55 G30 ◇

KNIGHT, J.F. & WOODROW, R.E. *A complete theory with arbitrarily large minimality ranks* ◇ C15 C40 C50 ◇

KNIGHT, J.F. *Additive structure in uncountable models for a fixed completion of P* ◇ C50 C57 C62 C75 D30 ◇

KOPPELBERG, S. *Groups of permutations with few fixed points* ◇ C07 C50 E05 G05 ◇

KOSSAK, R. *A certain class of models of Peano arithmetic* ◇ C50 C57 C62 ◇

KOTLARSKI, H. *On cofinal extensions of models of arithmetic* ◇ C50 C62 ◇

KOTLARSKI, H. *On elementary cuts in models of arithmetic* ◇ C50 C57 C62 ◇

KUDAJBERGENOV, K.ZH. *The number of constructive homogeneous models of a complete decidable theory (Russian)* ◇ C50 C57 ◇

LUO, LIBO *On the number of countable homogeneous models* ◇ C15 C50 ◇

MACPHERSON, H.D. *The action of an infinite permutation group on unordered subsets of a set* ◇ C50 C60 E05 ◇

MILLAR, T.S. *Persistently finite theories with hyperarithmetic models* ◇ C15 C50 C57 D30 ◇

MLCEK, J. *Compactness and homogeneity of saturated structures I,II* ◇ C35 C50 C60 C62 ◇

OMAROV, B. *A question of V.Harnik (Russian)* ◇ C15 C35 C50 ◇

PARIGOT, M. *Le modele compagnon de la theorie des arbres* ◇ C10 C25 C35 C50 C65 ◇

PILLAY, A. *A note on finitely generated models* ◇ C45 C50 ◇

POIZAT, B. *Beaucoup de modeles a peu de frais* ◇ C45 C50 C52 C55 ◇

SAFFE, J. *The number of uncountable models of ω-stable theories* ◇ C45 C50 ◇

SCHMITT, P.H. *The L^t-theory of profinite abelian groups* ◇ B25 C50 C60 C80 C90 ◇

SHELAH, S. *Construction of many complicated uncountable structures and boolean algebras* ◇ C50 C55 G05 ◇

SHELAH, S. *Models with second order properties IV. A general method and eliminating diamonds* ◇ C50 C55 C60 C65 C80 E65 G05 ◇

STEINHORN, C.I. *A new omitting types theorem* ◇ C07 C15 C45 C50 C57 ◇

TSUBOI, A. *On M-recursively saturated models of Peano arithmetic (Japanese)* ◇ C50 C57 C62 ◇

1984

BALDWIN, JOHN T. *Strong saturation and the foundations of stability theory* ◇ C45 C50 ◇

BUECHLER, S. *Expansions of models of ω-stable theories* ◇ C45 C50 ◇

BUECHLER, S. *Kueker's conjecture for superstable theories*
⋄ C35 C45 C50 ⋄

BUECHLER, S. *Resplendency and recursive definability in ω-stable theories* ⋄ C45 C50 C57 ⋄

CHIKHACHEV, S.A. *An example of a theory without weakly (Σ,Σ)-atomic models (Russian)* ⋄ B20 C50 ⋄

DELON, F. *Espaces ultrametriques*
⋄ C25 C35 C45 C50 C65 ⋄

DENISOV, A.S. *Constructive homogeneous extensions (Russian)* ⋄ C50 C57 ⋄

FLEISCHER, I. *One-variable equationally compact distributive lattices (Russian summary)*
⋄ C05 C50 G10 ⋄

FUNK, M. & KEGEL, O.H. & STRAMBACH, K. *On group universality and homogenity* ⋄ C50 C60 ⋄

HODGES, W. *Finite extensions of finite groups*
⋄ C13 C50 C60 D35 ⋄

HOOVER, D.N. & KEISLER, H.J. *Adapted probability distributions* ⋄ C50 C65 H10 ⋄

KAUFMANN, M. & SCHMERL, J.H. *Saturation and simple extensions of models of Peano-arithmetic*
⋄ C50 C57 C62 ⋄

KOSSAK, R. $L_{\infty\omega_1}$-*elementary equivalence of ω_1-like models of PA* ⋄ C50 C57 C62 C75 ⋄

KOSSAK, R. *Remarks on free sets*
⋄ C50 C57 C62 E05 ⋄

KOTLARSKI, H. *On elementary cuts in recursively saturated models of Peano arithmetic* ⋄ C50 C57 C62 ⋄

KUDAJBERGENOV, K.ZH. *Autostability and extensions of constructivizations (Russian)* ⋄ C50 C57 F60 ⋄

KUDAJBERGENOV, K.ZH. *Constructivizability of a prime model (Russian)* ⋄ C50 C57 F60 ⋄

LACHLAN, A.H. *Binary homogeneous structures. I*
⋄ C10 C13 C15 C45 C50 C57 ⋄

LACHLAN, A.H. *Countable homogeneous tournaments*
⋄ C15 C50 C65 ⋄

LACHLAN, A.H. *On countable stable structures which are homogeneous for a finite relational language*
⋄ C10 C13 C15 C45 C50 ⋄

LACHLAN, A.H. & SHELAH, S. *Stable structures homogeneous for a binary language*
⋄ C10 C13 C15 C45 C50 ⋄

LASCAR, D. *Sous-groupes d'automorphismes d'une structure saturee* ⋄ C07 C35 C45 C50 ⋄

MACINTYRE, A. & MARKER, D. *Degrees of recursively saturated models* ⋄ C50 C57 C62 D30 D45 ⋄

NEGRI, M. *An application of recursive saturation*
⋄ C50 C57 C62 F30 ⋄

OGER, F. *The model theory of finitely generated finite-by-abelian groups* ⋄ C50 C60 ⋄

OMAROV, A.I. *Elementary theory of D-degrees (Russian)*
⋄ C20 C30 C50 D30 G10 ⋄

PILLAY, A. *Regular types in nonmultidimensional ω-stable theories* ⋄ C45 C50 ⋄

POIZAT, B. *Deux remarques a propos de la propriete de recouvrement fini* ⋄ C45 C50 C60 C62 ⋄

PUDLAK, P. & SOCHOR, A. *Models of the alternative set theory* ⋄ C50 C62 E50 E70 ⋄

SARACINO, D.H. & WOOD, C. *Nonexistence of a universal countable commutative ring* ⋄ C50 C60 ⋄

SARACINO, D.H. & WOOD, C. *QE commutative nilrings*
⋄ C10 C50 C60 ⋄

SEREMET, Z. *On automorphisms of resplendent models of arithmetic* ⋄ C07 C50 C62 ⋄

SHELAH, S. *On universal graphs without instances of CH*
⋄ C50 E05 E35 E50 ⋄

1985

BANKSTON, P. & SCHUTT, R. *On minimally free algebras*
⋄ C05 C50 C60 C65 ⋄

BONNET, R. & SHELAH, S. *Narrow Boolean algebras*
⋄ C50 E75 G05 ⋄

CHIKHACHEV, S.A. *An example of an atomic model which is not algebraically prime (Russian)* ⋄ C15 C50 ⋄

DOW, A. *Saturated boolean algebras and their Stone spaces* ⋄ C50 C55 E35 E50 E75 G05 ⋄

DROSTE, M. *Classes of universal words for the infinite symmetric groups* ⋄ C05 C50 C60 ⋄

FRANEK, F. *Isomorphisms of trees* ⋄ C50 C65 E05 ⋄

FUNK, M. & KEGEL, O.H. & STRAMBACH, K. *Gruppenuniversalitaet und Homogenisierbarkeit (English summary)* ⋄ C50 C60 ⋄

HARRINGTON, L.A. & MAKKAI, M. *An exposition of Shelah's "main gap": counting uncountable models of ω-stable and superstable theories*
⋄ C45 C50 C52 ⋄

KASHIWAGI, T. *Algebraically closed structures and a certain kind of saturated structure* ⋄ C50 ⋄

KOPPELBERG, S. *Homogeneous boolean algebras may have non-simple automorphism groups*
⋄ C07 C50 E50 E75 G05 ⋄

KOSSAK, R. *A note on satisfaction classes*
⋄ C50 C57 C62 F30 ⋄

KOTLARSKI, H. *Bounded induction and satisfaction classes* ⋄ C15 C50 C57 C62 ⋄

MIJAJLOVIC, Z. *On a proof on the Erdoes-Monk theorem*
⋄ C50 E05 G05 ⋄

PILLAY, A. & SHELAH, S. *Classification theory over a predicate I* ⋄ C40 C45 C50 C55 ⋄

SARACINO, D.H. & WOOD, C. *Finite QE rings in characteristic p^2* ⋄ C13 C50 C60 ⋄

SCHMERL, J.H. *Recursively saturated models generated by indiscernibles* ⋄ C30 C50 C57 ⋄

SHELAH, S. *Remarks in abstract model theory*
⋄ C40 C50 C52 C55 C80 C95 ⋄

TRUSS, J.K. *The group of the countable universal graph*
⋄ C07 C50 C65 ⋄

C52 Properties of classes of models

1951
ROBINSON, A. *On axiomatic systems which possess finite models* ⋄ C13 C52 C60 ⋄

1952
PEREMANS, W. *Some theorems on free algebras and on direct products of algebras* ⋄ C05 C52 ⋄
TARSKI, A. *Some notions and methods on the borderline of algebra and metamathematics*
⋄ C07 C35 C52 C60 ⋄

1954
TARSKI, A. *Contributions to the theory of models I,II*
⋄ C13 C30 C52 ⋄
VAUGHT, R.L. *Remarks on universal classes of relational systems* ⋄ C52 ⋄

1955
BING, K. *On arithmetical classes not closed under direct union* ⋄ C05 C40 C52 ⋄
LOS, J. *On the extending of models I* ⋄ C30 C52 ⋄
LOS, J. & SUSZKO, R. *On the extending of models II*
⋄ C05 C30 C52 ⋄
LOS, J. & SUSZKO, R. *On the infinite sums of models*
⋄ C30 C40 C52 ⋄
LOS, J. *Quelques remarques, theoremes et problemes sur les classes definissables d'algebres*
⋄ C05 C20 C30 C52 ⋄
TARSKI, A. *Contributions to the theory of models III*
⋄ C05 C52 G15 ⋄

1956
HENKIN, L. *Two concepts from the theory of models*
⋄ C13 C52 C60 ⋄
ROBINSON, A. *Completeness and persistence in the theory of models* ⋄ C30 C35 C52 ⋄
ROBINSON, A. *Note on a problem of L.Henkin*
⋄ C30 C52 ⋄

1957
CHANG, C.C. *Unions of chains of models and direct products of models* ⋄ C30 C52 ⋄
LIGHTSTONE, A.H. & ROBINSON, A. *Syntactical transforms* ⋄ B10 C52 C60 ⋄
LOS, J. & SUSZKO, R. *On the extending of models IV: Infinite sums of models* ⋄ C30 C52 ⋄
LYNDON, R.C. *Sentences preserved under homomorphisms; sentences preserved under subdirect products* ⋄ C07 C30 C40 C52 ⋄
MAL'TSEV, A.I. *Classes of models with an operation of generation (Russian)* ⋄ C52 ⋄
MAL'TSEV, A.I. *Derived operations and predicates (Russian)* ⋄ C52 ⋄

MYCIELSKI, J. *A characterisation of arithmetical classes (Russian summary)* ⋄ C52 ⋄
ZUBIETA RUSSI, G. *Clases aritmeticas definidas sin igualdad* ⋄ B10 B20 C07 C52 ⋄

1958
CRAIG, W. & VAUGHT, R.L. *Finite axiomatizability using additional predicates* ⋄ B10 C52 ⋄
KOGALOVSKIJ, S.R. *On universal classes of algebras (Russian)* ⋄ C52 ⋄
MAL'TSEV, A.I. *Certain classes of models (Russian)*
⋄ C05 C52 G30 ⋄
MAL'TSEV, A.I. *Defining relations in categories (Russian)*
⋄ C05 C52 G30 ⋄
MAL'TSEV, A.I. *The structural characterization of certain classes of algebras (Russian)* ⋄ C05 C52 G30 ⋄
OBERSCHELP, A. *Ueber die Axiome produkt-abgeschlossener arithmetischer Klassen*
⋄ B20 B25 C30 C40 C52 ⋄
SCOTT, D.S. *Convergent sequences of complete theories*
⋄ C35 C52 ⋄
SCOTT, D.S. & SUPPES, P. *Foundational aspects of theories of measurement* ⋄ B30 C13 C52 ⋄
TAJMANOV, A.D. *On classes of models, closed under direct product (Russian)* ⋄ C30 C52 D35 ⋄

1959
APPEL, K.I. *Horn sentences in identity theory*
⋄ C13 C30 C52 ⋄
CHANG, C.C. *On unions of chains of models*
⋄ C40 C52 ⋄
KOGALOVSKIJ, S.R. *On universal classes of algebras, closed under direct product (Russian)* ⋄ C05 C52 ⋄
KOGALOVSKIJ, S.R. *Universal classes of models (Russian)*
⋄ C05 C52 ⋄
LYNDON, R.C. *Existential Horn sentences*
⋄ B20 C30 C40 C52 ⋄
LYNDON, R.C. *Properties preserved under homomorphism*
⋄ C07 C40 C52 ⋄
LYNDON, R.C. *Properties preserved under algebraic constructions* ⋄ C07 C30 C40 C52 ⋄
LYNDON, R.C. *Properties preserved in subdirect products*
⋄ C30 C40 C52 ⋄
MAL'TSEV, A.I. *Model correspondences (Russian)*
⋄ B15 C05 C07 C52 C60 C85 ⋄
MAL'TSEV, A.I. *Small models (Russian)* ⋄ C52 C55 ⋄
OBERSCHELP, A. *Ueber die Axiome arithmetischer Klassen mit Abgeschlossenheitsbedingungen*
⋄ B25 C40 C52 ⋄
ROBINSON, A. *Obstructions to arithmetical extension and the theorem of Los and Suszko* ⋄ C40 C52 ⋄
TAJMANOV, A.D. *On a class of models closed with respect to direct products (Russian)* ⋄ C30 C52 ⋄

1960

KEISLER, H.J. *Theory of models with generalized atomic formulas* ⋄ C07 C30 C52 ⋄

LOS, J. & SLOMINSKI, J. & SUSZKO, R. *On the extending of models V: Embedding theorems for relational models* ⋄ C07 C30 C52 ⋄

SVENONIUS, L. *Some problems in logical model-theory* ⋄ C07 C52 C95 ⋄

TAJMANOV, A.D. *On a class of models closed with respect to direct products (Russian)* ⋄ C30 C52 ⋄

1961

EHRENFEUCHT, A. & MOSTOWSKI, ANDRZEJ *A compact space of models of first order theories* ⋄ C07 C52 ⋄

KEISLER, H.J. *Ultraproducts and elementary classes* ⋄ C20 C40 C50 C52 C55 E05 E50 ⋄

KOCHEN, S. *Ultraproducts in the theory of models* ⋄ C20 C35 C52 C60 ⋄

KOGALOVSKIJ, S.R. *Categorial characteristics of axiomatized classes (Russian)* ⋄ C52 G30 ⋄

KOGALOVSKIJ, S.R. *On a general method of obtaining structural characterizations of axiomatized classes (Russian)* ⋄ C52 ⋄

KOGALOVSKIJ, S.R. *On multiplicative semigroups of rings (Russian)* ⋄ C52 C60 ⋄

TAJMANOV, A.D. *Characteristic properties of axiomatizable classes of models I,II (Russian)* ⋄ C52 ⋄

TAJMANOV, A.D. *Characterization of finitely axiomatizable classes of models (Russian)* ⋄ C52 ⋄

VAUGHT, R.L. *The elementary character of two notions from general algebra* ⋄ C05 C40 C52 C60 ⋄

ZAKHAROV, D.A. *On the Los-Suszko theorem (Russian)* ⋄ C52 ⋄

1962

ADDISON, J.W. *The theory of hierarchies* ⋄ C40 C52 D55 E15 ⋄

ERSHOV, YU.L. *Axiomatizable classes of models with infinite signature (Russian)* ⋄ C52 ⋄

JONSSON, B. *Algebraic extensions of relational systems* ⋄ C05 C52 C60 ⋄

LOS, J. *Common extension in equational classes* ⋄ C30 C52 ⋄

MOSTOWSKI, ANDRZEJ *A problem in the theory of models* ⋄ C15 C52 ⋄

RABIN, M.O. *Classes of models and sets of sentences with the intersection property* ⋄ C07 C52 ⋄

TAJMANOV, A.D. *On a theorem of Beth and Kochen (Russian)* ⋄ C20 C40 C52 ⋄

TAJMANOV, A.D. *The characteristics of axiomatizable classes of models (Russian)* ⋄ C20 C52 ⋄

1963

GAO, HENGSHAN *Remarks on Los and Suszko's paper "On the extending of models. IV." (Chinese)* ⋄ C30 C52 ⋄

KOGALOVSKIJ, S.R. *Quasiprojective classes of models (Russian)* ⋄ C20 C52 C60 ⋄

KROM, MELVEN R. *Separation principles in the hierarchy theory of pure first-order logic* ⋄ B10 C40 C52 D55 ⋄

MERZLYAKOV, YU.I. *On the theory of generalized solvable and generalized nilpotent groups (Russian)* ⋄ C52 C60 ⋄

ZAKHAROV, D.A. *Convex classes of models (Russian)* ⋄ C52 ⋄

1964

KEISLER, H.J. *Unions of relational systems* ⋄ C30 C52 ⋄

KOGALOVSKIJ, S.R. *A remark on ultraproducts (Russian)* ⋄ C20 C52 ⋄

KOGALOVSKIJ, S.R. *The relation between finitely-projective and finitely-reductive classes of models (Russian)* ⋄ C13 C52 ⋄

MAKKAI, M. *On PC_Δ-classes in the theory of models* ⋄ C20 C40 C52 ⋄

MAKKAI, M. *Remarks on my paper "On PC_Δ-classes in the theory of models"* ⋄ C40 C52 ⋄

ONYSZKIEWICZ, J. *On the different topologies in the space of models* ⋄ C15 C52 C85 ⋄

PICKETT, H.E. *Subdirect representations of relational systems* ⋄ C05 C52 ⋄

1965

ALMAGAMBETOV, ZH.A. *On classes of axioms closed with respect to the given reduced products and powers (Russian)* ⋄ B25 C20 C52 D35 ⋄

BOUVERE DE, K.L. *Logical synonymity* ⋄ A05 C40 C52 ⋄

JONSSON, B. *Extensions of relational structures* ⋄ C30 C50 C52 ⋄

KEISLER, H.J. *Reduced products and Horn classes* ⋄ B20 C20 C52 C55 ⋄

KOCHEN, S. *Topics in the theory of definition* ⋄ C20 C40 C52 ⋄

KOGALOVSKIJ, S.R. *Certain remarks on classes axiomatizable in a nonelementary way (Russian)* ⋄ C20 C52 C85 ⋄

KOGALOVSKIJ, S.R. *Generalized quasiuniversal classes of models (Russian)* ⋄ C52 C85 ⋄

KOGALOVSKIJ, S.R. *On finitely reduced classes of models (Russian)* ⋄ C52 ⋄

KOGALOVSKIJ, S.R. *On multiplicative semigroups of rings (Russian)* ⋄ C52 C60 ⋄

KOGALOVSKIJ, S.R. *Some remarks on ultraproducts (Russian)* ⋄ C20 C52 E55 ⋄

KOGALOVSKIJ, S.R. *Two questions concerning finite projective classes (Russian)* ⋄ C52 ⋄

MAKKAI, M. *A compactness result concerning direct products of models* ⋄ C05 C30 C52 ⋄

MONK, J.D. *Model-theoretic methods and results in the theory of cylindric algebras* ⋄ C05 C20 C52 G15 ⋄

SCHEIN, B.M. *Relation algebras* ⋄ C05 C20 C52 ⋄

SVENONIUS, L. *On the denumerable models of theories with extra predicates* ⋄ C15 C40 C52 ⋄

SZCZERBA, L.W. & TARSKI, A. *Metamathematical properties of some affine geometries* ⋄ B25 B30 C52 C65 D35 ⋄

1966

FEFERMAN, S. & KREISEL, G. *Persistent and invariant formulas relative to theories of higher order*
⋄ B15 C40 C52 C85 ⋄

KHISAMIEV, N.G. *Certain applications of the method of extension of mappings (Russian) (Kazakh summary)*
⋄ C35 C52 ⋄

KOGALOVSKIJ, S.R. *On the properties preserved under algebraic constructions (Russian)*
⋄ B25 C30 C52 C60 D35 D45 ⋄

LINDSTROEM, P. *On relations between structures*
⋄ C07 C40 C52 ⋄

VAUGHT, R.L. *Elementary classes closed under descending intersection* ⋄ C52 ⋄

1967

CHANG, C.C. *Omitting types of prenex formulas*
⋄ C07 C52 ⋄

FRAISSE, R. *La classe universelle: Une notion a la limite de la logique et de la theorie des relations*
⋄ C07 C30 C52 ⋄

JEAN, M. *Relations monomorphes et classes universelles*
⋄ C07 C30 C52 ⋄

KOGALOVSKIJ, S.R. *On compactness-preserving algebraic constructions (Russian)* ⋄ C20 C30 C52 ⋄

MACZYNSKI, M.J. & TRACZYK, T. *The m-amalgamation property for m-distributive boolean algebras*
⋄ C52 G05 ⋄

MAL'TSEV, A.I. *Universally axiomatizable subclasses of locally finite classes of models (Russian)* ⋄ C52 ⋄

OMAROV, A.I. *Filtered products of models (Russian) (English summary)* ⋄ C20 C30 C40 C50 C52 ⋄

OMAROV, A.I. *On compact classes of models (Russian) (English summary)* ⋄ C20 C52 ⋄

RAUTENBERG, W. *Elementare Schemata nichtelementarer Axiome* ⋄ B25 C52 C60 C65 C85 D35 ⋄

TAJMANOV, A.D. *Remarks to the Mostowski-Ehrenfeucht theorem (Russian) (English summary)*
⋄ C07 C20 C30 C52 E50 ⋄

VAUGHT, R.L. *Axiomatizability by a schema*
⋄ C07 C52 ⋄

WEGLORZ, B. *Completeness and compactness of lattices*
⋄ C05 C52 G10 ⋄

1968

FELSCHER, W. *On criteria of definability* ⋄ C20 C52 ⋄

GEIGER, D. *Closed systems of functions and predicates*
⋄ C05 C52 ⋄

KOGALOVSKIJ, S.R. *On compact classes of algebraic systems (Russian)* ⋄ C30 C52 ⋄

KOGALOVSKIJ, S.R. *Remarks on compact classes of algebraic systems (Russian)* ⋄ C30 C52 ⋄

SAYEKI, H. *Some consequences of the normality of the space of models* ⋄ C07 C40 C52 ⋄

WEGLORZ, B. & WOJCIECHOWSKA, A. *Summability of pure extensions of relational structures*
⋄ C05 C07 C52 ⋄

WEINSTEIN, J.M. *(ω_1, ω) properties of unions of models*
⋄ C30 C40 C52 C75 ⋄

1969

CHEREMISIN, A.I. *Structural characterization of certain classes of models (Russian)* ⋄ C05 C30 C52 ⋄

CLEAVE, J.P. *Local properties of systems*
⋄ C52 C60 C85 ⋄

GRAETZER, G. & LAKSER, H. & PLONKA, J. *Joins and direct products of equational classes* ⋄ C05 C52 ⋄

MACZYNSKI, M.J. *The amalgamation of α-distributive α-complete boolean algebras (Russian summary)*
⋄ C52 G05 ⋄

MALITZ, J. *Universal classes in infinitary languages*
⋄ C40 C52 C75 ⋄

MENDELSOHN, E. *An elementary characterization of the category of (free) relational systems* ⋄ C52 G30 ⋄

PRELLER, A. *Interpolation et amalgamation*
⋄ C40 C52 C75 G15 G25 ⋄

TABATA, H. *Free structures and universal Horn sentences*
⋄ B20 C05 C52 ⋄

1970

BENEJAM, J.-P. *Quelques proprietes algebriques des isomorphismes locaux* ⋄ C07 C52 E07 G15 ⋄

CHUDNOVSKY, G.V. *Problems of the theory of models, related to categoricity (Russian)*
⋄ C35 C45 C50 C52 C55 C75 ⋄

DRABBE, J. *Sur une propriete des formules negatives*
⋄ C30 C40 C52 ⋄

EKLOF, P.C. & SABBAGH, G. *Model-completions and modules* ⋄ C25 C35 C52 C60 ⋄

GECSEG, F. *On certain classes of Σ-structures*
⋄ C05 C52 ⋄

HAUSCHILD, K. & WOLTER, H. *Einige Anwendungen des Koenigschen Graphensatzes in der mathematischen Logik* ⋄ C52 ⋄

HERRE, H. & RAUTENBERG, W. *Das Basistheorem und einige Anwendungen in der Modelltheorie*
⋄ C40 C52 G05 ⋄

KAISER, KLAUS *Zur Einbettungsbedingung im Zusammenhang mit einer Programmiersprache von Engeler* ⋄ B75 C20 C35 C52 ⋄

KAISER, KLAUS *Zur Theorie algorithmisch abgeschlossener Modellklassen* ⋄ C20 C52 D75 ⋄

MAREK, W. & ONYSZKIEWICZ, J. *The existence of constructible elements in some classes of models (Russian summary)* ⋄ C52 E45 ⋄

OMAROV, A.I. *A certain property of Frechet filters (Russian)* ⋄ C20 C30 C50 C52 ⋄

SUZUKI, Y. *Orbits of denumerable models of complete theories* ⋄ C15 C35 C52 C62 ⋄

VOL'VACHEV, R.T. *Linear groups as axiomatizable classes of models (Russian)* ⋄ C20 C52 C60 ⋄

1971

ARMBRUST, M. & KAISER, KLAUS *Remarks on model classes acting on one another*
⋄ C05 C20 C30 C52 ⋄

BENEJAM, J.-P. *La methode des s-ludions et les classes logiques* ⋄ C07 C52 ⋄

CARPINTERO ORGANERO, P. *Numero de tipos diferentes de algebras de Boole de cardinal m finito que poseen 2^m ideales primos* ⋄ C52 G05 ⋄

EKLOF, P.C. & SABBAGH, G. *Definability problems for modules and rings* ⋄ C20 C52 C60 C75 ⋄

FUJIWARA, T. *On the construction of the least universal Horn class containing a given class* ⋄ B20 C05 C52 ⋄

HAUSCHILD, K. *Nichtaxiomatisierbarkeit von Satzmengen durch Ausdruecke spezieller Gestalt* ⋄ B28 C20 C40 C52 F30 ⋄

IGOSHIN, V.I. *Characterizable varieties of lattices (Russian)* ⋄ C05 C52 G10 ⋄

MAKKAI, M. *Algebraic operations on classes of structures and their connections with logical formulas I,II (Hungarian) (English summary)* ⋄ C30 C40 C52 C75 ⋄

MCKENZIE, R. *Cardinal multiplication of structures with a reflexive relation* ⋄ C05 C30 C52 ⋄

ROBINSON, A. *Forcing in model theory* ⋄ C25 C35 C52 ⋄

ROBINSON, A. *Infinite forcing in model theory* ⋄ C25 C35 C52 ⋄

SABBAGH, G. *A note on the embedding property* ⋄ C20 C52 ⋄

SABBAGH, G. *On properties of countable character* ⋄ C20 C50 C52 ⋄

SCHWARTZ, DIETRICH *Die Nichtkompaktheit der Klasse der archimedisch geordneten Koerper* ⋄ C52 C60 ⋄

SHELAH, S. *Two cardinal compactness* ⋄ C30 C52 C55 C80 ⋄

TABATA, H. *A generalized free structure and several properties of universal Horn classes* ⋄ B20 C05 C52 ⋄

VINOGRADOV, A.A. *Nonaxiomatizability of lattice-orderable groups (Russian)* ⋄ C52 C60 ⋄

1972

ABAKUMOV, A.I. & PALYUTIN, E.A. & SHISHMAREV, YU.E. & TAJTSLIN, M.A. *Categorial quasivarieties (Russian)* ⋄ C05 C35 C52 ⋄

ARMBRUST, M. & KAISER, KLAUS *On quasi-universal model classes* ⋄ B15 C05 C20 C52 C85 ⋄

ARMBRUST, M. & KAISER, KLAUS *Some remarks on projective model classes and the interpolation theorem* ⋄ C40 C52 ⋄

CLARK, D.M. & KRAUSS, P.H. *Monomorphic relational systems* ⋄ C05 C30 C52 ⋄

COLE, J.C. & DICKMANN, M.A. *Non-axiomatizability results in infinitary languages for higher-order structures* ⋄ C52 C65 C75 C85 ⋄

FITTLER, R. *Some categories of models* ⋄ C30 C52 G30 ⋄

FRAISSE, R. *Chaines compatibles pour un groupe de permutations. Application aux relations enchainables et monomorphes* ⋄ C07 C30 C52 C65 E07 ⋄

HAUSCHILD, K. & MELIS, E. *Zusammenhaenge zwischen Axiomatisierbarkeit und Definierbarkeit in elementaren Theorien mit eingeschraenktem Wohlordnungsschema* ⋄ C40 C52 ⋄

KASHIWAGI, T. *On (m, A)-implicational classes* ⋄ C20 C52 C75 ⋄

KOTAS, J. *About the equivalent theories of algebras with relations (Polish and Russian summaries)* ⋄ C52 ⋄

MACZYNSKI, M.J. *Strong m-extension property and m-amalgamation in boolean algebras (Russian summary)* ⋄ C52 G05 ⋄

MAKKAI, M. *Svenonius sentences and Lindstroem's theory on preservation theorems* ⋄ C40 C52 C75 ⋄

MAKOWSKY, J.A. *Note on almost strongly minimal theories (Russian summary)* ⋄ C35 C52 ⋄

MANSFIELD, R. *Horn classes and reduced direct products* ⋄ C20 C52 C90 ⋄

PACHOLSKI, L. & WASZKIEWICZ, J. *On compact classes of models* ⋄ C20 C52 ⋄

PIGOZZI, D. *Amalgamation, congruence-extension, and interpolation properties in algebras* ⋄ C05 C40 C52 G15 ⋄

RINGEL, C.M. *The intersection property of amalgamations* ⋄ C52 G30 ⋄

ROBINSON, A. *Generic categories* ⋄ C25 C52 C60 G30 ⋄

TULIPANI, S. *Proprieta metamatematiche di alcune classi di algebre* ⋄ C05 C52 ⋄

VAZHENIN, YU.M. *The elementary definability and elementary characterizability of classes of reflexive graphs (Russian)* ⋄ C52 C65 ⋄

YASUHARA, M. *The generic structures as a variety (Russian summary)* ⋄ C25 C52 ⋄

1973

BACSICH, P.D. *Defining algebraic elements* ⋄ C40 C52 C75 ⋄

BALDWIN, JOHN T. *A sufficient condition for a variety to have the amalgamation property* ⋄ C05 C35 C52 ⋄

CHEPURNOV, B.A. *Some remarks on quasiuniversal classes (Russian)* ⋄ C30 C52 ⋄

DICKMANN, M.A. *The problem of non-finite axiomatizability of $\aleph_1$-categorical theories* ⋄ C35 C52 C98 ⋄

GRAETZER, G. & LAKSER, H. *A note on the implicational class generated by a class of structures* ⋄ C05 C52 ⋄

HENRARD, P. *Classes of cotheories* ⋄ C52 ⋄

HICKIN, K.K. *Countable type local theorems in algebra* ⋄ C05 C52 ⋄

HICKIN, K.K. *Countable type local theorems in algebra* ⋄ C05 C52 C60 ⋄

KAISER, KLAUS *Diagonal-like embeddings* ⋄ C20 C52 ⋄

KAISER, KLAUS *On generalized projective classes* ⋄ C20 C52 ⋄

KAISER, KLAUS *On reductive and projective classes* ⋄ C20 C52 ⋄

KOKORIN, A.I. & MART'YANOV, V.I. *Universal extended theories (Russian)* ⋄ B25 C52 D35 ⋄

MAKKAI, M. *Preservation theorems for pseudo-elementary classes (Russian)* ⋄ C40 C52 ⋄

MAKKAI, M. *Vaught sentences and Lindstroem's regular relations* ⋄ C40 C52 C75 D55 D70 ⋄

MARCHINI, C. *Funtori che conservano e riflettono le formule di Horn* ⋄ C40 C52 G30 ⋄

MARCHINI, C. *Funtori rappresentabili e formule di Horn* ⋄ C52 G30 ⋄

MORLEY, M.D. *Countable models with standard part* ⋄ C15 C50 C52 ⋄

PACHOLSKI, L. *On countable universal boolean algebras and compact classes of models*
⋄ C20 C50 C52 G05 ⋄

POUZET, M. *Extensions completes d'une theorie forcing complete* ⋄ C25 C52 ⋄

SHAGIDULLIN, R.R. *The axiomatizability of one projection of a universally axiomatizable class of algebraic systems (Russian)* ⋄ C52 ⋄

TAJTSLIN, M.A. *Existentially closed regular commutative semigroups (Russian)* ⋄ C25 C35 C52 C60 ⋄

TAYLOR, W. *Characterizing Mal'cev conditions*
⋄ C05 C52 ⋄

VINOGRADOV, A.A. *Nonaxiomatizability of directionally ordered groups in the class of nontrivially partially ordered groups* ⋄ C52 C60 ⋄

1974

ARMBRUST, M. & KAISER, KLAUS *On some properties a projective model class passes on to the generated axiomatic class* ⋄ C20 C52 ⋄

BELYAEV, V.YA. *Theories with the Ehrenfeucht property (Russian)* ⋄ C52 ⋄

FITTLER, R. *Saturated models of incomplete theories*
⋄ C20 C50 C52 ⋄

FRENKIN, B.R. *Σ-free models (Russian)* ⋄ C05 C52 ⋄

FRODA-SCHECHTER, M. & KWATINETZ, M. *Some remarks on a paper of R. C. Lyndon: "Properties preserved under homomorphism"* ⋄ C05 C40 C52 ⋄

HERRE, H. *Nonfinitely axiomatizable theories of graphs*
⋄ C52 C65 ⋄

JAKUBOWICZ, C.A. *On generalized products preserving compactness (Russian)* ⋄ C30 C52 G05 ⋄

KAISER, KLAUS *Direct limits in quasi-universal model classes* ⋄ B15 C30 C52 C85 ⋄

KASHIWAGI, T. *Note on locally definable classes of structures* ⋄ C52 C75 ⋄

LASCAR, D. *Sur les theories convexes "modeles completes"*
⋄ C15 C35 C52 ⋄

MAKOWSKY, J.A. *On some conjectures connected with complete sentences* ⋄ C35 C45 C52 C98 ⋄

MONRO, G.P. *The strong amalgamation property for complete boolean algebras* ⋄ C52 E40 G05 ⋄

ONYSZKIEWICZ, J. *Dimensions of PC space of models (Russian summary)* ⋄ C15 C52 C85 ⋄

TAYLOR, W. *Uniformity of congruences* ⋄ C05 C52 ⋄

YASUHARA, M. *The amalgamation property, the universal-homogeneous models, and the generic models*
⋄ C25 C50 C52 ⋄

1975

BACSICH, P.D. *Amalgamation properties and interpolation theorems for equational theories*
⋄ B20 C05 C40 C52 ⋄

BACSICH, P.D. *The strong amalgamation property*
⋄ C05 C52 C60 ⋄

BALDWIN, JOHN T. & BERMAN, J. *The number of subdirectly irreducible algebras in a variety*
⋄ C05 C52 ⋄

BELYAEV, V.YA. *On theories with the Ehrenfeucht property (Russian)* ⋄ C52 ⋄

CHIKHACHEV, S.A. *Categorical universal classes of semigroups (Russian)* ⋄ C05 C35 C52 C60 ⋄

CLARE, F. *Operations on elementary classes of groups*
⋄ C05 C52 C60 ⋄

COLLET, J.-C. *Theories stables par isotopie (English summary)* ⋄ C52 ⋄

FITTLER, R. *Closed models and hulls of theories*
⋄ C25 C52 C60 ⋄

GIRI, R.D. *On a varietal structure of algebras*
⋄ C05 C52 ⋄

HENRARD, P. *Weak elimination of quantifiers and cotheories* ⋄ C10 C52 ⋄

KAISER, KLAUS *Various remarks on quasi-universal model classes* ⋄ B15 C05 C52 C85 ⋄

KUEKER, D.W. *Core structures for theories*
⋄ C40 C50 C52 ⋄

MACINTYRE, A. & SIMMONS, H. *Algebraic properties of number theories*
⋄ C25 C40 C52 C62 D25 H15 ⋄

MURSKIJ, V.L. *A finite basis of identities and other properties of "almost all" finite algebras (Russian)*
⋄ C05 C13 C52 ⋄

SHELAH, S. *Existence of rigid-like families of abelian p-groups* ⋄ C52 C55 C60 ⋄

TAYLOR, W. *Continuum many Mal'cev conditions*
⋄ C05 C52 ⋄

TAYLOR, W. *The fine spectrum of a variety*
⋄ C05 C13 C52 ⋄

WEISPFENNING, V. *Model-completeness and elimination of quantifiers for subdirect products of structures*
⋄ C10 C30 C35 C52 C60 C90 ⋄

1976

AMIT, R. & SHELAH, S. *The complete finitely axiomatized theories of order are dense* ⋄ C52 C65 ⋄

BURRIS, S. *Subdirect representations in axiomatic classes*
⋄ C05 C52 ⋄

CHEPURNOV, B.A. & KOGALOVSKIJ, S.R. *Some criteria of heredity and locality for formulas of higher order (Russian)* ⋄ B15 C52 C85 ⋄

EKLOF, P.C. *Infinitary model theory of abelian groups*
⋄ C52 C60 C75 E35 E55 E75 ⋄

FRIEDRICHSDORF, U. *Einige Bemerkungen zur Peano-Arithmetik* ⋄ C25 C52 C62 F30 ⋄

GNANVO, C. *Γ-cotheories, test de Robinson (English summary)* ⋄ C35 C52 ⋄

GNANVO, C. *Γ-extension logique (English summary)*
⋄ C30 C52 ⋄

HARNIK, V. & MAKKAI, M. *Applications of Vaught sentences and the covering theorem*
⋄ C15 C40 C45 C50 C52 C75 D70 E15 ⋄

HOFMANN, K.H. & MISLOVE, M.W. *Amalgamation in categories with concrete duals* ⋄ C05 C52 G30 ⋄

KAISER, KLAUS *Quasi-axiomatic classes*
⋄ B15 C20 C52 C75 C85 ⋄

KASHIWAGI, T. *A generalizaton of a certain compactness property* ⋄ C20 C52 C75 ⋄

KASHIWAGI, T. *On certain compactness properties for classes of structures* ⋄ C20 C52 ⋄

KOGALOVSKIJ, S.R. *On local properties (Russian)*
⋄ C52 C75 ⋄

MACINTYRE, A. & SHELAH, S. *Uncountable universal locally finite groups* ⋄ C25 C50 C52 C55 C60 ⋄

MARCUS, L. *The ≺-order on submodels* ⋄ C50 C52 ⋄
OLIN, P. *Homomorphisms of elementary types of boolean algebras* ⋄ C05 C52 G05 ⋄
PIGOZZI, D. *The universality of the variety of quasigroups* ⋄ B25 C05 C52 D35 ⋄
SWIJTINK, Z.G. *Eliminability in a cardinal* ⋄ C52 ⋄
WEGLORZ, B. *A note on: "Atomic compactness in $\aleph_1$-categorical Horn theories" by John T. Baldwin* ⋄ C05 C35 C50 C52 ⋄

1977

BALDWIN, JOHN T. & BERMAN, J. *A model theoretic approach to Mal'cev conditions* ⋄ C05 C52 ⋄
BENEJAM, J.-P. *Algebraic characterizations of the satisfiability of first-order logical formulas and the halting of programs* ⋄ C52 D10 G05 ⋄
BONNET, R. *Sur le type d'isomorphie d'algebres de Boole dispersees* ⋄ C52 C55 E50 G05 ⋄
CARSTENS, H.G. *The theorem of Matijasevic is provable in Peano's arithmetic by finitely many axioms* ⋄ C52 D25 D35 F30 H15 ⋄
DIXON, P.G. *Classes of algebraic systems defined by universal Horn sentences* ⋄ C05 C52 ⋄
EKLOF, P.C. *Classes closed under substructures and direct limits* ⋄ C52 C60 C75 ⋄
FORREST, W.K. *Model theory for universal classes with the amalgamation property: A study in the foundations of model theory and algebra* ⋄ C05 C25 C30 C35 C45 C52 C60 ⋄
GNANVO, C. *Γ-cotheories et elimination des quantificateurs (English summary)* ⋄ C10 C25 C35 C52 ⋄
GORBUNOV, V.A. *Covers in lattices of quasivarieties and independent axiomatizability (Russian)* ⋄ C05 C52 ⋄
KRAUSS, P.H. *Homogeneous universal models of universal theories* ⋄ C05 C35 C50 C52 ⋄
MIJAJLOVIC, Z. *Homogeneous-universal models of theories which have model completions* ⋄ C35 C50 C52 ⋄
NELSON, EVELYN *Classes defined by implications* ⋄ B20 C05 C52 ⋄
PALYUTIN, E.A. *Number of models in $L_{\infty \omega_1}$-theories I,II (Russian)* ⋄ C35 C52 C55 C75 E45 E65 ⋄
PINTER, C. *Some theorems on omitting types, with applications to model completeness, amalgamation, and related properties* ⋄ C07 C25 C35 C52 C75 ⋄
SCHMERL, J.H. *Theories with only a finite number of existentially complete models* ⋄ C25 C52 ⋄
TAJTSLIN, M.A. *Existentially closed commutative semigroups (Russian) (English summary)* ⋄ C25 C52 C60 ⋄
TWER VON DER, T. *Pseudoelementare Relationen und Aussagen vom Typ des Bernstein'schen Aequivalenzsatzes* ⋄ C07 C52 ⋄
VINOGRADOV, A.A. *Nonaxiomatizability of lattice-orderable rings (Russian)* ⋄ C52 C60 ⋄
ZIL'BER, B.I. *The structure of a model of a categorical theory and finite-axiomatizability problem (Russian)* ⋄ C35 C52 ⋄

1978

ADLER, A. *On the multiplicative semigroups of rings* ⋄ C52 C60 ⋄
BALDWIN, JOHN T. *Some EC_Σ classes of rings* ⋄ C52 C60 ⋄
FORREST, W.K. *A note on universal classes with applications to the theory of graphs* ⋄ C05 C50 C52 C65 ⋄
HAUSCHILD, K. *Gruppengraphen und endliche Axiomatisierbarkeit* ⋄ C35 C52 C60 ⋄
HERRERA MIRANDA, J. *Les theories convexes de Horn (English summary)* ⋄ B20 C05 C52 G30 ⋄
HULE, H. *Relations between the amalgamation property and algebraic equations* ⋄ C05 C52 ⋄
McKEE, T.A. *Forbidden subgraphs in terms of forbidden quantifiers* ⋄ C52 C65 ⋄
McKENZIE, R. *A finite algebra $\underline{A}$ with $SP(\underline{A})$ not elementary* ⋄ C05 C52 ⋄
MIKENBERG, I. *From total to partial algebras* ⋄ C05 C40 C52 ⋄
NEMETI, I. *From hereditary classes to varieties in abstract model theory and partial algebra* ⋄ C05 C20 C52 C90 C95 ⋄
PINTER, C. *A note on the decomposition of theories with respect to amalgamation, convexity, and related properties* ⋄ C07 C52 ⋄
SKORNYAKOV, L.A. *The axiomatizability of a class of injective polygons (Russian)* ⋄ C05 C52 C60 ⋄
TOKARENKO, A.I. *The group-theoretic properties N_0, N, and $\tilde{N}$ are not axiomatizable (Russian)* ⋄ C20 C52 C60 ⋄
WEISPFENNING, V. *A note on $\aleph_0$-categorical model-companions* ⋄ C25 C35 C50 C52 C65 G05 G10 ⋄

1979

BAJZHANOV, B.S. & OMAROV, B. *On finite diagrams (Russian)* ⋄ C52 ⋄
BENENSON, I.E. *The structure of quasivarieties of models (Russian)* ⋄ C05 C52 ⋄
FRASNAY, C. *Sur quelques classes universelles de relations m-aires* ⋄ C07 C30 C52 E07 ⋄
GLEBSKIJ, YU.V. & GORDON, E.I. *The elementary theories of certain distributive lattices with an additive measure (Russian)* ⋄ C52 G10 ⋄
LASCAR, D. *Les modeles denombrables d'une theorie superstable ayant des fonctions de Skolem (English summary)* ⋄ C15 C45 C52 ⋄
LOULLIS, G. *Sheaves and boolean valued model theory* ⋄ C35 C52 C60 C90 ⋄
MILLER, DOUGLAS E. *On classes closed under unions of chains* ⋄ C52 C75 ⋄
MUNDICI, D. *Compattezza = JEP in ogni logica* ⋄ C40 C52 C95 ⋄
OIKKONEN, J. *On PC- and RPC-classes in generalized model theory* ⋄ C13 C52 C75 C80 C95 ⋄
PALYUTIN, E.A. *Categorical positive Horn theories (Russian)* ⋄ C05 C35 C52 ⋄
PERMINOV, E.A. & VAZHENIN, YU.M. *On rigid lattices and graphs (Russian)* ⋄ C50 C52 C65 G10 ⋄

POUZET, M. *Chaines de theories universelles*
⋄ B20   C52 ⋄

SKORNYAKOV, L.A. *Finite axiomatizability of a class of point modules (Russian)* ⋄ C52   C60 ⋄

SZCZERBA, L.W. & TARSKI, A. *Metamathematical discussion of some affine geometries*
⋄ B25   B30   C35   C52   C65   D35 ⋄

TOKARENKO, A.I. *Axiomatizability of the group-theoretic property $\overline{Z}$ (Russian)* ⋄ C20   C52   C60 ⋄

TUMANOV, V.I. *Lattices of equational theories of models (Russian)* ⋄ C05   C52   G10 ⋄

TUSCHIK, H.-P. *A reduction of a problem about ω_1-categoricity* ⋄ C35   C52 ⋄

WHEELER, W.H. *Amalgamation and elimination of quantifiers for theories of fields*
⋄ C10   C25   C52   C60 ⋄

1980

BAJRAMOV, R.A. & MAMEDOV, O.M. *Σ-subdirect representations in axiomatic classes (Russian) (English summary)* ⋄ C05   C30   C52 ⋄

BEKENOV, M.I. *Theories with a basis (Russian)*
⋄ C45   C52 ⋄

BURRIS, S. & WERNER, H. *Remarks on boolean products* ⋄ C05   C30   C52   G05 ⋄

CAICEDO, X. *The subdirect decomposition theorem for classes of structures closed under direct limits*
⋄ C05   C52 ⋄

COHN, P.M. *On semifir constructions*
⋄ C05   C20   C52   C60 ⋄

EDA, K. *A note on the hereditary properties in the product space* ⋄ C52   C65 ⋄

KAISER, KLAUS *On a lattice of relational reducts*
⋄ C52   C75 ⋄

MARKOVSKI, S. *On quasivarieties of generalized subalgebras* ⋄ C05   C52 ⋄

MCNULTY, G.F. *Classes which generate the variety of all lattice-ordered groups* ⋄ C05   C52   C60 ⋄

MEKLER, A.H. *On residual properties* ⋄ C52   C75 ⋄

MUSTAFIN, T.G. *Theories with a two-cardinal formula (Russian)* ⋄ C45   C50   C52   C55 ⋄

PALYUTIN, E.A. *Categorical Horn classes I (Russian)*
⋄ C05   C35   C45   C52 ⋄

PALYUTIN, E.A. *Description of categorical positive Horn classes (Russian)* ⋄ C05   C35   C52 ⋄

SAPIR, M.V. *On finite and independent axiomatizability of certain quasi-varieties of semigroups (Russian)*
⋄ C05   C52 ⋄

TAYLOR, W. *Mal'tsev conditions and spectra*
⋄ C05   C13   C52 ⋄

ZIL'BER, B.I. *Solution of the problem of finite axiomatizability for theories that are categorical in all infinite powers (Russian)* ⋄ C35   C45   C52 ⋄

ZIL'BER, B.I. *Strongly minimal countably categorical theories (Russian)* ⋄ C35   C45   C52 ⋄

ZIL'BER, B.I. *Totally categorical theories: Structural properties and the non-finite axiomatizability*
⋄ C35   C45   C52 ⋄

1981

BALDWIN, JOHN T. & BERMAN, J. *Elementary classes of varieties* ⋄ C05   C52 ⋄

BLATTER, C. & SPECKER, E. *Le nombre de structures finies d'une theorie a caractere fini* ⋄ C13   C52   C85 ⋄

DZIOBIAK, W. *Concerning axiomatizability of the quasivariety generated by a finite Heyting or topological boolean algebra* ⋄ C05   C13   C52 ⋄

FISHBURN, P.C. *Restricted thresholds for interval orders: a case of nonaxiomatizability by a universal sentence*
⋄ B70   B75   C52 ⋄

FISHER, J.R. *Axiomatic radical and semisimple classes of rings* ⋄ C05   C52   C60 ⋄

KAISER, KLAUS *Lattices acting on universal classes*
⋄ C52   G10 ⋄

KAISER, KLAUS *On complementedly normal lattices*
⋄ C52   G10 ⋄

KASHIWAGI, T. *On classes which have the weak amalgamation property* ⋄ C25   C52 ⋄

KEPKA, T. *Strong amalgamation property in some groupoid varieties* ⋄ C05   C52 ⋄

KOGALOVSKIJ, S.R. & SOLDATOVA, V.V. *Remarks on countably local classes (Russian)* ⋄ C15   C52 ⋄

LABLANQUIE, J.-C. *Modeles generiques et compacite (English summary)* ⋄ C25   C52 ⋄

LAWRENCE, J. *Primitive rings do not form an elementary class* ⋄ C20   C52   C60 ⋄

MUNDICI, D. *A group-theoretical invariant for elementary equivalence and its role in representations of elementary classes* ⋄ B10   C07   C52 ⋄

OIKKONEN, J. *Finite and infinite versions of the operations PC and RPC in infinitary languages* ⋄ C52   C75 ⋄

PRESTEL, A. *Pseudo real closed fields*
⋄ C10   C25   C35   C52   C60 ⋄

ZEITLER, H. *Die Amalgamierungseigenschaft und reine Ringe* ⋄ C52   C60 ⋄

ZIL'BER, B.I. *Solution of the problem of finite axiomatizability for theories that are categorical in all infinite powers (Russian)* ⋄ B30   C35   C45   C52 ⋄

1982

ARAGANE, K. *Characterization of elementary classes of structures whose every substructure is a congruence class*
⋄ C05   C52 ⋄

BALDWIN, JOHN T. & MCKENZIE, R. *Counting models in universal Horn classes* ⋄ C05   C52 ⋄

BALDWIN, JOHN T. & BERMAN, J. *Definable principal congruence relations: kith and kin*
⋄ C05   C52   E07 ⋄

BEKENOV, M.I. *On the spectrum of quasitranscendental theories (Russian)* ⋄ C35   C45   C52 ⋄

BUYS, A. & HEIDEMA, J. *Radicals and subdirect products in models* ⋄ C05   C52 ⋄

COOPER, GLEN R. *On complexity of complete first-order theories* ⋄ C35   C45   C52 ⋄

CZELAKOWSKI, J. *Logical matrices and the amalgamation property* ⋄ B22   C40   C52   C90 ⋄

DZIOBIAK, W. *Concerning axiomatizability of the quasivariety generated by a finite Heyting or topological boolean algebra* ⋄ C05   C13   C52   G10 ⋄

ERSHOV, YU.L. *Totally real field extensions (Russian)*
⋄ B25 C52 C60 ⋄

FUJIWARA, T. *Characterization of axiomatic classes of groupoids closed under identity element extensions*
⋄ C05 C52 ⋄

GORBACHUK, E.L. & KOMARNITS'KIJ, N.YA. *On the axiomatizability of radical and semisimple classes of modules and abelian groups (Russian)* ⋄ C52 C60 ⋄

HAUSCHILD, K. *Model-theoretic properties of cause-and-effect structures* ⋄ B45 C52 D35 ⋄

JARDEN, M. *The elementary theory of large e-fold ordered fields* ⋄ C52 C60 ⋄

JONSSON, B. *Varieties of relation algebras* ⋄ C05 C52 ⋄

KLEIMAN, J.G. *On identities in groups (Russian)*
⋄ C52 C60 D40 ⋄

KRAPEZ, A. *Some nonaxiomatizable classes of semigroups*
⋄ C05 C52 ⋄

LIPPARINI, P. *Locally finite theories with model companion (Italian summary)* ⋄ C25 C35 C52 ⋄

MUNDICI, D. *Interpolation, compactness and JEP in soft model theory* ⋄ C40 C52 C95 ⋄

MYCIELSKI, J. *An essay about old model theory*
⋄ C07 C52 ⋄

PALYUTIN, E.A. *Number of models in complete varieties*
⋄ C05 C45 C52 ⋄

SAIN, I. *On classes of algebraic systems closed with respect to quotients* ⋄ C05 C20 C30 C52 ⋄

SHELAH, S. *The spectrum problem I: $\aleph_\varepsilon$-saturated models, the main gap. II: Totally transcendental and the infinite depth* ⋄ C45 C50 C52 ⋄

SMIRNOV, D.M. *Universal definability of Mal'tsev classes (Russian)* ⋄ C05 C52 ⋄

TULIPANI, S. *On classes of algebras with the definability of congruences* ⋄ C05 C20 C52 ⋄

WEISPFENNING, V. *Model theory and lattices of formulas*
⋄ C25 C35 C52 C75 G10 ⋄

1983

ARAGANE, K. & FUJIWARA, T. *Elementary classes whose every model is algebraically closed* ⋄ C25 C52 ⋄

DANKO, W. *Algorithmic properties of finitely generated structures* ⋄ B75 C40 C52 C75 ⋄

DELFRATE, M.G. *Oriented 2-graphs and their generalization to dimension n (Italian) (English summary)* ⋄ C52 C65 ⋄

FLEISCHER, I. *Subdirect decomposability into irreducibles*
⋄ C05 C52 ⋄

FUJIWARA, T. *Characterization of axiomatic classes of groupoids closed under identity extensions*
⋄ C05 C52 ⋄

HANF, W.P. & MYERS, D.L. *Boolean sentence algebras: isomorphism constructions* ⋄ B10 C52 ⋄

MAKOWSKY, J.A. & SHELAH, S. *Positive results in abstract model theory: a theory of compact logics*
⋄ C20 C52 C95 E05 E55 ⋄

MARONGIU, G. *Definibilita al primo ordine delle congruenze principali e delle congruenze compatte*
⋄ C05 C20 C52 ⋄

MCKENZIE, R. *The number of non-isomorphic models in quasi-varieties of semigroups* ⋄ C05 C52 ⋄

MIJAJLOVIC, Z. *On the embedding property of models*
⋄ C25 C35 C52 G05 G10 ⋄

MUNDICI, D. *Compactness = JEP in any logic*
⋄ C40 C52 C95 ⋄

PALASINSKI, M. *Varieties of commutative BCK-algebras not generated by their finite members*
⋄ C52 G10 G25 ⋄

PITTS, A.M. *Amalgamation and interpolation in the category of Heyting algebras* ⋄ C40 C52 G30 ⋄

PITTS, A.M. *An application of open maps to categorical logic* ⋄ C40 C52 G30 ⋄

POIZAT, B. *Beaucoup de modeles a peu de frais*
⋄ C45 C50 C52 C55 ⋄

POIZAT, B. *Proprietes "modele-theoriques" d'une theorie du premier ordre* ⋄ C45 C52 C95 ⋄

SHELAH, S. *Classification theory of non-elementary classes, Part I: The number of uncountable models of $\psi \in L_{\omega_1,\omega}$* ⋄ C45 C52 C55 C75 C80 E50 ⋄

1984

ALVES, E.H. *Paraconsistent logic and model theory*
⋄ B53 C40 C52 C90 ⋄

ARAGANE, K. *Characterization of elementary classes of structures whose every substructure is a congruence class*
⋄ C05 C52 ⋄

BAJRAMOV, R.A. & MAMEDOV, O.M. *Nonstandard subdirect representations in axiomatizable classes (Russian)* ⋄ C05 C30 C52 ⋄

BAUDISCH, A. *Tensor products of modules and elementary equivalence* ⋄ C30 C52 C60 ⋄

BLATTER, C. & SPECKER, E. *Recurrence relations for the number of labeled structures on a finite set*
⋄ C13 C52 C85 ⋄

CHERLIN, G.L. & VOLGER, H. *Convexity properties and algebraic closure operators* ⋄ C52 ⋄

ECSEDI-TOTH, P. *A characterization of quasi-varieties in equality-free languages* ⋄ C05 C52 ⋄

GIORGETTA, D. & SHELAH, S. *Existentially closed structures in the power of the continuum*
⋄ C15 C25 C30 C52 C55 C60 C75 ⋄

GORBUNOV, V.A. *Axiomatizability of replete classes (Russian)* ⋄ C05 C20 C52 ⋄

GRAETZER, G. & WHITNEY, S. *Infinitary varieties of structures closed under the formation of complex structures* ⋄ C05 C52 C75 ⋄

HEBERT, M. *Un theoreme de preservation pour les factorisations* ⋄ C07 C40 C52 ⋄

HODGES, W. *Constructing many non-isomorphic models*
⋄ C52 ⋄

HODGES, W. *On constructing many non-isomorphic algebras* ⋄ C05 C52 C55 C75 E50 E55 E75 ⋄

HUMBERSTONE, I.L. *Monadic representability of certain binary relations* ⋄ C52 D60 E07 ⋄

KALFA, C. *Decidable properties of finite sets of equations in trivial languages* ⋄ C05 C35 C52 ⋄

KASHIWAGI, T. *A certain compactness property and free structures* ⋄ C52 ⋄

PALASINSKI, M. *On BCK-algebras with the operation (S)*
⋄ C52 G10 G25 ⋄

PILLAY, A. *Countable modules*
⋄ C15 C35 C45 C52 C60 ⋄

TULIPANI, S. *On the universal theory of classes of finite models* ⋄ C05 C13 C52 C60 ⋄

WRONSKI, A. *Interpolation and amalgamation properties of BCK-algebras* ⋄ C40 C52 G10 G25 ⋄

WRONSKI, A. *On varieties of commutative BCK-algebras not generated by their finite members* ⋄ C05 C52 G25 ⋄

1985

ARAGANE, K. & FUJIWARA, T. *Elementary classes whose every model is α-existentially closed* ⋄ C25 C52 ⋄

ARAGANE, K. & FUJIWARA, T. *Elementary classes of α-existentially closed structures* ⋄ C25 C52 ⋄

CHERLIN, G.L. & HARRINGTON, L.A. & LACHLAN, A.H. *$\aleph_0$-categorical, $\aleph_0$-stable structures* ⋄ C35 C45 C52 ⋄

CZELAKOWSKI, J. *Sentential logics and Maehara interpolation property* ⋄ B22 C05 C40 C52 ⋄

D'OTTAVIANO, I.M.L. *The model extension theorems for Π_3-theories* ⋄ B50 C07 C52 ⋄

GONCHAROV, S.S. *Axioms for classes with strong epimorphisms (Russian) (English summary)* ⋄ C52 ⋄

HARRINGTON, L.A. & MAKKAI, M. *An exposition of Shelah's "main gap": counting uncountable models of ω-stable and superstable theories* ⋄ C45 C50 C52 ⋄

MAKOWSKY, J.A. *Why Horn formulas matter in computer science: Initial structure and generic examples* ⋄ B75 C52 ⋄

MEDVEDEV, N.YA. *Quasivarieties of l-groups and groups* ⋄ C05 C52 C60 ⋄

NEMETI, I. *Cylindric-relativised set algebras have strong amalgamation* ⋄ C05 C52 G15 G25 ⋄

PRESTEL, A. *On the axiomatization of PRC-fields* ⋄ C52 C60 ⋄

SHELAH, S. *Classification of first order theories which have a structure theorem* ⋄ C35 C45 C52 C75 ⋄

SHELAH, S. *Remarks in abstract model theory* ⋄ C40 C50 C52 C55 C80 C95 ⋄

C10 ∪ B25 Elimination of quantifiers and decidability

1908
LOEWENHEIM, L. *Ueber das Aufloesungsproblem im logischen Klassenkalkuel* ⋄ B20 B25 G05 ⋄

1915
LOEWENHEIM, L. *Ueber Moeglichkeiten im Relativkalkuel* ⋄ B25 C07 C10 C85 E07 G15 ⋄

1919
SKOLEM, T.A. *Untersuchungen ueber die Axiome des Klassenkalkuels und ueber Produktions- und Summationsprobleme, welche gewisse Klassen von Aussagen betreffen* ⋄ B20 B25 C10 D35 G05 ⋄

1920
SKOLEM, T.A. *Logisch-kombinatorische Untersuchungen ueber die Erfuellbarkeit oder Beweisbarkeit mathematischer Saetze nebst einem Theorem ueber dichte Mengen* ⋄ B10 B25 C07 C65 C75 ⋄

1922
BEHMANN, H. *Beitraege zur Algebra der Logik, insbesondere zum Entscheidungsproblem* ⋄ B20 B25 D35 ⋄

1923
BEHMANN, H. *Algebra der Logik und Entscheidungsproblem* ⋄ B20 B25 D35 ⋄

1926
LANGFORD, C.H. *Some theorems on deducibility* ⋄ B30 C10 C35 C65 ⋄
LANGFORD, C.H. *Theorems on deducibility (second paper)* ⋄ B30 C10 C35 C65 ⋄

1927
BEHMANN, H. *Entscheidungsproblem und Logik der Beziehungen* ⋄ B25 D35 ⋄

1928
ACKERMANN, W. *Ueber die Erfuellbarkeit gewisser Zaehlausdruecke* ⋄ B25 D35 ⋄
BERNAYS, P. & SCHOENFINKEL, M. *Zum Entscheidungsproblem der mathematischen Logik* ⋄ B20 B25 D35 ⋄
HILBERT, D. & ACKERMANN, W. *Grundzuege der theoretischen Logik* ⋄ B25 B98 D35 ⋄
KALMAR, L. *Eine Bemerkung zur Entscheidungstheorie* ⋄ B20 B25 F30 ⋄
SKOLEM, T.A. *Ueber die mathematische Logik* ⋄ A05 B25 B98 C07 C10 C35 ⋄

1929
GOEDEL, K. *Ein Spezialfall des Entscheidungsproblems der theoretischen Logik* ⋄ B20 B25 ⋄

1930
HERBRAND, J. *Recherches sur la theorie de la demonstration* ⋄ B10 B25 F05 F07 F25 F30 ⋄
KALMAR, L. *Ein Beitrag zum Entscheidungsproblem* ⋄ B25 D35 ⋄
PRESBURGER, M. *Ueber die Vollstaendigkeit eines gewissen Systems der Arithmetik ganzer Zahlen, in welchem die Addition als einzige Operation hervortritt* ⋄ B25 B28 C10 C35 F30 ⋄
RAMSEY, F.P. *On a problem of formal logic* ⋄ B10 B25 E05 E20 E75 ⋄
WAERDEN VAN DER, B.L. *Eine Bemerkung ueber die Unzerlegbarkeit von Polynomen* ⋄ B25 C60 D45 F55 ⋄

1931
HERBRAND, J. *Sur le probleme fondamental de la logique mathematique* ⋄ A05 B25 B30 F30 F99 ⋄
SKOLEM, T.A. *Ueber einige Satzfunktionen in der Arithmetik* ⋄ B25 B28 C10 F30 ⋄
TARSKI, A. *Sur les ensembles definissables de nombres reels I* ⋄ B25 B28 C10 C40 C60 C65 D55 E15 E47 F35 ⋄

1932
KALMAR, L. *Zum Entscheidungsproblem der mathematischen Logik* ⋄ B20 B25 D35 ⋄
LANGFORD, C.H. & LEWIS, C.I. *Symbolic logic* ⋄ B25 B98 C10 C98 ⋄

1933
GOEDEL, K. *Zum Entscheidungsproblem des logischen Funktionenkalkuels* ⋄ B10 B20 B25 C13 D35 ⋄
KALMAR, L. *Ueber die Erfuellbarkeit derjenigen Zaehlausdruecke, welche in der Normalform zwei benachbarte Allzeichen enthalten* ⋄ B20 B25 ⋄
SCHUETTE, K. *Untersuchungen zum Entscheidungsproblem der mathematischen Logik* ⋄ B20 B25 D35 ⋄
SKOLEM, T.A. *Ein kombinatorischer Satz mit Anwendung auf ein logisches Entscheidungsproblem* ⋄ B20 B25 E05 ⋄

1934
KIREEVSKIJ, N.N. *On the problem of the solvability of the decision problem (Russian)* ⋄ B25 D35 ⋄
SCHUETTE, K. *Ueber die Erfuellbarkeit einer Klasse von logischen Formeln* ⋄ B20 B25 ⋄

1935
GENTZEN, G. *Untersuchungen ueber das logische Schliessen I,II* ⋄ B10 B25 F05 F07 F30 F50 ⋄

SKOLEM, T.A. *Ueber die Erfuellbarkeit gewisser Zaehlausdruecke* ◊ B25 B28 ◊

1936

ACKERMANN, W. *Beitraege zum Entscheidungsproblem der mathematischen Logik* ◊ B25 D35 ◊

SKOLEM, T.A. *Ein Satz ueber die Erfuellbarkeit von einigen Zaehlausdruecken der Form* $(x)(\exists y_1,\ldots,y_n)K_1(x,y_1,\ldots,y_n) \& (x_1,x_2,x_3)K_2(x_1,x_2,x_3)$ ◊ B10 B25 ◊

1937

PEPIS, J. *Ueber das Entscheidungsproblem des engeren logischen Funktionskalkuels (Polish) (German summary)* ◊ B20 B25 D35 ◊

SKOLEM, T.A. *Eine Bemerkung zum Entscheidungsproblem* ◊ B20 B25 D35 ◊

1938

WAJSBERG, M. *Untersuchungen ueber den Aussagenkalkuel von A.Heyting* ◊ B25 F50 ◊

1939

ZHEGALKIN, I.I. *Sur l'Entscheidungsproblem (Russian) (French summary)* ◊ B20 B25 D35 ◊

1941

MCKINSEY, J.C.C. *A solution of the decision problem for the Lewis systems S2 and S4, with an application to topology* ◊ B25 B45 C65 ◊

1943

MCKINSEY, J.C.C. *The decision problem for some classes of sentences without quantifiers* ◊ B20 B25 C05 C60 G10 ◊

1945

QUINE, W.V.O. *On the logic of quantification* ◊ B10 B25 ◊

1946

MCKINSEY, J.C.C. & TARSKI, A. *On closed elements in closure algebras* ◊ B25 C05 C65 G05 G25 ◊

ZHEGALKIN, I.I. *Sur le probleme de la resolubilite pour les classes finies (Russian) (French summary)* ◊ B20 B25 C13 ◊

1948

TARSKI, A. *A decision method for elementary algebra and geometry* ◊ B25 B30 C10 C35 C60 ◊

WRIGHT VON, G.H. *On the idea of logical truth I* ◊ A05 B05 B20 B25 ◊

1949

HALLDEN, S. *On the decision problem of Lewis' calculus S5* ◊ B25 B45 ◊

HALLDEN, S. *Results concerning the decision problem of Lewis's calculi S3 and S6* ◊ B25 B45 ◊

NOVIKOV, P.S. *On the axiom of complete induction (Russian)* ◊ B25 B28 D35 F30 ◊

QUINE, W.V.O. *On decidability and completeness* ◊ B25 D35 F30 ◊

SZMIELEW, W. *Decision problem in group theory* ◊ B25 C10 C60 ◊

1950

BEHMANN, H. *Das Aufloesungsproblem in der Klassenlogik* ◊ A05 B20 B25 ◊

BETH, E.W. *Decision problems of logic and mathematics* ◊ B20 B25 C10 D35 ◊

JANICZAK, A. *A remark conerning decidability of complete theories* ◊ B25 C35 ◊

PIL'CHAK, B.YU. *On the decision problem for the calculus of problems (Russian)* ◊ B25 F50 ◊

SCHUETTE, K. *Schlussweisen-Kalkuele der Praedikatenlogik* ◊ B10 B25 F05 F07 F50 ◊

WRIGHT VON, G.H. *On the idea of logical truth II* ◊ A05 B05 B20 B25 ◊

1951

CHURCH, A. *Special cases of the decision problem* ◊ B25 ◊

SHEPHERDSON, J.C. *Inverses and zero divisors in matrix rings* ◊ B25 C60 D40 ◊

WRIGHT VON, G.H. *An essay in modal logic* ◊ B25 B45 ◊

WRIGHT VON, G.H. *Deontic logic* ◊ A05 B25 B45 ◊

1952

GEACH, P.T. & WRIGHT VON, G.H. *On an extended logic of relations* ◊ B20 B25 ◊

KALICKI, J. *On comparison of finite algebras* ◊ B25 C05 C13 ◊

MOSTOWSKI, ANDRZEJ *On direct products of theories* ◊ B25 C10 C30 ◊

THOMAS, I. *A new decision procedure for Aristotle's syllogistic* ◊ A10 B20 B25 ◊

WRIGHT VON, G.H. *On double quantification* ◊ A05 B20 B25 ◊

1953

ISSMANN, S. *Une methode de decision pour certaines formules du calcul des predicats* ◊ B25 ◊

JANICZAK, A. *Undecidability of some simple formalized theories* ◊ B25 D35 ◊

KOLMOGOROV, A.N. *On the concept of algorithm (Russian)* ◊ B25 D20 ◊

RASIOWA, H. & SIKORSKI, R. *On satisfiability and decidability in non-classical functional calculi* ◊ B10 B25 B45 C90 F50 ◊

WANG, HAO *Quelques notions d'axiomatique* ◊ B25 B30 C07 C35 E30 F30 ◊

1954

ACKERMANN, W. *Solvable cases of the decision problem* ◊ B25 ◊

ANDERSON, A.R. *Improved decision procedures for Lewis's calculus S4 and von Wright's calculus M* ◊ B25 B45 ◊

BETH, E.W. *Observations metamathematiques sur les structures simplement ordonnees* ◊ B25 C65 ◊

HARROP, R. *An investigation of the propositional calculus used in a particular system of logic* ◊ B20 B25 ◊

HENRY, DESMOND PAUL *Expressions trivially decidable* ◊ B25 ◊

SEIDENBERG, A. *A new decision method for elementary algebra* ◊ B25 C10 C60 ◊

TURING, A.M. *Solvable and unsolvable problems*
    ⋄ B25 D10 D35 ⋄
VAUGHT, R.L. *Applications of the Loewenheim-Skolem-Tarski theorem to problems of completeness and decidability* ⋄ B25 C07 C35 ⋄
VAUGHT, R.L. *On sentences holding in direct products of relational systems* ⋄ B20 B25 C30 ⋄

1955

GOODSTEIN, R.L. *On non-constructive theorems of analysis and the decision problem*
    ⋄ B25 B28 F30 F60 ⋄
HENKIN, L. *On a theorem of Vaught* ⋄ B25 C35 ⋄
HERMES, H. *Vorlesung ueber Entscheidungsprobleme in Mathematik und Logik*
    ⋄ B25 B40 B65 D10 D20 D35 D98 ⋄
KLAUA, D. *Systematische Behandlung der loesbaren Faelle des Entscheidungsproblems fuer den Praedikatenkalkuel der ersten Stufe* ⋄ B25 ⋄
SZMIELEW, W. *Elementary properties of abelian groups*
    ⋄ B25 C10 C60 ⋄

1956

MEREDITH, D. *A correction to von Wright's decision procedure for the deontic system P* ⋄ B25 B45 ⋄
ROBINSON, A. *Complete theories*
    ⋄ B25 C35 C60 C98 ⋄

1957

ADYAN, S.I. *Finitely presented groups and algorithms (Russian)* ⋄ B25 C60 D40 ⋄
BAR-HILLEL, Y. *New light on the liar* ⋄ B25 ⋄
COBHAM, A. *Effectively decidable theories* ⋄ B25 F30 ⋄
COLLINS, G.E. *Tarski's decision method for elementary algebra* ⋄ C10 C60 ⋄
DREBEN, B. *Systematic treatment of the decision problem*
    ⋄ B20 B25 D35 ⋄
EHRENFEUCHT, A. *Two theories with axioms built by means of pleonasmus* ⋄ B25 C35 D35 ⋄
EICHHOLZ, T. *Semantische Untersuchungen zur Entscheidbarkeit im Praedikatenkalkuel mit Funktionsvariablen* ⋄ B25 ⋄
GRZEGORCZYK, A. *Decision problems*
    ⋄ B25 C10 D35 D98 F30 ⋄
MATSUMOTO, K. & OHNISHI, M. *Gentzen method in modal calculi. I* ⋄ B25 B45 ⋄
MOSTOWSKI, ANDRZEJ *On a generalisation of quantifiers*
    ⋄ B25 C55 C80 C95 ⋄
PUTNAM, H. *Decidability and essential undecidability*
    ⋄ B25 D35 ⋄
ROBINSON, A. *Relative model-completeness and the elimination of quantifiers* ⋄ C10 C35 ⋄

1958

BORKOWSKI, L. *On proper quantifiers I (Polish and Russian summaries)* ⋄ B20 B25 ⋄
GAL, L.N. *A note on direct products* ⋄ B25 C30 D35 ⋄
HARROP, R. *On the existence of finite models and decision procedures for propositional calculi* ⋄ B22 B25 ⋄
KIREEVSKIJ, N.N. *Ueber die Allgemeingueltigkeit gewisser Zaehlausdruecke (Russian)* ⋄ B25 F30 ⋄
OBERSCHELP, A. *Ueber die Axiome produkt-abgeschlossener arithmetischer Klassen*
    ⋄ B20 B25 C30 C40 C52 ⋄

REICHBACH, J. *On the first-order functional calculus and the truncation of models (Polish and Russian summaries)* ⋄ B25 C07 C90 ⋄
ROBINSON, A. *Relative model-completeness and the elimination of quantifiers (English, French, and German summaries)* ⋄ C10 C35 C98 ⋄
SCHMIDT, H.A. *Un procede maniable de decision pour la logique propositionnelle intuitionniste* ⋄ B25 F50 ⋄

1959

EHRENFEUCHT, A. *Decidability of the theory of one function* ⋄ B25 ⋄
ELGOT, C.C. & WRIGHT, J.B. *Quantifier elimination in a problem of logical design* ⋄ B25 B70 C10 ⋄
FEFERMAN, S. & VAUGHT, R.L. *The first order properties of products of algebraic systems*
    ⋄ B25 C10 C20 C30 G05 ⋄
KUZNETSOV, A.V. *On the equivalence and functional completeness problems (Russian)* ⋄ B25 D35 ⋄
MATSUMOTO, K. & OHNISHI, M. *Gentzen method in modal calculi II* ⋄ B25 B45 ⋄
OBERSCHELP, A. *Ueber die Axiome arithmetischer Klassen mit Abgeschlossenheitsbedingungen*
    ⋄ B25 C40 C52 ⋄
ROBINSON, A. *Solution of a problem of Tarski*
    ⋄ B25 C35 C60 ⋄
SCHWABHAEUSER, W. *Entscheidbarkeit und Vollstaendigkeit der elementaren hyperbolischen Geometrie* ⋄ B25 B30 C35 C65 ⋄

1960

BORKOWSKI, L. *On proper quantifiers II (Polish and Russian summaries)* ⋄ B20 B25 ⋄
BUECHI, J.R. *Weak second-order arithmetic and finite automata* ⋄ B15 B25 C85 D05 F35 ⋄
MATSUMOTO, K. *Decision procedure for modal sentential calculus S3* ⋄ B25 B45 ⋄
ROBINSON, A. & ZAKON, E. *Elementary properties of ordered abelian groups* ⋄ B25 C35 C60 C65 ⋄
SEIDENBERG, A. *An elimination theory for differential algebra* ⋄ C10 C60 ⋄
TRAKHTENBROT, B.A. *Algorithms and machine solutions of problems (Russian)* ⋄ B25 B35 D20 ⋄

1961

BORKOWSKI, L. *A didactical approach to the zero-one decision procedure of the expressions of the first order monadic predicate calculus (Polish) (Russian and English summaries)* ⋄ B20 B25 C10 ⋄
ELGOT, C.C. *Decision problems of finite automata design and related arithmetics*
    ⋄ B25 D03 D05 D35 F30 ⋄
MAL'TSEV, A.I. *On the elementary theories of locally free universal algebras (Russian)* ⋄ B25 C05 C60 ⋄
OGLESBY, F.C. *Report: an examination of a decision procedure* ⋄ B25 F07 ⋄
OHNISHI, M. *Gentzen decision procedures for Lewis's systems S2 and S3* ⋄ B25 B45 ⋄
POLIFERNO, M.J. *Decision algorithms for some functional calculi with modality* ⋄ B25 B45 ⋄

1962

BLANCHE, R. *Axiomatics* ◇ A10 B25 B30 ◇

BUECHI, J.R. *On a decision method in restricted second order arithmetic* ◇ B25 C85 D05 F35 ◇

DREBEN, B. & KAHR, A.S. & WANG, HAO *Classification of AEA formulas by letter atoms* ◇ B20 B25 D35 ◇

DREBEN, B. *Solvable Suranyi subclasses: an introduction to the Herbrand theory* ◇ B20 B25 ◇

ERSHOV, YU.L. *Decidability of elementary theories of certain classes of abelian groups (Russian)* ◇ B25 C60 ◇

KOSTYRKO, V.F. *On an error in the paper of I.I.Zhegalkin "Sur le probleme de la resolubilite pour les classes finies" (Russian)* ◇ B25 C13 ◇

MAL'TSEV, A.I. *Axiomatisable classes of locally free algebras of various types (Russian)* ◇ B25 C05 C10 ◇

McCALL, S. *A simple decision procedure for one-variable implication/negation formulae in intuitionistic logic* ◇ B25 F50 ◇

OGLESBY, F.C. *An examination of a decision procedure* ◇ B25 ◇

PETRESCO, J. *Algorithmes de decision et de construction dans les groupes libres* ◇ B25 D40 ◇

SCHWABHAEUSER, W. *On completeness and decidability and some non-definable notions of elementary hyperbolic geometry* ◇ B25 C35 C65 ◇

1963

CHURCH, A. *Logic, arithmetic, and automata* ◇ A10 B25 D05 D98 ◇

FRIEDMAN, JOYCE *A semi-decision procedure for the functional calculus* ◇ B25 B35 ◇

GOLDBERG, R. *On the solvability of a subclass of the Suranyi reduction class* ◇ B25 ◇

GOODSTEIN, R.L. *A decidable fragment of recursive arithmetic* ◇ B20 B25 F30 ◇

NERODE, A. *A decision method for p-adic integral zeros of diophantine equations* ◇ B25 C10 ◇

ROGERS JR., H. *An example in mathematical logic* ◇ B25 C10 C65 ◇

TAJMANOV, A.D. *Decidability of the elementary theory of sphere inclusion (Russian)* ◇ B25 C65 ◇

WANG, HAO *Dominoes and the AEA case of the decision problem* ◇ B20 B25 D05 D35 ◇

1964

ERSHOV, YU.L. *Decidability of certain non-elementary theories (Russian)* ◇ B25 C30 C85 G05 ◇

ERSHOV, YU.L. *Decidability of the elementary theory of relatively complemented distributive lattices and of the theory of filters (Russian)* ◇ B25 G10 ◇

GUREVICH, Y. *Elementary properties of ordered abelian groups (Russian)* ◇ B25 C60 ◇

HUBER-DYSON, V. *On the decision problem for theories of finite models* ◇ B25 C13 D35 ◇

KEISLER, H.J. *Complete theories of algebraically closed fields with distinguished subfields* ◇ B25 C35 C60 ◇

KROM, MELVEN R. *A decision procedure for a class of formulas of first order predicate calculus* ◇ B20 B25 ◇

MAKANIN, G.S. *A new solvable case of the decision problem for first-order predicate calculus (Russian)* ◇ B03 B25 ◇

TAJTSLIN, M.A. *Decidability of certain elementary theories (Russian)* ◇ B25 C60 ◇

TAJTSLIN, M.A. *On elementary theories of free nilpotent algebras (Russian)* ◇ B25 C60 D35 ◇

TAUTS, A. *Solution of logical equations in the first-order one-place predicate calculus (Russian) (English summary)* ◇ B20 B25 ◇

1965

ALMAGAMBETOV, ZH.A. *On classes of axioms closed with respect to the given reduced products and powers (Russian)* ◇ B25 C20 C52 D35 ◇

ALMAGAMBETOV, ZH.A. *Solvability of the elementary theory of certain classes of free nilpotent algebras (Russian)* ◇ B25 C60 ◇

AX, J. & KOCHEN, S. *Diophantine problems over local fields I,II (II: A complete set of axioms for p-adic number theory)* ◇ B25 C20 C35 C60 ◇

BUECHI, J.R. *Decision methods in the theory of ordinals* ◇ B15 B25 C85 D05 E10 ◇

BUECHI, J.R. *Transfinite automata recursions and weak second order theory of ordinals* ◇ B15 B25 C85 D05 E10 ◇

ERSHOV, YU.L. & LAVROV, I.A. & TAJMANOV, A.D. & TAJTSLIN, M.A. *Elementary theories (Russian)* ◇ B25 C98 D35 D98 ◇

ERSHOV, YU.L. *On elementary theories of local fields (Russian)* ◇ B25 C35 C60 ◇

ERSHOV, YU.L. *On the elementary theory of maximal normed fields (Russian)* ◇ B25 C35 C60 ◇

ERSHOV, YU.L. *On the elementary theory of maximal normed fields (Russian)* ◇ B25 C60 ◇

GUREVICH, Y. *Existential interpretation (Russian)* ◇ B20 B25 C60 D35 F25 ◇

HANF, W.P. *Model-theoretic methods in the study of elementary logic* ◇ B25 C07 D20 D25 D35 F25 ◇

HARROP, R. *Some generalizations and applications of a relativization procedure for propositional calculi* ◇ B22 B25 D35 ◇

ITO, MAKOTO *Boolean recursive functions and closure algebra* ◇ B25 B45 D20 ◇

JENSEN, R.B. *Ein neuer Beweis fuer die Entscheidbarkeit des einstelligen Praedikatenkalkuels mit Identitaet* ◇ B25 ◇

KOSTYRKO, V.F. *On the decidability problem for Ackermann's case (Russian)* ◇ B25 ◇

McNAUGHTON, R. *Undefinability of addition from one unary operator* ◇ B28 C10 ◇

ROBINSON, JULIA *The decision problem for fields* ◇ B25 C60 D35 ◇

ROBITASHVILI, N.G. *On the problem of reducing a pp-formula of the pure first-order functional calculus to some special case of the decision problem (Georgian) (Russian summary)* ◇ B20 B25 ◇

SCHWABHAEUSER, W. *On models of elementary elliptic geometry* ◇ B25 C65 ◇

SZCZERBA, L.W. & TARSKI, A. *Metamathematical properties of some affine geometries*
⋄ B25 B30 C52 C65 D35 ⋄

WANG, HAO *Remarks on machines, sets, and the decision problem* ⋄ B20 B25 D03 D05 D10 D35 E30 ⋄

WANG, SHIQIANG *Some decidable cases of elementary propositions on partial order sets I (Chinese)*
⋄ B25 E07 ⋄

1966

AX, J. & KOCHEN, S. *Diophantine problems over local fields, III: Decidable fields*
⋄ B25 C10 C20 C35 C60 C65 ⋄

ELGOT, C.C. & RABIN, M.O. *Decidability and undecidability of extensions of second (first) order theory of (generalized) successor*
⋄ B15 B25 C85 D05 D35 F30 F35 ⋄

FRIEDMAN, JOYCE *Computer realization of a decision procedure in logic* ⋄ B25 B35 ⋄

GLADSTONE, M.D. *Finite models for inequations*
⋄ B25 C05 C13 ⋄

GOODSTEIN, R.L. & LEE, R.D. *A decidable class of equations in recursive arithmetic* ⋄ B20 B25 F30 ⋄

GUREVICH, Y. *Certain algorithmic questions of the theory of classes of algebraic systems (Russian)*
⋄ B25 D35 ⋄

GUREVICH, Y. *Effective recognition of satisfiability of formulae of the restricted predicate calculus (Russian)*
⋄ B20 B25 D35 ⋄

HANSON, W.H. *On some alleged decision procedures for S4* ⋄ B25 B45 ⋄

KHISAMIEV, N.G. *Universal theory of lattice-ordered abelian groups (Russian)* ⋄ B25 C60 ⋄

KOGALOVSKIJ, S.R. *On the properties preserved under algebraic constructions (Russian)*
⋄ B25 C30 C52 C60 D35 D45 ⋄

LAEUCHLI, H. & LEONARD, J. *On the elementary theory of linear order* ⋄ B25 C65 ⋄

MASLOV, S.YU. *Application of the inverse method for establishing deducibility to the theory of decidable fragments in the classical predicate calculus (Russian)*
⋄ B20 B25 ⋄

MINTS, G.E. *Skolem's method of elimination of positive quantifiers in sequential calculi (Russian)*
⋄ B25 C10 F50 ⋄

ROEDDING, D. *Der Entscheidbarkeitsbegriff in der mathematischen Logik*
⋄ B25 D10 D20 D25 D35 ⋄

ROUSSEAU, G. *A decidable class of number theoretic equations* ⋄ B20 B25 F30 ⋄

TAJTSLIN, M.A. *On elementary theories of commutative semigroups with cancellation (Russian)*
⋄ B25 C60 D40 ⋄

TAJTSLIN, M.A. *On elementary theories of commutative semigroups (Russian)* ⋄ B25 C05 C60 ⋄

1967

AX, J. *Solving diophantine problems modulo every prime*
⋄ B25 C13 C20 C60 ⋄

BONDI, I.L. *The decidability or undecidability of certain formal logical calculi (Russian)* ⋄ B25 D35 ⋄

ERSHOV, YU.L. *Elementary theories of Post varieties (Russian)* ⋄ B25 C05 C13 G20 ⋄

ERSHOV, YU.L. *Fields with a solvable theory (Russian)*
⋄ B25 C35 C60 ⋄

GALVIN, F. *Reduced products, Horn sentences, and decision problems*
⋄ B20 B25 C20 C30 C40 D35 ⋄

GUREVICH, Y. *A contribution to the elementary theory of lattice ordered abelian groups and K-lineals (Russian)*
⋄ B25 C60 D35 ⋄

GUREVICH, Y. *A new decision procedure for the theory of ordered abelian groups (Russian)* ⋄ B25 C60 ⋄

JOHNSTONE JR., H.W. *An inductive decision-procedure for the monadic predicate calculus* ⋄ B25 ⋄

KROM, MELVEN R. *The decision problem for segregated formulas in first-order logic* ⋄ B20 B25 D35 ⋄

LIFSCHITZ, V. *The decision problem for some constructive theories of equality (Russian)* ⋄ B25 D35 F50 ⋄

LUCKHARDT, H. *Aussagenlogisch fundierte Theorien*
⋄ B25 F05 ⋄

MCCALL, S. *Connexive class logic* ⋄ B25 E70 G05 ⋄

MIHAILESCU, E.G. *Decision problem in the classical logic*
⋄ B25 D35 ⋄

NIELAND, J.J.F. *Beth's tableau-method*
⋄ B05 B25 B50 F07 ⋄

PAIR, C. *Sur des algorithmes pour les problemes de cheminement dans les graphes finis (English summary)*
⋄ B25 ⋄

PAOLA DI, R.A. *A survey of Soviet work in the theory of computer programming* ⋄ B25 B75 D35 G30 ⋄

RAUTENBERG, W. *Elementare Schemata nichtelementarer Axiome* ⋄ B25 C52 C60 C65 C85 D35 ⋄

TAJMANOV, A.D. *Systems with a solvable universal theory (Russian)* ⋄ B20 B25 C20 ⋄

TAJTSLIN, M.A. *Two remarks on isomorphisms of commutative semigroups (Russian) (English summary)*
⋄ B25 C05 C60 ⋄

TARSKI, A. *The completeness of elementary algebra and geometry* ⋄ B25 B30 C10 C35 C60 C65 ⋄

THOMAS, I. *Decision for K4* ⋄ B25 ⋄

THOMASON, R.H. *A decision procedure for Fitch's propositional calculus* ⋄ B25 B45 ⋄

1968

AX, J. *The elementary theory of finite fields*
⋄ B25 B30 C13 C20 C60 ⋄

ERSHOV, YU.L. *Restricted theories of totally ordered sets (Russian)* ⋄ B25 C65 D35 E07 ⋄

FOL'K, N.F. & SHESTOPAL, G.A. *The solvability of elementary theories of integral and natural numbers with addition (Russian)* ⋄ B25 C10 F30 ⋄

HODES, L. & SPECKER, E. *Lengths of formulas and elimination of quantifiers I* ⋄ B35 C10 C40 ⋄

KOKORIN, A.I. & KOZLOV, G.T. *Extended elementary and universal theories of lattice-ordered abelian groups with a finite number of threads (Russian)* ⋄ B25 C60 ⋄

LACHLAN, A.H. *On the lattice of recursively enumerable sets* ⋄ B25 C10 D25 ⋄

LACHLAN, A.H. *The elementary theory of recursively enumerable sets* ⋄ B25 D25 ⋄

LAEUCHLI, H. *A decision procedure for the weak second order theory of linear order* ⋄ B15 B25 C65 C85 ⋄
LOPEZ-ESCOBAR, E.G.K. *A decision method for the intuitionistic theory of successor* ⋄ B25 C10 F30 F50 ⋄
MAL'TSEV, A.I. *Some questions bordering on algebra and mathematical logic (Russian)* ⋄ B25 C05 ⋄
MCKAY, C.G. *The decidability of certain intermediate propositional logics* ⋄ B25 B55 ⋄
MINTS, G.E. *Solvability of the problem of deducibility in LJ for a class of formulas which do not contain negative occurences of quantors (Russian)* ⋄ B25 F50 ⋄
RABIN, M.O. *Decidability of second-order theories and automata on infinite trees* ⋄ B15 B25 C85 D05 ⋄
RENNIE, M.K. *A function which bounds truth-tabular calculations in S5* ⋄ B25 B45 D15 ⋄
ROUSSEAU, G. *Note on the decidability of a certain class of number theoretic equations* ⋄ B20 B25 F30 ⋄
ROUTLEY, R. *Decision procedures and semantics for C1, E1, and S0.5^0* ⋄ B25 B45 ⋄
ROUTLEY, R. *The decidability and semantical incompleteness of Lemmon's system S0.5* ⋄ B25 B45 ⋄
SAADE, M. *A decision problem for finite groupoids (Russian summary)* ⋄ B25 C13 ⋄
SEGERBERG, K. *Decidability of S4.1* ⋄ B25 B45 ⋄
SEGERBERG, K. *Decidability of four modal logics* ⋄ B25 B45 ⋄
SIEFKES, D. *Recursion theory and the theorem of Ramsey in one-place second order successor arithmetic* ⋄ B15 B25 C85 ⋄
SLOMSON, A. *The monadic fragment of predicate calculus with the Chang quantifier and equality* ⋄ B25 C20 C55 C80 ⋄
TAJTSLIN, M.A. *Algorithmic problems for commutative semigroups (Russian)* ⋄ B25 C05 C60 ⋄
THATCHER, J.W. & WRIGHT, J.B. *Generalized finite automata theory with an application to a decision problem of second order logic* ⋄ B15 B25 D05 ⋄

1969

BUECHI, J.R. & LANDWEBER, L.H. *Definability in the monadic second-order theory of successor* ⋄ B15 B25 C40 C85 D05 F35 ⋄
COHEN, P.J. *Decision procedures for real and p-adic fields* ⋄ B25 C10 C60 ⋄
GOLOTA, YA.YA. *Some techniques for simplifying the construction of nets of marks (Russian)* ⋄ B25 B75 F50 ⋄
GOODSTEIN, R.L. *Decision methods in recursive arithmetic* ⋄ B25 F30 ⋄
GUREVICH, Y. *A decision problem for decision problems (Russian)* ⋄ B25 D35 ⋄
GUREVICH, Y. *The decision problem for the logic of predicates and operations (Russian)* ⋄ B25 D35 ⋄
HEATH, I.J. *Decidable classes of number-theoretic sentences* ⋄ B25 ⋄
HUBER-DYSON, V. *On the decision problem for extensions of a decidable theory* ⋄ B25 C13 C60 D35 ⋄
KALLICK, B. *A decision procedure based on the resolution method* ⋄ B25 B35 ⋄

KOKORIN, A.I. & KOZLOV, G.T. *Elementary theory of abelian groups without torsion, with a predicate for selecting subgroups (Russian)* ⋄ B25 C35 C60 ⋄
LUCKHARDT, H. *Kodifikation und Aussagenlogik* ⋄ B05 B25 F50 ⋄
MATERNA, P. *Identity, equivalence and isomorphism of problems* ⋄ B25 ⋄
OREVKOV, V.P. *On nonlengthening applications of equality rules (Russian)* ⋄ B25 B35 F05 F07 ⋄
RABIN, M.O. *Decidability of second order theories and automata on infinite trees* ⋄ B15 B25 C85 D05 ⋄
SHAFAAT, A. *A remark on a paper of Taimanov* ⋄ B25 C20 ⋄
STASZEK, W. *Investigations on the classical logic of names (Polish) (Russian and English summaries)* ⋄ A10 B25 ⋄
TSINMAN, L.L. *Certain algorithms in a formal arithmetic system (Russian)* ⋄ B25 F30 ⋄
ZEMAN, J.J. *Decision procedures for S3^0 and S4^0* ⋄ B25 B45 ⋄

1970

DONER, J.E. *Tree acceptors and some of their applications* ⋄ B25 C85 D05 ⋄
GABBAY, D.M. *The decidability of the Kreisel-Putnam system* ⋄ B25 B55 F50 ⋄
GRZEGORCZYK, A. *Decision procedures for theories categorical in $\aleph_0$* ⋄ B25 C35 ⋄
KANGER, S. *Equational calculi and automatic demonstration* ⋄ B25 B35 C05 ⋄
LIOGON'KIJ, M.I. *On the question of quantitative characteristics of logical formulae (Russian) (English summary)* ⋄ B20 B25 C13 ⋄
RABIN, M.O. *Weakly definable relations and special automata* ⋄ B15 B25 C40 C85 D05 ⋄
ROUTLEY, R. *Decision procedures and semantics for Feys' system S2^0 and surrounding systems* ⋄ B25 B45 ⋄
SCHUPP, P.E. *A note on recursively enumerable predicates in groups* ⋄ B25 D25 D40 ⋄
SHUKLA, A. *Decision procedures for Lewis system S1 and related modal systems* ⋄ B25 B45 ⋄
SIEFKES, D. *Decidable theories I. Buechi's monadic second order successor arithmetic* ⋄ B15 B25 C85 D05 F35 F98 ⋄
SIEFKES, D. *Decidable extensions of monadic second order successor arithmetic* ⋄ B15 B25 C85 D05 F35 ⋄
SIMMONS, H. *The solution of a decision problem for several classes of rings* ⋄ B25 C60 ⋄
TAJTSLIN, M.A. *On elementary theories of lattices of subgroups (Russian)* ⋄ B25 C60 D35 G10 ⋄

1971

BAKER, A. *Effective methods in the theory of numbers* ⋄ B25 B28 ⋄
BAKER, A. *Effective methods in Diophantine problems* ⋄ B25 ⋄
BONDI, I.L. *The solvability problem for certain formal-logical calculi (Russian)* ⋄ B25 D35 ⋄
CHAPIN JR., E.W. *The strong decidability of cut-logics I: Partial propositional calculi. II: Generalizations* ⋄ B20 B25 ⋄

DULAC, M.-H. *Decidabilite et operations entre theories* ⋄ B10 B25 D35 ⋄

GABBAY, D.M. *Decidability results in non-classical logic, III: Systems with stability operators* ⋄ B25 B45 C90 ⋄

GABBAY, D.M. *On decidable, finitely axiomatizable, modal and tense logics without the finite model property I, II* ⋄ B25 B45 C90 ⋄

GREENLEAF, N. *Fields in which varieties have rational points: a note on a problem of Ax* ⋄ B25 C60 ⋄

HAUSCHILD, K. *Entscheidbarkeit in der Theorie bewerteter Graphen (Russian and English summaries)* ⋄ B25 C65 ⋄

HAUSCHILD, K. & RAUTENBERG, W. *Interpretierbarkeit und Entscheidbarkeit in der Graphentheorie I* ⋄ B25 C65 D35 F25 ⋄

HERRE, H. *Entscheidungsprobleme in der elementaren Theorie einer zweistelligen Relation* ⋄ B25 C10 ⋄

LEWIS, D. *Completeness and decidability of three logics of counterfactual conditionals* ⋄ B25 B45 ⋄

MCKAY, C.G. *A class of decidable intermediate propositional logics* ⋄ B25 B55 ⋄

PACHOLSKI, L. *On countably compact reduced products III* ⋄ C10 C20 C50 G05 ⋄

PINUS, A.G. *A remark on the paper by Ju.M.Vazhenin: "Elementary properties of transformation semigroups of ordered sets" (Russian)* ⋄ B25 C65 D35 E07 ⋄

RABIN, M.O. *Decidability and definability in second-order theories* ⋄ B15 B25 C40 C85 D05 ⋄

REICHBACH, J. *A note about generalized models and the origin of forcing in the theory of models* ⋄ B25 C25 C90 ⋄

REICHBACH, J. *Generalized models and decidability - weak and strong statistical decidability of theorems* ⋄ B25 C25 C90 ⋄

SABBAGH, G. *Logique mathematique V. Decidabilite et fonctions recursives* ⋄ B25 D20 D35 ⋄

SHOENFIELD, J.R. *A theorem on quantifier elimination* ⋄ C10 C60 ⋄

SIEFKES, D. *Undecidable extensions of monadic second order successor arithmetic* ⋄ B15 B25 C85 D35 F35 ⋄

T"RKALANOV, K.D. *A solution of the inequality in a certain class of partially ordered semigroups by the method of Osipova (Russian and French summmaries)* ⋄ B25 D40 ⋄

T"RKALANOV, K.D. & ZHELEVA, S. *On the problem of inequality for partially ordered semigroups (Bulgarian) (Russian and German summaries)* ⋄ B25 D40 ⋄

TURAN, P. *On the work of Alan Baker* ⋄ B25 D35 ⋄

VILLE, F. *Decidabilite des formules existentielles en theorie des ensembles* ⋄ B25 E30 ⋄

1972

BURRIS, S. *Models in equational theories of unary algebras* ⋄ B25 C05 C13 C35 ⋄

COLLET, J.-C. *Multirelation homogene et elimination de quanteurs* ⋄ C10 C50 ⋄

COOPER, D.C. *Theorem proving in arithmetic without multiplication* ⋄ B25 B35 F30 ⋄

EKLOF, P.C. & FISHER, E.R. *The elementary theory of abelian groups* ⋄ B25 C10 C35 C50 C60 ⋄

ELLENTUCK, E. & MANASTER, A.B. *The decidability of a class of ∀∃ sentences in the isols* ⋄ B25 D50 ⋄

ERSHOV, YU.L. *Elementary group theories (Russian)* ⋄ B25 C05 C13 C60 D35 ⋄

FRIDMAN, EH.I. & PENZIN, YU.G. *The elementary and the universal theory of the ordered group of integers with maximal subgroups (Russian)* ⋄ B20 B25 D35 ⋄

FRIEDRICH, U. *Entscheidbarkeit der monadischen Nachfolgerarithmetik mit endlichen Automaten ohne Analysetheorem (Russian, English and French summaries)* ⋄ B25 C85 D05 F35 ⋄

GABBAY, D.M. *Decidability of some intuitionistic predicate theories* ⋄ B25 C90 F50 ⋄

GRABOWSKI, M. *The set of all tautologies of the zero-order algorithmic logic is decidable (Russian summary)* ⋄ B25 B70 ⋄

HAUSCHILD, K. & HERRE, H. & RAUTENBERG, W. *Entscheidbarkeit der elementaren Theorien der endlichen Baeume und verwandter Klassen endlicher Strukturen (Russian, English and French summaries)* ⋄ B25 C13 ⋄

HAUSCHILD, K. & HERRE, H. & RAUTENBERG, W. *Entscheidbarkeit der monadischen Theorie 2. Stufe der n-separierten Graphen (Russian, English and French summaries)* ⋄ B15 B25 ⋄

HAUSCHILD, K. & HERRE, H. & RAUTENBERG, W. *Interpretierbarkeit und Entscheidbarkeit in der Graphentheorie II* ⋄ B25 C65 D35 F25 ⋄

HENSON, C.W. *Countable homogeneous relational structures and $\aleph_0$-categorical theories* ⋄ C10 C15 C35 C50 D35 ⋄

HERRE, H. *Die Entscheidbarkeit der elementaren Theorie der n-separierten symmetrischen Graphen endlicher Valenz* ⋄ B25 C65 ⋄

MART'YANOV, V.I. *Decidability of the theories of some classes of abelian groups with an automorphism predicate and a group predicate (Russian)* ⋄ B25 C60 ⋄

MASLOV, S.YU. & OREVKOV, V.P. *Decidable classes that reduce to a single quantifier class (Russian)* ⋄ B20 B25 ⋄

MCKENZIE, R. *Equational bases and nonmodular lattice varieties* ⋄ B25 C05 G10 ⋄

MUELLER, HORST *Entscheidungsprobleme im Bereich der Semantik von Programmiersprachen* ⋄ B25 B75 ⋄

PENZIN, YU.G. *Decidability of the theory of algebraic integers with addition and congruences according to divisors (Russian)* ⋄ B25 C60 ⋄

PINUS, A.G. *On the theory of convex subsets (Russian)* ⋄ B25 C65 C85 D35 ⋄

RABIN, M.O. *Automata on infinite objects and Church's problem* ⋄ B15 B25 C85 D05 ⋄

REICHBACH, J. *Generalized models, probability of formulas, decisions and statistical decidability of theorems* ⋄ B25 C25 C90 ⋄

REICHBACH, J. *Notes about decidablility of mathematics with probability 1 and generalized models* ⋄ B25 C90 ⋄

SLOMSON, A. *Generalized quantifiers and well orderings*
⋄ B25 C55 C65 C80 E07 ⋄

SLUPECKI, J. *L-decidability and decidability*
⋄ B22 B25 ⋄

STILLWELL, J.C. *Decidability of the "almost all" theory of degrees* ⋄ B25 C80 D30 ⋄

VAKARELOV, D. *Extensional logics (Russian)*
⋄ B05 B25 ⋄

VINNER, S. *A generalization of Ehrenfeucht's game and some applications* ⋄ B25 C07 C55 C80 E60 ⋄

WEESE, M. *Zur Modellvollstaendigkeit und Entscheidbarkeit gewisser topologischer Raeume (Russian, English and French summaries)*
⋄ B25 C35 C65 D80 ⋄

ZIEGLER, M. *Die elementare Theorie der Henselschen Koerper* ⋄ B25 C10 C35 C60 ⋄

ZINOV'EV, A.A. *Semantische Entscheidbarkeit der Quantorentheorie* ⋄ A05 B25 ⋄

1973

AANDERAA, S.O. & LEWIS, H.R. *Prefix classes of Krom formulas* ⋄ B20 B25 D35 ⋄

BALDWIN, JOHN T. & LACHLAN, A.H. *On universal Horn classes categorical in some infinite power*
⋄ C05 C10 C35 C45 ⋄

BENOIS, M. *Application de l'etude de certaines congruences a un probleme de decidabilite* ⋄ B25 ⋄

BOERGER, E. *Eine entscheidbare Klasse von Kromformeln*
⋄ B25 ⋄

BONDI, I.L. *The decision problem for normed linear spaces and for Hilbert spaces (Russian)* ⋄ B25 C65 D35 ⋄

BRYLL, G. & SLUPECKI, J. *Syntactic proof of the L-decidability of classical logic and Lukasiewicz's three-valued logic (Polish) (English summary)*
⋄ B25 B50 ⋄

BUECHI, J.R. & SIEFKES, D. *Axiomatization of the monadic second order theory of ω_1* ⋄ B15 B25 C85 E10 ⋄

BUECHI, J.R. & SIEFKES, D. *The monadic second order theory of all countable ordinals*
⋄ B15 B25 C85 E10 ⋄

BUECHI, J.R. *The monadic second order theory of ω_1*
⋄ B15 B25 C85 E10 ⋄

CUSIN, R. *Sur les structures generiques en theorie des modeles* ⋄ C10 C25 C35 C50 ⋄

DORACZYNSKI, R. *Elimination of bound variables in logic with an arbitrary quantifier (Polish and Russian summaries)* ⋄ C10 C80 ⋄

FRIDMAN, EH.I. *Positive element-subgroup theories (Russian)* ⋄ B25 D35 ⋄

FRIDMAN, EH.I. & SLOBODSKOJ, A.M. *The theory of the additive group of the integers with an arbitrary number of predicates that define maximal subgroups (Russian)*
⋄ C10 F30 ⋄

GABBAY, D.M. *A survey of decidability results for modal, tense and intermediate logics*
⋄ B25 B45 B55 C98 F50 ⋄

GABBAY, D.M. *The undecidability of intuitionistic theories of algebraically closed fields and real closed fields*
⋄ B25 C60 C90 D35 F50 ⋄

GENSLER, H.J. *A simplified decision procedure for categorical syllogisms* ⋄ B20 B25 ⋄

GOLDBLATT, R.I. *A model-theoretic study of some systems containing S3* ⋄ B25 B45 C90 ⋄

GUREVICH, Y. *Formulas with a single $\forall$ (Russian)*
⋄ B20 B25 C13 ⋄

HAKEN, W. *Connections between topological and group theoretical decision problems*
⋄ B25 C60 C65 D35 D40 D80 ⋄

HAMBLIN, C.L. *A felicitous fragment of the predicate calculus* ⋄ B20 B25 ⋄

HAUSCHILD, K. & RAUTENBERG, W. *Entscheidungsprobleme der Theorie zweier Aequivalenzrelationen mit beschraenkter Zahl von Elementen in den Klassen* ⋄ B25 ⋄

IWANUS, B. *On Lesniewski's elementary ontology (Polish and Russian summaries)* ⋄ B25 B60 E70 ⋄

IWANUS, B. *Proof of decidability of the traditional calculus of names (Polish and Russian summaries)* ⋄ B25 ⋄

JOYNER JR., W.M. *Automatic theorem-proving and the decision problem* ⋄ B25 B35 ⋄

KOKORIN, A.I. & MART'YANOV, V.I. *Universal extended theories (Russian)* ⋄ B25 C52 D35 ⋄

LIPSHITZ, L. & SARACINO, D.H. *The model companion of the theory of commutative rings without nilpotent elements* ⋄ B25 C25 C35 C60 ⋄

LIVCHAK, A.B. *A resolving procedure for the elementary theory of a torsion-free abelian group with a distinguished sub-group (Russian)*
⋄ B25 C10 C60 ⋄

MART'YANOV, V.I. *Decidability of the universal element-subgroup theory of the integers (Russian)*
⋄ B25 ⋄

MOSTERIN, J. *El problema de la decision en la logica de predicados* ⋄ B25 ⋄

OPPEN, D.C. *Elementary bounds for Presburger arithmetic* ⋄ B25 C10 D15 F30 ⋄

PAOLA DI, R.A. *The solvability of the decision problem for classes of proper formulas and related results*
⋄ B20 B25 ⋄

PARIKH, R. *Some results on the length of proofs*
⋄ B25 F20 F30 ⋄

PENZIN, YU.G. *Decidability of certain theories of integers (Russian)* ⋄ B25 F30 ⋄

PENZIN, YU.G. *Decidability of the theory of integers with addition, order and multiplication by an arbitrary number (Russian)* ⋄ B25 F30 ⋄

PENZIN, YU.G. *Decidability of a theory of the integers with addition, order and predicates that distinguish a chain of subgroups (Russian)* ⋄ B25 F30 ⋄

ROBINSON, JULIA *Solving diophantine equations*
⋄ B25 D35 D80 H15 ⋄

SMORYNSKI, C.A. *Elementary intuitionistic theories*
⋄ B25 C90 D35 F50 ⋄

STIHI, T. *A case of decidability in the first order predicate calculus (Romanian)* ⋄ B20 B25 ⋄

VASIL'EV, EH.S. *Elementary theories of two-base models of abelian groups (Russian)* ⋄ B25 C60 C85 ⋄

VASIL'EV, EH.S. *The elementary theories of complete torsion-free abelian groups with p-adic topology (Russian)* ⋄ B25 C20 C60 D35 ⋄

VIDAL-NAQUET, G. *Quelques applications des automates a arbres infinis* ⋄ B25 D05 D35 ⋄

WEAVER, G.E. *A note on compactness and decidability* ⋄ B25 C95 ⋄

WEISPFENNING, V. *Infinitary model-theoretic properties of κ-saturated structures*
⋄ C10 C25 C35 C50 C60 C65 C75 ⋄

WOLTER, H. *Eine Erweiterung der elementaren Praedikatenlogik: Anwendungen in der Arithmetik und anderen mathematischen Theorien*
⋄ B28 C10 C80 F30 ⋄

ZAMYATIN, A.P. *A prevariety of semigroups whose elementary theory is solvable (Russian)* ⋄ B25 C05 ⋄

1974

AANDERAA, S.O. & GOLDFARB, W.D. *The finite controllability of the Maslov case* ⋄ B25 ⋄

ANDREWS, P.B. *Provability in elementary type theory*
⋄ B15 B25 D35 F35 ⋄

ASH, C.J. *Reduced powers and boolean extensions*
⋄ B25 C20 C30 G20 ⋄

BOERGER, E. *Beitrag zur Reduktion des Entscheidungsproblems auf Klassen von Hornformeln mit kurzen Alternationen* ⋄ B20 B25 D35 ⋄

COLLET, J.-C. *Multirelation (k,p)-homogene et elimination de quanteurs* ⋄ C10 C40 C50 ⋄

COMER, S.D. *Elementary properties of structures of sections* ⋄ B25 C90 ⋄

COSTA DA, N.C.A. *Remarques sur les calculs $\mathscr{C}_n$, $\mathscr{C}_n^*$, $\mathscr{C}_n^=$ et $\mathscr{D}_n$* ⋄ B25 B53 ⋄

ERSHOV, YU.L. *Theory of numerations, III: Constructive models (Russian)* ⋄ B25 C57 C98 D45 F50 ⋄

FISCHER, MICHAEL J. & RABIN, M.O. *Super-exponential complexity of Presburger arithmetic*
⋄ B25 D15 F20 F30 ⋄

GABBAY, D.M. & JONGH DE, D.H.J. *A sequence of decidable finitely axiomatizable intermediate logics with the disjunction property* ⋄ B22 B25 B55 ⋄

GALDA, K. & PASSOS, E.P. *Application of quantifier elimination to mechanical theorem proving*
⋄ B35 C10 ⋄

GOLDBLATT, R.I. *Decidability of some extensions of J*
⋄ B25 B55 ⋄

GOLOTA, YA.YA. *A matrix notation for nets of marks (Russian) (English summary)* ⋄ B25 F50 ⋄

GUREVICH, Y. *A resolving procedure for the extended theory of ordered abelian groups (Russian)*
⋄ B25 C10 C60 C85 ⋄

LONGO, G. *I problemi di decisione e la loro complessita*
⋄ B25 B35 D15 F20 ⋄

LUCAS, T. *Universal classes of monadic algebras*
⋄ B25 C05 G15 ⋄

MARCJA, A. & TULIPANI, S. *Questioni di teoria dei modelli per linguaggi universali positivi (English summary)*
⋄ B20 C07 C10 ⋄

MARKOV, A.A. *On the language A_0 (Russian)*
⋄ B25 F50 F65 ⋄

MEHLHORN, K. *The "almost all" theory of subrecursive degrees is decidable* ⋄ B25 C80 D20 ⋄

MIJAJLOVIC, Z. *On decidability of one class of boolean formulas* ⋄ B05 B25 ⋄

PALYUTIN, E.A. *The algebras of formulae of countably categorical theories (Russian)* ⋄ B25 C35 ⋄

PIGOZZI, D. *The join of equational theories*
⋄ B25 C05 D35 ⋄

RACKOFF, C.W. *On the complexity of the theories of weak direct products: a preliminary report*
⋄ B25 C13 C30 C60 D15 ⋄

REICHBACH, J. *About decidability, generalized models and main theorems of many valued predicate calculi*
⋄ B25 B50 C90 ⋄

ROUTLEY, R. *Semantical analyses of propositional systems of Fitch and Nelson* ⋄ B25 B46 C90 F50 ⋄

SMIRNOVA, O.S. *Completeness and decidability of the axiomatic theory SG for spherical geometry (Russian)*
⋄ B25 C35 C65 ⋄

VAZHENIN, YU.M. *On the elementary theory of free inverse semigroups* ⋄ B25 C60 D35 ⋄

WEAVER, G.E. *Finite partitions and their generators*
⋄ B20 B25 C07 C40 ⋄

WOOD, C. *Prime model extensions for differential fields of characteristic $p \neq 0$* ⋄ C10 C35 C60 ⋄

1975

ALTON, D.A. *Embedding relations in the lattice of recursively enumerable sets* ⋄ B25 D25 ⋄

BASARAB, S.A. *The models of the elementary theory of finite abelian groups (Romanian) (English summary)*
⋄ B25 C13 C60 ⋄

BAUDISCH, A. *Die elementare Theorie der Gruppe vom Typ p^∞ mit Untergruppen* ⋄ B25 C60 C85 ⋄

BAUDISCH, A. *Entscheidbarkeitsprobleme elementarer Theorien von Klassen abelscher Gruppen mit Untergruppen (English and Russian summaries)*
⋄ B25 C60 ⋄

BAUR, W. *Decidability and undecidability of theories of abelian groups with predicates for subgroups*
⋄ B25 C60 D35 ⋄

BERNARDI, C. *On the equational class of diagonalizable algebras (the algebraization of the theories which express Theor. VI)* ⋄ B25 F30 G05 G25 ⋄

BURRIS, S. *Boolean powers* ⋄ B25 C05 C30 C90 ⋄

BURRIS, S. & SANKAPPANAVAR, H.P. *Lattice-theoretic decision problems in universal algebra*
⋄ B25 C05 D35 ⋄

COLLINS, G.E. *Quantifier elimination for real closed fields by cylindrical algebraic decomposition*
⋄ C10 C60 G15 ⋄

COMER, S.D. *Monadic algebras with finite degree*
⋄ B25 C05 C90 G15 ⋄

EHRENFEUCHT, A. *Practical decidability* ⋄ B25 D15 ⋄

ENGELER, E. *Algorithmic logic* ⋄ B25 B75 D15 ⋄

FERRANTE, J. & RACKOFF, C.W. *A decision procedure for the first order theory of real addition with order*
⋄ B25 C10 C60 D15 F35 ⋄

FRANEK, M. *On some relations in boolean algebras*
⋄ B25 G05 ⋄

GABBAY, D.M. *Decidability results in non-classical logics I*
⋄ B25 B45 C90 F50 ⋄

GABBAY, D.M. *The decision problem for some finite extensions of the intuitionistic theory of abelian groups* ◊ B25 F50 ◊

GAERDENFORS, P. & HANSSON, B. *Filtrations and the finite frame property in boolean semantics* ◊ B25 B45 C90 G05 G25 ◊

GOLD, A.J. *Sur les arbres logiques de M. Krasner* ◊ B25 E70 ◊

HAUSCHILD, K. *Entscheidbarkeit in der Theorie baumartiger Graphen (Russian, English and French summaries)* ◊ B25 ◊

HAUSCHILD, K. *Ueber zwei Spiele und ihre Anwendung in der Modelltheorie (Russian, English and French summaries)* ◊ B25 C07 C30 C35 C85 ◊

HAUSCHILD, K. *Zur Uebertragbarkeit von Entscheidbarkeitsresultaten elementarer Theorien (Russian, English and French summaries)* ◊ B15 B25 C65 C85 ◊

HENRARD, P. *Weak elimination of quantifiers and cotheories* ◊ C10 C52 ◊

HERRE, H. & WOLTER, H. *Entscheidbarkeit von Theorien in Logiken mit verallgemeinerten Quantoren* ◊ B25 C10 C55 C80 D35 ◊

JARDEN, M. & KIEHNE, U. *The elementary theory of algebraic fields of finite corank* ◊ B25 C10 C20 C60 ◊

JENSEN, F.V. *On completeness in cardinality logics* ◊ C10 C35 C55 C60 C80 ◊

JONSSON, B. & MCNULTY, G.F. & QUACKENBUSH, R.W. *The ascending and descending varietal chains of a variety* ◊ B25 C05 D35 ◊

KOCHEN, S. *The model theory of local fields* ◊ B25 C20 C35 C60 ◊

KOPIEKI, R. & SARALSKI, B. & WALIGORA, G. *The research of Jaskowski on decidability theory of first order sentences I* ◊ A10 B20 B25 D35 ◊

KRAUSS, P.H. *Quantifier elimination* ◊ C10 C35 C50 ◊

KUZNETSOV, A.V. *On superintuitionistic logics* ◊ B25 B55 ◊

LACAVA, F. *Teoria dei campi differenziali ordinati* ◊ C10 C35 C60 ◊

MACINTYRE, A. *Dense embeddings. I. A theorem of Robinson in a general setting* ◊ B25 C10 C35 C60 D40 ◊

MARINI, D. & MIGLIOLI, P.A. & ORNAGHI, M. *First order logic as a tool to solve and classify problems* ◊ B10 B25 F50 ◊

MAZOYER, J. *Sur les theories categoriques finiment axiomatisables (English summary)* ◊ C10 C35 C45 ◊

MEYER, A.R. *The inherent computational complexity of theories of ordered sets* ◊ B25 D15 ◊

MEYER, A.R. *Weak monadic second order theory of successor is not elementary recursive* ◊ B25 D10 D15 D20 ◊

MONK, L.G. *Elementary-recursive decision procedures* ◊ B25 C60 D20 ◊

MOSTOWSKI, T. *A note on a decision procedure for rings of formal power series and its applications* ◊ B25 C60 ◊

NUTE, D.E. *Counterfactuals* ◊ A05 B25 ◊

REICHBACH, J. *Generalized models for classical and intuitionistic predicate calculi* ◊ B10 B25 C90 F50 ◊

REICHBACH, J. *Infinite proof methods in decidability problems with Ulam's models and classification of predicate calculi* ◊ B25 C25 C90 ◊

SAELI, D. *Problemi di decisione per algebre connesse a logiche a piu valori (English summary)* ◊ B25 B50 C60 G20 ◊

SARACINO, D.H. & WEISPFENNING, V. *On algebraic curves over commutative regular rings* ◊ C10 C25 C35 C60 ◊

SEESE, D.G. *Zur Entscheidbarkeit der monadischen Theorie 2. Stufe baumartiger Graphen (Russian, English and French summaries)* ◊ B15 B25 ◊

SHELAH, S. *The monadic theory of order* ◊ B25 C65 C85 D35 E50 ◊

SHOENFIELD, J.R. *The decision problem for recursively enumerable degrees* ◊ B25 D25 ◊

TENNEY, R.L. *Second-order Ehrenfeucht games and the decidability of the second-order theory of an equivalence relation* ◊ B15 B25 C85 ◊

THOMASON, R.H. *Decidability in the logic of conditionals* ◊ B25 B45 ◊

WEISPFENNING, V. *Model-completeness and elimination of quantifiers for subdirect products of structures* ◊ C10 C30 C35 C52 C60 C90 ◊

WOLTER, H. *Entscheidbarkeit der Arithmetik mit Addition und Ordnung in Logiken mit verallgemeinerten Quantoren* ◊ B25 C10 C55 C80 F30 ◊

1976

ANSHEL, M. & STEBE, P. *Conjugate powers in free products with amalgamation* ◊ B25 D40 ◊

AUSIELLO, G. *Difficult logical theories and their computer approximations* ◊ B25 B35 D15 ◊

BAUDISCH, A. *Elimination of the quantifier Q_α in the theory of abelian groups (Russian summary)* ◊ B25 C10 C60 C80 ◊

BAUR, W. *Elimination of quantifiers for modules* ◊ C10 C60 ◊

BEL'TYUKOV, A.P. *Decidability of the universal theory of natural numbers with addition and divisibility (Russian)* ◊ B25 B28 F30 ◊

BOERGER, E. *Ueber einige Interpretationen von Registermaschinen mit Anwendungen auf Entscheidungsprobleme in der Logik, der Algorithmentheorie und der Theorie formaler Sprachen* ◊ B25 D10 ◊

BOOLOS, G. *On deciding the truth of certain statements involving the notion of consistency* ◊ B25 B45 F30 ◊

CADEVALL SOLER, M. *Decidability of the logic of monadic predicates by the method of rejection (Spanish)* ◊ B25 ◊

COMER, S.D. *Complete and model-complete theories of monadic algebras* ◊ B25 C05 C25 C35 C60 G15 ◊

FLEISCHER, I. *Quantifier elimination for modules and ordered groups* ◊ B25 C10 C60 ◊

FRIED, M. & SACERDOTE, G.S. *Solving diophantine problems over all residue class fields of a number field and all finite fields* ⋄ B25 C10 C13 C60 D35 ⋄

FRIEDMAN, H.M. *On decidability of equational theories* ⋄ B25 C05 D35 ⋄

GABBAY, D.M. *Investigations in modal and tense logics with applications to problems in philosophy and linguistics* ⋄ A05 B25 B46 C90 ⋄

GONCHAROV, S.S. *Restricted theories of constructive boolean algebras (Russian)* ⋄ B25 C57 D35 F60 G05 ⋄

GUREVICH, Y. *The decision problem for standard classes* ⋄ B20 B25 D35 ⋄

HARROP, R. *Some results concerning finite model separability of propositional calculi* ⋄ B22 B25 ⋄

HARTMANIS, J. *On effective speed-up and long proofs of trivial theorems in formal theories (French summary)* ⋄ B25 B35 D15 D20 F20 ⋄

HERRMANN, E. *On Lindenbaum functions of $\aleph_0$-categorical theories of finite similarity type* ⋄ B25 C35 C57 ⋄

JOYNER JR., W.M. *Resolution strategies as decision procedures* ⋄ B25 B35 ⋄

KASSLER, M. *The decidability of languages that assert music* ⋄ B25 B80 ⋄

KOREC, I. & RAUTENBERG, W. *Model-interpretability into trees and applications* ⋄ B25 C45 C60 C65 F25 ⋄

LITMAN, A. *On the monadic theory of ω_1 without AC* ⋄ B25 C85 D05 E10 E25 ⋄

LIVCHAK, A.B. *Elimination of identity in the theory of abelian groups with predicates for subgroups (Russian)* ⋄ B25 C60 ⋄

LUCAS, T. *Universal classes of monadic algebras* ⋄ B25 C05 G15 ⋄

MACINTYRE, A. *On definable subsets of p-adic fields* ⋄ C10 C60 ⋄

MAKOWSKY, J.A. & MARCJA, A. *Problemi di decidibilita in logica topologica (English summary)* ⋄ B25 C65 C80 ⋄

MART'YANOV, V.I. *Decidability of the theory of closed torsion free abelian groups of finite rank with predicates for addition and automorphism (Russian)* ⋄ B25 C60 ⋄

MONTGOMERY, H. & ROUTLEY, R. *Algebraic semantics for $S2^0$ and necessitated extensions* ⋄ B25 B45 ⋄

MOSTOWSKI, T. *Analytic applications of decidability theorems* ⋄ B25 C60 C65 D80 ⋄

NORGELA, S.A. *On approximating reduction classes of CPC by decidable classes (Russian) (English summary)* ⋄ B20 B25 D35 ⋄

NORGELA, S.A. *On frequency decidability of some classes of predicate calculus (Russian)* ⋄ B25 ⋄

OREVKOV, V.P. *Solvable classes of pseudoprenex formulas (Russian) (English summary)* ⋄ B25 F50 ⋄

PIGOZZI, D. *The universality of the variety of quasigroups* ⋄ B25 C05 C52 D35 ⋄

PINUS, A.G. *Theories of boolean algebras in a calculus with the quantifier "infinitely many exist" (Russian)* ⋄ B25 C10 C35 C57 C80 G05 ⋄

RACKOFF, C.W. *On the complexity of the theories of weak direct powers* ⋄ B25 C13 C30 C60 D15 ⋄

REICHBACH, J. *Graphs, generalized models, games and decidability* ⋄ B25 C65 C90 ⋄

SCHMERL, J.H. *The decidability of some $\aleph_0$-categorical theories* ⋄ B25 C35 ⋄

SEGERBERG, K. *Discrete linear future time without axioms* ⋄ B25 B45 C90 ⋄

SLOMSON, A. *Decision problems for generalized quantifiers - a survey* ⋄ B25 C80 C98 ⋄

VASIL'EV, EH.S. *On elementary theories of abelian p-groups with heights of elements (Russian)* ⋄ B25 C60 ⋄

WASILEWSKA, A. *A sequence formalization for SCI* ⋄ B25 F07 ⋄

WASILEWSKA, A. *On the decidability theorems* ⋄ B25 F07 ⋄

WASSERMAN, H.C. *An analysis of the counterfactual conditional* ⋄ B25 B45 ⋄

WEESE, M. *Entscheidbarkeit in speziellen uniformen Strukturen bezueglich Sprachen mit Maechtigkeitsquantoren* ⋄ B25 C55 C65 C80 ⋄

WEISPFENNING, V. *On the elementary theory of Hensel fields* ⋄ B25 C10 C35 C60 ⋄

WUETHRICH, H.R. *Ein Entscheidungsverfahren fuer die Theorie der reell abgeschlossenen Koerper* ⋄ B25 C60 D15 ⋄

ZAMYATIN, A.P. *Varieties of associative rings whose elementary theory is decidable (Russian)* ⋄ B25 C05 C60 ⋄

ZIEGLER, M. *A language for topological structures which satisfies a Lindstroem-theorem* ⋄ B25 C40 C50 C65 C80 C90 C95 D35 ⋄

1977

BASARAB, S.A. *On the elementary theories of abelian profinite groups and Abelian torsion groups* ⋄ B25 C60 ⋄

BAUDISCH, A. *A note on the elementary theory of torsion free Abelian groups with one predicate for subgroups* ⋄ B25 C60 ⋄

BAUDISCH, A. *Decidability of the theory of abelian groups with Ramsey quantifiers* ⋄ B25 C10 C60 C80 ⋄

BAUDISCH, A. *The theory of abelian groups with the quantifier $(\leq x)$* ⋄ B25 C10 C60 C80 ⋄

BERMAN, L. *Precise bounds for Presburger arithmetic and the reals with addition: preliminary report* ⋄ B25 D15 F30 ⋄

BOOLOS, G. *On deciding the provability of certain fixed point statements* ⋄ B25 B45 F30 ⋄

BUECHI, J.R. *Using determinancy of games to eliminate quantifiers* ⋄ B15 C10 C85 E60 ⋄

CLARK, D.M. & KRAUSS, P.H. *Relatively homogeneous structures* ⋄ C10 C35 C50 C60 C65 G05 ⋄

DAVIDSON, B. & JACKSON, F.C. & PARGETTER, R. *Modal trees for T and S5* ⋄ B25 B45 ⋄

DRIES VAN DEN, L. *Artin-Schreier theory for commutative regular rings* ⋄ B25 C10 C25 C60 ⋄

DROBOTUN, B.N. *On countable models of decidable almost categorical theories (Russian)* ⋄ B25 C15 C35 C57 ⋄

FERENCZI, M. *On valid assertions - in probability logic*
⋄ B25 B48 C90 ⋄

FERRANTE, J. & GEISER, J.R. *An efficient decision procedure for the theory of rational order*
⋄ B25 D15 ⋄

FIDEL, M.M. *The decidability of the calculi $\mathscr{C}_n$*
⋄ B25 B53 G25 ⋄

FLUM, J. *Distributive normal forms* ⋄ B25 C07 ⋄

GNANVO, C. *Γ-cotheories et elimination des quantificateurs (English summary)*
⋄ C10 C25 C35 C52 ⋄

GUREVICH, Y. *Expanded theory of ordered abelian groups*
⋄ B25 C10 C60 C85 ⋄

HAJEK, P. *Generalized quantifiers and finite sets*
⋄ B25 B50 C13 C40 C80 ⋄

HAUCK, J. *Die Entscheidbarkeit der n-Beweisbarkeit aus Axiomenschemata (English, Russian and French summaries)* ⋄ B25 ⋄

HAVRANEK, T. *Towards a model theory of statistical theories* ⋄ B25 C90 ⋄

HENSON, C.W. & JOCKUSCH JR., C.G. & RUBEL, L.A. & TAKEUTI, G. *First-order topology*
⋄ B25 C65 C75 D35 H05 ⋄

HERRE, H. *Decidability of theories in logics with additional monadic quantifiers* ⋄ B25 C55 C80 ⋄

HERRE, H. & WOLTER, H. *Entscheidbarkeit der Theorie der linearen Ordnung in L_{Q_1}*
⋄ B25 C55 C65 C80 ⋄

HERRMANN, E. *Ueber Lindenbaumfunktionen von $\aleph_0$-kategorischen Theorien endlicher Signatur*
⋄ B25 C35 C57 ⋄

KLUPP, H. & SCHNORR, C.-P. *A universally hard set of formulae with respect to non-deterministic Turing acceptors* ⋄ B25 D10 D15 ⋄

KRYNICKI, M. *Henkin quantifier and decidability*
⋄ B25 C10 C80 D35 ⋄

LADNER, R.E. *Application of model theoretic games to discrete linear orders and finite automata*
⋄ B15 B25 C07 C13 C65 C85 D05 E07 E60 ⋄

LEWIS, H.R. *A measure of complexity for combinatorial decision problems of the tiling variety*
⋄ B25 D10 D15 ⋄

LOPARIC, A. *Une etude semantique de quelques calculs propositionnels (English summary)* ⋄ B05 B25 ⋄

MAKANIN, G.S. *The problem of the solvability of equations in a free semigroup (Russian)* ⋄ B03 B25 D40 ⋄

MAKANIN, G.S. *The problem of solvability of equations in a free semigroup (Russian)* ⋄ B03 B25 D40 ⋄

MART'YANOV, V.I. *Extended universal theories of the integers (Russian)* ⋄ B25 D35 F30 ⋄

MASLOV, S.YU. & NORGELA, S.A. *Herbrand strategies and the "greater deducibility" relation (Russian) (English summary)* ⋄ B25 B35 F05 ⋄

MIZUTANI, C. *Monadic second order logic with an added quantifier Q* ⋄ B15 B25 C40 C55 C80 ⋄

RABIN, M.O. *Decidable theories* ⋄ B25 C98 D20 ⋄

SCHIEMANN, I. *Eine Axiomatisierung des monadischen Praedikatenkalkuels mit verallgemeinerten Quantoren (Russian, English and French summaries)*
⋄ B25 C40 C55 C80 ⋄

SCHMERL, J.H. *On $\aleph_0$-categoricity and the theory of trees*
⋄ B25 C35 C65 ⋄

SEESE, D.G. & TUSCHIK, H.-P. *Construction of nice trees*
⋄ B25 C55 C65 C80 ⋄

SEESE, D.G. *Decidability of ω-trees with bounded sets - a survey* ⋄ B25 C65 C80 C85 ⋄

SEESE, D.G. *Second order logic, generalized quantifiers and decidability*
⋄ B15 B25 C55 C65 C80 C85 D35 ⋄

SHELAH, S. *Decidability of a portion of the predicate calculus* ⋄ B20 B25 ⋄

SHOENFIELD, J.R. *Quantifier elimination in fields*
⋄ C10 C60 C98 ⋄

SHOSTAK, R.E. *On the SUP-INF method for proving Presburger formulas* ⋄ B25 B35 F30 ⋄

TUSCHIK, H.-P. *Elimination verallgemeinerter Quantoren in ω_1-kategorischen Theorien (Russian, English and French summaries)* ⋄ B25 C10 C35 C55 C80 ⋄

TUSCHIK, H.-P. *On the decidability of the theory of linear orderings in the language $L(Q_1)$*
⋄ B25 C55 C65 C80 ⋄

VASIL'EV, EH.S. *Elementary theories of certain classes of two-base models of abelian groups (Russian)*
⋄ B25 C60 C85 ⋄

WEESE, M. *Ein neuer Beweis fuer die Entscheidbarkeit der Theorie der booleschen Algebren (Russian, English and French summaries)* ⋄ B25 G05 ⋄

WEESE, M. *Entscheidbarkeit der Theorie der booleschen Algebren in Sprachen mit Maechtigkeitsquantoren (Russian and English summaries)*
⋄ B25 C55 C80 G05 ⋄

WEESE, M. *The decidability of the theory of boolean algebras with cardinality quantifiers (Russian summary)* ⋄ B25 C10 C55 C80 G05 ⋄

WEESE, M. *The decidability of the theory of boolean algebras with the quantifier "there exist infinitely many"*
⋄ B25 C80 ⋄

WEISPFENNING, V. *Nullstellensaetze - a model theoretic framework* ⋄ C10 C25 C60 ⋄

WERNER, H. *Varieties generated by quasi-primal algebras have decidable theories* ⋄ B25 C05 ⋄

WIRSING, M. *Das Entscheidungsproblem der Klasse von Formeln, die hoechstens zwei Primformeln enthalten*
⋄ B20 B25 D35 ⋄

WOLTER, H. *Entscheidbarkeit der Theorie der Wohlordnung mit Elimination der Quantoren (Russian, English and French summaries)*
⋄ B25 C10 C65 C80 E07 ⋄

1978

ALVES, E.H. *On decidability of a system of dialectical propositional logic* ⋄ B25 B53 ⋄

ANSHEL, M. *Decision problems for HNN groups and commutative semigroups* ⋄ B25 D40 ⋄

BROWN, S.S. *Bounds on transfer principles for algebraically closed and complete discretely valued fields* ⋄ B25 C10 C60 ⋄

BRUSS, A.R. & MEYER, A.R. *On time-space classes and their relation to the theory of real addition*
⋄ B25 D15 ⋄

DEGTEV, A.N. *Solvability of the ∀∃-theory of a certain factor-lattice of recursively enumerable sets (Russian)*
⋄ B25 D25 D50 ⋄

DONER, J.E. & MOSTOWSKI, ANDRZEJ & TARSKI, A. *The elementary theory of well-ordering - a metamathematical study* ⋄ B25 C10 C65 E07 ⋄

DRIES VAN DEN, L. *Model theory of fields*
⋄ B25 C10 C25 C35 C60 H20 ⋄

FERRO, A. & OMODEO, E.G. *An efficient validity test for formulae in extensional two-level syllogistic*
⋄ B20 B25 B35 ⋄

GOAD, C.A. *Monadic infinitary propositional logic: a special operator* ⋄ B25 B60 C75 F50 ⋄

GOGOL, D. *The $\forall_n \exists$-completeness of Zermelo-Fraenkel set theory* ⋄ B25 C35 E30 ⋄

GURARI, E.M. & IBARRA, O.H. *An NP-complete number-theoretic problem* ⋄ B25 D15 D80 ⋄

HERRE, H. & WOLTER, H. *Entscheidbarkeit der Theorie der linearen Ordnung in L_{Q_κ} fuer regulaeres ω_κ*
⋄ B25 C55 C65 C80 E07 E50 ⋄

HERRE, H. & PINUS, A.G. *Zum Entscheidungsproblem fuer Theorien in Logiken mit monadischen verallgemeinerten Quantoren*
⋄ B25 C10 C55 C80 D35 ⋄

HERRMANN, E. *Der Verband der rekursiv aufzaehlbaren Mengen (Entscheidungs-Problem) (Russian and English summaries)* ⋄ B25 D25 D98 ⋄

KOKORIN, A.I. & PINUS, A.G. *Decidability problems of extended theories (Russian)*
⋄ B25 C85 C98 D35 ⋄

KRAMOSIL, I. & SINDELAR, J. *Statistical deducibility testing with stochastic parameters* ⋄ B25 B48 ⋄

KRASNER, M. *Abstract Galois theory*
⋄ C07 C10 C60 C75 ⋄

KRON, A. *Decision procedures for two positive relevance logics* ⋄ B25 B46 ⋄

KURKE, H. & MOSTOWSKI, T. & PFISTER, G. & POPESCU, D. & ROCZEN, M. *Die Approximationseigenschaft lokaler Ringe*
⋄ C10 C60 ⋄

LERMAN, M. *Lattices of α-recursively enumerable sets*
⋄ B25 D60 ⋄

LERMAN, M. *On elementary theories of some lattices of α-recursively enumerable sets* ⋄ B25 D60 E45 ⋄

LEWIS, H.R. *Complexity of solvable cases of the decision problem for the predicate calculus* ⋄ B20 B25 D15 ⋄

LIPSHITZ, L. *The diophantine problem for addition and divisibility* ⋄ B25 D35 ⋄

LOVELAND, D.W. & REDDY, C.R. *Presburger arithmetic with bounded quantifier alternation*
⋄ B25 C10 D15 F20 F30 ⋄

MEREDITH, D. *Positive logic and λ-constants*
⋄ B20 B25 B40 ⋄

MUZALEWSKI, M. *Restricted decision problems in some classes of algebraic systems* ⋄ B25 ⋄

NORGELA, S.A. *Herbrand strategies of deduction-search in predicate calculus I (Russian) (English and Lithuanian summaries)* ⋄ B25 B35 D35 F07 ⋄

OPPEN, D.C. *A $2^{2^{2^{pn}}}$ upper bound on the complexity of Presburger arithmetic* ⋄ B25 C10 D15 F30 ⋄

PARIKH, R. *A decidability result for a second-order process logic* ⋄ B25 ⋄

ROSE, B.I. *Model theory of alternative rings*
⋄ C10 C35 C45 C60 ⋄

ROSE, B.I. *Rings which admit elimination of quantifiers*
⋄ C10 C60 ⋄

RYBAKOV, V.V. *A decidable noncompact extension of the logic S4 (Russian)* ⋄ B25 B45 ⋄

RYBAKOV, V.V. *Modal logics with LM-axioms (Russian)*
⋄ B25 B45 ⋄

SANKAPPANAVAR, H.P. *Decision problems: History and methods* ⋄ A10 B25 D35 D98 ⋄

SCHMERL, J.H. *A decidable $\aleph_0$-categorical theory with a non-recursive Ryll-Nardzewski function*
⋄ B25 C35 C57 D30 ⋄

SCHUETTE, K. *Ein Ansatz zum Entscheidungsverfahren fuer eine Formelklasse der Praedikatenlogik mit Identitaet* ⋄ B10 B25 ⋄

SEESE, D.G. *Decidability of ω-trees with bounded sets (German and Russian summaries)*
⋄ B25 C65 C85 D05 ⋄

SEESE, D.G. *Decidability and generalized quantifiers*
⋄ B25 C80 C98 ⋄

SEJTENOV, S.M. *Theory of finite fields with an additional predicate distinguishing a subfield (Russian)*
⋄ B25 C13 C60 ⋄

SEMENOV, A.L. *Some algorithmic problems for systems of algorithmic algebras (Russian)* ⋄ B25 B75 ⋄

SHELAH, S. *A weak generalization of MA to higher cardinals* ⋄ B25 C55 E35 E50 E75 ⋄

SIEFKES, D. *An axiom system for the weak monadic second order theory of two successors*
⋄ B15 B25 C85 D05 ⋄

SINGER, M.F. *The model theory of ordered differential fields* ⋄ B25 C25 C35 C60 C65 D80 ⋄

THOMAS, WOLFGANG *The theory of successor with an extra predicate* ⋄ B25 D35 ⋄

TOFFALORI, C. *Eliminazione dei quantificatori per certe teorie di coppie di campi (Teoria dei modelli) (English summary)* ⋄ C10 C60 ⋄

WASILEWSKA, A. *Machines, logics and decidability*
⋄ B25 B35 ⋄

WEESE, M. *Definierbare Praedikate in booleschen Algebren. II* ⋄ B25 C10 C40 C80 G05 ⋄

WERNER, H. *Discriminator-algebras. Algebraic representation and model theoretic properties*
⋄ B25 C05 C25 C35 C90 ⋄

ZAMYATIN, A.P. *Prevarieties of associative rings whose elementary theory is decidable (Russian)*
⋄ B25 C05 C60 ⋄

1979

ADAMSON, A. & GILES, R. *A game-based formal system for L_∞* ⋄ B25 B50 D15 ⋄

BERMAN, P. *Complexity of the theory of atomless boolean algebras* ⋄ B25 D15 ⋄

BLASS, A.R. & HARARY, F. *Properties of almost all graphs and complexes* ⋄ C10 C13 C65 ⋄

BOERGER, E. *A new general approach to the theory of the many-one equivalence of decision problems for algorithmic systems*
⋄ B25 D03 D05 D10 D25 D30 D40 ⋄

BOERGER, E. & KLEINE BUENING, H. *The reachability problem for Petri nets and decision problems for Skolem arithmetic* ⋄ B25 D15 D35 D80 ⋄

BOOLOS, G. *The unprovability of consistency. An essay in modal logic* ⋄ B25 B28 B45 B98 F30 F98 ⋄

BRITTON, J.L. *Integer solutions of systems of quadratic equations* ⋄ B25 D35 D80 ⋄

BURRIS, S. & WERNER, H. *Sheaf constructions and their elementary properties*
⋄ B25 C05 C20 C25 C30 C60 C90 ⋄

CHLEBUS, B.S. *On decidability of propositional algorithmic logic (Polish) (English summary)*
⋄ B25 B75 D10 D15 ⋄

COWLES, J.R. *The relative expressive power of some logics extending first-order logic*
⋄ C10 C75 C80 C85 E15 ⋄

COWLES, J.R. *The theory of Archimedean real closed fields in logics with Ramsey quantifiers*
⋄ C10 C35 C60 C80 ⋄

DREBEN, B. & GOLDFARB, W.D. *The decision problem. Solvable classes of quantificational formulas*
⋄ B25 C98 ⋄

DRIES VAN DEN, L. *New decidable fields of algebraic numbers* ⋄ B25 C60 ⋄

EKLOF, P.C. & MEKLER, A.H. *Stationary logic of finitely determinate structures*
⋄ B25 C10 C55 C60 C65 C80 ⋄

FERRANTE, J. & RACKOFF, C.W. *The computational complexity of logical theories*
⋄ B25 C98 D10 D15 ⋄

GAO, HENGSHAN & MO, SHAOKUI *The decidability of the $\aleph_0$-validity of formulas in the monadic predicate calculus with enumerative quantifiers and its applications (Chinese)* ⋄ B25 C80 ⋄

GLADSTONE, M.D. *The decidability of one-variable propositional calculi* ⋄ B20 B25 ⋄

GOGOL, D. *Sentences with three quantifiers are decidable in set theory* ⋄ B25 C35 E30 ⋄

GURARI, E.M. & IBARRA, O.H. *An NP-complete number-theoretic problem*
⋄ B25 D10 D15 D25 D80 ⋄

GUREVICH, Y. *Modest theory of short chains I*
⋄ B25 C65 C85 D35 E07 ⋄

GUREVICH, Y. & SHELAH, S. *Modest theory of short chains II* ⋄ B25 C65 C85 D35 E07 E50 ⋄

HEINTZ, JOOS *Definability bounds of first order theories of algebraically closed fields*
⋄ B25 C40 C60 D10 D15 ⋄

HERRE, H. & WOLTER, H. *Entscheidbarkeit der Theorie der linearen Ordnung in Logiken mit Maechtigkeitsquantoren bzw. mit Chang-Quantor*
⋄ B25 C55 C80 E07 E50 E55 ⋄

HERRE, H. & WOLTER, H. *The decision problem for the theory of linear orderings in extended logics*
⋄ B25 C55 C80 C98 ⋄

JUKNA, S. *On decidability of equivalence problem in some algebras of recursive functions (Russian)*
⋄ B25 D20 D75 ⋄

KOPPEL, M. *Some decidable diophantine problems: positive solution to a problem of Davis, Matijasevich and Robinson* ⋄ B25 ⋄

KRYNICKI, M. *On the expressive power of the language using the Henkin quantifier* ⋄ B15 B25 C80 C85 ⋄

KRYNICKI, M. & LACHLAN, A.H. *On the semantics of the Henkin quantifier*
⋄ B15 B25 C40 C55 C80 C85 ⋄

LEWIS, H.R. *Unsolvable classes of quantificational formulas* ⋄ B20 B25 D35 D98 ⋄

MANDERS, K.L. *The theory of all substructures of a structure: Characterisation and decision problems*
⋄ B20 B25 C57 C60 ⋄

MOSTOWSKI, A.WLODZIMIERZ *A note concerning the complexity of a decision problem for positive formulas in SkS* ⋄ B15 B25 D15 ⋄

MOSTOWSKI, T. & PFISTER, G. *Der Cohensche Eliminationssatz in positiver Charakteristik*
⋄ B25 C10 C60 ⋄

PINUS, A.G. *A calculus of one-place predicates (Russian)*
⋄ B20 B25 C55 C80 C85 ⋄

PINUS, A.G. *Elimination of the quantifiers Q_0 and Q_1 on symmetric groups (Russian)*
⋄ C10 C55 C60 C80 ⋄

PRATT, V.R. *Axioms or algorithms* ⋄ B25 B75 D15 ⋄

PRESIC, M.D. *On the greatest congruence of relations*
⋄ B25 E07 ⋄

PRESTEL, A. *Entscheidbarkeit mathematischer Theorien*
⋄ B25 C60 C98 D35 ⋄

ROGAVA, M.G. *A new decision procedure for SCI (Russian)* ⋄ B25 B60 F05 ⋄

ROMAN'KOV, V.A. *Universal theory of nilpotent groups (Russian)* ⋄ B25 C60 D35 ⋄

SCHWABHAEUSER, W. *Modellvollstaendigkeit der Mittelpunktsgeometrie und der Theorie der Vektorgruppen* ⋄ B25 C35 C65 ⋄

SEESE, D.G. *Some graph-theoretical operations and decidability* ⋄ B15 B25 C65 C85 ⋄

SEESE, D.G. *Stationary logic and ordinals*
⋄ B25 C55 C65 C80 ⋄

SEMENOV, A.L. *On certain extensions of the arithmetic of addition of natural numbers (Russian)* ⋄ B25 F30 ⋄

SHANK, H.S. *The rational case of a matrix problem of Harrison* ⋄ B25 ⋄

SHOSTAK, R.E. *A practical decision procedure for arithmetic with function symbols* ⋄ B25 B35 F30 ⋄

SZCZERBA, L.W. & TARSKI, A. *Metamathematical discussion of some affine geometries*
⋄ B25 B30 C35 C52 C65 D35 ⋄

VOLGER, H. *Sheaf representations of algebras and transfer theorems for model theoretic properties*
⋄ B25 C05 C30 C35 C90 ⋄

WEISPFENNING, V. *Lattice products*
⋄ C10 C25 C30 C35 C90 ⋄

WHEELER, W.H. *Amalgamation and elimination of quantifiers for theories of fields*
⋄ C10 C25 C52 C60 ⋄

WHEELER, W.H. *The first order theory of N-colorable graphs* ⋄ B30 C10 C25 C35 C65 ⋄

WOEHL, K. *Zur Komplexitaet der Presburger Arithmetik und des Aequivalenz-Problems einfacher Programme* ⋄ B25 B75 D15 F20 F30 ⋄

WOODROW, R.E. *There are four countable ultrahomogeneous graphs without triangles* ⋄ C10 C50 C65 E05 ⋄

ZUBENKO, V.V. *Standard program algebras (Ukrainain) (English and Russian summaries)* ⋄ B25 B75 G25 ⋄

1980

ABRAMSKY, M. *The classical decision problem and partial functions* ⋄ B25 ⋄

BALDWIN, JOHN T. & KUEKER, D.W. *Ramsey quantifiers and the finite cover property* ⋄ C10 C35 C45 C80 ⋄

BAUDISCH, A. *Application of the theory of Shirshov-Witt in the theory of groups and Lie algebras* ⋄ B25 C45 C60 ⋄

BAUDISCH, A. & SEESE, D.G. & TUSCHIK, H.-P. & WEESE, M. *Decidability and generalized quantifiers* ⋄ B25 B98 C10 C60 C80 C98 ⋄

BAUDISCH, A. *The theory of abelian p-groups with the quantifier I is decidable* ⋄ B25 C10 C80 ⋄

BAUR, W. *On the elementary theory of quadruples of vector spaces* ⋄ B25 C60 D80 ⋄

BELEGRADEK, O.V. *Decidable fragments of universal theories and existentially closed models (Russian)* ⋄ B25 C25 D40 ⋄

BERLINE, C. *Elimination of quantifiers for non semi-simple rings of characteristic p* ⋄ C10 C60 ⋄

BERMAN, L. *The complexity of logical theories* ⋄ B25 D10 D15 F30 ⋄

BOERGER, E. *Entscheidbarkeit* ⋄ B25 C98 ⋄

BOERGER, E. & KLEINE BUENING, H. *The reachability problem for Petri nets and decision problems for Skolem arithmetic* ⋄ B25 D15 D35 D80 ⋄

BOFFA, M. & CHERLIN, G.L. *Elimination des quantificateurs dans les faisceaux (English summary)* ⋄ C10 C60 C90 ⋄

BOFFA, M. *Elimination des quantificateurs en algebre* ⋄ C10 C98 ⋄

BOFFA, M. & MACINTYRE, A. & POINT, F. *The quantifier elimination problem for rings without nilpotent elements and for semi-simple rings* ⋄ C10 C60 ⋄

BOZZI, S. & MELONI, G.C. *Representation of Heyting algebras with covering and propositional intuitionistic logic with local operator* ⋄ B25 B55 C90 F50 G10 ⋄

BRUSS, A.R. & MEYER, A.R. *On time-space classes and their relation to the theory of real addition* ⋄ B25 D15 ⋄

BURGESS, J.P. *Decidability for branching time* ⋄ B25 B45 C85 ⋄

CEGIELSKI, P. *La theorie elementaire de la multiplication (English summary)* ⋄ B25 C10 C80 F30 ⋄

CHERLIN, G.L. & DICKMANN, M.A. *Anneaux reels clos et anneaux de fonctions continues (English summary)* ⋄ C10 C35 C60 C65 ⋄

CHERLIN, G.L. *Rings of continuous functions: decision problems* ⋄ B25 C60 C65 D35 ⋄

CHLEBUS, B.S. *Decidability and definability results concerning well-orderings and some extensions of first order logic* ⋄ B15 B25 C40 C65 C80 D35 E07 ⋄

CUTLAND, N.J. *Computability. An introduction to recursive function theory* ⋄ B25 D98 ⋄

DICHEV, KH.V. *Representation of data in calculation of logical formulas on finite models (Russian)* ⋄ B25 C13 ⋄

DRIES VAN DEN, L. *A linearly ordered ring whose theory admits elimination of quantifiers is a real closed field* ⋄ C10 C60 ⋄

DRIES VAN DEN, L. *Some model theory and number theory for models of weak systems of arithmetic* ⋄ B25 C62 F30 H15 ⋄

DROBOTUN, B.N. & GONCHAROV, S.S. *Numerations of saturated and homogeneous models (Russian)* ⋄ B25 C50 C57 D45 ⋄

ERSHOV, YU.L. *Decision problems and constructivizable models (Russian)* ⋄ B25 C57 C60 C98 D45 D98 ⋄

ERSHOV, YU.L. *Regularly closed fields (Russian)* ⋄ B25 C35 C60 D35 ⋄

FAJARDO, S. *Compacidad y decidibilidad de logicas monadicas con cuantificadores cardinales* ⋄ B25 C55 C80 ⋄

FERRO, A. & OMODEO, E.G. & SCHWARTZ, J.T. *Decision procedures for some fragments of set theory* ⋄ B25 E30 ⋄

FERRO, A. & OMODEO, E.G. & SCHWARTZ, J.T. *Decision procedures for elementary sublanguages of set theory. I: Multi-level syllogistic and some extensions* ⋄ B25 E30 ⋄

FIDEL, M.M. *An algebraic study of logic with constructible negation* ⋄ B25 B55 G25 ⋄

GARAVAGLIA, S. *Decomposition of totally transcendental modules* ⋄ C10 C35 C45 C60 ⋄

GOLTZ, H.-J. *Untersuchungen zur Elimination verallgemeinerter Quantoren in Koerpertheorien* ⋄ C10 C55 C60 C80 ⋄

GUREVICH, Y. *Ordered abelian groups* ⋄ B25 C10 C60 C85 ⋄

HEINDORF, L. *The decidability of the L_t - theory of boolean spaces* ⋄ B25 C65 C80 C90 ⋄

HENDRY, H.E. *Two remarks on the atomistic calculus of individuals* ⋄ C10 G05 ⋄

HERRE, H. & SEESE, D.G. *Concerning the monadic theory of the topology of well-orderings and scattered spaces* ⋄ B25 C55 C65 C85 E07 E35 E50 E55 ⋄

JACOB, B. *The model theory of "R-formal" fields* ⋄ B25 C10 C25 C35 C60 ⋄

JARDEN, M. *An analogue of Chebotarev density theorem for fields of finite corank* ⋄ B25 C10 C60 ⋄

KHMELEVSKIJ, YU.I. & TAJMANOV, A.D. *Decidability of the universal theory of a free semigroup (Russian)* ⋄ B25 C05 C60 ⋄

KIM, KUKJIN *On the structure of Hensel fields* ⋄ B25 C60 ⋄

KOMORI, Y. *Completeness of two theories on ordered abelian groups and embedding relations*
⋄ B25 C10 C35 C60 ⋄

KOZEN, D. *Complexity of boolean algebras*
⋄ B25 D15 G05 ⋄

LADNER, R.E. *Complexity theory with emphasis on the complexity of logical theories*
⋄ B25 C98 D05 D10 D15 ⋄

LERMAN, M. & SOARE, R.I. *A decidable fragment of the elementary theory of the lattice of recursively enumerable sets* ⋄ B25 C10 D25 ⋄

LEWIS, H.R. *Complexity results for classes of quantificational formulas* ⋄ B20 B25 D10 D15 ⋄

LIVCHAK, A.B. *A letter to the editors (Russian)*
⋄ B25 C10 C60 ⋄

LOPARIC, Z. *Decidability and cognitive significance in Carnap* ⋄ A05 A10 B25 ⋄

LUCAS, T. *Reduction d'ensembles de formules* ⋄ C10 ⋄

LUE, YIZHONG *Research on a decision problem (Chinese) (English summary)* ⋄ B25 ⋄

MAKANIN, G.S. *Equations in a free semigroup (Russian)*
⋄ B03 B25 D40 ⋄

MARCONI, D. *A decision method for the calculus C_1*
⋄ A05 B25 B53 ⋄

MCKEE, T.A. *Monadic characterizations in nonstandard topology* ⋄ C10 H05 H20 ⋄

MCROBBIE, M.A. & MEYER, R.K. & THISTLEWAITE, P.B. *A mechanized decision procedure for non-classical logics: The program KRIPKE* ⋄ B25 B35 ⋄

MOSTOWSKI, A.WLODZIMIERZ *Finite automata on infinite trees and subtheories of SkS* ⋄ B25 C85 D05 ⋄

NAKAMURA, A. & ONO, H. *Decidability results on a query language for data bases with incomplete information*
⋄ B25 B52 B75 ⋄

NOAH, A. *Predicate-functors and the limits of decidability in logic* ⋄ B25 G15 ⋄

NORGELA, S.A. *On approximation of some classes of classical predicate calculus by decidable classes. I,II (Russian) (English and Lithuanian summaries)*
⋄ B10 B20 B25 ⋄

OPPEN, D.C. *Complexity, convexity and combinations of theories* ⋄ B25 B35 D15 ⋄

OPPEN, D.C. *Reasoning about recursively defined data structures* ⋄ B25 D15 ⋄

REICHBACH, J. *Decidability of mathematical sciences and their undecidability* ⋄ B25 D35 ⋄

ROSE, B.I. *On the model theory of finite-dimensional algebras* ⋄ B25 C35 C45 C60 ⋄

ROSE, B.I. *Prime quantifier eliminable rings*
⋄ C10 C60 ⋄

SCHMERL, J.H. *Decidability and $\aleph_0$-categoricity of theories of partially ordered sets*
⋄ B25 C15 C35 C65 D35 ⋄

SMORYNSKI, C.A. *Skolem's solution to a problem of Frobenius* ⋄ C10 ⋄

SWART DE, H.C.M. *Gentzen-type systems for C, K and several extensions of C and K; constructive completeness proofs and effective decision procedures for these systems* ⋄ B25 B45 ⋄

SZCZERBA, L.W. *Interpretations with parameters*
⋄ B25 C07 C35 C45 F25 ⋄

THOMAS, WOLFGANG *On the bounded monadic theory of well-ordered structures* ⋄ B15 B25 C85 E07 ⋄

TOFFALORI, C. *Fasci di coppie di campi algebricamente chiusi* ⋄ B25 C35 C60 C90 ⋄

TOFFALORI, C. *Sul model-completamento di certe teorie di coppie di anelli (English summary)*
⋄ C10 C25 C35 C60 ⋄

TUCKER, J.V. *Computability and the algebra of fields: some affine constructions*
⋄ B25 C57 C60 C65 D45 ⋄

TULIPANI, S. *Invarianti per l'equivalenza elementare per una classe assiomatica di spazi generali di misura (English summary)* ⋄ B25 C10 C65 G05 ⋄

TUSCHIK, H.-P. *On the decidability of the theory of linear orderings with generalized quantifiers*
⋄ B25 C55 C65 C80 ⋄

VALIEV, M.K. *Decision complexity of variants of propositional dynamic logic* ⋄ B25 B75 D15 ⋄

WASILEWSKA, A. *On the Gentzen-type formalizations*
⋄ B25 D05 F07 ⋄

1981

ANAPOLITANOS, D.A. & VAEAENAENEN, J. *Decidability of some logics with free quantifier variables*
⋄ B25 C80 ⋄

BAUDISCH, A. *The elementary theory of abelian groups with m-chains of pure subgroups*
⋄ B25 C55 C60 C80 ⋄

BAUDISCH, A. *There is no module having the finite cover property* ⋄ C10 C45 C60 C80 ⋄

BERLINE, C. & CHERLIN, G.L. *QE nilrings of prime characteristic* ⋄ C10 C60 ⋄

BERLINE, C. & CHERLIN, G.L. *QE rings in characteristic p*
⋄ C10 C60 ⋄

BERLINE, C. *Rings which admit elimination of quantifiers*
⋄ C10 C60 ⋄

BEZVERKHNIJ, V.N. *Solvability of the inclusion problem in a class of HNN-groups (Russian)* ⋄ B25 C60 D40 ⋄

BEZVERKHNIJ, V.N. & GRINBLAT, V.A. *The root problem in Artin Groups (Russian)* ⋄ B25 C60 D40 ⋄

BOERGER, E. *Logical description of computation processes*
⋄ B25 D05 D15 D35 ⋄

BOFFA, M. *Quantifier elimination in boolean sheaves*
⋄ C10 C90 G30 ⋄

BREBAN, M. & FERRO, A. & OMODEO, E.G. & SCHWARTZ, J.T. *Decision procedures for elementary sublanguages of set theory. II. Formulas involving restricted quantifiers, together with ordinal, integer, map, and domain notions* ⋄ B25 E30 ⋄

BURRIS, S. & MCKENZIE, R. *Decidability and boolean representations* ⋄ B25 C05 C30 C60 ⋄

BURRIS, S. & MCKENZIE, R. *Decidable varieties with modular congruence lattices* ⋄ B25 C05 ⋄

CEGIELSKI, P. *Theorie elementaire de la multiplication des entiers naturels* ⋄ B25 C10 C80 F30 ⋄

CHATZIDAKIS, Z. *La representation en termes de faisceaux des modeles de la theorie elementaire de la multiplication des entiers naturels*
⋄ C10 C30 C62 C90 ⋄

CHERLIN, G.L. & DRIES VAN DEN, L. & MACINTYRE, A. *Decidability and undecidability theorems for PAC-fields* ⋄ B25 C60 D35 ⋄

CLOTE, P. *A note on decidable model theory* ⋄ B25 C15 C50 C57 D30 ⋄

COMER, S.D. *The decision problem for certain nilpotent closed varieties* ⋄ B25 C05 D35 ⋄

COWLES, J.R. *Generalized archimedean fields and logics with Malitz quantifiers* ⋄ C10 C60 C80 ⋄

CUPPLES, B. *On Copi's misapplication of a decision procedure* ⋄ B25 ⋄

DELON, F. *Types sur $C((X))$* ⋄ C10 C45 C60 ⋄

DRIES VAN DEN, L. *Quantifier elimination for linear formulas over ordered and valued fields* ⋄ C10 C60 ⋄

DRIES VAN DEN, L. *Which curves over Z have points with coordinates in a discrete ordered ring?* ⋄ B25 C60 C62 ⋄

ERSHOV, YU.L. *Eliminability of quantifiers in regularly closed fields (Russian)* ⋄ B25 C10 C60 ⋄

ERSHOV, YU.L. *On elementary theories of regularly closed fields (Russian)* ⋄ B25 C60 ⋄

FUERER, M. *Alternation and the Ackermann case of the decision problem* ⋄ B20 B25 D15 ⋄

GOLDBLATT, R.I. *Grothendieck topology as geometric modality* ⋄ B25 B45 C90 F50 G30 ⋄

GOLDFARB, W.D. *On the Goedel class with identity* ⋄ B20 B25 D35 ⋄

GRUNEWALD, F. & SEGAL, D. *How to solve a quadratic equation in integers* ⋄ B25 ⋄

GURARI, E.M. & IBARRA, O.H. *The complexity of the equivalence problem for two characterizations of Presburger sets* ⋄ B25 D05 D15 F30 ⋄

HAJEK, P. *Decision problems of some statistically motivated monadic modal calculi* ⋄ B25 B45 ⋄

HAUSCHILD, K. *Zum Vergleich von Haertigquantor und Rescherquantor* ⋄ B25 C10 C55 C80 D35 ⋄

HAUSSLER, D. *Model completeness of an algebra of languages* ⋄ B25 C35 D05 ⋄

HEINDORF, L. *Comparing the expressive power of some languages for boolean algebras* ⋄ B25 C10 C80 C85 C90 G05 ⋄

HERRE, H. *Miscellaneous results and problems in extended model theory* ⋄ B25 C65 C80 C98 D35 ⋄

HERRE, H. & WOLTER, H. *Untersuchungen zur Theorie der linearen Ordnung in Logiken mit Maechtigkeitsquantoren* ⋄ B25 C55 C65 C80 E07 ⋄

HUET, G. *A complete proof of correctness of the Knuth-Bendix completion algorithm* ⋄ B25 B75 C05 ⋄

JACOB, B. *The model theory of generalized real closed fields* ⋄ B25 C35 C60 ⋄

JAFFAR, J. *Presburger arithmetic with array segments* ⋄ B25 ⋄

JUKNA, S. *On the decidability of Hoare's system* ⋄ B25 B75 ⋄

KOSTYRKO, V.F. *On elementary ∃-theory of semigroups (Russian)* ⋄ B25 C05 ⋄

KOZEN, D. *Positive first-order logic is NP-complete* ⋄ B25 D15 ⋄

LIPSHITZ, L. *Some remarks on the diophantine problem for addition and divisibility* ⋄ B25 D15 ⋄

LISOVIK, L.P. *Sets that are expressible in theories of commutative semigroups (Russian) (English Summary)* ⋄ B25 C05 C40 ⋄

MACINTYRE, A. *The laws of exponentiation* ⋄ B25 C65 ⋄

MANASTER, A.B. & REMMEL, J.B. *Some decision problems for subtheories of two-dimensional partial orderings* ⋄ B25 C10 C65 D35 ⋄

MICHEL, P. *Borne superieure de la complexite de la theorie de N muni de la relation de divisibilite* ⋄ B25 D15 ⋄

MILLAR, T.S. *Counterexamples via model completions* ⋄ C10 C35 C50 C57 ⋄

MOSTOWSKI, A.WLODZIMIERZ *The complexity of automata and subtheories of monadic second order arithmetics* ⋄ B15 B25 D05 D15 F35 ⋄

MULLER, D.E. & SCHUPP, P.E. *Context-free languages, groups, the theory of ends, second-order logic, tiling problems, cellular automata, and vector addition systems* ⋄ B15 B25 D05 D80 ⋄

NAKAMURA, A. & ONO, H. *Undecidability of extensions of the monadic first-order theory of successor and two-dimensional finite automata* ⋄ B15 B25 D05 D35 ⋄

POINT, F. *Elimination des quantificateurs dans les L-anneaux ∗-reguliers et application aux anneaux bireguliers* ⋄ C10 C60 C90 ⋄

PREST, M. *Quantifier elimination for modules* ⋄ C10 C60 ⋄

PRESTEL, A. *Pseudo real closed fields* ⋄ C10 C25 C35 C52 C60 ⋄

ROSE, B.I. & WOODROW, R.E. *Ultrahomogeneous structures* ⋄ C10 C50 ⋄

ROTHMALER, P. *Q_0 is eliminable in every complete theory of modules* ⋄ C10 C60 C80 ⋄

ROZENBLAT, B.V. & VAZHENIN, YU.M. *Decidability of the positive theory of a free countably generated semigroup (Russian)* ⋄ B25 C05 D35 ⋄

ROZENBLAT, B.V. *Positive theories of some varieties of semigroups (Russian)* ⋄ B25 C05 ⋄

SCHMERL, J.H. *Arborescent structures II: Interpretability in the theory of trees* ⋄ B25 C50 C65 C75 ⋄

SCHMERL, J.H. *Decidability and finite axiomatizability of theories of $\aleph_0$-categorical partially ordered sets* ⋄ B25 C35 C65 ⋄

SCHMIDT, K. *Ein Rechenverfahren fuer die elementare Logik (English summary)* ⋄ B25 B35 ⋄

SEESE, D.G. *Elimination of second-order quantifiers for well-founded trees in stationary logic and finitely determinate structures* ⋄ C10 C55 C65 C80 ⋄

SEESE, D.G. *Stationary logic and ordinals* ⋄ B25 C55 C65 C80 ⋄

THOMAS, WOLFGANG *A combinatorial approach to the theory of ω-automata* ⋄ B25 D05 ⋄

TULIPANI, S. *Model-completions of theories of finitely additive measures with values in an ordered field* ⋄ C10 C25 C35 C65 G05 ⋄

TWER VON DER, T. *On the strength of several versions of Dirichlets ("pigeon-hole"-) principle in the sense of first-order logic* ⋄ C10 H15 ⋄

URQUHART, A.I.F. *Decidability and the finite model property* ◇ B22 B25 D35 ◇

URQUHART, A.I.F. *The decision problem for equational theories* ◇ B25 C05 D35 ◇

VAZHENIN, YU.M. *Sur la liaison entre problemes combinatoires et algorithmiques*
◇ B25 D05 D15 D35 ◇

WEESE, M. *Decidability with respect to Haertig quantifier and Rescher quantifier* ◇ B25 C55 C80 D35 ◇

WEISPFENNING, V. *Elimination of quantifiers for certain ordered and lattice-ordered abelian groups*
◇ B25 C10 C60 ◇

ZYGMUNT, J. *Notes on decidability and finite approximability of sentential logics*
◇ A10 B22 B25 ◇

1982

AANDERAA, S.O. & BOERGER, E. & GUREVICH, Y. *Prefix classes of Krom formulae with identity (German summary)* ◇ B20 B25 D35 ◇

ARNON, D.S. & MCCALLUM, S. *Cylindrical algebraic decomposition by quantifier elimination*
◇ C10 C40 ◇

BAUDISCH, A. *Decidability and stability of free nilpotent Lie algebras and free nilpotent p-groups of finite exponent* ◇ B25 C45 C60 ◇

BAUR, W. *Die Theorie der Paare reell abgeschlossener Koerper* ◇ B25 C60 ◇

BAUR, W. *On the elementary theory of pairs of real closed fields II* ◇ B25 C60 ◇

BURRIS, S. *The first order theory of Boolean algebras with a distinguished group of automorphisms*
◇ B25 C05 C60 G05 ◇

BUSZKOWSKI, W. *Some decision problems in the theory of syntactic categories* ◇ B25 D05 ◇

CHERLIN, G.L. & FELGNER, U. *Quantifier eliminable groups* ◇ C10 C13 C60 ◇

CHLEBUS, B.S. *On the decidability of propositional algorithmic logic* ◇ B25 B75 ◇

COLLINS, G.E. *Quantifier elimination for real closed fields: a guide to the literature* ◇ C10 C60 C98 ◇

DRIES VAN DEN, L. *Some applications of a model theoretic fact to (semi-)algebraic geometry* ◇ C10 C60 C65 ◇

ERSHOV, YU.L. *Algorithmic problems in the theory of fields (positive aspects) (Russian)* ◇ B25 C10 C60 C98 ◇

ERSHOV, YU.L. *Multiply valued fields (Russian)*
◇ B25 C25 C35 C60 C98 ◇

ERSHOV, YU.L. *Regularly r-closed fields (Russian)*
◇ B25 C25 C35 C60 ◇

ERSHOV, YU.L. *Totally real field extensions (Russian)*
◇ B25 C52 C60 ◇

FUERER, M. *The complexity of Presburger arithmetic with bounded quantifier alternation depth* ◇ B25 D15 ◇

GUREVICH, Y. & HARRINGTON, L.A. *Automata, trees, and games* ◇ B25 C85 D05 ◇

GUREVICH, Y. *Crumbly spaces* ◇ B25 C65 C85 ◇

GUREVICH, Y. & SHELAH, S. *Monadic theory of order and topology in ZFC* ◇ B25 C65 C85 D35 E07 ◇

HARAN, D. & JARDEN, M. *Bounded statements in the theory of algebraically closed fields with distinguished automorphisms* ◇ B25 C60 ◇

HARAN, D. & LUBOTZKY, A. *Embedding covers and the theory of Frobenius fields* ◇ B25 C60 ◇

HODGSON, B.R. *On direct products of automaton decidable theories* ◇ B25 D05 ◇

HUBER-DYSON, V. *Decision problems in group theory*
◇ B25 D40 ◇

JURIE, P.-F. *Decidabilite de la theorie elementaire des anneaux booleiens a operateurs dans un group fini (English summary)* ◇ B25 C60 ◇

KESEL'MAN, D.YA. *Decidability of theories of certain classes of elimination graphs* ◇ B25 C65 D35 ◇

KREJNOVICH, V.YA. & OSWALD, U. *A decision method for the universal theorems of Quine's new foundations*
◇ B20 B25 E70 ◇

LOMECKY, Z. *Algorithms for the computation of free lattices* ◇ B25 G10 ◇

MOLZAN, B. *How to eliminate quantifiers in the elementary theory of p-rings* ◇ B25 C60 ◇

MOLZAN, B. *The theory of superatomic boolean algebras in the logic with the binary Ramsey quantifier*
◇ B25 C10 C80 G05 ◇

MYASNIKOV, A.G. & REMESLENNIKOV, V.N. *Definability of the set of Mal'tsev bases and elementary theories of finite-dimensional algebras I (Russian)* ◇ B25 C60 ◇

OSWALD, U. *A decision method for the existential theorems of NF_2* ◇ B25 E70 ◇

PRESTEL, A. *Decidable theories of preordered fields*
◇ B25 C25 C35 C60 ◇

ROSENSTEIN, J.G. *Linear orderings*
◇ B25 C65 C98 E07 E98 ◇

ROTHMALER, P. & TUSCHIK, H.-P. *A two cardinal theorem for homogeneous sets and the elimination of Malitz quantifiers* ◇ C10 C45 C80 ◇

SARACINO, D.H. & WOOD, C. *QE nil-2 groups of exponent 4* ◇ C10 C15 C60 ◇

SCHMERL, J.H. & SIMPSON, S.G. *On the role of Ramsey quantifiers in first order arithmetic*
◇ B28 C10 C62 C80 ◇

SCHMITT, P.H. *The elementary theory of torsion-free abelian groups with a predicate specifying a subgroup*
◇ B25 C60 ◇

SCHOENHAGE, A. *Random access machines and Presburger arithmetic* ◇ B25 D15 ◇

SEESE, D.G. & WEESE, M. *L(aa)-elementary types of well-orderings* ◇ B25 C55 C65 C80 E10 ◇

SHOSTAK, R.E. *Deciding combinations of theories*
◇ B25 B35 ◇

SIEKMANN, J. & SZABO, P. *Universal unification and a classification of equational theories*
◇ B25 B35 C05 ◇

TOFFALORI, C. *Stabilita, categoricita ed eliminazione dei quantificatori per una classe di anelli locali*
◇ C10 C35 C45 C60 ◇

TOFFALORI, C. *Teoria dei modelli per alcune classi di anelli locali* ◇ C10 C25 C35 C60 ◇

TULIPANI, S. *A use of the method of interpretations for decidability or undecidability of measure spaces*
◇ B25 C65 D35 ◇

TUSCHIK, H.-P. *Elimination of cardinality quantifiers*
◇ C10 C45 C55 C60 C80 ◇

WEISPFENNING, V. *Valuation rings and Boolean products*
⋄ C10 C30 C60 ⋄

WERNER, H. *Sheaf constructions in universal algebra and model theory*
⋄ B25 C05 C25 C30 C35 C90 G05 ⋄

1983

BAUDISCH, A. & SEESE, D.G. & TUSCHIK, H.-P. *ω-trees in stationary logic* ⋄ B25 C55 C65 C80 ⋄

BELYAKOV, E.B. & MART'YANOV, V.I. *Universal theories of integers and the extended Bliznetsov hypothesis (Russian)* ⋄ B25 ⋄

BERLINE, C. & CHERLIN, G.L. *QE rings in characteristic p^n* ⋄ C10 C60 ⋄

BOERGER, E. *From decision problems to problems of complexity* ⋄ B25 D15 ⋄

BORICIC, B.R. *A decision procedure for certain disjunction-free intermediate propositional calculi*
⋄ B25 B55 ⋄

BUECHI, J.R. & ZAIONTZ, C. *Deterministic automata and the monadic theory of ordinals $< \omega_2$*
⋄ B15 B25 C85 D05 E10 ⋄

BURRIS, S. *Boolean constructions*
⋄ A10 B25 C05 C30 C98 D35 G05 ⋄

CHERLIN, G.L. & SCHMITT, P.H. *Locally pure topological abelian groups: Elementary invariants*
⋄ B25 C60 C65 C80 C90 ⋄

CHERLIN, G.L. & DICKMANN, M.A. *Real closed rings II: Model theory* ⋄ B25 C10 C60 ⋄

CZEDLI, G. & FREESE, R. *On congruence distributivity and modularity* ⋄ B25 C05 ⋄

DANKO, W. *Interpretability of algorithmic theories*
⋄ B25 B75 C35 F25 ⋄

DENEF, J. & JARDEN, M. & LEWIS, D. *On Ax-fields which are C_i* ⋄ B25 C60 ⋄

DOSHITA, S. & YAMASAKI, S. *The satisfiability problem for a class consisting of Horn sentences and some non-Horn sentences in propositional logic* ⋄ B20 B25 D15 ⋄

DRIES VAN DEN, L. & MACINTYRE, A. & MCKENNA, K. *Elimination of quantifiers in algebraic structures*
⋄ C10 C60 ⋄

ERSHOV, YU.L. *Regularly r-closed fields (Russian)*
⋄ B25 C25 C35 C60 ⋄

FERRO, A. *Some decidability questions in set theory (Italian)* ⋄ B25 E30 ⋄

GRANDJEAN, E. *Complexity of the first-order theory of almost all finite structures* ⋄ B25 C13 D15 ⋄

GUREVICH, Y. & SHELAH, S. *Random models and the Goedel case of the decision problem*
⋄ B20 B25 C13 D35 ⋄

GUREVICH, Y. & MAGIDOR, M. & SHELAH, S. *The monadic theory of ω_2*
⋄ B15 B25 C65 C85 D35 E10 E35 ⋄

HAREL, D. *Recurring dominoes: making the highly undecidable highly understandable*
⋄ B25 B75 D15 D55 ⋄

HEINTZ, JOOS *Definability and fast quantifier elimination in algebraically closed fields*
⋄ B25 C10 C40 C60 D10 D15 ⋄

HODGSON, B.R. *Decidabilite par automate fini*
⋄ B25 D05 D35 ⋄

IVANOV, A.A. *Decidability of theories in a certain calculus (Russian)* ⋄ B25 C10 C55 C60 C80 D35 ⋄

IVANOV, A.A. *Some theories in generalized calculi (Russian)* ⋄ B25 C10 C55 C60 C80 ⋄

JARDEN, M. *On the model companion of the theory of e-fold ordered fields* ⋄ B25 C25 C35 C60 ⋄

KAPETANOVIC, M. *Some simple decidability proofs*
⋄ B20 B25 ⋄

LEHMANN, D.J. & SHELAH, S. *Reasoning with time and chance* ⋄ B25 B45 ⋄

LOLLI, G. *Complessita delle teorie* ⋄ B25 D15 ⋄

MACINTYRE, A. *Decision problems for exponential rings: the p-adic case* ⋄ B25 C60 ⋄

MUNDICI, D. *Natural limitations of decision procedures for arithmetic with bounded quantifiers*
⋄ B25 D10 D15 F30 ⋄

NOSKOV, G.A. *The elementary theory of a finitely generated almost solvable group (Russian)*
⋄ B25 C60 D35 ⋄

NURTAZIN, A.T. *An elementary classification and some properties of complete orders (Russian)*
⋄ B25 C35 C65 C75 E10 ⋄

OMODEO, E.G. *Decidability and validity in the presence of choice operators (Italian)* ⋄ B25 ⋄

ONO, H. *Equational theories and universal theories of fields* ⋄ B25 C05 C60 ⋄

PARIGOT, M. *Le modele compagnon de la theorie des arbres* ⋄ C10 C25 C35 C50 C65 ⋄

PINUS, A.G. *The operation of Cartesian product*
⋄ B25 C05 D35 ⋄

PRESTEL, A. & ROQUETTE, P. *Lectures on formally p-adic fields* ⋄ B25 C10 C35 C60 C98 ⋄

REMESLENNIKOV, V.N. & ROMAN'KOV, V.A. *Model-theoretic and algorithmic questions of group theory (Russian)*
⋄ B25 C60 C98 D15 D30 D40 ⋄

ROZENBLAT, B.V. & VAZHENIN, YU.M. *On positive theories of free algebraic systems (Russian)*
⋄ B20 B25 C05 ⋄

SCHMITT, P.H. *The L^1-theory of profinite abelian groups*
⋄ B25 C50 C60 C80 C90 ⋄

SEMENOV, A.L. *Logical theories of one-place functions on the set of natural numbers (Russian)*
⋄ B25 C10 C62 C85 ⋄

TOFFALORI, C. *Questioni di teoria dei modelli per coppie di campi* ⋄ C10 C35 C45 C60 ⋄

ULRICH, D. *The finite model property and recursive bounds on the size of countermodels* ⋄ B22 B25 C13 ⋄

VAZHENIN, YU.M. *Semigroups with one defining relation whose elementary theories are decidable (Russian)*
⋄ B25 C05 C60 ⋄

VOJSHVILLO, E.K. *A decision procedure for the system E (of entailment) I* ⋄ B25 B46 ⋄

VOLGER, H. *A new hierarchy of elementary recursive decision problems* ⋄ B25 D15 D20 ⋄

VOLGER, H. *Turing machines with linear alternation, theories of bounded concatenation and the decision problem of first order theories* ⋄ B25 D10 D15 ⋄

ZAIONTZ, C. *Axiomatization of the monadic theory of ordinals $<\omega_2$* ⋄ B25 B28 C65 C85 ⋄

ZAMYATIN, A.P. *Decidability of the elementary theories of certain varieties of rings (Russian)*
⋄ B25 C05 C60 ⋄

ZHALDOKAS, R. *An approach to the construction of complete term rewriting systems (Russian) (English and Lithuanian summaries)* ⋄ B25 B35 B75 C05 ⋄

ZYGMUNT, J. *On decidability and finite approximability of sentential logics* ⋄ B22 B25 ⋄

1984

AANDERAA, S.O. *On the solvability of the extended $\forall\exists\wedge\exists\forall^*$-Ackermann class with identity*
⋄ B20 B25 ⋄

BASARAB, S.A. *On some classes of Hilbertian fields*
⋄ B25 C10 C25 C35 C60 ⋄

BAUDISCH, A. *Magidor-Malitz quantifiers in modules*
⋄ C10 C60 C80 ⋄

BECKER, E. *Extended Artin-Schreier theory of fields*
⋄ B25 C35 C60 ⋄

BOERGER, E. *Decision problems in predicate logic*
⋄ B20 B25 D35 ⋄

BOERGER, E. *Spektralproblem and completeness of logical decision problems* ⋄ B25 C13 D15 ⋄

BREBAN, M. & FERRO, A. *Decision procedures for elementary sublanguages of set theory. III. Restricted classes of formulas involving the power set operator and the general set union operator* ⋄ B25 E30 ⋄

BURRIS, S. *Model companions for finitely generated universal Horn classes*
⋄ B25 C05 C10 C35 D20 ⋄

CANTOR, DAVID G. & ROQUETTE, P. *On diophantine equations over the ring of all algebraic integers*
⋄ B25 C60 D35 ⋄

CAVALLI, A.R. & FARINAS DEL CERRO, L. *A decision method for linear temporal logic* ⋄ B25 B45 ⋄

CEGIELSKI, P. *La theorie elementaire de la divisibilite est finiment axiomatisable (English summary)*
⋄ B25 B28 F30 ⋄

CHEN, JIYUAN *The satisfiability problem for simple boolean expressions belongs to P (Chinese) (English summary)*
⋄ B05 B25 D15 ⋄

CHERLIN, G.L. *Decidable theories of pseudo-algebraically closed fields* ⋄ B25 C60 ⋄

CHISTOV, A.L. & GRIGOR'EV, D.YU. *Complexity of quantifier elimination in theory of algebraically closed fields* ⋄ C10 C60 D15 ⋄

DENEF, J. & LIPSHITZ, L. *Power series solutions of algebraic differential equations*
⋄ B25 C60 C65 D35 D80 ⋄

DENEF, J. *The rationality of the Poincare series associated to the p-adic points on a variety* ⋄ C10 C60 ⋄

DENENBERG, L. & LEWIS, H.R. *Logical syntax and computational complexity* ⋄ B25 D15 ⋄

DENENBERG, L. & LEWIS, H.R. *The complexity of the satisfiability problem for Krom formulas*
⋄ B20 B25 D15 ⋄

DRIES VAN DEN, L. *Algebraic theories with definable Skolem functions* ⋄ C10 C40 C60 ⋄

DRIES VAN DEN, L. *Remarks on Tarski's problem concerning* $(\mathbb{R},+,\cdot,exp)$ ⋄ C10 C60 C65 ⋄

DRUGUSH, YA.M. *Finite approximability of forest superintuitionistic logics (Russian)* ⋄ B25 B55 ⋄

EMERSON, E.A. & SISTLA, A.P. *Deciding full branching time logic* ⋄ B25 B45 D15 ⋄

ERSHOV, YU.L. *Regularly r-closed fields with weakly universal Galois group (Russian)* ⋄ B25 C60 D35 ⋄

FRIED, M. & HARAN, D. & JARDEN, M. *Galois stratification over Frobenius fields* ⋄ B25 C60 ⋄

GECSEG, F. & STEINBY, M. *Tree automata*
⋄ A05 B25 D05 D98 ⋄

GOLDFARB, W.D. & GUREVICH, Y. & SHELAH, S. *A decidable subclass of the minimal Goedel class with identity* ⋄ B20 B25 ⋄

GUREVICH, R. *Decidability of the equational theory of positive numbers with raising to a power (Russian)*
⋄ B25 B28 C65 ⋄

HAILPERIN, T. *Probability logic* ⋄ B25 B48 C90 ⋄

HARAN, D. *The undecidability of pseudo-real-closed fields*
⋄ B25 C60 D35 ⋄

HEINDORF, L. *Beitraege zur Modelltheorie der Booleschen Algebren (English and Russian summaries)*
⋄ B25 C10 C35 C55 C80 G05 ⋄

HODES, H.T. *The modal theory of pure identity and some related decision problems* ⋄ B25 B45 D35 ⋄

IVANOV, A.A. *Complete theories of unars (Russian)*
⋄ B25 C35 ⋄

IVANOV, A.A. *Decidability of extended theories of addition of the natural numbers and the integers (Russian)*
⋄ B25 C65 F30 ⋄

KETONEN, J. & WEYHRAUCH, R.W. *A decidable fragment of predicate calculus* ⋄ B20 B25 ⋄

KOZEN, D. & PARIKH, R. *A decision procedure for the propositional μ-calculus* ⋄ B25 B75 ⋄

LACHLAN, A.H. *Binary homogeneous structures. I*
⋄ C10 C13 C15 C45 C50 C57 ⋄

LACHLAN, A.H. *On countable stable structures which are homogeneous for a finite relational language*
⋄ C10 C13 C15 C45 C50 ⋄

LACHLAN, A.H. & SHELAH, S. *Stable structures homogeneous for a binary language*
⋄ C10 C13 C15 C45 C50 ⋄

LENSKI, W. *Elimination of quantifiers for the theory of archimedean ordered divisible groups in a logic with Ramsey quantifiers* ⋄ C10 C55 C60 C80 ⋄

LISOVIK, L.P. *Construction of decidable singular theories of two successor functions with an extra predicate (Russian)* ⋄ B25 D05 ⋄

LISOVIK, L.P. *Monadic second-order theories of two successor functions with an additional predicate (Russian)* ⋄ B15 B25 C10 C85 D05 ⋄

MAKANIN, G.S. *Decidability of the universal and positive theories of a free group (Russian)* ⋄ B25 C60 ⋄

MEKLER, A.H. *Stationary logic of ordinals*
⋄ B25 C10 C55 C65 C80 E10 ⋄

NELSON, GREG *Combining satisfiability procedures by equality-sharing* ⋄ B25 B35 ⋄

ROQUETTE, P. *Some tendencies in contemporary algebra*
⋄ C10 C35 C60 C98 ⋄

ROTHMALER, P. *Some model theory of modules III. On infiniteness of sets definable in modules*
⋄ C10 C35 C60 C80 ⋄

RUYER, H. *Deux resultats concernant les modules sur un anneau de Dedekind (English summary)*
⋄ C10 C60 C75 ⋄

RYBAKOV, V.V. *Decidability of the admissibility problem in layer-finite modal logics (Russian)* ⋄ B25 B45 ⋄

SARACINO, D.H. & WOOD, C. *QE commutative nilrings*
⋄ C10 C50 C60 ⋄

SCARPELLINI, B. *Complexity of subcases of Presburger arithmetic* ⋄ B25 D15 F20 F30 ⋄

SCHMERL, J.H. $\aleph_0$-*categorical partially ordered sets (French summary)*
⋄ B25 C15 C35 C65 C98 E07 ⋄

SCHMITT, P.H. *Model- and substructure complete theories of ordered abelian groups* ⋄ C10 C35 C60 ⋄

SCOWCROFT, P. *The real algebraic structure of Scott's model of intuitionistic analysis*
⋄ B25 C90 F35 F50 ⋄

SEMENOV, A.L. *Decidability of monadic theories*
⋄ B25 C85 C98 ⋄

TETRUASHVILI, M.R. *The computational complexity of the theory of abelian groups with a given number of generators* ⋄ B25 D15 ⋄

UMIRBAEV, U.U. *Equality problem for center-by-metabelian Lie-algebras (Russian)*
⋄ B25 C60 D40 ⋄

VOLGER, H. *Filtered and stable boolean powers are relativized full boolean powers*
⋄ B25 C20 C30 C90 ⋄

WEISPFENNING, V. *Aspects of quantifier elimination in algebra* ⋄ C10 C60 C98 ⋄

WEISPFENNING, V. *Quantifier elimination and decision procedures for valued fields* ⋄ B25 C10 C35 C60 ⋄

WOJTYLAK, P. *A proof of Herbrand's theorem*
⋄ C10 F05 ⋄

WOLTER, H. *Some results about exponential fields (survey)*
⋄ B25 C60 C65 C98 ⋄

ZIEGLER, M. *Model theory of modules*
⋄ B25 C10 C45 C60 ⋄

1985

BAUDISCH, A. & SEESE, D.G. & TUSCHIK, H.-P. & WEESE, M. *Decidability and quantifier elimination*
⋄ B25 C10 C60 C65 C80 C98 ⋄

BOFFA, M. & MICHAUX, C. & POINT, F. & PRAAG VAN, P. *L'elimination lineaire dans les corps* ⋄ C10 C60 ⋄

BURGESS, J.P. & GUREVICH, Y. *The decision problem for linear temporal logic* ⋄ B25 B45 ⋄

CANTONE, D. & FERRO, A. & SCHWARTZ, J.T. *Decision procedures for elementary sublanguages of set theory VI: Multi-level syllogistic extended by the powerset operator* ⋄ B25 E30 ⋄

COMER, S.D. *The elementary theory of interval real numbers* ⋄ B25 B28 C65 ⋄

CRESSWELL, M.J. *The decidable normal modal logics are not recursively enumerable*
⋄ B25 B45 D25 D35 D80 ⋄

DEUTSCH, H. *A note on the decidability of a strong relevant logic* ⋄ B25 B46 ⋄

DICKMANN, M.A. *Applications of model theory to real algebraic geometry; a survey* ⋄ C10 C60 C98 ⋄

DICKMANN, M.A. *Elimination of quantifiers for ordered valuation rings* ⋄ C10 C60 ⋄

DRIES VAN DEN, L. & SMITH, RICK L. *Decidable regularly closed fields of algebraic numbers*
⋄ B25 C57 C60 ⋄

DRIES VAN DEN, L. *The field of reals with a predicate for the powers of two* ⋄ B25 C60 C65 ⋄

FERRO, A. *A note on the decidability of MLS extended with the powerset operator* ⋄ B25 E30 ⋄

GIAMBRONE, S. TW_+ *and* RW_+ *are decidable*
⋄ B25 B46 ⋄

GUREVICH, R. *Equational theory of positive numbers with exponentiation* ⋄ B25 B28 C65 ⋄

GUREVICH, Y. *Monadic second-order theories*
⋄ B25 C65 C85 C98 D35 D98 ⋄

GUREVICH, Y. & SHELAH, S. *The decision problem for branching time logic* ⋄ B25 B45 C65 D05 D25 ⋄

HAREL, D. *Recurring dominoes: Making the highly undecidable highly understandable*
⋄ B25 B75 D05 D15 D55 ⋄

HUHN, A.P. *On nonmodular n-distributive lattices: the decision problem for identities in finite n-distributive lattices* ⋄ B25 C13 G10 ⋄

KORKINA, E.I. & KUSHNIRENKO, A.G. *Another proof of the Tarski-Seidenberg theorem (Russian)* ⋄ B25 ⋄

LEVITZ, H. *Decidability of some problems pertaining to base 2 exponential diophantine equations* ⋄ B25 ⋄

MACINTYRE, A. *Effective determination of the zeros of p-adic exponential functions* ⋄ B25 ⋄

MAKANIN, G.S. *On the decidability of the theory of a free group (Russian)* ⋄ B25 C60 C98 D35 ⋄

MILLAR, T.S. *Decidable Ehrenfeucht theories*
⋄ B25 C15 C57 C98 D30 D35 D55 ⋄

MOLZAN, B. *On the theory of Boolean algebras in the logic with Ramsey quantifiers* ⋄ B25 C10 C80 G05 ⋄

MUCHNIK, A.A. *Games on infinite trees and automata with dead ends. A new proof of decidability for the monadic theory with two successor functions (Russian)*
⋄ B15 B25 C85 D05 ⋄

MULLER, D.E. & SCHUPP, P.E. *Alternating automata on infinite objects, determinacy and Rabin's Theorem*
⋄ B25 C85 D05 E60 ⋄

MULLER, D.E. & SCHUPP, P.E. *The theory of ends, pushdown automata, and second-order logic*
⋄ B15 B25 D10 D80 ⋄

PAPPAS, P. *The model-theoretic structure of abelian group rings* ⋄ B25 C20 C60 C90 ⋄

POINT, F. *Transfer properties of discriminator varieties*
⋄ C10 C25 C30 C35 ⋄

RAPP, A. *Elimination of Malitz quantifiers in stable theories* ⋄ C10 C45 C55 C80 ⋄

RAPP, A. *The ordered field of real numbers and logics with Malitz quantifiers* ⋄ B25 C10 C55 C65 C80 ⋄

SAWAMURA, H. *Axiomatization of computer-oriented modal logic and decision procedure* ⋄ B25 B45 ⋄

TOURAILLE, A. *Elimination des quantificateurs dans la theorie elementaire des algebres de Boole munies d'une famille d'ideaux distingues (English summary)*
⋄ C10 G05 ⋄

URSINI, A. *Decision problems for classes of diagonalizable algebras* ⋄ B25 B45 C05 D35 ⋄

WEISPFENNING, V. *Quantifier elimination for distributive lattices and measure algebras*
⋄ B25 B50 C10 C65 G05 G10 ⋄

WEISPFENNING, V. *Quantifier elimination for modules*
⋄ C10 C35 C57 C60 ⋄

WEISPFENNING, V. *The complexity of elementary problems in archimedian ordered groups*
⋄ B25 C10 D15 D40 ⋄

C13 Finite structures

1917
HUNTINGTON, E.V. *Complete existential theory of the postulates for serial order* ⋄ B30 C13 C35 E07 ⋄

HUNTINGTON, E.V. *Complete existential theory of the postulates for well ordered sets* ⋄ B30 C13 C35 E07 ⋄

1933
GOEDEL, K. *Zum Entscheidungsproblem des logischen Funktionenkalkuels* ⋄ B10 B20 B25 C13 D35 ⋄

WAJSBERG, M. *Untersuchungen ueber den Funktionenkalkuel fuer endliche Individuenbereiche* ⋄ B10 C13 ⋄

1946
ZHEGALKIN, I.I. *Sur le probleme de la resolubilite pour les classes finies (Russian) (French summary)* ⋄ B20 B25 C13 ⋄

1947
JONSSON, B. & TARSKI, A. *Direct decomposition of finite algebraic systems* ⋄ C05 C13 ⋄

1949
KUSTAANHEIMA, P. *Ueber die Vollstaendigkeit der Axiomensysteme mit einem endlichen Individuenbereich* ⋄ C13 C35 C85 ⋄

1950
TRAKHTENBROT, B.A. *The impossibility of an algorithm for the decision problem in finite domains (Russian)* ⋄ C13 D35 ⋄

1951
ROBINSON, A. *On axiomatic systems which possess finite models* ⋄ C13 C52 C60 ⋄

1952
CRAIG, W. *Incompletability, with respect to validity in every finite nonempty domain of first order functional calculus* ⋄ C13 D35 ⋄

KALICKI, J. *On comparison of finite algebras* ⋄ B25 C05 C13 ⋄

SCHOLZ, H. *Ein ungeloestes Problem in der symbolischen Logik (problem 1)* ⋄ B10 C13 ⋄

1953
DAVIS, R.L. *The number of structures of finite relations* ⋄ C13 ⋄

TRAKHTENBROT, B.A. *On recursively separability (Russian)* ⋄ B10 C13 D35 ⋄

ZYKOV, A.A. *The spectrum problem in the extended predicate calculus (Russian)* ⋄ B15 C13 ⋄

1954
LYNDON, R.C. *Identities in finite algebras* ⋄ C05 C13 ⋄

TARSKI, A. *Contributions to the theory of models I,II* ⋄ C13 C30 C52 ⋄

1955
ASSER, G. *Das Repraesentantenproblem im Praedikatenkalkuel der ersten Stufe mit Identitaet* ⋄ C13 ⋄

1956
HENKIN, L. *Two concepts from the theory of models* ⋄ C13 C52 C60 ⋄

MOSTOWSKI, ANDRZEJ *Concerning a problem of H.Scholz* ⋄ C13 D20 ⋄

SCOTT, D.S. *Equationally complete extensions of finite algebras* ⋄ C05 C13 ⋄

1957
VAUGHT, R.L. *Sentences true in all constructive models* ⋄ C13 C57 D35 ⋄

1958
MAL'TSEV, A.I. *Homomorphisms onto finite groups (Russian)* ⋄ C07 C13 C60 ⋄

SCOTT, D.S. & SUPPES, P. *Foundational aspects of theories of measurement* ⋄ B30 C13 C52 ⋄

1959
APPEL, K.I. *Horn sentences in identity theory* ⋄ C13 C30 C52 ⋄

TAIT, W.W. *A counterexample to a conjecture of Scott and Suppes* ⋄ C13 ⋄

1960
BUTLER, J.W. *On complete and independent sets of operations in finite algebras* ⋄ C05 C13 ⋄

VAUGHT, R.L. *Sentences true in all constructive models* ⋄ C13 C57 D35 ⋄

1961
HAILPERIN, T. *A complete set of axioms for logical formulas invalid in some finite domain* ⋄ B20 C13 ⋄

MAL'TSEV, A.I. *Undecidability of the elementary theory of finite groups (Russian)* ⋄ C13 D35 ⋄

1962
KOSTYRKO, V.F. *On an error in the paper of I.I.Zhegalkin "Sur le probleme de la resolubilite pour les classes finies" (Russian)* ⋄ B25 C13 ⋄

TAJTSLIN, M.A. *Effective inseparability of the set of identically true and the set of finitely refutable formulas of the elementary theory of lattices (Russian)* ⋄ C13 D25 D35 G10 ⋄

1963

LAVROV, I.A. *Effective inseparability of the sets of identically true formulae and finitely refutable formulae for certain elementary theories (Russian)*
⋄ C13 D25 D35 ⋄

TAJTSLIN, M.A. *Undecidability of elementary theories of certain classes of finite commutative associative rings (Russian)* ⋄ C13 D35 ⋄

1964

ERSHOV, YU.L. *Unsolvability of theories of symmetric and simple finite groups (Russian)* ⋄ C13 C60 D35 ⋄

HUBER-DYSON, V. *On the decision problem for theories of finite models* ⋄ B25 C13 D35 ⋄

KOGALOVSKIJ, S.R. *The relation between finitely-projective and finitely-reductive classes of models (Russian)*
⋄ C13 C52 ⋄

1965

ASTROMOFF, A. *Some structure theorems for primal and categorical algebras* ⋄ C05 C13 ⋄

FRAISSE, R. *A hypothesis concerning the extension of finite relations and its verification for certain special cases*
⋄ C07 C13 E07 ⋄

TAJTSLIN, M.A. *On the theory of finite rings with division (Russian)* ⋄ C13 D35 ⋄

1966

GLADSTONE, M.D. *Finite models for inequations*
⋄ B25 C05 C13 ⋄

GUREVICH, Y. *On the decision problem for pure restricted predicate logic (Russian)* ⋄ B20 C13 D35 ⋄

GUREVICH, Y. *The decision problem for the restricted predicate calculus (Russian)* ⋄ B20 C13 D35 ⋄

JONSSON, B. *The unique factorization problem for finite relational structures* ⋄ C05 C13 ⋄

1967

AX, J. *Solving diophantine problems modulo every prime*
⋄ B25 C13 C20 C60 ⋄

CHANG, C.C. *Cardinal factorization of finite relational structures* ⋄ C05 C13 C30 ⋄

ERSHOV, YU.L. *Elementary theories of Post varieties (Russian)* ⋄ B25 C05 C13 G20 ⋄

EVANS, T. *The spectrum of a variety* ⋄ C05 C13 ⋄

FRAISSE, R. *Une hypothese sur l'extension des relations finies et sa verification dans certaines classes particulieres. II* ⋄ C07 C13 E07 ⋄

GRAETZER, G. *On the spectra of classes of algebras*
⋄ C05 C13 ⋄

KEISLER, H.J. *Ultraproducts of finite sets*
⋄ C13 C20 C55 E05 ⋄

LOVASZ, L. *Operations with structures*
⋄ C07 C13 C30 ⋄

YANKOV, V.A. *Finite validity of formulas of a special form (Russian)* ⋄ B55 C13 F50 ⋄

1968

AX, J. *The elementary theory of finite fields*
⋄ B25 B30 C13 C20 C60 ⋄

MCKENZIE, R. *On finite groupoids and $\aleph$-prime algebras*
⋄ C05 C13 ⋄

OBERSCHELP, W. *Strukturzahlen in endlichen Relationssystemen* ⋄ C13 ⋄

RAUTENBERG, W. *Unterscheidbarkeit endlicher geordneter Mengen mit gegebener Anzahl von Quantoren*
⋄ B20 C07 C13 C65 D35 ⋄

SAADE, M. *A decision problem for finite groupoids (Russian summary)* ⋄ B25 C13 ⋄

1969

BALDI, G. *Modelli finiti nella teoria degli insiemi (English summary)* ⋄ C13 C62 E70 ⋄

GLEBSKIJ, YU.V. & KOGAN, D.I. & LIOGON'KIJ, M.I. & TALANOV, V.A. *Range and degree of realizability of formulas in the restricted predicate calculus (Russian)*
⋄ B10 C07 C13 ⋄

HUBER-DYSON, V. *On the decision problem for extensions of a decidable theory* ⋄ B25 C13 C60 D35 ⋄

SAADE, M. *A comment on a paper by Evans*
⋄ C05 C13 ⋄

1970

LIOGON'KIJ, M.I. *On the question of quantitative characteristics of logical formulae (Russian) (English summary)* ⋄ B20 B25 C13 ⋄

SHELAH, S. *On the cardinality of ultraproduct of finite sets*
⋄ C13 C20 C55 E05 ⋄

1971

GARFUNKEL, S. & SHANK, H.S. *On the undecidability of finite planar graphs* ⋄ C13 D35 ⋄

HENSON, C.W. *A family of countable homogeneous graphs* ⋄ C13 C15 C50 C65 ⋄

KOSTYRKO, V.F. *The reduction class*
$\forall x \forall y \exists z F(x,y,z) \wedge \forall^m \mathfrak{A}(F)$ *(Russian) (English summary)* ⋄ B20 C13 D35 ⋄

LINDNER, C.C. *Finite partial cyclic triple systems can be finitely embedded* ⋄ C05 C13 D35 ⋄

LINDNER, C.C. *Identities preserved by singular direct product* ⋄ C05 C13 C40 ⋄

ROMOV, B.A. *Definability by formulas for predicates on a finite model (Russian) (English summary)*
⋄ C13 C40 C60 ⋄

TAYLOR, W. *Atomic compactness and elementary equivalence* ⋄ C05 C13 C50 C65 ⋄

VANCKO, R.M. *The spectrum of some classes of free universal algebras* ⋄ C05 C13 ⋄

YASUHARA, M. *On a problem of Mostowski on finite spectra* ⋄ C13 ⋄

1972

BULLOCK, A.M. & SCHNEIDER, H.H. *A calculus for finitely satisfiable formulas with identity* ⋄ B10 C07 C13 ⋄

BURRIS, S. *Models in equational theories of unary algebras* ⋄ B25 C05 C13 C35 ⋄

ERSHOV, YU.L. *Elementary group theories (Russian)*
⋄ B25 C05 C13 C60 D35 ⋄

GARFUNKEL, S. & SHANK, H.S. *On the indecidability of finite planar cubic graphs* ⋄ C13 D35 ⋄

HAUSCHILD, K. & HERRE, H. & RAUTENBERG, W. *Entscheidbarkeit der elementaren Theorien der endlichen Baeume und verwandter Klassen endlicher Strukturen (Russian, English and French summaries)*
⋄ B25 C13 ⋄

JONES, N.D. & SELMAN, A.L. *Turing machines and the spectra of first-order formulas with equality*
◇ C13 D10 D15 ◇

KOPFERMANN, K. *Ultraprodukte endlicher Koerper*
◇ C13 C20 C60 ◇

KRAUSS, P.H. *On primal algebras* ◇ C05 C13 ◇

LINDNER, C.C. *Identities preserved by the singular direct product II* ◇ C05 C13 C40 ◇

ROEDDING, D. & SCHWICHTENBERG, H. *Bemerkungen zum Spektralproblem* ◇ B15 C13 D10 D15 ◇

SMOLIN, V.P. *On the question of definability in the lower predicate calculus on a finite model without equality (Russian) (English summary)* ◇ C13 C40 ◇

VANCKO, R.M. *The family of locally independent sets in finite algebras* ◇ C05 C13 ◇

1973

BULLOCK, A.M. & SCHNEIDER, H.H. *On generating the finitely satisfiable formulas* ◇ B10 C07 C13 D25 ◇

DAHN, B.I. $\aleph_0$-*kategorische zyklenbeschraenkte Graphen*
◇ C13 C35 C65 ◇

DAHN, B.I. $\aleph_0$-*kategorische zyklenbeschraenkte Graphen (English summary)* ◇ C13 C35 C65 ◇

GUREVICH, Y. *Formulas with a single* $\forall$ *(Russian)*
◇ B20 B25 C13 ◇

HAJEK, P. *Automatic listing of important observational statements. I,II* ◇ B35 C13 C80 C90 ◇

KEISLER, H.J. & WALKOE JR., W.J. *The diversity of quantifier prefixes* ◇ B10 C07 C13 D55 ◇

KOPFERMANN, K. *Ultraprodukte endlicher Koerper*
◇ C13 C20 C60 ◇

KRAUSS, P.H. *On quasi primal algebras* ◇ C05 C13 ◇

KUEKER, D.W. *A note on the elementary theory of finite abelian groups* ◇ C13 C20 C60 ◇

MCLAUGHLIN, T.G. *A non-enumerability theorem for infinite classes of finite structures* ◇ C13 D25 D55 ◇

SELMAN, A.L. *Sets of formulas valid in finite structures*
◇ C13 D30 ◇

WASZKIEWICZ, J. *On cardinalities of algebras of formulas for* ω_0-*categorical theories* ◇ C13 C35 G05 ◇

WOLENSKI, J. *Wajsberg on the first-order predicate calculus for the finite models* ◇ A10 C13 ◇

1974

BORSHCHEV, V.B. & KHOMYAKOV, M.V. *Schemes for functions and relations (Russian)* ◇ B75 C13 ◇

FAGIN, R. *Generalized first-order spectra and polynomial-time recognizable sets*
◇ C13 C85 D15 ◇

HAJEK, P. *Automatic listing of important observational statements. III* ◇ B35 C13 C80 C90 ◇

JONES, N.D. & SELMAN, A.L. *Turing machines and the spectra of first-order formulas* ◇ C13 D10 D15 ◇

KEIMEL, K. & WERNER, H. *Stone duality for varieties generated by quasi-primal algebras*
◇ C05 C13 C90 ◇

MASHURYAN, A.S. *Recursive definitions on induction models (Russian) (Armenian summary)*
◇ C13 D20 ◇

RACKOFF, C.W. *On the complexity of the theories of weak direct products: a preliminary report*
◇ B25 C13 C30 C60 D15 ◇

1975

ASH, C.J. *Sentences with finite models* ◇ C13 ◇

BASARAB, S.A. *The models of the elementary theory of finite abelian groups (Romanian) (English summary)*
◇ B25 C13 C60 ◇

BAUDISCH, A. *Endliche n-aequivalente Gruppen*
◇ C13 C60 ◇

DEUTSCH, M. *Zur Theorie der spektralen Darstellung von Praedikaten durch Ausdruecke der Praedikatenlogik 1.Stufe* ◇ B10 C13 D25 D35 ◇

FAGIN, R. *A spectrum hierarchy* ◇ C13 ◇

FAGIN, R. *A two-cardinal characterization of double spectra* ◇ C13 ◇

FAGIN, R. *Monadic generalized spectra* ◇ C13 C85 ◇

FROEMKE, J. & QUACKENBUSH, R.W. *The spectrum of an equational class of groupoids* ◇ C05 C13 ◇

GOULD, M. *An equational spectrum giving cardinalities of endomorphism monoids* ◇ C05 C13 ◇

HAY, L. *Spectra and halting problems*
◇ C13 D10 D25 D30 ◇

LOPEZ, G. *Le probleme de l'isomorphie des restrictions strictes, pour les extensions a un element de relations enchainables non constantes (English summary)*
◇ C13 C30 E07 ◇

MCKENZIE, R. *On spectra, and the negative solution of the decision problem for identities having a finite nontrivial model* ◇ B20 C05 C13 D35 ◇

MURSKIJ, V.L. *A finite basis of identities and other properties of "almost all" finite algebras (Russian)*
◇ C05 C13 C52 ◇

RAKHMATULIN, N.A. *Automata-theoretic characteristics for the spectra of formulas of finite levels (Russian)*
◇ C13 D05 D10 ◇

RAUTENBERG, W. & ZIEGLER, M. *Recursive inseparability in graph theory* ◇ C13 D35 ◇

TAYLOR, W. *The fine spectrum of a variety*
◇ C05 C13 C52 ◇

1976

APPLESON, R.R. *Zero-divisors among finite structures of a fixed type* ◇ C05 C13 C30 ◇

BALDWIN, JOHN T. & BERMAN, J. *Varieties and finite closure conditions* ◇ C05 C13 ◇

CHRISTEN, C. *Spektralproblem und Komplexitaetstheorie*
◇ B10 C13 D10 D15 D25 D35 ◇

CLARK, D.M. & KRAUSS, P.H. *Para primal algebras*
◇ C05 C13 ◇

FAGIN, R. *Probabilities on finite models* ◇ C13 ◇

FRIED, M. & SACERDOTE, G.S. *Solving diophantine problems over all residue class fields of a number field and all finite fields* ◇ B25 C10 C13 C60 D35 ◇

GAO, HENGSHAN *The decision problem of modal predicate calculi. II: On Slomson's reductions (Chinese)*
◇ B45 C13 D35 ◇

HAJEK, P. *Observationsfunktorenkalkuele und die Logik der automatisierten Forschung*
◇ B60 C13 C80 C90 ◇

HAJEK, P. *Some remarks on observational model-theoretic languages* ◇ C13 C95 ◇

KIEFE, C. *Sets definable over finite fields: their zeta-functions* ◇ C13 C60 ◇

MYCIELSKI, J. *A 01-law in a finite space* ◊ C13 ◊
RACKOFF, C.W. *On the complexity of the theories of weak direct powers* ◊ B25 C13 C30 C60 D15 ◊
WILKIE, A.J. *A note on products of finite structures with an application to graphs* ◊ C13 ◊

1977

CLARK, D.M. & KRAUSS, P.H. *Varieties generated by para primal algebras* ◊ C05 C13 ◊
DEUTSCH, M. *Zur Axiomatisierung zyklischer Gruppen* ◊ C13 C60 ◊
FAGIN, R. *The number of finite relational structures* ◊ C07 C13 ◊
GACS, P. & LOVASZ, L. *Some remarks on generalized spectra* ◊ B15 C13 D15 ◊
HAJEK, P. *Generalized quantifiers and finite sets* ◊ B25 B50 C13 C40 C80 ◊
LADNER, R.E. *Application of model theoretic games to discrete linear orders and finite automata* ◊ B15 B25 C07 C13 C65 C85 D05 E07 E60 ◊
OBERSCHELP, W. *Monotonicity for structure numbers in theories without identity* ◊ C13 ◊

1978

HAJEK, P. & HAVRANEK, T. *Mechanizing hypothesis formation. Mathematical foundations for a general theory* ◊ A05 B35 C13 C80 C90 D80 ◊
LOPEZ, G. *L'indeformabilite des relations et multirelations binaires* ◊ C07 C13 E07 ◊
NEUMANN, W.D. *Mal'cev conditions, spectra and Kronecker product* ◊ C05 C13 ◊
SEJTENOV, S.M. *Theory of finite fields with an additional predicate distinguishing a subfield (Russian)* ◊ B25 C13 C60 ◊
TINHOFER, G. *On the simultaneous isomorphism of special relations (isomorphism of automata)* ◊ C13 D05 ◊

1979

BLASS, A.R. & HARARY, F. *Properties of almost all graphs and complexes* ◊ C10 C13 C65 ◊
MASHURYAN, A.S. *On a class of primitive recursive functions (Russian)* ◊ C13 D20 ◊
OIKKONEN, J. *On PC- and RPC-classes in generalized model theory* ◊ C13 C52 C75 C80 C95 ◊
POTTHOFF, K. *Ultraproduits des groupes finis et applications a la theorie de Galois* ◊ C13 C20 C60 ◊

1980

BERMAN, J. *A proof of Lyndon's finite basis theorem* ◊ C05 C13 ◊
CLARK, D.M. & KRAUSS, P.H. *Plain para primal algebras* ◊ C05 C13 ◊
DICHEV, KH.V. *Representation of data in calculation of logical formulas on finite models (Russian)* ◊ B25 C13 ◊
EBBINGHAUS, H.-D. *Mathematische Logik und Informatik* ◊ C13 D10 D15 D98 E70 H15 ◊
GORALCIK, P. & GORALCIKOVA, A. & KOUBEK, V. *Testing of properties of finite algebras* ◊ C13 D15 D40 ◊
KOPPELBERG, S. *Cardinalities of ultraproducts of finite sets* ◊ C13 C20 C55 E05 E10 ◊

LOVASZ, L. *Efficient algorithms: an approach by formal logic* ◊ B35 C13 D15 ◊
LYNCH, J.F. *Almost sure theories* ◊ C13 C15 C30 ◊
MASHURYAN, A.S. *On the class of primitive recursive functions definable on finite models (Russian) (Armenian summary)* ◊ C13 D20 ◊
QUACKENBUSH, R.W. *Algebras with minimal spectrum* ◊ C05 C13 ◊
TAYLOR, W. *Mal'tsev conditions and spectra* ◊ C05 C13 C52 ◊

1981

BANASCHEWSKI, B. *When are divisible abelian groups injective?* ◊ C13 C60 G10 ◊
BLATTER, C. & SPECKER, E. *Le nombre de structures finies d'une theorie a caractere fini* ◊ C13 C52 C85 ◊
CSAKANY, B. & GAVALCOVA, T. *Three-element quasi-primal algebras* ◊ C05 C13 ◊
DZIOBIAK, W. *Concerning axiomatizability of the quasivariety generated by a finite Heyting or topological boolean algebra* ◊ C05 C13 C52 ◊
HODGES, W. *Encoding orders and trees in binary relations* ◊ C13 C45 ◊
HUBER-DYSON, V. *A reduction of the open sentence problem for finite groups* ◊ C13 C60 D40 ◊
MCKENZIE, R. *Residually small varieties of semigroups* ◊ C05 C13 ◊
POLAK, L. & ROSICKY, J. *Implicit operations on finite algebras* ◊ C05 C13 ◊
SHABANOV-KUSHNARENKO, YU.P. *Modeling of finite mathematical structures (Russian)* ◊ C13 ◊
SLOBODSKOJ, A.M. *Unsolvability of the universal theory of finite groups (Russian)* ◊ C13 C60 D35 D40 ◊
TALANOV, V.A. *Asymptotic solvability of logical formulas (Russian)* ◊ C13 ◊

1982

AANDERAA, S.O. & BOERGER, E. & LEWIS, H.R. *Conservative reduction classes of Krom formulas* ◊ B20 C13 D35 ◊
APPS, A.B. *Boolean powers of groups* ◊ C13 C30 C60 ◊
CALL, R.L. *Systems with parity quantifiers* ◊ C13 C80 ◊
CHERLIN, G.L. & FELGNER, U. *Quantifier eliminable groups* ◊ C10 C13 C60 ◊
DENECKE, K. *Preprimal algebras* ◊ B50 C05 C13 ◊
DZIOBIAK, W. *Concerning axiomatizability of the quasivariety generated by a finite Heyting or topological boolean algebra* ◊ C05 C13 C52 G10 ◊
GAIFMAN, H. *On local and non-local properties* ◊ C07 C13 C65 E47 ◊
HUBER-DYSON, V. *Symmetric groups and the open sentence problem* ◊ C13 C60 D35 ◊
LACHLAN, A.H. *Finite homogeneous simple digraphs* ◊ C13 C50 C65 ◊
LYNCH, J.F. *Complexity classes and theories of finite models* ◊ C13 D15 ◊
POIZAT, B. *Deux ou trois choses que je sais de L_n* ◊ B20 C07 C13 C50 ◊

QUACKENBUSH, R.W. *Enumeration in classes of ordered structures* ◊ C13 C98 ◊

REITERMAN, J. *The Birkhoff theorem for finite algebras* ◊ C05 C13 ◊

SMORYNSKI, C.A. *The finite inseparability of the first-order theory of diagonalisable algebras* ◊ B45 C13 D35 F30 ◊

1983

AJTAI, M. *Σ_1^1-formulae on finite structures* ◊ C13 C62 D15 H15 ◊

BELLISSIMA, F. & MIROLLI, M. *On the axiomatization of finite K-frames* ◊ B45 C13 C90 ◊

FURMANOWSKI, T. *The logic of algebraic rules as a generalization of equational logic* ◊ C05 C07 C13 ◊

GRANDJEAN, E. *Complexity of the first-order theory of almost all finite structures* ◊ B25 C13 D15 ◊

GUREVICH, Y. & SHELAH, S. *Random models and the Goedel case of the decision problem* ◊ B20 B25 C13 D35 ◊

HALPERN, M. *Inductive inference in finite algebraic structures* ◊ B48 C13 ◊

LOPEZ, G. *Reconstruction d'une S-expansion* ◊ C07 C13 E07 ◊

MCKENZIE, R. *Finite forbidden lattices* ◊ C13 G10 ◊

ULRICH, D. *The finite model property and recursive bounds on the size of countermodels* ◊ B22 B25 C13 ◊

1984

BLATTER, C. & SPECKER, E. *Recurrence relations for the number of labeled structures on a finite set* ◊ C13 C52 C85 ◊

BOERGER, E. *Spektralproblem and completeness of logical decision problems* ◊ B25 C13 D15 ◊

BOOLOS, G. *Trees and finite satisfiability: proof of a conjecture of Burgess* ◊ B10 C13 F07 ◊

COMPTON, K.J. *An undecidable problem in finite combinatorics* ◊ C13 D80 ◊

DAHLHAUS, E. *Reduction to NP-complete problems by interpretations* ◊ C13 D15 ◊

FRASNAY, C. *Relations enchaineables, rangements et pseudo-rangements (English summary)* ◊ C07 C13 C30 E07 ◊

FRIEDMAN, H.M. *On the spectra of universal relational sentences* ◊ C13 D15 ◊

GRANDJEAN, E. *Spectre des formules du premier order et complexite algorithmique* ◊ C13 C98 D15 ◊

GRANDJEAN, E. *The spectra of first-order sentences and computational complexity* ◊ C13 D15 ◊

GRANDJEAN, E. *Universal quantifiers and time complexity of random access machines* ◊ C13 D15 ◊

GUREVICH, Y. *Toward logic tailored for computational complexity* ◊ B75 C13 D15 ◊

HODGES, W. *Finite extensions of finite groups* ◊ C13 C50 C60 D35 ◊

LACHLAN, A.H. *Binary homogeneous structures. I* ◊ C10 C13 C15 C45 C50 C57 ◊

LACHLAN, A.H. *On countable stable structures which are homogeneous for a finite relational language* ◊ C10 C13 C15 C45 C50 ◊

LACHLAN, A.H. & SHELAH, S. *Stable structures homogeneous for a binary language* ◊ C10 C13 C15 C45 C50 ◊

MAKOWSKY, J.A. *Model theoretic issues in theoretical computer science, part I: Relational data bases and abstract data types* ◊ B75 C13 C95 ◊

SCARPELLINI, B. *Complete second order spectra* ◊ B15 C13 ◊

SCARPELLINI, B. *Second order spectra* ◊ B15 C13 ◊

TULIPANI, S. *On the universal theory of classes of finite models* ◊ C05 C13 C52 C60 ◊

TUMANOV, V.I. *Finite lattices with no independent basis of quasiidentities (Russian)* ◊ C05 C13 G10 ◊

TURAN, G. *On the definability of properties of finite graphs* ◊ C13 C65 ◊

1985

ASH, C.J. *Pseudovarieties, generalized varieties and similiarly described classes* ◊ C05 C13 ◊

COMPTON, K.J. *Application of a Tauberian theorem to finite model theory* ◊ C13 ◊

ECSEDI-TOTH, P. *A partial solution of the finite spectrum problem* ◊ C13 ◊

GILMAN, R.H. *An application of ultraproducts to finite groups* ◊ C13 C20 C60 ◊

GRANDJEAN, E. *Universal quantifiers and time complexity of random access machines* ◊ C13 D10 D15 ◊

HUHN, A.P. *On nonmodular n-distributive lattices: the decision problem for identities in finite n-distributive lattices* ◊ B25 C13 G10 ◊

KAUFMANN, M. & SHELAH, S. *On random models of finite power and monadic logic* ◊ C13 C85 ◊

LYNCH, J.F. *Probability of first-order sentences about unary functions* ◊ C13 ◊

MCKENZIE, R. *The structure of finite algebras* ◊ C05 C13 C98 ◊

PLONKA, J. *On identities satisfied in finite algebras* ◊ C05 C13 ◊

SACKS, G.E. *Some open questions in recursion theory* ◊ C13 D98 E60 ◊

SARACINO, D.H. & WOOD, C. *Finite QE rings in characteristic p^2* ◊ C13 C50 C60 ◊

SCARPELLINI, B. *Lower bound results on lengths of second-order formulas* ◊ B15 C13 D10 F20 F35 ◊

SZWAST, W. *On some properties of Horn's spectra (Polish) (English summary)* ◊ C13 D15 ◊

TALJA, J. *Semantic games on finite trees* ◊ C13 ◊

C57 Recursion-theoretic model theory

1926
HERMANN, G. *Die Frage der endlich vielen Schritte in der Theorie der Polynomideale* ⋄ C57 C60 D45 F55 ⋄

1950
KREISEL, G. *Note on arithmetic models for consistent formulae of the predicate calculus I*
⋄ B10 C57 D45 D55 F30 ⋄

1953
HASENJAEGER, G. *Eine Bemerkung zu Henkin's Beweis fuer die Vollstaendigkeit des Praedikatenkalkuels der ersten Stufe* ⋄ B10 C07 C57 D55 ⋄
KREISEL, G. *Note on arithmetic models for consistent formulae of the predicate calculus II*
⋄ B10 C57 D45 D55 F30 ⋄
MOSTOWSKI, ANDRZEJ *On a system of axioms which has no recursively enumerable arithmetic model*
⋄ C57 D45 E30 E70 ⋄

1955
FROEHLICH, A. & SHEPHERDSON, J.C. *Effective procedures in field theory* ⋄ C57 C60 D45 ⋄
MOSTOWSKI, ANDRZEJ *A formula with no recursively enumerable model* ⋄ C57 D45 ⋄

1956
MOSTOWSKI, ANDRZEJ *Development and applications of the "projective" classification of sets of integers*
⋄ C57 C80 D55 E15 E45 ⋄

1957
HENKIN, L. *Sums of squares* ⋄ C57 C60 ⋄
KREISEL, G. *Sums of squares*
⋄ C57 C60 D20 F07 F99 ⋄
MOSTOWSKI, ANDRZEJ *On recursive models of formalised arithmetic* ⋄ C57 C62 D45 ⋄
RABIN, M.O. *Computable algebraic systems*
⋄ C57 C60 D40 D45 D80 ⋄
VAUGHT, R.L. *Sentences true in all constructive models*
⋄ C13 C57 D35 ⋄

1958
KREISEL, G. *Mathematical significance of consistency proofs* ⋄ C57 C60 F05 F25 F50 ⋄
RABIN, M.O. *On recursively enumerable and arithmetic models of set theory* ⋄ C57 C62 D45 ⋄

1960
RABIN, M.O. *Computable algebra, general theory and theory of computable fields* ⋄ C57 C60 D45 ⋄
SCOTT, D.S. *On a theorem of Rabin*
⋄ C35 C57 C62 D35 ⋄
SHOENFIELD, J.R. *Degrees of models*
⋄ C57 D30 D35 ⋄

VAUGHT, R.L. *Sentences true in all constructive models*
⋄ C13 C57 D35 ⋄

1961
MAL'TSEV, A.I. *Constructive algebra I (Russian)*
⋄ C57 C98 F60 F98 ⋄
ROBINSON, A. *Model theory and non-standard arithmetic*
⋄ C35 C50 C57 C60 C62 H15 ⋄
SCOTT, D.S. *On constructing models for arithmetic*
⋄ C20 C57 C62 H15 ⋄

1962
GRZEGORCZYK, A. *A theory without recursive models*
⋄ C57 C62 D55 ⋄
KENT, C.F. *Constructive analogues of the group of permutations of the natural numbers* ⋄ C57 D45 ⋄
MAL'TSEV, A.I. *On recursive abelian groups (Russian)*
⋄ C57 C60 D45 ⋄
MAL'TSEV, A.I. *Strongly related models and recursively complete algebras (Russian)*
⋄ C57 C60 C62 D45 ⋄
MOSTOWSKI, ANDRZEJ *L'espace des modeles d'une theorie formalisee et quelques-unes de ses applications*
⋄ B50 C07 C40 C57 C80 C85 C90 D45 ⋄

1963
FRENKEL', V.I. *Algorithmic problems in partially ordered groups (Russian)* ⋄ C57 C60 D40 ⋄

1964
CELLUCCI, C. *Categorie ricorsive* ⋄ C57 D20 G30 ⋄
FRENKEL', V.I. *On the effective partial ordering of finitely defined groups (Russian)* ⋄ C57 C60 ⋄
MOSCHOVAKIS, Y.N. *Recursive metric spaces*
⋄ C57 C65 D45 F60 ⋄

1965
PUTNAM, H. *Trial and error predicates and the solution to a problem of Mostowski* ⋄ C57 D20 D55 ⋄

1966
EHRENFEUCHT, A. & KREISEL, G. *Strong models of arithmetic* ⋄ C57 C62 ⋄
MOSCHOVAKIS, Y.N. *Notation systems and recursive ordered fields* ⋄ C57 C60 D45 F60 ⋄
NOGINA, E.YU. *On effective topological spaces*
⋄ C57 C65 ⋄

1967
MAYOH, B.H. *Groups and semigroups with solvable word problems* ⋄ C57 C60 D40 ⋄
VUCKOVIC, V. *Recursive models for three-valued propositional calculi with classical implication*
⋄ B50 C57 C90 F30 ⋄

1968

CLEAVE, J.P. *Hyperarithmetic ultrafilters*
⋄ C57 D55 E05 ⋄

ERSHOV, YU.L. *Numbered fields* ⋄ C57 C60 D45 ⋄

MADISON, E.W. *Computable algebraic structures and nonstandard arithmetic* ⋄ C57 C60 C62 D45 ⋄

MADISON, E.W. *Structures elementarily closed relative to a model for arithmetic* ⋄ C57 C60 C62 ⋄

1969

DEKKER, J.C.E. *Countable vector spaces with recursive operations. Part I* ⋄ C57 C60 D45 D50 ⋄

HENSEL, G. & PUTNAM, H. *Normal models and the field Σ_1^** ⋄ C07 C30 C57 D55 ⋄

NOGINA, E.YU. *Correlations between certain classes of effectively topological spaces (Russian)*
⋄ C57 C65 D45 ⋄

1970

APPLEBAUM, C.H. & DEKKER, J.C.E. *Partial recursive functions and ω-functions* ⋄ C57 D50 ⋄

COLLINS, D.J. *A universal semigroup (Russian)*
⋄ C05 C57 D05 D40 ⋄

FEINER, L. *Hierarchies of boolean algebras*
⋄ C57 D45 G05 ⋄

HAMILTON, A.G. *Bases and α-dimensions of countable vector spaces with recursive operations*
⋄ C57 C60 D45 D50 ⋄

KOSOVSKIJ, N.K. *Some questions in the constructive theory of normed boolean algebras (Russian)*
⋄ C57 F60 G05 ⋄

LACHLAN, A.H. & MADISON, E.W. *Computable fields and arithmetically definable ordered fields*
⋄ C57 C60 D45 ⋄

MADISON, E.W. *A note on computable real fields*
⋄ C57 C60 D45 F60 ⋄

SEIDENBERG, A. *Construction of the integral closure of a finite integral domain* ⋄ C57 C60 F55 ⋄

1971

DEKKER, J.C.E. *Countable vector spaces with recursive operations. Part II* ⋄ C57 C60 D45 D50 ⋄

DEKKER, J.C.E. *Two notes on vector spaces with recursive operations* ⋄ C57 C60 D45 ⋄

KHISAMIEV, N.G. *Strongly constructive models (Russian) (Kazakh summary)* ⋄ C30 C57 ⋄

MADISON, E.W. *Some remarks on computable (non-archimedean) ordered fields*
⋄ C57 C60 D45 ⋄

PERETYAT'KIN, M.G. *Strongly constructive models and numerations of the boolean algebra of recursive sets (Russian)* ⋄ C57 D20 D45 G05 ⋄

VILLE, F. *Complexite des structures rigidement contenues dans une theorie du premier ordre* ⋄ C50 C57 ⋄

1972

CUTLAND, N.J. *Π_1^1-models and Π_1^1-categoricity*
⋄ C57 C70 D55 ⋄

DENISOV, S.D. *Models of a noncontradictory formulas and the Ershov hierarchy (Russian)* ⋄ C07 C57 D55 ⋄

ERSHOV, YU.L. *Existence of constructivizations (Russian)*
⋄ C57 D45 ⋄

HOWARD, P.E. *A proof of a theorem of Tennenbaum*
⋄ C57 C62 H15 ⋄

MACINTYRE, A. *Omitting quantifier-free types in generic structures* ⋄ C25 C57 C60 C75 D30 D40 ⋄

MACINTYRE, A. *On algebraically closed groups*
⋄ C25 C57 C60 D40 ⋄

MADISON, E.W. *Real fields with characterization of the natural numbers* ⋄ C57 H15 ⋄

PARIKH, R. *A note on rigid substructures* ⋄ C50 C57 ⋄

PIXLEY, A.F. *Local Mal'cev conditions* ⋄ C05 C57 ⋄

1973

ALTON, D.A. & MADISON, E.W. *Computability of boolean algebras and their extensions* ⋄ C57 D45 G05 ⋄

EMDE BOAS VAN, P. *Mostowski's universal set-algebra*
⋄ C50 C57 C65 D45 E07 E20 G05 ⋄

ERSHOV, YU.L. *Constructive models (Russian)*
⋄ C57 D45 ⋄

ERSHOV, YU.L. *Skolem functions and constructive models (Russian)* ⋄ C57 C65 F50 ⋄

FEINER, L. *Degrees of nonrecursive presentability*
⋄ C57 D30 D45 ⋄

GONCHAROV, S.S. *Constructivizability of superatomic boolean algebras (Russian)* ⋄ C57 D45 G05 ⋄

GONCHAROV, S.S. & NURTAZIN, A.T. *Constructive models of complete solvable theories (Russian)*
⋄ C45 C50 C57 G05 ⋄

PERETYAT'KIN, M.G. *A strongly constructive model without elementary submodels and extensions (Russian)*
⋄ C35 C57 D45 ⋄

PERETYAT'KIN, M.G. *Every recursively enumerable extension of a theory of linear order has a constructive model (Russian)* ⋄ C57 C65 ⋄

PERETYAT'KIN, M.G. *On complete theories with a finite number of denumerable models (Russian)* ⋄ C57 ⋄

SHEVYAKOV, V.S. *Formulas of the restricted predicate calculus which distinguish certain classes of models with simply computable predicates (Russian)* ⋄ C57 ⋄

SUTER, G.H. *Recursive elements and constructive extensions of computable local integral domains*
⋄ C57 C60 D45 ⋄

1974

BAUR, W. *Ueber rekursive Strukturen*
⋄ C50 C57 D45 ⋄

BOONE, W.W. *Between logic and group theory*
⋄ C57 C60 D40 ⋄

CROSSLEY, J.N. & NERODE, A. *Combinatorial functors*
⋄ C57 C62 D45 D50 G30 ⋄

ERSHOV, YU.L. *Theory of numerations, III: Constructive models (Russian)* ⋄ B25 C57 C98 D45 F50 ⋄

HARRINGTON, L.A. *Recursively presentable prime models*
⋄ C35 C50 C57 C60 ⋄

KHISAMIEV, N.G. *Strongly constructive models of a decidable theory (Russian)* ⋄ C15 C57 F60 ⋄

MEJTUS, V.YU. & VERSHININ, K.P. *On some unsolvable problems in computable categories (Russian)*
⋄ C57 D80 G30 ⋄

NURTAZIN, A.T. *Strong and weak constructivization and computable families (Russian)* ⋄ C57 D45 ⋄

TAJMANOV, A.D. *On the elementary theory of topological algebras (Russian)* ◊ C57 C65 ◊

1975

BARWISE, J. & SCHLIPF, J.S. *On recursively saturated models of arithmetic* ◊ C50 C57 C62 D80 H15 ◊

CARSTENS, H.G. *Reducing hyperarithmetic sequences* ◊ C57 D55 ◊

DOBRITSA, V.P. *Recursively numbered classes of constructive extensions and autostability of algebras (Russian)* ◊ C57 D45 ◊

ELLENTUCK, E. *Semigroups, Horn sentences and isolic structures* ◊ C05 C57 D50 E10 ◊

FEFERMAN, S. *Impredicativity of the existence of the largest divisible subgroup of an abelian p-group* ◊ C57 C60 F65 ◊

FOWLER III, N. *Intersections of α-spaces* ◊ C57 C60 D50 ◊

FOWLER III, N. *Sum of α-spaces* ◊ C57 C60 D45 ◊

GONCHAROV, S.S. *Autostability and computable families of constructivizations (Russian)* ◊ C50 C57 D45 ◊

GONCHAROV, S.S. *Certain properties of the constructivization of boolean algebras (Russian)* ◊ C57 D45 G05 ◊

GUHL, R. *A theorem on recursively enumerable vector spaces* ◊ C57 C60 D45 ◊

KREISEL, G. *Observations on a recent generalization of completeness theorems due to Schuette* ◊ A05 C07 C57 F05 F20 F35 F50 ◊

KREISEL, G. & MINTS, G.E. & SIMPSON, S.G. *The use of abstract language in elementary metamathematics: Some pedagogic examples* ◊ A05 C07 C57 C75 F05 F07 F20 F50 ◊

MADISON, E.W. & NELSON, GEORGE C. *Some examples of constructive and non-constructive extension of the countable atomless boolean algebra* ◊ C57 F60 G05 ◊

METAKIDES, G. & NERODE, A. *Recursion theory and algebra* ◊ C57 C60 D25 D45 ◊

PINUS, A.G. *Effective linear orders (Russian)* ◊ C57 D45 E07 F15 ◊

ROBINSON, A. *Algorithms in algebra* ◊ C40 C57 C60 D45 D75 ◊

SEIDENBERG, A. *Construction of the integral closure of a finite integral domain. II* ◊ C57 C60 F55 ◊

1976

BARWISE, J. & SCHLIPF, J.S. *An introduction to recursively saturated and resplendent models* ◊ C15 C40 C50 C57 ◊

BOERGER, E. *On the construction of simple first-order formulae without recursive models* ◊ C57 D25 ◊

CROSSLEY, J.N. & NERODE, A. *Effective dimension* ◊ C57 D45 D50 ◊

DOBRITSA, V.P. & GONCHAROV, S.S. *An example of a constructive abelian group with non-constructivizable reduced subgroup (Russian)* ◊ C57 C60 D45 ◊

DOBRITSA, V.P. *On computable and strictly computable classes of constructive algebras (Russian)* ◊ C57 C60 D45 ◊

FOWLER III, N. *α-decompositions of α-spaces* ◊ C57 C60 D45 D50 ◊

FRIEDMAN, H.M. *The complexity of explicit definitions* ◊ C40 C57 ◊

GONCHAROV, S.S. *Non-self-equivalent constructivization of atomic boolean algebras (Russian)* ◊ C57 D45 G05 ◊

GONCHAROV, S.S. *Restricted theories of constructive boolean algebras (Russian)* ◊ B25 C57 D35 F60 G05 ◊

HERRMANN, E. *On Lindenbaum functions of $\aleph_0$-categorical theories of finite similarity type* ◊ B25 C35 C57 ◊

MORLEY, M.D. *Decidable models* ◊ C50 C57 ◊

PINUS, A.G. *Theories of boolean algebras in a calculus with the quantifier "infinitely many exist" (Russian)* ◊ B25 C10 C35 C57 C80 G05 ◊

SCHMERL, J.H. *Effectiveness and Vaught's gap ω two-cardinal theorem* ◊ C55 C57 ◊

SHANIN, N.A. *On the quantifier of limiting realizability (Russian)* ◊ C57 C80 F50 ◊

1977

APT, K.R. *Recursive embeddings of partial orderings* ◊ C57 D20 G05 ◊

BIENENSTOCK, E. *Sets of degrees of computable fields* ◊ C57 C60 D30 D45 ◊

DOBRITSA, V.P. *Computability of certain classes of constructive algebras (Russian)* ◊ C57 C60 D45 ◊

DROBOTUN, B.N. *Enumerations of simple models (Russian)* ◊ C57 D45 ◊

DROBOTUN, B.N. *On countable models of decidable almost categorical theories (Russian)* ◊ B25 C15 C35 C57 ◊

GONCHAROV, S.S. *The quantity of nonautoequivalent constructivizations (Russian)* ◊ C50 C57 D45 ◊

GUHL, R. *Two notes on recursively enumerable vector spaces* ◊ C57 C60 D45 ◊

HAY, L. & MANASTER, A.B. & ROSENSTEIN, J.G. *Concerning partial recursive similarity transformations of linearly ordered sets* ◊ C57 C65 D20 D25 D30 D45 E07 ◊

HERRMANN, E. *Ueber Lindenbaumfunktionen von $\aleph_0$-kategorischen Theorien endlicher Signatur* ◊ B25 C35 C57 ◊

KALANTARI, I. & RETZLAFF, A.T. *Maximal vector spaces under automorphisms of the lattice of recursively enumerable vector spaces* ◊ C57 C60 D45 ◊

KAUFMANN, M. *A rather classless model* ◊ C50 C57 C62 E65 ◊

KHISAMIEV, N.G. *On the periodical part of a strongly constructivizable abelian group (Russian)* ◊ C57 C60 D45 ◊

METAKIDES, G. *A return to constructive algebra via recursive function theory* ◊ C57 D45 ◊

METAKIDES, G. & NERODE, A. *Recursively enumerable vector spaces* ◊ C57 C60 D45 ◊

REMMEL, J.B. *Maximal and cohesive vector spaces* ◊ C57 D25 D45 ◊

SHANIN, N.A. *On the quantifier of limiting realizability* ◊ B55 C57 C80 F50 ◊

VUCKOVIC, V. *Recursive and recursive enumerable manifolds I,II* ⋄ C57 D45 D80 ⋄

1978

BLOSHCHITSYN, V.YA. & ZAKIR'YANOV, K.KH. *Constructive abelian groups (Russian)* ⋄ C57 C60 D45 ⋄

DOBRITSA, V.P. & KHISAMIEV, N.G. & NURTAZIN, A.T. *Constructive periodic abelian groups (Russian)* ⋄ C57 C60 F60 ⋄

ELLENTUCK, E. *Model theoretic methods in the theory of isols* ⋄ C57 D50 ⋄

FOWLER III, N. *Effective inner product spaces* ⋄ C57 C60 D45 D80 ⋄

GONCHAROV, S.S. *Constructive models of $\aleph_1$-categorical theories (Russian)* ⋄ C35 C57 ⋄

GONCHAROV, S.S. *Strong constructivizability of homogeneous models (Russian)* ⋄ C50 C57 D45 ⋄

KALANTARI, I. *Major subspaces of recursively enumerable vector spaces* ⋄ C57 C60 D25 D45 ⋄

KHISAMIEV, N.G. *Strongly constructive periodic abelian groups (Russian) (Kazakh summary)* ⋄ C57 C60 D45 ⋄

LASCAR, D. *Caractere effectif des theoremes d'approximation d'Artin (English summary)* ⋄ C57 ⋄

LIPSHITZ, L. & NADEL, M.E. *The additive structure of models of arithmetic* ⋄ C15 C50 C57 C62 ⋄

MCALOON, K. *Diagonal methods and strong cuts in models of arithmetic* ⋄ C57 C62 D80 F30 ⋄

METAKIDES, G. *Constructive algebra in a new frame* ⋄ C57 C60 D45 ⋄

MILLAR, T.S. *Foundations of recursive model theory* ⋄ C50 C57 ⋄

NOGINA, E.YU. *Numerierte topologische Raeume (Russisch)* ⋄ C57 C60 D45 ⋄

PERETYAT'KIN, M.G. *Criterion for strong constructivizability of a homogeneous model (Russian)* ⋄ C35 C50 C57 D45 ⋄

REMMEL, J.B. *A r-maximal vector space not contained in any maximal vector space* ⋄ C57 D25 D45 ⋄

REMMEL, J.B. *Recursively enumerable boolean algebras* ⋄ C57 D25 D45 G05 ⋄

RETZLAFF, A.T. *Simple and hyperhypersimple vector spaces* ⋄ C57 D25 D45 ⋄

SCHLIPF, J.S. *Toward model theory through recursive saturation* ⋄ C15 C40 C50 C57 C70 ⋄

SCHMERL, J.H. *A decidable $\aleph_0$-categorical theory with a non-recursive Ryll-Nardzewski function* ⋄ B25 C35 C57 D30 ⋄

1979

BERTONI, A. & MAURI, G. & MIGLIOLI, P.A. *A characterization of abstract data as model-theoretic invariants* ⋄ B75 C57 D45 D80 ⋄

DEMILLO, R.A. & LIPTON, R.J. *Some connections between mathematical logic and complexity theory* ⋄ C57 D15 F30 H15 ⋄

DRIES VAN DEN, L. *Algorithms and bounds for polynomial rings* ⋄ C57 C60 ⋄

DZGOEV, V.D. *Recursive automorphisms of constructive models (Russian)* ⋄ C57 D45 ⋄

GIRSTMAIR, K. *Ueber konstruktive Methoden der Galoistheorie* ⋄ C57 C60 F55 ⋄

KALANTARI, I. *Automorphisms of the lattice of recursively enumerable vector spaces* ⋄ C57 C60 D45 ⋄

KALANTARI, I. & RETZLAFF, A.T. *Recursive constructions in topological spaces* ⋄ C57 C65 D45 ⋄

KHISAMIEV, N.G. *Criterion of the constructivizability of a direct sum of cyclic p-groups (Russian)* ⋄ C57 C60 D45 ⋄

KHISAMIEV, N.G. *On subgroups of finite index of abelian groups (Russian) (Kazakh summary)* ⋄ C57 C60 D45 F60 ⋄

KUDAJBERGENOV, K.ZH. *A theory with two strongly constructivizable models (Russian)* ⋄ C57 D45 ⋄

KUKIN, G.P. *Subalgebras of finitely defined Lie algebras (Russian)* ⋄ C05 C57 ⋄

LERMAN, M. & SCHMERL, J.H. *Theories with recursive models* ⋄ C35 C57 C65 ⋄

MANDERS, K.L. *The theory of all substructures of a structure: Characterisation and decision problems* ⋄ B20 B25 C57 C60 ⋄

MEAD, J. *Recursive prime models for boolean algebras* ⋄ C50 C57 G05 ⋄

METAKIDES, G. & NERODE, A. *Effective content of field theory* ⋄ C57 C60 D45 ⋄

METAKIDES, G. & REMMEL, J.B. *Recursion theory on orderings I. A model theoretic setting* ⋄ C57 D45 ⋄

MILLAR, T.S. *A complete, decidable theory with two decidable models* ⋄ C57 ⋄

REMMEL, J.B. *R-maximal boolean algebras* ⋄ C57 D45 G05 ⋄

RETZLAFF, A.T. *Direct summands of recursively enumerable vector spaces* ⋄ C57 C60 D45 ⋄

1980

BERTONI, A. & MAURI, G. & MIGLIOLI, P.A. *Towards a theory of abstract data types: a discussion on problems and tools* ⋄ B75 C57 D80 ⋄

DROBOTUN, B.N. & GONCHAROV, S.S. *Numerations of saturated and homogeneous models (Russian)* ⋄ B25 C50 C57 D45 ⋄

DZGOEV, V.D. & GONCHAROV, S.S. *Autostability of models (Russian)* ⋄ C57 D45 G10 ⋄

DZGOEV, V.D. *On the constructivization of certain structures (Russian)* ⋄ C57 D45 ⋄

ERSHOV, YU.L. *Decision problems and constructivizable models (Russian)* ⋄ B25 C57 C60 C98 D45 D98 ⋄

GONCHAROV, S.S. *A totally transcendental decidable theory without constructivizable homogeneous models (Russian)* ⋄ C45 C50 C57 ⋄

GONCHAROV, S.S. *Autostability of models and abelian groups (Russian)* ⋄ C50 C57 C60 D45 ⋄

GONCHAROV, S.S. *Problem of the number of non-self-equivalent constructivizations (Russian)* ⋄ C57 D45 ⋄

GONCHAROV, S.S. *The problem of the number of nonautoequivalent constructivizations (Russian)* ⋄ C57 D45 ⋄

GONCHAROV, S.S. *Totally transcendental theory with non-constructivizable prime model (Russian)* ⋄ C45 C50 C57 D45 ⋄

HARNIK, V. *Game sentences, recursive saturation and definability* ◊ C40 C50 C57 C70 ◊
KUDAJBERGENOV, K.ZH. *On constructive models of undecidable theories (Russian)* ◊ C15 C35 C57 D35 ◊
MANASTER, A.B. & ROSENSTEIN, J.G. *Two-dimensional partial orderings: Recursive model theory* ◊ C57 C65 E07 ◊
MANASTER, A.B. & ROSENSTEIN, J.G. *Two-dimensional partial orderings: Undecidability* ◊ C57 C65 D35 E07 G10 ◊
METAKIDES, G. & NERODE, A. *Recursion theory on fields and abstract dependence* ◊ C57 C60 D45 ◊
MILLAR, T.S. *Homogeneous models and decidability* ◊ C50 C57 ◊
MYERS, R.W. *Complexity of model-theoretic notions* ◊ C07 C35 C57 D55 ◊
REMMEL, J.B. *On r.e. and co-r.e. vector spaces with nonextendible bases* ◊ C57 C60 D45 ◊
REMMEL, J.B. *Recursion theory on orderings II* ◊ C57 D45 ◊
REMMEL, J.B. *Recursion theory on algebraic structures with independent sets* ◊ C57 C60 D45 ◊
SCHLIPF, J.S. *Recursively saturated models of set theory* ◊ C50 C57 C62 E70 H20 ◊
SMORYNSKI, C.A. & STAVI, J. *Cofinal extension preserves recursive saturation* ◊ C50 C57 C62 ◊
TUCKER, J.V. *Computability and the algebra of fields: some affine constructions* ◊ B25 C57 C60 C65 D45 ◊

1981

BERTONI, A. & MAURI, G. & MIGLIOLI, P.A. *Model theoretic aspects of abstract data specification* ◊ B75 C50 C57 ◊
CLOTE, P. *A note on decidable model theory* ◊ B25 C15 C50 C57 D30 ◊
CROSSLEY, J.N. (ED.) *Aspects of effective algebra. Proceedings of a conference at Monash University, Australia, 1-4 August, 1979* ◊ C57 C97 D45 D97 ◊
DOBRITSA, V.P. *On constructivizable abelian groups (Russian)* ◊ C57 C60 D40 D45 ◊
ELLENTUCK, E. *Galois theorems for isolated fields* ◊ C57 C60 D45 D50 ◊
GONCHAROV, S.S. *Groups with a finite number of constructivizations (Russian)* ◊ C57 C60 D45 ◊
HINGSTON, P. *Effective decomposition in Noetherian rings* ◊ C57 C60 D45 ◊
KARR, M. *Summation in finite terms* ◊ C57 ◊
KHISAMIEV, N.G. *Criterion for constructivizability of a direct sum of cyclic p-groups (Russian) (Kazakh summary)* ◊ C57 C60 D45 ◊
KIRBY, L.A.S. & MCALOON, K. & MURAWSKI, R. *Indicators, recursive saturation and expandability* ◊ C50 C57 C62 H15 ◊
KOTLARSKI, H. *On elementary cuts in models of arithmetic* ◊ C50 C57 C62 ◊
LACHLAN, A.H. *Full satisfaction classes and recursive saturation* ◊ C50 C57 C62 ◊

LERMAN, M. *On recursive linear orderings* ◊ C57 C65 D30 D45 D55 E07 ◊
LIN, C. *Recursively presented abelian groups: effective p-group theory. I* ◊ C57 C60 D45 ◊
LIN, C. *The effective content of Ulm's theorem* ◊ C57 C60 D45 ◊
MACINTYRE, A. *The complexity of types in field theory* ◊ C45 C50 C57 C60 ◊
MILLAR, T.S. *Counterexamples via model completions* ◊ C10 C35 C50 C57 ◊
MILLAR, T.S. *Vaught's theorem recursively revisited* ◊ C15 C57 D45 ◊
MURAWSKI, R. *A simple remark on satisfaction classes, indiscernibles and recursive saturation* ◊ C30 C50 C57 H15 ◊
REMMEL, J.B. *Effective structures not contained in recursively enumerable structures* ◊ C57 D45 ◊
REMMEL, J.B. *Recursive isomorphism types of recursive boolean algebras* ◊ C57 D45 G05 ◊
REMMEL, J.B. *Recursive boolean algebras with recursive atoms* ◊ C57 D45 G05 ◊
REMMEL, J.B. *Recursively categorical linear orderings* ◊ C35 C57 C65 ◊
RICHTER, L.J. *Degrees of structures* ◊ C57 D30 D45 ◊
ROCHE LA, P. *Effective Galois theory* ◊ C57 C60 D45 F60 ◊
RODRIGUEZ ARTALEJO, M. *Eine syntaktisch-algebraische Methode zur Konstruktion von Modellen* ◊ B10 C07 C30 C57 ◊
SCHMERL, J.H. *Arborescent structures I: Recursive models* ◊ C50 C57 C65 ◊
SCHMERL, J.H. *Recursively saturated, rather classless models of Peano arithmetic* ◊ C50 C57 C62 E45 E55 ◊
SMITH, RICK L. *Effective valuation theory* ◊ C57 C60 D45 F55 ◊
SMITH, RICK L. *Two theorems on autostability in p-groups* ◊ C57 C60 ◊
SMORYNSKI, C.A. *Cofinal extensions of nonstandard models of arithmetic* ◊ C50 C57 C62 ◊
SMORYNSKI, C.A. *Elementary extensions of recursively saturated models of arithmetic* ◊ C50 C57 C62 ◊
SMORYNSKI, C.A. *Recursively saturated nonstandard models of arithmetic* ◊ C50 C57 C62 ◊
URZYCZYN, P. *The unwind property in certain algebras* ◊ C05 C57 ◊

1982

BALDWIN, JOHN T. *Recursion theory and abstract dependence* ◊ C57 C60 D25 D45 ◊
CEGIELSKI, P. & MCALOON, K. & WILMERS, G.M. *Modeles recursivement satures de l'addition et de la multiplication des entiers naturels (English summary)* ◊ C50 C57 C62 F30 H15 ◊
COMYN, G. *Arbres infinitaires. Approximations et proprietes de calculabilite* ◊ C57 ◊
CROSSLEY, J.N. *The given* ◊ C57 D45 ◊
DZGOEV, V.D. *Constructivizations of direct products of algebraic systems (Russian)* ◊ C30 C57 D45 G05 G10 ◊

EISENBERG, E.F. & REMMEL, J.B. *Effective isomorphisms of algebraic structures* ⋄ C57 D45 ⋄

GONCHAROV, S.S. *Limiting equivalent constructivizations (Russian)* ⋄ C57 D45 F60 ⋄

KALANTARI, I. *Major subsets in effective topology* ⋄ C57 C65 D25 D45 ⋄

KALANTARI, I. & LEGGETT, A. *Simplicity in effective topology* ⋄ C57 C65 D25 ⋄

KNIGHT, J.F. & NADEL, M.E. *Expansions of models and Turing degrees* ⋄ C50 C57 D30 ⋄

LERMAN, M. & ROSENSTEIN, J.G. *Recursive linear orderings* ⋄ C57 C65 D45 D55 E07 ⋄

MACINTYRE, A. *Residue fields of models of P* ⋄ C57 C60 C62 H15 ⋄

MARKER, D. *Degrees of models of true arithmetic* ⋄ C57 C62 D30 H15 ⋄

MCALOON, K. *On the complexity of models of arithmetic* ⋄ C57 C62 ⋄

METAKIDES, G. & NERODE, A. *The introduction of non-recursive methods into mathematics* ⋄ C57 D45 D80 F60 ⋄

MILLAR, T.S. *Type structure complexity and decidability* ⋄ C50 C57 ⋄

MOROZOV, A.S. *Countable homogeneous boolean algebras (Russian)* ⋄ C15 C50 C57 G05 ⋄

MOROZOV, A.S. *Strong constructivizability of countable saturated boolean algebras (Russian)* ⋄ C50 C57 G05 ⋄

NERODE, A. & REMMEL, J.B. *Recursion theory on matroids* ⋄ C57 D25 D45 ⋄

NOSKOV, G.A. *On conjugacy in metabelian groups (Russian)* ⋄ C57 C60 ⋄

PERETYAT'KIN, M.G. *Finitely axiomatizable totally transcendental theories (Russian)* ⋄ C45 C57 ⋄

PERETYAT'KIN, M.G. *Turing machine computations in finitely axiomatizable theories (Russian)* ⋄ C57 D45 ⋄

SHI, NIANDONG *Creative pairs of subalgebras of recursively enumerable boolean algebras (Chinese)* ⋄ C57 D25 D45 G05 ⋄

SMORYNSKI, C.A. *A note on initial segment constructions in recursively saturated models of arithmetic* ⋄ C50 C57 C62 ⋄

SMORYNSKI, C.A. *Back-and-forth inside a recursively saturated model of arithmetic* ⋄ C15 C50 C57 C62 ⋄

TSUBOI, A. *On M-recursively saturated models of arithmetic* ⋄ C50 C57 C62 ⋄

TVERSKOJ, A.A. *Investigation of recursiveness and arithmeticity of signature functions in nonstandard models of arithmetics (Russian)* ⋄ C57 C62 D45 H15 ⋄

1983

ASH, C.J. & MILLAR, T.S. *Persistently finite, persistently arithmetic theories* ⋄ C57 ⋄

BERTONI, A. & MAURI, G. & MIGLIOLI, P.A. *On the power of model theory in specifying abstract data types and in capturing their recursiveness* ⋄ B75 C50 C57 D20 ⋄

DOBRITSA, V.P. *Complexity of the index set of a constructive model (Russian)* ⋄ C57 D45 ⋄

DOBRITSA, V.P. *Some constructivizations of abelian groups (Russian)* ⋄ C57 C60 D45 ⋄

DOWNEY, R.G. *On a question of A.Retzlaff* ⋄ C57 C60 D25 D45 ⋄

GONCHAROV, S.S. *Universal recursively enumerable boolean algebras (Russian)* ⋄ C50 C57 D45 G05 ⋄

GONCHAROV, V.A. *A recursively representable Boolean algebra (Russian)* ⋄ C57 D45 G05 ⋄

GUICHARD, D.R. *Automorphisms of substructure lattices in recursive algebra* ⋄ C57 C60 D45 G05 ⋄

HICKIN, K.K. & PHILLIPS, R.E. *Isomorphism types in wreath products and effective embeddings of periodic groups* ⋄ C57 C60 D40 ⋄

KALANTARI, I. & REMMEL, J.B. *Degrees of recursively enumarable topological spaces* ⋄ C57 C65 D25 D45 ⋄

KALANTARI, I. & LEGGETT, A. *Maximality in effective topology* ⋄ C57 C65 D25 D45 ⋄

KHISAMIEV, N.G. *Strongly constructive abelian p-groups (Russian)* ⋄ C57 C60 D45 F60 ⋄

KIERSTEAD, H.A. *An effective version of Hall's theorem* ⋄ C57 D80 ⋄

KIERSTEAD, H.A. & REMMEL, J.B. *Indiscernibles and decidable models* ⋄ C30 C45 C57 C80 ⋄

KNIGHT, J.F. *Additive structure in uncountable models for a fixed completion of P* ⋄ C50 C57 C62 C75 D30 ⋄

KOSSAK, R. *A certain class of models of Peano arithmetic* ⋄ C50 C57 C62 ⋄

KOTLARSKI, H. *On elementary cuts in models of arithmetic* ⋄ C50 C57 C62 ⋄

KUDAJBERGENOV, K.ZH. *The number of constructive homogeneous models of a complete decidable theory (Russian)* ⋄ C50 C57 ⋄

MADISON, E.W. *The existence of countable totally nonconstructive extensions of the countable atomless boolean algebras* ⋄ C57 D45 G05 ⋄

MILLAR, T.S. *Omitting types, type spectrums, and decidability* ⋄ C07 C57 ⋄

MILLAR, T.S. *Persistently finite theories with hyperarithmetic models* ⋄ C15 C50 C57 D30 ⋄

MOROZOV, A.S. *Groups of recursive automorphisms of constructive boolean algebras (Russian)* ⋄ C07 C57 D45 F60 G05 ⋄

MOSES, M. *Recursive properties of isomorphism types* ⋄ C57 D45 ⋄

ROY, D.K. *R.e. presented linear orders* ⋄ C57 C65 D45 E07 ⋄

STEINHORN, C.I. *A new omitting types theorem* ⋄ C07 C15 C45 C50 C57 ⋄

TSUBOI, A. *On M-recursively saturated models of Peano arithmetic (Japanese)* ⋄ C50 C57 C62 ⋄

1984

ABDRAZAKOV, K.T. & KHISAMIEV, N.G. *A criterion for strong constructivizability of a class of abelian p-groups (Russian)* ⋄ C57 C60 D45 ⋄

ASH, C.J. & DOWNEY, R.G. *Decidable subspaces and recursively enumerable subspaces*
⋄ C57 C60 D35 D45 ⋄

BALDWIN, JOHN T. *First-order theories of abstract dependence relations*
⋄ C35 C45 C57 C60 D25 D45 ⋄

BUECHLER, S. *Resplendency and recursive definability in ω-stable theories* ⋄ C45 C50 C57 ⋄

DENISOV, A.S. *Constructive homogeneous extensions (Russian)* ⋄ C50 C57 ⋄

DOWNEY, R.G. *A note on decompositions of recursively enumerable subspaces* ⋄ C57 C60 D25 D45 ⋄

DOWNEY, R.G. *Bases of supermaximal subspaces and Steinitz systems. I* ⋄ C57 C60 D25 D45 ⋄

DOWNEY, R.G. *Co-immune subspaces and complementation in V_∞* ⋄ C57 C60 D45 D50 ⋄

DOWNEY, R.G. & REMMEL, J.B. *The universal complementation property* ⋄ C57 C60 D25 D45 ⋄

DYMENT, E.Z. *Recursive metrizability of numbered topological spaces and bases of effective linear topological spaces (Russian)* ⋄ C57 C65 D45 ⋄

GUICHARD, D.R. *A note on r-maximal subspaces of V_∞*
⋄ C57 C60 D45 ⋄

GUPTA, N. *Recursively presented two generated infinite p-groups* ⋄ C57 C60 ⋄

JOCKUSCH JR., C.G. & KALANTARI, I. *Recursively enumerable sets and van der Waerden's theorem on arithmetic progressions* ⋄ C57 D25 D80 ⋄

KAUFMANN, M. & SCHMERL, J.H. *Saturation and simple extensions of models of Peano-arithmetic*
⋄ C50 C57 C62 ⋄

KHISAMIEV, N.G. *Connection between constructivizability and strong constructivizability for different classes of abelian groups (Russian)* ⋄ C57 C60 D45 ⋄

KNIGHT, J.F. & LACHLAN, A.H. & SOARE, R.I. *Two theorems on degrees of models of true arithmetic*
⋄ C57 C62 D30 ⋄

KOSSAK, R. $L_{\infty\omega_1}$-*elementary equivalence of ω_1-like models of PA* ⋄ C50 C57 C62 C75 ⋄

KOSSAK, R. *Remarks on free sets*
⋄ C50 C57 C62 E05 ⋄

KOTLARSKI, H. *On elementary cuts in recursively saturated models of Peano arithmetic* ⋄ C50 C57 C62 ⋄

KUDAJBERGENOV, K.ZH. *Autostability and extensions of constructivizations (Russian)* ⋄ C50 C57 F60 ⋄

KUDAJBERGENOV, K.ZH. *Constructivizability of a prime model (Russian)* ⋄ C50 C57 F60 ⋄

LACHLAN, A.H. *Binary homogeneous structures. I*
⋄ C10 C13 C15 C45 C50 C57 ⋄

MACINTYRE, A. & MARKER, D. *Degrees of recursively saturated models* ⋄ C50 C57 C62 D30 D45 ⋄

MILLAR, T.S. *Decidability and the number of countable models* ⋄ C15 C57 ⋄

MOROZOV, A.S. *Group $Aut_r(Q, \leq)$ is not constructivizable (Russian)* ⋄ C07 C57 F60 ⋄

MOSES, M. *Recursive linear orders with recursive successivities* ⋄ C57 D45 E07 ⋄

MOSES, M. *Recursive properties of isomorphism types*
⋄ C57 D45 ⋄

NEGRI, M. *An application of recursive saturation*
⋄ C50 C57 C62 F30 ⋄

NELSON, GEORGE C. *Boolean powers, recursive models, and the Horn theory of a structure*
⋄ C05 C20 C30 C57 G05 ⋄

NIKITIN, A.A. *Some algorithmic problems for projective planes (Russian)* ⋄ C57 C65 D80 ⋄

ODINTSOV, S.P. *Atomless ideals of constructive Boolean algebras (Russian)* ⋄ C57 G05 ⋄

ROSENSTEIN, J.G. *Recursive linear orderings (French summary)* ⋄ C57 C65 D45 ⋄

SCHRAM, J.M. *Recursively prime trees*
⋄ C57 C65 D45 ⋄

SCHWARZ, S. *Recursive automorphisms of recursive linear orderings* ⋄ C57 C65 D45 ⋄

TVERSKOJ, A.A. *Constructivizability of formal arithmetical structures (Russian)* ⋄ C57 D45 H15 ⋄

WATNICK, R. *A generalization of Tennenbaum's theorem on effectively finite recursive linear orderings*
⋄ C57 C65 D25 D45 ⋄

1985

ADAMOWICZ, Z. & MORALES-LUNA, G. *A recursive model for arithmetic with weak induction*
⋄ C57 F30 H15 ⋄

DOWNEY, R.G. & HIRD, G.R. *Automorphisms of supermaximal subspaces* ⋄ C57 C60 D25 D45 ⋄

DOWNEY, R.G. & KALANTARI, I. *Effective extensions of linear forms on a recursive vector space over a recursive field* ⋄ C57 C60 D45 ⋄

DRIES VAN DEN, L. & SMITH, RICK L. *Decidable regularly closed fields of algebraic numbers*
⋄ B25 C57 C60 ⋄

GLASS, A.M.W. *Effective extensions of lattices-ordered groups that preserve the degree of the conjugacy and the word problems* ⋄ C57 C60 D40 ⋄

GLASS, A.M.W. *The word and isomorphism problems in universal algebra* ⋄ C05 C57 D40 ⋄

GONCHAROV, S.S. *Stongly constructive models (Russian) (English summary)* ⋄ C57 ⋄

KALANTARI, I. & WEITKAMP, G. *Effective topological spaces I: A definability theory. II: A hierarchy*
⋄ C40 C57 C65 D45 D75 D80 ⋄

KAMIMURA, T. *An effective given initial semigroup*
⋄ C57 D45 ⋄

KANDA, A. *Numeration models of λ-calculus*
⋄ B40 B60 C57 D45 ⋄

KHISAMIEV, N.G. & KHISAMIEV, Z.G. *Nonconstructivizability of the reduced part of a strongly constructive torsion-free abelian group (Russian)*
⋄ C57 C60 D45 ⋄

KIERSTEAD, H.A. & REMMEL, J.B. *Degrees of indiscernibles in decidable models*
⋄ C30 C35 C57 D30 ⋄

KOSSAK, R. *A note on satisfaction classes*
⋄ C50 C57 C62 F30 ⋄

KOSSAK, R. *Recursively saturated ω_1-like models of arithmetic* ⋄ C55 C57 C62 C75 C80 E65 ⋄

KOTLARSKI, H. *Bounded induction and satisfaction classes* ⋄ C15 C50 C57 C62 ⋄

MADISON, E.W. *On boolean algebras and their recursive completions* ⋄ C57 D45 G05 ⋄
MILLAR, T.S. *Decidable Ehrenfeucht theories* ⋄ B25 C15 C57 C98 D30 D35 D55 ⋄
MONTAGNA, F. & SORBI, A. *Universal recursion theoretic properties of r.e. preordered structures* ⋄ C57 D45 G10 ⋄
MOROZOV, A.S. *Constructive Boolean algebras with almost identical automorphisms (Russian)* ⋄ C07 C57 G05 ⋄
NERODE, A. & REMMEL, J.B. *A survey of lattices of R.E. substructures* ⋄ C57 C98 D45 D98 ⋄
PINUS, A.G. *Applications of boolean powers of algebraic systems (Russian)* ⋄ C05 C30 C55 C57 C80 C85 D35 E50 G05 ⋄
ROY, D.K. *Linear order types of nonrecursive presentability* ⋄ C57 C65 D45 ⋄
SCHMERL, J.H. *Recursively saturated models generated by indiscernibles* ⋄ C30 C50 C57 ⋄
SCHRIEBER, L. *Recursive properties of euclidean domains* ⋄ C57 C60 D45 ⋄
SOLOVAY, R.M. *Infinite fixed-point algebras* ⋄ C57 C62 G25 ⋄
TVERSKOJ, A.A. *Constructivizable and nonconstructivizable formal arithmetic structures (Russian)* ⋄ C57 C62 D45 H15 ⋄
WEISPFENNING, V. *Quantifier elimination for modules* ⋄ C10 C35 C57 C60 ⋄

C05 Equational classes, universal algebra

1933
BIRKHOFF, GARRETT *On the combination of subalgebras*
◇ C05 C60 G10 ◇

1935
BIRKHOFF, GARRETT *On the structure of abstract algebras*
◇ C05 C07 G25 ◇

1939
MAL'TSEV, A.I. *On the embedding of associative systems into groups (Russian) (German summary)*
◇ C05 C60 ◇

1940
MAL'TSEV, A.I. *On the embedding of associative systems into groups II (Russian) (German summary)*
◇ C05 C60 ◇

1943
MCKINSEY, J.C.C. *The decision problem for some classes of sentences without quantifiers*
◇ B20 B25 C05 C60 G10 ◇

1944
BIRKHOFF, GARRETT *Subdirect unions in universal algebra*
◇ C05 ◇
MCKINSEY, J.C.C. & TARSKI, A. *The algebra of topology*
◇ C05 C50 C65 G05 G25 ◇

1946
BIRKHOFF, GARRETT *Sobre los grupos de automorfismos*
◇ C05 C07 G10 ◇
MCKINSEY, J.C.C. & TARSKI, A. *On closed elements in closure algebras* ◇ B25 C05 C65 G05 G25 ◇
TARSKI, A. *A remark on functionally free algebras*
◇ C05 C30 G25 ◇

1947
JONSSON, B. & TARSKI, A. *Direct decomposition of finite algebraic systems* ◇ C05 C13 ◇

1949
JONSSON, B. & TARSKI, A. *Cardinal products of isomorphism types* ◇ C05 C30 E75 ◇
JORDAN, P. *Zur Axiomatik der Verknuepfungsbereiche*
◇ C05 C75 G10 ◇

1951
HORN, A. *On sentences which are true of direct unions of algebras* ◇ C05 C30 C40 ◇
MARCZEWSKI, E. *Sur les congruences et les proprietes positives d'algebres abstraites* ◇ C05 C40 ◇

1952
HIGMAN, G. *Ordering by divisibility in abstract algebras*
◇ C05 C65 ◇

KALICKI, J. *On comparison of finite algebras*
◇ B25 C05 C13 ◇
LOS, J. *Recherches algebriques sur les operations analytiques et quasi-analytiques* ◇ C05 ◇
PEREMANS, W. *Some theorems on free algebras and on direct products of algebras* ◇ C05 C52 ◇
SHODA, K. *Zur Theorie der algebraischen Erweiterungen*
◇ C05 C60 ◇
SIKORSKI, R. *Products of abstract algebras*
◇ C05 C75 ◇

1953
FOSTER, A.L. *Generalized "boolean" theory of universal algebras, part I: subdirect sums and normal representation theorem* ◇ C05 C30 ◇
FOSTER, A.L. *Generalized "boolean" theory of universal algebras, II: identities and subdirect sums of functionally complete algebras* ◇ C05 C30 ◇
GERICKE, H. *Ueber den Begriff der algebraischen Struktur*
◇ C05 ◇
KRISHNAN, V.S. *Closure operations on c-structures*
◇ C05 C30 E75 ◇

1954
CHANG, C.C. *Some general theorems on direct products and their applications in the theory of models*
◇ C05 C75 ◇
LYNDON, R.C. *Identities in finite algebras*
◇ C05 C13 ◇
MAL'TSEV, A.I. *On the general theory of algebraic systems (Russian)* ◇ C05 C65 ◇
NEUMANN, B.H. *An embedding theorem for algebraic systems* ◇ C05 ◇

1955
BING, K. *On arithmetical classes not closed under direct union* ◇ C05 C40 C52 ◇
KALICKI, J. & SCOTT, D.S. *Equational completeness of abstract algebras* ◇ C05 ◇
KALICKI, J. *The number of equationally complete classes of equations* ◇ C05 ◇
LOS, J. & SUSZKO, R. *On the extending of models II*
◇ C05 C30 C52 ◇
LOS, J. *Quelques remarques, theoremes et problemes sur les classes definissables d'algebres*
◇ C05 C20 C30 C52 ◇
SHODA, K. *Bemerkungen ueber die Existenz der algebraisch abgeschlossenen Erweiterung*
◇ C05 C60 ◇
TARSKI, A. *Contributions to the theory of models III*
◇ C05 C52 G15 ◇

1956

FUJIWARA, T. *On the existence of algebraically closed algebraic systems* ⋄ C05 ⋄

MAL'TSEV, A.I. *Quasiprimitive classes of abstract algebras (Russian)* ⋄ C05 ⋄

MAL'TSEV, A.I. *Subdirect products of models (Russian)* ⋄ C05 ⋄

RIBEIRO, H. *The notion of universal completeness* ⋄ C05 C35 ⋄

SCOTT, D.S. *Equationally complete extensions of finite algebras* ⋄ C05 C13 ⋄

SLOMINSKI, J. *On the extending of models III. Extensions in equationally definable classes of algebras* ⋄ C05 C30 C60 G05 ⋄

TARSKI, A. *Equationally complete rings and relation algebras* ⋄ C05 G15 ⋄

1957

HANF, W.P. *On some fundamental problems concerning isomorphism of boolean algebras* ⋄ C05 C15 C30 G05 ⋄

JONSSON, B. *On direct decompositions of torsion-free abelian groups* ⋄ C05 C30 C60 ⋄

JONSSON, B. *On isomorphism types of groups and other algebraic systems* ⋄ C05 C30 C60 ⋄

NERODE, A. *Composita, equations, and freely generated algebras* ⋄ C05 ⋄

RIBEIRO, H. *Universal completeness* ⋄ C05 C35 ⋄

TARSKI, A. & VAUGHT, R.L. *Arithmetical extensions of relational systems* ⋄ C05 C07 C55 C60 ⋄

TARSKI, A. & VAUGHT, R.L. *Elementary (arithmetical) extensions* ⋄ C05 C07 C35 C55 C60 ⋄

TARSKI, A. *Remarks on direct products of commutative semigroups* ⋄ C05 C30 G05 ⋄

YAQUB, A. *On the identities of certain algebras* ⋄ C05 ⋄

1958

MAL'TSEV, A.I. *Certain classes of models (Russian)* ⋄ C05 C52 G30 ⋄

MAL'TSEV, A.I. *Defining relations in categories (Russian)* ⋄ C05 C52 G30 ⋄

MAL'TSEV, A.I. *The structural characterization of certain classes of algebras (Russian)* ⋄ C05 C52 G30 ⋄

SLOMINSKI, J. *Theory of models with infinitary operations and relations* ⋄ C05 C75 ⋄

1959

CLARKE, A.B. *On the representation of cardinal algebras by directed sums* ⋄ C05 C30 ⋄

KOGALOVSKIJ, S.R. *On universal classes of algebras, closed under direct product (Russian)* ⋄ C05 C52 ⋄

KOGALOVSKIJ, S.R. *Universal classes of models (Russian)* ⋄ C05 C52 ⋄

MAL'TSEV, A.I. *Algebraic systems (Russian)* ⋄ C05 C60 ⋄

MAL'TSEV, A.I. *Model correspondences (Russian)* ⋄ B15 C05 C07 C52 C60 C85 ⋄

NERODE, A. *Composita, equations, and freely generated algebras* ⋄ C05 ⋄

SLOMINSKI, J. *The theory of abstract algebras with infinitary operations* ⋄ C05 ⋄

1960

BRYANT, S.J. & MARICA, J.G. *Unary algebras* ⋄ C05 C30 ⋄

BUTLER, J.W. *On complete and independent sets of operations in finite algebras* ⋄ C05 C13 ⋄

1961

JONSSON, B. & TARSKI, A. *On two properties of free algebras* ⋄ C05 ⋄

KEISLER, H.J. *On some results of Jonsson and Tarski concerning free algebras* ⋄ C05 ⋄

MAL'TSEV, A.I. *On the elementary theories of locally free universal algebras (Russian)* ⋄ B25 C05 C60 ⋄

MAURY, G. *La condition "integralement clos" dans quelques structures algébriques* ⋄ C05 C60 ⋄

RIBEIRO, H. *On the universal completeness of classes of relational systems* ⋄ C05 C35 ⋄

SCHMIDT, J. *Algebraic operations and algebraic independence in algebras with infinitary operations* ⋄ C05 ⋄

VAUGHT, R.L. *The elementary character of two notions from general algebra* ⋄ C05 C40 C52 C60 ⋄

1962

JONSSON, B. *Algebraic extensions of relational systems* ⋄ C05 C52 C60 ⋄

MAL'TSEV, A.I. *Axiomatisable classes of locally free algebras of various types (Russian)* ⋄ B25 C05 C10 ⋄

MONK, J.D. *On pseudo-simple universal algebras* ⋄ C05 ⋄

NEUMANN, B.H. *Special topics in algebra: Universal algebra* ⋄ C05 C98 ⋄

SIKORSKI, R. *Algebra of formalized languages* ⋄ C05 C07 G15 G25 ⋄

YANOV, YU.I. *On systems of identities for algebras (Russian)* ⋄ C05 ⋄

1963

FLEISCHER, I. *Elementary properties of inductive limits* ⋄ C05 C20 C30 ⋄

GRAETZER, G. *A theorem on doubly transitive permutation groups with application to universal algebras* ⋄ C05 C50 ⋄

KOGALOVSKIJ, S.R. *Structural characteristics of universal classes (Russian)* ⋄ C05 ⋄

PIERCE, R.S. *A note on free products of abstract algebras* ⋄ C05 ⋄

SCHMIDT, E.T. *Universale Algebren mit gegebenen Automorphismengruppen und Unteralgebrenverbaenden* ⋄ C05 ⋄

SIKORSKI, R. *Products of generalized algebras and products of realizations* ⋄ C05 C07 G25 ⋄

1964

ARMBRUST, M. & SCHMIDT, J. *Zum Cayleyschen Darstellungssatz* ⋄ C05 ⋄

CHANG, C.C. & JONSSON, B. & TARSKI, A. *Refinement properties for relational structures* ⋄ C05 C30 ⋄

CRAWLEY, P. & JONSSON, B. *Refinements for infinite direct decompositions of algebraic systems* ⋄ C05 C30 C60 ⋄

CSAKANY, B. *Abelian properties of primitive classes of universal algebras (Russian)* ◊ C05 C60 ◊
DAIGNEAULT, A. *Freedom in polyadic algebras and two theorems of Beth and Craig* ◊ C05 C40 G15 ◊
GRAETZER, G. *On the class of subdirect powers of a finite algebra* ◊ C05 C30 ◊
LOS, J. *Normal subalgebras in general algebras* ◊ C05 ◊
MYCIELSKI, J. *Some compactifications of general algebras* ◊ C05 C50 ◊
PICKETT, H.E. *Subdirect representations of relational systems* ◊ C05 C52 ◊
SCHMIDT, E.T. *Universale Algebren mit gegebenen Automorphismengruppen und Kongruenzverbaenden* ◊ C05 ◊

1965

ASTROMOFF, A. *Some structure theorems for primal and categorical algebras* ◊ C05 C13 ◊
COHN, P.M. *Universal Algebra* ◊ C05 C98 ◊
CSAKANY, B. *Inner automorphisms of universal algebras* ◊ C05 ◊
FELSCHER, W. *Zur Algebra unendlich langer Zeichenreihen* ◊ C05 C75 E10 ◊
HAJEK, P. *The concept of a primitive class of algebras (Birkhoff theorem) (Czech) (Russian and German summaries)* ◊ C05 ◊
KOGALOVSKIJ, S.R. *On the theorem of Birkhoff (Russian)* ◊ C05 ◊
LOS, J. *Free product in general algebras* ◊ C05 C30 ◊
MAKKAI, M. *A compactness result concerning direct products of models* ◊ C05 C30 C52 ◊
MONK, J.D. *Model-theoretic methods and results in the theory of cylindric algebras* ◊ C05 C20 C52 G15 ◊
RIBEIRO, H. & SCHWABAUER, R. *A remark on equational completeness* ◊ C05 C35 ◊
SCHEIN, B.M. *Relation algebras* ◊ C05 C20 C52 ◊
WEGLORZ, B. *Compactness of algebraic systems (Russian summary)* ◊ C05 C50 ◊

1966

AUSTIN, A.K. *Varieties of groups defined by a single law* ◊ C05 C60 ◊
CHANG, C.C. & KEISLER, H.J. *Continuous model theory* ◊ B50 C05 C20 C50 C90 C98 ◊
ERDOES, P. & HAJNAL, A. *On a problem of B.Jonsson* ◊ C05 C55 E05 E50 E55 ◊
GLADSTONE, M.D. *Finite models for inequations* ◊ B25 C05 C13 ◊
HOEHNKE, H.-J. *Zur Strukturgleichheit axiomatischer Klassen* ◊ C05 G15 ◊
ISBELL, J.R. *Epimorphisms and dominions* ◊ C05 C50 ◊
JONSSON, B. *The unique factorization problem for finite relational structures* ◊ C05 C13 ◊
KOSTINSKY, A. *Recent results on Jonsson algebras* ◊ C05 C55 E05 E55 ◊
LOS, J. *Direct sums in general algebras* ◊ C05 C30 ◊
LYNDON, R.C. *Dependence in groups* ◊ C05 C60 ◊
MACZYNSKI, M.J. *A property of retracts of α^+-homogeneous α^+-universal boolean algebras* ◊ C05 C50 G05 ◊

MAL'TSEV, A.I. *A few remarks on quasivarieties of algebraic systems (Russian)* ◊ C05 ◊
MAL'TSEV, A.I. *On standard notations and terminology in the theory of algebraic systems (Russian)* ◊ C05 C07 ◊
NEUMANN, B.H. & WIEGOLD, E.C. *A semigroup representation of varieties of algebras* ◊ C05 ◊
OHKUMA, T. *Ultrapowers in categories* ◊ C05 C20 G30 ◊
SHEVRIN, L.N. *Elementary lattice properties of semigroups (Russian)* ◊ C05 C60 ◊
TAJTSLIN, M.A. *On elementary theories of commutative semigroups (Russian)* ◊ B25 C05 C60 ◊
WEGLORZ, B. *Equationally compact algebras I* ◊ C05 C20 C50 ◊

1967

BOL'BOT, A.D. *Equationally complete varieties of totally symmetric quasigroups (Russian) (English summary)* ◊ C05 ◊
CHANG, C.C. *Cardinal factorization of finite relational structures* ◊ C05 C13 C30 ◊
ERSHOV, YU.L. *Elementary theories of Post varieties (Russian)* ◊ B25 C05 C13 G20 ◊
EVANS, T. *The spectrum of a variety* ◊ C05 C13 ◊
GRAETZER, G. *On the endomorphism semigroup of simple algebras* ◊ C05 ◊
GRAETZER, G. *On the spectra of classes of algebras* ◊ C05 C13 ◊
HOWIE, J.M. & ISBELL, J.R. *Epimorphisms and dominions. II* ◊ C05 C50 ◊
JOHNSON, J. & SEIFERT, R.L. *A survey of multiunary algebras* ◊ C05 C98 ◊
JONSSON, B. *Algebras whose congruence lattices are distributive* ◊ C05 G10 ◊
MAL'TSEV, A.I. *Multiplication of classes of algebraic systems (Russian)* ◊ C05 ◊
NEUMANN, H. *Varieties of groups* ◊ C05 C60 C98 ◊
PLONKA, J. *On a method of construction of abstract algebras* ◊ C05 C30 ◊
TAJTSLIN, M.A. *Two remarks on isomorphisms of commutative semigroups (Russian) (English summary)* ◊ B25 C05 C60 ◊
WEGLORZ, B. *Completeness and compactness of lattices* ◊ C05 C52 G10 ◊
WEGLORZ, B. *Equationally compact algebras III* ◊ C05 C20 C50 ◊

1968

BRYCE, R.A. *A note on free products with normal amalgamation* ◊ C05 C30 ◊
BURMEISTER, P. *Ueber die Maechtigkeiten und Unabhaengigkeitsgrade der Basen freier Algebren. I* ◊ C05 C75 ◊
FAJTLOWICZ, S. & HOLSZTYNSKI, W. & MYCIELSKI, J. & WEGLORZ, B. *On powers of bases in some compact algebras* ◊ C05 C50 ◊
FLEISCHER, I. *On Cohn's theorem* ◊ C05 ◊
FOSTER, A.L. *Pre-fields and universal algebraic extensions; equational precessions* ◊ C05 C60 ◊
GEIGER, D. *Closed systems of functions and predicates* ◊ C05 C52 ◊

GRAETZER, G. *On the existence of free structures over universal classes* ⋄ C05 ⋄
GRAETZER, G. *Universal algebra* ⋄ C05 C98 ⋄
IMREH, B. *On a theorem of G.Birkhoff* ⋄ C05 ⋄
JONSSON, B. *Algebraic structures with prescribed automorphism groups* ⋄ C05 C07 ⋄
JONSSON, B. & OLIN, P. *Almost direct products and saturation* ⋄ C05 C20 C50 G05 ⋄
MAL'TSEV, A.I. *Some questions bordering on algebra and mathematical logic (Russian)* ⋄ B25 C05 ⋄
MCKENZIE, R. *On finite groupoids and $\aleph$-prime algebras* ⋄ C05 C13 ⋄
MYCIELSKI, J. & RYLL-NARDZEWSKI, C. *Equationally compact algebras II* ⋄ C05 C20 C50 C65 ⋄
NEUMANN, B.H. *Lectures on topics in the theory of infinite groups* ⋄ C05 C60 C98 D40 ⋄
PIERCE, R.S. *Introduction to the theory of abstract algebras* ⋄ C05 C98 ⋄
PLONKA, E. *On a problem of Bjarni Jonsson concerning automorphisms of a general algebra* ⋄ C05 ⋄
SCHMIDT, J. *Universelle Halbgruppe, Kategorien, freies Produkt* ⋄ C05 C30 G30 ⋄
TAJTSLIN, M.A. *Algorithmic problems for commutative semigroups (Russian)* ⋄ B25 C05 C60 ⋄
TAJTSLIN, M.A. *On the isomorphism problem for commutative semigroups (Russian)* ⋄ C05 C07 C60 D40 ⋄
TARSKI, A. *Equational logic and equational theories of algebras* ⋄ C05 ⋄
WEGLORZ, B. & WOJCIECHOWSKA, A. *Summability of pure extensions of relational structures* ⋄ C05 C07 C52 ⋄

1969

CHEREMISIN, A.I. *Structural characterization of certain classes of models (Russian)* ⋄ C05 C30 C52 ⋄
FELSCHER, W. *On the algebra of quantifiers (Russian summary)* ⋄ B10 C05 C07 G15 ⋄
GRAETZER, G. *Free Σ-structures* ⋄ C05 ⋄
GRAETZER, G. & LAKSER, H. & PLONKA, J. *Joins and direct products of equational classes* ⋄ C05 C52 ⋄
KAISER, KLAUS *Ueber eine Verallgemeinerung der Robinsonschen Modellvervollstaendigung I,II (II: Abgeschlossene Algebren)* ⋄ C05 C25 C35 ⋄
KLEIN, ABRAHAM A. *Necessary conditions for embedding rings into fields* ⋄ C05 C60 ⋄
PADMANABHAN, R. *On single equational-axiom systems for abelian groups* ⋄ C05 ⋄
SAADE, M. *A comment on a paper by Evans* ⋄ C05 C13 ⋄
SHAFAAT, A. *On implicationally defined classes of algebras* ⋄ C05 C75 ⋄
TABATA, H. *Free structures and universal Horn sentences* ⋄ B20 C05 C52 ⋄
TAYLOR, W. *Atomic compactness and graph theory* ⋄ C05 C50 C65 ⋄
WHALEY, T.P. *Algebras satisfying the descending chain condition for subalgebras* ⋄ C05 C50 ⋄
YANOV, YU.I. *On directed transformations of formulas (Russian)* ⋄ C05 ⋄

1970

BURMEISTER, P. *Ueber die Maechtigkeiten und Unabhaengigkeitsgrade der Basen freier Algebren II* ⋄ C05 ⋄
COLLINS, D.J. *A universal semigroup (Russian)* ⋄ C05 C57 D05 D40 ⋄
FUJIWARA, T. & NAKANO, Y. *On the relation between free structures and direct limits* ⋄ C05 C30 ⋄
GALVIN, F. & HORN, A. *Operations preserving all equivalence relations* ⋄ C05 C75 ⋄
GECSEG, F. *On certain classes of Σ-structures* ⋄ C05 C52 ⋄
HALEY, D.K. *Equationally compact noetherian rings* ⋄ C05 C60 ⋄
HALEY, D.K. *On compact commutative noetherian rings* ⋄ C05 C60 ⋄
KANGER, S. *Equational calculi and automatic demonstration* ⋄ B25 B35 C05 ⋄
KIRCH, A.M. *Generalized restricted direct products* ⋄ C05 C30 ⋄
KULIK, V.T. *Sentences preserved under homomorphisms that are one-to-one at zero (Russian)* ⋄ C05 C40 C60 ⋄
MAL'TSEV, A.I. *Algebraic systems (Russian)* ⋄ C05 C98 ⋄
MCKENZIE, R. *Equational bases for lattice theories* ⋄ C05 G10 ⋄
NAKANO, Y. *On the finite characters in general algebras* ⋄ C05 ⋄
NEUMANN, W.D. *On the quasivariety of convex subsets of affine spaces* ⋄ C05 C60 ⋄
SENFT, J.R. *On weak autormorphisms of universal algebras* ⋄ C05 C07 ⋄
TAYLOR, W. *Compactness and chromatic number* ⋄ C05 C50 C65 ⋄
WENZEL, G.H. *Subdirect irreducibility and equational compactness in unary algebras $\langle A;f \rangle$* ⋄ C05 C50 ⋄
WOJDYLO, B. *On some problems of J.Slominski concerning equations in quasi-algebras* ⋄ C05 ⋄
ZIL'BER, B.I. *On the isomorphism and elementary equivalence of commutative semigroups (Russian)* ⋄ C05 C60 ⋄

1971

ANUSIAK, J. & WEGLORZ, B. *Remarks on c-independence in cartesian products of abstract algebras* ⋄ C05 ⋄
ARMBRUST, M. & KAISER, KLAUS *Remarks on model classes acting on one another* ⋄ C05 C20 C30 C52 ⋄
BALBES, R. & DWINGER, P. *Uniqueness of representations of a distributive lattice as a free product of a boolean algebra and a chain* ⋄ C05 C30 G10 ⋄
BAUMSLAG, B. & BAUMSLAG, G. *On ascending chain conditions* ⋄ C05 C60 ⋄
CHUAQUI, R.B. *A representation theorem for linearly ordered cardinal algebras* ⋄ C05 C65 E10 G25 ⋄
COMER, S.D. *Representations by algebras of sections over boolean spaces* ⋄ C05 C90 ⋄
FELSCHER, W. *An algebraic approach to first order logic* ⋄ B10 C05 C07 C20 C90 G15 ⋄

FUHRKEN, G. & TAYLOR, W. *Weakly atomic-compact relational structures* ⋄ C05 C50 ⋄

FUJIWARA, T. *Freely generable classes of structures* ⋄ C05 ⋄

FUJIWARA, T. *On the construction of the least universal Horn class containing a given class* ⋄ B20 C05 C52 ⋄

GOETZ, A. *A generalization of the direct product of universal algebras* ⋄ C05 C30 ⋄

IGOSHIN, V.I. *Characterizable varieties of lattices (Russian)* ⋄ C05 C52 G10 ⋄

KRATOCHVIL, P. *Note on a representation of universal algebras as subdirect powers* ⋄ C05 C30 ⋄

KRAUSS, P.H. *Universally complete universal theories* ⋄ C05 C30 C35 ⋄

LINDNER, C.C. *Finite partial cyclic triple systems can be finitely embedded* ⋄ C05 C13 D35 ⋄

LINDNER, C.C. *Identities preserved by singular direct product* ⋄ C05 C13 C40 ⋄

LUSTIG, M. *On isotone and homomorphic maps of ordered unary algebras* ⋄ C05 C07 ⋄

MCKENZIE, R. $\aleph_1$-*incompactness of Z* ⋄ C05 C60 ⋄

MCKENZIE, R. *Cardinal multiplication of structures with a reflexive relation* ⋄ C05 C30 C52 ⋄

MCKENZIE, R. *Definability in lattices of equational theories* ⋄ C05 C40 G10 ⋄

MCKENZIE, R. *On semigroups whose proper subsemigroups have lesser power* ⋄ C05 E50 E75 ⋄

PIERCE, R.S. *The submodule lattice of a cyclic module* ⋄ C05 C60 ⋄

PLATT, C.R. *Iterated limits of universal algebras* ⋄ C05 C30 ⋄

SCHEIN, B.M. *Linearly fundamentally ordered semigroups* ⋄ C05 C60 E20 ⋄

SEIFERT, R.L. *On prime binary relational structures* ⋄ C05 C50 ⋄

SMITH JR., D.B. *Universal homogeneous algebras* ⋄ C05 C50 ⋄

TABATA, H. *A generalized free structure and several properties of universal Horn classes* ⋄ B20 C05 C52 ⋄

TAYLOR, W. *Atomic compactness and elementary equivalence* ⋄ C05 C13 C50 C65 ⋄

TAYLOR, W. *Some constructions of compact algebras* ⋄ C05 C50 C55 C65 ⋄

TULIPANI, S. *Cardinali misurabili e algebre semplici in una varieta generata da un'algebra infinitaria m-funzionalmente completa (English summary)* ⋄ C05 C75 E55 ⋄

VANCKO, R.M. *The spectrum of some classes of free universal algebras* ⋄ C05 C13 ⋄

WENZEL, G.H. *On ($\mathfrak{S}, \mathfrak{A}, \mathfrak{m}$)-atomic compact relational systems* ⋄ C05 C50 ⋄

1972

ABAKUMOV, A.I. & PALYUTIN, E.A. & SHISHMAREV, YU.E. & TAJTSLIN, M.A. *Categorial quasivarieties (Russian)* ⋄ C05 C35 C52 ⋄

ARMBRUST, M. & KAISER, KLAUS *On quasi-universal model classes* ⋄ B15 C05 C20 C52 C85 ⋄

BACSICH, P.D. *Cofinal simplicity and algebraic closedness* ⋄ C05 C25 C50 C60 C75 G10 ⋄

BACSICH, P.D. *Injectivity in model theory* ⋄ C05 C35 C50 G05 G10 G30 ⋄

BANASCHEWSKI, B. & NELSON, EVELYN *Equational compactness in equational classes of algebras* ⋄ C05 C50 ⋄

BELEGRADEK, O.V. & TAJTSLIN, M.A. *Two remarks on the varieties $\mathfrak{A}_{m,n}$ (Russian)* ⋄ C05 C50 ⋄

BULMAN-FLEMING, S. & TAYLOR, W. *On a question of G.H. Wenzel* ⋄ C05 E50 ⋄

BULMAN-FLEMING, S. *On equationally compact semilattices* ⋄ C05 C50 ⋄

BURRIS, S. *Models in equational theories of unary algebras* ⋄ B25 C05 C13 C35 ⋄

CLARK, D.M. & KRAUSS, P.H. *Monomorphic relational systems* ⋄ C05 C30 C52 ⋄

ERSHOV, YU.L. *Elementary group theories (Russian)* ⋄ B25 C05 C13 C60 D35 ⋄

FELSCHER, W. *Modelltheorie als universelle Algebra* ⋄ C05 C07 ⋄

HEIDEMA, J. *Metamathematical representation of radicals in universal algebra* ⋄ C05 C60 ⋄

JONSSON, B. *Topics in universal algebra* ⋄ C05 ⋄

KELLY, DAVID *A note on equationally compact lattices* ⋄ C05 C50 G10 ⋄

KRAUSS, P.H. *Extending congruence relations* ⋄ C05 C07 C50 ⋄

KRAUSS, P.H. *On primal algebras* ⋄ C05 C13 ⋄

LINDNER, C.C. *Identities preserved by the singular direct product II* ⋄ C05 C13 C40 ⋄

LUCAS, T. *Equations in the theory of monadic algebras* ⋄ C05 G15 ⋄

MCKENZIE, R. *A method for obtaining refinement theorems with an application to direct products of semigroups* ⋄ C05 C30 ⋄

MCKENZIE, R. *Equational bases and nonmodular lattice varieties* ⋄ B25 C05 G10 ⋄

PALYUTIN, E.A. *On complete quasimanifolds (Russian)* ⋄ C05 C35 ⋄

PIGOZZI, D. *Amalgamation, congruence-extension, and interpolation properties in algebras* ⋄ C05 C40 C52 G15 ⋄

PIXLEY, A.F. *Local Mal'cev conditions* ⋄ C05 C57 ⋄

SELMAN, A.L. *Completeness of calculi for axiomatically defined classes of algebras* ⋄ B55 C05 C07 ⋄

SUSZKO, R. *Equational logic and theories in sentential language* ⋄ B05 C05 C07 ⋄

TAYLOR, W. *Fixed points of endomorphisms* ⋄ C05 C50 ⋄

TAYLOR, W. *Note on pure-essential extensions* ⋄ C05 ⋄

TAYLOR, W. *On equationally compact semigroups* ⋄ C05 C50 ⋄

TAYLOR, W. *Residually small varieties* ⋄ C05 C50 C55 ⋄

TULIPANI, S. *Proprieta metamatematiche di alcune classi di algebre* ⋄ C05 C52 ⋄

TULIPANI, S. *Sulla completezza e sulla categoricita della teoria delle W-algebre semplici (English summary)* ⋄ C05 C35 C50 ⋄

VANCKO, R.M. *The family of locally independent sets in finite algebras* ⋄ C05 C13 ⋄

VAZHENIN, YU.M. *On finitely approximable semigroups of idempotents (Russian)* ⋄ C05 C60 ⋄

1973

ARMBRUST, M. *A first order approach to correspondence and congruence systems* ⋄ C05 C75 ⋄

ARMBRUST, M. *Direct limits of congruence systems* ⋄ C05 C85 ⋄

BACSICH, P.D. *An epi-reflector for universal theories* ⋄ C05 C30 C60 G30 ⋄

BACSICH, P.D. *Primality and model-completion* ⋄ C05 C25 C35 ⋄

BALDWIN, JOHN T. *A sufficient condition for a variety to have the amalgamation property* ⋄ C05 C35 C52 ⋄

BALDWIN, JOHN T. & LACHLAN, A.H. *On universal Horn classes categorical in some infinite power* ⋄ C05 C10 C35 C45 ⋄

BANASCHEWSKI, B. & NELSON, EVELYN *Equational compactness in infinitary algebras* ⋄ C05 C50 ⋄

BUDKIN, A.I. & GORBUNOV, V.A. *Implicative classes of algebras (Russian)* ⋄ C05 ⋄

BURRIS, S. *Scott sentences and a problem of Vaught for mono-unary algebras* ⋄ C05 C15 C75 ⋄

COMER, S.D. *Arithmetic properties of relatively free products* ⋄ C05 ⋄

DAVEY, B.A. *Sheaf spaces and sheaves of universal algebras* ⋄ C05 C90 ⋄

EDGAR, G.A. *The class of topological spaces is equationally definable* ⋄ C05 ⋄

FAKIR, S. *Monomorphismes et objects algebriquement et existentiellement clos dans les categories localement presentables. Applications aux monoides, categories et groupoides* ⋄ C05 C25 C30 G30 ⋄

FUJIWARA, T. *A generalization of Lyndon's theorem characterizing sentences preserved in subdirect products* ⋄ C05 C30 C40 ⋄

FUJIWARA, T. *Note on generalized atomic sets of formulas* ⋄ B20 C05 C40 ⋄

GANTER, B. & PLONKA, J. & WERNER, H. *Homogeneous algebras are simple* ⋄ C05 C50 ⋄

GRAETZER, G. & LAKSER, H. *A note on the implicational class generated by a class of structures* ⋄ C05 C52 ⋄

HAIMO, F. *The index of an algebra automorphism group* ⋄ C05 C07 ⋄

HALEY, D.K. *Equationally compact artinian rings* ⋄ C05 C60 ⋄

HICKIN, K.K. *Countable type local theorems in algebra* ⋄ C05 C52 ⋄

HICKIN, K.K. *Countable type local theorems in algebra* ⋄ C05 C52 C60 ⋄

ISBELL, J.R. *Functorial implicit operations* ⋄ C05 C40 ⋄

JOHNSON, J.S. *Axiom systems for first order logic with finitely many variables* ⋄ B20 C05 G15 ⋄

KRAUSS, P.H. *On quasi primal algebras* ⋄ C05 C13 ⋄

MAKKAI, M. *A proof of Baker's finite-base theorem on equational classes generated by finite elements on congruence distributive varieties* ⋄ C05 ⋄

MALCOLM, W.G. *Application of higher-order ultraproducts to the theory of local properties in universal algebras and relational systems* ⋄ C05 C20 C85 E40 ⋄

MCKENZIE, R. *Some unsolvable problems between lattice theory and equational logic* ⋄ C05 C98 G10 ⋄

MOREL, A.C. *Algebras with a maximal C-dependence property* ⋄ C05 ⋄

PALYUTIN, E.A. *Categorical quasivarieties of arbitrary signature (Russian)* ⋄ C05 C35 ⋄

PLONKA, J. *On the sum of a direct system of relational systems* ⋄ C05 C30 ⋄

ROSENBERG, I.G. *Strongly rigid relations* ⋄ B50 C05 C50 ⋄

SALIJ, V.N. *Boolean valued algebras (Russian)* ⋄ C05 C90 G05 ⋄

SHAFAAT, A. *Remarks on quasivarieties of algebras* ⋄ C05 ⋄

SICHLER, J. *Weak automorphisms of universal algebras* ⋄ C05 E50 ⋄

TAYLOR, W. *Characterizing Mal'cev conditions* ⋄ C05 C52 ⋄

WENZEL, G.H. *Eine Charakterisierung gleichungskompakter universeller Algebren* ⋄ C05 C50 ⋄

ZAMYATIN, A.P. *A prevariety of semigroups whose elementary theory is solvable (Russian)* ⋄ B25 C05 ⋄

1974

ADAMEK, J. & REITERMAN, J. *Fixed-point property of unary algebras* ⋄ C05 C07 ⋄

BACSICH, P.D. & ROWLANDS-HUGHES, D. *Syntactic characterizations of amalgamation, convexity and related properties* ⋄ C05 C07 C60 C75 ⋄

BANASCHEWSKI, B. *Equational compactness of G-sets* ⋄ C05 C50 C60 ⋄

BANASCHEWSKI, B. *On equationally compact extensions of algebras* ⋄ C05 C50 ⋄

BLASS, A.R. & NEUMANN, P.M. *An application of universal algebra in group theory* ⋄ C05 C60 ⋄

BULMAN-FLEMING, S. *A note on equationally compact algebras* ⋄ C05 C50 ⋄

BULMAN-FLEMING, S. *Algebraic compactness and its relations to topology* ⋄ C05 C50 ⋄

COMER, S.D. *Restricted direct products and sectional representations* ⋄ C05 C30 C90 ⋄

EKLOF, P.C. *Algebraic closure operators and strong amalgamation bases* ⋄ C05 ⋄

FRENKIN, B.R. *Σ-free models (Russian)* ⋄ C05 C52 ⋄

FRENKIN, B.R. *Reducibility and uniform reducibility of algebraic operations (Russian)* ⋄ C05 ⋄

FRODA-SCHECHTER, M. & KWATINETZ, M. *Some remarks on a paper of R. C. Lyndon: "Properties preserved under homomorphism"* ⋄ C05 C40 C52 ⋄

GRAETZER, G. & SICHLER, J. *Agassiz sum of algebras* ⋄ C05 C30 C40 ⋄

HALEY, D.K. *A note on equational compactness* ⋄ C05 C60 ⋄

JONSSON, B. *Some recent trends in general algebra* ⋄ C05 C98 ⋄

KEIMEL, K. *Representation des algebres universelles par des faisceaux* ⋄ C05 C35 C60 C90 ⋄

KEIMEL, K. & WERNER, H. *Stone duality for varieties generated by quasi-primal algebras* ⋄ C05 C13 C90 ⋄

LAGRANGE, R. *Amalgamation and epimorphisms in m-complete boolean algebras* ◇ C05 G05 ◇

LUCAS, T. *Universal classes of monadic algebras* ◇ B25 C05 G15 ◇

MCKENZIE, R. & SHELAH, S. *The cardinals of simple models for universal theories* ◇ C05 C50 C55 E05 ◇

NELSON, EVELYN *Infinitary equational compactness* ◇ C05 C50 C75 ◇

NELSON, EVELYN *Not every equational class of infinitary algebras contains a simple algebra* ◇ C05 C50 ◇

NEUMANN, W.D. *On Malcev conditions* ◇ C05 ◇

OLIN, P. *Free products and elementary equivalence* ◇ C05 C60 ◇

PIGOZZI, D. *The join of equational theories* ◇ B25 C05 D35 ◇

PLONKA, J. *On connections between the decomposition of an algebra into sums of direct systems of subalgebras* ◇ C05 C30 ◇

PLONKA, J. *On the subdirect product of some equational classes of algebras* ◇ C05 C30 ◇

SUSZKO, R. *Equational logic and theories in sentential languages* ◇ B05 C05 C07 ◇

TAYLOR, W. *Pure-irreducible mono-unary algebras* ◇ C05 C50 G10 ◇

TAYLOR, W. *Uniformity of congruences* ◇ C05 C52 ◇

VAZHENIN, YU.M. *On global including semigroups of symmetric semigroups (Russian)* ◇ C05 C60 ◇

WOLF, A. *Sheaf representations of arithmetical algebras* ◇ C05 C90 G30 ◇

ZYGMUNT, J. *A note on direct products and ultraproducts of logical matrices* ◇ B50 C05 C20 ◇

1975

AMER, M.A. *On a theorem of Garrett Birkhoff (Arabic summary)* ◇ C05 ◇

ANDREKA, H. & NEMETI, I. *Subalgebra systems of algebras with finite and infinite, regular and irregular arity* ◇ C05 C75 ◇

BACSICH, P.D. *Amalgamation properties and interpolation theorems for equational theories* ◇ B20 C05 C40 C52 ◇

BACSICH, P.D. *The strong amalgamation property* ◇ C05 C52 C60 ◇

BALDWIN, JOHN T. & BERMAN, J. *The number of subdirectly irreducible algebras in a variety* ◇ C05 C52 ◇

BAUDISCH, A. *Elementare Theorien von Halbgruppen mit Kuerzungsregeln mit einem einstelligen Praedikat* ◇ C05 C60 D35 ◇

BELEGRADEK, O.V. & TAJTSLIN, M.A. *Categorical varieties of groupoids (Russian)* ◇ C05 C35 ◇

BELESOV, A.U. *α-atomic compactness of filtered products (Russian)* ◇ C05 C20 C50 ◇

BELKIN, V.P. & GORBUNOV, V.A. *Filters in lattices of quasivarieties of algebraic systems (Russian)* ◇ C05 C60 D40 G10 ◇

BLOOM, S.L. *Some theorems on structural consequence operations* ◇ B22 C05 C20 ◇

BUDKIN, A.I. & GORBUNOV, V.A. *Quasivarieties of algebraic systems (Russian)* ◇ C05 ◇

BURRIS, S. *Boolean powers* ◇ B25 C05 C30 C90 ◇

BURRIS, S. & SANKAPPANAVAR, H.P. *Lattice-theoretic decision problems in universal algebra* ◇ B25 C05 D35 ◇

CHIKHACHEV, S.A. *Categorical universal classes of semigroups (Russian)* ◇ C05 C35 C52 C60 ◇

CLARE, F. *Operations on elementary classes of groups* ◇ C05 C52 C60 ◇

COMER, S.D. *Monadic algebras with finite degree* ◇ B25 C05 C90 G15 ◇

ELLENTUCK, E. *Semigroups, Horn sentences and isolic structures* ◇ C05 C57 D50 E10 ◇

FROEMKE, J. & QUACKENBUSH, R.W. *The spectrum of an equational class of groupoids* ◇ C05 C13 ◇

GARCIA, O.C. *Injectivity in categories of groups and algebras* ◇ C05 C60 G30 ◇

GIRI, R.D. *On a varietal structure of algebras* ◇ C05 C52 ◇

GOULD, M. *An equational spectrum giving cardinalities of endomorphism monoids* ◇ C05 C13 ◇

GOULD, M. *Endomorphism and automorphism structure of direct squares of universal algebras* ◇ C05 C07 ◇

HATCHER, W.S. & SHAFAAT, A. *Categorical languages for algebraic structures* ◇ C05 G30 ◇

HUTCHINSON, G. *On the representation of lattices by modules* ◇ C05 C60 G10 ◇

INDERMARK, K. *Gleichungsdefinierbarkeit in Relationstrukturen* ◇ C05 C40 ◇

JONSSON, B. & MCNULTY, G.F. & QUACKENBUSH, R.W. *The ascending and descending varietal chains of a variety* ◇ B25 C05 D35 ◇

KAISER, KLAUS *Various remarks on quasi-universal model classes* ◇ B15 C05 C52 C85 ◇

MCKENZIE, R. *On spectra, and the negative solution of the decision problem for identities having a finite nontrivial model* ◇ B20 C05 C13 D35 ◇

MONK, J.D. *Some cardinal functions on algebras I,II* ◇ C05 E50 E75 ◇

MOVSISYAN, YU.M. *Algebraic systems of the second level (Russian)* ◇ B15 C05 ◇

MURSKIJ, V.L. *A finite basis of identities and other properties of "almost all" finite algebras (Russian)* ◇ C05 C13 C52 ◇

NELSON, EVELYN *Injectivity and equational compactness in the class of $\aleph_0$-semilattices* ◇ C05 C50 G10 ◇

NELSON, EVELYN *On the adjointness between operations and relations and its impact on atomic compactness* ◇ C05 C50 G30 ◇

NELSON, EVELYN *Semilattices do not have equationally compact hulls* ◇ C05 C50 G10 ◇

NELSON, EVELYN *Some functorial aspects of atomic compactness* ◇ C05 C50 E50 G30 ◇

OLIN, P. *$\aleph_0$-categoricity of products of 1-unary algebras* ◇ C05 C35 ◇

PALYUTIN, E.A. *Description of categorical quasivarieties (Russian)* ◇ C05 C35 ◇

PIGOZZI, D. *Equational logic and equational theories of algebras* ◇ C05 C98 ◇

PLONKA, J. *On relations between decompositions of an algebra into the sums of direct systems of subalgebras* ◇ C05 C30 ◇

PLONKA, J. *Remark on direct products and the sums of direct systems of algebras (Russian summary)*
⋄ C05 C30 ⋄

TARSKI, A. *An interpolation theorem for irredundant bases of closure structures* ⋄ C05 G15 ⋄

TAYLOR, W. *Continuum many Mal'cev conditions*
⋄ C05 C52 ⋄

TAYLOR, W. *The fine spectrum of a variety*
⋄ C05 C13 C52 ⋄

1976

APPLESON, R.R. *Zero-divisors among finite structures of a fixed type* ⋄ C05 C13 C30 ⋄

BALDWIN, JOHN T. & BERMAN, J. *Varieties and finite closure conditions* ⋄ C05 C13 ⋄

BELESOV, A.U. *Injective and equationally compact modules (Russian)* ⋄ C05 C50 C60 ⋄

BIRKHOFF, GARRETT *Note on universal topological algebra* ⋄ C05 ⋄

BLOOM, S.L. *Varieties of ordered algebras*
⋄ C05 C07 ⋄

BUDKIN, A.I. *Quasi-identities in a free group (Russian)*
⋄ C05 ⋄

BURRIS, S. *Subdirect representations in axiomatic classes*
⋄ C05 C52 ⋄

CLARK, D.M. & KRAUSS, P.H. *Para primal algebras*
⋄ C05 C13 ⋄

COMER, S.D. *Complete and model-complete theories of monadic algebras*
⋄ B25 C05 C25 C35 C60 G15 ⋄

CORNISH, W.H. *A sheaf representation for generalized Stone lattices* ⋄ C05 C90 G10 ⋄

CSAKANY, B. *Conditions involving universally quantified function variables* ⋄ C05 C40 C85 ⋄

FRIEDMAN, H.M. *On decidability of equational theories*
⋄ B25 C05 D35 ⋄

GALVIN, F. & PRIKRY, K. *Infinitary Jonsson algebras and partition relations* ⋄ C05 C50 C55 C75 E05 ⋄

HALEY, D.K. *Equational compactness and compact topologies in rings satisfying the a.c.c.* ⋄ C05 C60 ⋄

HOFMANN, K.H. & MISLOVE, M.W. *Amalgamation in categories with concrete duals* ⋄ C05 C52 G30 ⋄

ISBELL, J.R. *Compatibility and extensions of algebraic theories* ⋄ C05 G05 G30 ⋄

JEZEK, J. *Universal algebra and model theory (Czech)*
⋄ C05 C98 ⋄

JONSSON, B. & OLIN, P. *Elementary equivalence and relatively free products of lattices* ⋄ C05 C30 G10 ⋄

KRASNER, M. *Polytheorie de Galois abstraite dans le cas infini general* ⋄ C05 C07 C60 ⋄

LINTON, F.E.J. *Companions for a lonely result of Isbell*
⋄ C05 ⋄

LUCAS, T. *Universal classes of monadic algebras*
⋄ B25 C05 G15 ⋄

MYCIELSKI, J. & TAYLOR, W. *A compactification of the algebra of terms* ⋄ C05 C75 ⋄

OLIN, P. *Homomorphisms of elementary types of boolean algebras* ⋄ C05 C52 G05 ⋄

PIGOZZI, D. *The universality of the variety of quasigroups*
⋄ B25 C05 C52 D35 ⋄

SCHEIN, B.M. *Injective commutative semigroups*
⋄ C05 C60 ⋄

SCHUMACHER, D. *Absolutely free algebras in a topos containing an infinite object* ⋄ C05 C30 G30 ⋄

SHAFAAT, A. *Constructions preserving the associative and the commutative laws* ⋄ C05 ⋄

SZENDREI, A. *The operation ISKP on classes of algebras*
⋄ C05 ⋄

TAYLOR, W. *Pure compactifications in quasi-primal varieties* ⋄ C05 ⋄

WEGLORZ, B. *A note on: "Atomic compactness in $\aleph_1$-categorical Horn theories" by John T. Baldwin*
⋄ C05 C35 C50 C52 ⋄

ZAMYATIN, A.P. *Varieties of associative rings whose elementary theory is decidable (Russian)*
⋄ B25 C05 C60 ⋄

1977

BALDWIN, JOHN T. & BERMAN, J. *A model theoretic approach to Mal'cev conditions* ⋄ C05 C52 ⋄

BANASCHEWSKI, B. & NELSON, EVELYN *Elementary properties of limit reduced powers with applications to boolean powers* ⋄ C05 C20 C30 C90 ⋄

BELESOV, A.U. *Equationally compact systems (Russian)*
⋄ C05 C50 C60 ⋄

BOZOVIC, N.B. *On Markov properties of finitely presented groups* ⋄ C05 C50 D45 ⋄

BULMAN-FLEMING, S. & WERNER, H. *Equational compactness in quasi-primal varieties* ⋄ C05 C60 ⋄

BURRIS, S. *An example concerning definable principal congruences* ⋄ C05 C40 ⋄

CLARK, D.M. & KRAUSS, P.H. *Varieties generated by para primal algebras* ⋄ C05 C13 ⋄

DIXON, P.G. *Classes of algebraic systems defined by universal Horn sentences* ⋄ C05 C52 ⋄

FORREST, W.K. *Model theory for universal classes with the amalgamation property: A study in the foundations of model theory and algebra*
⋄ C05 C25 C30 C35 C45 C52 C60 ⋄

FUJIWARA, T. *A remark on existence of maximal homomorphisms* ⋄ C05 ⋄

FUJIWARA, T. *On implicational classes of structures*
⋄ C05 C75 ⋄

GORBUNOV, V.A. *Covers in lattices of quasivarieties and independent axiomatizability (Russian)*
⋄ C05 C52 ⋄

HENKIN, L. *The logic of equality* ⋄ B20 C05 ⋄

HICKIN, K.K. & PLOTKIN, J.M. *Some algebraic properties of weakly compact and compact cardinals*
⋄ C05 C55 E55 ⋄

ISBELL, J.R. & KLUN, M.I. & SCHANUEL, S.H. *Affine parts of algebraic theories. I* ⋄ C05 G30 ⋄

KRAUSS, P.H. *Homogeneous universal models of universal theories* ⋄ C05 C35 C50 C52 ⋄

LATYSHEV, V.N. *Complexity of nonmatrix varieties of associative algebras I,II (Russian)* ⋄ C05 C60 ⋄

MAKKAI, M. & McNULTY, G.F. *Universal Horn axiom systems for lattices of submodules*
⋄ C05 C60 G10 ⋄

MAKSIMOVA, L.L. *Craig's interpolation theorem and amalgamable varieties (Russian)*
⋄ B22 B55 C05 C40 C90 G10 ⋄

MAKSIMOVA, L.L. *Craig's theorem in superintuitionistic logics and amalgamable varieties of pseudo-boolean algebras (Russian)*
◊ B55 C05 C40 C90 G05 G10 ◊

NELSON, EVELYN *Classes defined by implications*
◊ B20 C05 C52 ◊

OLIN, P. *Elementary properties of V-free products of groups* ◊ C05 C30 C60 ◊

PRESIC, M.D. & PRESIC, S.B. *On the embedding of Ω-algebras in groupoids* ◊ C05 ◊

SAUER, N. & STONE, MICHAEL G. *The algebraic closure of a semigroup of functions* ◊ C05 ◊

TAYLOR, W. *Equational logic* ◊ C05 C98 D40 ◊

WERNER, H. *Varieties generated by quasi-primal algebras have decidable theories* ◊ B25 C05 ◊

1978

ANDREKA, H. & NEMETI, I. *Neat reducts of varieties*
◊ C05 C20 G15 G25 G30 ◊

BERMAN, J. & GRAETZER, G. *Uniform representations of congruence schemes* ◊ C05 C40 ◊

BURRIS, S. & JEFFERS, E. *On the simplicity and subdirect irreducibility of boolean ultrapowers*
◊ C05 C20 C50 ◊

BURRIS, S. *Rigid boolean powers*
◊ C05 C30 C50 G05 ◊

FABER, V. & LAVER, R. & MCKENZIE, R. *Coverings of groups by abelian subgroups*
◊ C05 C60 E50 E75 ◊

FORREST, W.K. *A note on universal classes with applications to the theory of graphs*
◊ C05 C50 C52 C65 ◊

FUJIWARA, T. *Algebraically closed algebraic extensions in universal classes* ◊ C05 C25 ◊

GARAVAGLIA, S. *Homology with equationally compact coeffcients* ◊ C05 C20 ◊

GIVANT, S. *Universal Horn classes categorical or free in power* ◊ B20 C05 C35 ◊

HATCHER, W.S. *A language for type-free algebra*
◊ C05 G30 ◊

HERRERA MIRANDA, J. *Les theories convexes de Horn (English summary)* ◊ B20 C05 C52 G30 ◊

HULE, H. *Relations between the amalgamation property and algebraic equations* ◊ C05 C52 ◊

MCKENZIE, R. *A finite algebra $\underline{A}$ with $SP(\underline{A})$ not elementary* ◊ C05 C52 ◊

MCKENZIE, R. *Para primal varieties: A study of finite axiomatizability and definable principal congruences in locally finite varieties* ◊ C05 ◊

MIKENBERG, I. *From total to partial algebras*
◊ C05 C40 C52 ◊

NEMETI, I. *From hereditary classes to varieties in abstract model theory and partial algebra*
◊ C05 C20 C52 C90 C95 ◊

NEUMANN, W.D. *Mal'cev conditions, spectra and Kronecker product* ◊ C05 C13 ◊

REGGIANI, C. *Le SH nei limiti diretti di strutture e di interpretazioni (English summary)*
◊ C05 C30 C40 ◊

SHELAH, S. *Jonsson algebras in successor cardinals*
◊ C05 C55 E05 E10 E75 ◊

SKORNYAKOV, L.A. *The axiomatizability of a class of injective polygons (Russian)* ◊ C05 C52 C60 ◊

SZABO, L. *Concrete representation of related structures of universal algebras I* ◊ C05 C40 ◊

TAJMANOV, A.D. *On topologizing countable algebras (Russian)* ◊ C05 C15 C65 ◊

WERNER, H. *Discriminator-algebras. Algebraic representation and model theoretic properties*
◊ B25 C05 C25 C35 C90 ◊

ZAMYATIN, A.P. *A non-abelian variety of groups has an undecidable elementary theory(Russian)*
◊ C05 C60 D35 ◊

ZAMYATIN, A.P. *Prevarieties of associative rings whose elementary theory is decidable (Russian)*
◊ B25 C05 C60 ◊

1979

ANDREKA, H. & NEMETI, I. *Formulas and ultraproducts in categories* ◊ C05 C20 G30 ◊

BENENSON, I.E. *The structure of quasivarieties of models (Russian)* ◊ C05 C52 ◊

BURRIS, S. & LAWRENCE, J. *Definable principle congruences in varieties of groups and rings*
◊ C05 C40 ◊

BURRIS, S. & WERNER, H. *Sheaf constructions and their elementary properties*
◊ B25 C05 C20 C25 C30 C60 C90 ◊

CLARK, D.M. & KRAUSS, P.H. *Global subdirect products*
◊ C05 C30 C90 ◊

CZELAKOWSKI, J. *A characterization of Matr(C)*
◊ B22 C05 ◊

GIVANT, S. *A representation theorem for universal Horn classes categorical in power* ◊ B20 C05 C35 ◊

HALEY, D.K. *Equational compactness in rings*
◊ C05 C60 ◊

HULE, H. *Solutionally complete varieties* ◊ C05 ◊

JONSSON, B. *On finitely based varieties of algebras*
◊ C05 ◊

KUKIN, G.P. *Subalgebras of finitely defined Lie algebras (Russian)* ◊ C05 C57 ◊

MAKSIMOVA, L.L. *Interpolation theorems in modal logics and amalgamable varieties of topological boolean algebras (Russian)* ◊ B45 C05 C40 C90 G05 ◊

PALYUTIN, E.A. *Categorical positive Horn theories (Russian)* ◊ C05 C35 C52 ◊

PLONKA, J. *On automorphism groups of relational systems and universal algebras* ◊ C05 C07 E20 ◊

PREST, M. *Torsion and universal Horn classes of modules*
◊ C05 C60 ◊

REGGIANI, C. *Conservazione delle SH di tipo $\forall\exists$ nei limiti diretti (English summary)* ◊ C05 C30 C40 ◊

ROZENBLAT, B.V. *Positive theories of free inverse semigroups (Russian)* ◊ B20 C05 D35 D40 ◊

SCHMID, J. *Algebraically and existentially closed distributive lattices* ◊ C05 C25 G10 ◊

SHISHMAREV, YU.E. *Categorical quasivarieties of quasigroups (Russian)* ◊ C05 C35 ◊

TAYLOR, W. *Equational logic* ◊ C05 C98 ◊

TULIPANI, S. *The Hanf number for classes of algebras whose largest congruence is always finitely generated*
◊ C05 C55 ◊

TUMANOV, V.I. *Lattices of equational theories of models (Russian)* ⋄ C05 C52 G10 ⋄

VOLGER, H. *Sheaf representations of algebras and transfer theorems for model theoretic properties*
⋄ B25 C05 C30 C35 C90 ⋄

ZLATOS, P. *A characterization of some boolean powers*
⋄ C05 C20 C30 C75 ⋄

1980

ANDREKA, H. & NEMETI, I. *On systems of varieties definable by schemes of equations* ⋄ C05 G15 ⋄

ANDREKA, H. & BURMEISTER, P. & NEMETI, I. *Quasi-equational logic of partial algebras*
⋄ C05 C90 ⋄

BAJRAMOV, R.A. & MAMEDOV, O.M. *Σ-subdirect representations in axiomatic classes (Russian) (English summary)* ⋄ C05 C30 C52 ⋄

BALDWIN, JOHN T. *The number of subdirectly irreducible algebras in a variety. II* ⋄ C05 C75 ⋄

BANASCHEWSKI, B. & NELSON, EVELYN *Boolean powers as algebras of continuous functions*
⋄ C05 C20 C30 C50 ⋄

BERMAN, J. *A proof of Lyndon's finite basis theorem*
⋄ C05 C13 ⋄

BURRIS, S. & WERNER, H. *Remarks on boolean products*
⋄ C05 C30 C52 G05 ⋄

BUYS, A. & HEIDEMA, J. *Model-theoretical aspects of radicals and subdirect decomposability* ⋄ C05 C60 ⋄

CAICEDO, X. *The subdirect decomposition theorem for classes of structures closed under direct limits*
⋄ C05 C52 ⋄

CLARK, D.M. & KRAUSS, P.H. *Plain para primal algebras*
⋄ C05 C13 ⋄

COHN, P.M. *On semifir constructions*
⋄ C05 C20 C52 C60 ⋄

CZELAKOWSKI, J. *Model-theoretic methods in methodology of propositional calculi*
⋄ B22 C05 G99 ⋄

CZELAKOWSKI, J. *Reduced products of logical matrices*
⋄ B22 C05 C20 ⋄

DAY, A. & FREESE, R. *A characterization of identities implying congruence modularity. I* ⋄ C05 ⋄

GIVANT, S. *Union decompositions and universal classes categorical in power* ⋄ C05 C35 ⋄

HEIDEMA, J. & MALAN, S.W. *Congruences on relational structures* ⋄ C05 C07 ⋄

JAKUBIKOVA-STUDENOVSKA, D. *On weakly rigid monounary algebras* ⋄ C05 C55 E10 E75 ⋄

JONSSON, B. *Congruence varieties* ⋄ C05 ⋄

KHMELEVSKIJ, YU.I. & TAJMANOV, A.D. *Decidability of the universal theory of a free semigroup (Russian)*
⋄ B25 C05 C60 ⋄

LAMPE, W.A. & MYERS, D.L. *Elementary properties of free extensions* ⋄ C05 ⋄

MADDUX, R. *The equational theory of CA_3 is undecidable*
⋄ C05 D35 G15 ⋄

MARKOVSKI, S. *On quasivarieties of generalized subalgebras* ⋄ C05 C52 ⋄

McNULTY, G.F. *Classes which generate the variety of all lattice-ordered groups* ⋄ C05 C52 C60 ⋄

OLIN, P. *Varietal free products of bands*
⋄ C05 C30 G10 ⋄

PALYUTIN, E.A. *Categorical Horn classes I (Russian)*
⋄ C05 C35 C45 C52 ⋄

PALYUTIN, E.A. *Description of categorical positive Horn classes (Russian)* ⋄ C05 C35 C52 ⋄

QUACKENBUSH, R.W. *Algebras with minimal spectrum*
⋄ C05 C13 ⋄

SAPIR, M.V. *On finite and independent axiomatizability of certain quasi-varieties of semigroups (Russian)*
⋄ C05 C52 ⋄

SHAFAAT, A. *Consistency in categorical languages for algebras* ⋄ C05 G30 ⋄

TAYLOR, W. *Mal'tsev conditions and spectra*
⋄ C05 C13 C52 ⋄

UROSU, C. *On filter products of congruences (Romanian) (French summary)* ⋄ C05 C20 ⋄

1981

ANDREKA, H. & NEMETI, I. *HSP K is equational class, without the axiom of choice* ⋄ C05 E25 ⋄

ANDREKA, H. & BURMEISTER, P. & NEMETI, I. *Quasivarieties of partial algebras. A unifying approach towards a two-valued model theory for partial algebras*
⋄ C05 C90 ⋄

ANDREKA, H. & NEMETI, I. *Some universal algebraic and model theoretic results in computer science*
⋄ B75 C05 C98 ⋄

BAKER, K.A. *Definable normal closures in locally finite varieties of groups* ⋄ C05 C40 C60 ⋄

BALDWIN, JOHN T. & BERMAN, J. *Elementary classes of varieties* ⋄ C05 C52 ⋄

BAUDISCH, A. *Subgroups of semifree groups*
⋄ C05 C60 ⋄

BELEGRADEK, O.V. *Classes of algebras with inner mappings (Russian)* ⋄ C05 C25 C50 ⋄

BURMEISTER, P. *Quasi-equational logic for partial algebras* ⋄ C05 C90 ⋄

BURRIS, S. & SANKAPPANAVAR, H.P. *A course in universal algebra* ⋄ C05 C98 ⋄

BURRIS, S. & McKENZIE, R. *Decidability and boolean representations* ⋄ B25 C05 C30 C60 ⋄

BURRIS, S. & McKENZIE, R. *Decidable varieties with modular congruence lattices* ⋄ B25 C05 ⋄

COMER, S.D. *The decision problem for certain nilpotent closed varieties* ⋄ B25 C05 D35 ⋄

CSAKANY, B. & GAVALCOVA, T. *Three-element quasi-primal algebras* ⋄ C05 C13 ⋄

DZIOBIAK, W. *Concerning axiomatizability of the quasivariety generated by a finite Heyting or topological boolean algebra* ⋄ C05 C13 C52 ⋄

FISHER, J.R. *Axiomatic radical and semisimple classes of rings* ⋄ C05 C52 C60 ⋄

FRANZEN, B. *Algebraic compactness of filter quotients*
⋄ C05 C30 C60 ⋄

GIVANT, S. *The number of nonisomorphic denumerable models of certain universal Horn classes*
⋄ C05 C15 C35 ⋄

HODGES, W. *In singular cardinality, locally free algebras are free* ⋄ C05 C55 C60 E05 E60 E75 ⋄

HUET, G. *A complete proof of correctness of the Knuth-Bendix completion algorithm* ⋄ B25 B75 C05 ⋄

HUTCHINSON, G. *A complete logic for n-permutable congruence lattices* ⋄ C05 C35 G10 ⋄

JEZEK, J. *The lattice of equational theories I: Modular elements. II: The lattice of full sets of terms* ⋄ C05 C40 G10 ⋄

KEPKA, T. *Strong amalgamation property in some groupoid varieties* ⋄ C05 C52 ⋄

KOSTYRKO, V.F. *On elementary ∃-theory of semigroups (Russian)* ⋄ B25 C05 ⋄

LISOVIK, L.P. *Sets that are expressible in theories of commutative semigroups (Russian) (English Summary)* ⋄ B25 C05 C40 ⋄

MCKENZIE, R. *Residually small varieties of semigroups* ⋄ C05 C13 ⋄

MYCIELSKI, J. & PERLMUTTER, P. *Model completeness of some metric completions of absolutely free algebras* ⋄ C05 C35 C50 ⋄

PINUS, A.G. *The spectrum of rigid systems of Horn classes (Russian)* ⋄ C05 C30 C50 ⋄

POLAK, L. & ROSICKY, J. *Implicit operations on finite algebras* ⋄ C05 C13 ⋄

PROTASOV, I.V. & SIDORCHUK, A.D. *On varieties of topological algebraic systems (Russian)* ⋄ C05 C90 ⋄

RICHTER, M.M. & TRIESCH, E. *Universelle Algebra* ⋄ C05 C98 ⋄

ROZENBLAT, B.V. & VAZHENIN, YU.M. *Decidability of the positive theory of a free countably generated semigroup (Russian)* ⋄ B25 C05 D35 ⋄

ROZENBLAT, B.V. *Positive theories of some varieties of semigroups (Russian)* ⋄ B25 C05 ⋄

TAYLOR, W. *Some universal sets of terms* ⋄ C05 D80 ⋄

URQUHART, A.I.F. *The decision problem for equational theories* ⋄ B25 C05 D35 ⋄

URZYCZYN, P. *The unwind property in certain algebras* ⋄ C05 C57 ⋄

1982

ANDREKA, H. & NEMETI, I. *A general axiomatizability theorem formulated in terms of cone-injective subcategories* ⋄ C05 C20 C90 G30 ⋄

ARAGANE, K. *Characterization of elementary classes of structures whose every substructure is a congruence class* ⋄ C05 C52 ⋄

BALDWIN, JOHN T. & BERMAN, J. & GLASS, A.M.W. & HODGES, W. *A combinatorial fact about free algebras* ⋄ C05 ⋄

BALDWIN, JOHN T. & MCKENZIE, R. *Counting models in universal Horn classes* ⋄ C05 C52 ⋄

BALDWIN, JOHN T. & BERMAN, J. *Definable principal congruence relations: kith and kin* ⋄ C05 C52 E07 ⋄

BANKSTON, P. *Some obstacles to duality in topological algebra* ⋄ C05 C20 C65 G30 ⋄

BELYAEV, V.YA. *On algebraic closure and amalgamation of semigroups* ⋄ C05 C25 C60 ⋄

BLOOM, S.L. *A note on the logic of signed equations* ⋄ B60 C05 C07 ⋄

BUDKIN, A.I. *Independent axiomatizability of quasivarieties of groups (Russian)* ⋄ C05 C60 ⋄

BURMEISTER, P. *Partial algebras - survey of a unifying approach towards a two-valued model theory for partial algebras* ⋄ C05 C90 C98 ⋄

BURRIS, S. *Remarks on reducts of varieties* ⋄ C05 ⋄

BURRIS, S. *The first order theory of Boolean algebras with a distinguished group of automorphisms* ⋄ B25 C05 C60 G05 ⋄

BURRIS, S. & LAWRENCE, J. *Two undecidability results using modified boolean powers* ⋄ C05 C60 D35 G05 ⋄

BUYS, A. & HEIDEMA, J. *Radicals and subdirect products in models* ⋄ C05 C52 ⋄

CZEDLI, G. *A note on the compactness of the consequence relation for congruence varieties* ⋄ C05 ⋄

CZELAKOWSKI, J. & DZIOBIAK, W. *Another proof that $ISP_r(K)$ is the least quasivariety containing K* ⋄ C05 ⋄

DATUASHVILI, N.I. *Characterization of varieties of B-valued models (Russian) (English and Georgian summaries)* ⋄ C05 C90 E40 ⋄

DENECKE, K. *Preprimal algebras* ⋄ B50 C05 C13 ⋄

DOBBERTIN, H. *On Vaught's criterion for isomorphisms of countable boolean algebras* ⋄ C05 C15 G05 ⋄

DZIOBIAK, W. *Concerning axiomatizability of the quasivariety generated by a finite Heyting or topological boolean algebra* ⋄ C05 C13 C52 G10 ⋄

FUJIWARA, T. *Characterization of axiomatic classes of groupoids closed under identity element extensions* ⋄ C05 C52 ⋄

FURMANOWSKI, T. *The logic of algebraic rules as a generalization of equational logic* ⋄ C05 C07 ⋄

ISBELL, J.R. *Generating the algebraic theory of $C(X)$* ⋄ C05 C75 ⋄

JEZEK, J. *The lattice of equational theories III: definability and automorphisms* ⋄ C05 C40 G10 ⋄

JONSSON, B. *Varieties of relation algebras* ⋄ C05 C52 ⋄

KASHIWAGI, T. *Note on algebraic extensions in universal classes* ⋄ C05 C25 C30 ⋄

KOMATSU, H. *On the equational definability of addition in rings* ⋄ C05 C60 ⋄

KRAPEZ, A. *Some nonaxiomatizable classes of semigroups* ⋄ C05 C52 ⋄

MESKHI, S. *Complete boolean powers of quasiprimal algebras with lattice reducts* ⋄ C05 C30 ⋄

NEMETI, I. & SAIN, I. *Cone-implicational subcategories and some Birkhoff-type theorems* ⋄ C05 C20 C90 G30 ⋄

PALYUTIN, E.A. *Number of models in complete varieties* ⋄ C05 C45 C52 ⋄

PROTASOV, I.V. *Varieties of topological spaces (Russian) (English summary)* ⋄ C05 C65 ⋄

REITERMAN, J. *The Birkhoff theorem for finite algebras* ⋄ C05 C13 ⋄

SAIN, I. *On classes of algebraic systems closed with respect to quotients* ⋄ C05 C20 C30 C52 ⋄

SAIN, I. *Weak products for universal algebra and model theory* ⋄ C05 C30 ⋄

SCHMID, J. *Algebraically closed distributive p-algebras*
  ◇ C05 C25 G05 G10 G20 ◇
SCHMID, J. *Model companions of distributive p-algebras*
  ◇ C05 C25 C35 G10 ◇
SIEKMANN, J. & SZABO, P. *Universal unification and a classification of equational theories*
  ◇ B25 B35 C05 ◇
SMIRNOV, D.M. *Universal definability of Mal'tsev classes (Russian)* ◇ C05 C52 ◇
TAYLOR, W. *Some interesting identities* ◇ C05 C98 ◇
TULIPANI, S. *On classes of algebras with the definability of congruences* ◇ C05 C20 C52 ◇
TULIPANI, S. *Sui reticolo di congruenze di strutture algebriche modelli di teorie universali positive (English summary)* ◇ C05 C50 C55 ◇
WERNER, H. *Sheaf constructions in universal algebra and model theory*
  ◇ B25 C05 C25 C30 C35 C90 G05 ◇

1983

ANDREKA, H. & NEMETI, I. *Generalization of the concept of variety and quasivariety to partial algebras through category theory*
  ◇ C05 C40 C75 C90 C95 G30 ◇
BAKER, K.A. *Nondefinability of projectivity in lattice varieties* ◇ C05 C30 C40 G10 ◇
BALDWIN, JOHN T. & GIVANT, S. *Model complete universal Horn classes* ◇ C05 C35 ◇
BALDWIN, JOHN T. & SHELAH, S. *The structure of saturated free algebras* ◇ C05 C30 C45 C50 ◇
BANKSTON, P. *Obstacles to duality between classes of relational structures* ◇ C05 C20 E55 G30 ◇
BANKSTON, P. & FOX, R. *On categories of algebras equivalent to a quasivariety* ◇ C05 C20 G30 ◇
BANKSTON, P. *On productive classes of function rings*
  ◇ C05 C20 C65 E75 G30 ◇
BENECKE, K. & REICHEL, HORST *Equational partiality*
  ◇ C05 C90 ◇
BERMAN, J. *Free spectra of 3-element algebras*
  ◇ C05 C98 ◇
BLASS, A.R. *Words, free algebras, and coequalizers*
  ◇ C05 C75 E25 E35 F50 F55 G30 ◇
BURRIS, S. *Boolean constructions*
  ◇ A10 B25 C05 C30 C98 D35 G05 ◇
COMER, S.D. *Extension of polygroups by polygroups and their representations using color schemes*
  ◇ C05 C90 ◇
CZEDLI, G. & FREESE, R. *On congruence distributivity and modularity* ◇ B25 C05 ◇
FLEISCHER, I. *Subdirect decomposability into irreducibles*
  ◇ C05 C52 ◇
FUJIWARA, T. *Characterization of axiomatic classes of groupoids closed under identity extensions*
  ◇ C05 C52 ◇
FURMANOWSKI, T. *The logic of algebraic rules as a generalization of equational logic*
  ◇ C05 C07 C13 ◇
HOLLAND, W.C. *A survey of varieties of lattice ordered groups* ◇ C05 C60 C98 ◇
ISBELL, J.R. *Generic algebras* ◇ C05 C50 E55 G30 ◇

LIPPARINI, P. *On Hilbert's Nullstellensatz (Italian)*
  ◇ C05 C60 ◇
MARONGIU, G. *Definibilita al primo ordine delle congruenze principali e delle congruenze compatte*
  ◇ C05 C20 C52 ◇
MCKENZIE, R. *The number of non-isomorphic models in quasi-varieties of semigroups* ◇ C05 C52 ◇
ONO, H. *Equational theories and universal theories of fields* ◇ B25 C05 C60 ◇
PINUS, A.G. *The operation of Cartesian product*
  ◇ B25 C05 D35 ◇
QUACKENBUSH, R.W. *Minimal paraprimal algebras*
  ◇ C05 C98 ◇
ROZENBLAT, B.V. & VAZHENIN, YU.M. *On positive theories of free algebraic systems (Russian)*
  ◇ B20 B25 C05 ◇
RUDAK, W. *A completeness theorem for weak equational logic* ◇ C05 C07 ◇
TULIPANI, S. *On the congruence lattice of stable algebras with definability of compact congruences*
  ◇ C05 C45 ◇
TULIPANI, S. *On the size of congruence lattices for models of theories with definability of congruences*
  ◇ C05 C35 G10 ◇
VAZHENIN, YU.M. *Semigroups with one defining relation whose elementary theories are decidable (Russian)*
  ◇ B25 C05 C60 ◇
ZAMYATIN, A.P. *Decidability of the elementary theories of certain varieties of rings (Russian)*
  ◇ B25 C05 C60 ◇
ZHALDOKAS, R. *An approach to the construction of complete term rewriting systems (Russian) (English and Lithuanian summaries)* ◇ B25 B35 B75 C05 ◇

1984

ALIEV, V.G. *Model companions of theories of semigroup classes (Russian) (English and Azerbaijani summaries)*
  ◇ C05 C25 C35 ◇
ANDREKA, H. & NEMETI, I. *Importance of universal algebra for computer science*
  ◇ C05 C20 C95 G15 ◇
ARAGANE, K. *Characterization of elementary classes of structures whose every substructure is a congruence class*
  ◇ C05 C52 ◇
BAJRAMOV, R.A. & MAMEDOV, O.M. *Nonstandard subdirect representations in axiomatizable classes (Russian)* ◇ C05 C30 C52 ◇
BELYAEV, V.YA. *Uncountable extensions of countable algebraically closed semigroups (Russian)*
  ◇ C05 C55 C60 C75 ◇
BUDKIN, A.I. *Quasiidentities and direct wreaths of groups (Russian)* ◇ C05 C30 C60 ◇
BURRIS, S. *Model companions for finitely generated universal Horn classes*
  ◇ B25 C05 C10 C35 D20 ◇
CLARK, D.M. *$\aleph_0$-categoricity in infra primal varieties*
  ◇ C05 C30 C35 E05 ◇
CZEDLI, G. & DAY, A. *Horn sentences with (W) and weak Mal'cev conditions* ◇ C05 ◇
CZEDLI, G. *Mal'cev conditions for Horn sentences with congruence permutability* ◇ C05 ◇

ECSEDI-TOTH, P. *A characterization of quasi-varieties in equality-free languages* ⋄ C05 C52 ⋄

FLEISCHER, I. *One-variable equationally compact distributive lattices (Russian summary)* ⋄ C05 C50 G10 ⋄

GALLIER, J.H. *n-rational algebras I: Basic properties and free algebras. II: Varieties and logic of inequalities* ⋄ C05 C75 ⋄

GORBUNOV, V.A. *Axiomatizability of replete classes (Russian)* ⋄ C05 C20 C52 ⋄

GRAETZER, G. & WHITNEY, S. *Infinitary varieties of structures closed under the formation of complex structures* ⋄ C05 C52 C75 ⋄

HODGES, W. *On constructing many non-isomorphic algebras* ⋄ C05 C52 C55 C75 E50 E55 E75 ⋄

HOLLAND, W.C. *Classification of lattice ordered groups (French summary)* ⋄ C05 C60 C98 ⋄

ISKANDER, A.A. *Extensions of algebraic systems* ⋄ C05 C30 ⋄

KALFA, C. *Decidable properties of finite sets of equations in trivial languages* ⋄ C05 C35 C52 ⋄

MANCA, V. & SALIBRA, A. *First-order theories as many-sorted algebras* ⋄ C05 C07 ⋄

MCKENZIE, R. *A new product of algebras and a type reduction theorem* ⋄ C05 C30 ⋄

NELSON, GEORGE C. *Boolean powers, recursive models, and the Horn theory of a structure* ⋄ C05 C20 C30 C57 G05 ⋄

RICHTER, M.M. *Some aspects of nonstandard methods in general algebra* ⋄ C05 C30 H05 ⋄

STEPANOVA, A.A. *Categorical quasivarieties of abelian groupoids and quasigroups (Russian)* ⋄ C05 C35 ⋄

STEPANOVA, A.A. *Groupoids of rank 2 generating categorical Horn classes (Russian)* ⋄ C05 C35 ⋄

TULIPANI, S. *On the universal theory of classes of finite models* ⋄ C05 C13 C52 C60 ⋄

TUMANOV, V.I. *Finite lattices with no independent basis of quasiidentities (Russian)* ⋄ C05 C13 G10 ⋄

WRONSKI, A. *On varieties of commutative BCK-algebras not generated by their finite members* ⋄ C05 C52 G25 ⋄

1985

ASH, C.J. *Pseudovarieties, generalized varieties and similiarly described classes* ⋄ C05 C13 ⋄

BANKSTON, P. & SCHUTT, R. *On minimally free algebras* ⋄ C05 C50 C60 C65 ⋄

BARANSKIJ, V.A. *Independence of equational theories and automorphism groups of lattices* ⋄ C05 C07 G10 ⋄

BURRIS, S. *A simple proof of the hereditary undecidability of the theory of lattice ordered abelian groups* ⋄ C05 C60 D35 G10 ⋄

CZELAKOWSKI, J. *Sentential logics and Maehara interpolation property* ⋄ B22 C05 C40 C52 ⋄

DROSTE, M. *Classes of universal words for the infinite symmetric groups* ⋄ C05 C50 C60 ⋄

GARCIA, O.C. & TAYLOR, W. *Generalized commutativity* ⋄ C05 C60 ⋄

GLASS, A.M.W. *The word and isomorphism problems in universal algebra* ⋄ C05 C57 D40 ⋄

GOGUEN, J.A. & MESEGUER, J. *Completeness of many-sorted equational logic* ⋄ B20 C05 C07 ⋄

HUYNH, D.T. *The complexity of equivalence problems for commutative grammars* ⋄ C05 C15 ⋄

MAKOWSKY, J.A. *Vopenka's principle and compact logics* ⋄ C05 C55 C95 E55 ⋄

MCKENZIE, R. *The structure of finite algebras* ⋄ C05 C13 C98 ⋄

MEDVEDEV, N.YA. *Quasivarieties of l-groups and groups* ⋄ C05 C52 C60 ⋄

MONK, J.D. *On endomorphism bases* ⋄ C05 C55 E10 G05 ⋄

NEMETI, I. *Cylindric-relativised set algebras have strong amalgamation* ⋄ C05 C52 G15 G25 ⋄

NOVAK, VITEZSLAV *On a power of relational stuctures* ⋄ C05 C30 ⋄

PALYUTIN, E.A. & SAFFE, J. & STARCHENKO, S.S. *Models of superstable Horn theories* ⋄ C05 C35 C45 ⋄

PAUL, E. *Equational methods in first order predicate calculus* ⋄ B75 C05 C07 ⋄

PINUS, A.G. *Applications of boolean powers of algebraic systems (Russian)* ⋄ C05 C30 C55 C57 C80 C85 D35 E50 G05 ⋄

PIXLEY, A.F. *Principal congruence formulas in arithmetical varieties* ⋄ C05 ⋄

PLONKA, J. *On identities satisfied in finite algebras* ⋄ C05 C13 ⋄

STEPANOVA, A.A. *Semigroups generating categorical Horn classes (Russian)* ⋄ C05 C35 ⋄

TULIPANI, S. *Algebre sottodirettamente irriducibili e assiomatizzazione di varieta* ⋄ C05 C98 ⋄

URSINI, A. *Decision problems for classes of diagonalizable algebras* ⋄ B25 B45 C05 D35 ⋄

WRONSKI, A. *On a form of equational interpolation property* ⋄ C05 C40 ⋄

ZLATOS, P. *On conceptual completeness of syntactic-semantical systems* ⋄ C05 C20 C90 G15 G30 ⋄

C60 Model-theoretic algebra

1926
HERMANN, G. *Die Frage der endlich vielen Schritte in der Theorie der Polynomideale* ◊ C57 C60 D45 F55 ◊

1930
WAERDEN VAN DER, B.L. *Eine Bemerkung ueber die Unzerlegbarkeit von Polynomen*
◊ B25 C60 D45 F55 ◊

1931
TARSKI, A. *Sur les ensembles definissables de nombres reels I*
◊ B25 B28 C10 C40 C60 C65 D55 E15 E47 F35 ◊

1933
BIRKHOFF, GARRETT *On the combination of subalgebras*
◊ C05 C60 G10 ◊

1938
KRASNER, M. *Une generalisation de la notion de corps*
◊ C07 C60 C75 ◊
TARSKI, A. *Ein Beitrag zur Axiomatik der abelschen Gruppen* ◊ B30 C60 ◊

1939
MAL'TSEV, A.I. *On the embedding of associative systems into groups (Russian) (German summary)*
◊ C05 C60 ◊

1940
MAL'TSEV, A.I. *On the embedding of associative systems into groups II (Russian) (German summary)*
◊ C05 C60 ◊

1941
MAL'TSEV, A.I. *A general method for obtaining local theorems in group theory (Russian)* ◊ C07 C60 ◊

1943
MCKINSEY, J.C.C. *The decision problem for some classes of sentences without quantifiers*
◊ B20 B25 C05 C60 G10 ◊
NEUMANN, B.H. *Adjunction of elements to groups*
◊ C60 ◊

1945
KRASNER, M. *Generalisation et analogues de la theorie de Galois* ◊ C07 C60 C75 ◊

1948
HEWITT, E. *Rings of real-valued continuous functions*
◊ C20 C50 C60 C65 E75 ◊
TARSKI, A. *A decision method for elementary algebra and geometry* ◊ B25 B30 C10 C35 C60 ◊

1949
HIGMAN, G. & NEUMANN, B.H. & NEUMANN, H. *Embedding theorems for groups* ◊ C60 D40 ◊
SZMIELEW, W. *Decision problem in group theory*
◊ B25 C10 C60 ◊

1950
GRENIEWSKI, H. *Groups and fields definable in the propositional calculus* ◊ B05 C60 ◊
KRASNER, M. *Generalisation abstraite de la theorie de Galois* ◊ C07 C60 C75 ◊
SPECKER, E. *Additive Gruppen von Folgen ganzer Zahlen*
◊ C60 ◊
SZELE, T. *Ein Analogon der Koerpertheorie fuer abelsche Gruppen* ◊ C25 C60 ◊

1951
MAL'TSEV, A.I. *On the completing of group order (Russian)* ◊ C60 ◊
ROBINSON, A. *On axiomatic systems which possess finite models* ◊ C13 C52 C60 ◊
ROBINSON, A. *On the metamathematics of algebra*
◊ C07 C60 C98 ◊
ROBINSON, R.M. *Arithmetical definability of field elements* ◊ C60 ◊
ROBINSON, R.M. *Undecidable rings* ◊ C60 D35 ◊
SCOTT, W.R. *Algebraically closed groups* ◊ C25 C60 ◊
SHEPHERDSON, J.C. *Inverses and zero divisors in matrix rings* ◊ B25 C60 D40 ◊

1952
NEUMANN, B.H. *A note on algebraically closed groups*
◊ C25 C60 ◊
ROBINSON, A. *On the application of symbolic logic to algebra* ◊ C60 ◊
SHODA, K. *Zur Theorie der algebraischen Erweiterungen*
◊ C05 C60 ◊
TARSKI, A. *Some notions and methods on the borderline of algebra and metamathematics*
◊ C07 C35 C52 C60 ◊

1953
HENKIN, L. *Some interconnections between modern algebra and mathematical logic* ◊ C07 C60 C85 ◊
MAEDA, F. *A lattice formulation for algebraic and transcendental extensions in abstract algebras*
◊ C60 G10 ◊
ROBINSON, A. *Les rapports entre le calcul deductif et l'interpretation semantique d'un systeme axiomatique*
◊ C60 ◊

1954

EHRENFEUCHT, A. & LOS, J. *Sur les produits cartesiens des groupes cycliques infinis* ⋄ C30 C60 ⋄

NEUMANN, B.H. *An essay on free products of groups with amalgamations* ⋄ C60 ⋄

ROBINSON, A. *L'application de la logique formelle aux mathematiques* ⋄ B30 C60 ⋄

ROBINSON, A. *On predicates in algebraically closed fields* ⋄ C60 ⋄

SEIDENBERG, A. *A new decision method for elementary algebra* ⋄ B25 C10 C60 ⋄

1955

BERNAYS, P. *Betrachtungen ueber das Vollstaendigkeitsaxiom und verwandte Axiome* ⋄ B28 C60 C65 ⋄

BUECHI, J.R. & WRIGHT, J.B. *The theory of proportionality as an abstraction of group theory* ⋄ C60 ⋄

ERDOES, P. & GILLMAN, L. & HENRIKSEN, M. *An isomorphism theorem for real-closed fields* ⋄ C60 E05 E50 E75 ⋄

FROEHLICH, A. & SHEPHERDSON, J.C. *Effective procedures in field theory* ⋄ C57 C60 D45 ⋄

GILMORE, P.C. & ROBINSON, A. *Metamathematical considerations on the relative irreducibility of polynomials* ⋄ C60 ⋄

HORN, A. *A characterisation of unions of linearly independent sets* ⋄ C60 E75 ⋄

ROBINSON, A. *Further remarks on ordered fields and definite functions* ⋄ C60 ⋄

ROBINSON, A. *Note on an embedding theorem for algebraic systems* ⋄ C60 ⋄

ROBINSON, A. *On ordered fields and definite functions* ⋄ C30 C60 ⋄

ROBINSON, A. *Ordered structures and related concepts* ⋄ C30 C35 C50 C60 ⋄

ROBINSON, A. *Theorie metamathematique des ideaux* ⋄ C60 C98 ⋄

SHODA, K. *Bemerkungen ueber die Existenz der algebraisch abgeschlossenen Erweiterung* ⋄ C05 C60 ⋄

SZMIELEW, W. *Elementary properties of abelian groups* ⋄ B25 C10 C60 ⋄

1956

FUCHS, L. & KERTESZ, A. & SZELE, T. *On abelian groups in which every homomorphic image can be embedded* ⋄ C60 ⋄

HENKIN, L. *Two concepts from the theory of models* ⋄ C13 C52 C60 ⋄

MAL'TSEV, A.I. *Remark on densely ordered groups (Russian)* ⋄ C60 ⋄

MAL'TSEV, A.I. *Representations of models (Russian)* ⋄ C07 C60 ⋄

ROBINSON, A. *Apercu metamathematique sur les nombres reels* ⋄ C60 C65 ⋄

ROBINSON, A. *Complete theories* ⋄ B25 C35 C60 C98 ⋄

ROBINSON, A. *Solution of a problem by Erdoes-Gillman-Henriksen* ⋄ C60 ⋄

SEIDENBERG, A. *Some remarks on Hilbert's Nullstellensatz* ⋄ C60 ⋄

SLOMINSKI, J. *On the extending of models III. Extensions in equationally definable classes of algebras* ⋄ C05 C30 C60 G05 ⋄

1957

ADYAN, S.I. *Finitely presented groups and algorithms (Russian)* ⋄ B25 C60 D40 ⋄

BUECHI, J.R. & WRIGHT, J.B. *Invariants of the anti-automorphisms of a group* ⋄ C60 ⋄

COLLINS, G.E. *Tarski's decision method for elementary algebra* ⋄ C10 C60 ⋄

HENKIN, L. *Sums of squares* ⋄ C57 C60 ⋄

JONSSON, B. *On direct decompositions of torsion-free abelian groups* ⋄ C05 C30 C60 ⋄

JONSSON, B. *On isomorphism types of groups and other algebraic systems* ⋄ C05 C30 C60 ⋄

KOCHEN, S. *Completeness of algebraic systems in higher order calculi* ⋄ B15 C35 C60 C85 ⋄

KREISEL, G. *Sums of squares* ⋄ C57 C60 D20 F07 F99 ⋄

LIGHTSTONE, A.H. & ROBINSON, A. *On the representation of Herbrand functions in algebraically closed fields* ⋄ C60 ⋄

LIGHTSTONE, A.H. & ROBINSON, A. *Syntactical transforms* ⋄ B10 C52 C60 ⋄

RABIN, M.O. *Computable algebraic systems* ⋄ C57 C60 D40 D45 D80 ⋄

ROBINSON, A. *Applications to field theory* ⋄ C60 ⋄

ROBINSON, A. *Some problems of definability in the lower predicate calculus* ⋄ C35 C40 C60 ⋄

TARSKI, A. & VAUGHT, R.L. *Arithmetical extensions of relational systems* ⋄ C05 C07 C55 C60 ⋄

TARSKI, A. & VAUGHT, R.L. *Elementary (arithmetical) extensions* ⋄ C05 C07 C35 C55 C60 ⋄

1958

ADYAN, S.I. *On algorithmic problems in effectively complete classes of groups (Russian)* ⋄ C60 D40 ⋄

KRASNER, M. *Les algebres cylindriques* ⋄ C07 C60 G15 ⋄

KREISEL, G. *Mathematical significance of consistency proofs* ⋄ C57 C60 F05 F25 F50 ⋄

MAL'TSEV, A.I. *Homomorphisms onto finite groups (Russian)* ⋄ C07 C13 C60 ⋄

SEIDENBERG, A. *Comments on Lefschetz' principle* ⋄ C60 ⋄

1959

COHN, P.M. *On the free product of associative rings* ⋄ C30 C60 ⋄

ERDELYI, M. *Systems of equations over non-commutative groups (Ungarisch) (Engl. Zusammenfassung)* ⋄ C25 C60 ⋄

FUCHS, L. *The existence of indecomposable abelian groups of arbitrary power* ⋄ C55 C60 ⋄

HALL, P. *On the finiteness of certain soluble groups* ⋄ C60 ⋄

HALL, P. *Some constructions for locally finite groups* ⋄ C25 C60 ⋄

MAL'TSEV, A.I. *Algebraic systems (Russian)* ⋄ C05 C60 ⋄

MAL'TSEV, A.I. *Model correspondences (Russian)*
⋄ B15 C05 C07 C52 C60 C85 ⋄
ROBINSON, A. *On the concept of a differentially closed field* ⋄ C35 C60 ⋄
ROBINSON, A. *Solution of a problem of Tarski*
⋄ B25 C35 C60 ⋄
ROBINSON, JULIA *The undecidability of algebraic rings and fields* ⋄ C60 D35 ⋄
SHIKHANOVICH, YU.A. *Examples of application of mathematical logic to algebra (Russian)* ⋄ C60 ⋄

1960

ALLING, N.L. *On ordered divisible groups* ⋄ C55 C60 ⋄
COHN, P.M. *On the free product of associative rings. II: the case of (skew) fields* ⋄ C30 C60 ⋄
JOHNSON, D.G. *A structure theory for a class of lattice-ordered rings* ⋄ C60 G10 ⋄
MAL'TSEV, A.I. *On free soluble groups (Russian)*
⋄ C60 D35 ⋄
MAL'TSEV, A.I. *Some correspondences between rings and groups (Russian)* ⋄ C60 D35 ⋄
RABIN, M.O. *Computable algebra, general theory and theory of computable fields* ⋄ C57 C60 D45 ⋄
ROBINSON, A. & ZAKON, E. *Elementary properties of ordered abelian groups* ⋄ B25 C35 C60 C65 ⋄
SEIDENBERG, A. *An elimination theory for differential algebra* ⋄ C10 C60 ⋄

1961

ALLING, N.L. *A characterization of abelian η_α-groups in terms of their natural valuation* ⋄ C55 C60 ⋄
COHN, P.M. *On the embedding of rings in skew fields*
⋄ C60 ⋄
HIGMAN, G. *Subgroups of finitely presented groups*
⋄ C60 D25 D45 ⋄
KOCHEN, S. *Ultraproducts in the theory of models*
⋄ C20 C35 C52 C60 ⋄
KOGALOVSKIJ, S.R. *On multiplicative semigroups of rings (Russian)* ⋄ C52 C60 ⋄
MAL'TSEV, A.I. *Elementary properties of linear groups (Russian)* ⋄ C60 ⋄
MAL'TSEV, A.I. *On the elementary theories of locally free universal algebras (Russian)* ⋄ B25 C05 C60 ⋄
MAURY, G. *La condition "integralement clos" dans quelques structures algébriques* ⋄ C05 C60 ⋄
ROBINSON, A. *Model theory and non-standard arithmetic*
⋄ C35 C50 C57 C60 C62 H15 ⋄
VAUGHT, R.L. *The elementary character of two notions from general algebra* ⋄ C05 C40 C52 C60 ⋄
ZAKON, E. *Generalized archimedean groups* ⋄ C60 ⋄

1962

ALLING, N.L. *On the existence of real-closed fields that are η_α-sets of power $\aleph_\alpha$* ⋄ C55 C60 C65 ⋄
BAER, REINHOLD *Abzaehlbar erkennbare gruppentheoretische Eigenschaften* ⋄ C60 ⋄
EISEN, M. *Ideal theory and difference algebra* ⋄ C60 ⋄
ERSHOV, YU.L. *Decidability of elementary theories of certain classes of abelian groups (Russian)*
⋄ B25 C60 ⋄

JONSSON, B. *Algebraic extensions of relational systems*
⋄ C05 C52 C60 ⋄
KARGAPOLOV, M.I. *On the elementary theory of lattices of subgroups (Russian)* ⋄ C60 G10 ⋄
KARGAPOLOV, M.I. *On the elementary theory of abelian groups (Russian)* ⋄ C35 C60 ⋄
KLIMOVSKY, G. *El axioma de eleccion y la existencia de subgrupos commutativos maximales*
⋄ C60 E25 E75 ⋄
KOKORIN, A.I. & KOPYTOV, V.M. *Certain classes of ordered groups (Russian)* ⋄ C60 ⋄
LAEUCHLI, H. *Auswahlaxiom in der Algebra*
⋄ C60 E25 E75 ⋄
LAVROV, I.A. *Undecidability of elementary theories of certain rings (Russian)* ⋄ C60 D35 ⋄
LORENZEN, P. *Metamathematik*
⋄ A05 B98 C60 D20 D35 D98 F98 ⋄
LYNDON, R.C. *Metamathematics and algebras: an example* ⋄ C40 C60 ⋄
MAL'TSEV, A.I. *On recursive abelian groups (Russian)*
⋄ C57 C60 D45 ⋄
MAL'TSEV, A.I. *Strongly related models and recursively complete algebras (Russian)*
⋄ C57 C60 C62 D45 ⋄
ROBINSON, A. *A note on embedding problems*
⋄ C07 C60 ⋄
ROBINSON, JULIA *On the decision problem for algebraic rings* ⋄ C60 D35 ⋄
TAJTSLIN, M.A. *Relatively elementary subspaces in compact Lie algebras (Russian)* ⋄ C60 ⋄

1963

FRENKEL', V.I. *Algorithmic problems in partially ordered groups (Russian)* ⋄ C57 C60 D40 ⋄
GUREVICH, Y. & KOKORIN, A.I. *Universal equivalence of ordered abelian groups (Russian)* ⋄ C60 ⋄
KARGAPOLOV, M.I. *Classification of ordered abelian groups by elementary properties (Russian)* ⋄ C60 ⋄
KOGALOVSKIJ, S.R. *Quasiprojective classes of models (Russian)* ⋄ C20 C52 C60 ⋄
MERZLYAKOV, YU.I. *On the theory of generalized solvable and generalized nilpotent groups (Russian)*
⋄ C52 C60 ⋄
ROBINSON, A. *Introduction to model theory and to the metamathematics of algebra*
⋄ B98 C60 C98 H15 ⋄
ZAKHAROV, D.A. *The ring of p-adic integers is elementary in the field of p-padic numbers (Russian)* ⋄ C60 ⋄

1964

CRAWLEY, P. & JONSSON, B. *Refinements for infinite direct decompositions of algebraic systems*
⋄ C05 C30 C60 ⋄
CSAKANY, B. *Abelian properties of primitive classes of universal algebras (Russian)* ⋄ C05 C60 ⋄
ERSHOV, YU.L. *Unsolvability of theories of symmetric and simple finite groups (Russian)* ⋄ C13 C60 D35 ⋄
FRENKEL', V.I. *On the effective partial ordering of finitely defined groups (Russian)* ⋄ C57 C60 ⋄
GUREVICH, Y. *Elementary properties of ordered abelian groups (Russian)* ⋄ B25 C60 ⋄

KEISLER, H.J. *Complete theories of algebraically closed fields with distinguished subfields* ⋄ B25 C35 C60 ⋄
ROBINSON, A. *Between logic and mathematics* ⋄ C60 ⋄
TAJTSLIN, M.A. *Decidability of certain elementary theories (Russian)* ⋄ B25 C60 ⋄
TAJTSLIN, M.A. *On elementary theories of free nilpotent algebras (Russian)* ⋄ B25 C60 D35 ⋄

1965

ALLING, N.L. *Rings of continuous integer-valued functions and nonstandard arithmetic* ⋄ C20 C60 H15 ⋄
ALMAGAMBETOV, ZH.A. *Solvability of the elementary theory of certain classes of free nilpotent algebras (Russian)* ⋄ B25 C60 ⋄
AX, J. & KOCHEN, S. *Diophantine problems over local fields I,II (II: A complete set of axioms for p-adic number theory)* ⋄ B25 C20 C35 C60 ⋄
BRIGNOLE, D. & RIBEIRO, H. *On the universal equivalence for ordered abelian groups* ⋄ C60 ⋄
BUECHI, J.R. *Relatively categorical and normal theories* ⋄ C35 C40 C60 ⋄
ERSHOV, YU.L. *On elementary theories of local fields (Russian)* ⋄ B25 C35 C60 ⋄
ERSHOV, YU.L. *On the elementary theory of maximal normed fields (Russian)* ⋄ B25 C35 C60 ⋄
ERSHOV, YU.L. *On the elementary theory of maximal normed fields (Russian)* ⋄ B25 C60 ⋄
GUREVICH, Y. *Existential interpretation (Russian)* ⋄ B20 B25 C60 D35 F25 ⋄
JACOBSON, R.A. & YOCOM, K.L. *Absolutely independent group axioms* ⋄ C60 ⋄
KOGALOVSKIJ, S.R. *On multiplicative semigroups of rings (Russian)* ⋄ C52 C60 ⋄
MAKOWIECKA, H. *On a primitive notion of one-dimensional geometries over some fields* ⋄ B30 C60 ⋄
RAUTENBERG, W. *Beweis des Kommutativgesetzes in elementar-archimedisch geordneten Gruppen* ⋄ C60 ⋄
RIBENBOIM, P. *On the existence of totally ordered abelian groups which are η_α-sets* ⋄ C55 C60 E07 E75 ⋄
ROBINSON, JULIA *The decision problem for fields* ⋄ B25 C60 D35 ⋄
TAJTSLIN, M.A. *On the elementary theory of classical Lie algebras (Russian)* ⋄ C60 D35 ⋄

1966

AMITSUR, S.A. *Rational identities and applications to algebra and geometry* ⋄ C50 C60 C65 ⋄
AUSTIN, A.K. *Varieties of groups defined by a single law* ⋄ C05 C60 ⋄
AX, J. & KOCHEN, S. *Diophantine problems over local fields, III: Decidable fields* ⋄ B25 C10 C20 C35 C60 C65 ⋄
BROWKIN, J. *On forms over p-adic fields (Russian summary)* ⋄ C60 ⋄
ERSHOV, YU.L. *On the elementary theory of maximal normed fields II (Russian)* ⋄ C35 C60 ⋄
KENZHEBAEV, S. *Certain undecidable rings (Russian) (Kazakh summary)* ⋄ C60 D35 ⋄
KHISAMIEV, N.G. & KOKORIN, A.I. *Elementary classification of structurally ordered abelian groups with finite number of threads (Russian)* ⋄ C60 ⋄
KHISAMIEV, N.G. *Universal theory of lattice-ordered abelian groups (Russian)* ⋄ B25 C60 ⋄
KOGALOVSKIJ, S.R. *On the properties preserved under algebraic constructions (Russian)* ⋄ B25 C30 C52 C60 D35 D45 ⋄
LYNDON, R.C. *Dependence in groups* ⋄ C05 C60 ⋄
MERZLYAKOV, YU.I. *Positive formulae on free groups (Russian)* ⋄ C60 ⋄
MOSCHOVAKIS, Y.N. *Notation systems and recursive ordered fields* ⋄ C57 C60 D45 F60 ⋄
ROBINSON, A. *A new approach to the theory of algebraic numbers I,II (Italian summary)* ⋄ C60 H15 H20 ⋄
ROBINSON, A. *On some applications of model theory to algebra and analysis* ⋄ C60 C98 H05 H98 ⋄
SHEVRIN, L.N. *Elementary lattice properties of semigroups (Russian)* ⋄ C05 C60 ⋄
STRUNKOV, S.P. *Subgroups of periodic groups (Russian)* ⋄ C60 E75 ⋄
TAJTSLIN, M.A. *On elementary theories of commutative semigroups with cancellation (Russian)* ⋄ B25 C60 D40 ⋄
TAJTSLIN, M.A. *On elementary theories of commutative semigroups (Russian)* ⋄ B25 C05 C60 ⋄
TANG, C.Y. *An existence theorem for generalized direct products with amalgamated subgroups* ⋄ C30 C60 ⋄

1967

AX, J. *Solving diophantine problems modulo every prime* ⋄ B25 C13 C20 C60 ⋄
BUNYATOV, M.R. *Dirac spaces (Russian)* ⋄ C20 C60 H05 H10 ⋄
ERSHOV, YU.L. *Fields with a solvable theory (Russian)* ⋄ B25 C35 C60 ⋄
ERSHOV, YU.L. *On the elementary theory of maximal normed fields III (Russian) (English summary)* ⋄ C35 C60 ⋄
GUREVICH, Y. *A contribution to the elementary theory of lattice ordered abelian groups and K-lineals (Russian)* ⋄ B25 C60 D35 ⋄
GUREVICH, Y. *A new decision procedure for the theory of ordered abelian groups (Russian)* ⋄ B25 C60 ⋄
HAUSCHILD, K. *Cauchyfolgen hoeheren Typus' in angeordneten Koerpern* ⋄ C60 ⋄
HAUSCHILD, K. *Ueber ein Definierbarkeitsproblem in der Koerpertheorie (English and French summaries)* ⋄ C40 C60 ⋄
HEIDEMA, J. & WALT VAN DER, A.P.J. *Contributions to the metamathematical theory of ideals I. Domains with dense kernel* ⋄ C07 C60 ⋄
MAYOH, B.H. *Groups and semigroups with solvable word problems* ⋄ C57 C60 D40 ⋄
MEGIBBEN, C. *On mixed groups of torsion-free rank one* ⋄ C60 ⋄
MEISTERS, G.H. *Nonarchimedean integers* ⋄ C60 H15 ⋄
NEUMANN, H. *Varieties of groups* ⋄ C05 C60 C98 ⋄
RAUTENBERG, W. *Elementare Schemata nichtelementarer Axiome* ⋄ B25 C52 C60 C65 C85 D35 ⋄
ROBINSON, A. *Non-standard theory of Dedekind rings* ⋄ C20 C60 H15 H20 ⋄

ROBINSON, A. *Nonstandard arithmetic*
  ⋄ C20 C60 H15 H20 ⋄
TAJTSLIN, M.A. *Two remarks on isomorphisms of commutative semigroups (Russian) (English summary)*
  ⋄ B25 C05 C60 ⋄
TARSKI, A. *The completeness of elementary algebra and geometry* ⋄ B25 B30 C10 C35 C60 C65 ⋄
VAMOS, P. *On ring classes defined by modules* ⋄ C60 ⋄
VOL'VACHEV, R.T. *The elementary theory of modules (Russian)* ⋄ C60 ⋄

1968

AX, J. *The elementary theory of finite fields*
  ⋄ B25 B30 C13 C20 C60 ⋄
BARTOCCI, U. *Sulla generalizzazione di un teorema di Kochen (English summary)* ⋄ C20 C60 ⋄
ERSHOV, YU.L. *Numbered fields* ⋄ C57 C60 D45 ⋄
FOSTER, A.L. *Pre-fields and universal algebraic extensions; equational precessions* ⋄ C05 C60 ⋄
HEIDEMA, J. *Contributions to the metamathematical theory of ideals II. Metamathematical prime ideals and radicals* ⋄ C07 C60 ⋄
KOKORIN, A.I. & KOZLOV, G.T. *Extended elementary and universal theories of lattice-ordered abelian groups with a finite number of threads (Russian)* ⋄ B25 C60 ⋄
KOPPERMAN, R.D. & MATHIAS, A.R.D. *Some problems in group theory* ⋄ C60 C75 ⋄
KUZ'MIN, E.N. *Algebraic sets in Mal'cev algebras (Russian)* ⋄ C60 ⋄
MADISON, E.W. *Computable algebraic structures and nonstandard arithmetic* ⋄ C57 C60 C62 D45 ⋄
MADISON, E.W. *Structures elementarily closed relative to a model for arithmetic* ⋄ C57 C60 C62 ⋄
MOREL, A.C. *Structure and order structure in Abelian groups* ⋄ C60 E07 ⋄
NEUMANN, B.H. *Lectures on topics in the theory of infinite groups* ⋄ C05 C60 C98 D40 ⋄
RAUTENBERG, W. *Nichtdefinierbarkeit der Multiplikation in dividierbaren Ringen* ⋄ C60 ⋄
SABBAGH, G. *Un theoreme de plongement en algebre* ⋄ C60 ⋄
TAJTSLIN, M.A. *Algorithmic problems for commutative semigroups (Russian)* ⋄ B25 C05 C60 ⋄
TAJTSLIN, M.A. *Elementary lattice theories for ideals in polynomial rings (Russian)* ⋄ C60 D35 G10 ⋄
TAJTSLIN, M.A. *On the isomorphism problem for commutative semigroups (Russian)*
  ⋄ C05 C07 C60 D40 ⋄
TIMOSHENKO, E.I. *Preservation of elementary and universal equivalence under the wreath product (Russian)* ⋄ C30 C60 ⋄
VALIEV, M.K. *A theorem of G.Higman (Russian)*
  ⋄ C60 D35 D40 ⋄
VOL'VACHEV, R.T. *Positive and elementary linear groups (Russian)* ⋄ C60 ⋄

1969

BARWISE, J. & EKLOF, P.C. *Lefschetz's principle*
  ⋄ B15 C60 C75 C85 ⋄
CLEAVE, J.P. *Local properties of systems*
  ⋄ C52 C60 C85 ⋄

COHEN, P.J. *Decision procedures for real and p-adic fields*
  ⋄ B25 C10 C60 ⋄
DEKKER, J.C.E. *Countable vector spaces with recursive operations. Part I* ⋄ C57 C60 D45 D50 ⋄
EKLOF, P.C. *Resolutions of singularities in prime characteristic for almost all primes* ⋄ C20 C60 ⋄
ERSHOV, YU.L. *The number of linear orders on a field (Russian)* ⋄ C55 C60 ⋄
GUPTA, H.N. *On some axioms in foundations of cartesian spaces* ⋄ C60 ⋄
HAUSCHILD, K. & WOLTER, H. *Ueber die Kategorisierbarkeit gewisser Koerper in nicht-elementaren Logiken* ⋄ C60 C80 C85 ⋄
HUBER-DYSON, V. *On the decision problem for extensions of a decidable theory* ⋄ B25 C13 C60 D35 ⋄
JARDEN, M. *Rational points on algebraic varieties over large number fields* ⋄ C60 ⋄
KLEIN, ABRAHAM A. *Necessary conditions for embedding rings into fields* ⋄ C05 C60 ⋄
KOCHEN, S. *Integer valued rational functions over the p-adic numbers: a p-adic analogue of the theory of real fields* ⋄ C60 ⋄
KOKORIN, A.I. & KOZLOV, G.T. *Elementary theory of abelian groups without torsion, with a predicate for selecting subgroups (Russian)* ⋄ B25 C35 C60 ⋄
KOPPERMAN, R.D. *Applications of infinitary languages to analysis* ⋄ C35 C60 C65 C75 ⋄
REILLY, N.R. *Some applications of wreath products and ultraproducts in the theory of lattice ordered groups*
  ⋄ C20 C30 C60 ⋄
ROBINSON, A. *Compactification of groups and rings and non-standard analysis* ⋄ C40 C60 H05 H20 ⋄
ROBINSON, A. *Problems and methods of model theory*
  ⋄ C60 C98 ⋄
ROBINSON, A. *Topics in nonstandard algebraic number theory* ⋄ C60 H05 H15 H20 ⋄
SABBAGH, G. *How not to characterize the multiplicative groups of fields* ⋄ C60 ⋄
SCARPELLINI, B. *On the metamathematics of rings and integral domains* ⋄ C60 ⋄
SCOTT, D.S. *On completing ordered fields* ⋄ C60 ⋄

1970

AOYAMA, K. *On the structure space of a direct product of rings* ⋄ C30 C60 ⋄
BARWISE, J. & EKLOF, P.C. *Infinitary properties of abelian torsion groups* ⋄ C60 C75 ⋄
BAZHAEV, YU.V. *Universal groups (Russian)*
  ⋄ C50 C60 ⋄
BREDIKHIN, S.V. & ERSHOV, YU.L. & KAL'NEJ, V.E. *Fields with two linear orderings (Russian)* ⋄ C35 C60 ⋄
CHARNOW, A. *The automorphisms of an algebraically closed field* ⋄ C60 ⋄
COLLINS, D.J. *On recognizing properties of groups which have solvable word problem* ⋄ C60 D40 ⋄
DELIYANNIS, P.C. *Group representations and cardinal algebras* ⋄ C60 E10 ⋄
DURNEV, V.G. *Positive formulae on groups (Russian)*
  ⋄ C60 ⋄
EKLOF, P.C. & SABBAGH, G. *Model-completions and modules* ⋄ C25 C35 C52 C60 ⋄

HALEY, D.K. *Equationally compact noetherian rings*
⋄ C05 C60 ⋄

HALEY, D.K. *On compact commutative noetherian rings*
⋄ C05 C60 ⋄

HAMILTON, A.G. *Bases and α-dimensions of countable vector spaces with recursive operations*
⋄ C57 C60 D45 D50 ⋄

KARGAPOLOV, M.I. *Universal groups (Russian)*
⋄ C50 C60 ⋄

KOZLOV, G.T. *Unsolvability of the elementary theory of lattices of subgroups of finite abelian p-groups (Russian)*
⋄ C60 D35 ⋄

KRASNER, M. *Endotheorie de Galois abstraite*
⋄ C07 C60 C75 ⋄

KROM, MELVEN R. & KROM, MYREN *Groups with free nonabelian subgroups* ⋄ C60 ⋄

KULIK, V.T. *Sentences preserved under homomorphisms that are one-to-one at zero (Russian)*
⋄ C05 C40 C60 ⋄

LACHLAN, A.H. & MADISON, E.W. *Computable fields and arithmetically definable ordered fields*
⋄ C57 C60 D45 ⋄

LUXEMBURG, W.A.J. & TAYLOR, R.F. *Almost commuting matrices are near commuting matrices* ⋄ C60 H05 ⋄

MADISON, E.W. *A note on computable real fields*
⋄ C57 C60 D45 F60 ⋄

NEUMANN, W.D. *On the quasivariety of convex subsets of affine spaces* ⋄ C05 C60 ⋄

OLIN, P. *Some first order properties of direct sums of modules* ⋄ C30 C40 C60 ⋄

REMESLENNIKOV, V.N. & SOKOLOV, V.G. *Some properties of a Magnus embedding (Russian)* ⋄ C60 D40 ⋄

ROBINSON, A. *Elementary embeddings of fields of power series* ⋄ C20 C60 H20 ⋄

SABBAGH, G. *Aspects logiques de la purete dans les modules* ⋄ C20 C30 C60 ⋄

SEIDENBERG, A. *Construction of the integral closure of a finite integral domain* ⋄ C57 C60 F55 ⋄

SIMMONS, H. *The solution of a decision problem for several classes of rings* ⋄ B25 C60 ⋄

SZCZERBA, L.W. *Filling of gaps in ordered fields*
⋄ C60 ⋄

TAJTSLIN, M.A. *On elementary theories of lattices of subgroups (Russian)* ⋄ B25 C60 D35 G10 ⋄

TAMURA, S. *Axioms for commutative rings* ⋄ C60 ⋄

VUL'VACHEV, R.T. *Linear groups as axiomatizable classes of models (Russian)* ⋄ C20 C52 C60 ⋄

ZIL'BER, B.I. *On the isomorphism and elementary equivalence of commutative semigroups (Russian)*
⋄ C05 C60 ⋄

1971

ARMBRUST, M. *On the theorem of Ax and Kochen*
⋄ C60 ⋄

AX, J. *A metamathematical approach to some problems in number theory* ⋄ C20 C60 C98 ⋄

BAUMSLAG, B. & BAUMSLAG, G. *On ascending chain conditions* ⋄ C05 C60 ⋄

DEKKER, J.C.E. *Countable vector spaces with recursive operations. Part II* ⋄ C57 C60 D45 D50 ⋄

DEKKER, J.C.E. *Two notes on vector spaces with recursive operations* ⋄ C57 C60 D45 ⋄

EKLOF, P.C. & SABBAGH, G. *Definability problems for modules and rings* ⋄ C20 C52 C60 C75 ⋄

GREENLEAF, N. *Fields in which varieties have rational points: a note on a problem of Ax* ⋄ B25 C60 ⋄

HAUSCHILD, K. & RAUTENBERG, W. *Interpretierbarkeit in der Gruppentheorie* ⋄ C60 D35 F25 ⋄

HILL, P. *Classes of abelian groups closed under taking subgroups and direct limits* ⋄ C30 C60 ⋄

LARSON, L.C. *Nonstandard theory of Zariski rings*
⋄ C60 H20 ⋄

MACINTYRE, A. *On ω_1-categorical theories of abelian groups* ⋄ C35 C45 C60 ⋄

MACINTYRE, A. *On ω_1-categorical theories of fields*
⋄ C35 C60 ⋄

MACINTYRE, A. *On the elementary theory of Banach algebras* ⋄ C60 C65 D35 ⋄

MADISON, E.W. *Some remarks on computable (non-archimedean) ordered fields*
⋄ C57 C60 D45 ⋄

MARCO DE, G. & RICHTER, M.M. *Rings of continuous functions with values in a non-archimedean ordered field* ⋄ C20 C60 C65 E75 ⋄

MCKENZIE, R. $\aleph_1$-*incompactness of Z* ⋄ C05 C60 ⋄

MCKENZIE, R. *A note on subgroups of infinite symmetric groups* ⋄ C60 ⋄

MCKENZIE, R. *On elementary types of symmetric groups*
⋄ C60 E10 E47 ⋄

OLIN, P. *Direct multiples and powers of modules*
⋄ C30 C40 C60 ⋄

PABION, J.F. *Problemes d'immersions en algebre*
⋄ C07 C60 ⋄

PIERCE, R.S. *The submodule lattice of a cyclic module*
⋄ C05 C60 ⋄

POIZAT, B. *Theorie de Galois des relations*
⋄ C07 C60 C75 E07 ⋄

ROBINSON, A. *On the notion of algebraic closedness for noncommutative groups and fields* ⋄ C25 C60 ⋄

ROMOV, B.A. *Definability by formulas for predicates on a finite model (Russian) (English summary)*
⋄ C13 C40 C60 ⋄

ROQUETTE, P. *Bemerkungen zur Theorie der formal p-adischen Koerper* ⋄ C60 ⋄

SABBAGH, G. *Embedding problems for modules and rings with application to model-companions*
⋄ C25 C35 C60 ⋄

SABBAGH, G. *Sous-modules purs, existentiellement clos et elementaires* ⋄ C25 C35 C60 ⋄

SCHEIN, B.M. *Linearly fundamentally ordered semigroups*
⋄ C05 C60 E20 ⋄

SCHWARTZ, DIETRICH *Die Nichtkompaktheit der Klasse der archimedisch geordneten Koerper* ⋄ C52 C60 ⋄

SEIDENBERG, A. *On the length of a Hilbert ascending chain* ⋄ C60 F55 ⋄

SHOENFIELD, J.R. *A theorem on quantifier elimination*
⋄ C10 C60 ⋄

VINOGRADOV, A.A. *Nonaxiomatizability of lattice-orderable groups (Russian)* ⋄ C52 C60 ⋄

ZIL'BER, B.I. *An example of two elementarily equivalent but not isomorphic finitely generated metabelian groups (Russian)* ⋄ C60 ⋄

1972

ABIAN, A. *On solvability of infinite systems of polynomial equations over a finite ring* ⋄ C60 E25 ⋄

BACSICH, P.D. *Cofinal simplicity and algebraic closedness* ⋄ C05 C25 C50 C60 C75 G10 ⋄

BACSICH, P.D. *Diagrammatic construction of homogenous universal models* ⋄ C50 C60 ⋄

BELEGRADEK, O.V. *On unstable theories of groups (Russian)* ⋄ C45 C60 ⋄

BOFFA, M. *Corps λ-clos* ⋄ C60 ⋄

BOFFA, M. *Sur l'existence des corps universels-homogenes* ⋄ C50 C55 C60 ⋄

BOFFA, M. & PRAAG VAN, P. *Sur les corps generiques* ⋄ C25 C60 ⋄

BOFFA, M. & PRAAG VAN, P. *Sur les sous-champs maximaux des corps generiques denombrables* ⋄ C15 C25 C60 ⋄

BOFFA, M. *Ultraproduits et application a l'algebre* ⋄ C20 C60 ⋄

DURNEV, V.G. *The positive theory of a free semigroup (Russian)* ⋄ C60 ⋄

EKLOF, P.C. *Homogeneous universal modules* ⋄ C50 C60 ⋄

EKLOF, P.C. *Some model theory of abelian groups* ⋄ C35 C60 ⋄

EKLOF, P.C. & FISHER, E.R. *The elementary theory of abelian groups* ⋄ B25 C10 C35 C50 C60 ⋄

ERSHOV, YU.L. *Elementary group theories (Russian)* ⋄ B25 C05 C13 C60 D35 ⋄

FRIDMAN, EH.I. *Undecidability of the elementary theory of abelian torsion-free groups with a finite set of serving subgroups (Russian)* ⋄ C60 D35 ⋄

GLASS, A.M.W. *An application of ultraproducts to lattice-ordered groups* ⋄ C20 C60 G10 ⋄

GOEBEL, R. & RICHTER, M.M. *Cartesian closed classes of perfect groups* ⋄ C60 ⋄

GUPTA, H.N. & PRESTEL, A. *On a class of Pasch-free Euclidean planes* ⋄ C60 C65 ⋄

HAUSCHILD, K. *Universalitaet in der Ringtheorie* ⋄ C60 D35 F25 ⋄

HEIDEMA, J. *Metamathematical representation of radicals in universal algebra* ⋄ C05 C60 ⋄

HICKIN, K.K. & PLOTKIN, J.M. *On the equivalence of three local theorem techniques* ⋄ C60 ⋄

HIRSCHELMANN, A. *An application of ultra-products to prime rings with polynomial identities* ⋄ C20 C60 ⋄

HOFMANN, K.H. *Representation of algebras by continuous sections* ⋄ C60 C90 C98 ⋄

JANSSEN, G. *Restricted ultraproducts of finite von Neumann algebras* ⋄ C20 C60 H05 ⋄

JARDEN, M. *Elementary statements over large algebraic fields* ⋄ C20 C60 ⋄

KOPFERMANN, K. *Ultraprodukte endlicher Koerper* ⋄ C13 C20 C60 ⋄

KOPPERMAN, R.D. *Model theory and its applications* ⋄ C60 C98 ⋄

MACINTYRE, A. *Omitting quantifier-free types in generic structures* ⋄ C25 C57 C60 C75 D30 D40 ⋄

MACINTYRE, A. *On algebraically closed groups* ⋄ C25 C57 C60 D40 ⋄

MART'YANOV, V.I. *Decidability of the theories of some classes of abelian groups with an automorphism predicate and a group predicate (Russian)* ⋄ B25 C60 ⋄

MAY, W. *Multiplicative groups of fields* ⋄ C60 ⋄

OLIN, P. *$\aleph_0$-categoricity of two-sorted structures* ⋄ C20 C30 C35 C60 ⋄

OLIN, P. *First order preservation theorems in two-sorted languages* ⋄ C30 C40 C60 ⋄

PENZIN, YU.G. *Decidability of the theory of algebraic integers with addition and congruences according to divisors (Russian)* ⋄ B25 C60 ⋄

PHILLIPS, R.G. & SPERRY, P.L. *Elementary extensions of linear topological abelian groups* ⋄ C60 ⋄

PIERCE, K.R. *Amalgamating abelian ordered groups* ⋄ C60 ⋄

ROBINSON, A. *Algebraic function fields and non-standard arithmetic* ⋄ C60 H15 ⋄

ROBINSON, A. *Generic categories* ⋄ C25 C52 C60 G30 ⋄

ROBINSON, A. *On the real closure of a Hardy field* ⋄ C60 C65 ⋄

ROSENSTEIN, J.G. *On $GL_2(R)$ where R is a boolean ring* ⋄ C35 C60 G05 ⋄

SACERDOTE, G.S. *On a problem of Boone* ⋄ C60 D30 D35 D40 ⋄

SACKS, G.E. *The differential closure of a differential field* ⋄ C60 C98 ⋄

SIMSON, D. *$\aleph$-flat and $\aleph$-projective modules (Russian summary)* ⋄ C60 ⋄

SIMSON, D. *On the structure of flat modules (Russian summary)* ⋄ C60 ⋄

TIMOSHENKO, E.I. *On the question of the elementary equivalence of groups (Russian)* ⋄ C30 C60 ⋄

VAZHENIN, YU.M. *On finitely approximable semigroups of idempotents (Russian)* ⋄ C05 C60 ⋄

WATANABE, T. *Ultraproducts and diophantine problems* ⋄ C20 C60 ⋄

WHEELER, W.H. *Algebraically closed division rings, forcing, and the analytic hierarchy* ⋄ C25 C60 D55 ⋄

ZIEGLER, M. *Die elementare Theorie der Henselschen Koerper* ⋄ B25 C10 C35 C60 ⋄

1973

AANDERAA, S.O. *A proof of Higman's embedding theorem using Britton extensions of groups* ⋄ C60 D40 D80 ⋄

BACSICH, P.D. *An epi-reflector for universal theories* ⋄ C05 C30 C60 G30 ⋄

BASARAB, S.A. *Certain metamathematical aspects of the theory of henselian fields (Romanian) (English summary)* ⋄ C60 ⋄

BERNSTEIN, A.R. *Non-standard analysis* ⋄ C60 C65 H05 ⋄

BERTRAND, D. *Le theoreme d'Ax et Kochen* ⋄ C60 ⋄

BRITTON, J.L. *The existence of infinite Burnside groups*
⋄ C60 D40 ⋄

CARSON, A.B. *The model-completion of the theory of commutative regular rings* ⋄ C25 C35 C60 ⋄

CHARRETTON, C. & RICHARD, D. *Elements d'une theorie non standard des groupes topologiques* ⋄ C60 H05 ⋄

CHERLIN, G.L. *Algebraically closed commutative rings*
⋄ C25 C60 ⋄

EKLOF, P.C. *Lefschetz's principle and local functors*
⋄ C60 C75 ⋄

EKLOF, P.C. *The structure of ultraproducts of abelian groups* ⋄ C20 C60 ⋄

GABBAY, D.M. *The undecidability of intuitionistic theories of algebraically closed fields and real closed fields*
⋄ B25 C60 C90 D35 F50 ⋄

HAKEN, W. *Connections between topological and group theoretical decision problems*
⋄ B25 C60 C65 D35 D40 D80 ⋄

HALEY, D.K. *Equationally compact artinian rings*
⋄ C05 C60 ⋄

HICKIN, K.K. *Countable type local theorems in algebra*
⋄ C05 C52 C60 ⋄

HICKIN, K.K. & PHILLIPS, R.E. *Local theorems and group extensions* ⋄ C60 ⋄

HILL, P. *New criteria for freeness in abelian groups*
⋄ C60 C75 ⋄

JURIE, P.-F. *Une extension de la theorie de Galois abstraite finitaire* ⋄ C60 G15 ⋄

KASAHARA, S. *A characterization of nonstandard real fields* ⋄ C20 C60 H05 ⋄

KEGEL, O.H. & WEHRFRITZ, B.A.F. *Locally finite groups*
⋄ C60 C98 ⋄

KOPFERMANN, K. *Ultraprodukte endlicher Koerper*
⋄ C13 C20 C60 ⋄

KUEKER, D.W. *A note on the elementary theory of finite abelian groups* ⋄ C13 C20 C60 ⋄

LIPSHITZ, L. & SARACINO, D.H. *The model companion of the theory of commutative rings without nilpotent elements* ⋄ B25 C25 C35 C60 ⋄

LIVCHAK, A.B. *A resolving procedure for the elementary theory of a torsion-free abelian group with a distinguished sub-group (Russian)*
⋄ B25 C10 C60 ⋄

MACINTYRE, A. *Martin's axiom applied to existentially closed groups* ⋄ C25 C55 C60 C75 C80 E50 ⋄

MELONI, G.C. *Analisi non standard di anelli e corpi topologici (English summary)* ⋄ C60 H05 H20 ⋄

MLCEK, J. *A representation of models of Peano arithmetic*
⋄ C60 C62 ⋄

NEUMANN, B.H. *The isomorphism problem for algebraically closed groups* ⋄ C25 C60 D40 ⋄

PINUS, A.G. *Elementary definability of symmetry groups*
⋄ C60 ⋄

ROBINSON, A. *Function theory on some nonarchimedian fields* ⋄ C60 C65 H05 ⋄

ROBINSON, A. *Metamathematical problems*
⋄ A05 B30 C60 C98 H05 ⋄

ROBINSON, A. *Model theory as a framework for algebra*
⋄ C25 C35 C60 C98 ⋄

ROBINSON, A. *Nonstandard arithmetic and generic arithmetic* ⋄ C25 C60 C62 C98 H15 ⋄

ROBINSON, A. *Nonstandard points on algebraic curves*
⋄ C20 C60 H20 ⋄

ROBINSON, A. *On bounds in the theory of polynomial ideals (Russian)* ⋄ C60 H20 ⋄

ROBINSON, A. *Standard and nonstandard number systems* ⋄ C25 C60 C98 H05 H15 H20 H98 ⋄

ROSENSTEIN, J.G. $\aleph_0$-*categoricity of groups*
⋄ C35 C60 ⋄

SACERDOTE, G.S. *Almost all free products of groups have the same positive theory* ⋄ C60 ⋄

SACERDOTE, G.S. *Elementary properties of free groups*
⋄ C60 ⋄

SARACINO, D.H. *Model companions for $\aleph_0$-categorical theories* ⋄ C25 C35 C60 ⋄

SCHUPP, P.E. *A survey of small cancellation theory*
⋄ C60 D40 ⋄

SEIDENBERG, A. *On the impossibility of some constructions in polynomial rings* ⋄ C60 F55 ⋄

SHELAH, S. *Differentially closed fields* ⋄ C45 C60 ⋄

SHELAH, S. *First order theory of permutation groups*
⋄ B15 C55 C60 C85 ⋄

SUTER, G.H. *Recursive elements and constructive extensions of computable local integral domains*
⋄ C57 C60 D45 ⋄

TAJTSLIN, M.A. *Existentially closed regular commutative semigroups (Russian)* ⋄ C25 C35 C52 C60 ⋄

VASIL'EV, EH.S. *Elementary theories of two-base models of abelian groups (Russian)* ⋄ B25 C60 C85 ⋄

VASIL'EV, EH.S. *The elementary theories of complete torsion-free abelian groups with p-adic topology (Russian)* ⋄ B25 C20 C60 D35 ⋄

VASIL'EV, V.G. *Solvable elementarily equivalent groups (Russian)* ⋄ C60 ⋄

VINOGRADOV, A.A. *Nonaxiomatizability of directionally ordered groups in the class of nontrivially partially ordered groups* ⋄ C52 C60 ⋄

WEISPFENNING, V. *Infinitary model-theoretic properties of κ-saturated structures*
⋄ C10 C25 C35 C50 C60 C65 C75 ⋄

WHITELEY, W. *Homogeneous sets and homogeneous formulas* ⋄ C40 C60 ⋄

WHITELEY, W. *Logic and invariant theory I. Invariant theory of projective properties* ⋄ B30 C60 ⋄

WOOD, C. *The model theory of differential fields of characteristic $p \neq 0$* ⋄ C25 C35 C45 C60 ⋄

1974

BACSICH, P.D. & ROWLANDS-HUGHES, D. *Syntactic characterizations of amalgamation, convexity and related properties* ⋄ C05 C07 C60 C75 ⋄

BANASCHEWSKI, B. *Equational compactness of G-sets*
⋄ C05 C50 C60 ⋄

BAUDISCH, A. *Theorien abelscher Gruppen mit einem einstelligen Praedikat* ⋄ C60 D35 ⋄

BELEGRADEK, O.V. *Algebraically closed groups (Russian)*
⋄ C25 C60 D40 ⋄

BELEGRADEK, O.V. *Definability in algebraically closed groups (Russian)* ⋄ C25 C40 C60 ⋄

BELEGRADEK, O.V. *On algebraically closed groups (Russian)* ⋄ C25 C60 ⋄

BELYAEV, V.YA. *Elementary types of algebraically closed semigroups (Russian)* ⋄ C25 C60 ⋄

BLASS, A.R. & NEUMANN, P.M. *An application of universal algebra in group theory* ⋄ C05 C60 ⋄

BOONE, W.W. *Between logic and group theory* ⋄ C57 C60 D40 ⋄

CARSON, A.B. *Algebraically closed regular rings* ⋄ C25 C60 ⋄

COHN, P.M. *The class of rings imbeddable in skew fields* ⋄ C60 ⋄

COZART, D. & MOORE JR., L.C. *The nonstandard hull of a Riesz space* ⋄ C60 H05 ⋄

DORACZYNSKA, E. *The complete theory of the midpoint operation (Russian summary)* ⋄ C35 C60 ⋄

DUBROVSKY, D.L. *Some subfields of $\mathbb{Q}_p$ and their non-standard analogues* ⋄ C60 H15 H20 ⋄

EHDEL'MAN, G.S. *Certain radicals of metaideals (Russian)* ⋄ B10 C07 C60 ⋄

EKLOF, P.C. *Infinitary equivalence of abelian groups* ⋄ C60 C75 ⋄

FELGNER, U. *$\aleph_1$-kategorische Theorien nicht-kommutativer Ringe* ⋄ C35 C45 C60 ⋄

FUCHS, L. *Indecomposable abelian groups of measurable cardinalities* ⋄ C55 C60 E55 ⋄

GUREVICH, Y. *A resolving procedure for the extended theory of ordered abelian groups (Russian)* ⋄ B25 C10 C60 C85 ⋄

HALEY, D.K. *A note on equational compactness* ⋄ C05 C60 ⋄

HARRINGTON, L.A. *Recursively presentable prime models* ⋄ C35 C50 C57 C60 ⋄

HERRMANN, C. & POGUNTKE, W. *The class of sublattices of normal subgroup lattices is not elementary* ⋄ C60 G10 ⋄

JARDEN, M. *An injective rational map of an abstract variety into itself* ⋄ C20 C60 ⋄

KEIMEL, K. *Representation des algebres universelles par des faisceaux* ⋄ C05 C35 C60 C90 ⋄

KEISLER, H.J. *Monotone complete fields* ⋄ C60 C65 H05 ⋄

KERTESZ, A. *Transfinite Methoden in der Algebra (Russian summary)* ⋄ C60 E25 E75 ⋄

KLINGEN, N. *Elementar aequivalente Koerper und ihre absolute Galoisgruppe* ⋄ C20 C60 H20 ⋄

KOLCHIN, E.R. *Constrained extensions of differential fields* ⋄ C60 ⋄

KOPPELBERG, S. & TITS, J. *Une propriete des produits directs infinis de groupes finis isomorphes* ⋄ C30 C60 E75 ⋄

KOREC, I. & PERETYAT'KIN, M.G. & RAUTENBERG, W. *Definability in structures of finite valency* ⋄ C60 C65 F25 ⋄

KRIVINE, J.-L. *Langages a valeurs reelles et applications* ⋄ B50 C40 C60 C90 ⋄

MACINTYRE, A. *A note on axioms for infinite-generic structures* ⋄ C25 C60 C62 C75 E07 ⋄

MALCOLM, W.G. *Some results and algebraic applications in the theory of higher-order ultraproducts* ⋄ C20 C60 C85 G05 ⋄

MULVEY, C.J. *Intuitionistic algebra and representations of rings* ⋄ C60 C90 F50 F55 G30 ⋄

MYERS, D.L. *The boolean algebras of abelian groups and well-orders* ⋄ C60 C65 G05 ⋄

OLIN, P. *Free products and elementary equivalence* ⋄ C05 C60 ⋄

PODEWSKI, K.-P. & REINEKE, J. *An ω_1-categorical ring which is not almost strongly minimal* ⋄ C35 C50 C60 ⋄

RACKOFF, C.W. *On the complexity of the theories of weak direct products: a preliminary report* ⋄ B25 C13 C30 C60 D15 ⋄

ROBINSON, A. *A decision method for elementary algebra and geometry - revisited* ⋄ C25 C60 ⋄

ROSENLICHT, M. *The nonminimality of the differential closure* ⋄ C60 ⋄

SARACINO, D.H. *Wreath products and existentially complete solvable groups* ⋄ C25 C35 C60 ⋄

SCHWABHAEUSER, W. & SZCZERBA, L.W. *An affine space as union of spaces of higher dimension* ⋄ C60 C65 ⋄

SHELAH, S. *Infinite abelian groups, Whitehead problem and some constructions* ⋄ C55 C60 E05 E35 E45 E50 E75 ⋄

SOPRUNOV, S.F. *On the power of a real-closed field (Russian) (English summary)* ⋄ C20 C60 D35 ⋄

SPIVAKOV, YU.L. *Algebraic extensions of the field of formal power series $F_D((t))$ (Russian) (Uzbek summary)* ⋄ C20 C60 H05 ⋄

TAJTSLIN, M.A. *Existentially closed commutative rings* ⋄ C25 C60 ⋄

VAZHENIN, YU.M. *On global including semigroups of symmetric semigroups (Russian)* ⋄ C05 C60 ⋄

VAZHENIN, YU.M. *On the elementary theory of free inverse semigroups* ⋄ B25 C60 D35 ⋄

WOOD, C. *Prime model extensions for differential fields of characteristic $p \ne 0$* ⋄ C10 C35 C60 ⋄

ZAKON, E. *Model-completeness and elementary properties of torsion free abelian groups* ⋄ C35 C60 ⋄

ZIL'BER, B.I. *Rings with $\aleph_1$-categorical theories (Russian)* ⋄ C35 C60 ⋄

1975

BACSICH, P.D. *The strong amalgamation property* ⋄ C05 C52 C60 ⋄

BASARAB, S.A. *The models of the elementary theory of finite abelian groups (Romanian) (English summary)* ⋄ B25 C13 C60 ⋄

BAUDISCH, A. *Die elementare Theorie der Gruppe vom Typ p^∞ mit Untergruppen* ⋄ B25 C60 C85 ⋄

BAUDISCH, A. *Elementare Theorien von Halbgruppen mit Kuerzungsregeln mit einem einstelligen Praedikat* ⋄ C05 C60 D35 ⋄

BAUDISCH, A. *Endliche n-aequivalente Gruppen* ⋄ C13 C60 ⋄

BAUDISCH, A. *Entscheidbarkeitsprobleme elementarer Theorien von Klassen abelscher Gruppen mit Untergruppen (English and Russian summaries)* ⋄ B25 C60 ⋄

BAUR, W. *$\aleph_0$-categorical modules* ⋄ C35 C60 ⋄

BAUR, W. *Decidability and undecidability of theories of abelian groups with predicates for subgroups*
⋄ B25 C60 D35 ⋄

BELKIN, V.P. & GORBUNOV, V.A. *Filters in lattices of quasivarieties of algebraic systems (Russian)*
⋄ C05 C60 D40 G10 ⋄

BERLINE, C. *Categoricite en $\aleph_0$ du groupe lineaire d'un anneau de Boole (English summary)* ⋄ C35 C60 ⋄

BERTHIER, D. *Stability of non-model-complete theories; products, groups* ⋄ C20 C30 C45 C60 ⋄

BOFFA, M. *A note on existentially complete division rings*
⋄ C25 C60 ⋄

CHERLIN, G.L. *Ideals in some nonstandard Dedekind rings* ⋄ C60 H15 H20 ⋄

CHERLIN, G.L. *Ideals of integers in nonstandard number fields* ⋄ C60 H15 H20 ⋄

CHIKHACHEV, S.A. *Categorical universal classes of semigroups (Russian)* ⋄ C05 C35 C52 C60 ⋄

CLARE, F. *Embedding theorems for infinite symmetric groups* ⋄ C30 C60 ⋄

CLARE, F. *Operations on elementary classes of groups*
⋄ C05 C52 C60 ⋄

COHN, P.M. *Presentation of skew fields. I. Existentially closed skew fields and the "Nullstellensatz"*
⋄ C25 C60 ⋄

COLLINS, G.E. *Quantifier elimination for real closed fields by cylindrical algebraic decomposition*
⋄ C10 C60 G15 ⋄

DENEF, J. *Hilbert's tenth problem for quadratic rings*
⋄ C60 D35 ⋄

EKLOF, P.C. *Categories of local functors*
⋄ C60 C75 G30 ⋄

EKLOF, P.C. *On the existence of κ-free abelian groups*
⋄ C60 C75 E10 E45 E75 ⋄

ENGELER, E. *On the solvability of algorithmic problems*
⋄ B75 C60 C75 D75 ⋄

FEFERMAN, S. *Impredicativity of the existence of the largest divisible subgroup of an abelian p-group*
⋄ C57 C60 F65 ⋄

FELGNER, U. *On $\aleph_0$-categorical extra-special p-groups*
⋄ C35 C60 ⋄

FERRANTE, J. & RACKOFF, C.W. *A decision procedure for the first order theory of real addition with order*
⋄ B25 C10 C60 D15 F35 ⋄

FITTLER, R. *Closed models and hulls of theories*
⋄ C25 C52 C60 ⋄

FOWLER III, N. *Intersections of α-spaces*
⋄ C57 C60 D50 ⋄

FOWLER III, N. *Sum of α-spaces* ⋄ C57 C60 D45 ⋄

FRIDMAN, EH.I. & SLOBODSKOJ, A.M. *Theories of Abelian groups with predicates specifying a subgroup (Russian)*
⋄ C60 D35 ⋄

GARCIA, O.C. *Injectivity in categories of groups and algebras* ⋄ C05 C60 G30 ⋄

GRAINGER, A.D. *Invariant subspaces of compact operators on topological vector spaces* ⋄ C60 C65 H05 ⋄

GUETING, R. *Subtractive abelian groups* ⋄ C60 ⋄

GUHL, R. *A theorem on recursively enumerable vector spaces* ⋄ C57 C60 D45 ⋄

HICKMAN, J.L. & NEUMANN, B.H. *A question of Babai on groups* ⋄ C60 E25 E35 E75 ⋄

HIRSCHFELD, J. *Finite forcing and generic filters in arithmetic* ⋄ C25 C60 C62 ⋄

HIRSCHFELD, J. & WHEELER, W.H. *Forcing, arithmetic, division rings* ⋄ C25 C60 C62 C98 ⋄

HUTCHINSON, G. *On the representation of lattices by modules* ⋄ C05 C60 G10 ⋄

JARDEN, M. & KIEHNE, U. *The elementary theory of algebraic fields of finite corank*
⋄ B25 C10 C20 C60 ⋄

JENSEN, F.V. *On completeness in cardinality logics*
⋄ C10 C35 C55 C60 C80 ⋄

JOYAL, A. *Les theoremes de Chevalley-Tarski et remarque sur l'algebre constructive* ⋄ C60 F50 G30 ⋄

KLINGEN, N. *Zur Idealstruktur in Nichtstandardmodellen von Dedekindringen* ⋄ C20 C60 H15 H20 ⋄

KOCHEN, S. *The model theory of local fields*
⋄ B25 C20 C35 C60 ⋄

KOKORIN, A.I. & KOZLOV, G.T. *Proof of a lemma on model completeness (Russian)* ⋄ C35 C60 ⋄

LACAVA, F. *Teoria dei campi differenziali ordinati*
⋄ C10 C35 C60 ⋄

LIPSHITZ, L. *Commutative regular rings with integral closure* ⋄ C60 ⋄

MACINTYRE, A. *Dense embeddings. I. A theorem of Robinson in a general setting*
⋄ B25 C10 C35 C60 D40 ⋄

MAGID, A.R. *Ultrafunctors* ⋄ C20 C60 ⋄

MAKOWIECKA, H. *An elementary geometry in connection with decomposition of a plane* ⋄ B30 C60 ⋄

MAKOWIECKA, H. *On primitive notions in n-dimensional elementary geometries* ⋄ B30 C60 ⋄

MAKOWIECKA, H. *Quartenary relations in weak euclidean geometries (Russian summary)* ⋄ B30 C60 ⋄

MAKOWIECKA, H. *The norm relation in weak euclidean geometry with the axiom on two circles (Russian summary)* ⋄ B30 C60 ⋄

MAKOWIECKA, H. *The theory of bi-proportionality as a geometry* ⋄ B30 C60 ⋄

MART'YANOV, V.I. *The theory of abelian groups with predicates specifying a subgroup, and with endomorphism operations (Russian)* ⋄ C60 ⋄

MCKENNA, K. *New facts about Hilbert's seventeenth problem* ⋄ C60 ⋄

METAKIDES, G. & NERODE, A. *Recursion theory and algebra* ⋄ C57 C60 D25 D45 ⋄

MONK, L.G. *Elementary-recursive decision procedures*
⋄ B25 C60 D20 ⋄

MOSTOWSKI, T. *A note on a decision procedure for rings of formal power series and its applications* ⋄ B25 C60 ⋄

NEBRES, B.F. *Elementary invariants for abelian groups (uses of saturated model theory in algebra)*
⋄ C50 C60 ⋄

PHILLIPS, R.E. & PLOTKIN, J.M. *On factor coverings of groups* ⋄ C60 ⋄

PODEWSKI, K.-P. *Minimale Ringe* ⋄ C60 ⋄

PRAAG VAN, P. *Sur les centralisateurs des corps de type fini dans les corps existentiellement clos (English summary)*
⋄ C25 C60 ⋄

PRESTEL, A. & ZIEGLER, M. *Erblich euklidische Koerper*
⋄ C60 D35 ⋄
PRESTEL, A. *Lectures on formally real fields*
⋄ C60 C98 ⋄
REINEKE, J. *Minimale Gruppen* ⋄ C50 C60 ⋄
ROBINSON, A. *Algorithms in algebra*
⋄ C40 C57 C60 D45 D75 ⋄
ROBINSON, A. & ROQUETTE, P. *On the finiteness theorem of Siegel and Mahler concerning diophantine equations*
⋄ C60 H05 H15 H20 ⋄
ROQUETTE, P. *Nonstandard aspects of Hilbert's irreducibility theorem* ⋄ C60 H15 H20 ⋄
SABBAGH, G. *Categoricite en $\aleph_0$ et stabilite: constructions les preservant et conditions de chaine (English summary)* ⋄ C35 C45 C60 ⋄
SABBAGH, G. *Categoricite et stabilite: quelques examples parmi les groupes et anneaux (English summary)*
⋄ C35 C45 C60 ⋄
SABBAGH, G. *Sur les groupes qui ne sont pas reunion d'une suite croissante de sous-groupes propres (English summary)* ⋄ C25 C60 ⋄
SACERDOTE, G.S. *Infinite coforcing in model theory*
⋄ C25 C30 C60 ⋄
SACERDOTE, G.S. *Projective model theory and coforcing*
⋄ C25 C30 C35 C40 C50 C60 ⋄
SAELI, D. *Problemi di decisione per algebre connesse a logiche a piu valori (English summary)*
⋄ B25 B50 C60 G20 ⋄
SANERIB JR., R.A. *Ultraproducts and elementary types of some groups related to infinite symmetric groups*
⋄ C20 C60 ⋄
SARACINO, D.H. & WEISPFENNING, V. *Commutative regular rings without prime model extensions*
⋄ C25 C35 C60 ⋄
SARACINO, D.H. & WEISPFENNING, V. (EDS.) *Model theory and algebra. A memorial tribute to Abraham Robinson*
⋄ C60 C97 ⋄
SARACINO, D.H. & WEISPFENNING, V. *On algebraic curves over commutative regular rings*
⋄ C10 C25 C35 C60 ⋄
SCHMITT, P.H. *Categorical lattices* ⋄ C35 C60 G10 ⋄
SEIDENBERG, A. *Construction of the integral closure of a finite integral domain. II* ⋄ C57 C60 F55 ⋄
SHELAH, S. *A compactness theorem for singular cardinals, free algebras, Whitehead problem and transversals*
⋄ C60 C75 E05 E75 ⋄
SHELAH, S. *Existence of rigid-like families of abelian p-groups* ⋄ C52 C55 C60 ⋄
SHELAH, S. *The lazy model theoretician's guide to stability*
⋄ C25 C35 C45 C50 C60 C98 ⋄
SHELAH, S. *Why there are many nonisomorphic models for unsuperstable theories*
⋄ C45 C50 C55 C60 C65 G05 ⋄
SOBOCINSKI, B. *Concerning the postulate-systems of subtractive abelian groups* ⋄ C60 ⋄
TROFIMOV, M.YU. *Definability in algebraically closed systems (Russian)* ⋄ C25 C40 C60 C85 ⋄
TYSHKEVICH, R.I. *Models and groups (Russian)*
⋄ C60 ⋄
WEISPFENNING, V. *Model-completeness and elimination of quantifiers for subdirect products of structures*
⋄ C10 C30 C35 C52 C60 C90 ⋄
WINKLER, P.M. *Model-completeness and Skolem expansions* ⋄ C25 C30 C35 C60 ⋄
ZIEGLER, M. *A counterexample in the theory of definable automorphisms* ⋄ C07 C60 ⋄

1976

ADLER, A. & KIEFE, C. *Pseudofinite fields, procyclic fields and model-completion* ⋄ C35 C60 ⋄
BALDWIN, JOHN T. & SAXL, J. *Logical stability in group theory* ⋄ C45 C60 ⋄
BARWISE, J. *Some applications of Henkin quantifiers*
⋄ C40 C50 C60 C75 C80 ⋄
BAUDISCH, A. *Elimination of the quantifier Q_α in the theory of abelian groups (Russian summary)*
⋄ B25 C10 C60 C80 ⋄
BAUMSLAG, B. & LEVIN, F. *Algebraically closed torsion-free nilpotent groups of class 2* ⋄ C25 C60 ⋄
BAUR, W. *Elimination of quantifiers for modules*
⋄ C10 C60 ⋄
BELESOV, A.U. *Injective and equationally compact modules (Russian)* ⋄ C05 C50 C60 ⋄
BERNARD, C.-L. *Sous-groupes nilpotents des groupes multiplicatifs de corps (English summary)* ⋄ C60 ⋄
CHERLIN, G.L. *Amalgamation bases for commutative rings without nilpotent elements* ⋄ C60 ⋄
CHERLIN, G.L. & REINEKE, J. *Categoricity and stability of commutative rings* ⋄ C45 C60 ⋄
CHERLIN, G.L. *Model theoretic algebra* ⋄ C25 C60 ⋄
CHERLIN, G.L. *Model theoretic algebra. Selected topics*
⋄ C25 C60 C98 ⋄
CLARE, F. *On elementary equivalence of abelian groups*
⋄ C60 ⋄
COMER, S.D. *Complete and model-complete theories of monadic algebras*
⋄ B25 C05 C25 C35 C60 G15 ⋄
DOBRITSA, V.P. & GONCHAROV, S.S. *An example of a constructive abelian group with non-constructivizable reduced subgroup (Russian)* ⋄ C57 C60 D45 ⋄
DOBRITSA, V.P. *On computable and strictly computable classes of constructive algebras (Russian)*
⋄ C57 C60 D45 ⋄
EKLOF, P.C. *Infinitary model theory of abelian groups*
⋄ C52 C60 C75 E35 E55 E75 ⋄
FELGNER, U. *Einige gruppentheoretische Aequivalente zum Auswahlaxiom* ⋄ C60 E25 E75 ⋄
FLEISCHER, I. *Quantifier elimination for modules and ordered groups* ⋄ B25 C10 C60 ⋄
FLUM, J. *Algebra y logica* ⋄ C60 C98 ⋄
FLUM, J. *El problema de las palabras en la teoria de grupos* ⋄ C25 C60 D40 ⋄
FOWLER III, N. *α-decompositions of α-spaces*
⋄ C57 C60 D45 D50 ⋄
FRIED, M. & SACERDOTE, G.S. *Solving diophantine problems over all residue class fields of a number field and all finite fields* ⋄ B25 C10 C13 C60 D35 ⋄
FRIED, M. & JARDEN, M. *Stable extensions and fields with the global density property* ⋄ C60 ⋄

GROSS, W.F. *Models with dimension*
  ◇ C35 C50 C60 E25 E35 ◇
HALEY, D.K. *Equational compactness and compact topologies in rings satisfying the a.c.c.* ◇ C05 C60 ◇
HICKMAN, J.L. *The construction of groups in models of set theory that fail the axiom of choice*
  ◇ C60 E25 E35 ◇
HIRSCHFELD, J. *Non standard analysis and the compactification of groups* ◇ C60 H05 H20 ◇
HODGES, W. *On the effectivity of some field constructions*
  ◇ C60 C75 D65 E35 E47 E75 ◇
JARDEN, M. *Algebraically closed fields with distinguished subfields* ◇ C20 C60 ◇
JARDEN, M. *The elementary theory of ω-free Ax fields*
  ◇ C20 C60 ◇
KENNISON, J.F. *Integral domain type representations in sheaves and other topoi* ◇ C60 C90 G30 ◇
KIEFE, C. *Sets definable over finite fields: their zeta-functions* ◇ C13 C60 ◇
KOREC, I. & RAUTENBERG, W. *Model-interpretability into trees and applications* ◇ B25 C45 C60 C65 F25 ◇
KRASNER, M. *Endotheorie de Galois abstraite et son theoreme d'homomorphie (English summary)*
  ◇ C07 C60 C75 ◇
KRASNER, M. *Polytheorie de Galois abstraite dans le cas infini general* ◇ C05 C07 C60 ◇
LIPSHITZ, L. *Integral closures of uncountable commutative regular rings* ◇ C60 ◇
LIVCHAK, A.B. *Definable sets of integers (Russian)*
  ◇ C60 C62 F30 ◇
LIVCHAK, A.B. *Elimination of identity in the theory of abelian groups with predicates for subgroups (Russian)*
  ◇ B25 C60 ◇
LIVCHAK, A.B. *Formula subsets of abelian groups*
  ◇ C60 ◇
LUXEMBURG, W.A.J. *On a class of valuation fields introduced by A. Robinson* ◇ C60 H20 ◇
MACINTYRE, A. & ROSENSTEIN, J.G. $\aleph_0$-*categoricity for rings without nilpotent elements and for boolean structures* ◇ C35 C60 ◇
MACINTYRE, A. *Existentially closed structures and Jensen's principle* ◇ ◇ C25 C60 E45 E65 ◇
MACINTYRE, A. *On definable subsets of p-adic fields*
  ◇ C10 C60 ◇
MACINTYRE, A. & SHELAH, S. *Uncountable universal locally finite groups* ◇ C25 C50 C52 C55 C60 ◇
MAKOWIECKA, H. *A general property of ternary relations in elementary geometry (Russian summary)*
  ◇ B30 C60 ◇
MART'YANOV, V.I. *Decidability of the theory of closed torsion free abelian groups of finite rank with predicates for addition and automorphism (Russian)*
  ◇ B25 C60 ◇
MOSTOWSKI, T. *Analytic applications of decidability theorems* ◇ B25 C60 C65 D80 ◇
NITA, A. *Quelques observations sur les ultraproduits d'anneaux* ◇ C20 C60 ◇
PERLIS, D. *Group algebras and model theory* ◇ C60 ◇
RACKOFF, C.W. *On the complexity of the theories of weak direct powers* ◇ B25 C13 C30 C60 D15 ◇

ROSENSTEIN, J.G. $\aleph_0$-*categoricity is not inherited by factor groups* ◇ C35 C60 ◇
RYATOVA, N.A. *Conditions for elementary equivalence of modules over discretely normed rings* ◇ C60 ◇
SACERDOTE, G.S. *A characterization of the subgroups of finitely presented groups* ◇ C60 D40 ◇
SACERDOTE, G.S. *Some logical problems concerning free and free product groups* ◇ C60 C98 ◇
SARACINO, D.H. *Existentially complete nilpotent groups*
  ◇ C20 C25 C35 C60 ◇
SCHEIN, B.M. *Injective commutative semigroups*
  ◇ C05 C60 ◇
SCHMITT, P.H. *The model-completion of Stone algebras*
  ◇ C25 C35 C60 G10 ◇
VASIL'EV, EH.S. *On elementary theories of abelian p-groups with heights of elements (Russian)*
  ◇ B25 C60 ◇
WEISPFENNING, V. *Negative-existentially complete structures and definability in free extensions*
  ◇ C25 C40 C60 C75 D40 G05 ◇
WEISPFENNING, V. *On the elementary theory of Hensel fields* ◇ B25 C10 C35 C60 ◇
WHEELER, W.H. *Model-companions and definability in existentially complete structures*
  ◇ C25 C35 C60 C98 ◇
WOOD, C. *The model theory of differential fields revisited*
  ◇ C60 C98 ◇
WUETHRICH, H.R. *Ein Entscheidungsverfahren fuer die Theorie der reell abgeschlossenen Koerper*
  ◇ B25 C60 D15 ◇
ZAMYATIN, A.P. *Varieties of associative rings whose elementary theory is decidable (Russian)*
  ◇ B25 C05 C60 ◇

1977

BALDWIN, JOHN T. & ROSE, B.I. $\aleph_0$-*categoricity and stability of rings* ◇ C35 C45 C60 ◇
BASARAB, S.A. *On the elementary theories of abelian profinite groups and Abelian torsion groups*
  ◇ B25 C60 ◇
BAUDISCH, A. *A note on the elementary theory of torsion free Abelian groups with one predicate for subgroups*
  ◇ B25 C60 ◇
BAUDISCH, A. *Decidability of the theory of abelian groups with Ramsey quantifiers* ◇ B25 C10 C60 C80 ◇
BAUDISCH, A. *The theory of abelian groups with the quantifier $(\leq x)$* ◇ B25 C10 C60 C80 ◇
BAUDISCH, A. & WEESE, M. *The Lindenbaum-algebra of the theory of well-orders and abelian groups with the quantifier Q_α* ◇ C60 C80 ◇
BELESOV, A.U. *Equationally compact systems (Russian)*
  ◇ C05 C50 C60 ◇
BELYAEV, V.YA. *Algebraically closed semigroups (Russian)* ◇ C25 C60 ◇
BIENENSTOCK, E. *Sets of degrees of computable fields*
  ◇ C57 C60 D30 D45 ◇
BLUM, L. *Differentially closed fields: a model-theoretic tour* ◇ C60 ◇
BOZZI, S. & MELONI, G.C. *Ideal properties and intuitionistic algebra* ◇ C60 C90 F50 F55 ◇

BULMAN-FLEMING, S. & WERNER, H. *Equational compactness in quasi-primal varieties* ◇ C05 C60 ◇

CHIKHACHEV, S.A. *$\aleph_1$-categorical commutative rings (Russian)* ◇ C35 C60 ◇

CLARK, D.M. & KRAUSS, P.H. *Relatively homogeneous structures* ◇ C10 C35 C50 C60 C65 G05 ◇

COWLES, J.R. *The theory of differentially closed fields in logics with cardinal quantifiers* ◇ C60 C80 ◇

DEUTSCH, M. *Zur Axiomatisierung zyklischer Gruppen* ◇ C13 C60 ◇

DIARRA, B. *Ultra-produits de corps munis d'une valeur absolue ultra-metrique (English summary)* ◇ C20 C60 ◇

DICKMANN, M.A. *Deux applications de la methode de va-et-vient* ◇ C60 C75 ◇

DOBRITSA, V.P. *Computability of certain classes of constructive algebras (Russian)* ◇ C57 C60 D45 ◇

DRIES VAN DEN, L. *Artin-Schreier theory for commutative regular rings* ◇ B25 C10 C25 C60 ◇

DURET, J.-L. *Instabilite des corps formellement reels* ◇ C45 C60 ◇

EKLOF, P.C. *Applications of logic to the problem of splitting abelian groups* ◇ C55 C60 C75 C80 E35 E45 E75 ◇

EKLOF, P.C. *Classes closed under substructures and direct limits* ◇ C52 C60 C75 ◇

EKLOF, P.C. *Methods of logic in abelian group theory* ◇ C60 C75 E35 E45 E55 E75 ◇

EKLOF, P.C. & MEKLER, A.H. *On constructing indecomposable groups in L* ◇ C55 C60 C75 E35 E45 E55 ◇

EKLOF, P.C. *Ultraproducts for algebraists* ◇ C20 C60 C98 ◇

ELLENTUCK, E. *Boolean valued rings* ◇ C30 C60 C90 D50 ◇

ESTERLE, J. *Solution d'un probleme d'Erdoes, Gillman et Henriksen et application a l'etude des homomorphismes de $\mathscr{C}(K)$* ◇ C50 C60 E50 E75 ◇

FELGNER, U. *Stability and $\aleph_0$-categoricity of nonabelian groups* ◇ C35 C45 C60 ◇

FISHER, E.R. *Abelian structures I* ◇ C60 G30 ◇

FORREST, W.K. *Model theory for universal classes with the amalgamation property: A study in the foundations of model theory and algebra* ◇ C05 C25 C30 C35 C45 C52 C60 ◇

FRIED, E. *Automorphism group of integral domains fixing a given subring* ◇ C07 C60 ◇

GROSS, W.F. *Dimension and finite closure* ◇ C35 C50 C60 E25 ◇

GUHL, R. *Two notes on recursively enumerable vector spaces* ◇ C57 C60 D45 ◇

GUREVICH, Y. *Expanded theory of ordered abelian groups* ◇ B25 C10 C60 C85 ◇

HUBER, M. *Sur le probleme de Whitehead concernant les groupes abeliens libres (English summary)* ◇ C60 E65 E75 ◇

HUBER-DYSON, V. *Talking about free groups in naturally enriched languages* ◇ C60 D40 ◇

JEHNE, W. & KLINGEN, N. *Superprimes and a generalized Frobenius symbol* ◇ C20 C60 E50 E75 ◇

KALANTARI, I. & RETZLAFF, A.T. *Maximal vector spaces under automorphisms of the lattice of recursively enumerable vector spaces* ◇ C57 C60 D45 ◇

KHISAMIEV, N.G. *On the periodical part of a strongly constructivizable abelian group (Russian)* ◇ C57 C60 D45 ◇

KOPPELBERG, S. *Groups cannot be Souslin ordered* ◇ C60 E07 ◇

KRECZMAR, A. *On infinite sets of polynomial relations* ◇ C60 C75 ◇

LACAVA, F. & SAELI, D. *Sul model-completamento della teoria delle L-catene* ◇ C35 C60 G20 ◇

LATYSHEV, V.N. *Complexity of nonmatrix varieties of associative algebras I,II (Russian)* ◇ C05 C60 ◇

LIPSHITZ, L. *The real closure of a commutative regular f-ring* ◇ C60 ◇

LIVCHAK, A.B. *Expressible subsets of the ordered group of integers (Russian)* ◇ C60 C62 F30 ◇

MACINTYRE, A. *Model completeness* ◇ C25 C35 C60 C98 ◇

MAKKAI, M. & MCNULTY, G.F. *Universal Horn axiom systems for lattices of submodules* ◇ C05 C60 G10 ◇

METAKIDES, G. & NERODE, A. *Recursively enumerable vector spaces* ◇ C57 C60 D45 ◇

OLIN, P. *Elementary properties of V-free products of groups* ◇ C05 C30 C60 ◇

PAILLET, J.L. *Resultats relatifs a l'existence de structures n-elementaires (English summary)* ◇ C50 C60 ◇

PETRY, A. *A propos des centralisateurs de certains sous-corps des corps generiques (English summary)* ◇ C25 C60 ◇

PRIDE, S.J. *The isomorphism problem for two-generator one-relator groups with torsion is solvable* ◇ C60 D40 ◇

SHELAH, S. *Existentially-closed groups in $\aleph_1$ with special properties* ◇ C15 C25 C60 C75 ◇

SHELAH, S. *Whitehead groups may be not free, even assuming CH. I* ◇ C60 E05 E35 E50 E75 ◇

SHOENFIELD, J.R. *Quantifier elimination in fields* ◇ C10 C60 C98 ◇

STEFANESCU, M. *A correspondence between a class of near-rings and a class of groups (Italian summary)* ◇ C60 ◇

STUETZER, H. *Die elementare Formulierbarkeit von Dimension und Grad algebraischer Mannigfaltigkeiten* ◇ C20 C60 ◇

TAJTSLIN, M.A. *Existentially closed commutative semigroups (Russian) (English summary)* ◇ C25 C52 C60 ◇

TOFFALORI, C. *Alcune proprieta di teorie di campi con un sottocampo privilegiato (English summary)* ◇ C25 C35 C60 ◇

VASIL'EV, EH.S. *Elementary theories of certain classes of two-base models of abelian groups (Russian)* ◇ B25 C60 C85 ◇

VINOGRADOV, A.A. *Nonaxiomatizability of lattice-orderable rings (Russian)* ◇ C52 C60 ◇

WEISPFENNING, V. *Nullstellensaetze - a model theoretic framework* ◇ C10 C25 C60 ◇

WHITELEY, W. *Logic and invariant theory III. Axiom systems and basic syzygies* ⋄ B30 C60 ⋄

ZIL'BER, B.I. *Groups and rings, the theory of which is categorical (Russian) (English summary)*
⋄ C35 C45 C60 ⋄

1978

ADLER, A. *On the multiplicative semigroups of rings*
⋄ C52 C60 ⋄

AMERBAEV, V.M. & KASIMOV, YU.F. *Formal expressibility of order relations in algebraic systems (Russian) (Kazakh summary)* ⋄ C40 C60 ⋄

AOE, H. *On the Whitehead problem* ⋄ C60 E35 ⋄

BALDWIN, JOHN T. *Some EC_Σ classes of rings*
⋄ C52 C60 ⋄

BASARAB, S.A. *Quelques proprietes modele-theoriques des corps values henseliens (English summary)*
⋄ C25 C35 C60 ⋄

BASARAB, S.A. *Some model theory for henselian valued fields* ⋄ C35 C60 ⋄

BELEGRADEK, O.V. *Elementary properties of algebraically closed groups (Russian) (English summary)*
⋄ C25 C60 D40 ⋄

BELEGRADEK, O.V. *On nonstable group theories (Russian)*
⋄ C45 C60 ⋄

BELYAEV, V.YA. *On algebraic closure in classes of semigroups and groups (Russian)* ⋄ C25 C60 ⋄

BELYAEV, V.YA. *On maximal subgroups of an algebraically closed semigroup (Russian)*
⋄ C25 C60 ⋄

BLOSHCHITSYN, V.YA. & ZAKIR'YANOV, K.KH. *Constructive abelian groups (Russian)* ⋄ C57 C60 D45 ⋄

BOKUT', L.A. *On algebraically closed and simple Lie algebras (Russian)* ⋄ C25 C60 ⋄

BROWN, S.S. *Bounds on transfer principles for algebraically closed and complete discretely valued fields* ⋄ B25 C10 C60 ⋄

BRYANT, R.M. & GROVES, J.R.J. *Wreath products and ultraproducts of groups* ⋄ C20 C30 C60 ⋄

CAMERON, P.J. *Orbits of permutation groups on unordered sets* ⋄ C30 C60 E05 ⋄

CARSON, A.B. *Coherent polynomial rings over regular rings of finite index* ⋄ C60 ⋄

CHERLIN, G.L. & ROSENSTEIN, J.G. *On $\aleph_0$-categorical abelian-by-finite groups* ⋄ C35 C60 ⋄

CHERLIN, G.L. *Superstable division rings* ⋄ C45 C60 ⋄

DELON, F. *Definition de l'arithmetique dans la theorie des anneaux de series formelles (English summary)*
⋄ C60 D35 F25 F35 ⋄

DENEF, J. *Diophantine sets over $Z[T]$*
⋄ C60 D25 D35 ⋄

DENEF, J. & LIPSHITZ, L. *Diophantine sets over some rings of algebraic integers* ⋄ C60 D35 ⋄

DENEF, J. *The Diophantine problem for polynomial rings and fields of rational functions* ⋄ C60 D35 ⋄

DOBRITSA, V.P. & KHISAMIEV, N.G. & NURTAZIN, A.T. *Constructive periodic abelian groups (Russian)*
⋄ C57 C60 F60 ⋄

DRIES VAN DEN, L. *Model theory of fields*
⋄ B25 C10 C25 C35 C60 H20 ⋄

ENJALBERT, P. *Epimorphismes: quelques resultats algebriques et logiques (English summary)* ⋄ C60 ⋄

FABER, V. & LAVER, R. & MCKENZIE, R. *Coverings of groups by abelian subgroups*
⋄ C05 C60 E50 E75 ⋄

FELGNER, U. *$\aleph_0$-categorical stable groups*
⋄ C35 C45 C60 ⋄

FOWLER III, N. *Effective inner product spaces*
⋄ C57 C60 D45 D80 ⋄

HAUSCHILD, K. *Gruppengraphen und endliche Axiomatisierbarkeit* ⋄ C35 C52 C60 ⋄

HICKIN, K.K. *Complete universal locally finite groups*
⋄ C50 C60 ⋄

JARDEN, M. *Intersections of conjugate fields of finite corank over Hilbertian fields* ⋄ C60 ⋄

KAISER, KLAUS *On relational selections for complete theories* ⋄ C20 C35 C50 C60 ⋄

KALANTARI, I. *Major subspaces of recursively enumerable vector spaces* ⋄ C57 C60 D25 D45 ⋄

KHISAMIEV, N.G. *Strongly constructive periodic abelian groups (Russian) (Kazakh summary)*
⋄ C57 C60 D45 ⋄

KLINGEN, N. *Zur Modelltheorie lokaler und globaler Koerper* ⋄ C60 H20 ⋄

KRASNER, M. *Abstract Galois theory*
⋄ C07 C10 C60 C75 ⋄

KURKE, H. & MOSTOWSKI, T. & PFISTER, G. & POPESCU, D. & ROCZEN, M. *Die Approximationseigenschaft lokaler Ringe*
⋄ C10 C60 ⋄

LOLLI, G. *Alcune applicazioni della compattezza*
⋄ C20 C60 H05 ⋄

MACDONALD, J.L. *Categorical shape theory and the back and forth property* ⋄ C60 C90 G05 G30 ⋄

MARZI LI, E.M. *An elementary class of hypergroups (Italian)* ⋄ C60 ⋄

METAKIDES, G. *Constructive algebra in a new frame*
⋄ C57 C60 D45 ⋄

NOGINA, E.YU. *Numerierte topologische Raeume (Russisch)* ⋄ C57 C60 D45 ⋄

POIZAT, B. *Rangs des types dans les corps differentiels*
⋄ C45 C60 ⋄

PREST, M. *Some model-theoretic aspects of torsion theories* ⋄ C25 C35 C60 ⋄

PRESTEL, A. *Artin's conjecture on p-adic number fields*
⋄ C60 C98 ⋄

PRESTEL, A. & ZIEGLER, M. *Model theoretic methods in the theory of topological fields* ⋄ C60 C65 C80 ⋄

ROSE, B.I. *Model theory of alternative rings*
⋄ C10 C35 C45 C60 ⋄

ROSE, B.I. *Rings which admit elimination of quantifiers*
⋄ C10 C60 ⋄

ROSE, B.I. *The $\aleph_1$-categoricity of strictly upper triangular matrix rings over algebraically closed fields*
⋄ C35 C45 C60 ⋄

ROTHMALER, P. *Total transzendente abelsche Gruppen und Morley-Rang (English and Russian summaries)*
⋄ C35 C45 C60 ⋄

SARACINO, D.H. *Existentially complete torsion-free nilpotent groups* ⋄ C25 C60 ⋄

SCHUPPAR, B. *Modelltheoretische Untersuchungen zur Galoistheorie von Funktionenkoerpern (Dissertation)* ⋄ C60 ⋄

SEJTENOV, S.M. *Theory of finite fields with an additional predicate distinguishing a subfield (Russian)* ⋄ B25 C13 C60 ⋄

SINGER, M.F. *A class of differential fields with minimal differential closures* ⋄ C60 ⋄

SINGER, M.F. *The model theory of ordered differential fields* ⋄ B25 C25 C35 C60 C65 D80 ⋄

SKORNYAKOV, L.A. *The axiomatizability of a class of injective polygons (Russian)* ⋄ C05 C52 C60 ⋄

TOFFALORI, C. *Eliminazione dei quantificatori per certe teorie di coppie di campi (Teoria dei modelli) (English summary)* ⋄ C10 C60 ⋄

TOKARENKO, A.I. *The group-theoretic properties N_0, N, and $\tilde{N}$ are not axiomatizable (Russian)* ⋄ C20 C52 C60 ⋄

WHEELER, W.H. *A characterization of companionable, universal theories* ⋄ C25 C35 C60 C65 ⋄

WHITELEY, W. *Logic and invariant theory II: Homogeneous coordinates, the introduction of higher quantities, and structural geometry* ⋄ B30 C60 ⋄

ZAMYATIN, A.P. *A non-abelian variety of groups has an undecidable elementary theory (Russian)* ⋄ C05 C60 D35 ⋄

ZAMYATIN, A.P. *Prevarieties of associative rings whose elementary theory is decidable (Russian)* ⋄ B25 C05 C60 ⋄

1979

ARIBAUD, F. *Un ideal maximal de l'anneau des endomorphismes d'un espace vectoriel de dimension infinie* ⋄ C60 C62 E40 E75 ⋄

BALDWIN, JOHN T. *Stability theory and algebra* ⋄ C45 C60 ⋄

BASARAB, S.A. *A model-theoretic transfer theorem for henselian valued fields* ⋄ C25 C35 C60 ⋄

BASARAB, S.A. *Model-theoretic methods in the theory of henselian valued fields. I,II (Romanian) (English summaries)* ⋄ C25 C35 C60 ⋄

BAUR, W. & CHERLIN, G.L. & MACINTYRE, A. *Totally categorical groups and rings* ⋄ C35 C45 C60 ⋄

BECKER, J.A. & DENEF, J. & DRIES VAN DEN, L. & LIPSHITZ, L. *Ultraproducts and approximation in local rings. I* ⋄ C20 C60 ⋄

BELYAEV, V.YA. & TAJTSLIN, M.A. *On elementary properties of existentially closed systems (Russian)* ⋄ C25 C40 C60 C75 C85 D55 ⋄

BERNAL, E. *On classes of groups closed under ultraproducts* ⋄ C20 C60 ⋄

BURRIS, S. & WERNER, H. *Sheaf constructions and their elementary properties* ⋄ B25 C05 C20 C25 C30 C60 C90 ⋄

BUSZKOWSKI, W. *Undecidability of the theory of lattice-orderable groups* ⋄ C60 D35 ⋄

CHERLIN, G.L. *Groups of small Morley rank* ⋄ C45 C60 ⋄

CHERLIN, G.L. *Stable algebraic theories* ⋄ C45 C60 ⋄

COSTE, M. & COSTE, M.-F. *The generic model of an ε-stable geometric extension of the theory of rings is of line type* ⋄ C60 C90 G30 ⋄

COSTE, M. & COSTE, M.-F. *Topologies for real algebraic geometry* ⋄ C60 G30 ⋄

COWLES, J.R. *The theory of Archimedean real closed fields in logics with Ramsey quantifiers* ⋄ C10 C35 C60 C80 ⋄

DEISSLER, R. *Minimal and prime models of complete theories of torsion free abelian groups* ⋄ C50 C60 ⋄

DENEF, J. *The Diophantine problem for polynomial rings of positive characteristic* ⋄ C60 D35 ⋄

DRIES VAN DEN, L. *Algorithms and bounds for polynomial rings* ⋄ C57 C60 ⋄

DRIES VAN DEN, L. & RIBENBOIM, P. *Application de la theorie des modeles aux groupes de Galois de corps de fonctions (English summary)* ⋄ C60 ⋄

DRIES VAN DEN, L. *New decidable fields of algebraic numbers* ⋄ B25 C60 ⋄

DUBUC, E.J. & REYES, G.E. *Subtoposes of the ring classifier* ⋄ C60 C75 C90 G30 ⋄

EKLOF, P.C. & MEKLER, A.H. *Stationary logic of finitely determinate structures* ⋄ B25 C10 C55 C60 C65 C80 ⋄

GARAVAGLIA, S. *Direct product decomposition of theories of modules* ⋄ C30 C45 C60 ⋄

GIRSTMAIR, K. *Ueber konstruktive Methoden der Galoistheorie* ⋄ C57 C60 F55 ⋄

GREITHER, C. *Struktur und Transzendenzgrad von $\eta_{\alpha+1}$-Koerpern und Ultrapotenzen von Koerpern* ⋄ C20 C60 ⋄

HALEY, D.K. *Equational compactness in rings* ⋄ C05 C60 ⋄

HEINTZ, JOOS *Definability bounds of first order theories of algebraically closed fields* ⋄ B25 C40 C60 D10 D15 ⋄

HICKIN, K.K. & PHILLIPS, R.E. *A construction of locally finite simple groups* ⋄ C30 C60 ⋄

HUBER, M. *On cartesian powers of a rational group* ⋄ C60 E45 E55 E75 ⋄

HUBER-DYSON, V. *An inductive theory for free products of groups* ⋄ C60 ⋄

JENSEN, C.U. & VAMOS, P. *On the axiomatizability of certain classes of modules* ⋄ C60 ⋄

JOHNSTONE, P.T. *Another condition equivalent to De Morgan's law* ⋄ C60 C90 F50 G05 G30 ⋄

JOHNSTONE, P.T. *Remarks on the previous paper. Extracts from two letters to Anders Kock* ⋄ C60 C90 G30 ⋄

KALANTARI, I. *Automorphisms of the lattice of recursively enumerable vector spaces* ⋄ C57 C60 D45 ⋄

KHISAMIEV, N.G. *Criterion of the constructivizability of a direct sum of cyclic p-groups (Russian)* ⋄ C57 C60 D45 ⋄

KHISAMIEV, N.G. *On subgroups of finite index of abelian groups (Russian) (Kazakh summary)* ⋄ C57 C60 D45 F60 ⋄

KIEHNE, U. *Bounded products in the theory of valued fields* ⋄ C20 C60 ⋄

LACAVA, F. *Alcune proprieta delle L-algebre e delle L-algebre esistenzialmente chiuse*
◇ C25 C35 C60 G20 ◇

LEAVITT, W.G. & LEEUWEN VAN, L.C.A. *Rings isomorphic with all proper factor-rings* ◇ C60 E10 E75 ◇

LOULLIS, G. *Sheaves and boolean valued model theory*
◇ C35 C52 C60 C90 ◇

MANDERS, K.L. *The theory of all substructures of a structure: Characterisation and decision problems*
◇ B20 B25 C57 C60 ◇

MEKLER, A.H. *Model complete theories with a distinguished substructure* ◇ C35 C45 C60 ◇

METAKIDES, G. & NERODE, A. *Effective content of field theory* ◇ C57 C60 D45 ◇

MOSTOWSKI, T. & PFISTER, G. *Der Cohensche Eliminationssatz in positiver Charakteristik*
◇ B25 C10 C60 ◇

NITA, A. *The behavior of ultraproducts of modules under passage to fractions (Romanian) (English summary)*
◇ C20 C60 ◇

NITA, A. *Ultraproducts of rings which retain factor properties (Romanian) (French summary)*
◇ C20 C60 ◇

OIKKONEN, J. *Mathematical applications of model theory (Finnish) (English summary)* ◇ C07 C60 ◇

PINUS, A.G. *Elimination of the quantifiers Q_0 and Q_1 on symmetric groups (Russian)*
◇ C10 C55 C60 C80 ◇

PODEWSKI, K.-P. & REINEKE, J. *Algebraically closed commutative local rings* ◇ C25 C35 C60 ◇

POPESCU, D. *Algebraically pure morphisms*
◇ C20 C60 ◇

POTTHOFF, K. *Ultraproduits des groupes finis et applications a la theorie de Galois*
◇ C13 C20 C60 ◇

PREST, M. *Model-completions of some theories of modules*
◇ C25 C35 C50 C60 ◇

PREST, M. *Torsion and universal Horn classes of modules*
◇ C05 C60 ◇

PRESTEL, A. *Entscheidbarkeit mathematischer Theorien*
◇ B25 C60 C98 D35 ◇

RETZLAFF, A.T. *Direct summands of recursively enumerable vector spaces* ◇ C57 C60 D45 ◇

RICKERT, U. *Ueber Summen in abelschen Gruppen*
◇ C60 ◇

ROBBIN, J.W. *Evaluation fields for power series I: The nullstellensatz. II: The reellnullstellensatz* ◇ C60 ◇

ROBINSON, A. *Selected papers of Abraham Robinson. Vol.I: Model theory and algebra* ◇ C60 C96 ◇

ROMAN'KOV, V.A. *On definable subgroups of solvable groups* ◇ C40 C60 ◇

ROMAN'KOV, V.A. *Universal theory of nilpotent groups (Russian)* ◇ B25 C60 D35 ◇

SARACINO, D.H. & WOOD, C. *Periodic existentially closed nilpotent groups* ◇ C25 C35 C45 C60 ◇

SHELAH, S. & ZIEGLER, M. *Algebraically closed groups of large cardinality* ◇ C25 C60 C75 ◇

SHELAH, S. *On uncountable abelian groups*
◇ C60 E35 E45 E50 E75 ◇

SKORNYAKOV, L.A. *Finite axiomatizability of a class of point modules (Russian)* ◇ C52 C60 ◇

SMITH, B.D. *An application of non-standard model theory*
◇ C60 H20 ◇

STEGBAUER, W. *A generalized model companion for a theory of partially ordered fields* ◇ C25 C35 C60 ◇

SUKHORUTCHENKO, I.I. *Programmably categorical abelian p-groups I,II (Russian)* ◇ B75 C35 C60 ◇

TAJTSLIN, M.A. *Program theories of periodic abelian groups (Russian)* ◇ B75 C60 ◇

TIMOSHENKO, E.I. *On elementary theories of wreath products (Russian)* ◇ C30 C60 ◇

TOFFALORI, C. *Alcune osservazioni sugli anelli commutativi esistenzialmente chiusi (English summary)*
◇ C25 C60 ◇

TOFFALORI, C. *Semigruppi archimedei esistenzialmente chiusi (English summary)* ◇ C25 C60 ◇

TOKARENKO, A.I. *Axiomatizability of the group-theoretic property $\bar{Z}$ (Russian)* ◇ C20 C52 C60 ◇

TRANSIER, R. *Verallgemeinerte formal p-adische Koerper*
◇ C25 C60 ◇

VOLGER, H. *Preservation theorems for limits of structures and global sections of sheaves of structures*
◇ C40 C60 C90 ◇

WHEELER, W.H. *Amalgamation and elimination of quantifiers for theories of fields*
◇ C10 C25 C52 C60 ◇

WHEELER, W.H. *Model-complete theories of pseudo-algebraically closed fields*
◇ C25 C35 C60 ◇

WHITELEY, W. *Logic and invariant theory IV. Invariants and syzygies in combinatorial geometry* ◇ C60 ◇

WOOD, C. *Notes on the stability of separably closed fields*
◇ C25 C35 C45 C60 ◇

WRAITH, G.C. *Generic Galois theory of local rings*
◇ C60 G30 ◇

1980

ASH, C.J. & ROSENTHAL, J.W. *Some theories associated with algebraically closed fields* ◇ C60 D35 ◇

BASARAB, S.A. & NICA, V. & POPESCU, D. *Approximation properties and existential completeness for ring morphisms* ◇ C25 C60 H15 ◇

BASARAB, S.A. *Towards a general theory of formally p-adic fields* ◇ C60 ◇

BAUDISCH, A. *Application of the theory of Shirshov-Witt in the theory of groups and Lie algebras*
◇ B25 C45 C60 ◇

BAUDISCH, A. & SEESE, D.G. & TUSCHIK, H.-P. & WEESE, M. *Decidability and generalized quantifiers*
◇ B25 B98 C10 C60 C80 C98 ◇

BAUMSLAG, G. & DYER, E. & HELLER, A. *The topology of discrete groups* ◇ C25 C60 ◇

BAUR, W. *On the elementary theory of quadruples of vector spaces* ◇ B25 C60 D80 ◇

BECKER, J.A. & DENEF, J. & LIPSHITZ, L. *Further remarks on the elementary theory of formal power series rings*
◇ C60 D35 ◇

BECKER, J.A. & LIPSHITZ, L. *Remarks of the elementary theories of formal and convergent power series*
◇ C60 ◇

BERLINE, C. *Elimination of quantifiers for non semi-simple rings of characteristic p* ◊ C10 C60 ◊

BOFFA, M. & CHERLIN, G.L. *Elimination des quantificateurs dans les faisceaux (English summary)* ◊ C10 C60 C90 ◊

BOFFA, M. & MACINTYRE, A. & POINT, F. *The quantifier elimination problem for rings without nilpotent elements and for semi-simple rings* ◊ C10 C60 ◊

BOUSCAREN, E. *Existentially closed modules: Types and prime models* ◊ C25 C60 ◊

BUYS, A. & HEIDEMA, J. *Model-theoretical aspects of radicals and subdirect decomposability* ◊ C05 C60 ◊

CHERLIN, G.L. & DICKMANN, M.A. *Anneaux reels clos et anneaux de fonctions continues (English summary)* ◊ C10 C35 C60 C65 ◊

CHERLIN, G.L. *On $\aleph_0$-categorical nilrings* ◊ C35 C60 ◊

CHERLIN, G.L. *On $\aleph_0$-categorical nilrings. II* ◊ C35 C60 ◊

CHERLIN, G.L. *Rings of continuous functions: decision problems* ◊ B25 C60 C65 D35 ◊

CHERLIN, G.L. & SHELAH, S. *Superstable fields and groups* ◊ C45 C60 ◊

CHITI, E. *Modelli saturati per particolari teorie di gruppi abeliani ordinati (English summary)* ◊ C50 C60 ◊

COHN, P.M. *On semifir constructions* ◊ C05 C20 C52 C60 ◊

COHN, P.M. *Problems* ◊ C60 ◊

DELON, F. *Hensel fields in equal characteristic $p>0$* ◊ C60 ◊

DENEF, J. & LIPSHITZ, L. *Ultraproducts and approximation in local rings. II* ◊ C20 C60 ◊

DICKMANN, M.A. *On polynomials over real closed rings* ◊ C60 ◊

DICKMANN, M.A. *Sur les anneaux de polynomes a coefficients dans un anneau reel clos (English summary)* ◊ C60 ◊

DRIES VAN DEN, L. *A linearly ordered ring whose theory admits elimination of quantifiers is a real closed field* ◊ C10 C60 ◊

DURET, J.-L. *Les corps pseudo-finis ont la propriete d'independance (English summary)* ◊ C45 C60 ◊

DURET, J.-L. *Les corps faiblement algebriquement clos non separablement clos ont la propriete d'independance* ◊ C45 C60 ◊

EKLOF, P.C. & HUBER, M. *On the rank of Ext* ◊ C60 E35 E45 E50 E75 ◊

ERSHOV, YU.L. *Decision problems and constructivizable models (Russian)* ◊ B25 C57 C60 C98 D45 D98 ◊

ERSHOV, YU.L. *Multiply valued fields (Russian)* ◊ C25 C35 C60 ◊

ERSHOV, YU.L. *Regularly closed fields (Russian)* ◊ B25 C35 C60 D35 ◊

FELGNER, U. *Horn-theories of Abelian groups* ◊ C20 C60 ◊

FELGNER, U. *Kategorizitaet* ◊ C35 C45 C60 C98 ◊

FELGNER, U. *The model theory of FC-groups* ◊ C35 C45 C60 D35 ◊

GARAVAGLIA, S. *Decomposition of totally transcendental modules* ◊ C10 C35 C45 C60 ◊

GIORGETTA, D. *Embeddings in groups of permutations with supports of bounded cardinality* ◊ C60 E05 ◊

GLASS, A.M.W. & PIERCE, K.R. *Equations and inequations in lattice-ordered groups* ◊ C25 C60 G10 ◊

GLASS, A.M.W. & PIERCE, K.R. *Existentially complete abelian lattice-ordered groups* ◊ C25 C60 G10 ◊

GLASS, A.M.W. & PIERCE, K.R. *Existentially complete lattice-ordered groups* ◊ C25 C35 C60 G10 ◊

GOLTZ, H.-J. *Untersuchungen zur Elimination verallgemeinerter Quantoren in Koerpertheorien* ◊ C10 C55 C60 C80 ◊

GONCHAROV, S.S. *Autostability of models and abelian groups (Russian)* ◊ C50 C57 C60 D45 ◊

GUREVICH, Y. *Ordered abelian groups* ◊ B25 C10 C60 C85 ◊

HICKIN, K.K. & MACINTYRE, A. *Algebraically closed groups: embeddings and centralizers* ◊ C25 C60 D40 ◊

HODGES, W. *Interpreting number theory in nilpotent groups* ◊ C25 C60 C62 F25 F35 ◊

HOLLAND, W.C. *Trying to recognize the real line* ◊ C07 C60 C65 ◊

JACOB, B. *A Nullstellensatz for $\mathbb{R}((t))$* ◊ C60 ◊

JACOB, B. *The model theory of "R-formal" fields* ◊ B25 C10 C25 C35 C60 ◊

JAMBU-GIRAUDET, M. *Theorie des modeles de groupes d'automorphismes d'ensembles totalement ordonnes 2-homogenes (English summary)* ◊ C07 C60 C62 C65 D35 E07 F25 F35 ◊

JARDEN, M. *An analogue of Chebotarev density theorem for fields of finite corank* ◊ B25 C10 C60 ◊

JARDEN, M. & ROQUETTE, P. *The Nullstellensatz over $\mathfrak{p}$-adically closed fields* ◊ C60 ◊

JARDEN, M. *Transfer principles for finite and p-adic fields* ◊ C60 ◊

JENSEN, C.U. *Applications logiques en theorie des anneaux et des modules* ◊ C60 ◊

JENSEN, C.U. & LENZING, H. *Model theory and representations of algebras* ◊ C60 ◊

JENSEN, C.U. *Peano rings of arbitrary global dimension* ◊ C60 E50 E75 ◊

JENSEN, C.U. *Proprietes homologiques et logiques des anneaux de fonctions entieres (English summary)* ◊ C60 C65 E50 E75 ◊

KHMELEVSKIJ, YU.I. & TAJMANOV, A.D. *Decidability of the universal theory of a free semigroup (Russian)* ◊ B25 C05 C60 ◊

KIM, KUKJIN *On the structure of Hensel fields* ◊ B25 C60 ◊

KOMARNITS'KIJ, N.YA. *Axiomatizability of certain classes of modules connected with a torsion (Russian)* ◊ C60 ◊

KOMORI, Y. *Completeness of two theories on ordered abelian groups and embedding relations* ◊ B25 C10 C35 C60 ◊

KRYAZHOVSKIKH, G.V. *Approximability of finitely presented algebras (Russian)* ◊ C60 D40 ◊

LACAVA, F. *Osservazioni sulla teoria dei gruppi abeliani reticolari* ⋄ C25 C35 C60 ⋄
LIVCHAK, A.B. *A letter to the editors (Russian)* ⋄ B25 C10 C60 ⋄
MACINTYRE, A. *Nonstandard number theory* ⋄ C60 H15 ⋄
MCKEE, T.A. *Generalized equivalence and the foundations of quasigroups* ⋄ C60 ⋄
MCKENNA, K. *Some diophantine Nullstellensaetze* ⋄ C60 ⋄
MCNULTY, G.F. *Classes which generate the variety of all lattice-ordered groups* ⋄ C05 C52 C60 ⋄
MEJREMBEKOV, K.A. *Remarks on theories with a bounded spectrum (Russian)* ⋄ C35 C45 C60 ⋄
MEKLER, A.H. *How to construct almost free groups* ⋄ C60 C75 E35 E45 E75 ⋄
METAKIDES, G. & NERODE, A. *Recursion theory on fields and abstract dependence* ⋄ C57 C60 D45 ⋄
MURZIN, F.A. *Elementary equivalence of rings of continuous functions (Russian)* ⋄ C60 ⋄
NITA, A. *Sur les ultraproduits des modules* ⋄ C20 C60 ⋄
NITA, A. *Ultraproducts for certain classes of rings (Romanian) (English summary)* ⋄ C20 C60 ⋄
PACHOLSKI, L. & WIERZEJEWSKI, J. & WILKIE, A.J. (EDS.) *Model theory of algebra and arithmetic. Proceedings, Karpacz, Poland, September 1-7, 1979* ⋄ C60 C62 C97 ⋄
POIZAT, B. *Les equations differentielles vues par un logicien* ⋄ C45 C60 ⋄
POIZAT, B. *Une preuve d'un theoreme de James Ax sur les extensions algebriques de corps (English summary)* ⋄ C60 ⋄
PREST, M. *Elementary torsion theories and locally finitely presented categories* ⋄ C60 G30 ⋄
PRESTEL, A. & ZIEGLER, M. *Non-axiomatizable classes of V-topological fields* ⋄ C60 C65 C80 ⋄
REMMEL, J.B. *On r.e. and co-r.e. vector spaces with nonextendible bases* ⋄ C57 C60 D45 ⋄
REMMEL, J.B. *Recursion theory on algebraic structures with independent sets* ⋄ C57 C60 D45 ⋄
ROMANOVSKIJ, N.S. *The elementary theory of an almost polycyclic group (Russian)* ⋄ C60 D35 ⋄
ROSE, B.I. *On the model theory of finite-dimensional algebras* ⋄ B25 C35 C45 C60 ⋄
ROSE, B.I. *Prime quantifier eliminable rings* ⋄ C10 C60 ⋄
RUMELY, R. *Undecidability and definability for the theory of global fields* ⋄ C40 C60 D35 ⋄
RYCHKOV, S.V. *On direct products of abelian groups (Russian)* ⋄ C30 C60 E55 E75 ⋄
SCHUPPAR, B. *Elementare Aussagen zur Arithmetik und Galoistheorie von Funktionenkoerpern* ⋄ C60 ⋄
SHELAH, S. *On a problem of Kurosh, Jonsson groups, and applications* ⋄ C60 ⋄
SHELAH, S. *Whitehead groups may not be free even assuming CH. II* ⋄ C60 E05 E35 E50 E75 ⋄
TOFFALORI, C. *Fasci di coppie di campi algebricamente chiusi* ⋄ B25 C35 C60 C90 ⋄
TOFFALORI, C. *Sheaves of pairs of real closed fields* ⋄ C25 C35 C60 C90 ⋄
TOFFALORI, C. *Sugli anelli nilreali (English summary)* ⋄ C20 C25 C60 ⋄
TOFFALORI, C. *Sul model-completamento di certe teorie di coppie di anelli (English summary)* ⋄ C10 C25 C35 C60 ⋄
TUCKER, J.V. *Computability and the algebra of fields: some affine constructions* ⋄ B25 C57 C60 C65 D45 ⋄
WHEELER, W.H. *Model theory of strictly upper triangular matrix rings* ⋄ C25 C35 C60 ⋄
ZIEGLER, M. *Algebraisch abgeschlossene Gruppen (English summary)* ⋄ C25 C60 D30 D40 ⋄

1981

ANTONOVSKIJ, M.YA. & CHUDNOVSKY, D.V. & CHUDNOVSKY, G.V. & HEWITT, E. *Rings of real-valued continuous functions II* ⋄ C20 C50 C60 C65 E35 E50 E75 ⋄
ASH, C.J. & NERODE, A. *Functorial properties of algebraic closure and Skolemization* ⋄ C60 G30 ⋄
BAKER, K.A. *Definable normal closures in locally finite varieties of groups* ⋄ C05 C40 C60 ⋄
BANASCHEWSKI, B. *When are divisible abelian groups injective?* ⋄ C13 C60 G10 ⋄
BASARAB, S.A. *Extension of places and contraction properties for function fields over $\mathfrak{p}$-adically closed fields* ⋄ C60 H20 ⋄
BAUDISCH, A. *Subgroups of semifree groups* ⋄ C05 C60 ⋄
BAUDISCH, A. *The elementary theory of abelian groups with m-chains of pure subgroups* ⋄ B25 C55 C60 C80 ⋄
BAUDISCH, A. *There is no module having the finite cover property* ⋄ C10 C45 C60 C80 ⋄
BERLINE, C. *Ideaux des anneaux de Peano (d'apres Cherlin)* ⋄ C60 ⋄
BERLINE, C. & CHERLIN, G.L. *QE nilrings of prime characteristic* ⋄ C10 C60 ⋄
BERLINE, C. & CHERLIN, G.L. *QE rings in characteristic p* ⋄ C10 C60 ⋄
BERLINE, C. *Rings which admit elimination of quantifiers* ⋄ C10 C60 ⋄
BERLINE, C. *Stabilite et algebre II. Groupes Abeliens et modules* ⋄ C35 C45 C60 C98 ⋄
BERLINE, C. *Stabilite et algebre III. Anneaux* ⋄ C45 C60 C98 ⋄
BERLINE, C. *Stabilite et algebre IV. Groupes* ⋄ C45 C60 C98 ⋄
BEZVERKHNIJ, V.N. *Solvability of the inclusion problem in a class of HNN-groups (Russian)* ⋄ B25 C60 D40 ⋄
BEZVERKHNIJ, V.N. & GRINBLAT, V.A. *The root problem in Artin Groups (Russian)* ⋄ B25 C60 D40 ⋄
BUNGE, MARTA C. *Sheaves and prime model extensions* ⋄ C35 C60 C90 ⋄
BURRIS, S. & MCKENZIE, R. *Decidability and boolean representations* ⋄ B25 C05 C30 C60 ⋄
CAMERON, P.J. *Orbits of permutation groups on unordered sets II* ⋄ C30 C60 E05 ⋄
CHERLIN, G.L. & DRIES VAN DEN, L. & MACINTYRE, A. *Decidability and undecidability theorems for PAC-fields* ⋄ B25 C60 D35 ⋄

CHERLIN, G.L. & DICKMANN, M.A. *Note on a Nullstellensatz* ◊ C60 ◊

CHERLIN, G.L. & SCHMITT, P.H. *Undecidable L^t theories of topological abelian groups* ◊ C60 C90 D35 ◊

COWLES, J.R. *Generalized archimedean fields and logics with Malitz quantifiers* ◊ C10 C60 C80 ◊

COWLES, J.R. *The Henkin quantifier and real closed fields* ◊ C60 C80 ◊

DELON, F. *Indecidabilite de la theorie des anneaux de series formelles a plusieurs indeterminees (English summary)* ◊ C60 D35 ◊

DELON, F. *Types sur C((X))* ◊ C10 C45 C60 ◊

DICKMANN, M.A. *Sur les ouverts semi-algebriques d'une cloture reelle* ◊ C60 ◊

DOBRITSA, V.P. *On constructivizable abelian groups (Russian)* ◊ C57 C60 D40 D45 ◊

DRIES VAN DEN, L. *Quantifier elimination for linear formulas over ordered and valued fields* ◊ C10 C60 ◊

DRIES VAN DEN, L. *Which curves over Z have points with coordinates in a discrete ordered ring?* ◊ B25 C60 C62 ◊

EHDEL'MAN, G.S. *π-decomposition of metaideals (Russian)* ◊ C07 C60 ◊

EKLOF, P.C. & MEKLER, A.H. *Infinitary stationary logic and abelian groups* ◊ C60 C75 C80 ◊

ELLENTUCK, E. *Galois theorems for isolated fields* ◊ C57 C60 D45 D50 ◊

ENGELER, E. *Generalized Galois theory and its application to complexity* ◊ C60 D15 ◊

ERIMBETOV, M.M. *A fragment of a logic with infinite formulas (Russian)* ◊ C60 C75 ◊

ERSHOV, YU.L. *Eliminability of quantifiers in regularly closed fields (Russian)* ◊ B25 C10 C60 ◊

ERSHOV, YU.L. *On elementary theories of regularly closed fields (Russian)* ◊ B25 C60 ◊

ERSHOV, YU.L. *Undecidability of regularly closed fields (Russian)* ◊ C60 D35 ◊

FISHER, J.R. *Axiomatic radical and semisimple classes of rings* ◊ C05 C52 C60 ◊

FRANZEN, B. *Algebraic compactness of filter quotients* ◊ C05 C30 C60 ◊

GARAVAGLIA, S. *Forking in modules* ◊ C45 C60 ◊

GLASS, A.M.W. *Ordered permutation groups* ◊ C07 C60 C65 C98 ◊

GONCHAROV, S.S. *Groups with a finite number of constructivizations (Russian)* ◊ C57 C60 D45 ◊

HERRMANN, C. & JENSEN, C.U. & LENZING, H. *Applications of model theory to representations of finite-dimensional algebras* ◊ C60 ◊

HINGSTON, P. *Effective decomposition in Noetherian rings* ◊ C57 C60 D45 ◊

HODGES, W. *In singular cardinality, locally free algebras are free* ◊ C05 C55 C60 E05 E60 E75 ◊

HUBER-DYSON, V. *A reduction of the open sentence problem for finite groups* ◊ C13 C60 D40 ◊

JACOB, B. *The model theory of generalized real closed fields* ◊ B25 C35 C60 ◊

KHISAMIEV, N.G. *Criterion for constructivizability of a direct sum of cyclic p-groups (Russian) (Kazakh summary)* ◊ C57 C60 D45 ◊

LAWRENCE, J. *Primitive rings do not form an elementary class* ◊ C20 C52 C60 ◊

LIN, C. *Recursively presented abelian groups: effective p-group theory. I* ◊ C57 C60 D45 ◊

LIN, C. *The effective content of Ulm's theorem* ◊ C57 C60 D45 ◊

MACINTYRE, A. *The complexity of types in field theory* ◊ C45 C50 C57 C60 ◊

MAIER, B.J. *Existenziell abgeschlossene lokal endliche p-Gruppen* ◊ C25 C60 ◊

MANSFIELD, R. *How many slopes in a polygon?* ◊ C60 ◊

MEJREMBEKOV, K.A. *Spectra of superstable abelian groups (Russian)* ◊ C45 C60 ◊

MEKLER, A.H. *Stability of nilpotent groups of class 2 and prime exponent* ◊ C45 C60 ◊

MOCKOR, J. *The completion of valued fields and nonstandard models* ◊ C60 H20 ◊

MYASNIKOV, A.G. & REMESLENNIKOV, V.N. *Isomorphisms and elementary properties of nilpotent power groups (Russian)* ◊ C60 ◊

OGER, F. *Equivalence elementaire et genre de groupes finis-par-abeliens de type fini (English summary)* ◊ C60 ◊

PLOTKIN, J.M. *ZF and locally finite groups* ◊ C60 C62 E25 E35 E75 ◊

PODEWSKI, K.-P. & REINEKE, J. *On algebraically closed commutative indecomposable rings* ◊ C25 C35 C60 ◊

POINT, F. *Elimination des quantificateurs dans les L-anneaux *-reguliers et application aux anneaux bireguliers* ◊ C10 C60 C90 ◊

POIZAT, B. *Sous-groupes definissables d'un groupe stable* ◊ C45 C60 ◊

PREST, M. *Quantifier elimination for modules* ◊ C10 C60 ◊

PRESTEL, A. *Pseudo real closed fields* ◊ C10 C25 C35 C52 C60 ◊

PRESTEL, A. *Zur Axiomatisierung gewisser affiner Geometrien* ◊ C60 C65 ◊

RABINOVICH, E.B. *Inductive limits of symmetric groups and universal groups (Russian) (English summary)* ◊ C15 C50 C60 ◊

ROCHE LA, P. *Effective Galois theory* ◊ C57 C60 D45 F60 ◊

ROGERS, P. *Preservation of saturation and stability in a variety of nilpotent groups* ◊ C45 C50 C60 ◊

ROMO SANTOS, C. *Two methods for the resolution of singularities in any characteristic: ultraproducts and maximal contact (Spanish)* ◊ C20 C60 ◊

ROTHMALER, P. *Q_0 is eliminable in every complete theory of modules* ◊ C10 C60 C80 ◊

SEBEL'DIN, A.M. *Definability of a torsion-free unreduced abelian group by its group of endomorphisms (Russian)* ◊ C60 E50 E75 ◊

SHELAH, S. *On endo-rigid strongly $\aleph_1$-free abelian groups in $\aleph_1$* ◊ C50 C55 C60 E05 E50 E75 ◊

SLOBODSKOJ, A.M. *Unsolvability of the universal theory of finite groups (Russian)* ◊ C13 C60 D35 D40 ◊

SMITH, RICK L. *Effective valuation theory* ◊ C57 C60 D45 F55 ◊

SMITH, RICK L. *Two theorems on autostability in p-groups* ⋄ C57 C60 ⋄

TYUKAVKIN, L.V. *Axiomatizability of the class of completely reducible modules (Russian)* ⋄ C60 ⋄

TYUKAVKIN, L.V. *Axiomatizability of the class of irreducible modules (Russian)* ⋄ C60 ⋄

WEISPFENNING, V. *Elimination of quantifiers for certain ordered and lattice-ordered abelian groups* ⋄ B25 C10 C60 ⋄

WILKIE, A.J. *On discretely ordered rings in which every definable ideal is principal* ⋄ C60 C62 H15 ⋄

ZEITLER, H. *Die Amalgamierungseigenschaft und reine Ringe* ⋄ C52 C60 ⋄

ZYAPKOV, N.P. *Reduction theorems for universally consistent fields (Russian)* ⋄ C60 ⋄

1982

ANELLIS, I.H. *Boolean groups* ⋄ C30 C60 G05 ⋄

APPS, A.B. *Boolean powers of groups* ⋄ C13 C30 C60 ⋄

BALDWIN, JOHN T. *Recursion theory and abstract dependence* ⋄ C57 C60 D25 D45 ⋄

BAUDISCH, A. *Decidability and stability of free nilpotent Lie algebras and free nilpotent p-groups of finite exponent* ⋄ B25 C45 C60 ⋄

BAUR, W. *Die Theorie der Paare reell abgeschlossener Koerper* ⋄ B25 C60 ⋄

BAUR, W. *On the elementary theory of pairs of real closed fields II* ⋄ B25 C60 ⋄

BAUVAL, A. *Une condition necessaire d'equivalence elementaire entre anneaux de polynomes sur des corps (English summary)* ⋄ C60 ⋄

BELYAEV, V.YA. *On algebraic closure and amalgamation of semigroups* ⋄ C05 C25 C60 ⋄

BUDKIN, A.I. *Independent axiomatizability of quasivarieties of groups (Russian)* ⋄ C05 C60 ⋄

BUNGE, MARTA C. *On the transfer of an abstract Nullstellensatz* ⋄ C35 C60 C90 G30 ⋄

BURRIS, S. *The first order theory of Boolean algebras with a distinguished group of automorphisms* ⋄ B25 C05 C60 G05 ⋄

BURRIS, S. & LAWRENCE, J. *Two undecidability results using modified boolean powers* ⋄ C05 C60 D35 G05 ⋄

CAMERON, P.J. *Orbits and enumeration* ⋄ C30 C60 ⋄

CHERLIN, G.L. & FELGNER, U. *Quantifier eliminable groups* ⋄ C10 C13 C60 ⋄

COLLINS, G.E. *Quantifier elimination for real closed fields: a guide to the literature* ⋄ C10 C60 C98 ⋄

DAHN, B.I. *The limit behaviour of exponential terms* ⋄ C60 C65 ⋄

DAI, ZHIZHONG *Henselian valued fields (Chinese) (English summary)* ⋄ C60 ⋄

DELZELL, C.N. *Case distinctions are necessary for representing polynomials as sums of squares* ⋄ A10 C60 C65 F55 ⋄

DELZELL, C.N. *Continuous sums of squares of forms* ⋄ C60 C65 F55 ⋄

DRIES VAN DEN, L. & GLASS, A.M.W. & MACINTYRE, A. & MEKLER, A.H. & POLAND, J. *Elementary equivalence and the commutator subgroup* ⋄ C60 ⋄

DRIES VAN DEN, L. *Some applications of a model theoretic fact to (semi-)algebraic geometry* ⋄ C10 C60 C65 ⋄

DUGAS, M. & GOEBEL, R. *Every cotorsion-free algebra is an endomorphism algebra* ⋄ C55 C60 E75 ⋄

DUGAS, M. & GOEBEL, R. *Every cotorsion-free ring is an endomorphism ring* ⋄ C55 C60 E45 E65 E75 ⋄

EDA, K. *A boolean power and a direct product of abelian groups* ⋄ C30 C60 C90 E40 E75 ⋄

EKLOF, P.C. *Homological dimension and stationary sets* ⋄ C60 E05 E75 ⋄

ERSHOV, YU.L. *Absolute irreducibility and Henselization properties (Russian)* ⋄ C60 ⋄

ERSHOV, YU.L. *Algorithmic problems in the theory of fields (positive aspects) (Russian)* ⋄ B25 C10 C60 C98 ⋄

ERSHOV, YU.L. *Multiply valued fields (Russian)* ⋄ B25 C25 C35 C60 C98 ⋄

ERSHOV, YU.L. *Regularly r-closed fields (Russian)* ⋄ B25 C25 C35 C60 ⋄

ERSHOV, YU.L. *Totally real field extensions (Russian)* ⋄ B25 C52 C60 ⋄

GONDARD, D. *Theorie des modeles et fonctions definies positives sur les varietes algebriques reelles (English summary)* ⋄ C35 C60 ⋄

GORBACHUK, E.L. & KOMARNITS'KIJ, N.YA. *On the axiomatizability of radical and semisimple classes of modules and abelian groups (Russian)* ⋄ C52 C60 ⋄

HARAN, D. & JARDEN, M. *Bounded statements in the theory of algebraically closed fields with distinguished automorphisms* ⋄ B25 C60 ⋄

HARAN, D. & LUBOTZKY, A. *Embedding covers and the theory of Frobenius fields* ⋄ B25 C60 ⋄

HICKMAN, J.L. *Automorphisms of medial fields* ⋄ C07 C60 E25 E75 ⋄

HOEPPNER, M. *On the freeness of Whitehead-diagrams* ⋄ C60 ⋄

HUBER-DYSON, V. *Symmetric groups and the open sentence problem* ⋄ C13 C60 D35 ⋄

JARDEN, M. *The elementary theory of large e-fold ordered fields* ⋄ C52 C60 ⋄

JENSEN, C.U. & LENZING, H. *Homological dimensions and representation type of algebras under base field extension* ⋄ C20 C60 ⋄

JENSEN, C.U. *La dimension globale de l'anneau des fonctions entieres (English summary)* ⋄ C60 C65 E35 E75 ⋄

JENSEN, C.U. *Sur une classe de corps indecidables (English summary)* ⋄ C60 D35 ⋄

JURIE, P.-F. *Decidabilite de la theorie elementaire des anneaux booleiens a operateurs dans un group fini (English summary)* ⋄ B25 C60 ⋄

KLEIMAN, J.G. *On identities in groups (Russian)* ⋄ C52 C60 D40 ⋄

KOMATSU, H. *On the equational definability of addition in rings* ⋄ C05 C60 ⋄

KOPYTOV, V.M. *Nilpotent lattice ordered groups (Russian)* ⋄ C60 ⋄

KUSRAEV, A.G. *Boolean-valued analysis of duality of extended modules (Russian)* ⋄ C60 C65 C90 E40 E75 ⋄

LEONARDO DI, M.V. *On formal axiomatization of algebraic structures (Italian)* ⋄ C60 ⋄

MACINTYRE, A. *Residue fields of models of P*
⋄ C57 C60 C62 H15 ⋄

MANN, A. *A note on recursively presented and co-recursively presented groups* ⋄ C60 D40 D45 ⋄

MCKENZIE, R. *Subdirect powers of non-abelian groups*
⋄ C30 C60 ⋄

MEKLER, A.H. *Primitive rings are not definable in $L_{\infty\infty}$*
⋄ C60 C75 ⋄

MEZ, H.-C. *Existentially closed linear groups*
⋄ C25 C35 C60 ⋄

MIJAJLOVIC, Z. *Completions of models and Galois theory*
⋄ C60 ⋄

MOLZAN, B. *How to eliminate quantifiers in the elementary theory of p-rings* ⋄ B25 C60 ⋄

MYASNIKOV, A.G. & REMESLENNIKOV, V.N. *Classification of nilpotent powered groups according to elementary properties (Russian)* ⋄ C60 D40 ⋄

MYASNIKOV, A.G. & REMESLENNIKOV, V.N. *Definability of the set of Mal'tsev bases and elementary theories of finite-dimensional algebras I (Russian)* ⋄ B25 C60 ⋄

NOSKOV, G.A. *On conjugacy in metabelian groups (Russian)* ⋄ C57 C60 ⋄

OGER, F. *Des groupes nilpotents de classe 2 sans toraion de type fini ayant les memes images finies peuvent ne pas etre elementairement equivalents* ⋄ C60 ⋄

OGER, F. *Equivalence elementaire entre groupes finis-par-abeliens de type fini* ⋄ C60 ⋄

PASINI, L. *Sulla chiusura reale dei campi generalizzati di Hardy (English summary)* ⋄ C60 C65 H20 ⋄

PELC, A. *Semiregular invariant measures on abelian groups* ⋄ C60 E55 E75 ⋄

POINT, F. *Sur l'elimination lineaire (English summary)*
⋄ C60 ⋄

PREST, M. *Elementary equivalence of Σ-injective modules*
⋄ C60 ⋄

PRESTEL, A. *Decidable theories of preordered fields*
⋄ B25 C25 C35 C60 ⋄

REINEKE, J. *On algebraically closed models of theories of commutative rings* ⋄ C25 C60 ⋄

RICHTER, M.M. *Ideale Punkte, Monaden und Nichtstandard-Methoden*
⋄ C20 C60 E70 H05 H98 ⋄

RIPS, E. *Subgroups of small cancellation groups* ⋄ C60 ⋄

ROLLAND, R. *Etude des coupures dans les groupes et corps ordonnes* ⋄ C60 C65 ⋄

ROMO SANTOS, C. *Ultraproducts of nonsingular varietes (Spanish)* ⋄ C20 C60 ⋄

ROSE, B.I. *Preservation of elementary equivalence under scalar extension* ⋄ C60 ⋄

SARACINO, D.H. & WOOD, C. *QE nil-2 groups of exponent 4* ⋄ C10 C15 C60 ⋄

SCHMITT, P.H. *The elementary theory of torsion-free abelian groups with a predicate specifying a subgroup*
⋄ B25 C60 ⋄

TAHA, F. *Algebres simples centrales sur les corps ultra-produits de corps p-adiques* ⋄ C20 C60 ⋄

TAKAHASHI, MICHIHIRO *Extensions of semimodules. I*
⋄ C60 G30 ⋄

TAMAGAWA, T. *On regularly closed fields* ⋄ C60 ⋄

TOFFALORI, C. *Stabilita, categoricita ed eliminazione dei quantificatori per una classe di anelli locali*
⋄ C10 C35 C45 C60 ⋄

TOFFALORI, C. *Strutture esistenzialmente complete per certe classi di anelli (English summary)*
⋄ C20 C25 C60 ⋄

TOFFALORI, C. *Teoria dei modelli per alcune classi di anelli locali* ⋄ C10 C25 C35 C60 ⋄

TUSCHIK, H.-P. *Elimination of cardinality quantifiers*
⋄ C10 C45 C55 C60 C80 ⋄

TYUKAVKIN, L.V. *On the model completion of certain theories of modules (Russian)* ⋄ C35 C60 ⋄

WANG, SHIQIANG *A class of commutative rings with Goldbach property (Chinese) (English summary)*
⋄ C60 ⋄

WANG, SHIQIANG *A class of commutative rings with prime formulas (Chinese)* ⋄ C60 ⋄

WANG, SHIQIANG & WU, TAO *Extensions of quadratic rings of algebraic integers without the Goldbach property (Chinese) (English summary)* ⋄ C60 ⋄

WEISPFENNING, V. *Valuation rings and Boolean products*
⋄ C10 C30 C60 ⋄

WEISSAUER, R. *Der Hilbertsche Irreduzibilitaetssatz*
⋄ C60 H20 ⋄

WILSON, J.S. *The algebraic structure of $\aleph_0$-categorical groups* ⋄ C15 C35 C60 ⋄

ZAYED, M. *Characterisation des algebres de representation finie sur des corps algebriquement clos* ⋄ C20 C60 ⋄

ZIEGLER, M. *Einige unentscheidbare Koerpertheorien*
⋄ C60 D35 ⋄

ZIL'BER, B.I. *Uncountable categorical nilpotent groups and Lie algebras (Russian)* ⋄ C35 C60 ⋄

1983

APPS, A.B. *$\aleph_0$-categorical finite extensions of boolean powers* ⋄ C30 C35 C60 ⋄

APPS, A.B. *On $\aleph_0$-categorical class two groups*
⋄ C35 C60 ⋄

APPS, A.B. *On the structure of $\aleph_0$-categorical groups*
⋄ C35 C60 ⋄

BAUDISCH, A. *A remark on $\aleph_0$-categorical stable groups*
⋄ C35 C45 C60 ⋄

BECKER, T. *Real closed rings and ordered valuation rings*
⋄ C60 ⋄

BERLINE, C. *Deviation des types dans les corps algebriquement clos* ⋄ C45 C60 ⋄

BERLINE, C. & CHERLIN, G.L. *QE rings in characteristic p^n* ⋄ C10 C60 ⋄

CAMERON, P.J. *Orbits of permutation groups on unordered sets. III: Imprimitive groups. IV: Homogeneity and transitivity* ⋄ C30 C60 E05 ⋄

CHERLIN, G.L. & SCHMITT, P.H. *Locally pure topological abelian groups: Elementary invariants*
⋄ B25 C60 C65 C80 C90 ⋄

CHERLIN, G.L. & DICKMANN, M.A. *Real closed rings II: Model theory* ⋄ B25 C10 C60 ⋄

COWLES, J.R. & LAGRANGE, R. *Generalized archimedean fields* ⋄ C60 ⋄

DAHN, B.I. & GEHNE, J. *A theory of ordered exponential fields which has no model completion*
⋄ C35 C60 C65 ⋄

DAHN, B.I. & WOLTER, H. *On the theory of exponential fields* ⋄ C25 C60 C65 ⋄

DENEF, J. & JARDEN, M. & LEWIS, D. *On Ax-fields which are C_i* ⋄ B25 C60 ⋄

DOBRITSA, V.P. *Some constructivizations of abelian groups (Russian)* ⋄ C57 C60 D45 ⋄

DOWNEY, R.G. *On a question of A.Retzlaff*
⋄ C57 C60 D25 D45 ⋄

DRIES VAN DEN, L. & MACINTYRE, A. & MCKENNA, K. *Elimination of quantifiers in algebraic structures*
⋄ C10 C60 ⋄

DRIES VAN DEN, L. *Reducing to prime characteristic by means of Artin approximation and constructible properties, and applied to Hochster algebras* ⋄ C60 ⋄

DUGAS, M. & HERDEN, G. *Abitrary torsion classes and almost free abelian groups*
⋄ C55 C60 E50 E55 E65 E75 ⋄

DUGAS, M. & HERDEN, G. *Arbitrary torsion classes of abelian groups* ⋄ C55 C60 E55 E65 E75 ⋄

DURET, J.-L. *Stabilite des corps separablement clos*
⋄ C35 C45 C60 ⋄

EDA, K. *On a boolean power of a torsion free abelian group* ⋄ C30 C60 C90 E40 E75 G05 ⋄

EKLOF, P.C. & MEKLER, A.H. *On endomorphism rings of ω_1-separable primary groups*
⋄ C60 E45 E50 E75 ⋄

EKLOF, P.C. *Set theory and structure theorems*
⋄ C60 C98 E50 E65 E75 ⋄

EKLOF, P.C. *The structure of ω_1-separable groups*
⋄ C60 E35 E50 E75 ⋄

ERSHOV, YU.L. *Algebraic properties of regularly closed fields (Russian)* ⋄ C60 ⋄

ERSHOV, YU.L. *Regularly r-closed fields (Russian)*
⋄ B25 C25 C35 C60 ⋄

ESPANOL, L. *First-order aspects of *-rings and constructive version of the Gauss lemma (Spanish) (English summary)* ⋄ C60 ⋄

FORREST, W.K. *The theory of affine constructible sets*
⋄ C60 C65 ⋄

FREY, G. *Nonstandard arithmetic and application to height functions* ⋄ C60 H10 H15 H20 ⋄

GRAMAIN, F. *Non-minimalite de la cloture differentielle I: La preuve de S.Shelah* ⋄ C50 C60 C65 ⋄

GRAMAIN, F. *Non-minimalite de la cloture differentielle II: La preuve de M.Rosenlicht* ⋄ C50 C60 C65 ⋄

GROSSBERG, R. & SHELAH, S. *On universal locally finite groups* ⋄ C60 ⋄

GUICHARD, D.R. *Automorphisms of substructure lattices in recursive algebra* ⋄ C57 C60 D45 G05 ⋄

HEINTZ, JOOS *Definability and fast quantifier elimination in algebraically closed fields*
⋄ B25 C10 C40 C60 D10 D15 ⋄

HICKIN, K.K. & PHILLIPS, R.E. *Isomorphism types in wreath products and effective embeddings of periodic groups* ⋄ C57 C60 D40 ⋄

HOLLAND, W.C. *A survey of varieties of lattice ordered groups* ⋄ C05 C60 C98 ⋄

HUBER, M. *Methods of set theory and the abundance of separable abelian p-groups* ⋄ C55 C60 E05 E75 ⋄

HUBER, M. *On reflexive modules and abelian groups*
⋄ C55 C60 E50 E55 E75 G30 ⋄

IVANOV, A.A. *Decidability of theories in a certain calculus (Russian)* ⋄ B25 C10 C55 C60 C80 D35 ⋄

IVANOV, A.A. *Some theories in generalized calculi (Russian)* ⋄ B25 C10 C55 C60 C80 ⋄

JARDEN, M. & WHEELER, W.H. *Model-complete theories of e-free Ax fields* ⋄ C35 C60 ⋄

JARDEN, M. *On the model companion of the theory of e-fold ordered fields* ⋄ B25 C25 C35 C60 ⋄

JARDEN, M. & SHELAH, S. *Pseudo-algebraically closed fields over rational function fields* ⋄ C60 ⋄

JARDEN, M. *The elementary theory of normal Frobenius fields* ⋄ C60 ⋄

KHARLAMPOVICH, O.G. *The universal theory of the class of finite nilpotent groups is undecidable (Russian)*
⋄ C60 D35 ⋄

KHISAMIEV, N.G. *Strongly constructive abelian p-groups (Russian)* ⋄ C57 C60 D45 F60 ⋄

LASCAR, D. *Les corps differentiellement clos denombrables* ⋄ C15 C45 C60 ⋄

LAWRENCE, J. *The theory of $\forall\exists$ elementary conditions on rings* ⋄ C60 ⋄

LIPPARINI, P. *On Hilbert's Nullstellensatz (Italian)*
⋄ C05 C60 ⋄

LUO, LIBO *The τ-theory for free groups is undecidable*
⋄ C35 C60 D35 ⋄

MACINTYRE, A. *Decision problems for exponential rings: the p-adic case* ⋄ B25 C60 ⋄

MACPHERSON, H.D. *The action of an infinite permutation group on unordered subsets of a set*
⋄ C50 C60 E05 ⋄

MAIER, B.J. *On existentially closed and generic nilpotent groups* ⋄ C25 C60 ⋄

MEGIBBEN, C. *Crawley's problem on the unique ω-elongation of p-groups is undecidable*
⋄ C55 C60 E35 E45 E50 E65 E75 ⋄

MEKLER, A.H. *Proper forcing and abelian groups*
⋄ C55 C60 E35 E50 E75 ⋄

MLCEK, J. *Compactness and homogeneity of saturated structures I,II* ⋄ C35 C50 C60 C62 ⋄

MYASNIKOV, A.G. & REMESLENNIKOV, V.N. *Definability of sets of Mal'tsev bases and elementary theories of finite-dimensional algebras II (Russian)* ⋄ C60 ⋄

NOSKOV, G.A. *Elementary theory of a finitely generated commutative ring (Russian)* ⋄ C60 D35 ⋄

NOSKOV, G.A. *The elementary theory of a finitely generated almost solvable group (Russian)*
⋄ B25 C60 D35 ⋄

OGER, F. *Cancellation and elementary equivalence of groups* ⋄ C60 ⋄

ONO, H. *Equational theories and universal theories of fields* ⋄ B25 C05 C60 ⋄

OZHIGOV, YU.I. *Equations with two unknowns in a free group (Russian)* ⋄ C60 ⋄

PILLAY, A. & PREST, M. *Forking and pushouts in modules*
⋄ C45 C60 ⋄

POIZAT, B. *C'est beau et chaud* ⋄ C45 C60 C98 ⋄
POIZAT, B. *Groupes stables, avec types generiques reguliers* ⋄ C45 C60 ⋄
POIZAT, B. *Une theorie de Galois imaginaire* ⋄ C45 C60 ⋄
POTTHOFF, K. *Quelques applications des methodes non-standard a la theorie des groupes* ⋄ C60 H05 H20 ⋄
PREST, M. *Existentially complete prime rings* ⋄ C25 C60 ⋄
PRESTEL, A. & ROQUETTE, P. *Lectures on formally p-adic fields* ⋄ B25 C10 C35 C60 C98 ⋄
REMESLENNIKOV, V.N. & ROMAN'KOV, V.A. *Model-theoretic and algorithmic questions of group theory (Russian)* ⋄ B25 C60 C98 D15 D30 D40 ⋄
REPIN, N.N. *Equations with one unknown in nilpotent groups (Russian)* ⋄ C60 ⋄
ROTHMALER, P. *Some model theory of modules. I: On total transcendence of modules. II: On stability and categoricity of flat modules* ⋄ C35 C45 C60 ⋄
ROTHMALER, P. *Stationary types in modules* ⋄ C35 C45 C60 ⋄
SARACINO, D.H. *Amalgamation bases for nil-2 groups* ⋄ C60 ⋄
SARACINO, D.H. & WOOD, C. *Finitely generic abelian lattice-ordered groups* ⋄ C25 C60 ⋄
SCHMITT, P.H. *The L^t-theory of profinite abelian groups* ⋄ B25 C50 C60 C80 C90 ⋄
SHELAH, S. *Models with second order properties IV. A general method and eliminating diamonds* ⋄ C50 C55 C60 C65 C80 E65 G05 ⋄
SHELAH, S. *On the number of non conjugate subgroups* ⋄ C55 C60 E50 E75 ⋄
TOFFALORI, C. *Differentially closed rings for some classes of differential rings (Italian) (English summary)* ⋄ C60 ⋄
TOFFALORI, C. *Questioni di teoria dei modelli per coppie di campi* ⋄ C10 C35 C45 C60 ⋄
TOFFALORI, C. *Teoria dei modelli per una classe di anelli differenziali* ⋄ C25 C60 ⋄
VAZHENIN, YU.M. *Semigroups with one defining relation whose elementary theories are decidable (Russian)* ⋄ B25 C05 C60 ⋄
WALD, B. *On κ-products modulo μ-products* ⋄ C20 C60 E45 E75 ⋄
WHEELER, W.H. *Model complete theories of formally real fields and formally p-adic fields* ⋄ C35 C60 ⋄
WOLTER, H. *On the "problem of the last root" for exponential terms* ⋄ C60 C65 ⋄
ZAMYATIN, A.P. *Decidability of the elementary theories of certain varieties of rings (Russian)* ⋄ B25 C05 C60 ⋄

1984

ABDRAZAKOV, K.T. & KHISAMIEV, N.G. *A criterion for strong constructivizability of a class of abelian p-groups (Russian)* ⋄ C57 C60 D45 ⋄
ASH, C.J. & DOWNEY, R.G. *Decidable subspaces and recursively enumerable subspaces* ⋄ C57 C60 D35 D45 ⋄
BALDWIN, JOHN T. *First-order theories of abstract dependence relations* ⋄ C35 C45 C57 C60 D25 D45 ⋄
BASARAB, S.A. *Axioms for pseudo-real-closed fields* ⋄ C25 C60 ⋄
BASARAB, S.A. *Definite functions on algebraic varieties over ordered fields* ⋄ C25 C60 ⋄
BASARAB, S.A. *On some classes of Hilbertian fields* ⋄ B25 C10 C25 C35 C60 ⋄
BAUDISCH, A. *Magidor-Malitz quantifiers in modules* ⋄ C10 C60 C80 ⋄
BAUDISCH, A. *On elementary properties of free Lie algebras* ⋄ C45 C60 ⋄
BAUDISCH, A. *Tensor products of modules and elementary equivalence* ⋄ C30 C52 C60 ⋄
BAUDISCH, A. & ROTHMALER, P. *The stratified order in modules* ⋄ C45 C60 ⋄
BAUVAL, A. *La theorie d'un anneau de polynomes* ⋄ C60 C85 ⋄
BECKER, E. *Extended Artin-Schreier theory of fields* ⋄ B25 C35 C60 ⋄
BELYAEV, V.V. *Locally finite Chevalley groups (Russian)* ⋄ C60 ⋄
BELYAEV, V.YA. *Uncountable extensions of countable algebraically closed semigroups (Russian)* ⋄ C05 C55 C60 C75 ⋄
BILLINGTON, N. *Growth of groups and graded algebras* ⋄ C60 D40 ⋄
BUDKIN, A.I. *Quasiidentities and direct wreaths of groups (Russian)* ⋄ C05 C30 C60 ⋄
BURNS, R.G. & OKOH, F. & SMITH, HOWARD & WIEGOLD, J. *On the number of normal subgroups of an uncountable soluble group* ⋄ C55 C60 E35 E50 E75 ⋄
CAMERON, P.J. *Aspects of random graph* ⋄ C07 C60 C65 ⋄
CANTOR, DAVID G. & ROQUETTE, P. *On diophantine equations over the ring of all algebraic integers* ⋄ B25 C60 D35 ⋄
CHERLIN, G.L. *Decidable theories of pseudo-algebraically closed fields* ⋄ B25 C60 ⋄
CHERLIN, G.L. *Definability in power series rings of nonzero characteristic* ⋄ C40 C60 D35 ⋄
CHERLIN, G.L. *Undecidability of rational function fields in nonzero characteristic* ⋄ C60 C85 D35 ⋄
CHISTOV, A.L. & GRIGOR'EV, D.YU. *Complexity of quantifier elimination in theory of algebraically closed fields* ⋄ C10 C60 D15 ⋄
CONTESSA, M. *Ultraproducts of pm-rings and mp-rings* ⋄ C20 C60 G05 ⋄
COQUAND, T. *Le theoreme de representation d'Arens et Kaplansky* ⋄ C60 C98 ⋄
DAHN, B.I. & GOERING, P. *Notes on exponential-logarithmic terms* ⋄ C60 C65 ⋄
DAHN, B.I. *The limit behaviour of exponential terms* ⋄ C60 C65 D15 ⋄
DELON, F. *Corps equivalents a leurs corps de series* ⋄ C25 C60 ⋄
DELON, F. *Theories completes de corps* ⋄ C35 C60 C98 ⋄

DELZELL, C.N. *A continuous, constructive solution to Hilbert's 17th problem* ⋄ C60 F55 ⋄

DENEF, J. & LIPSHITZ, L. *Power series solutions of algebraic differential equations* ⋄ B25 C60 C65 D35 D80 ⋄

DENEF, J. *The rationality of the Poincare series associated to the p-adic points on a variety* ⋄ C10 C60 ⋄

DIARRA, B. *Ultraproduits ultrametriques de corps values* ⋄ C20 C60 ⋄

DOWNEY, R.G. *A note on decompositions of recursively enumerable subspaces* ⋄ C57 C60 D25 D45 ⋄

DOWNEY, R.G. *Bases of supermaximal subspaces and Steinitz systems. I* ⋄ C57 C60 D25 D45 ⋄

DOWNEY, R.G. *Co-immune subspaces and complementation in V_∞* ⋄ C57 C60 D45 D50 ⋄

DOWNEY, R.G. & REMMEL, J.B. *The universal complementation property* ⋄ C57 C60 D25 D45 ⋄

DRIES VAN DEN, L. *Algebraic theories with definable Skolem functions* ⋄ C10 C40 C60 ⋄

DRIES VAN DEN, L. & SCHMIDT, K. *Bounds in the theory of polynomial rings over fields. A nonstandard approach* ⋄ C60 H10 H20 ⋄

DRIES VAN DEN, L. *Exponential rings, exponential polynomials and exponential functions* ⋄ C60 C65 ⋄

DRIES VAN DEN, L. & WILKIE, A.J. *Gromov's theorem on groups of polynomial growth and elementary logic* ⋄ C60 H20 ⋄

DRIES VAN DEN, L. *Remarks on Tarski's problem concerning $(\mathbb{R}, +, \cdot, exp)$* ⋄ C10 C60 C65 ⋄

DUBOIS, DONALD WARD & RECIO MUNIZ, T.J. *A note on Robinson's non-negativity criterion* ⋄ C60 ⋄

DUGAS, M. & GOEBEL, R. *Almost Σ-cyclic abelian p-groups in L* ⋄ C55 C60 E45 E75 ⋄

DUGAS, M. & GOEBEL, R. & GOLDSMITH, B. *Representation of algebras over a complete discrete valuation ring* ⋄ C55 C60 E75 ⋄

EDA, K. *Maximal quotient rings and boolean extensions (Japanese)* ⋄ C60 C90 E40 E75 ⋄

EDA, K. & HIBINO, K. *On boolean powers of the group Z and (ω, ω)-weak distributivity* ⋄ C30 C60 C90 E35 E50 E75 G05 ⋄

EKLOF, P.C. & MEZ, H.-C. *Additive groups of existentially closed rings* ⋄ C25 C60 ⋄

EKLOF, P.C. & MEKLER, A.H. & SHELAH, S. *Almost disjoint abelian groups* ⋄ C55 C60 E45 E50 E75 ⋄

ERSHOV, YU.L. *Regularly r-closed fields with weakly universal Galois group (Russian)* ⋄ B25 C60 D35 ⋄

ERSHOV, YU.L. *Two theorems on regularly r-closed fields* ⋄ C60 ⋄

FELGNER, U. & SCHULZ, KLAUS *Algebraische Konsequenzen des Determiniertheits-Axioms* ⋄ C60 E60 E75 ⋄

FRIED, M. & HARAN, D. & JARDEN, M. *Galois stratification over Frobenius fields* ⋄ B25 C60 ⋄

FUNK, M. & KEGEL, O.H. & STRAMBACH, K. *On group universality and homogenity* ⋄ C50 C60 ⋄

GILMAN, R.H. *Characteristically simple $\aleph_0$-categorical groups* ⋄ C35 C60 ⋄

GIORGETTA, D. & SHELAH, S. *Existentially closed structures in the power of the continuum* ⋄ C15 C25 C30 C52 C55 C60 C75 ⋄

GLASS, A.M.W. *The isomorphism problem and undecidable properties for finitely presented lattice-ordered groups* ⋄ C60 D40 G10 ⋄

GOEBEL, R. *The existence of rigid systems of maximal size* ⋄ C60 E75 ⋄

GUICHARD, D.R. *A note on r-maximal subspaces of V_∞* ⋄ C57 C60 D45 ⋄

GUPTA, N. *Recursively presented two generated infinite p-groups* ⋄ C57 C60 ⋄

GUREVICH, Y. & SCHMITT, P.H. *The theory of ordered abelian groups does not have the independence property* ⋄ C45 C60 ⋄

HARAN, D. *The undecidability of pseudo-real-closed fields* ⋄ B25 C60 D35 ⋄

HEINEMANN, B. & PRESTEL, A. *Fields regularly closed with respect to finitely many valuations and orderings* ⋄ C35 C60 ⋄

HODGES, W. *Finite extensions of finite groups* ⋄ C13 C50 C60 D35 ⋄

HODGES, W. *Groupes nilpotents existentiellement clos de classe fixee* ⋄ C25 C60 ⋄

HOLLAND, W.C. *Classification of lattice ordered groups (French summary)* ⋄ C05 C60 C98 ⋄

JENSEN, C.U. *Theorie des modeles pour des anneaux de fonctions entieres et des corps de fonctions meromorphes* ⋄ C60 C65 D35 F35 ⋄

KHARLAMPOVICH, O.G. *The universal theory of certain classes of Lie rings (Russian)* ⋄ C60 ⋄

KHISAMIEV, N.G. *Connection between constructivizability and strong constructivizability for different classes of abelian groups (Russian)* ⋄ C57 C60 D45 ⋄

KREMER, E.M. *Modules with an almost categorical theory (Russian)* ⋄ C35 C45 C60 ⋄

KREMER, E.M. *Rings over which all modules of a given type are almost categorical (Russian)* ⋄ C35 C45 C60 ⋄

KRYNICKI, M. & LACHLAN, A.H. & VAEAENAENEN, J. *Vector spaces and binary quantifiers* ⋄ C60 C75 C80 ⋄

KUHLMANN, F.V. & PRESTEL, A. *Places of algebraic function fields* ⋄ C60 ⋄

LENSKI, W. *Elimination of quantifiers for the theory of archimedean ordered divisible groups in a logic with Ramsey quantifiers* ⋄ C10 C55 C60 C80 ⋄

MAIER, B.J. *Amalgame nilpotenter Guppen der Klasse zwei* ⋄ C60 ⋄

MAIER, B.J. *Amalgams of torsion-free nilpotent groups of class three* ⋄ C60 ⋄

MAIER, B.J. *Existentially closed torsion-free nilpotent groups of class three* ⋄ C25 C60 ⋄

MAKANIN, G.S. *Decidability of the universal and positive theories of a free group (Russian)* ⋄ B25 C60 ⋄

MATEESCU, C. & POPESCU, D. *Ultraproducts and big Cohen-Macaulay modules* ⋄ C20 C60 ⋄

MYASNIKOV, A.G. & REMESLENNIKOV, V.N. *Finite dimensional algebras and k-groups of finite rank* ⋄ C60 C98 ⋄

OGER, F. *Elementary equivalence and isomorphism of finitely generated nilpotent groups* ⋄ C60 ⋄

OGER, F. *The model theory of finitely generated finite-by-abelian groups* ⋄ C50 C60 ⋄

PHILLIPS, R.E. *Existentially closed locally finite central extensions; multipliers and local systems* ⋄ C25 C60 ⋄

PILLAY, A. *Countable modules* ⋄ C15 C35 C45 C52 C60 ⋄

PILLAY, A. & STEINHORN, C.I. *Definable sets in ordered structures* ⋄ C40 C60 C65 ⋄

POIZAT, B. *Deux remarques a propos de la propriete de recouvrement fini* ⋄ C45 C50 C60 C62 ⋄

POIZAT, B. *La structure geometrique des groupes stables* ⋄ C45 C60 C98 ⋄

PREST, M. *Rings of finite representation type and modules of finite Morley rank* ⋄ C45 C60 ⋄

PRESTEL, A. *Model theory of fields: An application to positive semidefinite polynomials* ⋄ C60 ⋄

RIBENBOIM, P. *Remarks on existentially closed fields and diophantine equations* ⋄ C25 C60 ⋄

ROGGENKAMP, K.W. & WIEDEMANN, A. *Auslander - Reiten quivers of Schurian orders* ⋄ C60 ⋄

ROQUETTE, P. *Some tendencies in contemporary algebra* ⋄ C10 C35 C60 C98 ⋄

ROSENTHAL, D. *The order indiscernibles of divisible ordered abelian groups* ⋄ C30 C60 ⋄

ROTHMALER, P. *Some model theory of modules III. On infiniteness of sets definable in modules* ⋄ C10 C35 C60 C80 ⋄

RUYER, H. *Deux resultats concernant les modules sur un anneau de Dedekind (English summary)* ⋄ C10 C60 C75 ⋄

SARACINO, D.H. & WOOD, C. *An example in the model theory of abelian lattice-ordered groups* ⋄ C25 C60 G10 ⋄

SARACINO, D.H. & WOOD, C. *Nonexistence of a universal countable commutative ring* ⋄ C50 C60 ⋄

SARACINO, D.H. & WOOD, C. *QE commutative nilrings* ⋄ C10 C50 C60 ⋄

SCHMITT, P.H. *Model- and substructure complete theories of ordered abelian groups* ⋄ C10 C35 C60 ⋄

SCHMITT, P.H. *Undecidable theories of valued Abelian groups* ⋄ C60 D35 ⋄

SHELAH, S. *A combinatorial principle and endomorphism rings. I. On p-groups* ⋄ C60 E65 E75 ⋄

SHELAH, S. *A combinatorial theorem and endomorphism rings of abelian groups II* ⋄ C60 E05 E65 E75 ⋄

SHEN, FUXING & WANG, SHIQIANG & YUE, QIJING *Number-theoretic properties of some Goldbach and non-Goldbach commutative rings (Chinese) (English summary)* ⋄ C60 ⋄

SMITH, KAY *Commutative regular rings and boolean valued fields* ⋄ C60 C90 E40 E75 ⋄

STEPRANS, J. *The number of submodules* ⋄ C60 E35 E50 E75 ⋄

TOFFALORI, C. *Anelli regolari separabilmente chiusi (English summary)* ⋄ C35 C45 C60 ⋄

TOFFALORI, C. *On a class of differential local rings (Italian) (English summary)* ⋄ C25 C60 ⋄

TULIPANI, S. *On the universal theory of classes of finite models* ⋄ C05 C13 C52 C60 ⋄

UMIRBAEV, U.U. *Equality problem for center-by-metabelian Lie-algebras (Russian)* ⋄ B25 C60 D40 ⋄

WANG, CHUSHENG *On ultraproducts in the variety of fields (Chinese)* ⋄ C20 C60 ⋄

WANG, SHIQIANG *A class of commutative rings without Goldbach property (Chinese)* ⋄ C60 ⋄

WANG, SHIQIANG *Extensions with and without Goldbach property of some cubic rings of integers (Chinese)* ⋄ C60 ⋄

WANG, SHIQIANG *Some number-theoretic properties of a kind of Goldbach commutative rings (Chinese)* ⋄ C60 ⋄

WEISPFENNING, V. *Aspects of quantifier elimination in algebra* ⋄ C10 C60 C98 ⋄

WEISPFENNING, V. *Quantifier elimination and decision procedures for valued fields* ⋄ B25 C10 C35 C60 ⋄

WOLTER, H. *Some results about exponential fields (survey)* ⋄ B25 C60 C65 C98 ⋄

ZIEGLER, M. *Model theory of modules* ⋄ B25 C10 C45 C60 ⋄

ZIL'BER, B.I. *Some model theory of simple algebraic groups over algebraically closed fields* ⋄ C60 ⋄

1985

ABIAN, A. *Solvability of infinite systems of polynomial equations over the field of complex numbers* ⋄ C60 ⋄

ADAMOWICZ, Z. *Algebraic approach to $\exists_1$ induction* ⋄ C60 C62 F30 ⋄

APPS, A.B. *Two counterexamples in $\aleph_0$-categorical groups* ⋄ C35 C60 ⋄

BANKSTON, P. & SCHUTT, R. *On minimally free algebras* ⋄ C05 C50 C60 C65 ⋄

BASARAB, S.A. *The absolute Galois group of pseudoreal closed field with finitely many orders* ⋄ C25 C60 ⋄

BASARAB, S.A. *Transfer principles for PRC fields* ⋄ C25 C60 ⋄

BAUDISCH, A. & SEESE, D.G. & TUSCHIK, H.-P. & WEESE, M. *Decidability and quantifier elimination* ⋄ B25 C10 C60 C65 C80 C98 ⋄

BAUDISCH, A. *On Lascar rank in nonmultidimensional ω-stable theories* ⋄ C45 C60 ⋄

BAUVAL, A. *Polynomial rings and weak second-order logic* ⋄ C60 C85 F35 ⋄

BECKER, E. & JACOB, B. *Rational points on algebraic varieties over a generalized real closed field: a model theoretic approach* ⋄ C60 ⋄

BOERGER, R. & RAJAGOPALAN, M. *When do all ring homomorphisms depend on only one co-ordinate?* ⋄ C60 E55 E75 ⋄

BOFFA, M. & MICHAUX, C. & POINT, F. & PRAAG VAN, P. *L'elimination lineaire dans les corps* ⋄ C10 C60 ⋄

BOZOVIC, N.B. *First-order classes of groups having no groups with a given property* ⋄ C60 ⋄

BURRIS, S. *A simple proof of the hereditary undecidability of the theory of lattice ordered abelian groups* ⋄ C05 C60 D35 G10 ⋄

CORNER, A.L.S. & GOEBEL, R. *Prescribing endomorphism algebras, a unified treatment* ◊ C60 ◊

DICKMANN, M.A. *Applications of model theory to real algebraic geometry; a survey* ◊ C10 C60 C98 ◊

DICKMANN, M.A. *Elimination of quantifiers for ordered valuation rings* ◊ C10 C60 ◊

DOWNEY, R.G. & HIRD, G.R. *Automorphisms of supermaximal subspaces* ◊ C57 C60 D25 D45 ◊

DOWNEY, R.G. & KALANTARI, I. *Effective extensions of linear forms on a recursive vector space over a recursive field* ◊ C57 C60 D45 ◊

DRIES VAN DEN, L. & SMITH, RICK L. *Decidable regularly closed fields of algebraic numbers* ◊ B25 C57 C60 ◊

DRIES VAN DEN, L. *The field of reals with a predicate for the powers of two* ◊ B25 C60 C65 ◊

DROSTE, M. & SHELAH, S. *A construction of all normal subgroup lattices of 2-transitive automorphism groups of linearly ordered sets* ◊ C07 C60 C65 E07 ◊

DROSTE, M. *Classes of universal words for the infinite symmetric groups* ◊ C05 C50 C60 ◊

DROSTE, M. *The normal subgroup lattice of 2-transitive automorphism groups of linearly ordered sets* ◊ C07 C60 C65 ◊

DUGAS, M. *On reduced products of abelian groups* ◊ C20 C60 E75 ◊

EKLOF, P.C. *Applications to algebra* ◊ C60 C75 C98 ◊

EKLOF, P.C. & HUBER, M. *On ω-filtered vector spaces and their applications to abelian p-groups* ◊ C60 E75 ◊

EKLOF, P.C. & MEZ, H.-C. *The ideal structure of existentially closed algebras* ◊ C25 C60 ◊

FELGNER, U. *Die Definierbarkeit von K in K(t) im Falle eines pseudo-endlichen Koerpers K (English summary)* ◊ C60 ◊

FUCHS, P.H. & PILZ, G. *Ultraproducts and ultra-limits of near rings* ◊ C20 C60 ◊

FUNK, M. & KEGEL, O.H. & STRAMBACH, K. *Gruppenuniversalitaet und Homogenisierbarkeit (English summary)* ◊ C50 C60 ◊

GARCIA, O.C. & TAYLOR, W. *Generalized commutativity* ◊ C05 C60 ◊

GILMAN, R.H. *An application of ultraproducts to finite groups* ◊ C13 C20 C60 ◊

GLASS, A.M.W. *Effective extensions of lattices-ordered groups that preserve the degree of the conjugacy and the word problems* ◊ C57 C60 D40 ◊

GOEBEL, R. & SHELAH, S. *Semi-rigid classes of cotorsion-free abelian groups* ◊ C55 C60 E05 E45 E55 E65 E75 ◊

HERFORT, W.N. & MANEVITZ, L.M. *Topological Frobenius groups* ◊ C60 H20 ◊

HODGES, W. *Building models by games* ◊ C25 C55 C60 C80 C98 E60 ◊

HODKINSON, I.M. *A construction of many uncountable rings* ◊ C55 C60 E05 E50 E75 ◊

KHISAMIEV, N.G. & KHISAMIEV, Z.G. *Nonconstructivizability of the reduced part of a strongly constructive torsion-free abelian group (Russian)* ◊ C57 C60 D45 ◊

LASCAR, D. *Les groupes ω-stables de rang fini* ◊ C35 C45 C60 ◊

LASLANDES, B. *Modele-compagnons de theories de corps munis de n ordres* ◊ C25 C35 C60 ◊

LAWRENCE, J. *The definability of the commutator subgroup in a variety generated by a finite group* ◊ C60 ◊

LEINEN, F. *Existentially closed groups in locally finite group classes* ◊ C25 C60 ◊

LINDSAY, P.A. *On recognizing cyclic modules effectively* ◊ C60 D45 ◊

LOEB, P.A. *A nonstandard functional approach to Fubini's theorem* ◊ C60 C65 H05 H20 ◊

LYUBETSKIJ, V.A. *Some algebraic questions of nonstandard analysis (Russian)* ◊ C60 C90 H05 ◊

MACPHERSON, H.D. *Growth rates in infinite graphs and permutation groups* ◊ C60 C65 ◊

MACPHERSON, H.D. *Orbits of infinite permutation groups* ◊ C60 ◊

MAKANIN, G.S. *On the decidability of the theory of a free group (Russian)* ◊ B25 C60 C98 D35 ◊

MEDVEDEV, N.YA. *Quasivarieties of l-groups and groups* ◊ C05 C52 C60 ◊

NEUMANN, P.M. *Some primitive permutation groups* ◊ C60 ◊

NIEFIELD, S.B. & ROSENTHAL, K.I. *Strong de Morgan's law and the spectrum of a commutative ring* ◊ C60 C90 G30 ◊

NITA, A. & NITA, C. *Sur l'ultraproduit des anneaux de matrices* ◊ C20 C60 ◊

PAPPAS, P. *A Diophantine problem for Laurent polynomial rings* ◊ C60 D35 D80 ◊

PAPPAS, P. *The model-theoretic structure of abelian group rings* ◊ B25 C20 C60 C90 ◊

POINT, F. *Finitely generic models of T_{UH}, for certain model companionable theories T* ◊ C25 C60 ◊

PREST, M. *The generalised RK-order, orthogonality and regular types for modules* ◊ C45 C60 ◊

PRESTEL, A. *On the axiomatization of PRC-fields* ◊ C52 C60 ◊

RYCHKOV, S.V. *The problem of splitting pure extensions of abelian groups and axiomatic set theory (Russian)* ◊ C60 E35 E45 E50 E75 ◊

SAGEEV, G. & SHELAH, S. *On the structure of $Ext(A, \mathbb{Z})$ in ZFC^+* ◊ C60 E05 E45 E50 E55 E65 E75 ◊

SARACINO, D.H. & WOOD, C. *Finite QE rings in characteristic p^2* ◊ C13 C50 C60 ◊

SCHMERL, J.H. *Models of Peano arithmetic and a question of Sikorski on ordered fields* ◊ C55 C60 C62 E05 E75 ◊

SCHMIDT, K. *Nonstandard methods in field theory. Bounds and definability, and specialization properties in prime characteristic* ◊ C25 C60 H20 ◊

SCHRIEBER, L. *Recursive properties of euclidean domains* ◊ C57 C60 D45 ◊

SHELAH, S. *Incompactness in regular cardinals* ◊ C55 C60 E05 ◊

SHELAH, S. *Uncountable constructions for B.A.,e.c. groups and Banach spaces*
◊ C25 C60 C65 E05 E65 E75 G05 ◊

SHEN, FUXING *Quadratic residues in a sort of commutative rings with Goldbach property (Chinese)* ⋄ C60 ⋄

THOMAS, S. *Complete existentially closed locally finite groups* ⋄ C25 C60 ⋄

THOMAS, S. *The automorphism tower problem* ⋄ C07 C60 E75 ⋄

TOFFALORI, C. *Teorie p-$\aleph_0$-categoriche* ⋄ C35 C45 C60 ⋄

VARADARAJAN, K. *Pseudo-mitotic groups* ⋄ C25 C60 ⋄

WEISPFENNING, V. *Quantifier elimination for modules* ⋄ C10 C35 C57 C60 ⋄

WOLTER, H. *On the "problem of the last root" for exponential terms* ⋄ C60 C65 ⋄

C62 ∪ H15 Models of arithmetic and set theory

1923
SKOLEM, T.A. *Einige Bemerkungen zur axiomatischen Begruendung der Mengenlehre*
⋄ A05 B30 C07 C62 E30 ⋄

1933
SKOLEM, T.A. *Ueber die Unmoeglichkeit einer Charakterisierung der Zahlenreihe mittels eines endlichen Axiomensystems*
⋄ B28 C20 C62 F30 H15 ⋄
TARSKI, A. *Einige Betrachtungen ueber die Begriffe der ω-Widerspruchsfreiheit und der ω-Vollstaendigkeit*
⋄ B28 C62 F30 ⋄

1934
SKOLEM, T.A. *Ueber die Nichtcharakterisierbarkeit der Zahlenreihe mittels endlich oder abzaehlbar unendlich vieler Aussagen mit ausschliesslich Zahlenvariablen*
⋄ B28 C20 C62 F30 H15 ⋄

1941
SKOLEM, T.A. *Sur la portee du theoreme de Loewenheim-Skolem* ⋄ C07 C62 E30 ⋄

1950
ROSSER, J.B. & WANG, HAO *Non-standard models for formal logics*
⋄ B15 C62 E30 E70 F25 H15 H20 ⋄

1951
SHEPHERDSON, J.C. *Inner models for set theory I*
⋄ C62 E35 E55 ⋄

1952
MOLODSHIJ, V.N. *On interrelations of certain assertions of generality with the induction axiom in Peano's system of axioms (Russian)* ⋄ C62 F30 ⋄
MOSTOWSKI, ANDRZEJ *On models of axiomatic systems*
⋄ C62 F25 F30 F35 ⋄
SHEPHERDSON, J.C. *Inner models for set theory II*
⋄ C62 E35 E55 ⋄

1953
SHEPHERDSON, J.C. *Inner models for set theory III*
⋄ C62 E35 E55 ⋄

1954
MCNAUGHTON, R. *A non-standard truth definition*
⋄ B28 C62 E30 H10 H20 ⋄

1955
SKOLEM, T.A. *Peano's axioms and models of arithmetic*
⋄ B28 C20 C62 F30 H15 ⋄

1956
MOSTOWSKI, ANDRZEJ *On models of axiomatic set theory*
⋄ C62 E30 E47 ⋄

1957
MAEHARA, S. *Remark on Skolem's theorem concerning the impossibility of characterization of the natural number sequence* ⋄ B28 C62 F30 ⋄
MENDELSON, E. *Non standard models* ⋄ C62 H15 ⋄
MONTAGUE, R. & VAUGHT, R.L. *Models of set theory*
⋄ C62 E30 E45 ⋄
MOSTOWSKI, ANDRZEJ *On recursive models of formalised arithmetic* ⋄ C57 C62 D45 ⋄
RIEGER, L. *A contribution to Goedel's axiomatic set-theory, I* ⋄ C62 E30 E35 E45 E50 E70 ⋄
SCOTT, D.S. *The notion of rank in set-theory*
⋄ C62 E10 E20 E30 ⋄

1958
GANDY, R.O. *Note on a paper of Kemeny's* ⋄ H15 ⋄
GRZEGORCZYK, A. & MOSTOWSKI, ANDRZEJ & RYLL-NARDZEWSKI, C. *The classical and the ω-complete arithmetic* ⋄ C62 D55 D70 F30 ⋄
KEMENY, J.G. *Undecidable problems of elementary number theory* ⋄ F30 H15 ⋄
MOSTOWSKI, ANDRZEJ *Quelques observations sur l'usage des methodes non finitistes dans la meta-mathematique*
⋄ A05 C30 C62 C80 E30 E35 F99 ⋄
RABIN, M.O. *On recursively enumerable and arithmetic models of set theory* ⋄ C57 C62 D45 ⋄

1959
MONTAGUE, R. & VAUGHT, R.L. *Natural models of set theories* ⋄ C62 E47 E55 E65 ⋄
MOSTOWSKI, ANDRZEJ *A class of models for second order arithmetic* ⋄ C62 ⋄
MOSTOWSKI, ANDRZEJ *Notes concerning an existence proof for standard models (Russian)*
⋄ C62 E30 E45 ⋄
RIEGER, L. *A contribution to Goedel's axiomatic set theory II* ⋄ C62 E30 E70 ⋄
WETTE, E. *Von operativen Modellen der axiomatischen Mengenlehre* ⋄ C62 E35 F99 ⋄

1960
GANDY, R.O. *On a problem of Kleene's*
⋄ C62 D30 D55 ⋄
GANDY, R.O. & KREISEL, G. & TAIT, W.W. *Set existence*
⋄ C62 D55 F35 ⋄
NISHIMURA, T. *Note on axiomatic set theory I. The independence of Zermelo's "Aussonderungsaxiom" from other axioms of set theory* ⋄ C62 E30 E35 ⋄
NISHIMURA, T. *Note on axiomatic set theory II. A construction of a model satisfying the axioms of set*

theory without Zermelo's Aussonderungsaxiom in a certain axiom system of ordinal numbers ◇ C62 E10 E30 E35 ◇

SCOTT, D.S. *On a theorem of Rabin* ◇ C35 C57 C62 D35 ◇

SKOLEM, T.A. *Investigations on a comprehension axiom without negation in the defining propositional functions* ◇ C62 E30 ◇

1961

CHAUVIN, A. *Deux modeles verifiant certains axiomes de la theorie des ensembles de Goedel, et construits dans la theorie des ensembles arithmetiques de Kleene. Construction des modeles* ◇ C62 D55 E30 E70 ◇

CHAUVIN, A. *Deux modeles verifiant certains axiomes de la theorie des ensembles de Goedel, et construits dans la theorie des ensembles arithmetiques de Kleene. Validite des axiomes* ◇ C62 D55 E30 E70 ◇

GANDY, R.O. & KREISEL, G. & TAIT, W.W. *Set existence II* ◇ C62 D55 F35 ◇

KREISEL, G. *Set theoretic problems suggested by the notion of potential totality* ◇ A05 C62 D20 D55 D65 ◇

MACDOWELL, R. & SPECKER, E. *Modelle der Arithmetik* ◇ B28 C20 C55 C62 ◇

MENDELSON, E. *On non-standard models for number theory* ◇ C62 H15 ◇

MOSTOWSKI, ANDRZEJ *Formal systems of analysis based on an infinitistic rule of proof* ◇ C62 E45 F35 ◇

MUELLER, GERT H. *Nicht-Standardmodelle der Zahlentheorie* ◇ C62 F30 ◇

OREY, S. *Relative interpretations* ◇ C62 F25 ◇

RABIN, M.O. *Non-standard models and independence of the induction axiom* ◇ B28 C62 F30 H15 ◇

ROBINSON, A. *Model theory and non-standard arithmetic* ◇ C35 C50 C57 C60 C62 H15 ◇

ROBINSON, A. *Non-standard analysis* ◇ H05 H15 ◇

SCOTT, D.S. *On constructing models for arithmetic* ◇ C20 C57 C62 H15 ◇

1962

GRZEGORCZYK, A. *A kind of categoricity* ◇ C35 C50 C62 C65 ◇

GRZEGORCZYK, A. *A theory without recursive models* ◇ C57 C62 D55 ◇

KRUSE, A.H. *Completion of mathematical systems* ◇ C25 C35 C62 ◇

MAL'TSEV, A.I. *Strongly related models and recursively complete algebras (Russian)* ◇ C57 C60 C62 D45 ◇

RABIN, M.O. *Diophantine equations and non-standard models of arithmetic* ◇ C62 D25 H15 ◇

SCOTT, D.S. *Algebras of sets binumerable in complete extensions of arithmetic* ◇ C40 C62 F30 ◇

VOPENKA, P. *A method of constructing a nonstandard model of the Bernays-Goedel axiomatic theory of sets (Russian)* ◇ C20 C62 E30 E70 H20 ◇

VOPENKA, P. *Construction of models for set theory by the method of ultra products (Russian) (German summary)* ◇ C20 C62 E05 E30 E55 E70 ◇

VOPENKA, P. *Models for set theory (Russian) (German summary)* ◇ C62 E30 E70 ◇

1963

COHEN, P.J. *A minimal model for set theory* ◇ C62 E35 E45 ◇

GRINDLINGER, M.D. *Inductive definition of a function (Russian)* ◇ C40 C62 ◇

HAUSCHILD, K. *Modelle der Mengenlehre, die aus endlichen Mengen bestehen* ◇ C62 E30 E35 ◇

HAUSCHILD, K. *Ueber die Charakterisierbarkeit der Zahlenreihe in gewissen Nichtstandardmodellen der Arithmetik* ◇ H15 ◇

KREISEL, G. *Axiomatic results of second-order arithmetic* ◇ C62 F35 ◇

MONTAGUE, R. *Syntactical treatments of modality, with corollaries on reflexion principles and finite axiomatizability* ◇ B45 B65 C62 F30 ◇

ROBINSON, A. *Introduction to model theory and to the metamathematics of algebra* ◇ B98 C60 C98 H15 ◇

ROBINSON, A. *On languages which are based on non-standard arithmetic* ◇ C75 H05 H15 ◇

VOPENKA, P. *Construction of models of set theory by the spectrum method (Russian) (German summary)* ◇ C62 E30 E70 ◇

VOPENKA, P. *Construction of non-standard, non-regular models in set theory (Russian) (German summary)* ◇ C20 C62 E30 E70 ◇

VOPENKA, P. *Elementary concepts in set theory (Russian) (German summary)* ◇ C62 E47 ◇

WANG, HAO *Partial systems of number theory* ◇ C62 F30 ◇

1964

HAJEK, P. *Die durch die schwach inneren Relationen gegebenen Modelle der Mengenlehre* ◇ C62 E30 E70 ◇

KURATA, R. *On the existence of a proper complete model of set theory* ◇ C62 E30 E70 ◇

MYCIELSKI, J. *The definition of arithmetic operations in the Ackermann model (Russian)* ◇ C62 E47 ◇

SHEPHERDSON, J.C. *A non-standard model for a free variable fragment of number theory* ◇ C62 F30 ◇

VOPENKA, P. *Submodels of models of set theory (Russian) (German summary)* ◇ C20 C62 E30 E55 E70 ◇

1965

ALLING, N.L. *Rings of continuous integer-valued functions and nonstandard arithmetic* ◇ C20 C60 H15 ◇

BALCAR, B. & JECH, T.J. *Models of the theory of sets generated by a perfect relation (Russian) (Czech and German summaries)* ◇ C62 E30 E55 E70 ◇

FEFERMAN, S. *Some applications of the notions of forcing and generic sets* ◇ C62 D55 E25 E35 E40 E45 F35 ◇

GOODSTEIN, R.L. *Multiple successor arithmetics* ◇ C62 F30 ◇

HAJEK, P. *Eine Bemerkung ueber standarde nichtregulaere Modelle der Mengenlehre* ◇ C62 E30 E55 E70 ◇

HAJEK, P. *Modelle der Mengenlehre, in denen Mengen gegebener Gestalt existieren* ◇ C62 E30 E35 ◇

HAJEK, P. & VOPENKA, P. *Permutation submodels of the model* ▽ ◇ C62 E25 E35 ◇

KREISEL, G. *Model-theoretic invariants: applications to recursive and hyperarithmetic operations*
⋄ C40 C62 D55 D60 ⋄
LEVY, A. *The Fraenkel-Mostowski method for independence proofs in set theory* ⋄ C62 E25 E35 ⋄
MOLLER, S. *Induction models (Danish)* ⋄ C62 F30 ⋄
MONTAGUE, R. *Set theory and higher-order logic*
⋄ B15 C62 E30 ⋄
MOSTOWSKI, ANDRZEJ *On models of Zermelo-Fraenkel set theory satisfying the axiom of constructibility*
⋄ C62 E45 ⋄
ROBINSON, A. *Topics in non-archimedean mathematics*
⋄ H05 H15 H20 ⋄
SHEPHERDSON, J.C. *Non-standard models for fragments of number theory* ⋄ C62 F30 ⋄
UMEZAWA, T. *On a model in the ordinal numbers*
⋄ C62 E30 E35 E45 E70 ⋄
VOPENKA, P. *Construction of a model for Goedel-Bernays set theory for which the class of natural numbers is a set of the model and a proper class in the theory*
⋄ C62 E30 E70 ⋄
VOPENKA, P. *On ∇-model of set theory*
⋄ C20 C62 C90 E30 E35 E40 E70 ⋄
VOPENKA, P. *Properties of ∇-model*
⋄ C62 E10 E35 E40 ⋄
VOPENKA, P. *The limits of sheaves and applications on constructions of models*
⋄ C20 C62 C90 E25 E30 E35 E40 E70 ⋄

1966

BUKOVSKY, L. & HAJEK, P. *On the standardness and regularity of normal syntactic models of the set theory*
⋄ C62 E30 E55 ⋄
DRABBE, J. *Les ordinaux dans les modeles internes de la theorie des ensembles* ⋄ C62 E10 E45 ⋄
EHRENFEUCHT, A. & KREISEL, G. *Strong models of arithmetic* ⋄ C57 C62 ⋄
HAJEK, P. & VOPENKA, P. *Some permutation submodels of the the model ∇* ⋄ C62 E25 E35 ⋄
KRUSE, A.H. *Grothendieck universes and the super-complete models of Shepherdson* ⋄ C62 E55 ⋄
LEBLANC, L. *Representabilite et definissabilite dans les algebres transformationnelles et dans les algebres polyadiques* ⋄ C40 C62 C90 G15 ⋄
MAREK, W. & ONYSZKIEWICZ, J. *Representation of partial ordering in cardinals of model of Zermelo-Fraenkel set theory I,II (II. Sets of incomparable powers) (Russian summaries)* ⋄ C62 E10 E25 E35 E55 ⋄
NAGASHIMA, T. *A remark on a comprehension axiom without negation* ⋄ C62 E30 ⋄
NERODE, A. *Diophantine correct non-standard models in the isols* ⋄ C62 D50 ⋄
ROBINSON, A. *A new approach to the theory of algebraic numbers I,II (Italian summary)* ⋄ C60 H15 H20 ⋄
ROBINSON, A. *Non-standard analysis*
⋄ A10 B98 H05 H10 H15 H98 ⋄
SCARPELLINI, B. *On a family of models of Zermelo-Fraenkel set theory*
⋄ C62 E30 E35 E45 ⋄
STERN, S.T. *The seminatural numbers* ⋄ E10 H15 ⋄

UMEZAWA, T. *On a model in the ordinal numbers. II*
⋄ C62 E30 E35 E45 E70 ⋄
VOPENKA, P. *∇-models in which the generalized continuum hypothesis does not hold*
⋄ C62 E35 E50 ⋄

1967

BALCAR, B. & VOPENKA, P. *On complete models of the set theory* ⋄ C62 E10 E25 ⋄
BOFFA, M. *Modeles de la theorie des ensembles, associes aux permutations de l'univers*
⋄ C62 E30 E35 E70 ⋄
BOFFA, M. *Remarques concernant les modeles associes aux permutations de l'univers* ⋄ C62 E30 E35 ⋄
ENDERTON, H.B. *An infinitistic rule of proof*
⋄ C62 F35 ⋄
GAIFMAN, H. *Uniform extension operators for models and their applications* ⋄ C20 C30 C62 ⋄
HAJEK, P. & VOPENKA, P. *Concerning the ∇-models of set theory (Russian summary)* ⋄ C62 E35 E40 E45 ⋄
HAUSCHILD, K. *Die Nichtexistenz starkregulaerer standarder Modelle in der Mengenlehre von Zermelo-Fraenkel und ihren widerspruchsfreien Erweiterungen* ⋄ C62 E30 ⋄
HERMES, H. *Ein mengentheoretisches Modell des Peanoschen Axiomensystems* ⋄ C62 ⋄
JENSEN, R.B. *Concrete models of set theory*
⋄ C62 E25 E35 E45 E50 ⋄
JENSEN, R.B. *Modelle der Mengenlehre*
⋄ C62 D55 E25 E35 E40 E45 E50 E98 ⋄
MAREK, W. & ONYSZKIEWICZ, J. *Some results in the foundations of the set theory*
⋄ C62 E25 E35 E45 E50 ⋄
MEISTERS, G.H. *Nonarchimedean integers*
⋄ C60 H15 ⋄
MOSTOWSKI, ANDRZEJ *Modeles transitifs de la theorie des ensembles de Zermelo-Fraenkel*
⋄ C62 E35 E45 E98 ⋄
PABION, J.F. *Modeles elementaires pour la theorie des ensembles* ⋄ C62 E30 ⋄
PABION, J.F. *Une approche de la construction d'un modele denombrable pour la theorie de Zermelo-Fraenkel*
⋄ C62 E30 ⋄
ROBINSON, A. *Non-standard theory of Dedekind rings*
⋄ C20 C60 H15 H20 ⋄
ROBINSON, A. *Nonstandard arithmetic*
⋄ C20 C60 H15 H20 ⋄
SHEPHERDSON, J.C. *The rule of induction in the free variable arithmetic based on + and ∗* ⋄ C62 F30 ⋄
SUZUKI, Y. *Applications of the theory of β-models*
⋄ C62 D65 ⋄
VOPENKA, P. *General theory of ∇-models*
⋄ C62 E35 E40 ⋄
YASUGI, M. *Interpretations of set theory and ordinal number theory* ⋄ C62 E10 E30 ⋄

1968

BENI, J. & BOFFA, M. *Elimination de cycles d'appartenance par permutation de l'univers* ⋄ C62 E35 E70 ⋄
ELLENTUCK, E. *The representation of cardinals in models of set theory* ⋄ C62 E10 E25 E35 ⋄

FJELSTAD, P. *Set-theories as algebras*
  ⋄ C62 E30 G25 ⋄
GEISER, J.R. *Nonstandard logic* ⋄ C20 H10 H15 ⋄
KEISLER, H.J. & MORLEY, M.D. *Elementary extensions of models of set theory* ⋄ C62 ⋄
MADISON, E.W. *Computable algebraic structures and nonstandard arithmetic* ⋄ C57 C60 C62 D45 ⋄
MADISON, E.W. *Structures elementarily closed relative to a model for arithmetic* ⋄ C57 C60 C62 ⋄
MORLEY, M.D. *Partitions and models*
  ⋄ C30 C50 C55 C62 C75 E05 E45 E55 ⋄
OLIN, P. *An almost everywhere direct power*
  ⋄ C30 C62 E50 ⋄
PHILLIPS, N.C.K. *A model for urelements* ⋄ C62 E70 ⋄
SUZUKI, Y. $\aleph_0$-*standard models for set theory (Russian summary)* ⋄ C62 D55 ⋄
WILMERS, G.M. *Some properties of constructible models for set theory related to elementary equivalence*
  ⋄ C62 E45 ⋄

1969

ADLER, A. *Extensions of non-standard models of number theory* ⋄ C62 D25 D35 H15 ⋄
BALCAR, B. & SOCHOR, A. *The general theory of semisets, syntactic models of axiomatic set theory*
  ⋄ C62 E35 E70 ⋄
BALDI, G. *Modelli finiti nella teoria degli insiemi (English summary)* ⋄ C13 C62 E70 ⋄
BOFFA, M. & RIBEAUFOSSE, P. *Modeles de la theorie generale des ensembles, construits sur les nombres-ε*
  ⋄ C62 E10 E30 ⋄
CHUDNOVSKY, G.V. *Certain properties of inaccessible cardinals in ZF models of set theory (Russian)*
  ⋄ C62 E35 E55 ⋄
DOETS, H.C. *Novak's result by Henkin's method*
  ⋄ C62 E35 E70 ⋄
GREWE, R. *Natural models of Ackermann's set theory*
  ⋄ C62 E30 E70 ⋄
HART, J. $\mathscr{E}^0$-*arithmetic* ⋄ C62 D20 ⋄
KRIVINE, J.-L. *Modeles de ZF+AC dans lesquels tout ensemble de reels definissable en termes d'ordinaux est mesurable-Lebesgue* ⋄ C62 E35 E45 E75 ⋄
MOSTOWSKI, ANDRZEJ *Constructible sets with applications*
  ⋄ C62 E25 E35 E45 E50 E70 E98 ⋄
MOSTOWSKI, ANDRZEJ *Models of set theory*
  ⋄ C62 E98 ⋄
MOSTOWSKI, ANDRZEJ & SUZUKI, Y. *On ω-models which are not β-models* ⋄ C62 ⋄
OLIN, P. *Indefinability in the arithmetic of isolic integers*
  ⋄ D50 H15 ⋄
PACHOLSKI, L. *Elementary substructures of direct powers (Russian summary)* ⋄ C30 C62 ⋄
PLOTKIN, J.M. *Generic embeddings* ⋄ C62 E25 E35 ⋄
POTTHOFF, K. *Ueber Nichtstandardmodelle der Arithmetik und der rationalen Zahlen* ⋄ H15 H20 ⋄
ROBINSON, A. & ZAKON, E. *A set-theoretical characterization of enlargements*
  ⋄ C20 C62 E75 H05 H20 ⋄

ROBINSON, A. *Topics in nonstandard algebraic number theory* ⋄ C60 H05 H15 H20 ⋄
SHEPHERDSON, J.C. *Weak and strong induction*
  ⋄ C62 F30 ⋄
WEGLORZ, B. *A model of set theory ⅁ over a given Boolean algebra (Russian summary)*
  ⋄ C62 E25 E35 G05 ⋄

1970

BARROS DE, C.M. *Sur la classe des ordinaux appartenant a un univers et univers bien ordonnes*
  ⋄ C62 E10 E55 E70 ⋄
BARWISE, J. & FISHER, E.R. *The Shoenfield absoluteness lemma* ⋄ C62 D55 E45 ⋄
BELYAKIN, N.V. *Generalized computations and second order arithmetic (Russian)*
  ⋄ C62 D10 D45 D65 D70 F35 ⋄
CONNES, A. *Determination de modeles minimaux en analyse non standard et application* ⋄ H05 H15 ⋄
FENSTAD, J.E. *Non-standard models for arithmetic and analysis* ⋄ H05 H15 ⋄
FRIEDMAN, H.M. *Iterated inductive definitions and* Σ_2^1-AC ⋄ C62 D55 E25 F35 ⋄
GAIFMAN, H. *On local arithmetical functions and their application for constructing types of Peano's arithmetic*
  ⋄ C30 C62 H15 ⋄
JENSEN, R.B. & SOLOVAY, R.M. *Some applications of almost disjoint sets*
  ⋄ C62 D55 E05 E35 E45 E55 ⋄
KEISLER, H.J. *Logic with the quantifier "there exist uncountably many"* ⋄ C30 C55 C62 C75 C80 ⋄
KUNEN, K. & PARIS, J.B. *Boolean extensions and measurable cardinals*
  ⋄ C20 C62 E35 E40 E50 E55 ⋄
KUNEN, K. *Some applications of iterated ultrapowers in set theory*
  ⋄ C20 C55 C62 E05 E35 E45 E50 E55 ⋄
LOEB, M.H. *A model theoretic characterization of effective operations* ⋄ C62 D20 ⋄
NAGASHIMA, T. *A model of the comprehension axiom without negation* ⋄ C62 E30 ⋄
POTTHOFF, K. *Ideale in Nichtstandardmodellen der ganzen Zahlen* ⋄ H15 H20 ⋄
SCHULTZ, KONRAD *Modelle modaler Mengenlehren*
  ⋄ B45 C62 C90 E35 E70 ⋄
SUZUKI, Y. *Orbits of denumerable models of complete theories* ⋄ C15 C35 C52 C62 ⋄
TAKAHASHI, MOTO-O *On models of axiomatic set theory (Japanese)* ⋄ C62 ⋄

1971

BARWISE, J. *Infinitary methods in the model theory of set theory* ⋄ C62 C70 E45 E47 ⋄
BARWISE, J. & GANDY, R.O. & MOSCHOVAKIS, Y.N. *The next admissible set*
  ⋄ C40 C62 D55 D60 D70 E45 E47 ⋄
BOFFA, M. *Sentences universelles en theorie des ensembles*
  ⋄ C62 E30 ⋄
CASARI, E. *Regular models for second order set theory*
  ⋄ C62 C85 E55 E70 ⋄

DRAGALIN, A.G. & FUKSON, V.I. & LYUBETSKIJ, V.A. *Definable sequences of countable ordinals (Russian)*
⋄ C40 C62 E10 E35 E45 E47 ⋄

ELLENTUCK, E. *A minimal model for strong analysis*
⋄ C50 C62 C85 E35 ⋄

ENDERTON, H.B. & FRIEDMAN, H.M. *Approximating the standard model of analysis* ⋄ C62 E45 F35 ⋄

FELGNER, U. *Models of ZF set theory*
⋄ C62 E25 E35 E45 E50 E98 ⋄

FRIEDMAN, H.M. *A more explicit set theory*
⋄ A05 C62 E30 E35 E45 F30 F35 ⋄

GRZEGORCZYK, A. *An unfinitizability proof by means of restricted reduced power* ⋄ C20 C62 ⋄

GRZEGORCZYK, A. *Outline of theoretical arithmetic (Polish)* ⋄ B28 C62 F30 F35 F98 ⋄

HAJEK, P. *Sets, semisets, models*
⋄ A10 C62 E35 E40 E70 ⋄

JECH, T.J. *On models for the set theory without AC*
⋄ C62 E10 E25 E35 E40 ⋄

JECH, T.J. & POWELL, W.C. *Standard models of set theory with predication* ⋄ C62 E30 E47 E55 E65 E70 ⋄

JECH, T.J. *Two remarks on elementary embeddings of the universe* ⋄ C62 E40 E45 E55 ⋄

JEROSLOW, R. *Non-effectiveness in S.Orey's arithmetical compactness theorem* ⋄ C62 ⋄

KEISLER, H.J. & SILVER, J.H. *End extensions of models of set theory* ⋄ C62 E45 ⋄

KUNEN, K. *Elementary embeddings and infinitary combinatorics* ⋄ C62 E05 E55 ⋄

MOSCHOVAKIS, Y.N. *Predicative classes*
⋄ C62 D55 D60 D65 D70 E70 ⋄

PARIKH, R. *Existence and feasibility in arithmetic*
⋄ A05 D15 F30 F65 H15 ⋄

PHILLIPS, R.G. *On the structure of nonstandard models of arithmetic* ⋄ H15 ⋄

PURITZ, C.W. *Ultrafilters and standard functions in non-standard arithmetic* ⋄ C20 E05 H15 ⋄

ROSENTHAL, J.W. *Relations not determining the structure of L* ⋄ C35 C62 E45 ⋄

TOMASIK, J. *On the number of elementary theories of some hierarchies of sets (Russian summary)* ⋄ C62 E45 ⋄

WILMERS, G.M. *Internally standard set theories*
⋄ C62 C70 E45 E47 ⋄

ZBIERSKI, P. *Models for higher order arithmetics (Russian summary)* ⋄ C62 E30 F35 ⋄

1972

AABERG, C. *Some results on models for set theory*
⋄ C62 E40 ⋄

ABIAN, A. *Set theoretical canonical models (Armenian and Russian summaries)* ⋄ C62 E35 ⋄

ABIAN, A. *The method of synergistic models*
⋄ C62 E25 E35 E45 ⋄

APT, K.R. *ω-models in analytical hierarchy (Russian summary)* ⋄ C62 D30 D55 ⋄

BLASS, A.R. *The intersection of nonstandard models of arithmetic* ⋄ C62 ⋄

BOFFA, M. *Isomorphismes des systemes extensionnels superuniversels* ⋄ C62 E30 ⋄

BUKOVSKY, L. *Models of set theory with axiom of constructibility which contain non-constructible sets*
⋄ C20 C62 E45 ⋄

COSTA DA, N.C.A. *Modeles et univers de Dedecker*
⋄ C62 E55 E70 G30 ⋄

DAGUENET, M. *Un modele non standard de l'arithmetique*
⋄ C62 E05 H15 ⋄

DOEPP, K. *Bemerkung zu Henkins Beweis fuer die Nichtstandard-Vollstaendigkeit der Typentheorie*
⋄ B15 C62 C85 ⋄

GAIFMAN, H. *A note on models and submodels of arithmetic* ⋄ C62 ⋄

GRILLIOT, T.J. *Omitting types; applications to recursion theory* ⋄ C07 C62 D30 D55 D60 D75 ⋄

HINNION, R. *Sur les modeles de NF ("New Foundations") de Quine)* ⋄ C62 E70 ⋄

HOWARD, P.E. *A proof of a theorem of Tennenbaum*
⋄ C57 C62 H15 ⋄

MADISON, E.W. *Real fields with characterization of the natural numbers* ⋄ C57 H15 ⋄

MLCEK, J. & SOCHOR, A. *Contributions to the theory of semisets. II: The theory of semisets and end-extensions in a syntactic setting* ⋄ C62 E70 ⋄

MOSTOWSKI, ANDRZEJ *A transfinite sequence of ω-models* ⋄ C62 D30 ⋄

MOSTOWSKI, ANDRZEJ *Models of second order arithmetic with definable Skolem functions* ⋄ C62 ⋄

NADEL, M.E. *Some Loewenheim-Skolem results for admissible sets* ⋄ C62 C70 D60 E55 ⋄

PARIS, J.B. *On models of arithmetic* ⋄ C62 ⋄

PHILLIPS, R.G. *Addition in nonstandard models of arithmetic* ⋄ H15 ⋄

POTTHOFF, K. *Ordnungseigenschaften von Nichtstandardmodellen* ⋄ C20 C62 H05 H15 ⋄

POTTHOFF, K. *Ueber Enderweiterungen von Nichtstandardmodellen der Arithmetik und anderer Strukturen* ⋄ C62 H15 H20 ⋄

PURITZ, C.W. *Skies, constellations and monads*
⋄ H15 H20 ⋄

ROBINSON, A. *Algebraic function fields and non-standard arithmetic* ⋄ C60 H15 ⋄

SABBAGH, G. *The epimorphisms of the ∈-relation*
⋄ C62 E20 E35 G30 ⋄

SANMARTIN ESPLUGUES, J. *Standard models of axiomatic systems of sets I (Spanish)* ⋄ C62 E35 ⋄

SHILLETO, J.R. *Minimum models of analysis*
⋄ C50 C62 D55 E45 ⋄

TAKAHASHI, MOTO-O $\tilde{\Delta}_1$-definability in set theory
⋄ C62 E35 E45 E47 ⋄

1973

APT, K.R. *Infinitistic rules of proof and their semantics (Russian summary)* ⋄ C62 F07 F35 ⋄

ARTIGUE, M. & ISAMBERT, E. & PERRIN, M.J. & ZALC, A. *A notion of extension for models of full second order arithmetic (Russian summary)* ⋄ C62 E45 ⋄

BALDWIN, JOHN T. & BLASS, A.R. & GLASS, A.M.W. & KUEKER, D.W. *A "natural" theory without a prime model* ⋄ C15 C50 C62 C65 ⋄

BOFFA, M. *Structures extensionnelles generiques*
⋄ C25 C62 E70 ⋄

BOOLOS, G. *A note on Beth's theorem (Russian summary)*
⋄ C40 C62 ⋄

BOWEN, K.A. *Infinite forcing and natural models of set theory (Russian summary)*
⋄ C25 C55 C62 C75 C80 E55 G30 ⋄

BUKOVSKY, L. *Characterization of generic extensions of models of set theory* ⋄ C62 E40 G05 ⋄

COLE, J.C. *Categories of sets and models of set theory*
⋄ C62 E30 G30 ⋄

DUBIEL, M. *Nonstandard well-orderings in ω-models of A_2 (Russian summary)* ⋄ C62 ⋄

EHRENFEUCHT, A. *Discernible elements in models for Peano arithmetic* ⋄ C50 C62 ⋄

ELLENTUCK, E. *The structure of Dedekind cardinals*
⋄ C62 E10 E25 E35 ⋄

ENDERTON, H.B. *Constructible β-models* ⋄ C62 E45 ⋄

FRIEDMAN, H.M. *Countable models of set theories*
⋄ C15 C62 E45 F35 H20 ⋄

GOLDREI, D.C. & MACINTYRE, A. & SIMMONS, H. *The forcing companions of number theories* ⋄ C25 C62 ⋄

HAJEK, P. & VOPENKA, P. *Existence of a generalized model of the Goedel-Bernays set theory* ⋄ C62 E70 ⋄

KNIGHT, J.F. *Generic expansions of structures*
⋄ C25 C62 C85 ⋄

KRIVINE, J.-L. & MCALOON, K. *Some true unprovable formulas for set theory*
⋄ C62 C75 E35 E45 E55 ⋄

LINDSTROEM, P. *A note on weak second order logic with variables for elementarily definable relations*
⋄ B15 C55 C62 ⋄

MAREK, W. *Observations concerning elementary extensions of ω-models II* ⋄ C62 E45 ⋄

MAREK, W. *On the metamathematics of impredicative set theory* ⋄ C62 E35 E47 E70 ⋄

MAREK, W. & SREBRNY, M. *On transitive models for fragments of set theory (Russian summary)*
⋄ C62 E30 E45 ⋄

MLCEK, J. *A representation of models of Peano arithmetic*
⋄ C60 C62 ⋄

MOSTOWSKI, ANDRZEJ *A contribution to teratology (Russian)* ⋄ C20 C62 ⋄

MOSTOWSKI, ANDRZEJ *Partial orderings of the family of ω-models* ⋄ C62 ⋄

NELSON, GEORGE C. *Nonconstructivity of models of the reals (Russian summary)*
⋄ C62 C65 D55 H05 H15 ⋄

PARIS, J.B. *Minimal models of ZF* ⋄ C50 C62 E45 ⋄

ROBINSON, A. *Nonstandard arithmetic and generic arithmetic* ⋄ C25 C60 C62 C98 H15 ⋄

ROBINSON, A. *Standard and nonstandard number systems* ⋄ C25 C60 C98 H05 H15 H20 H98 ⋄

ROBINSON, JULIA *Axioms for number theoretic functions (Russian)* ⋄ C62 D20 D75 F30 ⋄

ROBINSON, JULIA *Solving diophantine equations*
⋄ B25 D35 D80 H15 ⋄

SCHMERL, J.H. *Peano models with many generic classes*
⋄ C55 C62 E40 ⋄

SCHOCK, R. *Set theory can provide for all accessible ordinals* ⋄ C62 E30 E35 E55 ⋄

SUZUKI, Y. & WILMERS, G.M. *Non-standard models for set theory* ⋄ C62 E35 ⋄

TAKAHASHI, S. *A non-standard treatment of infinitely near points* ⋄ D35 H15 H20 ⋄

WILMERS, G.M. *An $\aleph_1$-standard model for ZF which is an element of the minimal model for ZF*
⋄ C62 C70 E30 E35 E45 H20 ⋄

1974

AABERG, C. *Relativity phenomena in set theory* ⋄ C62 ⋄

ABIAN, A. *Nonstandard models for arithmetics and analysis* ⋄ C20 H05 H15 ⋄

APT, K.R. & MAREK, W. *Second order arithmetic and related topics* ⋄ C62 D55 E45 F25 F35 ⋄

BARNES, D.W. & MONRO, G.P. *A simple model for a weak system of arithmetic* ⋄ C62 F30 ⋄

BARWISE, J. *Admissible sets over models of set theory*
⋄ C62 D60 E30 ⋄

BILY, J. & BUKOVSKY, L. *On expansion of β-models*
⋄ C62 E25 E40 E70 F35 ⋄

BLASS, A.R. *On certain types and models for arithmetic*
⋄ C62 E05 ⋄

COHEN, P.E. *Models of set theory with more real numbers than ordinals* ⋄ C62 E25 E35 ⋄

COHEN, P.J. *Automorphisms of set theory*
⋄ C62 E25 E35 ⋄

COSTA DA, N.C.A. *α-models and systems T and T^**
⋄ C62 E70 G30 ⋄

CROSSLEY, J.N. & NERODE, A. *Combinatorial functors*
⋄ C57 C62 D45 D50 G30 ⋄

DUBROVSKY, D.L. *Some subfields of $\mathbf{Q}_p$ and their non-standard analogues* ⋄ C60 H15 H20 ⋄

ELLENTUCK, E. *A model of arithmetic*
⋄ C62 E10 E25 ⋄

GAIFMAN, H. *Elementary embeddings of models of set theory and certain subtheories*
⋄ C62 E30 E45 E55 ⋄

GEISER, J.R. *A formalization of Essenin-Volpin's proof theoretical studies by means of nonstandard analysis*
⋄ C75 F50 H10 H15 ⋄

GREEN, J. *Σ_1-compactness for next admissible sets*
⋄ C62 C70 D60 ⋄

GUZICKI, W. *Elementary extensions of Levy's model of A_2^-* ⋄ C55 C62 C80 E25 E35 E40 F35 ⋄

GUZICKI, W. *Uncountable β-models with countable height* ⋄ C62 E40 E50 ⋄

HIRSCHFELD, J. *Models of arithmetic and the semiring of recursive functions* ⋄ C62 D20 H15 ⋄

HUTCHINSON, J.E. *Models of set theories and their applications* ⋄ C55 C62 C80 ⋄

ISAMBERT, E. *Extensions elementaires non-standard de modeles de theorie des ensembles* ⋄ C62 C80 ⋄

ISAMBERT, E. *Theories syntaxiquement non-standard*
⋄ C62 ⋄

KANOVEJ, V.G. *On degrees of constructibility and descriptive properties of the set of real numbers in an initial model and in its extensions (Russian)*
⋄ C62 D30 E40 E45 ⋄

KEISLER, H.J. *Models with tree structures*
⋄ C55 C62 E65 ⋄

KOGALOVSKIJ, S.R. *Certain simple consequences of the axiom of constructibility (Russian) (English summary)*
⋄ C40 C62 C85 E45 ⋄

KRAJEWSKI, S. *A remark on automorphisms and nonstandard properties (Russian summary)*
⋄ C62 H15 ⋄

KRAJEWSKI, S. *Mutually inconsistent satisfaction classes*
⋄ C62 F30 ⋄

KRAJEWSKI, S. *Predicative expansions of axiomatic theories* ⋄ C30 C62 E30 E70 F25 ⋄

LEEDS, S. & PUTNAM, H. *Solution to a problem of Gandy's*
⋄ C62 D55 ⋄

MACINTYRE, A. *A note on axioms for infinite-generic structures* ⋄ C25 C60 C62 C75 E07 ⋄

MAREK, W. & SREBRNY, M. *Gaps in the constructible universe* ⋄ C62 E10 E30 E45 ⋄

MAREK, W. & ZBIERSKI, P. *On the size of the family of β-models* ⋄ C62 E15 E45 ⋄

MAREK, W. *Stable sets, a characterization of β_2-models of full second order arithmetic and some related facts*
⋄ C62 E45 E47 ⋄

MCALOON, K. *On the sequence of models HOD_n*
⋄ C62 E35 E45 ⋄

MONTAGNA, F. *Sui limiti di certe teorie (English summary)*
⋄ C30 C35 C50 C62 H15 ⋄

MOSTOWSKI, ANDRZEJ *Observations concerning elementary extensions of ω-models* ⋄ C62 C80 ⋄

NADEL, M.E. *More Loewenheim-Skolem results for admissible sets* ⋄ C62 C70 D60 ⋄

PHILLIPS, R.G. *A minimal extension that is not conservative* ⋄ C62 ⋄

PHILLIPS, R.G. *Omitting types in arithmetic and conservative extensions* ⋄ C62 ⋄

POWELL, W.C. *Variations of Keisler's theorem for complete embeddings* ⋄ C07 C62 ⋄

SCHOCK, R. *The consistency of the axiom of constructibility in ZF with non-predicative ultimate classes*
⋄ C62 E35 E45 E70 ⋄

SIMPSON, S.G. *Forcing and models of arithmetic*
⋄ C15 C25 C62 E40 E47 ⋄

SREBRNY, M. *Note on a theorem of Mostowski (Russian summary)* ⋄ C62 E50 ⋄

STEPANEK, P. *Contributions to the theory of semisets. IV*
⋄ C20 C62 E35 E70 ⋄

TANAKA, H. *Some analytical rules of inference in the second-order arithmetic* ⋄ C62 D55 F35 ⋄

VILLE, F. *More on set existence* ⋄ C62 C70 E47 ⋄

WAID, C.C. *On Dirichlet's theorem and finite primes*
⋄ H15 ⋄

1975

AABERG, C. *Generic extensions and elementary embeddings* ⋄ C62 E40 ⋄

ABIAN, A. *On the standard-model hypothesis of ZF*
⋄ C62 E30 ⋄

ABIAN, A. *On the use of more than two-element boolean valued models* ⋄ C62 C90 E35 ⋄

BARWISE, J. & SCHLIPF, J.S. *On recursively saturated models of arithmetic* ⋄ C50 C57 C62 D80 H15 ⋄

CHERLIN, G.L. *Ideals in some nonstandard Dedekind rings* ⋄ C60 H15 H20 ⋄

CHERLIN, G.L. *Ideals of integers in nonstandard number fields* ⋄ C60 H15 H20 ⋄

DEHORNOY, P. *Solution d'une conjecture de Bukovsky (English summary)* ⋄ C20 C62 E40 E55 ⋄

FLEISSNER, W.G. *Lemma on measurable cardinals*
⋄ C20 C62 E45 E55 ⋄

FRIEDMAN, H.M. *Large models of countable height*
⋄ C15 C55 C62 E25 E40 ⋄

GREEN, J. *A note on $\mathscr{P}$-admissible sets with urelements*
⋄ C62 E30 ⋄

GREGORY, J. *On a finiteness condition for infinitary languages* ⋄ C40 C62 C70 C75 E40 ⋄

HARRINGTON, L.A. & KECHRIS, A.S. *On characterizing Spector classes*
⋄ C62 D55 D60 D65 D75 E45 E55 E60 ⋄

HIRSCHFELD, J. *Finite forcing and generic filters in arithmetic* ⋄ C25 C60 C62 ⋄

HIRSCHFELD, J. & WHEELER, W.H. *Forcing, arithmetic, division rings* ⋄ C25 C60 C62 C98 ⋄

HIRSCHFELD, J. *Models of arithmetic and recursive functions* ⋄ C62 D20 H15 ⋄

HIRSCHFELD, J. *The model companion of ZF*
⋄ C25 C62 E30 ⋄

KECHRIS, A.S. *Countable ordinals and the analytical hierarchy I* ⋄ C62 D55 E10 E15 E60 ⋄

KLINGEN, N. *Zur Idealstruktur in Nichtstandardmodellen von Dedekindringen* ⋄ C20 C60 H15 H20 ⋄

KNIGHT, J.F. *Types omitted in uncountable models of arithmetic* ⋄ C07 C62 ⋄

LAKE, J. *Natural models and Ackermann-type set theories*
⋄ C62 E30 E45 E55 E70 ⋄

LEVITZ, H. *An ordered set of arithmetic functions representing the least ε-number* ⋄ C62 F15 F30 ⋄

LOLLI, G. *Sui modelli di certe teorie delle classi (English summary)* ⋄ C62 E30 E70 ⋄

MACINTYRE, A. & SIMMONS, H. *Algebraic properties of number theories*
⋄ C25 C40 C52 C62 D25 H15 ⋄

MAREK, W. & SREBRNY, M. *No minimal transitive model of Z^-* ⋄ C62 E30 E70 F35 ⋄

MAREK, W. & MOSTOWSKI, ANDRZEJ *On extendability of models of ZF set theory to the models of Kelley-Morse theory of classes* ⋄ C62 E30 E45 E70 ⋄

MCALOON, K. *Applications alternees de theoremes d'incompletude et des theoremes de completude (English summary)* ⋄ C62 E35 F30 ⋄

MCALOON, K. *Formules de Rosser pour ZF (English summary)* ⋄ C62 C70 E30 E35 E45 F25 ⋄

METAKIDES, G. & PLOTKIN, J.M. *An algebraic characterization of power set in countable standard models of ZF* ⋄ C15 C62 E35 G05 ⋄

PERRIN, M.J. & ZALC, A. *Un theoreme de compacite en analyse. Application aux β_n-modeles et β_n-extensions (English summary)* ⋄ B15 C62 ⋄

ROBINSON, A. & ROQUETTE, P. *On the finiteness theorem of Siegel and Mahler concerning diophantine equations*
⋄ C60 H05 H15 H20 ⋄

ROQUETTE, P. *Nonstandard aspects of Hilbert's irreducibility theorem* ⋄ C60 H15 H20 ⋄

ROSENTHAL, J.W. *Truth in all of certain well-founded countable models arising in set theory* ⋄ C40 C62 C70 E45 E47 ⋄

SCHLIPF, J.S. *Some hyperelementary aspects of model theory* ⋄ C50 C62 C70 D55 ⋄

SOCHOR, A. & VOPENKA, P. *Contributions to the theory of semisets. V: On axiom of general collapse* ⋄ C20 C62 E35 E70 ⋄

SOCHOR, A. *Contributions to the theory of semisets. VI: Non-existence of the class of all absolute natural numbers* ⋄ C20 C62 E35 E70 ⋄

SOCHOR, A. *Real classes in the ultrapower of hereditarily finite sets* ⋄ C20 E70 H15 ⋄

SOPRUNOV, S.F. *Initial segments of nonstandard arithmetics (Russian)* ⋄ C62 ⋄

SOPRUNOV, S.F. *Strong nonstandard models of arithmetic (Russian)* ⋄ C62 ⋄

STEEL, J.R. *Descending sequences of degrees* ⋄ C62 D30 ⋄

STOJAKOVIC, M. *Inductive and Peano models (Serbo-Croatian) (English summary)* ⋄ B28 C62 ⋄

WAUW-DE KINDER VAN DE, G. *Arithmetique de premier ordre dans les topos (English summary)* ⋄ F30 F50 G30 H15 ⋄

WEINGARTNER, P. *A finite approximation to models of set theory* ⋄ C62 ⋄

WILKIE, A.J. *On models of arithmetic – answers to two problems raised by H.Gaifman* ⋄ C62 ⋄

WILMERS, G.M. *Non-standard models and their application to model theory* ⋄ C50 C62 H20 ⋄

1976

ADAMOWICZ, Z. *One more aspect of forcing and omitting types* ⋄ C62 C75 E35 E40 ⋄

APT, K.R. *Semantics of the infinitistic rules of proof* ⋄ C62 D55 E45 F07 F35 ⋄

BELL, J.L. *Uncountable standard models of ZFC + $V \neq L$* ⋄ C62 E40 E45 ⋄

BENIOFF, P.A. *Models of Zermelo Fraenkel set theory as carriers for the mathematics of physics I,II* ⋄ C62 D80 E75 ⋄

BEZBORUAH, A. & SHEPHERDSON, J.C. *Goedel's second incompleteness theorem for Q* ⋄ C62 F30 ⋄

BUKOVSKY, L. *Changing cofinality of $\aleph_2$* ⋄ C62 E35 E40 ⋄

CONWAY, J.H. *On numbers and games* ⋄ B28 E10 E70 E75 H15 H20 ⋄

COWEN, R.H. *Elementary equivalence and constructible models of Zermelo-Fraenkel set theory* ⋄ C62 E45 ⋄

DEHORNOY, P. *Intersections d'ultrapuissances iterees de modeles de la theorie des ensembles (English summary)* ⋄ C20 C62 E25 E55 ⋄

DRABBE, J. *Extensions Cohen-generiques des ensembles admissibles (English summary)* ⋄ C62 E30 E40 ⋄

DUBIEL, M. *Elementary extensions of β-models of A_2 with fixed height* ⋄ C62 ⋄

EHRENFEUCHT, A. & JENSEN, D.C. *Some problems in elementary arithmetic* ⋄ C62 F30 H15 ⋄

FRIEDRICHSDORF, U. *Einige Bemerkungen zur Peano-Arithmetik* ⋄ C25 C52 C62 F30 ⋄

GAIFMAN, H. *Models and types of Peano's arithmetic* ⋄ C30 C62 H15 ⋄

HINNION, R. *Modeles de fragments de la theorie des ensembles de Zermelo-Fraenkel dans les "New foundations" de Quine (English summary)* ⋄ C62 E30 E35 E70 ⋄

HUTCHINSON, J.E. *Elementary extensions of countable models of set theory* ⋄ C15 C62 ⋄

HUTCHINSON, J.E. *Model theory via set theory* ⋄ C40 C55 C62 C70 C75 C80 C95 E75 ⋄

HUTCHINSON, J.E. *Order types of ordinals in models of set theory* ⋄ C30 C62 E10 ⋄

KNIGHT, J.F. *Hanf numbers for omitting types over particular theories* ⋄ C07 C55 C62 ⋄

KNIGHT, J.F. *Omitting types in set theory and arithmetic* ⋄ C07 C62 E40 ⋄

KRAJEWSKI, S. *Non-standard satisfaction classes* ⋄ C62 C85 H15 H20 ⋄

LIVCHAK, A.B. *Definable sets of integers (Russian)* ⋄ C60 C62 F30 ⋄

MANEVITZ, L.M. *Internal end-extensions of Peano arithmetic and a problem of Gaifman* ⋄ C62 ⋄

MAREK, W. & SREBRNY, M. *Urelements and extendability* ⋄ C62 E30 E70 F25 F30 F35 ⋄

MIJOULE, R. *Theorie des types et modeles de la theorie des ensembles (English summary)* ⋄ B15 C62 G30 ⋄

MOSTOWSKI, ANDRZEJ *A remark on models of the Goedel-Bernays axioms for set theory* ⋄ C62 E30 E70 ⋄

MOSTOWSKI, ANDRZEJ *Two remarks on the models of Morse' set theory* ⋄ C62 E25 E35 E40 E70 ⋄

MURAWSKI, R. *Expandability of models for elementary arithmetic* ⋄ C62 H15 ⋄

MURAWSKI, R. *On expandability of models of Peano arithmetic. I,II* ⋄ C62 C98 ⋄

POTTHOFF, K. *A simple tree lemma and its application to a counterexample of Philips* ⋄ C20 C62 H15 ⋄

ROGUSKI, S. *Extensions of models for ZFC to models for $ZF+V=HOD$ with applications* ⋄ C62 E35 E45 E47 ⋄

SIMMONS, H. *Each regular number structure is biregular* ⋄ C25 C62 ⋄

SOPRUNOV, S.F. *Denumerable non-standard models of arithmetic (Russian)* ⋄ C15 C62 ⋄

VETULANI, Z. *Categoricity relative to ordinals for models of set theory and the nonabsoluteness of L* ⋄ C35 C62 E45 H20 ⋄

ZARACH, A. *Generic extension of admissible sets* ⋄ C62 E25 E35 F35 ⋄

1977

ABIAN, A. *A preliminary to forcing in set theory. On the standard and minimal standard model axioms of ZF* ⋄ C62 E30 ⋄

BIELINSKI, K. *Extendability of structures as infinitary property* ⋄ C62 C70 E30 E70 ⋄

BLASS, A.R. *Amalgamation of non-standard models of arithmetic* ⋄ C62 E05 ⋄

BLASS, A.R. *End extensions, conservative extensions, and the Rudin-Frolik ordering* ◊ C62 E05 E50 ◊

BOFFA, M. *Modeles cumulatifs de la theorie des types* ◊ B15 C62 E30 ◊

BUKOVSKY, L. *Iterated ultrapower and Prikry's forcing* ◊ C20 C62 E40 E55 ◊

BURGIN, M.S. *Nonclassical models of the natural numbers (Russian)* ◊ H15 ◊

CARSTENS, H.G. *The theorem of Matijasevic is provable in Peano's arithmetic by finitely many axioms* ◊ C52 D25 D35 F30 H15 ◊

CUDA, K. *The nonabsolute boundedness model of the theory of semisets* ◊ C62 E35 E70 ◊

DAMYANOV, B.P. & DANOV, D.L. & KHRISTOV, KH.YA. *Sets of asymptotic numbers closed with respect to each of the four algebraic operations (Russian)* ◊ H15 ◊

DAWES, A.M. *End extensions which are models of a given theory* ◊ C07 C15 C30 C62 ◊

GREEN, J. *Next $\mathscr{P}$-admissible sets are of cofinality ω* ◊ C62 C70 ◊

GUZICKI, W. *The number of β-models of Kelley-Morse set theory* ◊ C62 E50 E70 ◊

HAJEK, P. *Experimental logics and Π_3^0 theories* ◊ B60 C62 D55 F30 ◊

HARRINGTON, L.A. & PARIS, J.B. *A mathematical incompleteness in Peano arithmetic* ◊ C30 C62 E05 F30 ◊

HARRINGTON, L.A. *Long projective wellorderings* ◊ C62 D55 E15 E35 E50 ◊

KAUFMANN, M. *A rather classless model* ◊ C50 C57 C62 E65 ◊

KIRBY, L.A.S. & PARIS, J.B. *Initial segments of models of Peano's axioms* ◊ C62 F30 ◊

KRAJEWSKI, S. *A note on expansion of models of set theories* ◊ C62 E30 E70 ◊

LIVCHAK, A.B. *Expressible subsets of the ordered group of integers (Russian)* ◊ C60 C62 F30 ◊

MAREK, W. & ZBIERSKI, P. *On the number of models of Kelley-Morse theory of classes with prescribed V* ◊ C62 E70 ◊

MAREK, W. *On the standard part problem* ◊ C62 E45 ◊

MCALOON, K. *Consistency statements and number theories* ◊ C62 F30 ◊

MIJAJLOVIC, Z. *A note on elementary end extension* ◊ C30 C62 ◊

MURAWSKI, R. *On expandability of models of Peano arithmetic. III* ◊ C62 C98 ◊

NADEL, M.E. & STAVI, J. *The pure part of HYP($\mathfrak{M}$)* ◊ C62 C70 ◊

PARIS, J.B. *Models of arithmetic and the 1-3-1 lattice* ◊ C50 C62 G10 ◊

RICHARD, D. *On external properties of nonstandard models of arithmetic (French summary)* ◊ C62 ◊

SAKAROVITCH, JOEL *Systemes normaux d'ensembles de conditions de "forcing" et applications aux suites decroissantes de modeles de ZFC (English summary)* ◊ C62 E40 ◊

SCHLIPF, J.S. *Ordinal spectra of first-order theories* ◊ C35 C50 C55 C62 C70 D15 D30 D70 ◊

SIMMONS, H. *Existentially closed models of basic number theory* ◊ C25 C62 ◊

SMORYNSKI, C.A. *The incompleteness theorems* ◊ C62 F25 F30 F98 ◊

SREBRNY, M. *Relatively constructible transitive models* ◊ C62 D55 D60 E10 E40 E45 E55 ◊

SREBRNY, M. *Second-order theories of countable transitive models* ◊ C62 C85 ◊

URSINI, A. *A sequence of theories for arithmetic whose union is complete* ◊ C62 F30 ◊

VETULANI, Z. *Some remarks on ultrapowers of extendable models of ZF* ◊ C20 C62 E30 E70 H20 ◊

WILKIE, A.J. *On models of arithmetic having non-modular substructure lattices* ◊ C20 C62 ◊

WILKIE, A.J. *On the theories of end extensions of models of arithmetic* ◊ C62 ◊

ZARACH, A. *Extension of ZF-models to models with the scheme of choice* ◊ C15 C62 E25 E30 E35 ◊

1978

ABRAMSON, F.G. & HARRINGTON, L.A. *Models without indiscernibles* ◊ C30 C50 C55 C62 C65 C75 E05 E40 H15 ◊

ANAPOLITANOS, D.A. *Absolute indiscernibles and standard models of ZFC* ◊ C30 C62 E40 ◊

APT, K.R. *Inductive definitions, models of comprehension and invariant definability* ◊ C40 C62 D55 D70 F35 ◊

ARTIGUE, M. & ISAMBERT, E. & PERRIN, M.J. & ZALC, A. *Some remarks on bicommutability* ◊ C62 E30 F25 F35 ◊

BLASS, A.R. *A model-theoretic view of some special ultrafilters* ◊ C20 E05 H15 ◊

BUKOVSKY, L. *Cogeneric extensions* ◊ C62 E40 ◊

CHARRETTON, C. *Type d'ordre des ordinaux de modeles non denombrables de la theorie des ensembles* ◊ C20 C50 C62 C70 E10 E50 H20 ◊

DEHORNOY, P. *Iterated ultrapowers and Prikry forcing* ◊ C20 C62 E25 E40 E55 ◊

FENSTAD, J.E. & GANDY, R.O. & SACKS, G.E. (EDS.) *Generalized recursion theory II* ◊ C62 C70 C80 D97 ◊

FRIEDMAN, H.M. *Categoricity with respect to ordinals* ◊ C35 C62 E45 ◊

GLOEDE, K. *Hierarchies of sets definable by means of infinitary languages* ◊ C62 C70 C75 E45 ◊

KELLER, J.P. & RICHARD, D. *Remarques sur les structures additives des modeles de l'arithmetique (English summary)* ◊ C62 H15 ◊

KIRBY, L.A.S. & PARIS, J.B. *Σ_n-collection schemas in arithmetic* ◊ C62 F30 ◊

KLEINBERG, E.M. *A combinatorial characterization of normal M-ultrafilters* ◊ C20 C62 E05 E40 E55 ◊

KUEHNRICH, M. *Superclasses in a finite extension of Zermelo set theory* ◊ C62 E30 E55 E70 G30 ◊

LIPSHITZ, L. & NADEL, M.E. *The additive structure of models of arithmetic* ◊ C15 C50 C57 C62 ◊

LIPTON, R.J. *Model theoretic aspects of computational complexity* ◊ C62 D15 F30 F65 H15 ◊

MAREK, W. ω-models of second order arithmetic and admissible sets ◊ C62 E30 E45 F35 ◊

MAREK, W. & NYBERG, A.M. Extendability of ZF models in the von Neumann hierarchy to models of KM theory of classes ◊ C62 C70 E30 E70 ◊

MAREK, W. & ZBIERSKI, P. On a class of models of the N-th order arithmetic ◊ C62 ◊

MAREK, W. Some comments on the paper by Artigue, Isambert, Perrin and Zalc: "Some remarks on bicommutability" ◊ C62 E30 E45 F25 F35 ◊

MCALOON, K. Completeness theorems, incompleteness theorems and models of arithmetic
◊ C25 C62 F30 ◊

MCALOON, K. Diagonal methods and strong cuts in models of arithmetic ◊ C57 C62 D80 F30 ◊

MILLS, G. A model of Peano arithmetic with no elementary end extension ◊ C25 C62 H15 ◊

MLCEK, J. A note on cofinal extensions and segments
◊ C30 C62 C65 ◊

MLCEK, J. End-extensions of countable structures and the induction schema ◊ C15 C62 F30 ◊

MURAWSKI, R. Indicators, satisfaction classes and expandability ◊ C62 H15 ◊

PARIS, J.B. Note on an induction axiom
◊ B28 C62 F30 ◊

PARIS, J.B. Some independence results for Peano arithmetic ◊ C62 F30 ◊

POTTHOFF, K. Orderings of types of countable arithmetic
◊ C20 C62 H15 ◊

SCHMERL, J.H. Extending models of arithmetic ◊ C62 ◊

SHELAH, S. End extensions and numbers of countable models ◊ C15 C30 C55 C62 E50 ◊

SHELAH, S. Models with second order properties. II: Trees with no undefined branches
◊ C50 C55 C62 C75 C80 C95 E05 E35 E65 ◊

STEEL, J.R. Forcing with tagged trees
◊ C62 D30 D55 E40 F35 ◊

WILKIE, A.J. Some results and problems on weak systems of arithmetic ◊ C62 F30 H15 ◊

YASUMOTO, M. Applications of Cherlin chains to set theory (Japanese) ◊ C62 ◊

ZBIERSKI, P. Axiomatizability of second order arithmetic with ω-rule ◊ B15 C62 E45 E70 F35 ◊

1979

ABRAMSON, F.G. Σ_1-separation
◊ C62 C70 D60 E45 E47 ◊

ALKOR, C. The problem of extendability (Italian summary) ◊ C50 C62 E70 ◊

ANAPOLITANOS, D.A. Automorphisms of finite order
◊ C07 C62 E25 E35 ◊

ANAPOLITANOS, D.A. Automorphisms of finite order and the axiom of choice ◊ C62 E25 E35 ◊

ARIBAUD, F. Un ideal maximal de l'anneau des endomorphismes d'un espace vectoriel de dimension infinie ◊ C60 C62 E40 E75 ◊

BELYAKIN, N.V. Nonstandardly finite sets (Russian)
◊ E30 H05 H15 ◊

CSIRMAZ, L. On definability in Peano arithmetic
◊ C40 C62 F30 ◊

DAWSON JR., J.W. The Goedel incompleteness theorem from a length-of-proof perspective
◊ A10 F20 F30 H15 ◊

DEMILLO, R.A. & LIPTON, R.J. Some connections between mathematical logic and complexity theory
◊ C57 D15 F30 H15 ◊

FRIEDMAN, H.M. On the naturalness of definable operations ◊ C30 C62 E35 E47 ◊

GUASPARI, D. Partially conservative extensions of arithmetic ◊ C62 F25 F30 ◊

HINNION, R. Modele constructible de la theorie des ensembles de Zermelo dans la theorie des types
◊ B15 C62 E30 E35 E45 ◊

KANOVEJ, V.G. The set of all analytically definable sets of natural numbers can be defined analytically (Russian)
◊ C62 D55 E15 E35 E45 ◊

LIPSHITZ, L. Diophantine correct models of arithmetic
◊ C15 C62 H15 ◊

LOLLI, G. & MAREK, W. On elementary theories of v. Neumann's levels ◊ C62 E30 ◊

MARTIN, D.A. & MITCHELL, W.J. On the ultrafilter of closed, unbounded sets
◊ C20 C62 E05 E35 E55 E60 ◊

MCALOON, K. Formes combinatoires du theoreme d'incompletude (d'apres J. Paris et d'autres) ◊ C62 ◊

MILLS, G. & PARIS, J.B. Closure properties of countable nonstandard integers ◊ H15 ◊

MILLS, G. Substructure lattices of models of arithmetic
◊ C62 H15 ◊

MURAWSKI, R. Indicators and the structure of expansions
◊ C62 H15 ◊

PENZIN, YU.G. Twins problem in formal arithmetic (Russian) ◊ F30 H15 ◊

RATAJCZYK, Z. On the number of expansions of the models of ZFC-set theory to models of KM-theory of classes
◊ C62 E70 ◊

RESL, M. On models in the alternative set theory
◊ C62 E70 ◊

ROSENTHAL, J.W. Truth in all of certain well-founded countable models arising in set theory II
◊ C40 C62 C70 E45 E47 ◊

SCHWABHAEUSER, W. Non-finitizability of a weak second-order theory ◊ B15 C62 C65 C85 F35 ◊

SOPRUNOV, S.F. Lattices of non-standard arithmetics (Russian) ◊ C62 E50 G10 H15 ◊

VAEAENAENEN, J. A new incompleteness in arithmetic (Finnish) (English summary)
◊ A10 C62 D25 E05 F30 ◊

ZHANG, JINWEN A nonstandard model of arithmetic constructed by means of a forcing method (Chinese) (English summary) ◊ E40 H15 ◊

1980

BASARAB, S.A. & NICA, V. & POPESCU, D. Approximation properties and existential completeness for ring morphisms ◊ C25 C60 H15 ◊

BLASS, A.R. Conservative extensions of models of arithmetic ◊ C20 C62 E05 ◊

CLOTE, P. Weak partition relations, finite games, and independence results in Peano arithmetic
◊ C62 E05 F30 ◊

CORRADA, M. *Remarks on extendability of set-theoretical models preserving cardinality* ⋄ C62 E30 E70 ⋄

CUDA, K. *An elimination of infinitely small quantities and infinitely large numbers (within the framework of AST)* ⋄ E70 H05 H15 ⋄

DICKMANN, M.A. *Types remarquables et extensions de modeles dans l'arithmetique de Peano I* ⋄ C25 C30 C62 ⋄

DRIES VAN DEN, L. *Some model theory and number theory for models of weak systems of arithmetic* ⋄ B25 C62 F30 H15 ⋄

DUBIEL, M. *Generalized quantifiers in models of set theory* ⋄ C62 C80 E35 ⋄

EBBINGHAUS, H.-D. *Mathematische Logik und Informatik* ⋄ C13 D10 D15 D98 E70 H15 ⋄

FLANNAGAN, T.B. *Expansions of models of ZFC* ⋄ C62 E25 ⋄

FOURMAN, M.P. *Sheaf models for set theory* ⋄ C62 C90 E25 E35 E70 F50 G30 ⋄

FREY, G. *Cyclic isogenies and nonstandard arithmetic* ⋄ H15 H20 ⋄

GOSTANIAN, R. *Constructible models of subsystems of ZF* ⋄ C62 E45 ⋄

GUASPARI, D. *Definability in models of set theory* ⋄ C62 D55 E45 E47 ⋄

HAJEK, P. & PUDLAK, P. *Two orderings of the class of all countable models of Peano arithmetic* ⋄ C15 C62 ⋄

HINNION, R. *Contraction de structures et application a NFU (les "New Foundations" des Quine avec extensionnalite pour les ensembles non vides). Definition du "degre de nonextensionnalite" d'une relation quelconque (English summary)* ⋄ C30 C62 E35 E70 ⋄

HODGES, W. *Interpreting number theory in nilpotent groups* ⋄ C25 C60 C62 F25 F35 ⋄

JAMBU-GIRAUDET, M. *Theorie des modeles de groupes d'automorphismes d'ensembles totalement ordonnes 2-homogenes (English summary)* ⋄ C07 C60 C62 C65 D35 E07 F25 F35 ⋄

KAKUDA, Y. *Set theory based on the language with the additional quantifier "for almost all" I* ⋄ C55 C62 C80 E70 ⋄

KANI, E. *Nonstandard diophantine geometry* ⋄ H15 H20 ⋄

KIRBY, L.A.S. *La methode des indicatrices et le theoreme d'incompletude* ⋄ C62 F30 ⋄

KOSSAK, R. *An application of definable types of Peano's arithmetic* ⋄ C62 ⋄

KOTLARSKI, H. *On Skolem ultrapowers and their non-standard variant* ⋄ C20 C62 E45 ⋄

KUEHNRICH, M. & SCHULTZ, KONRAD *A hierarchy of models for Skala's set theory* ⋄ C62 E35 E70 ⋄

LASCAR, D. *Une indicatrice de type "Ramsey" pour l'arithmetique de Peano et la formule de Paris-Harrington* ⋄ C62 F30 ⋄

MACINTYRE, A. *Nonstandard number theory* ⋄ C60 H15 ⋄

MACINTYRE, A. *Ramsey quantifiers in arithmetic* ⋄ C62 C80 F30 F35 ⋄

MANEVITZ, L.M. & STAVI, J. Δ_0^2 *operators and alternating sentences in arithmetic* ⋄ C62 F30 H15 ⋄

MAREK, W. & ZBIERSKI, P. *On the number of models of the Kelley-Morse theory of classes* ⋄ C62 E70 ⋄

MCALOON, K. *Les rapports entre la methode des indicatrices et la methode de Goedel pour obtenir des resultats d'independance* ⋄ C62 E35 F30 ⋄

MCALOON, K. (ED.) *Modeles de l'arithmetique* ⋄ C62 C97 F30 ⋄

MCALOON, K. *Progressions transfinies de theories axiomatiques, formes combinatoires du theoreme d'incompletude et fonctions recursives a croissance rapide* ⋄ C62 D20 F15 F30 ⋄

MURAWSKI, R. *Some remarks on the structure of expansions* ⋄ C50 C62 E45 ⋄

NADEL, M.E. *On a problem of MacDowell and Specker* ⋄ C50 C62 E50 ⋄

NERODE, A. & SHORE, R.A. *Reducibility orderings: theories, definability and automorphisms* ⋄ C62 D30 D35 ⋄

PACHOLSKI, L. & WIERZEJEWSKI, J. & WILKIE, A.J. (EDS.) *Model theory of algebra and arithmetic. Proceedings, Karpacz, Poland, September 1-7, 1979* ⋄ C60 C62 C97 ⋄

PARIS, J.B. *A hierarchy of cuts in models of arithmetic* ⋄ C15 C62 F30 ⋄

RESSAYRE, J.-P. *Types remarquables et extensions de modeles dans l'arithmetique de Peano. II (Avec un appendice de M.A.Dickmann)* ⋄ C30 C62 E05 ⋄

RICHARD, D. *Saturation des modeles de Peano* ⋄ C50 C62 H15 ⋄

SCHLIPF, J.S. *Recursively saturated models of set theory* ⋄ C50 C57 C62 E70 H20 ⋄

SMORYNSKI, C.A. & STAVI, J. *Cofinal extension preserves recursive saturation* ⋄ C50 C57 C62 ⋄

TVERSKOJ, A.A. *A sequence of combinatorial judgements which are independent of Peano arithmetic (Russian) (English summary)* ⋄ C62 E05 F30 ⋄

WILKIE, A.J. *Applications of complexity theory to Σ_0-definability problems in arithmetic* ⋄ C62 D15 F30 ⋄

YASUMOTO, M. *Classes on ZF models* ⋄ C62 E30 E47 ⋄

ZBIERSKI, P. *Indicators and incompleteness of Peano arithmetic* ⋄ C62 F30 ⋄

ZHANG, JINWEN *A family of models of lattice-valued sets (Chinese) (English summary)* ⋄ C62 C90 E70 ⋄

1981

ABRAMSON, F.G. *Locally countable models of Σ_1-separation* ⋄ C62 C70 E30 E47 ⋄

ACZEL, P. *Two notes on the Paris independence result I: A generalization of Ramsey's theorem. II: The ordinal height of a density* ⋄ C62 E05 F15 F30 ⋄

BERLINE, C. & MCALOON, K. & RESSAYRE, J.-P. (EDS.) *Model theory and arithmetic. Comptes rendus d'une action thematique programmee du C.N.R.S. sur la theorie des modeles et l'arithmetique* ⋄ C62 C97 F30 ⋄

BLASS, A.R. *The model of set theory generated by countably many generic reals* ⋄ C62 E25 E40 ⋄

BUKOVSKY, L. *Cogeneric extensions II* ⋄ C62 E40 ⋄

CHATZIDAKIS, Z. *La representation en termes de faisceaux des modeles de la theorie elementaire de la multiplication des entiers naturels*
⋄ C10 C30 C62 C90 ⋄

CLOTE, P. *Anti-basis theorems and their relation to independence results in Peano arithmetic*
⋄ C62 C98 E05 F30 ⋄

DRIES VAN DEN, L. *Which curves over Z have points with coordinates in a discrete ordered ring?*
⋄ B25 C60 C62 ⋄

ELLENTUCK, E. *Hyper-torre isols* ⋄ C62 D50 ⋄

GUZICKI, W. *The equivalence of definable quantifiers in second order arithmetic*
⋄ C30 C55 C62 C80 E40 E45 F35 ⋄

HAJEK, P. *Completion closed algebras and models of Peano arithmetic* ⋄ C62 F30 ⋄

HINNION, R. *Extensional quotients of structures and applications to the study of the axiom of extensionality*
⋄ C30 C62 E30 E35 E70 ⋄

JAMBU-GIRAUDET, M. *Interpretations d'arithmetiques dans des groupes et des treillis*
⋄ C07 C62 C65 D35 E07 F25 F30 G10 ⋄

KAMO, S. *Nonstandard natural number systems and nonstandard models* ⋄ H15 ⋄

KAUFMANN, M. *On existence of Σ_n-end extensions*
⋄ C62 ⋄

KIRBY, L.A.S. & MCALOON, K. & MURAWSKI, R. *Indicators, recursive saturation and expandability*
⋄ C50 C57 C62 H15 ⋄

KOSSAK, R. & PARIS, J.B. *Subsets of models of arithmetic*
⋄ C62 F30 ⋄

KOTLARSKI, H. & KRAJEWSKI, S. & LACHLAN, A.H. *Construction of satisfaction classes for nonstandard models* ⋄ C50 C62 ⋄

KOTLARSKI, H. *On elementary cuts in models of arithmetic* ⋄ C50 C57 C62 ⋄

KURATA, R. *The reflection principle, transfinite induction, and the Paris-Harrington principle (Japanese)*
⋄ C62 F30 ⋄

LACHLAN, A.H. *Full satisfaction classes and recursive saturation* ⋄ C50 C57 C62 ⋄

MACINTYRE, A. *A theorem of Rabin in a general setting*
⋄ C25 C62 ⋄

MCALOON, K. & RESSAYRE, J.-P. *Les methodes de Kirby-Paris et la theorie des ensembles*
⋄ C62 E05 E30 E55 F30 ⋄

MISERCQUE, D. *Solutions de deux problemes poses par H. Simmons* ⋄ C62 F30 H15 ⋄

MURAWSKI, R. *A note on inner interpretations of models of Peano arithmetic* ⋄ C62 ⋄

MURAWSKI, R. *A simple remark on satisfaction classes, indiscernibles and recursive saturation*
⋄ C30 C50 C57 H15 ⋄

MURAWSKI, R. *Incompleteness of Σ_n^0-definable theories via indicators* ⋄ C62 ⋄

MURAWSKI, R. *Some more remarks on expandability of initial segments* ⋄ C62 ⋄

PABION, J.F. *Saturation en arithmetique et en analyse*
⋄ C50 C62 ⋄

PABION, J.F. & RICHARD, D. *Synonymy and re-interpretation for some sublanguages of Peano arithmetic* ⋄ B28 C62 F30 ⋄

PARIS, J.B. & WILKIE, A.J. *Δ_0 sets and induction*
⋄ C62 F30 ⋄

PARIS, J.B. & WILKIE, A.J. *Models of arithmetic and the rudimentary sets* ⋄ C62 D20 ⋄

PARIS, J.B. *Some conservation results for fragments of arithmetic* ⋄ C62 F30 ⋄

PILLAY, A. *Cuts in models of arithmetic* ⋄ C62 E05 ⋄

PILLAY, A. *Models of Peano arithmetic (a survey of basic results)* ⋄ C62 C98 H15 ⋄

PILLAY, A. *Partition properties and definable types in Peano arithmetic* ⋄ C30 C62 E05 ⋄

PLOTKIN, J.M. *ZF and locally finite groups*
⋄ C60 C62 E25 E35 E75 ⋄

RASKOVIC, M. *On extendability of models of ZF set theory to models of alternative set theory* ⋄ C62 E70 ⋄

RATAJCZYK, Z. *A characterization of expandability of models for ZF to models for KM*
⋄ C62 C70 E45 E70 ⋄

RATAJCZYK, Z. *Note on submodels for A_2* ⋄ C62 ⋄

RICHARD, D. *De la structure additive a la saturation des modeles de Peano et a une classification des sous-langages de l'arithmetique* ⋄ C50 C62 ⋄

SAGEEV, G. *A model of ZF + there exists an inaccessible, in which the Dedekind cardinals constitute a natural nonstandard model of arithmetic* ⋄ E25 E35 H15 ⋄

SCHMERL, J.H. *Recursively saturated, rather classless models of Peano arithmetic*
⋄ C50 C57 C62 E45 E55 ⋄

SMORYNSKI, C.A. *Cofinal extensions of nonstandard models of arithmetic* ⋄ C50 C57 C62 ⋄

SMORYNSKI, C.A. *Elementary extensions of recursively saturated models of arithmetic* ⋄ C50 C57 C62 ⋄

SMORYNSKI, C.A. *Recursively saturated nonstandard models of arithmetic* ⋄ C50 C57 C62 ⋄

TWER VON DER, T. *On the strength of several versions of Dirichlets ("pigeon-hole"-) principle in the sense of first-order logic* ⋄ C10 H15 ⋄

TWER VON DER, T. *Some remarks on the mathematical incompleteness of Peano's arithmetic found by Paris and Harrington* ⋄ C62 E05 F30 ⋄

VAEAENAENEN, J. *Generalized quantfiers in models of set theory* ⋄ C55 C62 C80 C85 C95 E35 ⋄

WILKIE, A.J. *On discretely ordered rings in which every definable ideal is principal* ⋄ C60 C62 H15 ⋄

WILSON, T.P. *General models of set theory*
⋄ C62 E55 E70 ⋄

ZBIERSKI, P. *Nonstandard interpretations of higher order theories* ⋄ C62 F35 H15 ⋄

1982

BOFFA, M. *Algebres de Boole atomiques et modeles de la theorie des types* ⋄ B15 C62 E70 ⋄

CEGIELSKI, P. & MCALOON, K. & WILMERS, G.M. *Modeles recursivement satures de l'addition et de la multiplication des entiers naturels (English summary)*
⋄ C50 C57 C62 F30 H15 ⋄

CUDA, K. *An elimination of the predicate "to be a standard member" in nonstandard models of arithmetic*
  ⋄ H15 ⋄
DAVID, R. *Some applications of Jensen's coding theorem*
  ⋄ C62 D55 E40 E45 ⋄
DIMITRACOPOULOS, C. & PARIS, J.B. *Truth definitions for Δ_0 formulae* ⋄ C40 C62 H15 ⋄
FORSTER, T.E. *Axiomatizing set theory with a universal set*
  ⋄ C62 E70 ⋄
GORDEEV, L.N. *Constructive models for set theory with extensionality* ⋄ C62 E70 F50 ⋄
KIRBY, L.A.S. & PARIS, J.B. *Accessible independence results for Peano arithmetic* ⋄ C62 F30 ⋄
KIRBY, L.A.S. *Flipping properties in arithmetic*
  ⋄ C62 E05 F30 ⋄
KNIGHT, J.F. & NADEL, M.E. *Models of arithmetic and closed ideals* ⋄ C62 D30 ⋄
KOCHEN, S. & KRIPKE, S.A. *Nonstandard models of Peano arithmetic* ⋄ C20 C62 ⋄
KRANAKIS, E. *Definable ultrafilters and end extensions of constructible sets* ⋄ C62 E05 E45 E47 ⋄
LYNCH, J.F. *On sets of relations definable by addition*
  ⋄ C40 C62 ⋄
MACINTYRE, A. *Residue fields of models of P*
  ⋄ C57 C60 C62 H15 ⋄
MAREK, W. & PRISCO DI, C.A. *Models closed under projective operations (Russian summary)*
  ⋄ C62 E15 E45 ⋄
MARKER, D. *Degrees of models of true arithmetic*
  ⋄ C57 C62 D30 H15 ⋄
McALOON, K. *On the complexity of models of arithmetic*
  ⋄ C57 C62 ⋄
MISERCQUE, D. *The nonhomogeneity of the E-tree - answer to a problem raised by D. Jensen and A. Ehrenfeucht* ⋄ C62 F30 ⋄
MORGENSTERN, C.F. *On generalized quantifiers in arithmetic* ⋄ C62 C80 F30 ⋄
PABION, J.F. Π_2-*theorie des ensembles*
  ⋄ C50 C62 E30 E45 ⋄
PABION, J.F. *Saturated models of Peano arithmetic*
  ⋄ C50 C62 ⋄
PETRY, A. *Une characterisation algebrique des structures satisfaisant les memes sentences stratifees*
  ⋄ C20 C62 E70 ⋄
RATAJCZYK, Z. *Satisfaction classes and combinatorial sentences independent from PA* ⋄ C62 E05 F30 ⋄
SCHMERL, J.H. & SIMPSON, S.G. *On the role of Ramsey quantifiers in first order arithmetic*
  ⋄ B28 C10 C62 C80 ⋄
SIMPSON, S.G. *Set theoretic aspects of ATR_0*
  ⋄ C62 E30 F35 ⋄
SMORYNSKI, C.A. *A note on initial segment constructions in recursively saturated models of arithmetic*
  ⋄ C50 C57 C62 ⋄
SMORYNSKI, C.A. *Back-and-forth inside a recursively saturated model of arithmetic*
  ⋄ C15 C50 C57 C62 ⋄
SMORYNSKI, C.A. *Nonstandard models and constructivity*
  ⋄ C62 F30 F50 ⋄

SOCHOR, A. *Metamathematics of the alternative set theory II* ⋄ C20 C62 E35 E70 ⋄
SUN, WENZHI *A property of the upper bounding ordinal of a model of ZFC (Chinese) (English summary)*
  ⋄ C62 E10 ⋄
TSUBOI, A. *On M-recursively saturated models of arithmetic* ⋄ C50 C57 C62 ⋄
TVERSKOJ, A.A. *Investigation of recursiveness and arithmeticity of signature functions in nonstandard models of arithmetics (Russian)*
  ⋄ C57 C62 D45 H15 ⋄
VAEAENAENEN, J. *Generalized quantifiers in models of set theory* ⋄ C55 C62 C80 C85 C95 E35 ⋄
WILKIE, A.J. *On core structures for Peano arithmetic*
  ⋄ C50 C62 F30 H15 ⋄
ZARACH, A. *Unions of ZF^--models which are themselves ZF^--models* ⋄ C62 E30 E35 ⋄

1983

ABE, Y. *Strongly compact cardinals and the fixed points of elementary embeddings* ⋄ C62 E55 ⋄
AJTAI, M. Σ_1^1-*formulae on finite structures*
  ⋄ C13 C62 D15 H15 ⋄
BASARAB, S.A. *Roth's theorem: non-standard aspects (Romanian)* ⋄ H10 H15 ⋄
BELYAKIN, N.V. *A means of modeling a classical second-order arithmetic (Russian)*
  ⋄ B15 C62 D55 F35 ⋄
BOFFA, M. *Arithmetic and theory of types*
  ⋄ B15 C62 E70 F35 ⋄
CANTINI, A. Σ-*models of subtheories of KPI (Italian)*
  ⋄ C62 C70 E30 ⋄
CANTINI, A. *A note on the theory of admissible sets with ε-induction restricted to formulas with one quantifier and related systems (Italian summary)*
  ⋄ C62 E30 E45 E70 F10 F15 F35 ⋄
CLOTE, P. & McALOON, K. *Two further combinatorial theorems equivalent to the 1-consistency of Peano arithmetic* ⋄ B28 C62 F30 ⋄
CUDA, K. *Nonstandard models of arithmetic as an alternative basis for continuum considerations*
  ⋄ E70 F30 H15 ⋄
DIMITRACOPOULOS, C. & PARIS, J.B. *A note on the undefinability of cuts* ⋄ C40 C62 F30 H15 ⋄
ERSHOV, YU.L. *The Σ-enumeration principle (Russian)*
  ⋄ C62 C70 D60 E30 E65 ⋄
FOREMAN, M. *More saturated ideals*
  ⋄ C55 C62 E05 E35 E55 ⋄
FORTI, M. & HONSELL, F. *Formalizzazioni del "Principio di libera costruzione"* ⋄ C62 E35 E70 ⋄
FREY, G. *Nonstandard arithmetic and application to height functions* ⋄ C60 H10 H15 H20 ⋄
GUZICKI, W. *Definable quantifiers in second order arithmetic and elementary extensions of ω-models*
  ⋄ C30 C55 C62 C80 E40 E50 F35 ⋄
KAUFMANN, M. *Blunt and topless end extensions of models of set theory* ⋄ C62 E55 ⋄
KAUFMANN, M. *Set theory with a filter quantifier*
  ⋄ C62 C80 E30 E70 ⋄

KNIGHT, J.F. *Additive structure in uncountable models for a fixed completion of P*
⋄ C50 C57 C62 C75 D30 ⋄

KOSSAK, R. *A certain class of models of Peano arithmetic*
⋄ C50 C57 C62 ⋄

KOTLARSKI, H. *On cofinal extensions of models of arithmetic* ⋄ C50 C62 ⋄

KOTLARSKI, H. *On elementary cuts in models of arithmetic* ⋄ C50 C57 C62 ⋄

MAGIDOR, M. & SHELAH, S. & STAVI, J. *On the standard part of nonstandard models of set theory*
⋄ C62 E45 H05 ⋄

MAREK, W. & RASMUSSEN, K. *Spectrum of L*
⋄ C55 C62 E45 ⋄

MIJAJLOVIC, Z. *On Σ_1^0-extensions of ω*
⋄ C15 C30 C62 ⋄

MIJAJLOVIC, Z. *Submodels and definable points in models of Peano arithmetics* ⋄ C62 ⋄

MLCEK, J. *Compactness and homogeneity of saturated structures I,II* ⋄ C35 C50 C60 C62 ⋄

MONTAGNA, F. *ZFC-models as Kripke-models*
⋄ B45 C62 E30 F30 ⋄

PUDLAK, P. *A definition of exponentiation by a bounded arithmetical formula* ⋄ B28 C62 F30 ⋄

RESSAYRE, J.-P. *Modeles isomorphes a leurs propres segments initiaux en theorie des ensembles (English summary)* ⋄ C62 E30 ⋄

RIMSCHA VON, M. *Bases of ZF^0-models*
⋄ C62 E47 E70 ⋄

SEMENOV, A.L. *Logical theories of one-place functions on the set of natural numbers (Russian)*
⋄ B25 C10 C62 C85 ⋄

SOCHOR, A. *Metamathematics of the alternative set theory III* ⋄ C62 E35 E70 ⋄

TSUBOI, A. *On M-recursively saturated models of Peano arithmetic (Japanese)* ⋄ C50 C57 C62 ⋄

YASUMOTO, M. *Nonstandard arithmetic of function fields over H-convex subfields of $^*Q$ (Japanese)*
⋄ H05 H15 ⋄

YASUMOTO, M. *Nonstandard arithmetic of function fields over H-convex subfields of $^*Q$* ⋄ H05 H15 ⋄

ZADROZNY, W. *Iterating ordinal definability*
⋄ C62 E35 E45 ⋄

1984

ABRAHAM, U. *A minimal model for $\neg$ CH: iteration of Jensen's reals*
⋄ C62 D30 E35 E40 E45 E50 E65 ⋄

ADAMOWICZ, Z. *Axiomatization of the forcing relation with an application to Peano arithmetic*
⋄ C62 E40 F30 ⋄

BAETEN, J. *Filters and ultrafilters over definable subsets of admissible ordinals* ⋄ C62 D60 E05 E10 E47 ⋄

CSIRMAZ, L. & PARIS, J.B. *A property of 2-sorted Peano models and program verification* ⋄ B75 H15 ⋄

CUDA, K. & VOJTASKOVA, B. *Models of AST without choice* ⋄ C62 E25 E35 E70 E75 H15 ⋄

ENAYAT, A. *On certain elementary extensions of models of set theory* ⋄ C30 C62 E55 ⋄

GIRARD, J.-Y. *The Ω-rule* ⋄ B15 C62 C75 F35 ⋄

GITIK, M. *The nonstationary ideal on $\aleph_2$*
⋄ C62 E05 E35 E55 ⋄

GUREVICH, Y. & SHELAH, S. *The monadic theory and the "next world"* ⋄ C62 C65 C85 E40 F25 ⋄

HAJEK, P. *On a new notion of partial conservativity*
⋄ C62 F30 ⋄

HARRINGTON, L.A. *Admissible ordinals are not immortal*
⋄ C62 C70 E30 E45 E47 ⋄

HENLE, J.M. *An extravagant partition relation for a model of arithmetic* ⋄ C62 E05 E07 E25 E35 E40 ⋄

HENSON, C.W. & KAUFMANN, M. & KEISLER, H.J. *The strength of nonstandard methods in arithmetic*
⋄ C62 E30 F30 F35 H05 H15 ⋄

KAUFMANN, M. & KRANAKIS, E. *Definable ultrapowers and ultrafilters over admissible ordinals*
⋄ C20 C62 D60 E05 E45 ⋄

KAUFMANN, M. *Mutually generic classes and incompatible expansions* ⋄ C62 E40 ⋄

KAUFMANN, M. *On expandability of models of arithmetic and set theory to models of weak second-order theories*
⋄ C15 C30 C62 C70 C75 E30 E70 ⋄

KAUFMANN, M. & SCHMERL, J.H. *Saturation and simple extensions of models of Peano-arithmetic*
⋄ C50 C57 C62 ⋄

KIRBY, L.A.S. *Ultrafilters and types on models of arithmetic* ⋄ C20 C62 E05 H15 ⋄

KNIGHT, J.F. & LACHLAN, A.H. & SOARE, R.I. *Two theorems on degrees of models of true arithmetic*
⋄ C57 C62 D30 ⋄

KOSSAK, R. *$L_{\infty \omega_1}$-elementary equivalence of ω_1-like models of PA* ⋄ C50 C57 C62 C75 ⋄

KOSSAK, R. *Remarks on free sets*
⋄ C50 C57 C62 E05 ⋄

KOTLARSKI, H. *On elementary cuts in recursively saturated models of Peano arithmetic* ⋄ C50 C57 C62 ⋄

KOTLARSKI, H. *Some remarks on initial segments in models of Peano arithmetic* ⋄ C62 ⋄

KURATA, R. *Paris-Harrington theory and reflection principles* ⋄ C62 F30 ⋄

MACINTYRE, A. & MARKER, D. *Degrees of recursively saturated models* ⋄ C50 C57 C62 D30 D45 ⋄

MAGIDOR, M. & SHELAH, S. & STAVI, J. *Countably decomposable admissible sets*
⋄ C40 C62 C70 E30 E45 E47 ⋄

MILLS, G. & PARIS, J.B. *Regularity in models of arithmetic*
⋄ C62 C80 F30 ⋄

MOTOHASHI, N. *A normal form theorem for first order formulas and its application to Gaifman's splitting theorem* ⋄ C07 C62 ⋄

MOTOHASHI, N. *A normal form theorem for first order formulas and its application to Gaifman's splitting theorem* ⋄ B10 C07 C62 ⋄

MURAWSKI, R. *A contribution to nonstandard teratology*
⋄ H15 ⋄

MURAWSKI, R. *Expandability of models of arithmetic*
⋄ C62 H15 ⋄

MURAWSKI, R. *Trace expansions of initial segments*
⋄ C62 ⋄

NEGRI, M. *An application of recursive saturation*
  ◇ C50 C57 C62 F30 ◇
PARIS, J.B. *On the structure of models with restricted E_1-induction (Czech) (Russian and English summaries)*
  ◇ C62 F30 ◇
POIZAT, B. *Deux remarques a propos de la propriete de recouvrement fini* ◇ C45 C50 C60 C62 ◇
PUDLAK, P. & SOCHOR, A. *Models of the alternative set theory* ◇ C50 C62 E50 E70 ◇
RESSAYRE, J.-P. *Deux fantaisies d'univers non standard (English summary)* ◇ C62 E30 ◇
RICHARD, D. *The arithmetics as theories of two orders (English and French summaries)*
  ◇ B28 C62 D35 F30 ◇
SCHMERL, U.R. *Diophantine equations in a fragment of number theory* ◇ C62 F30 ◇
SEREMET, Z. *On automorphisms of resplendent models of arithmetic* ◇ C07 C50 C62 ◇
SHELAH, S. *On logical sentences in PA*
  ◇ C62 C80 E05 F25 F30 ◇
SMORYNSKI, C.A. *Lectures on nonstandard models of arithmetic* ◇ A10 C62 C98 F30 F98 H15 ◇
TVERSKOJ, A.A. *Constructivizability of formal arithmetical structures (Russian)* ◇ C57 D45 H15 ◇
VETULANI, Z. *Ramified analysis and the minimal β-models of higher order arithmetics*
  ◇ C62 D55 E45 F35 F65 ◇
VISSER, A. *The provability logic of recursively enumerable theories extending Peano arithmetic and arbitrary theories extending Peano arithmetic*
  ◇ B45 C62 F30 ◇
ZADROZNY, W. *Ordinal definability in Jensen's model*
  ◇ C62 E45 ◇

1985

ABRAHAM, U. & RUBIN, M. & SHELAH, S. *On the consistency of some partition theorems for continuous colorings, and structure of $\aleph_1$-dense real order types*
  ◇ C55 C62 C65 C80 E05 E07 E35 ◇
ADAMOWICZ, Z. & MORALES-LUNA, G. *A recursive model for arithmetic with weak induction*
  ◇ C57 F30 H15 ◇
ADAMOWICZ, Z. *Algebraic approach to $\exists_1$ induction*
  ◇ C60 C62 F30 ◇
ALLING, N.L. *Conway's field of surreal numbers*
  ◇ H10 H15 H20 ◇
APTER, A.W. *Successors of singular cardinals and measurability* ◇ C62 E25 E35 E55 ◇
CLOTE, P. *Applications of the low-basis theorem in arithmetic* ◇ C62 D20 D30 D55 F30 ◇
CLOTE, P. *Partition relations in arithmetic*
  ◇ C62 E05 F30 ◇
DAVID, R. & FRIEDMAN, S.D. *Uncountable ZF-ordinals*
  ◇ C62 D60 E40 E45 ◇

DIMITRACOPOULOS, C. *A generalization of a theorem of H.Friedman* ◇ C62 D15 F30 ◇
ENAYAT, A. *Weakly compact cardinals in models of set theory* ◇ C30 C62 E50 E55 E65 ◇
FORTI, M. *f-admissible models of set theory (Italian)*
  ◇ C62 ◇
GAVALEC, M. *Local properties of complete Boolean products* ◇ C30 C62 G05 ◇
GRILLIOT, T.J. *Disturbing arithmetic*
  ◇ C07 C30 C40 C62 ◇
KOSSAK, R. *A note on satisfaction classes*
  ◇ C50 C57 C62 F30 ◇
KOSSAK, R. *Recursively saturated ω_1-like models of arithmetic* ◇ C55 C57 C62 C75 C80 E65 ◇
KOTLARSKI, H. *Bounded induction and satisfaction classes* ◇ C15 C50 C57 C62 ◇
MCCARTHY, T. *Abstraction and definability in semantically closed structures*
  ◇ A05 C40 C62 C90 ◇
MLCEK, J. *Some automorphisms of natural numbers in the alternative set theory* ◇ C07 C30 E70 H15 ◇
PARIS, J.B. & WILKIE, A.J. *Counting problems in bounded arithmetic* ◇ B28 C62 F30 ◇
PINO, R. & RESSAYRE, J.-P. *Definable ultrafilters and elementary end extensions*
  ◇ C20 C30 C62 E05 E45 E47 ◇
PUDLAK, P. *Cuts, consistency statements and interpretations* ◇ C62 E30 F25 ◇
PUDLAK, P. & SOCHOR, A. *Elementary extensions of models of the alternative set theory*
  ◇ C62 E70 H15 ◇
SCHMERL, J.H. *Models of Peano arithmetic and a question of Sikorski on ordered fields*
  ◇ C55 C60 C62 E05 E75 ◇
SHEARD, M. *Co-critical points of elementary embeddings*
  ◇ C62 E45 E55 ◇
SIMPSON, S.G. *Friedman's research on subsystems of second order arithmetic* ◇ C62 C98 F35 F98 ◇
SMORYNSKI, C.A. *Nonstandard models and related developments* ◇ C62 C98 F30 F98 ◇
SOCHOR, A. *Constructibility and shiftings of view*
  ◇ E45 E70 H15 ◇
SOCHOR, A. *Notes on revealed classes* ◇ E70 H15 ◇
SOLOVAY, R.M. *Infinite fixed-point algebras*
  ◇ C57 C62 G25 ◇
SURESON, C. *Non-closure of the image model and absence of fixed points* ◇ C62 E05 E45 E55 ◇
TVERSKOJ, A.A. *Constructivizable and nonconstructivizable formal arithmetic structures (Russian)* ◇ C57 C62 D45 H15 ◇
TZOUVARAS, A.D. *Minimal ultrafilters and maximal endomorphic universes* ◇ C62 E70 ◇
WILMERS, G.M. *Bounded existential induction*
  ◇ C62 F30 ◇

C65 Models of other mathematical theories

1920
SKOLEM, T.A. *Logisch-kombinatorische Untersuchungen ueber die Erfuellbarkeit oder Beweisbarkeit mathematischer Saetze nebst einem Theorem ueber dichte Mengen* ⋄ B10 B25 C07 C65 C75 ⋄

1925
URYSOHN, P. *Sur un espace metrique universel*
⋄ C50 C65 E75 ⋄

1926
LANGFORD, C.H. *Analytic completeness of postulate sets*
⋄ B20 C35 C65 E07 ⋄
LANGFORD, C.H. *Some theorems on deducibility*
⋄ B30 C10 C35 C65 ⋄
LANGFORD, C.H. *Theorems on deducibility (second paper)*
⋄ B30 C10 C35 C65 ⋄

1927
LANGFORD, C.H. *On a type of completeness characterizing the general laws for separation of point pairs*
⋄ C35 C65 E07 ⋄
URYSOHN, P. *Sur un espace metrique universel I,II*
⋄ C50 C65 E75 ⋄

1931
FOLLEY, K.W. *Sets of the first and second α-category and α-residual sets* ⋄ C55 C65 E07 E75 ⋄
TARSKI, A. *Sur les ensembles definissables de nombres reels I*
⋄ B25 B28 C10 C40 C60 C65 D55 E15 E47 F35 ⋄

1935
LINDENBAUM, A. & TARSKI, A. *Ueber die Beschraenktheit der Ausdrucksmittel deduktiver Theorien*
⋄ B30 C07 C35 C65 ⋄

1939
LANGFORD, C.H. *A theorem on deducibility for second-order functions* ⋄ B15 C35 C65 C85 ⋄

1941
MCKINSEY, J.C.C. *A solution of the decision problem for the Lewis systems S2 and S4, with an application to topology* ⋄ B25 B45 C65 ⋄

1944
MCKINSEY, J.C.C. & TARSKI, A. *The algebra of topology*
⋄ C05 C50 C65 G05 G25 ⋄

1946
MCKINSEY, J.C.C. & TARSKI, A. *On closed elements in closure algebras* ⋄ B25 C05 C65 G05 G25 ⋄

1948
HEWITT, E. *Rings of real-valued continuous functions*
⋄ C20 C50 C60 C65 E75 ⋄

1951
ROBINSON, R.M. *Arithmetical definitions in the ring of integers* ⋄ C65 F30 ⋄

1952
HIGMAN, G. *Ordering by divisibility in abstract algebras*
⋄ C05 C65 ⋄

1953
DENJOY, A. *Les matrices d'ordination de toutes puissances*
⋄ C65 E07 ⋄

1954
BETH, E.W. *Observations metamathematiques sur les structures simplement ordonnees* ⋄ B25 C65 ⋄
KREISEL, G. *Applications of mathematical logic to various branches of mathematics* ⋄ B98 C65 ⋄
MAL'TSEV, A.I. *On the general theory of algebraic systems (Russian)* ⋄ C05 C65 ⋄

1955
BERNAYS, P. *Betrachtungen ueber das Vollstaendigkeitsaxiom und verwandte Axiome*
⋄ B28 C60 C65 ⋄
FRAISSE, R. *Sur quelques classifications des relations basees sur des isomorphismes restreints I: Etude generale. II: Applications aux relations d'ordre, et construction d'exemples montrant que ces classifications sont distinctes* ⋄ C07 C20 C65 E07 ⋄

1956
FRAISSE, R. *Application des γ-operateurs au calcul logique du premier echelon* ⋄ B10 C07 C65 ⋄
JOHNSTON, J.B. *Universal infinite partially ordered sets*
⋄ C50 C65 E07 E50 ⋄
ROBINSON, A. *Apercu metamathematique sur les nombres reels* ⋄ C60 C65 ⋄
SCHWABHAEUSER, W. *Ueber die Vollstaendigkeit der elementaren euklidischen Geometrie*
⋄ B30 C35 C65 ⋄

1957
COHN, P.M. *Groups of order automorphisms of ordered sets* ⋄ C07 C65 E07 ⋄
EHRENFEUCHT, A. *Application of games to some problems of mathematical logic (Russian summary)*
⋄ C07 C35 C65 E60 ⋄
FEFERMAN, S. *Some recent work of Ehrenfeucht and Fraisse* ⋄ B10 C07 C65 E10 ⋄

1958

LAUGWITZ, D. & SCHMIEDEN, C. *Eine Erweiterung der Infinitesimalrechnung* ⋄ B28 C65 E20 H05 ⋄

1959

HENKIN, L. & SUPPES, P. & TARSKI, A. (EDS.) *The axiomatic method, with special reference to geometry and physics* ⋄ B30 B97 C65 D35 ⋄

SCHWABHAEUSER, W. *Entscheidbarkeit und Vollstaendigkeit der elementaren hyperbolischen Geometrie* ⋄ B25 B30 C35 C65 ⋄

1960

ROBINSON, A. & ZAKON, E. *Elementary properties of ordered abelian groups* ⋄ B25 C35 C60 C65 ⋄

1961

EHRENFEUCHT, A. *An application of games to the completeness problem for formalized theories* ⋄ C07 C65 C85 E10 E60 ⋄

1962

ALLING, N.L. *On the existence of real-closed fields that are η_α-sets of power $\aleph_\alpha$* ⋄ C55 C60 C65 ⋄

GRZEGORCZYK, A. *A kind of categoricity* ⋄ C35 C50 C62 C65 ⋄

RAUTENBERG, W. *Ueber metatheoretische Eigenschaften einiger geometrischer Theorien* ⋄ B30 C65 D35 ⋄

SCHWABHAEUSER, W. *On completeness and decidability and some non-definable notions of elementary hyperbolic geometry* ⋄ B25 C35 C65 ⋄

SZCZERBA, L.W. *On the models of the affine geometry* ⋄ C65 ⋄

SZMIELEW, W. *New foundations of absolute geometry* ⋄ B30 C65 ⋄

1963

ROGERS JR., H. *An example in mathematical logic* ⋄ B25 C10 C65 ⋄

SCARPELLINI, B. *Eine Anwendung der unendlichwertigen Logik auf topologische Raeume* ⋄ B80 C65 ⋄

TAJMANOV, A.D. *Decidability of the elementary theory of sphere inclusion (Russian)* ⋄ B25 C65 ⋄

1964

FENSTAD, J.E. *Model theory, ultraproducts and topology* ⋄ C20 C65 E75 ⋄

MOREL, A.C. *Ordering relations admitting automorphisms* ⋄ C07 C65 E07 ⋄

MOSCHOVAKIS, Y.N. *Recursive metric spaces* ⋄ C57 C65 D45 F60 ⋄

1965

RAUTENBERG, W. *Hilberts Schnittpunktsaetze und einige modelltheoretische Aspekte der formalisierten Geometrie* ⋄ C65 ⋄

SCHWABHAEUSER, W. *On models of elementary elliptic geometry* ⋄ B25 C65 ⋄

STONE, A.L. *Extensive ultraproducts and Haar measures* ⋄ C20 C65 E75 H05 ⋄

SZCZERBA, L.W. & TARSKI, A. *Metamathematical properties of some affine geometries* ⋄ B25 B30 C52 C65 D35 ⋄

1966

AMITSUR, S.A. *Rational identities and applications to algebra and geometry* ⋄ C50 C60 C65 ⋄

AX, J. & KOCHEN, S. *Diophantine problems over local fields, III: Decidable fields* ⋄ B25 C10 C20 C35 C60 C65 ⋄

KROM, MELVEN R. *A property of sentences that define quasi-order* ⋄ C65 ⋄

LAEUCHLI, H. & LEONARD, J. *On the elementary theory of linear order* ⋄ B25 C65 ⋄

LOPEZ-ESCOBAR, E.G.K. *On defining well-orderings* ⋄ C55 C65 C75 E10 ⋄

NOGINA, E.YU. *On effective topological spaces* ⋄ C57 C65 ⋄

RAUTENBERG, W. *Ueber Hilberts Schnittpunktsaetze* ⋄ B30 C65 ⋄

1967

FEFFERMAN, C. *Cardinally maximal sets of non-equivalent order types* ⋄ C65 C75 E07 ⋄

GRZEGORCZYK, A. *Some relational systems and the associated topological spaces* ⋄ B45 C65 G25 ⋄

KOPPERMAN, R.D. *Application of infinitary languages to metric spaces* ⋄ C65 C75 ⋄

KOPPERMAN, R.D. *On the axiomatizability of uniform spaces* ⋄ C65 C75 ⋄

KOPPERMAN, R.D. *The L_{ω_1,ω_1}-theory of Hilbert spaces* ⋄ C65 C75 ⋄

RAUTENBERG, W. *Elementare Schemata nichtelementarer Axiome* ⋄ B25 C52 C60 C65 C85 D35 ⋄

TARSKI, A. *The completeness of elementary algebra and geometry* ⋄ B25 B30 C10 C35 C60 C65 ⋄

1968

ERSHOV, YU.L. *Restricted theories of totally ordered sets (Russian)* ⋄ B25 C65 D35 E07 ⋄

LAEUCHLI, H. *A decision procedure for the weak second order theory of linear order* ⋄ B15 B25 C65 C85 ⋄

LOPEZ-ESCOBAR, E.G.K. *Well-orderings and finite quantifiers* ⋄ C65 C75 ⋄

MYCIELSKI, J. & RYLL-NARDZEWSKI, C. *Equationally compact algebras II* ⋄ C05 C20 C50 C65 ⋄

RAUTENBERG, W. *Unterscheidbarkeit endlicher geordneter Mengen mit gegebener Anzahl von Quantoren* ⋄ B20 C07 C13 C65 D35 ⋄

ROSENSTEIN, J.G. *Theories which are not $\aleph_0$-categorical* ⋄ C35 C65 ⋄

1969

BONNET, R. & POUZET, M. *Extensions et stratifications d'ensembles disperses* ⋄ C65 E07 ⋄

BONNET, R. *Stratifications et extension des genres de chaines denombrables* ⋄ C15 C65 E07 ⋄

KEISLER, H.J. *Infinite quantifiers and continuous games* ⋄ C65 C75 C80 ⋄

KOPPERMAN, R.D. *Applications of infinitary languages to analysis* ⋄ C35 C60 C65 C75 ⋄

NOGINA, E.YU. *Correlations between certain classes of effectively topological spaces (Russian)* ⋄ C57 C65 D45 ⋄

PHILLIPS, R.G. *Liouville's theorem* ⋄ C65 H05 ⋄
ROSENSTEIN, J.G. $\aleph_0$*-categoricity of linear orderings*
  ⋄ C35 C65 E07 ⋄
TAYLOR, W. *Atomic compactness and graph theory*
  ⋄ C05 C50 C65 ⋄
ZAKON, E. *Remarks on the nonstandard real axis*
  ⋄ C65 H05 ⋄

1970

BAKER, K.A. & FISHBURN, P.C. & ROBERTS, F.S. *A new characterization of partial orders of dimension two*
  ⋄ C65 E07 ⋄
BARANSKIJ, V.A. & TRAKHTMAN, A.N. *Subsemigroup graphs (Russian) (English summary)* ⋄ C65 D05 ⋄
KONCEWICZ, L. *The simplest expressions of the predicate calculus that define classes of symmetric graphs without loops I (Polish) (English and Russian summaries)*
  ⋄ C65 ⋄
TAYLOR, W. *Compactness and chromatic number*
  ⋄ C05 C50 C65 ⋄
VAZHENIN, YU.M. *Elementary properties of semigroups of transformations of ordered sets (Russian)*
  ⋄ C65 E07 ⋄

1971

BONNET, R. *Categorie et α-categorie des ensembles ordonnes* ⋄ C65 E07 ⋄
CHUAQUI, R.B. *A representation theorem for linearly ordered cardinal algebras* ⋄ C05 C65 E10 G25 ⋄
FRAISSE, R. *Abritement entre relations et specialement entre chaines* ⋄ C65 E07 ⋄
FRAISSE, R. & POUZET, M. *Interpretabilite d'une relation par une chaine* ⋄ C07 C30 C65 E07 ⋄
FRAISSE, R. & POUZET, M. *Sur une classe de relations n'ayant qu'un nombre fini de bornes*
  ⋄ C07 C30 C65 E07 ⋄
HAUSCHILD, K. *Entscheidbarkeit in der Theorie bewerteter Graphen (Russian and English summaries)*
  ⋄ B25 C65 ⋄
HAUSCHILD, K. & RAUTENBERG, W. *Interpretierbarkeit und Entscheidbarkeit in der Graphentheorie I*
  ⋄ B25 C65 D35 F25 ⋄
HENSON, C.W. *A family of countable homogeneous graphs* ⋄ C13 C15 C50 C65 ⋄
MACINTYRE, A. *On the elementary theory of Banach algebras* ⋄ C60 C65 D35 ⋄
MARCO DE, G. & RICHTER, M.M. *Rings of continuous functions with values in a non-archimedean ordered field* ⋄ C20 C60 C65 E75 ⋄
PINUS, A.G. *A remark on the paper by Ju.M.Vazhenin: "Elementary properties of transformation semigroups of ordered sets" (Russian)* ⋄ B25 C65 D35 E07 ⋄
ROTMAN, B. *Remarks on some theorems of Rado on universal graphs* ⋄ C50 C65 ⋄
SZCZERBA, L.W. *Incompleteness degree of elementary Pasch-free geometry* ⋄ C35 C65 ⋄
TAYLOR, W. *Atomic compactness and elementary equivalence* ⋄ C05 C13 C50 C65 ⋄
TAYLOR, W. *Some constructions of compact algebras*
  ⋄ C05 C50 C55 C65 ⋄

1972

BAKER, K.A. & FISHBURN, P.C. & ROBERTS, F.S. *Partial orders of dimension 2* ⋄ C65 E07 ⋄
CALAIS, J.-P. *Partial isomorphisms and infinitary languages* ⋄ C07 C65 C75 ⋄
COLE, J.C. & DICKMANN, M.A. *Non-axiomatizability results in infinitary languages for higher-order structures* ⋄ C52 C65 C75 C85 ⋄
DACUNHA-CASTELLE, D. & KRIVINE, J.-L. *Applications des ultraproduits a l'etude des espaces et des algebres de Banach* ⋄ C20 C65 E75 ⋄
DACUNHA-CASTELLE, D. *Applications des ultraproduits a la theorie des plongements des espaces de Banach* ⋄ C20 C65 E75 ⋄
DACUNHA-CASTELLE, D. *Ultraproduits d'espaces de Banach* ⋄ C20 C65 E75 ⋄
DACUNHA-CASTELLE, D. *Ultraproduits d'espaces L^p et d'espaces d'Orlicz* ⋄ C20 C65 E75 ⋄
FRAISSE, R. *Chaines compatibles pour un groupe de permutations. Application aux relations enchainables et monomorphes* ⋄ C07 C30 C52 C65 E07 ⋄
GUPTA, H.N. & PRESTEL, A. *On a class of Pasch-free Euclidean planes* ⋄ C60 C65 ⋄
HAGENDORF, J.G. *Extensions immediates de chaines et de relations* ⋄ C65 E07 ⋄
HAUSCHILD, K. & HERRE, H. & RAUTENBERG, W. *Interpretierbarkeit und Entscheidbarkeit in der Graphentheorie II* ⋄ B25 C65 D35 F25 ⋄
HENSON, C.W. *The nonstandard hulls of a uniform space*
  ⋄ C65 H05 ⋄
HENSON, C.W. & MOORE JR., L.C. *The nonstandard theory of topological vector spaces* ⋄ C65 H05 ⋄
HERRE, H. *Die Entscheidbarkeit der elementaren Theorie der n-separierten symmetrischen Graphen endlicher Valenz* ⋄ B25 C65 ⋄
KONCEWICZ, L. *The simplest expressions of the predicate calculus that define classes of symmetric graphs without loops II (Polish) (English and Russian summaries)*
  ⋄ C65 ⋄
KRIVINE, J.-L. *Theorie des modeles et espaces L^p*
  ⋄ C65 ⋄
PINUS, A.G. *On the theory of convex subsets (Russian)*
  ⋄ B25 C65 C85 D35 ⋄
ROBINSON, A. *On the real closure of a Hardy field*
  ⋄ C60 C65 ⋄
SCHREIBER, M. *Quelques remarques sur les caracterisations des espaces L^p, $0 \leqslant p < i$ (English summary)* ⋄ C20 C65 ⋄
SLOMSON, A. *Generalized quantifiers and well orderings*
  ⋄ B25 C55 C65 C80 E07 ⋄
TITIEV, R.J. *Multidimensional measurement and universal axiomatizability* ⋄ C65 ⋄
VAZHENIN, YU.M. *The elementary definability and elementary characterizability of classes of reflexive graphs (Russian)* ⋄ C52 C65 ⋄
WEESE, M. *Zur Modellvollstaendigkeit und Entscheidbarkeit gewisser topologischer Raeume (Russian, English and French summaries)*
  ⋄ B25 C35 C65 D80 ⋄

WOLFF, M. *Nonstandard-Komplettierung von Cauchy-Algebren* ⋄ C65 H05 ⋄

YOUNG, L. *Functional analysis - a non-standard treatment with semifields* ⋄ C65 H05 ⋄

1973

BALDWIN, JOHN T. & BLASS, A.R. & GLASS, A.M.W. & KUEKER, D.W. *A "natural" theory without a prime model* ⋄ C15 C50 C62 C65 ⋄

BERNSTEIN, A.R. *Non-standard analysis*
⋄ C60 C65 H05 ⋄

BONDI, I.L. *The decision problem for normed linear spaces and for Hilbert spaces (Russian)* ⋄ B25 C65 D35 ⋄

DAHN, B.I. $\aleph_0$-*kategorische zyklenbeschraenkte Graphen*
⋄ C13 C35 C65 ⋄

DAHN, B.I. $\aleph_0$-*kategorische zyklenbeschraenkte Graphen (English summary)* ⋄ C13 C35 C65 ⋄

EMDE BOAS VAN, P. *Mostowski's universal set-algebra*
⋄ C50 C57 C65 D45 E07 E20 G05 ⋄

ERSHOV, YU.L. *Skolem functions and constructive models (Russian)* ⋄ C57 C65 F50 ⋄

HAKEN, W. *Connections between topological and group theoretical decision problems*
⋄ B25 C60 C65 D35 D40 D80 ⋄

HECHLER, S.H. *Large superuniversal metric spaces*
⋄ C50 C65 E75 ⋄

HENSON, C.W. & MOORE JR., L.C. *Invariance of the nonstandard hulls of locally convex spaces*
⋄ C65 H05 ⋄

KONCEWICZ, L. *Definability of classes of graphs in the first order predicate calculus with identity (Polish and Russian summaries)* ⋄ C65 ⋄

LILLO DE, N.J. *A formal characterization of ordinal numbers* ⋄ C65 E10 ⋄

LUCE, R. DUNCAN *Three axiom systems for additive semiordered structures* ⋄ C65 ⋄

MAREK, W. *Consistance d'une hypothese de Fraisse sur la definissabilite dans un langage du second ordre*
⋄ B15 C65 E10 E35 E45 ⋄

MAREK, W. *Sur la consistance d'une hypothese de Fraisse sur la definissabilite dans un langage du second ordre*
⋄ B15 C65 E10 E35 E45 ⋄

NELSON, GEORGE C. *Nonconstructivity of models of the reals (Russian summary)*
⋄ C62 C65 D55 I105 H15 ⋄

OSTASZEWSKI, A.J. *Saturated structures and a theorem of Arhangelskij* ⋄ C50 C65 E75 ⋄

PERETYAT'KIN, M.G. *Every recursively enumerable extension of a theory of linear order has a constructive model (Russian)* ⋄ C57 C65 ⋄

POL, R. *There is no universal totally disconnected space*
⋄ C50 C65 ⋄

ROBINSON, A. *Function theory on some nonarchimedian fields* ⋄ C60 C65 H05 ⋄

WEISPFENNING, V. *Infinitary model-theoretic properties of κ-saturated structures*
⋄ C10 C25 C35 C50 C60 C65 C75 ⋄

1974

DENISOV, S.D. *Three theorems on elementary theories and tt-reducibility (Russian)* ⋄ C65 D30 ⋄

HARTVIGSON, Z.R. *A non-desarguesian space geometry*
⋄ C65 ⋄

HENSON, C.W. & MOORE JR., L.C. *Invariance of the nonstandard hulls of a uniform space* ⋄ C65 H05 ⋄

HENSON, C.W. & MOORE JR., L.C. *Nonstandard hulls of the classical Banach spaces* ⋄ C65 H05 ⋄

HENSON, C.W. & MOORE JR., L.C. *Semi-reflexivity of the nonstandard hulls of a locally convex space*
⋄ C65 H05 ⋄

HENSON, C.W. & MOORE JR., L.C. *Subspaces of the nonstandard hull of a normed space* ⋄ C65 H05 ⋄

HENSON, C.W. *The isomorphism property in nonstandard analysis and its use in the theory of Banach spaces*
⋄ C65 H05 ⋄

HERRE, H. *Nonfinitely axiomatizable theories of graphs*
⋄ C52 C65 ⋄

KEISLER, H.J. *Monotone complete fields*
⋄ C60 C65 H05 ⋄

KESEL'MAN, D.YA. *The elementary theories of graphs and abelian loops (Russian)* ⋄ C65 D35 ⋄

KOREC, I. & PERETYAT'KIN, M.G. & RAUTENBERG, W. *Definability in structures of finite valency*
⋄ C60 C65 F25 ⋄

MYERS, D.L. *The boolean algebras of abelian groups and well-orders* ⋄ C60 C65 G05 ⋄

NADEL, M.E. *Scott sentences and admissible sets*
⋄ C15 C65 C70 C75 ⋄

NOVAK, VITEZSLAV & NOVOTNY, M. *Abstrakte Dimension von Strukturen* ⋄ C30 C65 ⋄

PIETSCH, A. *Ultraprodukte von Operatoren in Banachraeumen* ⋄ C20 C65 E75 H20 ⋄

ROBINSON, A. *A note on topological model theory*
⋄ C65 C90 ⋄

ROSENTHAL, J.W. *Models of Th(ω^ω, <)* ⋄ C65 ⋄

RUBIN, M. *Theories of linear order*
⋄ C15 C35 C50 C65 E07 ⋄

SCHWABHAEUSER, W. & SZCZERBA, L.W. *An affine space as union of spaces of higher dimension* ⋄ C60 C65 ⋄

SMIRNOVA, O.S. *Completeness and decidability of the axiomatic theory SG for spherical geometry (Russian)*
⋄ B25 C35 C65 ⋄

STERN, J. *Sur certaines classes d'espaces de Banach caracterisees par des formules* ⋄ C20 C65 ⋄

SZMIELEW, W. *The order and the semi-order of n-dimensional euclidean space in the axiomatic and model-theoretic aspects* ⋄ B30 C65 ⋄

SZMIELEW, W. *The role of the Pasch axiom in the foundations of euclidean geometry* ⋄ C65 ⋄

TAJMANOV, A.D. *On the elementary theory of topological algebras (Russian)* ⋄ C57 C65 ⋄

1975

BEAUCHEMIN, P. & REYES, G.E. *Espaces de Baire et espaces de probabilite de structures relationnelles*
⋄ C65 C75 E25 E55 E75 ⋄

DACUNHA-CASTELLE, D. & KRIVINE, J.-L. *Sous-espaces de L^1* ⋄ C20 C65 E75 ⋄

GRAINGER, A.D. *Invariant subspaces of compact operators on topological vector spaces* ⋄ C60 C65 H05 ⋄

HANAZAWA, M. *A remark on ordered structures with unary predicates* ⋄ C65 E07 ⋄

HAUSCHILD, K. *Zur Uebertragbarkeit von Entscheidbarkeitsresultaten elementarer Theorien (Russian, English and French summaries)* ⋄ B15 B25 C65 C85 ⋄

HENSON, C.W. *The monad system of the finest compatible uniform structure* ⋄ C65 H05 ⋄

HENSON, C.W. *When do two Banach spaces have isometrically isomorphic nonstandard hulls?* ⋄ C65 H05 ⋄

LAKE, J. *Characterising the largest, countable partial ordering* ⋄ C15 C50 C65 E07 ⋄

MAKOWSKY, J.A. *Topological model theory* ⋄ C65 C80 ⋄

MCKEE, T.A. *A formal analogy between Baer subplanes and their complements* ⋄ C65 ⋄

MCKEE, T.A. *Infinitary logic and topological homeomorphisms* ⋄ C65 C75 C80 C90 ⋄

SCHWABHAEUSER, W. & SZCZERBA, L.W. *Relations on lines as primitive notions for euclidean geometry* ⋄ C65 ⋄

SHELAH, S. *The monadic theory of order* ⋄ B25 C65 C85 D35 E50 ⋄

SHELAH, S. *Why there are many nonisomorphic models for unsuperstable theories* ⋄ C45 C50 C55 C60 C65 G05 ⋄

STERN, J. *Le probleme des enveloppes d'espace de Banach (English summary)* ⋄ C65 ⋄

STERN, J. *Proprietes locales et ultrapuissances d'espaces de Banach* ⋄ C20 C65 ⋄

TERRA, A. *Questions on the axiomatization of n-dimensional order (Russian)* ⋄ C65 ⋄

WEISPFENNING, V. *Two model theoretic proofs of Rueckert's Nullstellensatz* ⋄ C65 H20 ⋄

1976

AMIT, R. & SHELAH, S. *The complete finitely axiomatized theories of order are dense* ⋄ C52 C65 ⋄

BECKER, J.A. & LIPSHITZ, L. *An application of logic to analysis* ⋄ B80 C65 ⋄

BLANC, G. & RAMBAUD, C. *Theories formelles sur graphe (English summary)* ⋄ C07 C65 G30 ⋄

HENSON, C.W. *Nonstandard hulls of Banach spaces* ⋄ C65 H05 ⋄

HENSON, C.W. *Ultraproducts of Banach spaces* ⋄ C20 C65 H05 ⋄

KOREC, I. & RAUTENBERG, W. *Model-interpretability into trees and applications* ⋄ B25 C45 C60 C65 F25 ⋄

KRIVINE, J.-L. *Sous-espaces de dimension finie des espaces de Banach reticules* ⋄ C20 C65 ⋄

MAKOWSKY, J.A. & MARCJA, A. *Problemi di decidibilita in logica topologica (English summary)* ⋄ B25 C65 C80 ⋄

MCKEE, T.A. *Sentences preserved between equivalent topological bases* ⋄ C65 C80 C90 ⋄

MOSTOWSKI, T. *Analytic applications of decidability theorems* ⋄ B25 C60 C65 D80 ⋄

REICHBACH, J. *Graphs, generalized models, games and decidability* ⋄ B25 C65 C90 ⋄

SGRO, J. *Completeness theorems for continuous functions and product topologies* ⋄ C40 C65 C80 C90 E75 ⋄

STERN, J. *Some applications of model theory in Banach space theory* ⋄ C20 C65 ⋄

STERN, J. *The problem of envelopes for Banach spaces* ⋄ C20 C65 ⋄

WEESE, M. *Entscheidbarkeit in speziellen uniformen Strukturen bezueglich Sprachen mit Maechtigkeitsquantoren* ⋄ B25 C55 C65 C80 ⋄

ZIEGLER, M. *A language for topological structures which satisfies a Lindstroem-theorem* ⋄ B25 C40 C50 C65 C80 C90 C95 D35 ⋄

1977

BANKSTON, P. *Topological reduced products and the GCH* ⋄ C20 C65 E50 E75 ⋄

BANKSTON, P. *Ultraproducts in topology* ⋄ C20 C65 E75 ⋄

CHADWICK, J.J.M. & WICKSTEAD, A.W. *A quotient of ultrapowers of Banach spaces and semi-Fredholm operators* ⋄ C20 C65 H05 ⋄

CLARK, D.M. & KRAUSS, P.H. *Relatively homogeneous structures* ⋄ C10 C35 C50 C60 C65 G05 ⋄

DACUNHA-CASTELLE, D. & KRIVINE, J.-L. *Sous-espaces de L^1* ⋄ C20 C65 E75 ⋄

DALES, H.G. & ESTERLE, J. *Discontinuous homomorphisms from C(X)* ⋄ C65 E50 E75 ⋄

DONNADIEU, M.-R. & RAMBAUD, C. *Theoremes de completude dans les theories sur graphes orientes (English summary)* ⋄ C07 C65 G30 ⋄

DUBIEL, M. *Generalized quantifiers and elementary extensions of countable models* ⋄ C15 C65 C80 ⋄

ESTERLE, J. *Proprietes universelles de certains semi-groupes et algebres de convolution* ⋄ C65 E35 E50 ⋄

GUREVICH, Y. *Monadic theory of order and topology I* ⋄ B15 C65 C85 E07 E50 E75 ⋄

HAY, L. & MANASTER, A.B. & ROSENSTEIN, J.G. *Concerning partial recursive similarity transformations of linearly ordered sets* ⋄ C57 C65 D20 D25 D30 D45 E07 ⋄

HENSON, C.W. & JOCKUSCH JR., C.G. & RUBEL, L.A. & TAKEUTI, G. *First-order topology* ⋄ B25 C65 C75 D35 H05 ⋄

HERRE, H. & WOLTER, H. *Entscheidbarkeit der Theorie der linearen Ordnung in L_{Q_1}* ⋄ B25 C55 C65 C80 ⋄

HIGMAN, G. *Homogeneous relations* ⋄ C07 C30 C65 E05 ⋄

KEISLER, H.J. *Hyperfinite model theory* ⋄ B50 C65 C75 C80 C90 H20 ⋄

LADNER, R.E. *Application of model theoretic games to discrete linear orders and finite automata* ⋄ B15 B25 C07 C13 C65 C85 D05 E07 E60 ⋄

LANDRAITIS, C.K. *Definability in well quasi-ordered sets of structures* ⋄ C15 C65 C75 ⋄

SCHMERL, J.H. $\aleph_0$-categoricity of partially ordered sets of width 2 ⋄ C35 C65 ⋄

SCHMERL, J.H. On $\aleph_0$-categoricity and the theory of trees ⋄ B25 C35 C65 ⋄

SEESE, D.G. & TUSCHIK, H.-P. Construction of nice trees ⋄ B25 C55 C65 C80 ⋄

SEESE, D.G. Decidability of ω-trees with bounded sets - a survey ⋄ B25 C65 C80 C85 ⋄

SEESE, D.G. Partitions for trees ⋄ C55 C65 E05 ⋄

SEESE, D.G. Second order logic, generalized quantifiers and decidability
⋄ B15 B25 C55 C65 C80 C85 D35 ⋄

STERN, J. Examples d'application de la theorie des modeles a la theorie des espaces de Banach ⋄ C20 C65 ⋄

TUSCHIK, H.-P. On the decidability of the theory of linear orderings in the language $L(Q_1)$
⋄ B25 C55 C65 C80 ⋄

WACHS, E.A. Models of well-orderings ⋄ C65 E07 ⋄

WEESE, M. The undecidability of well-ordering with the Haertig quantifier (Russian summary)
⋄ C55 C65 C80 D35 E07 ⋄

WOLTER, H. Entscheidbarkeit der Theorie der Wohlordnung mit Elimination der Quantoren (Russian, English and French summaries)
⋄ B25 C10 C65 C80 E07 ⋄

1978

ABRAMSON, F.G. & HARRINGTON, L.A. Models without indiscernibles
⋄ C30 C50 C55 C62 C65 C75 E05 E40 H15 ⋄

ANAPOLITANOS, D.A. A theorem on absolute indiscernibles ⋄ C30 C65 ⋄

DONER, J.E. & MOSTOWSKI, ANDRZEJ & TARSKI, A. The elementary theory of well-ordering - a metamathematical study ⋄ B25 C10 C65 E07 ⋄

FORREST, W.K. A note on universal classes with applications to the theory of graphs
⋄ C05 C50 C52 C65 ⋄

GARAVAGLIA, S. Model theory of topological structures
⋄ C20 C40 C65 C80 C90 ⋄

HERRE, H. & WOLTER, H. Entscheidbarkeit der Theorie der linearen Ordnung in L_{Q_κ} fuer regulaeres ω_κ
⋄ B25 C55 C65 C80 E07 E50 ⋄

HERRMANN, R.A. The nonstandard theory of semi-uniform spaces ⋄ C65 H05 H20 ⋄

JANOS, L. Topological dimension as a first order property (Italian summary) ⋄ C65 ⋄

LACZKOVICH, M. Solvability and consistency for infinite systems of linear inequalities ⋄ C65 ⋄

MCKEE, T.A. Forbidden subgraphs in terms of forbidden quantifiers ⋄ C52 C65 ⋄

MLCEK, J. A note on cofinal extensions and segments
⋄ C30 C62 C65 ⋄

MURZIN, F.A. A question by Jokusch, Rubel and Takeuti (Russian) ⋄ C65 ⋄

PODEWSKI, K.-P. & ZIEGLER, M. Stable graphs
⋄ C45 C65 ⋄

PRESTEL, A. & ZIEGLER, M. Model theoretic methods in the theory of topological fields ⋄ C60 C65 C80 ⋄

SCHMERL, J.H. $\aleph_0$-categoricity and comparability graphs
⋄ C15 C35 C65 ⋄

SCHWARTZ, N. η_α-Strukturen
⋄ C50 C55 C65 E07 ⋄

SEESE, D.G. A remark to the undecidability of well-orderings with the Haertig quantifier
⋄ C55 C65 C80 D35 ⋄

SEESE, D.G. Decidability of ω-trees with bounded sets (German and Russian summaries)
⋄ B25 C65 C85 D05 ⋄

SHELAH, S. & STERN, J. The Hanf number of the first order theory of Banach spaces ⋄ C55 C65 H10 ⋄

SINGER, M.F. The model theory of ordered differential fields ⋄ B25 C25 C35 C60 C65 D80 ⋄

SONENBERG, E.A. On the elementary theory of inductive order ⋄ C65 E07 G05 ⋄

STEEL, J.R. On Vaught's conjecture
⋄ C15 C65 C70 C75 ⋄

STERN, J. Ultrapowers and local properties of Banach spaces ⋄ C20 C65 ⋄

TAJMANOV, A.D. On topologizing countable algebras (Russian) ⋄ C05 C15 C65 ⋄

TRIPPEL, J.R. Die Algebra der Saetze der monadischen Theorie schwach zweiter Stufe der linearen Ordnung ist atomar ⋄ C65 C85 ⋄

WEISPFENNING, V. A note on $\aleph_0$-categorical model-companions
⋄ C25 C35 C50 C52 C65 G05 G10 ⋄

WHEELER, W.H. A characterization of companionable, universal theories ⋄ C25 C35 C60 C65 ⋄

1979

BANKSTON, P. Note on "Ultraproducts in topology"
⋄ C20 C65 E75 ⋄

BANKSTON, P. Topological reduced products via good ultrafilters ⋄ C20 C65 E35 E50 E75 ⋄

BLASS, A.R. & HARARY, F. Properties of almost all graphs and complexes ⋄ C10 C13 C65 ⋄

BOYD, M.W. & TRANSUE, W.R. Properties of ultraproducts ⋄ C20 C65 ⋄

EKLOF, P.C. & MEKLER, A.H. Stationary logic of finitely determinate structures
⋄ B25 C10 C55 C60 C65 C80 ⋄

FOURMAN, M.P. & HYLAND, J.M.E. Sheaf models for analysis ⋄ C65 C90 F35 F50 G10 G30 ⋄

GUREVICH, Y. Modest theory of short chains I
⋄ B25 C65 C85 D35 E07 ⋄

GUREVICH, Y. & SHELAH, S. Modest theory of short chains II ⋄ B25 C65 C85 D35 E07 E50 ⋄

GUREVICH, Y. Monadic theory of order and topology II
⋄ B15 C65 C85 E07 E45 E50 ⋄

GUTIERREZ-NOVOA, L. Non-Euclidean real numbers
⋄ B28 C65 ⋄

JOHNSTONE, P.T. Conditions related to de Morgan's law
⋄ B55 C65 C90 F50 G30 ⋄

KALANTARI, I. & RETZLAFF, A.T. Recursive constructions in topological spaces ⋄ C57 C65 D45 ⋄

KAWAI, T. An application of nonstandard analysis to characters of groups of continuous functions
⋄ C65 H10 ⋄

LERMAN, M. & SCHMERL, J.H. *Theories with recursive models* ⋄ C35 C57 C65 ⋄

LILLO DE, N.J. *Models of an extension of the theory ORD* ⋄ B28 C40 C65 E10 E30 ⋄

MAKOWSKY, J.A. *The reals cannot be characterized topologically with strictly local properties and countability axioms* ⋄ C65 C80 C90 ⋄

PELLER, V. *Applications of ultraproducts in operator theory. A simple proof of E. Bishop's theorem (Russian) (English summary)* ⋄ C20 C65 ⋄

PERMINOV, E.A. & VAZHENIN, YU.M. *On rigid lattices and graphs (Russian)* ⋄ C50 C52 C65 G10 ⋄

PINUS, A.G. *Elementary equivalence of topological spaces (Russian)* ⋄ C55 C65 C85 G05 G10 ⋄

RAMBAUD, C. *Completude en theorie sur graphes orientes* ⋄ C07 C65 ⋄

SCHMERL, J.H. *Countable homogeneous partially ordered sets* ⋄ C15 C50 C65 E07 ⋄

SCHWABHAEUSER, W. *Modellvollstaendigkeit der Mittelpunktsgeometrie und der Theorie der Vektorgruppen* ⋄ B25 C35 C65 ⋄

SCHWABHAEUSER, W. *Non-finitizability of a weak second-order theory* ⋄ B15 C62 C65 C85 F35 ⋄

SEESE, D.G. *Some graph-theoretical operations and decidability* ⋄ B15 B25 C65 C85 ⋄

SEESE, D.G. *Stationary logic and ordinals* ⋄ B25 C55 C65 C80 ⋄

SONENBERG, E.A. *Non-standard models of ordinal arithmetics* ⋄ C65 H20 ⋄

STERN, J. *Le groupe des isometries d'un espace de Banach* ⋄ C65 ⋄

SZCZERBA, L.W. & TARSKI, A. *Metamathematical discussion of some affine geometries* ⋄ B25 B30 C35 C52 C65 D35 ⋄

WHEELER, W.H. *The first order theory of N-colorable graphs* ⋄ B30 C10 C25 C35 C65 ⋄

WOODROW, R.E. *There are four countable ultrahomogeneous graphs without triangles* ⋄ C10 C50 C65 E05 ⋄

1980

ANAPOLITANOS, D.A. & VAEAENAENEN, J. *On the axiomatizability of the notion of an automorphism of a finite order* ⋄ C07 C65 ⋄

BECKER, J.A. & HENSON, C.W. & RUBEL, L.A. *First-order conformal invariants* ⋄ C35 C65 D35 E35 E75 ⋄

CHERLIN, G.L. & DICKMANN, M.A. *Anneaux reels clos et anneaux de fonctions continues (English summary)* ⋄ C10 C35 C60 C65 ⋄

CHERLIN, G.L. *Rings of continuous functions: decision problems* ⋄ B25 C60 C65 D35 ⋄

CHLEBUS, B.S. *Decidability and definability results concerning well-orderings and some extensions of first order logic* ⋄ B15 B25 C40 C65 C80 D35 E07 ⋄

DAHN, B.I. *First order logics for metric structures* ⋄ C65 C80 C90 ⋄

EDA, K. *A note on the hereditary properties in the product space* ⋄ C52 C65 ⋄

FLUM, J. & ZIEGLER, M. *Topological model theory* ⋄ C65 C80 C90 C98 ⋄

GUREVICH, Y. *Two notes on formalized topology* ⋄ C65 C85 F25 ⋄

HEINDORF, L. *The decidability of the L_t - theory of boolean spaces* ⋄ B25 C65 C80 C90 ⋄

HEINRICH, S. *The isomorphic problem of envelopes* ⋄ C20 C65 H05 ⋄

HEINRICH, S. *Ultraproducts in Banach space theory* ⋄ C20 C65 C98 H05 ⋄

HERRE, H. & SEESE, D.G. *Concerning the monadic theory of the topology of well-orderings and scattered spaces* ⋄ B25 C55 C65 C85 E07 E35 E50 E55 ⋄

HERRE, H. *Modelltheoretische Eigenschaften endlichvalenter Graphen* ⋄ C35 C45 C50 C65 ⋄

HOLLAND, W.C. *Trying to recognize the real line* ⋄ C07 C60 C65 ⋄

IMAZEKI, K. *Characterizations of compact operators and semi-Fredholm operators* ⋄ C20 C65 ⋄

JAMBU-GIRAUDET, M. *Theorie des modeles de groupes d'automorphismes d'ensembles totalement ordonnes 2-homogenes (English summary)* ⋄ C07 C60 C62 C65 D35 E07 F25 F35 ⋄

JENSEN, C.U. *Proprietes homologiques et logiques des anneaux de fonctions entieres (English summary)* ⋄ C60 C65 E50 E75 ⋄

KULUNKOV, P.A. *Questions of definability in ordered sets (Russian) (English summary)* ⋄ C15 C65 E07 ⋄

LACHLAN, A.H. & WOODROW, R.E. *Countable ultrahomogeneous undirected graphs* ⋄ C15 C50 C65 ⋄

LANDRAITIS, C.K. *$L_{\omega_1\omega}$ equivalence between countable and uncountable linear orderings* ⋄ C65 C75 E07 ⋄

MANASTER, A.B. & ROSENSTEIN, J.G. *Two-dimensional partial orderings: Recursive model theory* ⋄ C57 C65 E07 ⋄

MANASTER, A.B. & ROSENSTEIN, J.G. *Two-dimensional partial orderings: Undecidability* ⋄ C57 C65 D35 E07 G10 ⋄

MCKEE, T.A. *Generalized equivalence and the phraseology of configuration theorems* ⋄ C65 ⋄

MYERS, D.L. *The boolean algebra of the theory of linear orders* ⋄ C65 G05 ⋄

PRESTEL, A. & ZIEGLER, M. *Non-axiomatizable classes of V-topological fields* ⋄ C60 C65 C80 ⋄

RIMSCHA VON, M. *Mengentheoretische Modelle des λK-Kalkuels* ⋄ B40 C65 E65 E70 ⋄

SCHMERL, J.H. *Decidability and $\aleph_0$-categoricity of theories of partially ordered sets* ⋄ B25 C15 C35 C65 D35 ⋄

TUCKER, J.V. *Computability and the algebra of fields: some affine constructions* ⋄ B25 C57 C60 C65 D45 ⋄

TULIPANI, S. *Invarianti per l'equivalenza elementare per una classe assiomatica di spazi generali di misura (English summary)* ⋄ B25 C10 C65 G05 ⋄

TUSCHIK, H.-P. *On the decidability of the theory of linear orderings with generalized quantifiers* ⋄ B25 C55 C65 C80 ⋄

WEINBERG, E.C. *Automorphism groups of minimal η_α-sets* ⋄ C07 C65 E07 ⋄

1981

ABBATI, M.C. & MANIA, A. *A spectral theory for order unit spaces* ◊ C65 ◊

ANTONOVSKIJ, M.YA. & CHUDNOVSKY, D.V. & CHUDNOVSKY, G.V. & HEWITT, E. *Rings of real-valued continuous functions II*
◊ C20 C50 C60 C65 E35 E50 E75 ◊

ASSOUS, R. & POUZET, M. *Structures invariantes et classes de meilleur ordre* ◊ C65 E07 ◊

BABAI, L. *On the abstract group of automorphisms*
◊ C07 C65 ◊

BOTHE, H.G. *A first-order characterization of 3-dimensional manifolds* ◊ C65 G10 ◊

GLASS, A.M.W. & GUREVICH, Y. & HOLLAND, W.C. & JAMBU-GIRAUDET, M. *Elementary theory of automorphism groups of doubly homogeneous chains*
◊ C07 C50 C65 E07 ◊

GLASS, A.M.W. *Elementary types of automorphisms of linearly ordered sets -- a survey* ◊ C07 C65 C98 ◊

GLASS, A.M.W. *Ordered permutation groups*
◊ C07 C60 C65 C98 ◊

GLASS, A.M.W. & GUREVICH, Y. & HOLLAND, W.C. & SHELAH, S. *Rigid homogeneous chains*
◊ C07 C50 C65 E07 E35 E50 ◊

GOLDBLATT, R.I. *"Locally-at" as a topological quantifier-former* ◊ C65 C80 ◊

GUREVICH, Y. & HOLLAND, W.C. *Recognizing the real line* ◊ C07 C65 ◊

HEINRICH, S. *Ultraproducts of L_1-predual spaces*
◊ C20 C65 H05 ◊

HERRE, H. *Miscellaneous results and problems in extended model theory* ◊ B25 C65 C80 C98 D35 ◊

HERRE, H. & WOLTER, H. *Untersuchungen zur Theorie der linearen Ordnung in Logiken mit Maechtigkeitsquantoren*
◊ B25 C55 C65 C80 E07 ◊

JAMBU-GIRAUDET, M. *Interpretations d'arithmetiques dans des groupes et des treillis*
◊ C07 C62 C65 D35 E07 F25 F30 G10 ◊

KEGEL, O.H. *Existentially closed projective planes*
◊ C25 C65 ◊

KOPPERMAN, R.D. *First-order topological axioms*
◊ C65 ◊

LERMAN, M. *On recursive linear orderings*
◊ C57 C65 D30 D45 D55 E07 ◊

MACINTYRE, A. *The laws of exponentiation*
◊ B25 C65 ◊

MANASTER, A.B. & REMMEL, J.B. *Partial orderings of fixed finite dimension: Model companions and density*
◊ C25 C65 D35 E07 ◊

MANASTER, A.B. & REMMEL, J.B. *Some decision problems for subtheories of two-dimensional partial orderings*
◊ B25 C10 C65 D35 ◊

PASINI, A. *On the 1°-order translations of the theory of the geometric closure structures* ◊ C65 ◊

PRESTEL, A. *Zur Axiomatisierung gewisser affiner Geometrien* ◊ C60 C65 ◊

RAYNAUD, Y. *Deux nouveaux exemples d'espaces de Banach stables (English summary)* ◊ C20 C65 ◊

RAYNAUD, Y. *Espaces de Banach superstables (English summary)* ◊ C20 C65 ◊

REMMEL, J.B. *Recursively categorical linear orderings*
◊ C35 C57 C65 ◊

SCHMERL, J.H. *Arborescent structures I: Recursive models*
◊ C50 C57 C65 ◊

SCHMERL, J.H. *Arborescent structures II: Interpretability in the theory of trees* ◊ B25 C50 C65 C75 ◊

SCHMERL, J.H. *Decidability and finite axiomatizability of theories of $\aleph_0$-categorical partially ordered sets*
◊ B25 C35 C65 ◊

SEESE, D.G. *Elimination of second-order quantifiers for well-founded trees in stationary logic and finitely determinate structures* ◊ C10 C55 C65 C80 ◊

SEESE, D.G. *Stationary logic and ordinals*
◊ B25 C55 C65 C80 ◊

SHELAH, S. *On saturation for a predicate*
◊ C45 C50 C55 C65 E50 ◊

TULIPANI, S. *Model-completions of theories of finitely additive measures with values in an ordered field*
◊ C10 C25 C35 C65 G05 ◊

1982

BANKSTON, P. *Some obstacles to duality in topological algebra* ◊ C05 C20 C65 G30 ◊

BONNET, R. & POUZET, M. *Linear extensions of ordered sets* ◊ C65 C98 E07 ◊

DAHN, B.I. *The limit behaviour of exponential terms*
◊ C60 C65 ◊

DELZELL, C.N. *Case distinctions are necessary for representing polynomials as sums of squares*
◊ A10 C60 C65 F55 ◊

DELZELL, C.N. *Continuous sums of squares of forms*
◊ C60 C65 F55 ◊

DRIES VAN DEN, L. *Some applications of a model theoretic fact to (semi-)algebraic geometry* ◊ C10 C60 C65 ◊

FACENDA AGUIRRE, J.A. *Ultraproducts of locally convex spaces (Spanish) (English summary)* ◊ C20 C65 ◊

GAIFMAN, H. *On local and non-local properties*
◊ C07 C13 C65 E47 ◊

GUREVICH, Y. *Crumbly spaces* ◊ B25 C65 C85 ◊

GUREVICH, Y. & SHELAH, S. *Monadic theory of order and topology in ZFC* ◊ B25 C65 C85 D35 E07 ◊

HAGENDORF, J.G. *Extensions immediates de chaines*
◊ C65 E07 ◊

HEINRICH, S. & MANKIEWICZ, P. *Applications of ultrapowers to the uniform and Lipschitz classification of Banach spaces* ◊ C20 C65 H05 ◊

HEINRICH, S. *The isomorphic problem of envelopes*
◊ C20 C65 H05 ◊

JENSEN, C.U. *La dimension globale de l'anneau des fonctions entieres (English summary)*
◊ C60 C65 E35 E75 ◊

JONSSON, B. *Arithmetic of ordered sets*
◊ C65 C98 E07 ◊

KALANTARI, I. *Major subsets in effective topology*
◊ C57 C65 D25 D45 ◊

KALANTARI, I. & LEGGETT, A. *Simplicity in effective topology* ◊ C57 C65 D25 ◊

KESEL'MAN, D.YA. *Decidability of theories of certain classes of elimination graphs* ◊ B25 C65 D35 ◊

KUSRAEV, A.G. *Boolean-valued analysis of duality of extended modules (Russian)*
◊ C60 C65 C90 E40 E75 ◊
LACHLAN, A.H. *Finite homogeneous simple digraphs*
◊ C13 C50 C65 ◊
LERMAN, M. & ROSENSTEIN, J.G. *Recursive linear orderings* ◊ C57 C65 D45 D55 E07 ◊
MALLIAVIN, M.P. *Ultra-produits d'algebre de Lie*
◊ C20 C65 ◊
MEISSNER, W. *A Loewenheim-Skolem theorem for inner product spaces* ◊ C55 C65 ◊
MOORE, S.M. *Nonstandard analysis applied to path integrals and generalized functions (Italian and Russian summaries)* ◊ C65 H10 H20 ◊
MOSCATELLI, V.B. *Ultraprodotti e super-riflessivita in teoria degli spazi di Banach* ◊ C20 C65 ◊
PARIGOT, M. *Theories d'arbres* ◊ C45 C65 ◊
PASINI, L. *Sulla chiusura reale dei campi generalizzati di Hardy (English summary)* ◊ C60 C65 H20 ◊
PROTASOV, I.V. *Varieties of topological spaces (Russian) (English summary)* ◊ C05 C65 ◊
ROLLAND, R. *Etude des coupures dans les groupes et corps ordonnes* ◊ C60 C65 ◊
ROSENSTEIN, J.G. *Linear orderings*
◊ B25 C65 C98 E07 E98 ◊
SEESE, D.G. & WEESE, M. *L(aa)-elementary types of well-orderings* ◊ B25 C55 C65 C80 E10 ◊
SEESE, D.G. & TUSCHIK, H.-P. & WEESE, M. *Undecidable theories in stationary logic*
◊ C55 C65 C80 D35 E05 E75 ◊
TULIPANI, S. *A use of the method of interpretations for decidability or undecidability of measure spaces*
◊ B25 C65 D35 ◊
WAGNER, C.M. *On Martin's conjecture*
◊ C15 C35 C45 C65 C75 ◊

1983

BANKSTON, P. *On productive classes of function rings*
◊ C05 C20 C65 E75 G30 ◊
BARANSKIJ, V.A. *Algebraic systems whose elementary theory is compatible with an arbitrary group (Russian)*
◊ C07 C65 G10 ◊
BAUDISCH, A. & SEESE, D.G. & TUSCHIK, H.-P. *ω-trees in stationary logic* ◊ B25 C55 C65 C80 ◊
BUGAJSKI, S. *Semantics in Banach spaces*
◊ B51 C65 C90 ◊
CHARRETTON, C. & POUZET, M. *Comparaison des structures engendrees par des chaines*
◊ C30 C65 G05 ◊
CHERLIN, G.L. & SCHMITT, P.H. *Locally pure topological abelian groups: Elementary invariants*
◊ B25 C60 C65 C80 C90 ◊
COMPTON, K.J. *Some useful preservation theorems*
◊ C30 C40 C65 ◊
DAHN, B.I. & GEHNE, J. *A theory of ordered exponential fields which has no model completion*
◊ C35 C60 C65 ◊
DAHN, B.I. & WOLTER, H. *On the theory of exponential fields* ◊ C25 C60 C65 ◊
DELFRATE, M.G. *Oriented 2-graphs and their generalization to dimension n (Italian) (English summary)* ◊ C52 C65 ◊

EDA, K. *A note on subgroups of Z^N*
◊ C65 E35 E50 E75 ◊
FORREST, W.K. *The theory of affine constructible sets*
◊ C60 C65 ◊
GORDON, E.I. & LYUBETSKIJ, V.A. *Boolean extensions of uniform structures (Russian)* ◊ C65 C90 E40 ◊
GRAMAIN, F. *Non-minimalite de la cloture differentielle I: La preuve de S.Shelah* ◊ C50 C60 C65 ◊
GRAMAIN, F. *Non-minimalite de la cloture differentielle II: La preuve de M.Rosenlicht* ◊ C50 C60 C65 ◊
GUERRE, S. & LEVY, M. *Espaces l^p dans les sous-espaces de L^1* ◊ C20 C65 ◊
GUREVICH, Y. & SHELAH, S. *Interpreting second-order logic in the monadic theory of order*
◊ B15 C65 C85 D35 E07 E50 F25 ◊
GUREVICH, Y. & SHELAH, S. *Rabin's uniformization problem* ◊ B15 C40 C65 C85 E40 ◊
GUREVICH, Y. & MAGIDOR, M. & SHELAH, S. *The monadic theory of ω_2*
◊ B15 B25 C65 C85 D35 E10 E35 ◊
HEINRICH, S. & HENSON, C.W. & MOORE JR., L.C. *Elementary equivalence of L_1-preduals* ◊ C20 C65 ◊
HEINRICH, S. & MANKIEWICZ, P. *Some open problems in the nonlinear classification of Banach spaces*
◊ C20 C65 ◊
HEINRICH, S. *Ultrapowers of locally convex spaces and applications I* ◊ C20 C65 H10 ◊
HEINRICH, S. *Ultrapowers of locally convex spaces and applications II* ◊ C20 C65 H10 ◊
HENSON, C.W. & MOORE JR., L.C. *Nonstandard analysis and the theory of Banach spaces* ◊ C65 H05 ◊
HERRE, H. & MEKLER, A.H. & SMITH, KENNETH W. *Superstable graphs*
◊ C45 C65 ◊
JAMBU-GIRAUDET, M. *Bi-interpretable groups and lattices*
◊ C07 C65 G10 ◊
KALANTARI, I. & REMMEL, J.B. *Degrees of recursively enumarable topological spaces*
◊ C57 C65 D25 D45 ◊
KALANTARI, I. & LEGGETT, A. *Maximality in effective topology* ◊ C57 C65 D25 D45 ◊
KATZBERG, H. *Complex exponential terms with only finitely many zeros* ◊ C65 ◊
KUERSTEN, K.-D. *Local duality of ultraproducts of Banach lattices* ◊ C20 C65 ◊
KUSRAEV, A.G. *On some categories and functors of boolean-valued analysis (Russian)*
◊ C65 C90 E40 G30 ◊
KUSRAEV, A.G. & KUTATELADZE, S.S. *Subdifferentials in Boolean-valued models of set theory (Russian)*
◊ C65 C90 E40 E75 ◊
LEISS, H. *Implizit definierte Mengensysteme*
◊ C40 C65 C90 ◊
NURTAZIN, A.T. *An elementary classification and some properties of complete orders (Russian)*
◊ B25 C35 C65 C75 E10 ◊
ODIFREDDI, P. *On the first order theory of the arithmetical degrees* ◊ C65 D30 D55 ◊

PARIGOT, M. *Le modele compagnon de la theorie des arbres* ⋄ C10 C25 C35 C50 C65 ⋄
RICHARDSON, D.B. *Roots of real exponential functions* ⋄ C35 C65 ⋄
ROY, D.K. *R.e. presented linear orders* ⋄ C57 C65 D45 E07 ⋄
RUBEL, L.A. *Some research problems about algebraic differential equations* ⋄ C65 E47 ⋄
SCHWABHAEUSER, W. & SZMIELEW, W. & TARSKI, A. *Metamathematische Methoden in der Geometrie* ⋄ B30 C65 C98 D35 ⋄
SCHWABHAEUSER, W. *Metamathematische Methoden in der Geometrie. Teil II: Metamathematische Betrachtungen* ⋄ B30 C65 C98 ⋄
SHELAH, S. *Models with second order properties IV. A general method and eliminating diamonds* ⋄ C50 C55 C60 C65 C80 E65 G05 ⋄
TUSCHIK, H.-P. *The Ramsey theorem for additive colourings* ⋄ C55 C65 E05 ⋄
WOLTER, H. *On the "problem of the last root" for exponential terms* ⋄ C60 C65 ⋄
ZAIONTZ, C. *Axiomatization of the monadic theory of ordinals $<\omega_2$* ⋄ B25 B28 C65 C85 ⋄

1984

BANKSTON, P. *Expressive power in first order topology* ⋄ C65 C90 ⋄
BANKSTON, P. *First order representations of compact Hausdorff spaces* ⋄ C65 ⋄
CAMERON, P.J. *Aspects of random graph* ⋄ C07 C60 C65 ⋄
DAHN, B.I. & GOERING, P. *Notes on exponential-logarithmic terms* ⋄ C60 C65 ⋄
DAHN, B.I. & WOLTER, H. *Ordered fields with several exponential functions* ⋄ C25 C35 C65 D35 ⋄
DAHN, B.I. *The limit behaviour of exponential terms* ⋄ C60 C65 D15 ⋄
DALEN VAN, D. *How to glue analysis models* ⋄ C65 C90 F35 F50 ⋄
DELON, F. *Espaces ultrametriques* ⋄ C25 C35 C45 C50 C65 ⋄
DENEF, J. & LIPSHITZ, L. *Power series solutions of algebraic differential equations* ⋄ B25 C60 C65 D35 D80 ⋄
DRIES VAN DEN, L. *Exponential rings, exponential polynomials and exponential functions* ⋄ C60 C65 ⋄
DRIES VAN DEN, L. *Remarks on Tarski's problem concerning $(\mathbb{R}, +, \cdot, exp)$* ⋄ C10 C60 C65 ⋄
DYMENT, E.Z. *Recursive metrizability of numbered topological spaces and bases of effective linear topological spaces (Russian)* ⋄ C57 C65 D45 ⋄
FLUM, J. *Modelltheorie - topologische Modelltheorie* ⋄ C65 C90 C98 ⋄
GOLDBLATT, R.I. *Orthomodularity is not elementary* ⋄ B51 C65 G10 ⋄
GUREVICH, R. *Decidability of the equational theory of positive numbers with raising to a power (Russian)* ⋄ B25 B28 C65 ⋄
GUREVICH, Y. & SHELAH, S. *The monadic theory and the "next world"* ⋄ C62 C65 C85 E40 F25 ⋄

HEINDORF, L. *Continuous functions on countable ordinals* ⋄ C65 E10 E45 ⋄
HENSON, C.W. & RUBEL, L.A. *Some applications of Nevanlinna theory to mathematical logic:identities of exponential functions* ⋄ B20 C65 F30 ⋄
HOOVER, D.N. & KEISLER, H.J. *Adapted probability distributions* ⋄ C50 C65 H10 ⋄
IVANOV, A.A. *Decidability of extended theories of addition of the natural numbers and the integers (Russian)* ⋄ B25 C65 F30 ⋄
JENSEN, C.U. *Theorie des modeles pour des anneaux de fonctions entieres et des corps de fonctions meromorphes* ⋄ C60 C65 D35 F35 ⋄
KAUFMANN, M. *Some remarks on equivalence in infinitary and stationary logic* ⋄ C55 C65 C75 C80 E45 ⋄
KIERSTEAD, H.A. & MCNULTY, G.F. & TROTTER JR., W.T. *A theory of recursive dimension of ordered sets* ⋄ C65 D45 ⋄
KRIVINE, J.-L. *Methodes de theorie des modeles en geometrie des espaces de Banach* ⋄ C65 C98 E75 ⋄
KRUPA, A. & ZAWISZA, B. *Applications of ultrapowers in analysis of unbounded selfadjoint operators* ⋄ C20 C65 H10 ⋄
KUERSTEN, K.-D. *Lokale Reflexivitaet und lokale Dualitaet von Ultraprodukten fuer halbgeordnete Banachraeume (English and Russian summaries)* ⋄ C20 C65 ⋄
KUSRAEV, A.G. *Order-continuous functionals in Boolean-valued models of set theory (Russian)* ⋄ C65 C90 E40 ⋄
LACHLAN, A.H. *Countable homogeneous tournaments* ⋄ C15 C50 C65 ⋄
LEVY, M. & RAYNAUD, Y. *Ultrapuissances des espaces $L^p(L^q)$ (English summary)* ⋄ C20 C65 H10 H20 ⋄
MANDERS, K.L. *Interpretations and the model theory of the classical geometries* ⋄ C25 C35 C65 ⋄
MARTINEZ, JUAN CARLOS *Accessible sets and $(L_{\omega_1\omega})_t$-equivalence for T_3 spaces* ⋄ C65 C75 C80 C90 ⋄
MCKEE, T.A. *Relative expressiveness of the edge/adjacency language for graph theory* ⋄ C65 ⋄
MEKLER, A.H. *Stationary logic of ordinals* ⋄ B25 C10 C55 C65 C80 E10 ⋄
NIKITIN, A.A. *Some algorithmic problems for projective planes (Russian)* ⋄ C57 C65 D80 ⋄
OZAWA, M. *A classification of type I AW^*-algebras and Boolean valued analysis* ⋄ C65 C90 E40 E75 ⋄
PILLAY, A. & STEINHORN, C.I. *Definable sets in ordered structures* ⋄ C40 C60 C65 ⋄
POUZET, M. & RICHARD, D. (EDS.) *Orders: description and roles in set theory, lattices, ordered groups, topology, theory of models and relations, combinatorics, effectiveness, social sciences* ⋄ C65 C97 ⋄
ROBERTS, I. *Ultrapowers in the Lipschitz and uniform classification of Banach spaces* ⋄ C20 C65 ⋄
ROSENSTEIN, J.G. *Recursive linear orderings (French summary)* ⋄ C57 C65 D45 ⋄

SCHMERL, J.H. $\aleph_0$-categorical partially ordered sets (French summary)
  ◊ B25 C15 C35 C65 C98 E07 ◊

SCHRAM, J.M. *Recursively prime trees*
  ◊ C57 C65 D45 ◊

SCHWARZ, S. *Recursive automorphisms of recursive linear orderings* ◊ C57 C65 D45 ◊

TURAN, G. *On the definability of properties of finite graphs*
  ◊ C13 C65 ◊

WATNICK, R. *A generalization of Tennenbaum's theorem on effectively finite recursive linear orderings*
  ◊ C57 C65 D25 D45 ◊

WOLFF, M. *Spectral theory of group representations and their nonstandard hull* ◊ C65 H10 ◊

WOLTER, H. *Some remarks on exponential functions in ordered fields* ◊ C25 C65 D35 ◊

WOLTER, H. *Some results about exponential fields (survey)*
  ◊ B25 C60 C65 C98 ◊

1985

ABRAHAM, U. & RUBIN, M. & SHELAH, S. *On the consistency of some partition theorems for continuous colorings, and structure of $\aleph_1$-dense real order types*
  ◊ C55 C62 C65 C80 E05 E07 E35 ◊

BALDWIN, JOHN T. & SHELAH, S. *Second-order quantifiers and the complexity of theories*
  ◊ C40 C55 C65 C75 C85 ◊

BANKSTON, P. & SCHUTT, R. *On minimally free algebras*
  ◊ C05 C50 C60 C65 ◊

BAUDISCH, A. & SEESE, D.G. & TUSCHIK, H.-P. & WEESE, M. *Decidability and quantifier elimination*
  ◊ B25 C10 C60 C65 C80 C98 ◊

BESLAGIC, A. & RUDIN, M.E. *Set-theoretic constructions of nonshrinking open covers* ◊ C55 C65 E05 E75 ◊

COMER, S.D. *The elementary theory of interval real numbers* ◊ B25 B28 C65 ◊

DRIES VAN DEN, L. *The field of reals with a predicate for the powers of two* ◊ B25 C60 C65 ◊

DROSTE, M. & SHELAH, S. *A construction of all normal subgroup lattices of 2-transitive automorphism groups of linearly ordered sets* ◊ C07 C60 C65 E07 ◊

DROSTE, M. *The normal subgroup lattice of 2-transitive automorphism groups of linearly ordered sets*
  ◊ C07 C60 C65 ◊

FAJARDO, S. *Completeness theorems for the general theory of stochastic processes* ◊ B48 C65 C90 ◊

FRANEK, F. *Isomorphisms of trees* ◊ C50 C65 E05 ◊

GOLTZ, H.-J. *The boolean sentence algebra of the theory of linear ordering is atomic with respect to logics with a Malitz quantifier* ◊ C55 C65 C80 G05 ◊

GUREVICH, R. *Equational theory of positive numbers with exponentiation* ◊ B25 B28 C65 ◊

GUREVICH, Y. *Monadic second-order theories*
  ◊ B25 C65 C85 C98 D35 D98 ◊

GUREVICH, Y. & SHELAH, S. *The decision problem for branching time logic* ◊ B25 B45 C65 D05 D25 ◊

HATCHER, W.S. *Elementary extension and the hyperreal numbers* ◊ C65 H05 ◊

HOLLAND, W.C. & MEKLER, A.H. & SHELAH, S. *Lawless order* ◊ C65 E05 E07 E75 ◊

JAMBU-GIRAUDET, M. *Quelques remarques sur l'equivalence elementaire entre groupes ou treillis d'automorphismes de chaines 2-homogenes*
  ◊ C07 C65 ◊

KALANTARI, I. & WEITKAMP, G. *Effective topological spaces I: A definability theory. II: A hierarchy*
  ◊ C40 C57 C65 D45 D75 D80 ◊

LIN, PEIKEE *Unconditional bases and fixed points of nonexpansive mappings* ◊ C20 C65 H05 ◊

LOEB, P.A. *A nonstandard functional approach to Fubini's theorem* ◊ C60 C65 H05 H20 ◊

MACPHERSON, H.D. *Growth rates in infinite graphs and permutation groups* ◊ C60 C65 ◊

MARONGIU, G. *Alcune osservazioni sulla $\aleph_0$-categoricita degli ordini circolari* ◊ C35 C65 ◊

PAETZOLD, P. *On fields and terms with arctan function*
  ◊ C65 ◊

RAPP, A. *The ordered field of real numbers and logics with Malitz quantifiers* ◊ B25 C10 C55 C65 C80 ◊

ROY, D.K. *Linear order types of nonrecursive presentability* ◊ C57 C65 D45 ◊

SHELAH, S. *Monadic logic and Loewenheim numbers*
  ◊ C45 C55 C65 C75 C85 E45 ◊

SHELAH, S. *Uncountable constructions for B.A.,e.c. groups and Banach spaces*
  ◊ C25 C60 C65 E05 E65 E75 G05 ◊

TOKAREV, E.V. *The notion of indistinguishability and the p-Mazur property in Banach spaces (Russian)*
  ◊ C65 E75 ◊

TRUSS, J.K. *The group of the countable universal graph*
  ◊ C07 C50 C65 ◊

WEISPFENNING, V. *Quantifier elimination for distributive lattices and measure algebras*
  ◊ B25 B50 C10 C65 G05 G10 ◊

WOLTER, H. *On the "problem of the last root" for exponential terms* ◊ C60 C65 ◊

C55 Set-theoretic model theory

1931
FOLLEY, K.W. *Sets of the first and second α-category and α-residual sets* ◊ C55 C65 E07 E75 ◊

1957
MOSTOWSKI, ANDRZEJ *On a generalisation of quantifiers* ◊ B25 C55 C80 C95 ◊

TARSKI, A. & VAUGHT, R.L. *Arithmetical extensions of relational systems* ◊ C05 C07 C55 C60 ◊

TARSKI, A. & VAUGHT, R.L. *Elementary (arithmetical) extensions* ◊ C05 C07 C35 C55 C60 ◊

1959
FUCHS, L. *The existence of indecomposable abelian groups of arbitrary power* ◊ C55 C60 ◊

MAL'TSEV, A.I. *Small models (Russian)* ◊ C52 C55 ◊

RABIN, M.O. *Arithmetical extensions with prescribed cardinality* ◊ C20 C30 C55 ◊

1960
ALLING, N.L. *On ordered divisible groups* ◊ C55 C60 ◊

HANF, W.P. *Models of languages with infinitely long expressions* ◊ C55 C75 ◊

1961
ALLING, N.L. *A characterization of abelian η_α-groups in terms of their natural valuation* ◊ C55 C60 ◊

KEISLER, H.J. *Ultraproducts and elementary classes* ◊ C20 C40 C50 C52 C55 E05 E50 ◊

MACDOWELL, R. & SPECKER, E. *Modelle der Arithmetik* ◊ B28 C20 C55 C62 ◊

1962
ALLING, N.L. *On the existence of real-closed fields that are η_α-sets of power $\aleph_\alpha$* ◊ C55 C60 C65 ◊

CHANG, C.C. & KEISLER, H.J. *Applications of ultraproducts of pairs of cardinals to the theory of models* ◊ C20 C55 E05 E50 ◊

FRAYNE, T.E. & MOREL, A.C. & SCOTT, D.S. *Reduced direct products* ◊ B20 C20 C55 ◊

KEISLER, H.J. *Some applications of the theory of models to set theory* ◊ C20 C55 E55 ◊

RESCHER, N. *Plurality-quantification* ◊ C55 C80 ◊

1963
KEISLER, H.J. *Limit ultrapowers* ◊ C20 C55 ◊

1964
FUHRKEN, G. *Skolem-type normal forms for first-order languages with a generalized quantifier* ◊ C55 C80 ◊

KEISLER, H.J. *Good ideals in fields of sets* ◊ C20 C55 E05 ◊

KEISLER, H.J. *On cardinalities of ultraproducts* ◊ C20 C55 E05 ◊

KEISLER, H.J. *Ultraproducts and saturated models* ◊ C20 C50 C55 ◊

PARK, D. *Set theoretic constructions in model theory* ◊ C55 ◊

VAUGHT, R.L. *The completeness of logic with the added quantifier "there are uncountable many"* ◊ C55 C80 ◊

1965
CHANG, C.C. *A note on the two cardinal problem* ◊ C55 E50 ◊

FUHRKEN, G. *Languages with added quantifier "There exist at least $\aleph_\alpha$"* ◊ C55 C80 ◊

HAERTIG, K. *Ueber einen Quantifikator mit zwei Wirkungsbereichen* ◊ C55 C80 ◊

KEISLER, H.J. *A survey of ultraproducts* ◊ C20 C55 C98 E05 ◊

KEISLER, H.J. *Ideals with prescribed degree of goodness* ◊ C20 C55 E05 ◊

KEISLER, H.J. *Limit ultraproducts* ◊ C20 C55 E55 ◊

KEISLER, H.J. *Reduced products and Horn classes* ◊ B20 C20 C52 C55 ◊

MORLEY, M.D. *Omitting classes of elements* ◊ C30 C55 C75 ◊

RIBENBOIM, P. *On the existence of totally ordered abelian groups which are η_α-sets* ◊ C55 C60 E07 E75 ◊

VAUGHT, R.L. *A Loewenheim-Skolem theorem for cardinals far apart* ◊ C55 C80 ◊

VAUGHT, R.L. *The Loewenheim-Skolem theorem* ◊ C07 C55 C80 ◊

1966
ERDOES, P. & HAJNAL, A. *On a problem of B.Jonsson* ◊ C05 C55 E05 E50 E55 ◊

HELLING, M. *Model-theoretic problems for some extensions of first-order languages* ◊ C55 C75 C80 E55 ◊

KEISLER, H.J. *First order properties of pairs of cardinals* ◊ C55 ◊

KEISLER, H.J. *Some model theoretic results for ω-logic* ◊ C35 C50 C55 C75 ◊

KOSTINSKY, A. *Recent results on Jonsson algebras* ◊ C05 C55 E05 E55 ◊

LOPEZ-ESCOBAR, E.G.K. *On defining well-orderings* ◊ C55 C65 C75 E10 ◊

YASUHARA, M. *An axiomatic system for the first order language with an equi-cardinality quantifier* ◊ C55 C80 ◊

YASUHARA, M. *Syntactical and semantical properties of generalized quantifiers* ◊ C40 C55 C80 ◊

1967

HASENJAEGER, G. *On Loewenheim-Skolem-type insufficiencies of second order logic*
⋄ B15 C55 C85 ⋄

KEISLER, H.J. & MORLEY, M.D. *On the number of homogeneous models of a given power*
⋄ C50 C55 C75 E50 ⋄

KEISLER, H.J. *Ultraproducts which are not saturated*
⋄ C20 C50 C55 ⋄

KEISLER, H.J. *Ultraproducts of finite sets*
⋄ C13 C20 C55 E05 ⋄

SHAFAAT, A. *Principle of localization for a more general type of languages* ⋄ B10 C55 C75 C90 ⋄

1968

CHANG, C.C. *Some remarks on the model theory of infinitary languages* ⋄ C07 C30 C55 C75 ⋄

KEISLER, H.J. *Models with orderings* ⋄ C55 C80 ⋄

MALITZ, J. *The Hanf number for complete $L_{\omega_1,\omega}$-sentences* ⋄ C55 C75 ⋄

MORLEY, M.D. *Partitions and models*
⋄ C30 C50 C55 C62 C75 E05 E45 E55 ⋄

SLOMSON, A. *The monadic fragment of predicate calculus with the Chang quantifier and equality*
⋄ B25 C20 C55 C80 ⋄

VAUGHT, R.L. *Model theory and set theory*
⋄ C55 C80 C98 ⋄

YASUHARA, M. *On minimal ordinal types of theories*
⋄ C55 ⋄

1969

ERSHOV, YU.L. *The number of linear orders on a field (Russian)* ⋄ C55 C60 ⋄

ISSEL, W. *Semantische Untersuchungen ueber Quantoren I* ⋄ B10 C55 C80 ⋄

PABION, J.F. *Extensions du theoreme de Loewenheim-Skolem en logique infinitaire*
⋄ C55 C75 ⋄

RESSAYRE, J.-P. *Sur les theories du premier ordre categorique en un cardinal* ⋄ C30 C35 C55 ⋄

WOJCIECHOWSKA, A. *Generalized products for Q_α-languages (Russian summary)*
⋄ C30 C55 C80 ⋄

1970

CHUDNOVSKY, G.V. *Problems of the theory of models, related to categoricity (Russian)*
⋄ C35 C45 C50 C52 C55 C75 ⋄

FUHRKEN, G. *On the degree of compactness of the languages Q_α* ⋄ C55 C80 ⋄

ISSEL, W. *Semantische Untersuchungen ueber Quantoren II,III* ⋄ C40 C55 C80 ⋄

KEISLER, H.J. *Logic with the quantifier "there exist uncountably many"* ⋄ C30 C55 C62 C75 C80 ⋄

KUNEN, K. *Some applications of iterated ultrapowers in set theory*
⋄ C20 C55 C62 E05 E35 E45 E50 E55 ⋄

LIPNER, L.D. *Some aspects of generalized quantifiers*
⋄ C55 C80 ⋄

MAKKAI, M. *Structures elementarily equivalent relative to infinitary languages to models of higher power*
⋄ C55 C75 ⋄

REYES, G.E. *Local definability theory*
⋄ C07 C15 C40 C50 C55 C75 ⋄

SHELAH, S. *A note on Hanf numbers*
⋄ C35 C55 C75 ⋄

SHELAH, S. *Finite diagrams stable in power*
⋄ C35 C45 C50 C55 ⋄

SHELAH, S. *On the cardinality of ultraproduct of finite sets*
⋄ C13 C20 C55 E05 ⋄

SHELAH, S. *Remark to "Local definability theory" of Reyes* ⋄ C40 C55 ⋄

SLOMSON, A. *An algebraic characterization of indistinguishable cardinals* ⋄ C55 C85 E10 ⋄

WOJCIECHOWSKA, A. *A note on models with orderings (Russian summary)* ⋄ C55 E05 E55 ⋄

YASUHARA, M. *Incompleteness of L_p-languages*
⋄ C55 C80 ⋄

1971

BARWISE, J. & KUNEN, K. *Hanf numbers for fragments of $L_{\infty\omega}$* ⋄ C55 C75 ⋄

EBBINGHAUS, H.-D. *On models with large automorphism groups* ⋄ C30 C55 C80 ⋄

KEISLER, H.J. *On theories categorical in their own power*
⋄ C35 C55 ⋄

KUNEN, K. & PRIKRY, K. *On descendingly incomplete ultrafilters* ⋄ C55 E05 ⋄

LOPEZ, G. *Un exemple d'une χ-classe M ne contenant pas de multirelation (M-χ)-universelle de cardinal χ (χ nombre beth)* ⋄ C50 C55 ⋄

MAGIDOR, M. *On the role of supercompact and extendible cardinals in logic* ⋄ C55 C75 C85 E47 E55 ⋄

MAGIDOR, M. *There are many normal ultrafilters corresponding to a supercompact cardinal*
⋄ C20 C55 E05 E55 ⋄

MALITZ, J. *Infinitary analogs of theorems from first order model theory* ⋄ C30 C40 C55 C75 ⋄

MORLEY, M.D. *The Loewenheim-Skolem theorem for models with standard part* ⋄ C30 C55 ⋄

SHELAH, S. *On the number of non-almost isomorphic models of T in a power* ⋄ C55 C75 ⋄

SHELAH, S. *Stability, the f.c.p., and superstability: model theoretic properties of formulas in first order theory*
⋄ C35 C45 C55 ⋄

SHELAH, S. *The number of non-isomorphic models of an unstable first-order theory* ⋄ C45 C55 ⋄

SHELAH, S. *Two cardinal compactness*
⋄ C30 C52 C55 C80 ⋄

SILVER, J.H. *Some applications of model theory in set theory* ⋄ C30 C55 D55 E05 E45 E55 ⋄

SILVER, J.H. *The independence of Kurepa's conjecture and two-cardinal conjectures in model theory*
⋄ C55 E05 E35 E55 ⋄

TAYLOR, W. *Some constructions of compact algebras*
⋄ C05 C50 C55 C65 ⋄

1972

BARWISE, J. *The Hanf number of second order logic*
⋄ B15 C55 C85 E30 ⋄

BELL, J.L. *On the relationship between weak compactness in $L_{\omega_1,\omega}, L_{\omega_1,\omega_1}$ and restricted second order languages* ⋄ C55 C75 C80 C85 E05 E55 ⋄

BENDA, M. *On reduced products and filters*
⋄ C20 C55 E05 E50 ⋄

BOFFA, M. *Sur l'existence des corps universels-homogenes*
⋄ C50 C55 C60 ⋄

CHUDNOVSKY, D.V. *Topological properties of products of discrete spaces and set theory (Russian)*
⋄ C55 C75 E35 E55 E75 ⋄

FLUM, J. *Die Automorphismenmengen der Modelle einer L_{Q_κ}-Theorie* ⋄ C30 C55 C80 ⋄

FLUM, J. *Hanf numbers and well-ordering numbers*
⋄ C30 C55 C75 C80 ⋄

FUHRKEN, G. *A remark on the Haertig quantifier*
⋄ C55 C80 ⋄

HODGES, W. *On order-types of models* ⋄ C30 C55 ⋄

JENSEN, R.B. *The fine structure of the constructible hierarchy* ⋄ C55 E05 E35 E45 E47 E55 E65 ⋄

KETONEN, J. *On nonregular ultrafilters*
⋄ C20 C55 E05 E55 G05 ⋄

KUEKER, D.W. *Loewenheim-Skolem and interpolation theorems in infinitary languages*
⋄ C40 C55 C75 E05 ⋄

KUNEN, K. *Ultrafilters and independent sets*
⋄ C55 E05 ⋄

LACHLAN, A.H. *A property of stable theories*
⋄ C45 C55 ⋄

LOPEZ-ESCOBAR, E.G.K. *The infinitary language $L_{\theta\theta}$ is not local (Russian summary)* ⋄ C55 C75 E55 ⋄

MALITZ, J. & REINHARDT, W.N. *A complete countable $L^Q_{\omega_1}$-theory with maximal models of many cardinalities*
⋄ C55 C80 ⋄

MALITZ, J. & REINHARDT, W.N. *Maximal models in the language with quantifier "there exist uncountably many"* ⋄ C50 C55 C80 E55 ⋄

MITCHELL, W.J. *Aronszajn trees and the independence of the transfer property* ⋄ C55 E05 E35 E55 ⋄

REYES, G.E. *$L_{\omega_1\omega}$ is enough: a reduction theorem for some infinitary languages* ⋄ C55 C75 E55 ⋄

ROSENTHAL, J.W. *A new proof of a theorem of Shelah*
⋄ C35 C45 C55 E45 E50 ⋄

SCHMERL, J.H. *An elementary sentence which has ordered models* ⋄ C55 E55 ⋄

SCHMERL, J.H. & SHELAH, S. *On power-like models for hyperinaccessible cardinals* ⋄ C30 C55 E55 ⋄

SHELAH, S. *A combinatorial problem; stability and order for models and theories in infinitary languages*
⋄ C45 C55 C75 E05 ⋄

SHELAH, S. *For what filters is every reduced product saturated ?* ⋄ C20 C50 C55 ⋄

SHELAH, S. *On models with power-like orderings*
⋄ C55 ⋄

SHELAH, S. *Saturation of ultrapowers and Keisler's order*
⋄ C20 C35 C45 C50 C55 ⋄

SLOMSON, A. *Generalized quantifiers and well orderings*
⋄ B25 C55 C65 C80 E07 ⋄

TAYLOR, W. *Residually small varieties*
⋄ C05 C50 C55 ⋄

VINNER, S. *A generalization of Ehrenfeucht's game and some applications* ⋄ B25 C07 C55 C80 E60 ⋄

WOJCIECHOWSKA, A. *On Helling cardinals*
⋄ C55 E05 E55 ⋄

1973

BELEGRADEK, O.V. *Almost categorical theories (Russian)*
⋄ C35 C45 C55 ⋄

BOWEN, K.A. *Infinite forcing and natural models of set theory (Russian summary)*
⋄ C25 C55 C62 C75 C80 E55 G30 ⋄

CHANG, C.C. *What's so special about saturated models ?*
⋄ C40 C50 C55 ⋄

CUTLAND, N.J. *Model theory on admissible sets*
⋄ C45 C55 C70 ⋄

FRIEDMAN, H.M. *Beth's theorem in cardinality logics*
⋄ C40 C55 C80 ⋄

GOGOL, D. *Low cardinality models for a type of infinitary theory* ⋄ C55 C75 ⋄

HODGES, W. *Models in which all long indiscernible sequences are indiscernible sets* ⋄ C30 C55 ⋄

KLEINBERG, E.M. *Rowbottom cardinals and Jonsson cardinals are almost the same* ⋄ C55 E05 E55 ⋄

LASCAR, D. *Types definissables et produits de types*
⋄ C45 C55 ⋄

LINDSTROEM, P. *A note on weak second order logic with variables for elementarily definable relations*
⋄ B15 C55 C62 ⋄

MACINTYRE, A. *Martin's axiom applied to existentially closed groups* ⋄ C25 C55 C60 C75 C80 E50 ⋄

PRIKRY, K. *On descendingly complete ultrafilters*
⋄ C55 E05 E55 ⋄

RESSAYRE, J.-P. *Boolean models and infinitary first order languages*
⋄ C40 C50 C55 C70 C75 C80 C90 ⋄

SCHMERL, J.H. *Peano models with many generic classes*
⋄ C55 C62 E40 ⋄

SHELAH, S. *First order theory of permutation groups*
⋄ B15 C55 C60 C85 ⋄

SHELAH, S. *Weak definability in infinitary languages*
⋄ C40 C55 C75 ⋄

WOLTER, H. *Untersuchungen zum Spektralproblem gewisser Logiken 2. Stufe* ⋄ C55 C85 E10 ⋄

1974

BARWISE, J. *Mostowski's collapsing function and the closed unbounded filter* ⋄ C15 C55 C75 E47 ⋄

BAUMGARTNER, J.E. *The Hanf number for complete $L_{\omega_1,\omega}$-sentences (without GCH)*
⋄ C35 C50 C55 C75 ⋄

BELL, J.L. *On compact cardinals*
⋄ C55 C75 E45 E55 ⋄

COMFORT, W.W. & NEGREPONTIS, S. *The theory of ultrafilters*
⋄ C20 C55 C98 E05 E55 E75 E98 G05 ⋄

DEVLIN, K.J. *An introduction to the fine structure of the constructible hierarchy*
⋄ C55 D55 E45 E47 E55 E98 ⋄

ERIMBETOV, M.M. *Categoricity in power and the invalidity of the two-cardinal theorem for a formula of finite rank (Russian)* ⋄ C35 C45 C55 ⋄

FRIEDMAN, H.M. *On existence proofs of Hanf numbers*
⋄ C55 C75 E30 E70 F50 ⋄

FUCHS, L. *Indecomposable abelian groups of measurable cardinalities* ⋄ C55 C60 E55 ⋄

GARLAND, S.J. *Second-order cardinal characterizability* ⋄ B15 C40 C55 C85 D55 E10 E47 E55 ⋄

GRILLIOT, T.J. *Model theory for dissecting recursion theory* ⋄ C55 C70 C75 C80 D60 D75 ⋄

GUZICKI, W. *Elementary extensions of Levy's model of A_2^-* ⋄ C55 C62 C80 E25 E35 E40 F35 ⋄

HUTCHINSON, J.E. *Models of set theories and their applications* ⋄ C55 C62 C80 ⋄

KEISLER, H.J. & PRIKRY, K. *A result concerning cardinalities of ultraproducts* ⋄ C20 C55 E05 ⋄

KEISLER, H.J. *Models with tree structures* ⋄ C55 C62 E65 ⋄

KOTLARSKI, H. *On the existence of well-ordered models* ⋄ C50 C55 ⋄

MCKENZIE, R. & SHELAH, S. *The cardinals of simple models for universal theories* ⋄ C05 C50 C55 E05 ⋄

REBHOLZ, J.R. *Some consequences of the Morass and Diamond* ⋄ C55 E05 E45 E65 ⋄

SCHMERL, J.H. *Generalizing special Aronszajn trees* ⋄ C55 E05 E35 E55 ⋄

SHELAH, S. *Infinite abelian groups, Whitehead problem and some constructions* ⋄ C55 C60 E05 E35 E45 E50 E75 ⋄

SHELAH, S. *The Hanf number of omitting complete types* ⋄ C55 C75 ⋄

1975

ANDLER, D. *Semi-minimal theories and categoricity* ⋄ C30 C35 C55 ⋄

BALDWIN, JOHN T. *Conservative extensions and the two cardinal theorems for stable theories* ⋄ C45 C55 ⋄

BENDA, M. *Some properties of mirrored orders* ⋄ C55 E07 ⋄

EBBINGHAUS, H.-D. *Zum L(Q)-Interpolationsproblem* ⋄ C40 C55 C80 ⋄

ERIMBETOV, M.M. *Complete theories with 1-cardinal formulas (Russian)* ⋄ C35 C45 C55 ⋄

FLUM, J. *First-order logic and its extensions* ⋄ C40 C55 C75 C95 C98 ⋄

FLUM, J. *L(Q)-preservation theorems* ⋄ C20 C40 C55 C80 E60 ⋄

FRIEDMAN, H.M. *Large models of countable height* ⋄ C15 C55 C62 E25 E40 ⋄

HARNIK, V. *A two cardinal theorem for sets of formulas in a stable theory* ⋄ C45 C50 C55 ⋄

HERRE, H. & WOLTER, H. *Entscheidbarkeit von Theorien in Logiken mit verallgemeinerten Quantoren* ⋄ B25 C10 C55 C80 D35 ⋄

JENSEN, F.V. *On completeness in cardinality logics* ⋄ C10 C35 C55 C60 C80 ⋄

MAKOWSKY, J.A. *Securable quantifiers, κ-unions and admissible sets* ⋄ C55 C70 C75 C80 C95 ⋄

SHELAH, S. *A two-cardinal theorem* ⋄ C55 E05 ⋄

SHELAH, S. *Categoricity in $\aleph_1$ of sentences in $L_{\omega_1\omega}(Q)$* ⋄ C35 C55 C75 C80 ⋄

SHELAH, S. *Existence of rigid-like families of abelian p-groups* ⋄ C52 C55 C60 ⋄

SHELAH, S. *Generalized quantifiers and compact logic* ⋄ C55 C80 C95 E55 ⋄

SHELAH, S. *Why there are many nonisomorphic models for unsuperstable theories* ⋄ C45 C50 C55 C60 C65 G05 ⋄

STAVI, J. *Extensions of Kripke's embedding theorem* ⋄ C55 C75 C90 G05 ⋄

VINNER, S. *Model-completeness in a first order language with a generalized quantifier* ⋄ C35 C55 C80 ⋄

VINNER, S. *On two complete sets in the analytical and the arithmetical hierarchies* ⋄ C55 C80 D55 ⋄

WEAVER, G.E. *Tarski-Vaught and Loewenheim-Skolem numbers* ⋄ C55 C95 ⋄

WOLTER, H. *Entscheidbarkeit der Arithmetik mit Addition und Ordnung in Logiken mit verallgemeinerten Quantoren* ⋄ B25 C10 C55 C80 F30 ⋄

WOLTER, H. *Untersuchungen zu Logiken mit Maechtigkeitsquantoren (Russian, English and French summaries)* ⋄ C55 C80 ⋄

1976

FERRO, R. *Consistency property and model existence theorem for second order negative languages with conjunctions and quantifications over sets of cardinality smaller than a strong limit cardinal of denumerable cofinality* ⋄ C55 C75 C85 ⋄

GALVIN, F. & PRIKRY, K. *Infinitary Jonsson algebras and partition relations* ⋄ C05 C50 C55 C75 E05 ⋄

GOLD, J.M. *A reflection property for saturated models* ⋄ C50 C55 ⋄

HUTCHINSON, J.E. *Model theory via set theory* ⋄ C40 C55 C62 C70 C75 C80 C95 E75 ⋄

KAUFMANN, M. *Some results in stationary logic. Doctoral Dissertation, University of Wisconsin, Madison* ⋄ C55 C80 ⋄

KEISLER, H.J. *Six classes of theories* ⋄ C45 C55 ⋄

KNIGHT, J.F. *Hanf numbers for omitting types over particular theories* ⋄ C07 C55 C62 ⋄

MACINTYRE, A. & SHELAH, S. *Uncountable universal locally finite groups* ⋄ C25 C50 C52 C55 C60 ⋄

MAKOWSKY, J.A. & SHELAH, S. & STAVI, J. *Δ-logics and generalized quantifiers* ⋄ C40 C55 C80 C95 ⋄

OKAZAKI, H. *An application of Fuhrken's reduction technique* ⋄ C55 C80 ⋄

SCHMERL, J.H. *Effectiveness and Vaught's gap ω two-cardinal theorem* ⋄ C55 C57 ⋄

SCHMERL, J.H. *On κ-like structures which embed stationary and closed unbounded subsets* ⋄ C55 E05 E55 ⋄

SCHMERL, J.H. *Remarks on self-extending models* ⋄ C30 C55 E55 ⋄

SHELAH, S. *Interpreting set theory in the endomorphism semi-group of a free algebra or in a category* ⋄ C55 E05 F25 G30 ⋄

SHELAH, S. *Refuting Ehrenfeucht conjecture on rigid models* ⋄ C50 C55 E50 ⋄

WEESE, M. *Entscheidbarkeit in speziellen uniformen Strukturen bezueglich Sprachen mit Maechtigkeitsquantoren* ⋄ B25 C55 C65 C80 ⋄

WEESE, M. *The universality of boolean algebras with the Haertig quantifier* ⋄ C55 C80 D35 F25 G05 ⋄

1977

BARWISE, J. *Some eastern two-cardinal theorems*
⋄ C55 C80 ⋄

BAUMGARTNER, J.E. *Ineffability properties of cardinals. II*
⋄ C55 E05 E45 E55 ⋄

BONNET, R. *Sur le type d'isomorphie d'algebres de Boole dispersees* ⋄ C52 C55 E50 G05 ⋄

EKLOF, P.C. *Applications of logic to the problem of splitting abelian groups*
⋄ C55 C60 C75 C80 E35 E45 E75 ⋄

EKLOF, P.C. & MEKLER, A.H. *On constructing indecomposable groups in L*
⋄ C55 C60 C75 E35 E45 E55 ⋄

GILLAM, D.W.H. *Morley numbers for generalized languages* ⋄ C55 C75 C80 ⋄

HERRE, H. *Decidability of theories in logics with additional monadic quantifiers* ⋄ B25 C55 C80 ⋄

HERRE, H. & WOLTER, H. *Entscheidbarkeit der Theorie der linearen Ordnung in L_{Q_1}*
⋄ B25 C55 C65 C80 ⋄

HICKIN, K.K. & PLOTKIN, J.M. *Some algebraic properties of weakly compact and compact cardinals*
⋄ C05 C55 E55 ⋄

KLEINBERG, E.M. *AD ⊢ "The $\aleph_n$ are Jonsson cardinals and $\aleph_\omega$ is a Rowbottom cardinal"*
⋄ C55 E05 E55 E60 ⋄

KNIGHT, J.F. *A complete $L_{\omega_1\omega}$-sentence characterizing $\aleph_1$* ⋄ C55 C75 ⋄

KRAWCZYK, A. & MAREK, W. *On the rules of proof generated by hierarchies* ⋄ B15 C55 C85 E45 ⋄

KUEKER, D.W. *Countable approximations and Loewenheim-Skolem theorems*
⋄ C15 C55 C75 E05 ⋄

LITMAN, A. & SHELAH, S. *Models with few isomorphic expansions* ⋄ C15 C35 C50 C55 ⋄

MAGIDOR, M. *Chang's conjecture and powers of singular cardinals* ⋄ C55 E40 E50 E55 E65 ⋄

MAGIDOR, M. & MALITZ, J. *Compact extensions of $L(Q)$ (part Ia)* ⋄ C55 C80 E55 E65 ⋄

MAGIDOR, M. & MALITZ, J. *Compactness and transfer for a fragment of L^2* ⋄ C55 C80 E05 E65 ⋄

MEKLER, A.H. *Theories with models of prescribed cardinalities* ⋄ C55 ⋄

MIZUTANI, C. *Monadic second order logic with an added quantifier Q* ⋄ B15 B25 C40 C55 C80 ⋄

MORLEY, M.D. *Homogenous sets*
⋄ C30 C55 C98 E05 E55 ⋄

MUSTAFIN, T.G. *On the two-cardinal problem (Russian) (Kazakh summary)* ⋄ C55 E15 E50 ⋄

PACHOLSKI, L. *Saturated limit reduced powers*
⋄ C20 C50 C55 ⋄

PALYUTIN, E.A. *Number of models in $L_{\infty\omega_1}$-theories I,II (Russian)* ⋄ C35 C52 C55 C75 E45 E65 ⋄

RESSAYRE, J.-P. *Models with compactness properties relative to an admissible language*
⋄ C50 C55 C70 C80 D55 D60 E15 ⋄

SCHIEMANN, I. *Eine Axiomatisierung des monadischen Praedikatenkalkuels mit verallgemeinerten Quantoren (Russian, English and French summaries)*
⋄ B25 C40 C55 C80 ⋄

SCHLIPF, J.S. *Ordinal spectra of first-order theories*
⋄ C35 C50 C55 C62 C70 D15 D30 D70 ⋄

SCHMERL, J.H. *An axiomatization for a class of two-cardinal models* ⋄ C55 ⋄

SEESE, D.G. & TUSCHIK, H.-P. *Construction of nice trees*
⋄ B25 C55 C65 C80 ⋄

SEESE, D.G. *Partitions for trees* ⋄ C55 C65 E05 ⋄

SEESE, D.G. *Second order logic, generalized quantifiers and decidability*
⋄ B15 B25 C55 C65 C80 C85 D35 ⋄

SHELAH, S. *A two cardinal theorem and a combinatorial theorem* ⋄ C55 E05 ⋄

TUSCHIK, H.-P. *Elimination verallgemeinerter Quantoren in ω_1-kategorischen Theorien (Russian, English and French summaries)* ⋄ B25 C10 C35 C55 C80 ⋄

TUSCHIK, H.-P. *On the decidability of the theory of linear orderings in the language $L(Q_1)$*
⋄ B25 C55 C65 C80 ⋄

WEESE, M. *Entscheidbarkeit der Theorie der booleschen Algebren in Sprachen mit Maechtigkeitsquantoren (Russian and English summaries)*
⋄ B25 C55 C80 G05 ⋄

WEESE, M. *The decidability of the theory of boolean algebras with cardinality quantifiers (Russian summary)* ⋄ B25 C10 C55 C80 G05 ⋄

WEESE, M. *The undecidability of well-ordering with the Haertig quantifier (Russian summary)*
⋄ C55 C65 C80 D35 E07 ⋄

1978

ABRAMSON, F.G. & HARRINGTON, L.A. *Models without indiscernibles*
⋄ C30 C50 C55 C62 C65 C75 E05 E40 H15 ⋄

BARWISE, J. & KAUFMANN, M. & MAKKAI, M. *Stationary logic* ⋄ C55 C75 C80 ⋄

BAUMGARTNER, J.E. & GALVIN, F. *Generalized Erdoes cardinals and $0^{\#}$* ⋄ C55 C75 E05 E45 E55 ⋄

BEN-DAVID, S. *On Shelah's compactness of cardinals*
⋄ C55 C80 E05 E35 E50 E55 ⋄

BENDA, M. *Compactness for omitting of types*
⋄ C20 C30 C55 C70 C75 E55 ⋄

BOUSCAREN, E. & CHATZIDAKIS, Z. *Modeles premiers et atomiques: theoreme des deux cardinaux dans les theories totalement transcendantes*
⋄ C45 C50 C55 ⋄

BURGESS, J.P. *Consistency proofs in model theory: A contribution to "Jensenlehre"*
⋄ C55 E05 E35 E50 ⋄

BURGESS, J.P. *On the Hanf number of Souslin logic*
⋄ C55 C75 E35 ⋄

DELON, F. *Types localement isoles et theoreme des deux cardinaux dans les theories stables denombrables*
⋄ C45 C50 C55 ⋄

GALVIN, F. & HESSE, G. & STEFFENS, K. *On the number of automorphisms of structures* ⋄ C55 E05 ⋄

GARAVAGLIA, S. *Relative strength of Malitz quantifiers*
⋄ C55 C80 ⋄

GILLAM, D.W.H. *Refined Morley numbers*
⋄ C55 C75 ⋄

HERRE, H. & WOLTER, H. *Entscheidbarkeit der Theorie der linearen Ordnung in L_{Q_κ} fuer regulaeres ω_κ*
⋄ B25 C55 C65 C80 E07 E50 ⋄

HERRE, H. & PINUS, A.G. *Zum Entscheidungsproblem fuer Theorien in Logiken mit monadischen verallgemeinerten Quantoren*
⋄ B25 C10 C55 C80 D35 ⋄

JOVANOVIC, A. *A note on ultrapower cardinality*
⋄ C20 C55 E05 E10 ⋄

KEISLER, H.J. *The stability function of a theory*
⋄ C45 C55 ⋄

KNIGHT, J.F. *Prime and atomic models* ⋄ C50 C55 ⋄

KOTLARSKI, H. *A note on countable well-ordered models*
⋄ C50 C55 E47 ⋄

KOTLARSKI, H. *Some remarks on well-ordered models*
⋄ C20 C50 C55 E07 ⋄

KUEKER, D.W. *Uniform theorems in infinitary logic*
⋄ C40 C55 C75 ⋄

KUNEN, K. *Saturated ideals*
⋄ C55 E05 E35 E40 E55 ⋄

MAKOWSKY, J.A. *Quantifying over countable sets: Positive vs. stationary logic* ⋄ C55 C80 ⋄

MEKLER, A.H. *The size of epimorphic extensions*
⋄ C55 C75 C95 E55 G30 ⋄

NADEL, M.E. & STAVI, J. *$L_{\infty\lambda}$-equivalence, isomorphism and potential isomorphism*
⋄ C55 C75 E07 E40 E65 ⋄

PINUS, A.G. *Cardinality of models for theories in a calculus with a Haertig quantifier (Russian)* ⋄ C55 C80 ⋄

PINUS, A.G. *The Loewenheim-Skolem property for second-order logics (Russian)*
⋄ C55 C80 C85 E45 ⋄

SCHWARTZ, N. *η_α-Strukturen*
⋄ C50 C55 C65 E07 ⋄

SEESE, D.G. *A remark to the undecidability of well-orderings with the Haertig quantifier*
⋄ C55 C65 C80 D35 ⋄

SHELAH, S. *A weak generalization of MA to higher cardinals* ⋄ B25 C55 E35 E50 E75 ⋄

SHELAH, S. *End extensions and numbers of countable models* ⋄ C15 C30 C55 C62 E50 ⋄

SHELAH, S. *Jonsson algebras in successor cardinals*
⋄ C05 C55 E05 E10 E75 ⋄

SHELAH, S. *Models with second order properties. I: Boolean algebras with no definable automorphisms*
⋄ C07 C30 C50 C55 E50 G05 ⋄

SHELAH, S. *Models with second order properties. II: Trees with no undefined branches*
⋄ C50 C55 C62 C75 C80 C95 E05 E35 E65 ⋄

SHELAH, S. & STERN, J. *The Hanf number of the first order theory of Banach spaces* ⋄ C55 C65 H10 ⋄

VAEAENAENEN, J. *Two axioms of set theory with applications to logic*
⋄ C40 C55 C80 C85 E35 E47 E65 ⋄

1979

AJTAI, M. *Isomorphism and higher order equivalence*
⋄ C15 C55 E35 E45 ⋄

ASSER, G. *Ueber die Charakterisierbarkeit transfiniter Maechtigkeiten im Praedikatenkalkuel der zweiten Stufe* ⋄ B15 C55 C80 C85 ⋄

BOURGAIN, J. & ROSENTHAL, H.P. & SCHECHTMAN, G. *Uniform validity and stationary logic* ⋄ C55 C80 ⋄

EKLOF, P.C. & MEKLER, A.H. *Stationary logic of finitely determinate structures*
⋄ B25 C10 C55 C60 C65 C80 ⋄

FEFERMAN, S. *Generalizing set-theoretical model theory and an analogue theory on admissible sets*
⋄ C50 C55 C70 E70 ⋄

GREEN, J. *Some model theory for game logics*
⋄ C55 C75 E60 ⋄

HERRE, H. & WOLTER, H. *Entscheidbarkeit der Theorie der linearen Ordnung in Logiken mit Maechtigkeitsquantoren bzw. mit Chang-Quantor*
⋄ B25 C55 C80 E07 E50 E55 ⋄

HERRE, H. & WOLTER, H. *The decision problem for the theory of linear orderings in extended logics*
⋄ B25 C55 C80 C98 ⋄

KAUFMANN, M. *A new omitting types theorem for L(Q)*
⋄ C55 C75 C80 ⋄

KRYNICKI, M. & LACHLAN, A.H. *On the semantics of the Henkin quantifier*
⋄ B15 B25 C40 C55 C80 C85 ⋄

MONK, J.D. & RASSBACH, W. *The number of rigid boolean algebras* ⋄ C50 C55 E75 G05 ⋄

MORGENSTERN, C.F. *On amalgamations of languages with Magidor-Malitz quantifiers* ⋄ C55 C80 E55 ⋄

MORGENSTERN, C.F. *The measure quantifier*
⋄ C55 C75 C80 E55 ⋄

MUNDICI, D. *Robinson consistency theorem in soft model theory (Italian summary)* ⋄ C40 C55 C80 C95 ⋄

OIKKONEN, J. *A generalization of the infinitely deep languages of Hintikka and Rantala*
⋄ C55 C75 C80 ⋄

PINUS, A.G. *A calculus of one-place predicates (Russian)*
⋄ B20 B25 C55 C80 C85 ⋄

PINUS, A.G. *Elementary equivalence of topological spaces (Russian)* ⋄ C55 C65 C85 G05 G10 ⋄

PINUS, A.G. *Elimination of the quantifiers Q_0 and Q_1 on symmetric groups (Russian)*
⋄ C10 C55 C60 C80 ⋄

PINUS, A.G. *Hanf number for the calculus with the Haertig quantifier (Russian)* ⋄ C55 C80 C85 E45 ⋄

SEESE, D.G. *Stationary logic and ordinals*
⋄ B25 C55 C65 C80 ⋄

SHELAH, S. *Hanf number of omitting type for simple first-order theories* ⋄ C20 C50 C55 C75 ⋄

TULIPANI, S. *The Hanf number for classes of algebras whose largest congruence is always finitely generated*
⋄ C05 C55 ⋄

VAEAENAENEN, J. *Abstract logic and set theory. I. Definability* ⋄ C40 C55 C95 E75 ⋄

VAEAENAENEN, J. *On Hanf numbers of unbounded logics*
⋄ C55 C75 C80 C85 C95 E55 ⋄

VAEAENAENEN, J. *Remarks on free quantifier variables*
◇ C55 C80 C85 C95 E45 ◇

1980

BRUCE, KIM B. *Model constructions in stationary logic I. Forcing* ◇ C25 C55 C70 C75 C80 ◇

CAICEDO, X. *Back-and-forth systems for arbitrary quantifiers* ◇ C55 C75 C80 ◇

CHARRETTON, C. & POUZET, M. *Les chaines dans les modeles d'Ehrenfeucht-Mostowski (English summary)*
◇ C30 C55 ◇

DECIU, E. *On the cardinal number and on the structure of the set of all subrings of the rational numbers field*
◇ C55 ◇

FAJARDO, S. *Compacidad y decidibilidad de logicas monadicas con cuantificadores cardinales*
◇ B25 C55 C80 ◇

GOLD, J.M. *Some results on the structure of the φ-spectrum* ◇ C50 C55 E55 ◇

GOLTZ, H.-J. *Untersuchungen zur Elimination verallgemeinerter Quantoren in Koerpertheorien*
◇ C10 C55 C60 C80 ◇

GOSTANIAN, R. & HRBACEK, K. *Propositional extensions of $L_{\omega_1\omega}$* ◇ C55 C75 E15 E50 ◇

HARRINGTON, L.A. *Extensions of countable infinitary logic which preserve most of its nice properties*
◇ C40 C55 C70 C75 C95 ◇

HAUSCHILD, K. *Generalized Haertig quantifiers*
◇ C55 C80 ◇

HERRE, H. & SEESE, D.G. *Concerning the monadic theory of the topology of well-orderings and scattered spaces*
◇ B25 C55 C65 C85 E07 E35 E50 E55 ◇

JAKUBIKOVA-STUDENOVSKA, D. *On weakly rigid monounary algebras* ◇ C05 C55 E10 E75 ◇

JOVANOVIC, A. *A note on the two-cardinal problem*
◇ C20 C55 E10 E55 ◇

JOVANOVIC, A. *Ultraproducts of well orders*
◇ C20 C55 E05 E07 ◇

KAKUDA, Y. *Set theory based on the language with the additional quantifier "for almost all" I*
◇ C55 C62 C80 E70 ◇

KOPPELBERG, S. *Cardinalities of ultraproducts of finite sets* ◇ C13 C20 C55 E05 E10 ◇

MALITZ, J. & RUBIN, M. *Compact fragments of higher order logic* ◇ B15 C55 C80 C85 ◇

MUSTAFIN, T.G. *A non-two-cardinal set of stable types (Russian)* ◇ C35 C45 C55 ◇

MUSTAFIN, T.G. *Theories with a two-cardinal formula (Russian)* ◇ C45 C50 C52 C55 ◇

PRZYMUSINSKA, H. *Craig interpolation theorem and Hanf number for v-valued infinitary predicate calculi (Russian summary)* ◇ B50 C40 C55 C75 G20 ◇

SGRO, J. *The interior operator logic and product topologies*
◇ C40 C55 C80 C90 ◇

SHELAH, S. *A note on cardinal exponentiation*
◇ C55 E05 E50 E65 ◇

SHELAH, S. *Independence results*
◇ C45 C55 E07 E35 E40 ◇

SHELAH, S. *Simple unstable theories*
◇ C45 C50 C55 ◇

TUSCHIK, H.-P. *On the decidability of the theory of linear orderings with generalized quantifiers*
◇ B25 C55 C65 C80 ◇

TZOUVARAS, A.D. *A non-standard characterization of the norm of free ultrafilters* ◇ C55 E05 H05 ◇

VAEAENAENEN, J. *Boolean-valued models and generalized quantifiers* ◇ C55 C80 C85 C90 E35 E40 ◇

VAEAENAENEN, J. *The Hanf number of $L_{\omega_1\omega_1}$*
◇ C55 C75 E35 E55 ◇

1981

BAUDISCH, A. *Formulas of $L(aa)$ where aa is not in the scope of "$\neg$"* ◇ C55 C80 ◇

BAUDISCH, A. *The elementary theory of abelian groups with m-chains of pure subgroups*
◇ B25 C55 C60 C80 ◇

CAICEDO, X. *On extensions of $L_{\omega\omega}(Q_1)$*
◇ C55 C75 C80 ◇

GUZICKI, W. *The equivalence of definable quantifiers in second order arithmetic*
◇ C30 C55 C62 C80 E40 E45 F35 ◇

HAUSCHILD, K. *Zum Vergleich von Haertigquantor und Rescherquantor* ◇ B25 C10 C55 C80 D35 ◇

HERRE, H. & WOLTER, H. *Untersuchungen zur Theorie der linearen Ordnung in Logiken mit Maechtigkeitsquantoren*
◇ B25 C55 C65 C80 E07 ◇

HODGES, W. *In singular cardinality, locally free algebras are free* ◇ C05 C55 C60 E05 E60 E75 ◇

KAKUDA, Y. *Completeness theorem for the system ST with the quantifier "aa"* ◇ C55 C80 E70 ◇

KAUFMANN, M. *Filter logics: Filters on ω_1*
◇ C55 C80 E05 ◇

KLEINBERG, E.M. *Producing measurable cardinals beyond κ* ◇ C20 C55 E05 E60 ◇

KNIGHT, J.F. *Algebraic independence*
◇ C40 C55 C75 ◇

SEESE, D.G. *Elimination of second-order quantifiers for well-founded trees in stationary logic and finitely determinate structures* ◇ C10 C55 C65 C80 ◇

SEESE, D.G. *Stationary logic and ordinals*
◇ B25 C55 C65 C80 ◇

SHELAH, S. *Models with second order properties. III. Omitting types for $L(Q)$* ◇ C55 C75 C80 E05 ◇

SHELAH, S. *On endo-rigid strongly $\aleph_1$-free abelian groups in $\aleph_1$* ◇ C50 C55 C60 E05 E50 E75 ◇

SHELAH, S. *On saturation for a predicate*
◇ C45 C50 C55 C65 E50 ◇

SHELAH, S. *On the number of nonisomorphic models of cardinality λ, $L_{\infty\lambda}$-equivalent to a fixed model*
◇ C55 C75 E45 E55 E75 ◇

TIURYN, J. *Logic of effective definitions*
◇ B60 C55 C75 ◇

VAEAENAENEN, J. *Generalized quantfiers in models of set theory* ◇ C55 C62 C80 C85 C95 E35 ◇

WEESE, M. *Decidability with respect to Haertig quantifier and Rescher quantifier* ◇ B25 C55 C80 D35 ◇

1982

DUGAS, M. & GOEBEL, R. *Every cotorsion-free algebra is an endomorphism algebra* ◇ C55 C60 E75 ◇

DUGAS, M. & GOEBEL, R. *Every cotorsion-free ring is an endomorphism ring* ◇ C55 C60 E45 E65 E75 ◇

FERRO, R. ω-*satisfiability,* ω-*consistency property and the downward Loewenheim-Skolem theorem for* $L_{\kappa\kappa}$ *(Italian summary)* ◇ C55 C75 ◇

FOREMAN, M. *Large cardinals and strong model theoretic transfer properties* ◇ C55 E35 E55 ◇

GORETTI, A. *Alcune osservazioni sul linguaggio* $L(Q_1)$ *(English summary)* ◇ C50 C55 C80 ◇

GUREVICH, R. *Loewenheim-Skolem problem for functors* ◇ C55 C75 G30 ◇

JA'JA', J. *The computational complexity of a set of quadratic functions* ◇ C55 ◇

LAVER, R. *Saturated ideals and nonregular ultrafilters* ◇ C20 C55 E05 E55 E65 ◇

MEISSNER, W. *A Loewenheim-Skolem theorem for inner product spaces* ◇ C55 C65 ◇

SEESE, D.G. & WEESE, M. *L(aa)-elementary types of well-orderings* ◇ B25 C55 C65 C80 E10 ◇

SEESE, D.G. & TUSCHIK, H.-P. & WEESE, M. *Undecidable theories in stationary logic* ◇ C55 C65 C80 D35 E05 E75 ◇

SHELAH, S. *On the number of nonisomorphic models in* $L_{\infty,\kappa}$ *when* κ *is weakly compact* ◇ C55 C75 E55 ◇

TULIPANI, S. *Sui reticolo di congruenze di strutture algebriche modelli di teorie universali positive (English summary)* ◇ C05 C50 C55 ◇

TUSCHIK, H.-P. *Elimination of cardinality quantifiers* ◇ C10 C45 C55 C60 C80 ◇

VAEAENAENEN, J. *Abstract logic and set theory. II. Large cardinals* ◇ C55 C95 E35 E55 ◇

VAEAENAENEN, J. *Generalized quantifiers in models of set theory* ◇ C55 C62 C80 C85 C95 E35 ◇

1983

ADAMSON, A. *A note on two-cardinal models* ◇ C55 ◇

BAUDISCH, A. & SEESE, D.G. & TUSCHIK, H.-P. ω-*trees in stationary logic* ◇ B25 C55 C65 C80 ◇

COMBASE, J. *On the existence of finitely determinate models for some theories in stationary logic* ◇ C55 C80 E55 ◇

DONDER, H.-D. & KOEPKE, P. *On the consistency strength of "accessible" Jonsson cardinals and of the weak Chang conjecture* ◇ C55 E05 E35 E55 ◇

DUGAS, M. & HERDEN, G. *Abitrary torsion classes and almost free abelian groups* ◇ C55 C60 E50 E55 E65 E75 ◇

DUGAS, M. & HERDEN, G. *Arbitrary torsion classes of abelian groups* ◇ C55 C60 E55 E65 E75 ◇

FOREMAN, M. *More saturated ideals* ◇ C55 C62 E05 E35 E55 ◇

GUZICKI, W. *Definable quantifiers in second order arithmetic and elementary extensions of* ω-*models* ◇ C30 C55 C62 C80 E40 E50 F35 ◇

HARIZANOV, V. & MIJAJLOVIC, Z. *Regular relations and the quantifier "there exist uncountably many"* ◇ C55 C80 ◇

HUBER, M. *Methods of set theory and the abundance of separable abelian p-groups* ◇ C55 C60 E05 E75 ◇

HUBER, M. *On reflexive modules and abelian groups* ◇ C55 C60 E50 E55 E75 G30 ◇

IVANOV, A.A. *Decidability of theories in a certain calculus (Russian)* ◇ B25 C10 C55 C60 C80 D35 ◇

IVANOV, A.A. *Some theories in generalized calculi (Russian)* ◇ B25 C10 C55 C60 C80 ◇

JOVANOVIC, A. *On the two-cardinal problem* ◇ C20 C55 E05 ◇

KAKUDA, Y. *Reflection principles via filter quantifier (Japanese)* ◇ C55 C80 E30 ◇

KUBOTA, N. *A splitting property on a set of ideals and a weak saturation hypothesis* ◇ C55 E05 ◇

MALITZ, J. *Downward transfer of satisfiability for sentences of* $L^{1,1}$ ◇ C55 C80 ◇

MAREK, W. & RASMUSSEN, K. *Spectrum of L* ◇ C55 C62 E45 ◇

MEGIBBEN, C. *Crawley's problem on the unique* ω-*elongation of p-groups is undecidable* ◇ C55 C60 E35 E45 E50 E65 E75 ◇

MEKLER, A.H. *Proper forcing and abelian groups* ◇ C55 C60 E35 E50 E75 ◇

MITCHELL, W.J. *Sets constructed from sequences of measures: revisited* ◇ C55 E45 E55 ◇

PINUS, A.G. *Calculus with the quantifier of elementary equivalence (Russian)* ◇ C40 C55 C80 D35 ◇

POIZAT, B. *Beaucoup de modeles a peu de frais* ◇ C45 C50 C52 C55 ◇

RUBIN, M. *A Boolean algebra with few subalgebras, interval algebras and retractiveness* ◇ C55 C80 D35 E50 E65 G05 ◇

RUBIN, M. & SHELAH, S. *On the expressibility hierarchy of Magidor-Malitz quantifiers* ◇ C55 C80 E40 E65 ◇

SHELAH, S. *Classification theory of non-elementary classes, Part I: The number of uncountable models of* $\psi \in L_{\omega_1,\omega}$ ◇ C45 C52 C55 C75 C80 E50 ◇

SHELAH, S. *Construction of many complicated uncountable structures and boolean algebras* ◇ C50 C55 G05 ◇

SHELAH, S. *Models with second order properties IV. A general method and eliminating diamonds* ◇ C50 C55 C60 C65 C80 E65 G05 ◇

SHELAH, S. *On the number of non conjugate subgroups* ◇ C55 C60 E50 E75 ◇

TUSCHIK, H.-P. *The Ramsey theorem for additive colourings* ◇ C55 C65 E05 ◇

VAEAENAENEN, J. Δ-*extension and Hanf-numbers* ◇ C40 C55 C80 C95 ◇

ZAWADOWSKI, M. *The Skolem-Loewenheim theorem in toposes* ◇ C55 C80 E10 G30 ◇

1984

BELYAEV, V.YA. *Uncountable extensions of countable algebraically closed semigroups (Russian)* ◇ C05 C55 C60 C75 ◇

BURNS, R.G. & OKOH, F. & SMITH, HOWARD & WIEGOLD, J. *On the number of normal subgroups of an uncountable soluble group* ◇ C55 C60 E35 E50 E75 ◇

COMBASE, J. *Introduction a la logique stationnaire*
◇ C55 C80 C98 ◇

DUGAS, M. & GOEBEL, R. *Almost Σ-cyclic abelian p-groups in L* ◇ C55 C60 E45 E75 ◇

DUGAS, M. & GOEBEL, R. & GOLDSMITH, B. *Representation of algebras over a complete discrete valuation ring* ◇ C55 C60 E75 ◇

EKLOF, P.C. & MEKLER, A.H. & SHELAH, S. *Almost disjoint abelian groups*
◇ C55 C60 E45 E50 E75 ◇

GIORGETTA, D. & SHELAH, S. *Existentially closed structures in the power of the continuum*
◇ C15 C25 C30 C52 C55 C60 C75 ◇

HEINDORF, L. *Beitraege zur Modelltheorie der Booleschen Algebren (English and Russian summaries)*
◇ B25 C10 C35 C55 C80 G05 ◇

HENLE, J.M. & KLEINBERG, E.M. & WATRO, R.J. *On the ultrafilters and ultrapowers of strong partition cardinals*
◇ C20 C55 E05 E60 ◇

HODGES, W. *Models built on linear orderings (French summary)* ◇ C30 C55 C98 E07 ◇

HODGES, W. *On constructing many non-isomorphic algebras* ◇ C05 C52 C55 C75 E50 E55 E75 ◇

HORT, C. & OSSWALD, H. *On nonstandard models in higher order logic*
◇ B15 C55 C85 E55 H05 H20 ◇

KANAMORI, A. & TAYLOR, A.D. *Separating ultrafilters on uncountable cardinals* ◇ C20 C55 E05 E55 ◇

KAUFMANN, M. *Some remarks on equivalence in infinitary and stationary logic* ◇ C55 C65 C75 C80 E45 ◇

KOEPKE, P. *The consistency strength of the free-subset property for* ω_ω ◇ C55 E05 E35 E45 E55 ◇

LENSKI, W. *Elimination of quantifiers for the theory of archimedean ordered divisible groups in a logic with Ramsey quantifiers* ◇ C10 C55 C60 C80 ◇

LEVINSKI, J.-P. *Instances of the conjecture of Chang*
◇ C55 E05 E45 E55 ◇

MEKLER, A.H. *Stationary logic of ordinals*
◇ B25 C10 C55 C65 C80 E10 ◇

RAPP, A. *On the expressive power of the logics* $L(Q_\alpha^{(n_1,...,n_m)})$ ◇ C55 C80 ◇

SHELAH, S. *A pair of non-isomorphic* $\equiv_{\infty\lambda}$ *models of power* λ *for* λ *singular with* $\lambda^\omega = \lambda$ ◇ C55 C75 ◇

SHELAH, S. *Can you take Solovay's inaccessible away?*
◇ C55 C75 C80 E15 E25 E35 E55 E75 ◇

TRYBA, J. *On Jonsson cardinals with uncountable cofinality* ◇ C55 E05 E55 ◇

1985

ABRAHAM, U. & RUBIN, M. & SHELAH, S. *On the consistency of some partition theorems for continuous colorings, and structure of* $\aleph_1$-*dense real order types*
◇ C55 C62 C65 C80 E05 E07 E35 ◇

BALDWIN, JOHN T. & SHELAH, S. *Second-order quantifiers and the complexity of theories*
◇ C40 C55 C65 C75 C85 ◇

BESLAGIC, A. & RUDIN, M.E. *Set-theoretic constructions of nonshrinking open covers* ◇ C55 C65 E05 E75 ◇

DOW, A. *Saturated boolean algebras and their Stone spaces* ◇ C50 C55 E35 E50 E75 G05 ◇

EBBINGHAUS, H.-D. *Extended logics: The general framework* ◇ C40 C55 C80 C95 C98 ◇

FLUM, J. *Maximale monadische Logiken*
◇ C55 C80 C95 ◇

GOEBEL, R. & SHELAH, S. *Semi-rigid classes of cotorsion-free abelian groups*
◇ C55 C60 E05 E45 E55 E65 E75 ◇

GOLTZ, H.-J. *The boolean sentence algebra of the theory of linear ordering is atomic with respect to logics with a Malitz quantifier* ◇ C55 C65 C80 G05 ◇

HODGES, W. *Building models by games*
◇ C25 C55 C60 C80 C98 E60 ◇

HODKINSON, I.M. *A construction of many uncountable rings* ◇ C55 C60 E05 E50 E75 ◇

KAUFMANN, M. *A note on the Hanf number of second-order logic* ◇ C55 C80 C85 ◇

KAUFMANN, M. *The quantifier "There exist uncountable many" and some of its relatives* ◇ C55 C80 C98 ◇

KOSSAK, R. *Recursively saturated* ω_1-*like models of arithmetic* ◇ C55 C57 C62 C75 C80 E65 ◇

MAKOWSKY, J.A. *Vopenka's principle and compact logics*
◇ C05 C55 C95 E55 ◇

MEKLER, A.H. & SHELAH, S. *Stationary logic and its friends I* ◇ C40 C55 C75 C80 E35 E50 ◇

MIJAJLOVIC, Z. *On the definability of the quantifier "there exist uncountably many"* ◇ C55 C80 ◇

MONK, J.D. *On endomorphism bases*
◇ C05 C55 E10 G05 ◇

PILLAY, A. & SHELAH, S. *Classification theory over a predicate I* ◇ C40 C45 C50 C55 ◇

PINUS, A.G. *Applications of boolean powers of algebraic systems (Russian)*
◇ C05 C30 C55 C57 C80 C85 D35 E50 G05 ◇

RAPP, A. *Elimination of Malitz quantifiers in stable theories* ◇ C10 C45 C55 C80 ◇

RAPP, A. *The ordered field of real numbers and logics with Malitz quantifiers* ◇ B25 C10 C55 C65 C80 ◇

SCHMERL, J.H. *Models of Peano arithmetic and a question of Sikorski on ordered fields*
◇ C55 C60 C62 E05 E75 ◇

SCHMERL, J.H. *Transfer theorems and their applications to logic* ◇ C55 C80 C98 ◇

SHELAH, S. *Incompactness in regular cardinals*
◇ C55 C60 E05 ◇

SHELAH, S. *Monadic logic and Loewenheim numbers*
◇ C45 C55 C65 C75 C85 E45 ◇

SHELAH, S. *On the possible number* $no(M) =$ *the number of nonisomorphic models* $L_{\infty,\lambda}$-*equivalent to M of power* λ, *for* λ *singular* ◇ C55 C75 ◇

SHELAH, S. *Remarks in abstract model theory*
◇ C40 C50 C52 C55 C80 C95 ◇

SPECTOR, M. *Model theory under the axiom of determinateness* ◇ C20 C55 C75 E60 ◇

VAEAENAENEN, J. *Set-theoretical definability of logics*
◇ C40 C55 C80 C95 C98 E47 ◇

ZAWADOWSKI, M. *The Skolem-Loewenheim theorem in toposes II* ◇ C55 C90 G30 ◇

C70 ∪ C75 Admissible sets and infinitary languages

1920
SKOLEM, T.A. *Logisch-kombinatorische Untersuchungen ueber die Erfuellbarkeit oder Beweisbarkeit mathematischer Saetze nebst einem Theorem ueber dichte Mengen* ⋄ B10 B25 C07 C65 C75 ⋄

1931
ZERMELO, E. *Ueber Stufen der Quantifikation und die Logik des Unendlichen*
 ⋄ A05 C75 D35 E07 E30 ⋄

1934
ITO, MAKOTO *Einige Anwendungen der Theorie des Entscheidungsproblems zur Axiomatik I*
 ⋄ B30 C75 F25 ⋄

1935
ZERMELO, E. *Grundlagen einer allgemeinen Theorie der mathematischen Satzsysteme (erste Mitteilung)*
 ⋄ B05 C75 E20 ⋄

1937
KURATOWSKI, K. *Les types d'ordre definissables et les ensembles boreliens* ⋄ C75 D55 E07 E15 ⋄

1938
HELMER, O. *Languages with expressions of infinite length*
 ⋄ C75 ⋄
KRASNER, M. *Une generalisation de la notion de corps*
 ⋄ C07 C60 C75 ⋄

1939
NOVIKOV, P.S. *On some existence theorems (Russian)*
 ⋄ B05 C75 ⋄

1940
BOCHVAR, D.A. *Ueber einen Aussagenkalkuel mit abzaehlbaren logischen Summen und Produkten (Russian summary)* ⋄ B05 C75 ⋄

1943
NOVIKOV, P.S. *On the consistency of certain logical calculus (Russian summary)* ⋄ B05 B28 C75 ⋄

1945
KRASNER, M. *Generalisation et analogues de la theorie de Galois* ⋄ C07 C60 C75 ⋄

1947
MOSTOWSKI, ANDRZEJ *On absolute properties of relations*
 ⋄ B15 C75 C85 E15 E47 ⋄

1949
JORDAN, P. *Zur Axiomatik der Verknuepfungsbereiche*
 ⋄ C05 C75 G10 ⋄

1950
KRASNER, M. *Generalisation abstraite de la theorie de Galois* ⋄ C07 C60 C75 ⋄

1952
SIKORSKI, R. *Products of abstract algebras*
 ⋄ C05 C75 ⋄

1954
CHANG, C.C. *Some general theorems on direct products and their applications in the theory of models*
 ⋄ C05 C75 ⋄

1957
SCOTT, D.S. & TARSKI, A. *The sentential calculus with infinitely long expressions* ⋄ C75 ⋄
TARSKI, A. *Remarks on predicate logic with infinitely long expressions* ⋄ C75 ⋄

1958
ENGELER, E. *Untersuchungen zur Modelltheorie*
 ⋄ C07 C75 ⋄
SLOMINSKI, J. *Theory of models with infinitary operations and relations* ⋄ C05 C75 ⋄
TARSKI, A. *Remarks on predicate logic with infinitely long expressions* ⋄ C75 ⋄

1960
HANF, W.P. *Models of languages with infinitely long expressions* ⋄ C55 C75 ⋄

1961
ENGELER, E. *Unendliche Formeln in der Modell-Theorie*
 ⋄ C15 C35 C40 C75 ⋄
ENGELER, E. *Zur Beweistheorie von Sprachen mit unendlich langen Formeln* ⋄ C75 F07 ⋄
HENKIN, L. *Some remarks on infinitely long formulas*
 ⋄ C75 C80 ⋄
MACHOVER, M. *The theory of transfinite recursion*
 ⋄ C75 D60 E45 ⋄
MAEHARA, S. & TAKEUTI, G. *A formal system of first-order predicate calculus with infinitely long expressions*
 ⋄ C40 C75 F05 ⋄
NOVIKOV, P.S. *On the consistency of certain logical calculi (Russian)* ⋄ C75 ⋄

1962
FUHRKEN, G. *Bemerkung zu einer Arbeit E.Engelers*
 ⋄ C15 C40 C75 ⋄
KARP, C.R. *Independence proofs in predicate logic with infinitely long expressions* ⋄ C75 ⋄
TARSKI, A. *Some problems and results relevant to the foundations of set theory*
 ⋄ C75 E05 E10 E55 G05 ⋄

1963

ENGELER, E. *A reduction-principle for infinite formulas*
  ⋄ C75 F07 ⋄
KARP, C.R. *A note on the representation of α-complete boolean algebras* ⋄ C75 G05 ⋄
KEISLER, H.J. *A complete first-order logic with infinitary predicates* ⋄ C75 G15 ⋄
KINO, A. & TAKEUTI, G. *On predicates with constructive infinitely long expressions* ⋄ C70 F15 ⋄
ROBINSON, A. *On languages which are based on non-standard arithmetic* ⋄ C75 H05 H15 ⋄

1964

HANF, W.P. *Incompactness in languages with infinitely long expressions* ⋄ C75 E55 ⋄
KARP, C.R. *Languages with expressions of infinite length*
  ⋄ C75 C98 G05 ⋄
KEISLER, H.J. & TARSKI, A. *From accessible to inaccessible cardinals* ⋄ C75 E05 E55 ⋄
RYLL-NARDZEWSKI, C. *On Borel measurability of orbits*
  ⋄ C75 E15 E75 ⋄
SCOTT, D.S. *Invariant Borel sets* ⋄ C75 D55 E15 ⋄

1965

FELSCHER, W. *Zur Algebra unendlich langer Zeichenreihen* ⋄ C05 C75 E10 ⋄
KARP, C.R. *Finite-quantifier equivalence* ⋄ C75 ⋄
KEISLER, H.J. *Finite approximations of infinitely long formulas* ⋄ C40 C50 C75 ⋄
KEISLER, H.J. *Some application of infinitely long formulas*
  ⋄ C40 C50 C75 ⋄
LOPEZ-ESCOBAR, E.G.K. *An interpolation theorem for denumerably long sentences* ⋄ C40 C75 ⋄
LOPEZ-ESCOBAR, E.G.K. *Universal formulas in the infinitary language $L_{\alpha,\beta}$* ⋄ C75 ⋄
MALITZ, J. *Problems in the model theory of infinite languages* ⋄ C40 C75 ⋄
MORLEY, M.D. *Omitting classes of elements*
  ⋄ C30 C55 C75 ⋄
NISHIMURA, T. & TANAKA, H. *On two systems for arithmetic* ⋄ C75 F30 F35 ⋄
SCOTT, D.S. *Logic with denumerably long formulas and finite strings of quantifiers* ⋄ C15 C40 C75 C98 ⋄
TAIT, W.W. *Infinitely long terms of transfinite type*
  ⋄ C75 F10 F15 F50 ⋄

1966

HELLING, M. *Model-theoretic problems for some extensions of first-order languages*
  ⋄ C55 C75 C80 E55 ⋄
KEISLER, H.J. *Some model theoretic results for ω-logic*
  ⋄ C35 C50 C55 C75 ⋄
KINO, A. *On definability of ordinals in logic with infinitely long expressions* ⋄ C40 C75 E10 E47 ⋄
KRAUSS, P.H. & SCOTT, D.S. *Assigning probabilities to logical formulas* ⋄ B48 C75 C90 ⋄
LINDSTROEM, P. *On characterizability in L_{ω_1,ω_0}*
  ⋄ C75 ⋄
LOPEZ-ESCOBAR, E.G.K. *On defining well-orderings*
  ⋄ C55 C65 C75 E10 ⋄

1967

BARWISE, J. *Infinitary logic and admissible sets*
  ⋄ C70 D60 ⋄
ENGELER, E. *Algorithmic properties of structures*
  ⋄ B75 C75 D10 D75 ⋄
FEFFERMAN, C. *Cardinally maximal sets of non-equivalent order types* ⋄ C65 C75 E07 ⋄
KARP, C.R. *Nonaxiomatizability results for infinitary systems* ⋄ C75 G05 ⋄
KEISLER, H.J. & MORLEY, M.D. *On the number of homogeneous models of a given power*
  ⋄ C50 C55 C75 E50 ⋄
KOPPERMAN, R.D. *Application of infinitary languages to metric spaces* ⋄ C65 C75 ⋄
KOPPERMAN, R.D. *On the axiomatizability of uniform spaces* ⋄ C65 C75 ⋄
KOPPERMAN, R.D. *The L_{ω_1,ω_1}-theory of Hilbert spaces*
  ⋄ C65 C75 ⋄
LOPEZ-ESCOBAR, E.G.K. *A complete, infinitary axiomatization of weak second-order logic*
  ⋄ B15 C75 C85 F35 ⋄
LOPEZ-ESCOBAR, E.G.K. *Remarks on an infinitary language with constructive formulas*
  ⋄ C40 C70 C75 F50 ⋄
RYLL-NARDZEWSKI, C. & WEGLORZ, B. *Compactness and homomorphisms of algebraic structures*
  ⋄ C07 C50 C75 ⋄
SHAFAAT, A. *Principle of localization for a more general type of languages* ⋄ B10 C55 C75 C90 ⋄

1968

BARWISE, J. *Implicit definability and compactness in infinitary languages* ⋄ C40 C70 ⋄
BARWISE, J. (ED.) *The syntax and semantics of infinitary languages* ⋄ C70 C75 C97 ⋄
BENDA, M. *Ultraproducts and non-standard logics (Russian summary)* ⋄ C20 C50 C75 ⋄
BURMEISTER, P. *Ueber die Maechtigkeiten und Unabhaengigkeitsgrade der Basen freier Algebren. I*
  ⋄ C05 C75 ⋄
CHANG, C.C. *Infinitary properties of models generated from indiscernibles* ⋄ C30 C75 ⋄
CHANG, C.C. *Some remarks on the model theory of infinitary languages* ⋄ C07 C30 C55 C75 ⋄
CHUDNOVSKY, G.V. *Some results in the theory of infinitely long expressions (Russian)* ⋄ C40 C75 ⋄
ENGELER, E. *Remarks on the theory of geometrical constructions* ⋄ C75 D80 ⋄
FEFERMAN, S. *Lectures on proof theory*
  ⋄ C40 C70 C75 F05 F15 F98 ⋄
FEFERMAN, S. *Persistent and invariant formulas for outer extensions* ⋄ C30 C40 C70 C75 F35 ⋄
FRIEDMAN, H.M. & JENSEN, R.B. *Note on admissible ordinals* ⋄ C70 D60 ⋄
KARP, C.R. *An algebraic proof of the Barwise compactness theorem* ⋄ C70 C90 ⋄
KEISLER, H.J. *Formulas with linearly ordered quantifiers*
  ⋄ C40 C50 C75 C80 ⋄
KOPPERMAN, R.D. & MATHIAS, A.R.D. *Some problems in group theory* ⋄ C60 C75 ⋄

KREISEL, G. *Choice of infinitary languages by means of definability criteria: generalized recursion theory*
  ⋄ C40 C70 C75 D60 D75 ⋄

KUEKER, D.W. *Definability, automormophisms and infinitary languages* ⋄ C07 C40 C75 ⋄

KUNEN, K. *Implicit definability and infinitary languages*
  ⋄ C40 C70 C75 D70 E47 E55 ⋄

LABLANQUIE, J.-C. *Sur les approximations finies des formules infiniment longues* ⋄ C75 ⋄

LOPEZ-ESCOBAR, E.G.K. *Well-orderings and finite quantifiers* ⋄ C65 C75 ⋄

MALITZ, J. *The Hanf number for complete $L_{\omega_1,\omega}$-sentences* ⋄ C55 C75 ⋄

MORLEY, M.D. *Partitions and models*
  ⋄ C30 C50 C55 C62 C75 E05 E45 E55 ⋄

PRELLER, A. *Quantified algebras* ⋄ C75 G15 G25 ⋄

TAIT, W.W. *Normal derivability in classical logic*
  ⋄ C75 F05 F15 F30 F35 F60 ⋄

TAKEUTI, G. *A determinate logic* ⋄ C40 C75 E60 ⋄

WEINSTEIN, J.M. *(ω_1,ω) properties of unions of models*
  ⋄ C30 C40 C52 C75 ⋄

1969

BARWISE, J. *Applications of strict Π_1^1 predicates to infinitary logic* ⋄ C40 C70 D60 ⋄

BARWISE, J. *Infinitary logic and admissible sets*
  ⋄ C70 E45 E47 ⋄

BARWISE, J. & EKLOF, P.C. *Lefschetz's principle*
  ⋄ B15 C60 C75 C85 ⋄

BARWISE, J. *Remarks on universal sentences of $L_{\omega_1,\omega}$*
  ⋄ C15 C40 C75 ⋄

BENDA, M. *Reduced products and nonstandard logics*
  ⋄ C20 C50 C75 ⋄

CALAIS, J.-P. *La methode de Fraisse dans les langages infinis* ⋄ C75 ⋄

KEISLER, H.J. *Infinite quantifiers and continuous games*
  ⋄ C65 C75 C80 ⋄

KOPPERMAN, R.D. *Applications of infinitary languages to analysis* ⋄ C35 C60 C65 C75 ⋄

KRAUSS, P.H. *Representation of symmetric probability models* ⋄ C75 C90 ⋄

LOPEZ-ESCOBAR, E.G.K. *A non-interpolation theorem (Russian summary)* ⋄ C40 C75 ⋄

MAKKAI, M. *An application of a method of Smullyan to logics on admissible sets* ⋄ C70 ⋄

MAKKAI, M. *On the model theory of denumerable long formulas with finite strings of quantifiers*
  ⋄ C40 C70 C75 ⋄

MALITZ, J. *Universal classes in infinitary languages*
  ⋄ C40 C52 C75 ⋄

PABION, J.F. *Extensions du theoreme de Loewenheim-Skolem en logique infinitaire*
  ⋄ C55 C75 ⋄

PRELLER, A. *Interpolation et amalgamation*
  ⋄ C40 C52 C75 G15 G25 ⋄

PRELLER, A. *Logique algebrique infinitaire. Completude des calculs $L_{\alpha\beta}$* ⋄ C75 G25 ⋄

SHAFAAT, A. *On implicationally defined classes of algebras* ⋄ C05 C75 ⋄

UESU, T. *On logics with infinitely long expressions (Japanese)* ⋄ C75 ⋄

YASUGI, M. *An evaluation of the ω-complexity of first order arithmetic with the constructive ω-rule* ⋄ C75 ⋄

1970

BARWISE, J. & EKLOF, P.C. *Infinitary properties of abelian torsion groups* ⋄ C60 C75 ⋄

CHUDNOVSKY, G.V. *Problems of the theory of models, related to categoricity (Russian)*
  ⋄ C35 C45 C50 C52 C55 C75 ⋄

CUTLAND, N.J. *The theory of hyperarithmetic and Π_1^1 models* ⋄ C70 D55 ⋄

DICKMANN, M.A. *Model theory of infinitary languages. Vol. I, II* ⋄ C75 C98 E55 ⋄

EKLOF, P.C. *On the existence of $L_{\infty,\kappa}$ indiscernibles*
  ⋄ C30 C50 C75 ⋄

ENGELER, E. *Geometry and language* ⋄ B30 C75 ⋄

GALVIN, F. & HORN, A. *Operations preserving all equivalence relations* ⋄ C05 C75 ⋄

KEISLER, H.J. *Logic with the quantifier "there exist uncountably many"* ⋄ C30 C55 C62 C75 C80 ⋄

KRASNER, M. *Endotheorie de Galois abstraite*
  ⋄ C07 C60 C75 ⋄

KUEKER, D.W. *Generalized interpolation and definability*
  ⋄ C40 C50 C75 ⋄

MAKKAI, M. *Structures elementarily equivalent relative to infinitary languages to models of higher power*
  ⋄ C55 C75 ⋄

MORLEY, M.D. *The number of countable models*
  ⋄ C15 C75 ⋄

REYES, G.E. *Local definability theory*
  ⋄ C07 C15 C40 C50 C55 C75 ⋄

SHELAH, S. *A note on Hanf numbers*
  ⋄ C35 C55 C75 ⋄

SHELAH, S. *On languages with non-homogeneous strings of quantifiers* ⋄ C75 C80 ⋄

SURMA, S.J. *On the axiomatic treatment of the theory of models II: Syntactical characterization of a fragment of the theory of models (Polish summary)*
  ⋄ B22 C07 C75 ⋄

TAKEUTI, G. *A determinate logic* ⋄ C40 C75 E60 ⋄

UESU, T. *On the recursively restricted rules*
  ⋄ C75 F30 ⋄

1971

BARWISE, J. & KUNEN, K. *Hanf numbers for fragments of $L_{\infty\omega}$* ⋄ C55 C75 ⋄

BARWISE, J. *Infinitary methods in the model theory of set theory* ⋄ C62 C70 E45 E47 ⋄

CAMPBELL, P.J. *Suslin logic*
  ⋄ C75 D55 E15 G05 G25 ⋄

CARSTENGERDES, W. *Mehrsortige logische Systeme mit unendlich langen Formeln I, II* ⋄ C40 C75 F05 ⋄

CHANG, C.C. *Sets constructible using $L_{\kappa\kappa}$*
  ⋄ C75 E25 E45 E50 E55 ⋄

CHANG, C.C. *Two interpolation theorems*
  ⋄ C40 C50 C75 ⋄

COVEN, C. *Forcing in infinitary languages*
  ⋄ C25 C75 ⋄

EKLOF, P.C. & SABBAGH, G. *Definability problems for modules and rings* ◊ C20 C52 C60 C75 ◊

FLUM, J. *A remark on infinitary languages* ◊ C75 ◊

GREGORY, J. *Incompleteness of a formal system for infinitary finite quantifier formulas* ◊ C75 ◊

KEISLER, H.J. *Model theory for infinitary logic. Logic with countable conjunctions and finite quantifiers*
◊ C70 C75 C98 ◊

LASSAIGNE, R. *Produits reduits et langages infinis*
◊ C20 C75 E05 E55 ◊

MAGIDOR, M. *On the role of supercompact and extendible cardinals in logic* ◊ C55 C75 C85 E47 E55 ◊

MAKKAI, M. *Algebraic operations on classes of structures and their connections with logical formulas I, II (Hungarian) (English summary)*
◊ C30 C40 C52 C75 ◊

MALITZ, J. *Infinitary analogs of theorems from first order model theory* ◊ C30 C40 C55 C75 ◊

POIZAT, B. *Theorie de Galois des relations*
◊ C07 C60 C75 E07 ◊

SHELAH, S. *On the number of non-almost isomorphic models of T in a power* ◊ C55 C75 ◊

TULIPANI, S. *Cardinali misurabili e algebre semplici in una varieta generata da un'algebra infinitaria m-funzionalmente completa (English summary)*
◊ C05 C75 E55 ◊

UESU, T. *Simple type theory with constructive infinitely long expressions* ◊ B15 C75 F05 F35 ◊

WILMERS, G.M. *Internally standard set theories*
◊ C62 C70 E45 E47 ◊

1972

BACSICH, P.D. *Cofinal simplicity and algebraic closedness*
◊ C05 C25 C50 C60 C75 G10 ◊

BARWISE, J. *Absolute logics and $L_{\infty\omega}$* ◊ C75 C95 ◊

BELL, J.L. *On the relationship between weak compactness in $L_{\omega_1,\omega}, L_{\omega_1,\omega_1}$ and restricted second order languages* ◊ C55 C75 C80 C85 E05 E55 ◊

CALAIS, J.-P. *Partial isomorphisms and infinitary languages* ◊ C07 C65 C75 ◊

CHUDNOVSKY, D.V. *Topological properties of products of discrete spaces and set theory (Russian)*
◊ C55 C75 E35 E55 E75 ◊

COLE, J.C. & DICKMANN, M.A. *Non-axiomatizability results in infinitary languages for higher-order structures* ◊ C52 C65 C75 C85 ◊

CUTLAND, N.J. *Π_1^1-models and Π_1^1-categoricity*
◊ C57 C70 D55 ◊

CUTLAND, N.J. *Σ_1-compactness and ultraproducts*
◊ C20 C70 ◊

FEFERMAN, S. *Infinitary properties, local functors, and systems of ordinal functions*
◊ C30 C40 C75 E10 F15 G30 ◊

FLUM, J. *Hanf numbers and well-ordering numbers*
◊ C30 C55 C75 C80 ◊

GARLAND, S.J. *Generalized interpolation theorems*
◊ C40 C75 C85 D55 E15 ◊

JERVELL, H.R. *Herbrand and Skolem theorems in infinitary languages* ◊ C75 F05 ◊

KASHIWAGI, T. *On (m, A)-implicational classes*
◊ C20 C52 C75 ◊

KUEKER, D.W. *Loewenheim-Skolem and interpolation theorems in infinitary languages*
◊ C40 C55 C75 E05 ◊

LASSAIGNE, R. & RESSAYRE, J.-P. *Algebres de Boole et langages infinis* ◊ C35 C75 G05 ◊

LOPEZ-ESCOBAR, E.G.K. *The infinitary language $L_{\theta\theta}$ is not local (Russian summary)* ◊ C55 C75 E55 ◊

MACINTYRE, A. *Omitting quantifier-free types in generic structures* ◊ C25 C57 C60 C75 D30 D40 ◊

MAKKAI, M. *Svenonius sentences and Lindstroem's theory on preservation theorems* ◊ C40 C52 C75 ◊

MALCOLM, W.G. *On a compactness theorem of A. Shafaat*
◊ C20 C75 ◊

MANSFIELD, R. *The completeness theorem for infinitary logic* ◊ C75 C90 ◊

MARCUS, L. *A minimal prime model with an infinite set of indiscernibles* ◊ C30 C50 C75 ◊

MARTIN-LOEF, P. *Infinite terms and a system of natural deduction* ◊ C75 F05 F50 ◊

MOSCHOVAKIS, Y.N. *The game quantifier*
◊ C75 C80 D55 D70 D75 E60 ◊

MOTOHASHI, N. *A new theorem on definability in a positive second order logic with countable conjunctions and disjunctions* ◊ B15 C40 C75 C85 ◊

MOTOHASHI, N. *Countable structures for uncountable infinitary languages* ◊ C15 C40 C75 ◊

MOTOHASHI, N. *Interpolation theorem and characterization theorem* ◊ C40 C75 F05 ◊

NADEL, M.E. *An application of set theory to model theory*
◊ C75 E47 E75 ◊

NADEL, M.E. *Some Loewenheim-Skolem results for admissible sets* ◊ C62 C70 D60 E55 ◊

NEBRES, B.F. *Herbrand uniformity theorems for infinitary languages* ◊ C40 C70 C75 F05 ◊

NEBRES, B.F. *Infinitary formulas preserved under unions of models* ◊ C40 C70 C75 ◊

NGUYEN CAT HO *Generalized algebras of Post and their applications to many-valued logics with infinitely long formulas* ◊ B50 C75 G20 ◊

OLIN, P. *Products of two-sorted structures*
◊ C30 C40 C75 ◊

REYES, G.E. *$L_{\omega_1\omega}$ is enough: a reduction theorem for some infinitary languages* ◊ C55 C75 E55 ◊

SHELAH, S. *A combinatorial problem; stability and order for models and theories in infinitary languages*
◊ C45 C55 C75 E05 ◊

WOOD, C. *Forcing for infinitary languages*
◊ C25 C75 ◊

1973

ACZEL, P. *Infinitary logic and the Barwise compactness theorem* ◊ C70 C98 ◊

ARMBRUST, M. *A first order approach to correspondence and congruence systems* ◊ C05 C75 ◊

BACSICH, P.D. *Defining algebraic elements*
◊ C40 C52 C75 ◊

BARWISE, J. *Back and forth through infinitary logics*
◊ C15 C75 ◊

BOWEN, K.A. *Infinite forcing and natural models of set theory (Russian summary)*
⋄ C25 C55 C62 C75 C80 E55 G30 ⋄

BUECHI, J.R. & DANHOF, K.J. *Variations on a theme of Cantor in the theory of relational structures*
⋄ C07 C35 C50 C75 ⋄

BURRIS, S. *Scott sentences and a problem of Vaught for mono-unary algebras* ⋄ C05 C15 C75 ⋄

CUTLAND, N.J. *Model theory on admissible sets*
⋄ C45 C55 C70 ⋄

EKLOF, P.C. *Lefschetz's principle and local functors*
⋄ C60 C75 ⋄

GOGOL, D. *Low cardinality models for a type of infinitary theory* ⋄ C55 C75 ⋄

GRANT, J. *Automorphisms definable by formulas*
⋄ C07 C40 C75 ⋄

GREGORY, J. *Uncountable models and infinitary elementary extensions* ⋄ C30 C70 ⋄

GRILLIOT, T.J. *Implicit definability and hyperprojectivity*
⋄ C40 C75 D75 ⋄

HILL, P. *New criteria for freeness in abelian groups*
⋄ C60 C75 ⋄

KEISLER, H.J. *Forcing and the omitting types theorem*
⋄ C25 C75 ⋄

KNIGHT, J.F. *Complete types and the natural numbers*
⋄ C07 C75 ⋄

KRIVINE, J.-L. & MCALOON, K. *Some true unprovable formulas for set theory*
⋄ C62 C75 E35 E45 E55 ⋄

MACINTYRE, A. *Martin's axiom applied to existentially closed groups* ⋄ C25 C55 C60 C75 C80 E50 ⋄

MAKKAI, M. *Global definability theory in $L_{\omega_1,\omega}$*
⋄ C40 C75 ⋄

MAKKAI, M. *Vaught sentences and Lindstroem's regular relations* ⋄ C40 C52 C75 D55 D70 ⋄

MAKOWSKY, J.A. *Langages engendres a partir des formules de Scott* ⋄ C75 C95 ⋄

MOTOHASHI, N. *An extended relativization theorem*
⋄ C40 C75 ⋄

MOTOHASHI, N. *Model theory on a positive second order logic with countable conjunctions and disjunctions*
⋄ B15 C40 C75 C85 ⋄

MOTOHASHI, N. *Two theorems on mix-relativization*
⋄ B10 C40 C75 ⋄

NGUYEN CAT HO *Generalized Post algebras and their application to some infinitary many-valued logics*
⋄ B50 C75 G20 ⋄

RESSAYRE, J.-P. *Boolean models and infinitary first order languages*
⋄ C40 C50 C55 C70 C75 C80 C90 ⋄

SHELAH, S. *Weak definability in infinitary languages*
⋄ C40 C55 C75 ⋄

STAVI, J. *A converse of the Barwise completeness theorem*
⋄ C70 D60 D70 ⋄

TAYLOR, W. *A heterogeneous interpolant* ⋄ C40 C75 ⋄

VAUGHT, R.L. *A Borel invariantization*
⋄ C75 D55 E15 ⋄

VAUGHT, R.L. *Descriptive set theory in $L_{\omega_1,\omega}$*
⋄ C15 C70 C75 D55 D70 E15 ⋄

WEISPFENNING, V. *Infinitary model-theoretic properties of κ-saturated structures*
⋄ C10 C25 C35 C50 C60 C65 C75 ⋄

WILMERS, G.M. *An $\aleph_1$-standard model for ZF which is an element of the minimal model for ZF*
⋄ C62 C70 E30 E35 E45 H20 ⋄

1974

AFRICK, H. *Scott's interpolation theorem fails for $L_{\omega_1,\omega}$*
⋄ C40 C75 ⋄

BACSICH, P.D. & ROWLANDS-HUGHES, D. *Syntactic characterizations of amalgamation, convexity and related properties* ⋄ C05 C07 C60 C75 ⋄

BARWISE, J. *Mostowski's collapsing function and the closed unbounded filter* ⋄ C15 C55 C75 E47 ⋄

BAUMGARTNER, J.E. *The Hanf number for complete $L_{\omega_1,\omega}$-sentences (without GCH)*
⋄ C35 C50 C55 C75 ⋄

BEAUCHEMIN, P. *Quelques consequences en logique d'une loi de probabilite 0-1 pour les multirelations denombrables* ⋄ C75 E25 E55 E75 ⋄

BELL, J.L. *On compact cardinals*
⋄ C55 C75 E45 E55 ⋄

BOWEN, K.A. *Forcing in a general setting*
⋄ C25 C75 C80 C85 E40 ⋄

BOWEN, K.A. *Primitive recursive notations for infinitary formulas* ⋄ C75 F15 ⋄

CHUDNOVSKY, G.V. *Transversals and properties of compactness type (Russian)* ⋄ C20 C75 E05 E55 ⋄

EKLOF, P.C. *Infinitary equivalence of abelian groups*
⋄ C60 C75 ⋄

FRAISSE, R. *Isomorphisme local et equivalence associes a un ordinal; utilite en calcul des formules infinies a quanteurs finis* ⋄ B15 C07 C75 C85 ⋄

FRIEDMAN, H.M. *On existence proofs of Hanf numbers*
⋄ C55 C75 E30 E70 F50 ⋄

FUKUYAMA, M. & NAGAOKA, K. *Strong representability of α-recursive functions* ⋄ C70 D60 ⋄

GEISER, J.R. *A formalization of Essenin-Volpin's proof theoretical studies by means of nonstandard analysis*
⋄ C75 F50 H10 H15 ⋄

GORDON, C.E. *Prime and search computability, characterized as definability in certain sublanguages of constructible $L_{\omega_1,\omega}$* ⋄ C75 D75 ⋄

GREEN, J. *Σ_1-compactness for next admissible sets*
⋄ C62 C70 D60 ⋄

GREGORY, J. *Beth definability in infinitary languages*
⋄ C40 C75 ⋄

GRILLIOT, T.J. *Model theory for dissecting recursion theory* ⋄ C55 C70 C75 C80 D60 D75 ⋄

HODGES, W. *A normal form for algebraic constructions*
⋄ C40 C75 E47 ⋄

KARP, C.R. *Infinite-quantifier languages and ω-chains of models* ⋄ C40 C75 ⋄

KASHIWAGI, T. *Note on locally definable classes of structures* ⋄ C52 C75 ⋄

MACINTYRE, A. *A note on axioms for infinite-generic structures* ⋄ C25 C60 C62 C75 E07 ⋄

MAKKAI, M. *A remark on a paper of J.P. Ressayre*
⋄ C70 ⋄

MAKKAI, M. *Generalizing Vaught sentences from ω to strong cofinality ω* ⋄ C40 C75 ⋄

MORLEY, M.D. *Applications of topology to $L_{\omega_1,\omega}$*
⋄ C75 ⋄

MOTOHASHI, N. *Atomic structures and Scott sentences (Japanese)* ⋄ C50 C75 ⋄

NADEL, M.E. *More Loewenheim-Skolem results for admissible sets* ⋄ C62 C70 D60 ⋄

NADEL, M.E. *Scott sentences and admissible sets*
⋄ C15 C65 C70 C75 ⋄

NELSON, EVELYN *Infinitary equational compactness*
⋄ C05 C50 C75 ⋄

NISHIMURA, T. *Gentzen-style formulation of systems of set-calculus* ⋄ C75 E30 F05 ⋄

NISHIMURA, T. *Gentzen-style formal system representing properties of functions* ⋄ C75 E30 F05 ⋄

SHELAH, S. *The Hanf number of omitting complete types*
⋄ C55 C75 ⋄

TAUTS, A. *A game for the construction of a propositional semantics in generalized Beth models (Russian) (Estonian and German summaries)*
⋄ C75 C90 E60 F50 ⋄

VAUGHT, R.L. *Invariant sets in topology and logic*
⋄ C75 D55 D70 E15 ⋄

VILLE, F. *More on set existence* ⋄ C62 C70 E47 ⋄

1975

ACZEL, P. *Quantifiers, games and inductive definitions*
⋄ C75 C80 D70 ⋄

ANDREKA, H. & NEMETI, I. *Subalgebra systems of algebras with finite and infinite, regular and irregular arity* ⋄ C05 C75 ⋄

ARZARELLO, F. *Un sistema di deduzione naturale per linguaggi infinitari (English summary)* ⋄ C75 F05 ⋄

BARWISE, J. *Admissible sets and the interaction of model theory, recursion theory and set theory*
⋄ C70 C98 D55 D60 D70 E30 E35 E45 E47 ⋄

BARWISE, J. *Admissible sets and structures. An approach to definability theory*
⋄ B98 C40 C70 C98 D60 D98 E30 E98 ⋄

BEAUCHEMIN, P. & REYES, G.E. *Espaces de Baire et espaces de probabilité de structures relationnelles*
⋄ C65 C75 E25 E55 E75 ⋄

BURGESS, J.P. & MILLER, DOUGLAS E. *Remarks on invariant descriptive set theory* ⋄ C75 D55 E15 ⋄

CHUDNOVSKY, G.V. *Combinatorial properties of compact cardinals* ⋄ C75 E05 E55 ⋄

CUNNINGHAM, E. *Chain models: applications of consistency properties and back-and-forth techniques in infinite-quantifier languages* ⋄ C40 C75 ⋄

DICKMANN, M.A. *Large infinitary languages; model theory* ⋄ C75 C98 E55 ⋄

EKLOF, P.C. *Categories of local functors*
⋄ C60 C75 G30 ⋄

EKLOF, P.C. *On the existence of κ-free abelian groups*
⋄ C60 C75 E10 E45 E75 ⋄

ELLENTUCK, E. *The foundations of Suslin logic*
⋄ C75 E75 ⋄

ELLENTUCK, E. & HALPERN, F. *Theories having many extensions* ⋄ C07 C75 C80 E75 ⋄

ENGELER, E. *On the solvability of algorithmic problems*
⋄ B75 C60 C75 D75 ⋄

FEFERMAN, S. *Two notes on abstract model theory II. Languages for which the set of valid sentences is semi-invariantly implicitly definable*
⋄ C40 C70 C95 ⋄

FERRO, R. *Limits to some interpolation theorems*
⋄ C40 C75 ⋄

FLUM, J. *First-order logic and its extensions*
⋄ C40 C55 C75 C95 C98 ⋄

FRIEDMAN, H.M. *Adding propositional connectives to countable infinitary logic* ⋄ C70 C75 ⋄

GLOEDE, K. *Set theory in infinitary languages*
⋄ C70 C75 E30 E65 ⋄

GREEN, J. *Consistency properties for finite quantifier languages* ⋄ C30 C40 C70 C75 ⋄

GREGORY, J. *On a finiteness condition for infinitary languages* ⋄ C40 C62 C70 C75 E40 ⋄

HICKIN, K.K. & PLOTKIN, J.M. *Compactness and ρ-valued logics* ⋄ B50 C75 C90 ⋄

HODGES, W. *A normal form for algebraic constructions II*
⋄ C40 C75 E25 G30 ⋄

KEISLER, H.J. *Constructions in model theory*
⋄ C25 C30 C50 C75 C95 C98 ⋄

KINO, A. & MYHILL, J.R. *A hierarchy of languages with infinitely long expressions* ⋄ C40 C75 E10 F15 ⋄

KRAWCZYK, A. *On a class of models for sentences of languages α_A* ⋄ C70 E45 ⋄

KREISEL, G. & MINTS, G.E. & SIMPSON, S.G. *The use of abstract language in elementary metamathematics: Some pedagogic examples*
⋄ A05 C07 C57 C75 F05 F07 F20 F50 ⋄

KREISEL, G. *Was hat die Logik in den letzten 25 Jahren für die Mathematik geleistet?* ⋄ A05 B98 C75 D40 ⋄

KUEKER, D.W. *Back-and-forth arguments and infinitary logics* ⋄ C75 C98 ⋄

KUEKER, D.W. (ED.) *Infinitary logic: In memoriam Carol Karp* ⋄ C70 C75 C97 ⋄

LEE, V. & NADEL, M.E. *On the number of generic models*
⋄ C25 C70 ⋄

MAKOWSKY, J.A. *Securable quantifiers, κ-unions and admissible sets* ⋄ C55 C70 C75 C80 C95 ⋄

MANSFIELD, R. *Omitting types: application to descriptive set theory* ⋄ C75 D55 E15 E45 ⋄

McALOON, K. *Formules de Rosser pour ZF (English summary)* ⋄ C62 C70 E30 E35 E45 F25 ⋄

McKEE, T.A. *Infinitary logic and topological homeomorphisms* ⋄ C65 C75 C80 C90 ⋄

MOTOHASHI, N. *Some proof-theoretic properties of dense linear orderings and countable well-orderings*
⋄ C40 C75 ⋄

ROSENTHAL, J.W. *Truth in all of certain well-founded countable models arising in set theory*
⋄ C40 C62 C70 E45 E47 ⋄

SCHLIPF, J.S. *Some hyperelementary aspects of model theory* ⋄ C50 C62 C70 D55 ⋄

SCHWICHTENBERG, H. *Elimination of higher type levels in definitions of primitive recursive functionals by means of transfinite recursion* ⋄ C75 F10 F15 F35 F50 ⋄

SCHWICHTENBERG, H. & WAINER, S.S. *Infinite terms and recursion in higher types* ◇ C75 D65 F10 ◇

SHELAH, S. *A compactness theorem for singular cardinals, free algebras, Whitehead problem and transversals* ◇ C60 C75 E05 E75 ◇

SHELAH, S. *Categoricity in $\aleph_1$ of sentences in $L_{\omega_1\omega}(Q)$* ◇ C35 C55 C75 C80 ◇

STAVI, J. *Applications of a theorem of Levy to boolean terms and algebras* ◇ C75 E47 E75 G05 ◇

STAVI, J. *Extensions of Kripke's embedding theorem* ◇ C55 C75 C90 G05 ◇

STERN, J. *A new look at the interpolation problem* ◇ C25 C40 C70 ◇

TAUTS, A. *A semantic model for infinite formulas (Russian) (Estonian and German summaries)* ◇ C75 C90 F50 G10 ◇

TAUTS, A. *Search for deduction by means of a semantic model (Russian) (Estonian and German summaries)* ◇ B55 C75 C90 F50 ◇

TAUTS, A. *Strong tautology and construction of a countermodel for nonderivable propositions (Russian) (Estonian and German summaries)* ◇ C75 C90 F50 ◇

WAINER, S.S. *Some hierachies based on higher type quantification* ◇ C75 D55 D65 E45 ◇

1976

ADAMOWICZ, Z. *One more aspect of forcing and omitting types* ◇ C62 C75 E35 E40 ◇

ARZARELLO, F. *Alcune questioni riguardanti l'aritmetica e i linguaggi infinitari (English summary)* ◇ C75 F05 ◇

BARWISE, J. *Some applications of Henkin quantifiers* ◇ C40 C50 C60 C75 C80 ◇

BOOS, W. *Infinitary compactness without strong inaccessibility* ◇ C75 E35 E55 ◇

COSTA DA, N.C.A. & PINTER, C. *α-logic and infinitary languages* ◇ B60 C75 ◇

EKLOF, P.C. *Infinitary model theory of abelian groups* ◇ C52 C60 C75 E35 E55 E75 ◇

ELLENTUCK, E. *Categoricity regained* ◇ C35 C75 C90 E40 ◇

FELSCHER, W. *On interpolation when function symbols are present* ◇ C40 C75 F50 ◇

FERRO, R. *Consistency property and model existence theorem for second order negative languages with conjunctions and quantifications over sets of cardinality smaller than a strong limit cardinal of denumerable cofinality* ◇ C55 C75 C85 ◇

GALVIN, F. & PRIKRY, K. *Infinitary Jonsson algebras and partition relations* ◇ C05 C50 C55 C75 E05 ◇

GOSTANIAN, R. & HRBACEK, K. *On the failure of the weak Beth property* ◇ C40 C75 C80 ◇

HARNIK, V. & MAKKAI, M. *Applications of Vaught sentences and the covering theorem* ◇ C15 C40 C45 C50 C52 C75 D70 E15 ◇

HARNIK, V. *Approximation theorems and model theoretic forcing* ◇ C25 C75 ◇

HINTIKKA, K.J.J. & RANTALA, V. *A new approach to infinitary languages* ◇ C40 C75 E60 ◇

HODGES, W. *On the effectivity of some field constructions* ◇ C60 C75 D65 E35 E47 E75 ◇

HUTCHINSON, J.E. *Model theory via set theory* ◇ C40 C55 C62 C70 C75 C80 C95 E75 ◇

KAISER, KLAUS *Quasi-axiomatic classes* ◇ B15 C20 C52 C75 C85 ◇

KASHIWAGI, T. *A generalizaton of a certain compactness property* ◇ C20 C52 C75 ◇

KASHIWAGI, T. *On locally definable classes closed under reduced products* ◇ C20 C75 ◇

KOGALOVSKIJ, S.R. *On local properties (Russian)* ◇ C52 C75 ◇

KRASNER, M. *Endotheorie de Galois abstraite et son theoreme d'homomorphie (English summary)* ◇ C07 C60 C75 ◇

MAKKAI, M. & REYES, G.E. *Model-theoretical methods in the theory of topoi and related categories I,II (Russian summaries)* ◇ C75 C90 G30 ◇

MANSOUX, A. *Logique sans egalite et (k,p)-quasivalence (English summary)* ◇ C07 C40 C75 ◇

MORAIS, R. *Projective logic* ◇ B60 C75 G05 ◇

MYCIELSKI, J. & TAYLOR, W. *A compactification of the algebra of terms* ◇ C05 C75 ◇

MYERS, D.L. *Invariant uniformization* ◇ C75 D55 E15 E35 ◇

NADEL, M.E. *A transfer principle for simple properties of theories* ◇ C50 C70 C75 ◇

NADEL, M.E. *On models $\equiv_{\infty\omega}$ to an uncountable model* ◇ C15 C70 C75 ◇

NYBERG, A.M. *Uniform inductive definability and infinitary languages* ◇ C70 D70 ◇

SLONNEGER, K. *A complete infinitary logic* ◇ C75 F05 ◇

VINNER, S. *Implicit axioms, ω-rule and the axiom of induction in high school mathematics* ◇ B28 C75 ◇

WEISPFENNING, V. *Negative-existentially complete structures and definability in free extensions* ◇ C25 C40 C60 C75 D40 G05 ◇

1977

BIELINSKI, K. *Extendability of structures as infinitary property* ◇ C62 C70 E30 E70 ◇

BURGESS, J.P. *Descriptive set theory and infinitary languages* ◇ C75 E15 ◇

CELLUCCI, C. *Proprieta di coerenza e completezza in $L_{\omega_1\omega}$* ◇ C75 ◇

CICHON, J. *Some remarks on selectors (Russian summary)* ◇ C75 E05 E40 E45 E55 ◇

DEISSLER, R. *Minimal models* ◇ C15 C40 C50 C75 ◇

DICKMANN, M.A. *Deux applications de la methode de va-et-vient* ◇ C60 C75 ◇

DICKMANN, M.A. *Structures Σ-saturees* ◇ C40 C50 C70 ◇

EKLOF, P.C. *Applications of logic to the problem of splitting abelian groups* ◇ C55 C60 C75 C80 E35 E45 E75 ◇

EKLOF, P.C. *Classes closed under substructures and direct limits* ◇ C52 C60 C75 ◇

EKLOF, P.C. *Methods of logic in abelian group theory* ◇ C60 C75 E35 E45 E55 E75 ◇

EKLOF, P.C. & MEKLER, A.H. *On constructing indecomposable groups in L*
⋄ C55 C60 C75 E35 E45 E55 ⋄

FUJIWARA, T. *On implicational classes of structures*
⋄ C05 C75 ⋄

GILLAM, D.W.H. *Morley numbers for generalized languages* ⋄ C55 C75 C80 ⋄

GLOEDE, K. *The metamathematics of infinitary set theoretical systems* ⋄ C70 C75 E30 E70 ⋄

GRANT, P.W. *Strict Π_1^1-predicates on countable and cofinality ω transitive sets*
⋄ C15 C40 C70 D55 D60 D70 E47 E60 ⋄

GREEN, J. *Next $\mathcal{P}$-admissible sets are of cofinality ω*
⋄ C62 C70 ⋄

HARNIK, V. & MAKKAI, M. *A tree argument in infinitary model theory* ⋄ C15 C75 ⋄

HENSON, C.W. & JOCKUSCH JR., C.G. & RUBEL, L.A. & TAKEUTI, G. *First-order topology*
⋄ B25 C65 C75 D35 H05 ⋄

KAMO, S. *Reduced powers and models for infinitely logic*
⋄ C20 C75 C90 ⋄

KEISLER, H.J. *Hyperfinite model theory*
⋄ B50 C65 C75 C80 C90 H20 ⋄

KEISLER, H.J. *The monotone class theorem in infinitary logic* ⋄ C70 C75 ⋄

KNIGHT, J.F. *A complete $L_{\omega_1\omega}$-sentence characterizing $\aleph_1$* ⋄ C55 C75 ⋄

KRAWCZYK, A. *On a class of models for sentences of language $\mathcal{L}_A$* ⋄ C70 ⋄

KRECZMAR, A. *On infinite sets of polynomial relations*
⋄ C60 C75 ⋄

KRYNICKI, M. *Some consequences of the omitting and realizing types property* ⋄ C75 C95 ⋄

KUEKER, D.W. *Countable approximations and Loewenheim-Skolem theorems*
⋄ C15 C55 C75 E05 ⋄

LABLANQUIE, J.-C. *Proprietes de forcing et modeles generiques* ⋄ C15 C25 C70 ⋄

LANDRAITIS, C.K. *Definability in well quasi-ordered sets of structures* ⋄ C15 C65 C75 ⋄

LASCAR, D. *Sur la hauteur des structures denombrables (English summary)* ⋄ C15 C50 C75 ⋄

LEE, V & NADEL, M.E. *Remarks on generic models*
⋄ C25 C70 ⋄

MAKKAI, M. *Admissible sets and infinitary logic*
⋄ C70 C98 D55 D60 D70 D98 ⋄

MAKKAI, M. & MYCIELSKI, J. *An $L_{\omega_1\omega}$-complete and consistent theory without models* ⋄ C35 C75 E75 ⋄

MAKKAI, M. *An "admissible" generalization of a theorem on countable Σ_1^1 sets of reals with applications*
⋄ C15 C40 C50 C70 D55 E15 ⋄

MAKKAI, M. & REYES, G.E. *First order categorical logic. Model-theoretical methods in the theory of topoi and related categories* ⋄ C75 C90 F50 G30 ⋄

MISAWA, T. & YASUDA, Y. *Suslin logics of Gentzen style*
⋄ C75 F07 ⋄

MORAIS, R. *Projective logics and projective boolean algebras* ⋄ B60 C75 E15 G05 G25 ⋄

MOTOHASHI, N. *A remark on Scott's interpolation theorem for $L_{\omega_1\omega}$* ⋄ C40 C75 ⋄

NADEL, M.E. & STAVI, J. *The pure part of $HYP(\mathfrak{M})$*
⋄ C62 C70 ⋄

NEPEJVODA, L.K. & NEPEJVODA, N.N. *Languages without set-variables for the description of sets (Russian)*
⋄ C75 D55 ⋄

NYBERG, A.M. *Inductive operators on resolvable structures* ⋄ C70 D60 D70 D75 ⋄

ONYSZKIEWICZ, J. *Omitting types in some extended system of logic* ⋄ C75 C95 ⋄

PALYUTIN, E.A. *Number of models in $L_{\infty\omega_1}$-theories I,II (Russian)* ⋄ C35 C52 C55 C75 E45 E65 ⋄

PARK, D. *Finiteness is μ-ineffable*
⋄ B75 C75 C80 C85 ⋄

PINTER, C. *Some theorems on omitting types, with applications to model completeness, amalgamation, and related properties* ⋄ C07 C25 C35 C52 C75 ⋄

RANTALA, V. *Aspects of definability*
⋄ A05 C07 C40 C75 ⋄

RESSAYRE, J.-P. *Models with compactness properties relative to an admissible language*
⋄ C50 C55 C70 C80 D55 D60 E15 ⋄

REYES, G.E. *Sheaves and concepts: a model-theoretic interpretation of Grothendieck topoi*
⋄ C75 C90 F50 G30 ⋄

SCHLIPF, J.S. *A guide to the identification of admissible sets above structures* ⋄ C15 C50 C70 ⋄

SCHLIPF, J.S. *Ordinal spectra of first-order theories*
⋄ C35 C50 C55 C62 C70 D15 D30 D70 ⋄

SGRO, J. *Completeness theorems for topological models*
⋄ C20 C40 C75 C80 C90 ⋄

SHELAH, S. *Existentially-closed groups in $\aleph_1$ with special properties* ⋄ C15 C25 C60 C75 ⋄

SUNDHOLM, G. *A completeness proof for an infinitary tense-logic* ⋄ B45 C75 ⋄

TIURYN, J. *Fixed-points and algebras with infinitely long expressions. I* ⋄ B75 C75 ⋄

TIURYN, J. *Fixed-points and algebras with infinitely long expressions. II* ⋄ B75 C75 ⋄

1978

ABRAMSON, F.G. & HARRINGTON, L.A. *Models without indiscernibles*
⋄ C30 C50 C55 C62 C65 C75 E05 E40 H15 ⋄

ADAMSON, A. *Admissible sets and the saturation of structures* ⋄ C50 C70 C75 E40 ⋄

BARWISE, J. & MOSCHOVAKIS, Y.N. *Global inductive definability* ⋄ C40 C70 D70 ⋄

BARWISE, J. *Monotone quantifiers and admissible sets*
⋄ C70 C80 D60 D70 ⋄

BARWISE, J. & KAUFMANN, M. & MAKKAI, M. *Stationary logic* ⋄ C55 C75 C80 ⋄

BAUMGARTNER, J.E. & GALVIN, F. *Generalized Erdoes cardinals and $0^{\#}$* ⋄ C55 C75 E05 E45 E55 ⋄

BENDA, M. *Compactness for omitting of types*
⋄ C20 C30 C55 C70 C75 E55 ⋄

BRUCE, KIM B. *Model-theoretic forcing in logic with a generalized quantifier* ⋄ C25 C70 C75 C80 ⋄

BURGESS, J.P. *On the Hanf number of Souslin logic* ⋄ C55 C75 E35 ⋄

CHARRETTON, C. *Type d'ordre des ordinaux de modeles non denombrables de la theorie des ensembles* ⋄ C20 C50 C62 C70 E10 E50 H20 ⋄

COHEN, P.E. *Some continuity properties for ultraproducts* ⋄ C20 C75 ⋄

CUTLAND, N.J. Σ_1-*compactness in languages stronger than* $\mathscr{L}_A$ ⋄ C70 C75 C80 ⋄

CUTLAND, N.J. *On the form of countable admissible ordinals* ⋄ C70 D60 ⋄

FENSTAD, J.E. & GANDY, R.O. & SACKS, G.E. (EDS.) *Generalized recursion theory II* ⋄ C62 C70 C80 D97 ⋄

FERRO, R. *Interpolation theorem for* $L_{k,k}^{2+}$ ⋄ B15 C40 C75 C85 ⋄

FITTING, M. *Elementary formal systems for hyperarithmetical relations* ⋄ C75 D55 D75 ⋄

GILLAM, D.W.H. *Refined Morley numbers* ⋄ C55 C75 ⋄

GLOEDE, K. *Hierarchies of sets definable by means of infinitary languages* ⋄ C62 C70 C75 E45 ⋄

GOAD, C.A. *Monadic infinitary propositional logic: a special operator* ⋄ B25 B60 C75 F50 ⋄

GOLD, B. *Compact and* ω-*compact formulas in* $L_{\omega_1,\omega}$ ⋄ C75 ⋄

GRANT, P.W. *The completeness of* $L_{\omega_1\omega}(P)$ ⋄ C75 C80 ⋄

GREEN, J. κ-*Suslin logic* ⋄ C75 ⋄

KRASNER, M. *Abstract Galois theory* ⋄ C07 C10 C60 C75 ⋄

KUEKER, D.W. *Uniform theorems in infinitary logic* ⋄ C40 C55 C75 ⋄

LINDSTROEM, P. *Omitting uncountable types and extensions of elementary logic* ⋄ C75 C80 C95 ⋄

MAKOWSKY, J.A. *Some observations on uniform reduction for properties invariant on the range of definable relations* ⋄ C30 C40 C75 C95 ⋄

MAREK, W. & NYBERG, A.M. *Extendability of ZF models in the von Neumann hierarchy to models of KM theory of classes* ⋄ C62 C70 E30 E70 ⋄

MEKLER, A.H. *The size of epimorphic extensions* ⋄ C55 C75 C95 E55 G30 ⋄

MILLER, DOUGLAS E. *The invariant* Π_α^0 *separation principle* ⋄ C15 C70 C75 D55 E15 ⋄

NADEL, M.E. & STAVI, J. $L_{\infty\lambda}$-*equivalence, isomorphism and potential isomorphism* ⋄ C55 C75 E07 E40 E65 ⋄

NADEL, M.E. *Infinitary intuitionistic logic from a classical point of view* ⋄ C75 C90 F50 ⋄

OIKKONEN, J. *Second-order definability, game quantifiers and related expressions* ⋄ C40 C75 C80 C85 E60 ⋄

ROSICKY, J. *Categories of models of infinitary Horn theories* ⋄ C75 G30 ⋄

SCHLIPF, J.S. *Toward model theory through recursive saturation* ⋄ C15 C40 C50 C57 C70 ⋄

SHELAH, S. *Models with second order properties. II: Trees with no undefined branches* ⋄ C50 C55 C62 C75 C80 C95 E05 E35 E65 ⋄

STARK, W.R. *A forcing approach to the strict-*Π_1^1 *reflection and strict-*$\Pi_1^1 = \Sigma_1^0$ ⋄ C70 D60 E40 ⋄

STAVI, J. *Compactness properties of infinitary and abstract languages. I. General results* ⋄ C70 C75 C95 ⋄

STEEL, J.R. *On Vaught's conjecture* ⋄ C15 C65 C70 C75 ⋄

TAUTS, A. *Ambiguity of the axiom of normal functions (Russian) (Estonian and German summaries)* ⋄ B15 C75 E10 E55 E65 ⋄

TIURYN, J. *Fixed-points and algebras with infinitely long expressions I: Regular algebras* ⋄ B75 C75 ⋄

1979

ABRAMSON, F.G. Σ_1-*separation* ⋄ C62 C70 D60 E45 E47 ⋄

BELYAEV, V.YA. & TAJTSLIN, M.A. *On elementary properties of existentially closed systems (Russian)* ⋄ C25 C40 C60 C75 C85 D55 ⋄

BRUCE, KIM B. & KEISLER, H.J. $L_A(\exists)$ ⋄ C25 C70 C75 C80 ⋄

BURGESS, J.P. *A reflection phenomenon in descriptive set theory* ⋄ C75 D55 E15 ⋄

COWLES, J.R. *The relative expressive power of some logics extending first-order logic* ⋄ C10 C75 C80 C85 E15 ⋄

DUBUC, E.J. & REYES, G.E. *Subtoposes of the ring classifier* ⋄ C60 C75 C90 G30 ⋄

EYTAN, M. *Engeler algorithms in a topos* ⋄ B75 C75 G30 ⋄

FEFERMAN, S. *Generalizing set-theoretical model theory and an analogue theory on admissible sets* ⋄ C50 C55 C70 E70 ⋄

FRIEDMAN, S.D. *HC of an admissible set* ⋄ C70 D60 E30 ⋄

GREEN, J. *Some model theory for game logics* ⋄ C55 C75 E60 ⋄

HARNIK, V. & MAKKAI, M. *New axiomatizations for logics with generalized quantifiers* ⋄ C70 C80 ⋄

HARNIK, V. *Refinements of Vaught's normal form theorem* ⋄ C40 C75 ⋄

KARTTUNEN, M. *Infinitary languages* $N_{\infty\lambda}$ *and generalized partial isomorphisms* ⋄ C75 ⋄

KAUFMANN, M. *A new omitting types theorem for* $L(Q)$ ⋄ C55 C75 C80 ⋄

LABLANQUIE, J.-C. *Proprietes de consistance et forcing* ⋄ C25 C40 C75 ⋄

LASCAR, D. *La conjecture de Vaught* ⋄ C15 C50 C75 ⋄

MAKKAI, M. *There are atomic Nadel structures* ⋄ C15 C70 ⋄

MAKOWSKY, J.A. & SHELAH, S. *The theorems of Beth and Craig in abstract model theory I: The abstract setting* ⋄ C40 C75 C80 C95 ⋄

MILLER, DOUGLAS E. *An application of invariant sets to global definability* ◇ C40 C70 C75 ◇

MILLER, DOUGLAS E. *On classes closed under unions of chains* ◇ C52 C75 ◇

MORGENSTERN, C.F. *The measure quantifier* ◇ C55 C75 C80 E55 ◇

MOTOHASHI, N. *A remark on Africk's paper on Scott's interpolation theorem for $L_{\omega_1,\omega}$* ◇ C40 C75 ◇

OIKKONEN, J. *A generalization of the infinitely deep languages of Hintikka and Rantala* ◇ C55 C75 C80 ◇

OIKKONEN, J. *Hierarchies of model-theoretic definability - an approach to second-order logics* ◇ B15 C40 C70 C75 C80 C85 ◇

OIKKONEN, J. *On PC- and RPC-classes in generalized model theory* ◇ C13 C52 C75 C80 C95 ◇

RANTALA, V. *Game-theoretical semantics and back-and-forth* ◇ C75 ◇

ROSENTHAL, J.W. *Truth in all of certain well-founded countable models arising in set theory II* ◇ C40 C62 C70 E45 E47 ◇

SACKS, G.E. *Effective bounds on Morley rank* ◇ C15 C35 C45 C70 ◇

SHELAH, S. & ZIEGLER, M. *Algebraically closed groups of large cardinality* ◇ C25 C60 C75 ◇

SHELAH, S. *Hanf number of omitting type for simple first-order theories* ◇ C20 C50 C55 C75 ◇

TIURYN, J. *Fixed-points and algebras with infinitely long expressions II: μ-clones of regular algebras* ◇ B75 C75 ◇

VAEAENAENEN, J. *On Hanf numbers of unbounded logics* ◇ C55 C75 C80 C85 C95 E55 ◇

ZLATOS, P. *A characterization of some boolean powers* ◇ C05 C20 C30 C75 ◇

1980

ADAMSON, A. *Saturated structures, unions of chains, and preservation theorems* ◇ C40 C50 C70 ◇

BALDWIN, JOHN T. *The number of subdirectly irreducible algebras in a variety. II* ◇ C05 C75 ◇

BRUCE, KIM B. *Model constructions in stationary logic I. Forcing* ◇ C25 C55 C70 C75 C80 ◇

CAICEDO, X. *Back-and-forth systems for arbitrary quantifiers* ◇ C55 C75 C80 ◇

CUTLAND, N.J. & KAUFMANN, M. *Σ_1-well-founded compactness* ◇ C70 C80 ◇

FUJIWARA, T. *Existentially closed structures and algebraically closed structures* ◇ C25 C75 ◇

GOSTANIAN, R. & HRBACEK, K. *Propositional extensions of $L_{\omega_1\omega}$* ◇ C55 C75 E15 E50 ◇

GUICHARD, D.R. *A many-sorted interpolation theorem for L(Q)* ◇ C40 C75 C80 ◇

HARNIK, V. *Game sentences, recursive saturation and definability* ◇ C40 C50 C57 C70 ◇

HARRINGTON, L.A. *Extensions of countable infinitary logic which preserve most of its nice properties* ◇ C40 C55 C70 C75 C95 ◇

HODGES, W. *Functorial uniform reducibility* ◇ C20 C40 C75 ◇

JECH, T.J. & MAGIDOR, M. & MITCHELL, W.J. & PRIKRY, K. *Precipitous ideals* ◇ C70 E05 E35 E40 E50 E55 ◇

JONGH DE, D.H.J. *A class of intuitionistic connectives* ◇ B55 C75 C90 F50 ◇

KAISER, KLAUS *On a lattice of relational reducts* ◇ C52 C75 ◇

KALICKI, C. *Infinitary propositional intuitionistic logic* ◇ B55 C75 F50 ◇

KIERSTEAD, H.A. *Countable models of ω_1-categorical theories in admissible languages* ◇ C15 C35 C70 ◇

LANDRAITIS, C.K. *$L_{\omega_1\omega}$ equivalence between countable and uncountable linear orderings* ◇ C65 C75 E07 ◇

LUO, LIBO *The interpolation theorem for the $P(\kappa)$ of a strongly inaccessible cardinal (Chinese)* ◇ C40 C75 E50 E55 ◇

MEKLER, A.H. *How to construct almost free groups* ◇ C60 C75 E35 E45 E75 ◇

MEKLER, A.H. *On residual properties* ◇ C52 C75 ◇

MISAWA, T. *Consistency properties for Hausdorff logic* ◇ C75 ◇

MITSCHKE, G. *Infinite terms and infinite reductions* ◇ B40 C75 ◇

NADEL, M.E. *An arbitrary equivalence relation as elementary equivalence in an abstract logic* ◇ C75 C80 C95 ◇

PILLAY, A. *$\aleph_0$-categoricity in regular subsets of $L_{\omega_1\omega}$* ◇ C35 C75 ◇

PRZYMUSINSKA, H. *Craig interpolation theorem and Hanf number for ν-valued infinitary predicate calculi (Russian summary)* ◇ B50 C40 C55 C75 G20 ◇

PRZYMUSINSKA, H. *Gentzen-type semantics for ν-valued infinitary predicate calculi (Russian summary)* ◇ B50 C75 G20 ◇

PRZYMUSINSKA, H. *On ν-valued infinitary predicate calculi (Russian summary)* ◇ C75 G20 ◇

PRZYMUSINSKA, H. *Some results concerning ν-valued infinitary predicate calculi (Russian summary)* ◇ C20 C75 ◇

RUBIN, M. & SHELAH, S. *On the elementary equivalence of automorphism groups of boolean algebras; downward Skolem Loewenheim theorems and compactness of related quantifiers* ◇ C07 C30 C75 C80 C95 G05 ◇

STAHL, S.H. *Intensional sets* ◇ C75 E35 E45 E50 ◇

STARK, W.R. *Martin's axiom in the model theory of L_A* ◇ C70 E50 ◇

VAEAENAENEN, J. *The Hanf number of $L_{\omega_1\omega_1}$* ◇ C55 C75 E35 E55 ◇

VAUGHT, R.L. *On $PC_d(\mathfrak{A})$-classes for an admissible set $\mathfrak{A}$* ◇ C70 ◇

1981

ABRAMSON, F.G. *Locally countable models of Σ_1-separation* ◇ C62 C70 E30 E47 ◇

BACK, R.J.R. *Proving total correctness of nondeterministic programs in infinitary logic* ◇ B75 C75 ◇

BARWISE, J. *Infinitary logics* ⋄ C75 C98 ⋄
BARWISE, J. *The role of the omitting types theorem in infinitary logic* ⋄ C75 C80 C95 ⋄
BELL, J.L. *Isomorphism of structures in S-toposes* ⋄ C75 C90 ⋄
BERGSTRA, J.A. & TIURYN, J. *Implicit definability of algebraic structures by means of program properties* ⋄ B75 C40 C70 ⋄
CAICEDO, X. *Independent sets of axioms in $L_{\kappa\alpha}$* ⋄ C40 C75 ⋄
CAICEDO, X. *On extensions of $L_{\omega\omega}(Q_1)$* ⋄ C55 C75 C80 ⋄
DAHN, B.I. *Partial isomorphisms and intuitionistic logic* ⋄ C75 C90 F50 ⋄
EKLOF, P.C. & MEKLER, A.H. *Infinitary stationary logic and abelian groups* ⋄ C60 C75 C80 ⋄
ERIMBETOV, M.M. *A fragment of a logic with infinite formulas (Russian)* ⋄ C60 C75 ⋄
FERRO, R. *An analysis of Karp's interpolation theorem and the notion of k-consistency property (Italian summary)* ⋄ C40 C75 ⋄
FRIEDMAN, S.D. *Uncountable admissibles II: Compactness* ⋄ C70 D60 E45 E47 ⋄
HODGES, W. & SHELAH, S. *Infinite games and reduced products* ⋄ C20 C40 C75 E55 E60 ⋄
KARTTUNEN, M. *Model theoretic results for infinitely deep languages* ⋄ C20 C75 ⋄
KNIGHT, J.F. *Algebraic independence* ⋄ C40 C55 C75 ⋄
KUEKER, D.W. *$L_{\infty\omega_1}$ elementarily equivalent models of power ω_1* ⋄ C75 ⋄
MAKKAI, M. *An example concerning Scott heights* ⋄ C70 ⋄
MAKOWSKY, J.A. & SHELAH, S. *The theorems of Beth and Craig in abstract model theory II: Compact logics* ⋄ C40 C75 C80 C95 ⋄
MAKOWSKY, J.A. & ZIEGLER, M. *Topological model theory with an interior operator: consistency properties and back-and-forth arguments* ⋄ C40 C75 C80 ⋄
MILLER, A.W. *Vaught's conjecture for theories of one unary operation* ⋄ C15 C75 ⋄
MUNDICI, D. *Variations on Friedman's third and fourth problem* ⋄ C40 C75 C80 C95 ⋄
OIKKONEN, J. *Finite and infinite versions of the operations PC and RPC in infinitary languages* ⋄ C52 C75 ⋄
POHLERS, W. *Cut-elimination for impredicative infinitary systems I. Ordinal-analysis for ID_1*
⋄ C75 F05 F15 F35 ⋄
POIZAT, B. *Theorie de Galois pour les algebres de Post infinitaires* ⋄ C75 G20 ⋄
RATAJCZYK, Z. *A characterization of expandability of models for ZF to models for KM* ⋄ C62 C70 E45 E70 ⋄
ROSICKY, J. *Concrete categories and infinitary languages* ⋄ C75 G30 ⋄
ROTHMALER, P. *A note on I-types* ⋄ C07 C45 C75 ⋄
SCHMERL, J.H. *Arborescent structures II: Interpretability in the theory of trees* ⋄ B25 C50 C65 C75 ⋄

SHELAH, S. *Models with second order properties. III. Omitting types for L(Q)* ⋄ C55 C75 C80 E05 ⋄
SHELAH, S. *On the number of nonisomorphic models of cardinality λ, $L_{\infty\lambda}$-equivalent to a fixed model* ⋄ C55 C75 E45 E55 E75 ⋄
TAUTS, A. *The connection between semantic models and pseudo-boolean algebras (Russian) (Estonian and German summaries)* ⋄ C75 C90 F50 G10 ⋄
TIURYN, J. *Logic of effective definitions* ⋄ B60 C55 C75 ⋄
WILMERS, G.M. *An observation concerning the relationship between finite and infinitary Σ^1_1-sentences* ⋄ C70 ⋄

1982

FEDYSZAK-KOSZELA, A. *Some remarks on infinitary languages $N_{\kappa\alpha}$* ⋄ C75 ⋄
FERRO, R. *ω-satisfiability, ω-consistency property and the downward Loewenheim-Skolem theorem for $L_{\kappa\kappa}$ (Italian summary)* ⋄ C55 C75 ⋄
FRIEDMAN, S.D. *Steel forcing and Barwise compactness* ⋄ C70 E40 ⋄
FRIEDMAN, S.D. *Uncountable admissibles I: Forcing* ⋄ C70 D60 E40 ⋄
GEORGESCU, G. *A generalized quantifier for the hereditary σ-rings* ⋄ C75 C80 ⋄
GEORGESCU, G. *Une generalisation du theoreme d'omission des types dans les algebres polyadiques* ⋄ C75 C90 G15 ⋄
GRULOVIC, M. *A note on forcing and weak interpolation theorem for infinitary logics (Serbo-Croatian summary)* ⋄ C25 C40 C75 ⋄
GUREVICH, R. *Loewenheim-Skolem problem for functors* ⋄ C55 C75 G30 ⋄
HOOVER, D.N. *A normal form theorem for $L_{\omega_1 p}$, with applications* ⋄ C75 C90 ⋄
ISBELL, J.R. *Generating the algebraic theory of $C(X)$* ⋄ C05 C75 ⋄
JAEGER, G. *Zur Beweistheorie der Kripke-Platek-Mengenlehre ueber den natuerlichen Zahlen* ⋄ C70 E30 F15 F35 ⋄
KARTTUNEN, M. *Model theoretic results for infinitely deep languages* ⋄ C20 C75 ⋄
KAWAI, H. *Eine Logik erster Stufe mit einem infinitaeren Zeitoperator* ⋄ B45 C40 C75 C90 ⋄
MAGIDOR, M. *The simplest counterexample to compactness in the constructible universe* ⋄ C70 E45 ⋄
MEKLER, A.H. *Primitive rings are not definable in $L_{\infty\infty}$* ⋄ C60 C75 ⋄
MOTOHASHI, N. *An axiomatization theorem* ⋄ B10 C07 C40 C75 ⋄
MUNDICI, D. *Compactness, interpolation and Friedman's third problem* ⋄ C40 C75 C95 ⋄
POHLERS, W. *Admissibility in proof theory; a survey* ⋄ C70 F15 F35 F98 ⋄
POHLERS, W. *Cut elimination for impredicative infinitary systems II. Ordinal analysis for iterated inductive definitions* ⋄ C75 F05 F15 F35 ⋄

RANTALA, V. *Infinitely deep game sentences and interpolations* ◇ C40 C75 ◇

RASIOWA, H. *Lectures on infinitary logic and logics of programs* ◇ B75 C75 C98 ◇

SHELAH, S. *On the number of nonisomorphic models in $L_{\infty,\kappa}$ when κ is weakly compact*
◇ C55 C75 E55 ◇

WAGNER, C.M. *On Martin's conjecture*
◇ C15 C35 C45 C65 C75 ◇

WANG, SHIQIANG *An omitting types theorem in lattice-valued model theory (Chinese)*
◇ C07 C75 C90 ◇

WEISPFENNING, V. *Model theory and lattices of formulas*
◇ C25 C35 C52 C75 G10 ◇

1983

ANDREKA, H. & NEMETI, I. *Generalization of the concept of variety and quasivariety to partial algebras through category theory*
◇ C05 C40 C75 C90 C95 G30 ◇

ASH, C.J. *Model-theoretic forms of the axiom of choice*
◇ C75 E25 E35 ◇

BALDO, A. *Complete interpolation theorems for L_{kk}^{2+} (Italian summary)* ◇ C40 C75 C85 ◇

BLASS, A.R. *Words, free algebras, and coequalizers*
◇ C05 C75 E25 E35 F50 F55 G30 ◇

CANTINI, A. *Σ-models of subtheories of KPI (Italian)*
◇ C62 C70 E30 ◇

DANKO, W. *Algorithmic properties of finitely generated structures* ◇ B75 C40 C52 C75 ◇

DIENER, K.-H. *On constructing infinitary languages $L_{\alpha\beta}$ without the axiom of choice* ◇ C75 E25 ◇

ERSHOV, YU.L. *Dynamic logic over admissible sets (Russian)* ◇ B75 C70 ◇

ERSHOV, YU.L. *The Σ-enumeration principle (Russian)*
◇ C62 C70 D60 E30 E65 ◇

FERRO, R. *Seq-consistency property and interpolation theorems* ◇ C40 C75 ◇

FRIEDMAN, S.D. & SHELAH, S. *Tall α-recursive structures*
◇ C70 C75 E45 ◇

GAO, HENGSHAN *Comments on "The interpolation theorem for the propositional calculus $P(\kappa)$ when κ is a strongly inaccessible cardinal" by Luo, Libo (Chinese)*
◇ B05 C40 C75 E55 ◇

GERLA, G. *Model theory and modal logic (Italian)*
◇ B45 C25 C75 C90 ◇

KARTTUNEN, M. *Model theoretic results for infinitely deep languages* ◇ C20 C75 ◇

KNIGHT, J.F. *Additive structure in uncountable models for a fixed completion of P*
◇ C50 C57 C62 C75 D30 ◇

MOTOHASHI, N. *Some remarks on Barwise approximation theorem on Henkin quantifiers* ◇ C75 C80 ◇

NURTAZIN, A.T. *An elementary classification and some properties of complete orders (Russian)*
◇ B25 C35 C65 C75 E10 ◇

OIKKONEN, J. *Logical operations and iterated infinitely deep languages* ◇ C75 C95 ◇

SACKS, G.E. *On the number of countable models*
◇ C15 C70 C75 ◇

SHELAH, S. *Classification theory of non-elementary classes, Part I: The number of uncountable models of $\psi \in L_{\omega_1,\omega}$* ◇ C45 C52 C55 C75 C80 E50 ◇

1984

BELYAEV, V.YA. *Uncountable extensions of countable algebraically closed semigroups (Russian)*
◇ C05 C55 C60 C75 ◇

BLAIR, H.A. *The intractability of validity in logic programming and dynamic logic*
◇ B75 C75 D35 D55 ◇

BURD, B. *Decomposable collections of sets*
◇ C75 E05 ◇

FERBUS, M.-C. *Functorial bounds for cut elimination in $L_{\beta\omega}I$* ◇ C75 F05 F15 G30 ◇

FRIEDMAN, S.D. *Infinitary logic and $0^\#$*
◇ C75 E15 E45 E55 ◇

FRIEDMAN, S.D. *Model theory for $L_{\infty\omega_1}$* ◇ C75 ◇

GALLIER, J.H. *n-rational algebras I: Basic properties and free algebras. II: Varieties and logic of inequalities*
◇ C05 C75 ◇

GERLA, G. & VACCARO, V. *Modal logic and model theory*
◇ B45 C25 C75 C90 ◇

GIORGETTA, D. & SHELAH, S. *Existentially closed structures in the power of the continuum*
◇ C15 C25 C30 C52 C55 C60 C75 ◇

GIRARD, J.-Y. *The Ω-rule* ◇ B15 C62 C75 F35 ◇

GOLDBLATT, R.I. *An abstract setting for Henkin proofs*
◇ C07 C75 ◇

GRAETZER, G. & WHITNEY, S. *Infinitary varieties of structures closed under the formation of complex structures* ◇ C05 C52 C75 ◇

HARRINGTON, L.A. *Admissible ordinals are not immortal*
◇ C62 C70 E30 E45 E47 ◇

HODGES, W. *On constructing many non-isomorphic algebras* ◇ C05 C52 C55 C75 E50 E55 E75 ◇

HUGLY, P. & SAYWARD, C.W. *Do we need quantification?*
◇ A05 C75 ◇

KARTTUNEN, M. *Model theory for infinitely deep languages* ◇ C20 C75 ◇

KAUFMANN, M. *Filter logics on ω*
◇ C40 C75 C80 C95 ◇

KAUFMANN, M. *On expandability of models of arithmetic and set theory to models of weak second-order theories*
◇ C15 C30 C62 C70 C75 E30 E70 ◇

KAUFMANN, M. *Some remarks on equivalence in infinitary and stationary logic* ◇ C55 C65 C75 C80 E45 ◇

KOSSAK, R. *$L_{\infty\omega_1}$-elementary equivalence of ω_1-like models of PA* ◇ C50 C57 C62 C75 ◇

KRYNICKI, M. & LACHLAN, A.H. & VAEAENAENEN, J. *Vector spaces and binary quantifiers*
◇ C60 C75 C80 ◇

MAGIDOR, M. & SHELAH, S. & STAVI, J. *Countably decomposable admissible sets*
◇ C40 C62 C70 E30 E45 E47 ◇

MARTINEZ, JUAN CARLOS *Accessible sets and $(L_{\omega_1\omega})_t$-equivalence for T_3 spaces*
◇ C65 C75 C80 C90 ◇

MOTOHASHI, N. *Approximation theory of uniqueness conditions by existence conditions*
⋄ C40 C75 C80 F07 F50 ⋄

MOTOHASHI, N. *Equality and Lyndon's interpolation theorem* ⋄ B20 C40 C75 ⋄

PARIGOT, M. *Omission des types dans les theories $\Pi_2^0(\Delta)$*
⋄ C40 C75 ⋄

RADEV, S.R. *Interpolation in the infinitary propositional normal modal logic KL_{ω_1}* ⋄ B45 C40 C75 ⋄

RUYER, H. *Deux resultats concernant les modules sur un anneau de Dedekind (English summary)*
⋄ C10 C60 C75 ⋄

SHELAH, S. *A pair of non-isomorphic $\equiv_{\infty\lambda}$ models of power λ for λ singular with $\lambda^\omega = \lambda$* ⋄ C55 C75 ⋄

SHELAH, S. *Can you take Solovay's inaccessible away?*
⋄ C55 C75 C80 E15 E25 E35 E55 E75 ⋄

1985

BALDWIN, JOHN T. & SHELAH, S. *Second-order quantifiers and the complexity of theories*
⋄ C40 C55 C65 C75 C85 ⋄

BARWISE, J. & FEFERMAN, S. (EDS.) *Model-theoretic logics*
⋄ C75 C80 C85 C90 C95 C98 ⋄

CAICEDO, X. *Failure of interpolation for quantifiers of monadic type* ⋄ C40 C75 C80 C95 ⋄

DICKMANN, M.A. *Larger infinitary languages*
⋄ C75 C98 ⋄

EKLOF, P.C. *Applications to algebra*
⋄ C60 C75 C98 ⋄

ERSHOV, YU.L. *Σ-definability in admissible sets (Russian)*
⋄ C70 E47 ⋄

FLUM, J. & MARTINEZ, JUAN CARLOS *κ-local operations for topological structures* ⋄ C40 C75 C80 C90 ⋄

FLUM, J. *Characterizing logics*
⋄ C75 C80 C95 C98 ⋄

HOOVER, D.N. *A probabilistic interpolation theorem*
⋄ C40 C70 C90 ⋄

KOLAITIS, P.G. *Game quantification*
⋄ C40 C70 C75 C98 D60 D65 D70 E60 ⋄

KOSSAK, R. *Recursively saturated ω_1-like models of arithmetic* ⋄ C55 C57 C62 C75 C80 E65 ⋄

LASCAR, D. *Quelques precisions sur la d.o.p. et la profondeur d'une theorie (English summary)*
⋄ C45 C75 ⋄

LASCAR, D. *Why some people are excited by Vaught's conjecture* ⋄ C15 C70 C75 ⋄

MEKLER, A.H. & SHELAH, S. *Stationary logic and its friends I* ⋄ C40 C55 C75 C80 E35 E50 ⋄

NADEL, M.E. *$L_{\omega_1\omega}$ and admissible fragments*
⋄ C70 C75 C98 ⋄

SHELAH, S. *Classification of first order theories which have a structure theorem* ⋄ C35 C45 C52 C75 ⋄

SHELAH, S. *Monadic logic and Loewenheim numbers*
⋄ C45 C55 C65 C75 C85 E45 ⋄

SHELAH, S. *On the possible number $no(M)$ = the number of nonisomorphic models $L_{\infty,\lambda}$-equivalent to M of power λ, for λ singular* ⋄ C55 C75 ⋄

SPECTOR, M. *Model theory under the axiom of determinateness* ⋄ C20 C55 C75 E60 ⋄

C80 ∪ C85 ∪ C95 Other extended logics and generalized model theory

1915
LOEWENHEIM, L. *Ueber Moeglichkeiten im Relativkalkuel*
⋄ B25 C07 C10 C85 E07 G15 ⋄

1939
LANGFORD, C.H. *A theorem on deducibility for second-order functions* ⋄ B15 C35 C65 C85 ⋄

1947
MOSTOWSKI, ANDRZEJ *On absolute properties of relations*
⋄ B15 C75 C85 E15 E47 ⋄

1949
KUSTAANHEIMA, P. *Ueber die Vollstaendigkeit der Axiomensysteme mit einem endlichen Individuenbereich* ⋄ C13 C35 C85 ⋄

1950
HENKIN, L. *Completeness in the theory of types*
⋄ A05 B15 C85 F35 ⋄

1953
HENKIN, L. *Some interconnections between modern algebra and mathematical logic* ⋄ C07 C60 C85 ⋄

1954
MARKWALD, W. *Zur Theorie der konstruktiven Wohlordnungen* ⋄ C80 D55 F15 ⋄

1955
HINTIKKA, K.J.J. *Reductions in the theory of types*
⋄ B15 C85 ⋄

1956
MOSTOWSKI, ANDRZEJ *Development and applications of the "projective" classification of sets of integers*
⋄ C57 C80 D55 E15 E45 ⋄

1957
KOCHEN, S. *Completeness of algebraic systems in higher order calculi* ⋄ B15 C35 C60 C85 ⋄
MOSTOWSKI, ANDRZEJ *On a generalisation of quantifiers*
⋄ B25 C55 C80 C95 ⋄

1958
FRAISSE, R. *Sur une extension de la polyrelation et des parentes tirant son origine du calcul logique du $k^{ème}$-echelon* ⋄ B15 C07 C85 E07 ⋄
MOSTOWSKI, ANDRZEJ *Quelques observations sur l'usage des methodes non finitistes dans la meta-mathematique*
⋄ A05 C30 C62 C80 E30 E35 F99 ⋄

1959
MAL'TSEV, A.I. *Model correspondences (Russian)*
⋄ B15 C05 C07 C52 C60 C85 ⋄
OREY, S. *Model theory for the higher order predicate calculus* ⋄ B15 C85 ⋄

1960
BUECHI, J.R. *Weak second-order arithmetic and finite automata* ⋄ B15 B25 C85 D05 F35 ⋄
SVENONIUS, L. *Some problems in logical model-theory*
⋄ C07 C52 C95 ⋄

1961
EHRENFEUCHT, A. *An application of games to the completeness problem for formalized theories*
⋄ C07 C65 C85 E10 E60 ⋄
HENKIN, L. *Some remarks on infinitely long formulas*
⋄ C75 C80 ⋄
LACHLAN, A.H. *The U-quantifier* ⋄ C80 ⋄
MOSTOWSKI, ANDRZEJ *Concerning the problem of axiomatizability of the field of real numbers in the weak second order logic* ⋄ B15 B28 C85 ⋄

1962
BAAYEN, P.C. *Partial ordering of quantifiers and of clopen equivalence relations* ⋄ C80 C99 E07 G15 G99 ⋄
BUECHI, J.R. *On a decision method in restricted second order arithmetic* ⋄ B25 C85 D05 F35 ⋄
MOSTOWSKI, ANDRZEJ *L'espace des modeles d'une theorie formalisee et quelques-unes de ses applications*
⋄ B50 C07 C40 C57 C80 C85 C90 D45 ⋄
MOSTOWSKI, ANDRZEJ *On invariant, dual invariant and absolute formulas* ⋄ B15 C40 C85 ⋄
RESCHER, N. *Plurality-quantification* ⋄ C55 C80 ⋄

1964
ERSHOV, YU.L. *Decidability of certain non-elementary theories (Russian)* ⋄ B25 C30 C85 G05 ⋄
FUHRKEN, G. *Skolem-type normal forms for first-order languages with a generalized quantifier*
⋄ C55 C80 ⋄
MO, SHAOKUI *Enumeration quantifiers and predicate calculus (Chinese)* ⋄ B10 C07 C80 ⋄
OHKUMA, T. *Closed classes of models* ⋄ C95 G30 ⋄
ONYSZKIEWICZ, J. *On the different topologies in the space of models* ⋄ C15 C52 C85 ⋄
SCHOCK, R. *On the logic of variable binders*
⋄ B10 C80 ⋄
VAUGHT, R.L. *The completeness of logic with the added quantifier "there are uncountable many"*
⋄ C55 C80 ⋄

1965
BUECHI, J.R. *Decision methods in the theory of ordinals*
⋄ B15 B25 C85 D05 E10 ⋄
BUECHI, J.R. *Transfinite automata recursions and weak second order theory of ordinals*
⋄ B15 B25 C85 D05 E10 ⋄
ENGELER, E. *Combinatorial theorems for the construction of models* ⋄ C80 C85 C95 ⋄

FUHRKEN, G. *Languages with added quantifier "There exist at least $\aleph_\alpha$"* ⋄ C55 C80 ⋄

HAERTIG, K. *Ueber einen Quantifikator mit zwei Wirkungsbereichen* ⋄ C55 C80 ⋄

KOGALOVSKIJ, S.R. *Certain remarks on classes axiomatizable in a nonelementary way (Russian)* ⋄ C20 C52 C85 ⋄

KOGALOVSKIJ, S.R. *Generalized quasiuniversal classes of models (Russian)* ⋄ C52 C85 ⋄

SCHREIBER, P. *Untersuchungen ueber die Modelle der Typentheorie* ⋄ B15 C85 ⋄

VAUGHT, R.L. *A Loewenheim-Skolem theorem for cardinals far apart* ⋄ C55 C80 ⋄

VAUGHT, R.L. *The Loewenheim-Skolem theorem* ⋄ C07 C55 C80 ⋄

1966

APELT, H. *Axiomatische Untersuchungen ueber einige mit der Presburgerschen Arithmetik verwandte Systeme* ⋄ B28 C35 C80 F30 ⋄

ELGOT, C.C. & RABIN, M.O. *Decidability and undecidability of extensions of second (first) order theory of (generalized) successor* ⋄ B15 B25 C85 D05 D35 F30 F35 ⋄

FEFERMAN, S. & KREISEL, G. *Persistent and invariant formulas relative to theories of higher order* ⋄ B15 C40 C52 C85 ⋄

HELLING, M. *Model-theoretic problems for some extensions of first-order languages* ⋄ C55 C75 C80 E55 ⋄

KOGALOVSKIJ, S.R. *On higher order logic (Russian)* ⋄ B15 C40 C85 ⋄

KOGALOVSKIJ, S.R. *On the semantics of the theory of types (Russian)* ⋄ B15 C40 C85 ⋄

LINDSTROEM, P. *First order predicate logic with generalized quantifiers* ⋄ C80 C95 ⋄

ROEDDING, D. *Anzahlquantoren in der Kleene-Hierarchie* ⋄ C80 D55 ⋄

ROEDDING, D. *Anzahlquantoren in der Praedikatenlogik* ⋄ C07 C80 ⋄

YASUHARA, M. *An axiomatic system for the first order language with an equi-cardinality quantifier* ⋄ C55 C80 ⋄

YASUHARA, M. *Syntactical and semantical properties of generalized quantifiers* ⋄ C40 C55 C80 ⋄

1967

HASENJAEGER, G. *On Loewenheim-Skolem-type insufficiencies of second order logic* ⋄ B15 C55 C85 ⋄

LOPEZ-ESCOBAR, E.G.K. *A complete, infinitary axiomatization of weak second-order logic* ⋄ B15 C75 C85 F35 ⋄

LOPEZ-ESCOBAR, E.G.K. *On a theorem of J.I.Malitz* ⋄ C30 C40 C85 ⋄

RAUTENBERG, W. *Elementare Schemata nichtelementarer Axiome* ⋄ B25 C52 C60 C65 C85 D35 ⋄

SCHWABHAEUSER, W. *Zur Axiomatisierbarkeit von Theorien in der schwachen Logik der zweiten Stufe* ⋄ B15 C85 ⋄

SLOMSON, A. *Some problems in mathematical logic* ⋄ C80 ⋄

1968

KEISLER, H.J. *Formulas with linearly ordered quantifiers* ⋄ C40 C50 C75 C80 ⋄

KEISLER, H.J. *Models with orderings* ⋄ C55 C80 ⋄

KOGALOVSKIJ, S.R. *Some remarks on higher order logic (Russian)* ⋄ B15 C40 C85 F35 ⋄

LAEUCHLI, H. *A decision procedure for the weak second order theory of linear order* ⋄ B15 B25 C65 C85 ⋄

MOSTOWSKI, ANDRZEJ *Craig's interpolation theorem in some extended systems of logic* ⋄ C40 C95 ⋄

RABIN, M.O. *Decidability of second-order theories and automata on infinite trees* ⋄ B15 B25 C85 D05 ⋄

SIEFKES, D. *Recursion theory and the theorem of Ramsey in one-place second order successor arithmetic* ⋄ B15 B25 C85 ⋄

SLOMSON, A. *The monadic fragment of predicate calculus with the Chang quantifier and equality* ⋄ B25 C20 C55 C80 ⋄

VAUGHT, R.L. *Model theory and set theory* ⋄ C55 C80 C98 ⋄

1969

BARWISE, J. & EKLOF, P.C. *Lefschetz's principle* ⋄ B15 C60 C75 C85 ⋄

BUECHI, J.R. & LANDWEBER, L.H. *Definability in the monadic second-order theory of successor* ⋄ B15 B25 C40 C85 D05 F35 ⋄

CLEAVE, J.P. *Local properties of systems* ⋄ C52 C60 C85 ⋄

EBBINGHAUS, H.-D. *Ueber fuer-fast-alle-Quantoren* ⋄ C80 ⋄

HAUSCHILD, K. & WOLTER, H. *Ueber die Kategorisierbarkeit gewisser Koerper in nicht-elementaren Logiken* ⋄ C60 C80 C85 ⋄

ISSEL, W. *Semantische Untersuchungen ueber Quantoren I* ⋄ B10 C55 C80 ⋄

JOHNSON, D.R. & THOMASON, R.H. *Predicate calculus with free quantifier variables* ⋄ B15 C80 ⋄

KEISLER, H.J. *Infinite quantifiers and continuous games* ⋄ C65 C75 C80 ⋄

LINDSTROEM, P. *On extensions of elementary logic* ⋄ C95 ⋄

RABIN, M.O. *Decidability of second order theories and automata on infinite trees* ⋄ B15 B25 C85 D05 ⋄

WOJCIECHOWSKA, A. *Generalized products for Q_α-languages (Russian summary)* ⋄ C30 C55 C80 ⋄

1970

BELL, J.L. *Weak compactness in restricted second-order languages* ⋄ C85 E55 ⋄

CANTY, J.T. & KUENG, G. *Substitutional quantification and Lesniewskian quantifiers* ⋄ A05 B60 C80 E70 ⋄

DONER, J.E. *Tree acceptors and some of their applications* ⋄ B25 C85 D05 ⋄

ENDERTON, H.B. *Finite partially-ordered quantifiers* ⋄ B15 C80 C85 ⋄

ENDERTON, H.B. *The unique existential quantifier*
⋄ C80 D55 ⋄

FUHRKEN, G. *On the degree of compactness of the languages Q_α* ⋄ C55 C80 ⋄

ISSEL, W. *Semantische Untersuchungen ueber Quantoren II,III* ⋄ C40 C55 C80 ⋄

KEISLER, H.J. *Logic with the quantifier "there exist uncountably many"* ⋄ C30 C55 C62 C75 C80 ⋄

LIPNER, L.D. *Some aspects of generalized quantifiers*
⋄ C55 C80 ⋄

RABIN, M.O. *Weakly definable relations and special automata* ⋄ B15 B25 C40 C85 D05 ⋄

SHELAH, S. *On languages with non-homogeneous strings of quantifiers* ⋄ C75 C80 ⋄

SIEFKES, D. *Decidable theories I. Buechi's monadic second order successor arithmetic*
⋄ B15 B25 C85 D05 F35 F98 ⋄

SIEFKES, D. *Decidable extensions of monadic second order successor arithmetic* ⋄ B15 B25 C85 D05 F35 ⋄

SLOMSON, A. *An algebraic characterization of indistinguishable cardinals* ⋄ C55 C85 E10 ⋄

WALKOE JR., W.J. *Finite partially-ordered quantification*
⋄ B15 C80 ⋄

YASUHARA, M. *Incompleteness of L_p-languages*
⋄ C55 C80 ⋄

1971

ALTHAM, J.E.J. *The logic of plurality*
⋄ A05 B98 C80 ⋄

CASARI, E. *Regular models for second order set theory*
⋄ C62 C85 E55 E70 ⋄

CORCORAN, J. & HERRING, J.M. *Notes on a semantic analysis of variable binding term operators*
⋄ A05 A10 B10 C80 ⋄

EBBINGHAUS, H.-D. *On models with large automorphism groups* ⋄ C30 C55 C80 ⋄

ELLENTUCK, E. *A minimal model for strong analysis*
⋄ C50 C62 C85 E35 ⋄

FLUM, J. *Ganzgeschlossene und praedikatengeschlossene Logiken I,II* ⋄ B10 C85 C95 ⋄

MAGIDOR, M. *On the role of supercompact and extendible cardinals in logic* ⋄ C55 C75 C85 E47 E55 ⋄

PALYUTIN, E.A. *Boolean algebras with a categorical theory in a weak second order logic (Russian)*
⋄ B15 C35 C85 D35 G05 ⋄

RABIN, M.O. *Decidability and definability in second-order theories* ⋄ B15 B25 C40 C85 D05 ⋄

SHELAH, S. *Two cardinal compactness*
⋄ C30 C52 C55 C80 ⋄

SIEFKES, D. *Undecidable extensions of monadic second order successor arithmetic*
⋄ B15 B25 C85 D35 F35 ⋄

1972

ANDREWS, P.B. *General models, descriptions, and choice in type theory* ⋄ B15 C85 E25 E35 F35 ⋄

ANDREWS, P.B. *General models and extensionality*
⋄ B15 C85 E35 F35 ⋄

ARMBRUST, M. & KAISER, KLAUS *On quasi-universal model classes* ⋄ B15 C05 C20 C52 C85 ⋄

BARWISE, J. *Absolute logics and $L_{\infty\omega}$* ⋄ C75 C95 ⋄

BARWISE, J. *The Hanf number of second order logic*
⋄ B15 C55 C85 E30 ⋄

BELL, J.L. *On the relationship between weak compactness in $L_{\omega_1,\omega}, L_{\omega_1,\omega_1}$ and restricted second order languages* ⋄ C55 C75 C80 C85 E05 E55 ⋄

COLE, J.C. & DICKMANN, M.A. *Non-axiomatizability results in infinitary languages for higher-order structures* ⋄ C52 C65 C75 C85 ⋄

CORCORAN, J. & HATCHER, W.S. & HERRING, J.M. *Variable binding term operators* ⋄ B10 C80 ⋄

DOEPP, K. *Bemerkung zu Henkins Beweis fuer die Nichtstandard-Vollstaendigkeit der Typentheorie*
⋄ B15 C62 C85 ⋄

DONER, J.E. *Definability in the extended arithmetic of ordinal numbers* ⋄ C85 E10 E47 ⋄

FINE, K. *For so many individuals* ⋄ C07 C80 ⋄

FLUM, J. *Die Automorphismenmengen der Modelle einer L_{Q_κ}-Theorie* ⋄ C30 C55 C80 ⋄

FLUM, J. *Hanf numbers and well-ordering numbers*
⋄ C30 C55 C75 C80 ⋄

FRIEDRICH, U. *Entscheidbarkeit der monadischen Nachfolgerarithmetik mit endlichen Automaten ohne Analysetheorem (Russian, English and French summaries)* ⋄ B25 C85 D05 F35 ⋄

FUHRKEN, G. *A remark on the Haertig quantifier*
⋄ C55 C80 ⋄

GARLAND, S.J. *Generalized interpolation theorems*
⋄ C40 C75 C85 D55 E15 ⋄

MALITZ, J. & REINHARDT, W.N. *A complete countable $L_{\omega_1}^Q$-theory with maximal models of many cardinalities*
⋄ C55 C80 ⋄

MALITZ, J. & REINHARDT, W.N. *Maximal models in the language with quantifier "there exist uncountably many"* ⋄ C50 C55 C80 E55 ⋄

MOSCHOVAKIS, Y.N. *The game quantifier*
⋄ C75 C80 D55 D70 D75 E60 ⋄

MOTOHASHI, N. *A new theorem on definability in a positive second order logic with countable conjunctions and disjunctions* ⋄ B15 C40 C75 C85 ⋄

PINUS, A.G. *On the theory of convex subsets (Russian)*
⋄ B25 C65 C85 D35 ⋄

RABIN, M.O. *Automata on infinite objects and Church's problem* ⋄ B15 B25 C85 D05 ⋄

SEESE, D.G. *Entscheidbarkeits- und Definierbarkeitsfragen der Theorie "netzartiger" Graphen I (Russian) (English and French summaries)*
⋄ C80 C85 D35 ⋄

SLOMSON, A. *Generalized quantifiers and well orderings*
⋄ B25 C55 C65 C80 E07 ⋄

STILLWELL, J.C. *Decidability of the "almost all" theory of degrees* ⋄ B25 C80 D30 ⋄

VINNER, S. *A generalization of Ehrenfeucht's game and some applications* ⋄ B25 C07 C55 C80 E60 ⋄

WOLTER, H. *Ueber Mengen von Ausdruecken der Logik hoeherer Stufe, fuer die der Endlichkeitssatz und der Satz von Loewenheim-Skolem gelten*
⋄ B15 C20 C85 ⋄

WOLTER, H. *Untersuchung ueber Algebren und formalisierte Sprachen hoeherer Stufen (Russian, English and French summaries)*
⋄ B15 C20 C30 C40 C85 ⋄

1973

ANDREKA, H. & GERGELY, T. & NEMETI, I. *Some questions on languages of order n (Hungarian)*
⋄ B15 C07 C85 ⋄

ARMBRUST, M. *Direct limits of congruence systems*
⋄ C05 C85 ⋄

BOWEN, K.A. *Infinite forcing and natural models of set theory (Russian summary)*
⋄ C25 C55 C62 C75 C80 E55 G30 ⋄

BUECHI, J.R. & SIEFKES, D. *Axiomatization of the monadic second order theory of ω_1* ⋄ B15 B25 C85 E10 ⋄

BUECHI, J.R. & SIEFKES, D. *The monadic second order theory of all countable ordinals*
⋄ B15 B25 C85 E10 ⋄

BUECHI, J.R. *The monadic second order theory of ω_1*
⋄ B15 B25 C85 E10 ⋄

DORACZYNSKI, R. *Elimination of bound variables in logic with an arbitrary quantifier (Polish and Russian summaries)* ⋄ C10 C80 ⋄

FRIEDMAN, H.M. *Beth's theorem in cardinality logics*
⋄ C40 C55 C80 ⋄

HAJEK, P. *Automatic listing of important observational statements. I,II* ⋄ B35 C13 C80 C90 ⋄

KNIGHT, J.F. *Generic expansions of structures*
⋄ C25 C62 C85 ⋄

KOGALOVSKIJ, S.R. & RORER, M.A. *On the question of the definability of the concept of definability (Russian)*
⋄ B15 C40 C85 ⋄

KRIVINE, J.-L. & MCALOON, K. *Forcing and generalized quantifiers* ⋄ C25 C80 ⋄

LINDSTROEM, P. *A characterization of elementary logic*
⋄ C95 ⋄

MACINTYRE, A. *Martin's axiom applied to existentially closed groups* ⋄ C25 C55 C60 C75 C80 E50 ⋄

MAKOWSKY, J.A. *Langages engendres a partir des formules de Scott* ⋄ C75 C95 ⋄

MALCOLM, W.G. *Application of higher-order ultraproducts to the theory of local properties in universal algebras and relational systems* ⋄ C05 C20 C85 E40 ⋄

MOTOHASHI, N. *Model theory on a positive second order logic with countable conjunctions and disjunctions*
⋄ B15 C40 C75 C85 ⋄

REICHBACH, J. *Generalized models for intuitionistic and classical predicate calculi with ultraproducts*
⋄ C20 C25 C80 C90 F50 ⋄

RESSAYRE, J.-P. *Boolean models and infinitary first order languages*
⋄ C40 C50 C55 C70 C75 C80 C90 ⋄

SHELAH, S. *First order theory of permutation groups*
⋄ B15 C55 C60 C85 ⋄

SHELAH, S. *There are just four second-order quantifiers*
⋄ B15 C80 C85 ⋄

THARP, L.H. *The characterization of monadic logic*
⋄ B15 B20 C95 ⋄

VASIL'EV, EH.S. *Elementary theories of two-base models of abelian groups (Russian)* ⋄ B25 C60 C85 ⋄

WEAVER, G.E. *A note on compactness and decidability*
⋄ B25 C95 ⋄

WOLTER, H. *Eine Erweiterung der elementaren Praedikatenlogik: Anwendungen in der Arithmetik und anderen mathematischen Theorien*
⋄ B28 C10 C80 F30 ⋄

WOLTER, H. *Untersuchungen zum Spektralproblem gewisser Logiken 2. Stufe* ⋄ C55 C85 E10 ⋄

1974

ADAMS, E.W. *The logic of "almost all"* ⋄ C80 ⋄

ANDREKA, H. & GERGELY, T. & NEMETI, I. *Some questions on languages of order n I,II (Russian) (English summaries)* ⋄ B15 C07 C85 ⋄

BARWISE, J. *Axioms for abstract model theory* ⋄ C95 ⋄

BOWEN, K.A. *Forcing in a general setting*
⋄ C25 C75 C80 C85 E40 ⋄

EBBINGHAUS, H.-D. & FLUM, J. *Eine Maximalitaetseigenschaft der Praedikatenlogik erster Stufe* ⋄ C95 ⋄

FAGIN, R. *Generalized first-order spectra and polynomial-time recognizable sets*
⋄ C13 C85 D15 ⋄

FEFERMAN, S. *Applications of many-sorted interpolation theorems* ⋄ C40 C95 ⋄

FEFERMAN, S. *Two notes on abstract model theory I. Properties invariant on the range of definable relations between structures* ⋄ C40 C95 ⋄

FLUM, J. *On Horn theories* ⋄ B20 C20 C80 ⋄

FRAISSE, R. *Isomorphisme local et equivalence associes a un ordinal; utilite en calcul des formules infinies a quanteurs finis* ⋄ B15 C07 C75 C85 ⋄

GABBAY, D.M. & MORAVCSIK, J.M.E. *Branching quantifiers and Montague semantics* ⋄ B65 C80 ⋄

GARLAND, S.J. *Second-order cardinal characterizability*
⋄ B15 C40 C55 C85 D55 E10 E47 E55 ⋄

GRILLIOT, T.J. *Model theory for dissecting recursion theory* ⋄ C55 C70 C75 C80 D60 D75 ⋄

GUREVICH, Y. *A resolving procedure for the extended theory of ordered abelian groups (Russian)*
⋄ B25 C10 C60 C85 ⋄

GUZICKI, W. *Elementary extensions of Levy's model of A_2^-* ⋄ C55 C62 C80 E25 E35 E40 F35 ⋄

HAJEK, P. *Automatic listing of important observational statements. III* ⋄ B35 C13 C80 C90 ⋄

HINTIKKA, K.J.J. *Quantifiers vs. quantification theory*
⋄ A05 B10 C80 ⋄

HUTCHINSON, J.E. *Models of set theories and their applications* ⋄ C55 C62 C80 ⋄

ISAMBERT, E. *Extensions elementaires non-standard de modeles de theorie des ensembles* ⋄ C62 C80 ⋄

JENSEN, F.V. *Interpolation and definability in abstract logics* ⋄ C40 C95 ⋄

KAISER, KLAUS *Direct limits in quasi-universal model classes* ⋄ B15 C30 C52 C85 ⋄

KOGALOVSKIJ, S.R. *Certain criteria for localness for higher order formulas (Russian)* ⋄ B15 C85 ⋄

KOGALOVSKIJ, S.R. *Certain simple consequences of the axiom of constructibility (Russian) (English summary)* ⋄ C40 C62 C85 E45 ⋄

KOGALOVSKIJ, S.R. *Reductions in second-order logic (Russian)* ⋄ B15 C85 ⋄

LINDSTROEM, P. *On characterizing elementary logic* ⋄ B10 C95 ⋄

MALCOLM, W.G. *Some results and algebraic applications in the theory of higher-order ultraproducts* ⋄ C20 C60 C85 G05 ⋄

MEHLHORN, K. *The "almost all" theory of subrecursive degrees is decidable* ⋄ B25 C80 D20 ⋄

MOSTOWSKI, ANDRZEJ *Observations concerning elementary extensions of ω-models* ⋄ C62 C80 ⋄

ONYSZKIEWICZ, J. *Dimensions of PC space of models (Russian summary)* ⋄ C15 C52 C85 ⋄

THARP, L.H. *Continuity and elementary logic* ⋄ C07 C80 C95 D30 D55 ⋄

1975

ACZEL, P. *Quantifiers, games and inductive definitions* ⋄ C75 C80 D70 ⋄

ALTHAM, J.E.J. & TENNANT, N. *Sortal quantification* ⋄ A05 C80 ⋄

ANDREKA, H. & GERGELY, T. & NEMETI, I. *Many-sorted languages and their connection with nth order languages (Russian) (English summary)* ⋄ B15 C85 ⋄

BAUDISCH, A. *Die elementare Theorie der Gruppe vom Typ p^∞ mit Untergruppen* ⋄ B25 C60 C85 ⋄

BROESTERHUIZEN, G. *A generalized Łoś ultraproduct theorem* ⋄ C20 C80 ⋄

BROESTERHUIZEN, G. *Structures for a logic with additional generalized quantifier* ⋄ C80 ⋄

BRUIJN DE, N.G. *Set theory with type restrictions* ⋄ C80 E30 E70 ⋄

CARLSTROM, I.F. *Truth and entailment for a vague quantifier* ⋄ B52 C80 ⋄

CHERLIN, G.L. *Second order forcing, algebraically closed structures, and large cardinals* ⋄ C25 C85 E55 ⋄

COCCHIARELLA, N.B. *A second order logic of variable-binding operators* ⋄ A05 B15 C80 ⋄

COSTA DA, N.C.A. & DRUCK, I.F. *Sur les "VBTOs" selon M.Hatcher (English summary)* ⋄ C07 C40 C80 ⋄

EBBINGHAUS, H.-D. *Modelltheorie und Logik* ⋄ C95 C98 ⋄

EBBINGHAUS, H.-D. *Zum L(Q)-Interpolationsproblem* ⋄ C40 C55 C80 ⋄

ELLENTUCK, E. & HALPERN, F. *Theories having many extensions* ⋄ C07 C75 C80 E75 ⋄

FAGIN, R. *Monadic generalized spectra* ⋄ C13 C85 ⋄

FEFERMAN, S. *Two notes on abstract model theory II. Languages for which the set of valid sentences is semi-invariantly implicitly definable* ⋄ C40 C70 C95 ⋄

FLUM, J. *First-order logic and its extensions* ⋄ C40 C55 C75 C95 C98 ⋄

FLUM, J. *L(Q)-preservation theorems* ⋄ C20 C40 C55 C80 E60 ⋄

GALLIN, D. *Intensional and higher-order modal logic. With applications to Montague semantics* ⋄ B45 C85 C90 E35 E50 E70 ⋄

GOGOL, D. *Formulas with two generalized quantifiers* ⋄ C80 ⋄

GUENTHNER, F. & HOEPELMAN, J.P. *A note on the representation of "branching quantifiers"* ⋄ B65 C80 ⋄

HAUSCHILD, K. *Ueber zwei Spiele und ihre Anwendung in der Modelltheorie (Russian, English and French summaries)* ⋄ B25 C07 C30 C35 C85 ⋄

HAUSCHILD, K. *Zur Uebertragbarkeit von Entscheidbarkeitsresultaten elementarer Theorien (Russian, English and French summaries)* ⋄ B15 B25 C65 C85 ⋄

HAVRANEK, T. *Statistical quantifiers in observational calculi* ⋄ B50 C80 C90 ⋄

HERRE, H. & WOLTER, H. *Entscheidbarkeit von Theorien in Logiken mit verallgemeinerten Quantoren* ⋄ B25 C10 C55 C80 D35 ⋄

JENSEN, F.V. *On completeness in cardinality logics* ⋄ C10 C35 C55 C60 C80 ⋄

JERVELL, H.R. *Conservative endextensions and the quantifier "there exist uncountably many"* ⋄ C30 C80 ⋄

KAISER, KLAUS *Various remarks on quasi-universal model classes* ⋄ B15 C05 C52 C85 ⋄

KEISLER, H.J. *Constructions in model theory* ⋄ C25 C30 C50 C75 C95 C98 ⋄

MAKOWSKY, J.A. *Securable quantifiers, κ-unions and admissible sets* ⋄ C55 C70 C75 C80 C95 ⋄

MAKOWSKY, J.A. *Topological model theory* ⋄ C65 C80 ⋄

MCKEE, T.A. *Infinitary logic and topological homeomorphisms* ⋄ C65 C75 C80 C90 ⋄

PINTER, C. *Algebraic logic with generalized quantifiers* ⋄ C80 G15 ⋄

SHELAH, S. *Categoricity in $\aleph_1$ of sentences in $L_{\omega_1\omega}(Q)$* ⋄ C35 C55 C75 C80 ⋄

SHELAH, S. *Generalized quantifiers and compact logic* ⋄ C55 C80 C95 E55 ⋄

SHELAH, S. *The monadic theory of order* ⋄ B25 C65 C85 D35 E50 ⋄

TENNEY, R.L. *Second-order Ehrenfeucht games and the decidability of the second-order theory of an equivalence relation* ⋄ B15 B25 C85 ⋄

TROFIMOV, M.YU. *Definability in algebraically closed systems (Russian)* ⋄ C25 C40 C60 C85 ⋄

VINNER, S. *Model-completeness in a first order language with a generalized quantifier* ⋄ C35 C55 C80 ⋄

VINNER, S. *On two complete sets in the analytical and the arithmetical hierarchies* ⋄ C55 C80 D55 ⋄

VOLGER, H. *Logical categories, semantical categories and topoi* ⋄ C20 C85 G30 ⋄

WEAVER, G.E. *Tarski-Vaught and Loewenheim-Skolem numbers* ⋄ C55 C95 ⋄

WOLTER, H. *Entscheidbarkeit der Arithmetik mit Addition und Ordnung in Logiken mit verallgemeinerten Quantoren* ⋄ B25 C10 C55 C80 F30 ⋄

WOLTER, H. *Untersuchungen zu Logiken mit Maechtigkeitsquantoren (Russian, English and French summaries)* ⋄ C55 C80 ⋄

1976

BARWISE, J. *Some applications of Henkin quantifiers*
⋄ C40 C50 C60 C75 C80 ⋄

BAUDISCH, A. *Elimination of the quantifier Q_α in the theory of abelian groups (Russian summary)*
⋄ B25 C10 C60 C80 ⋄

CHEPURNOV, B.A. & KOGALOVSKIJ, S.R. *Some criteria of heredity and locality for formulas of higher order (Russian)* ⋄ B15 C52 C85 ⋄

CSAKANY, B. *Conditions involving universally quantified function variables* ⋄ C05 C40 C85 ⋄

EBBINGHAUS, H.-D. *General logic systems* ⋄ C95 ⋄

FERRO, R. *Consistency property and model existence theorem for second order negative languages with conjunctions and quantifications over sets of cardinality smaller than a strong limit cardinal of denumerable cofinality* ⋄ C55 C75 C85 ⋄

GIRARD, J.-Y. *Three-valued logic and cut-elimination: The actual meaning of Takeuti's conjecture*
⋄ B50 C85 C90 F05 F35 F50 ⋄

GOSTANIAN, R. & HRBACEK, K. *On the failure of the weak Beth property* ⋄ C40 C75 C80 ⋄

HAJEK, P. *Observationsfunktorenkalkuele und die Logik der automatisierten Forschung*
⋄ B60 C13 C80 C90 ⋄

HAJEK, P. *Some remarks on observational model-theoretic languages* ⋄ C13 C95 ⋄

HINTIKKA, K.J.J. *Partially ordered quantifiers vs. partially ordered ideas* ⋄ A05 C80 ⋄

HINTIKKA, K.J.J. *Quantifiers in logic and quantifiers in natural languages* ⋄ A05 B65 C80 ⋄

HUTCHINSON, J.E. *Model theory via set theory*
⋄ C40 C55 C62 C70 C75 C80 C95 E75 ⋄

JENSEN, F.V. *Souslin-Kleene does not imply Beth*
⋄ C40 C95 ⋄

KAISER, KLAUS *Quasi-axiomatic classes*
⋄ B15 C20 C52 C75 C85 ⋄

KAUFMANN, M. *Some results in stationary logic. Doctoral Dissertation, University of Wisconsin, Madison*
⋄ C55 C80 ⋄

KRAJEWSKI, S. *Non-standard satisfaction classes*
⋄ C62 C85 H15 H20 ⋄

KRAWCZYK, A. & KRYNICKI, M. *Ehrenfeucht games for generalized quantifiers* ⋄ C80 ⋄

LITMAN, A. *On the monadic theory of ω_1 without AC*
⋄ B25 C85 D05 E10 E25 ⋄

MAKOWSKY, J.A. & SHELAH, S. & STAVI, J. *Δ-logics and generalized quantifiers* ⋄ C40 C55 C80 C95 ⋄

MAKOWSKY, J.A. & MARCJA, A. *Problemi di decidibilita in logica topologica (English summary)*
⋄ B25 C65 C80 ⋄

MCKEE, T.A. *Sentences preserved between equivalent topological bases* ⋄ C65 C80 C90 ⋄

OKAZAKI, H. *An application of Fuhrken's reduction technique* ⋄ C55 C80 ⋄

PAULOS, J.A. *Noncharacterizability of the syntax set*
⋄ C40 C95 ⋄

PINUS, A.G. *Theories of boolean algebras in a calculus with the quantifier "infinitely many exist" (Russian)*
⋄ B25 C10 C35 C57 C80 G05 ⋄

SGRO, J. *Completeness theorems for continuous functions and product topologies*
⋄ C40 C65 C80 C90 E75 ⋄

SHANIN, N.A. *On the quantifier of limiting realizability (Russian)* ⋄ C57 C80 F50 ⋄

SLOMSON, A. *Decision problems for generalized quantifiers - a survey* ⋄ B25 C80 C98 ⋄

STENIUS, E. *Comments on Jaakko Hintikka's paper "Quantifiers vs. quantification theory"*
⋄ A05 B65 C80 ⋄

WALKOE JR., W.J. *A small step backwards*
⋄ B10 C80 ⋄

WEESE, M. *Entscheidbarkeit in speziellen uniformen Strukturen bezueglich Sprachen mit Maechtigkeitsquantoren* ⋄ B25 C55 C65 C80 ⋄

WEESE, M. *The universality of boolean algebras with the Haertig quantifier* ⋄ C55 C80 D35 F25 G05 ⋄

WESTERSTAAHL, D. *Some philosophical aspects of abstract model theory* ⋄ A05 C95 ⋄

ZIEGLER, M. *A language for topological structures which satisfies a Lindstroem-theorem*
⋄ B25 C40 C50 C65 C80 C90 C95 D35 ⋄

1977

BADGER, L.W. *An Ehrenfeucht game for the multivariable quantifiers of Malitz and some applications*
⋄ C40 C80 E60 ⋄

BARWISE, J. *Some eastern two-cardinal theorems*
⋄ C55 C80 ⋄

BAUDISCH, A. *Decidability of the theory of abelian groups with Ramsey quantifiers* ⋄ B25 C10 C60 C80 ⋄

BAUDISCH, A. *The theory of abelian groups with the quantifier $(\leq x)$* ⋄ B25 C10 C60 C80 ⋄

BAUDISCH, A. & WEESE, M. *The Lindenbaum-algebra of the theory of well-orders and abelian groups with the quantifier Q_α* ⋄ C60 C80 ⋄

BUECHI, J.R. *Using determinancy of games to eliminate quantifiers* ⋄ B15 C10 C85 E60 ⋄

BUTRICK, R. *A deduction rule for $VBTO()_{i=1}^n$* ⋄ C80 ⋄

CAICEDO, X. *A back-and-forth characterization of elementary equivalence in stationary logic* ⋄ C80 ⋄

COWLES, J.R. *The theory of differentially closed fields in logics with cardinal quantifiers* ⋄ C60 C80 ⋄

DANIELS, C.B. & FREEMAN, J.B. *Classical second-order intensional logic with maximal propositions*
⋄ B15 C85 C90 ⋄

DUBIEL, M. *Generalized quantifiers and elementary extensions of countable models* ⋄ C15 C65 C80 ⋄

EKLOF, P.C. *Applications of logic to the problem of splitting abelian groups*
⋄ C55 C60 C75 C80 E35 E45 E75 ⋄

FRAISSE, R. *Deux relations denombrables, logiquement equivalentes pour le second ordre, sont isomorphes*
⋄ B15 C15 C85 E07 E45 ⋄

GILLAM, D.W.H. *Morley numbers for generalized languages* ⋄ C55 C75 C80 ⋄

GUREVICH, Y. *Expanded theory of ordered abelian groups*
⋄ B25 C10 C60 C85 ⋄

GUREVICH, Y. *Monadic theory of order and topology I*
⋄ B15 C65 C85 E07 E50 E75 ⋄

HAJEK, P. *Generalized quantifiers and finite sets*
◇ B25 B50 C13 C40 C80 ◇

HERRE, H. *Decidability of theories in logics with additional monadic quantifiers* ◇ B25 C55 C80 ◇

HERRE, H. & WOLTER, H. *Entscheidbarkeit der Theorie der linearen Ordnung in L_{Q_1}*
◇ B25 C55 C65 C80 ◇

HINTIKKA, K.J.J. *Quantifiers in natural languages: Some logical problems. II* ◇ A05 C80 ◇

KEISLER, H.J. *Hyperfinite model theory*
◇ B50 C65 C75 C80 C90 H20 ◇

KRAWCZYK, A. & MAREK, W. *On the rules of proof generated by hierarchies* ◇ B15 C55 C85 E45 ◇

KRYNICKI, M. *Henkin quantifier and decidability*
◇ B25 C10 C80 D35 ◇

KRYNICKI, M. *Some consequences of the omitting and realizing types property* ◇ C75 C95 ◇

LADNER, R.E. *Application of model theoretic games to discrete linear orders and finite automata*
◇ B15 B25 C07 C13 C65 C85 D05 E07 E60 ◇

MAGIDOR, M. & MALITZ, J. *Compact extensions of $L(Q)$ (part Ia)* ◇ C55 C80 E55 E65 ◇

MAGIDOR, M. & MALITZ, J. *Compactness and transfer for a fragment of L^2* ◇ C55 C80 E05 E65 ◇

MAKOWSKY, J.A. & TULIPANI, S. *Some model theory for monotone quantifiers* ◇ C20 C40 C80 ◇

MIETTINEN, S. *Some remarks on definability* ◇ C95 ◇

MIZUTANI, C. *Monadic second order logic with an added quantifier Q* ◇ B15 B25 C40 C55 C80 ◇

MOLDESTAD, J. *On monotone second order relations*
◇ C40 C85 ◇

ONYSZKIEWICZ, J. *Omitting types in some extended system of logic* ◇ C75 C95 ◇

PARK, D. *Finiteness is μ-ineffable*
◇ B75 C75 C80 C85 ◇

PUDLAK, P. *Generalized quantifiers and semisets*
◇ B10 C80 E70 ◇

QVARNSTROEM, B.O. *On the concept of formalization and partially ordered quantifiers* ◇ A05 C80 ◇

RESSAYRE, J.-P. *Models with compactness properties relative to an admissible language*
◇ C50 C55 C70 C80 D55 D60 E15 ◇

SCHIEMANN, I. *Eine Axiomatisierung des monadischen Praedikatenkalkuels mit verallgemeinerten Quantoren (Russian, English and French summaries)*
◇ B25 C40 C55 C80 ◇

SEESE, D.G. & TUSCHIK, H.-P. *Construction of nice trees*
◇ B25 C55 C65 C80 ◇

SEESE, D.G. *Decidability of ω-trees with bounded sets – a survey* ◇ B25 C65 C80 C85 ◇

SEESE, D.G. *Second order logic, generalized quantifiers and decidability*
◇ B15 B25 C55 C65 C80 C85 D35 ◇

SGRO, J. *Completeness theorems for topological models*
◇ C20 C40 C75 C80 C90 ◇

SGRO, J. *Maximal logics*
◇ C20 C40 C80 C90 C95 ◇

SHANIN, N.A. *On the quantifier of limiting realizability*
◇ B55 C57 C80 F50 ◇

SHREJDER, YU.A. & VILENKIN, N.YA. *Majorant spaces and "majority" quantifier (Russian) (English summary)*
◇ B45 C80 E20 ◇

SOBOLEV, S.K. *The intuitionistic propositional calculus with quantifiers (Russian)* ◇ B55 C80 C90 D35 F50 ◇

SREBRNY, M. *Second-order theories of countable transitive models* ◇ C62 C85 ◇

TUSCHIK, H.-P. *Elimination verallgemeinerter Quantoren in ω_1-kategorischen Theorien (Russian, English and French summaries)* ◇ B25 C10 C35 C55 C80 ◇

TUSCHIK, H.-P. *On the decidability of the theory of linear orderings in the language $L(Q_1)$*
◇ B25 C55 C65 C80 ◇

VAEAENAENEN, J. *On the compactness theorem*
◇ C20 C40 C95 ◇

VAEAENAENEN, J. *Remarks on generalized quantifiers and second order logics* ◇ B15 C40 C80 C85 C95 ◇

VASIL'EV, EH.S. *Elementary theories of certain classes of two-base models of abelian groups (Russian)*
◇ B25 C60 C85 ◇

WEESE, M. *Definierbare Praedikate in booleschen Algebren. I* ◇ C40 C80 G05 ◇

WEESE, M. *Entscheidbarkeit der Theorie der booleschen Algebren in Sprachen mit Maechtigkeitsquantoren (Russian and English summaries)*
◇ B25 C55 C80 G05 ◇

WEESE, M. *The decidability of the theory of boolean algebras with cardinality quantifiers (Russian summary)* ◇ B25 C10 C55 C80 G05 ◇

WEESE, M. *The decidability of the theory of boolean algebras with the quantifier "there exist infinitely many"*
◇ B25 C80 ◇

WEESE, M. *The undecidability of well-ordering with the Haertig quantifier (Russian summary)*
◇ C55 C65 C80 D35 E07 ◇

WOLTER, H. *Entscheidbarkeit der Theorie der Wohlordnung mit Elimination der Quantoren (Russian, English and French summaries)*
◇ B25 C10 C65 C80 E07 ◇

1978

BARWISE, J. *Monotone quantifiers and admissible sets*
◇ C70 C80 D60 D70 ◇

BARWISE, J. & KAUFMANN, M. & MAKKAI, M. *Stationary logic* ◇ C55 C75 C80 ◇

BEN-DAVID, S. *On Shelah's compactness of cardinals*
◇ C55 C80 E05 E35 E50 E55 ◇

BRUCE, KIM B. *Ideal models and some not so ideal problems in the model theory of $L(Q)$* ◇ C40 C80 ◇

BRUCE, KIM B. *Model-theoretic forcing in logic with a generalized quantifier* ◇ C25 C70 C75 C80 ◇

CUTLAND, N.J. *Σ_1-compactness in languages stronger than $\mathscr{L}_A$* ◇ C70 C75 C80 ◇

DANIELS, C.B. & FREEMAN, J.B. *A logic of generalized quantification* ◇ B15 C80 C90 ◇

EBBINGHAUS, H.-D. & FLUM, J. & THOMAS, WOLFGANG *Einfuehrung in die mathematische Logik* ◇ B98 C95 C98 ◇

FENSTAD, J.E. & GANDY, R.O. & SACKS, G.E. (EDS.) *Generalized recursion theory II*
◇ C62 C70 C80 D97 ◇

FERRO, R. *Interpolation theorem for $L_{k,k}^{2+}$*
  ⋄ B15 C40 C75 C85 ⋄
GARAVAGLIA, S. *Model theory of topological structures*
  ⋄ C20 C40 C65 C80 C90 ⋄
GARAVAGLIA, S. *Relative strength of Malitz quantifiers*
  ⋄ C55 C80 ⋄
GRANT, P.W. *The completeness of $L_{\omega_1\omega}(P)$*
  ⋄ C75 C80 ⋄
HAJEK, P. & HAVRANEK, T. *Mechanizing hypothesis formation. Mathematical foundations for a general theory* ⋄ A05 B35 C13 C80 C90 D80 ⋄
HAVRANEK, T. & VOSAHLO, J. *A GUHA procedure with correlational quantifiers* ⋄ B35 C80 ⋄
HERRE, H. & WOLTER, H. *Entscheidbarkeit der Theorie der linearen Ordnung in L_{Q_κ} fuer regulaeres ω_κ*
  ⋄ B25 C55 C65 C80 E07 E50 ⋄
HERRE, H. & PINUS, A.G. *Zum Entscheidungsproblem fuer Theorien in Logiken mit monadischen verallgemeinerten Quantoren*
  ⋄ B25 C10 C55 C80 D35 ⋄
HOOVER, D.N. *Probability logic* ⋄ B48 C80 C90 ⋄
KOKORIN, A.I. & PINUS, A.G. *Decidability problems of extended theories (Russian)*
  ⋄ B25 C85 C98 D35 ⋄
LINDSTROEM, P. *Omitting uncountable types and extensions of elementary logic* ⋄ C75 C80 C95 ⋄
MAKOWSKY, J.A. *Quantifying over countable sets: Positive vs. stationary logic* ⋄ C55 C80 ⋄
MAKOWSKY, J.A. *Some observations on uniform reduction for properties invariant on the range of definable relations* ⋄ C30 C40 C75 C95 ⋄
MEKLER, A.H. *The size of epimorphic extensions*
  ⋄ C55 C75 C95 E55 G30 ⋄
NEMETI, I. *From hereditary classes to varieties in abstract model theory and partial algebra*
  ⋄ C05 C20 C52 C90 C95 ⋄
OIKKONEN, J. *Second-order definability, game quantifiers and related expressions*
  ⋄ C40 C75 C80 C85 E60 ⋄
PINUS, A.G. *Cardinality of models for theories in a calculus with a Haertig quantifier (Russian)* ⋄ C55 C80 ⋄
PINUS, A.G. *The Loewenheim-Skolem property for second-order logics (Russian)*
  ⋄ C55 C80 C85 E45 ⋄
PRESTEL, A. & ZIEGLER, M. *Model theoretic methods in the theory of topological fields* ⋄ C60 C65 C80 ⋄
SEESE, D.G. *A remark to the undecidability of well-orderings with the Haertig quantifier*
  ⋄ C55 C65 C80 D35 ⋄
SEESE, D.G. *Decidability of ω-trees with bounded sets (German and Russian summaries)*
  ⋄ B25 C65 C85 D05 ⋄
SEESE, D.G. *Decidability and generalized quantifiers*
  ⋄ B25 C80 C98 ⋄
SHELAH, S. *Models with second order properties. II: Trees with no undefined branches*
  ⋄ C50 C55 C62 C75 C80 C95 E05 E35 E65 ⋄

SIEFKES, D. *An axiom system for the weak monadic second order theory of two successors*
  ⋄ B15 B25 C85 D05 ⋄
STAVI, J. *Compactness properties of infinitary and abstract languages. I. General results* ⋄ C70 C75 C95 ⋄
TRIPPEL, J.R. *Die Algebra der Saetze der monadischen Theorie schwach zweiter Stufe der linearen Ordnung ist atomar* ⋄ C65 C85 ⋄
VAEAENAENEN, J. *Two axioms of set theory with applications to logic*
  ⋄ C40 C55 C80 C85 E35 E47 E65 ⋄
WEESE, M. *Definierbare Praedikate in booleschen Algebren. II* ⋄ B25 C10 C40 C80 G05 ⋄
ZIEGLER, M. *Definable bases of monotone systems*
  ⋄ C40 C80 ⋄

1979

ANDREKA, H. & NEMETI, I. & SAIN, I. *Completeness problems in verification of programs and program schemes* ⋄ B75 C90 C95 ⋄
ASSER, G. *Ueber die Charakterisierbarkeit transfiniter Maechtigkeiten im Praedikatenkalkuel der zweiten Stufe* ⋄ B15 C55 C80 C85 ⋄
BAUDISCH, A. *Uncountable n-cubes in models of $\aleph_0$-categorical theories* ⋄ C35 C80 ⋄
BELYAEV, V.YA. & TAJTSLIN, M.A. *On elementary properties of existentially closed systems (Russian)*
  ⋄ C25 C40 C60 C75 C85 D55 ⋄
BOUDREAUX, J.C. *Defining general structures*
  ⋄ B15 C85 ⋄
BOURGAIN, J. & ROSENTHAL, H.P. & SCHECHTMAN, G. *Uniform validity and stationary logic* ⋄ C55 C80 ⋄
BRUCE, KIM B. & KEISLER, H.J. $L_A(\exists)$
  ⋄ C25 C70 C75 C80 ⋄
COWLES, J.R. *The relative expressive power of some logics extending first-order logic*
  ⋄ C10 C75 C80 C85 E15 ⋄
COWLES, J.R. *The theory of Archimedean real closed fields in logics with Ramsey quantifiers*
  ⋄ C10 C35 C60 C80 ⋄
EKLOF, P.C. & MEKLER, A.H. *Stationary logic of finitely determinate structures*
  ⋄ B25 C10 C55 C60 C65 C80 ⋄
GABBAY, D.M. & KASHER, A. *On the quantifier "there is a certain X"* ⋄ C80 ⋄
GAO, HENGSHAN & MO, SHAOKUI *The decidability of the $\aleph_0$-validity of formulas in the monadic predicate calculus with enumerative quantifiers and its applications (Chinese)* ⋄ B25 C80 ⋄
GOMBOCZ, W.L. *Lesniewski and Mally*
  ⋄ A05 A10 C80 ⋄
GUREVICH, Y. *Modest theory of short chains I*
  ⋄ B25 C65 C85 D35 E07 ⋄
GUREVICH, Y. & SHELAH, S. *Modest theory of short chains II* ⋄ B25 C65 C85 D35 E07 E50 ⋄
GUREVICH, Y. *Monadic theory of order and topology II*
  ⋄ B15 C65 C85 E07 E45 E50 ⋄
HAREL, D. *Characterizing second order logic with first order quantifiers* ⋄ B15 C80 C85 ⋄

HARNIK, V. & MAKKAI, M. *New axiomatizations for logics with generalized quantifiers* ⋄ C70 C80 ⋄

HERRE, H. & WOLTER, H. *Entscheidbarkeit der Theorie der linearen Ordnung in Logiken mit Maechtigkeitsquantoren bzw. mit Chang-Quantor*
⋄ B25 C55 C80 E07 E50 E55 ⋄

HERRE, H. & WOLTER, H. *The decision problem for the theory of linear orderings in extended logics*
⋄ B25 C55 C80 C98 ⋄

HINTIKKA, K.J.J. *Quantifiers in natural languages: Some logical problems. I* ⋄ A05 B65 C80 ⋄

KAUFMANN, M. *A new omitting types theorem for L(Q)*
⋄ C55 C75 C80 ⋄

KISELEV, A.A. & KOGALOVSKIJ, S.R. *Some remarks on higher order monadic languages (Russian)*
⋄ B15 C85 ⋄

KRYNICKI, M. *On the expressive power of the language using the Henkin quantifier* ⋄ B15 B25 C80 C85 ⋄

KRYNICKI, M. & LACHLAN, A.H. *On the semantics of the Henkin quantifier*
⋄ B15 B25 C40 C55 C80 C85 ⋄

MAKOWSKY, J.A. *The reals cannot be characterized topologically with strictly local properties and countability axioms* ⋄ C65 C80 C90 ⋄

MAKOWSKY, J.A. & SHELAH, S. *The theorems of Beth and Craig in abstract model theory I: The abstract setting*
⋄ C40 C75 C80 C95 ⋄

MORGENSTERN, C.F. *On amalgamations of languages with Magidor-Malitz quantifiers* ⋄ C55 C80 E55 ⋄

MORGENSTERN, C.F. *The measure quantifier*
⋄ C55 C75 C80 E55 ⋄

MUNDICI, D. *Compactness + Craig interpolation = Robinson consistency in any logic* ⋄ C40 C95 ⋄

MUNDICI, D. *Compattezza = JEP in ogni logica*
⋄ C40 C52 C95 ⋄

MUNDICI, D. *Robinson consistency theorem in soft model theory (Italian summary)* ⋄ C40 C55 C80 C95 ⋄

OIKKONEN, J. *A generalization of the infinitely deep languages of Hintikka and Rantala*
⋄ C55 C75 C80 ⋄

OIKKONEN, J. *Hierarchies of model-theoretic definability - an approach to second-order logics*
⋄ B15 C40 C70 C75 C80 C85 ⋄

OIKKONEN, J. *On PC- and RPC-classes in generalized model theory* ⋄ C13 C52 C75 C80 C95 ⋄

PETERSON, P.L. *On the logic of "few", "many", and "most"*
⋄ B60 C80 ⋄

PINUS, A.G. *A calculus of one-place predicates (Russian)*
⋄ B20 B25 C55 C80 C85 ⋄

PINUS, A.G. *Elementary equivalence of topological spaces (Russian)* ⋄ C55 C65 C85 G05 G10 ⋄

PINUS, A.G. *Elimination of the quantifiers Q_0 and Q_1 on symmetric groups (Russian)*
⋄ C10 C55 C60 C80 ⋄

PINUS, A.G. *Hanf number for the calculus with the Haertig quantifier (Russian)* ⋄ C55 C80 C85 E45 ⋄

ROWLANDS-HUGHES, D. *Back and forth methods in abstract model theory* ⋄ C40 C80 C95 ⋄

RUBIN, M. *On the automorphism groups of homogeneous and saturated boolean algebras*
⋄ C07 C50 C85 G05 ⋄

SAIN, I. *There are general rules for specifying semantics: Observations of abstract model theory*
⋄ B75 C90 C95 E75 ⋄

SCHWABHAEUSER, W. *Non-finitizability of a weak second-order theory* ⋄ B15 C62 C65 C85 F35 ⋄

SEESE, D.G. *Some graph-theoretical operations and decidability* ⋄ B15 B25 C65 C85 ⋄

SEESE, D.G. *Stationary logic and ordinals*
⋄ B25 C55 C65 C80 ⋄

TAJTSLIN, M.A. *A description of algebraic systems in weak logic of order ω and in program logic (Russian)*
⋄ B15 B75 C35 C85 ⋄

VAEAENAENEN, J. *Abstract logic and set theory. I. Definability* ⋄ C40 C55 C95 E75 ⋄

VAEAENAENEN, J. *On Hanf numbers of unbounded logics*
⋄ C55 C75 C80 C85 C95 E55 ⋄

VAEAENAENEN, J. *Remarks on free quantifier variables*
⋄ C55 C80 C85 C95 E45 ⋄

1980

BADGER, L.W. *Beth's property fails in $L^{<\omega}$*
⋄ C40 C80 ⋄

BALDWIN, JOHN T. & KUEKER, D.W. *Ramsey quantifiers and the finite cover property*
⋄ C10 C35 C45 C80 ⋄

BAUDISCH, A. & SEESE, D.G. & TUSCHIK, H.-P. & WEESE, M. *Decidability and generalized quantifiers*
⋄ B25 B98 C10 C60 C80 C98 ⋄

BAUDISCH, A. *The theory of abelian p-groups with the quantifier I is decidable* ⋄ B25 C10 C80 ⋄

BOUDREAUX, J.C. *Frames versus minimally restricted structures* ⋄ B15 C85 ⋄

BRUCE, KIM B. *Model constructions in stationary logic I. Forcing* ⋄ C25 C55 C70 C75 C80 ⋄

BURGESS, J.P. *Decidability for branching time*
⋄ B25 B45 C85 ⋄

CAICEDO, X. *Back-and-forth systems for arbitrary quantifiers* ⋄ C55 C75 C80 ⋄

CEGIELSKI, P. *La theorie elementaire de la multiplication (English summary)* ⋄ B25 C10 C80 F30 ⋄

CHLEBUS, B.S. *Decidability and definability results concerning well-orderings and some extensions of first order logic*
⋄ B15 B25 C40 C65 C80 D35 E07 ⋄

COCCHIARELLA, N.B. *Nominalism and conceptualism as predicative second-order theories of predication*
⋄ A05 B15 C85 ⋄

CORCORAN, J. *Categoricity* ⋄ A05 B15 C35 C85 ⋄

COSTA DA, N.C.A. *A model-theoretical approach to variable binding term operators* ⋄ C07 C80 ⋄

CUTLAND, N.J. & KAUFMANN, M. *Σ_1-well-founded compactness* ⋄ C70 C80 ⋄

DAHN, B.I. *First order logics for metric structures*
⋄ C65 C80 C90 ⋄

DUBIEL, M. *Generalized quantifiers in models of set theory*
⋄ C62 C80 E35 ⋄

FAJARDO, S. *Compacidad y decidibilidad de logicas monadicas con cuantificadores cardinales*
⋄ B25 C55 C80 ⋄

FLUM, J. & ZIEGLER, M. *Topological model theory*
  ⋄ C65 C80 C90 C98 ⋄
GEORGESCU, G. *Generalized quantifiers on polyadic algebras* ⋄ C80 G15 ⋄
GOLTZ, H.-J. *Untersuchungen zur Elimination verallgemeinerter Quantoren in Koerpertheorien*
  ⋄ C10 C55 C60 C80 ⋄
GUICHARD, D.R. *A many-sorted interpolation theorem for $L(Q)$* ⋄ C40 C75 C80 ⋄
GUREVICH, Y. *Ordered abelian groups*
  ⋄ B25 C10 C60 C85 ⋄
GUREVICH, Y. *Two notes on formalized topology*
  ⋄ C65 C85 F25 ⋄
HARRINGTON, L.A. *Extensions of countable infinitary logic which preserve most of its nice properties*
  ⋄ C40 C55 C70 C75 C95 ⋄
HAUSCHILD, K. *Generalized Haertig quantifiers*
  ⋄ C55 C80 ⋄
HEINDORF, L. *The decidability of the L_t - theory of boolean spaces* ⋄ B25 C65 C80 C90 ⋄
HERRE, H. & SEESE, D.G. *Concerning the monadic theory of the topology of well-orderings and scattered spaces*
  ⋄ B25 C55 C65 C85 E07 E35 E50 E55 ⋄
HERRMANN, E. & WOLTER, H. *Untersuchungen zu schwachen Logiken der zweiten Stufe*
  ⋄ B15 C15 C85 ⋄
JANKOWSKI, A.W. *Embeddings of logical structures. I,II*
  ⋄ C95 ⋄
KAKUDA, Y. *Set theory based on the language with the additional quantifier "for almost all" I*
  ⋄ C55 C62 C80 E70 ⋄
LOPEZ-ESCOBAR, E.G.K. *Semantical models for intuitionistic logics* ⋄ C90 C95 F50 ⋄
MACINTYRE, A. *Ramsey quantifiers in arithmetic*
  ⋄ C62 C80 F30 F35 ⋄
MAKOWSKY, J.A. *Measuring the expressive power of dynamic logics: an application of abstract model theory* ⋄ B75 C95 ⋄
MALITZ, J. & RUBIN, M. *Compact fragments of higher order logic* ⋄ B15 C55 C80 C85 ⋄
MANDERS, K.L. *First-order logical systems and set-theoretical definability* ⋄ C95 E45 ⋄
MONTAGNA, F. *Some modal logics with quantifiers (Italian summary)* ⋄ B45 C80 C90 ⋄
MOSTOWSKI, A.WLODZIMIERZ *Finite automata on infinite trees and subtheories of SkS* ⋄ B25 C85 D05 ⋄
NADEL, M.E. *An arbitrary equivalence relation as elementary equivalence in an abstract logic*
  ⋄ C75 C80 C95 ⋄
PRESTEL, A. & ZIEGLER, M. *Non-axiomatizable classes of V-topological fields* ⋄ C60 C65 C80 ⋄
RUBIN, M. *On the automorphism groups of countable boolean algebras* ⋄ C07 C15 C85 E45 G05 ⋄
RUBIN, M. & SHELAH, S. *On the elementary equivalence of automorphism groups of boolean algebras; downward Skolem Loewenheim theorems and compactness of related quantifiers*
  ⋄ C07 C30 C75 C80 C95 G05 ⋄
SCHWARTZ, DIETRICH *Cylindric algebras with filter quantifiers* ⋄ C80 G15 ⋄

SGRO, J. *Interpolation fails for the Souslin-Kleene closure of the open set quantifier logic*
  ⋄ C40 C80 C90 C95 ⋄
SGRO, J. *The interior operator logic and product topologies*
  ⋄ C40 C55 C80 C90 ⋄
SGRO, J. *Ultraproduct invariant logics*
  ⋄ C20 C40 C80 C95 ⋄
THOMAS, WOLFGANG *On the bounded monadic theory of well-ordered structures* ⋄ B15 B25 C85 E07 ⋄
TUSCHIK, H.-P. *On the decidability of the theory of linear orderings with generalized quantifiers*
  ⋄ B25 C55 C65 C80 ⋄
VAEAENAENEN, J. *A quantifier for isomorphisms*
  ⋄ C40 C80 C85 C95 ⋄
VAEAENAENEN, J. *Boolean-valued models and generalized quantifiers* ⋄ C55 C80 C85 C90 E35 E40 ⋄
WEESE, M. *Generalized Ehrenfeucht games* ⋄ C80 ⋄

1981

ANAPOLITANOS, D.A. & VAEAENAENEN, J. *Decidability of some logics with free quantifier variables*
  ⋄ B25 C80 ⋄
BARWISE, J. & COOPER, R. *Generalized quantifiers and natural language* ⋄ A05 B60 C80 ⋄
BARWISE, J. *The role of the omitting types theorem in infinitary logic* ⋄ C75 C80 C95 ⋄
BAUDISCH, A. *Formulas of $L(aa)$ where aa is not in the scope of "¬"* ⋄ C55 C80 ⋄
BAUDISCH, A. *The elementary theory of abelian groups with m-chains of pure subgroups*
  ⋄ B25 C55 C60 C80 ⋄
BAUDISCH, A. *There is no module having the finite cover property* ⋄ C10 C45 C60 C80 ⋄
BLATTER, C. & SPECKER, E. *Le nombre de structures finies d'une theorie a caractere fini* ⋄ C13 C52 C85 ⋄
CAICEDO, X. *On extensions of $L_{\omega\omega}(Q_1)$*
  ⋄ C55 C75 C80 ⋄
CEGIELSKI, P. *Theorie elementaire de la multiplication des entiers naturels* ⋄ B25 C10 C80 F30 ⋄
COWLES, J.R. *Generalized archimedean fields and logics with Malitz quantifiers* ⋄ C10 C60 C80 ⋄
COWLES, J.R. *The Henkin quantifier and real closed fields*
  ⋄ C60 C80 ⋄
EKLOF, P.C. & MEKLER, A.H. *Infinitary stationary logic and abelian groups* ⋄ C60 C75 C80 ⋄
GOLDBLATT, R.I. *"Locally-at" as a topological quantifier-former* ⋄ C65 C80 ⋄
GUZICKI, W. *The equivalence of definable quantifiers in second order arithmetic*
  ⋄ C30 C55 C62 C80 E40 E45 F35 ⋄
HAJEK, P. & KURKA, P. *A second-order dynamic logic with array assignments* ⋄ B75 C85 ⋄
HAUSCHILD, K. *Zum Vergleich von Haertigquantor und Rescherquantor* ⋄ B25 C10 C55 C80 D35 ⋄
HEINDORF, L. *Comparing the expressive power of some languages for boolean algebras*
  ⋄ B25 C10 C80 C85 C90 G05 ⋄
HERRE, H. *Miscellaneous results and problems in extended model theory* ⋄ B25 C65 C80 C98 D35 ⋄
HERRE, H. & WOLTER, H. *Untersuchungen zur Theorie der linearen Ordnung in Logiken mit*

Maechtigkeitsquantoren
◇ B25 C55 C65 C80 E07 ◇

KAKUDA, Y. *Completeness theorem for the system ST with the quantifier "aa"* ◇ C55 C80 E70 ◇

KAUFMANN, M. *Filter logics: Filters on ω_1*
◇ C55 C80 E05 ◇

MAKOWSKY, J.A. & SHELAH, S. *The theorems of Beth and Craig in abstract model theory II: Compact logics*
◇ C40 C75 C80 C95 ◇

MAKOWSKY, J.A. & ZIEGLER, M. *Topological model theory with an interior operator: consistency properties and back-and-forth arguments* ◇ C40 C75 C80 ◇

MOLZAN, B. *On the number of different theories of boolean algebras in several logics* ◇ C80 G05 ◇

MUNDICI, D. *An algebraic result about soft model theoretical equivalence relations with an application to H. Friedman's fourth problem* ◇ C15 C40 C95 ◇

MUNDICI, D. *Applications of many-sorted Robinson consistency theorem* ◇ C40 C80 C95 ◇

MUNDICI, D. *Robinson's consistency theorem in soft model theory* ◇ C40 C80 C95 ◇

MUNDICI, D. *Variations on Friedman's third and fourth problem* ◇ C40 C75 C80 C95 ◇

ROTHMALER, P. *Q_0 is eliminable in every complete theory of modules* ◇ C10 C60 C80 ◇

SEESE, D.G. *Elimination of second-order quantifiers for well-founded trees in stationary logic and finitely determinate structures* ◇ C10 C55 C65 C80 ◇

SEESE, D.G. *Stationary logic and ordinals*
◇ B25 C55 C65 C80 ◇

SHELAH, S. *Models with second order properties. III. Omitting types for L(Q)* ◇ C55 C75 C80 E05 ◇

TURAN, G. *On cellular graph-automata and second-order definable graph-properties* ◇ C85 D05 D80 ◇

VAEAENAENEN, J. *Generalized quantfiers in models of set theory* ◇ C55 C62 C80 C85 C95 E35 ◇

WEESE, M. *Decidability with respect to Haertig quantifier and Rescher quantifier* ◇ B25 C55 C80 D35 ◇

1982

BALDWIN, JOHN T. & MILLER, DOUGLAS E. *Some contributions to definability theory for languages with generalized quantifiers* ◇ C40 C80 ◇

CALL, R.L. *Systems with parity quantifiers*
◇ C13 C80 ◇

CARLSON, L. *Plural quantifiers and informational independence* ◇ A05 C80 ◇

DAPUETO, C. *Interpretation in an elementary topos (Italian summary)* ◇ C85 G30 ◇

EBBINGHAUS, H.-D. & ZIEGLER, M. *Interpolation in Logiken monotoner Systeme* ◇ C40 C80 ◇

GEORGESCU, G. *A generalized quantifier for the hereditary σ-rings* ◇ C75 C80 ◇

GEORGESCU, G. *Algebraic analysis of the topological logic L(I)* ◇ C80 G15 ◇

GORETTI, A. *Alcune osservazioni sul linguaggio $L(Q_1)$ (English summary)* ◇ C50 C55 C80 ◇

GUREVICH, Y. & HARRINGTON, L.A. *Automata, trees, and games* ◇ B25 C85 D05 ◇

GUREVICH, Y. *Crumbly spaces* ◇ B25 C65 C85 ◇

GUREVICH, Y. & SHELAH, S. *Monadic theory of order and topology in ZFC* ◇ B25 C65 C85 D35 E07 ◇

HINTIKKA, K.J.J. *Game-theoretical semantics: Insights and prospects* ◇ A05 B60 C80 ◇

JANKOWSKI, A.W. *An alternative characterization of elementary logic (Russian summary)* ◇ C95 ◇

KRYNICKI, M. & VAEAENAENEN, J. *On orderings of the family of all logics* ◇ C95 ◇

LIPPARINI, P. *Some results about compact logics (Italian summary)* ◇ C95 ◇

LYONS, R. *Measure-theoretic quantifiers and Haar measure* ◇ C80 ◇

MANNILA, H. *Restricted compactness notions in abstract logic and topology* ◇ C95 ◇

MOLZAN, B. *The theory of superatomic boolean algebras in the logic with the binary Ramsey quantifier*
◇ B25 C10 C80 G05 ◇

MORGENSTERN, C.F. *On generalized quantifiers in arithmetic* ◇ C62 C80 F30 ◇

MUNDICI, D. *Compactness, interpolation and Friedman's third problem* ◇ C40 C75 C95 ◇

MUNDICI, D. *Duality between logics and equivalence relations* ◇ C40 C95 ◇

MUNDICI, D. *Interpolation, compactness and JEP in soft model theory* ◇ C40 C52 C95 ◇

MUNDICI, D. *L-embedding, amalgamation and L-elementary equivalence (Italian summary)* ◇ C95 ◇

MUNDICI, D. *Lectures on abstract model theory I,II,III*
◇ C95 C98 ◇

NEPEJVODA, N.N. *Constructive logics (Russian)*
◇ B60 C90 C95 F50 ◇

ROTHMALER, P. & TUSCHIK, H.-P. *A two cardinal theorem for homogeneous sets and the elimination of Malitz quantifiers* ◇ C10 C45 C80 ◇

SCHMERL, J.H. & SIMPSON, S.G. *On the role of Ramsey quantifiers in first order arithmetic*
◇ B28 C10 C62 C80 ◇

SEESE, D.G. & WEESE, M. *L(aa)-elementary types of well-orderings* ◇ B25 C55 C65 C80 E10 ◇

SEESE, D.G. & TUSCHIK, H.-P. & WEESE, M. *Undecidable theories in stationary logic*
◇ C55 C65 C80 D35 E05 E75 ◇

THOMPSON, B. *Syllogisms using "few", "many", and "most"* ◇ B60 C80 ◇

TUSCHIK, H.-P. *Elimination of cardinality quantifiers*
◇ C10 C45 C55 C60 C80 ◇

VAEAENAENEN, J. *Abstract logic and set theory. II. Large cardinals* ◇ C55 C95 E35 E55 ◇

VAEAENAENEN, J. *Generalized quantifiers in models of set theory* ◇ C55 C62 C80 C85 C95 E35 ◇

1983

ANDREKA, H. & NEMETI, I. *Generalization of the concept of variety and quasivariety to partial algebras through category theory*
◇ C05 C40 C75 C90 C95 G30 ◇

BALDO, A. *Complete interpolation theorems for L_{kk}^{2+} (Italian summary)* ◇ C40 C75 C85 ◇

BAUDISCH, A. & SEESE, D.G. & TUSCHIK, H.-P. ω-trees in stationary logic ◊ B25 C55 C65 C80 ◊

BUECHI, J.R. & ZAIONTZ, C. Deterministic automata and the monadic theory of ordinals $< \omega_2$
◊ B15 B25 C85 D05 E10 ◊

BUECHI, J.R. & SIEFKES, D. The complete extensions of the monadic second order theory of countable ordinals
◊ B15 C35 C85 D05 E10 ◊

CHERLIN, G.L. & SCHMITT, P.H. Locally pure topological abelian groups: Elementary invariants
◊ B25 C60 C65 C80 C90 ◊

COMBASE, J. On the existence of finitely determinate models for some theories in stationary logic
◊ C55 C80 E55 ◊

GOL'DSHTEJN, B.G. Decomposition of logics (Russian)
◊ C95 H05 ◊

GRABOWSKI, M. Some model-theoretical properties of logic for programs with random control ◊ B75 C85 ◊

GUREVICH, Y. & SHELAH, S. Interpreting second-order logic in the monadic theory of order
◊ B15 C65 C85 D35 E07 E50 F25 ◊

GUREVICH, Y. & SHELAH, S. Rabin's uniformization problem ◊ B15 C40 C65 C85 E40 ◊

GUREVICH, Y. & MAGIDOR, M. & SHELAH, S. The monadic theory of ω_2
◊ B15 B25 C65 C85 D35 E10 E35 ◊

GUZICKI, W. Definable quantifiers in second order arithmetic and elementary extensions of ω-models
◊ C30 C55 C62 C80 E40 E50 F35 ◊

HARIZANOV, V. & MIJAJLOVIC, Z. Regular relations and the quantifier "there exist uncountably many"
◊ C55 C80 ◊

IVANOV, A.A. Decidability of theories in a certain calculus (Russian) ◊ B25 C10 C55 C60 C80 D35 ◊

IVANOV, A.A. Some theories in generalized calculi (Russian) ◊ B25 C10 C55 C60 C80 ◊

KAKUDA, Y. Reflection principles via filter quantifier (Japanese) ◊ C55 C80 E30 ◊

KAUFMANN, M. Set theory with a filter quantifier
◊ C62 C80 E30 E70 ◊

KIERSTEAD, H.A. & REMMEL, J.B. Indiscernibles and decidable models ◊ C30 C45 C57 C80 ◊

KNYAZEV, V.V. Asymptotic solvability of a language with integral quantifier (Russian) ◊ B75 C80 ◊

MAKOWSKY, J.A. & SHELAH, S. Positive results in abstract model theory: a theory of compact logics
◊ C20 C52 C95 E05 E55 ◊

MALITZ, J. Downward transfer of satisfiability for sentences of $L^{1,1}$ ◊ C55 C80 ◊

MANNILA, H. A topological characterization of $(\lambda, \mu)^*$-compactness ◊ C95 E75 ◊

MOTOHASHI, N. Some remarks on Barwise approximation theorem on Henkin quantifiers ◊ C75 C80 ◊

MUNDICI, D. Compactness = JEP in any logic
◊ C40 C52 C95 ◊

MUNDICI, D. Inverse topological systems and compactness in abstract model theory ◊ C40 C80 C95 ◊

OIKKONEN, J. Logical operations and iterated infinitely deep languages ◊ C75 C95 ◊

PINUS, A.G. Calculus with the quantifier of elementary equivalence (Russian) ◊ C40 C55 C80 D35 ◊

POIZAT, B. Proprietes "modele-theoriques" d'une theorie du premier ordre ◊ C45 C52 C95 ◊

RUBIN, M. A Boolean algebra with few subalgebras, interval algebras and retractiveness
◊ C55 C80 D35 E50 E65 G05 ◊

RUBIN, M. & SHELAH, S. On the expressibility hierarchy of Magidor-Malitz quantifiers
◊ C55 C80 E40 E65 ◊

SCHMITT, P.H. The L^t-theory of profinite abelian groups
◊ B25 C50 C60 C80 C90 ◊

SEMENOV, A.L. Logical theories of one-place functions on the set of natural numbers (Russian)
◊ B25 C10 C62 C85 ◊

SHELAH, S. Classification theory of non-elementary classes, Part I: The number of uncountable models of $\psi \in L_{\omega_1, \omega}$ ◊ C45 C52 C55 C75 C80 E50 ◊

SHELAH, S. Models with second order properties IV. A general method and eliminating diamonds
◊ C50 C55 C60 C65 C80 E65 G05 ◊

SKVORTSOV, D.P. On some ways of construction logical languages with quantifiers over finite sequences (Russian) ◊ B15 C80 ◊

VAEAENAENEN, J. Δ-extension and Hanf-numbers
◊ C40 C55 C80 C95 ◊

ZAIONTZ, C. Axiomatization of the monadic theory of ordinals $< \omega_2$ ◊ B25 B28 C65 C85 ◊

ZAWADOWSKI, M. The Skolem-Loewenheim theorem in toposes ◊ C55 C80 E10 G30 ◊

1984

ANDREKA, H. & NEMETI, I. Importance of universal algebra for computer science
◊ C05 C20 C95 G15 ◊

BAUDISCH, A. Magidor-Malitz quantifiers in modules
◊ C10 C60 C80 ◊

BAUVAL, A. La theorie d'un anneau de polynomes
◊ C60 C85 ◊

BENTHEM VAN, J.F.A.K. Questions about quantifiers
◊ A05 B65 C80 ◊

BLASS, A.R. & GUREVICH, Y. & KOZEN, D. A 0-1 law for logic with a fixed-point operator
◊ B10 B75 C80 D70 ◊

BLASS, A.R. & GUREVICH, Y. Henkin quantifiers and complete problems ◊ C80 D15 ◊

BLATTER, C. & SPECKER, E. Recurrence relations for the number of labeled structures on a finite set
◊ C13 C52 C85 ◊

BROWN, M.A. Generalized quantifiers and the square of opposition ◊ B65 C80 ◊

CHERLIN, G.L. Undecidability of rational function fields in nonzero characteristic ◊ C60 C85 D35 ◊

COMBASE, J. Introduction a la logique stationnaire
◊ C55 C80 C98 ◊

COMPTON, K.J. On rich words ◊ C85 D05 D45 ◊

GEORGESCU, G. Algebraic analysis of the logic with the quantifier "there exist uncountably many"
◊ C80 G15 ◊

GUREVICH, Y. & SHELAH, S. *The monadic theory and the "next world"* ⋄ C62 C65 C85 E40 F25 ⋄

HEINDORF, L. *Beitraege zur Modelltheorie der Booleschen Algebren (English and Russian summaries)*
⋄ B25 C10 C35 C55 C80 G05 ⋄

HEINDORF, L. *Regular ideals and boolean pairs*
⋄ C30 C85 D35 G05 ⋄

HORT, C. & OSSWALD, H. *On nonstandard models in higher order logic*
⋄ B15 C55 C85 E55 H05 H20 ⋄

KANOVICH, M.I. *Restricted quantors in a consecutive semantical system (Russian)* ⋄ C80 ⋄

KAUFMANN, M. & SHELAH, S. *A nonconservativity result on global choice* ⋄ C80 E25 E35 E55 ⋄

KAUFMANN, M. *Filter logics on ω*
⋄ C40 C75 C80 C95 ⋄

KAUFMANN, M. *Some remarks on equivalence in infinitary and stationary logic* ⋄ C55 C65 C75 C80 E45 ⋄

KRYNICKI, M. & LACHLAN, A.H. & VAEAENAENEN, J. *Vector spaces and binary quantifiers*
⋄ C60 C75 C80 ⋄

LENSKI, W. *Elimination of quantifiers for the theory of archimedean ordered divisible groups in a logic with Ramsey quantifiers* ⋄ C10 C55 C60 C80 ⋄

LISOVIK, L.P. *Monadic second-order theories of two successor functions with an additional predicate (Russian)* ⋄ B15 B25 C10 C85 D05 ⋄

MAKOWSKY, J.A. *Model theoretic issues in theoretical computer science, part I: Relational data bases and abstract data types* ⋄ B75 C13 C95 ⋄

MARTINEZ, JUAN CARLOS *Accessible sets and $(L_{\omega_1\omega})_t$-equivalence for T_3 spaces*
⋄ C65 C75 C80 C90 ⋄

MEKLER, A.H. *Stationary logic of ordinals*
⋄ B25 C10 C55 C65 C80 E10 ⋄

MILLS, G. & PARIS, J.B. *Regularity in models of arithmetic*
⋄ C62 C80 F30 ⋄

MOTOHASHI, N. *Approximation theory of uniqueness conditions by existence conditions*
⋄ C40 C75 C80 F07 F50 ⋄

MUNDICI, D. *A generalization of abstract model theory*
⋄ C90 C95 G30 ⋄

MUNDICI, D. *Abstract model theory and nets of C_*-algebras: noncommutative interpolation and preservation properties* ⋄ C40 C95 ⋄

MUNDICI, D. *Abstract model theory of many-valued logics and K-theory of certain C^*-algebras*
⋄ B50 C90 C95 ⋄

MUNDICI, D. *Embeddings, amalgamations, and elementary equivalence: the representation of compact logics* ⋄ C95 ⋄

PARIGOT, M. *Extensions de la logique du premier order. Theoreme de Lindstroem* ⋄ C95 C98 ⋄

RAPP, A. *On the expressive power of the logics $L(Q_\alpha^{\langle n_1,\ldots,n_m\rangle})$* ⋄ C55 C80 ⋄

ROTHMALER, P. *Some model theory of modules III. On infiniteness of sets definable in modules*
⋄ C10 C35 C60 C80 ⋄

SEMENOV, A.L. *Decidability of monadic theories*
⋄ B25 C85 C98 ⋄

SHELAH, S. *Can you take Solovay's inaccessible away?*
⋄ C55 C75 C80 E15 E25 E35 E55 E75 ⋄

SHELAH, S. *On logical sentences in PA*
⋄ C62 C80 E05 F25 F30 ⋄

WESTERSTAAHL, D. *Some results on quantifiers*
⋄ B65 C80 ⋄

1985

ABRAHAM, U. & RUBIN, M. & SHELAH, S. *On the consistency of some partition theorems for continuous colorings, and structure of $\aleph_1$-dense real order types*
⋄ C55 C62 C65 C80 E05 E07 E35 ⋄

BALDWIN, JOHN T. *Definable second-order quantifiers*
⋄ C30 C80 C85 C98 ⋄

BALDWIN, JOHN T. & SHELAH, S. *Second-order quantifiers and the complexity of theories*
⋄ C40 C55 C65 C75 C85 ⋄

BARWISE, J. & FEFERMAN, S. (EDS.) *Model-theoretic logics*
⋄ C75 C80 C85 C90 C95 C98 ⋄

BAUDISCH, A. & SEESE, D.G. & TUSCHIK, H.-P. & WEESE, M. *Decidability and quantifier elimination*
⋄ B25 C10 C60 C65 C80 C98 ⋄

BAUVAL, A. *Polynomial rings and weak second-order logic*
⋄ C60 C85 F35 ⋄

CAICEDO, X. *Failure of interpolation for quantifiers of monadic type* ⋄ C40 C75 C80 C95 ⋄

EBBINGHAUS, H.-D. *Extended logics: The general framework* ⋄ C40 C55 C80 C95 C98 ⋄

FLUM, J. & MARTINEZ, JUAN CARLOS *κ-local operations for topological structures* ⋄ C40 C75 C80 C90 ⋄

FLUM, J. *Characterizing logics*
⋄ C75 C80 C95 C98 ⋄

FLUM, J. *Maximale monadische Logiken*
⋄ C55 C80 C95 ⋄

FRAISSE, R. *Deux relations denombrables, logiquement equivalentes pour le second ordre, sont isomorphes (modulo un axiome de constructibilite)*
⋄ B15 C15 C85 E45 ⋄

GOLTZ, H.-J. *The boolean sentence algebra of the theory of linear ordering is atomic with respect to logics with a Malitz quantifier* ⋄ C55 C65 C80 G05 ⋄

GUREVICH, Y. *Monadic second-order theories*
⋄ B25 C65 C85 C98 D35 D98 ⋄

HINMAN, P.G. & ZACHOS, S. *Probabilistic machines, oracles, and quantifiers* ⋄ C80 D15 ⋄

HODGES, W. *Building models by games*
⋄ C25 C55 C60 C80 C98 E60 ⋄

JANKOWSKI, A.W. *Galois structures* ⋄ B22 C95 ⋄

KAUFMANN, M. *A note on the Hanf number of second-order logic* ⋄ C55 C80 C85 ⋄

KAUFMANN, M. & SHELAH, S. *On random models of finite power and monadic logic* ⋄ C13 C85 ⋄

KAUFMANN, M. *The quantifier "There exist uncountable many" and some of its relatives* ⋄ C55 C80 C98 ⋄

KEISLER, H.J. *Probability quantifiers*
⋄ C40 C80 C90 C98 ⋄

KOSSAK, R. *Recursively saturated ω_1-like models of arithmetic* ⋄ C55 C57 C62 C75 C80 E65 ⋄

KRYNICKI, M. *Interpretations in nonelementary languages* ⋄ C80 C95 F25 ⋄

LIPPARINI, P. *Compactness properties of logics generated by monadic quantifiers and cardinalities of limit ultrapowers* ⋄ C20 C80 C95 ⋄

LIPPARINI, P. *Duality for compact logics and substitution in abstract model theory* ⋄ C95 ⋄

LIPPARINI, P. *Robinson equivalence relations through limit ultrapowers (Italian summary)* ⋄ C20 C95 ⋄

MAKOWSKY, J.A. & MUNDICI, D. *Abstract equivalence relations* ⋄ C95 ⋄

MAKOWSKY, J.A. *Abstract embedding relations* ⋄ C95 ⋄

MAKOWSKY, J.A. *Compactness, embeddings and definability* ⋄ C40 C95 C98 ⋄

MAKOWSKY, J.A. *Vopenka's principle and compact logics* ⋄ C05 C55 C95 E55 ⋄

MEKLER, A.H. & SHELAH, S. *Stationary logic and its friends I* ⋄ C40 C55 C75 C80 E35 E50 ⋄

MIJAJLOVIC, Z. *On the definability of the quantifier "there exist uncountably many"* ⋄ C55 C80 ⋄

MOLZAN, B. *On the theory of Boolean algebras in the logic with Ramsey quantifiers* ⋄ B25 C10 C80 G05 ⋄

MUCHNIK, A.A. *Games on infinite trees and automata with dead ends. A new proof of decidability for the monadic theory with two successor functions (Russian)* ⋄ B15 B25 C85 D05 ⋄

MULLER, D.E. & SCHUPP, P.E. *Alternating automata on infinite objects, determinacy and Rabin's Theorem* ⋄ B25 C85 D05 E60 ⋄

MUNDICI, D. *Other quantifiers: On overview* ⋄ C80 C98 ⋄

PINUS, A.G. *Applications of boolean powers of algebraic systems (Russian)* ⋄ C05 C30 C55 C57 C80 C85 D35 E50 G05 ⋄

RAPP, A. *Elimination of Malitz quantifiers in stable theories* ⋄ C10 C45 C55 C80 ⋄

RAPP, A. *The ordered field of real numbers and logics with Malitz quantifiers* ⋄ B25 C10 C55 C65 C80 ⋄

SCHMERL, J.H. *Transfer theorems and their applications to logic* ⋄ C55 C80 C98 ⋄

SHAPIRO, S. *Second-order languages and mathematical practice* ⋄ A05 C85 ⋄

SHELAH, S. *Monadic logic and Loewenheim numbers* ⋄ C45 C55 C65 C75 C85 E45 ⋄

SHELAH, S. *Remarks in abstract model theory* ⋄ C40 C50 C52 C55 C80 C95 ⋄

STEINHORN, C.I. *Borel structures for first-order and extended logics* ⋄ C80 C90 C98 ⋄

TARLECKI, A. *On the existence of free models in abstract algebraic institutions* ⋄ B75 C95 ⋄

VAEAENAENEN, J. *Set-theoretical definability of logics* ⋄ C40 C55 C80 C95 C98 E47 ⋄

C90 Nonclassical models

1949
MOSTOWSKI, ANDRZEJ *Sur l'interpretation geometrique et topologique des notions logiques* ⋄ B10 C90 ⋄

1953
RASIOWA, H. & SIKORSKI, R. *Algebraic treatment of the notion of satisfiability* ⋄ B10 C90 G05 ⋄
RASIOWA, H. & SIKORSKI, R. *On satisfiability and decidability in non-classical functional calculi* ⋄ B10 B25 B45 C90 F50 ⋄

1955
RASIOWA, H. *Algebraic models of axiomatic theories* ⋄ B30 C90 F50 G05 G10 G25 ⋄
RASIOWA, H. & SIKORSKI, R. *An application of lattices to logic* ⋄ B45 C90 G10 ⋄
RASIOWA, H. & SIKORSKI, R. *On existential theorems in non-classical functional calculi* ⋄ B45 C90 F50 G05 ⋄

1958
MOSTOWSKI, ANDRZEJ *Quelques applications de la topologie a la logique mathematique* ⋄ C90 C98 E15 ⋄
REICHBACH, J. *On the first-order functional calculus and the truncation of models (Polish and Russian summaries)* ⋄ B25 C07 C90 ⋄

1960
GAIFMAN, H. *Probability models and the completeness theorem* ⋄ B48 C90 ⋄

1962
CHANG, C.C. & KEISLER, H.J. *Model theories with truth values in a uniform space* ⋄ B50 C20 C50 C90 ⋄
KRIPKE, S.A. *"Flexible" predicates of formal number theory* ⋄ C90 F30 ⋄
MOSTOWSKI, ANDRZEJ *L'espace des modeles d'une theorie formalisee et quelques-unes de ses applications* ⋄ B50 C07 C40 C57 C80 C85 C90 D45 ⋄

1963
REZNIKOFF, I. *Chaines de formules* ⋄ B05 C40 C90 F50 ⋄

1964
GAIFMAN, H. *Concerning measures on first order calculi* ⋄ B48 C90 ⋄
MELTER, R.A. *Boolean valued rings and boolean metric spaces* ⋄ C90 ⋄

1965
CHANG, C.C. & KEISLER, H.J. *Continuous model theory* ⋄ C90 ⋄
EHRESMANN, C. *Categories et structures* ⋄ C90 C98 G30 ⋄
KRIPKE, S.A. *Semantical analysis of intuitionistic logic I* ⋄ B55 C90 F50 ⋄
VOPENKA, P. *On ∇-model of set theory* ⋄ C20 C62 C90 E30 E35 E40 E70 ⋄
VOPENKA, P. *The limits of sheaves and applications on constructions of models* ⋄ C20 C62 C90 E25 E30 E35 E40 E70 ⋄

1966
BAR-HILLEL, Y. *Do natural languages contain paradoxes?* ⋄ A05 B65 C90 ⋄
CHANG, C.C. & KEISLER, H.J. *Continuous model theory* ⋄ B50 C05 C20 C50 C90 C98 ⋄
FOLLESDAL, D. *A model theoretic approach to causal logic* ⋄ A05 B45 C90 ⋄
KRAUSS, P.H. & SCOTT, D.S. *Assigning probabilities to logical formulas* ⋄ B48 C75 C90 ⋄
LEBLANC, L. *Representabilite et definissabilite dans les algebres transformationnelles et dans les algebres polyadiques* ⋄ C40 C62 C90 G15 ⋄
POLLOCK, J.L. *Model theory and modal logic* ⋄ B45 C90 ⋄
REZNIKOFF, I. *Sur les ensembles denombrables de formules en logique intuitionniste* ⋄ C40 C90 F50 ⋄

1967
SHAFAAT, A. *Principle of localization for a more general type of languages* ⋄ B10 C55 C75 C90 ⋄
SKORDEV, D.G. *Certain algebraic aspects of mathematical logic (Bulgarian) (German summary)* ⋄ C25 C90 G10 ⋄
SUSZKO, R. *A proposal concerning the formulation of the infinitistic axiom in the theory of logical probability* ⋄ B48 C90 ⋄
VUCKOVIC, V. *Recursive models for three-valued propositional calculi with classical implication* ⋄ B50 C57 C90 F30 ⋄

1968
CHUDNOVSKY, D.V. *Probability on first order logics (Ukrainian) (Russian and English summaries)* ⋄ B48 C90 ⋄
FOLLESDAL, D. *Interpretation of quantifiers* ⋄ A05 B10 C90 ⋄
KARP, C.R. *An algebraic proof of the Barwise compactness theorem* ⋄ C70 C90 ⋄
SCHUETTE, K. *Vollstaendige Systeme modaler und intuitionistischer Logik* ⋄ B45 C90 F50 F98 ⋄

1969
ANDERSON, J.G. *An application of Kripke's completeness theorem for intuitionism to superconstructive propositional calculi* ⋄ B55 C90 ⋄

EBBINGHAUS, H.-D. *Ueber eine Praedikatenlogik mit partiell definierten Praedikaten und Funktionen* ⋄ B60 C90 ⋄

FITTING, M. *Intuitionistic logic, model theory, and forcing* ⋄ B98 C90 C98 E25 E35 E45 E50 F50 F98 ⋄

KRAUSS, P.H. *Representation of symmetric probability models* ⋄ C75 C90 ⋄

RESCHER, N. *Nonstandard quantificational logic* ⋄ A05 B45 C90 ⋄

RESCHER, N. *Topics in philosophical logic* ⋄ A05 B45 B98 C90 ⋄

RUZSA, I. *Random models of some quasi-intuitionistic logics* ⋄ B55 C90 ⋄

RUZSA, I. *Random models of logical systems I: Models of valuing logics* ⋄ B50 C90 ⋄

WASZKIEWICZ, J. & WEGLORZ, B. *On products of structures for generalized logics (Polish and Russian summaries)* ⋄ B50 C20 C30 C90 ⋄

WATANABE, S. *Modified concepts of logic, probability, and information based on generalized continuous characteristic function* ⋄ C90 ⋄

1970

CHUDNOVSKY, D.V. *Logical probability and conditional probability on Boolean algebras (Russian) (English summary)* ⋄ B48 C90 G05 ⋄

FITTING, M. *Intuitionistic model theory and the Cohen independence proofs* ⋄ A05 C90 E35 E40 F50 ⋄

LUCKHARDT, H. *Ein Henkin-Vollstaendigkeitsbeweis fuer die intuitionistische Praedikatenlogik bezueglich der Kripke-Semantik* ⋄ C90 F50 ⋄

LUCKHARDT, H. *Kripke-Semantik der derivativen Praedikatenlogik* ⋄ C90 F50 ⋄

MAEHARA, S. *A general theory of completeness proofs* ⋄ C07 C90 F50 ⋄

MANSFIELD, R. *The theory of boolean ultrapowers* ⋄ C20 C90 G05 ⋄

OSSWALD, H. *Modelltheoretische Untersuchungen in der Kripke-Semantik* ⋄ C20 C90 F50 ⋄

RASIOWA, H. *Ultraproducts of m-valued models and a generalization of the Loewenheim-Skolem-Goedel-Mal'cev theorem for theories based on m-valued logics (Russian summary)* ⋄ B50 C20 C90 ⋄

REICHBACH, J. *On statistical tests in generalizations of Herbrand's theorems and asymptotic probabilistic models (main theorems of mathematics)* ⋄ B10 C40 C90 ⋄

SCHULTZ, KONRAD *Modelle modaler Mengenlehren* ⋄ B45 C62 C90 E35 E70 ⋄

SEGERBERG, K. *Modal logics with linear alternative relations* ⋄ B45 C90 ⋄

1971

COMER, S.D. *Representations by algebras of sections over boolean spaces* ⋄ C05 C90 ⋄

ELLERMAN, D.P. *Sheaves or relation structures and ultraproducts* ⋄ C20 C90 ⋄

FELSCHER, W. *An algebraic approach to first order logic* ⋄ B10 C05 C07 C20 C90 G15 ⋄

GABBAY, D.M. *Decidability results in non-classical logic, III: Systems with statability operators* ⋄ B25 B45 C90 ⋄

GABBAY, D.M. *On decidable, finitely axiomatizable, modal and tense logics without the finite model property I, II* ⋄ B25 B45 C90 ⋄

GABBAY, D.M. *Semantic proof of the Craig interpolation theorem for intuitionistic logic and extensions I,II* ⋄ B55 C40 C90 F50 ⋄

GOERNEMANN, S. *A logic stronger than intuitionism* ⋄ B55 C90 F50 ⋄

KLEMKE, D. *Ein Henkin-Beweis fuer die Vollstaendigkeit eines Kalkuels relativ zur Grzegorczyk-Semantik* ⋄ B55 C90 F50 ⋄

KOCK, A. & WRAITH, G.C. *Elementary toposes* ⋄ C90 G30 ⋄

NADIU, G.S. *Sur la logique de Heyting* ⋄ C90 F50 G10 G30 ⋄

REICHBACH, J. *A note about generalized models and the origin of forcing in the theory of models* ⋄ B25 C25 C90 ⋄

REICHBACH, J. *Generalized models and decidability – weak and strong statistical decidability of theorems* ⋄ B25 C25 C90 ⋄

RUTKOWSKI, A. *Some remarks about boolean-valued models (Russian summary)* ⋄ C90 G05 ⋄

RUZSA, I. *Random models of logical systems II: Models of some modal logics* ⋄ B45 C90 ⋄

RUZSA, I. *Random models of logical systems III: Models of quantified logics* ⋄ B50 C90 ⋄

WASZKIEWICZ, J. *The notions of isomorphism and identity for many-valued relational structures (Polish and Russian summaries)* ⋄ B50 C90 ⋄

1972

COMER, S.D. *A sheaf-theoretic duality theory for cylindric algebras* ⋄ C90 G15 ⋄

ERSHOV, YU.L. *Relationship between sheaf spaces and numbered sets with the C_2^* property (Russian) (English summary)* ⋄ C90 D25 D45 D65 ⋄

FITTING, M. *Non-classical logics and the independence results of set theory* ⋄ B45 C90 E35 ⋄

GABBAY, D.M. *Application of trees to intermediate logics* ⋄ B55 C90 ⋄

GABBAY, D.M. *Craig's interpolation theorem for modal logics* ⋄ B45 C40 C90 ⋄

GABBAY, D.M. *Decidability of some intuitionistic predicate theories* ⋄ B25 C90 F50 ⋄

GABBAY, D.M. *Model theory for intuitionistic logic* ⋄ B55 C20 C90 F50 ⋄

HOFMANN, K.H. *Representation of algebras by continuous sections* ⋄ C60 C90 C98 ⋄

MANSFIELD, R. *Horn classes and reduced direct products* ⋄ C20 C52 C90 ⋄

MANSFIELD, R. *The completeness theorem for infinitary logic* ⋄ C75 C90 ⋄

MARTIN-LOEF, P. *About models for intuitionistic type theories and the notion of definitional equality* ⋄ C90 F50 ⋄

MIURA, S. *Probabilistic models of modal logics*
  ⋄ B45 C90 ⋄
OSSWALD, H. *Homomorphie-invariante Formeln in der intuitionistischen Logik* ⋄ C40 C90 F50 ⋄
RASIOWA, H. *The Craig interpolation theorem for m-valued predicate calculi (Russian summary)*
  ⋄ B50 C40 C90 ⋄
REICHBACH, J. *Generalized models, probability of formulas, decisions and statistical decidability of theorems* ⋄ B25 C25 C90 ⋄
REICHBACH, J. *Notes about decidablility of mathematics with probability 1 and generalized models*
  ⋄ B25 C90 ⋄
SEGERBERG, K. *Post completeness in modal logic*
  ⋄ B45 C90 ⋄
SHORB, A.M. *Filter constructions in boolean valued model theory* ⋄ C90 ⋄
TAUTS, A. *A connection between generalized Beth models and topological pseudo-Boolean algebras (Russian) (Estonian and German summaries)*
  ⋄ C90 F50 G10 ⋄
TAUTS, A. *A semantic interpretation of formulas in generalized Beth models and in pseudo-boolean algebras (Russian) (Estonian and German summaries)*
  ⋄ C90 F50 G10 ⋄
TIERNEY, M. *Sheaf theory and the continuum hypothesis*
  ⋄ C90 E35 E50 G30 ⋄

1973

CHANG, C.C. *Modal model theory* ⋄ B45 C90 ⋄
COSTE, M. *Logique du 1er ordre dans les topos elementaires* ⋄ C90 G30 ⋄
DAHN, B.I. *Generalized Kripke-models* ⋄ B45 C90 ⋄
DAVEY, B.A. *Sheaf spaces and sheaves of universal algebras* ⋄ C05 C90 ⋄
DRAGALIN, A.G. *Intuitionistic model theory (Russian)*
  ⋄ C90 F50 ⋄
GABBAY, D.M. *The undecidability of intuitionistic theories of algebraically closed fields and real closed fields*
  ⋄ B25 C60 C90 D35 F50 ⋄
GOBLE, L.F. *A new modal model* ⋄ B45 C90 ⋄
GOLDBLATT, R.I. *A model-theoretic study of some systems containing S3* ⋄ B25 B45 C90 ⋄
HAJEK, P. *Automatic listing of important observational statements. I,II* ⋄ B35 C13 C80 C90 ⋄
LESAFFRE, B. *Structures algebriques dans les topos elementaires* ⋄ C90 G30 ⋄
OSIUS, G. *The internal and external aspect of logic and set theory in elementary topoi*
  ⋄ C90 E70 E75 F50 G30 ⋄
REICHBACH, J. *Generalized models for intuitionistic and classical predicate calculi with ultraproducts*
  ⋄ C20 C25 C80 C90 F50 ⋄
RESSAYRE, J.-P. *Boolean models and infinitary first order languages*
  ⋄ C40 C50 C55 C70 C75 C80 C90 ⋄
SALIJ, V.N. *Boolean valued algebras (Russian)*
  ⋄ C05 C90 G05 ⋄
SANCHIS, L.E. *Formally defined operations in Kripke models* ⋄ C90 F50 ⋄

SANMARTIN ESPLUGUES, J. *Syllogistics, many-valued logic and model theory (Spanish)* ⋄ A10 B20 B50 C90 ⋄
SCOTT, D.S. *Models for various type-free calculi*
  ⋄ B40 C90 ⋄
SMORYNSKI, C.A. *Elementary intuitionistic theories*
  ⋄ B25 C90 D35 F50 ⋄
SMORYNSKI, C.A. *Investigations of intuitionistic formal systems by means of Kripke models*
  ⋄ B55 C90 F50 ⋄
SZCZERBA, L.W. *The notion of an elementary subsystem for a boolean-valued relational system* ⋄ C07 C90 ⋄

1974

BENDA, M. *Note on boolean ultrapowers*
  ⋄ C20 C50 C90 ⋄
COMER, S.D. *Elementary properties of structures of sections* ⋄ B25 C90 ⋄
COMER, S.D. *Restricted direct products and sectional representations* ⋄ C05 C30 C90 ⋄
DAHN, B.I. *A note on generalized Kripke-models*
  ⋄ B45 C90 ⋄
DAHN, B.I. *Contributions to the model theory for non-classical logics* ⋄ B45 C90 ⋄
DAHN, B.I. *Generalized Kripke models* ⋄ C90 ⋄
DAHN, B.I. *Meta-mathematics of some many-valued calculi* ⋄ B50 C40 C90 ⋄
ELLERMAN, D.P. *Sheaves of structures and generalized ultraproducts* ⋄ C20 C25 C90 G30 ⋄
GRANT, J. *Incomplete models* ⋄ B53 C90 ⋄
HAJEK, P. *Automatic listing of important observational statements. III* ⋄ B35 C13 C80 C90 ⋄
KEIMEL, K. *Representation des algebres universelles par des faisceaux* ⋄ C05 C35 C60 C90 ⋄
KEIMEL, K. & WERNER, H. *Stone duality for varieties generated by quasi-primal algebras*
  ⋄ C05 C13 C90 ⋄
KRIVINE, J.-L. *Langages a valeurs reelles et applications*
  ⋄ B50 C40 C60 C90 ⋄
MACINTYRE, A. *Model-completeness for sheaves of structures* ⋄ C35 C90 ⋄
MORTIMER, M. *Some results in modal model theory*
  ⋄ B45 C90 ⋄
MULVEY, C.J. *Intuitionistic algebra and representations of rings* ⋄ C60 C90 F50 F55 G30 ⋄
PASHENKOV, V.V. *Duality of topological models (Russian)*
  ⋄ C30 C90 ⋄
PETRESCU, I. *A theorem of Loewenheim-Skolem type for topological models (Romanian) (English summary)*
  ⋄ C20 C90 ⋄
POTTHOFF, K. *Boolean ultrapowers* ⋄ C20 C90 ⋄
REICHBACH, J. *About decidability, generalized models and main theorems of many valued predicate calculi*
  ⋄ B25 B50 C90 ⋄
REYES, G.E. *From sheaves to logic* ⋄ C90 F50 G30 ⋄
ROBINSON, A. *A note on topological model theory*
  ⋄ C65 C90 ⋄
ROUTLEY, R. *Semantical analyses of propositional systems of Fitch and Nelson* ⋄ B25 B46 C90 F50 ⋄
TAUTS, A. *A formal deduction of tautological formulas in pseudo-boolean algebras (Russian) (Estonian and German summaries)* ⋄ B60 C90 F50 G05 G10 ⋄

TAUTS, A. *A game for the construction of a propositional semantics in generalized Beth models (Russian) (Estonian and German summaries)*
⋄ C75 C90 E60 F50 ⋄

WOLF, A. *Sheaf representations of arithmetical algebras*
⋄ C05 C90 G30 ⋄

1975

ABIAN, A. *On the use of more than two-element boolean valued models* ⋄ C62 C90 E35 ⋄

BOWEN, K.A. *Normal modal model theory*
⋄ B45 C90 ⋄

BURRIS, S. *Boolean powers* ⋄ B25 C05 C30 C90 ⋄

COMER, S.D. *Monadic algebras with finite degree*
⋄ B25 C05 C90 G15 ⋄

DAHN, B.I. *Eine Anwendung des Basistheorems in der nichtklassischen Logik* ⋄ B55 C40 C90 ⋄

DAHN, B.I. *On models with variable universe* ⋄ C90 ⋄

FINE, K. *Some connections between elementary and modal logic* ⋄ B45 C07 C90 ⋄

GABBAY, D.M. *Decidability results in non-classical logics I*
⋄ B25 B45 C90 F50 ⋄

GABBAY, D.M. *Model theory for tense logics*
⋄ B45 C20 C90 ⋄

GAERDENFORS, P. & HANSSON, B. *Filtrations and the finite frame property in boolean semantics*
⋄ B25 B45 C90 G05 G25 ⋄

GALLIN, D. *Intensional and higher-order modal logic. With applications to Montague semantics*
⋄ B45 C85 C90 E35 E50 E70 ⋄

GOLDBLATT, R.I. *First order definability in modal logic*
⋄ B45 C20 C90 ⋄

HALPERN, F. *Transfer theorems for topological structures*
⋄ C20 C40 C90 ⋄

HAVRANEK, T. *Statistical quantifiers in observational calculi* ⋄ B50 C80 C90 ⋄

HICKIN, K.K. & PLOTKIN, J.M. *Compactness and ρ-valued logics* ⋄ B50 C75 C90 ⋄

KAMO, S. *Completeness theorem of a logical system on a complete boolean algebra* ⋄ C90 G05 ⋄

LOPEZ-ESCOBAR, E.G.K. & VELDMAN, W. *Intuitionistic completeness of a restricted second-order logic*
⋄ B15 C90 F50 ⋄

MCKEE, T.A. *Infinitary logic and topological homeomorphisms* ⋄ C65 C75 C80 C90 ⋄

MLEZIVA, M. *Semantics based on states of affairs*
⋄ C90 ⋄

OSIUS, G. *A note on Kripke-Joyal semantics for the internal language of topoi* ⋄ C90 F50 G30 ⋄

OSIUS, G. *Logical and set theoretical tools in elementary topoi* ⋄ C90 E70 F50 G30 ⋄

PASHENKOV, V.V. *Boolean models and some of their applications (Russian)*
⋄ C90 E07 G05 G10 G30 ⋄

PASHENKOV, V.V. *On dualities (Russian)* ⋄ C30 C90 ⋄

RAMA RAO, V.V. & SWAMY, U.M. *Triple and sheaf representation of Stone lattices* ⋄ C90 G10 G30 ⋄

RANTALA, V. *Urn models: A new kind of non-standard model for first-order logic* ⋄ A05 B10 C07 C90 ⋄

REICHBACH, J. *Generalized models for classical and intuitionistic predicate calculi*
⋄ B10 B25 C90 F50 ⋄

REICHBACH, J. *Infinite proof methods in decidability problems with Ulam's models and classification of predicate calculi* ⋄ B25 C25 C90 ⋄

REYES, G.E. *Faisceaux et concepts* ⋄ C90 F50 G30 ⋄

SONOBE, O. *Bi-relational frameworks for minimal and intuitionistic logics* ⋄ B45 C90 F50 ⋄

STAVI, J. *Extensions of Kripke's embedding theorem*
⋄ C55 C75 C90 G05 ⋄

TAUTS, A. *A semantic model for infinite formulas (Russian) (Estonian and German summaries)*
⋄ C75 C90 F50 G10 ⋄

TAUTS, A. *Search for deduction by means of a semantic model (Russian) (Estonian and German summaries)*
⋄ B55 C75 C90 F50 ⋄

TAUTS, A. *Strong tautology and construction of a countermodel for nonderivable propositions (Russian) (Estonian and German summaries)*
⋄ C75 C90 F50 ⋄

THOMASON, S.K. *Categories of frames for modal logic*
⋄ B45 C90 G30 ⋄

WEISPFENNING, V. *Model-completeness and elimination of quantifiers for subdirect products of structures*
⋄ C10 C30 C35 C52 C60 C90 ⋄

1976

AMER, M.A. *Typed boolean structures I (Arabic summary)*
⋄ B15 B40 C90 E40 G05 ⋄

ANDREKA, H. & DAHN, B.I. & NEMETI, I. *On a proof of Shelah (Russian summary)* ⋄ C07 C20 C90 ⋄

BERNHARDT, K. *Semantics for finitary predicate calculi*
⋄ B20 C90 ⋄

BIGELOW, J.C. *Possible worlds foundations for probability*
⋄ B48 C90 ⋄

BORSHCHEV, V.B. & KHOMYAKOV, M.V. *Club systems (a formal calculus for the description of composed systems) (Russian) (English and German summaries)* ⋄ C90 ⋄

CORNISH, W.H. *A sheaf representation for generalized Stone lattices* ⋄ C05 C90 G10 ⋄

DIACONESCU, R. *Grothendieck toposes have boolean points - a new proof* ⋄ C90 G30 ⋄

ELLENTUCK, E. *Categoricity regained*
⋄ C35 C75 C90 E40 ⋄

GABBAY, D.M. *Investigations in modal and tense logics with applications to problems in philosophy and linguistics* ⋄ A05 B25 B46 C90 ⋄

GIRARD, J.-Y. *Three-valued logic and cut-elimination: The actual meaning of Takeuti's conjecture*
⋄ B50 C85 C90 F05 F35 F50 ⋄

HAJEK, P. *Observationsfunktorenkalkuele und die Logik der automatisierten Forschung*
⋄ B60 C13 C80 C90 ⋄

JONGH DE, D.H.J. & SMORYNSKI, C.A. *Kripke models and the intuitionistic theory of species* ⋄ C90 F35 F50 ⋄

KENNISON, J.F. *Integral domain type representations in sheaves and other topoi* ⋄ C60 C90 G30 ⋄

KROL', M.D. *The topological models of intuitionistic analysis. One counter-example (Russian)*
⋄ C90 F35 F50 ⋄

LAITA, L.M. *Un estudio de logica algebraica desde el punto de vista de la teoria de categorias*
⋄ C90 G15 G30 ⋄

MAKKAI, M. & REYES, G.E. *Model-theoretical methods in the theory of topoi and related categories I,II (Russian summaries)* ⋄ C75 C90 G30 ⋄

MANGANI, P. *Alcune questioni di teoria dei modelli per linguaggi predicativi generali (English summary)* ⋄ B50 C20 C90 ⋄

MCKEE, T.A. *Sentences preserved between equivalent topological bases* ⋄ C65 C80 C90 ⋄

MURZIN, F.A. *Two arbitrary elementarily equivalent models of continuous logic have isomorphic ultra-powers (Russian)* ⋄ C20 C90 ⋄

PAULOS, J.A. *A model-theoretic semantics for modal logic* ⋄ B45 C07 C90 ⋄

POWELL, W.C. *A completeness theorem for Zermelo-Fraenkel set theory* ⋄ C90 E70 ⋄

REICHBACH, J. *Graphs, generalized models, games and decidability* ⋄ B25 C65 C90 ⋄

SEGERBERG, K. *Discrete linear future time without axioms* ⋄ B25 B45 C90 ⋄

SGRO, J. *Completeness theorems for continuous functions and product topologies* ⋄ C40 C65 C80 C90 E75 ⋄

VOLGER, H. *The Feferman-Vaught theorem revisited* ⋄ C20 C30 C90 ⋄

ZIEGLER, M. *A language for topological structures which satisfies a Lindstroem-theorem* ⋄ B25 C40 C50 C65 C80 C90 C95 D35 ⋄

1977

AMER, M.A. *Typed boolean structures II (Arabic summary)* ⋄ B15 B40 C90 E40 G05 ⋄

BANASCHEWSKI, B. & NELSON, EVELYN *Elementary properties of limit reduced powers with applications to boolean powers* ⋄ C05 C20 C30 C90 ⋄

BERNHARDT, K. *Abschliessbarkeit von Kripke-Semantiken* ⋄ B55 C90 ⋄

BOSSCHE VAN DEN, G. *Algebres de Heyting completes etalees et faisceaux* ⋄ C90 G10 ⋄

BOZZI, S. & MELONI, G.C. *Ideal properties and intuitionistic algebra* ⋄ C60 C90 F50 F55 ⋄

CHUAQUI, R.B. *A semantical definition of probability* ⋄ C90 ⋄

DANIELS, C.B. & FREEMAN, J.B. *Classical second-order intensional logic with maximal propositions* ⋄ B15 B85 C90 ⋄

ELLENTUCK, E. *Boolean valued rings* ⋄ C30 C60 C90 D50 ⋄

ELLENTUCK, E. *Free Suslin algebras* ⋄ C90 G25 ⋄

FERENCZI, M. *On valid assertions in probability logic* ⋄ B25 B48 C90 ⋄

GABBAY, D.M. *A new version of Beth semantics for intuitionistic logic* ⋄ C90 F50 ⋄

GABBAY, D.M. *Craig interpolation theorem for intuitionistic logic and extensions part III* ⋄ B55 C40 C90 F50 ⋄

GIRARD, J.-Y. *Functional interpretation and Kripke models* ⋄ C90 F10 F30 F50 ⋄

HAJEK, P. & HAVRANEK, T. *On generation of inductive hypotheses* ⋄ B35 C90 ⋄

HAVRANEK, T. *Towards a model theory of statistical theories* ⋄ B25 C90 ⋄

JOHNSTONE, P.T. *Rings, fields, and spectra* ⋄ C90 F50 G30 ⋄

KAISER, KLAUS *On generic stalks of sheaves* ⋄ C25 C90 ⋄

KAMO, S. *Reduced powers and models for infinitely logic* ⋄ C20 C75 C90 ⋄

KEISLER, H.J. *Hyperfinite model theory* ⋄ B50 C65 C75 C80 C90 H20 ⋄

KROL', M.D. *Disjunctive and existential properties of intuitionistic analysis with Kripke's scheme (Russian)* ⋄ C90 F35 F50 ⋄

MACNAB, D.S. *Some applications of double-negation sheafification* ⋄ C30 C90 F50 G30 ⋄

MAKKAI, M. & REYES, G.E. *First order categorical logic. Model-theoretical methods in the theory of topoi and related categories* ⋄ C75 C90 F50 G30 ⋄

MAKOWSKY, J.A. & MARCJA, A. *Completeness theorems for modal model theory with the Montague-Chang semantics I* ⋄ B45 C90 ⋄

MAKSIMOVA, L.L. *Craig's interpolation theorem and amalgamable varieties (Russian)* ⋄ B22 B55 C05 C40 C90 G10 ⋄

MAKSIMOVA, L.L. *Craig's theorem in superintuitionistic logics and amalgamable varieties of pseudo-boolean algebras (Russian)* ⋄ B55 C05 C40 C90 G05 G10 ⋄

MANSFIELD, R. *Sheaves and normal submodels* ⋄ C40 C90 E15 G05 G10 ⋄

MARCJA, A. *Prodotti booleani e prodotti booleani ridotti* ⋄ C20 C30 C90 ⋄

MARKOVIC, Z. *Reduced products of saturated intuitionistic theories* ⋄ C20 C90 F50 ⋄

MOTOHASHI, N. *Partially ordered interpretations* ⋄ C90 E40 F25 F50 ⋄

RAUSZER, C. *Applications of Kripke models to Heyting-Brouwer logic* ⋄ B55 C90 F50 ⋄

RAUSZER, C. *Model theory for an extension of intuitionistic logic* ⋄ B55 C90 F50 ⋄

REYES, G.E. *Sheaves and concepts: a model-theoretic interpretation of Grothendieck topoi* ⋄ C75 C90 F50 G30 ⋄

ROMANSKI, J. *On connections between algebraic and Kripke semantics* ⋄ C90 F50 G10 G25 ⋄

RYBAKOV, V.V. *Noncompact extensions of the logic S4 (Russian)* ⋄ B45 C90 ⋄

SATO, M. *A study of Kripke-type models for some modal logics by Gentzen's sequential method* ⋄ B45 C90 ⋄

SGRO, J. *Completeness theorems for topological models* ⋄ C20 C40 C75 C80 C90 ⋄

SGRO, J. *Maximal logics* ⋄ C20 C40 C80 C90 C95 ⋄

SMORYNSKI, C.A. *On axiomatizing fragments* ⋄ C90 F50 ⋄

SOBOLEV, S.K. *The intuitionistic propositional calculus with quantifiers (Russian)* ⋄ B55 C80 C90 D35 F50 ⋄

SWART DE, H.C.M. *An intuitionistically plausible interpretation of intuitionistic logic* ⋄ C90 F50 ⋄

TROELSTRA, A.S. *Completeness and validity for intuitionistic predicate logic* ⋄ C90 F50 ⋄

YASHIN, A.L. *On an extension of intuitionistic logic and its interpretations (Russian)* ⋄ B55 C90 ⋄

1978

BOWEN, K.A. *Model theory for modal logic. Kripke models for modal predicate calculi* ⋄ B45 C90 ⋄

DALEN VAN, D. *An interpretation of intuitionistic analysis* ⋄ C90 F35 F50 ⋄

DANIELS, C.B. & FREEMAN, J.B. *A logic of generalized quantification* ⋄ B15 C80 C90 ⋄

DZIOBIAK, W. *A note on incompleteness of modal logics with respect to neighbourhood semantics* ⋄ B45 C90 ⋄

ELLERMAN, D.P. & ROTA, G.-C. *A measure theoretic approach to logical quantification* ⋄ B10 C90 G25 ⋄

FINE, K. *Model theory for modal logic. part I: the de re/de dicto distinction. part II: the elimination of de re modality* ⋄ A05 B45 C40 C90 ⋄

GARAVAGLIA, S. *Model theory of topological structures* ⋄ C20 C40 C65 C80 C90 ⋄

HAJEK, P. & HAVRANEK, T. *Mechanizing hypothesis formation. Mathematical foundations for a general theory* ⋄ A05 B35 C13 C80 C90 D80 ⋄

HOOVER, D.N. *Probability logic* ⋄ B48 C80 C90 ⋄

KEARNS, J.T. *Intuitionist logic, a logic of justification* ⋄ A05 C90 F50 ⋄

KROL', M.D. *A topological model for intuitionistic analysis with Kripke's scheme* ⋄ C90 F35 F50 ⋄

KROL', M.D. *Distinct variants of Kripke's scheme in intuitionistic analysis (Russian)* ⋄ C90 F35 F50 ⋄

LOULLIS, G. *Infinite forcing for boolean valued models* ⋄ C25 C35 C90 ⋄

MACDONALD, J.L. *Categorical shape theory and the back and forth property* ⋄ C60 C90 G05 G30 ⋄

MULVEY, C.J. *A remark on the prime stalk theorem* ⋄ C90 ⋄

NADEL, M.E. *Infinitary intuitionistic logic from a classical point of view* ⋄ C75 C90 F50 ⋄

NEMETI, I. *From hereditary classes to varieties in abstract model theory and partial algebra* ⋄ C05 C20 C52 C90 C95 ⋄

OLIN, P. *Urn models and categoricity* ⋄ A05 C35 C90 ⋄

REYES, G.E. *Theorie des modeles et faisceaux* ⋄ C90 F50 G30 ⋄

ROUTLEY, R. *An inadequacy in Kripke-semantics for intuitionistic quantificational logic* ⋄ B55 C90 F50 ⋄

SCHUMM, G.F. *An incomplete nonnormal extension of S3* ⋄ B45 C90 ⋄

SCHUMM, G.F. *Putting ℵ in its place* ⋄ B45 C90 ⋄

SHEKHTMAN, V.B. *Rieger-Nishimura lattices (Russian)* ⋄ B55 C90 G10 G25 ⋄

SHEKHTMAN, V.B. *Two-dimensional modal logic (Russian)* ⋄ B45 C90 ⋄

SWART DE, H.C.M. *First steps in intuitionistic model theory* ⋄ C90 F50 ⋄

TAKAHASHI, MAKOTO *Sheaves of B-valued structures (Japanese)* ⋄ C90 G30 ⋄

TAKEUTI, G. *Two applications of logic to mathematics* ⋄ B98 C90 E40 E75 F05 F30 F35 G12 ⋄

TROELSTRA, A.S. *Some remarks on the complexity of Henkin-Kripke models* ⋄ C90 D55 F50 ⋄

VEIT RICCIOLI, B. *Il forcing come principio logico per la costruzione dei fasci. I,II (French summaries)* ⋄ C25 C90 E40 G30 ⋄

VOJVODIC, G. *Some theorems for model theory of mixed-valued predicate calculi* ⋄ B50 C20 C35 C90 ⋄

WERNER, H. *Discriminator-algebras. Algebraic representation and model theoretic properties* ⋄ B25 C05 C25 C35 C90 ⋄

ZACHOROWSKI, S. *Dummett's LC has the interpolation property* ⋄ B55 C40 C90 ⋄

1979

ANDREKA, H. & NEMETI, I. & SAIN, I. *Completeness problems in verification of programs and program schemes* ⋄ B75 C90 C95 ⋄

ANDREKA, H. & NEMETI, I. & SAIN, I. *Program verification within and without logic* ⋄ B75 C90 ⋄

BENDA, M. *Modeloids I* ⋄ C90 ⋄

BENTHEM VAN, J.F.A.K. *Canonical modal logics and ultrafilter extensions* ⋄ B45 C90 G25 ⋄

BURRIS, S. & WERNER, H. *Sheaf constructions and their elementary properties* ⋄ B25 C05 C20 C25 C30 C60 C90 ⋄

CLARK, D.M. & KRAUSS, P.H. *Global subdirect products* ⋄ C05 C30 C90 ⋄

COLE, J.C. *Classifying topoi* ⋄ C90 G30 ⋄

COSTE, M. *Localisation, spectra and sheaf representation* ⋄ C30 C90 ⋄

COSTE, M. & COSTE, M.-F. *The generic model of an ε-stable geometric extension of the theory of rings is of line type* ⋄ C60 C90 G30 ⋄

DAHN, B.I. *Constructions of classical models by means of Kripke models (survey)* ⋄ C07 C20 C90 E35 E45 ⋄

DRAGALIN, A.G. *Algebraic approach to the analysis of models of non-standard logics (Russian)* ⋄ C90 ⋄

DUBUC, E.J. & REYES, G.E. *Subtoposes of the ring classifier* ⋄ C60 C75 C90 G30 ⋄

FOURMAN, M.P. & MULVEY, C.J. & SCOTT, D.S. (EDS.) *Applications of sheaves* ⋄ C90 C97 G97 ⋄

FOURMAN, M.P. & HYLAND, J.M.E. *Sheaf models for analysis* ⋄ C65 C90 F35 F50 G10 G30 ⋄

FOURMAN, M.P. & SCOTT, D.S. *Sheaves and logic* ⋄ C90 C98 E70 F35 F50 G30 ⋄

HEIJENOORT VAN, J. *Introduction a la semantique des logiques non-classiques* ⋄ B45 C90 F50 ⋄

HIROSE, K. & TAKAHASHI, S. *Sentences preserved by sheaves of structures (Japanese)* ⋄ C40 C90 G30 ⋄

HOOGEWIJS, A. *On a formalization of the non-definedness notion* ⋄ B60 C90 ⋄

JOHNSTONE, P.T. *Another condition equivalent to De Morgan's law* ⋄ C60 C90 F50 G05 G30 ⋄

JOHNSTONE, P.T. *Conditions related to de Morgan's law* ⋄ B55 C65 C90 F50 G30 ⋄

JOHNSTONE, P.T. *Remarks on the previous paper. Extracts from two letters to Anders Kock* ⋄ C60 C90 G30 ⋄

KENNISON, J.F. & LEDBETTER, C.S. *Sheaf representations and the Dedekind reals* ⋄ C90 G30 ⋄

KIRK, R.E. *Some classes of Kripke frames characteristic for the intuitionistic logic* ⋄ B55 C90 F50 ⋄

LOULLIS, G. *Sheaves and boolean valued model theory*
  ◇ C35 C52 C60 C90 ◇
MAKOWSKY, J.A. *The reals cannot be characterized topologically with strictly local properties and countability axioms* ◇ C65 C80 C90 ◇
MAKSIMOVA, L.L. *Interpolation theorems in modal logics and amalgamable varieties of topological boolean algebras (Russian)* ◇ B45 C05 C40 C90 G05 ◇
MARKOVIC, Z. *An intuitionistic omitting types theorem*
  ◇ C07 C90 F50 ◇
NAT VANDER, A. *Beyond non-normal possible worlds*
  ◇ B45 C90 ◇
NEPEJVODA, N.N. *Stable truth and computability (Russian)* ◇ C90 F35 F50 ◇
RAUTENBERG, W. *Klassische und nichtklassische Aussagenlogik* ◇ B45 B50 B98 C90 ◇
SAIN, I. *There are general rules for specifying semantics: Observations of abstract model theory*
  ◇ B75 C90 C95 E75 ◇
SCHWARTZ, DIETRICH *Untersuchungen ueber die Banach-Logik* ◇ C90 ◇
SHEKHTMAN, V.B. *Topological semantics of superintuitionistic logics (Russian)* ◇ B55 C90 ◇
SHIMODA, M. *Sheaves over cHa and the Heyting valued model $V^{(H)}$ (Japanese)* ◇ C90 E40 G30 ◇
SKVORTSOV, D.P. *Logic of infinite problems and Kripke models on atomic semilattices of sets (Russian)*
  ◇ B55 C90 F50 G20 ◇
SWEET, A.M. *Intended model theory* ◇ A05 C90 ◇
URSINI, A. *Two remarks on models in modal logics (Italian summary)* ◇ B45 C90 ◇
VOLGER, H. *Preservation theorems for limits of structures and global sections of sheaves of structures*
  ◇ C40 C60 C90 ◇
VOLGER, H. *Sheaf representations of algebras and transfer theorems for model theoretic properties*
  ◇ B25 C05 C30 C35 C90 ◇
WEINSTEIN, S. *Some applications of Kripke models to formal systems of intuitionistic analysis*
  ◇ C90 F35 F50 ◇
WEISPFENNING, V. *Lattice products*
  ◇ C10 C25 C30 C35 C90 ◇

1980

ANDREKA, H. & BURMEISTER, P. & NEMETI, I. *Quasi-equational logic of partial algebras*
  ◇ C05 C90 ◇
ARTEMOV, S.N. *Arithmetically complete modal theories (Russian)* ◇ B45 C90 F50 ◇
BOFFA, M. & CHERLIN, G.L. *Elimination des quantificateurs dans les faisceaux (English summary)*
  ◇ C10 C60 C90 ◇
BOZZI, S. & MELONI, G.C. *Representation of Heyting algebras with covering and propositional intuitionistic logic with local operator*
  ◇ B25 B55 C90 F50 G10 ◇
CHUAQUI, R.B. *Foundations of statistical methods using a semantical definition of probability* ◇ B48 C90 ◇
DAHN, B.I. *First order logics for metric structures*
  ◇ C65 C80 C90 ◇
FLUM, J. & ZIEGLER, M. *Topological model theory*
  ◇ C65 C80 C90 C98 ◇
FOURMAN, M.P. *Sheaf models for set theory*
  ◇ C62 C90 E25 E35 E70 F50 G30 ◇
GUITART, R. & LAIR, C. *Calcul syntaxique des modeles et calcul des formules internes* ◇ C90 G30 ◇
HEINDORF, L. *The decidability of the L_t - theory of boolean spaces* ◇ B25 C65 C80 C90 ◇
JOHNSTONE, P.T. *The Gleason cover of a topos. I*
  ◇ C90 G30 ◇
JONGH DE, D.H.J. *A class of intuitionistic connectives*
  ◇ B55 C75 C90 F50 ◇
KIRK, R.E. *A characterization of the classes of finite tree frames which are adequate for the intuitionistic logic*
  ◇ B55 C90 F50 ◇
LAVENDHOMME, R. & LUCAS, T. *Logique locale*
  ◇ C90 G30 ◇
LOPEZ-ESCOBAR, E.G.K. *Semantical models for intuitionistic logics* ◇ C90 C95 F50 ◇
MIRAGLIA, F. *Relations between structures of stable continuous functions and filtered powers*
  ◇ C30 C35 C90 ◇
MONTAGNA, F. *Some modal logics with quantifiers (Italian summary)* ◇ B45 C80 C90 ◇
NISHIMURA, H. *A preservation theorem for tense logic*
  ◇ B45 C20 C40 C90 ◇
NISHIMURA, H. *Saturated and special models in modal model theory with applications to the modal and de re hierarchies* ◇ B45 C40 C50 C90 ◇
POPOV, S.V. & ZAKHAR'YASHCHEV, M.V. *On the power of countermodels in intuitionistic calculus (Russian)*
  ◇ C90 F50 ◇
PROTASOV, I.V. *On the theory of topological algebraic systems (Russian) (English summary)* ◇ C90 ◇
RAUSZER, C. *An algebraic and Kripke-style approach to a certain extension of intuitionistic logic*
  ◇ B55 C20 C90 F50 G25 ◇
REITERMAN, J. & TRNKOVA, V. *Dynamic algebras which are not Kripke structures* ◇ B75 C90 G25 ◇
SCHULTZ, KONRAD *A topological model for Troelstra's system CS of intuitionistic analysis*
  ◇ C90 E70 F35 F50 ◇
SCOTT, D.S. *Lambda calculus: some models, some philosophy* ◇ A05 B40 C90 ◇
SCOTT, D.S. *Relating theories of the λ-calculus*
  ◇ B40 C90 F50 G30 ◇
SETTE, A.-M. *Functorial approach to interpretability*
  ◇ C90 F25 G30 ◇
SGRO, J. *Interpolation fails for the Souslin-Kleene closure of the open set quantifier logic*
  ◇ C40 C80 C90 C95 ◇
SGRO, J. *The interior operator logic and product topologies*
  ◇ C40 C55 C80 C90 ◇
SHEKHTMAN, V.B. *Topological models of propositional logics (Russian)* ◇ B45 C90 ◇
TAKAHASHI, MAKOTO *Topological powers and reduced powers* ◇ C20 C30 C90 ◇
TOFFALORI, C. *Fasci di coppie di campi algebricamente chiusi* ◇ B25 C35 C60 C90 ◇

TOFFALORI, C. *Sheaves of pairs of real closed fields*
⋄ C25 C35 C60 C90 ⋄

VAEAENAENEN, J. *Boolean-valued models and generalized quantifiers* ⋄ C55 C80 C85 C90 E35 E40 ⋄

VOJVODIC, G. *The Craig interpolation theorem for mixed-valued predicate calculi (Serbo-Croatian summary)* ⋄ B50 C40 C90 ⋄

WANG, SHIQIANG *A proof of the compactness theorem in lattice-valued model theory (Chinese) (English summary)* ⋄ C90 ⋄

ZHANG, JINWEN *A family of models of lattice-valued sets (Chinese) (English summary)* ⋄ C62 C90 E70 ⋄

1981

ANDREKA, H. & BURMEISTER, P. & NEMETI, I. *Quasivarieties of partial algebras. A unifying approach towards a two-valued model theory for partial algebras*
⋄ C05 C90 ⋄

BELL, J.L. *Isomorphism of structures in S-toposes*
⋄ C75 C90 ⋄

BOFFA, M. *Quantifier elimination in boolean sheaves*
⋄ C10 C90 G30 ⋄

BUNGE, MARTA C. & REYES, G.E. *Boolean spectra and model completions* ⋄ C35 C90 G30 ⋄

BUNGE, MARTA C. *Sheaves and prime model extensions*
⋄ C35 C60 C90 ⋄

BURMEISTER, P. *Quasi-equational logic for partial algebras* ⋄ C05 C90 ⋄

CHATZIDAKIS, Z. *La representation en termes de faisceaux des modeles de la theorie elementaire de la multiplication des entiers naturels*
⋄ C10 C30 C62 C90 ⋄

CHERLIN, G.L. & SCHMITT, P.H. *Undecidable L^t theories of topological abelian groups* ⋄ C60 C90 D35 ⋄

DAHN, B.I. *Partial isomorphisms and intuitionistic logic*
⋄ C75 C90 F50 ⋄

FINE, K. *Model theory for modal logic. III. Existence and predication* ⋄ A05 B45 C90 ⋄

GOLDBLATT, R.I. *Grothendieck topology as geometric modality* ⋄ B25 B45 C90 F50 G30 ⋄

GRAYSON, R.J. *Concepts of general topology in constructive mathematics and in sheaves*
⋄ C90 E35 E70 E75 F35 F50 F55 G30 ⋄

HEINDORF, L. *Comparing the expressive power of some languages for boolean algebras*
⋄ B25 C10 C80 C85 C90 G05 ⋄

HORIGUCHI, H. *A definition of the category of boolean-valued models* ⋄ C20 C30 C90 G30 ⋄

KEARNS, J.T. *A more satisfactory description of the semantics of justification* ⋄ A05 C90 ⋄

KOPPELBERG, S. *A lattice structure on the isomorphism types of complete Boolean algebras*
⋄ C50 C90 G05 G10 ⋄

KOZEN, D. *On the duality of dynamic algebras and Kripke models* ⋄ B75 C90 ⋄

LAIR, C. *Categories modelables et categories esquissables*
⋄ C90 G30 ⋄

LASCAR, D. *Dans quelle mesure la categorie des modeles d'une theorie complete determine-t-elle cette theorie?*
⋄ C35 C90 G30 ⋄

LAVENDHOMME, R. & LUCAS, T. *A propos d'un theoreme de Macintyre* ⋄ C35 C90 ⋄

LOPEZ-ESCOBAR, E.G.K. *On the interpolation theorem for the logic of constant domains*
⋄ B55 C40 C90 F50 ⋄

LU, JINGBO & WANG, SHIQIANG *On the fundamental theorem of ultraproducts for lattice-valued models (Chinese)* ⋄ C20 C90 ⋄

MAKKAI, M. *The topos of types* ⋄ C35 C90 G30 ⋄

MIGLIOLI, P.A. & MOSCATO, U. & ORNAGHI, M. *Trees in Kripke models and in an intuitionistic refutation system*
⋄ C90 F50 ⋄

MURAVITSKIJ, A.YU. *Strong equivalence on an intuitionistic Kripke model and assertorically equivolumetric logics (Russian)* ⋄ B55 C90 F50 ⋄

NISHIMURA, H. *Model theory for tense logic: Saturated and special models with applications to the tense hierarchy* ⋄ B45 C40 C50 C90 ⋄

NISHIMURA, H. *The semantical characterization of de dicto in continuous modal model theory*
⋄ B45 C40 C90 ⋄

POINT, F. *Elimination des quantificateurs dans les L-anneaux *-reguliers et application aux anneaux bireguliers* ⋄ C10 C60 C90 ⋄

PROTASOV, I.V. & SIDORCHUK, A.D. *On varieties of topological algebraic systems (Russian)* ⋄ C05 C90 ⋄

SHIMODA, M. *Categorical aspects of Heyting-valued models for intuitionistic set theory*
⋄ C90 E40 E70 F50 G30 ⋄

STACHNIAK, Z. *Introduction to model theory for Lesniewski's ontology* ⋄ C20 C90 E70 ⋄

TAKEUTI, G. & TITANI, S. *Heyting-valued universes of intuitionistic set theory* ⋄ C90 E40 E70 G30 ⋄

TAUTS, A. *The connection between semantic models and pseudo-boolean algebras (Russian) (Estonian and German summaries)* ⋄ C75 C90 F50 G10 ⋄

WANG, SHIQIANG *Lattice-valued atomic and countable saturated models (Chinese)* ⋄ C50 C90 ⋄

WANG, SHIQIANG *Two cases of application of the method of construction by constants in lattice-valued model theory (Chinese)* ⋄ C30 C90 ⋄

ZLATOS, P. *Two-levelled logic and model theory*
⋄ B50 C90 G25 ⋄

1982

ANDREKA, H. & NEMETI, I. & SAIN, I. *A complete logic for reasoning about programs via nonstandard model theory. I,II* ⋄ B75 C90 H10 H20 ⋄

ANDREKA, H. & NEMETI, I. *A general axiomatizability theorem formulated in terms of cone-injective subcategories* ⋄ C05 C20 C90 G30 ⋄

BOZZI, S. *Formule attuali ed ereditarie nella logica intuizionista* ⋄ C90 F50 ⋄

BUNGE, MARTA C. *On the transfer of an abstract Nullstellensatz* ⋄ C35 C60 C90 G30 ⋄

BURMEISTER, P. *Partial algebras – survey of a unifying approach towards a two-valued model theory for partial algebras* ⋄ C05 C90 C98 ⋄

COMER, S.D. *Boolean combinations of monadic formulas*
⋄ C07 C90 G15 ⋄

CZELAKOWSKI, J. *Logical matrices and the amalgamation property* ◇ B22 C40 C52 C90 ◇

DATUASHVILI, N.I. *Characterization of varieties of B-valued models (Russian) (English and Georgian summaries)* ◇ C05 C90 E40 ◇

EDA, K. *A boolean power and a direct product of abelian groups* ◇ C30 C60 C90 E40 E75 ◇

GABBAY, D.M. *Intuitionistic basis for non-monotonic logic* ◇ B55 C90 F50 ◇

GEORGESCU, G. *Une generalisation du theoreme d'omission des types dans les algebres polyadiques* ◇ C75 C90 G15 ◇

GUITART, R. *From where do figurative algebras come?* ◇ C90 G30 ◇

HOOVER, D.N. *A normal form theorem for $L_{\omega_1 p}$, with applications* ◇ C75 C90 ◇

HORIGUCHI, H. *Limits in the category of Boolean-valued models* ◇ C90 G30 ◇

HORIGUCHI, H. *Strict Boolean-valued models* ◇ C90 G30 ◇

KATZ, M. *Real-valued models with metric equality and uniformly continuous predicates* ◇ C90 ◇

KAWAI, H. *Eine Logik erster Stufe mit einem infinitaeren Zeitoperator* ◇ B45 C40 C75 C90 ◇

KUSRAEV, A.G. *Boolean-valued analysis of duality of extended modules (Russian)* ◇ C60 C65 C90 E40 E75 ◇

LAMBEK, J. & MOERDIJK, I. *Two sheaf representations of elementary toposes* ◇ C90 G30 ◇

LASCAR, D. *On the category of models of a complete theory* ◇ C35 C90 G30 ◇

LU, JINGBO *A note on infinite-lattice-valued model theory (Chinese) (English summary)* ◇ C90 ◇

LU, JINGBO & SHEN, CHENGMIN *Saturated models in lattice-valued model theory (Chinese) (English summary)* ◇ C50 C90 ◇

MAKKAI, M. *Full continuous embeddings of toposes* ◇ C35 C90 G30 ◇

MAKSIMOVA, L.L. *Absence of the interpolation property in modal companions of a Dummett logic (Russian)* ◇ B45 C40 C90 ◇

MAKSIMOVA, L.L. *Interpolation properties of superintuitionistic, positive and modal logics* ◇ B55 C40 C90 ◇

MAKSIMOVA, L.L. *Lyndon's interpolation theorem in modal logics (Russian)* ◇ B45 C40 C90 ◇

MARCHINI, C. *Realizations and witnesses for Kripke models* ◇ C90 F50 G30 ◇

MOENS, J.L. *Forcing et semantique de Kripke-Joyal* ◇ C25 C90 E40 G30 ◇

MORGAN, C.G. *Simple probabilistic semantics for propositional K, T, B, S4, and S5* ◇ B45 C90 ◇

MORGAN, C.G. *There is a probabilistic semantics for every extension of classical sentence logic* ◇ B45 C90 ◇

NEMETI, I. & SAIN, I. *Cone-implicational subcategories and some Birkhoff-type theorems* ◇ C05 C20 C90 G30 ◇

NEMETI, I. *On notions of factorization systems and their applications to cone-injective subcategories* ◇ C90 G10 G30 ◇

NEPEJVODA, N.N. *Constructive logics (Russian)* ◇ B60 C90 C95 F50 ◇

PROTASOV, I.V. *The local method and the compactness theorem for topological algebraic systems (Russian)* ◇ C90 ◇

RANTALA, V. *Quantified modal logic: non-normal worlds and propositional attitudes* ◇ B45 C90 ◇

SEGERBERG, K. *A completeness theorem in the modal logic of programs* ◇ B75 C90 ◇

SHEN, FUXING *Elementary extensions and elementary chains of lattice-valued models* ◇ C35 C40 C90 ◇

WANG, SHIQIANG *An omitting types theorem in lattice-valued model theory (Chinese)* ◇ C07 C75 C90 ◇

WERNER, H. *Sheaf constructions in universal algebra and model theory* ◇ B25 C05 C25 C30 C35 C90 G05 ◇

1983

ADAMEK, J. *Theory of mathematical structures* ◇ C90 C98 G30 ◇

ANDREKA, H. & NEMETI, I. *Generalization of the concept of variety and quasivariety to partial algebras through category theory* ◇ C05 C40 C75 C90 C95 G30 ◇

BELL, J.L. *Orthologic, forcing, and the manifestation of attributes* ◇ B51 C90 ◇

BELLISSIMA, F. & MIROLLI, M. *On the axiomatization of finite K-frames* ◇ B45 C13 C90 ◇

BENECKE, K. & REICHEL, HORST *Equational partiality* ◇ C05 C90 ◇

BENTHEM VAN, J.F.A.K. & HUMBERSTONE, I.L. *Hallden-completeness by gluing of Kripke frames* ◇ B45 C30 C90 ◇

BLOK, W.J. & KOEHLER, P. *Algebraic semantics for quasiclassical modal logics* ◇ B45 C90 G25 ◇

BORCEUX, F. & BOSSCHE VAN DEN, G. *Recovering a frame from its sheaves of algebras* ◇ C90 G30 ◇

BUGAJSKI, S. *Semantics in Banach spaces* ◇ B51 C65 C90 ◇

CHERLIN, G.L. & SCHMITT, P.H. *Locally pure topological abelian groups: Elementary invariants* ◇ B25 C60 C65 C80 C90 ◇

COMER, S.D. *Extension of polygroups by polygroups and their representations using color schemes* ◇ C05 C90 ◇

DIERS, Y. *Une description axiomatique des categories de faisceaux de structures algebriques sur les espaces topologiques booleens* ◇ C90 G30 ◇

EDA, K. *On a boolean power of a torsion free abelian group* ◇ C30 C60 C90 E40 E75 G05 ◇

EISENMENGER, H. *Some local definability results on countable topological structures* ◇ C15 C40 C90 ◇

GERLA, G. *Model theory and modal logic (Italian)* ◇ B45 C25 C75 C90 ◇

GORDON, E.I. & LYUBETSKIJ, V.A. *Boolean extensions of uniform structures (Russian)* ◇ C65 C90 E40 ◇

GORDON, E.I. & LYUBETSKIJ, V.A. *Embedding of sheaves in a Heyting-valued universe (Russian)* ◇ C90 E40 E70 G30 ◇

HIEN, BUIHUY & SAIN, I. *Category theoretic notions of ultraproducts* ◇ C20 C90 G30 ◇

HIEN, BUIHUY & SAIN, I. *In which categories are first-order axiomatizable hulls characterizable by ultraproducts?* ◇ C20 C90 G30 ◇

HOOGEWIJS, A. *A partial predicate calculus in a two-valued logic* ◇ B60 C90 ◇

HOOGEWIJS, A. *The non-definedness notion and forcing* ◇ C25 C90 ◇

JANKOWSKI, A.W. & RAUSZER, C. *Logical foundation approach to users' domain restriction in data bases* ◇ C90 ◇

KATZ, M. *Deduction systems and valuation spaces* ◇ C25 C90 G25 ◇

KLEINKNECHT, R. *Modal logic without possible worlds* ◇ A05 C90 ◇

KUSRAEV, A.G. *On some categories and functors of boolean-valued analysis (Russian)* ◇ C65 C90 E40 G30 ◇

KUSRAEV, A.G. & KUTATELADZE, S.S. *Subdifferentials in Boolean-valued models of set theory (Russian)* ◇ C65 C90 E40 E75 ◇

LEBLANC, H. & MORGAN, C.G. *Probability theory, intuitionism, semantics, and the Dutch book argument* ◇ B48 C90 F50 ◇

LEISS, H. *Implizit definierte Mengensysteme* ◇ C40 C65 C90 ◇

LOPEZ-ESCOBAR, E.G.K. *A second paper on the interpolation theorem for the logic of constant domains* ◇ B55 C40 C90 F05 F50 ◇

MARKOVIC, Z. *On reduced products of Kripke models* ◇ C20 C90 ◇

MARKOVIC, Z. *Some preservation results for classical and intuitionistic satisfiability in Kripke models* ◇ B20 C25 C90 F50 ◇

MINARI, P. *Completeness theorems for some intermediate predicate calculi* ◇ B55 C90 ◇

MURAVITSKIJ, A.YU. *Comparison of the topological and relational semantics of superintuitionistic logics (Russian)* ◇ B55 C90 ◇

ONO, H. *Model extension theorem and Craig's interpolation theorem for intermediate predicate logics* ◇ B55 C20 C40 C90 ◇

SCHMITT, P.H. *The L^t-theory of profinite abelian groups* ◇ B25 C50 C60 C80 C90 ◇

SIDORCHUK, A.D. *On the completeness theorem for topological models (Russian) (English summary)* ◇ C90 ◇

STEPANOV, V.I. *On model theory for intuitionistic logic (Russian)* ◇ C90 F50 ◇

ZAKHAR'YASHCHEV, M.V. *On intermediate logics (Russian)* ◇ B55 C90 ◇

1984

ALVES, E.H. *Paraconsistent logic and model theory* ◇ B53 C40 C52 C90 ◇

ANSHAKOV, O.M. & RYCHKOV, S.V. *Axiomatization of finite-valued logical calculi (Russian)* ◇ B50 C90 ◇

ANSHAKOV, O.M. & RYCHKOV, S.V. *On a way of the formalization and classification of many valued logics (Russian)* ◇ B50 C90 ◇

BANKSTON, P. *Expressive power in first order topology* ◇ C65 C90 ◇

BOZIC, M. & DOSEN, K. *Models for normal intuitionistic modal logics* ◇ B45 C90 ◇

BRADY, R.T. *Natural deduction systems for some quantified relevant logics* ◇ B46 C90 ◇

BUNGE, MARTA C. *Toposes in logic and logic in toposes* ◇ C90 E70 G30 ◇

CZELAKOWSKI, J. *Remarks on finitely based logics* ◇ B22 C90 ◇

DALEN VAN, D. *How to glue analysis models* ◇ C65 C90 F35 F50 ◇

EDA, K. *Maximal quotient rings and boolean extensions (Japanese)* ◇ C60 C90 E40 E75 ◇

EDA, K. & HIBINO, K. *On boolean powers of the group Z and (ω, ω)-weak distributivity* ◇ C30 C60 C90 E35 E50 E75 G05 ◇

FLEISCHER, I. *"Kripke semantics" = algebra + poetry (French summary)* ◇ C90 ◇

FLUM, J. *Modelltheorie - topologische Modelltheorie* ◇ C65 C90 C98 ◇

GABBAY, D.M. & GUENTHNER, F. (EDS.) *Handbook of philosophical logic Vol II. Extensions of classical logic* ◇ B98 C90 C98 ◇

GARSON, J.W. *Quantification in modal logic* ◇ B45 C90 C98 ◇

GERLA, G. & VACCARO, V. *Modal logic and model theory* ◇ B45 C25 C75 C90 ◇

GRAYSON, R.J. *Heyting-valued semantics* ◇ B15 C90 F35 F50 G30 ◇

GUMB, R.D. *"Conservative" Kripke-closures* ◇ C40 C90 ◇

HAILPERIN, T. *Probability logic* ◇ B25 B48 C90 ◇

HERRERA MIRANDA, J. *Sur la logique de premier ordre et la theorie des categories* ◇ C90 G30 ◇

HIEN, BUIHUY *Elementary classes in the injective subcategories approach to abstract model theory* ◇ C20 C90 G30 ◇

HOEVEN VAN DER, G.F. & MOERDIJK, I. *Constructing choice sequences from lawless sequences of neighbourhood functions* ◇ C90 F35 F50 ◇

HOEVEN VAN DER, G.F. & MOERDIJK, I. *Sheaf models for choice sequences* ◇ C90 F35 F50 G30 ◇

JOZSA, R. *Sheaf models and massless fields* ◇ B30 C90 F50 G30 ◇

KUSRAEV, A.G. *Order-continuous functionals in Boolean-valued models of set theory (Russian)* ◇ C65 C90 E40 ◇

MAKKAI, M. *A Stone-type representation theory for first order logic* ◇ C90 G30 ◇

MARKOVIC, Z. *Kripke models for intuitionistic theories with decidable atomic formulas* ◇ C90 F50 ◇

MARTINEZ, JUAN CARLOS *Accessible sets and $(L_{\omega_1 \omega})_t$-equivalence for T_3 spaces* ◇ C65 C75 C80 C90 ◇

MOERDIJK, I. *Heine-Borel does not imply the Fan theorem* ◇ C90 E35 F35 F50 G30 ◇

MONTAGNA, F. *The predicate modal logic of provability* ◇ B45 C90 F30 ◇

MUNDICI, D. *A generalization of abstract model theory*
⋄ C90 C95 G30 ⋄

MUNDICI, D. *Abstract model theory of many-valued logics and K-theory of certain C^*-algebras*
⋄ B50 C90 C95 ⋄

MURAVITSKIJ, A.YU. *A result on the completeness of superintuitionistic logics* ⋄ B55 C90 ⋄

OZAWA, M. *A classification of type I AW^*-algebras and Boolean valued analysis* ⋄ C65 C90 E40 E75 ⋄

SCEDROV, A. *Forcing and classifying topoi*
⋄ C90 E35 E40 F50 G30 ⋄

SCOWCROFT, P. *The real algebraic structure of Scott's model of intuitionistic analysis*
⋄ B25 C90 F35 F50 ⋄

SHEN, FUXING *Forcing in lattice-valued model theory (Chinese)* ⋄ C25 C90 ⋄

SMITH, KAY *Commutative regular rings and boolean valued fields* ⋄ C60 C90 E40 E75 ⋄

SZATKOWSKI, M. *Concerning Kripke semantics for intermediate predicate logics* ⋄ B55 C90 ⋄

TAKEUTI, G. *Quantum logic and quantization*
⋄ B51 C90 E40 E75 ⋄

VOLGER, H. *Filtered and stable boolean powers are relativized full boolean powers*
⋄ B25 C20 C30 C90 ⋄

YASHIN, A.D. *Completeness of the intuitionistic predicate calculus with the concept of "bar" (Russian)*
⋄ C90 F05 F50 ⋄

YASHIN, A.D. *Nishimura's formulas as one-place logical connectives in the elementary theory of Kripke models (Russian)* ⋄ C90 F50 ⋄

1985

BARWISE, J. & FEFERMAN, S. (EDS.) *Model-theoretic logics*
⋄ C75 C80 C85 C90 C95 C98 ⋄

BERTOSSI, L. & CHUAQUI, R.B. *Approximation to truth and theory of errors* ⋄ C90 H05 ⋄

BOZIC, M. *Semantics for some intermediate logics*
⋄ B55 C90 ⋄

DOSEN, K. *A completeness theorem for the Lambek calculus of syntactic categories* ⋄ B65 C90 D05 ⋄

FAJARDO, S. *Completeness theorems for the general theory of stochastic processes* ⋄ B48 C65 C90 ⋄

FAJARDO, S. *Probability logic with conditional expectation*
⋄ B48 C40 C90 ⋄

FLUM, J. & MARTINEZ, JUAN CARLOS *κ-local operations for topological structures* ⋄ C40 C75 C80 C90 ⋄

GEORGESCU, G. & VOICULESCU, I. *Eastern model-theory for boolean-valued theories* ⋄ C25 C35 C40 C90 ⋄

HOOVER, D.N. *A probabilistic interpolation theorem*
⋄ C40 C70 C90 ⋄

HORIGUCHI, H. *The category of boolean-valued models and its applications* ⋄ C30 C90 E40 G30 ⋄

JANKOWSKI, A.W. & ZAWADOWSKI, M. *Sheaves over Heyting lattices* ⋄ C90 F35 F50 G10 G30 ⋄

KEISLER, H.J. *Probability quantifiers*
⋄ C40 C80 C90 C98 ⋄

KOPPELBERG, S. *Booleschwertige Logik*
⋄ B50 C90 C98 E35 E45 E50 E65 G05 ⋄

LAVENDHOMME, R. & LUCAS, T. *A non-boolean version of Feferman-Vaught's theorem* ⋄ C30 C90 ⋄

LEONE, M. *On temporal invariance of menu systems*
⋄ B05 B75 C90 ⋄

LYUBETSKIJ, V.A. *Some algebraic questions of nonstandard analysis (Russian)* ⋄ C60 C90 H05 ⋄

MCCARTHY, T. *Abstraction and definability in semantically closed structures*
⋄ A05 C40 C62 C90 ⋄

MINARI, P. *Kripke-definable ordinals* ⋄ C90 E47 ⋄

NIEFIELD, S.B. & ROSENTHAL, K.I. *Strong de Morgan's law and the spectrum of a commutative ring*
⋄ C60 C90 G30 ⋄

ONO, H. *Semantical analysis of predicate logics without the contraction rule* ⋄ B10 C90 F50 ⋄

PAPPAS, P. *The model-theoretic structure of abelian group rings* ⋄ B25 C20 C60 C90 ⋄

RASIOWA, H. *Topological representations of Post algebras of order ω^+ and open theories based on ω^+-valued Post logic* ⋄ B50 C90 G20 ⋄

RASKOVIC, M. *Model theory for $L_{\mathfrak{A}M}$ logic* ⋄ C90 ⋄

RAUSZER, C. *Formalizations of certain intermediate logics. I* ⋄ B55 C40 C90 ⋄

SHEN, FUXING *Lattice-valued model complete theory (Chinese) (English summary)* ⋄ C35 C90 ⋄

STEINHORN, C.I. *Borel structures and measure and category logics* ⋄ C90 C98 ⋄

STEINHORN, C.I. *Borel structures for first-order and extended logics* ⋄ C80 C90 C98 ⋄

ZAWADOWSKI, M. *The Skolem-Loewenheim theorem in toposes II* ⋄ C55 C90 G30 ⋄

ZIEGLER, M. *Topological model theory* ⋄ C90 C98 ⋄

ZLATOS, P. *On conceptual completeness of syntactic-semantical systems*
⋄ C05 C20 C90 G15 G30 ⋄

C97 Proceedings

1965

ADDISON, J.W. & HENKIN, L. & TARSKI, A. (EDS.) *The theory of models. Proceedings of the 1963 International Symposium at Berkeley* ◊ C97 ◊

1967

CROSSLEY, J.N. (ED.) *Sets, models and recursion theory* ◊ B97 C97 D97 ◊

1968

BARWISE, J. (ED.) *The syntax and semantics of infinitary languages* ◊ C70 C75 C97 ◊

LOEB, M.H. (ED.) *Proceedings of the summer school in logic, Leeds, 1967* ◊ B97 C97 F97 ◊

1969

DEJON, B. & HENRICI, P. (EDS.) *Constructive aspects of the fundamental theorem of algebra: proceedings of a symposium, IBM* ◊ B28 C97 F97 ◊

LUXEMBURG, W.A.J. (ED.) *Applications of model theory to algebra, analysis and probability* ◊ C97 H97 ◊

1972

HODGES, W. (ED.) *Conference in Mathematical Logic -- London '70* ◊ B97 C97 ◊

1973

BELL, J.L. & COLE, J. & PRIEST, G. & SLOMSON, A. *Proceedings of the Bertrand Russell Memorial Logic Conference* ◊ C97 ◊

1974

ADDISON, J.W. & CHANG, C.C. & CRAIG, W. & HENKIN, L. & SCOTT, D.S. & VAUGHT, R.L. (EDS.) *Proceedings of the Tarski Symposium* ◊ B97 C97 ◊

1975

CROSSLEY, J.N. (ED.) *Algebra and logic* ◊ B97 C97 G97 ◊

DILLER, J. & MUELLER, GERT H. (EDS.) ⊦ *ISILC proof theory symposium* ◊ C97 F97 ◊

KUEKER, D.W. (ED.) *Infinitary logic: In memoriam Carol Karp* ◊ C70 C75 C97 ◊

LAWVERE, F.W. & MAURER, C. & WRAITH, G.C. (EDS.) *Model theory and topoi* ◊ C97 F97 G97 ◊

MANGANI, P. (ED.) *Model Theory and Applications. Centro Internazionale Matematico Estivo (CIME) II Ciclo* ◊ C97 ◊

SARACINO, D.H. & WEISPFENNING, V. (EDS.) *Model theory and algebra. A memorial tribute to Abraham Robinson* ◊ C60 C97 ◊

1977

ARRUDA, A.I. & CHUAQUI, R.B. & COSTA DA, N.C.A. (EDS.) *Nonclassical logics, model theory and computability. Proceedings of the Third Latin-American Symposium on Mathematical Logic* ◊ B97 C97 ◊

GANDY, R.O. & HYLAND, J.M.E. (EDS.) *Logic Colloquium 76. Proceedings of a conference held in Oxford in July 1976* ◊ B97 C97 D97 ◊

HENRARD, P. (ED.) *Six days of model theory. Proceedings of a conference held at Louvain-la-Neuve in March 1975* ◊ C97 ◊

MIETTINEN, S. & VAEAENAENEN, J. (EDS.) *Proceedings of the Symposium on Mathematical Logic in Oulu 1974 and in Helsinki 1975* ◊ C97 ◊

1978

MACINTYRE, A. & PACHOLSKI, L. & PARIS, J.B. (EDS.) *Logic colloquium '77. Proceedings of the colloquium held in Wroclaw, august 1977* ◊ B97 C97 ◊

POIZAT, B. (ED.) *Groupe d'etude de theories stables. 1re annee: 1977/78. Exposes 1 a 9* ◊ C45 C97 C98 ◊

1979

FOURMAN, M.P. & MULVEY, C.J. & SCOTT, D.S. (EDS.) *Applications of sheaves* ◊ C90 C97 G97 ◊

1980

BAJZHANOV, B.S. & TAJTSLIN, M.A. (EDS.) *Theory of models and its applications (Russian)* ◊ C97 ◊

MCALOON, K. (ED.) *Modeles de l'arithmetique* ◊ C62 C97 F30 ◊

PACHOLSKI, L. & WIERZEJEWSKI, J. & WILKIE, A.J. (EDS.) *Model theory of algebra and arithmetic. Proceedings, Karpacz, Poland, September 1-7, 1979* ◊ C60 C62 C97 ◊

1981

BERLINE, C. & MCALOON, K. & RESSAYRE, J.-P. (EDS.) *Model theory and arithmetic. Comptes rendus d'une action thematique programmee du C.N.R.S. sur la theorie des modeles et l'arithmetique* ◊ C62 C97 F30 ◊

CROSSLEY, J.N. (ED.) *Aspects of effective algebra. Proceedings of a conference at Monash University, Australia, 1-4 August, 1979* ◊ C57 C97 D45 D97 ◊

GOEBEL, R. & WALKER, E.A. (EDS.) *Abelian group theory* ◊ C97 ◊

GUZICKI, W. & MAREK, W. & PELC, A. & RAUSZER, C. (EDS.) *Proceedings of the International Conference "Open Days in Model Theory and Set Theory"* ◊ C97 ◊

HERRE, H. (ED.) *Workshop on extended model theory* ◊ C97 G05 ◊

JENSEN, R.B. & PRESTEL, A. (EDS.) *Set theory and model theory. Proceedings of an informal symposium, held at Bonn, June 1-3, 1979* ◊ C97 E97 ◊

LERMAN, M. & SCHMERL, J.H. & SOARE, R.I. (EDS.) *Logic year 1979-80, The University of Connecticut, USA*
⋄ B97 C97 ⋄

POIZAT, B. (ED.) *Groupe d'etude de theories stables. 2e annee 1978/1979. Exposes 1 a 11*
⋄ C45 C97 C98 ⋄

1982

METAKIDES, G. (ED.) *Patras logic symposium*
⋄ B97 C97 ⋄

SOBOLEV, S.L. (ED.) *Mathematical logic and the theory of algorithms (Russian)* ⋄ B97 C97 D97 ⋄

TRACZYK, T. (ED.) *Universal algebra and applications. Papers presented at Stefan Banach International Center at the Semester "Universal Algebra and Applications" held February 15 - June 9, 1978* ⋄ C97 ⋄

1983

BERNARDI, C. (ED.) *Atti degli incontri di logica matematica* ⋄ B97 C97 ⋄

CUPONA, G. & MIJAJLOVIC, Z. & MILIC, S. & PERIC, V. & PRESIC, S.B. (EDS.) *Proceedings of the third algebraic conference* ⋄ C97 G05 ⋄

DAHN, B.I. (ED.) *Proceedings of the first Easter conference on model theory* ⋄ C97 ⋄

MEULEN TER, A.G.B. (ED.) *Studies in modeltheoretic semantics* ⋄ B97 C97 ⋄

POIZAT, B. (ED.) *Groupe d'etude de theories stables. 3e annee 1980-1982* ⋄ C45 C97 C98 ⋄

1984

DAHN, B.I. & WOLTER, H. (EDS.) *Proceedings of the second Easter conference on model theory, Wittenberg (GDR), April 23-27,1984* ⋄ C97 ⋄

DELON, F. & LASCAR, D. & PARIGOT, M. & SABBAGH, G. (EDS.) *Logique, Octobre 1983, Paris. Compte rendu de la table ronde de logique des 15 et 16 octobre 1983, Paris* ⋄ B97 C97 ⋄

LOLLI, G. & LONGO, G. & MARCJA, A. (EDS.) *Logic colloquium '82. Proceedings of the colloquium held in Florence, 23-28 August, 1982*
⋄ B97 C97 D97 F97 ⋄

LOLLI, G. & LONGO, G. & MARCJA, A. (EDS.) *Proceedings of the Logic Colloqium '82* ⋄ C97 ⋄

MUELLER, GERT H. & RICHTER, M.M. (EDS.) *Models and sets* ⋄ B97 C97 D97 H97 ⋄

POUZET, M. & RICHARD, D. (EDS.) *Orders: description and roles in set theory, lattices, ordered groups, topology, theory of models and relations, combinatorics, effectiveness, social sciences* ⋄ C65 C97 ⋄

WECHSUNG, G. (ED.) *Frege conference 1984. Proceedings of the International Conference held at Schwerin, Sept.10-14, 1984* ⋄ A05 A10 B97 C97 ⋄

1985

HARRINGTON, L.A. & MORLEY, M.D. & SCEDROV, A. & SIMPSON, S.G. *Harvey Friedman's research on the foundations of mathematics*
⋄ B97 C97 D97 E97 F97 ⋄

PRISCO DI, C.A. (ED.) *Methods in mathematical logic. Proceedings of the 6th Latin American Symposium on Mathematical Logic held in Caracas, Venezuela, Aug. 1-6, 1983* ⋄ B97 C97 ⋄

C98 Textbooks, surveys

1932
LANGFORD, C.H. & LEWIS, C.I. *Symbolic logic*
⋄ B25 B98 C10 C98 ⋄

1951
ROBINSON, A. *On the metamathematics of algebra*
⋄ C07 C60 C98 ⋄

1954
FUENTES MIRAS, J.R. & JOSE, R. *"Deductive truth" according to Tarski (Spanish)* ⋄ C07 C98 ⋄

1955
ROBINSON, A. *Theorie metamathematique des ideaux*
⋄ C60 C98 ⋄

1956
KREISEL, G. *Some uses of metamathematics*
⋄ A05 C98 F99 ⋄
ROBINSON, A. *Complete theories*
⋄ B25 C35 C60 C98 ⋄

1957
TARSKI, A. *Introductory remarks on the theory of models*
⋄ C98 ⋄

1958
MOSTOWSKI, ANDRZEJ *Quelques applications de la topologie a la logique mathematique*
⋄ C90 C98 E15 ⋄
ROBINSON, A. *Relative model-completeness and the elimination of quantifiers (English, French, and German summaries)* ⋄ C10 C35 C98 ⋄

1959
ROBINSON, A. *Outline of an introduction to mathematical logic IV* ⋄ B10 B98 C07 C98 G05 ⋄

1961
MAL'TSEV, A.I. *Constructive algebra I (Russian)*
⋄ C57 C98 F60 F98 ⋄

1962
NEUMANN, B.H. *Special topics in algebra: Universal algebra* ⋄ C05 C98 ⋄
ROBINSON, A. *Recent developments in model theory*
⋄ C98 ⋄

1963
DAIGNEAULT, A. *Theorie des modeles en logique mathematique* ⋄ C98 G05 G15 ⋄
MAL'TSEV, A.I. *Some problems of the theory of classes of models (Russian)* ⋄ C98 ⋄
RASIOWA, H. & SIKORSKI, R. *The mathematics of metamathematics*
⋄ B98 C98 F50 F98 G05 G10 G98 ⋄

ROBINSON, A. *Introduction to model theory and to the metamathematics of algebra*
⋄ B98 C60 C98 H15 ⋄
VAUGHT, R.L. *Models of complete theories* ⋄ C98 ⋄

1964
KARP, C.R. *Languages with expressions of infinite length*
⋄ C75 C98 G05 ⋄

1965
ADDISON, J.W. & BOUVERE DE, K.L. & PITT, W.B. *A bibliography of the theory of models* ⋄ A10 C98 ⋄
BELL, J.L. & SLOMSON, A. *Introduction to model theory*
⋄ C98 ⋄
COHN, P.M. *Universal Algebra* ⋄ C05 C98 ⋄
EHRESMANN, C. *Categories et structures*
⋄ C90 C98 G30 ⋄
ERSHOV, YU.L. & LAVROV, I.A. & TAJMANOV, A.D. & TAJTSLIN, M.A. *Elementary theories (Russian)*
⋄ B25 C98 D35 D98 ⋄
KEISLER, H.J. *A survey of ultraproducts*
⋄ C20 C55 C98 E05 ⋄
MOSTOWSKI, ANDRZEJ *Thirty years of foundational studies. Lectures on the development of mathematical logic and the studies of the foundations of mathematics in 1930-1964* ⋄ A10 B98 C98 D98 E98 F98 ⋄
MUELLER, GERT H. *Der Modellbegriff in der Mathematik*
⋄ C07 C98 ⋄
SCOTT, D.S. *Logic with denumerably long formulas and finite strings of quantifiers* ⋄ C15 C40 C75 C98 ⋄

1966
CHANG, C.C. & KEISLER, H.J. *Continuous model theory*
⋄ B50 C05 C20 C50 C90 C98 ⋄
KOPPERMAN, R.D. *Introductory notes on model theory (Spanish summary)* ⋄ C98 ⋄
LYNDON, R.C. *Notes on logic* ⋄ C98 G98 ⋄
ROBINSON, A. *On some applications of model theory to algebra and analysis* ⋄ C60 C98 H05 H98 ⋄

1967
FRAISSE, R. *Cours de logique mathematique. Tome I. Relation, formule logique, compacite, completude*
⋄ B98 C98 ⋄
JOHNSON, J. & SEIFERT, R.L. *A survey of multiunary algebras* ⋄ C05 C98 ⋄
KREISEL, G. & KRIVINE, J.-L. *Elements de logique mathematique. Theorie des modeles* ⋄ B98 C98 ⋄
NEUMANN, H. *Varieties of groups* ⋄ C05 C60 C98 ⋄
SHOENFIELD, J.R. *Mathematical logic*
⋄ B98 C98 D98 E98 F98 ⋄

1968

DALLA CHIARA SCABIA, M.L. *Modelli sintattici e semantici delle teorie elementari*
⋄ C07 C98 E35 E45 F25 F98 ⋄

GRAETZER, G. *Universal algebra* ⋄ C05 C98 ⋄

NEUMANN, B.H. *Lectures on topics in the theory of infinite groups* ⋄ C05 C60 C98 D40 ⋄

PIERCE, R.S. *Introduction to the theory of abstract algebras* ⋄ C05 C98 ⋄

ROBINSON, A. *Model theory* ⋄ C98 ⋄

VAUGHT, R.L. *Model theory and set theory*
⋄ C55 C80 C98 ⋄

1969

BELL, J.L. & SLOMSON, A. *Models and ultraproducts: An introduction* ⋄ C20 C98 ⋄

FITTING, M. *Intuitionistic logic, model theory, and forcing*
⋄ B98 C90 C98 E25 E35 E45 E50 F50 F98 ⋄

ROBINSON, A. *Problems and methods of model theory*
⋄ C60 C98 ⋄

1970

DICKMANN, M.A. *Model theory of infinitary languages. Vol. I, II* ⋄ C75 C98 E55 ⋄

MAL'TSEV, A.I. *Algebraic systems (Russian)*
⋄ C05 C98 ⋄

SPERANZA, F. *Relazioni e strutture* ⋄ C98 E98 ⋄

TAJTSLIN, M.A. *Model theory (Russian)* ⋄ B98 C98 ⋄

1971

AX, J. *A metamathematical approach to some problems in number theory* ⋄ C20 C60 C98 ⋄

DRABBE, J. *Quelques notions de la theorie des modeles*
⋄ C07 C98 ⋄

GRZEGORCZYK, A. *Lezioni sulla teoria dei modelli (raccolte a cura di Marisa Venturini Zilli)* ⋄ C98 ⋄

HENKIN, L. *Mathematical foundations for mathematics*
⋄ A10 C98 E30 E55 F55 G05 G15 ⋄

KEISLER, H.J. *A short course in model theory* ⋄ C98 ⋄

KEISLER, H.J. *Model theory* ⋄ C98 ⋄

KEISLER, H.J. *Model theory for infinitary logic. Logic with countable conjunctions and finite quantifiers*
⋄ C70 C75 C98 ⋄

ROGERS, R. *Mathematical logic and formalized theories. A survey of basic concepts and results* ⋄ B98 C98 ⋄

SABBAGH, G. *Logique mathematique III. Theorie des modeles* ⋄ C98 ⋄

SCHWABHAEUSER, W. *Modelltheorie I* ⋄ C98 ⋄

ZYGMUNT, J. *A survey of the methods of proof of the Goedel-Malcev's completeness theorem*
⋄ A10 C07 C98 ⋄

1972

ENDERTON, H.B. *A mathematical introduction to logic*
⋄ B98 C98 F30 ⋄

FRAISSE, R. *Cours de logique mathematique. Tome 2: Theorie des modeles* ⋄ B98 C98 ⋄

HEIDEMA, J. *Aspects of model theory (Africaans) (English summary)* ⋄ C98 ⋄

HOFMANN, K.H. *Representation of algebras by continuous sections* ⋄ C60 C90 C98 ⋄

KOPPERMAN, R.D. *Model theory and its applications*
⋄ C60 C98 ⋄

MAREK, W. & ONYSZKIEWICZ, J. *Elements of logic and foundations of mathematics in problems (Polish)*
⋄ B98 C98 D98 E98 ⋄

SACKS, G.E. *Saturated model theory* ⋄ C98 ⋄

SACKS, G.E. *The differential closure of a differential field*
⋄ C60 C98 ⋄

SCHWABHAEUSER, W. *Modelltheorie II* ⋄ C98 ⋄

SMIRNOV, V.A. *Formal derivation and logical calculi (Russian)* ⋄ B98 C07 C40 C98 F05 F07 F98 ⋄

1973

ACZEL, P. *Infinitary logic and the Barwise compactness theorem* ⋄ C70 C98 ⋄

BLAQUIER, M. *Formal systems and their models (Spanish)*
⋄ C98 ⋄

CHANG, C.C. & KEISLER, H.J. *Model theory* ⋄ C98 ⋄

DICKMANN, M.A. *The problem of non-finite axiomatizability of $\aleph_1$-categorical theories*
⋄ C35 C52 C98 ⋄

ERSHOV, YU.L. & PALYUTIN, E.A. & TAJTSLIN, M.A. *Mathematical logic (Russian)* ⋄ B98 C98 ⋄

GABBAY, D.M. *A survey of decidability results for modal, tense and intermediate logics*
⋄ B25 B45 B55 C98 F50 ⋄

HEYTING, A. *Address to Professor A. Robinson*
⋄ A10 C98 ⋄

KEGEL, O.H. & WEHRFRITZ, B.A.F. *Locally finite groups*
⋄ C60 C98 ⋄

McKENZIE, R. *Some unsolvable problems between lattice theory and equational logic* ⋄ C05 C98 G10 ⋄

MORLEY, M.D. *Studies in model theory* ⋄ C98 ⋄

ROBINSON, A. *Metamathematical problems*
⋄ A05 B30 C60 C98 H05 ⋄

ROBINSON, A. *Model theory as a framework for algebra*
⋄ C25 C35 C60 C98 ⋄

ROBINSON, A. *Nonstandard arithmetic and generic arithmetic* ⋄ C25 C60 C62 C98 H15 ⋄

ROBINSON, A. *Standard and nonstandard number systems* ⋄ C25 C60 C98 H05 H15 H20 H98 ⋄

VAUGHT, R.L. *Some aspects of the theory of models*
⋄ C98 ⋄

1974

CHANG, C.C. *Model theory 1945-1971* ⋄ A10 C98 ⋄

COMFORT, W.W. & NEGREPONTIS, S. *The theory of ultrafilters*
⋄ C20 C55 C98 E05 E55 E75 E98 G05 ⋄

CROSSLEY, J.N. *Satisfaction (a brief survey of model theory)* ⋄ C07 C98 ⋄

ERSHOV, YU.L. *Theory of numerations, III: Constructive models (Russian)* ⋄ B25 C57 C98 D45 F50 ⋄

GOSTANIAN, R. *Lectures on model theory. Part I*
⋄ C98 ⋄

JONSSON, B. *Some recent trends in general algebra*
⋄ C05 C98 ⋄

MAKOWSKY, J.A. *On some conjectures connected with complete sentences* ⋄ C35 C45 C52 C98 ⋄

SCHWABHAEUSER, W. *Modelltheorie* ⋄ C98 ⋄

SIMMONS, H. *Companion theories (Forcing in model theory)* ⋄ C25 C98 ⋄
TAKAHASHI, S. *Methodes logiques en geometrie diophantienne* ⋄ B28 C98 D98 G30 H98 ⋄
VAUGHT, R.L. *Model theory before 1945* ⋄ A10 C98 ⋄

1975

BARWISE, J. *Admissible sets and the interaction of model theory, recursion theory and set theory*
⋄ C70 C98 D55 D60 D70 E30 E35 E45 E47 ⋄
BARWISE, J. *Admissible sets and structures. An approach to definability theory*
⋄ B98 C40 C70 C98 D60 D98 E30 E98 ⋄
CROSSLEY, J.N. *A brief survey of model theory*
⋄ C07 C98 ⋄
DICKMANN, M.A. *Large infinitary languages; model theory* ⋄ C75 C98 E55 ⋄
EBBINGHAUS, H.-D. *Modelltheorie und Logik*
⋄ C95 C98 ⋄
FLUM, J. *First-order logic and its extensions*
⋄ C40 C55 C75 C95 C98 ⋄
FRIEDMAN, H.M. *One hundred and two problems in mathematical logic* ⋄ B98 C98 D98 E98 F98 ⋄
HIRSCHFELD, J. & WHEELER, W.H. *Forcing, arithmetic, division rings* ⋄ C25 C60 C62 C98 ⋄
KEISLER, H.J. *Constructions in model theory*
⋄ C25 C30 C50 C75 C95 C98 ⋄
KUEKER, D.W. *Back-and-forth arguments and infinitary logics* ⋄ C75 C98 ⋄
PIGOZZI, D. *Equational logic and equational theories of algebras* ⋄ C05 C98 ⋄
PRESTEL, A. *Lectures on formally real fields*
⋄ C60 C98 ⋄
SHELAH, S. *The lazy model theoretician's guide to stability*
⋄ C25 C35 C45 C50 C60 C98 ⋄
SIMMONS, H. *Counting countable e.c. structures*
⋄ C15 C25 C50 C98 ⋄

1976

CHERLIN, G.L. *Model theoretic algebra. Selected topics*
⋄ C25 C60 C98 ⋄
FLUM, J. *Algebra y logica* ⋄ C60 C98 ⋄
HALKOWSKA, K. & PIROG-RZEPECKA, K. & SLUPECKI, J. *Mathematical logic (Polish)* ⋄ B98 C07 C98 E98 ⋄
JEZEK, J. *Universal algebra and model theory (Czech)*
⋄ C05 C98 ⋄
MAKKAI, M. *Review of Chang and Keisler: model theory*
⋄ C98 ⋄
MURAWSKI, R. *On expandability of models of Peano arithmetic. I,II* ⋄ C62 C98 ⋄
SACERDOTE, G.S. *Some logical problems concerning free and free product groups* ⋄ C60 C98 ⋄
SLOMSON, A. *Decision problems for generalized quantifiers - a survey* ⋄ B25 C80 C98 ⋄
WHEELER, W.H. *Model-companions and definability in existentially complete structures*
⋄ C25 C35 C60 C98 ⋄
WOOD, C. *The model theory of differential fields revisited*
⋄ C60 C98 ⋄
WRAITH, G.C. *Logic from topology: a survey of topoi*
⋄ C98 F98 G30 ⋄

1977

BARWISE, J. *An introduction to first-order logic*
⋄ B10 C07 C98 ⋄
BARWISE, J. & KEISLER, H.J. *Guide to part A: Model theory* ⋄ C98 ⋄
BARWISE, J. (ED.) *Handbook of mathematical logic*
⋄ B98 C98 D98 E98 F98 H98 ⋄
BENDA, M. *Some directions in model theory* ⋄ C98 ⋄
BRIDGE, J. *Beginning model theory. The completeness theorem and some consequences* ⋄ C07 C98 ⋄
EKLOF, P.C. *Ultraproducts for algebraists*
⋄ C20 C60 C98 ⋄
KEISLER, H.J. *Fundamentals of model theory* ⋄ C98 ⋄
LASCAR, D. *La theorie des modeles: histoire et orientations actuelles* ⋄ C98 ⋄
LOLLI, G. *Indiscernibili in teoria dei modelli e in teoria degli insiemi (English summary)*
⋄ C30 C98 E05 E45 E55 ⋄
MACINTYRE, A. *Model completeness*
⋄ C25 C35 C60 C98 ⋄
MAKKAI, M. *Admissible sets and infinitary logic*
⋄ C70 C98 D55 D60 D70 D98 ⋄
MORLEY, M.D. *Homogenous sets*
⋄ C30 C55 C98 E05 E55 ⋄
MURAWSKI, R. *On expandability of models of Peano arithmetic. III* ⋄ C62 C98 ⋄
PRESTEL, A. *An introduction to ultraproducts*
⋄ C20 C98 ⋄
RABIN, M.O. *Decidable theories* ⋄ B25 C98 D20 ⋄
SHOENFIELD, J.R. *Quantifier elimination in fields*
⋄ C10 C60 C98 ⋄
STILLWELL, J.C. *Concise survey of mathematical logic*
⋄ B98 C98 D10 D35 ⋄
TAYLOR, W. *Equational logic* ⋄ C05 C98 D40 ⋄

1978

EBBINGHAUS, H.-D. & FLUM, J. & THOMAS, WOLFGANG *Einfuehrung in die mathematische Logik* ⋄ B98 C95 C98 ⋄
EHRSAM, S. *Expose elementaire de la theorie de la stabilite*
⋄ C45 C98 ⋄
EHRSAM, S. *L'ordre fondamental, le theoreme de la borne, le theoreme de la relation d'equivalence finie*
⋄ C20 C45 C98 ⋄
KOKORIN, A.I. & PINUS, A.G. *Decidability problems of extended theories (Russian)*
⋄ B25 C85 C98 D35 ⋄
LIGHTSTONE, A.H. *Mathematical logic. An introduction to model theory. Edited by H. B. Enderton*
⋄ B98 C98 H05 ⋄
POIZAT, B. (ED.) *Groupe d'etude de theories stables. 1re annee: 1977/78. Exposes 1 a 9* ⋄ C45 C97 C98 ⋄
PRESTEL, A. *Artin's conjecture on p-adic number fields*
⋄ C60 C98 ⋄
SEESE, D.G. *Decidability and generalized quantifiers*
⋄ B25 C80 C98 ⋄
SHELAH, S. *Classification theory and the number of nonisomorphic models* ⋄ C45 C98 ⋄

1979

DREBEN, B. & GOLDFARB, W.D. *The decision problem. Solvable classes of quantificational formulas*
⋄ B25 C98 ⋄

FERRANTE, J. & RACKOFF, C.W. *The computational complexity of logical theories*
⋄ B25 C98 D10 D15 ⋄

FOURMAN, M.P. & SCOTT, D.S. *Sheaves and logic*
⋄ C90 C98 E70 F35 F50 G30 ⋄

GOLDBLATT, R.I. *Topoi. The categorial analysis of logic*
⋄ B98 C98 E98 F35 F50 F98 G30 ⋄

HERRE, H. & WOLTER, H. *The decision problem for the theory of linear orderings in extended logics*
⋄ B25 C55 C80 C98 ⋄

MALITZ, J. *Introduction to mathematical logic. Set theory, computable functions, model theory*
⋄ B98 C98 D98 ⋄

PRESTEL, A. *Entscheidbarkeit mathematischer Theorien*
⋄ B25 C60 C98 D35 ⋄

TAYLOR, W. *Equational logic* ⋄ C05 C98 ⋄

1980

BAUDISCH, A. & SEESE, D.G. & TUSCHIK, H.-P. & WEESE, M. *Decidability and generalized quantifiers*
⋄ B25 B98 C10 C60 C80 C98 ⋄

BOERGER, E. *Entscheidbarkeit* ⋄ B25 C98 ⋄

BOFFA, M. *Elimination des quantificateurs en algebre*
⋄ C10 C98 ⋄

BUCUR, I. *Special chapters of algebra (Romanian)*
⋄ B80 C98 D98 ⋄

DALEN VAN, D. *Logic and structure* ⋄ B98 C98 ⋄

DOUWEN VAN, E.K. & MONK, J.D. & RUBIN, M. *Some questions about boolean algebras*
⋄ C98 G05 G98 ⋄

ERSHOV, YU.L. *Decision problems and constructivizable models (Russian)*
⋄ B25 C57 C60 C98 D45 D98 ⋄

FELGNER, U. *Kategorizitaet* ⋄ C35 C45 C60 C98 ⋄

FLUM, J. & ZIEGLER, M. *Topological model theory*
⋄ C65 C80 C90 C98 ⋄

HEINRICH, S. *Ultraproducts in Banach space theory*
⋄ C20 C65 C98 H05 ⋄

KREISEL, G. *Modell* ⋄ C98 ⋄

KREISEL, G. *Modelltheorie* ⋄ C98 ⋄

LADNER, R.E. *Complexity theory with emphasis on the complexity of logical theories*
⋄ B25 C98 D05 D10 D15 ⋄

1981

ANDREKA, H. & NEMETI, I. *Some universal algebraic and model theoretic results in computer science*
⋄ B75 C05 C98 ⋄

BARWISE, J. *Infinitary logics* ⋄ C75 C98 ⋄

BERLINE, C. *Stabilite et algebre I. Pratique de la stabilite et de la categoricite* ⋄ C35 C45 C98 ⋄

BERLINE, C. *Stabilite et algebre II. Groupes Abeliens et modules* ⋄ C35 C45 C60 C98 ⋄

BERLINE, C. *Stabilite et algebre III. Anneaux*
⋄ C45 C60 C98 ⋄

BERLINE, C. *Stabilite et algebre IV. Groupes*
⋄ C45 C60 C98 ⋄

BURRIS, S. & SANKAPPANAVAR, H.P. *A course in universal algebra* ⋄ C05 C98 ⋄

CLOTE, P. *Anti-basis theorems and their relation to independence results in Peano arithmetic*
⋄ C62 C98 E05 F30 ⋄

GLASS, A.M.W. *Elementary types of automorphisms of linearly ordered sets -- a survey* ⋄ C07 C65 C98 ⋄

GLASS, A.M.W. *Ordered permutation groups*
⋄ C07 C60 C65 C98 ⋄

HERRE, H. *Miscellaneous results and problems in extended model theory* ⋄ B25 C65 C80 C98 D35 ⋄

KRAJEWSKI, S. *Kurt Goedel and his work (Polish)*
⋄ A10 C98 D98 E98 F99 ⋄

MACINTYRE, A. *Model theory* ⋄ C98 ⋄

PILLAY, A. *Models of Peano arithmetic (a survey of basic results)* ⋄ C62 C98 H15 ⋄

POIZAT, B. (ED.) *Groupe d'etude de theories stables. 2e annee 1978/1979. Exposes 1 a 11*
⋄ C45 C97 C98 ⋄

POIZAT, B. *Le rang selon Lascar* ⋄ C45 C98 ⋄

POTTHOFF, K. *Einfuehrung in die Modelltheorie und ihre Anwendungen* ⋄ C98 ⋄

RICHTER, M.M. & SCHLOESSER, B. & SCHWARZ, D. *Modelltheorie* ⋄ C98 ⋄

RICHTER, M.M. & TRIESCH, E. *Universelle Algebra*
⋄ C05 C98 ⋄

WANG, HAO *Popular lectures on mathematical logic (Chinese)* ⋄ A05 B98 C98 E98 F30 F98 ⋄

1982

BONNET, R. & POUZET, M. *Linear extensions of ordered sets* ⋄ C65 C98 E07 ⋄

BURMEISTER, P. *Partial algebras - survey of a unifying approach towards a two-valued model theory for partial algebras* ⋄ C05 C90 C98 ⋄

COLLINS, G.E. *Quantifier elimination for real closed fields: a guide to the literature* ⋄ C10 C60 C98 ⋄

ERSHOV, YU.L. *Algorithmic problems in the theory of fields (positive aspects) (Russian)* ⋄ B25 C10 C60 C98 ⋄

ERSHOV, YU.L. *Multiply valued fields (Russian)*
⋄ B25 C25 C35 C60 C98 ⋄

JONSSON, B. *Arithmetic of ordered sets*
⋄ C65 C98 E07 ⋄

MUNDICI, D. *Lectures on abstract model theory I,II,III*
⋄ C95 C98 ⋄

QUACKENBUSH, R.W. *Enumeration in classes of ordered structures* ⋄ C13 C98 ⋄

RASIOWA, H. *Lectures on infinitary logic and logics of programs* ⋄ B75 C75 C98 ⋄

ROSENSTEIN, J.G. *Linear orderings*
⋄ B25 C65 C98 E07 E98 ⋄

TAYLOR, W. *Some interesting identities* ⋄ C05 C98 ⋄

1983

ADAMEK, J. *Theory of mathematical structures*
⋄ C90 C98 G30 ⋄

BAUDISCH, A. & ROTHMALER, P. *Vorlesungen zur Einfuehrung in die Theorie des Shelah'schen "forking"*
⋄ C45 C98 ⋄

BERMAN, J. *Free spectra of 3-element algebras*
⋄ C05 C98 ⋄

BURRIS, S. *Boolean constructions*
  ⋄ A10 B25 C05 C30 C98 D35 G05 ⋄
EKLOF, P.C. *Set theory and structure theorems*
  ⋄ C60 C98 E50 E65 E75 ⋄
GABBAY, D.M. & GUENTHNER, F. (EDS.) *Handbook of philosophical logic Vol. I: Elements of classical logic*
  ⋄ B98 C98 ⋄
HODGES, W. *Elementary predicate logic* ⋄ B10 C98 ⋄
HOLLAND, W.C. *A survey of varieties of lattice ordered groups* ⋄ C05 C60 C98 ⋄
MANGIONE, C. *Stages of the relation between algebra and logic (Italian)* ⋄ C98 ⋄
MARCJA, A. *Categoricita e algebra* ⋄ C35 C45 C98 ⋄
PILLAY, A. *An introduction to stability theory*
  ⋄ C45 C98 ⋄
POIZAT, B. *C'est beau et chaud* ⋄ C45 C60 C98 ⋄
POIZAT, B. (ED.) *Groupe d'etude de theories stables. 3e annee 1980-1982* ⋄ C45 C97 C98 ⋄
POIZAT, B. *Travaux publies par les membres du groupe "theories stables"* ⋄ C45 C98 ⋄
PRESTEL, A. & ROQUETTE, P. *Lectures on formally p-adic fields* ⋄ B25 C10 C35 C60 C98 ⋄
QUACKENBUSH, R.W. *Minimal paraprimal algebras*
  ⋄ C05 C98 ⋄
REMESLENNIKOV, V.N. & ROMAN'KOV, V.A. *Model-theoretic and algorithmic questions of group theory (Russian)*
  ⋄ B25 C60 C98 D15 D30 D40 ⋄
SCHWABHAEUSER, W. & SZMIELEW, W. & TARSKI, A. *Metamathematische Methoden in der Geometrie*
  ⋄ B30 C65 C98 D35 ⋄
SCHWABHAEUSER, W. *Metamathematische Methoden in der Geometrie. Teil II: Metamathematische Betrachtungen* ⋄ B30 C65 C98 ⋄

1984

BAUDISCH, A. & ROTHMALER, P. *Vorlesungen zur Einfuehrung in die Theorie des Shelah'schen "forking" 2. Semester* ⋄ C45 C98 ⋄
CHERLIN, G.L. *Totally categorial structures*
  ⋄ C35 C45 C98 ⋄
COMBASE, J. *Introduction a la logique stationnaire*
  ⋄ C55 C80 C98 ⋄
COQUAND, T. *Le theoreme de representation d'Arens et Kaplansky* ⋄ C60 C98 ⋄
DELON, F. *Theories completes de corps*
  ⋄ C35 C60 C98 ⋄
FLUM, J. *Modelltheorie - topologische Modelltheorie*
  ⋄ C65 C90 C98 ⋄
GABBAY, D.M. & GUENTHNER, F. (EDS.) *Handbook of philosophical logic Vol II. Extensions of classical logic*
  ⋄ B98 C90 C98 ⋄
GARSON, J.W. *Quantification in modal logic*
  ⋄ B45 C90 C98 ⋄
GRANDJEAN, E. *Spectre des formules du premier order et complexite algorithmique* ⋄ C13 C98 D15 ⋄
HODGES, W. *Models built on linear orderings (French summary)* ⋄ C30 C55 C98 E07 ⋄
HOLLAND, W.C. *Classification of lattice ordered groups (French summary)* ⋄ C05 C60 C98 ⋄

KEISLER, H.J. *Lyndon's research in mathematical logic*
  ⋄ A10 C98 G98 ⋄
KRIVINE, J.-L. *Methodes de theorie des modeles en geometrie des espaces de Banach* ⋄ C65 C98 E75 ⋄
MAKKAI, M. *A survey of basic stability theory, with particular emphasis on orthogonality and regular types*
  ⋄ C45 C98 ⋄
MANEVITZ, L.M. *Applied model theory and metamathematics: an Abraham Robinson memorial problem list* ⋄ C98 H98 ⋄
MYASNIKOV, A.G. & REMESLENNIKOV, V.N. *Finite dimensional algebras and k-groups of finite rank*
  ⋄ C60 C98 ⋄
PARIGOT, M. *Extensions de la logique du premier order. Theoreme de Lindstroem* ⋄ C95 C98 ⋄
POIZAT, B. *La structure geometrique des groupes stables*
  ⋄ C45 C60 C98 ⋄
ROQUETTE, P. *Some tendencies in contemporary algebra*
  ⋄ C10 C35 C60 C98 ⋄
SCHMERL, J.H. $\aleph_0$-*categorical partially ordered sets (French summary)*
  ⋄ B25 C15 C35 C65 C98 E07 ⋄
SEMENOV, A.L. *Decidability of monadic theories*
  ⋄ B25 C85 C98 ⋄
SMORYNSKI, C.A. *Lectures on nonstandard models of arithmetic* ⋄ A10 C62 C98 F30 F98 H15 ⋄
STEGMUELLER, W. & VARGA VON KIBED, M. *Strukturtypen der Logik. Probleme und Resultate der Wissenschaftstheorie und analytischen Philosophie, Band III: Strukturtypen der Logik* ⋄ A05 C98 ⋄
WEISPFENNING, V. *Aspects of quantifier elimination in algebra* ⋄ C10 C60 C98 ⋄
WOLTER, H. *Some results about exponential fields (survey)*
  ⋄ B25 C60 C65 C98 ⋄
ZIL'BER, B.I. *The structure of models of uncountably categorical theories* ⋄ C35 C98 ⋄

1985

BALDWIN, JOHN T. *Definable second-order quantifiers*
  ⋄ C30 C80 C85 C98 ⋄
BARR, M. & WELLS, C. *Toposes, triples and theories*
  ⋄ C98 E98 F35 F50 G30 ⋄
BARWISE, J. & FEFERMAN, S. (EDS.) *Model-theoretic logics*
  ⋄ C75 C80 C85 C90 C95 C98 ⋄
BARWISE, J. *Model-theoretic logics: background and aims*
  ⋄ A05 C98 ⋄
BAUDISCH, A. & SEESE, D.G. & TUSCHIK, H.-P. & WEESE, M. *Decidability and quantifier elimination*
  ⋄ B25 C10 C60 C65 C80 C98 ⋄
DELON, F. & LASCAR, D. & LOUVEAU, A. & SABBAGH, G. (EDS.) *Seminaire general de logique 1982-83* ⋄ B97 C98 D97 ⋄
DICKMANN, M.A. *Applications of model theory to real algebraic geometry; a survey* ⋄ C10 C60 C98 ⋄
DICKMANN, M.A. *Larger infinitary languages*
  ⋄ C75 C98 ⋄
EBBINGHAUS, H.-D. *Extended logics: The general framework* ⋄ C40 C55 C80 C95 C98 ⋄
EKLOF, P.C. *Applications to algebra*
  ⋄ C60 C75 C98 ⋄

FLUM, J. *Characterizing logics*
⋄ C75 C80 C95 C98 ⋄
GUREVICH, Y. *Monadic second-order theories*
⋄ B25 C65 C85 C98 D35 D98 ⋄
HODGES, W. *Building models by games*
⋄ C25 C55 C60 C80 C98 E60 ⋄
KAUFMANN, M. *The quantifier "There exist uncountable many" and some of its relatives* ⋄ C55 C80 C98 ⋄
KEISLER, H.J. *Probability quantifiers*
⋄ C40 C80 C90 C98 ⋄
KOLAITIS, P.G. *Game quantification*
⋄ C40 C70 C75 C98 D60 D65 D70 E60 ⋄
KOPPELBERG, S. *Booleschwertige Logik*
⋄ B50 C90 C98 E35 E45 E50 E65 G05 ⋄
LEHMANN, G. *Modell- und rekursionstheoretische Grundlagen psychologischer Theorienbildung*
⋄ B10 C98 D80 D99 ⋄
MAKANIN, G.S. *On the decidability of the theory of a free group (Russian)* ⋄ B25 C60 C98 D35 ⋄
MAKOWSKY, J.A. *Compactness, embeddings and definability* ⋄ C40 C95 C98 ⋄
MCKENZIE, R. *The structure of finite algebras*
⋄ C05 C13 C98 ⋄
MILLAR, T.S. *Decidable Ehrenfeucht theories*
⋄ B25 C15 C57 C98 D30 D35 D55 ⋄

MUNDICI, D. *Other quantifiers: On overview*
⋄ C80 C98 ⋄
NADEL, M.E. $L_{\omega_1\omega}$ *and admissible fragments*
⋄ C70 C75 C98 ⋄
NERODE, A. & REMMEL, J.B. *A survey of lattices of R.E. substructures* ⋄ C57 C98 D45 D98 ⋄
POIZAT, B. *Cours de theorie des modeles* ⋄ C98 ⋄
SCHMERL, J.H. *Transfer theorems and their applications to logic* ⋄ C55 C80 C98 ⋄
SCOTT, D.S. & MCCARTY, C. & HORTY, J.F. *Bibliography*
⋄ A10 C98 ⋄
SIMPSON, S.G. *Friedman's research on subsystems of second order arithmetic* ⋄ C62 C98 F35 F98 ⋄
SMORYNSKI, C.A. *Nonstandard models and related developments* ⋄ C62 C98 F30 F98 ⋄
STEINHORN, C.I. *Borel structures and measure and category logics* ⋄ C90 C98 ⋄
STEINHORN, C.I. *Borel structures for first-order and extended logics* ⋄ C80 C90 C98 ⋄
TULIPANI, S. *Algebre sottodirettamente irriducibili e assiomatizzazione di varieta* ⋄ C05 C98 ⋄
VAEAENAENEN, J. *Set-theoretical definability of logics*
⋄ C40 C55 C80 C95 C98 E47 ⋄
ZIEGLER, M. *Topological model theory* ⋄ C90 C98 ⋄

Author Index

AABERG, C. [1972] *Some results on models for set theory* (**X** 0882) Univ Filos Foeren: Uppsala 53pp
⋄ C62 E40 ⋄ REV MR 50 #96 Zbl 343 #02046 • ID 17143

AABERG, C. [1974] *Relativity phenomena in set theory* (**J** 0154) Synthese 27*189-198
⋄ C62 ⋄ REV MR 58 #5202 Zbl 322 #02055 • ID 31611

AABERG, C. [1975] *Generic extensions and elementary embeddings* (**J** 0105) Theoria (Lund) 41*96-104
⋄ C62 E40 ⋄ REV Zbl 343 #02041 • ID 31610

AABERG, C. see Vol. V for further entries

AANDERAA, S.O. [1973] *A proof of Higman's embedding theorem using Britton extensions of groups* (**P** 0678) Word Probl: Decis & Burnside Probl in Group Th;1969 Irvine 1-18
⋄ C60 D40 D80 ⋄ REV MR 54 #412 Zbl 265 #20033 JSL 41.785 • ID 23990

AANDERAA, S.O. & LEWIS, H.R. [1973] *Prefix classes of Krom formulas* (**J** 0036) J Symb Logic 38*628-642
⋄ B20 B25 D35 ⋄ REV MR 49 #2326 Zbl 326 #02035
• ID 00010

AANDERAA, S.O. & GOLDFARB, W.D. [1974] *The finite controllability of the Maslov case* (**J** 0036) J Symb Logic 39*509-518
⋄ B25 ⋄ REV MR 51 #113 Zbl 306 #02042 • ID 03801

AANDERAA, S.O. & BOERGER, E. & LEWIS, H.R. [1982] *Conservative reduction classes of Krom formulas* (**J** 0036) J Symb Logic 47*110-130
⋄ B20 C13 D35 ⋄ REV MR 83e:03021 Zbl 487 #03005
• ID 35210

AANDERAA, S.O. & BOERGER, E. & GUREVICH, Y. [1982] *Prefix classes of Krom formulae with identity (German summary)* (**J** 0009) Arch Math Logik Grundlagenforsch 22*43-49
⋄ B20 B25 D35 ⋄ REV MR 83m:03019 Zbl 494 #03007
• ID 33756

AANDERAA, S.O. [1984] *On the solvability of the extended ∀∃ ∧ ∃∀*-Ackermann class with identity* (**P** 2342) Symp Rek Kombin;1983 Muenster 270-284
⋄ B20 B25 ⋄ REV MR 86h:03016 Zbl 575 #03005
• ID 41791

AANDERAA, S.O. see Vol. I, IV, VI for further entries

ABAKUMOV, A.I. & PALYUTIN, E.A. & SHISHMAREV, YU.E. & TAJTSLIN, M.A. [1972] *Categorial quasivarieties (Russian)* (**J** 0003) Algebra i Logika 11*3-38,121
• TRANSL [1972] (**J** 0069) Algeb and Log 11*1-20
⋄ C05 C35 C52 ⋄ REV MR 50 #12857 Zbl 249 #02024
• ID 32034

ABBATI, M.C. & MANIA, A. [1981] *A spectral theory for order unit spaces* (**J** 2658) Ann Inst Henri Poincare, Sect A 35*259-285
⋄ C65 ⋄ REV MR 83d:46005 Zbl 482 #46006 • ID 37700

ABBATI, M.C. see Vol. II for further entries

ABDRAZAKOV, K.T. & KHISAMIEV, N.G. [1984] *A criterion for strong constructivizability of a class of abelian p-groups (Russian)* (**J** 0092) Sib Mat Zh 25/4*3-8
• TRANSL [1984] (**J** 0475) Sib Math J 25*511-515
⋄ C57 C60 D45 ⋄ REV MR 86g:03073 • ID 41873

ABE, Y. [1983] *Strongly compact cardinals and the fixed points of elementary embeddings* (**P** 4113) Found of Math;1982 Kyoto 1-11
⋄ C62 E55 ⋄ ID 47672

ABE, Y. see Vol. V for further entries

ABIAN, A. [1972] *On solvability of infinite systems of polynomial equations over a finite ring* (**J** 0061) Simon Stevin 46*33-37
⋄ C60 E25 ⋄ REV MR 47 #3353 Zbl 253 #12018
• ID 00069

ABIAN, A. [1972] *Set theoretical canonical models (Armenian and Russian summaries)* (**J** 0312) Izv Akad Nauk Armyan SSR, Ser Mat 7*399-404
⋄ C62 E35 ⋄ REV MR 50 #100 Zbl 255 #02073
• ID 04126

ABIAN, A. [1972] *The method of synergistic models* (**J** 0051) Commentat Math, Ann Soc Math Pol, Ser 1 16*265-270
⋄ C62 E25 E35 E45 ⋄ REV MR 48 #3739
Zbl 239 #02027 • ID 00071

ABIAN, A. [1974] *Nonstandard models for arithmetics and analysis* (**J** 0063) Studia Logica 33*11-22
⋄ C20 H05 H15 ⋄ REV MR 50 #92 Zbl 287 #02038
• ID 03810

ABIAN, A. [1975] *On the standard-model hypothesis of ZF* (**J** 0068) Z Math Logik Grundlagen Math 21*87-88
⋄ C62 E30 ⋄ REV MR 51 #5301 Zbl 307 #02043
• ID 03813

ABIAN, A. [1975] *On the use of more than two-element boolean valued models* (**J** 0047) Notre Dame J Formal Log 16*555-564
⋄ C62 C90 E35 ⋄ REV MR 52 #2877 Zbl 254 #02048
• ID 21301

ABIAN, A. [1977] *A preliminary to forcing in set theory. On the standard and minimal standard model axioms of ZF* (**P** 2994) Nat Math Conf (8);1977 Tehran 1-11
⋄ C62 E30 ⋄ REV MR 80c:03052 • ID 70262

ABIAN, A. [1985] *Solvability of infinite systems of polynomial equations over the field of complex numbers* (**J** 0005) Amer Math Mon 92*94-98
⋄ C60 ⋄ ID 39859

ABIAN, A. also published under the name ABIAN, S.

ABIAN, A. see Vol. I, II, IV, V for further entries

ABRAHAM, U. [1984] *A minimal model for ¬ CH: iteration of Jensen's reals* (**J** 0064) Trans Amer Math Soc 281*657-674
⋄ C62 D30 E35 E40 E45 E50 E65 ⋄ REV
MR 85f:03044 Zbl 541 #03029 • ID 40744

ABRAHAM, U. & RUBIN, M. & SHELAH, S. [1985] *On the consistency of some partition theorems for continuous colorings, and structure of $\aleph_1$-dense real order types* (**J** 0073) Ann Pure Appl Logic 29*123-206
⋄ C55 C62 C65 C80 E05 E07 E35 ⋄ ID 47480

ABRAHAM, U. see Vol. V for further entries

ABRAMSKY, M. [1980] *The classical decision problem and partial functions* (**J** 0009) Arch Math Logik Grundlagenforsch 20*3-12
⋄ B25 ⋄ REV MR 81i:03014 Zbl 431 # 03007 • ID 53913

ABRAMSON, F.G. [1978] *Interpolation theorems for program schemata* (**J** 0194) Inform & Control 36*217-233
⋄ B70 C40 ⋄ REV MR 57 # 14564 Zbl 367 # 68019
• ID 31648

ABRAMSON, F.G. & HARRINGTON, L.A. [1978] *Models without indiscernibles* (**J** 0036) J Symb Logic 43*572-600
⋄ C30 C50 C55 C62 C65 C75 E05 E40 H15 ⋄ REV MR 80a:03045 Zbl 391 # 03027 JSL 48.484 • ID 29285

ABRAMSON, F.G. [1979] *Σ_1-separation* (**J** 0036) J Symb Logic 44*374-382
⋄ C62 C70 D60 E45 E47 ⋄ REV MR 82j:03057
Zbl 428 # 03029 • ID 53788

ABRAMSON, F.G. [1981] *Locally countable models of Σ_1-separation* (**J** 0036) J Symb Logic 46*96-100
⋄ C62 C70 E30 E47 ⋄ REV MR 83i:03058
Zbl 481 # 03023 • ID 33267

ABRAMSON, F.G. see Vol. IV, V for further entries

ACKERMANN, W. [1928] see HILBERT, D.

ACKERMANN, W. [1928] *Ueber die Erfuellbarkeit gewisser Zaehlausdruecke* (**J** 0043) Math Ann 100*638-649
⋄ B25 D35 ⋄ REV FdM 54.57 • ID 00109

ACKERMANN, W. [1936] *Beitraege zum Entscheidungsproblem der mathematischen Logik* (**J** 0043) Math Ann 112*419-432
⋄ B25 D35 ⋄ REV Zbl 13.241 JSL 1.43 FdM 62.41
• ID 00112

ACKERMANN, W. [1954] *Solvable cases of the decision problem* (**X** 0809) North Holland: Amsterdam viii + 114pp
⋄ B25 ⋄ REV MR 16.323 Zbl 56.245 JSL 22.68 • ID 00126

ACKERMANN, W. see Vol. I, II, IV, V, VI for further entries

ACZEL, P. [1973] *Infinitary logic and the Barwise compactness theorem* (**P** 0710) Russell Mem Logic Conf;1971 Uldum 234-277
⋄ C70 C98 ⋄ REV MR 51 # 50 • ID 24788

ACZEL, P. [1975] *Quantifiers, games and inductive definitions* (**P** 0757) Scand Logic Symp (3);1973 Uppsala 1-14
⋄ C75 C80 D70 ⋄ REV MR 54 # 12477 Zbl 324 # 02009 JSL 43.373 • ID 30616

ACZEL, P. [1981] *Two notes on the Paris independence result I: A generalization of Ramsey's theorem. II: The ordinal height of a density* (**P** 3404) Model Th & Arithm;1979/80 Paris 21-31
⋄ C62 E05 F15 F30 ⋄ REV MR 84i:03071
Zbl 487 # 03025 • ID 34552

ACZEL, P. see Vol. II, IV, V, VI for further entries

ADAMEK, J. & REITERMAN, J. [1974] *Fixed-point property of unary algebras* (**J** 0004) Algeb Universalis 4*163-165
⋄ C05 C07 ⋄ REV MR 50 # 9746 Zbl 296 # 08006
• ID 00152

ADAMEK, J. [1983] *Theory of mathematical structures* (**X** 0835) Reidel: Dordrecht x + 317pp
⋄ C90 C98 G30 ⋄ REV MR 85c:18001 Zbl 523 # 18001
• ID 36993

ADAMEK, J. see Vol. IV, V for further entries

ADAMOWICZ, Z. [1976] *One more aspect of forcing and omitting types* (**J** 0036) J Symb Logic 41*73-80
⋄ C62 C75 E35 E40 ⋄ REV MR 57 # 16050
Zbl 406 # 03063 • ID 14749

ADAMOWICZ, Z. [1984] *Axiomatization of the forcing relation with an application to Peano arithmetic* (**J** 0027) Fund Math 120*167-186
⋄ C62 E40 F30 ⋄ REV MR 86h:03090 • ID 45406

ADAMOWICZ, Z. & MORALES-LUNA, G. [1985] *A recursive model for arithmetic with weak induction* (**J** 0036) J Symb Logic 50*49-54
⋄ C57 F30 H15 ⋄ REV MR 86d:03058 Zbl 576 # 03045
• ID 39780

ADAMOWICZ, Z. [1985] *Algebraic approach to $\exists_1$ induction* (**P** 4310) Easter Conf on Model Th (3);1985 Gross Koeris 5-15
⋄ C60 C62 F30 ⋄ ID 49053

ADAMOWICZ, Z. also published under the name SMOLSKA-ADAMOWICZ, Z.

ADAMOWICZ, Z. see Vol. IV, V for further entries

ADAMS, E.W. [1974] *The logic of "almost all"* (**J** 0122) J Philos Logic 3*3-17
⋄ C80 ⋄ REV MR 54 # 12471 Zbl 278 # 02022 • ID 29058

ADAMS, E.W. see Vol. I, II for further entries

ADAMSON, A. [1978] *Admissible sets and the saturation of structures* (**J** 0007) Ann Math Logic 14*111-157
⋄ C50 C70 C75 E40 ⋄ REV MR 80h:03052
Zbl 395 # 03025 • ID 29149

ADAMSON, A. & GILES, R. [1979] *A game-based formal system for L_∞* (**J** 0063) Studia Logica 38*49-73
⋄ B25 B50 D15 ⋄ REV MR 80m:03049 Zbl 417 # 03008
• ID 53245

ADAMSON, A. [1980] *A note on the realization of types* (**J** 0018) Canad Math Bull 23*95-98
⋄ C07 C50 ⋄ REV MR 81i:03039 Zbl 463 # 03018
• ID 54558

ADAMSON, A. [1980] *Saturated structures, unions of chains, and preservation theorems* (**J** 0007) Ann Math Logic 19*67-96
⋄ C40 C50 C70 ⋄ REV MR 83i:03055 Zbl 471 # 03043
• ID 55238

ADAMSON, A. [1983] *A note on two-cardinal models* (**J** 0068) Z Math Logik Grundlagen Math 29*193-196
⋄ C55 ⋄ REV MR 84m:03051 Zbl 526 # 03020 • ID 35752

ADDISON, J.W. [1962] *Some problems in hierarchy theory* (**P** 0613) Rec Fct Th;1961 New York 123-130
⋄ C40 D55 E15 ⋄ REV MR 25 # 4997 Zbl 143.255
JSL 29.61 • ID 00176

ADDISON, J.W. [1962] *The theory of hierarchies* (**P** 0612) Int Congr Log, Meth & Phil of Sci (1,Proc);1960 Stanford 26-37
⋄ C40 C52 D55 E15 ⋄ REV MR 28 # 1127 Zbl 133.251
JSL 29.61 • ID 00175

ADDISON, J.W. & BOUVERE DE, K.L. & PITT, W.B. [1965] *A bibliography of the theory of models* (P 0614) Th Models;1963 Berkeley 438-492
⋄ A10 C98 ⋄ REV MR 33 # 3879 • ID 00178

ADDISON, J.W. [1965] *The method of alternating chains* (P 0614) Th Models;1963 Berkeley 1-16
⋄ C40 D55 D75 E15 ⋄ REV MR 34 # 1197 Zbl 199.12
• ID 16215

ADDISON, J.W. & HENKIN, L. & TARSKI, A. (EDS.) [1965] *The theory of models. Proceedings of the 1963 International Symposium at Berkeley* (X 0809) North Holland: Amsterdam xv+494pp
⋄ C97 ⋄ REV MR 33 # 3878 Zbl 148.1 • REM 3rd ed. 1972
• ID 29690

ADDISON, J.W. & CHANG, C.C. & CRAIG, W. & HENKIN, L. & SCOTT, D.S. & VAUGHT, R.L. (EDS.) [1974] *Proceedings of the Tarski Symposium* (S 3304) Proc Symp Pure Math 25*xxi+498pp
⋄ B97 C97 ⋄ REV MR 50 # 1829 • REM Corr. ed. 1979; xx+498pp. Contains bibliography of A.Tarski with a supplement in the corr. ed. • ID 70206

ADDISON, J.W. see Vol. I, IV, V for further entries

ADLER, A. [1969] *Extensions of non-standard models of number theory* (J 0068) Z Math Logik Grundlagen Math 15*289-290
⋄ C62 D25 D35 H15 ⋄ REV MR 42 # 54 Zbl 188.322 JSL 40.244 • ID 00184

ADLER, A. [1971] *The cardinality of ultrapowers - an example* (J 0053) Proc Amer Math Soc 28*311-312
⋄ C20 E05 E55 ⋄ REV MR 43 # 6081 Zbl 217.16
• ID 00185

ADLER, A. & JORGENSEN, M. [1972] *Descendingly incomplete ultrafilters and the cardinality of ultrapowers* (J 0017) Canad J Math 24*830-834
⋄ C20 E05 E50 ⋄ REV MR 46 # 8829 Zbl 251 # 02051
• ID 00186

ADLER, A. [1972] *Representation of models of full theories* (J 0068) Z Math Logik Grundlagen Math 18*183-188
⋄ C07 C20 ⋄ REV MR 46 # 1582 Zbl 268 # 02035
• ID 00188

ADLER, A. & HAMILTON, J. [1973] *Invariant means via the ultrapower* (J 0043) Math Ann 202*71-76
⋄ C20 H05 ⋄ REV MR 48 # 2664 Zbl 235 # 43001
• ID 00190

ADLER, A. [1974] *An application of elementary model theory to topological boolean algebras* (P 1083) Victoria Symp Nonstand Anal;1972 Victoria 1-4
⋄ C20 G05 H20 ⋄ REV MR 58 # 206 Zbl 272 # 54041
• ID 26608

ADLER, A. & KIEFE, C. [1976] *Pseudofinite fields, procyclic fields and model-completion* (J 0048) Pac J Math 62*305-309
⋄ C35 C60 ⋄ REV MR 54 # 7245 Zbl 339 # 02048
• ID 18475

ADLER, A. [1978] *On the multiplicative semigroups of rings* (J 0394) Commun Algeb 6*1751-1753
⋄ C52 C60 ⋄ REV MR 58 # 22166 Zbl 399 # 03021
• ID 52817

ADLER, A. see Vol. I, IV, V for further entries

ADYAN, S.I. [1957] *Finitely presented groups and algorithms (Russian)* (J 0023) Dokl Akad Nauk SSSR 117*9-12
⋄ B25 C60 D40 ⋄ REV MR 20 # 2371 Zbl 85.251
• ID 00193

ADYAN, S.I. [1958] *On algorithmic problems in effectively complete classes of groups (Russian)* (J 0023) Dokl Akad Nauk SSSR 123*13-16
⋄ C60 D40 ⋄ REV MR 21 # 1998 Zbl 90.11 • ID 00195

ADYAN, S.I. see Vol. IV, V for further entries

AFRICK, H. [1972] *A proof theoretic proof of Scott's general interpolation theorem* (J 0036) J Symb Logic 37*683-695
⋄ C40 F07 ⋄ REV MR 51 # 7833 Zbl 258 # 02037
• ID 00199

AFRICK, H. [1974] *Scott's interpolation theorem fails for $L_{\omega_1,\omega}$* (J 0036) J Symb Logic 39*124-126
⋄ C40 C75 ⋄ REV MR 50 # 1836 Zbl 344 # 02010
• ID 00200

AJTAI, M. [1979] *Isomorphism and higher order equivalence* (J 0007) Ann Math Logic 16*181-203
⋄ C15 C55 E35 E45 ⋄ REV MR 81j:03052
Zbl 415 # 03044 • ID 53146

AJTAI, M. [1983] *Σ_1^1-formulae on finite structures* (J 0073) Ann Pure Appl Logic 24*1-48
⋄ C13 C62 D15 H15 ⋄ REV MR 85b:03048
Zbl 519 # 03021 • ID 37538

AJTAI, M. see Vol. IV, V for further entries

ALCANTARA DE, L.P. & CORRADA, M. [1980] *Notes on many-sorted systems* (P 3006) Brazil Conf Math Log (3);1979 Recife 83-108
⋄ B10 C07 ⋄ REV MR 83k:03015 Zbl 455 # 03005
• ID 54266

ALCANTARA DE, L.P. see Vol. I, II, V for further entries

ALIEV, V.G. [1984] *Model companions of theories of semigroup classes (Russian) (English and Azerbaijani summaries)* (J 0135) Izv Akad Nauk Azerb SSR, Ser Fiz-Tekh Mat 5/4*3-7
⋄ C05 C25 C35 ⋄ ID 45594

ALKOR, C. [1979] *The problem of extendability (Italian summary)* (J 3285) Boll Unione Mat Ital, V Ser, A 16*85-91
⋄ C50 C62 E70 ⋄ REV MR 80h:03078 Zbl 395 # 03036
• ID 52587

ALKOR, C. see Vol. V for further entries

ALLING, N.L. [1960] *On ordered divisible groups* (J 0064) Trans Amer Math Soc 94*498-514
⋄ C55 C60 ⋄ REV MR 25 # 4013 Zbl 94.247 • ID 00274

ALLING, N.L. [1961] *A characterization of abelian η_α-groups in terms of their natural valuation* (J 0054) Proc Nat Acad Sci USA 47*711-713
⋄ C55 C60 ⋄ REV MR 31 # 259 Zbl 108.257 • ID 00275

ALLING, N.L. [1962] *On the existence of real-closed fields that are η_α-sets of power $\aleph_\alpha$* (J 0064) Trans Amer Math Soc 103*341-352
⋄ C55 C60 C65 ⋄ REV MR 26 # 3615 Zbl 108.257
• ID 00276

ALLING, N.L. [1965] *Rings of continuous integer-valued functions and nonstandard arithmetic* (J 0064) Trans Amer Math Soc 118*498-525
 ⋄ C20 C60 H15 ⋄ REV MR 32 # 2431 Zbl 143.10
 • ID 00277

ALLING, N.L. [1985] *Conway's field of surreal numbers* (J 0064) Trans Amer Math Soc 287*365-386
 ⋄ H10 H15 H20 ⋄ REV MR 86f:04002 • ID 44331

ALMAGAMBETOV, ZH.A. [1965] *On classes of axioms closed with respect to the given reduced products and powers (Russian)* (J 0003) Algebra i Logika 4/3*71-77
 ⋄ B25 C20 C52 D35 ⋄ REV MR 33 # 7255 Zbl 223 # 02052 • ID 00278

ALMAGAMBETOV, ZH.A. [1965] *Solvability of the elementary theory of certain classes of free nilpotent algebras (Russian)* (J 0003) Algebra i Logika 4/6*5-14
 ⋄ B25 C60 ⋄ REV MR 34 # 59 Zbl 199.30 • ID 00279

ALTHAM, J.E.J. [1971] *The logic of plurality* (X 0816) Methuen: London & New York ix+84pp
 ⋄ A05 B98 C80 ⋄ REV MR 47 # 12 Zbl 276 # 02008
 • ID 00285

ALTHAM, J.E.J. & TENNANT, N. [1975] *Sortal quantification* (P 1967) Form Semant of Nat Lang;1973 Cambridge GB 46-58
 ⋄ A05 C80 ⋄ REV Zbl 342 # 02009 • ID 60108

ALTON, D.A. & MADISON, E.W. [1973] *Computability of boolean algebras and their extensions* (J 0007) Ann Math Logic 6*95-128
 ⋄ C57 D45 G05 ⋄ REV MR 49 # 4766 Zbl 272 # 02061
 • ID 00289

ALTON, D.A. [1975] *Embedding relations in the lattice of recursively enumerable sets* (J 0009) Arch Math Logik Grundlagenforsch 17*37-41
 ⋄ B25 D25 ⋄ REV MR 52 # 5391 Zbl 315 # 02040
 • ID 17945

ALTON, D.A. see Vol. IV for further entries

ALVES, E.H. [1978] *On decidability of a system of dialectical propositional logic* (J 0387) Bull Sect Logic, Pol Acad Sci 7*179-184
 ⋄ B25 B53 ⋄ REV MR 80b:03028 Zbl 412 # 03012
 • ID 52943

ALVES, E.H. [1984] *Paraconsistent logic and model theory* (J 0063) Studia Logica 43*17-32
 ⋄ B53 C40 C52 C90 ⋄ ID 39893

ALVES, E.H. see Vol. II for further entries

AMER, M.A. [1975] *On a theorem of Garrett Birkhoff (Arabic summary)* (J 0397) Proc Math Phys Soc Egypt 40*19-21
 ⋄ C05 ⋄ REV MR 51 # 12627 Zbl 426 # 06005 • ID 30620

AMER, M.A. [1976] *Typed boolean structures I (Arabic summary)* (J 0397) Proc Math Phys Soc Egypt 41*15-22
 ⋄ B15 B40 C90 E40 G05 ⋄ REV MR 52 # 7860 Zbl 418 # 03009 • REM Part II 1977 • ID 17946

AMER, M.A. [1977] *Typed boolean structures II (Arabic summary)* (J 0397) Proc Math Phys Soc Egypt 44*11-15
 ⋄ B15 B40 C90 E40 G05 ⋄ REV MR 56 # 2784 Zbl 427 # 03011 • REM Part I 1976 • ID 30619

AMER, M.A. see Vol. II, VI for further entries

AMERBAEV, V.M. & KASIMOV, YU.F. [1978] *Formal expressibility of order relations in algebraic systems (Russian) (Kazakh summary)* (J 0429) Vest Akad Nauk Kazak SSR 1978/10*55-59
 ⋄ C40 C60 ⋄ REV MR 58 # 21582 Zbl 413 # 06009
 • ID 70388

AMIT, R. & SHELAH, S. [1976] *The complete finitely axiomatized theories of order are dense* (J 0029) Israel J Math 23*200-208
 ⋄ C52 C65 ⋄ REV MR 58 # 5162 Zbl 341 # 02041
 • ID 17947

AMITSUR, S.A. [1966] *Rational identities and applications to algebra and geometry* (J 0032) J Algeb 3*304-359
 ⋄ C50 C60 C65 ⋄ REV MR 33 # 139 Zbl 203.40
 • ID 00301

ANAPOLITANOS, D.A. [1978] *A new kind of indiscernibles* (J 3133) Bull Soc Math Belg, Ser B 30*135-153
 ⋄ C07 C30 C50 ⋄ REV MR 81j:03050 Zbl 423 # 03032
 • ID 53544

ANAPOLITANOS, D.A. [1978] *A theorem on absolute indiscernibles* (J 0063) Studia Logica 37*291-295
 ⋄ C30 C65 ⋄ REV MR 80e:03031 Zbl 398 # 03017
 • ID 52745

ANAPOLITANOS, D.A. [1978] *Absolute indiscernibles and standard models of ZFC* (J 0465) Bull Greek Math Soc (NS) 19*265-273
 ⋄ C30 C62 E40 ⋄ REV MR 80d:03053 Zbl 449 # 03050
 • ID 56717

ANAPOLITANOS, D.A. [1979] *Automorphisms of finite order* (J 0068) Z Math Logik Grundlagen Math 25*565-575
 ⋄ C07 C62 E25 E35 ⋄ REV MR 81b:03055 Zbl 421 # 03040 • ID 53439

ANAPOLITANOS, D.A. [1979] *Automorphisms of finite order and the axiom of choice* (J 3133) Bull Soc Math Belg, Ser B 31*39-44
 ⋄ C62 E25 E35 ⋄ REV MR 82k:03049 Zbl 448 # 03038
 • ID 56638

ANAPOLITANOS, D.A. & VAEAENAENEN, J. [1980] *On the axiomatizability of the notion of an automorphism of a finite order* (J 0068) Z Math Logik Grundlagen Math 26*433-437
 ⋄ C07 C65 ⋄ REV MR 82b:03070 Zbl 463 # 03019
 • ID 54559

ANAPOLITANOS, D.A. [1981] *Cyclic indiscernibles and Skolem functions* (J 0068) Z Math Logik Grundlagen Math 27*353-362
 ⋄ C07 C30 ⋄ REV MR 82k:03043 Zbl 474 # 03014
 • ID 55418

ANAPOLITANOS, D.A. & VAEAENAENEN, J. [1981] *Decidability of some logics with free quantifier variables* (J 0068) Z Math Logik Grundlagen Math 27*17-22
 ⋄ B25 C80 ⋄ REV MR 82d:03059 Zbl 483 # 03018
 • ID 70402

ANDERSON, A.R. [1954] *Improved decision procedures for Lewis's calculus S4 and von Wright's calculus M* (J 0036) J Symb Logic 19*201-214 • ERR/ADD ibid 20*150
 ⋄ B25 B45 ⋄ REV MR 16.103 Zbl 57.8 JSL 20.302
 • ID 00309

ANDERSON, A.R. see Vol. I, II, VI for further entries

ANDERSON, J.G. [1969] *An application of Kripke's completeness theorem for intuitionism to superconstructive propositional calculi* (**J** 0068) Z Math Logik Grundlagen Math 15∗259-288
  ⋄ B55 C90 ⋄ REV MR 40 # 1261 Zbl 216.288 • ID 00344

ANDERSON, J.G. see Vol. I, II for further entries

ANDLER, D. [1975] *Finite-dimensional models of categorical semi-minimal theories* (**J** 0079) Logique & Anal, NS 18∗359-378
  • REPR [1977] (**P** 1625) Six Days of Model Th;1975 Louvain-la-Neuve 127-146
  ⋄ C35 C50 ⋄ REV MR 58 # 5168 Zbl 355 # 02037 Zbl 402 # 03027 • ID 50235

ANDLER, D. [1975] *Semi-minimal theories and categoricity* (**J** 0036) J Symb Logic 40∗419-438
  ⋄ C30 C35 C55 ⋄ REV MR 55 # 2548 Zbl 331 # 02028
  • ID 14817

ANDREKA, H. & NEMETI, I. [1973] *On the equivalence of sets definable by satisfaction and ultrafilters* (**J** 0411) Studia Sci Math Hung 8∗463-467
  ⋄ C07 C20 ⋄ REV MR 51 # 46 Zbl 296 # 02028 • ID 17949

ANDREKA, H. & GERGELY, T. & NEMETI, I. [1973] *Some questions on languages of order n (Hungarian)* (**J** 0396) Mat Lapok 24∗63-94
  ⋄ B15 C07 C85 ⋄ REV MR 53 # 5249 • ID 22879

ANDREKA, H. & GERGELY, T. & NEMETI, I. [1974] *Some questions on languages of order n I,II (Russian) (English summaries)* (**J** 0040) Kibernetika, Akad Nauk Ukr SSR 1974/5∗61-67, 1974/6∗79-83
  • TRANSL [1974] (**J** 0021) Cybernetics 10∗804-812, 1003-1008
  ⋄ B15 C07 C85 ⋄ REV MR 56 # 1811 MR 56 # 1812 Zbl 316 # 68048 Zbl 316 # 68056 • ID 60135

ANDREKA, H. & GERGELY, T. & NEMETI, I. [1974] *Sufficient and necessary condition for the completeness of a calculus* (**J** 0068) Z Math Logik Grundlagen Math 20∗433-434
  ⋄ B10 C07 G05 ⋄ REV MR 51 # 10023 Zbl 305 # 02063
  • ID 03829

ANDREKA, H. & NEMETI, I. [1975] *A simple, purely algebraic proof of the completeness of some first order logics* (**J** 0004) Algeb Universalis 5∗8-15
  ⋄ C07 G15 ⋄ REV MR 52 # 2875 Zbl 306 # 02059
  • ID 17659

ANDREKA, H. & GERGELY, T. & NEMETI, I. [1975] *Many-sorted languages and their connection with nth order languages (Russian) (English summary)* (**J** 0040) Kibernetika, Akad Nauk Ukr SSR 1975/4∗86-92
  • TRANSL [1975] (**J** 0021) Cybernetics 11∗605-612
  ⋄ B15 C85 ⋄ REV MR 56 # 11742 Zbl 436 # 03004
  • ID 55844

ANDREKA, H. & NEMETI, I. [1975] *Subalgebra systems of algebras with finite and infinite, regular and irregular arity* (**J** 0006) Ann Univ Budapest, Sect Math 17∗103-118
  ⋄ C05 C75 ⋄ REV MR 52 # 226 Zbl 271 # 08006
  • ID 17951

ANDREKA, H. & DAHN, B.I. & NEMETI, I. [1976] *On a proof of Shelah (Russian summary)* (**J** 0014) Bull Acad Pol Sci, Ser Math Astron Phys 24∗1-7
  ⋄ C07 C20 C90 ⋄ REV MR 53 # 2666 Zbl 339 # 02046
  • ID 17953

ANDREKA, H. & NEMETI, I. [1978] *Los lemma holds in every category* (**J** 0411) Studia Sci Math Hung 13∗361-376
  ⋄ C20 G30 ⋄ REV MR 82i:03049 Zbl 502 # 03015
  • ID 70424

ANDREKA, H. & NEMETI, I. [1978] *Neat reducts of varieties* (**J** 0411) Studia Sci Math Hung 13∗47-51
  ⋄ C05 C20 G15 G25 G30 ⋄ REV MR 82j:08013 Zbl 421 # 03047 • ID 54384

ANDREKA, H. & NEMETI, I. & SAIN, I. [1979] *Completeness problems in verification of programs and program schemes* (**P** 2059) Math Founds of Comput Sci (8);1979 Olomouc 208-218
  ⋄ B75 C90 C95 ⋄ REV MR 81e:68011 Zbl 411 # 03017
  • ID 52871

ANDREKA, H. & NEMETI, I. [1979] *Formulas and ultraproducts in categories* (**S** 0410) Math Beitr Univ Halle-Wittenberg 12∗133-151
  • REPR [1979] (**J** 3124) Beitr Algebra Geom 8∗133-151
  ⋄ C05 C20 G30 ⋄ REV MR 81m:18004 Zbl 531 # 03042
  • ID 80546

ANDREKA, H. & NEMETI, I. [1979] *Not all representable cylindric algebras are neat reducts* (**J** 0387) Bull Sect Logic, Pol Acad Sci 8∗145-147
  ⋄ C20 G15 ⋄ REV MR 80m:03100 Zbl 446 # 03046
  • ID 56575

ANDREKA, H. & NEMETI, I. & SAIN, I. [1979] *Program verification within and without logic* (**J** 0387) Bull Sect Logic, Pol Acad Sci 8∗124-129
  ⋄ B75 C90 ⋄ REV MR 82b:68005 Zbl 441 # 68024
  • ID 80545

ANDREKA, H. & MAKAI JR., E. & MARKI, L. & NEMETI, I. [1979] *Reduced products in categories* (**P** 2920) Contrib to Gen Algeb (1);1978 Klagenfurt 25-45
  ⋄ C20 G30 ⋄ REV MR 80h:03049 Zbl 409 # 18001
  • ID 56347

ANDREKA, H. & NEMETI, I. [1980] *On systems of varieties definable by schemes of equations* (**J** 0004) Algeb Universalis 11∗105-116
  ⋄ C05 G15 ⋄ REV MR 82d:08007 Zbl 387 # 03025
  • ID 56040

ANDREKA, H. & BURMEISTER, P. & NEMETI, I. [1980] *Quasi-equational logic of partial algebras* (**J** 0387) Bull Sect Logic, Pol Acad Sci 9∗193-199
  ⋄ C05 C90 ⋄ REV MR 83d:03036 Zbl 491 # 08007 Zbl 491 # 08008 • ID 35182

ANDREKA, H. & NEMETI, I. [1981] *HSP K is equational class, without the axiom of choice* (**J** 0004) Algeb Universalis 13∗164-166
  ⋄ C05 E25 ⋄ REV MR 82k:03081 Zbl 492 # 03011
  • ID 70419

ANDREKA, H. & BURMEISTER, P. & NEMETI, I. [1981] *Quasivarieties of partial algebras. A unifying approach towards a two-valued model theory for partial algebras* (**J** 0411) Studia Sci Math Hung 16∗325-372
  ⋄ C05 C90 ⋄ REV MR 85i:08003 Zbl 537 # 08004
  • ID 44339

ANDREKA, H. & NEMETI, I. [1981] *Some universal algebraic and model theoretic results in computer science* (**P** 3165) FCT'81 Fund of Comput Th;1981 Szeged 16-23
  ⋄ B75 C05 C98 ⋄ REV MR 84g:08018 Zbl 483.68050
  • ID 39432

ANDREKA, H. & NEMETI, I. & SAIN, I. [1982] *A complete logic for reasoning about programs via nonstandard model theory. I,II* (**J** 1426) Theor Comput Sci 17*193-212,259-278
⋄ B75 C90 H10 H20 ⋄ REV MR 84d:68007 Zbl 475 # 68009 Zbl 475 # 68010 • ID 55505

ANDREKA, H. & NEMETI, I. [1982] *A general axiomatizability theorem formulated in terms of cone-injective subcategories* (**P** 3237) Colloq Universal Algeb;1977 Esztergom 13-35
⋄ C05 C20 C90 G30 ⋄ REV MR 84a:03031a Zbl 489 # 03006 • ID 35571

ANDREKA, H. & NEMETI, I. [1983] *Generalization of the concept of variety and quasivariety to partial algebras through category theory* (**J** 0202) Diss Math (Warsaw) 204*51pp
⋄ C05 C40 C75 C90 C95 G30 ⋄ REV MR 85c:08006 Zbl 518 # 08007 • ID 38372

ANDREKA, H. [1983] *Sharpening the characterization of the power of Floyd method* (**P** 3830) Logics of Progr & Appl;1980 Poznan 1-26
⋄ B75 C20 ⋄ REV MR 84e:68006a Zbl 528 # 68007 • ID 36759

ANDREKA, H. & NEMETI, I. [1984] *Importance of universal algebra for computer science* (**P** 3088) Univer Alg & Link Log, Alg, Combin, Comp Sci;1983 Darmstadt 204-215
⋄ C05 C20 C95 G15 ⋄ REV Zbl 548 # 68022 • ID 43245

ANDREKA, H. see Vol. V for further entries

ANDREWS, P.B. [1972] *General models, descriptions, and choice in type theory* (**J** 0036) J Symb Logic 37*385-394
⋄ B15 C85 E25 E35 F35 ⋄ REV MR 47 # 6433 Zbl 264 # 02049 • ID 00367

ANDREWS, P.B. [1972] *General models and extensionality* (**J** 0036) J Symb Logic 37*395-397
⋄ B15 C85 E35 F35 ⋄ REV MR 47 # 6434 Zbl 264 # 02050 • ID 00366

ANDREWS, P.B. [1974] *Provability in elementary type theory* (**J** 0068) Z Math Logik Grundlag Math 20*411-418
⋄ B15 B25 D35 F35 ⋄ REV MR 52 # 7867 Zbl 306 # 02017 • ID 03832

ANDREWS, P.B. see Vol. I, VI for further entries

ANELLIS, I.H. [1982] *Boolean groups* (**J** 0535) Abh Braunschweig Wiss Ges 33*85-97
⋄ C30 C60 G05 ⋄ REV MR 84g:03102 Zbl 504 # 06016 • ID 34206

ANELLIS, I.H. see Vol. V for further entries

ANSHAKOV, O.M. & RYCHKOV, S.V. [1984] *Axiomatization of finite-valued logical calculi (Russian)* (**J** 0142) Mat Sb, Akad Nauk SSSR, NS 123(165)*477-495
• TRANSL [1984] (**J** 0349) Math of USSR, Sbor 51*473-491
⋄ B50 C90 ⋄ REV MR 85i:03065 Zbl 551 # 03015 Zbl 566 # 03010 • ID 44158

ANSHAKOV, O.M. & RYCHKOV, S.V. [1984] *On a way of the formalization and classification of many valued logics (Russian)* (**S** 2582) Semiotika & Inf, Akad Nauk SSSR 23*78-106,156
⋄ B50 C90 ⋄ REV Zbl 562 # 03012 • ID 47425

ANSHAKOV, O.M. see Vol. II for further entries

ANSHEL, M. & STEBE, P. [1976] *Conjugate powers in free products with amalgamation* (**J** 1447) Houston J Math 2*139-147
⋄ B25 D40 ⋄ REV MR 53 # 13416 Zbl 344 # 20032 • ID 30627

ANSHEL, M. [1978] *Decision problems for HNN groups and commutative semigroups* (**J** 1447) Houston J Math 4*137-142
⋄ B25 D40 ⋄ REV MR 58 # 16889 Zbl 391 # 20024 • ID 30625

ANSHEL, M. see Vol. IV, VI for further entries

ANTONOVSKIJ, M.YA. & CHUDNOVSKY, D.V. & CHUDNOVSKY, G.V. & HEWITT, E. [1981] *Rings of real-valued continuous functions II* (**J** 0044) Math Z 176*151-186
⋄ C20 C50 C60 C65 E35 E50 E75 ⋄ REV MR 83f:54011 Zbl 432 # 03030 • REM Part I 1948 by Hewitt,E. • ID 53987

ANTONOVSKIJ, M.YA. see Vol. V for further entries

ANUSIAK, J. & WEGLORZ, B. [1971] *Remarks on c-independence in cartesian products of abstract algebras* (**S** 0019) Colloq Math (Warsaw) 22*161-165
⋄ C05 ⋄ REV MR 43 # 3195 Zbl 245 # 08004 • ID 00394

ANUSIAK, J. see Vol. V for further entries

AOE, H. [1978] *On the Whitehead problem* (**J** 1508) Math Sem Notes, Kobe Univ 6*363-367
⋄ C60 E35 ⋄ REV MR 80a:03066 Zbl 405 # 20048 • ID 70444

AOYAMA, K. [1970] *On the structure space of a direct product of rings* (**J** 0102) Hiroshima Univ J Sci, Ser A Math 34*339-353
⋄ C30 C60 ⋄ REV MR 45 # 1890 Zbl 215.77 • ID 00396

APELT, H. [1966] *Axiomatische Untersuchungen ueber einige mit der Presburgerschen Arithmetik verwandte Systeme* (**J** 0068) Z Math Logik Grundlagen Math 12*131-168
⋄ B28 C35 C80 F30 ⋄ REV MR 36 # 6279 Zbl 149.7 • ID 00403

APPEL, K.I. [1959] *Horn sentences in identity theory* (**J** 0036) J Symb Logic 24*306-310
⋄ C13 C30 C52 ⋄ REV MR 24 # A2528 Zbl 114.245 JSL 31.131 • ID 00406

APPEL, K.I. see Vol. IV for further entries

APPLEBAUM, C.H. & DEKKER, J.C.E. [1970] *Partial recursive functions and ω-functions* (**J** 0036) J Symb Logic 35*559-568
⋄ C57 D50 ⋄ REV MR 43 # 3116 Zbl 217.13 • ID 00414

APPLEBAUM, C.H. see Vol. IV for further entries

APPLESON, R.R. [1976] *Zero-divisors among finite structures of a fixed type* (**J** 0004) Algeb Universalis 6*25-35
⋄ C05 C13 C30 ⋄ REV MR 53 # 2796 Zbl 356 # 08002 • ID 21642

APPS, A.B. [1982] *Boolean powers of groups* (**J** 0332) Math Proc Cambridge Phil Soc 91*375-395
⋄ C13 C30 C60 ⋄ ID 48361

APPS, A.B. [1983] $\aleph_0$-*categorical finite extensions of boolean powers* (**J** 3240) Proc London Math Soc, Ser 3 47*385-410
⋄ C30 C35 C60 ⋄ REV MR 85g:03051 Zbl 493 # 03007 • ID 41227

APPS, A.B. [1983] *On $\aleph_0$-categorical class two groups* (**J** 0032) J Algeb 82*516-538
⋄ C35 C60 ⋄ REV MR 85d:20031 Zbl 533 # 20001 • ID 36573

APPS, A.B. [1983] *On the structure of $\aleph_0$-categorical groups*
(J 0032) J Algeb 81*320-339
⋄ C35 C60 ⋄ REV MR 84h:03079 Zbl 512 # 20013
• ID 34275

APPS, A.B. [1985] *Two counterexamples in $\aleph_0$-categorical groups* (J 0004) Algeb Universalis 20*51-56
⋄ C35 C60 ⋄ REV Zbl 569 # 20003 • ID 46360

APT, K.R. [1972] *ω-models in analytical hierarchy (Russian summary)* (J 0014) Bull Acad Pol Sci, Ser Math Astron Phys 20*901-904
⋄ C62 D30 D55 ⋄ REV MR 47 # 4777 Zbl 252 # 02045
• ID 00429

APT, K.R. [1973] *Infinitistic rules of proof and their semantics (Russian summary)* (J 0014) Bull Acad Pol Sci, Ser Math Astron Phys 21*879-886
⋄ C62 F07 F35 ⋄ REV MR 51 # 5291 Zbl 272 # 02077
• ID 00431

APT, K.R. & MAREK, W. [1974] *Second order arithmetic and related topics* (J 0007) Ann Math Logic 6*177-229
⋄ C62 D55 E45 F25 F35 ⋄ REV MR 51 # 12512
Zbl 299 # 02066 • ID 00432

APT, K.R. [1976] *Semantics of the infinitistic rules of proof* (J 0036) J Symb Logic 41*121-138
⋄ C62 D55 E45 F07 F35 ⋄ REV MR 55 # 10230
Zbl 328 # 02016 • ID 14789

APT, K.R. [1977] *Recursive embeddings of partial orderings* (J 0017) Canad J Math 29*349-359
⋄ C57 D20 G05 ⋄ REV MR 55 # 2673 Zbl 331 # 06002
• ID 60167

APT, K.R. [1978] *Inductive definitions, models of comprehension and invariant definability* (J 0029) Israel J Math 29*221-238
⋄ C40 C62 D55 D70 F35 ⋄ REV MR 81e:03048
Zbl 379 # 02014 • ID 29131

APT, K.R. see Vol. I, II, IV, VI for further entries

APTER, A.W. [1985] *A cardinal structure theorem for an ultrapower* (J 0018) Canad Math Bull 28*472-473
⋄ C20 E35 E55 ⋄ ID 48362

APTER, A.W. [1985] *Successors of singular cardinals and measurability* (J 0345) Adv Math 55*228-241
⋄ C62 E25 E35 E55 ⋄ ID 39959

APTER, A.W. see Vol. V for further entries

ARAGANE, K. & FUJIWARA, T. [1980] *Free product decompositions in a regular class* (P 3010) Symp Semigroups (3);1979 Kobe 1-6
⋄ C30 ⋄ REV MR 81h:08007 Zbl 445 # 08005 • ID 56509

ARAGANE, K. [1982] *Characterization of elementary classes of structures whose every substructure is a congruence class* (P 3833) Symp Semigroups (6);1982 Kyoto 65-67
⋄ C05 C52 ⋄ REV Zbl 509 # 08006 • ID 36792

ARAGANE, K. & FUJIWARA, T. [1983] *Elementary classes whose every model is algebraically closed* (P 4006) Symp Semigroups (7);1983 Tokyo 31-40
⋄ C25 C52 ⋄ REV MR 85d:03066 Zbl 558 # 03015
• ID 40437

ARAGANE, K. [1984] *Characterization of elementary classes of structures whose every substructure is a congruence class* (J 0352) Math Jap 29*23-28
⋄ C05 C52 ⋄ REV MR 85g:03050 Zbl 539 # 03014
• ID 39982

ARAGANE, K. & FUJIWARA, T. [1985] *Elementary classes whose every model is α-existentially closed* (P 4423) Semigroups (8);1984 Matsue 52-56
⋄ C25 C52 ⋄ ID 48347

ARAGANE, K. & FUJIWARA, T. [1985] *Elementary classes of α-existentially closed structures* (J 0352) Math Jap 30*989-1007
⋄ C25 C52 ⋄ ID 49043

ARIBAUD, F. [1979] *Un ideal maximal de l'anneau des endomorphismes d'un espace vectoriel de dimension infinie* (S 2250) Semin Dubreil: Algebre 31(1977/78)*184-189
⋄ C60 C62 E40 E75 ⋄ REV MR 81b:16015
Zbl 416 # 16010 • ID 53226

ARMBRUST, M. & SCHMIDT, J. [1964] *Zum Cayleyschen Darstellungssatz* (J 0043) Math Ann 154*70-72
⋄ C05 ⋄ REV MR 28 # 3109 Zbl 122.267 • ID 30819

ARMBRUST, M. [1971] *On the theorem of Ax and Kochen* (J 0008) Arch Math (Basel) 22*55-58
⋄ C60 ⋄ REV MR 45 # 3375 Zbl 221 # 12056 • ID 00464

ARMBRUST, M. & KAISER, KLAUS [1971] *Remarks on model classes acting on one another* (J 0043) Math Ann 193*136-138
⋄ C05 C20 C30 C52 ⋄ REV MR 44 # 6466 Zbl 209.13
• ID 31802

ARMBRUST, M. & KAISER, KLAUS [1972] *On quasi-universal model classes* (J 0068) Z Math Logik Grundlagen Math 18*403-406
⋄ B15 C05 C20 C52 C85 ⋄ REV MR 47 # 4784
Zbl 249 # 02025 • ID 00468

ARMBRUST, M. & KAISER, KLAUS [1972] *Some remarks on projective model classes and the interpolation theorem* (J 0043) Math Ann 197*5-8
⋄ C40 C52 ⋄ REV MR 45 # 6605 Zbl 222 # 02059
• ID 00465

ARMBRUST, M. [1973] *A first order approach to correspondence and congruence systems* (J 0068) Z Math Logik Grundlagen Math 19*215-222
⋄ C05 C75 ⋄ REV MR 51 # 12659 Zbl 298 # 08005
• ID 00470

ARMBRUST, M. [1973] *Direct limits of congruence systems* (S 0019) Colloq Math (Warsaw) 27*177-185
⋄ C05 C85 ⋄ REV MR 51 # 5445 Zbl 231 # 08003
• ID 17424

ARMBRUST, M. & KAISER, KLAUS [1974] *On some properties a projective model class passes on to the generated axiomatic class* (J 0009) Arch Math Logik Grundlagenforsch 16*133-136
⋄ C20 C52 ⋄ REV MR 50 # 9570 Zbl 286 # 02057
• ID 00471

ARMBRUST, M. see Vol. V for further entries

ARNON, D.S. & MCCALLUM, S. [1982] *Cylindrical algebraic decomposition by quantifier elimination* (P 3826) Proc Europ Conf Comput Algeb;1982 Marseille 215-222
⋄ C10 C40 ⋄ REV MR 84m:03044 Zbl 552 # 68046
• ID 35747

ARNON, D.S. see Vol. I for further entries

ARRUDA, A.I. & CHUAQUI, R.B. & COSTA DA, N.C.A. (EDS.) [1977] *Nonclassical logics, model theory and computability. Proceedings of the Third Latin-American Symposium on Mathematical Logic* (X 0809) North Holland: Amsterdam xviii+307pp
◇ B97 C97 ◇ REV MR 55#12445 Zbl 348#00003
• ID 16589

ARRUDA, A.I. see Vol. I, II, V, VI for further entries

ARTEMOV, S.N. [1980] *Arithmetically complete modal theories (Russian)* (S 2582) Semiotika & Inf, Akad Nauk SSSR 14*115-133
◇ B45 C90 F50 ◇ REV MR 82a:03056 Zbl 463#03006
• ID 54546

ARTEMOV, S.N. see Vol. II, IV, VI for further entries

ARTIGUE, M. & ISAMBERT, E. & PERRIN, M.J. & ZALC, A. [1973] *A notion of extension for models of full second order arithmetic (Russian summary)* (J 0014) Bull Acad Pol Sci, Ser Math Astron Phys 21*1087-1091
◇ C62 E45 ◇ REV MR 50#1880 Zbl 276#02038
• ID 03836

ARTIGUE, M. & ISAMBERT, E. & PERRIN, M.J. & ZALC, A. [1978] *Some remarks on bicommutability* (J 0027) Fund Math 101*207-226
◇ C62 E30 F25 F35 ◇ REV MR 80j:03019a Zbl 401#03024 • REM Comments have been published by Marek ibid 101*227-228 • ID 70505

ARTIGUE, M. see Vol. IV, V, VI for further entries

ARZARELLO, F. [1975] *Un sistema di deduzione naturale per linguaggi infinitari (English summary)* (J 0220) Atti Accad Sci Torino, Fis Mat Nat 109*633-641
◇ C75 F05 ◇ REV MR 55#7744 Zbl 367#02006
• ID 51166

ARZARELLO, F. [1976] *Alcune questioni riguardanti l'aritmetica e i linguaggi infinitari (English summary)* (J 2038) Rend Sem Mat, Torino 34*369-380
◇ C75 F05 ◇ REV MR 55#12463 Zbl 359#02008
• ID 50556

ARZARELLO, F. see Vol. V, VI for further entries

ASH, C.J. [1974] *Reduced powers and boolean extensions* (J 3172) J London Math Soc, Ser 2 9*429-432
◇ B25 C20 C30 G20 ◇ REV MR 58#10410 Zbl 304#02021 • ID 60199

ASH, C.J. [1975] *Sentences with finite models* (J 0068) Z Math Logik Grundlagen Math 21*401-404
◇ C13 ◇ REV MR 52#5402 Zbl 337#02031 • ID 03840

ASH, C.J. & ROSENTHAL, J.W. [1980] *Some theories associated with algebraically closed fields* (J 0036) J Symb Logic 45*359-362
◇ C60 D35 ◇ REV MR 81e:03030 Zbl 487#03017
• ID 28370

ASH, C.J. & NERODE, A. [1981] *Functorial properties of algebraic closure and Skolemization* (J 3194) J Austral Math Soc, Ser A 31*136-141
◇ C60 G30 ◇ REV MR 82m:12014 Zbl 463#18001
• ID 54583

ASH, C.J. [1983] *Model-theoretic forms of the axiom of choice* (P 3669) SE Asian Conf on Log;1981 Singapore 1-12
◇ C75 E25 E35 ◇ REV MR 84m:03068 Zbl 542#03026
• ID 35766

ASH, C.J. & MILLAR, T.S. [1983] *Persistently finite, persistently arithmetic theories* (J 0053) Proc Amer Math Soc 89*487-492
◇ C57 ◇ REV Zbl 527#03014 • ID 37509

ASH, C.J. & DOWNEY, R.G. [1984] *Decidable subspaces and recursively enumerable subspaces* (J 0036) J Symb Logic 49*1137-1145
◇ C57 C60 D35 D45 ◇ ID 39867

ASH, C.J. [1985] *Pseudovarieties, generalized varieties and similarly described classes* (J 0032) J Algeb 92*104-115
◇ C05 C13 ◇ REV Zbl 548#08007 • ID 43219

ASH, C.J. see Vol. I, II, IV, V for further entries

ASSER, G. [1955] *Das Repraesentantenproblem im Praedikatenkalkuel der ersten Stufe mit Identitaet* (J 0068) Z Math Logik Grundlagen Math 1*252-263
◇ C13 ◇ REV MR 17.1038 Zbl 66.257 JSL 23.38
• ID 00541

ASSER, G. [1955] *Eine semantische Charakterisierung der deduktiv abgeschlossenen Mengen des Praedikatenkalkuels erster Stufe* (J 0068) Z Math Logik Grundlagen Math 1*3-28
◇ C07 ◇ REV MR 16.1079 Zbl 64.11 JSL 20.282
• ID 00540

ASSER, G. [1979] *Ueber die Charakterisierbarkeit transfiniter Maechtigkeiten im Praedikatenkalkuel der zweiten Stufe* (P 2539) Frege Konferenz (1);1979 Jena 33-42
◇ B15 C55 C80 C85 ◇ REV MR 82d:03017 • ID 70532

ASSER, G. see Vol. I, II, IV, V for further entries

ASSOUS, R. & POUZET, M. [1981] *Structures invariantes et classes de meilleur ordre* (P 2614) Open Days in Model Th & Set Th;1981 Jadwisin 5-14
◇ C65 E07 ◇ ID 33709

ASSOUS, R. see Vol. V for further entries

ASTROMOFF, A. [1965] *Some structure theorems for primal and categorical algebras* (J 0044) Math Z 87*365-377
◇ C05 C13 ◇ REV MR 32#2355 Zbl 136.262 • ID 00559

AUSIELLO, G. [1976] *Difficult logical theories and their computer approximations* (J 1620) Asterisque 38-39*3-21
◇ B25 B35 D15 ◇ REV MR 57#16032 Zbl 365#02022
• ID 51023

AUSIELLO, G. see Vol. I, IV, VI for further entries

AUSTIN, A.K. [1966] *Varieties of groups defined by a single law* (J 0057) Publ Math (Univ Debrecen) 13*89-93
◇ C05 C60 ◇ REV MR 35#5497 Zbl 168.269 • ID 00570

AUSTIN, A.K. see Vol. IV for further entries

AX, J. & KOCHEN, S. [1965] *Diophantine problems over local fields I,II (II: A complete set of axioms for p-adic number theory)* (J 0100) Amer J Math 87*605-630,631-648
◇ B25 C20 C35 C60 ◇ REV MR 32#2401 MR 32#2402 Zbl 136.328 JSL 36.683 • REM Part III 1966
• ID 00574

Ax, J. & Kochen, S. [1966] *Diophantine problems over local fields, III: Decidable fields* (J 0120) Ann of Math, Ser 2 83*437-456
⋄ B25 C10 C20 C35 C60 C65 ⋄ REV MR 34 # 1262 Zbl 223 # 02050 • REM Parts I,II 1965 • ID 90140

Ax, J. [1967] *Solving diophantine problems modulo every prime* (J 0120) Ann of Math, Ser 2 85*161-183
⋄ B25 C13 C20 C60 ⋄ REV MR 35 # 126 Zbl 239 # 10032 JSL 38.161 • ID 00579

Ax, J. [1968] *The elementary theory of finite fields* (J 0120) Ann of Math, Ser 2 88*239-271
⋄ B25 B30 C13 C20 C60 ⋄ REV MR 37 # 5187 Zbl 195.57 JSL 38.162 • ID 00580

Ax, J. [1971] *A metamathematical approach to some problems in number theory* (P 1469) AMS Numb Th Summer Inst;1969 Stony Brook 161-190
⋄ C20 C60 C98 ⋄ REV MR 47 # 4966 Zbl 218 # 10075
• ID 26296

Ax, J. see Vol. IV for further entries

Baayen, P.C. [1962] *Partial ordering of quantifiers and of clopen equivalence relations* (X 1121) Math Centr: Amsterdam ZW-025*15pp
⋄ C80 C99 E07 G15 G99 ⋄ REV Zbl 109.240 • ID 33458

Babai, L. [1981] *On the abstract group of automorphisms* (S 3306) Lond Math Soc Lect Note Ser 52*1-40
⋄ C07 C65 ⋄ REV MR 83a:05064 Zbl 467 # 05031
• ID 38915

Babai, L. see Vol. V for further entries

Back, R.J.R. [1981] *Proving total correctness of nondeterministic programs in infinitary logic* (J 1431) Acta Inf 15*233-249
⋄ B75 C75 ⋄ REV MR 82h:68010 Zbl 437 # 68008 Zbl 467 # 03028 • ID 55026

Back, R.J.R. see Vol. I, VI for further entries

Bacsich, P.D. [1972] *Cofinal simplicity and algebraic closedness* (J 0004) Algeb Universalis 2*354-360
⋄ C05 C25 C50 C60 C75 G10 ⋄ REV MR 48 # 5962 Zbl 259 # 08005 • ID 00618

Bacsich, P.D. [1972] *Diagrammatic construction of homogenous universal models* (J 3172) J London Math Soc, Ser 2 5*762-768
⋄ C50 C60 ⋄ REV MR 47 # 34 Zbl 257 # 02043 • ID 00621

Bacsich, P.D. [1972] *Injectivity in model theory* (S 0019) Colloq Math (Warsaw) 25*165-176
⋄ C05 C35 C50 G05 G10 G30 ⋄ REV MR 48 # 3840 Zbl 214.14 JSL 40.88 • ID 00619

Bacsich, P.D. [1973] *An epi-reflector for universal theories* (J 0018) Canad Math Bull 16*167-171
⋄ C05 C30 C60 G30 ⋄ REV MR 49 # 370 Zbl 265 # 08008 • ID 03849

Bacsich, P.D. [1973] *Defining algebraic elements* (J 0036) J Symb Logic 38*93-101
⋄ C40 C52 C75 ⋄ REV MR 53 # 115 Zbl 259 # 02041
• ID 00623

Bacsich, P.D. [1973] *Primality and model-completion* (J 0004) Algeb Universalis 3*265-270
⋄ C05 C25 C35 ⋄ REV MR 50 # 4441 Zbl 308 # 02052
• ID 03850

Bacsich, P.D. [1974] *Model theory of epimorphisms* (J 0018) Canad Math Bull 17*471-477
⋄ C07 G30 ⋄ REV MR 51 # 10081 Zbl 313 # 02032
• ID 17531

Bacsich, P.D. & Rowlands-Hughes, D. [1974] *Syntactic characterizations of amalgamation, convexity and related properties* (J 0036) J Symb Logic 39*433-451
⋄ C05 C07 C60 C75 ⋄ REV MR 52 # 71 Zbl 302 # 02019
• ID 03851

Bacsich, P.D. [1975] *Amalgamation properties and interpolation theorems for equational theories* (J 0004) Algeb Universalis 5*45-55
⋄ B20 C05 C40 C52 ⋄ REV MR 52 # 2873 Zbl 324 # 02036 • ID 17658

Bacsich, P.D. [1975] *The strong amalgamation property* (S 0019) Colloq Math (Warsaw) 33*13-23
⋄ C05 C52 C60 ⋄ REV MR 53 # 116 Zbl 331 # 02033
• ID 16580

Bacsich, P.D. see Vol. I, V for further entries

Badger, L.W. [1977] *An Ehrenfeucht game for the multivariable quantifiers of Malitz and some applications* (J 0048) Pac J Math 72*293-304
⋄ C40 C80 E60 ⋄ REV MR 57 # 2877 Zbl 351 # 90107
• ID 27166

Badger, L.W. [1980] *Beth's property fails in $L^{<\omega}$* (J 0036) J Symb Logic 45*284-290
⋄ C40 C80 ⋄ REV MR 81h:03077 Zbl 439 # 03013
• ID 56005

Baer, Reinhold [1962] *Abzaehlbar erkennbare gruppentheoretische Eigenschaften* (J 0044) Math Z 79*344-363
⋄ C60 ⋄ REV MR 26 # 6246 Zbl 105.259 • ID 00633

Baer, Reinhold see Vol. I, V for further entries

Baeten, J. [1984] *Filters and ultrafilters over definable subsets of admissible ordinals* (P 2153) Logic Colloq;1983 Aachen 1*1-8
⋄ C62 D60 E05 E10 E47 ⋄ REV MR 86g:03075 Zbl 562 # 03024 • ID 45395

Baeten, J. see Vol. VI for further entries

Bagemihl, F. & Gillman, L. [1956] *Some cofinality theorems on ordered sets* (J 0027) Fund Math 43*178-184
⋄ C50 E07 E10 ⋄ REV MR 18.551 Zbl 72.43 • ID 00653

Bagemihl, F. see Vol. V for further entries

Bajramov, R.A. & Mamedov, O.M. [1980] *Σ-subdirect representations in axiomatic classes (Russian) (English summary)* (J 0134) Dokl Akad Nauk Azerb SSR 36/4*3-7
⋄ C05 C30 C52 ⋄ REV MR 82c:08011 Zbl 445 # 08002
• ID 32583

Bajramov, R.A. & Mamedov, O.M. [1984] *Nonstandard subdirect representations in axiomatizable classes (Russian)* (J 0092) Sib Mat Zh 25/3*14-29
• TRANSL [1984] (J 0475) Sib Math J 25*347-360
⋄ C05 C30 C52 ⋄ REV MR 85m:08016 • ID 40263

Bajramov, R.A. see Vol. II for further entries

Bajzhanov, B.S. & Omarov, B. [1979] *On finite diagrams (Russian)* (C 2065) Teor Nereg Kriv Raz Geom Post 11-15
⋄ C52 ⋄ REV MR 81f:03040 • ID 70594

BAJZHANOV, B.S. & OMAROV, B. [1979] *Restriction of a theory to a formula (Russian)* (C 2065) Teor Nereg Kriv Raz Geom Post 9-11
 ⋄ C30 C45 ⋄ REV MR 81f:03039 • ID 70595

BAJZHANOV, B.S. & OMAROV, B. [1980] *On omitting models (Russian) (Kazakh summary)* (J 0403) Izv Akad Nauk Kazak SSR, Ser Fiz-Mat 1980/1*57-59,93
 ⋄ C15 C50 ⋄ REV MR 81g:03035 Zbl 428 # 03024
 • ID 53783

BAJZHANOV, B.S. [1980] *Questions on the spectra of totally transcendental theories of finite rank (Russian)* (C 2620) Teor Model & Primen 25-44
 ⋄ C35 C45 ⋄ REV MR 82f:03026 Zbl 574 # 03015
 • ID 90344

BAJZHANOV, B.S. [1980] *Some properties of totally transcendental theories (Russian)* (C 2620) Teor Model & Primen 14-24 • ERR/ADD [1981] (C 3806) Issl Teor Progr 102
 ⋄ C35 C45 ⋄ REV MR 82e:03031 MR 84k:03096 Zbl 574 # 03016 • ID 90343

BAJZHANOV, B.S. & TAJTSLIN, M.A. (EDS.) [1980] *Theory of models and its applications (Russian)* (X 2769) Kazakh Gos Univ: Alma-Ata 83pp
 ⋄ C97 ⋄ REV MR 81k:03002 Zbl 492 # 00003 • ID 37726

BAJZHANOV, B.S. see Vol. V for further entries

BAKER, A. [1971] *Effective methods in the theory of numbers* (P 0743) Int Congr Math (II,11,Proc);1970 Nice 1*19-26
 ⋄ B25 B28 ⋄ REV MR 54 # 10162 Zbl 222 # 10001 JSL 37.606 • ID 42680

BAKER, A. [1971] *Effective methods in Diophantine problems* (S 3304) Proc Symp Pure Math 20*195-205
 ⋄ B25 ⋄ REV MR 47 # 3324 • ID 42681

BAKER, K.A. & FISHBURN, P.C. & ROBERTS, F.S. [1970] *A new characterization of partial orders of dimension two* (J 1377) Ann New York Acad Sci 175*23-24
 ⋄ C65 E07 ⋄ REV MR 42 # 140 Zbl 241 # 06002
 • ID 28844

BAKER, K.A. & FISHBURN, P.C. & ROBERTS, F.S. [1972] *Partial orders of dimension 2* (J 0138) Networks 2*11-28
 ⋄ C65 E07 ⋄ REV MR 46 # 104 Zbl 247 # 06002
 • ID 00687

BAKER, K.A. [1981] *Definable normal closures in locally finite varieties of groups* (J 1447) Houston J Math 7*467-471
 ⋄ C05 C40 C60 ⋄ REV MR 83g:20026 Zbl 494 # 20012
 • ID 36651

BAKER, K.A. [1983] *Nondefinability of projectivity in lattice varieties* (J 0004) Algeb Universalis 17*267-274
 ⋄ C05 C30 C40 G10 ⋄ REV MR 85k:06006 Zbl 535 # 06005 • ID 45226

BAKER, K.A. see Vol. II for further entries

BALBES, R. & DWINGER, P. [1971] *Uniqueness of representations of a distributive lattice as a free product of a boolean algebra and a chain* (S 0019) Colloq Math (Warsaw) 24*27-35
 ⋄ C05 C30 G10 ⋄ REV MR 46 # 3389 Zbl 231 # 06017
 • ID 00716

BALBES, R. see Vol. IV, V for further entries

BALCAR, B. & JECH, T.J. [1965] *Models of the theory of sets generated by a perfect relation (Russian) (Czech and German summaries)* (J 0086) Cas Pestovani Mat, Ceskoslov Akad Ved 90*413-434
 ⋄ C62 E30 E55 E70 ⋄ REV MR 35 # 48 Zbl 137.15
 • ID 00722

BALCAR, B. & VOPENKA, P. [1967] *On complete models of the set theory* (J 0014) Bull Acad Pol Sci, Ser Math Astron Phys 15*839-841
 ⋄ C62 E10 E25 ⋄ REV MR 39 # 3989 Zbl 177.14
 • ID 13883

BALCAR, B. & SOCHOR, A. [1969] *The general theory of semisets, syntactic models of axiomatic set theory* (P 0630) Aspects Math Log;1968 Varenna 267-285
 ⋄ C62 E35 E70 ⋄ REV MR 41 # 50 Zbl 212.324
 • ID 00726

BALCAR, B. see Vol. IV, V for further entries

BALDI, G. [1969] *Modelli finiti nella teoria degli insiemi (English summary)* (J 0012) Boll Unione Mat Ital, IV Ser 2*326-340
 ⋄ C13 C62 E70 ⋄ REV MR 40 # 34 Zbl 188.23 • ID 00734

BALDO, A. [1983] *Complete interpolation theorems for L_{kk}^{2+} (Italian summary)* (J 2100) Boll Unione Mat Ital, VI Ser, B 2*759-777
 ⋄ C40 C75 C85 ⋄ REV MR 85i:03099 Zbl 558 # 03013
 • ID 41844

BALDWIN, JOHN T. & LACHLAN, A.H. [1971] *On strongly minimal sets* (J 0036) J Symb Logic 36*79-96
 ⋄ C35 C45 C50 ⋄ REV MR 44 # 3851 Zbl 217.304 JSL 40.636 • ID 00736

BALDWIN, JOHN T. [1972] *Almost strongly minimal theories I,II* (J 0036) J Symb Logic 37*487-493,657-660
 ⋄ C35 C45 ⋄ REV MR 48 # 89 Zbl 266 # 02028 Zbl 266 # 02029 • ID 00738

BALDWIN, JOHN T. [1973] *α_T is finite for $\aleph_1$-categorical T* (J 0064) Trans Amer Math Soc 181*37-51
 ⋄ C35 C45 ⋄ REV MR 47 # 8289 Zbl 265 # 02034
 • ID 00740

BALDWIN, JOHN T. & BLASS, A.R. & GLASS, A.M.W. & KUEKER, D.W. [1973] *A "natural" theory without a prime model* (J 0004) Algeb Universalis 3*152-155
 ⋄ C15 C50 C62 C65 ⋄ REV MR 50 # 86 Zbl 273 # 02033
 • ID 03854

BALDWIN, JOHN T. [1973] *A sufficient condition for a variety to have the amalgamation property* (S 0019) Colloq Math (Warsaw) 28*181-183,329
 ⋄ C05 C35 C52 ⋄ REV MR 48 # 5957 Zbl 241 # 08005
 • ID 00741

BALDWIN, JOHN T. & LACHLAN, A.H. [1973] *On universal Horn classes categorical in some infinite power* (J 0004) Algeb Universalis 3*98-111
 ⋄ C05 C10 C35 C45 ⋄ REV MR 50 # 4273 Zbl 272 # 02078 • ID 03858

BALDWIN, JOHN T. [1973] *The number of automorphisms of models of $\aleph_1$-categorical theories* (J 0027) Fund Math 83*1-6
 ⋄ C07 C35 C50 ⋄ REV MR 48 # 10794 Zbl 287 # 02034
 • ID 00742

BALDWIN, JOHN T. & PLOTKIN, J.M. [1974] *A topology for the space of countable models of a first order theory* (J 0068) Z Math Logik Grundlagen Math 20*173-178
◇ C15 ◇ REV MR 50 # 4276   Zbl 327 # 02043 • ID 28363

BALDWIN, JOHN T. & BLASS, A.R. [1974] *An axiomatic approach to rank in model theory* (J 0007) Ann Math Logic 7*295-324
◇ C45   C50 ◇ REV MR 51 # 130   Zbl 298 # 02055
• ID 00743

BALDWIN, JOHN T. [1974] *Atomic compactness in $\aleph_1$-categorical Horn theories* (J 0027) Fund Math 83*263-268
◇ C35   C50 ◇ REV MR 49 # 2334   Zbl 286 # 02058
• ID 60275

BALDWIN, JOHN T. [1975] *Conservative extensions and the two cardinal theorems for stable theories* (J 0027) Fund Math 88*7-9
◇ C45   C55 ◇ REV MR 52 # 72   Zbl 307 # 02033 • ID 00745

BALDWIN, JOHN T. & BERMAN, J. [1975] *The number of subdirectly irreducible algebras in a variety* (J 0004) Algeb Universalis 5*379-389
◇ C05   C52 ◇ REV MR 52 # 13578   Zbl 348 # 08002 • REM Part I. Part II 1980 • ID 21862

BALDWIN, JOHN T. & SAXL, J. [1976] *Logical stability in group theory* (J 3194) J Austral Math Soc, Ser A 21*267-276
◇ C45   C60 ◇ REV MR 53 # 10934   Zbl 342 # 02036 JSL 49.317 • ID 23100

BALDWIN, JOHN T. & BERMAN, J. [1976] *Varieties and finite closure conditions* (S 0019) Colloq Math (Warsaw) 35*15-20
◇ C05   C13 ◇ REV MR 53 # 5427   Zbl 328 # 08004
• ID 30660

BALDWIN, JOHN T. & ROSE, B.I. [1977] *$\aleph_0$-categoricity and stability of rings* (J 0032) J Algeb 45*1-16
◇ C35   C45   C60 ◇ REV MR 55 # 12507   Zbl 368 # 16001
• ID 28375

BALDWIN, JOHN T. & BERMAN, J. [1977] *A model theoretic approach to Mal'cev conditions* (J 0036) J Symb Logic 42*277-288
◇ C05   C52 ◇ REV MR 58 # 207   Zbl 412 # 03016 Zbl 446 # 03025 • ID 26458

BALDWIN, JOHN T. [1978] *Some EC_Σ classes of rings* (J 0068) Z Math Logik Grundlagen Math 24*489-492
◇ C52   C60 ◇ REV MR 80a:03042   Zbl 402 # 03034
• ID 54681

BALDWIN, JOHN T. [1979] *Stability theory and algebra* (J 0036) J Symb Logic 44*599-608
◇ C45   C60 ◇ REV MR 81i:03051   Zbl 437 # 03013
• ID 55881

BALDWIN, JOHN T. & KUEKER, D.W. [1980] *Ramsey quantifiers and the finite cover property* (J 0048) Pac J Math 90*11-19
◇ C10   C35   C45   C80 ◇ REV MR 83e:03054
Zbl 471 # 03022 • ID 55217

BALDWIN, JOHN T. [1980] *The number of subdirectly irreducible algebras in a variety. II* (J 0004) Algeb Universalis 11*1-6
◇ C05   C75 ◇ REV MR 82c:08008   Zbl 469 # 08004 • REM Part I 1975 by Baldwin,J.T. & Berman,J. • ID 55181

BALDWIN, JOHN T. & KUEKER, D.W. [1981] *Algebraically prime models* (J 0007) Ann Math Logic 20*289-330
◇ C50 ◇ REV MR 83d:03042   Zbl 494 # 03024 • ID 90342

BALDWIN, JOHN T. [1981] *Definability and the hierarchy of stable theories* (P 2628) Log Year;1979/80 Storrs 1-15
◇ C40   C45 ◇ REV MR 82i:03040   Zbl 465 # 03014
• ID 54917

BALDWIN, JOHN T. & BERMAN, J. [1981] *Elementary classes of varieties* (J 1447) Houston J Math 7*473-492
◇ C05   C52 ◇ REV MR 83h:08011   Zbl 487 # 08007
• ID 33334

BALDWIN, JOHN T. & BERMAN, J. & GLASS, A.M.W. & HODGES, W. [1982] *A combinatorial fact about free algebras* (J 0004) Algeb Universalis 15*145-152
◇ C05 ◇ REV MR 84e:08009   Zbl 535 # 08005 • ID 33742

BALDWIN, JOHN T. & MCKENZIE, R. [1982] *Counting models in universal Horn classes* (J 0004) Algeb Universalis 15*359-384
◇ C05   C52 ◇ REV MR 84m:03042   Zbl 525 # 03028
• ID 33741

BALDWIN, JOHN T. & BERMAN, J. [1982] *Definable principal congruence relations: kith and kin* (J 0002) Acta Sci Math (Szeged) 44*255-270
◇ C05   C52   E07 ◇ REV MR 84e:08002   Zbl 503 # 08003
• ID 33740

BALDWIN, JOHN T. [1982] *Recursion theory and abstract dependence* (P 3634) Patras Logic Symp;1980 Patras 67-76
◇ C57   C60   D25   D45 ◇ REV MR 85i:03147
Zbl 519 # 03035 • ID 33739

BALDWIN, JOHN T. & MILLER, DOUGLAS E. [1982] *Some contributions to definability theory for languages with generalized quantifiers* (J 0036) J Symb Logic 47*572-586
◇ C40   C80 ◇ REV MR 83j:03060   Zbl 499 # 03013
• ID 33738

BALDWIN, JOHN T. & GIVANT, S. [1983] *Model complete universal Horn classes* (J 0004) Algeb Universalis 17*110-119
◇ C05   C35 ◇ REV MR 84m:03045   Zbl 549 # 03025
• ID 33743

BALDWIN, JOHN T. & SHELAH, S. [1983] *The structure of saturated free algebras* (J 0004) Algeb Universalis 17*191-199
◇ C05   C30   C45   C50 ◇ REV MR 85h:03032
Zbl 537 # 03020 • ID 33745

BALDWIN, JOHN T. [1984] *First-order theories of abstract dependence relations* (J 0073) Ann Pure Appl Logic 26*215-243
◇ C35   C45   C57   C60   D25   D45 ◇ REV MR 85g:03056
• ID 33746

BALDWIN, JOHN T. [1984] *Strong saturation and the foundations of stability theory* (P 3710) Logic Colloq;1982 Firenze 71-84
◇ C45   C50 ◇ REV MR 85j:03043   Zbl 554 # 03020
• ID 33744

BALDWIN, JOHN T. [1985] *Definable second-order quantifiers* (C 4183) Model-Theor Log 445-477
◇ C30   C80   C85   C98 ◇ ID 48329

BALDWIN, JOHN T. & SHELAH, S. [1985] *Second-order quantifiers and the complexity of theories* (J 0047) Notre Dame J Formal Log 26*229-303
◇ C40   C55   C65   C75   C85 ◇ ID 47544

BALL, R.N. [1980] *Cauchy completions are homomorphic images of submodels of ultrapowers* (P 2992) Conf Convergence Structs;1980 Lawton 1-7
⋄ C20 ⋄ REV MR 82c:03041   Zbl 471 # 03026 • ID 55221

BANASCHEWSKI, B. & NELSON, EVELYN [1972] *Equational compactness in equational classes of algebras* (J 0004) Algeb Universalis 2*152-165
⋄ C05 C50 ⋄ REV MR 46 # 7125   Zbl 258 # 08005
• ID 00763

BANASCHEWSKI, B. & NELSON, EVELYN [1973] *Equational compactness in infinitary algebras* (S 0019) Colloq Math (Warsaw) 27*197-205
⋄ C05 C50 ⋄ REV MR 48 # 5958   Zbl 255 # 08001
• ID 00765

BANASCHEWSKI, B. [1974] *Equational compactness of G-sets* (J 0018) Canad Math Bull 17*11-18
⋄ C05 C50 C60 ⋄ REV MR 50 # 4442   Zbl 294 # 08003
• ID 03860

BANASCHEWSKI, B. [1974] *On equationally compact extensions of algebras* (J 0004) Algeb Universalis 4*20-35
⋄ C05 C50 ⋄ REV MR 50 # 6974   Zbl 314 # 08005
• ID 00766

BANASCHEWSKI, B. & NELSON, EVELYN [1977] *Elementary properties of limit reduced powers with applications to boolean powers* (P 2112) Contrib to Universal Algeb;1975 Szeged 21-25
⋄ C05 C20 C30 C90 ⋄ REV MR 58 # 5178
Zbl 384 # 03020 • ID 52062

BANASCHEWSKI, B. & NELSON, EVELYN [1980] *Boolean powers as algebras of continuous functions* (J 0202) Diss Math (Warsaw) 179*51pp
⋄ C05 C20 C30 C50 ⋄ REV MR 81i:03040
Zbl 439 # 08004 • ID 56043

BANASCHEWSKI, B. [1981] *When are divisible abelian groups injective?* (J 2823) Quaest Math, S Africa 4*285-307
⋄ C13 C60 G10 ⋄ REV MR 83e:20063   Zbl 484 # 18009
• ID 36612

BANASCHEWSKI, B. see Vol. I, IV, V for further entries

BANKSTON, P. [1977] *Topological reduced products and the GCH* (S 2848) Topology Proc 1*261-267
⋄ C20 C65 E50 E75 ⋄ REV MR 56 # 13141
Zbl 401 # 54005 • ID 33381

BANKSTON, P. [1977] *Ultraproducts in topology* (J 0254) Gen Topology Appl 7*283-308
⋄ C20 C65 E75 ⋄ REV MR 56 # 16554   Zbl 364 # 54005
• ID 31009

BANKSTON, P. [1979] *Note on "Ultraproducts in topology"* (J 0254) Gen Topology Appl 10*231-232
⋄ C20 C65 E75 ⋄ REV MR 80h:54007   Zbl 405 # 54004
• ID 33384

BANKSTON, P. [1979] *Topological reduced products via good ultrafilters* (J 0254) Gen Topology Appl 10*121-137
⋄ C20 C65 E35 E50 E75 ⋄ REV MR 82g:54017
Zbl 398 # 54003 • ID 52789

BANKSTON, P. [1982] *Some obstacles to duality in topological algebra* (J 0017) Canad J Math 34*80-90 • ERR/ADD ibid 37*82-83
⋄ C05 C20 C65 G30 ⋄ REV MR 83i:18012
Zbl 452 # 18003 • ID 39343

BANKSTON, P. [1983] *Obstacles to duality between classes of relational structures* (J 0004) Algeb Universalis 17*87-91
⋄ C05 C20 E55 G30 ⋄ REV MR 84j:18005
Zbl 522 # 08008 • ID 37051

BANKSTON, P. & FOX, R. [1983] *On categories of algebras equivalent to a quasivariety* (J 0004) Algeb Universalis 16*153-158
⋄ C05 C20 G30 ⋄ REV MR 84g:08026   Zbl 517 # 18011
• ID 39434

BANKSTON, P. [1983] *On productive classes of function rings* (J 0053) Proc Amer Math Soc 87*11-14
⋄ C05 C20 C65 E75 G30 ⋄ REV MR 84c:03056
Zbl 509 # 08014 • ID 34000

BANKSTON, P. [1984] *Expressive power in first order topology* (J 0036) J Symb Logic 49*478-487
⋄ C65 C90 ⋄ REV MR 85h:03038   Zbl 576 # 03023
• ID 42418

BANKSTON, P. [1984] *First order representations of compact Hausdorff spaces* (P 4286) Categor Topol;1983 Toledo 23-28
⋄ C65 ⋄ REV MR 86e:54021   Zbl 548 # 54016 • ID 48911

BANKSTON, P. & SCHUTT, R. [1985] *On minimally free algebras* (J 0017) Canad J Math 37*963-978
⋄ C05 C50 C60 C65 ⋄ ID 48366

BANKSTON, P. see Vol. I, V for further entries

BAR-HILLEL, Y. [1957] *New light on the liar* (J 0103) Analysis (Oxford) 18*1-6
• REPR [1970] (C 4689) Bar-Hillel: Aspects of Lang 253-257
⋄ B25 ⋄ REV JSL 34.645 • ID 37306

BAR-HILLEL, Y. [1966] *Do natural languages contain paradoxes?* (J 0178) Stud Gen 19*391-397
• REPR [1970] (C 4689) Bar-Hillel: Aspects of Lang 273-285
⋄ A05 B65 C90 ⋄ REV Zbl 147.246 JSL 34.645
• ID 37309

BAR-HILLEL, Y. see Vol. I, II, IV, V, VI for further entries

BARANSKIJ, V.A. & TRAKHTMAN, A.N. [1970] *Subsemigroup graphs (Russian) (English summary)* (J 0143) Mat Chasopis (Slov Akad Ved) 20*135-140
⋄ C65 D05 ⋄ REV MR 47 # 3233   Zbl 196.41 • ID 00771

BARANSKIJ, V.A. [1983] *Algebraic systems whose elementary theory is compatible with an arbitrary group (Russian)* (J 0003) Algebra i Logika 22*599-607,719
• TRANSL [1983] (J 0069) Algeb and Log 22*425-431
⋄ C07 C65 G10 ⋄ REV Zbl 556 # 20005 • ID 44790

BARANSKIJ, V.A. [1985] *Independence of equational theories and automorphism groups of lattices* (J 0092) Sib Mat Zh 26/4*3-10
• TRANSL [1985] (J 0475) Sib Math J 26*479-485
⋄ C05 C07 G10 ⋄ ID 49812

BARNES, D.W. & MONRO, G.P. [1974] *A simple model for a weak system of arithmetic* (J 0016) Bull Austral Math Soc 11*321-323
⋄ C62 F30 ⋄ REV MR 50 # 9571   Zbl 284 # 02030
• ID 03863

BARNES, D.W. see Vol. I, II for further entries

BARONE, E. & GIANNONE, A. & SCOZZAFAVA, R. [1980] *On some aspects of the theory and applications of finitely additive probability measures* (J 2821) Pubbl Ist Mat App Univ Stud Roma 16∗43-53
  ⋄ C20 H05 ⋄ REV MR 82f:28001   Zbl 466 # 28002
  • ID 80631

BARR, M. & WELLS, C. [1985] *Toposes, triples and theories* (X 0811) Springer: Heidelberg & New York xiii + 345pp
  ⋄ C98 E98 F35 F50 G30 ⋄ REV MR 86f:18001   Zbl 567 # 18001 • ID 40111

BARR, M. see Vol. V for further entries

BARROS DE, C.M. [1970] *Sur la classe des ordinaux appartenant a un univers et univers bien ordonnes* (J 0068) Z Math Logik Grundlagen Math 16∗475-488
  ⋄ C62 E10 E55 E70 ⋄ REV MR 42 # 7501   Zbl 246 # 02048 • ID 00818

BARROS DE, C.M. see Vol. I, V for further entries

BARTOCCI, U. [1968] *Sulla generalizzazione di un teorema di Kochen (English summary)* (J 0149) Atti Accad Naz Lincei Fis Mat Nat, Ser 8 45∗221-230
  ⋄ C20 C60 ⋄ REV MR 45 # 309   Zbl 179.60 • ID 02018

BARWISE, J. [1967] *Infinitary logic and admissible sets* (0000) Diss., Habil. etc 124pp
  ⋄ C70 D60 ⋄ REM Doctorial diss., Stanford University
  • ID 16732

BARWISE, J. [1968] *Implicit definability and compactness in infinitary languages* (P 0637) Syntax & Semant Infinitary Lang;1967 Los Angeles 1-35
  ⋄ C40 C70 ⋄ REV MR 38 # 3141   Zbl 195.303   JSL 37.201
  • ID 00909

BARWISE, J. (ED.) [1968] *The syntax and semantics of infinitary languages* (S 3301) Lect Notes Math 72∗iv + 268pp
  ⋄ C70 C75 C97 ⋄ REV MR 38 # 3141   Zbl 165.1
  • ID 25069

BARWISE, J. [1969] *Applications of strict Π_1^1 predicates to infinitary logic* (J 0036) J Symb Logic 34∗409-423
  ⋄ C40 C70 D60 ⋄ REV MR 41 # 5218   Zbl 216.3   JSL 39.335 • ID 00825

BARWISE, J. [1969] *Infinitary logic and admissible sets* (J 0036) J Symb Logic 34∗226-252
  ⋄ C70 E45 E47 ⋄ REV MR 53 # 10546   Zbl 215.318   JSL 36.156 • ID 00827

BARWISE, J. & EKLOF, P.C. [1969] *Lefschetz's principle* (J 0032) J Algeb 13∗554-570
  ⋄ B15 C60 C75 C85 ⋄ REV MR 41 # 5207   Zbl 194.517
  • ID 00822

BARWISE, J. [1969] *Remarks on universal sentences of $L_{\omega_1,\omega}$* (J 0025) Duke Math J 36∗631-637
  ⋄ C15 C40 C75 ⋄ REV MR 39 # 6744   Zbl 215.319
  • ID 00824

BARWISE, J. & ROBINSON, A. [1970] *Completing theories by forcing* (J 0007) Ann Math Logic 2∗119-142
  • REPR [1979] (C 4594) Sel Pap Robinson 1∗219-242
  ⋄ C25 C35 ⋄ REV MR 42 # 7494   Zbl 222 # 02058   JSL 40.633 • ID 00832

BARWISE, J. & EKLOF, P.C. [1970] *Infinitary properties of abelian torsion groups* (J 0007) Ann Math Logic 2∗25-68
  ⋄ C60 C75 ⋄ REV MR 43 # 4906   Zbl 222 # 02014
  • ID 00830

BARWISE, J. & FISHER, E.R. [1970] *The Shoenfield absoluteness lemma* (J 0029) Israel J Math 8∗329-339
  ⋄ C62 D55 E45 ⋄ REV MR 43 # 4660   Zbl 206.11
  • ID 00828

BARWISE, J. & KUNEN, K. [1971] *Hanf numbers for fragments of $L_{\infty\omega}$* (J 0029) Israel J Math 10∗306-320
  ⋄ C55 C75 ⋄ REV MR 47 # 4758   Zbl 253 # 02010   JSL 49.315 • ID 30134

BARWISE, J. [1971] *Infinitary methods in the model theory of set theory* (P 0638) Logic Colloq;1969 Manchester 53-66
  ⋄ C62 C70 E45 E47 ⋄ REV MR 43 # 3103   Zbl 219 # 02046 • ID 00910

BARWISE, J. & GANDY, R.O. & MOSCHOVAKIS, Y.N. [1971] *The next admissible set* (J 0036) J Symb Logic 36∗108-120
  ⋄ C40 C62 D55 D60 D70 E45 E47 ⋄ REV MR 46 # 36   Zbl 236 # 02033 • ID 00836

BARWISE, J. [1972] *Absolute logics and $L_{\infty\omega}$* (J 0007) Ann Math Logic 4∗309-340
  ⋄ C75 C95 ⋄ REV MR 49 # 2252   Zbl 248 # 02061
  • ID 00840

BARWISE, J. [1972] *The Hanf number of second order logic* (J 0036) J Symb Logic 37∗588-594
  ⋄ B15 C55 C85 E30 ⋄ REV MR 58 # 21603   Zbl 281 # 02020 • ID 00839

BARWISE, J. [1973] *A preservation theorem for interpretations* (P 0713) Cambridge Summer School Math Log;1971 Cambridge GB 618-621
  ⋄ C40 F25 ⋄ REV MR 51 # 5285   Zbl 276 # 02034
  • ID 04141

BARWISE, J. [1973] *Back and forth through infinitary logics* (C 0654) Stud in Model Th 5-34
  ⋄ C15 C75 ⋄ REV MR 49 # 7116 • ID 00912

BARWISE, J. [1974] *Admissible sets over models of set theory* (P 0602) Generalized Recursion Th (1);1972 Oslo 97-122
  ⋄ C62 D60 E30 ⋄ REV MR 53 # 2670   Zbl 355 # 02031
  • ID 00911

BARWISE, J. [1974] *Axioms for abstract model theory* (J 0007) Ann Math Logic 7∗221-265
  ⋄ C95 ⋄ REV MR 51 # 12513   Zbl 324 # 02034 • ID 00841

BARWISE, J. [1974] *Mostowski's collapsing function and the closed unbounded filter* (J 0027) Fund Math 82∗95-103
  ⋄ C15 C55 C75 E47 ⋄ REV MR 50 # 12724   Zbl 306 # 02019   JSL 51.232 • ID 00913

BARWISE, J. [1975] *Admissible sets and the interaction of model theory, recursion theory and set theory* (P 1521) Int Congr Math (II,12);1974 Vancouver 1∗229-234
  ⋄ C70 C98 D55 D60 D70 E30 E35 E45 E47 ⋄ REV MR 55 # 2570   Zbl 342 # 02029 • ID 31017

BARWISE, J. [1975] *Admissible sets and structures. An approach to definability theory* (X 0811) Springer: Heidelberg & New York xiii + 394pp
  ⋄ B98 C40 C70 C98 D60 D98 E30 E98 ⋄ REV MR 54 # 12519   Zbl 316 # 02047   JSL 43.139 • ID 60316

BARWISE, J. & SCHLIPF, J.S. [1975] *On recursively saturated models of arithmetic* (C 0782) Model Th & Algeb (A. Robinson) 42-55
⋄ C50 C57 C62 D80 H15 ⋄ REV MR 53 # 12934 Zbl 343 # 02031 • ID 23183

BARWISE, J. & SCHLIPF, J.S. [1976] *An introduction to recursively saturated and resplendent models* (J 0036) J Symb Logic 41*531-536
⋄ C15 C40 C50 C57 ⋄ REV MR 53 # 7761 Zbl 343 # 02032 JSL 47.440 • ID 14694

BARWISE, J. [1976] *Some applications of Henkin quantifiers* (J 0029) Israel J Math 25*47-63
⋄ C40 C50 C60 C75 C80 ⋄ REV MR 57 # 5676 Zbl 347 # 02007 • ID 26072

BARWISE, J. [1977] *An introduction to first-order logic* (C 1523) Handb of Math Logic 5-46
⋄ B10 C07 C98 ⋄ REV MR 58 # 10395 JSL 49.968 • ID 24196

BARWISE, J. & KEISLER, H.J. [1977] *Guide to part A: Model theory* (C 1523) Handb of Math Logic 3-4
⋄ C98 ⋄ ID 27660

BARWISE, J. (ED.) [1977] *Handbook of mathematical logic* (X 0809) North Holland: Amsterdam xi+1165pp
• TRANSL [1982] (X 2027) Nauka: Moskva
⋄ B98 C98 D98 E98 F98 H98 ⋄ REV MR 56 # 15351 MR 84g:03004 MR 84j:03006 Zbl 443 # 03001 JSL 49.968 JSL 49.971 JSL 49.975 JSL 49.980 • REM 3rd ed 1982. Transl. in 4 parts. Russian suppl. by Mints,G.E. & Orevkov,V.P • ID 70117

BARWISE, J. [1977] *Some eastern two-cardinal theorems* (P 1704) Int Congr Log, Meth & Phil of Sci (5);1975 London ON 1*11-31
⋄ C55 C80 ⋄ REV MR 58 # 21570 Zbl 393 # 03023 • ID 32255

BARWISE, J. & MOSCHOVAKIS, Y.N. [1978] *Global inductive definability* (J 0036) J Symb Logic 43*521-534
⋄ C40 C70 D70 ⋄ REV MR 81g:03059 Zbl 395 # 03021 • ID 29280

BARWISE, J. [1978] *Monotone quantifiers and admissible sets* (P 1628) Generalized Recursion Th (2);1977 Oslo 1-38
⋄ C70 C80 D60 D70 ⋄ REV MR 81d:03037 Zbl 453 # 03047 • ID 70698

BARWISE, J. & KAUFMANN, M. & MAKKAI, M. [1978] *Stationary logic* (J 0007) Ann Math Logic 13*171-224 • ERR/ADD ibid 20*231-232
⋄ C55 C75 C80 ⋄ REV MR 82f:03031a MR 82f:03031b Zbl 372 # 02031 JSL 46.867 • ID 31872

BARWISE, J. & COOPER, R. [1981] *Generalized quantifiers and natural language* (J 2130) Linguist Philos 4*159-219
⋄ A05 B60 C80 ⋄ REV Zbl 473 # 03033 • ID 55362

BARWISE, J. [1981] *Infinitary logics* (C 2617) Modern Log Survey 93-112
⋄ C75 C98 ⋄ REV MR 82f:03002 Zbl 464 # 03001 • ID 90137

BARWISE, J. [1981] *The role of the omitting types theorem in infinitary logic* (J 0009) Arch Math Logik Grundlagenforsch 21*55-68
⋄ C75 C80 C95 ⋄ REV MR 83h:03052 Zbl 467 # 03034 • ID 55032

BARWISE, J. & FEFERMAN, S. (EDS.) [1985] *Model-theoretic logics* (X 0811) Springer: Heidelberg & New York xviii+893pp
⋄ C75 C80 C85 C90 C95 C98 ⋄ ID 48323

BARWISE, J. [1985] *Model-theoretic logics: background and aims* (C 4183) Model-Theor Log 3-23
⋄ A05 C98 ⋄ ID 48324

BARWISE, J. see Vol. I, II, IV, V for further entries

BASARAB, S.A. [1973] *Certain metamathematical aspects of the theory of henselian fields (Romanian) (English summary)* (J 0197) Stud Cercet Mat Acad Romana 25*1449-1459
⋄ C60 ⋄ REV MR 52 # 8104 Zbl 295 # 12106 • ID 17968

BASARAB, S.A. [1975] *The models of the elementary theory of finite abelian groups (Romanian) (English summary)* (J 0197) Stud Cercet Mat Acad Romana 27*381-386
⋄ B25 C13 C60 ⋄ REV MR 53 # 7770 Zbl 335 # 02036 • ID 23010

BASARAB, S.A. [1977] *On the elementary theories of abelian profinite groups and Abelian torsion groups* (J 0060) Rev Roumaine Math Pures Appl 22*299-309
⋄ B25 C60 ⋄ REV MR 55 # 12508 Zbl 388 # 03013 • ID 52267

BASARAB, S.A. [1978] *Quelques proprietes modele-theoriques des corps values henseliens (English summary)* (J 2313) C R Acad Sci, Paris, Ser A-B 287*A189-A191,A291-A293
⋄ C25 C35 C60 ⋄ REV MR 58 # 21583 MR 80a:12028 Zbl 394 # 12008 • ID 90147

BASARAB, S.A. [1978] *Some model theory for henselian valued fields* (J 0032) J Algeb 55*191-212
⋄ C35 C60 ⋄ REV MR 82m:03046 Zbl 424 # 03016 • ID 70730

BASARAB, S.A. [1979] *A model-theoretic transfer theorem for henselian valued fields* (J 0127) J Reine Angew Math 311-312*1-30
⋄ C25 C35 C60 ⋄ REV MR 81h:03071 Zbl 409 # 12030 • ID 56346

BASARAB, S.A. [1979] *Model-theoretic methods in the theory of henselian valued fields. I,II (Romanian) (English summaries)* (J 0197) Stud Cercet Mat Acad Romana 31*3-39,617-656
⋄ C25 C35 C60 ⋄ REV MR 82e:12034 MR 82e:12035 Zbl 424 # 03015 Zbl 446 # 03028 Zbl 454 # 03013 • ID 80641

BASARAB, S.A. & NICA, V. & POPESCU, D. [1980] *Approximation properties and existential completeness for ring morphisms* (J 0504) Manuscr Math 33*227-282
⋄ C25 C60 H15 ⋄ REV MR 82k:03047 Zbl 472 # 13013 • ID 70736

BASARAB, S.A. [1980] *Towards a general theory of formally p-adic fields* (J 0504) Manuscr Math 30*279-327
⋄ C60 ⋄ REV MR 81e:12028 Zbl 451 # 12016 • ID 80642

BASARAB, S.A. [1981] *Extension of places and contraction properties for function fields over $\mathfrak{p}$-adically closed fields* (J 0127) J Reine Angew Math 326*54-78
⋄ C60 H20 ⋄ REV MR 82j:03040 Zbl 491 # 12025 • ID 37757

BASARAB, S.A. [1983] *Roth's theorem: non-standard aspects (Romanian)* (J 0197) Stud Cercet Mat Acad Romana 35*105-113
⋄ H10 H15 ⋄ REV MR 85b:11116 Zbl 534 # 10050 • ID 38348

BASARAB, S.A. [1984] *Axioms for pseudo-real-closed fields* (J 0060) Rev Roumaine Math Pures Appl 29*449-456
⋄ C25 C60 ⋄ REV Zbl 555 # 12009 • ID 44236

BASARAB, S.A. [1984] *Definite functions on algebraic varieties over ordered fields* (J 0060) Rev Roumaine Math Pures Appl 29*527-535
⋄ C25 C60 ⋄ REV MR 85k:12006 Zbl 578 # 12019 • ID 47120

BASARAB, S.A. [1984] *On some classes of Hilbertian fields* (J 2563) Result Math 7*1-34
⋄ B25 C10 C25 C35 C60 ⋄ REV MR 86c:12002 Zbl 547 # 12016 • ID 43270

BASARAB, S.A. [1985] *The absolute Galois group of pseudoreal closed field with finitely many orders* (J 0326) J Pure Appl Algebra 38*1-18
⋄ C25 C60 ⋄ ID 48978

BASARAB, S.A. [1985] *Transfer principles for PRC fields* (P 4310) Easter Conf on Model Th (3);1985 Gross Koeris 16-20
⋄ C25 C60 ⋄ ID 49044

BAUDISCH, A. [1974] *Theorien abelscher Gruppen mit einem einstelligen Praedikat* (J 0027) Fund Math 83*121-127
⋄ C60 D35 ⋄ REV MR 54 # 10001 Zbl 289 # 02035 • ID 00867

BAUDISCH, A. [1975] *Die elementare Theorie der Gruppe vom Typ p^∞ mit Untergruppen* (J 0068) Z Math Logik Grundlagen Math 21*347-352
⋄ B25 C60 C85 ⋄ REV MR 53 # 7753 Zbl 319 # 02038 • ID 03870

BAUDISCH, A. [1975] *Elementare Theorien von Halbgruppen mit Kuerzungsregeln mit einem einstelligen Praedikat* (J 0014) Bull Acad Pol Sci, Ser Math Astron Phys 23*107-109
⋄ C05 C60 D35 ⋄ REV MR 51 # 10067 Zbl 307 # 02029 • ID 00868

BAUDISCH, A. [1975] *Endliche n-aequivalente Gruppen* (J 0115) Wiss Z Humboldt-Univ Berlin, Math-Nat Reihe 24*757-764
⋄ C13 C60 ⋄ REV MR 58 # 5181 Zbl 325 # 02034 • ID 30445

BAUDISCH, A. [1975] *Entscheidbarkeitsprobleme elementarer Theorien von Klassen abelscher Gruppen mit Untergruppen (English and Russian summaries)* (J 0014) Bull Acad Pol Sci, Ser Math Astron Phys 23*111-115
⋄ B25 C60 ⋄ REV MR 53 # 7752 Zbl 307 # 02030 • ID 00869

BAUDISCH, A. [1976] *Elimination of the quantifier Q_α in the theory of abelian groups (Russian summary)* (J 0014) Bull Acad Pol Sci, Ser Math Astron Phys 24*543-549
⋄ B25 C10 C60 C80 ⋄ REV MR 54 # 7246 Zbl 341 # 02040 • ID 25798

BAUDISCH, A. [1977] *A note on the elementary theory of torsion free Abelian groups with one predicate for subgroups* (J 0115) Wiss Z Humboldt-Univ Berlin, Math-Nat Reihe 26*611-612
⋄ B25 C60 ⋄ REV MR 80d:03012 Zbl 423 # 03036 • ID 53548

BAUDISCH, A. [1977] *Decidability of the theory of abelian groups with Ramsey quantifiers* (J 0014) Bull Acad Pol Sci, Ser Math Astron Phys 25*733-739 • ERR/ADD ibid 31*99-105
⋄ B25 C10 C60 C80 ⋄ REV MR 58 # 21556 MR 85j:03011 Zbl 375 # 02045 Zbl 533 # 03002 • ID 27133

BAUDISCH, A. & TUSCHIK, H.-P. [1977] *Eine Bemerkung ueber Vermeidung von Typen in ueberabzaehlbaren Modellen* (J 0115) Wiss Z Humboldt-Univ Berlin, Math-Nat Reihe 26*613-614
⋄ C07 ⋄ REV MR 80e:03025 Zbl 423 # 03043 • ID 53555

BAUDISCH, A. [1977] *The theory of abelian groups with the quantifier $(\leqslant x)$* (J 0068) Z Math Logik Grundlagen Math 23*447-462
⋄ B25 C10 C60 C80 ⋄ REV MR 58 # 16262 Zbl 441 # 03009 • ID 56062

BAUDISCH, A. & WEESE, M. [1977] *The Lindenbaum-algebra of the theory of well-orders and abelian groups with the quantifier Q_α* (P 1695) Set Th & Hierarch Th (3);1976 Bierutowice 59-73
⋄ C60 C80 ⋄ REV MR 58 # 10290 Zbl 364 # 02033 • ID 50963

BAUDISCH, A. [1979] *Uncountable n-cubes in models of $\aleph_0$-categorical theories* (J 3293) Bull Acad Pol Sci, Ser Math 27*1-9
⋄ C35 C80 ⋄ REV MR 81h:03069 Zbl 423 # 03039 • ID 53551

BAUDISCH, A. [1980] *Application of the theory of Shirshov-Witt in the theory of groups and Lie algebras* (S 3415) Prepr Akad Wiss DDR ZI Math Mech P-25/80*44pp • ERR/ADD ibid 30/80*9pp
⋄ B25 C45 C60 ⋄ REV Zbl 436 # 03028 Zbl 454 # 03015 • ID 54227

BAUDISCH, A. & SEESE, D.G. & TUSCHIK, H.-P. & WEESE, M. [1980] *Decidability and generalized quantifiers* (X 0911) Akademie Verlag: Berlin xii + 235pp
⋄ B25 B98 C10 C60 C80 C98 ⋄ REV MR 82i:03048 Zbl 442 # 03011 JSL 47.907 • ID 56368

BAUDISCH, A. [1980] *The theory of abelian p-groups with the quantifier I is decidable* (J 0027) Fund Math 108*183-197
⋄ B25 C10 C80 ⋄ REV MR 82g:03068 Zbl 369 # 02024 • ID 51324

BAUDISCH, A. [1981] *Formulas of L(aa) where aa is not in the scope of "¬"* (J 0068) Z Math Logik Grundlagen Math 27*249-254
⋄ C55 C80 ⋄ REV MR 84c:03072 Zbl 467 # 03031 • ID 55029

BAUDISCH, A. [1981] *Subgroups of semifree groups* (J 0001) Acta Math Acad Sci Hung 38*19-28
⋄ C05 C60 ⋄ REV MR 82k:20059 • ID 42906

BAUDISCH, A. [1981] *The elementary theory of abelian groups with m-chains of pure subgroups* (J 0027) Fund Math 112*147-157
⋄ B25 C55 C60 C80 ⋄ REV MR 82h:03026 Zbl 477 # 20005 • ID 55606

BAUDISCH, A. [1981] *There is no module having the finite cover property* (S 3414) Prepr, Akad Wiss DDR, Inst Math 20/81*16pp
⋄ C10 C45 C60 C80 ⋄ REV Zbl 462 # 03008 • ID 54515

BAUDISCH, A. [1982] *Decidability and stability of free nilpotent Lie algebras and free nilpotent p-groups of finite exponent* (J 0007) Ann Math Logic 23*1-25
⋄ B25 C45 C60 ⋄ REV MR 84e:03019 Zbl 499 # 03016 • ID 34358

BAUDISCH, A. & SEESE, D.G. & TUSCHIK, H.-P. [1983] ω-trees in stationary logic (J 0027) Fund Math 119*205-215
⋄ B25 C55 C65 C80 ⋄ REV MR 86a:03036 Zbl 574 # 03018 • ID 40630

BAUDISCH, A. [1983] A remark on $\aleph_0$-categorical stable groups (P 1601) Easter Conf on Model Th (1);1983 Diedrichshagen 4-9
⋄ C35 C45 C60 ⋄ REV MR 86a:03032 Zbl 528 # 03017 • ID 37635

BAUDISCH, A. & ROTHMALER, P. [1983] Vorlesungen zur Einfuehrung in die Theorie des Shelah'schen "forking" (S 3382) Sem-ber, Humboldt-Univ Berlin, Sekt Math 47*96pp
⋄ C45 C98 ⋄ REV MR 84f:03030 Zbl 526 # 03015 • REM Part I. Part II 1984 • ID 34450

BAUDISCH, A. [1984] Magidor-Malitz quantifiers in modules (J 0036) J Symb Logic 49*1-8
⋄ C10 C60 C80 ⋄ REV MR 85e:03070 • ID 40351

BAUDISCH, A. [1984] On elementary properties of free Lie algebras (S 3414) Prepr, Akad Wiss DDR, Inst Math 28pp
⋄ C45 C60 ⋄ REV Zbl 535 # 03012 • ID 38316

BAUDISCH, A. [1984] Tensor products of modules and elementary equivalence (J 0004) Algeb Universalis 19*120-127
⋄ C30 C52 C60 ⋄ REV MR 85j:03048 • ID 42907

BAUDISCH, A. & ROTHMALER, P. [1984] The stratified order in modules (P 3621) Frege Konferenz (2);1984 Schwerin 44-56
⋄ C45 C60 ⋄ REV MR 85m:03006 Zbl 551 # 03020 • ID 40187

BAUDISCH, A. & ROTHMALER, P. [1984] Vorlesungen zur Einfuehrung in die Theorie des Shelah'schen "forking" 2. Semester (S 3382) Sem-ber, Humboldt-Univ Berlin, Sekt Math 59*117pp
⋄ C45 C98 ⋄ REV MR 86f:03054 • REM Part II. Part I 1983 • ID 42908

BAUDISCH, A. & SEESE, D.G. & TUSCHIK, H.-P. & WEESE, M. [1985] Decidability and quantifier elimination (C 4183) Model-Theor Log 236-268
⋄ B25 C10 C60 C65 C80 C98 ⋄ ID 48327

BAUDISCH, A. [1985] On Lascar rank in nonmultidimensional ω-stable theories (P 4310) Easter Conf on Model Th (3);1985 Gross Koeris 21-44
⋄ C45 C60 ⋄ ID 49045

BAUMGARTNER, J.E. [1974] The Hanf number for complete $L_{\omega_1,\omega}$-sentences (without GCH) (J 0036) J Symb Logic 39*575-578
⋄ C35 C50 C55 C75 ⋄ REV MR 51 # 51 Zbl 299 # 02063 • ID 03871

BAUMGARTNER, J.E. [1977] Ineffability properties of cardinals. II (P 1704) Int Congr Log, Meth & Phil of Sci (5);1975 London ON 1*87-106
⋄ C55 E05 E45 E55 ⋄ REV MR 58 # 27487 Zbl 373 # 04002 • REM Part I 1975 • ID 51513

BAUMGARTNER, J.E. & GALVIN, F. [1978] Generalized Erdoes cardinals and $0^{\#}$ (J 0007) Ann Math Logic 15*289-313
⋄ C55 C75 E05 E45 E55 ⋄ REV MR 80f:03051 Zbl 437 # 03028 • ID 55896

BAUMGARTNER, J.E. see Vol. V for further entries

BAUMSLAG, B. & BAUMSLAG, G. [1971] On ascending chain conditions (J 3240) Proc London Math Soc, Ser 3 22*681-704
⋄ C05 C60 ⋄ REV MR 45 # 351 Zbl 226 # 20023 • ID 00879

BAUMSLAG, B. & LEVIN, F. [1976] Algebraically closed torsion-free nilpotent groups of class 2 (J 0394) Commun Algeb 4*533-560
⋄ C25 C60 ⋄ REV MR 53 # 5749 Zbl 379 # 20032 • ID 90152

BAUMSLAG, G. [1971] see BAUMSLAG, B.

BAUMSLAG, G. & DYER, E. & HELLER, A. [1980] The topology of discrete groups (J 0326) J Pure Appl Algebra 16*1-47
⋄ C25 C60 ⋄ ID 48367

BAUMSLAG, G. see Vol. IV for further entries

BAUR, W. [1974] Ueber rekursive Strukturen (J 0305) Invent Math 23*89-95
⋄ C50 C57 D45 ⋄ REV MR 49 # 2335 Zbl 285 # 02050 • ID 04142

BAUR, W. [1975] $\aleph_0$-categorical modules (J 0036) J Symb Logic 40*213-220
⋄ C35 C60 ⋄ REV MR 51 # 5283 Zbl 309 # 02059 JSL 40.213 • ID 04246

BAUR, W. [1975] Decidability and undecidability of theories of abelian groups with predicates for subgroups (J 0020) Compos Math 31*23-30
⋄ B25 C60 D35 ⋄ REV MR 52 # 5399 Zbl 335 # 02032 • ID 03874

BAUR, W. [1976] Elimination of quantifiers for modules (J 0029) Israel J Math 25*64-70
⋄ C10 C60 ⋄ REV MR 56 # 15409 Zbl 354 # 02043 • ID 26073

BAUR, W. & CHERLIN, G.L. & MACINTYRE, A. [1979] Totally categorical groups and rings (J 0032) J Algeb 57*407-440
⋄ C35 C45 C60 ⋄ REV MR 80e:03034 Zbl 401 # 03012 JSL 49.317 • ID 70792

BAUR, W. [1980] On the elementary theory of quadruples of vector spaces (J 0007) Ann Math Logic 19*243-262
⋄ B25 C60 D80 ⋄ REV MR 82g:03056 Zbl 453 # 03010 • ID 54140

BAUR, W. [1982] Die Theorie der Paare reell abgeschlossener Koerper (P 3482) Logic & Algor (Specker);1980 Zuerich 25-34
⋄ B25 C60 ⋄ REV MR 83c:03028 Zbl 505 # 03009 • REM Part I. Part II 1982. Title: On the elementary theory of pairs of real closed fields • ID 35134

BAUR, W. [1982] On the elementary theory of pairs of real closed fields II (J 0036) J Symb Logic 47*669-679
⋄ B25 C60 ⋄ REV MR 84j:03072 Zbl 529 # 03002 • REM Part I 1982. Title: Die Theorie der Paare reell abgeschlossener Koerper • ID 34664

BAUR, W. see Vol. IV, V for further entries

BAUVAL, A. [1982] Une condition necessaire d'equivalence elementaire entre anneaux de polynomes sur des corps (English summary) (J 3364) C R Acad Sci, Paris, Ser 1 295*31-33
⋄ C60 ⋄ REV MR 83j:03052 Zbl 505 # 03017 • ID 34865

BAUVAL, A. [1984] *La theorie d'un anneau de polynomes* (S 3521) Mem Soc Math Fr 16*77-84
⋄ C60 C85 ⋄ REV Zbl 563 # 12027 • ID 39726

BAUVAL, A. [1985] *Polynomial rings and weak second-order logic* (J 0036) J Symb Logic 50*953-972
⋄ C60 C85 F35 ⋄ ID 48368

BAZHAEV, YU.V. [1970] *Universal groups (Russian)* (C 2601) Vlozhenie Grupp: Algor Vopr 23-31
⋄ C50 C60 ⋄ REV MR 54 # 7622 Zbl 251 # 20041 • ID 90358

BEAUCHEMIN, P. [1974] *Quelques consequences en logique d'une loi de probabilite 0-1 pour les multirelations denombrables* (J 2313) C R Acad Sci, Paris, Ser A-B 278*A1155-A1157
⋄ C75 E25 E55 E75 ⋄ REV MR 52 # 7899 Zbl 291 # 02042 • ID 17974

BEAUCHEMIN, P. & REYES, G.E. [1975] *Espaces de Baire et espaces de probabilite de structures relationnelles* (J 0345) Adv Math 17*14-24
⋄ C65 C75 E25 E55 E75 ⋄ REV MR 54 # 7265 Zbl 339 # 02047 • ID 25817

BEAUCHEMIN, P. [1975] *Un theoreme de Loewenheim-Skolem-Tarski pour les structures homogenes (English summary)* (J 2313) C R Acad Sci, Paris, Ser A-B 280*A1249-A1251
⋄ C07 C50 E50 ⋄ REV MR 51 # 7851 Zbl 306 # 02056 • ID 17975

BECKER, E. [1984] *Extended Artin-Schreier theory of fields* (J 0308) Rocky Mountain J Math 14*881-897
⋄ B25 C35 C60 ⋄ REV MR 86f:12001 Zbl 563 # 12023 • ID 47467

BECKER, E. & JACOB, B. [1985] *Rational points on algebraic varieties over a generalized real closed field: a model theoretic approach* (J 0127) J Reine Angew Math 357*77-95
⋄ C60 ⋄ REV Zbl 548 # 14007 • ID 43221

BECKER, J.A. & LIPSHITZ, L. [1976] *An application of logic to analysis* (J 0017) Canad J Math 28*83-91
⋄ B80 C65 ⋄ REV MR 55 # 8397 Zbl 319 # 02051 • ID 29684

BECKER, J.A. & DENEF, J. & DRIES VAN DEN, L. & LIPSHITZ, L. [1979] *Ultraproducts and approximation in local rings. I* (J 0305) Invent Math 51*189-203
⋄ C20 C60 ⋄ REV MR 80k:14009 Zbl 416 # 13004 • REM Part II 1980 by Denef,J. & Lipshitz,L. • ID 31827

BECKER, J.A. & HENSON, C.W. & RUBEL, L.A. [1980] *First-order conformal invariants* (J 0120) Ann of Math, Ser 2 112*123-178
⋄ C35 C65 D35 E35 E75 ⋄ REV MR 83a:30011 Zbl 459 # 03019 • ID 54459

BECKER, J.A. & DENEF, J. & LIPSHITZ, L. [1980] *Further remarks on the elementary theory of formal power series rings* (P 2625) Model Th of Algeb & Arithm;1979 Karpacz 1-9
⋄ C60 D35 ⋄ REV MR 83a:13013 Zbl 452 # 12013 JSL 50.853 • ID 47356

BECKER, J.A. & LIPSHITZ, L. [1980] *Remarks of the elementary theories of formal and convergent power series* (J 0027) Fund Math 105*229-239 • ERR/ADD ibid 112*241
⋄ C60 ⋄ REV MR 82i:14004 Zbl 465 # 03015 • ID 54918

BECKER, T. [1983] *Real closed rings and ordered valuation rings* (J 0068) Z Math Logik Grundlagen Math 29*417-425
⋄ C60 ⋄ REV MR 86a:03034 Zbl 528 # 03019 • ID 37637

BEHMANN, H. [1922] *Beitraege zur Algebra der Logik, insbesondere zum Entscheidungsproblem* (J 0043) Math Ann 86*163-229
⋄ B20 B25 D35 ⋄ REV FdM 48.1119 • ID 00942

BEHMANN, H. [1923] *Algebra der Logik und Entscheidungsproblem* (J 0157) Jbuchber Dtsch Math-Ver 32*66-67,2.Abteilung
⋄ B20 B25 D35 ⋄ ID 00943

BEHMANN, H. [1927] *Entscheidungsproblem und Logik der Beziehungen* (J 0157) Jbuchber Dtsch Math-Ver 36*17-18,2.Abteilung
⋄ B25 D35 ⋄ REV FdM 53.41 • ID 00946

BEHMANN, H. [1950] *Das Aufloesungsproblem in der Klassenlogik* (J 0009) Arch Math Logik Grundlagenforsch 1*17-29
⋄ A05 B20 B25 ⋄ REV MR 14.122 Zbl 41.348 JSL 18.74 • ID 01559

BEHMANN, H. see Vol. I, II, V, VI for further entries

BEKENOV, M.I. & HERRE, H. [1979] *A remark concerning a generalized Morley-rank* (X 2888) ZI Math Mech Akad Wiss DDR: Berlin 13pp
⋄ C45 ⋄ REV Zbl 402 # 03031 • ID 54678

BEKENOV, M.I. & MUSTAFIN, T.G. [1979] *On rank functions, definable and indecomposable types in stable theories (Russian)* (J 0023) Dokl Akad Nauk SSSR 245*777-780
• TRANSL [1979] (J 0062) Sov Math, Dokl 20*333-336
⋄ C45 ⋄ REV MR 81j:03049 Zbl 435 # 03027 • ID 55787

BEKENOV, M.I. [1980] *Theories with a basis (Russian)* (C 2620) Teor Model & Primen 45-47
⋄ C45 C52 ⋄ REV MR 82h:03025 Zbl 556 # 03028 • ID 70840

BEKENOV, M.I. & MUSTAFIN, T.G. [1981] *Properties of m-types in stable theories (Russian)* (J 0092) Sib Mat Zh 22/4*27-34,228
• TRANSL [1981] (J 0475) Sib Math J 22*19-25
⋄ C45 ⋄ REV MR 82i:03041 Zbl 527 # 03012 • ID 70841

BEKENOV, M.I. [1982] *On the spectrum of quasitranscendental theories (Russian)* (J 0003) Algebra i Logika 21*3-12,124
• TRANSL [1982] (J 0069) Algeb and Log 21*1-7
⋄ C35 C45 C52 ⋄ REV MR 84i:03069 Zbl 519 # 03023 • ID 34550

BELEGRADEK, O.V. [1972] *Categoricity in nondenumerable powers and $\aleph_1$-homogeneous models (Russian)* (J 0003) Algebra i Logika 11*125-129,237
• TRANSL [1972] (J 0069) Algeb and Log 11*71-73
⋄ C35 C50 ⋄ REV MR 47 # 8283 Zbl 252 # 02051 • ID 27762

BELEGRADEK, O.V. [1972] *On unstable theories of groups (Russian)* (P 2585) All-Union Conf Math Log (2);1972 Moskva 5-6
⋄ C45 C60 ⋄ ID 43223

BELEGRADEK, O.V. & TAJTSLIN, M.A. [1972] *Two remarks on the varieties $\mathfrak{A}_{m,n}$ (Russian)* (J 0003) Algebra i Logika 11*501-508,614
• TRANSL [1972] (J 0069) Algeb and Log 11*275-279
⋄ C05 C50 ⋄ REV MR 47 # 8396 Zbl 272 # 08004 • ID 32035

BELEGRADEK, O.V. [1973] *Almost categorical theories (Russian)* (J 0092) Sib Mat Zh 14*277-288
• TRANSL [1973] (J 0475) Sib Math J 14*191-198
◇ C35 C45 C55 ◇ REV MR 49#2332 Zbl 262#02047
• ID 30368

BELEGRADEK, O.V. [1973] *Certain nonelementary properties of models of categorical theories (Russian)* (J 0031) Izv Vyssh Ucheb Zaved, Mat (Kazan) 1973/4(131)*3-7
◇ C35 ◇ REV MR 48#98 Zbl 264#08004 • ID 00965

BELEGRADEK, O.V. [1974] *Algebraically closed groups (Russian)* (J 0003) Algebra i Logika 13*239-255,363
• TRANSL [1974] (J 0069) Algeb and Log 13*135-143
◇ C25 C60 D40 ◇ REV MR 52#2859 Zbl 304#20019 Zbl 319#20039 • ID 17646

BELEGRADEK, O.V. [1974] *Definability in algebraically closed groups (Russian)* (J 0087) Mat Zametki (Akad Nauk SSSR) 16*375-380
• TRANSL [1974] (J 1044) Math Notes, Acad Sci USSR 16*813-816
◇ C25 C40 C60 ◇ REV MR 50#12710 Zbl 334#02025
• ID 03883

BELEGRADEK, O.V. [1974] *On algebraically closed groups (Russian)* (P 1590) All-Union Conf Math Log (3);1974 Novosibirsk 12-14
◇ C25 C60 ◇ ID 43238

BELEGRADEK, O.V. & ZIL'BER, B.I. [1974] *The model companion of an $\aleph_1$-categorical theory (Russian)* (P 1590) All-Union Conf Math Log (3);1974 Novosibirsk 10-11
◇ C25 C35 ◇ ID 27341

BELEGRADEK, O.V. & TAJTSLIN, M.A. [1975] *Categorical varieties of groupoids (Russian)* (P 4208) All-Union Symp Algeb;1975 Gomel 377
◇ C05 C35 ◇ ID 43240

BELEGRADEK, O.V. [1978] *Elementary properties of algebraically closed groups (Russian) (English summary)* (J 0027) Fund Math 98*83-101
◇ C25 C60 D40 ◇ REV MR 57#9530 Zbl 389#20030
• ID 29209

BELEGRADEK, O.V. [1978] *On nonstable group theories (Russian)* (J 0031) Izv Vyssh Ucheb Zaved, Mat (Kazan) 1978/8(195)*41-44
• TRANSL [1978] (J 3449) Sov Math 22/8*31-33
◇ C45 C60 ◇ REV MR 82f:03027 Zbl 398#20001
• ID 52786

BELEGRADEK, O.V. [1980] *Decidable fragments of universal theories and existentially closed models (Russian)* (J 0092) Sib Mat Zh 21/6*196-201,223
• TRANSL [1980] (J 0475) Sib Math J 21*898-902
◇ B25 C25 D40 ◇ REV MR 82d:20033 Zbl 498#20024
• ID 80668

BELEGRADEK, O.V. [1981] *Classes of algebras with inner mappings (Russian)* (C 3806) Issl Teor Progr 3-10
◇ C05 C25 C50 ◇ REV MR 85e:08004 • ID 39911

BELEGRADEK, O.V. see Vol. IV for further entries

BELESOV, A.U. [1975] *α-atomic compactness of filtered products (Russian)* (J 0403) Izv Akad Nauk Kazak SSR, Ser Fiz-Mat 1975/3*75-77,93
◇ C05 C20 C50 ◇ REV MR 52#5403 Zbl 316#08003
• ID 70859

BELESOV, A.U. [1976] *Injective and equationally compact modules (Russian)* (J 0403) Izv Akad Nauk Kazak SSR, Ser Fiz-Mat 1976/1*85-86,94
◇ C05 C50 C60 ◇ REV MR 56#15410 Zbl 362#16012
• ID 70857

BELESOV, A.U. [1977] *Equationally compact systems (Russian)* (J 0403) Izv Akad Nauk Kazak SSR, Ser Fiz-Mat 1977/1*61-62,92
◇ C05 C50 C60 ◇ REV MR 56#5275 Zbl 352#08004
• ID 70858

BELKIN, V.P. & GORBUNOV, V.A. [1975] *Filters in lattices of quasivarieties of algebraic systems (Russian)* (J 0003) Algebra i Logika 14*373-392
• TRANSL [1975] (J 0069) Algeb and Log 14*229-239
◇ C05 C60 D40 G10 ◇ REV MR 53#5428 Zbl 328#08005 • ID 22948

BELL, J.L. & SLOMSON, A. [1965] *Introduction to model theory* (X 1518) Oxford Univ Math Inst: Oxford
◇ C98 ◇ ID 30565

BELL, J.L. & SLOMSON, A. [1969] *Models and ultraproducts: An introduction* (X 0809) North Holland: Amsterdam 322pp
◇ C20 C98 ◇ REV MR 42#4381 Zbl 179.314 JSL 37.763
• ID 00990

BELL, J.L. [1970] *Weak compactness in restricted second-order languages* (J 0014) Bull Acad Pol Sci, Ser Math Astron Phys 18*111-114
◇ C85 E55 ◇ REV MR 41#5202 Zbl 195.16 • ID 00992

BELL, J.L. [1972] *On the relationship between weak compactness in $L_{\omega_1,\omega}, L_{\omega_1,\omega_1}$ and restricted second order languages* (J 0009) Arch Math Logik Grundlagenforsch 15*74-78
◇ C55 C75 C80 C85 E05 E55 ◇ REV MR 48#1878 Zbl 253#02009 • ID 00999

BELL, J.L. & COLE, J. & PRIEST, G. & SLOMSON, A. [1973] *Proceedings of the Bertrand Russell Memorial Logic Conference* (P 0710) Russell Mem Logic Conf;1971 Uldum VI+404pp
◇ C97 ◇ REV MR 49#4747 • ID 49817

BELL, J.L. [1974] *On compact cardinals* (J 0068) Z Math Logik Grundlagen Math 20*389-393
◇ C55 C75 E45 E55 ◇ REV MR 52#2885 Zbl 302#02028 • ID 03885

BELL, J.L. [1976] *Uncountable standard models of ZFC + V ≠ L* (P 1476) Set Th & Hierarch Th (2) (Mostowski);1975 Bierutowice 29-36
◇ C62 E40 E45 ◇ REV MR 55#2565 Zbl 362#02073
• ID 23794

BELL, J.L. [1981] *Isomorphism of structures in S-toposes* (J 0036) J Symb Logic 46*449-459
◇ C75 C90 ◇ REV MR 83a:03068 Zbl 465#03019
• ID 54922

BELL, J.L. [1983] *Orthologic, forcing, and the manifestation of attributes* (P 3669) SE Asian Conf on Log;1981 Singapore 13-36
◇ B51 C90 ◇ REV MR 85m:03043 Zbl 553#03040
• ID 43055

BELL, J.L. see Vol. I, II, V for further entries

BELLISSIMA, F. & MIROLLI, M. [1983] *On the axiomatization of finite K-frames* (J 0063) Studia Logica 42*383-388
⋄ B45 C13 C90 ⋄ REV MR 86a:03014 Zbl 556 # 03017
• ID 42314

BELLISSIMA, F. see Vol. II, VI for further entries

BEL'TYUKOV, A.P. [1976] *Decidability of the universal theory of natural numbers with addition and divisibility (Russian)* (S 0228) Zap Nauch Sem Leningrad Otd Mat Inst Steklov 60*15-28,221
• TRANSL [1980] (J 1531) J Sov Math 14*1436-1444
⋄ B25 B28 F30 ⋄ REV MR 58 # 27419 Zbl 345 # 02035
• ID 29789

BEL'TYUKOV, A.P. see Vol. IV for further entries

BELYAEV, V.V. [1984] *Locally finite Chevalley groups (Russian)* (C 4674) Stud on Group Th (Sverdlovsk) 39-50
⋄ C60 ⋄ ID 49811

BELYAEV, V.YA. [1974] *Elementary types of algebraically closed semigroups (Russian)* (P 1590) All-Union Conf Math Log (3);1974 Novosibirsk 18-19
⋄ C25 C60 ⋄ ID 32618

BELYAEV, V.YA. [1974] *Theories with the Ehrenfeucht property (Russian)* (P 1590) All-Union Conf Math Log (3);1974 Novosibirsk 15-17
⋄ C52 ⋄ ID 32617

BELYAEV, V.YA. [1975] *On theories with the Ehrenfeucht property (Russian)* (J 0092) Sib Mat Zh 16*175-177,198
• TRANSL [1975] (J 0475) Sib Math J 16*141-143
⋄ C52 ⋄ REV MR 51 # 124 Zbl 317 # 02057 • ID 17440

BELYAEV, V.YA. [1977] *Algebraically closed semigroups (Russian)* (J 0092) Sib Mat Zh 18*32-39,237
• TRANSL [1977] (J 0475) Sib Math J 18*23-29
⋄ C25 C60 ⋄ REV MR 58 # 939 Zbl 368 # 20051
• ID 32619

BELYAEV, V.YA. [1978] *On algebraic closure in classes of semigroups and groups (Russian)* (P 4599) All-Union Conf Stud & Sci-Tech Prog; 95-105
⋄ C25 C60 ⋄ ID 48370

BELYAEV, V.YA. [1978] *On maximal subgroups of an algebraically closed semigroup (Russian)* (P 2559) All-Union Symp on Th of Semigroups (2);1978 Sverdlovsk 9
⋄ C25 C60 ⋄ ID 32620

BELYAEV, V.YA. & TAJTSLIN, M.A. [1979] *On elementary properties of existentially closed systems (Russian)* (J 0067) Usp Mat Nauk 34/2(206)*39-94
• TRANSL [1979] (J 1399) Russ Math Surv 34/2*43-107
⋄ C25 C40 C60 C75 C85 D55 ⋄ REV MR 82a:03028 Zbl 413 # 03025 • ID 32621

BELYAEV, V.YA. [1982] *On algebraic closure and amalgamation of semigroups* (J 0136) Semigroup Forum 25*223-267
⋄ C05 C25 C60 ⋄ REV MR 84d:20059 Zbl 528 # 20047
• ID 39702

BELYAEV, V.YA. [1984] *Uncountable extensions of countable algebraically closed semigroups (Russian)* (J 0092) Sib Mat Zh 25/1*30-38
• TRANSL [1984] (J 0475) Sib Math J 25*24-30
⋄ C05 C55 C60 C75 ⋄ REV MR 85m:03025 Zbl 541 # 20039 • ID 48174

BELYAEV, V.YA. see Vol. IV for further entries

BELYAKIN, N.V. [1970] *Generalized computations and second order arithmetic (Russian)* (J 0003) Algebra i Logika 9*375-405
• TRANSL [1970] (J 0069) Algeb and Log 9*225-243
⋄ C62 D10 D45 D65 D70 F35 ⋄ REV MR 46 # 32 Zbl 278 # 02036 • ID 29065

BELYAKIN, N.V. [1979] *Nonstandardly finite sets (Russian)* (C 2967) Metodol Probl Mat 48-54
⋄ E30 H05 H15 ⋄ REV MR 81f:03004 • ID 70865

BELYAKIN, N.V. [1983] *A means of modeling a classical second-order arithmetic (Russian)* (J 0003) Algebra i Logika 22*3-25
• TRANSL [1983] (J 0069) Algeb and Log 22*1-18
⋄ B15 C62 D55 F35 ⋄ REV MR 85h:03046 Zbl 538 # 03040 • ID 41479

BELYAKIN, N.V. see Vol. IV, VI for further entries

BELYAKOV, E.B. & MART'YANOV, V.I. [1983] *Universal theories of integers and the extended Bliznetsov hypothesis (Russian)* (J 0003) Algebra i Logika 22*26-34
• TRANSL [1983] (J 0069) Algeb and Log 22*19-26
⋄ B25 ⋄ REV MR 86g:03017 Zbl 564 # 03013 • ID 45833

BEN-DAVID, S. [1978] *On Shelah's compactness of cardinals* (J 0029) Israel J Math 31*34-56 • ERR/ADD ibid 31*394
⋄ C55 C80 E05 E35 E50 E55 ⋄ REV MR 80f:03060 Zbl 384 # 03036 • ID 29139

BENDA, M. [1968] *Ultraproducts and non-standard logics (Russian summary)* (J 0014) Bull Acad Pol Sci, Ser Math Astron Phys 16*453-456
⋄ C20 C50 C75 ⋄ REV MR 40 # 5429 Zbl 184.14
• ID 01020

BENDA, M. [1969] *Reduced products and nonstandard logics* (J 0036) J Symb Logic 34*424-436
⋄ C20 C50 C75 ⋄ REV MR 40 # 4092 Zbl 184.14
• ID 01021

BENDA, M. [1970] *Reduced products filters and Boolean ultrapowers* (0000) Diss., Habil. etc vii+60pp
⋄ C20 ⋄ REM Ph.d.thesis, University of Wisconsin, Madison, WI • ID 16738

BENDA, M. [1971] *On saturated reduced products* (J 0048) Pac J Math 39*557-571
⋄ C20 C50 E05 ⋄ REV MR 46 # 7019 Zbl 262 # 02050
JSL 40.546 • ID 01023

BENDA, M. [1972] *On reduced products and filters* (J 0007) Ann Math Logic 4*1-29
⋄ C20 C55 E05 E50 ⋄ REV MR 45 # 1741
Zbl 235 ⫫ 02049 • ID 01024

BENDA, M. [1974] *Note on boolean ultrapowers* (J 0053) Proc Amer Math Soc 46*289-293
⋄ C20 C50 C90 ⋄ REV MR 50 # 9575 Zbl 299 # 02061
• ID 03887

BENDA, M. [1974] *Remarks on countable models* (J 0027) Fund Math 81*107-119
⋄ C15 C50 ⋄ REV MR 51 # 7852 Zbl 289 # 02038
• ID 01025

BENDA, M. [1975] *Construction of models from groups of permutations* (J 0036) J Symb Logic 40*383-388
⋄ C30 ⋄ REV MR 52 # 5404 Zbl 325 # 02037 • ID 14821

BENDA, M. [1975] *Some properties of mirrored orders* (J 0132) Math Scand 37*5-12
⋄ C55 E07 ⋄ REV MR 56 # 15411 Zbl 318 # 02050
• ID 60439

BENDA, M. [1977] *Some directions in model theory* (P 1076) Latin Amer Symp Math Log (3);1976 Campinas 117-133
⋄ C98 ⋄ REV MR 57 # 16062 Zbl 361 # 02063 • ID 16601

BENDA, M. [1978] *Compactness for omitting of types* (J 0007) Ann Math Logic 14*39-56
⋄ C20 C30 C55 C70 C75 E55 ⋄ REV MR 80k:03032 Zbl 384 # 03037 • ID 29145

BENDA, M. [1979] *Modeloids I* (J 0064) Trans Amer Math Soc 250*47-90
⋄ C90 ⋄ REV MR 82g:08002 Zbl 408 # 03034 • ID 56273

BENDA, M. see Vol. V, VI for further entries

BENECKE, K. & REICHEL, HORST [1983] *Equational partiality* (J 0004) Algeb Universalis 16*219-232
⋄ C05 C90 ⋄ REV MR 84g:08016 Zbl 524 # 08005
• ID 38227

BENEJAM, J.-P. [1970] *Quelques proprietes algebriques des isomorphismes locaux* (J 2313) C R Acad Sci, Paris, Ser A-B 271*A972-A975
⋄ C07 C52 E07 G15 ⋄ REV MR 42 # 5784 Zbl 298 # 02065 • ID 01028

BENEJAM, J.-P. [1970] *Une classification des isomorphismes locaux et son application a la logique* (J 2313) C R Acad Sci, Paris, Ser A-B 271*A529-A532
⋄ C07 G15 ⋄ REV MR 42 # 5783 Zbl 277 # 02015
• ID 01027

BENEJAM, J.-P. [1971] *La methode des s-ludions et les classes logiques* (J 2313) C R Acad Sci, Paris, Ser A-B 272*A981-A984
⋄ C07 C52 ⋄ REV MR 43 # 7312 Zbl 277 # 02016
• ID 01029

BENEJAM, J.-P. [1977] *Algebraic characterizations of the satisfiability of first-order logical formulas and the halting of programs* (J 0068) Z Math Logik Grundlagen Math 23*111-120
⋄ C52 D10 G05 ⋄ REV MR 58 # 10396 Zbl 383 # 03008
• ID 26472

BENEJAM, J.-P. see Vol. I, IV, V for further entries

BENENSON, I.E. [1979] *The structure of quasivarieties of models (Russian)* (J 0031) Izv Vyssh Ucheb Zaved, Mat (Kazan) 1979/11*14-20
• TRANSL [1979] (J 3449) Sov Math 23/12*13-21
⋄ C05 C52 ⋄ REV MR 81e:08011 Zbl 432 # 08007
• ID 80681

BENI, J. & BOFFA, M. [1968] *Elimination de cycles d'appartenance par permutation de l'univers* (J 2313) C R Acad Sci, Paris, Ser A-B 266*A545-A546 • ERR/ADD ibid 266*A1027
⋄ C62 E35 E70 ⋄ REV MR 38 # 4301 Zbl 169.307
• ID 01035

BENIAMINOV, E.M. [1980] *An algebraic approach to models of data bases of the relational type (Russian)* (S 2582) Semiotika & Inf, Akad Nauk SSSR 14*44-80
⋄ B75 C30 ⋄ REV MR 82h:68127a Zbl 509 # 68114
• ID 36800

BENIOFF, P.A. [1976] *Models of Zermelo Fraenkel set theory as carriers for the mathematics of physics I,II* (J 0209) J Math Phys 17*618-628,629-640
⋄ C62 D80 E75 ⋄ REV MR 57 # 12210 Zbl 331 # 02047 Zbl 331 # 02048 • ID 30647

BENIOFF, P.A. see Vol. I, II, IV for further entries

BENNETT, D.W. [1973] *An elementary completeness proof for a system of natural deduction* (J 0047) Notre Dame J Formal Log 14*430-432
⋄ B10 C07 F07 ⋄ REV MR 47 # 8284 Zbl 258 # 02048
• ID 01038

BENNETT, D.W. [1977] *A note on the completeness proof for natural deduction* (J 0047) Notre Dame J Formal Log 18*145-146
⋄ B10 C07 F07 ⋄ REV MR 56 # 8355 Zbl 338 # 02014
• ID 21957

BENNETT, D.W. see Vol. I for further entries

BENOIS, M. [1973] *Application de l'etude de certaines congruences a un probleme de decidabilite* (S 2250) Semin Dubreil: Algebre 7*1pp
⋄ B25 ⋄ REV Zbl 319 # 02040 • ID 29682

BENTHEM VAN, J.F.A.K. [1974] *Semantic tableaus* (J 3077) Nieuw Arch Wisk, Ser 3 22*44-59
⋄ B10 C07 F07 ⋄ REV MR 49 # 2282 Zbl 285 # 02048
• ID 03889

BENTHEM VAN, J.F.A.K. [1978] *Ramsey eliminability* (J 0063) Studia Logica 37*321-336
⋄ A05 C40 ⋄ REV MR 81c:03022 Zbl 402 # 03014
• ID 54661

BENTHEM VAN, J.F.A.K. [1979] *Canonical modal logics and ultrafilter extensions* (J 0036) J Symb Logic 44*1-8
⋄ B45 C90 G25 ⋄ REV MR 80i:03028 Zbl 405 # 03011 JSL 47.441 • ID 28157

BENTHEM VAN, J.F.A.K. & HUMBERSTONE, I.L. [1983] *Hallden-completeness by gluing of Kripke frames* (J 0047) Notre Dame J Formal Log 24*426-430
⋄ B45 C30 C90 ⋄ REV MR 85i:03040 Zbl 487 # 03008
• ID 37277

BENTHEM VAN, J.F.A.K. & PEARCE, D.A. [1984] *A mathematical characterization of interpretation between theories* (J 0063) Studia Logica 43*295-303
⋄ C20 F25 ⋄ REV Zbl 573 # 03009 • ID 39892

BENTHEM VAN, J.F.A.K. [1984] *Questions about quantifiers* (J 0036) J Symb Logic 49*443-466
⋄ A05 B65 C80 ⋄ REV MR 85e:03064 Zbl 573 # 03008
• ID 40307

BENTHEM VAN, J.F.A.K. see Vol. II, V for further entries

BERESTOVOY, S. [1974] *Modeles homogenes d'une theorie forcee* (J 2313) C R Acad Sci, Paris, Ser A-B 278*A397-A399
⋄ C25 C35 C50 ⋄ REV MR 49 # 35 Zbl 279 # 02030
• ID 03892

BERGSTRA, J.A. & TIURYN, J. [1981] *Implicit definability of algebraic structures by means of program properties* (J 2095) Fund Inform, Ann Soc Math Pol, Ser 4 4*661-674
⋄ B75 C40 C70 ⋄ REV MR 84c:03069 Zbl 456 # 68025
• ID 34037

BERGSTRA, J.A. see Vol. I, II, IV, VI for further entries

BERLINE, C. [1975] *Categoricite en $\aleph_0$ du groupe lineaire d'un anneau de Boole (English summary)* (J 2313) C R Acad Sci, Paris, Ser A-B 280∗A753-A754
  ⋄ C35 C60 ⋄ REV MR 51 # 7847   Zbl 299 # 02058
  • ID 17978

BERLINE, C. [1980] *Elimination of quantifiers for non semi-simple rings of characteristic p* (P 2625) Model Th of Algeb & Arith;1979 Karpacz 10-19
  ⋄ C10 C60 ⋄ REV MR 82k:03034   Zbl 446 # 03027
  JSL 50.1079 • ID 56556

BERLINE, C. [1981] *Ideaux des anneaux de Peano (d'apres Cherlin)* (P 3404) Model Th & Arith;1979/80 Paris 32-43
  ⋄ C60 ⋄ REV MR 83f:03032   Zbl 479 # 03030 • ID 55692

BERLINE, C. & MCALOON, K. & RESSAYRE, J.-P. (EDS.) [1981] *Model theory and arithmetic. Comptes rendus d'une action thematique programmee du C.N.R.S. sur la theorie des modeles et l'arithmetique* (S 3301) Lect Notes Math 890∗vi+306pp
  ⋄ C62 C97 F30 ⋄ REV MR 82m:03004   Zbl 465 # 00004
  • ID 54902

BERLINE, C. & CHERLIN, G.L. [1981] *QE nilrings of prime characteristic* (J 3133) Bull Soc Math Belg, Ser B 33∗3-17
  ⋄ C10 C60 ⋄ REV MR 84a:03033   Zbl 479 # 03020
  JSL 50.1080 • ID 55682

BERLINE, C. & CHERLIN, G.L. [1981] *QE rings in characteristic p* (P 2628) Log Year;1979/80 Storrs 16-31
  ⋄ C10 C60 ⋄ REV MR 84a:03032   Zbl 471 # 03028
  JSL 50.1080 • ID 55223

BERLINE, C. [1981] *Rings which admit elimination of quantifiers* (J 0036) J Symb Logic 46∗56-58
  ⋄ C10 C60 ⋄ REV MR 82d:03046   Zbl 462 # 03007
  JSL 50.1079 • ID 54514

BERLINE, C. [1981] *Stabilite et algebre I. Pratique de la stabilite et de la categoricite* (C 2619) Groupe Etude Th Stables (2) 1978/79 1∗13pp
  ⋄ C35 C45 C98 ⋄ REV MR 83j:03049a   Zbl 454 # 03012
  • REM Part II 1981 • ID 35351

BERLINE, C. [1981] *Stabilite et algebre II. Groupes Abeliens et modules* (C 2619) Groupe Etude Th Stables (2) 1978/79 2∗8pp
  ⋄ C35 C45 C60 C98 ⋄ REV MR 83j:03049b
  Zbl 454 # 03012 • REM Parts I,III 1981 • ID 35352

BERLINE, C. [1981] *Stabilite et algebre III. Anneaux* (C 2619) Groupe Etude Th Stables (2) 1978/79 3∗8pp
  ⋄ C45 C60 C98 ⋄ REV MR 83j:03049c   Zbl 454 # 03012
  • REM Parts II,IV 1981 • ID 35353

BERLINE, C. [1981] *Stabilite et algebre IV. Groupes* (C 2619) Groupe Etude Th Stables (2) 1978/79 4∗8pp
  ⋄ C45 C60 C98 ⋄ REV MR 83j:03049d   Zbl 454 # 03012
  • REM Part III 1981 • ID 35354

BERLINE, C. [1983] *Deviation des types dans les corps algebriquement clos* (C 3822) Groupe Etude Th Stables (3) 1980/82 10pp
  ⋄ C45 C60 ⋄ REV MR 85d:03069   Zbl 512 # 03016
  • ID 36511

BERLINE, C. & CHERLIN, G.L. [1983] *QE rings in characteristic p^n* (J 0036) J Symb Logic 48∗140-162
  ⋄ C10 C60 ⋄ REV MR 84f:03027   Zbl 524 # 03016
  JSL 50.1080 • ID 34447

BERMAN, J. [1975] see BALDWIN, JOHN T.

BERMAN, J. [1976] see BALDWIN, JOHN T.

BERMAN, J. [1977] see BALDWIN, JOHN T.

BERMAN, J. & GRAETZER, G. [1978] *Uniform representations of congruence schemes* (J 0048) Pac J Math 76∗301-311
  ⋄ C05 C40 ⋄ REV MR 58 # 10667   Zbl 435 # 08005
  • ID 69185

BERMAN, J. [1980] *A proof of Lyndon's finite basis theorem* (J 0193) Discr Math 29∗229-233
  ⋄ C05 C13 ⋄ REV MR 81g:03033   Zbl 449 # 08004
  • ID 70970

BERMAN, J. [1981] see BALDWIN, JOHN T.

BERMAN, J. [1982] see BALDWIN, JOHN T.

BERMAN, J. [1983] *Free spectra of 3-element algebras* (P 3841) Universal Algeb & Lattice Th (4);1982 Puebla 10-53
  ⋄ C05 C98 ⋄ REV MR 85d:08099   Zbl 518 # 08010
  • ID 38374

BERMAN, J. see Vol. II for further entries

BERMAN, L. [1977] *Precise bounds for Presburger arithmetic and the reals with addition: preliminary report* (P 3572) IEEE Symp Found of Comput Sci (18);1977 Providence 95-99
  ⋄ B25 D15 F30 ⋄ REV MR 58 # 5116 • ID 70972

BERMAN, L. [1980] *The complexity of logical theories* (J 1426) Theor Comput Sci 11∗71-77
  ⋄ B25 D10 D15 F30 ⋄ REV MR 82c:03061b
  Zbl 475 # 03017 • ID 55471

BERMAN, L. see Vol. IV for further entries

BERMAN, P. [1979] *Complexity of the theory of atomless boolean algebras* (P 2935) FCT'79 Fund of Comput Th;1979 Berlin/Wendisch-Rietz 64-70
  ⋄ B25 D15 ⋄ REV MR 81k:03033   Zbl 421 # 03028
  • ID 53427

BERMAN, P. see Vol. II, IV for further entries

BERNAL, E. [1979] *On classes of groups closed under ultraproducts* (P 4097) Jorn Mat Luso-Espanol (6);1979 Santander 719-732
  ⋄ C20 C60 ⋄ ID 45473

BERNARD, C.-L. [1976] *Sous-groupes nilpotents des groupes multiplicatifs de corps (English summary)* (J 2313) C R Acad Sci, Paris, Ser A-B 282∗A1129-A1130
  ⋄ C60 ⋄ REV MR 53 # 8279   Zbl 338 # 20032 • ID 27261

BERNARDI, C. [1975] *On the equational class of diagonalizable algebras (the algebraization of the theories which express Theor. VI)* (J 0063) Studia Logica 34∗321-331
  ⋄ B25 F30 G05 G25 ⋄ REV MR 57 # 109
  Zbl 322 # 02033 • REM Part V 1975 by Montagna,F. Part VII 1976 by Magari,R. • ID 31012

BERNARDI, C. [1980] *Lo spazio duale di un prodotto di algebre di Boole e le compattificazioni di Stone (English summary)* (J 3526) Ann Mat Pura Appl, Ser 4 126∗253-266
  ⋄ C20 G05 ⋄ REV MR 82i:06018   Zbl 481 # 54022
  • ID 80697

BERNARDI, C. (ED.) [1983] *Atti degli incontri di logica matematica* (X 3812) Univ Siena, Dip Mat: Siena 398pp
  ⋄ B97 C97 ⋄ REV MR 84k:03006   Zbl 505 # 00007
  • ID 34945

BERNARDI, C. see Vol. II, IV, VI for further entries

BERNAYS, P. & SCHOENFINKEL, M. [1928] *Zum Entscheidungsproblem der mathematischen Logik* (J 0043) Math Ann 99*342-372
- ◇ B20 C25 D35 ◇ REV FdM 54.56 • ID 01076

BERNAYS, P. [1955] *Betrachtungen ueber das Vollstaendigkeitsaxiom und verwandte Axiome* (J 0044) Math Z 63*219-229
- ◇ B28 C60 C65 ◇ REV MR 17.447 Zbl 68.269 • ID 01090

BERNAYS, P. [1957] *Betrachtungen zum Paradoxon von Thoralf Skolem* (J 0752) Avh Norske Vid-Akad Oslo I 5*9pp
- ◇ A05 C07 ◇ REV MR 20#6362 • ID 42821

BERNAYS, P. [1972] *Bemerkung ueber eine Ordnung der zahlentheoretischen Funktionen mittels eines additiven 0-1-Masses fuer die Menge der natuerlichen Zahlen* (C 0648) Th of Sets and Topology (Hausdorff) 39-43
- ◇ C20 E05 E50 ◇ REV MR 52#5429 Zbl 298#02072 • ID 01265

BERNAYS, P. see Vol. I, II, IV, V, VI for further entries

BERNHARDT, K. [1976] *Semantics for finitary predicate calculi* (J 0063) Studia Logica 35*227-241
- ◇ B20 C90 ◇ REV MR 56#5228 Zbl 344#02017 • ID 60487

BERNHARDT, K. [1977] *Abschliessbarkeit von Kripke-Semantiken* (J 0115) Wiss Z Humboldt-Univ Berlin, Math-Nat Reihe 26*615-622
- ◇ B55 C90 ◇ REV MR 80e:03039 Zbl 423#03040 • ID 53552

BERNHARDT, K. see Vol. V for further entries

BERNSTEIN, A.R. [1973] *Non-standard analysis* (C 0654) Stud in Model Th 35-58
- ◇ C60 C65 H05 ◇ REV MR 49#2366 • ID 03898

BERNSTEIN, A.R. see Vol. I, V for further entries

BERRY, GEORGE D.W. [1953] *Symposium: On the ontological significance of the Loewenheim-Skolem theorem* (C 4191) Acad Freedom, Log & Religion 39-55
- ◇ A05 C07 ◇ REV JSL 20.63 • ID 42420

BERRY, GEORGE D.W. see Vol. I for further entries

BERTHIER, D. [1975] *Stability of non-model-complete theories; products, groups* (J 3172) J London Math Soc, Ser 2 11*453-464
- ◇ C20 C30 C45 C60 ◇ REV MR 52#2870 Zbl 328#02034 • ID 17655

BERTONI, A. & MAURI, G. & MIGLIOLI, P.A. [1979] *A characterization of abstract data as model-theoretic invariants* (P 1873) Automata, Lang & Progr (6);1979 Graz 26-37
- ◇ B75 C57 D45 D80 ◇ REV MR 82a:68027 Zbl 411#68033 • ID 52925

BERTONI, A. & MAURI, G. & MIGLIOLI, P.A. [1980] *Towards a theory of abstract data types: a discussion on problems and tools* (P 2946) Int Symp Progr (4);1980 Paris 44-58
- ◇ B75 C57 D80 ◇ REV MR 82c:68013 Zbl 435#68022 • ID 80710

BERTONI, A. & MAURI, G. & MIGLIOLI, P.A. [1981] *Model theoretic aspects of abstract data specification* (P 3642) Colloq Math Log in Computer Sci;1978 Salgotarjan 181-193
- ◇ B75 C50 C57 ◇ REV MR 83g:68007 Zbl 503#68013 • ID 36666

BERTONI, A. & MAURI, G. & MIGLIOLI, P.A. [1983] *On the power of model theory in specifying abstract data types and in capturing their recursiveness* (J 2095) Fund Inform, Ann Soc Math Pol, Ser 4 6*127-170
- ◇ B75 C50 C57 D20 ◇ REV MR 84f:68014 Zbl 529#68008 • ID 38487

BERTONI, A. see Vol. IV, VI for further entries

BERTOSSI, L. & CHUAQUI, R.B. [1985] *Approximation to truth and theory of errors* (P 2160) Latin Amer Symp Math Log (6);1983 Caracas 13-31
- ◇ C90 H05 ◇ REV Zbl 569#03011 • ID 41802

BERTRAND, D. [1973] *Le theoreme d'Ax et Kochen* (S 2348) Semin Th Nombres Bordeaux 1972-73/10*2pp
- ◇ C60 ◇ REV MR 52#13754 Zbl 287#12022 • ID 30542

BESLAGIC, A. & RUDIN, M.E. [1985] *Set-theoretic constructions of nonshrinking open covers* (J 2635) Topology Appl 20*167-177
- ◇ C55 C65 E05 E75 ◇ REV Zbl 574#54020 • ID 48251

BESLAGIC, A. see Vol. V for further entries

BETH, E.W. [1950] *Decision problems of logic and mathematics* (C 0643) Philosophie 1946-48 3-18
- ◇ B20 B25 C10 D35 ◇ REV JSL 22.359 • ID 01156

BETH, E.W. [1951] *A topological proof of the theorem of Loewenheim-Skolem-Goedel* (J 0028) Indag Math 13*436-444
- ◇ C07 ◇ REV MR 13.614 Zbl 44.2 JSL 19.61 • ID 01157

BETH, E.W. [1953] *On Padoa's method in the theory of definition* (J 0028) Indag Math 15*330-339
- ◇ B10 C40 ◇ REV MR 15.385 Zbl 53.344 JSL 21.194 • ID 01159

BETH, E.W. [1953] *Some consequences of the theorem of Loewenheim-Skolem-Goedel-Malcev* (J 0028) Indag Math 15*66-71
- ◇ C07 ◇ REV MR 14.714 Zbl 51.6 JSL 19.61 • ID 01161

BETH, E.W. [1953] *Sur la description de certains modeles d'un systeme formel* (P 0645) Int Congr Philos (11);1953 Bruxelles 5*64-69
- ◇ C07 ◇ REV MR 15.189 Zbl 53.2 JSL 19.224 • ID 01160

BETH, E.W. [1954] *Observations metamathematiques sur les structures simplement ordonnees* (P 0646) Appl Sci de Log Math;1952 Paris 29-35
- ◇ B25 C65 ◇ REV MR 16.556 Zbl 58.246 JSL 23.34 • ID 01162

BETH, E.W. [1955] *Semantic entailment and formal derivability* (J 0182) Kon Nederl Akad Wetensch Afd Let Med N S 18/13*309-342
- • TRANSL [1978] (X 2698) Univ Valencia Dept Log Filos Cienc: Valencia xvi+54pp • REPR [1969] (C 0569) Phil of Math Oxford Readings 9-41
- ◇ B10 C07 ◇ REV MR 19.625 JSL 22.360 • ID 01269

BETH, E.W. [1960] *Completeness results for formal systems* (P 0660) Int Congr Math (II, 8);1958 Edinburgh 281-288
- ◇ B10 C07 ◇ REV MR 23#A42 Zbl 119.251 JSL 27.110 • ID 01534

BETH, E.W. [1962] *Observations concernant la theorie de la definition* (P 1606) Colloq Math (Pascal);1962 Clermont-Ferrand 83-87
- ◇ C40 ◇ REV MR 44#1553 JSL 40.457 • ID 27940

BETH, E.W. see Vol. I, II, V, VI for further entries

BEZBORUAH, A. & SHEPHERDSON, J.C. [1976] *Goedel's second incompleteness theorem for Q* (**J** 0036) J Symb Logic 41*503-512
⋄ C62 F30 ⋄ REV MR 53 # 7756 Zbl 328 # 02017
• ID 14785

BEZVERKHNIJ, V.N. [1981] *Solvability of the inclusion problem in a class of HNN-groups (Russian)* (**C** 2264) Algor Probl Teor Grupp & Polygrupp 20-62
⋄ B25 C60 D40 ⋄ REV MR 84h:20020 • ID 39526

BEZVERKHNIJ, V.N. & GRINBLAT, V.A. [1981] *The root problem in Artin Groups (Russian)* (**C** 2264) Algor Probl Teor Grupp & Polygrupp 72-81
⋄ B25 C60 D40 ⋄ REV MR 83j:20043 • ID 39904

BEZVERKHNIJ, V.N. see Vol. IV for further entries

BIELINSKI, K. [1977] *Extendability of structures as infinitary property* (**P** 1695) Set Th & Hierarch Th (3);1976 Bierutowice 75-93
⋄ C62 C70 E30 E70 ⋄ REV MR 57 # 9475 Zbl 385 # 03043 • ID 52151

BIENENSTOCK, E. [1977] *Sets of degrees of computable fields* (**J** 0029) Israel J Math 27*348-356
⋄ C57 C60 D30 D45 ⋄ REV MR 58 # 16233 Zbl 359 # 02030 • ID 50578

BIGELOW, J.C. [1976] *Possible worlds foundations for probability* (**J** 0122) J Philos Logic 5*299-320
⋄ B48 C90 ⋄ REV Zbl 347 # 02020 • ID 60510

BIGELOW, J.C. see Vol. II for further entries

BILLINGTON, N. [1984] *Growth of groups and graded algebras* (**J** 0394) Commun Algeb 12*2579-2588 • ERR/ADD ibid 13/3*753-755
⋄ C60 D40 ⋄ REV MR 86e:20039ab Zbl 547 # 20026
• ID 45773

BILY, J. & BUKOVSKY, L. [1974] *On expansion of β-models* (**J** 0027) Fund Math 82*239-244
⋄ C62 E25 E40 E70 F35 ⋄ REV MR 51 # 149 Zbl 311 # 02062 • ID 01205

BING, K. [1955] *On arithmetical classes not closed under direct union* (**J** 0053) Proc Amer Math Soc 6*836-846
⋄ C05 C40 C52 ⋄ REV MR 17.226 Zbl 68.247 JSL 21.321 • ID 01207

BING, K. see Vol. I, VI for further entries

BIRKHOFF, GARRETT [1933] *On the combination of subalgebras* (**J** 0171) Proc Cambridge Phil Soc Math Phys 29*441-464
⋄ C05 C60 G10 ⋄ REV Zbl 7.395 FdM 59.154 • ID 33575

BIRKHOFF, GARRETT [1935] *On the structure of abstract algebras* (**J** 0171) Proc Cambridge Phil Soc Math Phys 31*433-454
⋄ C05 C07 G25 ⋄ REV Zbl 13.1 FdM 61.1026 • ID 01219

BIRKHOFF, GARRETT [1944] *Subdirect unions in universal algebra* (**J** 0015) Bull Amer Math Soc 50*764-768
⋄ C05 ⋄ REV MR 6.33 Zbl 60.58 • ID 01228

BIRKHOFF, GARRETT [1946] *Sobre los grupos de automorfismos* (**J** 0188) Rev Union Mat Argentina 11*155-157
⋄ C05 C07 G10 ⋄ REV MR 7.411 • ID 01527

BIRKHOFF, GARRETT [1976] *Note on universal topological algebra* (**J** 0004) Algeb Universalis 6*21-23
⋄ C05 ⋄ REV MR 53 # 2800 Zbl 375 # 06006 • ID 21645

BIRKHOFF, GARRETT see Vol. I, II, IV, V for further entries

BITTER-RUCKER VON, R. & ELLENTUCK, E. [1972] *Martin's axiom and saturated models* (**J** 0053) Proc Amer Math Soc 34*243-249
⋄ C20 C50 E05 E50 ⋄ REV MR 45 # 54 Zbl 246 # 02035
• ID 03312

BITTER-RUCKER VON, R. see Vol. V for further entries

BLACK, M. [1967] *Craig's theorem* (**C** 0601) Encycl of Philos 2*249-251
⋄ A05 C40 ⋄ REV JSL 35.304 • ID 01247

BLACK, M. see Vol. I, II for further entries

BLAIR, H.A. [1984] *The intractability of validity in logic programming and dynamic logic* (**P** 2989) Log of Progr; 1983 Pittsburgh 57-67
⋄ B75 C75 D35 D55 ⋄ REV Zbl 547 # 68036 • ID 43300

BLAIR, H.A. see Vol. I, IV for further entries

BLANC, G. & RAMBAUD, C. [1976] *Theories formelles sur graphe (English summary)* (**J** 2313) C R Acad Sci, Paris, Ser A-B 282*A1185-A1188
⋄ C07 C65 G30 ⋄ REV MR 54 # 9983 Zbl 339 # 02023
• ID 27263

BLANC, G. [1978] *Equivalence naturelle et formules logiques en theorie des categories* (**J** 0009) Arch Math Logik Grundlagenforsch 19*131-137
⋄ C40 G30 ⋄ REV MR 80m:03109 Zbl 407 # 03035
• ID 56201

BLANC, G. see Vol. V for further entries

BLANCHE, R. [1962] *Axiomatics* (**X** 0866) Routledge & Kegan Paul: Henley on Thames 65pp
⋄ A10 B25 B30 ⋄ REV MR 26 # 1259 Zbl 115.4 JSL 30.105 • ID 22464

BLANCHE, R. see Vol. I, II, VI for further entries

BLAQUIER, M. [1973] *Formal systems and their models (Spanish)* (**J** 0388) Human, Ser 4 Logica Mat 3*19-33
⋄ C98 ⋄ REV MR 52 # 35 • ID 17208

BLASS, A.R. [1972] *The intersection of nonstandard models of arithmetic* (**J** 0036) J Symb Logic 37*103-106
⋄ C62 ⋄ REV MR 48 # 1916 Zbl 246 # 02039 • ID 01296

BLASS, A.R. [1972] *Theories without countable models* (**J** 0036) J Symb Logic 37*562-568
⋄ C15 E50 ⋄ REV MR 47 # 6470 Zbl 258 # 02052
• ID 01294

BLASS, A.R. [1973] see BALDWIN, JOHN T.

BLASS, A.R. & NEUMANN, P.M. [1974] *An application of universal algebra in group theory* (**J** 0133) Michigan Math J 21*167-169
⋄ C05 C60 ⋄ REV MR 51 # 729 Zbl 292 # 20033
• ID 17980

BLASS, A.R. [1974] see BALDWIN, JOHN T.

BLASS, A.R. [1974] *On certain types and models for arithmetic* (**J** 0036) J Symb Logic 39*151-162
⋄ C62 E05 ⋄ REV MR 51 # 5286 Zbl 296 # 02031
• ID 01299

BLASS, A.R. [1974] *Ultrafilter mappings and their Dedekind cuts* (**J** 0064) Trans Amer Math Soc 188*327-340
⋄ C20 E05 E50 ⋄ REV MR 50 # 4310 Zbl 305 # 02065
• ID 01298

BLASS, A.R. [1977] *Amalgamation of non-standard models of arithmetic* (J 0036) J Symb Logic 42*372-386
⋄ C62 E05 ⋄ REV MR 57 # 9531 Zbl 381 # 03050
• ID 28164

BLASS, A.R. [1977] *End extensions, conservative extensions, and the Rudin-Frolik ordering* (J 0064) Trans Amer Math Soc 225*325-340
⋄ C62 E05 E50 ⋄ REV MR 54 # 12515 Zbl 362 # 02052
• ID 26228

BLASS, A.R. [1978] *A model-theoretic view of some special ultrafilters* (P 1897) Logic Colloq;1977 Wroclaw 96*79-90
⋄ C20 E05 H15 ⋄ REV MR 80d:03046 Zbl 439 # 03031
• ID 56022

BLASS, A.R. & HARARY, F. [1979] *Properties of almost all graphs and complexes* (J 2753) J Graph Th 3*225-240
⋄ C10 C13 C65 ⋄ REV MR 81c:05081 Zbl 418 # 05050
• ID 40313

BLASS, A.R. [1980] *Conservative extensions of models of arithmetic* (J 0009) Arch Math Logik Grundlagenforsch 20*85-94
⋄ C20 C62 E05 ⋄ REV MR 82c:03051 Zbl 453 # 03072
• ID 54202

BLASS, A.R. [1981] *Some initial segments of the Rudin-Keisler ordering* (J 0036) J Symb Logic 46*147-157
⋄ C20 E05 E50 ⋄ REV MR 82c:04002 Zbl 463 # 03027
• ID 54567

BLASS, A.R. [1981] *The model of set theory generated by countably many generic reals* (J 0036) J Symb Logic 46*732-752
⋄ C62 E25 E40 ⋄ REV MR 83e:03050 Zbl 482 # 03022
• ID 35229

BLASS, A.R. & SCEDROV, A. [1983] *Boolean classifying topoi* (J 0326) J Pure Appl Algebra 28*15-30
⋄ C35 G30 ⋄ REV MR 84e:03078 Zbl 504 # 03030
• ID 34410

BLASS, A.R. & SCEDROV, A. [1983] *Classifying topoi and finite forcing* (J 0326) J Pure Appl Algebra 28*111-140
⋄ C25 G30 ⋄ REV MR 84g:03035 Zbl 516 # 03042
• ID 34149

BLASS, A.R. [1983] *Words, free algebras, and coequalizers* (J 0027) Fund Math 117*117-160
⋄ C05 C75 E25 E35 F50 F55 G30 ⋄ REV MR 85d:03127 Zbl 542 # 08005 • ID 40316

BLASS, A.R. & GUREVICH, Y. & KOZEN, D. [1984] *A 0-1 law for logic with a fixed-point operator* (1111) Preprints, Manuscr., Techn. Reports etc. CRL-TR-38-84
⋄ B10 B75 C80 D70 ⋄ REM Univ. of Michigan, Computing Research Lab. • ID 47716

BLASS, A.R. & GUREVICH, Y. [1984] *Henkin quantifiers and complete problems* (1111) Preprints, Manuscr., Techn. Reports etc.
⋄ C80 D15 ⋄ REM Univ. of Michigan Technical Report
• ID 47720

BLASS, A.R. [1984] *There are not exactly five objects* (J 0036) J Symb Logic 49*467-469 • ERR/ADD ibid 50*781
⋄ B20 C07 ⋄ REV MR 85e:03021 • ID 40306

BLASS, A.R. [1985] *Kleene degrees of ultrafilters* (P 3342) Rec Th Week;1984 Oberwolfach 29-48
⋄ C20 D65 E05 ⋄ REV Zbl 573 # 03020 • ID 45296

BLASS, A.R. see Vol. IV, V for further entries

BLATTER, C. & SPECKER, E. [1981] *Le nombre de structures finies d'une theorie a caractere fini* (S 4311) Groupes de Contact FNRS, Sci Math 41-44
⋄ C13 C52 C85 ⋄ ID 48371

BLATTER, C. & SPECKER, E. [1984] *Recurrence relations for the number of labeled structures on a finite set* (P 2342) Symp Rek Kombin;1983 Muenster 43-61
⋄ C13 C52 C85 ⋄ ID 45381

BLOK, W.J. & KOEHLER, P. [1983] *Algebraic semantics for quasiclassical modal logics* (J 0036) J Symb Logic 48*941-964
⋄ B45 C90 G25 ⋄ REV MR 85f:03014 Zbl 562 # 03011
• ID 40651

BLOK, W.J. see Vol. II for further entries

BLOOM, S.L. [1973] *Extensions of Goedel's completeness theorem and the Loewenheim-Skolem theorem* (J 0047) Notre Dame J Formal Log 14*408-410
⋄ C07 ⋄ REV MR 48 # 92 Zbl 225 # 02033 • ID 01334

BLOOM, S.L. [1975] *Some theorems on structural consequence operations* (J 0063) Studia Logica 34*1-9
⋄ B22 C05 C20 ⋄ REV MR 53 # 2675 Zbl 311 # 02016
• ID 21621

BLOOM, S.L. & SUSZKO, R. [1975] *Ultraproducts of SCI models* (J 0387) Bull Sect Logic, Pol Acad Sci 4*9-14
⋄ C20 ⋄ REV MR 53 # 119 • ID 16583

BLOOM, S.L. [1976] *Varieties of ordered algebras* (J 0119) J Comp Syst Sci 13*200-212
⋄ C05 C07 ⋄ REV MR 55 # 239 Zbl 337 # 06008
• ID 28145

BLOOM, S.L. [1982] *A note on the logic of signed equations* (J 0063) Studia Logica 41*75-81
⋄ B60 C05 C07 ⋄ REV MR 85e:03069 Zbl 509 # 03015
• ID 36525

BLOOM, S.L. see Vol. I, II, IV, V, VI for further entries

BLOSHCHITSYN, V.YA. & ZAKIR'YANOV, K.KH. [1978] *Constructive abelian groups (Russian)* (C 3848) Algeb & Teor Chisel ('78) 18-25
⋄ C57 C60 D45 ⋄ REV Zbl 485 # 03020 • ID 36848

BLUM, L. [1977] *Differentially closed fields: a model-theoretic tour* (C 2919) Contrib Algeb (Kolchin) 37-61
⋄ C60 ⋄ REV MR 58 # 10415 Zbl 368 # 12013 • ID 51285

BLUM, L. see Vol. II, IV for further entries

BOCHVAR, D.A. [1940] *Ueber einen Aussagenkalkuel mit abzaehlbaren logischen Summen und Produkten (Russian summary)* (J 0142) Mat Sb, Akad Nauk SSSR, NS 7(49)*65-100
⋄ B05 C75 ⋄ REV MR 1.321 Zbl 23.99 JSL 5.119 FdM 66.29 • ID 01348

BOCHVAR, D.A. see Vol. I, II, V for further entries

BOERGER, E. [1973] *Eine entscheidbare Klasse von Kromformeln* (J 0068) Z Math Logik Grundlagen Math 19*117-120
⋄ B25 ⋄ REV MR 49 # 2327 Zbl 298 # 02048 • ID 01367

BOERGER, E. [1974] *Beitrag zur Reduktion des Entscheidungsproblems auf Klassen von Hornformeln mit kurzen Alternationen* (J 0009) Arch Math Logik Grundlagenforsch 16*67-84
⋄ B20 B25 D35 ⋄ REV MR 49 # 10535 Zbl 277 # 02009
• ID 01368

BOERGER, E. [1976] *On the construction of simple first-order formulae without recursive models* (P 1619) Coloq Log Simb;1975 Madrid 7-24
⋄ C57 D25 ⋄ REV MR 56 # 8348 Zbl 357 # 02010
• ID 50358

BOERGER, E. [1976] *Ueber einige Interpretationen von Registermaschinen mit Anwendungen auf Entscheidungsprobleme in der Logik, der Algorithmentheorie und der Theorie formaler Sprachen* (P 3198) Incont Compl Calc, Cod & Ling Form;1975 Napoli 27-46
⋄ B25 D10 ⋄ REV Zbl 411 # 68045 • ID 52927

BOERGER, E. [1979] *A new general approach to the theory of the many-one equivalence of decision problems for algorithmic systems* (J 0068) Z Math Logik Grundlagen Math 25*135-162
⋄ B25 D03 D05 D10 D25 D30 D40 ⋄ REV MR 80f:03045 Zbl 429 # 03016 • ID 53847

BOERGER, E. & KLEINE BUENING, H. [1979] *The reachability problem for Petri nets and decision problems for Skolem arithmetic* (P 2615) Scand Logic Symp (5);1979 Aalborg 59-96
⋄ B25 D15 D35 D80 ⋄ REV MR 82b:03079 Zbl 453 # 03013 • ID 54143

BOERGER, E. [1980] *Entscheidbarkeit* (C 4082) Handb Wiss Begriffe 159-160
⋄ B25 C98 ⋄ ID 40022

BOERGER, E. & KLEINE BUENING, H. [1980] *The reachability problem for Petri nets and decision problems for Skolem arithmetic* (J 1426) Theor Comput Sci 11*123-143
⋄ B25 D15 D35 D80 ⋄ REV MR 81h:68034 Zbl 453 # 03012 • ID 80766

BOERGER, E. [1981] *Logical description of computation processes* (P 3165) FCT'81 Fund of Comput Th;1981 Szeged 410-424
⋄ B25 D05 D15 D35 ⋄ REV MR 83e:03066 Zbl 467 # 03037 • ID 55035

BOERGER, E. [1982] see AANDERAA, S.O.

BOERGER, E. [1983] *From decision problems to problems of complexity* (P 3091) Conv Int Storica Logica;1982 San Gimignano 211-215
⋄ B25 D15 ⋄ ID 40032

BOERGER, E. [1984] *Decision problems in predicate logic* (P 3710) Logic Colloq;1982 Firenze 263-301
⋄ B20 B25 D35 ⋄ REV MR 85III:03008 Zbl 556 # 03012
• ID 40047

BOERGER, E. [1984] *Spektralproblem and completeness of logical decision problems* (P 2342) Symp Rek Kombin;1983 Muenster 333-356
⋄ B25 C13 D15 ⋄ REV Zbl 564 # 03008 • ID 40054

BOERGER, E. see Vol. I, IV, VI for further entries

BOERGER, R. & RAJAGOPALAN, M. [1985] *When do all ring homomorphisms depend on only one co-ordinate?* (J 0008) Arch Math (Basel) 45*223-228
⋄ C60 E55 E75 ⋄ REV Zbl 568 # 03026 • ID 49224

BOERGER, R. see Vol. V for further entries

BOFFA, M. [1967] *Modeles de la theorie des ensembles, associes aux permutations de l'univers* (J 2313) C R Acad Sci, Paris, Ser A-B 264*A221-A222
⋄ C62 E30 E35 E70 ⋄ REV MR 36 # 42 Zbl 168.250
• ID 01404

BOFFA, M. [1967] *Remarques concernant les modeles associes aux permutations de l'univers* (J 2313) C R Acad Sci, Paris, Ser A-B 265*A205-A206
⋄ C62 E30 E35 ⋄ REV MR 36 # 4980 Zbl 183.12
• ID 31019

BOFFA, M. [1968] see BENI, J.

BOFFA, M. & RIBEAUFOSSE, P. [1969] *Modeles de la theorie generale des ensembles, construits sur les nombres-ε* (J 0068) Z Math Logik Grundlagen Math 15*239-240
⋄ C62 E10 E30 ⋄ REV MR 40 # 2532 Zbl 181.11
• ID 01371

BOFFA, M. [1971] *Forcing and reflection (Russian summary)* (J 0014) Bull Acad Pol Sci, Ser Math Astron Phys 19*181-183
⋄ C25 E65 ⋄ REV MR 44 # 59 Zbl 245 # 02057 • ID 01390

BOFFA, M. [1971] *Sentences universelles en theorie des ensembles* (J 2313) C R Acad Sci, Paris, Ser A-B 273*A69-A70
⋄ C62 E30 ⋄ REV MR 45 # 6612 Zbl 218 # 02059
• ID 01387

BOFFA, M. [1972] *Corps λ-clos* (J 2313) C R Acad Sci, Paris, Ser A-B 275*A881-A882
⋄ C60 ⋄ REV MR 47 # 4789 Zbl 248 # 02062 • ID 01393

BOFFA, M. [1972] *Isomorphismes des systemes extensionnels superuniversels* (J 2313) C R Acad Sci, Paris, Ser A-B 274*A369-A370
⋄ C62 E30 ⋄ REV MR 46 # 3309 Zbl 301 # 02071
• ID 01395

BOFFA, M. [1972] *Modeles universels homogenes et modeles generiques* (J 2313) C R Acad Sci, Paris, Ser A-B 274*A693-A694
⋄ C25 C50 ⋄ REV MR 45 # 6606 Zbl 235 # 02048
• ID 26187

BOFFA, M. [1972] *Sur l'existence des corps universels-homogenes* (J 2313) C R Acad Sci, Paris, Ser A-B 275*A1267-A1268
⋄ C50 C55 C60 ⋄ REV MR 47 # 1598 Zbl 257 # 02044
• ID 01394

BOFFA, M. & PRAAG VAN, P. [1972] *Sur les corps generiques* (J 2313) C R Acad Sci, Paris, Ser A-B 274*A1325-A1327
⋄ C25 C60 ⋄ REV MR 45 # 6607 Zbl 291 # 02038
• ID 01397

BOFFA, M. & PRAAG VAN, P. [1972] *Sur les sous-champs maximaux des corps generiques denombrables* (J 2313) C R Acad Sci, Paris, Ser A-B 275*A945-A947
⋄ C15 C25 C60 ⋄ REV MR 48 # 101 Zbl 286 # 02061
• ID 01391

BOFFA, M. [1972] *Ultraproduits et application a l'algebre* (J 0082) Bull Soc Math Belg 24*124-132
⋄ C20 C60 ⋄ REV MR 48 # 99 Zbl 255 # 02056 • ID 01396

BOFFA, M. [1973] *Structures extensionnelles generiques* (J 0082) Bull Soc Math Belg 25*3-10
⋄ C25 C62 E70 ⋄ REV MR 50 # 101 Zbl 286 # 02069
• ID 03930

BOFFA, M. [1975] *A note on existentially complete division rings* (C 0782) Model Th & Algeb (A. Robinson) 56-59
  ⋄ C25 C60 ⋄ REV MR 53 # 121 Zbl 317 # 02066
  • ID 16586

BOFFA, M. [1977] *Modeles cumulatifs de la theorie des types* (J 0056) Publ Dep Math, Lyon 14/2∗9-12
  ⋄ B15 C62 E30 ⋄ REV MR 58 # 27449 Zbl 432 # 03026
  • ID 53983

BOFFA, M. & CHERLIN, G.L. [1980] *Elimination des quantificateurs dans les faisceaux (English summary)* (J 2313) C R Acad Sci, Paris, Ser A-B 290∗A355-A357
  ⋄ C10 C60 C90 ⋄ REV MR 81i:03037 Zbl 423 # 03033
  • ID 53545

BOFFA, M. [1980] *Elimination des quantificateurs en algebre* (J 3133) Bull Soc Math Belg, Ser B 32∗107-133
  ⋄ C10 C98 ⋄ REV MR 84e:03041 Zbl 469 # 03017
  • ID 55145

BOFFA, M. & MACINTYRE, A. & POINT, F. [1980] *The quantifier elimination problem for rings without nilpotent elements and for semi-simple rings* (P 2625) Model Th of Algeb & Arithm;1979 Karpacz 20-30
  ⋄ C10 C60 ⋄ REV MR 82d:03054 Zbl 446 # 03026 JSL 50.1079 • ID 56555

BOFFA, M. [1981] *Quantifier elimination in boolean sheaves* (P 2614) Open Days in Model Th & Set Th;1981 Jadwisin 35-44
  ⋄ C10 C90 G30 ⋄ ID 33712

BOFFA, M. [1982] *Algebres de Boole atomiques et modeles de la theorie des types* (P 3774) Th d'Ensembl de Quine;1981 Louvain-la-Neuve 1-5
  ⋄ B15 C62 E70 ⋄ REV MR 84g:03085 Zbl 536 # 03036
  • ID 34190

BOFFA, M. [1983] *Arithmetic and theory of types* (P 1601) Easter Conf on Model Th (1);1983 Diedrichshagen 10-16
  ⋄ B15 C62 E70 F35 ⋄ REV MR 84i:03008 Zbl 528 # 03020 • ID 37638

BOFFA, M. & MICHAUX, C. & POINT, F. & PRAAG VAN, P. [1985] *L'elimination lineaire dans les corps* (J 3364) C R Acad Sci, Paris, Ser 1 300∗323-325
  ⋄ C10 C60 ⋄ ID 45595

BOFFA, M. see Vol. I, IV, V, VI for further entries

BOKUT', L.A. [1978] *On algebraically closed and simple Lie algebras (Russian)* (S 0066) Tr Mat Inst Steklov 148∗30-42,271
  • TRANSL [1978] (S 0055) Proc Steklov Inst Math 148∗27-33
  ⋄ C25 C60 ⋄ REV MR 82c:17009 • ID 48372

BOKUT', L.A. see Vol. IV for further entries

BOL'BOT, A.D. [1967] *Equationally complete varieties of totally symmetric quasigroups (Russian) (English summary)* (J 0003) Algebra i Logika 6/2∗13-19
  ⋄ C05 ⋄ REV MR 35 # 6536 Zbl 153.345 • ID 01418

BONDI, I.L. [1967] *The decidability or undecidability of certain formal logical calculi (Russian)* (J 0023) Dokl Akad Nauk SSSR 172∗1001-1002
  • TRANSL [1967] (J 0062) Sov Math, Dokl 8∗205-206
  ⋄ B25 D35 ⋄ REV MR 35 # 1464 Zbl 162.21 • ID 01424

BONDI, I.L. [1971] *The solvability problem for certain formal-logical calculi (Russian)* (S 0208) Uch Zap, Ped Inst, Moskva 277∗205-213
  ⋄ B25 D35 ⋄ REV MR 46 # 5100 • ID 02009

BONDI, I.L. [1973] *The decision problem for normed linear spaces and for Hilbert spaces (Russian)* (J 0031) Izv Vyssh Ucheb Zaved, Mat (Kazan) 1973/5(132)∗3-10
  ⋄ B25 C65 D35 ⋄ REV MR 48 # 8218 Zbl 275 # 02044
  • ID 01425

BONDI, I.L. see Vol. IV for further entries

BONNET, R. & POUZET, M. [1969] *Extensions et stratifications d'ensembles disperses* (J 2313) C R Acad Sci, Paris, Ser A-B 268∗A1512-1515
  ⋄ C65 E07 ⋄ REV MR 39 # 4055 Zbl 188.42 • ID 33172

BONNET, R. [1969] *Stratifications et extension des genres de chaines denombrables* (J 2313) C R Acad Sci, Paris, Ser A-B 269∗A880-A882
  ⋄ C15 C65 E07 ⋄ REV MR 40 # 5503 Zbl 206.280
  • ID 01426

BONNET, R. [1971] *Categorie et α-categorie des ensembles ordonnes* (J 0056) Publ Dep Math, Lyon 8/1∗69-90
  ⋄ C65 E07 ⋄ REV MR 48 # 186 Zbl 261 # 06002
  • ID 30471

BONNET, R. [1977] *Sur le type d'isomorphie d'algebres de Boole dispersees* (P 1729) Colloq Int Log;1975 Clermont-Ferrand 107-122
  ⋄ C52 C55 E50 G05 ⋄ REV MR 58 # 27682 Zbl 455 # 06007 • ID 54295

BONNET, R. [1980] *Very strongly rigid boolean algebras, continuum discrete set condition, countable antichain condition I* (J 0004) Algeb Universalis 11∗341-364
  ⋄ C50 E50 E75 G05 ⋄ REV MR 82c:06025 Zbl 467 # 06008 • ID 55060

BONNET, R. & POUZET, M. [1982] *Linear extensions of ordered sets* (P 3780) Ordered Sets;1981 Banff 125-170
  ⋄ C65 C98 E07 ⋄ REV MR 83h:06004 Zbl 499 # 06002
  • ID 38291

BONNET, R. & SHELAH, S. [1985] *Narrow Boolean algebras* (J 0073) Ann Pure Appl Logic 28∗1-12
  ⋄ C50 E75 G05 ⋄ REV Zbl 563 # 03035 • ID 38773

BONNET, R. see Vol. V for further entries

BOOLOS, G. [1970] *A proof of the Loewenheim-Skolem theorem* (J 0047) Notre Dame J Formal Log 11∗76-78
  ⋄ C07 ⋄ REV MR 48 # 93 Zbl 209.304 JSL 38.519
  • ID 01434

BOOLOS, G. [1973] *A note on Beth's theorem (Russian summary)* (J 0014) Bull Acad Pol Sci, Ser Math Astron Phys 21∗1-2
  ⋄ C40 C62 ⋄ REV MR 47 # 6469 Zbl 255 # 02057
  • ID 01436

BOOLOS, G. [1976] *On deciding the truth of certain statements involving the notion of consistency* (J 0036) J Symb Logic 41∗779-781
  ⋄ B25 B45 F30 ⋄ REV MR 56 # 8353 Zbl 359 # 02050
  • ID 14578

BOOLOS, G. [1977] *On deciding the provability of certain fixed point statements* (J 0036) J Symb Logic 42∗191-193
  ⋄ B25 B45 F30 ⋄ REV MR 58 # 192 Zbl 381 # 03013
  • ID 26442

BOOLOS, G. [1979] *The unprovability of consistency. An essay in modal logic* (X 0805) Cambridge Univ Pr: Cambridge, GB viii + 184pp
 ⋄ B25 B28 B45 B98 F30 F98 ⋄ REV MR 81c:03013 Zbl 409 # 03009 JSL 46.871 • ID 56312

BOOLOS, G. [1984] *Trees and finite satisfiability: proof of a conjecture of Burgess* (J 0047) Notre Dame J Formal Log 25*193-197
 ⋄ B10 C13 F07 ⋄ REV MR 85f:03005 Zbl 561 # 03004 • ID 40442

BOOLOS, G. see Vol. I, II, IV, V, VI for further entries

BOONE, W.W. [1974] *Between logic and group theory* (P 0709) Int Conf Th of Groups (2);1973 Canberra 90-102
 ⋄ C57 C60 D40 ⋄ REV MR 50 # 7357 Zbl 298 # 20027 • ID 04158

BOONE, W.W. see Vol. IV for further entries

BOOS, W. [1976] *Infinitary compactness without strong inaccessibility* (J 0036) J Symb Logic 41*33-38
 ⋄ C75 E35 E55 ⋄ REV MR 53 # 12946 Zbl 345 # 02049 • ID 14754

BOOS, W. see Vol. IV, V for further entries

BOOTH, D. [1970] *Ultrafilters on a countable set* (J 0007) Ann Math Logic 2*1-24
 ⋄ C20 E05 ⋄ REV MR 43 # 3104 Zbl 231 # 02067 • ID 01459

BOOTH, D. see Vol. V for further entries

BORCEUX, F. & BOSSCHE VAN DEN, G. [1983] *Recovering a frame from its sheaves of algebras* (J 0326) J Pure Appl Algebra 28*141-154
 ⋄ C90 G30 ⋄ REV MR 85e:18009 Zbl 517 # 18014 Zbl 519 # 32018 • ID 39930

BORCEUX, F. see Vol. V for further entries

BORICIC, B.R. [1983] *A decision procedure for certain disjunction-free intermediate propositional calculi* (J 0400) Publ Inst Math, NS (Belgrade) 34(48)*19-26
 ⋄ B25 B55 ⋄ REV MR 86d:03027 Zbl 556 # 03025 • ID 44659

BORICIC, B.R. see Vol. II, VI for further entries

BORKOWSKI, L. [1958] *On proper quantifiers I (Polish and Russian summaries)* (J 0063) Studia Logica 8*65-130
 ⋄ B20 B25 ⋄ REV MR 21 # 4095 Zbl 198.13 JSL 32.262 • REM Part II 1960 • ID 01482

BORKOWSKI, L. [1960] *On proper quantifiers II (Polish and Russian summaries)* (J 0063) Studia Logica 10*7-28
 • ERR/ADD ibid 15*272
 ⋄ B20 B25 ⋄ REV MR 24 # A681 Zbl 198.13 JSL 32.263 • REM Part I 1958 • ID 01483

BORKOWSKI, L. [1961] *A didactical approach to the zero-one decision procedure of the expressions of the first order monadic predicate calculus (Polish) (Russian and English summaries)* (J 0063) Studia Logica 11*57-76 • ERR/ADD ibid 15*271-272
 ⋄ B20 B25 C10 ⋄ REV MR 23 # A3076 Zbl 121.253 • ID 01484

BORKOWSKI, L. [1968] *Some remarks about the notion of definition (Polish) (Russian and English summaries)* (J 0063) Studia Logica 23*59-70
 ⋄ A05 B10 C40 ⋄ REV MR 39 # 1310 Zbl 305 # 02010 JSL 35.468 • ID 01485

BORKOWSKI, L. see Vol. I, II, V, VI for further entries

BORSHCHEV, V.B. & KHOMYAKOV, M.V. [1974] *Schemes for functions and relations (Russian)* (C 2577) Issl Formaliz Yazyk & Neklass Log 23-49
 ⋄ B75 C13 ⋄ REV MR 57 # 16027 • ID 71249

BORSHCHEV, V.B. & KHOMYAKOV, M.V. [1976] *Club systems (a formal calculus for the description of composed systems) (Russian) (English and German summaries)* (J 0338) Nauch-Tekh Inf, Ser 2, Akad Nauk SSSR 1976/8*3-6,39-40
 ⋄ C90 ⋄ REV MR 55 # 5249 Zbl 357 # 02042 • ID 50390

BOSSCHE VAN DEN, G. [1977] *Algebres de Heyting completes etalees et faisceaux* (J 0962) Ann Soc Sci Bruxelles, Ser 1 91*107-116
 ⋄ C90 G10 ⋄ REV MR 58 # 5854 Zbl 402 # 18010 • ID 54708

BOSSCHE VAN DEN, G. [1983] see BORCEUX, F.

BOTHE, H.G. [1981] *A first-order characterization of 3-dimensional manifolds* (P 2623) Worksh Extended Model Th;1980 Berlin 1-19
 ⋄ C65 G10 ⋄ REV MR 83e:54052 Zbl 485 # 57004 • ID 38502

BOUDREAUX, J.C. [1979] *Defining general structures* (J 0047) Notre Dame J Formal Log 20*465-488
 ⋄ B15 C85 ⋄ REV MR 80i:03021 Zbl 368 # 02051 • ID 56096

BOUDREAUX, J.C. [1980] *Frames versus minimally restricted structures* (J 0047) Notre Dame J Formal Log 21*251-262
 ⋄ B15 C85 ⋄ REV MR 81h:03025 Zbl 368 # 02052 • ID 53790

BOURGAIN, J. & ROSENTHAL, H.P. & SCHECHTMAN, G. [1979] *Uniform validity and stationary logic* (1111) Preprints, Manuscr., Techn. Reports etc. 25pp
 ⋄ C55 C80 ⋄ REM Manuscript, Univ. of Maryland • ID 33340

BOURGAIN, J. see Vol. V for further entries

BOURTOT, B. & CUSIN, R. [1966] *Une demonstration du theoreme de Loewenheim-Skolem* (J 0056) Publ Dep Math, Lyon 3/1*9-16
 ⋄ C07 ⋄ REV MR 33 # 7234 Zbl 219 # 02005 • ID 02646

BOUSCAREN, E. & CHATZIDAKIS, Z. [1978] *Modeles premiers et atomiques: theoreme des deux cardinaux dans les theories totalement transcendantes* (C 3093) Groupe Etude Th Stables (1) 1977/78 1/2*11pp
 ⋄ C45 C50 C55 ⋄ REV MR 80d:03030a Zbl 408 # 03026 • ID 71271

BOUSCAREN, E. [1980] *Existentially closed modules: Types and prime models* (P 2625) Model Th of Algeb & Arithm;1979 Karpacz 31-43
 ⋄ C25 C60 ⋄ REV MR 82j:03038 Zbl 471 # 03030 • ID 55225

BOUSCAREN, E. & LASCAR, D. [1983] *Countable models of nonmultidimensional $\aleph_0$-stable theories* (J 0036) J Symb Logic 48*197-205
⋄ C15 C45 C50 ⋄ REV MR 85b:03049 Zbl 517 # 03008
• ID 39997

BOUSCAREN, E. [1983] *Countable models of multidimensional $\aleph_0$-stable theories* (J 0036) J Symb Logic 48*377-383
⋄ C15 C45 C50 ⋄ REV MR 84k:03097 Zbl 527 # 03010
• ID 36114

BOUSCAREN, E. [1984] *Martin's conjecture for ω-stable theories* (J 0029) Israel J Math 49*15-25
⋄ C15 C45 ⋄ ID 45596

BOUVERE DE, K.L. [1959] *A method in proofs of undefinability, with applications to functions in the arithmetic of natural numbers* (X 0809) North Holland: Amsterdam xvi + 64pp
⋄ C40 F30 ⋄ REV MR 21 # 3327 Zbl 231 # 02061 JSL 25.271 • ID 01501

BOUVERE DE, K.L. [1963] *A mathematical characterization of explicit definability* (J 0028) Indag Math 25*264-274
⋄ C40 ⋄ REV MR 28 # 28 Zbl 122.245 JSL 36.687
• ID 02029

BOUVERE DE, K.L. [1965] see ADDISON, J.W.

BOUVERE DE, K.L. [1965] *Logical synonymity* (J 0028) Indag Math 27*622-629
• REPR [1968] (C 0684) Automation in Lang Transl & Theorem Prov 129-136
⋄ A05 C40 C52 ⋄ REV MR 32 # 2316 Zbl 221 # 02042
• ID 01502

BOUVERE DE, K.L. [1965] *Synonymous theories* (P 0614) Th Models;1963 Berkeley 402-406
• REPR [1968] (C 0684) Automation in Lang Transl & Theorem Prov 123-127
⋄ B30 C40 ⋄ REV Zbl 221 # 02041 • ID 04261

BOUVERE DE, K.L. [1967] *Some remarks about synonymity and the theorem of Beth* (P 0683) Log & Founds of Sci (Beth);1964 Paris 10-18
⋄ C40 G15 ⋄ REV MR 39 # 3979 Zbl 221 # 02043
• ID 28110

BOUVERE DE, K.L. [1968] *On formal ambiguity* (C 0684) Automation in Lang Transl & Theorem Prov 101-109
⋄ C40 ⋄ REV MR 41 # 5206 Zbl 198.317 • ID 02708

BOUVERE DE, K.L. [1969] *Remarks on classification of theories by their complete extensions* (J 0047) Notre Dame J Formal Log 10*1-17
⋄ C07 C35 ⋄ REV MR 40 # 5434 Zbl 184.12 • ID 01503

BOUVERE DE, K.L. see Vol. I for further entries

BOWEN, K.A. [1972] *A note on cut elimination and completeness in first order theories* (J 0068) Z Math Logik Grundlagen Math 18*173-176
⋄ B10 C07 F05 ⋄ REV MR 46 # 8809 Zbl 243 # 02020
• ID 01505

BOWEN, K.A. [1973] *Infinite forcing and natural models of set theory (Russian summary)* (J 0014) Bull Acad Pol Sci, Ser Math Astron Phys 21*195-199
⋄ C25 C55 C62 C75 C80 E55 G30 ⋄ REV MR 48 # 5859 Zbl 257 # 02042 • ID 01506

BOWEN, K.A. [1974] *Forcing in a general setting* (J 0027) Fund Math 81*315-329
⋄ C25 C75 C80 C85 E40 ⋄ REV MR 51 # 5292 Zbl 346 # 02036 • ID 01508

BOWEN, K.A. [1974] *Primitive recursive notations for infinitary formulas* (S 0019) Colloq Math (Warsaw) 30*1-5
⋄ C75 F15 ⋄ REV MR 50 # 6785 Zbl 262 # 02014
• ID 03942

BOWEN, K.A. [1975] *Normal modal model theory* (J 0122) J Philos Logic 4*97-131
⋄ B45 C90 ⋄ REV MR 57 # 5733 Zbl 317 # 02015
• ID 29636

BOWEN, K.A. [1978] *Model theory for modal logic. Kripke models for modal predicate calculi* (X 0835) Reidel: Dordrecht x + 128pp
⋄ B45 C90 ⋄ REV MR 80j:03023 Zbl 395 # 03022 JSL 46.415 • ID 30659

BOWEN, K.A. [1980] *Interpolation in loop-free logic* (J 0063) Studia Logica 39*297-310
⋄ B75 C40 ⋄ REV MR 82c:03020 Zbl 457 # 03012
• ID 54337

BOWEN, K.A. see Vol. I, II, V, VI for further entries

BOYD, M.W. & TRANSUE, W.R. [1979] *Properties of ultraproducts* (J 3522) Rend Circ Mat Palermo, Ser 2 28*387-397
⋄ C20 C65 ⋄ REV MR 82b:46021 Zbl 455 # 46015
• ID 54299

BOZIC, M. & DOSEN, K. [1984] *Models for normal intuitionistic modal logics* (J 0063) Studia Logica 43*217-245
⋄ B45 C90 ⋄ ID 40033

BOZIC, M. [1985] *Semantics for some intermediate logics* (J 0400) Publ Inst Math, NS (Belgrade) 37*7-15
⋄ B55 C90 ⋄ ID 48301

BOZIC, M. see Vol. II, V, VI for further entries

BOZOVIC, N.B. [1977] *On Markov properties of finitely presented groups* (J 0400) Publ Inst Math, NS (Belgrade) 21(35)*29-32
⋄ C05 C50 D45 ⋄ REV MR 57 # 422 Zbl 374 # 02010
• ID 51534

BOZOVIC, N.B. [1985] *First-order classes of groups having no groups with a given property* (J 0400) Publ Inst Math, NS (Belgrade) 37(51)*51-55
⋄ C60 ⋄ ID 49533

BOZOVIC, N.B. see Vol. IV for further entries

BOZZI, S. & MELONI, G.C. [1977] *Ideal properties and intuitionistic algebra* (J 0207) Ist Lombardo Accad Sci Rend, A (Milano) 111*145-150
⋄ C60 C90 F50 F55 ⋄ REV MR 58 # 16827 Zbl 393 # 03026 • ID 52446

BOZZI, S. & MELONI, G.C. [1980] *Representation of Heyting algebras with covering and propositional intuitionistic logic with local operator* (J 3285) Boll Unione Mat Ital, V Ser, A 17*436-442
⋄ B25 B55 C90 F50 G10 ⋄ REV MR 82k:03099 Zbl 454 # 03008 • ID 54221

BOZZI, S. [1982] *Formule attuali ed ereditarie nella logica intuizionista* (J 3746) Note Math (Lecce) 2*159-166
⋄ C90 F50 ⋄ REV MR 85e:03024 Zbl 547 # 03038
• ID 40429

BOZZI, S. see Vol. II, VI for further entries

BRADY, R.T. [1977] *Unspecified constants in predicate calculus and first-order theories* (J 0079) Logique & Anal, NS 20*229-243
 ⋄ B10 C07 ⋄ REV MR 57 # 5672 Zbl 368 # 02013
 • ID 30639

BRADY, R.T. [1984] *Natural deduction systems for some quantified relevant logics* (J 0079) Logique & Anal, NS 27*355-377
 ⋄ B46 C90 ⋄ REV Zbl 559 # 03011 • ID 45522

BRADY, R.T. see Vol. I, II, V for further entries

BREBAN, M. & FERRO, A. & OMODEO, E.G. & SCHWARTZ, J.T. [1981] *Decision procedures for elementary sublanguages of set theory. II. Formulas involving restricted quantifiers, together with ordinal, integer, map, and domain notions* (J 0155) Commun Pure Appl Math 34*177-195
 ⋄ B25 E30 ⋄ REV MR 82i:03018b Zbl 465 # 03003 • REM Part I 1980 by Ferro,A & Omodeo,E.G. & Schwartz,J.T. Part III 1984 by Breban,M. & Ferro,A. Common author of all parts: Ferro,A. • ID 54906

BREBAN, M. & FERRO, A. [1984] *Decision procedures for elementary sublanguages of set theory. III. Restricted classes of formulas involving the power set operator and the general set union operator* (J 2650) Adv Appl Math 5*147-215
 ⋄ B25 E30 ⋄ REV MR 86b:03018 Zbl 542 # 03003 • REM Part II 1981 by Breban,M. & Ferro,A. & Omodeo,E.G. & Schwartz,J.T. Part IV to appear in J0155, part V in J0119. Part VI 1985 by Cantone,D. & Ferro,A. & Schwartz,J.T. Common author of all parts: Ferro,A. • ID 40365

BREDIKHIN, S.V. & ERSHOV, YU.L. & KAL'NEJ, V.E. [1970] *Fields with two linear orderings (Russian)* (J 0087) Mat Zametki (Akad Nauk SSSR) 7*525-536
 • TRANSL [1970] (J 1044) Math Notes, Acad Sci USSR 7*319-325
 ⋄ C35 C60 ⋄ REV JSL 51.235 • ID 48373

BRIDGE, J. [1977] *Beginning model theory. The completeness theorem and some consequences* (X 0815) Clarendon Pr: Oxford vii+143pp
 ⋄ C07 C98 ⋄ REV MR 58 # 27171 Zbl 361 # 02064 JSL 44.283 • ID 50699

BRIDGE, J. see Vol. V, VI for further entries

BRIGNOLE, D. & RIBEIRO, H. [1965] *On the universal equivalence for ordered abelian groups* (J 0003) Algebra i Logika 4/2*51-55
 ⋄ C60 ⋄ REV MR 31 # 3520 • ID 31431

BRIGNOLE, D. see Vol. V for further entries

BRITTON, J.L. [1973] *The existence of infinite Burnside groups* (P 0678) Word Probl: Decis & Burnside Probl in Group Th;1969 Irvine 67-348
 ⋄ C60 D40 ⋄ REV MR 53 # 10945 Zbl 264 # 20028 JSL 41.786 • ID 44434

BRITTON, J.L. [1979] *Integer solutions of systems of quadratic equations* (J 0332) Math Proc Cambridge Phil Soc 86*385-389
 ⋄ B25 D35 D80 ⋄ REV MR 80i:10080 Zbl 418 # 10019
 • ID 53330

BRITTON, J.L. see Vol. IV for further entries

BROESTERHUIZEN, G. [1975] *A generalized Łoś ultraproduct theorem* (S 0019) Colloq Math (Warsaw) 33*161-173
 ⋄ C20 C80 ⋄ REV MR 53 # 120 Zbl 282 # 02023
 • ID 16585

BROESTERHUIZEN, G. [1975] *Structures for a logic with additional generalized quantifier* (S 0019) Colloq Math (Warsaw) 33*1-12
 ⋄ C80 ⋄ REV MR 53 # 92 Zbl 278 # 02014 • ID 16556

BROESTERHUIZEN, G. & WEGLORZ, B. & WIERZEJEWSKI, J. [1977] *Remarks on rigid structures* (J 0028) Indag Math 39*383-388
 ⋄ C50 ⋄ REV MR 57 # 93 Zbl 367 # 02026 • ID 31530

BROESTERHUIZEN, G. [1979] *Extensions of automorphisms* (J 0028) Indag Math 41*235-244
 ⋄ C30 C50 ⋄ REV MR 83j:03051 Zbl 421 # 03022
 • ID 53421

BROESTERHUIZEN, G. [1982] *Locally orderable structures* (J 0068) Z Math Logik Grundlagen Math 28*7-14
 ⋄ C30 ⋄ REV MR 83e:03046 Zbl 497 # 03024 • ID 35226

BROWKIN, J. [1966] *On forms over p-adic fields (Russian summary)* (J 0014) Bull Acad Pol Sci, Ser Math Astron Phys 14*489-492
 ⋄ C60 ⋄ REV MR 34 # 2522 Zbl 139.282 • ID 01643

BROWN, M.A. [1984] *Generalized quantifiers and the square of opposition* (J 0047) Notre Dame J Formal Log 25*303-322
 ⋄ B65 C80 ⋄ REV MR 85j:03034 Zbl 559 # 03023
 • ID 42575

BROWN, M.A. see Vol. II for further entries

BROWN, S.S. [1978] *Bounds on transfer principles for algebraically closed and complete discretely valued fields* (S 0167) Mem Amer Math Soc 15(204)*iv+92pp
 ⋄ B25 C10 C60 ⋄ REV MR 80j:03045 Zbl 383 # 03022
 • ID 52004

BRUCE, KIM B. [1978] *Ideal models and some not so ideal problems in the model theory of L(Q)* (J 0036) J Symb Logic 43*304-321
 ⋄ C40 C80 ⋄ REV MR 80a:03048 Zbl 385 # 03029
 • ID 29263

BRUCE, KIM B. [1978] *Model-theoretic forcing in logic with a generalized quantifier* (J 0007) Ann Math Logic 13*225-265
 ⋄ C25 C70 C75 C80 ⋄ REV MR 80c:03033 Zbl 436 # 03025 • ID 27965

BRUCE, KIM B. & KEISLER, H.J. [1979] $L_A(\exists)$ (J 0036) J Symb Logic 44*15-28
 ⋄ C25 C70 C75 C80 ⋄ REV MR 80f:03037 Zbl 413 # 03024 • ID 53018

BRUCE, KIM B. [1980] *Model constructions in stationary logic I. Forcing* (J 0036) J Symb Logic 45*439-454
 ⋄ C25 C55 C70 C75 C80 ⋄ REV MR 82f:03032 Zbl 479 # 03021 • ID 55683

BRUCE, KIM B. see Vol. VI for further entries

BRUIJN DE, N.G. [1975] *Set theory with type restrictions* (P 0759) Infinite & Finite Sets (Erdoes);1973 Keszthely 1*205-214
 ⋄ C80 E30 E70 ⋄ REV MR 53 # 2615 Zbl 302 # 02002
 • ID 21509

BRUIJN DE, N.G. see Vol. I, II, V, VI for further entries

BRUSS, A.R. & MEYER, A.R. [1978] *On time-space classes and their relation to the theory of real addition* (P 1740) ACM Symp Th of Comput (10);1978 San Diego 233-239
⋄ B25 D15 ⋄ REV MR 81i:68056 • ID 80801

BRUSS, A.R. & MEYER, A.R. [1980] *On time-space classes and their relation to the theory of real addition* (J 1426) Theor Comput Sci 11*59-69
⋄ B25 D15 ⋄ REV MR 82c:03061a Zbl 467 # 03038 • ID 55036

BRYANT, R.M. & GROVES, J.R.J. [1978] *Wreath products and ultraproducts of groups* (J 0131) Quart J Math, Oxford Ser 2 29*301-308
⋄ C20 C30 C60 ⋄ REV MR 80d:20027 Zbl 388 # 20022 • ID 90163

BRYANT, S.J. & MARICA, J.G. [1960] *Unary algebras* (J 0048) Pac J Math 10*1347-1353
⋄ C05 C30 ⋄ REV MR 22 # 9463 • ID 08765

BRYCE, R.A. [1968] *A note on free products with normal amalgamation* (J 0038) J Austral Math Soc 8*631-637
⋄ C05 C30 ⋄ REV MR 37 # 5303 Zbl 164.19 • ID 01677

BRYLL, G. & SLUPECKI, J. [1973] *Syntactic proof of the L-decidability of classical logic and Lukasiewicz's three-valued logic (Polish) (English summary)* (S 1454) Zesz Nauk Wyz Szk Ped Mat, Opole 13*131-152
⋄ B25 B50 ⋄ REV MR 54 # 4954 Zbl 291 # 02007 • ID 24728

BRYLL, G. [1975] *Classes of asymmetric and bounded graphs defined by the simplest expressions of the predicate calculus (Polish. English translation)* (S 1454) Zesz Nauk Wyz Szk Ped Mat, Opole 15*45-48,49-52
⋄ C40 ⋄ REV MR 54 # 7313 Zbl 319 # 02013 • ID 25820

BRYLL, G. see Vol. I, II, IV for further entries

BUCUR, I. [1980] *Special chapters of algebra (Romanian)* (X 0871) Acad Rep Soc Romania: Bucharest 335pp
• TRANSL [1984] (X 0835) Reidel: Dordrecht viii+406pp
⋄ B80 C98 D98 ⋄ REV MR 84c:14001 MR 86f:14001 • ID 46245

BUDKIN, A.I. & GORBUNOV, V.A. [1973] *Implicative classes of algebras (Russian)* (J 0003) Algebra i Logika 12*249-268,363
• TRANSL [1973] (J 0069) Algeb and Log 12*139-149
⋄ C05 ⋄ REV MR 54 # 5085 Zbl 287 # 08005 • ID 24758

BUDKIN, A.I. & GORBUNOV, V.A. [1975] *Quasivarieties of algebraic systems (Russian)* (J 0003) Algebra i Logika 14*123-142,240
• TRANSL [1975] (J 0069) Algeb and Log 14*73-84
⋄ C05 ⋄ REV MR 53 # 240 Zbl 317 # 08003 • ID 16676

BUDKIN, A.I. [1976] *Quasi-identities in a free group (Russian)* (J 0003) Algebra i Logika 15*39-52,112
• TRANSL [1976] (J 0069) Algeb and Log 15*25-33
⋄ C05 ⋄ REV MR 58 # 21584 Zbl 364 # 20035 • ID 71386

BUDKIN, A.I. [1982] *Independent axiomatizability of quasivarieties of groups (Russian)* (J 0087) Mat Zametki (Akad Nauk SSSR) 31*817-826,956
• TRANSL [1982] (J 1044) Math Notes, Acad Sci USSR 31*413-417
⋄ C05 C60 ⋄ REV MR 84b:20029 Zbl 508 # 20014 • ID 38802

BUDKIN, A.I. [1984] *Quasiidentities and direct wreaths of groups (Russian)* (J 0003) Algebra i Logika 23*367-382
• TRANSL [1984] (J 0069) Algeb and Log 23*253-264
⋄ C05 C30 C60 ⋄ REV MR 86d:20030 • ID 48300

BUECHI, J.R. & WRIGHT, J.B. [1955] *The theory of proportionality as an abstraction of group theory* (J 0043) Math Ann 130*102-108
⋄ C60 ⋄ REV MR 17.455 Zbl 65.9 JSL 22.369 • ID 01684

BUECHI, J.R. & WRIGHT, J.B. [1957] *Invariants of the anti-automorphisms of a group* (J 0053) Proc Amer Math Soc 8*1134-1140
⋄ C60 ⋄ REV MR 20 # 2389 Zbl 83.246 • ID 01688

BUECHI, J.R. [1960] *Weak second-order arithmetic and finite automata* (J 0068) Z Math Logik Grundlagen Math 6*66-92
⋄ B15 B25 C85 D05 F35 ⋄ REV MR 23 # A2317 Zbl 103.247 JSL 28.100 • ID 01690

BUECHI, J.R. [1962] *On a decision method in restricted second order arithmetic* (P 0612) Int Congr Log, Meth & Phil of Sci (1,Proc);1960 Stanford 1-11
• TRANSL [1964] (J 1048) Kiber Sb Perevodov 1964/8*78-90 (Russian)
⋄ B25 C85 D05 F35 ⋄ REV MR 32 # 1116 Zbl 147.251 JSL 28.100 • ID 01693

BUECHI, J.R. [1965] *Decision methods in the theory of ordinals* (J 0015) Bull Amer Math Soc 71*767-770
⋄ B15 B25 C85 D05 E10 ⋄ REV MR 32 # 7413 • ID 01697

BUECHI, J.R. [1965] *Relatively categorical and normal theories* (P 0614) Th Models;1963 Berkeley 424-426
⋄ C35 C40 C60 ⋄ REV Zbl 163.247 • ID 01696

BUECHI, J.R. [1965] *Transfinite automata recursions and weak second order theory of ordinals* (P 0623) Int Congr Log, Meth & Phil of Sci (2,Proc);1964 Jerusalem 3-23
⋄ B15 B25 C85 D05 E10 ⋄ REV MR 35 # 1480 • ID 01694

BUECHI, J.R. & LANDWEBER, L.H. [1969] *Definability in the monadic second-order theory of successor* (J 0036) J Symb Logic 34*166-170
⋄ B15 B25 C40 C85 D05 F35 ⋄ REV MR 42 # 4387 Zbl 209.22 • ID 01700

BUECHI, J.R. & DANHOF, K.J. [1972] *Model theoretic approaches to definability* (J 0068) Z Math Logik Grundlagen Math 18*61-70
⋄ C20 C40 ⋄ REV MR 49 # 2330 Zbl 241 # 02015 • ID 01704

BUECHI, J.R. & SIEFKES, D. [1973] *Axiomatization of the monadic second order theory of* ω_1 (C 3046) Decid Theories II - Monad 2nd Ord Th Count Ordinals 129-217
⋄ B15 B25 C85 E10 ⋄ REV Zbl 298 # 02050 • ID 71381

BUECHI, J.R. & DANHOF, K.J. [1973] *Definability in normal theories* (J 0029) Israel J Math 14*248-256
⋄ C40 ⋄ REV MR 49 # 10537 Zbl 267 # 02038 • ID 01710

BUECHI, J.R. & SIEFKES, D. [1973] *The monadic second order theory of all countable ordinals* (C 3046) Decid Theories II - Monad 2nd Ord Th Count Ordinals vi+217pp
⋄ B15 B25 C85 E10 ⋄ REV MR 49 # 10534 Zbl 298 # 02050 • ID 01708

BUECHI, J.R. [1973] *The monadic second order theory of ω_1* (C 3046) Decid Theories II - Monad 2nd Ord Th Count Ordinals 1-127
⋄ B15 B25 C85 E10 ⋄ REV MR 57 # 16033 • ID 71377

BUECHI, J.R. & DANHOF, K.J. [1973] *Variations on a theme of Cantor in the theory of relational structures* (J 0068) Z Math Logik Grundlagen Math 19*411-426
⋄ C07 C35 C50 C75 ⋄ REV MR 49 # 2336 Zbl 317 # 02063 • ID 01706

BUECHI, J.R. [1977] *Using determinancy of games to eliminate quantifiers* (P 2588) FCT'77 Fund of Comput Th;1977 Poznan 367-378
⋄ B15 C10 C85 E60 ⋄ REV MR 58 # 16244 Zbl 367 # 02005 • ID 51165

BUECHI, J.R. & ZAIONTZ, C. [1983] *Deterministic automata and the monadic theory of ordinals $<\omega_2$* (J 0068) Z Math Logik Grundlagen Math 29*313-336
⋄ B15 B25 C85 D05 E10 ⋄ REV MR 85j:03008 Zbl 541 # 03004 • ID 40515

BUECHI, J.R. & SIEFKES, D. [1983] *The complete extensions of the monadic second order theory of countable ordinals* (J 0068) Z Math Logik Grundlagen Math 29*289-312
⋄ B15 C35 C85 D05 E10 ⋄ REV MR 85g:03058 Zbl 541 # 03003 • ID 33901

BUECHI, J.R. see Vol. I, IV, V for further entries

BUECHLER, S. [1984] *Expansions of models of ω-stable theories* (J 0036) J Symb Logic 49*470-477
⋄ C45 C50 ⋄ REV MR 86b:03037 • ID 42425

BUECHLER, S. [1984] *Kueker's conjecture for superstable theories* (J 0036) J Symb Logic 49*930-934
⋄ C35 C45 C50 ⋄ REV MR 86a:03033 • ID 42426

BUECHLER, S. [1984] *Resplendency and recursive definability in ω-stable theories* (J 0029) Israel J Math 49*26-33
⋄ C45 C50 C57 ⋄ REV Zbl 568 # 03015 • ID 45597

BUECHLER, S. [1985] *Coordinatization in superstable theories I. Stationary types* (J 0064) Trans Amer Math Soc 288*101-114
⋄ C45 ⋄ REV MR 86c:03031 • ID 40304

BUECHLER, S. [1985] *Invariantes for ω-categorical, ω-stable theories* (J 0029) Israel J Math 52*65-81
⋄ C35 C45 ⋄ ID 49403

BUECHLER, S. [1985] *One theorem of Zil'ber's on strongly minimal sets* (J 0036) J Symb Logic 50*1054-1061
⋄ C35 C45 ⋄ ID 48374

BUECHLER, S. [1985] *The geometry of weakly minimal types* (J 0036) J Symb Logic 50*1044-1053
⋄ C45 ⋄ ID 48375

BUGAJSKI, S. [1983] *Semantics in Banach spaces* (J 0063) Studia Logica 42*81-88
⋄ B51 C65 C90 ⋄ REV MR 86c:03036 Zbl 559 # 03017 • ID 42320

BUGAJSKI, S. see Vol. I, II for further entries

BUKOVSKY, L. & HAJEK, P. [1966] *On the standardness and regularity of normal syntactic models of the set theory* (J 0014) Bull Acad Pol Sci, Ser Math Astron Phys 14*101-105
⋄ C62 E30 E55 ⋄ REV MR 34 # 1164 Zbl 147.260 • ID 01719

BUKOVSKY, L. & PRIKRY, K. [1966] *Some metamathematical properties of measurable cardinals* (J 0014) Bull Acad Pol Sci, Ser Math Astron Phys 14*9-14
⋄ C20 E47 E55 ⋄ REV MR 36 # 6285 Zbl 141.6 JSL 33.476 • ID 01721

BUKOVSKY, L. [1972] *Models of set theory with axiom of constructibility which contain non-constructible sets* (J 0014) Bull Acad Pol Sci, Ser Math Astron Phys 20*969-972
⋄ C20 C62 E45 ⋄ REV MR 47 # 1609 Zbl 251 # 02062 • ID 01728

BUKOVSKY, L. [1973] *Characterization of generic extensions of models of set theory* (J 0027) Fund Math 83*35-46
⋄ C62 E40 G05 ⋄ REV MR 48 # 10804 Zbl 344 # 02043 • ID 01729

BUKOVSKY, L. [1974] see BILY, J.

BUKOVSKY, L. [1976] *Changing cofinality of $\aleph_2$* (P 1476) Set Th & Hierarch Th (2) (Mostowski);1975 Bierutowice 37-49
⋄ C62 E35 E40 ⋄ REV MR 55 # 5441 Zbl 365 # 02049 • ID 31650

BUKOVSKY, L. [1977] *Iterated ultrapower and Prikry's forcing* (J 0140) Comm Math Univ Carolinae (Prague) 18*77-85
⋄ C20 C62 E40 E55 ⋄ REV MR 56 # 5295 Zbl 358 # 02069 • ID 31651

BUKOVSKY, L. [1978] *Cogeneric extensions* (P 1897) Logic Colloq;1977 Wroclaw 91-98
⋄ C62 E40 ⋄ REV MR 80e:03062 Zbl 444 # 03027 • REM Part I. Part II 1981 • ID 71397

BUKOVSKY, L. [1979] *Any partition into Lebesgue measure zero sets produces a non-measurable set* (J 3293) Bull Acad Pol Sci, Ser Math 27*431-435
⋄ C20 E40 E75 ⋄ REV MR 81h:03095 Zbl 429 # 28001 • ID 53889

BUKOVSKY, L. [1981] *Cogeneric extensions II* (P 2614) Open Days in Model Th & Set Th;1981 Jadwisin 49-60
⋄ C62 E40 ⋄ REM Part I 1978 • ID 33714

BUKOVSKY, L. see Vol. IV, V for further entries

BULLOCK, A.M. & SCHNEIDER, H.H. [1972] *A calculus for finitely satisfiable formulas with identity* (J 0009) Arch Math Logik Grundlagenforsch 15*158-163
⋄ B10 C07 C13 ⋄ REV MR 48 # 8219 Zbl 262 # 02013 • ID 01746

BULLOCK, A.M. & SCHNEIDER, H.H. [1973] *On generating the finitely satisfiable formulas* (J 0047) Notre Dame J Formal Log 14*373-376
⋄ B10 C07 C13 D25 ⋄ REV MR 47 # 8280 Zbl 236 # 02014 • ID 01748

BULMAN-FLEMING, S. & TAYLOR, W. [1972] *On a question of G.H.Wenzel* (J 0004) Algeb Universalis 2*142-145
⋄ C05 E50 ⋄ REV MR 47 # 8397 Zbl 268 # 08004 • ID 01751

BULMAN-FLEMING, S. [1972] *On equationally compact semilattices* (J 0004) Algeb Universalis 2*146-151
⋄ C05 C50 ⋄ REV MR 46 # 7118 Zbl 267 # 08006 JSL 40.88 • ID 01753

BULMAN-FLEMING, S. [1974] *A note on equationally compact algebras* (J 0004) Algeb Universalis 4*41-43
⋄ C05 C50 ⋄ REV MR 50 # 12858 Zbl 296 # 08015 • ID 01757

BULMAN-FLEMING, S. [1974] *Algebraic compactness and its relations to topology* (**P** 0706) Gen Topol & Appl (de Groot);1972 Pittsburgh 89-94
⋄ C05 C50 ⋄ REV MR 50#12868 Zbl 286#08008
• ID 04168

BULMAN-FLEMING, S. & WERNER, H. [1977] *Equational compactness in quasi-primal varieties* (**J** 0004) Algeb Universalis 7*33-46
⋄ C05 C60 ⋄ REV MR 55#2705 Zbl 367#08006
• ID 24172

BUNGE, MARTA C. & REYES, G.E. [1981] *Boolean spectra and model completions* (**J** 0027) Fund Math 113*165-173
⋄ C35 C90 G30 ⋄ REV MR 83a:03069 Zbl 481#03021
• ID 35077

BUNGE, MARTA C. [1981] *Sheaves and prime model extensions* (**J** 0032) J Algeb 68*79-96
⋄ C35 C60 C90 ⋄ REV MR 83d:03043 Zbl 465#03016
• ID 54919

BUNGE, MARTA C. [1982] *On the transfer of an abstract Nullstellensatz* (**J** 0394) Commun Algeb 10*1891-1906
⋄ C35 C60 C90 G30 ⋄ REV MR 84c:03074 Zbl 494#18005 • ID 34013

BUNGE, MARTA C. [1984] *Toposes in logic and logic in toposes* (**J** 3781) Topoi 3*13-22
⋄ C90 E70 G30 ⋄ REV MR 86e:03060 • ID 44974

BUNGE, MARTA C. see Vol. V for further entries

BUNYATOV, M.R. [1967] *Dirac spaces (Russian)* (**S** 0223) Nauch Trud NS, Politekh Inst Tashkent 43*24-38
⋄ C20 C60 H05 H10 ⋄ REV MR 43#32 • ID 02032

BURD, B. [1984] *Decomposable collections of sets* (**J** 0047) Notre Dame J Formal Log 25*17-26
⋄ C75 E05 ⋄ REV MR 85i:04005 Zbl 541#03030
• ID 41389

BURD, B. see Vol. V for further entries

BURGESS, J.P. & MILLER, DOUGLAS E. [1975] *Remarks on invariant descriptive set theory* (**J** 0027) Fund Math 90*53-75
⋄ C75 D55 E15 ⋄ REV MR 53#7784 Zbl 342#02049
• ID 03956

BURGESS, J.P. [1977] *Descriptive set theory and infinitary languages* (**P** 3269) Set Th Found Math (Kurepa);1977 Beograd 9-30
⋄ C75 E15 ⋄ REV MR 58#16131 Zbl 408#03030
• ID 56269

BURGESS, J.P. [1978] *Consistency proofs in model theory: A contribution to "Jensenlehre"* (**J** 0007) Ann Math Logic 14*1-12
⋄ C55 E05 E35 E50 ⋄ REV MR 58#5206 Zbl 384#03035 • ID 29143

BURGESS, J.P. [1978] *On the Hanf number of Souslin logic* (**J** 0036) J Symb Logic 43*568-571
⋄ C55 C75 E35 ⋄ REV MR 81c:03026 Zbl 386#03016
• ID 29284

BURGESS, J.P. [1979] *A reflection phenomenon in descriptive set theory* (**J** 0027) Fund Math 104*127-139
⋄ C75 D55 E15 ⋄ REV MR 81i:04002 Zbl 449#03048
• ID 56715

BURGESS, J.P. [1980] *Decidability for branching time* (**J** 0063) Studia Logica 39*203-218
⋄ B25 B45 C85 ⋄ REV MR 82e:03020 Zbl 467#03006
• ID 55004

BURGESS, J.P. & GUREVICH, Y. [1985] *The decision problem for linear temporal logic* (**J** 0047) Notre Dame J Formal Log 26*115-128
⋄ B25 B45 ⋄ REV Zbl 573#03004 • ID 42591

BURGESS, J.P. see Vol. II, IV, V, VI for further entries

BURGIN, M.S. [1977] *Nonclassical models of the natural numbers (Russian)* (**J** 0067) Usp Mat Nauk 32/6*209-210
⋄ H15 ⋄ REV MR 58#5187 Zbl 432#40002 • ID 71469

BURGIN, M.S. see Vol. IV, V for further entries

BURMEISTER, P. [1968] *Ueber die Maechtigkeiten und Unabhaengigkeitsgrade der Basen freier Algebren. I* (**J** 0027) Fund Math 62*165-189
⋄ C05 C75 ⋄ REV MR 37#2658 Zbl 159.19 • REM Part II 1970 • ID 48809

BURMEISTER, P. [1970] *Ueber die Maechtigkeiten und Unabhaengigkeitsgrade der Basen freier Algebren II* (**J** 0027) Fund Math 67*323-336
⋄ C05 ⋄ REV MR 42#3007 Zbl 199.325 • REM Part I 1968
• ID 01786

BURMEISTER, P. [1980] see ANDREKA, H.

BURMEISTER, P. [1981] *Quasi-equational logic for partial algebras* (**P** 3165) FCT'81 Fund of Comput Th;1981 Szeged 71-80
⋄ C05 C90 ⋄ REV MR 83e:08011 Zbl 491#08007
• ID 40147

BURMEISTER, P. [1981] see ANDREKA, H.

BURMEISTER, P. [1982] *Partial algebras - survey of a unifying approach towards a two-valued model theory for partial algebras* (**J** 0004) Algeb Universalis 15*306-358
⋄ C05 C90 C98 ⋄ REV MR 84g:03040 Zbl 511#03014
• ID 34153

BURMEISTER, P. see Vol. V for further entries

BURNS, R.G. & OKOH, F. & SMITH, HOWARD & WIEGOLD, J. [1984] *On the number of normal subgroups of an uncountable soluble group* (**J** 0008) Arch Math (Basel) 42*289-295
⋄ C55 C60 E35 E50 E75 ⋄ REV MR 85k:20112 Zbl 528#20001 • ID 47165

BURRIS, S. [1972] *Models in equational theories of unary algebras* (**J** 0004) Algeb Universalis 1*386-392
⋄ B25 C05 C13 C35 ⋄ REV MR 45#3287 Zbl 238#08003 • ID 27910

BURRIS, S. [1973] *Scott sentences and a problem of Vaught for mono-unary algebras* (**J** 0027) Fund Math 80*111-115
⋄ C05 C15 C75 ⋄ REV MR 48#5817 Zbl 284#02026
• ID 01792

BURRIS, S. [1975] *An existence theorem for model companions* (**P** 1679) Lattice Th;1975 Ulm 33-37
⋄ C25 C35 ⋄ REV MR 55#2557 Zbl 365#02040
• ID 30657

BURRIS, S. [1975] *Boolean powers* (**J** 0004) Algeb Universalis 5*341-360
⋄ B25 C05 C30 C90 ⋄ REV MR 56#5393 Zbl 328#08003 • ID 23872

BURRIS, S. & SANKAPPANAVAR, H.P. [1975] *Lattice-theoretic decision problems in universal algebra* (J 0004) Algeb Universalis 5*163-177
⋄ B25 C05 D35 ⋄ REV MR 52#13359 Zbl 322#02045
• ID 21812

BURRIS, S. [1976] *Subdirect representations in axiomatic classes* (S 0019) Colloq Math (Warsaw) 34*191-196
⋄ C05 C52 ⋄ REV Zbl 326#08005 • ID 30654

BURRIS, S. [1977] *An example concerning definable principal congruences* (J 0004) Algeb Universalis 7*403-404
⋄ C05 C40 ⋄ REV MR 56#216 Zbl 364#08005
• ID 26605

BURRIS, S. [1978] *Bounded boolean powers and* $\equiv_n$ (J 0004) Algeb Universalis 8*137-138
⋄ C30 G05 ⋄ REV MR 56#8467 Zbl 369#02028
• ID 30655

BURRIS, S. & JEFFERS, E. [1978] *On the simplicity and subdirect irreducibility of boolean ultrapowers* (S 0019) Colloq Math (Warsaw) 39*215-218
⋄ C05 C20 C50 ⋄ REV MR 80d:03036 Zbl 397#03014
• ID 52680

BURRIS, S. [1978] *Rigid boolean powers* (J 0004) Algeb Universalis 8*264-265
⋄ C05 C30 C50 G05 ⋄ REV MR 57#3042
Zbl 382#06014 • ID 30656

BURRIS, S. & LAWRENCE, J. [1979] *Definable principle congruences in varieties of groups and rings* (J 0004) Algeb Universalis 9*152-164 • ERR/ADD ibid 13*264-267
⋄ C05 C40 ⋄ REV MR 80c:08004 Zbl 407#08006
• ID 56227

BURRIS, S. & WERNER, H. [1979] *Sheaf constructions and their elementary properties* (J 0064) Trans Amer Math Soc 248*269-309
⋄ B25 C05 C20 C25 C30 C60 C90 ⋄ REV
MR 82d:03049 Zbl 411#03022 • ID 52876

BURRIS, S. & WERNER, H. [1980] *Remarks on boolean products* (J 0004) Algeb Universalis 10*333-344
⋄ C05 C30 C52 G05 ⋄ REV MR 81f:08009
Zbl 454#08007 • ID 69284

BURRIS, S. & SANKAPPANAVAR, H.P. [1981] *A course in universal algebra* (X 0811) Springer: Heidelberg & New York xvi + 276pp
⋄ C05 C98 ⋄ REV MR 83k:08001 Zbl 478#08001
• ID 69286

BURRIS, S. & MCKENZIE, R. [1981] *Decidability and boolean representations* (S 0167) Mem Amer Math Soc 32/246*viii + 106pp
⋄ B25 C05 C30 C60 ⋄ REV MR 83j:03024
Zbl 483#03019 • ID 90164

BURRIS, S. & MCKENZIE, R. [1981] *Decidable varieties with modular congruence lattices* (J 0589) Bull Amer Math Soc (NS) 4*350-352
⋄ B25 C05 ⋄ REV MR 82m:08006 Zbl 468#08008
• ID 55113

BURRIS, S. [1982] *Remarks on reducts of varieties* (P 3237) Colloq Universal Algeb;1977 Esztergom 161-168
⋄ C05 ⋄ REV MR 83f:08016 Zbl 503#08010 • ID 36663

BURRIS, S. [1982] *The first order theory of Boolean algebras with a distinguished group of automorphisms* (J 0004) Algeb Universalis 15*156-161
⋄ B25 C05 C60 G05 ⋄ REV MR 84f:03008
Zbl 465#08004 • ID 38205

BURRIS, S. & LAWRENCE, J. [1982] *Two undecidability results using modified boolean powers* (J 0017) Canad J Math 34*500-505
⋄ C05 C60 D35 G05 ⋄ REV MR 83k:03051
Zbl 499#03029 • ID 35402

BURRIS, S. [1983] *Boolean constructions* (P 3841) Universal Algeb & Lattice Th (4);1982 Puebla 67-90
⋄ A10 B25 C05 C30 C98 D35 G05 ⋄ REV
MR 85d:08010 Zbl 517#08001 • ID 36689

BURRIS, S. [1984] *Model companions for finitely generated universal Horn classes* (J 0036) J Symb Logic 49*68-74
⋄ B25 C05 C10 C35 D20 ⋄ REV MR 85g:03052
• ID 42428

BURRIS, S. [1985] *A simple proof of the hereditary undecidability of the theory of lattice ordered abelian groups* (J 0004) Algeb Universalis 20*400-401
⋄ C05 C60 D35 G10 ⋄ REV Zbl 575#06017 • ID 48302

BURRIS, S. see Vol. V for further entries

BUSZKOWSKI, W. [1979] *Undecidability of the theory of lattice-orderable groups* (J 2718) Fct Approximatio, Comment Math, Poznan 7*23-28
⋄ C60 D35 ⋄ REV MR 80m:03078 Zbl 408#03010
• ID 56249

BUSZKOWSKI, W. [1982] *Some decision problems in the theory of syntactic categories* (J 0068) Z Math Logik Grundlagen Math 28*539-548
⋄ B25 D05 ⋄ REV MR 84m:03014 Zbl 499#03010
• ID 35719

BUSZKOWSKI, W. see Vol. IV, V, VI for further entries

BUTLER, J.W. [1960] *On complete and independent sets of operations in finite algebras* (J 0048) Pac J Math 10*1169-1179
⋄ C05 C13 ⋄ REV MR 27#3543 JSL 30.246 • ID 01800

BUTLER, J.W. see Vol. II for further entries

BUTRICK, R. [1977] *A deduction rule for* $VBTO()_{i=1}^{n}$ (J 0047) Notre Dame J Formal Log 18*510-512
⋄ C80 ⋄ REV MR 58#16121 Zbl 231#02018 • ID 60801

BUTRICK, R. see Vol. I, V, VI for further entries

BUTTON, R.W. [1979] *When do two topologies have the same monads?* (J 0254) Gen Topology Appl 10*7-11
⋄ C20 H05 ⋄ REV MR 81g:54070 Zbl 408#54033
• ID 80839

BUTTON, R.W. see Vol. I for further entries

BUYS, A. & HEIDEMA, J. [1980] *Model-theoretical aspects of radicals and subdirect decomposability* (P 3853) Algeb Symp (2);1980 Pretoria 21-25
⋄ C05 C60 ⋄ REV Zbl 505#13001 • ID 38206

BUYS, A. & HEIDEMA, J. [1982] *Radicals and subdirect products in models* (J 0394) Commun Algeb 10*1315-1330
⋄ C05 C52 ⋄ REV MR 84k:03094 Zbl 504#03017
• ID 36111

CADEVALL SOLER, M. [1976] *Decidability of the logic of monadic predicates by the method of rejection (Spanish)* (J 0162) Teorema (Valencia) 6*455-484
⋄ B25 ⋄ REV MR 55#12498 • ID 71502

CAICEDO, X. [1977] *A back-and-forth characterization of elementary equivalence in stationary logic* (1111) Preprints, Manuscr., Techn. Reports etc. TR77-46*22pp
⋄ C80 ⋄ REM University of Maryland • ID 31668

CAICEDO, X. [1980] *Back-and-forth systems for arbitrary quantifiers* (P 2958) Latin Amer Symp Math Log (4);1978 Santiago 83-102
⋄ C55 C75 C80 ⋄ REV MR 83j:03061 Zbl 434#03025 • ID 55729

CAICEDO, X. [1980] *The subdirect decomposition theorem for classes of structures closed under direct limits* (J 3194) J Austral Math Soc, Ser A 30*171-179
⋄ C05 C52 ⋄ REV MR 82c:08001 Zbl 462#08004 • ID 31669

CAICEDO, X. [1981] *Independent sets of axioms in $L_{\kappa\alpha}$* (J 0018) Canad Math Bull 24*219-223
⋄ C40 C75 ⋄ REV MR 84d:03043 Zbl 457#03035 • ID 31673

CAICEDO, X. [1981] *On extensions of $L_{\omega\omega}(Q_1)$* (J 0047) Notre Dame J Formal Log 22*85-93
⋄ C55 C75 C80 ⋄ REV MR 82b:03071 Zbl 438#03040 • ID 56626

CAICEDO, X. [1985] *Failure of interpolation for quantifiers of monadic type* (P 2160) Latin Amer Symp Math Log (6);1983 Caracas 1-12
⋄ C40 C75 C80 C95 ⋄ REV Zbl 571#03015 • ID 41803

CAICEDO, X. see Vol. I, IV for further entries

CALAIS, J.-P. [1967] *Relation et multirelation pseudohomogene* (J 2313) C R Acad Sci, Paris, Ser A-B 265*A2-A4
⋄ C07 C50 ⋄ REV MR 38#1048 Zbl 165.311 • ID 01813

CALAIS, J.-P. [1969] *Isomorphismes locaux generalises* (J 2313) C R Acad Sci, Paris, Ser A-B 268*A761-A764
⋄ C07 ⋄ REV MR 40#7096 Zbl 179.14 • ID 01815

CALAIS, J.-P. [1969] *La methode de Fraisse dans les langages infinis* (J 2313) C R Acad Sci, Paris, Ser A-B 268*A785-A788,A845-A848
⋄ C75 ⋄ REV Zbl 179.14 • ID 01814

CALAIS, J.-P. [1972] *Partial isomorphisms and infinitary languages* (J 0068) Z Math Logik Grundlagen Math 18*435-456
⋄ C07 C65 C75 ⋄ REV MR 47#8240 Zbl 257#02014 • ID 01816

CALL, R.L. [1972] *The Goedel-Herbrand theorems* (J 0047) Notre Dame J Formal Log 13*131-134
⋄ B10 C07 ⋄ REV MR 46#5097 Zbl 227#02014 • ID 71509

CALL, R.L. [1982] *Systems with parity quantifiers* (J 0068) Z Math Logik Grundlagen Math 28*297-309
⋄ C13 C80 ⋄ REV MR 84c:03070 Zbl 503#03004 • ID 34011

CALL, R.L. see Vol. I, II, VI for further entries

CAMERON, P.J. [1978] *Orbits of permutation groups on unordered sets* (J 3172) J London Math Soc, Ser 2 17*410-414
⋄ C30 C60 E05 ⋄ REV MR 58#11136 Zbl 385#20002
• REM Part I. Part II 1981 • ID 90363

CAMERON, P.J. [1981] *Orbits of permutation groups on unordered sets II* (J 3172) J London Math Soc, Ser 2 23*249-264
⋄ C30 C60 E05 ⋄ REV MR 82k:20005 Zbl 498#20003
• REM Part I 1978. Parts III,IV 1983 • ID 49857

CAMERON, P.J. [1982] *Orbits and enumeration* (P 3856) Combin Th;1982 Rauischholzhausen 86-99
⋄ C30 C60 ⋄ REV MR 85b:05010 Zbl 497#05006
• ID 36632

CAMERON, P.J. [1983] *Orbits of permutation groups on unordered sets. III: Imprimitive groups. IV: Homogeneity and transitivity* (J 3172) J London Math Soc, Ser 2 27*229-237,238-247
⋄ C30 C60 E05 ⋄ REV MR 84f:20076 Zbl 519#20004
• REM Part II 1981 • ID 48376

CAMERON, P.J. [1984] *Aspects of random graph* (P 3085) Graph Th & Combin (Erdoes);1983 Cambridge 65-79
⋄ C07 C60 C65 ⋄ REV Zbl 561#05047 • ID 49813

CAMERON, P.J. see Vol. V for further entries

CAMPBELL, P.J. [1971] *Suslin logic* (0000) Diss., Habil. etc
⋄ C75 D55 E15 G05 G25 ⋄ REM Ph.D.thesis, Cornell University • ID 16762

CAMPBELL, P.J. see Vol. I, V for further entries

CANTINI, A. [1980] *A note on three-valued logic and Tarski theorem on truth definitions* (J 0063) Studia Logica 39*405-414
⋄ B50 C40 F30 ⋄ REV MR 83i:03019 Zbl 457#03024
• ID 54349

CANTINI, A. [1983] *Σ-models of subtheories of KPI (Italian)* (P 3829) Atti Incontri Log Mat (1);1982 Siena 201-204
⋄ C62 C70 E30 ⋄ REV MR 84k:03006 • ID 44727

CANTINI, A. [1983] *A note on the theory of admissible sets with ε-induction restricted to formulas with one quantifier and related systems (Italian summary)* (J 2100) Boll Unione Mat Ital, VI Ser, B 2*721-737
⋄ C62 E30 E45 E70 F10 F15 F35 ⋄ REV Zbl 558#03027 • ID 41846

CANTINI, A. see Vol. I, IV, V, VI for further entries

CANTONE, D. & FERRO, A. & SCHWARTZ, J.T. [1985] *Decision procedures for elementary sublanguages of set theory VI: Multi-level syllogistic extended by the powerset operator* (J 0155) Commun Pure Appl Math 38*549-571
⋄ B25 E30 ⋄ REM Part IV to appear in J0155; part V in J0119 Part III 1984 by Breban,M. & Ferro,A.. Common author of all parts: Ferro,A. • ID 48303

CANTOR, DAVID G. & ROQUETTE, P. [1984] *On diophantine equations over the ring of all algebraic integers* (J 0401) J Number Th 18*1-26
⋄ B25 C60 D35 ⋄ REV MR 85j:11036 Zbl 538#12014
• ID 46787

CANTOR, DAVID G. see Vol. IV for further entries

CANTY, J.T. [1963] *Completeness of Copi's method of deduction* (J 0047) Notre Dame J Formal Log 4*142-144
⋄ B05 C07 ⋄ REV MR 29#3361 Zbl 154.4 JSL 30.366
• ID 01832

CANTY, J.T. & KUENG, G. [1970] *Substitutional quantification and Lesniewskian quantifiers* (J 0105) Theoria (Lund) 36*165-182
⋄ A05 B60 C80 E70 ⋄ REV Zbl 213.14 • ID 27980

CANTY, J.T. see Vol. II for further entries

CARLSON, L. [1982] *Plural quantifiers and informational independence* (P 3800) Intens Log: Th & Appl;1979 Moskva 163-174
⋄ A05 C80 ⋄ REV MR 84h:03091 • ID 34340

CARLSTROM, I.F. [1975] *Truth and entailment for a vague quantifier* (J 0154) Synthese 30*461-495
⋄ B52 C80 ⋄ REV Zbl 309 # 02023 • ID 60828

CARPINTERO ORGANERO, P. [1971] *Numero de tipos diferentes de algebras de Boole de cardinal m finito que poseen 2^m ideales primos* (J 0236) Rev Mat Hisp-Amer, Ser 4 31*93-97
⋄ C52 G05 ⋄ REV MR 46 # 5199 Zbl 252 # 06010
• ID 02236

CARPINTERO ORGANERO, P. [1971] *Numero de tipos diferentes de algebras de Boole de cardinal m infinito (English summary)* (J 0296) Acta Salamantica Cienc 38*54pp • ERR/ADD [1972] (J 0236) Rev Mat Hisp-Amer, Ser 4 32*239-240
⋄ C15 G05 ⋄ REV MR 48 # 5942 MR 49 # 4893
Zbl 222 # 06007 Zbl 252 # 06009 • ID 16767

CARPINTERO ORGANERO, P. see Vol. V for further entries

CARSON, A.B. [1973] *The model-completion of the theory of commutative regular rings* (J 0032) J Algeb 27*136-146
⋄ C25 C35 C60 ⋄ REV MR 52 # 7893 Zbl 279 # 02037
• ID 01871

CARSON, A.B. [1974] *Algebraically closed regular rings* (J 0017) Canad J Math 26*1036-1049
⋄ C25 C60 ⋄ REV MR 50 # 13124 Zbl 285 # 13003
• ID 29975

CARSON, A.B. [1978] *Coherent polynomial rings over regular rings of finite index* (J 0048) Pac J Math 74*327-332
⋄ C60 ⋄ REV MR 57 # 12207 Zbl 349 # 16005 • ID 71551

CARSTENGERDES, W. [1971] *Mehrsortige logische Systeme mit unendlich langen Formeln I, II* (J 0009) Arch Math Logik Grundlagenforsch 14*38-53,108-126
⋄ C40 C75 F05 ⋄ REV MR 44 # 2583 MR 48 # 5818
Zbl 235 # 02023 • ID 01873

CARSTENS, H.G. [1975] *Reducing hyperarithmetic sequences* (J 0027) Fund Math 89*5-11
⋄ C57 D55 ⋄ REV MR 52 # 2855 Zbl 334 # 02023
• ID 03968

CARSTENS, H.G. [1977] *The theorem of Matijasevic is provable in Peano's arithmetic by finitely many axioms* (J 0079) Logique & Anal, NS 20*116-121
⋄ C52 D25 D35 F30 H15 ⋄ REV MR 58 # 27400
Zbl 382 # 03037 • ID 51961

CARSTENS, H.G. see Vol. IV, V for further entries

CASARI, E. [1971] *Regular models for second order set theory* (P 0669) Conv Teor Modelli & Geom;1969/70 Roma 177-188
⋄ C62 C85 E55 E70 ⋄ REV MR 43 # 34 Zbl 226 # 02049
• ID 01884

CASARI, E. [1981] *Positively omitting types* (C 3515) Ital Studies in Phil of Sci 3-11
⋄ B10 C07 ⋄ REV MR 82i:03017 Zbl 449 # 03008
• ID 56675

CASARI, E. see Vol. I, II, V for further entries

CASCANTE DAVILA, J.M. & PLA I CARRERA, J. [1977] *A precise proof of criterion CST.6 in Chapter IV of Nicolas Bourbaki's Theorie des ensembles (Spanish)* (P 3547) Jorn Mat Hispano-Lusitanas (1);1973 Madrid 291-294
⋄ C07 E30 ⋄ REV MR 82h:04001 • ID 71563

CAVALLI, A.R. & FARINAS DEL CERRO, L. [1984] *A decision method for linear temporal logic* (P 2633) Autom Deduct (7);1984 Napa 113-127
⋄ B25 B45 ⋄ REV MR 86g:03050 Zbl 547 # 03008
• ID 43192

CAVALLI, A.R. see Vol. II for further entries

CEGIELSKI, P. [1980] *La theorie elementaire de la multiplication (English summary)* (J 2313) C R Acad Sci, Paris, Ser A-B 290*A935-A938
⋄ B25 C10 C80 F30 ⋄ REV MR 82d:03020
Zbl 435 # 03009 • ID 55770

CEGIELSKI, P. [1981] *Theorie elementaire de la multiplication des entiers naturels* (P 3404) Model Th & Arithm;1979/80 Paris 44-89
⋄ B25 C10 C80 F30 ⋄ REV MR 83e:03094
Zbl 478 # 03011 • ID 55626

CEGIELSKI, P. & MCALOON, K. & WILMERS, G.M. [1982] *Modeles recursivement satures de l'addition et de la multiplication des entiers naturels (English summary)* (P 3623) Logic Colloq;1980 Prague 57-68
⋄ C50 C57 C62 F30 H15 ⋄ REV MR 84i:03072
Zbl 527 # 03043 • ID 34553

CEGIELSKI, P. [1984] *La theorie elementaire de la divisibilite est finiment axiomatisable (English summary)* (J 3364) C R Acad Sci, Paris, Ser 1 299*367-369
⋄ B25 B28 F30 ⋄ REV MR 85j:03042 • ID 46773

CEGIELSKI, P. see Vol. I for further entries

CELLUCCI, C. [1964] *Categorie ricorsive* (J 4408) Boll Unione Mat Ital, III Ser 19*300-305
⋄ C57 D20 G30 ⋄ REV MR 30 # 1928 Zbl 149.8
• ID 02198

CELLUCCI, C. [1977] *Proprieta di coerenza e completezza in $L_{\omega_1\omega}$* (J 0319) Matematiche (Sem Mat Catania) 32*153-174
⋄ C75 ⋄ REV MR 81m:03047 Zbl 438 # 03039 • ID 55951

CELLUCCI, C. see Vol. IV, V, VI for further entries

CHADWICK, J.J.M. & WICKSTEAD, A.W. [1977] *A quotient of ultrapowers of Banach spaces and semi-Fredholm operators* (J 0161) Bull London Math Soc 9*321-325
⋄ C20 C65 H05 ⋄ REV MR 57 # 1080 Zbl 388 # 46028
• ID 80867

CHADWICK, J.J.M. see Vol. I for further entries

CHANG, C.C. [1954] *Some general theorems on direct products and their applications in the theory of models* (J 0028) Indag Math 16*592-598
⋄ C05 C75 ⋄ REV MR 16.555 Zbl 58.248 JSL 21.406
• ID 01931

CHANG, C.C. [1957] *Unions of chains of models and direct products of models* (**P** 1675) Summer Inst Symb Log;1957 Ithaca 141-143
⋄ C30 C52 ⋄ ID 29333

CHANG, C.C. & MOREL, A.C. [1958] *On closure under direct product* (**J** 0036) J Symb Logic 23∗149-154
⋄ C30 C40 ⋄ REV MR 21#3359 Zbl 92.6 JSL 27.234 • ID 01936

CHANG, C.C. [1959] *On unions of chains of models* (**J** 0053) Proc Amer Math Soc 10∗120-127
⋄ C40 C52 ⋄ REV MR 21#2576 Zbl 96.242 JSL 25.487 • ID 01939

CHANG, C.C. & KEISLER, H.J. [1962] *Applications of ultraproducts of pairs of cardinals to the theory of models* (**J** 0048) Pac J Math 12∗835-845
⋄ C20 C55 E05 E50 ⋄ REV MR 26#3606 Zbl 109.8 JSL 36.338 • ID 01950

CHANG, C.C. & KEISLER, H.J. [1962] *Model theories with truth values in a uniform space* (**J** 0015) Bull Amer Math Soc 68∗107-109
⋄ B50 C20 C50 C90 ⋄ REV MR 25#12 Zbl 104.241 • ID 01946

CHANG, C.C. & JONSSON, B. & TARSKI, A. [1964] *Refinement properties for relational structures* (**J** 0027) Fund Math 55∗249-281
⋄ C05 C30 ⋄ REV MR 30#3029 Zbl 171.258 • ID 01962

CHANG, C.C. [1964] *Some new results in definability* (**J** 0015) Bull Amer Math Soc 70∗808-813
⋄ C40 ⋄ REV MR 30#1053 • ID 01965

CHANG, C.C. [1965] *A note on the two cardinal problem* (**J** 0053) Proc Amer Math Soc 16∗1148-1155
⋄ C55 E50 ⋄ REV MR 33#1238 Zbl 137.8 JSL 36.339 • ID 01969

CHANG, C.C. [1965] *A simple proof of the Rabin-Keisler theorem* (**J** 0015) Bull Amer Math Soc 71∗642-643
⋄ C30 E05 ⋄ REV MR 31#2154 Zbl 158.265 JSL 32.277 • ID 01970

CHANG, C.C. & KEISLER, H.J. [1965] *Continuous model theory* (**P** 0614) Th Models;1963 Berkeley 25-38
⋄ C90 ⋄ REV MR 38#36 Zbl 199.8 • ID 16213

CHANG, C.C. & KEISLER, H.J. [1966] *Continuous model theory* (**X** 0857) Princeton Univ Pr: Princeton xi+165pp
• TRANSL [1971] (**X** 0885) Mir: Moskva 184pp
⋄ B50 C05 C20 C50 C90 C98 ⋄ REV MR 38#36 Zbl 149.4 • ID 02410

CHANG, C.C. [1967] *Cardinal factorization of finite relational structures* (**J** 0027) Fund Math 60∗251-269
⋄ C05 C13 C30 ⋄ REV MR 37#126 Zbl 183.10 • ID 01973

CHANG, C.C. [1967] *Omitting types of prenex formulas* (**J** 0036) J Symb Logic 32∗61-74
⋄ C07 C52 ⋄ REV MR 35#4094 Zbl 161.5 JSL 39.182 • ID 01972

CHANG, C.C. [1967] *Ultraproducts and other methods of constructing models* (**P** 0691) Sets, Models & Recursion Th;1965 Leicester 85-121
⋄ C20 C30 ⋄ REV MR 36#1311 Zbl 157.27 • ID 02677

CHANG, C.C. [1968] *Infinitary properties of models generated from indiscernibles* (**P** 0627) Int Congr Log, Meth & Phil of Sci (3,Proc);1967 Amsterdam 9-21
⋄ C30 C75 ⋄ REV MR 40#4091 Zbl 205.307 • ID 01975

CHANG, C.C. [1968] *Some remarks on the model theory of infinitary languages* (**P** 0637) Syntax & Semant Infinitary Lang;1967 Los Angeles 36-63
⋄ C07 C30 C55 C75 ⋄ REV Zbl 175.268 • ID 01929

CHANG, C.C. [1971] *Sets constructible using $L_{\kappa\kappa}$* (**P** 0693) Axiomatic Set Th;1967 Los Angeles 1∗1-8
⋄ C75 E25 E45 E50 E55 ⋄ REV MR 43#6077 Zbl 218#02061 • ID 02676

CHANG, C.C. [1971] *Two interpolation theorems* (**P** 0669) Conv Teor Modelli & Geom;1969/70 Roma 5-19
⋄ C40 C50 C75 ⋄ REV MR 44#53 Zbl 222#02008 • ID 01976

CHANG, C.C. [1973] *Modal model theory* (**P** 0713) Cambridge Summer School Math Log;1971 Cambridge GB 599-617
⋄ B45 C90 ⋄ REV MR 50#9531 Zbl 276#02012 • ID 03978

CHANG, C.C. & KEISLER, H.J. [1973] *Model theory* (**X** 0809) North Holland: Amsterdam xii+550pp
• TRANSL [1977] (**X** 0885) Mir: Moskva 614pp
⋄ C98 ⋄ REV MR 53#12927 MR 57#9519 MR 58#27177 Zbl 276#02032 JSL 41.697 • ID 01981

CHANG, C.C. [1973] *What's so special about saturated models?* (**C** 0654) Stud in Model Th 59-95
⋄ C40 C50 C55 ⋄ REV MR 49#2356 • ID 03979

CHANG, C.C. [1974] *Model theory 1945-1971* (**P** 0610) Tarski Symp;1971 Berkeley 173-186
⋄ A10 C98 ⋄ REV MR 57#12200 Zbl 306#02006 • ID 01983

CHANG, C.C. [1974] see ADDISON, J.W.

CHANG, C.C. see Vol. I, II, IV, V for further entries

CHAPIN JR., E.W. [1971] *The strong decidability of cut-logics I: Partial propositional calculi. II: Generalizations* (**J** 0047) Notre Dame J Formal Log 12∗322-328,429-434
⋄ B20 B25 ⋄ REV MR 45#6571 MR 45#6572 Zbl 188.28 Zbl 214.18 • ID 01987

CHAPIN JR., E.W. see Vol. I, V for further entries

CHARNOW, A. [1970] *The automorphisms of an algebraically closed field* (**J** 0018) Canad Math Bull 13∗95-97
⋄ C60 ⋄ REV MR 41#5337 • ID 15245

CHARRETTON, C. & RICHARD, D. [1973] *Elements d'une theorie non standard des groupes topologiques* (**J** 0056) Publ Dep Math, Lyon 9/2∗1-29
⋄ C60 H05 ⋄ REV MR 50#2385 Zbl 281#22004 • ID 03982

CHARRETTON, C. [1978] *Type d'ordre des ordinaux de modeles non denombrables de la theorie des ensembles* (**J** 0056) Publ Dep Math, Lyon 15/3∗37-52
⋄ C20 C50 C62 C70 E10 E50 H20 ⋄ REV MR 81d:03036 Zbl 429#03031 • ID 53862

CHARRETTON, C. & POUZET, M. [1980] *Les chaines dans les modeles d'Ehrenfeucht-Mostowski (English summary)* (**J** 2313) C R Acad Sci, Paris, Ser A-B 290∗A715-A717
⋄ C30 C55 ⋄ REV MR 81d:03034 Zbl 434#03024 • ID 55728

CHARRETTON, C. & POUZET, M. [1983] *Chains in Ehrenfeucht-Mostowski models* (J 0027) Fund Math 118*109-122
⋄ C30 ⋄ REV MR 85j:03045   Zbl 545 # 03012 • ID 41226

CHARRETTON, C. & POUZET, M. [1983] *Comparaison des structures engendrees par des chaines* (P 1601) Easter Conf on Model Th (1);1983 Diedrichshagen 17-27
⋄ C30 C65 G05 ⋄ REV MR 84i:03008   Zbl 533 # 03019 • ID 36543

CHARRETTON, C. see Vol. I for further entries

CHATZIDAKIS, Z. [1978] see BOUSCAREN, E.

CHATZIDAKIS, Z. [1981] *Forking et rangs locaux selon Shelah* (C 2619) Groupe Etude Th Stables (2) 1978/79 2/10*12pp
⋄ C45 ⋄ REV MR 82g:03052   Zbl 464 # 03012 • ID 71657

CHATZIDAKIS, Z. [1981] *La representation en termes de faisceaux des modeles de la theorie elementaire de la multiplication des entiers naturels* (P 3404) Model Th & Arithm;1979/80 Paris 90-110
⋄ C10 C30 C62 C90 ⋄ REV MR 83h:03049   Zbl 496 # 03018 • ID 36063

CHAUBARD, A. [1970] *Definissabilite explicite* (J 0247) Bull Sci Math, Ser 2 94*33-39
⋄ C40 ⋄ REV MR 43 # 1824   Zbl 216.293 • ID 02412

CHAUBARD, A. [1970] *Ultraproduit de suites generiques et application* (J 2313) C R Acad Sci, Paris, Ser A-B 270*A497-A500
⋄ C20 C25 ⋄ REV MR 41 # 3256   Zbl 231 # 02068 • ID 01993

CHAUBARD, A. [1971] *Puissances ultrafiltrees* (J 2313) C R Acad Sci, Paris, Ser A-B 273*A1190-A1193
⋄ C20 ⋄ REV MR 44 # 6467   Zbl 231 # 02069 • ID 01994

CHAUBARD, A. [1972] *Sur les puissances ultrafiltrees* (J 2313) C R Acad Sci, Paris, Ser A-B 274*A1-A4
⋄ C20 ⋄ REV MR 44 # 6468   Zbl 231 # 02070 • ID 01995

CHAUVIN, A. [1961] *Deux modeles verifiant certains axiomes de la theorie des ensembles de Goedel, et construits dans la theorie des ensembles arithmetiques de Kleene. Construction des modeles* (J 0109) C R Acad Sci, Paris 253*1394-1396
⋄ C62 D55 E30 E70 ⋄ REV MR 24 # A1218   Zbl 112.11 • ID 01999

CHAUVIN, A. [1961] *Deux modeles verifiant certains axiomes de la theorie des ensembles de Goedel, et construits dans la theorie des ensembles arithmetiques de Kleene. Validite des axiomes* (J 0109) C R Acad Sci, Paris 253*1519-1521
⋄ C62 D55 E30 E70 ⋄ REV MR 24 # A1835   Zbl 112.11 • ID 25473

CHAUVIN, A. see Vol. I, II, IV, V, VI for further entries

CHEN, JIYUAN [1984] *The satisfiability problem for simple boolean expressions belongs to P (Chinese) (English summary)* (J 2521) Beijing Shifan Daxue Xuebao, Ziran Kexue 1984/1*14-21
⋄ B05 B25 D15 ⋄ REV MR 85j:03059 • ID 44191

CHEPURNOV, B.A. [1973] *Some remarks on quasiuniversal classes (Russian)* (J 0226) Uch Zap Ped Inst, Ivanovo 125*175-190
⋄ C30 C52 ⋄ REV MR 57 # 209 • ID 80894

CHEPURNOV, B.A. & KOGALOVSKIJ, S.R. [1976] *Some criteria of heredity and locality for formulas of higher order (Russian)* (C 3271) Issl Teor Mnozh & Neklass Logik 127-156
⋄ B15 C52 C85 ⋄ REV MR 58 # 21423   Zbl 412 # 03018 • ID 52949

CHEREMISIN, A.I. [1969] *Structural characterization of certain classes of models (Russian)* (J 0226) Uch Zap Ped Inst, Ivanovo 61*271-286
⋄ C05 C30 C52 ⋄ REV MR 47 # 1721 • ID 02199

CHERLIN, G.L. [1971] *A new approach to the theory of infinitely generic structures* (0000) Diss., Habil. etc
⋄ C25 ⋄ REM Doctorial diss., Yale University • ID 16772

CHERLIN, G.L. [1972] *The model-companion of a class of structures* (J 0036) J Symb Logic 37*546-556
⋄ C25 ⋄ REV MR 52 # 73   Zbl 259 # 02043   JSL 48.496 • ID 02077

CHERLIN, G.L. & HIRSCHFELD, J. [1972] *Ultrafilters and ultraproducts in non-standard analysis* (P 0732) Contrib to Non-Standard Anal;1970 Oberwolfach 261-279
⋄ C20 E05 H05 ⋄ REV MR 58 # 5191   Zbl 262 # 02055   JSL 40.634 • ID 14684

CHERLIN, G.L. [1973] *Algebraically closed commutative rings* (J 0036) J Symb Logic 38*493-499
⋄ C25 C60 ⋄ REV MR 49 # 36   Zbl 274 # 02028 • ID 02078

CHERLIN, G.L. [1975] *Ideals in some nonstandard Dedekind rings* (J 0079) Logique & Anal, NS 18*379-406
• REPR [1977] (P 1625) Six Days of Model Th;1975 Louvain-la-Neuve 147-174
⋄ C60 H15 H20 ⋄ REV MR 58 # 16263   Zbl 362 # 02053   Zbl 396 # 03048 • ID 50775

CHERLIN, G.L. [1975] *Ideals of integers in nonstandard number fields* (C 0782) Model Th & Algeb (A. Robinson) 60-90
⋄ C60 H15 H20 ⋄ REV MR 55 # 7768   Zbl 323 # 02061 • ID 31377

CHERLIN, G.L. [1975] *Second order forcing, algebraically closed structures, and large cardinals* (J 0027) Fund Math 87*141-160
⋄ C25 C85 E55 ⋄ REV MR 51 # 12515   Zbl 308 # 02063 • ID 02079

CHERLIN, G.L. [1976] *Amalgamation bases for commutative rings without nilpotent elements* (J 0029) Israel J Math 25*87-96
⋄ C60 ⋄ REV MR 56 # 11779   Zbl 352 # 02041 • ID 26076

CHERLIN, G.L. & REINEKE, J. [1976] *Categoricity and stability of commutative rings* (J 0007) Ann Math Logic 9*367-399
⋄ C45 C60 ⋄ REV MR 58 ## 208   Zbl 326 # 02041 • ID 17996

CHERLIN, G.L. [1976] *Model theoretic algebra* (J 0036) J Symb Logic 41*537-545
⋄ C25 C60 ⋄ REV MR 54 # 90   Zbl 338 # 02029 • ID 14842

CHERLIN, G.L. [1976] *Model theoretic algebra. Selected topics* (S 3301) Lect Notes Math 521*iv+234pp
⋄ C25 C60 C98 ⋄ REV MR 58 # 27455   Zbl 332 # 02056   JSL 47.222 • ID 60940

CHERLIN, G.L. & ROSENSTEIN, J.G. [1978] *On $\aleph_0$-categorical abelian-by-finite groups* (J 0032) J Algeb 53*188-226
⋄ C35 C60 ⋄ REV MR 58 # 11146   Zbl 381 # 20003 • ID 30795

CHERLIN, G.L. [1978] *Superstable division rings* (P 1897) Logic Colloq;1977 Wroclaw 99-111
  ◇ C45 C60 ◇ REV MR 80g:03034 Zbl 466 # 03011
  • ID 54962

CHERLIN, G.L. [1979] *Groups of small Morley rank* (J 0007) Ann Math Logic 17*1-28
  ◇ C45 C60 ◇ REV MR 81h:03072 Zbl 427 # 20001 JSL 49.317 • ID 53747

CHERLIN, G.L. [1979] *Lindenbaum algebras and model companions* (J 0027) Fund Math 104*213-219
  ◇ C25 C35 ◇ REV MR 81i:03046 Zbl 428 # 03022
  • ID 53781

CHERLIN, G.L. [1979] *Stable algebraic theories* (P 2627) Logic Colloq;1978 Mons 53-74
  ◇ C45 C60 ◇ REV MR 81i:03052 Zbl 444 # 03015
  • ID 56463

CHERLIN, G.L. [1979] see BAUR, W.

CHERLIN, G.L. & DICKMANN, M.A. [1980] *Anneaux reels clos et anneaux de fonctions continues (English summary)* (J 2313) C R Acad Sci, Paris, Ser A-B 290*A1-A4
  ◇ C10 C35 C60 C65 ◇ REV MR 80m:03069 Zbl 423 # 03034 • ID 53546

CHERLIN, G.L. [1980] see BOFFA, M.

CHERLIN, G.L. [1980] *On $\aleph_0$-categorical nilrings* (J 0004) Algeb Universalis 10*27-30
  ◇ C35 C60 ◇ REV MR 81e:03031 Zbl 438 # 13002 • REM Part I. Part II 1980 • ID 55973

CHERLIN, G.L. [1980] *On $\aleph_0$-categorical nilrings. II* (J 0036) J Symb Logic 45*291-301
  ◇ C35 C60 ◇ REV MR 81j:03053 Zbl 447 # 16003 • REM Part I 1980 • ID 56585

CHERLIN, G.L. [1980] *Rings of continuous functions: decision problems* (P 2625) Model Th of Algeb & Arithm;1979 Karpacz 44-91
  ◇ B25 C60 C65 D35 ◇ REV MR 82e:03019 Zbl 454 # 03004 • ID 54217

CHERLIN, G.L. & SHELAH, S. [1980] *Superstable fields and groups* (J 0007) Ann Math Logic 18*227-270
  ◇ C45 C60 ◇ REV MR 82c:03045 Zbl 467 # 03025 Zbl 475 # 03012 JSL 49.317 • ID 55023

CHERLIN, G.L. & DRIES VAN DEN, L. & MACINTYRE, A. [1981] *Decidability and undecidability theorems for PAC-fields* (J 0589) Bull Amer Math Soc (NS) 4*101-104
  ◇ B25 C60 D35 ◇ REV MR 82g:03057 Zbl 466 # 12017
  • ID 54982

CHERLIN, G.L. & DICKMANN, M.A. [1981] *Note on a Nullstellensatz* (P 3404) Model Th & Arithm;1979/80 Paris 111-114
  ◇ C60 ◇ REV MR 83h:12032 Zbl 484 # 13019 • ID 36610

CHERLIN, G.L. [1981] see BERLINE, C.

CHERLIN, G.L. & SCHMITT, P.H. [1981] *Undecidable L^t theories of topological abelian groups* (J 0036) J Symb Logic 46*761-772
  ◇ C60 C90 D35 ◇ REV MR 83e:03047 Zbl 482 # 03016
  • ID 35227

CHERLIN, G.L. & FELGNER, U. [1982] *Quantifier eliminable groups* (P 3623) Logic Colloq;1980 Prague 69-81
  ◇ C10 C13 C60 ◇ REV MR 84d:20007 Zbl 496 # 03016
  • ID 36863

CHERLIN, G.L. & SCHMITT, P.H. [1983] *Locally pure topological abelian groups: Elementary invariants* (J 0073) Ann Pure Appl Logic 24*49-85
  ◇ B25 C60 C65 C80 C90 ◇ REV MR 85a:03050 Zbl 522 # 03023 • ID 33342

CHERLIN, G.L. [1983] see BERLINE, C.

CHERLIN, G.L. & DICKMANN, M.A. [1983] *Real closed rings II: Model theory* (J 0073) Ann Pure Appl Logic 25*213-231
  ◇ B25 C10 C60 ◇ REV Zbl 538 # 03028 • ID 41472

CHERLIN, G.L. & VOLGER, H. [1984] *Convexity properties and algebraic closure operators* (P 2153) Logic Colloq;1983 Aachen 1*113-146
  ◇ C52 ◇ ID 39709

CHERLIN, G.L. [1984] *Decidable theories of pseudo-algebraically closed fields* (P 2153) Logic Colloq;1983 Aachen 1*83-101
  ◇ B25 C60 ◇ ID 41753

CHERLIN, G.L. [1984] *Definability in power series rings of nonzero characteristic* (P 2153) Logic Colloq;1983 Aachen 1*102-112
  ◇ C40 C60 D35 ◇ REV MR 86h:03059 Zbl 574 # 03017
  • ID 41760

CHERLIN, G.L. [1984] *Totally categorial structures* (P 4313) Int Congr Math (II,14);1983 Warsaw 1*301-306
  ◇ C35 C45 C98 ◇ ID 48348

CHERLIN, G.L. [1984] *Undecidability of rational function fields in nonzero characteristic* (P 3710) Logic Colloq;1982 Firenze 85-95
  ◇ C60 C85 D35 ◇ REV MR 86f:03068 Zbl 551 # 03027
  • ID 43903

CHERLIN, G.L. & HARRINGTON, L.A. & LACHLAN, A.H. [1985] *$\aleph_0$-categorical, $\aleph_0$-stable structures* (J 0073) Ann Pure Appl Logic 28*103-135
  ◇ C35 C45 C52 ◇ REV MR 86g:03054 Zbl 566 # 03022
  • ID 39899

CHIKHACHEV, S.A. [1975] *A question of Simmons (Russian)* (J 0092) Sib Mat Zh 16*876-880,887
  • TRANSL [1975] (J 0475) Sib Math J 16*674-677
  ◇ C25 C50 ◇ REV MR 52 # 5405 Zbl 342 # 02037
  • ID 18104

CHIKHACHEV, S.A. [1975] *Categorical universal classes of semigroups (Russian)* (J 0092) Sib Mat Zh 16*149-157,197
  • TRANSL [1975] (J 0475) Sib Math J 16*120-126
  ◇ C05 C35 C52 C60 ◇ REV MR 54 # 7239 Zbl 305 # 20032 • ID 25029

CHIKHACHEV, S.A. [1975] *Generic models (Russian)* (J 0003) Algebra i Logika 14*345-353,370
  • TRANSL [1975] (J 0069) Algeb and Log 14*214-218
  ◇ C25 C35 ◇ REV MR 58 # 27442 Zbl 328 # 02035
  • ID 26007

CHIKHACHEV, S.A. [1975] *Generic models of countable theories (Russian)* (J 0003) Algebra i Logika 14*704-721
  • TRANSL [1975] (J 0069) Algeb and Log 14*421-429
  ◇ C15 C25 ◇ REV MR 56 # 11777 Zbl 352 # 02039
  • ID 26033

CHIKHACHEV, S.A. [1977] $\aleph_1$-categorical commutative rings (Russian) (J 0092) Sib Mat Zh 18∗908-914,958
• TRANSL [1977] (J 0475) Sib Math J 18∗644-649
⋄ C35 C60 ⋄ REV MR 56 # 15401 Zbl 365 # 02046
• ID 71706

CHIKHACHEV, S.A. [1984] An example of a theory without weakly (Σ, Σ)-atomic models (Russian) (J 0003) Algebra i Logika 23∗336-340
• TRANSL [1984] (J 0069) Algeb and Log 23∗233-237
⋄ B20 C50 ⋄ REV MR 86h:03049 • ID 42695

CHIKHACHEV, S.A. [1985] An example of an atomic model which is not algebraically prime (Russian) (J 0092) Sib Mat Zh 26/3∗192-196
• TRANSL [1985] (J 0475) Sib Math J 26∗460-464
⋄ C15 C50 ⋄ REV Zbl 576 # 03020 Zbl 578 # 03017
• ID 46506

CHISTOV, A.L. & GRIGOR'EV, D.YU. [1984] Complexity of quantifier elimination in theory of algebraically closed fields (P 3658) Math Founds of Comput Sci (11);1984 Prague 17-31
⋄ C10 C60 D15 ⋄ REV Zbl 562 # 03015 • ID 44792

CHITI, E. [1980] Modelli saturati per particolari teorie di gruppi abeliani ordinati (English summary) (J 3128) Boll Unione Mat Ital, Suppl 2∗145-155
⋄ C50 C60 ⋄ REV MR 84c:03061 Zbl 448 # 03020
• ID 56620

CHLEBUS, B.S. [1979] On decidability of propositional algorithmic logic (Polish) (English summary) (S 3270) Spraw Inst Inf, Uniw Warsaw 83∗28pp
⋄ B25 B75 D10 D15 ⋄ REV Zbl 454 # 03006 • ID 54219

CHLEBUS, B.S. [1980] Decidability and definability results concerning well-orderings and some extensions of first order logic (J 0068) Z Math Logik Grundlagen Math 26∗529-536
⋄ B15 B25 C40 C65 C80 D35 E07 ⋄ REV MR 82b:03072 Zbl 445 # 03018 • ID 56482

CHLEBUS, B.S. [1982] On the decidability of propositional algorithmic logic (J 0068) Z Math Logik Grundlagen Math 28∗247-261
⋄ B25 B75 ⋄ REV MR 84e:03022 Zbl 502 # 03012
• ID 34361

CHLEBUS, B.S. see Vol. II, IV for further entries

CHOBANOV, I. [1967] A model-axiomatizing scheme (Bulgarian) (English summary) (J 0255) God Fak Mat & Mekh, Univ Sofiya 62∗157-159
⋄ C07 ⋄ REV MR 40 # 7101 Zbl 193.301 • ID 02446

CHOQUET, G. [1968] Construction d'ultrafiltres sur N (J 0247) Bull Sci Math, Ser 2 92∗41-48
⋄ C20 E05 E50 ⋄ REV MR 38 # 2722 Zbl 157.531
• ID 02686

CHOQUET, G. [1968] Deux classes remarquables d'ultrafiltres sur N (J 0247) Bull Sci Math, Ser 2 92∗143-153
⋄ C20 E05 E50 ⋄ REV MR 38 # 5154 Zbl 162.262
• ID 02687

CHOQUET, G. see Vol. IV, V for further entries

CHRISTEN, C. [1976] Spektralproblem und Komplexitaetstheorie (P 3196) Kompl von Entscheid Probl;1973/74 Zuerich 102-126
⋄ B10 C13 D10 D15 D25 D35 ⋄ REV MR 57 # 18232 Zbl 391 # 03021 • ID 52348

CHUAQUI, R.B. [1971] A representation theorem for linearly ordered cardinal algebras (J 0027) Fund Math 71∗85-91
⋄ C05 C65 E10 G25 ⋄ REV MR 44 # 5264
Zbl 227 # 06009 • ID 10843

CHUAQUI, R.B. [1977] A semantical definition of probability (P 1076) Latin Amer Symp Math Log (3);1976 Campinas 135-167
⋄ C90 ⋄ REV MR 58 # 11333 Zbl 394 # 03037 • ID 16602

CHUAQUI, R.B. [1977] see ARRUDA, A.I.

CHUAQUI, R.B. [1980] Foundations of statistical methods using a semantical definition of probability (P 2958) Latin Amer Symp Math Log (4);1978 Santiago 103-120
⋄ B48 C90 ⋄ REV MR 82i:62010 Zbl 469 # 03024
• ID 55151

CHUAQUI, R.B. [1985] see BERTOSSI, L.

CHUAQUI, R.B. see Vol. I, II, V for further entries

CHUDNOVSKY, D.V. [1968] Probability on first order logics (Ukrainian) (Russian and English summaries) (J 0270) Dokl Akad Nauk Ukr SSR, Ser A 1968∗914-916
⋄ B48 C90 ⋄ REV MR 39 # 5338 • ID 02732

CHUDNOVSKY, D.V. [1970] Logical probability and conditional probability on Boolean algebras (Russian) (English summary) (J 0295) Teor Veroyat i Mat Stat (Kiev) 2∗221-225
• TRANSL [1974] (J 3456) Th Probab & Math Stat 2∗225-229
⋄ B48 C90 G05 ⋄ REV MR 44 # 1063 Zbl 222 # 02066
• ID 26322

CHUDNOVSKY, D.V. [1972] Topological properties of products of discrete spaces and set theory (Russian) (J 0023) Dokl Akad Nauk SSSR 204∗298-301
• TRANSL [1972] (J 0062) Sov Math, Dokl 13∗661-665
⋄ C55 C75 E35 E55 E75 ⋄ REV MR 46 # 62
Zbl 265 # 02040 • ID 02580

CHUDNOVSKY, D.V. [1981] see ANTONOVSKIJ, M.YA.

CHUDNOVSKY, D.V. see Vol. I, V for further entries

CHUDNOVSKY, G.V. [1968] Some results in the theory of infinitely long expressions (Russian) (J 0023) Dokl Akad Nauk SSSR 179∗1286-1288
• TRANSL [1968] (J 0062) Sov Math, Dokl 9∗556-559
⋄ C40 C75 ⋄ REV MR 37 # 6167 Zbl 315 # 02016
JSL 37.202 • ID 02575

CHUDNOVSKY, G.V. [1969] Certain properties of inaccessible cardinals in ZF models of set theory (Russian) (P 3546) Teor Avtom (2);1969 Kiev 3-16
⋄ C62 E35 E55 ⋄ REV MR 45 # 8521 • ID 71792

CHUDNOVSKY, G.V. [1970] Problems of the theory of models, related to categoricity (Russian) (J 0003) Algebra i Logika 9∗80-120
• TRANSL [1970] (J 0069) Algeb and Log 9∗50-74
⋄ C35 C45 C50 C52 C55 C75 ⋄ REV MR 44 # 1549
Zbl 231 # 02066 • ID 02578

CHUDNOVSKY, G.V. [1974] Transversals and properties of compactness type (Russian) (J 0023) Dokl Akad Nauk SSSR 216∗748-750
• TRANSL [1974] (J 0062) Sov Math, Dokl 15∗891-894
⋄ C20 C75 E05 E55 ⋄ REV MR 51 # 7874
Zbl 309 # 02072 • ID 17593

CHUDNOVSKY, G.V. [1975] *Combinatorial properties of compact cardinals* (P 0759) Infinite & Finite Sets (Erdoes);1973 Keszthely 1*289-306
⋄ C75 E05 E55 ⋄ REV MR 51 #7873 Zbl 324 #02066
• ID 17594

CHUDNOVSKY, G.V. [1981] see ANTONOVSKIJ, M.YA.

CHUDNOVSKY, G.V. see Vol. IV, V for further entries

CHURCH, A. [1951] *Special cases of the decision problem* (J 0252) Rev Philos Louvain 49*203-221 • ERR/ADD ibid 50*270-272
⋄ B25 ⋄ REV JSL 17.73 • ID 02459

CHURCH, A. [1963] *Logic, arithmetic, and automata* (P 0677) Int Congr Math (II, 9,Proc);1962 Djursholm 23-35
⋄ A10 B25 D05 D98 ⋄ REV MR 31 #65 Zbl 116.336 JSL 29.210 • ID 02403

CHURCH, A. see Vol. I, II, IV, V, VI for further entries

CICHON, J. [1977] *Some remarks on selectors (Russian summary)* (J 0014) Bull Acad Pol Sci, Ser Math Astron Phys 25*611-614
⋄ C75 E05 E40 E45 E55 ⋄ REV MR 57 #9548 Zbl 362 #02076 • ID 27131

CICHON, J. see Vol. V for further entries

CLARE, F. [1975] *Embedding theorems for infinite symmetric groups* (J 0028) Indag Math 37*155-163
⋄ C30 C60 ⋄ REV MR 51 #5786 Zbl 324 #20039
• ID 02241

CLARE, F. [1975] *Operations on elementary classes of groups* (J 0004) Algeb Universalis 5*120-124
⋄ C05 C52 C60 ⋄ REV MR 52 #5413 Zbl 328 #20045
• ID 18105

CLARE, F. [1976] *On elementary equivalence of abelian groups* (S 0019) Colloq Math (Warsaw) 34*199-207
⋄ C60 ⋄ REV MR 53 #649 Zbl 339 #02049 • ID 16689

CLARK, D.M. & KRAUSS, P.H. [1972] *Monomorphic relational systems* (J 0068) Z Math Logik Grundlagen Math 18*229-235
⋄ C05 C30 C52 ⋄ REV MR 46 #8830 Zbl 252 #02052
• ID 02242

CLARK, D.M. & KRAUSS, P.H. [1976] *Para primal algebras* (J 0004) Algeb Universalis 6*165-192
⋄ C05 C13 ⋄ REV MR 54 #10105 Zbl 368 #08004
• ID 25869

CLARK, D.M. & KRAUSS, P.H. [1977] *Relatively homogeneous structures* (P 1075) Logic Colloq;1976 Oxford 87*255-285
⋄ C10 C35 C50 C60 C65 G05 ⋄ REV MR 58 #10416 Zbl 423 #03037 • ID 16624

CLARK, D.M. & KRAUSS, P.H. [1977] *Varieties generated by para primal algebras* (J 0004) Algeb Universalis 7*93-114
⋄ C05 C13 ⋄ REV MR 55 #2707 Zbl 435 #08004
• ID 24176

CLARK, D.M. & KRAUSS, P.H. [1979] *Global subdirect products* (S 0167) Mem Amer Math Soc 17/210*ii+109pp
⋄ C05 C30 C90 ⋄ REV MR 80b:08001 Zbl 421 #08001
• ID 90275

CLARK, D.M. & KRAUSS, P.H. [1980] *Plain para primal algebras* (J 0004) Algeb Universalis 11*365-388
⋄ C05 C13 ⋄ REV MR 82g:08007 Zbl 455 #08005
• ID 69335

CLARK, D.M. [1984] $\aleph_0$-*categoricity in infra primal varieties* (J 0004) Algeb Universalis 19*160-176
⋄ C05 C30 C35 E05 ⋄ REV MR 86e:08009 • ID 44212

CLARKE, A.B. [1959] *On the representation of cardinal algebras by directed sums* (J 0064) Trans Amer Math Soc 91*161-192
⋄ C05 C30 ⋄ REV MR 21 #4933 Zbl 85.259 • ID 02245

CLARKE, A.B. see Vol. V for further entries

CLEAVE, J.P. [1968] *Hyperarithmetic ultrafilters* (P 0692) Summer School in Logic;1967 Leeds 223-240
⋄ C57 D55 E05 ⋄ REV MR 39 #3999 Zbl 217.14
• ID 02681

CLEAVE, J.P. [1969] *Local properties of systems* (J 3172) J London Math Soc, Ser 2 44*121-130 • ERR/ADD ibid 1*384
⋄ C52 C60 C85 ⋄ REV MR 38 #4293 Zbl 169.7
• ID 02267

CLEAVE, J.P. see Vol. I, II, IV, VI for further entries

CLOTE, P. [1980] *Weak partition relations, finite games, and independence results in Peano arithmetic* (P 2625) Model Th of Algeb & Arith;1979 Karpacz 92-107
⋄ C62 E05 F30 ⋄ REV MR 84h:03144 Zbl 445 #03029
• ID 56493

CLOTE, P. [1981] *A note on decidable model theory* (P 3404) Model Th & Arithm;1979/80 Paris 134-142
⋄ B25 C15 C50 C57 D30 ⋄ REV MR 83d:03046 Zbl 474 #03020 • ID 55424

CLOTE, P. [1981] *Anti-basis theorems and their relation to independence results in Peano arithmetic* (P 3404) Model Th & Arithm;1979/80 Paris 115-133
⋄ C62 C98 E05 F30 ⋄ REV MR 83e:03095 Zbl 474 #03021 • ID 55425

CLOTE, P. & MCALOON, K. [1983] *Two further combinatorial theorems equivalent to the 1-consistency of Peano arithmetic* (J 0036) J Symb Logic 48*1090-1104
⋄ B28 C62 F30 ⋄ REV MR 85e:03087 Zbl 545 #03033
• ID 40361

CLOTE, P. [1985] *Applications of the low-basis theorem in arithmetic* (P 3342) Rec Th Week;1984 Oberwolfach 65-88
⋄ C62 D20 D30 D55 F30 ⋄ REV Zbl 574 #03053
• ID 45298

CLOTE, P. [1985] *Partition relations in arithmetic* (P 2160) Latin Amer Symp Math Log (6);1983 Caracas 32-68
⋄ C62 E05 F30 ⋄ REV Zbl 567 #03029 • ID 41801

CLOTE, P. see Vol. IV, V for further entries

COBHAM, A. [1957] *Effectively decidable theories* (P 1675) Summer Inst Symb Log;1957 Ithaca 391-395
⋄ B25 F30 ⋄ REV JSL 31.653 • ID 29385

COBHAM, A. see Vol. I, IV for further entries

COCCHIARELLA, N.B. [1975] *A second order logic of variable-binding operators* (J 0302) Rep Math Logic, Krakow & Katowice 5*3-18
⋄ A05 B15 C80 ⋄ REV MR 57 #5675 Zbl 348 #02019
• ID 21896

COCCHIARELLA, N.B. [1980] *Nominalism and conceptualism as predicative second-order theories of predication* (J 0047) Notre Dame J Formal Log 21*481-500
⋄ A05 B15 C85 ⋄ REV MR 81j:03026 Zbl 416 #03013
• ID 53911

COCCHIARELLA, N.B. see Vol. I, II, V, VI for further entries

COHEN, P.E. [1974] *Models of set theory with more real numbers than ordinals* (J 0036) J Symb Logic 39*579-583
⋄ C62 E25 E35 ⋄ REV MR 53 #7779 Zbl 301 #02069 • ID 04012

COHEN, P.E. [1978] *Some continuity properties for ultraproducts* (J 0068) Z Math Logik Grundlagen Math 24*319-321
⋄ C20 C75 ⋄ REV MR 80b:03042 Zbl 396 #03028 • ID 52638

COHEN, P.E. see Vol. IV, V for further entries

COHEN, P.J. [1963] *A minimal model for set theory* (J 0015) Bull Amer Math Soc 69*537-540
⋄ C62 E35 E45 ⋄ REV MR 27 #41 Zbl 112.10 JSL 30.250 • ID 02303

COHEN, P.J. [1969] *Decision procedures for real and p-adic fields* (J 0155) Commun Pure Appl Math 22*131-151
⋄ B25 C10 C60 ⋄ REV MR 39 #5342 Zbl 167.15 • ID 16781

COHEN, P.J. [1974] *Automorphisms of set theory* (P 0610) Tarski Symp;1971 Berkeley 325-330
⋄ C62 E25 E35 ⋄ REV MR 51 #10086 Zbl 317 #02074 • ID 02309

COHEN, P.J. see Vol. V for further entries

COHN, P.M. [1957] *Groups of order automorphisms of ordered sets* (J 0303) Mathematika (Univ Coll London) 4*41-50
⋄ C07 C65 E07 ⋄ REV MR 19.940 Zbl 88.27 • ID 41136

COHN, P.M. [1959] *On the free product of associative rings* (J 0044) Math Z 71*380-398
⋄ C30 C60 ⋄ REV MR 21 #5648 Zbl 87.263 • REM Part I. Part II 1960 • ID 02310

COHN, P.M. [1960] *On the free product of associative rings. II: the case of (skew) fields* (J 0044) Math Z 73*433-456
⋄ C30 C60 ⋄ REV MR 22 #4747 Zbl 95.257 • REM Part I 1959. Part II 1968 • ID 48887

COHN, P.M. [1961] *On the embedding of rings in skew fields* (J 3240) Proc London Math Soc, Ser 3 11*511-530
⋄ C60 ⋄ REV MR 25 #100 Zbl 104.32 • ID 02311

COHN, P.M. [1965] *Universal Algebra* (X 0837) Harper & Row: New York xv+333pp
• LAST ED [1981] (X 0835) Reidel: Dordrecht xv+412pp
⋄ C05 C98 ⋄ REV MR 31 #224 MR 82j:08001 Zbl 461 #08001 JSL 34.113 • ID 02312

COHN, P.M. [1974] *The class of rings imbeddable in skew fields* (J 0161) Bull London Math Soc 6*147-148
⋄ C60 ⋄ REV MR 51 #3216 • ID 48379

COHN, P.M. [1975] *Presentation of skew fields. I. Existentially closed skew fields and the "Nullstellensatz"* (J 0332) Math Proc Cambridge Phil Soc 77*7-19
⋄ C25 C60 ⋄ REV MR 50 #13132 Zbl 297 #16012 • ID 17131

COHN, P.M. [1980] *On semifir constructions* (P 2634) Word Problems II;1976 Oxford 73-80
⋄ C05 C20 C52 C60 ⋄ REV MR 81m:16009 Zbl 433 #16003 • ID 80942

COHN, P.M. [1980] *Problems* (P 2634) Word Problems II;1976 Oxford 577-578
⋄ C60 ⋄ REV MR 81m:16021 Zbl 433 #16017 • ID 80941

COHN, P.M. see Vol. IV for further entries

COLE, J. [1973] see BELL, J.L.

COLE, J.C. & DICKMANN, M.A. [1972] *Non-axiomatizability results in infinitary languages for higher-order structures* (P 2080) Conf Math Log;1970 London 29-41
⋄ C52 C65 C75 C85 ⋄ REV MR 49 #2331 Zbl 234 #02009 • ID 04014

COLE, J.C. [1973] *Categories of sets and models of set theory* (P 0710) Russell Mem Logic Conf;1971 Uldum 351-399
⋄ C62 E30 G30 ⋄ REV MR 50 #9584 • ID 04015

COLE, J.C. [1979] *Classifying topoi* (C 2897) Algeb Model, Kateg & Gruppoide 5-20
⋄ C90 G30 ⋄ REV MR 81e:18008 Zbl 429 #18005 • ID 53887

COLE, J.C. see Vol. I for further entries

COLLET, J.-C. [1972] *Multirelation homogene et elimination de quanteurs* (J 2313) C R Acad Sci, Paris, Ser A-B 275*A1211-A1213
⋄ C10 C50 ⋄ REV MR 47 #1608 Zbl 379 #02017 • ID 02319

COLLET, J.-C. [1974] *Multirelation (k,p)-homogene et elimination de quanteurs* (J 2313) C R Acad Sci, Paris, Ser A-B 278*A197-A198
⋄ C10 C40 C50 ⋄ REV MR 48 #10795 Zbl 379 #02018 • ID 02320

COLLET, J.-C. [1974] *Multirelations (k,p)-homogenes et (k,p)-ages* (J 2313) C R Acad Sci, Paris, Ser A-B 278*A1473-A1474
⋄ C07 C50 ⋄ REV MR 50 #4277 Zbl 296 #02025 • ID 04016

COLLET, J.-C. [1975] *Theories stables par isotopie (English summary)* (J 2313) C R Acad Sci, Paris, Ser A-B 281*A677-A679
⋄ C52 ⋄ REV MR 53 #12928 Zbl 317 #02058 • ID 23177

COLLINS, D.J. [1970] *A universal semigroup (Russian)* (J 0003) Algebra i Logika 9*731-740
• TRANSL [1970] (J 0069) Algeb and Log 9*442-446
⋄ C05 C57 D05 D40 ⋄ REV MR 44 #1564 Zbl 225 #02030 Zbl 234 #02030 • ID 02327

COLLINS, D.J. [1970] *On recognizing properties of groups which have solvable word problem* (J 0008) Arch Math (Basel) 21*31-39
⋄ C60 D40 ⋄ REV MR 42 #351 Zbl 195.20 JSL 39.340 • ID 02448

COLLINS, D.J. see Vol. IV for further entries

COLLINS, G.E. [1957] *Tarski's decision method for elementary algebra* (P 1675) Summer Inst Symb Log;1957 Ithaca 64-70
⋄ C10 C60 ⋄ ID 02331

COLLINS, G.E. [1975] *Quantifier elimination for real closed fields by cylindrical algebraic decomposition* (P 1449) Automata Th & Formal Lang;1975 Kaiserslautern 134-183
⋄ C10 C60 G15 ⋄ REV MR 53 #7771 Zbl 318 #02051 • ID 23011

COLLINS, G.E. [1982] *Quantifier elimination for real closed fields: a guide to the literature* (C 3861) Comput Algeb. Symb & Algeb Comput 79-81
⋄ C10 C60 C98 ⋄ REV Zbl 495 #03016 • ID 36882

COLLINS, G.E. see Vol. I, IV, V, VI for further entries

COMBASE, J. [1983] *On the existence of finitely determinate models for some theories in stationary logic* (**P** 3829) Atti Incontri Log Mat (1);1982 Siena 137-158
⋄ C55 C80 E55 ⋄ REV MR 84k:03006 Zbl 519 # 03027
• ID 37540

COMBASE, J. [1984] *Introduction a la logique stationnaire* (**C** 4356) Gen Log Semin Paris 1982/83 13-34
⋄ C55 C80 C98 ⋄ ID 48059

COMBASE, J. see Vol. V for further entries

COMER, S.D. & TOURNEAU LE, J.J. [1969] *Isomorphism types of infinite algebras* (**J** 0053) Proc Amer Math Soc 21*635-639
⋄ C07 ⋄ REV MR 39 # 116 Zbl 175.15 • ID 31056

COMER, S.D. [1971] *Representations by algebras of sections over boolean spaces* (**J** 0048) Pac J Math 38*29-38
⋄ C05 C90 ⋄ REV MR 46 # 3412 Zbl 219 # 08002
• ID 02337

COMER, S.D. [1972] *A sheaf-theoretic duality theory for cylindric algebras* (**J** 0064) Trans Amer Math Soc 169*75-87
⋄ C90 G15 ⋄ REV MR 46 # 7023 Zbl 264 # 02052
• ID 02338

COMER, S.D. [1973] *Arithmetic properties of relatively free products* (**P** 0457) Lattice Th;1973 Houston 180-193
⋄ C05 ⋄ REV MR 53 # 5429 Zbl 294 # 08002 • ID 22950

COMER, S.D. [1974] *Elementary properties of structures of sections* (**J** 3127) Bol Soc Mat Mexicana, Ser 2 19*78-85
⋄ B25 C90 ⋄ REV MR 55 # 10265 Zbl 338 # 02032
• ID 27342

COMER, S.D. [1974] *Restricted direct products and sectional representations* (**J** 0114) Math Nachr 64*333-344
⋄ C05 C30 C90 ⋄ REV MR 51 # 5447 Zbl 302 # 08001
• ID 17426

COMER, S.D. [1975] *Monadic algebras with finite degree* (**J** 0004) Algeb Universalis 5*313-327
⋄ B25 C05 C90 G15 ⋄ REV MR 53 # 7774
Zbl 338 # 02033 • ID 23014

COMER, S.D. [1976] *Complete and model-complete theories of monadic algebras* (**S** 0019) Colloq Math (Warsaw) 34*183-190
⋄ B25 C05 C25 C35 C60 G15 ⋄ REV MR 58 # 209
Zbl 332 # 02062 • ID 31052

COMER, S.D. [1981] *The decision problem for certain nilpotent closed varieties* (**J** 0068) Z Math Logik Grundlagen Math 27*557-560
⋄ B25 C05 D35 ⋄ REV MR 83g:08008 Zbl 474 # 08003
• ID 55444

COMER, S.D. [1982] *Boolean combinations of monadic formulas* (**J** 0004) Algeb Universalis 15*299-305
⋄ C07 C90 G15 ⋄ REV MR 84m:03046 Zbl 521 # 03048
• ID 35748

COMER, S.D. [1983] *Extension of polygroups by polygroups and their representations using color schemes* (**P** 3841) Universal Algeb & Lattice Th (4);1982 Puebla 91-103
⋄ C05 C90 ⋄ REV MR 85d:20086 Zbl 521 # 20052
• ID 37471

COMER, S.D. [1985] *The elementary theory of interval real numbers* (**J** 0068) Z Math Logik Grundlagen Math 31*89-95
⋄ B25 B28 C65 ⋄ REV MR 86h:03069 Zbl 547 # 03007
Zbl 565 # 03008 • ID 42290

COMER, S.D. see Vol. IV, V for further entries

COMFORT, W.W. & NEGREPONTIS, S. [1974] *The theory of ultrafilters* (**X** 0811) Springer: Heidelberg & New York x+482pp
⋄ C20 C55 C98 E05 E55 E75 E98 G05 ⋄ REV
MR 53 # 135 Zbl 298 # 02004 JSL 41.782 • ID 16659

COMFORT, W.W. see Vol. V for further entries

COMPTON, K.J. [1983] *Some useful preservation theorems* (**J** 0036) J Symb Logic 48*427-440
⋄ C30 C40 C65 ⋄ REV MR 84j:03068 Zbl 523 # 03021
• ID 34659

COMPTON, K.J. [1984] *An undecidable problem in finite combinatorics* (**J** 0036) J Symb Logic 49*842-850
⋄ C13 D80 ⋄ REV MR 85i:03094 • ID 42434

COMPTON, K.J. [1984] *On rich words* (**P** 4064) Prog Graph Th;1982 Waterloo 39-61
⋄ C85 D05 D45 ⋄ REV Zbl 563 # 03022 • ID 47457

COMPTON, K.J. [1985] *Application of a Tauberian theorem to finite model theory* (**J** 0009) Arch Math Logik Grundlagenforsch 25*91-98
⋄ C13 ⋄ ID 48380

COMYN, G. [1982] *Arbres infinitaires. Approximations et proprietes de calculabilite* (**P** 4004) CAAP'82 Arbres en Algeb & Progr (7);1982 Lille 65-81
⋄ C57 ⋄ REV Zbl 539 # 68057 • ID 41258

COMYN, G. see Vol. IV, VI for further entries

CONNES, A. [1970] *Determination de modeles minimaux en analyse non standard et application* (**J** 2313) C R Acad Sci, Paris, Ser A-B 271*A969-A971
⋄ H05 H15 ⋄ REV MR 44 # 3854 Zbl 255 # 02059
• ID 02352

CONNES, A. [1970] *Ultrapuissances et applications dans le cadre de l'analyse non standard* (**C** 1525) Semin Init Analyse (9-10) Paris 1969/71 9/8*25pp
⋄ C20 H05 ⋄ REV MR 44 # 3855 Zbl 216.12 • ID 26249

CONTESSA, M. [1984] *Ultraproducts of pm-rings and mp-rings* (**J** 0326) J Pure Appl Algebra 32*11-20
⋄ C20 C60 G05 ⋄ REV MR 85i:03096 Zbl 539 # 13003
• REM See also 1985 • ID 44190

CONTESSA, M. [1985] *A note on ultraproducts of complete Boolean algebras* (**J** 0326) J Pure Appl Algebra 36*217
⋄ C20 G05 ⋄ REV MR 86h:03051 Zbl 558 # 06017 • REM
See also 1984 • ID 45598

CONWAY, J.H. [1976] *On numbers and games* (**X** 0801) Academic Pr: New York ix+238pp
• TRANSL [1983] (**X** 0900) Vieweg: Wiesbaden vii+205pp
⋄ B28 E10 E70 E75 H15 H20 ⋄ REV MR 56 # 8365
MR 84m:03086 Zbl 334 # 00004 Zbl 491 # 00024
• ID 71909

CONWAY, J.H. see Vol. IV, V for further entries

COOPER, D.C. [1972] *Theorem proving in arithmetic without multiplication* (**J** 0508) Machine Intelligence 7*91-99
⋄ B25 B35 F30 ⋄ REV Zbl 258 # 68046 • ID 61080

COOPER, D.C. see Vol. I, IV for further entries

COOPER, GLEN R. [1982] *On complexity of complete first-order theories* (J 0068) Z Math Logik Grundlagen Math 28*93-136
◇ C35 C45 C52 ◇ REV MR 84e:03047 Zbl 496 # 03019 • ID 34383

COOPER, R. [1981] see BARWISE, J.

COOPER, R. see Vol. V for further entries

COQUAND, T. [1984] *Le theoreme de representation d'Arens et Kaplansky* (C 4356) Gen Log Semin Paris 1982/83 97-112
◇ C60 C98 ◇ ID 48081

COQUAND, T. see Vol. I for further entries

CORCORAN, J. & HERRING, J.M. [1971] *Notes on a semantic analysis of variable binding term operators* (J 0079) Logique & Anal, NS 14*644-657
◇ A05 A10 B10 C80 ◇ REV MR 46 # 6989 Zbl 239 # 02007 • ID 27917

CORCORAN, J. & HATCHER, W.S. & HERRING, J.M. [1972] *Variable binding term operators* (J 0068) Z Math Logik Grundlagen Math 18*177-182
◇ B10 C80 ◇ REV MR 46 # 5098 Zbl 257 # 02013 • ID 02392

CORCORAN, J. [1980] *Categoricity* (J 2028) Hist & Phil Log 1*187-207
◇ A05 B15 C35 C85 ◇ REV MR 82j:03034 Zbl 504 # 03014 • ID 71934

CORCORAN, J. [1980] *On definitional equivalence and related topics* (J 2028) Hist & Phil Log 1*231-234
◇ A10 C40 F25 ◇ REV MR 83g:03014 • ID 35994

CORCORAN, J. [1981] *From categoricity to completeness* (J 2028) Hist & Phil Log 2*113-119
◇ A10 C35 ◇ REV MR 83d:03040 • ID 34825

CORCORAN, J. see Vol. I, II, VI for further entries

CORNER, A.L.S. & GOEBEL, R. [1985] *Prescribing endomorphism algebras, a unified treatment* (J 3240) Proc London Math Soc, Ser 3 50*447-479
◇ C60 ◇ REV MR 86h:16031 Zbl 562 # 20030 • ID 44687

CORNISH, W.H. [1976] *A sheaf representation for generalized Stone lattices* (J 0014) Bull Acad Pol Sci, Ser Math Astron Phys 24*33-36
◇ C05 C90 G10 ◇ REV MR 53 # 2786 Zbl 325 # 06006 • ID 18108

CORNMAN, J.W. [1972] *Craig's theorem, Ramsey-sentences, and scientific instrumentalism* (J 0154) Synthese 25*82-128
◇ A05 C40 ◇ REV Zbl 262 # 02006 • ID 27472

CORNMAN, J.W. see Vol. I, II for further entries

CORRADA, M. [1980] see ALCANTARA DE, L.P.

CORRADA, M. [1980] *Remarks on extendability of set-theoretical models preserving cardinality* (P 3006) Brazil Conf Math Log (3);1979 Recife 71-82
◇ C62 E30 E70 ◇ REV MR 82e:03036 Zbl 451 # 03022 • ID 54037

CORRADA, M. see Vol. V for further entries

COSTA DA, N.C.A. [1972] *Modeles et univers de Dedecker* (J 2313) C R Acad Sci, Paris, Ser A-B 275*A483-A486
◇ C62 E55 E70 G30 ◇ REV MR 46 # 3310 Zbl 243 # 02047 • ID 02485

COSTA DA, N.C.A. [1974] *α-models and systems T and T*** (J 0047) Notre Dame J Formal Log 15*443-454
◇ C62 E70 G30 ◇ REV MR 52 # 2881 Zbl 246 # 02049 • ID 02486

COSTA DA, N.C.A. [1974] *Remarques sur les calculs $\mathscr{C}_n$, $\mathscr{C}_n^*$, $\mathscr{C}_n^=$ et $\mathscr{D}_n$* (J 2313) C R Acad Sci, Paris, Ser A-B 278*A819-A821
◇ B25 B53 ◇ REV MR 53 # 5291 Zbl 278 # 02021 • ID 22926

COSTA DA, N.C.A. & DRUCK, I.F. [1975] *Sur les "VBTOs" selon M.Hatcher (English summary)* (J 2313) C R Acad Sci, Paris, Ser A-B 281*A741-A743
◇ C07 C40 C80 ◇ REV MR 57 # 94 Zbl 316 # 02056 • ID 27258

COSTA DA, N.C.A. & PINTER, C. [1976] *α-logic and infinitary languages* (J 0068) Z Math Logik Grundlagen Math 22*105-112
◇ B60 C75 ◇ REV MR 58 # 21426 Zbl 336 # 02039 • ID 14829

COSTA DA, N.C.A. [1977] see ARRUDA, A.I.

COSTA DA, N.C.A. [1980] *A model-theoretical approach to variable binding term operators* (P 2958) Latin Amer Symp Math Log (4);1978 Santiago 133-162
◇ C07 C80 ◇ REV MR 81i:03056 Zbl 424 # 03018 • ID 72065

COSTA DA, N.C.A. see Vol. I, II, V, VI for further entries

COSTE, M.-F. [1979] see COSTE, M.

COSTE, M.-F. also published under the name COSTE-ROY, M.-F.

COSTE, M.-F. see Vol. V for further entries

COSTE, M. [1973] *Logique du 1er ordre dans les topos elementaires* (1111) Preprints, Manuscr., Techn. Reports etc.
◇ C90 G30 ◇ REM Seminaire Benabou, Paris • ID 27665

COSTE, M. [1979] *Localisation, spectra and sheaf representation* (P 2901) Appl Sheaves;1977 Durham 212-238
◇ C30 C90 ◇ REV MR 83c:18002 Zbl 422 # 18007 • ID 53503

COSTE, M. & COSTE, M.-F. [1979] *The generic model of an ε-stable geometric extension of the theory of rings is of line type* (C 3055) Topos-Theor Meth in Geom 29-36
◇ C60 C90 G30 ◇ REV MR 80m:03112 Zbl 427 # 18008 • ID 71955

COSTE, M. & COSTE, M.-F. [1979] *Topologies for real algebraic geometry* (C 3055) Topos-Theor Meth in Geom 37-100
◇ C60 G30 ◇ REV MR 80m:14002 Zbl 417 # 14002 • ID 80976

COSTE, M. see Vol. IV, V, VI for further entries

COVEN, C. [1971] *Forcing in infinitary languages* (0000) Diss., Habil. etc 91pp
◇ C25 C75 ◇ REM Doctorial diss., Yale University, New Haven, CT • ID 16782

COWEN, R.H. [1976] *Elementary equivalence and constructible models of Zermelo-Fraenkel set theory* (J 0068) Z Math Logik Grundlagen Math 22*333-338
◇ C62 E45 ◇ REV MR 57 # 2915 Zbl 346 # 02037 • ID 18443

COWEN, R.H. see Vol. I, V for further entries

COWLES, J.R. [1977] *The theory of differentially closed fields in logics with cardinal quantifiers* (J 0009) Arch Math Logik Grundlagenforsch 18*105-114
⋄ C60 C80 ⋄ REV MR 57 # 5738 Zbl 371 # 02024
• ID 24334

COWLES, J.R. [1979] *The relative expressive power of some logics extending first-order logic* (J 0036) J Symb Logic 44*129-146
⋄ C10 C75 C80 C85 E15 ⋄ REV MR 80f:03039 Zbl 408 # 03032 • ID 56271

COWLES, J.R. [1979] *The theory of Archimedean real closed fields in logics with Ramsey quantifiers* (J 0027) Fund Math 103*65-76
⋄ C10 C35 C60 C80 ⋄ REV MR 80f:03032 Zbl 412 # 03017 • ID 52948

COWLES, J.R. [1981] *Generalized archimedean fields and logics with Malitz quantifiers* (J 0027) Fund Math 112*45-59
⋄ C10 C60 C80 ⋄ REV MR 82h:03032 Zbl 533 # 03020
• ID 71964

COWLES, J.R. [1981] *The Henkin quantifier and real closed fields* (J 0068) Z Math Logik Grundlagen Math 27*549-555
⋄ C60 C80 ⋄ REV MR 84g:03048 Zbl 505 # 03015
• ID 33197

COWLES, J.R. & LAGRANGE, R. [1983] *Generalized archimedean fields* (J 0047) Notre Dame J Formal Log 24*133-140
⋄ C60 ⋄ REV MR 84m:03054 Zbl 488 # 03027 • ID 35755

COZART, D. & MOORE JR., L.C. [1974] *The nonstandard hull of a Riesz space* (J 0025) Duke Math J 41*263-276
⋄ C60 H05 ⋄ REV MR 50 # 10747 Zbl 293 # 46012
• ID 02503

CRAIG, W. [1952] *Incompletability, with respect to validity in every finite nonempty domain of first order functional calculus* (P 0593) Int Congr Math (II, 6);1950 Cambridge MA 1*721
⋄ C13 D35 ⋄ ID 28058

CRAIG, W. & QUINE, W.V.O. [1952] *On reduction to a symmetric relation* (J 0036) J Symb Logic 17*188
⋄ B10 C40 ⋄ REV MR 14.233 Zbl 47.9 JSL 18.269
• ID 02505

CRAIG, W. [1953] *On axiomatizability within a system* (J 0036) J Symb Logic 18*30-32
⋄ C07 D20 ⋄ REV MR 14.1051 Zbl 53.201 JSL 19.62
• ID 02507

CRAIG, W. [1957] *Linear reasoning. A new form of the Herbrand-Gentzen theorem* (J 0036) J Symb Logic 22*250-268
⋄ B10 C07 C40 F05 ⋄ REV MR 21 # 3317 Zbl 81.244 JSL 24.243 • ID 02510

CRAIG, W. [1957] *Three uses of the Herbrand-Gentzen theorem in relating model theory and proof theory* (J 0036) J Symb Logic 22*269-285
⋄ B10 C07 C40 F05 ⋄ REV MR 21 # 3318 Zbl 79.245 JSL 24.243 • ID 02508

CRAIG, W. & VAUGHT, R.L. [1958] *Finite axiomatizability using additional predicates* (J 0036) J Symb Logic 23*289-308
⋄ B10 C52 ⋄ REV MR 21 # 4909 Zbl 85.246 JSL 36.334
• ID 02511

CRAIG, W. [1965] *Satisfaction for n-th order languages defined in n-th order languages* (J 0036) J Symb Logic 30*13-25
⋄ B15 C40 ⋄ REV MR 33 # 3883 Zbl 199.3 • ID 02514

CRAIG, W. [1974] see ADDISON, J.W.

CRAIG, W. see Vol. I, IV, V, VI for further entries

CRAWLEY, P. & JONSSON, B. [1964] *Refinements for infinite direct decompositions of algebraic systems* (J 0048) Pac J Math 14*797-855
⋄ C05 C30 C60 ⋄ REV MR 30 # 49 Zbl 134.255
• ID 02526

CRESSWELL, M.J. [1985] *The decidable normal modal logics are not recursively enumerable* (J 0122) J Philos Logic 14*231-233
⋄ B25 B45 D25 D35 D80 ⋄ ID 42657

CRESSWELL, M.J. see Vol. I, II, IV, V for further entries

CROSSLEY, J.N. [1966] *Some theorems in logic* (J 0248) Math Student 34*125-129
⋄ B10 C07 ⋄ ID 31222

CROSSLEY, J.N. (ED.) [1967] *Sets, models and recursion theory* (X 0809) North Holland: Amsterdam 331pp
⋄ B97 C97 D97 ⋄ REV MR 36 # 24 Zbl 158.2 • ID 31684

CROSSLEY, J.N. & NERODE, A. [1974] *Combinatorial functors* (X 0811) Springer: Heidelberg & New York viii+146pp
⋄ C57 C62 D45 D50 G30 ⋄ REV MR 52 # 10397 Zbl 283 # 02036 JSL 42.586 • ID 21692

CROSSLEY, J.N. [1974] *Satisfaction (a brief survey of model theory)* (J 0329) Math Chron (Auckland) 3*1-8
⋄ C07 C98 ⋄ REV MR 50 # 1869 Zbl 289 # 02003
• ID 17136

CROSSLEY, J.N. [1975] *A brief survey of model theory* (P 2052) Math Educ & New Areas in Math;1974 Penang 1-10
⋄ C07 C98 ⋄ ID 31687

CROSSLEY, J.N. (ED.) [1975] *Algebra and logic* (X 0811) Springer: Heidelberg & New York viii+307pp
⋄ B97 C97 G97 ⋄ REV MR 51 # 2851 Zbl 293 # 00006
• ID 31725

CROSSLEY, J.N. [1975] *Saturated models* (J 1480) Bull Malaysian Math Soc 1975*1-7
⋄ C50 ⋄ REV MR 54 # 4963 • ID 24132

CROSSLEY, J.N. & NERODE, A. [1976] *Effective dimension* (J 0032) J Algeb 41*398-412
⋄ C57 D45 D50 ⋄ REV MR 54 # 7234 Zbl 344 # 02036
• ID 25025

CROSSLEY, J.N. (ED.) [1981] *Aspects of effective algebra. Proceedings of a conference at Monash University, Australia, 1-4 August, 1979* (X 2863) Upside Down A Book: Yarra Glen x+290pp
⋄ C57 C97 D45 D97 ⋄ REV MR 82h:03003 Zbl 456 # 00004 • ID 54309

CROSSLEY, J.N. [1982] *The given* (J 0063) Studia Logica 41*131-139
⋄ C57 D45 ⋄ REV MR 85i:03149 Zbl 536 # 03024
• ID 37106

CROSSLEY, J.N. see Vol. I, II, IV, V, VI for further entries

CSAKANY, B. [1964] *Abelian properties of primitive classes of universal algebras (Russian)* (J 0002) Acta Sci Math (Szeged) 25*202-208
⋄ C05 C60 ⋄ REV MR 30 # 48 JSL 37.189 • ID 01811

CSAKANY, B. [1965] *Inner automorphisms of universal algebras* (J 0057) Publ Math (Univ Debrecen) 12∗331-333
  ⋄ C05 ⋄ REV MR 32 # 2356 Zbl 135.25 • ID 02562

CSAKANY, B. [1976] *Conditions involving universally quantified function variables* (J 0002) Acta Sci Math (Szeged) 38∗7-11
  ⋄ C05 C40 C85 ⋄ REV MR 53 # 13074 Zbl 335 # 08003
  • ID 27233

CSAKANY, B. & GAVALCOVA, T. [1981] *Three-element quasi-primal algebras* (J 0411) Studia Sci Math Hung 16∗237-248
  ⋄ C05 C13 ⋄ REV MR 84j:08008 Zbl 527 # 08005
  • ID 38555

CSAKANY, B. see Vol. II for further entries

CSIRMAZ, L. [1979] *On definability in Peano arithmetic* (J 0387) Bull Sect Logic, Pol Acad Sci 8∗148-153
  ⋄ C40 C62 F30 ⋄ REV MR 81b:03001 Zbl 424 # 03008
  • ID 72011

CSIRMAZ, L. & PARIS, J.B. [1984] *A property of 2-sorted Peano models and program verification* (J 0068) Z Math Logik Grundlagen Math 30∗325-334
  ⋄ B75 H15 ⋄ REV Zbl 564 # 03049 • ID 42265

CSIRMAZ, L. see Vol. I, IV, V for further entries

CUDA, K. [1977] *The nonabsolute boundedness model of the theory of semisets* (J 0140) Comm Math Univ Carolinae (Prague) 18∗763-769
  ⋄ C62 E35 E70 ⋄ REV MR 58 # 21610 Zbl 367 # 02033
  • ID 32281

CUDA, K. [1980] *An elimination of infinitely small quantities and infinitely large numbers (within the framework of AST)* (J 0140) Comm Math Univ Carolinae (Prague) 21∗433-445
  ⋄ E70 H05 H15 ⋄ REV MR 83i:03104 Zbl 445 # 03042
  • ID 56506

CUDA, K. [1982] *An elimination of the predicate "to be a standard member" in nonstandard models of arithmetic* (J 0140) Comm Math Univ Carolinae (Prague) 23∗785-803
  ⋄ H15 ⋄ REV MR 84g:03108 Zbl 522 # 03059 • ID 34210

CUDA, K. [1983] *Nonstandard models of arithmetic as an alternative basis for continuum considerations* (J 0140) Comm Math Univ Carolinae (Prague) 24∗415-430
  ⋄ E70 F30 H15 ⋄ REV MR 85g:03095 Zbl 531 # 03045
  • ID 37690

CUDA, K. & VOJTASKOVA, B. [1984] *Models of AST without choice* (J 0140) Comm Math Univ Carolinae (Prague) 25∗555-589
  ⋄ C62 E25 E35 E70 E75 H15 ⋄ REV Zbl 561 # 03028
  • ID 39637

CUDA, K. see Vol. I, V for further entries

CUNNINGHAM, E. [1975] *Chain models: applications of consistency properties and back-and-forth techniques in infinite-quantifier languages* (C 0781) Infinitary Logic (Karp) 125-142
  ⋄ C40 C75 ⋄ REV MR 57 # 16047 Zbl 316 # 02018
  • ID 72013

CUPONA, G. & MIJAJLOVIC, Z. & MILIC, S. & PERIC, V. & PRESIC, S.B. (EDS.) [1983] *Proceedings of the third algebraic conference* (X 4030) Univ Novom Sadu, Inst Mat: Novi Sad iii + 157pp
  ⋄ C97 G05 ⋄ REV MR 85i:20004 Zbl 521 # 00011
  • ID 44322

CUPONA, G. see Vol. IV, V for further entries

CUPPLES, B. [1981] *On Copi's misapplication of a decision procedure* (J 0079) Logique & Anal, NS 24∗421-424
  ⋄ B25 ⋄ REV MR 83j:03026 • ID 35338

CUSIN, R. [1966] see BOURTOT, B.

CUSIN, R. [1966] *Une demonstration concernant le theoreme d'interpolation generalise aux ensembles d'enonces* (J 0056) Publ Dep Math, Lyon 3/2∗62-70
  ⋄ C40 ⋄ REV MR 33 # 7235 Zbl 178.312 • ID 02647

CUSIN, R. [1970] *Theories quasi-completes* (J 2313) C R Acad Sci, Paris, Ser A-B 270∗A297-A299
  ⋄ C35 G05 ⋄ REV MR 41 # 1519 Zbl 211.14 • ID 02652

CUSIN, R. [1971] *Recherche du forcing-compagnon et du modele-compagnon d'une theorie liee a l'existence de modeles $\aleph_\alpha$-universels* (J 2313) C R Acad Sci, Paris, Ser A-B 273∗A956-A959
  ⋄ C25 C35 C50 ⋄ REV MR 50 # 88 Zbl 235 # 02047
  • ID 02654

CUSIN, R. & PABION, J.F. [1971] *Structures generiques associees a une classe de theories* (J 2313) C R Acad Sci, Paris, Ser A-B 272∗A1620-A1623
  ⋄ C25 C50 ⋄ REV MR 44 # 47 Zbl 234 # 02037 • ID 02656

CUSIN, R. [1971] *Structures prehomogenes et structures generiques* (J 2313) C R Acad Sci, Paris, Ser A-B 273∗A137-A140
  ⋄ C25 C50 ⋄ REV MR 45 # 1742 Zbl 241 # 02017
  • ID 02655

CUSIN, R. [1971] *Sur le "forcing" en theorie des modeles* (J 2313) C R Acad Sci, Paris, Ser A-B 272∗A845-A848
  ⋄ C25 ⋄ REV MR 44 # 2589 Zbl 235 # 02046 • ID 02653

CUSIN, R. [1972] *L'algebre des formules d'une theorie complete et "forcing-compagnon"* (J 2313) C R Acad Sci, Paris, Ser A-B 275∗A1269-A1272
  ⋄ C25 C35 ⋄ REV MR 47 # 3162 Zbl 254 # 02042
  • ID 02659

CUSIN, R. [1972] *Quasi-complete theories* (P 2080) Conf Math Log;1970 London 337-338
  ⋄ C35 ⋄ REV Zbl 227 # 02031 • ID 27361

CUSIN, R. [1973] *Sur les structures generiques en theorie des modeles* (J 0060) Rev Roumaine Math Pures Appl 18∗519-541
  ⋄ C10 C25 C35 C50 ⋄ REV MR 53 # 7762 Zbl 272 # 02073 • ID 23004

CUSIN, R. [1974] *The number of countable generic models for finite forcing* (J 0027) Fund Math 84∗265-270
  ⋄ C15 C25 ⋄ REV MR 49 # 2349 Zbl 287 # 02035
  • ID 02658

CUSIN, R. see Vol. V for further entries

CUTLAND, N.J. [1970] *The theory of hyperarithmetic and Π_1^1 models* (0000) Diss., Habil. etc
  ⋄ C70 D55 ⋄ REM Ph.D. Thesis, University of Bristol
  • ID 16784

CUTLAND, N.J. [1972] *Π_1^1-models and Π_1^1-categoricity* (P 2080) Conf Math Log;1970 London 42-62
  ⋄ C57 C70 D55 ⋄ REV MR 51 # 133 Zbl 242 # 02057
  • ID 02662

CUTLAND, N.J. [1972] Σ_1-*compactness and ultraproducts*
(**J** 0036) J Symb Logic 37*668-672
- ◇ C20 C70 ◇ REV MR 47 # 4759 Zbl 259 # 02010
- • ID 02661

CUTLAND, N.J. [1973] *Model theory on admissible sets* (**J** 0007) Ann Math Logic 5*257-289
- ◇ C45 C55 C70 ◇ REV MR 51 # 12504 Zbl 293 # 02037
- • ID 02663

CUTLAND, N.J. [1976] *Compactness without languages* (**J** 0068) Z Math Logik Grundlagen Math 22*113-115
- ◇ C07 C20 ◇ REV MR 56 # 5271 Zbl 328 # 02037
- • ID 18499

CUTLAND, N.J. [1977] *Some theories having countably many countable models* (**J** 0068) Z Math Logik Grundlagen Math 23*105-110
- ◇ C15 C35 ◇ REV MR 58 # 5169 Zbl 446 # 03024
- • ID 26471

CUTLAND, N.J. [1978] Σ_1-*compactness in languages stronger than* $\mathscr{L}_A$ (**J** 0036) J Symb Logic 43*508-520
- ◇ C70 C75 C80 ◇ REV MR 80b:03041 Zbl 393 # 03024
- • ID 29279

CUTLAND, N.J. [1978] *On the form of countable admissible ordinals* (**S** 0019) Colloq Math (Warsaw) 38*173-174
- ◇ C70 D60 ◇ REV MR 58 # 21535 Zbl 381 # 03025
- • ID 51893

CUTLAND, N.J. & KAUFMANN, M. [1980] Σ_1-*well-founded compactness* (**J** 0007) Ann Math Logic 18*271-296
- ◇ C70 C80 ◇ REV MR 82d:03060 Zbl 473 # 03030
- • ID 72037

CUTLAND, N.J. [1980] *Computability. An introduction to recursive function theory* (**X** 0805) Cambridge Univ Pr: Cambridge, GB x+251pp
- • TRANSL [1983] (**X** 0885) Mir: Moskva 256pp
- ◇ B25 D98 ◇ REV MR 81i:03001 MR 84f:03001 Zbl 448 # 03029 • ID 56629

CUTLAND, N.J. see Vol. I, II, IV, V for further entries

CZEDLI, G. [1982] *A note on the compactness of the consequence relation for congruence varieties* (**J** 0004) Algeb Universalis 15*142-143
- ◇ C05 ◇ REV MR 83i:08002 Zbl 509 # 08012 • ID 39325

CZEDLI, G. & FREESE, R. [1983] *On congruence distributivity and modularity* (**J** 0004) Algeb Universalis 17*216-219
- ◇ B25 C05 ◇ REV MR 85e:08011 • ID 39922

CZEDLI, G. & DAY, A. [1984] *Horn sentences with (W) and weak Mal'cev conditions* (**J** 0004) Algeb Universalis 19*217-230
- ◇ C05 ◇ REV Zbl 549 # 08003 • ID 44241

CZEDLI, G. [1984] *Mal'cev conditions for Horn sentences with congruence permutability* (**J** 4729) Acta Math Hung 44*115-124
- ◇ C05 ◇ REV Zbl 541 # 08005 • ID 44321

CZELAKOWSKI, J. [1979] ω-*saturated matrices* (**J** 0387) Bull Sect Logic, Pol Acad Sci 8*120-123
- ◇ C50 ◇ REV MR 82b:03062 • ID 53305

CZELAKOWSKI, J. [1979] *A characterization of Matr(C)* (**J** 0387) Bull Sect Logic, Pol Acad Sci 8*83-86
- ◇ B22 C05 ◇ REV MR 80g:03024 Zbl 409 # 03016
- • ID 31660

CZELAKOWSKI, J. [1979] *A purely algebraic proof of the omitting types theorem* (**J** 0387) Bull Sect Logic, Pol Acad Sci 8*7-9
- ◇ C07 C15 C30 G99 ◇ REV MR 80m:03067 Zbl 418 # 03023 • ID 31658

CZELAKOWSKI, J. [1979] *A remark on countable algebraic models* (**J** 0387) Bull Sect Logic, Pol Acad Sci 8*2-6
- ◇ C15 C20 G05 G99 ◇ REV MR 80i:03042 • ID 31657

CZELAKOWSKI, J. [1979] *Large matrices which induce finite consequence operations* (**J** 0387) Bull Sect Logic, Pol Acad Sci 8*79-82
- ◇ B22 C20 ◇ REV MR 80g:03023 • ID 31659

CZELAKOWSKI, J. [1980] *Model-theoretic methods in methodology of propositional calculi* (**X** 2733) Acad Sci Inst Phi Soc: Wroclaw 75pp
- ◇ B22 C05 G99 ◇ REV MR 82c:03046 Zbl 445 # 03008
- • ID 56472

CZELAKOWSKI, J. [1980] *Reduced products of logical matrices* (**J** 0063) Studia Logica 39*19-43
- ◇ B22 C05 C20 ◇ REV MR 81h:03060 Zbl 445 # 03009
- • ID 56473

CZELAKOWSKI, J. & DZIOBIAK, W. [1982] *Another proof that* $ISP_r(K)$ *is the least quasivariety containing K* (**J** 0063) Studia Logica 41*343-345
- ◇ C05 ◇ REV Zbl 542 # 08006 • ID 42300

CZELAKOWSKI, J. [1982] *Logical matrices and the amalgamation property* (**J** 0063) Studia Logica 41*329-341
- ◇ B22 C40 C52 C90 ◇ REV MR 85i:03091 Zbl 549 # 03014 • ID 42743

CZELAKOWSKI, J. [1984] *Remarks on finitely based logics* (**P** 2153) Logic Colloq;1983 Aachen 1*147-168
- ◇ B22 C90 ◇ ID 41764

CZELAKOWSKI, J. [1985] *Sentential logics and Maehara interpolation property* (**J** 0063) Studia Logica 44*265-284
- ◇ B22 C05 C40 C52 ◇ ID 47502

CZELAKOWSKI, J. see Vol. I, II for further entries

DACUNHA-CASTELLE, D. & KRIVINE, J.-L. [1972] *Applications des ultraproduits a l'etude des espaces et des algebres de Banach* (**J** 0343) Studia Math, Pol Akad Nauk 41*315-334
- ◇ C20 C65 E75 ◇ REV MR 46 # 4165 Zbl 275 # 46023
- • ID 17036

DACUNHA-CASTELLE, D. [1972] *Applications des ultraproduits a la theorie des plongements des espaces de Banach* (**P** 0773) Colloq Anal Fonctionelle;1971 Bordeaux 117-125
- ◇ C20 C65 E75 ◇ REV MR 51 # 13647 Zbl 254 # 46013
- • ID 17234

DACUNHA-CASTELLE, D. [1972] *Ultraproduits d'espaces de Banach* (**P** 3030) Semin Goulaouic-Schwartz;1972 Paris 9*11pp
- ◇ C20 C65 E75 ◇ REV MR 53 # 1232 Zbl 243 # 46030
- • ID 81005

DACUNHA-CASTELLE, D. [1972] *Ultraproduits d'espaces* L^p *et d'espaces d'Orlicz* (**P** 3030) Semin Goulaouic-Schwartz;1972 Paris 10*9pp
- ◇ C20 C65 E75 ◇ REV MR 53 # 1233 Zbl 244 # 46032
- • ID 81004

DACUNHA-CASTELLE, D. & KRIVINE, J.-L. [1975] *Sous-espaces de* L^1 (**X** 2854) Univ Paris XI UER Math: Paris 51pp
- ◇ C20 C65 E75 ◇ REV MR 55 # 11018 • ID 81007

DACUNHA-CASTELLE, D. & KRIVINE, J.-L. [1977] *Sous-espaces de L^1* (**J** 0029) Israel J Math 26∗320-351
⋄ C20 C65 E75 ⋄ REV MR 58 # 30117 Zbl 344 # 46051
• ID 81006

DACUNHA-CASTELLE, D. see Vol. I for further entries

DAGUENET, M. [1972] *Un modele non standard de l'arithmetique* (**J** 2313) C R Acad Sci, Paris, Ser A-B 274∗A685-A688
⋄ C62 E05 H15 ⋄ REV MR 47 # 36 Zbl 242 # 02061
• ID 02744

DAGUENET, M. [1975] *Rapport entre l'ensemble des ultrafiltres admettant un ultrafiltre donne pour image et l'ensemble des images de cet ultrafiltre* (**J** 0140) Comm Math Univ Carolinae (Prague) 16∗99-113
⋄ C20 E05 ⋄ REV MR 51 # 6729 Zbl 302 # 54006
• ID 17438

DAGUENET, M. also published under the name DAGUENET-TEISSIER, M.

DAGUENET, M. see Vol. V for further entries

DAGUENET-TEISSIER, M. [1979] *Ultrafiltres a la facon de Ramsey* (**J** 0064) Trans Amer Math Soc 250∗91-120
⋄ C20 E05 E50 ⋄ REV MR 81b:04004 Zbl 426 # 03049
• ID 53645

DAGUENET-TEISSIER, M. also published under the name DAGUENET, M.

DAHLHAUS, E. [1984] *Reduction to NP-complete problems by interpretations* (**P** 2342) Symp Rek Kombin;1983 Muenster 357-365
⋄ C13 D15 ⋄ REV MR 86f:03060 Zbl 558 # 03019
• ID 41789

DAHN, B.I. [1973] $\aleph_0$-*kategorische zyklenbeschraenkte Graphen* (**J** 0014) Bull Acad Pol Sci, Ser Math Astron Phys 21∗293-298
⋄ C13 C35 C65 ⋄ REV MR 53 # 7757 Zbl 258 # 02053
• ID 61209

DAHN, B.I. [1973] $\aleph_0$-*kategorische zyklenbeschraenkte Graphen (English summary)* (**J** 0027) Fund Math 80∗117-131
⋄ C13 C35 C65 ⋄ REV MR 48 # 3728 Zbl 269 # 02021
• ID 61214

DAHN, B.I. [1973] *Generalized Kripke-models* (**J** 0014) Bull Acad Pol Sci, Ser Math Astron Phys 21∗1073-1077
⋄ B45 C90 ⋄ REV MR 50 # 12663 Zbl 279 # 02034
• ID 04057

DAHN, B.I. [1974] *A note on generalized Kripke-models* (**J** 0387) Bull Sect Logic, Pol Acad Sci 3/1∗8-12
⋄ B45 C90 ⋄ REV MR 53 # 5262 • ID 22893

DAHN, B.I. [1974] *Contributions to the model theory for non-classical logics* (**J** 0068) Z Math Logik Grundlagen Math 20∗473-479
⋄ B45 C90 ⋄ REV MR 54 # 12491a Zbl 306 # 02052
• ID 04058

DAHN, B.I. [1974] *Generalized Kripke models* (**J** 0387) Bull Sect Logic, Pol Acad Sci 3/1∗3-7
⋄ C90 ⋄ REV MR 53 # 5261 • ID 22892

DAHN, B.I. [1974] *Meta-mathematics of some many-valued calculi* (**J** 0014) Bull Acad Pol Sci, Ser Math Astron Phys 22∗747-750
⋄ B50 C40 C90 ⋄ REV MR 50 # 6789 Zbl 293 # 02019
• ID 02746

DAHN, B.I. [1975] *Eine Anwendung des Basistheorems in der nichtklassischen Logik* (**J** 0115) Wiss Z Humboldt-Univ Berlin, Math-Nat Reihe 24∗794-796
⋄ B55 C40 C90 ⋄ REV MR 58 # 5048 Zbl 339 # 02015
• ID 61216

DAHN, B.I. [1975] *On models with variable universe* (**J** 0063) Studia Logica 34∗11-23
⋄ C90 ⋄ REV MR 54 # 12491b Zbl 307 # 02035 • ID 61213

DAHN, B.I. [1976] see ANDREKA, H.

DAHN, B.I. [1979] *Constructions of classical models by means of Kripke models (survey)* (**J** 0063) Studia Logica 38∗401-405
⋄ C07 C20 C90 E35 E45 ⋄ REV MR 82g:03065
Zbl 439 # 03014 • ID 56006

DAHN, B.I. [1980] *First order logics for metric structures* (**J** 0068) Z Math Logik Grundlagen Math 26∗77-88
⋄ C65 C80 C90 ⋄ REV MR 81g:03006 Zbl 433 # 03021
• ID 72091

DAHN, B.I. [1981] *Partial isomorphisms and intuitionistic logic* (**J** 0063) Studia Logica 40∗405-413
⋄ C75 C90 F50 ⋄ REV MR 84m:03012 Zbl 496 # 03011
• ID 35717

DAHN, B.I. [1982] *The limit behaviour of exponential terms* (**S** 3231) Prepr NF, Sekt Math, Humboldt-Univ Berlin 48∗30pp
⋄ C60 C65 ⋄ REV MR 86f:03058 Zbl 516 # 03019
• ID 37254

DAHN, B.I. & GEHNE, J. [1983] *A theory of ordered exponential fields which has no model completion* (**P** 1601) Easter Conf on Model Th (1);1983 Diedrichshagen 39-41
⋄ C35 C60 C65 ⋄ REV MR 84i:03008 Zbl 571 # 03013
• ID 41836

DAHN, B.I. & WOLTER, H. [1983] *On the theory of exponential fields* (**J** 0068) Z Math Logik Grundlagen Math 29∗465-480
⋄ C25 C60 C65 ⋄ REV MR 85a:12010 Zbl 531 # 03017
• ID 37678

DAHN, B.I. (ED.) [1983] *Proceedings of the first Easter conference on model theory* (**S** 3382) Sem-ber, Humboldt-Univ Berlin, Sekt Math 49∗154pp
⋄ C97 ⋄ REV MR 84i:03008 Zbl 509 # 00040 • ID 48635

DAHN, B.I. & GOERING, P. [1984] *Notes on exponential-logarithmic terms* (**P** 1545) Easter Conf on Model Th (2);1984 Wittenberg 11-19
⋄ C60 C65 ⋄ REV MR 86h:03060 Zbl 564 # 03033
• ID 44630

DAHN, B.I. & WOLTER, H. [1984] *Ordered fields with several exponential functions* (**J** 0068) Z Math Logik Grundlagen Math 30∗341-348
⋄ C25 C35 C65 D35 ⋄ REV Zbl 504 # 12024
Zbl 527 # 03016 Zbl 548 # 03013 • ID 42268

DAHN, B.I. & WOLTER, H. (EDS.) [1984] *Proceedings of the second Easter conference on model theory, Wittenberg (GDR), April 23-27,1984* (**S** 3382) Sem-ber, Humboldt-Univ Berlin, Sekt Math 60∗243pp
⋄ C97 ⋄ REV Zbl 542 # 00001 • ID 43707

DAHN, B.I. [1984] *The limit behaviour of exponential terms* (J 0027) Fund Math 124*169-186
  ◊ C60 C65 D15 ◊ REV MR 86f:03058 • ID 45383

DAHN, B.I. see Vol. II, IV for further entries

DAI, ZHIZHONG [1982] *Henselian valued fields (Chinese) (English summary)* (J 3742) Shuxue Yanjiu yu Pinglun 2/1*153-161
  ◊ C60 ◊ REV MR 83k:12024 Zbl 498 # 12028 • ID 40276

DAIGNEAULT, A. [1963] *Theorie des modeles en logique mathematique* (X 0893) Pr Univ Montreal: Montreal 184pp
  ◊ C98 G05 G15 ◊ REV MR 40 # 31 Zbl 164.310
  • ID 23589

DAIGNEAULT, A. [1964] *Freedom in polyadic algebras and two theorems of Beth and Craig* (J 0133) Michigan Math J 11*129-135
  ◊ C05 C40 G15 ◊ REV MR 29 # 3408 JSL 36.337
  • ID 02751

DAIGNEAULT, A. [1971] *Boolean powers in algebraic logic* (J 0068) Z Math Logik Grundlagen Math 17*411-420
  ◊ C20 C30 G05 G15 ◊ REV MR 45 # 8512
  Zbl 241 # 02022 • ID 02756

DALEN VAN, D. [1978] *An interpretation of intuitionistic analysis* (J 0007) Ann Math Logic 13*1-43
  ◊ C90 F35 F50 ◊ REV MR 80i:03067 Zbl 399 # 03049
  • ID 27959

DALEN VAN, D. [1980] *Logic and structure* (X 0811) Springer: Heidelberg & New York viii + 172pp
  ◊ B98 C98 ◊ REV MR 81f:03001 MR 84k:03002
  Zbl 434 # 03001 • REM 2nd ed. 1983; X + 207pp • ID 55705

DALEN VAN, D. [1984] *How to glue analysis models* (J 0036) J Symb Logic 49*1339-1349
  ◊ C65 C90 F35 F50 ◊ REV MR 86e:03056
  Zbl 576 # 03038 • ID 39878

DALEN VAN, D. see Vol. I, II, IV, V, VI for further entries

DALES, H.G. & ESTERLE, J. [1977] *Discontinuous homomorphisms from C(X)* (J 0015) Bull Amer Math Soc 83*257-259
  ◊ C65 E50 E75 ◊ REV MR 55 # 3791 Zbl 341 # 46037
  • ID 81016

DALES, H.G. see Vol. V for further entries

DALLA CHIARA SCABIA, M.L. [1968] *Modelli sintattici e semantici delle teorie elementari* (X 0844) Feltrinelli: Milano 240pp
  ◊ C07 C98 E35 E45 F25 F98 ◊ REV MR 39 # 5308
  Zbl 202.8 JSL 39.959 JSL 40.236 • ID 04285

DALLA CHIARA SCABIA, M.L. see Vol. I, II, V for further entries

DAMYANOV, B.P. & DANOV, D.L. & KHRISTOV, KH.YA. [1977] *Sets of asymptotic numbers closed with respect to each of the four algebraic operations (Russian)* (J 2547) Serdica, Bulgar Math Publ 3*136-151
  ◊ H15 ◊ REV MR 58 # 11260 Zbl 459 # 10036 • ID 54472

DAMYANOV, B.P. see Vol. I for further entries

DANHOF, K.J. [1972] see BUECHI, J.R.

DANHOF, K.J. [1973] see BUECHI, J.R.

DANHOF, K.J. [1974] *A proof of the compactness theorem* (J 0068) Z Math Logik Grundlagen Math 20*179-182
  ◊ C07 ◊ REV MR 50 # 4278 Zbl 317 # 02064 • ID 02770

DANIELS, C.B. & FREEMAN, J.B. [1977] *Classical second-order intensional logic with maximal propositions* (J 0122) J Philos Logic 6*1-31
  ◊ B15 C85 C90 ◊ REV MR 58 # 10310 Zbl 353 # 02005
  • ID 30696

DANIELS, C.B. & FREEMAN, J.B. [1978] *A logic of generalized quantification* (J 0302) Rep Math Logic, Krakow & Katowice 10*9-42
  ◊ B15 C80 C90 ◊ REV MR 81m:03022 Zbl 446 # 03008
  • ID 56537

DANIELS, C.B. see Vol. II for further entries

DANILOVA, S.A. [1979] *The group of automorphisms of a model (Russian)* (J 1516) Vest Ser Fiz Mat Mekh, Univ Minsk 1979*69-70,88
  ◊ C07 C30 ◊ REV MR 83d:03049 • ID 35189

DANKO, W. [1983] *Algorithmic properties of finitely generated structures* (P 3830) Logics of Progr & Appl;1980 Poznan 118-131
  ◊ B75 C40 C52 C75 ◊ REV MR 84b:68001
  Zbl 508 # 03012 • ID 36935

DANKO, W. [1983] *Interpretability of algorithmic theories* (J 2095) Fund Inform, Ann Soc Math Pol, Ser 4 6*217-233
  ◊ B25 B75 C35 F25 ◊ REV MR 84h:68022
  Zbl 559 # 68048 • ID 39522

DANKO, W. see Vol. II, IV for further entries

DANOV, D.L. [1977] see DAMYANOV, B.P.

DAPUETO, C. [1982] *Interpretation in an elementary topos (Italian summary)* (J 3741) Rend Mat Appl, Ser 7 2*531-545
  ◊ C85 G30 ◊ REV MR 84f:03056 Zbl 509 # 03035
  • ID 34475

DAPUETO, C. see Vol. I, VI for further entries

DATUASHVILI, N.I. [1982] *Characterization of varieties of B-valued models (Russian) (English and Georgian summaries)* (J 0954) Tr Inst Prikl Mat, Tbilisi 11*5-14
  ◊ C05 C90 E40 ◊ REV MR 85k:03022 Zbl 576 # 03022
  • ID 49878

DAVEY, B.A. [1973] *Sheaf spaces and sheaves of universal algebras* (J 0044) Math Z 134*275-290
  ◊ C05 C90 ◊ REV MR 48 # 8345 Zbl 266 # 08002
  • ID 02791

DAVEY, B.A. & WERNER, H. [1979] *Injectivity and boolean powers* (J 0044) Math Z 166*205-223
  ◊ C30 G05 ◊ REV MR 80b:08004 Zbl 402 # 08010
  • ID 90184

DAVEY, B.A. see Vol. V for further entries

DAVID, R. [1982] *Some applications of Jensen's coding theorem* (J 0007) Ann Math Logic 22*177-196
  ◊ C62 D55 E40 E45 ◊ REV MR 83j:03084
  Zbl 489 # 03021 • ID 35373

DAVID, R. & FRIEDMAN, S.D. [1985] *Uncountable ZF-ordinals* (P 4046) Rec Th;1982 Ithaca 217-222
  ◊ C62 D60 E40 E45 ◊ REV Zbl 566 # 03033 • ID 46391

DAVID, R. see Vol. IV, V for further entries

DAVIDSON, B. & JACKSON, F.C. & PARGETTER, R. [1977] *Modal trees for T and S5* (J 0047) Notre Dame J Formal Log 18*602-606
⋄ B25 B45 ⋄ REV MR 58 # 16145 Zbl 299 # 02022
• ID 23697

DAVIDSON, B. see Vol. II for further entries

DAVIS, R.L. [1953] *The number of structures of finite relations* (J 0053) Proc Amer Math Soc 4*486-495
⋄ C13 ⋄ REV MR 14.1053 Zbl 51.247 JSL 39.341
• ID 02839

DAVIS, R.L. see Vol. V for further entries

DAWES, A.M. [1977] *End extensions which are models of a given theory* (J 0068) Z Math Logik Grundlagen Math 23*463-467
⋄ C07 C15 C30 C62 ⋄ REV MR 58 # 10411
Zbl 441 # 03005 • ID 28190

DAWES, A.M. see Vol. IV, V, VI for further entries

DAWSON JR., J.W. [1979] *The Goedel incompleteness theorem from a length-of-proof perspective* (J 0005) Amer Math Mon 86*740-748
⋄ A10 F20 F30 H15 ⋄ REV MR 80i:03001
Zbl 426 # 03061 • ID 53657

DAWSON JR., J.W. see Vol. IV, V for further entries

DAY, A. & FREESE, R. [1980] *A characterization of identities implying congruence modularity. I* (J 0017) Canad J Math 32*1140-1167
⋄ C05 ⋄ REV MR 82b:08009 Zbl 414 # 08003 • ID 69428

DAY, A. [1984] see CZEDLI, G.

DAY, A. see Vol. IV for further entries

DECIU, E. [1980] *On the cardinal number and on the structure of the set of all subrings of the rational numbers field* (J 0440) Bul Inst Politeh Bucuresti, Ser Mec 42/2*3-8
⋄ C55 ⋄ REV MR 82a:04003 Zbl 462 # 04003 • ID 72214

DEGTEV, A.N. [1978] *Solvability of the ∀∃-theory of a certain factor-lattice of recursively enumerable sets (Russian)* (J 0003) Algebra i Logika 17*134-143,241
• TRANSL [1978] (J 0069) Algeb and Log 17*94-101
⋄ B25 D25 D50 ⋄ REV MR 80j:03055 Zbl 445 # 03020
• ID 72221

DEGTEV, A.N. see Vol. IV, V for further entries

DEHORNOY, P. [1975] *Solution d'une conjecture de Bukovsky (English summary)* (J 2313) C R Acad Sci, Paris, Ser A-B 281*A821-A824
⋄ C20 C62 E40 E55 ⋄ REV MR 53 # 12947
Zbl 365 # 02059 • ID 23194

DEHORNOY, P. [1976] *Intersections d'ultrapuissances iterees de modeles de la theorie des ensembles (English summary)* (J 2313) C R Acad Sci, Paris, Ser A-B 282*A935-A937
⋄ C20 C62 E25 E55 ⋄ REV MR 53 # 7786
Zbl 397 # 03015 • ID 23024

DEHORNOY, P. [1978] *Iterated ultrapowers and Prikry forcing* (J 0007) Ann Math Logic 15*109-160
⋄ C20 C62 E25 E40 E55 ⋄ REV MR 80k:03057
Zbl 417 # 03025 • ID 29158

DEHORNOY, P. [1983] *An application of ultrapowers to changing cofinality* (J 0036) J Symb Logic 48*225-235
⋄ C20 E40 E55 ⋄ REV MR 85b:03088 Zbl 526 # 03033
• ID 38177

DEHORNOY, P. see Vol. V for further entries

DEISSLER, R. [1977] *Minimal models* (J 0036) J Symb Logic 42*254-260
⋄ C15 C40 C50 C75 ⋄ REV MR 58 # 27443
Zbl 393 # 03019 • ID 26453

DEISSLER, R. [1979] *Minimal and prime models of complete theories of torsion free abelian groups* (J 0004) Algeb Universalis 9*250-265
⋄ C50 C60 ⋄ REV MR 80e:03035 Zbl 428 # 03026
• ID 53785

DEJON, B. & HENRICI, P. (EDS.) [1969] *Constructive aspects of the fundamental theorem of algebra: proceedings of a symposium, IBM* (X 0827) Wiley & Sons: New York 337pp
⋄ B28 C97 F97 ⋄ REV Zbl 175.1 • ID 23438

DEKKER, J.C.E. [1969] *Countable vector spaces with recursive operations. Part I* (J 0036) J Symb Logic 34*363-387
⋄ C57 C60 D45 D50 ⋄ REV MR 40 # 5449 Zbl 185.20
• REM Part II 1971 • ID 02900

DEKKER, J.C.E. [1970] see APPLEBAUM, C.H.

DEKKER, J.C.E. [1971] *Countable vector spaces with recursive operations. Part II* (J 0036) J Symb Logic 36*477-493
⋄ C57 C60 D45 D50 ⋄ REV MR 45 # 4967
Zbl 231 # 02052 • REM Part I 1969 • ID 02901

DEKKER, J.C.E. [1971] *Two notes on vector spaces with recursive operations* (J 0047) Notre Dame J Formal Log 12*329-334
⋄ C57 C60 D45 ⋄ REV MR 47 # 3160 Zbl 205.308
• ID 02902

DEKKER, J.C.E. see Vol. IV for further entries

DELFRATE, M.G. [1983] *Oriented 2-graphs and their generalization to dimension n (Italian) (English summary)* (J 0549) Riv Mat Univ Parma, Ser 4 9*153-160
⋄ C52 C65 ⋄ REV MR 85f:05066 Zbl 549 # 05058
• ID 39944

DELIYANNIS, P.C. [1970] *Group representations and cardinal algebras* (J 0017) Canad J Math 22*759-772
⋄ C60 E10 ⋄ REV MR 46 # 3625 Zbl 238 # 06015
• ID 27908

DELIYANNIS, P.C. see Vol. II for further entries

DELON, F. [1978] *Definition de l'arithmetique dans la theorie des anneaux de series formelles (English summary)* (J 2313) C R Acad Sci, Paris, Ser A-B 286*A87-A89
⋄ C60 D35 F25 F35 ⋄ REV MR 82a:12018
Zbl 382 # 13010 • ID 27313

DELON, F. [1978] *Types localement isoles et theoreme des deux cardinaux dans les theories stables denombrables* (C 3093) Groupe Etude Th Stables (1) 1977/78 1/7*10pp
⋄ C45 C50 C55 ⋄ REV MR 80d:03030b Zbl 408 # 03026
• ID 72262

DELON, F. [1980] *Hensel fields in equal characteristic p>0* (P 2625) Model Th of Algeb & Arith;1979 Karpacz 108-116
⋄ C60 ⋄ REV MR 82g:03058 Zbl 477 # 12028 JSL 50.853
• ID 55602

DELON, F. [1981] *Indecidabilite de la theorie des anneaux de series formelles a plusieurs indeterminees (English summary)* (J 0027) Fund Math 112*215-229
⋄ C60 D35 ⋄ REV MR 83e:03048 Zbl 515 # 12020
• ID 35228

DELON, F. [1981] *Types sur C((X))* (**C** 2619) Groupe Etude Th Stables (2) 1978/79 2/5*29pp
- ⋄ C10 C45 C60 ⋄ REV MR 82i:03044   Zbl 454 # 03012
- • ID 72260

DELON, F. [1984] *Corps equivalents a leurs corps de series* (**S** 3521) Mem Soc Math Fr 16*95-103
- ⋄ C25 C60 ⋄ REV MR 86f:03003 • ID 39723

DELON, F. [1984] *Espaces ultrametriques* (**J** 0036) J Symb Logic 49*405-424
- ⋄ C25 C35 C45 C50 C65 ⋄ REV MR 85i:03116
- • ID 42437

DELON, F. & LASCAR, D. & PARIGOT, M. & SABBAGH, G. (EDS.) [1984] *Logique, Octobre 1983, Paris. Compte rendu de la table ronde de logique des 15 et 16 octobre 1983, Paris* (**S** 3521) Mem Soc Math Fr 16*iii+103pp
- ⋄ B97 C97 ⋄ REV MR 86f:03003   Zbl 547 # 00009
- • ID 43258

DELON, F. [1984] *Theories completes de corps* (**C** 4356) Gen Log Semin Paris 1982/83 123-137
- ⋄ C35 C60 C98 ⋄ ID 47518

DELON, F. & LASCAR, D. & LOUVEAU, A. & SABBAGH, G. (EDS.) [1985] *Seminaire general de logique 1982-83* (**X** 4643) Univ Paris VII, UER Math: Paris 186pp
- ⋄ B97 C98 D97 ⋄ REV MR 86f:03005   Zbl 567 # 00003
- • ID 47516

DELZELL, C.N. [1982] *Case distinctions are necessary for representing polynomials as sums of squares* (**P** 3708) Herbrand Symp Logic Colloq;1981 Marseille 87-103
- ⋄ A10 C60 C65 F55 ⋄ REV Zbl 502 # 03032 • ID 36898

DELZELL, C.N. [1982] *Continuous sums of squares of forms* (**P** 3638) Brouwer Centenary Symp;1981 Noordwijkerhout 65-75
- ⋄ C60 C65 F55 ⋄ REV MR 85g:03086   Zbl 527 # 10016
- • ID 38556

DELZELL, C.N. [1984] *A continuous, constructive solution to Hilbert's 17th problem* (**J** 0305) Invent Math 76*365-384
- ⋄ C60 F55 ⋄ REV MR 86e:12003   Zbl 547 # 12017
- • ID 43271

DEMILLO, R.A. & LIPTON, R.J. [1979] *Some connections between mathematical logic and complexity theory* (**P** 3542) ACM Symp Th of Comput (11);1979 Atlanta 153-159
- ⋄ C57 D15 F30 H15 ⋄ REV MR 81h:03084 • ID 72280

DEMILLO, R.A. see Vol. IV, VI for further entries

DENECKE, K. [1982] *Preprimal algebras* (**X** 0911) Akademie Verlag: Berlin 162pp
- ⋄ B50 C05 C13 ⋄ REV MR 85a:08003   Zbl 506 # 08003
- • ID 37824

DENEF, J. [1975] *Hilbert's tenth problem for quadratic rings* (**J** 0053) Proc Amer Math Soc 48*214-220
- ⋄ C60 D35 ⋄ REV MR 50 # 12961   Zbl 324 # 02032
- • ID 04079

DENEF, J. [1978] *Diophantine sets over Z[T]* (**J** 0053) Proc Amer Math Soc 69*148-150
- ⋄ C60 D25 D35 ⋄ REV MR 57 # 2899   Zbl 393 # 03035
- • ID 52455

DENEF, J. & LIPSHITZ, L. [1978] *Diophantine sets over some rings of algebraic integers* (**J** 3172) J London Math Soc, Ser 2 18*385-391
- ⋄ C60 D35 ⋄ REV MR 80a:12030   Zbl 399 # 10049
- • ID 52854

DENEF, J. [1978] *The Diophantine problem for polynomial rings and fields of rational functions* (**J** 0064) Trans Amer Math Soc 242*391-399
- ⋄ C60 D35 ⋄ REV MR 58 # 10809   Zbl 399 # 10048
- • ID 52853

DENEF, J. [1979] *The Diophantine problem for polynomial rings of positive characteristic* (**P** 2627) Logic Colloq;1978 Mons 131-145
- ⋄ C60 D35 ⋄ REV MR 81h:03090   Zbl 457 # 12011
- • ID 54394

DENEF, J. [1979] see BECKER, J.A.

DENEF, J. [1980] see BECKER, J.A.

DENEF, J. & LIPSHITZ, L. [1980] *Ultraproducts and approximation in local rings. II* (**J** 0043) Math Ann 253*1-28
- ⋄ C20 C60 ⋄ REV MR 82g:13021   Zbl 439 # 13018 • REM Part I 1979 by Becker,J. & Denef,J. & Dries van den,L. & Lipshitz,L. • ID 81085

DENEF, J. & JARDEN, M. & LEWIS, D. [1983] *On Ax-fields which are C_i* (**J** 0131) Quart J Math, Oxford Ser 2 34*21-36
- ⋄ B25 C60 ⋄ REV MR 84j:12028   Zbl 519 # 12015
- • ID 36723

DENEF, J. & LIPSHITZ, L. [1984] *Power series solutions of algebraic differential equations* (**J** 0043) Math Ann 267*213-238
- ⋄ B25 C60 C65 D35 D80 ⋄ REV MR 85j:12010 Zbl 518 # 12015 • ID 38480

DENEF, J. [1984] *The rationality of the Poincare series associated to the p-adic points on a variety* (**J** 0305) Invent Math 77*1-23
- ⋄ C10 C60 ⋄ REV MR 86c:11043   Zbl 537 # 12011
- • ID 43852

DENENBERG, L. & LEWIS, H.R. [1984] *Logical syntax and computational complexity* (**P** 2153) Logic Colloq;1983 Aachen 2*101-116
- ⋄ B25 D15 ⋄ REV MR 86e:03013   Zbl 564 # 03009
- • ID 43003

DENENBERG, L. & LEWIS, H.R. [1984] *The complexity of the satisfiability problem for Krom formulas* (**J** 1426) Theor Comput Sci 30*319-341
- ⋄ B20 B25 D15 ⋄ ID 43373

DENISOV, A.S. [1984] *Constructive homogeneous extensions (Russian)* (**J** 0092) Sib Mat Zh 25/6*60-69
- • TRANSL [1984] (**J** 0475) Sib Math J 25*879-888
- ⋄ C50 C57 ⋄ REV Zbl 571 # 03012 • ID 42732

DENISOV, S.D. [1972] *Models of a noncontradictory formulas and the Ershov hierarchy (Russian)* (**J** 0003) Algebra i Logika 11*648-655,736
- • TRANSL [1972] (**J** 0069) Algeb and Log 11*359-362
- ⋄ C07 C57 D55 ⋄ REV MR 52 # 5406   Zbl 268 # 02028
- • ID 61324

DENISOV, S.D. [1974] *Three theorems on elementary theories and tt-reducibility (Russian)* (J 0003) Algebra i Logika 13*5-8,120
- TRANSL [1974] (J 0069) Algeb and Log 13*1-2
⋄ C65 D30 ⋄ REV MR 50 # 9561 Zbl 289 # 02036
• ID 25962

DENISOV, S.D. see Vol. IV for further entries

DENJOY, A. [1953] *Les matrices d'ordination de toutes puissances* (J 0109) C R Acad Sci, Paris 236*345-348
⋄ C65 E07 ⋄ REV MR 15.408 Zbl 50.53 • ID 43585

DENJOY, A. see Vol. V for further entries

DESUA, F. [1956] *Consistency and completeness - a resume* (J 0005) Amer Math Mon 63*295-305
⋄ C07 ⋄ REV Zbl 71.7 • ID 37353

DEUTSCH, H. [1985] *A note on the decidability of a strong relevant logic* (J 0063) Studia Logica 44*159-164
⋄ B25 B46 ⋄ ID 47508

DEUTSCH, H. see Vol. II for further entries

DEUTSCH, M. [1975] *Zur Theorie der spektralen Darstellung von Praedikaten durch Ausdruecke der Praedikatenlogik 1.Stufe* (J 0009) Arch Math Logik Grundlagenforsch 17*9-16
⋄ B10 C13 D25 D35 ⋄ REV MR 54 # 2436 Zbl 337 # 02025 • ID 02972

DEUTSCH, M. [1977] *Zur Axiomatisierung zyklischer Gruppen* (J 1514) Praxis Math 19*169-172
⋄ C13 C60 ⋄ REV MR 56 # 5261 Zbl 361 # 20002 • ID 72328

DEUTSCH, M. see Vol. I, IV, V, VI for further entries

DEVLIN, K.J. [1974] *An introduction to the fine structure of the constructible hierarchy* (P 0602) Generalized Recursion Th (1);1972 Oslo 123-163
⋄ C55 D55 E45 E47 E55 E98 ⋄ REV MR 53 # 2659 Zbl 295 # 02037 • ID 15073

DEVLIN, K.J. [1983] *Reduced powers of $\aleph_2$-trees* (J 0027) Fund Math 118*129-134
⋄ C20 E05 E45 ⋄ REV MR 85i:03156 Zbl 536 # 03028 • ID 37114

DEVLIN, K.J. see Vol. IV, V for further entries

DIACONESCU, R. [1976] *Grothendieck toposes have boolean points - a new proof* (J 0394) Commun Algeb 4*723-729
⋄ C90 G30 ⋄ REV MR 54 # 2757 Zbl 358 # 18011 • ID 24062

DIACONESCU, R. see Vol. V, VI for further entries

DIARRA, B. [1977] *Ultra-produits de corps munis d'une valeur absolue ultra-metrique (English summary)* (J 2313) C R Acad Sci, Paris, Ser A-B 284*A1261-A1263
⋄ C20 C60 ⋄ REV MR 56 # 2816 Zbl 351 # 12103 • ID 72375

DIARRA, B. [1984] *Ultraproduits ultrametriques de corps values* (J 1934) Ann Sci Univ Clermont Math 22*1-37
⋄ C20 C60 ⋄ REV Zbl 566 # 12020 • ID 46361

DICHEV, KH.V. [1980] *Representation of data in calculation of logical formulas on finite models (Russian)* (C 3048) Teor Vopr Proekt Vych Sist 88-96,102
⋄ B25 C13 ⋄ REV MR 83a:68020 • ID 38854

DICKMANN, M.A. [1970] *Model theory of infinitary languages. Vol. I, II* (X 1599) Aarhus Univ Mat Inst: Aarhus viii + 390pp
⋄ C75 C98 E55 ⋄ REV MR 42 # 7495 Zbl 255 # 02006 • ID 28954

DICKMANN, M.A. [1972] see COLE, J.C.

DICKMANN, M.A. [1973] *The problem of non-finite axiomatizability of $\aleph_1$-categorical theories* (P 0710) Russell Mem Logic Conf;1971 Uldum 141-216
⋄ C35 C52 C98 ⋄ REV MR 50 # 4272 • ID 17034

DICKMANN, M.A. [1975] *Large infinitary languages; model theory* (X 0809) North Holland: Amsterdam xv + 464pp
⋄ C75 C98 E55 ⋄ REV MR 58 # 27450 Zbl 324 # 02010 JSL 43.144 • ID 26233

DICKMANN, M.A. [1977] *Deux applications de la methode de va-et-vient* (J 0056) Publ Dep Math, Lyon 14/2*63-92
⋄ C60 C75 ⋄ REV MR 58 # 27456 Zbl 402 # 03035 • ID 54682

DICKMANN, M.A. [1977] *Structures Σ-saturees* (P 1695) Set Th & Hierarch Th (3);1976 Bierutowice 153-168
⋄ C40 C50 C70 ⋄ REV MR 58 # 21574 Zbl 393 # 03022 • ID 31711

DICKMANN, M.A. [1980] see CHERLIN, G.L.

DICKMANN, M.A. [1980] *On polynomials over real closed rings* (P 2625) Model Th of Algeb & Arithm;1979 Karpacz 117-135
⋄ C60 ⋄ REV MR 83h:13037 Zbl 452 # 13001 • ID 54118

DICKMANN, M.A. [1980] *Sur les anneaux de polynomes a coefficients dans un anneau reel clos (English summary)* (J 2313) C R Acad Sci, Paris, Ser A-B 290*A57-A60
⋄ C60 ⋄ REV MR 81e:03032 Zbl 423 # 03035 • ID 53547

DICKMANN, M.A. [1980] *Types remarquables et extensions de modeles dans l'arithmetique de Peano I* (J 1620) Asterisque 73*59-117
⋄ C25 C30 C62 ⋄ REV MR 82h:03029 Zbl 463 # 03040 JSL 48.483 • REM Part II 1980 by Ressayre,J.-P. • ID 54580

DICKMANN, M.A. [1981] see CHERLIN, G.L.

DICKMANN, M.A. [1981] *Sur les ouverts semi-algebriques d'une cloture reelle* (J 0050) Port Math 40*231-238
⋄ C60 ⋄ ID 48285

DICKMANN, M.A. [1983] see CHERLIN, G.L.

DICKMANN, M.A. [1985] *Applications of model theory to real algebraic geometry; a survey* (P 2160) Latin Amer Symp Math Log (6);1983 Caracas 76-150
⋄ C10 C60 C98 ⋄ ID 41800

DICKMANN, M.A. [1985] *Elimination of quantifiers for ordered valuation rings* (P 4310) Easter Conf on Model Th (3);1985 Gross Koeris 64-88
⋄ C10 C60 ⋄ REV Zbl 573 # 03010 • ID 49046

DICKMANN, M.A. [1985] *Larger infinitary languages* (C 4183) Model-Theor Log 317-363
⋄ C75 C98 ⋄ ID 48286

DICKMANN, M.A. see Vol. V for further entries

DIENER, K.-H. [1983] *On constructing infinitary languages $L_{\alpha\beta}$ without the axiom of choice* (J 0068) Z Math Logik Grundlagen Math 29*357-376
- ◊ C75 E25 ◊ REV MR 85a:03078 Zbl 549 # 03050
- ID 34750

DIERS, Y. [1983] *Une description axiomatique des categories de faisceaux de structures algebriques sur les espaces topologiques booleens* (J 0345) Adv Math 47*258-299
- ◊ C90 G30 ◊ REV MR 84m:18013 Zbl 513 # 18008
- ID 37808

DILLER, J. & MUELLER, GERT H. (EDS.) [1975] ⊢ *ISILC proof theory symposium* (S 3301) Lect Notes Math 500*viii+383pp
- ◊ C97 F97 ◊ REV MR 52 # 13304 Zbl 309 # 00006 • REM This volume contains only the proof theory part of the conference • ID 21771

DILLER, J. see Vol. I, II, IV, VI for further entries

DIMITRACOPOULOS, C. & PARIS, J.B. [1982] *Truth definitions for Δ_0 formulae* (P 3482) Logic & Algor (Specker);1980 Zuerich 317-329
- ◊ C40 C62 H15 ◊ REV MR 84d:03041 Zbl 475 # 03033
- ID 55487

DIMITRACOPOULOS, C. & PARIS, J.B. [1983] *A note on the undefinability of cuts* (J 0036) J Symb Logic 48*564-569
- ◊ C40 C62 F30 H15 ◊ REV MR 84k:03151 Zbl 529 # 03040 • ID 35043

DIMITRACOPOULOS, C. [1985] *A generalization of a theorem of H.Friedman* (J 0068) Z Math Logik Grundlagen Math 31*221-225
- ◊ C62 D15 F30 ◊ ID 47569

DIMITRACOPOULOS, C. see Vol. VI for further entries

DIXON, P.G. [1977] *Classes of algebraic systems defined by universal Horn sentences* (J 0004) Algeb Universalis 7*315-339
- ◊ C05 C52 ◊ REV MR 56 # 114 Zbl 364 # 08004
- ID 26599

DOBBERTIN, H. [1982] *On Vaught's criterion for isomorphisms of countable boolean algebras* (J 0004) Algeb Universalis 15*95-114
- ◊ C05 C15 G05 ◊ REV MR 83m:06017 Zbl 443 # 08002
- ID 40383

DOBRITSA, V.P. [1975] *Recursively numbered classes of constructive extensions and autostability of algebras (Russian)* (J 0092) Sib Mat Zh 16*1148-1154,1369
- • TRANSL [1975] (J 0475) Sib Math J 16*879-883
- ◊ C57 D45 ◊ REV MR 53 # 12917 Zbl 328 # 02030
- ID 23167

DOBRITSA, V.P. & GONCHAROV, S.S. [1976] *An example of a constructive abelian group with non-constructivizable reduced subgroup (Russian)* (P 2064) All-Union Conf Math Log (4);1976 Kishinev 33
- ◊ C57 C60 D45 ◊ ID 32652

DOBRITSA, V.P. [1976] *On computable and strictly computable classes of constructive algebras (Russian)* (P 2572) Material Respub Konf Molod Uchen;1976 Alma Ata 187
- ◊ C57 C60 D45 ◊ ID 32653

DOBRITSA, V.P. [1977] *Computability of certain classes of constructive algebras (Russian)* (J 0092) Sib Mat Zh 18*570-579
- • TRANSL [1977] (J 0475) Sib Math J 18*406-413
- ◊ C57 C60 D45 ◊ REV MR 56 # 5244 Zbl 361 # 02065
- ID 32655

DOBRITSA, V.P. & KHISAMIEV, N.G. & NURTAZIN, A.T. [1978] *Constructive periodic abelian groups (Russian)* (J 0092) Sib Mat Zh 19*1260-1265
- • TRANSL [1978] (J 0475) Sib Math J 19*886-890
- ◊ C57 C60 F60 ◊ REV MR 80b:20067 Zbl 421 # 20021
- ID 32600

DOBRITSA, V.P. [1981] *On constructivizable abelian groups (Russian)* (J 0092) Sib Mat Zh 22/3*208-213,239
- ◊ C57 C60 D40 D45 ◊ REV MR 82i:03058 Zbl 473 # 03037 • ID 55366

DOBRITSA, V.P. [1983] *Complexity of the index set of a constructive model (Russian)* (J 0003) Algebra i Logika 22*372-381
- • TRANSL [1983] (J 0069) Algeb and Log 22*269-276
- ◊ C57 D45 ◊ REV MR 86d:03042 Zbl 537 # 03022
- ID 39619

DOBRITSA, V.P. [1983] *Some constructivizations of abelian groups (Russian)* (J 0092) Sib Mat Zh 24/2*18-25
- • TRANSL [1983] (J 0475) Sib Math J 24*167-173
- ◊ C57 C60 D45 ◊ REV MR 85d:20062 Zbl 528 # 20038
- ID 36756

DOBRITSA, V.P. see Vol. IV for further entries

DOEPP, K. [1972] *Bemerkung zu Henkins Beweis fuer die Nichtstandard-Vollstaendigkeit der Typentheorie* (J 0047) Notre Dame J Formal Log 13*561-562
- ◊ B15 C62 C85 ◊ REV MR 50 # 12700 Zbl 232 # 02013
- ID 04094

DOEPP, K. see Vol. I, IV for further entries

DOETS, H.C. [1969] *Novak's result by Henkin's method* (J 0027) Fund Math 64*329-333
- ◊ C62 E35 E70 ◊ REV MR 39 # 3983 Zbl 293 # 02047
- ID 03072

DOETS, H.C. [1971] *A theorem on the existence of expansions (Russian summary)* (J 0014) Bull Acad Pol Sci, Ser Math Astron Phys 19*1-3
- ◊ C07 C20 E25 ◊ REV MR 44 # 6469 Zbl 208.13
- ID 03075

DOETS, H.C. see Vol. V for further entries

DOKAS, L. [1974] *The notion of mathematical structure (Greek) (French summary)* (C 1503) In Honor of Plakides 317-324
- ◊ C07 ◊ REV MR 54 # 4964 Zbl 289 # 00001 • ID 24133

DOKAS, L. see Vol. V for further entries

DONDER, H.-D. & KOEPKE, P. [1983] *On the consistency strength of "accessible" Jonsson cardinals and of the weak Chang conjecture* (J 0073) Ann Pure Appl Logic 25*233-261
- ◊ C55 E05 E35 E55 ◊ REV MR 85j:03084 Zbl 556 # 03041 • ID 39341

DONDER, H.-D. see Vol. V for further entries

DONER, J.E. [1970] *Tree acceptors and some of their applications* (J 0119) J Comp Syst Sci 4*406-451 • ERR/ADD ibid 5*453(1971))
- ◊ B25 C85 D05 ◊ REV MR 44 # 5179ab Zbl 212.29 JSL 37.619 • ID 03082

DONER, J.E. [1972] *Definability in the extended arithmetic of ordinal numbers* (J 0202) Diss Math (Warsaw) 96∗50pp
⋄ C85 E10 E47 ⋄ REV MR 49 # 2396 Zbl 248 # 02056
• ID 04095

DONER, J.E. & MOSTOWSKI, ANDRZEJ & TARSKI, A. [1978] *The elementary theory of well-ordering - a metamathematical study* (P 1897) Logic Colloq;1977 Wroclaw 1-54
⋄ B25 C10 C65 E07 ⋄ REV MR 80d:03027a
Zbl 461 # 03003 • ID 54480

DONER, J.E. see Vol. IV, V for further entries

DONNADIEU, M.-R. & RAMBAUD, C. [1977] *Theoremes de completude dans les theories sur graphes orientes (English summary)* (J 2313) C R Acad Sci, Paris, Ser A-B 284∗A355-A358
⋄ C07 C65 G30 ⋄ REV MR 56 # 112 Zbl 359 # 02051
• ID 27294

DONNADIEU, M.-R. see Vol. II, V for further entries

DORACZYNSKA, E. [1974] *The complete theory of the midpoint operation (Russian summary)* (J 0014) Bull Acad Pol Sci, Ser Math Astron Phys 22∗1195-1200
⋄ C35 C60 ⋄ REV MR 51 # 5281 Zbl 298 # 20066
• ID 17459

DORACZYNSKI, R. [1973] *Elimination of bound variables in logic with an arbitrary quantifier (Polish and Russian summaries)* (J 0063) Studia Logica 32∗117-129
⋄ C10 C80 ⋄ REV MR 50 # 4279 Zbl 322 # 02009
• ID 04096

DOSEN, K. [1984] see BOZIC, M.

DOSEN, K. [1985] *A completeness theorem for the Lambek calculus of syntactic categories* (J 0068) Z Math Logik Grundlagen Math 31∗235-241
⋄ B65 C90 D05 ⋄ REV Zbl 547 # 03027 • ID 47571

DOSEN, K. see Vol. II, VI for further entries

DOSHITA, S. & YAMASAKI, S. [1983] *The satisfiability problem for a class consisting of Horn sentences and some non-Horn sentences in propositional logic* (J 0194) Inform & Control 59∗1-12 • ERR/ADD ibid 60∗174
⋄ B20 B25 D15 ⋄ REV Zbl 564 # 03010 • ID 44209

DOSHITA, S. see Vol. I, IV for further entries

D'OTTAVIANO, I.M.L. [1985] *The model extension theorems for Π_3-theories* (P 2160) Latin Amer Symp Math Log (6);1983 Caracas 157-173
⋄ B50 C07 C52 ⋄ ID 41799

D'OTTAVIANO, I.M.L. see Vol. II for further entries

DOUWEN VAN, E.K. & MONK, J.D. & RUBIN, M. [1980] *Some questions about boolean algebras* (J 0004) Algeb Universalis 11∗220-243
⋄ C98 G05 G98 ⋄ REV MR 82a:06024 Zbl 451 # 06014
• ID 54046

DOUWEN VAN, E.K. see Vol. V for further entries

DOW, A. [1985] *Saturated boolean algebras and their Stone spaces* (J 2635) Topology Appl 21∗193-207
⋄ C50 C55 E35 E50 E75 G05 ⋄ ID 48304

DOW, A. see Vol. V for further entries

DOWNEY, R.G. [1983] *On a question of A.Retzlaff* (J 0068) Z Math Logik Grundlagen Math 29∗379-384
⋄ C57 C60 D25 D45 ⋄ REV MR 85b:03075
Zbl 526 # 03028 • ID 38174

DOWNEY, R.G. [1984] *A note on decompositions of recursively enumerable subspaces* (J 0068) Z Math Logik Grundlagen Math 30∗465-470
⋄ C57 C60 D25 D45 ⋄ REV MR 86b:03052
Zbl 535 # 03022 • ID 43527

DOWNEY, R.G. [1984] *Bases of supermaximal subspaces and Steinitz systems. I* (J 0036) J Symb Logic 49∗1146-1159
⋄ C57 C60 D25 D45 ⋄ ID 39869

DOWNEY, R.G. [1984] *Co-immune subspaces and complementation in V_∞* (J 0036) J Symb Logic 49∗528-538
⋄ C57 C60 D45 D50 ⋄ REV MR 86d:03043 • ID 42438

DOWNEY, R.G. [1984] see ASH, C.J.

DOWNEY, R.G. & REMMEL, J.B. [1984] *The universal complementation property* (J 0036) J Symb Logic 49∗1125-1136
⋄ C57 C60 D25 D45 ⋄ ID 39864

DOWNEY, R.G. & HIRD, G.R. [1985] *Automorphisms of supermaximal subspaces* (J 0036) J Symb Logic 50∗1-9
⋄ C57 C60 D25 D45 ⋄ REV Zbl 572 # 03024 • ID 39783

DOWNEY, R.G. & KALANTARI, I. [1985] *Effective extensions of linear forms on a recursive vector space over a recursive field* (J 0068) Z Math Logik Grundlagen Math 31∗193-200
⋄ C57 C60 D45 ⋄ ID 47566

DOWNEY, R.G. see Vol. IV for further entries

DRABBE, J. [1966] *Les ordinaux dans les modeles internes de la theorie des ensembles* (J 0150) Acad Roy Belg Bull Cl Sci (5) 52∗808-815
⋄ C62 E10 E45 ⋄ REV MR 35 # 43 Zbl 192.42 • ID 03098

DRABBE, J. [1970] *Sur une propriete des formules negatives* (J 2313) C R Acad Sci, Paris, Ser A-B 271∗A629
⋄ C30 C40 C52 ⋄ REV MR 42 # 4382 • ID 03106

DRABBE, J. [1971] *Quelques notions de la theorie des modeles* (J 0082) Bull Soc Math Belg 23∗274-282
⋄ C07 C98 ⋄ REV MR 54 # 12469 Zbl 323 # 02002
• ID 30401

DRABBE, J. [1971] *Sur une propriete de preservation* (J 0047) Notre Dame J Formal Log 12∗505-506
⋄ C40 ⋄ REV MR 46 # 44 Zbl 224 # 02041 • ID 03108

DRABBE, J. [1974] *Un concept de forcing* (J 0082) Bull Soc Math Belg 26∗381-391
⋄ C25 ⋄ REV MR 56 # 8359 Zbl 336 # 02042 • ID 61390

DRABBE, J. [1976] *Extensions Cohen-generiques des ensembles admissibles (English summary)* (J 2313) C R Acad Sci, Paris, Ser A-B 283∗A267-A268
⋄ C62 E30 E40 ⋄ REV MR 55 # 2566 Zbl 377 # 02028
• ID 27288

DRABBE, J. see Vol. I, II, IV, V, VI for further entries

DRAGALIN, A.G. & FUKSON, V.I. & LYUBETSKIJ, V.A. [1971] *Definable sequences of countable ordinals (Russian)* (J 0023) Dokl Akad Nauk SSSR 196∗1263-1265
• TRANSL [1971] (J 0062) Sov Math, Dokl 12∗341-343
⋄ C40 C62 E10 E35 E45 E47 ⋄ REV MR 43 # 35
Zbl 245 ∗ 02061 • ID 03116

DRAGALIN, A.G. [1973] *Intuitionistic model theory (Russian)* (S 2337) Istor Metodol Estest Nauk (Moskva) 14*106-126
◇ C90 F50 ◇ REV MR 58 # 27451   Zbl 275 # 02026
• ID 29696

DRAGALIN, A.G. [1979] *Algebraic approach to the analysis of models of non-standard logics (Russian)* (S 2582) Semiotika & Inf, Akad Nauk SSSR 53-55
◇ C90 ◇ ID 90008

DRAGALIN, A.G. see Vol. I, IV, V, VI for further entries

DREBEN, B. [1952] *On the completeness of quantification theory* (J 0054) Proc Nat Acad Sci USA 38*1047-1052
◇ C07 ◇ REV MR 14.526 Zbl 49.292 JSL 18.339
• ID 03125

DREBEN, B. [1957] *Systematic treatment of the decision problem* (P 1675) Summer Inst Symb Log;1957 Ithaca 363
◇ B20 B25 D35 ◇ ID 03611

DREBEN, B. & KAHR, A.S. & WANG, HAO [1962] *Classification of AEA formulas by letter atoms* (J 0015) Bull Amer Math Soc 68*528-532
◇ B20 B25 D35 ◇ REV MR 30 # 22 Zbl 112.5 JSL 29.101
• ID 03127

DREBEN, B. [1962] *Solvable Suranyi subclasses: an introduction to the Herbrand theory* (P 0698) Harvard Symp Digit Comput & Appl;1961 Cambridge MA 32-47
◇ B20 B25 ◇ REV MR 36 # 1318 JSL 30.390 • ID 03612

DREBEN, B. & PUTNAM, H. [1967] *The Craig interpolation lemma* (J 0047) Notre Dame J Formal Log 8*229-233
◇ C40 ◇ REV MR 39 # 1290 Zbl 203.9 • ID 03130

DREBEN, B. & GOLDFARB, W.D. [1979] *The decision problem. Solvable classes of quantificational formulas* (X 0832) Addison-Wesley: Reading xii + 271pp
◇ B25 C98 ◇ REV MR 81i:03015 Zbl 457 # 03005 JSL 47.452 • ID 54330

DREBEN, B. see Vol. I, II, VI for further entries

DRIES VAN DEN, L. [1977] *Artin-Schreier theory for commutative regular rings* (J 0007) Ann Math Logic 12*113-150
◇ B25 C10 C25 C60 ◇ REV MR 58 # 16264 Zbl 376 # 13012 • ID 23674

DRIES VAN DEN, L. [1978] *Model theory of fields* (0000) Diss., Habil. etc 157pp
◇ B25 C10 C25 C35 C60 H20 ◇ REM Rijksuniversiteit Utrecht • ID 45429

DRIES VAN DEN, L. [1979] *Algorithms and bounds for polynomial rings* (P 2627) Logic Colloq;1978 Mons 147-157
◇ C57 C60 ◇ REV MR 81f:03045 Zbl 461 # 13015
• ID 54499

DRIES VAN DEN, L. & RIBENBOIM, P. [1979] *Application de la theorie des modeles aux groupes de Galois de corps de fonctions (English summary)* (J 2313) C R Acad Sci, Paris, Ser A-B 288*A789-A792
◇ C60 ◇ REV MR 80b:12013 Zbl 426 # 12004 • ID 53671

DRIES VAN DEN, L. [1979] *New decidable fields of algebraic numbers* (J 0053) Proc Amer Math Soc 77*251-256
◇ B25 C60 ◇ REV MR 82h:03027 Zbl 408 # 12026
• ID 79616

DRIES VAN DEN, L. [1979] see BECKER, J.A.

DRIES VAN DEN, L. [1980] *A linearly ordered ring whose theory admits elimination of quantifiers is a real closed field* (J 0053) Proc Amer Math Soc 79*97-100
◇ C10 C60 ◇ REV MR 81e:03033 Zbl 397 # 06017 JSL 50.1079 • ID 52716

DRIES VAN DEN, L. [1980] *Some model theory and number theory for models of weak systems of arithmetic* (P 2625) Model Th of Algeb & Arithm;1979 Karpacz 346-362
◇ B25 C62 F30 H15 ◇ REV MR 82f:03029 Zbl 454 # 03034 • ID 54246

DRIES VAN DEN, L. [1981] see CHERLIN, G.L.

DRIES VAN DEN, L. [1981] *Quantifier elimination for linear formulas over ordered and valued fields* (J 3133) Bull Soc Math Belg, Ser B 33*19-31
◇ C10 C60 ◇ REV MR 82k:03035 Zbl 479 # 03018
• ID 55680

DRIES VAN DEN, L. [1981] *Which curves over Z have points with coordinates in a discrete ordered ring?* (J 0064) Trans Amer Math Soc 264*181-189
◇ B25 C60 C62 ◇ REV MR 82i:03046 Zbl 464 # 10047
• ID 79620

DRIES VAN DEN, L. & GLASS, A.M.W. & MACINTYRE, A. & MEKLER, A.H. & POLAND, J. [1982] *Elementary equivalence and the commutator subgroup* (J 0126) Glasgow Math J 23*115-117
◇ C60 ◇ REV MR 84g:03041 Zbl 504 # 03006 • ID 34154

DRIES VAN DEN, L. [1982] *Some applications of a model theoretic fact to (semi-)algebraic geometry* (J 0028) Indag Math 44*397-401
◇ C10 C60 C65 ◇ REV MR 84m:14029 Zbl 538 # 14017
• ID 39472

DRIES VAN DEN, L. & MACINTYRE, A. & MCKENNA, K. [1983] *Elimination of quantifiers in algebraic structures* (J 0345) Adv Math 47*74-87
◇ C10 C60 ◇ REV MR 84f:03028 Zbl 531 # 03016 JSL 50.1079 • ID 34448

DRIES VAN DEN, L. [1983] *Reducing to prime characteristic by means of Artin approximation and constructible properties, and applied to Hochster algebras* (X 1051) Univ Utrecht Math Inst: Utrecht v + 65pp
◇ C60 ◇ REV MR 85f:13025 Zbl 513 # 13010 • ID 39947

DRIES VAN DEN, L. [1984] *Algebraic theories with definable Skolem functions* (J 0036) J Symb Logic 49*625-629
◇ C10 C40 C60 ◇ REV MR 85e:03076 • ID 40354

DRIES VAN DEN, L. & SCHMIDT, K. [1984] *Bounds in the theory of polynomial rings over fields. A nonstandard approach* (J 0305) Invent Math 76*77-91
◇ C60 H10 H20 ◇ REV MR 85i:12016 Zbl 539 # 13011
• ID 41246

DRIES VAN DEN, L. [1984] *Exponential rings, exponential polynomials and exponential functions* (J 0048) Pac J Math 113*51-66
◇ C60 C65 ◇ REV MR 85j:13040 • ID 45151

DRIES VAN DEN, L. & WILKIE, A.J. [1984] *Gromov's theorem on groups of polynomial growth and elementary logic* (J 0032) J Algeb 89*349-374
◇ C60 H20 ◇ REV MR 85k:20101 Zbl 552 # 20017
• ID 45851

DRIES VAN DEN, L. [1984] *Remarks on Tarski's problem concerning* ($\mathbb{R}, +, \cdot, exp$) (P 3710) Logic Colloq;1982 Firenze 97-121
⋄ C10 C60 C65 ⋄ REV MR 86g:03052 • ID 41839

DRIES VAN DEN, L. & SMITH, RICK L. [1985] *Decidable regularly closed fields of algebraic numbers* (J 0036) J Symb Logic 50*468-475
⋄ B25 C57 C60 ⋄ REV Zbl 574 # 12023 • ID 41808

DRIES VAN DEN, L. [1985] *The field of reals with a predicate for the powers of two* (J 0504) Manuscr Math 54*187-195
⋄ B25 C60 C65 ⋄ ID 48383

DRIES VAN DEN, L. see Vol. V, VI for further entries

DROBOTUN, B.N. [1977] *Enumerations of simple models (Russian)* (J 0092) Sib Mat Zh 18*1002-1014,1205
• TRANSL [1977] (J 0475) Sib Math J 18*707-716
⋄ C57 D45 ⋄ REV MR 58 # 16254 Zbl 382 # 03032 • ID 51956

DROBOTUN, B.N. [1977] *On countable models of decidable almost categorical theories (Russian)* (J 0403) Izv Akad Nauk Kazak SSR, Ser Fiz-Mat 1977/5*22-25,91
⋄ B25 C15 C35 C57 ⋄ REV MR 58 # 27444 Zbl 374 # 02032 • ID 51556

DROBOTUN, B.N. & GONCHAROV, S.S. [1980] *Numerations of saturated and homogeneous models (Russian)* (J 0092) Sib Mat Zh 21/2*25-41,236
• TRANSL [1980] (J 0475) Sib Math J 21*164-176
⋄ B25 C50 C57 D45 ⋄ REV MR 82k:03068 Zbl 441 # 03016 • ID 56069

DROSTE, M. & SHELAH, S. [1985] *A construction of all normal subgroup lattices of 2-transitive automorphism groups of linearly ordered sets* (J 0029) Israel J Math 51*223-261
⋄ C07 C60 C65 E07 ⋄ ID 48384

DROSTE, M. [1985] *Classes of universal words for the infinite symmetric groups* (J 0004) Algeb Universalis 20*205-216
⋄ C05 C50 C60 ⋄ ID 48321

DROSTE, M. [1985] *The normal subgroup lattice of 2-transitive automorphism groups of linearly ordered sets* (J 3457) Order 2*291-319
⋄ C07 C60 C65 ⋄ ID 49798

DROSTE, M. see Vol. V for further entries

DRUCK, I.F. [1975] see COSTA DA, N.C.A.

DRUGUSH, YA.M. [1984] *Finite approximability of forest superintuitionistic logics (Russian)* (J 0087) Mat Zametki (Akad Nauk SSSR) 36*755-764
⋄ B25 B55 ⋄ REV MR 86h:03041 • ID 45347

DRUGUSH, YA.M. see Vol. II, VI for further entries

DUBIEL, M. [1973] *Nonstandard well-orderings in ω-models of A_2 (Russian summary)* (J 0014) Bull Acad Pol Sci, Ser Math Astron Phys 21*887-892
⋄ C62 ⋄ REV MR 51 # 2903 Zbl 273 # 02040 • ID 17385

DUBIEL, M. [1976] *Elementary extensions of β-models of A_2 with fixed height* (P 1476) Set Th & Hierarch Th (2) (Mostowski);1975 Bierutowice 77-81
⋄ C62 ⋄ REV MR 56 # 15402 Zbl 342 # 02041 • ID 23798

DUBIEL, M. [1977] *Generalized quantifiers and elementary extensions of countable models* (J 0036) J Symb Logic 42*341-348
⋄ C15 C65 C80 ⋄ REV MR 57 # 9526 Zbl 399 # 03022 • ID 31694

DUBIEL, M. [1980] *Generalized quantifiers in models of set theory* (J 0027) Fund Math 106*153-161
⋄ C62 C80 E35 ⋄ REV MR 81j:03057 Zbl 452 # 03029 • ID 54093

DUBOIS, DONALD WARD & RECIO MUNIZ, T.J. [1984] *A note on Robinson's non-negativity criterion* (J 0027) Fund Math 122*71-76
⋄ C60 ⋄ REV MR 85j:12003 Zbl 562 # 14007 • ID 47436

DUBOIS, DONALD WARD see Vol. IV for further entries

DUBROVSKY, D.L. [1974] *Some subfields of $\mathbb{Q}_p$ and their non-standard analogues* (J 0017) Canad J Math 26*473-491
⋄ C60 H15 H20 ⋄ REV MR 50 # 9845 Zbl 273 # 12105 • ID 04100

DUBUC, E.J. & REYES, G.E. [1979] *Subtoposes of the ring classifier* (S 3462) Var Publ Ser, Aarhus Univ 30*101-122
⋄ C60 C75 C90 G30 ⋄ REV MR 81j:03098 Zbl 403 # 18003 • ID 54769

DUGAS, M. & GOEBEL, R. [1982] *Every cotorsion-free algebra is an endomorphism algebra* (J 0044) Math Z 181*451-470
⋄ C55 C60 E75 ⋄ REV MR 84h:13008 Zbl 506 # 16022 • ID 45871

DUGAS, M. & GOEBEL, R. [1982] *Every cotorsion-free ring is an endomorphism ring* (J 3240) Proc London Math Soc, Ser 3 45*319-336
⋄ C55 C60 E45 E65 E75 ⋄ REV MR 84b:20064 Zbl 506 # 16022 • ID 37825

DUGAS, M. & HERDEN, G. [1983] *Abitrary torsion classes and almost free abelian groups* (J 0029) Israel J Math 44*322-334
⋄ C55 C60 E50 E55 E65 E75 ⋄ REV MR 84k:16034 Zbl 525 # 20037 • ID 37486

DUGAS, M. & HERDEN, G. [1983] *Arbitrary torsion classes of abelian groups* (J 0394) Commun Algeb 11*1455-1472
⋄ C55 C60 E55 E65 E75 ⋄ REV MR 84f:20058 Zbl 543 # 20038 • ID 48322

DUGAS, M. & GOEBEL, R. [1984] *Almost Σ-cyclic abelian p-groups in L* (P 4185) Abel Group & Modul;1984 Udine 87-105
⋄ C55 C60 E45 E75 ⋄ REV MR 86g:20072 Zbl 574 # 20041 • ID 46405

DUGAS, M. & GOEBEL, R. & GOLDSMITH, B. [1984] *Representation of algebras over a complete discrete valuation ring* (J 0131) Quart J Math, Oxford Ser 2 35*131-146
⋄ C55 C60 E75 ⋄ REV MR 85k:16052 Zbl 536 # 16031 • ID 36775

DUGAS, M. [1985] *On reduced products of abelian groups* (J 0144) Rend Sem Mat Univ Padova 73*41-47
⋄ C20 C60 E75 ⋄ ID 42704

DUGAS, M. see Vol. V for further entries

DULAC, M.-H. [1971] *Decidabilite et operations entre theories* (J 2313) C R Acad Sci, Paris, Ser A-B 273*A1113-A1114
⋄ B10 B25 D35 ⋄ REV MR 45 # 3204 Zbl 224 # 02038 • ID 03165

DURET, J.-L. [1977] *Instabilite des corps formellement reels*
(J 0018) Canad Math Bull 20*385-387
⋄ C45 C60 ⋄ REV MR 57 # 5739 Zbl 385 # 03027
• ID 31027

DURET, J.-L. [1980] *Les corps pseudo-finis ont la propriete d'independance (English summary)* (J 2313) C R Acad Sci, Paris, Ser A-B 290*A981-A983
⋄ C45 C60 ⋄ REV MR 81m:03043 Zbl 469 # 03020
• ID 55148

DURET, J.-L. [1980] *Les corps faiblement algebriquement clos non separablement clos ont la propriete d'independance* (P 2625) Model Th of Algeb & Arithm;1979 Karpacz 136-162
⋄ C45 C60 ⋄ REV MR 83i:12024 Zbl 489 # 03009
• ID 90362

DURET, J.-L. [1983] *Stabilite des corps separablement clos*
(C 3822) Groupe Etude Th Stables (3) 1980/82 Exp.8*13pp
⋄ C35 C45 C60 ⋄ REV MR 85f:03032 Zbl 515 # 03015
• ID 38140

DURNEV, V.G. [1970] *Positive formulae on groups (Russian)*
(J 0789) Uch Zap Mat Ped Inst, Tula 2*215-241
⋄ C60 ⋄ REV MR 57 # 16053 • REM Geometr. i Algebra
• ID 72512

DURNEV, V.G. [1972] *The positive theory of a free semigroup (Russian)* (C 1435) Vopr Teor Grupp & Polugrupp 122-172
⋄ C60 ⋄ REV MR 52 # 13360 • ID 21814

DURNEV, V.G. see Vol. I, II, IV, V for further entries

DWINGER, P. [1971] see BALBES, R.

DWINGER, P. see Vol. V for further entries

DYER, E. [1980] see BAUMSLAG, G.

DYER, E. see Vol. IV for further entries

DYMENT, E.Z. [1984] *Recursive metrizability of numbered topological spaces and bases of effective linear topological spaces (Russian)* (J 0031) Izv Vyssh Ucheb Zaved, Mat (Kazan) 1984/8*59-61
• TRANSL [1984] (J 3449) Sov Math 28/8*74-78
⋄ C57 C65 D45 ⋄ REV MR 86d:03044 • ID 44722

DYMENT, E.Z. see Vol. IV, VI for further entries

DYWAN, Z. & STEPIEN, T. [1980] *Every two-valued propositional calculus has the interpolation property* (J 0387) Bull Sect Logic, Pol Acad Sci 9*152-153
⋄ B20 C40 ⋄ REV MR 82b:03026 Zbl 453 # 03028
• ID 54158

DYWAN, Z. see Vol. I, II, V, VI for further entries

DZGOEV, V.D. [1979] *Recursive automorphisms of constructive models (Russian)* (P 2564) All-Union Algeb Conf (15);1979 Novosibirsk
⋄ C57 D45 ⋄ ID 32626

DZGOEV, V.D. & GONCHAROV, S.S. [1980] *Autostability of models (Russian)* (J 0003) Algebra i Logika 19*45-58,132
• TRANSL [1980] (J 0069) Algeb and Log 19*28-37
⋄ C57 D45 G10 ⋄ REV MR 82g:03082 Zbl 468 # 03023
• ID 55088

DZGOEV, V.D. [1980] *On the constructivization of certain structures (Russian)* (J 0092) Sib Mat Zh 21/1*231
⋄ C57 D45 ⋄ ID 32623

DZGOEV, V.D. [1982] *Constructivizations of direct products of algebraic systems (Russian)* (J 0003) Algebra i Logika 21*138-148
• TRANSL [1982] (J 0069) Algeb and Log 21*88-96
⋄ C30 C57 D45 G05 G10 ⋄ REV MR 85b:03076 Zbl 507 # 03013 • ID 37236

DZGOEV, V.D. see Vol. IV for further entries

DZIOBIAK, W. [1978] *A note on incompleteness of modal logics with respect to neighbourhood semantics* (J 0387) Bull Sect Logic, Pol Acad Sci 7*185-190
⋄ B45 C90 ⋄ REV MR 80b:03018 Zbl 415 # 03020
• ID 53122

DZIOBIAK, W. [1981] *Concerning axiomatizability of the quasivariety generated by a finite Heyting or topological boolean algebra* (J 0387) Bull Sect Logic, Pol Acad Sci 10*177-180
⋄ C05 C13 C52 ⋄ REV Zbl 492 # 03029 • ID 38099

DZIOBIAK, W. [1982] see CZELAKOWSKI, J.

DZIOBIAK, W. [1982] *Concerning axiomatizability of the quasivariety generated by a finite Heyting or topological boolean algebra* (J 0063) Studia Logica 41*415-428
⋄ C05 C13 C52 G10 ⋄ REV MR 86c:08012 • ID 42297

DZIOBIAK, W. see Vol. I, II for further entries

EBBINGHAUS, H.-D. [1969] *Ueber eine Praedikatenlogik mit partiell definierten Praedikaten und Funktionen* (J 0009) Arch Math Logik Grundlagenforsch 12*39-53
⋄ B60 C90 ⋄ REV MR 40 # 7092 JSL 37.617 • ID 03219

EBBINGHAUS, H.-D. [1969] *Ueber fuer-fast-alle-Quantoren*
(J 0009) Arch Math Logik Grundlagenforsch 12*179-193
⋄ C80 ⋄ REV MR 43 # 23 Zbl 259 # 02008 • ID 03218

EBBINGHAUS, H.-D. [1971] *On models with large automorphism groups* (J 0009) Arch Math Logik Grundlagenforsch 14*179-197
⋄ C30 C55 C80 ⋄ REV MR 47 # 6471 Zbl 237 # 02015
• ID 03220

EBBINGHAUS, H.-D. & FLUM, J. [1974] *Algebraische Charakterisierung der elementaren Aequivalenz* (J 0160) Math-Phys Sem-ber, NS 21*80-95
⋄ C07 C35 ⋄ REV MR 51 # 10077 Zbl 287 # 02032
• ID 17532

EBBINGHAUS, H.-D. & FLUM, J. [1974] *Eine Maximalitaetseigenschaft der Praedikatenlogik erster Stufe* (J 0160) Math-Phys Sem-ber, NS 21*182-202
⋄ C95 ⋄ REV MR 51 # 10076 Zbl 312 # 02038 • ID 17534

EBBINGHAUS, H.-D. [1975] *Modelltheorie und Logik* (J 0157) Jbuchber Dtsch Math-Ver 76*165-182
⋄ C95 C98 ⋄ REV MR 58 # 10394 Zbl 349 # 02042
• ID 28199

EBBINGHAUS, H.-D. [1975] *Zum L(Q)-Interpolationsproblem*
(J 0044) Math Z 142*271-279
⋄ C40 C55 C80 ⋄ REV MR 52 # 7844 Zbl 288 # 02012
• ID 18125

EBBINGHAUS, H.-D. [1976] *General logic systems* (P 1619) Coloq Log Simb;1975 Madrid 43-49
⋄ C95 ⋄ REV MR 56 # 5210 Zbl 362 # 02002 • ID 50724

EBBINGHAUS, H.-D. & FLUM, J. & THOMAS, WOLFGANG [1978] *Einfuehrung in die mathematische Logik* (**X** 0890) Wiss Buchges: Darmstadt ix+288pp
• TRANSL [1984] (**X** 0811) Springer: Heidelberg & New York ix+216pp
⋄ B98 C95 C98 ⋄ REV MR 81h:03001 Zbl 399 # 03001 Zbl 556 # 03001 • ID 28201

EBBINGHAUS, H.-D. [1980] *Mathematische Logik und Informatik* (**P** 3608) Wechselwirk Inform Math;1979 Wien 119-145
⋄ C13 D10 D15 D98 E70 H15 ⋄ ID 41134

EBBINGHAUS, H.-D. & ZIEGLER, M. [1982] *Interpolation in Logiken monotoner Systeme* (**J** 0009) Arch Math Logik Grundlagenforsch 22∗1-17
⋄ C40 C80 ⋄ REV MR 83m:03044 Zbl 508 # 03015
• ID 90113

EBBINGHAUS, H.-D. [1985] *Extended logics: The general framework* (**C** 4183) Model-Theor Log 25-76
⋄ C40 C55 C80 C95 C98 ⋄ ID 48325

EBBINGHAUS, H.-D. see Vol. I, IV, V, VI for further entries

ECSEDI-TOTH, P. & TURI, L. [1982] *On enumerability of interpolants* (**P** 3758) Algeb Conf (2);1981 Novi Sad 71-78
⋄ B10 C40 ⋄ REV MR 84b:03021 Zbl 536 # 03002
• ID 34907

ECSEDI-TOTH, P. [1984] *A characterization of quasi-varieties in equality-free languages* (**J** 0002) Acta Sci Math (Szeged) 47∗41-54
⋄ C05 C52 ⋄ ID 45760

ECSEDI-TOTH, P. & TURI, L. [1984] *On the number of zero order interpolants* (**J** 0380) Acta Cybern (Szeged) 6∗425-436
⋄ C40 ⋄ REV MR 85i:03027 Zbl 561 # 03002 • ID 44065

ECSEDI-TOTH, P. [1985] *A partial solution of the finite spectrum problem* (**J** 0380) Acta Cybern (Szeged) 7∗211-215
⋄ C13 ⋄ REV MR 86f:03020 • ID 47689

ECSEDI-TOTH, P. see Vol. I for further entries

EDA, K. [1980] *A note on the hereditary properties in the product space* (**J** 2606) Tsukuba J Math 4∗157-159
⋄ C52 C65 ⋄ REV MR 83d:54010 Zbl 499 # 54005
• ID 37159

EDA, K. [1980] *Limit systems and chain condition* (**J** 2606) Tsukuba J Math 4∗147-155
⋄ C30 E40 G05 ⋄ REV MR 82k:06018 Zbl 466 # 06012
• ID 37158

EDA, K. [1982] *A boolean power and a direct product of abelian groups* (**J** 2606) Tsukuba J Math 6∗187-194
⋄ C30 C60 C90 E40 E75 ⋄ REV MR 85b:20078 Zbl 533 # 20026 • ID 37160

EDA, K. [1983] *A note on subgroups of Z^N* (**P** 3802) Abel Group Th;1983 Honolulu 371-374
⋄ C65 E35 E50 E75 ⋄ REV MR 85h:20065 Zbl 521 # 20036 • ID 37470

EDA, K. [1983] *Infinite products of algebras and measurable cardinals (Japanese)* (**P** 4113) Found of Math;1982 Kyoto 12-19
⋄ C30 E55 E75 ⋄ ID 47673

EDA, K. [1983] *On a boolean power of a torsion free abelian group* (**J** 0032) J Algeb 82∗84-93
⋄ C30 C60 C90 E40 E75 G05 ⋄ REV MR 84i:20055 Zbl 538 # 20027 • ID 37161

EDA, K. [1984] *Maximal quotient rings and boolean extensions (Japanese)* (**P** 3668) Log & Founds of Math;1983 Kyoto 142-179
⋄ C60 C90 E40 E75 ⋄ ID 42939

EDA, K. & HIBINO, K. [1984] *On boolean powers of the group Z and (ω,ω)-weak distributivity* (**J** 0090) J Math Soc Japan 36∗619-628
⋄ C30 C60 C90 E35 E50 E75 G05 ⋄ REV MR 85m:20081 Zbl 556 # 20039 • ID 44306

EDA, K. see Vol. V for further entries

EDGAR, G.A. [1973] *The class of topological spaces is equationally definable* (**J** 0004) Algeb Universalis 3∗139-146
⋄ C05 ⋄ REV MR 51 # 3026 Zbl 273 # 08005 • ID 17365

EHDEL'MAN, G.S. [1974] *Certain radicals of metaideals (Russian)* (**J** 3536) Uch Zap Univ, Ivanovo 130∗44-50
⋄ B10 C07 C60 ⋄ REV MR 58 # 16265 • ID 72569

EHDEL'MAN, G.S. [1981] *π-decomposition of metaideals (Russian)* (**C** 3865) Algeb Sistemy (Ivanovo) 218-232
⋄ C07 C60 ⋄ REV MR 85h:03012 • ID 43393

EHRENFEUCHT, A. & LOS, J. [1954] *Sur les produits cartesiens des groupes cycliques infinis* (**J** 0014) Bull Acad Pol Sci, Ser Math Astron Phys 2∗261-263
⋄ C30 C60 ⋄ REV MR 16.110 Zbl 55.253 • ID 03230

EHRENFEUCHT, A. & MOSTOWSKI, ANDRZEJ [1956] *Models of axiomatic theories admitting automorphisms* (**J** 0027) Fund Math 43∗50-68
⋄ C07 C30 E75 ⋄ REV MR 18.863 Zbl 73.7 JSL 31.644
• ID 03232

EHRENFEUCHT, A. [1957] *Application of games to some problems of mathematical logic (Russian summary)* (**J** 0014) Bull Acad Pol Sci, Ser Math Astron Phys 5∗35-37
⋄ C07 C35 C65 E60 ⋄ REV MR 19.4 Zbl 105.9
• ID 03234

EHRENFEUCHT, A. [1957] *On theories categorical in power* (**J** 0027) Fund Math 44∗241-248
⋄ C30 C35 ⋄ REV MR 20 # 3089 Zbl 105.6 JSL 31.645
• ID 03236

EHRENFEUCHT, A. [1957] *Two theories with axioms built by means of pleonasmus* (**J** 0036) J Symb Logic 22∗36-38
⋄ B25 C35 D35 ⋄ REV MR 19.933 Zbl 78.244 JSL 23.445 • ID 03235

EHRENFEUCHT, A. [1959] *Decidability of the theory of one function* (**J** 0369) Notices Amer Math Soc 6∗268
⋄ B25 ⋄ ID 49906

EHRENFEUCHT, A. & MOSTOWSKI, ANDRZEJ [1961] *A compact space of models of first order theories* (**J** 0014) Bull Acad Pol Sci, Ser Math Astron Phys 9∗369-373
⋄ C07 C52 ⋄ REV MR 26 # 6043 Zbl 106.5 JSL 35.586
• ID 03239

EHRENFEUCHT, A. [1961] *An application of games to the completeness problem for formalized theories* (**J** 0027) Fund Math 49∗129-141
⋄ C07 C65 C85 E10 E60 ⋄ REV MR 23 # A3666 Zbl 96.243 JSL 32.281 • ID 24896

EHRENFEUCHT, A. [1965] *Elementary theories with models without automorphisms* (P 0614) Th Models;1963 Berkeley 70-76
⋄ C50 ⋄ REV MR 35 # 37 Zbl 163.7 JSL 39.338 • ID 03244

EHRENFEUCHT, A. & KREISEL, G. [1966] *Strong models of arithmetic* (J 0014) Bull Acad Pol Sci, Ser Math Astron Phys 14*107-110
⋄ C57 C62 ⋄ REV MR 36 # 2497 Zbl 137.7 • ID 03245

EHRENFEUCHT, A. & FUHRKEN, G. [1971] *A finitely axiomatizable complete theory with atomless $F_1(T)$* (J 0009) Arch Math Logik Grundlagenforsch 14*162-166
⋄ C07 ⋄ REV MR 48 # 1910 Zbl 254 # 02043 • ID 03255

EHRENFEUCHT, A. & FUHRKEN, G. [1971] *On models with undefinable elements* (J 0132) Math Scand 28*325-328
⋄ C40 C50 ⋄ REV MR 46 # 8828 Zbl 236 # 02041
• ID 03251

EHRENFEUCHT, A. [1972] *There are continuum ω_0-categorical theories* (J 0014) Bull Acad Pol Sci, Ser Math Astron Phys 20*425-427
⋄ C35 ⋄ REV MR 47 # 1596 Zbl 293 # 02033 • ID 03257

EHRENFEUCHT, A. [1973] *Discernible elements in models for Peano arithmetic* (J 0036) J Symb Logic 38*291-292
⋄ C50 C62 ⋄ REV MR 49 # 2352 Zbl 279 # 02036
• ID 03258

EHRENFEUCHT, A. [1975] *Practical decidability* (J 0119) J Comp Syst Sci 11*392-396
⋄ B25 D15 ⋄ REV MR 52 # 9683 Zbl 329 # 02020
• ID 18126

EHRENFEUCHT, A. & JENSEN, D.C. [1976] *Some problems in elementary arithmetics* (J 0027) Fund Math 92*223-245
⋄ C62 F30 H15 ⋄ REV MR 54 # 7244 Zbl 362 # 02049
• ID 25797

EHRENFEUCHT, A. see Vol. I, II, IV, V, VI for further entries

EHRESMANN, C. [1965] *Categories et structures* (X 0856) Dunod: Paris 358pp
⋄ C90 C98 G30 ⋄ REV MR 35 # 4274 • ID 23592

EHRSAM, S. [1978] *Expose elementaire de la theorie de la stabilite* (C 3093) Groupe Etude Th Stables (1) 1977/78 1/1*30pp
⋄ C45 C98 ⋄ REV MR 80e:03029 Zbl 408 # 03026
• ID 72591

EHRSAM, S. [1978] *L'ordre fondamental, le theoreme de la borne, le theoreme de la relation d'equivalence finie* (C 3093) Groupe Etude Th Stables (1) 1977/78 1/4*16pp
⋄ C20 C45 C98 ⋄ REV MR 80c:03034 Zbl 408 # 03026
• ID 72593

EHRSAM, S. [1978] *Theories ω_1-categoriques* (C 3093) Groupe Etude Th Stables (1) 1977/78 1/3*6pp
⋄ C35 C45 ⋄ REV MR 80e:03027 Zbl 408 # 03026
• ID 72592

EICHHOLZ, T. [1957] *Semantische Untersuchungen zur Entscheidbarkeit im Praedikatenkalkuel mit Funktionsvariablen* (J 0009) Arch Math Logik Grundlagenforsch 3*19-28
⋄ B25 ⋄ REV MR 19.522 Zbl 80.9 JSL 24.241 • ID 03262

EISEN, M. [1962] *Ideal theory and difference algebra* (J 0352) Math Jap 7*159-180
⋄ C60 ⋄ REV MR 25 # 3832 Zbl 111.7 • ID 17927

EISENBERG, E.F. & REMMEL, J.B. [1982] *Effective isomorphisms of algebraic structures* (P 3634) Patras Logic Symp;1980 Patras 95-122
⋄ C57 D45 ⋄ REV MR 84m:03052 Zbl 525 # 03039
• ID 35753

EISENMENGER, H. [1983] *Some local definability results on countable topological structures* (J 0036) J Symb Logic 48*683-692
⋄ C15 C40 C90 ⋄ REV MR 85i:03100 Zbl 549 # 03024
• ID 42405

EKLOF, P.C. [1969] see BARWISE, J.

EKLOF, P.C. [1969] *Resolutions of singularities in prime characteristic for almost all primes* (J 0064) Trans Amer Math Soc 146*429-438
⋄ C20 C60 ⋄ REV MR 41 # 8417 Zbl 191.195 • ID 90195

EKLOF, P.C. [1970] see BARWISE, J.

EKLOF, P.C. & SABBAGH, G. [1970] *Model-completions and modules* (J 0007) Ann Math Logic 2*251-295
⋄ C25 C35 C52 C60 ⋄ REV MR 43 # 3105
Zbl 227 # 02029 • ID 03274

EKLOF, P.C. [1970] *On the existence of $L_{\infty,\kappa}$ indiscernibles* (J 0053) Proc Amer Math Soc 25*798-800
⋄ C30 C50 C75 ⋄ REV MR 41 # 5203 Zbl 231 # 02019
• ID 03273

EKLOF, P.C. & SABBAGH, G. [1971] *Definability problems for modules and rings* (J 0036) J Symb Logic 36*623-649
⋄ C20 C52 C60 C75 ⋄ REV MR 47 # 1605
Zbl 251 # 02052 • ID 03276

EKLOF, P.C. [1972] *Homogeneous universal modules* (J 0132) Math Scand 29*187-196
⋄ C50 C60 ⋄ REV MR 47 # 3168 Zbl 235 # 16017
• ID 03281

EKLOF, P.C. [1972] *Some model theory of abelian groups* (J 0036) J Symb Logic 37*335-342
⋄ C35 C60 ⋄ REV MR 52 # 10411 Zbl 259 # 02042
• ID 03278

EKLOF, P.C. & FISHER, E.R. [1972] *The elementary theory of abelian groups* (J 0007) Ann Math Logic 4*115-171
⋄ B25 C10 C35 C50 C60 ⋄ REV MR 58 # 27457
Zbl 248 # 02049 JSL 39.603 • ID 03279

EKLOF, P.C. [1973] *Lefschetz's principle and local functors* (J 0053) Proc Amer Math Soc 37*333-339
⋄ C60 C75 ⋄ REV MR 48 # 3736 Zbl 254 # 14004
• ID 03282

EKLOF, P.C. [1973] *The structure of ultraproducts of abelian groups* (J 0048) Pac J Math 47*67-79
⋄ C20 C60 ⋄ REV MR 48 # 6281 Zbl 244 # 20063
• ID 03283

EKLOF, P.C. [1974] *Algebraic closure operators and strong amalgamation bases* (J 0004) Algeb Universalis 4*89-98
⋄ C05 ⋄ REV MR 52 # 5521 Zbl 296 # 08009 • ID 03285

EKLOF, P.C. [1974] *Infinitary equivalence of abelian groups* (J 0027) Fund Math 81*305-314
⋄ C60 C75 ⋄ REV MR 50 # 6829 Zbl 327 # 02050
• ID 03284

EKLOF, P.C. [1975] *Categories of local functors* (C 0782) Model Th & Algeb (A. Robinson) 91-116
⋄ C60 C75 G30 ⋄ REV MR 55 # 7766 Zbl 317 # 02067
• ID 29659

EKLOF, P.C. [1975] *On the existence of κ-free abelian groups* (J 0053) Proc Amer Math Soc 47*65-72
⋄ C60 C75 E10 E45 E75 ⋄ REV MR 52 # 599
Zbl 293 # 20036 • ID 18127

EKLOF, P.C. [1976] *Infinitary model theory of abelian groups* (J 0029) Israel J Math 25*97-107
⋄ C52 C60 C75 E35 E55 E75 ⋄ REV MR 56 # 11778
Zbl 347 # 02035 • ID 26077

EKLOF, P.C. [1977] *Applications of logic to the problem of splitting abelian groups* (P 1075) Logic Colloq;1976 Oxford 87*287-299
⋄ C55 C60 C75 C80 E35 E45 E75 ⋄ REV
MR 58 # 27458 Zbl 436 # 03029 • ID 16626

EKLOF, P.C. [1977] *Classes closed under substructures and direct limits* (J 0068) Z Math Logik Grundlagen Math 23*427-430
⋄ C52 C60 C75 ⋄ REV MR 80m:03068 Zbl 448 # 03024
• ID 56624

EKLOF, P.C. [1977] *Methods of logic in abelian group theory* (P 1682) Abel Group Th;1976 Las Cruces 251-269
⋄ C60 C75 E35 E45 E55 E75 ⋄ REV MR 57 # 5740
Zbl 381 # 03041 • ID 30690

EKLOF, P.C. & MEKLER, A.H. [1977] *On constructing indecomposable groups in L* (J 0032) J Algeb 49*96-103
⋄ C55 C60 C75 E35 E45 E55 ⋄ REV MR 56 # 15412
Zbl 372 # 20042 • ID 51455

EKLOF, P.C. [1977] *Ultraproducts for algebraists* (C 1523) Handb of Math Logic 105-137
⋄ C20 C60 C98 ⋄ REV MR 58 # 10395 JSL 49.968
• ID 24198

EKLOF, P.C. & MEKLER, A.H. [1979] *Stationary logic of finitely determinate structures* (J 0007) Ann Math Logic 17*227-269
⋄ B25 C10 C55 C60 C65 C80 ⋄ REV MR 82f:03033
Zbl 448 # 03025 • ID 56625

EKLOF, P.C. & HUBER, M. [1980] *On the rank of Ext* (J 0044) Math Z 174*159-185
⋄ C60 E35 E45 E50 E75 ⋄ REV MR 82a:20061
Zbl 443 # 20046 • ID 56438

EKLOF, P.C. & MEKLER, A.H. [1981] *Infinitary stationary logic and abelian groups* (J 0027) Fund Math 112*1-15
⋄ C60 C75 C80 ⋄ REV MR 82j:03042 Zbl 467 # 03029
• ID 55027

EKLOF, P.C. [1982] *Homological dimension and stationary sets* (J 0044) Math Z 180*1-9 • ERR/ADD ibid 183*575-576
⋄ C60 E05 E75 ⋄ REV MR 84c:16025a MR 84c:16025b
Zbl 504 # 16019 Zbl 509 # 16013 • ID 38408

EKLOF, P.C. & MEKLER, A.H. [1983] *On endomorphism rings of ω_1-separable primary groups* (P 3802) Abel Group Th;1983 Honolulu 320-339
⋄ C60 E45 E50 E75 ⋄ REV MR 85i:20052
Zbl 536 # 20036 • ID 41825

EKLOF, P.C. [1983] *Set theory and structure theorems* (P 3802) Abel Group Th;1983 Honolulu 275-284
⋄ C60 C98 E50 E65 E75 ⋄ REV MR 85a:03061
Zbl 519 # 20040 • ID 34809

EKLOF, P.C. [1983] *The structure of ω_1-separable groups* (J 0064) Trans Amer Math Soc 279*497-523
⋄ C60 E35 E50 E75 ⋄ REV MR 84k:03124
Zbl 578 # 03027 • ID 36142

EKLOF, P.C. & MEZ, H.-C. [1984] *Additive groups of existentially closed rings* (P 4185) Abel Group & Modul;1984 Udine 243-252
⋄ C25 C60 ⋄ REV Zbl 568 # 20057 • ID 46356

EKLOF, P.C. & MEKLER, A.H. & SHELAH, S. [1984] *Almost disjoint abelian groups* (J 0029) Israel J Math 49*34-54
⋄ C55 C60 E45 E50 E75 ⋄ REV Zbl 558 # 20031
• ID 38751

EKLOF, P.C. [1985] *Applications to algebra* (C 4183) Model-Theor Log 423-441
⋄ C60 C75 C98 ⋄ ID 48328

EKLOF, P.C. & HUBER, M. [1985] *On ω-filtered vector spaces and their applications to abelian p-groups* (J 2022) Comm Math Helvetici 60*145-170
⋄ C60 E75 ⋄ REV Zbl 573 # 20054 • ID 40414

EKLOF, P.C. & MEZ, H.-C. [1985] *The ideal structure of existentially closed algebras* (J 0036) J Symb Logic 50*1025-1043
⋄ C25 C60 ⋄ ID 48386

EKLOF, P.C. see Vol. V for further entries

ELGOT, C.C. & WRIGHT, J.B. [1959] *Quantifier elimination in a problem of logical design* (J 0133) Michigan Math J 6*65-69
⋄ B25 B70 C10 ⋄ REV MR 21 # 3287 Zbl 85.340
• ID 03286

ELGOT, C.C. [1961] *Decision problems of finite automata design and related arithmetics* (J 0064) Trans Amer Math Soc 98*21-51 • ERR/ADD ibid 103*558-559
⋄ B25 D03 D05 D35 F30 ⋄ REV MR 25 # 2962
Zbl 111.11 JSL 34.509 • ID 03288

ELGOT, C.C. & RABIN, M.O. [1966] *Decidability and undecidability of extensions of second (first) order theory of (generalized) successor* (J 0036) J Symb Logic 31*169-181
⋄ B15 B25 C85 D05 D35 F30 F35 ⋄ REV Zbl 144.245
• ID 03291

ELGOT, C.C. see Vol. IV, VI for further entries

ELLENTUCK, E. [1968] *The representation of cardinals in models of set theory* (J 0068) Z Math Logik Grundlagen Math 14*143-158
⋄ C62 E10 E25 E35 ⋄ REV MR 37 # 2597 Zbl 169.306
• ID 03300

ELLENTUCK, E. [1971] *A minimal model for strong analysis* (J 0027) Fund Math 73*125-131
⋄ C50 C62 C85 E35 ⋄ REV MR 58 # 5188
Zbl 243 # 02042 JSL 38.524 • ID 03314

ELLENTUCK, E. [1972] see BITTER-RUCKER VON, R.

ELLENTUCK, E. & MANASTER, A.B. [1972] *The decidability of a class of ∀∃ sentences in the isols* (J 0048) Pac J Math 43*573-584
⋄ B25 D50 ⋄ REV MR 48 # 5841 Zbl 242 # 02051
• ID 03309

ELLENTUCK, E. [1973] *The structure of Dedekind cardinals* (J 0064) Trans Amer Math Soc 180*109-125
⋄ C62 E10 E25 E35 ⋄ REV MR 48 # 3743
Zbl 274 # 02038 • ID 03316

ELLENTUCK, E. [1974] *A model of arithmetic* (J 0014) Bull Acad Pol Sci, Ser Math Astron Phys 22*353-355
⋄ C62 E10 E25 ⋄ REV MR 50#93 Zbl 299#02079
• ID 03323

ELLENTUCK, E. [1975] *Semigroups, Horn sentences and isolic structures* (J 0048) Pac J Math 61*87-101
⋄ C05 C57 D50 E10 ⋄ REV MR 53#5279 Zbl 324#02029 • ID 22911

ELLENTUCK, E. [1975] *The foundations of Suslin logic* (J 0036) J Symb Logic 40*567-575
⋄ C75 E75 ⋄ REV MR 52#13319 Zbl 339#02014
• ID 14810

ELLENTUCK, E. & HALPERN, F. [1975] *Theories having many extensions* (J 0302) Rep Math Logic, Krakow & Katowice 4*25-29
⋄ C07 C75 C80 E75 ⋄ REV MR 52#75 Zbl 315#02048
• ID 18129

ELLENTUCK, E. [1976] *Categoricity regained* (J 0036) J Symb Logic 41*639-643
⋄ C35 C75 C90 E40 ⋄ REV MR 58#5163 Zbl 344#02037 • ID 14586

ELLENTUCK, E. [1977] *Boolean valued rings* (J 0027) Fund Math 96*67-86
⋄ C30 C60 C90 D50 ⋄ REV MR 58#16237 Zbl 365#02044 • ID 27196

ELLENTUCK, E. [1977] *Free Suslin algebras* (J 0022) Cheskoslov Mat Zh 27(102)*201-219
⋄ C90 G25 ⋄ REV MR 56#2818 Zbl 382#03045
• ID 51969

ELLENTUCK, E. [1978] *Model theoretic methods in the theory of isols* (J 0007) Ann Math Logic 14*273-285
⋄ C57 D50 ⋄ REV MR 81h:03093 Zbl 393#03032
• ID 29153

ELLENTUCK, E. [1981] *Galois theorems for isolated fields* (J 0068) Z Math Logik Grundlagen Math 27*1-9
⋄ C57 C60 D45 D50 ⋄ REV MR 84a:03048 Zbl 523#03032 • ID 33198

ELLENTUCK, E. [1981] *Hyper-torre isols* (J 0036) J Symb Logic 46*1-5
⋄ C62 D50 ⋄ REV MR 82f:03038 Zbl 503.03020
• ID 72617

ELLENTUCK, E. see Vol. IV, V, VI for further entries

ELLERMAN, D.P. [1971] *Sheaves or relation structures and ultraproducts* (S 1133) Res Rep Boston Univ 1971*71-79
⋄ C20 C90 ⋄ ID 22105

ELLERMAN, D.P. [1974] *Sheaves of structures and generalized ultraproducts* (J 0007) Ann Math Logic 7*163-195
⋄ C20 C25 C90 G30 ⋄ REV MR 51#12516 Zbl 295#02031 • ID 03325

ELLERMAN, D.P. & ROTA, G.-C. [1978] *A measure theoretic approach to logical quantification* (J 0144) Rend Sem Mat Univ Padova 59*227-246
⋄ B10 C90 G25 ⋄ REV MR 81c:03027 Zbl 449#03028
• ID 56695

EMDE BOAS VAN, P. [1973] *Mostowski's universal set-algebra* (X 1121) Math Centr: Amsterdam ZW14/73*24pp
⋄ C50 C57 C65 D45 E07 E20 G05 ⋄ REV Zbl 257#02022 • ID 29008

EMDE BOAS VAN, P. see Vol. II, IV, VI for further entries

EMERSON, E.A. & SISTLA, A.P. [1984] *Deciding full branching time logic* (J 0194) Inform & Control 61*175-201
⋄ B25 B45 D15 ⋄ REV MR 86h:03021 • ID 45349

EMERSON, E.A. see Vol. II, IV for further entries

ENAYAT, A. [1984] *On certain elementary extensions of models of set theory* (J 0064) Trans Amer Math Soc 283*705-715
⋄ C30 C62 E55 ⋄ REV MR 85j:03050 Zbl 563#03033
• ID 41809

ENAYAT, A. [1985] *Weakly compact cardinals in models of set theory* (J 0036) J Symb Logic 50*476-486
⋄ C30 C62 E50 E55 E65 ⋄ ID 41810

ENDERTON, H.B. [1967] *An infinitistic rule of proof* (J 0036) J Symb Logic 32*447-451
⋄ C62 F35 ⋄ REV MR 37#66 Zbl 157.27 • ID 03347

ENDERTON, H.B. [1970] *Finite partially-ordered quantifiers* (J 0068) Z Math Logik Grundlagen Math 16*393-397
⋄ B15 C80 C85 ⋄ REV MR 44#1546 Zbl 193.294
• ID 03351

ENDERTON, H.B. [1970] *The unique existential quantifier* (J 0009) Arch Math Logik Grundlagenforsch 13*52-54
⋄ C80 D55 ⋄ REV MR 44#1567 Zbl 272#02019 JSL 40.627 • ID 03352

ENDERTON, H.B. & FRIEDMAN, H.M. [1971] *Approximating the standard model of analysis* (J 0027) Fund Math 72*175-188
⋄ C62 E45 F35 ⋄ REV MR 45#6609 Zbl 235#02053 JSL 39.600 • ID 03353

ENDERTON, H.B. [1972] *A mathematical introduction to logic* (X 0801) Academic Pr: New York xiii+295pp
⋄ B98 C98 F30 ⋄ REV MR 49#2239 Zbl 298#02002 JSL 38.340 • ID 03355

ENDERTON, H.B. [1973] *Constructible β-models* (J 0068) Z Math Logik Grundlagen Math 19*277-282
⋄ C62 E45 ⋄ REV MR 48#3723 Zbl 301#02053
• ID 03357

ENDERTON, H.B. see Vol. IV, V, VI for further entries

ENGELER, E. [1958] *Untersuchungen zur Modelltheorie* (0000) Diss., Habil. etc 28pp
⋄ C07 C75 ⋄ REV MR 21#5552 • REM Promotionsarbeit, Art Inst Orell Fuessli AG, Zuerich • ID 21316

ENGELER, E. [1959] *Aequivalenzklassen von n-Tupeln* (J 0068) Z Math Logik Grundlagen Math 5*340-345
⋄ C07 ⋄ REV MR 25#4996 Zbl 92.249 JSL 31.504
• ID 03360

ENGELER, E. [1959] *Eine Konstruktion von Modellerweiterungen* (J 0068) Z Math Logik Grundlagen Math 5*126-131
⋄ C30 ⋄ REV MR 22#12 Zbl 87.9 • ID 03359

ENGELER, E. [1961] *Unendliche Formeln in der Modell-Theorie* (J 0068) Z Math Logik Grundlagen Math 7*154-160
⋄ C15 C35 C40 C75 ⋄ REV MR 27#1371 Zbl 109.7 JSL 35.343 • ID 03361

ENGELER, E. [1961] *Zur Beweistheorie von Sprachen mit unendlich langen Formeln* (J 0068) Z Math Logik Grundlagen Math 7*213-218
⋄ C75 F07 ⋄ REV MR 30#1030 Zbl 124.248 JSL 36.685
• ID 03362

ENGELER, E. [1963] *A reduction-principle for infinite formulas* (J 0043) Math Ann 151*296-303
  ⋄ C75 F07 ⋄ REV MR 28 # 2974 Zbl 114.245 JSL 33.123
  • ID 03364

ENGELER, E. [1965] *Combinatorial theorems for the construction of models* (P 0614) Th Models;1963 Berkeley 77-88
  ⋄ C80 C85 C95 ⋄ REV MR 34 # 1186 Zbl 199.9
  • ID 03363

ENGELER, E. [1966] *On structures defined by mapping filters* (J 0043) Math Ann 167*105-112
  ⋄ C20 E75 G30 ⋄ REV MR 35 # 1455 Zbl 149.246
  • ID 03365

ENGELER, E. [1967] *Algorithmic properties of structures* (J 0041) Math Syst Theory 1*183-195
  ⋄ B75 C75 D10 D75 ⋄ REV MR 37 # 72 Zbl 202.8 JSL 37.197 • ID 03366

ENGELER, E. [1967] *On the structure of elementary maps* (J 0068) Z Math Logik Grundlagen Math 13*323-328
  ⋄ C07 C20 ⋄ REV MR 36 # 4972 Zbl 174.14 • ID 03367

ENGELER, E. [1968] *Remarks on the theory of geometrical constructions* (P 0637) Syntax & Semant Infinitary Lang;1967 Los Angeles 64-76
  ⋄ C75 D80 ⋄ REV Zbl 179.14 • ID 29303

ENGELER, E. [1970] *Geometry and language* (J 0076) Dialectica 24*77-85
  ⋄ B30 C75 ⋄ REV MR 44 # 5216 Zbl 256 # 02008
  • ID 03368

ENGELER, E. [1975] *Algorithmic logic* (P 1430) Adv Course Founds Computer Sci;1974 Amsterdam 55-85
  ⋄ B25 B75 D15 ⋄ REV MR 52 # 12380 Zbl 314 # 68009 JSL 42.420 • ID 21738

ENGELER, E. [1975] *On the solvability of algorithmic problems* (P 0775) Logic Colloq;1973 Bristol 231-251
  ⋄ B75 C60 C75 D75 ⋄ REV MR 52 # 2842 Zbl 317 # 02051 • ID 17633

ENGELER, E. [1976] *Lower bounds by Galois theory* (J 1620) Asterisque 38-39*45-52
  ⋄ B75 C50 ⋄ REV MR 56 # 7315 Zbl 355 # 02027
  • ID 50225

ENGELER, E. [1981] *Generalized Galois theory and its application to complexity* (J 1426) Theor Comput Sci 13*271-293
  ⋄ C60 D15 ⋄ REV MR 82j:68018 Zbl 468 # 12014
  • ID 81166

ENGELER, E. see Vol. I, II, IV, V, VI for further entries

ENJALBERT, P. [1978] *Epimorphismes: quelques resultats algebriques et logiques (English summary)* (J 2313) C R Acad Sci, Paris, Ser A-B 287*A175-A177
  ⋄ C60 ⋄ REV MR 80j:18002 Zbl 386 # 18001 • ID 81173

ENJALBERT, P. see Vol. II for further entries

ERDELYI, M. [1959] *Systems of equations over non-commutative groups (Ungarisch) (Engl. Zusammenfassung)* (S 1596) Acta Univ Debrecen 6*43-54
  ⋄ C25 C60 ⋄ REV MR 26 # 178 Zbl 234 # 20019
  • ID 27813

ERDOES, P. & GILLMAN, L. & HENRIKSEN, M. [1955] *An isomorphism theorem for real-closed fields* (J 0120) Ann of Math, Ser 2 61*542-554
  ⋄ C60 E05 E50 E75 ⋄ REV MR 16.993 Zbl 65.23
  • ID 03401

ERDOES, P. & HAJNAL, A. [1966] *On a problem of B.Jonsson* (J 0014) Bull Acad Pol Sci, Ser Math Astron Phys 14*19-23
  ⋄ C05 C55 E05 E50 E55 ⋄ REV MR 35 # 64 Zbl 171.265 • ID 03455

ERDOES, P. & HAJNAL, A. & MILNER, E.C. [1972] *Partition relations for η_α and for $\aleph_\alpha$-saturated models* (C 0648) Th of Sets and Topology (Hausdorff) 95-108
  ⋄ C50 E05 E07 E50 ⋄ REV MR 49 # 7143 Zbl 277 # 04006 • ID 17190

ERDOES, P. see Vol. V, VI for further entries

ERIMBETOV, M.M. [1974] *Categoricity in power and the invalidity of the two-cardinal theorem for a formula of finite rank (Russian)* (J 0003) Algebra i Logika 13*493-500,605
  • TRANSL [1974] (J 0069) Algeb and Log 13*283-286
  ⋄ C35 C45 C55 ⋄ REV MR 52 # 13366 Zbl 324 # 02037
  • ID 31746

ERIMBETOV, M.M. [1974] *On almost categorical theories (Russian)* (P 1590) All-Union Conf Math Log (3);1974 Novosibirsk 70-72
  ⋄ C35 C45 ⋄ REV MR 52 # 2867 • ID 31748

ERIMBETOV, M.M. [1975] *Complete theories with 1-cardinal formulas (Russian)* (J 0003) Algebra i Logika 14*245-257,368
  • TRANSL [1975] (J 0069) Algeb and Log 14*151-158
  ⋄ C35 C45 C55 ⋄ REV MR 54 # 10005 Zbl 331 # 02034
  • ID 25846

ERIMBETOV, M.M. [1975] *On almost categorical theories (Russian)* (J 0092) Sib Mat Zh 16*279-292,420
  • TRANSL [1975] (J 0475) Sib Math J 16*215-225
  ⋄ C35 C45 ⋄ REV MR 52 # 2867 Zbl 311 # 02058
  • ID 17653

ERIMBETOV, M.M. [1981] *A fragment of a logic with infinite formulas (Russian)* (C 3806) Issl Teor Progr 30-48
  ⋄ C60 C75 ⋄ REV MR 84m:03059 • ID 35759

ERIMBETOV, M.M. [1981] *A property of countable stable theories (Russian)* (J 0092) Sib Mat Zh 22/1*81-86,229
  • TRANSL [1981] (J 0475) Sib Math J 22*59-63
  ⋄ C45 ⋄ REV MR 82e:03032 Zbl 522 # 03022 • ID 72703

ERIMBETOV, M.M. see Vol. II, IV for further entries

ERSHOV, YU.L. [1962] *Axiomatizable classes of models with infinite signature (Russian)* (J 0003) Algebra i Logika 1/4*32-44
  ⋄ C52 ⋄ REV MR 33 # 3930 • ID 03515

ERSHOV, YU.L. [1962] *Decidability of elementary theories of certain classes of abelian groups (Russian)* (J 0003) Algebra i Logika 1/6*37-41
  ⋄ B25 C60 ⋄ REV MR 31 # 1192 Zbl 227 # 02027
  • ID 03518

ERSHOV, YU.L. [1964] *Decidability of certain non-elementary theories (Russian)* (J 0003) Algebra i Logika 3/2*45-47
  ⋄ B25 C30 C85 G05 ⋄ REV MR 30 # 27 Zbl 276 # 02030 • ID 03521

ERSHOV, YU.L. [1964] *Decidability of the elementary theory of relatively complemented distributive lattices and of the theory of filters (Russian)* (J 0003) Algebra i Logika 3/3*17-38
  ⋄ B25 G10 ⋄ REV MR 31 # 4725 JSL 34.126 • ID 06607

ERSHOV, YU.L. [1964] *Unsolvability of theories of symmetric and simple finite groups (Russian)* (**J** 0023) Dokl Akad Nauk SSSR 158*777-779
- TRANSL [1964] (**J** 0062) Sov Math, Dokl 5*1309-1311
- ⋄ C13 C60 D35 ⋄ REV MR 30 # 3019 Zbl 199.30
- ID 03520

ERSHOV, YU.L. & LAVROV, I.A. & TAJMANOV, A.D. & TAJTSLIN, M.A. [1965] *Elementary theories (Russian)* (**J** 0067) Usp Mat Nauk 20/4*37-108
- TRANSL [1965] (**J** 1399) Russ Math Surv 20/4*35-105
- ⋄ B25 C98 D35 D98 ⋄ REV MR 32 # 4012 Zbl 199.30 JSL 39.603
- ID 03525

ERSHOV, YU.L. [1965] *On elementary theories of local fields (Russian)* (**J** 0003) Algebra i Logika 4/2*5-30
- ⋄ B25 C35 C60 ⋄ REV MR 34 # 7501 Zbl 274 # 02021
- ID 61591

ERSHOV, YU.L. [1965] *On the elementary theory of maximal normed fields (Russian)* (**J** 0003) Algebra i Logika 4/3*31-70,4/6*47-48
- ⋄ B25 C35 C60 ⋄ REV MR 33 # 1307 MR 33 # 5619 Zbl 274 # 02022 Zbl 274 # 02023
- REM Part I. Part II 1966
- ID 03522

ERSHOV, YU.L. [1965] *On the elementary theory of maximal normed fields (Russian)* (**J** 0023) Dokl Akad Nauk SSSR 165*21-23
- TRANSL [1965] (**J** 0062) Sov Math, Dokl 6*1390-1393
- ⋄ B25 C60 ⋄ REV MR 32 # 7554 Zbl 152.24
- ID 42959

ERSHOV, YU.L. [1966] *On the elementary theory of maximal normed fields II (Russian)* (**J** 0003) Algebra i Logika 5/1*8-40
- ⋄ C35 C60 ⋄ REV MR 34 # 4134 Zbl 283 # 02039
- REM Part I 1965. Part III 1967
- ID 17052

ERSHOV, YU.L. [1967] *Elementary theories of Post varieties (Russian)* (**J** 0003) Algebra i Logika 6/5*7-15
- ⋄ B25 C05 C13 G20 ⋄ REV MR 38 # 34 Zbl 207.298
- ID 03527

ERSHOV, YU.L. [1967] *Fields with a solvable theory (Russian)* (**J** 0023) Dokl Akad Nauk SSSR 174*19-20
- TRANSL [1967] (**J** 0062) Sov Math, Dokl 8*575-576
- ⋄ B25 C35 C60 ⋄ REV MR 35 # 5424 Zbl 153.372
- ID 03529

ERSHOV, YU.L. [1967] *On the elementary theory of maximal normed fields III (Russian) (English summary)* (**J** 0003) Algebra i Logika 6/3*31-38
- ⋄ C35 C60 ⋄ REV MR 36 # 6280
- REM Part II 1966
- ID 90199

ERSHOV, YU.L. [1968] *Numbered fields* (**P** 0627) Int Congr Log, Meth & Phil of Sci (3,Proc);1967 Amsterdam 31-34
- ⋄ C57 C60 D45 ⋄ REV MR 41 # 55
- ID 37357

ERSHOV, YU.L. [1968] *Restricted theories of totally ordered sets (Russian)* (**J** 0003) Algebra i Logika 7/3*38-47
- TRANSL [1968] (**J** 0069) Algeb and Log 7*153-159
- ⋄ B25 C65 D35 E07 ⋄ REV MR 41 # 44 Zbl 191.297
- ID 03532

ERSHOV, YU.L. [1969] *The number of linear orders on a field (Russian)* (**J** 0087) Mat Zametki (Akad Nauk SSSR) 6*201-211
- ⋄ C55 C60 ⋄ REV MR 40 # 2587
- ID 03534

ERSHOV, YU.L. [1970] see BREDIKHIN, S.V.

ERSHOV, YU.L. [1972] *Elementary group theories (Russian)* (**J** 0023) Dokl Akad Nauk SSSR 203*1240-1243
- TRANSL [1972] (**J** 0062) Sov Math, Dokl 13*528-532
- ⋄ B25 C05 C13 C60 D35 ⋄ REV MR 45 # 6892 Zbl 264 # 02047
- ID 03551

ERSHOV, YU.L. [1972] *Existence of constructivizations (Russian)* (**J** 0023) Dokl Akad Nauk SSSR 204*1041-1044
- TRANSL [1972] (**J** 0062) Sov Math, Dokl 13*779-783
- ⋄ C57 D45 ⋄ REV MR 47 # 8288 Zbl 261 # 02017
- ID 03552

ERSHOV, YU.L. [1972] *Relationship between sheaf spaces and numbered sets with the C_2^* property (Russian) (English summary)* (**S** 0228) Zap Nauch Sem Leningrad Otd Mat Inst Steklov 32*18-20,154
- TRANSL [1976] (**J** 1531) J Sov Math 6*358-360
- ⋄ C90 D25 D45 D65 ⋄ REV MR 52 # 5382 Zbl 349 # 02030
- ID 18137

ERSHOV, YU.L. [1973] *Constructive models (Russian)* (**C** 0733) Izbr Vopr Algeb & Log (Mal'tsev) 111-130
- ⋄ C57 D45 ⋄ REV MR 50 # 4291 Zbl 293 # 02038
- ID 17197

ERSHOV, YU.L. & PALYUTIN, E.A. & TAJTSLIN, M.A. [1973] *Mathematical logic (Russian)* (**X** 0913) Novosibirsk Gos Univ: Novosibirsk 159pp
- ⋄ B98 C98 ⋄ REV MR 57 # 9448
- ID 32036

ERSHOV, YU.L. [1973] *Skolem functions and constructive models (Russian)* (**J** 0003) Algebra i Logika 12*644-654,735
- TRANSL [1973] (**J** 0069) Algeb and Log 12*368-373
- ⋄ C57 C65 F50 ⋄ REV MR 57 # 5727 Zbl 287 # 02024
- ID 30533

ERSHOV, YU.L. [1974] *Theory of numerations, III: Constructive models (Russian)* (**X** 0913) Novosibirsk Gos Univ: Novosibirsk 139pp
- TRANSL [1977] (**J** 0068) Z Math Logik Grundlagen Math 23*289-371
- ⋄ B25 C57 C98 D45 F50 ⋄ REV MR 54 # 9995 Zbl 374 # 02028
- REM Part II 1973
- ID 25764

ERSHOV, YU.L. [1979] *Relatively complemented, distributive lattices (Russian)* (**J** 0003) Algebra i Logika 18*680-722
- TRANSL [1979] (**J** 0069) Algeb and Log 18*431-459
- ⋄ C20 G05 G10 ⋄ REV MR 82b:06012 Zbl 451 # 06013
- ID 69534

ERSHOV, YU.L. [1980] *Decision problems and constructivizable models (Russian)* (**X** 2027) Nauka: Moskva 416pp
- ⋄ B25 C57 C60 C98 D45 D98 ⋄ REV MR 82h:03009 Zbl 495 # 03009
- ID 72718

ERSHOV, YU.L. [1980] *Multiply valued fields (Russian)* (**J** 0023) Dokl Akad Nauk SSSR 253*274-277
- TRANSL [1980] (**J** 0062) Sov Math, Dokl 22*63-66
- ⋄ C25 C35 C60 ⋄ REV MR 81m:03044 Zbl 499 # 03017
- ID 72720

ERSHOV, YU.L. [1980] *Regularly closed fields (Russian)* (**J** 0023) Dokl Akad Nauk SSSR 251*783-785
- TRANSL [1980] (**J** 0062) Sov Math, Dokl 21*510-512
- ⋄ B25 C35 C60 D35 ⋄ REV MR 82g:03050 Zbl 467 # 03026
- ID 55024

ERSHOV, YU.L. [1981] *Eliminability of quantifiers in regularly closed fields (Russian)* (J 0023) Dokl Akad Nauk SSSR 258*16-20
- TRANSL [1981] (J 0062) Sov Math, Dokl 23*463-467
- ⋄ B25 C10 C60 ⋄ REV MR 82k:03036 Zbl 521 # 03020
- ID 72716

ERSHOV, YU.L. [1981] *On elementary theories of regularly closed fields (Russian)* (J 0023) Dokl Akad Nauk SSSR 257*271-274
- TRANSL [1981] (J 0062) Sov Math, Dokl 23*259-262
- ⋄ B25 C60 ⋄ REV MR 82h:03010 Zbl 481 # 03008
- ID 72717

ERSHOV, YU.L. [1981] *Undecidability of regularly closed fields (Russian)* (J 0003) Algebra i Logika 20*389-394,484
- TRANSL [1981] (J 0069) Algeb and Log 20*257-260
- ⋄ C60 D35 ⋄ REV MR 84e:03052 Zbl 499 # 03018
- ID 34387

ERSHOV, YU.L. [1982] *Absolute irreducibility and Henselization properties (Russian)* (J 0003) Algebra i Logika 21*530-536
- TRANSL [1982] (J 0069) Algeb and Log 21*353-357
- ⋄ C60 ⋄ REV MR 85c:12007 Zbl 507 # 12021 • ID 39026

ERSHOV, YU.L. [1982] *Algorithmic problems in the theory of fields (positive aspects) (Russian)* (C 1920) Spravochnaya Kniga po Mat Logike, Chast 1-4 3*269-353
- ⋄ B25 C10 C60 C98 ⋄ ID 48305

ERSHOV, YU.L. [1982] *Multiply valued fields (Russian)* (J 0067) Usp Mat Nauk 37/3*55-93,224
- TRANSL [1982] (J 1399) Russ Math Surv 37/3*63-107
- ⋄ B25 C25 C35 C60 C98 ⋄ REV MR 83j:03053 Zbl 523 # 12024 • ID 36990

ERSHOV, YU.L. [1982] *Regularly r-closed fields (Russian)* (J 0023) Dokl Akad Nauk SSSR 266*538-540
- TRANSL [1982] (J 0062) Sov Math, Dokl 26*363-366
- ⋄ B25 C25 C35 C60 ⋄ REV MR 84c:03060 Zbl 555 # 12010 JSL 51.235 • ID 34004

ERSHOV, YU.L. [1982] *Totally real field extensions (Russian)* (J 0023) Dokl Akad Nauk SSSR 263*1047-1049
- TRANSL [1982] (J 0062) Sov Math, Dokl 25*477-480
- ⋄ B25 C52 C60 ⋄ REV MR 83m:12045 Zbl 514 # 03019
- ID 37406

ERSHOV, YU.L. [1983] *Algebraic properties of regularly closed fields (Russian)* (S 0066) Tr Mat Inst Steklov 158*80-86
- TRANSL [1983] (S 0055) Proc Steklov Inst Math 158*85-91
- ⋄ C60 ⋄ REV MR 83k:12021 Zbl 523 # 12019 • ID 36989

ERSHOV, YU.L. [1983] *Dynamic logic over admissible sets (Russian)* (J 0023) Dokl Akad Nauk SSSR 273*1045-1048
- TRANSL [1983] (J 0062) Sov Math, Dokl 28*739-742
- ⋄ B75 C70 ⋄ REV MR 85d:03075 • ID 40496

ERSHOV, YU.L. [1983] *Regularly r-closed fields (Russian)* (J 0003) Algebra i Logika 22*382-402
- TRANSL [1983] (J 0069) Algeb and Log 22*277-291
- ⋄ B25 C25 C35 C60 ⋄ REV Zbl 544 # 12017 • ID 41033

ERSHOV, YU.L. [1983] *The Σ-enumeration principle (Russian)* (J 0023) Dokl Akad Nauk SSSR 270*786-788
- TRANSL [1983] (J 0062) Sov Math, Dokl 27*670-672
- ⋄ C62 C70 D60 E30 E65 ⋄ REV MR 85d:03074 Zbl 539 # 03016 • ID 40494

ERSHOV, YU.L. [1984] *Regularly r-closed fields with weakly universal Galois group (Russian)* (J 0003) Algebra i Logika 23*637-669
- TRANSL [1984] (J 0069) Algeb and Log 23*426-449
- ⋄ B25 C60 D35 ⋄ REV Zbl 574 # 12024 • ID 48388

ERSHOV, YU.L. [1984] *Two theorems on regularly r-closed fields* (J 0127) J Reine Angew Math 347*154-167
- ⋄ C60 ⋄ REV MR 85i:12006 Zbl 514 # 12022 • ID 37442

ERSHOV, YU.L. [1985] *Σ-definability in admissible sets (Russian)* (J 0023) Dokl Akad Nauk SSSR 285*792-795
- ⋄ C70 E47 ⋄ ID 49047

ERSHOV, YU.L. see Vol. I, II, IV, V, VI for further entries

ESPANOL, L. [1983] *First-order aspects of *-rings and constructive version of the Gauss lemma (Spanish) (English summary)* (J 1951) Rev Acad Cienc Exact Fis-Quim Zaragoza, Ser 2 38*11-14
- ⋄ C60 ⋄ REV MR 86b:03041 • ID 45153

ESPANOL, L. see Vol. V, VI for further entries

ESTERLE, J. [1977] see DALES, H.G.

ESTERLE, J. [1977] *Proprietes universelles de certains semi-groupes et algebres de convolution* (P 2910) Colloq Anal Harmon & Complexe;1977 Les Aludes 1977/9*6pp
- ⋄ C65 E35 E50 ⋄ REV MR 81e:46036 Zbl 479 # 46038
- ID 55696

ESTERLE, J. [1977] *Solution d'un probleme d'Erdoes, Gillman et Henriksen et application a l'etude des homomorphismes de $\mathscr{C}(K)$* (J 0001) Acta Math Acad Sci Hung 30*113-127
- ⋄ C50 C60 E50 E75 ⋄ REV MR 58 # 2298 Zbl 411 # 46037 • ID 52920

ESTERLE, J. see Vol. I, V for further entries

EVANS, T. [1967] *The spectrum of a variety* (J 0068) Z Math Logik Grundlagen Math 13*213-218
- ⋄ C05 C13 ⋄ REV MR 35 # 6539 Zbl 199.327 • ID 03567

EVANS, T. see Vol. II, IV for further entries

EYTAN, M. [1979] *Engeler algorithms in a topos* (P 2935) FCT'79 Fund of Comput Th;1979 Berlin/Wendisch-Rietz 128-137
- ⋄ B75 C75 G30 ⋄ REV MR 81g:03081 Zbl 415 # 03055
- ID 53157

EYTAN, M. see Vol. II, V for further entries

FABER, V. & LAVER, R. & MCKENZIE, R. [1978] *Coverings of groups by abelian subgroups* (J 0017) Canad J Math 30*933-945
- ⋄ C05 C60 E50 E75 ⋄ REV MR 80f:20042 Zbl 383 # 20026 • ID 31943

FACENDA AGUIRRE, J.A. [1982] *Ultraproducts of locally convex spaces (Spanish) (English summary)* (J 0296) Acta Salamantica Cienc 46*277-280
- ⋄ C20 C65 ⋄ REV MR 85j:46001 • ID 45448

FACENDA AGUIRRE, J.A. see Vol. I, V for further entries

FAGIN, R. [1974] *Generalized first-order spectra and polynomial-time recognizable sets* (P 0761) Compl of Computation;1973 New York 43-73
- ⋄ C13 C85 D15 ⋄ REV MR 51 # 7840 Zbl 303 # 68035
- ID 18140

FAGIN, R. [1975] *A spectrum hierarchy* (J 0068) Z Math Logik Grundlagen Math 21*123-134
⋄ C13 ⋄ REV MR 51 # 7842 Zbl 311 # 02020 • ID 04121

FAGIN, R. [1975] *A two-cardinal characterization of double spectra* (J 0068) Z Math Logik Grundlagen Math 21*121-122
⋄ C13 ⋄ REV MR 51 # 12505 Zbl 309 # 02055 • ID 17291

FAGIN, R. [1975] *Monadic generalized spectra* (J 0068) Z Math Logik Grundlagen Math 21*89-96
⋄ C13 C85 ⋄ REV MR 51 # 7841 Zbl 317 # 02054 • ID 04122

FAGIN, R. [1976] *Probabilities on finite models* (J 0036) J Symb Logic 41*50-58
⋄ C13 ⋄ REV MR 57 # 16042 Zbl 341 # 02044 JSL 50.1073 • ID 14751

FAGIN, R. [1977] *The number of finite relational structures* (J 0193) Discr Math 17*27-45
⋄ C07 C13 ⋄ REV MR 56 # 15444 • ID 21321

FAGIN, R. see Vol. I, IV, V for further entries

FAJARDO, S. [1980] *Compacidad y decidibilidad de logicas monadicas con cuantificadores cardinales* (J 0307) Rev Colomb Mat 14*173-196
⋄ B25 C55 C80 ⋄ REV MR 82e:03038 Zbl 467 # 03033 • ID 90114

FAJARDO, S. [1985] *Completeness theorems for the general theory of stochastic processes* (P 2160) Latin Amer Symp Math Log (6);1983 Caracas 174-194
⋄ B48 C65 C90 ⋄ ID 41798

FAJARDO, S. [1985] *Probability logic with conditional expectation* (J 0073) Ann Pure Appl Logic 28*137-161
⋄ B48 C40 C90 ⋄ REV Zbl 564 # 03019 • ID 39901

FAJTLOWICZ, S. & HOLSZTYNSKI, W. & MYCIELSKI, J. & WEGLORZ, B. [1968] *On powers of bases in some compact algebras* (S 0019) Colloq Math (Warsaw) 19*43-46
⋄ C05 C50 ⋄ REV MR 37 # 127 Zbl 156.258 • ID 32040

FAJTLOWICZ, S. see Vol. IV, V for further entries

FAKIR, S. [1972] *Categories localement coherentes et objets α-injectifs* (J 2313) C R Acad Sci, Paris, Ser A-B 274*A1392-A1395
⋄ C20 G30 ⋄ REV MR 45 # 5188 Zbl 238 # 18008 • ID 27912

FAKIR, S. & HADDAD, L. [1972] *Objets coherents et ultraproduits dans les categories* (J 0032) J Algeb 21*410-421
⋄ C20 G30 ⋄ REV MR 47 # 304 Zbl 248 # 18021 • ID 03649

FAKIR, S. [1973] *Monomorphismes et objects algebriquement et existentiellement clos dans les categories localement presentables. Applications aux monoides, categories et groupoides* (J 2313) C R Acad Sci, Paris, Ser A-B 276*A723-A726
⋄ C05 C25 C30 G30 ⋄ REV MR 47 # 8651 Zbl 267 # 18001 • ID 29906

FARINAS DEL CERRO, L. [1984] see CAVALLI, A.R.

FARINAS DEL CERRO, L. see Vol. I, II for further entries

FEDYSZAK-KOSZELA, A. [1982] *Some remarks on infinitary languages $N_{\kappa\alpha}$* (S 3291) Zesz Nauk Mat Fiz Chem (Uniw Gdansk) 5*19-35
⋄ C75 ⋄ REV Zbl 527 # 03018 • ID 37569

FEFERMAN, S. [1957] *Some recent work of Ehrenfeucht and Fraisse* (P 1675) Summer Inst Symb Log;1957 Ithaca 201-209
⋄ B10 C07 C65 E10 ⋄ REV JSL 32.282 • ID 03683

FEFERMAN, S. & VAUGHT, R.L. [1959] *The first order properties of products of algebraic systems* (J 0027) Fund Math 47*57-103
⋄ B25 C10 C20 C30 G05 ⋄ REV MR 21 # 7171 Zbl 88.248 JSL 32.276 • ID 03684

FEFERMAN, S. [1965] *Some applications of the notions of forcing and generic sets* (J 0027) Fund Math 56*325-345
• REPR [1965] (P 0614) Th Models;1963 Berkeley 89-95
⋄ C62 D55 E25 E35 E40 E45 F35 ⋄ REV MR 31 # 1193 MR 34 # 2439 Zbl 129.264 JSL 37.612
• REM Reprint is a summary • ID 03696

FEFERMAN, S. & KREISEL, G. [1966] *Persistent and invariant formulas relative to theories of higher order* (J 0015) Bull Amer Math Soc 72*480-485
⋄ B15 C40 C52 C85 ⋄ REV MR 33 # 1229 Zbl 234 # 02038 JSL 37.764 • ID 03697

FEFERMAN, S. [1968] *Lectures on proof theory* (P 0692) Summer School in Logic;1967 Leeds 1-107
⋄ C40 C70 C75 F05 F15 F98 ⋄ REV MR 38 # 4294 Zbl 248 # 02033 • ID 03700

FEFERMAN, S. [1968] *Persistent and invariant formulas for outer extensions* (J 0020) Compos Math 20*29-52
• REPR [1968] (C 0727) Logic Found of Math (Heyting) 29-52
⋄ C30 C40 C70 C75 F35 ⋄ REV MR 37 # 6170 Zbl 162.16 JSL 37.764 • ID 03703

FEFERMAN, S. [1972] *Infinitary properties, local functors, and systems of ordinal functions* (P 2080) Conf Math Log;1970 London 63-97
⋄ C30 C40 C75 E10 F15 G30 ⋄ REV MR 50 # 12646 Zbl 302 # 02018 • ID 03706

FEFERMAN, S. [1974] *Applications of many-sorted interpolation theorems* (P 0610) Tarski Symp;1971 Berkeley 205-223
⋄ C40 C95 ⋄ REV MR 53 # 10558 Zbl 311 # 02060 • ID 03707

FEFERMAN, S. [1974] *Two notes on abstract model theory I. Properties invariant on the range of definable relations between structures* (J 0027) Fund Math 82*153-165
⋄ C40 C95 ⋄ REV MR 51 # 134 MR 58 # 10404 Zbl 296 # 02026 • REM Part II 1975 • ID 03708

FEFERMAN, S. [1975] *Impredicativity of the existence of the largest divisible subgroup of an abelian p-group* (C 0782) Model Th & Algeb (A. Robinson) 117-130
⋄ C57 C60 F65 ⋄ REV MR 53 # 5274 Zbl 316 # 02058 • ID 22905

FEFERMAN, S. [1975] *Two notes on abstract model theory II. Languages for which the set of valid sentences is semi-invariantly implicitly definable* (J 0027) Fund Math 89*111-130
⋄ C40 C70 C95 ⋄ REV MR 58 # 10405 Zbl 335 # 02034 • REM Part I 1974 • ID 04236

FEFERMAN, S. [1979] *Generalizing set-theoretical model theory and an analogue theory on admissible sets* (P 1705) Scand Logic Symp (4);1976 Jyvaeskylae 171-195
⋄ C50 C55 C70 E70 ⋄ REV MR 80m:03084 Zbl 439 # 03010 • ID 56002

FEFERMAN, S. [1985] see BARWISE, J.

FEFERMAN, S. see Vol. I, II, IV, V, VI for further entries

FEFFERMAN, C. [1967] *Cardinally maximal sets of non-equivalent order types* (J 0068) Z Math Logik Grundlagen Math 13*205-212
⋄ C65 C75 E07 ⋄ REV MR 36 # 1313   Zbl 281 # 02054
• ID 03675

FEINER, L. [1970] *Hierarchies of boolean algebras* (J 0036) J Symb Logic 35*365-374
⋄ C57 D45 G05 ⋄ REV MR 44 # 39   Zbl 222 # 02048
• ID 03711

FEINER, L. [1973] *Degrees of nonrecursive presentability* (J 0053) Proc Amer Math Soc 38*621-624
⋄ C57 D30 D45 ⋄ REV MR 48 # 5836   Zbl 266 # 02021
• ID 03714

FEINER, L. see Vol. IV for further entries

FELGNER, U. [1971] *Models of ZF set theory* (X 0811) Springer: Heidelberg & New York vi+173pp
⋄ C62 E25 E35 E45 E50 E98 ⋄ REV MR 50 # 4298
Zbl 269 # 02029   JSL 40.92 • ID 03721

FELGNER, U. [1974] $\aleph_1$-*kategorische Theorien nicht-kommutativer Ringe* (J 0027) Fund Math 82*331-346
⋄ C35 C45 C60 ⋄ REV MR 51 # 10078   Zbl 301 # 02059
• ID 03726

FELGNER, U. [1975] *On $\aleph_0$-categorical extra-special p-groups* (J 0079) Logique & Anal, NS 18*407-428
• REPR [1977] (P 1625) Six Days of Model Th;1975 Louvain-la-Neuve 175-196
⋄ C35 C60 ⋄ REV MR 57 # 16054   Zbl 355 # 02038
Zbl 412 # 03015 • ID 50236

FELGNER, U. [1976] *Einige gruppentheoretische Aequivalente zum Auswahlaxiom* (J 0001) Acta Math Acad Sci Hung 28*13-18
⋄ C60 E25 E75 ⋄ REV MR 54 # 10017   Zbl 342 # 02048
• ID 25860

FELGNER, U. [1977] *Stability and $\aleph_0$-categoricity of nonabelian groups* (P 1075) Logic Colloq;1976 Oxford 301-324
⋄ C35 C45 C60 ⋄ REV MR 58 # 16266   Zbl 418 # 03024
• ID 16627

FELGNER, U. [1978] $\aleph_0$-*categorical stable groups* (J 0044) Math Z 160*27-49
⋄ C35 C45 C60 ⋄ REV MR 58 # 10417   Zbl 357 # 02048
• ID 51327

FELGNER, U. [1980] *Horn-theories of Abelian groups* (P 2625) Model Th of Algeb & Arithm;1979 Karpacz 163-173
⋄ C20 C60 ⋄ REV MR 82d:03055   Zbl 457 # 03033
• ID 54358

FELGNER, U. [1980] *Kategorizitaet* (J 0157) Jbuchber Dtsch Math-Ver 82*12-32
⋄ C35 C45 C60 C98 ⋄ REV MR 81c:03020
Zbl 453 # 03033 • ID 54163

FELGNER, U. [1980] *The model theory of FC-groups* (P 2958) Latin Amer Symp Math Log (4);1978 Santiago 163-190
⋄ C35 C45 C60 D35 ⋄ REV MR 81m:03045
Zbl 437 # 03014 • ID 55882

FELGNER, U. [1982] see CHERLIN, G.L.

FELGNER, U. & SCHULZ, KLAUS [1984] *Algebraische Konsequenzen des Determiniertheits-Axioms* (J 0008) Arch Math (Basel) 42*557-563
⋄ C60 E60 E75 ⋄ REV MR 85k:03034 • ID 39998

FELGNER, U. [1985] *Die Definierbarkeit von K in K(t) im Falle eines pseudo-endlichen Koerpers K (English summary)* (P 4310) Easter Conf on Model Th (3);1985 Gross Koeris 89-98
⋄ C60 ⋄ ID 49054

FELGNER, U. see Vol. V for further entries

FELSCHER, W. [1965] *Zur Algebra unendlich langer Zeichenreihen* (J 0068) Z Math Logik Grundlagen Math 11*5-16
⋄ C05 C75 E10 ⋄ REV MR 32 # 2358   Zbl 156.5
JSL 36.157 • ID 03734

FELSCHER, W. [1968] *On criteria of definability* (J 0053) Proc Amer Math Soc 19*834-836
⋄ C20 C52 ⋄ REV MR 37 # 5087   Zbl 165.310 JSL 34.300
• ID 03736

FELSCHER, W. [1969] *On the algebra of quantifiers (Russian summary)* (J 0014) Bull Acad Pol Sci, Ser Math Astron Phys 17*327-332
⋄ B10 C05 C07 G15 ⋄ REV MR 43 # 7313 • ID 03737

FELSCHER, W. [1971] *An algebraic approach to first order logic* (P 0669) Conv Teor Modelli & Geom;1969/70 Roma 133-148
⋄ B10 C05 C07 C20 C90 G15 ⋄ REV MR 43 # 1822
Zbl 209.12 • ID 03738

FELSCHER, W. [1972] *Modelltheorie als universelle Algebra* (C 0722) Stud Algeb & Anwendgn 129-130
⋄ C05 C07 ⋄ REV Zbl 258 # 02051 • ID 61666

FELSCHER, W. [1976] *On interpolation when function symbols are present* (J 0009) Arch Math Logik Grundlagenforsch 17*145-157
⋄ C40 C75 F50 ⋄ REV MR 54 # 12476   Zbl 339 # 02011
• ID 04239

FELSCHER, W. see Vol. I, V, VI for further entries

FENSTAD, J.E. [1964] *Model theory, ultraproducts and topology* (1111) Preprints, Manuscr., Techn. Reports etc. 21pp
⋄ C20 C65 E75 ⋄ REM Seminar reports. Institute of mathematics Oslo • ID 33314

FENSTAD, J.E. [1970] *Non-standard models for arithmetic and analysis* (P 0724) Skand Mat Kongr (15);1968 Oslo 30-47
⋄ H05 H15 ⋄ REV MR 41 # 1523   Zbl 191.298 • ID 17153

FENSTAD, J.E. & GANDY, R.O. & SACKS, G.E. (EDS.) [1978] *Generalized recursion theory II* (X 0809) North Holland: Amsterdam vii+466pp
⋄ C62 C70 C80 D97 ⋄ REV MR 80a:03004
Zbl 453 # 03047 • ID 54177

FENSTAD, J.E. see Vol. I, II, IV, V, VI for further entries

FERBUS, M.-C. [1984] *Functorial bounds for cut elimination in $L_{\beta\omega} I$* (J 0009) Arch Math Logik Grundlagenforsch 24*141-158
⋄ C75 F05 F15 G30 ⋄ REV MR 86c:03049
Zbl 558 # 03026 • ID 42605

FERBUS, M.-C. see Vol. VI for further entries

FERENCZI, M. [1977] *On valid assertions - in probability logic*
(J 0411) Studia Sci Math Hung 12*101-116
⋄ B25 B48 C90 ⋄ REV MR 82c:03026 Zbl 435 # 03022
• ID 55783

FERENCZI, M. see Vol. I for further entries

FERRANTE, J. & RACKOFF, C.W. [1975] *A decision procedure for the first order theory of real addition with order* (J 1428) SIAM J Comp 4*69-76
⋄ B25 C10 C60 D15 F35 ⋄ REV MR 52 # 10403
Zbl 277 # 02010 • ID 21702

FERRANTE, J. & GEISER, J.R. [1977] *An efficient decision procedure for the theory of rational order* (J 1426) Theor Comput Sci 4*227-233
⋄ B25 D15 ⋄ REV MR 56 # 13777 Zbl 372 # 02024
• ID 51432

FERRANTE, J. & RACKOFF, C.W. [1979] *The computational complexity of logical theories* (S 3301) Lect Notes Math 718*x+243pp
⋄ B25 C98 D10 D15 ⋄ REV MR 81d:03013
Zbl 404 # 03028 JSL 49.670 • ID 54815

FERRO, A. & OMODEO, E.G. [1978] *An efficient validity test for formulae in extensional two-level syllogistic* (J 0319) Matematiche (Sem Mat Catania) 33*130-137
⋄ B20 B25 B35 ⋄ REV MR 83i:03025 Zbl 448 # 68021
• ID 56666

FERRO, A. & MICALE, B. [1978] *Sui modelli di una teoria del 1° ordine nei quali certe formule definiscono un modello di un'altra teoria* (J 0319) Matematiche (Sem Mat Catania) 33*203-224
⋄ C07 ⋄ REV MR 82i:03043 Zbl 448 # 03018 • ID 56618

FERRO, A. & OMODEO, E.G. & SCHWARTZ, J.T. [1980] *Decision procedures for some fragments of set theory* (P 3063) Autom Deduct (5);1980 Les Arcs 88-96
⋄ B25 E30 ⋄ REV Zbl 457 # 03009 • ID 54334

FERRO, A. & OMODEO, E.G. & SCHWARTZ, J.T. [1980] *Decision procedures for elementary sublanguages of set theory. I: Multi-level syllogistic and some extensions* (J 0155) Commun Pure Appl Math 33*599-608
⋄ B25 E30 ⋄ REV MR 82i:03018a Zbl 453 # 03009 • REM Part II 1981 by Breban,M. & Ferro,A. & Omodeo,E.G. & Schwartz,J.T. Common author of all parts: Ferro,A.
• ID 54139

FERRO, A. [1981] see BREBAN, M.

FERRO, A. [1983] *Some decidability questions in set theory (Italian)* (P 3829) Atti Incontri Log Mat (1);1982 Siena 205-207
⋄ B25 E30 ⋄ REV MR 84k:03006 • ID 44643

FERRO, A. [1984] see BREBAN, M.

FERRO, A. [1985] *A note on the decidability of MLS extended with the powerset operator* (J 0155) Commun Pure Appl Math 38*367-374
⋄ B25 E30 ⋄ ID 47612

FERRO, A. [1985] see CANTONE, D.

FERRO, A. see Vol. IV for further entries

FERRO, R. [1975] *Limits to some interpolation theorems* (J 0144) Rend Sem Mat Univ Padova 54*201-213
⋄ C40 C75 ⋄ REV MR 56 # 2800 Zbl 354 # 02019
• ID 31071

FERRO, R. [1976] *Consistency property and model existence theorem for second order negative languages with conjunctions and quantifications over sets of cardinality smaller than a strong limit cardinal of denumerable cofinality* (J 0144) Rend Sem Mat Univ Padova 55*123-141
⋄ C55 C75 C85 ⋄ REV MR 57 # 61 Zbl 365 # 02006
• ID 31072

FERRO, R. [1978] *Interpolation theorem for $L^{2+}_{k,k}$* (J 0036) J Symb Logic 43*535-549
⋄ B15 C40 C75 C85 ⋄ REV MR 81a:03028
Zbl 397 # 03019 • ID 29281

FERRO, R. [1981] *An analysis of Karp's interpolation theorem and the notion of k-consistency property (Italian summary)* (J 0144) Rend Sem Mat Univ Padova 65*111-118
⋄ C40 C75 ⋄ REV MR 83k:03041 Zbl 485 # 03014
• ID 33346

FERRO, R. [1982] *ω-satisfiability, ω-consistency property and the downward Loewenheim-Skolem theorem for L_{KK} (Italian summary)* (J 0144) Rend Sem Mat Univ Padova 66*7-19
⋄ C55 C75 ⋄ REV MR 84i:03074 Zbl 531 # 03018
• ID 33345

FERRO, R. [1983] *Seq-consistency property and interpolation theorems* (J 0144) Rend Sem Mat Univ Padova 70*133-145
⋄ C40 C75 ⋄ REV MR 85j:03054 Zbl 534 # 03014
• ID 36559

FERRO, R. see Vol. I for further entries

FIDEL, M.M. [1977] *The decidability of the calculi $\mathscr{C}_n$* (J 0302) Rep Math Logic, Krakow & Katowice 8*31-40
⋄ B25 B53 G25 ⋄ REV MR 58 # 158 Zbl 378 # 02011
• ID 24188

FIDEL, M.M. [1980] *An algebraic study of logic with constructible negation* (P 3006) Brazil Conf Math Log (3);1979 Recife 119-129
⋄ B25 B55 G25 ⋄ REV MR 82i:03075 Zbl 453 # 03024
• ID 54154

FIDEL, M.M. see Vol. II, VI for further entries

FINE, K. [1972] *For so many individuals* (J 0047) Notre Dame J Formal Log 13*569-572
⋄ C07 C80 ⋄ REV MR 55 # 7736 Zbl 205.303 • ID 26337

FINE, K. [1975] *Some connections between elementary and modal logic* (P 0757) Scand Logic Symp (3);1973 Uppsala 15-31
⋄ B45 C07 C90 ⋄ REV MR 53 # 5265 Zbl 316 # 02021
• ID 22897

FINE, K. [1978] *Model theory for modal logic. part I: the de re/de dicto distinction. part II: the elimination of de re modality* (J 0122) J Philos Logic 7*125-156,277-306
⋄ A05 B45 C40 C90 ⋄ REV MR 80c:03021
MR 81f:03024 Zbl 375 # 02008 Zbl 409 # 03007
JSL 50.1083 • REM Part III 1981 • ID 30694

FINE, K. [1981] *Model theory for modal logic. III. Existence and predication* (J 0122) J Philos Logic 10*293-307
⋄ A05 B45 C90 ⋄ REV MR 83c:03016 Zbl 464 # 03017
JSL 50.1083 • REM Parts I,II 1978 • ID 54607

FINE, K. see Vol. II, V, VI for further entries

FISCHER, MICHAEL J. & RABIN, M.O. [1974] *Super-exponential complexity of Presburger arithmetic* (**P** 0761) Compl of Computation;1973 New York 27-41
⋄ B25 D15 F20 F30 ⋄ REV MR 51 # 2893 Zbl 319 # 68024 • ID 17344

FISCHER, MICHAEL J. see Vol. I, II, IV for further entries

FISHBURN, P.C. [1970] see BAKER, K.A.

FISHBURN, P.C. [1972] see BAKER, K.A.

FISHBURN, P.C. [1981] *Restricted thresholds for interval orders: a case of nonaxiomatizability by a universal sentence* (**J** 0035) J Math Psychol 24*276-283
⋄ B70 B75 C52 ⋄ REV MR 83m:03015 Zbl 484 # 03001 • ID 35423

FISHBURN, P.C. see Vol. I, II, V for further entries

FISHER, E.R. [1970] see BARWISE, J.

FISHER, E.R. & ROBINSON, A. [1972] *Inductive theories and their forcing companions* (**J** 0029) Israel J Math 12*95-107
• REPR [1979] (**C** 4594) Sel Pap Robinson 1*267-279
⋄ C25 G05 G10 ⋄ REV MR 47 # 3163 Zbl 268 # 02036 • ID 04311

FISHER, E.R. [1972] see EKLOF, P.C.

FISHER, E.R. & SIMMONS, H. & WHEELER, W.H. [1975] *Elementary equivalence classes of generic structures and existentially complete structures* (**C** 0782) Model Th & Algeb (A. Robinson) 131-169
⋄ C25 ⋄ REV MR 53 # 7763 Zbl 319 # 02046 • ID 23005

FISHER, E.R. [1977] *Abelian structures I* (**P** 1682) Abel Group Th;1976 Las Cruces 270-322
⋄ C60 G30 ⋄ REV MR 58 # 27459 Zbl 414 # 18001 • ID 72925

FISHER, J.R. [1981] *Axiomatic radical and semisimple classes of rings* (**J** 0048) Pac J Math 97*81-91
⋄ C05 C52 C60 ⋄ REV MR 83d:16008 Zbl 451 # 16004 • ID 54048

FITTING, M. [1969] *Intuitionistic logic, model theory, and forcing* (**X** 0809) North Holland: Amsterdam 191pp
⋄ B98 C90 C98 E25 E35 E45 E50 F50 F98 ⋄ REV MR 41 # 6666 Zbl 188.320 JSL 36.166 • ID 04349

FITTING, M. [1970] *Intuitionistic model theory and the Cohen independence proofs* (**P** 0603) Intuitionism & Proof Th;1968 Buffalo 219-226
⋄ A05 C90 E35 E40 F50 ⋄ REV MR 42 # 4390 Zbl 205.306 • ID 14547

FITTING, M. [1972] *Non-classical logics and the independence results of set theory* (**J** 0105) Theoria (Lund) 38*133-142
⋄ B45 C90 E35 ⋄ REV MR 58 # 16285 Zbl 255 # 02070 • ID 31074

FITTING, M. [1978] *Elementary formal systems for hyperarithmetical relations* (**J** 0068) Z Math Logik Grundlagen Math 24*25-30
⋄ C75 D55 D75 ⋄ REV MR 80a:03059 Zbl 374 # 02024 • ID 31075

FITTING, M. [1984] *Linear reasoning in modal logic* (**J** 0036) J Symb Logic 49*1363-1378
⋄ B45 C40 ⋄ REV MR 86f:03027 • ID 42443

FITTING, M. see Vol. I, II, IV, VI for further entries

FITTLER, R. [1969] *Categories of models with initial objects* (**P** 0723) Categ Th, Homol Th & Appl;1968 Seattle 1*33-45
⋄ C40 G30 ⋄ REV MR 40 # 5692 Zbl 198.20 • ID 16202

FITTLER, R. [1970] *Direct limits of models* (**J** 0068) Z Math Logik Grundlagen Math 16*377-384
⋄ C30 ⋄ REV MR 43 # 6066 Zbl 211.307 • ID 04358

FITTLER, R. [1971] *Generalized prime models* (**J** 0036) J Symb Logic 36*593-606
⋄ C50 ⋄ REV MR 46 # 1583 Zbl 241 # 02019 • ID 04359

FITTLER, R. [1972] *A characteristic direct limit property* (**J** 0068) Z Math Logik Grundlagen Math 18*131-133
⋄ C30 ⋄ REV MR 46 # 1584 Zbl 261 # 02036 • ID 04360

FITTLER, R. [1972] *Some categories of models* (**J** 0009) Arch Math Logik Grundlagenforsch 15*179-189 • ERR/ADD ibid 16*14
⋄ C30 C52 G30 ⋄ REV MR 48 # 8225 Zbl 272 # 02072 • ID 04361

FITTLER, R. [1972] *Verallgemeinerte Prim-Modelle (Zusammenfassung)* (**C** 0722) Stud Algeb & Anwendgn 1-5
⋄ C50 ⋄ REV MR 49 # 10538 Zbl 266 # 02030 • ID 17121

FITTLER, R. [1973] *Categories with ultraproducts* (**J** 2022) Comm Math Helvetici 48*100-115
⋄ C20 G30 ⋄ REV MR 48 # 3734 Zbl 262 # 18002 • ID 04362

FITTLER, R. [1974] *Saturated models of incomplete theories* (**J** 0009) Arch Math Logik Grundlagenforsch 16*3-14
⋄ C20 C50 C52 ⋄ REV MR 50 # 4280 Zbl 281 # 02051 • ID 04363

FITTLER, R. [1975] *Closed models and hulls of theories* (**P** 1442) ⊦ ISILC Logic Conf;1974 Kiel 169-189
⋄ C25 C52 C60 ⋄ REV MR 55 # 5431 Zbl 317 # 02056 • ID 27212

FITTLER, R. see Vol. I for further entries

FJELSTAD, P. [1968] *Set-theories as algebras* (**J** 0068) Z Math Logik Grundlagen Math 14*383-411
⋄ C62 E30 G25 ⋄ REV MR 39 # 1322 Zbl 172.295 • ID 04367

FLANNAGAN, T.B. [1980] *Expansions of models of ZFC* (**J** 0009) Arch Math Logik Grundlagenforsch 20*173-180
⋄ C62 E25 ⋄ REV MR 82f:03041 Zbl 483 # 03029 • ID 72958

FLANNAGAN, T.B. see Vol. V, VI for further entries

FLEISCHER, I. [1963] *Elementary properties of inductive limits* (**J** 0068) Z Math Logik Grundlagen Math 9*347-350
⋄ C05 C20 C30 ⋄ REV MR 28 # 1130 Zbl 119.17 • ID 04374

FLEISCHER, I. [1966] *A note on universal homogenous models* (**J** 0132) Math Scand 19*183-184
⋄ C50 ⋄ REV MR 35 # 4095 Zbl 147.258 • ID 04375

FLEISCHER, I. [1968] *On Cohn's theorem* (**J** 0039) J London Math Soc 43*237-238
⋄ C05 ⋄ REV MR 37 # 128 • ID 04377

FLEISCHER, I. [1976] *Concerning my note on universal homogeneous models* (**J** 0132) Math Scand 38*24
⋄ C50 ⋄ REV MR 53 # 5292 Zbl 319 # 02042 • ID 22927

FLEISCHER, I. [1976] *Quantifier elimination for modules and ordered groups* (J 0014) Bull Acad Pol Sci, Ser Math Astron Phys 24*9-15
⋄ B25 C10 C60 ⋄ REV MR 58 # 27689 Zbl 327 # 02048
• ID 18143

FLEISCHER, I. [1983] *Subdirect decomposability into irreducibles* (J 0004) Algeb Universalis 16*261-263
⋄ C05 C52 ⋄ REV MR 84g:08003 Zbl 519 # 08004
• ID 46257

FLEISCHER, I. [1984] *"Kripke semantics" = algebra + poetry (French summary)* (J 0079) Logique & Anal, NS 27*283-295
⋄ C90 ⋄ REV MR 86g:03032 Zbl 571 # 03005 • ID 44445

FLEISCHER, I. [1984] *One-variable equationally compact distributive lattices (Russian summary)* (J 1522) Math Slovaca 34*385-386
⋄ C05 C50 G10 ⋄ REV MR 86d:06015 • ID 48881

FLEISCHER, I. see Vol. I, V for further entries

FLEISSNER, W.G. [1975] *Lemma on measurable cardinals* (J 0053) Proc Amer Math Soc 49*517-518
⋄ C20 C62 E45 E55 ⋄ REV MR 51 # 2916
Zbl 331 # 02046 • ID 17356

FLEISSNER, W.G. see Vol. V for further entries

FLETCHER, T.J. [1963] *Models of many valued logics* (J 0005) Amer Math Mon 70*381-391
⋄ C07 ⋄ REV MR 27 # 3507 Zbl 118.13 • ID 04380

FLUM, J. [1971] *A remark on infinitary languages* (J 0036) J Symb Logic 36*461-462
⋄ C75 ⋄ REV MR 45 # 6598 Zbl 228 # 02034 JSL 37.764
• ID 04389

FLUM, J. [1971] *Ganzgeschlossene und praedikatengeschlossene Logiken I,II* (J 0009) Arch Math Logik Grundlagenforsch 14*24-37,99-107
⋄ B10 C85 C95 ⋄ REV MR 44 # 38 MR 51 # 12506
Zbl 219 # 02007 Zbl 232 # 02012 • ID 04388

FLUM, J. [1972] *Bemerkungen ueber minimale Modelle* (J 0009) Arch Math Logik Grundlagenforsch 15*79-82
⋄ C50 ⋄ REV MR 51 # 12507 Zbl 246 # 02034 • ID 04393

FLUM, J. [1972] *Die Automorphismenmengen der Modelle einer L_{Q_κ}-Theorie* (J 0009) Arch Math Logik Grundlagenforsch 15*83-85
⋄ C30 C55 C80 ⋄ REV MR 48 # 5816 Zbl 248 # 02060
• ID 04392

FLUM, J. [1972] *Hanf numbers and well-ordering numbers* (J 0009) Arch Math Logik Grundlagenforsch 15*164-178
⋄ C30 C55 C75 C80 ⋄ REV MR 48 # 1879
Zbl 267 # 02039 • ID 04394

FLUM, J. [1974] see EBBINGHAUS, H.-D.

FLUM, J. [1974] *On Horn theories* (J 0044) Math Z 138*205-212
⋄ B20 C20 C80 ⋄ REV MR 51 # 7853 Zbl 275 # 02046
• ID 15020

FLUM, J. [1975] *First-order logic and its extensions* (P 1442) ⊢ ISILC Logic Conf;1974 Kiel 248-310
⋄ C40 C55 C75 C95 C98 ⋄ REV MR 53 # 5293
Zbl 342 # 02007 • ID 22928

FLUM, J. [1975] *L(Q)-preservation theorems* (J 0036) J Symb Logic 40*410-418
⋄ C20 C40 C55 C80 E60 ⋄ REV MR 52 # 40
Zbl 345 # 02008 • ID 17211

FLUM, J. [1976] *Algebra y logica* (P 1619) Coloq Log Simb;1975 Madrid 91-101
⋄ C60 C98 ⋄ REV MR 56 # 5276 Zbl 357 # 02054
• ID 31079

FLUM, J. [1976] *El problema de las palabras en la teoria de grupos* (P 1619) Coloq Log Simb;1975 Madrid 81-90
⋄ C25 C60 D40 ⋄ REV MR 56 # 8701 Zbl 362 # 02032
• ID 31078

FLUM, J. [1977] *Distributive normal forms* (P 1629) Symp Math Log;1974 Oulo;1975 Helsinki 71-76
⋄ B25 C07 ⋄ ID 48389

FLUM, J. [1978] see EBBINGHAUS, H.-D.

FLUM, J. & ZIEGLER, M. [1980] *Topological model theory* (S 3301) Lect Notes Math 769*x + 151pp
⋄ C65 C80 C90 C98 ⋄ REV MR 81j:03059
Zbl 421 # 03024 • ID 53423

FLUM, J. [1984] *Modelltheorie - topologische Modelltheorie* (J 0157) Jbuchber Dtsch Math-Ver 86*69-82
⋄ C65 C90 C98 ⋄ REV MR 86a:03035 Zbl 538 # 03033
• ID 41476

FLUM, J. & MARTINEZ, JUAN CARLOS [1985] *κ-local operations for topological structures* (J 0068) Z Math Logik Grundlagen Math 31*345-349
⋄ C40 C75 C80 C90 ⋄ ID 47561

FLUM, J. [1985] *Characterizing logics* (C 4183) Model-Theor Log 77-120
⋄ C75 C80 C95 C98 ⋄ ID 48326

FLUM, J. [1985] *Maximale monadische Logiken* (J 0009) Arch Math Logik Grundlagenforsch 25*145-152
⋄ C55 C80 C95 ⋄ ID 48390

FLUM, J. see Vol. I for further entries

FOELDES, S. [1980] *On intervals in relational structures* (J 0068) Z Math Logik Grundlagen Math 26*97-101
⋄ C07 E07 ⋄ REV MR 81i:03048 • ID 72989

FOELDES, S. see Vol. IV, V for further entries

FOL'K, N.F. & SHESTOPAL, G.A. [1968] *The solvability of elementary theories of integral and natural numbers with addition (Russian)* (J 0323) Uch Zap Ped Inst, Ryazan 67*141-155
⋄ B25 C10 F30 ⋄ REV MR 41 # 1522 • ID 17126

FOLLESDAL, D. [1966] *A model theoretic approach to causal logic* (J 0239) Skr, K Nor Vidensk Selsk 2*13pp
⋄ A05 B45 C90 ⋄ REV Zbl 192.32 • ID 32328

FOLLESDAL, D. [1968] *Interpretation of quantifiers* (P 0627) Int Congr Log, Meth & Phil of Sci (3,Proc);1967 Amsterdam 271-281
⋄ A05 B10 C90 ⋄ REV MR 40 # 2525 Zbl 182.4
• ID 04443

FOLLESDAL, D. see Vol. II for further entries

FOLLEY, K.W. [1931] *Sets of the first and second α-category and α-residual sets* (J 3998) Proc & Trans Roy Soc Canada, Ser 3 25*137-143
⋄ C55 C65 E07 E75 ⋄ REV Zbl 4.55 FdM 57.1331
• ID 39801

FOLLEY, K.W. see Vol. V for further entries

FOO, N.Y. [1979] *Homomorphisms in the theory of modelling* (J 1743) Int J Gen Syst 5*13-16
⋄ C07 C40 ⋄ REV Zbl 394 # 93005 • ID 52551

FOREMAN, M. [1982] *Large cardinals and strong model theoretic transfer properties* (J 0064) Trans Amer Math Soc 272*427-463
⋄ C55 E35 E55 ⋄ REV MR 84d:03038 Zbl 494 # 03034 • ID 34078

FOREMAN, M. [1983] *More saturated ideals* (C 3875) Cabal Seminar Los Angeles 1979-81 1-27
⋄ C55 C62 E05 E35 E55 ⋄ REV Zbl 536 # 03032 • ID 37111

FOREMAN, M. see Vol. V for further entries

FORREST, W.K. [1977] *Model theory for universal classes with the amalgamation property: A study in the foundations of model theory and algebra* (J 0007) Ann Math Logic 11*263-366
⋄ C05 C25 C30 C35 C45 C52 C60 ⋄ REV MR 58 # 202 Zbl 409 # 03020 • ID 23667

FORREST, W.K. [1978] *A note on universal classes with applications to the theory of graphs* (J 0068) Z Math Logik Grundlagen Math 24*335-346
⋄ C05 C50 C52 C65 ⋄ REV MR 80f:03031 Zbl 416 # 03036 • ID 53202

FORREST, W.K. [1979] *Some basic results in the theory of ω-stable theories* (J 0068) Z Math Logik Grundlagen Math 25*513-520
⋄ C45 ⋄ REV MR 80m:03065 Zbl 421 # 03020 • ID 53419

FORREST, W.K. [1983] *The theory of affine constructible sets* (J 0068) Z Math Logik Grundlagen Math 29*97-135
⋄ C60 C65 ⋄ REV MR 85e:14003 Zbl 543 # 14001 • ID 39929

FORSTER, T.E. [1982] *Axiomatizing set theory with a universal set* (P 3774) Th d'Ensembl de Quine;1981 Louvain-la-Neuve 61-76
⋄ C62 E70 ⋄ REV MR 84g:03086 Zbl 536 # 03037 • ID 37116

FORSTER, T.E. see Vol. V for further entries

FORTI, M. & HONSELL, F. [1983] *Formalizzazioni del "Principio di libera costruzione"* (P 3829) Atti Incontri Log Mat (1);1982 Siena 209-213
⋄ C62 E35 E70 ⋄ REV MR 84k:03006 Zbl 514 # 03033 • ID 37413

FORTI, M. [1985] *f-admissible models of set theory (Italian)* (P 4646) Atti Incontri Log Mat (2);1983/84 Siena 479-485
⋄ C62 ⋄ ID 49795

FORTI, M. see Vol. V for further entries

FOSTER, A.L. [1953] *Generalized "boolean" theory of universal algebras, part I: subdirect sums and normal representation theorem* (J 0044) Math Z 58*306-336
⋄ C05 C30 ⋄ REV MR 15.194 Zbl 51.22 • REM Part II 1953 • ID 04455

FOSTER, A.L. [1953] *Generalized "boolean" theory of universal algebras, II: identities and subdirect sums of functionally complete algebras* (J 0044) Math Z 59*191-199
⋄ C05 C30 ⋄ REV MR 15.194 Zbl 51.262 • REM Part I 1953 • ID 42163

FOSTER, A.L. [1968] *Pre-fields and universal algebraic extensions; equational precessions* (J 0045) Monatsh Math 72*315-324
⋄ C05 C60 ⋄ REV MR 38 # 100 Zbl 162.325 • ID 04464

FOSTER, A.L. see Vol. II for further entries

FOURMAN, M.P. & MULVEY, C.J. & SCOTT, D.S. (EDS.) [1979] *Applications of sheaves* (S 3301) Lect Notes Math 753*xiv+779pp
⋄ C90 C97 G97 ⋄ REV MR 80j:18001 Zbl 407 # 00001 • ID 56166

FOURMAN, M.P. & HYLAND, J.M.E. [1979] *Sheaf models for analysis* (P 2901) Appl Sheaves;1977 Durham 280-301
⋄ C65 C90 F35 F50 G10 G30 ⋄ REV MR 82g:03069 Zbl 427 # 03028 • ID 53715

FOURMAN, M.P. & SCOTT, D.S. [1979] *Sheaves and logic* (P 2901) Appl Sheaves;1977 Durham 302-401
⋄ C90 C98 E70 F35 F50 G30 ⋄ REV MR 82d:03061 Zbl 415 # 03053 JSL 48.1201 • ID 53155

FOURMAN, M.P. [1980] *Sheaf models for set theory* (J 0326) J Pure Appl Algebra 19*91-101
⋄ C62 C90 E25 E35 E70 F50 G30 ⋄ REV MR 82i:03078 Zbl 446 # 03041 • ID 56570

FOURMAN, M.P. see Vol. V, VI for further entries

FOWLER III, N. [1975] *Intersections of α-spaces* (J 0038) J Austral Math Soc 20*398-418
⋄ C57 C60 D50 ⋄ REV MR 52 # 5397 Zbl 317 # 02049 • ID 15238

FOWLER III, N. [1975] *Sum of α-spaces* (J 0047) Notre Dame J Formal Log 16*379-388
⋄ C57 C60 D45 ⋄ REV MR 51 # 10065 Zbl 271 # 02030 • ID 17130

FOWLER III, N. [1976] *α-decompositions of α-spaces* (J 0036) J Symb Logic 41*483-488
⋄ C57 C60 D45 D50 ⋄ REV MR 55 # 92 Zbl 344 # 02035 • ID 14783

FOWLER III, N. [1978] *Effective inner product spaces* (J 0047) Notre Dame J Formal Log 19*693-701
⋄ C57 C60 D45 D80 ⋄ REV MR 80a:03056 Zbl 368 # 02044 • ID 52144

FOX, R. [1983] see BANKSTON, P.

FRAASSEN VAN, B.C. [1968] *A topological proof of the Loewenheim-Skolem, compactness, and strong completeness theorems for free logic* (J 0068) Z Math Logik Grundlagen Math 14*245-254
⋄ B60 C07 ⋄ REV MR 38 # 35 Zbl 165.15 • ID 04480

FRAASSEN VAN, B.C. see Vol. I, II for further entries

FRAISSE, R. [1949] *Sur une classification des systemes de relations faisant intervenir les ordinaux transfinis* (J 0109) C R Acad Sci, Paris 228*1682-1684
⋄ C07 E07 ⋄ REV MR 11.17 Zbl 32.337 • ID 04508

FRAISSE, R. [1950] *Sur les types de polyrelation et sur une hypothese d'origine logistique* (J 0109) C R Acad Sci, Paris 230*1557-1559
⋄ B10 C07 E07 E10 ⋄ REV MR 12.14 Zbl 40.164 • ID 04510

FRAISSE, R. [1950] *Sur une nouvelle classification des systemes de relations* (J 0109) C R Acad Sci, Paris 230*1022-1024
⋄ C07 E07 ⋄ REV MR 11.585 Zbl 40.164 • ID 04509

FRAISSE, R. & POSSEL DE, R. [1954] *Sur certaines suites d'equivalences dans une classe ordonnee, et sur leur application a la definition des parentes entre relations* (J 0109) C R Acad Sci, Paris 239∗940-942
⋄ C07 E07 ⋄ REV MR 16.227 Zbl 59.25 JSL 32.280
• ID 21995

FRAISSE, R. [1954] *Sur l'extension aux relations de quelques proprietes des ordres* (J 0204) Ann Sci Ecole Norm Sup, Ser 3 71∗363-388
⋄ C07 C30 C50 E07 ⋄ REV MR 16.1006 Zbl 57.42 JSL 32.280 • ID 04515

FRAISSE, R. [1954] *Sur quelques classifications des systemes de relations (English summary)* (J 0364) Publ Sci Univ Alger Ser A Math 1∗35-182
⋄ C07 E07 ⋄ REV MR 15.296 JSL 22.371 • ID 17070

FRAISSE, R. [1955] *La construction des γ-operateurs et leur application au calcul logique du premier ordre* (J 0109) C R Acad Sci, Paris 240∗2191-2193
⋄ B10 C07 E07 ⋄ REV MR 16.1006 Zbl 64.288 JSL 32.280 • ID 04516

FRAISSE, R. [1955] *Sur certains operateurs dans les classes de relations* (J 0109) C R Acad Sci, Paris 240∗2109-2110
⋄ C07 E07 ⋄ REV MR 16.1005 Zbl 64.287 JSL 32.280
• ID 04517

FRAISSE, R. [1955] *Sur quelques classifications des systemes de relations* (X 1259) Durand: Paris 156pp
⋄ C07 E07 ⋄ REV MR 15.296 Zbl 68.243 • ID 23594

FRAISSE, R. [1955] *Sur quelques classifications des relations basees sur des isomorphismes restreints I: Etude generale. II: Applications aux relations d'ordre, et construction d'exemples montrant que ces classifications sont distinctes* (J 0364) Publ Sci Univ Alger Ser A Math 2∗15-60,273-295
⋄ C07 C20 C65 E07 ⋄ REV MR 18.139 Zbl 126.13 Zbl 126.14 JSL 32.280 • REM Part III 1956 • ID 33347

FRAISSE, R. [1956] *Application des γ-operateurs au calcul logique du premier echelon* (J 0068) Z Math Logik Grundlagen Math 2∗76-92
⋄ B10 C07 C65 ⋄ REV MR 19.829 Zbl 126.17 JSL 32.280 • ID 04518

FRAISSE, R. [1956] *Etude de certains operateurs dans les classes de relations, definis a partir d'isomorphismes restreints* (J 0068) Z Math Logik Grundlagen Math 2∗59-75
⋄ C07 E07 ⋄ REV MR 18.456 Zbl 126.16 JSL 32.280
• ID 04519

FRAISSE, R. [1956] *Sur quelques classifications des relations, basees sur des isomorphismes restreints. III. Comparaison des parentes introduites dans la premiere partie avec des parentes precedemment etudiees* (J 0364) Publ Sci Univ Alger Ser A Math 3∗143-159
⋄ C07 E07 ⋄ REV MR 20 # 5145 Zbl 126.15 JSL 32.280
• REM Parts I,II 1955 • ID 17067

FRAISSE, R. [1957] *Obtention en theorie des relations de certaines classes d'origine logique* (P 0575) Int Congr Math (II, 7);1954 Amsterdam 1∗537-538
⋄ C07 ⋄ ID 29460

FRAISSE, R. [1958] *Sur une extension de la polyrelation et des parentes tirant son origine du calcul logique du $k^{\grave{e}me}$-echelon* (P 0576) Raisonn en Math & Sci Exper;1955 Paris 45-50
⋄ B15 C07 C85 E07 ⋄ REV MR 21 # 4110 Zbl 126.18 JSL 25.285 • ID 04520

FRAISSE, R. [1960] *Les modeles et l'algebre logique* (J 0154) Synthese 12∗197-201
• REPR [1961] (P 0711) Concept & Role of Model in Math & Sci;1960 Utrecht 73-77
⋄ C07 ⋄ REV MR 27 # 3545 Zbl 148.243 • ID 04522

FRAISSE, R. [1965] *A hypothesis concerning the extension of finite relations and its verification for certain special cases* (P 0614) Th Models;1963 Berkeley 96-106
⋄ C07 C13 E07 ⋄ REV MR 33 # 3942 Zbl 168.7 • REM Part I. Part II 1967 • ID 04526

FRAISSE, R. [1966] *Une generalisation de l'ultraproduit* (J 0036) J Symb Logic 31∗235-244
⋄ C20 ⋄ REV MR 34 # 1187 Zbl 145.242 JSL 40.457
• ID 04527

FRAISSE, R. [1967] *Cours de logique mathematique. Tome I. Relation, formule logique, compacite, completude* (X 0834) Gauthier-Villars: Paris xii+187pp
• TRANSL [1973] (X 0835) Reidel: Dordrecht xvi+186pp
⋄ B98 C98 ⋄ REV MR 37 # 3902 Zbl 247 # 02001 Zbl 247 # 02002 JSL 35.580 • REM Tome II 1972 • ID 04529

FRAISSE, R. [1967] *L'algebre logique et ses rapports avec la theorie des relations* (X 0893) Pr Univ Montreal: Montreal 78pp
⋄ C07 C50 E07 ⋄ REV MR 40 # 5430 Zbl 162.15
• ID 17072

FRAISSE, R. [1967] *La classe universelle: Une notion a la limite de la logique et de la theorie des relations* (J 0056) Publ Dep Math, Lyon 4/3∗55-61
⋄ C07 C30 C52 ⋄ REV MR 39 # 47 Zbl 189.8 • ID 04528

FRAISSE, R. [1967] *Une hypothese sur l'extension des relations finies et sa verification dans certaines classes particulieres. II* (P 0683) Log & Founds of Sci (Beth);1964 Paris 46-58
⋄ C07 C13 E07 ⋄ REV MR 45 # 3180 Zbl 174.15 • REM Part I 1965 • ID 04525

FRAISSE, R. [1970] *Aspects du theoreme de completude selon Herbrand* (P 0625) Symp Autom Demonst;1968 Versailles 73-86
⋄ B10 C07 ⋄ REV MR 42 # 5793 Zbl 205.305 • ID 04531

FRAISSE, R. [1971] *Abritement entre relations et specialement entre chaines* (P 0669) Conv Teor Modelli & Geom;1969/70 Roma 203-251
⋄ C65 E07 ⋄ REV MR 43 # 110 Zbl 211.18 • ID 04530

FRAISSE, R. & POUZET, M. [1971] *Interpretabilite d'une relation par une chaine* (J 2313) C R Acad Sci, Paris, Ser A-B 272∗A1624-A1627
⋄ C07 C30 C65 E07 ⋄ REV MR 44 # 2616 Zbl 223 # 04001 • ID 31976

FRAISSE, R. & POUZET, M. [1971] *Sur une classe de relations n'ayant qu'un nombre fini de bornes* (J 2313) C R Acad Sci, Paris, Ser A-B 273∗A275-A278
⋄ C07 C30 C65 E07 ⋄ REV MR 44 # 2617 Zbl 223 # 04002 • ID 31977

FRAISSE, R. [1972] *Chaines compatibles pour un groupe de permutations. Application aux relations enchainables et monomorphes* (C 0648) Th of Sets and Topology (Hausdorff) 129-149
⋄ C07 C30 C52 C65 E07 ⋄ REV Zbl 344 # 06001
• ID 61774

FRAISSE, R. [1972] *Cours de logique mathematique. Tome 2: Theorie des modeles* (**X** 0834) Gauthier-Villars: Paris xiv+177pp
• TRANSL [1974] (**X** 0835) Reidel: Dordrecht xix+191pp
◊ B98 C98 ◊ REV MR 49#10514 Zbl 247#02003 • REM Tome I 1967. Tome III 1975. • ID 04535

FRAISSE, R. [1972] *Reflexions sur la completude selon Herbrand* (**J** 0286) Int Logic Rev 5*86-98
◊ B10 C07 ◊ REV MR 51#125 Zbl 341#02042 JSL 40.238 • ID 04294

FRAISSE, R. [1974] *Isomorphisme local et equivalence associes a un ordinal; utilite en calcul des formules infinies a quanteurs finis* (**P** 0610) Tarski Symp;1971 Berkeley 241-254
◊ B15 C07 C75 C85 ◊ REV MR 50#9591 Zbl 327#02015 • ID 04537

FRAISSE, R. [1974] *Multirelation et age 1-extensifs* (**J** 0027) Fund Math 82*323-330
◊ C07 E07 ◊ REV MR 51#12529 Zbl 348#02045 • ID 17244

FRAISSE, R. [1974] *Problematique apportee par les relations en theorie des permutations* (**P** 0776) Permutations;1972 Paris 211-228
◊ C07 C30 E07 ◊ REV MR 52#539 Zbl 296#04002 • ID 61769

FRAISSE, R. [1977] *Deux relations denombrables, logiquement equivalentes pour le second ordre, sont isomorphes* (**J** 0056) Publ Dep Math, Lyon 14/2*41-62
◊ B15 C15 C85 E07 E45 ◊ REV MR 58#27297 Zbl 397#03008 • ID 52674

FRAISSE, R. [1977] *Present problems about intervals in relation-theory and logic* (**P** 1076) Latin Amer Symp Math Log (3);1976 Campinas 179-200
◊ C07 E07 ◊ REV MR 57#16082 Zbl 362#02054 • ID 16604

FRAISSE, R. [1979] *Le forcing faible; son utilisation pour caracteriser les relations unaires generales pour la chaine ou la consecutivite des entiers naturels (resultat de R. Solovay, 1976)* (**J** 0056) Publ Dep Math, Lyon 16/3-4*89-99
◊ C25 E40 ◊ REV MR 82b:03098 Zbl 449#03024 • ID 56691

FRAISSE, R. [1982] *La zerologie, une recherche aux frontieres de la logique et de l'art; application a la logique des relations de base vide* (**J** 0286) Int Logic Rev 26*67-79
◊ B60 C07 ◊ REV MR 85g:03039 • ID 43874

FRAISSE, R. [1984] *L'intervalle en theorie des relations; ses generalisations, filtre intervallaire et cloture d'une relation (English summary)* (**P** 2167) Orders: Descr & Roles;1982 L'Arbresle 313-341
◊ C07 E07 ◊ REV Zbl 574#04001 • ID 41777

FRAISSE, R. [1985] *Deux relations denombrables, logiquement equivalentes pour le second ordre, sont isomorphes (modulo un axiome de constructibilite)* (**C** 4181) Math Log & Formal Syst (Costa da) 161-182
◊ B15 C15 C85 E45 ◊ ID 48204

FRAISSE, R. see Vol. I, IV, V for further entries

FRANCI, R. [1977] *Un'osservazione sulle connessioni di Galois (English summary)* (**J** 0319) Matematiche (Sem Mat Catania) 32*175-182
◊ C07 E20 ◊ REV MR 81e:06004 Zbl 441#06005 • ID 81255

FRANEK, F. [1985] *Certain values of completeness and saturatedness of a uniform ideal rule out certain sizes of the underlying index set* (**J** 0018) Canad Math Bull 28*501-504
◊ C20 E05 ◊ ID 49474

FRANEK, F. [1985] *Isomorphisms of trees* (**J** 0053) Proc Amer Math Soc 95*95-100
◊ C50 C65 E05 ◊ ID 47703

FRANEK, F. see Vol. V for further entries

FRANEK, M. [1975] *On some relations in boolean algebras* (**J** 0143) Mat Chasopis (Slov Akad Ved) 25*111-127
◊ B25 G05 ◊ REV MR 53#5284 Zbl 306#02057 • ID 22919

FRANEK, M. see Vol. I for further entries

FRANZEN, B. [1981] *Algebraic compactness of filter quotients* (**P** 2622) Abel Group Th;1981 Oberwolfach 228-241
◊ C05 C30 C60 ◊ REV MR 83f:20040 Zbl 501.20034 • ID 90312

FRASNAY, C. [1962] *Relations invariantes dans un ensemble totalement ordonne* (**J** 0109) C R Acad Sci, Paris 255*2878-2879
◊ C07 C30 E07 ◊ REV MR 27#57 Zbl 122.20 • ID 31766

FRASNAY, C. [1963] *Groupes de permutations finies et familles d'ordres totaux, application a la theorie des relations* (**J** 0109) C R Acad Sci, Paris 257*2944-2947
◊ C07 C30 E07 ◊ REV MR 28#1142 Zbl 119.28 • ID 31768

FRASNAY, C. [1964] *Groupes de compatibilite de deux ordres totaux; application aux n-morphismes d'une relation m-aire* (**J** 0109) C R Acad Sci, Paris 259*3910-3913
◊ C07 C30 E07 ◊ REV MR 30#3035 Zbl 129.13 • ID 31770

FRASNAY, C. [1964] *Partages d'ensembles de parties et de produits d'ensembles, applications a la theorie des relations* (**J** 0109) C R Acad Sci, Paris 258*1373-1376
◊ C07 C30 E07 ◊ REV MR 32#5532 Zbl 137.16 • ID 31769

FRASNAY, C. [1965] *Quelques problemes combinatoires concernant les ordres totaux et les relations monomorphes* (**J** 0240) Ann Inst Fourier 15*415-524
◊ C07 C30 E07 ◊ REV MR 33#54 • ID 04556

FRASNAY, C. [1971] *Relations interpretables par une chaine, relations monomorphes* (**P** 3600) Ordres Totaux Finis;1967 Aix-en-Provence 163-169
◊ C07 E07 ◊ REV MR 51#12625 • ID 31771

FRASNAY, C. [1973] *Interpretation relationniste du seuil de recollement d'un groupe naturel* (**J** 2313) C R Acad Sci, Paris, Ser A-B 277*A865-A868
◊ C07 C30 E07 ◊ REV MR 51#5448 Zbl 274#04002 • ID 61790

FRASNAY, C. [1974] *Theoreme de G-recollement (d'une famille d'ordres totaux). Application aux relations monomorphes, extension aux multirelations* (**P** 0776) Permutations;1972 Paris 229-237
◊ C07 C30 E07 ◊ REV MR 50#112 Zbl 296#04003 • ID 32060

FRASNAY, C. [1979] *Sur quelques classes universelles de relations m-aires* (**J** 0056) Publ Dep Math, Lyon 16/3-4*21-32
◊ C07 C30 C52 E07 ◊ REV MR 82h:04002 Zbl 451#04002 • ID 73033

FRASNAY, C. [1984] *Relations enchaineables, rangements et pseudo-rangements (English summary)* (P 2167) Orders: Descr & Roles;1982 L'Arbresle 235-268
⋄ C07 C13 C30 E07 ⋄ REV Zbl 574#04002 • ID 41782

FRASNAY, C. see Vol. V for further entries

FRAYNE, T.E. & MOREL, A.C. & SCOTT, D.S. [1962] *Reduced direct products* (J 0027) Fund Math 51*195-228 • ERR/ADD ibid 53*117
⋄ B20 C20 C55 ⋄ REV MR 26#28 MR 27#4751 Zbl 108.5 JSL 31.506 • ID 04557

FREEMAN, J.B. [1977] see DANIELS, C.B.

FREEMAN, J.B. [1978] see DANIELS, C.B.

FREEMAN, J.B. see Vol. I, II for further entries

FREESE, R. [1980] see DAY, A.

FREESE, R. [1983] see CZEDLI, G.

FREESE, R. see Vol. IV for further entries

FRENKEL', V.I. [1963] *Algorithmic problems in partially ordered groups (Russian)* (J 0023) Dokl Akad Nauk SSSR 152*67-70
• TRANSL [1963] (J 0062) Sov Math, Dokl 4*1266-1269
⋄ C57 C60 D40 ⋄ REV MR 27#3715 MR 31#1194
• ID 20969

FRENKEL', V.I. [1964] *On the effective partial ordering of finitely defined groups (Russian)* (J 0092) Sib Mat Zh 5*651-670
⋄ C57 C60 ⋄ REV MR 29#2290 • ID 04602

FRENKEL', V.I. see Vol. IV for further entries

FRENKIN, B.R. [1974] *Σ-free models (Russian)* (J 0092) Sib Mat Zh 15*580-597,702
• TRANSL [1974] (J 0475) Sib Math J 15*416-428
⋄ C05 C52 ⋄ REV MR 54#4965 Zbl 282#08002
• ID 24134

FRENKIN, B.R. [1974] *Reducibility and uniform reducibility of algebraic operations (Russian)* (J 0142) Mat Sb, Akad Nauk SSSR, NS 95(137)*384-395,471
• TRANSL [1974] (J 0349) Math of USSR, Sbor 24*373-384
⋄ C05 ⋄ REV MR 52#3014 Zbl 319#20067 • ID 15239

FREY, G. [1980] *Cyclic isogenies and nonstandard arithmetic* (J 0401) J Number Th 12*343-363
⋄ H15 H20 ⋄ REV MR 82e:14048 Zbl 439#14004
• ID 56046

FREY, G. [1983] *Nonstandard arithmetic and application to height functions* (P 4230) Groupe Etude Anal Ultrametrique (9);1982 Marseille 3/J8*2pp
⋄ C60 H10 H15 H20 ⋄ REV MR 85m:11035 Zbl 513#12021 • ID 37806

FREY, G. see Vol. II for further entries

FREYD, P. [1965] *The theories of functors and models* (P 0614) Th Models;1963 Berkeley 107-120
⋄ C30 G30 ⋄ REV MR 34#1371 Zbl 192.39 JSL 36.336
• ID 04617

FREYD, P. [1976] *Properties invariant within equivalence types of categories* (P 1463) Algeb, Topol & Categ Th (Eilenberg);1974 New York 55-61
⋄ C40 G30 ⋄ REV MR 54#376 Zbl 342#18001
• ID 23989

FREYD, P. see Vol. I, V for further entries

FRIDMAN, EH.I. & PENZIN, YU.G. [1972] *The elementary and the universal theory of the ordered group of integers with maximal subgroups (Russian)* (C 3549) Algebra, Vyp 1 (Irkutsk) 80-86
⋄ B20 B25 D35 ⋄ REV MR 56#11776 • ID 77209

FRIDMAN, EH.I. [1972] *Undecidability of the elementary theory of abelian torsion-free groups with a finite set of serving subgroups (Russian)* (C 3549) Algebra, Vyp 1 (Irkutsk) 97-100
⋄ C60 D35 ⋄ REV MR 57#2901 • ID 73069

FRIDMAN, EH.I. [1973] *Positive element-subgroup theories (Russian)* (C 1443) Algebra, Vyp 2 (Irkutsk) 161-163
⋄ B25 D35 ⋄ REV MR 53#7754 • ID 22999

FRIDMAN, EH.I. & SLOBODSKOJ, A.M. [1973] *The theory of the additive group of the integers with an arbitrary number of predicates that define maximal subgroups (Russian)* (C 1443) Algebra, Vyp 2 (Irkutsk) 130-137
⋄ C10 F30 ⋄ REV MR 53#10575 • ID 23081

FRIDMAN, EH.I. & SLOBODSKOJ, A.M. [1975] *Theories of Abelian groups with predicates specifying a subgroup (Russian)* (J 0003) Algebra i Logika 14*572-575
• TRANSL [1975] (J 0069) Algeb and Log 14*353-355
⋄ C60 D35 ⋄ REV MR 55#12512 Zbl 347#02037
• ID 52191

FRIDMAN, EH.I. see Vol. IV for further entries

FRIED, E. [1977] *Automorphism group of integral domains fixing a given subring* (J 0004) Algeb Universalis 7*373-387
⋄ C07 C60 ⋄ REV MR 56#2985 Zbl 374#13003
• ID 26603

FRIED, M. & SACERDOTE, G.S. [1976] *Solving diophantine problems over all residue class fields of a number field and all finite fields* (J 0120) Ann of Math, Ser 2 104*203-233
⋄ B25 C10 C13 C60 D35 ⋄ REV MR 58#10722 Zbl 376#02042 • ID 51684

FRIED, M. & JARDEN, M. [1976] *Stable extensions and fields with the global density property* (J 0017) Canad J Math 28*774-787
⋄ C60 ⋄ ID 48391

FRIED, M. & HARAN, D. & JARDEN, M. [1984] *Galois stratification over Frobenius fields* (J 0345) Adv Math 51*1-35
⋄ B25 C60 ⋄ REV MR 86c:12007 Zbl 554#12016
• ID 44952

FRIEDMAN, H.M. & JENSEN, R.B. [1968] *Note on admissible ordinals* (P 0637) Syntax & Semant Infinitary Lang;1967 Los Angeles 77-79
⋄ C70 D60 ⋄ REV Zbl 199.305 • ID 04643

FRIEDMAN, H.M. [1970] *Iterated inductive definitions and Σ_2^1-AC* (P 0603) Intuitionism & Proof Th;1968 Buffalo 435-442
⋄ C62 D55 E25 F35 ⋄ REV MR 44#1555 Zbl 216.6
• ID 14546

FRIEDMAN, H.M. [1971] *A more explicit set theory* (P 0693) Axiomatic Set Th;1967 Los Angeles 1*49-66
⋄ A05 C62 E30 E35 E45 F30 F35 ⋄ REV MR 43#4658 Zbl 233#02026 • ID 17088

FRIEDMAN, H.M. [1971] see ENDERTON, H.B.

FRIEDMAN, H.M. [1973] *Beth's theorem in cardinality logics* (J 0029) Israel J Math 14*205-212
⋄ C40 C55 C80 ⋄ REV MR 47 # 6472   Zbl 263 # 02010
• ID 04655

FRIEDMAN, H.M. [1973] *Countable models of set theories* (P 0713) Cambridge Summer School Math Log;1971 Cambridge GB 539-573
⋄ C15 C62 E45 F35 H20 ⋄ REV MR 50 # 102
Zbl 271 # 02036 • ID 04654

FRIEDMAN, H.M. [1974] *On existence proofs of Hanf numbers* (J 0036) J Symb Logic 39*318-324
⋄ C55 C75 E30 E70 F50 ⋄ REV MR 53 # 12943
Zbl 293 # 02039 • ID 04656

FRIEDMAN, H.M. [1975] *Adding propositional connectives to countable infinitary logic* (J 0332) Math Proc Cambridge Phil Soc 77*1-6
⋄ C70 C75 ⋄ REV MR 50 # 6786   Zbl 323 # 02022
• ID 17089

FRIEDMAN, H.M. [1975] *Large models of countable height* (J 0064) Trans Amer Math Soc 201*227-239
⋄ C15 C55 C62 E25 E40 ⋄ REV MR 54 # 4966
Zbl 296 # 02036 • ID 04661

FRIEDMAN, H.M. [1975] *One hundred and two problems in mathematical logic* (J 0036) J Symb Logic 40*113-129
⋄ B98 C98 D98 E98 F98 ⋄ REV MR 51 # 5254
Zbl 318 # 02002 • ID 04296

FRIEDMAN, H.M. [1976] *On decidability of equational theories* (J 0326) J Pure Appl Algebra 7*1-3
⋄ B25 C05 D35 ⋄ REV MR 53 # 5285   Zbl 323 # 02056
• ID 22920

FRIEDMAN, H.M. [1976] *The complexity of explicit definitions* (J 0345) Adv Math 20*18-29
⋄ C40 C57 ⋄ REV MR 53 # 5287   Zbl 355 # 02010
• ID 22924

FRIEDMAN, H.M. [1978] *Categoricity with respect to ordinals* (P 1864) Higher Set Th;1977 Oberwolfach 669*17-20
⋄ C35 C62 E45 ⋄ REV MR 80m:03089   Zbl 395 # 03035
• ID 31764

FRIEDMAN, H.M. [1979] *On the naturalness of definable operations* (J 1447) Houston J Math 5*325-330
⋄ C30 C62 E35 E47 ⋄ REV MR 81d:03055
Zbl 453 # 03051 • ID 54181

FRIEDMAN, H.M. [1984] *On the spectra of universal relational sentences* (J 0194) Inform & Control 62*205-209
⋄ C13 D15 ⋄ ID 46357

FRIEDMAN, H.M. see Vol. I, IV, V, VI for further entries

FRIEDMAN, JOYCE [1963] *A semi-decision procedure for the functional calculus* (J 0037) ACM J 10*1-24
• REPR [1983] (C 4659) Autom of Reasoning 1*331-354
⋄ B25 B35 ⋄ REV MR 26 # 4892   Zbl 144.244   JSL 29.101
• ID 04667

FRIEDMAN, JOYCE [1966] *Computer realization of a decision procedure in logic* (P 0573) Inform Processing (3);1965 New York 2*327-328
⋄ B25 B35 ⋄ REV Zbl 173.17 • ID 29439

FRIEDMAN, JOYCE see Vol. I, IV, VI for further entries

FRIEDMAN, S.D. [1979] *HC of an admissible set* (J 0036) J Symb Logic 44*95-102
⋄ C70 D60 E30 ⋄ REV MR 81i:03054   Zbl 406 # 03056
• ID 56139

FRIEDMAN, S.D. [1981] *Uncountable admissibles II: Compactness* (J 0029) Israel J Math 40*129-149
⋄ C70 D60 E45 E47 ⋄ REV MR 83k:03058b
Zbl 522 # 03035 • REM Part I 1982 • ID 33348

FRIEDMAN, S.D. [1982] *Steel forcing and Barwise compactness* (J 0007) Ann Math Logic 22*31-46
⋄ C70 E40 ⋄ REV MR 83i:03061   Zbl 501 # 03032
• ID 33349

FRIEDMAN, S.D. [1982] *Uncountable admissibles I: Forcing* (J 0064) Trans Amer Math Soc 270*61-73
⋄ C70 D60 E40 ⋄ REV MR 83k:03058a   Zbl 493 # 03025
• REM Part II 1981 • ID 33350

FRIEDMAN, S.D. & SHELAH, S. [1983] *Tall α-recursive structures* (J 0053) Proc Amer Math Soc 88*672-678
⋄ C70 C75 E45 ⋄ REV MR 85g:03057   Zbl 528 # 03022
• ID 37639

FRIEDMAN, S.D. [1984] *Infinitary logic and $0^{\#}$* (P 3823) Axiomatic Set Th;1983 Boulder 99-107
⋄ C75 E15 E45 E55 ⋄ REV MR 86h:03091
Zbl 552 # 03033 • ID 42671

FRIEDMAN, S.D. [1984] *Model theory for $L_{\infty \omega_1}$* (J 0073) Ann Pure Appl Logic 26*103-122
⋄ C75 ⋄ REV MR 86g:03062   Zbl 558 # 03018 • ID 45237

FRIEDMAN, S.D. [1985] see DAVID, R.

FRIEDMAN, S.D. see Vol. IV, V for further entries

FRIEDRICH, U. [1972] *Entscheidbarkeit der monadischen Nachfolgerarithmetik mit endlichen Automaten ohne Analysetheorem (Russian, English and French summaries)* (J 0115) Wiss Z Humboldt-Univ Berlin, Math-Nat Reihe 21*503-504
⋄ B25 C85 D05 F35 ⋄ REV MR 48 # 3724
Zbl 248 # 02051 • ID 04668

FRIEDRICHSDORF, U. [1976] *Einige Bemerkungen zur Peano-Arithmetik* (J 0068) Z Math Logik Grundlagen Math 22*431-436
⋄ C25 C52 C62 F30 ⋄ REV MR 56 # 119
Zbl 356 # 02044 • ID 24288

FRIEDRICHSDORF, U. see Vol. V for further entries

FRODA-SCHECHTER, M. & KWATINETZ, M. [1974] *Some remarks on a paper of R. C. Lyndon; "Properties preserved under homomorphism"* (J 0517) Mathematica (Cluj) 16(39)*255-257
⋄ C05 C40 C52 ⋄ REV MR 58 # 27463   Zbl 382 # 03020
• REM Lyndon's paper was published in J0048 9*143-154 (1959) • ID 51944

FRODA-SCHECHTER, M. [1974] *Sur les homomorphismes des structures relationnelles I* (J 3451) Stud Univ Cluj, Ser Math-Mech 19/2*20-25
⋄ C07 ⋄ REV MR 50 # 9744   Zbl 296 # 08001 • ID 61846

FRODA-SCHECHTER, M. see Vol. V for further entries

FROEHLICH, A. & SHEPHERDSON, J.C. [1955] *Effective procedures in field theory* (J 0354) Phil Trans Roy Soc London, Ser A 248*407-432
⋄ C57 C60 D45 ⋄ REV MR 17.570 Zbl 70.35 JSL 24.169
• ID 17092

FROEHLICH, A. see Vol. IV for further entries

FROEMKE, J. & QUACKENBUSH, R.W. [1975] *The spectrum of an equational class of groupoids* (J 0048) Pac J Math 58*381-386
⋄ C05 C13 ⋄ REV MR 52 # 5522 Zbl 326 # 08007
• ID 14832

FUCHS, L. & KERTESZ, A. & SZELE, T. [1956] *On abelian groups in which every homomorphic image can be embedded* (J 0001) Acta Math Acad Sci Hung 7*467-474
⋄ C60 ⋄ REV MR 19.12 Zbl 75.241 • ID 04701

FUCHS, L. [1959] *The existence of indecomposable abelian groups of arbitrary power* (J 0001) Acta Math Acad Sci Hung 10*453-457
⋄ C55 C60 ⋄ REV MR 23 # A3176 • ID 90209

FUCHS, L. [1974] *Indecomposable abelian groups of measurable cardinalities* (P 2632) Conv Gruppi Abel & Conv Gruppi e loro Rappresent;1972 Roma 233-244
⋄ C55 C60 E55 ⋄ REV MR 52 # 5828 Zbl 328 # 20044
• ID 18150

FUCHS, P.H. & PILZ, G. [1985] *Ultraproducts and ultra-limits of near rings* (J 0045) Monatsh Math 100*105-112
⋄ C20 C60 ⋄ REV Zbl 571 # 16018 • ID 48306

FUCHS, P.H. see Vol. IV for further entries

FUENTES MIRAS, J.R. & JOSE, R. [1954] *"Deductive truth" according to Tarski (Spanish)* (J 4443) Gac Mat (Madrid) 6*110-120
⋄ C07 C98 ⋄ REV MR 18.270 Zbl 56.13 • ID 47948

FUENTES MIRAS, J.R. see Vol. I, VI for further entries

FUERER, M. [1981] *Alternation and the Ackermann case of the decision problem* (J 3370) Enseign Math, Ser 2 27*137-162
• REPR [1982] (P 3482) Logic & Algor (Specker);1980 Zuerich 161-186
⋄ B20 B25 D15 ⋄ REV MR 83d:03015a Zbl 479 # 03005 Zbl 502 # 03025 • ID 35169

FUERER, M. [1982] *The complexity of Presburger arithmetic with bounded quantifier alternation depth* (J 1426) Theor Comput Sci 18*105-111
⋄ B25 D15 ⋄ REV MR 83b:03049 Zbl 484 # 03003
• ID 35102

FUERER, M. see Vol. I, IV, V for further entries

FUHRKEN, G. [1962] *Bemerkung zu einer Arbeit E.Engelers* (J 0068) Z Math Logik Grundlagen Math 8*277-279
⋄ C15 C40 C75 ⋄ REV MR 26 # 3600 Zbl 109.8 JSL 35.343 • ID 04705

FUHRKEN, G. [1964] *Skolem-type normal forms for first-order languages with a generalized quantifier* (J 0027) Fund Math 54*291-302
⋄ C55 C80 ⋄ REV MR 29 # 3363 Zbl 166.260 JSL 33.121
• ID 04706

FUHRKEN, G. [1965] *Languages with added quantifier "There exist at least $\aleph_\alpha$"* (P 0614) Th Models;1963 Berkeley 121-131
⋄ C55 C80 ⋄ REV MR 39 # 2624 Zbl 166.260 JSL 35.342
• ID 04707

FUHRKEN, G. [1966] *Minimal- und Primmodelle* (J 0009) Arch Math Logik Grundlagenforsch 9*3-11
⋄ C50 ⋄ REV MR 35 # 38 Zbl 168.7 • ID 17100

FUHRKEN, G. [1968] *A model with exactly one undefinable element* (S 0019) Colloq Math (Warsaw) 19*183-185
⋄ C40 C50 ⋄ REV MR 37 # 5088 Zbl 176.275 • ID 04708

FUHRKEN, G. [1970] *On the degree of compactness of the languages Q_α* (1111) Preprints, Manuscr., Techn. Reports etc. 3pp
⋄ C55 C80 ⋄ REM Univ. Colorado, Univ. Minnesota
• ID 21342

FUHRKEN, G. [1971] see EHRENFEUCHT, A.

FUHRKEN, G. & TAYLOR, W. [1971] *Weakly atomic-compact relational structures* (J 0036) J Symb Logic 36*129-140
⋄ C05 C50 ⋄ REV MR 44 # 2590 Zbl 217.13 • ID 04710

FUHRKEN, G. [1972] *A remark on the Haertig quantifier* (J 0068) Z Math Logik Grundlagen Math 18*227-228
⋄ C55 C80 ⋄ REV MR 46 # 3265 Zbl 168.7 Zbl 246 # 02013 • ID 04712

FUJIWARA, T. [1956] *On the existence of algebraically closed algebraic systems* (J 1770) Osaka Math J 8*23-33
⋄ C05 ⋄ REV MR 18.11 Zbl 71.26 • ID 17101

FUJIWARA, T. & NAKANO, Y. [1970] *On the relation between free structures and direct limits* (J 0352) Math Jap 15*19-23
⋄ C05 C30 ⋄ REV MR 43 # 7388 Zbl 209.42 • ID 17102

FUJIWARA, T. [1971] *Freely generable classes of structures* (J 0081) Proc Japan Acad 47*761-764
⋄ C05 ⋄ REV MR 46 # 5127 Zbl 243 # 08003 • ID 04713

FUJIWARA, T. [1971] *On the construction of the least universal Horn class containing a given class* (J 0351) Osaka J Math 8*425-436
⋄ B20 C05 C52 ⋄ REV MR 47 # 108 Zbl 264 # 08003
• ID 17104

FUJIWARA, T. [1973] *A generalization of Lyndon's theorem characterizing sentences preserved in subdirect products* (J 0004) Algeb Universalis 3*280-287
⋄ C05 C30 C40 ⋄ REV MR 50 # 4281 Zbl 312 # 02040
• ID 04715

FUJIWARA, T. [1973] *Note on generalized atomic sets of formulas* (J 0081) Proc Japan Acad 49*443-448
⋄ B20 C05 C40 ⋄ REV MR 49 # 2337 Zbl 276 # 02033
• ID 04714

FUJIWARA, T. [1975] *Universal sentences preserved under certain extensions* (J 0081) Proc Japan Acad 51*29-33
⋄ B20 C40 ⋄ REV MR 52 # 13367 Zbl 331 # 02030
• ID 21821

FUJIWARA, T. [1977] *A remark on existence of maximal homomorphisms* (P 2997) Symp Semigroups (1);1977 Matsue 35-43
⋄ C05 ⋄ REV MR 57 # 16043 Zbl 411 # 18005 • ID 52918

FUJIWARA, T. [1977] *On implicational classes of structures* (P 3262) Semin Semigroup Th;1976 Kyoto 138-159
⋄ C05 C75 ⋄ REV MR 57 # 5734 Zbl 362 # 02048
• ID 50770

FUJIWARA, T. [1978] *A variation of Lyndon-Keisler's homomorphism theorem and its applications to interpolation theorems* (J 0090) J Math Soc Japan 30*287-302
⋄ C40 ⋄ REV MR 80b:03043 Zbl 369 # 02030 • ID 51805

FUJIWARA, T. [1978] *Algebraically closed algebraic extensions in universal classes* (P 3009) Symp Semigroups (2);1978 Tokyo 55-67
⋄ C05 C25 ⋄ REV MR 81g:03037 Zbl 408 # 03022
• ID 56261

FUJIWARA, T. [1980] *Existentially closed structures and algebraically closed structures* (P 3011) Symp Semigroups (4);1980 Yamaguchi 1-6
⋄ C25 C75 ⋄ REV MR 82f:20008 Zbl 476 # 03037
• ID 55549

FUJIWARA, T. [1980] see ARAGANE, K.

FUJIWARA, T. [1982] *Characterization of axiomatic classes of groupoids closed under identity element extensions* (P 3833) Symp Semigroups (6);1982 Kyoto 62-64
⋄ C05 C52 ⋄ REV Zbl 504 # 20038 • ID 38410

FUJIWARA, T. [1983] *Characterization of axiomatic classes of groupoids closed under identity extensions* (J 0352) Math Jap 28∗633-637
⋄ C05 C52 ⋄ REV MR 85h:08015 Zbl 523 # 20035
• ID 44257

FUJIWARA, T. [1983] see ARAGANE, K.

FUJIWARA, T. [1985] see ARAGANE, K.

FUKSON, V.I. [1971] see DRAGALIN, A.G.

FUKSON, V.I. see Vol. I, V for further entries

FUKUYAMA, M. & NAGAOKA, K. [1974] *Strong representability of α -recursive functions* (S 1459) Mem School Sci & Engin, Waseda Univ 38∗119-128
⋄ C70 D60 ⋄ REV MR 53 # 12907 Zbl 351 # 02032
• ID 23159

FUKUYAMA, M. see Vol. IV for further entries

FUNK, M. & KEGEL, O.H. & STRAMBACH, K. [1984] *On group universality and homogenity* (P 3088) Univer Alg & Link Log, Alg, Combin, Comp Sci;1983 Darmstadt 173-182
⋄ C50 C60 ⋄ REV Zbl 552 # 08002 • ID 47254

FUNK, M. & KEGEL, O.H. & STRAMBACH, K. [1985] *Gruppenuniversalitaet und Homogenisierbarkeit (English summary)* (J 3526) Ann Mat Pura Appl, Ser 4 141∗1-126
⋄ C50 C60 ⋄ ID 49538

FURMANOWSKI, T. [1982] *The logic of algebraic rules as a generalization of equational logic* (J 0387) Bull Sect Logic, Pol Acad Sci 11∗6-12
⋄ C05 C07 ⋄ REV Zbl 528 # 03015 • ID 37633

FURMANOWSKI, T. [1983] *The logic of algebraic rules as a generalization of equational logic* (J 0063) Studia Logica 42∗251-257
⋄ C05 C07 C13 ⋄ REV MR 85j:03040 Zbl 528 # 03015
• ID 42328

FURMANOWSKI, T. see Vol. I, II, VI for further entries

GABBAY, D.M. [1970] *The decidability of the Kreisel-Putnam system* (J 0036) J Symb Logic 35∗431-436
⋄ B25 B55 F50 ⋄ REV MR 58 # 16167 Zbl 228 # 02013
• ID 04721

GABBAY, D.M. [1971] *Decidability results in non-classical logic, III: Systems with stability operators* (J 0029) Israel J Math 10∗135-146
⋄ B25 B45 C90 ⋄ REV MR 46 # 1553 Zbl 289 # 02033
• REM Part I 1975. Parts II,IV have not yet appeared
• ID 30000

GABBAY, D.M. [1971] *On decidable, finitely axiomatizable, modal and tense logics without the finite model property I, II* (J 0029) Israel J Math 10∗478-495,496-503
⋄ B25 B45 C90 ⋄ REV MR 45 # 4961 Zbl 231 # 02023
• ID 04723

GABBAY, D.M. [1971] *Semantic proof of the Craig interpolation theorem for intuitionistic logic and extensions I,II* (P 0638) Logic Colloq;1969 Manchester 391-401,403-410
⋄ B55 C40 C90 F50 ⋄ REV MR 43 # 3094a
MR 43 ∗ 3094b Zbl 234 # 02017 • REM Part III 1977
• ID 04726

GABBAY, D.M. [1972] *Application of trees to intermediate logics* (J 0036) J Symb Logic 37∗135-138
⋄ B55 C90 ⋄ REV MR 47 # 8251 Zbl 243 # 02019
• ID 04732

GABBAY, D.M. [1972] *Craig's interpolation theorem for modal logics* (P 2080) Conf Math Log;1970 London 111-127
⋄ B45 C40 C90 ⋄ REV MR 49 # 8815 Zbl 233 # 02009
• ID 04733

GABBAY, D.M. [1972] *Decidability of some intuitionistic predicate theories* (J 0036) J Symb Logic 37∗579-587
⋄ B25 C90 F50 ⋄ REV MR 48 # 86 Zbl 266 # 02025
• ID 04729

GABBAY, D.M. [1972] *Model theory for intuitionistic logic* (J 0068) Z Math Logik Grundlagen Math 18∗49-54
⋄ B55 C20 C90 F50 ⋄ REV MR 45 # 4959
Zbl 242 # 02059 • ID 04730

GABBAY, D.M. [1973] *A survey of decidability results for modal, tense and intermediate logics* (P 0793) Int Congr Log, Meth & Phil of Sci (4,Proc);1971 Bucharest 29-43
⋄ B25 B45 B55 C98 F50 ⋄ REV MR 55 # 12470
• ID 17911

GABBAY, D.M. [1973] *The undecidability of intuitionistic theories of algebraically closed fields and real closed fields* (J 0036) J Symb Logic 38∗86-92
⋄ B25 C60 C90 D35 F50 ⋄ REV MR 47 # 6466
Zbl 266 # 02027 • ID 61880

GABBAY, D.M. & JONGH DE, D.H.J. [1974] *A sequence of decidable finitely axiomatizable intermediate logics with the disjunction property* (J 0036) J Symb Logic 39∗67-78
⋄ B22 B25 B55 ⋄ REV MR 51 # 10038 Zbl 289 # 02032
• ID 17555

GABBAY, D.M. & MORAVCSIK, J.M.E. [1974] *Branching quantifiers and Montague semantics* (J 1517) Theor Linguist 1∗139-157
⋄ B65 C80 ⋄ REV MR 58 # 21427 • ID 32157

GABBAY, D.M. [1975] *Decidability results in non-classical logics I* (J 0007) Ann Math Logic 8∗237-295
⋄ B25 B45 C90 F50 ⋄ REV MR 52 # 68 Zbl 309 # 02053
• REM Parts II,IV have not yet appeared. Part III 1971
• ID 04738

GABBAY, D.M. [1975] *Model theory for tense logics* (J 0007) Ann Math Logic 8∗185-236
⋄ B45 C20 C90 ⋄ REV MR 51 # 7826 Zbl 307 # 02014
• ID 04739

GABBAY, D.M. [1975] *The decision problem for some finite extensions of the intuitionistic theory of abelian groups* (J 0063) Studia Logica 34∗59-67
⋄ B25 F50 ⋄ REV MR 51 # 12500 Zbl 306 # 02027
• ID 17286

GABBAY, D.M. [1976] *Investigations in modal and tense logics with applications to problems in philosophy and linguistics* (**X** 0835) Reidel: Dordrecht xi+306pp
⋄ A05 B25 B46 C90 ⋄ REV MR 58#27328 Zbl 374#02013 JSL 44.656 • ID 32148

GABBAY, D.M. [1977] *A new version of Beth semantics for intuitionistic logic* (**J** 0036) J Symb Logic 42∗306-308
⋄ C90 F50 ⋄ REV MR 58#16169 Zbl 379#02007 • ID 26463

GABBAY, D.M. [1977] *Craig interpolation theorem for intuitionistic logic and extensions part III* (**J** 0036) J Symb Logic 42∗269-271
⋄ B55 C40 C90 F50 ⋄ REV MR 58#16168 Zbl 372#02016 • REM Parts I,II 1971 • ID 26456

GABBAY, D.M. & KASHER, A. [1979] *On the quantifier "there is a certain X"* (**P** 4597) Cognitive Viewpoint;1977 Ghent 71-78
⋄ C80 ⋄ ID 48392

GABBAY, D.M. [1982] *Intuitionistic basis for non-monotonic logic* (**P** 3840) Autom Deduct (6);1982 New York 260-273
⋄ B55 C90 F50 ⋄ REV MR 85e:03062 Zbl 481#68091 • ID 38464

GABBAY, D.M. & GUENTHNER, F. (EDS.) [1983] *Handbook of philosophical logic Vol. I: Elements of classical logic* (**S** 3307) Synth Libr 164∗xi+493pp
⋄ B98 C98 ⋄ REV Zbl 538#03001 • ID 41457

GABBAY, D.M. & GUENTHNER, F. (EDS.) [1984] *Handbook of philosophical logic Vol II. Extensions of classical logic* (**S** 3307) Synth Libr 165∗x+776pp
⋄ B98 C90 C98 ⋄ REV Zbl 572#03003 • REM Part I 1983 • ID 41831

GABBAY, D.M. see Vol. I, II, IV, VI for further entries

GACS, P. & LOVASZ, L. [1977] *Some remarks on generalized spectra* (**J** 0068) Z Math Logik Grundlagen Math 23∗547-554
⋄ B15 C13 D15 ⋄ REV MR 58#10398 Zbl 398#03025 • ID 52753

GACS, P. see Vol. II, IV for further entries

GAERDENFORS, P. & HANSSON, B. [1975] *Filtrations and the finite frame property in boolean semantics* (**P** 0757) Scand Logic Symp (3);1973 Uppsala 32-39
⋄ B25 B45 C90 G05 G25 ⋄ REV MR 53#2633 Zbl 311#02031 JSL 43.373 • ID 21522

GAERDENFORS, P. see Vol. II for further entries

GAIFMAN, H. [1960] *Probability models and the completeness theorem* (**P** 1953) Int Congr Log, Meth & Phil of Sci (1;Abstr);1960 Stanford 77-78
⋄ B48 C90 ⋄ ID 16922

GAIFMAN, H. [1964] *Concerning measures on first order calculi* (**J** 0029) Israel J Math 2∗1-18
⋄ B48 C90 ⋄ REV MR 31#31 Zbl 192.33 • ID 33352

GAIFMAN, H. [1967] *Uniform extension operators for models and their applications* (**P** 0691) Sets, Models & Recursion Th;1965 Leicester 122-155
⋄ C20 C30 C62 ⋄ REV MR 36#3640 Zbl 174.15 • ID 17108

GAIFMAN, H. [1970] *On local arithmetical functions and their application for constructing types of Peano's arithmetic* (**P** 1072) Math Log & Founds of Set Th;1968 Jerusalem 105-121
⋄ C30 C62 H15 ⋄ REV MR 44#1550 Zbl 209.308 • ID 28477

GAIFMAN, H. [1972] *A note on models and submodels of arithmetic* (**P** 2080) Conf Math Log;1970 London 128-144
⋄ C62 ⋄ REV MR 54#7247 Zbl 255#02058 JSL 40.244 • ID 04749

GAIFMAN, H. [1974] *Elementary embeddings of models of set theory and certain subtheories* (**P** 0693) Axiomatic Set Th;1967 Los Angeles 2∗33-101
⋄ C62 E30 E45 E55 ⋄ REV MR 51#12523 Zbl 342#02042 • ID 17241

GAIFMAN, H. [1974] *Finiteness is not a Σ_0-property* (**J** 0029) Israel J Math 19∗359-368
⋄ C40 E47 ⋄ REV MR 55#10266 Zbl 307#02052 • ID 04751

GAIFMAN, H. [1974] *Operations on relational structures, functors and classes I* (**P** 0610) Tarski Symp;1971 Berkeley 21-39
⋄ C20 C30 C40 G30 ⋄ REV MR 51#7866 Zbl 319#02048 • ID 04750

GAIFMAN, H. [1976] *Models and types of Peano's arithmetic* (**J** 0007) Ann Math Logic 9∗223-306
⋄ C30 C62 H15 ⋄ REV MR 53#10577 Zbl 332#02058 JSL 48.484 • ID 18152

GAIFMAN, H. [1982] *On local and non-local properties* (**P** 3708) Herbrand Symp Logic Colloq;1981 Marseille 105-135
⋄ C07 C13 C65 E47 ⋄ REV MR 85k:03020 Zbl 518#03008 • ID 37513

GAIFMAN, H. see Vol. I, II, IV, V, VI for further entries

GAL, L.N. [1958] *A note on direct products* (**J** 0036) J Symb Logic 23∗1-6
⋄ B25 C30 D35 ⋄ REV MR 20#5121 Zbl 92.6 JSL 36.541 • ID 04757

GALDA, K. & PASSOS, E.P. [1974] *Application of quantifier elimination to mechanical theorem proving* (**P** 3103) Adv Cybern Syst;1972 Oxford 1∗335-348
⋄ B35 C10 ⋄ REV Zbl 332#68064 • ID 61899

GALLIER, J.H. [1984] *n-rational algebras I: Basic properties and free algebras. II: Varieties and logic of inequalities* (**J** 1428) SIAM J Comp 13∗750-775,776-794
⋄ C05 C75 ⋄ REV Zbl 554#68017 Zbl 554#68019 • ID 44119

GALLIER, J.H. see Vol. I, IV for further entries

GALLIN, D. [1975] *Intensional and higher-order modal logic. With applications to Montague semantics* (**X** 0809) North Holland: Amsterdam vii+148pp
⋄ B45 C85 C90 E35 E50 E70 ⋄ REV MR 58#21470 Zbl 341#02014 JSL 42.581 • ID 29739

GALLIN, D. see Vol. I for further entries

GALVIN, F. [1967] *Reduced products, Horn sentences, and decision problems* (**J** 0015) Bull Amer Math Soc 73∗59-64
⋄ B20 B25 C20 C30 C40 D35 ⋄ REV MR 35#39 Zbl 155.349 JSL 33.477 • ID 04765

GALVIN, F. [1970] *Horn sentences* (J 0007) Ann Math Logic 1*389-422
⋄ B20 C20 C30 C40 D35 ⋄ REV MR 48#3729 Zbl 206.278 JSL 38.651 • ID 04766

GALVIN, F. & HORN, A. [1970] *Operations preserving all equivalence relations* (J 0053) Proc Amer Math Soc 24*521-523
⋄ C05 C75 ⋄ REV MR 41#3359 Zbl 209.42 • ID 28246

GALVIN, F. & PRIKRY, K. [1976] *Infinitary Jonsson algebras and partition relations* (J 0004) Algeb Universalis 6*367-376
⋄ C05 C50 C55 C75 E05 ⋄ REV MR 55#7786 Zbl 363#08001 • ID 23921

GALVIN, F. [1978] see BAUMGARTNER, J.E.

GALVIN, F. & HESSE, G. & STEFFENS, K. [1978] *On the number of automorphisms of structures* (J 0193) Discr Math 24*161-166
⋄ C55 E05 ⋄ REV MR 80b:03038 Zbl 405#05030 • ID 73205

GALVIN, F. see Vol. IV, V for further entries

GANDY, R.O. [1958] *Note on a paper of Kemeny's* (J 0043) Math Ann 136*466
⋄ H15 ⋄ REV MR 20#5141 Zbl 81.248 • ID 04773

GANDY, R.O. [1960] *On a problem of Kleene's* (J 0015) Bull Amer Math Soc 66*501-502
⋄ C62 D30 D55 ⋄ REV MR 23#A64 Zbl 97.246 JSL 29.104 • ID 04774

GANDY, R.O. & KREISEL, G. & TAIT, W.W. [1960] *Set existence* (J 0014) Bull Acad Pol Sci, Ser Math Astron Phys 8*577-583
⋄ C62 D55 F35 ⋄ REV MR 28#2964a JSL 27.232 • REM Part I. Part II 1961 • ID 15061

GANDY, R.O. & KREISEL, G. & TAIT, W.W. [1961] *Set existence II* (J 0014) Bull Acad Pol Sci, Ser Math Astron Phys 9*881-882
⋄ C62 D55 F35 ⋄ REV MR 28#2964b JSL 27.232 • REM Part I 1960 • ID 22330

GANDY, R.O. [1971] see BARWISE, J.

GANDY, R.O. & HYLAND, J.M.E. (EDS.) [1977] *Logic Colloquium 76. Proceedings of a conference held in Oxford in July 1976* (S 3303) Stud Logic Found Math 87*x+612pp
⋄ B97 C97 D97 ⋄ REV MR 57#53 Zbl 409#00002 • ID 16612

GANDY, R.O. [1978] see FENSTAD, J.E.

GANDY, R.O. see Vol. I, IV, V, VI for further entries

GANTER, B. & PLONKA, J. & WERNER, H. [1973] *Homogeneous algebras are simple* (J 0027) Fund Math 79*217-220
⋄ C05 C50 ⋄ REV MR 47#8400 Zbl 264#08006 • ID 04785

GAO, HENGSHAN [1963] *Remarks on Los and Suszko's paper "On the extending of models. IV."* (Chinese) (J 0420) Shuxue Jinzhan 6*388-390
⋄ C30 C52 ⋄ REV MR 33#3931 JSL 36.339 • ID 06908

GAO, HENGSHAN [1976] *The decision problem of modal predicate calculi. II: On Slomson's reductions* (Chinese) (J 0418) Shuxue Xuebao 19*276-280
⋄ B45 C13 D35 ⋄ REV MR 58#27330 Zbl 349#02016 • REM Part I 1973 • ID 61914

GAO, HENGSHAN & MO, SHAOKUI [1979] *The decidability of the $\aleph_0$-validity of formulas in the monadic predicate calculus with enumerative quantifiers and its applications* (Chinese) (J 2771) Kexue Tongbao 24*101-102
⋄ B25 C80 ⋄ REV MR 82f:03010 Zbl 421#03010 • ID 53409

GAO, HENGSHAN [1983] *Comments on "The interpolation theorem for the propositional calculus $P(\kappa)$ when κ is a strongly inaccessible cardinal" by Luo, Libo* (Chinese) (J 3742) Shuxue Yanjiu yu Pinglun 3/4*133-134
⋄ B05 C40 C75 E55 ⋄ REV MR 86a:03007 • REM Comments to Luo, Libo 1980 • ID 45159

GAO, HENGSHAN see Vol. II, IV, V, VI for further entries

GARAVAGLIA, S. [1978] *Homology with equationally compact coefficients* (J 0027) Fund Math 100*89-95
⋄ C05 C20 ⋄ REV MR 58#12999 Zbl 377#55006 • ID 29201

GARAVAGLIA, S. [1978] *Model theory of topological structures* (J 0007) Ann Math Logic 14*13-37
⋄ C20 C40 C65 C80 C90 ⋄ REV MR 58#10406 Zbl 409#03041 • ID 29144

GARAVAGLIA, S. [1978] *Relative strength of Malitz quantifiers* (J 0047) Notre Dame J Formal Log 19*495-503
⋄ C55 C80 ⋄ REV MR 58#5018 Zbl 351#02012 • ID 51483

GARAVAGLIA, S. [1979] *Direct product decomposition of theories of modules* (J 0036) J Symb Logic 44*77-88
⋄ C30 C45 C60 ⋄ REV MR 80c:03038 Zbl 438#03038 • ID 55950

GARAVAGLIA, S. [1980] *Decomposition of totally transcendental modules* (J 0036) J Symb Logic 45*155-164
⋄ C10 C35 C45 C60 ⋄ REV MR 81a:03032 Zbl 453#03036 • ID 54166

GARAVAGLIA, S. [1981] *Forking in modules* (J 0047) Notre Dame J Formal Log 22*155-162
⋄ C45 C60 ⋄ REV MR 82g:03059 Zbl 438#03037 • ID 54092

GARAVAGLIA, S. & PLOTKIN, J.M. [1984] *Separation properties and boolean powers* (S 0019) Colloq Math (Warsaw) 48*167-172
⋄ C30 E50 G05 G10 ⋄ ID 44219

GARCIA, O.C. [1975] *Injectivity in categories of groups and algebras* (J 1507) An Inst Mat, Univ Nac Aut Mexico 14*95-115
⋄ C05 C60 G30 ⋄ REV MR 57#16421 Zbl 323#08006 • ID 61915

GARCIA, O.C. & TAYLOR, W. [1985] *Generalized commutativity* (P 4178) Universal Algeb & Lattice Th; 1984 Charleston 101-122
⋄ C05 C60 ⋄ ID 47862

GARFUNKEL, S. & SHANK, H.S. [1971] *On the undecidability of finite planar graphs* (J 0036) J Symb Logic 36*121-125
⋄ C13 D35 ⋄ REV MR 44#2607 Zbl 222#02056 • ID 04793

GARFUNKEL, S. & SHANK, H.S. [1972] *On the indecidability of finite planar cubic graphs* (J 0036) J Symb Logic 37*595-597
⋄ C13 D35 ⋄ REV MR 47#4781 Zbl 269#02020 • ID 04795

GARFUNKEL, S. see Vol. IV for further entries

GARLAND, S.J. [1972] *Generalized interpolation theorems* (J 0036) J Symb Logic 37*343-351
⋄ C40 C75 C85 D55 E15 ⋄ REV MR 51 # 10045 Zbl 267 # 02037 • ID 04799

GARLAND, S.J. [1974] *Second-order cardinal characterizability* (P 0693) Axiomatic Set Th;1967 Los Angeles 2*127-146
⋄ B15 C40 C55 C85 D55 E10 E47 E55 ⋄ REV MR 54 # 4982 Zbl 319 # 02065 • ID 24144

GARSON, J.W. [1984] *Quantification in modal logic* (C 4085) Handb Philos Log 2*249-307
⋄ B45 C90 C98 ⋄ ID 41834

GARSON, J.W. see Vol. II for further entries

GAVALCOVA, T. [1981] see CSAKANY, B.

GAVALEC, M. [1985] *Local properties of complete Boolean products* (S 0019) Colloq Math (Warsaw) 50*19-37
⋄ C30 C62 G05 ⋄ REM CMP 18/6 • ID 49542

GAVALEC, M. see Vol. V for further entries

GEACH, P.T. & WRIGHT VON, G.H. [1952] *On an extended logic of relations* (J 0990) Soc Sci Fennicae Comment Phys-Math 16/1*37pp
⋄ B20 B25 ⋄ REV Zbl 49.149 JSL 22.72 • ID 16851

GEACH, P.T. see Vol. I, II, V for further entries

GECSEG, F. [1970] *On certain classes of Σ-structures* (J 0002) Acta Sci Math (Szeged) 31*191-195
⋄ C05 C52 ⋄ REV MR 43 # 4747 Zbl 211.30 • ID 04821

GECSEG, F. & STEINBY, M. [1984] *Tree automata* (X 0928) Akad Kiado: Budapest 235pp
⋄ A05 B25 D05 D98 ⋄ REV MR 86c:68061 Zbl 537 # 68056 • ID 41334

GECSEG, F. see Vol. IV for further entries

GEHNE, J. [1983] see DAHN, B.I.

GEIGER, D. [1968] *Closed systems of functions and predicates* (J 0048) Pac J Math 27*95-100
⋄ C05 C52 ⋄ REV MR 38 # 3207 Zbl 186.25 • ID 04826

GEISER, J.R. [1968] *A result on limit ultra powers* (1111) Preprints, Manuscr., Techn. Reports etc.
⋄ C20 ⋄ REM Notes Dartmouth College • ID 21345

GEISER, J.R. [1968] *Nonstandard logic* (J 0036) J Symb Logic 33*236-250
⋄ C20 H10 H15 ⋄ REV MR 42 # 55 Zbl 157.17 • ID 04827

GEISER, J.R. [1974] *A formalization of Essenin-Volpin's proof theoretical studies by means of nonstandard analysis* (J 0036) J Symb Logic 39*81-87
⋄ C75 F50 H10 H15 ⋄ REV MR 55 # 5406 Zbl 296 # 02018 • ID 04831

GEISER, J.R. [1977] see FERRANTE, J.

GEISER, J.R. see Vol. I, VI for further entries

GENSLER, H.J. [1973] *A simplified decision procedure for categorical syllogisms* (J 0047) Notre Dame J Formal Log 14*457-466
⋄ B20 B25 ⋄ REV MR 52 # 2814 Zbl 265 # 02009 • ID 18156

GENSLER, H.J. see Vol. II for further entries

GENTZEN, G. [1935] *Untersuchungen ueber das logische Schliessen I,II* (J 0044) Math Z 39*176-210,405-431
• TRANSL [1955] (X 0840) Pr Univ France: Paris xi+170pp (French) [1964] (J 0325) Amer Phil Quart 1*288-306, 2*204-218 (English) [1969] (C 1409) Gentzen: Collected Papers 69-131 (English) [1978] (J 2751) J Fac Lib Art Yamaguchi Univ 12*1-20, 13*21-40 (Japanese)
• REPR [1969] (X 0890) Wiss Buchges: Darmstadt ii+62pp
⋄ B10 B25 F05 F07 F30 F50 ⋄ REV MR 17.3 MR 50 # 4228 MR 82m:01080 Zbl 10.145 Zbl 10.146 JSL 1.209 JSL 22.350 JSL 35.144 FdM 60.20 • REM 1969 transl. contains additional notes • ID 24221

GENTZEN, G. see Vol. I, VI for further entries

GEORGE, A. [1985] *Skolem and the Loewenheim-Skolem theorem: a case study of the philosophical significance of mathematical results* (J 2028) Hist & Phil Log 6*75-89
⋄ A05 A10 C07 ⋄ REV Zbl 572 # 03004 • ID 46273

GEORGESCU, G. & PETRESCU, I. [1975] *Model-completion in categories* (J 0149) Atti Accad Naz Lincei Fis Mat Nat, Ser 8 59*627-634
⋄ C25 C35 G30 ⋄ REV MR 58 # 21576 Zbl 357 # 02052 • ID 50400

GEORGESCU, G. [1976] *Algebraic structures introduced by forcing (Romanian) (English summary)* (J 0197) Stud Cercet Mat Acad Romana 28*161-168
⋄ C25 G25 ⋄ REV MR 53 # 7776 Zbl 357 # 02055 • ID 23016

GEORGESCU, G. [1977] *Forcing infini en theorie des modeles booleens* (J 1738) Rev Roumaine Sci Soc 19*197-203
⋄ C25 ⋄ ID 31093

GEORGESCU, G. [1977] *Structures algebriques introduites par forcing* (P 1729) Colloq Int Log;1975 Clermont-Ferrand 221-224
⋄ C25 G25 ⋄ REV Zbl 438 # 03036 • ID 55948

GEORGESCU, G. [1980] *Generalized quantifiers on polyadic algebras* (J 0060) Rev Roumaine Math Pures Appl 25*1027-1032
⋄ C80 G15 ⋄ REV MR 82d:03100 Zbl 454 # 03033 • ID 54245

GEORGESCU, G. [1982] *A generalized quantifier for the hereditary σ-rings* (J 0070) Bull Soc Sci Math Roumanie, NS 26(74)*237-240
⋄ C75 C80 ⋄ REV MR 84g:03049 Zbl 501 # 03019 • ID 34159

GEORGESCU, G. [1982] *Algebraic analysis of the topological logic L(I)* (J 0068) Z Math Logik Grundlagen Math 28*447-454
⋄ C80 G15 ⋄ REV MR 85d:03079 Zbl 496 # 03043 • ID 36870

GEORGESCU, G. [1982] *Une generalisation du theoreme d'omission des types dans les algebres polyadiques* (J 1934) Ann Sci Univ Clermont Math 21*67-74
⋄ C75 C90 G15 ⋄ REV MR 85i:03189 Zbl 523 # 03051 • ID 37038

GEORGESCU, G. [1984] *Algebraic analysis of the logic with the quantifier "there exist uncountably many"* (J 0004) Algeb Universalis 19*99-105
⋄ C80 G15 ⋄ REV MR 85j:03114 • ID 45289

GEORGESCU, G. & VOICULESCU, I. [1985] *Eastern model-theory for boolean-valued theories* (J 0068) Z Math Logik Grundlagen Math 31*79-88
⋄ C25 C35 C40 C90 ⋄ ID 42289

GEORGESCU, G. see Vol. I, II, V for further entries

GEORGIEVA, N.V. [1973] *On the logical signs in Schuette's interpolation formulae (Bulgarian) (Russian and English summaries)* (J 1156) Izv Bulgar Akad Nauk Mat Inst 14*99-103
⋄ C40 F50 ⋄ REV MR 50 # 9543 Zbl 334 # 02013 • ID 20996

GEORGIEVA, N.V. see Vol. I, II, IV, VI for further entries

GERGELY, T. [1973] see ANDREKA, H.

GERGELY, T. [1974] see ANDREKA, H.

GERGELY, T. [1975] see ANDREKA, H.

GERGELY, T. & VERSHININ, K.P. [1978] *Model theoretical investigation of theorem proving methods* (J 0047) Notre Dame J Formal Log 19*523-542
⋄ B35 C07 ⋄ REV MR 80b:03015 Zbl 333 # 68060 • ID 52253

GERGELY, T. see Vol. I, II, IV for further entries

GERICKE, H. [1953] *Ueber den Begriff der algebraischen Struktur* (J 0008) Arch Math (Basel) 4*163-171
⋄ C05 ⋄ REV MR 15.280 Zbl 50.259 • ID 04902

GERICKE, H. see Vol. I, V for further entries

GERLA, G. [1983] *Model theory and modal logic (Italian)* (P 3829) Atti Incontri Log Mat (1);1982 Siena 87-89
⋄ B45 C25 C75 C90 ⋄ REV MR 84k:03006 • ID 44834

GERLA, G. & VACCARO, V. [1984] *Modal logic and model theory* (J 0063) Studia Logica 43*203-216
⋄ B45 C25 C75 C90 ⋄ REV MR 86h:03023 • ID 39889

GERLA, G. see Vol. II, IV, V for further entries

GERMANO, G. [1971] *Incompleteness and truth definitions* (J 0105) Theoria (Lund) 37*86-90
⋄ C40 D35 F30 ⋄ REV MR 50 # 4274 Zbl 243 # 02038 • ID 04907

GERMANO, G. [1973] *Incompleteness theorem via weak definability of truth: a short proof* (J 0047) Notre Dame J Formal Log 14*377-380
⋄ C40 F30 ⋄ REV MR 48 # 90 Zbl 232 # 02035 • ID 04910

GERMANO, G. see Vol. I, IV, VI for further entries

GIAMBRONE, S. [1985] TW_+ and RW_+ *are decidable* (J 0122) J Philos Logic 14*235-254
⋄ B25 B46 ⋄ ID 42658

GIAMBRONE, S. see Vol. II, VI for further entries

GIANNONE, A. [1980] see BARONE, E.

GIANNONE, A. [1982] *Probabilita non σ-additive e analisi non-standard (English summary)* (J 3741) Rend Mat Appl, Ser 7 2*47-58
⋄ C20 H10 ⋄ REV MR 83i:03105 Zbl 492 # 60002 • ID 34858

GIANNONE, A. see Vol. I for further entries

GILES, R. [1979] see ADAMSON, A.

GILES, R. see Vol. I, II, V for further entries

GILLAM, D.W.H. [1976] *Relatively complete theories* (J 0068) Z Math Logik Grundlagen Math 22*245-250
⋄ C35 ⋄ REV MR 56 # 15403 Zbl 343 # 02036 • ID 18451

GILLAM, D.W.H. [1976] *Saturation and omitting types* (J 0009) Arch Math Logik Grundlagenforsch 18*5-18
⋄ C07 C30 C50 ⋄ REV MR 57 # 5728 Zbl 345 # 02039 • ID 24315

GILLAM, D.W.H. [1977] *Morley numbers for generalized languages* (J 0027) Fund Math 94*145-153
⋄ C55 C75 C80 ⋄ REV MR 56 # 115 Zbl 359 # 02009 • ID 26518

GILLAM, D.W.H. [1978] *Refined Morley numbers* (S 0019) Colloq Math (Warsaw) 39*201-209
⋄ C55 C75 ⋄ REV MR 81i:03045 Zbl 416 # 03039 • ID 53205

GILLAM, D.W.H. see Vol. V for further entries

GILLMAN, L. [1955] see ERDOES, P.

GILLMAN, L. [1955] *Remarque sur les ensembles* η_α (J 0109) C R Acad Sci, Paris 241*12-13
⋄ C50 E07 ⋄ REV MR 17.135 Zbl 68.43 • ID 04937

GILLMAN, L. [1956] see BAGEMIHL, F.

GILLMAN, L. [1956] *Some remarks on* η_α-*sets* (J 0027) Fund Math 43*77-82
⋄ C50 E07 ⋄ REV Zbl 70.51 • ID 04938

GILLMAN, L. see Vol. V for further entries

GILMAN, R.H. [1984] *Characteristically simple* $\aleph_0$-*categorical groups* (J 0036) J Symb Logic 49*900-907
⋄ C35 C60 ⋄ REV MR 86b:03036 • ID 42445

GILMAN, R.H. [1985] *An application of ultraproducts to finite groups* (C 4631) Rutgers Group Th Year New Brunswick 1983/84 409-412
⋄ C13 C20 C60 ⋄ ID 49545

GILMORE, P.C. & ROBINSON, A. [1955] *Metamathematical considerations on the relative irreducibility of polynomials* (J 0017) Canad J Math 7*483-489
• REPR [1979] (C 4594) Sel Pap Robinson 1*348-354
⋄ C60 ⋄ REV MR 17.226 Zbl 66.269 • ID 04940

GILMORE, P.C. [1962] *Some forms of completeness* (J 0036) J Symb Logic 27*344-352
⋄ C35 ⋄ REV MR 30 # 1925 Zbl 118.248 • ID 04945

GILMORE, P.C. see Vol. I, II, V, VI for further entries

GIORGETTA, D. [1980] *Embeddings in groups of permutations with supports of bounded cardinality* (J 0032) J Algeb 65*361-372
⋄ C60 E05 ⋄ REV MR 81j:20011 Zbl 443 # 20006 • ID 56436

GIORGETTA, D. & SHELAH, S. [1984] *Existentially closed structures in the power of the continuum* (J 0073) Ann Pure Appl Logic 26*123-148
⋄ C15 C25 C30 C52 C55 C60 C75 ⋄ REV MR 86e:03035 Zbl 561 # 03018 • ID 38741

GIRARD, J.-Y. [1976] *Three-valued logic and cut-elimination: The actual meaning of Takeuti's conjecture* (J 0202) Diss Math (Warsaw) 136*45pp
⋄ B50 C85 C90 F05 F35 F50 ⋄ REV MR 56 # 5235 Zbl 357 # 02027 • ID 50375

GIRARD, J.-Y. [1977] *Functional interpretation and Kripke models* (P 1704) Int Congr Log, Meth & Phil of Sci (5);1975 London ON 1∗33-57
  ⋄ C90 F10 F30 F50 ⋄ REV MR 58#159 Zbl 417#03029 • ID 53265

GIRARD, J.-Y. [1984] *The Ω-rule* (P 4313) Int Congr Math (II,14);1983 Warsaw 1∗307-321
  ⋄ B15 C62 C75 F35 ⋄ REV Zbl 569#03023 • ID 48565

GIRARD, J.-Y. see Vol. IV, V, VI for further entries

GIRI, R.D. [1975] *On a varietal structure of algebras* (J 0064) Trans Amer Math Soc 213∗53-60
  ⋄ C05 C52 ⋄ REV MR 52#3008 Zbl 317#08002 • ID 17682

GIRSTMAIR, K. [1979] *Ueber konstruktive Methoden der Galoistheorie* (J 0504) Manuscr Math 26∗423-441
  ⋄ C57 C60 F55 ⋄ REV MR 80b:12016 Zbl 408#12024 • ID 56289

GITIK, M. [1984] *The nonstationary ideal on $\aleph_2$* (J 0029) Israel J Math 48∗257-288
  ⋄ C62 E05 E35 E55 ⋄ REV Zbl 563#03016 • ID 39952

GITIK, M. see Vol. V for further entries

GIVANT, S. [1978] *Universal Horn classes categorical or free in power* (J 0007) Ann Math Logic 15∗1-53
  ⋄ B20 C05 C35 ⋄ REV MR 80c:03032 Zbl 401#03009 • ID 29155

GIVANT, S. [1979] *A representation theorem for universal Horn classes categorical in power* (J 0007) Ann Math Logic 17∗91-116
  ⋄ B20 C05 C35 ⋄ REV MR 81b:03038 Zbl 436#03020 • ID 55860

GIVANT, S. [1980] *Union decompositions and universal classes categorical in power* (J 0004) Algeb Universalis 10∗155-175
  ⋄ C05 C35 ⋄ REV MR 82c:03042 Zbl 433#03016 • ID 73391

GIVANT, S. [1981] *The number of nonisomorphic denumerable models of certain universal Horn classes* (J 0004) Algeb Universalis 13∗56-68
  ⋄ C05 C15 C35 ⋄ REV MR 83f:03029 Zbl 497#03026 • ID 35303

GIVANT, S. [1983] see BALDWIN, JOHN T.

GLADSTONE, M.D. [1966] *Finite models for inequations* (J 0036) J Symb Logic 31∗581-592
  ⋄ B25 C05 C13 ⋄ REV MR 35#1466 Zbl 192.41 JSL 33.479 • ID 05011

GLADSTONE, M.D. [1979] *The decidability of one-variable propositional calculi* (J 0047) Notre Dame J Formal Log 20∗438-450
  ⋄ B20 B25 ⋄ REV MR 80f:03016 Zbl 351#02010 • ID 52622

GLADSTONE, M.D. see Vol. I, IV for further entries

GLASS, A.M.W. [1972] *An application of ultraproducts to lattice-ordered groups* (J 0017) Canad J Math 24∗1063-1064
  ⋄ C20 C60 G10 ⋄ REV MR 47#96 Zbl 246#06005 • ID 05020

GLASS, A.M.W. [1973] see BALDWIN, JOHN T.

GLASS, A.M.W. & PIERCE, K.R. [1980] *Equations and inequations in lattice-ordered groups* (P 2983) Ordered Groups;1978 Boise 141-171
  ⋄ C25 C60 G10 ⋄ REV MR 82c:06030 Zbl 478#06006 • ID 55647

GLASS, A.M.W. & PIERCE, K.R. [1980] *Existentially complete abelian lattice-ordered groups* (J 0064) Trans Amer Math Soc 261∗255-270
  ⋄ C25 C60 G10 ⋄ REV MR 81k:03032 Zbl 482#06011 • ID 41679

GLASS, A.M.W. & PIERCE, K.R. [1980] *Existentially complete lattice-ordered groups* (J 0029) Israel J Math 36∗257-272
  ⋄ C25 C35 C60 G10 ⋄ REV MR 82b:06016 Zbl 454#06007 • ID 54247

GLASS, A.M.W. & GUREVICH, Y. & HOLLAND, W.C. & JAMBU-GIRAUDET, M. [1981] *Elementary theory of automorphism groups of doubly homogeneous chains* (P 2628) Log Year;1979/80 Storrs 67-82
  ⋄ C07 C50 C65 E07 ⋄ REV MR 82i:03047 Zbl 508.06001 • ID 73400

GLASS, A.M.W. [1981] *Elementary types of automorphisms of linearly ordered sets -- a survey* (P 2630) Lie Algeb, Group Th, Part Ord Algeb Struct;1980 Carbondale 218-229
  ⋄ C07 C65 C98 ⋄ REV MR 82e:06001 Zbl 466#06016 • ID 81347

GLASS, A.M.W. [1981] *Ordered permutation groups* (S 3306) Lond Math Soc Lect Note Ser 55∗x+266pp+xi-xlix
  ⋄ C07 C60 C65 C98 ⋄ REV MR 83j:06004 Zbl 473#06010 • ID 90194

GLASS, A.M.W. & GUREVICH, Y. & HOLLAND, W.C. & SHELAH, S. [1981] *Rigid homogeneous chains* (J 0332) Math Proc Cambridge Phil Soc 89∗7-17
  ⋄ C07 C50 C65 E07 E35 E50 ⋄ REV MR 82c:06001 Zbl 462#06011 • ID 54528

GLASS, A.M.W. [1982] see BALDWIN, JOHN T.

GLASS, A.M.W. [1982] see DRIES VAN DEN, L.

GLASS, A.M.W. [1984] *The isomorphism problem and undecidable properties for finitely presented lattice-ordered groups* (P 2167) Orders: Descr & Roles;1982 L'Arbresle 157-170
  ⋄ C60 D40 G10 ⋄ REV Zbl 549#06010 • ID 44713

GLASS, A.M.W. [1985] *Effective extensions of lattices-ordered groups that preserve the degree of the conjugacy and the word problems* (P 4663) Ordered Algeb Struct;1982 Cincinnati 89-98
  ⋄ C57 C60 D40 ⋄ ID 49075

GLASS, A.M.W. [1985] *The word and isomorphism problems in universal algebra* (P 4178) Universal Algeb & Lattice Th;1984 Charleston 123-128
  ⋄ C05 C57 D40 ⋄ ID 47863

GLASS, A.M.W. see Vol. IV for further entries

GLASSMIRE, W. [1971] *There are $2^{\aleph_0}$ countable categorical theories* (J 0014) Bull Acad Pol Sci, Ser Math Astron Phys 19∗185-190
  ⋄ C35 ⋄ REV MR 45#8513 Zbl 216.11 • ID 05021

GLEBSKIJ, YU.V. & KOGAN, D.I. & LIOGON'KIJ, M.I. & TALANOV, V.A. [1969] *Range and degree of realizability of formulas in the restricted predicate calculus (Russian)* (J 0040) Kibernetika, Akad Nauk Ukr SSR 1969/2*17-27
- TRANSL [1969] (J 0021) Cybernetics 5*142-154
⋄ B10 C07 C13 ⋄ REV MR 46 # 42 Zbl 209.308 JSL 50.1073 • ID 05032

GLEBSKIJ, YU.V. & GORDON, E.I. [1979] *The elementary theories of certain distributive lattices with an additive measure (Russian)* (S 3911) Komb-Algeb Met Prikl Mat (Gor'kij) 1979*24-35,121
⋄ C52 G10 ⋄ ID 46342

GLEBSKIJ, YU.V. see Vol. IV for further entries

GLOEDE, K. [1975] *Set theory in infinitary languages* (P 1442) ⊦ ISILC Logic Conf;1974 Kiel 311-362
⋄ C70 C75 E30 E65 ⋄ REV MR 53 # 12866 Zbl 316 # 02065 • ID 27214

GLOEDE, K. [1977] *The metamathematics of infinitary set theoretical systems* (J 0068) Z Math Logik Grundlagen Math 23*19-44
⋄ C70 C75 E30 E70 ⋄ REV MR 58 # 16290 Zbl 359 # 02019 • ID 26465

GLOEDE, K. [1978] *Hierarchies of sets definable by means of infinitary languages* (P 1864) Higher Set Th;1977 Oberwolfach 29-54
⋄ C62 C70 C75 E45 ⋄ REV MR 80g:03053 Zbl 384 # 03040 • ID 52082

GLOEDE, K. see Vol. V for further entries

GLUBRECHT, J.-M. [1982] *Ein Vollstaendigkeitsbeweis fuer schnittfreie Kalkuele mit der Maximalisierungsmethode von Henkin* (J 0009) Arch Math Logik Grundlagenforsch 22*159-166
⋄ B10 C07 F05 ⋄ REV MR 84b:03022 Zbl 496 # 03002 • ID 35619

GLUBRECHT, J.-M. see Vol. I, II, V for further entries

GNANVO, C. [1974] *Theorie forcee et "forcing-compagnon"* (J 2313) C R Acad Sci, Paris, Ser A-B 278*A305-A306
⋄ C25 ⋄ REV MR 49 # 10539 Zbl 279 # 02032 • ID 05054

GNANVO, C. [1976] *Γ-cotheories, test de Robinson (English summary)* (J 2313) C R Acad Sci, Paris, Ser A-B 282*A1183-A1184
⋄ C35 C52 ⋄ REV MR 57 # 9520 Zbl 354 # 02040 • ID 27262

GNANVO, C. [1976] *Γ-extension logique (English summary)* (J 2313) C R Acad Sci, Paris, Ser A-B 282*A309-A311
⋄ C30 C52 ⋄ REV MR 56 # 2814 Zbl 366 # 02036 • ID 27223

GNANVO, C. [1977] *Γ-cotheories et elimination des quantificateurs (English summary)* (J 2313) C R Acad Sci, Paris, Ser A-B 284*A129-A131
⋄ C10 C25 C35 C52 ⋄ REV MR 55 # 12503 Zbl 354 # 02039 • ID 27292

GOAD, C.A. [1978] *Monadic infinitary propositional logic: a special operator* (J 0302) Rep Math Logic, Krakow & Katowice 10*43-50
⋄ B25 B60 C75 F50 ⋄ REV MR 81e:03024 Zbl 424 # 03029 • ID 73424

GOAD, C.A. see Vol. I, VI for further entries

GOBLE, L.F. [1973] *A new modal model* (J 0079) Logique & Anal, NS 16*301-309
⋄ B45 C90 ⋄ REV MR 50 # 9532 Zbl 291 # 02012 • ID 05058

GOBLE, L.F. see Vol. II, VI for further entries

GOEBEL, R. & RICHTER, M.M. [1972] *Cartesian closed classes of perfect groups* (J 0032) J Algeb 23*370-381
⋄ C60 ⋄ REV MR 46 # 3631 • ID 33425

GOEBEL, R. & WALKER, E.A. (EDS.) [1981] *Abelian group theory* (S 3301) Lect Notes Math 874*xxi+447pp
⋄ C97 ⋄ REV MR 83a:20069 Zbl 451 # 00003 • ID 48645

GOEBEL, R. [1982] see DUGAS, M.

GOEBEL, R. [1984] see DUGAS, M.

GOEBEL, R. [1984] *The existence of rigid systems of maximal size* (P 4185) Abel Group & Modul;1984 Udine 189-202
⋄ C60 E75 ⋄ REV MR 86f:20001 Zbl 564 # 16030 • ID 48235

GOEBEL, R. [1985] see CORNER, A.L.S.

GOEBEL, R. & SHELAH, S. [1985] *Semi-rigid classes of cotorsion-free abelian groups* (J 0032) J Algeb 93*136-150
⋄ C55 C60 E05 E45 E55 E65 E75 ⋄ REV MR 86d:20061 Zbl 554 # 20018 • ID 44953

GOEBEL, R. see Vol. V for further entries

GOEDEL, K. [1929] *Ein Spezialfall des Entscheidungsproblems der theoretischen Logik* (J 1124) Ergebn Math Kolloquium 2*27-28
⋄ B20 B25 ⋄ REV FdM 57.1321 • ID 22123

GOEDEL, K. [1930] *Die Vollstaendigkeit der Axiome des logischen Funktionenkalkuels* (J 0124) Monatsh Math-Phys 37*349-360
- TRANSL [1967] (C 0675) From Frege to Goedel 583-591
⋄ B10 C07 ⋄ REV FdM 56.46 • ID 20884

GOEDEL, K. [1933] *Zum Entscheidungsproblem des logischen Funktionenkalkuels* (J 0124) Monatsh Math-Phys 40*433-443
⋄ B10 B20 B25 C13 D35 ⋄ REV Zbl 8.289 FdM 59.865 • ID 20885

GOEDEL, K. see Vol. I, II, IV, V, VI for further entries

GOERING, P. [1984] see DAHN, B.I.

GOERNEMANN, S. [1971] *A logic stronger than intuitionism* (J 0036) J Symb Logic 36*249-261
⋄ B55 C90 F50 ⋄ REV MR 45 # 27 Zbl 276 # 02013 • ID 39863

GOERNEMANN, S. also published under the name KOPPELBERG, S.

GOETZ, A. [1971] *A generalization of the direct product of universal algebras* (S 0019) Colloq Math (Warsaw) 22*167-176
⋄ C05 C30 ⋄ REV MR 44 # 6588 Zbl 236 # 08003 • ID 05087

GOGOL, D. [1973] *Low cardinality models for a type of infinitary theory* (J 0027) Fund Math 81*29-34
⋄ C55 C75 ⋄ REV MR 49 # 2251 Zbl 317 # 02061 • ID 29657

GOGOL, D. [1975] *Formulas with two generalized quantifiers* (J 0047) Notre Dame J Formal Log 16*133-136
⋄ C80 ⋄ REV MR 50 # 12645 Zbl 271 # 02014 • ID 05092

GOGOL, D. [1978] *The $\forall_n \exists$-completeness of Zermelo-Fraenkel set theory* (J 0068) Z Math Logik Grundlagen Math 24*289-290
⋄ B25 C35 E30 ⋄ REV MR 58 # 16286 Zbl 387 # 03016
• ID 52232

GOGOL, D. [1979] *Sentences with three quantifiers are decidable in set theory* (J 0027) Fund Math 102*1-8
⋄ B25 C35 E30 ⋄ REV MR 80j:03077 Zbl 397 # 03013
• ID 52679

GOGUEN, J.A. & MESEGUER, J. [1985] *Completeness of many-sorted equational logic* (J 1447) Houston J Math 11*307-334
⋄ B20 C05 C07 ⋄ REV Zbl 498 # 03018 • ID 36903

GOGUEN, J.A. see Vol. II, IV, V for further entries

GOLD, A.J. [1975] *Sur les arbres logiques de M. Krasner* (J 1934) Ann Sci Univ Clermont Math 11*17-72
⋄ B25 E70 ⋄ REV MR 55 # 5430 Zbl 318 # 02058
• ID 62082

GOLD, B. [1978] *Compact and ω-compact formulas in $L_{\omega_1,\omega}$* (J 0009) Arch Math Logik Grundlagenforsch 19*51-64
⋄ C75 ⋄ REV MR 81a:03037 Zbl 397 # 03020 • ID 29164

GOLD, J.M. [1976] *A reflection property for saturated models* (J 0068) Z Math Logik Grundlagen Math 22*425-430
⋄ C50 C55 ⋄ REV MR 56 # 2815 Zbl 355 # 02033
• ID 24287

GOLD, J.M. [1980] *Some results on the structure of the φ-spectrum* (J 0068) Z Math Logik Grundlagen Math 26*503-507
⋄ C50 C55 E55 ⋄ REV MR 82e:03037 Zbl 468 # 03032
• ID 55097

GOLDBERG, R. [1963] *On the solvability of a subclass of the Suranyi reduction class* (J 0036) J Symb Logic 28*237-244
⋄ B25 ⋄ REV MR 35 # 23 JSL 30.391 • ID 05102

GOLDBLATT, R.I. [1973] *A model-theoretic study of some systems containing S3* (J 0068) Z Math Logik Grundlagen Math 19*75-82
⋄ B25 B45 C90 ⋄ REV MR 47 # 6441 Zbl 255 # 02016
• ID 05104

GOLDBLATT, R.I. [1974] *Decidability of some extensions of J* (J 0068) Z Math Logik Grundlagen Math 20*203-205
⋄ B25 B55 ⋄ REV MR 51 # 60 Zbl 306 # 02045 • ID 05106

GOLDBLATT, R.I. [1975] *First order definability in modal logic* (J 0036) J Symb Logic 40*35-40
⋄ B45 C20 C90 ⋄ REV MR 55 # 5389 Zbl 311 # 02029
JSL 47.440 • ID 05108

GOLDBLATT, R.I. [1979] *Topoi. The categorial analysis of logic* (S 3303) Stud Logic Found Math 98*xv + 486pp
• TRANSL [1983] (X 0885) Mir: Moskva 488pp
⋄ B98 C98 E98 F35 F50 F98 G30 ⋄ REV
MR 81a:03063 Zbl 434 # 03050 JSL 47.445 • REM 2nd rev. ed. 1984; xvi + 552pp • ID 55754

GOLDBLATT, R.I. [1981] *"Locally-at" as a topological quantifier-former* (P 3113) Aspects Philos Logic;1977 Tuebingen 119-127
⋄ C65 C80 ⋄ REV MR 83d:03051 Zbl 474 # 03017
• ID 55421

GOLDBLATT, R.I. [1981] *Grothendieck topology as geometric modality* (J 0068) Z Math Logik Grundlagen Math 27*495-529
⋄ B25 B45 C90 F50 G30 ⋄ REV MR 83d:03069
Zbl 474 # 03018 • ID 55422

GOLDBLATT, R.I. [1984] *An abstract setting for Henkin proofs* (J 3781) Topoi 3*37-41
⋄ C07 C75 ⋄ REV MR 86f:03021 • ID 44668

GOLDBLATT, R.I. [1984] *Orthomodularity is not elementary* (J 0036) J Symb Logic 49*401-404
⋄ B51 C65 G10 ⋄ REV MR 85e:03154 • ID 40768

GOLDBLATT, R.I. [1985] *On the role of the Baire category theorem and dependent choice in the foundations of logic* (J 0036) J Symb Logic 50*412-422
⋄ C07 E25 E40 E75 G30 ⋄ REV Zbl 567 # 03023
• ID 42555

GOLDBLATT, R.I. see Vol. I, II, VI for further entries

GOLDFARB, W.D. [1974] see AANDERAA, S.O.

GOLDFARB, W.D. [1979] see DREBEN, B.

GOLDFARB, W.D. [1981] *On the Goedel class with identity* (J 0036) J Symb Logic 46*354-364
⋄ B20 B25 D35 ⋄ REV MR 82g:03079 Zbl 472 # 03009
• ID 55275

GOLDFARB, W.D. & GUREVICH, Y. & SHELAH, S. [1984] *A decidable subclass of the minimal Goedel class with identity* (J 0036) J Symb Logic 49*1253-1261
⋄ B20 B25 ⋄ REV MR 86g:03015b Zbl 576 # 03009
• ID 38765

GOLDFARB, W.D. see Vol. I, IV, VI for further entries

GOLDREI, D.C. & MACINTYRE, A. & SIMMONS, H. [1973] *The forcing companions of number theories* (J 0029) Israel J Math 14*317-337
⋄ C25 C62 ⋄ REV MR 48 # 5853 Zbl 301 # 02054
• ID 22294

GOL'DSHTEJN, B.G. [1983] *Decomposition of logics (Russian)* (J 0030) Izv Akad Nauk Uzb SSR, Ser Fiz-Mat 1983/6*10-12
⋄ C95 H05 ⋄ REV MR 85f:81007 Zbl 565 # 03034
• ID 45227

GOLDSMITH, B. [1984] see DUGAS, M.

GOLOTA, YA.YA. [1969] *Some techniques for simplifying the construction of nets of marks (Russian)* (S 0228) Zap Nauch Sem Leningrad Otd Mat Inst Steklov 16*44-53 • ERR/ADD ibid 20*292-294
• TRANSL [1969] (J 0521) Semin Math, Inst Steklov 16*20-25
⋄ B25 B75 F50 ⋄ REV MR 41 # 6668 Zbl 237 # 02004
• ID 05124

GOLOTA, YA.YA. [1974] *A matrix notation for nets of marks (Russian) (English summary)* (S 0228) Zap Nauch Sem Leningrad Otd Mat Inst Steklov 40*4-9,155
• TRANSL [1977] (J 1531) J Sov Math 8*247-251
⋄ B25 F50 ⋄ REV MR 51 # 82 Zbl 356 # 02021 • ID 15236

GOLOTA, YA.YA. see Vol. II, VI for further entries

GOLTZ, H.-J. [1980] *Untersuchungen zur Elimination verallgemeinerter Quantoren in Koerpertheorien* (J 0115) Wiss Z Humboldt-Univ Berlin, Math-Nat Reihe 29*391-397
 ⋄ C10 C55 C60 C80 ⋄ REV MR 82j:03032 Zbl 465 # 03018 • ID 54921

GOLTZ, H.-J. [1985] *The boolean sentence algebra of the theory of linear ordering is atomic with respect to logics with a Malitz quantifier* (J 0068) Z Math Logik Grundlagen Math 31*131-162
 ⋄ C55 C65 C80 G05 ⋄ ID 41814

GOMBOCZ, W.L. [1979] *Lesniewski and Mally* (J 0047) Notre Dame J Formal Log 20*934-946
 ⋄ A05 A10 C80 ⋄ REV MR 80h:03010 Zbl 363 # 02002
 • ID 56172

GONCHAROV, S.S. [1973] *Constructivizability of superatomic boolean algebras (Russian)* (J 0003) Algebra i Logika 12*31-40,120
 • TRANSL [1973] (J 0069) Algeb and Log 12*17-22
 ⋄ C57 D45 G05 ⋄ REV MR 47 # 8265 Zbl 278 # 02039
 • ID 05125

GONCHAROV, S.S. & NURTAZIN, A.T. [1973] *Constructive models of complete solvable theories (Russian)* (J 0003) Algebra i Logika 12*125-142,243
 • TRANSL [1973] (J 0069) Algeb and Log 12*67-77
 ⋄ C45 C50 C57 G05 ⋄ REV MR 53 # 2667
 Zbl 278 # 02038 • ID 21613

GONCHAROV, S.S. [1975] *Autostability and computable families of constructivizations (Russian)* (J 0003) Algebra i Logika 14*647-680,727
 • TRANSL [1975] (J 0069) Algeb and Log 14*392-409
 ⋄ C50 C57 D45 ⋄ REV MR 55 # 10267 Zbl 367 # 02023
 • ID 26031

GONCHAROV, S.S. [1975] *Certain properties of the constructivization of boolean algebras (Russian)* (J 0092) Sib Mat Zh 16*264-278,420
 • TRANSL [1975] (J 0475) Sib Math J 16*203-214
 ⋄ C57 D45 G05 ⋄ REV MR 52 # 2846 Zbl 309 # 02052
 • ID 17637

GONCHAROV, S.S. [1976] see DOBRITSA, V.P.

GONCHAROV, S.S. [1976] *Non-self-equivalent constructivization of atomic boolean algebras (Russian)* (J 0087) Mat Zametki (Akad Nauk SSSR) 19*853-858
 • TRANSL [1976] (J 1044) Math Notes, Acad Sci USSR 19*500-503
 ⋄ C57 D45 G05 ⋄ REV MR 55 # 12490 Zbl 357 # 02043
 • ID 50391

GONCHAROV, S.S. [1976] *Restricted theories of constructive boolean algebras (Russian)* (J 0092) Sib Mat Zh 17*797-812
 • TRANSL [1976] (J 0475) Sib Math J 17*601-611
 ⋄ B25 C57 D35 F60 G05 ⋄ REV MR 55 # 93
 Zbl 361 # 02066 • ID 52189

GONCHAROV, S.S. [1977] *The quantity of nonautoequivalent constructivizations (Russian)* (J 0003) Algebra i Logika 16*257-282
 • TRANSL [1977] (J 0069) Algeb and Log 16*169-185
 ⋄ C50 C57 D45 ⋄ REV MR 81h:03067 Zbl 405 # 03016
 • ID 54891

GONCHAROV, S.S. [1978] *Constructive models of $\aleph_1$-categorical theories (Russian)* (J 0087) Mat Zametki (Akad Nauk SSSR) 23*885-888
 • TRANSL [1978] (J 1044) Math Notes, Acad Sci USSR 23*486-487
 ⋄ C35 C57 ⋄ REV MR 80g:03029 Zbl 385 # 03025
 • ID 52133

GONCHAROV, S.S. [1978] *Strong constructivizability of homogeneous models (Russian)* (J 0003) Algebra i Logika 17*363-388
 • TRANSL [1978] (J 0069) Algeb and Log 17*247-263
 ⋄ C50 C57 D45 ⋄ REV MR 82a:03026 Zbl 441 # 03015
 • ID 56068

GONCHAROV, S.S. [1980] *A totally transcendental decidable theory without constructivizable homogeneous models (Russian)* (J 0003) Algebra i Logika 19*137-149
 • TRANSL [1980] (J 0069) Algeb and Log 19*85-93
 ⋄ C45 C50 C57 ⋄ REV MR 83d:03047 Zbl 468 # 03024
 • ID 55089

GONCHAROV, S.S. [1980] *Autostability of models and abelian groups (Russian)* (J 0003) Algebra i Logika 19*23-44
 • TRANSL [1980] (J 0069) Algeb and Log 19*13-27
 ⋄ C50 C57 C60 D45 ⋄ REV MR 82g:03081
 Zbl 468 # 03022 • ID 55087

GONCHAROV, S.S. [1980] see DZGOEV, V.D.

GONCHAROV, S.S. [1980] see DROBOTUN, B.N.

GONCHAROV, S.S. [1980] *Problem of the number of non-self-equivalent constructivizations (Russian)* (J 0003) Algebra i Logika 19*621-639
 • TRANSL [1980] (J 0069) Algeb and Log 19*401-414
 ⋄ C57 D45 ⋄ REV MR 82k:03046 Zbl 476 # 03046
 • ID 55558

GONCHAROV, S.S. [1980] *The problem of the number of nonautoequivalent constructivizations (Russian)* (J 0023) Dokl Akad Nauk SSSR 251*271-274
 • TRANSL [1980] (J 0062) Sov Math, Dokl 21*411-414
 ⋄ C57 D45 ⋄ REV MR 81i:03070 Zbl 476 # 03045
 • ID 73488

GONCHAROV, S.S. [1980] *Totally transcendental theory with non-constructivizable prime model (Russian)* (J 0092) Sib Mat Zh 21/1*44-51
 • TRANSL [1980] (J 0475) Sib Math J 21*32-37
 ⋄ C45 C50 C57 D45 ⋄ REV MR 81b:03039
 Zbl 463 # 03005 • ID 54545

GONCHAROV, S.S. [1981] *Groups with a finite number of constructivizations (Russian)* (J 0023) Dokl Akad Nauk SSSR 256*269-272
 • TRANSL [1981] (J 0062) Sov Math, Dokl 23*58-61
 ⋄ C57 C60 D45 ⋄ REV MR 82g:03083 Zbl 496 # 20021
 • ID 37746

GONCHAROV, S.S. [1982] *Limiting equivalent constructivizations (Russian)* (C 3953) Mat Log & Teor Algor 4-12
 ⋄ C57 D45 F60 ⋄ REV MR 85d:03122 Zbl 543 # 03017
 • ID 40426

GONCHAROV, S.S. [1983] *Universal recursively enumerable boolean algebras (Russian)* (J 0092) Sib Mat Zh 24/6*36-43
 • TRANSL [1983] (J 0475) Sib Math J 24*852-858
 ⋄ C50 C57 D45 G05 ⋄ REV MR 86b:03054
 Zbl 522 # 03026 • ID 43384

GONCHAROV, S.S. [1985] *Axioms for classes with strong epimorphisms (Russian) (English summary)* (P 4310) Easter Conf on Model Th (3);1985 Gross Koeris 99-102
  ⋄ C52 ⋄ REV Zbl 575 # 03023 • ID 48824

GONCHAROV, S.S. [1985] *Stongly constructive models (Russian) (English summary)* (P 4310) Easter Conf on Model Th (3);1985 Gross Koeris 103-114
  ⋄ C57 ⋄ REV Zbl 575 # 03024 • ID 48825

GONCHAROV, S.S. see Vol. IV for further entries

GONCHAROV, V.A. [1983] *A recursively representable Boolean algebra (Russian)* (C 4019) Kraevye Zadach Dlya Diff Uravnenij & Priloz Mekh & Tekh 43-46
  ⋄ C57 D45 G05 ⋄ REV MR 86b:03054 Zbl 565 # 03020
  • ID 45205

GONDARD, D. [1982] *Theorie des modeles et fonctions definies positives sur les varietes algebriques reelles (English summary)* (P 2294) Ordered Fields & Real Algeb Geom;1981 San Francisco 119-139
  ⋄ C35 C60 ⋄ REV MR 84a:12026 Zbl 503 # 14011
  • ID 38833

GOODSTEIN, R.L. [1955] *On non-constructive theorems of analysis and the decision problem* (J 0132) Math Scand 3*261-263
  ⋄ B25 B28 F30 F60 ⋄ REV MR 17.816 Zbl 67.248
  • ID 05177

GOODSTEIN, R.L. [1963] *A decidable fragment of recursive arithmetic* (J 0068) Z Math Logik Grundlagen Math 9*199-201
  ⋄ B20 B25 F30 ⋄ REV MR 27 # 2403 Zbl 114.248
  JSL 33.618 • ID 05189

GOODSTEIN, R.L. [1965] *Multiple successor arithmetics* (P 0688) Logic Colloq;1963 Oxford 265-271
  ⋄ C62 F30 ⋄ REV MR 35 # 6553 Zbl 166.5 • ID 05187

GOODSTEIN, R.L. & LEE, R.D. [1966] *A decidable class of equations in recursive arithmetic* (J 0068) Z Math Logik Grundlagen Math 12*235-239
  ⋄ B20 B25 F30 ⋄ REV MR 33 # 5491 Zbl 148.246
  JSL 33.618 • ID 05190

GOODSTEIN, R.L. [1969] *Decision methods in recursive arithmetic* (P 1841) Fct Recurs & Appl;1967 Tihany 99-104
  ⋄ B25 F30 ⋄ ID 32559

GOODSTEIN, R.L. [1972] *A new proof of completeness* (J 0047) Notre Dame J Formal Log 13*563-564
  ⋄ B05 C07 ⋄ REV MR 48 # 52 Zbl 225 # 02010 • ID 05195

GOODSTEIN, R.L. [1974] *Satisfiable in a larger domain* (J 0047) Notre Dame J Formal Log 15*598-600
  ⋄ B10 C07 ⋄ REV MR 51 # 47 Zbl 271 # 02013 • ID 05197

GOODSTEIN, R.L. see Vol. I, II, IV, V, VI for further entries

GORALCIK, P. & GORALCIKOVA, A. & KOUBEK, V. [1980] *Testing of properties of finite algebras* (P 2904) Automata, Lang & Progr (7);1980 Noordwijkerhout 273-281
  ⋄ C13 D15 D40 ⋄ REV MR 82b:03088 Zbl 449 # 68033
  • ID 73522

GORALCIKOVA, A. [1980] see GORALCIK, P.

GORANKO, V.F. [1985] *Propositional logics with strong negation and the Craig interpolation theorem* (J 0137) C R Acad Bulgar Sci 38*825-827
  ⋄ B55 C40 ⋄ ID 49398

GORANKO, V.F. [1985] *The Craig interpolation theorem for propositional logics with strong negation* (J 0063) Studia Logica 44*291-317
  ⋄ B55 C40 ⋄ ID 47504

GORBACHUK, E.L. & KOMARNITS'KIJ, N.YA. [1982] *On the axiomatizability of radical and semisimple classes of modules and abelian groups (Russian)* (J 0265) Ukr Mat Zh, Akad Nauk Ukr SSR 34*151-157,267
  • TRANSL [1982] (J 3281) Ukr Math J 34*124-129
  ⋄ C52 C60 ⋄ REV MR 83i:20049 Zbl 503 # 20021
  • ID 39345

GORBUNOV, V.A. [1973] see BUDKIN, A.I.

GORBUNOV, V.A. [1975] see BELKIN, V.P.

GORBUNOV, V.A. [1975] see BUDKIN, A.I.

GORBUNOV, V.A. [1977] *Covers in lattices of quasivarieties and independent axiomatizability (Russian)* (J 0003) Algebra i Logika 16*507-549
  • TRANSL [1977] (J 0069) Algeb and Log 16*340-369
  ⋄ C05 C52 ⋄ REV MR 58 # 27424 Zbl 403 # 08009
  • ID 54768

GORBUNOV, V.A. [1984] *Axiomatizability of replete classes (Russian)* (J 0087) Mat Zametki (Akad Nauk SSSR) 35*641-645
  • TRANSL [1984] (J 1044) Math Notes, Acad Sci USSR 35*337-339
  ⋄ C05 C20 C52 ⋄ REV MR 85i:08006 Zbl 559 # 08006
  • ID 44345

GORDEEV, L.N. [1982] *Constructive models for set theory with extensionality* (P 3638) Brouwer Centenary Symp;1981 Noordwijkerhout 123-147
  ⋄ C62 E70 F50 ⋄ REV MR 85k:03037 Zbl 524 # 03051
  • ID 37615

GORDEEV, L.N. see Vol. IV, V, VI for further entries

GORDON, C.E. [1974] *Prime and search computability, characterized as definability in certain sublanguages of constructible $L_{\omega_1,\omega}$* (J 0064) Trans Amer Math Soc 197*391-408
  ⋄ C75 D75 ⋄ REV MR 54 # 4945 Zbl 341 # 02030
  • ID 05199

GORDON, C.E. see Vol. IV, V, VI for further entries

GORDON, E.I. [1979] see GLEBSKIJ, YU.V.

GORDON, E.I. & LYUBETSKIJ, V.A. [1983] *Boolean extensions of uniform structures (Russian)* (C 3807) Issl Neklass Log & Formal Sist 82-153
  ⋄ C65 C90 E40 ⋄ REV MR 86d:03050 • ID 44349

GORDON, E.I. & LYUBETSKIJ, V.A. [1983] *Embedding of sheaves in a Heyting-valued universe (Russian)* (J 0023) Dokl Akad Nauk SSSR 268*794-798
  • TRANSL [1983] (J 0062) Sov Math, Dokl 27*159-164
  ⋄ C90 E40 E70 G30 ⋄ REV MR 84f:03034
  Zbl 546 # 03032 • ID 34454

GORDON, E.I. see Vol. I, IV, V for further entries

GORETTI, A. [1982] *Alcune osservazioni sul linguaggio $L(Q_1)$ (English summary)* (J 0149) Atti Accad Naz Lincei Fis Mat Nat, Ser 8 72*109-114
  ⋄ C50 C55 C80 ⋄ REV MR 85i:03119 Zbl 517 # 03010
  • ID 37281

GOSTANIAN, R. [1974] *Lectures on model theory. Part I* (X 1599) Aarhus Univ Mat Inst: Aarhus ii+186pp
⋄ C98 ⋄ REV MR 50#9572 Zbl 298#02052 • ID 28719

GOSTANIAN, R. & HRBACEK, K. [1976] *On the failure of the weak Beth property* (J 0053) Proc Amer Math Soc 58*245-249
⋄ C40 C75 C80 ⋄ REV MR 54#9975 Zbl 343#02011 • ID 25601

GOSTANIAN, R. [1980] *Constructible models of subsystems of ZF* (J 0036) J Symb Logic 45*237-250
⋄ C62 E45 ⋄ REV MR 81e:03052 Zbl 436#03048 • ID 73541

GOSTANIAN, R. & HRBACEK, K. [1980] *Propositional extensions of $L_{\omega_1\omega}$* (J 0202) Diss Math (Warsaw) 169*5-54
⋄ C55 C75 E15 E50 ⋄ REV MR 81i:03057 Zbl 436#03032 • ID 73544

GOSTANIAN, R. see Vol. IV, V for further entries

GOULD, M. [1975] *An equational spectrum giving cardinalities of endomorphism monoids* (J 0018) Canad Math Bull 18*427-429
⋄ C05 C13 ⋄ REV MR 52#13571 Zbl 321#08005 • ID 21859

GOULD, M. [1975] *Endomorphism and automorphism structure of direct squares of universal algebras* (J 0048) Pac J Math 59*69-84
⋄ C05 C07 ⋄ REV MR 53#13079 Zbl 295#08010 • ID 23212

GRABOWSKI, M. [1972] *The set of all tautologies of the zero-order algorithmic logic is decidable (Russian summary)* (J 0014) Bull Acad Pol Sci, Ser Math Astron Phys 20*575-581
⋄ B25 B70 ⋄ REV MR 48#5819 Zbl 256#02024 • ID 05228

GRABOWSKI, M. [1983] *Some model-theoretical properties of logic for programs with random control* (P 3830) Logics of Progr & Appl;1980 Poznan 148-155
⋄ B75 C85 ⋄ REV Zbl 519#68048 • ID 36731

GRABOWSKI, M. see Vol. I, IV for further entries

GRAETZER, G. [1963] *A theorem on doubly transitive permutation groups with application to universal algebras* (J 0027) Fund Math 53*25-41
⋄ C05 C50 ⋄ REV MR 28#53 Zbl 218#08005 • ID 26295

GRAETZER, G. [1964] *On the class of subdirect powers of a finite algebra* (J 0002) Acta Sci Math (Szeged) 25*160-168
⋄ C05 C30 ⋄ REV MR 29#5772 Zbl 192.96 JSL 37.189 • ID 05245

GRAETZER, G. [1967] *On the endomorphism semigroup of simple algebras* (J 0043) Math Ann 170*334-338
⋄ C05 ⋄ REV MR 35#106 Zbl 146.21 • ID 05250

GRAETZER, G. [1967] *On the spectra of classes of algebras* (J 0053) Proc Amer Math Soc 18*729-735
⋄ C05 C13 ⋄ REV MR 36#5055 Zbl 156.257 • ID 05249

GRAETZER, G. [1968] *On the existence of free structures over universal classes* (J 0114) Math Nachr 36*135-140
⋄ C05 ⋄ REV MR 37#6230 Zbl 155.35 • ID 05253

GRAETZER, G. [1968] *Universal algebra* (X 0864) Van Nostrand: New York xvi+368pp
• LAST ED [1979] (X 0811) Springer: Heidelberg & New York xviii+581pp
⋄ C05 C98 ⋄ REV MR 40#1320 MR 80g:08001 Zbl 182.342 Zbl 412#08001 JSL 38.643 JSL 47.450 • ID 26191

GRAETZER, G. [1969] *Free Σ-structures* (J 0064) Trans Amer Math Soc 135*517-542
⋄ C05 ⋄ REV MR 38#1047 Zbl 179.34 • ID 05261

GRAETZER, G. & LAKSER, H. & PLONKA, J. [1969] *Joins and direct products of equational classes* (J 0018) Canad Math Bull 12*741-744
⋄ C05 C52 ⋄ REV MR 43#1908 Zbl 188.49 • ID 05254

GRAETZER, G. & LAKSER, H. [1973] *A note on the implicational class generated by a class of structures* (J 0018) Canad Math Bull 16*603-605
⋄ C05 C52 ⋄ REV MR 51#5451 Zbl 299#08007 • ID 17430

GRAETZER, G. & SICHLER, J. [1974] *Agassiz sum of algebras* (S 0019) Colloq Math (Warsaw) 30*57-59
⋄ C05 C30 C40 ⋄ REV MR 51#5450 Zbl 296#08011 • ID 17428

GRAETZER, G. [1978] see BERMAN, J.

GRAETZER, G. & KELLY, DAVID [1984] *Free m-products of lattices I* (S 0019) Colloq Math (Warsaw) 48*181-192
⋄ C30 G10 ⋄ REV MR 86a:06009 Zbl 555#06007 • ID 46131

GRAETZER, G. & WHITNEY, S. [1984] *Infinitary varieties of structures closed under the formation of complex structures* (S 0019) Colloq Math (Warsaw) 48*1-5
⋄ C05 C52 C75 ⋄ REV MR 85j:08016 Zbl 545#08001 • ID 41201

GRAINGER, A.D. [1975] *Invariant subspaces of compact operators on topological vector spaces* (J 0048) Pac J Math 56*477-493
⋄ C60 C65 H05 ⋄ REV MR 51#11133 Zbl 325#47004 • ID 17507

GRAINGER, A.D. see Vol. I for further entries

GRAMAIN, F. [1983] *Non-minimalite de la cloture differentielle I: La preuve de S.Shelah* (C 3822) Groupe Etude Th Stables (3) 1980/82 4*9pp
⋄ C50 C60 C65 ⋄ REV MR 85h:03033a Zbl 509#03016 • REM Part II 1983 • ID 36526

GRAMAIN, F. [1983] *Non-minimalite de la cloture differentielle II: La preuve de M.Rosenlicht* (C 3822) Groupe Etude Th Stables (3) 1980/82 5*7pp
⋄ C50 C60 C65 ⋄ REV MR 85h:03033b Zbl 509#03017 • REM Part I 1983 • ID 36527

GRANDJEAN, E. [1983] *Complexity of the first-order theory of almost all finite structures* (J 0194) Inform & Control 57*180-204
⋄ B25 C13 D15 ⋄ REV MR 85j:03061 Zbl 542#03018 • ID 40384

GRANDJEAN, E. [1984] *Spectre des formules du premier order et complexite algorithmique* (C 4356) Gen Log Semin Paris 1982/83 139-152
⋄ C13 C98 D15 ⋄ ID 48072

GRANDJEAN, E. [1984] *The spectra of first-order sentences and computational complexity* (**J** 1428) SIAM J Comp 13*356-373
⋄ C13 D15 ⋄ REV MR 86g:68055 Zbl 535 # 03014
• ID 45011

GRANDJEAN, E. [1984] *Universal quantifiers and time complexity of random access machines* (**P** 2342) Symp Rek Kombin;1983 Muenster 366-379
⋄ C13 D15 ⋄ REV Zbl 548 # 03017 • ID 41788

GRANDJEAN, E. [1985] *Universal quantifiers and time complexity of random access machines* (**J** 0041) Math Syst Theory 18*171-187
⋄ C13 D10 D15 ⋄ REV Zbl 578 # 03022 • ID 42738

GRANDJEAN, E. see Vol. IV for further entries

GRANDY, R.E. [1976] *On the relation between free description theories and standard quantification theory* (**J** 0047) Notre Dame J Formal Log 17*149-152
⋄ B60 C20 ⋄ REV MR 58 # 21428 Zbl 292 # 02041
• ID 18176

GRANDY, R.E. see Vol. I, II, IV, VI for further entries

GRANT, J. [1972] *Recognizable algebras of formulas* (**J** 0047) Notre Dame J Formal Log 13*521-526 • ERR/ADD ibid 16*132
⋄ C07 C40 G05 G15 G25 ⋄ REV MR 48 # 5854 MR 50 # 9579 Zbl 214.14 • ID 05304

GRANT, J. [1973] *Automorphisms definable by formulas* (**J** 0048) Pac J Math 44*107-115 • ERR/ADD ibid 55*639
⋄ C07 C40 C75 ⋄ REV MR 47 # 3164 MR 51 # 5287 Zbl 241 # 02021 Zbl 326 # 02039 • ID 22287

GRANT, J. [1974] *Incomplete models* (**J** 0047) Notre Dame J Formal Log 15*601-607
⋄ B53 C90 ⋄ REV MR 50 # 4282 Zbl 254 # 02045
• ID 05305

GRANT, J. [1978] *Classifications for inconsistent theories* (**J** 0047) Notre Dame J Formal Log 19*435-444
⋄ B53 C07 ⋄ REV MR 58 # 10392 Zbl 305 # 02040
• ID 28232

GRANT, J. see Vol. I, II for further entries

GRANT, P.W. [1977] *Strict Π_1^1-predicates on countable and cofinality ω transitive sets* (**J** 0036) J Symb Logic 42*161-173
⋄ C15 C40 C70 D55 D60 D70 E47 E60 ⋄ REV MR 58 # 21537 Zbl 369 # 02020 • ID 26439

GRANT, P.W. [1978] *The completeness of $L_{\omega_1\omega}(P)$* (**J** 0068) Z Math Logik Grundlagen Math 24*357-364
⋄ C75 C80 ⋄ REV MR 80a:03049 Zbl 397 # 03021
• ID 31081

GRANT, P.W. see Vol. IV for further entries

GRAYSON, R.J. [1981] *Concepts of general topology in constructive mathematics and in sheaves* (**J** 0007) Ann Math Logic 20*1-41 • ERR/ADD ibid 23*99
⋄ C90 E35 E70 E75 F35 F50 F55 G30 ⋄ REV MR 82g:03105 MR 84d:03075 Zbl 458 # 03015 • REM Part I. Part II 1982 • ID 54421

GRAYSON, R.J. [1984] *Heyting-valued semantics* (**P** 3710) Logic Colloq;1982 Firenze 181-208
⋄ B15 C90 F35 F50 G30 ⋄ REV MR 86f:03099 Zbl 574 # 03048 • ID 41840

GRAYSON, R.J. see Vol. V, VI for further entries

GREEN, J. [1974] *Σ_1-compactness for next admissible sets* (**J** 0036) J Symb Logic 39*105-116
⋄ C62 C70 D60 ⋄ REV MR 51 # 5262 Zbl 286 # 02059
• ID 05312

GREEN, J. [1975] *A note on $\mathscr{P}$-admissible sets with urelements* (**J** 0047) Notre Dame J Formal Log 16*415-417
⋄ C62 E30 ⋄ REV MR 54 # 2459 Zbl 304 # 02032
• ID 05313

GREEN, J. [1975] *Consistency properties for finite quantifier languages* (**C** 0781) Infinitary Logic (Karp) 73-123
⋄ C30 C40 C70 C75 ⋄ REV MR 57 # 9527 • ID 73609

GREEN, J. [1977] *Next $\mathscr{P}$-admissible sets are of cofinality ω* (**J** 0047) Notre Dame J Formal Log 18*175-176
⋄ C62 C70 ⋄ REV MR 56 # 2826 Zbl 314 # 02051
• ID 21961

GREEN, J. [1978] *κ-Suslin logic* (**J** 0036) J Symb Logic 43*659-666
⋄ C75 ⋄ REV MR 80k:03033 Zbl 406 # 03057 • ID 56140

GREEN, J. [1979] *Some model theory for game logics* (**J** 0036) J Symb Logic 44*147-152
⋄ C55 C75 E60 ⋄ REV MR 80h:03053 Zbl 418 # 03025
• ID 53310

GREENLEAF, N. [1971] *Fields in which varieties have rational points: a note on a problem of Ax* (**J** 0053) Proc Amer Math Soc 27*139-140
⋄ B25 C60 ⋄ REV MR 42 # 4551 Zbl 213.471 JSL 38.163
• ID 05316

GREENLEAF, N. see Vol. V, VI for further entries

GREGORY, J. [1971] *Incompleteness of a formal system for infinitary finite quantifier formulas* (**J** 0036) J Symb Logic 36*445-455
⋄ C75 ⋄ REV MR 48 # 10758 Zbl 239 # 02005 • ID 05319

GREGORY, J. [1973] *Uncountable models and infinitary elementary extensions* (**J** 0036) J Symb Logic 38*460-470
⋄ C30 C70 ⋄ REV MR 51 # 12514 Zbl 296 # 02029 JSL 47.438 • ID 05320

GREGORY, J. [1974] *Beth definability in infinitary languages* (**J** 0036) J Symb Logic 39*22-26
⋄ C40 C75 ⋄ REV MR 50 # 6787 Zbl 327 # 02014
• ID 05321

GREGORY, J. [1975] *On a finiteness condition for infinitary languages* (**C** 0781) Infinitary Logic (Karp) 143-206
⋄ C40 C62 C70 C75 E40 ⋄ REV MR 58 # 16132
• ID 73614

GREGORY, J. see Vol. V for further entries

GREITHER, C. [1979] *Struktur und Transzendenzgrad von $\eta_{\alpha+1}$-Koerpern und Ultrapotenzen von Koerpern* (**J** 0032) J Algeb 58*70-93
⋄ C20 C60 ⋄ REV MR 80f:12028 Zbl 417 # 12009
• ID 81400

GRENIEWSKI, H. [1950] *Groups and fields definable in the propositional calculus* (**J** 0459) C R Soc Sci Lett Varsovie Cl 3 43*53-68
⋄ B05 C60 ⋄ REV MR 14.834 JSL 33.304 • ID 22000

GRENIEWSKI, H. see Vol. I, II, VI for further entries

GREWE, R. [1969] *Natural models of Ackermann's set theory* (J 0036) J Symb Logic 34*481-488
⋄ C62 E30 E70 ⋄ REV MR 40 # 5437 Zbl 191.301
• ID 05335

GRIGOR'EV, D.YU. [1984] see CHISTOV, A.L.

GRIGOR'EV, D.YU. see Vol. I, IV, V for further entries

GRILLIOT, T.J. [1972] *Omitting types; applications to recursion theory* (J 0036) J Symb Logic 37*81-89
⋄ C07 C62 D30 D55 D60 D75 ⋄ REV MR 49 # 8839 Zbl 244 # 02016 JSL 40.87 • ID 05347

GRILLIOT, T.J. [1973] *Implicit definability and hyperprojectivity* (J 0287) Scripta Math 29*151-155
⋄ C40 C75 D75 ⋄ REV MR 50 # 9558 Zbl 259 # 02032
• ID 05348

GRILLIOT, T.J. [1974] *Model theory for dissecting recursion theory* (P 0602) Generalized Recursion Th (1);1972 Oslo 421-428
⋄ C55 C70 C75 C80 D60 D75 ⋄ REV MR 53 # 12929 Zbl 281 # 02046 • ID 23178

GRILLIOT, T.J. [1985] *Disturbing arithmetic* (J 0036) J Symb Logic 50*375-379
⋄ C07 C30 C40 C62 ⋄ ID 41805

GRILLIOT, T.J. see Vol. IV for further entries

GRINBLAT, V.A. [1981] see BEZVERKHNIJ, V.N.

GRINBLAT, V.A. see Vol. IV for further entries

GRINDLINGER, M.D. [1963] *Inductive definition of a function (Russian)* (J 0226) Uch Zap Ped Inst, Ivanovo 31*56-58
⋄ C40 C62 ⋄ REV MR 48 # 1911 • ID 05358

GRINDLINGER, M.D. see Vol. IV for further entries

GROSS, W.F. [1976] *Models with dimension* (J 0016) Bull Austral Math Soc 14*153-154
⋄ C35 C50 C60 E25 E35 ⋄ REV Zbl 322 # 02052
• ID 62204

GROSS, W.F. [1977] *Dimension and finite closure* (J 3194) J Austral Math Soc, Ser A 23*421-430
⋄ C35 C50 C60 E25 ⋄ REV MR 57 # 16044 Zbl 376 # 02041 • ID 51683

GROSS, W.F. see Vol. IV, VI for further entries

GROSSBERG, R. & SHELAH, S. [1983] *On universal locally finite groups* (J 0029) Israel J Math 44*289-302
⋄ C60 ⋄ REV MR 85c:20001 Zbl 525 # 20025 • ID 33564

GROVES, J.R.J. [1978] see BRYANT, R.M.

GRULOVIC, M. [1980] *A comment on forcing (Serbo-Croatian) (English summary)* (J 2855) Zbor Rad, Prir-Mat Fak, Ser Mat (Novi Sad) 10*161-171
⋄ C25 ⋄ REV MR 84c:03057 Zbl 488 # 03017 • ID 34001

GRULOVIC, M. [1982] *A note on forcing and weak interpolation theorem for infinitary logics (Serbo-Croatian summary)* (J 2887) Zbor Radova, NS 12*327-348
⋄ C25 C40 C75 ⋄ REV MR 85i:03098 Zbl 538 # 03032 Zbl 559 # 03021 • ID 41475

GRULOVIC, M. [1983] *On n-finite forcing* (J 2855) Zbor Rad, Prir-Mat Fak, Ser Mat (Novi Sad) 13*405-421
⋄ C25 E40 ⋄ REV MR 86h:03052 • ID 45085

GRULOVIC, M. see Vol. V for further entries

GRUNEWALD, F. & SEGAL, D. [1981] *How to solve a quadratic equation in integers* (J 0332) Math Proc Cambridge Phil Soc 89*1-5
⋄ B25 ⋄ REV MR 82j:10029 Zbl 471 # 10012 • ID 55255

GRUNEWALD, F. see Vol. IV for further entries

GRZEGORCZYK, A. [1957] *Decision problems* (X 1034) PWN: Warsaw 142pp
⋄ B25 C10 D35 D98 F30 ⋄ REV MR 22 # 7936 Zbl 89.246 • ID 24864

GRZEGORCZYK, A. & MOSTOWSKI, ANDRZEJ & RYLL-NARDZEWSKI, C. [1958] *The classical and the ω-complete arithmetic* (J 0036) J Symb Logic 23*188-206
⋄ C62 D55 D70 F30 ⋄ REV MR 21 # 4908 Zbl 84.248 JSL 27.80 • ID 05398

GRZEGORCZYK, A. & MOSTOWSKI, ANDRZEJ & RYLL-NARDZEWSKI, C. [1961] *Definability of sets in models of axiomatic theories* (J 0014) Bull Acad Pol Sci, Ser Math Astron Phys 9*163-167
⋄ C40 D55 ⋄ REV MR 29 # 1138 Zbl 99.9 JSL 34.126
• ID 05402

GRZEGORCZYK, A. [1962] *A kind of categoricity* (S 0019) Colloq Math (Warsaw) 9*183-187
⋄ C35 C50 C62 C65 ⋄ REV MR 26 # 3589 Zbl 108.5 JSL 30.387 • ID 05408

GRZEGORCZYK, A. [1962] *A theory without recursive models* (J 0014) Bull Acad Pol Sci, Ser Math Astron Phys 10*63-69
⋄ C57 C62 D55 ⋄ REV MR 25 # 3820 Zbl 113.6 JSL 28.102 • ID 05406

GRZEGORCZYK, A. [1962] *On the concept of categoricity (Polish and Russian summaries)* (J 0063) Studia Logica 13*39-66
⋄ C35 ⋄ REV MR 33 # 2545 Zbl 192.36 JSL 30.387
• ID 20930

GRZEGORCZYK, A. [1967] *Some relational systems and the associated topological spaces* (J 0027) Fund Math 60*223-231
⋄ B45 C65 G25 ⋄ REV MR 36 # 1304 Zbl 207.296 JSL 34.652 • ID 05412

GRZEGORCZYK, A. [1968] *Logical uniformity by decomposition and categoricity in $\aleph_0$* (J 0014) Bull Acad Pol Sci, Ser Math Astron Phys 16*687-692
⋄ C35 ⋄ REV MR 39 # 1317 Zbl 184.12 • ID 05413

GRZEGORCZYK, A. [1970] *Decision procedures for theories categorical in $\aleph_0$* (P 0625) Symp Autom Demonst;1968 Versailles 87-100
⋄ B25 C35 ⋄ REV MR 42 # 7496 Zbl 202.13 • ID 05416

GRZEGORCZYK, A. [1971] *An unfinitizability proof by means of restricted reduced power* (J 0027) Fund Math 73*37-49
⋄ C20 C62 ⋄ REV MR 58 # 5160 Zbl 225 # 02034 JSL 38.159 • ID 05417

GRZEGORCZYK, A. [1971] *Lezioni sulla teoria dei modelli (raccolte a cura di Marisa Venturini Zilli)* (J 3434) Pubbl Ist Appl Calcolo, Ser 3 68*53pp
⋄ C98 ⋄ REV Zbl 307 # 02032 • ID 62213

GRZEGORCZYK, A. [1971] *Outline of theoretical arithmetic (Polish)* (X 1034) PWN: Warsaw 314pp
⋄ B28 C62 F30 F35 F98 ⋄ REV MR 53 # 12856 MR 85e:03002 Zbl 251 # 02002 • REM 2nd ed. 1983
• ID 23119

GRZEGORCZYK, A. see Vol. I, II, IV, V, VI for further entries

GUASPARI, D. [1979] *Partially conservative extensions of arithmetic* (J 0064) Trans Amer Math Soc 254*47-68
◇ C62 F25 F30 ◇ REV MR 80j:03083 Zbl 417#03030
• ID 53266

GUASPARI, D. [1980] *Definability in models of set theory* (J 0036) J Symb Logic 45*9-19
◇ C62 D55 E45 E47 ◇ REV MR 83k:03065 Zbl 453#03053 • ID 54183

GUASPARI, D. see Vol. II, IV, V, VI for further entries

GUDJONSSON, H. [1970] *Remarks on limitultrapowers* (J 0009) Arch Math Logik Grundlagenforsch 13*154-157
◇ C20 E05 ◇ REV MR 43#6067 Zbl 217.304 • ID 05428

GUENTHNER, F. & HOEPELMAN, J.P. [1975] *A note on the representation of "branching quantifiers"* (J 1517) Theor Linguist 2*285-289
◇ B65 C80 ◇ REV MR 54#9976 • ID 25603

GUENTHNER, F. [1983] see GABBAY, D.M.

GUENTHNER, F. [1984] see GABBAY, D.M.

GUENTHNER, F. see Vol. II for further entries

GUERRE, S. & LEVY, M. [1983] *Espaces l^p dans les sous-espaces de L^1* (C 4256) Semin Geom Espace Banach 1982 Paris 103-105
◇ C20 C65 ◇ REV MR 84m:46006 Zbl 556#46012
• ID 46162

GUETING, R. [1975] *Subtractive abelian groups* (J 0047) Notre Dame J Formal Log 16*425-428
◇ C60 ◇ REV MR 51#10070 Zbl 304#20039 • ID 05431

GUHL, R. [1975] *A theorem on recursively enumerable vector spaces* (J 0047) Notre Dame J Formal Log 16*357-362
◇ C57 C60 D45 ◇ REV MR 52#2857 Zbl 254#02034
• ID 05433

GUHL, R. [1977] *Two notes on recursively enumerable vector spaces* (J 0047) Notre Dame J Formal Log 18*295-298
◇ C57 C60 D45 ◇ REV MR 56#5252 Zbl 305#02057
• ID 24226

GUICHARD, D.R. [1980] *A many-sorted interpolation theorem for L(Q)* (J 0053) Proc Amer Math Soc 80*469-474
◇ C40 C75 C80 ◇ REV MR 81i:03055 Zbl 464#03030
• ID 54620

GUICHARD, D.R. [1983] *Automorphisms of substructure lattices in recursive algebra* (J 0073) Ann Pure Appl Logic 25*47-58
◇ C57 C60 D45 G05 ◇ REV MR 85e:03104 • ID 40309

GUICHARD, D.R. [1984] *A note on r-maximal subspaces of V_∞* (J 0073) Ann Pure Appl Logic 26*1-9
◇ C57 C60 D45 ◇ REV MR 85j:03075 Zbl 522#03027
• ID 43386

GUITART, R. & LAIR, C. [1980] *Calcul syntaxique des modeles et calcul des formules internes* (J 3797) Diagrammes 4*106pp
◇ C90 G30 ◇ REV MR 84h:18012 Zbl 508#03030
• ID 36944

GUITART, R. [1982] *From where do figurative algebras come?* (J 3797) Diagrammes 7*RG1-RG7
◇ C90 G30 ◇ REV MR 84f:18011 Zbl 515#18004
• ID 39675

GUITART, R. see Vol. IV, V for further entries

GUMB, R.D. [1979] *An extended joint consistency theorem for free logic with equality* (J 0047) Notre Dame J Formal Log 20*321-335
◇ B60 C40 ◇ REV MR 80i:03020 Zbl 321#02012
• ID 62230

GUMB, R.D. [1984] *"Conservative" Kripke-closures* (J 0154) Synthese 60*39-49
◇ C40 C90 ◇ REV MR 86f:03029 • ID 44025

GUMB, R.D. see Vol. I, II, V for further entries

GUPTA, H.N. [1969] *On some axioms in foundations of cartesian spaces* (J 0018) Canad Math Bull 12*831-836
◇ C60 ◇ REV MR 40#7103 Zbl 192.573 • ID 05445

GUPTA, H.N. & PRESTEL, A. [1972] *On a class of Pasch-free Euclidean planes* (J 0014) Bull Acad Pol Sci, Ser Math Astron Phys 20*17-23
◇ C60 C65 ◇ REV MR 46.432 MR 47#942 Zbl 243.50001 • ID 33806

GUPTA, H.N. see Vol. I, VI for further entries

GUPTA, N. [1984] *Recursively presented two generated infinite p-groups* (J 0044) Math Z 188*89-90
◇ C57 C60 ◇ REV MR 86f:20032 Zbl 553#20015
• ID 39718

GURARI, E.M. & IBARRA, O.H. [1978] *An NP-complete number-theoretic problem* (P 1740) ACM Symp Th of Comput (10);1978 San Diego 205-215
◇ B25 D15 D80 ◇ REV MR 81c:68034 • ID 42545

GURARI, E.M. & IBARRA, O.H. [1979] *An NP-complete number-theoretic problem* (J 0037) ACM J 26*567-581
◇ B25 D10 D15 D25 D80 ◇ REV Zbl 407#68053
• ID 56232

GURARI, E.M. & IBARRA, O.H. [1981] *The complexity of the equivalence problem for two characterizations of Presburger sets* (J 1426) Theor Comput Sci 13*295-314
◇ B25 D05 D15 F30 ◇ REV MR 82m:68084 Zbl 454#03005 • ID 54218

GURARI, E.M. see Vol. IV for further entries

GUREVICH, R. [1982] *Loewenheim-Skolem problem for functors* (J 0029) Israel J Math 42*273-276
◇ C55 C75 G30 ◇ REV MR 84g:08027 Zbl 495#03018
• ID 33565

GUREVICH, R. [1984] *Decidability of the equational theory of positive numbers with raising to a power (Russian)* (J 0092) Sib Mat Zh 25/2(144)*216-219
◇ B25 B28 C65 ◇ REV MR 86f:03023 Zbl 555#03004
• ID 45002

GUREVICH, R. [1985] *Equational theory of positive numbers with exponentiation* (J 0053) Proc Amer Math Soc 94*135-141
◇ B25 B28 C65 ◇ REV Zbl 572#03014 • ID 44633

GUREVICH, Y. & KOKORIN, A.I. [1963] *Universal equivalence of ordered abelian groups (Russian)* (J 0003) Algebra i Logika 2/1*37-39
◇ C60 ◇ REV MR 27#5687 • ID 05449

GUREVICH, Y. [1964] *Elementary properties of ordered abelian groups (Russian)* (J 0003) Algebra i Logika 3/1*5-39
• TRANSL [1965] (J 0225) Amer Math Soc, Transl, Ser 2 46*165-192
◇ B25 C60 ◇ REV MR 28#5004 Zbl 178.314 • ID 05451

GUREVICH, Y. [1965] *Existential interpretation (Russian)* (J 0003) Algebra i Logika 4/4*71-85
⋄ B20 B25 C60 D35 F25 ⋄ REV MR 34 # 46 Zbl 294 # 02023 • REM Part II 1982 • ID 05452

GUREVICH, Y. [1966] *Certain algorithmic questions of the theory of classes of algebraic systems (Russian)* (C 1549) Tartu Mezhvuz Nauch Simp Obshchej Algeb 38-41
⋄ B25 D35 ⋄ REV MR 34 # 29 Zbl 248 # 02048 • ID 05457

GUREVICH, Y. [1966] *Effective recognition of satisfiability of formulae of the restricted predicate calculus (Russian)* (J 0003) Algebra i Logika 5/2*25-55
⋄ B20 B25 D35 ⋄ REV MR 35 # 4096 Zbl 198.324 • ID 05455

GUREVICH, Y. [1966] *On the decision problem for pure restricted predicate logic (Russian)* (J 0023) Dokl Akad Nauk SSSR 166*1032-1034
• TRANSL [1966] (J 0062) Sov Math, Dokl 7*217-219
⋄ B20 C13 D35 ⋄ REV MR 34 # 47 Zbl 158.252 • ID 05458

GUREVICH, Y. [1966] *The decision problem for the restricted predicate calculus (Russian)* (J 0023) Dokl Akad Nauk SSSR 168*510-511
• TRANSL [1966] (J 0062) Sov Math, Dokl 7*669-670
⋄ B20 C13 D35 ⋄ REV MR 34 # 60 Zbl 163.253 • ID 05456

GUREVICH, Y. [1967] *A contribution to the elementary theory of lattice ordered abelian groups and K-lineals (Russian)* (J 0023) Dokl Akad Nauk SSSR 175*1213-1215
• TRANSL [1967] (J 0062) Sov Math, Dokl 8*987-989
⋄ B25 C60 D35 ⋄ REV MR 36 # 1373 Zbl 189.326 • ID 33753

GUREVICH, Y. [1967] *A new decision procedure for the theory of ordered abelian groups (Russian)* (J 0003) Algebra i Logika 6/5*5-6
⋄ B25 C60 ⋄ ID 33754

GUREVICH, Y. [1969] *A decision problem for decision problems (Russian)* (J 0003) Algebra i Logika 8*640-642
• TRANSL [1969] (J 0069) Algeb and Log 8*362-363
⋄ B25 D35 ⋄ REV MR 44 # 71 Zbl 211.313 • ID 05461

GUREVICH, Y. [1969] *The decision problem for the logic of predicates and operations (Russian)* (J 0003) Algebra i Logika 8*284-308
• TRANSL [1969] (J 0069) Algeb and Log 8*160-174
⋄ B25 D35 ⋄ REV MR 41 # 8205 Zbl 198.325 • ID 31104

GUREVICH, Y. [1973] *Formulas with a single ∀ (Russian)* (C 0733) Izbr Vopr Algeb & Log (Mal'tsev) 97-110
⋄ B20 B25 C13 ⋄ REV MR 49 # 7115 Zbl 298 # 02049 • ID 32346

GUREVICH, Y. [1974] *A resolving procedure for the extended theory of ordered abelian groups (Russian)* (J 0092) Sib Mat Zh 15*230
• TRANSL [1974] (J 0475) Sib Math J 15*166
⋄ B25 C10 C60 C85 ⋄ REV Zbl 283 # 06008 • ID 62239

GUREVICH, Y. [1976] *The decision problem for standard classes* (J 0036) J Symb Logic 41*460-464
⋄ B20 B25 D35 ⋄ REV MR 53 # 10572 Zbl 339 # 02045 • ID 14777

GUREVICH, Y. [1977] *Expanded theory of ordered abelian groups* (J 0007) Ann Math Logic 12*193-228
⋄ B25 C10 C60 C85 ⋄ REV MR 58 # 16245 Zbl 368 # 06016 • ID 24282

GUREVICH, Y. [1977] *Monadic theory of order and topology I* (J 0029) Israel J Math 27*299-319
⋄ B15 C65 C85 E07 E50 E75 ⋄ REV MR 56 # 5259 Zbl 359 # 02061 • REM Part II 1979 • ID 50608

GUREVICH, Y. [1979] *Modest theory of short chains I* (J 0036) J Symb Logic 44*481-490
⋄ B25 C65 C85 D35 E07 ⋄ REV MR 81a:03038a Zbl 464 # 03013 • REM Part II 1979 by Gurevich,Y. & Shelah,S. • ID 54603

GUREVICH, Y. & SHELAH, S. [1979] *Modest theory of short chains II* (J 0036) J Symb Logic 44*491-502
⋄ B25 C65 C85 D35 E07 E50 ⋄ REV MR 81a:03038b Zbl 464 # 03014 • REM Part I 1979 by Gurevich,Y. • ID 54604

GUREVICH, Y. [1979] *Monadic theory of order and topology II* (J 0029) Israel J Math 34*45-71
⋄ B15 C65 C85 E07 E45 E50 ⋄ REV MR 81f:03049 Zbl 428 # 03034 • REM Part I 1977 • ID 53793

GUREVICH, Y. [1980] *Ordered abelian groups* (P 2983) Ordered Groups;1978 Boise 173-174
⋄ B25 C10 C60 C85 ⋄ REV Zbl 445 # 06014 • ID 56508

GUREVICH, Y. [1980] *Two notes on formalized topology* (J 0027) Fund Math 107*145-148
⋄ C65 C85 F25 ⋄ REV MR 82c:03057 Zbl 362 # 54004 • ID 53951

GUREVICH, Y. [1981] see GLASS, A.M.W.

GUREVICH, Y. & HOLLAND, W.C. [1981] *Recognizing the real line* (J 0064) Trans Amer Math Soc 265*527-534
⋄ C07 C65 ⋄ REV MR 82g:03060 Zbl 471 # 03033 • ID 55228

GUREVICH, Y. & HARRINGTON, L.A. [1982] *Automata, trees, and games* (P 3714) ACM Symp Th of Comput (14);1982 60-65
⋄ B25 C85 D05 ⋄ ID 33759

GUREVICH, Y. [1982] *Crumbly spaces* (P 3622) Int Congr Log, Meth & Phil of Sci (6,Proc);1979 Hannover 179-191
⋄ B25 C65 C85 ⋄ REV MR 84g:03044 Zbl 515 # 03019 • ID 33755

GUREVICH, Y. & SHELAH, S. [1982] *Monadic theory of order and topology in ZFC* (J 0007) Ann Math Logic 23*179-198
⋄ B25 C65 C85 D35 E07 ⋄ REV MR 85d:03080 Zbl 516 # 03007 • ID 33758

GUREVICH, Y. [1982] see AANDERAA, S.O.

GUREVICH, Y. & SHELAH, S. [1983] *Interpreting second-order logic in the monadic theory of order* (J 0036) J Symb Logic 48*816-828
⋄ B15 C65 C85 D35 E07 E50 F25 ⋄ REV MR 85f:03007 Zbl 559 # 03008 • ID 33765

GUREVICH, Y. & SHELAH, S. [1983] *Rabin's uniformization problem* (J 0036) J Symb Logic 48*1105-1119
⋄ B15 C40 C65 C85 E40 ⋄ REV MR 85g:03055 Zbl 537 # 03007 • ID 33768

GUREVICH, Y. & SHELAH, S. [1983] *Random models and the Goedel case of the decision problem* (J 0036) J Symb Logic 48*1120-1124
⋄ B20 B25 C13 D35 ⋄ REV MR 85d:03019 Zbl 534 # 03006 • ID 33766

GUREVICH, Y. & MAGIDOR, M. & SHELAH, S. [1983] *The monadic theory of ω_2* (J 0036) J Symb Logic 48*387-398
⋄ B15 B25 C65 C85 D35 E10 E35 ⋄ REV MR 84i:03076 Zbl 549 # 03010 • ID 33764

GUREVICH, Y. [1984] see GOLDFARB, W.D.

GUREVICH, Y. [1984] see BLASS, A.R.

GUREVICH, Y. & SHELAH, S. [1984] *The monadic theory and the "next world"* (J 0029) Israel J Math 49*55-68
⋄ C62 C65 C85 E40 F25 ⋄ REV Zbl 575 # 03028 • ID 38749

GUREVICH, Y. & SCHMITT, P.H. [1984] *The theory of ordered abelian groups does not have the independence property* (J 0064) Trans Amer Math Soc 284*171-182
⋄ C45 C60 ⋄ REV MR 85g:03053 Zbl 536 # 03018 • ID 33770

GUREVICH, Y. [1984] *Toward logic tailored for computational complexity* (P 2153) Logic Colloq;1983 Aachen 2*175-216
⋄ B75 C13 D15 ⋄ REV MR 86f:03061 • ID 43005

GUREVICH, Y. [1985] *Monadic second-order theories* (C 4183) Model-Theor Log 479-509
⋄ B25 C65 C85 C98 D35 D98 ⋄ ID 48332

GUREVICH, Y. [1985] see BURGESS, J.P.

GUREVICH, Y. & SHELAH, S. [1985] *The decision problem for branching time logic* (J 0036) J Symb Logic 50*668-681
⋄ B25 B45 C65 D05 D25 ⋄ ID 47372

GUREVICH, Y. see Vol. I, II, IV, V, VI for further entries

GUTIERREZ-NOVOA, L. [1979] *Non-Euclidean real numbers* (J 3495) Boll Unione Mat Ital, V Ser, B 16*390-404
⋄ B28 C65 ⋄ REV MR 80j:51020 Zbl 452 # 12011 • ID 54116

GUZICKI, W. [1974] *Elementary extensions of Levy's model of A_2^-* (J 0154) Synthese 27*265-270
⋄ C55 C62 C80 E25 E35 E40 F35 ⋄ REV MR 58 # 16268 Zbl 344 # 02045 • ID 62249

GUZICKI, W. [1974] *Uncountable β-models with countable height* (J 0027) Fund Math 82*143-152
⋄ C62 E40 E50 ⋄ REV MR 50 # 4283 Zbl 294 # 02030 • ID 05464

GUZICKI, W. [1977] *The number of β-models of Kelley-Morse set theory* (P 1639) Set Th & Hierarch Th (1);1974 Karpacz 23-26
⋄ C62 E50 E70 ⋄ REV MR 58 # 5225 Zbl 403 # 03042 • ID 54755

GUZICKI, W. & MAREK, W. & PELC, A. & RAUSZER, C. (EDS.) [1981] *Proceedings of the International Conference "Open Days in Model Theory and Set Theory"* (P 2614) Open Days in Model Th & Set Th;1981 Jadwisin 333pp
⋄ C97 ⋄ ID 49818

GUZICKI, W. [1981] *The equivalence of definable quantifiers in second order arithmetic* (J 0027) Fund Math 113*59-65
⋄ C30 C55 C62 C80 E40 E45 F35 ⋄ REV MR 84g:03050 Zbl 493 # 03011 • ID 34160

GUZICKI, W. [1983] *Definable quantifiers in second order arithmetic and elementary extensions of ω-models* (J 0202) Diss Math (Warsaw) 208*51pp
⋄ C30 C55 C62 C80 E40 E50 F35 ⋄ REV MR 85f:03035 Zbl 518 # 03022 • ID 37520

GUZICKI, W. see Vol. IV, V, VI for further entries

HADDAD, L. [1972] see FAKIR, S.

HADDAD, L. see Vol. I, V for further entries

HAERLEN, H. [1928] *Ueber die Vollstaendigkeit und Entscheidbarkeit* (J 0157) Jbuchber Dtsch Math-Ver 37*226-230
⋄ A05 C35 ⋄ REV FdM 54.54 • ID 05484

HAERLEN, H. see Vol. I, V for further entries

HAERTIG, K. [1958] *Einbettung elementarer Theorien in endlich axiomatisierbare* (J 0068) Z Math Logik Grundlagen Math 4*249-292
⋄ C07 ⋄ REV MR 21 # 7153 Zbl 99.9 JSL 30.397 • ID 05491

HAERTIG, K. [1965] *Ueber einen Quantifikator mit zwei Wirkungsbereichen* (P 0797) Fonds des Math, Machines Math & Appl;1962 Tihany 31-36
⋄ C55 C80 ⋄ REV MR 32 # 7396 • ID 05494

HAERTIG, K. see Vol. I for further entries

HAGENDORF, J.G. [1972] *Extensions immediates de chaines et de relations* (J 2313) C R Acad Sci, Paris, Ser A-B 274*A607-A609
⋄ C65 E07 ⋄ REV MR 48 # 5927 Zbl 241 # 06001 • ID 28843

HAGENDORF, J.G. [1982] *Extensions immediates de chaines* (J 0068) Z Math Logik Grundlagen Math 28*15-44
⋄ C65 E07 ⋄ REV MR 83f:06001 Zbl 486 # 06001 • ID 42208

HAGENDORF, J.G. see Vol. V for further entries

HAILPERIN, T. [1961] *A complete set of axioms for logical formulas invalid in some finite domain* (J 0068) Z Math Logik Grundlagen Math 7*84-96
⋄ B20 C13 ⋄ REV MR 26 # 3588 Zbl 111.8 JSL 27.108 • ID 05508

HAILPERIN, T. [1984] *Probability logic* (J 0047) Notre Dame J Formal Log 25*198-212
⋄ B25 B48 C90 ⋄ REV MR 85m:03013 Zbl 547 # 03018 • ID 41842

HAILPERIN, T. see Vol. I, II, V, VI for further entries

HAIMO, F. [1973] *The index of an algebra automorphism group* (P 0779) Conf Group Th;1972 Racine 85-90
⋄ C05 C07 ⋄ REV MR 54 # 5090 Zbl 275 # 20076 • ID 24155

HAIMO, F. see Vol. IV for further entries

HAJEK, P. [1964] *Die durch die schwach inneren Relationen gegebenen Modelle der Mengenlehre* (J 0068) Z Math Logik Grundlagen Math 10*151-157
⋄ C62 E30 E70 ⋄ REV MR 29 # 2179 Zbl 158.265 JSL 32.412 • ID 05512

HAJEK, P. [1965] *Eine Bemerkung ueber standarde nichtregulaere Modelle der Mengenlehre* (J 0140) Comm Math Univ Carolinae (Prague) 6*1-6
⋄ C62 E30 E55 E70 ⋄ REV MR 31 # 36 Zbl 131.247 • ID 05514

HAJEK, P. [1965] *Modelle der Mengenlehre, in denen Mengen gegebener Gestalt existieren* (J 0068) Z Math Logik Grundlagen Math 11*103-115
⋄ C62 E30 E35 ⋄ REV MR 31 #2155 Zbl 171.264 JSL 33.474 • ID 05515

HAJEK, P. & VOPENKA, P. [1965] *Permutation submodels of the model* ∇ (J 0014) Bull Acad Pol Sci, Ser Math Astron Phys 13*611-614
⋄ C62 E25 E35 ⋄ REV MR 33 #2530 Zbl 143.259 JSL 34.515 • ID 13878

HAJEK, P. [1965] *The concept of a primitive class of algebras (Birkhoff theorem) (Czech) (Russian and German summaries)* (J 0086) Cas Pestovani Mat, Ceskoslov Akad Ved 90*477-486
⋄ C05 ⋄ REV MR 35 #40 Zbl 139.13 JSL 35.489
• ID 05516

HAJEK, P. [1966] *Generalized interpretability in terms of models. Note to a paper of R.Montague (Czech and Russian summaries)* (J 0086) Cas Pestovani Mat, Ceskoslov Akad Ved 91*352-357
⋄ C07 F25 ⋄ REV MR 35 #1456 Zbl 144.245 • REM Montague's paper appeared 1965 in J0028 27*467-476
• ID 05517

HAJEK, P. [1966] see BUKOVSKY, L.

HAJEK, P. & VOPENKA, P. [1966] *Some permutation submodels of the the model* ∇ (J 0014) Bull Acad Pol Sci, Ser Math Astron Phys 14*1-7
⋄ C62 E25 E35 ⋄ REV MR 33 #2531 Zbl 143.258 JSL 34.515 • ID 05519

HAJEK, P. & VOPENKA, P. [1967] *Concerning the* ∇ *-models of set theory (Russian summary)* (J 0014) Bull Acad Pol Sci, Ser Math Astron Phys 15*113-117
⋄ C62 E35 E40 E45 ⋄ REV MR 37 #5094 Zbl 164.315
• ID 13881

HAJEK, P. [1970] *Logische Kategorien* (J 0009) Arch Math Logik Grundlagenforsch 13*168-193
⋄ B10 C40 E40 F25 G30 ⋄ REV MR 43 #7386 Zbl 226 #02043 • ID 05521

HAJEK, P. [1971] *Sets, semisets, models* (P 0693) Axiomatic Set Th;1967 Los Angeles 1*67-82
⋄ A10 C62 E35 E40 E70 ⋄ REV MR 43 #3110 Zbl 233 #02025 JSL 40.505 • ID 05523

HAJEK, P. [1973] *Automatic listing of important observational statements. I,II* (J 0156) Kybernetika (Prague) 9*187-205,251-271
⋄ B35 C13 C80 C90 ⋄ REV MR 53 #5288 Zbl 289 #68046 Zbl 289 #68047 • REM Part III 1974
• ID 22921

HAJEK, P. & VOPENKA, P. [1973] *Existence of a generalized model of the Goedel-Bernays set theory* (J 0014) Bull Acad Pol Sci, Ser Math Astron Phys 21*1079-1086
⋄ C62 E70 ⋄ REV MR 54 #10016 Zbl 274 #02036
• ID 31134

HAJEK, P. [1974] *Automatic listing of important observational statements. III* (J 0156) Kybernetika (Prague) 10*95-124
⋄ B35 C13 C80 C90 ⋄ REV MR 54 #4960 Zbl 289 #68047 • REM Parts I,II 1973 • ID 24128

HAJEK, P. [1976] *Observationsfunktorenkalkuele und die Logik der automatisierten Forschung* (J 0129) Elektr Informationsverarbeitung & Kybern 12*181-186
⋄ B60 C13 C80 C90 ⋄ REV MR 58 #27413 Zbl 356 #68095 • ID 31127

HAJEK, P. [1976] *Some remarks on observational model-theoretic languages* (P 1476) Set Th & Hierarch Th (2) (Mostowski);1975 Bierutowice 335-345
⋄ C13 C95 ⋄ REV MR 58 #16258 Zbl 438 #03044
• ID 23817

HAJEK, P. [1977] *Experimental logics and Π_3^0 theories* (J 0036) J Symb Logic 42*515-522
⋄ B60 C62 D55 F30 ⋄ REV MR 58 #16243 Zbl 428 #03043 • ID 26853

HAJEK, P. [1977] *Generalized quantifiers and finite sets* (P 1639) Set Th & Hierarch Th (1);1974 Karpacz 14*91-104
⋄ B25 B50 C13 C40 C80 ⋄ REV MR 58 #5086 Zbl 386 #03017 • ID 31126

HAJEK, P. & HAVRANEK, T. [1977] *On generation of inductive hypotheses* (J 1741) Int J Man-Mach Stud 9*415-438
⋄ B35 C90 ⋄ REV Zbl 372 #68026 • ID 31130

HAJEK, P. & HAVRANEK, T. [1978] *Mechanizing hypothesis formation. Mathematical foundations for a general theory* (X 0811) Springer: Heidelberg & New York xv+396pp
• TRANSL [1984] (X 2027) Nauka: Moskva 278pp
⋄ A05 B35 C13 C80 C90 D80 ⋄ REV MR 82f:03017 MR 86e:03022 Zbl 371 #02002 • ID 31131

HAJEK, P. & PUDLAK, P. [1980] *Two orderings of the class of all countable models of Peano arithmetic* (P 2625) Model Th of Algeb & Arithm;1979 Karpacz 174-185
⋄ C15 C62 ⋄ REV MR 82g:03061 Zbl 471 #03024
• ID 55219

HAJEK, P. & KURKA, P. [1981] *A second-order dynamic logic with array assignments* (J 2095) Fund Inform, Ann Soc Math Pol, Ser 4 4*919-933
⋄ B75 C85 ⋄ REV MR 84c:03035 Zbl 491 #68033
• ID 35692

HAJEK, P. [1981] *Completion closed algebras and models of Peano arithmetic* (J 0140) Comm Math Univ Carolinae (Prague) 22*585-594
⋄ C62 F30 ⋄ REV MR 83f:03071 Zbl 499 #03022
• ID 35314

HAJEK, P. [1981] *Decision problems of some statistically motivated monadic modal calculi* (J 1741) Int J Man-Mach Stud 15*351-358
⋄ B25 B45 ⋄ REV MR 83j:03039 Zbl 464 #03015
• ID 54605

HAJEK, P. [1984] *On a new notion of partial conservativity* (P 2153) Logic Colloq;1983 Aachen 2*217-232
⋄ C62 F30 ⋄ REV MR 86d:03054 • ID 43006

HAJEK, P. see Vol. I, II, IV, V, VI for further entries

HAJNAL, A. [1966] see ERDOES, P.

HAJNAL, A. [1972] see ERDOES, P.

HAJNAL, A. see Vol. V for further entries

HAKEN, W. [1973] *Connections between topological and group theoretical decision problems* (P 0678) Word Probl: Decis & Burnside Probl in Group Th;1969 Irvine 427–441
⋄ B25 C60 C65 D35 D40 D80 ⋄ REV MR 53 #1594 Zbl 265 #02033 JSL 41.786 • ID 29825

HAKEN, W. see Vol. IV for further entries

HALEY, D.K. [1970] *Equationally compact noetherian rings* (P 0629) Conf Universal Algeb;1969 Kingston 258–267
⋄ C05 C60 ⋄ REV MR 41 #5342 Zbl 232 #13022
• ID 05575

HALEY, D.K. [1970] *On compact commutative noetherian rings* (J 0043) Math Ann 189*272–274
⋄ C05 C60 ⋄ REV Zbl 193.7 JSL 40.88 • ID 05574

HALEY, D.K. [1973] *Equationally compact artinian rings* (J 0017) Canad J Math 25*273–283
⋄ C05 C60 ⋄ REV MR 47 #3445 Zbl 258 #16016
• ID 05576

HALEY, D.K. [1974] *A note on equational compactness* (J 0004) Algeb Universalis 4*36–40
⋄ C05 C60 ⋄ REV MR 50 #6977 Zbl 296 #08014
• ID 05577

HALEY, D.K. [1976] *Equational compactness and compact topologies in rings satisfying the a.c.c.* (J 0048) Pac J Math 62*99–115
⋄ C05 C60 ⋄ REV MR 54 #2564 Zbl 319 #16030
• ID 18471

HALEY, D.K. [1979] *Equational compactness in rings* (S 3301) Lect Notes Math 745*iii+167pp
⋄ C05 C60 ⋄ REV MR 81e:16036 Zbl 463 #16001
• ID 81466

HALKOWSKA, K. & PIROG-RZEPECKA, K. [1975] *On proofs in the theories containig conditional definition (Polish) (English summary)* (S 1454) Zesz Nauk Wyz Szk Ped Mat, Opole 15*63–76
⋄ B20 C07 ⋄ REV MR 54 #7200 Zbl 315 #02014
• ID 29801

HALKOWSKA, K. & PIROG-RZEPECKA, K. & SLUPECKI, J. [1976] *Mathematical logic (Polish)* (X 1034) PWN: Warsaw 289pp
⋄ B98 C07 C98 E98 ⋄ REV MR 57 #15933 Zbl 393 #03001 • ID 52421

HALKOWSKA, K. & PLONKA, J. [1978] *On formulas preserved by the sum of a direct system of relational systems* (J 0114) Math Nachr 86*117–120
⋄ C30 C40 ⋄ REV MR 80g:03033 Zbl 402 #03029
• ID 54676

HALKOWSKA, K. see Vol. I, II, V for further entries

HALL, P. [1959] *On the finiteness of certain soluble groups* (J 3240) Proc London Math Soc, Ser 3 9*595–622
⋄ C60 ⋄ REV MR 22 #1618 • ID 48397

HALL, P. [1959] *Some constructions for locally finite groups* (J 0039) J London Math Soc 34*305–319
⋄ C25 C60 ⋄ REV Zbl 88.23 • ID 40052

HALL, P. see Vol. IV for further entries

HALLDEN, S. [1949] *On the decision problem of Lewis' calculus S5* (J 4510) Norsk Mat Tidsskr 31*89–94
⋄ B25 B45 ⋄ REV MR 11.411 Zbl 40.146 JSL 15.224
• ID 05587

HALLDEN, S. [1949] *Results concerning the decision problem of Lewis's calculi S3 and S6* (J 0036) J Symb Logic 14*230–236
⋄ B25 B45 ⋄ REV MR 11.303 Zbl 36.7 JSL 15.70
• ID 05586

HALLDEN, S. see Vol. I, II for further entries

HALPERN, F. [1975] see ELLENTUCK, E.

HALPERN, F. [1975] *Transfer theorems for topological structures* (J 0048) Pac J Math 61*427–440
⋄ C20 C40 C90 ⋄ REV MR 53 #5294 Zbl 336 #02040
• ID 22929

HALPERN, J.D. [1975] *Nonstandard combinatorics* (J 3240) Proc London Math Soc, Ser 3 30*40–54
⋄ C30 E05 G05 ⋄ REV MR 52 #10436 Zbl 313 #05001
• ID 21724

HALPERN, J.D. see Vol. I, IV, V, VI for further entries

HALPERN, M. [1983] *Inductive inference in finite algebraic structures* (P 3858) Adequate Modeling of Syst;1982 Bad Honnef 167–176
⋄ B48 C13 ⋄ REV Zbl 499 #68035 • ID 45211

HAMBLIN, C.L. [1973] *A felicitous fragment of the predicate calculus* (J 0047) Notre Dame J Formal Log 14*433–447
⋄ B20 B25 ⋄ REV MR 51 #115 Zbl 265 #02008
• ID 17398

HAMBLIN, C.L. see Vol. I, II, IV for further entries

HAMILTON, A.G. [1970] *Bases and α-dimensions of countable vector spaces with recursive operations* (J 0036) J Symb Logic 35*85–96
⋄ C57 C60 D45 D50 ⋄ REV MR 44 #2604 Zbl 195.18
• ID 05625

HAMILTON, A.G. see Vol. I, II, IV for further entries

HAMILTON, J. [1973] see ADLER, A.

HANAZAWA, M. [1975] *A remark on ordered structures with unary predicates* (J 0090) J Math Soc Japan 27*345–349
⋄ C65 E07 ⋄ REV MR 52 #109 Zbl 304 #02023
• ID 18182

HANAZAWA, M. [1979] *An interpretation of Skolem's paradox in the predicate calculus with ε-symbol* (J 1472) Sci Rep Saitama Univ, Ser A 9/2*11–13
⋄ B10 C07 F05 F25 ⋄ REV MR 81h:03110 Zbl 457 #03004 • ID 54329

HANAZAWA, M. see Vol. I, V, VI for further entries

HANF, W.P. [1957] *On some fundamental problems concerning isomorphism of boolean algebras* (J 0132) Math Scand 5*205–217
⋄ C05 C15 C30 G05 ⋄ REV MR 21 #7167 Zbl 81.261 JSL 24.251 • ID 05633

HANF, W.P. [1960] *Models of languages with infinitely long expressions* (P 1953) Int Congr Log, Meth & Phil of Sci (1;Abstr);1960 Stanford 24
⋄ C55 C75 ⋄ ID 41156

HANF, W.P. [1964] *Incompactness in languages with infinitely long expressions* (J 0027) Fund Math 53*309–324
⋄ C75 E55 ⋄ REV MR 28 #3943 Zbl 207.302 JSL 30.95
• ID 33566

HANF, W.P. [1965] *Model-theoretic methods in the study of elementary logic* (P 0614) Th Models;1963 Berkeley 132-145
⋄ B25 C07 D20 D25 D35 F25 ⋄ REV MR 35#1457 Zbl 166.258 JSL 34.127 • ID 05635

HANF, W.P. & MYERS, D.L. [1983] *Boolean sentence algebras: isomorphism constructions* (J 0036) J Symb Logic 48*329-338
⋄ B10 C52 ⋄ REV MR 84m:03096 Zbl 511#03005 • ID 35790

HANF, W.P. see Vol. IV, V for further entries

HANSON, W.H. [1966] *On some alleged decision procedures for S4* (J 0036) J Symb Logic 31*641-643
⋄ B25 B45 ⋄ REV MR 35#5298 Zbl 166.3 JSL 35.326 • ID 05648

HANSON, W.H. see Vol. II, VI for further entries

HANSSON, B. [1975] see GAERDENFORS, P.

HANSSON, B. see Vol. I, II for further entries

HARAN, D. & JARDEN, M. [1982] *Bounded statements in the theory of algebraically closed fields with distinguished automorphisms* (J 0127) J Reine Angew Math 337*1-17
⋄ B25 C60 ⋄ REV MR 84m:03015 Zbl 486#12010 • ID 35720

HARAN, D. & LUBOTZKY, A. [1982] *Embedding covers and the theory of Frobenius fields* (J 0029) Israel J Math 41*181-202
⋄ B25 C60 ⋄ REV MR 83m:12035 Zbl 502#20013 • ID 40356

HARAN, D. [1984] see FRIED, M.

HARAN, D. [1984] *The undecidability of pseudo-real-closed fields* (J 0504) Manuscr Math 49*91-108
⋄ B25 C60 D35 ⋄ REV MR 85j:03072 Zbl 578#12020 • ID 44141

HARARY, F. [1979] see BLASS, A.R.

HARARY, F. see Vol. I, II, V for further entries

HAREL, D. [1979] *Characterizing second order logic with first order quantifiers* (J 0068) Z Math Logik Grundlagen Math 25*419-422
⋄ B15 C80 C85 ⋄ REV MR 80k:03034 Zbl 432#03007 • ID 53965

HAREL, D. [1983] *Recurring dominoes: making the highly undecidable highly understandable* (P 3864) FCT'83 Found of Comput Th;1983 Borgholm 177-194
⋄ B25 B75 D15 D55 ⋄ REV MR 85e:68003 Zbl 531#68002 • ID 38572

HAREL, D. [1985] *Recurring dominoes: Making the highly undecidable highly understandable* (S 3358) Ann Discrete Math 24*51-72
⋄ B25 B75 D05 D15 D55 ⋄ ID 41208

HAREL, D. see Vol. II, IV for further entries

HARIZANOV, V. & MIJAJLOVIC, Z. [1983] *Regular relations and the quantifier "there exist uncountably many"* (J 0068) Z Math Logik Grundlagen Math 29*151-161
⋄ C55 C80 ⋄ REV MR 85d:03078 Zbl 519#03029 • ID 37541

HARIZANOV, V. see Vol. V for further entries

HARNIK, V. & RESSAYRE, J.-P. [1971] *Prime extensions and categoricity in power* (J 0029) Israel J Math 10*172-185
⋄ C35 C50 ⋄ REV MR 46#45 Zbl 224#02044 • ID 05661

HARNIK, V. [1975] *A two cardinal theorem for sets of formulas in a stable theory* (J 0029) Israel J Math 21*7-23
⋄ C45 C50 C55 ⋄ REV MR 55#12504 Zbl 329#02024 • ID 05663

HARNIK, V. [1975] *On the existence of saturated models of stable theories* (J 0053) Proc Amer Math Soc 52*361-367
⋄ C45 C50 ⋄ REV MR 52#7890 Zbl 316#02055 • ID 18183

HARNIK, V. & MAKKAI, M. [1976] *Applications of Vaught sentences and the covering theorem* (J 0036) J Symb Logic 41*171-187
⋄ C15 C40 C45 C50 C52 C75 D70 E15 ⋄ REV MR 56#5265 Zbl 333#02013 • ID 14797

HARNIK, V. [1976] *Approximation theorems and model theoretic forcing* (J 0036) J Symb Logic 41*59-72
⋄ C25 C75 ⋄ REV MR 55#5432 Zbl 355#02011 • ID 14750

HARNIK, V. & MAKKAI, M. [1977] *A tree argument in infinitary model theory* (J 0053) Proc Amer Math Soc 67*309-314
⋄ C15 C75 ⋄ REV MR 57#12204 Zbl 384#03019 • ID 31139

HARNIK, V. & MAKKAI, M. [1979] *New axiomatizations for logics with generalized quantifiers* (J 0029) Israel J Math 32*257-281
⋄ C70 C80 ⋄ REV MR 80f:03040 Zbl 404#03023 • ID 54810

HARNIK, V. [1979] *Refinements of Vaught's normal form theorem* (J 0036) J Symb Logic 44*289-306
⋄ C40 C75 ⋄ REV MR 80i:03044 Zbl 436#03033 • ID 73849

HARNIK, V. [1980] *Game sentences, recursive saturation and definability* (J 0036) J Symb Logic 45*35-46
⋄ C40 C50 C57 C70 ⋄ REV MR 81h:03076 Zbl 445#03014 • ID 56478

HARNIK, V. & HARRINGTON, L.A. [1984] *Fundamentals of forking* (J 0073) Ann Pure Appl Logic 26*245-286
⋄ C45 ⋄ REV MR 86c:03032 • ID 45277

HARNIK, V. [1985] *Stability theory and set existence axioms* (J 0036) J Symb Logic 50*123-137
⋄ C45 F35 ⋄ ID 39778

HARRINGTON, L.A. [1974] *Recursively presentable prime models* (J 0036) J Symb Logic 39*305-309
⋄ C35 C50 C57 C60 ⋄ REV MR 50#4292 Zbl 332#02055 JSL 49.671 • ID 05664

HARRINGTON, L.A. & KECHRIS, A.S. [1975] *On characterizing Spector classes* (J 0036) J Symb Logic 40*19-24
⋄ C62 D55 D60 D65 D75 E45 E55 E60 ⋄ REV MR 55#5420 Zbl 312#02033 • ID 05665

HARRINGTON, L.A. & PARIS, J.B. [1977] *A mathematical incompleteness in Peano arithmetic* (C 1523) Handb of Math Logic 1133-1142
⋄ C30 C62 E05 F30 ⋄ REV MR 58#10343 JSL 49.980 • ID 27334

HARRINGTON, L.A. [1977] *Long projective wellorderings* (J 0007) Ann Math Logic 12*1-24
⋄ C62 D55 E15 E35 E50 ⋄ REV MR 57 # 5752 Zbl 384 # 03033 • ID 24274

HARRINGTON, L.A. [1978] see ABRAMSON, F.G.

HARRINGTON, L.A. [1980] *Extensions of countable infinitary logic which preserve most of its nice properties* (J 0009) Arch Math Logik Grundlagenforsch 20*95-102
⋄ C40 C55 C70 C75 C95 ⋄ REV MR 82g:03066 Zbl 473 # 03031 • ID 55360

HARRINGTON, L.A. [1982] see GUREVICH, Y.

HARRINGTON, L.A. & MAKKAI, M. & SHELAH, S. [1984] *A proof of Vaught's conjecture for ω-stable theories* (J 0029) Israel J Math 49*259-280
⋄ C15 C35 C45 ⋄ ID 38754

HARRINGTON, L.A. [1984] *Admissible ordinals are not immortal* (J 0073) Ann Pure Appl Logic 26*358-361
⋄ C62 C70 E30 E45 E47 ⋄ REM Appendix to an article (ibid 26*287-361) • ID 48307

HARRINGTON, L.A. [1984] see HARNIK, V.

HARRINGTON, L.A. [1985] see CHERLIN, G.L.

HARRINGTON, L.A. & MAKKAI, M. [1985] *An exposition of Shelah's "main gap": counting uncountable models of ω-stable and superstable theories* (J 0047) Notre Dame J Formal Log 26*139-177
⋄ C45 C50 C52 ⋄ ID 42593

HARRINGTON, L.A. & MORLEY, M.D. & SCEDROV, A. & SIMPSON, S.G. [1985] *Harvey Friedman's research on the foundations of mathematics* (X 0809) North Holland: Amsterdam xvi+408pp
⋄ B97 C97 D97 E97 F97 ⋄ ID 49810

HARRINGTON, L.A. see Vol. IV, V for further entries

HARROP, R. [1954] *An investigation of the propositional calculus used in a particular system of logic* (J 0171) Proc Cambridge Phil Soc Math Phys 50*495-512
⋄ B20 B25 ⋄ REV MR 16.661 Zbl 56.8 JSL 23.65 • ID 05676

HARROP, R. [1958] *On the existence of finite models and decision procedures for propositional calculi* (J 0171) Proc Cambridge Phil Soc Math Phys 54*1-13
⋄ B22 B25 ⋄ REV MR 20 # 6 Zbl 80.8 JSL 25.180 • ID 05678

HARROP, R. [1965] *Some generalizations and applications of a relativization procedure for propositional calculi* (P 0688) Logic Colloq;1963 Oxford 12-41
⋄ B22 B25 D35 ⋄ REV MR 35 # 2737 Zbl 158.252 JSL 32.125 • ID 05684

HARROP, R. [1976] *Some results concerning finite model separability of propositional calculi* (J 0063) Studia Logica 35*179-189
⋄ B22 B25 ⋄ REV MR 55 # 7729 Zbl 361 # 02019 • ID 32206

HARROP, R. see Vol. I, IV, V, VI for further entries

HART, J. [1969] $\mathscr{E}^0$-*arithmetic* (J 0068) Z Math Logik Grundlagen Math 15*237
⋄ C62 D20 ⋄ REV MR 40 # 40 Zbl 157.336 • ID 05689

HARTMANIS, J. [1976] *On effective speed-up and long proofs of trivial theorems in formal theories (French summary)* (J 4698) Rev Franc Autom, Inf & Rech Operat, Ser Rouge Inf Th 10/R-1*29-38
⋄ B25 B35 D15 D20 F20 ⋄ REV MR 54 # 6550 Zbl 399 # 03042 • ID 24160

HARTMANIS, J. see Vol. IV, VI for further entries

HARTVIGSON, Z.R. [1974] *A non-desarguesian space geometry* (J 0027) Fund Math 86*144-147
⋄ C65 ⋄ REV MR 50 # 8259 Zbl 292 # 50003 • ID 05706

HASENJAEGER, G. [1952] *Topologische Untersuchungen zur Semantik und Syntax eines erweiterten Praedikatenkalkuels* (J 0009) Arch Math Logik Grundlagenforsch 1*99-129
⋄ B10 C07 G05 ⋄ REV MR 15.668 Zbl 49.6 • ID 05722

HASENJAEGER, G. [1953] *Eine Bemerkung zu Henkin's Beweis fuer die Vollstaendigkeit des Praedikatenkalkuels der ersten Stufe* (J 0036) J Symb Logic 18*42-48
⋄ B10 C07 C57 D55 ⋄ REV MR 14.1052 Zbl 51.5 JSL 31.268 • ID 05723

HASENJAEGER, G. [1955] *On definability and derivability* (P 1589) Math Interpr of Formal Systs;1954 Amsterdam 15-25
⋄ B10 B15 C07 D55 ⋄ REV MR 17.699 Zbl 68.246 JSL 24.171 • ID 27719

HASENJAEGER, G. [1967] *On Loewenheim-Skolem-type insufficiencies of second order logic* (P 0691) Sets, Models & Recursion Th;1965 Leicester 173-182
⋄ B15 C55 C85 ⋄ REV MR 36 # 3637 Zbl 175.267 • ID 05731

HASENJAEGER, G. see Vol. I, II, IV, V, VI for further entries

HATCHER, W.S. [1968] *Remarques sur l'equivalence des systemes logiques* (J 2313) C R Acad Sci, Paris, Ser A-B 267*A529-A530
⋄ C07 G05 ⋄ REV MR 38 # 3125 Zbl 169.6 • ID 05747

HATCHER, W.S. [1972] see CORCORAN, J.

HATCHER, W.S. & SHAFAAT, A. [1975] *Categorical languages for algebraic structures* (J 0068) Z Math Logik Grundlagen Math 21*433-438
⋄ C05 G30 ⋄ REV MR 52 # 10408 Zbl 408 # 03029 • ID 05749

HATCHER, W.S. [1978] *A language for type-free algebra* (J 0068) Z Math Logik Grundlagen Math 24*385-397
⋄ C05 G30 ⋄ REV MR 80f:18005 Zbl 396 # 03015 • ID 32205

HATCHER, W.S. [1985] *Elementary extension and the hyperreal numbers* (C 4181) Math Log & Formal Syst (Costa da) 205-219
⋄ C65 H05 ⋄ ID 48208

HATCHER, W.S. see Vol. I, II, IV, V, VI for further entries

HAUCK, J. & HERRE, H. & POSEGGA, M. [1972] *Zur Metatheorie formaler Systeme* (C 1533) Quantoren, Modal, Paradox 107-122
⋄ B10 C07 F30 ⋄ REV Zbl 256 # 02006 • ID 62383

HAUCK, J. [1977] *Die Entscheidbarkeit der n-Beweisbarkeit aus Axiomenschemata (English, Russian and French summaries)* (J 0115) Wiss Z Humboldt-Univ Berlin, Math-Nat Reihe 26*629-634
⋄ B25 ⋄ REV MR 80e:03009 Zbl 423 # 03004 • ID 53516

HAUCK, J. see Vol. IV, VI for further entries

HAUSCHILD, K. [1963] *Modelle der Mengenlehre, die aus endlichen Mengen bestehen* (J 0068) Z Math Logik Grundlagen Math 9*7-12
⋄ C62 E30 E35 ⋄ REV MR 26 # 2358 Zbl 112.10
• ID 05759

HAUSCHILD, K. [1963] *Ueber die Charakterisierbarkeit der Zahlenreihe in gewissen Nichtstandardmodellen der Arithmetik* (J 0068) Z Math Logik Grundlagen Math 9*113-116
⋄ H15 ⋄ REV MR 26 # 4887 Zbl 112.8 • ID 05758

HAUSCHILD, K. [1967] *Cauchyfolgen hoeheren Typus' in angeordneten Koerpern* (J 0068) Z Math Logik Grundlagen Math 13*55-66
⋄ C60 ⋄ REV MR 35 # 1527 Zbl 184.295 • ID 05767

HAUSCHILD, K. [1967] *Die Nichtexistenz starkregulaerer standarder Modelle in der Mengenlehre von Zermelo-Fraenkel und ihren widerspruchsfreien Erweiterungen* (J 0140) Comm Math Univ Carolinae (Prague) 8*249-255
⋄ C62 E30 ⋄ REV MR 37 # 6175 Zbl 174.18 • ID 05765

HAUSCHILD, K. [1967] *Ueber ein Definierbarkeitsproblem in der Koerpertheorie (English and French summaries)* (J 0115) Wiss Z Humboldt-Univ Berlin, Math-Nat Reihe 16*673-674
⋄ C40 C60 ⋄ REV MR 40 # 7230 Zbl 301 # 02058
• ID 62393

HAUSCHILD, K. [1967] *Verallgemeinerung eines Satzes von Cantor* (J 0115) Wiss Z Humboldt-Univ Berlin, Math-Nat Reihe 16*669-671
⋄ C07 C15 C35 ⋄ REV MR 38 # 5598 Zbl 313 # 02031
• ID 05766

HAUSCHILD, K. [1968] *Metatheoretische Eigenschaften gewisser Klassen von elementaren Theorien* (J 0068) Z Math Logik Grundlagen Math 14*205-244
⋄ B28 C15 C35 E30 E70 ⋄ REV MR 38 # 3146
Zbl 185.12 • ID 05770

HAUSCHILD, K. & WOLTER, H. [1968] *Zur Charakterisierung von Theorien mit syntaktisch beschreibbarer Auswahlfunktion* (J 0068) Z Math Logik Grundlagen Math 14*263-266
⋄ C07 C20 ⋄ REV MR 37 # 2588 Zbl 186.9 • ID 05768

HAUSCHILD, K. & WOLTER, H. [1969] *Ueber die Kategorisierbarkeit gewisser Koerper in nicht-elementaren Logiken* (J 0068) Z Math Logik Grundlagen Math 15*157-162
⋄ C60 C80 C85 ⋄ REV MR 39 # 5341 Zbl 293 # 02040
• ID 05771

HAUSCHILD, K. & WOLTER, H. [1970] *Einige Anwendungen des Koenigschen Graphensatzes in der mathematischen Logik* (J 0068) Z Math Logik Grundlagen Math 16*265-269
⋄ C52 ⋄ REV MR 46 # 7013 Zbl 206.279 • ID 05776

HAUSCHILD, K. [1971] *Entscheidbarkeit in der Theorie bewerteter Graphen (Russian and English summaries)* (J 0129) Elektr Informationsverarbeitung & Kybern 7*485-492
⋄ B25 C65 ⋄ REV MR 47 # 8279 Zbl 238 # 02042
• ID 05778

HAUSCHILD, K. & RAUTENBERG, W. [1971] *Interpretierbarkeit in der Gruppentheorie* (J 0004) Algeb Universalis 1*136-151
⋄ C60 D35 F25 ⋄ REV MR 46 # 39 Zbl 242 # 02052
• ID 05780

HAUSCHILD, K. & RAUTENBERG, W. [1971] *Interpretierbarkeit und Entscheidbarkeit in der Graphentheorie I* (J 0068) Z Math Logik Grundlagen Math 17*47-55
⋄ B25 C65 D35 F25 ⋄ REV MR 44 # 6491
Zbl 231 # 02058 • REM Part II 1972 by Hauschild,K. & Herre,H. & Rautenberg,W • ID 27450

HAUSCHILD, K. [1971] *Nichtaxiomatisierbarkeit von Satzmengen durch Ausdruecke spezieller Gestalt* (J 0027) Fund Math 72*245-253
⋄ B28 C20 C40 C52 F30 ⋄ REV MR 48 # 88
Zbl 215.48 JSL 38.161 • ID 05779

HAUSCHILD, K. & HERRE, H. & RAUTENBERG, W. [1972] *Entscheidbarkeit der elementaren Theorien der endlichen Baeume und verwandter Klassen endlicher Strukturen (Russian, English and French summaries)* (J 0115) Wiss Z Humboldt-Univ Berlin, Math-Nat Reihe 21*497-502
⋄ B25 C13 ⋄ REV MR 50 # 6818 Zbl 252 # 02050
• ID 05785

HAUSCHILD, K. & HERRE, H. & RAUTENBERG, W. [1972] *Entscheidbarkeit der monadischen Theorie 2. Stufe der n-separierten Graphen (Russian, English and French summaries)* (J 0115) Wiss Z Humboldt-Univ Berlin, Math-Nat Reihe 21*507-511
⋄ B15 B25 ⋄ REV MR 50 # 79 Zbl 264 # 02046 • ID 05782

HAUSCHILD, K. & HERRE, H. & RAUTENBERG, W. [1972] *Interpretierbarkeit und Entscheidbarkeit in der Graphentheorie II* (J 0068) Z Math Logik Grundlagen Math 18*457-480
⋄ B25 C65 D35 F25 ⋄ REV MR 48 # 3737
Zbl 284 # 02023 • REM Part I 1971 • ID 05787

HAUSCHILD, K. [1972] *Universalitaet in der Ringtheorie* (J 0115) Wiss Z Humboldt-Univ Berlin, Math-Nat Reihe 21*505-506
⋄ C60 D35 F25 ⋄ REV MR 48 # 8221 Zbl 245 # 02041
• ID 05790

HAUSCHILD, K. & MELIS, E. [1972] *Zusammenhaenge zwischen Axiomatisierbarkeit und Definierbarkeit in elementaren Theorien mit eingeschraenktem Wohlordnungsschema* (J 0115) Wiss Z Humboldt-Univ Berlin, Math-Nat Reihe 21*519-521
⋄ C40 C52 ⋄ REV MR 48 # 5843 Zbl 215.49 • ID 05791

HAUSCHILD, K. & RAUTENBERG, W. [1973] *Entscheidungsprobleme der Theorie zweier Aequivalenzrelationen mit beschraenkter Zahl von Elementen in den Klassen* (J 0027) Fund Math 81*35-41
⋄ B25 ⋄ REV MR 49 # 2338 Zbl 319 # 02035 • ID 05794

HAUSCHILD, K. [1975] *Entscheidbarkeit in der Theorie baumartiger Graphen (Russian, English and French summaries)* (J 0115) Wiss Z Humboldt-Univ Berlin, Math-Nat Reihe 24*764-768
⋄ B25 ⋄ REV MR 58 # 5395 Zbl 328 # 02031 • ID 62392

HAUSCHILD, K. [1975] *Ueber zwei Spiele und ihre Anwendung in der Modelltheorie (Russian, English and French summaries)* (J 0115) Wiss Z Humboldt-Univ Berlin, Math-Nat Reihe 24*783-788
⋄ B25 C07 C30 C35 C85 ⋄ REV MR 58 # 5152
Zbl 347 # 02031 • ID 27134

HAUSCHILD, K. [1975] *Zur Uebertragbarkeit von Entscheidbarkeitsresultaten elementarer Theorien (Russian, English and French summaries)* (J 0115) Wiss Z Humboldt-Univ Berlin, Math-Nat Reihe 24*780-783
◊ B15 B25 C65 C85 ◊ REV MR 58 # 5151 Zbl 341 # 02039 • ID 29757

HAUSCHILD, K. [1978] *Gruppengraphen und endliche Axiomatisierbarkeit* (X 2888) ZI Math Mech Akad Wiss DDR: Berlin 14pp
◊ C35 C52 C60 ◊ REV Zbl 402 # 03028 • ID 54675

HAUSCHILD, K. [1980] *Generalized Haertig quantifiers* (J 3293) Bull Acad Pol Sci, Ser Math 28*523-528
◊ C55 C80 ◊ REV MR 83a:03030 Zbl 503 # 03010 • ID 90122

HAUSCHILD, K. [1981] *Zum Vergleich von Haertigquantor und Rescherquantor* (J 0068) Z Math Logik Grundlagen Math 27*255-264
◊ B25 C10 C55 C80 D35 ◊ REV MR 82h:03033 Zbl 503.03011 • ID 73915

HAUSCHILD, K. [1982] *Model-theoretic properties of cause-and-effect structures* (J 0140) Comm Math Univ Carolinae (Prague) 23*541-555
◊ B45 C52 D35 ◊ REV MR 84e:03023 Zbl 522 # 03005 • ID 34362

HAUSCHILD, K. see Vol. I, IV, V, VI for further entries

HAUSDORFF, F. [1908] *Grundzuege einer Theorie der geordneten Mengen* (J 0043) Math Ann 65*435-505
◊ C50 E07 E10 ◊ REV FdM 39.99 • ID 05796

HAUSDORFF, F. see Vol. IV, V for further entries

HAUSSLER, D. [1981] *Model completeness of an algebra of languages* (J 0053) Proc Amer Math Soc 83*371-374
◊ B25 C35 D05 ◊ REV MR 82k:03048 Zbl 478 # 03014 • ID 55629

HAUSSLER, D. see Vol. IV for further entries

HAVRANEK, T. [1975] *Statistical quantifiers in observational calculi* (J 0472) Theory Decis 6*213-230
◊ B50 C80 C90 ◊ REV MR 57 # 12187 Zbl 313 # 68070 • ID 31144

HAVRANEK, T. [1977] see HAJEK, P.

HAVRANEK, T. [1977] *Towards a model theory of statistical theories* (J 0154) Synthese 36*441-458
◊ B25 C90 ◊ REV MR 58 # 27464 Zbl 393 # 62001 • ID 32356

HAVRANEK, T. & VOSAHLO, J. [1978] *A GUHA procedure with correlational quantifiers* (J 1741) Int J Man-Mach Stud 10*67-74
◊ B35 C80 ◊ REV MR 80g:68118b Zbl 404 # 68095 • ID 54868

HAVRANEK, T. [1978] see HAJEK, P.

HAVRANEK, T. see Vol. I, II, IV for further entries

HAY, L. [1975] *Spectra and halting problems* (J 0068) Z Math Logik Grundlagen Math 21*167-176
◊ C13 D10 D25 D30 ◊ REV MR 51 # 10056 Zbl 309 # 02049 • ID 05822

HAY, L. & MANASTER, A.B. & ROSENSTEIN, J.G. [1977] *Concerning partial recursive similarity transformations of linearly ordered sets* (J 0048) Pac J Math 71*57-70
◊ C57 C65 D20 D25 D30 D45 E07 ◊ REV MR 56 # 2806 Zbl 409 # 03027 • ID 30705

HAY, L. see Vol. I, II, IV, V for further entries

HEATH, I.J. [1969] *Decidable classes of number-theoretic sentences* (J 0068) Z Math Logik Grundlagen Math 15*411-420
◊ B25 ◊ REV MR 42 # 72 Zbl 199.309 • ID 05834

HEATH, I.J. see Vol. VI for further entries

HEBERT, M. [1984] *Un theoreme de preservation pour les factorisations* (J 2128) C R Math Acad Sci, Soc Roy Canada 6*39-42
◊ C07 C40 C52 ◊ REV MR 85h:03031 Zbl 532 # 03011 • ID 39621

HECHLER, S.H. [1973] *Large superuniversal metric spaces* (J 0029) Israel J Math 14*115-148
◊ C50 C65 E75 ◊ REV MR 48 # 2967 Zbl 258 # 54004 • ID 05842

HECHLER, S.H. see Vol. I, V for further entries

HEDRLIN, Z. [1969] *On universal partly ordered sets and classes* (J 0032) J Algeb 11*503-509
◊ C50 E07 G30 ◊ REV MR 39 # 279 Zbl 209.16 • ID 05863

HEDRLIN, Z. see Vol. V for further entries

HEIDEMA, J. & WALT VAN DER, A.P.J. [1967] *Contributions to the metamathematical theory of ideals I. Domains with dense kernel* (J 0028) Indag Math 29*156-162
◊ C07 C60 ◊ REV MR 35 # 5301 Zbl 232 # 02038 JSL 36.158 • REM Part II 1968 • ID 05866

HEIDEMA, J. [1968] *Contributions to the metamathematical theory of ideals II. Metamathematical prime ideals and radicals* (J 0028) Indag Math 30*280-285
◊ C07 C60 ◊ REV MR 37 # 6171 Zbl 232 # 02039 JSL 36.158 • REM Part I 1967 by Heidema,J. & Walt van der,A.P.J. • ID 05868

HEIDEMA, J. [1972] *Aspects of model theory (Africaans) (English summary)* (J 1069) Tydskr Natuurwetenskap (Pretoria) 12*37-49
◊ C98 ◊ REV MR 51 # 5288 Zbl 237 # 02014 • ID 24807

HEIDEMA, J. [1972] *Metamathematical representation of radicals in universal algebra* (P 2080) Conf Math Log;1970 London 342-343
◊ C05 C60 ◊ REV Zbl 227 # 08009 • ID 27381

HEIDEMA, J. & MALAN, S.W. [1980] *Congruences on relational structures* (P 3882) Algeb Symp (1);1979 Potchefstroom 55-68
◊ C05 C07 ◊ REV MR 82i:16002 Zbl 491 # 08004 • ID 38474

HEIDEMA, J. [1980] see BUYS, A.

HEIDEMA, J. [1982] see BUYS, A.

HEIDEMA, J. see Vol. VI for further entries

HEIJENOORT VAN, J. [1979] *Introduction a la semantique des logiques non-classiques* (X 3562) Ecole Norm Sup Jeunes Filles: Paris ix + 108pp
◊ B45 C90 F50 ◊ REV MR 81f:03025 • ID 79641

HEIJENOORT VAN, J. see Vol. I, VI for further entries

HEINDORF, L. [1980] *The decidability of the L_t - theory of boolean spaces* (J 0115) Wiss Z Humboldt-Univ Berlin, Math-Nat Reihe 29*413-419
⋄ B25 C65 C80 C90 ⋄ REV MR 83b:03013 Zbl 467 # 03032 • ID 55030

HEINDORF, L. [1981] *Comparing the expressive power of some languages for boolean algebras* (J 0068) Z Math Logik Grundlagen Math 27*419-434
⋄ B25 C10 C80 C85 C90 G05 ⋄ REV MR 82m:03050 Zbl 472 # 03008 • ID 55274

HEINDORF, L. [1984] *Beitraege zur Modelltheorie der Booleschen Algebren (English and Russian summaries)* (S 3382) Sem-ber, Humboldt-Univ Berlin, Sekt Math 53*112pp
⋄ B25 C10 C35 C55 C80 G05 ⋄ REV MR 85e:03077 Zbl 532 # 03016 • ID 38279

HEINDORF, L. [1984] *Continuous functions on countable ordinals* (J 0068) Z Math Logik Grundlagen Math 30*339-340
⋄ C65 E10 E45 ⋄ REV MR 86h:03061 Zbl 526 # 03018 • ID 42267

HEINDORF, L. [1984] *Regular ideals and boolean pairs* (J 0068) Z Math Logik Grundlagen Math 30*547-560
⋄ C30 C85 D35 G05 ⋄ REV MR 86g:03063 Zbl 558 # 03016 • ID 39661

HEINEMANN, B. & PRESTEL, A. [1984] *Fields regularly closed with respect to finitely many valuations and orderings* (S 3382) Sem-ber, Humboldt-Univ Berlin, Sekt Math 60*20-59
⋄ C35 C60 ⋄ REV MR 86i:12013 Zbl 566 # 03023 • ID 48736

HEINEMANN, B. & PRESTEL, A. [1984] *Fields regularly closed with respect to finitely many valuations and orderings* (P 4285) Quad & Hermitian Forms;1983 Hamilton 297-336
⋄ C35 C60 ⋄ REV MR 86g:12012 Zbl 566 # 03023 • ID 40505

HEINRICH, S. [1980] *The isomorphic problem of envelopes* (S 3230) Prepr, Inst Math, Pol Acad Sci 213*17pp
⋄ C20 C65 H05 ⋄ REV MR 84h:46025 Zbl 448 # 46020 • ID 56660

HEINRICH, S. [1980] *Ultraproducts in Banach space theory* (J 0127) J Reine Angew Math 313*72-104
⋄ C20 C65 C98 H05 ⋄ REV MR 82b:46013 Zbl 412 # 46017 • ID 81516

HEINRICH, S. [1981] *Ultraproducts of L_1-predual spaces* (J 0027) Fund Math 113*221-234
⋄ C20 C65 H05 ⋄ REV MR 83m:46020 Zbl 427 # 46009 Zbl 472 # 46022 • ID 53749

HEINRICH, S. & MANKIEWICZ, P. [1982] *Applications of ultrapowers to the uniform and Lipschitz classification of Banach spaces* (J 0343) Studia Math, Pol Akad Nauk 73*225-251
⋄ C20 C65 H05 ⋄ REV MR 84h:46026 Zbl 448 # 46021 Zbl 506 # 46008 • ID 37826

HEINRICH, S. [1982] *The isomorphic problem of envelopes* (J 0343) Studia Math, Pol Akad Nauk 73*41-49
⋄ C20 C65 H05 ⋄ REV MR 84h:46025 Zbl 519 # 46015 • ID 46460

HEINRICH, S. & HENSON, C.W. & MOORE JR., L.C. [1983] *Elementary equivalence of L_1-preduals* (P 3883) Banach Space Th & Appl (1);1981 Bucharest 79-90
⋄ C20 C65 ⋄ REV MR 84j:46021 Zbl 514 # 46052 • ID 37446

HEINRICH, S. & MANKIEWICZ, P. [1983] *Some open problems in the nonlinear classification of Banach spaces* (P 3883) Banach Space Th & Appl (1);1981 Bucharest 91-95
⋄ C20 C65 ⋄ REV MR 84h:46001 Zbl 514 # 46053 • ID 37447

HEINRICH, S. [1983] *Ultrapowers of locally convex spaces and applications I* (S 3414) Prepr, Akad Wiss DDR, Inst Math 10/83*58pp
⋄ C20 C65 H10 ⋄ REV Zbl 522 # 46002 • REM Part II 1983 • ID 37053

HEINRICH, S. [1983] *Ultrapowers of locally convex spaces and applications II* (S 3414) Prepr, Akad Wiss DDR, Inst Math 11/83*37pp
• REPR [1985] (J 0114) Math Nachr 121*211-229
⋄ C20 C65 H10 ⋄ REV Zbl 522 # 46003 • REM Part I 1983 • ID 37054

HEINTZ, JOOS [1979] *Definability bounds of first order theories of algebraically closed fields* (P 2935) FCT'79 Fund of Comput Th;1979 Berlin/Wendisch-Rietz 160-166
⋄ B25 C40 C60 D10 D15 ⋄ REV MR 81j:03064 Zbl 439 # 03003 • ID 55995

HEINTZ, JOOS [1983] *Definability and fast quantifier elimination in algebraically closed fields* (J 1426) Theor Comput Sci 24*239-277 • ERR/ADD ibid 39*343
⋄ B25 C10 C40 C60 D10 D15 ⋄ REV MR 85a:68062 Zbl 546 # 03017 • ID 38863

HEINTZ, JOOS see Vol. IV for further entries

HELLER, A. [1980] see BAUMSLAG, G.

HELLER, A. see Vol. IV for further entries

HELLING, M. [1966] *Model-theoretic problems for some extensions of first-order languages* (0000) Diss., Habil. etc
⋄ C55 C75 C80 E55 ⋄ REM Diss., University of California, Berkeley • ID 20920

HELMER, O. [1938] *Languages with expressions of infinite length* (J 0748) Erkenntnis (Leipzig) 7*138-141
⋄ C75 ⋄ REV JSL 4.25 FdM 64.930 • ID 05884

HELMER, O. see Vol. I, II, VI for further entries

HENDRY, H.E. [1980] *Two remarks on the atomistic calculus of individuals* (J 0097) Nous, Quart J Phil 14*235-237
⋄ C10 G05 ⋄ REV MR 82b:03109 • ID 74015

HENDRY, H.E. [1982] *Complete extensions of the calculus of individuals* (J 0097) Nous, Quart J Phil 16*453-460
⋄ C35 G05 ⋄ REV MR 83m:03042 • ID 35444

HENDRY, H.E. see Vol. I, II, VI for further entries

HENKIN, L. [1949] *The completeness of the first-order functional calculus* (J 0036) J Symb Logic 14*159-166
• REPR [1969] (C 0569) Phil of Math Oxford Readings 42-50
⋄ B10 C07 ⋄ REV MR 11.487 Zbl 34.6 JSL 15.68 • ID 05892

HENKIN, L. [1950] *Completeness in the theory of types* (J 0036) J Symb Logic 15*81-91
• REPR [1969] (C 0569) Phil of Math Oxford Readings 51-63
⋄ A05 B15 C85 F35 ⋄ REV MR 12.70 Zbl 39.8 JSL 16.72 • ID 05894

HENKIN, L. [1953] *Some interconnections between modern algebra and mathematical logic* (J 0064) Trans Amer Math Soc 74*410-427
⋄ C07 C60 C85 ⋄ REV MR 14.1052 Zbl 50.6 JSL 20.183 • ID 05898

HENKIN, L. [1954] *A generalization of the notion of ω-consistency* (J 0036) J Symb Logic 19*183-196
⋄ B10 C07 F30 ⋄ REV MR 16.103 Zbl 56.11 JSL 23.40 • ID 05900

HENKIN, L. [1955] *On a theorem of Vaught* (J 0028) Indag Math 17*326-328
⋄ B25 C35 ⋄ REV MR 16.1080 Zbl 68.12 JSL 24.58 • ID 05904

HENKIN, L. [1956] *La structure algebrique des theories mathematiques* (X 0834) Gauthier-Villars: Paris 52pp
⋄ C07 F50 G05 G10 G15 ⋄ REV MR 18.272 Zbl 87.250 JSL 22.215 • ID 05908

HENKIN, L. [1956] *Two concepts from the theory of models* (J 0036) J Symb Logic 21*28-32
⋄ C13 C52 C60 ⋄ REV MR 17.816 Zbl 71.7 JSL 27.95 • ID 05909

HENKIN, L. [1957] *A generalization of the concept of ω-completeness* (J 0036) J Symb Logic 22*1-14
⋄ B10 C07 F30 ⋄ REV MR 20#1626 Zbl 81.12 JSL 24.172 • ID 05910

HENKIN, L. [1957] *Sums of squares* (P 1675) Summer Inst Symb Log;1957 Ithaca 284-291
⋄ C57 C60 ⋄ REV JSL 31.128 • ID 05913

HENKIN, L. & SUPPES, P. & TARSKI, A. (EDS.) [1959] *The axiomatic method, with special reference to geometry and physics* (X 0809) North Holland: Amsterdam xi+488pp
⋄ B30 B97 C65 D35 ⋄ REV Zbl 88.244 • ID 22387

HENKIN, L. [1961] *Some remarks on infinitely long formulas* (P 0633) Infinitist Meth;1959 Warsaw 167-183
⋄ C75 C80 ⋄ REV MR 26#1244 Zbl 121.253 JSL 30.96 • ID 05917

HENKIN, L. [1963] *An extension of the Craig-Lyndon interpolation theorem* (J 0036) J Symb Logic 28*201-216
⋄ C40 ⋄ REV MR 30#1927 Zbl 219#02006 JSL 30.98 • ID 05925

HENKIN, L. [1965] see ADDISON, J.W.

HENKIN, L. [1971] *Mathematical foundations for mathematics* (J 0005) Amer Math Mon 78*463-487
⋄ A10 C98 E30 E55 F55 G05 G15 ⋄ REV MR 44#5 Zbl 217.6 JSL 39.333 • ID 05928

HENKIN, L. [1974] see ADDISON, J.W.

HENKIN, L. [1977] *The logic of equality* (J 0005) Amer Math Mon 84*597-612
⋄ B20 C05 ⋄ REV MR 57#12345 Zbl 376#02017 • ID 27268

HENKIN, L. see Vol. I, II, V for further entries

HENLE, J.M. [1984] *An extravagant partition relation for a model of arithmetic* (P 3823) Axiomatic Set Th;1983 Boulder 109-113
⋄ C62 E05 E07 E25 E35 E40 ⋄ REV MR 86g:03080 Zbl 549#03027 • ID 42672

HENLE, J.M. & KLEINBERG, E.M. & WATRO, R.J. [1984] *On the ultrafilters and ultrapowers of strong partition cardinals* (J 0036) J Symb Logic 49*1268-1272
⋄ C20 C55 E05 E60 ⋄ REV MR 86g:03076 • ID 39876

HENLE, J.M. see Vol. I, V for further entries

HENNESSY, M. [1980] *A proof system for the first-order relational calculus* (J 0119) J Comp Syst Sci 20*96-110
⋄ B35 C07 ⋄ REV MR 81h:03023 Zbl 431#68084 • ID 53954

HENNESSY, M. see Vol. VI for further entries

HENRARD, P. [1971] *Une theorie sans modele generique* (J 2313) C R Acad Sci, Paris, Ser A-B 272*A293-A294
⋄ C25 ⋄ REV MR 44#48 Zbl 205.307 • ID 26192

HENRARD, P. [1972] *Forcing with infinite conditions* (P 2080) Conf Math Log;1970 London 343-345
⋄ C25 ⋄ REV Zbl 228#02031 • ID 29127

HENRARD, P. [1973] *Classes of cotheories* (P 0710) Russell Mem Logic Conf;1971 Uldum 217-220
⋄ C52 ⋄ REV MR 54#4967 • ID 24135

HENRARD, P. [1973] *Le "forcing-compagnon" sans "forcing"* (J 2313) C R Acad Sci, Paris, Ser A-B 276*A821-A822
⋄ C25 ⋄ REV MR 47#3165 Zbl 314#02059 • ID 05938

HENRARD, P. [1975] *Weak elimination of quantifiers and cotheories* (P 0775) Logic Colloq;1973 Bristol 395-398
⋄ C10 C52 ⋄ REV MR 52#76 Zbl 315#02049 • ID 18189

HENRARD, P. (ED.) [1977] *Six days of model theory. Proceedings of a conference held at Louvain-la-Neuve in March 1975* (X 2706) Ed Castella: Albeuve 275pp
⋄ C97 ⋄ REV MR 57#9453 Zbl 391#00002 • ID 52327

HENRICI, P. [1969] see DEJON, B.

HENRIKSEN, M. [1955] see ERDOES, P.

HENRY, DESMOND PAUL [1954] *Expressions trivially decidable* (J 0093) J Comp Syst 1*221-224
⋄ B25 ⋄ REV MR 16.555 Zbl 57.8 JSL 23.63 • ID 05940

HENRY, DESMOND PAUL see Vol. I for further entries

HENSEL, G. & PUTNAM, H. [1969] *Normal models and the field Σ_1^** (J 0027) Fund Math 64*231-240
⋄ C07 C30 C57 D55 ⋄ REV MR 39#5357 Zbl 193.302 • ID 05948

HENSEL, G. see Vol. IV, VI for further entries

HENSON, C.W. [1971] *A family of countable homogeneous graphs* (J 0048) Pac J Math 38*69-83
⋄ C13 C15 C50 C65 ⋄ REV MR 46#3377 Zbl 204.241 • ID 05951

HENSON, C.W. [1972] *Countable homogeneous relational structures and $\aleph_0$-categorical theories* (J 0036) J Symb Logic 37*494-500
⋄ C10 C15 C35 C50 D35 ⋄ REV MR 48#94 Zbl 259#02040 • ID 05954

HENSON, C.W. [1972] *The nonstandard hulls of a uniform space* (J 0048) Pac J Math 43*115-137
⋄ C65 H05 ⋄ REV MR 47 # 2559 Zbl 245 # 54046
• ID 05952

HENSON, C.W. & MOORE JR., L.C. [1972] *The nonstandard theory of topological vector spaces* (J 0064) Trans Amer Math Soc 172*405-435 • ERR/ADD ibid 184*509
⋄ C65 H05 ⋄ REV MR 46 # 7836 Zbl 254 # 46001
• ID 22286

HENSON, C.W. & MOORE JR., L.C. [1973] *Invariance of the nonstandard hulls of locally convex spaces* (J 0025) Duke Math J 40*193-206
⋄ C65 H05 ⋄ REV MR 51 # 8775 Zbl 256 # 46001
• ID 05956

HENSON, C.W. & MOORE JR., L.C. [1974] *Invariance of the nonstandard hulls of a uniform space* (P 1083) Victoria Symp Nonstand Anal;1972 Victoria 85-98
⋄ C65 H05 ⋄ REV MR 58 # 2784 Zbl 272 # 54040
• ID 26633

HENSON, C.W. & MOORE JR., L.C. [1974] *Nonstandard hulls of the classical Banach spaces* (J 0025) Duke Math J 41*277-284
⋄ C65 H05 ⋄ REV MR 51 # 3867 Zbl 298 # 46028
• ID 05962

HENSON, C.W. & MOORE JR., L.C. [1974] *Semi-reflexivity of the nonstandard hulls of a locally convex space* (P 1083) Victoria Symp Nonstand Anal;1972 Victoria 71-84
⋄ C65 H05 ⋄ REV MR 57 # 12159 Zbl 276 # 46006
• ID 26632

HENSON, C.W. & MOORE JR., L.C. [1974] *Subspaces of the nonstandard hull of a normed space* (J 0064) Trans Amer Math Soc 197*131-143
⋄ C65 H05 ⋄ REV MR 51 # 1351 Zbl 307 # 46008
• ID 05960

HENSON, C.W. [1974] *The isomorphism property in nonstandard analysis and its use in the theory of Banach spaces* (J 0036) J Symb Logic 39*717-731
⋄ C65 H05 ⋄ REV MR 50 # 12713 Zbl 306 # 02055
• ID 05959

HENSON, C.W. [1975] *The monad system of the finest compatible uniform structure* (J 0053) Proc Amer Math Soc 51*163-170
⋄ C65 H05 ⋄ REV MR 51 # 6779 Zbl 307 # 54023
• ID 17439

HENSON, C.W. [1975] *When do two Banach spaces have isometrically isomorphic nonstandard hulls ?* (J 0029) Israel J Math 22*57-67
⋄ C65 H05 ⋄ REV MR 52 # 6386 Zbl 314 # 46023
• ID 05963

HENSON, C.W. [1976] *Nonstandard hulls of Banach spaces* (J 0029) Israel J Math 25*108-144
⋄ C65 H05 ⋄ REV MR 57 # 1089 Zbl 348 # 46014
• ID 26078

HENSON, C.W. [1976] *Ultraproducts of Banach spaces* (C 4232) Altgeld Book
⋄ C20 C65 H05 ⋄ ID 46653

HENSON, C.W. & JOCKUSCH JR., C.G. & RUBEL, L.A. & TAKEUTI, G. [1977] *First-order topology* (J 0202) Diss Math (Warsaw) 143*40pp
⋄ B25 C65 C75 D35 H05 ⋄ REV MR 55 # 5434 Zbl 399 # 03019 • ID 30718

HENSON, C.W. [1980] see BECKER, J.A.

HENSON, C.W. [1983] see HEINRICH, S.

HENSON, C.W. & MOORE JR., L.C. [1983] *Nonstandard analysis and the theory of Banach spaces* (C 3884) Nonstandard Anal - Recent Develop 27-112
⋄ C65 H05 ⋄ REV MR 85f:46033 Zbl 511 # 46070
• ID 38543

HENSON, C.W. & RUBEL, L.A. [1984] *Some applications of Nevanlinna theory to mathematical logic:identities of exponential functions* (J 0064) Trans Amer Math Soc 282*1-32 • ERR/ADD ibid 294*381
⋄ B20 C65 F30 ⋄ REV MR 85h:03015 Zbl 533 # 03015
• ID 36539

HENSON, C.W. & KAUFMANN, M. & KEISLER, H.J. [1984] *The strength of nonstandard methods in arithmetic* (J 0036) J Symb Logic 49*1039-1058
⋄ C62 E30 F30 F35 H05 H15 ⋄ REV MR 86h:03115
• ID 39860

HENSON, C.W. see Vol. I, V for further entries

HERBRAND, J. [1930] *Recherches sur la theorie de la demonstration* (J 0459) C R Soc Sci Lett Varsovie Cl 3 33*128pp
• TRANSL [1967] (C 0675) From Frege to Goedel 525-581 [1971] (C 0745) Herbrand: Log Writings 46-188,272-276
• REPR [1968] (C 2486) Herbrand: Ecrits Logiques 35-153 [1931] (J 1092) Ann Univ Paris 6*186-189
⋄ B10 B25 F05 F07 F25 F30 ⋄ REV JSL 36.523 JSL 40.94 FdM 56.824 • ID 20866

HERBRAND, J. [1931] *Sur le probleme fondamental de la logique mathematique* (J 0459) C R Soc Sci Lett Varsovie Cl 3 24*12-56
• TRANSL [1971] (C 0745) Herbrand: Log Writings 215-271
⋄ A05 B25 B30 F30 F99 ⋄ REV Zbl 3.290 JSL 36.523 FdM 57.1320 • ID 16812

HERBRAND, J. see Vol. I, VI for further entries

HERDEN, G. [1983] see DUGAS, M.

HERFORT, W.N. & MANEVITZ, L.M. [1985] *Topological Frobenius groups* (J 0032) J Algeb 92*16-32
⋄ C60 H20 ⋄ ID 44483

HERMANN, G. [1926] *Die Frage der endlich vielen Schritte in der Theorie der Polynomideale* (J 0043) Math Ann 95*736-788
⋄ C57 C60 D45 F55 ⋄ REV FdM 52.127 • ID 05997

HERMES, H. [1955] *Vorlesung ueber Entscheidungsprobleme in Mathematik und Logik* (X 0910) Aschendorffsche Verlagsbuchh: Muenster ii + 140pp
⋄ B25 B40 B65 D10 D20 D35 D98 ⋄ REV MR 17.569 Zbl 67.249 • ID 06008

HERMES, H. [1967] *Ein mengentheoretisches Modell des Peanoschen Axiomensystems* (J 0487) Math Unterricht 13/3*5-31
⋄ C62 ⋄ ID 32177

HERMES, H. see Vol. I, II, IV, VI for further entries

HERNANDEZ, R. [1983] *Espaces L^p, factorisation et produits tensoriels dans les espaces de Banach* (J 3364) C R Acad Sci, Paris, Ser 1 296*385-388
⋄ C20 ⋄ REV MR 84e:46014 Zbl 552 # 46039 • ID 43436

HERRE, H. & RAUTENBERG, W. [1970] *Das Basistheorem und einige Anwendungen in der Modelltheorie* (J 0115) Wiss Z Humboldt-Univ Berlin, Math-Nat Reihe 19∗579-583
⋄ C40 C52 G05 ⋄ REV MR 48 # 10796 Zbl 275 # 02049 • ID 06024

HERRE, H. [1971] *Entscheidungsprobleme in der elementaren Theorie einer zweistelligen Relation* (J 0068) Z Math Logik Grundlagen Math 17∗301-313
⋄ B25 C10 ⋄ REV MR 45 # 1749 Zbl 255 # 02051 • ID 06026

HERRE, H. [1972] *Die Entscheidbarkeit der elementaren Theorie der n-separierten symmetrischen Graphen endlicher Valenz* (J 0068) Z Math Logik Grundlagen Math 18∗249-254
⋄ B25 C65 ⋄ REV MR 46 # 3289 Zbl 273 # 02031 • ID 06027

HERRE, H. [1972] see HAUSCHILD, K.

HERRE, H. [1972] see HAUCK, J.

HERRE, H. [1974] *Nonfinitely axiomatizable theories of graphs* (J 0014) Bull Acad Pol Sci, Ser Math Astron Phys 22∗1187-1190
⋄ C52 C65 ⋄ REV MR 51 # 119 Zbl 325 # 02035 • ID 06029

HERRE, H. & WOLTER, H. [1975] *Entscheidbarkeit von Theorien in Logiken mit verallgemeinerten Quantoren* (J 0068) Z Math Logik Grundlagen Math 21∗229-246
⋄ B25 C10 C55 C80 D35 ⋄ REV MR 53 # 2623 Zbl 318 # 02049 • ID 14283

HERRE, H. [1977] *Decidability of theories in logics with additional monadic quantifiers* (P 1629) Symp Math Log;1974 Oulo;1975 Helsinki 77-80
⋄ B25 C55 C80 ⋄ ID 48399

HERRE, H. & WOLTER, H. [1977] *Entscheidbarkeit der Theorie der linearen Ordnung in L_{Q_1}* (J 0068) Z Math Logik Grundlagen Math 23∗273-282
⋄ B25 C55 C65 C80 ⋄ REV MR 56 # 5260 Zbl 368 # 02015 • ID 26487

HERRE, H. & WOLTER, H. [1978] *Entscheidbarkeit der Theorie der linearen Ordnung in L_{Q_K} fuer regulaeres ω_K* (J 0068) Z Math Logik Grundlagen Math 24∗73-78
⋄ B25 C55 C65 C80 E07 E50 ⋄ REV MR 57 # 16034 Zbl 382 # 03011 • ID 51935

HERRE, H. & PINUS, A.G. [1978] *Zum Entscheidungsproblem fuer Theorien in Logiken mit monadischen verallgemeinerten Quantoren* (J 0068) Z Math Logik Grundlagen Math 24∗375-384
⋄ B25 C10 C55 C80 D35 ⋄ REV MR 58 # 16246 Zbl 397 # 03010 • ID 52676

HERRE, H. [1979] see BEKENOV, M.I.

HERRE, H. & WOLTER, H. [1979] *Entscheidbarkeit der Theorie der linearen Ordnung in Logiken mit Maechtigkeitsquantoren bzw. mit Chang-Quantor* (J 0068) Z Math Logik Grundlagen Math 25∗345-358
⋄ B25 C55 C80 E07 E50 E55 ⋄ REV MR 81a:03010 Zbl 429 # 03008 • ID 53839

HERRE, H. & WOLTER, H. [1979] *The decision problem for the theory of linear orderings in extended logics* (P 2539) Frege Konferenz (1);1979 Jena 162-165
⋄ B25 C55 C80 C98 ⋄ REV MR 82c:03055 • ID 74070

HERRE, H. [1979] *Transzendente Theorien* (X 2888) ZI Math Mech Akad Wiss DDR: Berlin 23pp
⋄ C45 C50 E75 ⋄ REV Zbl 402 # 03032 • ID 54679

HERRE, H. & SEESE, D.G. [1980] *Concerning the monadic theory of the topology of well-orderings and scattered spaces* (J 3293) Bull Acad Pol Sci, Ser Math 28∗1-6
⋄ B25 C55 C65 C85 E07 E35 E50 E55 ⋄ REV MR 82f:03030 Zbl 467 # 03009 • ID 55007

HERRE, H. [1980] *Modelltheoretische Eigenschaften endlichvalenter Graphen* (J 0068) Z Math Logik Grundlagen Math 26∗51-58
⋄ C35 C45 C50 C65 ⋄ REV MR 81k:03031 Zbl 445 # 03015 • ID 56479

HERRE, H. [1981] *Miscellaneous results and problems in extended model theory* (P 2623) Worksh Extended Model Th;1980 Berlin 20-65
⋄ B25 C65 C80 C98 D35 ⋄ REV MR 84h:03095 Zbl 522 # 03026 • ID 34285

HERRE, H. & WOLTER, H. [1981] *Untersuchungen zur Theorie der linearen Ordnung in Logiken mit Maechtigkeitsquantoren* (J 0068) Z Math Logik Grundlagen Math 27∗73-94
⋄ B25 C55 C65 C80 E07 ⋄ REV MR 82j:03008 Zbl 469 # 03005 • ID 55133

HERRE, H. (ED.) [1981] *Workshop on extended model theory* (X 2655) Akad Wiss DDR Inst Math: Berlin ii + 160pp
⋄ C97 G05 ⋄ REV MR 82m:03006 • ID 70002

HERRE, H. & MEKLER, A.H. & SMITH, KENNETH W. [1983] *Superstable graphs* (J 0027) Fund Math 118∗75-79
⋄ C45 C65 ⋄ REV MR 85i:03102 • ID 44192

HERRE, H. see Vol. IV for further entries

HERRERA MIRANDA, J. [1978] *Les theories convexes de Horn (English summary)* (J 2313) C R Acad Sci, Paris, Ser A-B 287∗A593-A594
⋄ B20 C05 C52 G30 ⋄ REV MR 83k:03038 Zbl 407 # 03038 • ID 36172

HERRERA MIRANDA, J. [1984] *Sur la logique de premier ordre et la theorie des categories* (J 0307) Rev Colomb Mat 18∗41-82
⋄ C90 G30 ⋄ REV Zbl 555 # 03033 • ID 44709

HERRING, J.M. [1971] see CORCORAN, J.

HERRING, J.M. [1972] see CORCORAN, J.

HERRING, J.M. [1976] *Equivalence of several notions of theory completeness in a free logic* (J 0302) Rep Math Logic, Krakow & Katowice 6∗87-91
⋄ B20 C35 ⋄ REV MR 57 # 9517 Zbl 384 # 03005 • ID 21916

HERRMANN, C. & POGUNTKE, W. [1974] *The class of sublattices of normal subgroup lattices is not elementary* (J 0004) Algeb Universalis 4∗280-286
⋄ C60 G10 ⋄ REV MR 50 # 6936 Zbl 303 # 20031 • ID 06035

HERRMANN, C. & JENSEN, C.U. & LENZING, H. [1981] *Applications of model theory to representations of finite-dimensional algebras* (J 0044) Math Z 178∗83-98
⋄ C60 ⋄ REV MR 83c:16023 Zbl 445 # 16020 • ID 54500

HERRMANN, C. see Vol. IV for further entries

HERRMANN, E. [1976] *On Lindenbaum functions of* $\aleph_0$-*categorical theories of finite similarity type* (J 0014) Bull Acad Pol Sci, Ser Math Astron Phys 24*17-21
◊ B25 C35 C57 ◊ REV MR 53 # 7758 Zbl 333 # 02042
• ID 18190

HERRMANN, E. [1977] *Ueber Lindenbaumfunktionen von* $\aleph_0$-*kategorischen Theorien endlicher Signatur* (J 0115) Wiss Z Humboldt-Univ Berlin, Math-Nat Reihe 26*637-646
◊ B25 C35 C57 ◊ REV MR 80d:03028 Zbl 423 # 03005
• ID 53517

HERRMANN, E. [1978] *Der Verband der rekursiv aufzaehlbaren Mengen (Entscheidungs-Problem) (Russian and English summaries)* (S 3382) Sem-ber, Humboldt-Univ Berlin, Sekt Math 10*v+304pp
◊ B25 D25 D98 ◊ REV MR 81e:03041 Zbl 415 # 03029
• ID 53131

HERRMANN, E. & WOLTER, H. [1980] *Untersuchungen zu schwachen Logiken der zweiten Stufe* (J 0068) Z Math Logik Grundlagen Math 26*59-68
◊ B15 C15 C85 ◊ REV MR 81k:03013 Zbl 445 # 03019
• ID 56483

HERRMANN, E. see Vol. IV for further entries

HERRMANN, R.A. [1978] *The nonstandard theory of semi-uniform spaces* (J 0068) Z Math Logik Grundlagen Math 24*237-256
◊ C65 H05 H20 ◊ REV MR 58 # 12992 Zbl 468 # 54037
• ID 55117

HERRMANN, R.A. see Vol. I for further entries

HESSE, G. [1978] see GALVIN, F.

HESSE, G. see Vol. V for further entries

HEWITT, E. [1948] *Rings of real-valued continuous functions* (J 0064) Trans Amer Math Soc 64*45-99
◊ C20 C50 C60 C65 E75 ◊ REV MR 10.126 Zbl 32.286
• REM Part I. Part II 1981 by Antonovskij,M.Ya. & Chudnovskij,D.V. & Chudnovskij,G.V. & Hewitt,E.
• ID 42963

HEWITT, E. [1981] see ANTONOVSKIJ, M.YA.

HEWITT, E. see Vol. V for further entries

HEYTING, A. [1973] *Address to Professor A. Robinson* (J 3077) Nieuw Arch Wisk, Ser 3 21*134-137
◊ A10 C98 ◊ REV MR 55 # 7720 • ID 73999

HEYTING, A. see Vol. I, II, IV, VI for further entries

HIBINO, K. [1984] see EDA, K.

HICKIN, K.K. & PLOTKIN, J.M. [1972] *On the equivalence of three local theorem techniques* (J 0053) Proc Amer Math Soc 35*389-392
◊ C60 ◊ REV MR 46 # 1908 Zbl 258 # 20038 • ID 90233

HICKIN, K.K. [1973] *Countable type local theorems in algebra* (P 0779) Conf Group Th;1972 Racine 123-125
◊ C05 C52 ◊ REV MR 51 # 7989 Zbl 285 # 08005
• ID 17583

HICKIN, K.K. [1973] *Countable type local theorems in algebra* (J 0032) J Algeb 27*523-537
◊ C05 C52 C60 ◊ REV MR 49 # 5179 Zbl 275 # 08005
• ID 29727

HICKIN, K.K. & PHILLIPS, R.E. [1973] *Local theorems and group extensions* (J 0171) Proc Cambridge Phil Soc Math Phys 73*7-20
◊ C60 ◊ REV MR 46 # 9167 Zbl 248 # 20038 • ID 90231

HICKIN, K.K. & PLOTKIN, J.M. [1975] *Compactness and* ρ *-valued logics* (P 1805) Int Symp Multi-Val Log (5,Proc);1975 Bloomington 400-405
◊ B50 C75 C90 ◊ REV MR 58 # 10299 • ID 28362

HICKIN, K.K. & PLOTKIN, J.M. [1977] *Some algebraic properties of weakly compact and compact cardinals* (J 0027) Fund Math 97*177-185
◊ C05 C55 E55 ◊ REV MR 56 # 15424 Zbl 365 # 02058
• ID 27199

HICKIN, K.K. [1978] *Complete universal locally finite groups* (J 0064) Trans Amer Math Soc 239*213-227
◊ C50 C60 ◊ REV MR 58 # 902 Zbl 386 # 20014
• ID 52213

HICKIN, K.K. & PHILLIPS, R.E. [1979] *A construction of locally finite simple groups* (J 0008) Arch Math (Basel) 32*97-100
◊ C30 C60 ◊ REV MR 84k:20014 Zbl 394 # 20024
• ID 90232

HICKIN, K.K. & MACINTYRE, A. [1980] *Algebraically closed groups: embeddings and centralizers* (P 2634) Word Problems II;1976 Oxford 141-155
◊ C25 C60 D40 ◊ REV MR 82c:20065 Zbl 444 # 20025
• ID 81551

HICKIN, K.K. & PHILLIPS, R.E. [1983] *Isomorphism types in wreath products and effective embeddings of periodic groups* (J 0064) Trans Amer Math Soc 277*765-778
◊ C57 C60 D40 ◊ REV MR 85i:20034 Zbl 516 # 20015
• ID 38580

HICKIN, K.K. see Vol. IV, V for further entries

HICKMAN, J.L. & NEUMANN, B.H. [1975] *A question of Babai on groups* (J 0016) Bull Austral Math Soc 13*355-368
◊ C60 E25 E35 E75 ◊ REV MR 53 # 3129 Zbl 314 # 20030 • ID 21655

HICKMAN, J.L. [1976] *The construction of groups in models of set theory that fail the axiom of choice* (J 0016) Bull Austral Math Soc 14*199-232
◊ C60 E25 E35 ◊ REV MR 53 # 12940 Zbl 324 # 02055
• ID 23189

HICKMAN, J.L. [1982] *Automorphisms of medial fields* (J 0068) Z Math Logik Grundlagen Math 28*263-267
◊ C07 C60 E25 E75 ◊ REV MR 84c:12019 Zbl 501 # 03037 • ID 36952

HICKMAN, J.L. see Vol. V, VI for further entries

HIEN, BUIHUY & NEMETI, I. [1981] *Problems with the category theoretic notions of ultraproducts* (J 0387) Bull Sect Logic, Pol Acad Sci 10*122-127
◊ C20 G30 ◊ REV MR 83a:18004 Zbl 489 # 03026
• ID 37210

HIEN, BUIHUY [1982] *On a problem in algebraic model theory* (J 0387) Bull Sect Logic, Pol Acad Sci 11*103-108
◊ C20 G30 ◊ REV MR 85d:03128 Zbl 531 # 03043
• ID 40281

HIEN, BUIHUY & SAIN, I. [1983] *Category theoretic notions of ultraproducts* (J 0411) Studia Sci Math Hung 18*309-317
◊ C20 C90 G30 ◊ REV Zbl 574 # 18002 • ID 45605

HIEN, BUIHUY & SAIN, I. [1983] *In which categories are first-order axiomatizable hulls characterizable by ultraproducts?* (**J** 0306) Cah Topol & Geom Differ 24*215-222
- ◊ C20 C90 G30 ◊ REV MR 85d:03081 Zbl 519 # 18003
- ID 36725

HIEN, BUIHUY [1984] *Elementary classes in the injective subcategories approach to abstract model theory* (**J** 0049) Period Math Hung 15*87-99
- ◊ C20 C90 G30 ◊ REV MR 86b:03044 • ID 45150

HIGMAN, G. & NEUMANN, B.H. & NEUMANN, H. [1949] *Embedding theorems for groups* (**J** 0039) J London Math Soc 24*247-254
- ◊ C60 D40 ◊ REV MR 11.322 Zbl 34.301 • ID 90239

HIGMAN, G. [1952] *Ordering by divisibility in abstract algebras* (**J** 3240) Proc London Math Soc, Ser 3 2*326-336
- ◊ C05 C65 ◊ REV MR 14.238 Zbl 47.34 • ID 20911

HIGMAN, G. [1961] *Subgroups of finitely presented groups* (**J** 1150) Proc Roy Soc London, Ser A 262*455-475
- ◊ C60 D25 D45 ◊ REV MR 24 # A152 Zbl 104.21 JSL 29.204 • ID 22083

HIGMAN, G. [1977] *Homogeneous relations* (**J** 0131) Quart J Math, Oxford Ser 2 28(2)*31-39
- ◊ C07 C30 C65 E05 ◊ REV MR 55 # 3090 Zbl 349 # 20017 • ID 81557

HIGMAN, G. see Vol. IV, V for further entries

HILBERT, D. & ACKERMANN, W. [1928] *Grundzuege der theoretischen Logik* (**X** 0811) Springer: Heidelberg & New York viii+120pp
- TRANSL [1950] (**X** 0848) Chelsea: New York xii+172pp (English) [1950] (**X** 1876) Kexue Chubanshe: Beijing
- ◊ B25 B98 D35 ◊ REV MR 50 # 4230 Zbl 239 # 02001 JSL 15.59 JSL 16.52 JSL 25.158 JSL 3.83 FdM 54.55
- REM 4th ed. 1959;viii+188pp. • ID 00107

HILBERT, D. see Vol. I, II, IV, V, VI for further entries

HILL, P. [1971] *Classes of abelian groups closed under taking subgroups and direct limits* (**J** 0004) Algeb Universalis 1*63-70
- ◊ C30 C60 ◊ REV MR 45 # 2016 Zbl 236 # 20042
- ID 06089

HILL, P. [1973] *New criteria for freeness in abelian groups* (**J** 0064) Trans Amer Math Soc 182*201-209
- ◊ C60 C75 ◊ REV MR 48 # 4151 Zbl 275 # 20096
- ID 30475

HINGSTON, P. [1981] *Effective decomposition in Noetherian rings* (**P** 2902) Aspects Effective Algeb;1979 Clayton 122-127
- ◊ C57 C60 D45 ◊ REV MR 83a:03040 Zbl 473 # 03036
- ID 55365

HINMAN, P.G. & ZACHOS, S. [1985] *Probabilistic machines, oracles, and quantifiers* (**P** 3342) Rec Th Week;1984 Oberwolfach 159-192
- ◊ C80 D15 ◊ ID 45303

HINMAN, P.G. see Vol. IV, V, VI for further entries

HINNION, R. [1972] *Sur les modeles de NF ("New Foundations" de Quine)* (**J** 2313) C R Acad Sci, Paris, Ser A-B 275*A567
- ◊ C62 E70 ◊ REV MR 46 # 5131 Zbl 257 # 02052
- ID 06109

HINNION, R. [1976] *Modeles de fragments de la theorie des ensembles de Zermelo-Fraenkel dans les "New foundations" de Quine (English summary)* (**J** 2313) C R Acad Sci, Paris, Ser A-B 282*A1-A3
- ◊ C62 E30 E35 E70 ◊ REV MR 53 # 7781 Zbl 324 # 02056 • ID 23019

HINNION, R. [1979] *Modele constructible de la theorie des ensembles de Zermelo dans la theorie des types* (**J** 3133) Bull Soc Math Belg, Ser B 31*3-11
- ◊ B15 C62 E30 E35 E45 ◊ REV MR 82e:03048 Zbl 439 # 03032 • ID 56023

HINNION, R. [1980] *Contraction de structures et application a NFU (les "New Foundations" des Quine avec extensionnalite pour les ensembles non vides). Definition du "degre de nonextensionnalite" d'une relation quelconque (English summary)* (**J** 2313) C R Acad Sci, Paris, Ser A-B 290*A677-A680
- ◊ C30 C62 E35 E70 ◊ REV MR 81i:03082 Zbl 468 # 03035 • ID 55100

HINNION, R. [1981] *Extensional quotients of structures and applications to the study of the axiom of extensionality* (**J** 3133) Bull Soc Math Belg, Ser B 33*173-206
- ◊ C30 C62 E30 E35 E70 ◊ REV MR 84c:03089 Zbl 484 # 03029 • ID 34023

HINNION, R. see Vol. V for further entries

HINTIKKA, K.J.J. [1953] *Distributive normal forms in the calculus of predicates* (**J** 0096) Acta Philos Fenn 6*71pp
- ◊ B10 C07 ◊ REV MR 16.1079 Zbl 50.246 JSL 20.75
- ID 06111

HINTIKKA, K.J.J. [1955] *Reductions in the theory of types* (**J** 0096) Acta Philos Fenn 8*57-115
- ◊ B15 C85 ◊ REV MR 17.119 Zbl 67.2 JSL 31.660
- ID 06112

HINTIKKA, K.J.J. [1964] *Distributive normal forms and deductive interpolation* (**J** 0068) Z Math Logik Grundlagen Math 10*185-191
- ◊ B10 C07 C40 ◊ REV MR 34 # 5641 JSL 31.267
- ID 06122

HINTIKKA, K.J.J. [1965] *Distributive normal forms in first-order logic* (**P** 0688) Logic Colloq;1963 Oxford 48-91
- TRANSL [1980] (**C** 4675) Hintikka: Logiko-Epist Issled 105-157
- ◊ B10 C07 ◊ REV MR 35 # 2726 JSL 31.267 • ID 06123

HINTIKKA, K.J.J. & TUOMELA, R. [1970] *Toward a general theory of auxiliary concepts and definability in first-order theories* (**P** 0785) Scand Logic Symp (1);1968 Aabo 19-60
- REPR [1970] (**C** 1203) Inform & Infer 298-330
- ◊ A05 C40 ◊ REV MR 43 # 7280 MR 58 # 16098 Zbl 337 # 02030 • ID 74176

HINTIKKA, K.J.J. [1972] *Constituents and finite identifiability* (**J** 0122) J Philos Logic 1*45-52
- ◊ C40 ◊ REV MR 46 # 7014 Zbl 231 # 02062 • ID 06127

HINTIKKA, K.J.J. [1973] *Logic, language-games and information. Kantian themes in the philosophy of logic* (**X** 0815) Clarendon Pr: Oxford x+291pp
- ◊ A05 C07 ◊ REV MR 54 # 4911 Zbl 253 # 02005
- ID 24087

HINTIKKA, K.J.J. & NIINILUOTO, I. [1973] *On the surface semantics of quantificational proof procedures* (J 0963) Ajatus (Helsinki) 35*197-215
⋄ B35 C07 F07 ⋄ REV Zbl 291 # 02005 • ID 32405

HINTIKKA, K.J.J. [1974] *Quantifiers vs. quantification theory* (J 0076) Dialectica 27*329-358
• REPR [1979] (C 4695) Game-Th Semantics 49-79
⋄ A05 B10 C80 ⋄ REV Zbl 362 # 02008 JSL 51.240
• ID 50730

HINTIKKA, K.J.J. & RANTALA, V. [1975] *Systematizing definability theory* (P 0757) Scand Logic Symp (3);1973 Uppsala 40-62
⋄ C40 ⋄ REV MR 53 # 10578 Zbl 315 # 02047 JSL 43.373
• ID 23083

HINTIKKA, K.J.J. & RANTALA, V. [1976] *A new approach to infinitary languages* (J 0007) Ann Math Logic 10*95-115
⋄ C40 C75 E60 ⋄ REV MR 55 # 12462 Zbl 339 # 02013
• ID 18195

HINTIKKA, K.J.J. [1976] *Partially ordered quantifiers vs. partially ordered ideas* (J 0076) Dialectica 30*89-99
⋄ A05 C80 ⋄ REV Zbl 428 # 03006 • ID 53766

HINTIKKA, K.J.J. [1976] *Quantifiers in logic and quantifiers in natural languages* (P 1502) Phil of Logic;1976 Bristol 208-270
• REPR [1979] (C 4695) Game-Th Semantics 27-47
⋄ A05 B65 C80 ⋄ REV MR 54 # 7193 JSL 51.240
• ID 24984

HINTIKKA, K.J.J. [1977] *Quantifiers in natural languages: Some logical problems. II* (J 2130) Linguist Philos 1*153-172
• REPR [1979] (C 4695) Game-Th Semantics 81-117
⋄ A05 C80 ⋄ REV Zbl 366 # 02010 JSL 51.240 • REM Part I 1979. Reprinted together with part I. • ID 51093

HINTIKKA, K.J.J. [1979] *Quantifiers in natural languages: Some logical problems. I* (P 1705) Scand Logic Symp (4);1976 Jyvaeskylae 295-314
• REPR [1979] (C 4695) Game-Th Semantics 81-117
⋄ A05 B65 C80 ⋄ REV MR 80j:03008 Zbl 406 # 03007 JSL 51.240 • REM Part II 1977. Reprinted together with part II. • ID 56090

HINTIKKA, K.J.J. [1982] *Game-theoretical semantics: Insights and prospects* (J 0047) Notre Dame J Formal Log 23*219-241
⋄ A05 B60 C80 ⋄ REV MR 84a:03029a Zbl 464 # 03008
• ID 55406

HINTIKKA, K.J.J. see Vol. I, II, V, VI for further entries

HIRD, G.R. [1985] see DOWNEY, R.G.

HIROSE, K. & TAKAHASHI, S. [1979] *Sentences preserved by sheaves of structures (Japanese)* (P 4108) Found of Math;1979 Kyoto 157-165
⋄ C40 C90 G30 ⋄ ID 47665

HIROSE, K. see Vol. II, IV, V for further entries

HIRSCHELMANN, A. [1972] *An application of ultra-products to prime rings with polynomial identities* (P 2080) Conf Math Log;1970 London 255*145-148
⋄ C20 C60 ⋄ REV MR 49 # 343 Zbl 238 # 02051
• ID 19023

HIRSCHELMANN, A. see Vol. IV for further entries

HIRSCHFELD, J. [1972] see CHERLIN, G.L.

HIRSCHFELD, J. [1974] *Models of arithmetic and the semiring of recursive functions* (P 1083) Victoria Symp Nonstand Anal;1972 Victoria 369*99-105
⋄ C62 D20 H15 ⋄ REV MR 58 # 5189 Zbl 281 # 02056
• ID 21188

HIRSCHFELD, J. [1975] *Finite forcing and generic filters in arithmetic* (C 0782) Model Th & Algeb (A. Robinson) 172-199
⋄ C25 C60 C62 ⋄ REV MR 58 # 21577 Zbl 317 # 02062
• ID 29658

HIRSCHFELD, J. & WHEELER, W.H. [1975] *Forcing, arithmetic, division rings* (S 3301) Lect Notes Math 454*vii+266pp
⋄ C25 C60 C62 C98 ⋄ REV MR 52 # 10412
Zbl 304 # 02024 JSL 45.188 • ID 21708

HIRSCHFELD, J. [1975] *Models of arithmetic and recursive functions* (J 0029) Israel J Math 20*111-126
⋄ C62 D20 H15 ⋄ REV MR 52 # 2858 Zbl 311 # 02050
• ID 06142

HIRSCHFELD, J. [1975] *The model companion of ZF* (J 0053) Proc Amer Math Soc 50*369-374
⋄ C25 C62 E30 ⋄ REV MR 51 # 12508 Zbl 311 # 02068
• ID 17294

HIRSCHFELD, J. [1976] *Another approach to infinite forcing* (J 1934) Ann Sci Univ Clermont Math 13*81-86
⋄ C25 ⋄ REV MR 57 # 16045 Zbl 426 # 03033 • ID 53629

HIRSCHFELD, J. [1976] *Non standard analysis and the compactification of groups* (J 0029) Israel J Math 25*145-153
⋄ C60 H05 H20 ⋄ REV MR 58 # 22374 Zbl 348 # 22002
• ID 26079

HIRSCHFELD, J. [1978] *Examples in the theory of existential completeness* (J 0036) J Symb Logic 43*650-658
⋄ C25 C35 C50 ⋄ REV MR 81i:03043 Zbl 408 # 03025
• ID 31774

HIRSCHFELD, J. [1980] *Finite forcing, existential types and complete types* (J 0036) J Symb Logic 45*93-102
⋄ C25 C35 ⋄ REV MR 81b:03037 Zbl 436 # 03026
• ID 55866

HIRSCHFELD, J. [1980] *Generalized ultrapowers* (J 0009) Arch Math Logik Grundlagenforsch 20*13-26
⋄ C20 ⋄ REV MR 82b:03067 Zbl 453 # 03032 • ID 54162

HIRSCHFELD, J. see Vol. I, II, VI for further entries

HODES, H.T. [1984] *The modal theory of pure identity and some related decision problems* (J 0068) Z Math Logik Grundlagen Math 30*415-423
⋄ B25 B45 D35 ⋄ REV MR 86d:03018 Zbl 534 # 03007
• ID 41142

HODES, H.T. see Vol. II, IV, V for further entries

HODES, L. & SPECKER, E. [1968] *Lengths of formulas and elimination of quantifiers I* (P 0608) Logic Colloq;1966 Hannover 175-188
• TRANSL [1973] (J 3079) Kiber Sb Perevodov, NS 10*99-113
⋄ B35 C10 C40 ⋄ REV MR 39 # 32 MR 48 # 5714
Zbl 191.285 JSL 40.503 • ID 24939

HODES, L. [1970] *The logical complexity of geometric properties on the plane* (J 0037) ACM J 17*339-347
⋄ B35 C40 ⋄ REV MR 45 # 8527 Zbl 198.35 • ID 06168

HODGES, W. (ED.) [1972] *Conference in Mathematical Logic -- London '70* (S 3301) Lect Notes Math 255∗viii+351pp
◇ B97 C97 ◇ REV MR 48#5804 Zbl 227#02011
• ID 70226

HODGES, W. [1972] *On order-types of models* (J 0036) J Symb Logic 37∗69-70
◇ C30 C55 ◇ REV MR 48#95 Zbl 245#02051 • ID 06170

HODGES, W. [1973] *Models in which all long indiscernible sequences are indiscernible sets* (J 0027) Fund Math 78∗1-6
◇ C30 C55 ◇ REV MR 47#8285 Zbl 246#02036
• ID 06169

HODGES, W. [1974] *A normal form for algebraic constructions* (J 0161) Bull London Math Soc 6∗57-60
◇ C40 C75 E47 ◇ REV MR 51#10198 Zbl 276#02040
• REM Part I. Part II 1975 • ID 17513

HODGES, W. [1975] *A normal form for algebraic constructions II* (J 0079) Logique & Anal, NS 18∗429-487
• REPR [1977] (P 1625) Six Days of Model Th;1975 Louvain-la-Neuve 197-255
◇ C40 C75 E25 G30 ◇ REV MR 58#21903 Zbl 346#02029 Zbl 398#03015 • REM Part I 1974
• ID 30712

HODGES, W. [1976] *On the effectivity of some field constructions* (J 3240) Proc London Math Soc, Ser 3 32∗133-162
◇ C60 C75 D65 E35 E47 E75 ◇ REV MR 55#10252 Zbl 325#12105 • ID 30711

HODGES, W. & LACHLAN, A.H. & SHELAH, S. [1977] *Possible orderings of an indiscernible sequence* (J 0161) Bull London Math Soc 9∗212-215
◇ C30 E07 ◇ REV MR 57#16085 Zbl 361#20006
• ID 30713

HODGES, W. [1980] *Functorial uniform reducibility* (J 0027) Fund Math 108∗77-81
◇ C20 C40 C75 ◇ REV MR 82e:03030 Zbl 363#02058
• ID 55785

HODGES, W. [1980] *Interpreting number theory in nilpotent groups* (J 0009) Arch Math Logik Grundlagenforsch 20∗103-111
◇ C25 C60 C62 F25 F35 ◇ REV MR 82c:03047 Zbl 454#03014 • ID 54226

HODGES, W. [1981] *Encoding orders and trees in binary relations* (J 0303) Mathematika (Univ Coll London) 28∗67-71
◇ C13 C45 ◇ REV MR 82k:03040 Zbl 449#03025
• ID 54357

HODGES, W. [1981] *In singular cardinality, locally free algebras are free* (J 0004) Algeb Universalis 12∗205-220 • ERR/ADD ibid 19∗135
◇ C05 C55 C60 E05 E60 E75 ◇ REV MR 82i:08005 Zbl 476#03039 • ID 55551

HODGES, W. & SHELAH, S. [1981] *Infinite games and reduced products* (J 0007) Ann Math Logic 20∗77-108
◇ C20 C40 C75 E55 E60 ◇ REV MR 82f:03025 Zbl 501.03014 • ID 74221

HODGES, W. [1982] see BALDWIN, JOHN T.

HODGES, W. [1983] *Elementary predicate logic* (C 4085) Handb Philos Log 1∗1-131
◇ B10 C98 ◇ REV Zbl 538#03001 • ID 39725

HODGES, W. [1984] *Constructing many non-isomorphic models* (S 3382) Sem-ber, Humboldt-Univ Berlin, Sekt Math 60∗73-75
◇ C52 ◇ REV Zbl 551#03016 • ID 42736

HODGES, W. [1984] *Finite extensions of finite groups* (P 2153) Logic Colloq;1983 Aachen 1∗193-206
◇ C13 C50 C60 D35 ◇ REV MR 86g:03059 • ID 39733

HODGES, W. [1984] *Groupes nilpotents existentiellement clos de classe fixee* (S 3521) Mem Soc Math Fr 16∗1-10
◇ C25 C60 ◇ REV Zbl 554#20007 • ID 39722

HODGES, W. [1984] *Models built on linear orderings (French summary)* (P 2167) Orders: Descr & Roles;1982 L'Arbresle 207-234
◇ C30 C55 C98 E07 ◇ REV MR 86h:03053 Zbl 553#03020 • ID 39727

HODGES, W. [1984] *On constructing many non-isomorphic algebras* (P 3088) Univer Alg & Link Log, Alg, Combin, Comp Sci;1983 Darmstadt 67-77
◇ C05 C52 C55 C75 E50 E55 E75 ◇ REV Zbl 545#03015 • ID 39731

HODGES, W. [1985] *Building models by games* (X 0805) Cambridge Univ Pr: Cambridge, GB vi+311pp
◇ C25 C55 C60 C80 C98 E60 ◇ REV Zbl 569#03015
• ID 39750

HODGES, W. see Vol. I, II, V for further entries

HODGSON, B.R. [1982] *On direct products of automaton decidable theories* (J 1426) Theor Comput Sci 19∗331-335
◇ B25 D05 ◇ REV MR 83m:03046 Zbl 493#03002
• ID 35446

HODGSON, B.R. [1983] *Decidabilite par automate fini* (J 2660) Ann Sci Math Quebec 7∗39-57
◇ B25 D05 D35 ◇ REV MR 84m:03016 Zbl 531#03007
• ID 35721

HODGSON, B.R. see Vol. IV, VI for further entries

HODKINSON, I.M. [1985] *A construction of many uncountable rings* (P 4310) Easter Conf on Model Th (3);1985 Gross Koeris 134-142
◇ C55 C60 E05 E50 E75 ◇ REV Zbl 575#03026
• ID 48826

HOEHNKE, H.-J. [1966] *Zur Strukturgleichheit axiomatischer Klassen* (J 0068) Z Math Logik Grundlagen Math 12∗69-83
◇ C05 G15 ◇ REV MR 32#5499 Zbl 163.245 • ID 06177

HOEHNKE, H.-J. see Vol. IV for further entries

HOEPELMAN, J.P. [1975] see GUENTHNER, F.

HOEPELMAN, J.P. see Vol. II for further entries

HOEPPNER, M. [1982] *On the freeness of Whitehead-diagrams* (P 3885) Categor Th.Appl to Algeb Log & Topol;1981 Gummersbach 133-137
◇ C60 ◇ REV MR 84c:18025 Zbl 513#18011 • ID 37809

HOEVEN VAN DER, G.F. & MOERDIJK, I. [1984] *Constructing choice sequences from lawless sequences of neighbourhood functions* (P 2153) Logic Colloq;1983 Aachen 1∗207-234
◇ C90 F35 F50 ◇ REV MR 86f:03100 • ID 41766

HOEVEN VAN DER, G.F. & MOERDIJK, I. [1984] *Sheaf models for choice sequences* (J 0073) Ann Pure Appl Logic 27∗63-107
◇ C90 F35 F50 G30 ◇ REV Zbl 546#03018 • ID 41155

HOEVEN VAN DER, G.F. see Vol. IV, V, VI for further entries

HOFMANN, K.H. [1972] *Representation of algebras by continuous sections* (J 0015) Bull Amer Math Soc 78*291-373
 ◊ C60 C90 C98 ◊ REV MR 50 # 415 Zbl 237 # 16018
 • ID 06185

HOFMANN, K.H. & MISLOVE, M.W. [1976] *Amalgamation in categories with concrete duals* (J 0004) Algeb Universalis 6*327-347
 ◊ C05 C52 G30 ◊ REV MR 56 # 5676 Zbl 381 # 18008
 • ID 24522

HOFMANN, K.H. see Vol. V for further entries

HOLLAND, W.C. [1980] *Trying to recognize the real line* (P 2983) Ordered Groups;1978 Boise 131-134
 ◊ C07 C60 C65 ◊ REV MR 82i:06022 Zbl 447 # 06014
 • ID 56584

HOLLAND, W.C. [1981] see GLASS, A.M.W.

HOLLAND, W.C. [1981] see GUREVICH, Y.

HOLLAND, W.C. [1983] *A survey of varieties of lattice ordered groups* (P 3841) Universal Algeb & Lattice Th (4);1982 Puebla 153-158
 ◊ C05 C60 C98 ◊ REV MR 84j:06020 Zbl 513 # 06008
 • ID 43082

HOLLAND, W.C. [1984] *Classification of lattice ordered groups (French summary)* (P 2167) Orders: Descr & Roles;1982 L'Arbresle 151-155
 ◊ C05 C60 C98 ◊ REV MR 86c:06021 Zbl 547 # 06012
 • ID 47038

HOLLAND, W.C. & MEKLER, A.H. & SHELAH, S. [1985] *Lawless order* (J 3457) Order 1*383-397
 ◊ C65 E05 E07 E75 ◊ REV Zbl 567 # 06012 • ID 45604

HOLLAND, W.C. see Vol. IV for further entries

HOLSZTYNSKI, W. [1968] see FAJTLOWICZ, S.

HOLSZTYNSKI, W. see Vol. IV, V for further entries

HOMMA, K. [1969] *Some results on saturated models* (J 0352) Math Jap 14*1-5
 ◊ C20 C50 ◊ REV MR 41 # 3257 Zbl 194.309 • ID 06204

HOMMA, K. see Vol. V for further entries

HONSELL, F. [1983] see FORTI, M.

HONSELL, F. see Vol. V, VI for further entries

HOOGEWIJS, A. [1979] *On a formalization of the non-definedness notion* (J 0068) Z Math Logik Grundlagen Math 25*213-217
 ◊ B60 C90 ◊ REV MR 80f:03017 Zbl 415 # 03019
 • ID 53121

HOOGEWIJS, A. [1983] *A partial predicate calculus in a two-valued logic* (J 0068) Z Math Logik Grundlagen Math 29*239 243
 ◊ B60 C90 ◊ REV MR 85d:03017 Zbl 521 # 03006
 • ID 37062

HOOGEWIJS, A. [1983] *The non-definedness notion and forcing* (J 0230) An Univ Iasi, NS, Sect Ia 29*5-11
 ◊ C25 C90 ◊ REV MR 85d:03063 Zbl 513 # 03026
 • ID 37227

HOOK, J.L. [1985] *A note on interpretations of many-sorted theories* (J 0036) J Symb Logic 50*372-374
 ◊ B10 C07 F25 ◊ REV Zbl 571 # 03026 • ID 41804

HOOVER, D.N. [1978] *Probability logic* (J 0007) Ann Math Logic 14*287-313
 ◊ B48 C80 C90 ◊ REV MR 80b:03044 Zbl 394 # 03033
 • ID 29154

HOOVER, D.N. [1982] *A normal form theorem for $L_{\omega_1 p}$, with applications* (J 0036) J Symb Logic 47*605-624
 ◊ C75 C90 ◊ REV MR 84j:03079 Zbl 496 # 03017
 • ID 34671

HOOVER, D.N. & KEISLER, H.J. [1984] *Adapted probability distributions* (J 0064) Trans Amer Math Soc 286*159-201
 ◊ C50 C65 H10 ◊ REV Zbl 548 # 60019 • ID 43234

HOOVER, D.N. [1985] *A probabilistic interpolation theorem* (J 0036) J Symb Logic 50*708-713
 ◊ C40 C70 C90 ◊ ID 47375

HOOVER, D.N. see Vol. I, II, IV for further entries

HORIGUCHI, H. [1981] *A definition of the category of boolean-valued models* (J 0407) Comm Math Univ St Pauli (Tokyo) 30*135-147
 ◊ C20 C30 C90 G30 ◊ REV MR 83b:03044
 Zbl 476 # 03041 • ID 55553

HORIGUCHI, H. [1982] *Limits in the category of Boolean-valued models* (J 0407) Comm Math Univ St Pauli (Tokyo) 31*9-13
 ◊ C90 G30 ◊ REV MR 84d:03046a Zbl 501 # 03020
 • ID 34081

HORIGUCHI, H. [1982] *Strict Boolean-valued models* (J 0407) Comm Math Univ St Pauli (Tokyo) 31*15-18
 ◊ C90 G30 ◊ REV MR 84d:03046b Zbl 501 # 03021
 • ID 34082

HORIGUCHI, H. [1985] *The category of boolean-valued models and its applications* (J 0407) Comm Math Univ St Pauli (Tokyo) 34*71-89
 ◊ C30 C90 E40 G30 ◊ ID 41746

HORN, A. [1951] *On sentences which are true of direct unions of algebras* (J 0036) J Symb Logic 16*14-21
 ◊ C05 C30 C40 ◊ REV MR 12.662 Zbl 43.248
 JSL 16.216 • ID 06226

HORN, A. [1955] *A characterisation of unions of linearly independent sets* (J 0039) J London Math Soc 30*494-496
 ◊ C60 E75 ◊ REV MR 17.135 Zbl 66.41 • ID 43634

HORN, A. [1970] see GALVIN, F.

HORN, A. see Vol. I, II, VI for further entries

HORT, C. & OSSWALD, H. [1984] *On nonstandard models in higher order logic* (J 0036) J Symb Logic 49*204-219
 ◊ B15 C55 C85 E55 H05 H20 ◊ REV MR 85i:03122
 • ID 44206

HORTY, J.F. [1985] see SCOTT, D.S.

HOWARD, P.E. [1972] *A proof of a theorem of Tennenbaum* (J 0068) Z Math Logik Grundlagen Math 18*111-112
 ◊ C57 C62 H15 ◊ REV MR 46 # 3293 Zbl 251 # 02053
 • ID 06260

HOWARD, P.E. [1975] *Los' theorem and the boolean prime ideal theorem imply the axiom of choice* (J 0053) Proc Amer Math Soc 49*426-428
 ◊ C20 E25 G05 ◊ REV MR 52 # 5422 Zbl 312 # 02052
 • ID 18198

HOWARD, P.E. see Vol. V for further entries

HOWIE, J.M. & ISBELL, J.R. [1967] *Epimorphisms and dominions. II* (**J** 0032) J Algeb 6*7-21
◇ C05 C50 ◇ REV MR 35 # 105b Zbl 211.333 • REM Part I 1966 by Isbell,J.R. • ID 48988

HRBACEK, K. & VOPENKA, P. [1966] *On strongly measurable cardinals* (**J** 0014) Bull Acad Pol Sci, Ser Math Astron Phys 14*587-591
◇ C20 E35 E45 E55 ◇ REV MR 35 # 2747 Zbl 158.262
• ID 15070

HRBACEK, K. [1976] see GOSTANIAN, R.

HRBACEK, K. [1980] see GOSTANIAN, R.

HRBACEK, K. see Vol. I, IV, V for further entries

HUBER, M. [1977] *Sur le probleme de Whitehead concernant les groupes abeliens libres (English summary)* (**J** 2313) C R Acad Sci, Paris, Ser A-B 284*A471-A472
◇ C60 E65 E75 ◇ REV MR 55 # 98 Zbl 349 # 20023
• ID 74296

HUBER, M. [1979] *On cartesian powers of a rational group* (**J** 0044) Math Z 169*253-259
◇ C60 E45 E55 E75 ◇ REV MR 81i:20068
Zbl 403 # 20034 • ID 53089

HUBER, M. [1980] see EKLOF, P.C.

HUBER, M. [1983] *Methods of set theory and the abundance of separable abelian p-groups* (**P** 3802) Abel Group Th;1983 Honolulu 304-319
◇ C55 C60 E05 E75 ◇ REV MR 85h:20059
Zbl 519 # 20039 • ID 36726

HUBER, M. [1983] *On reflexive modules and abelian groups* (**J** 0032) J Algeb 82*469-487
◇ C55 C60 E50 E55 E75 G30 ◇ REV MR 85f:16022
Zbl 517 # 16020 • ID 36692

HUBER, M. [1985] see EKLOF, P.C.

HUBER, M. see Vol. V for further entries

HUBER-DYSON, V. [1964] *On the decision problem for theories of finite models* (**J** 0029) Israel J Math 2*55-70
◇ B25 C13 D35 ◇ REV MR 31 # 2149 Zbl 143.249
• ID 03214

HUBER-DYSON, V. [1969] *On the decision problem for extensions of a decidable theory* (**J** 0027) Fund Math 64*7-40
◇ B25 C13 C60 D35 ◇ REV MR 40 # 5431 Zbl 193.312
• ID 03215

HUBER-DYSON, V. [1977] *Talking about free groups in naturally enriched languages* (**J** 0394) Commun Algeb 5*1163-1191
◇ C60 D40 ◇ REV MR 56 # 11775 Zbl 365 # 02045
• ID 51046

HUBER-DYSON, V. [1979] *An inductive theory for free products of groups* (**J** 0004) Algeb Universalis 9*35-44
◇ C60 ◇ REV MR 81i:20042 Zbl 412 # 20003 • ID 52976

HUBER-DYSON, V. [1981] *A reduction of the open sentence problem for finite groups* (**J** 0161) Bull London Math Soc 13*331-338
◇ C13 C60 D40 ◇ REV MR 82j:03052 Zbl 441 # 20002
• ID 54396

HUBER-DYSON, V. [1982] *Decision problems in group theory* (**P** 3808) Rect Trends in Math;1982 Reinhardsbrunn 174-182
◇ B25 D40 ◇ REV MR 84d:20032 Zbl 508 # 20018
• ID 39680

HUBER-DYSON, V. [1982] *Symmetric groups and the open sentence problem* (**P** 3634) Patras Logic Symp;1980 Patras 159-169
◇ C13 C60 D35 ◇ REV MR 84j:20032 Zbl 514 # 20025
• ID 37444

HUBER-DYSON, V. see Vol. IV, VI for further entries

HUET, G. [1981] *A complete proof of correctness of the Knuth-Bendix completion algorithm* (**J** 0119) J Comp Syst Sci 23*11-21
◇ B25 B75 C05 ◇ REV MR 83e:68012 Zbl 465 # 68014
• ID 54947

HUET, G. see Vol. I, IV, VI for further entries

HUGLY, P. & SAYWARD, C.W. [1982] *Indenumerability and substitutional quantification* (**J** 0047) Notre Dame J Formal Log 23*358-366
◇ A05 B10 C07 ◇ REV MR 83k:03016 Zbl 452 # 03004
• ID 54068

HUGLY, P. & SAYWARD, C.W. [1984] *Do we need quantification?* (**J** 0047) Notre Dame J Formal Log 25*289-302
◇ A05 C75 ◇ REV MR 86d:03008 Zbl 566 # 03001
• ID 42574

HUGLY, P. see Vol. I, II for further entries

HUHN, A.P. [1985] *On nonmodular n-distributive lattices: the decision problem for identities in finite n-distributive lattices* (**J** 0002) Acta Sci Math (Szeged) 48*215-219
◇ B25 C13 G10 ◇ ID 49445

HUHN, A.P. see Vol. IV for further entries

HULE, H. [1978] *Relations between the amalgamation property and algebraic equations* (**J** 3194) J Austral Math Soc, Ser A 25*257-263
◇ C05 C52 ◇ REV MR 58 # 5454 Zbl 405 # 08002
• ID 90252

HULE, H. [1979] *Solutionally complete varieties* (**J** 3194) J Austral Math Soc, Ser A 28*82-86
◇ C05 ◇ REV MR 81i:08007 Zbl 419 # 08007 • ID 81609

HULE, H. see Vol. IV for further entries

HUMBERSTONE, I.L. [1983] see BENTHEM VAN, J.F.A.K.

HUMBERSTONE, I.L. [1984] *Monadic representability of certain binary relations* (**J** 0016) Bull Austral Math Soc 29*365-376
◇ C52 D60 E07 ◇ REV MR 85f:04001 Zbl 531 # 04004
• ID 37693

HUMBERSTONE, I.L. see Vol. II for further entries

HUNTINGTON, E.V. [1917] *Complete existential theory of the postulates for serial order* (**J** 0015) Bull Amer Math Soc 23*276-280
◇ B30 C13 C35 E07 ◇ REV FdM 46.308 • ID 41774

HUNTINGTON, E.V. [1917] *Complete existential theory of the postulates for well ordered sets* (**J** 0015) Bull Amer Math Soc 23*280-282
◇ B30 C13 C35 E07 ◇ REV FdM 46.308 • ID 41776

HUNTINGTON, E.V. see Vol. I, II, V, VI for further entries

HUTCHINSON, G. [1975] *On the representation of lattices by modules* (**J** 0064) Trans Amer Math Soc 209*311-351
◇ C05 C60 G10 ◇ REV MR 51 # 12637 Zbl 328 # 06002
• ID 17255

HUTCHINSON, G. [1981] *A complete logic for n-permutable congruence lattices* (J 0004) Algeb Universalis 13*206-224
⋄ C05 C35 G10 ⋄ REV MR 83a:03067 Zbl 479 # 08007
• ID 55695

HUTCHINSON, G. see Vol. IV for further entries

HUTCHINSON, J.E. [1974] *Models of set theories and their applications* (0000) Diss., Habil. etc
⋄ C55 C62 C80 ⋄ REM Thesis, Stanford University, Stanford • ID 21363

HUTCHINSON, J.E. [1976] *Elementary extensions of countable models of set theory* (J 0036) J Symb Logic 41*139-145
⋄ C15 C62 ⋄ REV MR 53 # 12944 Zbl 364 # 02029
• ID 14790

HUTCHINSON, J.E. [1976] *Model theory via set theory* (J 0029) Israel J Math 24*286-304
⋄ C40 C55 C62 C70 C75 C80 C95 E75 ⋄ REV MR 55 # 10268 Zbl 373 # 02038 • ID 26068

HUTCHINSON, J.E. [1976] *Order types of ordinals in models of set theory* (J 0036) J Symb Logic 41*489-502
⋄ C30 C62 E10 ⋄ REV MR 53 # 7767 Zbl 338 # 02034
• ID 14784

HUYNH, D.T. [1985] *The complexity of equivalence problems for commutative grammars* (J 0194) Inform & Control 66*103-121
⋄ C05 C15 ⋄ ID 49572

HUYNH, D.T. see Vol. IV for further entries

HYLAND, J.M.E. [1977] see GANDY, R.O.

HYLAND, J.M.E. [1979] see FOURMAN, M.P.

HYLAND, J.M.E. see Vol. IV, V, VI for further entries

IBARRA, O.H. [1978] see GURARI, E.M.

IBARRA, O.H. [1979] see GURARI, E.M.

IBARRA, O.H. [1981] see GURARI, E.M.

IBARRA, O.H. see Vol. IV, VI for further entries

IGOSHIN, V.I. [1971] *Characterizable varieties of lattices (Russian)* (S 0778) Uporyad Mnozhestva Reshetki (Saratov) 1*22-30
⋄ C05 C52 G10 ⋄ REV MR 51 # 7970 Zbl 244 # 06004
• ID 17587

IMAZEKI, K. [1980] *Characterizations of compact operators and semi-Fredholm operators* (J 2853) TRU Math (Tokyo) 16*1-8
⋄ C20 C65 ⋄ REV MR 82g:47011 Zbl 457 # 47015
• ID 81636

IMMERMAN, N. [1981] *Number of quantifiers is better than number of tape cells* (J 0119) J Comp Syst Sci 22*384-406
⋄ C07 D15 ⋄ REV MR 84e:68039 Zbl 486 # 03019
• ID 38079

IMMERMAN, N. see Vol. IV for further entries

IMREH, B. [1968] *On a theorem of G.Birkhoff* (J 0057) Publ Math (Univ Debrecen) 15*147-148
⋄ C05 ⋄ REV MR 38 # 3209 Zbl 172.21 • ID 06386

INDERMARK, K. [1975] *Gleichungsdefinierbarkeit in Relationstrukturen* (X 0817) Ges Math Datenverarbeit: Bonn 49pp
⋄ C05 C40 ⋄ REV MR 53 # 7745 Zbl 306 # 02041
• ID 22993

INDERMARK, K. see Vol. IV, VI for further entries

ISAMBERT, E. [1973] see ARTIGUE, M.

ISAMBERT, E. [1974] *Extensions elementaires non-standard de modeles de theorie des ensembles* (J 2313) C R Acad Sci, Paris, Ser A-B 279*A83-A86
⋄ C62 C80 ⋄ REV MR 51 # 5306 Zbl 289 # 02047
• ID 17407

ISAMBERT, E. [1974] *Theories syntaxiquement non-standard* (J 2313) C R Acad Sci, Paris, Ser A-B 278*A1469-A1471
⋄ C62 ⋄ REV MR 50 # 103 Zbl 291 # 02046 • ID 06403

ISAMBERT, E. [1978] see ARTIGUE, M.

ISAMBERT, E. see Vol. V, VI for further entries

ISBELL, J.R. [1966] *Epimorphisms and dominions* (P 1100) Conf Categor Algeb;1965 La Jolla 232-246
⋄ C05 C50 ⋄ REV MR 35 # 105a Zbl 194.16 • REM Part I. Part II 1967 by Howie,J.M. & Isbell,J.R. • ID 48985

ISBELL, J.R. [1967] see HOWIE, J.M.

ISBELL, J.R. [1973] *Functorial implicit operations* (J 0029) Israel J Math 15*185-188
⋄ C05 C40 ⋄ REV MR 48 # 2027 Zbl 266 # 02032
• ID 06418

ISBELL, J.R. [1976] *Compatibility and extensions of algebraic theories* (J 0004) Algeb Universalis 6*37-51
⋄ C05 G05 G30 ⋄ REV MR 54 # 212 Zbl 335 # 08005
• ID 24590

ISBELL, J.R. & KLUN, M.I. & SCHANUEL, S.H. [1977] *Affine parts of algebraic theories. I* (J 0032) J Algeb 44*1-8
⋄ C05 G30 ⋄ REV MR 55 # 2555 Zbl 353 # 18005
• ID 74372

ISBELL, J.R. [1982] *Generating the algebraic theory of C(X)* (J 0004) Algeb Universalis 15*153-155
⋄ C05 C75 ⋄ REV MR 84e:08004 Zbl 516 # 18008
• ID 39556

ISBELL, J.R. [1983] *Generic algebras* (J 0064) Trans Amer Math Soc 275*497-510
⋄ C05 C50 E55 G30 ⋄ REV MR 84h:18010 Zbl 513 # 08001 • ID 37803

ISBELL, J.R. see Vol. V for further entries

ISKANDER, A.A. [1984] *Extensions of algebraic systems* (J 0064) Trans Amer Math Soc 281*309-327
⋄ C05 C30 ⋄ REV MR 85h:08012 Zbl 537 # 08005
• ID 43851

ISKANDER, A.A. see Vol. IV for further entries

ISSEL, W. [1969] *Semantische Untersuchungen ueber Quantoren I* (J 0068) Z Math Logik Grundlagen Math 15*353-358
⋄ B10 C55 C80 ⋄ REV MR 41 # 6675 Zbl 188.16 • REM Parts II,III 1970 • ID 06459

ISSEL, W. [1970] *Semantische Untersuchungen ueber Quantoren II,III* (J 0068) Z Math Logik Grundlagen Math 16*281-296,421-438
⋄ C40 C55 C80 ⋄ REV MR 43 # 1821 Zbl 205.9 Zbl 211.305 • REM Part I 1969 • ID 06460

ISSMANN, S. [1953] *Une methode de decision pour certaines formules du calcul des predicats* (P 0645) Int Congr Philos (11);1953 Bruxelles 19*35-38
⋄ B25 ⋄ REV MR 15.90 Zbl 53.1 JSL 19.132 • ID 16865

ITHIER, P. [1971] *Sur les formules stables par produit reduit suivant certains filtres* (J 2313) C R Acad Sci, Paris, Ser A-B 272*A1429-A1432
  ⋄ C20 ⋄ REV MR 44#6470 Zbl 216.12 • ID 06462

ITO, MAKOTO [1934] *Einige Anwendungen der Theorie des Entscheidungsproblems zur Axiomatik I* (J 0261) Tohoku Math J 37*222-235
  ⋄ B30 C75 F25 ⋄ REV Zbl 7.385 FdM 61.51 • REM Part II 1935 • ID 32185

ITO, MAKOTO [1965] *Boolean recursive functions and closure algebra* (P 0614) Th Models;1963 Berkeley 431-432
  ⋄ B25 B45 D20 ⋄ REV Zbl 147.256 • ID 27530

ITO, MAKOTO see Vol. I, II, VI for further entries

IVANOV, A.A. [1983] *Decidability of theories in a certain calculus (Russian)* (J 0087) Mat Zametki (Akad Nauk SSSR) 33*617-625
  • TRANSL [1983] (J 1044) Math Notes, Acad Sci USSR 33*317-321
  ⋄ B25 C10 C55 C60 C80 D35 ⋄ REV MR 85c:03013 Zbl 524#03023 • ID 37599

IVANOV, A.A. [1983] *Some theories in generalized calculi (Russian)* (J 0092) Sib Mat Zh 24/6*56-65
  • TRANSL [1983] (J 0475) Sib Math J 24*868-876
  ⋄ B25 C10 C55 C60 C80 ⋄ REV MR 85i:03120 Zbl 543#03021 • ID 40394

IVANOV, A.A. [1984] *Complete theories of unars (Russian)* (J 0003) Algebra i Logika 23*48-73
  • TRANSL [1984] (J 0069) Algeb and Log 23*36-55
  ⋄ B25 C35 ⋄ REV Zbl 546#08006 • ID 39642

IVANOV, A.A. [1984] *Decidability of extended theories of addition of the natural numbers and the integers (Russian)* (J 0092) Sib Mat Zh 25/4*78-81
  • TRANSL [1984] (J 0475) Sib Math J 25*572-574
  ⋄ B25 C65 F30 ⋄ REV MR 86f:03024 • ID 41869

IWANUS, B. [1973] *On Lesniewski's elementary ontology (Polish and Russian summaries)* (J 0063) Studia Logica 31*73-125
  ⋄ B25 B60 E70 ⋄ REV MR 51#10041 Zbl 275#02019 • ID 17553

IWANUS, B. [1973] *Proof of decidability of the traditional calculus of names (Polish and Russian summaries)* (J 0063) Studia Logica 32*131-147
  ⋄ B25 ⋄ REV MR 50#6819 Zbl 362#02010 • ID 06474

IWANUS, B. see Vol. I, II for further entries

IYANAGA, S. [1983] *Problems in foundation of mathematics: Complete and uncomplete theories (Japanese)* (J 4439) Kagaku 53*393-398
  ⋄ C35 ⋄ ID 46957

IYANAGA, S. see Vol. VI for further entries

JACKSON, F.C. [1977] see DAVIDSON, B.

JACKSON, F.C. see Vol. II for further entries

JACOB, B. [1980] *A Nullstellensatz for $\mathbb{R}((t))$* (J 0394) Commun Algeb 8*1083-1094
  ⋄ C60 ⋄ REV MR 82e:13006 Zbl 503#14013 • ID 81681

JACOB, B. [1980] *The model theory of "R-formal" fields* (J 0007) Ann Math Logic 19*263-282
  ⋄ B25 C10 C25 C35 C60 ⋄ REV MR 82b:03068 Zbl 463#03021 • ID 54561

JACOB, B. [1981] *The model theory of generalized real closed fields* (J 0127) J Reine Angew Math 323*213-220
  ⋄ B25 C35 C60 ⋄ REV MR 82m:03047 Zbl 446#12021 • ID 74440

JACOB, B. [1985] see BECKER, E.

JACOBSON, R.A. & YOCOM, K.L. [1965] *Absolutely independent group axioms* (J 0005) Amer Math Mon 72*756-758
  ⋄ C60 ⋄ REV MR 32#5517 Zbl 132.20 • ID 06491

JAEGER, G. [1982] *Zur Beweistheorie der Kripke-Platek-Mengenlehre ueber den natuerlichen Zahlen* (J 0009) Arch Math Logik Grundlagenforsch 22*121-139
  ⋄ C70 E30 F15 F35 ⋄ REV MR 84j:03110 Zbl 503#03014 • ID 34700

JAEGER, G. see Vol. V, VI for further entries

JAFFAR, J. [1981] *Presburger arithmetic with array segments* (J 0232) Inform Process Lett 12*79-82
  ⋄ B25 ⋄ REV MR 83g:68072 Zbl 474#03006 • ID 55410

JA'JA', J. [1982] *The computational complexity of a set of quadratic functions* (J 0119) J Comp Syst Sci 24*209-223
  ⋄ C55 ⋄ REV MR 83m:68077 Zbl 484#68033 • ID 40413

JA'JA', J. see Vol. IV for further entries

JAKUBIKOVA-STUDENOVSKA, D. [1980] *On weakly rigid monounary algebras* (J 1522) Math Slovaca 30*197-206
  ⋄ C05 C55 E10 E75 ⋄ REV MR 82f:08007 Zbl 439#08003 • ID 56042

JAKUBOWICZ, C.A. [1974] *On generalized products preserving compactness (Russian)* (J 0003) Algebra i Logika 13*227-231,235 • ERR/ADD ibid 13*721
  • TRANSL [1974] (J 0069) Algeb and Log 13*131-133
  ⋄ C30 C52 G05 ⋄ REV MR 51#7854 MR 53#117 Zbl 305#08002 • ID 17700

JAKUBOWICZ, C.A. see Vol. V for further entries

JAMBU-GIRAUDET, M. [1980] *Theorie des modeles de groupes d'automorphismes d'ensembles totalement ordonnes 2-homogenes (English summary)* (J 2313) C R Acad Sci, Paris, Ser A-B 290*A1037-A1039
  ⋄ C07 C60 C62 C65 D35 E07 F25 F35 ⋄ REV MR 81f:03046 Zbl 453#03037 • ID 54167

JAMBU-GIRAUDET, M. [1981] see GLASS, A.M.W.

JAMBU-GIRAUDET, M. [1981] *Interpretations d'arithmetiques dans des groupes et des treillis* (P 3404) Model Th & Arithm;1979/80 Paris 143-153
  ⋄ C07 C62 C65 D35 E07 F25 F30 G10 ⋄ REV MR 83g:03064 Zbl 476#03023 • ID 55535

JAMBU-GIRAUDET, M. [1983] *Bi-interpretable groups and lattices* (J 0064) Trans Amer Math Soc 278*253-269
  ⋄ C07 C65 G10 ⋄ REV MR 84g:06028 Zbl 522#03024 • ID 37786

JAMBU-GIRAUDET, M. [1985] *Quelques remarques sur l'equivalence elementaire entre groupes ou treillis d'automorphismes de chaines 2-homogenes* (J 0193) Discr Math 53*117-124
  ⋄ C07 C65 ⋄ REV Zbl 561#06001 • ID 47491

JANICZAK, A. [1950] *A remark concerning decidability of complete theories* (J 0036) J Symb Logic 15*277-279
  ⋄ B25 C35 ⋄ REV MR 12.790 Zbl 41.4 JSL 16.146 • ID 06507

JANICZAK, A. [1953] *Undecidability of some simple formalized theories* (J 0027) Fund Math 40*131-139
⋄ B25 D35 ⋄ REV MR 15.669 Zbl 52.252 JSL 22.217
• ID 06508

JANICZAK, A. see Vol. IV for further entries

JANKOWSKI, A.W. [1980] *Embeddings of logical structures. I,II* (S 3230) Prepr, Inst Math, Pol Acad Sci 10*1-125,126-237
⋄ C95 ⋄ REV Zbl 432 # 03022 • ID 53978

JANKOWSKI, A.W. [1982] *An alternative characterization of elementary logic (Russian summary)* (J 3293) Bull Acad Pol Sci, Ser Math 30*9-13
⋄ C95 ⋄ REV MR 84a:03039 Zbl 501 # 03022 • ID 35576

JANKOWSKI, A.W. & RAUSZER, C. [1983] *Logical foundation approach to users' domain restriction in data bases* (J 1426) Theor Comput Sci 23*11-36
⋄ C90 ⋄ REV MR 84c:68016 Zbl 529 # 68072 • ID 39571

JANKOWSKI, A.W. [1985] *Galois structures* (J 0063) Studia Logica 44*109-124
⋄ B22 C95 ⋄ ID 47505

JANKOWSKI, A.W. & ZAWADOWSKI, M. [1985] *Sheaves over Heyting lattices* (J 0063) Studia Logica 44*237-256
⋄ C90 F35 F50 G10 G30 ⋄ ID 47500

JANKOWSKI, A.W. see Vol. II, V for further entries

JANOS, L. [1978] *Topological dimension as a first order property (Italian summary)* (J 0149) Atti Accad Naz Lincei Fis Mat Nat, Ser 8 64*572-577
⋄ C65 ⋄ REV MR 81a:03039 Zbl 429 # 54018 • ID 53893

JANSSEN, G. [1972] *Restricted ultraproducts of finite von Neumann algebras* (P 0732) Contrib to Non-Standard Anal;1970 Oberwolfach 101-114
⋄ C20 C60 H05 ⋄ REV MR 58 # 2335 Zbl 248 # 46049
• ID 06540

JARDEN, M. [1969] *Rational points on algebraic varieties over large number fields* (J 0015) Bull Amer Math Soc 75*603-606
⋄ C60 ⋄ REV MR 39 # 1456 Zbl 174.527 • ID 06541

JARDEN, M. [1972] *Elementary statements over large algebraic fields* (J 0064) Trans Amer Math Soc 164*67-91
⋄ C20 C60 ⋄ REV MR 46 # 1795 Zbl 235 # 12104
• ID 27830

JARDEN, M. [1974] *An injective rational map of an abstract variety into itself* (J 0127) J Reine Angew Math 265*23-30
⋄ C20 C60 ⋄ REV MR 49 # 2730 Zbl 275 # 14001
• ID 06542

JARDEN, M. & KIEHNE, U. [1975] *The elementary theory of algebraic fields of finite corank* (J 0305) Invent Math 30*275-294
⋄ B25 C10 C20 C60 ⋄ REV MR 55 # 8012
Zbl 315 # 12107 • ID 27344

JARDEN, M. [1976] *Algebraically closed fields with distinguished subfields* (J 0008) Arch Math (Basel) 27*502-505
⋄ C20 C60 ⋄ REV MR 54 # 12517 Zbl 339 # 12107
• ID 26562

JARDEN, M. [1976] see FRIED, M.

JARDEN, M. [1976] *The elementary theory of ω-free Ax fields* (J 0305) Invent Math 38*187-206
⋄ C20 C60 ⋄ REV MR 55 # 8013 Zbl 342 # 12104
• ID 81699

JARDEN, M. [1978] *Intersections of conjugate fields of finite corank over Hilbertian fields* (J 3172) J London Math Soc, Ser 2 17*393-396
⋄ C60 ⋄ REV MR 58 # 16608 Zbl 402 # 12004 • ID 81698

JARDEN, M. [1980] *An analogue of Chebotarev density theorem for fields of finite corank* (J 2631) J Math Kyoto Univ 20*141-147
⋄ B25 C10 C60 ⋄ REV MR 81d:12010 Zbl 433 # 12013
• ID 81697

JARDEN, M. & ROQUETTE, P. [1980] *The Nullstellensatz over p-adically closed fields* (J 0090) J Math Soc Japan 32*425-460
⋄ C60 ⋄ REV MR 82g:14027 Zbl 446 # 12016 • ID 81702

JARDEN, M. [1980] *Transfer principles for finite and p-adic fields* (J 3077) Nieuw Arch Wisk, Ser 3 28*139-158
⋄ C60 ⋄ REV MR 83h:12041 Zbl 446 # 12022 • ID 39202

JARDEN, M. [1982] see HARAN, D.

JARDEN, M. [1982] *The elementary theory of large e-fold ordered fields* (J 0118) Acta Math 149*239-260
⋄ C52 C60 ⋄ REV MR 84e:12029 Zbl 513 # 12020
• ID 37805

JARDEN, M. & WHEELER, W.H. [1983] *Model-complete theories of e-free Ax fields* (J 0036) J Symb Logic 48*1125-1129
⋄ C35 C60 ⋄ REV MR 85f:03034 Zbl 572 # 03015
• ID 40718

JARDEN, M. [1983] *On the model companion of the theory of e-fold ordered fields* (J 0118) Acta Math 150*243-253
⋄ B25 C25 C35 C60 ⋄ REV MR 85e:03078
Zbl 518 # 12018 JSL 51.235 • ID 38375

JARDEN, M. [1983] see DENEF, J.

JARDEN, M. & SHELAH, S. [1983] *Pseudo-algebraically closed fields over rational function fields* (J 0053) Proc Amer Math Soc 87*223-228
⋄ C60 ⋄ REV MR 84c:12015 Zbl 508 # 12021 • ID 38756

JARDEN, M. [1983] *The elementary theory of normal Frobenius fields* (J 0133) Michigan Math J 30*155-163
⋄ C60 ⋄ REV MR 85e:12004 Zbl 535 # 12016 • ID 38333

JARDEN, M. [1984] see FRIED, M.

JEAN, M. [1967] *Relations monomorphes et classes universelles* (J 2313) C R Acad Sci, Paris, Ser A-B 264*A591-A593
⋄ C07 C30 C52 ⋄ REV MR 38 # 54 Zbl 153.18
• ID 06562

JECH, T.J. [1965] see BALCAR, B.

JECH, T.J. [1971] *On models for the set theory without AC* (P 0693) Axiomatic Set Th;1967 Los Angeles 1*135-142
⋄ C62 E10 E25 E35 E40 ⋄ REV MR 43 # 6078
Zbl 263 # 02031 JSL 40.505 • ID 06575

JECH, T.J. & POWELL, W.C. [1971] *Standard models of set theory with predication* (J 0015) Bull Amer Math Soc 77*808-813
⋄ C62 E30 E47 E55 E65 E70 ⋄ REV MR 43 # 6074
Zbl 263 # 02030 • ID 29869

JECH, T.J. [1971] *Two remarks on elementary embeddings of the universe* (J 0048) Pac J Math 39*395-400
⋄ C62 E40 E45 E55 ⋄ REV MR 46 # 7032
Zbl 225 # 02037 JSL 38.335 • ID 06576

JECH, T.J. & PRIKRY, K. [1976] *On ideals of sets and the power set operation* (J 0015) Bull Amer Math Soc 82*593-596
⋄ C20 E05 E40 E50 E55 ⋄ REV MR 58 # 21618 Zbl 339 # 02060 • ID 30721

JECH, T.J. & PRIKRY, K. [1979] *Ideals over uncountable sets: Application of almost disjoint functions and generic ultrapowers* (S 0167) Mem Amer Math Soc 214*iii+71pp
⋄ C20 E05 E40 E50 E55 ⋄ REV MR 80f:03059 Zbl 398 # 03044 • ID 52772

JECH, T.J. & MAGIDOR, M. & MITCHELL, W.J. & PRIKRY, K. [1980] *Precipitous ideals* (J 0036) J Symb Logic 45*1-8
⋄ C70 E05 E35 E40 E50 E55 ⋄ REV MR 81h:03097 Zbl 437 # 03026 • ID 55894

JECH, T.J. see Vol. IV, V for further entries

JEFFERS, E. [1978] see BURRIS, S.

JEHNE, W. & KLINGEN, N. [1977] *Superprimes and a generalized Frobenius symbol* (J 0399) Acta Arith, Pol Akad Nauk 32*209-232
⋄ C20 C60 E50 E75 ⋄ REV MR 55 # 12694 Zbl 316 # 12004 • ID 81712

JENSEN, A. [1965] *Some results concerning a general set-theoretical approach to logic* (J 0132) Math Scand 16*5-24
⋄ C07 ⋄ REV MR 33 # 27 • ID 06589

JENSEN, A. see Vol. I, V for further entries

JENSEN, C.U. & VAMOS, P. [1979] *On the axiomatizability of certain classes of modules* (J 0044) Math Z 167*227-237
⋄ C60 ⋄ REV MR 81h:03073 Zbl 387 # 03012 • ID 52228

JENSEN, C.U. [1980] *Applications logiques en theorie des anneaux et des modules* (P 3140) Colloq d'Algeb;1980 Rennes 29-37
⋄ C60 ⋄ REV MR 84h:16017 Zbl 478 # 03013 • ID 55628

JENSEN, C.U. & LENZING, H. [1980] *Model theory and representations of algebras* (P 3023) Repr Algeb (2);1979 Ottawa 302-310
⋄ C60 ⋄ REV MR 82c:16029 Zbl 445 # 16019 • ID 56511

JENSEN, C.U. [1980] *Peano rings of arbitrary global dimension* (J 3172) J London Math Soc, Ser 2 21*39-44
⋄ C60 E50 E75 ⋄ REV MR 81i:16037 Zbl 416 # 03038 • ID 53472

JENSEN, C.U. [1980] *Proprietes homologiques et logiques des anneaux de fonctions entieres (English summary)* (J 2313) C R Acad Sci, Paris, Ser A-B 291*A515-A517
⋄ C60 C65 E50 E75 ⋄ REV MR 82k:30057 Zbl 456 # 18007 • ID 54316

JENSEN, C.U. [1981] see HERRMANN, C.

JENSEN, C.U. & LENZING, H. [1982] *Homological dimensions and representation type of algebras under base field extension* (J 0504) Manuscr Math 39*1-13
⋄ C20 C60 ⋄ REV MR 83k:16019 Zbl 498 # 16023 • ID 38526

JENSEN, C.U. [1982] *La dimension globale de l'anneau des fonctions entieres (English summary)* (J 3364) C R Acad Sci, Paris, Ser 1 294*385-386
⋄ C60 C65 E35 E75 ⋄ REV MR 83f:16060 Zbl 536 # 18006 • ID 46674

JENSEN, C.U. [1982] *Sur une classe de corps indecidables (English summary)* (J 3364) C R Acad Sci, Paris, Ser 1 295*507-509
⋄ C60 D35 ⋄ REV MR 83m:12046 Zbl 514 # 03020 • ID 37407

JENSEN, C.U. [1984] *Theorie des modeles pour des anneaux de fonctions entieres et des corps de fonctions meromorphes* (S 3521) Mem Soc Math Fr 16*23-40
⋄ C60 C65 D35 F35 ⋄ REV Zbl 562 # 12025 • ID 39754

JENSEN, C.U. see Vol. IV for further entries

JENSEN, D.C. [1976] see EHRENFEUCHT, A.

JENSEN, D.C. see Vol. I, VI for further entries

JENSEN, F.V. [1974] *Interpolation and definability in abstract logics* (J 0154) Synthese 27*251-257
⋄ C40 C95 ⋄ REV MR 58 # 21565 Zbl 311 # 02024 • ID 15059

JENSEN, F.V. [1975] *On completeness in cardinality logics* (J 0014) Bull Acad Pol Sci, Ser Math Astron Phys 23*117-122
⋄ C10 C35 C55 C60 C80 ⋄ REV MR 55 # 2489 Zbl 306 # 02018 • ID 06593

JENSEN, F.V. [1976] *Souslin-Kleene does not imply Beth* (J 0027) Fund Math 90*269-273
⋄ C40 C95 ⋄ REV MR 54 # 2452 Zbl 326 # 02014 • ID 18210

JENSEN, F.V. see Vol. I, II, IV for further entries

JENSEN, R.B. [1965] *Ein neuer Beweis fuer die Entscheidbarkeit des einstelligen Praedikatenkalkuels mit Identitaet* (J 0009) Arch Math Logik Grundlagenforsch 7*128-138
⋄ B25 ⋄ REV MR 34 # 1193 Zbl 158.252 • ID 06594

JENSEN, R.B. [1967] *Concrete models of set theory* (P 0691) Sets, Models & Recursion Th;1965 Leicester 44-74
⋄ C62 E25 E35 E45 E50 ⋄ REV MR 37 # 53 Zbl 204.6 JSL 35.472 • ID 06597

JENSEN, R.B. [1967] *Modelle der Mengenlehre* (S 3301) Lect Notes Math 37*viii+176pp
⋄ C62 D55 E25 E35 E40 E45 E50 E98 ⋄ REV MR 36 # 4982 Zbl 191.299 JSL 40.92 • ID 06596

JENSEN, R.B. [1968] see FRIEDMAN, H.M.

JENSEN, R.B. & SOLOVAY, R.M. [1970] *Some applications of almost disjoint sets* (P 1072) Math Log & Founds of Set Th;1968 Jerusalem 84-104
⋄ C62 D55 E05 E35 E45 E55 ⋄ REV MR 44 # 6482 Zbl 222 # 02077 • ID 22245

JENSEN, R.B. [1972] *The fine structure of the constructible hierarchy* (J 0007) Ann Math Logic 4*229-308,443
⋄ C55 E05 E35 E45 E47 E55 E65 ⋄ REV MR 46 # 8834 Zbl 257 # 02035 JSL 40.632 • ID 19016

JENSEN, R.B. & PRESTEL, A. (EDS.) [1981] *Set theory and model theory. Proceedings of an informal symposium, held at Bonn, June 1-3, 1979* (S 3301) Lect Notes Math 872*v+174pp
⋄ C97 E97 ⋄ REV MR 83b:03005 Zbl 453 # 00008 • ID 54128

JENSEN, R.B. see Vol. IV, V, VI for further entries

JEROSLOW, R. [1971] *Non-effectiveness in S.Orey's arithmetical compactness theorem* (J 0068) Z Math Logik Grundlagen Math 17*285-289
⋄ C62 ⋄ REV MR 45 # 1762 Zbl 234 # 02036 • ID 06604

JEROSLOW, R. see Vol. II, IV, VI for further entries

JERVELL, H.R. [1972] *Herbrand and Skolem theorems in infinitary languages* (S 1626) Oslo Preprint Ser 3
⋄ C75 F05 ⋄ ID 33262

JERVELL, H.R. [1973] *Skolem and Herbrand theorems in first order logic* (S 1626) Oslo Preprint Ser 6*77pp
⋄ B10 C07 F05 ⋄ ID 33265

JERVELL, H.R. [1975] *Conservative endextensions and the quantifier "there exist uncountably many"* (P 0757) Scand Logic Symp (3);1973 Uppsala 63-80
⋄ C30 C80 ⋄ REV MR 51 # 10027 Zbl 309 # 02011 JSL 43.373 • ID 17561

JERVELL, H.R. see Vol. II, IV, V, VI for further entries

JEZEK, J. [1976] *Universal algebra and model theory (Czech)* (X 1906) SNTL: Prague 227pp
⋄ C05 C98 ⋄ REV MR 80f:08001 • ID 81717

JEZEK, J. [1981] *The lattice of equational theories I: Modular elements. II: The lattice of full sets of terms* (J 0022) Cheskoslov Mat Zh 31(106)*127-152,573-603
⋄ C05 C40 G10 ⋄ REV MR 84e:08007 Zbl 477 # 08006 Zbl 486 # 08009 • REM Part III 1982 • ID 55600

JEZEK, J. [1982] *The lattice of equational theories III: definability and automorphisms* (J 0022) Cheskoslov Mat Zh 32(107)*129-165
⋄ C05 C40 G10 ⋄ REV MR 84e:08007c Zbl 499 # 08005 • REM Parts I,II 1981 • ID 38294

JEZEK, J. [1985] *Elementarily nonequivalent infinite partition lattices* (J 0004) Algeb Universalis 20*132-133
⋄ C07 E20 G10 ⋄ REV Zbl 565 # 06006 • ID 46363

JEZEK, J. see Vol. V for further entries

JOCKUSCH JR., C.G. [1977] see HENSON, C.W.

JOCKUSCH JR., C.G. & KALANTARI, I. [1984] *Recursively enumerable sets and van der Waerden's theorem on arithmetic progressions* (J 0048) Pac J Math 115*143-153
⋄ C57 D25 D80 ⋄ REV MR 86d:03045 Zbl 571 # 03017 • ID 39945

JOCKUSCH JR., C.G. see Vol. I, IV, V, VI for further entries

JOHNSON, D.G. [1960] *A structure theory for a class of lattice-ordered rings* (J 0118) Acta Math 104*163-215
⋄ C60 G10 ⋄ REV MR 23 # A2447 Zbl 94.253 JSL 35.347 • ID 06647

JOHNSON, D.R. & THOMASON, R.H. [1969] *Predicate calculus with free quantifier variables* (J 0036) J Symb Logic 34*1-7
⋄ B15 C80 ⋄ REV MR 39 # 5328 Zbl 188.314 • ID 06648

JOHNSON, D.R. see Vol. I for further entries

JOHNSON, J. & SEIFERT, R.L. [1967] *A survey of multiunary algebras* (1111) Preprints, Manuscr., Techn. Reports etc. 27pp
⋄ C05 C98 ⋄ REM Seminar Notes, Berkeley • ID 19037

JOHNSON, J.S. [1973] *Axiom systems for first order logic with finitely many variables* (J 0036) J Symb Logic 38*576-578
⋄ B20 C05 G15 ⋄ REV MR 49 # 8847 Zbl 286 # 02017 • ID 06681

JOHNSTON, J.B. [1956] *Universal infinite partially ordered sets* (J 0053) Proc Amer Math Soc 7*507-514
⋄ C50 C65 E07 E50 ⋄ REV MR 20 # 3090 Zbl 71.50 • ID 42818

JOHNSTONE, P.T. [1977] *Rings, fields, and spectra* (J 0032) J Algeb 49*238-260
⋄ C90 F50 G30 ⋄ REV MR 56 # 8335 Zbl 369 # 13019 • ID 51354

JOHNSTONE, P.T. [1979] *Another condition equivalent to De Morgan's law* (J 0394) Commun Algeb 7*1309-1312
⋄ C60 C90 F50 G05 G30 ⋄ REV MR 81d:03067 Zbl 417 # 18002 • ID 74552

JOHNSTONE, P.T. [1979] *Conditions related to de Morgan's law* (P 2901) Appl Sheaves;1977 Durham 479-491
⋄ B55 C65 C90 F50 G30 ⋄ REV MR 81e:03062 Zbl 445 # 03041 • ID 56505

JOHNSTONE, P.T. [1979] *Remarks on the previous paper. Extracts from two letters to Anders Kock* (S 3462) Var Publ Ser, Aarhus Univ 30*137-138
⋄ C60 C90 G30 ⋄ REV Zbl 428 # 03057 • ID 53816

JOHNSTONE, P.T. [1980] *The Gleason cover of a topos. I* (J 0326) J Pure Appl Algebra 19*171-192
⋄ C90 G30 ⋄ REV MR 82a:18002 Zbl 445 # 18004 • REM Part II 1981 • ID 56515

JOHNSTONE, P.T. see Vol. V, VI for further entries

JOHNSTONE JR., H.W. [1967] *An inductive decision-procedure for the monadic predicate calculus* (J 0079) Logique & Anal, NS 10*324-327
⋄ B25 ⋄ REV MR 37 # 49 Zbl 155.341 • ID 06688

JOHNSTONE JR., H.W. see Vol. I, II, VI for further entries

JOKISZ-ZABILSKA, I. [1984] *On a construction of many-sorted structures* (J 1008) Demonstr Math (Warsaw) 17*757-765
⋄ C30 ⋄ REV Zbl 576 # 03019 • ID 48024

JONES, N.D. & SELMAN, A.L. [1972] *Turing machines and the spectra of first-order formulas with equality* (P 1901) ACM Symp Th of Comput (4);1972 Denver 157-167
⋄ C13 D10 D15 ⋄ REV Zbl 381 # 03026 • ID 51894

JONES, N.D. & SELMAN, A.L. [1974] *Turing machines and the spectra of first-order formulas* (J 0036) J Symb Logic 39*139-150
⋄ C13 D10 D15 ⋄ REV MR 58 # 5164 Zbl 288 # 02021 • ID 06700

JONES, N.D. see Vol. IV for further entries

JONGH DE, D.H.J. [1974] see GABBAY, D.M.

JONGH DE, D.H.J. & SMORYNSKI, C.A. [1976] *Kripke models and the intuitionistic theory of species* (J 0007) Ann Math Logic 9*157-186
⋄ C90 F35 F50 ⋄ REV MR 53 # 5266 Zbl 299 # 02037 Zbl 317 # 02037 • ID 18214

JONGH DE, D.H.J. [1980] *A class of intuitionistic connectives* (P 2058) Kleene Symp;1978 Madison 103-111
⋄ B55 C75 C90 F50 ⋄ REV MR 81m:03013 Zbl 479 # 03015 • ID 55677

JONGH DE, D.H.J. see Vol. IV, V, VI for further entries

JONSSON, B. & TARSKI, A. [1947] *Direct decomposition of finite algebraic systems* (X 0845) Univ Notre Dame Pr: Notre Dame v+64pp
⋄ C05 C13 ⋄ REV MR 8.560 Zbl 41.345 • ID 25199

JONSSON, B. & TARSKI, A. [1949] *Cardinal products of isomorphism types* (**X** 0894) Oxford Univ Pr: Oxford 253-312
⋄ C05 C30 E75 ⋄ REV Zbl 41.345 JSL 14.188 • REM Appendix to Tarski,A.: Cardinal algebras; Oxford Univ. Press 1949 • ID 41606

JONSSON, B. [1956] *Universal relational systems* (**J** 0132) Math Scand 4∗193-208
⋄ C50 ⋄ REV MR 20#3091 Zbl 77.253 JSL 32.534 • ID 06710

JONSSON, B. [1957] *On direct decompositions of torsion-free abelian groups* (**J** 0132) Math Scand 5∗230-235
⋄ C05 C30 C60 ⋄ REV MR 21#7170 Zbl 81.262 • ID 06712

JONSSON, B. [1957] *On isomorphism types of groups and other algebraic systems* (**J** 0132) Math Scand 5∗224-229
⋄ C05 C30 C60 ⋄ REV MR 21#7169 Zbl 81.262 • ID 06711

JONSSON, B. [1960] *Homogeneous universal relational systems* (**J** 0132) Math Scand 8∗137-142
⋄ C50 ⋄ REV MR 23#A2328 JSL 32.534 • ID 06715

JONSSON, B. & TARSKI, A. [1961] *On two properties of free algebras* (**J** 0132) Math Scand 9∗95-101
⋄ C05 ⋄ REV MR 23#A3695 • ID 06717

JONSSON, B. [1962] *Algebraic extensions of relational systems* (**J** 0132) Math Scand 11∗197-205
⋄ C05 C52 C60 ⋄ REV MR 27#4777 • ID 06721

JONSSON, B. [1964] see CHANG, C.C.

JONSSON, B. [1964] see CRAWLEY, P.

JONSSON, B. [1965] *Extensions of relational structures* (**P** 0614) Th Models;1963 Berkeley 146-157
⋄ C30 C50 C52 ⋄ REV MR 34#2463 Zbl 158.263 • ID 22313

JONSSON, B. [1966] *The unique factorization problem for finite relational structures* (**S** 0019) Colloq Math (Warsaw) 14∗1-32
⋄ C05 C13 ⋄ REV MR 33#87 • ID 06723

JONSSON, B. [1967] *Algebras whose congruence lattices are distributive* (**J** 0132) Math Scand 21∗110-121
⋄ C05 G10 ⋄ REV MR 38#5689 Zbl 167.284 • ID 06724

JONSSON, B. [1968] *Algebraic structures with prescribed automorphism groups* (**S** 0019) Colloq Math (Warsaw) 19∗1-4
⋄ C05 C07 ⋄ REV MR 36#6336 Zbl 153.336 • ID 06725

JONSSON, B. & OLIN, P. [1968] *Almost direct products and saturation* (**J** 0020) Compos Math 20∗125-132
• REPR [1968] (**C** 0727) Logic Found of Math (Heyting) 125-132
⋄ C05 C20 C50 G05 ⋄ REV MR 37#2589 Zbl 155.35 • ID 06727

JONSSON, B. [1972] *Topics in universal algebra* (**X** 0811) Springer: Heidelberg & New York vi+220pp
⋄ C05 ⋄ REV MR 49#10625 Zbl 225#08001 • ID 21375

JONSSON, B. [1974] *Some recent trends in general algebra* (**P** 0610) Tarski Symp;1971 Berkeley 1-19
⋄ C05 C98 ⋄ REV MR 51#7988 Zbl 311#08002 • ID 06734

JONSSON, B. & MCNULTY, G.F. & QUACKENBUSH, R.W. [1975] *The ascending and descending varietal chains of a variety* (**J** 0017) Canad J Math 27∗25-31
⋄ B25 C05 D35 ⋄ REV MR 50#12860 Zbl 305#08003 • ID 06740

JONSSON, B. & OLIN, P. [1976] *Elementary equivalence and relatively free products of lattices* (**J** 0004) Algeb Universalis 6∗313-325
⋄ C05 C30 G10 ⋄ REV MR 55#215 Zbl 381#06007 • ID 24520

JONSSON, B. [1979] *On finitely based varieties of algebras* (**S** 0019) Colloq Math (Warsaw) 42∗255-261
⋄ C05 ⋄ REV MR 81g:08015 Zbl 427#08003 • ID 81735

JONSSON, B. [1980] *Congruence varieties* (**J** 0004) Algeb Universalis 10∗355-394
⋄ C05 ⋄ REV MR 81e:08004 Zbl 438#08003 • ID 69646

JONSSON, B. [1982] *Arithmetic of ordered sets* (**P** 3780) Ordered Sets;1981 Banff 3-41
⋄ C65 C98 E07 ⋄ REV MR 83i:06004 Zbl 499#06001 • ID 38290

JONSSON, B. [1982] *Varieties of relation algebras* (**J** 0004) Algeb Universalis 15∗273-298
⋄ C05 C52 ⋄ REV MR 84g:08023 Zbl 545#08009 • ID 39444

JONSSON, B. see Vol. V for further entries

JORDAN, P. [1949] *Zur Axiomatik der Verknuepfungsbereiche* (**J** 0107) Abh Math Sem Univ Hamburg 16∗54-70
⋄ C05 C75 G10 ⋄ REV MR 11.75 Zbl 37.156 JSL 23.361 • ID 06743

JORDAN, P. see Vol. II for further entries

JORGENSEN, M. [1972] see ADLER, A.

JORGENSEN, M. [1975] *Regular ultrafilters and long ultrapowers* (**J** 0018) Canad Math Bull 18∗41-43
⋄ C20 E05 ⋄ REV MR 52#13374 Zbl 319#02043 • ID 21825

JORGENSEN, M. see Vol. V for further entries

JOSE, R. [1954] see FUENTES MIRAS, J.R.

JOVANOVIC, A. [1978] *A note on ultrapower cardinality* (**J** 0400) Publ Inst Math, NS (Belgrade) 24(38)∗79-81
⋄ C20 C55 E05 E10 ⋄ REV MR 80d:04005 Zbl 401#03010 • ID 74570

JOVANOVIC, A. [1980] *A note on the two-cardinal problem* (**J** 0400) Publ Inst Math, NS (Belgrade) 28(42)∗101-103
⋄ C20 C55 E10 E55 ⋄ REV MR 83g:03030 Zbl 483#03021 • ID 36003

JOVANOVIC, A. [1980] *Ultraproducts of well orders* (**J** 0400) Publ Inst Math, NS (Belgrade) 27(41)∗99-102
⋄ C20 C55 E05 E07 ⋄ REV MR 82i:03038 Zbl 483#03020 • ID 74569

JOVANOVIC, A. [1983] *On the two-cardinal problem* (**P** 3855) Algeb Conf (3);1982 Beograd 73-83
⋄ C20 C55 E05 ⋄ REV MR 85j:03047 Zbl 538#03027 • ID 41471

JOVANOVIC, A. see Vol. V for further entries

JOYAL, A. [1975] *Les theoremes de Chevalley-Tarski et remarque sur l'algebre constructive* (**J** 0306) Cah Topol & Geom Differ 16∗256-258
⋄ C60 F50 G30 ⋄ REV Zbl 354#02038 • ID 27668

JOYAL, A. see Vol. V, VI for further entries

JOYNER JR., W.M. [1973] *Automatic theorem-proving and the decision problem* (P 3062) IEEE Symp Switch & Automata Th (14);1973 Iowa City 159-166
⋄ B25 B35 ⋄ REV MR 55 # 6989 • ID 81740

JOYNER JR., W.M. [1976] *Resolution strategies as decision procedures* (J 0037) ACM J 23*398-417
⋄ B25 B35 ⋄ REV MR 53 # 15007 Zbl 335 # 68062
• ID 23231

JOZSA, R. [1984] *Sheaf models and massless fields* (J 2736) Int J Theor Phys 23*67-97
⋄ B30 C90 F50 G30 ⋄ REV MR 85m:81076 • ID 45334

JUKNA, S. [1979] *On decidability of equivalence problem in some algebras of recursive functions (Russian)* (J 2574) Litov Mat Sb (Vil'nyus) 19/3*133-136
⋄ B25 D20 D75 ⋄ ID 33258

JUKNA, S. [1981] *On the decidability of Hoare's system* (P 3642) Colloq Math Log in Computer Sci;1978 Salgotarjan
⋄ B25 B75 ⋄ ID 33909

JUKNA, S. see Vol. IV, VI for further entries

JURIE, P.-F. [1973] *Une extension de la theorie de Galois abstraite finitaire* (J 2313) C R Acad Sci, Paris, Ser A-B 276*A81-A84
⋄ C60 G15 ⋄ REV MR 47 # 6481 Zbl 259 # 02045
• ID 06778

JURIE, P.-F. [1982] *Decidabilite de la theorie elementaire des anneaux booleiens a operateurs dans un group fini (English summary)* (J 3364) C R Acad Sci, Paris, Ser 1 295*215-217
⋄ B25 C60 ⋄ REV MR 84a:03016 Zbl 519 # 03007
• ID 35562

JURIE, P.-F. & TOURAILLE, A. [1984] *Ideaux elementairement equivalents dans une algebre Booleenne* (J 3364) C R Acad Sci, Paris, Ser 1 299*415-418
⋄ C07 G05 ⋄ REV MR 85j:03111 • ID 39636

KAHR, A.S. [1962] see DREBEN, B.

KAHR, A.S. see Vol. I, IV for further entries

KAISER, KLAUS [1969] *Ueber eine Verallgemeinerung der Robinsonschen Modellvervollstaendigung I,II (II: Abgeschlossene Algebren)* (J 0068) Z Math Logik Grundlagen Math 15*37-48,123-134
⋄ C05 C25 C35 ⋄ REV MR 39 # 48 MR 41 # 1520
Zbl 188.321 • ID 06797

KAISER, KLAUS [1970] *Zur Einbettungsbedingung im Zusammenhang mit einer Programmiersprache von Engeler* (J 0044) Math Z 113*363-368
⋄ B75 C20 C35 C52 ⋄ REV MR 42 # 4385 Zbl 188.20
• ID 06799

KAISER, KLAUS [1970] *Zur Theorie algorithmisch abgeschlossener Modellklassen* (J 0041) Math Syst Theory 4*160-167
⋄ C20 C52 D75 ⋄ REV MR 42 # 4384 Zbl 202.8
• ID 06800

KAISER, KLAUS [1971] see ARMBRUST, M.

KAISER, KLAUS [1972] see ARMBRUST, M.

KAISER, KLAUS [1973] *Diagonal-like embeddings* (S 0019) Colloq Math (Warsaw) 27*171-174
⋄ C20 C52 ⋄ REV MR 49 # 2339 Zbl 238 # 02043
• ID 06803

KAISER, KLAUS [1973] *On generalized projective classes* (S 0019) Colloq Math (Warsaw) 28*1-5
⋄ C20 C52 ⋄ REV MR 48 # 3731 Zbl 259 # 02037
• ID 06801

KAISER, KLAUS [1973] *On reductive and projective classes* (S 0019) Colloq Math (Warsaw) 28*7-9
⋄ C20 C52 ⋄ REV MR 48 # 3730 Zbl 259 # 02038
• ID 06802

KAISER, KLAUS [1974] *Direct limits in quasi-universal model classes* (J 3172) J London Math Soc, Ser 2 9*585-588
⋄ B15 C30 C52 C85 ⋄ REV MR 52 # 77 Zbl 309 # 02054
• ID 06804

KAISER, KLAUS [1974] see ARMBRUST, M.

KAISER, KLAUS [1975] *Various remarks on quasi-universal model classes* (J 0009) Arch Math Logik Grundlagenforsch 17*91-95
⋄ B15 C05 C52 C85 ⋄ REV MR 52 # 5407
Zbl 331 # 02031 • ID 06805

KAISER, KLAUS [1976] *Quasi-axiomatic classes* (J 0009) Arch Math Logik Grundlagenforsch 17*129-134
⋄ B15 C20 C52 C75 C85 ⋄ REV MR 54 # 4968
Zbl 331 # 02032 • ID 06806

KAISER, KLAUS [1977] *On generic stalks of sheaves* (J 3172) J London Math Soc, Ser 2 16(2)*385-392
⋄ C25 C90 ⋄ REV MR 57 # 9521 Zbl 373 # 02037
• ID 31803

KAISER, KLAUS [1978] *On relational selections for complete theories* (J 0009) Arch Math Logik Grundlagenforsch 19*79-87
⋄ C20 C35 C50 C60 ⋄ REV MR 80c:03037
Zbl 398 # 03022 • ID 31804

KAISER, KLAUS [1980] *On a lattice of relational reducts* (J 0068) Z Math Logik Grundlagen Math 26*429-432
⋄ C52 C75 ⋄ REV MR 82d:06010 Zbl 453 # 03041
• ID 54171

KAISER, KLAUS [1981] *Lattices acting on universal classes* (J 0068) Z Math Logik Grundlagen Math 27*127-130
⋄ C52 G10 ⋄ REV MR 82j:03080 Zbl 472 # 03026
• ID 55291

KAISER, KLAUS [1981] *On complementedly normal lattices* (J 0008) Arch Math (Basel) 37*278-284
⋄ C52 G10 ⋄ REV MR 83c:06014 Zbl 474 # 06013 • REM
Part I. Part II 1984 • ID 55442

KAKUDA, Y. [1980] *Set theory based on the language with the additional quantifier "for almost all" I* (J 1508) Math Sem Notes, Kobe Univ 8*603-609
⋄ C55 C62 C80 E70 ⋄ REV MR 82f:03044
Zbl 497 # 03042 • ID 74609

KAKUDA, Y. [1981] *Completeness theorem for the system ST with the quantifier "aa"* (P 4153) B-Val Anal & Nonstand Anal;1981 Kyoto 122-128
⋄ C55 C80 E70 ⋄ ID 47772

KAKUDA, Y. [1983] *Reflection principles via filter quantifier (Japanese)* (P 4113) Found of Math;1982 Kyoto 53-63
⋄ C55 C80 E30 ⋄ ID 47676

KAKUDA, Y. see Vol. I, V for further entries

KALANTARI, I. & RETZLAFF, A.T. [1977] *Maximal vector spaces under automorphisms of the lattice of recursively enumerable vector spaces* (J 0036) J Symb Logic 42*481-491
⋄ C57 C60 D45 ⋄ REV MR 58 # 21539 Zbl 383 # 03033 • ID 26850

KALANTARI, I. [1978] *Major subspaces of recursively enumerable vector spaces* (J 0036) J Symb Logic 43*293-303
⋄ C57 C60 D25 D45 ⋄ REV MR 81f:03058 Zbl 403 # 03034 • ID 29262

KALANTARI, I. [1979] *Automorphisms of the lattice of recursively enumerable vector spaces* (J 0068) Z Math Logik Grundlagen Math 25*385-401
⋄ C57 C60 D45 ⋄ REV MR 81b:03047 Zbl 424 # 03023 • ID 74612

KALANTARI, I. & RETZLAFF, A.T. [1979] *Recursive constructions in topological spaces* (J 0036) J Symb Logic 44*609-625
⋄ C57 C65 D45 ⋄ REV MR 81f:03059 Zbl 427 # 03035 • ID 53722

KALANTARI, I. [1982] *Major subsets in effective topology* (P 3634) Patras Logic Symp;1980 Patras 77-94
⋄ C57 C65 D25 D45 ⋄ REV MR 84g:03064 Zbl 521 # 03029 • ID 34171

KALANTARI, I. & LEGGETT, A. [1982] *Simplicity in effective topology* (J 0036) J Symb Logic 47*169-183
⋄ C57 C65 D25 ⋄ REV MR 83m:03053 Zbl 516 # 03023 • ID 35452

KALANTARI, I. & REMMEL, J.B. [1983] *Degrees of recursively enumarable topological spaces* (J 0036) J Symb Logic 48*610-622
⋄ C57 C65 D25 D45 ⋄ REV MR 85i:03151 Zbl 532 # 03021 • ID 33595

KALANTARI, I. & LEGGETT, A. [1983] *Maximality in effective topology* (J 0036) J Symb Logic 48*100-112
⋄ C57 C65 D25 D45 ⋄ REV MR 84k:03112 Zbl 532 # 03020 • ID 36130

KALANTARI, I. [1984] see JOCKUSCH JR., C.G.

KALANTARI, I. & WEITKAMP, G. [1985] *Effective topological spaces I: A definability theory. II: A hierarchy* (J 0073) Ann Pure Appl Logic 29*1-27,207-224
⋄ C40 C57 C65 D45 D75 D80 ⋄ REV Zbl 569 # 03018 • ID 47472

KALANTARI, I. [1985] see DOWNEY, R.G.

KALANTARI, I. see Vol. IV for further entries

KALFA, C. [1984] *Decidable properties of finite sets of equations in trivial languages* (J 0036) J Symb Logic 49*1333-1338
⋄ C05 C35 C52 ⋄ REV MR 86h:03017 Zbl 575 # 03006 • ID 40045

KALFA, C. see Vol. IV for further entries

KALICKI, C. [1980] *Infinitary propositional intuitionistic logic* (J 0047) Notre Dame J Formal Log 21*216-228
⋄ B55 C75 F50 ⋄ REV MR 81e:03025 Zbl 336 # 02021 • ID 53776

KALICKI, J. [1952] *On comparison of finite algebras* (J 0053) Proc Amer Math Soc 3*36-40
⋄ B25 C05 C13 ⋄ REV MR 13.898 Zbl 46.31 JSL 22.378 • ID 06814

KALICKI, J. & SCOTT, D.S. [1955] *Equational completeness of abstract algebras* (J 0028) Indag Math 17*650-659
⋄ C05 ⋄ REV MR 17.571 Zbl 73.245 JSL 23.56 • ID 06817

KALICKI, J. [1955] *The number of equationally complete classes of equations* (J 0028) Indag Math 17*660-662
⋄ C05 ⋄ REV MR 17.571 Zbl 73.246 JSL 23.56 • ID 06816

KALICKI, J. see Vol. I, II, IV for further entries

KALLICK, B. [1969] *A decision procedure based on the resolution method* (P 0594) Inform Processing (4);1968 Edinburgh 1*269-275
⋄ B25 B35 ⋄ REV Zbl 213.19 JSL 38.656 • ID 22190

KALMAR, L. [1928] *Eine Bemerkung zur Entscheidungstheorie* (J 0460) Acta Univ Szeged, Sect Mat 4*248-252
⋄ B20 B25 F30 ⋄ REV FdM 55.32 • ID 06835

KALMAR, L. [1930] *Ein Beitrag zum Entscheidungsproblem* (J 0460) Acta Univ Szeged, Sect Mat 5*222-236
⋄ B25 D35 ⋄ REV Zbl 4.146 FdM 57.1321 • ID 06836

KALMAR, L. [1932] *Zum Entscheidungsproblem der mathematischen Logik* (P 0653) Int Congr Math (II, 4);1932 Zuerich 2*337-338
⋄ B20 B25 D35 ⋄ REV FdM 58.70 • ID 06837

KALMAR, L. [1933] *Ueber die Erfuellbarkeit derjenigen Zaehlausdruecke, welche in der Normalform zwei benachbarte Allzeichen enthalten* (J 0043) Math Ann 108*466-484
⋄ B20 B25 ⋄ REV Zbl 6.385 FdM 59.864 • ID 06838

KALMAR, L. [1934] *Ueber einen Loewenheimschen Satz* (J 0460) Acta Univ Szeged, Sect Mat 7*112-121
⋄ C07 ⋄ REV Zbl 11.3 FdM 60.851 • ID 06839

KALMAR, L. see Vol. I, II, IV, V, VI for further entries

KAL'NEJ, V.E. [1970] see BREDIKHIN, S.V.

KAMIMURA, T. [1985] *An effective given initial semigroup* (J 1431) Acta Inf 203-227
⋄ C57 D45 ⋄ REV Zbl 571 # 20057 • ID 48019

KAMIMURA, T. see Vol. IV for further entries

KAMO, S. [1975] *Completeness theorem of a logical system on a complete boolean algebra* (J 0350) Sci Rep Tokyo Kyoiku Daigaku Sect A 13*23-27
⋄ C90 G05 ⋄ REV MR 52 # 10405 Zbl 347 # 02039 • ID 21705

KAMO, S. [1977] *Reduced powers and models for infinitely logic* (J 0407) Comm Math Univ St Pauli (Tokyo) 26*187-194
⋄ C20 C75 C90 ⋄ REV MR 57 # 5736 Zbl 378 # 02024 • ID 51806

KAMO, S. [1981] *Nonstandard natural number systems and nonstandard models* (J 0036) J Symb Logic 46*365-376
⋄ H15 ⋄ REV MR 82g:03112 Zbl 471 # 03052 • ID 55247

KAMO, S. see Vol. V for further entries

KANAMORI, A. [1978] see SOLOVAY, R.M.

KANAMORI, A. & TAYLOR, A.D. [1984] *Separating ultrafilters on uncountable cardinals* (J 0029) Israel J Math 47*131-138
⋄ C20 C55 E05 E55 ⋄ REV MR 85h:03058 Zbl 541 # 03031 • ID 40364

KANAMORI, A. see Vol. V for further entries

KANDA, A. [1985] *Numeration models of λ-calculus* (J 0068) Z Math Logik Grundlagen Math 31*209-220
⋄ B40 B60 C57 D45 ⋄ ID 47568

KANDA, A. see Vol. IV, V, VI for further entries

KANGER, S. [1968] *Equivalent theories* (**J** 0105) Theoria (Lund) 34*1-6
⋄ C07 G05 ⋄ REV MR 39 # 6742 • ID 06889

KANGER, S. [1970] *Equational calculi and automatic demonstration* (**C** 0735) Logic & Value (Dahlquist) 220-226
⋄ B25 B35 C05 ⋄ REV MR 53 # 12897 • ID 23148

KANGER, S. see Vol. I, II, VI for further entries

KANI, E. [1980] *Nonstandard diophantine geometry* (**P** 3241) Queen's Numb Th Conf;1979 Kingston 129-172
⋄ H15 H20 ⋄ REV MR 83a:10020 Zbl 454 # 14010
• ID 54248

KANOVEJ, V.G. [1974] *On degrees of constructibility and descriptive properties of the set of real numbers in an initial model and in its extensions (Russian)* (**J** 0023) Dokl Akad Nauk SSSR 216*728-729
• TRANSL [1974] (**J** 0062) Sov Math, Dokl 15*866-868
⋄ C62 D30 E40 E45 ⋄ REV MR 51 # 143
Zbl 327 # 02051 • ID 17445

KANOVEJ, V.G. [1979] *The set of all analytically definable sets of natural numbers can be defined analytically (Russian)* (**J** 0216) Izv Akad Nauk SSSR, Ser Mat 43*1259-1293
• TRANSL [1980] (**J** 0448) Math of USSR, Izv 15*469-500
⋄ C62 D55 E15 E35 E45 ⋄ REV MR 82a:03043
Zbl 427 # 03045 • ID 54105

KANOVEJ, V.G. see Vol. I, IV, V, VI for further entries

KANOVICH, M.I. [1975] *A hierachical semantic system with set variables (Russian)* (**J** 0023) Dokl Akad Nauk SSSR 221*1256-1259
• TRANSL [1975] (**J** 0062) Sov Math, Dokl 16*504-509
⋄ B20 B65 C40 D05 F50 ⋄ REV MR 52 # 5385
Zbl 325 # 02030 • ID 18218

KANOVICH, M.I. [1984] *Restricted quantors in a consecutive semantical system (Russian)* (**P** 4392) Mat Logika (Markova);1980 Sofia 63-64
⋄ C80 ⋄ ID 46961

KANOVICH, M.I. see Vol. I, IV, VI for further entries

KAPETANOVIC, M. [1983] *Some simple decidability proofs* (**J** 0042) Mat Vesn, Drust Mat Fiz Astron Serb 7(20)(35)*27-29
⋄ B20 B25 ⋄ REV MR 85f:03008 Zbl 553 # 03004
• ID 40445

KAPETANOVIC, M. see Vol. II for further entries

KARGAPOLOV, M.I. [1962] *On the elementary theory of lattices of subgroups (Russian)* (**J** 0003) Algebra i Logika 1/3*46-53
⋄ C60 G10 ⋄ REV MR 27 # 4754 JSL 32.279 • ID 19059

KARGAPOLOV, M.I. [1962] *On the elementary theory of abelian groups (Russian)* (**J** 0003) Algebra i Logika 1/6*26-36
⋄ C35 C60 ⋄ REV MR 31 # 251 JSL 32.535 • ID 19058

KARGAPOLOV, M.I. [1963] *Classification of ordered abelian groups by elementary properties (Russian)* (**J** 0003) Algebra i Logika 2/2*31-46
⋄ C60 ⋄ REV MR 27 # 5823 • ID 90263

KARGAPOLOV, M.I. [1970] *Universal groups (Russian)* (**J** 0003) Algebra i Logika 9*428-435
• TRANSL [1970] (**J** 0069) Algeb and Log 9*257-261
⋄ C50 C60 ⋄ REV MR 43 # 6312 Zbl 217.358 • ID 06926

KARGAPOLOV, M.I. see Vol. I, IV for further entries

KARP, C.R. [1962] *Independence proofs in predicate logic with infinitely long expressions* (**J** 0036) J Symb Logic 27*171-188
⋄ C75 ⋄ REV MR 28 # 19 Zbl 114.245 JSL 30.96
• ID 06927

KARP, C.R. [1963] *A note on the representation of α-complete boolean algebras* (**J** 0053) Proc Amer Math Soc 14*705-707
⋄ C75 G05 ⋄ REV MR 27 # 3570 JSL 30.393 • ID 06928

KARP, C.R. [1964] *Languages with expressions of infinite length* (**X** 0809) North Holland: Amsterdam xix+183pp
⋄ C75 C98 G05 ⋄ REV MR 31 # 1178 Zbl 127.9
JSL 33.477 • ID 06929

KARP, C.R. [1965] *Finite-quantifier equivalence* (**P** 0614) Th Models;1963 Berkeley 407-412
⋄ C75 ⋄ REV MR 35 # 36 Zbl 253 # 02052 JSL 36.158
• ID 06930

KARP, C.R. [1967] *Nonaxiomatizability results for infinitary systems* (**J** 0036) J Symb Logic 32*367-384
⋄ C75 G05 ⋄ REV MR 36 # 2484 Zbl 203.10 JSL 33.478
• ID 06931

KARP, C.R. [1968] *An algebraic proof of the Barwise compactness theorem* (**P** 0637) Syntax & Semant Infinitary Lang;1967 Los Angeles 80-95
⋄ C70 C90 ⋄ REV Zbl 195.303 JSL 39.335 • ID 06933

KARP, C.R. [1974] *Infinite-quantifier languages and ω-chains of models* (**P** 0610) Tarski Symp;1971 Berkeley 225-232
⋄ C40 C75 ⋄ REV MR 51 # 12517 Zbl 308 # 02016
• ID 06936

KARP, C.R. see Vol. IV, V for further entries

KARR, M. [1981] *Summation in finite terms* (**J** 0037) ACM J 28*305-350
⋄ C57 ⋄ REV MR 82m:12018 Zbl 494 # 68044 • ID 81790

KARR, M. see Vol. VI for further entries

KARTTUNEN, M. [1979] *Infinitary languages $N_{\infty\lambda}$ and generalized partial isomorphisms* (**P** 1705) Scand Logic Symp (4);1976 Jyvaeskylae 153-168
⋄ C75 ⋄ REV MR 82k:03051 Zbl 413 # 03023 • ID 53017

KARTTUNEN, M. [1981] *Model theoretic results for infinitely deep languages* (**J** 3437) Rep, Akad Wiss DDR, Inst Math R-MATH-03/81*66-101
⋄ C20 C75 ⋄ REV MR 84h:03088 Zbl 478 # 03017
• ID 55632

KARTTUNEN, M. [1982] *Model theoretic results for infinitely deep languages* (**J** 0387) Bull Sect Logic, Pol Acad Sci 11*21-23
⋄ C20 C75 ⋄ REV Zbl 527 # 03017 • ID 37568

KARTTUNEN, M. [1983] *Model theoretic results for infinitely deep languages* (**J** 0063) Studia Logica 42*223-241
⋄ C20 C75 ⋄ REV MR 85i:03117 Zbl 568 # 03016
• ID 42335

KARTTUNEN, M. [1984] *Model theory for infinitely deep languages* (**J** 0446) Ann Acad Sci Fennicae, Ser A I, Diss 50*96pp
⋄ C20 C75 ⋄ REV MR 86h:03071 Zbl 568 # 03017
• ID 44045

KASAHARA, S. [1973] *A characterization of nonstandard real fields* (**J** 1508) Math Sem Notes, Kobe Univ 1*vi+8pp
• REPR [1974] (**J** 0081) Proc Japan Acad 50*53-56
⋄ C20 C60 H05 ⋄ REV MR 51 # 12804 Zbl 287 # 02039
• ID 30535

KASAHARA, S. see Vol. I, V for further entries

KASHER, A. [1979] see GABBAY, D.M.

KASHER, A. see Vol. I, II for further entries

KASHIWAGI, T. [1972] *On (m,A)-implicational classes* (J 0352) Math Jap 17*1-12
  ⋄ C20 C52 C75 ⋄ REV MR 48 # 5846 Zbl 251 # 02018 • ID 06945

KASHIWAGI, T. [1974] *Note on locally definable classes of structures* (J 0081) Proc Japan Acad 50*436-439
  ⋄ C52 C75 ⋄ REV MR 52 # 78 Zbl 311 # 08007 • ID 18220

KASHIWAGI, T. [1976] *A generalizaton of a certain compactness property* (J 2751) J Fac Lib Art Yamaguchi Univ 10*215-217
  ⋄ C20 C52 C75 ⋄ REV MR 83c:03031 • ID 35136

KASHIWAGI, T. [1976] *On certain compactness properties for classes of structures* (J 0352) Math Jap 21*327-332
  ⋄ C20 C52 ⋄ REV MR 58 # 27453 Zbl 363 # 02062 • ID 50888

KASHIWAGI, T. [1976] *On locally definable classes closed under reduced products* (J 2751) J Fac Lib Art Yamaguchi Univ 10*211-213
  ⋄ C20 C75 ⋄ REV MR 83c:03029 • ID 35135

KASHIWAGI, T. [1981] *On classes which have the weak amalgamation property* (P 2998) Symp Semigroups (5);1981 Sakado 10-17
  ⋄ C25 C52 ⋄ REV MR 83d:03045 Zbl 489 # 03010 • ID 34826

KASHIWAGI, T. [1982] *Note on algebraic extensions in universal classes* (J 2751) J Fac Lib Art Yamaguchi Univ 16*7-12
  ⋄ C05 C25 C30 ⋄ REV MR 85c:03011 • ID 39831

KASHIWAGI, T. [1984] *A certain compactness property and free structures* (J 2751) J Fac Lib Art Yamaguchi Univ 18*13-16
  ⋄ C52 ⋄ ID 45087

KASHIWAGI, T. [1985] *Algebraically closed structures and a certain kind of saturated structure* (P 4423) Semigroups (8);1984 Matsue 40-44
  ⋄ C50 ⋄ ID 48349

KASIMOV, YU.F. [1978] see AMERBAEV, V.M.

KASSLER, M. [1976] *The decidability of languages that assert music* (J 1023) Perspectives of New Music 14*249-251
  ⋄ B25 B80 ⋄ ID 32363

KASSLER, M. see Vol. I for further entries

KATZ, M. [1982] *Real-valued models with metric equality and uniformly continuous predicates* (J 0036) J Symb Logic 47*772-792
  ⋄ C90 ⋄ REV MR 84c:03048 Zbl 498 # 03022 • ID 35700

KATZ, M. [1983] *Deduction systems and valuation spaces* (J 0079) Logique & Anal, NS 26*157-175
  ⋄ C25 C90 G25 ⋄ REV MR 85i:03193 Zbl 539 # 03017 • ID 39792

KATZ, M. see Vol. II for further entries

KATZBERG, H. [1983] *Complex exponential terms with only finitely many zeros* (S 3382) Sem-ber, Humboldt-Univ Berlin, Sekt Math 49*68-72
  ⋄ C65 ⋄ REV MR 84i:03008 Zbl 572 # 03013 • ID 49231

KAUFMANN, M. [1976] *Some results in stationary logic. Doctoral Dissertation, University of Wisconsin, Madison* (0000) Diss., Habil. etc iii+141pp
  ⋄ C55 C80 ⋄ REV JSL 43.369 • REM Doct. diss., University of Wisconsin, Madison • ID 31498

KAUFMANN, M. [1977] *A rather classless model* (J 0053) Proc Amer Math Soc 62*330-333
  ⋄ C50 C57 C62 E65 ⋄ REV MR 57 # 16058 Zbl 359 # 02054 • ID 31400

KAUFMANN, M. [1978] see BARWISE, J.

KAUFMANN, M. [1979] *A new omitting types theorem for L(Q)* (J 0036) J Symb Logic 44*507-521
  ⋄ C55 C75 C80 ⋄ REV MR 80m:03073 Zbl 427 # 03026 JSL 50.1076 • ID 53713

KAUFMANN, M. [1980] see CUTLAND, N.J.

KAUFMANN, M. [1981] *Filter logics: Filters on ω_1* (J 0007) Ann Math Logic 20*155-200
  ⋄ C55 C80 E05 ⋄ REV MR 84d:03044 Zbl 467 # 03030 • ID 55028

KAUFMANN, M. [1981] *On existence of Σ_n-end extensions* (P 2628) Log Year;1979/80 Storrs 92-103
  ⋄ C62 ⋄ REV MR 82h:03031 Zbl 467 # 03021 • ID 55019

KAUFMANN, M. [1983] *Blunt and topless end extensions of models of set theory* (J 0036) J Symb Logic 48*1053-1073
  ⋄ C62 E55 ⋄ REV MR 85f:03036 Zbl 537 # 03024 • ID 39848

KAUFMANN, M. [1983] *Set theory with a filter quantifier* (J 0036) J Symb Logic 48*263-287
  ⋄ C62 C80 E30 E70 ⋄ REV MR 85d:03076 Zbl 518 # 03007 • ID 37512

KAUFMANN, M. & SHELAH, S. [1984] *A nonconservativity result on global choice* (J 0073) Ann Pure Appl Logic 27*209-214
  ⋄ C80 E25 E35 E55 ⋄ REV MR 86a:03037 Zbl 554 # 03025 • ID 38764

KAUFMANN, M. & KRANAKIS, E. [1984] *Definable ultrapowers and ultrafilters over admissible ordinals* (J 0068) Z Math Logik Grundlagen Math 30*97-118
  ⋄ C20 C62 D60 E05 E45 ⋄ REV MR 86e:03037 Zbl 519 # 03022 • ID 41225

KAUFMANN, M. [1984] *Filter logics on ω* (J 0036) J Symb Logic 49*241-256
  ⋄ C40 C75 C80 C95 ⋄ REV MR 86a:03038 • ID 42468

KAUFMANN, M. [1984] *Mutually generic classes and incompatible expansions* (J 0027) Fund Math 121*213-218
  ⋄ C62 E40 ⋄ REV MR 86d:03032 • ID 39887

KAUFMANN, M. [1984] *On expandability of models of arithmetic and set theory to models of weak second-order theories* (J 0027) Fund Math 122*57-60
  ⋄ C15 C30 C62 C70 C75 E30 E70 ⋄ REV MR 85f:03030 Zbl 551 # 03021 • ID 40711

KAUFMANN, M. & SCHMERL, J.H. [1984] *Saturation and simple extensions of models of Peano-arithmetic* (J 0073) Ann Pure Appl Logic 27*109-136
  ⋄ C50 C57 C62 ⋄ REV MR 85j:03051 Zbl 557 # 03021 • ID 39638

KAUFMANN, M. [1984] *Some remarks on equivalence in infinitary and stationary logic* (J 0047) Notre Dame J Formal Log 25*383-389
⋄ C55 C65 C75 C80 E45 ⋄ REV MR 86d:03033 Zbl 554 # 03022 • ID 46625

KAUFMANN, M. [1984] see HENSON, C.W.

KAUFMANN, M. [1985] *A note on the Hanf number of second-order logic* (J 0047) Notre Dame J Formal Log 26*305-308
⋄ C55 C80 C85 ⋄ ID 47527

KAUFMANN, M. & SHELAH, S. [1985] *On random models of finite power and monadic logic* (J 0193) Discr Math 54*285-293
⋄ C13 C85 ⋄ ID 46358

KAUFMANN, M. [1985] *The quantifier "There exist uncountable many" and some of its relatives* (C 4183) Model-Theor Log 123-176
⋄ C55 C80 C98 ⋄ ID 48330

KAWAI, H. [1982] *Eine Logik erster Stufe mit einem infinitaeren Zeitoperator* (J 0068) Z Math Logik Grundlagen Math 28*173-180
⋄ B45 C40 C75 C90 ⋄ REV MR 83h:03023 Zbl 511 # 03006 • ID 36043

KAWAI, H. see Vol. II for further entries

KAWAI, T. [1979] *An application of nonstandard analysis to characters of groups of continuous functions* (J 2826) Rep Fac Sci, Kagoshima Univ, Math Phys Chem 12*43-45
⋄ C65 H10 ⋄ REV MR 81i:22002 Zbl 499 # 22003 • ID 38297

KAWAI, T. see Vol. I, V for further entries

KEARNS, J.T. [1978] *Intuitionist logic, a logic of justification* (J 0063) Studia Logica 37*243-260
⋄ A05 C90 F50 ⋄ REV MR 80d:03025 Zbl 408 # 03045 • ID 31151

KEARNS, J.T. [1981] *A more satisfactory description of the semantics of justification* (J 0047) Notre Dame J Formal Log 22*109-119
⋄ A05 C90 ⋄ REV MR 82d:03034 Zbl 419 # 03039 • ID 54109

KEARNS, J.T. see Vol. I, II, VI for further entries

KECHRIS, A.S. [1975] *Countable ordinals and the analytical hierarchy I* (J 0048) Pac J Math 60*223-227
⋄ C62 D55 E10 E15 E60 ⋄ REV MR 52 # 7900 Zbl 287 # 02042 • REM Part II 1978 • ID 14831

KECHRIS, A.S. [1975] see HARRINGTON, L.A.

KECHRIS, A.S. see Vol. IV, V for further entries

KEGEL, O.H. & WEHRFRITZ, B.A.F. [1973] *Locally finite groups* (X 0809) North Holland: Amsterdam xi+210pp
⋄ C60 C98 ⋄ REV MR 57 # 9848 Zbl 259 # 20001 • ID 49809

KEGEL, O.H. [1981] *Existentially closed projective planes* (P 3894) Geom-von Staudt's Point of View; 1980 Bad Windsheim 1-49
⋄ C25 C65 ⋄ REV MR 83k:51002 Zbl 512 # 51009 • ID 37559

KEGEL, O.H. [1984] see FUNK, M.

KEGEL, O.H. [1985] see FUNK, M.

KEIMEL, K. [1974] *Representation des algebres universelles par des faisceaux* (S 2250) Semin Dubreil: Algebre 27/12(2)*9pp
⋄ C05 C35 C60 C90 ⋄ REV MR 53 # 3062 Zbl 335 # 08008 • ID 21652

KEIMEL, K. & WERNER, H. [1974] *Stone duality for varieties generated by quasi-primal algebras* (S 0167) Mem Amer Math Soc 148*59-85
⋄ C05 C13 C90 ⋄ REV MR 50 # 12861 Zbl 274 # 08002 • ID 28667

KEIMEL, K. see Vol. V for further entries

KEISLER, H.J. [1960] *Theory of models with generalized atomic formulas* (J 0036) J Symb Logic 25*1-26
⋄ C07 C30 C52 ⋄ REV MR 24 # A36 Zbl 107.8 JSL 34.651 • ID 06999

KEISLER, H.J. [1961] *On some results of Jonsson and Tarski concerning free algebras* (J 0132) Math Scand 9*102-106
⋄ C05 ⋄ REV MR 23 # A3696 Zbl 111.21 • ID 06997

KEISLER, H.J. [1961] *Ultraproducts and elementary classes* (J 0028) Indag Math 23*477-495
⋄ C20 C40 C50 C52 C55 E05 E50 ⋄ REV MR 25 # 3816 JSL 27.357 • ID 06998

KEISLER, H.J. [1962] see CHANG, C.C.

KEISLER, H.J. [1962] *Some applications of the theory of models to set theory* (P 0612) Int Congr Log, Meth & Phil of Sci (1,Proc);1960 Stanford 80-86
⋄ C20 C55 E55 ⋄ REV MR 32 # 7418 Zbl 199.19 JSL 32.410 • ID 07000

KEISLER, H.J. [1963] *A complete first-order logic with infinitary predicates* (J 0027) Fund Math 52*177-203
⋄ C75 G15 ⋄ REV MR 27 # 2399 Zbl 115.6 JSL 31.269 • ID 07002

KEISLER, H.J. [1963] *Limit ultrapowers* (J 0064) Trans Amer Math Soc 107*382-408
⋄ C20 C55 ⋄ REV MR 26 # 6054 Zbl 122.245 JSL 32.277 • ID 07001

KEISLER, H.J. [1964] *Complete theories of algebraically closed fields with distinguished subfields* (J 0133) Michigan Math J 11*71-81
⋄ B25 C35 C60 ⋄ REV MR 31 # 3331 Zbl 134.8 • ID 07006

KEISLER, H.J. & TARSKI, A. [1964] *From accessible to inaccessible cardinals* (J 0027) Fund Math 53*225-308 • ERR/ADD ibid 57*119
⋄ C75 E05 E55 ⋄ REV MR 29 # 3385 MR 31 # 3340 Zbl 173.8 JSL 32.411 • ID 19046

KEISLER, H.J. [1964] *Good ideals in fields of sets* (J 0120) Ann of Math, Ser 2 79*338-359
⋄ C20 C55 E05 ⋄ REV MR 29 # 3383 Zbl 137.8 JSL 39.332 • ID 07003

KEISLER, H.J. [1964] *On cardinalities of ultraproducts* (J 0015) Bull Amer Math Soc 70*644-647
⋄ C20 C55 E05 ⋄ REV MR 29 # 3384 Zbl 129.258 JSL 38.650 • ID 07004

KEISLER, H.J. [1964] *Ultraproducts and saturated models* (J 0028) Indag Math 26*178-186
⋄ C20 C50 C55 ⋄ REV MR 29 # 5745 Zbl 199.11 JSL 35.584 • ID 07005

KEISLER, H.J. [1964] *Unions of relational systems* (J 0053) Proc Amer Math Soc 15*540-545
⋄ C30 C52 ⋄ REV MR 29 # 2185 Zbl 199.11 JSL 33.287
• ID 07007

KEISLER, H.J. [1965] *A survey of ultraproducts* (P 0623) Int Congr Log, Meth & Phil of Sci (2,Proc);1964 Jerusalem 112-126
⋄ C20 C55 C98 E05 ⋄ REV MR 34 # 5678 Zbl 156.321 JSL 35.585 • ID 07014

KEISLER, H.J. [1965] see CHANG, C.C.

KEISLER, H.J. [1965] *Finite approximations of infinitely long formulas* (P 0614) Th Models;1963 Berkeley 158-169
⋄ C40 C50 C75 ⋄ REV MR 34 # 2464 JSL 34.129
• ID 07011

KEISLER, H.J. [1965] *Ideals with prescribed degree of goodness* (J 0120) Ann of Math, Ser 2 81*112-116
⋄ C20 C55 E05 ⋄ REV MR 30 # 1063 Zbl 147.257 JSL 39.332 • ID 07012

KEISLER, H.J. [1965] *Limit ultraproducts* (J 0036) J Symb Logic 30*212-234
⋄ C20 C55 E55 ⋄ REV MR 33 # 46 Zbl 147.256 JSL 32.277 • ID 07013

KEISLER, H.J. [1965] *Reduced products and Horn classes* (J 0064) Trans Amer Math Soc 117*307-328
⋄ B20 C20 C52 C55 ⋄ REV MR 30 # 1047 Zbl 199.11 JSL 31.507 • ID 07008

KEISLER, H.J. [1965] *Some application of infinitely long formulas* (J 0036) J Symb Logic 30*339-349
⋄ C40 C50 C75 ⋄ REV MR 33 # 7256 Zbl 158.13 JSL 34.129 • ID 07009

KEISLER, H.J. [1966] see CHANG, C.C.

KEISLER, H.J. [1966] *First order properties of pairs of cardinals* (J 0015) Bull Amer Math Soc 72*141-144
⋄ C55 ⋄ REV MR 32 # 1117 Zbl 143.260 JSL 33.122
• ID 07017

KEISLER, H.J. [1966] *Some model theoretic results for ω-logic* (J 0029) Israel J Math 4*249-261
⋄ C35 C50 C55 C75 ⋄ REV MR 36 # 4974 Zbl 167.14
• ID 07015

KEISLER, H.J. [1966] *Universal homogeneous boolean algebras* (J 0133) Michigan Math J 13*129-132
⋄ C50 G05 ⋄ REV MR 33 # 3968 Zbl 168.266 JSL 33.123
• ID 07016

KEISLER, H.J. & MORLEY, M.D. [1967] *On the number of homogeneous models of a given power* (J 0029) Israel J Math 5*73-78
⋄ C50 C55 C75 E50 ⋄ REV MR 36 # 6274 Zbl 153.321
• ID 07018

KEISLER, H.J. [1967] *Ultraproducts which are not saturated* (J 0036) J Symb Logic 32*23-46
⋄ C20 C50 C55 ⋄ REV MR 36 # 1312 Zbl 162.15 JSL 35.585 • ID 90266

KEISLER, H.J. [1967] *Ultraproducts of finite sets* (J 0036) J Symb Logic 32*47-57
⋄ C13 C20 C55 E05 ⋄ REV MR 38 # 4296 Zbl 153.17 JSL 35.586 • ID 07021

KEISLER, H.J. & MORLEY, M.D. [1968] *Elementary extensions of models of set theory* (J 0029) Israel J Math 6*49-65
⋄ C62 ⋄ REV MR 38 # 5611 Zbl 172.295 • ID 07023

KEISLER, H.J. [1968] *Formulas with linearly ordered quantifiers* (P 0637) Syntax & Semant Infinitary Lang;1967 Los Angeles 96-130
⋄ C40 C50 C75 C80 ⋄ REV Zbl 182.10 • ID 07025

KEISLER, H.J. [1968] *Models with orderings* (P 0627) Int Congr Log, Meth & Phil of Sci (3,Proc);1967 Amsterdam 35-62
⋄ C55 C80 ⋄ REV MR 42 # 4386 Zbl 191.295 JSL 39.334
• ID 07022

KEISLER, H.J. [1969] *Infinite quantifiers and continuous games* (P 0649) Appl Model Th to Algeb, Anal & Probab;1967 Pasadena 228-264
⋄ C65 C75 C80 ⋄ REV MR 40 # 1143 Zbl 217.305 JSL 38.523 • ID 07026

KEISLER, H.J. [1970] *Logic with the quantifier "there exist uncountably many"* (J 0007) Ann Math Logic 1*1-93
⋄ C30 C55 C62 C75 C80 ⋄ REV MR 41 # 8217 Zbl 206.273 JSL 36.685 • ID 24832

KEISLER, H.J. [1971] *A short course in model theory* (1111) Preprints, Manuscr., Techn. Reports etc.
⋄ C98 ⋄ REM Mimeographed Notes, Cambridge, GB
• ID 21380

KEISLER, H.J. & SILVER, J.H. [1971] *End extensions of models of set theory* (P 0693) Axiomatic Set Th;1967 Los Angeles 1*177-188
⋄ C62 E45 ⋄ REV MR 48 # 96 Zbl 225 # 02039 JSL 40.505 • ID 07029

KEISLER, H.J. [1971] *Model theory* (P 0743) Int Congr Math (II,11,Proc);1970 Nice 1*141-150
⋄ C98 ⋄ REV MR 57 # 5729 Zbl 225 # 02036 JSL 38.648
• ID 07031

KEISLER, H.J. [1971] *Model theory for infinitary logic. Logic with countable conjunctions and finite quantifiers* (X 0809) North Holland: Amsterdam x + 208pp
⋄ C70 C75 C98 ⋄ REV MR 49 # 8855 Zbl 222 # 02064 JSL 38.522 • ID 74760

KEISLER, H.J. [1971] *On theories categorical in their own power* (J 0036) J Symb Logic 36*240-243
⋄ C35 C55 ⋄ REV MR 45 # 3184 Zbl 233 # 02022
• ID 07028

KEISLER, H.J. [1973] *Forcing and the omitting types theorem* (C 0654) Stud in Model Th 96-133
⋄ C25 C75 ⋄ REV MR 49 # 2340 • ID 07032

KEISLER, H.J. [1973] see CHANG, C.C.

KEISLER, H.J. & WALKOE JR., W.J. [1973] *The diversity of quantifier prefixes* (J 0036) J Symb Logic 38*79-85
⋄ B10 C07 C13 D55 ⋄ REV MR 51 # 12472 Zbl 259 # 02007 • ID 07033

KEISLER, H.J. & PRIKRY, K. [1974] *A result concerning cardinalities of ultraproducts* (J 0036) J Symb Logic 39*43-48
⋄ C20 C55 E05 ⋄ REV MR 49 # 8862 Zbl 291 # 02036
• ID 07035

KEISLER, H.J. [1974] *Models with tree structures* (P 0610) Tarski Symp;1971 Berkeley 331-348
⋄ C55 C62 E65 ⋄ REV MR 50 # 9576 Zbl 323 # 02065
• ID 07037

KEISLER, H.J. [1974] *Monotone complete fields* (**P** 1083) Victoria Symp Nonstand Anal;1972 Victoria 113-115
⋄ C60 C65 H05 ⋄ REV MR 58 # 21585 Zbl 289 # 02040
• ID 30004

KEISLER, H.J. [1975] *Constructions in model theory* (**P** 2021) Model Th & Appl;1975 Bressanone 56-108
⋄ C25 C30 C50 C75 C95 C98 ⋄ ID 90125

KEISLER, H.J. [1976] *Foundations of infinitesimal calculus* (**X** 1337) Prindle Weber Schmidt: Boston x+214pp
⋄ C20 H05 H98 ⋄ REV Zbl 333 # 26001 JSL 46.643
• ID 33913

KEISLER, H.J. [1976] *Six classes of theories* (**J** 3194) J Austral Math Soc, Ser A 21*257-266
⋄ C45 C55 ⋄ REV MR 53 # 12930 Zbl 342 # 02035
• ID 23179

KEISLER, H.J. [1977] *Fundamentals of model theory* (**C** 1523) Handb of Math Logic 47-103
⋄ C98 ⋄ REV MR 58 # 10395 JSL 49.968 • ID 24197

KEISLER, H.J. [1977] see BARWISE, J.

KEISLER, H.J. [1977] *Hyperfinite model theory* (**P** 1075) Logic Colloq;1976 Oxford 5-110
⋄ B50 C65 C75 C80 C90 H20 ⋄ REV MR 58 # 10421 Zbl 423 # 03041 • ID 16614

KEISLER, H.J. [1977] *The monotone class theorem in infinitary logic* (**J** 0053) Proc Amer Math Soc 64*129-134
⋄ C70 C75 ⋄ REV MR 56 # 85 Zbl 364 # 02030 • ID 50960

KEISLER, H.J. [1978] *The stability function of a theory* (**J** 0036) J Symb Logic 43*481-486
⋄ C45 C55 ⋄ REV MR 80a:03040 Zbl 434 # 03023 JSL 48.496 • ID 29274

KEISLER, H.J. [1979] see BRUCE, KIM B.

KEISLER, H.J. [1984] see HOOVER, D.N.

KEISLER, H.J. [1984] *Lyndon's research in mathematical logic* (**C** 3510) Contrib Group Theory (Lyndon) 29-32
⋄ A10 C98 G98 ⋄ REV MR 85i:03003 Zbl 554 # 03001
• ID 44024

KEISLER, H.J. [1984] see HENSON, C.W.

KEISLER, H.J. [1985] *Probability quantifiers* (**C** 4183) Model-Theor Log 509-556
⋄ C40 C80 C90 C98 ⋄ ID 48333

KEISLER, H.J. see Vol. I, II for further entries

KELLER, J.P. & RICHARD, D. [1978] *Remarques sur les structures additives des modeles de l'arithmetique (English summary)* (**J** 2313) C R Acad Sci, Paris, Ser A-B 287*A101-A104
⋄ C62 H15 ⋄ REV MR 58 # 10422 Zbl 385 # 03058
• ID 52165

KELLER, J.P. see Vol. I for further entries

KELLY, DAVID [1972] *A note on equationally compact lattices* (**J** 0004) Algeb Universalis 2*80-84
⋄ C05 C50 G10 ⋄ REV MR 46 # 7102 Zbl 268 # 06005
• ID 07045

KELLY, DAVID [1984] see GRAETZER, G.

KELLY, DAVID see Vol. V for further entries

KEMENY, J.G. [1958] *Undecidable problems of elementary number theory* (**J** 0043) Math Ann 135*160-169
⋄ F30 H15 ⋄ REV MR 20 # 5140 Zbl 81.247 JSL 23.359
• ID 07053

KEMENY, J.G. see Vol. I, II for further entries

KENNISON, J.F. [1976] *Integral domain type representations in sheaves and other topoi* (**J** 0044) Math Z 151*35-56
⋄ C60 C90 G30 ⋄ REV MR 54 # 2758 Zbl 346 # 18012
• ID 24063

KENNISON, J.F. & LEDBETTER, C.S. [1979] *Sheaf representations and the Dedekind reals* (**P** 2901) Appl Sheaves;1977 Durham 500-513
⋄ C90 G30 ⋄ REV MR 83j:18012 Zbl 419 # 18002
• ID 39875

KENNISON, J.F. see Vol. V for further entries

KENT, C.F. [1962] *Constructive analogues of the group of permutations of the natural numbers* (**J** 0064) Trans Amer Math Soc 104*347-362
⋄ C57 D45 ⋄ REV MR 25 # 3826 Zbl 105.248 JSL 34.517
• ID 07059

KENT, C.F. see Vol. I, IV, VI for further entries

KENZHEBAEV, S. [1966] *Certain undecidable rings (Russian) (Kazakh summary)* (**J** 0403) Izv Akad Nauk Kazak SSR, Ser Fiz-Mat 1966/1*25-30
⋄ C60 D35 ⋄ REV MR 34 # 1346 Zbl 199.31 • ID 16220

KENZHEBAEV, S. see Vol. IV for further entries

KEPKA, T. [1981] *Strong amalgamation property in some groupoid varieties* (**P** 2552) Conf Finite Algeb & Multi-Val Log;1979 Szeged 373-385
⋄ C05 C52 ⋄ REV MR 83f:08020 Zbl 478 # 08013
• ID 40224

KERTESZ, A. [1956] see FUCHS, L.

KERTESZ, A. [1974] *Transfinite Methoden in der Algebra (Russian summary)* (**J** 0667) Scripta Fac Sci Math, Brno 4*35-37
⋄ C60 E25 E75 ⋄ REV MR 53 # 140 Zbl 367 # 04002
• ID 16665

KERTESZ, A. see Vol. V for further entries

KESEL'MAN, D.YA. [1974] *The elementary theories of graphs and abelian loops (Russian)* (**J** 0087) Mat Zametki (Akad Nauk SSSR) 16*957-968
• TRANSL [1974] (**J** 1044) Math Notes, Acad Sci USSR 16*1167-1171
⋄ C65 D35 ⋄ REV MR 51 # 7959 Zbl 311 # 02063
• ID 17588

KESEL'MAN, D.YA. [1982] *Decidability of theories of certain classes of elimination graphs* (**J** 0040) Kibernetika, Akad Nauk Ukr SSR 1982/3i*34-37,133
• TRANSL [1982] (**J** 0021) Cybernetics 18*305-310
⋄ B25 C65 D35 ⋄ REV MR 84m:05076 • ID 39437

KETONEN, J. [1972] *On nonregular ultrafilters* (**J** 0036) J Symb Logic 37*71-74
⋄ C20 C55 E05 E55 G05 ⋄ REV MR 48 # 1926 Zbl 248 # 02058 • ID 07071

KETONEN, J. [1978] *The structure of countable boolean algebras* (**J** 0120) Ann of Math, Ser 2 108*41-89
⋄ C15 G05 ⋄ REV MR 58 # 10647 Zbl 418 # 06006
• ID 81833

KETONEN, J. & WEYHRAUCH, R.W. [1984] *A decidable fragment of predicate calculus* (**J** 1426) Theor Comput Sci 32*297-307
⋄ B20   B25   ⋄ REV   MR 86g:03018   • ID 44002

KETONEN, J. see Vol. I, V, VI for further entries

KFOURY, D. [1973] *Comparing algebraic structures up to algorithmic equivalence* (**P** 0763) Automata, Lang & Progr (1);1972 Rocquencourt 253-263
⋄ B75   C20   C35   ⋄ REV   MR 51 # 9582   Zbl 281 # 68007
• ID 62950

KHARLAMPOVICH, O.G. [1983] *The universal theory of the class of finite nilpotent groups is undecidable (Russian)* (**J** 0087) Mat Zametki (Akad Nauk SSSR) 33*499-516
• TRANSL [1983] (**J** 1044) Math Notes, Acad Sci USSR 33*254-263
⋄ C60   D35   ⋄ REV   MR 85b:20046   Zbl 516 # 20018
• ID 39367

KHARLAMPOVICH, O.G. [1984] *The universal theory of certain classes of Lie rings (Russian)* (**J** 0340) Mat Zap (Univ Sverdlovsk) 13/4*156-164
⋄ C60   ⋄ ID 49471

KHARLAMPOVICH, O.G. see Vol. IV for further entries

KHISAMIEV, N.G. [1966] *Certain applications of the method of extension of mappings (Russian) (Kazakh summary)* (**J** 0429) Vest Akad Nauk Kazak SSR 22*68-72
⋄ C35   C52   ⋄ REV   MR 34 # 2462   Zbl 212.321 • ID 06144

KHISAMIEV, N.G. & KOKORIN, A.I. [1966] *Elementary classification of structurally ordered abelian groups with finite number of threads (Russian)* (**J** 0003) Algebra i Logika 5/1*41-50
⋄ C60   ⋄ REV   MR 33 # 7433   Zbl 254 # 06015 • ID 32596

KHISAMIEV, N.G. [1966] *Universal theory of lattice-ordered abelian groups (Russian)* (**J** 0003) Algebra i Logika 5/3*71-76
⋄ B25   C60   ⋄ REV   MR 34 # 2727   Zbl 275 # 02050
• ID 29709

KHISAMIEV, N.G. [1971] *Strongly constructive models (Russian) (Kazakh summary)* (**J** 0403) Izv Akad Nauk Kazak SSR, Ser Fiz-Mat 1971/3*59-63
⋄ C30   C57   ⋄ REV   MR 45 # 4960   Zbl 305 # 02062
• ID 06146

KHISAMIEV, N.G. [1974] *Strongly constructive models of a decidable theory (Russian)* (**J** 0403) Izv Akad Nauk Kazak SSR, Ser Fiz-Mat 1974/1*83-84,94
⋄ C15   C57   F60   ⋄ REV   MR 50 # 6824   Zbl 275 # 02051
• ID 06147

KHISAMIEV, N.G. [1977] *On the periodical part of a strongly constructivizable abelian group (Russian)* (**C** 2557) Teor & Priklad Zad Mat & Mekh 299-303
⋄ C57   C60   D45   ⋄ REV   MR 80c:03046   • ID 32598

KHISAMIEV, N.G. [1978] see DOBRITSA, V.P.

KHISAMIEV, N.G. [1978] *Strongly constructive periodic abelian groups (Russian) (Kazakh summary)* (**J** 0403) Izv Akad Nauk Kazak SSR, Ser Fiz-Mat 1978/1*58-62,92
⋄ C57   C60   D45   ⋄ REV   MR 80b:03040   Zbl 396 # 20034
• ID 32599

KHISAMIEV, N.G. [1979] *Criterion of the constructivizability of a direct sum of cyclic p-groups (Russian)* (**P** 2558) All-Union Conf Math Log (5) (Mal'tsev);1979 Novosibirsk 157
⋄ C57   C60   D45   ⋄ REV   MR 82h:20042   • ID 32602

KHISAMIEV, N.G. [1979] *On subgroups of finite index of abelian groups (Russian) (Kazakh summary)* (**J** 0403) Izv Akad Nauk Kazak SSR, Ser Fiz-Mat 1979/3*43-47,89
⋄ C57   C60   D45   F60   ⋄ REV   MR 81b:20037
Zbl 416 # 20049 • ID 32601

KHISAMIEV, N.G. [1981] *Criterion for constructivizability of a direct sum of cyclic p-groups (Russian) (Kazakh summary)* (**J** 0403) Izv Akad Nauk Kazak SSR, Ser Fiz-Mat 1981/1*51-55,86
⋄ C57   C60   D45   ⋄ REV   MR 82h:20042   Zbl 457 # 20001
• ID 81571

KHISAMIEV, N.G. [1983] *Strongly constructive abelian p-groups (Russian)* (**J** 0003) Algebra i Logika 22*198-217
• TRANSL [1983] (**J** 0069) Algeb and Log 22*142-158
⋄ C57   C60   D45   F60   ⋄ REV   MR 85e:03105
Zbl 568 # 20052 • ID 40239

KHISAMIEV, N.G. [1984] see ABDRAZAKOV, K.T.

KHISAMIEV, N.G. [1984] *Connection between constructivizability and strong constructivizability for different classes of abelian groups (Russian)* (**J** 0003) Algebra i Logika 23*319-335,363
• TRANSL [1984] (**J** 0069) Algeb and Log 23*220-233
⋄ C57   C60   D45   ⋄ REV   MR 86h:03080 • ID 42697

KHISAMIEV, N.G. & KHISAMIEV, Z.G. [1985] *Nonconstructivizability of the reduced part of a strongly constructive torsion-free abelian group (Russian)* (**J** 0003) Algebra i Logika 24*108-118,123
• TRANSL [1985] (**J** 0069) Algeb and Log 24*69-76
⋄ C57   C60   D45   ⋄ ID 49409

KHISAMIEV, N.G. see Vol. IV for further entries

KHISAMIEV, Z.G. [1985] see KHISAMIEV, N.G.

KHISAMIEV, Z.G. see Vol. IV for further entries

KHMELEVSKIJ, YU.I. & TAJMANOV, A.D. [1980] *Decidability of the universal theory of a free semigroup (Russian)* (**J** 0092) Sib Mat Zh 21/1*228-230,240
⋄ B25   C05   C60   ⋄ REV   MR 81b:03012   Zbl 438 # 20046
• ID 79193

KHMELEVSKIJ, YU.I. see Vol. IV for further entries

KHOMYAKOV, M.V. [1974] see BORSHCHEV, V.B.

KHOMYAKOV, M.V. [1976] see BORSHCHEV, V.B.

KHRISTOV, KH.YA. [1977] see DAMYANOV, B.P.

KHRISTOV, KH.YA. see Vol. I for further entries

KIEFE, C. [1976] see ADLER, A.

KIEFE, C. [1976] *Sets definable over finite fields: their zeta-functions* (**J** 0064) Trans Amer Math Soc 223*45-59
⋄ C13   C60   ⋄ REV   MR 54 # 10272   Zbl 372 # 02032
• ID 25877

KIEHNE, U. [1975] see JARDEN, M.

KIEHNE, U. [1979] *Bounded products in the theory of valued fields* (**J** 0127) J Reine Angew Math 305*9-36
⋄ C20   C60   ⋄ REV   MR 80i:12018   Zbl 423 # 12022
• ID 53577

KIERSTEAD, H.A. [1980] *Countable models of ω_1-categorical theories in admissible languages* (**J** 0007) Ann Math Logic 19*127-175
⋄ C15   C35   C70   ⋄ REV   MR 82c:03054   Zbl 478 # 03016
• ID 55631

KIERSTEAD, H.A. [1983] *An effective version of Hall's theorem* (J 0053) Proc Amer Math Soc 88*124-128
◇ C57 D80 ◇ REV MR 84g:03065 Zbl 533 # 03028
• ID 34172

KIERSTEAD, H.A. & REMMEL, J.B. [1983] *Indiscernibles and decidable models* (J 0036) J Symb Logic 48*21-32
◇ C30 C45 C57 C80 ◇ REV MR 84k:03103 Zbl 523 # 03030 • ID 36120

KIERSTEAD, H.A. & MCNULTY, G.F. & TROTTER JR., W.T. [1984] *A theory of recursive dimension of ordered sets* (J 3457) Order 1*67-82
◇ C65 D45 ◇ REV MR 86a:06003 Zbl 562 # 06001
• ID 45164

KIERSTEAD, H.A. & REMMEL, J.B. [1985] *Degrees of indiscernibles in decidable models* (J 0064) Trans Amer Math Soc 289*41-57
◇ C30 C35 C57 D30 ◇ ID 44599

KIERSTEAD, H.A. see Vol. IV, V for further entries

KIM, KUKJIN [1980] *On the structure of Hensel fields* (J 0546) J Korean Math Soc 16*155-160
◇ B25 C60 ◇ REV MR 82a:03030 Zbl 453 # 12011
• ID 54204

KINO, A. & TAKEUTI, G. [1963] *On predicates with constructive infinitely long expressions* (J 0090) J Math Soc Japan 15*176-190
◇ C70 F15 ◇ REV Zbl 118.250 JSL 30.97 • ID 07100

KINO, A. [1966] *On definability of ordinals in logic with infinitely long expressions* (J 0036) J Symb Logic 31*365-375
• ERR/ADD ibid 32*343-344
◇ C40 C75 E10 E47 ◇ REV MR 34 # 4113 Zbl 183.11 JSL 35.341 • ID 07101

KINO, A. & MYHILL, J.R. [1975] *A hierarchy of languages with infinitely long expressions* (P 0775) Logic Colloq;1973 Bristol 55-71
◇ C40 C75 E10 F15 ◇ REV MR 52 # 2817 Zbl 357 # 02013 • ID 17616

KINO, A. see Vol. IV, VI for further entries

KIRBY, L.A.S. & PARIS, J.B. [1977] *Initial segments of models of Peano's axioms* (P 1695) Set Th & Hierarch Th (3);1976 Bierutowice 211-226
◇ C62 F30 ◇ REV MR 58 # 10423 Zbl 364 # 02032 JSL 48.482 • ID 50962

KIRBY, L.A.S. & PARIS, J.B. [1978] Σ_n-*collection schemas in arithmetic* (P 1897) Logic Colloq;1977 Wroclaw 199-209
◇ C62 F30 ◇ REV MR 81e:03056 Zbl 442 # 03042
• ID 56399

KIRBY, L.A.S. [1980] *La methode des indicatrices et le theoreme d'incompletude* (C 2969) Modeles de l'Arithm; Paris 1977 5-18
◇ C62 F30 ◇ REV MR 84d:03040 Zbl 458 # 03023 JSL 48.483 • ID 54429

KIRBY, L.A.S. & MCALOON, K. & MURAWSKI, R. [1981] *Indicators, recursive saturation and expandability* (J 0027) Fund Math 114*127-139
◇ C50 C57 C62 H15 ◇ REV MR 84i:03117 Zbl 488 # 03038 • ID 34593

KIRBY, L.A.S. & PARIS, J.B. [1982] *Accessible independence results for Peano arithmetic* (J 0161) Bull London Math Soc 14*285-293
◇ C62 F30 ◇ REV MR 83j:03096 Zbl 501 # 03017
• ID 35381

KIRBY, L.A.S. [1982] *Flipping properties in arithmetic* (J 0036) J Symb Logic 47*416-422
◇ C62 E05 F30 ◇ REV MR 84e:03084 Zbl 488 # 03031
• ID 34416

KIRBY, L.A.S. [1984] *Ultrafilters and types on models of arithmetic* (J 0073) Ann Pure Appl Logic 27*215-252
◇ C20 C62 E05 H15 ◇ REV MR 86f:03107 Zbl 565 # 03014 • ID 39686

KIRCH, A.M. [1970] *Generalized restricted direct products* (J 0025) Duke Math J 37*171-176
◇ C05 C30 ◇ REV MR 41 # 6755 Zbl 204.225 • ID 07111

KIREEVSKIJ, N.N. [1934] *On the problem of the solvability of the decision problem (Russian)* (J 4717) Izv Akad Nauk SSSR 1934*1493-1499
◇ B25 D35 ◇ REV FdM 60.850 • ID 07112

KIREEVSKIJ, N.N. [1958] *Ueber die Allgemeingueltigkeit gewisser Zaehlausdruecke (Russian)* (J 0142) Mat Sb, Akad Nauk SSSR, NS 42*669-678
◇ B25 F30 ◇ REV Zbl 14.2 FdM 61.973 • ID 40811

KIRK, R.E. [1979] *Some classes of Kripke frames characteristic for the intuitionistic logic* (J 0068) Z Math Logik Grundlagen Math 25*409-410
◇ B55 C90 F50 ◇ REV MR 81b:03011 Zbl 423 # 03042
• ID 53554

KIRK, R.E. [1980] *A characterization of the classes of finite tree frames which are adequate for the intuitionistic logic* (J 0068) Z Math Logik Grundlagen Math 26*497-501
◇ B55 C90 F50 ◇ REV MR 81m:03014 Zbl 446 # 03017
• ID 56546

KIRK, R.E. see Vol. I for further entries

KISELEV, A.A. & KOGALOVSKIJ, S.R. [1979] *Some remarks on higher order monadic languages (Russian)* (P 2554) All-Union Symp Th Log Infer;1974 Moskva 274-283
◇ B15 C85 ◇ REV MR 84k:03033 • ID 32589

KISELEV, A.A. see Vol. IV, V for further entries

KLAUA, D. [1955] *Systematische Behandlung der loesbaren Faelle des Entscheidungsproblems fuer den Praedikatenkalkuel der ersten Stufe* (J 0068) Z Math Logik Grundlagen Math 1*264-270
◇ B25 ◇ REV MR 17.1038 Zbl 66.257 JSL 36.168
• ID 07136

KLAUA, D. see Vol. I, II, V, VI for further entries

KLEIMAN, J.G. [1982] *On identities in groups (Russian)* (J 0065) Tr Moskva Mat Obshch 44*62-108
◇ C52 C60 D40 ◇ REV MR 84e:20040 • ID 39318

KLEIMAN, J.G. see Vol. IV for further entries

KLEIN, ABRAHAM A. [1969] *Necessary conditions for embedding rings into fields* (J 0064) Trans Amer Math Soc 137*141-151
◇ C05 C60 ◇ REV MR 38 # 4510 Zbl 176.312 • ID 07200

KLEIN-BARMEN, F. [1956] *Zur Theorie der Strukturen und Algebren* (J 0352) Math Jap 4*83-94
◇ C07 ◇ REV MR 20 # 6995 Zbl 72.261 JSL 24.254
• ID 07209

KLEIN-BARMEN, F. see Vol. I, V for further entries

KLEINBERG, E.M. [1973] *Rowbottom cardinals and Jonsson cardinals are almost the same* (**J** 0036) J Symb Logic 38*423-427
⋄ C55 E05 E55 ⋄ REV MR 49 # 2385   Zbl 275 # 02059
• ID 07222

KLEINBERG, E.M. [1977] $AD \vdash$ *"The $\aleph_n$ are Jonsson cardinals and $\aleph_\omega$ is a Rowbottom cardinal"* (**J** 0007) Ann Math Logic 12*229-248
⋄ C55 E05 E55 E60 ⋄ REV MR 57 # 9550   Zbl 378 # 02032   • ID 27947

KLEINBERG, E.M. [1977] *Infinitary combinatorics and the axiom of determinateness* (**X** 0811) Springer: Heidelberg & New York iii + 50pp
⋄ C20 E05 E25 E55 E60 E98 ⋄ REV MR 58 # 109   Zbl 362 # 02067 • ID 50789

KLEINBERG, E.M. [1978] *A combinatorial characterization of normal M-ultrafilters* (**J** 0345) Adv Math 30*77-84
⋄ C20 C62 E05 E40 E55 ⋄ REV MR 80e:03064   Zbl 402 # 03049 • ID 54696

KLEINBERG, E.M. [1981] *Producing measurable cardinals beyond κ* (**J** 0036) J Symb Logic 46*643-648
⋄ C20 C55 E05 E60 ⋄ REV MR 83a:03052   Zbl 483 # 03028 • ID 35069

KLEINBERG, E.M. [1984] see HENLE, J.M.

KLEINBERG, E.M. see Vol. I, IV, V, VI for further entries

KLEINE BUENING, H. [1979] see BOERGER, E.

KLEINE BUENING, H. [1980] see BOERGER, E.

KLEINE BUENING, H. see Vol. I, IV for further entries

KLEINKNECHT, R. [1983] *Modal logic without possible worlds* (**J** 0260) Ann Jap Ass Phil Sci 6*161-171
⋄ A05 C90 ⋄ REV MR 85i:03051   Zbl 536 # 03005
• ID 37093

KLEINKNECHT, R. see Vol. I, V for further entries

KLEJNER, M.L. [1976] *On one question of Simmons (Russian)* (**P** 2064) All-Union Conf Math Log (4);1976 Kishinev 61
⋄ C25 ⋄ ID 48403

KLEMKE, D. [1971] *Ein Henkin-Beweis fuer die Vollstaendigkeit eines Kalkuels relativ zur Grzegorczyk-Semantik* (**J** 0009) Arch Math Logik Grundlagenforsch 14*148-161
⋄ B55 C90 F50 ⋄ REV MR 47 # 8253   Zbl 233 # 02020
• ID 07229

KLENK, V. [1976] *Intended models and the Loewenheim-Skolem theorem* (**J** 0122) J Philos Logic 5*475-489
⋄ A05 C07 ⋄ REV MR 58 # 10274   Zbl 337 # 02011
• ID 62997

KLIMOVSKY, G. [1962] *El axioma de eleccion y la existencia de subgrupos commutativos maximales* (**J** 0188) Rev Union Mat Argentina 20*267-287
⋄ C60 E25 E75 ⋄ REV MR 26 # 4922   Zbl 112.10
• ID 07236

KLIMOVSKY, G. see Vol. I, V for further entries

KLINGEN, N. [1974] *Elementar aequivalente Koerper und ihre absolute Galoisgruppe* (**J** 0008) Arch Math (Basel) 25*604-612
⋄ C20 C60 H20 ⋄ REV MR 50 # 9844   Zbl 408 # 12022
• ID 07238

KLINGEN, N. [1975] *Zur Idealstruktur in Nichtstandardmodellen von Dedekindringen* (**J** 0127) J Reine Angew Math 274/275*38-60
⋄ C20 C60 H15 H20 ⋄ REV MR 52 # 8120   Zbl 305 # 02066 • ID 18226

KLINGEN, N. [1977] see JEHNE, W.

KLINGEN, N. [1978] *Zur Modelltheorie lokaler und globaler Koerper* (**J** 0068) Z Math Logik Grundlagen Math 24*509-522
⋄ C60 H20 ⋄ REV MR 80a:12015   Zbl 422 # 12017
• ID 53502

KLUN, M.I. [1977] see ISBELL, J.R.

KLUPP, H. & SCHNORR, C.-P. [1977] *A universally hard set of formulae with respect to non-deterministic Turing acceptors* (**J** 0232) Inform Process Lett 6*35-37
⋄ B25 D10 D15 ⋄ REV MR 56 # 4241   Zbl 353 # 68064
• ID 50134

KNIGHT, J.F. [1973] *Complete types and the natural numbers* (**J** 0036) J Symb Logic 38*413-415
⋄ C07 C75 ⋄ REV MR 49 # 2359   Zbl 276 # 02035
• ID 17321

KNIGHT, J.F. [1973] *Generic expansions of structures* (**J** 0036) J Symb Logic 38*561-570
⋄ C25 C62 C85 ⋄ REV MR 49 # 2353   Zbl 285 # 02044
• ID 07253

KNIGHT, J.F. [1975] *Types omitted in uncountable models of arithmetic* (**J** 0036) J Symb Logic 40*317-320
⋄ C07 C62 ⋄ REV MR 52 # 2869   Zbl 319 # 02045
• ID 14824

KNIGHT, J.F. [1976] *Hanf numbers for omitting types over particular theories* (**J** 0036) J Symb Logic 41*583-588
⋄ C07 C55 C62 ⋄ REV MR 58 # 5170   Zbl 343 # 02039
JSL 48.484 • ID 14581

KNIGHT, J.F. [1976] *Omitting types in set theory and arithmetic* (**J** 0036) J Symb Logic 41*25-32
⋄ C07 C62 E40 ⋄ REV MR 53 # 10579   Zbl 328 # 02039
JSL 48.484 • ID 14755

KNIGHT, J.F. [1977] *A complete $L_{\omega_1\omega}$-sentence characterizing $\aleph_1$* (**J** 0036) J Symb Logic 42*59-62
⋄ C55 C75 ⋄ REV MR 58 # 10407   Zbl 426 # 03037
• ID 24351

KNIGHT, J.F. [1977] *Skolem functions and elementary embeddings* (**J** 0036) J Symb Logic 42*94-98
⋄ C15 C30 ⋄ REV MR 58 # 5179   Zbl 381 # 03022
• ID 23751

KNIGHT, J.F. [1978] *An inelastic model with indiscernibles* (**J** 0036) J Symb Logic 43*331-334
⋄ C15 C30 C50 ⋄ REV MR 58 # 10412   Zbl 385 # 03024
• ID 29265

KNIGHT, J.F. [1978] *Prime and atomic models* (**J** 0036) J Symb Logic 43*385-393
⋄ C50 C55 ⋄ REV MR 58 # 21578   Zbl 388 # 03014
• ID 29267

KNIGHT, J.F. [1981] *Algebraic independence* (**J** 0036) J Symb Logic 46*377-384
⋄ C40 C55 C75 ⋄ REV MR 82d:03050   Zbl 484 # 03014
• ID 74894

KNIGHT, J.F. & NADEL, M.E. [1982] *Expansions of models and Turing degrees* (J 0036) J Symb Logic 47*587-604
⋄ C50 C57 D30 ⋄ REV MR 83k:03039 Zbl 527 # 03013
• ID 36173

KNIGHT, J.F. & NADEL, M.E. [1982] *Models of arithmetic and closed ideals* (J 0036) J Symb Logic 47*833-840
⋄ C62 D30 ⋄ REV MR 85d:03072 Zbl 518 # 03032
• ID 37529

KNIGHT, J.F. [1982] *Theories whose resplendent models are homogeneous* (J 0029) Israel J Math 42*151-161
⋄ C50 ⋄ REV MR 84g:03039 Zbl 497 # 03023 • ID 34152

KNIGHT, J.F. & WOODROW, R.E. [1983] *A complete theory with arbitrarily large minimality ranks* (J 0036) J Symb Logic 48*321-328
⋄ C15 C40 C50 ⋄ REV MR 84m:03050 Zbl 538 # 03026
• ID 35751

KNIGHT, J.F. [1983] *Additive structure in uncountable models for a fixed completion of P* (J 0036) J Symb Logic 48*623-628
⋄ C50 C57 C62 C75 D30 ⋄ REV MR 85h:03034 Zbl 549 # 03026 • ID 42733

KNIGHT, J.F. [1983] *Degrees of types and independent sequences* (J 0036) J Symb Logic 48*1074-1081
⋄ C35 C45 D30 ⋄ REV MR 86b:03039 Zbl 541 # 03015
• ID 40363

KNIGHT, J.F. & LACHLAN, A.H. & SOARE, R.I. [1984] *Two theorems on degrees of models of true arithmetic* (J 0036) J Symb Logic 49*425-436
⋄ C57 C62 D30 ⋄ REV MR 85i:03144 Zbl 576 # 03044
• ID 33597

KNYAZEV, V.V. [1983] *Asymptotic solvability of a language with integral quantifier (Russian)* (S 3911) Komb-Algeb Met Prikl Mat (Gor'kij) 1983*134-150
⋄ B75 C80 ⋄ REV Zbl 567 # 03011 • ID 49211

KOCHEN, S. [1957] *Completeness of algebraic systems in higher order calculi* (P 1675) Summer Inst Symb Log;1957 Ithaca 370-376
⋄ B15 C35 C60 C85 ⋄ REV JSL 27.97 • ID 07272

KOCHEN, S. [1961] *Ultraproducts in the theory of models* (J 0120) Ann of Math, Ser 2 74*221-261
⋄ C20 C35 C52 C60 ⋄ REV MR 25 # 1992 Zbl 132.246 JSL 27.355 • ID 07273

KOCHEN, S. [1965] see AX, J.

KOCHEN, S. [1965] *Topics in the theory of definition* (P 0614) Th Models;1963 Berkeley 170-176
⋄ C20 C40 C52 ⋄ REV MR 38 # 991 Zbl 156.25 JSL 35.300 • ID 07274

KOCHEN, S. [1966] see AX, J.

KOCHEN, S. [1969] *Integer valued rational functions over the p-adic numbers: a p-adic analogue of the theory of real fields* (P 1061) Numb Th;1969 Houston 57-73
⋄ C60 ⋄ REV MR 41 # 1685 • ID 20950

KOCHEN, S. [1975] *The model theory of local fields* (P 1442)
⊢ ISILC Logic Conf;1974 Kiel 384-425
⋄ B25 C20 C35 C60 ⋄ REV MR 58 # 27918 Zbl 319 # 12009 • ID 27216

KOCHEN, S. & KRIPKE, S.A. [1982] *Nonstandard models of Peano arithmetic* (J 3370) Enseign Math, Ser 2 28*211-231
• REPR [1982] (P 3482) Logic & Algor (Specker);1980 Zuerich 275-295
⋄ C20 C62 ⋄ REV MR 83i:03059b Zbl 514 # 03045
• ID 35522

KOCHEN, S. see Vol. II for further entries

KOCK, A. & WRAITH, G.C. [1971] *Elementary toposes* (X 1599) Aarhus Univ Mat Inst: Aarhus i+118pp
⋄ C90 G30 ⋄ REV MR 49 # 7324 Zbl 251 # 18015
• ID 28673

KOCK, A. see Vol. I, V, VI for further entries

KOEHLER, P. [1983] see BLOK, W.J.

KOEPKE, P. [1983] see DONDER, H.-D.

KOEPKE, P. [1984] *The consistency strength of the free-subset property for ω_ω* (J 0036) J Symb Logic 49*1198-1204
⋄ C55 E05 E35 E45 E55 ⋄ REV MR 86f:03091
• ID 39394

KOEPKE, P. see Vol. V for further entries

KOGALOVSKIJ, S.R. [1958] *On universal classes of algebras (Russian)* (J 0023) Dokl Akad Nauk SSSR 122*759-762
⋄ C52 ⋄ REV MR 21 # 1281 Zbl 89.246 • ID 33855

KOGALOVSKIJ, S.R. [1959] *On universal classes of algebras, closed under direct product (Russian)* (J 0031) Izv Vyssh Ucheb Zaved, Mat (Kazan) 1959/3(10)*88-96
⋄ C05 C52 ⋄ REV MR 27 # 84 • ID 20973

KOGALOVSKIJ, S.R. [1959] *Universal classes of models (Russian)* (J 0023) Dokl Akad Nauk SSSR 124*260-263
⋄ C05 C52 ⋄ REV MR 21 # 1932 Zbl 89.246 • ID 07285

KOGALOVSKIJ, S.R. [1961] *Categorial characteristics of axiomatized classes (Russian)* (J 0067) Usp Mat Nauk 16/2*209
⋄ C52 G30 ⋄ ID 33856

KOGALOVSKIJ, S.R. [1961] *On a general method of obtaining structural characterizations of axiomatized classes (Russian)* (J 0023) Dokl Akad Nauk SSSR 136*1291-1294
• TRANSL [1961] (J 0062) Sov Math, Dokl 2*196-199
⋄ C52 ⋄ REV MR 22 # 12045 Zbl 192.40 • ID 19082

KOGALOVSKIJ, S.R. [1961] *On multiplicative semigroups of rings (Russian)* (J 0023) Dokl Akad Nauk SSSR 140*1005-1007
• TRANSL [1961] (J 0062) Sov Math, Dokl 2*1299-1301
⋄ C52 C60 ⋄ REV MR 25 # 140 Zbl 118.253 • ID 07286

KOGALOVSKIJ, S.R. [1963] *Quasiprojective classes of models (Russian)* (J 0023) Dokl Akad Nauk SSSR 148*505-507
• TRANSL [1963] (J 0062) Sov Math, Dokl 4*114-116
⋄ C20 C52 C60 ⋄ REV MR 27 # 2424 Zbl 192.39 JSL 37.401 • ID 19080

KOGALOVSKIJ, S.R. [1963] *Structural characteristics of universal classes (Russian)* (J 0092) Sib Mat Zh 4*97-119
⋄ C05 ⋄ REV MR 26 # 4909 Zbl 199.10 JSL 37.401
• ID 07287

KOGALOVSKIJ, S.R. [1964] *A remark on ultraproducts (Russian)* (C 3551) Tr Molodykh Uchen Vyp Mat 52-54
⋄ C20 C52 ⋄ REV MR 32 # 7408 Zbl 304 # 02020
• ID 22320

KOGALOVSKIJ, S.R. [1964] *The relation between finitely-projective and finitely-reductive classes of models (Russian)* (J 0023) Dokl Akad Nauk SSSR 155*1255-1257
• TRANSL [1964] (J 0062) Sov Math, Dokl 5*562-565
◊ C13 C52 ◊ REV MR 30 # 1048 Zbl 137.8 • ID 07288

KOGALOVSKIJ, S.R. [1965] *Certain remarks on classes axiomatizable in a nonelementary way (Russian)* (J 0340) Mat Zap (Univ Sverdlovsk) 5/1*54-60
◊ C20 C52 C85 ◊ REV MR 34 # 1188 Zbl 298 # 02058 • ID 07290

KOGALOVSKIJ, S.R. [1965] *Generalized quasiuniversal classes of models (Russian)* (J 0216) Izv Akad Nauk SSSR, Ser Mat 29*1273-1282
• TRANSL [1970] (J 0225) Amer Math Soc, Transl, Ser 2 94*107-118
◊ C52 C85 ◊ REV MR 33 # 3932 Zbl 202.302 • ID 19077

KOGALOVSKIJ, S.R. [1965] *On finitely reduced classes of models (Russian)* (J 0092) Sib Mat Zh 6*1021-1025
◊ C52 ◊ REV MR 33 # 47 Zbl 209.14 • ID 07291

KOGALOVSKIJ, S.R. [1965] *On multiplicative semigroups of rings (Russian)* (S 1544) Teor Polugrupp & Prilozh 1*251-261
◊ C52 C60 ◊ REV MR 33 # 3933 Zbl 118.253 • ID 42972

KOGALOVSKIJ, S.R. [1965] *On the theorem of Birkhoff (Russian)* (J 0067) Usp Mat Nauk 20/5*206-207
◊ C05 ◊ REV MR 34 # 1249 Zbl 201.346 • ID 33857

KOGALOVSKIJ, S.R. [1965] *Some remarks on ultraproducts (Russian)* (J 0216) Izv Akad Nauk SSSR, Ser Mat 29*997-1004
• TRANSL [1970] (J 0225) Amer Math Soc, Transl, Ser 2 94*119-126
◊ C20 C52 E55 ◊ REV MR 34 # 4135 Zbl 202.302 • ID 24972

KOGALOVSKIJ, S.R. [1965] *Two questions concerning finite projective classes (Russian)* (J 0092) Sib Mat Zh 6*1429-1431
◊ C52 ◊ REV MR 34 # 36 Zbl 209.13 • ID 07289

KOGALOVSKIJ, S.R. [1966] *On higher order logic (Russian)* (J 0023) Dokl Akad Nauk SSSR 171*1272-1274
• TRANSL [1966] (J 0062) Sov Math, Dokl 7*1642-1645
◊ B15 C40 C85 ◊ REV MR 35 # 35 Zbl 204.1 • ID 07292

KOGALOVSKIJ, S.R. [1966] *On the properties preserved under algebraic constructions (Russian)* (C 1549) Tartu Mezhvuz Nauch Simp Obshchej Algeb 44-51
◊ B25 C30 C52 C60 D35 D45 ◊ REV MR 34 # 5679 Zbl 248 # 02050 • ID 63026

KOGALOVSKIJ, S.R. [1966] *On the semantics of the theory of types (Russian)* (J 0031) Izv Vyssh Ucheb Zaved, Mat (Kazan) 1966/1*89-98
• TRANSL [1970] (J 0225) Amer Math Soc, Transl, Ser 2 94*93-104
◊ B15 C40 C85 ◊ REV MR 34 # 2442 Zbl 202.301 JSL 37.193 • ID 24967

KOGALOVSKIJ, S.R. [1967] *On compactness-preserving algebraic constructions (Russian)* (J 0092) Sib Mat Zh 8*1202-1205
• TRANSL [1967] (J 0475) Sib Math J 8*920-922
◊ C20 C30 C52 ◊ REV MR 37 # 1236 Zbl 162.313 JSL 39.338 • ID 07293

KOGALOVSKIJ, S.R. [1968] *On compact classes of algebraic systems (Russian)* (J 0003) Algebra i Logika 7/2*27-41
• TRANSL [1968] (J 0069) Algeb and Log 7*86-94
◊ C30 C52 ◊ REV MR 40 # 1271 Zbl 191.294 JSL 39.339 • ID 07297

KOGALOVSKIJ, S.R. [1968] *Remarks on compact classes of algebraic systems (Russian)* (J 0023) Dokl Akad Nauk SSSR 180*1029-1032
• TRANSL [1968] (J 0062) Sov Math, Dokl 9*710-713
◊ C30 C52 ◊ REV MR 37 # 6172 Zbl 191.294 • ID 07298

KOGALOVSKIJ, S.R. [1968] *Some remarks on higher order logic (Russian)* (J 0023) Dokl Akad Nauk SSSR 178*1007-1009
• TRANSL [1968] (J 0062) Sov Math, Dokl 9*227-229
◊ B15 C40 C85 F35 ◊ REV MR 39 # 43 Zbl 182.321 • ID 07299

KOGALOVSKIJ, S.R. [1972] *On the degrees of indefinability (Russian)* (P 2585) All-Union Conf Math Log (2);1972 Moskva 22
◊ C40 ◊ ID 33859

KOGALOVSKIJ, S.R. & RORER, M.A. [1973] *On the question of the definability of the concept of definability (Russian)* (J 0226) Uch Zap Ped Inst, Ivanovo 125*46-72
◊ B15 C40 C85 ◊ REV MR 58 # 10289 • ID 74934

KOGALOVSKIJ, S.R. [1974] *Certain criteria for localness for higher order formulas (Russian)* (J 3536) Uch Zap Univ, Ivanovo 130*3-18
◊ B15 C85 ◊ REV MR 57 # 5730 • ID 74930

KOGALOVSKIJ, S.R. [1974] *Certain simple consequences of the axiom of constructibility (Russian) (English summary)* (J 0027) Fund Math 82*245-267
◊ C40 C62 C85 E45 ◊ REV MR 51 # 150 Zbl 299 # 02082 • ID 17448

KOGALOVSKIJ, S.R. [1974] *Reductions in second-order logic (Russian)* (C 2578) Filos & Logika 363-397
◊ B15 C85 ◊ REV MR 58 # 27298 • ID 74929

KOGALOVSKIJ, S.R. [1976] *On local properties (Russian)* (J 0023) Dokl Akad Nauk SSSR 230*1275-1278 • ERR/ADD ibid 18*IV-VIII
• TRANSL [1976] (J 0062) Sov Math, Dokl 17*1465-1469
◊ C52 C75 ◊ REV MR 54 # 7240 Zbl 359 # 02055 • ID 25030

KOGALOVSKIJ, S.R. [1976] see CHEPURNOV, B.A.

KOGALOVSKIJ, S.R. & SOLDATOVA, V.V. [1977] *On definability by intension (Russian)* (C 3625) Metod Log Anal 41-45
◊ C40 ◊ ID 33888

KOGALOVSKIJ, S.R. [1979] see KISELEV, A.A.

KOGALOVSKIJ, S.R. & SOLDATOVA, V.V. [1981] *Remarks on countably local classes (Russian)* (C 3865) Algeb Sistemy (Ivanovo) 158-168
◊ C15 C52 ◊ REV MR 85g:20036 Zbl 526 # 08004 • ID 38186

KOGALOVSKIJ, S.R. see Vol. I, VI for further entries

KOGAN, D.I. [1969] see GLEBSKIJ, YU.V.

KOGAN, D.I. see Vol. IV for further entries

KOKORIN, A.I. & KOPYTOV, V.M. [1962] *Certain classes of ordered groups (Russian)* (J 0003) Algebra i Logika 1/3*21-23
◊ C60 ◊ REV MR 27 # 5840 • ID 16274

KOKORIN, A.I. [1963] see GUREVICH, Y.

KOKORIN, A.I. [1966] see KHISAMIEV, N.G.

KOKORIN, A.I. & KOZLOV, G.T. [1968] *Extended elementary and universal theories of lattice-ordered abelian groups with a finite number of threads (Russian)* (J 0003) Algebra i Logika 7/1*91-103
• TRANSL [1968] (J 0069) Algeb and Log 7*54-61
◊ B25 C60 ◊ REV MR 38 # 4298 Zbl 206.37 • ID 07306

KOKORIN, A.I. & KOZLOV, G.T. [1969] *Elementary theory of abelian groups without torsion, with a predicate for selecting subgroups (Russian)* (J 0003) Algebra i Logika 8*320-334
• TRANSL [1969] (J 0069) Algeb and Log 8*182-190
◊ B25 C35 C60 ◊ REV MR 41 # 3263 • ID 07420

KOKORIN, A.I. & MART'YANOV, V.I. [1973] *Universal extended theories (Russian)* (C 1443) Algebra, Vyp 2 (Irkutsk) 107-113
◊ B25 C52 D35 ◊ REV MR 53 # 5286 • ID 22922

KOKORIN, A.I. & KOZLOV, G.T. [1975] *Proof of a lemma on model completeness (Russian)* (J 0003) Algebra i Logika 14*533-535,606
• TRANSL [1975] (J 0069) Algeb and Log 14*328-330
◊ C35 C60 ◊ REV MR 56 # 11729 Zbl 359 # 02056
• ID 26023

KOKORIN, A.I. & PINUS, A.G. [1978] *Decidability problems of extended theories (Russian)* (J 0067) Usp Mat Nauk 33/2(200)*49-84
• TRANSL [1978] (J 1399) Russ Math Surv 33/2*53-96
◊ B25 C85 C98 D35 ◊ REV MR 58 # 5153 Zbl 395 # 03008 Zbl 434 # 03014 • ID 55718

KOLAITIS, P.G. [1985] *Game quantification* (C 4183) Model-Theor Log 365-421
◊ C40 C70 C75 C98 D60 D65 D70 E60 ◊ ID 48331

KOLAITIS, P.G. see Vol. IV for further entries

KOLCHIN, E.R. [1974] *Constrained extensions of differential fields* (J 0345) Adv Math 12*141-170
◊ C60 ◊ REV MR 49 # 4982 Zbl 279 # 12103 • ID 07309

KOLMOGOROV, A.N. [1953] *On the concept of algorithm (Russian)* (J 0067) Usp Mat Nauk 8/4(56)*175-176
◊ B25 D20 ◊ REV Zbl 51.245 • ID 43578

KOLMOGOROV, A.N. see Vol. I, IV, V, VI for further entries

KOMARNITS'KIJ, N.YA. [1980] *Axiomatizability of certain classes of modules connected with a torsion (Russian)* (S 0166) Mat Issl, Mold SSR 56*92-109,160-161
◊ C60 ◊ REV MR 82a:03031 Zbl 446 # 16024 • ID 74952

KOMARNITS'KIJ, N.YA. [1982] see GORBACHUK, E.L.

KOMATSU, H. [1982] *On the equational definability of addition in rings* (J 0439) Math J Okayama Univ 24*133-136
◊ C05 C60 ◊ REV MR 84d:16021 Zbl 512 # 16016
• ID 37561

KOMORI, Y. [1978] *Logics without Craig's interpolation property* (J 3239) Proc Japan Acad, Ser A 54*46-48
◊ B55 C40 ◊ REV MR 58 # 10328 Zbl 399 # 03048
• ID 52844

KOMORI, Y. [1980] *Completeness of two theories on ordered abelian groups and embedding relations* (J 0111) Nagoya Math J 77*33-39
◊ B25 C10 C35 C60 ◊ REV MR 82c:03053 Zbl 404 # 03014 • ID 53112

KOMORI, Y. see Vol. I, II for further entries

KONCEWICZ, L. [1970] *The simplest expressions of the predicate calculus that define classes of symmetric graphs without loops I (Polish) (English and Russian summaries)* (S 4733) Prace Inst Mat, Politech Wroclaw, Ser Stud Mater 3*23-42
◊ C65 ◊ REV MR 55 # 2486a Zbl 254 # 05102 • REM Part II 1972 • ID 74961

KONCEWICZ, L. [1972] *The simplest expressions of the predicate calculus that define classes of symmetric graphs without loops II (Polish) (English and Russian summaries)* (S 4733) Prace Inst Mat, Politech Wroclaw, Ser Stud Mater 4*47-59
◊ C65 ◊ REV MR 55 # 2486b Zbl 254 # 05103 • REM Part I 1970 • ID 74960

KONCEWICZ, L. [1973] *Definability of classes of graphs in the first order predicate calculus with identity (Polish and Russian summaries)* (J 0063) Studia Logica 32*159-190 • ERR/ADD ibid 33*119
◊ C65 ◊ REV MR 50 # 4284 Zbl 312 # 02044 • ID 19132

KOPFERMANN, K. [1972] *Ultraprodukte endlicher Koerper* (J 0127) J Reine Angew Math 253*138-140
◊ C13 C20 C60 ◊ REV MR 45 # 6792 Zbl 228 # 12109
• ID 07354

KOPFERMANN, K. [1973] *Ultraprodukte endlicher Koerper* (J 0127) J Reine Angew Math 259*211-218
◊ C13 C20 C60 ◊ REV MR 47 # 193 Zbl 249 # 12110
• ID 07355

KOPIEKI, R. & SARALSKI, B. & WALIGORA, G. [1975] *The research of Jaskowski on decidability theory of first order sentences I* (J 0063) Studia Logica 34*201-214
◊ A10 B20 B25 D35 ◊ REV MR 54 # 86 Zbl 315 # 02044
• ID 23964

KOPPEL, M. [1979] *Some decidable diophantine problems: positive solution to a problem of Davis, Matijasevich and Robinson* (J 0053) Proc Amer Math Soc 77*319-323
◊ B25 ◊ REV MR 81a:10069 Zbl 424 # 03020 • ID 81911

KOPPEL, M. see Vol. II for further entries

KOPPELBERG, B. & KOPPELBERG, S. [1976] *A boolean ultrapower which is not an ultrapower* (J 0036) J Symb Logic 41*245-249
◊ C20 E05 ◊ REV MR 53 # 7768 Zbl 368 # 02053
• ID 14805

KOPPELBERG, B. [1980] *Ultrapowers and boolean ultrapowers of ω and ω_1* (J 0009) Arch Math Logik Grundlagenforsch 20*147-153
◊ C20 E05 E50 E55 E65 ◊ REV MR 82d:03048 Zbl 474 # 03030 • ID 55434

KOPPELBERG, B. see Vol. V for further entries

KOPPELBERG, S. [1973] *Injective hulls of chains* (J 0008) Arch Math (Basel) 24*225-229
◊ C30 G05 ◊ REV MR 48 # 194 Zbl 262 # 06008
• ID 07358

KOPPELBERG, S. & TITS, J. [1974] *Une propriete des produits directs infinis de groupes finis isomorphes* (J 2313) C R Acad Sci, Paris, Ser A-B 279*A583-A585
◊ C30 C60 E75 ◊ REV MR 51 # 13058 Zbl 302 # 20027
• ID 39872

KOPPELBERG, S. [1976] see KOPPELBERG, B.

KOPPELBERG, S. [1977] *Groups cannot be Souslin ordered* (J 0008) Arch Math (Basel) 29*315-317
⋄ C60 E07 ⋄ REV MR 57#5741 Zbl 369#06011
• ID 27169

KOPPELBERG, S. [1980] *Cardinalities of ultraproducts of finite sets* (J 0036) J Symb Logic 45*574-584
⋄ C13 C20 C55 E05 E10 ⋄ REV MR 81i:04006 Zbl 497#03016 • ID 74974

KOPPELBERG, S. [1981] *A lattice structure on the isomorphism types of complete Boolean algebras* (P 3268) Set Th & Model Th;1979 Bonn 98-126
⋄ C50 C90 G05 G10 ⋄ REV Zbl 497#03017 • ID 38110

KOPPELBERG, S. [1983] *Groups of permutations with few fixed points* (J 0004) Algeb Universalis 17*50-64
⋄ C07 C50 E05 G05 ⋄ REV MR 85k:20009 Zbl 529#20002 • ID 39874

KOPPELBERG, S. [1985] *Booleschwertige Logik* (J 0157) Jbuchber Dtsch Math-Ver 87*19-38
⋄ B50 C90 C98 E35 E45 E50 E65 G05 ⋄ REV MR 86e:03039 • ID 39673

KOPPELBERG, S. [1985] *Homogeneous boolean algebras may have non-simple automorphism groups* (J 2635) Topology Appl 21*103-120
⋄ C07 C50 E50 E75 G05 ⋄ ID 48309

KOPPELBERG, S. also published under the name GOERNEMANN, S.

KOPPELBERG, S. see Vol. V for further entries

KOPPERMAN, R.D. [1966] *Introductory notes on model theory (Spanish summary)* (J 0348) Rev Mat Elementales 8*19-31
⋄ C98 ⋄ REV MR 35#5302 • ID 16809

KOPPERMAN, R.D. [1967] *Application of infinitary languages to metric spaces* (J 0048) Pac J Math 23*299-310
⋄ C65 C75 ⋄ REV MR 36#1307 Zbl 222#02013
• ID 07359

KOPPERMAN, R.D. [1967] *On the axiomatizability of uniform spaces* (J 0036) J Symb Logic 32*289-294
⋄ C65 C75 ⋄ REV MR 35#6535 Zbl 222#02011
• ID 07360

KOPPERMAN, R.D. [1967] *The L_{ω_1,ω_1}-theory of Hilbert spaces* (J 0036) J Symb Logic 32*295-304
⋄ C65 C75 ⋄ REV MR 35#6534 Zbl 222#02012
• ID 07361

KOPPERMAN, R.D. & MATHIAS, A.R.D. [1968] *Some problems in group theory* (P 0637) Syntax & Semant Infinitary Lang;1967 Los Angeles 131-138
⋄ C60 C75 ⋄ REV Zbl 182.326 • ID 29305

KOPPERMAN, R.D. [1969] *Applications of infinitary languages to analysis* (P 0649) Appl Model Th to Algeb, Anal & Probab;1967 Pasadena 265-273
⋄ C35 C60 C65 C75 ⋄ REV MR 38#5596 Zbl 248#02065 • ID 07362

KOPPERMAN, R.D. [1972] *Model theory and its applications* (X 0802) Allyn & Bacon: London x+333pp
⋄ C60 C98 ⋄ REV MR 51#128 Zbl 233#02021 JSL 38.647 • ID 07363

KOPPERMAN, R.D. [1981] *First-order topological axioms* (J 0036) J Symb Logic 46*475-489
⋄ C65 ⋄ REV MR 83a:54004 Zbl 472#54002 • ID 55326

KOPPERMAN, R.D. see Vol. V for further entries

KOPYTOV, V.M. [1962] see KOKORIN, A.I.

KOPYTOV, V.M. [1982] *Nilpotent lattice ordered groups (Russian)* (J 0092) Sib Mat Zh 23/5*127-131
• TRANSL [1982] (J 0475) Sib Math J 23*690-693
⋄ C60 ⋄ REV MR 84b:06016 Zbl 511#06011 • ID 38906

KOPYTOV, V.M. see Vol. IV for further entries

KOREC, I. & PERETYAT'KIN, M.G. & RAUTENBERG, W. [1974] *Definability in structures of finite valency* (J 0027) Fund Math 81*173-181
⋄ C60 C65 F25 ⋄ REV MR 49#2360 Zbl 276#02036
• ID 07371

KOREC, I. & RAUTENBERG, W. [1976] *Model-interpretability into trees and applications* (J 0009) Arch Math Logik Grundlagenforsch 17*97-104
⋄ B25 C45 C60 C65 F25 ⋄ REV MR 54#2443 Zbl 324#02041 • ID 07374

KOREC, I. see Vol. IV, V for further entries

KORKINA, E.I. & KUSHNIRENKO, A.G. [1985] *Another proof of the Tarski-Seidenberg theorem (Russian)* (J 0092) Sib Mat Zh 26/5*94-98,205
• TRANSL [1985] (J 0475) Sib Math J 26/5*703-707
⋄ B25 ⋄ ID 49259

KOSOVSKIJ, N.K. [1970] *Some questions in the constructive theory of normed boolean algebras (Russian)* (S 0066) Tr Mat Inst Steklov 113*2-38
• TRANSL [1970] (S 0055) Proc Steklov Inst Math 113*1-41
⋄ C57 F60 G05 ⋄ REV MR 44#5988 Zbl 229#02032
• ID 28600

KOSOVSKIJ, N.K. see Vol. I, II, IV, VI for further entries

KOSSAK, R. [1980] *An application of definable types of Peano's arithmetic* (J 3293) Bull Acad Pol Sci, Ser Math 28*213-217
⋄ C62 ⋄ REV MR 83b:03038 Zbl 457#03031 • ID 54356

KOSSAK, R. & PARIS, J.B. [1981] *Subsets of models of arithmetic* (P 2614) Open Days in Model Th & Set Th;1981 Jadwisin 159-174
⋄ C62 F30 ⋄ ID 33723

KOSSAK, R. [1983] *A certain class of models of Peano arithmetic* (J 0036) J Symb Logic 48*311-320
⋄ C50 C57 C62 ⋄ REV MR 84j:03076 Zbl 514#03036
• ID 34668

KOSSAK, R. [1984] $L_{\infty\omega_1}$-*elementary equivalence of ω_1-like models of PA* (J 0027) Fund Math 123*123-131
⋄ C50 C57 C62 C75 ⋄ REV MR 86f:03108 Zbl 545#03018 • ID 41820

KOSSAK, R. [1984] *Remarks on free sets* (S 3382) Sem-ber, Humboldt-Univ Berlin, Sekt Math 60*78-86
⋄ C50 C57 C62 E05 ⋄ REV MR 86h:03066 Zbl 562#03017 • ID 44634

KOSSAK, R. [1985] *A note on satisfaction classes* (J 0047) Notre Dame J Formal Log 26*1-8
⋄ C50 C57 C62 F30 ⋄ REV MR 86c:03055 Zbl 562#03040 • ID 42582

KOSSAK, R. [1985] *Recursively saturated ω_1-like models of arithmetic* (J 0047) Notre Dame J Formal Log 26*413-422
⋄ C55 C57 C62 C75 C80 E65 ⋄ REV Zbl 552#03021 Zbl 571#03014 • ID 47534

KOSSOWSKI, P. [1970] *On the axiomatic treatment of the theory of models IV: Independence of some axiom system of model theory (Polish summary)* (S 0458) Zesz Nauk, Prace Log, Uniw Krakow 5*15-24
⋄ B22 C07 ⋄ REV MR 45 # 1748 • REM Part III 1970 by Surma,S.J. • ID 07392

KOSTINSKY, A. [1966] *Recent results on Jonsson algebras* (1111) Preprints, Manuscr., Techn. Reports etc.
⋄ C05 C55 E05 E55 ⋄ REM Seminar Notes, Dept. Math., Univ. California, Berkley, CA • ID 21393

KOSTYRKO, V.F. [1962] *On an error in the paper of I.I.Zhegalkin "Sur le probleme de la resolubilite pour les classes finies" (Russian)* (J 0003) Algebra i Logika 1/5*31-36
⋄ B25 C13 ⋄ REV MR 27 # 2421 Zbl 178.324 JSL 30.254 • ID 19125

KOSTYRKO, V.F. [1965] *On the decidability problem for Ackermann's case (Russian)* (J 0092) Sib Mat Zh 6*342-363
⋄ B25 ⋄ REV MR 31 # 2150 Zbl 178.323 • ID 07394

KOSTYRKO, V.F. [1971] *The reduction class $\forall x \forall y \exists z F(x,y,z) \wedge \forall^m \mathfrak{A}(F)$ (Russian) (English summary)* (J 0040) Kibernetika, Akad Nauk Ukr SSR 1971/5*1-3
⋄ B20 C13 D35 ⋄ REV MR 46 # 1550 Zbl 231 # 02055 • ID 19122

KOSTYRKO, V.F. [1981] *On elementary ∃-theory of semigroups (Russian)* (J 0040) Kibernetika, Akad Nauk Ukr SSR 1981/5*8-11
• TRANSL [1981] (J 0021) Cybernetics 17*578-582
⋄ B25 C05 ⋄ REV MR 84k:20029 Zbl 498 # 03007 • ID 36901

KOSTYRKO, V.F. see Vol. I, IV for further entries

KOTAS, J. [1972] *About the equivalent theories of algebras with relations (Polish and Russian summaries)* (J 0063) Studia Logica 30*79-96
⋄ C52 ⋄ REV MR 47 # 4783 Zbl 274 # 02025 • ID 07408

KOTAS, J. see Vol. I, II, VI for further entries

KOTLARSKI, H. [1973] *Some simple results on automorphisms of models (Russian summary)* (J 0014) Bull Acad Pol Sci, Ser Math Astron Phys 21*503-507
⋄ C07 E25 ⋄ REV MR 49 # 8851 Zbl 271 # 02033 • ID 07412

KOTLARSKI, H. [1974] *On the existence of well-ordered models* (J 0014) Bull Acad Pol Sci, Ser Math Astron Phys 22*459-462
⋄ C50 C55 ⋄ REV MR 50 # 6825 Zbl 288 # 02029 • ID 07413

KOTLARSKI, H. [1978] *A note on countable well-ordered models* (J 0014) Bull Acad Pol Sci, Ser Math Astron Phys 26*573-575
⋄ C50 C55 E47 ⋄ REV MR 80a:03046 Zbl 406 # 03061 • ID 29176

KOTLARSKI, H. [1978] *Some remarks on well-ordered models* (J 0027) Fund Math 99*123-132
⋄ C20 C50 C55 E07 ⋄ REV MR 57 # 12201 Zbl 395 # 03024 • ID 29196

KOTLARSKI, H. [1980] *On Skolem ultrapowers and their non-standard variant* (J 0068) Z Math Logik Grundlagen Math 26*227-236
⋄ C20 C62 E45 ⋄ REV MR 81m:03040 Zbl 436 # 03023 • ID 55863

KOTLARSKI, H. & KRAJEWSKI, S. & LACHLAN, A.H. [1981] *Construction of satisfaction classes for nonstandard models* (J 0018) Canad Math Bull 24*283-293
⋄ C50 C62 ⋄ REV MR 84c:03066a Zbl 471 # 03054 • ID 55249

KOTLARSKI, H. [1981] *On elementary cuts in models of arithmetic* (J 3293) Bull Acad Pol Sci, Ser Math 29*419-423
⋄ C50 C57 C62 ⋄ REV MR 83b:03075 Zbl 471 # 03053 • ID 55248

KOTLARSKI, H. [1983] *On cofinal extensions of models of arithmetic* (J 0036) J Symb Logic 48*253-262
⋄ C50 C62 ⋄ REV MR 85d:03129 Zbl 537 # 03051 • ID 40024

KOTLARSKI, H. [1983] *On elementary cuts in models of arithmetic* (J 0027) Fund Math 115*27-31
⋄ C50 C57 C62 ⋄ REV MR 84f:03031 Zbl 515 # 03038 • ID 34451

KOTLARSKI, H. [1984] *On elementary cuts in recursively saturated models of Peano arithmetic* (J 0027) Fund Math 120*205-222
⋄ C50 C57 C62 ⋄ REV MR 86f:03056 Zbl 572 # 03016 • ID 45761

KOTLARSKI, H. [1984] *Some remarks on initial segments in models of Peano arithmetic* (J 0036) J Symb Logic 49*955-960
⋄ C62 ⋄ REV MR 85h:03076 Zbl 574 # 03052 • ID 42469

KOTLARSKI, H. [1985] *Bounded induction and satisfaction classes* (P 4310) Easter Conf on Model Th (3);1985 Gross Koeris 143-167
⋄ C15 C50 C57 C62 ⋄ ID 49911

KOUBEK, V. [1980] see GORALCIK, P.

KOUBEK, V. see Vol. IV, V for further entries

KOZEN, D. [1980] *Complexity of boolean algebras* (J 1426) Theor Comput Sci 10*221-247
⋄ B25 D15 G05 ⋄ REV MR 81e:03008 Zbl 428 # 03036 • ID 53795

KOZEN, D. [1981] *On the duality of dynamic algebras and Kripke models* (P 3497) Log of Progr;1979 Zuerich 1-11
⋄ B75 C90 ⋄ REV MR 85b:03042 Zbl 482 # 03008 • ID 36840

KOZEN, D. [1981] *Positive first-order logic is NP-complete* (J 0284) IBM J Res Dev 25*327-332
⋄ B25 D15 ⋄ REV MR 83b:03050 Zbl 481 # 03026 • ID 35103

KOZEN, D. & PARIKH, R. [1984] *A decision procedure for the propositional μ-calculus* (P 2989) Log of Progr; 1983 Pittsburgh 313-325
⋄ B25 B75 ⋄ REV MR 86a:68004 Zbl 564 # 03012 • ID 44553

KOZEN, D. [1984] see BLASS, A.R.

KOZEN, D. see Vol. II, IV, VI for further entries

KOZLOV, G.T. [1968] see KOKORIN, A.I.

KOZLOV, G.T. [1969] see KOKORIN, A.I.

KOZLOV, G.T. [1970] *Unsolvability of the elementary theory of lattices of subgroups of finite abelian p-groups (Russian)* (J 0003) Algebra i Logika 9*167-171
- TRANSL [1970] (J 0069) Algeb and Log 9*104-107
◇ C60 D35 ◇ REV MR 38 #4298 MR 43 #7329 Zbl 222 #02055 • ID 07424

KOZLOV, G.T. [1975] see KOKORIN, A.I.

KOZLOV, G.T. see Vol. IV for further entries

KRABBE, E.C.W. [1978] *The adequacy of material dialogue-games* (J 0047) Notre Dame J Formal Log 19*321-330
◇ A05 B60 C07 ◇ REV MR 58 # 10341 Zbl 316 # 02011 • ID 51481

KRABBE, E.C.W. see Vol. II, VI for further entries

KRAJEWSKI, S. [1974] *A remark on automorphisms and nonstandard properties (Russian summary)* (J 0014) Bull Acad Pol Sci, Ser Math Astron Phys 22*989-991
◇ C62 H15 ◇ REV MR 52 # 86 Zbl 298 # 02062 • ID 07446

KRAJEWSKI, S. [1974] *Mutually inconsistent satisfaction classes* (J 0014) Bull Acad Pol Sci, Ser Math Astron Phys 22*983-987
◇ C62 F30 ◇ REV MR 52 # 10417 Zbl 298 # 02069 • ID 07447

KRAJEWSKI, S. [1974] *Predicative expansions of axiomatic theories* (J 0068) Z Math Logik Grundlagen Math 20*435-452
◇ C30 C62 E30 E70 F25 ◇ REV MR 51 # 10082 Zbl 298 # 02068 • ID 07445

KRAJEWSKI, S. [1976] *Non-standard satisfaction classes* (P 1476) Set Th & Hierarch Th (2) (Mostowski);1975 Bierutowice 121-144
◇ C62 C85 H15 H20 ◇ REV MR 56 # 5280 Zbl 338 # 02030 • ID 23801

KRAJEWSKI, S. [1977] *A note on expansion of models of set theories* (P 1639) Set Th & Hierarch Th (1);1974 Karpacz 14*63-67
◇ C62 E30 E70 ◇ REV MR 58 # 10447 Zbl 407 # 03037 • ID 31406

KRAJEWSKI, S. [1981] see KOTLARSKI, H.

KRAJEWSKI, S. [1981] *Kurt Goedel and his work (Polish)* (J 0519) Wiad Mat, Ann Soc Math Pol, Ser 2 23*161-187
◇ A10 C98 D98 E98 F99 ◇ REV MR 84j:01068 Zbl 535 # 01011 • ID 38331

KRAJEWSKI, S. see Vol. II, IV for further entries

KRAMOSIL, I. & SINDELAR, J. [1978] *Statistical deducibility testing with stochastic parameters* (J 0156) Kybernetika (Prague) 14*385-396
◇ B25 B48 ◇ REV MR 81b:68114 Zbl 403 # 62014 • ID 54777

KRAMOSIL, I. see Vol. I, II, IV, VI for further entries

KRANAKIS, E. [1982] *Definable ultrafilters and end extensions of constructible sets* (J 0068) Z Math Logik Grundlagen Math 28*395-412
◇ C62 E05 E45 E47 ◇ REV MR 84g:03082 Zbl 494 # 03031 • ID 34187

KRANAKIS, E. [1983] *Definable Ramsey and definable Erdoes ordinals* (J 0009) Arch Math Logik Grundlagenforsch 23*115-128
◇ C30 D60 E05 E45 E55 ◇ REV MR 86b:03067 Zbl 535 # 03028 • ID 38326

KRANAKIS, E. [1984] see KAUFMANN, M.

KRANAKIS, E. see Vol. IV, V for further entries

KRAPEZ, A. [1982] *Some nonaxiomatizable classes of semigroups* (P 3758) Algeb Conf (2);1981 Novi Sad 101-105
◇ C05 C52 ◇ REV MR 84j:03073 • ID 34665

KRASNER, M. [1938] *Une generalisation de la notion de corps* (J 3941) J Math Pures Appl, Ser 9 17*367-385 • ERR/ADD ibid 18*
◇ C07 C60 C75 ◇ REV Zbl 20.200 FdM 64.86 • ID 33444

KRASNER, M. [1945] *Generalisation et analogues de la theorie de Galois* (J 0109) C R Acad Sci, Paris
◇ C07 C60 C75 ◇ ID 33674

KRASNER, M. [1950] *Generalisation abstraite de la theorie de Galois* (P 4593) Algeb & Th Nombres;1949 Paris 163-168
◇ C07 C60 C75 ◇ REV MR 12.796 Zbl 39.261 • ID 33445

KRASNER, M. [1958] *Les algebres cylindriques* (J 0353) Bull Soc Math Fr 86*315-319
◇ C07 C60 G15 ◇ REV MR 21 # 3356 Zbl 89.18 JSL 36.337 • ID 33446

KRASNER, M. [1970] *Endotheorie de Galois abstraite* (S 1057) Semin Dubreil: Alg Th Nombr 22/6*19pp
◇ C07 C60 C75 ◇ REV MR 44 # 45 Zbl 222 # 02068 • ID 26324

KRASNER, M. [1976] *Endotheorie de Galois abstraite et son theoreme d'homomorphie (English summary)* (J 2313) C R Acad Sci, Paris, Ser A-B 282*A683-A686
◇ C07 C60 C75 ◇ REV MR 53 # 5550 Zbl 344 # 02040 • ID 63120

KRASNER, M. [1976] *Polytheorie de Galois abstraite dans le cas infini general* (J 1934) Ann Sci Univ Clermont Math 13*87-91
◇ C05 C07 C60 ◇ REV MR 58 # 16468 Zbl 376 # 08001 • ID 33447

KRASNER, M. [1978] *Abstract Galois theory* (J 2713) Eleutheria, Math J Sem Zervos (Athens) 1978*15-34
◇ C07 C10 C60 C75 ◇ REV MR 82j:03041 • ID 75059

KRASNER, M. see Vol. V, VI for further entries

KRATOCHVIL, P. [1971] *Note on a representation of universal algebras as subdirect powers* (S 0019) Colloq Math (Warsaw) 24*11-14
◇ C05 C30 ◇ REV MR 46 # 3415 Zbl 245 # 08005 • ID 07455

KRAUSS, P.H. & SCOTT, D.S. [1966] *Assigning probabilities to logical formulas* (C 1107) Aspects Inductive Log 219-264
◇ B48 C75 C90 ◇ REV Zbl 202.299 • ID 27709

KRAUSS, P.H. [1969] *Representation of symmetric probability models* (J 0036) J Symb Logic 34*183-193
◇ C75 C90 ◇ REV MR 43 # 1236 Zbl 218 # 02022 • ID 07460

KRAUSS, P.H. [1971] *Universally complete universal theories* (J 0068) Z Math Logik Grundlagen Math 17*351-370
◇ C05 C30 C35 ◇ REV MR 45 # 50 Zbl 231 # 02063 • ID 07461

KRAUSS, P.H. [1972] *Extending congruence relations* (J 0053) Proc Amer Math Soc 31∗517-520
⋄ C05 C07 C50 ⋄ REV MR 44 # 2688 Zbl 255 # 08006
• ID 07463

KRAUSS, P.H. [1972] see CLARK, D.M.

KRAUSS, P.H. [1972] *On primal algebras* (J 0004) Algeb Universalis 2∗62-67
⋄ C05 C13 ⋄ REV MR 46 # 5213 Zbl 265 # 08001
• ID 07462

KRAUSS, P.H. [1973] *On quasi primal algebras* (J 0044) Math Z 134∗85-89
⋄ C05 C13 ⋄ REV MR 48 # 10952 Zbl 257 # 08004
• ID 07464

KRAUSS, P.H. [1975] *Quantifier elimination* (P 1442) ⊢ ISILC Logic Conf;1974 Kiel 426-444
⋄ C10 C35 C50 ⋄ REV MR 54 # 4969 Zbl 345 # 02041
• ID 27217

KRAUSS, P.H. [1976] see CLARK, D.M.

KRAUSS, P.H. [1977] *Homogeneous universal models of universal theories* (J 0068) Z Math Logik Grundlagen Math 23∗415-426
⋄ C05 C35 C50 C52 ⋄ REV MR 80g:03028 Zbl 453 # 03035 • ID 54165

KRAUSS, P.H. [1977] see CLARK, D.M.

KRAUSS, P.H. [1979] see CLARK, D.M.

KRAUSS, P.H. [1980] see CLARK, D.M.

KRAWCZYK, A. [1975] *On a class of models for sentences of languages α_A* (J 0014) Bull Acad Pol Sci, Ser Math Astron Phys 23∗381-385 • ERR/ADD ibid 23∗1 loose page
⋄ C70 E45 ⋄ REV MR 52 # 5347 MR 53 # 12867 Zbl 356 # 02011 Zbl 357 # 02038 • ID 23128

KRAWCZYK, A. & KRYNICKI, M. [1976] *Ehrenfeucht games for generalized quantifiers* (P 1476) Set Th & Hierarch Th (2) (Mostowski);1975 Bierutowice 145-152
⋄ C80 ⋄ REV MR 55 # 79 Zbl 338 # 02008 • ID 24408

KRAWCZYK, A. [1977] *On a class of models for sentences of language $\mathcal{L}_A$* (P 1629) Symp Math Log;1974 Oulo;1975 Helsinki 81-88
⋄ C70 ⋄ ID 48404

KRAWCZYK, A. & MAREK, W. [1977] *On the rules of proof generated by hierarchies* (P 1695) Set Th & Hierarch Th (3);1976 Bierutowice 227-239
⋄ B15 C55 C85 E45 ⋄ REV MR 58 # 21621 Zbl 424 # 03026 • ID 31404

KRAWCZYK, A. see Vol. V for further entries

KRECZMAR, A. [1977] *On infinite sets of polynomial relations* (J 0014) Bull Acad Pol Sci, Ser Math Astron Phys 25∗1-6
⋄ C60 C75 ⋄ REV MR 56 # 5277 Zbl 357 # 02014
• ID 26550

KRECZMAR, A. see Vol. II, IV for further entries

KREISEL, G. [1950] *Note on arithmetic models for consistent formulae of the predicate calculus I* (J 0027) Fund Math 37∗265-285
⋄ B10 C57 D45 D55 F30 ⋄ REV MR 12.790 JSL 18.180
• REM Part II 1953 • ID 07481

KREISEL, G. [1953] *Note on arithmetic models for consistent formulae of the predicate calculus II* (P 0645) Int Congr Philos (11);1953 Bruxelles 14∗39-49
⋄ B10 C57 D45 D55 F30 ⋄ REV MR 15.668 Zbl 53.200 JSL 21.403 • REM Part I 1951 • ID 20815

KREISEL, G. [1954] *Applications of mathematical logic to various branches of mathematics* (P 0646) Appl Sci de Log Math;1952 Paris 37-49
⋄ B98 C65 ⋄ REV MR 16.782 Zbl 57.245 JSL 24.236
• ID 16973

KREISEL, G. [1956] *Some uses of metamathematics* (J 0013) Brit J Phil Sci 7∗161-173
⋄ A05 C98 F99 ⋄ ID 90276

KREISEL, G. [1957] *Sums of squares* (P 1675) Summer Inst Symb Log;1957 Ithaca 313-320
⋄ C57 C60 D20 F07 F99 ⋄ REV JSL 31.128 • ID 29369

KREISEL, G. [1958] *Mathematical significance of consistency proofs* (J 0036) J Symb Logic 23∗155-182
⋄ C57 C60 F05 F25 F50 ⋄ REV MR 22 # 6710 Zbl 88.15 JSL 31.129 • ID 07514

KREISEL, G. [1960] see GANDY, R.O.

KREISEL, G. [1961] see GANDY, R.O.

KREISEL, G. [1961] *Set theoretic problems suggested by the notion of potential totality* (P 0633) Infinitist Meth;1959 Warsaw 103-140
⋄ A05 C62 D20 D55 D65 ⋄ REV MR 26 # 3599
• ID 07523

KREISEL, G. [1963] *Axiomatic results of second-order arithmetic* (C 4220) Rep Sem Found Anal 1∗1.1-1.48
⋄ C62 F35 ⋄ ID 42740

KREISEL, G. [1965] *Model-theoretic invariants: applications to recursive and hyperarithmetic operations* (P 0614) Th Models;1963 Berkeley 190-205
⋄ C40 C62 D55 D60 ⋄ REV MR 33 # 7257 Zbl 225 # 02040 • ID 33546

KREISEL, G. [1966] see FEFERMAN, S.

KREISEL, G. [1966] see EHRENFEUCHT, A.

KREISEL, G. & KRIVINE, J.-L. [1967] *Elements de logique mathematique. Theorie des modeles* (X 0856) Dunod: Paris viii + 213pp
• TRANSL [1972] (X 0811) Springer: Heidelberg & New York xv + 274pp [1967] (X 0809) North Holland: Amsterdam xi + 222pp (English)
⋄ B98 C98 ⋄ REV MR 34 # 7331 MR 36 # 2463 MR 50 # 4231 Zbl 146.7 Zbl 238 # 02003 JSL 34.112
• ID 22411

KREISEL, G. [1967] *Informal rigour and completeness proofs* (P 2268) Int Colloq Philos of Sci;1965 London 1∗138-186
• TRANSL [1978] (C 4649) Paradiso di Cantor 59-93 • REPR [1969] (C 0569) Phil of Math Oxford Readings 78-94
⋄ A05 B30 C07 E30 E50 F50 ⋄ ID 07533

KREISEL, G. [1968] *Choice of infinitary languages by means of definability criteria: generalized recursion theory* (P 0637) Syntax & Semant Infinitary Lang;1967 Los Angeles 139-151
⋄ C40 C70 C75 D60 D75 ⋄ REV Zbl 177.10 • ID 07540

KREISEL, G. [1975] *Observations on a recent generalization of completeness theorems due to Schuette* (P 1440) ⊢ ISILC Proof Th Symp (Schuette);1974 Kiel 164-181
  ⋄ A05 C07 C57 F05 F20 F35 F50 ⋄ REV MR 54#73 Zbl 324#02020 • ID 23955

KREISEL, G. & MINTS, G.E. & SIMPSON, S.G. [1975] *The use of abstract language in elementary metamathematics: Some pedagogic examples* (C 0758) Logic Colloq Boston 1972-73 38-131
  ⋄ A05 C07 C57 C75 F05 F07 F20 F50 ⋄ REV Zbl 318#02003 • ID 27698

KREISEL, G. [1975] *Was hat die Logik in den letzten 25 Jahren für die Mathematik geleistet?* (J 2688) Conceptus (Wien) 9*40-45
  ⋄ A05 B98 C75 D40 ⋄ REV MR 58#21356 • ID 75092

KREISEL, G. [1980] *Modell* (C 4082) Handb Wiss Begriffe 437-440
  ⋄ C98 ⋄ ID 48850

KREISEL, G. [1980] *Modelltheorie* (C 4082) Handb Wiss Begriffe 440-443
  ⋄ C98 ⋄ ID 48851

KREISEL, G. see Vol. I, II, IV, V, VI for further entries

KREJNOVICH, V.YA. & OSWALD, U. [1982] *A decision method for the universal theorems of Quine's new foundations* (J 0068) Z Math Logik Grundlagen Math 28*181-187
  ⋄ B20 B25 E70 ⋄ REV MR 84b:03071 Zbl 515#03030 • ID 35653

KREJNOVICH, V.YA. see Vol. VI for further entries

KREMER, E.M. [1984] *Modules with an almost categorical theory (Russian)* (J 0092) Sib Mat Zh 25/6*70-75
  • TRANSL [1984] (J 0475) Sib Math J 25*888-892
  ⋄ C35 C45 C60 ⋄ REV MR 86h:03062 • ID 42706

KREMER, E.M. [1984] *Rings over which all modules of a given type are almost categorical (Russian)* (J 0003) Algebra i Logika 23*159-174
  • TRANSL [1984] (J 0069) Algeb and Log 23*113-124
  ⋄ C35 C45 C60 ⋄ REV MR 86h:16023 • ID 41867

KRIPKE, S.A. [1962] *"Flexible" predicates of formal number theory* (J 0053) Proc Amer Math Soc 13*647-650
  ⋄ C90 F30 ⋄ REV MR 25#3827 Zbl 109.9 • ID 07556

KRIPKE, S.A. [1965] *Semantical analysis of intuitionistic logic I* (P 0688) Logic Colloq;1963 Oxford 92-130
  ⋄ B55 C90 F50 ⋄ REV MR 34#1184 Zbl 137.7 JSL 35.330 • REM Part II never appeared • ID 07557

KRIPKE, S.A. [1982] see KOCHEN, S.

KRIPKE, S.A. see Vol. I, II, IV, VI for further entries

KRISHNAN, V.S. [1953] *Closure operations on c-structures* (J 0028) Indag Math 15*317-329
  ⋄ C05 C30 E75 ⋄ REV MR 15.675 Zbl 53.123 • ID 41162

KRISHNAN, V.S. see Vol. V for further entries

KRIVINE, J.-L. [1967] see KREISEL, G.

KRIVINE, J.-L. [1969] *Modeles de ZF+AC dans lesquels tout ensemble de reels definissable en termes d'ordinaux est mesurable-Lebesgue* (J 2313) C R Acad Sci, Paris, Ser A-B 269*A549-A552
  ⋄ C62 E35 E45 E75 ⋄ REV MR 40#7107 Zbl 182.328 • ID 07561

KRIVINE, J.-L. [1972] see DACUNHA-CASTELLE, D.

KRIVINE, J.-L. [1972] *Theorie des modeles et espaces L^p* (J 2313) C R Acad Sci, Paris, Ser A-B 275*A1207-A1210
  ⋄ C65 ⋄ REV MR 46#7879 Zbl 246#02037 • ID 07563

KRIVINE, J.-L. & MCALOON, K. [1973] *Forcing and generalized quantifiers* (J 0007) Ann Math Logic 5*199-255 • ERR/ADD ibid 6*93
  ⋄ C25 C80 ⋄ REV MR 56#5209 Zbl 259#02009 • ID 07564

KRIVINE, J.-L. & MCALOON, K. [1973] *Some true unprovable formulas for set theory* (P 0710) Russell Mem Logic Conf;1971 Uldum 332-341
  ⋄ C62 C75 E35 E45 E55 ⋄ REV MR 50#9580 • ID 24214

KRIVINE, J.-L. [1974] *Langages a valeurs reelles et applications* (J 0027) Fund Math 81*213-253
  ⋄ B50 C40 C60 C90 ⋄ REV MR 50#1873 Zbl 292#02019 • ID 07566

KRIVINE, J.-L. [1975] see DACUNHA-CASTELLE, D.

KRIVINE, J.-L. [1976] *Sous-espaces de dimension finie des espaces de Banach reticules* (J 0120) Ann of Math, Ser 2 104*1-29
  ⋄ C20 C65 ⋄ REV MR 53#11341 Zbl 329#46008 • ID 23105

KRIVINE, J.-L. [1977] see DACUNHA-CASTELLE, D.

KRIVINE, J.-L. [1984] *Methodes de theorie des modeles en geometrie des espaces de Banach* (C 4356) Gen Log Semin Paris 1982/83 179-186
  ⋄ C65 C98 E75 ⋄ ID 48084

KRIVINE, J.-L. see Vol. I, V for further entries

KROL', M.D. [1976] *The topological models of intuitionistic analysis. One counter-example (Russian)* (J 0087) Mat Zametki (Akad Nauk SSSR) 19*859-862
  • TRANSL [1976] (J 1044) Math Notes, Acad Sci USSR 19*503-504
  ⋄ C90 F35 F50 ⋄ REV MR 55#2524 Zbl 355#02025 JSL 46.660 • ID 50223

KROL', M.D. [1977] *Disjunctive and existential properties of intuitionistic analysis with Kripke's scheme (Russian)* (J 0023) Dokl Akad Nauk SSSR 234*750-753
  • TRANSL [1977] (J 0062) Sov Math, Dokl 18*755-758
  ⋄ C90 F35 F50 ⋄ REV MR 57#9494 Zbl 384#03043 • ID 52085

KROL', M.D. [1978] *A topological model for intuitionistic analysis with Kripke's scheme* (J 0068) Z Math Logik Grundlagen Math 24*427-436
  ⋄ C90 F35 F50 ⋄ REV MR 80b:03096 Zbl 418#03039 JSL 46.660 • ID 53324

KROL', M.D. [1978] *Distinct variants of Kripke's scheme in intuitionistic analysis (Russian)* (J 0023) Dokl Akad Nauk SSSR 239*1048-1051
  • TRANSL [1978] (J 0062) Sov Math, Dokl 19*474-477
  ⋄ C90 F35 F50 ⋄ REV MR 58#5096 Zbl 397#03037 JSL 46.660 • ID 52703

KROL', M.D. see Vol. VI for further entries

KROM, MELVEN R. [1963] *Separation principles in the hierarchy theory of pure first-order logic* (J 0036) J Symb Logic 28*222-236
⋄ B10 C40 C52 D55 ⋄ REV MR 31 # 2133 Zbl 137.9 JSL 31.503 • ID 07571

KROM, MELVEN R. [1964] *A decision procedure for a class of formulas of first order predicate calculus* (J 0048) Pac J Math 14*1305-1319
⋄ B20 B25 ⋄ REV MR 31 # 3316 Zbl 171.270 • ID 07570

KROM, MELVEN R. [1966] *A property of sentences that define quasi-order* (J 0047) Notre Dame J Formal Log 7*349-352
⋄ C65 ⋄ REV MR 38 # 4297 Zbl 172.8 • ID 07572

KROM, MELVEN R. [1967] *The decision problem for segregated formulas in first-order logic* (J 0132) Math Scand 21*233-240
⋄ B20 B25 D35 ⋄ REV MR 39 # 1286 Zbl 169.310 • ID 07574

KROM, MELVEN R. [1968] *Some interpolation theorems for first-order formulas in which all disjunctions are binary* (J 0079) Logique & Anal, NS 11*403-412
⋄ B20 C40 ⋄ REV MR 39 # 2603 • ID 07575

KROM, MELVEN R. & KROM, MYREN [1970] *Groups with free nonabelian subgroups* (J 0048) Pac J Math 35*425
⋄ C60 ⋄ REV MR 43 # 4904 • ID 28492

KROM, MELVEN R. see Vol. I, IV, V for further entries

KROM, MYREN [1970] see KROM, MELVEN R.

KRON, A. [1978] *Decision procedures for two positive relevance logics* (J 0302) Rep Math Logic, Krakow & Katowice 10*61-78
⋄ B25 B46 ⋄ REV MR 81b:03020 Zbl 432 # 03013 • ID 53971

KRON, A. see Vol. II for further entries

KRUPA, A. & ZAWISZA, B. [1984] *Applications of ultrapowers in analysis of unbounded selfadjoint operators* (J 3417) Bull Pol Acad Sci, Math 32*581-588
⋄ C20 C65 H10 ⋄ REV MR 86h:47029 • ID 44797

KRUSE, A.H. [1962] *Completion of mathematical systems* (J 0048) Pac J Math 12*589-605
⋄ C25 C35 C62 ⋄ REV MR 26 # 1273 Zbl 121.13 • ID 07580

KRUSE, A.H. [1966] *Grothendieck universes and the super-complete models of Shepherdson* (J 0020) Compos Math 17*96-101
⋄ C62 E55 ⋄ REV MR 31 # 4716 Zbl 129.5 JSL 37.613 • ID 07586

KRUSE, A.H. see Vol. IV, V for further entries

KRYAZHOVSKIKH, G.V. [1980] *Approximability of finitely presented algebras (Russian)* (J 0092) Sib Mat Zh 21/5*58-62
• TRANSL [1980] (J 0475) Sib Math J 21*688-691
⋄ C60 D40 ⋄ REV MR 82f:08004 Zbl 457 # 17001 • ID 54395

KRYNICKI, M. [1976] see KRAWCZYK, A.

KRYNICKI, M. [1977] *Henkin quantifier and decidability* (P 1629) Symp Math Log;1974 Oulo;1975 Helsinki 2*89-90
⋄ B25 C10 C80 D35 ⋄ ID 33547

KRYNICKI, M. [1977] *Some consequences of the omitting and realizing types property* (P 1629) Symp Math Log;1974 Oulo;1975 Helsinki 33-43
⋄ C75 C95 ⋄ ID 90115

KRYNICKI, M. [1979] *On the expressive power of the language using the Henkin quantifier* (P 1705) Scand Logic Symp (4);1976 Jyvaeskylae 259-265
⋄ B15 B25 C80 C85 ⋄ REV MR 82c:03056 Zbl 411 # 03027 • ID 52881

KRYNICKI, M. & LACHLAN, A.H. [1979] *On the semantics of the Henkin quantifier* (J 0036) J Symb Logic 44*184-200
⋄ B15 B25 C40 C55 C80 C85 ⋄ REV MR 80h:03054 Zbl 418 # 03026 • ID 53311

KRYNICKI, M. & VAEAENAENEN, J. [1982] *On orderings of the family of all logics* (J 0009) Arch Math Logik Grundlagenforsch 22*141-158
⋄ C95 ⋄ REV MR 84g:03053 Zbl 499 # 03024 • ID 34163

KRYNICKI, M. & LACHLAN, A.H. & VAEAENAENEN, J. [1984] *Vector spaces and binary quantifiers* (J 0047) Notre Dame J Formal Log 25*72-78
⋄ C60 C75 C80 ⋄ REV MR 85d:03077 Zbl 543 # 03020 • ID 40288

KRYNICKI, M. [1985] *Interpretations in nonelementary languages* (P 4310) Easter Conf on Model Th (3);1985 Gross Koeris 168-182
⋄ C80 C95 F25 ⋄ ID 49048

KRZYSTEK, P.S. & ZACHOROWSKI, S. [1977] *Lukasiewicz logics have not the interpolation property* (J 0302) Rep Math Logic, Krakow & Katowice 9*39-40
⋄ B50 C40 ⋄ REV MR 58 # 21440 Zbl 384 # 03011 • ID 52053

KRZYSTEK, P.S. see Vol. II for further entries

KUBOTA, N. [1983] *A splitting property on a set of ideals and a weak saturation hypothesis* (J 0407) Comm Math Univ St Pauli (Tokyo) 32*209-216
⋄ C55 E05 ⋄ REV MR 85e:04005 Zbl 524 # 03032 • ID 37604

KUBOTA, N. see Vol. V for further entries

KUCIA, A. [1977] *On saturating ultrafilters on N* (S 0019) Colloq Math (Warsaw) 37*23-27
⋄ C20 E05 E50 ⋄ REV MR 58 # 5240 Zbl 446 # 03039 • ID 56568

KUCIA, A. see Vol. V for further entries

KUDAJBERGENOV, K.ZH. [1979] *A theory with two strongly constructivizable models (Russian)* (J 0003) Algebra i Logika 18*176-185,253
• TRANSL [1979] (J 0069) Algeb and Log 18*111-117
⋄ C57 D45 ⋄ REV MR 81f:03038 Zbl 448 # 03028 • ID 75139

KUDAJBERGENOV, K.ZH. [1980] *On constructive models of undecidable theories (Russian)* (J 0092) Sib Mat Zh 21/5*155-158,192
⋄ C15 C35 C57 D35 ⋄ REV MR 82h:03040 Zbl 454 # 03011 • ID 54224

KUDAJBERGENOV, K.ZH. [1982] *A remark on omitting models (Russian)* (J 0429) Vest Akad Nauk Kazak SSR 1982/10*70-71
⋄ C15 C50 ⋄ REV MR 85a:03047 Zbl 525 # 03021 • ID 34798

KUDAJBERGENOV, K.ZH. [1983] *The number of constructive homogeneous models of a complete decidable theory (Russian)* (J 0087) Mat Zametki (Akad Nauk SSSR) 34*135-143
- TRANSL [1983] (J 1044) Math Notes, Acad Sci USSR 34*552-557
  ⋄ C50 C57 ⋄ REV MR 85e:03075 Zbl 537 # 03021
  • ID 40506

KUDAJBERGENOV, K.ZH. [1984] *Autostability and extensions of constructivizations (Russian)* (J 0092) Sib Mat Zh 25/5*72-78
- TRANSL [1984] (J 0475) Sib Math J 25*743-749
  ⋄ C50 C57 F60 ⋄ REV MR 86b:03040 • ID 41816

KUDAJBERGENOV, K.ZH. [1984] *Constructivizability of a prime model (Russian)* (J 0092) Sib Mat Zh 25/4*93-98
- TRANSL [1984] (J 0475) Sib Math J 25*584-588
  ⋄ C50 C57 F60 ⋄ REV MR 86d:03030 • ID 41871

KUEHNRICH, M. [1971] *Ein Beweis der wesentlichen Nichtdefinierbarkeit* (J 0342) Monatsber Dt Akad Wiss 13*255-257
  ⋄ C40 E47 E70 ⋄ REV MR 47 # 3174 Zbl 233 # 02028
  • ID 07616

KUEHNRICH, M. [1978] *On the Hermes term logic* (X 2888) ZI Math Mech Akad Wiss DDR: Berlin 14/78*26pp
  ⋄ C07 C20 ⋄ REV Zbl 406 # 03053 • ID 56136

KUEHNRICH, M. [1978] *Superclasses in a finite extension of Zermelo set theory* (J 0068) Z Math Logik Grundlagen Math 24*539-552
  ⋄ C62 E30 E55 E70 G30 ⋄ REV MR 80f:03056 Zbl 428 # 03048 • ID 53807

KUEHNRICH, M. & SCHULTZ, KONRAD [1980] *A hierarchy of models for Skala's set theory* (J 0068) Z Math Logik Grundlagen Math 26*555-559
  ⋄ C62 E35 E70 ⋄ REV MR 82e:03046 Zbl 473 # 03043
  • ID 55372

KUEHNRICH, M. [1983] *On the Hermes term logic* (J 0302) Rep Math Logic, Krakow & Katowice 16*3-16
  ⋄ B10 C07 C20 ⋄ REV MR 86f:03051 Zbl 406 # 03053
  • ID 36540

KUEHNRICH, M. see Vol. I, V for further entries

KUEKER, D.W. [1968] *Definability, automormophisms and infinitary languages* (P 0637) Syntax & Semant Infinitary Lang;1967 Los Angeles 152-165
  ⋄ C07 C40 C75 ⋄ REV Zbl 235 # 02018 • ID 16801

KUEKER, D.W. [1970] *Generalized interpolation and definability* (J 0007) Ann Math Logic 1*423-468
  ⋄ C40 C50 C75 ⋄ REV MR 44 # 54 JSL 39.337
  • ID 07619

KUEKER, D.W. [1972] *Loewenheim-Skolem and interpolation theorems in infinitary languages* (J 0015) Bull Amer Math Soc 78*211-215
  ⋄ C40 C55 C75 E05 ⋄ REV MR 45 # 36 Zbl 264 # 02017 JSL 51.232 • ID 07620

KUEKER, D.W. [1973] see BALDWIN, JOHN T.

KUEKER, D.W. [1973] *A note on the elementary theory of finite abelian groups* (J 0004) Algeb Universalis 3*156-159
  ⋄ C13 C20 C60 ⋄ REV MR 50 # 87 Zbl 315 # 02051
  • ID 07621

KUEKER, D.W. [1975] *Back-and-forth arguments and infinitary logics* (C 0781) Infinitary Logic (Karp) 17-71
  ⋄ C75 C98 ⋄ REV MR 57 # 2905 • ID 75152

KUEKER, D.W. [1975] *Core structures for theories* (J 0027) Fund Math 89*155-171
  ⋄ C40 C50 C52 ⋄ REV MR 52 # 13368 Zbl 327 # 02041
  • ID 07622

KUEKER, D.W. (ED.) [1975] *Infinitary logic: In memoriam Carol Karp* (X 0811) Springer: Heidelberg & New York vi+206pp
  ⋄ C70 C75 C97 ⋄ REV MR 52 # 7837 Zbl 316 # 02018
  • ID 66270

KUEKER, D.W. [1977] *Countable approximations and Loewenheim-Skolem theorems* (J 0007) Ann Math Logic 11*57-103
  ⋄ C15 C55 C75 E05 ⋄ REV MR 56 # 15406 Zbl 364 # 02009 JSL 51.232 • ID 24265

KUEKER, D.W. [1978] *Uniform theorems in infinitary logic* (P 1897) Logic Colloq;1977 Wroclaw 161-170
  ⋄ C40 C55 C75 ⋄ REV MR 80f:03038 Zbl 473 # 03032
  • ID 55361

KUEKER, D.W. [1980] see BALDWIN, JOHN T.

KUEKER, D.W. [1981] $L_{\infty \omega_1}$ *elementarily equivalent models of power* ω_1 (P 2628) Log Year;1979/80 Storrs 120-131
  ⋄ C75 ⋄ REV MR 82k:03052 Zbl 486 # 03018 • ID 75150

KUEKER, D.W. [1981] see BALDWIN, JOHN T.

KUENG, G. [1970] see CANTY, J.T.

KUENG, G. see Vol. II for further entries

KUERSTEN, K.-D. [1983] *Local duality of ultraproducts of Banach lattices* (P 3883) Banach Space Th & Appl (1);1981 Bucharest 137-142
  ⋄ C20 C65 ⋄ REV MR 85c:46014 Zbl 515 # 46018
  • ID 39057

KUERSTEN, K.-D. [1984] *Lokale Reflexivitaet und lokale Dualitaet von Ultraprodukten fuer halbgeordnete Banachraeume (English and Russian summaries)* (J 4231) Z Anal & Anwendungen 3*245-262
  ⋄ C20 C65 ⋄ REV MR 85k:46016 Zbl 544 # 46007
  • ID 45474

KUHLMANN, F.V. & PRESTEL, A. [1984] *Places of algebraic function fields* (J 0127) J Reine Angew Math 353*181-195
  ⋄ C60 ⋄ REV MR 86d:12014 Zbl 535 # 12015 • ID 38332

KUKIN, G.P. [1979] *Subalgebras of finitely defined Lie algebras (Russian)* (J 0003) Algebra i Logika 18*311-327
- TRANSL [1979] (J 0069) Algeb and Log 18*190-201
  ⋄ C05 C57 ⋄ REV MR 81k:17010 Zbl 445 # 17011
  • ID 56514

KUKIN, G.P. see Vol. IV for further entries

KULIK, V.T. [1970] *Sentences preserved under homomorphisms that are one-to-one at zero (Russian)* (J 0031) Izv Vyssh Ucheb Zaved, Mat (Kazan) 1970/4(95)*56-63
  ⋄ C05 C40 C60 ⋄ REV MR 43 # 1911 Zbl 205.319
  • ID 07629

KULUNKOV, P.A. [1980] *Questions of definability in ordered sets (Russian) (English summary)* (J 0288) Vest Ser Mat Mekh, Univ Moskva 1980/1*48-51,92-93
• TRANSL [1980] (J 0510) Moscow Univ Math Bull 35/1*51-55
◊ C15 C65 E07 ◊ REV MR 81g:03038 Zbl 451 # 06001
• ID 54045

KULUNKOV, P.A. see Vol. V for further entries

KUNEN, K. [1968] *Implicit definability and infinitary languages* (J 0036) J Symb Logic 33*446-451
◊ C40 C70 C75 D70 E47 E55 ◊ REV MR 38 # 5597 Zbl 195.302 JSL 35.341 • ID 07631

KUNEN, K. & PARIS, J.B. [1970] *Boolean extensions and measurable cardinals* (J 0007) Ann Math Logic 2*359-377
◊ C20 C62 E35 E40 E50 E55 ◊ REV MR 43 # 3114 Zbl 216.14 • ID 07632

KUNEN, K. [1970] *Some applications of iterated ultrapowers in set theory* (J 0007) Ann Math Logic 1*179-227
◊ C20 C55 C62 E05 E35 E45 E50 E55 ◊ REV MR 43 # 3080 Zbl 236 # 02053 • ID 07637

KUNEN, K. [1971] *Elementary embeddings and infinitary combinatorics* (J 0036) J Symb Logic 36*407-413
◊ C62 E05 E55 ◊ REV MR 47 # 40 Zbl 272 # 02087 JSL 39.331 • ID 07638

KUNEN, K. [1971] see BARWISE, J.

KUNEN, K. & PRIKRY, K. [1971] *On descendingly incomplete ultrafilters* (J 0036) J Symb Logic 36*650-652
◊ C55 E05 ◊ REV MR 46 # 1585 Zbl 259 # 02053
• ID 07635

KUNEN, K. [1972] *Ultrafilters and independent sets* (J 0064) Trans Amer Math Soc 172*299-306
◊ C55 E05 ◊ REV MR 47 # 3170 Zbl 263 # 02033
• ID 07640

KUNEN, K. [1978] *Saturated ideals* (J 0036) J Symb Logic 43*65-76
◊ C55 E05 E35 E40 E55 ◊ REV MR 80a:03068 Zbl 395 # 03031 • ID 29242

KUNEN, K. see Vol. I, II, V for further entries

KURATA, R. [1964] *On the existence of a proper complete model of set theory* (J 0407) Comm Math Univ St Pauli (Tokyo) 13*35-43
◊ C62 E30 E70 ◊ REV MR 31 # 4723 Zbl 229 # 02049
• ID 07649

KURATA, R. [1981] *The reflection principle, transfinite induction, and the Paris-Harrington principle (Japanese)* (P 4153) B-Val Anal & Nonstand Anal;1981 Kyoto 1-14
◊ C62 F30 ◊ ID 47747

KURATA, R. [1984] *Paris-Harrington theory and reflection principles* (P 3668) Log & Founds of Math;1983 Kyoto 123-131
• REPR [1984] (J 3940) Saitama Math J 2*33-45
◊ C62 F30 ◊ REV MR 86g:03095 Zbl 552 # 03038
• ID 42937

KURATA, R. see Vol. V, VI for further entries

KURATOWSKI, K. [1937] *Les types d'ordre définissables et les ensembles boreliens* (J 0027) Fund Math 29*97-100
◊ C75 D55 E07 E15 ◊ REV Zbl 17.49 JSL 3.48 FdM 64.833 • ID 07668

KURATOWSKI, K. see Vol. I, IV, V for further entries

KURKA, P. [1981] see HAJEK, P.

KURKA, P. see Vol. II, VI for further entries

KURKE, H. & MOSTOWSKI, T. & PFISTER, G. & POPESCU, D. & ROCZEN, M. [1978] *Die Approximationseigenschaft lokaler Ringe* (S 3301) Lect Notes Math 634*iv+204pp
◊ C10 C60 ◊ REV MR 58 # 5653 Zbl 401 # 13013
• ID 54633

KUSHNIRENKO, A.G. [1985] see KORKINA, E.I.

KUSRAEV, A.G. [1982] *Boolean-valued analysis of duality of extended modules (Russian)* (J 0023) Dokl Akad Nauk SSSR 267*1049-1052
• TRANSL [1982] (J 0062) Sov Math, Dokl 26*732-735
◊ C60 C65 C90 E40 E75 ◊ REV MR 84f:03045 Zbl 538 # 46059 • ID 34464

KUSRAEV, A.G. [1983] *On some categories and functors of boolean-valued analysis (Russian)* (J 0023) Dokl Akad Nauk SSSR 271*283-286
• TRANSL [1983] (J 0062) Sov Math, Dokl 28*73-77
◊ C65 C90 E40 G30 ◊ REV MR 85d:03103 Zbl 545 # 03021 • ID 40491

KUSRAEV, A.G. & KUTATELADZE, S.S. [1983] *Subdifferentials in Boolean-valued models of set theory (Russian)* (J 0092) Sib Mat Zh 24/5*109-122
• TRANSL [1983] (J 0475) Sib Math J 24*735-746
◊ C65 C90 E40 E75 ◊ REV MR 85m:49029 Zbl 573 # 03022 • ID 44867

KUSRAEV, A.G. [1984] *Order-continuous functionals in Boolean-valued models of set theory (Russian)* (J 0092) Sib Mat Zh 25/1*69-79
• TRANSL [1984] (J 0475) Sib Math J 25*57-65
◊ C65 C90 E40 ◊ REV MR 86e:03048 Zbl 573 # 03023
• ID 45341

KUSRAEV, A.G. see Vol. V for further entries

KUSTAANHEIMA, P. [1949] *Ueber die Vollstaendigkeit der Axiomensysteme mit einem endlichen Individuenbereich* (J 0990) Soc Sci Fennicae Comment Phys-Math 63*44pp
◊ C13 C35 C85 ◊ REV MR 12.2 Zbl 34.294 JSL 16.282
• ID 07705

KUTATELADZE, S.S. [1983] see KUSRAEV, A.G.

KUTATELADZE, S.S. see Vol. I, V for further entries

KUZ'MIN, E.N. [1968] *Algebraic sets in Mal'cev algebras (Russian)* (J 0003) Algebra i Logika 7/2*42-47
• TRANSL [1968] (J 0069) Algeb and Log 7*95-97
◊ C60 ◊ REV MR 38 # 4533 Zbl 204.361 • ID 07725

KUZNETSOV, A.V. [1959] *On the equivalence and functional completeness problems (Russian)* (P 0607) All-Union Math Conf (3);1956 Moskva 2*145-146
◊ B25 D35 ◊ ID 28393

KUZNETSOV, A.V. [1975] *On superintuitionistic logics* (P 1521) Int Congr Math (II,12);1974 Vancouver 1*243-249
◊ B25 B55 ◊ REV MR 55 # 2509 Zbl 342 # 02015
• ID 63246

KUZNETSOV, A.V. see Vol. I, II, IV, VI for further entries

KWATINETZ, M. [1974] see FRODA-SCHECHTER, M.

LABLANQUIE, J.-C. [1968] *Sur les approximations finies des formules infiniment longues* (J 2313) C R Acad Sci, Paris, Ser A-B 267*A65-A68
⋄ C75 ⋄ REV MR 38 # 2018 Zbl 199.304 • ID 48135

LABLANQUIE, J.-C. [1974] *Quelques proprietes des enonces universellement forces* (J 0068) Z Math Logik Grundlagen Math 20*31-36
⋄ C25 ⋄ REV MR 49 # 8852 Zbl 308 # 02048 • ID 07733

LABLANQUIE, J.-C. [1976] *Le nombre des structures generiques d'une propriete de "forcing" (English summary)* (J 2313) C R Acad Sci, Paris, Ser A-B 282*A817-A819
⋄ C15 C25 ⋄ REV MR 53 # 7769 Zbl 343 # 02042 • ID 23009

LABLANQUIE, J.-C. [1977] *Le nombre de classes d'equivalence elementaire des modeles generiques d'une propriete de "forcing" (English summary)* (J 2313) C R Acad Sci, Paris, Ser A-B 284*A787-A790
⋄ C15 C25 C35 ⋄ REV MR 55 # 2550 Zbl 358 # 02065 • ID 27297

LABLANQUIE, J.-C. [1977] *Proprietes de forcing et modeles generiques* (J 0056) Publ Dep Math, Lyon 14/2*13-20
⋄ C15 C25 C70 ⋄ REV MR 58 # 27454 Zbl 396 # 03029 • ID 52639

LABLANQUIE, J.-C. [1978] *Forcing infini et theoreme d'omission des types de Chang* (J 1934) Ann Sci Univ Clermont Math 16*27-31
⋄ C07 C25 ⋄ REV MR 80e:03026 Zbl 397 # 03016 • ID 52682

LABLANQUIE, J.-C. [1979] *Proprietes de consistance et forcing* (J 1934) Ann Sci Univ Clermont Math 18*37-45
⋄ C25 C40 C75 ⋄ REV MR 81j:03047 Zbl 427 # 03020 • ID 53707

LABLANQUIE, J.-C. [1980] *Modeles premiers et modeles generiques (English summary)* (J 2313) C R Acad Sci, Paris, Ser A-B 290*A931-A933
⋄ C25 C50 ⋄ REV MR 82e:03028 Zbl 435 # 03024 • ID 55784

LABLANQUIE, J.-C. [1981] *Modeles generiques et compacite (English summary)* (J 3364) C R Acad Sci, Paris, Ser 1 293*227-230
⋄ C25 C52 ⋄ REV MR 83b:03033 Zbl 481 # 03019 • ID 35095

LABLANQUIE, J.-C. [1982] *Extensions existentielles forcing-completes de theories forcing-completes (English summary)* (J 3364) C R Acad Sci, Paris, Ser 1 294*669-672
⋄ C25 ⋄ REV MR 83i:03053 Zbl 523 # 03019 • ID 35519

LACAVA, F. [1975] *Teoria dei campi differenziali ordinati* (J 0149) Atti Accad Naz Lincei Fis Mat Nat, Ser 8 59*322-327
⋄ C10 C35 C60 ⋄ REV MR 58 # 210 Zbl 353 # 02032 • ID 50115

LACAVA, F. & SAELI, D. [1976] *Proprieta e model-completamento di alcune varieta di algebre di Lukasiewicz (English summary)* (J 0149) Atti Accad Naz Lincei Fis Mat Nat, Ser 8 60*359-367
⋄ C35 G20 ⋄ REV MR 57 # 97 Zbl 362 # 02038 • ID 50760

LACAVA, F. & SAELI, D. [1977] *Sul model-completamento della teoria delle L-catene* (J 3285) Boll Unione Mat Ital, V Ser, A 14*107-110
⋄ C35 C60 G20 ⋄ REV MR 55 # 12509 Zbl 363 # 02056 • ID 50882

LACAVA, F. [1979] *Alcune proprieta delle L-algebre e delle L-algebre esistenzialmente chiuse* (J 3285) Boll Unione Mat Ital, V Ser, A 16*360-366
⋄ C25 C35 C60 G20 ⋄ REV MR 81a:03062 Zbl 427 # 03024 • ID 53711

LACAVA, F. [1980] *Osservazioni sulla teoria dei gruppi abeliani reticolari* (J 3285) Boll Unione Mat Ital, V Ser, A 17*319-322
⋄ C25 C35 C60 ⋄ REV MR 82a:03032 Zbl 439 # 03012 • ID 56004

LACAVA, F. see Vol. IV for further entries

LACHLAN, A.H. [1961] *The U-quantifier* (J 0068) Z Math Logik Grundlagen Math 7*171-174
⋄ C80 ⋄ REV MR 25 # 2955 Zbl 100.248 JSL 27.244 • ID 07734

LACHLAN, A.H. [1968] *On the lattice of recursively enumerable sets* (J 0064) Trans Amer Math Soc 130*1-37
⋄ B25 C10 D25 ⋄ REV MR 37 # 2594 Zbl 281 # 02042 JSL 35.153 • ID 07751

LACHLAN, A.H. [1968] *The elementary theory of recursively enumerable sets* (J 0025) Duke Math J 35*123-146
⋄ B25 D25 ⋄ REV MR 37 # 2593 Zbl 281 # 02043 JSL 35.153 • ID 07752

LACHLAN, A.H. & MADISON, E.W. [1970] *Computable fields and arithmetically definable ordered fields* (J 0053) Proc Amer Math Soc 24*803-807
⋄ C57 C60 D45 ⋄ REV MR 40 # 7110 Zbl 229 # 02039 • ID 07756

LACHLAN, A.H. [1971] see BALDWIN, JOHN T.

LACHLAN, A.H. [1971] *The transcendental rank of a theory* (J 0048) Pac J Math 37*119-122
⋄ C45 ⋄ REV MR 46 # 7016 Zbl 231 # 02064 • ID 22280

LACHLAN, A.H. [1972] *A property of stable theories* (J 0027) Fund Math 77*9-20
⋄ C45 C55 ⋄ REV MR 48 # 5847 Zbl 268 # 02034 • ID 07759

LACHLAN, A.H. [1973] *On the number of countable models of a countable superstable theory* (P 0793) Int Congr Log, Meth & Phil of Sci (4,Proc);1971 Bucharest 45-56
⋄ C15 C45 ⋄ REV MR 56 # 5266 • ID 75277

LACHLAN, A.H. [1973] see BALDWIN, JOHN T.

LACHLAN, A.H. [1974] *Two conjectures regarding the stability of ω-categorical theories* (J 0027) Fund Math 81*133-145
⋄ C35 C45 ⋄ REV MR 49 # 2341 Zbl 284 # 02025 • ID 07762

LACHLAN, A.H. [1975] *A remark on the strict order property* (J 0068) Z Math Logik Grundlagen Math 21*69-70
⋄ C45 ⋄ REV MR 51 # 131 Zbl 308 # 02050 • ID 07763

LACHLAN, A.H. [1975] *Theories with a finite number of models in an uncountable power are categorical* (J 0048) Pac J Math 61*465-481
⋄ C35 C45 ⋄ REV MR 55 # 7762 Zbl 338 # 02028 • ID 63274

LACHLAN, A.H. [1976] *Dimension and totally transcendental theories of rank 2* (P 1476) Set Th & Hierarch Th (2) (Mostowski);1975 Bierutowice 153-183
⋄ C35 C45 ⋄ REV MR 56 # 15404 Zbl 345 # 02040
• ID 23804

LACHLAN, A.H. [1977] see HODGES, W.

LACHLAN, A.H. [1978] *Skolem functions and elementary extensions* (J 3172) J London Math Soc, Ser 2 18*1-6
⋄ C30 C50 ⋄ REV MR 80a:03036 Zbl 402 # 03033
• ID 54680

LACHLAN, A.H. [1978] *Spectra of ω-stable theories* (J 0068) Z Math Logik Grundlagen Math 24*129-139
⋄ C35 C45 ⋄ REV MR 80a:03041 Zbl 401 # 03013
• ID 75273

LACHLAN, A.H. [1979] see KRYNICKI, M.

LACHLAN, A.H. & WOODROW, R.E. [1980] *Countable ultrahomogeneous undirected graphs* (J 0064) Trans Amer Math Soc 262*51-94
⋄ C15 C50 C65 ⋄ REV MR 82c:05083 Zbl 471 # 03025
• ID 55220

LACHLAN, A.H. [1980] *Singular properties of Morley rank* (J 0027) Fund Math 108*145-157
⋄ C35 C45 ⋄ REV MR 82e:03033 Zbl 473 # 03026
• ID 55356

LACHLAN, A.H. [1981] see KOTLARSKI, H.

LACHLAN, A.H. [1981] *Full satisfaction classes and recursive saturation* (J 0018) Canad Math Bull 24*295-297
⋄ C50 C57 C62 ⋄ REV MR 84c:03066b Zbl 471 # 03055
• ID 55250

LACHLAN, A.H. [1982] *Finite homogeneous simple digraphs* (P 3708) Herbrand Symp Logic Colloq;1981 Marseille 189-208
⋄ C13 C50 C65 ⋄ REV MR 85h:05049 Zbl 518 # 05037
• ID 41881

LACHLAN, A.H. [1984] *Binary homogeneous structures. I* (S 3382) Sem-ber, Humboldt-Univ Berlin, Sekt Math 60*100-156
⋄ C10 C13 C15 C45 C50 C57 ⋄ REV MR 86h:03048 Zbl 552 # 03020 • ID 43365

LACHLAN, A.H. [1984] *Countable homogeneous tournaments* (J 0064) Trans Amer Math Soc 284*431-461
⋄ C15 C50 C65 ⋄ REV MR 85i:05118 Zbl 562 # 05025
• ID 44333

LACHLAN, A.H. [1984] *On countable stable structures which are homogeneous for a finite relational language* (J 0029) Israel J Math 49*69-153
⋄ C10 C13 C15 C45 C50 ⋄ ID 45599

LACHLAN, A.H. & SHELAH, S. [1984] *Stable structures homogeneous for a binary language* (J 0029) Israel J Math 49*155-180
⋄ C10 C13 C15 C45 C50 ⋄ ID 38753

LACHLAN, A.H. [1984] see KNIGHT, J.F.

LACHLAN, A.H. [1984] see KRYNICKI, M.

LACHLAN, A.H. [1985] see CHERLIN, G.L.

LACHLAN, A.H. see Vol. II, IV, V, VI for further entries

LACZKOVICH, M. [1978] *Solvability and consistency for infinite systems of linear inequalities* (J 0049) Period Math Hung 9*63-70
⋄ C65 ⋄ REV MR 80e:03024 Zbl 342 # 15012 • ID 75294

LACZKOVICH, M. see Vol. V for further entries

LADNER, R.E. [1977] *Application of model theoretic games to discrete linear orders and finite automata* (J 0194) Inform & Control 33*281-303
• TRANSL [1980] (J 3079) Kiber Sb Perevodov, NS 17*164-191
⋄ B15 B25 C07 C13 C65 C85 D05 E07 E60 ⋄ REV MR 58 # 10387 Zbl 387 # 68037 • ID 52252

LADNER, R.E. [1980] *Complexity theory with emphasis on the complexity of logical theories* (P 3021) Logic Colloq;1979 Leeds 286-319
⋄ B25 C98 D05 D10 D15 ⋄ REV MR 82h:03037 Zbl 444 # 03018 • ID 75296

LADNER, R.E. see Vol. II, IV for further entries

LADRIERE, J. [1969] *Le theoreme de Loewenheim-Skolem* (J 1030) Formalisation 10*108-130
⋄ C07 ⋄ ID 15162

LADRIERE, J. see Vol. I, II, IV, VI for further entries

LAEUCHLI, H. [1962] *Auswahlaxiom in der Algebra* (J 2022) Comm Math Helvetici 37*1-18
⋄ C60 E25 E75 ⋄ REV MR 26 # 1258 Zbl 108.10
• ID 07783

LAEUCHLI, H. & LEONARD, J. [1966] *On the elementary theory of linear order* (J 0027) Fund Math 59*109-116
⋄ B25 C65 ⋄ REV MR 33 # 7258 Zbl 156.253 JSL 33.287
• ID 07785

LAEUCHLI, H. [1968] *A decision procedure for the weak second order theory of linear order* (P 0608) Logic Colloq;1966 Hannover 189-197
⋄ B15 B25 C65 C85 ⋄ REV MR 39 # 5343 • ID 07787

LAEUCHLI, H. see Vol. II, V, VI for further entries

LAGRANGE, R. [1974] *Amalgamation and epimorphisms in m-complete boolean algebras* (J 0004) Algeb Universalis 4*277-279
⋄ C05 G05 ⋄ REV MR 51 # 300 Zbl 303 # 06010
• ID 07792

LAGRANGE, R. [1983] see COWLES, J.R.

LAIR, C. [1980] see GUITART, R.

LAIR, C. [1981] *Categories modelables et categories esquissables* (J 3797) Diagrammes 6*L1-L20
⋄ C90 G30 ⋄ REV MR 84j:18004 Zbl 522 # 18008
• ID 39149

LAIR, C. see Vol. V for further entries

LAITA, L.M. [1976] *Un estudio de logica algebraica desde el punto de vista de la teoria de categorias* (J 0047) Notre Dame J Formal Log 17*89-118
⋄ C90 G15 G30 ⋄ REV MR 55 # 7772 Zbl 299 # 02070
• ID 18251

LAKE, J. [1975] *Characterising the largest, countable partial ordering* (J 0068) Z Math Logik Grundlagen Math 21*353-354
⋄ C15 C50 C65 E07 ⋄ REV MR 52 # 2888 Zbl 307 # 06001 • ID 07801

LAKE, J. [1975] *Natural models and Ackermann-type set theories*
(J 0036) J Symb Logic 40*151-158
⋄ C62 E30 E45 E55 E70 ⋄ REV MR 51 # 5307
Zbl 307 # 02048 • ID 07800

LAKE, J. see Vol. I, V for further entries

LAKSER, H. [1969] see GRAETZER, G.

LAKSER, H. [1973] see GRAETZER, G.

LAMBEK, J. & MOERDIJK, I. [1982] *Two sheaf representations of elementary toposes* (P 3638) Brouwer Centenary Symp;1981 Noordwijkerhout 275-295
⋄ C90 G30 ⋄ REV MR 85h:03070 Zbl 511 # 03028
• ID 36913

LAMBEK, J. see Vol. I, IV, V, VI for further entries

LAMPE, W.A. & MYERS, D.L. [1980] *Elementary properties of free extensions* (J 0004) Algeb Universalis 11*269-284
⋄ C05 ⋄ REV MR 82j:03033 Zbl 455 # 03010 • ID 54271

LAMPE, W.A. see Vol. V for further entries

LANDRAITIS, C.K. [1977] *Definability in well quasi-ordered sets of structures* (J 0036) J Symb Logic 42*289-291
⋄ C15 C65 C75 ⋄ REV MR 58 # 134 Zbl 371 # 02023
• ID 26460

LANDRAITIS, C.K. [1980] $L_{\omega_1\omega}$ *equivalence between countable and uncountable linear orderings* (J 0027) Fund Math 107*99-112
⋄ C65 C75 E07 ⋄ REV MR 81j:03056 Zbl 453 # 03038
• ID 54168

LANDRAITIS, C.K. see Vol. V for further entries

LANDWEBER, L.H. [1969] see BUECHI, J.R.

LANDWEBER, L.H. see Vol. IV for further entries

LANGFORD, C.H. [1926] *Analytic completeness of postulate sets* (J 1910) Proc London Math Soc, Ser 2 25*115-142
⋄ B20 C35 C65 E07 ⋄ REV FdM 52.49 • ID 07849

LANGFORD, C.H. [1926] *Some theorems on deducibility* (J 0120) Ann of Math, Ser 2 28*16-40
⋄ B30 C10 C35 C65 ⋄ REV FdM 52.48 • REM Part I. Part II 1926 • ID 07850

LANGFORD, C.H. [1926] *Theorems on deducibility (second paper)* (J 0120) Ann of Math, Ser 2 28*459-471
⋄ B30 C10 C35 C65 ⋄ REV FdM 53.43 • REM Part II. Part I 1926 • ID 07851

LANGFORD, C.H. [1927] *On a type of completeness characterizing the general laws for separation of point pairs* (J 0064) Trans Amer Math Soc 29*96-110
⋄ C35 C65 E07 ⋄ REV FdM 53.539 • ID 07852

LANGFORD, C.H. & LEWIS, C.I. [1932] *Symbolic logic* (X 1228) Appleton-Century-Crofts: New York xi+506pp
• LAST ED [1959] (X 0813) Dover: New York ix+518pp
⋄ B25 B98 C10 C98 ⋄ REV MR 21 # 4091 Zbl 87.8
FdM 58.56 • ID 22592

LANGFORD, C.H. [1939] *A theorem on deducibility for second-order functions* (J 0036) J Symb Logic 4*77-79
⋄ B15 C35 C65 C85 ⋄ REV Zbl 21.98 JSL 5.30
FdM 65.28 • ID 07855

LANGFORD, C.H. see Vol. I for further entries

LARSON, L.C. [1971] *Nonstandard theory of Zariski rings*
(J 0053) Proc Amer Math Soc 29*23-29
⋄ C60 H20 ⋄ REV MR 43 # 4808 Zbl 197.314 • ID 07867

LASCAR, D. [1973] *Types definissables et produits de types*
(J 2313) C R Acad Sci, Paris, Ser A-B 276*A1253-A1256
⋄ C45 C55 ⋄ REV MR 49 # 33 Zbl 259 # 02039 • ID 07868

LASCAR, D. [1974] *Sur les theories convexes "modeles completes"*
(J 2313) C R Acad Sci, Paris, Ser A-B 278*A1001-A1004
⋄ C15 C35 C52 ⋄ REV MR 50 # 1870 Zbl 285 # 02047
• ID 07869

LASCAR, D. [1975] *Definissabilite dans les theories stables*
(J 0079) Logique & Anal, NS 18*489-507
• REPR [1977] (P 1625) Six Days of Model Th;1975
Louvain-la-Neuve 257-275
⋄ C45 ⋄ REV MR 58 # 10397 Zbl 354 # 02041
Zbl 399 # 03020 • ID 50181

LASCAR, D. [1975] *Quelques remarques sur la notion de bifurcation (forking) (English summary)* (J 2313) C R Acad Sci, Paris, Ser A-B 280*A607-A610
⋄ C45 ⋄ REV MR 51 # 7855 Zbl 306 # 02049 • ID 17699

LASCAR, D. [1976] *Ranks and definability in superstable theories*
(J 0029) Israel J Math 23*53-87
⋄ C15 C45 ⋄ REV MR 53 # 12931 Zbl 326 # 02038
JSL 47.215 • ID 07870

LASCAR, D. [1977] *Generalisation de l'ordre de Rudin-Keisler aux types d'une theorie* (P 1729) Colloq Int Log;1975
Clermont-Ferrand 73-81
⋄ C15 C35 C45 E05 ⋄ REV MR 82a:03024
Zbl 469 # 03023 • ID 75356

LASCAR, D. [1977] *La theorie des modeles: histoire et orientations actuelles* (S 3677) Fund Scientiae
⋄ C98 ⋄ ID 47588

LASCAR, D. [1977] *Sur la hauteur des structures denombrables (English summary)* (J 2313) C R Acad Sci, Paris, Ser A-B 285*A415-A418
⋄ C15 C50 C75 ⋄ REV MR 56 # 5211 Zbl 371 # 02025
• ID 27312

LASCAR, D. [1978] *Caractere effectif des theoremes d'approximation d'Artin (English summary)* (J 2313) C R Acad Sci, Paris, Ser A-B 287*A907-A910
⋄ C57 ⋄ REV MR 80c:14006 Zbl 424 # 13005 • ID 82025

LASCAR, D. & POIZAT, B. [1979] *An introduction to forking*
(J 0036) J Symb Logic 44*330-350
⋄ C45 ⋄ REV MR 80k:03030 Zbl 424 # 03013 JSL 51.234
• ID 75364

LASCAR, D. [1979] *La conjecture de Vaught* (C 2050) Semin Th des Ensembles GMS Paris 1976/78 13-20
⋄ C15 C50 C75 ⋄ REV MR 84e:03044 Zbl 521 # 03022
• ID 34381

LASCAR, D. [1979] *Les modeles denombrables d'une theorie superstable ayant des fonctions de Skolem (English summary)*
(J 2313) C R Acad Sci, Paris, Ser A-B 289*A655-A658
⋄ C15 C45 C52 ⋄ REV MR 80m:03064 Zbl 428 # 03023
• ID 53782

LASCAR, D. [1980] *Une indicatrice de type "Ramsey" pour l'arithmetique de Peano et la formule de Paris-Harrington*
(C 2969) Modeles de l'Arithm; Paris 1977 19-30
⋄ C62 F30 ⋄ REV MR 83g:03074 Zbl 458 # 03024
JSL 48.483 • ID 54430

LASCAR, D. [1981] *Dans quelle mesure la categorie des modeles d'une theorie complete determine-t-elle cette theorie?* (J 3133) Bull Soc Math Belg, Ser B 33∗33-40
 ⋄ C35 C90 G30 ⋄ REV MR 83e:03052 Zbl 485 # 03036
 • ID 90316

LASCAR, D. [1981] *Les modeles denombrables d'une theorie ayant des fonctions de Skolem (English summary)* (J 0064) Trans Amer Math Soc 268∗345-366
 ⋄ C15 C45 C50 ⋄ REV MR 83f:03030 Zbl 527 # 03011
 • ID 35275

LASCAR, D. [1982] *On the category of models of a complete theory* (J 0036) J Symb Logic 47∗249-266
 ⋄ C35 C90 G30 ⋄ REV MR 84g:03055 Zbl 498 # 03019
 • ID 34164

LASCAR, D. [1982] *Ordre de Rudin-Keisler et poids dans les theories stables* (J 0068) Z Math Logik Grundlagen Math 28∗413-430
 ⋄ C45 ⋄ REV MR 84d:03037 Zbl 497 # 03020 • ID 34077

LASCAR, D. [1983] see BOUSCAREN, E.

LASCAR, D. [1983] *Les corps differentiellement clos denombrables* (C 3822) Groupe Etude Th Stables (3) 1980/82 6∗9pp
 ⋄ C15 C45 C60 ⋄ REV MR 85b:03053 Zbl 512 # 03015
 • ID 36510

LASCAR, D. [1984] see DELON, F.

LASCAR, D. [1984] *Relation entre le rang U et le poids (English summary)* (J 0027) Fund Math 121∗117-123
 ⋄ C45 ⋄ REV MR 86g:03057 Zbl 551 # 03019 • ID 42734

LASCAR, D. [1984] *Sous-groupes d'automorphismes d'une structure saturee* (P 3710) Logic Colloq;1982 Firenze 123-134
 ⋄ C07 C35 C45 C50 ⋄ REV Zbl 556 # 03029 • ID 41838

LASCAR, D. [1984] *Theorie de la classification* (C 4356) Gen Log Semin Paris 1982/83 35-43
 ⋄ C45 ⋄ REV MR 86f:03005 Zbl 575 # 03020 • ID 48073

LASCAR, D. [1985] *Les groupes ω-stables de rang fini* (J 0064) Trans Amer Math Soc 292∗451-462
 ⋄ C35 C45 C60 ⋄ ID 48405

LASCAR, D. [1985] *Quelques precisions sur la d.o.p. et la profondeur d'une theorie (English summary)* (J 0036) J Symb Logic 50∗316-330
 ⋄ C45 C75 ⋄ ID 41811

LASCAR, D. [1985] see DELON, F.

LASCAR, D. [1985] *Why some people are excited by Vaught's conjecture* (J 0036) J Symb Logic 50∗973-982
 ⋄ C15 C70 C75 ⋄ ID 48406

LASCAR, D. see Vol. I, IV for further entries

LASLANDES, B. [1985] *Modele-compagnons de theories de corps munis de n ordres* (J 3364) C R Acad Sci, Paris, Ser 1 300∗411-414
 ⋄ C25 C35 C60 ⋄ ID 47271

LASSAIGNE, R. [1971] *Produits reduits et langages infinis* (J 2313) C R Acad Sci, Paris, Ser A-B 273∗A791-A794
 ⋄ C20 C75 E05 E55 ⋄ REV MR 44 # 2591 Zbl 238 # 02050 • ID 07871

LASSAIGNE, R. & RESSAYRE, J.-P. [1972] *Algebres de Boole et langages infinis* (J 2313) C R Acad Sci, Paris, Ser A-B 274∗A689-A692
 ⋄ C35 C75 G05 ⋄ REV MR 45 # 6599 Zbl 236 # 02045
 • ID 07872

LATYSHEV, V.N. [1977] *Complexity of nonmatrix varieties of associative algebras I,II (Russian)* (J 0003) Algebra i Logika 16∗149-183,184-199,249,250
 • TRANSL [1977] (J 0069) Algeb and Log 16∗98-122,122-133
 ⋄ C05 C60 ⋄ REV MR 58 # 27695 Zbl 395 # 16010 Zbl 395 # 16011 • ID 29220

LAUGWITZ, D. & SCHMIEDEN, C. [1958] *Eine Erweiterung der Infinitesimalrechnung* (J 0044) Math Z 69∗1-39
 ⋄ B28 C65 E20 H05 ⋄ REV MR 20 # 2404 Zbl 82.42
 • ID 48699

LAUGWITZ, D. see Vol. I for further entries

LAVENDHOMME, R. & LUCAS, T. [1980] *Logique locale* (J 1650) Rapp, Sem Math Pure, Univ Cathol Louvain 101∗38pp
 ⋄ C90 G30 ⋄ REV Zbl 452 # 03031 • ID 54095

LAVENDHOMME, R. & LUCAS, T. [1981] *A propos d'un theoreme de Macintyre* (J 0306) Cah Topol & Geom Differ 22∗387-398
 ⋄ C35 C90 ⋄ REV MR 83b:03045 Zbl 485 # 03037
 • ID 35098

LAVENDHOMME, R. & LUCAS, T. [1985] *A non-boolean version of Feferman-Vaught's theorem* (J 0068) Z Math Logik Grundlagen Math 31∗299-308
 ⋄ C30 C90 ⋄ REV Zbl 553 # 03022 Zbl 567 # 03012
 • ID 47556

LAVENDHOMME, R. see Vol. II, V, VI for further entries

LAVER, R. [1978] see FABER, V.

LAVER, R. [1982] *Saturated ideals and nonregular ultrafilters* (P 3634) Patras Logic Symp;1980 Patras 297-305
 ⋄ C20 C55 E05 E55 E65 ⋄ REV MR 85e:03110 Zbl 534 # 03024 Zbl 536 # 03031 • ID 36563

LAVER, R. see Vol. V for further entries

LAVROV, I.A. [1962] *Undecidability of elementary theories of certain rings (Russian)* (J 0003) Algebra i Logika 1/3∗39-45
 ⋄ C60 D35 ⋄ REV MR 28 # 3934 • ID 07884

LAVROV, I.A. [1963] *Effective inseparability of the sets of identically true formulae and finitely refutable formulae for certain elementary theories (Russian)* (J 0003) Algebra i Logika 2/1∗5-18
 ⋄ C13 D25 D35 ⋄ REV MR 28 # 1132 Zbl 199.32
 • ID 07885

LAVROV, I.A. [1965] see ERSHOV, YU.L.

LAVROV, I.A. see Vol. I, II, IV, V, VI for further entries

LAWRENCE, J. [1979] see BURRIS, S.

LAWRENCE, J. [1981] *Primitive rings do not form an elementary class* (J 0394) Commun Algeb 9∗397-400
 ⋄ C20 C52 C60 ⋄ REV MR 82d:16005 Zbl 454 # 16001
 • ID 54249

LAWRENCE, J. [1982] see BURRIS, S.

LAWRENCE, J. [1983] *The theory of ∀∃ elementary conditions on rings* (**J** 2128) C R Math Acad Sci, Soc Roy Canada 5*41-45
⋄ C60 ⋄ REV MR 84c:16019 Zbl 505 # 16006 • ID 38210

LAWRENCE, J. [1985] *The definability of the commutator subgroup in a variety generated by a finite group* (**J** 0018) Canad Math Bull 28*505-507
⋄ C60 ⋄ ID 48407

LAWVERE, F.W. & MAURER, C. & WRAITH, G.C. (EDS.) [1975] *Model theory and topoi* (**X** 0811) Springer: Heidelberg & New York 354pp
⋄ C97 F97 G97 ⋄ REV MR 51 # 10016 Zbl 299 # 00014 JSL 46.158 • ID 70191

LAWVERE, F.W. see Vol. V, VI for further entries

LEAVITT, W.G. & LEEUWEN VAN, L.C.A. [1979] *Rings isomorphic with all proper factor-rings* (**P** 2624) Ring Th;1978 Antwerpen 783-798
⋄ C60 E10 E75 ⋄ REV MR 81i:16010 Zbl 427 # 16019 • ID 53745

LEBLANC, H. [1969] *Three generalizations of a theorem of Beth's* (**J** 0079) Logique & Anal, NS 12*205-220
⋄ B15 C07 ⋄ REV MR 44 # 3832 Zbl 209.302 • ID 07936

LEBLANC, H. & MORGAN, C.G. [1983] *Probability theory, intuitionism, semantics, and the Dutch book argument* (**J** 0047) Notre Dame J Formal Log 24*289-304
⋄ B48 C90 F50 ⋄ REV MR 85d:03048 Zbl 531 # 03037 • ID 37685

LEBLANC, H. [1984] *A new semantics for first-order logic, multivalent and mostly intensional* (**J** 3781) Topoi 3*55-62
⋄ B10 C30 ⋄ REV MR 86f:03038 • ID 44672

LEBLANC, H. see Vol. I, II, VI for further entries

LEBLANC, L. [1966] *Representabilite et definissabilite dans les algebres transformationnelles et dans les algebres polyadiques* (**X** 0893) Pr Univ Montreal: Montreal 124pp
⋄ C40 C62 C90 G15 ⋄ REV MR 40 # 4090 Zbl 199.6 • ID 07949

LEBLANC, L. see Vol. I for further entries

LEDBETTER, C.S. [1979] see KENNISON, J.F.

LEE, R.D. [1966] see GOODSTEIN, R.L.

LEE, R.D. see Vol. VI for further entries

LEE, V. & NADEL, M.E. [1975] *On the number of generic models* (**J** 0027) Fund Math 90*105-114
⋄ C25 C70 ⋄ REV MR 53 # 7723 Zbl 322 # 02051 • ID 07965

LEE, V. & NADEL, M.E. [1977] *Remarks on generic models* (**J** 0027) Fund Math 95*73-84
⋄ C25 C70 ⋄ REV MR 57 # 12205 Zbl 402 # 03026 • ID 26522

LEEDS, S. & PUTNAM, H. [1974] *Solution to a problem of Gandy's* (**J** 0027) Fund Math 81*99-106
⋄ C62 D55 ⋄ REV MR 55 # 91 Zbl 325 # 02028 • ID 07969

LEEDS, S. see Vol. IV, V for further entries

LEEUWEN VAN, L.C.A. [1979] see LEAVITT, W.G.

LEGGETT, A. [1982] see KALANTARI, I.

LEGGETT, A. [1983] see KALANTARI, I.

LEGGETT, A. see Vol. IV for further entries

LEHMANN, D.J. & SHELAH, S. [1983] *Reasoning with time and chance* (**P** 3851) Automata, Lang & Progr (10);1983 Barcelona 445-457
⋄ B25 B45 ⋄ REV MR 84k:68004 Zbl 546 # 03010 • ID 41143

LEHMANN, D.J. see Vol. II, IV for further entries

LEHMANN, G. [1985] *Modell- und rekursionstheoretische Grundlagen psychologischer Theorienbildung* (**X** 0811) Springer: Heidelberg & New York xxii+297pp
⋄ B10 C98 D80 D99 ⋄ ID 42737

LEINEN, F. [1985] *Existentially closed groups in locally finite group classes* (**J** 0394) Commun Algeb 13*1991-2024
⋄ C25 C60 ⋄ ID 48310

LEISS, H. [1983] *Implizit definierte Mengensysteme* (**S** 0478) Bonn Math Schr ii+159pp
⋄ C40 C65 C90 ⋄ REV MR 84f:03029 Zbl 506 # 03006 • ID 34449

LENSKI, W. [1984] *Elimination of quantifiers for the theory of archimedean ordered divisible groups in a logic with Ramsey quantifiers* (**P** 2153) Logic Colloq;1983 Aachen 1*261-280
⋄ C10 C55 C60 C80 ⋄ REV Zbl 565 # 03012 • ID 41767

LENZING, H. [1980] see JENSEN, C.U.

LENZING, H. [1981] see HERRMANN, C.

LENZING, H. [1982] see JENSEN, C.U.

LEONARD, J. [1966] see LAEUCHLI, H.

LEONARDO DI, M.V. [1982] *On formal axiomatization of algebraic structures (Italian)* (**J** 0104) Atti Accad Sci Lett Arti Palermo, Ser 4/I 39*445-475
⋄ C60 ⋄ REV Zbl 523 # 03025 • ID 37027

LEONE, M. [1985] *On temporal invariance of menu systems* (**P** 2999) Proc Conf Databasis (Calzone);1985 Heidelberg 1-12
⋄ B05 B75 C90 ⋄ ID 49912

LERMAN, M. [1978] *Lattices of α-recursively enumerable sets* (**P** 1628) Generalized Recursion Th (2);1977 Oslo 223-238
⋄ B25 D60 ⋄ REV MR 80e:03054 Zbl 453 # 03047 • ID 30739

LERMAN, M. [1978] *On elementary theories of some lattices of α-recursively enumerable sets* (**J** 0007) Ann Math Logic 14*227-272
⋄ B25 D60 E45 ⋄ REV MR 80i:03056 Zbl 391 # 03022 • ID 29152

LERMAN, M. & SCHMERL, J.H. [1979] *Theories with recursive models* (**J** 0036) J Symb Logic 44*59-76
⋄ C35 C57 C65 ⋄ REV MR 81g:03036 Zbl 423 # 03038 • ID 53550

LERMAN, M. & SOARE, R.I. [1980] *A decidable fragment of the elementary theory of the lattice of recursively enumerable sets* (**J** 0064) Trans Amer Math Soc 257*1-37
⋄ B25 C10 D25 ⋄ REV MR 81c:03034 Zbl 439 # 03023 • ID 56014

LERMAN, M. & SCHMERL, J.H. & SOARE, R.I. (EDS.) [1981] *Logic year 1979-80, The University of Connecticut, USA* (**S** 3301) Lect Notes Math 859*viii+326pp
⋄ B97 C97 ⋄ REV MR 82e:03007 Zbl 456 # 00006 • ID 54311

LERMAN, M. [1981] *On recursive linear orderings* (**P** 2628) Log Year;1979/80 Storrs 132-142
⋄ C57 C65 D30 D45 D55 E07 ⋄ REV MR 82i:03059 Zbl 467 # 03045 • ID 55043

LERMAN, M. & ROSENSTEIN, J.G. [1982] *Recursive linear orderings* (**P** 3634) Patras Logic Symp;1980 Patras 123-126
⋄ C57 C65 D45 D55 E07 ⋄ REV MR 84j:03092 Zbl 511 # 03018 • ID 34683

LERMAN, M. & SHORE, R.A. & SOARE, R.I. [1984] *The elementary theory of the recursively enumerable degrees is not $\aleph_0$ categorical* (**J** 0345) Adv Math 53∗301-320
⋄ C35 D25 ⋄ REV MR 86d:03040 Zbl 559 # 03027
• ID 33603

LERMAN, M. see Vol. IV for further entries

LESAFFRE, B. [1973] *Structures algebriques dans les topos elementaires* (**J** 2313) C R Acad Sci, Paris, Ser A-B 277∗A663-A666
⋄ C90 G30 ⋄ REV MR 48 # 8597 Zbl 364 # 18006
• ID 82053

LEVIN, F. [1976] see BAUMSLAG, B.

LEVINSKI, J.-P. [1984] *Instances of the conjecture of Chang* (**J** 0029) Israel J Math 48∗225-243
⋄ C55 E05 E45 E55 ⋄ ID 39720

LEVITZ, H. [1975] *An ordered set of arithmetic functions representing the least ε-number* (**J** 0068) Z Math Logik Grundlagen Math 21∗115-120
⋄ C62 F15 F30 ⋄ REV MR 51 # 7845 Zbl 325 # 04002
• ID 18259

LEVITZ, H. [1985] *Decidability of some problems pertaining to base 2 exponential diophantine equations* (**J** 0068) Z Math Logik Grundlagen Math 31∗109-115
⋄ B25 ⋄ ID 41815

LEVITZ, H. see Vol. I, IV, V, VI for further entries

LEVY, A. [1958] *Comparison of subtheories* (**J** 0053) Proc Amer Math Soc 9∗942-945 • ERR/ADD ibid 10∗1000
⋄ C07 C40 E30 F30 ⋄ REV MR 21 # 1934 JSL 24.226
• ID 24860

LEVY, A. [1961] *Axiomatization of induced theories* (**J** 0053) Proc Amer Math Soc 12∗251-253
⋄ C07 ⋄ REV MR 23 # A43 JSL 34.302 • ID 08067

LEVY, A. [1965] *The Fraenkel-Mostowski method for independence proofs in set theory* (**P** 0614) Th Models;1963 Berkeley 221-228
⋄ C62 E25 E35 ⋄ REV MR 34 # 1166 JSL 40.431
• ID 22314

LEVY, A. see Vol. I, II, IV, V, VI for further entries

LEVY, M. [1983] see GUERRE, S.

LEVY, M. & RAYNAUD, Y. [1984] *Ultrapuissances des espaces $L^p(L^q)$ (English summary)* (**J** 3364) C R Acad Sci, Paris, Ser 1 299∗81-84
⋄ C20 C65 H10 H20 ⋄ REV MR 85i:46044 Zbl 569 # 46008 • ID 44350

LEVY, P. [1970] *Un paradoxe relatif a l'ensemble des nombres definissables* (**J** 2313) C R Acad Sci, Paris, Ser A-B 270∗A693-A695
⋄ C40 E47 ⋄ REV MR 41 # 3237 Zbl 188.312 • ID 08094

LEVY, P. see Vol. II, V, VI for further entries

LEWIS, C.I. [1932] see LANGFORD, C.H.

LEWIS, C.I. see Vol. I, II for further entries

LEWIS, D. [1971] *Completeness and decidability of three logics of counterfactual conditionals* (**J** 0105) Theoria (Lund) 37∗74-85
⋄ B25 B45 ⋄ REV MR 49 # 2278 Zbl 229 # 02023
• ID 08107

LEWIS, D. [1983] see DENEF, J.

LEWIS, D. see Vol. II, V for further entries

LEWIS, H.R. [1973] see AANDERAA, S.O.

LEWIS, H.R. [1977] *A measure of complexity for combinatorial decision problems of the tiling variety* (**P** 3238) Conf Theoret Comput Sci;1977 Waterloo ON 94-99
⋄ B25 D10 D15 ⋄ REV MR 58 # 10388 Zbl 414 # 03008
• ID 53054

LEWIS, H.R. [1978] *Complexity of solvable cases of the decision problem for the predicate calculus* (**P** 3578) IEEE Symp Found of Comput Sci (19);1978 Ann Arbor 35-47
⋄ B20 B25 D15 ⋄ REV MR 80e:03041 • ID 75527

LEWIS, H.R. [1979] *Unsolvable classes of quantificational formulas* (**X** 0832) Addison-Wesley: Reading xvii+198pp
⋄ B20 B25 D35 D98 ⋄ REV MR 81i:03069 Zbl 423 # 03003 JSL 47.221 • ID 53515

LEWIS, H.R. [1980] *Complexity results for classes of quantificational formulas* (**J** 0119) J Comp Syst Sci 21∗317-353
• TRANSL [1983] (**J** 3079) Kiber Sb Perevodov, NS 20∗64-106
⋄ B20 B25 D10 D15 ⋄ REV MR 82m:03052 Zbl 471 # 03034 • ID 36546

LEWIS, H.R. [1982] see AANDERAA, S.O.

LEWIS, H.R. [1984] see DENENBERG, L.

LEWIS, H.R. see Vol. I, IV for further entries

LIFSCHITZ, V. [1967] *The decision problem for some constructive theories of equality (Russian)* (**S** 0228) Zap Nauch Sem Leningrad Otd Mat Inst Steklov 4∗78-85
• TRANSL [1967] (**J** 0521) Semin Math, Inst Steklov 4∗29-31
⋄ B25 D35 F50 ⋄ REV MR 39 # 65 Zbl 165.19 • ID 33665

LIFSCHITZ, V. see Vol. I, IV, VI for further entries

LIGHTSTONE, A.H. & ROBINSON, A. [1957] *On the representation of Herbrand functions in algebraically closed fields* (**J** 0036) J Symb Logic 22∗187-204
• REPR [1979] (**C** 4594) Sel Pap Robinson 1∗396-413
⋄ C60 ⋄ REV MR 20 # 4555 Zbl 208.12 JSL 35.148
• ID 08147

LIGHTSTONE, A.H. & ROBINSON, A. [1957] *Syntactical transforms* (**J** 0064) Trans Amer Math Soc 86∗220-245
• REPR [1979] (**C** 4594) Sel Pap Robinson 1∗120-145
⋄ B10 C52 C60 ⋄ REV MR 19.935 Zbl 208.10 JSL 24.244 • ID 08149

LIGHTSTONE, A.H. [1978] *Mathematical logic. An introduction to model theory. Edited by H. B. Enderton* (**X** 1332) Plenum Publ: New York xiii+338pp
⋄ B98 C98 H05 ⋄ REV MR 80i:03002 Zbl 382 # 03002
• ID 51926

LIGHTSTONE, A.H. see Vol. I, II for further entries

LILLO DE, N.J. [1973] *A formal characterization of ordinal numbers* (J 0047) Notre Dame J Formal Log 14*397-400
• ERR/ADD ibid 15*648
⋄ C65 E10 ⋄ REV MR 48#8223 MR 50#1866 Zbl 258#02036 • ID 02906

LILLO DE, N.J. [1979] *Models of an extension of the theory ORD* (J 0047) Notre Dame J Formal Log 20*729-734
⋄ B28 C40 C65 E10 E30 ⋄ REV MR 81f:03043 Zbl 349#02009 • ID 56184

LILLO DE, N.J. see Vol. IV for further entries

LIN, C. [1981] *Recursively presented abelian groups: effective p-group theory. I* (J 0036) J Symb Logic 46*617-624
⋄ C57 C60 D45 ⋄ REV MR 82j:03054 Zbl 499#03032
• ID 75552

LIN, C. [1981] *The effective content of Ulm's theorem* (P 2902) Aspects Effective Algeb;1979 Clayton 147-160
⋄ C57 C60 D45 ⋄ REV MR 83f:03038 Zbl 499#03033
• ID 35279

LIN, PEIKEE [1985] *Unconditional bases and fixed points of nonexpansive mappings* (J 0048) Pac J Math 116*69-76
⋄ C20 C65 H05 ⋄ REV MR 86c:47075 Zbl 566#47038
• ID 48745

LINDENBAUM, A. & TARSKI, A. [1927] *Sur l'independance des notions primitives dans les systemes mathematiques* (J 0283) Ann Soc Pol Math 5*111-113
⋄ B30 C40 E75 ⋄ ID 30918

LINDENBAUM, A. & TARSKI, A. [1935] *Ueber die Beschraenktheit der Ausdrucksmittel deduktiver Theorien* (J 1124) Ergebn Math Kolloquium 7*15-22
• TRANSL [1956] (C 1159) Tarski: Logic, Semantics, Metamathematics 384-392
⋄ B30 C07 C35 C65 ⋄ REV JSL 1.115 FdM 62.39
• ID 30921

LINDENBAUM, A. see Vol. I, V, VI for further entries

LINDNER, C.C. [1971] *Finite partial cyclic triple systems can be finitely embedded* (J 0004) Algeb Universalis 1*93-96
⋄ C05 C13 D35 ⋄ REV MR 46#1617 Zbl 219#05011
• ID 08168

LINDNER, C.C. [1971] *Identities preserved by singular direct product* (J 0004) Algeb Universalis 1*86-89
⋄ C05 C13 C40 ⋄ REV MR 44#5401 Zbl 221#20099
• REM Part I. Part II 1972 • ID 08167

LINDNER, C.C. [1972] *Identities preserved by the singular direct product II* (J 0004) Algeb Universalis 2*113-117
⋄ C05 C13 C40 ⋄ REV Zbl 249#20038 • REM Part I 1971
• ID 08169

LINDNER, C.C. see Vol. IV for further entries

LINDSAY, P.A. [1985] *On recognizing cyclic modules effectively* (J 0394) Commun Algeb 13*1579-1595
⋄ C60 D45 ⋄ ID 46372

LINDSTROEM, P. [1964] *On model-completeness* (J 0105) Theoria (Lund) 30*183-196
⋄ C35 ⋄ REV MR 31#3317 JSL 35.587 • ID 08171

LINDSTROEM, P. [1966] *First order predicate logic with generalized quantifiers* (J 0105) Theoria (Lund) 32*186-195
⋄ C80 C95 ⋄ REV MR 39#5329 JSL 34.650 • ID 08174

LINDSTROEM, P. [1966] *On characterizability in L_{ω_1,ω_0}* (J 0105) Theoria (Lund) 32*165-171
⋄ C75 ⋄ REV MR 39#45 JSL 35.587 • ID 08172

LINDSTROEM, P. [1966] *On relations between structures* (J 0105) Theoria (Lund) 32*172-185
⋄ C07 C40 C52 ⋄ REV MR 39#5339 JSL 34.515
• ID 08173

LINDSTROEM, P. [1968] *Remarks on some theorems of Keisler* (J 0036) J Symb Logic 33*571-576
⋄ C07 C20 C30 ⋄ REV MR 42#4388 Zbl 181.302 JSL 36.339 • ID 08175

LINDSTROEM, P. [1969] *On extensions of elementary logic* (J 0105) Theoria (Lund) 35*1-11
⋄ C95 ⋄ REV MR 39#5330 Zbl 206.272 JSL 39.183
• ID 08176

LINDSTROEM, P. [1973] *A characterization of elementary logic* (C 1389) Modality, Morality, Probl of Sense & Nonsense (Hallden) 189-191
⋄ C95 ⋄ REV MR 56#8322 • ID 75561

LINDSTROEM, P. [1973] *A note on weak second order logic with variables for elementarily definable relations* (P 0710) Russell Mem Logic Conf;1971 Uldum 221-233
⋄ B15 C55 C62 ⋄ REV MR 50#4238 • ID 08177

LINDSTROEM, P. [1974] *On characterizing elementary logic* (C 1936) Log Th & Semant Anal (Kanger) 129-146
⋄ B10 C95 ⋄ REV Zbl 289#02001 • ID 29976

LINDSTROEM, P. [1978] *Omitting uncountable types and extensions of elementary logic* (J 0105) Theoria (Lund) 44*152-156
⋄ C75 C80 C95 ⋄ REV MR 82d:03062 Zbl 437#03016
• ID 55884

LINDSTROEM, P. see Vol. VI for further entries

LINTON, F.E.J. [1976] *Companions for a lonely result of Isbell* (J 0004) Algeb Universalis 6*91
⋄ C05 ⋄ REV MR 54#213 Zbl 335#08006 • ID 24591

LINTON, F.E.J. see Vol. V for further entries

LIOGON'KIJ, M.I. [1969] see GLEBSKIJ, YU.V.

LIOGON'KIJ, M.I. [1970] *On the question of quantitative characteristics of logical formulae (Russian) (English summary)* (J 0040) Kibernetika, Akad Nauk Ukr SSR 1970/3*16-22
• TRANSL [1970] (J 0021) Cybernetics 6*205-212
⋄ B20 B25 C13 ⋄ REV MR 47#1597 Zbl 229#02040
• ID 08182

LIOGON'KIJ, M.I. see Vol. I, IV for further entries

LIPNER, L.D. [1970] *Some aspects of generalized quantifiers* (0000) Diss., Habil. etc 97pp
⋄ C55 C80 ⋄ REM Thesis, University of California, Berkeley, CA • ID 21414

LIPPARINI, P. [1982] *Existentially complete closure algebras (Italian summary)* (J 3768) Boll Unione Mat Ital, VI Ser, D 1*13-19
⋄ C25 G25 ⋄ REV MR 84h:03087 Zbl 514#03022
• ID 34280

LIPPARINI, P. [1982] *Locally finite theories with model companion (Italian summary)* (J 0149) Atti Accad Naz Lincei Fis Mat Nat, Ser 8 72*6-11
⋄ C25 C35 C52 ⋄ REV MR 85g:03048 Zbl 524#03021
• ID 37598

LIPPARINI, P. [1982] *Some results about compact logics (Italian summary)* (**J** 0149) Atti Accad Naz Lincei Fis Mat Nat, Ser 8 72*308-311
◇ C95 ◇ REV MR 85b:03059 Zbl 548 # 03015 • ID 33568

LIPPARINI, P. [1983] *On Hilbert's Nullstellensatz (Italian)* (**P** 3829) Atti Incontri Log Mat (1);1982 Siena 91-92
◇ C05 C60 ◇ REV MR 84k:03006 Zbl 515 # 08001
• ID 48408

LIPPARINI, P. [1985] *Compactness properties of logics generated by monadic quantifiers and cardinalities of limit ultrapowers* (**P** 4310) Easter Conf on Model Th (3);1985 Gross Koeris 183-185
◇ C20 C80 C95 ◇ ID 49049

LIPPARINI, P. [1985] *Duality for compact logics and substitution in abstract model theory* (**J** 0068) Z Math Logik Grundlagen Math 31*517-532
◇ C95 ◇ REV Zbl 572 # 03017 • ID 47802

LIPPARINI, P. [1985] *Robinson equivalence relations through limit ultrapowers (Italian summary)* (**J** 2100) Boll Unione Mat Ital, VI Ser, B 4*569-583
◇ C20 C95 ◇ ID 48311

LIPSHITZ, L. & SARACINO, D.H. [1973] *The model companion of the theory of commutative rings without nilpotent elements* (**J** 0053) Proc Amer Math Soc 38*381-387
◇ B25 C25 C35 C60 ◇ REV MR 55 # 12510
Zbl 267 # 02040 JSL 48.496 • ID 08186

LIPSHITZ, L. [1975] *Commutative regular rings with integral closure* (**J** 0064) Trans Amer Math Soc 211*161-170
◇ C60 ◇ REV MR 53 # 122 Zbl 315 # 13006 • ID 16587

LIPSHITZ, L. [1976] see BECKER, J.A.

LIPSHITZ, L. [1976] *Integral closures of uncountable commutative regular rings* (**J** 0053) Proc Amer Math Soc 56*19-24
◇ C60 ◇ REV MR 58 # 5627 Zbl 339 # 13010 • ID 31819

LIPSHITZ, L. [1977] *The real closure of a commutative regular f-ring* (**J** 0027) Fund Math 94*173-176
◇ C60 ◇ REV MR 55 # 5620 Zbl 359 # 02057 • ID 26520

LIPSHITZ, L. [1978] see DENEF, J.

LIPSHITZ, L. & NADEL, M.E. [1978] *The additive structure of models of arithmetic* (**J** 0053) Proc Amer Math Soc 68*331-336
◇ C15 C50 C57 C62 ◇ REV MR 58 # 10424
Zbl 383 # 03047 • ID 31825

LIPSHITZ, L. [1978] *The diophantine problem for addition and divisibility* (**J** 0064) Trans Amer Math Soc 235*271-283
◇ B25 D35 ◇ REV MR 57 # 9666 Zbl 374 # 02025
• ID 31822

LIPSHITZ, L. [1979] *Diophantine correct models of arithmetic* (**J** 0053) Proc Amer Math Soc 73*107-108
◇ C15 C62 H15 ◇ REV MR 80a:03092 Zbl 402 # 03055
• ID 31826

LIPSHITZ, L. [1979] see BECKER, J.A.

LIPSHITZ, L. [1980] see BECKER, J.A.

LIPSHITZ, L. [1980] see DENEF, J.

LIPSHITZ, L. [1981] *Some remarks on the diophantine problem for addition and divisibility* (**J** 3133) Bull Soc Math Belg, Ser B 33*41-52
◇ B25 D15 ◇ REV MR 82k:03011 Zbl 497 # 03007
• ID 75573

LIPSHITZ, L. [1984] see DENEF, J.

LIPSHITZ, L. see Vol. IV for further entries

LIPTON, R.J. [1978] *Model theoretic aspects of computational complexity* (**P** 3578) IEEE Symp Found of Comput Sci (19);1978 Ann Arbor 193-200
◇ C62 D15 F30 F65 H15 ◇ REV MR 80e:03042
• ID 75578

LIPTON, R.J. [1979] see DEMILLO, R.A.

LIPTON, R.J. see Vol. IV, VI for further entries

LIS, Z. [1960] *Logical consequence, semantic and formal* (**J** 0063) Studia Logica 10*39-60
◇ B10 C07 ◇ REV MR 24 # A1813 Zbl 121.252 • ID 08189

LIS, Z. see Vol. I for further entries

LISOVIK, L.P. [1981] *Sets that are expressible in theories of commutative semigroups (Russian) (English Summary)* (**J** 0270) Dokl Akad Nauk Ukr SSR, Ser A 1981/8*80-83
◇ B25 C05 C40 ◇ REV MR 82j:03039 Zbl 484 # 03002
• ID 75583

LISOVIK, L.P. [1984] *Construction of decidable singular theories of two successor functions with an extra predicate (Russian)* (**J** 0003) Algebra i Logika 23*266-277
• TRANSL [1984] (**J** 0069) Algeb and Log 23*181-189
◇ B25 D05 ◇ ID 42703

LISOVIK, L.P. [1984] *Monadic second-order theories of two successor functions with an additional predicate (Russian)* (**J** 0270) Dokl Akad Nauk Ukr SSR, Ser A 8*80-82
◇ B15 B25 C10 C85 D05 ◇ REV MR 86c:03008
• ID 48409

LISOVIK, L.P. see Vol. IV for further entries

LITMAN, A. [1976] *On the monadic theory of ω_1 without AC* (**J** 0029) Israel J Math 23*251-266
◇ B25 C85 D05 E10 E25 ◇ REV MR 54 # 12526
Zbl 348 # 02041 • ID 18261

LITMAN, A. & SHELAH, S. [1977] *Models with few isomorphic expansions* (**J** 0029) Israel J Math 28*331-338
◇ C15 C35 C50 C55 ◇ REV MR 57 # 9522
Zbl 365 # 02036 • ID 27204

LITMAN, A. see Vol. V for further entries

LIVCHAK, A.B. [1973] *A resolving procedure for the elementary theory of a torsion-free abelian group with a distinguished sub-group (Russian)* (**J** 0340) Mat Zap (Univ Sverdlovsk) 8/3*73-81,141
◇ B25 C10 C60 ◇ REV MR 48 # 1906 Zbl 322 # 02046
Zbl 476 # 03024 • REM A comment was published ibid 12*161
• ID 08212

LIVCHAK, A.B. [1976] *Definable sets of integers (Russian)* (**J** 0340) Mat Zap (Univ Sverdlovsk) 10/1*54-59,151
◇ C60 C62 F30 ◇ REV MR 58 # 10418 Zbl 435 # 03028
• ID 55788

LIVCHAK, A.B. [1976] *Elimination of identity in the theory of abelian groups with predicates for subgroups (Russian)* (**C** 2555) Algeb Sistemy (Irkutsk) 79-93
◇ B25 C60 ◇ ID 90160

LIVCHAK, A.B. [1976] *Formula subsets of abelian groups* (**P** 2064) All-Union Conf Math Log (4);1976 Kishinev 77
◇ C60 ◇ ID 33960

LIVCHAK, A.B. [1977] *Expressible subsets of the ordered group of integers (Russian)* (J 0031) Izv Vyssh Ucheb Zaved, Mat (Kazan) 1977/8*105-107
• TRANSL [1977] (J 3449) Sov Math 21/8*81-83
◇ C60 C62 F30 ◇ REV MR 57#5742 Zbl 378#02025
• ID 33959

LIVCHAK, A.B. [1980] *A letter to the editors (Russian)* (J 0340) Mat Zap (Univ Sverdlovsk) 12/2*161
◇ B25 C10 C60 ◇ REV MR 83e:03041 Zbl 322#02046 Zbl 476#03024 • REM Concerns Livchak,A.B. ibid 8/3*73-81(1973) • ID 55536

LIVCHAK, A.B. see Vol. IV for further entries

LOATS, J.T. & RUBIN, M. [1978] *Boolean algebras without nontrivial onto endomorphisms exist in every uncountable cardinality* (J 0053) Proc Amer Math Soc 72*346-351
◇ C50 G05 ◇ REV MR 80f:03071 Zbl 401#06007
• ID 54630

LOATS, J.T. see Vol. V for further entries

LOEB, M.H. (ED.) [1968] *Proceedings of the summer school in logic, Leeds, 1967* (S 3301) Lect Notes Math 70*iv+331pp
◇ B97 C97 F97 ◇ ID 37293

LOEB, M.H. [1970] *A model theoretic characterization of effective operations* (J 0036) J Symb Logic 35*217-222 • ERR/ADD ibid 39*225
◇ C62 D20 ◇ REV MR 43#48 MR 50#74 Zbl 223#02044 Zbl 289#02029 • ID 22235

LOEB, M.H. see Vol. I, II, IV, VI for further entries

LOEB, P.A. [1985] *A nonstandard functional approach to Fubini's theorem* (J 0053) Proc Amer Math Soc 93*343-346
◇ C60 C65 H05 H20 ◇ REV Zbl 566#28004 • ID 44489

LOEB, P.A. see Vol. I, V for further entries

LOEWEN, K. [1969] *Some examples of ultraproducts (Polish and Russian summaries)* (J 0063) Studia Logica 24*47-53
◇ C20 ◇ REV MR 40#1272 Zbl 254#02040 • ID 08240

LOEWEN, K. see Vol. VI for further entries

LOEWENHEIM, L. [1908] *Ueber das Aufloesungsproblem im logischen Klassenkalkuel* (J 0366) Sitzber Berlin Math Ges 7*89-94
◇ B20 B25 G05 ◇ REV FdM 39.89 • ID 08241

LOEWENHEIM, L. [1915] *Ueber Moeglichkeiten im Relativkalkuel* (J 0043) Math Ann 76*447-470
• TRANSL [1967] (C 0675) From Frege to Goedel 232-251
◇ B25 C07 C10 C85 E07 G15 ◇ REV FdM 45.108
• ID 08246

LOEWENHEIM, L. see Vol. I, V, VI for further entries

LOLLI, G. [1975] *Sui modelli di certe teorie delle classi (English summary)* (J 0549) Riv Mat Univ Parma, Ser 4 1*159-176
◇ C62 E30 E70 ◇ REV MR 55#12520 Zbl 359#02064
• ID 31828

LOLLI, G. [1977] *Indiscernibili in teoria dei modelli e in teoria degli insiemi (English summary)* (J 3285) Boll Unione Mat Ital, V Ser, A A14*10-24
◇ C30 C98 E05 E45 E55 ◇ REV MR 58#21605 Zbl 353#02029 • ID 31830

LOLLI, G. [1978] *Alcune applicazioni della compattezza* (J 0319) Matematiche (Sem Mat Catania) 33*321-332
◇ C20 C60 H05 ◇ REV MR 84i:03115 Zbl 469#03050
• ID 55177

LOLLI, G. & MAREK, W. [1979] *On elementary theories of v. Neumann's levels* (J 3285) Boll Unione Mat Ital, V Ser, A 16*406-411
◇ C62 E30 ◇ REV MR 80j:03074 Zbl 417#03023
• ID 53259

LOLLI, G. [1983] *Complessita delle teorie* (P 3829) Atti Incontri Log Mat (1);1982 Siena 159-182
◇ B25 D15 ◇ REV MR 84k:03006 Zbl 525#03030
• ID 38263

LOLLI, G. & LONGO, G. & MARCJA, A. (EDS.) [1984] *Logic colloquium '82. Proceedings of the colloquium held in Florence, 23-28 August, 1982* (S 3303) Stud Logic Found Math 112*viii+358pp
◇ B97 C97 D97 F97 ◇ REV MR 85g:03006 Zbl 538#00003 • ID 41493

LOLLI, G. & LONGO, G. & MARCJA, A. (EDS.) [1984] *Proceedings of the Logic Colloqium '82* (S 3303) Stud Logic Found Math 112*viii+358pp
◇ C97 ◇ REV MR 85g:03006 Zbl 538#00003 • ID 49819

LOLLI, G. see Vol. I, II, IV, V, VI for further entries

LOMECKY, Z. [1982] *Algorithms for the computation of free lattices* (P 3826) Proc Europ Conf Comput Algeb;1982 Marseille 223-230
◇ B25 G10 ◇ REV MR 84a:06001 • ID 38823

LONGO, G. [1974] *I problemi di decisione e la loro complessita* (J 3436) Quad, Ist Appl Calcolo, Ser 3 8*87-108
◇ B25 B35 D15 F20 ◇ REV Zbl 427#03010 • ID 53697

LONGO, G. [1984] see LOLLI, G.

LONGO, G. see Vol. IV, V, VI for further entries

LOPARIC, A. [1977] *Une etude semantique de quelques calculs propositionnels (English summary)* (J 2313) C R Acad Sci, Paris, Ser A-B 284*A835-A838
◇ B05 B25 ◇ REV MR 55#5399 Zbl 378#02010
• ID 27299

LOPARIC, A. see Vol. II for further entries

LOPARIC, Z. [1980] *Decidability and cognitive significance in Carnap* (P 3006) Brazil Conf Math Log (3);1979 Recife 173-197
◇ A05 A10 B25 ◇ REV MR 82b:03014 Zbl 448#03008
• ID 56608

LOPEZ, G. [1971] *Un exemple d'une χ-classe M ne contenant pas de multirelation (M-χ)-universelle de cardinal χ (χ nombre beth)* (J 2313) C R Acad Sci, Paris, Ser A-B 272*A1241
◇ C50 C55 ◇ REV MR 44#49 Zbl 215.325 • ID 28023

LOPEZ, G. [1975] *Le probleme de l'isomorphie des restrictions strictes, pour les extensions a un element de relations enchainables non constantes (English summary)* (J 2313) C R Acad Sci, Paris, Ser A-B 281*A593-A595
◇ C13 C30 E07 ◇ REV MR 52#2889 Zbl 322#04003
• ID 63507

LOPEZ, G. [1978] *L'indeformabilite des relations et multirelations binaires* (J 0068) Z Math Logik Grundlagen Math 24*303-317
◇ C07 C13 E07 ◇ REV MR 81i:03049 Zbl 397#04002
• ID 75653

LOPEZ, G. [1983] *Reconstruction d'une S-expansion* (J 0068) Z Math Logik Grundlagen Math 29*11-24
◇ C07 C13 E07 ◇ REV MR 84f:04002 Zbl 536#04004
• ID 37133

LOPEZ, G. see Vol. V for further entries

LOPEZ-ESCOBAR, E.G.K. [1965] *An interpolation theorem for denumerably long sentences* (J 0027) Fund Math 57*253-272
 ⋄ C40 C75 ⋄ REV MR 32#5500 Zbl 137.7 JSL 34.301 JSL 35.301 • ID 08260

LOPEZ-ESCOBAR, E.G.K. [1965] *Universal formulas in the infinitary language $L_{\alpha,\beta}$* (J 0014) Bull Acad Pol Sci, Ser Math Astron Phys 13*383-388
 ⋄ C75 ⋄ REV MR 33#2534 Zbl 137.6 JSL 34.301 JSL 35.301 • ID 08259

LOPEZ-ESCOBAR, E.G.K. [1966] *On defining well-orderings* (J 0027) Fund Math 59*13-21,299-300
 ⋄ C55 C65 C75 E10 ⋄ REV MR 34#30 Zbl 166.4 Zbl 199.13 JSL 33.123 • ID 22322

LOPEZ-ESCOBAR, E.G.K. [1967] *A complete, infinitary axiomatization of weak second-order logic* (J 0027) Fund Math 61*93-103
 ⋄ B15 C75 C85 F35 ⋄ REV MR 36#2482 Zbl 174.13 JSL 35.467 • ID 08262

LOPEZ-ESCOBAR, E.G.K. [1967] *On a theorem of J.I.Malitz* (J 0014) Bull Acad Pol Sci, Ser Math Astron Phys 15*739-743
 ⋄ C30 C40 C85 ⋄ REV MR 37#1237 Zbl 188.21 JSL 35.586 • ID 08261

LOPEZ-ESCOBAR, E.G.K. [1967] *Remarks on an infinitary language with constructive formulas* (J 0036) J Symb Logic 32*305-318
 ⋄ C40 C70 C75 F50 ⋄ REV MR 37#6168 Zbl 221#02005 • ID 08263

LOPEZ-ESCOBAR, E.G.K. [1968] *A decision method for the intuitionistic theory of successor* (J 0028) Indag Math 30*466-467
 ⋄ B25 C10 F30 F50 ⋄ REV MR 40#4081 Zbl 169.310 • ID 08264

LOPEZ-ESCOBAR, E.G.K. [1968] *Well-orderings and finite quantifiers* (J 0090) J Math Soc Japan 20*477-489
 ⋄ C65 C75 ⋄ REV MR 37#6169 Zbl 169.7 • ID 08265

LOPEZ-ESCOBAR, E.G.K. [1969] *A non-interpolation theorem (Russian summary)* (J 0014) Bull Acad Pol Sci, Ser Math Astron Phys 17*109-112
 ⋄ C40 C75 ⋄ REV MR 39#3978 Zbl 181.301 JSL 40.457-458 • ID 08266

LOPEZ-ESCOBAR, E.G.K. [1972] *The infinitary language $L_{\theta\theta}$ is not local (Russian summary)* (J 0014) Bull Acad Pol Sci, Ser Math Astron Phys 20*527-528
 ⋄ C55 C75 E55 ⋄ REV MR 47#4760 Zbl 286#02022 • ID 19200

LOPEZ-ESCOBAR, E.G.K. & VELDMAN, W. [1975] *Intuitionistic completeness of a restricted second-order logic* (P 1440) ⊢ ISILC Proof Th Symp (Schuette);1974 Kiel 500*198-232
 ⋄ B15 C90 F50 ⋄ REV MR 54#67 Zbl 334#02015 • ID 24555

LOPEZ-ESCOBAR, E.G.K. [1980] *Semantical models for intuitionistic logics* (P 2958) Latin Amer Symp Math Log (4);1978 Santiago 191-207
 ⋄ C90 C95 F50 ⋄ REV MR 81j:03090 Zbl 469#03022 • ID 55150

LOPEZ-ESCOBAR, E.G.K. [1981] *On the interpolation theorem for the logic of constant domains* (J 0036) J Symb Logic 46*87-88
 ⋄ B55 C40 C90 F50 ⋄ REV MR 82i:03068 Zbl 469#03015 • REM See also 1983 • ID 55143

LOPEZ-ESCOBAR, E.G.K. [1983] *A second paper on the interpolation theorem for the logic of constant domains* (J 0036) J Symb Logic 48*595-599
 ⋄ B55 C40 C90 F05 F50 ⋄ REV MR 85b:03099 Zbl 547#03022 • REM See also 1981 • ID 39786

LOPEZ-ESCOBAR, E.G.K. see Vol. I, II, V, VI for further entries

LORENZEN, P. [1962] *Metamathematik* (X 0876) Bibl Inst: Mannheim 173pp
 • TRANSL [1967] (X 0834) Gauthier-Villars: Paris 162pp (French) [1971] (X 1781) Tecnos: Madrid (Spanish)
 ⋄ A05 B98 C60 D20 D35 D98 F98 ⋄ REV MR 28#3932 Zbl 105.246 JSL 31.106 • ID 08303

LORENZEN, P. see Vol. I, II, IV, V, VI for further entries

LOS, J. [1952] *Recherches algebriques sur les operations analytiques et quasi-analytiques* (J 0283) Ann Soc Pol Math 25*131-139
 ⋄ C05 ⋄ REV MR 15.204 Zbl 48.284 • ID 08319

LOS, J. [1954] *On the categoricity in power of elementary deductive systems and some related problems* (S 0019) Colloq Math (Warsaw) 3*58-62
 ⋄ C35 ⋄ REV MR 15.845 Zbl 55.5 JSL 23.360 • ID 08323

LOS, J. [1954] *Sur le theoreme de Goedel pour les theories indenombrables* (J 0014) Bull Acad Pol Sci, Ser Math Astron Phys 2*319-320
 ⋄ C07 E25 G05 ⋄ REV MR 16.103 Zbl 56.10 • ID 08322

LOS, J. [1954] see EHRENFEUCHT, A.

LOS, J. [1955] *On the extending of models I* (J 0027) Fund Math 42*38-54
 ⋄ C30 C52 ⋄ REV MR 17.224 Zbl 65.4 JSL 27.93 • REM Part II 1955 by Los,J. & Suszko,R. • ID 08326

LOS, J. & SUSZKO, R. [1955] *On the extending of models II* (J 0027) Fund Math 42*343-347
 ⋄ C05 C30 C52 ⋄ REV MR 17.815 Zbl 66.255 JSL 27.94 • REM Part I 1955. Part III 1956 by Slominski,J. • ID 08330

LOS, J. & SUSZKO, R. [1955] *On the infinite sums of models* (J 0014) Bull Acad Pol Sci, Ser Math Astron Phys 3*201-202
 ⋄ C30 C40 C52 ⋄ REV MR 17.224 Zbl 66.255 • ID 08324

LOS, J. [1955] *Quelques remarques, theoremes et problemes sur les classes definissables d'algebres* (P 1589) Math Interpr of Formal Systs;1954 Amsterdam 98-113
 ⋄ C05 C20 C30 C52 ⋄ REV MR 17.700 Zbl 68.244 JSL 25.168 • ID 27338

LOS, J. [1955] *The algebraic treatment of the methodology of elementary deductive system(Polish and Russian summaries)* (J 0063) Studia Logica 2*151-212,328
 ⋄ B22 C07 ⋄ REV MR 18.785 Zbl 67.251 JSL 21.193 • ID 08332

LOS, J. & MOSTOWSKI, ANDRZEJ & RASIOWA, H. [1956] *A proof of Herbrand's theorem* (J 3941) J Math Pures Appl, Ser 9 35*19-24 • ERR/ADD ibid 40*129-134
 ⋄ B10 C07 F05 F50 ⋄ REV MR 17.699 MR 28#22 Zbl 199.7 Zbl 73.7 JSL 36.168 • ID 08333

Los, J. & Suszko, R. [1957] *On the extending of models IV: Infinite sums of models* (J 0027) Fund Math 44*52-60
◊ C30 C52 ◊ REV MR 19.724 Zbl 78.6 JSL 27.94 • REM Part III 1956 by Slominski,J. Part V 1960 by Los,J. & Slominski,J. & Suszko,R. • ID 21238

Los, J. & Slominski, J. & Suszko, R. [1960] *On the extending of models V: Embedding theorems for relational models* (J 0027) Fund Math 48*113-121
◊ C07 C30 C52 ◊ REV MR 22 # 3676 Zbl 129.259 • REM Part IV 1957 • ID 08341

Los, J. [1962] *Common extension in equational classes* (P 0612) Int Congr Log, Meth & Phil of Sci (1,Proc);1960 Stanford 136-142
◊ C30 C52 ◊ REV MR 27 # 2400 • ID 24929

Los, J. [1964] *Normal subalgebras in general algebras* (S 0019) Colloq Math (Warsaw) 12*151-153
◊ C05 ◊ REV MR 30 # 3046 Zbl 128.23 • ID 31416

Los, J. [1965] *Free product in general algebras* (P 0614) Th Models;1963 Berkeley 229-237
◊ C05 C30 ◊ REV MR 33 # 3976 Zbl 171.258 • ID 08349

Los, J. [1966] *Direct sums in general algebras* (S 0019) Colloq Math (Warsaw) 14*33-38
◊ C05 C30 ◊ REV MR 35 # 1530 Zbl 197.291 • ID 08350

Los, J. see Vol. I, II, IV, V for further entries

Loullis, G. [1978] *Infinite forcing for boolean valued models* (J 0465) Bull Greek Math Soc (NS) 19*155-182
◊ C25 C35 C90 ◊ REV MR 80f:03042 Zbl 418 # 03027 • ID 53312

Loullis, G. [1979] *Sheaves and boolean valued model theory* (J 0036) J Symb Logic 44*153-183
◊ C35 C52 C60 C90 ◊ REV MR 80k:03037 Zbl 411 # 03028 • ID 52882

Louveau, A. [1985] see Delon, F.

Louveau, A. see Vol. IV, V for further entries

Lovasz, L. [1967] *Operations with structures* (J 0001) Acta Math Acad Sci Hung 18*321-328
◊ C07 C13 C30 ◊ REV MR 35 # 5379 Zbl 174.14 • ID 08353

Lovasz, L. [1977] see Gacs, P.

Lovasz, L. [1980] *Efficient algorithms: an approach by formal logic* (C 3494) Stud on Math Progr. Math Meth Oper Res, Vol 1 1*119-126
◊ B35 C13 D15 ◊ REV MR 82e:68042 Zbl 427 # 68050 • ID 53755

Lovasz, L. see Vol. V for further entries

Loveland, D.W. & Reddy, C.R. [1978] *Presburger arithmetic with bounded quantifier alternation* (P 1740) ACM Symp Th of Comput (10);1978 San Diego 320-325
◊ B25 C10 D15 F20 F30 ◊ REV MR 80d:68057 • ID 31164

Loveland, D.W. see Vol. I, IV for further entries

Loveys, J. [1984] *The uniqueness of envelopes in $\aleph_0$-categorical, $\aleph_0$-stable structures* (J 0036) J Symb Logic 49*1171-1184
◊ C35 C45 ◊ REV MR 86g:03055 • ID 39871

Lu, Jingbo & Wang, Shiqiang [1981] *On the fundamental theorem of ultraproducts for lattice-valued models (Chinese)* (J 2771) Kexue Tongbao 26*71-74
• TRANSL [1982] (J 3769) Sci Bull, Foreign Lang Ed 27*241-245
◊ C20 C90 ◊ REV MR 85i:03123 Zbl 516 # 03016 • ID 37252

Lu, Jingbo [1982] *A note on infinite-lattice-valued model theory (Chinese) (English summary)* (J 2521) Beijing Shifan Daxue Xuebao, Ziran Kexue 1982/2*1-8
◊ C90 ◊ REV MR 84k:03105 • ID 36122

Lu, Jingbo & Shen, Chengmin [1982] *Saturated models in lattice-valued model theory (Chinese) (English summary)* (J 3770) Dongbei Shifan Daxue Xuebao, Ziran Kexue 1982/2*33-38
◊ C50 C90 ◊ REV MR 84d:03047 Zbl 566 # 03025 • ID 34083

Lubotzky, A. [1982] see Haran, D.

Lucas, F. [1978] *Equivalence elementaire et produits (English summary)* (J 2313) C R Acad Sci, Paris, Ser A-B 287*A41-A42
◊ C30 ◊ REV MR 58 # 10419 Zbl 397 # 03017 • ID 52683

Lucas, T. [1972] *Equations in the theory of monadic algebras* (J 0053) Proc Amer Math Soc 31*239-244
◊ C05 G15 ◊ REV MR 45 # 1740 Zbl 237 # 02024 • ID 08364

Lucas, T. [1974] *Universal classes of monadic algebras* (J 1650) Rapp, Sem Math Pure, Univ Cathol Louvain 47*ii+23pp
◊ B25 C05 G15 ◊ REV Zbl 325 # 02042 • ID 30448

Lucas, T. [1976] *Une utilisation du langage des categories pour la presentation de theoremes de theorie des modeles. I* (J 1650) Rapp, Sem Math Pure, Univ Cathol Louvain 62*18pp
◊ C40 G30 ◊ REV Zbl 353 # 02031 • ID 50114

Lucas, T. [1976] *Universal classes of monadic algebras* (J 0068) Z Math Logik Grundlagen Math 22*35-44
◊ B25 C05 G15 ◊ REV MR 53 # 7764 Zbl 341 # 02049 • ID 18490

Lucas, T. [1980] see Lavendhomme, R.

Lucas, T. [1980] *Reduction d'ensembles de formules* (J 1650) Rapp, Sem Math Pure, Univ Cathol Louvain 94*13pp
◊ C10 ◊ REV Zbl 455 # 03011 • ID 54272

Lucas, T. [1981] see Lavendhomme, R.

Lucas, T. [1985] see Lavendhomme, R.

Lucas, T. see Vol. I, II, V, VI for further entries

Luce, R. Duncan [1973] *Three axiom systems for additive semiordered structures* (J 0374) SIAM J Appl Math 25*41-53
◊ C65 ◊ REV MR 47 # 8365 Zbl 256 # 06009 • ID 08366

Luckhardt, H. [1967] *Aussagenlogisch fundierte Theorien* (J 0009) Arch Math Logik Grundlagenforsch 10*37-58
◊ B25 F05 ◊ REV MR 35 # 4083 Zbl 165.305 • ID 08378

Luckhardt, H. [1969] *Kodifikation und Aussagenlogik* (J 0009) Arch Math Logik Grundlagenforsch 12*18-38
◊ B05 B25 F50 ◊ REV MR 41 # 3239 Zbl 182.6 • ID 08380

LUCKHARDT, H. [1970] *Ein Henkin-Vollstaendigkeitsbeweis fuer die intuitionistische Praedikatenlogik bezueglich der Kripke-Semantik* (J 0009) Arch Math Logik Grundlagenforsch 13*55-59
⋄ C90 F50 ⋄ REV MR 44 #32 Zbl 212.17 • ID 08381

LUCKHARDT, H. [1970] *Kripke-Semantik der derivativen Praedikatenlogik* (J 0009) Arch Math Logik Grundlagenforsch 13*134-135
⋄ C90 F50 ⋄ REV MR 45 #8499 Zbl 218 #02019
• ID 08382

LUCKHARDT, H. see Vol. I, II, IV, VI for further entries

LUE, QICI [1983] *Counterexample to Bowen's theorem and failures of interpolation theorem in modal logic (Chinese)* (J 2771) Kexue Tongbao 28*897-899
• TRANSL [1984] (J 3769) Sci Bull, Foreign Lang Ed 29*433-436
⋄ B45 C40 ⋄ REV MR 86b:03022 Zbl 543 #03012
• ID 40907

LUE, QICI see Vol. I, II for further entries

LUE, YIZHONG [1980] *Research on a decision problem (Chinese) (English summary)* (J 0298) Nanyang Univ J 1980/2*1-5
⋄ B25 ⋄ REV Zbl 444 #03005 • ID 56453

LUE, YIZHONG see Vol. IV for further entries

LUO, LIBO [1980] *The interpolation theorem for the $P(\kappa)$ of a strongly inaccessible cardinal (Chinese)* (J 0418) Shuxue Xuebao 23*177-182
⋄ C40 C75 E50 E55 ⋄ REV MR 82i:03014 • REM For comments see Gao,Hongshan 1983 • ID 55737

LUO, LIBO [1980] *The union and product of models, and homogeneous models (Chinese) (English summary)* (J 2521) Beijing Shifan Daxue Xuebao, Ziran Kexue 1980/3-4*31-39
⋄ C15 C30 C50 ⋄ REV MR 84k:03104 • ID 36121

LUO, LIBO [1983] *On the number of countable homogeneous models* (J 0036) J Symb Logic 48*539-541
⋄ C15 C50 ⋄ REV MR 84i:03070 Zbl 527 #03009
• ID 34551

LUO, LIBO [1983] *The τ-theory for free groups is undecidable* (J 0036) J Symb Logic 48*700-703
⋄ C35 C60 D35 ⋄ REV MR 85j:03073 Zbl 527 #03021
• ID 44346

LUSTIG, M. [1971] *On isotone and homomorphic maps of ordered unary algebras* (J 0001) Acta Math Acad Sci Hung 22*431-439
⋄ C05 C07 ⋄ REV MR 45 #1813 Zbl 246 #06007
• ID 08424

LUXEMBURG, W.A.J. [1962] *A remark on a paper by N.G.de Bruijn and P.Erdoes* (J 0028) Indag Math 24*343-345
⋄ C07 C20 E05 E25 ⋄ ID 27624

LUXEMBURG, W.A.J. [1962] *Nonstandard analysis. Lectures on A. Robinson's theory of infinitesimals and infinitely large numbers* (X 1246) Caltech: Pasadena
⋄ C20 H05 H98 ⋄ ID 21419

LUXEMBURG, W.A.J. [1962] *Two applications of the method of construction by ultrapowers to analysis* (J 0015) Bull Amer Math Soc 68*416-419
⋄ C20 E05 E75 H05 ⋄ REV MR 25 #3837 Zbl 109.8
• ID 08426

LUXEMBURG, W.A.J. [1969] *A general theory of monads* (P 0649) Appl Model Th to Algeb, Anal & Probab;1967 Pasadena 18-86
⋄ B15 C20 H05 ⋄ REV MR 39 #6244 Zbl 207.524 JSL 36.541 • ID 21147

LUXEMBURG, W.A.J. (ED.) [1969] *Applications of model theory to algebra, analysis and probability* (X 0818) Holt Rinehart & Winston: New York vii+307pp
⋄ C97 H97 ⋄ REV Zbl 174.285 • ID 25073

LUXEMBURG, W.A.J. [1969] *Reduced powers of the real number system and equivalents of the Hahn-Banach extension theorem* (P 0649) Appl Model Th to Algeb, Anal & Probab;1967 Pasadena 123-137
⋄ C20 E25 E75 H05 ⋄ REV MR 38 #5616 Zbl 181.401
• ID 24816

LUXEMBURG, W.A.J. & TAYLOR, R.F. [1970] *Almost commuting matrices are near commuting matrices* (J 0028) Indag Math 32*96-98
⋄ C60 H05 ⋄ REV MR 41 #3502 Zbl 195.50 • ID 08429

LUXEMBURG, W.A.J. [1976] *On a class of valuation fields introduced by A. Robinson* (J 0029) Israel J Math 25*189-201
⋄ C60 H20 ⋄ REV MR 57 #98 Zbl 341 #12106 • ID 26081

LUXEMBURG, W.A.J. see Vol. I, V for further entries

LYNCH, J.F. [1980] *Almost sure theories* (J 0007) Ann Math Logic 18*91-135
⋄ C13 C15 C30 ⋄ REV MR 81i:03059 Zbl 433 #03020 JSL 50.1073 • ID 75731

LYNCH, J.F. [1982] *Complexity classes and theories of finite models* (J 0041) Math Syst Theory 15*127-144
⋄ C13 D15 ⋄ REV MR 84e:03042 Zbl 484 #03020
• ID 34379

LYNCH, J.F. [1982] *On sets of relations definable by addition* (J 0036) J Symb Logic 47*659-668
⋄ C40 C62 ⋄ REV MR 84i:03073 Zbl 504 #03013
• ID 34554

LYNCH, J.F. [1985] *Probability of first-order sentences about unary functions* (J 0064) Trans Amer Math Soc 287*543-568
⋄ C13 ⋄ REV MR 86g:03012 Zbl 526 #03012 • ID 42693

LYNDON, R.C. [1954] *Identities in finite algebras* (J 0053) Proc Amer Math Soc 5*8-9
⋄ C05 C13 ⋄ REV MR 15.676 Zbl 55.27 JSL 22.379
• ID 08437

LYNDON, R.C. [1957] *Sentences preserved under homomorphisms; sentences preserved under subdirect products* (P 1675) Summer Inst Symb Log;1957 Ithaca 122-124
⋄ C07 C30 C40 C52 ⋄ REV JSL 32.533 • ID 08443

LYNDON, R.C. [1959] *An interpolation theorem in the predicate calculus* (J 0048) Pac J Math 9*129-142
⋄ C40 ⋄ REV MR 21 #5555 Zbl 93.10 JSL 25.273
• ID 08438

LYNDON, R.C. [1959] *Existential Horn sentences* (J 0053) Proc Amer Math Soc 10*994-998
⋄ B20 C30 C40 C52 ⋄ REV MR 22 #6712 Zbl 114.249 JSL 30.253 • ID 08439

LYNDON, R.C. [1959] *Properties preserved under homomorphism* (J 0048) Pac J Math 9*143-154
⋄ C07 C40 C52 ⋄ REV MR 21 #7157 Zbl 93.11 JSL 32.533 • ID 08441

LYNDON, R.C. [1959] *Properties preserved under algebraic constructions* (J 0015) Bull Amer Math Soc 65*287-299
  ◊ C07 C30 C40 C52 ◊ REV MR 22 # 2549 Zbl 87.9 JSL 32.533 • ID 08442

LYNDON, R.C. [1959] *Properties preserved in subdirect products* (J 0048) Pac J Math 9*155-164
  ◊ C30 C40 C52 ◊ REV MR 21 # 6331 Zbl 93.11 • ID 37294

LYNDON, R.C. [1962] *Metamathematics and algebras: an example* (P 0612) Int Congr Log, Meth & Phil of Sci (1,Proc);1960 Stanford 143-150
  ◊ C40 C60 ◊ REV MR 28 # 2052 Zbl 142.247 JSL 34.653 • ID 08445

LYNDON, R.C. [1966] *Dependence in groups* (S 0019) Colloq Math (Warsaw) 14*275-283
  ◊ C05 C60 ◊ REV MR 32 # 5717 Zbl 141.21 JSL 34.151 • ID 08446

LYNDON, R.C. [1966] *Notes on logic* (X 0864) Van Nostrand: New York vi+97pp
  • TRANSL [1968] (X 0885) Mir: Moskva 128pp [1968] (X 1034) PWN: Warsaw 111pp
  ◊ C98 G98 ◊ REV MR 33 # 2521 MR 52 # 13301 Zbl 168.3 JSL 38.644 • ID 21767

LYNDON, R.C. see Vol. I, IV, V for further entries

LYONS, R. [1982] *Measure-theoretic quantifiers and Haar measure* (J 0053) Proc Amer Math Soc 86*67-70
  • ERR/ADD ibid 91/2*329-330
  ◊ C80 ◊ REV MR 83i:43001 MR 86d:43001 Zbl 468 # 28013 • ID 39347

LYUBETSKIJ, V.A. [1971] see DRAGALIN, A.G.

LYUBETSKIJ, V.A. [1983] see GORDON, E.I.

LYUBETSKIJ, V.A. [1985] *Some algebraic questions of nonstandard analysis (Russian)* (J 0023) Dokl Akad Nauk SSSR 280*38-41
  • TRANSL [1985] (J 0062) Sov Math, Dokl 31*30-34
  ◊ C60 C90 H05 ◊ REV MR 86e:03063 • ID 45424

LYUBETSKIJ, V.A. see Vol. I, IV, V for further entries

MACDONALD, J.L. [1978] *Categorical shape theory and the back and forth property* (J 0326) J Pure Appl Algebra 12*79-92
  ◊ C60 C90 G05 G30 ◊ REV MR 80j:55013 Zbl 394 # 18002 • ID 82132

MACDOWELL, R. & SPECKER, E. [1961] *Modelle der Arithmetik* (P 0633) Infinitist Meth;1959 Warsaw 257-263
  ◊ B28 C20 C55 C62 ◊ REV MR 27 # 2425 Zbl 126.11 JSL 38.651 • ID 19285

MACHOVER, M. [1960] *A note on sentences preserved under direct products and powers* (J 0014) Bull Acad Pol Sci, Ser Math Astron Phys 8*519-523
  ◊ C30 C40 ◊ REV MR 28 # 4997 Zbl 95.8 JSL 32.533 • ID 08456

MACHOVER, M. [1961] *The theory of transfinite recursion* (J 0015) Bull Amer Math Soc 67*575-578
  ◊ C75 D60 E45 ◊ REV MR 26 # 17 Zbl 103.246 JSL 35.335 • ID 08457

MACHOVER, M. see Vol. I, II, IV, V for further entries

MACINTYRE, A. [1971] *On ω_1-categorical theories of abelian groups* (J 0027) Fund Math 70*253-270
  ◊ C35 C45 C60 ◊ REV MR 44 # 6471 Zbl 234 # 02035 JSL 49.317 • ID 27801

MACINTYRE, A. [1971] *On ω_1-categorical theories of fields* (J 0027) Fund Math 71*1-25
  ◊ C35 C60 ◊ REV MR 45 # 48 Zbl 228 # 02033 JSL 49.317 • ID 08465

MACINTYRE, A. [1971] *On the elementary theory of Banach algebras* (J 0007) Ann Math Logic 3*239-269
  ◊ C60 C65 D35 ◊ REV MR 48 # 87 Zbl 286 # 02049 • ID 08464

MACINTYRE, A. [1972] *Direct powers with distinguished diagonal* (P 2080) Conf Math Log;1970 London 178-203
  ◊ C30 ◊ REV MR 49 # 2342 Zbl 256 # 02026 • ID 08466

MACINTYRE, A. [1972] *Omitting quantifier-free types in generic structures* (J 0036) J Symb Logic 37*512-520
  ◊ C25 C57 C60 C75 D30 D40 ◊ REV MR 49 # 37 Zbl 273 # 02038 • ID 08467

MACINTYRE, A. [1972] *On algebraically closed groups* (J 0120) Ann of Math, Ser 2 96*53-97
  ◊ C25 C57 C60 D40 ◊ REV MR 47 # 6477 Zbl 254 # 20021 • ID 08468

MACINTYRE, A. [1973] *Martin's axiom applied to existentially closed groups* (J 0132) Math Scand 32*46-56
  ◊ C25 C55 C60 C75 C80 E50 ◊ REV MR 49 # 4770 Zbl 268 # 02038 • ID 08469

MACINTYRE, A. [1973] see GOLDREI, D.C.

MACINTYRE, A. [1974] *A note on axioms for infinite-generic structures* (J 3172) J London Math Soc, Ser 2 9*581-584
  ◊ C25 C60 C62 C75 E07 ◊ REV MR 51 # 121 Zbl 308 # 02055 • ID 15208

MACINTYRE, A. [1974] *Model-completeness for sheaves of structures* (J 0027) Fund Math 81*73-89
  ◊ C35 C90 ◊ REV MR 49 # 2361 Zbl 317 # 02065 JSL 48.496 • ID 08473

MACINTYRE, A. & SIMMONS, H. [1975] *Algebraic properties of number theories* (J 0029) Israel J Math 22*7-27
  ◊ C25 C40 C52 C62 D25 H15 ◊ REV MR 53 # 2671 Zbl 356 # 02043 • ID 08474

MACINTYRE, A. [1975] *Dense embeddings. I. A theorem of Robinson in a general setting* (C 0782) Model Th & Algeb (A. Robinson) 200-219
  ◊ B25 C10 C35 C60 D40 ◊ REV MR 53 # 10574 Zbl 327 # 02049 • ID 23079

MACINTYRE, A. & ROSENSTEIN, J.G. [1976] *$\aleph_0$-categoricity for rings without nilpotent elements and for boolean structures* (J 0032) J Algeb 43*129-154
  ◊ C35 C60 ◊ REV MR 55 # 2558 Zbl 346 # 02031 • ID 30793

MACINTYRE, A. [1976] *Existentially closed structures and Jensen's principle* ◊ (J 0029) Israel J Math 25*202-210
  ◊ C25 C60 E45 E65 ◊ REV MR 58 # 211 Zbl 374 # 02034 • ID 26082

MACINTYRE, A. [1976] *On definable subsets of p-adic fields* (J 0036) J Symb Logic 41*605-610
  ◊ C10 C60 ◊ REV MR 58 # 5182 Zbl 362 # 02046 • ID 14583

MACINTYRE, A. & SHELAH, S. [1976] *Uncountable universal locally finite groups* (J 0032) J Algeb 43*168-175
◇ C25 C50 C52 C55 C60 ◇ REV MR 55 # 12511 Zbl 363 # 20032 • ID 31465

MACINTYRE, A. [1977] *Model completeness* (C 1523) Handb of Math Logic 139-180
◇ C25 C35 C60 C98 ◇ REV MR 58 # 10395 JSL 49.968
• ID 24199

MACINTYRE, A. & PACHOLSKI, L. & PARIS, J.B. (EDS.) [1978] *Logic colloquium '77. Proceedings of the colloquium held in Wroclaw, august 1977* (S 3303) Stud Logic Found Math 96*x+311pp
◇ B97 C97 ◇ REV MR 80a:03003 Zbl 426 # 00004
• ID 53596

MACINTYRE, A. [1979] see BAUR, W.

MACINTYRE, A. [1980] see HICKIN, K.K.

MACINTYRE, A. [1980] *Nonstandard number theory* (P 1959) Int Congr Math (II,13);1978 Helsinki 1*253-262
◇ C60 H15 ◇ REV MR 82h:03071 Zbl 453 # 03071
• ID 54201

MACINTYRE, A. [1980] *Ramsey quantifiers in arithmetic* (P 2625) Model Th of Algeb & Arithm;1979 Karpacz 186-210
◇ C62 C80 F30 F35 ◇ REV MR 83j:03099
Zbl 464 # 03031 JSL 50.1078 • ID 54621

MACINTYRE, A. [1980] see BOFFA, M.

MACINTYRE, A. [1981] *A theorem of Rabin in a general setting* (J 3133) Bull Soc Math Belg, Ser B 33*53-63
◇ C25 C62 ◇ REV MR 82k:03106 Zbl 502.03034
• ID 75760

MACINTYRE, A. [1981] see CHERLIN, G.L.

MACINTYRE, A. [1981] *Model theory* (C 2617) Modern Log Survey 45-65
◇ C98 ◇ REV MR 82f:03002 Zbl 464 # 03001 • ID 42764

MACINTYRE, A. [1981] *The complexity of types in field theory* (P 2628) Log Year;1979/80 Storrs 143-156
◇ C45 C50 C57 C60 ◇ REV MR 83g:03044
Zbl 499 # 03015 • ID 90377

MACINTYRE, A. [1981] *The laws of exponentiation* (P 3404) Model Th & Arithm;1979/80 Paris 185-197
◇ B25 C65 ◇ REV MR 83f:03037 Zbl 503 # 08008
• ID 35278

MACINTYRE, A. [1982] see DRIES VAN DEN, L.

MACINTYRE, A. [1982] *Residue fields of models of P* (P 3622) Int Congr Log, Meth & Phil of Sci (6,Proc);1979 Hannover 193-206
◇ C57 C60 C62 H15 ◇ REV MR 86b:03042
Zbl 514 # 03021 • ID 37408

MACINTYRE, A. [1983] *Decision problems for exponential rings: the p-adic case* (P 3864) FCT'83 Found of Comput Th;1983 Borgholm 285-289
◇ B25 C60 ◇ REV MR 85f:12014 Zbl 533 # 13013
• ID 36572

MACINTYRE, A. [1983] see DRIES VAN DEN, L.

MACINTYRE, A. & MARKER, D. [1984] *Degrees of recursively saturated models* (J 0064) Trans Amer Math Soc 282*539-554
◇ C50 C57 C62 D30 D45 ◇ REV MR 85e:03106
Zbl 557 # 03046 • ID 40310

MACINTYRE, A. [1985] *Effective determination of the zeros of p-adic exponential functions* (P 3083) FCT'83 Found of Comput Th (Sel Pap);1983 Borgholm 103-109
◇ B25 ◇ REV Zbl 554 # 12009 • ID 44951

MACINTYRE, A. see Vol. IV, VI for further entries

MACNAB, D.S. [1977] *Some applications of double-negation sheafification* (J 3420) Proc Edinburgh Math Soc, Ser 2 279-285
◇ C30 C90 F50 G30 ◇ REV MR 57 # 412
Zbl 371 # 18010 • ID 51394

MACNAB, D.S. see Vol. II for further entries

MACPHERSON, H.D. [1983] *The action of an infinite permutation group on unordered subsets of a set* (J 3240) Proc London Math Soc, Ser 3 46*471-486
◇ C50 C60 E05 ◇ REV MR 84m:20006 Zbl 509 # 20003
• ID 39487

MACPHERSON, H.D. [1985] *Growth rates in infinite graphs and permutation groups* (J 3240) Proc London Math Soc, Ser 3 51*285-294
◇ C60 C65 ◇ ID 48410

MACPHERSON, H.D. [1985] *Orbits of infinite permutation groups* (J 3240) Proc London Math Soc, Ser 3 51*246-284
◇ C60 ◇ ID 48411

MACZYNSKI, M.J. [1966] *A property of retracts of α^+-homogeneous α^+-universal boolean algebras* (J 0014) Bull Acad Pol Sci, Ser Math Astron Phys 14*607-608
◇ C05 C50 G05 ◇ REV MR 34 # 7423 Zbl 151.11
• ID 08488

MACZYNSKI, M.J. [1966] *Generalized free m-products of m-distributive boolean algebras with an m-amalgamated subalgebra* (J 0014) Bull Acad Pol Sci, Ser Math Astron Phys 14*539-542
◇ C30 G05 ◇ REV MR 34 # 4177 Zbl 143.26 JSL 35.346
• ID 08489

MACZYNSKI, M.J. & TRACZYK, T. [1967] *The m-amalgamation property for m-distributive boolean algebras* (J 0014) Bull Acad Pol Sci, Ser Math Astron Phys 15*57-60
◇ C52 G05 ◇ REV MR 39 # 110 Zbl 168.20 JSL 35.346
• ID 08490

MACZYNSKI, M.J. [1969] *The amalgamation of α-distributive α-complete boolean algebras (Russian summary)* (J 0014) Bull Acad Pol Sci, Ser Math Astron Phys 17*415-418
◇ C52 G05 ◇ REV MR 40 # 7166 Zbl 186.307 • ID 08492

MACZYNSKI, M.J. [1972] *Strong m-extension property and m-amalgamation in boolean algebras (Russian summary)* (J 0014) Bull Acad Pol Sci, Ser Math Astron Phys 20*259-263
◇ C52 G05 ◇ REV MR 46 # 3396 Zbl 241 # 06012
• ID 08493

MACZYNSKI, M.J. see Vol. II for further entries

MADDUX, R. [1980] *The equational theory of CA_3 is undecidable* (J 0036) J Symb Logic 45*311-316
◇ C05 D35 G15 ◇ REV MR 81e:03060 Zbl 435 # 03010
• ID 55771

MADDUX, R. see Vol. V, VI for further entries

MADISON, E.W. [1968] *Computable algebraic structures and nonstandard arithmetic* (**J** 0064) Trans Amer Math Soc 130*38-54
⋄ C57 C60 C62 D45 ⋄ REV MR 36 # 2498 Zbl 176.275
• ID 08500

MADISON, E.W. [1968] *Structures elementarily closed relative to a model for arithmetic* (**J** 0036) J Symb Logic 33*101-104
⋄ C57 C60 C62 ⋄ REV MR 38 # 5599 Zbl 191.296
• ID 08499

MADISON, E.W. [1970] *A note on computable real fields* (**J** 0036) J Symb Logic 35*239-241
⋄ C57 C60 D45 F60 ⋄ REV MR 42 # 7500 Zbl 291 # 02031 • ID 24826

MADISON, E.W. [1970] see LACHLAN, A.H.

MADISON, E.W. [1971] *Some remarks on computable (non-archimedean) ordered fields* (**J** 3172) J London Math Soc, Ser 2 4*304-308
⋄ C57 C60 D45 ⋄ REV MR 45 # 6624 Zbl 258 # 02044
• ID 63572

MADISON, E.W. [1972] *Real fields with characterization of the natural numbers* (**J** 0047) Notre Dame J Formal Log 13*211-218
⋄ C57 H15 ⋄ REV MR 45 # 4963 Zbl 197.278 • ID 08502

MADISON, E.W. [1973] see ALTON, D.A.

MADISON, E.W. & NELSON, GEORGE C. [1975] *Some examples of constructive and non-constructive extension of the countable atomless boolean algebra* (**J** 3172) J London Math Soc, Ser 2 11*325-336
⋄ C57 F60 G05 ⋄ REV MR 52 # 89 Zbl 375 # 06009
• ID 18269

MADISON, E.W. [1983] *The existence of countable totally nonconstructive extensions of the countable atomless boolean algebras* (**J** 0036) J Symb Logic 48*167-170
⋄ C57 D45 G05 ⋄ REV MR 84g:03066 Zbl 523 # 06020
• ID 34173

MADISON, E.W. [1985] *On boolean algebras and their recursive completions* (**J** 0068) Z Math Logik Grundlagen Math 31*481-486
⋄ C57 D45 G05 ⋄ ID 47796

MADISON, E.W. see Vol. IV for further entries

MAEDA, F. [1953] *A lattice formulation for algebraic and transcendental extensions in abstract algebras* (**J** 0102) Hiroshima Univ J Sci, Ser A Math 16*383-397
⋄ C60 G10 ⋄ REV MR 15.675 Zbl 51.22 • ID 08505

MAEHARA, S. [1957] *Remark on Skolem's theorem concerning the impossibility of characterization of the natural number sequence* (**J** 0081) Proc Japan Acad 33*588-590
⋄ B28 C62 F30 ⋄ REV MR 20 # 6364 Zbl 79.8
JSL 31.659 • ID 08516

MAEHARA, S. [1958] *Another proof of Takeuti's theorems on Skolem's paradox* (**J** 0434) J Fac Sci Univ Tokyo, Sect 1 7*541-556
⋄ C07 F25 F30 ⋄ REV MR 20 # 2283 Zbl 102.7
JSL 31.659 • ID 08519

MAEHARA, S. [1960] *On the interpolation theorem of Craig (Japanese)* (**J** 0091) Sugaku 12*235-237
⋄ B10 C40 F05 ⋄ REV MR 29 # 3356 Zbl 123.246
• ID 08520

MAEHARA, S. & TAKEUTI, G. [1961] *A formal system of first-order predicate calculus with infinitely long expressions* (**J** 0090) J Math Soc Japan 13*357-370
⋄ C40 C75 F05 ⋄ REV MR 29 # 2178 Zbl 108.2
JSL 27.468 • ID 08524

MAEHARA, S. [1970] *A general theory of completeness proofs* (**J** 0260) Ann Jap Ass Phil Sci 3*242-256
⋄ C07 C90 F50 ⋄ REV MR 44 # 6443 Zbl 226 # 02032
• ID 08455

MAEHARA, S. & TAKEUTI, G. [1971] *Two interpolation theorems for a Π_1^1 predicate calculus* (**J** 0036) J Symb Logic 36*262-270
⋄ B15 C40 F35 ⋄ REV MR 46 # 6991 Zbl 278 # 02013
• ID 08528

MAEHARA, S. see Vol. I, II, IV, VI for further entries

MAGAJNA, B. [1983] *Infinitesimals (Slovenian) (English summary)* (**J** 2310) Obz Mat Fiz, Ljubljana 30*33-41
⋄ C20 H05 ⋄ REV MR 84i:03116 Zbl 512 # 26010
• ID 34592

MAGID, A.R. [1975] *Ultrafunctors* (**J** 0017) Canad J Math 27*372-375
⋄ C20 C60 ⋄ REV MR 52 # 3285 Zbl 259 # 13006
• ID 17688

MAGIDOR, M. [1971] *On the role of supercompact and extendible cardinals in logic* (**J** 0029) Israel J Math 10*147-157
⋄ C55 C75 C85 E47 E55 ⋄ REV MR 45 # 4966
Zbl 263 # 02034 • ID 08543

MAGIDOR, M. [1971] *There are many normal ultrafilters corresponding to a supercompact cardinal* (**J** 0029) Israel J Math 9*186-192
⋄ C20 C55 E05 E55 ⋄ REV MR 50 # 110 Zbl 211.309
• ID 08542

MAGIDOR, M. [1977] *Chang's conjecture and powers of singular cardinals* (**J** 0036) J Symb Logic 42*272-276
⋄ C55 E40 E50 E55 E65 ⋄ REV MR 58 # 5221
Zbl 394 # 03046 • ID 26457

MAGIDOR, M. & MALITZ, J. [1977] *Compact extensions of $L(Q)$ (part Ia)* (**J** 0007) Ann Math Logic 11*217-261
⋄ C55 C80 E55 E65 ⋄ REV MR 56 # 11746
Zbl 356 # 02012 JSL 50.1076 • ID 23666

MAGIDOR, M. & MALITZ, J. [1977] *Compactness and transfer for a fragment of L^2* (**J** 0036) J Symb Logic 42*261-268
⋄ C55 C80 E05 E65 ⋄ REV MR 58 # 16129
Zbl 394 # 03016 • ID 26454

MAGIDOR, M. [1980] see JECH, T.J.

MAGIDOR, M. [1982] *The simplest counterexample to compactness in the constructible universe* (**P** 3622) Int Congr Log, Meth & Phil of Sci (6,Proc);1979 Hannover 279-288
⋄ C70 E45 ⋄ REV MR 84a:03038 Zbl 521 # 03035
• ID 35575

MAGIDOR, M. & SHELAH, S. & STAVI, J. [1983] *On the standard part of nonstandard models of set theory* (**J** 0036) J Symb Logic 48*33-38
⋄ C62 E45 H05 ⋄ REV MR 84m:03058 Zbl 522 # 03060
• ID 35758

MAGIDOR, M. [1983] see GUREVICH, Y.

MAGIDOR, M. & SHELAH, S. & STAVI, J. [1984] *Countably decomposable admissible sets* (J 0073) Ann Pure Appl Logic 26*287-361
⋄ C40 C62 C70 E30 E45 E47 ⋄ REV MR 86i:03048 Zbl 558 # 03017 • ID 38750

MAGIDOR, M. see Vol. IV, V for further entries

MAIER, B.J. [1981] *Existenziell abgeschlossene lokal endliche p-Gruppen* (J 0008) Arch Math (Basel) 37*113-128
⋄ C25 C60 ⋄ REV MR 83e:20005 Zbl 451 # 20032
• ID 54051

MAIER, B.J. [1983] *On existentially closed and generic nilpotent groups* (J 0029) Israel J Math 46*170-188
⋄ C25 C60 ⋄ REV MR 85m:03026 MR 86a:20030 Zbl 539 # 03015 • ID 39974

MAIER, B.J. [1984] *Amalgame nilpotenter Guppen der Klasse zwei* (J 0057) Publ Math (Univ Debrecen) 31*57-70
⋄ C60 ⋄ REV MR 85k:20117 • ID 48412

MAIER, B.J. [1984] *Amalgams of torsion-free nilpotent groups of class three* (J 0144) Rend Sem Mat Univ Padova 72*231-247
⋄ C60 ⋄ REV MR 86d:20027 Zbl 559 # 20020 • ID 44465

MAIER, B.J. [1984] *Existentially closed torsion-free nilpotent groups of class three* (J 0036) J Symb Logic 49*220-230
⋄ C25 C60 ⋄ REV MR 85m:03026 • ID 42477

MAKAI JR., E. [1979] see ANDREKA, H.

MAKANIN, G.S. [1964] *A new solvable case of the decision problem for first-order predicate calculus (Russian)* (C 0570) Form Log & Metodol Nauk 125-153
⋄ B03 B25 ⋄ REV Zbl 148.8 • ID 14845

MAKANIN, G.S. [1977] *The problem of the solvability of equations in a free semigroup (Russian)* (J 0142) Mat Sb, Akad Nauk SSSR, NS 103(145)*147-236,319
• TRANSL [1977] (J 0349) Math of USSR, Sbor 32*129-198
⋄ B03 B25 D40 ⋄ REV MR 57 # 9874 Zbl 371 # 20047
• ID 51395

MAKANIN, G.S. [1977] *The problem of solvability of equations in a free semigroup (Russian)* (J 0023) Dokl Akad Nauk SSSR 233/2*287-290
• TRANSL [1977] (J 0062) Sov Math, Dokl 18*330-334
⋄ B03 B25 D40 ⋄ REV MR 58 # 5997 Zbl 379 # 20046
• ID 51863

MAKANIN, G.S. [1980] *Equations in a free semigroup (Russian)* (P 1959) Int Congr Math (II,13);1978 Helsinki 1*263-268
• TRANSL [1981] (J 0225) Amer Math Soc, Transl, Ser 2 117*1-6
⋄ B03 B25 D40 ⋄ REV MR 82b:20081 Zbl 436 # 20039
• ID 82143

MAKANIN, G.S. [1984] *Decidability of the universal and positive theories of a free group (Russian)* (J 0216) Izv Akad Nauk SSSR, Ser Mat 48*735-749
• TRANSL [1985] (J 0448) Math of USSR, Izv 25*75-88
⋄ B25 C60 ⋄ REV MR 86c:03009 Zbl 578 # 20001
• ID 45751

MAKANIN, G.S. [1985] *On the decidability of the theory of a free group (Russian)* (P 4647) FCT'85 Fund of Comput Th;1985 Cottbus 279-284
⋄ B25 C60 C98 D35 ⋄ REV Zbl 574 # 20001 • ID 48835

MAKANIN, G.S. see Vol. IV for further entries

MAKKAI, M. [1964] *On PC_Δ-classes in the theory of models* (J 1487) Publ Math Inst Acad Sci Hung 9*159-194
⋄ C20 C40 C52 ⋄ REV MR 30 # 1052 Zbl 133.245 JSL 36.335 • ID 31861

MAKKAI, M. [1964] *On a generalization of a theorem of E.W.Beth* (J 0001) Acta Math Acad Sci Hung 15*227-235
⋄ C40 C50 ⋄ REV MR 28 # 5001 • ID 08577

MAKKAI, M. [1964] *Remarks on my paper "On PC_Δ-classes in the theory of models"* (J 1487) Publ Math Inst Acad Sci Hung 9*601-602
⋄ C40 C52 ⋄ REV MR 33 # 48 Zbl 133.245 JSL 36.335
• ID 31862

MAKKAI, M. [1965] *A compactness result concerning direct products of models* (J 0027) Fund Math 57*313-325
⋄ C05 C30 C52 ⋄ REV MR 32 # 5512 Zbl 133.245 JSL 33.477 • ID 08578

MAKKAI, M. [1969] *An application of a method of Smullyan to logics on admissible sets* (J 0014) Bull Acad Pol Sci, Ser Math Astron Phys 17*341-346
⋄ C70 ⋄ REV MR 42 # 1649 Zbl 191.297 • ID 08580

MAKKAI, M. [1969] *On the model theory of denumerable long formulas with finite strings of quantifiers* (J 0036) J Symb Logic 34*437-459
⋄ C40 C70 C75 ⋄ REV MR 41 # 45 Zbl 235 # 02050 JSL 38.337 • ID 08579

MAKKAI, M. [1970] *Structures elementarily equivalent relative to infinitary languages to models of higher power* (J 0001) Acta Math Acad Sci Hung 21*283-295
⋄ C55 C75 ⋄ REV MR 42 # 7486 Zbl 235 # 02051
• ID 08581

MAKKAI, M. [1971] *Algebraic operations on classes of structures and their connections with logical formulas I,II (Hungarian) (English summary)* (J 0462) Mat Fiz Oszt Koezlem, Acad Sci Hung 20*45-83,221-276
⋄ C30 C40 C52 C75 ⋄ REV MR 44 # 50 MR 56 # 5278 Zbl 219 # 02039 Zbl 243 # 02041 • ID 28091

MAKKAI, M. [1972] *Svenonius sentences and Lindstroem's theory on preservation theorems* (J 0027) Fund Math 73*219-233
⋄ C40 C52 C75 ⋄ REV MR 47 # 3166 Zbl 248 # 02021 JSL 40.635 • ID 08582

MAKKAI, M. [1973] *A proof of Baker's finite-base theorem on equational classes generated by finite elements on congruence distributive varieties* (J 0004) Algeb Universalis 3*174-181
⋄ C05 ⋄ REV MR 50 # 4444 Zbl 288 # 08007 • ID 08584

MAKKAI, M. [1973] *Global definability theory in $L_{\omega_1,\omega}$*
(J 0015) Bull Amer Math Soc 79*916-921
⋄ C40 C75 ⋄ REV MR 55 # 2490 Zbl 281 # 02053
• ID 14521

MAKKAI, M. [1973] *Preservation theorems for pseudo-elementary classes (Russian)* (C 0733) Izbr Vopr Algeb & Log (Mal'tsev) 161-183
⋄ C40 C52 ⋄ REV MR 49 # 10540 Zbl 277 # 02013
• ID 08583

MAKKAI, M. [1973] *Vaught sentences and Lindstroem's regular relations* (P 0713) Cambridge Summer School Math Log;1971 Cambridge GB 622-660
⋄ C40 C52 C75 D55 D70 ⋄ REV MR 50 # 4285 Zbl 268 # 02009 • ID 20987

MAKKAI, M. [1974] *A remark on a paper of J.P. Ressayre* (J 0007) Ann Math Logic 7*157-162
⋄ C70 ⋄ REV MR 50 # 12709 Zbl 299 # 02013 • ID 08586

MAKKAI, M. [1974] *Generalizing Vaught sentences from ω to strong cofinality ω* (J 0027) Fund Math 82*105-119
• ERR/ADD ibid 82*385
⋄ C40 C75 ⋄ REV MR 53 # 5295a MR 53 # 5295b Zbl 301 # 02051 Zbl 301 # 02052 • ID 08585

MAKKAI, M. [1976] see HARNIK, V.

MAKKAI, M. & REYES, G.E. [1976] *Model-theoretical methods in the theory of topoi and related categories I,II (Russian summaries)* (J 0014) Bull Acad Pol Sci, Ser Math Astron Phys 24*379-384,385-392
⋄ C75 C90 G30 ⋄ REV MR 54 # 10009 Zbl 337 # 18004 Zbl 337 # 18005 • ID 25850

MAKKAI, M. [1976] *Review of Chang and Keisler: model theory* (J 0015) Bull Amer Math Soc 82*433-446
⋄ C98 ⋄ ID 31867

MAKKAI, M. [1977] see HARNIK, V.

MAKKAI, M. [1977] *Admissible sets and infinitary logic* (C 1523) Handb of Math Logic 233-281
⋄ C70 C98 D55 D60 D70 D98 ⋄ REV MR 58 # 10395 JSL 49.968 • ID 24202

MAKKAI, M. & MYCIELSKI, J. [1977] *An $L_{\omega_1\omega}$-complete and consistent theory without models* (J 0053) Proc Amer Math Soc 62*131-133
⋄ C35 C75 E75 ⋄ REV MR 55 # 12464 Zbl 383 # 03009 • ID 31868

MAKKAI, M. [1977] *An "admissible" generalization of a theorem on countable Σ_1^1 sets of reals with applications* (J 0007) Ann Math Logic 11*1-30
⋄ C15 C40 C50 C70 D55 E15 ⋄ REV MR 58 # 10408 Zbl 376 # 02031 • ID 23657

MAKKAI, M. & REYES, G.E. [1977] *First order categorical logic. Model-theoretical methods in the theory of topoi and related categories* (X 0811) Springer: Heidelberg & New York viii + 301pp
⋄ C75 C90 F50 G30 ⋄ REV MR 58 # 21600 Zbl 357 # 18002 • ID 50433

MAKKAI, M. & MCNULTY, G.F. [1977] *Universal Horn axiom systems for lattices of submodules* (J 0004) Algeb Universalis 7*25-31
⋄ C05 C60 G10 ⋄ REV MR 55 # 2682 Zbl 375 # 06003 • ID 31900

MAKKAI, M. [1978] see BARWISE, J.

MAKKAI, M. [1979] see HARNIK, V.

MAKKAI, M. [1979] *There are atomic Nadel structures* (J 2128) C R Math Acad Sci, Soc Roy Canada 1*157-160
⋄ C15 C70 ⋄ REV MR 80e:03038 Zbl 421 # 03023 • ID 53422

MAKKAI, M. [1980] *On full embeddings I* (J 0326) J Pure Appl Algebra 16*183-195
⋄ C40 G30 ⋄ REV MR 83g:18007 Zbl 449 # 18001 • ID 56746

MAKKAI, M. [1981] *An example concerning Scott heights* (J 0036) J Symb Logic 46*301-318
⋄ C70 ⋄ REV MR 82m:03049 Zbl 501.03018 • ID 75837

MAKKAI, M. [1981] *The topos of types* (P 2628) Log Year;1979/80 Storrs 157-201
⋄ C35 C90 G30 ⋄ REV MR 82k:03103 Zbl 527 # 03042 • ID 75838

MAKKAI, M. [1982] *Full continuous embeddings of toposes* (J 0064) Trans Amer Math Soc 269*167-196
⋄ C35 C90 G30 ⋄ REV MR 83c:03058 Zbl 487 # 18004 • ID 35151

MAKKAI, M. [1982] *Stone duality for first order logic* (P 3708) Herbrand Symp Logic Colloq;1981 Marseille 217-232
⋄ C07 G30 ⋄ REV MR 85h:03071 Zbl 522 # 03006 • ID 37439

MAKKAI, M. [1984] see HARRINGTON, L.A.

MAKKAI, M. [1984] *A survey of basic stability theory, with particular emphasis on orthogonality and regular types* (J 0029) Israel J Math 49*181-238
⋄ C45 C98 ⋄ REV MR 86h:03055 • ID 45600

MAKKAI, M. [1984] *A Stone-type representation theory for first order logic* (P 2180) Math Appl Categ Th;1983 Denver 175-243
⋄ C90 G30 ⋄ REV MR 85i:03197 Zbl 561 # 03034 • ID 44303

MAKKAI, M. [1985] see HARRINGTON, L.A.

MAKKAI, M. [1985] *Ultraproducts and categorical logic* (P 2160) Latin Amer Symp Math Log (6);1983 Caracas 222-309
⋄ C20 G30 ⋄ ID 41797

MAKKAI, M. see Vol. V, VI for further entries

MAKOWIECKA, H. [1965] *On a primitive notion of one-dimensional geometries over some fields* (J 0014) Bull Acad Pol Sci, Ser Math Astron Phys 13*43-47
⋄ B30 C60 ⋄ REV MR 30 # 4752 Zbl 132.408 • ID 08587

MAKOWIECKA, H. [1975] *An elementary geometry in connection with decomposition of a plane* (J 0014) Bull Acad Pol Sci, Ser Math Astron Phys 23*665-674
⋄ B30 C60 ⋄ REV MR 52 # 5410 Zbl 312 # 50003 • ID 08589

MAKOWIECKA, H. [1975] *On primitive notions in n-dimensional elementary geometries* (J 0014) Bull Acad Pol Sci, Ser Math Astron Phys 23*675-682
⋄ B30 C60 ⋄ REV MR 52 # 5411 Zbl 312 # 50004 • ID 08588

MAKOWIECKA, H. [1975] *Quarternary relations in weak euclidean geometries (Russian summary)* (J 0014) Bull Acad Pol Sci, Ser Math Astron Phys 23*683-692
⋄ B30 C60 ⋄ REV MR 52 # 5412 Zbl 312 # 50005 • ID 18272

MAKOWIECKA, H. [1975] *The norm relation in weak euclidean geometry with the axiom on two circles (Russian summary)* (J 0014) Bull Acad Pol Sci, Ser Math Astron Phys 23*1181-1187
⋄ B30 C60 ⋄ REV MR 52 # 13363 Zbl 325 # 50002 • ID 21816

MAKOWIECKA, H. [1975] *The theory of bi-proportionality as a geometry* (J 0014) Bull Acad Pol Sci, Ser Math Astron Phys 23*657-664
⋄ B30 C60 ⋄ REV MR 52 # 5409 Zbl 312 # 50006 • ID 08590

MAKOWIECKA, H. [1976] *A general property of ternary relations in elementary geometry (Russian summary)* (J 0014) Bull Acad Pol Sci, Ser Math Astron Phys 24*163-169
  ⋄ B30 C60 ⋄ REV MR 54 # 2448 Zbl 325 # 50003
  • ID 24642

MAKOWIECKA, H. see Vol. I for further entries

MAKOWSKY, J.A. [1972] *Note on almost strongly minimal theories (Russian summary)* (J 0014) Bull Acad Pol Sci, Ser Math Astron Phys 20*529-534
  ⋄ C35 C52 ⋄ REV MR 47 # 35 Zbl 267 # 02036 • ID 08594

MAKOWSKY, J.A. [1973] *Langages engendres a partir des formules de Scott* (J 2313) C R Acad Sci, Paris, Ser A-B 276*A1585-A1587
  ⋄ C75 C95 ⋄ REV MR 49 # 2357 Zbl 273 # 02011
  • ID 30483

MAKOWSKY, J.A. [1974] *On some conjectures connected with complete sentences* (J 0027) Fund Math 81*193-202
  ⋄ C35 C45 C52 C98 ⋄ REV MR 51 # 2894
  Zbl 285 # 02042 • ID 17339

MAKOWSKY, J.A. [1975] *Securable quantifiers, κ-unions and admissible sets* (P 0775) Logic Colloq;1973 Bristol 409-428
  ⋄ C55 C70 C75 C80 C95 ⋄ REV MR 52 # 5345
  Zbl 311 # 02023 • ID 18273

MAKOWSKY, J.A. [1975] *Topological model theory* (P 2021) Model Th & Appl;1975 Bressanone 122-150
  ⋄ C65 C80 ⋄ ID 90127

MAKOWSKY, J.A. & SHELAH, S. & STAVI, J. [1976] *Δ-logics and generalized quantifiers* (J 0007) Ann Math Logic 10*155-192
  ⋄ C40 C55 C80 C95 ⋄ REV MR 56 # 15362
  Zbl 346 # 02007 • ID 23650

MAKOWSKY, J.A. & MARCJA, A. [1976] *Problemi di decidibilita in logica topologica (English summary)* (J 0144) Rend Sem Mat Univ Padova 56*67-78
  ⋄ B25 C65 C80 ⋄ REV MR 57 # 16035 Zbl 402 # 03020
  • ID 30757

MAKOWSKY, J.A. & MARCJA, A. [1977] *Completeness theorems for modal model theory with the Montague-Chang semantics I* (J 0068) Z Math Logik Grundlagen Math 23*97-104
  ⋄ B45 C90 ⋄ REV MR 58 # 5057 Zbl 362 # 02043
  • ID 31849

MAKOWSKY, J.A. & TULIPANI, S. [1977] *Some model theory for monotone quantifiers* (J 0009) Arch Math Logik Grundlagenforsch 18*115-134
  ⋄ C20 C40 C80 ⋄ REV MR 57 # 9474 Zbl 365 # 02042
  • ID 31850

MAKOWSKY, J.A. [1978] *Quantifying over countable sets: Positive vs. stationary logic* (P 1897) Logic Colloq;1977 Wroclaw 183-193
  ⋄ C55 C80 ⋄ REV MR 80j:03052 Zbl 469 # 03021
  • ID 55149

MAKOWSKY, J.A. [1978] *Some observations on uniform reduction for properties invariant on the range of definable relations* (J 0027) Fund Math 99*199-203
  ⋄ C30 C40 C75 C95 ⋄ REV MR 81e:03029
  Zbl 383 # 03021 • ID 29205

MAKOWSKY, J.A. [1979] *The reals cannot be characterized topologically with strictly local properties and countability axioms* (P 1705) Scand Logic Symp (4);1976 Jyvaeskylae 251-257
  ⋄ C65 C80 C90 ⋄ REV MR 80h:03055 Zbl 402 # 03037
  • ID 54684

MAKOWSKY, J.A. & SHELAH, S. [1979] *The theorems of Beth and Craig in abstract model theory I: The abstract setting* (J 0064) Trans Amer Math Soc 256*215-239
  ⋄ C40 C75 C80 C95 ⋄ REV MR 81b:03041
  Zbl 428 # 03032 • REM Part II 1981 • ID 53791

MAKOWSKY, J.A. [1980] *Measuring the expressive power of dynamic logics: an application of abstract model theory* (P 2904) Automata, Lang & Progr (7);1980 Noordwijkerhout 409-421 • ERR/ADD [1981] (P 2903) Automata, Lang & Progr (8);1981 Akko 551
  ⋄ B75 C95 ⋄ REV MR 82c:03022 MR 82j:03021
  Zbl 465 # 68012 • ID 54945

MAKOWSKY, J.A. & SHELAH, S. [1981] *The theorems of Beth and Craig in abstract model theory II: Compact logics* (J 0009) Arch Math Logik Grundlagenforsch 21*13-35
  ⋄ C40 C75 C80 C95 ⋄ REV MR 83g:03034
  Zbl 472 # 03028 • REM Part I 1979 • ID 55293

MAKOWSKY, J.A. & ZIEGLER, M. [1981] *Topological model theory with an interior operator: consistency properties and back-and-forth arguments* (J 0009) Arch Math Logik Grundlagenforsch 21*37-54
  ⋄ C40 C75 C80 ⋄ REV MR 83h:03051 Zbl 472 # 03027
  • ID 55292

MAKOWSKY, J.A. & SHELAH, S. [1983] *Positive results in abstract model theory: a theory of compact logics* (J 0073) Ann Pure Appl Logic 25*263-299
  ⋄ C20 C52 C95 E05 E55 ⋄ REV MR 85i:03125
  Zbl 544 # 03013 • ID 38744

MAKOWSKY, J.A. [1984] *Model theoretic issues in theoretical computer science, part I: Relational data bases and abstract data types* (P 3710) Logic Colloq;1982 Firenze 303-343
  ⋄ B75 C13 C95 ⋄ REV MR 86c:68056 Zbl 553 # 68028
  • ID 39796

MAKOWSKY, J.A. & MUNDICI, D. [1985] *Abstract equivalence relations* (C 4183) Model-Theor Log 717-746
  ⋄ C95 ⋄ ID 48334

MAKOWSKY, J.A. [1985] *Abstract embedding relations* (C 4183) Model-Theor Log 747-791
  ⋄ C95 ⋄ ID 48336

MAKOWSKY, J.A. [1985] *Compactness, embeddings and definability* (C 4183) Model-Theor Log 645-716
  ⋄ C40 C95 C98 ⋄ ID 48337

MAKOWSKY, J.A. [1985] *Vopenka's principle and compact logics* (J 0036) J Symb Logic 50*42-48
  ⋄ C05 C55 C95 E55 ⋄ REV MR 86h:03073 • ID 39781

MAKOWSKY, J.A. [1985] *Why Horn formulas matter in computer science: Initial structure and generic examples* (P 4627) CAAP'85 Arbres en Algeb & Progr (10);1985 Berlin 374-385
  ⋄ B75 C52 ⋄ ID 39812

MAKOWSKY, J.A. see Vol. II for further entries

MAKSIMOVA, L.L. [1977] *Craig's interpolation theorem and amalgamable varieties (Russian)* (**J** 0023) Dokl Akad Nauk SSSR 237*1281-1284
- TRANSL [1977] (**J** 0062) Sov Math, Dokl 18*1550-1553
⋄ B22 B55 C05 C40 C90 G10 ⋄ REV MR 57 # 5698 Zbl 393 # 03013 • ID 52433

MAKSIMOVA, L.L. [1977] *Craig's theorem in superintuitionistic logics and amalgamable varieties of pseudo-boolean algebras (Russian)* (**J** 0003) Algebra i Logika 16*643-681,741
- TRANSL [1977] (**J** 0069) Algeb and Log 16*427-455
⋄ B55 C05 C40 C90 G05 G10 ⋄ REV MR 80c:03028 Zbl 403 # 03047 • ID 54760

MAKSIMOVA, L.L. [1979] *Interpolation theorems in modal logics and amalgamable varieties of topological boolean algebras (Russian)* (**J** 0003) Algebra i Logika 18*556-586,632
- TRANSL [1979] (**J** 0069) Algeb and Log 18*348-370
⋄ B45 C05 C40 C90 G05 ⋄ REV MR 81j:03031 Zbl 436 # 03011 • ID 55851

MAKSIMOVA, L.L. [1982] *Absence of the interpolation property in modal companions of a Dummett logic (Russian)* (**J** 0003) Algebra i Logika 21*690-694
- TRANSL [1982] (**J** 0069) Algeb and Log 21*460-462
⋄ B45 C40 C90 ⋄ REV MR 85e:03043 • ID 40502

MAKSIMOVA, L.L. [1982] *Interpolation properties of superintuitionistic, positive and modal logics* (**P** 3800) Intens Log: Th & Appl;1979 Moskva 70-78
- TRANSL [1984] (**C** 4366) Modal & Intens Log & Primen Probl Metodol Nauk 81-88
⋄ B55 C40 C90 ⋄ REV MR 85f:03017 Zbl 521 # 03014 • ID 37067

MAKSIMOVA, L.L. [1982] *Lyndon's interpolation theorem in modal logics (Russian)* (**C** 3953) Mat Log & Teor Algor 45-55
⋄ B45 C40 C90 ⋄ REV MR 85e:03031 Zbl 543 # 03011 • ID 40416

MAKSIMOVA, L.L. see Vol. I, II, IV, V for further entries

MALAN, S.W. [1980] see HEIDEMA, J.

MALCOLM, W.G. [1972] *On a compactness theorem of A.Shafaat* (**J** 3172) J London Math Soc, Ser 2 5*719-725
⋄ C20 C75 ⋄ REV MR 48 # 1913 Zbl 263 # 02026 • ID 08618

MALCOLM, W.G. [1972] *Variations in definition of ultraproducts of a family of first order relational structures* (**J** 0047) Notre Dame J Formal Log 13*394-398
⋄ C20 ⋄ REV MR 47 # 3167 Zbl 238 # 02046 • ID 08619

MALCOLM, W.G. [1973] *Application of higher-order ultraproducts to the theory of local properties in universal algebras and relational systems* (**J** 3240) Proc London Math Soc, Ser 3 27*617-637
⋄ C05 C20 C85 E40 ⋄ REV MR 49 # 38 Zbl 292 # 02044 • ID 08620

MALCOLM, W.G. [1974] *Some results and algebraic applications in the theory of higher-order ultraproducts* (**J** 0047) Notre Dame J Formal Log 15*1-15
⋄ C20 C60 C85 G05 ⋄ REV MR 49 # 2362 Zbl 212.18 • ID 08621

MALINOWSKI, G. & MICHALCZYK, M. [1981] *Interpolation properties for a class of many-valued propositional calculi* (**J** 0387) Bull Sect Logic, Pol Acad Sci 10*9-16
⋄ B50 C40 ⋄ REV MR 82g:03037 Zbl 457 # 03023 • ID 54348

MALINOWSKI, G. & MICHALCZYK, M. [1982] *That SCI has the interpolation property* (**J** 0063) Studia Logica 41*375-380
⋄ B60 C40 ⋄ REV MR 86e:03006 Zbl 549 # 03008 • ID 42305

MALINOWSKI, G. see Vol. I, II for further entries

MALITZ, J. [1965] *Problems in the model theory of infinite languages* (0000) Diss., Habil. etc
⋄ C40 C75 ⋄ REM Doct. diss., University of California, Berkeley, CA • ID 21428

MALITZ, J. [1968] *The Hanf number for complete $L_{\omega_1,\omega}$-sentences* (**P** 0637) Syntax & Semant Infinitary Lang;1967 Los Angeles 166-181
⋄ C55 C75 ⋄ REV Zbl 212.322 • ID 08626

MALITZ, J. [1969] *Universal classes in infinitary languages* (**J** 0025) Duke Math J 36*621-630
⋄ C40 C52 C75 ⋄ REV MR 39 # 6745 Zbl 249 # 02008 JSL 39.336 • ID 08627

MALITZ, J. [1971] *Infinitary analogs of theorems from first order model theory* (**J** 0036) J Symb Logic 36*216-228
⋄ C30 C40 C55 C75 ⋄ REV MR 45 # 37 Zbl 232 # 02037 • ID 08628

MALITZ, J. & REINHARDT, W.N. [1972] *A complete countable $L_{\omega_1}^Q$-theory with maximal models of many cardinalities* (**J** 0048) Pac J Math 43*691-700
⋄ C55 C80 ⋄ REV MR 47 # 4788 Zbl 274 # 02027 JSL 40.635 • ID 08629

MALITZ, J. & REINHARDT, W.N. [1972] *Maximal models in the language with quantifier "there exist uncountably many"* (**J** 0048) Pac J Math 40*139-155
⋄ C50 C55 C80 E55 ⋄ REV MR 47 # 1574 Zbl 242 # 02058 JSL 40.635 • ID 08631

MALITZ, J. [1975] *Complete theories with countably many rigid nonisomorphic models* (**J** 0036) J Symb Logic 40*389-392
⋄ C15 C35 C50 ⋄ REV MR 52 # 10406 Zbl 332 # 02054 • ID 14822

MALITZ, J. [1977] see MAGIDOR, M.

MALITZ, J. [1979] *Introduction to mathematical logic. Set theory, computable functions, model theory* (**X** 0811) Springer: Heidelberg & New York xii+198pp
⋄ B98 C98 D98 ⋄ REV MR 81h:03002 Zbl 407 # 03001 JSL 49.672 • ID 56167

MALITZ, J. & RUBIN, M. [1980] *Compact fragments of higher order logic* (**P** 2958) Latin Amer Symp Math Log (4);1978 Santiago 219-238
⋄ B15 C55 C80 C85 ⋄ REV MR 82k:03055 Zbl 434 # 03013 • ID 55717

MALITZ, J. [1983] *Downward transfer of satisfiability for sentences of $L^{1,1}$* (**J** 0036) J Symb Logic 48*1146-1150
⋄ C55 C80 ⋄ REV MR 85e:03089 Zbl 537 # 03025 • ID 39844

MALITZ, J. see Vol. I, IV, V for further entries

MALLIAVIN, M.P. [1982] *Ultra-produits d'algebre de Lie*
(S 2250) Semin Dubreil: Algebre 34*157-166
⋄ C20 C65 ⋄ REV MR 84k:17007 Zbl 482 # 17003
• ID 37697

MAL'TSEV, A.I. [1936] *Untersuchungen aus dem Gebiete der mathematischen Logik (Russian summary)* (J 0142) Mat Sb, Akad Nauk SSSR, NS 1(43)*323-336
• TRANSL [1971] (C 2621) Mal'tsev: Metamath of Algeb Syst 1-14
⋄ B10 C07 ⋄ REV Zbl 14.385 JSL 2.84 FdM 62.42
• ID 08602

MAL'TSEV, A.I. [1939] *On the embedding of associative systems into groups (Russian) (German summary)* (J 0142) Mat Sb, Akad Nauk SSSR, NS 6(48)*331-336
⋄ C05 C60 ⋄ REV MR 2.7 FdM 65.79 • REM Part I. Part II 1940 • ID 41173

MAL'TSEV, A.I. [1940] *On the embedding of associative systems into groups II (Russian) (German summary)* (J 0142) Mat Sb, Akad Nauk SSSR, NS 8(50)*251-264
⋄ C05 C60 ⋄ REV MR 2.128 FdM 66.97 • REM Part I 1939
• ID 41174

MAL'TSEV, A.I. [1941] *A general method for obtaining local theorems in group theory (Russian)* (J 0226) Uch Zap Ped Inst, Ivanovo 1*3-9
• TRANSL [1971] (C 2621) Mal'tsev: Metamath of Algeb Syst 15-21
⋄ C07 C60 ⋄ REV MR 17.823 JSL 24.55 • ID 19267

MAL'TSEV, A.I. [1951] *On the completing of group order (Russian)* (S 0066) Tr Mat Inst Steklov 38*173-175
⋄ C60 ⋄ REV MR 14.13 • ID 43748

MAL'TSEV, A.I. [1954] *On the general theory of algebraic systems (Russian)* (J 0142) Mat Sb, Akad Nauk SSSR, NS 35(77)*3-20
⋄ C05 C65 ⋄ REV MR 16.440 • ID 41175

MAL'TSEV, A.I. [1956] *Quasiprimitive classes of abstract algebras (Russian)* (J 0023) Dokl Akad Nauk SSSR 108*187-189
• TRANSL [1971] (C 2621) Mal'tsev: Metamath of Algeb Syst 27-31
⋄ C05 ⋄ REV MR 18.107 Zbl 231 # 02002 Zbl 73.258 JSL 24.57 • ID 19263

MAL'TSEV, A.I. [1956] *Remark on densely ordered groups (Russian)* (J 0226) Uch Zap Ped Inst, Ivanovo 10*3-5
⋄ C60 ⋄ REV MR 19.530 • ID 43750

MAL'TSEV, A.I. [1956] *Representations of models (Russian)* (J 0023) Dokl Akad Nauk SSSR 108*27-29
• TRANSL [1971] (C 2621) Mal'tsev: Metamath of Algeb Syst 22-26
⋄ C07 C60 ⋄ REV MR 18.370 Zbl 74.11 JSL 24.55
• ID 19266

MAL'TSEV, A.I. [1956] *Subdirect products of models (Russian)* (J 0023) Dokl Akad Nauk SSSR 109*264-266
• TRANSL [1971] (C 2621) Mal'tsev: Metamath of Algeb Syst 32-36
⋄ C05 ⋄ REV MR 19.240 Zbl 73.251 JSL 24.57 • ID 19264

MAL'TSEV, A.I. [1957] *Classes of models with an operation of generation (Russian)* (J 0023) Dokl Akad Nauk SSSR 116*738-741
• TRANSL [1971] (C 2621) Mal'tsev: Metamath of Algeb Syst 44-50
⋄ C52 ⋄ REV MR 20 # 2271 Zbl 79.7 JSL 40.640
• ID 08605

MAL'TSEV, A.I. [1957] *Derived operations and predicates (Russian)* (J 0023) Dokl Akad Nauk SSSR 116*24-27
• TRANSL [1971] (C 2621) Mal'tsev: Metamath of Algeb Syst 37-43
⋄ C52 ⋄ REV MR 20 # 1647 Zbl 79.245 JSL 40.640
• ID 08606

MAL'TSEV, A.I. [1958] *Certain classes of models (Russian)* (J 0023) Dokl Akad Nauk SSSR 120*245-248
• TRANSL [1971] (C 2621) Mal'tsev: Metamath of Algeb Syst 61-65
⋄ C05 C52 G30 ⋄ REV MR 20 # 5155 Zbl 80.253 JSL 40.640 • ID 90269

MAL'TSEV, A.I. [1958] *Defining relations in categories (Russian)* (J 0023) Dokl Akad Nauk SSSR 119*1095-1098
• TRANSL [1971] (C 2621) Mal'tsev: Metamath of Algeb Syst
⋄ C05 C52 G30 ⋄ REV MR 20 # 3805 Zbl 80.252
• ID 41176

MAL'TSEV, A.I. [1958] *Homomorphisms onto finite groups (Russian)* (J 0226) Uch Zap Ped Inst, Ivanovo 18*49-60
⋄ C07 C13 C60 ⋄ ID 42983

MAL'TSEV, A.I. [1958] *The structural characterization of certain classes of algebras (Russian)* (J 0023) Dokl Akad Nauk SSSR 120*29-32
• TRANSL [1971] (C 2621) Mal'tsev: Metamath of Algeb Syst 56-60
⋄ C05 C52 G30 ⋄ REV MR 20 # 5154 Zbl 80.252 JSL 40.640 • ID 08607

MAL'TSEV, A.I. [1959] *Algebraic systems (Russian)* (P 0607) All-Union Math Conf (3);1956 Moskva 2*8
⋄ C05 C60 ⋄ ID 28388

MAL'TSEV, A.I. [1959] *Model correspondences (Russian)* (J 0216) Izv Akad Nauk SSSR, Ser Mat 23*313-336
• TRANSL [1971] (C 2621) Mal'tsev: Metamath of Algeb Syst 66-94
⋄ B15 C05 C07 C52 C60 C85 ⋄ REV MR 22 # 10909 Zbl 100.12 JSL 34.299 JSL 35.299 • ID 19260

MAL'TSEV, A.I. [1959] *Regular products of models (Russian)* (J 0216) Izv Akad Nauk SSSR, Ser Mat 23*489-502
• TRANSL [1964] (J 0225) Amer Math Soc, Transl, Ser 2 39*193-206 [1971] (C 2621) Mal'tsev: Metamath of Algeb Syst 95-113
⋄ C30 ⋄ REV MR 23 # A1536 Zbl 117.12 JSL 34.651
• ID 41268

MAL'TSEV, A.I. [1959] *Small models (Russian)* (J 0023) Dokl Akad Nauk SSSR 127*258-261
• TRANSL [1971] (C 2621) Mal'tsev: Metamath of Algeb Syst 114-118
⋄ C52 C55 ⋄ REV MR 21 # 5553 Zbl 117.12 JSL 34.513
• ID 19259

MAL'TSEV, A.I. [1960] *On free soluble groups (Russian)* (J 0023) Dokl Akad Nauk SSSR 130*495-498
• TRANSL [1960] (J 0062) Sov Math, Dokl 1*65-68 [1971] (C 2621) Mal'tsev: Metamath of Algeb Syst 119-123
⋄ C60 D35 ⋄ REV MR 22 # 8056 Zbl 97.248 JSL 30.99
• ID 19257

MAL'TSEV, A.I. [1960] *Some correspondences between rings and groups (Russian)* (J 0142) Mat Sb, Akad Nauk SSSR, NS 50(92)*257-266
• TRANSL [1965] (J 0225) Amer Math Soc, Transl, Ser 2 45*221-231 [1971] (C 2621) Mal'tsev: Metamath of Algeb Syst 124-137
◊ C60 D35 ◊ REV MR 22 # 9448 Zbl 100.14 JSL 30.393 • ID 08609

MAL'TSEV, A.I. [1961] *Constructive algebra I (Russian)* (J 0067) Usp Mat Nauk 16/3*3-60
• TRANSL [1961] (J 1399) Russ Math Surv 16/3*77-129 [1971] (C 2621) Mal'tsev: Metamath of Algeb Syst 148-214
◊ C57 C98 F60 F98 ◊ REV MR 27 # 1362 Zbl 129.259 JSL 31.647 • ID 28706

MAL'TSEV, A.I. [1961] *Elementary properties of linear groups (Russian)* (C 0566) Nekotorye Prob Mat & Mekh (Lavrent'ev) 110-132
• TRANSL [1971] (C 2621) Mal'tsev: Metamath of Algeb Syst 221-247
◊ C60 ◊ REV MR 42 # 389 JSL 40.640 • ID 08610

MAL'TSEV, A.I. [1961] *On the elementary theories of locally free universal algebras (Russian)* (J 0023) Dokl Akad Nauk SSSR 138*1009-1012
• TRANSL [1961] (J 0062) Sov Math, Dokl 2*768-771
◊ B25 C05 C60 ◊ REV MR 24 # A3055 Zbl 119.252 JSL 32.278 • ID 19253

MAL'TSEV, A.I. [1961] *Undecidability of the elementary theory of finite groups (Russian)* (J 0023) Dokl Akad Nauk SSSR 138*771-774
• TRANSL [1961] (J 0062) Sov Math, Dokl 2*714-717 [1971] (C 2621) Mal'tsev: Metamath of Algeb Syst 215-220
◊ C13 D35 ◊ REV MR 27 # 3550 Zbl 119.252 JSL 30.394 • ID 19251

MAL'TSEV, A.I. [1962] *Axiomatisable classes of locally free algebras of various types (Russian)* (J 0092) Sib Mat Zh 3*729-743
• TRANSL [1971] (C 2621) Mal'tsev: Metamath of Algeb Syst 262-281
◊ B25 C05 C10 ◊ REV MR 26 # 59 Zbl 142.247 JSL 32.278 • ID 19245

MAL'TSEV, A.I. [1962] *On recursive abelian groups (Russian)* (J 0023) Dokl Akad Nauk SSSR 146*1009-1012
• TRANSL [1962] (J 0062) Sov Math, Dokl 3*1431-1434 [1971] (C 2621) Mal'tsev: Metamath of Algeb Syst 282-286
◊ C57 C60 D45 ◊ REV MR 27 # 1363 Zbl 156.11 JSL 31.649 • ID 41272

MAL'TSEV, A.I. [1962] *Strongly related models and recursively complete algebras (Russian)* (J 0023) Dokl Akad Nauk SSSR 145*276-279
• TRANSL [1962] (J 0062) Sov Math, Dokl 3*987-991 [1971] (C 2621) Mal'tsev: Metamath of Algeb Syst 255-261
◊ C57 C60 C62 D45 ◊ REV MR 26 # 1254 Zbl 132.247 JSL 31.649 • ID 19247

MAL'TSEV, A.I. [1963] *Some problems of the theory of classes of models (Russian)* (P 0754) All-Union Math Conf (4);1961 Leningrad 1*169-198
• TRANSL [1969] (J 0225) Amer Math Soc, Transl, Ser 2 83*1-48 [1971] (C 2621) Mal'tsev: Metamath of Algeb Syst 313-326
◊ C98 ◊ REV MR 27 # 5693 Zbl 191.295 • ID 20972

MAL'TSEV, A.I. [1966] *A few remarks on quasivarieties of algebraic systems (Russian)* (J 0003) Algebra i Logika 5/3*3-9
• TRANSL [1971] (C 2621) Mal'tsev: Metamath of Algeb Syst 416-421
◊ C05 ◊ REV MR 34 # 5728 • ID 90294

MAL'TSEV, A.I. [1966] *On standard notations and terminology in the theory of algebraic systems (Russian)* (J 0003) Algebra i Logika 5/1*71-77
◊ C05 C07 ◊ REV MR 33 # 2585 Zbl 243 # 08001 • ID 28852

MAL'TSEV, A.I. [1967] *Multiplication of classes of algebraic systems (Russian)* (J 0092) Sib Mat Zh 8*346-365
• TRANSL [1967] (J 0475) Sib Math J 8*254-266 [1971] (C 2621) Mal'tsev: Metamath of Algeb Syst 422-446
◊ C05 ◊ REV MR 35 # 4140 Zbl 228 # 08007 JSL 40.640 • ID 41274

MAL'TSEV, A.I. [1967] *Universally axiomatizable subclasses of locally finite classes of models (Russian)* (J 0092) Sib Mat Zh 8*1005-1014
• TRANSL [1967] (J 0475) Sib Math J 8*764-771 [1971] (C 2621) Mal'tsev: Metamath of Algeb Syst 447-454
◊ C52 ◊ REV MR 36 # 6275 Zbl 153.16 JSL 40.640 • ID 14651

MAL'TSEV, A.I. [1968] *Some questions bordering on algebra and mathematical logic (Russian)* (P 0657) Int Congr Math (II,10);1966 Moskva 217-231
• TRANSL [1968] (J 0225) Amer Math Soc, Transl, Ser 2 70*89-100 [1971] (C 2621) Mal'tsev: Metamath of Algeb Syst 460-473
◊ B25 C05 ◊ REV MR 38 # 2072 Zbl 205.306 JSL 40.641 • ID 08611

MAL'TSEV, A.I. [1970] *Algebraic systems (Russian)* (X 2027) Nauka: Moskva 392pp
• TRANSL [1973] (X 0911) Akademie Verlag: Berlin xii+317pp [1973] (X 0811) Springer: Heidelberg & New York xii+317pp
◊ C05 C98 ◊ REV MR 44 # 142 Zbl 223 # 08001 Zbl 266 # 08001 • ID 21828

MAL'TSEV, A.I. [1971] *The metamathematics of algebraic systems. Collected papers: 1936-1967* (S 3303) Stud Logic Found Math 66*xviii+494pp
◊ C96 ◊ REV MR 50 # 1877 Zbl 231 # 02002 • REM Transl. from several Russian articles. Edited and provided with suppl. notes by Wells III,B.F. • ID 27419

MAL'TSEV, A.I. [1976] *Selected works. I: Classical algebra. II: Mathematical logic and the general theory of algebraic systems (Russian)* (X 2027) Nauka: Moskva 484pp,388pp
◊ A10 C96 G96 ◊ REV MR 54 # 12464 MR 54 # 12465 Zbl 422 # 01024 Zbl 422 # 01025 • REM Edited by Lavrent'ev, M.A. and Shirshov, A.I. • ID 53460

MAL'TSEV, A.I. see Vol. I, II, IV for further entries

MAMEDOV, O.M. [1980] see BAJRAMOV, R.A.

MAMEDOV, O.M. [1984] see BAJRAMOV, R.A.

MANASTER, A.B. [1972] see ELLENTUCK, E.

MANASTER, A.B. [1975] *Completeness, compactness, and undecidability: an introduction to mathematical logic* (X 0819) Prentice Hall: Englewood Cliffs vi+154pp
◊ B98 C07 D35 D98 ◊ REV MR 53 # 12857 Zbl 306 # 02001 JSL 42.320 • ID 23120

MANASTER, A.B. [1977] see HAY, L.

MANASTER, A.B. & ROSENSTEIN, J.G. [1980] *Two-dimensional partial orderings: Recursive model theory* (**J** 0036) J Symb Logic 45*121-132
⋄ C57 C65 E07 ⋄ REV MR 81d:03047a Zbl 468 # 03009
• ID 55074

MANASTER, A.B. & ROSENSTEIN, J.G. [1980] *Two-dimensional partial orderings: Undecidability* (**J** 0036) J Symb Logic 45*133-143
⋄ C57 C65 D35 E07 G10 ⋄ REV MR 81d:03047b Zbl 468 # 03008 • ID 55073

MANASTER, A.B. & REMMEL, J.B. [1981] *Partial orderings of fixed finite dimension: Model companions and density* (**J** 0036) J Symb Logic 46*789-802
⋄ C25 C65 D35 E07 ⋄ REV MR 83b:06002 Zbl 491 # 03012 • ID 33275

MANASTER, A.B. & REMMEL, J.B. [1981] *Some decision problems for subtheories of two-dimensional partial orderings* (**P** 2628) Log Year;1979/80 Storrs 202-214
⋄ B25 C10 C65 D35 ⋄ REV MR 82h:03011 Zbl 486 # 03010 • ID 75934

MANASTER, A.B. see Vol. IV, V for further entries

MANCA, V. & SALIBRA, A. [1984] *First-order theories as many-sorted algebras* (**J** 0047) Notre Dame J Formal Log 25*86-94
⋄ C05 C07 ⋄ REV MR 86a:03069 Zbl 543 # 03016
• ID 40386

MANCA, V. see Vol. IV, VI for further entries

MANDERS, K.L. [1979] *The theory of all substructures of a structure: Characterisation and decision problems* (**J** 0036) J Symb Logic 44*583-598
⋄ B20 B25 C57 C60 ⋄ REV MR 81i:03047 Zbl 429 # 03007 • ID 53838

MANDERS, K.L. [1980] *First-order logical systems and set-theoretical definability* (**X** 1331) Univ Pittsburgh Pr: Pittsburgh 35pp
⋄ C95 E45 ⋄ ID 33515

MANDERS, K.L. [1980] *Theories with the existential substructure property* (**J** 0068) Z Math Logik Grundlagen Math 26*89-92
⋄ C25 C35 ⋄ REV MR 81c:03021 Zbl 514 # 03016
• ID 75942

MANDERS, K.L. [1984] *Interpretations and the model theory of the classical geometries* (**P** 2153) Logic Colloq;1983 Aachen 1*297-330
⋄ C25 C35 C65 ⋄ REV MR 86h:03070 Zbl 561 # 03019
• ID 41768

MANDERS, K.L. see Vol. I, II, IV for further entries

MANEVITZ, L.M. [1976] *Internal end-extensions of Peano arithmetic and a problem of Gaifman* (**J** 3172) J London Math Soc, Ser 2 13*80-82
⋄ C62 ⋄ REV MR 56 # 120 Zbl 362 # 02039 • ID 50761

MANEVITZ, L.M. [1976] *Robinson forcing is not absolute* (**J** 0029) Israel J Math 25*211-232
⋄ C25 E35 E45 E50 ⋄ REV MR 56 # 15407 Zbl 376 # 02050 • ID 26083

MANEVITZ, L.M. & STAVI, J. [1980] Δ_0^2 *operators and alternating sentences in arithmetic* (**J** 0036) J Symb Logic 45*144-154
⋄ C62 F30 H15 ⋄ REV MR 81a:03064 Zbl 458 # 03022
• ID 54428

MANEVITZ, L.M. [1984] *Applied model theory and metamathematics: an Abraham Robinson memorial problem list* (**J** 0029) Israel J Math 49*3-14
⋄ C98 H98 ⋄ ID 45601

MANEVITZ, L.M. [1985] see HERFORT, W.N.

MANEVITZ, L.M. see Vol. V for further entries

MANGANI, P. (ED.) [1975] *Model Theory and Applications. Centro Internazionale Matematico Estivo (CIME) II Ciclo* (**X** 0860) Cremonese: Firenze 151pp
⋄ C97 ⋄ ID 49820

MANGANI, P. [1976] *Alcune questioni di teoria dei modelli per linguaggi predicativi generali (English summary)* (**J** 0149) Atti Accad Naz Lincei Fis Mat Nat, Ser 8 60*368-376
⋄ B50 C20 C90 ⋄ REV MR 57 # 2906 Zbl 367 # 02027
• ID 51187

MANGANI, P. & MARCJA, A. [1980] *Shelah rank for boolean algebras and some application to elementary theories I* (**J** 0004) Algeb Universalis 10*247-257
⋄ C35 C40 C45 C50 G05 ⋄ REV MR 81i:03044 Zbl 429 # 03043 • ID 53874

MANGANI, P. & MARCJA, A. [1982] $\aleph_1$-*boolean spectrum and stability (Italian summary)* (**J** 0149) Atti Accad Naz Lincei Fis Mat Nat, Ser 8 72*269-272
⋄ C40 C45 ⋄ REV MR 85f:03033 Zbl 531 # 03013
• ID 37676

MANGANI, P. see Vol. I, II for further entries

MANGIONE, C. [1983] *Stages of the relation between algebra and logic (Italian)* (**P** 3829) Atti Incontri Log Mat (1);1982 Siena 65-80
⋄ C98 ⋄ REV MR 84k:03006 • ID 44819

MANGIONE, C. see Vol. I, II, VI for further entries

MANIA, A. [1981] see ABBATI, M.C.

MANIA, A. see Vol. II for further entries

MANIN, YU.I. [1977] *A course in mathematical logic* (**X** 0811) Springer: Heidelberg & New York xiii+286pp
⋄ B98 C07 D98 E35 E50 F30 G12 ⋄ REV MR 56 # 15345 Zbl 383 # 03002 • REM Translated from Russian • ID 51984

MANIN, YU.I. see Vol. I, II, IV, V, VI for further entries

MANKIEWICZ, P. [1982] see HEINRICH, S.

MANKIEWICZ, P. [1983] see HEINRICH, S.

MANN, A. [1982] *A note on recursively presented and co-recursively presented groups* (**J** 0161) Bull London Math Soc 14*112-118
⋄ C60 D40 D45 ⋄ REV MR 84d:20033 Zbl 483 # 20020
• ID 39712

MANNILA, H. [1982] *Restricted compactness notions in abstract logic and topology* (1111) Preprints, Manuscr., Techn. Reports etc. 31pp
⋄ C95 ⋄ REM University of Helsinki • ID 33518

MANNILA, H. [1983] *A topological characterization of $(\lambda,\mu)^*$-compactness* (J 0073) Ann Pure Appl Logic 25*301-305
⋄ C95 E75 ⋄ REV MR 85m:03027 Zbl 544 # 03014
• ID 40998

MANNILA, H. see Vol. I for further entries

MANSFIELD, R. [1970] *The theory of boolean ultrapowers* (J 0007) Ann Math Logic 2*297-323
⋄ C20 C90 G05 ⋄ REV MR 46 # 47 Zbl 216.294
• ID 08669

MANSFIELD, R. [1972] *Horn classes and reduced direct products* (J 0064) Trans Amer Math Soc 172*279-286
⋄ C20 C52 C90 ⋄ REV MR 47 # 6473 Zbl 261 # 02034
• ID 08671

MANSFIELD, R. [1972] *The completeness theorem for infinitary logic* (J 0036) J Symb Logic 37*31-32
⋄ C75 C90 ⋄ REV MR 53 # 2624 Zbl 244 # 02005
• ID 08672

MANSFIELD, R. [1975] *Omitting types: application to descriptive set theory* (J 0053) Proc Amer Math Soc 47*198-200
⋄ C75 D55 E15 E45 ⋄ REV MR 50 # 6851 Zbl 302 # 02027 • ID 08674

MANSFIELD, R. [1977] *Sheaves and normal submodels* (J 0036) J Symb Logic 42*241-250
⋄ C40 C90 E15 G05 G10 ⋄ REV MR 58 # 21580 Zbl 393 # 03025 • ID 26451

MANSFIELD, R. [1981] *How many slopes in a polygon?* (J 0029) Israel J Math 39*265-272
⋄ C60 ⋄ REV MR 84j:03074 Zbl 473 # 10007 • ID 34666

MANSFIELD, R. see Vol. I, IV, V for further entries

MANSOUX, A. [1976] *Logique sans egalite et (k,p)-quasivalence (English summary)* (J 2313) C R Acad Sci, Paris, Ser A-B 283*A137-A140
⋄ C07 C40 C75 ⋄ REV MR 54 # 7241 Zbl 356 # 02045
• ID 25793

MANSOUX, A. [1976] *Theories de Galois locales (English summary)* (J 2313) C R Acad Sci, Paris, Ser A-B 282*A759-A762
⋄ C07 C50 E07 ⋄ REV MR 53 # 13072 Zbl 339 # 02050
• ID 27225

MARCHINI, C. [1973] *Funtori che conservano e riflettono le formule di Horn* (J 0385) Atti Sem Mat Fis Univ Modena 22*60-65
⋄ C40 C52 G30 ⋄ REV MR 51 # 3251 Zbl 346 # 18001
• ID 17383

MARCHINI, C. [1973] *Funtori rappresentabili e formule di Horn* (J 0385) Atti Sem Mat Fis Univ Modena 22*335-347
⋄ C52 G30 ⋄ REV MR 52 # 7891 Zbl 346 # 18002
• ID 18276

MARCHINI, C. [1982] *Realizations and witnesses for Kripke models* (P 3845) Conf Math Service of Man (2,Proc)(Feriet);1982 Las Palmas 471-476
⋄ C90 F50 G30 ⋄ REV Zbl 538 # 03046 • ID 41484

MARCHINI, C. see Vol. II, V for further entries

MARCJA, A. & MARONGIU, G. [1969] *Estensioni booleane di strutture relazionali (English summary)* (J 0088) Ann Univ Ferrara, NS, Sez 7 14*53-76
⋄ C20 C30 G05 ⋄ REV MR 42 # 50 Zbl 204.311
• ID 08696

MARCJA, A. [1970] *Alcune questioni intorno alle ultraestensioni di strutture realizionali (English summary)* (J 0088) Ann Univ Ferrara, NS, Sez 7 15*1-12
⋄ C20 G05 ⋄ REV MR 44 # 5212 Zbl 216.307 • ID 08699

MARCJA, A. [1970] *Tipi di ultrafiltri in algebre di Boole complete con alcune applicazioni alle ultraestensioni* (J 0319) Matematiche (Sem Mat Catania) 25*290-297
⋄ C20 G05 ⋄ REV MR 46 # 7021 Zbl 222 # 02061
• ID 08698

MARCJA, A. [1972] *Una generalizzazione del concetto di prodotto diretto* (P 0694) Conv Algeb Assoc (Scorza);1970 Roma 363-377
⋄ C20 C30 C40 G05 ⋄ REV MR 50 # 9745 Zbl 243 # 02040 • ID 08700

MARCJA, A. & TULIPANI, S. [1974] *Questioni di teoria dei modelli per linguaggi universali positivi (English summary)* (J 0149) Atti Accad Naz Lincei Fis Mat Nat, Ser 8 56*915-923
⋄ B20 C07 C10 ⋄ REV MR 52 # 10409 Zbl 317 # 02055
• REM Part I. Part II 1976 by Tulipani,S. • ID 21706

MARCJA, A. [1976] see MAKOWSKY, J.A.

MARCJA, A. [1977] see MAKOWSKY, J.A.

MARCJA, A. [1977] *Prodotti booleani e prodotti booleani ridotti* (J 0319) Matematiche (Sem Mat Catania) 32*23-34
⋄ C20 C30 C90 ⋄ REV MR 81m:03041 Zbl 438 # 03042
• ID 55954

MARCJA, A. [1980] see MANGANI, P.

MARCJA, A. [1982] see MANGANI, P.

MARCJA, A. [1982] *An algebraic approach to superstability (Italian summary)* (J 2099) Boll Unione Mat Ital, VI Ser, A 1*71-76
⋄ C40 C45 G05 ⋄ REV MR 83j:03050 Zbl 497 # 03021
• ID 35355

MARCJA, A. [1983] *Categoricita e algebra* (P 3829) Atti Incontri Log Mat (1);1982 Siena 35-64
⋄ C35 C45 C98 ⋄ REV MR 84k:03006 Zbl 527 # 03015
• ID 37510

MARCJA, A. [1984] see LOLLI, G.

MARCJA, A. & TOFFALORI, C. [1984] *On pseudo-$\aleph_0$-categorical theories* (J 0068) Z Math Logik Grundlagen Math 30*533-540
⋄ C35 C40 C45 G05 ⋄ REV MR 86c:03030 Zbl 561 # 03017 • ID 39667

MARCJA, A. & TOFFALORI, C. [1984] *On Cantor-Bendixson spectra containing (1,1) I* (P 2153) Logic Colloq;1983 Aachen 1*331-350
⋄ C35 C40 C45 G05 ⋄ REV Zbl 564 # 03030 • REM Part II 1985 • ID 41769

MARCJA, A. & TOFFALORI, C. [1985] *On Cantor-Bendixon spectra containing (1,1) II* (J 0036) J Symb Logic 50*611-618
⋄ C35 C40 C45 G05 ⋄ REM Part I 1984 • ID 47368

MARCO DE, G. & RICHTER, M.M. [1971] *Rings of continuous functions with values in a non-archimedean ordered field* (J 0144) Rend Sem Mat Univ Padova 45*327-336
⋄ C20 C60 C65 E75 ⋄ REV MR 46 # 655 • ID 33905

MARCONI, D. [1980] *A decision method for the calculus C_1* (P 3006) Brazil Conf Math Log (3);1979 Recife 211-223
⋄ A05 B25 B53 ⋄ REV MR 83b:03026 Zbl 448 # 03014
• ID 56614

MARCONI, D. see Vol. II for further entries

MARCUS, L. [1972] *A minimal prime model with an infinite set of indiscernibles* (J 0029) Israel J Math 11*180-183
⋄ C30 C50 C75 ⋄ REV MR 47 # 8241 Zbl 299 # 02062
• ID 08701

MARCUS, L. [1974] *Minimal models of theories of one function symbol* (J 0029) Israel J Math 18*117-131
⋄ C50 ⋄ REV MR 50 # 4286 Zbl 298 # 02059 • ID 08702

MARCUS, L. [1975] *A theory with only non-homogenous minimal models* (J 0004) Algeb Universalis 5*147
⋄ C50 ⋄ REV MR 52 # 5408 Zbl 316 # 02054 • ID 18277

MARCUS, L. [1975] *A type-open minimal model* (J 0009) Arch Math Logik Grundlagenforsch 17*17-24
⋄ C50 ⋄ REV MR 52 # 10407 Zbl 323 # 02058 • ID 08703

MARCUS, L. [1976] *Elementary separation by elementary expansion* (J 0053) Proc Amer Math Soc 59*144-145
⋄ C15 C50 ⋄ REV MR 55 # 2551 Zbl 345 # 02038
• ID 29790

MARCUS, L. [1976] *The $\prec$-order on submodels* (J 0036) J Symb Logic 41*215-221
⋄ C50 C52 ⋄ REV MR 53 # 7765 Zbl 332 # 02057
• ID 14801

MARCUS, L. [1979] *An expansion of an $\aleph_0$-categorical model* (J 0027) Fund Math 103*183-188
⋄ C35 ⋄ REV MR 81b:03035 Zbl 417 # 03013 • ID 53250

MARCUS, L. [1980] *The number of countable models of a theory of one unary function* (J 0027) Fund Math 108*171-181
⋄ C15 ⋄ REV MR 81m:03039 Zbl 363 # 02055 • ID 76008

MARCZEWSKI, E. [1951] *Sur les congruences et les proprietes positives d'algebres abstraites* (S 0019) Colloq Math (Warsaw) 2*220-228
⋄ C05 C40 ⋄ REV MR 14.347 Zbl 45.1 JSL 32.533
• ID 08724

MARCZEWSKI, E. also published under the name SZPILRAJN-MARCZEWSKI, E. and SZPILRAJN, E.

MARCZEWSKI, E. see Vol. V for further entries

MAREK, W. & ONYSZKIEWICZ, J. [1966] *Representation of partial ordering in cardinals of model of Zermelo-Fraenkel set theory I,II (II: Sets of incomparable powers) (Russian summaries)* (J 0014) Bull Acad Pol Sci, Ser Math Astron Phys 14*357-358,479-481
⋄ C62 E10 E25 E35 E55 ⋄ REV MR 35 # 46
Zbl 143.259 • ID 19298

MAREK, W. & ONYSZKIEWICZ, J. [1967] *Some results in the foundations of the set theory* (J 0014) Bull Acad Pol Sci, Ser Math Astron Phys 15*51-52
⋄ C62 E25 E35 E45 E50 ⋄ REV MR 38 # 5614
Zbl 157.26 • ID 28474

MAREK, W. & ONYSZKIEWICZ, J. [1970] *The existence of constructible elements in some classes of models (Russian summary)* (J 0014) Bull Acad Pol Sci, Ser Math Astron Phys 18*43-46
⋄ C52 E45 ⋄ REV MR 42 # 60 Zbl 197.277 • ID 08741

MAREK, W. & ONYSZKIEWICZ, J. [1972] *Elements of logic and foundations of mathematics in problems (Polish)* (X 1034) PWN: Warsaw 278pp
• TRANSL [1982] (X 0835) Reidel: Dordrecht viii + 276pp
⋄ B98 C98 D98 E98 ⋄ REV MR 84h:03001
Zbl 288 # 02001 Zbl 574 # 03001 • ID 34216

MAREK, W. [1973] *Consistance d'une hypothese de Fraisse sur la definissabilite dans un langage du second ordre* (J 2313) C R Acad Sci, Paris, Ser A-B 276*A1147-A1150
⋄ B15 C65 E10 E35 E45 ⋄ REV MR 52 # 2879a
Zbl 259 # 02048 • ID 17662

MAREK, W. [1973] *Observations concerning elementary extensions of ω-models II* (J 0036) J Symb Logic 38*227-231
⋄ C62 E45 ⋄ REV MR 49 # 2381 Zbl 287 # 02037 • REM
Part I 1974 by Mostowski,Andrzej • ID 08752

MAREK, W. [1973] *On the metamathematics of impredicative set theory* (J 0202) Diss Math (Warsaw) 98*40pp
⋄ C62 E35 E47 E70 ⋄ REV MR 48 # 107
Zbl 273 # 02046 • ID 08748

MAREK, W. & SREBRNY, M. [1973] *On transitive models for fragments of set theory (Russian summary)* (J 0014) Bull Acad Pol Sci, Ser Math Astron Phys 21*389-392
⋄ C62 E30 E45 ⋄ REV MR 55 # 7773 Zbl 262 # 02054
• ID 08747

MAREK, W. [1973] *Sur la consistance d'une hypothese de Fraisse sur la definissabilite dans un langage du second ordre* (J 2313) C R Acad Sci, Paris, Ser A-B 276*A1169-A1172
⋄ B15 C65 E10 E35 E45 ⋄ REV MR 52 # 2879b
Zbl 257 # 02050 • ID 17663

MAREK, W. & SREBRNY, M. [1974] *Gaps in the constructible universe* (J 0007) Ann Math Logic 6*359-394
⋄ C62 E10 E30 E45 ⋄ REV MR 51 # 10102
Zbl 279 # 02049 • ID 08755

MAREK, W. & ZBIERSKI, P. [1974] *On the size of the family of β-models* (J 0014) Bull Acad Pol Sci, Ser Math Astron Phys 22*779-781
⋄ C62 E15 E45 ⋄ REV MR 50 # 6852 Zbl 289 # 02046
• ID 08753

MAREK, W. [1974] see APT, K.R.

MAREK, W. [1974] *Stable sets, a characterization of β_2-models of full second order arithmetic and some related facts* (J 0027) Fund Math 82*175-189
⋄ C62 E45 E47 ⋄ REV MR 51 # 10097 Zbl 294 # 02031
• ID 08757

MAREK, W. & SREBRNY, M. [1975] *No minimal transitive model of Z^-* (J 0068) Z Math Logik Grundlagen Math 21*225-228
⋄ C62 E30 E70 F35 ⋄ REV MR 52 # 2884
Zbl 315 # 02055 • ID 08758

MAREK, W. & MOSTOWSKI, ANDRZEJ [1975] *On extendability of models of ZF set theory to the models of Kelley-Morse theory of classes* (P 1442) ⊢ ISILC Logic Conf;1974 Kiel 460-542
⋄ C62 E30 E45 E70 ⋄ REV MR 53 # 7782 Zbl 339 # 02053 • ID 23020

MAREK, W. & SREBRNY, M. [1976] *Urelements and extendability* (P 1476) Set Th & Hierarch Th (2) (Mostowski);1975 Bierutowice 203-219
⋄ C62 E30 E70 F25 F30 F35 ⋄ REV MR 56 # 15415 Zbl 358 # 02067 • ID 24412

MAREK, W. & ZBIERSKI, P. [1977] *On the number of models of Kelley-Morse theory of classes with prescribed V* (P 1639) Set Th & Hierarch Th (1);1974 Karpacz 1*13-21
⋄ C62 E70 ⋄ REV MR 58 # 10452 Zbl 415 # 03045
• ID 53147

MAREK, W. [1977] see KRAWCZYK, A.

MAREK, W. [1977] *On the standard part problem* (P 1639) Set Th & Hierarch Th (1);1974 Karpacz 1*51-55
⋄ C62 E45 ⋄ REV MR 58 # 5226 Zbl 404 # 03039
• ID 54826

MAREK, W. [1978] *ω-models of second order arithmetic and admissible sets* (J 0027) Fund Math 98*103-120
⋄ C62 E30 E45 F35 ⋄ REV MR 57 # 16051 Zbl 385 # 03030 • ID 29210

MAREK, W. & NYBERG, A.M. [1978] *Extendability of ZF models in the von Neumann hierarchy to models of KM theory of classes* (P 1628) Generalized Recursion Th (2);1977 Oslo 271-282
⋄ C62 C70 E30 E70 ⋄ REV MR 80f:03034 Zbl 453 # 03047 • ID 28323

MAREK, W. & ZBIERSKI, P. [1978] *On a class of models of the N-th order arithmetic* (P 1864) Higher Set Th;1977 Oberwolfach 669*361-374
⋄ C62 ⋄ REV MR 81i:03104 Zbl 383 # 03046 • ID 31572

MAREK, W. [1978] *Some comments on the paper by Artigue, Isambert, Perrin and Zalc: "Some remarks on bicommutability"* (J 0027) Fund Math 101*227-228
⋄ C62 E30 E45 F25 F35 ⋄ REV MR 80j:03019b Zbl 401 # 03025 • REM The paper was published ibid 101*207-226 • ID 76016

MAREK, W. [1979] see LOLLI, G.

MAREK, W. & ZBIERSKI, P. [1980] *On the number of models of the Kelley-Morse theory of classes* (J 0027) Fund Math 109*169-173
⋄ C62 E70 ⋄ REV MR 82b:03069 Zbl 464 # 03048
• ID 76030

MAREK, W. [1981] *On cores of iterated ultrapowers* (P 2614) Open Days in Model Th & Set Th;1981 Jadwisin 175-180
⋄ C20 E45 E55 ⋄ ID 33724

MAREK, W. [1981] see GUZICKI, W.

MAREK, W. & PRISCO DI, C.A. [1982] *Models closed under projective operations (Russian summary)* (J 3293) Bull Acad Pol Sci, Ser Math 30*15-20
⋄ C62 E15 E45 ⋄ REV MR 83j:03085 Zbl 513 # 03027
• ID 35374

MAREK, W. & RASMUSSEN, K. [1983] *Spectrum of L* (J 0202) Diss Math (Warsaw) 211*38pp
⋄ C55 C62 E45 ⋄ REV MR 84j:03098 Zbl 535 # 03027
• ID 34688

MAREK, W. see Vol. II, IV, V for further entries

MARICA, J.G. [1960] see BRYANT, S.J.

MARINI, D. & MIGLIOLI, P.A. & ORNAGHI, M. [1975] *First order logic as a tool to solve and classify problems* (P 0784) GI Jahrestag (5);1975 Dortmund 669-679
⋄ B10 B25 F50 ⋄ REV MR 53 # 5282 Zbl 329 # 68079
• ID 22915

MARKER, D. [1982] *Degrees of models of true arithmetic* (P 3708) Herbrand Symp Logic Colloq;1981 Marseille 233-242
⋄ C57 C62 D30 H15 ⋄ REV MR 85i:03109 Zbl 522 # 03058 • ID 37802

MARKER, D. [1984] *A model theoretic proof of Feferman's preservation theorem* (J 0047) Notre Dame J Formal Log 25*213-216
⋄ C40 ⋄ REV MR 85i:03101 Zbl 566 # 03021 • ID 42568

MARKER, D. [1984] see MACINTYRE, A.

MARKI, L. [1979] see ANDREKA, H.

MARKOV, A.A. [1974] *On the language $Я_0$ (Russian)* (J 0023) Dokl Akad Nauk SSSR 214*40-43
• TRANSL [1974] (J 0062) Sov Math, Dokl 15*38-42
⋄ B25 F50 F65 ⋄ REV MR 49 # 2287 Zbl 308 # 02032
• ID 63716

MARKOV, A.A. [1974] *The completeness of the classical predicate calculus in constructive logic (Russian)* (J 0023) Dokl Akad Nauk SSSR 215*266-269
• TRANSL [1974] (J 0062) Sov Math, Dokl 15*476-481
⋄ C07 F50 ⋄ REV MR 49 # 2294 Zbl 306 # 02046
• ID 63722

MARKOV, A.A. see Vol. I, II, IV, V, VI for further entries

MARKOVIC, Z. [1977] *Reduced products of saturated intuitionistic theories* (J 0400) Publ Inst Math, NS (Belgrade) 21(35)*131-133
⋄ C20 C90 F50 ⋄ REV MR 57 # 16048 Zbl 367 # 02009
• ID 51169

MARKOVIC, Z. [1979] *An intuitionistic omitting types theorem* (J 0400) Publ Inst Math, NS (Belgrade) 26(40)*167-169
⋄ C07 C90 F50 ⋄ REV MR 82g:03103 Zbl 442 # 03027
• ID 56384

MARKOVIC, Z. [1983] *On reduced products of Kripke models* (J 0400) Publ Inst Math, NS (Belgrade) 34(48)*117-120
⋄ C20 C90 ⋄ REV MR 86g:03101 • ID 44692

MARKOVIC, Z. [1983] *Some preservation results for classical and intuitionistic satisfiability in Kripke models* (J 0047) Notre Dame J Formal Log 24*395-398
⋄ B20 C25 C90 F50 ⋄ REV MR 85a:03016 Zbl 487 # 03015 • ID 37386

MARKOVIC, Z. [1984] *Kripke models for intuitionistic theories with decidable atomic formulas* (J 0400) Publ Inst Math, NS (Belgrade) 36(50)*3-7
⋄ C90 F50 ⋄ REV MR 86h:03105 • ID 45080

MARKOVIC, Z. see Vol. I for further entries

MARKOVSKI, S. [1980] *On quasivarieties of generalized subalgebras* (P 3355) Algeb Conf (1);1980 Skopje 125-129
⋄ C05 C52 ⋄ REV MR 84j:08006b Zbl 482 # 03010
• ID 36841

MARKOVSKI, S. see Vol. IV for further entries

MARKUSZ, Z. [1983] *On first order many-sorted logic* (J 2845) Tanulmanyok 151*85pp
⋄ B10 C07 ⋄ REV MR 86e:03009 • ID 44560

MARKWALD, W. [1954] *Zur Theorie der konstruktiven Wohlordnungen* (J 0043) Math Ann 127*135-149
⋄ C80 D55 F15 ⋄ REV MR 15.771 Zbl 56.47 JSL 20.283 • ID 08779

MARKWALD, W. see Vol. I, II, IV for further entries

MARONGIU, G. [1969] see MARCJA, A.

MARONGIU, G. [1982] *Alcuni risultati sulla $\aleph_0$-categoricita per strutture con un numero finito di relazioni di equivalenza (English summary)* (J 2100) Boll Unione Mat Ital, VI Ser, B 1*809-825
⋄ C35 E07 ⋄ REV MR 84c:03058 Zbl 511 # 03012 • ID 34002

MARONGIU, G. [1983] *Definibilita al primo ordine delle congruenze principali e delle congruenze compatte* (J 2099) Boll Unione Mat Ital, VI Ser, A 2*47-53 • ERR/ADD ibid 4*337-338
⋄ C05 C20 C52 ⋄ REV MR 85b:08002 Zbl 521 # 08004 Zbl 566 # 08001 • ID 37464

MARONGIU, G. [1985] *Alcune osservazioni sulla $\aleph_0$-categoricita degli ordini circolari* (J 2100) Boll Unione Mat Ital, VI Ser, B 4*883-900
⋄ C35 C65 ⋄ ID 48413

MARONGIU, G. see Vol. VI for further entries

MARTIN, D.A. & MITCHELL, W.J. [1979] *On the ultrafilter of closed, unbounded sets* (J 0036) J Symb Logic 44*503-506
⋄ C20 C62 E05 E35 E55 E60 ⋄ REV MR 81a:03051 Zbl 439 # 03039 • ID 56030

MARTIN, D.A. see Vol. IV, V for further entries

MARTIN-LOEF, P. [1972] *About models for intuitionistic type theories and the notion of definitional equality* (1111) Preprints, Manuscr., Techn. Reports etc.
⋄ C90 F50 ⋄ REM Univ. Stockholm, Dep. Math., Report No. 4 • ID 21229

MARTIN-LOEF, P. [1972] *Infinite terms and a system of natural deduction* (J 0020) Compos Math 24*93-103
⋄ C75 F05 F50 ⋄ REV MR 46 # 20 Zbl 237 # 02006 • ID 08833

MARTIN-LOEF, P. see Vol. I, IV, V, VI for further entries

MARTINEZ, JUAN CARLOS [1984] *Accessible sets and $(L_{\omega_1\omega})_t$-equivalence for T_3 spaces* (J 0036) J Symb Logic 49*961-967
⋄ C65 C75 C80 C90 ⋄ REV MR 86e:03038 Zbl 576 # 03024 • ID 42479

MARTINEZ, JUAN CARLOS [1985] see FLUM, J.

MART'YANOV, V.I. [1972] *Decidability of the theories of some classes of abelian groups with an automorphism predicate and a group predicate (Russian)* (C 3549) Algebra, Vyp 1 (Irkutsk) 55-67
⋄ B25 C60 ⋄ REV MR 57 # 9514 • ID 76102

MART'YANOV, V.I. [1973] *Decidability of the universal element-subgroup theory of the integers (Russian)* (C 1443) Algebra, Vyp 2 (Irkutsk) 114-121
⋄ B25 ⋄ REV MR 53 # 5296 • ID 22931

MART'YANOV, V.I. [1973] see KOKORIN, A.I.

MART'YANOV, V.I. [1975] *The theory of abelian groups with predicates specifying a subgroup, and with endomorphism operations (Russian)* (J 0003) Algebra i Logika 14*536-542,606
• TRANSL [1975] (J 0069) Algeb and Log 14*330-334
⋄ C60 ⋄ REV MR 55 # 12513 Zbl 347 # 02036 • ID 26025

MART'YANOV, V.I. [1976] *Decidability of the theory of closed torsion free abelian groups of finite rank with predicates for addition and automorphism (Russian)* (C 2555) Algeb Sistemy (Irkutsk) 94-106
⋄ B25 C60 ⋄ ID 90341

MART'YANOV, V.I. [1977] *Extended universal theories of the integers (Russian)* (J 0003) Algebra i Logika 16*588-602,624
• TRANSL [1977] (J 0069) Algeb and Log 16*395-405
⋄ B25 D35 F30 ⋄ REV MR 58 # 27421 Zbl 394 # 03038 • ID 29230

MART'YANOV, V.I. [1983] see BELYAKOV, E.B.

MART'YANOV, V.I. see Vol. I, IV for further entries

MARZI LI, E.M. [1978] *An elementary class of hypergroups (Italian)* (P 4243) Conf Binary Syst & Appl;1978 Taormina 175-182
⋄ C60 ⋄ ID 45953

MASHURYAN, A.S. [1974] *Recursive definitions on induction models (Russian) (Armenian summary)* (J 0346) Dokl Akad Nauk Armyan SSR 59*199-204
⋄ C13 D20 ⋄ REV MR 51 # 2899 Zbl 309 # 02060 • ID 17343

MASHURYAN, A.S. [1979] *On a class of primitive recursive functions (Russian)* (J 0346) Dokl Akad Nauk Armyan SSR 69*209-212
⋄ C13 D20 ⋄ REV Zbl 431 # 03028 • ID 53934

MASHURYAN, A.S. [1980] *On the class of primitive recursive functions definable on finite models (Russian) (Armenian summary)* (J 0346) Dokl Akad Nauk Armyan SSR 71*209-211
⋄ C13 D20 ⋄ REV MR 82i:03052 Zbl 494 # 03026 • ID 76106

MASHURYAN, A.S. see Vol. II for further entries

MASLOV, S.YU. [1966] *Application of the inverse method for establishing deducibility to the theory of decidable fragments in the classical predicate calculus (Russian)* (J 0023) Dokl Akad Nauk SSSR 171*1282-1285
• TRANSL [1966] (J 0062) Sov Math, Dokl 7*1653-1657
⋄ B20 B25 ⋄ REV MR 35 # 25 Zbl 162.21 • ID 19334

MASLOV, S.YU. & OREVKOV, V.P. [1972] *Decidable classes that reduce to a single quantifier class (Russian)* (S 0066) Tr Mat Inst Steklov 121*57-66,165
• TRANSL [1972] (S 0055) Proc Steklov Inst Math 121*61-72
⋄ B20 B25 ⋄ REV MR 50 # 12678 Zbl 301 # 02041 • ID 08848

MASLOV, S.YU. & NORGELA, S.A. [1977] *Herbrand strategies and the "greater deducibility" relation (Russian) (English summary)* (S 0228) Zap Nauch Sem Leningrad Otd Mat Inst Steklov 68*51-61,144
• TRANSL [1981] (J 1531) J Sov Math 15*28-33
⋄ B25 B35 F05 ⋄ REV MR 58 # 21504 Zbl 358 # 02030 • ID 32672

MASLOV, S.YU. see Vol. I, II, IV, VI for further entries

MATEESCU, C. & POPESCU, D. [1984] *Ultraproducts and big Cohen-Macaulay modules* (**J** 0197) Stud Cercet Mat Acad Romana 36*424-428
⋄ C20 C60 ⋄ ID 48025

MATERNA, P. [1969] *Identity, equivalence and isomorphism of problems* (**J** 0036) J Symb Logic 34*24-34
⋄ B25 ⋄ REV MR 40 # 5421 Zbl 209.8 • ID 08870

MATERNA, P. see Vol. I, II for further entries

MATHIAS, A.R.D. [1968] see KOPPERMAN, R.D.

MATHIAS, A.R.D. [1972] *Solution of problems of Choquet and Puritz* (**P** 2080) Conf Math Log;1970 London 204-210
⋄ C20 D55 E05 E15 ⋄ REV MR 51 # 166 Zbl 232 # 02036 • ID 17477

MATHIAS, A.R.D. see Vol. I, IV, V for further entries

MATSUMOTO, K. & OHNISHI, M. [1957] *Gentzen method in modal calculi. I* (**J** 1770) Osaka Math J 9*113-130
⋄ B25 B45 ⋄ REV MR 22 # 9441a Zbl 80.7 JSL 40.466
• REM Part II 1959 • ID 16393

MATSUMOTO, K. & OHNISHI, M. [1959] *Gentzen method in modal calculi II* (**J** 1770) Osaka Math J 11*115-120
⋄ B25 B45 ⋄ REV MR 22 # 9441b Zbl 89.6 JSL 40.466
• REM Part I 1957 • ID 16395

MATSUMOTO, K. [1960] *Decision procedure for modal sentential calculus S3* (**J** 1770) Osaka Math J 12*167-175
⋄ B25 B45 ⋄ REV MR 25 # 2954 Zbl 99.8 JSL 40.468
• ID 08893

MATSUMOTO, K. see Vol. I, II, IV, VI for further entries

MAURER, C. [1975] see LAWVERE, F.W.

MAURER, C. see Vol. V for further entries

MAURI, G. [1979] see BERTONI, A.

MAURI, G. [1980] see BERTONI, A.

MAURI, G. [1981] see BERTONI, A.

MAURI, G. [1983] see BERTONI, A.

MAURI, G. see Vol. IV, VI for further entries

MAURY, G. [1961] *La condition "integralement clos" dans quelques structures algebriques* (**J** 0204) Ann Sci Ecole Norm Sup, Ser 3 78*31-100
⋄ C05 C60 ⋄ REV MR 25 # 5019 • ID 08909

MAY, W. [1972] *Multiplicative groups of fields* (**J** 3240) Proc London Math Soc, Ser 3 24*295-306
⋄ C60 ⋄ REV MR 45 # 3560 Zbl 232 # 20075 • ID 08917

MAYOH, B.H. [1967] *Groups and semigroups with solvable word problems* (**J** 0053) Proc Amer Math Soc 18*1038-1039
⋄ C57 C60 D40 ⋄ REV MR 37 # 4151 Zbl 153.349 JSL 36.541 • ID 08925

MAYOH, B.H. see Vol. I, II, IV, VI for further entries

MAZOYER, J. [1975] *Sur les theories categoriques finiment axiomatisables (English summary)* (**J** 2313) C R Acad Sci, Paris, Ser A-B 281*A403-A406
⋄ C10 C35 C45 ⋄ REV MR 52 # 2864 Zbl 364 # 02028
• ID 17650

McALOON, K. [1973] see KRIVINE, J.-L.

McALOON, K. [1974] *On the sequence of models HOD_n* (**J** 0027) Fund Math 82*85-93
⋄ C62 E35 E45 ⋄ REV MR 50 # 98 Zbl 295 # 02033
• ID 08939

McALOON, K. [1975] *Applications alternees de theoremes d'incompletude et des theoremes de completude (English summary)* (**J** 2313) C R Acad Sci, Paris, Ser A-B 280*A849-A852
⋄ C62 E35 F30 ⋄ REV MR 51 # 5295 Zbl 307 # 02036
• ID 17403

McALOON, K. [1975] *Formules de Rosser pour ZF (English summary)* (**J** 2313) C R Acad Sci, Paris, Ser A-B 281*A669-A672
⋄ C62 C70 E30 E35 E45 F25 ⋄ REV MR 52 # 5417 Zbl 333 # 02046 • ID 18286

McALOON, K. [1977] *Consistency statements and number theories* (**P** 1729) Colloq Int Log;1975 Clermont-Ferrand 199-207
⋄ C62 F30 ⋄ REV MR 80b:03110 Zbl 435 # 03037
• ID 55797

McALOON, K. [1978] *Completeness theorems, incompleteness theorems and models of arithmetic* (**J** 0064) Trans Amer Math Soc 239*253-277
⋄ C25 C62 F30 ⋄ REV MR 81e:03066 Zbl 388 # 03028
• ID 52282

McALOON, K. [1978] *Diagonal methods and strong cuts in models of arithmetic* (**P** 1897) Logic Colloq;1977 Wroclaw 171-181
⋄ C57 C62 D80 F30 ⋄ REV MR 80k:03068 Zbl 458 # 03021 • ID 54427

McALOON, K. [1979] *Formes combinatoires du theoreme d'incompletude (d'apres J. Paris et d'autres)* (**S** 1567) Semin Bourbaki Exp.521*263-276
⋄ C62 ⋄ REV MR 83e:03051 Zbl 416 # 03054 • REM Springer: Heidelberg & New York; Lecture Notes Math 710
• ID 53220

McALOON, K. [1980] *Les rapports entre la methode des indicatrices et la methode de Goedel pour obtenir des resultats d'independance* (**J** 1620) Asterisque 73*31-39
⋄ C62 E35 F30 ⋄ REV MR 82g:03098a Zbl 462 # 03016 JSL 48.483 • ID 54523

McALOON, K. (ED.) [1980] *Modeles de l'arithmetique* (**X** 2244) Soc Math France: Paris 155pp
⋄ C62 C97 F30 ⋄ REV MR 81j:03006 Zbl 424 # 00006
• ID 70045

McALOON, K. [1980] *Progressions transfinies de theories axiomatiques, formes combinatoires du theoreme d'incompletude et fonctions recursives a croissance rapide* (**J** 1620) Asterisque 73*41-58
⋄ C62 D20 F15 F30 ⋄ REV MR 82g:03098b Zbl 462 # 03017 JSL 48.483 • ID 54524

McALOON, K. [1981] see KIRBY, L.A.S.

McALOON, K. & RESSAYRE, J.-P. [1981] *Les methodes de Kirby-Paris et la theorie des ensembles* (**P** 3404) Model Th & Arith;1979/80 Paris 154-184
⋄ C62 E05 E30 E55 F30 ⋄ REV MR 84c:03067 Zbl 498 # 03041 • ID 34009

McALOON, K. [1981] see BERLINE, C.

McALOON, K. [1982] see CEGIELSKI, P.

MCALOON, K. [1982] *On the complexity of models of arithmetic* (J 0036) J Symb Logic 47*403-415
⋄ C57 C62 ⋄ REV MR 84h:03084   Zbl 519 # 03056
• ID 34279

MCALOON, K. [1983] see CLOTE, P.

MCALOON, K. see Vol. I, IV, V, VI for further entries

MCCALL, S. [1962] *A simple decision procedure for one-variable implication/negation formulae in intuitionistic logic* (J 0047) Notre Dame J Formal Log 3*120-122
⋄ B25 F50 ⋄ REV MR 26 # 6048   Zbl 118.248 JSL 29.212
• ID 08942

MCCALL, S. [1967] *Connexive class logic* (J 0036) J Symb Logic 32*83-90
⋄ B25 E70 G05 ⋄ REV MR 36 # 41   Zbl 174.8 • ID 08949

MCCALL, S. see Vol. II, VI for further entries

MCCALLUM, S. [1982] see ARNON, D.S.

MCCARTHY, T. [1985] *Abstraction and definability in semantically closed structures* (J 0122) J Philos Logic 14*255-266
⋄ A05 C40 C62 C90 ⋄ ID 42659

MCCARTHY, T. see Vol. I for further entries

MCCARTY, C. [1985] see SCOTT, D.S.

MCCARTY, C. see Vol. VI for further entries

MCKAY, C.G. [1968] *The decidability of certain intermediate propositional logics* (J 0036) J Symb Logic 33*258-264
⋄ B25 B55 ⋄ REV MR 39 # 1300   Zbl 175.271 • ID 08986

MCKAY, C.G. [1971] *A class of decidable intermediate propositional logics* (J 0036) J Symb Logic 36*127-128
⋄ B25 B55 ⋄ REV MR 43 # 7297   Zbl 216.292 • ID 08988

MCKAY, C.G. see Vol. II, VI for further entries

MCKEE, T.A. [1975] *A formal analogy between Baer subplanes and their complements* (J 0395) J Geom 6*97-104
⋄ C65 ⋄ REV MR 52 # 9072   Zbl 301 # 50009 • ID 30761

MCKEE, T.A. [1975] *Infinitary logic and topological homeomorphisms* (J 0068) Z Math Logik Grundlagen Math 21*405-408
⋄ C65 C75 C80 C90 ⋄ REV MR 52 # 7845
Zbl 341 # 02013 • ID 08989

MCKEE, T.A. [1976] *Sentences preserved between equivalent topological bases* (J 0068) Z Math Logik Grundlagen Math 22*79-84
⋄ C65 C80 C90 ⋄ REV MR 54 # 4970   Zbl 334 # 02026
• ID 18488

MCKEE, T.A. [1978] *Forbidden subgraphs in terms of forbidden quantifiers* (J 0047) Notre Dame J Formal Log 19*186-188
⋄ C52 C65 ⋄ REV MR 58 # 16269   Zbl 351 # 02036
• ID 27105

MCKEE, T.A. [1980] *Generalized equivalence and the foundations of quasigroups* (J 0047) Notre Dame J Formal Log 21*135-140
⋄ C60 ⋄ REV MR 81b:03007   Zbl 394 # 03035 • ID 53201

MCKEE, T.A. [1980] *Generalized equivalence and the phraseology of configuration theorems* (J 0047) Notre Dame J Formal Log 21*141-147
⋄ C65 ⋄ REV MR 81b:03008   Zbl 394 # 03034 • ID 53200

MCKEE, T.A. [1980] *Monadic characterizations in nonstandard topology* (J 0068) Z Math Logik Grundlagen Math 26*395-397
⋄ C10 H05 H20 ⋄ REV MR 81m:03074   Zbl 449 # 54046
• ID 76216

MCKEE, T.A. [1984] *Relative expressiveness of the edge/adjacency language for graph theory* (J 0027) Fund Math 123*163-167
⋄ C65 ⋄ REV MR 85k:05094   Zbl 556 # 05060 • ID 44038

MCKEE, T.A. see Vol. I, II for further entries

MCKENNA, K. [1975] *New facts about Hilbert's seventeenth problem* (C 0782) Model Th & Algeb (A. Robinson) 220-230
⋄ C60 ⋄ REV MR 53 # 5547   Zbl 357 # 12019 • ID 22955

MCKENNA, K. [1980] *Some diophantine Nullstellensaetze* (P 2625) Model Th of Algeb & Arithm;1979 Karpacz 228-247
⋄ C60 ⋄ REV MR 83i:12023   Zbl 452 # 13002 • ID 54119

MCKENNA, K. [1983] see DRIES VAN DEN, L.

MCKENZIE, R. [1968] *On finite groupoids and $\mathfrak{K}$-prime algebras* (J 0064) Trans Amer Math Soc 133*115-129
⋄ C05 C13 ⋄ REV MR 37 # 1293   Zbl 215.62 • ID 08990

MCKENZIE, R. [1970] *Equational bases for lattice theories* (J 0132) Math Scand 27*24-38
⋄ C05 G10 ⋄ REV MR 43 # 118   Zbl 307 # 08001
• ID 08992

MCKENZIE, R. [1971] *$\aleph_1$-incompactness of Z* (S 0019) Colloq Math (Warsaw) 23*199-202,325
⋄ C05 C60 ⋄ REV MR 46 # 3294   Zbl 253 # 08001
JSL 40.88 • ID 08995

MCKENZIE, R. [1971] *A note on subgroups of infinite symmetric groups* (J 0028) Indag Math 33*53-58
⋄ C60 ⋄ REV MR 45 # 3545 • ID 08999

MCKENZIE, R. [1971] *Cardinal multiplication of structures with a reflexive relation* (J 0027) Fund Math 70*59-101
⋄ C05 C30 C52 ⋄ REV MR 43 # 6150   Zbl 228 # 08002
• ID 08998

MCKENZIE, R. [1971] *Definability in lattices of equational theories* (J 0007) Ann Math Logic 3*197-237
⋄ C05 C40 G10 ⋄ REV MR 43 # 6069   Zbl 328 # 02038
JSL 39.601 • ID 08996

MCKENZIE, R. [1971] *On elementary types of symmetric groups* (J 0004) Algeb Universalis 1*13-20
⋄ C60 E10 E47 ⋄ REV MR 44 # 3852   Zbl 232 # 20057
• ID 09000

MCKENZIE, R. [1971] *On semigroups whose proper subsemigroups have lesser power* (J 0004) Algeb Universalis 1*21-25
⋄ C05 E50 E75 ⋄ REV MR 45 # 2060   Zbl 234 # 20030
• ID 08993

MCKENZIE, R. [1972] *A method for obtaining refinement theorems with an application to direct products of semigroups* (J 0004) Algeb Universalis 2*324-338
⋄ C05 C30 ⋄ REV MR 49 # 2493   Zbl 273 # 08003
• ID 09005

MCKENZIE, R. [1972] *Equational bases and nonmodular lattice varieties* (J 0064) Trans Amer Math Soc 174*1-43
⋄ B25 C05 G10 ⋄ REV MR 47 # 1696   Zbl 265 # 08006
• ID 09004

MCKENZIE, R. [1973] *Some unsolvable problems between lattice theory and equational logic* (P 0457) Lattice Th;1973 Houston 564-573
⋄ C05 C98 G10 ⋄ REV MR 53 # 2771   Zbl 329 # 06002
• ID 21639

MCKENZIE, R. & SHELAH, S. [1974] *The cardinals of simple models for universal theories* (P 0610) Tarski Symp;1971 Berkeley 53-74
⋄ C05 C50 C55 E05 ⋄ REV MR 50 # 12711   Zbl 316 # 02057 • ID 09006

MCKENZIE, R. [1975] *On spectra, and the negative solution of the decision problem for identities having a finite nontrivial model* (J 0036) J Symb Logic 40*186-196
⋄ B20 C05 C13 D35 ⋄ REV MR 51 # 12499   Zbl 316 # 02052 • ID 09009

MCKENZIE, R. [1978] *A finite algebra $\underline{A}$ with $SP(\underline{A})$ not elementary* (J 0004) Algeb Universalis 8*5-7
⋄ C05 C52 ⋄ REV MR 58 # 10665   Zbl 371 # 08005
• ID 31946

MCKENZIE, R. [1978] see FABER, V.

MCKENZIE, R. [1978] *Para primal varieties: A study of finite axiomatizability and definable principal congruences in locally finite varieties* (J 0004) Algeb Universalis 8*336-348
⋄ C05 ⋄ REV MR 57 # 9634   Zbl 383 # 08008 • ID 31945

MCKENZIE, R. [1981] see BURRIS, S.

MCKENZIE, R. [1981] *Residually small varieties of semigroups* (J 0004) Algeb Universalis 13*171-201
⋄ C05 C13 ⋄ REV MR 82k:20097   Zbl 475 # 20051
• ID 41678

MCKENZIE, R. [1982] see BALDWIN, JOHN T.

MCKENZIE, R. [1982] *Subdirect powers of non-abelian groups* (J 1447) Houston J Math 8*389-399
⋄ C30 C60 ⋄ REV MR 84d:20027   Zbl 514 # 20024
• ID 37443

MCKENZIE, R. [1983] *Finite forbidden lattices* (P 3841) Universal Algeb & Lattice Th (4);1982 Puebla 176-205
⋄ C13 G10 ⋄ REV MR 85b:06006   Zbl 523 # 06012
• ID 43084

MCKENZIE, R. [1983] *The number of non-isomorphic models in quasi-varieties of semigroups* (J 0004) Algeb Universalis 16*195-203
⋄ C05 C52 ⋄ REV MR 84f:08008   Zbl 525 # 20042
• ID 37487

MCKENZIE, R. [1984] *A new product of algebras and a type reduction theorem* (J 0004) Algeb Universalis 18*29-69
⋄ C05 C30 ⋄ REV MR 86h:08011 • ID 49882

MCKENZIE, R. [1985] *The structure of finite algebras* (P 4646) Atti Incontri Log Mat (2);1983/84 Siena 561-584
⋄ C05 C13 C98 ⋄ ID 49905

MCKENZIE, R. see Vol. I, IV, V, VI for further entries

MCKINSEY, J.C.C. [1941] *A solution of the decision problem for the Lewis systems S2 and S4, with an application to topology* (J 0036) J Symb Logic 6*117-134
⋄ B25 B45 C65 ⋄ REV MR 3.290   JSL 7.118 • ID 09020

MCKINSEY, J.C.C. [1943] *The decision problem for some classes of sentences without quantifiers* (J 0036) J Symb Logic 8*61-76
⋄ B20 B25 C05 C60 G10 ⋄ REV MR 5.85   JSL 9.30
• ID 09021

MCKINSEY, J.C.C. & TARSKI, A. [1944] *The algebra of topology* (J 0120) Ann of Math, Ser 2 45*141-191
⋄ C05 C50 C65 G05 G25 ⋄ REV MR 5.211   Zbl 60.62   JSL 9.96 • ID 09023

MCKINSEY, J.C.C. & TARSKI, A. [1946] *On closed elements in closure algebras* (J 0120) Ann of Math, Ser 2 47*122-162
⋄ B25 C05 C65 G05 G25 ⋄ REV MR 7.359   Zbl 60.62   JSL 11.83 • ID 09026

MCKINSEY, J.C.C. see Vol. I, II, V, VI for further entries

MCLAUGHLIN, T.G. [1973] *A non-enumerability theorem for infinite classes of finite structures* (P 0678) Word Probl: Decis & Burnside Probl in Group Th;1969 Irvine 479-481
⋄ C13 D25 D55 ⋄ REV MR 55 * 5418   Zbl 267 # 02031   JSL 41.786 • ID 29894

MCLAUGHLIN, T.G. see Vol. I, IV, V for further entries

MCNAUGHTON, R. [1954] *A non-standard truth definition* (J 0053) Proc Amer Math Soc 5*505-509
⋄ B28 C62 E30 H10 H20 ⋄ REV MR 15.925   Zbl 56.12
• ID 09068

MCNAUGHTON, R. [1965] *Undefinability of addition from one unary operator* (J 0064) Trans Amer Math Soc 117*329-337
⋄ B28 C10 ⋄ REV MR 31 # 1191   Zbl 143.250   JSL 31.270
• ID 09073

MCNAUGHTON, R. see Vol. I, II, IV, V, VI for further entries

MCNULTY, G.F. [1975] see JONSSON, B.

MCNULTY, G.F. [1977] see MAKKAI, M.

MCNULTY, G.F. [1980] *Classes which generate the variety of all lattice-ordered groups* (P 2983) Ordered Groups;1978 Boise 135-140
⋄ C05 C52 C60 ⋄ REV MR 82c:06031   Zbl 448 # 06017
• ID 56656

MCNULTY, G.F. [1984] see KIERSTEAD, H.A.

MCNULTY, G.F. see Vol. I, IV, V for further entries

MCROBBIE, M.A. & MEYER, R.K. & THISTLEWAITE, P.B. [1980] *A mechanized decision procedure for non-classical logics: The program KRIPKE* (J 0387) Bull Sect Logic, Pol Acad Sci 9*189-192
⋄ B25 B35 ⋄ REV MR 82b:03035   Zbl 453 # 03002
• ID 54132

MCROBBIE, M.A. see Vol. II for further entries

MEAD, J. [1979] *Recursive prime models for boolean algebras* (S 0019) Colloq Math (Warsaw) 41*25-33
⋄ C50 C57 G05 ⋄ REV MR 80j:03050   Zbl 445 # 03016
• ID 56480

MEAD, J. & NELSON, GEORGE C. [1980] *Model companions and k-model completeness for the complete theories of boolean algebras* (J 0036) J Symb Logic 45*47-55
⋄ C25 C35 G05 ⋄ REV MR 81f:03047   Zbl 445 # 03017
• ID 56481

MEDVEDEV, N.YA. [1985] *Quasivarieties of l-groups and groups* (J 0092) Sib Mat Zh 26*111-117
- TRANSL [1985] (J 0475) Sib Math J 26*717-723
⋄ C05 C52 C60 ⋄ ID 49883

MEGIBBEN, C. [1967] *On mixed groups of torsion-free rank one* (J 0316) Illinois J Math 11*134-144
⋄ C60 ⋄ REV MR 34 # 2691 Zbl 139.252 • ID 26155

MEGIBBEN, C. [1983] *Crawley's problem on the unique ω-elongation of p-groups is undecidable* (J 0048) Pac J Math 107*205-212
⋄ C55 C60 E35 E45 E50 E65 E75 ⋄ REV MR 84m:20058 Zbl 521 # 20035 • ID 37469

MEGIBBEN, C. see Vol. V for further entries

MEHLHORN, K. [1974] *The "almost all" theory of subrecursive degrees is decidable* (P 1869) Automata, Lang & Progr (2);1974 Saarbruecken 317-325
⋄ B25 C80 D20 ⋄ REV MR 55 # 7752 Zbl 284 # 68041 • ID 63853

MEHLHORN, K. see Vol. IV for further entries

MEISSNER, W. [1982] *A Loewenheim-Skolem theorem for inner product spaces* (J 0068) Z Math Logik Grundlagen Math 28*549-556
⋄ C55 C65 ⋄ REV MR 84j:03078 Zbl 513 # 03015 • ID 34670

MEISTERS, G.H. [1967] *Nonarchimedean integers* (J 0005) Amer Math Mon 74*434-436
⋄ C60 H15 ⋄ REV MR 35 # 5430 Zbl 159.336 • ID 09086

MEISTERS, G.H. & MONK, J.D. [1973] *Construction of the reals via ultrapowers* (J 0308) Rocky Mountain J Math 3*141-158
⋄ C20 H05 ⋄ REV MR 50 # 292 Zbl 259 # 12104 • ID 09087

MEISTERS, G.H. see Vol. V for further entries

MEJREMBEKOV, K.A. [1980] *Remarks on theories with a bounded spectrum (Russian)* (C 2620) Teor Model & Primen 65-72
⋄ C35 C45 C60 ⋄ REV MR 82c:03048 • ID 76265

MEJREMBEKOV, K.A. [1981] *Spectra of superstable abelian groups (Russian)* (C 3806) Issl Teor Progr 80-84
⋄ C45 C60 ⋄ REV MR 84k:03098 • ID 36115

MEJTUS, V.YU. & VERSHININ, K.P. [1974] *On some unsolvable problems in computable categories (Russian)* (J 0023) Dokl Akad Nauk SSSR 216*42-43
- TRANSL [1974] (J 0062) Sov Math, Dokl 15*752-754
⋄ C57 D80 G30 ⋄ REV MR 49 # 10536 Zbl 304 # 02019 • ID 09089

MEKLER, A.H. [1977] see EKLOF, P.C.

MEKLER, A.H. [1977] *Theories with models of prescribed cardinalities* (J 0036) J Symb Logic 42*251-253
⋄ C55 ⋄ REV MR 58 # 27445 Zbl 382 # 03023 • ID 26452

MEKLER, A.H. [1978] *The size of epimorphic extensions* (J 0004) Algeb Universalis 8*228-232
⋄ C55 C75 C95 E55 G30 ⋄ REV MR 57 # 3209 Zbl 411 # 03029 • ID 52883

MEKLER, A.H. [1979] *Model complete theories with a distinguished substructure* (J 0053) Proc Amer Math Soc 75*294-299
⋄ C35 C45 C60 ⋄ REV MR 80f:03033 Zbl 416 # 03037 • ID 53203

MEKLER, A.H. [1979] see EKLOF, P.C.

MEKLER, A.H. [1980] *How to construct almost free groups* (J 0017) Canad J Math 32*1206-1228
⋄ C60 C75 E35 E45 E75 ⋄ REV MR 82b:20038 Zbl 413 # 20023 • ID 53035

MEKLER, A.H. [1980] *On residual properties* (J 0053) Proc Amer Math Soc 78*187-188
⋄ C52 C75 ⋄ REV MR 80j:03044 Zbl 448 # 03021 • ID 56621

MEKLER, A.H. [1981] see EKLOF, P.C.

MEKLER, A.H. [1981] *Stability of nilpotent groups of class 2 and prime exponent* (J 0036) J Symb Logic 46*781-788
⋄ C45 C60 ⋄ REV MR 83b:03035 Zbl 482 # 03014 • ID 33276

MEKLER, A.H. [1982] see DRIES VAN DEN, L.

MEKLER, A.H. [1982] *Primitive rings are not definable in $L_{\infty\infty}$* (J 0394) Commun Algeb 10*1689-1690
⋄ C60 C75 ⋄ REV MR 83h:03048 Zbl 507 # 16008 • ID 36062

MEKLER, A.H. [1983] see EKLOF, P.C.

MEKLER, A.H. [1983] *Proper forcing and abelian groups* (P 3802) Abel Group Th;1983 Honolulu 285-303
⋄ C55 C60 E35 E50 E75 ⋄ REV MR 85h:03053 Zbl 528 # 03032 • ID 37643

MEKLER, A.H. [1983] see HERRE, H.

MEKLER, A.H. [1984] see EKLOF, P.C.

MEKLER, A.H. [1984] *Stationary logic of ordinals* (J 0073) Ann Pure Appl Logic 26*47-68
⋄ B25 C10 C55 C65 C80 E10 ⋄ REV MR 85i:03121 Zbl 574 # 03019 • ID 44204

MEKLER, A.H. [1985] see HOLLAND, W.C.

MEKLER, A.H. & SHELAH, S. [1985] *Stationary logic and its friends I* (J 0047) Notre Dame J Formal Log 26*129-138
⋄ C40 C55 C75 C80 E35 E50 ⋄ ID 42592

MEKLER, A.H. see Vol. V for further entries

MELIS, E. [1972] see HAUSCHILD, K.

MELONI, G.C. [1973] *Analisi non standard di anelli e corpi topologici (English summary)* (J 0207) Ist Lombardo Accad Sci Rend, A (Milano) 107*503-510
⋄ C60 H05 H20 ⋄ REV MR 52 # 5414 Zbl 282 # 54028 • ID 18293

MELONI, G.C. [1977] see BOZZI, S.

MELONI, G.C. [1980] see BOZZI, S.

MELONI, G.C. see Vol. I, II, V, VI for further entries

MELTER, R.A. [1964] *Boolean valued rings and boolean metric spaces* (J 0008) Arch Math (Basel) 15*354-363
⋄ C90 ⋄ REV MR 30 # 46 Zbl 126.274 JSL 35.488 • ID 09092

MELTER, R.A. see Vol. I for further entries

MENDELSOHN, E. [1969] *An elementary characterization of the category of (free) relational systems* (J 0140) Comm Math Univ Carolinae (Prague) 10*571-588
- REPR [1970] (J 0044) Math Z 113*224-232
⋄ C52 G30 ⋄ REV MR 42 # 3005 MR 42 # 3147 Zbl 206.300 Zbl 241 # 08001 • ID 09101

MENDELSOHN, E. see Vol. V for further entries

MENDELSON, E. [1957] *Non standard models* (P 1675) Summer Inst Symb Log;1957 Ithaca 167-168
⋄ C62 H15 ⋄ ID 29340

MENDELSON, E. [1958] *On a class of universal ordered sets* (J 0053) Proc Amer Math Soc 9*712-713
⋄ C50 E07 ⋄ REV MR 20 # 3075 Zbl 87.270 • ID 09112

MENDELSON, E. [1961] *On non-standard models for number theory* (C 0622) Essays Found of Math (Fraenkel) 259-268
⋄ C62 H15 ⋄ REV MR 29 # 1141 Zbl 148.255 JSL 32.128 • ID 09115

MENDELSON, E. see Vol. I, II, IV, V, VI for further entries

MEREDITH, D. [1956] *A correction to von Wright's decision procedure for the deontic system P* (J 0094) Mind 65*548-550
⋄ B25 B45 ⋄ REV JSL 22.92 • ID 09147

MEREDITH, D. [1978] *Positive logic and λ-constants* (J 0063) Studia Logica 37*269-285
⋄ B20 B25 B40 ⋄ REV MR 80c:03019 Zbl 393 # 03008 • ID 31932

MEREDITH, D. see Vol. I, II, VI for further entries

MERZLYAKOV, YU.I. [1963] *On the theory of generalized solvable and generalized nilpotent groups (Russian)* (J 0003) Algebra i Logika 2/5*29-36
⋄ C52 C60 ⋄ REV MR 28 # 4030 Zbl 136.277 • ID 41677

MERZLYAKOV, YU.I. [1966] *Positive formulae on free groups (Russian)* (J 0003) Algebra i Logika 5/4*25-42
⋄ C60 ⋄ REV MR 36 # 5201 Zbl 216.294 • ID 26265

MERZLYAKOV, YU.I. see Vol. I for further entries

MESEGUER, J. [1985] see GOGUEN, J.A.

MESEGUER, J. see Vol. IV, V for further entries

MESKHI, S. [1982] *Complete boolean powers of quasiprimal algebras with lattice reducts* (J 0004) Algeb Universalis 14*388-390
⋄ C05 C30 ⋄ REV MR 84g:08012 Zbl 456 # 08007 • ID 39430

MESKHI, S. see Vol. II for further entries

METAKIDES, G. & PLOTKIN, J.M. [1975] *An algebraic characterization of power set in countable standard models of ZF* (J 0036) J Symb Logic 40*167-170
⋄ C15 C62 E35 G05 ⋄ REV MR 52 # 99 Zbl 307 # 02044 • ID 09167

METAKIDES, G. & NERODE, A. [1975] *Recursion theory and algebra* (P 0765) Algeb & Log;1974 Clayton 209-219
⋄ C57 C60 D25 D45 ⋄ REV MR 51 # 7798 Zbl 306 # 02038 • ID 17296

METAKIDES, G. [1977] *A return to constructive algebra via recursive function theory* (J 4135) Dialexeis 2*112-126
⋄ C57 D45 ⋄ ID 33933

METAKIDES, G. & NERODE, A. [1977] *Recursively enumerable vector spaces* (J 0007) Ann Math Logic 11*147-171
⋄ C57 C60 D45 ⋄ REV MR 56 # 5253 Zbl 389 # 03019 JSL 48.880 • ID 23661

METAKIDES, G. [1978] *Constructive algebra in a new frame* (J 4373) GMS Math Inst Patras 4*13-25
⋄ C57 C60 D45 ⋄ ID 33934

METAKIDES, G. & NERODE, A. [1979] *Effective content of field theory* (J 0007) Ann Math Logic 17*289-320
⋄ C57 C60 D45 ⋄ REV MR 82b:03082 Zbl 469 # 03028 JSL 48.880 • ID 55155

METAKIDES, G. & REMMEL, J.B. [1979] *Recursion theory on orderings I. A model theoretic setting* (J 0036) J Symb Logic 44*383-402
⋄ C57 D45 ⋄ REV MR 80m:03080 Zbl 471 # 03035 • REM Part II 1980 by Remmel,J.B. • ID 55230

METAKIDES, G. & NERODE, A. [1980] *Recursion theory on fields and abstract dependence* (J 0032) J Algeb 65*36-59
⋄ C57 C60 D45 ⋄ REV MR 81k:03041 Zbl 469 # 03029 JSL 48.880 • ID 55156

METAKIDES, G. (ED.) [1982] *Patras logic symposium* (S 3303) Stud Logic Found Math 109*ix+391pp
⋄ B97 C97 ⋄ REV MR 84b:03008 Zbl 504 # 00001 • ID 35609

METAKIDES, G. & NERODE, A. [1982] *The introduction of non-recursive methods into mathematics* (P 3638) Brouwer Centenary Symp;1981 Noordwijkerhout 319-335
⋄ C57 D45 D80 F60 ⋄ REV MR 85d:03121 Zbl 511 # 03020 • ID 37391

METAKIDES, G. see Vol. IV for further entries

MEULEN TER, A.G.B. (ED.) [1983] *Studies in modeltheoretic semantics* (X 4217) Foris: Dordrecht x+206pp
⋄ B97 C97 ⋄ REV Zbl 563 # 00003 • ID 47464

MEYER, A.R. [1975] *The inherent computational complexity of theories of ordered sets* (P 1521) Int Congr Math (II,12);1974 Vancouver 2*477-482
⋄ B25 D15 ⋄ REV MR 56 # 103 Zbl 361 # 02061 • ID 27549

MEYER, A.R. [1975] *Weak monadic second order theory of successor is not elementary recursive* (C 0758) Logic Colloq Boston 1972-73 132-154
• TRANSL [1975] (J 3079) Kiber Sb Perevodov, NS 12*62-77
⋄ B25 D10 D15 D20 ⋄ REV MR 52 # 13358 Zbl 326 # 02036 • ID 21811

MEYER, A.R. [1978] see BRUSS, A.R.

MEYER, A.R. [1980] see BRUSS, A.R.

MEYER, A.R. see Vol. I, II, IV, VI for further entries

MEYER, R.K. [1980] see MCROBBIE, M.A.

MEYER, R.K. see Vol. I, II, IV, V, VI for further entries

MEZ, H.-C. [1982] *Existentially closed linear groups* (J 0032) J Algeb 76*84-98
⋄ C25 C35 C60 ⋄ REV MR 84a:20050 Zbl 487 # 20030 • ID 36707

MEZ, H.-C. [1984] see EKLOF, P.C.

MEZ, H.-C. [1985] see EKLOF, P.C.

MICALE, B. [1978] see FERRO, A.

MICALE, B. see Vol. IV for further entries

MICHALCZYK, M. [1981] see MALINOWSKI, G.

MICHALCZYK, M. [1982] see MALINOWSKI, G.

MICHAUX, C. [1985] see BOFFA, M.

MICHEL, P. [1981] *Borne superieure de la complexite de la theorie de N muni de la relation de divisibilite* (P 3404) Model Th & Arithm;1979/80 Paris 242-250
⋄ B25 D15 ⋄ REV MR 83e:03063 Zbl 488 # 03021
• ID 35234

MIETTINEN, S. & VAEAENAENEN, J. (EDS.) [1977] *Proceedings of the Symposium on Mathematical Logic in Oulu 1974 and in Helsinki 1975* (1111) Preprints, Manuscr., Techn. Reports etc. iv+103pp
⋄ C97 ⋄ REM University of Helsinki, Department of Philosophy • ID 48417

MIETTINEN, S. [1977] *Some remarks on definability* (P 1629) Symp Math Log;1974 Oulo;1975 Helsinki 69-70
⋄ C95 ⋄ ID 49821

MIGLIOLI, P.A. [1975] see MARINI, D.

MIGLIOLI, P.A. [1979] see BERTONI, A.

MIGLIOLI, P.A. [1980] see BERTONI, A.

MIGLIOLI, P.A. [1981] see BERTONI, A.

MIGLIOLI, P.A. & MOSCATO, U. & ORNAGHI, M. [1981] *Trees in Kripke models and in an intuitionistic refutation system* (P 2923) CAAP'81 Arbres en Algeb & Progr (6);1981 Genova 316-331
⋄ C90 F50 ⋄ REV MR 83h:03019 Zbl 472 # 03048
• ID 55311

MIGLIOLI, P.A. [1983] see BERTONI, A.

MIGLIOLI, P.A. see Vol. II, IV, VI for further entries

MIHAILESCU, E.G. [1967] *Decision problem in the classical logic* (J 0047) Notre Dame J Formal Log 8*239-253
⋄ B25 D35 ⋄ REV MR 38 # 3124 Zbl 174.21 • ID 09234

MIHAILESCU, E.G. see Vol. I, II for further entries

MIJAJLOVIC, Z. [1974] *On decidability of one class of boolean formulas* (J 0042) Mat Vesn, Drust Mat Fiz Astron Serb 11(26)*48-54
⋄ B05 B25 ⋄ REV MR 50 # 80 Zbl 286 # 02065 • ID 09242

MIJAJLOVIC, Z. [1977] *A note on elementary end extension* (J 0400) Publ Inst Math, NS (Belgrade) 21(35)*141-144
⋄ C30 C62 ⋄ REV MR 57 # 2904 Zbl 365 # 02043
• ID 51044

MIJAJLOVIC, Z. [1977] *Homogeneous-universal models of theories which have model completions* (P 3269) Set Th Found Math (Kurepa);1977 Beograd 87-97
⋄ C35 C50 C52 ⋄ REV MR 58 # 16256 Zbl 407 # 03036
• ID 56202

MIJAJLOVIC, Z. [1977] *Some remarks on boolean terms - model theoretic approach* (J 0400) Publ Inst Math, NS (Belgrade) 21(35)*135-140
⋄ B05 C50 G05 ⋄ REV MR 57 # 2909 Zbl 362 # 02055
• ID 50777

MIJAJLOVIC, Z. [1979] *Saturated boolean algebras with ultrafilters* (J 0400) Publ Inst Math, NS (Belgrade) 26(40)*175-197
⋄ C20 C35 C50 G05 ⋄ REV MR 81i:03053
Zbl 436 # 03022 • ID 55862

MIJAJLOVIC, Z. [1982] *Completions of models and Galois theory* (P 3758) Algeb Conf (2);1981 Novi Sad 19-26
⋄ C60 ⋄ REV MR 84c:03062 Zbl 537 # 12016 • ID 34005

MIJAJLOVIC, Z. [1983] *On Σ_1^0-extensions of ω* (J 0400) Publ Inst Math, NS (Belgrade) 34(48)*121-124
⋄ C15 C30 C62 ⋄ REV MR 86f:03057 Zbl 554 # 03037
• ID 44948

MIJAJLOVIC, Z. [1983] *On the embedding property of models* (P 3855) Algeb Conf (3);1982 Beograd 103-110
⋄ C25 C35 C52 G05 G10 ⋄ REV MR 85j:03046
Zbl 533 # 03018 • ID 36542

MIJAJLOVIC, Z. [1983] see CUPONA, G.

MIJAJLOVIC, Z. [1983] see HARIZANOV, V.

MIJAJLOVIC, Z. [1983] *Submodels and definable points in models of Peano arithmetics* (J 0047) Notre Dame J Formal Log 24*417-425
⋄ C62 ⋄ REV MR 84k:03150 Zbl 545 # 03019 • ID 35042

MIJAJLOVIC, Z. [1985] *On ε-bounded ultraproducts* (P 4661) Algeb & Log;1984 Zagreb 105-110
⋄ C20 ⋄ ID 49050

MIJAJLOVIC, Z. [1985] *On a proof on the Erdoes-Monk theorem* (J 0400) Publ Inst Math, NS (Belgrade) 37(51)*25-28
⋄ C50 E05 G05 ⋄ REV Zbl 578 # 03018 • ID 49547

MIJAJLOVIC, Z. [1985] *On the definability of the quantifier "there exist uncountably many"* (J 0063) Studia Logica 44*257-264
⋄ C55 C80 ⋄ ID 47501

MIJOULE, R. [1976] *Theorie des types et modeles de la theorie des ensembles (English summary)* (J 2313) C R Acad Sci, Paris, Ser A-B 283*A733-A735
⋄ B15 C62 G30 ⋄ REV MR 54 # 10010 Zbl 355 # 02050
• ID 25854

MIJOULE, R. see Vol. IV, VI for further entries

MIKENBERG, I. [1978] *From total to partial algebras* (P 1800) Brazil Conf Math Log (1);1977 Campinas 203-223
⋄ C05 C40 C52 ⋄ REV MR 80f:08003 Zbl 384 # 03017
• ID 52059

MILIC, S. [1983] see CUPONA, G.

MILLAR, T.S. [1978] *Foundations of recursive model theory* (J 0007) Ann Math Logic 13*45-72
⋄ C50 C57 ⋄ REV MR 80a:03051 Zbl 432 # 03018
JSL 49.671 • ID 27960

MILLAR, T.S. [1979] *A complete, decidable theory with two decidable models* (J 0036) J Symb Logic 44*307-312
⋄ C57 ⋄ REV MR 81b:03040 Zbl 421 # 03026 JSL 49.671
• ID 53425

MILLAR, T.S. [1980] *Homogeneous models and decidability* (J 0048) Pac J Math 91*407-418
⋄ C50 C57 ⋄ REV MR 83i:03056 Zbl 467 # 03007
• ID 55005

MILLAR, T.S. [1981] *Counterexamples via model completions* (P 2628) Log Year;1979/80 Storrs 215-229
⋄ C10 C35 C50 C57 ⋄ REV MR 84b:03052
Zbl 493 # 03009 • ID 35640

MILLAR, T.S. [1981] *Stability, complete extensions, and the number of countable models* (P 2902) Aspects Effective Algeb;1979 Clayton 196-205
⋄ C15 C35 C45 ⋄ REV MR 83d:03041 Zbl 483 # 03022
• ID 35186

MILLAR, T.S. [1981] *Vaught's theorem recursively revisited* (J 0036) J Symb Logic 46*397-411
⋄ C15 C57 D45 ⋄ REV MR 82d:03053 Zbl 493.03008
• ID 76379

MILLAR, T.S. [1982] *Type structure complexity and decidability* (J 0064) Trans Amer Math Soc 271*73-81
⋄ C50 C57 ⋄ REV MR 83h:03046 Zbl 493 # 03010
• ID 36061

MILLAR, T.S. [1983] *Omitting types, type spectrums, and decidability* (J 0036) J Symb Logic 48*171-181
⋄ C07 C57 ⋄ REV MR 85a:03049 Zbl 516 # 03017
• ID 34800

MILLAR, T.S. [1983] *Persistently finite theories with hyperarithmetic models* (J 0064) Trans Amer Math Soc 278*91-99
⋄ C15 C50 C57 D30 ⋄ REV MR 84m:03053 Zbl 525 # 03027 • ID 35754

MILLAR, T.S. [1983] see ASH, C.J.

MILLAR, T.S. [1984] *Decidability and the number of countable models* (J 0073) Ann Pure Appl Logic 27*137-153
⋄ C15 C57 ⋄ REV MR 86f:03052 Zbl 563 # 03014
• ID 39678

MILLAR, T.S. [1985] *Decidable Ehrenfeucht theories* (P 4046) Rec Th;1982 Ithaca 311-321
⋄ B25 C15 C57 C98 D30 D35 D55 ⋄ REV Zbl 573 # 03003 • ID 46374

MILLER, A.W. [1981] *Vaught's conjecture for theories of one unary operation* (J 0027) Fund Math 111*135-141
⋄ C15 C75 ⋄ REV MR 82m:03044 Zbl 492 # 03012
• ID 76382

MILLER, A.W. [1983] *On the Borel classification of the isomorphism class of a countable model* (J 0047) Notre Dame J Formal Log 24*22-34
⋄ C15 D55 E15 ⋄ REV MR 84c:03055 Zbl 487 # 03013
• ID 35705

MILLER, A.W. see Vol. IV, V for further entries

MILLER, DOUGLAS E. [1975] see BURGESS, J.P.

MILLER, DOUGLAS E. [1978] *The invariant Π^0_α separation principle* (J 0064) Trans Amer Math Soc 242*185-204
⋄ C15 C70 C75 D55 E15 ⋄ REV MR 81b:03054 Zbl 409 # 03030 • ID 33521

MILLER, DOUGLAS E. [1979] *An application of invariant sets to global definability* (J 0036) J Symb Logic 44*9-14
⋄ C40 C70 C75 ⋄ REV MR 80m:03072 Zbl 404 # 03022
• ID 54809

MILLER, DOUGLAS E. [1979] *On classes closed under unions of chains* (J 0036) J Symb Logic 44*29-31
⋄ C52 C75 ⋄ REV MR 80e:03032 Zbl 415 # 03022
• ID 53124

MILLER, DOUGLAS E. [1981] *The metamathematics of model theory: Discovering language in action* (J 0036) J Symb Logic 46*490-498
⋄ C15 E15 ⋄ REV MR 83a:03028 Zbl 485 # 03015
• ID 33277

MILLER, DOUGLAS E. [1982] see BALDWIN, JOHN T.

MILLER, DOUGLAS E. see Vol. IV, V for further entries

MILLS, G. [1978] *A model of Peano arithmetic with no elementary end extension* (J 0036) J Symb Logic 43*563-567
⋄ C25 C62 H15 ⋄ REV MR 58 # 10425 Zbl 388 # 03029
• ID 29283

MILLS, G. & PARIS, J.B. [1979] *Closure properties of countable nonstandard integers* (J 0027) Fund Math 103*205-215
⋄ H15 ⋄ REV MR 81j:03099 Zbl 421 # 03051 • ID 53450

MILLS, G. [1979] *Substructure lattices of models of arithmetic* (J 0007) Ann Math Logic 16*145-180
⋄ C62 H15 ⋄ REV MR 81i:03105 Zbl 427 # 03057
• ID 53744

MILLS, G. & PARIS, J.B. [1984] *Regularity in models of arithmetic* (J 0036) J Symb Logic 49*272-280
⋄ C62 C80 F30 ⋄ REV MR 85h:03035 • ID 42484

MILLS, G. see Vol. VI for further entries

MILNER, E.C. [1972] see ERDOES, P.

MILNER, E.C. see Vol. V for further entries

MINARI, P. [1983] *Completeness theorems for some intermediate predicate calculi* (J 0063) Studia Logica 42*431-441
⋄ B55 C90 ⋄ REV MR 86f:03045 Zbl 571 # 03011
• ID 42336

MINARI, P. [1985] *Kripke-definable ordinals* (P 4646) Atti Incontri Log Mat (2);1983/84 Siena 185-188
⋄ C90 E47 ⋄ ID 49720

MINARI, P. see Vol. II for further entries

MINTS, G.E. [1966] *Skolem's method of elimination of positive quantifiers in sequential calculi (Russian)* (J 0023) Dokl Akad Nauk SSSR 169*24-27
• TRANSL [1966] (J 0062) Sov Math, Dokl 7*861-864
⋄ B25 C10 F50 ⋄ REV MR 34 # 5648 Zbl 186.5 JSL 36.526 • ID 19428

MINTS, G.E. [1968] *Solvability of the problem of deducibility in LJ for a class of formulas which do not contain negative occurences of quantors (Russian)* (S 0066) Tr Mat Inst Steklov 98*121-130
• TRANSL [1968] (S 0055) Proc Steklov Inst Math 98*135-145
⋄ B25 F50 ⋄ REV MR 41 # 6669 Zbl 179.11 • ID 19418

MINTS, G.E. [1975] see KREISEL, G.

MINTS, G.E. see Vol. I, II, IV, V, VI for further entries

MIRAGLIA, F. [1980] *Relations between structures of stable continuous functions and filtered powers* (P 3006) Brazil Conf Math Log (3);1979 Recife 225-244
⋄ C30 C35 C90 ⋄ REV MR 82g:03049 Zbl 448 # 03027
• ID 56627

MIRAGLIA, F. [1982] *On the preservation of elementary equivalence and embedding by bounded filtered powers and structures of stable continuous functions* (P 3860) Coll Papers to Farah on Retirement;1981 Sao Paulo 57-65
⋄ C20 C40 ⋄ REV MR 85g:03049 Zbl 548 # 54027
• ID 42744

MIRAGLIA, F. [1983] *On the preservation of elementary equivalence and embedding by bounded filtered powers and structures of stable continuous functions* (J 3250) Bol Soc Brasil Mat 14*81-86
⋄ C20 C40 ⋄ REV MR 85d:03067 Zbl 569 # 03014
• ID 40584

MIROLLI, M. [1983] see BELLISSIMA, F.

MIROLLI, M. see Vol. I, II, VI for further entries

MISAWA, T. & YASUDA, Y. [1977] *Suslin logics of Gentzen style* (J 3630) Bull Dept of Lib Arts (Numazu) 4*1-12
⋄ C75 F07 ⋄ ID 33405

MISAWA, T. [1980] *Consistency properties for Hausdorff logic* (J 0407) Comm Math Univ St Pauli (Tokyo) 29*169-181
⋄ C75 ⋄ REV MR 82g:03067 Zbl 453 # 03039 • ID 54169

MISAWA, T. see Vol. V for further entries

MISERCQUE, D. [1981] *Solutions de deux problemes poses par H. Simmons* (J 3133) Bull Soc Math Belg, Ser B 33*65-72
⋄ C62 F30 H15 ⋄ REV MR 82k:03091 Zbl 482 # 03015
• ID 76440

MISERCQUE, D. [1982] *The nonhomogeneity of the E-tree - answer to a problem raised by D. Jensen and A. Ehrenfeucht* (J 0053) Proc Amer Math Soc 84*573-575
⋄ C62 F30 ⋄ REV MR 83h:03085 Zbl 495 # 03048
• ID 36085

MISERCQUE, D. see Vol. I, IV, VI for further entries

MISLOVE, M.W. [1976] see HOFMANN, K.H.

MISLOVE, M.W. see Vol. V for further entries

MITCHELL, W.J. [1972] *Aronszajn trees and the independence of the transfer property* (J 0007) Ann Math Logic 5*21-46
⋄ C55 E05 E35 E55 ⋄ REV MR 47 # 1612
Zbl 255 # 02069 • ID 09322

MITCHELL, W.J. [1979] see MARTIN, D.A.

MITCHELL, W.J. [1980] see JECH, T.J.

MITCHELL, W.J. [1983] *Sets constructed from sequences of measures: revisited* (J 0036) J Symb Logic 48*600-609
⋄ C55 E45 E55 ⋄ REV MR 85j:03052 Zbl 527 # 03032
• ID 37578

MITCHELL, W.J. see Vol. V for further entries

MITSCHKE, G. [1980] *Infinite terms and infinite reductions* (C 3050) Essays Combin Log, Lambda Calc & Formalism (Curry) 243-257
⋄ B40 C75 ⋄ REV MR 82g:03021 Zbl 469 # 03006
• ID 76450

MITSCHKE, G. see Vol. IV, VI for further entries

MIURA, S. [1972] *Probabilistic models of modal logics* (J 0523) Bull Nagoya Inst Tech 24*67-72
⋄ B45 C90 ⋄ REV MR 48 # 1885 • ID 09336

MIURA, S. see Vol. I, II for further entries

MIYAMA, T. [1974] *The interpolation theorem and Beth's theorem in many-valued logics* (J 0352) Math Jap 19*341-355
⋄ B50 C40 ⋄ REV MR 54 # 60 Zbl 319 # 02015 • ID 29672

MIYAMA, T. see Vol. II for further entries

MIZUTANI, C. [1977] *Monadic second order logic with an added quantifier Q* (J 2606) Tsukuba J Math 1*45-76
⋄ B15 B25 C40 C55 C80 ⋄ REV MR 58 # 10291
Zbl 414 # 03023 • ID 53069

MIZUTANI, C. see Vol. VI for further entries

MLCEK, J. & SOCHOR, A. [1972] *Contributions to the theory of semisets. II: The theory of semisets and end-extensions in a syntactic setting* (J 0068) Z Math Logik Grundlagen Math 18*407-417
⋄ C62 E70 ⋄ REV MR 54 # 2460 Zbl 296 # 02038 • REM Part I 1972 by Hajek,P. Part III 1973 by Cuda,K. • ID 09207

MLCEK, J. [1973] *A representation of models of Peano arithmetic* (J 0140) Comm Math Univ Carolinae (Prague) 14*553-558
⋄ C60 C62 ⋄ REV MR 48 # 1917 Zbl 284 # 02029
• ID 09338

MLCEK, J. [1978] *A note on cofinal extensions and segments* (J 0140) Comm Math Univ Carolinae (Prague) 19*727-742
⋄ C30 C62 C65 ⋄ REV MR 80i:03045 Zbl 398 # 03020
• ID 31965

MLCEK, J. [1978] *End-extensions of countable structures and the induction schema* (J 0140) Comm Math Univ Carolinae (Prague) 19*291-308
⋄ C15 C62 F30 ⋄ REV MR 58 # 10413 Zbl 372 # 02030
• ID 31964

MLCEK, J. [1983] *Compactness and homogeneity of saturated structures I,II* (J 0140) Comm Math Univ Carolinae (Prague) 24*701-716,717-729
⋄ C35 C50 C60 C62 ⋄ REV MR 85i:03108
Zbl 575 # 03021 Zbl 575 # 03022 • ID 45239

MLCEK, J. [1985] *Some automorphisms of natural numbers in the alternative set theory* (J 0140) Comm Math Univ Carolinae (Prague) 26*467-475
⋄ C07 C30 E70 H15 ⋄ ID 48312

MLCEK, J. see Vol. I, V, VI for further entries

MLEZIVA, M. [1975] *Semantics based on states of affairs* (J 0156) Kybernetika (Prague) 11*319-335
⋄ C90 ⋄ REV MR 54 # 7242 Zbl 317 # 02053 • ID 25794

MLEZIVA, M. see Vol. I, II for further entries

MO, SHAOKUI [1964] *Enumeration quantifiers and predicate calculus (Chinese)* (J 0418) Shuxue Xuebao 14*218-230
• TRANSL [1964] (J 0419) Chinese Math Acta 5*239-253 (English)
⋄ B10 C07 C80 ⋄ REV Zbl 192.28 • ID 47141

MO, SHAOKUI [1979] see GAO, HENGSHAN

MO, SHAOKUI see Vol. I, II, IV, V, VI for further entries

MOCKOR, J. [1981] *The completion of valued fields and nonstandard models* (J 0407) Comm Math Univ St Pauli (Tokyo) 30*1-16
⋄ C60 H20 ⋄ REV MR 84d:12024 Zbl 476 # 12020
• ID 39739

MOENS, J.L. [1982] *Forcing et semantique de Kripke-Joyal* (S 3935) Cah Cent Log (Louvain) 3*28pp
⋄ C25 C90 E40 G30 ⋄ REV Zbl 565 # 03013 • ID 48346

MOERDIJK, I. [1982] see LAMBEK, J.

MOERDIJK, I. [1984] see HOEVEN VAN DER, G.F.

MOERDIJK, I. [1984] *Heine-Borel does not imply the Fan theorem* (J 0036) J Symb Logic 49*514
⋄ C90 E35 F35 F50 G30 ⋄ ID 42485

MOERDIJK, I. see Vol. IV, V, VI for further entries

MOLDESTAD, J. [1977] *On monotone second order relations* (P 1629) Symp Math Log;1974 Oulo;1975 Helsinki 44-50
⋄ C40 C85 ⋄ ID 48419

MOLDESTAD, J. see Vol. IV for further entries

MOLLER, S. [1965] *Induction models (Danish)* (J 0311) Nordisk Mat Tidskr 13∗127-137
⋄ C62 F30 ⋄ REV MR 34 # 38 Zbl 131.6 • ID 48096

MOLODSHIJ, V.N. [1952] *On interrelations of certain assertions of generality with the induction axiom in Peano's system of axioms (Russian)* (S 1498) Uch Zap Univ, Moskva 155∗168-173
⋄ C62 F30 ⋄ REV MR 17.1040 • ID 09388

MOLODSHIJ, V.N. see Vol. V, VI for further entries

MOLZAN, B. [1981] *On the number of different theories of boolean algebras in several logics* (J 3437) Rep, Akad Wiss DDR, Inst Math R-MATH-03/81∗102-113
⋄ C80 G05 ⋄ REV MR 84h:03092 Zbl 476 # 03040 • ID 55552

MOLZAN, B. [1982] *How to eliminate quantifiers in the elementary theory of p-rings* (J 0068) Z Math Logik Grundlagen Math 28∗82-92
⋄ B25 C60 ⋄ REV MR 83e:03042 Zbl 517 # 03009 • ID 90320

MOLZAN, B. [1982] *The theory of superatomic boolean algebras in the logic with the binary Ramsey quantifier* (J 0068) Z Math Logik Grundlagen Math 28∗365-376
⋄ B25 C10 C80 G05 ⋄ REV MR 84d:03045 Zbl 459 # 03018 • ID 54458

MOLZAN, B. [1985] *On the theory of Boolean algebras in the logic with Ramsey quantifiers* (P 4310) Easter Conf on Model Th (3);1985 Gross Koeris 186-192
⋄ B25 C10 C80 G05 ⋄ ID 49051

MONK, J.D. [1962] *On pseudo-simple universal algebras* (J 0053) Proc Amer Math Soc 13∗543-546
⋄ C05 ⋄ REV MR 26 # 2381 Zbl 106.17 • ID 09393

MONK, J.D. [1965] *Model-theoretic methods and results in the theory of cylindric algebras* (P 0614) Th Models;1963 Berkeley 238-250
⋄ C05 C20 C52 G15 ⋄ REV MR 34 # 58 Zbl 199.6 • ID 41180

MONK, J.D. [1973] see MEISTERS, G.H.

MONK, J.D. [1975] *Some cardinal functions on algebras I,II* (J 0004) Algeb Universalis 5∗76-81,361-366
⋄ C05 E50 E75 ⋄ REV MR 52 # 13588 MR 52 # 13589 Zbl 311 # 08005 Zbl 345 # 08004 • ID 21872

MONK, J.D. [1978] *Omitting types algebraically* (J 1934) Ann Sci Univ Clermont Math 16∗101-105
⋄ C07 G15 ⋄ REV MR 80a:03077 Zbl 396 # 03046 • ID 52656

MONK, J.D. & RASSBACH, W. [1979] *The number of rigid boolean algebras* (J 0004) Algeb Universalis 9∗207-210
⋄ C50 C55 E75 G05 ⋄ REV MR 80c:06020 Zbl 416 # 06016 • ID 53222

MONK, J.D. [1980] see DOUWEN VAN, E.K.

MONK, J.D. [1985] *On endomorphism bases* (J 0004) Algeb Universalis 20∗264-266
⋄ C05 C55 E10 G05 ⋄ REV Zbl 568 # 08003 • ID 49230

MONK, J.D. see Vol. I, II, V for further entries

MONK, L.G. [1975] *Elementary-recursive decision procedures* (0000) Diss., Habil. etc 89pp
⋄ B25 C60 D20 ⋄ REM Univ. Berkeley • ID 90301

MONRO, G.P. [1974] see BARNES, D.W.

MONRO, G.P. [1974] *The strong amalgamation property for complete boolean algebras* (J 0068) Z Math Logik Grundlagen Math 20∗499-502
⋄ C52 E40 G05 ⋄ REV MR 55 # 2563 Zbl 317 # 02084 • ID 09414

MONRO, G.P. see Vol. V for further entries

MONTAGNA, F. [1974] *Sui limiti di certe teorie (English summary)* (J 0319) Matematiche (Sem Mat Catania) 29∗182-194
⋄ C30 C35 C50 C62 H15 ⋄ REV MR 51 # 5293 Zbl 308 # 02049 • ID 17401

MONTAGNA, F. [1980] *Some modal logics with quantifiers (Italian summary)* (J 3495) Boll Unione Mat Ital, V Ser, B 17∗1395-1410
⋄ B45 C80 C90 ⋄ REV MR 85m:03011 Zbl 452 # 03012 • ID 54076

MONTAGNA, F. [1983] *ZFC-models as Kripke-models* (J 0068) Z Math Logik Grundlagen Math 29∗163-168
⋄ B45 C62 E30 F30 ⋄ REV MR 84j:03043 Zbl 519 # 03013 • ID 34635

MONTAGNA, F. [1984] *The predicate modal logic of provability* (J 0047) Notre Dame J Formal Log 25∗179-192
⋄ B45 C90 F30 ⋄ REV MR 86b:03023 Zbl 549 # 03013 • ID 42567

MONTAGNA, F. & SORBI, A. [1985] *Universal recursion theoretic properties of r.e. preordered structures* (J 0036) J Symb Logic 50∗397-406
⋄ C57 D45 G10 ⋄ ID 41807

MONTAGNA, F. see Vol. II, IV, VI for further entries

MONTAGUE, R. & VAUGHT, R.L. [1957] *Models of set theory* (P 1675) Summer Inst Symb Log;1957 Ithaca 353-354
⋄ C62 E30 E45 ⋄ ID 29377

MONTAGUE, R. & VAUGHT, R.L. [1959] *A note on theories with selectors* (J 0027) Fund Math 47∗243-247
⋄ C07 ⋄ REV MR 23 # A1521 Zbl 94.8 JSL 25.177 • ID 16982

MONTAGUE, R. & VAUGHT, R.L. [1959] *Natural models of set theories* (J 0027) Fund Math 47∗219-242
⋄ C62 E47 E55 E65 ⋄ REV MR 23 # A1520 Zbl 94.8 JSL 25.177 • ID 24781

MONTAGUE, R. [1962] *Theories incomparable with respect to relative interpretability* (J 0036) J Symb Logic 27∗195-211
⋄ C07 F25 F30 ⋄ REV MR 27 # 5684 Zbl 112.246 JSL 36.688 • ID 09425

MONTAGUE, R. [1963] *Syntactical treatments of modality, with corollaries on reflexion principles and finite axiomatizability* (J 0096) Acta Philos Fenn 16∗153-167
• REPR [1974] (C 4062) Montague: Formal Philos 286-302
⋄ B45 B65 C62 F30 ⋄ REV MR 29 # 1140 Zbl 117.13 JSL 40.600 JSL 47.210 • ID 14478

MONTAGUE, R. [1965] *Interpretability in terms of models* (J 0028) Indag Math 27∗467-476
⋄ C07 F25 ⋄ REV MR 31 # 4724 Zbl 151.11 • ID 09431

MONTAGUE, R. [1965] *Set theory and higher-order logic* (P 0688) Logic Colloq;1963 Oxford 131-148
⋄ B15 C62 E30 ⋄ REV MR 36 # 3638 Zbl 129.259 JSL 40.459 • ID 24776

MONTAGUE, R. [1968] *Recursion theory as a branch of model theory* (P 0627) Int Congr Log, Meth & Phil of Sci (3,Proc);1967 Amsterdam 63-86
⋄ C40 D75 ⋄ REV MR 42 # 7505 Zbl 247 # 02040 JSL 38.158 • ID 09432

MONTAGUE, R. see Vol. I, II, IV, V, VI for further entries

MONTALI, T. [1971] *Pseudolimiti e formule SH* (J 1526) Riv Mat Univ Parma, Ser 2 12*245-258
⋄ B20 C30 G30 ⋄ REV MR 52 # 39 Zbl 293 # 02036
• ID 17210

MONTEVERDI, D. [1972] *SH relative e pseudolimiti (English summary)* (J 3254) Riv Mat Univ Parma, Ser 3 1*189-194
⋄ C30 G30 ⋄ REV MR 51 # 135 Zbl 304 # 02022
• ID 15219

MONTEVERDI, D. [1973] *SH a piu sorte di variabili* (J 3254) Riv Mat Univ Parma, Ser 3 2*67-77
⋄ B20 C30 G30 ⋄ REV MR 52 # 13372 Zbl 341 # 02043
• ID 21823

MONTGOMERY, H. & ROUTLEY, R. [1976] *Algebraic semantics for $S2^0$ and necessitated extensions* (J 0047) Notre Dame J Formal Log 17*44-58
⋄ B25 B45 ⋄ REV MR 55 # 5392 Zbl 313 # 02010
• ID 18366

MONTGOMERY, H. see Vol. I, II for further entries

MOORE, A.W. [1985] *Set theory, Skolem's paradox and the "Tractatus"* (J 0103) Analysis (Oxford) 45*13-20
⋄ A05 C07 E30 ⋄ ID 45047

MOORE, S.M. [1982] *Nonstandard analysis applied to path integrals and generalized functions (Italian and Russian summaries)* (J 2775) Nuovo Cimento B, Ser 2 70/2*277-290
⋄ C65 H10 H20 ⋄ REV MR 84a:81011 • ID 38856

MOORE, S.M. see Vol. I for further entries

MOORE JR., L.C. [1972] see HENSON, C.W.

MOORE JR., L.C. [1973] see HENSON, C.W.

MOORE JR., L.C. [1974] see HENSON, C.W.

MOORE JR., L.C. [1974] see COZART, D.

MOORE JR., L.C. [1983] see HEINRICH, S.

MOORE JR., L.C. [1983] see HENSON, C.W.

MOORE JR., L.C. see Vol. I, V for further entries

MORAIS, R. [1976] *Projective logic* (J 0110) Anais Acad Bras Cienc 48*627-661
⋄ B60 C75 G05 ⋄ REV MR 58 # 135 Zbl 403 # 03028
• ID 54742

MORAIS, R. [1977] *Projective logics and projective boolean algebras* (P 1076) Latin Amer Symp Math Log (3);1976 Campinas 201-221
⋄ B60 C75 E15 G05 G25 ⋄ REV MR 57 # 16065 Zbl 366 # 02038 • ID 16605

MORALES-LUNA, G. [1985] see ADAMOWICZ, Z.

MORAVCSIK, J.M.E. [1974] see GABBAY, D.M.

MORAVCSIK, J.M.E. see Vol. II for further entries

MOREL, A.C. [1958] see CHANG, C.C.

MOREL, A.C. [1962] see FRAYNE, T.E.

MOREL, A.C. [1964] *Ordering relations admitting automorphisms* (J 0027) Fund Math 54*279-284
⋄ C07 C65 E07 ⋄ REV MR 30 # 1061 Zbl 146.15
• ID 09473

MOREL, A.C. [1968] *Structure and order structure in Abelian groups* (S 0019) Colloq Math (Warsaw) 19*199-209
⋄ C60 E07 ⋄ REV MR 38 # 1166 • ID 09475

MOREL, A.C. [1973] *Algebras with a maximal C-dependence property* (J 0004) Algeb Universalis 3*160-173
⋄ C05 ⋄ REV MR 51 # 312 Zbl 301 # 08007 • ID 17487

MOREL, A.C. also published under the name DAVIS, A.C.

MOREL, A.C. see Vol. V for further entries

MORGAN, C.G. [1982] *Simple probabilistic semantics for propositional K, T, B, S4, and S5* (J 0122) J Philos Logic 11*443-458
⋄ B45 C90 ⋄ REV MR 85d:03056 Zbl 512 # 03011
• ID 36507

MORGAN, C.G. [1982] *There is a probabilistic semantics for every extension of classical sentence logic* (J 0122) J Philos Logic 11*431-442
⋄ B45 C90 ⋄ REV MR 85d:03055 Zbl 515 # 60003
• ID 37860

MORGAN, C.G. [1983] see LEBLANC, H.

MORGAN, C.G. see Vol. I, II, IV, VI for further entries

MORGENSTERN, C.F. [1979] *On amalgamations of languages with Magidor-Malitz quantifiers* (J 0036) J Symb Logic 44*549-558
⋄ C55 C80 E55 ⋄ REV MR 80k:03035 Zbl 431 # 03023
• ID 53929

MORGENSTERN, C.F. [1979] *The measure quantifier* (J 0036) J Symb Logic 44*103-108
⋄ C55 C75 C80 E55 ⋄ REV MR 80f:03041
Zbl 404 # 03025 • ID 54812

MORGENSTERN, C.F. [1982] *On generalized quantifiers in arithmetic* (J 0036) J Symb Logic 47*187-190
⋄ C62 C80 F30 ⋄ REV MR 84g:03051 Zbl 487 # 03019 JSL 50.1078 • ID 34161

MORGENSTERN, C.F. see Vol. V for further entries

MORLEY, M.D. & VAUGHT, R.L. [1962] *Homogeneous universal models* (J 0132) Math Scand 11*37-57
⋄ C50 ⋄ REV MR 27 # 37 Zbl 112.6 JSL 32.535 • ID 09489

MORLEY, M.D. [1963] *On theories categorical in uncountable powers* (J 0054) Proc Nat Acad Sci USA 49*213-216
⋄ C35 C45 ⋄ REV MR 32 # 7410 Zbl 118.14 JSL 31.646
• ID 09491

MORLEY, M.D. [1965] *Categoricity in power* (J 0064) Trans Amer Math Soc 114*514-538
⋄ C35 C45 ⋄ REV MR 31 # 58 Zbl 151.11 JSL 31.646
• ID 09492

MORLEY, M.D. [1965] *Omitting classes of elements* (P 0614) Th Models;1963 Berkeley 265-273
⋄ C30 C55 C75 ⋄ REV MR 34 # 1189 Zbl 168.249 JSL 33.286 • ID 24850

MORLEY, M.D. [1967] *Countable models of $\aleph_1$-categorical theories* (J 0029) Israel J Math 5*65-72
⋄ C15 C35 ⋄ REV MR 36 # 2488 Zbl 173.6 JSL 40.636
• ID 09493

MORLEY, M.D. [1967] see KEISLER, H.J.

MORLEY, M.D. [1968] see KEISLER, H.J.

MORLEY, M.D. [1968] *Partitions and models* (P 0692) Summer School in Logic;1967 Leeds 109–158
⋄ C30 C50 C55 C62 C75 E05 E45 E55 ⋄ REV MR 40#1273 Zbl 191.295 JSL 39.182 • ID 09494

MORLEY, M.D. [1970] *The number of countable models* (J 0036) J Symb Logic 35*14–18
⋄ C15 C75 ⋄ REV MR 44#5213 Zbl 196.10 JSL 49.314 • ID 09495

MORLEY, M.D. [1971] *The Loewenheim-Skolem theorem for models with standard part* (P 0669) Conv Teor Modelli & Geom;1969/70 Roma 43–52
⋄ C30 C55 ⋄ REV MR 44#1551 Zbl 215.324 • ID 09496

MORLEY, M.D. [1973] *Countable models with standard part* (P 0793) Int Congr Log, Meth & Phil of Sci (4,Proc);1971 Bucharest 57–62
⋄ C15 C50 C52 ⋄ REV MR 56#8357 • ID 31895

MORLEY, M.D. [1973] *Studies in model theory* (C 0654) Stud in Model Th vii+197pp
⋄ C98 ⋄ REV MR 48#5845 Zbl 298#02051 • ID 31896

MORLEY, M.D. [1974] *A remark on a paper by Abian* (J 0001) Acta Math Acad Sci Hung 25*413
⋄ C35 G05 ⋄ REV MR 50#4287 Zbl 308#02051 • ID 09498

MORLEY, M.D. [1974] *Applications of topology to $L_{\omega_1,\omega}$* (P 0610) Tarski Symp;1971 Berkeley 233–240
⋄ C75 ⋄ REV MR 51#10079 Zbl 306#02051 • ID 09497

MORLEY, M.D. [1976] *Decidable models* (J 0029) Israel J Math 25*233–240
⋄ C50 C57 ⋄ REV MR 56#15405 Zbl 361#02067 • ID 26084

MORLEY, M.D. [1977] *Homogenous sets* (C 1523) Handb of Math Logic 181–196
⋄ C30 C55 C98 E05 E55 ⋄ REV MR 58#10395 JSL 49.968 • ID 24200

MORLEY, M.D. [1985] see HARRINGTON, L.A.

MORLEY, M.D. see Vol. IV for further entries

MOROZOV, A.S. [1982] *Countable homogeneous boolean algebras (Russian)* (J 0003) Algebra i Logika 21*269–282
• TRANSL [1982] (J 0069) Algeb and Log 21*181–190
⋄ C15 C50 C57 G05 ⋄ REV MR 85a:06023 Zbl 539#03011 • ID 39816

MOROZOV, A.S. [1982] *Strong constructivizability of countable saturated boolean algebras (Russian)* (J 0003) Algebra i Logika 21*193–203
• TRANSL [1982] (J 0069) Algeb and Log 21*130–137
⋄ C50 C57 G05 ⋄ REV MR 85b:03052 Zbl 526#03016 • ID 38169

MOROZOV, A.S. [1983] *Groups of recursive automorphisms of constructive boolean algebras (Russian)* (J 0003) Algebra i Logika 22*138–158
• TRANSL [1983] (J 0069) Algeb and Log 22*95–112
⋄ C07 C57 D45 F60 G05 ⋄ REV Zbl 549#03031 • ID 43123

MOROZOV, A.S. [1984] *Group $Aut_r(Q, \leq)$ is not constructivizable (Russian)* (J 0087) Mat Zametki (Akad Nauk SSSR) 36*473–478
• TRANSL [1984] (J 1044) Math Notes, Acad Sci USSR 36*733–736
⋄ C07 C57 F60 ⋄ REV MR 86g:20001 Zbl 574#03028 • ID 41757

MOROZOV, A.S. [1985] *Constructive Boolean algebras with almost identical automorphisms (Russian)* (J 0087) Mat Zametki (Akad Nauk SSSR) 37*478–482,599
• TRANSL [1985] (J 1044) Math Notes, Acad Sci USSR 37*266–268
⋄ C07 C57 G05 ⋄ ID 46375

MOROZOV, A.S. see Vol. IV for further entries

MORTIMER, M. [1974] *Some results in modal model theory* (J 0036) J Symb Logic 39*496–508
⋄ B45 C90 ⋄ REV MR 50#9537 Zbl 306#02021 • ID 09506

MORTIMER, M. see Vol. I for further entries

MOSCATELLI, V.B. [1982] *Ultraprodotti e super-riflessivita in teoria degli spazi di Banach* (J 3601) Boll Unione Mat Ital, VI Ser, C 1/1*81–89
⋄ C20 C65 ⋄ REV MR 84d:46013 Zbl 514#46054 • ID 39703

MOSCATO, U. [1981] see MIGLIOLI, P.A.

MOSCATO, U. see Vol. II, VI for further entries

MOSCHOVAKIS, Y.N. [1964] *Recursive metric spaces* (J 0027) Fund Math 55*215–238
⋄ C57 C65 D45 F60 ⋄ REV MR 32#45 Zbl 221#02015 JSL 31.651 • ID 09514

MOSCHOVAKIS, Y.N. [1966] *Notation systems and recursive ordered fields* (J 0020) Compos Math 17*40–71
⋄ C57 C60 D45 F60 ⋄ REV MR 31#5798 Zbl 143.13 JSL 31.650 • ID 21265

MOSCHOVAKIS, Y.N. [1969] *Abstract computability and invariant definability* (J 0036) J Symb Logic 34*605–633
⋄ C40 D70 D75 ⋄ REV MR 42#5791 Zbl 218#02039 • ID 09517

MOSCHOVAKIS, Y.N. [1970] *The Suslin-Kleene theorem for countable structures* (J 0025) Duke Math J 37*341–352
⋄ C15 C40 D55 D70 D75 ⋄ REV MR 42#7509 Zbl 207.12 • ID 09518

MOSCHOVAKIS, Y.N. [1971] *Predicative classes* (P 0693) Axiomatic Set Th;1967 Los Angeles 1*247–264
⋄ C62 D55 D60 D65 D70 E70 ⋄ REV MR 43#7314 Zbl 234#02029 JSL 40.506 • ID 09520

MOSCHOVAKIS, Y.N. [1971] see BARWISE, J.

MOSCHOVAKIS, Y.N. [1972] *The game quantifier* (J 0053) Proc Amer Math Soc 31*245–250
⋄ C75 C80 D55 D70 D75 E60 ⋄ REV MR 44#3871 Zbl 243#02035 JSL 38.653 • ID 09521

MOSCHOVAKIS, Y.N. [1978] see BARWISE, J.

MOSCHOVAKIS, Y.N. see Vol. IV, V for further entries

MOSES, M. [1983] *Recursive properties of isomorphism types* (J 3194) J Austral Math Soc, Ser A 34*269–286
⋄ C57 D45 ⋄ REV MR 84g:03067 Zbl 532#03022 Zbl 546#03025 • ID 34174

MOSES, M. [1984] *Recursive linear orders with recursive successivities* (J 0073) Ann Pure Appl Logic 27∗253-264
⋄ C57 D45 E07 ⋄ REV Zbl 572 # 03025 • ID 39682

MOSES, M. [1984] *Recursive properties of isomorphism types* (J 0016) Bull Austral Math Soc 29∗419-421
⋄ C57 D45 ⋄ REV Zbl 532 # 03022 • ID 46602

MOSTERIN, J. [1973] *El problema de la decision en la logica de predicados* (J 1798) Convivium (Barcelona) 39∗4-11
⋄ B25 ⋄ ID 31172

MOSTERIN, J. see Vol. I, II, V for further entries

MOSTOWSKI, A.WLODZIMIERZ [1979] *A note concerning the complexity of a decision problem for positive formulas in SkS* (P 2952) CAAP'79 Arbres en Algeb & Progr (4);1979 Lille 173-180
⋄ B15 B25 D15 ⋄ REV MR 80m:03028 • ID 76601

MOSTOWSKI, A.WLODZIMIERZ [1980] *Finite automata on infinite trees and subtheories of SkS* (P 3057) CAAP'80 Arbres en Algeb & Progr (5);1980 Lille 228-240
⋄ B25 C85 D05 ⋄ REV MR 83e:03061 Zbl 489 # 03012 • ID 34829

MOSTOWSKI, A.WLODZIMIERZ [1981] *The complexity of automata and subtheories of monadic second order arithmetics* (P 3165) FCT'81 Fund of Comput Th;1981 Szeged 453-466
⋄ B15 B25 D05 D15 F35 ⋄ REV MR 83g:03036 Zbl 469 # 03025 • ID 55152

MOSTOWSKI, A.WLODZIMIERZ see Vol. I, II, IV for further entries

MOSTOWSKI, ANDRZEJ [1937] *Abzaehlbare Boolesche Koerper und ihre Anwendung auf die allgemeine Metamamathematik* (J 0027) Fund Math 29∗34-53
⋄ C15 C50 G05 ⋄ REV Zbl 16.337 JSL 3.47 FdM 63.829 • ID 09527

MOSTOWSKI, ANDRZEJ [1938] *Ueber gewisse universelle Relationen* (J 0283) Ann Soc Pol Math 17∗117-118
⋄ C50 ⋄ ID 09529

MOSTOWSKI, ANDRZEJ [1947] *On absolute properties of relations* (J 0036) J Symb Logic 12∗33-42
⋄ B15 C75 C85 E15 E47 ⋄ REV MR 10.93 Zbl 29.100 JSL 13.46 • ID 09538

MOSTOWSKI, ANDRZEJ [1949] *Sur l'interpretation geometrique et topologique des notions logiques* (P 0682) Int Congr Philos (10);1948 Amsterdam 767-769
⋄ B10 C90 ⋄ REV MR 10.423 Zbl 31.193 JSL 14.184 • ID 09545

MOSTOWSKI, ANDRZEJ [1952] *On direct products of theories* (J 0036) J Symb Logic 17∗1 31
⋄ B25 C10 C30 ⋄ REV MR 13.897 Zbl 47.7 JSL 17.203 • ID 09549

MOSTOWSKI, ANDRZEJ [1952] *On models of axiomatic systems* (J 0027) Fund Math 39∗133-158
⋄ C62 F25 F30 F35 ⋄ REV MR 14.938 Zbl 53.201 JSL 19.220 • ID 09550

MOSTOWSKI, ANDRZEJ [1953] *On a system of axioms which has no recursively enumerable arithmetic model* (J 0027) Fund Math 40∗56-61
⋄ C57 D45 E30 E70 ⋄ REV MR 15.667 Zbl 53.3 JSL 23.45 • ID 09553

MOSTOWSKI, ANDRZEJ [1955] *A formula with no recursively enumerable model* (J 0027) Fund Math 42∗125-140
⋄ C57 D45 ⋄ REV MR 17.225 Zbl 67.251 JSL 23.45 • ID 09558

MOSTOWSKI, ANDRZEJ [1956] see LOS, J.

MOSTOWSKI, ANDRZEJ [1956] *Concerning a problem of H.Scholz* (J 0068) Z Math Logik Grundlagen Math 2∗210-214
⋄ C13 D20 ⋄ REV MR 19.240 Zbl 74.249 JSL 24.241 • ID 09561

MOSTOWSKI, ANDRZEJ [1956] *Development and applications of the "projective" classification of sets of integers* (P 0575) Int Congr Math (II, 7);1954 Amsterdam 3∗280-288
⋄ C57 C80 D55 E15 E45 ⋄ REV MR 19.238 Zbl 75.235 JSL 23.44 • ID 09560

MOSTOWSKI, ANDRZEJ [1956] see EHRENFEUCHT, A.

MOSTOWSKI, ANDRZEJ [1956] *On models of axiomatic set theory* (J 0014) Bull Acad Pol Sci, Ser Math Astron Phys 4∗663-667
⋄ C62 E30 E47 ⋄ REV MR 18.711 Zbl 75.8 JSL 32.531 • ID 09559

MOSTOWSKI, ANDRZEJ [1957] *On a generalisation of quantifiers* (J 0027) Fund Math 44∗12-36
⋄ B25 C55 C80 C95 ⋄ REV MR 19.724 Zbl 78.244 JSL 23.217 JSL 25.365 • ID 16980

MOSTOWSKI, ANDRZEJ [1957] *On recursive models of formalised arithmetic* (J 0014) Bull Acad Pol Sci, Ser Math Astron Phys 5∗705-710
⋄ C57 C62 D45 ⋄ REV MR 20 # 7 Zbl 81.12 JSL 23.45 • ID 09562

MOSTOWSKI, ANDRZEJ [1958] *Quelques applications de la topologie a la logique mathematique* (2222) See Remarks 470-477
⋄ C90 C98 E15 ⋄ REV JSL 36.688 • REM Appendix to the 4th ed. of Kuratowski,C.: Topologie I, 1933 • ID 43873

MOSTOWSKI, ANDRZEJ [1958] *Quelques observations sur l'usage des methodes non finitistes dans la meta-mathematique* (P 0576) Raisonn en Math & Sci Exper;1955 Paris 19-29 (Disc 29-32)
⋄ A05 C30 C62 C80 E30 E35 F99 ⋄ REV MR 21 # 4896 Zbl 105.5 JSL 24.234 • ID 09566

MOSTOWSKI, ANDRZEJ [1958] see GRZEGORCZYK, A.

MOSTOWSKI, ANDRZEJ [1959] *A class of models for second order arithmetic* (J 0014) Bull Acad Pol Sci, Ser Math Astron Phys 7∗401-404
⋄ C62 ⋄ REV MR 22 # 6706 Zbl 100.11 JSL 34.128 • ID 09567

MOSTOWSKI, ANDRZEJ [1959] *Notes concerning an existence proof for standard models (Russian)* (P 0607) All-Union Math Conf (3);1956 Moskva 4∗232-236
⋄ C62 E30 E45 ⋄ REV Zbl 135.6 • ID 31287

MOSTOWSKI, ANDRZEJ [1961] see EHRENFEUCHT, A.

MOSTOWSKI, ANDRZEJ [1961] *Concerning the problem of axiomatizability of the field of real numbers in the weak second order logic* (C 0622) Essays Found of Math (Fraenkel) 269-286
⋄ B15 B28 C85 ⋄ REV MR 28 # 5003 JSL 32.130 • ID 09570

MOSTOWSKI, ANDRZEJ [1961] see GRZEGORCZYK, A.

MOSTOWSKI, ANDRZEJ [1961] *Formal systems of analysis based on an infinitistic rule of proof* (P 0633) Infinitist Meth;1959 Warsaw 141-166
⋄ C62 E45 F35 ⋄ REV MR 36#3631 Zbl 121.15 JSL 34.128 • ID 24858

MOSTOWSKI, ANDRZEJ [1962] *A problem in the theory of models* (J 0014) Bull Acad Pol Sci, Ser Math Astron Phys 10*121-126
⋄ C15 C52 ⋄ REV MR 32#5513 Zbl 108.4 JSL 39.600 • ID 09573

MOSTOWSKI, ANDRZEJ [1962] *L'espace des modeles d'une theorie formalisee et quelques-unes de ses applications* (P 1606) Colloq Math (Pascal);1962 Clermont-Ferrand 7*107-116
⋄ B50 C07 C40 C57 C80 C85 C90 D45 ⋄ REV MR 47#4785 JSL 40.501 • ID 09574

MOSTOWSKI, ANDRZEJ [1962] *On invariant, dual invariant and absolute formulas* (J 0202) Diss Math (Warsaw) 29*1-38
⋄ B15 C40 C85 ⋄ REV MR 29#5721 Zbl 106.5 • ID 22070

MOSTOWSKI, ANDRZEJ [1965] *On models of Zermelo-Fraenkel set theory satisfying the axiom of constructibility* (J 0096) Acta Philos Fenn 18*135-144
⋄ C62 E45 ⋄ REV MR 37#58 Zbl 143.258 JSL 36.542 • ID 09579

MOSTOWSKI, ANDRZEJ [1965] *Thirty years of foundational studies. Lectures on the development of mathematical logic and the studies of the foundations of mathematics in 1930-1964* (J 0096) Acta Philos Fenn 17*1-180
• REPR [1966] (X 1096) Blackwell: Oxford 180pp
⋄ A10 B98 C98 D98 E98 F98 ⋄ REV MR 33#18 MR 33#5445 Zbl 146.245 JSL 33.111 • ID 09578

MOSTOWSKI, ANDRZEJ [1967] *Modeles transitifs de la theorie des ensembles de Zermelo-Fraenkel* (X 0893) Pr Univ Montreal: Montreal 170pp
⋄ C62 E35 E45 E98 ⋄ REV MR 45#3197 Zbl 189.286 • ID 22257

MOSTOWSKI, ANDRZEJ [1968] *Craig's interpolation theorem in some extended systems of logic* (P 0627) Int Congr Log, Meth & Phil of Sci (3,Proc);1967 Amsterdam 87-103
⋄ C40 C95 ⋄ REV MR 40#4093 Zbl 182.9 • ID 09582

MOSTOWSKI, ANDRZEJ [1969] *Constructible sets with applications* (X 0809) North Holland: Amsterdam ix+269pp
• TRANSL [1973] (X 0885) Mir: Moskva 256pp
⋄ C62 E25 E35 E45 E50 E70 E98 ⋄ REV MR 41#52 MR 49#10548 Zbl 185.14 JSL 40.631 • ID 21062

MOSTOWSKI, ANDRZEJ [1969] *Models of set theory* (P 0630) Aspects Math Log;1968 Varenna 65-179
⋄ C62 E98 ⋄ REV MR 40#7106 Zbl 228#02037 • ID 09584

MOSTOWSKI, ANDRZEJ & SUZUKI, Y. [1969] *On ω-models which are not β-models* (J 0027) Fund Math 65*83-93
⋄ C62 ⋄ REV MR 40#7104 Zbl 294#02029 • ID 09586

MOSTOWSKI, ANDRZEJ [1972] *A transfinite sequence of ω-models* (J 0036) J Symb Logic 37*96-102
⋄ C62 D30 ⋄ REV MR 48#3721 Zbl 246#02038 • ID 09588

MOSTOWSKI, ANDRZEJ [1972] *Models of second order arithmetic with definable Skolem functions* (J 0027) Fund Math 75*223-234 • ERR/ADD ibid 84*173
⋄ C62 ⋄ REV MR 47#8286 MR 50#89 Zbl 245#02049 Zbl 275#02047 JSL 38.652 • ID 19461

MOSTOWSKI, ANDRZEJ [1973] *A contribution to teratology* (Russian) (C 0733) Izbr Vopr Algeb & Log (Mal'tsev) 184-196
⋄ C20 C62 ⋄ REV MR 48#5851 Zbl 279#02038 • ID 29548

MOSTOWSKI, ANDRZEJ [1973] *Partial orderings of the family of ω-models* (P 0793) Int Congr Log, Meth & Phil of Sci (4,Proc);1971 Bucharest 13-28
⋄ C62 ⋄ REV MR 56#8358 • ID 14836

MOSTOWSKI, ANDRZEJ [1974] *Observations concerning elementary extensions of ω-models* (P 0610) Tarski Symp;1971 Berkeley 349-355
⋄ C62 C80 ⋄ REV MR 51#12509 Zbl 329#02023 • REM Part I. Part II 1973 by Marek,W. • ID 09589

MOSTOWSKI, ANDRZEJ [1975] see MAREK, W.

MOSTOWSKI, ANDRZEJ [1976] *A remark on models of the Goedel-Bernays axioms for set theory* (C 1468) Sets & Classes (Bernays) 325-340
⋄ C62 E30 E70 ⋄ REV MR 56#5293 Zbl 338#02037 • ID 23738

MOSTOWSKI, ANDRZEJ [1976] *Two remarks on the models of Morse' set theory* (P 1476) Set Th & Hierarch Th (2) (Mostowski);1975 Bierutowice 13-21
• REPR [1977] (P 1639) Set Th & Hierarch Th (1);1974 Karpacz 1*5-11
⋄ C62 E25 E35 E40 E70 ⋄ REV MR 56#11789 MR 58#5219 Zbl 338#02036 Zbl 378#02030 • ID 23791

MOSTOWSKI, ANDRZEJ [1978] see DONER, J.E.

MOSTOWSKI, ANDRZEJ [1979] *Foundational studies. Selected works Vol. I, II* (X 0809) North Holland: Amsterdam xlvi+635pp,viii+605pp
⋄ B96 C96 D96 E96 ⋄ REV MR 81i:01018 Zbl 425#01021 • ID 82308

MOSTOWSKI, ANDRZEJ see Vol. I, II, IV, V, VI for further entries

MOSTOWSKI, T. [1975] *A note on a decision procedure for rings of formal power series and its applications* (J 0014) Bull Acad Pol Sci, Ser Math Astron Phys 23*1229-1232
⋄ B25 C60 ⋄ REV MR 55#2899 Zbl 345#13022 Zbl 357#13012 • ID 50432

MOSTOWSKI, T. [1976] *Analytic applications of decidability theorems* (J 0519) Wiad Mat, Ann Soc Math Pol, Ser 2 20*1-6
⋄ B25 C60 C65 D80 ⋄ REV MR 56#5518 • ID 82309

MOSTOWSKI, T. [1978] see KURKE, H.

MOSTOWSKI, T. & PFISTER, G. [1979] *Der Cohensche Eliminationssatz in positiver Charakteristik* (S 0410) Math Beitr Univ Halle-Wittenberg 12*115-123
• REPR [1979] (J 3124) Beitr Algebra Geom 8*115-123
⋄ B25 C10 C60 ⋄ REV MR 81h:13015 Zbl 489#13011 • ID 36597

MOTOHASHI, N. [1969] *On normal operations on models* (J 0090) J Math Soc Japan 21*564-573
⋄ C30 C40 C50 ⋄ REV MR 41#3258 Zbl 191.296 • ID 09594

MOTOHASHI, N. [1970] *A theorem in the theory of definition* (J 0090) J Math Soc Japan 22*490-494
⋄ C40 C50 ⋄ REV MR 42 # 7499 Zbl 198.323 • ID 09595

MOTOHASHI, N. [1972] *A new theorem on definability in a positive second order logic with countable conjunctions and disjunctions* (J 0081) Proc Japan Acad 48*153-156
⋄ B15 C40 C75 C85 ⋄ REV MR 47 # 4761 Zbl 252 # 02006 • ID 09599

MOTOHASHI, N. [1972] *Countable structures for uncountable infinitary languages* (J 0081) Proc Japan Acad 48*716-718
⋄ C15 C40 C75 ⋄ REV MR 51 # 7817 Zbl 268 # 02008 • ID 17315

MOTOHASHI, N. [1972] *Interpolation theorem and characterization theorem* (J 0260) Ann Jap Ass Phil Sci 4*85-150
⋄ C40 C75 F05 ⋄ REV MR 49 # 10530 Zbl 241 # 02005 • ID 09596

MOTOHASHI, N. [1973] *An extended relativization theorem* (J 0090) J Math Soc Japan 25*250-256
⋄ C40 C75 ⋄ REV MR 49 # 2343 Zbl 254 # 02015 • ID 09601

MOTOHASHI, N. [1973] *Model theory on a positive second order logic with countable conjunctions and disjunctions* (J 0090) J Math Soc Japan 25*27-42
⋄ B15 C40 C75 C85 ⋄ REV MR 47 # 6476 Zbl 244 # 02019 • ID 09602

MOTOHASHI, N. [1973] *Two theorems on mix-relativization* (J 0081) Proc Japan Acad 49*161-163
⋄ B10 C40 C75 ⋄ REV MR 49 # 2344 Zbl 289 # 02011 • ID 09600

MOTOHASHI, N. [1974] *Atomic structures and Scott sentences (Japanese)* (J 0091) Sugaku 26*256-257
⋄ C50 C75 ⋄ REV MR 58 # 27479 • ID 90093

MOTOHASHI, N. [1975] *On a theorem of Schoenfield (Japanese)* (J 0091) Sugaku 27*368-371
⋄ C40 ⋄ REV MR 58 # 27295 • ID 90094

MOTOHASHI, N. [1975] *Some proof-theoretic properties of dense linear orderings and countable well-orderings* (J 0081) Proc Japan Acad 51*301-303
⋄ C40 C75 ⋄ REV MR 51 # 7836 Zbl 343 # 02019 • ID 18305

MOTOHASHI, N. [1977] *A remark on Scott's interpolation theorem for* $L_{\omega_1\omega}$ (J 0036) J Symb Logic 42*63
⋄ C40 C75 ⋄ REV MR 58 # 10293 Zbl 368 # 02018 • ID 23746

MOTOHASHI, N. [1977] *Partially ordered interpretations* (J 0036) J Symb Logic 42*83-93
⋄ C90 E40 F25 F50 ⋄ REV MR 58 # 201 Zbl 411 # 03030 • ID 23750

MOTOHASHI, N. [1978] *An elimination theorem of uniqueness condition* (P 2607) IBM Symp Math Log & Comput Sci (3);
⋄ B10 C40 ⋄ ID 90092

MOTOHASHI, N. [1979] *A remark on Africk's paper on Scott's interpolation theorem for* $L_{\omega_1,\omega}$ (J 0036) J Symb Logic 44*32
⋄ C40 C75 ⋄ REV MR 80c:03040 Zbl 404 # 03024 • ID 54811

MOTOHASHI, N. [1981] *Homogeneous formulas and definability theorems* (P 3201) Logic Symposia;1979/80 Hakone 109-116
⋄ C40 ⋄ REV MR 83k:03036 Zbl 504 # 03016 • ID 33417

MOTOHASHI, N. [1982] *ε-theorems and elimination theorems of uniqueness conditions* (P 3634) Patras Logic Symp;1980 Patras 373-387
⋄ B10 C40 F05 ⋄ REV MR 84h:03027 Zbl 535 # 03009 • ID 33418

MOTOHASHI, N. [1982] *An axiomatization theorem* (J 0090) J Math Soc Japan 34*551-560
⋄ B10 C07 C40 C75 ⋄ REV MR 83h:03018 Zbl 476 # 03022 • ID 55534

MOTOHASHI, N. [1983] *Some remarks on Barwise approximation theorem on Henkin quantifiers* (P 3669) SE Asian Conf on Log;1981 Singapore 107-114
⋄ C75 C80 ⋄ REV MR 85d:03073 Zbl 553 # 03003 • ID 40405

MOTOHASHI, N. [1984] *A normal form theorem for first order formulas and its application to Gaifman's splitting theorem* (P 3668) Log & Founds of Math;1983 Kyoto 40-52
⋄ C07 C62 ⋄ ID 42932

MOTOHASHI, N. [1984] *A normal form theorem for first order formulas and its application to Gaifman's splitting theorem* (J 0036) J Symb Logic 49*1262-1267
⋄ B10 C07 C62 ⋄ ID 42487

MOTOHASHI, N. [1984] *Approximation theory of uniqueness conditions by existence conditions* (J 0027) Fund Math 120*127-142
⋄ C40 C75 C80 F07 F50 ⋄ ID 45359

MOTOHASHI, N. [1984] *Equality and Lyndon's interpolation theorem* (J 0036) J Symb Logic 49*123-128
⋄ B20 C40 C75 ⋄ REV MR 86f:03053 Zbl 574 # 03014 • ID 42488

MOTOHASHI, N. see Vol. I, II, VI for further entries

MOVSISYAN, YU.M. [1975] *Algebraic systems of the second level (Russian)* (S 0166) Mat Issl, Mold SSR 10/2*182-191,285-286
⋄ B15 C05 ⋄ REV MR 53 # 235 Zbl 347 # 08005 • ID 16671

MUCHNIK, A.A. [1985] *Games on infinite trees and automata with dead ends. A new proof of decidability for the monadic theory with two successor functions (Russian)* (S 2582) Semiotika & Inf, Akad Nauk SSSR 24*16-40,142
⋄ B15 B25 C85 D05 ⋄ REV Zbl 576 # 03010 • ID 47256

MUCHNIK, A.A. see Vol. I, II, IV, VI for further entries

MUELLER, GERT H. [1961] *Nicht-Standardmodelle der Zahlentheorie* (J 0044) Math Z 77*414-438
⋄ C62 F30 ⋄ REV MR 31 # 3333 Zbl 268 # 02039 JSL 37.405 • ID 09619

MUELLER, GERT H. [1965] *Der Modellbegriff in der Mathematik* (J 0178) Stud Gen 18*154-166
⋄ C07 C98 ⋄ REV Zbl 128.11 • ID 33930

MUELLER, GERT H. [1975] see DILLER, J.

MUELLER, GERT H. & RICHTER, M.M. (EDS.) [1984] *Models and sets* (S 3301) Lect Notes Math 1103*viii+484pp
⋄ B97 C97 D97 H97 ⋄ REV MR 85k:03002a Zbl 547 # 00008 • REM Logic Colloquium;1983 Aachen, Vol.I. Vol.II 1984 by Boerger,E. • ID 41750

MUELLER, GERT H. see Vol. I, II, IV, V, VI for further entries

MUELLER, HORST [1972] *Entscheidungsprobleme im Bereich der Semantik von Programmiersprachen* (J 0487) Math Unterricht 18*58-72
⋄ B25 B75 ⋄ ID 28299

MUELLER, HORST see Vol. IV, V for further entries

MULLER, D.E. & SCHUPP, P.E. [1981] *Context-free languages, groups, the theory of ends, second-order logic, tiling problems, cellular automata, and vector addition systems* (J 0589) Bull Amer Math Soc (NS) 4*331-334
⋄ B15 B25 D05 D80 ⋄ REV MR 82m:03051 Zbl 484 # 03019 • ID 76648

MULLER, D.E. & SCHUPP, P.E. [1985] *Alternating automata on infinite objects, determinacy and Rabin's Theorem* (P 4595) Autom Infinite Words;1984 Le Mont Dore 100-107
⋄ B25 C85 D05 E60 ⋄ ID 48421

MULLER, D.E. & SCHUPP, P.E. [1985] *The theory of ends, pushdown automata, and second-order logic* (J 1426) Theor Comput Sci 37*51-75
⋄ B15 B25 D10 D80 ⋄ ID 47646

MULLER, D.E. see Vol. I, IV for further entries

MULVEY, C.J. [1974] *Intuitionistic algebra and representations of rings* (S 0167) Mem Amer Math Soc 148*3-57
⋄ C60 C90 F50 F55 G30 ⋄ REV MR 53 # 2650 Zbl 274 # 18012 • ID 21534

MULVEY, C.J. [1978] *A remark on the prime stalk theorem* (J 0326) J Pure Appl Algebra 10*253-256
⋄ C90 ⋄ REV MR 57 # 397 Zbl 403 # 18007 • ID 54770

MULVEY, C.J. [1979] see FOURMAN, M.P.

MULVEY, C.J. see Vol. V for further entries

MUNDICI, D. [1979] *Compactness + Craig interpolation = Robinson consistency in any logic* (1111) Preprints, Manuscr., Techn. Reports etc. 8pp
⋄ C40 C95 ⋄ REM Math. Institut, Univ. of Florence
• ID 33522

MUNDICI, D. [1979] *Compattezza = JEP in ogni logica* (J 3639) Notiz Unione Mat Ital 8-9*19
⋄ C40 C52 C95 ⋄ REV MR 85c:03014 • ID 33527

MUNDICI, D. [1979] *Robinson consistency theorem in soft model theory (Italian summary)* (J 0149) Atti Accad Naz Lincei Fis Mat Nat, Ser 8 67*383-386
⋄ C40 C55 C80 C95 ⋄ REV MR 83e:03056 Zbl 464 # 03032 • ID 90129

MUNDICI, D. [1980] *Natural limitations of algorithmic procedures in logic (Italian summary)* (J 0149) Atti Accad Naz Lincei Fis Mat Nat, Ser 8 69*101-105
⋄ C40 D05 D80 ⋄ REV MR 84e:03007 Zbl 518 # 03005
• ID 33528

MUNDICI, D. [1981] *A group-theoretical invariant for elementary equivalence and its role in representations of elementary classes* (J 0063) Studia Logica 40*253-267
⋄ B10 C07 C52 ⋄ REV MR 84e:03045 Zbl 482 # 03013
• ID 90124

MUNDICI, D. [1981] *An algebraic result about soft model theoretical equivalence relations with an application to H. Friedman's fourth problem* (J 0036) J Symb Logic 46*523-530
⋄ C15 C40 C95 ⋄ REV MR 82k:03056 Zbl 465 # 03020
• ID 54923

MUNDICI, D. [1981] *Applications of many-sorted Robinson consistency theorem* (J 0068) Z Math Logik Grundlagen Math 27*181-188
⋄ C40 C80 C95 ⋄ REV MR 82k:03057 Zbl 463 # 03023
• ID 54563

MUNDICI, D. [1981] *Complexity of Craig's interpolation* (P 2614) Open Days in Model Th & Set Th;1981 Jadwisin 185-204
⋄ C40 D15 ⋄ REV Zbl 507 # 03025 • ID 33726

MUNDICI, D. [1981] *Craig's interpolation theorem in computation theory* (J 0149) Atti Accad Naz Lincei Fis Mat Nat, Ser 8 70*6-11
⋄ C40 D15 ⋄ REV Zbl 523 # 03027 • ID 33529

MUNDICI, D. [1981] *Robinson's consistency theorem in soft model theory* (J 0064) Trans Amer Math Soc 263*231-241
⋄ C40 C80 C95 ⋄ REV MR 82d:03063 • ID 76664

MUNDICI, D. [1981] *Variations on Friedman's third and fourth problem* (P 2614) Open Days in Model Th & Set Th;1981 Jadwisin 205-220
⋄ C40 C75 C80 C95 ⋄ ID 90134

MUNDICI, D. [1982] *Compactness, interpolation and Friedman's third problem* (J 0007) Ann Math Logic 22*197-211
⋄ C40 C75 C95 ⋄ REV MR 84a:03040 Zbl 495 # 03020
• ID 35577

MUNDICI, D. [1982] *Complexity of Craig's interpolation* (J 2095) Fund Inform, Ann Soc Math Pol, Ser 4 5*261-278
⋄ C40 D15 ⋄ REV MR 84h:03081 Zbl 507 # 03025
• ID 34339

MUNDICI, D. [1982] *Duality between logics and equivalence relations* (J 0064) Trans Amer Math Soc 270*111-129
⋄ C40 C95 ⋄ REV MR 84g:03054 Zbl 497 # 03018
• ID 90118

MUNDICI, D. [1982] *Interpolation, compactness and JEP in soft model theory* (J 0009) Arch Math Logik Grundlagenforsch 22*61-67
⋄ C40 C52 C95 ⋄ REV MR 84c:03075 Zbl 495 # 03019
• ID 90130

MUNDICI, D. [1982] *L-embedding, amalgamation and L-elementary equivalence (Italian summary)* (J 0149) Atti Accad Naz Lincei Fis Mat Nat, Ser 8 72*312-314
⋄ C95 ⋄ REV MR 85e:03091 Zbl 527 # 03019 • ID 33523

MUNDICI, D. [1982] *Lectures on abstract model theory I,II,III* (1111) Preprints, Manuscr., Techn. Reports etc. 170pp
⋄ C95 C98 ⋄ REM University of Florence, Mathematical Institute • ID 48422

MUNDICI, D. [1983] *A lower bound for the complexity of Craig's interpolants in sentential logic* (J 0009) Arch Math Logik Grundlagenforsch 23*27-36
⋄ B20 C40 D15 ⋄ REV MR 85c:03003 Zbl 511 # 03004
• ID 37383

MUNDICI, D. [1983] *Compactness = JEP in any logic* (J 0027) Fund Math 116*99-108
⋄ C40 C52 C95 ⋄ REV MR 85c:03014 Zbl 564 # 03034
• ID 44362

MUNDICI, D. [1983] *Inverse topological systems and compactness in abstract model theory* (P 1601) Easter Conf on Model Th (1);1983 Diedrichshagen 73-98
⋄ C40 C80 C95 ⋄ REV MR 84i:03008   Zbl 537 # 03027
• ID 40143

MUNDICI, D. [1983] *Natural limitations of decision procedures for arithmetic with bounded quantifiers* (J 0009) Arch Math Logik Grundlagenforsch 23∗37-54
⋄ B25 D10 D15 F30 ⋄ REV MR 84k:03040   Zbl 523 # 03028 • ID 34976

MUNDICI, D. [1984] *Δ-tautologies, uniform and nonuniform upper bounds in computation theory (Italian summary)* (J 0149) Atti Accad Naz Lincei Fis Mat Nat, Ser 8 75∗99-101
⋄ B75 C40 D15 ⋄ REV MR 86d:03037   Zbl 568 # 03018   Zbl 568 # 03019 • ID 44693

MUNDICI, D. [1984] *A generalization of abstract model theory* (J 0027) Fund Math 124∗1-25
⋄ C90 C95 G30 ⋄ ID 41817

MUNDICI, D. [1984] *Abstract model theory and nets of C_*-algebras: noncommutative interpolation and preservation properties* (P 2153) Logic Colloq;1983 Aachen 1∗351-377
⋄ C40 C95 ⋄ ID 41770

MUNDICI, D. [1984] *Abstract model theory of many-valued logics and K-theory of certain C^*-algebras* (P 1545) Easter Conf on Model Th (2);1984 Wittenberg 157-204
⋄ B50 C90 C95 ⋄ REV Zbl 561 # 03011 • ID 44636

MUNDICI, D. [1984] *Embeddings, amalgamations, and elementary equivalence: the representation of compact logics* (J 0027) Fund Math 124∗109-122
⋄ C95 ⋄ REV MR 86e:03040 • ID 41819

MUNDICI, D. [1984] *NP and Craig's interpolation theorem* (P 3710) Logic Colloq;1982 Firenze 345-358
⋄ C40 D15 ⋄ REV MR 86e:03042 • ID 41841

MUNDICI, D. [1984] *Tautologies with a unique Craig interpolant, uniform vs. nonuniform complexity* (J 0073) Ann Pure Appl Logic 27∗265-273
⋄ C40 D15 ⋄ REV MR 86g:03056 • ID 44069

MUNDICI, D. [1985] see MAKOWSKY, J.A.

MUNDICI, D. [1985] *Other quantifiers: On overview* (C 4183) Model-Theor Log 211-233
⋄ C80 C98 ⋄ ID 48335

MUNDICI, D. see Vol. IV, V for further entries

MURAVITSKIJ, A.YU. [1981] *Strong equivalence on an intuitionistic Kripke model and assertorically equivolumetric logics (Russian)* (J 0003) Algebra i Logika 20∗165-182,250
• TRANSL [1981] (J 0069) Algeb and Log 20∗112-123
⋄ B55 C90 F50 ⋄ REV MR 83h:03087   Zbl 486 # 03015
• ID 36087

MURAVITSKIJ, A.YU. [1983] *Comparison of the topological and relational semantics of superintuitionistic logics (Russian)* (J 0003) Algebra i Logika 22∗276-296
• TRANSL [1983] (J 0069) Algeb and Log 22∗197-213
⋄ B55 C90 ⋄ REV MR 85j:03031   Zbl 543 # 03022
• ID 40399

MURAVITSKIJ, A.YU. [1984] *A result on the completeness of superintuitionistic logics* (J 0087) Mat Zametki (Akad Nauk SSSR) 36∗765-776,799
• TRANSL [1984] (J 1044) Math Notes, Acad Sci USSR 36∗883-889
⋄ B55 C90 ⋄ REV MR 86f:03046   Zbl 566 # 03017
• ID 45360

MURAVITSKIJ, A.YU. see Vol. II, VI for further entries

MURAWSKI, R. [1976] *Expandability of models for elementary arithmetic* (X 1916) Komunikaty i Rozprawy Inst: Poznan 26pp
⋄ C62 H15 ⋄ ID 39649

MURAWSKI, R. [1976] *On expandability of models of Peano arithmetic. I,II* (J 0063) Studia Logica 35∗409-419,421-431
• ERR/ADD ibid 36∗237
⋄ C62 C98 ⋄ REV MR 56 # 121   Zbl 353 # 02035   Zbl 353 # 02036 • REM Part III 1977 • ID 50118

MURAWSKI, R. [1977] *On expandability of models of Peano arithmetic. III* (J 0063) Studia Logica 36∗181-188
⋄ C62 C98 ⋄ REV MR 56 # 121c   Zbl 359 # 02058 • REM Parts I,II 1976 • ID 50605

MURAWSKI, R. [1978] *Indicators, satisfaction classes and expandability* (X 1916) Komunikaty i Rozprawy Inst: Poznan 17pp
⋄ C62 H15 ⋄ ID 39652

MURAWSKI, R. [1979] *Indicators and the structure of expansions* (X 1916) Komunikaty i Rozprawy Inst: Poznan 18pp
⋄ C62 H15 ⋄ ID 35976

MURAWSKI, R. [1980] *Some remarks on the structure of expansions* (J 0068) Z Math Logik Grundlagen Math 26∗537-546
⋄ C50 C62 E45 ⋄ REV MR 82c:03052   Zbl 444 # 03037
• ID 76666

MURAWSKI, R. [1981] *A note on inner interpretations of models of Peano arithmetic* (J 0302) Rep Math Logic, Krakow & Katowice 13∗53-57
⋄ C62 ⋄ REV MR 83i:03060   Zbl 483 # 03037 • ID 35523

MURAWSKI, R. [1981] *A simple remark on satisfaction classes, indiscernibles and recursive saturation* (J 2718) Fct Approximatio, Comment Math, Poznan 11∗149-151
⋄ C30 C50 C57 H15 ⋄ REV MR 84f:03059   Zbl 481 # 03022 • ID 34478

MURAWSKI, R. [1981] *Incompleteness of Σ_n^0-definable theories via indicators* (J 2718) Fct Approximatio, Comment Math, Poznan 11∗57-63
⋄ C62 ⋄ REV MR 84g:03109   Zbl 475 # 03040 • ID 55494

MURAWSKI, R. [1981] see KIRBY, L.A.S.

MURAWSKI, R. [1981] *Some more remarks on expandability of initial segments* (P 2614) Open Days in Model Th & Set Th;1981 Jadwisin 221-230
⋄ C62 ⋄ ID 33727

MURAWSKI, R. [1984] *A contribution to nonstandard teratology* (P 2153) Logic Colloq;1983 Aachen 1∗379-388
⋄ H15 ⋄ REV MR 86h:03067   Zbl 565 # 03036 • ID 39688

MURAWSKI, R. [1984] *Expandability of models of arithmetic* (P 3621) Frege Konferenz (2);1984 Schwerin 87-93
⋄ C62 H15 ⋄ REV MR 85m:03006   Zbl 565 # 03035
• ID 39693

MURAWSKI, R. [1984] *Trace expansions of initial segments* (**J** 0068) Z Math Logik Grundlagen Math 30∗471-476
  ⋄ C62 ⋄ REV MR 86g:03111 Zbl 536 # 03051 • ID 39659

MURAWSKI, R. see Vol. IV, V, VI for further entries

MURSKIJ, V.L. [1975] *A finite basis of identities and other properties of "almost all" finite algebras (Russian)* (**J** 0052) Probl Kibern 30∗43-56
  ⋄ C05 C13 C52 ⋄ REV JSL 50.1073 • ID 48343

MURSKIJ, V.L. see Vol. I, II, IV for further entries

MURZIN, F.A. [1974] *Totally transcendental and $\aleph_1$-categorical theories (Russian)* (**P** 1590) All-Union Conf Math Log (3);1974 Novosibirsk 145-147
  ⋄ C35 C45 ⋄ ID 32632

MURZIN, F.A. [1976] *$\aleph_1$-categorical theories (Russian)* (**X** 2235) VINITI: Moskva 3095-76∗28pp
  ⋄ C35 C45 ⋄ REV Zbl 352 # 02037 • ID 32635

MURZIN, F.A. [1976] *Two arbitrary elementarily equivalent models of continuous logic have isomorphic ultra-powers (Russian)* (**C** 2555) Algeb Sistemy (Irkutsk) 107-121
  ⋄ C20 C90 ⋄ ID 32631

MURZIN, F.A. [1977] *$\aleph_1$-categorical theories (Russian)* (**J** 0092) Sib Mat Zh 18∗232
  • TRANSL [1977] (**J** 0475) Sib Math J 18∗171
  ⋄ C35 C45 ⋄ REV Zbl 352 # 02037 • REM Summary of Murzin; 1976 • ID 32630

MURZIN, F.A. [1978] *A question by Jokusch, Rubel and Takeuti (Russian)* (**J** 0092) Sib Mat Zh 19∗353-359,479
  • TRANSL [1978] (**J** 0475) Sib Math J 19∗247-252
  ⋄ C65 ⋄ REV MR 58 # 10420 Zbl 409 # 54024 • ID 32629

MURZIN, F.A. [1980] *Elementary equivalence of rings of continuous functions (Russian)* (**C** 2620) Teor Model & Primen 73-74
  ⋄ C60 ⋄ REV MR 82c:03049 • ID 76675

MURZIN, F.A. see Vol. VI for further entries

MUSTAFIN, T.G. & TAJMANOV, A.D. [1970] *Countable models of theories which are categorical in power $\aleph_1$ but not in power $\aleph_0$ (Russian)* (**J** 0003) Algebra i Logika 9∗559-565
  • TRANSL [1970] (**J** 0069) Algeb and Log 9∗338-341
  ⋄ C15 C35 ⋄ REV MR 45 # 41 Zbl 222 # 02063 • ID 76686

MUSTAFIN, T.G. [1970] *Models of totally transcendental theories (Russian)(Kazakh summary)* (**J** 0403) Izv Akad Nauk Kazak SSR, Ser Fiz-Mat 1970/1∗37-43
  ⋄ C45 C50 ⋄ REV MR 43 # 6068 Zbl 188.20 • ID 09659

MUSTAFIN, T.G. [1977] *A strong base of the elementary types of theories (Russian)* (**J** 0092) Sib Mat Zh 18∗1356-1366,1437
  • TRANSL [1977] (**J** 0475) Sib Math J 18∗961-969
  ⋄ C45 C50 ⋄ REV MR 57 # 5731 Zbl 387 # 03011
  • ID 52227

MUSTAFIN, T.G. [1977] *On the two-cardinal problem (Russian)(Kazakh summary)* (**J** 0403) Izv Akad Nauk Kazak SSR, Ser Fiz-Mat 1977/3∗40-44,89
  ⋄ C55 E15 E50 ⋄ REV MR 58 # 5171 Zbl 364 # 02039
  • ID 50969

MUSTAFIN, T.G. [1979] see BEKENOV, M.I.

MUSTAFIN, T.G. [1980] *A non-two-cardinal set of stable types (Russian)* (**J** 0087) Mat Zametki (Akad Nauk SSSR) 27∗515-525,668
  • TRANSL [1980] (**J** 1044) Math Notes, Acad Sci USSR 27∗253-259
  ⋄ C35 C45 C55 ⋄ REV MR 81f:03041 Zbl 448 # 03019
  • ID 76683

MUSTAFIN, T.G. [1980] *Rank functions in stable theories (Russian)* (**J** 0092) Sib Mat Zh 21/6∗84-95
  • TRANSL [1980] (**J** 0475) Sib Math J 21∗815-824
  ⋄ C45 ⋄ REV MR 83c:03030 Zbl 478 # 03012 • ID 55627

MUSTAFIN, T.G. [1980] *Theories with a two-cardinal formula (Russian)* (**J** 0003) Algebra i Logika 19∗676-682,746
  • TRANSL [1980] (**J** 0069) Algeb and Log 19∗438-442
  ⋄ C45 C50 C52 C55 ⋄ REV MR 82k:03045
  Zbl 473 # 03028 • ID 90302

MUSTAFIN, T.G. [1981] *On the number of countable models of a countable complete theory (Russian)* (**J** 0003) Algebra i Logika 20∗69-91,124
  • TRANSL [1981] (**J** 0069) Algeb and Log 20∗48-65
  ⋄ C15 C45 ⋄ REV MR 84m:03047 Zbl 473 # 03025
  • ID 55355

MUSTAFIN, T.G. [1981] *Principles of normalization of formulas (Russian)* (**J** 0092) Sib Mat Zh 22/2∗158-169,237
  • TRANSL [1981] (**J** 0475) Sib Math J 22∗291-299
  ⋄ C45 ⋄ REV MR 82e:03034 Zbl 529 # 03010 • ID 76682

MUSTAFIN, T.G. [1981] see BEKENOV, M.I.

MUSTAFIN, T.G. & NURMAGAMBETOV, T.A. [1982] *Separable types and rank functions in stable theories (Russian)* (**J** 0003) Algebra i Logika 21∗204-218
  • TRANSL [1982] (**J** 0069) Algeb and Log 21∗138-148
  ⋄ C45 ⋄ REV MR 84k:03099 Zbl 544 # 03012 • ID 36116

MUSTAFIN, T.G. [1985] *Classification of superstable theories by rank functions (Russian)* (**J** 0003) Algebra i Logika 24∗42-64,123
  • TRANSL [1985] (**J** 0069) Algeb and Log 24∗27-40
  ⋄ C45 ⋄ ID 49414

MUZALEWSKI, M. [1978] *Restricted decision problems in some classes of algebraic systems* (**J** 0068) Z Math Logik Grundlagen Math 24∗279-287
  ⋄ B25 ⋄ REV MR 81g:03010 Zbl 396 # 03014 • ID 52624

MUZALEWSKI, M. see Vol. IV for further entries

MYASNIKOV, A.G. & REMESLENNIKOV, V.N. [1981] *Isomorphisms and elementary properties of nilpotent power groups (Russian)* (**J** 0023) Dokl Akad Nauk SSSR 258∗1056-1059
  • TRANSL [1981] (**J** 0062) Sov Math, Dokl 23∗637-640
  ⋄ C60 ⋄ REV MR 83d:20026 Zbl 512 # 20019 • ID 39605

MYASNIKOV, A.G. & REMESLENNIKOV, V.N. [1982] *Classification of nilpotent powered groups according to elementary properties (Russian)* (**C** 3953) Mat Log & Teor Algor 56-87
  ⋄ C60 D40 ⋄ REV MR 85h:20042 Zbl 516 # 20021
  • ID 38582

MYASNIKOV, A.G. & REMESLENNIKOV, V.N. [1982] *Definability of the set of Mal'tsev bases and elementary theories of finite-dimensional algebras I (Russian)* (J 0092) Sib Mat Zh 23/5∗152-167,224
- TRANSL [1982] (J 0475) Sib Math J 23∗711-724
⋄ B25 C60 ⋄ REV MR 84d:03039 Zbl 516 # 20022 • REM Part II 1983 • ID 34079

MYASNIKOV, A.G. & REMESLENNIKOV, V.N. [1983] *Definability of sets of Mal'tsev bases and elementary theories of finite-dimensional algebras II (Russian)* (J 0092) Sib Mat Zh 24/2∗97-113
- TRANSL [1983] (J 0475) Sib Math J 24∗231-246
⋄ C60 ⋄ REV MR 85b:03054 Zbl 531 # 20018 • REM Part I 1982 • ID 38567

MYASNIKOV, A.G. & REMESLENNIKOV, V.N. [1984] *Finite dimensional algebras and k-groups of finite rank* (C 3510) Contrib Group Theory (Lyndon) 436-454
⋄ C60 C98 ⋄ REV MR 86c:20036 Zbl 551 # 20025
• ID 44148

MYCIELSKI, J. [1957] *A characterisation of arithmetical classes (Russian summary)* (J 0014) Bull Acad Pol Sci, Ser Math Astron Phys 5∗1025-1027
⋄ C52 ⋄ REV MR 20 # 5 Zbl 79.245 • ID 09675

MYCIELSKI, J. [1964] *Some compactifications of general algebras* (S 0019) Colloq Math (Warsaw) 13∗1-9 • ERR/ADD [1968] (J 0027) Fund Math 61∗281
⋄ C05 C50 ⋄ REV MR 37 # 3986 Zbl 136.261 JSL 40.88
• ID 19447

MYCIELSKI, J. [1964] *The definition of arithmetic operations in the Ackermann model (Russian)* (J 0003) Algebra i Logika 3/5-6∗64-65
⋄ C62 E47 ⋄ REV MR 31 # 2134 Zbl 171.264 • ID 09682

MYCIELSKI, J. [1965] *On unions of denumerable models (English)* (J 0003) Algebra i Logika 4/2∗57-58
⋄ C07 C15 ⋄ REV MR 31 # 5799 Zbl 163.245 JSL 33.287
• ID 09688

MYCIELSKI, J. & RYLL-NARDZEWSKI, C. [1968] *Equationally compact algebras II* (J 0027) Fund Math 61∗271-281
• ERR/ADD ibid 62∗309
⋄ C05 C20 C50 C65 ⋄ REV MR 37 # 1238
Zbl 263 # 08001 JSL 40.88 • REM Part I 1966 by Weglorz,B. 1966. Part III 1967 by Weglorz,B. • ID 09691

MYCIELSKI, J. [1968] see FAJTLOWICZ, S.

MYCIELSKI, J. & TAYLOR, W. [1976] *A compactification of the algebra of terms* (J 0004) Algeb Universalis 6∗159-163
⋄ C05 C75 ⋄ REV MR 55 # 7886 Zbl 358 # 08001
• ID 24503

MYCIELSKI, J. [1976] *A 01-law in a finite space* (J 0396) Mat Lapok 23∗288
⋄ C13 ⋄ ID 15416

MYCIELSKI, J. [1977] see MAKKAI, M.

MYCIELSKI, J. & PERLMUTTER, P. [1981] *Model completeness of some metric completions of absolutely free algebras* (J 0004) Algeb Universalis 12∗137-144
⋄ C05 C35 C50 ⋄ REV MR 82k:03037 Zbl 467 # 03024
• ID 55022

MYCIELSKI, J. [1982] *An essay about old model theory* (J 0302) Rep Math Logic, Krakow & Katowice 14∗49-58
⋄ C07 C52 ⋄ REV MR 84f:03032 Zbl 487 # 03020
• ID 34452

MYCIELSKI, J. see Vol. I, IV, V, VI for further entries

MYERS, D.L. [1974] *The back-and-forth isomorphism construction* (J 0048) Pac J Math 55∗521-529
⋄ C07 C15 G15 G30 ⋄ REV MR 51 # 5289
Zbl 309 # 02057 • ID 17400

MYERS, D.L. [1974] *The boolean algebras of abelian groups and well-orders* (J 0036) J Symb Logic 39∗452-458
⋄ C60 C65 G05 ⋄ REV MR 51 # 138 Zbl 301 # 02067
• ID 09696

MYERS, D.L. [1976] *Invariant uniformization* (J 0027) Fund Math 91∗65-72
⋄ C75 D55 E15 E35 ⋄ REV MR 53 # 10582
Zbl 358 # 02068 • ID 30751

MYERS, D.L. [1980] see LAMPE, W.A.

MYERS, D.L. [1980] *The boolean algebra of the theory of linear orders* (J 0029) Israel J Math 35∗234-256
⋄ C65 G05 ⋄ REV MR 81i:03038 Zbl 437 # 03032
• ID 55900

MYERS, D.L. [1983] see HANF, W.P.

MYERS, D.L. see Vol. IV for further entries

MYERS, R.W. [1980] *Complexity of model-theoretic notions* (J 0047) Notre Dame J Formal Log 21∗656-658
⋄ C07 C35 C57 D55 ⋄ REV MR 81j:03070
Zbl 416 # 03043 • ID 55957

MYHILL, J.R. [1975] see KINO, A.

MYHILL, J.R. see Vol. I, II, IV, V, VI for further entries

NABEBIN, A.A. [1977] *Expressibility in restricted second-order arithmetic (Russian)* (J 0092) Sib Mat Zh 18∗830-837,957
- TRANSL [1977] (J 0475) Sib Math J 18∗588-593
⋄ B15 C40 D05 F35 ⋄ REV MR 58 # 16250
Zbl 385 # 03047 • ID 52155

NABEBIN, A.A. see Vol. IV for further entries

NADEL, M.E. [1972] *An application of set theory to model theory* (J 0029) Israel J Math 11∗386-393
⋄ C75 E47 E75 ⋄ REV MR 46 # 3298 Zbl 301 # 02049
• ID 09744

NADEL, M.E. [1972] *Some Loewenheim-Skolem results for admissible sets* (J 0029) Israel J Math 12∗427-432
⋄ C62 C70 D60 E55 ⋄ REV MR 47 # 3143
Zbl 262 # 02053 • ID 09743

NADEL, M.E. [1974] *More Loewenheim-Skolem results for admissible sets* (J 0029) Israel J Math 18∗53-64
⋄ C62 C70 D60 ⋄ REV MR 51 # 2901 Zbl 309 # 02058
• ID 16930

NADEL, M.E. [1974] *Scott sentences and admissible sets* (J 0007) Ann Math Logic 7∗267-294
⋄ C15 C65 C70 C75 ⋄ REV MR 52 # 5348
Zbl 301 # 02050 • ID 09746

NADEL, M.E. [1975] see LEE, V.

NADEL, M.E. [1976] *A transfer principle for simple properties of theories* (J 0053) Proc Amer Math Soc 59∗353-357
⋄ C50 C70 C75 ⋄ REV MR 54 # 2453 Zbl 368 # 02050
• ID 24039

NADEL, M.E. [1976] *On models $\equiv_{\infty\omega}$ to an uncountable model* (J 0053) Proc Amer Math Soc 54∗307-310
⋄ C15 C70 C75 ⋄ REV MR 52 # 13373 Zbl 328 # 02036
• ID 21824

NADEL, M.E. [1977] see LEE, V.

NADEL, M.E. & STAVI, J. [1977] *The pure part of HYP($\mathfrak{M}$)* (J 0036) J Symb Logic 42*33-46
⋄ C62 C70 ⋄ REV MR 58 # 27465 Zbl 382 # 03022
• ID 23742

NADEL, M.E. & STAVI, J. [1978] $L_{\infty\lambda}$-*equivalence, isomorphism and potential isomorphism* (J 0064) Trans Amer Math Soc 236*51-74
⋄ C55 C75 E07 E40 E65 ⋄ REV MR 57 # 2907 Zbl 381 # 03024 • ID 51892

NADEL, M.E. [1978] *Infinitary intuitionistic logic from a classical point of view* (J 0007) Ann Math Logic 14*159-191
⋄ C75 C90 F50 ⋄ REV MR 80f:03027 Zbl 406 # 03055
• ID 29150

NADEL, M.E. [1978] see LIPSHITZ, L.

NADEL, M.E. [1980] *An arbitrary equivalence relation as elementary equivalence in an abstract logic* (J 0068) Z Math Logik Grundlagen Math 26*103-109
⋄ C75 C80 C95 ⋄ REV MR 82g:03070 Zbl 444 # 03016
• ID 56464

NADEL, M.E. [1980] *On a problem of MacDowell and Specker* (J 0036) J Symb Logic 45*612-622
⋄ C50 C62 E50 ⋄ REV MR 82d:03056 Zbl 522 # 03021
• ID 76724

NADEL, M.E. [1982] see KNIGHT, J.F.

NADEL, M.E. [1985] $L_{\omega_1\omega}$ *and admissible fragments* (C 4183) Model-Theor Log 271-316
⋄ C70 C75 C98 ⋄ ID 48340

NADEL, M.E. see Vol. VI for further entries

NADIU, G.S. [1969] *The interpolation theorem in strict positive logic* (J 0070) Bull Soc Sci Math Roumanie, NS 13(61)*185-193
⋄ B20 C40 ⋄ REV MR 47 # 23 Zbl 204.310 • ID 09749

NADIU, G.S. [1971] *Elementary logic and models* (J 0070) Bull Soc Sci Math Roumanie, NS 15(63)*95-111
⋄ C07 C20 C30 ⋄ REV MR 51 # 136 Zbl 324 # 02039
• ID 17443

NADIU, G.S. [1971] *Sur la logique de Heyting* (C 0640) Log, Autom, Inform 41-70
⋄ C90 F50 G10 G30 ⋄ REV MR 49 # 21 Zbl 255 # 02018 • ID 09750

NADIU, G.S. see Vol. IV for further entries

NAGAOKA, K. [1974] see FUKUYAMA, M.

NAGAOKA, K. see Vol. II for further entries

NAGASHIMA, T. [1966] *A remark on a comprehension axiom without negation* (J 0081) Proc Japan Acad 42*425-426
⋄ C62 E30 ⋄ REV MR 34 # 4115 Zbl 147.254 • ID 09756

NAGASHIMA, T. [1966] *An extension of the Craig-Schuette interpolation theorem* (J 0260) Ann Jap Ass Phil Sci 3*12-18
⋄ C40 F50 ⋄ REV MR 33 # 30 Zbl 148.6 JSL 33.291
• ID 09755

NAGASHIMA, T. [1970] *A model of the comprehension axiom without negation* (J 0531) Hitotsubashi J Arts Sci (Tokyo) 11*50-52
⋄ C62 E30 ⋄ REV MR 43 # 41 • ID 09758

NAGASHIMA, T. see Vol. II, IV for further entries

NAKAMURA, A. & ONO, H. [1980] *Decidability results on a query language for data bases with incomplete information* (P 3210) Math Founds of Comput Sci (9);1980 Rydzyna 452-459
⋄ B25 B52 B75 ⋄ REV MR 82e:68102 Zbl 452 # 68095
• ID 54126

NAKAMURA, A. & ONO, H. [1981] *Undecidability of extensions of the monadic first-order theory of successor and two-dimensional finite automata* (P 3201) Logic Symposia;1979/80 Hakone 155-174
⋄ B15 B25 D05 D35 ⋄ REV MR 83k:03054 Zbl 474 # 03023 • ID 55427

NAKAMURA, A. see Vol. I, II, IV, VI for further entries

NAKANO, Y. [1970] *On the finite characters in general algebras* (J 2751) J Fac Lib Art Yamaguchi Univ
⋄ C05 ⋄ ID 33389

NAKANO, Y. [1970] see FUJIWARA, T.

NAKANO, Y. [1971] *An application of A.Robinson's proof of the completeness theorem* (J 0081) Proc Japan Acad 47*929-931
⋄ C07 ⋄ REV MR 47 # 113 Zbl 256 # 08002 • ID 09794

NAKANO, Y. [1982] *On the theories with countably many unary predicate symbols* (J 2751) J Fac Lib Art Yamaguchi Univ 16*1-6
⋄ C07 ⋄ REV MR 84g:03045 • ID 33392

NAKANO, Y. see Vol. V for further entries

NAMBA, K. [1970] *On arithmetical extension operators* (J 0260) Ann Jap Ass Phil Sci 3*216-230
⋄ C20 E47 E55 ⋄ REV MR 46 # 7033 Zbl 262 # 02036
• ID 09802

NAMBA, K. [1970] *On arithmetical extension operators (Japanese)* (J 0091) Sugaku 22*92-105
⋄ C20 E47 E55 ⋄ REV MR 44 # 3859 • ID 09803

NAMBA, K. see Vol. V, VI for further entries

NARENS, L. [1974] *Minimal conditions for additive conjoint measurement and qualitative probability* (J 0035) J Math Psychol 11*404-430
⋄ C20 ⋄ REV MR 50 # 15979 Zbl 307 # 02038 • ID 64160

NARENS, L. [1981] *On the scales of measurements* (J 0035) J Math Psychol 24*249-275
⋄ C50 ⋄ REV MR 83d:03044 Zbl 496 # 92020 • ID 35187

NARENS, L. see Vol. I, II for further entries

NAT VANDER, A. [1979] *Beyond non-normal possible worlds* (J 0047) Notre Dame J Formal Log 20*631-635
⋄ B45 C90 ⋄ REV MR 80k:03022 Zbl 368 # 02022
• ID 56121

NAT VANDER, A. see Vol. II for further entries

NEBRES, B.F. [1972] *Herbrand uniformity theorems for infinitary languages* (J 0090) J Math Soc Japan 24*1-19
⋄ C40 C70 C75 F05 ⋄ REV MR 46 # 21 Zbl 226 # 02001
• ID 30771

NEBRES, B.F. [1972] *Infinitary formulas preserved under unions of models* (J 0036) J Symb Logic 37*449-465
⋄ C40 C70 C75 ⋄ REV MR 53 # 7724 Zbl 262 # 02015
• ID 09831

NEBRES, B.F. [1975] *Elementary invariants for abelian groups (uses of saturated model theory in algebra)* (J 1480) Bull Malaysian Math Soc 1975*8-16
⋄ C50 C60 ⋄ REV MR 54 # 2455 • ID 24040

NEGREPONTIS, S. [1969] *The Stone space of the saturated boolean algebras* (J 0064) Trans Amer Math Soc 141*515-527
⋄ C50 G05 ⋄ REV MR 40 # 1311 Zbl 223 # 06002
• ID 09837

NEGREPONTIS, S. [1971] *A property of the saturated boolean algebras* (J 0028) Indag Math 33*117-120
⋄ C50 G05 ⋄ REV MR 44 # 6570 Zbl 251 # 06022
• ID 09838

NEGREPONTIS, S. [1972] *Adequate ultrafilters of special boolean algebras* (J 0064) Trans Amer Math Soc 174*345-367
⋄ C20 E05 G05 ⋄ REV MR 47 # 1607 Zbl 261 # 02035
• ID 09839

NEGREPONTIS, S. [1974] see COMFORT, W.W.

NEGREPONTIS, S. see Vol. V for further entries

NEGRI, M. [1984] *An application of recursive saturation* (J 2099) Boll Unione Mat Ital, VI Ser, A 3*449-451
⋄ C50 C57 C62 F30 ⋄ REV MR 86f:03109
Zbl 562 # 03039 • ID 44715

NEGRI, M. see Vol. VI for further entries

NELSON, EVELYN [1972] see BANASCHEWSKI, B.

NELSON, EVELYN [1973] see BANASCHEWSKI, B.

NELSON, EVELYN [1974] *Infinitary equational compactness* (J 0004) Algeb Universalis 4*1-13
⋄ C05 C50 C75 ⋄ REV MR 50 # 12863 Zbl 326 # 08009
• ID 09851

NELSON, EVELYN [1974] *Not every equational class of infinitary algebras contains a simple algebra* (S 0019) Colloq Math (Warsaw) 30*27-30
⋄ C05 C50 ⋄ REV MR 50 # 4445 Zbl 254 # 08007
• ID 09850

NELSON, EVELYN [1975] *Injectivity and equational compactness in the class of $\aleph_0$-semilattices* (J 0018) Canad Math Bull 18*387-392
⋄ C05 C50 G10 ⋄ REV MR 52 # 13532 Zbl 323 # 06006
• ID 21848

NELSON, EVELYN [1975] *On the adjointness between operations and relations and its impact on atomic compactness* (S 0019) Colloq Math (Warsaw) 33*33-40
⋄ C05 C50 G30 ⋄ REV MR 52 # 13573 Zbl 312 # 08001
• ID 29106

NELSON, EVELYN [1975] *Semilattices do not have equationally compact hulls* (S 0019) Colloq Math (Warsaw) 34*1-5
⋄ C05 C50 G10 ⋄ REV MR 54 # 211 Zbl 364 # 06003
• ID 24589

NELSON, EVELYN [1975] *Some functorial aspects of atomic compactness* (J 0004) Algeb Universalis 5*367-378
⋄ C05 C50 E50 G30 ⋄ REV MR 52 # 13590
Zbl 342 # 08003 • ID 21874

NELSON, EVELYN [1977] *Classes defined by implications* (J 0004) Algeb Universalis 7*405-407
⋄ B20 C05 C52 ⋄ REV MR 56 # 218 Zbl 378 # 08004
• ID 26606

NELSON, EVELYN [1977] see BANASCHEWSKI, B.

NELSON, EVELYN [1980] see BANASCHEWSKI, B.

NELSON, EVELYN see Vol. IV, V for further entries

NELSON, GEORGE C. [1973] *Nonconstructivity of models of the reals (Russian summary)* (J 0014) Bull Acad Pol Sci, Ser Math Astron Phys 21*1067-1071
⋄ C62 C65 D55 H05 H15 ⋄ REV MR 51 # 10066
Zbl 293 # 02042 • ID 17539

NELSON, GEORGE C. [1975] see MADISON, E.W.

NELSON, GEORGE C. [1980] see MEAD, J.

NELSON, GEORGE C. [1982] *The periodic power of $\mathfrak{A}$ and complete Horn theories* (J 0004) Algeb Universalis 14*349-356
⋄ C20 C30 C35 ⋄ REV MR 83h:03047 Zbl 449 # 03023
• ID 56690

NELSON, GEORGE C. [1983] *Logic of reduced power structures* (J 0036) J Symb Logic 48*53-59 • ERR/ADD ibid 1145
⋄ C20 ⋄ REV MR 85i:03097a MR 85i:03097b
Zbl 511 # 03013 • ID 37387

NELSON, GEORGE C. [1984] *Boolean powers, recursive models, and the Horn theory of a structure* (J 0048) Pac J Math 114*207-220
⋄ C05 C20 C30 C57 G05 ⋄ REV MR 85k:03021
Zbl 505 # 03014 • ID 39839

NELSON, GEORGE C. see Vol. IV, VI for further entries

NELSON, GREG [1984] *Combining satisfiability procedures by equality-sharing* (P 3084) Autom Theor Prov After 25 Yea;1983 Denver 201-211
⋄ B25 B35 ⋄ REV MR 85d:68005 Zbl 564 # 03011
• ID 45270

NELSON, GREG see Vol. IV for further entries

NEMETI, I. [1973] see ANDREKA, H.

NEMETI, I. [1974] see ANDREKA, H.

NEMETI, I. [1975] see ANDREKA, H.

NEMETI, I. [1976] see ANDREKA, H.

NEMETI, I. [1978] *From hereditary classes to varieties in abstract model theory and partial algebra* (S 0410) Math Beitr Univ Halle-Wittenberg 11*69-78
⋄ C05 C20 C52 C90 C95 ⋄ REV MR 82a:03025
Zbl 415 # 08002 • ID 76804

NEMETI, I. [1978] see ANDREKA, H.

NEMETI, I. [1979] see ANDREKA, H.

NEMETI, I. [1980] see ANDREKA, H.

NEMETI, I. [1981] see ANDREKA, H.

NEMETI, I. [1981] see HIEN, BUIHUY

NEMETI, I. [1982] see ANDREKA, H.

NEMETI, I. & SAIN, I. [1982] *Cone-implicational subcategories and some Birkhoff-type theorems* (P 3237) Colloq Universal Algeb;1977 Esztergom 535-578
⋄ C05 C20 C90 G30 ⋄ REV MR 84a:03031b
Zbl 495 # 18001 • ID 35572

NEMETI, I. [1982] *On notions of factorization systems and their applications to cone-injective subcategories* (J 0049) Period Math Hung 13*229-235
⋄ C90 G10 G30 ⋄ REV MR 84a:18002 Zbl 516 # 18001
• ID 38578

NEMETI, I. [1983] see ANDREKA, H.

NEMETI, I. [1984] see ANDREKA, H.

NEMETI, I. [1985] *Cylindric-relativised set algebras have strong amalgamation* (J 0036) J Symb Logic 50*689-700
⋄ C05 C52 G15 G25 ⋄ ID 47373

NEMETI, I. see Vol. II, V for further entries

NEPEJVODA, L.K. & NEPEJVODA, N.N. [1977] *Languages without set-variables for the description of sets (Russian)* (S 2579) Teor Mnozhestv & Topol (Izhevsk) 1*70-75
⋄ C75 D55 ⋄ REV Zbl 479 # 03027 • ID 55689

NEPEJVODA, N.N. [1977] see NEPEJVODA, L.K.

NEPEJVODA, N.N. [1979] *Stable truth and computability (Russian)* (S 0554) Issl Teor Algor & Mat Logik (Moskva) 3*78-89,133
⋄ C90 F35 F50 ⋄ REV MR 82f:03052 Zbl 422 # 03036 • ID 53497

NEPEJVODA, N.N. [1982] *Constructive logics (Russian)* (C 3849) Modal & Relevant Log, Vyp 1 91-106
⋄ B60 C90 C95 F50 ⋄ REV Zbl 557 # 03039 • ID 46228

NEPEJVODA, N.N. see Vol. I, II, IV, V, VI for further entries

NERODE, A. [1957] *Composita, equations, and freely generated algebras* (P 1675) Summer Inst Symb Log;1957 Ithaca 321-325
⋄ C05 ⋄ REV Zbl 115.12 JSL 35.347 • ID 09877

NERODE, A. [1959] *Composita, equations, and freely generated algebras* (J 0064) Trans Amer Math Soc 91*139-151
⋄ C05 ⋄ REV MR 21 # 3362 Zbl 115.12 JSL 35.347 • ID 09876

NERODE, A. [1963] *A decision method for p-adic integral zeros of diophantine equations* (J 0015) Bull Amer Math Soc 69*513-517
⋄ B25 C10 ⋄ REV MR 29 # 5723 JSL 30.391 • ID 09881

NERODE, A. [1966] *Diophantine correct non-standard models in the isols* (J 0120) Ann of Math, Ser 2 84*421-432
⋄ C62 D50 ⋄ REV MR 34 # 2465 Zbl 158.251 JSL 33.619 • ID 09884

NERODE, A. [1974] see CROSSLEY, J.N.

NERODE, A. [1975] see METAKIDES, G.

NERODE, A. [1976] see CROSSLEY, J.N.

NERODE, A. [1977] see METAKIDES, G.

NERODE, A. [1979] see METAKIDES, G.

NERODE, A. [1980] see METAKIDES, G.

NERODE, A. & SHORE, R.A. [1980] *Reducibility orderings: theories, definability and automorphisms* (J 0007) Ann Math Logic 18*61-89
⋄ C62 D30 D35 ⋄ REV MR 81k:03040 Zbl 494 # 03028 • ID 76822

NERODE, A. [1981] see ASH, C.J.

NERODE, A. & REMMEL, J.B. [1982] *Recursion theory on matroids* (P 3634) Patras Logic Symp;1980 Patras 41-65
⋄ C57 D25 D45 ⋄ REV MR 86b:03055a Zbl 526 # 03026 • REM Part I. Part II 1983 • ID 38172

NERODE, A. [1982] see METAKIDES, G.

NERODE, A. & REMMEL, J.B. [1985] *A survey of lattices of R.E. substructures* (P 4046) Rec Th;1982 Ithaca 323-375
⋄ C57 C98 D45 D98 ⋄ REV Zbl 573 # 03015 • ID 46402

NERODE, A. see Vol. I, IV, V, VI for further entries

NEUMANN, B.H. [1943] *Adjunction of elements to groups* (J 0039) J London Math Soc 18*4-11
⋄ C60 ⋄ REV MR 5.58 • ID 48424

NEUMANN, B.H. [1949] see HIGMAN, G.

NEUMANN, B.H. [1952] *A note on algebraically closed groups* (J 0039) J London Math Soc 27*247-249
⋄ C25 C60 ⋄ REV Zbl 46.248 • ID 28768

NEUMANN, B.H. [1954] *An embedding theorem for algebraic systems* (J 3240) Proc London Math Soc, Ser 3 4*138-153
⋄ C05 ⋄ REV MR 17.448 Zbl 56.29 • ID 09895

NEUMANN, B.H. [1954] *An essay on free products of groups with amalgamations* (J 0354) Phil Trans Roy Soc London, Ser A 246*503-554
⋄ C60 ⋄ REV MR 16.10 Zbl 57.17 • ID 28783

NEUMANN, B.H. [1962] *Special topics in algebra: Universal algebra* (X 1214) New York Univ Cour Inst Math: New York ii+78pp
⋄ C05 C98 ⋄ ID 25046

NEUMANN, B.H. & WIEGOLD, E.C. [1966] *A semigroup representation of varieties of algebras* (S 0019) Colloq Math (Warsaw) 14*111-114
⋄ C05 ⋄ REV MR 33 # 211 Zbl 192.96 • ID 16252

NEUMANN, B.H. [1968] *Lectures on topics in the theory of infinite groups* (X 0925) Tata Inst Fund Res: Bombay iii+267+ivpp
⋄ C05 C60 C98 D40 ⋄ REV MR 42 # 1881 Zbl 237 # 20001 • ID 24829

NEUMANN, B.H. [1973] *The isomorphism problem for algebraically closed groups* (P 0678) Word Probl: Decis & Burnside Probl in Group Th;1969 Irvine 553-562
⋄ C25 C60 D40 ⋄ REV MR 54 # 2767 Zbl 262 # 20046 JSL 41.786 • ID 24064

NEUMANN, B.H. [1975] see HICKMAN, J.L.

NEUMANN, B.H. see Vol. IV, V for further entries

NEUMANN, H. [1949] see HIGMAN, G.

NEUMANN, H. [1967] *Varieties of groups* (X 0811) Springer: Heidelberg & New York x+192pp
⋄ C05 C60 C98 ⋄ REV MR 35 # 6733 • ID 24952

NEUMANN, H. see Vol. I, II for further entries

NEUMANN, P.M. [1974] see BLASS, A.R.

NEUMANN, P.M. [1985] *Some primitive permutation groups* (J 3240) Proc London Math Soc, Ser 3 50*265-281
⋄ C60 ⋄ ID 48425

NEUMANN, W.D. [1970] *On the quasivariety of convex subsets of affine spaces* (J 0008) Arch Math (Basel) 21*11-16
⋄ C05 C60 ⋄ REV MR 41 # 7516 Zbl 194.15 • ID 16235

NEUMANN, W.D. [1974] *On Malcev conditions* (J 0038) J Austral Math Soc 17*376-384
⋄ C05 ⋄ REV MR 51 # 7998 Zbl 294 # 08004 • ID 17582

NEUMANN, W.D. [1978] *Mal'cev conditions, spectra and Kronecker product* (J 3194) J Austral Math Soc, Ser A 25*103-117 • ERR/ADD ibid 28*510
⋄ C05 C13 ⋄ REV MR 58 # 449 Zbl 387 # 08004 Zbl 419 # 08009 • ID 69735

NGUYEN CAT HO [1972] *Generalized algebras of Post and their applications to many-valued logics with infinitely long formulas* (J 0387) Bull Sect Logic, Pol Acad Sci 1/1∗4-12
◇ B50   C75   G20   ◇ REV   MR 50 # 12650   • ID 09939

NGUYEN CAT HO [1973] *Generalized Post algebras and their application to some infinitary many-valued logics* (J 0202) Diss Math (Warsaw) 107∗72pp
◇ B50   C75   G20   ◇ REV   Zbl 281 # 02057   • ID 29577

NICA, V. [1980] see BASARAB, S.A.

NIEFIELD, S.B. & ROSENTHAL, K.I. [1985] *Strong de Morgan's law and the spectrum of a commutative ring* (J 0032) J Algeb 93∗169-181
◇ C60   C90   G30   ◇ REV   MR 86f:18009   Zbl 562 # 18005
• ID 47437

NIELAND, J.J.F. [1967] *Beth's tableau-method* (P 0683) Log & Founds of Sci (Beth);1964 Paris 19-38
◇ B05   B25   B50   F07   ◇ REV   MR 39 # 1287   Zbl 203.9
• ID 09951

NIELAND, J.J.F. see Vol. II for further entries

NIINILUOTO, I. [1973] see HINTIKKA, K.J.J.

NIINILUOTO, I. see Vol. I, II, V for further entries

NIKITIN, A.A. [1984] *Some algorithmic problems for projective planes (Russian)* (J 0003) Algebra i Logika 23∗512-529
• TRANSL [1984] (J 0069) Algeb and Log 23∗347-358
◇ C57   C65   D80   ◇ ID 48426

NISHIMURA, H. [1980] *A preservation theorem for tense logic* (J 0068) Z Math Logik Grundlagen Math 26∗331-335
◇ B45   C20   C40   C90   ◇ REV   MR 81i:03024
Zbl 452 # 03020   • ID 54084

NISHIMURA, H. [1980] *Saturated and special models in modal model theory with applications to the modal and de re hierarchies* (J 0068) Z Math Logik Grundlagen Math 26∗481-490
◇ B45   C40   C50   C90   ◇ REV   MR 81k:03023
Zbl 443 # 03018   • ID 56424

NISHIMURA, H. [1981] *Model theory for tense logic: Saturated and special models with applications to the tense hierarchy* (J 0063) Studia Logica 40∗89-98
◇ B45   C40   C50   C90   ◇ REV   MR 83g:03035
Zbl 469 # 03019   • ID 55147

NISHIMURA, H. [1981] *The semantical characterization of de dicto in continuous modal model theory* (J 0068) Z Math Logik Grundlagen Math 27∗233-240
◇ B45   C40   C90   ◇ REV   MR 84c:03038   Zbl 464 # 03021
• ID 54611

NISHIMURA, H. see Vol. II, V for further entries

NISHIMURA, T. [1960] *Note on axiomatic set theory I. The independence of Zermelo's "Aussonderungsaxiom" from other axioms of set theory* (J 1770) Osaka Math J 12∗319-329
◇ C62   E30   E35   ◇ REV   MR 27 # 1378   Zbl 103.246
JSL 29.107   • REM   Part II 1960   • ID 09970

NISHIMURA, T. [1960] *Note on axiomatic set theory II. A construction of a model satisfying the axioms of set theory without Zermelo's Aussonderungsaxiom in a certain axiom system of ordinal numbers* (J 0407) Comm Math Univ St Pauli (Tokyo) 9∗29-37
◇ C62   E10   E30   E35   ◇ REV   MR 27 # 1379   Zbl 126.21
JSL 29.107   • REM   Part I 1960   • ID 09972

NISHIMURA, T. & TANAKA, H. [1965] *On two systems for arithmetic* (J 0090) J Math Soc Japan 17∗244-280
◇ C75   F30   F35   ◇ REV   MR 32 # 37   Zbl 133.251
• ID 09975

NISHIMURA, T. [1974] *Gentzen-style formulation of systems of set-calculus* (J 0407) Comm Math Univ St Pauli (Tokyo) 23∗29-36
◇ C75   E30   F05   ◇ REV   MR 52 # 100   Zbl 361 # 02041
• ID 18312

NISHIMURA, T. [1974] *Gentzen-style formal system representing properties of functions* (J 0407) Comm Math Univ St Pauli (Tokyo) 23∗37-44
◇ C75   E30   F05   ◇ REV   MR 52 # 101   Zbl 361 # 02042
• ID 18313

NISHIMURA, T. see Vol. I, II, IV, V, VI for further entries

NITA, A. [1976] *Quelques observations sur les ultraproduits d'anneaux* (J 0070) Bull Soc Sci Math Roumanie, NS 20(68)∗325-328
◇ C20   C60   ◇ REV   MR 58 # 22136   Zbl 366 # 13006
• ID 51141

NITA, A. [1979] *The behavior of ultraproducts of modules under passage to fractions (Romanian) (English summary)* (J 3131) Bul Inst Politeh Bucuresti, Ser Chim-Metal 41/1∗3-5
◇ C20   C60   ◇ REV   MR 81d:16024   • ID 82360

NITA, A. [1979] *Ultraproducts of rings which retain factor properties (Romanian) (French summary)* (J 0197) Stud Cercet Mat Acad Romana 31∗209-211
◇ C20   C60   ◇ REV   MR 80i:16048   Zbl 423 # 16014
• ID 82361

NITA, A. [1980] *Sur les ultraproduits des modules* (J 0070) Bull Soc Sci Math Roumanie, NS 24(72)∗167-170
◇ C20   C60   ◇ REV   MR 82d:16038   Zbl 467 # 03022
• ID 55020

NITA, A. [1980] *Ultraproducts for certain classes of rings (Romanian) (English summary)* (J 0440) Bul Inst Politeh Bucuresti, Ser Mec 42/1∗13-17
◇ C20   C60   ◇ REV   MR 83b:13015   Zbl 484 # 13007
• ID 38965

NITA, A. & NITA, C. [1985] *Sur l'ultraproduit des anneaux de matrices* (J 0070) Bull Soc Sci Math Roumanie, NS 29(77)∗275-278
◇ C20   C60   ◇ REV   Zbl 578 # 16008   • ID 49884

NITA, C. [1985] see NITA, A.

NOAH, A. [1980] *Predicate-functors and the limits of decidability in logic* (J 0047) Notre Dame J Formal Log 21∗701-707
◇ B25   G15   ◇ REV   MR 82b:03042   Zbl 416 # 03014
• ID 55931

NOGINA, E.Yu. [1966] *On effective topological spaces* (J 0023) Dokl Akad Nauk SSSR 169∗28-31
◇ C57   C65   ◇ REV   Zbl 154.7   • ID 49889

NOGINA, E.Yu. [1969] *Correlations between certain classes of effectively topological spaces (Russian)* (J 0087) Mat Zametki (Akad Nauk SSSR) 5∗483-495
• TRANSL [1969] (J 1044) Math Notes, Acad Sci USSR 5∗288-294
◇ C57   C65   D45   ◇ REV   MR 40 # 4082   Zbl 188.328
• ID 09983

NOGINA, E.YU. [1978] *Numerierte topologische Raeume (Russisch)* (J 0068) Z Math Logik Grundlagen Math 24*141-176
- ◇ C57 C60 D45 ◇ REV MR 80e:03051 Zbl 415 # 03034
- ID 53136

NOGINA, E.YU. see Vol. IV for further entries

NORGELA, S.A. [1976] *On approximating reduction classes of CPC by decidable classes (Russian) (English summary)* (S 0228) Zap Nauch Sem Leningrad Otd Mat Inst Steklov 60*103-108,224
- TRANSL [1980] (J 1531) J Sov Math 14*1493-1496
- ◇ B20 B25 D35 ◇ REV MR 58 # 27422 Zbl 342 # 02033
- ID 32671

NORGELA, S.A. [1976] *On frequency decidability of some classes of predicate calculus (Russian)* (J 2574) Litov Mat Sb (Vil'nyus) 16/2*234-235
- ◇ B25 ◇ ID 32679

NORGELA, S.A. [1977] see MASLOV, S.YU.

NORGELA, S.A. [1978] *Herbrand strategies of deduction-search in predicate calculus I (Russian) (English and Lithuanian summaries)* (J 2574) Litov Mat Sb (Vil'nyus) 18/4*95-100,201
- TRANSL [1978] (J 3283) Lith Math J 18*513-517
- ◇ B25 B35 D35 F07 ◇ REV MR 80e:03050 Zbl 394 # 03017 • REM Part II 1979 • ID 32674

NORGELA, S.A. [1980] *On approximation of some classes of classical predicate calculus by decidable classes. I,II (Russian) (English and Lithuanian summaries)* (J 2574) Litov Mat Sb (Vil'nyus) 20/1*135-143,219,4*89-96,210
- ◇ B10 B20 B25 ◇ REV MR 81i:03016 MR 82i:03015 Zbl 432 # 03006 Zbl 455 # 03006 • ID 32676

NORGELA, S.A. see Vol. I, IV, VI for further entries

NOSKOV, G.A. [1982] *On conjugacy in metabelian groups (Russian)* (J 0087) Mat Zametki (Akad Nauk SSSR) 31*495-507,653
- TRANSL [1982] (J 1044) Math Notes, Acad Sci USSR 31*252-258
- ◇ C57 C60 ◇ REV MR 83i:20029 Zbl 492 # 20018
- ID 39350

NOSKOV, G.A. [1983] *Elementary theory of a finitely generated commutative ring (Russian)* (J 0087) Mat Zametki (Akad Nauk SSSR) 33*23-29,157
- TRANSL [1983] (J 1044) Math Notes, Acad Sci USSR 33*12-15
- ◇ C60 D35 ◇ REV MR 84g:03063 Zbl 524 # 13014
- ID 34170

NOSKOV, G.A. [1983] *The elementary theory of a finitely generated almost solvable group (Russian)* (J 0216) Izv Akad Nauk SSSR, Ser Mat 47*498-517
- TRANSL [1983] (J 0448) Math of USSR, Izv 22*465-482
- ◇ B25 C60 D35 ◇ REV MR 85d:20029 Zbl 521 # 20019
- ID 38972

NOVAK, I.L. [1950] *A construction for models of consistent systems* (J 0027) Fund Math 37*87-110
- ◇ B15 C07 E70 ◇ REV MR 12.791 JSL 16.273 • ID 09999

NOVAK, VITEZSLAV & NOVOTNY, M. [1974] *Abstrakte Dimension von Strukturen* (J 0068) Z Math Logik Grundlagen Math 20*207-220
- ◇ C30 C65 ◇ REV MR 53 # 2765 Zbl 324 # 08002
- ID 10012

NOVAK, VITEZSLAV & NOVOTNY, M. [1984] *On a power of cyclically ordered sets (Russian and Czech summaries)* (J 0086) Cas Pestovani Mat, Ceskoslov Akad Ved 109*421-424
- ◇ C30 E07 ◇ REV MR 86f:06009 • ID 46664

NOVAK, VITEZSLAV [1985] *On a power of relational stuctures* (J 0022) Cheskoslov Mat Zh 35(110)*167-172
- ◇ C05 C30 ◇ REV MR 86c:08002 • ID 44694

NOVAK, VITEZSLAV see Vol. V for further entries

NOVIKOV, P.S. [1939] *On some existence theorems (Russian)* (J 0023) Dokl Akad Nauk SSSR 23*438-440
- ◇ B05 C75 ◇ REV Zbl 21.290 JSL 5.69 FdM 65.1106
- ID 10016

NOVIKOV, P.S. [1943] *On the consistency of certain logical calculus (Russian summary)* (J 0142) Mat Sb, Akad Nauk SSSR, NS 12(54)*231-261
- ◇ B05 B28 C75 ◇ REV MR 5.197 Zbl 60.21 JSL 11.129
- ID 19479

NOVIKOV, P.S. [1949] *On the axiom of complete induction (Russian)* (J 0023) Dokl Akad Nauk SSSR 64*457-459
- ◇ B25 B28 D35 F30 ◇ REV MR 11.304 Zbl 38.151 JSL 14.256 • ID 19476

NOVIKOV, P.S. [1961] *On the consistency of certain logical calculi (Russian)* (P 0633) Infinitist Meth;1959 Warsaw 71-74
- ◇ C75 ◇ REV MR 25 # 4990 JSL 27.246 • ID 10024

NOVIKOV, P.S. [1979] *Selected works. Theory of sets and functions. Mathematical logic and algebra (Russian)* (X 2027) Nauka: Moskva 396pp
- ◇ B96 C96 E96 ◇ REV MR 80i:01017 Zbl 485 # 01027
- ID 38498

NOVIKOV, P.S. see Vol. I, II, IV, V, VI for further entries

NOVOTNY, M. [1974] see NOVAK, VITEZSLAV

NOVOTNY, M. [1984] see NOVAK, VITEZSLAV

NOVOTNY, M. see Vol. II, IV, V for further entries

NURMAGAMBETOV, T.A. [1981] *On almost ω-stable theories (Russian)(Kazakh summary)* (J 0403) Izv Akad Nauk Kazak SSR, Ser Fiz-Mat 1981/5*47-51,81-82
- ◇ C45 ◇ REV MR 83e:03043 Zbl 488 # 03018 • ID 35225

NURMAGAMBETOV, T.A. [1982] see MUSTAFIN, T.G.

NURTAZIN, A.T. [1973] see GONCHAROV, S.S.

NURTAZIN, A.T. [1974] *Strong and weak constructivization and computable families (Russian)* (J 0003) Algebra i Logika 13*311-323,364
- TRANSL [1974] (J 0069) Algeb and Log 13*177-184
- ◇ C57 D45 ◇ REV MR 52 # 2851 Zbl 302 # 02014
- ID 17642

NURTAZIN, A.T. [1974] *Theories categorical in an uncountable power without principal model-complete extensions (Russian)* (J 0403) Izv Akad Nauk Kazak SSR, Ser Fiz-Mat 1974/3*87-88,94
- ◇ C35 ◇ REV MR 50 # 6826 Zbl 306 # 02050 • ID 10041

NURTAZIN, A.T. [1978] see DOBRITSA, V.P.

NURTAZIN, A.T. [1983] *An elementary classification and some properties of complete orders (Russian)* (C 4019) Kraevye Zadach Dlya Diff Uravnenij & Priloz Mekh & Tekh 122-125
- ◇ B25 C35 C65 C75 E10 ◇ REV MR 85f:03037 Zbl 555 # 03014 • ID 40721

NUTE, D.E. [1975] *Counterfactuals* (J 0047) Notre Dame J Formal Log 16*476-482
  ⋄ A05 B25 ⋄ REV MR 52 # 2827 Zbl 283 # 02007
  • ID 18314

NUTE, D.E. see Vol. I, II for further entries

NYBERG, A.M. [1976] *Uniform inductive definability and infinitary languages* (J 0036) J Symb Logic 41*109-120
  ⋄ C70 D70 ⋄ REV MR 54 # 7201 Zbl 374 # 02023
  • ID 14788

NYBERG, A.M. [1977] *Inductive operators on resolvable structures* (P 1629) Symp Math Log;1974 Oulo;1975 Helsinki 91-100
  ⋄ C70 D60 D70 D75 ⋄ ID 28324

NYBERG, A.M. [1978] see MAREK, W.

NYBERG, A.M. see Vol. I, V for further entries

OBERSCHELP, A. [1958] *Ueber die Axiome produkt-abgeschlossener arithmetischer Klassen* (J 0009) Arch Math Logik Grundlagenforsch 4*95-123
  ⋄ B20 B25 C30 C40 C52 ⋄ REV MR 21 # 6330 Zbl 108.6 JSL 32.532 • ID 10042

OBERSCHELP, A. [1959] *Ueber die Axiome arithmetischer Klassen mit Abgeschlossenheitsbedingungen* (J 0009) Arch Math Logik Grundlagenforsch 5*26-36
  ⋄ B25 C40 C52 ⋄ REV MR 22 # 6711 Zbl 118.252 JSL 32.533 • ID 10043

OBERSCHELP, A. [1968] *On the Craig-Lyndon interpolation theorem* (J 0036) J Symb Logic 33*271-274
  ⋄ C40 ⋄ REV MR 39 # 1291 Zbl 182.9 • ID 10050

OBERSCHELP, A. see Vol. I, II, IV, V for further entries

OBERSCHELP, W. [1968] *Strukturzahlen in endlichen Relationssystemen* (P 0608) Logic Colloq;1966 Hannover 199-213
  ⋄ C13 ⋄ REV MR 38 # 5644 Zbl 216.306 JSL 40.503
  • ID 22209

OBERSCHELP, W. [1977] *Monotonicity for structure numbers in theories without identity* (P 2912) Combin & Repres Groupe Symetr;1976 Strasbourg 297-308
  ⋄ C13 ⋄ REV MR 57 # 5732 Zbl 365 # 02038 • ID 51039

OBERSCHELP, W. see Vol. I, IV, VI for further entries

ODIFREDDI, P. [1983] *On the first order theory of the arithmetical degrees* (J 0053) Proc Amer Math Soc 87*505-507
  ⋄ C65 D30 D55 ⋄ REV MR 84d:03056 Zbl 554 # 03023
  • ID 34091

ODIFREDDI, P. see Vol. IV, V, VI for further entries

ODINTSOV, S.P. [1984] *Atomless ideals of constructive Boolean algebras (Russian)* (J 0003) Algebra i Logika 23*278-295,362
  • TRANSL [1984] (J 0069) Algeb and Log 23*190-203
  ⋄ C57 G05 ⋄ REV MR 86f:03069 • ID 44901

OGER, F. [1981] *Equivalence elementaire et genre de groupes finis-par-abeliens de type fini (English summary)* (J 3364) C R Acad Sci, Paris, Ser 1 293*1-4
  ⋄ C60 ⋄ REV MR 83b:03043 Zbl 471 # 03032 • ID 90324

OGER, F. [1982] *Des groupes nilpotents de classe 2 sans torsion de type fini ayant les memes images finies peuvent ne pas etre elementairement equivalents* (J 3364) C R Acad Sci, Paris, Ser 1 294*1-4
  ⋄ C60 ⋄ REV MR 83c:20054 Zbl 489 # 20029 • ID 39038

OGER, F. [1982] *Equivalence elementaire entre groupes finis-par-abeliens de type fini* (J 2022) Comm Math Helvetici 57*469-480
  ⋄ C60 ⋄ REV MR 84g:03042 Zbl 512 # 20019 • ID 34155

OGER, F. [1983] *Cancellation and elementary equivalence of groups* (J 0326) J Pure Appl Algebra 30*293-299
  ⋄ C60 ⋄ REV MR 85e:03079 Zbl 524 # 20002 • ID 38230

OGER, F. [1984] *Elementary equivalence and isomorphism of finitely generated nilpotent groups* (J 0394) Commun Algeb 12*1899-1915
  ⋄ C60 ⋄ REV MR 86c:20037 Zbl 549 # 20024 • ID 45155

OGER, F. [1984] *The model theory of finitely generated finite-by-abelian groups* (J 0036) J Symb Logic 49*1115-1124
  ⋄ C50 C60 ⋄ REV MR 86g:03060 • ID 39862

OGLESBY, F.C. [1961] *Report: an examination of a decision procedure* (J 0015) Bull Amer Math Soc 67*300-304
  ⋄ B25 F07 ⋄ REV MR 23 # A795 Zbl 100.10 JSL 28.165
  • ID 10067

OGLESBY, F.C. [1962] *An examination of a decision procedure* (S 0167) Mem Amer Math Soc 44*148pp
  ⋄ B25 ⋄ REV MR 32 # 2317 Zbl 126.20 JSL 28.165
  • ID 10068

OGLESBY, F.C. see Vol. I for further entries

OHKUMA, T. [1964] *Closed classes of models* (J 0407) Comm Math Univ St Pauli (Tokyo) 13*63-80
  ⋄ C95 G30 ⋄ REV MR 31 # 3334 Zbl 199.9 • ID 10080

OHKUMA, T. [1966] *Ultrapowers in categories* (J 0537) Yokohama Math J 14*17-37
  ⋄ C05 C20 G30 ⋄ REV MR 35 # 5374 Zbl 168.268 JSL 37.402 JSL 38.402 • ID 10081

OHKUMA, T. [1973] *Finitary objects and ultrapowers* (J 0407) Comm Math Univ St Pauli (Tokyo) 21*73-82
  ⋄ C20 G30 ⋄ REV MR 48 # 2213 Zbl 262 # 18003
  • ID 10083

OHKUMA, T. see Vol. V for further entries

OHNISHI, M. [1957] see MATSUMOTO, K.

OHNISHI, M. [1959] see MATSUMOTO, K.

OHNISHI, M. [1961] *Gentzen decision procedures for Lewis's systems S2 and S3* (J 1770) Osaka Math J 13*125-137
  ⋄ B25 B45 ⋄ REV MR 24 # A1210 Zbl 106.4 JSL 40.468
  • ID 10085

OHNISHI, M. see Vol. II, VI for further entries

OIKKONEN, J. [1978] *Second-order definability, game quantifiers and related expressions* (J 0990) Soc Sci Fennicae Comment Phys-Math 48*39-101
  ⋄ C40 C75 C80 C85 E60 ⋄ REV MR 80k:03036 Zbl 402 # 03030 • ID 54677

OIKKONEN, J. [1979] *A generalization of the infinitely deep languages of Hintikka and Rantala* (C 1706) Essays Honour J. Hintikka 101-112
  ⋄ C55 C75 C80 ⋄ REV Zbl 432 # 03020 • ID 53977

OIKKONEN, J. [1979] *Hierarchies of model-theoretic definability - an approach to second-order logics* (P 1705) Scand Logic Symp (4);1976 Jyvaeskylae 197-225
  ⋄ B15 C40 C70 C75 C80 C85 ⋄ REV MR 80i:03048 Zbl 399 # 03035 • ID 52831

OIKKONEN, J. [1979] *Mathematical applications of model theory (Finnish) (English summary)* (J 1108) Arkhimedes (Helsinki) 31*216-226
⋄ C07 C60 ⋄ REV MR 80m:03070 Zbl 416 # 03055
• ID 53221

OIKKONEN, J. [1979] *On PC- and RPC-classes in generalized model theory* (P 2615) Scand Logic Symp (5);1979 Aalborg 257-270
⋄ C13 C52 C75 C80 C95 ⋄ REV MR 82d:03064 Zbl 431 # 03025 • ID 53931

OIKKONEN, J. [1981] *Finite and infinite versions of the operations PC and RPC in infinitary languages* (P 2623) Worksh Extended Model Th;1980 Berlin 114-135
⋄ C52 C75 ⋄ REV MR 84h:03089 Zbl 484 # 03016
• ID 34281

OIKKONEN, J. [1983] *Logical operations and iterated infinitely deep languages* (J 0063) Studia Logica 42*243-249
⋄ C75 C95 ⋄ REV MR 85i:03118 Zbl 548 # 03016
• ID 42338

OKAZAKI, H. [1976] *An application of Fuhrken's reduction technique* (J 4727) Kumamoto J Sci, Math 12*17-22
⋄ C55 C80 ⋄ REV MR 56 # 83 Zbl 362 # 02041 • ID 50763

OKAZAKI, H. see Vol. V for further entries

OKOH, F. [1984] see BURNS, R.G.

OLIN, P. [1968] see JONSSON, B.

OLIN, P. [1968] *An almost everywhere direct power* (J 0064) Trans Amer Math Soc 134*405-420
⋄ C30 C62 E50 ⋄ REV MR 38 # 38 Zbl 184.13 • ID 10091

OLIN, P. [1969] *Indefinability in the arithmetic of isolic integers* (J 0048) Pac J Math 29*175-186
⋄ D50 H15 ⋄ REV MR 43 # 49 Zbl 184.20 • ID 10092

OLIN, P. [1970] *Elementary extensions within reduced powers* (J 0025) Duke Math J 37*421-429
⋄ C20 ⋄ REV MR 41 # 6676 Zbl 206.278 • ID 10093

OLIN, P. [1970] *Some first order properties of direct sums of modules* (J 0068) Z Math Logik Grundlagen Math 16*405-416
⋄ C30 C40 C60 ⋄ REV MR 43 # 1825 Zbl 212.319
• ID 10094

OLIN, P. [1971] *Direct multiples and powers of modules* (J 0027) Fund Math 73*113-124
⋄ C30 C40 C60 ⋄ REV MR 45 # 6610 Zbl 235 # 02052
• ID 10095

OLIN, P. [1972] $\aleph_0$-*categoricity of two-sorted structures* (J 0004) Algeb Universalis 2*262-269 • ERR/ADD ibid 5*446
⋄ C20 C30 C35 C60 ⋄ REV MR 48 # 1909 Zbl 258 # 02054 • ID 10096

OLIN, P. [1972] *First order preservation theorems in two-sorted languages* (J 3172) J London Math Soc, Ser 2 4*631-637
⋄ C30 C40 C60 ⋄ REV MR 46 # 46 Zbl 244 # 02018
• ID 10097

OLIN, P. [1972] *Products of two-sorted structures* (J 0036) J Symb Logic 37*75-80
⋄ C30 C40 C75 ⋄ REV MR 47 # 8287 Zbl 238 # 02045
• ID 10098

OLIN, P. [1974] *Free products and elementary equivalence* (J 0048) Pac J Math 52*175-184
⋄ C05 C60 ⋄ REV MR 50 # 1874 Zbl 257 # 02040
• ID 10099

OLIN, P. [1975] $\aleph_0$-*categoricity of products of 1-unary algebras* (S 0019) Colloq Math (Warsaw) 33*25-31
⋄ C05 C35 ⋄ REV MR 52 # 13364 Zbl 312 # 02041
• ID 21817

OLIN, P. [1976] see JONSSON, B.

OLIN, P. [1976] *Free products and elementary types of boolean algebras* (J 0132) Math Scand 38*5-23
⋄ C30 C50 G05 ⋄ REV MR 56 # 2895 Zbl 339 # 06008
• ID 21599

OLIN, P. [1976] *Homomorphisms of elementary types of boolean algebras* (J 0004) Algeb Universalis 6*259-260
⋄ C05 C52 G05 ⋄ REV MR 55 # 7769 Zbl 355 # 02034
• ID 24512

OLIN, P. [1977] *Elementary properties of V-free products of groups* (J 0032) J Algeb 47*105-114
⋄ C05 C30 C60 ⋄ REV MR 58 # 22319 Zbl 365 # 02039
• ID 28309

OLIN, P. [1978] *Aleph-zero categorical Stone algebras* (J 3194) J Austral Math Soc, Ser A 26*337-347
⋄ C35 G10 ⋄ REV MR 80a:03043 Zbl 394 # 03036
• ID 52513

OLIN, P. [1978] *Urn models and categoricity* (J 0122) J Philos Logic 7*331-345
⋄ A05 C35 C90 ⋄ REV MR 81h:03081 Zbl 383 # 03006
• ID 51988

OLIN, P. [1980] *Varietal free products of bands* (J 0136) Semigroup Forum 21*83-87
⋄ C05 C30 G10 ⋄ REV MR 82a:20065 Zbl 453 # 20049
• ID 82391

OLIVEIRA DE, A.J.F. [1981] *A note on cyclic boolean algebras (Portuguese) (English summary)* (P 3751) Jorn Mat Luso-Espanol (8);1981 Coimbra I*153-161
⋄ C40 G05 ⋄ REV MR 83m:03043 • ID 34887

OLIVEIRA DE, A.J.F. & SEBASTIAO E SILVA, J. [1985] *On automorphisms of arbitrary mathematical systems* (J 2028) Hist & Phil Log 6*91-116
⋄ C07 ⋄ ID 46353

OLIVEIRA DE, A.J.F. see Vol. V for further entries

OMAROV, A.I. [1967] *Filtered products of models (Russian) (English summary)* (J 0003) Algebra i Logika 6/3*77-89
⋄ C20 C30 C40 C50 C52 ⋄ REV MR 37 # 2590 Zbl 165.310 • ID 10102

OMAROV, A.I. [1967] *On compact classes of models (Russian) (English summary)* (J 0003) Algebra i Logika 6/2*49-60
⋄ C20 C52 ⋄ REV MR 36 # 3641 Zbl 204.4 JSL 34.652
• ID 19542

OMAROV, A.I. [1968] *The $M(\alpha)$-class of models (Russian)* (J 0092) Sib Mat Zh 9*216-219
• TRANSL [1968] (J 0475) Sib Math J 9*163-165
⋄ C20 C50 ⋄ REV MR 36 # 6277 Zbl 165.311 • ID 33000

OMAROV, A.I. [1970] *A certain property of Frechet filters (Russian)* (J 0403) Izv Akad Nauk Kazak SSR, Ser Fiz-Mat 1970/3*66-68
⋄ C20 C30 C50 C52 ⋄ REV MR 45 # 42 Zbl 208.13
• ID 10104

OMAROV, A.I. [1971] *Products with respect to an ω-universal filter (Russian) (Kazakh summary)* (J 0403) Izv Akad Nauk Kazak SSR, Ser Fiz-Mat 1971/3*47-50
⋄ C20 C50 G05 ⋄ REV MR 46 # 3291 Zbl 224 # 02043
• ID 10105

OMAROV, A.I. [1973] *On subsystems of reduced powers (Russian)* (J 0003) Algebra i Logika 12*74-82,121
• TRANSL [1973] (J 0069) Algeb and Log 12*42-46
⋄ C20 G05 ⋄ REV MR 48 # 1914 Zbl 286 # 02055
• ID 10106

OMAROV, A.I. [1979] $\triangleleft_t$*-minimality and the finite cover property (Russian)* (C 2065) Teor Nereg Kriv Raz Geom Post 74-76
⋄ C20 C45 C50 ⋄ REV MR 81f:03044 • ID 76967

OMAROV, A.I. [1981] *A sufficient condition for instability of theories (Russian)* (C 3806) Issl Teor Progr 85-86
⋄ C45 ⋄ REV MR 84k:03100 • ID 36117

OMAROV, A.I. [1984] *Elementary theory of D-degrees (Russian)* (J 0003) Algebra i Logika 23*530-537
• TRANSL [1984] (J 0069) Algeb and Log 23*358-363
⋄ C20 C30 C50 D30 G10 ⋄ ID 46564

OMAROV, B. [1979] see BAJZHANOV, B.S.

OMAROV, B. [1980] see BAJZHANOV, B.S.

OMAROV, B. [1980] *Ranks and Deg of countable ω_0-categorical theories (Russian)* (C 2620) Teor Model & Primen 61-65
⋄ C35 C45 ⋄ REV MR 82e:03029 • ID 76971

OMAROV, B. [1981] *Properties of rank and Deg of complete countable stable theories (Russian)* (C 3806) Issl Teor Progr 87-93
⋄ C45 ⋄ REV MR 84k:03101 • ID 36118

OMAROV, B. [1983] *A question of V.Harnik (Russian)* (J 0429) Vest Akad Nauk Kazak SSR 10*66-67
⋄ C15 C35 C50 ⋄ REV MR 85e:03071 Zbl 559 # 03020
• ID 40619

OMAROV, B. [1983] *Nonessential extensions of complete theories (Russian)* (J 0003) Algebra i Logika 22*542-550
• TRANSL [1983] (J 0069) Algeb and Log 22*390-397
⋄ C15 C35 ⋄ REV MR 86h:03050 Zbl 561 # 03016
• ID 39617

OMODEO, E.G. [1978] see FERRO, A.

OMODEO, E.G. [1980] see FERRO, A.

OMODEO, E.G. [1981] see BREBAN, M.

OMODEO, E.G. [1983] *Decidability and validity in the presence of choice operators (Italian)* (P 3829) Atti Incontri Log Mat (1);1982 Siena 225-226
⋄ B25 ⋄ ID 44823

OMODEO, E.G. see Vol. I, II for further entries

ONO, H. [1980] see NAKAMURA, A.

ONO, H. [1981] see NAKAMURA, A.

ONO, H. [1983] *Equational theories and universal theories of fields* (J 0090) J Math Soc Japan 35*289-306
⋄ B25 C05 C60 ⋄ REV MR 84g:03046 Zbl 496 # 03015
• ID 34157

ONO, H. [1983] *Model extension theorem and Craig's interpolation theorem for intermediate predicate logics* (J 0302) Rep Math Logic, Krakow & Katowice 15*41-58
⋄ B55 C20 C40 C90 ⋄ REV MR 84j:03060
Zbl 519 # 03016 • ID 34652

ONO, H. [1985] *Semantical analysis of predicate logics without the contraction rule* (J 0063) Studia Logica 44*187-196
⋄ B10 C90 F50 ⋄ ID 47510

ONO, H. see Vol. I, II, IV, V, VI for further entries

ONYSZKIEWICZ, J. [1964] *On the different topologies in the space of models* (J 0014) Bull Acad Pol Sci, Ser Math Astron Phys 12*245-248
⋄ C15 C52 C85 ⋄ REV MR 34 # 2466 Zbl 137.250
• ID 10139

ONYSZKIEWICZ, J. [1966] see MAREK, W.

ONYSZKIEWICZ, J. [1967] see MAREK, W.

ONYSZKIEWICZ, J. [1970] see MAREK, W.

ONYSZKIEWICZ, J. [1972] see MAREK, W.

ONYSZKIEWICZ, J. [1974] *Dimensions of PC space of models (Russian summary)* (J 0014) Bull Acad Pol Sci, Ser Math Astron Phys 22*773-777
⋄ C15 C52 C85 ⋄ REV MR 51 # 5290 Zbl 291 # 02041
• ID 10140

ONYSZKIEWICZ, J. [1977] *Omitting types in some extended system of logic* (P 1629) Symp Math Log;1974 Oulo;1975 Helsinki 51-61
⋄ C75 C95 ⋄ ID 90131

ONYSZKIEWICZ, J. see Vol. V for further entries

OPPEN, D.C. [1973] *Elementary bounds for Presburger arithmetic* (P 1482) ACM Symp Th of Comput (5);1973 Austin 34-37
⋄ B25 C10 D15 F30 ⋄ REV MR 54 # 4187
Zbl 306 # 02044 • ID 24077

OPPEN, D.C. [1978] *A $2^{2^{2^{pn}}}$ upper bound on the complexity of Presburger arithmetic* (J 0119) J Comp Syst Sci 16*323-332
⋄ B25 C10 D15 F30 ⋄ REV MR 57 # 18224
Zbl 381 # 03021 • ID 51889

OPPEN, D.C. [1980] *Complexity, convexity and combinations of theories* (J 1426) Theor Comput Sci 12*291-302
⋄ B25 B35 D15 ⋄ REV MR 82a:03013 Zbl 437 # 03007
• ID 55875

OPPEN, D.C. [1980] *Reasoning about recursively defined data structures* (J 0037) ACM J 27*403-411
⋄ B25 D15 ⋄ REV MR 82a:68177 Zbl 477 # 68025
• ID 82395

OPPEN, D.C. see Vol. IV for further entries

OREVKOV, V.P. [1969] *On nonlengthening applications of equality rules (Russian)* (S 0228) Zap Nauch Sem Leningrad Otd Mat Inst Steklov 16*152-156 • ERR/ADD ibid 20*292-294
- TRANSL [1969] (J 0521) Semin Math, Inst Steklov 16*77-79
  ◇ B25 B35 F05 F07 ◇ REV MR 41#6665 Zbl 197.273
- ID 10154

OREVKOV, V.P. [1972] see MASLOV, S.YU.

OREVKOV, V.P. [1976] *Solvable classes of pseudoprenex formulas (Russian) (English summary)* (S 0228) Zap Nauch Sem Leningrad Otd Mat Inst Steklov 60*109-170,225 • ERR/ADD ibid 88*248-249
- TRANSL [1980] (J 1531) J Sov Math 14*1497-1538
  ◇ B25 F50 ◇ REV MR 58#27423 Zbl 342#02034
- ID 64291

OREVKOV, V.P. see Vol. I, II, IV, VI for further entries

OREY, S. [1956] *On ω-consistency and related properties* (J 0036) J Symb Logic 21*246-252
  ◇ C07 F30 ◇ REV MR 18.632 Zbl 71.8 JSL 23.40
- ID 10163

OREY, S. [1959] *Model theory for the higher order predicate calculus* (J 0064) Trans Amer Math Soc 92*72-84
  ◇ B15 C85 ◇ REV MR 21#6325 Zbl 88.10 JSL 27.96
- ID 10165

OREY, S. [1961] *Relative interpretations* (J 0068) Z Math Logik Grundlagen Math 7*146-153
  ◇ C62 F25 ◇ REV MR 26#3608 Zbl 121.255 JSL 40.627
- ID 10166

OREY, S. see Vol. I, V, VI for further entries

ORNAGHI, M. [1975] see MARINI, D.

ORNAGHI, M. [1981] see MIGLIOLI, P.A.

ORNAGHI, M. see Vol. II, IV, VI for further entries

OSDOL VAN, D.H. [1972] *Truth with respect to an ultrafilter or how to make intuition rigorous* (J 0005) Amer Math Mon 79*355-363
  ◇ C20 H05 ◇ REV MR 46#1545 Zbl 241#26004
- ID 22272

OSIUS, G. [1973] *The internal and external aspect of logic and set theory in elementary topoi* (J 0306) Cah Topol & Geom Differ 14*47-49
  ◇ C90 E70 E75 F50 G30 ◇ REV MR 52#7896 Zbl 362#18001 • ID 27669

OSIUS, G. [1975] *A note on Kripke-Joyal semantics for the internal language of topoi* (C 0772) Model Th & Topoi 349-354
  ◇ C90 F50 G30 ◇ REV MR 52#7898 Zbl 348#18003 JSL 46.158 • ID 18321

OSIUS, G. [1975] *Logical and set theoretical tools in elementary topoi* (C 0772) Model Th & Topoi 297-346
  ◇ C90 E70 F50 G30 ◇ REV MR 52#7897 Zbl 348#18002 JSL 46.158 • ID 18320

OSIUS, G. see Vol. V for further entries

OSSWALD, H. [1970] *Modelltheoretische Untersuchungen in der Kripke-Semantik* (J 0009) Arch Math Logik Grundlagenforsch 13*3-21
  ◇ C20 C90 F50 ◇ REV MR 43#7299 Zbl 216.12
- ID 10184

OSSWALD, H. [1972] *Homomorphie-invariante Formeln in der intuitionistischen Logik* (J 0009) Arch Math Logik Grundlagenforsch 15*86-96
  ◇ C40 C90 F50 ◇ REV MR 48#3707 Zbl 251#02031
- ID 10186

OSSWALD, H. [1984] see HORT, C.

OSSWALD, H. see Vol. VI for further entries

OSTASZEWSKI, A.J. [1973] *Saturated structures and a theorem of Arhangelskij* (J 3172) J London Math Soc, Ser 2 6*453-458
  ◇ C50 C65 E75 ◇ REV MR 48#5848 Zbl 257#06005
- ID 10189

OSTASZEWSKI, A.J. see Vol. IV, V for further entries

OSWALD, U. [1982] *A decision method for the existential theorems of NF_2* (P 3774) Th d'Ensembl de Quine;1981 Louvain-la-Neuve 23-43
  ◇ B25 E70 ◇ REV MR 84e:03062 Zbl 544#03028
- ID 34396

OSWALD, U. [1982] see KREJNOVICH, V.YA.

OSWALD, U. see Vol. V for further entries

OZAWA, M. [1984] *A classification of type I AW^*-algebras and Boolean valued analysis* (J 0090) J Math Soc Japan 36*589-608
  ◇ C65 C90 E40 E75 ◇ REV MR 85m:46068 • ID 44314

OZAWA, M. see Vol. V for further entries

OZHIGOV, YU.I. [1983] *Equations with two unknowns in a free group (Russian)* (J 0023) Dokl Akad Nauk SSSR 268*809-813
- TRANSL [1983] (J 0062) Sov Math, Dokl 27*177-181
  ◇ C60 ◇ REV MR 84j:20033 Zbl 525#20023 • ID 39151

PABION, J.F. [1967] *Modeles elementaires pour la theorie des ensembles* (J 0056) Publ Dep Math, Lyon 4/3*39-54
  ◇ C62 E30 ◇ REV MR 38#5612b Zbl 164.313 • ID 41363

PABION, J.F. [1967] *Une approche de la construction d'un modele denombrable pour la theorie de Zermelo-Fraenkel* (J 0056) Publ Dep Math, Lyon 4/3*15-31
  ◇ C62 E30 ◇ REV MR 38#5612a Zbl 164.313 • ID 10209

PABION, J.F. [1969] *Extensions du theoreme de Loewenheim-Skolem en logique infinitaire* (J 2313) C R Acad Sci, Paris, Ser A-B 268*A925-A927
  ◇ C55 C75 ◇ REV MR 39#5327 Zbl 182.326 • ID 10211

PABION, J.F. [1969] *Sur la methode du "forcing" de P.J.Cohen* (J 2313) C R Acad Sci, Paris, Ser A-B 268*A989-A991
  ◇ C25 E40 ◇ REV MR 40#35 Zbl 185.15 • ID 10210

PABION, J.F. [1970] *Theorie des modeles sans identite* (J 0056) Publ Dep Math, Lyon 7/1*47-85
  ◇ C07 ◇ REV MR 44#2592 Zbl 231#02071 • ID 27454

PABION, J.F. [1971] *Etude des bases dans les modeles des theories $\aleph_1$-categoriques* (J 2313) C R Acad Sci, Paris, Ser A-B 273*A537-A539
  ◇ C15 C35 C50 ◇ REV MR 45#1743 Zbl 235#02044
- ID 10213

PABION, J.F. [1971] *Problemes d'immersions en algebre* (C 3032) Semin Lefebvre Struct Algeb 1970/71 Exp.14*148-160
  ◇ C07 C60 ◇ REV MR 53#123 Zbl 383#03017
- ID 16588

PABION, J.F. [1971] *Remarques sur la definition d'un ensemble de parametres* (J 2313) C R Acad Sci, Paris, Ser A-B 272*A1073-A1075
⋄ C40 ⋄ REV MR 43 # 4650   Zbl 218 # 02050 • ID 10212

PABION, J.F. [1971] see CUSIN, R.

PABION, J.F. [1971] *Une notion de transcendance dans les structures relationnelles* (J 2313) C R Acad Sci, Paris, Ser A-B 272*A1541-A1544
⋄ C30 C45 ⋄ REV MR 44 # 51   Zbl 235 # 02043 • ID 27240

PABION, J.F. [1972] *Relations prehomogenes* (J 2313) C R Acad Sci, Paris, Ser A-B 274*A529-A531
⋄ C07 C35 C50 ⋄ REV MR 45 # 1744   Zbl 249 # 02026
• ID 10214

PABION, J.F. [1981] *Saturation en arithmetique et en analyse* (J 3133) Bull Soc Math Belg, Ser B 33*73-82
⋄ C50 C62 ⋄ REV MR 83b:03040   Zbl 499 # 03020
• ID 35096

PABION, J.F. & RICHARD, D. [1981] *Synonymy and re-interpretation for some sublanguages of Peano arithmetic* (P 2614) Open Days in Model Th & Set Th;1981 Jadwisin 231-236
⋄ B28 C62 F30 ⋄ ID 33728

PABION, J.F. [1982] Π_2-*theorie des ensembles* (J 1934) Ann Sci Univ Clermont Math 21*15-45
⋄ C50 C62 E30 E45 ⋄ REV MR 85e:03115
Zbl 574 # 03040 • ID 40714

PABION, J.F. [1982] *Saturated models of Peano arithmetic* (J 0036) J Symb Logic 47*625-637
⋄ C50 C62 ⋄ REV MR 83j:03054   Zbl 498 # 03021
• ID 35357

PABION, J.F. see Vol. I, II, V, VI for further entries

PACHOLSKI, L. & WEGLORZ, B. [1968] *Topologically compact structures and positive formulas* (S 0019) Colloq Math (Warsaw) 19*37-42
⋄ B20 C40 C50 ⋄ REV MR 37 # 871   Zbl 184.13
JSL 40.88 • ID 10215

PACHOLSKI, L. [1969] *Elementary substructures of direct powers (Russian summary)* (J 0014) Bull Acad Pol Sci, Ser Math Astron Phys 17*55-59
⋄ C30 C62 ⋄ REV MR 40 # 2530   Zbl 198.20 • ID 10217

PACHOLSKI, L. [1969] *On products of first order theories (Russian summary)* (J 0014) Bull Acad Pol Sci, Ser Math Astron Phys 17*793-796
⋄ C30 C50 ⋄ REV MR 41 # 6677   Zbl 199.303 • ID 10218

PACHOLSKI, L. & RYLL-NARDZEWSKI, C. [1970] *On countably compact reduced products I* (J 0027) Fund Math 67*155-161
⋄ C20 C50 ⋄ REV MR 42 # 157   Zbl 196.30 • REM Part II 1970 • ID 10219

PACHOLSKI, L. [1970] *On countably compact reduced products II (Russian summary)* (J 0014) Bull Acad Pol Sci, Ser Math Astron Phys 18*1-3
⋄ C20 C50 ⋄ REV MR 42 # 158   Zbl 204.311 • REM Part I 1970 by Pacholski,L. & Ryll-Nardzewski,C. Part III 1971
• ID 10221

PACHOLSKI, L. [1971] *On countably compact reduced products III* (S 0019) Colloq Math (Warsaw) 23*5-15
⋄ C10 C20 C50 G05 ⋄ REV MR 46 # 48
Zbl 242 # 02056 • REM Part II 1970 • ID 10222

PACHOLSKI, L. & WASZKIEWICZ, J. [1972] *On compact classes of models* (J 0027) Fund Math 76*139-147
⋄ C20 C52 ⋄ REV MR 46 # 7020   Zbl 255 # 08005
• ID 10224

PACHOLSKI, L. [1973] *On countable universal boolean algebras and compact classes of models* (J 0027) Fund Math 78*43-60
⋄ C20 C50 C52 G05 ⋄ REV MR 48 # 3733
Zbl 222 # 02060 • ID 10223

PACHOLSKI, L. [1976] *On limit reduced powers, saturatedness and universality* (P 1476) Set Th & Hierarch Th (2) (Mostowski);1975 Bierutowice 221-239
⋄ C20 C50 ⋄ REV MR 57 # 95   Zbl 346 # 02028 • ID 23808

PACHOLSKI, L. [1977] *Homogeneity, universality and saturatedness of limit reduced powers III* (J 0027) Fund Math 95*85-94
⋄ C20 C50 ⋄ REV MR 58 # 16261   Zbl 354 # 02042 • REM Part II 1977 by Weglorz,B. • ID 26524

PACHOLSKI, L. [1977] *Saturated limit reduced powers* (P 1639) Set Th & Hierarch Th (1);1974 Karpacz 1*79-89
⋄ C20 C50 C55 ⋄ REV MR 58 # 5172   Zbl 393 # 03020
• ID 28366

PACHOLSKI, L. [1978] see MACINTYRE, A.

PACHOLSKI, L. & WIERZEJEWSKI, J. & WILKIE, A.J. (EDS.) [1980] *Model theory of algebra and arithmetic. Proceedings, Karpacz, Poland, September 1-7, 1979* (S 3301) Lect Notes Math 834*vi+410pp
⋄ C60 C62 C97 ⋄ REV MR 82a:03003   Zbl 436 # 00008
• ID 55838

PACHOLSKI, L. [1981] *Homogeneous limit reduced powers* (J 0009) Arch Math Logik Grundlagenforsch 21*131-136
⋄ C20 C50 ⋄ REV MR 83b:03037   Zbl 497 # 03014
• ID 33311

PACHOLSKI, L. & TOMASIK, J. [1982] *Reduced products which are not saturated* (J 0004) Algeb Universalis 14*210-227
⋄ C20 C50 ⋄ REV MR 83d:03039   Zbl 497 # 03015
• ID 35185

PADMANABHAN, R. [1969] *On single equational-axiom systems for abelian groups* (J 0038) J Austral Math Soc 9*143-152
⋄ C05 ⋄ REV MR 39 # 4276   Zbl 204.340 • ID 10227

PADOA, A. [1902] *Essai d'une theorie algebrique des nombres entiers, precede d'une introduction logique a une theorie deductive quelconque* (P 1484) Int Congr Math (2);1900 Paris 3*309-365
• TRANSL [1967] (C 0675) From Frege to Goedel 119-123
⋄ A10 B30 C40 ⋄ REM Only a part is translated
• ID 14911

PADOA, A. see Vol. I, II, V for further entries

PAETZOLD, P. [1985] *On fields and terms with arctan function* (P 4310) Easter Conf on Model Th (3);1985 Gross Koeris 193-200
⋄ C65 ⋄ ID 49038

PAILLET, J.L. [1971] *Quelques rapports entre l'interpretabilite et l'extension logique* (J 2313) C R Acad Sci, Paris, Ser A-B 272*A189-A192
⋄ C07 C40 ⋄ REV MR 42 # 7493   Zbl 212.17 • ID 10254

PAILLET, J.L. [1972] *La methode des "systemes logiquement inductifs"* (J 2313) C R Acad Sci, Paris, Ser A-B 274*A1141-A1144
⋄ C07 C30 ⋄ REV MR 45 # 3185a Zbl 237 # 02017
• ID 64348

PAILLET, J.L. [1972] *La methode des "systemes logiquement inductifs"* (J 2313) C R Acad Sci, Paris, Ser A-B 274*A1189-A1192
⋄ C07 C30 ⋄ REV MR 45 # 3185b Zbl 246 # 02041
• ID 27245

PAILLET, J.L. [1973] *La methode des systemes logiquement inductifs* (J 0060) Rev Roumaine Math Pures Appl 18*1393-1412
⋄ B10 C07 C30 ⋄ REV MR 55 # 5386 Zbl 273 # 02037
• ID 30495

PAILLET, J.L. [1974] *Quelques rapports entre la fini-model-completabilite et des proprietes approchantes* (J 2313) C R Acad Sci, Paris, Ser A-B 278*A1237-A1240
⋄ C07 C35 ⋄ REV MR 50 # 4288 Zbl 289 # 02039
• ID 10256

PAILLET, J.L. [1974] *Une etude sur les theories "finiment model-completables"* (J 2313) C R Acad Sci, Paris, Ser A-B 278*A193-A196
⋄ C07 C35 ⋄ REV MR 49 # 10541 Zbl 279 # 02033
• ID 27255

PAILLET, J.L. [1976] *Une etude sur des structures ayant une certaine propriete de seuil pour les automorphismes elementaires (English summary)* (J 2313) C R Acad Sci, Paris, Ser A-B 283*A225-A228
⋄ C07 C35 C50 ⋄ REV MR 54 # 12514 Zbl 358 # 02066
• ID 27287

PAILLET, J.L. [1977] *Resultats relatifs a l'existence de structures n-elementaires (English summary)* (J 2313) C R Acad Sci, Paris, Ser A-B 284*A359-A362
⋄ C50 C60 ⋄ REV MR 55 # 95 Zbl 369 # 02031 • ID 27295

PAILLET, J.L. [1978] *Theories completes $\aleph_0$-categoriques et non finiment modele-completables (English summary)* (J 2313) C R Acad Sci, Paris, Ser A-B 286*A1167-A1170
⋄ C35 C45 ⋄ REV MR 80a:03039 Zbl 388 # 03011
• ID 52265

PAILLET, J.L. [1978] *Une etude sur des structures ayant une certaine propriete de seuil pour les automorphismes elementaires* (J 0068) Z Math Logik Grundlagen Math 24*7-24
⋄ C07 C35 C50 ⋄ REV MR 80e:03028 Zbl 374 # 02030
• ID 51554

PAILLET, J.L. see Vol. I, VI for further entries

PAIR, C. [1967] *Sur des algorithmes pour les problemes de cheminement dans les graphes finis (English summary)* (P 4680) Th of Graphs;1966 Roma 271-300
⋄ B25 ⋄ REV MR 36 # 2531 Zbl 203.266 • ID 24777

PAIR, C. see Vol. IV for further entries

PALASINSKI, M. [1983] *Varieties of commutative BCK-algebras not generated by their finite members* (J 0387) Bull Sect Logic, Pol Acad Sci 12*134-135
⋄ C52 G10 G25 ⋄ REV Zbl 548 # 03037 • ID 43205

PALASINSKI, M. [1984] *On BCK-algebras with the operation (S)* (J 0387) Bull Sect Logic, Pol Acad Sci 13*13-20
⋄ C52 G10 G25 ⋄ REV MR 85h:06030 Zbl 548 # 03038
• ID 43208

PALASINSKI, M. see Vol. IV for further entries

PALLADINO, D. [1976] *Questioni di categoricita delle teorie matematiche* (J 1515) Archimede 28*21-34
⋄ C35 ⋄ REV MR 57 # 90 Zbl 356 # 02042 • ID 50314

PALLADINO, D. see Vol. I, IV, V for further entries

PALYUTIN, E.A. [1971] *N^*-homogeneous models (Russian)* (J 0092) Sib Mat Zh 12*920-921
• TRANSL [1971] (J 0475) Sib Math J 12*665-666
⋄ C50 ⋄ REV MR 44 # 5215 Zbl 285 # 02043 • ID 10262

PALYUTIN, E.A. [1971] *Boolean algebras with a categorical theory in a weak second order logic (Russian)* (J 0003) Algebra i Logika 10*523-534
• TRANSL [1971] (J 0069) Algeb and Log 10*325-331
⋄ B15 C35 C85 D35 G05 ⋄ REV MR 46 # 3302 Zbl 248 # 02055 • ID 64365

PALYUTIN, E.A. [1971] *Models with countably categorical universal theories (Russian)* (J 0003) Algebra i Logika 10*23-32
• TRANSL [1971] (J 0069) Algeb and Log 10*15-20
⋄ C35 C45 G05 ⋄ REV MR 44 # 5214 Zbl 221 # 02040
• ID 28109

PALYUTIN, E.A. [1972] see ABAKUMOV, A.I.

PALYUTIN, E.A. [1972] *On complete quasimanifolds (Russian)* (J 0003) Algebra i Logika 11*689-693,737
• TRANSL [1972] (J 0069) Algeb and Log 11*384-386
⋄ C05 C35 ⋄ REV MR 48 # 10793 Zbl 264 # 08022
• ID 19527

PALYUTIN, E.A. [1973] *Categorical quasivarieties of arbitrary signature (Russian)* (J 0092) Sib Mat Zh 14*1285-1303,1367
• TRANSL [1973] (J 0475) Sib Math J 14*904-916
⋄ C05 C35 ⋄ REV MR 51 # 7857 Zbl 279 # 02029
• ID 17698

PALYUTIN, E.A. [1973] see ERSHOV, YU.L.

PALYUTIN, E.A. [1974] *The algebras of formulae of countably categorical theories (Russian)* (S 0019) Colloq Math (Warsaw) 31*157-159
⋄ B25 C35 ⋄ REV MR 51 # 2895 Zbl 295 # 02029
• ID 17340

PALYUTIN, E.A. [1975] *Description of categorical quasivarieties (Russian)* (J 0003) Algebra i Logika 14*145-185,240
• TRANSL [1975] (J 0069) Algeb and Log 14*86-111
⋄ C05 C35 ⋄ REV MR 53 # 2672 Zbl 319 # 08004
• ID 64358

PALYUTIN, E.A. [1977] *Number of models in $L_{\infty \omega_1}$-theories I,II (Russian)* (J 0003) Algebra i Logika 16*74-87,124,443-456,494
• TRANSL [1977] (J 0069) Algeb and Log 16*51-61,299-309
⋄ C35 C52 C55 C75 E45 E65 ⋄ REV MR 58 # 21425b Zbl 381 # 03023 Zbl 429 # 03014 • ID 29226

PALYUTIN, E.A. [1979] *Categorical positive Horn theories (Russian)* (J 0003) Algebra i Logika 18*47-72,122
• TRANSL [1979] (J 0069) Algeb and Log 18*31-49
⋄ C05 C35 C52 ⋄ REV MR 81f:03042 Zbl 448 # 03017
• ID 56617

PALYUTIN, E.A. [1980] *Categorical Horn classes I (Russian)* (J 0003) Algebra i Logika 19*582-614,617
- TRANSL [1980] (J 0069) Algeb and Log 19*377-400
⋄ C05 C35 C45 C52 ⋄ REV MR 82k:03038 Zbl 491 # 03011 • ID 77090

PALYUTIN, E.A. [1980] *Description of categorical positive Horn classes (Russian)* (J 0003) Algebra i Logika 19*683-700,746
- TRANSL [1980] (J 0069) Algeb and Log 19*443-455
⋄ C05 C35 C52 ⋄ REV MR 82k:03039 Zbl 491 # 03010
• ID 77089

PALYUTIN, E.A. [1982] *Number of models in complete varieties* (P 3622) Int Congr Log, Meth & Phil of Sci (6,Proc);1979 Hannover 207-221
⋄ C05 C45 C52 ⋄ REV MR 84g:03036 Zbl 523 # 03020
• ID 34150

PALYUTIN, E.A. & SAFFE, J. & STARCHENKO, S.S. [1985] *Models of superstable Horn theories* (J 0003) Algebra i Logika 24*278-326
- TRANSL [1985] (J 0069) Algeb and Log 24*171-210
⋄ C05 C35 C45 ⋄ ID 49888

PALYUTIN, E.A. see Vol. I, II, IV, V for further entries

PAOLA DI, R.A. [1967] *A survey of Soviet work in the theory of computer programming* (1111) Preprints, Manuscr., Techn. Reports etc. 146pp
⋄ B25 B75 D35 G30 ⋄ REM The Rand Corporation, RM-5424-PR, October 1967 • ID 30681

PAOLA DI, R.A. [1973] *The solvability of the decision problem for classes of proper formulas and related results* (J 0037) ACM J 20*112-126
⋄ B20 B25 ⋄ REV MR 48 # 7698 Zbl 285 # 68008
• ID 30684

PAOLA DI, R.A. see Vol. I, IV, VI for further entries

PAPPAS, P. [1985] *A Diophantine problem for Laurent polynomial rings* (J 0053) Proc Amer Math Soc 93*713-718
⋄ C60 D35 D80 ⋄ REV MR 86d:03041 Zbl 532 # 10032
• ID 44466

PAPPAS, P. [1985] *The model-theoretic structure of abelian group rings* (J 0073) Ann Pure Appl Logic 28*163-201
⋄ B25 C20 C60 C90 ⋄ REV Zbl 577 # 03020 • ID 39903

PARGETTER, R. [1977] see DAVIDSON, B.

PARGETTER, R. see Vol. II for further entries

PARIGOT, M. [1982] *Theories d'arbres* (J 0036) J Symb Logic 47*841-853
⋄ C45 C65 ⋄ REV MR 84g:03047 Zbl 523 # 03023 JSL 49.672 • ID 34158

PARIGOT, M. [1983] *Le modele compagnon de la theorie des arbres* (J 0068) Z Math Logik Grundlagen Math 29*137-150
⋄ C10 C25 C35 C50 C65 ⋄ REV MR 85e:03080 Zbl 539 # 03013 • ID 39807

PARIGOT, M. [1984] *Extensions de la logique du premier order. Theoreme de Lindstroem* (C 4356) Gen Log Semin Paris 1982/83 1-11
⋄ C95 C98 ⋄ REV MR 86f:03005 • ID 48078

PARIGOT, M. [1984] see DELON, F.

PARIGOT, M. [1984] *Omission des types dans les theories* $\Pi_2^0(\Delta)$ (J 3364) C R Acad Sci, Paris, Ser 1 299*195-198
⋄ C40 C75 ⋄ REV MR 86b:03038 Zbl 577 # 03017
• ID 44138

PARIKH, R. [1969] *A conservation result* (P 0649) Appl Model Th to Algeb, Anal & Probab;1967 Pasadena 107-108
⋄ C30 H20 ⋄ REV MR 38 # 3145 Zbl 188.21 • ID 24819

PARIKH, R. [1971] *Existence and feasibility in arithmetic* (J 0036) J Symb Logic 36*494-508
⋄ A05 D15 F30 F65 H15 ⋄ REV MR 46 # 3287 Zbl 243 # 02037 • ID 10282

PARIKH, R. [1972] *A note on rigid substructures* (J 0053) Proc Amer Math Soc 33*520-522
⋄ C50 C57 ⋄ REV MR 45 # 3186 Zbl 273 # 02034
• ID 10285

PARIKH, R. [1973] *Some results on the length of proofs* (J 0064) Trans Amer Math Soc 177*29-36
⋄ B25 F20 F30 ⋄ REV MR 55 # 5404 Zbl 269 # 02011
• ID 10286

PARIKH, R. [1975] *An $\aleph_0$-categorical theory whose language is countably infinite* (J 0053) Proc Amer Math Soc 49*216-218
⋄ C35 ⋄ REV MR 52 # 2868 Zbl 307 # 02037 • ID 17654

PARIKH, R. [1978] *A decidability result for a second-order process logic* (P 3578) IEEE Symp Found of Comput Sci (19);1978 Ann Arbor 177-183
⋄ B25 ⋄ REV MR 80e:68132 • ID 82433

PARIKH, R. [1984] see KOZEN, D.

PARIKH, R. see Vol. I, II, IV, V, VI for further entries

PARIS, J.B. [1970] see KUNEN, K.

PARIS, J.B. [1972] *On models of arithmetic* (P 2080) Conf Math Log;1970 London 251-280
⋄ C62 ⋄ REV MR 52 # 13369 Zbl 236 # 02042 • ID 14982

PARIS, J.B. [1973] *Minimal models of ZF* (P 0710) Russell Mem Logic Conf;1971 Uldum 327-331
⋄ C50 C62 E45 ⋄ REV MR 50 # 12704 • ID 23607

PARIS, J.B. [1974] *Patterns of indiscernibles* (J 0161) Bull London Math Soc 6*183-188
⋄ C30 E45 E55 ⋄ REV MR 51 # 10099 Zbl 287 # 04006
• ID 17522

PARIS, J.B. [1976] *Sets satisfying* $O^{\#}$ (J 0161) Bull London Math Soc 8*257-260
⋄ C30 E45 E55 E65 ⋄ REV MR 54 # 7264 Zbl 387 # 03023 • ID 25816

PARIS, J.B. [1977] see HARRINGTON, L.A.

PARIS, J.B. [1977] see KIRBY, L.A.S.

PARIS, J.B. [1977] *Models of arithmetic and the 1-3-1 lattice* (J 0027) Fund Math 95*195-199
⋄ C50 C62 G10 ⋄ REV MR 56 # 5270 Zbl 382 # 03048
• ID 26530

PARIS, J.B. [1978] see KIRBY, L.A.S.

PARIS, J.B. [1978] see MACINTYRE, A.

PARIS, J.B. [1978] *Note on an induction axiom* (J 0036) J Symb Logic 43*113-117
⋄ B28 C62 F30 ⋄ REV MR 81e:03057 Zbl 399 # 03009
• ID 29247

PARIS, J.B. [1978] *Some independence results for Peano arithmetic* (J 0036) J Symb Logic 43*725-731
  ⋄ C62 F30 ⋄ REV MR 82d:03095 Zbl 408#03048 JSL 48.482 • ID 31967

PARIS, J.B. [1979] see MILLS, G.

PARIS, J.B. [1980] *A hierarchy of cuts in models of arithmetic* (P 2625) Model Th of Algeb & Arithm;1979 Karpacz 312-337
  ⋄ C15 C62 F30 ⋄ REV MR 84e:03085 Zbl 448#03054 • ID 56654

PARIS, J.B. & WILKIE, A.J. [1981] Δ_0 *sets and induction* (P 2614) Open Days in Model Th & Set Th;1981 Jadwisin 237-248
  ⋄ C62 F30 ⋄ ID 33729

PARIS, J.B. & WILKIE, A.J. [1981] *Models of arithmetic and the rudimentary sets* (J 3133) Bull Soc Math Belg, Ser B 33*157-169
  ⋄ C62 D20 ⋄ REV MR 82k:03074 Zbl 499#03021 • ID 79972

PARIS, J.B. [1981] *Some conservation results for fragments of arithmetic* (P 3404) Model Th & Arithm;1979/80 Paris 251-262
  ⋄ C62 F30 ⋄ REV MR 83f:03060 Zbl 475#03041 • ID 55495

PARIS, J.B. [1981] see KOSSAK, R.

PARIS, J.B. [1982] see KIRBY, L.A.S.

PARIS, J.B. [1982] see DIMITRACOPOULOS, C.

PARIS, J.B. [1983] see DIMITRACOPOULOS, C.

PARIS, J.B. [1984] see CSIRMAZ, L.

PARIS, J.B. [1984] *On the structure of models with restricted E_1-induction (Czech) (Russian and English summaries)* (J 0086) Cas Pestovani Mat, Ceskoslov Akad Ved 109*372-379
  ⋄ C62 F30 ⋄ REV MR 86e:03036 Zbl 567#03028 • ID 46698

PARIS, J.B. [1984] see MILLS, G.

PARIS, J.B. & WILKIE, A.J. [1985] *Counting problems in bounded arithmetic* (P 2160) Latin Amer Symp Math Log (6);1983 Caracas 317-340
  ⋄ B28 C62 F30 ⋄ REV Zbl 572#03034 • ID 41796

PARIS, J.B. see Vol. IV, V, VI for further entries

PARK, D. [1964] *Set theoretic constructions in model theory* (0000) Diss., Habil. etc
  ⋄ C55 ⋄ REM Doct. diss., Massachusetts Institute of Technology • ID 19499

PARK, D. [1977] *Finiteness is μ-ineffable* (J 1426) Theor Comput Sci 3*173-181
  ⋄ B75 C75 C80 C85 ⋄ REV MR 55#10229 Zbl 353#02027 • ID 50110

PARK, D. see Vol. IV for further entries

PARMIGIANI BEDOGNA, D. [1980] *Le formule SH nelle potenze generalizzate ridotte (English summary)* (J 0549) Riv Mat Univ Parma, Ser 4 6*189-199
  ⋄ C20 C40 G30 ⋄ REV MR 82h:03023 • ID 77139

PASHENKOV, V.V. [1974] *Duality of topological models (Russian)* (J 0023) Dokl Akad Nauk SSSR 218*291-294
  • TRANSL [1974] (J 0062) Sov Math, Dokl 15*1336-1340
  ⋄ C30 C90 ⋄ REV MR 50#12705 Zbl 314#02066
  • ID 10315

PASHENKOV, V.V. [1975] *Boolean models and some of their applications (Russian)* (J 0142) Mat Sb, Akad Nauk SSSR, NS 97(139)*3-34,159
  • TRANSL [1976] (J 0349) Math of USSR, Sbor 26*1-30
  ⋄ C90 E07 G05 G10 G30 ⋄ REV MR 52#5498 Zbl 316#06009 • ID 18325

PASHENKOV, V.V. [1975] *On dualities (Russian)* (J 0023) Dokl Akad Nauk SSSR 222*1295-1298
  • TRANSL [1975] (J 0062) Sov Math, Dokl 16*777-781
  ⋄ C30 C90 ⋄ REV MR 52#5415 Zbl 401#03014
  • ID 18326

PASHENKOV, V.V. see Vol. V for further entries

PASINI, A. [1981] *On the 1°-order translations of the theory of the geometric closure structures* (J 3495) Boll Unione Mat Ital, V Ser, B 18*217-230
  ⋄ C65 ⋄ REV MR 83h:51017 Zbl 512#51007 • ID 37560

PASINI, A. see Vol. I for further entries

PASINI, L. [1982] *Sulla chiusura reale dei campi generalizzati di Hardy (English summary)* (J 3768) Boll Unione Mat Ital, VI Ser, D 1*147-159
  ⋄ C60 C65 H20 ⋄ REV MR 84e:26020 Zbl 502#03019
  • ID 36896

PASINI, L. see Vol. II for further entries

PASSOS, E.P. [1974] see GALDA, K.

PAUL, E. [1985] *Equational methods in first order predicate calculus* (J 4609) J Symb Comput 1*7-29
  ⋄ B75 C05 C07 ⋄ ID 49178

PAUL, E. see Vol. I for further entries

PAULOS, J.A. [1976] *A model-theoretic semantics for modal logic* (J 0047) Notre Dame J Formal Log 17*465-468
  ⋄ B45 C07 C90 ⋄ REV MR 56#2789 Zbl 314#02033
  • ID 18327

PAULOS, J.A. [1976] *Noncharacterizability of the syntax set* (J 0036) J Symb Logic 41*368-372
  ⋄ C40 C95 ⋄ REV MR 54#88 Zbl 333#02012 • ID 14768

PAULOS, J.A. see Vol. I, II for further entries

PAYNE, T.H. [1969] *An elementary submodel never preserved by Skolem expansions* (J 0068) Z Math Logik Grundlagen Math 15*435-436
  ⋄ C30 ⋄ REV MR 42#51 Zbl 195.14 • ID 10337

PAYNE, T.H. see Vol. IV, V for further entries

PEARCE, D.A. [1984] see BENTHEM VAN, J.F.A.K.

PELC, A. [1981] *Ideals on the real line and Ulam's problem* (J 0027) Fund Math 112*165-170
  ⋄ C20 E05 E40 ⋄ REV MR 83b:03057 Zbl 387#03019
  • ID 55433

PELC, A. [1981] see GUZICKI, W.

PELC, A. [1982] *Semiregular invariant measures on abelian groups* (J 0053) Proc Amer Math Soc 86*423-426
  ⋄ C60 E55 E75 ⋄ REV MR 84a:03061 Zbl 504#43001
  • ID 35588

PELC, A. see Vol. V for further entries

PELLER, V. [1979] *Applications of ultraproducts in operator theory. A simple proof of E. Bishop's theorem (Russian) (English summary)* (S 0228) Zap Nauch Sem Leningrad Otd Mat Inst Steklov 92*230-240,323-324
- ◊ C20 C65 ◊ REV MR 81k:47026 Zbl 429 # 47014
- • ID 53891

PENZIN, YU.G. [1972] *Decidability of the theory of algebraic integers with addition and congruences according to divisors (Russian)* (C 3549) Algebra, Vyp 1 (Irkutsk) 68-79
- ◊ B25 C60 ◊ REV MR 57 # 2910 • ID 32591

PENZIN, YU.G. [1972] see FRIDMAN, EH.I.

PENZIN, YU.G. [1973] *Decidability of certain theories of integers (Russian)* (J 0092) Sib Mat Zh 14*1139-1143,1160
- • TRANSL [1973] (J 0475) Sib Math J 14*796-799
- ◊ B25 F30 ◊ REV MR 48 # 8222 Zbl 272 # 02070
- • ID 10358

PENZIN, YU.G. [1973] *Decidability of the theory of integers with addition, order and multiplication by an arbitrary number (Russian)* (J 0087) Mat Zametki (Akad Nauk SSSR) 13*667-675
- • TRANSL [1973] (J 1044) Math Notes, Acad Sci USSR 13*401-405
- ◊ B25 F30 ◊ REV MR 48 # 1907 Zbl 267 # 02035
- • ID 10357

PENZIN, YU.G. [1973] *Decidability of a theory of the integers with addition, order and predicates that distinguish a chain of subgroups (Russian)* (C 1443) Algebra, Vyp 2 (Irkutsk) 138-153
- ◊ B25 F30 ◊ REV MR 53 # 7755 • ID 23000

PENZIN, YU.G. [1979] *Twins problem in formal arithmetic (Russian)* (J 0087) Mat Zametki (Akad Nauk SSSR) 26*505-511,653
- • TRANSL [1979] (J 1044) Math Notes, Acad Sci USSR 26*743-746
- ◊ F30 H15 ◊ REV MR 80m:03115 Zbl 422 # 03034
- • ID 32595

PENZIN, YU.G. see Vol. IV, VI for further entries

PEPIS, J. [1937] *Ueber das Entscheidungsproblem des engeren logischen Funktionskalkuels (Polish) (German summary)* (S 0281) Arch Towarz Nauk Lwow, Sect 3 7/8*1-172
- ◊ B20 B25 D35 ◊ REV Zbl 19.97 JSL 4.93 FdM 63.823
- • ID 19493

PEPIS, J. see Vol. I, IV for further entries

PEREMANS, W. [1952] *Some theorems on free algebras and on direct products of algebras* (J 0061) Simon Stevin 29*51-59
- ◊ C05 C52 ◊ REV MR 14.347 Zbl 47.15 JSL 20.184
- • ID 10366

PERETYAT'KIN, M.G. [1971] *Strongly constructive models and numerations of the boolean algebra of recursive sets (Russian)* (J 0003) Algebra i Logika 10*535-557
- • TRANSL [1971] (J 0069) Algeb and Log 10*332-345
- ◊ C57 D20 D45 G05 ◊ REV MR 46 # 5126 Zbl 311 # 02051 • ID 10368

PERETYAT'KIN, M.G. [1973] *A strongly constructive model without elementary submodels and extensions (Russian)* (J 0003) Algebra i Logika 12*312-322,364
- • TRANSL [1973] (J 0069) Algeb and Log 12*178-183
- ◊ C35 C57 D45 ◊ REV MR 52 # 7892 Zbl 298 # 02046
- • ID 18330

PERETYAT'KIN, M.G. [1973] *Every recursively enumerable extension of a theory of linear order has a constructive model (Russian)* (J 0003) Algebra i Logika 12*211-219,244
- • TRANSL [1973] (J 0069) Algeb and Log 12*120-124
- ◊ C57 C65 ◊ REV MR 54 # 7243 Zbl 298 # 02045
- • ID 25033

PERETYAT'KIN, M.G. [1973] *On complete theories with a finite number of denumerable models (Russian)* (J 0003) Algebra i Logika 12*550-576,618
- • TRANSL [1973] (J 0069) Algeb and Log 12*310-326
- ◊ C57 ◊ REV MR 50 # 6827 Zbl 298 # 02047 • ID 10370

PERETYAT'KIN, M.G. [1974] see KOREC, I.

PERETYAT'KIN, M.G. [1978] *Criterion for strong constructivizability of a homogeneous model (Russian)* (J 0003) Algebra i Logika 17*436-454,491
- • TRANSL [1978] (J 0069) Algeb and Log 17*290-301
- ◊ C35 C50 C57 D45 ◊ REV MR 81j:03051 Zbl 431 # 03021 • ID 53927

PERETYAT'KIN, M.G. [1980] *Example of an ω_1-categorical complete finitely axiomatizable theory (Russian)* (J 0003) Algebra i Logika 19*314-347,382-383
- • TRANSL [1980] (J 0069) Algeb and Log 19*202-229
- ◊ C35 ◊ REV MR 82j:03035 Zbl 468 # 03016 • ID 77212

PERETYAT'KIN, M.G. [1980] *Theories with three countable models (Russian)* (J 0003) Algebra i Logika 19*224-235,251
- • TRANSL [1980] (J 0069) Algeb and Log 19*139-147
- ◊ C15 ◊ REV MR 82g:03048 Zbl 481 # 03015 • ID 77213

PERETYAT'KIN, M.G. [1982] *Finitely axiomatizable totally transcendental theories (Russian)* (C 3953) Mat Log & Teor Algor 88-135
- ◊ C45 C57 ◊ REV MR 85i:03103 Zbl 524 # 03017
- • ID 37596

PERETYAT'KIN, M.G. [1982] *Turing machine computations in finitely axiomatizable theories (Russian)* (J 0003) Algebra i Logika 21*410-441
- • TRANSL [1982] (J 0069) Algeb and Log 21*272-295
- ◊ C57 D45 ◊ REV MR 85c:03015 Zbl 567 # 03014
- • ID 39943

PERIC, V. [1983] see CUPONA, G.

PERIC, V. see Vol. V for further entries

PERLIS, D. [1976] *Group algebras and model theory* (J 0316) Illinois J Math 20*298-305
- ◊ C60 ◊ REV MR 53 # 5297 Zbl 323 # 02059 • ID 22932

PERLIS, D. see Vol. V, VI for further entries

PERLMUTTER, P. [1981] see MYCIELSKI, J.

PERMINOV, E.A. & VAZHENIN, YU.M. [1979] *On rigid lattices and graphs (Russian)* (C 2536) Issl Sovrem Algeb 3-21
- ◊ C50 C52 C65 G10 ◊ REV MR 81g:05061 • ID 33025

PERRIN, M.J. [1973] see ARTIGUE, M.

PERRIN, M.J. & ZALC, A. [1975] *Un theoreme de compacite en analyse. Application aux β_n-modeles et β_n-extensions (English summary)* (J 2313) C R Acad Sci, Paris, Ser A-B 280*A177-A180
- ◊ B15 C62 ◊ REV MR 53 # 12933 Zbl 301 # 02056
- • ID 23181

PERRIN, M.J. [1978] see ARTIGUE, M.

PERRIN, M.J. see Vol. V, VI for further entries

PERYAZEV, N.A. [1985] *Positive indistinguishability of algebraic systems and completeness of positive theories (Russian)* (J 0087) Mat Zametki (Akad Nauk SSSR) 38*208-217,347
⋄ C35 ⋄ ID 49202

PETER, R. [1941] *The bounds of the axiomatic method (Hungarian) (German summary)* (J 0461) Mat Fiz Lapok 48*120-143
⋄ A05 C07 ⋄ REV MR 9.129 Zbl 24.241 JSL 6.110
• ID 10397

PETER, R. also published under the name POLITZER, R.

PETER, R. see Vol. I, IV, V, VI for further entries

PETERSON, P.L. [1979] *On the logic of "few", "many", and "most"* (J 0047) Notre Dame J Formal Log 20*155-179
⋄ B60 C80 ⋄ REV MR 80g:03019 Zbl 299 # 02012
• ID 52390

PETERSON, P.L. see Vol. II for further entries

PETRESCO, J. [1962] *Algorithmes de decision et de construction dans les groupes libres* (J 0044) Math Z 79*32-43
⋄ B25 D40 ⋄ REV MR 28 # 5106 Zbl 104.243 • ID 10432

PETRESCO, J. see Vol. IV for further entries

PETRESCU, I. [1974] *A theorem of Loewenheim-Skolem type for topological models (Romanian) (English summary)* (J 0197) Stud Cercet Mat Acad Romana 26*1237-1240
⋄ C20 C90 ⋄ REV MR 51 # 10083 Zbl 314 # 02067
• ID 17530

PETRESCU, I. [1975] see GEORGESCU, G.

PETRESCU, I. [1978] *Model-companions for logical functors* (J 0060) Rev Roumaine Math Pures Appl 23*413-417
⋄ C25 C35 G30 ⋄ REV MR 80a:03090 Zbl 378 # 02027
• ID 51809

PETRESCU, I. [1981] *Existential morphisms and existentially closed models of logical categories* (J 0068) Z Math Logik Grundlagen Math 27*363-370
⋄ C25 G30 ⋄ REV MR 82k:03104 Zbl 465 # 03032
• ID 54935

PETRY, A. [1977] *A propos des centralisateurs de certains sous-corps des corps generiques (English summary)* (J 0408) Bull Soc R Sci Liege 46*113-116
⋄ C25 C60 ⋄ REV MR 57 # 16055 Zbl 374 # 16012
• ID 31419

PETRY, A. [1982] *Une characterisation algebrique des structures satisfaisant les memes sentences stratifees* (P 3774) Th d'Ensembl de Quine;1981 Louvain-la-Neuve 7-16
⋄ C20 C62 E70 ⋄ REV MR 84e:03063 Zbl 543 # 03019
• ID 34397

PETRY, A. see Vol. V for further entries

PFISTER, G. [1978] see KURKE, H.

PFISTER, G. [1979] see MOSTOWSKI, T.

PHILLIPS, N.C.K. [1968] *A model for urelements* (J 0068) Z Math Logik Grundlagen Math 14*303-304
⋄ C62 E70 ⋄ REV MR 38 # 2021 Zbl 162.18 • ID 10454

PHILLIPS, N.C.K. see Vol. I, V for further entries

PHILLIPS, R.E. [1973] see HICKIN, K.K.

PHILLIPS, R.E. & PLOTKIN, J.M. [1975] *On factor coverings of groups* (S 0019) Colloq Math (Warsaw) 33*175-187
⋄ C60 ⋄ REV MR 52 # 8227 Zbl 311 # 20001 • ID 28361

PHILLIPS, R.E. [1979] see HICKIN, K.K.

PHILLIPS, R.E. [1983] see HICKIN, K.K.

PHILLIPS, R.E. [1984] *Existentially closed locally finite central extensions; multipliers and local systems* (J 0044) Math Z 187*383-392
⋄ C25 C60 ⋄ REV MR 85m:20060 Zbl 551 # 20021
• ID 48428

PHILLIPS, R.G. [1969] *Liouville's theorem* (J 0048) Pac J Math 28*397-405
⋄ C65 H05 ⋄ REV MR 42 # 4715 • ID 32232

PHILLIPS, R.G. [1971] *On the structure of nonstandard models of arithmetic* (J 0053) Proc Amer Math Soc 27*359-363
⋄ H15 ⋄ REV MR 43 # 33 Zbl 217.306 • ID 10456

PHILLIPS, R.G. [1972] *Addition in nonstandard models of arithmetic* (J 0036) J Symb Logic 37*483-486
⋄ H15 ⋄ REV MR 47 # 6478 Zbl 266 # 02033 • ID 10457

PHILLIPS, R.G. & SPERRY, P.L. [1972] *Elementary extensions of linear topological abelian groups* (J 0053) Proc Amer Math Soc 31*525-528
⋄ C60 ⋄ REV Zbl 242 # 02062 • ID 26364

PHILLIPS, R.G. [1974] *A minimal extension that is not conservative* (J 0133) Michigan Math J 21*27-32
⋄ C62 ⋄ REV MR 50 # 94 Zbl 288 # 02030 • ID 10458

PHILLIPS, R.G. [1974] *Omitting types in arithmetic and conservative extensions* (P 1083) Victoria Symp Nonstand Anal;1972 Victoria 369*195-202
⋄ C62 ⋄ REV MR 57 # 16059 Zbl 278 # 02043 • ID 29067

PICKETT, H.E. [1964] *Subdirect representations of relational systems* (J 0027) Fund Math 56*223-240 • ERR/ADD ibid 62*101
⋄ C05 C52 ⋄ REV MR 37 # 5134 Zbl 126.34 • ID 19600

PICKETT, H.E. see Vol. V for further entries

PIERCE, K.R. [1972] *Amalgamating abelian ordered groups* (J 0048) Pac J Math 43*711-723
⋄ C60 ⋄ REV MR 47 # 8389 Zbl 259 # 06018 • ID 10476

PIERCE, K.R. [1980] see GLASS, A.M.W.

PIERCE, R.S. [1963] *A note on free products of abstract algebras* (J 0028) Indag Math 25*401-407
⋄ C05 ⋄ REV MR 27 # 1400 Zbl 118.263 JSL 33.125
• ID 10483

PIERCE, R.S. [1968] *Introduction to the theory of abstract algebras* (X 0818) Holt Rinehart & Winston: New York viii + 148pp
⋄ C05 C98 ⋄ REV MR 37 # 2655 • ID 19507

PIERCE, R.S. [1971] *The submodule lattice of a cyclic module* (J 0004) Algeb Universalis 1*192-199
⋄ C05 C60 ⋄ REV MR 45 # 3460 Zbl 227 # 06003
• ID 10486

PIERCE, R.S. see Vol. V for further entries

PIETSCH, A. [1974] *Ultraprodukte von Operatoren in Banachraeumen* (J 0114) Math Nachr 61*123-132
⋄ C20 C65 E75 H20 ⋄ REV MR 50 # 8140
Zbl 288 # 47037 • ID 10492

PIGOZZI, D. [1972] *Amalgamation, congruence-extension, and interpolation properties in algebras* (J 0004) Algeb Universalis 1*269-349
⋄ C05 C40 C52 G15 ⋄ REV MR 46 # 57 Zbl 236 # 02047 • ID 10493

PIGOZZI, D. [1974] *The join of equational theories* (S 0019) Colloq Math (Warsaw) 30*15-25
⋄ B25 C05 D35 ⋄ REV MR 50 # 4270 Zbl 319 # 02037
• ID 10495

PIGOZZI, D. [1975] *Equational logic and equational theories of algebras* (1111) Preprints, Manuscr., Techn. Reports etc. 182pp
⋄ C05 C98 ⋄ REM Mimeographed Notes informally printed and bound by the Computer Sciences Dept. of Purdue University • ID 30774

PIGOZZI, D. [1976] *The universality of the variety of quasigroups* (J 3194) J Austral Math Soc, Ser A 21*194-219
⋄ B25 C05 C52 D35 ⋄ REV MR 52 # 13582 Zbl 323 # 20073 • ID 21868

PIGOZZI, D. see Vol. IV for further entries

PIL'CHAK, B.YU. [1950] *On the decision problem for the calculus of problems (Russian)* (J 0023) Dokl Akad Nauk SSSR 75*773-776
⋄ B25 F50 ⋄ REV MR 12.661 Zbl 39.7 JSL 16.226
• ID 16889

PIL'CHAK, B.YU. see Vol. VI for further entries

PILLAY, A. [1978] *Number of countable models* (J 0036) J Symb Logic 43*492-496
⋄ C15 C50 ⋄ REV MR 80a:03037 Zbl 393 # 03018
• ID 29276

PILLAY, A. [1980] $\aleph_0$-*categoricity in regular subsets of* $L_{\omega_1\omega}$ (J 3172) J London Math Soc, Ser 2 22*193-196
⋄ C35 C75 ⋄ REV MR 82d:03047 Zbl 435 # 03026
• ID 55786

PILLAY, A. [1980] *Instability and theories with few models* (J 0053) Proc Amer Math Soc 80*461-468
⋄ C15 C45 C50 ⋄ REV MR 81h:03070 Zbl 455 # 03012
• ID 54273

PILLAY, A. [1980] *Theories with exactly three countable models and theories with algebraic prime models* (J 0036) J Symb Logic 45*302-310
⋄ C15 C50 ⋄ REV MR 81e:03028 Zbl 441 # 03004
• ID 56057

PILLAY, A. [1981] *A class of* $\aleph_0$-*categorical theories* (J 0068) Z Math Logik Grundlagen Math 27*411-418
⋄ C35 C45 ⋄ REV MR 83g:03029 Zbl 472 # 03024
• ID 33208

PILLAY, A. [1981] *Cuts in models of arithmetic* (P 3404) Model Th & Arithm;1979/80 Paris 13-20
⋄ C62 E05 ⋄ REV MR 84h:03085 Zbl 475 # 03043
• ID 55497

PILLAY, A. [1981] *Finitely generated models of* ω-*stable theories* (J 3133) Bull Soc Math Belg, Ser B 33*83-91
⋄ C45 C50 ⋄ REV MR 84a:03035 Zbl 487 # 03016
• ID 90328

PILLAY, A. [1981] *Models of Peano arithmetic (a survey of basic results)* (P 3404) Model Th & Arithm;1979/80 Paris 1-12
⋄ C62 C98 H15 ⋄ REV MR 83j:03055 Zbl 475 # 03042
• ID 55496

PILLAY, A. [1981] *Number of models of theories with many types* (C 2619) Groupe Etude Th Stables (2) 1978/79 9*10pp
⋄ C45 ⋄ REV MR 82k:03042 Zbl 454 # 03012 • ID 77291

PILLAY, A. [1981] *Partition properties and definable types in Peano arithmetic* (P 3404) Model Th & Arithm;1979/80 Paris 263-269
⋄ C30 C62 E05 ⋄ REV MR 83f:03072 Zbl 475 # 03044
• ID 55498

PILLAY, A. [1981] *Prime models over subsets* (C 2619) Groupe Etude Th Stables (2) 1978/79 7*8pp
⋄ C45 C50 ⋄ REV MR 82k:03041 Zbl 454 # 03012
• ID 77292

PILLAY, A. [1982] *Dimension theory and homogeneity for elementary extensions of a model* (J 0036) J Symb Logic 47*147-160
⋄ C15 C45 C50 ⋄ REV MR 83d:03038 Zbl 502 # 03018
• ID 35184

PILLAY, A. [1982] *Weakly homogeneous models* (J 0053) Proc Amer Math Soc 86*126-132
⋄ C15 C45 C50 ⋄ REV MR 84c:03059 Zbl 496 # 03014
• ID 34003

PILLAY, A. [1983] $\aleph_0$-*categoricity over a predicate* (J 0047) Notre Dame J Formal Log 24*527-536
⋄ C35 ⋄ REV MR 85b:03050 Zbl 491 # 03009 • ID 37280

PILLAY, A. [1983] *A note on tight stable theories* (J 0009) Arch Math Logik Grundlagenforsch 23*147-152
⋄ C15 C35 C45 ⋄ REV MR 85a:03048 Zbl 536 # 03017
• ID 34799

PILLAY, A. [1983] *A note on finitely generated models* (J 0036) J Symb Logic 48*163-166
⋄ C45 C50 ⋄ REV MR 84d:03036 Zbl 519 # 03024
• ID 37539

PILLAY, A. [1983] *An introduction to stability theory* (X 0815) Clarendon Pr: Oxford xi+146pp
⋄ C45 C98 ⋄ REV MR 85i:03104 Zbl 526 # 03014 JSL 51.465 • ID 38168

PILLAY, A. [1983] *Countable models of stable theories* (J 0053) Proc Amer Math Soc 89*666-672
⋄ C15 C45 ⋄ REV MR 85i:03106 Zbl 545 # 03010
• ID 41224

PILLAY, A. & PREST, M. [1983] *Forking and pushouts in modules* (J 3240) Proc London Math Soc, Ser 3 46*365-384
⋄ C45 C60 ⋄ REV MR 84m:03048 Zbl 509 # 03018
• ID 35749

PILLAY, A. [1983] *The models of a non-multidimensional* ω-*stable theory* (C 3822) Groupe Etude Th Stables (3) 1980/82 10*22pp
⋄ C45 ⋄ REV MR 85i:03105 Zbl 523 # 03022 • ID 37025

PILLAY, A. [1983] *Well-definable types over subsets* (C 3822) Groupe Etude Th Stables (3) 1980/82 No.2*4pp
⋄ C45 ⋄ REV MR 85b:03051 Zbl 519 # 03025 • ID 40002

PILLAY, A. & SROUR, G. [1984] *Closed sets and chain conditions in stable theories* (J 0036) J Symb Logic 49*1350-1362
⋄ C35 C45 ⋄ REV MR 86h:03056 • ID 39881

PILLAY, A. [1984] *Countable modules* (J 0027) Fund Math 121*125-132
⋄ C15 C35 C45 C52 C60 ⋄ REV MR 86c:03029 Zbl 554 # 03021 • ID 39670

PILLAY, A. & STEINHORN, C.I. [1984] *Definable sets in ordered structures* (J 0589) Bull Amer Math Soc (NS) 11*159-162
⋄ C40 C60 C65 ⋄ REV MR 86c:03033 Zbl 542 # 03016
• ID 40378

PILLAY, A. [1984] *Regular types in nonmultidimensional ω-stable theories* (J 0036) J Symb Logic 49*880-891
⋄ C45 C50 ⋄ REV MR 86g:03058 • ID 42491

PILLAY, A. & STEINHORN, C.I. [1985] *A note on nonmultidimensional superstable theories* (J 0036) J Symb Logic 50*1020-1024
⋄ C15 C45 ⋄ ID 48430

PILLAY, A. & SHELAH, S. [1985] *Classification theory over a predicate I* (J 0047) Notre Dame J Formal Log 26*361-376
⋄ C40 C45 C50 C55 ⋄ ID 47531

PILZ, G. [1985] see FUCHS, P.H.

PINCUS, D. [1976] *Two model theoretic ideas in independence proofs* (J 0027) Fund Math 92*113-130
⋄ C30 E25 E35 ⋄ REV MR 56 # 124 Zbl 438 # 03051
• ID 26497

PINCUS, D. see Vol. I, V for further entries

PINO, R. & RESSAYRE, J.-P. [1985] *Definable ultrafilters and elementary end extensions* (P 2160) Latin Amer Symp Math Log (6);1983 Caracas 341-350
⋄ C20 C30 C62 E05 E45 E47 ⋄ ID 41795

PINTER, C. [1975] *Algebraic logic with generalized quantifiers* (J 0047) Notre Dame J Formal Log 16*511-516
⋄ C80 G15 ⋄ REV MR 52 # 13318 Zbl 311 # 02067
• ID 18337

PINTER, C. [1976] see COSTA DA, N.C.A.

PINTER, C. [1977] *Some theorems on omitting types, with applications to model completeness, amalgamation, and related properties* (P 1076) Latin Amer Symp Math Log (3);1976 Campinas 223-238
⋄ C07 C25 C35 C52 C75 ⋄ REV MR 58 # 21571
Zbl 361 # 02068 • ID 16606

PINTER, C. [1978] *A note on the decomposition of theories with respect to amalgamation, convexity, and related properties* (J 0047) Notre Dame J Formal Log 19*115-118
⋄ C07 C52 ⋄ REV MR 81a:03031 Zbl 351 # 02035
• ID 27092

PINTER, C. [1978] *Properties preserved under definitional equivalence and interpretations* (J 0068) Z Math Logik Grundlagen Math 24*481-488
⋄ C25 C35 C50 F25 ⋄ REV MR 80b:03039
Zbl 408 # 03028 • ID 56267

PINTER, C. [1980] *Topological duality theory in algebraic logic* (P 2958) Latin Amer Symp Math Log (4);1978 Santiago 255-266
⋄ C07 G25 ⋄ REV MR 81g:03077 Zbl 451 # 03028
• ID 54043

PINTER, C. see Vol. I, II, V for further entries

PINUS, A.G. [1971] *A remark on the paper by Ju.M.Vazhenin: "Elementary properties of transformation semigroups of ordered sets" (Russian)* (J 0003) Algebra i Logika 10*327-328
• TRANSL [1971] (J 0069) Algeb and Log 10*205
⋄ B25 C65 D35 E07 ⋄ REV MR 45 # 1804
Zbl 262 # 20079 • REM For the paper see Vazhenin 1970
• ID 19598

PINUS, A.G. [1972] *On the theory of convex subsets (Russian)* (J 0092) Sib Mat Zh 13*218-224
• TRANSL [1972] (J 0475) Sib Math J 13*157-161
⋄ B25 C65 C85 D35 ⋄ REV MR 45 # 3205
Zbl 255 # 02052 • ID 10513

PINUS, A.G. [1973] *Elementary definability of symmetry groups* (J 0004) Algeb Universalis 3*59-66
⋄ C60 ⋄ REV MR 49 # 2363 Zbl 272 # 02079 • ID 10514

PINUS, A.G. [1975] *Effective linear orders (Russian)* (J 0092) Sib Mat Zh 16*1246-1254,1371
• TRANSL [1975] (J 0475) Sib Math J 16*956-962
⋄ C57 D45 E07 F15 ⋄ REV MR 53 # 137
Zbl 333 # 02035 • ID 16662

PINUS, A.G. [1976] *Theories of boolean algebras in a calculus with the quantifier "infinitely many exist" (Russian)* (J 0092) Sib Mat Zh 17*1417-1421,1440
• TRANSL [1976] (J 0475) Sib Math J 17*1035-1038
⋄ B25 C10 C35 C57 C80 G05 ⋄ REV MR 56 # 84
Zbl 353 # 02006 • ID 50089

PINUS, A.G. [1978] *Cardinality of models for theories in a calculus with a Haertig quantifier (Russian)* (J 0092) Sib Mat Zh 19*1349-1356,1439
• TRANSL [1978] (J 0475) Sib Math J 19*949-955
⋄ C55 C80 ⋄ REV MR 80c:03041 Zbl 404 # 03027
• ID 54814

PINUS, A.G. [1978] see KOKORIN, A.I.

PINUS, A.G. [1978] *The Loewenheim-Skolem property for second-order logics (Russian)* (C 2595) Rekursiv Funktsii 55-60
⋄ C55 C80 C85 E45 ⋄ REV MR 82c:03058
Zbl 504 # 03020 • ID 77325

PINUS, A.G. [1978] see HERRE, H.

PINUS, A.G. [1979] *A calculus of one-place predicates (Russian)* (J 0031) Izv Vyssh Ucheb Zaved, Mat (Kazan) 1979/1(200)*54-60
• TRANSL [1979] (J 3449) Sov Math 23/1*43-47
⋄ B20 B25 C55 C80 C85 ⋄ REV MR 80j:03053
Zbl 408 # 03008 • ID 53695

PINUS, A.G. [1979] *Elementary equivalence of topological spaces (Russian)* (J 0092) Sib Mat Zh 20*433-439,464
• TRANSL [1979] (J 0475) Sib Math J 20*310-314
⋄ C55 C65 C85 G05 G10 ⋄ REV MR 80h:03050
Zbl 426 # 03038 • ID 53634

PINUS, A.G. [1979] *Elimination of the quantifiers Q_0 and Q_1 on symmetric groups (Russian)* (J 0031) Izv Vyssh Ucheb Zaved, Mat (Kazan) 1979/12(211)*45-47
• TRANSL [1979] (J 3449) Sov Math 23/12*47-49
⋄ C10 C55 C60 C80 ⋄ REV MR 81d:03039
Zbl 428 # 03030 • ID 53789

PINUS, A.G. [1979] *Hanf number for the calculus with the Haertig quantifier (Russian)* (J 0092) Sib Mat Zh 20*440-441,462
- TRANSL [1979] (J 0475) Sib Math J 20*315-316
- ◇ C55 C80 C85 E45 ◇ REV MR 80d:03034 Zbl 407 # 03039 • ID 56205

PINUS, A.G. [1980] *Weak commutativity of scattered summation of linear orders (Russian)* (J 0092) Sib Mat Zh 21/2*155-159,238
- TRANSL [1980] (J 0475) Sib Math J 21*262-272
- ◇ C30 E07 ◇ REV MR 81h:04003 Zbl 441 # 04003
- • ID 77327

PINUS, A.G. [1981] *The spectrum of rigid systems of Horn classes (Russian)* (J 0092) Sib Mat Zh 22/5*153-157,223
- TRANSL [1981] (J 0475) Sib Math J 22*769-772
- ◇ C05 C30 C50 ◇ REV MR 82k:03044 Zbl 481 # 03018
- • ID 36829

PINUS, A.G. [1982] *Complete embeddings of categories of algebraic systems and definability of a model for its semigroup of endomorphisms (Russian)* (J 0031) Izv Vyssh Ucheb Zaved, Mat (Kazan) 1982/1*80-83
- TRANSL [1982] (J 3449) Sov Math 26/1*97-101
- ◇ C07 F25 G30 ◇ REV MR 84a:03036 Zbl 514 # 03023
- • ID 35574

PINUS, A.G. [1983] *Calculus with the quantifier of elementary equivalence (Russian)* (J 0092) Sib Mat Zh 24/3(139)*136-141
- TRANSL [1983] (J 0475) Sib Math J 24*428-432
- ◇ C40 C55 C80 D35 ◇ REV MR 85b:03056 Zbl 536 # 03020 • ID 37103

PINUS, A.G. [1983] *The operation of Cartesian product* (J 0031) Izv Vyssh Ucheb Zaved, Mat (Kazan) 1983/8*51-53
- TRANSL [1983] (J 0062) Sov Math, Dokl 27*62-65
- ◇ B25 C05 D35 ◇ REV MR 85b:08009 Zbl 539 # 08004
- • ID 39357

PINUS, A.G. [1985] *Applications of boolean powers of algebraic systems (Russian)* (J 0092) Sib Mat Zh 26/3*117-125,225
- TRANSL [1985] (J 0475) Sib Math J 26*400-407
- ◇ C05 C30 C55 C57 C80 C85 D35 E50 G05 ◇ ID 47255

PINUS, A.G. see Vol. I, IV, V, VI for further entries

PIROG-RZEPECKA, K. [1975] see HALKOWSKA, K.

PIROG-RZEPECKA, K. [1976] see HALKOWSKA, K.

PIROG-RZEPECKA, K. see Vol. I, II, V for further entries

PITT, W.B. [1965] see ADDISON, J.W.

PITTS, A.M. [1983] *Amalgamation and interpolation in the category of Heyting algebras* (J 0326) J Pure Appl Algebra 29*155-165
- ◇ C40 C52 G30 ◇ REV MR 85c:03024 Zbl 517 # 18004
- • ID 36693

PITTS, A.M. [1983] *An application of open maps to categorical logic* (J 0326) J Pure Appl Algebra 29*313-326
- ◇ C40 C52 G30 ◇ REV MR 85c:03025 Zbl 521 # 03051
- • ID 37085

PITTS, A.M. see Vol. V, VI for further entries

PIXLEY, A.F. [1972] *Local Mal'cev conditions* (J 0018) Canad Math Bull 15*559-568
- ◇ C05 C57 ◇ REV MR 46 # 8942 Zbl 254 # 08009
- • ID 31207

PIXLEY, A.F. [1985] *Principal congruence formulas in arithmetical varieties* (P 4178) Universal Algeb & Lattice Th;1984 Charleston 238-254
- ◇ C05 ◇ ID 47872

PLA I CARRERA, J. [1977] see CASCANTE DAVILA, J.M.

PLA I CARRERA, J. see Vol. I, II, IV, V, VI for further entries

PLATEK, R.A. [1971] *The converse to a metatheorem in Goedel set theory* (J 0068) Z Math Logik Grundlagen Math 17*21-22
- ◇ C40 E30 E47 E70 ◇ REV MR 43 # 4656 Zbl 216.13
- • ID 10530

PLATEK, R.A. see Vol. IV, V for further entries

PLATT, C.R. [1971] *Iterated limits of universal algebras* (J 0004) Algeb Universalis 1*167-181
- ◇ C05 C30 ◇ REV MR 46 # 3409 Zbl 239 # 08005
- • ID 10533

PLATT, C.R. see Vol. V for further entries

PLONKA, E. [1968] *On a problem of Bjarni Jonsson concerning automorphisms of a general algebra* (S 0019) Colloq Math (Warsaw) 19*5-8
- ◇ C05 ◇ REV MR 36 # 6337 Zbl 153.337 • ID 16921

PLONKA, J. [1967] *On a method of construction of abstract algebras* (J 0027) Fund Math 61*183-189
- ◇ C05 C30 ◇ REV MR 37 # 1294 Zbl 168.267 • ID 10546

PLONKA, J. [1969] see GRAETZER, G.

PLONKA, J. [1969] *On sums of direct systems of boolean algebras* (S 0019) Colloq Math (Warsaw) 20*209-214
- ◇ C30 G05 ◇ REV MR 40 # 2590 Zbl 186.308 • ID 10549

PLONKA, J. [1973] see GANTER, B.

PLONKA, J. [1973] *On the sum of a direct system of relational systems* (J 0014) Bull Acad Pol Sci, Ser Math Astron Phys 21*595-597
- ◇ C05 C30 ◇ REV MR 49 # 10623 Zbl 267 # 08003
- • ID 29905

PLONKA, J. [1974] *On connections between the decomposition of an algebra into sums of direct systems of subalgebras* (J 0027) Fund Math 84*237-244
- ◇ C05 C30 ◇ REV MR 50 # 4439 Zbl 279 # 08003
- • ID 10561

PLONKA, J. [1974] *On the subdirect product of some equational classes of algebras* (J 0114) Math Nachr 63*303-305
- ◇ C05 C30 ◇ REV MR 50 # 12864 Zbl 289 # 08002
- • ID 10562

PLONKA, J. [1975] *On relations between decompositions of an algebra into the sums of direct systems of subalgebras* (J 0128) Acta Math Univ Comenianae (Bratislava) SPECIAL NO.*35-37
- ◇ C05 C30 ◇ REV MR 50 # 209 Zbl 301 # 08003
- • ID 10563

PLONKA, J. [1975] *Remark on direct products and the sums of direct systems of algebras (Russian summary)* (J 0014) Bull Acad Pol Sci, Ser Math Astron Phys 23*515-518
- ◇ C05 C30 ◇ REV MR 52 # 236 Zbl 314 # 08003
- • ID 10564

PLONKA, J. [1978] see HALKOWSKA, K.

PLONKA, J. [1979] *On automorphism groups of relational systems and universal algebras* (S 0019) Colloq Math (Warsaw) 42*341-344
◇ C05 C07 E20 ◇ REV MR 81e:04003 Zbl 428 # 08002
• ID 77370

PLONKA, J. [1985] *On identities satisfied in finite algebras* (P 4625) Contrib to Gen Algeb (3);1984 Wien 3*285-289
◇ C05 C13 ◇ REV Zbl 565 # 08007 • ID 49417

PLONKA, J. see Vol. V for further entries

PLOTKIN, J.M. [1969] *Generic embeddings* (J 0036) J Symb Logic 34*388-394
◇ C62 E25 E35 ◇ REV MR 40 # 5432 Zbl 182.329
• ID 10566

PLOTKIN, J.M. [1972] see HICKIN, K.K.

PLOTKIN, J.M. [1974] see BALDWIN, JOHN T.

PLOTKIN, J.M. [1975] see METAKIDES, G.

PLOTKIN, J.M. [1975] see HICKIN, K.K.

PLOTKIN, J.M. [1975] see PHILLIPS, R.E.

PLOTKIN, J.M. [1977] see HICKIN, K.K.

PLOTKIN, J.M. [1981] *ZF and locally finite groups* (J 0068) Z Math Logik Grundlagen Math 27*375-379
◇ C60 C62 E25 E35 E75 ◇ REV MR 83f:03044 Zbl 468 # 03037 • ID 55102

PLOTKIN, J.M. [1984] see GARAVAGLIA, S.

PLOTKIN, J.M. see Vol. I, IV, V, VI for further entries

PODEWSKI, K.-P. [1973] *Theorien mit relativ konstruktiblen saturierten Modellen* (J 0008) Arch Math (Basel) 24*449-455
◇ C45 C50 E45 ◇ REV MR 48 # 8226 Zbl 287 # 02036
• ID 10569

PODEWSKI, K.-P. [1973] *Zur Vergroesserung von Strukturen* (J 0068) Z Math Logik Grundlagen Math 19*265-270
◇ C15 C30 ◇ REV MR 49 # 2358 Zbl 307 # 02034
• ID 10570

PODEWSKI, K.-P. & REINEKE, J. [1974] *An ω_1-categorical ring which is not almost strongly minimal* (J 0036) J Symb Logic 39*665-668
◇ C35 C50 C60 ◇ REV MR 50 # 12706 Zbl 325 # 02040
• ID 10571

PODEWSKI, K.-P. [1975] *Minimale Ringe* (J 0160) Math-Phys Sem-ber, NS 22*193-197
◇ C60 ◇ REV MR 52 # 13775 Zbl 316 # 16001 • ID 21876

PODEWSKI, K.-P. & ZIEGLER, M. [1978] *Stable graphs* (J 0027) Fund Math 100*101-107
◇ C45 C65 ◇ REV MR 58 # 5183 Zbl 407 # 05072
• ID 77385

PODEWSKI, K.-P. & REINEKE, J. [1979] *Algebraically closed commutative local rings* (J 0036) J Symb Logic 44*89-94
◇ C25 C35 C60 ◇ REV MR 80c:03039 Zbl 431 # 03020
• ID 53926

PODEWSKI, K.-P. & REINEKE, J. [1981] *On algebraically closed commutative indecomposable rings* (J 0004) Algeb Universalis 12*123-131
◇ C25 C35 C60 ◇ REV MR 82c:03050 Zbl 471 # 03029
• ID 55224

PODEWSKI, K.-P. see Vol. I, II, IV, V for further entries

POGORZELSKI, W.A. & SLUPECKI, J. [1961] *A variant of the proof of the completeness of the first order functional calculus (Polish and Russian summaries)* (J 0063) Studia Logica 12*125-134
◇ B10 C07 ◇ REV MR 26 # 6040 Zbl 156.7 JSL 36.688
• ID 31969

POGORZELSKI, W.A. see Vol. I, II, VI for further entries

POGUNTKE, W. [1974] see HERRMANN, C.

POHLERS, W. [1981] *Cut-elimination for impredicative infinitary systems I. Ordinal-analysis for ID_1* (J 0009) Arch Math Logik Grundlagenforsch 21*113-129
◇ C75 F05 F15 F35 ◇ REV MR 84a:03070a Zbl 484 # 03030 JSL 48.879 • REM Part II 1982 • ID 33310

POHLERS, W. [1982] *Admissibility in proof theory; a survey* (P 3622) Int Congr Log, Meth & Phil of Sci (6,Proc);1979 Hannover 123-139
◇ C70 F15 F35 F98 ◇ REV MR 84h:03129 Zbl 506 # 03015 • ID 34316

POHLERS, W. [1982] *Cut elimination for impredicative infinitary systems II. Ordinal analysis for iterated inductive definitions* (J 0009) Arch Math Logik Grundlagenforsch 22*69-87
◇ C75 F05 F15 F35 ◇ REV MR 84a:03070b Zbl 497 # 03043 JSL 48.879 • REM Part I 1981 • ID 35591

POHLERS, W. see Vol. VI for further entries

POINT, F. [1980] see BOFFA, M.

POINT, F. [1981] *Elimination des quantificateurs dans les L-anneaux *-reguliers et application aux anneaux bireguliers* (J 3133) Bull Soc Math Belg, Ser B 33*93-107
◇ C10 C60 C90 ◇ REV MR 82h:03028 Zbl 475 # 03011
• ID 55465

POINT, F. [1982] *Sur l'elimination lineaire (English summary)* (J 3364) C R Acad Sci, Paris, Ser 1 295*211-213
◇ C60 ◇ REV MR 84a:03034 Zbl 518 # 03009 • ID 35573

POINT, F. [1985] *Finitely generic models of T_{UH}, for certain model companionable theories T* (J 0036) J Symb Logic 50*604-610
◇ C25 C60 ◇ REV Zbl 577 # 03014 • ID 47367

POINT, F. [1985] see BOFFA, M.

POINT, F. [1985] *Transfer properties of discriminator varieties* (P 4310) Easter Conf on Model Th (3);1985 Gross Koeris 201-214
◇ C10 C25 C30 C35 ◇ REV Zbl 575 # 03027 • ID 48827

POIZAT, B. [1971] *Theorie de Galois des relations* (J 2313) C R Acad Sci, Paris, Ser A-B 272*A645-A648
◇ C07 C60 C75 E07 ◇ REV MR 44 # 46 Zbl 209.11
• ID 48005

POIZAT, B. [1975] *Theoremes globaux (English summary)* (J 2313) C R Acad Sci, Paris, Ser A-B 280*A845-A847
◇ B20 C40 ◇ REV MR 52 # 2863 Zbl 301 # 02047
• ID 17649

POIZAT, B. [1976] *"Forcing" et relations generales sur une structure ω-categorique (English summary)* (J 2313) C R Acad Sci, Paris, Ser A-B 283*A223-A224
◇ C25 C35 ◇ REV MR 54 # 10007 Zbl 349 # 02048
• ID 25848

POIZAT, B. [1976] *Remarques sur le "forcing" en theorie des modeles (English summary)* (J 2313) C R Acad Sci, Paris, Ser A-B 283*A131-A134
◇ C25 ◇ REV MR 54 # 10006 Zbl 398 # 03014 • ID 25847

POIZAT, B. [1976] *Une relation particulierement rigide (English summary)* (J 2313) C R Acad Sci, Paris, Ser A-B 282*A671-A673
⋄ C50 E07 E55 ⋄ REV MR 53 # 12954   Zbl 327 # 04002
• ID 27224

POIZAT, B. [1978] *Etude d'un forcing en theorie des modeles* (J 0068) Z Math Logik Grundlagen Math 24*347-356
⋄ C25 C35 ⋄ REV MR 80g:03030   Zbl 408 # 03023
• ID 56262

POIZAT, B. (ED.) [1978] *Groupe d'etude de theories stables. 1re annee: 1977/78. Exposes 1 a 9* (X 1623) Univ Paris VI Inst Poincare: Paris 104pp
⋄ C45 C97 C98 ⋄ REV MR 80b:03003   Zbl 408 # 03026
• ID 56265

POIZAT, B. [1978] *Rangs des types dans les corps differentiels* (C 3093) Groupe Etude Th Stables (1) 1977/78 6*13pp
⋄ C45 C60 ⋄ REV MR 80c:03036   Zbl 408 # 03026
• ID 77418

POIZAT, B. [1978] *Suites d'indiscernables dans les theories stables* (C 3093) Groupe Etude Th Stables (1) 1977/78 5*7pp
⋄ C30 C45 ⋄ REV MR 80e:03030   Zbl 408 # 03026
• ID 77417

POIZAT, B. [1978] *Une preuve par la theorie de la deviation d'un theoreme de J. Baldwin (English summary)* (J 2313) C R Acad Sci, Paris, Ser A-B 287*A589-A591
⋄ C35 C45 ⋄ REV MR 83m:03039   Zbl 404 # 03021
• ID 54808

POIZAT, B. [1978] *Une theorie ω_1-categorique a α_T fini* (C 3093) Groupe Etude Th Stables (1) 1977/78 8*3pp
⋄ C35 C45 ⋄ REV MR 80j:03043   Zbl 408 # 03026
• ID 77414

POIZAT, B. [1979] see LASCAR, D.

POIZAT, B. [1979] *Le "forcing" de Fraisse dans un contexte arithmetique (English summary)* (J 2313) C R Acad Sci, Paris, Ser A-B 289*A409-A412
⋄ C25 ⋄ REV MR 80i:03043   Zbl 427 # 03021 • ID 53708

POIZAT, B. [1980] *Les equations differentielles vues par un logicien* (C 3591) Semin Equat Deriv Part Hyperb & Holom 1979/80 6*5pp
⋄ C45 C60 ⋄ REV MR 82g:12027 • ID 82511

POIZAT, B. [1980] *Une preuve d'un theoreme de James Ax sur les extensions algebriques de corps (English summary)* (J 2313) C R Acad Sci, Paris, Ser A-B 291*A245
⋄ C60 ⋄ REV MR 81m:12026   Zbl 469 # 12008 • ID 82512

POIZAT, B. [1981] *Degres de definissabilite arithmetique des generiques (English summary)* (J 3364) C R Acad Sci, Paris, Ser 1 293*289-291
⋄ C25 C40 D30 D55 ⋄ REV MR 83f:03031
Zbl 472 # 03022 • ID 55288

POIZAT, B. [1981] *Exercices en stabilite* (C 2619) Groupe Etude Th Stables (2) 1978/79 11*5pp
⋄ C45 ⋄ REV MR 82g:03053 • ID 77413

POIZAT, B. (ED.) [1981] *Groupe d'etude de theories stables. 2e annee 1978/1979. Exposes 1 a 11* (X 1623) Univ Paris VI Inst Poincare: Paris ii+115pp
⋄ C45 C97 C98 ⋄ REV MR 82f:03028   Zbl 454 # 03012
• ID 42002

POIZAT, B. [1981] *Le rang selon Lascar* (C 2619) Groupe Etude Th Stables (2) 1978/79 7pp
⋄ C45 C98 ⋄ REV Zbl 454 # 03012 • ID 48431

POIZAT, B. [1981] *Modeles premiers d'une theorie totalement transcendante* (C 2619) Groupe Etude Th Stables (2) 1978/79 8*7pp
⋄ C45 C50 ⋄ REV MR 82i:03042   Zbl 454 # 03012
• ID 77410

POIZAT, B. [1981] *Sous-groupes definissables d'un groupe stable* (J 0036) J Symb Logic 46*137-146
⋄ C45 C60 ⋄ REV MR 82g:03054   Zbl 476 # 03038
JSL 49.317 • ID 55550

POIZAT, B. [1981] *Theorie de Galois pour les algebres de Post infinitaires* (J 0068) Z Math Logik Grundlagen Math 27*31-44
⋄ C75 G20 ⋄ REV MR 82h:03070   Zbl 492 # 06010
• ID 77411

POIZAT, B. [1981] *Theories instables* (J 0036) J Symb Logic 46*513-522
⋄ C45 ⋄ REV MR 83e:03045   Zbl 473 # 03027 • REM
Postscript was published ibid 48*60-62 • ID 55357

POIZAT, B. [1982] *Deux ou trois choses que je sais de L_n* (J 0036) J Symb Logic 47*641-658
⋄ B20 C07 C13 C50 ⋄ REV MR 84b:03055
Zbl 507 # 03014 • ID 35642

POIZAT, B. [1983] *Beaucoup de modeles a peu de frais* (C 3822) Groupe Etude Th Stables (3) 1980/82 9*7pp
⋄ C45 C50 C52 C55 ⋄ REV MR 85d:03071
Zbl 525 # 03026 • ID 38261

POIZAT, B. [1983] *C'est beau et chaud* (C 3822) Groupe Etude Th Stables (3) 1980/82 7*11pp
⋄ C45 C60 C98 ⋄ REV MR 85e:03081   Zbl 525 # 03025
• ID 38260

POIZAT, B. (ED.) [1983] *Groupe d'etude de theories stables. 3e annee 1980-1982* (X 1623) Univ Paris VI Inst Poincare: Paris 115pp
⋄ C45 C97 C98 ⋄ REV MR 84i:03009 • ID 34493

POIZAT, B. [1983] *Groupes stables, avec types generiques reguliers* (J 0036) J Symb Logic 48*339-355
⋄ C45 C60 ⋄ REV MR 85e:03082   Zbl 525 # 03024
• ID 38259

POIZAT, B. [1983] *Paires de structures stables* (J 0036) J Symb Logic 48*239-249
⋄ C45 ⋄ REV MR 84h:03082   Zbl 525 # 03023 • ID 34276

POIZAT, B. [1983] *Post-scriptum a "Theories instables"* (J 0036) J Symb Logic 48*60-62
⋄ C45 ⋄ REV MR 83e:03045   MR 84g:03038
Zbl 533 # 03017 • REM The article was published ibid 46*513-522 • ID 34151

POIZAT, B. [1983] *Proprietes "modele-theoriques" d'une theorie du premier ordre* (P 1601) Easter Conf on Model Th (1);1983 Diedrichshagen 49*99-109
⋄ C45 C52 C95 ⋄ REV MR 84i:03008   Zbl 531 # 03014
• ID 37677

POIZAT, B. [1983] *Travaux publies par les membres du groupe "theories stables"* (C 3822) Groupe Etude Th Stables (3) 1980/82 11*7pp
⋄ C45 C98 ⋄ REV MR 84j:03069   Zbl 512 # 03014
• ID 34660

POIZAT, B. [1983] *Une theorie finiement axiomatisable et superstable* (C 3822) Groupe Etude Th Stables (3) 1980/82 1*9pp
- ⋄ C35 C45 ⋄ REV MR 85d:03070   Zbl 539 # 03012
- • ID 39768

POIZAT, B. [1983] *Une theorie de Galois imaginaire* (J 0036) J Symb Logic 48*1151-1170
- ⋄ C45 C60 ⋄ REV MR 85e:03083   Zbl 537 # 03023
- • ID 39846

POIZAT, B. [1984] *Deux remarques a propos de la propriete de recouvrement fini* (J 0036) J Symb Logic 49*803-807
- ⋄ C45 C50 C60 C62 ⋄ REV MR 85m:03023 • ID 42493

POIZAT, B. [1984] *La structure geometrique des groupes stables* (P 1545) Easter Conf on Model Th (2);1984 Wittenberg 205-217
- ⋄ C45 C60 C98 ⋄ REV MR 86h:03057   Zbl 564 # 03031
- • ID 44682

POIZAT, B. [1985] *Cours de theorie des modeles* (2222) See Remarks 584pp
- ⋄ C98 ⋄ REM Nur al-Mantiq wal-Ma'rifah, no. 1. Publ. by B.Poizat, 82, rue Racine, F-69100 Villeurbanne, France
- • ID 47654

POL, R. [1973] *There is no universal totally disconnected space* (J 0027) Fund Math 79*265-267
- ⋄ C50 C65 ⋄ REV MR 48 # 1139   Zbl 263 # 54006
- • ID 10607

POL, R. see Vol. V for further entries

POLAK, L. & ROSICKY, J. [1981] *Implicit operations on finite algebras* (P 2552) Conf Finite Algeb & Multi-Val Log;1979 Szeged 653-668
- ⋄ C05 C13 ⋄ REV MR 83e:08009   Zbl 478 # 08002
- • ID 40145

POLAND, J. [1982] see DRIES VAN DEN, L.

POLIFERNO, M.J. [1961] *Decision algorithms for some functional calculi with modality* (J 0079) Logique & Anal, NS 4*138-153
- ⋄ B25 B45 ⋄ REV JSL 32.244 • REM Cf. "Correction to a paper on modal logic", ibid 1964, 7*32-33 • ID 10611

POLLOCK, J.L. [1966] *Model theory and modal logic* (J 0079) Logique & Anal, NS 9*313-317
- ⋄ B45 C90 ⋄ REV MR 38 # 39   Zbl 192.31 • ID 10622

POLLOCK, J.L. see Vol. I, II, V for further entries

PONASSE, D. [1966] *Une demonstration du theoreme de completude de Goedel* (J 0056) Publ Dep Math, Lyon 3/1*2-8
- ⋄ C07 G05 ⋄ REV MR 33 # 5464   Zbl 207.296   JSL 32.418
- • ID 10635

PONASSE, D. [1967] *Quelques remarques topologiques sur l'espace des interpretations* (J 0056) Publ Dep Math, Lyon 4/3*33-37
- ⋄ C07 G05 ⋄ REV MR 40 # 5433   Zbl 165.16 • ID 10636

PONASSE, D. see Vol. I, II, V for further entries

POPESCU, D. [1978] see KURKE, H.

POPESCU, D. [1979] *Algebraically pure morphisms* (J 0060) Rev Roumaine Math Pures Appl 24*947-977
- ⋄ C20 C60 ⋄ REV MR 80j:13017   Zbl 416 # 13005
- • ID 53224

POPESCU, D. [1980] see BASARAB, S.A.

POPESCU, D. [1984] see MATEESCU, C.

POPOV, S.V. & ZAKHAR'YASHCHEV, M.V. [1980] *On the power of countermodels in intuitionistic calculus (Russian)* (S 2651) Prepr Inst Prikl Mat, Akad Nauk SSSR 80/45*28pp
- ⋄ C90 F50 ⋄ REV MR 82m:03022 • ID 80198

POPOV, S.V. see Vol. I, II, IV, VI for further entries

POREBSKA, M. [1985] *Interpolation for fragments of intermediate logics* (J 0387) Bull Sect Logic, Pol Acad Sci 14*79-83
- ⋄ B55 C40 ⋄ ID 48967

POREBSKA, M. see Vol. II, VI for further entries

POSEGGA, M. [1972] see HAUCK, J.

POSSEL DE, R. [1954] see FRAISSE, R.

POSSEL DE, R. see Vol. V for further entries

POTTHOFF, K. [1969] *Ueber Nichtstandardmodelle der Arithmetik und der rationalen Zahlen* (J 0068) Z Math Logik Grundlagen Math 15*223-236
- ⋄ H15 H20 ⋄ REV MR 40 # 4094   Zbl 187.270 • ID 10694

POTTHOFF, K. [1970] *Ideale in Nichtstandardmodellen der ganzen Zahlen* (J 0068) Z Math Logik Grundlagen Math 16*321-326
- ⋄ H15 H20 ⋄ REV MR 43 # 4654   Zbl 164.312 • ID 10695

POTTHOFF, K. [1971] *Representation of locally finite polyadic algebras and ultrapowers* (J 0068) Z Math Logik Grundlagen Math 17*91-96
- ⋄ C20 G15 ⋄ REV MR 43 # 7311   Zbl 221 # 02048
- • ID 10696

POTTHOFF, K. [1972] *Ordnungseigenschaften von Nichtstandardmodellen* (C 0648) Th of Sets and Topology (Hausdorff) 403-426
- ⋄ C20 C62 H05 H15 ⋄ REV MR 50 # 12712   Zbl 269 # 02024 • ID 10697

POTTHOFF, K. [1972] *Ueber Enderweiterungen von Nichtstandardmodellen der Arithmetik und anderer Strukturen* (0000) Diss., Habil. etc
- ⋄ C62 H15 H20 ⋄ REM Kiel • ID 40100

POTTHOFF, K. [1974] *Boolean ultrapowers* (J 0009) Arch Math Logik Grundlagenforsch 16*37-48
- ⋄ C20 C90 ⋄ REV MR 50 # 90   Zbl 285 # 02045 • ID 29963

POTTHOFF, K. [1976] *A simple tree lemma and its application to a counterexample of Philips* (J 0009) Arch Math Logik Grundlagenforsch 18*67-71
- ⋄ C20 C62 H15 ⋄ REV MR 57 # 2908   Zbl 402 # 03024
- • ID 23715

POTTHOFF, K. [1978] *Orderings of types of countable arithmetic* (J 0068) Z Math Logik Grundlagen Math 24*97-108
- ⋄ C20 C62 H15 ⋄ REV MR 80f:03035   Zbl 414 # 03040
- • ID 53086

POTTHOFF, K. [1979] *Ultraproduits des groupes finis et applications a la theorie de Galois* (J 0056) Publ Dep Math, Lyon 16/3-4*39-45
- ⋄ C13 C20 C60 ⋄ REV MR 83b:03032   Zbl 448 # 12005
- • ID 56658

POTTHOFF, K. [1981] *Einfuehrung in die Modelltheorie und ihre Anwendungen* (X 0890) Wiss Buchges: Darmstadt xii + 277pp
- ⋄ C98 ⋄ REV MR 84c:03001   Zbl 457 # 03028   JSL 48.219
- • ID 54353

POTTHOFF, K. [1983] *Quelques applications des methodes non-standard a la theorie des groupes* (J 4149) Publ Inst Rech Math Avancee 222/S-07*36-39
◇ C60 H05 H20 ◇ ID 40113

POTTHOFF, K. see Vol. I, II, V for further entries

POUR-EL, M.B. [1968] *Independent axiomatization and its relation to the hypersimple set* (J 0068) Z Math Logik Grundlagen Math 14*449-456
◇ C07 D25 ◇ REV MR 40 # 4105 Zbl 182.9 JSL 38.654
• ID 10704

POUR-EL, M.B. [1969] *A recursion-theoretic view of axiomatizable theories* (P 1841) Fct Recurs & Appl;1967 Tihany 26-42
◇ C07 D25 D35 ◇ ID 32554

POUR-EL, M.B. [1969] *Independent axiomatization and its relation to the hypersimple set* (P 1841) Fct Recurs & Appl;1967 Tihany 24-25
◇ C07 D25 D35 ◇ REV Zbl 182.9 • ID 32553

POUR-EL, M.B. [1970] *A recursion-theoretic view of axiomatizable theories* (J 0076) Dialectica 24*267-276
◇ C07 D25 D35 ◇ REV Zbl 274 # 02001 • ID 29012

POUR-EL, M.B. see Vol. I, IV, VI for further entries

POUZET, M. [1969] see BONNET, R.

POUZET, M. [1971] see FRAISSE, R.

POUZET, M. [1972] *Modele universel d'une theorie n-complete* (J 2313) C R Acad Sci, Paris, Ser A-B 274*A433-A436
◇ C15 C35 C50 ◇ REV MR 45 # 1745a Zbl 238 # 02048
• ID 64649

POUZET, M. [1972] *Modele universel d'une theorie n-complete: modele uniformement prehomogene* (J 2313) C R Acad Sci, Paris, Ser A-B 274*A695-A698
◇ C15 C35 C50 ◇ REV MR 45 # 1745b Zbl 238 # 02049
• ID 64652

POUZET, M. [1972] *Modele universel d'une theorie n-complete: modele prehomogene* (J 2313) C R Acad Sci, Paris, Ser A-B 274*A813-A816
◇ C15 C35 C50 ◇ REV MR 45 # 1745c Zbl 245 # 02050
• ID 19585

POUZET, M. [1973] *Extensions completes d'une theorie forcing complete* (J 0029) Israel J Math 16*212-215
◇ C25 C52 ◇ REV MR 48 # 10797 Zbl 279 # 02031
• ID 16928

POUZET, M. [1973] *Une remarque sur la libre interpretabilite d'une relation par une autre* (J 0004) Algeb Universalis 3*67-71
◇ C07 E07 ◇ REV MR 49 # 181 Zbl 304 # 04001
• ID 24428

POUZET, M. [1979] *Chaines de theories universelles* (J 0027) Fund Math 103*133-149
◇ B20 C52 ◇ REV MR 81j:03025 Zbl 365 # 02037
• ID 51038

POUZET, M. [1979] *Relation minimale pour son age* (J 0068) Z Math Logik Grundlagen Math 25*315-344
◇ C50 E07 ◇ REV MR 81a:04001 Zbl 481 # 04002
• ID 77496

POUZET, M. [1980] see CHARRETTON, C.

POUZET, M. [1981] *Application de la notion de relation presque-enchainable au denombrement des restrictions finies d'une relation* (J 0068) Z Math Logik Grundlagen Math 27*289-332
◇ C07 C30 E07 ◇ REV MR 82k:03050 Zbl 499 # 03019
• ID 77494

POUZET, M. [1981] see ASSOUS, R.

POUZET, M. [1982] see BONNET, R.

POUZET, M. [1983] see CHARRETTON, C.

POUZET, M. & RICHARD, D. (EDS.) [1984] *Orders: description and roles in set theory, lattices, ordered groups, topology, theory of models and relations, combinatorics, effectiveness, social sciences* (S 3358) Ann Discrete Math 23**xxvii + 548pp
◇ C65 C97 ◇ REV MR 85k:06001 Zbl 539 # 00003
• ID 41242

POUZET, M. see Vol. V for further entries

POWELL, W.C. [1971] see JECH, T.J.

POWELL, W.C. [1974] *Variations of Keisler's theorem for complete embeddings* (J 0027) Fund Math 81*121-132
◇ C07 C62 ◇ REV MR 49 # 7140 Zbl 363 # 02068
• ID 10717

POWELL, W.C. [1976] *A completeness theorem for Zermelo-Fraenkel set theory* (J 0036) J Symb Logic 41*323-327
◇ C90 E70 ◇ REV MR 54 # 4978 Zbl 332 # 02066
• ID 14762

POWELL, W.C. see Vol. V, VI for further entries

PRAAG VAN, P. [1972] see BOFFA, M.

PRAAG VAN, P. [1975] *Sur les centralisateurs des corps de type fini dans les corps existentiellement clos (English summary)* (J 2313) C R Acad Sci, Paris, Ser A-B 281*A891-A893
◇ C25 C60 ◇ REV MR 54 # 291 Zbl 321 # 16011
• ID 24594

PRAAG VAN, P. [1985] see BOFFA, M.

PRATT, V.R. [1979] *Axioms or algorithms* (P 2059) Math Founds of Comput Sci (8);1979 Olomouc 160-169
◇ B25 B75 D15 ◇ REV MR 81d:68023 Zbl 404 # 68044
• ID 54858

PRATT, V.R. see Vol. I, II, IV for further entries

PRAZMOWSKI, K. & SZCZERBA, L.W. [1976] *Interpretability and categoricity (Russian summary)* (J 0014) Bull Acad Pol Sci, Ser Math Astron Phys 24*309-312
◇ C35 F25 ◇ REV MR 56 # 8356 Zbl 342 # 02038
• ID 18347

PRAZMOWSKI, K. [1978] *Conciseness and hierarchy of theories (Russian summary)* (J 0014) Bull Acad Pol Sci, Ser Math Astron Phys 26*577-583
◇ C07 F25 ◇ REV MR 80j:03021 Zbl 408 # 03033
• ID 29177

PRELLER, A. [1968] *Quantified algebras* (P 0637) Syntax & Semant Infinitary Lang;1967 Los Angeles 182-203
◇ C75 G15 G25 ◇ REV Zbl 175.269 • ID 29306

PRELLER, A. [1969] *Interpolation et amalgamation* (J 0056) Publ Dep Math, Lyon 6/1*49-65
◇ C40 C52 C75 G15 G25 ◇ REV MR 39 # 6743 Zbl 182.11 • ID 10741

PRELLER, A. [1969] *Logique algebrique infinitaire. Completude des calculs $L_{\alpha\beta}$* (J 2313) C R Acad Sci, Paris, Ser A-B 268*A1509-A1511,A1589-A1592
- ⋄ C75 G25 ⋄ REV MR 39 # 5335 MR 39 # 5336 Zbl 182.326 • ID 21121

PRELLER, A. [1985] *A language for category theory, in which natural equivalence implies elementary equivalence of models* (J 0068) Z Math Logik Grundlagen Math 31*227-234
- ⋄ C07 G30 ⋄ REV Zbl 573 # 18001 • ID 47570

PRELLER, A. see Vol. V for further entries

PRESBURGER, M. [1930] *Ueber die Vollstaendigkeit eines gewissen Systems der Arithmetik ganzer Zahlen, in welchem die Addition als einzige Operation hervortritt* (P 0796) Congr Math Pays Slaves (1);1929 Warsaw 92-101,395
- ⋄ B25 B28 C10 C35 F30 ⋄ REV FdM 56.825 • ID 10749

PRESIC, M.D. & PRESIC, S.B. [1977] *On the embedding of Ω-algebras in groupoids* (J 0400) Publ Inst Math, NS (Belgrade) 21(35)*169-174
- ⋄ C05 ⋄ REV MR 56 # 5395 • ID 49902

PRESIC, M.D. [1979] *On the greatest congruence of relations* (J 0400) Publ Inst Math, NS (Belgrade) 26(40)*233-247
- ⋄ B25 E07 ⋄ REV MR 81h:03068 Zbl 449 # 03026 • ID 56693

PRESIC, M.D. [1980] *Some results on the extending of models* (P 3355) Algeb Conf (1);1980 Skopje 23-34
- ⋄ C30 ⋄ REV MR 84j:08006a Zbl 482 # 03011 • ID 36842

PRESIC, M.D. & PRESIC, S.B. [1981] *Some embedding theorems* (J 0400) Publ Inst Math, NS (Belgrade) 30(44)*161-167
- ⋄ C30 ⋄ REV MR 84j:03077 Zbl 521 # 03019 • ID 34669

PRESIC, M.D. see Vol. I, V for further entries

PRESIC, S.B. [1977] see PRESIC, M.D.

PRESIC, S.B. [1981] see PRESIC, M.D.

PRESIC, S.B. [1983] see CUPONA, G.

PRESIC, S.B. see Vol. I, II, IV for further entries

PREST, M. [1978] *Some model-theoretic aspects of torsion theories* (J 0326) J Pure Appl Algebra 12*295-310
- ⋄ C25 C35 C60 ⋄ REV MR 80a:18009 Zbl 398 # 03019 • ID 52747

PREST, M. [1979] *Model-completions of some theories of modules* (J 3172) J London Math Soc, Ser 2 20*369-372
- ⋄ C25 C35 C50 C60 ⋄ REV MR 81c:03023 Zbl 426 # 03035 • ID 53631

PREST, M. [1979] *Torsion and universal Horn classes of modules* (J 3172) J London Math Soc, Ser 2 19*411-416
- ⋄ C05 C60 ⋄ REV MR 80i:16037 Zbl 396 # 18005 • ID 52661

PREST, M. [1980] *Elementary torsion theories and locally finitely presented categories* (J 0326) J Pure Appl Algebra 18*205-212
- ⋄ C60 G30 ⋄ REV MR 82b:18015 Zbl 453 # 18009 • ID 82535

PREST, M. [1981] *Quantifier elimination for modules* (J 3133) Bull Soc Math Belg, Ser B 33*109-129
- ⋄ C10 C60 ⋄ REV MR 82i:03037 Zbl 479 # 03019 • ID 55681

PREST, M. [1982] *Elementary equivalence of Σ-injective modules* (J 3240) Proc London Math Soc, Ser 3 45*71-88
- ⋄ C60 ⋄ REV MR 83g:03031 Zbl 488 # 03019 • ID 36004

PREST, M. [1983] *Existentially complete prime rings* (J 3172) J London Math Soc, Ser 2 28*238-246
- ⋄ C25 C60 ⋄ REV MR 84m:03055 Zbl 538 # 03029 • ID 35756

PREST, M. [1983] see PILLAY, A.

PREST, M. [1984] *Rings of finite representation type and modules of finite Morley rank* (J 0032) J Algeb 88*502-533
- ⋄ C45 C60 ⋄ REV MR 85k:16030 Zbl 538 # 16025 • ID 41495

PREST, M. [1985] *The generalised RK-order, orthogonality and regular types for modules* (J 0036) J Symb Logic 50*202-219
- ⋄ C45 C60 ⋄ ID 39776

PRESTEL, A. [1972] see GUPTA, H.N.

PRESTEL, A. & ZIEGLER, M. [1975] *Erblich euklidische Koerper* (J 0127) J Reine Angew Math 274/275*196-205
- ⋄ C60 D35 ⋄ REV MR 52 # 363 Zbl 307 # 12103 • ID 18349

PRESTEL, A. [1975] *Lectures on formally real fields* (X 2233) Inst Mat Pura Apl: Rio Janeiro
- • REPR [1984] (S 3301) Lect Notes Math 1093*xi+125pp
- ⋄ C60 C98 ⋄ REV MR 86h:12013 • ID 33811

PRESTEL, A. [1977] *An introduction to ultraproducts* (C 1992) Scuola Algeb IMPA (3);1977 Rio de Janeiro 111-134
- ⋄ C20 C98 ⋄ ID 33810

PRESTEL, A. [1978] *Artin's conjecture on p-adic number fields* (P 2485) Attas Scuola Algeb IMPA (5);1978 Rio de Janeiro 79-109
- ⋄ C60 C98 ⋄ REV MR 81j:10029 • ID 40495

PRESTEL, A. & ZIEGLER, M. [1978] *Model theoretic methods in the theory of topological fields* (J 0127) J Reine Angew Math 299/300*318-341
- ⋄ C60 C65 C80 ⋄ REV MR 80f:54034 Zbl 367 # 12014 • ID 82537

PRESTEL, A. [1979] *Entscheidbarkeit mathematischer Theorien* (J 0157) Jbuchber Dtsch Math-Ver 81*177-188
- ⋄ B25 C60 C98 D35 ⋄ REV MR 80j:03001 Zbl 426 # 03013 • ID 53609

PRESTEL, A. & ZIEGLER, M. [1980] *Non-axiomatizable classes of V-topological fields* (J 0127) J Reine Angew Math 316*211-214
- ⋄ C60 C65 C80 ⋄ REV MR 82b:03073 Zbl 453 # 12012 • ID 54205

PRESTEL, A. [1981] *Pseudo real closed fields* (P 3268) Set Th & Model Th;1979 Bonn 127-156
- ⋄ C10 C25 C35 C52 C60 ⋄ REV MR 84b:12032 Zbl 466 # 12018 JSL 51.235 • ID 54983

PRESTEL, A. [1981] see JENSEN, R.B.

PRESTEL, A. [1981] *Zur Axiomatisierung gewisser affiner Geometrien* (J 3370) Enseign Math, Ser 2 27*125-136
- • REPR [1982] (P 3482) Logic & Algor (Specker);1980 Zuerich 341-352
- ⋄ C60 C65 ⋄ REV MR 83j:03025b Zbl 466 # 12020 • ID 40498

PRESTEL, A. [1982] *Decidable theories of preordered fields* (J 0043) Math Ann 258*481-492
⋄ B25 C25 C35 C60 ⋄ REV MR 83e:03049 Zbl 466 #12019 JSL 51.235 • ID 55648

PRESTEL, A. & ROQUETTE, P. [1983] *Lectures on formally p-adic fields* (S 3301) Lect Notes Math 1050*v+167pp
⋄ B25 C10 C35 C60 C98 ⋄ REV MR 85m:11090 Zbl 523 #12016 • ID 40503

PRESTEL, A. [1984] see HEINEMANN, B.

PRESTEL, A. [1984] *Model theory of fields: An application to positive semidefinite polynomials* (S 3521) Mem Soc Math Fr 16*53-65
⋄ C60 ⋄ ID 39729

PRESTEL, A. [1984] see KUHLMANN, F.V.

PRESTEL, A. [1985] *On the axiomatization of PRC-fields* (P 2160) Latin Amer Symp Math Log (6);1983 Caracas 351-359
⋄ C52 C60 ⋄ ID 41794

PRESTEL, A. see Vol. I, V for further entries

PREVIALE, F. [1971] *Applicazioni del concetto di diagramma nella teoria dei modelli di linguaggi con piu specie di variabili* (P 0669) Conv Teor Modelli & Geom;1969/70 Roma 21-41
⋄ C07 C40 ⋄ REV MR 43 #30 Zbl 212.19 • ID 22234

PREVIALE, F. see Vol. I, II, VI for further entries

PRIDE, S.J. [1977] *The isomorphism problem for two-generator one-relator groups with torsion is solvable* (J 0064) Trans Amer Math Soc 227*109-139
⋄ C60 D40 ⋄ REV MR 55 #3092 Zbl 356 #20037 • ID 26231

PRIEST, G. [1973] see BELL, J.L.

PRIEST, G. see Vol. I, II, V, VI for further entries

PRIJATELJ, N. [1978] *Formal construction of a logical language (Slovenian) (English summary)* (J 2310) Obz Mat Fiz, Ljubljana 25*20-33
⋄ C07 ⋄ REV MR 80m:03005 Zbl 366 #02003 • ID 51086

PRIJATELJ, N. see Vol. I, II, V for further entries

PRIKRY, K. [1966] see BUKOVSKY, L.

PRIKRY, K. [1971] see KUNEN, K.

PRIKRY, K. [1973] *On descendingly complete ultrafilters* (P 0713) Cambridge Summer School Math Log;1971 Cambridge GB 459-488
⋄ C55 E05 E55 ⋄ REV MR 50 #116 Zbl 268 #02050 • ID 10766

PRIKRY, K. [1974] see KEISLER, H.J.

PRIKRY, K. [1976] see GALVIN, F.

PRIKRY, K. [1976] see JECH, T.J.

PRIKRY, K. [1979] see JECH, T.J.

PRIKRY, K. [1980] see JECH, T.J.

PRIKRY, K. see Vol. IV, V for further entries

PRISCO DI, C.A. [1982] see MAREK, W.

PRISCO DI, C.A. (ED.) [1985] *Methods in mathematical logic. Proceedings of the 6th Latin American Symposium on Mathematical Logic held in Caracas, Venezuela, Aug. 1-6, 1983* (S 3301) Lect Notes Math 1130*vii+407pp
⋄ B97 C97 ⋄ REV MR 86d:03002 Zbl 556 #00007 • ID 41792

PRISCO DI, C.A. see Vol. V for further entries

PROTASOV, I.V. [1980] *On the theory of topological algebraic systems (Russian) (English summary)* (J 0270) Dokl Akad Nauk Ukr SSR, Ser A 1*15-17,92
⋄ C90 ⋄ REV MR 81e:08002 • ID 49885

PROTASOV, I.V. & SIDORCHUK, A.D. [1981] *On varieties of topological algebraic systems (Russian)* (J 0023) Dokl Akad Nauk SSSR 256*1314-1318
• TRANSL [1981] (J 0062) Sov Math, Dokl 23*184-187
⋄ C05 C90 ⋄ REV MR 82c:08009 Zbl 472 #08006 • ID 49870

PROTASOV, I.V. [1982] *The local method and the compactness theorem for topological algebraic systems (Russian)* (J 0092) Sib Mat Zh 23/1*136-143,221
• TRANSL [1982] (J 0475) Sib Math J 23*106-111
⋄ C90 ⋄ REV MR 84k:03106 Zbl 495 #03021 • ID 36123

PROTASOV, I.V. [1982] *Varieties of topological spaces (Russian) (English summary)* (J 0270) Dokl Akad Nauk Ukr SSR, Ser A 1982/6*21-23,88
⋄ C05 C65 ⋄ REV MR 83m:08014 • ID 47208

PRZELECKI, M. & WOJCICKI, R. [1971] *Inessential parts of extensions of first-order theories (Polish and Russian summaries)* (J 0063) Studia Logica 28*83-99
⋄ A05 C07 ⋄ REV MR 46 #8832 Zbl 247 #02048 • ID 10813

PRZELECKI, M. [1977] *On identifiability in extended domains* (P 1704) Int Congr Log, Meth & Phil of Sci (5);1975 London ON 3*81-89
⋄ C40 ⋄ REV MR 58 #21566 • ID 77577

PRZELECKI, M. see Vol. I for further entries

PRZYMUSINSKA, H. [1980] *Craig interpolation theorem and Hanf number for v-valued infinitary predicate calculi (Russian summary)* (J 3293) Bull Acad Pol Sci, Ser Math 28*207-211
⋄ B50 C40 C55 C75 G20 ⋄ REV MR 82k:03053d Zbl 465 #03012 • ID 54915

PRZYMUSINSKA, H. [1980] *Gentzen-type semantics for v-valued infinitary predicate calculi (Russian summary)* (J 3293) Bull Acad Pol Sci, Ser Math 28*203-206
⋄ B50 C75 G20 ⋄ REV MR 82k:03053c Zbl 465 #03011 • ID 54914

PRZYMUSINSKA, H. [1980] *On v-valued infinitary predicate calculi (Russian summary)* (J 3293) Bull Acad Pol Sci, Ser Math 28*193-198
⋄ C75 G20 ⋄ REV MR 82k:03053a Zbl 465 #03009 • ID 54912

PRZYMUSINSKA, H. [1980] *Some results concerning v-valued infinitary predicate calculi (Russian summary)* (J 3293) Bull Acad Pol Sci, Ser Math 28*199-202
⋄ C20 C75 ⋄ REV MR 82k:03053b Zbl 465 #03010 • ID 54913

PUDLAK, P. [1977] *Generalized quantifiers and semisets* (P 1639) Set Th & Hierarch Th (1);1974 Karpacz 109-116
⋄ B10 C80 E70 ⋄ REV MR 58 #5212 Zbl 411 #03043 • ID 32425

PUDLAK, P. [1980] see HAJEK, P.

PUDLAK, P. [1983] *A definition of exponentiation by a bounded arithmetical formula* (J 0140) Comm Math Univ Carolinae (Prague) 24*667-671
⋄ B28 C62 F30 ⋄ REV MR 85e:03142 Zbl 533 # 03032
• ID 36550

PUDLAK, P. & SOCHOR, A. [1984] *Models of the alternative set theory* (J 0036) J Symb Logic 49*570-585
⋄ C50 C62 E50 E70 ⋄ REV MR 86a:03060 Zbl 572 # 03032 • ID 42495

PUDLAK, P. [1985] *Cuts, consistency statements and interpretations* (J 0036) J Symb Logic 50*423-441
⋄ C62 E30 F25 ⋄ REV Zbl 569 # 03024 • ID 42556

PUDLAK, P. & SOCHOR, A. [1985] *Elementary extensions of models of the alternative set theory* (J 0068) Z Math Logik Grundlagen Math 31*309-316
⋄ C62 E70 H15 ⋄ REV Zbl 578 # 03028 • ID 47557

PUDLAK, P. see Vol. I, IV, V, VI for further entries

PUEHRINGER, C. [1982] *Ein Vollstaendigkeitsbeweis fuer die Theorie der atomlosen ∩ − ^{c}-partiellen Booleschen Algebren* (J 2688) Conceptus (Wien) 16*81-88
⋄ C30 G05 ⋄ REV MR 84c:03063 • ID 34006

PURITZ, C.W. [1971] *Ultrafilters and standard functions in non-standard arithmetic* (J 3240) Proc London Math Soc, Ser 3 22*705-733
⋄ C20 E05 H15 ⋄ REV MR 44 # 6474 Zbl 256 # 02028
• ID 10821

PURITZ, C.W. [1972] *Skies, constellations and monads* (P 0732) Contrib to Non-Standard Anal;1970 Oberwolfach 215-243
⋄ H15 H20 ⋄ REV MR 58 # 31038 Zbl 257 # 02047
• ID 29001

PURITZ, C.W. see Vol. I for further entries

PUTNAM, H. [1957] *Decidability and essential undecidability* (J 0036) J Symb Logic 22*39-54
⋄ B25 D35 ⋄ REV MR 19.626 Zbl 78.245 Zbl 79.454 JSL 23.446 • ID 10827

PUTNAM, H. [1964] *On families of sets represented in theories* (J 0009) Arch Math Logik Grundlagenforsch 6*66-70
⋄ C40 D25 F30 ⋄ REV MR 31 # 3329 Zbl 126.21
• ID 10838

PUTNAM, H. [1965] *Trial and error predicates and the solution to a problem of Mostowski* (J 0036) J Symb Logic 30*49-57
⋄ C57 D20 D55 ⋄ REV MR 33 # 3923 JSL 36.342
• ID 10840

PUTNAM, H. [1967] see DREBEN, B.

PUTNAM, H. [1969] see HENSEL, G.

PUTNAM, H. [1974] see LEEDS, S.

PUTNAM, H. see Vol. I, II, IV, V, VI for further entries

QUACKENBUSH, R.W. [1975] see JONSSON, B.

QUACKENBUSH, R.W. [1975] see FROEMKE, J.

QUACKENBUSH, R.W. [1980] *Algebras with minimal spectrum* (J 0004) Algeb Universalis 10*117-129
⋄ C05 C13 ⋄ REV MR 81c:08007 Zbl 437 # 08001
• ID 41676

QUACKENBUSH, R.W. [1982] *Enumeration in classes of ordered structures* (P 3780) Ordered Sets;1981 Banff 523-554
⋄ C13 C98 ⋄ REV MR 84h:06001 Zbl 531 # 06001
• ID 43070

QUACKENBUSH, R.W. [1983] *Minimal paraprimal algebras* (P 3895) Contrib to Gen Algeb (2);1982 Klagenfurt 2*291-304
⋄ C05 C98 ⋄ REV MR 85d:08005 Zbl 528 # 08001
• ID 38953

QUACKENBUSH, R.W. see Vol. V for further entries

QUINE, W.V.O. [1945] *On the logic of quantification* (J 0036) J Symb Logic 10*1-12
⋄ B10 B25 ⋄ REV MR 7.45 Zbl 60.22 JSL 12.17
• ID 10872

QUINE, W.V.O. [1949] *On decidability and completeness* (J 0154) Synthese 7*441-446
⋄ B25 D35 F30 ⋄ REV MR 12.70 JSL 16.76 • ID 10881

QUINE, W.V.O. [1952] see CRAIG, W.

QUINE, W.V.O. [1976] *Grades of discriminability* (J 0301) J Phil 73*113-116
⋄ A05 C40 ⋄ ID 30790

QUINE, W.V.O. see Vol. I, II, IV, V, VI for further entries

QUINN, C.F. [1974] *An analysis of the concept of constructive categoricity* (J 0047) Notre Dame J Formal Log 15*511-551
⋄ C35 ⋄ REV MR 51 # 7856 Zbl 212.17 • ID 10922

QVARNSTROEM, B.O. [1977] *On the concept of formalization and partially ordered quantifiers* (J 2130) Linguist Philos 1*307-319
⋄ A05 C80 ⋄ REV Zbl 385 # 03022 • ID 52130

RABIN, M.O. [1957] *Computable algebraic systems* (P 1675) Summer Inst Symb Log;1957 Ithaca 134-138
⋄ C57 C60 D40 D45 D80 ⋄ REV JSL 32.412 • ID 10933

RABIN, M.O. [1958] *On recursively enumerable and arithmetic models of set theory* (J 0036) J Symb Logic 23*408-416
⋄ C57 C62 D45 ⋄ REV MR 22 # 10908 Zbl 95.246 JSL 28.167 • ID 10930

RABIN, M.O. [1959] *Arithmetical extensions with prescribed cardinality* (J 0028) Indag Math 21*439-446
⋄ C20 C30 C55 ⋄ REV MR 21 # 5564 JSL 25.169
• ID 10929

RABIN, M.O. [1960] *Computable algebra, general theory and theory of computable fields* (J 0064) Trans Amer Math Soc 95*341-360
⋄ C57 C60 D45 ⋄ REV MR 22 # 4639 Zbl 156.12
• ID 33612

RABIN, M.O. [1961] *Non-standard models and independence of the induction axiom* (C 0622) Essays Found of Math (Fraenkel) 287-299
⋄ B28 C62 F30 H15 ⋄ REV MR 28 # 4999 Zbl 143.10 JSL 38.159 • ID 10934

RABIN, M.O. [1962] *Classes of models and sets of sentences with the intersection property* (P 1606) Colloq Math (Pascal);1962 Clermont-Ferrand 39-53
⋄ C07 C52 ⋄ REV MR 44 # 1554 JSL 32.413 • ID 28524

RABIN, M.O. [1962] *Diophantine equations and non-standard models of arithmetic* (P 0612) Int Congr Log, Meth & Phil of Sci (1,Proc);1960 Stanford 151-158
⋄ C62 D25 H15 ⋄ REV MR 27 # 3540 • ID 20907

RABIN, M.O. [1965] *Universal groups of automorphisms of models* (P 0614) Th Models;1963 Berkeley 274-284
⋄ C07 C30 ⋄ REV MR 34#1191 Zbl 163.247 • ID 10935

RABIN, M.O. [1966] see ELGOT, C.C.

RABIN, M.O. [1968] *Decidability of second-order theories and automata on infinite trees* (J 0015) Bull Amer Math Soc 74∗1025-1029
⋄ B15 B25 C85 D05 ⋄ REV MR 38#44 Zbl 313#02029 • ID 29625

RABIN, M.O. [1969] *Decidability of second order theories and automata on infinite trees* (J 0064) Trans Amer Math Soc 141∗1-35
• TRANSL [1971] (J 3079) Kiber Sb Perevodov, NS 8∗72-116
⋄ B15 B25 C85 D05 ⋄ REV MR 40#30
Zbl 221#02031 JSL 37.618 • ID 10942

RABIN, M.O. [1970] *Weakly definable relations and special automata* (P 1072) Math Log & Founds of Set Th;1968 Jerusalem 1-23
⋄ B15 B25 C40 C85 D05 ⋄ REV MR 43#3121
Zbl 214.22 JSL 40.622 • ID 22233

RABIN, M.O. [1971] *Decidability and definability in second-order theories* (P 0743) Int Congr Math (II,11,Proc);1970 Nice 1∗239-244
⋄ B15 B25 C40 C85 D05 ⋄ REV MR 54#12512
Zbl 226#02041 JSL 40.623 • ID 14690

RABIN, M.O. [1972] *Automata on infinite objects and Church's problem* (X 0803) Amer Math Soc: Providence iii+22pp
⋄ B15 B25 C85 D05 ⋄ REV MR 48#75
Zbl 315#02037 JSL 40.623 • ID 10943

RABIN, M.O. [1974] see FISCHER, MICHAEL J.

RABIN, M.O. [1977] *Decidable theories* (C 1523) Handb of Math Logic 595-629
⋄ B25 C98 D20 ⋄ REV MR 58#5109 JSL 49.975 • ID 27321

RABIN, M.O. see Vol. IV, VI for further entries

RABINOVICH, E.B. [1981] *Inductive limits of symmetric groups and universal groups (Russian) (English summary)* (J 0413) Izv Akad Nauk Belor SSR, Ser Fiz-Mat 1981/5∗39-42,138
⋄ C15 C50 C60 ⋄ REV MR 83c:20007 Zbl 497#20011 • ID 36636

RABINOVICH, E.B. see Vol. V for further entries

RACKOFF, C.W. [1974] *On the complexity of the theories of weak direct products: a preliminary report* (P 1464) ACM Symp Th of Comput (6);1974 Seattle 149-160
⋄ B25 C13 C30 C60 D15 ⋄ REV MR 54#4957
Zbl 358#68080 • ID 24125

RACKOFF, C.W. [1975] see FERRANTE, J.

RACKOFF, C.W. [1976] *On the complexity of the theories of weak direct powers* (J 0036) J Symb Logic 41∗561-573
⋄ B25 C13 C30 C60 D15 ⋄ REV MR 58#5155
Zbl 383#03011 • ID 14579

RACKOFF, C.W. [1979] see FERRANTE, J.

RACKOFF, C.W. see Vol. IV for further entries

RADEV, S.R. [1984] *Interpolation in the infinitary propositional normal modal logic KL_{ω_1}* (J 0137) C R Acad Bulgar Sci 37∗715-716
⋄ B45 C40 C75 ⋄ REV Zbl 561#03009 • ID 44072

RADEV, S.R. see Vol. I, II for further entries

RADU, A.G. & RADU, G.G. [1981] *Existential and universal quantifiers in an elementary topos* (J 0230) An Univ Iasi, NS, Sect Ia 27∗19-30
⋄ C07 G30 ⋄ REV MR 82k:18003 Zbl 465#03031 • ID 54934

RADU, A.G. see Vol. V for further entries

RADU, G.G. [1981] see RADU, A.G.

RADU, G.G. see Vol. V for further entries

RAJAGOPALAN, M. [1985] see BOERGER, R.

RAJAGOPALAN, M. see Vol. V for further entries

RAKHMATULIN, N.A. [1975] *Automata-theoretic characteristics for the spectra of formulas of finite levels (Russian)* (C 1050) Aktual Vopr Mat Log & Teor Mnozh 105-127
⋄ C13 D05 D10 ⋄ REV MR 58#5165 • ID 77649

RAKOV, S.A. [1977] *Ultraproducts and the "three spaces problem" (Russian)* (J 1207) Fkts Anal Prilozh 11∗88-89
• TRANSL [1977] (J 3163) Fct Anal & Appl 11∗236-237
⋄ C20 E75 ⋄ REV MR 57#1090 Zbl 394#46013 • ID 82565

RAMA RAO, V.V. & SWAMY, U.M. [1975] *Triple and sheaf representation of Stone lattices* (J 0004) Algeb Universalis 5∗104-113
⋄ C90 G10 G30 ⋄ REV MR 52#2998 Zbl 311#06008 • ID 24450

RAMBAUD, C. [1976] see BLANC, G.

RAMBAUD, C. [1977] see DONNADIEU, M.-R.

RAMBAUD, C. [1979] *Completude en theorie sur graphes orientes* (J 0056) Publ Dep Math, Lyon 16/3∗33-37
⋄ C07 C65 ⋄ REV MR 83j:03063 Zbl 464#03033 • ID 35361

RAMBAUD, C. see Vol. II for further entries

RAMSEY, F.P. [1930] *On a problem of formal logic* (J 1910) Proc London Math Soc, Ser 2 30∗264-286
⋄ B10 B25 E05 E20 E75 ⋄ ID 28781

RAMSEY, F.P. see Vol. I, II for further entries

RANTALA, V. [1973] *On the theory of definability in first order logic* (1111) Preprints, Manuscr., Techn. Reports etc.
⋄ C07 C40 ⋄ REM Univ. Helsinki, Inst. Philos. • ID 21228

RANTALA, V. [1975] see HINTIKKA, K.J.J.

RANTALA, V. [1975] *Urn models: A new kind of non-standard model for first-order logic* (J 0122) J Philos Logic 4∗455-474
⋄ A05 B10 C07 C90 ⋄ REV MR 57#12166
Zbl 324#02043 • ID 64743

RANTALA, V. [1976] see HINTIKKA, K.J.J.

RANTALA, V. [1977] *Aspects of definability* (J 0096) Acta Philos Fenn 29∗236pp
⋄ A05 C07 C40 C75 ⋄ REV MR 58#27435 • ID 77658

RANTALA, V. [1978] *The old and the new logic of metascience* (J 0154) Synthese 39∗233-247
⋄ A05 C07 ⋄ REV MR 80d:03007 Zbl 395#03002 • ID 52553

RANTALA, V. [1979] *Game-theoretical semantics and back-and-forth* (P 1705) Scand Logic Symp (4);1976 Jyvaeskylae 119-151
⋄ C75 ⋄ REV MR 80i:03049 Zbl 411 # 03026 • ID 52880

RANTALA, V. [1982] *Infinitely deep game sentences and interpolations* (P 3800) Intens Log: Th & Appl;1979 Moskva 211-219
⋄ C40 C75 ⋄ REV MR 84h:03090 Zbl 484 # 03015
• ID 34282

RANTALA, V. [1982] *Quantified modal logic: non-normal worlds and propositional attitudes* (J 0063) Studia Logica 41*41-65
⋄ B45 C90 ⋄ REV MR 85d:03037 Zbl 529 # 03004
• ID 37647

RANTALA, V. see Vol. I, II for further entries

RAO, A.P. [1977] *A more natural alternative to Mostowski's (MFL)* (J 0068) Z Math Logik Grundlagen Math 23*387-392
⋄ B60 C07 ⋄ REV MR 58 # 5011 Zbl 443 # 03014
• ID 56420

RAO, A.P. see Vol. II for further entries

RAPP, A. [1984] *On the expressive power of the logics $L(Q_\alpha^{\langle n_1,...,n_m \rangle})$* (J 0068) Z Math Logik Grundlagen Math 30*11-20
⋄ C55 C80 ⋄ REV MR 85h:03037 Zbl 541 # 03017
• ID 40605

RAPP, A. [1985] *Elimination of Malitz quantifiers in stable theories* (J 0048) Pac J Math 117*387-396
⋄ C10 C45 C55 C80 ⋄ REV Zbl 563 # 03013 • ID 41821

RAPP, A. [1985] *The ordered field of real numbers and logics with Malitz quantifiers* (J 0036) J Symb Logic 50*380-389
⋄ B25 C10 C55 C65 C80 ⋄ ID 41806

RASIOWA, H. & SIKORSKI, R. [1950] *A proof of the completeness theorem of Goedel* (J 0027) Fund Math 37*193-200
⋄ C07 G05 ⋄ REV MR 12.661 Zbl 40.293 JSL 17.72
• ID 10999

RASIOWA, H. & SIKORSKI, R. [1951] *A proof of the Skolem-Loewenheim theorem* (J 0027) Fund Math 38*230-232
⋄ C07 G05 ⋄ REV MR 15.385 Zbl 45.295 JSL 18.339
• ID 11002

RASIOWA, H. [1952] *A proof of the compactness theorem for arithmetical classes* (J 0027) Fund Math 39*8-14
⋄ C07 G05 ⋄ REV MR 14.938 Zbl 50.6 JSL 20.78
• ID 11008

RASIOWA, H. & SIKORSKI, R. [1953] *Algebraic treatment of the notion of satisfiability* (J 0027) Fund Math 40*62-95
⋄ B10 C90 G05 ⋄ REV MR 15.668 Zbl 53.2 JSL 20.78
• ID 11006

RASIOWA, H. & SIKORSKI, R. [1953] *On satisfiability and decidability in non-classical functional calculi* (J 0014) Bull Acad Pol Sci, Ser Math Astron Phys 1*229-231
⋄ B10 B25 B45 C90 F50 ⋄ REV MR 15.668 Zbl 51.245
• ID 11004

RASIOWA, H. [1955] *Algebraic models of axiomatic theories* (J 0027) Fund Math 41*291-310
⋄ B30 C90 F50 G05 G10 G25 ⋄ REV MR 19.111 Zbl 65.4 JSL 33.285 • ID 11010

RASIOWA, H. & SIKORSKI, R. [1955] *An application of lattices to logic* (J 0027) Fund Math 42*83-100
⋄ B45 C90 G10 ⋄ REV MR 19.240 Zbl 68.243 JSL 35.137 • ID 11014

RASIOWA, H. & SIKORSKI, R. [1955] *On existential theorems in non-classical functional calculi* (J 0027) Fund Math 41*21-28
⋄ B45 C90 F50 G05 ⋄ REV MR 16.987 Zbl 56.11 JSL 20.80 • ID 11011

RASIOWA, H. [1956] see LOS, J.

RASIOWA, H. [1957] *Sur la methode algebrique dans la methodologie des systemes deductifs elementaires* (J 0070) Bull Soc Sci Math Roumanie, NS 1*223-231
⋄ B10 C07 G05 G10 ⋄ REV MR 21 # 3329 • ID 11017

RASIOWA, H. & SIKORSKI, R. [1960] *On the Gentzen theorem* (J 0027) Fund Math 48*57-69
⋄ B10 C07 F05 ⋄ REV MR 22 # 1510 Zbl 99.6 • ID 11024

RASIOWA, H. & SIKORSKI, R. [1963] *The mathematics of metamathematics* (S 0257) Monograf Mat 41*522pp
• TRANSL [1972] (X 2027) Nauka: Moskva 591pp (Russian)
⋄ B98 C98 F50 F98 G05 G10 G98 ⋄ REV MR 29 # 1149 MR 50 # 4232 Zbl 122.243 JSL 32.274
• REM 3rd ed 1979 • ID 11028

RASIOWA, H. [1970] *Ultraproducts of m-valued models and a generalization of the Loewenheim-Skolem-Goedel-Mal'cev theorem for theories based on m-valued logics (Russian summary)* (J 0014) Bull Acad Pol Sci, Ser Math Astron Phys 18*415-420
⋄ B50 C20 C90 ⋄ REV MR 43 # 4638 Zbl 221 # 02044
• ID 11031

RASIOWA, H. [1972] *The Craig interpolation theorem for m-valued predicate calculi (Russian summary)* (J 0014) Bull Acad Pol Sci, Ser Math Astron Phys 20*341-346
⋄ B50 C40 C90 ⋄ REV MR 52 # 5355 Zbl 243 # 02014
• ID 11032

RASIOWA, H. [1982] *Lectures on infinitary logic and logics of programs* (J 3434) Pubbl Ist Appl Calcolo, Ser 3 142*122pp
⋄ B75 C75 C98 ⋄ REV MR 85i:03089 • ID 45056

RASIOWA, H. [1985] *Topological representations of Post algebras of order ω^+ and open theories based on ω^+-valued Post logic* (J 0063) Studia Logica 44*353-368
⋄ B50 C90 G20 ⋄ ID 48433

RASIOWA, H. see Vol. I, II, IV, V, VI for further entries

RASKOVIC, M. [1981] *On extendability of models of ZF set theory to models of alternative set theory* (P 2614) Open Days in Model Th & Set Th;1981 Jadwisin 259-264
⋄ C62 E70 ⋄ ID 33731

RASKOVIC, M. [1985] *Model theory for $L_{\mathfrak{U}M}$ logic* (J 0400) Publ Inst Math, NS (Belgrade) 37*17-22
⋄ C90 ⋄ ID 48313

RASKOVIC, M. see Vol. I, V for further entries

RASMUSSEN, K. [1983] see MAREK, W.

RASSBACH, W. [1979] see MONK, J.D.

RATAJCZYK, Z. [1979] *On the number of expansions of the models of ZFC-set theory to models of KM-theory of classes* (P 2627) Logic Colloq;1978 Mons 317-333
⋄ C62 E70 ⋄ REV MR 83j:03056 Zbl 445 # 03028
• ID 56492

RATAJCZYK, Z. [1981] *A characterization of expandability of models for ZF to models for KM* (J 0027) Fund Math 113*9-19
⋄ C62 C70 E45 E70 ⋄ REV MR 83b:03041 Zbl 478 # 03015 • ID 55630

RATAJCZYK, Z. [1981] *Note on submodels for A_2* (P 2614) Open Days in Model Th & Set Th;1981 Jadwisin 265-280
⋄ C62 ⋄ ID 33732

RATAJCZYK, Z. [1982] *Satisfaction classes and combinatorial sentences independent from PA* (J 0068) Z Math Logik Grundlagen Math 28*149-165
⋄ C62 E05 F30 ⋄ REV MR 83j:03097 Zbl 522 # 03043
• ID 35382

RATAJCZYK, Z. see Vol. V for further entries

RAUSZER, C. [1977] *Applications of Kripke models to Heyting-Brouwer logic* (J 0063) Studia Logica 36*61-71
⋄ B55 C90 F50 ⋄ REV MR 57 # 15977 Zbl 361 # 02033
• ID 50668

RAUSZER, C. [1977] *Model theory for an extension of intuitionistic logic* (J 0063) Studia Logica 36*73-87
⋄ B55 C90 F50 ⋄ REV MR 57 # 15978 Zbl 361 # 02034
• ID 50669

RAUSZER, C. [1980] *An algebraic and Kripke-style approach to a certain extension of intuitionistic logic* (J 0202) Diss Math (Warsaw) 167*62pp
⋄ B55 C20 C90 F50 G25 ⋄ REV MR 82k:03100 Zbl 442 # 03024 • ID 56381

RAUSZER, C. [1981] see GUZICKI, W.

RAUSZER, C. [1983] see JANKOWSKI, A.W.

RAUSZER, C. [1985] *Formalizations of certain intermediate logics. I* (P 2160) Latin Amer Symp Math Log (6);1983 Caracas 360-384
⋄ B55 C40 C90 ⋄ ID 41793

RAUSZER, C. see Vol. II, VI for further entries

RAUTENBERG, W. [1962] *Ueber metatheoretische Eigenschaften einiger geometrischer Theorien* (J 0068) Z Math Logik Grundlagen Math 8*5-41
⋄ B30 C65 D35 ⋄ REV MR 26 # 4883 Zbl 112.247
• ID 11041

RAUTENBERG, W. [1965] *Beweis des Kommutativgesetzes in elementar-archimedisch geordneten Gruppen* (J 0068) Z Math Logik Grundlagen Math 11*1-4
⋄ C60 ⋄ REV MR 30 # 1937 Zbl 135.249 • ID 11043

RAUTENBERG, W. [1965] *Hilberts Schnittpunktsaetze und einige modelltheoretische Aspekte der formalisierten Geometrie* (J 0115) Wiss Z Humboldt-Univ Berlin, Math-Nat Reihe 14*409-415
⋄ C65 ⋄ REV MR 33 # 3934 Zbl 168.250 • ID 11044

RAUTENBERG, W. [1966] *Ueber Hilberts Schnittpunktsaetze* (J 0068) Z Math Logik Grundlagen Math 12*57-59
⋄ B30 C65 ⋄ REV MR 34 # 3402 Zbl 168.250 • ID 11045

RAUTENBERG, W. [1967] *Elementare Schemata nichtelementarer Axiome* (J 0068) Z Math Logik Grundlagen Math 13*329-366
⋄ B25 C52 C60 C65 C85 D35 ⋄ REV MR 36 # 1316 Zbl 207.296 • ID 11046

RAUTENBERG, W. [1968] *Nichtdefinierbarkeit der Multiplikation in dividierbaren Ringen* (J 0068) Z Math Logik Grundlagen Math 14*59-60
⋄ C60 ⋄ REV MR 36 # 4977 Zbl 155.15 • ID 11049

RAUTENBERG, W. [1968] *Unterscheidbarkeit endlicher geordneter Mengen mit gegebener Anzahl von Quantoren* (J 0068) Z Math Logik Grundlagen Math 14*267-272
⋄ B20 C07 C13 C65 D35 ⋄ REV MR 37 # 2595 Zbl 169.7 • ID 11048

RAUTENBERG, W. [1970] see HERRE, H.

RAUTENBERG, W. [1971] see HAUSCHILD, K.

RAUTENBERG, W. [1972] see HAUSCHILD, K.

RAUTENBERG, W. [1973] see HAUSCHILD, K.

RAUTENBERG, W. [1974] see KOREC, I.

RAUTENBERG, W. & ZIEGLER, M. [1975] *Recursive inseparability in graph theory* (J 0369) Notices Amer Math Soc 22*523
⋄ C13 D35 ⋄ ID 49852

RAUTENBERG, W. [1976] see KOREC, I.

RAUTENBERG, W. [1979] *Klassische und nichtklassische Aussagenlogik* (X 0900) Vieweg: Wiesbaden xi + 361pp
⋄ B45 B50 B98 C90 ⋄ REV MR 81i:03002 Zbl 424 # 03007 • ID 77698

RAUTENBERG, W. [1983] *Modal tableau calculi and interpolation* (J 0122) J Philos Logic 12*403-423 • ERR/ADD ibid 14*229
⋄ B45 C40 ⋄ REV MR 85b:03029 Zbl 547 # 03015
• ID 39823

RAUTENBERG, W. see Vol. I, II, IV, VI for further entries

RAYNAUD, Y. [1981] *Deux nouveaux exemples d'espaces de Banach stables (English summary)* (J 3364) C R Acad Sci, Paris, Ser 1 292*715-717
⋄ C20 C65 ⋄ REV MR 82k:46026a Zbl 461 # 46017
• ID 82587

RAYNAUD, Y. [1981] *Espaces de Banach superstables (English summary)* (J 3364) C R Acad Sci, Paris, Ser 1 292*671-673
⋄ C20 C65 ⋄ REV MR 82k:46026b Zbl 461 # 46018
• ID 82586

RAYNAUD, Y. [1984] see LEVY, M.

REBHOLZ, J.R. [1974] *Some consequences of the Morass and Diamond* (J 0007) Ann Math Logic 7*361-385
⋄ C55 E05 E45 E65 ⋄ REV MR 52 # 10437 Zbl 362 # 02071 • ID 11057

RECIO MUNIZ, T.J. [1984] see DUBOIS, DONALD WARD

REDDY, C.R. [1978] see LOVELAND, D.W.

REDDY, C.R. see Vol. I for further entries

REGGIANI, C. [1978] *Le SH nei limiti diretti di strutture e di interpretazioni (English summary)* (J 0549) Riv Mat Univ Parma, Ser 4 4*269-275
⋄ C05 C30 C40 ⋄ REV MR 81i:03042 Zbl 416 # 18011
• ID 53227

REGGIANI, C. [1979] *Conservazione delle SH di tipo ∀∃ nei limiti diretti (English summary)* (J 3285) Boll Unione Mat Ital, V Ser, A 16*568-576
⋄ C05 C30 C40 ⋄ REV MR 81a:03029 Zbl 415 # 18001
• ID 77714

REICHBACH, J. [1955] *Completeness of the functional calculus of first order (Polish, Russian) (English summary)* (J 0063) Studia Logica 2*213-228,229-250 • ERR/ADD ibid 2*329
⋄ B10 C07 ⋄ REV MR 17.446 Zbl 67.250 JSL 21.194
• ID 33706

REICHBACH, J. [1958] *On the first-order functional calculus and the truncation of models (Polish and Russian summaries)* (J 0063) Studia Logica 7*181-220
⋄ B25 C07 C90 ⋄ REV MR 21 # 1264 JSL 23.61
• ID 11060

REICHBACH, J. [1969] *Propositional calculi and completeness theorem* (J 0537) Yokohama Math J 17*1-16
⋄ B05 C07 ⋄ REV MR 40 # 1259 Zbl 184.279 • ID 11073

REICHBACH, J. [1970] *On statistical tests in generalizations of Herbrand's theorems and asymptotic probabilistic models (main theorems of mathematics)* (J 1550) Creation Math 1*6-22
⋄ B10 C40 C90 ⋄ REV Zbl 237 # 02018 • ID 27880

REICHBACH, J. [1971] *A note about generalized models and the origin of forcing in the theory of models* (J 1550) Creation Math 2*9pp
⋄ B25 C25 C90 ⋄ REV Zbl 248 # 02064 • ID 32000

REICHBACH, J. [1971] *Generalized models and decidability - weak and strong statistical decidability of theorems* (J 1550) Creation Math 3*3-25
⋄ B25 C25 C90 ⋄ REV Zbl 237 # 02019 • ID 27881

REICHBACH, J. [1972] *Generalized models, probability of formulas, decisions and statistical decidability of theorems* (J 0537) Yokohama Math J 20*79-98
⋄ B25 C25 C90 ⋄ REV MR 46 # 6986 Zbl 257 # 02049
• ID 26655

REICHBACH, J. [1972] *Notes about decidablility of mathematics with probability 1 and generalized models* (J 1550) Creation Math 5*3-13
⋄ B25 C90 ⋄ REV Zbl 256 # 02029 • ID 32001

REICHBACH, J. [1973] *Generalized models for intuitionistic and classical predicate calculi with ultraproducts* (J 1550) Creation Math 6*9-44
⋄ C20 C25 C80 C90 F50 ⋄ REV Zbl 299 # 02068
• ID 31998

REICHBACH, J. [1974] *About decidability, generalized models and main theorems of many valued predicate calculi* (J 1550) Creation Math 7*11-20
⋄ B25 B50 C90 ⋄ REV Zbl 372 # 02034 • ID 31999

REICHBACH, J. [1975] *Generalized models for classical and intuitionistic predicate calculi* (J 0537) Yokohama Math J 23*5-30
⋄ B10 B25 C90 F50 ⋄ REV MR 55 # 2556
Zbl 343 # 02040 • ID 32003

REICHBACH, J. [1975] *Infinite proof methods in decidability problems with Ulam's models and classification of predicate calculi* (J 1550) Creation Math 8*3-14
⋄ B25 C25 C90 ⋄ REV Zbl 372 # 02035 • ID 32002

REICHBACH, J. [1976] *Graphs, generalized models, games and decidability* (J 1550) Creation Math 9*
⋄ B25 C65 C90 ⋄ ID 32004

REICHBACH, J. [1980] *Decidability of mathematical sciences and their undecidability* (J 1550) Creation Math 12*
⋄ B25 D35 ⋄ ID 32005

REICHBACH, J. see Vol. I, II, IV, VI for further entries

REICHEL, HORST [1983] see BENECKE, K.

REICHEL, HORST see Vol. IV for further entries

REILLY, N.R. [1969] *Some applications of wreath products and ultraproducts in the theory of lattice ordered groups* (J 0025) Duke Math J 36*825-834
⋄ C20 C30 C60 ⋄ REV MR 40 # 4184 Zbl 191.25
• ID 11089

REINEKE, J. [1974] see PODEWSKI, K.-P.

REINEKE, J. [1975] *Minimale Gruppen* (J 0068) Z Math Logik Grundlagen Math 21*357-359
⋄ C50 C60 ⋄ REV MR 52 # 85 Zbl 312 # 02045
JSL 49.317 • ID 11090

REINEKE, J. [1976] see CHERLIN, G.L.

REINEKE, J. [1979] see PODEWSKI, K.-P.

REINEKE, J. [1981] see PODEWSKI, K.-P.

REINEKE, J. [1982] *On algebraically closed models of theories of commutative rings* (P 3622) Int Congr Log, Meth & Phil of Sci (6,Proc);1979 Hannover 223-234
⋄ C25 C60 ⋄ REV MR 84c:03064 Zbl 504 # 03018
• ID 34007

REINHARDT, W.N. [1972] see MALITZ, J.

REINHARDT, W.N. [1974] *Remarks on reflection principles, large cardinals, and elementary embeddings* (P 0693) Axiomatic Set Th;1967 Los Angeles 2*189-205
⋄ C07 E47 E55 E65 ⋄ REV MR 53 # 5302
Zbl 325 # 02048 • ID 22936

REINHARDT, W.N. [1978] see SOLOVAY, R.M.

REINHARDT, W.N. see Vol. II, V, VI for further entries

REITERMAN, J. [1974] see ADAMEK, J.

REITERMAN, J. & TRNKOVA, V. [1980] *Dynamic algebras which are not Kripke structures* (P 3210) Math Founds of Comput Sci (9);1980 Rydzyna 528-538
⋄ B75 C90 G25 ⋄ REV MR 83f:68027 Zbl 451 # 03004
• ID 54019

REITERMAN, J. [1982] *The Birkhoff theorem for finite algebras* (J 0004) Algeb Universalis 14*1-10
⋄ C05 C13 ⋄ REV MR 84c:08008 Zbl 484 # 08007
• ID 39575

REITERMAN, J. see Vol. II, V for further entries

REMESLENNIKOV, V.N. & SOKOLOV, V.G. [1970] *Some properties of a Magnus embedding (Russian)* (J 0003) Algebra i Logika 9*566-578
• TRANSL [1970] (J 0069) Algeb and Log 9*342-349
⋄ C60 D40 ⋄ REV MR 45 # 2001 Zbl 247 # 20026
• ID 11095

REMESLENNIKOV, V.N. [1981] see MYASNIKOV, A.G.

REMESLENNIKOV, V.N. [1982] see MYASNIKOV, A.G.

REMESLENNIKOV, V.N. [1983] see MYASNIKOV, A.G.

REMESLENNIKOV, V.N. & ROMAN'KOV, V.A. [1983]
Model-theoretic and algorithmic questions of group theory (Russian) (J 1501) Itogi Nauki Tekh, Ser Algeb, Topol, Geom 21*3–79
- TRANSL [1985] (J 1531) J Sov Math 31*2887–2939
- ⋄ B25 C60 C98 D15 D30 D40 ⋄ REV MR 85f:20002 Zbl 563 # 20032 Zbl 573 # 20031 • ID 39954

REMESLENNIKOV, V.N. [1984] see MYASNIKOV, A.G.

REMESLENNIKOV, V.N. see Vol. IV for further entries

REMMEL, J.B. [1977] *Maximal and cohesive vector spaces* (J 0036) J Symb Logic 42*400–418
- ⋄ C57 D25 D45 ⋄ REV MR 57 # 5712 Zbl 429 # 03027
- • ID 53858

REMMEL, J.B. [1978] *A r-maximal vector space not contained in any maximal vector space* (J 0036) J Symb Logic 43*430–441
- ⋄ C57 D25 D45 ⋄ REV MR 58 # 16230 Zbl 409 # 03028
- • ID 29270

REMMEL, J.B. [1978] *Recursively enumerable boolean algebras* (J 0007) Ann Math Logic 15*75–107
- ⋄ C57 D25 D45 G05 ⋄ REV MR 80e:03052 Zbl 413 # 03027 • ID 29157

REMMEL, J.B. [1979] *R-maximal boolean algebras* (J 0036) J Symb Logic 44*533–548
- ⋄ C57 D45 G05 ⋄ REV MR 81b:03049 Zbl 439 # 03026
- • ID 56017

REMMEL, J.B. [1979] see METAKIDES, G.

REMMEL, J.B. [1980] *On r.e. and co-r.e. vector spaces with nonextendible bases* (J 0036) J Symb Logic 45*20–34
- ⋄ C57 C60 D45 ⋄ REV MR 81b:03048 Zbl 471 # 03038
- • ID 55233

REMMEL, J.B. [1980] *Recursion theory on orderings II* (J 0036) J Symb Logic 45*317–333
- ⋄ C57 D45 ⋄ REV MR 81e:03046 Zbl 471 # 03036
JSL 51.229 • REM Part I 1979 by Metakides,G. & Remmel,J.B.
- • ID 55231

REMMEL, J.B. [1980] *Recursion theory on algebraic structures with independent sets* (J 0007) Ann Math Logic 18*153–191
- ⋄ C57 C60 D45 ⋄ REV MR 81j:03076 Zbl 471 # 03037
- • ID 55232

REMMEL, J.B. [1981] *Effective structures not contained in recursively enumerable structures* (P 2902) Aspects Effective Algeb;1979 Clayton 206–225
- ⋄ C57 D45 ⋄ REV MR 83a:03029 Zbl 497 # 03035
- • ID 35064

REMMEL, J.B. [1981] see MANASTER, A.B.

REMMEL, J.B. [1981] *Recursive isomorphism types of recursive boolean algebras* (J 0036) J Symb Logic 46*572–594
- ⋄ C57 D45 G05 ⋄ REV MR 83a:03042 Zbl 543 # 03031
- • ID 33280

REMMEL, J.B. [1981] *Recursive boolean algebras with recursive atoms* (J 0036) J Symb Logic 46*595–616
- ⋄ C57 D45 G05 ⋄ REV MR 82j:03055 Zbl 543 # 03032
- • ID 77739

REMMEL, J.B. [1981] *Recursively categorical linear orderings* (J 0053) Proc Amer Math Soc 83*387–391
- ⋄ C35 C57 C65 ⋄ REV MR 82j:03037 Zbl 493 # 03022
- • ID 77729

REMMEL, J.B. [1982] see EISENBERG, E.F.

REMMEL, J.B. [1982] see NERODE, A.

REMMEL, J.B. [1983] see KALANTARI, I.

REMMEL, J.B. [1983] see KIERSTEAD, H.A.

REMMEL, J.B. [1984] see DOWNEY, R.G.

REMMEL, J.B. [1985] see NERODE, A.

REMMEL, J.B. [1985] see KIERSTEAD, H.A.

REMMEL, J.B. see Vol. IV, V for further entries

RENARDEL DE LAVALETTE, G.R. [1981] *The interpolation theorem in fragments of logics* (J 0028) Indag Math 43*71–86
- ⋄ B20 B55 C40 F50 ⋄ REV MR 82d:03019 Zbl 471 # 03012 • ID 55207

RENARDEL DE LAVALETTE, G.R. see Vol. I, VI for further entries

RENNIE, M.K. [1968] *A function which bounds truth-tabular calculations in S5* (J 0079) Logique & Anal, NS 11*425–439
- ⋄ B25 B45 D15 ⋄ REV MR 43 # 6062 Zbl 181.6
- • ID 11098

RENNIE, M.K. see Vol. I, II for further entries

REPIN, N.N. [1983] *Equations with one unknown in nilpotent groups (Russian)* (J 0087) Mat Zametki (Akad Nauk SSSR) 34*201–206
- TRANSL [1983] (J 1044) Math Notes, Acad Sci USSR 34*582–585
- ⋄ C60 ⋄ REV MR 85g:20045 Zbl 536 # 20020 • ID 43935

RESCHER, N. [1962] *Plurality-quantification* (J 0036) J Symb Logic 27*373–374
- ⋄ C55 C80 ⋄ ID 90119

RESCHER, N. [1969] *Nonstandard quantificational logic* (C 0555) Topic Philos Logic 162–181
- ⋄ A05 B45 C90 ⋄ ID 14969

RESCHER, N. [1969] *Topics in philosophical logic* (S 3307) Synth Libr xiv+347pp
- ⋄ A05 B45 B98 C90 ⋄ REV Zbl 175.264 • ID 25168

RESCHER, N. see Vol. I, II, VI for further entries

RESL, M. [1979] *On models in the alternative set theory* (J 0140) Comm Math Univ Carolinae (Prague) 20*723–736
- ⋄ C62 E70 ⋄ REV MR 81e:03053 Zbl 433 # 03032
- • ID 77752

RESL, M. see Vol. V, VI for further entries

RESSAYRE, J.-P. [1969] *Sur les theories du premier ordre categorique en un cardinal* (J 0064) Trans Amer Math Soc 142*481–505
- ⋄ C30 C35 C55 ⋄ REV MR 40 # 32 Zbl 209.304
JSL 36.684 • ID 11129

RESSAYRE, J.-P. [1971] see HARNIK, V.

RESSAYRE, J.-P. [1972] see LASSAIGNE, R.

RESSAYRE, J.-P. [1973] *Boolean models and infinitary first order languages* (J 0007) Ann Math Logic 6*41–92
- ⋄ C40 C50 C55 C70 C75 C80 C90 ⋄ REV MR 49 # 2368 Zbl 288 # 02013 JSL 47.439 • ID 23599

RESSAYRE, J.-P. [1977] *Models with compactness properties relative to an admissible language* (J 0007) Ann Math Logic 11∗31-55
　◇ C50　C55　C70　C80　D55　D60　E15　◇ REV MR 57 # 5735　Zbl 376 # 02032　JSL 47.439　• ID 23658

RESSAYRE, J.-P. [1980] *Types remarquables et extensions de modeles dans l'arithmetique de Peano. II (Avec un appendice de M.A.Dickmann)* (J 1620) Asterisque 73∗119-155
　◇ C30　C62　E05　◇ REV MR 82h:03030　Zbl 463 # 03041 JSL 48.483　• REM Part I 1980 by Dickmann,M.A.　• ID 54581

RESSAYRE, J.-P. [1981] see MCALOON, K.

RESSAYRE, J.-P. [1981] see BERLINE, C.

RESSAYRE, J.-P. [1983] *Modeles isomorphes a leurs propres segments initiaux en theorie des ensembles (English summary)* (J 3364) C R Acad Sci, Paris, Ser 1 296∗569-572
　◇ C62　E30　◇ REV MR 84g:03043　Zbl 538 # 03030
　• ID 41473

RESSAYRE, J.-P. [1984] *Deux fantaisies d'univers non standard (English summary)* (J 3364) C R Acad Sci, Paris, Ser 1 299∗583-586
　◇ C62　E30　◇ REV MR 86b:03043　• ID 44416

RESSAYRE, J.-P. [1985] see PINO, R.

RESSAYRE, J.-P. see Vol. IV, VI for further entries

RETZLAFF, A.T. [1977] see KALANTARI, I.

RETZLAFF, A.T. [1978] *Simple and hyperhypersimple vector spaces* (J 0036) J Symb Logic 43∗260-269
　◇ C57　D25　D45　◇ REV MR 81g:03053　Zbl 399 # 03033
　• ID 29259

RETZLAFF, A.T. [1979] *Direct summands of recursively enumerable vector spaces* (J 0068) Z Math Logik Grundlagen Math 25∗363-372
　◇ C57　C60　D45　◇ REV MR 81a:03049　Zbl 444 # 03024
　• ID 77762

RETZLAFF, A.T. [1979] see KALANTARI, I.

REYES, G.E. [1970] *Local definability theory* (J 0007) Ann Math Logic 1∗95-137
　◇ C07　C15　C40　C50　C55　C75　◇ REV MR 43 # 31 Zbl 217.305　• ID 22232

REYES, G.E. [1972] $L_{\omega_1\omega}$ *is enough: a reduction theorem for some infinitary languages* (J 0036) J Symb Logic 37∗705-710
　◇ C55　C75　E55　◇ REV MR 50 # 4240　Zbl 278 # 02015
　• ID 11130

REYES, G.E. [1974] *From sheaves to logic* (C 0768) Stud Algeb Logic 143-204
　◇ C90　F50　G30　◇ REV MR 50 # 13182　Zbl 344 # 02042 JSL 43.146　• ID 11131

REYES, G.E. [1975] see BEAUCHEMIN, P.

REYES, G.E. [1975] *Faisceaux et concepts* (J 0306) Cah Topol & Geom Differ 16∗307
　◇ C90　F50　G30　◇ REV Zbl 342 # 02039　• ID 64812

REYES, G.E. [1976] see MAKKAI, M.

REYES, G.E. [1977] see MAKKAI, M.

REYES, G.E. [1977] *Sheaves and concepts: a model-theoretic interpretation of Grothendieck topoi* (J 0306) Cah Topol & Geom Differ 18∗105-137
　◇ C75　C90　F50　G30　◇ REV MR 58 # 5184 Zbl 396 # 18002　• ID 52660

REYES, G.E. [1978] *Theorie des modeles et faisceaux* (J 0345) Adv Math 30∗156-170
　◇ C90　F50　G30　◇ REV MR 81f:03050　Zbl 409 # 03040
　• ID 56342

REYES, G.E. [1979] see DUBUC, E.J.

REYES, G.E. [1981] see BUNGE, MARTA C.

REYES, G.E. see Vol. V, VI for further entries

REZNIKOFF, I. [1963] *Chaines de formules* (J 0109) C R Acad Sci, Paris 256∗5021-5023
　◇ B05　C40　C90　F50　◇ REV MR 27 # 27　Zbl 143.7
　• ID 11135

REZNIKOFF, I. [1965] *Tout ensemble de formules de la logique classique est equivalent a un ensemble independant* (J 0109) C R Acad Sci, Paris 260∗2385-2388
　◇ B05　C40　◇ REV MR 31 # 2131　Zbl 143.7　• ID 11136

REZNIKOFF, I. [1966] *Sur les ensembles denombrables de formules en logique intuitionniste* (J 2313) C R Acad Sci, Paris, Ser A-B 262∗A415-A418
　◇ C40　C90　F50　◇ REV MR 33 # 2544　Zbl 143.7
　• ID 11137

REZNIKOFF, I. see Vol. VI for further entries

RIBEAUFOSSE, P. [1969] see BOFFA, M.

RIBEIRO, H. [1956] *The notion of universal completeness* (J 0050) Port Math 15∗83-86
　◇ C05　C35　◇ REV MR 18.785　Zbl 73.251　JSL 27.97
　• ID 11144

RIBEIRO, H. [1957] *Universal completeness* (P 1675) Summer Inst Symb Log;1957 Ithaca 81-82
　◇ C05　C35　◇ ID 29322

RIBEIRO, H. [1961] *On the universal completeness of classes of relational systems* (J 0009) Arch Math Logik Grundlagenforsch 5∗90-95
　◇ C05　C35　◇ REV MR 30 # 26　Zbl 121.12　• ID 11145

RIBEIRO, H. & SCHWABAUER, R. [1965] *A remark on equational completeness* (J 0009) Arch Math Logik Grundlagenforsch 7∗122-123
　◇ C05　C35　◇ REV MR 35 # 41　Zbl 154.4　JSL 36.161
　• ID 11146

RIBEIRO, H. [1965] see BRIGNOLE, D.

RIBEIRO, H. [1969] *Elementary languages and mathematical structures (Portuguese)* (J 0084) Gaz Mat (Lisboa) 30/113-116∗1-8
　◇ B10　C07　◇ REV MR 46 # 4798　Zbl 248 # 02004
　• ID 64815

RIBEIRO, H. see Vol. V for further entries

RIBENBOIM, P. [1965] *On the existence of totally ordered abelian groups which are η_α-sets* (J 0014) Bull Acad Pol Sci, Ser Math Astron Phys 13∗545-548
　◇ C55　C60　E07　E75　◇ REV MR 33 # 5756　Zbl 135.62
　• ID 11148

RIBENBOIM, P. [1969] *Boolean powers* (J 0027) Fund Math 65*243-268
⋄ C20 C30 G05 ⋄ REV MR 42#1732 Zbl 188.46
• ID 11149

RIBENBOIM, P. [1979] see DRIES VAN DEN, L.

RIBENBOIM, P. [1984] *Remarks on existentially closed fields and diophantine equations* (J 0144) Rend Sem Mat Univ Padova 71*229-237
⋄ C25 C60 ⋄ REV MR 86h:12009 • ID 49886

RICHARD, D. [1973] see CHARRETTON, C.

RICHARD, D. [1977] *On external properties of nonstandard models of arithmetic (French summary)* (J 0056) Publ Dep Math, Lyon 14/4*57-75
⋄ C62 ⋄ REV MR 80b:03111 Zbl 405#03036 • ID 77774

RICHARD, D. [1978] see KELLER, J.P.

RICHARD, D. [1980] *Saturation des modeles de Peano* (J 2313) C R Acad Sci, Paris, Ser A-B 290*A351-A353
⋄ C50 C62 H15 ⋄ REV MR 81f:03079 Zbl 429#03013
• ID 53844

RICHARD, D. [1981] *De la structure additive a la saturation des modeles de Peano et a une classification des sous-langages de l'arithmetique* (P 3404) Model Th & Arithm;1979/80 Paris 270-296
⋄ C50 C62 ⋄ REV MR 83j:03057 Zbl 498#03020
• ID 35358

RICHARD, D. [1981] see PABION, J.F.

RICHARD, D. [1984] see POUZET, M.

RICHARD, D. [1984] *The arithmetics as theories of two orders (English and French summaries)* (P 2167) Orders: Descr & Roles;1982 L'Arbresle 287-311
⋄ B28 C62 D35 F30 ⋄ REV MR 85k:06001
MR 86h:03102 Zbl 555#03026 • ID 41779

RICHARD, D. see Vol. I, IV, VI for further entries

RICHARDSON, D.B. [1983] *Roots of real exponential functions* (J 3172) J London Math Soc, Ser 2 28*46-56
⋄ C35 C65 ⋄ REV MR 85e:03072 • ID 40625

RICHARDSON, D.B. see Vol. IV, VI for further entries

RICHTER, L.J. [1981] *Degrees of structures* (J 0036) J Symb Logic 46*723-731
⋄ C57 D30 D45 ⋄ REV MR 83d:03048 Zbl 512#03024
• ID 33281

RICHTER, L.J. see Vol. IV for further entries

RICHTER, M.M. [1971] *Limites in Kategorien von Relationalsystemen* (J 0068) Z Math Logik Grundlagen Math 17*75-90
⋄ C20 C30 G30 ⋄ REV MR 43#4878 Zbl 227#02033
• ID 11168

RICHTER, M.M. [1971] see MARCO DE, G.

RICHTER, M.M. [1972] see GOEBEL, R.

RICHTER, M.M. & SCHLOESSER, B. & SCHWARZ, D. [1981] *Modelltheorie* (S 1642) Schr Inf Angew Math, Ber (Aachen) 73*113pp
⋄ C98 ⋄ ID 37695

RICHTER, M.M. & TRIESCH, E. [1981] *Universelle Algebra* (S 1642) Schr Inf Angew Math, Ber (Aachen) 74*72pp
⋄ C05 C98 ⋄ ID 37694

RICHTER, M.M. [1982] *Ideale Punkte, Monaden und Nichtstandard-Methoden* (X 0900) Vieweg: Wiesbaden vii + 264pp
⋄ C20 C60 E70 H05 H98 ⋄ REV MR 84g:03107
Zbl 487#03040 • ID 34209

RICHTER, M.M. [1984] see MUELLER, GERT H.

RICHTER, M.M. [1984] *Some aspects of nonstandard methods in general algebra* (P 3088) Univer Alg & Link Log, Alg, Combin, Comp Sci;1983 Darmstadt 78-84
⋄ C05 C30 H05 ⋄ REV Zbl 551#03038 • ID 43919

RICHTER, M.M. see Vol. I, II, IV, V, VI for further entries

RICKERT, U. [1979] *Ueber Summen in abelschen Gruppen* (J 0401) J Number Th 11*16-19
⋄ C60 ⋄ REV MR 80f:10059 Zbl 402#10049 • ID 54707

RICKEY, V.F. [1975] *Creative definitions in propositional calculi* (J 0047) Notre Dame J Formal Log 16*273-294
⋄ B05 C40 ⋄ REV MR 52#38 Zbl 232#02008 • ID 11178

RICKEY, V.F. [1975] *On creative definitions in the Principia Mathematica* (J 0079) Logique & Anal, NS 18*175-182
⋄ B15 C40 ⋄ REV MR 53#7721 Zbl 316#02003
• ID 22975

RICKEY, V.F. [1978] *On creative definitions in first order functional calculi* (J 0047) Notre Dame J Formal Log 19*307-309
⋄ B10 C40 ⋄ REV MR 58#5012 Zbl 368#02011
• ID 51658

RICKEY, V.F. see Vol. I, II, V for further entries

RIEGER, L. [1951] *On countable generalized σ-algebras, with a new proof of Goedel's completeness theorem (Czech)* (J 0022) Cheskoslov Mat Zh 1(76)*33-49
⋄ C07 G05 ⋄ REV MR 14.347 Zbl 45.150 JSL 20.281
• ID 28449

RIEGER, L. [1951] *On free $\aleph_\xi$-complete boolean algebras (with an application to logic)* (J 0027) Fund Math 38*35-52
⋄ C07 G05 ⋄ REV MR 14.347 Zbl 44.261 JSL 19.286
• ID 16866

RIEGER, L. [1955] *On a fundamental theorem of mathematical logic (Czech) (Russian and English summaries)* (J 0086) Cas Pestovani Mat, Ceskoslov Akad Ved 80*217-231
⋄ B10 C07 G05 G25 ⋄ REV MR 19.378 Zbl 68.242
JSL 35.489 • ID 19556

RIEGER, L. [1957] *A contribution to Goedel's axiomatic set-theory, I* (J 0022) Cheskoslov Mat Zh 7(82)*323-357
⋄ C62 E30 E35 E45 E50 E70 ⋄ REV MR 20#5739
Zbl 89.244 JSL 23.216 • REM Part II 1959 • ID 11188

RIEGER, L. [1959] *A contribution to Goedel's axiomatic set theory II* (J 0022) Cheskoslov Mat Zh 9(84)*1-49
⋄ C62 E30 E70 ⋄ REV MR 23#A1519 Zbl 95.9
JSL 40.242 • REM Part I 1957. Part III 1963 • ID 11189

RIEGER, L. [1964] *Zu den Strukturen der klassischen Praedikatenlogik* (J 0068) Z Math Logik Grundlagen Math 10*121-138
⋄ B10 C07 G05 ⋄ REV MR 29#2168 Zbl 124.247
JSL 35.440 • ID 11193

RIEGER, L. see Vol. I, II, IV, V, VI for further entries

RIMSCHA VON, M. [1980] *Mengentheoretische Modelle des λK-Kalkuels* (J 0009) Arch Math Logik Grundlagenforsch 20*65-73
⋄ B40 C65 E65 E70 ⋄ REV MR 81g:03012 Zbl 428 # 03045 • ID 53804

RIMSCHA VON, M. [1983] *Bases of ZF^0-models* (J 0009) Arch Math Logik Grundlagenforsch 23*11-19
⋄ C62 E47 E70 ⋄ REV MR 85h:03036 Zbl 541 # 03033 • ID 40517

RIMSCHA VON, M. see Vol. V for further entries

RINGEL, C.M. [1972] *The intersection property of amalgamations* (J 0326) J Pure Appl Algebra 2*341-342
⋄ C52 G30 ⋄ REV MR 48 # 11244 Zbl 263 # 18001 • ID 29872

RIPS, E. [1982] *Subgroups of small cancellation groups* (J 0161) Bull London Math Soc 14*45-47
⋄ C60 ⋄ REV MR 83c:20049 Zbl 481 # 20020 • ID 39042

RIPS, E. see Vol. IV for further entries

ROBBIN, J.W. [1979] *Evaluation fields for power series I: The nullstellensatz. II: The reellnullstellensatz* (J 0032) J Algeb 57*196-211,212-222
⋄ C60 ⋄ REV MR 80m:13006 Zbl 466 # 54022 • ID 82616

ROBBIN, J.W. see Vol. I, II for further entries

ROBERTS, F.S. [1970] see BAKER, K.A.

ROBERTS, F.S. [1972] see BAKER, K.A.

ROBERTS, F.S. see Vol. I, V for further entries

ROBERTS, I. [1984] *Ultrapowers in the Lipschitz and uniform classification of Banach spaces* (P 4225) Mini-Conf Nonlinear Anal;1984 Canberra 195-201
⋄ C20 C65 ⋄ ID 48118

ROBINSON, A. [1951] *On axiomatic systems which possess finite models* (J 0175) Methodos 3*140-149
• REPR [1979] (C 4594) Sel Pap Robinson 1*322-331
⋄ C13 C52 C60 ⋄ REV JSL 20.186 • ID 11223

ROBINSON, A. [1951] *On the metamathematics of algebra* (X 0809) North Holland: Amsterdam ix+195pp
⋄ C07 C60 C98 ⋄ REV MR 13.715 Zbl 43.247 JSL 17.205 • ID 11222

ROBINSON, A. [1952] *On the application of symbolic logic to algebra* (P 0593) Int Congr Math (II, 6);1950 Cambridge MA 1*686-694
• REPR [1979] (C 4594) Sel Pap Robinson 1*3-11
⋄ C60 ⋄ REV MR 13.716 Zbl 49.148 JSL 18.182 • ID 11224

ROBINSON, A. [1953] *Les rapports entre le calcul deductif et l'interpretation semantique d'un systeme axiomatique* (P 0644) Meth Form en Axiom;1950 Paris 35-51
⋄ C60 ⋄ REV MR 15.190 Zbl 50.246 JSL 20.185 • ID 11225

ROBINSON, A. [1954] *L'application de la logique formelle aux mathematiques* (P 0646) Appl Sci de Log Math;1952 Paris 51-64
⋄ B30 C60 ⋄ REV MR 16.782 Zbl 57.7 JSL 23.218 • ID 11226

ROBINSON, A. [1954] *On predicates in algebraically closed fields* (J 0036) J Symb Logic 19*103-114
• REPR [1979] (C 4594) Sel Pap Robinson 1*332-343
⋄ C60 ⋄ REV MR 15.925 Zbl 55.267 JSL 25.169 • ID 11227

ROBINSON, A. [1955] *Further remarks on ordered fields and definite functions* (J 0043) Math Ann 130*405-409
• REPR [1979] (C 4594) Sel Pap Robinson 1*370-374
⋄ C60 ⋄ REV MR 17.1180 Zbl 71.264 • ID 11236

ROBINSON, A. [1955] see GILMORE, P.C.

ROBINSON, A. [1955] *Note on an embedding theorem for algebraic systems* (J 0039) J London Math Soc 30*249-252
• REPR [1979] (C 4594) Sel Pap Robinson 1*344-347
⋄ C60 ⋄ REV MR 17.449 Zbl 64.8 • ID 11229

ROBINSON, A. [1955] *On ordered fields and definite functions* (J 0043) Math Ann 130*257-271
• REPR [1979] (C 4594) Sel Pap Robinson 1*355-369
⋄ C30 C60 ⋄ REV MR 17.822 Zbl 67.15 • ID 11230

ROBINSON, A. [1955] *Ordered structures and related concepts* (P 1589) Math Interpr of Formal Systs;1954 Amsterdam 51-56
• REPR [1979] (C 4594) Sel Pap Robinson 1*99-104
⋄ C30 C35 C50 C60 ⋄ REV MR 17.700 Zbl 68.245 JSL 25.170 • ID 27721

ROBINSON, A. [1955] *Theorie metamathematique des ideaux* (X 0834) Gauthier-Villars: Paris 186pp
⋄ C60 C98 ⋄ REV MR 16.1080 Zbl 64.243 JSL 20.279 • ID 11228

ROBINSON, A. [1956] *A result on consistency and its application to the theory of definition* (J 0028) Indag Math 18*47-58
• REPR [1979] (C 4594) Sel Pap Robinson 1*87-98
⋄ C40 ⋄ REV MR 17.1172 Zbl 75.7 JSL 25.174 • ID 11235

ROBINSON, A. [1956] *Apercu metamathematique sur les nombres reels* (P 4154) Deux Conf Prononcees;1956 Montreal 14pp
⋄ C60 C65 ⋄ ID 41675

ROBINSON, A. [1956] *Complete theories* (X 0809) North Holland: Amsterdam vii+129pp
⋄ B25 C35 C60 C98 ⋄ REV MR 17.817 MR 57 # 12202 Zbl 70.27 JSL 25.172 • ID 11232

ROBINSON, A. [1956] *Completeness and persistence in the theory of models* (J 0068) Z Math Logik Grundlagen Math 2*15-26
• REPR [1979] (C 4594) Sel Pap Robinson 1*108-119
⋄ C30 C35 C52 ⋄ REV MR 17.1173 Zbl 75.232 JSL 25.170 • ID 11233

ROBINSON, A. [1956] *Note on a problem of L.Henkin* (J 0036) J Symb Logic 21*33-35
• REPR [1979] (C 4594) Sel Pap Robinson 1*105-107
⋄ C30 C52 ⋄ REV MR 17.817 Zbl 71.7 JSL 27.96 • ID 11231

ROBINSON, A. [1956] *Solution of a problem by Erdoes-Gillman-Henriksen* (J 0053) Proc Amer Math Soc 7*908-909
• REPR [1979] (C 4594) Sel Pap Robinson 1*559-560
⋄ C60 ⋄ REV MR 18.374 Zbl 72.264 • ID 11234

ROBINSON, A. [1957] *Applications to field theory* (P 1675) Summer Inst Symb Log;1957 Ithaca 326-331
⋄ C60 ⋄ REV JSL 27.97 • ID 11245

ROBINSON, A. [1957] see LIGHTSTONE, A.H.

ROBINSON, A. [1957] *Relative model-completeness and the elimination of quantifiers* (P 1675) Summer Inst Symb Log;1957 Ithaca 155-159
- ⋄ C10 C35 ⋄ REV MR 21#1265 JSL 27.229 • REM Abbreviated version of Robinson,A. 1958 • ID 22159

ROBINSON, A. [1957] *Some problems of definability in the lower predicate calculus* (J 0027) Fund Math 44*309-329
- REPR [1979] (C 4594) Sel Pap Robinson 1*375-395
- ⋄ C35 C40 C60 ⋄ REV MR 19.1032 Zbl 79.6 JSL 25.171
- ID 11237

ROBINSON, A. [1958] *Outline of an introduction to mathematical logic I,II,III* (J 0018) Canad Math Bull 1*41-54,113-127,193-208
- ⋄ B98 C07 G05 ⋄ REV MR 20#5123 MR 20#5124 MR 21#6321 Zbl 80.5 Zbl 82.15 Zbl 84.4 • REM Part IV 1959 • ID 25442

ROBINSON, A. [1958] *Relative model-completeness and the elimination of quantifiers (English, French, and German summaries)* (J 0076) Dialectica 12*394-407
- REPR [1959] (C 1173) Logica (Bernays) 190-203 [1979] (C 4594) Sel Pap Robinson 1*146-159
- ⋄ C10 C35 C98 ⋄ REV MR 21#1265 Zbl 92.249 JSL 27.229 • ID 22158

ROBINSON, A. [1959] *Obstructions to arithmetical extension and the theorem of Los and Suszko* (J 0028) Indag Math 21*489-495
- REPR [1979] (C 4594) Sel Pap Robinson 1*160-166
- ⋄ C40 C52 ⋄ REV MR 22#2544 • ID 11242

ROBINSON, A. [1959] *On the concept of a differentially closed field* (J 0493) Bull Res Counc Israel Sect F 8*113-128
- REPR [1979] (C 4594) Sel Pap Robinson 1*440-455
- ⋄ C35 C60 ⋄ REV MR 23#A2323 Zbl 221#12054
- ID 11240

ROBINSON, A. [1959] *Outline of an introduction to mathematical logic IV* (J 0018) Canad Math Bull 2*33-42
- ⋄ B10 B98 C07 C98 G05 ⋄ REV MR 21#6321 Zbl 84.4 • REM Parts I-III 1958 • ID 19551

ROBINSON, A. [1959] *Solution of a problem of Tarski* (J 0027) Fund Math 47*179-204
- REPR [1979] (C 4594) Sel Pap Robinson 1*414-439
- ⋄ B25 C35 C60 ⋄ REV MR 22#3690 Zbl 93.13
- ID 11239

ROBINSON, A. & ZAKON, E. [1960] *Elementary properties of ordered abelian groups* (J 0064) Trans Amer Math Soc 96*222-236
- REPR [1979] (C 4594) Sel Pap Robinson 1*456-470
- ⋄ B25 C35 C60 C65 ⋄ REV MR 22#5673 Zbl 96.245 JSL 32.414 • ID 11247

ROBINSON, A. [1961] *Model theory and non-standard arithmetic* (P 0633) Infinitist Meth;1959 Warsaw 265-302
- REPR [1979] (C 4594) Sel Pap Robinson 1*167-204
- ⋄ C35 C50 C57 C60 C62 H15 ⋄ REV MR 26#32 Zbl 126.11 JSL 35.149 • ID 11249

ROBINSON, A. [1961] *Non-standard analysis* (J 0028) Indag Math 23*432-440
- REPR [1979] (C 4594) Sel Pap Robinson 2*3-11
- ⋄ H05 H15 ⋄ REV MR 26#33 Zbl 102.7 JSL 34.292
- ID 11251

ROBINSON, A. [1961] *On the construction of models* (C 0622) Essays Found of Math (Fraenkel) 207-217
- REPR [1979] (C 4594) Sel Pap Robinson 1*32-42
- ⋄ C07 C20 C30 E05 ⋄ REV MR 29#23 Zbl 221#02045
- ID 11250

ROBINSON, A. [1962] *A note on embedding problems* (J 0027) Fund Math 50*455-461
- REPR [1979] (C 4594) Sel Pap Robinson 1*471-477
- ⋄ C07 C60 ⋄ REV MR 25#2946 Zbl 224#02040
- ID 11252

ROBINSON, A. [1962] *Recent developments in model theory* (P 0612) Int Congr Log, Meth & Phil of Sci (1,Proc);1960 Stanford 60-79
- REPR [1979] (C 4594) Sel Pap Robinson 1*12-31
- ⋄ C98 ⋄ REV MR 29#4668 Zbl 221#02038 • ID 11253

ROBINSON, A. [1963] *Introduction to model theory and to the metamathematics of algebra* (X 0809) North Holland: Amsterdam ix+284pp
- TRANSL [1967] (X 2027) Nauka: Moskva 376pp
- ⋄ B98 C60 C98 H15 ⋄ REV MR 27#3533 MR 36#3642 JSL 25.172 • REM 2nd rev. ed. 1974 • ID 28804

ROBINSON, A. [1963] *On languages which are based on non-standard arithmetic* (J 0111) Nagoya Math J 22*83-117
- REPR [1979] (C 4594) Sel Pap Robinson 2*12-46
- ⋄ C75 H05 H15 ⋄ REV MR 27#3532 Zbl 166.261 JSL 34.516 • ID 11254

ROBINSON, A. [1964] *Between logic and mathematics* (J 1797) ICSU Review World Sci 6*218-224
- TRANSL [1966] (J 0519) Wiad Mat, Ann Soc Math Pol, Ser 2 9*89-96
- ⋄ C60 ⋄ REV Zbl 147.246 • ID 26136

ROBINSON, A. [1964] *On generalized limits and linear functionals* (J 0048) Pac J Math 14*269-283
- REPR [1979] (C 4594) Sel Pap Robinson 2*47-61
- ⋄ C20 H05 ⋄ REV MR 29#1534 Zbl 121.95 JSL 34.292
- ID 11258

ROBINSON, A. [1965] *Topics in non-archimedean mathematics* (P 0614) Th Models;1963 Berkeley 285-298
- REPR [1979] (C 4594) Sel Pap Robinson 2*99-112
- ⋄ H05 H15 H20 ⋄ REV MR 33#5489 Zbl 156.250 JSL 34.292 • ID 11259

ROBINSON, A. [1966] *A new approach to the theory of algebraic numbers I,II (Italian summary)* (J 0149) Atti Accad Naz Lincei Fis Mat Nat, Ser 8 40*222-225,770-774
- REPR [1979] (C 4594) Sel Pap Robinson 2*113-116,117-121
- ⋄ C60 I115 I120 ⋄ REV MR 35#5305 MR 35#5306 Zbl 173.8 • ID 11261

ROBINSON, A. [1966] *Non-standard analysis* (X 0809) North Holland: Amsterdam ix+293pp
- ⋄ A10 B98 H05 H10 H15 H98 ⋄ REV MR 34#5680 Zbl 151.8 JSL 35.292 • REM 2d edition 1968; 291pp
- ID 11263

ROBINSON, A. [1966] *On some applications of model theory to algebra and analysis* (J 0384) Rend Mat, Ser 5 25*562-592
- REPR [1979] (C 4594) Sel Pap Robinson 2*158-188
- ⋄ C60 C98 H05 H98 ⋄ REV MR 36#2489 Zbl 157.28
- ID 22170

ROBINSON, A. [1967] *Non-standard theory of Dedekind rings* (J 0028) Indag Math 29*444-452
- REPR [1979] (C 4594) Sel Pap Robinson 2*122-130
- ◇ C20 C60 H15 H20 ◇ REV MR 36 # 6399 Zbl 189.285
- ID 11265

ROBINSON, A. [1967] *Nonstandard arithmetic* (J 0015) Bull Amer Math Soc 73*818-843
- REPR [1979] (C 4594) Sel Pap Robinson 2*132-157
- ◇ C20 C60 H15 H20 ◇ REV MR 36 # 1319 Zbl 189.284
- ID 11264

ROBINSON, A. [1968] *Model theory* (C 0552) Phil Contemp - Chroniques 61-73
- REPR [1979] (C 4594) Sel Pap Robinson 2*524-536
- ◇ C98 ◇ ID 14648

ROBINSON, A. & ZAKON, E. [1969] *A set-theoretical characterization of enlargements* (P 0649) Appl Model Th to Algeb, Anal & Probab;1967 Pasadena 109-122
- REPR [1979] (C 4594) Sel Pap Robinson 2*206-219
- ◇ C20 C62 E75 H05 H20 ◇ REV MR 39 # 1319 Zbl 193.303 • ID 11267

ROBINSON, A. [1969] *Compactification of groups and rings and non-standard analysis* (J 0036) J Symb Logic 34*576-588
- REPR [1979] (C 4594) Sel Pap Robinson 2*243-255
- ◇ C40 C60 H05 H20 ◇ REV MR 44 # 1765 Zbl 196.11
- ID 11269

ROBINSON, A. [1969] *Problems and methods of model theory* (P 0630) Aspects Math Log;1968 Varenna 181-266
- ◇ C60 C98 ◇ REV MR 41 # 5208 Zbl 211.307 • ID 26144

ROBINSON, A. [1969] *Topics in nonstandard algebraic number theory* (P 0649) Appl Model Th to Algeb, Anal & Probab;1967 Pasadena 1-17
- REPR [1979] (C 4594) Sel Pap Robinson 2*189-205
- ◇ C60 H05 H15 H20 ◇ REV MR 42 # 3054 Zbl 182.327
- ID 21051

ROBINSON, A. [1970] see BARWISE, J.

ROBINSON, A. [1970] *Elementary embeddings of fields of power series* (J 0401) J Number Th 2*237-247
- REPR [1979] (C 4594) Sel Pap Robinson 2*232-242
- ◇ C20 C60 H20 ◇ REV MR 41 # 3451 Zbl 199.304
- ID 11271

ROBINSON, A. [1971] *Forcing in model theory* (P 0669) Conv Teor Modelli & Geom;1969/70 Roma 69-82
- REPR [1979] (C 4594) Sel Pap Robinson 1*205-218
- ◇ C25 C35 C52 ◇ REV MR 43 # 4651 Zbl 212.18 JSL 40.633 • ID 11270

ROBINSON, A. [1971] *Forcing in model theory* (P 0743) Int Congr Math (II,11,Proc);1970 Nice 1*245-250
- ◇ C25 ◇ REV MR 58 # 21581 Zbl 227 # 02028 JSL 40.633
- ID 21286

ROBINSON, A. [1971] *Infinite forcing in model theory* (P 0604) Scand Logic Symp (2);1970 Oslo 317-340
- REPR [1979] (C 4594) Sel Pap Robinson 1*243-266
- ◇ C25 C35 C52 ◇ REV MR 50 # 9574 Zbl 222 # 02057 JSL 40.633 • ID 11272

ROBINSON, A. [1971] *On the notion of algebraic closedness for noncommutative groups and fields* (J 0036) J Symb Logic 36*441-444
- REPR [1979] (C 4594) Sel Pap Robinson 1*478-481
- ◇ C25 C60 ◇ REV MR 45 # 43 Zbl 231 # 02065 • ID 11273

ROBINSON, A. [1972] *Algebraic function fields and non-standard arithmetic* (P 0732) Contrib to Non-Standard Anal;1970 Oberwolfach 1-14
- REPR [1979] (C 4594) Sel Pap Robinson 2*256-269
- ◇ C60 H15 ◇ REV MR 58 # 682 Zbl 247 # 12107 JSL 39.339 • ID 11274

ROBINSON, A. [1972] *Generic categories* (1111) Preprints, Manuscr., Techn. Reports etc. 20pp
- REPR [1979] (C 4594) Sel Pap Robinson 1*298-306
- ◇ C25 C52 C60 G30 ◇ REM Lect. pres. at the Logic Symp. Orleans, F, 1972 • ID 41674

ROBINSON, A. [1972] see FISHER, E.R.

ROBINSON, A. [1972] *On the real closure of a Hardy field* (C 0648) Th of Sets and Topology (Hausdorff) 427-433
- REPR [1979] (C 4594) Sel Pap Robinson 1*607-613
- ◇ C60 C65 ◇ REV MR 49 # 4980 Zbl 298 # 02061
- ID 11275

ROBINSON, A. [1973] *Function theory on some nonarchimedian fields* (J 0005) Amer Math Mon 80*87-109
- ◇ C60 C65 H05 ◇ REV MR 48 # 8464 Zbl 269 # 26020
- ID 11280

ROBINSON, A. [1973] *Metamathematical problems* (J 0036) J Symb Logic 38*500-516
- TRANSL [1973] (J 0396) Mat Lapok 24*1-17 • REPR [1979] (C 4594) Sel Pap Robinson 1*43-59
- ◇ A05 B30 C60 C98 H05 ◇ REV MR 49 # 2240 Zbl 289 # 02002 • ID 11276

ROBINSON, A. [1973] *Model theory as a framework for algebra* (C 0654) Stud in Model Th 134-157
- REPR [1979] (C 4594) Sel Pap Robinson 1*60-83
- ◇ C25 C35 C60 C98 ◇ REV MR 49 # 2365 • ID 11279

ROBINSON, A. [1973] *Nonstandard arithmetic and generic arithmetic* (P 0793) Int Congr Log, Meth & Phil of Sci (4,Proc);1971 Bucharest 137-154
- REPR [1979] (C 4594) Sel Pap Robinson 1*280-297
- ◇ C25 C60 C62 C98 H15 ◇ REV MR 56 # 5283
- ID 15099

ROBINSON, A. [1973] *Nonstandard points on algebraic curves* (J 0401) J Number Th 5*301-327
- REPR [1979] (C 4594) Sel Pap Robinson 2*306-332
- ◇ C20 C60 H20 ◇ REV MR 53 # 443 Zbl 272 # 12108
- ID 16679

ROBINSON, A. [1973] *On bounds in the theory of polynomial ideals (Russian)* (C 0733) Izbr Vopr Algeb & Log (Mal'tsev) 245-252
- REPR [1979] (C 4594) Sel Pap Robinson 1*482-489
- ◇ C60 H20 ◇ REV MR 49 # 2364 Zbl 296 # 13014
- ID 11278

ROBINSON, A. [1973] *Standard and nonstandard number systems* (J 3077) Nieuw Arch Wisk, Ser 3 21*115-133
- REPR [1979] (C 4594) Sel Pap Robinson 2*426-444
- ◇ C25 C60 C98 H05 H15 H20 H98 ◇ REV MR 55 # 7767 Zbl 267 # 02041 • ID 26149

ROBINSON, A. [1974] *A decision method for elementary algebra and geometry - revisited* (P 0610) Tarski Symp;1971 Berkeley 139-152
- REPR [1979] (C 4594) Sel Pap Robinson 1*490-503
- ◇ C25 C60 ◇ REV MR 51 # 2902 Zbl 325 # 02039
- ID 11282

ROBINSON, A. [1974] *A note on topological model theory* (J 0027) Fund Math 81*159-171
- REPR [1979] (C 4594) Sel Pap Robinson 1*307-319
- ◇ C65 C90 ◇ REV MR 49#7135 Zbl 276#02041
- ID 29043

ROBINSON, A. [1975] *Algorithms in algebra* (C 0782) Model Th & Algeb (A. Robinson) 14-40
- REPR [1979] (C 4594) Sel Pap Robinson 1*504-520
- ◇ C40 C57 C60 D45 D75 ◇ REV MR 53#5298 Zbl 318#02033 • ID 22933

ROBINSON, A. & ROQUETTE, P. [1975] *On the finiteness theorem of Siegel and Mahler concerning diophantine equations* (J 0401) J Number Th 7*121-176
- ◇ C60 H05 H15 H20 ◇ REV MR 51#10222 Zbl 299#12107 • ID 17511

ROBINSON, A. [1979] *Selected papers of Abraham Robinson. Vol.I: Model theory and algebra* (X 0875) Yale Univ Pr: New Haven xxxvii+694pp
- LAST ED [1979] (X 0809) North Holland: Amsterdam xxxvii+694pp
- ◇ C60 C96 ◇ REV MR 80h:01039a Zbl 424#01031 JSL 47.197 • REM Ed. by Keisler,H.J. & Koerner,S. & Luxemburg,W.A.J. & Young,A.D. Vol.II 1979 • ID 53899

ROBINSON, A. [1979] *Selected papers of Abraham Robinson. Vol.II: Nonstandard analysis and philosophy* (X 0875) Yale Univ Pr: New Haven xlv+582pp
- LAST ED [1979] (X 0809) North Holland: Amsterdam xlv+582pp
- ◇ A05 B96 C96 H05 H10 H96 ◇ REV MR 80h:01039b Zbl 424#01031 JSL 47.203 • REM Ed. by Keisler,H.J. & Koerner,S. & Luxemburg,W.A.J. & Young,A.D. Vol.II 1979 • ID 53900

ROBINSON, A. also published under the name ROBINSOHN, A.

ROBINSON, A. see Vol. I, IV for further entries

ROBINSON, JULIA [1959] *The undecidability of algebraic rings and fields* (J 0053) Proc Amer Math Soc 10*950-957
- ◇ C60 D35 ◇ REV MR 22#3691 Zbl 100.15 JSL 29.57
- ID 11295

ROBINSON, JULIA [1962] *On the decision problem for algebraic rings* (C 0595) Stud Math Anal & Rel Topics (Polya) 297-304
- ◇ C60 D35 ◇ REV MR 26#3609 Zbl 117.12 JSL 35.475
- ID 11296

ROBINSON, JULIA [1965] *The decision problem for fields* (P 0614) Th Models;1963 Berkeley 299-311
- ◇ B25 C60 D35 ◇ REV MR 34#62 Zbl 274#02020
- ID 11298

ROBINSON, JULIA [1973] *Axioms for number theoretic functions* (Russian) (C 0733) Izbr Vopr Algeb & Log (Mal'tsev) 253-263
- ◇ C62 D20 D75 F30 ◇ REV MR 48#8224 Zbl 279#02035 • ID 29547

ROBINSON, JULIA [1973] *Solving diophantine equations* (P 0793) Int Congr Log, Meth & Phil of Sci (4,Proc);1971 Bucharest 63-67
- ◇ B25 D35 D80 H15 ◇ REV MR 58#215 • ID 77822

ROBINSON, JULIA see Vol. IV, V, VI for further entries

ROBINSON, R.M. [1951] *Arithmetical definability of field elements* (J 0036) J Symb Logic 16*125-126
- ◇ C60 ◇ REV MR 13.97 Zbl 42.246 JSL 17.270 • ID 11310

ROBINSON, R.M. [1951] *Arithmetical definitions in the ring of integers* (J 0053) Proc Amer Math Soc 2*279-284
- ◇ C65 F30 ◇ REV MR 12.791 Zbl 54.7 JSL 17.269
- ID 11312

ROBINSON, R.M. [1951] *Undecidable rings* (J 0064) Trans Amer Math Soc 70*137-159
- ◇ C60 D35 ◇ REV MR 12.791 Zbl 42.245 JSL 17.268
- ID 11311

ROBINSON, R.M. see Vol. I, IV, V, VI for further entries

ROBITASHVILI, N.G. [1965] *On the problem of reducing a pp-formula of the pure first-order functional calculus to some special case of the decision problem (Georgian) (Russian summary)* (J 0233) Soobshch Akad Nauk Gruz SSR 39*269-276
- ◇ B20 B25 ◇ REV MR 32#5502 • ID 11330

ROBITASHVILI, N.G. see Vol. I, II, VI for further entries

ROCHE LA, P. [1981] *Effective Galois theory* (J 0036) J Symb Logic 46*385-392
- ◇ C57 C60 D45 F60 ◇ REV MR 82j:03053 Zbl 464#03039 • ID 75255

ROCKINGHAM GILL, R.R. [1970] *The Craig-Lyndon interpolation theorem in 3-valued logic* (J 0036) J Symb Logic 35*230-238
- ◇ B50 C40 ◇ REV MR 43#29 Zbl 212.315 • ID 22231

ROCKINGHAM GILL, R.R. [1975] *A note on the compactness theorem* (J 0068) Z Math Logik Grundlagen Math 21*377-378
- ◇ C07 C20 E25 ◇ REV MR 51#7858 Zbl 328#02033
- ID 18358

ROCKINGHAM GILL, R.R. see Vol. II, V for further entries

ROCZEN, M. [1978] see KURKE, H.

RODRIGUEZ ARTALEJO, M. [1981] *A method of constructing structures based on the distributive normal form (Spanish)* (J 0234) Rev Acad Cienc Exact Fis Nat Madrid 75*667-687
- ◇ C07 C30 ◇ REV MR 84i:03067 Zbl 499#03011
- ID 34548

RODRIGUEZ ARTALEJO, M. [1981] *Eine syntaktisch-algebraische Methode zur Konstruktion von Modellen* (J 0068) Z Math Logik Grundlagen Math 27*59-71
- ◇ B10 C07 C30 C57 ◇ REV MR 82e:03027 Zbl 481#03016 • ID 77840

RODRIGUEZ ARTALEJO, M. see Vol. I for further entries

ROEDDING, D. [1966] *Anzahlquantoren in der Kleene-Hierarchie* (J 0009) Arch Math Logik Grundlagenforsch 9*61-65
- ◇ C80 D55 ◇ REV MR 38#2023 Zbl 161.7 JSL 33.472
- ID 24963

ROEDDING, D. [1966] *Anzahlquantoren in der Praedikatenlogik* (J 0009) Arch Math Logik Grundlagenforsch 9*66-69
- ◇ C07 C80 ◇ REV MR 36#2485 Zbl 264#02016 JSL 33.473 • ID 11339

ROEDDING, D. [1966] *Der Entscheidbarkeitsbegriff in der mathematischen Logik* (J 0178) Stud Gen 19*516-522
- ◇ B25 D10 D20 D25 D35 ◇ REV Zbl 192.20 • ID 16282

ROEDDING, D. & SCHWICHTENBERG, H. [1972] *Bemerkungen zum Spektralproblem* (J 0068) Z Math Logik Grundlagen Math 18*1-12
- ◇ B15 C13 D10 D15 ◇ REV MR 46#5128 Zbl 242#02049 • ID 11343

ROEDDING, D. see Vol. I, IV for further entries

ROGAVA, M.G. [1979] *A new decision procedure for SCI (Russian)* (**P** 2539) Frege Konferenz (1);1979 Jena 361-364
⋄ B25 B60 F05 ⋄ REV MR 82h:03012 • ID 77850

ROGAVA, M.G. see Vol. I, II, VI for further entries

ROGERS, P. [1981] *Preservation of saturation and stability in a variety of nilpotent groups* (**J** 0036) J Symb Logic 46*499-512
⋄ C45 C50 C60 ⋄ REV MR 82k:20003 Zbl 497 # 03027 • ID 82634

ROGERS, R. [1971] *Mathematical logic and formalized theories. A survey of basic concepts and results* (**X** 0809) North Holland: Amsterdam xi+235pp
⋄ B98 C98 ⋄ REV MR 47 # 9 Zbl 245 # 02001 • ID 11365

ROGERS JR., H. [1963] *An example in mathematical logic* (**J** 0005) Amer Math Mon 70*929-945
⋄ B25 C10 C65 ⋄ REV JSL 37.616 • ID 11362

ROGERS JR., H. see Vol. I, IV, VI for further entries

ROGGENKAMP, K.W. & WIEDEMANN, A. [1984] *Auslander-Reiten quivers of Schurian orders* (**J** 0394) Commun Algeb 12*19-20,2525-2578
⋄ C60 ⋄ REV MR 86h:16030 • ID 49887

ROGUSKI, S. [1976] *Extensions of models for ZFC to models for ZF+V=HOD with applications* (**P** 1476) Set Th & Hierarch Th (2) (Mostowski);1975 Bierutowice 241-247
⋄ C62 E35 E45 E47 ⋄ REV MR 55 # 2567 Zbl 362 # 02061 • ID 23809

ROGUSKI, S. see Vol. IV, V for further entries

ROISIN, J.-R. [1977] *Introduction a une approche categorique en theorie des modeles* (**J** 3133) Bull Soc Math Belg, Ser B 29*161-174
⋄ C07 G30 ⋄ REV MR 57 # 16068 Zbl 433 # 03018 • ID 77863

ROISIN, J.-R. [1979] *On functorializing usual first-order model theory* (**P** 2901) Appl Sheaves;1977 Durham 612-622
⋄ C07 G30 ⋄ REV MR 81f:03051 Zbl 415 # 03054 • ID 53156

ROITMAN, J. [1982] *Non-isomorphic hyper-real fields from non-isomorphic ultrapowers* (**J** 0044) Math Z 181*93-96
⋄ C20 H05 H20 ⋄ REV MR 84a:54030 Zbl 474 # 54003 • ID 55446

ROITMAN, J. see Vol. V for further entries

ROLLAND, R. [1982] *Etude des coupures dans les groupes et corps ordonnes* (**P** 3852) Geom Algeb Reelle & Formes Quad;1981 Rennes 386-405
⋄ C60 C65 ⋄ REV MR 84c:06023 Zbl 508 # 06021 • ID 39572

ROMAN'KOV, V.A. [1979] *On definable subgroups of solvable groups* (**P** 2558) All-Union Conf Math Log (5) (Mal'tsev);1979 Novosibirsk 130
⋄ C40 C60 ⋄ ID 90353

ROMAN'KOV, V.A. [1979] *Universal theory of nilpotent groups (Russian)* (**J** 0087) Mat Zametki (Akad Nauk SSSR) 25*487-495,635
• TRANSL [1979] (**J** 1044) Math Notes, Acad Sci USSR 25*253-258
⋄ B25 C60 D35 ⋄ REV MR 80j:03058 Zbl 419 # 20031 • ID 53388

ROMAN'KOV, V.A. [1983] see REMESLENNIKOV, V.N.

ROMAN'KOV, V.A. see Vol. IV for further entries

ROMANOVSKIJ, N.S. [1980] *The elementary theory of an almost polycyclic group (Russian)* (**J** 0142) Mat Sb, Akad Nauk SSSR, NS 111(153)*135-143,160
• TRANSL [1981] (**J** 0349) Math of USSR, Sbor 39*125-132
⋄ C60 D35 ⋄ REV MR 81d:03035 Zbl 424 # 20031 • ID 77873

ROMANOVSKIJ, N.S. see Vol. IV for further entries

ROMANSKI, J. [1977] *On connections between algebraic and Kripke semantics* (**J** 1008) Demonstr Math (Warsaw) 10*123-127
⋄ C90 F50 G10 G25 ⋄ REV MR 56 # 5231 Zbl 369 # 02032 • ID 51332

ROMO SANTOS, C. [1981] *Two methods for the resolution of singularities in any characteristic: ultraproducts and maximal contact (Spanish)* (**P** 3751) Jorn Mat Luso-Espanol (8);1981 Coimbra 1*189-194
⋄ C20 C60 ⋄ REV MR 83m:14009 • ID 40404

ROMO SANTOS, C. [1982] *Ultraproducts of nonsingular varietes (Spanish)* (**J** 0236) Rev Mat Hisp-Amer, Ser 4 42*74-81
⋄ C20 C60 ⋄ ID 46364

ROMOV, B.A. [1971] *Definability by formulas for predicates on a finite model (Russian) (English summary)* (**J** 0040) Kibernetika, Akad Nauk Ukr SSR 1971/1*41-42
⋄ C13 C40 C60 ⋄ REV MR 45 # 3189 Zbl 253 # 02050 • ID 11388

ROMOV, B.A. see Vol. II, IV, V for further entries

ROQUETTE, P. [1971] *Bemerkungen zur Theorie der formal p-adischen Koerper* (**S** 0410) Math Beitr Univ Halle-Wittenberg 3*177-193
⋄ C60 ⋄ REV MR 45 # 8643 Zbl 245 # 12101 • ID 11400

ROQUETTE, P. [1975] *Nonstandard aspects of Hilbert's irreducibility theorem* (**C** 0782) Model Th & Algeb (A. Robinson) 231-275
⋄ C60 H15 H20 ⋄ REV MR 53 # 5598 Zbl 316 # 12103 • ID 22956

ROQUETTE, P. [1975] see ROBINSON, A.

ROQUETTE, P. [1980] see JARDEN, M.

ROQUETTE, P. [1983] see PRESTEL, A.

ROQUETTE, P. [1984] see CANTOR, DAVID G.

ROQUETTE, P. [1984] *Some tendencies in contemporary algebra* (**C** 3444) Perspectives in Math 393-422
⋄ C10 C35 C60 C98 ⋄ ID 44696

RORER, M.A. [1973] see KOGALOVSKIJ, S.R.

ROSE, B.I. [1977] see BALDWIN, JOHN T.

ROSE, B.I. [1978] *Model theory of alternative rings* (**J** 0047) Notre Dame J Formal Log 19*215-243
⋄ C10 C35 C45 C60 ⋄ REV MR 58 # 5185 Zbl 351 # 02037 • ID 28374

ROSE, B.I. [1978] *Rings which admit elimination of quantifiers* (**J** 0036) J Symb Logic 43*92-112 • ERR/ADD ibid 44*109-110
⋄ C10 C60 ⋄ REV MR 80e:03036 Zbl 388 # 03010 JSL 50.1079 • ID 28377

ROSE, B.I. [1978] *The $\aleph_1$-categoricity of strictly upper triangular matrix rings over algebraically closed fields* (J 0036) J Symb Logic 43*250-259
⋄ C35 C45 C60 ⋄ REV MR 80d:03031 Zbl 385 # 03026
• ID 28376

ROSE, B.I. [1980] *On the model theory of finite-dimensional algebras* (J 3240) Proc London Math Soc, Ser 3 40*21-39
⋄ B25 C35 C45 C60 ⋄ REV MR 81a:03033 Zbl 463 # 03020 • ID 54560

ROSE, B.I. [1980] *Prime quantifier eliminable rings* (J 3172) J London Math Soc, Ser 2 21*257-262
⋄ C10 C60 ⋄ REV MR 81g:03034 Zbl 439 # 03011
• ID 56003

ROSE, B.I. & WOODROW, R.E. [1981] *Ultrahomogeneous structures* (J 0068) Z Math Logik Grundlagen Math 27*23-30
⋄ C10 C50 ⋄ REV MR 82c:03044 Zbl 481 # 03017
• ID 77909

ROSE, B.I. [1982] *Preservation of elementary equivalence under scalar extension* (J 0036) J Symb Logic 47*734-738
⋄ C60 ⋄ REV MR 84c:03065 Zbl 514 # 03018 • ID 34008

ROSEN, N.I. [1983] *A characterization of 2-square ultrafilters* (J 0036) J Symb Logic 48*409-414
⋄ C20 E05 ⋄ REV MR 85e:04006 Zbl 564 # 04004
• ID 40800

ROSEN, N.I. see Vol. V for further entries

ROSENBERG, I.G. [1973] *Strongly rigid relations* (J 0308) Rocky Mountain J Math 3*631-639
⋄ B50 C05 C50 ⋄ REV MR 47 # 6588 Zbl 274 # 08001
• ID 29025

ROSENBERG, I.G. see Vol. I, II, V for further entries

ROSENLICHT, M. [1974] *The nonminimality of the differential closure* (J 0048) Pac J Math 52*529-537
⋄ C60 ⋄ REV MR 50 # 4556 Zbl 288 # 12103 • ID 11527

ROSENSTEIN, J.G. [1968] *Theories which are not $\aleph_0$-categorical* (P 0692) Summer School in Logic;1967 Leeds 273-278
⋄ C35 C65 ⋄ REV MR 38 # 5600 Zbl 182.324 • ID 11528

ROSENSTEIN, J.G. [1969] *$\aleph_0$-categoricity of linear orderings* (J 0027) Fund Math 64*1-5
⋄ C35 C65 E07 ⋄ REV MR 39 # 3982 Zbl 179.13
• ID 11530

ROSENSTEIN, J.G. [1971] *A note on a theorem of Vaught* (J 0036) J Symb Logic 36*439-440
⋄ C15 C35 C50 ⋄ REV MR 45 # 44 Zbl 228 # 02032
• ID 11531

ROSENSTEIN, J.G. [1972] *On $GL_2(R)$ where R is a boolean ring* (J 0018) Canad Math Bull 15*263-275
⋄ C35 C60 G05 ⋄ REV MR 46 # 3637 • ID 30829

ROSENSTEIN, J.G. [1973] *$\aleph_0$-categoricity of groups* (J 0032) J Algeb 25*435-467 • ERR/ADD ibid 48*236-240
⋄ C35 C60 ⋄ REV MR 47 # 6815 MR 58 # 16270 Zbl 256 # 20003 • ID 11532

ROSENSTEIN, J.G. [1976] *$\aleph_0$-categoricity is not inherited by factor groups* (J 0004) Algeb Universalis 6*93-95
⋄ C35 C60 ⋄ REV MR 53 # 3069 Zbl 342 # 02043
• ID 21653

ROSENSTEIN, J.G. [1976] see MACINTYRE, A.

ROSENSTEIN, J.G. [1977] see HAY, L.

ROSENSTEIN, J.G. [1978] see CHERLIN, G.L.

ROSENSTEIN, J.G. [1980] see MANASTER, A.B.

ROSENSTEIN, J.G. [1982] *Linear orderings* (X 0801) Academic Pr: New York xvii+487pp
⋄ B25 C65 C98 E07 E98 ⋄ REV MR 84m:06001 Zbl 488 # 04002 JSL 48.1207 • ID 38089

ROSENSTEIN, J.G. [1982] see LERMAN, M.

ROSENSTEIN, J.G. [1984] *Recursive linear orderings (French summary)* (P 2167) Orders: Descr & Roles;1982 L'Arbresle 465-475
⋄ C57 C65 D45 ⋄ REV MR 86d:06002 Zbl 554 # 06001
• ID 41775

ROSENSTEIN, J.G. see Vol. IV, V for further entries

ROSENTHAL, D. [1984] *The order indiscernibles of divisible ordered abelian groups* (J 0036) J Symb Logic 49*151-160
⋄ C30 C60 ⋄ REV MR 85c:03012 • ID 39836

ROSENTHAL, H.P. [1979] see BOURGAIN, J.

ROSENTHAL, H.P. see Vol. V for further entries

ROSENTHAL, J.W. [1971] *Relations not determining the structure of L* (J 0048) Pac J Math 37*497-514
⋄ C35 C62 E45 ⋄ REV MR 46 # 3295 Zbl 186.8
• ID 11534

ROSENTHAL, J.W. [1972] *A new proof of a theorem of Shelah* (J 0036) J Symb Logic 37*133-134
⋄ C35 C45 C55 E45 E50 ⋄ REV MR 48 # 1915 Zbl 254 # 02041 JSL 38.649 • ID 11533

ROSENTHAL, J.W. [1974] *Models of $Th(\omega^\omega, <)$* (J 0047) Notre Dame J Formal Log 15*122-132
⋄ C65 ⋄ REV MR 49 # 8853 Zbl 214.15 • ID 19604

ROSENTHAL, J.W. [1975] *Partial $\aleph_1$-homogeneity of the countable saturated model of an $\aleph_1$-categorical theory* (J 0068) Z Math Logik Grundlagen Math 21*307-308
⋄ C35 C50 ⋄ REV MR 52 # 70 Zbl 347 # 02032 • ID 11535

ROSENTHAL, J.W. [1975] *Truth in all of certain well-founded countable models arising in set theory* (J 0068) Z Math Logik Grundlagen Math 21*97-106
⋄ C40 C62 C70 E45 E47 ⋄ REV MR 54 # 2465 Zbl 315 # 02054 • REM Part I. Part II 1979 • ID 11536

ROSENTHAL, J.W. [1979] *On the dimension theory of $\aleph_1$-categorical theories with the nontrivial strong elementary intersection property* (J 0068) Z Math Logik Grundlagen Math 25*359-362
⋄ C35 ⋄ REV MR 82c:03035 Zbl 427 # 03023 • ID 28368

ROSENTHAL, J.W. [1979] *Truth in all of certain well-founded countable models arising in set theory II* (J 0068) Z Math Logik Grundlagen Math 25*403-405
⋄ C40 C62 C70 E45 E47 ⋄ REV MR 80i:03065 Zbl 422 # 03024 • REM Part I 1975 • ID 28371

ROSENTHAL, J.W. [1980] see ASH, C.J.

ROSENTHAL, J.W. see Vol. I, IV, V, VI for further entries

ROSENTHAL, K.I. [1985] see NIEFIELD, S.B.

ROSENTHAL, K.I. see Vol. V for further entries

ROSICKY, J. [1978] *Categories of models of infinitary Horn theories* (J 0322) Arch Math (Brno) 14∗219-226
⋄ C75 G30 ⋄ REV MR 80h:18004 Zbl 404 # 18003
• ID 54853

ROSICKY, J. [1981] *Concrete categories and infinitary languages* (J 0326) J Pure Appl Algebra 22∗309-339
⋄ C75 G30 ⋄ REV MR 82m:18004 Zbl 475 # 18001
• ID 55500

ROSICKY, J. [1981] see POLAK, L.

ROSICKY, J. see Vol. V for further entries

ROSSER, J.B. & WANG, HAO [1950] *Non-standard models for formal logics* (J 0036) J Symb Logic 15∗113-129
⋄ B15 C62 E30 E70 F25 H15 H20 ⋄ REV MR 12.384 Zbl 37.295 JSL 16.145 • ID 11558

ROSSER, J.B. [1955] *Les modeles des logiques formelles* (C 4157) Rosser:Deux Esquisses Log 33-65
⋄ C07 ⋄ REV JSL 22.293 • ID 41189

ROSSER, J.B. see Vol. I, II, IV, V, VI for further entries

ROTA, G.-C. [1978] see ELLERMAN, D.P.

ROTA, G.-C. see Vol. IV, VI for further entries

ROTHMALER, P. [1978] *Total transzendente abelsche Gruppen und Morley-Rang (English and Russian summaries)* (X 2888) ZI Math Mech Akad Wiss DDR: Berlin 45pp
⋄ C35 C45 C60 ⋄ REV MR 80c:03035 Zbl 388 # 03012
• ID 52266

ROTHMALER, P. [1981] Q_0 *is eliminable in every complete theory of modules* (P 2623) Worksh Extended Model Th;1980 Berlin 136-144
⋄ C10 C60 C80 ⋄ REV MR 84h:03078 Zbl 485 # 03012
• ID 33530

ROTHMALER, P. [1981] *A note on I-types* (J 0068) Z Math Logik Grundlagen Math 27∗95-96
⋄ C07 C45 C75 ⋄ REV MR 82d:03058 Zbl 475 # 03015
• ID 55469

ROTHMALER, P. & TUSCHIK, H.-P. [1982] *A two cardinal theorem for homogeneous sets and the elimination of Malitz quantifiers* (J 0064) Trans Amer Math Soc 269∗273-283
⋄ C10 C45 C80 ⋄ REV MR 83e:03055 Zbl 432 # 03017 Zbl 485 # 03011 • ID 53974

ROTHMALER, P. [1983] *Another treatment of the foundations of forking theory* (P 1601) Easter Conf on Model Th (1);1983 Diedrichshagen 110-128
⋄ C45 ⋄ REV MR 84i:03008 Zbl 532 # 03012 • ID 38277

ROTHMALER, P. [1983] *Some model theory of modules. I: On total transcendence of modules. II: On stability and categoricity of flat modules* (J 0036) J Symb Logic 48∗570-574,970-985
⋄ C35 C45 C60 ⋄ REV MR 84j:03075 MR 85e:03084 Zbl 524 # 03018 Zbl 524 # 03019 • REM Part III 1984
• ID 34667

ROTHMALER, P. [1983] *Stationary types in modules* (J 0068) Z Math Logik Grundlagen Math 29∗445-464
⋄ C35 C45 C60 ⋄ REV MR 85e:03086 Zbl 471 # 03031 Zbl 535 # 03011 • ID 38315

ROTHMALER, P. [1983] see BAUDISCH, A.

ROTHMALER, P. [1984] *Some model theory of modules III. On infiniteness of sets definable in modules* (J 0036) J Symb Logic 49∗32-46
⋄ C10 C35 C60 C80 ⋄ REV MR 85e:03085 Zbl 566 # 03020 • REM Parts I,II 1983 • ID 40402

ROTHMALER, P. [1984] see BAUDISCH, A.

ROTMAN, B. [1971] *Remarks on some theorems of Rado on universal graphs* (J 3172) J London Math Soc, Ser 2 4∗123-126
⋄ C50 C65 ⋄ REV MR 46 # 7124 Zbl 223 # 05131
• ID 11588

ROTMAN, B. see Vol. V for further entries

ROUSSEAU, G. [1966] *A decidable class of number theoretic equations* (J 0039) J London Math Soc 41∗737-741
⋄ B20 B25 F30 ⋄ REV MR 34 # 1182 Zbl 166.5
• ID 11593

ROUSSEAU, G. [1968] *Note on the decidability of a certain class of number theoretic equations* (J 0039) J London Math Soc 43∗385-386
⋄ B20 B25 F30 ⋄ REV MR 37 # 1244 Zbl 162.317
• ID 11598

ROUSSEAU, G. see Vol. II, V, VI for further entries

ROUTLEY, R. [1968] *Decision procedures and semantics for* C1, E1, *and* S0.5^0 (J 0079) Logique & Anal, NS 11∗468-471
⋄ B25 B45 ⋄ REV MR 40 # 1267 Zbl 195.297 JSL 38.329
• ID 11610

ROUTLEY, R. [1968] *The decidability and semantical incompleteness of Lemmon's system S0.5* (J 0079) Logique & Anal, NS 11∗413-421
⋄ B25 B45 ⋄ REV MR 38 # 4290 Zbl 172.292 JSL 38.328
• ID 11609

ROUTLEY, R. [1970] *Decision procedures and semantics for Feys' system* S2^0 *and surrounding systems* (J 0068) Z Math Logik Grundlagen Math 16∗165-174
⋄ B25 B45 ⋄ REV MR 43 # 1819 Zbl 206.275 • ID 11613

ROUTLEY, R. [1974] *Semantical analyses of propositional systems of Fitch and Nelson* (J 0063) Studia Logica 33∗283-298
⋄ B25 B46 C90 F50 ⋄ REV MR 51 # 70 Zbl 356 # 02022
• ID 15192

ROUTLEY, R. [1976] see MONTGOMERY, H.

ROUTLEY, R. [1978] *An inadequacy in Kripke-semantics for intuitionistic quantificational logic* (J 0387) Bull Sect Logic, Pol Acad Sci 7∗61-67
⋄ B55 C90 F50 ⋄ REV MR 58 # 16172 Zbl 414 # 03037
• ID 53083

ROUTLEY, R. see Vol. I, II, IV, V for further entries

ROWBOTTOM, F. [1964] *The Los conjecture for uncountable theories* (J 0369) Notices Amer Math Soc 11∗248
⋄ C35 C45 ⋄ ID 49814

ROWBOTTOM, F. see Vol. V for further entries

ROWLANDS-HUGHES, D. [1974] see BACSICH, P.D.

ROWLANDS-HUGHES, D. [1979] *Back and forth methods in abstract model theory* (0000) Diss., Habil. etc 149pp
⋄ C40 C80 C95 ⋄ REM Diss. Oxford University
• ID 48435

ROY, D.K. [1983] *R.e. presented linear orders* (J 0036) J Symb Logic 48*369-376
⋄ C57 C65 D45 E07 ⋄ REV MR 85b:03077 Zbl 525 # 03032 • ID 38265

ROY, D.K. [1985] *Linear order types of nonrecursive presentability* (J 0068) Z Math Logik Grundlagen Math 31*495-501
⋄ C57 C65 D45 ⋄ ID 47798

ROZENBLAT, B.V. [1979] *Positive theories of free inverse semigroups (Russian)* (J 0092) Sib Mat Zh 20*1282-1293,1408
• TRANSL [1979] (J 0475) Sib Math J 20*910-918
⋄ B20 C05 D35 D40 ⋄ REV MR 80m:03079 Zbl 431 # 20045 • ID 53949

ROZENBLAT, B.V. & VAZHENIN, YU.M. [1981] *Decidability of the positive theory of a free countably generated semigroup (Russian)* (J 0142) Mat Sb, Akad Nauk SSSR, NS 116(158)*120-127
• TRANSL [1983] (J 0349) Math of USSR, Sbor 44*109-116
⋄ B25 C05 D35 ⋄ REV MR 83b:03014 Zbl 472 # 20020
• ID 35085

ROZENBLAT, B.V. [1981] *Positive theories of some varieties of semigroups (Russian)* (J 0340) Mat Zap (Univ Sverdlovsk) 12/3*117-132,IV
⋄ B25 C05 ⋄ REV MR 84k:20032 Zbl 521 # 03008
• ID 37064

ROZENBLAT, B.V. & VAZHENIN, YU.M. [1983] *On positive theories of free algebraic systems (Russian)* (J 0031) Izv Vyssh Ucheb Zaved, Mat (Kazan) 1983/3*70-73
• TRANSL [1983] (J 3449) Sov Math 27/3*88-91
⋄ B20 B25 C05 ⋄ REV MR 84j:03028 Zbl 526 # 03010
• ID 34621

ROZHANSKAYA, YU.A. [1956] *On the equivalence of two definitions of completeness of a system of axioms of A. Kolmogorov and H. Weyl (Russian)* (S 1498) Uch Zap Univ, Moskva 181*197-198
⋄ C35 ⋄ REV MR 19.377 • ID 21019

ROZHANSKAYA, YU.A. [1959] *On the equivalence of two definitions of the completeness of a system of axioms (Russian)* (P 0607) All-Union Math Conf (3);1956 Moskva 2*147
⋄ C35 ⋄ ID 28396

RUBEL, L.A. [1977] see HENSON, C.W.

RUBEL, L.A. [1980] see BECKER, J.A.

RUBEL, L.A. [1983] *Some research problems about algebraic differential equations* (J 0064) Trans Amer Math Soc 280*43-52
⋄ C65 E47 ⋄ REV MR 84j:34005 Zbl 532 # 34009
• ID 39167

RUBEL, L.A. [1984] see HENSON, C.W.

RUBEL, L.A. see Vol. I for further entries

RUBIN, J.E. [1973] *The compactness theorem in mathematical logic* (J 0497) Math Mag 46*261-265
⋄ C07 ⋄ REV MR 48 # 5849 Zbl 274 # 02002 • ID 11655

RUBIN, J.E. see Vol. II, V for further entries

RUBIN, M. [1974] *Theories of linear order* (J 0029) Israel J Math 17*392-443
⋄ C15 C35 C50 C65 E07 ⋄ REV MR 50 # 1871 Zbl 304 # 02025 JSL 46.662 • ID 11657

RUBIN, M. [1978] see LOATS, J.T.

RUBIN, M. [1979] *On the automorphism groups of homogeneous and saturated boolean algebras* (J 0004) Algeb Universalis 9*54-86
⋄ C07 C50 C85 G05 ⋄ REV MR 80d:03032 Zbl 424 # 06012 • ID 77988

RUBIN, M. [1980] see MALITZ, J.

RUBIN, M. [1980] *On the automorphism groups of countable boolean algebras* (J 0029) Israel J Math 35*151-170
⋄ C07 C15 C85 E45 G05 ⋄ REV MR 81g:03039 Zbl 447 # 06010 • ID 56583

RUBIN, M. & SHELAH, S. [1980] *On the elementary equivalence of automorphism groups of boolean algebras; downward Skolem Loewenheim theorems and compactness of related quantifiers* (J 0036) J Symb Logic 45*265-283
⋄ C07 C30 C75 C80 C95 G05 ⋄ REV MR 81h:03078 Zbl 445 # 03012 • ID 56476

RUBIN, M. [1980] see DOUWEN VAN, E.K.

RUBIN, M. [1983] *A Boolean algebra with few subalgebras, interval algebras and retractiveness* (J 0064) Trans Amer Math Soc 278*65-89
⋄ C55 C80 D35 E50 E65 G05 ⋄ REV MR 85a:06024 Zbl 524 # 06020 • ID 33531

RUBIN, M. & SHELAH, S. [1983] *On the expressibility hierarchy of Magidor-Malitz quantifiers* (J 0036) J Symb Logic 48*542-557
⋄ C55 C80 E40 E65 ⋄ REV MR 85e:03090 Zbl 537 # 03026 • ID 38745

RUBIN, M. [1985] see ABRAHAM, U.

RUBIN, M. see Vol. IV for further entries

RUDAK, W. [1983] *A completeness theorem for weak equational logic* (J 0004) Algeb Universalis 16*331-337
⋄ C05 C07 ⋄ REV MR 85j:03041 • ID 45997

RUDIN, M.E. [1985] see BESLAGIC, A.

RUDIN, M.E. also published under the name ESTILL, M.E.

RUDIN, M.E. see Vol. I, V for further entries

RUDNIK, K. [1984] *A generalization of the interpolation theorem for the many-sorted calculus* (J 0387) Bull Sect Logic, Pol Acad Sci 13*2-12
⋄ C40 ⋄ REV MR 85i:03030 Zbl 551 # 03018 • ID 42735

RUMELY, R. [1980] *Undecidability and definability for the theory of global fields* (J 0064) Trans Amer Math Soc 262*195-217
⋄ C40 C60 D35 ⋄ REV MR 81m:03053 Zbl 472 # 03010
• ID 55276

RUTKOWSKI, A. [1971] *Some remarks about boolean-valued models (Russian summary)* (J 0014) Bull Acad Pol Sci, Ser Math Astron Phys 19*87-93
⋄ C90 G05 ⋄ REV MR 44 # 6480 Zbl 227 # 02037
• ID 11696

RUTKOWSKI, A. see Vol. I, II, IV, V for further entries

RUYER, H. [1984] *Deux resultats concernant les modules sur un anneau de Dedekind (English summary)* (J 3364) C R Acad Sci, Paris, Ser 1 298*1-3
⋄ C10 C60 C75 ⋄ REV MR 85b:13037 • ID 39360

RUZSA, I. [1969] *Random models of some quasi-intuitionistic logics* (J 0006) Ann Univ Budapest, Sect Math 12*77-93
⋄ B55 C90 ⋄ REV MR 47 # 1603 Zbl 193.302 • ID 11692

RUZSA, I. [1969] *Random models of logical systems I: Models of valuing logics* (J 0411) Studia Sci Math Hung 4*301-312
⋄ B50 C90 ⋄ REV MR 47 # 1600 Zbl 187.269 • REM Part II 1971 • ID 19613

RUZSA, I. [1971] *Random models of logical systems II: Models of some modal logics* (J 0411) Studia Sci Math Hung 5*255-265
⋄ B45 C90 ⋄ REV MR 47 # 1601 Zbl 237 # 02016 • REM Part I 1969. Part III 1971 • ID 19614

RUZSA, I. [1971] *Random models of logical systems III: Models of quantified logics* (J 0049) Period Math Hung 1*195-208
⋄ B50 C90 ⋄ REV MR 47 # 1602 Zbl 221 # 02046 • REM Part II 1971 • ID 11693

RUZSA, I. see Vol. II for further entries

RYATOVA, N.A. [1976] *Conditions for elementary equivalence of modules over discretely normed rings* (P 2064) All-Union Conf Math Log (4);1976 Kishinev 129
⋄ C60 ⋄ ID 90354

RYBAKOV, V.V. [1977] *Noncompact extensions of the logic S4 (Russian)* (J 0003) Algebra i Logika 16*472-490,494
• TRANSL [1977] (J 0069) Algeb and Log 16*321-334
⋄ B45 C90 ⋄ REV MR 58 # 27346 Zbl 406 # 03039
• ID 56122

RYBAKOV, V.V. [1978] *A decidable noncompact extension of the logic S4 (Russian)* (J 0003) Algebra i Logika 17*210-219
• TRANSL [1978] (J 0069) Algeb and Log 17*148-154
⋄ B25 B45 ⋄ REV MR 80m:03042 Zbl 415 # 03012
• ID 53114

RYBAKOV, V.V. [1978] *Modal logics with LM-axioms (Russian)* (J 0003) Algebra i Logika 17*455-467,491
• TRANSL [1978] (J 0069) Algeb and Log 17*302-310
⋄ B25 B45 ⋄ REV MR 80m:03043 Zbl 417 # 03007
• ID 53244

RYBAKOV, V.V. [1984] *Decidability of the admissibility problem in layer-finite modal logics (Russian)* (J 0003) Algebra i Logika 23*100-116,120
• TRANSL [1984] (J 0069) Algeb and Log 23*75-87
⋄ B25 B45 ⋄ REV Zbl 576 # 03012 • ID 39654

RYBAKOV, V.V. see Vol. II for further entries

RYCHKOV, S.V. [1980] *On direct products of abelian groups (Russian)* (J 0023) Dokl Akad Nauk SSSR 252*301-302
• TRANSL [1980] (J 0062) Sov Math, Dokl 21*747-748
⋄ C30 C60 E55 E75 ⋄ REV MR 82b:20075 Zbl 491 # 20043 • ID 82681

RYCHKOV, S.V. [1984] see ANSHAKOV, O.M.

RYCHKOV, S.V. [1985] *The problem of splitting pure extensions of abelian groups and axiomatic set theory (Russian)* (J 0067) Usp Mat Nauk 40/2(242)*195-196
• TRANSL [1985] (J 1399) Russ Math Surv 40/2*230-231
⋄ C60 E35 E45 E50 E75 ⋄ REV Zbl 578 # 20054
• ID 45118

RYCHKOV, S.V. see Vol. II, V for further entries

RYLL-NARDZEWSKI, C. [1958] see GRZEGORCZYK, A.

RYLL-NARDZEWSKI, C. [1959] *On the categoricity in power ≤ ℵ₀ (Russian summary)* (J 0014) Bull Acad Pol Sci, Ser Math Astron Phys 7*545-548
⋄ C35 ⋄ REV MR 22 # 2543 Zbl 117.11 JSL 31.505
• ID 11709

RYLL-NARDZEWSKI, C. [1961] see GRZEGORCZYK, A.

RYLL-NARDZEWSKI, C. [1964] *On Borel measurability of orbits* (J 0027) Fund Math 56*129-130
⋄ C75 E15 E75 ⋄ REV MR 30 # 3028 Zbl 152.214
• ID 11710

RYLL-NARDZEWSKI, C. & WEGLORZ, B. [1967] *Compactness and homomorphisms of algebraic structures* (S 0019) Colloq Math (Warsaw) 18*233-237
⋄ C07 C50 C75 ⋄ REV MR 37 # 129 Zbl 216.307
• ID 11711

RYLL-NARDZEWSKI, C. [1968] see MYCIELSKI, J.

RYLL-NARDZEWSKI, C. [1970] see PACHOLSKI, L.

RYLL-NARDZEWSKI, C. see Vol. I, IV, V for further entries

SAADE, M. [1968] *A decision problem for finite groupoids (Russian summary)* (J 0014) Bull Acad Pol Sci, Ser Math Astron Phys 16*819-820
⋄ B25 C13 ⋄ REV MR 38 # 4594 Zbl 167.15 • ID 11713

SAADE, M. [1969] *A comment on a paper by Evans* (J 0068) Z Math Logik Grundlagen Math 15*97-100
⋄ C05 C13 ⋄ REV MR 39 # 4309 Zbl 199.328 • ID 11714

SABBAGH, G. [1968] *Un theoreme de plongement en algebre* (J 0247) Bull Sci Math, Ser 2 92*49-52
⋄ C60 ⋄ REV MR 38 # 136 Zbl 159.338 • ID 11722

SABBAGH, G. [1969] *How not to characterize the multiplicative groups of fields* (J 3172) J London Math Soc, Ser 2 1*369-370
⋄ C60 ⋄ REV MR 39 # 5671 Zbl 184.13 • ID 11725

SABBAGH, G. [1970] *Aspects logiques de la purete dans les modules* (J 2313) C R Acad Sci, Paris, Ser A-B 271*A909-A912
⋄ C20 C30 C60 ⋄ REV MR 43 # 269 Zbl 202.9
• ID 11724

SABBAGH, G. [1970] see EKLOF, P.C.

SABBAGH, G. [1971] *A note on the embedding property* (J 0044) Math Z 121*239-242
⋄ C20 C52 ⋄ REV MR 45 # 45 Zbl 215.61 • ID 11728

SABBAGH, G. [1971] see EKLOF, P.C.

SABBAGH, G. [1971] *Embedding problems for modules and rings with application to model-companions* (J 0032) J Algeb 18*390-403
⋄ C25 C35 C60 ⋄ REV MR 43 # 6259 Zbl 218 # 02049
• ID 11726

SABBAGH, G. [1971] *Logique mathematique III. Theorie des modeles* (C 1495) Encycl Universalis 10*65-66
⋄ C98 ⋄ REV JSL 38.341 • REM Part II 1971 by Reznikoff,I. Part IV 1971 • ID 28608

SABBAGH, G. [1971] *Logique mathematique V. Decidabilite et fonctions recursives* (C 1495) Encycl Universalis 10*71-73
⋄ B25 D20 D35 ⋄ REV JSL 38.341 • REM Part IV 1971
• ID 28610

SABBAGH, G. [1971] *On properties of countable character* (J 0016) Bull Austral Math Soc 4*183-192
⋄ C20 C50 C52 ⋄ REV MR 43 # 7392 Zbl 209.321
• ID 11727

SABBAGH, G. [1971] *Sous-modules purs, existentiellement clos et elementaires* (J 2313) C R Acad Sci, Paris, Ser A-B 272*A1289-A1292
◇ C25 C35 C60 ◇ REV MR 46#3296 Zbl 215.325
• ID 11729

SABBAGH, G. [1972] *The epimorphisms of the ∈-relation* (J 0068) Z Math Logik Grundlagen Math 18*289-290
◇ C62 E20 E35 G30 ◇ REV MR 46#3306 Zbl 257#02059 • ID 11730

SABBAGH, G. [1975] *Categoricite en $\aleph_0$ et stabilite: constructions les preservant et conditions de chaine (English summary)* (J 2313) C R Acad Sci, Paris, Ser A-B 280*A531-A533
◇ C35 C45 C60 ◇ REV MR 51#7848 Zbl 327#02044
• ID 18367

SABBAGH, G. [1975] *Categoricite et stabilite: quelques examples parmi les groupes et anneaux (English summary)* (J 2313) C R Acad Sci, Paris, Ser A-B 280*A603-A606
◇ C35 C45 C60 ◇ REV MR 51#7849 Zbl 327#02045
• ID 18368

SABBAGH, G. [1975] *Sur les groupes qui ne sont pas reunion d'une suite croissante de sous-groupes propres (English summary)* (J 2313) C R Acad Sci, Paris, Ser A-B 280*A763-A766
◇ C25 C60 ◇ REV MR 51#3270 Zbl 315#20032
• ID 17367

SABBAGH, G. [1984] see DELON, F.

SABBAGH, G. [1985] see DELON, F.

SABBAGH, G. see Vol. I, II, IV, V for further entries

SACERDOTE, G.S. [1972] *On a problem of Boone* (J 0132) Math Scand 31*111-117
◇ C60 D30 D35 D40 ◇ REV MR 47#6871 Zbl 255#20022 • ID 30879

SACERDOTE, G.S. [1973] *Almost all free products of groups have the same positive theory* (J 0032) J Algeb 27*475-485
◇ C60 ◇ REV MR 48#11323 Zbl 284#02028 • ID 11737

SACERDOTE, G.S. [1973] *Elementary properties of free groups* (J 0064) Trans Amer Math Soc 178*127-138
◇ C60 ◇ REV MR 47#8686 Zbl 268#02037 • ID 11736

SACERDOTE, G.S. [1974] *Projective model completeness* (J 0036) J Symb Logic 39*117-123
◇ C35 ◇ REV MR 49#8854 Zbl 286#02056 • ID 11738

SACERDOTE, G.S. [1975] *Infinite coforcing in model theory* (J 0345) Adv Math 17*261-280
◇ C25 C30 C60 ◇ REV MR 57#9528 Zbl 314#02062
• ID 65001

SACERDOTE, G.S. [1975] *Projective model theory and coforcing* (C 0782) Model Th & Algeb (A. Robinson) 276-306
◇ C25 C30 C35 C40 C50 C60 ◇ REV MR 53#10580 Zbl 333#02041 • ID 23085

SACERDOTE, G.S. [1976] *A characterization of the subgroups of finitely presented groups* (J 0015) Bull Amer Math Soc 82*609-611
◇ C60 D40 ◇ REV MR 54#2814 Zbl 377#20028
• ID 24067

SACERDOTE, G.S. [1976] see FRIED, M.

SACERDOTE, G.S. [1976] *Some logical problems concerning free and free product groups* (J 0308) Rocky Mountain J Math 6*401-408
◇ C60 C98 ◇ REV MR 54#4958 Zbl 363#02061
• ID 24126

SACERDOTE, G.S. see Vol. IV for further entries

SACKS, G.E. [1972] *Saturated model theory* (X 0867) Benjamin: Reading xii+335pp
• TRANSL [1976] (X 0885) Mir: Moskva 190pp
◇ C98 ◇ REV MR 53#2668 Zbl 242#02054 JSL 40.637
• ID 21615

SACKS, G.E. [1972] *The differential closure of a differential field* (J 0015) Bull Amer Math Soc 78*629-634
◇ C60 C98 ◇ REV MR 45#8514 Zbl 276#02039
• ID 11756

SACKS, G.E. [1978] see FENSTAD, J.E.

SACKS, G.E. [1979] *Effective bounds on Morley rank* (J 0027) Fund Math 103*111-121
◇ C15 C35 C45 C70 ◇ REV MR 81a:03035 Zbl 437#03022 • ID 55890

SACKS, G.E. [1983] *On the number of countable models* (P 3669) SE Asian Conf on Log;1981 Singapore 185-196
◇ C15 C70 C75 ◇ REV MR 85i:03095 Zbl 563#03015
• ID 43058

SACKS, G.E. [1985] *Some open questions in recursion theory* (P 3342) Rec Th Week;1984 Oberwolfach 333-342
◇ C13 D98 E60 ◇ ID 42739

SACKS, G.E. see Vol. IV, V for further entries

SADOVSKIJ, V.N. & SMIRNOV, V.A. [1977] *Definability and identifiability: certain problems and hypotheses* (P 1704) Int Congr Log, Meth & Phil of Sci (5);1975 London ON 3*63-80
◇ C40 ◇ REV MR 58#21568 • ID 78069

SAELI, D. [1975] *Problemi di decisione per algebre connesse a logiche a piu valori (English summary)* (J 0149) Atti Accad Naz Lincei Fis Mat Nat, Ser 8 59*219-223
◇ B25 B50 C60 G20 ◇ REV MR 57#5722 Zbl 354#02037 • ID 78070

SAELI, D. [1976] see LACAVA, F.

SAELI, D. [1977] see LACAVA, F.

SAFFE, J. [1981] *Einige Ergebnisse ueber die Anzahl abzaehlbarer Modelle superstabiler Theorien (Dissertation)* (X 3158) Univ Hannover Fak Math Nat: Hannover 55pp
◇ C15 C45 ◇ REV Zbl 475#03008 • ID 55462

SAFFE, J. [1982] *A superstable theory with the dimensional order property has many models* (P 3708) Herbrand Symp Logic Colloq;1981 Marseille 281-286
◇ C45 ◇ REV MR 85i:03107 Zbl 499#03014 • ID 38119

SAFFE, J. [1983] *The number of uncountable models of ω-stable theories* (J 0073) Ann Pure Appl Logic 24*231-261
◇ C45 C50 ◇ REV MR 85j:03044 Zbl 518#03010
• ID 37514

SAFFE, J. [1984] *Categoricity and ranks* (J 0036) J Symb Logic 49*1379-1392
◇ C35 C45 ◇ REV MR 86e:03034 Zbl 578#03015
• ID 39884

SAFFE, J. [1985] see PALYUTIN, E.A.

SAGEEV, G. [1981] *A model of ZF + there exists an inaccessible, in which the Dedekind cardinals constitute a natural nonstandard model of arithmetic* (J 0007) Ann Math Logic 21*221-281
⋄ E25 E35 H15 ⋄ REV MR 83h:03074 Zbl 523 # 03038
• ID 36078

SAGEEV, G. & SHELAH, S. [1985] *On the structure of Ext$(A, \mathbb{Z})$ in ZFC^+* (J 0036) J Symb Logic 50*302-315
⋄ C60 E05 E45 E50 E55 E65 E75 ⋄ ID 41812

SAGEEV, G. see Vol. V for further entries

SAIN, I. [1979] see ANDREKA, H.

SAIN, I. [1979] *There are general rules for specifying semantics: Observations of abstract model theory* (J 2293) Comp Linguist & Comp Lang 13*195-250
⋄ B75 C90 C95 E75 ⋄ REV MR 81d:03040 Zbl 449 # 68007 • ID 56752

SAIN, I. [1982] see ANDREKA, H.

SAIN, I. [1982] see NEMETI, I.

SAIN, I. [1982] *On classes of algebraic systems closed with respect to quotients* (P 3831) Universal Algeb & Appl;1978 Warsaw 127-131
⋄ C05 C20 C30 C52 ⋄ REV MR 85d:00024 Zbl 509 # 08005 • ID 45243

SAIN, I. [1982] *Weak products for universal algebra and model theory* (J 3797) Diagrammes 8*S1-S15
⋄ C05 C30 ⋄ REV MR 86c:08009 Zbl 525 # 08002
• ID 44699

SAIN, I. [1983] *Applicability of category-theoretic notions of ultraproducts (Hungarian)* (J 0396) Mat Lapok 31*143-167
⋄ C20 G30 ⋄ REV MR 85j:03116 Zbl 553 # 18001
• ID 43417

SAIN, I. [1983] see HIEN, BUIHUY

SAIN, I. see Vol. I, II for further entries

SAITO, M. [1976] *Ultraproducts and nonstandard analysis (Japanese)* (X 3636) Tokyo Tosho: Tokyo vi + 160pp
⋄ C20 H05 ⋄ ID 33414

SAITO, M. see Vol. I, V for further entries

SAKAROVITCH, JOEL [1977] *Systemes normaux d'ensembles de conditions de "forcing" et applications aux suites decroissantes de modeles de ZFC (English summary)* (J 2313) C R Acad Sci, Paris, Ser A-B 285*A589-A592
⋄ C62 E40 ⋄ REV MR 56 # 5289 Zbl 372 # 02036
• ID 27314

SAKAROVITCH, JOEL see Vol. V for further entries

SALIBRA, A. [1984] see MANCA, V.

SALIJ, V.N. [1973] *Boolean valued algebras (Russian)* (J 0142) Mat Sb, Akad Nauk SSSR, NS 92(134)*550-563,647
• TRANSL [1973] (J 0349) Math of USSR, Sbor 21*544-557
⋄ C05 C90 G05 ⋄ REV MR 49 # 2345 Zbl 286 # 08006
• ID 11775

SALIJ, V.N. [1974] *Reduced extensions of models (Russian)* (S 0778) Uporyad Mnozhestva Reshetki (Saratov) 2*85-90
⋄ C20 C30 ⋄ REV MR 58 # 205 Zbl 299 # 02065
• ID 65030

SALIJ, V.N. see Vol. II, V for further entries

SAMI, R.L. [1982] *Sur la conjecture de Vaught en theorie descriptive (English summary)* (J 3364) C R Acad Sci, Paris, Ser 1 295*413-416
⋄ C15 E15 ⋄ REV MR 84a:03055 Zbl 509 # 03026
• ID 35583

SAMI, R.L. see Vol. IV, V for further entries

SANCHIS, L.E. [1973] *Formally defined operations in Kripke models* (J 0047) Notre Dame J Formal Log 14*467-480
⋄ C90 F50 ⋄ REV MR 48 # 10770 Zbl 226 # 02024
• ID 11828

SANCHIS, L.E. see Vol. I, IV, VI for further entries

SANERIB JR., R.A. [1975] *Ultraproducts and elementary types of some groups related to infinite symmetric groups* (J 0004) Algeb Universalis 5*24-38
⋄ C20 C60 ⋄ REV MR 52 # 2872 Zbl 333 # 20027
• ID 17657

SANERIB JR., R.A. see Vol. V for further entries

SANKAPPANAVAR, H.P. [1975] see BURRIS, S.

SANKAPPANAVAR, H.P. [1978] *Decision problems: History and methods* (P 1800) Brazil Conf Math Log (1);1977 Campinas 241-291
⋄ A10 B25 D35 D98 ⋄ REV MR 81a:03011 Zbl 385 # 03011 • ID 52119

SANKAPPANAVAR, H.P. [1981] see BURRIS, S.

SANKAPPANAVAR, H.P. see Vol. IV for further entries

SANMARTIN ESPLUGUES, J. [1972] *Standard models of axiomatic systems of sets I (Spanish)* (J 0162) Teorema (Valencia) 1972/5*55-60
⋄ C62 E35 ⋄ REV MR 47 # 6485 • ID 78126

SANMARTIN ESPLUGUES, J. [1973] *Syllogistics, many-valued logic and model theory (Spanish)* (J 0162) Teorema (Valencia) 3*355-365
⋄ A10 B20 B50 C90 ⋄ REV MR 50 # 9522 • ID 78125

SAPIR, M.V. [1980] *On finite and independent axiomatizability of certain quasi-varieties of semigroups (Russian)* (J 0031) Izv Vyssh Ucheb Zaved, Mat (Kazan) 1980/2*76-78
• TRANSL [1980] (J 3449) Sov Math 24/2*83-87
⋄ C05 C52 ⋄ REV MR 81h:20076 Zbl 429 # 08003
• ID 82707

SARACINO, D.H. [1973] *Model companions for $\aleph_0$-categorial theories* (J 0053) Proc Amer Math Soc 39*591-598
⋄ C25 C35 C60 ⋄ REV MR 47 # 4786 Zbl 272 # 02075
• ID 11845

SARACINO, D.H. [1973] see LIPSHITZ, L.

SARACINO, D.H. [1974] *m-existentially complete structures* (S 0019) Colloq Math (Warsaw) 30*7-13
⋄ C25 ⋄ REV MR 49 # 10542 Zbl 298 # 02054 • ID 11847

SARACINO, D.H. [1974] *Wreath products and existentially complete solvable groups* (J 0064) Trans Amer Math Soc 197*327-339
⋄ C25 C35 C60 ⋄ REV MR 49 # 7137 Zbl 311 # 02065
• ID 11846

SARACINO, D.H. [1975] *A counterexample in the theory of model companions* (J 0036) J Symb Logic 40*31-34
⋄ C25 C35 ⋄ REV MR 56 # 15408 Zbl 308 # 02056
• ID 11848

SARACINO, D.H. & WEISPFENNING, V. [1975] *Commutative regular rings without prime model extensions* (J 0053) Proc Amer Math Soc 47*201-207
⋄ C25 C35 C60 ⋄ REV MR 50 # 4293 Zbl 308 # 02057
• ID 11849

SARACINO, D.H. & WEISPFENNING, V. (EDS.) [1975] *Model theory and algebra. A memorial tribute to Abraham Robinson* (S 3301) Lect Notes Math 498*x+436pp
⋄ C60 C97 ⋄ REV MR 52 # 7889 Zbl 307 # 00005
• ID 70184

SARACINO, D.H. & WEISPFENNING, V. [1975] *On algebraic curves over commutative regular rings* (C 0782) Model Th & Algeb (A. Robinson) 307-383
⋄ C10 C25 C35 C60 ⋄ REV MR 53 # 5606
Zbl 318 # 13032 • ID 22958

SARACINO, D.H. [1976] *Existentially complete nilpotent groups* (J 0029) Israel J Math 25*241-248
⋄ C20 C25 C35 C60 ⋄ REV MR 56 # 11780
Zbl 347 # 02034 • ID 26085

SARACINO, D.H. [1978] *Existentially complete torsion-free nilpotent groups* (J 0036) J Symb Logic 43*126-134
⋄ C25 C60 ⋄ REV MR 58 # 212 Zbl 397 # 03018
• ID 29249

SARACINO, D.H. & WOOD, C. [1979] *Periodic existentially closed nilpotent groups* (J 0032) J Algeb 58*189-207
⋄ C25 C35 C45 C60 ⋄ REV MR 80m:03071 • ID 78134

SARACINO, D.H. & WOOD, C. [1982] *QE nil-2 groups of exponent 4* (J 0032) J Algeb 76*337-352
⋄ C10 C15 C60 ⋄ REV MR 83i:03052 Zbl 504 # 03012
• ID 35518

SARACINO, D.H. [1983] *Amalgamation bases for nil-2 groups* (J 0004) Algeb Universalis 16*47-62
⋄ C60 ⋄ REV MR 84i:20035 Zbl 517 # 20016 • ID 36694

SARACINO, D.H. & WOOD, C. [1983] *Finitely generic abelian lattice-ordered groups* (J 0064) Trans Amer Math Soc 277*113-123
⋄ C25 C60 ⋄ REV MR 85g:03054 Zbl 522 # 06016
• ID 37049

SARACINO, D.H. & WOOD, C. [1984] *An example in the model theory of abelian lattice-ordered groups* (J 0004) Algeb Universalis 19*34-37
⋄ C25 C60 G10 ⋄ REV MR 85k:06014 Zbl 549 # 06011
• ID 43151

SARACINO, D.H. & WOOD, C. [1984] *Nonexistence of a universal countable commutative ring* (J 0394) Commun Algeb 12*1171-1173
⋄ C50 C60 ⋄ REV MR 86d:13021 Zbl 542 # 13002
• ID 48436

SARACINO, D.H. & WOOD, C. [1984] *QE commutative nilrings* (J 0036) J Symb Logic 49*644-651
⋄ C10 C50 C60 ⋄ REV MR 85i:03093 JSL 50.1080
• ID 39894

SARACINO, D.H. & WOOD, C. [1985] *Finite QE rings in characteristic p^2* (J 0073) Ann Pure Appl Logic 28*13-31
⋄ C13 C50 C60 ⋄ REV MR 86g:03061 Zbl 557 # 03019
• ID 39906

SARALSKI, B. [1975] see KOPIEKI, R.

SATO, M. [1977] *A study of Kripke-type models for some modal logics by Gentzen's sequential method* (J 0390) Publ Res Inst Math Sci (Kyoto) 13*381-468
⋄ B45 C90 ⋄ REV MR 57 # 2884 Zbl 405 # 03013
• ID 30800

SATO, M. see Vol. I, II, VI for further entries

SAUER, N. & STONE, MICHAEL G. [1977] *The algebraic closure of a semigroup of functions* (J 0004) Algeb Universalis 7*219-233
⋄ C05 ⋄ ID 49901

SAUER, N. see Vol. V for further entries

SAWAMURA, H. [1985] *Axiomatization of computer-oriented modal logic and decision procedure* (J 3957) Bull Inf & Cybern (Kyushu Univ) 21/3-4*57-66
⋄ B25 B45 ⋄ REV Zbl 575 # 03011 • ID 47625

SAXL, J. [1976] see BALDWIN, JOHN T.

SAYEKI, H. [1968] *Some consequences of the normality of the space of models* (J 0027) Fund Math 61*243-251
⋄ C07 C40 C52 ⋄ REV MR 36 # 4975 Zbl 217.13
• ID 28047

SAYWARD, C.W. [1982] see HUGLY, P.

SAYWARD, C.W. [1984] see HUGLY, P.

SAYWARD, C.W. see Vol. I, II for further entries

SCARPELLINI, B. [1963] *Eine Anwendung der unendlichwertigen Logik auf topologische Raeume* (J 0027) Fund Math 52*129-150 • ERR/ADD ibid 53*345
⋄ B80 C65 ⋄ REV MR 34 # 8378a Zbl 149.193 • ID 11875

SCARPELLINI, B. [1966] *On a family of models of Zermelo-Fraenkel set theory* (J 0068) Z Math Logik Grundlagen Math 12*191-204
⋄ C62 E30 E35 E45 ⋄ REV MR 34 # 7362 Zbl 147.257
JSL 34.654 • ID 11880

SCARPELLINI, B. [1969] *On the metamathematics of rings and integral domains* (J 0064) Trans Amer Math Soc 138*71-96
⋄ C60 ⋄ REV MR 39 # 1295 Zbl 181.301 • ID 11882

SCARPELLINI, B. [1984] *Complete second order spectra* (J 0068) Z Math Logik Grundlagen Math 30*509-524
⋄ B15 C13 ⋄ REV MR 86f:03059 Zbl 564 # 03007
• ID 39663

SCARPELLINI, B. [1984] *Complexity of subcases of Presburger arithmetic* (J 0064) Trans Amer Math Soc 284*203-218
⋄ B25 D15 F20 F30 ⋄ REV MR 86c:03010
Zbl 548 # 03018 • ID 43196

SCARPELLINI, B. [1984] *Second order spectra* (P 2342) Symp Rek Kombin;1983 Muenster 171*380-389
⋄ B15 C13 ⋄ REV MR 85k:68004 Zbl 548 # 03004
• ID 41786

SCARPELLINI, B. [1985] *Lower bound results on lengths of second-order formulas* (J 0073) Ann Pure Appl Logic 29*29-58
⋄ B15 C13 D10 F20 F35 ⋄ ID 47473

SCARPELLINI, B. see Vol. II, IV, V, VI for further entries

SCEDROV, A. [1983] see BLASS, A.R.

SCEDROV, A. [1984] *Forcing and classifying topoi* (S 0167) Mem Amer Math Soc 295*93pp
⋄ C90 E35 E40 F50 G30 ⋄ REV MR 86d:03057
Zbl 536 # 03048 JSL 50.852 • ID 37128

SCEDROV, A. [1985] see HARRINGTON, L.A.

SCEDROV, A. see Vol. I, II, IV, V, VI for further entries

SCHANUEL, S.H. [1977] see ISBELL, J.R.

SCHECHTMAN, G. [1979] see BOURGAIN, J.

SCHECHTMAN, G. see Vol. V for further entries

SCHEIN, B.M. [1965] *Relation algebras* (J 0014) Bull Acad Pol Sci, Ser Math Astron Phys 13∗1-5
⋄ C05 C20 C52 ⋄ REV MR 31 # 2188 • ID 12865

SCHEIN, B.M. [1971] *Linearly fundamentally ordered semigroups* (S 0019) Colloq Math (Warsaw) 23∗17-23
⋄ C05 C60 E20 ⋄ REV MR 46 # 3398 Zbl 221 # 06004
• ID 28117

SCHEIN, B.M. [1976] *Injective commutative semigroups* (J 0004) Algeb Universalis 6∗395-397
⋄ C05 C60 ⋄ REV MR 55 # 8219 • ID 23927

SCHEIN, B.M. see Vol. IV, V for further entries

SCHIEMANN, I. [1977] *Eine Axiomatisierung des monadischen Praedikatenkalkuels mit verallgemeinerten Quantoren (Russian, English and French summaries)* (J 0115) Wiss Z Humboldt-Univ Berlin, Math-Nat Reihe 26∗647-657
⋄ B25 C40 C55 C80 ⋄ REV MR 80b:03045
Zbl 423 # 03006 • ID 53518

SCHLIPF, J.S. [1975] see BARWISE, J.

SCHLIPF, J.S. [1975] *Some hyperelementary aspects of model theory* (0000) Diss., Habil. etc 165pp
⋄ C50 C62 C70 D55 ⋄ REM Ph.D. thesis, University of Wisconsin, Madison, WI, USA • ID 19103

SCHLIPF, J.S. [1976] see BARWISE, J.

SCHLIPF, J.S. [1977] *A guide to the identification of admissible sets above structures* (J 0007) Ann Math Logic 12∗151-192
⋄ C15 C50 C70 ⋄ REV MR 58 # 5177 Zbl 374 # 02031
• ID 23675

SCHLIPF, J.S. [1977] *Ordinal spectra of first-order theories* (J 0036) J Symb Logic 42∗492-505
⋄ C35 C50 C55 C62 C70 D15 D30 D70 ⋄ REV
MR 58 # 27446 Zbl 411 # 03035 • ID 26851

SCHLIPF, J.S. [1978] *Toward model theory through recursive saturation* (J 0036) J Symb Logic 43∗183-206
⋄ C15 C40 C50 C57 C70 ⋄ REV MR 58 # 10399
Zbl 409 # 03019 • ID 29252

SCHLIPF, J.S. [1980] *Recursively saturated models of set theory* (J 0053) Proc Amer Math Soc 80∗135-142
⋄ C50 C57 C62 E70 H20 ⋄ REV MR 81h:03101
Zbl 455 # 03022 • ID 54283

SCHLIPF, J.S. see Vol. IV for further entries

SCHLOESSER, B. [1981] see RICHTER, M.M.

SCHMERL, J.H. [1972] *An elementary sentence which has ordered models* (J 0036) J Symb Logic 37∗521-530
⋄ C55 E55 ⋄ REV MR 55 # 96 Zbl 278 # 02041 • ID 12875

SCHMERL, J.H. & SHELAH, S. [1972] *On power-like models for hyperinaccessible cardinals* (J 0036) J Symb Logic 37∗531-537
⋄ C30 C55 E55 ⋄ REV MR 47 # 6474 Zbl 278 # 02042
• ID 12876

SCHMERL, J.H. [1973] *Peano models with many generic classes* (J 0048) Pac J Math 46∗523-536 • ERR/ADD ibid 92∗195-198
⋄ C55 C62 E40 ⋄ REV MR 50 # 6831 MR 83b:03076
Zbl 266 # 02031 Zbl 461 # 03005 • ID 12878

SCHMERL, J.H. [1974] *Generalizing special Aronszajn trees* (J 0036) J Symb Logic 39∗732-740
⋄ C55 E05 E35 E55 ⋄ REV MR 52 # 7905
Zbl 302 # 02029 • ID 12879

SCHMERL, J.H. [1975] *The number of equivalence classes of existentially complete structures* (C 0782) Model Th & Algeb (A. Robinson) 170-171
⋄ C25 ⋄ REV MR 54 # 2450 Zbl 319 # 02047 • ID 24038

SCHMERL, J.H. [1976] *Effectiveness and Vaught's gap ω two-cardinal theorem* (J 0053) Proc Amer Math Soc 58∗237-240
⋄ C55 C57 ⋄ REV MR 55 # 5433 Zbl 357 # 02049
• ID 50397

SCHMERL, J.H. [1976] *On κ-like structures which embed stationary and closed unbounded subsets* (J 0007) Ann Math Logic 10∗289-314
⋄ C55 E05 E55 ⋄ REV MR 55 # 7763 Zbl 357 # 02050
• ID 23655

SCHMERL, J.H. [1976] *Remarks on self-extending models* (J 0068) Z Math Logik Grundlagen Math 22∗509-512
⋄ C30 C55 E55 ⋄ REV MR 56 # 5267 Zbl 355 # 02035
• ID 50233

SCHMERL, J.H. [1976] *The decidability of some $\aleph_0$-categorical theories* (S 0019) Colloq Math (Warsaw) 36∗165-169
⋄ B25 C35 ⋄ REV MR 56 # 113 Zbl 362 # 02035
• ID 50757

SCHMERL, J.H. [1977] *$\aleph_0$-categoricity of partially ordered sets of width 2* (J 0053) Proc Amer Math Soc 63∗299-305
⋄ C35 C65 ⋄ REV MR 55 # 12501 Zbl 357 # 02047
• ID 50395

SCHMERL, J.H. [1977] *An axiomatization for a class of two-cardinal models* (J 0036) J Symb Logic 42∗174-178
⋄ C55 ⋄ REV MR 58 # 203 Zbl 374 # 02029 • ID 26440

SCHMERL, J.H. [1977] *On $\aleph_0$-categoricity and the theory of trees* (J 0027) Fund Math 94∗121-128
⋄ B25 C35 C65 ⋄ REV MR 55 # 10269 Zbl 352 # 02036
• ID 26516

SCHMERL, J.H. [1977] *Theories with only a finite number of existentially complete models* (J 0029) Israel J Math 28∗350-356
⋄ C25 C52 ⋄ REV MR 57 # 16052 Zbl 369 # 02026
• ID 27205

SCHMERL, J.H. [1978] *$\aleph_0$-categoricity and comparability graphs* (P 1897) Logic Colloq;1977 Wroclaw 221-228
⋄ C15 C35 C65 ⋄ REV MR 83i:03054 Zbl 441 # 03006
• ID 56059

SCHMERL, J.H. [1978] *A decidable $\aleph_0$-categorical theory with a non-recursive Ryll-Nardzewski function* (J 0027) Fund Math 98∗121-125
⋄ B25 C35 C57 D30 ⋄ REV MR 80g:03031
Zbl 372 # 02025 • ID 29211

SCHMERL, J.H. [1978] *Extending models of arithmetic* (J 0007) Ann Math Logic 14∗89-109
⋄ C62 ⋄ REV MR 80f:03036 Zbl 389 # 03028 • ID 29148

SCHMERL, J.H. [1978] *On $\aleph_0$-categoricity of filtered boolean extensions* (J 0004) Algeb Universalis 8∗159-161
⋄ C30 C35 ⋄ REV MR 57 # 9529 Zbl 426 # 03034
• ID 53630

SCHMERL, J.H. [1979] *Countable homogeneous partially ordered sets* (J 0004) Algeb Universalis 9∗317-321
⋄ C15 C50 C65 E07 ⋄ REV MR 81g:06001
Zbl 423 # 06002 • ID 69898

SCHMERL, J.H. [1979] see LERMAN, M.

SCHMERL, J.H. [1980] *Decidability and $\aleph_0$-categoricity of theories of partially ordered sets* (J 0036) J Symb Logic 45∗585-611
⋄ B25 C15 C35 C65 D35 ⋄ REV MR 84g:03037
Zbl 441 # 03007 • ID 56060

SCHMERL, J.H. [1981] *Arborescent structures I: Recursive models* (P 2902) Aspects Effective Algeb;1979 Clayton 226-231
⋄ C50 C57 C65 ⋄ REV MR 83g:03033a Zbl 475 # 03013
• REM Part II 1981 • ID 55467

SCHMERL, J.H. [1981] *Arborescent structures II: Interpretability in the theory of trees* (J 0064) Trans Amer Math Soc 266∗629-643
⋄ B25 C50 C65 C75 ⋄ REV MR 83g:03033b
Zbl 475 # 03014 • REM Part I 1981 • ID 55468

SCHMERL, J.H. [1981] *Decidability and finite axiomatizability of theories of $\aleph_0$-categorical partially ordered sets* (J 0036) J Symb Logic 46∗101-120
⋄ B25 C35 C65 ⋄ REV MR 82h:03024 Zbl 475 # 03010
• ID 55464

SCHMERL, J.H. [1981] see LERMAN, M.

SCHMERL, J.H. [1981] *Recursively saturated, rather classless models of Peano arithmetic* (P 2628) Log Year;1979/80 Storrs 268-282
⋄ C50 C57 C62 E45 E55 ⋄ REV MR 83b:03039
Zbl 469 # 03051 • ID 55178

SCHMERL, J.H. & SIMPSON, S.G. [1982] *On the role of Ramsey quantifiers in first order arithmetic* (J 0036) J Symb Logic 47∗423-435
⋄ B28 C10 C62 C80 ⋄ REV MR 83j:03062
Zbl 492 # 03015 JSL 50.1078 • ID 35360

SCHMERL, J.H. [1983] *$\aleph_0$-categorical distributive lattices of finite breadth* (J 0053) Proc Amer Math Soc 87∗707-713
⋄ C35 G10 ⋄ REV MR 85e:03073 Zbl 522 # 03020
• ID 37783

SCHMERL, J.II. [1984] *$\aleph_0$-categorical partially ordered sets (French summary)* (P 2167) Orders: Descr & Roles;1982 L'Arbresle 269-285
⋄ B25 C15 C35 C65 C98 E07 ⋄ REV MR 86h:03054
Zbl 551 # 03017 • ID 41781

SCHMERL, J.H. [1984] see KAUFMANN, M.

SCHMERL, J.H. [1985] *Models of Peano arithmetic and a question of Sikorski on ordered fields* (J 0029) Israel J Math 50∗145-159
⋄ C55 C60 C62 E05 E75 ⋄ REV Zbl 577 # 03016
• ID 45602

SCHMERL, J.H. [1985] *Recursively saturated models generated by indiscernibles* (J 0047) Notre Dame J Formal Log 26∗99-105
⋄ C30 C50 C57 ⋄ REV Zbl 556 # 03030 • ID 42589

SCHMERL, J.H. [1985] *Transfer theorems and their applications to logic* (C 4183) Model-Theor Log 177-209
⋄ C55 C80 C98 • ID 48338

SCHMERL, J.H. see Vol. IV, V for further entries

SCHMERL, U.R. [1984] *Diophantine equations in a fragment of number theory* (P 2153) Logic Colloq;1983 Aachen 2∗389-398
⋄ C62 F30 ⋄ REV MR 86g:03096 • ID 43011

SCHMERL, U.R. see Vol. IV, VI for further entries

SCHMID, J. [1979] *Algebraically and existentially closed distributive lattices* (J 0068) Z Math Logik Grundlagen Math 25∗525-530
⋄ C05 C25 G10 ⋄ REV MR 80j:03046 Zbl 425 # 06007
• ID 53585

SCHMID, J. [1982] *Algebraically closed distributive p-algebras* (J 0004) Algeb Universalis 15∗126-141
⋄ C05 C25 G05 G10 G20 ⋄ REV MR 83i:06012
Zbl 507 # 06009 • ID 39323

SCHMID, J. [1982] *Model companions of distributive p-algebras* (J 0036) J Symb Logic 47∗680-688
⋄ C05 C25 C35 G10 ⋄ REV MR 83i:03057
Zbl 494 # 03023 • ID 35520

SCHMID, J. [1985] *Algebraically closed p-semilattices* (J 0008) Arch Math (Basel) 45∗501-510
⋄ C25 G10 ⋄ REV Zbl 572 # 06002 • ID 48437

SCHMID, J. see Vol. I for further entries

SCHMIDT, E.T. [1963] *Universale Algebren mit gegebenen Automorphismengruppen und Unteralgebrenverbaenden* (J 0002) Acta Sci Math (Szeged) 24∗251-254
⋄ C05 ⋄ REV MR 28 # 2987 Zbl 113.249 • ID 12894

SCHMIDT, E.T. [1964] *Universale Algebren mit gegebenen Automorphismengruppen und Kongruenzverbaenden* (J 0001) Acta Math Acad Sci Hung 15∗37-45
⋄ C05 ⋄ REV MR 29 # 3411 Zbl 113.249 • ID 12895

SCHMIDT, H.A. [1958] *Un procede maniable de decision pour la logique propositionnelle intuitionniste* (P 0576) Raisonn en Math & Sci Exper;1955 Paris 57-65
⋄ B25 F50 ⋄ REV MR 21 # 2587 Zbl 88.248 JSL 25.286
• ID 12891

SCHMIDT, H.A. see Vol. I, II, VI for further entries

SCHMIDT, J. [1961] *Algebraic operations and algebraic independence in algebras with infinitary operations* (J 0352) Math Jap 6∗77-112
⋄ C05 ⋄ REV MR 28 # 54 Zbl 119.259 • ID 12913

SCHMIDT, J. [1964] see ARMBRUST, M.

SCHMIDT, J. [1968] *Universelle Halbgruppe, Kategorien, freies Produkt* (J 0114) Math Nachr 37∗345-358
⋄ C05 C30 G30 ⋄ REV MR 39 # 1384 Zbl 164.10
• ID 12920

SCHMIDT, J. see Vol. I, II, V, VI for further entries

SCHMIDT, K. [1981] *Ein Rechenverfahren fuer die elementare Logik (English summary)* (J 0989) Z Allg Wissth 12*110-115
 ⋄ B25 B35 ⋄ REV MR 83e:03024 • ID 35213

SCHMIDT, K. [1984] see DRIES VAN DEN, L.

SCHMIDT, K. [1985] *Nonstandard methods in field theory. Bounds and definability, and specialization properties in prime characteristic* (0000) Diss., Habil. etc xii + 118pp
 ⋄ C25 C60 H20 ⋄ REM Diss., Christian-Albrechts-Universitaet Kiel • ID 49816

SCHMIDT, K. see Vol. V for further entries

SCHMIEDEN, C. [1958] see LAUGWITZ, D.

SCHMITT, P.H. [1975] *Categorical lattices* (0000) Diss., Habil. etc
 ⋄ C35 C60 G10 ⋄ REM Diss., Universitaet Heidelberg • ID 19707

SCHMITT, P.H. [1976] *The model-completion of Stone algebras* (J 1934) Ann Sci Univ Clermont Math 13*135-155
 ⋄ C25 C35 C60 G10 ⋄ REV MR 57 # 5743 Zbl 352 # 02040 • ID 50039

SCHMITT, P.H. [1981] see CHERLIN, G.L.

SCHMITT, P.H. [1982] *The elementary theory of torsion-free abelian groups with a predicate specifying a subgroup* (J 0068) Z Math Logik Grundlagen Math 28*323-329
 ⋄ B25 C60 ⋄ REV MR 84b:03053 Zbl 495 # 03017 • ID 35641

SCHMITT, P.H. [1983] *Algebraically complete lattices* (J 0004) Algeb Universalis 17*135-142
 ⋄ C25 C35 G10 ⋄ REV Zbl 532 # 03017 • ID 38280

SCHMITT, P.H. [1983] see CHERLIN, G.L.

SCHMITT, P.H. [1983] *The L^t-theory of profinite abelian groups* (J 0027) Fund Math 119*135-150
 ⋄ B25 C50 C60 C80 C90 ⋄ REV MR 85j:03049 Zbl 545 # 03017 • ID 41228

SCHMITT, P.H. [1984] *Model- and substructure complete theories of ordered abelian groups* (P 2153) Logic Colloq;1983 Aachen 1*389-418
 ⋄ C10 C35 C60 ⋄ REV MR 86h:03063 Zbl 561 # 03020 • ID 39528

SCHMITT, P.H. [1984] see GUREVICH, Y.

SCHMITT, P.H. [1984] *Undecidable theories of valued Abelian groups* (S 3521) Mem Soc Math Fr 16*67-76
 ⋄ C60 D35 ⋄ REV Zbl 555 # 20034 • ID 46136

SCHMITT, P.H. see Vol. II for further entries

SCHNEIDER, H.H. [1972] see BULLOCK, A.M.

SCHNEIDER, H.H. [1973] see BULLOCK, A.M.

SCHNEIDER, H.H. see Vol. I, V for further entries

SCHNORR, C.-P. [1977] see KLUPP, H.

SCHNORR, C.-P. see Vol. I, IV, VI for further entries

SCHOCK, R. [1964] *On the logic of variable binders* (J 0009) Arch Math Logik Grundlagenforsch 6*71-90
 ⋄ B10 C80 ⋄ REV MR 31 # 34 Zbl 121.11 • ID 12942

SCHOCK, R. [1965] *Another proof of the strong Loewenheim-Skolem theorem* (J 0079) Logique & Anal, NS 8*30-38
 ⋄ C07 ⋄ REV MR 33 # 21 Zbl 146.10 • ID 12948

SCHOCK, R. [1965] *On definitions* (J 0009) Arch Math Logik Grundlagenforsch 8*28-44
 ⋄ A05 C40 ⋄ REV MR 31 # 3304 Zbl 173.5 • ID 12945

SCHOCK, R. [1973] *Set theory can provide for all accessible ordinals* (J 0079) Logique & Anal, NS 16*595-598
 ⋄ C62 E30 E35 E55 ⋄ REV MR 50 # 6847 Zbl 313 # 02041 • ID 12954

SCHOCK, R. [1974] *The consistency of the axiom of constructibility in ZF with non-predicative ultimate classes* (J 0079) Logique & Anal, NS 17*95-100
 ⋄ C62 E35 E45 E70 ⋄ REV MR 51 # 145 Zbl 313 # 02038 • ID 17446

SCHOCK, R. see Vol. I, II, V for further entries

SCHOENFELD, W. [1977] *Eine algebraische Konstruktion abzaehlbarer Modelle* (J 0009) Arch Math Logik Grundlagenforsch 18*135-144
 ⋄ C07 C15 ⋄ REV MR 57 # 12206 Zbl 411 # 03020 • ID 23721

SCHOENFELD, W. see Vol. I, IV, V, VI for further entries

SCHOENFINKEL, M. [1928] see BERNAYS, P.

SCHOENFINKEL, M. see Vol. I, VI for further entries

SCHOENHAGE, A. [1982] *Random access machines and Presburger arithmetic* (P 3482) Logic & Algor (Specker);1980 Zuerich 353-363
 ⋄ B25 D15 ⋄ REV MR 83c:03037 Zbl 496 # 03021 • ID 35139

SCHOENHAGE, A. see Vol. IV, V for further entries

SCHOLZ, H. [1952] *Ein ungeloestes Problem in der symbolischen Logik (problem 1)* (J 0036) J Symb Logic 17*160
 ⋄ B10 C13 ⋄ ID 33304

SCHOLZ, H. see Vol. I, II for further entries

SCHRAM, J.M. [1984] *Recursively prime trees* (J 2761) J Recreational Math 16*281-288
 ⋄ C57 C65 D45 ⋄ REV MR 85m:11008 • ID 44732

SCHREIBER, M. [1972] *Quelques remarques sur les caracterisations des espaces $L^p, 0 \leq p < i$ (English summary)* (J 0984) Ann Inst Henri Poincare, Sect B 8*83-92
 ⋄ C20 C65 ⋄ REV MR 47 # 762 Zbl 233 # 46040 • ID 82744

SCHREIBER, P. [1965] *Untersuchungen ueber die Modelle der Typentheorie* (J 0068) Z Math Logik Grundlagen Math 11*343-372
 ⋄ B15 C85 ⋄ REV MR 34 # 5654 Zbl 166.259 • ID 12976

SCHREIBER, P. [1978] *Kategorizitaetsbeweise in der Geometrie* (J 3124) Beitr Algebra Geom 7*79-90
 ⋄ C35 ⋄ REV MR 83b:03034 Zbl 405 # 03014 • ID 54889

SCHREIBER, P. see Vol. I, II, IV, VI for further entries

SCHRIEBER, L. [1985] *Recursive properties of euclidean domains* (J 0073) Ann Pure Appl Logic 29*59-77
 ⋄ C57 C60 D45 ⋄ REV Zbl 574 # 03029 • ID 47474

SCHROETER, K. [1953] *Theorie des mathematischen Schliessens* (P 0626) Ber Math-Tagung Berlin;1953 Berlin 5-12
◇ C07 ◇ REV MR 16.553 Zbl 52.9 • ID 12988

SCHROETER, K. [1958] *Theorie des logischen Schliessens II* (J 0068) Z Math Logik Grundlagen Math 4*10-65
◇ B10 C07 F05 F07 ◇ REV MR 20#6979 Zbl 80.8 JSL 32.418 • REM Part I 1955 • ID 24841

SCHROETER, K. see Vol. I, VI for further entries

SCHUETTE, K. [1933] *Untersuchungen zum Entscheidungsproblem der mathematischen Logik* (J 0043) Math Ann 109*572-603
◇ B20 B25 D35 ◇ REV Zbl 9.2 FdM 60.21 • ID 12997

SCHUETTE, K. [1934] *Ueber die Erfuellbarkeit einer Klasse von logischen Formeln* (J 0043) Math Ann 110*161-194
◇ B20 B25 ◇ REV Zbl 9.337 FdM 60.21 • ID 12998

SCHUETTE, K. [1950] *Schlussweisen-Kalkuele der Praedikatenlogik* (J 0043) Math Ann 122*47-65
◇ B10 B25 F05 F07 F50 ◇ REV MR 12.233 Zbl 36.148 JSL 16.155 • ID 12999

SCHUETTE, K. [1962] *Der Interpolationssatz der intuitionistischen Praedikatenlogik* (J 0043) Math Ann 148*192-200
◇ C40 F50 ◇ REV MR 26#24 Zbl 108.3 JSL 29.145 • ID 13012

SCHUETTE, K. [1968] *Vollstaendige Systeme modaler und intuitionischer Logik* (X 0811) Springer: Heidelberg & New York vii+87pp
• TRANSL [1974] (C 4138) Feys: Modal Logika
◇ B45 C90 F50 F98 ◇ REV MR 37#2587 Zbl 157.16 JSL 36.522 • ID 24822

SCHUETTE, K. [1978] *Ein Ansatz zum Entscheidungsverfahren fuer eine Formelklasse der Praedikatenlogik mit Identitaet* (J 0380) Acta Cybern (Szeged) 4*141-148
◇ B10 B25 ◇ REV MR 80b:03013 Zbl 406#03015 • ID 56098

SCHUETTE, K. see Vol. I, II, IV, V, VI for further entries

SCHULTE-MOENTING, J. [1976] *Interpolation formulae for predicates and terms which carry their own history* (J 0009) Arch Math Logik Grundlagenforsch 17*159-169
◇ C40 ◇ REV MR 55#5405 Zbl 339#02012 • ID 13030

SCHULTE-MOENTING, J. see Vol. I, VI for further entries

SCHULTZ, KONRAD [1970] *Modelle modaler Mengenlehren* (J 0068) Z Math Logik Grundlagen Math 16*327-339
◇ B45 C62 C90 E35 E70 ◇ REV MR 44#3849 Zbl 213.15 • ID 13035

SCHULTZ, KONRAD [1980] see KUEHNRICH, M.

SCHULTZ, KONRAD [1980] *A topological model for Troelstra's system CS of intuitionistic analysis* (J 0068) Z Math Logik Grundlagen Math 26*349-354
◇ C90 E70 F35 F50 ◇ REV MR 83d:03066 Zbl 457#03055 • ID 54380

SCHULTZ, KONRAD see Vol. II, V, VI for further entries

SCHULZ, KLAUS [1984] see FELGNER, U.

SCHUMACHER, D. [1976] *Absolutely free algebras in a topos containing an infinite object* (J 0018) Canad Math Bull 19*323-328
◇ C05 C30 G30 ◇ REV MR 55#12790 Zbl 363#08003 • ID 82749

SCHUMACHER, D. see Vol. V for further entries

SCHUMM, G.F. [1978] *An incomplete nonnormal extension of S3* (J 0036) J Symb Logic 43*211-212
◇ B45 C90 ◇ REV MR 58#5067 Zbl 386#03007 • ID 29254

SCHUMM, G.F. [1978] *Putting ℛ in its place* (J 0047) Notre Dame J Formal Log 19*623-628
◇ B45 C90 ◇ REV MR 80a:03029 Zbl 336#02019 • ID 52126

SCHUMM, G.F. see Vol. I, II, V for further entries

SCHUPP, P.E. [1970] *A note on recursively enumerable predicates in groups* (J 0027) Fund Math 66*61-63
◇ B25 D25 D40 ◇ REV MR 40#4345 Zbl 193.317 • ID 13048

SCHUPP, P.E. [1973] *A survey of small cancellation theory* (P 0678) Word Probl: Decis & Burnside Probl in Group Th;1969 Irvine 569-589
◇ C60 D40 ◇ REV MR 54#415 Zbl 292#20034 JSL 41.786 • ID 44402

SCHUPP, P.E. [1981] see MULLER, D.E.

SCHUPP, P.E. [1985] see MULLER, D.E.

SCHUPP, P.E. see Vol. IV for further entries

SCHUPPAR, B. [1978] *Modelltheoretische Untersuchungen zur Galoistheorie von Funktionenkoerpern (Dissertation)* (X 3157) Univ Essen Fachb Math: Essen 59pp
◇ C60 ◇ REV Zbl 477#12018 • ID 55601

SCHUPPAR, B. [1980] *Elementare Aussagen zur Arithmetik und Galoistheorie von Funktionenkoerpern* (J 0127) J Reine Angew Math 313*59-71
◇ C60 ◇ REV MR 80m:12011 Zbl 423#12015 • ID 53576

SCHUTT, R. [1985] see BANKSTON, P.

SCHWABAUER, R. [1965] see RIBEIRO, H.

SCHWABHAEUSER, W. [1956] *Ueber die Vollstaendigkeit der elementaren euklidischen Geometrie* (J 0068) Z Math Logik Grundlagen Math 2*137-165
◇ B30 C35 C65 ◇ REV MR 18.863 Zbl 74.12 JSL 36.156 • ID 13051

SCHWABHAEUSER, W. [1959] *Entscheidbarkeit und Vollstaendigkeit der elementaren hyperbolischen Geometrie* (J 0068) Z Math Logik Grundlagen Math 5*132-205
◇ B25 B30 C35 C65 ◇ REV MR 27#3549 Zbl 95.348 JSL 36.156 • ID 13052

SCHWABHAEUSER, W. [1962] *On completeness and decidability and some non-definable notions of elementary hyperbolic geometry* (P 0612) Int Congr Log, Meth & Phil of Sci (1,Proc);1960 Stanford 159-167
◇ B25 C35 C65 ◇ REV MR 27#4755 Zbl 135.396 JSL 36.156 • ID 13053

SCHWABHAEUSER, W. [1965] *On models of elementary elliptic geometry* (P 0614) Th Models;1963 Berkeley 312-328
◇ B25 C65 ◇ REV MR 33#7259 JSL 36.682 • ID 13056

SCHWABHAEUSER, W. [1967] *Zur Axiomatisierbarkeit von Theorien in der schwachen Logik der zweiten Stufe* (J 0009) Arch Math Logik Grundlagenforsch 10*60-96
◇ B15 C85 ◇ REV MR 37#3915 Zbl 165.15 • ID 13058

SCHWABHAEUSER, W. [1971] *Modelltheorie I* (X 0876) Bibl Inst: Mannheim vi+165pp
⋄ C98 ⋄ REV MR 48#3732 Zbl 286#02052 • REM Part II 1972 • ID 13060

SCHWABHAEUSER, W. [1972] *Modelltheorie II* (X 0876) Bibl Inst: Mannheim iii+123pp
⋄ C98 ⋄ REV MR 51#129 Zbl 286#02053 • REM Part I 1971 • ID 30556

SCHWABHAEUSER, W. & SZCZERBA, L.W. [1974] *An affine space as union of spaces of higher dimension* (P 0610) Tarski Symp;1971 Berkeley 133-138
⋄ C60 C65 ⋄ REV MR 53#5290 Zbl 309#50010 • ID 13061

SCHWABHAEUSER, W. [1974] *Modelltheorie* (J 0157) Jbuchber Dtsch Math-Ver 75∗114-129
⋄ C98 ⋄ REV MR 58#204 Zbl 274#02003 • ID 29013

SCHWABHAEUSER, W. & SZCZERBA, L.W. [1975] *Relations on lines as primitive notions for euclidean geometry* (J 0027) Fund Math 82∗347-355 • ERR/ADD ibid 87∗283
⋄ C65 ⋄ REV MR 50#12697 Zbl 296#50001 • ID 19689

SCHWABHAEUSER, W. [1979] *Modellvollstaendigkeit der Mittelpunktsgeometrie und der Theorie der Vektorgruppen* (J 0107) Abh Math Sem Univ Hamburg 48∗213-224
⋄ B25 C35 C65 ⋄ REV MR 80i:03022 Zbl 411#03025 • ID 52879

SCHWABHAEUSER, W. [1979] *Non-finitizability of a weak second-order theory* (J 0027) Fund Math 103∗83-102
⋄ B15 C62 C65 C85 F35 ⋄ REV MR 81h:03079 Zbl 422#03010 • ID 53471

SCHWABHAEUSER, W. & SZMIELEW, W. & TARSKI, A. [1983] *Metamathematische Methoden in der Geometrie* (X 0811) Springer: Heidelberg & New York viii+482pp
⋄ B30 C65 C98 D35 ⋄ REV MR 85e:03004 Zbl 564#51001 • ID 40225

SCHWABHAEUSER, W. [1983] *Metamathematische Methoden in der Geometrie. Teil II: Metamathematische Betrachtungen* (C 4602) Metamath Meth in Geom 173-452
⋄ B30 C65 C98 ⋄ REV MR 85e:03004 Zbl 564#51001 • ID 41875

SCHWABHAEUSER, W. see Vol. I, V for further entries

SCHWARTZ, DIETRICH [1970] *Nichtdefinierbarkeit der Stufenbeziehung durch die Elementbeziehung in der allgemeinen Mengenlehre von Klaua* (J 0342) Monatsber Dt Akad Wiss 12∗116-120
⋄ C40 C47 E70 ⋄ REV MR 42#4389 Zbl 216.14 • ID 13065

SCHWARTZ, DIETRICH [1971] *Die Nichtkompaktheit der Klasse der archimedisch geordneten Koerper* (J 0068) Z Math Logik Grundlagen Math 17∗189-191
⋄ C52 C60 ⋄ REV MR 44#6472 Zbl 221#02036 • ID 13066

SCHWARTZ, DIETRICH [1975] *Ultraprodukte in der Theorie der logischen Auswahlfunktionen* (J 0068) Z Math Logik Grundlagen Math 21∗385-394
⋄ B10 C20 ⋄ REV MR 52#13375 Zbl 333#02043 • ID 13068

SCHWARTZ, DIETRICH [1979] *Untersuchungen ueber die Banach-Logik* (J 0068) Z Math Logik Grundlagen Math 25∗111-118
⋄ C90 ⋄ REV MR 80e:03020 Zbl 438#03043 • ID 55955

SCHWARTZ, DIETRICH [1980] *Cylindric algebras with filter quantifiers* (J 0068) Z Math Logik Grundlagen Math 26∗251-254
⋄ C80 G15 ⋄ REV MR 81g:03078 Zbl 469#03043 • ID 55170

SCHWARTZ, DIETRICH see Vol. I, II, IV, V for further entries

SCHWARTZ, J.T. [1980] see FERRO, A.

SCHWARTZ, J.T. [1981] see BREBAN, M.

SCHWARTZ, J.T. [1985] see CANTONE, D.

SCHWARTZ, J.T. see Vol. I, II, IV, V for further entries

SCHWARTZ, N. [1978] *η_α-Strukturen* (J 0044) Math Z 158∗147-155
⋄ C50 C55 C65 E07 ⋄ REV MR 57#9523 Zbl 356#08001 • ID 50976

SCHWARZ, D. [1981] see RICHTER, M.M.

SCHWARZ, S. [1984] *Recursive automorphisms of recursive linear orderings* (J 0073) Ann Pure Appl Logic 26∗69-73
⋄ C57 C65 D45 ⋄ REV MR 85c:03018 Zbl 571#03018 • ID 39829

SCHWARZ, S. see Vol. IV, V for further entries

SCHWICHTENBERG, H. [1972] see ROEDDING, D.

SCHWICHTENBERG, H. [1975] *Elimination of higher type levels in definitions of primitive recursive functionals by means of transfinite recursion* (P 0775) Logic Colloq;1973 Bristol 279-303
⋄ C75 F10 F15 F35 F50 ⋄ REV MR 52#7866 Zbl 318#02047 • ID 18407

SCHWICHTENBERG, H. & WAINER, S.S. [1975] *Infinite terms and recursion in higher types* (P 1440) ⊢ ISILC Proof Th Symp (Schuette);1974 Kiel 341-364
⋄ C75 D65 F10 ⋄ REV MR 54#7231 Zbl 341#02033 • ID 25022

SCHWICHTENBERG, H. see Vol. I, IV, V, VI for further entries

SCOTT, D.S. [1955] see KALICKI, J.

SCOTT, D.S. [1956] *Equationally complete extensions of finite algebras* (J 0028) Indag Math 18∗35-38
⋄ C05 C13 ⋄ REV MR 18.636 Zbl 73.246 JSL 23.57 • ID 11899

SCOTT, D.S. [1957] *The notion of rank in set-theory* (P 1675) Summer Inst Symb Log;1957 Ithaca 267-269
⋄ C62 E10 E20 E30 ⋄ REV JSL 31.662 • ID 11911

SCOTT, D.S. & TARSKI, A. [1957] *The sentential calculus with infinitely long expressions* (P 1675) Summer Inst Symb Log;1957 Ithaca 83-89
• REPR [1958] (S 0019) Colloq Math (Warsaw) 6∗165-170
⋄ C75 ⋄ REV MR 20#6350 Zbl 119.250 JSL 30.95 • ID 21012

SCOTT, D.S. [1958] *Convergent sequences of complete theories* (0000) Diss., Habil. etc
⋄ C35 C52 ⋄ REM Ph.d.thesis, Princeton University • ID 19639

SCOTT, D.S. & SUPPES, P. [1958] *Foundational aspects of theories of measurement* (J 0036) J Symb Logic 23∗113-128
⋄ B30 C13 C52 ⋄ REV MR 22#6716 Zbl 84.246 JSL 33.287 • ID 11906

SCOTT, D.S. [1960] *On a theorem of Rabin* (J 0028) Indag Math 22*481-484
⋄ C35 C57 C62 D35 ⋄ REV MR 25#1985 • ID 11909

SCOTT, D.S. [1961] *Measurable cardinals and constructible sets* (J 0014) Bull Acad Pol Sci, Ser Math Astron Phys 9*521-524
⋄ C20 E45 E55 ⋄ REV MR 26#1263 Zbl 154.7 JSL 32.410 • ID 21996

SCOTT, D.S. [1961] *On constructing models for arithmetic* (P 0633) Infinitist Meth;1959 Warsaw 235-255
⋄ C20 C57 C62 H15 ⋄ REV MR 27#2423 Zbl 126.12 JSL 38.336 • ID 24930

SCOTT, D.S. [1962] *Algebras of sets binumerable in complete extensions of arithmetic* (P 0613) Rec Fct Th;1961 New York 117-121
⋄ C40 C62 F30 ⋄ REV MR 25#4993 Zbl 199.26 • ID 11914

SCOTT, D.S. [1962] see FRAYNE, T.E.

SCOTT, D.S. [1964] *Invariant Borel sets* (J 0027) Fund Math 56*117-128
⋄ C75 D55 E15 ⋄ REV MR 30#3027 Zbl 152.213 • ID 11916

SCOTT, D.S. [1965] *Logic with denumerably long formulas and finite strings of quantifiers* (P 0614) Th Models;1963 Berkeley 329-341
⋄ C15 C40 C75 C98 ⋄ REV MR 34#32 Zbl 166.260 JSL 36.157 • ID 11917

SCOTT, D.S. [1966] see KRAUSS, P.H.

SCOTT, D.S. [1969] *On completing ordered fields* (P 0649) Appl Model Th to Algeb, Anal & Probab;1967 Pasadena 274-278
⋄ C60 ⋄ REV MR 39#6866 Zbl 188.322 • ID 11920

SCOTT, D.S. [1973] *Models for various type-free calculi* (P 0793) Int Congr Log, Meth & Phil of Sci (4,Proc);1971 Bucharest 157-187
⋄ B40 C90 ⋄ REV MR 57#15987 • ID 78313

SCOTT, D.S. [1974] see ADDISON, J.W.

SCOTT, D.S. [1979] *A note on distributive normal forms* (C 1706) Essays Honour J. Hintikka 75-90
⋄ B10 C07 ⋄ REV Zbl 452#03005 • ID 54069

SCOTT, D.S. [1979] see FOURMAN, M.P.

SCOTT, D.S. [1980] *Lambda calculus: some models, some philosophy* (P 2058) Kleene Symp;1978 Madison 223-265
⋄ A05 B40 C90 ⋄ REV MR 82d:03024 Zbl 515#03004 • ID 78314

SCOTT, D.S. [1980] *Relating theories of the λ-calculus* (C 3050) Essays Combin Log, Lambda Calc & Formalism (Curry) 403-450
⋄ B40 C90 F50 G30 ⋄ REV MR 82d:03025 Zbl 469#03006 • ID 78323

SCOTT, D.S. & MCCARTY, C. & HORTY, J.F. [1985] *Bibliography* (C 4183) Model-Theor Log 793-893
⋄ A10 C98 ⋄ ID 48345

SCOTT, D.S. see Vol. I, II, IV, V, VI for further entries

SCOTT, W.R. [1951] *Algebraically closed groups* (J 0053) Proc Amer Math Soc 2*118-121
⋄ C25 C60 ⋄ REV MR 12.671 Zbl 43.23 • ID 11925

SCOWCROFT, P. [1984] *The real algebraic structure of Scott's model of intuitionistic analysis* (J 0073) Ann Pure Appl Logic 27*275-308
⋄ B25 C90 F35 F50 ⋄ REV MR 86f:03101 Zbl 566#03034 • ID 44277

SCOZZAFAVA, R. [1980] see BARONE, E.

SCOZZAFAVA, R. see Vol. II for further entries

SEBASTIAO E SILVA, J. [1985] see OLIVEIRA DE, A.J.F.

SEBEL'DIN, A.M. [1981] *Definability of a torsion-free unreduced abelian group by its group of endomorphisms* (Russian) (C 3910) Abel Gruppy & Moduli 102-108,137
⋄ C60 E50 E75 ⋄ REV MR 84b:20063 • ID 38919

SEESE, D.G. [1972] *Entscheidbarkeits- und Definierbarkeitsfragen der Theorie "netzartiger" Graphen I* (Russian) (English and French summaries) (J 0115) Wiss Z Humboldt-Univ Berlin, Math-Nat Reihe 21*513-517
⋄ C80 C85 D35 ⋄ REV MR 49#7133 Zbl 254#02036 • ID 11931

SEESE, D.G. [1975] *Zur Entscheidbarkeit der monadischen Theorie 2. Stufe baumartiger Graphen* (Russian, English and French summaries) (J 0115) Wiss Z Humboldt-Univ Berlin, Math-Nat Reihe 24*768-772
⋄ B15 B25 ⋄ REV MR 58#5157 Zbl 362#02037 • ID 50759

SEESE, D.G. & TUSCHIK, H.-P. [1977] *Construction of nice trees* (P 1695) Set Th & Hierarch Th (3);1976 Bierutowice 257-271
⋄ B25 C55 C65 C80 ⋄ REV MR 58#10292 Zbl 366#02027 • ID 51110

SEESE, D.G. [1977] *Decidability of ω-trees with bounded sets - a survey* (P 2588) FCT'77 Fund of Comput Th;1977 Poznan 511-515
⋄ B25 C65 C80 C85 ⋄ REV MR 58#16247 Zbl 375#02042 • ID 51619

SEESE, D.G. [1977] *Partitions for trees* (P 2921) Beitr Graphenth & Anw;1977 Oberhof 235-238
⋄ C55 C65 E05 ⋄ REV MR 82b:05056 Zbl 421#06003 • ID 82762

SEESE, D.G. [1977] *Second order logic, generalized quantifiers and decidability* (J 0014) Bull Acad Pol Sci, Ser Math Astron Phys 25*725-732
⋄ B15 B25 C55 C65 C80 C85 D35 ⋄ REV MR 57#87 Zbl 383#03010 • ID 27132

SEESE, D.G. [1978] *A remark to the undecidability of well-orderings with the Haertig quantifier* (J 0014) Bull Acad Pol Sci, Ser Math Astron Phys 26*951
⋄ C55 C65 C80 D35 ⋄ REV MR 80b:03014 Zbl 408#03031 • ID 56270

SEESE, D.G. [1978] *Decidability of ω-trees with bounded sets* (German and Russian summaries) (X 2888) ZI Math Mech Akad Wiss DDR: Berlin 52pp
⋄ B25 C65 C85 D05 ⋄ REV MR 58#194 Zbl 386#03004 • ID 52180

SEESE, D.G. [1978] *Decidability and generalized quantifiers* (P 1897) Logic Colloq;1977 Wroclaw 229-237
⋄ B25 C80 C98 ⋄ REV MR 80b:03046 Zbl 455#03007 • ID 54268

SEESE, D.G. [1979] *Some graph-theoretical operations and decidability* (J 0114) Math Nachr 87*15-21
  ⋄ B15 B25 C65 C85 ⋄ REV MR 81j:03028 Zbl 414 # 03007 • ID 53053

SEESE, D.G. [1979] *Stationary logic and ordinals* (S 3414) Prepr, Akad Wiss DDR, Inst Math 44*50pp
  ⋄ B25 C55 C65 C80 ⋄ REV MR 82f:03034 Zbl 416 # 03040 • ID 53206

SEESE, D.G. [1980] see HERRE, H.

SEESE, D.G. [1980] see BAUDISCH, A.

SEESE, D.G. [1981] *Elimination of second-order quantifiers for well-founded trees in stationary logic and finitely determinate structures* (P 3165) FCT'81 Fund of Comput Th;1981 Szeged 341-349
  ⋄ C10 C55 C65 C80 ⋄ REV MR 83h:03044 Zbl 474 # 03016 • ID 55420

SEESE, D.G. [1981] *Stationary logic and ordinals* (J 0064) Trans Amer Math Soc 263*111-124
  ⋄ B25 C55 C65 C80 ⋄ REV MR 82f:03034 Zbl 474 # 03015 • ID 55419

SEESE, D.G. & WEESE, M. [1982] *L(aa)-elementary types of well-orderings* (J 0068) Z Math Logik Grundlagen Math 28*557-564
  ⋄ B25 C55 C65 C80 E10 ⋄ REV MR 84c:03073 Zbl 519 # 03028 • ID 34012

SEESE, D.G. & TUSCHIK, H.-P. & WEESE, M. [1982] *Undecidable theories in stationary logic* (J 0053) Proc Amer Math Soc 84*563-567
  ⋄ C55 C65 C80 D35 E05 E75 ⋄ REV MR 84c:03071 Zbl 515 # 03002 • ID 33532

SEESE, D.G. [1983] see BAUDISCH, A.

SEESE, D.G. [1985] see BAUDISCH, A.

SEESE, D.G. see Vol. IV for further entries

SEGAL, D. [1981] see GRUNEWALD, F.

SEGAL, D. see Vol. IV for further entries

SEGERBERG, K. [1968] *Decidability of S4.1* (J 0105) Theoria (Lund) 34*7-20
  ⋄ B25 B45 ⋄ REV MR 39 # 1308 JSL 39.611 • ID 11935

SEGERBERG, K. [1968] *Decidability of four modal logics* (J 0105) Theoria (Lund) 34*21-25
  ⋄ B25 B45 ⋄ REV MR 39 # 1309 • ID 41673

SEGERBERG, K. [1970] *Modal logics with linear alternative relations* (J 0105) Theoria (Lund) 36*301-322
  ⋄ B45 C90 ⋄ REV MR 47 # 18 Zbl 235 # 02019 • ID 11938

SEGERBERG, K. [1972] *Post completeness in modal logic* (J 0036) J Symb Logic 37*711-715
  ⋄ B45 C90 ⋄ REV MR 47 # 4765 Zbl 268 # 02012 • ID 11939

SEGERBERG, K. [1976] *Discrete linear future time without axioms* (J 0063) Studia Logica 35*273-278,327
  ⋄ B25 B45 C90 ⋄ REV MR 55 # 10245 Zbl 343 # 02017 • ID 65185

SEGERBERG, K. [1982] *A completeness theorem in the modal logic of programs* (P 3831) Universal Algeb & Appl;1978 Warsaw 31-46
  ⋄ B75 C90 ⋄ REV MR 85k:03014 Zbl 546 # 03011 • ID 41144

SEGERBERG, K. see Vol. I, II, VI for further entries

SEIDENBERG, A. [1954] *A new decision method for elementary algebra* (J 0120) Ann of Math, Ser 2 60*365-374
  ⋄ B25 C10 C60 ⋄ REV MR 16.209 Zbl 56.18 JSL 22.295 • ID 11944

SEIDENBERG, A. [1956] *Some remarks on Hilbert's Nullstellensatz* (J 0008) Arch Math (Basel) 7*235-240
  ⋄ C60 ⋄ ID 49907

SEIDENBERG, A. [1958] *Comments on Lefschetz' principle* (J 0005) Amer Math Mon 65*685-690
  ⋄ C60 ⋄ REV MR 20 # 5201 Zbl 108.169 • ID 33533

SEIDENBERG, A. [1960] *An elimination theory for differential algebra* (S 0183) Publ Math Univ California 3*31-66
  ⋄ C10 C60 ⋄ ID 11945

SEIDENBERG, A. [1970] *Construction of the integral closure of a finite integral domain* (J 0059) Rend Sem Mat Fis Milano 40*100-120
  ⋄ C57 C60 F55 ⋄ REV MR 45 # 3396 Zbl 218 # 14023
  • REM Part II 1975 • ID 41695

SEIDENBERG, A. [1971] *On the length of a Hilbert ascending chain* (J 0053) Proc Amer Math Soc 29*443-450
  ⋄ C60 F55 ⋄ REV MR 43 # 6193 Zbl 216.328 • ID 41694

SEIDENBERG, A. [1973] *On the impossibility of some constructions in polynomial rings* (X 2121) Accad Naz Linc: Roma
  ⋄ C60 F55 ⋄ ID 41700

SEIDENBERG, A. [1975] *Construction of the integral closure of a finite integral domain. II* (J 0053) Proc Amer Math Soc 52*368-372
  ⋄ C57 C60 F55 ⋄ REV MR 54 # 12741 Zbl 333 # 13004
  • REM Part I 1970 • ID 41698

SEIDENBERG, A. see Vol. IV, VI for further entries

SEIFERT, R.L. [1967] see JOHNSON, J.

SEIFERT, R.L. [1971] *On prime binary relational structures* (J 0027) Fund Math 70*187-203
  ⋄ C05 C50 ⋄ REV MR 46 # 3407 Zbl 211.319 • ID 22274

SEJTENOV, S.M. [1978] *Theory of finite fields with an additional predicate distinguishing a subfield (Russian)* (J 0092) Sib Mat Zh 19*396-405
  • TRANSL [1978] (J 0475) Sib Math J 19*278-285
  ⋄ B25 C13 C60 ⋄ REV MR 81g:03040 Zbl 389 # 03009
  • ID 56248

SELMAN, A.L. [1972] *Completeness of calculi for axiomatically defined classes of algebras* (J 0004) Algeb Universalis 2*20-32
  ⋄ B55 C05 C07 ⋄ REV MR 47 # 1725 Zbl 251 # 08005
  • ID 11964

SELMAN, A.L. [1972] see JONES, N.D.

SELMAN, A.L. [1973] *Sets of formulas valid in finite structures* (J 0064) Trans Amer Math Soc 177*491-504
  ⋄ C13 D30 ⋄ REV MR 47 # 8272 Zbl 276 # 02025
  • ID 11966

SELMAN, A.L. [1974] see JONES, N.D.

SELMAN, A.L. see Vol. IV, V for further entries

SEMENOV, A.L. [1978] *Some algorithmic problems for systems of algorithmic algebras (Russian)* (J 0023) Dokl Akad Nauk SSSR 239∗1063-1066
- TRANSL [1978] (J 0062) Sov Math, Dokl 19∗490-493
- ◊ B25 B75 ◊ REV MR 80a:03055 Zbl 402 # 68036
- ID 54711

SEMENOV, A.L. [1979] *On certain extensions of the arithmetic of addition of natural numbers (Russian)* (J 0216) Izv Akad Nauk SSSR, Ser Mat 43∗1175-1195,1199
- TRANSL [1980] (J 0448) Math of USSR, Izv 15∗401-418
- ◊ B25 F30 ◊ REV MR 81e:03009 Zbl 417 # 03005
- ID 53242

SEMENOV, A.L. [1983] *Logical theories of one-place functions on the set of natural numbers (Russian)* (J 0216) Izv Akad Nauk SSSR, Ser Mat 47∗623-658
- TRANSL [1983] (J 0448) Math of USSR, Izv 22∗587-618
- ◊ B25 C10 C62 C85 ◊ REV MR 84i:03033 Zbl 541 # 03005 • ID 34517

SEMENOV, A.L. [1984] *Decidability of monadic theories* (P 3658) Math Founds of Comput Sci (11);1984 Prague 162-175
- ◊ B25 C85 C98 ◊ REV Zbl 553 # 03005 • ID 43294

SEMENOV, A.L. see Vol. I, IV, VI for further entries

SENFT, J.R. [1970] *On weak automorphisms of universal algebras* (J 0202) Diss Math (Warsaw) 74∗35pp
- ◊ C05 C07 ◊ REV MR 43 # 142 Zbl 211.30 • ID 11977

SEREMET, Z. [1984] *On automorphisms of resplendent models of arithmetic* (J 0068) Z Math Logik Grundlagen Math 30∗349-352
- ◊ C07 C50 C62 ◊ REV MR 85i:03115 Zbl 564 # 03032
- ID 42269

SERVI, M. [1971] *Una questione di teoria dei modelli nelle categorie con prodotti finiti (English summary)* (J 0319) Matematiche (Sem Mat Catania) 26∗307-324
- ◊ C40 G30 ◊ REV MR 52 # 79 Zbl 272 # 08008 • ID 18378

SERVI, M. [1974] *Su alcuni funtori che conservano le SH* (J 3254) Riv Mat Univ Parma, Ser 3 3∗291-308
- ◊ C40 G30 ◊ REV MR 55 # 10540 Zbl 379 # 18001
- ID 51862

SERVI, M. [1979] *A generalization of the exponential functor and its connections with the SH-formulas (Italian summary)* (J 0549) Riv Mat Univ Parma, Ser 4 5∗287-294
- ◊ C40 G30 ◊ REV MR 81h:18009 Zbl 454 # 18013
- ID 82777

SERVI, M. see Vol. V for further entries

SETTE, A.-M. [1973] *A propos d'une condition conduisant a une theorie forcee saturee* (J 2313) C R Acad Sci, Paris, Ser A-B 277∗A645-A646
- ◊ C25 C50 ◊ REV MR 49 # 34 Zbl 272 # 02074 • ID 11993

SETTE, A.-M. [1975] *Multirelation fortement homogene et ses rapports avec les theories forcees (English summary)* (J 2313) C R Acad Sci, Paris, Ser A-B 280∗A535-A537
- ◊ C25 C35 C50 ◊ REV MR 51 # 5284 Zbl 322 # 02050
- ID 17399

SETTE, A.-M. [1979] *Fraisse and Robinson's forcing (a comparative study)* (J 1573) Notas Commun Mat (Recife) 100∗19pp
- ◊ C25 ◊ REV MR 82i:03039 Zbl 427 # 03022 • ID 53709

SETTE, A.-M. [1980] *Functorial approach to interpretability* (P 2958) Latin Amer Symp Math Log (4);1978 Santiago 365-376
- ◊ C90 F25 G30 ◊ REV MR 83k:03043 Zbl 419 # 03020 Zbl 464 # 03029 • ID 53363

SETTE, A.-M. [1981] *Fraisse and Robinson's forcing* (J 0068) Z Math Logik Grundlagen Math 27∗225-231
- ◊ C25 ◊ REV MR 82i:03039 Zbl 475 # 03009 • ID 55463

SETTE, A.-M. [1984] *Partial isomorphism extension method and a representation theorem for Post-language* (J 0068) Z Math Logik Grundlagen Math 30∗289-293
- ◊ C07 G20 ◊ REV MR 85i:03192 • ID 42229

SETTE, A.-M. & SZCZERBA, L.W. [1985] *Characterizations of elementary interpretations in category theory* (C 4181) Math Log & Formal Syst (Costa da) 243-292
- ◊ C07 F25 G30 ◊ REV Zbl 574 # 03013 • ID 48209

SETTE, A.-M. see Vol. I, II, VI for further entries

SGRO, J. [1976] *Completeness theorems for continuous functions and product topologies* (J 0029) Israel J Math 25∗249-272
- ◊ C40 C65 C80 C90 E75 ◊ REV MR 58 # 27452 Zbl 344 # 54006 • ID 26086

SGRO, J. [1977] *Completeness theorems for topological models* (J 0007) Ann Math Logic 11∗173-193
- ◊ C20 C40 C75 C80 C90 ◊ REV MR 57 # 2878 Zbl 387 # 03010 • ID 23663

SGRO, J. [1977] *Maximal logics* (J 0053) Proc Amer Math Soc 63∗291-298
- ◊ C20 C40 C80 C90 C95 ◊ REV MR 58 # 10409 Zbl 415 # 03025 • ID 31509

SGRO, J. [1980] *Interpolation fails for the Souslin-Kleene closure of the open set quantifier logic* (J 0053) Proc Amer Math Soc 78∗568-572
- ◊ C40 C80 C90 C95 ◊ REV MR 81m:03048 Zbl 452 # 03030 • ID 54094

SGRO, J. [1980] *The interior operator logic and product topologies* (J 0064) Trans Amer Math Soc 258∗99-112
- ◊ C40 C55 C80 C90 ◊ REV MR 80m:03074 Zbl 426 # 03039 • ID 53635

SGRO, J. [1980] *Ultraproduct invariant logics* (J 0053) Proc Amer Math Soc 79∗635-638
- ◊ C20 C40 C80 C95 ◊ REV MR 81h:03066 Zbl 436 # 03034 • ID 78410

SHABANOV-KUSHNARENKO, YU.P. [1981] *Modeling of finite mathematical structures (Russian)* (J 2668) Avtom Sist Upravl & Prib Avtom, Khar'kov 60∗86-90
- ◊ C13 ◊ REV MR 84e:03043 • ID 34380

SHABANOV-KUSHNARENKO, YU.P. see Vol. I for further entries

SHAFAAT, A. [1967] *Principle of localization for a more general type of languages* (J 3240) Proc London Math Soc, Ser 3 17∗629-643
- ◊ B10 C55 C75 C90 ◊ REV MR 36 # 1308 Zbl 207.296
- ID 12002

SHAFAAT, A. [1969] *A remark on a paper of Taimanov* (J 0003) Algebra i Logika 8∗373-374
- TRANSL [1969] (J 0069) Algeb and Log 8∗216
- ◊ B25 C20 ◊ REV MR 41 # 3260 Zbl 195.18 Zbl 214.14
- ID 12003

SHAFAAT, A. [1969] *On implicationally defined classes of algebras* (J 3172) J London Math Soc, Ser 2 44∗137-140
◇ C05 C75 ◇ REV MR 38 # 5691   Zbl 182.31 • ID 14994

SHAFAAT, A. [1973] *On products of relational structures* (J 0001) Acta Math Acad Sci Hung 24∗13-19
◇ B15 C30 ◇ REV MR 52 # 80   Zbl 263 # 08003 • ID 18382

SHAFAAT, A. [1973] *Remarks on quasivarieties of algebras* (J 0322) Arch Math (Brno) 9∗67-71
◇ C05 ◇ REV MR 50 # 6981   Zbl 292 # 08003 • ID 12006

SHAFAAT, A. [1975] see HATCHER, W.S.

SHAFAAT, A. [1976] *Constructions preserving the associative and the commutative laws* (J 3194) J Austral Math Soc, Ser A 21∗112-117
◇ C05 ◇ REV MR 53 # 7910   Zbl 327 # 08005 • ID 23034

SHAFAAT, A. [1980] *Consistency in categorical languages for algebras* (J 0068) Z Math Logik Grundlagen Math 26∗205-207
◇ C05 G30 ◇ REV MR 81j:03061   Zbl 428 # 03027
• ID 53786

SHAGIDULLIN, R.R. [1973] *The axiomatizability of one projection of a universally axiomatizable class of algebraic systems (Russian)* (S 2738) Issl Prikl Mat (Univ Kazan') 1∗23-27
◇ C52 ◇ REV MR 57 # 9516   Zbl 328 # 08008 • ID 65022

SHANIN, N.A. [1976] *On the quantifier of limiting realizability (Russian)* (S 0228) Zap Nauch Sem Leningrad Otd Mat Inst Steklov 60∗209-220,227
• TRANSL [1980] (J 1531) J Sov Math 14∗1565-1572
◇ C57 C80 F50 ◇ REV MR 58 # 27418   Zbl 344 # 02025
• ID 65058

SHANIN, N.A. [1977] *On the quantifier of limiting realizability* (P 3269) Set Th Found Math (Kurepa);1977 Beograd 127
◇ B55 C57 C80 F50 ◇ REV MR 58 # 21516   Zbl 361 # 02046 • ID 53812

SHANIN, N.A. see Vol. I, IV, V, VI for further entries

SHANK, H.S. [1971] see GARFUNKEL, S.

SHANK, H.S. [1972] see GARFUNKEL, S.

SHANK, H.S. [1979] *The rational case of a matrix problem of Harrison* (J 0193) Discr Math 28∗207-212
◇ B25 ◇ REV MR 80k:15022   Zbl 426 # 15008 • ID 82783

SHANK, H.S. see Vol. IV for further entries

SHAPIRO, S. [1985] *Second-order languages and mathematical practice* (J 0036) J Symb Logic 50∗714-742
◇ A05 C85 ◇ ID 47376

SHAPIRO, S. see Vol. I, II, IV, VI for further entries

SHEARD, M. [1985] *Co-critical points of elementary embeddings* (J 0036) J Symb Logic 50∗220-226
◇ C62 E45 E55 ◇ ID 39762

SHEARD, M. see Vol. V for further entries

SHEKHTMAN, V.B. [1978] *Rieger-Nishimura lattices (Russian)* (J 0023) Dokl Akad Nauk SSSR 241∗1288-1291
• TRANSL [1978] (J 0062) Sov Math, Dokl 19∗1014-1018
◇ B55 C90 G10 G25 ◇ REV MR 80a:03075
Zbl 412 # 03010 • ID 52941

SHEKHTMAN, V.B. [1978] *Two-dimensional modal logic (Russian)* (J 0087) Mat Zametki (Akad Nauk SSSR) 23∗759-772
• TRANSL [1978] (J 1044) Math Notes, Acad Sci USSR 23∗417-424
◇ B45 C90 ◇ REV MR 58 # 5068   Zbl 384 # 03010
• ID 52052

SHEKHTMAN, V.B. [1979] *Topological semantics of superintuitionistic logics (Russian)* (S 2582) Semiotika & Inf, Akad Nauk SSSR 12∗62-67
◇ B55 C90 ◇ REV MR 82d:03040 • ID 78450

SHEKHTMAN, V.B. [1980] *Topological models of propositional logics (Russian)* (S 2582) Semiotika & Inf, Akad Nauk SSSR 15∗74-98
◇ B45 C90 ◇ REV MR 82d:03029   Zbl 455 # 03013
• ID 54274

SHEKHTMAN, V.B. see Vol. I, II, IV for further entries

SHELAH, S. [1969] *Categoricity of classes of models* (0000) Diss., Habil. etc
◇ C35 ◇ REM Ph.d.thesis, The Hebrew University Jerusalem
• ID 19628

SHELAH, S. [1969] *Stable theories* (J 0029) Israel J Math 7∗187-202
◇ C35 C45 ◇ REV MR 40 # 7102   Zbl 193.300   JSL 38.648
• ID 12032

SHELAH, S. [1970] *A note on Hanf numbers* (J 0048) Pac J Math 34∗541-545
◇ C35 C55 C75 ◇ REV MR 42 # 2932   Zbl 207.7
• ID 12036

SHELAH, S. [1970] *Finite diagrams stable in power* (J 0007) Ann Math Logic 2∗69-118
◇ C35 C45 C50 C55 ◇ REV MR 44 # 2593   Zbl 204.311
JSL 49.315 • ID 12038

SHELAH, S. [1970] *On languages with non-homogeneous strings of quantifiers* (J 0029) Israel J Math 8∗75-79
◇ C75 C80 ◇ REV MR 41 # 6674   Zbl 196.11 • ID 12037

SHELAH, S. [1970] *On the cardinality of ultraproduct of finite sets* (J 0036) J Symb Logic 35∗83-84
◇ C13 C20 C55 E05 ◇ REV MR 48 # 3735   Zbl 196.10
JSL 38.650 • ID 12033

SHELAH, S. [1970] *On theories T categorical in |T|* (J 0036) J Symb Logic 35∗73-82
◇ C35 C45 ◇ REV MR 44 # 52   Zbl 196.10 • ID 12034

SHELAH, S. [1970] *Remark to "Local definability theory" of Reyes* (J 0007) Ann Math Logic 2∗441-447
◇ C40 C55 ◇ REV MR 44 # 56   Zbl 217.305 • ID 12035

SHELAH, S. [1971] *Every two elementarily equivalent models have isomorphic ultrapowers* (J 0029) Israel J Math 10∗224-233
◇ C07 C20 E05 ◇ REV MR 45 # 6608   Zbl 224 # 02045
• ID 12040

SHELAH, S. [1971] *On the number of non-almost isomorphic models of T in a power* (J 0048) Pac J Math 36∗811-818
◇ C55 C75 ◇ REV MR 44 # 2594   Zbl 194.309 • ID 12043

SHELAH, S. [1971] *Stability, the f.c.p., and superstability: model theoretic properties of formulas in first order theory* (J 0007) Ann Math Logic 3∗271-362
◇ C35 C45 C55 ◇ REV MR 47 # 6475   Zbl 281 # 02052
JSL 38.648 • ID 12041

SHELAH, S. [1971] *The number of non-isomorphic models of an unstable first-order theory* (**J** 0029) Israel J Math 9∗473-487
  ⋄ C45 C55 ⋄ REV MR 43 # 4652 Zbl 226 # 02045
  JSL 47.436 • ID 12039

SHELAH, S. [1971] *Two cardinal compactness* (**J** 0029) Israel J Math 9∗193-198
  ⋄ C30 C52 C55 C80 ⋄ REV MR 46 # 1581 Zbl 212.20
  • ID 12042

SHELAH, S. [1972] *A combinatorial problem; stability and order for models and theories in infinitary languages* (**J** 0048) Pac J Math 41∗247-261
  ⋄ C45 C55 C75 E05 ⋄ REV MR 46 # 7018
  Zbl 239 # 02024 • ID 12048

SHELAH, S. [1972] *A note on model complete models and generic models* (**J** 0053) Proc Amer Math Soc 34∗509-514
  ⋄ C25 C35 ⋄ REV MR 45 # 3188 Zbl 262 # 02051
  • ID 12045

SHELAH, S. [1972] *For what filters is every reduced product saturated?* (**J** 0029) Israel J Math 12∗23-31
  ⋄ C20 C50 C55 ⋄ REV MR 46 # 3292 Zbl 247 # 02046
  JSL 40.456 • ID 12044

SHELAH, S. [1972] *On models with power-like orderings* (**J** 0036) J Symb Logic 37∗247-267
  ⋄ C55 ⋄ REV MR 56 # 5272 Zbl 273 # 02036 • ID 12046

SHELAH, S. [1972] see SCHMERL, J.H.

SHELAH, S. [1972] *Saturation of ultrapowers and Keisler's order* (**J** 0007) Ann Math Logic 4∗75-114
  ⋄ C20 C35 C45 C50 C55 ⋄ REV MR 45 # 3187
  Zbl 243 # 02039 • ID 12049

SHELAH, S. [1972] *Uniqueness and characterization of prime models over sets for totally transcendental first-order-theories* (**J** 0036) J Symb Logic 37∗107-113
  ⋄ C30 C45 C50 ⋄ REV MR 47 # 4787 Zbl 247 # 02047
  JSL 38.523 • ID 12047

SHELAH, S. [1973] *Differentially closed fields* (**J** 0029) Israel J Math 16∗314-328
  ⋄ C45 C60 ⋄ REV MR 49 # 8856 Zbl 306 # 12105
  • ID 12050

SHELAH, S. [1973] *First order theory of permutation groups* (**J** 0029) Israel J Math 14∗149-162 • ERR/ADD ibid 15∗437-441
  ⋄ B15 C55 C60 C85 ⋄ REV MR 54 # 4972
  Zbl 284 # 20003 • ID 19627

SHELAH, S. [1973] *There are just four second-order quantifiers* (**J** 0029) Israel J Math 15∗282-300
  ⋄ B15 C80 C85 ⋄ REV MR 49 # 20 Zbl 273 # 02009
  JSL 51.234 • ID 12052

SHELAH, S. [1973] *Weak definability in infinitary languages* (**J** 0036) J Symb Logic 38∗399-404
  ⋄ C40 C55 C75 ⋄ REV MR 51 # 5263 Zbl 284 # 02027
  • ID 12053

SHELAH, S. [1974] *Categoricity of uncountable theories* (**P** 0610) Tarski Symp;1971 Berkeley 187-203
  ⋄ C35 C45 C50 ⋄ REV MR 51 # 10074 Zbl 317 # 02059
  JSL 46.866 • ID 12055

SHELAH, S. [1974] *Infinite abelian groups, Whitehead problem and some constructions* (**J** 0029) Israel J Math 18∗243-256
  ⋄ C55 C60 E05 E35 E45 E50 E75 ⋄ REV
  MR 50 # 9582 Zbl 318 # 02053 • ID 12054

SHELAH, S. [1974] see MCKENZIE, R.

SHELAH, S. [1974] *The Hanf number of omitting complete types* (**J** 0048) Pac J Math 50∗163-168
  ⋄ C55 C75 ⋄ REV MR 51 # 132 Zbl 306 # 02048
  • ID 12056

SHELAH, S. [1975] *A compactness theorem for singular cardinals, free algebras, Whitehead problem and transversals* (**J** 0029) Israel J Math 21∗319-349
  ⋄ C60 C75 E05 E75 ⋄ REV MR 52 # 10410
  Zbl 343 # 02043 • ID 12058

SHELAH, S. [1975] *A two-cardinal theorem* (**J** 0053) Proc Amer Math Soc 48∗207-213
  ⋄ C55 E05 ⋄ REV MR 50 # 9573 Zbl 302 # 02017
  • ID 12057

SHELAH, S. [1975] *Categoricity in $\aleph_1$ of sentences in $L_{\omega_1\omega}(Q)$* (**J** 0029) Israel J Math 20∗127-148
  ⋄ C35 C55 C75 C80 ⋄ REV MR 52 # 83 Zbl 324 # 02038
  • ID 19626

SHELAH, S. [1975] *Existence of rigid-like families of abelian p-groups* (**C** 0782) Model Th & Algeb (A. Robinson) 384-402
  ⋄ C52 C55 C60 ⋄ REV MR 54 # 425 Zbl 329 # 20037
  • ID 23993

SHELAH, S. [1975] *Generalized quantifiers and compact logic* (**J** 0064) Trans Amer Math Soc 204∗342-364
  ⋄ C55 C80 C95 E55 ⋄ REV MR 51 # 12510
  Zbl 322 # 02010 • ID 12059

SHELAH, S. [1975] *The lazy model theoretician's guide to stability* (**J** 0079) Logique & Anal, NS 18∗241-308
  • REPR [1977] (**P** 1625) Six Days of Model Th;1975 Louvain-la-Neuve 9-76
  ⋄ C25 C35 C45 C50 C60 C98 ⋄ REV MR 58 # 27447
  Zbl 359 # 02052 Zbl 414 # 03022 • ID 31463

SHELAH, S. [1975] *The monadic theory of order* (**J** 0120) Ann of Math, Ser 2 102∗379-419
  ⋄ B25 C65 C85 D35 E50 ⋄ REV MR 58 # 10390
  Zbl 345 # 02034 • ID 15024

SHELAH, S. [1975] *Why there are many nonisomorphic models for unsuperstable theories* (**P** 1521) Int Congr Math (II,12);1974 Vancouver 1∗259-263
  ⋄ C45 C50 C55 C60 C65 G05 ⋄ REV MR 54 # 10008
  Zbl 367 # 02030 • ID 25849

SHELAH, S. [1976] see MAKOWSKY, J.A.

SHELAH, S. [1976] *Interpreting set theory in the endomorphism semi-group of a free algebra or in a category* (**J** 1934) Ann Sci Univ Clermont Math 13∗1-29
  ⋄ C55 E05 F25 G30 ⋄ REV MR 58 # 21622
  Zbl 372 # 02026 • ID 51434

SHELAH, S. [1976] *Refuting Ehrenfeucht conjecture on rigid models* (**J** 0029) Israel J Math 25∗273-286
  ⋄ C50 C55 E50 ⋄ REV MR 58 # 5173 Zbl 359 # 02053
  • ID 26087

SHELAH, S. [1976] see AMIT, R.

SHELAH, S. [1976] see MACINTYRE, A.

SHELAH, S. [1977] *A two cardinal theorem and a combinatorial theorem* (**J** 0053) Proc Amer Math Soc 62∗134-136
  ⋄ C55 E05 ⋄ REV MR 55 # 7764 Zbl 362 # 02040
  • ID 31462

SHELAH, S. [1977] *Decidability of a portion of the predicate calculus* (J 0029) Israel J Math 28*32-44
⋄ B20 B25 ⋄ REV MR 58#21562 Zbl 368#02014
• ID 27202

SHELAH, S. [1977] *Existentially-closed groups in $\aleph_1$ with special properties* (J 0465) Bull Greek Math Soc (NS) 18*17-27
⋄ C15 C25 C60 C75 ⋄ REV MR 80j:03047
Zbl 411#03024 • ID 52878

SHELAH, S. [1977] see LITMAN, A.

SHELAH, S. [1977] see HODGES, W.

SHELAH, S. [1977] *Whitehead groups may be not free, even assuming CH. I* (J 0029) Israel J Math 28*193-204
⋄ C60 E05 E35 E50 E75 ⋄ REV MR 57#9538
Zbl 369#02035 • REM Part II 1980 • ID 27203

SHELAH, S. [1978] *A weak generalization of MA to higher cardinals* (J 0029) Israel J Math 30*297-306
⋄ B25 C55 E35 E50 E75 ⋄ REV MR 58#21606
Zbl 384#03032 • ID 29138

SHELAH, S. [1978] *Classification theory and the number of nonisomorphic models* (X 0809) North Holland: Amsterdam xvi+544pp
⋄ C45 C98 ⋄ REV MR 81a:03030 JSL 47.694 • ID 78472

SHELAH, S. [1978] *End extensions and numbers of countable models* (J 0036) J Symb Logic 43*550-562
⋄ C15 C30 C55 C62 E50 ⋄ REV MR 80b:03037
Zbl 412#03043 JSL 46.663 • ID 29282

SHELAH, S. [1978] *Jonsson algebras in successor cardinals* (J 0029) Israel J Math 30*57-64
⋄ C05 C55 E05 E10 E75 ⋄ REV MR 58#21572
Zbl 384#03016 • ID 31288

SHELAH, S. [1978] *Models with second order properties. I: Boolean algebras with no definable automorphisms* (J 0007) Ann Math Logic 14*57-72
⋄ C07 C30 C50 C55 E50 G05 ⋄ REV MR 80b:03047a
Zbl 383#03018 • REM Part II 1978 • ID 52000

SHELAH, S. [1978] *Models with second order properties. II: Trees with no undefined branches* (J 0007) Ann Math Logic 14*73-87,Appendix:223-226
⋄ C50 C55 C62 C75 C80 C95 E05 E35 E65 ⋄ REV
MR 80b:03047bc Zbl 383#03019 Zbl 383#03020 • REM Part I 1978. Part III 1980 • ID 52002

SHELAH, S. [1978] *On the number of minimal models* (J 0036) J Symb Logic 43*475-480
⋄ C15 C50 ⋄ REV MR 58#10414 Zbl 389#03014
• ID 29273

SHELAH, S. & STERN, J. [1978] *The Hanf number of the first order theory of Banach spaces* (J 0064) Trans Amer Math Soc 244*147-171
⋄ C55 C65 H10 ⋄ REV MR 80a:03047 Zbl 396#03047
• ID 31519

SHELAH, S. & ZIEGLER, M. [1979] *Algebraically closed groups of large cardinality* (J 0036) J Symb Logic 44*522-532
⋄ C25 C60 C75 ⋄ REV MR 80j:03048 Zbl 427#03025
• ID 53712

SHELAH, S. [1979] *Hanf number of omitting type for simple first-order theories* (J 0036) J Symb Logic 44*319-324
⋄ C20 C50 C55 C75 ⋄ REV MR 80k:03031
Zbl 428#03025 • ID 53784

SHELAH, S. [1979] see GUREVICH, Y.

SHELAH, S. [1979] *On uncountable abelian groups* (J 0029) Israel J Math 32*311-330
⋄ C60 E35 E45 E50 E75 ⋄ REV MR 82h:03054
Zbl 412#20047 • ID 52977

SHELAH, S. [1979] *On uniqueness of prime models* (J 0036) J Symb Logic 44*215-220
⋄ C45 C50 ⋄ REV MR 80m:03066 Zbl 421#03021
JSL 48.497 • ID 53420

SHELAH, S. [1979] see MAKOWSKY, J.A.

SHELAH, S. [1980] *A note on cardinal exponentiation* (J 0036) J Symb Logic 45*56-66
⋄ C55 E05 E50 E65 ⋄ REV MR 82c:03070
Zbl 439#03038 • ID 56029

SHELAH, S. [1980] *Independence results* (J 0036) J Symb Logic 45*563-573
⋄ C45 C55 E07 E35 E40 ⋄ REV MR 82b:03099
Zbl 451#03017 • ID 54032

SHELAH, S. [1980] *On a problem of Kurosh, Jonsson groups, and applications* (P 2634) Word Problems II;1976 Oxford 373-394
⋄ C60 ⋄ REV MR 81j:20047 Zbl 438#20025 • ID 82788

SHELAH, S. [1980] see RUBIN, M.

SHELAH, S. [1980] *Simple unstable theories* (J 0007) Ann Math Logic 19*177-203
⋄ C45 C50 C55 ⋄ REV MR 82g:03055 Zbl 489#03008
• ID 78466

SHELAH, S. [1980] see CHERLIN, G.L.

SHELAH, S. [1980] *Whitehead groups may not be free even assuming CH. II* (J 0029) Israel J Math 35*257-285
⋄ C60 E05 E35 E50 E75 ⋄ REV MR 82h:03055
Zbl 467#03049 • REM Part I 1977 • ID 55047

SHELAH, S. [1981] see HODGES, W.

SHELAH, S. [1981] *Models with second order properties. III. Omitting types for L(Q)* (J 0009) Arch Math Logik Grundlagenforsch 21*1-11
⋄ C55 C75 C80 E05 ⋄ REV MR 83a:03031
Zbl 502#03016 • REM Part II 1978. Part IV 1983 • ID 33320

SHELAH, S. [1981] *On endo-rigid strongly $\aleph_1$-free abelian groups in $\aleph_1$* (J 0029) Israel J Math 40*291-295
⋄ C50 C55 C60 E05 E50 E75 ⋄ REV MR 83f:20042
Zbl 501#03015 • ID 38748

SHELAH, S. [1981] *On saturation for a predicate* (J 0047) Notre Dame J Formal Log 22*239-248
⋄ C45 C50 C55 C65 E50 ⋄ REV MR 83b:03036
Zbl 497#03022 • ID 33512

SHELAH, S. [1981] *On the number of nonisomorphic models of cardinality λ, $L_{\infty\lambda}$-equivalent to a fixed model* (J 0047) Notre Dame J Formal Log 22*5-10
⋄ C55 C75 E45 E55 E75 ⋄ REV MR 82k:03054
Zbl 453#03040 • ID 54170

SHELAH, S. [1981] see GLASS, A.M.W.

SHELAH, S. [1981] see MAKOWSKY, J.A.

SHELAH, S. [1982] see GUREVICH, Y.

SHELAH, S. [1982] *On the number of nonisomorphic models in $L_{\infty,\kappa}$ when κ is weakly compact* (J 0047) Notre Dame J Formal Log 23*21-26
⋄ C55 C75 E55 ⋄ REV MR 84h:03093 Zbl 481 # 03024
• ID 34283

SHELAH, S. [1982] *The spectrum problem I: $\aleph_\varepsilon$-saturated models, the main gap. II: Totally transcendental and the infinite depth* (J 0029) Israel J Math 43*324-356,357-364
⋄ C45 C50 C52 ⋄ REV MR 84j:03070 Zbl 532 # 03013 Zbl 532 # 03014 • ID 34661

SHELAH, S. [1983] *Classification theory of non-elementary classes, Part I: The number of uncountable models of $\psi \in L_{\omega_1,\omega}$* (J 0029) Israel J Math 46*212-273
⋄ C45 C52 C55 C75 C80 E50 ⋄ REV MR 85m:03024 Zbl 552 # 03019 • ID 33535

SHELAH, S. [1983] *Construction of many complicated uncountable structures and boolean algebras* (J 0029) Israel J Math 45*100-146
⋄ C50 C55 G05 ⋄ REV Zbl 552 # 03018 • ID 38746

SHELAH, S. [1983] see GUREVICH, Y.

SHELAH, S. [1983] *Models with second order properties IV. A general method and eliminating diamonds* (J 0073) Ann Pure Appl Logic 25*183-212
⋄ C50 C55 C60 C65 C80 E65 G05 ⋄ REV MR 85j:03056 Zbl 558 # 03014 • REM Part III 1980 • ID 38743

SHELAH, S. [1983] see RUBIN, M.

SHELAH, S. [1983] *On the number of non conjugate subgroups* (J 0004) Algeb Universalis 16*131-146
⋄ C55 C60 E50 E75 ⋄ REV MR 84i:20005 Zbl 521 # 20015 • ID 37468

SHELAH, S. [1983] see MAGIDOR, M.

SHELAH, S. [1983] see GROSSBERG, R.

SHELAH, S. [1983] see MAKOWSKY, J.A.

SHELAH, S. [1983] see JARDEN, M.

SHELAH, S. [1983] see LEHMANN, D.J.

SHELAH, S. [1983] see FRIEDMAN, S.D.

SHELAH, S. [1983] see BALDWIN, JOHN T.

SHELAH, S. [1984] *A combinatorial principle and endomorphism rings. I. On p-groups* (J 0029) Israel J Math 49*239-258
⋄ C60 E65 E75 ⋄ REV Zbl 559 # 20039 • REM Part II 1984 • ID 38759

SHELAH, S. [1984] *A combinatorial theorem and endomorphism rings of abelian groups II* (P 4185) Abel Group & Modul;1984 Udine 37-86
⋄ C60 E05 E65 E75 ⋄ REM Part I 1984 • ID 46365

SHELAH, S. [1984] see GOLDFARB, W.D.

SHELAH, S. [1984] see KAUFMANN, M.

SHELAH, S. [1984] *A pair of non-isomorphic $\equiv_{\infty\lambda}$ models of power λ for λ singular with $\lambda^\omega = \lambda$* (J 0047) Notre Dame J Formal Log 25*97-104
⋄ C55 C75 ⋄ REV MR 85j:03055 Zbl 559 # 03022
• ID 38768

SHELAH, S. [1984] see HARRINGTON, L.A.

SHELAH, S. [1984] see EKLOF, P.C.

SHELAH, S. [1984] *Can you take Solovay's inaccessible away?* (J 0029) Israel J Math 48*1-47
⋄ C55 C75 C80 E15 E25 E35 E55 E75 ⋄ REV MR 86g:03082a • ID 38762

SHELAH, S. [1984] see MAGIDOR, M.

SHELAH, S. [1984] see GIORGETTA, D.

SHELAH, S. [1984] *On logical sentences in PA* (P 3710) Logic Colloq;1982 Firenze 145-160
⋄ C62 C80 E05 F25 F30 ⋄ REV MR 86g:03097 Zbl 564 # 03045 • ID 38758

SHELAH, S. [1984] *On universal graphs without instances of CH* (J 0073) Ann Pure Appl Logic 26*75-87
⋄ C50 E05 E35 E50 ⋄ REV MR 85h:03054 Zbl 551 # 03032 • ID 38761

SHELAH, S. [1984] see LACHLAN, A.H.

SHELAH, S. [1984] see GUREVICH, Y.

SHELAH, S. [1985] see DROSTE, M.

SHELAH, S. [1985] *Classification of first order theories which have a structure theorem* (J 0589) Bull Amer Math Soc (NS) 12*227-232
⋄ C35 C45 C52 C75 ⋄ REV MR 86h:03058 Zbl 578 # 03016 • ID 41818

SHELAH, S. [1985] see PILLAY, A.

SHELAH, S. [1985] *Incompactness in regular cardinals* (J 0047) Notre Dame J Formal Log 26*195-228
⋄ C55 C60 E05 ⋄ REM Appendix by A.Mekler • ID 47543

SHELAH, S. [1985] see HOLLAND, W.C.

SHELAH, S. [1985] *Monadic logic and Loewenheim numbers* (J 0073) Ann Pure Appl Logic 28*203-216
⋄ C45 C55 C65 C75 C85 E45 ⋄ REV Zbl 565 # 03015
• ID 38772

SHELAH, S. [1985] see BONNET, R.

SHELAH, S. [1985] see KAUFMANN, M.

SHELAH, S. [1985] see ABRAHAM, U.

SHELAH, S. [1985] *On the possible number no(M) = the number of nonisomorphic models $L_{\infty,\lambda}$-equivalent to M of power λ, for λ singular* (J 0047) Notre Dame J Formal Log 26*36-50
⋄ C55 C75 ⋄ REV Zbl 567 # 03010 • ID 42584

SHELAH, S. [1985] see SAGEEV, G.

SHELAH, S. [1985] *Remarks in abstract model theory* (J 0073) Ann Pure Appl Logic 29*255-288
⋄ C40 C50 C52 C55 C80 C95 ⋄ ID 47478

SHELAH, S. [1985] see BALDWIN, JOHN T.

SHELAH, S. [1985] see GOEBEL, R.

SHELAH, S. [1985] see MEKLER, A.H.

SHELAH, S. [1985] see GUREVICH, Y.

SHELAH, S. [1985] *Uncountable constructions for B.A.,e.c. groups and Banach spaces* (J 0029) Israel J Math 51*273-297
⋄ C25 C60 C65 E05 E65 E75 G05 ⋄ ID 48344

SHELAH, S. see Vol. I, II, IV, V for further entries

SHEN, CHENGMIN [1982] see LU, JINGBO

SHEN, FUXING [1982] *Elementary extensions and elementary chains of lattice-valued models* (J 3769) Sci Bull, Foreign Lang Ed 27∗691-693
◇ C35 C40 C90 ◇ REV MR 85i:03124 Zbl 519#03030
• ID 37542

SHEN, FUXING [1984] *Forcing in lattice-valued model theory (Chinese)* (J 2771) Kexue Tongbao 29∗1286-1287
◇ C25 C90 ◇ ID 48207

SHEN, FUXING & WANG, SHIQIANG & YUE, QIJING [1984] *Number-theoretic properties of some Goldbach and non-Goldbach commutative rings (Chinese) (English summary)* (J 2521) Beijing Shifan Daxue Xuebao, Ziran Kexue 1984/4∗27-32
◇ C60 ◇ ID 48442

SHEN, FUXING [1985] *Lattice-valued model complete theory (Chinese) (English summary)* (J 2521) Beijing Shifan Daxue Xuebao, Ziran Kexue 1985/2∗9-14
◇ C35 C90 ◇ ID 48443

SHEN, FUXING [1985] *Quadratic residues in a sort of commutative rings with Goldbach property (Chinese)* (J 3742) Shuxue Yanjiu yu Pinglun 5/4∗117-118
◇ C60 ◇ ID 48444

SHEPHERDSON, J.C. [1951] *Inner models for set theory I* (J 0036) J Symb Logic 16∗161-190
◇ C62 E35 E55 ◇ REV MR 13.522 Zbl 43.53 JSL 18.342
• REM Part II 1952 • ID 28756

SHEPHERDSON, J.C. [1951] *Inverses and zero divisors in matrix rings* (J 3240) Proc London Math Soc, Ser 3 1∗71-85
• ERR/ADD ibid 1∗II
◇ B25 C60 D40 ◇ REV MR 13.7 Zbl 43.17 • ID 12066

SHEPHERDSON, J.C. [1952] *Inner models for set theory II* (J 0036) J Symb Logic 17∗225-237
◇ C62 E35 E55 ◇ REV MR 14.835 Zbl 48.281 JSL 18.342 • REM Part I 1951. Part III 1953 • ID 28757

SHEPHERDSON, J.C. [1953] *Inner models for set theory III* (J 0036) J Symb Logic 18∗145-167
◇ C62 E35 E55 ◇ REV MR 15.278 Zbl 51.38 JSL 18.342
• REM Part II 1952 • ID 28758

SHEPHERDSON, J.C. [1955] see FROEHLICH, A.

SHEPHERDSON, J.C. [1964] *A non-standard model for a free variable fragment of number theory* (J 0014) Bull Acad Pol Sci, Ser Math Astron Phys 12∗79-86
◇ C62 F30 ◇ REV MR 28#5002 Zbl 132.247 JSL 30.389
• ID 12075

SHEPHERDSON, J.C. [1965] *Non-standard models for fragments of number theory* (P 0614) Th Models;1963 Berkeley 342-358
◇ C62 F30 ◇ REV MR 33#5483 Zbl 154.262 • ID 12072

SHEPHERDSON, J.C. [1967] *The rule of induction in the free variable arithmetic based on $+$ and $\ast$* (J 0179) Ann Fac Sci Clermont 35∗25-31
◇ C62 F30 ◇ REV MR 43#1837 • ID 12077

SHEPHERDSON, J.C. [1969] *Weak and strong induction* (J 0005) Amer Math Mon 76∗989-1004
◇ C62 F30 ◇ REV MR 40#2518 Zbl 188.24 • ID 12078

SHEPHERDSON, J.C. [1976] see BEZBORUAH, A.

SHEPHERDSON, J.C. see Vol. I, II, IV, V, VI for further entries

SHESTOPAL, G.A. [1968] see FOL'K, N.F.

SHESTOPAL, G.A. see Vol. I for further entries

SHEVRIN, L.N. [1966] *Elementary lattice properties of semigroups (Russian)* (J 0023) Dokl Akad Nauk SSSR 167∗305-308
• TRANSL [1966] (J 0062) Sov Math, Dokl 7∗395-398
◇ C05 C60 ◇ REV MR 34#1430 Zbl 158.24 • ID 11996

SHEVYAKOV, V.S. [1973] *Formulas of the restricted predicate calculus which distinguish certain classes of models with simply computable predicates (Russian)* (J 0023) Dokl Akad Nauk SSSR 210∗285-287
• TRANSL [1973] (J 0062) Sov Math, Dokl 14∗743-745
◇ C57 ◇ REV MR 49#2346 Zbl 329#02004 • ID 11994

SHI, NIANDONG [1982] *Creative pairs of subalgebras of recursively enumerable boolean algebras (Chinese)* (J 0418) Shuxue Xuebao 25∗737-745
◇ C57 D25 D45 G05 ◇ REV MR 85b:03078 Zbl 524#03030 • ID 37603

SHI, NIANDONG see Vol. II for further entries

SHIKHANOVICH, YU.A. [1959] *Examples of application of mathematical logic to algebra (Russian)* (P 0607) All-Union Math Conf (3);1956 Moskva 2∗148-149
◇ C60 ◇ ID 45705

SHILLETO, J.R. [1972] *Minimum models of analysis* (J 0036) J Symb Logic 37∗48-54
◇ C50 C62 D55 E45 ◇ REV MR 51#7872
Zbl 247#02053 JSL 39.601 • ID 29513

SHIMODA, M. [1979] *Sheaves over cHa and the Heyting valued model $V^{(H)}$ (Japanese)* (P 4108) Found of Math;1979 Kyoto 92-105
◇ C90 E40 G30 ◇ ID 47660

SHIMODA, M. [1981] *Categorical aspects of Heyting-valued models for intuitionistic set theory* (J 0407) Comm Math Univ St Pauli (Tokyo) 30∗17-35
◇ C90 E40 E70 F50 G30 ◇ REV MR 83a:03070
Zbl 466#03020 • ID 54971

SHIRAI, K. [1975] *Intuitionistic version of the Los-Tarski-Robinson theorem* (J 0260) Ann Jap Ass Phil Sci 4∗323-332
◇ C40 F05 F50 ◇ REV MR 58#10332 Zbl 317#02024
• ID 29640

SHIRAI, K. see Vol. II, VI for further entries

SHISHMAREV, YU.E. [1972] see ABAKUMOV, A.I.

SHISHMAREV, YU.E. [1972] *Categorical theories of a function (Russian)* (J 0087) Mat Zametki (Akad Nauk SSSR) 11∗89-98
• TRANSL [1972] (J 1044) Math Notes, Acad Sci USSR 11∗58-63
◇ C35 ◇ REV MR 45#46 Zbl 241#02016 • ID 28823

SHISHMAREV, YU.E. [1979] *Categorical quasivarieties of quasigroups (Russian)* (J 0003) Algebra i Logika 18∗614-629
• TRANSL [1979] (J 0069) Algeb and Log 18∗389-400
◇ C05 C35 ◇ REV MR 81i:20104 Zbl 444#03014
• ID 54355

SHODA, K. [1952] *Zur Theorie der algebraischen Erweiterungen* (J 1770) Osaka Math J 4∗133-143
◇ C05 C60 ◇ REV MR 14.614 • ID 41672

SHODA, K. [1955] *Bemerkungen ueber die Existenz der algebraisch abgeschlossenen Erweiterung* (J 0081) Proc Japan Acad 31*128-130
⋄ C05   C60   ⋄ REV   MR 17.6   Zbl 67.12   • ID 12086

SHOENFIELD, J.R. [1960] *Degrees of models* (J 0036) J Symb Logic 25*233-237
⋄ C57   D30   D35   ⋄ REV   MR 25#2957   Zbl 105.248   JSL 22.623   JSL 33.623   • ID 12098

SHOENFIELD, J.R. [1965] *Applications of model theory to degrees of unsolvability* (P 0614) Th Models;1963 Berkeley 359-363
⋄ C50   D25   D30   ⋄ REV   MR 34#53   Zbl 192.52   JSL 37.610   • ID 12099

SHOENFIELD, J.R. [1967] *Mathematical logic* (X 0832) Addison-Wesley: Reading viii+344pp
• TRANSL [1975] (X 2027) Nauka: Moskva 527pp
⋄ B98   C98   D98   E98   F98   ⋄ REV   MR 37#1224   MR 53#87   Zbl 155.11   JSL 40.234   • ID 22384

SHOENFIELD, J.R. [1971] *A theorem on quantifier elimination* (P 0669) Conv Teor Modelli & Geom;1969/70 Roma 173-176
⋄ C10   C60   ⋄ REV   MR 42#7497   Zbl 222#02007   • ID 12104

SHOENFIELD, J.R. [1975] *The decision problem for recursively enumerable degrees* (J 0015) Bull Amer Math Soc 81*973-977
⋄ B25   D25   ⋄ REV   MR 52#7882   Zbl 339#02043   • ID 18383

SHOENFIELD, J.R. [1977] *Quantifier elimination in fields* (P 1076) Latin Amer Symp Math Log (3);1976 Campinas 243-252
⋄ C10   C60   C98   ⋄ REV   MR 57#100   Zbl 362#02045   • ID 16608

SHOENFIELD, J.R. see Vol. I, IV, V, VI for further entries

SHORB, A.M. [1972] *Filter constructions in boolean valued model theory* (J 0068) Z Math Logik Grundlagen Math 18*193-200
⋄ C90   ⋄ REV   MR 47#8291   Zbl 253#02053   • ID 12111

SHORB, A.M. see Vol. I for further entries

SHORE, R.A. [1980] see NERODE, A.

SHORE, R.A. [1981] *The degrees of unsolvability: global results* (P 2628) Log Year;1979/80 Storrs 283-301
⋄ C40   D30   D35   F35   ⋄ REV   MR 82k:03065   Zbl 474#03022   • ID 55426

SHORE, R.A. [1984] see LERMAN, M.

SHORE, R.A. see Vol. I, IV, V, VI for further entries

SHOSTAK, R.E. [1977] *On the SUP-INF method for proving Presburger formulas* (J 0037) ACM J 24*529-543
⋄ B25   B35   F30   ⋄ REV   MR 58#13992   Zbl 423#68052   • ID 69916

SHOSTAK, R.E. [1979] *A practical decision procedure for arithmetic with function symbols* (J 0037) ACM J 26*351-360
⋄ B25   B35   F30   ⋄ REV   MR 80f:68020   Zbl 496#03003   • ID 82799

SHOSTAK, R.E. [1982] *Deciding combinations of theories* (P 3840) Autom Deduct (6);1982 New York 209-222
⋄ B25   B35   ⋄ REV   MR 85h:68075   Zbl 481#68089   • ID 38462

SHOSTAK, R.E. see Vol. I for further entries

SHREJDER, YU.A. & VILENKIN, N.YA. [1977] *Majorant spaces and "majority" quantifier (Russian) (English summary)* (S 2582) Semiotika & Inf, Akad Nauk SSSR 8*45-82
• ERR/ADD ibid 212-213
⋄ B45   C80   E20   ⋄ REV   MR 57#5677   • ID 79715

SHREJDER, YU.A. see Vol. II, IV, V for further entries

SHTEJNBUK, V.B. [1982] *Semigroups of endomorphisms of filters (Russian)* (J 0031) Izv Vyssh Ucheb Zaved, Mat (Kazan) 1982/1*71-74
⋄ C07   E05   ⋄ REV   MR 83g:20075   Zbl 491#20054   • ID 39142

SHTEJNBUK, V.B. see Vol. V for further entries

SHUKLA, A. [1970] *Decision procedures for Lewis system S1 and related modal systems* (J 0047) Notre Dame J Formal Log 11*141-180   • ERR/ADD ibid 14*584
⋄ B25   B45   ⋄ REV   MR 44#1544   MR 48#8203   Zbl 182.6   JSL 37.754   • ID 19678

SHUKLA, A. see Vol. I, II for further entries

SICHLER, J. [1973] *Weak automorphisms of universal algebras* (J 0004) Algeb Universalis 3*1-7
⋄ C05   E50   ⋄ REV   MR 48#8354   Zbl 273#08008   • ID 12136

SICHLER, J. [1974] see GRAETZER, G.

SICHLER, J. [1975] *Note on free products of Hopfian lattices* (J 0004) Algeb Universalis 5*145-146
⋄ C30   G10   ⋄ REV   MR 53#5390   Zbl 321#06006   • ID 24458

SICHLER, J. see Vol. V for further entries

SIDORCHUK, A.D. [1981] see PROTASOV, I.V.

SIDORCHUK, A.D. [1983] *On the completeness theorem for topological models (Russian) (English summary)* (J 0270) Dokl Akad Nauk Ukr SSR, Ser A 1983/7*75-78
⋄ C90   ⋄ REV   MR 85b:03057   Zbl 513#03016   • ID 37222

SIEFKES, D. [1968] *Recursion theory and the theorem of Ramsey in one-place second order successor arithmetic* (P 0608) Logic Colloq;1966 Hannover 237-254
⋄ B15   B25   C85   ⋄ REV   MR 39#64   Zbl 186.11   JSL 40.504   • ID 21146

SIEFKES, D. [1970] *Decidable theories I. Buechi's monadic second order successor arithmetic* (S 3301) Lect Notes Math 120*xii+130pp
⋄ B15   B25   C85   D05   F35   F98   ⋄ REV   MR 44#6488   Zbl 399#03011   • ID 12145

SIEFKES, D. [1970] *Decidable extensions of monadic second order successor arithmetic* (P 0577) Automatenth & Formale Sprachen;1969 Oberwolfach 441-472
⋄ B15   B25   C85   D05   F35   ⋄ REV   MR 56#8354   Zbl 213.19   • ID 27984

SIEFKES, D. [1971] *Undecidable extensions of monadic second order successor arithmetic* (J 0068) Z Math Logik Grundlagen Math 17*385-394
⋄ B15   B25   C85   D35   F35   ⋄ REV   MR 45#1763   Zbl 193.312   • ID 12146

SIEFKES, D. [1973] see BUECHI, J.R.

SIEFKES, D. [1978] *An axiom system for the weak monadic second order theory of two successors* (J 0029) Israel J Math 30*264-284
  ⋄ B15 B25 C85 D05 ⋄ REV MR 80a:03015 Zbl 397 # 03009 • ID 29136

SIEFKES, D. [1983] see BUECHI, J.R.

SIEFKES, D. see Vol. IV, VI for further entries

SIEKMANN, J. & SZABO, P. [1982] *Universal unification and a classification of equational theories* (P 3840) Autom Deduct (6);1982 New York 369-389
  • TRANSL [1984] (J 3079) Kiber Sb Perevodov, NS 21*213-234
  ⋄ B25 B35 C05 ⋄ REV MR 85i:08005 Zbl 494 # 68087
  • ID 36659

SIEKMANN, J. see Vol. I, IV for further entries

SIKORSKI, R. [1950] see RASIOWA, H.

SIKORSKI, R. [1951] see RASIOWA, H.

SIKORSKI, R. [1952] *Products of abstract algebras* (J 0027) Fund Math 39*211-228
  ⋄ C05 C75 ⋄ REV MR 14.839 Zbl 50.27 • ID 12290

SIKORSKI, R. [1953] see RASIOWA, H.

SIKORSKI, R. [1955] see RASIOWA, H.

SIKORSKI, R. [1960] see RASIOWA, H.

SIKORSKI, R. [1961] *A topological characterization of open theories* (J 0014) Bull Acad Pol Sci, Ser Math Astron Phys 9*259-260
  ⋄ B20 C07 G05 ⋄ REV MR 25 # 2961 Zbl 124.249 JSL 38.163 • ID 12306

SIKORSKI, R. [1962] *Algebra of formalized languages* (S 0019) Colloq Math (Warsaw) 9*1-31
  ⋄ C05 C07 G15 G25 ⋄ REV MR 25 # 3814 Zbl 104.241 JSL 31.508 • ID 12309

SIKORSKI, R. [1962] *On open theories* (S 0019) Colloq Math (Warsaw) 9*171-182
  ⋄ B20 C07 G05 ⋄ REV MR 27 # 1370 Zbl 124.249 JSL 38.163 • ID 12308

SIKORSKI, R. [1963] *Products of generalized algebras and products of realizations* (S 0019) Colloq Math (Warsaw) 10*1-13
  ⋄ C05 C07 G25 ⋄ REV MR 33 # 2533 Zbl 122.9 JSL 36.682 • ID 12313

SIKORSKI, R. [1963] see RASIOWA, H.

SIKORSKI, R. see Vol. I, II, IV, V, VI for further entries

SILVER, J.H. [1971] see KEISLER, H.J.

SILVER, J.H. [1971] *Measurable cardinals and Δ^1_3 well-orderings* (J 0120) Ann of Math, Ser 2 94*414-446
  ⋄ C20 C30 D55 E05 E15 E35 E45 E55 ⋄ REV MR 45 # 8517 Zbl 259 # 02054 JSL 39.330 • ID 19669

SILVER, J.H. [1971] *Some applications of model theory in set theory* (J 0007) Ann Math Logic 3*45-110
  ⋄ C30 C55 D55 E05 E45 E55 ⋄ REV MR 53 # 12950 Zbl 215.324 JSL 39.597 • REM This paper is most of the author's 1966 Berkeley dissertation • ID 12319

SILVER, J.H. [1971] *The independence of Kurepa's conjecture and two-cardinal conjectures in model theory* (P 0693) Axiomatic Set Th;1967 Los Angeles 1*383-390
  ⋄ C55 E05 E35 E55 ⋄ REV MR 43 # 3112 Zbl 255 # 02068 JSL 40.506 • ID 12318

SILVER, J.H. [1973] *The bearing of large cardinals on constructibility* (C 0654) Stud in Model Th 158-182
  ⋄ C20 C30 E05 E45 E55 ⋄ REV MR 49 # 2375
  • ID 12321

SILVER, J.H. [1974] *Indecomposable ultrafilters and $O^\#$* (P 0610) Tarski Symp;1971 Berkeley 357-363
  ⋄ C20 E05 E45 E55 ⋄ REV MR 50 # 12726 Zbl 324 # 02063 • ID 12322

SILVER, J.H. see Vol. IV, V for further entries

SIMMONS, H. [1970] *The solution of a decision problem for several classes of rings* (J 0048) Pac J Math 34*547-557
  ⋄ B25 C60 ⋄ REV MR 42 # 1658 Zbl 193.313 • ID 12327

SIMMONS, H. [1972] *A possible characterization of generic structures* (J 0132) Math Scand 31*257-261
  ⋄ C25 ⋄ REV MR 52 # 81 Zbl 262 # 02052 • ID 18387

SIMMONS, H. [1972] *Existentially closed structures* (J 0036) J Symb Logic 37*293-310
  ⋄ C25 C35 C50 ⋄ REV MR 51 # 12518 Zbl 277 # 02011
  • ID 12328

SIMMONS, H. [1973] *An omitting types theorem with an application to the construction of generic structures* (J 0132) Math Scand 33*46-54
  ⋄ C07 C25 C30 ⋄ REV MR 48 # 10798 Zbl 277 # 02012
  • ID 12332

SIMMONS, H. [1973] *Le nombre de structures generiques d'une theorie* (J 2313) C R Acad Sci, Paris, Ser A-B 277*A487-A489
  ⋄ C25 ⋄ REV MR 49 # 2347 Zbl 314 # 02061 • ID 12330

SIMMONS, H. [1973] see GOLDREI, D.C.

SIMMONS, H. [1973] *Une construction du "forcing-compagnon" d'une theorie* (J 2313) C R Acad Sci, Paris, Ser A-B 277*A563-A566
  ⋄ C25 ⋄ REV MR 49 # 2348 Zbl 314 # 02060 • ID 12331

SIMMONS, H. [1974] *Companion theories (Forcing in model theory)* (X 1754) Univ Catholique: Louvain 61pp
  ⋄ C25 C98 ⋄ REV Zbl 355 # 02036 • ID 31240

SIMMONS, H. [1974] *The use of injective-like structures in model theory* (J 0020) Compos Math 28*113-142
  ⋄ C25 C35 C50 ⋄ REV MR 50 # 4289 Zbl 291 # 02035
  • ID 12333

SIMMONS, H. [1975] see MACINTYRE, A.

SIMMONS, H. [1975] *Counting countable e.c. structures* (J 0079) Logique & Anal, NS 18*309-357
  • REPR [1977] (P 1625) Six Days of Model Th;1975 Louvain-la-Neuve 77-125
  ⋄ C15 C25 C50 C98 ⋄ REV MR 57 # 12203 Zbl 352 # 02038 Zbl 396 # 03030 • ID 50037

SIMMONS, H. [1975] see FISHER, E.R.

SIMMONS, H. [1975] *The complexity of T^f and omitting types in F_T* (C 0782) Model Th & Algeb (A. Robinson) 403-407
  ⋄ C07 C25 ⋄ REV MR 53 # 7766 Zbl 322 # 02049
  • ID 23008

SIMMONS, H. [1976] *Each regular number structure is biregular* (J 0029) Israel J Math 23*347-352 • ERR/ADD ibid 26*95
⋄ C25 C62 ⋄ REV MR 54#12518 MR 54#2456 Zbl 367#02028 Zbl 367#02029 • ID 18389

SIMMONS, H. [1976] *Large and small existentially closed structures* (J 0036) J Symb Logic 41*379-390
⋄ C25 C35 C50 ⋄ REV MR 54#87 Zbl 372#02029
• ID 14770

SIMMONS, H. [1977] *Existentially closed models of basic number theory* (P 1075) Logic Colloq;1976 Oxford 325-369
⋄ C25 C62 ⋄ REV MR 58#5186 Zbl 433#03015
• ID 16628

SIMMONS, H. see Vol. II, IV, VI for further entries

SIMPSON, S.G. [1974] *Forcing and models of arithmetic* (J 0053) Proc Amer Math Soc 43*193-194
⋄ C15 C25 C62 E40 E47 ⋄ REV MR 55#7765 Zbl 291#02037 • ID 30799

SIMPSON, S.G. [1975] see KREISEL, G.

SIMPSON, S.G. [1982] see SCHMERL, J.H.

SIMPSON, S.G. [1982] *Set theoretic aspects of ATR_0* (P 3623) Logic Colloq;1980 Prague 255-271
⋄ C62 E30 F35 ⋄ REV MR 84b:03078b Zbl 497#03046
• ID 35660

SIMPSON, S.G. [1985] *Friedman's research on subsystems of second order arithmetic* (C 4699) Friedman's Res on Found of Math 137-159
⋄ C62 C98 F35 F98 ⋄ ID 49879

SIMPSON, S.G. [1985] see HARRINGTON, L.A.

SIMPSON, S.G. see Vol. I, IV, V, VI for further entries

SIMSON, D. [1972] *$\aleph$-flat and $\aleph$-projective modules (Russian summary)* (J 0014) Bull Acad Pol Sci, Ser Math Astron Phys 20*109-114
⋄ C60 ⋄ REV MR 47#3450 Zbl 232#16021 • ID 12343

SIMSON, D. [1972] *On the structure of flat modules (Russian summary)* (J 0014) Bull Acad Pol Sci, Ser Math Astron Phys 20*115-120
⋄ C60 ⋄ REV MR 47#3451 Zbl 232#16022 • ID 12344

SINDELAR, J. [1978] see KRAMOSIL, I.

SINDELAR, J. see Vol. I, II for further entries

SINGER, M.F. [1978] *A class of differential fields with minimal differential closures* (J 0053) Proc Amer Math Soc 69*319-322
⋄ C60 ⋄ REV MR 57#5737 Zbl 396#03032 • ID 52642

SINGER, M.F. [1978] *The model theory of ordered differential fields* (J 0036) J Symb Logic 43*82-91
⋄ B25 C25 C35 C60 C65 D80 ⋄ REV MR 80a:03044 Zbl 396#03031 • ID 29244

SINGER, M.F. see Vol. I, IV for further entries

SISTLA, A.P. [1984] see EMERSON, E.A.

SISTLA, A.P. see Vol. II, IV for further entries

SKOLEM, T.A. [1919] *Untersuchungen ueber die Axiome des Klassenkalkuels und ueber Produktions- und Summationsprobleme, welche gewisse Klassen von Aussagen betreffen* (J 0974) Norsk Vid-Akad Oslo Mat-Natur Kl Skr 3*37pp
• REPR [1970] (C 1098) Skolem: Select Works in Logic 67-101
⋄ B20 B25 C10 D35 G05 ⋄ ID 12383

SKOLEM, T.A. [1920] *Logisch-kombinatorische Untersuchungen ueber die Erfuellbarkeit oder Beweisbarkeit mathematischer Saetze nebst einem Theorem ueber dichte Mengen* (J 0974) Norsk Vid-Akad Oslo Mat-Natur Kl Skr 1920/4*1-36
• TRANSL [1967] (C 0675) From Frege to Goedel 254-263
• REPR [1970] (C 1098) Skolem: Select Works in Logic 103-136
⋄ B10 B25 C07 C65 C75 ⋄ REV FdM 48.1121
• ID 12382

SKOLEM, T.A. [1923] *Einige Bemerkungen zur axiomatischen Begruendung der Mengenlehre* (P 1086) Skand Mat Kongr (5);1922 Helsinki 217-232
• TRANSL [1967] (C 0675) From Frege to Goedel 290-301 (English) • REPR [1970] (C 1098) Skolem: Select Works in Logic 137-152
⋄ A05 B30 C07 C62 E30 ⋄ REV FdM 49.138 • ID 21221

SKOLEM, T.A. [1928] *Ueber die mathematische Logik* (J 4510) Norsk Mat Tidsskr 10*125-142
• TRANSL [1967] (C 0675) From Frege to Goedel 508-524
• REPR [1970] (C 1098) Skolem: Select Works in Logic 189-206
⋄ A05 B25 B98 C07 C10 C35 ⋄ REV FdM 54.58
• ID 12385

SKOLEM, T.A. [1929] *Ueber einige Grundlagenfragen der Mathematik* (J 0974) Norsk Vid-Akad Oslo Mat-Natur Kl Skr 4*1-49
• REPR [1970] (C 1098) Skolem: Select Works in Logic 227-273
⋄ A05 A10 B30 C07 E30 ⋄ REV FdM 55.31 • ID 12386

SKOLEM, T.A. [1931] *Ueber einige Satzfunktionen in der Arithmetik* (J 0974) Norsk Vid-Akad Oslo Mat-Natur Kl Skr 7*28pp
• REPR [1970] (C 1098) Skolem: Select Works in Logic 281-306
⋄ B25 B28 C10 F30 ⋄ REV Zbl 2.3 FdM 57.1320
• ID 12388

SKOLEM, T.A. [1933] *Ein kombinatorischer Satz mit Anwendung auf ein logisches Entscheidungsproblem* (J 0027) Fund Math 20*254-261
• REPR [1970] (C 1098) Skolem: Select Works in Logic 337-344
⋄ B20 B25 E05 ⋄ REV Zbl 7.97 FdM 59.54 • ID 12391

SKOLEM, T.A. [1933] *Ueber die Unmoeglichkeit einer Charakterisierung der Zahlenreihe mittels eines endlichen Axiomensystems* (J 0975) Norsk Mat Forenings Skr 10*73-82
• REPR [1970] (C 1098) Skolem: Select Works in Logic 345-354
⋄ B28 C20 C62 F30 H15 ⋄ REV Zbl 7.193 FdM 59.53
• ID 12392

SKOLEM, T.A. [1934] *Ueber die Nichtcharakterisierbarkeit der Zahlenreihe mittels endlich oder abzaehlbar unendlich vieler Aussagen mit ausschliesslich Zahlenvariablen* (**J** 0027) Fund Math 23*150-161
- REPR [1970] (**C** 1098) Skolem: Select Works in Logic 355-366
- ⋄ B28 C20 C62 F30 H15 ⋄ REV Zbl 10.49 FdM 60.25
- ID 12393

SKOLEM, T.A. [1935] *Ueber die Erfuellbarkeit gewisser Zaehlausdruecke* (**J** 0974) Norsk Vid-Akad Oslo Mat-Natur Kl Skr 6*14pp
- ⋄ B25 B28 ⋄ REV Zbl 13.97 FdM 61.973 • ID 12395

SKOLEM, T.A. [1936] *Ein Satz ueber die Erfuellbarkeit von einigen Zaehlausdruecken der Form* $(x)(\exists y_1,...,y_n) K_1(x,y_1,...,y_n)$ & $(x_1,x_2,x_3) K_2(x_1,x_2,x_3)$ (**J** 0752) Avh Norske Vid-Akad Oslo I 1935/8*1-10
- REPR [1970] (**C** 1098) Skolem: Select Works in Logic 375-394
- ⋄ B10 B25 ⋄ REV Zbl 13.242 JSL 1.111 FdM 62.42
- ID 40864

SKOLEM, T.A. [1937] *Eine Bemerkung zum Entscheidungsproblem* (**P** 1608) Int Congr Math (II, 5);1936 Oslo 2*268-270
- ⋄ B20 B25 D35 ⋄ REV JSL 3.57 FdM 63.31 • ID 27938

SKOLEM, T.A. [1941] *Sur la portee du theoreme de Loewenheim-Skolem* (**P** 0652) Entretiens Zuerich Fond & Method Sci Math;1938 Zuerich 25-52
- REPR [1970] (**C** 1098) Skolem: Select Works in Logic 455-482
- ⋄ C07 C62 E30 ⋄ REV MR 2.338 JSL 7.35 FdM 67.35
- ID 12399

SKOLEM, T.A. [1955] *Peano's axioms and models of arithmetic* (**P** 1589) Math Interpr of Formal Systs;1954 Amsterdam 1-14
- REPR [1970] (**C** 1098) Skolem: Select Works in Logic 587-600
- ⋄ B28 C20 C62 F30 H15 ⋄ REV MR 17.699 Zbl 68.246 JSL 22.306 • ID 27718

SKOLEM, T.A. [1960] *Investigations on a comprehension axiom without negation in the defining propositional functions* (**J** 0047) Notre Dame J Formal Log 1*13-22
- REPR [1970] (**C** 1098) Skolem: Select Works in Logic 6630672
- ⋄ C62 E30 ⋄ REV MR 26 # 4890 Zbl 116.10 • ID 12426

SKOLEM, T.A. [1970] *Selected works in logic* (**X** 1554) Universitesforlaget: Oslo 732pp
- ⋄ B96 C96 ⋄ REV MR 44 # 2562 Zbl 228 # 02001 • REM Edited by Fenstad,J.E. • ID 29109

SKOLEM, T.A. see Vol. I, II, IV, V, VI for further entries

SKORDEV, D.G. [1967] *Certain algebraic aspects of mathematical logic (Bulgarian) (German summary)* (**J** 0255) God Fak Mat & Mekh, Univ Sofiya 62*111-122
- ⋄ C25 C90 G10 ⋄ REV MR 43 # 4647 • ID 12441

SKORDEV, D.G. see Vol. II, IV, V, VI for further entries

SKORNYAKOV, L.A. [1978] *The axiomatizability of a class of injective polygons (Russian)* (**S** 2851) Tr Sem Petrovskogo, Univ Moskva 4*233-239
- ⋄ C05 C52 C60 ⋄ REV MR 80e:03037 Zbl 418 # 03021
- ID 53306

SKORNYAKOV, L.A. [1979] *Finite axiomatizability of a class of point modules (Russian)* (**S** 0019) Colloq Math (Warsaw) 42*365-366
- ⋄ C52 C60 ⋄ REV MR 81h:03075 Zbl 434 # 16018
- ID 55755

SKORNYAKOV, L.A. see Vol. IV, V for further entries

SKVORTSOV, D.P. [1979] *Logic of infinite problems and Kripke models on atomic semilattices of sets (Russian)* (**J** 0023) Dokl Akad Nauk SSSR 245*798-801
- TRANSL [1979] (**J** 0062) Sov Math, Dokl 20*360-363
- ⋄ B55 C90 F50 G20 ⋄ REV MR 80i:03040 Zbl 438 # 03028 • ID 55940

SKVORTSOV, D.P. [1983] *On some ways of construction logical languages with quantifiers over finite sequences (Russian)* (**S** 2582) Semiotika & Inf, Akad Nauk SSSR 20*102-126
- ⋄ B15 C80 ⋄ REV MR 84m:03011 Zbl 512 # 03027
- ID 40156

SKVORTSOV, D.P. see Vol. I, II, V, VI for further entries

SLAGLE, J.R. [1970] *Interpolation theorems for resolution in lower predicate calculus* (**J** 0037) ACM J 17*535-542
- ⋄ B35 C40 ⋄ REV MR 50 # 6804 Zbl 198.36 • ID 12452

SLAGLE, J.R. see Vol. I, VI for further entries

SLOBODSKOJ, A.M. [1973] see FRIDMAN, EH.I.

SLOBODSKOJ, A.M. [1975] see FRIDMAN, EH.I.

SLOBODSKOJ, A.M. [1981] *Unsolvability of the universal theory of finite groups (Russian)* (**J** 0003) Algebra i Logika 20*207-230,251
- TRANSL [1981] (**J** 0069) Algeb and Log 20*139-156
- ⋄ C13 C60 D35 D40 ⋄ REV MR 83h:03062 Zbl 519 # 03006 • ID 36070

SLOBODSKOJ, A.M. see Vol. IV for further entries

SLOMINSKI, J. [1956] *On the extending of models III. Extensions in equationally definable classes of algebras* (**J** 0027) Fund Math 43*69-76
- ⋄ C05 C30 C60 G05 ⋄ REV MR 18.2 Zbl 71.256 JSL 27.94 • REM Part II 1955 by Los,J. & Suszko,R. Part IV 1957 by Los,J. & Suszko,R. • ID 12466

SLOMINSKI, J. [1958] *Theory of models with infinitary operations and relations* (**J** 0014) Bull Acad Pol Sci, Ser Math Astron Phys 6*449-456
- ⋄ C05 C75 ⋄ REV MR 20 # 4479 Zbl 92.6 JSL 31.131
- ID 12467

SLOMINSKI, J. [1959] *The theory of abstract algebras with infinitary operations* (**J** 0202) Diss Math (Warsaw) 18*67pp
- ⋄ C05 ⋄ REV MR 21 # 7173 • ID 12470

SLOMINSKI, J. [1960] see LOS, J.

SLOMSON, A. [1965] see BELL, J.L.

SLOMSON, A. [1967] *Some problems in mathematical logic* (0000) Diss., Habil. etc viii+121pp
- ⋄ C80 ⋄ REM D.Phil.thesis, Oxford • ID 19657

SLOMSON, A. [1968] *The monadic fragment of predicate calculus with the Chang quantifier and equality* (**P** 0692) Summer School in Logic;1967 Leeds 279-301
- ⋄ B25 C20 C55 C80 ⋄ REV MR 39 # 1315 Zbl 195.302
- ID 28589

SLOMSON, A. [1969] see BELL, J.L.

SLOMSON, A. [1970] *An algebraic characterization of indistinguishable cardinals* (J 0036) J Symb Logic 35*97-104
⋄ C55 C85 E10 ⋄ REV MR 44#2595 Zbl 197.277
• ID 12474

SLOMSON, A. [1972] *Generalized quantifiers and well orderings* (J 0009) Arch Math Logik Grundlagenforsch 15*57-73
⋄ B25 C55 C65 C80 E07 ⋄ REV MR 48#3698 Zbl 247#02045 • ID 12475

SLOMSON, A. [1973] see BELL, J.L.

SLOMSON, A. [1976] *Decision problems for generalized quantifiers - a survey* (P 1476) Set Th & Hierarch Th (2) (Mostowski);1975 Bierutowice 249-258
⋄ B25 C80 C98 ⋄ REV MR 58#16248 Zbl 343#02034 • ID 23810

SLOMSON, A. see Vol. I, IV for further entries

SLONNEGER, K. [1976] *A complete infinitary logic* (J 0036) J Symb Logic 41*730-746
⋄ C75 F05 ⋄ REV MR 55#7739 Zbl 357#02015
• ID 14575

SLUPECKI, J. [1961] see POGORZELSKI, W.A.

SLUPECKI, J. [1972] *L-decidability and decidability* (J 0387) Bull Sect Logic, Pol Acad Sci 1/3*38-43
⋄ B22 B25 ⋄ REV MR 51#116 • ID 15183

SLUPECKI, J. [1973] see BRYLL, G.

SLUPECKI, J. [1976] see HALKOWSKA, K.

SLUPECKI, J. see Vol. I, II, V, VI for further entries

SMIRNOV, D.M. [1982] *Universal definability of Mal'tsev classes (Russian)* (J 0003) Algebra i Logika 21*721-738
• TRANSL [1982] (J 0069) Algeb and Log 21*481-494
⋄ C05 C52 ⋄ REV MR 85e:08010 Zbl 542#03015
• ID 40368

SMIRNOV, D.M. see Vol. I, V for further entries

SMIRNOV, V.A. [1972] *Formal derivation and logical calculi (Russian)* (X 2027) Nauka: Moskva 272pp
⋄ B98 C07 C40 C98 F05 F07 F98 ⋄ REV Zbl 269#02004 • ID 30359

SMIRNOV, V.A. [1974] *On the question of the definability of predicates introduced via bilateral reduction sentences (Russian)* (C 2578) Filos & Logika 165-167
⋄ C40 ⋄ REV MR 58#27436 • ID 78738

SMIRNOV, V.A. [1977] see SADOVSKIJ, V.N.

SMIRNOV, V.A. see Vol. I, II, V, VI for further entries

SMIRNOVA, O.S. [1974] *Completeness and decidability of the axiomatic theory SG for spherical geometry (Russian)* (J 3536) Uch Zap Univ, Ivanovo 130*25-37
⋄ B25 C35 C65 ⋄ REV MR 57#89 • ID 78746

SMIRNOVA, O.S. see Vol. I for further entries

SMITH, B.D. [1979] *An application of non-standard model theory* (J 0032) J Algeb 58*502-507
⋄ C60 H20 ⋄ REV MR 80i:17021 Zbl 417#17007
• ID 53275

SMITH, HOWARD [1984] see BURNS, R.G.

SMITH, KAY [1984] *Commutative regular rings and boolean valued fields* (J 0036) J Symb Logic 49*281-297
⋄ C60 C90 E40 E75 ⋄ REV MR 85i:03161 • ID 42504

SMITH, KENNETH W. [1981] *Stability and categoricity of lattices* (J 0017) Canad J Math 33*1380-1419
⋄ C35 C45 G10 ⋄ REV MR 83k:03037 Zbl 436#03027
• ID 55416

SMITH, KENNETH W. [1983] see HERRE, H.

SMITH, RICK L. [1981] *Effective valuation theory* (P 2902) Aspects Effective Algeb;1979 Clayton 232-245
⋄ C57 C60 D45 F55 ⋄ REV MR 83b:03053 Zbl 479#03023 • ID 55685

SMITH, RICK L. [1981] *Two theorems on autostability in p-groups* (P 2628) Log Year;1979/80 Storrs 302-311
⋄ C57 C60 ⋄ REV MR 83h:03064 Zbl 488#03024
• ID 36072

SMITH, RICK L. [1985] see DRIES VAN DEN, L.

SMITH, RICK L. see Vol. IV, V, VI for further entries

SMITH JR., D.B. [1971] *Universal homogeneous algebras* (J 0004) Algeb Universalis 1*254-260
⋄ C05 C50 ⋄ REV MR 45#3298 Zbl 239#08013
• ID 12525

SMOLIN, V.P. [1972] *On the question of definability in the lower predicate calculus on a finite model without equality (Russian) (English summary)* (J 0040) Kibernetika, Akad Nauk Ukr SSR 1972/4*147-148
⋄ C13 C40 ⋄ REV MR 46#7015 Zbl 248#02020
• ID 12534

SMOLIN, V.P. [1975] *Monadic algebras and the one-place predicate calculus* (J 2604) Progr Comput Software 1*455-462
⋄ B20 C07 G15 ⋄ REV Zbl 367#02032 • ID 51192

SMORYNSKI, C.A. [1973] *Elementary intuitionistic theories* (J 0036) J Symb Logic 38*102-134
⋄ B25 C90 D35 F50 ⋄ REV MR 48#5842 Zbl 261#02033 • ID 12535

SMORYNSKI, C.A. [1973] *Investigations of intuitionistic formal systems by means of Kripke models* (0000) Diss., Habil. etc
⋄ B55 C90 F50 ⋄ REM Diss., University of Chicago
• ID 27058

SMORYNSKI, C.A. [1976] see JONGH DE, D.H.J.

SMORYNSKI, C.A. [1977] *On axiomatizing fragments* (J 0036) J Symb Logic 42*530-544
⋄ C90 F50 ⋄ REV MR 57#15998 Zbl 381#03046
• ID 26856

SMORYNSKI, C.A. [1977] *The incompleteness theorems* (C 1523) Handb of Math Logic 821-865
⋄ C62 F25 F30 F98 ⋄ REV MR 58#10343 JSL 49.980
• ID 27327

SMORYNSKI, C.A. [1978] *Beth's theorem and self-referential sentences* (P 1897) Logic Colloq;1977 Wroclaw 253-261
⋄ B45 C40 F30 ⋄ REV MR 80c:03023 Zbl 453#03018
• ID 54148

SMORYNSKI, C.A. & STAVI, J. [1980] *Cofinal extension preserves recursive saturation* (P 2625) Model Th of Algeb & Arithm;1979 Karpacz 338-345
⋄ C50 C57 C62 ⋄ REV MR 82g:03062 Zbl 467#03059
• ID 55057

SMORYNSKI, C.A. [1980] *Skolem's solution to a problem of Frobenius* (J 2789) Math Intell 3*123-132
⋄ C10 ⋄ REV MR 83b:03031 Zbl 498#10013 • ID 35094

SMORYNSKI, C.A. [1981] *Cofinal extensions of nonstandard models of arithmetic* (J 0047) Notre Dame J Formal Log 22*133-144
- ◇ C50 C57 C62 ◇ REV MR 82g:03063   Zbl 481 # 03045
- • ID 78761

SMORYNSKI, C.A. [1981] *Elementary extensions of recursively saturated models of arithmetic* (J 0047) Notre Dame J Formal Log 22*193-203
- ◇ C50 C57 C62 ◇ REV MR 82g:03064   Zbl 503.03032
- • ID 78760

SMORYNSKI, C.A. [1981] *Recursively saturated nonstandard models of arithmetic* (J 0036) J Symb Logic 46*259-286
- • ERR/ADD ibid 47*493-494
- ◇ C50 C57 C62 ◇ REV MR 82i:03045 MR 84b:03054 Zbl 501 # 03044 Zbl 549 # 03059 • ID 78758

SMORYNSKI, C.A. [1982] *A note on initial segment constructions in recursively saturated models of arithmetic* (J 0047) Notre Dame J Formal Log 23*393-408
- ◇ C50 C57 C62 ◇ REV MR 83j:03058   Zbl 519 # 03055
- • ID 35359

SMORYNSKI, C.A. [1982] *Back-and-forth inside a recursively saturated model of arithmetic* (P 3623) Logic Colloq;1980 Prague 273-278
- ◇ C15 C50 C57 C62 ◇ REV MR 83m:03041 Zbl 503 # 03033 • ID 35443

SMORYNSKI, C.A. [1982] *Nonstandard models and constructivity* (P 3638) Brouwer Centenary Symp;1981 Noordwijkerhout 459-464
- ◇ C62 F30 F50 ◇ REV MR 85d:03114   Zbl 518 # 03026
- • ID 37523

SMORYNSKI, C.A. [1982] *The finite inseparability of the first-order theory of diagonalisable algebras* (J 0063) Studia Logica 41*347-349
- ◇ B45 C13 D35 F30 ◇ REV Zbl 542 # 03024 • ID 43687

SMORYNSKI, C.A. [1984] *Lectures on nonstandard models of arithmetic* (P 3710) Logic Colloq;1982 Firenze 1-70
- ◇ A10 C62 C98 F30 F98 H15 ◇ REV MR 86c:03035 Zbl 554 # 03036 • ID 41837

SMORYNSKI, C.A. [1985] *Nonstandard models and related developments* (C 4699) Friedman's Res on Found of Math 179-229
- ◇ C62 C98 F30 F98 ◇ ID 49880

SMORYNSKI, C.A. see Vol. II, IV, V, VI for further entries

SMULLYAN, R.M. [1970] *Abstract quantification theory* (P 0603) Intuitionism & Proof Th;1968 Buffalo 79-91
- ◇ B10 C07 F07 ◇ REV MR 42 # 2928 Zbl 206.272
- • ID 12557

SMULLYAN, R.M. see Vol. I, II, IV, V, VI for further entries

SNAPPER, E. [1977] *Omitting models* (J 0036) J Symb Logic 42*29-32
- ◇ C07 C15 C50 ◇ REV MR 58 # 21573 Zbl 369 # 02029
- • ID 24347

SOARE, R.I. [1980] see LERMAN, M.

SOARE, R.I. [1981] see LERMAN, M.

SOARE, R.I. [1984] see LERMAN, M.

SOARE, R.I. [1984] see KNIGHT, J.F.

SOARE, R.I. see Vol. IV, VI for further entries

SOBOCINSKI, B. [1975] *Concerning the postulate-systems of subtractive abelian groups* (J 0047) Notre Dame J Formal Log 16*429-444
- ◇ C60 ◇ REV MR 51 # 10071   Zbl 305 # 20003 • ID 12622

SOBOCINSKI, B. see Vol. I, II, V for further entries

SOBOLEV, S.K. [1977] *The intuitionistic propositional calculus with quantifiers (Russian)* (J 0087) Mat Zametki (Akad Nauk SSSR) 22*69-76
- • TRANSL [1977] (J 1044) Math Notes, Acad Sci USSR 22*528-532
- ◇ B55 C80 C90 D35 F50 ◇ REV MR 56 # 15371 Zbl 365 # 02013 • ID 51014

SOBOLEV, S.K. see Vol. II, IV, VI for further entries

SOBOLEV, S.L. (ED.) [1982] *Mathematical logic and the theory of algorithms (Russian)* (X 2642) Nauka: Novosibirsk 176pp
- ◇ B97 C97 D97 ◇ REV MR 84i:03007   Zbl 539 # 00002
- • ID 34492

SOCHOR, A. [1969] see BALCAR, B.

SOCHOR, A. [1972] see MLCEK, J.

SOCHOR, A. & VOPENKA, P. [1975] *Contributions to the theory of semisets. V: On axiom of general collapse* (J 0068) Z Math Logik Grundlagen Math 21*289-302
- ◇ C20 C62 E35 E70 ◇ REV MR 54 # 2463 Zbl 327 # 02053 • REM Part IV 1974 by Stepanek,P. Part VI 1975 • ID 13887

SOCHOR, A. [1975] *Contributions to the theory of semisets. VI: Non-existence of the class of all absolute natural numbers* (J 0068) Z Math Logik Grundlagen Math 21*439-442
- ◇ C20 C62 E35 E70 ◇ REV MR 54 # 2464 Zbl 344 # 02050 • REM Part V 1975 by Sochor,A. & Vopenka,P. • ID 12625

SOCHOR, A. [1975] *Real classes in the ultrapower of hereditarily finite sets* (J 0140) Comm Math Univ Carolinae (Prague) 16*637-640
- ◇ C20 E70 H15 ◇ REV MR 52 # 10419 Zbl 327 # 02047
- • ID 21715

SOCHOR, A. & VOPENKA, P. [1981] *The axiom of reflection* (J 0140) Comm Math Univ Carolinae (Prague) 22*87-111
- ◇ C20 E65 E70 H20 ◇ REV MR 83j:03087 Zbl 454 # 03027 • ID 54239

SOCHOR, A. [1982] *Metamathematics of the alternative set theory II* (J 0140) Comm Math Univ Carolinae (Prague) 23*55-79
- ◇ C20 C62 E35 E70 ◇ REV MR 83h:03080 Zbl 493 # 03030 • REM Part I 1979. Part III 1983 • ID 36081

SOCHOR, A. [1983] *Metamathematics of the alternative set theory III* (J 0140) Comm Math Univ Carolinae (Prague) 24*137-154
- ◇ C62 E35 E70 ◇ REV MR 85e:03127   Zbl 531 # 03031
- • REM Part II 1982 • ID 37683

SOCHOR, A. [1984] see PUDLAK, P.

SOCHOR, A. [1985] *Constructibility and shiftings of view* (J 0140) Comm Math Univ Carolinae (Prague) 26*477-498
- ◇ E45 E70 H15 ◇ ID 48318

SOCHOR, A. [1985] see PUDLAK, P.

SOCHOR, A. [1985] *Notes on revealed classes* (J 0140) Comm Math Univ Carolinae (Prague) 26*499-514
- ◇ E70 H15 ◇ ID 48319

SOCHOR, A. see Vol. I, V, VI for further entries

SOKOLOV, V.G. [1970] see REMESLENNIKOV, V.N.

SOKOLOV, V.G. see Vol. IV for further entries

SOLDATOVA, V.V. [1977] see KOGALOVSKIJ, S.R.

SOLDATOVA, V.V. [1981] see KOGALOVSKIJ, S.R.

SOLDATOVA, V.V. see Vol. IV for further entries

SOLOVAY, R.M. [1970] see JENSEN, R.B.

SOLOVAY, R.M. [1974] *Strongly compact cardinals and the GCH* (P 0610) Tarski Symp;1971 Berkeley 365-372
⋄ C20 E05 E50 E55 E65 ⋄ REV MR 52#106 Zbl 317#02083 • ID 12645

SOLOVAY, R.M. & REINHARDT, W.N. & KANAMORI, A. [1978] *Strong axioms of infinity and elementary embeddings* (J 0007) Ann Math Logic 13*73-116
⋄ C20 E35 E45 E55 E65 ⋄ REV MR 80h:03072 Zbl 376#02055 • ID 27961

SOLOVAY, R.M. [1985] *Infinite fixed-point algebras* (P 4046) Rec Th;1982 Ithaca 473-486
⋄ C57 C62 G25 ⋄ REV Zbl 573#03030 • ID 46419

SOLOVAY, R.M. see Vol. II, IV, V, VI for further entries

SONENBERG, E.A. [1978] *On the elementary theory of inductive order* (J 0009) Arch Math Logik Grundlagenforsch 19*13-22
⋄ C65 E07 G05 ⋄ REV MR 80d:03027b Zbl 395#03023
• ID 29161

SONENBERG, E.A. [1979] *Non-standard models of ordinal arithmetics* (J 0068) Z Math Logik Grundlagen Math 25*5-27
⋄ C65 H20 ⋄ REV MR 80h:03051 Zbl 437#03038
• ID 55906

SONOBE, O. [1975] *Bi-relational frameworks for minimal and intuitionistic logics* (P 3299) Progr Kiso Riron, Algor Okeru Shomei Ron;1973/74 Kyoto 174-189
⋄ B45 C90 F50 ⋄ REV MR 58#5075 Zbl 375#02009
• ID 51586

SONOBE, O. see Vol. II for further entries

SOPRUNOV, S.F. [1974] *On the power of a real-closed field (Russian) (English summary)* (J 0288) Vest Ser Mat Mekh, Univ Moskva 29/4*70-73
• TRANSL [1974] (J 0510) Moscow Univ Math Bull 29/3-4*98-101
⋄ C20 C60 D35 ⋄ REV MR 51#5294 Zbl 298#02060
• ID 17402

SOPRUNOV, S.F. [1975] *Initial segments of nonstandard arithmetics (Russian)* (J 0023) Dokl Akad Nauk SSSR 223*576-577
• TRANSL [1975] (J 0062) Sov Math, Dokl 16*968-970
⋄ C62 ⋄ REV MR 53#7772 Zbl 331#02035 • ID 23012

SOPRUNOV, S.F. [1975] *Strong nonstandard models of arithmetic (Russian)* (J 0023) Dokl Akad Nauk SSSR 220*293-296
• TRANSL [1975] (J 0062) Sov Math, Dokl 16*80-84
⋄ C62 ⋄ REV MR 53#12935 Zbl 362#02050 • ID 23185

SOPRUNOV, S.F. [1976] *Denumerable non-standard models of arithmetic (Russian)* (C 3271) Issl Teor Mnozh & Neklass Logik 157-173
⋄ C15 C62 ⋄ REV MR 57#5744 Zbl 412#03042
• ID 52973

SOPRUNOV, S.F. [1979] *Lattices of non-standard arithmetics (Russian)* (C 2581) Issl Neklass Log & Teor Mnozh 146-173
• TRANSL [1984] (S 3489) Sel Math Sov 4*203-223
⋄ C62 E50 G10 H15 ⋄ REV MR 81j:03100
Zbl 468#03045 • ID 55110

SORBI, A. [1985] see MONTAGNA, F.

SORBI, A. see Vol. IV, VI for further entries

SPECKER, E. [1950] *Additive Gruppen von Folgen ganzer Zahlen* (J 0050) Port Math 9*131-140
⋄ C60 ⋄ REV MR 12.587 Zbl 41.363 • ID 12659

SPECKER, E. [1958] *Dualitaet* (J 0076) Dialectica 12*451-465
⋄ B30 C07 E70 ⋄ REV MR 21#5551 Zbl 91.7
JSL 27.231 • ID 12669

SPECKER, E. [1961] see MACDOWELL, R.

SPECKER, E. [1968] see HODES, L.

SPECKER, E. [1981] see BLATTER, C.

SPECKER, E. [1984] see BLATTER, C.

SPECKER, E. see Vol. I, II, IV, V, VI for further entries

SPECTOR, M. [1980] *A measurable cardinal with a nonwellfounded ultrapower* (J 0036) J Symb Logic 45*623-628
⋄ C20 E25 E35 E40 ⋄ REV MR 81k:03053
Zbl 471#03047 • ID 55242

SPECTOR, M. [1985] *Model theory under the axiom of determinateness* (J 0036) J Symb Logic 50*773-780
⋄ C20 C55 C75 E60 ⋄ ID 47378

SPECTOR, M. see Vol. V for further entries

SPERANZA, F. [1970] *Relazioni e strutture* (X 1375) Zanichelli: Bologna 224pp
⋄ C98 E98 ⋄ REV MR 50#1882 • ID 78881

SPERRY, P.L. [1972] see PHILLIPS, R.G.

SPIVAKOV, YU.L. [1974] *Algebraic extensions of the field of formal power series $F_D((t))$ (Russian) (Uzbek summary)* (J 0030) Izv Akad Nauk Uzb SSR, Ser Fiz-Mat 18/6*15-21,78
⋄ C20 C60 H05 ⋄ REV MR 52#5639 Zbl 303#12102
• ID 18401

SREBRNY, M. [1973] see MAREK, W.

SREBRNY, M. [1974] see MAREK, W.

SREBRNY, M. [1974] *Note on a theorem of Mostowski (Russian summary)* (J 0014) Bull Acad Pol Sci, Ser Math Astron Phys 22*997-1000
⋄ C62 E50 ⋄ REV MR 54#89 Zbl 296#02035 • ID 23967

SREBRNY, M. [1975] see MAREK, W.

SREBRNY, M. [1976] see MAREK, W.

SREBRNY, M. [1977] *Relatively constructible transitive models* (J 0027) Fund Math 96*161-172
⋄ C62 D55 D60 E10 E40 E45 E55 ⋄ REV
MR 58#16273 Zbl 383#03036 • ID 26534

SREBRNY, M. [1977] *Second-order theories of countable transitive models* (P 1639) Set Th & Hierarch Th (1);1974 Karpacz 57-61
⋄ C62 C85 ⋄ REV MR 58#5228 Zbl 395#03026
• ID 52577

SREBRNY, M. see Vol. II, IV, V for further entries

SROUR, G. [1984] see PILLAY, A.

STACHNIAK, Z. [1981] *Introduction to model theory for Lesniewski's ontology* (J 0481) Acta Univ Wroclaw 586(Logika 9)*76pp
⋄ C20 C90 E70 ⋄ REV Zbl 474 # 03013 • ID 55417

STACHNIAK, Z. see Vol. I for further entries

STAHL, S.H. [1980] *Intensional sets* (J 0316) Illinois J Math 24*379-381
⋄ C75 E35 E45 E50 ⋄ REV MR 81g:03061 Zbl 411 # 03041 • ID 53487

STAHL, S.H. see Vol. IV for further entries

STARCHENKO, S.S. [1985] see PALYUTIN, E.A.

STARK, W.R. [1978] *A forcing approach to the strict-Π_1^1 reflection and strict-$\Pi_1^1 = \Sigma_1^0$* (J 0068) Z Math Logik Grundlagen Math 24*467-479
⋄ C70 D60 E40 ⋄ REV MR 81c:03025 Zbl 412 # 03033 • ID 52964

STARK, W.R. [1980] *Martin's axiom in the model theory of L_A* (J 0036) J Symb Logic 45*172-176
⋄ C70 E50 ⋄ REV MR 81a:03036 Zbl 436 # 03031 • ID 78943

STARK, W.R. see Vol. I, II for further entries

STASZEK, W. [1969] *Investigations on the classical logic of names (Polish) (Russian and English summaries)* (J 0063) Studia Logica 25*169-188
⋄ A10 B25 ⋄ REV MR 41 # 30 Zbl 248 # 02022 • ID 13130

STASZEK, W. see Vol. I for further entries

STAVI, J. [1973] *A converse of the Barwise completeness theorem* (J 0036) J Symb Logic 38*594-612
⋄ C70 D60 D70 ⋄ REV MR 51 # 12473 Zbl 308 # 02015 • ID 13133

STAVI, J. [1975] *Applications of a theorem of Levy to boolean terms and algebras* (J 0064) Trans Amer Math Soc 205*1-36
⋄ C75 E47 E75 G05 ⋄ REV MR 57 # 9476 Zbl 369 # 06010 • ID 13137

STAVI, J. [1975] *Extensions of Kripke's embedding theorem* (J 0007) Ann Math Logic 8*345-428
⋄ C55 C75 C90 G05 ⋄ REV MR 53 # 12937 Zbl 354 # 02044 • ID 50184

STAVI, J. [1976] see MAKOWSKY, J.A.

STAVI, J. [1977] see NADEL, M.E.

STAVI, J. [1978] see NADEL, M.E.

STAVI, J. [1978] *Compactness properties of infinitary and abstract languages. I. General results* (P 1897) Logic Colloq;1977 Wroclaw 263-275
⋄ C70 C75 C95 ⋄ REV MR 81d:03038 Zbl 448 # 03022 • ID 56622

STAVI, J. [1980] see MANEVITZ, L.M.

STAVI, J. [1980] see SMORYNSKI, C.A.

STAVI, J. [1983] see MAGIDOR, M.

STAVI, J. [1984] see MAGIDOR, M.

STAVI, J. see Vol. II, V for further entries

STEBE, P. [1976] see ANSHEL, M.

STEBE, P. see Vol. IV for further entries

STEEL, J.R. [1975] *Descending sequences of degrees* (J 0036) J Symb Logic 40*59-61
⋄ C62 D30 ⋄ REV MR 56 # 8347 Zbl 349 # 02036 • ID 13142

STEEL, J.R. [1978] *Forcing with tagged trees* (J 0007) Ann Math Logic 15*55-74
⋄ C62 D30 D55 E40 F35 ⋄ REV MR 81c:03044 Zbl 404 # 03020 • ID 29156

STEEL, J.R. [1978] *On Vaught's conjecture* (C 2908) Cabal Seminar Los Angeles 1976-77 193-208
⋄ C15 C65 C70 C75 ⋄ REV MR 81b:03036 Zbl 403 # 03027 • ID 54741

STEEL, J.R. see Vol. IV, V for further entries

STEFANESCU, M. [1977] *A correspondence between a class of near-rings and a class of groups (Italian summary)* (J 0149) Atti Accad Naz Lincei Fis Mat Nat, Ser 8 62*439-443
⋄ C60 ⋄ REV MR 58 # 11033 Zbl 375 # 17002 • ID 82870

STEFFENS, K. [1978] see GALVIN, F.

STEFFENS, K. see Vol. V for further entries

STEGBAUER, W. [1979] *A generalized model companion for a theory of partially ordered fields* (J 0036) J Symb Logic 44*643-652
⋄ C25 C35 C60 ⋄ REV MR 81a:03034 Zbl 436 # 03036 • ID 78974

STEGMUELLER, W. & VARGA VON KIBED, M. [1984] *Strukturtypen der Logik. Probleme und Resultate der Wissenschaftstheorie und analytischen Philosophie, Band III: Strukturtypen der Logik* (X 0811) Springer: Heidelberg & New York xv+524pp
⋄ A05 C98 ⋄ REV Zbl 548 # 03001 • REM Part II,2 1973. Part IV,1 1973 • ID 41822

STEGMUELLER, W. see Vol. I, II, IV, V, VI for further entries

STEINBY, M. [1984] see GECSEG, F.

STEINBY, M. see Vol. IV for further entries

STEINHORN, C.I. [1983] *A new omitting types theorem* (J 0053) Proc Amer Math Soc 89*480-486
⋄ C07 C15 C45 C50 C57 ⋄ REV MR 84j:03071 Zbl 529 # 03009 • ID 34663

STEINHORN, C.I. [1984] see PILLAY, A.

STEINHORN, C.I. [1985] see PILLAY, A.

STEINHORN, C.I. [1985] *Borel structures and measure and category logics* (C 4183) Model-Theor Log 579-596
⋄ C90 C98 ⋄ ID 48339

STEINHORN, C.I. [1985] *Borel structures for first-order and extended logics* (C 4699) Friedman's Res on Found of Math 161-178
⋄ C80 C90 C98 ⋄ ID 49881

STEINHORN, C.I. see Vol. V for further entries

STENIUS, E. [1976] *Comments on Jaakko Hintikka's paper "Quantifiers vs. quantification theory"* (J 0076) Dialectica 30*67-88
⋄ A05 B65 C80 ⋄ REV Zbl 413 # 03007 • ID 53001

STENIUS, E. see Vol. I, II, VI for further entries

STEPANEK, P. [1974] *Contributions to the theory of semisets. IV*
(J 0068) Z Math Logik Grundlagen Math 20∗373-384
⋄ C20 C62 E35 E70 ⋄ REV MR 54 # 2462
Zbl 342 # 02047 • REM Part III 1973 by Cuda,K. Part V 1975
by Sochor,A. & Vopenka,P. • ID 13172

STEPANEK, P. see Vol. V for further entries

STEPANOV, V.I. [1983] *On model theory for intuitionistic logic
(Russian)* (J 0142) Mat Sb, Akad Nauk SSSR, NS
120∗227-239,288
• TRANSL [1984] (J 0349) Math of USSR, Sbor 48∗223-235
⋄ C90 F50 ⋄ REV MR 85b:03058 Zbl 568 # 03018
• ID 39758

STEPANOV, V.I. see Vol. VI for further entries

STEPANOVA, A.A. [1984] *Categorical quasivarieties of abelian
groupoids and quasigroups (Russian)* (J 0092) Sib Mat Zh
25/4∗174-178
• TRANSL [1984] (J 0475) Sib Math J 25∗650-653
⋄ C05 C35 ⋄ REV MR 86a:20090 • ID 41872

STEPANOVA, A.A. [1984] *Groupoids of rank 2 generating
categorical Horn classes (Russian)* (J 0003) Algebra i Logika
23∗193-207
• TRANSL [1984] (J 0069) Algeb and Log 23∗136-146
⋄ C05 C35 ⋄ ID 41865

STEPANOVA, A.A. [1985] *Semigroups generating categorical
Horn classes (Russian)* (J 0087) Mat Zametki (Akad Nauk
SSSR) 38∗317-324,350
⋄ C05 C35 ⋄ ID 49258

STEPIEN, T. [1980] see DYWAN, Z.

STEPIEN, T. [1981] *Craig-Goedel-Lindenbaum's property and
Sobocinski-Tarski's property in propositional calculi* (J 0387)
Bull Sect Logic, Pol Acad Sci 10∗116-121
⋄ B22 C40 ⋄ REV MR 83b:03030 Zbl 481 # 03007
• ID 33327

STEPIEN, T. see Vol. I, II for further entries

STEPRANS, J. [1984] *The number of submodules* (J 3240) Proc
London Math Soc, Ser 3 49∗183-192
⋄ C60 E35 E50 E75 ⋄ REV MR 85d:03101
Zbl 539 # 16023 • ID 40628

STEPRANS, J. see Vol. V for further entries

STERN, J. [1972] *Theoremes d'interpolation obtenus par la
methode du "forcing"* (J 2313) C R Acad Sci, Paris, Ser A-B
275∗A237-A240
⋄ C25 C40 ⋄ REV MR 47 # 1604 Zbl 245 # 02044
• ID 13175

STERN, J. [1974] *Sur certaines classes d'espaces de Banach
caracterisees par des formules* (J 2313) C R Acad Sci, Paris,
Ser A-B 278∗A525-A528
⋄ C20 C65 ⋄ REV MR 49 # 9594 Zbl 293 # 02041
• ID 13178

STERN, J. [1975] *A new look at the interpolation problem*
(J 0036) J Symb Logic 40∗1-13
⋄ C25 C40 C70 ⋄ REV MR 56 # 2785 Zbl 319 # 02012
• ID 13179

STERN, J. [1975] *Le probleme des enveloppes d'espace de Banach
(English summary)* (J 2313) C R Acad Sci, Paris, Ser A-B
280∗A797-A799
⋄ C65 ⋄ REV MR 51 # 3869 Zbl 295 # 46033
Zbl 318 # 46026 • ID 17370

STERN, J. [1975] *Proprietes locales et ultrapuissances d'espaces de
Banach* (C 3031) Semin Maurey-Schwartz 1974/75
VI-VII∗30pp
⋄ C20 C65 ⋄ REV MR 56 # 6348 Zbl 318 # 46027
• ID 82881

STERN, J. [1976] *Some applications of model theory in Banach
space theory* (J 0007) Ann Math Logic 9∗49-121
⋄ C20 C65 ⋄ REV MR 52 # 6384 Zbl 378 # 02026
• ID 18409

STERN, J. [1976] *The problem of envelopes for Banach spaces*
(J 0029) Israel J Math 24∗1-15
⋄ C20 C65 ⋄ REV MR 53 # 14089 Zbl 348 # 46013
• ID 18408

STERN, J. [1977] *Examples d'application de la theorie des modeles
a la theorie des espaces de Banach* (P 1729) Colloq Int
Log;1975 Clermont-Ferrand 19-37
⋄ C20 C65 ⋄ REV MR 58 # 30086 Zbl 447 # 46013
• ID 31517

STERN, J. [1978] see SHELAH, S.

STERN, J. [1978] *Ultrapowers and local properties of Banach
spaces* (J 0064) Trans Amer Math Soc 240∗231-252
⋄ C20 C65 ⋄ REV MR 81f:46031 Zbl 402 # 03025
• ID 31516

STERN, J. [1979] *Le groupe des isometries d'un espace de Banach*
(J 0343) Studia Math, Pol Akad Nauk 64∗139-149
⋄ C65 ⋄ REV MR 80f:46022 Zbl 419 # 46019 • ID 31518

STERN, J. see Vol. I, IV, V for further entries

STERN, S.T. [1966] *The seminatural numbers* (J 0005) Amer
Math Mon 73∗589-603
⋄ E10 H15 ⋄ REV MR 33 # 5496 Zbl 192.45 • ID 13181

STERN, S.T. see Vol. I, V for further entries

STIHI, T. [1973] *A case of decidability in the first order predicate
calculus (Romanian)* (S 1613) Probl Logic (Bucharest)
5∗257-265
⋄ B20 B25 ⋄ REV MR 58 # 195 • ID 79027

STIHI, T. see Vol. I, V for further entries

STILLWELL, J.C. [1972] *Decidability of the "almost all" theory of
degrees* (J 0036) J Symb Logic 37∗501-506
⋄ B25 C80 D30 ⋄ REV MR 50 # 1863 Zbl 287 # 02029
• ID 13190

STILLWELL, J.C. [1977] *Concise survey of mathematical logic*
(J 3194) J Austral Math Soc, Ser A 24∗139-161
⋄ B98 C98 D10 D35 ⋄ REV MR 57 # 5655
Zbl 393 # 03002 • ID 52422

STILLWELL, J.C. see Vol. I, II, IV for further entries

STOJAKOVIC, M. [1975] *Inductive and Peano models
(Serbo-Croatian) (English summary)* (J 2855) Zbor Rad,
Prir-Mat Fak, Ser Mat (Novi Sad) 5∗1-8
⋄ B28 C62 ⋄ REV MR 56 # 15416 • ID 79037

STOJAKOVIC, M. see Vol. I, II, V for further entries

STONE, A.L. [1965] *Extensive ultraproducts and Haar measures*
(P 0614) Th Models;1963 Berkeley 419-423
⋄ C20 C65 E75 H05 ⋄ REV MR 34 # 1441 • ID 27645

STONE, A.L. see Vol. I for further entries

STONE, MICHAEL G. [1977] see SAUER, N.

STONE, MICHAEL G. see Vol. V for further entries

STRAMBACH, K. [1984] see FUNK, M.

STRAMBACH, K. [1985] see FUNK, M.

STRUNKOV, S.P. [1966] *Subgroups of periodic groups (Russian)* (J 0023) Dokl Akad Nauk SSSR 170*279-281
 • TRANSL [1966] (J 0062) Sov Math, Dokl 7*1201-1203
  ⋄ C60 E75 ⋄ REV MR 34 # 2705 Zbl 178.19 • ID 13237

STUETZER, H. [1977] *Die elementare Formulierbarkeit von Dimension und Grad algebraischer Mannigfaltigkeiten* (J 0008) Arch Math (Basel) 28*270-273
  ⋄ C20 C60 ⋄ REV MR 56 # 3003 Zbl 363 # 14001
 • ID 26565

STURM, T. [1980] *Universal coretracts in some categories of chains and boolean algebras* (J 1008) Demonstr Math (Warsaw) 13*693-702
  ⋄ C50 E07 G05 ⋄ REV MR 82d:06002 Zbl 447 # 06011
 • ID 69938

STURM, T. see Vol. V for further entries

SUKHORUTCHENKO, I.I. [1979] *Programmably categorical abelian p-groups I,II (Russian)* (C 2065) Teor Nereg Kriv Raz Geom Post 77-83,84-90
  ⋄ B75 C35 C60 ⋄ REV MR 81f:03048 • ID 79090

SUN, WENZHI [1982] *A property of the upper bounding ordinal of a model of ZFC (Chinese) (English summary)* (J 3742) Shuxue Yanjiu yu Pinglun 2/4*135-141
  ⋄ C62 E10 ⋄ REV MR 84d:03042 Zbl 505 # 03023
 • ID 34080

SUN, WENZHI see Vol. V for further entries

SUNDHOLM, G. [1977] *A completeness proof for an infinitary tense-logic* (J 0105) Theoria (Lund) 43*47-51
  ⋄ B45 C75 ⋄ REV MR 56 # 15370 Zbl 364 # 02011
 • ID 50941

SUNDHOLM, G. see Vol. VI for further entries

SUPPES, P. [1958] see SCOTT, D.S.

SUPPES, P. [1959] see HENKIN, L.

SUPPES, P. see Vol. I, II, V for further entries

SURESON, C. [1985] *Non-closure of the image model and absence of fixed points* (J 0073) Ann Pure Appl Logic 28*287-314
  ⋄ C62 E05 E45 E55 ⋄ ID 40297

SURESON, C. see Vol. V for further entries

SURMA, S.J. [1968] *Four studies in metamathematics (Polish) (Russian and English summaries)* (J 0063) Studia Logica 23*79-114
  ⋄ B22 C07 C35 E25 ⋄ REV MR 39 # 3965
Zbl 313 # 02028 • ID 12748

SURMA, S.J. [1969] *On the axiomatic treatment of the theory of models I: Theory of models as an extension of Tarski's consequence theory (Polish summary)* (S 0458) Zesz Nauk, Prace Log, Uniw Krakow 4*39-45
  ⋄ B22 C07 ⋄ REV MR 41 # 1521 • REM Part II 1970
 • ID 12752

SURMA, S.J. [1970] *On the axiomatic treatment of the theory of models II: Syntactical characterization of a fragment of the theory of models (Polish summary)* (S 0458) Zesz Nauk, Prace Log, Uniw Krakow 5*43-55
  ⋄ B22 C07 C75 ⋄ REV MR 45 # 1746 • REM Part I 1969. Part III 1970 • ID 79124

SURMA, S.J. [1970] *On the axiomatic treatment of the theory of models III: The function of model content (Polish summary)* (S 0458) Zesz Nauk, Prace Log, Uniw Krakow 5*57-63
  ⋄ B22 C07 ⋄ REV MR 45 # 1747 • REM Part II 1970. Part IV 1970 by Kossowski,P. • ID 12754

SURMA, S.J. see Vol. I, II, V, VI for further entries

SUSZKO, R. [1955] see LOS, J.

SUSZKO, R. [1957] see LOS, J.

SUSZKO, R. [1960] see LOS, J.

SUSZKO, R. [1967] *A proposal concerning the formulation of the infinitistic axiom in the theory of logical probability* (S 0019) Colloq Math (Warsaw) 17*347-349
  ⋄ B48 C90 ⋄ REV MR 36 # 4595 Zbl 218 # 02021
 • ID 26275

SUSZKO, R. [1971] *Quasi-completeness in non-Fregean logic (Polish and Russian summaries)* (J 0063) Studia Logica 29*7-16
  ⋄ B60 C07 ⋄ REV MR 46 # 5115 Zbl 272 # 02029
 • ID 12777

SUSZKO, R. [1972] *Equational logic and theories in sentential language* (J 0387) Bull Sect Logic, Pol Acad Sci 1/2*2-9
  ⋄ B05 C05 C07 ⋄ REV MR 50 # 12714 • ID 12781

SUSZKO, R. [1974] *Equational logic and theories in sentential languages* (S 0019) Colloq Math (Warsaw) 29*19-23
  ⋄ B05 C05 C07 ⋄ REV MR 49 # 39 Zbl 296 # 02027
 • ID 12782

SUSZKO, R. [1975] see BLOOM, S.L.

SUSZKO, R. see Vol. I, II, V, VI for further entries

SUTER, G.H. [1973] *Recursive elements and constructive extensions of computable local integral domains* (J 0036) J Symb Logic 38*272-290
  ⋄ C57 C60 D45 ⋄ REV MR 52 # 54 Zbl 276 # 02028
 • ID 12784

SUZUKI, Y. [1967] *Applications of the theory of β-models* (J 0407) Comm Math Univ St Pauli (Tokyo) 16*57-68
  ⋄ C62 D65 ⋄ REV MR 37 # 1240 Zbl 216.296 • ID 12790

SUZUKI, Y. [1968] *$\aleph_0$-standard models for set theory (Russian summary)* (J 0014) Bull Acad Pol Sci, Ser Math Astron Phys 16*265-267
  ⋄ C62 D55 ⋄ REV MR 39 # 51 Zbl 206.11 • ID 12791

SUZUKI, Y. [1969] see MOSTOWSKI, ANDRZEJ

SUZUKI, Y. [1970] *Orbits of denumerable models of complete theories* (J 0027) Fund Math 67*89-95
  ⋄ C15 C35 C52 C62 ⋄ REV MR 42 # 1651 Zbl 199.303
 • ID 12792

SUZUKI, Y. & WILMERS, G.M. [1973] *Non-standard models for set theory* (P 0710) Russell Mem Logic Conf;1971 Uldum 278-314
  ⋄ C62 E35 ⋄ REV MR 50 # 4302 • ID 24802

SUZUKI, Y. see Vol. IV, V for further entries

SVENONIUS, L. [1955] *Definability and simplicity* (J 0036) J Symb Logic 20*235-250
  ⋄ C40 ⋄ REV MR 18.270 Zbl 68.246 JSL 37.174
 • ID 12793

SVENONIUS, L. [1959] $\aleph_0$-categoricity in first-order predicate calculus (J 0105) Theoria (Lund) 25*82-94
⋄ C35 ⋄ REV MR 25 # 1986a JSL 31.504 • ID 12795

SVENONIUS, L. [1959] A theorem on permutations in models (J 0105) Theoria (Lund) 25*173-178
⋄ C30 C40 ⋄ REV MR 25 # 1986b JSL 31.505 • ID 12794

SVENONIUS, L. [1960] On minimal models of first-order systems (J 0105) Theoria (Lund) 26*44-52
⋄ C07 C50 ⋄ REV MR 25 # 1986c JSL 35.343 • ID 12796

SVENONIUS, L. [1960] Some problems in logical model-theory (X 1493) Gleerup: Lund 43pp
• LAST ED [1960] (X 1494) Munksgaard: Copenhagen
⋄ C07 C52 C95 ⋄ REV MR 25 # 1986d JSL 31.505
• ID 41194

SVENONIUS, L. [1965] On the denumerable models of theories with extra predicates (P 0614) Th Models;1963 Berkeley 376-389
⋄ C15 C40 C52 ⋄ REV MR 35 # 42 Zbl 156.25
• ID 12797

SVENONIUS, L. see Vol. I, IV, VI for further entries

SWAMY, U.M. [1975] see RAMA RAO, V.V.

SWART DE, H.C.M. [1977] An intuitionistically plausible interpretation of intuitionistic logic (J 0036) J Symb Logic 42*564-578
⋄ C90 F50 ⋄ REV MR 58 # 27354 Zbl 383 # 03040
• ID 26858

SWART DE, H.C.M. [1978] First steps in intuitionistic model theory (J 0036) J Symb Logic 43*3-12
⋄ C90 F50 ⋄ REV MR 80c:03059 Zbl 395 # 03037
• ID 29237

SWART DE, H.C.M. [1980] Gentzen-type systems for C, K and several extensions of C and K; constructive completeness proofs and effective decision procedures for these systems (J 0079) Logique & Anal, NS 23*263-284
⋄ B25 B45 ⋄ REV MR 83i:03037 Zbl 454 # 03007
• ID 54220

SWART DE, H.C.M. see Vol. II, V, VI for further entries

SWEET, A.M. [1979] Intended model theory (J 0047) Notre Dame J Formal Log 20*575-592
⋄ A05 C90 ⋄ REV MR 80i:03050 Zbl 368 # 02048
• ID 56141

SWEET, A.M. see Vol. II for further entries

SWIJTINK, Z.G. [1976] Eliminability in a cardinal (J 0063) Studia Logica 35*71-89
⋄ C52 ⋄ REV MR 56 # 5268 Zbl 353 # 02033 • ID 50116

SZABO, L. [1978] Concrete representation of related structures of universal algebras I (J 0002) Acta Sci Math (Szeged) 40*175-184
⋄ C05 C40 ⋄ REV MR 58 # 443 Zbl 388 # 08003
• ID 29187

SZABO, P. [1982] see SIEKMANN, J.

SZABO, P. see Vol. I, IV for further entries

SZATKOWSKI, M. [1984] Concerning Kripke semantics for intermediate predicate logics (S 1070) Probl Mat, Wyz Szk Ped, Bydgosszcz 4*43-62
⋄ B55 C90 ⋄ ID 44572

SZATKOWSKI, M. see Vol. II for further entries

SZCZERBA, L.W. [1962] On the models of the affine geometry (J 0014) Bull Acad Pol Sci, Ser Math Astron Phys 10*
⋄ C65 ⋄ REV MR 26 # 5461 Zbl 108.359 • ID 39480

SZCZERBA, L.W. & TARSKI, A. [1965] Metamathematical properties of some affine geometries (P 0623) Int Congr Log, Meth & Phil of Sci (2,Proc);1964 Jerusalem 166-178
⋄ B25 B30 C52 C65 D35 ⋄ REV MR 35 # 843
Zbl 149.385 JSL 36.333 • ID 12831

SZCZERBA, L.W. [1970] Filling of gaps in ordered fields (J 0014) Bull Acad Pol Sci, Ser Math Astron Phys 18*349-352
⋄ C60 ⋄ REV MR 42 # 3066 Zbl 199.321 • ID 12833

SZCZERBA, L.W. [1971] Incompleteness degree of elementary Pasch-free geometry (J 0014) Bull Acad Pol Sci, Ser Math Astron Phys 19*213-215
⋄ C35 C65 ⋄ REV MR 44 # 3171 Zbl 208.236 • ID 33243

SZCZERBA, L.W. [1973] The notion of an elementary subsystem for a boolean-valued relational system (J 0027) Fund Math 80*5-12
⋄ C07 C90 ⋄ REV MR 48 # 1919 Zbl 273 # 02035
• ID 12836

SZCZERBA, L.W. [1974] see SCHWABHAEUSER, W.

SZCZERBA, L.W. [1975] see SCHWABHAEUSER, W.

SZCZERBA, L.W. [1976] see PRAZMOWSKI, K.

SZCZERBA, L.W. & TARSKI, A. [1979] Metamathematical discussion of some affine geometries (J 0027) Fund Math 104*155-192
⋄ B25 B30 C35 C52 C65 D35 ⋄ REV MR 81a:03012
Zbl 497 # 03008 • ID 79177

SZCZERBA, L.W. [1980] Interpretations with parameters (J 0068) Z Math Logik Grundlagen Math 26*35-39
⋄ B25 C07 C35 C45 F25 ⋄ REV MR 81k:03062
Zbl 444 # 03017 • ID 79172

SZCZERBA, L.W. [1983] Interpretations (J 0387) Bull Sect Logic, Pol Acad Sci 12*208-213
⋄ B30 C07 F25 ⋄ ID 39510

SZCZERBA, L.W. [1985] see SETTE, A.-M.

SZCZERBA, L.W. see Vol. I, IV, VI for further entries

SZELE, T. [1950] Ein Analogon der Koerpertheorie fuer abelsche Gruppen (J 0127) J Reine Angew Math 188*167-192
⋄ C25 C60 ⋄ REV MR 14 # 132 Zbl 54.010 • ID 49815

SZELE, T. [1956] see FUCHS, L.

SZELE, T. see Vol. V for further entries

SZENDREI, A. [1976] The operation ISKP on classes of algebras (J 0004) Algeb Universalis 6*349-353
⋄ C05 ⋄ REV MR 55 # 242 Zbl 357 # 08008 • ID 23918

SZENDREI, A. see Vol. II for further entries

SZMIELEW, W. [1949] Decision problem in group theory (P 0682) Int Congr Philos (10);1948 Amsterdam 763-766
⋄ B25 C10 C60 ⋄ REV MR 10.500 Zbl 34.294 JSL 14.63
• ID 12846

SZMIELEW, W. [1955] Elementary properties of abelian groups (J 0027) Fund Math 41*203-271
⋄ B25 C10 C60 ⋄ REV MR 17.233 Zbl 64.8 JSL 24.59
• ID 12849

SZMIELEW, W. [1962] *New foundations of absolute geometry* (P 0612) Int Congr Log, Meth & Phil of Sci (1,Proc);1960 Stanford 168-175
⋄ B30 C65 ⋄ REV MR 28 # 1517 Zbl 135.207 JSL 37.201
• ID 12850

SZMIELEW, W. [1974] *The order and the semi-order of n-dimensional euclidean space in the axiomatic and model-theoretic aspects* (P 0777) Grundl Geom & Algeb Meth;1973 Potsdam 69-79
⋄ B30 C65 ⋄ REV MR 53 # 112 Zbl 291 # 50003
• ID 16578

SZMIELEW, W. [1974] *The role of the Pasch axiom in the foundations of euclidean geometry* (P 0610) Tarski Symp;1971 Berkeley 123-132
⋄ C65 ⋄ REV MR 51 # 10072 Zbl 312 # 50002 • ID 12853

SZMIELEW, W. [1983] see SCHWABHAEUSER, W.

SZMIELEW, W. see Vol. I, IV, V, VI for further entries

SZWAST, W. [1985] *On some properties of Horn's spectra (Polish) (English summary)* (S 1454) Zesz Nauk Wyz Szk Ped Mat, Opole 23*5-9
⋄ C13 D15 ⋄ REV Zbl 574 # 03021 • ID 48831

TABATA, H. [1969] *Free structures and universal Horn sentences* (J 0352) Math Jap 14*101-104
⋄ B20 C05 C52 ⋄ REV MR 44 # 2596 Zbl 199.327
• ID 13247

TABATA, H. [1971] *A generalized free structure and several properties of universal Horn classes* (J 0352) Math Jap 16*91-102
⋄ B20 C05 C52 ⋄ REV MR 46 # 120 Zbl 309 # 08003
• ID 13248

TABATA, H. [1973] *An application of a certain argument about isomorphisms of α-saturated structures* (J 0081) Proc Japan Acad 49*262-264
⋄ C07 C50 ⋄ REV MR 48 # 5850 Zbl 274 # 02026
• ID 29021

TAHA, F. [1982] *Algebres simples centrales sur les corps ultra-produits de corps p-adiques* (P 4031) Semin Algeb (34);1981 Paris 89-128
⋄ C20 C60 ⋄ REV MR 84a:16035 Zbl 489 # 12009
• ID 38839

TAIT, W.W. [1959] *A counterexample to a conjecture of Scott and Suppes* (J 0036) J Symb Logic 24*15-16
⋄ C13 ⋄ REV MR 22 # 6717 Zbl 92.249 JSL 33.288
• ID 13278

TAIT, W.W. [1960] see GANDY, R.O.

TAIT, W.W. [1961] see GANDY, R.O.

TAIT, W.W. [1965] *Infinitely long terms of transfinite type* (P 0688) Logic Colloq;1963 Oxford 176-185
⋄ C75 F10 F15 F50 ⋄ REV MR 33 # 3925 Zbl 154.5 JSL 40.623 • ID 13281

TAIT, W.W. [1968] *Normal derivability in classical logic* (P 0637) Syntax & Semant Infinitary Lang;1967 Los Angeles 204-236
⋄ C75 F05 F15 F30 F35 F60 ⋄ REV Zbl 206.5
• ID 13286

TAIT, W.W. see Vol. I, IV, V, VI for further entries

TAJMANOV, A.D. [1958] *On classes of models, closed under direct product (Russian)* (J 0067) Usp Mat Nauk 13/3*231-232
⋄ C30 C52 D35 ⋄ ID 13265

TAJMANOV, A.D. [1959] *On a class of models closed with respect to direct products (Russian)* (J 0023) Dokl Akad Nauk SSSR 127*1173-1175
⋄ C30 C52 ⋄ REV MR 21 # 5554 Zbl 123.246 JSL 30.253
• ID 19770

TAJMANOV, A.D. [1960] *On a class of models closed with respect to direct products (Russian)* (J 0216) Izv Akad Nauk SSSR, Ser Mat 24*493-510
• TRANSL [1966] (J 0225) Amer Math Soc, Transl, Ser 2 59*1-22
⋄ C30 C52 ⋄ REV MR 22 # 6713 Zbl 116.7 JSL 30.253
• ID 13267

TAJMANOV, A.D. [1961] *Characteristic properties of axiomatizable classes of models I,II (Russian)* (J 0216) Izv Akad Nauk SSSR, Ser Mat 25*601-620,755-764
⋄ C52 ⋄ REV MR 26 # 1255 MR 26 # 1256 Zbl 103.249 Zbl 119.12 JSL 38.164 JSL 38.165 • ID 13270

TAJMANOV, A.D. [1961] *Characterization of finitely axiomatizable classes of models (Russian)* (J 0023) Dokl Akad Nauk SSSR 138*67-69
• TRANSL [1961] (J 0062) Sov Math, Dokl 2*552-554
⋄ C52 ⋄ REV MR 23 # A1537 Zbl 113.243 JSL 38.165
• ID 19767

TAJMANOV, A.D. [1961] *Characterization of finitely axiomatizable classes of models (Russian)* (J 0092) Sib Mat Zh 2*759-766
⋄ C52 ⋄ REV MR 26 # 1257 Zbl 113.243 JSL 38.165
• ID 19768

TAJMANOV, A.D. [1962] *On a theorem of Beth and Kochen (Russian)* (J 0003) Algebra i Logika 1/6*4-16
⋄ C20 C40 C52 ⋄ REV MR 27 # 4752 Zbl 178.312
• ID 13272

TAJMANOV, A.D. [1962] *The characteristics of axiomatizable classes of models (Russian)* (J 0003) Algebra i Logika 1/4*5-31
⋄ C20 C52 ⋄ REV MR 33 # 5490 Zbl 178.313 JSL 38.164
• ID 13271

TAJMANOV, A.D. [1963] *Decidability of the elementary theory of sphere inclusion (Russian)* (J 0003) Algebra i Logika 2/3*23-27
⋄ B25 C65 ⋄ REV MR 27 # 5692 Zbl 199.32 • ID 13273

TAJMANOV, A.D. [1965] see ERSHOV, YU.L.

TAJMANOV, A.D. [1967] *Automorphisms of an algebra having one unary operation in the signature (Russian)* (J 0003) Algebra i Logika 6/5*21-31
⋄ C07 C20 C40 ⋄ REV MR 37 # 2591 Zbl 212.321
• ID 13276

TAJMANOV, A.D. [1967] *Remarks to the Mostowski-Ehrenfeucht theorem (Russian) (English summary)* (J 0003) Algebra i Logika 6/3*101-103
⋄ C07 C20 C30 C52 E50 ⋄ REV MR 36 # 1317 Zbl 165.310 • ID 13275

TAJMANOV, A.D. [1967] *Systems with a solvable universal theory (Russian)* (J 0003) Algebra i Logika 6/5*33-43
⋄ B20 B25 C20 ⋄ REV MR 37 # 6173 Zbl 165.318
• ID 13274

TAJMANOV, A.D. [1970] see MUSTAFIN, T.G.

TAJMANOV, A.D. [1974] *On the elementary theory of topological algebras (Russian)* (J 0027) Fund Math 81*331-342
- ◊ C57 C65 ◊ REV MR 50#4290 Zbl 292#13011
- ID 13277

TAJMANOV, A.D. [1978] *On topologizing countable algebras (Russian)* (J 0023) Dokl Akad Nauk SSSR 243*284-286
- TRANSL [1978] (J 0062) Sov Math, Dokl 19*1346-1349
- ◊ C05 C15 C65 ◊ REV MR 80d:08003 Zbl 415#03023
- ID 53125

TAJMANOV, A.D. [1980] see KHMELEVSKIJ, YU.I.

TAJMANOV, A.D. see Vol. I, IV, V for further entries

TAJTSLIN, M.A. [1962] *Effective inseparability of the set of identically true and the set of finitely refutable formulas of the elementary theory of lattices (Russian)* (J 0003) Algebra i Logika 1/3*24-38
- ◊ C13 D25 D35 G10 ◊ REV MR 28#1131 Zbl 199.31
- ID 19772

TAJTSLIN, M.A. [1962] *Relatively elementary subspaces in compact Lie algebras (Russian)* (J 0003) Algebra i Logika 1/2*30-46
- ◊ C60 ◊ REV MR 27#5869 Zbl 163.281 • ID 32023

TAJTSLIN, M.A. [1963] *Undecidability of elementary theories of certain classes of finite commutative associative rings (Russian)* (J 0003) Algebra i Logika 2/3*29-51
- ◊ C13 D35 ◊ REV MR 28#1128 Zbl 192.56 • ID 13254

TAJTSLIN, M.A. [1964] *Decidability of certain elementary theories (Russian)* (J 0003) Algebra i Logika 3/3*5-12
- ◊ B25 C60 ◊ REV MR 31#4726 • ID 13256

TAJTSLIN, M.A. [1964] *On elementary theories of free nilpotent algebras (Russian)* (J 0003) Algebra i Logika 3/5-6*57-63
- ◊ B25 C60 D35 ◊ REV MR 32#5519 Zbl 231#02059
- ID 13255

TAJTSLIN, M.A. [1965] see ERSHOV, YU.L.

TAJTSLIN, M.A. [1965] *On the elementary theory of classical Lie algebras (Russian)* (J 0023) Dokl Akad Nauk SSSR 164*1243-1245
- TRANSL [1965] (J 0062) Sov Math, Dokl 6*1373-1376
- ◊ C60 D35 ◊ REV MR 33#1338 Zbl 231#02060
- ID 65641

TAJTSLIN, M.A. [1965] *On the theory of finite rings with division (Russian)* (J 0003) Algebra i Logika 4/4*103-114
- ◊ C13 D35 ◊ REV MR 35#1641 Zbl 216.292 • ID 13257

TAJTSLIN, M.A. [1966] *On elementary theories of commutative semigroups with cancellation (Russian)* (J 0003) Algebra i Logika 5/1*51-69
- ◊ B25 C60 D40 ◊ REV MR 33#5486 Zbl 253#02047
- ID 13258

TAJTSLIN, M.A. [1966] *On elementary theories of commutative semigroups (Russian)* (J 0003) Algebra i Logika 5/4*55-89
- ◊ B25 C05 C60 ◊ REV MR 35#2985 Zbl 174.21
- ID 32027

TAJTSLIN, M.A. [1967] *Two remarks on isomorphisms of commutative semigroups (Russian) (English summary)* (J 0003) Algebra i Logika 6/1*95-110
- ◊ B25 C05 C60 ◊ REV MR 35#5531 Zbl 207.331
- ID 32028

TAJTSLIN, M.A. [1968] *Algorithmic problems for commutative semigroups (Russian)* (J 0023) Dokl Akad Nauk SSSR 178*786-789
- TRANSL [1968] (J 0062) Sov Math, Dokl 9*201-204
- ◊ B25 C05 C60 ◊ REV MR 37#2878 Zbl 174.22
- ID 32030

TAJTSLIN, M.A. [1968] *Elementary lattice theories for ideals in polynomial rings (Russian)* (J 0003) Algebra i Logika 7/2*94-97
- TRANSL [1968] (J 0069) Algeb and Log 7*127-129
- ◊ C60 D35 G10 ◊ REV MR 38#4312 Zbl 216.292
- ID 13260

TAJTSLIN, M.A. [1968] *On the isomorphism problem for commutative semigroups (Russian)* (J 0092) Sib Mat Zh 9*375-401
- TRANSL [1968] (J 0475) Sib Math J 9*286-304
- ◊ C05 C07 C60 D40 ◊ REV MR 37#330
Zbl 242#20069 • ID 32029

TAJTSLIN, M.A. [1970] *Model theory (Russian)* (X 0913) Novosibirsk Gos Univ: Novosibirsk 214pp
- ◊ B98 C98 ◊ REV MR 49#2350 • ID 28675

TAJTSLIN, M.A. [1970] *On elementary theories of lattices of subgroups (Russian)* (J 0003) Algebra i Logika 9*473-483
- TRANSL [1970] (J 0069) Algeb and Log 9*285-290
- ◊ B25 C60 D35 G10 ◊ REV MR 44#1739
Zbl 221#02033 • ID 28107

TAJTSLIN, M.A. [1972] see ABAKUMOV, A.I.

TAJTSLIN, M.A. [1972] see BELEGRADEK, O.V.

TAJTSLIN, M.A. [1973] *Existentially closed regular commutative semigroups (Russian)* (J 0003) Algebra i Logika 12*689-703,735-736
- TRANSL [1973] (J 0069) Algeb and Log 12*394-401
- ◊ C25 C35 C52 C60 ◊ REV MR 52#10413
Zbl 292#20056 • ID 21710

TAJTSLIN, M.A. [1973] see ERSHOV, YU.L.

TAJTSLIN, M.A. [1974] *Existentially closed commutative rings* (P 1590) All-Union Conf Math Log (3);1974 Novosibirsk 213-215
- ◊ C25 C60 ◊ ID 27350

TAJTSLIN, M.A. [1975] see BELEGRADEK, O.V.

TAJTSLIN, M.A. [1977] *Existentially closed commutative semigroups (Russian) (English summary)* (J 0027) Fund Math 94*231-243
- ◊ C25 C52 C60 ◊ REV MR 55#5436 Zbl 358#20068
- ID 79188

TAJTSLIN, M.A. [1979] *A description of algebraic systems in weak logic of order ω and in program logic (Russian)* (C 2065) Teor Nereg Kriv Raz Geom Post 91-98
- ◊ B15 B75 C35 C85 ◊ REV MR 82e:68031 • ID 82932

TAJTSLIN, M.A. [1979] see BELYAEV, V.YA.

TAJTSLIN, M.A. [1979] *Program theories of periodic abelian groups (Russian)* (C 2065) Teor Nereg Kriv Raz Geom Post 98-107
- ◊ B75 C60 ◊ REV MR 81j:03054 • ID 79187

TAJTSLIN, M.A. [1980] see BAJZHANOV, B.S.

TAJTSLIN, M.A. see Vol. II, IV for further entries

TAKAHASHI, MAKOTO [1978] *Sheaves of B-valued structures (Japanese)* (P 4109) B-Val Anal & Nonstand Anal;1978 Kyoto 87-100
⋄ C90 G30 ⋄ ID 47669

TAKAHASHI, MAKOTO [1980] *Topological powers and reduced powers* (J 2613) Tokyo J Math 3*141-147
⋄ C20 C30 C90 ⋄ REV MR 82c:03040 Zbl 436#03024 • ID 55864

TAKAHASHI, MAKOTO [1981] *Iterated boolean powers (Japanese)* (P 4153) B-Val Anal & Nonstand Anal;1981 Kyoto 57-65
⋄ C30 E40 ⋄ ID 47759

TAKAHASHI, MAKOTO [1985] *Iterated boolean powers* (J 0407) Comm Math Univ St Pauli (Tokyo) 34*59-66
⋄ C30 E40 ⋄ ID 46403

TAKAHASHI, MICHIHIRO [1982] *Extensions of semimodules. I* (J 1508) Math Sem Notes, Kobe Univ 10*563-592
⋄ C60 G30 ⋄ REV MR 84h:16024 • ID 46096

TAKAHASHI, MOTO-O [1970] *On models of axiomatic set theory (Japanese)* (J 0091) Sugaku 22*161-176
⋄ C62 ⋄ ID 90082

TAKAHASHI, MOTO-O [1972] $\tilde{\Delta}_1$-*definability in set theory* (P 2080) Conf Math Log;1970 London 255*281-304
⋄ C62 E35 E45 E47 ⋄ REV MR 54#97 Zbl 244#02027 • ID 24577

TAKAHASHI, MOTO-O see Vol. I, II, IV, V, VI for further entries

TAKAHASHI, S. [1973] *A non-standard treatment of infinitely near points* (C 2976) Numb Th, Algeb Geom & Comm Algeb (Akizuki) 231-241
⋄ D35 H15 H20 ⋄ REV MR 55#5632 Zbl 291#02040 • ID 65648

TAKAHASHI, S. [1974] *Methodes logiques en geometrie diophantienne* (X 0893) Pr Univ Montreal: Montreal 48*178pp
⋄ B28 C98 D98 G30 H98 ⋄ REV MR 51#424 Zbl 325#02038 • ID 17488

TAKAHASHI, S. [1979] see HIROSE, K.

TAKAHASHI, S. see Vol. I for further entries

TAKEUCHI, Y. [1977] *Construction of nonstandard numbers from the rationals (Spanish)* (J 0377) Bol Mat (Bogota) 11*185-207
⋄ C20 H05 ⋄ REV MR 80h:03096 Zbl 494#03053 • ID 79205

TAKEUCHI, Y. see Vol. I, V for further entries

TAKEUTI, G. [1957] *On Skolem's theorem* (J 0090) J Math Soc Japan 9*71-76
⋄ C07 F25 ⋄ REV MR 19.4 Zbl 86.8 JSL 24.66 • ID 13310

TAKEUTI, G. [1957] *Remark on my paper: On Skolem's theorem* (J 0090) J Math Soc Japan 9*192-194
⋄ C07 F25 F30 ⋄ REV MR 19.829 Zbl 105.6 JSL 24.66
• REM The paper was published ibid 9*71-76 • ID 13311

TAKEUTI, G. [1961] see MAEHARA, S.

TAKEUTI, G. [1962] *On the weak definability in set theory* (J 0081) Proc Japan Acad 38*43-46
⋄ C40 E25 E47 ⋄ REV MR 29#8 Zbl 104.242 • ID 13325

TAKEUTI, G. [1963] see KINO, A.

TAKEUTI, G. [1968] *A determinate logic* (P 0637) Syntax & Semant Infinitary Lang;1967 Los Angeles 237-264
⋄ C40 C75 E60 ⋄ REV Zbl 182.325 • ID 13332

TAKEUTI, G. [1970] *A determinate logic* (J 0111) Nagoya Math J 38*113-138
⋄ C40 C75 E60 ⋄ REV MR 46#3266 Zbl 194.309 • ID 13336

TAKEUTI, G. [1971] see MAEHARA, S.

TAKEUTI, G. [1977] see HENSON, C.W.

TAKEUTI, G. [1978] *Two applications of logic to mathematics* (X 3552) Iwanami Shoten: Tokyo viii+137pp
• LAST ED [1978] (X 0857) Princeton Univ Pr: Princeton viii+137pp
⋄ B98 C90 E40 E75 F05 F30 F35 G12 ⋄ REV MR 58#21591 Zbl 393#03027 • ID 52447

TAKEUTI, G. & TITANI, S. [1981] *Heyting-valued universes of intuitionistic set theory* (P 3201) Logic Symposia;1979/80 Hakone 189-306
⋄ C90 E40 E70 G30 ⋄ REV MR 84g:03100 Zbl 482#03023 • ID 34204

TAKEUTI, G. [1984] *Quantum logic and quantization* (P 4383) Found Quant Mech;1983 Kokubunji-shi 256-260
⋄ B51 C90 E40 E75 ⋄ REV MR 86g:03108 • ID 39581

TAKEUTI, G. see Vol. I, II, IV, V, VI for further entries

TAKHTADZHYAN, K.A. [1969] *Logical formulae that are preserved under a strong homomorphism (Armenian) (Russian and English summaries)* (J 0312) Izv Akad Nauk Armyan SSR, Ser Mat 4*66-68
⋄ C07 C40 ⋄ REV MR 42#52 Zbl 181.5 • ID 13252

TALANOV, V.A. [1969] see GLEBSKIJ, YU.V.

TALANOV, V.A. [1981] *Asymptotic solvability of logical formulas (Russian)* (S 3911) Komb-Algeb Met Prikl Mat (Gor'kij) 1981*118-126
⋄ C13 ⋄ REV MR 85i:03081 Zbl 538#03008 • ID 44184

TALJA, J. [1985] *Semantic games on finite trees* (P 4180) Int Congr Log, Meth & Phil of Sci (7,Pap);1983 Salzburg 153-180
⋄ C13 ⋄ ID 48121

TALJA, J. see Vol. II, IV for further entries

TAMAGAWA, T. [1982] *On regularly closed fields* (P 3817) Algeb Homage:Ring Th & Rel Top (Jacobson);1981 New Haven 325-334
⋄ C60 ⋄ REV MR 84m:14024 • ID 39469

TAMTHAI, M. [1978] *A note on conservative extensions* (J 1735) Bull South East Asian Soc 2*53-54
⋄ C07 ⋄ REV MR 80d:03033 Zbl 418#03018 • ID 53303

TAMURA, S. [1970] *Axioms for commutative rings* (J 0081) Proc Japan Acad 46*116-120
⋄ C60 ⋄ REV MR 42#6019 Zbl 215.383 • ID 37186

TAMURA, S. see Vol. I, II, IV for further entries

TANAKA, H. [1965] see NISHIMURA, T.

TANAKA, H. [1974] *Some analytical rules of inference in the second-order arithmetic* (J 0407) Comm Math Univ St Pauli (Tokyo) 23/1*71-81
⋄ C62 D55 F35 ⋄ REV MR 53#10559 Zbl 357#02012 • ID 23068

TANAKA, H. see Vol. I, II, IV, V for further entries

TANG, C.Y. [1966] *An existence theorem for generalized direct products with amalgamated subgroups* (J 0017) Canad J Math 18*75-82
⋄ C30 C60 ⋄ REV MR 32#4181 Zbl 178.350 • ID 13387

TARLECKI, A. [1985] *On the existence of free models in abstract algebraic institutions* (J 1426) Theor Comput Sci 37*269-304
⋄ B75 C95 ⋄ ID 48448

TARSKI, A. [1927] see LINDENBAUM, A.

TARSKI, A. [1931] *Sur les ensembles definissables de nombres reels I* (J 0027) Fund Math 17*210-239
• TRANSL [1956] (C 1159) Tarski: Logic, Semantics, Metamathematics 110-142 (English)
⋄ B25 B28 C10 C40 C60 C65 D55 E15 E47 F35 ⋄ REV Zbl 75.4 JSL 34.99 FdM 57.60 • ID 16924

TARSKI, A. [1932] *Der Wahrheitsbegriff in den Sprachen der deduktiven Disziplinen* (J 0931) Anz Oesterr Akad Wiss, Math-Nat Kl 69*23-25
⋄ A05 B30 C07 C40 F30 ⋄ REV Zbl 4.1 FdM 58.997
• REM Summary. The full article appeared in 1933 • ID 13406

TARSKI, A. [1933] *Einige Betrachtungen ueber die Begriffe der ω-Widerspruchsfreiheit und der ω-Vollstaendigkeit* (J 0124) Monatsh Math-Phys 40*97-112
• TRANSL [1956] (C 1159) Tarski: Logic, Semantics, Metamathematics 279-295 (English)
⋄ B28 C62 F30 ⋄ REV MR 17.1171 Zbl 7.97 JSL 34.99 FdM 59.53 • ID 13407

TARSKI, A. [1933] *On the notion of truth in reference to formalized deductive sciences (Polish)* (J 0459) C R Soc Sci Lett Varsovie Cl 3 34*vii+116pp
• TRANSL [1935] (J 4716) Stud Philos, Leopolis (Poznan) 1*261-405 (German) [1956] (C 1159) Tarski: Logic, Semantics, Metamathematics 152-278 (English) [1961] (C 0769) L'Antinom Ment nel Pensiero Contemp 391-677 (Italian)
⋄ A05 B15 B30 C07 C40 F30 ⋄ REV Zbl 13.289 Zbl 4.1 Zbl 75.7 JSL 34.99 FdM 58.997 FdM 62.1051
• REM Polish summary in J1093, 1930/31, 210-211; German summary in J0931, 1932 • ID 28816

TARSKI, A. [1934] *Some methodological investigations on the definability of terms (Polish)* (J 1125) Przeglad Filoz 37*438-460
• TRANSL [1956] (C 1159) Tarski: Logic, Semantics, Metamathematics 296-319 [1935] (J 0748) Erkenntnis (Leipzig) 5*80-100
⋄ B10 C35 C40 ⋄ REV Zbl 12.1 JSL 34.99 FdM 61.50
• ID 30917

TARSKI, A. [1935] *Grundzuege des Systemkalkuels. Teil I* (J 0027) Fund Math 25*503-526
• TRANSL [1956] (C 1159) Tarski: Logic, Semantics, Metamathematics 342-364
⋄ B22 C07 ⋄ REV Zbl 12.385 JSL 34.100 FdM 62.38
• REM Part II 1936 • ID 13409

TARSKI, A. [1935] see LINDENBAUM, A.

TARSKI, A. [1936] *Grundzuege des Systemkalkuels. Teil II* (J 0027) Fund Math 26*283-301
• TRANSL [1956] (C 1159) Tarski: Logic, Semantics, Metamathematics 364-383
⋄ A05 B22 C07 C35 G15 ⋄ REV Zbl 14.387 JSL 1.71 FdM 62.38 • REM Part I 1935 • ID 13415

TARSKI, A. [1936] *On the concept of logical consequences (Polish)* (J 1125) Przeglad Filoz 39*58-68
• TRANSL [1936] (P 0632) Congr Int Phil des Sci;1935 Paris 7*1-11 (German) [1956] (C 1159) Tarski: Logic, Semantics, Metamathematics 409-420
⋄ A05 C07 ⋄ REV JSL 2.83 JSL 34.100 FdM 62.1065
• ID 13414

TARSKI, A. [1936] *The establishment of scientific semantics (Polish)* (J 1125) Przeglad Filoz 39*50-57
• TRANSL [1936] (P 0632) Congr Int Phil des Sci;1935 Paris 3*1-8 (German) [1956] (C 1159) Tarski: Logic, Semantics, Metamathematics 401-408 (English)
⋄ A05 B10 B30 C07 ⋄ REV MR 17.1171 JSL 2.83 JSL 34.100 FdM 62.1065 • ID 13413

TARSKI, A. [1938] *Ein Beitrag zur Axiomatik der abelschen Gruppen* (J 0027) Fund Math 30*253-256
⋄ B30 C60 ⋄ REV Zbl 19.52 FdM 64.53 • ID 30925

TARSKI, A. [1944] see MCKINSEY, J.C.C.

TARSKI, A. [1944] *The semantic conception of truth and the foundations of semantics* (J 0075) Phil Phenom Research 4*341-376
• TRANSL [1955] (J 0290) Euclides 30*1-43 (Dutch) [1963] (X 1364) Vita e Pensiero: Milano 43pp
⋄ A05 B10 B30 C07 ⋄ REV MR 16.438 MR 6.31 Zbl 61.8 Zbl 64.4 JSL 9.68 • ID 13432

TARSKI, A. [1946] *A remark on functionally free algebras* (J 0120) Ann of Math, Ser 2 47*163-165
⋄ C05 C30 G25 ⋄ REV MR 7.360 Zbl 60.62 JSL 11.84
• ID 13434

TARSKI, A. [1946] see MCKINSEY, J.C.C.

TARSKI, A. [1947] see JONSSON, B.

TARSKI, A. [1948] *A decision method for elementary algebra and geometry* (X 4282) Rand Cor: Santa Monica iii+60pp
• LAST ED [1951] (X 0926) Univ Calif Pr: Berkeley iii+63pp
⋄ B25 B30 C10 C35 C60 ⋄ REV MR 10.499 MR 13.423 Zbl 44.251 JSL 14.188 JSL 17.207 • ID 19756

TARSKI, A. [1948] *A problem concerning the notion of definability* (J 0036) J Symb Logic 13*107-111
⋄ B10 B15 C40 ⋄ REV MR 10.176 Zbl 29.242 JSL 13.172 • ID 13437

TARSKI, A. [1948] *Axiomatic and algebraic aspects of two theorems on sums of cardinals* (J 0027) Fund Math 35*79-104
⋄ B28 C15 E10 E25 G05 ⋄ REV MR 10.687 Zbl 31.289 JSL 14.257 • ID 13436

TARSKI, A. [1949] see JONSSON, B.

TARSKI, A. [1952] *Some notions and methods on the borderline of algebra and metamathematics* (P 0593) Int Congr Math (II, 6);1950 Cambridge MA I*705-720
⋄ C07 C35 C52 C60 ⋄ REV MR 13.521 Zbl 49.7 JSL 18.182 • ID 13439

TARSKI, A. [1954] *Contributions to the theory of models I,II* (J 0028) Indag Math 16*572-581,582-588
⋄ C13 C30 C52 ⋄ REV MR 16.554 Zbl 58.247 JSL 21.405 • REM Part III 1955 • ID 13443

TARSKI, A. [1955] *Contributions to the theory of models III* (J 0028) Indag Math 17*56-64
⋄ C05 C52 G15 ⋄ REV MR 16.554 Zbl 58.247 JSL 21.405 • REM Parts I,II 1954 • ID 13445

TARSKI, A. [1956] *Equationally complete rings and relation algebras* (J 0028) Indag Math 18*39-46
⋄ C05 G15 ⋄ REV MR 18.636 Zbl 73.246 JSL 23.57
• ID 13448

TARSKI, A. [1956] *Logic, semantics, metamathematics. Papers from 1923 to 1938* (X 0815) Clarendon Pr: Oxford xiv + 471pp
• LAST ED [1983] (X 2725) Hackett Publ : Indianapolis xxx + 506pp
⋄ B96 C96 ⋄ REV MR 17.1171 MR 85e:01065 Zbl 75.7 JSL 34.99 • REM See also 1972 • ID 28568

TARSKI, A. & VAUGHT, R.L. [1957] *Arithmetical extensions of relational systems* (J 0020) Compos Math 13*81-102
⋄ C05 C07 C55 C60 ⋄ REV MR 20 # 1627 Zbl 91.12 JSL 32.131 • ID 13453

TARSKI, A. & VAUGHT, R.L. [1957] *Elementary (arithmetical) extensions* (P 1675) Summer Inst Symb Log;1957 Ithaca 51-55
⋄ C05 C07 C35 C55 C60 ⋄ REV JSL 32.131 • ID 13460

TARSKI, A. [1957] *Introductory remarks on the theory of models* (P 1675) Summer Inst Symb Log;1957 Ithaca 174
⋄ C98 ⋄ ID 29342

TARSKI, A. [1957] *Remarks on direct products of commutative semigroups* (J 0132) Math Scand 5*218-223
⋄ C05 C30 G05 ⋄ REV MR 21 # 7168 Zbl 81.261
• ID 13450

TARSKI, A. [1957] *Remarks on predicate logic with infinitely long expressions* (P 1675) Summer Inst Symb Log;1957 Ithaca 160-163
⋄ C75 ⋄ ID 29338

TARSKI, A. [1957] see SCOTT, D.S.

TARSKI, A. [1958] *Remarks on predicate logic with infinitely long expressions* (S 0019) Colloq Math (Warsaw) 6*171-176
⋄ C75 ⋄ REV MR 20 # 6351 Zbl 119.250 JSL 30.94
• ID 13457

TARSKI, A. [1959] see HENKIN, L.

TARSKI, A. [1961] see JONSSON, B.

TARSKI, A. [1962] *Some problems and results relevant to the foundations of set theory* (P 0612) Int Congr Log, Meth & Phil of Sci (1,Proc);1960 Stanford 125-135
• TRANSL [1965] (P 2251) Mat Log & Primen;1960 Stanford 146-158 (Russian)
⋄ C75 E05 E10 E55 G05 ⋄ REV MR 27 # 1382 Zbl 199.18 JSL 30.95 • ID 13461

TARSKI, A. [1964] see KEISLER, H.J.

TARSKI, A. [1964] see CHANG, C.C.

TARSKI, A. [1965] see SZCZERBA, L.W.

TARSKI, A. [1965] see ADDISON, J.W.

TARSKI, A. [1967] *The completeness of elementary algebra and geometry* (X 0999) CNRS Inst B Pascal: Paris iv + 50pp
⋄ B25 B30 C10 C35 C60 C65 ⋄ REV JSL 34.302
• ID 13463

TARSKI, A. [1968] *Equational logic and equational theories of algebras* (P 0608) Logic Colloq;1966 Hannover 275-288
⋄ C05 ⋄ REV MR 38 # 5692 Zbl 209.14 JSL 36.161
• ID 13464

TARSKI, A. [1972] *Logique, semantique, metamathematique 1923-1944. Tome I* (X 0850) Colin: Paris 276pp
⋄ B96 C96 ⋄ REV Zbl 259 # 02002 • REM See also 1956
• ID 30338

TARSKI, A. [1975] *An interpolation theorem for irredundant bases of closure structures* (J 0193) Discr Math 12*185-192
⋄ C05 G15 ⋄ REV MR 53 # 2806 Zbl 319 # 06002
• ID 21650

TARSKI, A. [1978] see DONER, J.E.

TARSKI, A. [1979] see SZCZERBA, L.W.

TARSKI, A. [1983] see SCHWABHAEUSER, W.

TARSKI, A. also published under the name TAJTELBAUM, A.

TARSKI, A. see Vol. I, II, IV, V, VI for further entries

TAUTS, A. [1964] *Solution of logical equations in the first-order one-place predicate calculus (Russian) (English summary)* (J 4701) Trudy Akad Nauk Ehston SSR, Fiz Astron (Tartu) 24*3-16
⋄ B20 B25 ⋄ REV MR 33 # 2526 • ID 13475

TAUTS, A. [1972] *A connection between generalized Beth models and topological pseudo-Boolean algebras (Russian) (Estonian and German summaries)* (S 3468) Tr Mat & Mekh (Tartu) 12(305)*3-8
⋄ C90 F50 G10 ⋄ REV MR 51 # 7862 Zbl 321 # 02013
• ID 65673

TAUTS, A. [1972] *A semantic interpretation of formulas in generalized Beth models and in pseudo-boolean algebras (Russian) (Estonian and German summaries)* (S 3468) Tr Mat & Mekh (Tartu) 12(305)*9-20
⋄ C90 F50 G10 ⋄ REV MR 51 # 7863 Zbl 321 # 02014
• ID 17695

TAUTS, A. [1974] *A formal deduction of tautological formulas in pseudo-boolean algebras (Russian) (Estonian and German summaries)* (S 3468) Tr Mat & Mekh (Tartu) 13(336)*3-30
⋄ B60 C90 F50 G05 G10 ⋄ REV MR 51 # 7864 Zbl 356 # 02010 • ID 17694

TAUTS, A. [1974] *A game for the construction of a propositional semantics in generalized Beth models (Russian) (Estonian and German summaries)* (S 3468) Tr Mat & Mekh (Tartu) 14(342)*13-28
⋄ C75 C90 E60 F50 ⋄ REV MR 51 # 7865 Zbl 362 # 02042 • ID 17693

TAUTS, A. [1975] *A semantic model for infinite formulas (Russian) (Estonian and German summaries)* (S 3468) Tr Mat & Mekh (Tartu) 15(355)*7-19
⋄ C75 C90 F50 G10 ⋄ REV MR 52 # 7846 Zbl 354 # 02018 • ID 18414

TAUTS, A. [1975] *Search for deduction by means of a semantic model (Russian) (Estonian and German summaries)* (S 3468) Tr Mat & Mekh (Tartu) 17(374)*3-28
⋄ B55 C75 C90 F50 ⋄ REV MR 58 # 5076 Zbl 393 # 03039 • ID 52459

TAUTS, A. [1975] *Strong tautology and construction of a countermodel for nonderivable propositions (Russian) (Estonian and German summaries)* (S 3468) Tr Mat & Mekh (Tartu) 17(374)*29-42
⋄ C75 C90 F50 ⋄ REV MR 57 # 16049 Zbl 393 # 03040
• ID 52460

TAUTS, A. [1978] *Ambiguity of the axiom of normal functions (Russian) (Estonian and German summaries)* (S 3468) Tr Mat & Mekh (Tartu) 22(464)*13-27
⋄ B15 C75 E10 E55 E65 ⋄ REV MR 80b:03086 Zbl 461 # 03006 • ID 54483

TAUTS, A. [1981] *The connection between semantic models and pseudo-boolean algebras (Russian) (Estonian and German summaries)* (S 0393) Uch Zap Univ, Tartu 556*3-10
⋄ C75 C90 F50 G10 ⋄ REV MR 83b:03070 Zbl 487 # 03011 • ID 38114

TAUTS, A. see Vol. I, II, IV for further entries

TAYLOR, A.D. [1984] see KANAMORI, A.

TAYLOR, A.D. see Vol. V for further entries

TAYLOR, R.F. [1970] see LUXEMBURG, W.A.J.

TAYLOR, R.F. see Vol. I for further entries

TAYLOR, W. [1969] *Atomic compactness and graph theory* (J 0027) Fund Math 65*139-145
⋄ C05 C50 C65 ⋄ REV MR 40 # 81 Zbl 182.344 • ID 13494

TAYLOR, W. [1970] *Compactness and chromatic number* (J 0027) Fund Math 67*147-153
⋄ C05 C50 C65 ⋄ REV MR 42 # 3009 Zbl 199.304 • ID 13495

TAYLOR, W. [1971] *Atomic compactness and elementary equivalence* (J 0027) Fund Math 71*103-112
⋄ C05 C13 C50 C65 ⋄ REV MR 45 # 8598 Zbl 238 # 02044 • ID 13498

TAYLOR, W. [1971] *Some constructions of compact algebras* (J 0007) Ann Math Logic 3*395-435
⋄ C05 C50 C55 C65 ⋄ REV MR 45 # 47 Zbl 239 # 08003 JSL 40.455 • ID 13497

TAYLOR, W. [1971] see FUHRKEN, G.

TAYLOR, W. [1972] *Fixed points of endomorphisms* (J 0004) Algeb Universalis 2*74-76
⋄ C05 C50 ⋄ REV MR 46 # 5217 Zbl 263 # 08004 • ID 13502

TAYLOR, W. [1972] *Note on pure-essential extensions* (J 0004) Algeb Universalis 2*234-237
⋄ C05 ⋄ REV MR 46 # 8936 Zbl 267 # 08005 • ID 13503

TAYLOR, W. [1972] see BULMAN-FLEMING, S.

TAYLOR, W. [1972] *On equationally compact semigroups* (J 0136) Semigroup Forum 5*81-88
⋄ C05 C50 ⋄ REV MR 47 # 1606 Zbl 276 # 20050 • ID 13500

TAYLOR, W. [1972] *Residually small varieties* (J 0004) Algeb Universalis 2*33-53
⋄ C05 C50 C55 ⋄ REV MR 47 # 3278 Zbl 263 # 08005 JSL 40.455 • ID 13501

TAYLOR, W. [1973] *A heterogeneous interpolant* (J 0111) Nagoya Math J 52*31-33
⋄ C40 C75 ⋄ REV MR 48 # 10759 Zbl 284 # 02006 • ID 13506

TAYLOR, W. [1973] *Characterizing Mal'cev conditions* (J 0004) Algeb Universalis 3*351-397
⋄ C05 C52 ⋄ REV MR 50 # 2030 Zbl 304 # 08003 • ID 13504

TAYLOR, W. [1974] *Pure-irreducible mono-unary algebras* (J 0004) Algeb Universalis 4*235-243
⋄ C05 C50 G10 ⋄ REV MR 50 # 6982 Zbl 306 # 08004 • ID 13507

TAYLOR, W. [1974] *Uniformity of congruences* (J 0004) Algeb Universalis 4*342-360
⋄ C05 C52 ⋄ REV MR 51 # 12658 Zbl 313 # 08001 • ID 13508

TAYLOR, W. [1975] *Continuum many Mal'cev conditions* (J 0004) Algeb Universalis 5*333-335
⋄ C05 C52 ⋄ REV MR 52 # 13583 Zbl 345 # 08001 • ID 21869

TAYLOR, W. [1975] *The fine spectrum of a variety* (J 0004) Algeb Universalis 5*263-303
⋄ C05 C13 C52 ⋄ REV MR 52 # 10547 Zbl 336 # 08004 • ID 21732

TAYLOR, W. [1976] see MYCIELSKI, J.

TAYLOR, W. [1976] *Pure compactifications in quasi-primal varieties* (J 0017) Canad J Math 28*50-62
⋄ C05 ⋄ REV MR 53 # 5435 Zbl 326 # 08006 • ID 22952

TAYLOR, W. [1977] *Equational logic* (P 2112) Contrib to Universal Algeb;1975 Szeged 465-501
⋄ C05 C98 D40 ⋄ REV MR 57 # 12341 Zbl 421 # 08003 • ID 69948

TAYLOR, W. [1979] *Equational logic* (J 1447) Houston J Math Survey 1979*iii+83pp
⋄ C05 C98 ⋄ REV MR 80j:03042 Zbl 421 # 08004 JSL 47.450 • ID 79270

TAYLOR, W. [1980] *Mal'tsev conditions and spectra* (J 3194) J Austral Math Soc, Ser A 29*143-152
⋄ C05 C13 C52 ⋄ REV MR 81h:08006 Zbl 439 # 08005 • ID 69949

TAYLOR, W. [1981] *Some universal sets of terms* (J 0064) Trans Amer Math Soc 267*595-607
⋄ C05 D80 ⋄ REV MR 82h:08005 Zbl 483 # 04003 • ID 82957

TAYLOR, W. [1982] *Some interesting identities* (J 1507) An Inst Mat, Univ Nac Aut Mexico 20*127-156
⋄ C05 C98 ⋄ REV MR 84c:08001 Zbl 501 # 08003 • ID 39573

TAYLOR, W. [1985] see GARCIA, O.C.

TENNANT, N. [1975] see ALTHAM, J.E.J.

TENNANT, N. see Vol. I, II, VI for further entries

TENNEY, R.L. [1975] *Second-order Ehrenfeucht games and the decidability of the second-order theory of an equivalence relation* (J 0038) J Austral Math Soc 20*323-331
⋄ B15 B25 C85 ⋄ REV MR 52 # 2861 Zbl 319 # 02036 • ID 17647

TENNEY, R.L. see Vol. I for further entries

TERRA, A. [1975] *Questions on the axiomatization of n-dimensional order (Russian)* (C 3170) Geom Sb Vyp 15 205-232,239-240
⋄ C65 ⋄ REV MR 54 # 10090 Zbl 473 # 06002 • ID 55394

TETRUASHVILI, M.R. [1984] *The computational complexity of the theory of abelian groups with a given number of generators* (P 3621) Frege Konferenz (2);1984 Schwerin 371-375
⋄ B25 D15 ⋄ ID 45394

TETRUASHVILI, M.R. see Vol. IV for further entries

THARP, L.H. [1973] *The characterization of monadic logic* (J 0036) J Symb Logic 38∗481-488
⋄ B15 B20 C95 ⋄ REV MR 49 # 2250 Zbl 275 # 02015
• ID 13527

THARP, L.H. [1974] *Continuity and elementary logic* (J 0036) J Symb Logic 39∗700-716
⋄ C07 C80 C95 D30 D55 ⋄ REV MR 51 # 49
Zbl 299 # 02011 • ID 13528

THARP, L.H. see Vol. I, V, VI for further entries

THATCHER, J.W. & WRIGHT, J.B. [1968] *Generalized finite automata theory with an application to a decision problem of second order logic* (J 0041) Math Syst Theory 2∗57-81
⋄ B15 B25 D05 ⋄ REV MR 37 # 75 Zbl 157.22
JSL 37.619 • ID 28596

THATCHER, J.W. see Vol. IV, V for further entries

THISTLEWAITE, P.B. [1980] see MCROBBIE, M.A.

THOMAS, I. [1952] *A new decision procedure for Aristotle's syllogistic* (J 0094) Mind 61∗564-566
⋄ A10 B20 B25 ⋄ REV JSL 21.315 • ID 13541

THOMAS, I. [1967] *Decision for K4* (J 0047) Notre Dame J Formal Log 8∗337-338
⋄ B25 ⋄ REV MR 45 # 1739 Zbl 189.282 JSL 37.182
• ID 13563

THOMAS, I. see Vol. I, II, V for further entries

THOMAS, S. [1985] *Complete existentially closed locally finite groups* (J 0008) Arch Math (Basel) 44∗97-109
⋄ C25 C60 ⋄ REV MR 86h:03064 Zbl 563 # 20037
• ID 39986

THOMAS, S. [1985] *The automorphism tower problem* (J 0053) Proc Amer Math Soc 95∗166-168
⋄ C07 C60 E75 ⋄ REV Zbl 575 # 20030 • ID 48871

THOMAS, WOLFGANG [1978] see EBBINGHAUS, H.-D.

THOMAS, WOLFGANG [1978] *The theory of successor with an extra predicate* (J 0043) Math Ann 237∗121-132
⋄ B25 D35 ⋄ REV MR 80b:03054 Zbl 369 # 02025
• ID 31271

THOMAS, WOLFGANG [1979] *Star-free regular sets of ω-sequences* (J 0194) Inform & Control 42∗148-156
⋄ C07 D05 ⋄ REV MR 80h:68052 Zbl 411 # 03031
• ID 52885

THOMAS, WOLFGANG [1980] *On the bounded monadic theory of well-ordered structures* (J 0036) J Symb Logic 45∗334-338
⋄ B15 B25 C85 E07 ⋄ REV MR 81h:03080
Zbl 437 # 03005 • ID 55873

THOMAS, WOLFGANG [1981] *A combinatorial approach to the theory of ω-automata* (J 0194) Inform & Control 48∗261-283
⋄ B25 D05 ⋄ REV MR 84d:68054 Zbl 478 # 03020
• ID 55635

THOMAS, WOLFGANG [1984] *An application of the Ehrenfeucht-Fraisse game in formal language theory* (S 3521) Mem Soc Math Fr 16∗11-21
⋄ C07 D05 ⋄ REV Zbl 558 # 68064 • ID 39756

THOMAS, WOLFGANG see Vol. I, IV, V, VI for further entries

THOMASON, R.H. [1967] *A decision procedure for Fitch's propositional calculus* (J 0047) Notre Dame J Formal Log 8∗101-117
⋄ B25 B45 ⋄ REV MR 38 # 3139 Zbl 183.15 • ID 13579

THOMASON, R.H. [1969] see JOHNSON, D.R.

THOMASON, R.H. [1975] *Decidability in the logic of conditionals* (C 1856) Log Enterprise 167-178
⋄ B25 B45 ⋄ REV Zbl 374 # 02014 • ID 51538

THOMASON, R.H. see Vol. I, II, VI for further entries

THOMASON, S.K. [1975] *Categories of frames for modal logic* (J 0036) J Symb Logic 40∗439-442
⋄ B45 C90 G30 ⋄ REV MR 52 # 2829 Zbl 317 # 02012
JSL 47.440 • ID 29635

THOMASON, S.K. see Vol. II, IV, V for further entries

THOMPSON, B. [1982] *Syllogisms using "few", "many", and "most"* (J 0047) Notre Dame J Formal Log 23∗75-84
⋄ B60 C80 ⋄ REV MR 84j:03067 Zbl 452 # 03025
• ID 55079

TIERNEY, M. [1972] *Sheaf theory and the continuum hypothesis* (P 0771) Toposes, Algeb Geom & Log;1971 Halifax 13-42
⋄ C90 E35 E50 G30 ⋄ REV MR 51 # 10088
Zbl 244 # 18005 • ID 17528

TIERNEY, M. see Vol. V for further entries

TIMOSHENKO, E.I. [1968] *Preservation of elementary and universal equivalence under the wreath product (Russian)* (J 0003) Algebra i Logika 7/4∗114-119
• TRANSL [1968] (J 0069) Algeb and Log 7∗273-276
⋄ C30 C60 ⋄ REV MR 41 # 352 Zbl 186.317 • ID 13609

TIMOSHENKO, E.I. [1972] *On the question of the elementary equivalence of groups (Russian)* (C 3549) Algebra, Vyp 1 (Irkutsk) 92-96
⋄ C30 C60 ⋄ REV MR 56 # 11781 • ID 79359

TIMOSHENKO, E.I. [1979] *On elementary theories of wreath products (Russian)* (S 2626) Vopr Teor Grupp Gomol Algeb 169-174
⋄ C30 C60 ⋄ REV Zbl 429 # 20002 • ID 90305

TIMOSHENKO, E.I. see Vol. IV for further entries

TINHOFER, G. [1978] *On the simultaneous isomorphism of special relations (isomorphism of automata)* (P 3148) Graphth Konzepte Inf (3);1977 Linz 205-213
⋄ C13 D05 ⋄ REV MR 58 # 20915 Zbl 389 # 68033
• ID 52324

TITANI, S. [1981] see TAKEUTI, G.

TITANI, S. see Vol. I, V, VI for further entries

TITIEV, R.J. [1972] *Multidimensional measurement and universal axiomatizability* (J 0105) Theoria (Lund) 38∗82-88
⋄ C65 ⋄ REV MR 47 # 6468 Zbl 264 # 92007 • ID 13620

TITIEV, R.J. see Vol. II for further entries

TITS, J. [1974] see KOPPELBERG, S.

TITS, J. see Vol. IV for further entries

TIURYN, J. [1977] *Fixed-points and algebras with infinitely long expressions. I* (P 1635) Math Founds of Comput Sci (6);1977 Tatranska Lomnica 513-522
⋄ B75 C75 ⋄ REV MR 58 # 3644a Zbl 397 # 68086 • REM Part II 1977 • ID 69955

TIURYN, J. [1977] *Fixed-points and algebras with infinitely long expressions. II* (P 2588) FCT'77 Fund of Comput Th;1977 Poznan 332-339
⋄ B75 C75 ⋄ REV MR 58 # 3644b Zbl 392 # 68013 • REM Part I 1977 • ID 69956

TIURYN, J. [1978] *Fixed-points and algebras with infinitely long expressions I: Regular algebras* (J 2095) Fund Inform, Ann Soc Math Pol, Ser 4 2*103-127
⋄ B75 C75 ⋄ REV MR 80a:68049 Zbl 401 # 68062 • REM Part II 1979 • ID 69957

TIURYN, J. [1979] *Fixed-points and algebras with infinitely long expressions II: μ-clones of regular algebras* (J 2095) Fund Inform, Ann Soc Math Pol, Ser 4 2*317-335
⋄ B75 C75 ⋄ REV MR 81e:08007 Zbl 436 # 68015 • REM Part I 1978 • ID 69958

TIURYN, J. [1981] see BERGSTRA, J.A.

TIURYN, J. [1981] *Logic of effective definitions* (J 2095) Fund Inform, Ann Soc Math Pol, Ser 4 4*629-659
⋄ B60 C55 C75 ⋄ REV MR 84c:03068 Zbl 486 # 68017 • ID 34010

TIURYN, J. [1984] *Implicit definability of finite binary trees by sets of equations* (P 2342) Symp Rek Kombin;1983 Muenster 320-332
⋄ C40 ⋄ ID 45378

TIURYN, J. see Vol. II, IV for further entries

TODORCEVIC, S.B. [1979] *Rigid boolean algebras* (J 0400) Publ Inst Math, NS (Belgrade) 25(39)*219-224
⋄ C50 E05 E75 G05 ⋄ REV MR 81j:06017 Zbl 414 # 06011 • ID 53088

TODORCEVIC, S.B. [1980] *Very strongly rigid boolean algebras* (J 0400) Publ Inst Math, NS (Belgrade) 27(41)*267-277
⋄ C50 E05 G05 ⋄ REV MR 82k:06017 Zbl 481 # 06010 • ID 82978

TODORCEVIC, S.B. see Vol. V for further entries

TOFFALORI, C. [1977] *Alcune proprieta di teorie di campi con un sottocampo privilegiato (English summary)* (J 3495) Boll Unione Mat Ital, V Ser, B 14*254-266
⋄ C25 C35 C60 ⋄ REV MR 58 # 16272 Zbl 362 # 02047 • ID 50769

TOFFALORI, C. [1978] *Eliminazione dei quantificatori per certe teorie di coppie di campi (Teoria dei modelli) (English summary)* (J 3285) Boll Unione Mat Ital, V Ser, A 15*159-166
⋄ C10 C60 ⋄ REV MR 58 # 213 Zbl 383 # 03016 • ID 51998

TOFFALORI, C. [1979] *Alcune osservazioni sugli anelli commutativi esistenzialimente chiusi (English summary)* (J 3495) Boll Unione Mat Ital, V Ser, B 16*1093-1102
⋄ C25 C60 ⋄ REV MR 82f:13008 Zbl 431 # 13002 • ID 82984

TOFFALORI, C. [1979] *Semigruppi archimedei esistenzialimente chiusi (English summary)* (J 0149) Atti Accad Naz Lincei Fis Mat Nat, Ser 8 67*162-167
⋄ C25 C60 ⋄ REV MR 82h:20080 Zbl 456 # 20031 • ID 82983

TOFFALORI, C. [1980] *Fasci di coppie di campi algebricamente chiusi* (J 3128) Boll Unione Mat Ital, Suppl 2*271-282
⋄ B25 C35 C60 C90 ⋄ REV MR 84e:12030 Zbl 467 # 03027 • ID 55025

TOFFALORI, C. [1980] *Sheaves of pairs of real closed fields* (J 2038) Rend Sem Mat, Torino 38/2*107-122 • ERR/ADD ibid 38*177
⋄ C25 C35 C60 C90 ⋄ REV MR 82m:03048a Zbl 465 # 03017 • ID 54920

TOFFALORI, C. [1980] *Sugli anelli nilreali (English summary)* (J 3128) Boll Unione Mat Ital, Suppl 2*127-144
⋄ C20 C25 C60 ⋄ REV MR 84c:13022 Zbl 466 # 13010 • ID 39578

TOFFALORI, C. [1980] *Sul model-completamento di certe teorie di coppie di anelli (English summary)* (J 3495) Boll Unione Mat Ital, V Ser, B 17*1439-1456
⋄ C10 C25 C35 C60 ⋄ REV MR 86d:03031 Zbl 464 # 13007 • ID 44417

TOFFALORI, C. [1982] *Stabilita, categoricita ed eliminazione dei quantificatori per una classe di anelli locali* (J 0088) Ann Univ Ferrara, NS, Sez 7 28*39-53
⋄ C10 C35 C45 C60 ⋄ REV MR 85b:13038 Zbl 522 # 03025 • ID 37787

TOFFALORI, C. [1982] *Strutture esistenzialmente complete per certe classi di anelli (English summary)* (J 0144) Rend Sem Mat Univ Padova 66*57-71
⋄ C20 C25 C60 ⋄ REV MR 83i:13024 Zbl 485 # 03013 • ID 36846

TOFFALORI, C. [1982] *Teoria dei modelli per alcune classi di anelli locali* (J 3768) Boll Unione Mat Ital, VI Ser, D 1*89-105
⋄ C10 C25 C35 C60 ⋄ REV MR 84i:13025 Zbl 501 # 03016 • ID 37776

TOFFALORI, C. [1983] *Differentially closed rings for some classes of differential rings (Italian) (English summary)* (J 3768) Boll Unione Mat Ital, VI Ser, D 2*51-70
⋄ C60 ⋄ REV MR 86a:13017 • ID 48787

TOFFALORI, C. [1983] *Questioni di teoria dei modelli per coppie di campi* (J 2100) Boll Unione Mat Ital, VI Ser, B 2*297-319
⋄ C10 C35 C45 C60 ⋄ REV MR 84i:12023 Zbl 532 # 03015 • ID 38278

TOFFALORI, C. [1983] *Teoria dei modelli per una classe di anelli differenziali* (P 3829) Atti Incontri Log Mat (1);1982 Siena 231-233
⋄ C25 C60 ⋄ REV MR 84k:03006 Zbl 516 # 03018 • ID 37253

TOFFALORI, C. [1984] *Anelli regolari separabilmente chiusi (English summary)* (J 0144) Rend Sem Mat Univ Padova 71*15-33
⋄ C35 C45 C60 ⋄ REV MR 86e:16022 • ID 48449

TOFFALORI, C. [1984] *On a class of differential local rings (Italian) (English summary)* (J 2100) Boll Unione Mat Ital, VI Ser, B 3*121-136
⋄ C25 C60 ⋄ REV MR 85f:13026 • ID 39950

TOFFALORI, C. [1984] see MARCJA, A.

TOFFALORI, C. [1985] see MARCJA, A.

TOFFALORI, C. [1985] *Teorie p-$\aleph_0$-categoriche* (P 4646) Atti Incontri Log Mat (2);1983/84 Siena 631-634
⋄ C35 C45 C60 ⋄ ID 49796

TOKARENKO, A.I. [1978] *The group-theoretic properties N_0, N, and $\tilde{N}$ are not axiomatizable (Russian)* (C 3467) Sb Rabot Algeb 329-331
⋄ C20 C52 C60 ⋄ REV MR 80i:03047 Zbl 477 # 20004 • ID 55605

TOKARENKO, A.I. [1979] *Axiomatizability of the group-theoretic property $\bar{Z}$ (Russian)* (J 0337) Mat Ezheg, Akad Nauk Latv SSR 23*155-157,275
⋄ C20 C52 C60 ⋄ REV MR 81c:20022 Zbl 432 # 20001
• ID 82985

TOKARENKO, A.I. [1984] *Two variants of the compactness theorem (Russian)* (J 0337) Mat Ezheg, Akad Nauk Latv SSR 28*208-211
⋄ C07 C20 ⋄ REV MR 86b:03035 Zbl 564 # 26006
• ID 44139

TOKAREV, E.V. [1985] *The notion of indistinguishability and the p-Mazur property in Banach spaces (Russian)* (J 0265) Ukr Mat Zh, Akad Nauk Ukr SSR 37*211-216
⋄ C65 E75 ⋄ ID 45606

TOMASIK, J. [1971] *On the number of elementary theories of some hierarchies of sets (Russian summary)* (J 0014) Bull Acad Pol Sci, Ser Math Astron Phys 19*271-274
⋄ C62 E45 ⋄ REV MR 45 # 4977 Zbl 295 # 02035
• ID 13634

TOMASIK, J. [1976] *On products of neat structures* (S 0019) Colloq Math (Warsaw) 36*13-16
⋄ C30 C50 ⋄ REV MR 55 # 10270 Zbl 363 # 02054
• ID 50881

TOMASIK, J. [1982] see PACHOLSKI, L.

TOURAILLE, A. [1984] see JURIE, P.-F.

TOURAILLE, A. [1985] *Elimination des quantificateurs dans la theorie elementaire des algebres de Boole munies d'une famille d'ideaux distingues (English summary)* (J 3364) C R Acad Sci, Paris, Ser 1 300*125-128
⋄ C10 G05 ⋄ REV MR 86c:03028 • ID 44684

TOURNEAU LE, J.J. [1969] see COMER, S.D.

TOURNEAU LE, J.J. see Vol. IV for further entries

TRACZYK, T. [1967] see MACZYNSKI, M.J.

TRACZYK, T. (ED.) [1982] *Universal algebra and applications. Papers presented at Stefan Banach International Center at the Semester "Universal Algebra and Applications" held February 15 - June 9, 1978* (X 1034) PWN: Warsaw 454pp
⋄ C97 ⋄ REV MR 85d:00024 Zbl 499 # 00007 • ID 38288

TRACZYK, T. see Vol. V for further entries

TRAKHTENBROT, B.A. [1950] *The impossibility of an algorithm for the decision problem in finite domains (Russian)* (J 0023) Dokl Akad Nauk SSSR 70*569-572
• TRANSL [1963] (J 0225) Amer Math Soc, Transl, Ser 2 23*1-5
⋄ C13 D35 ⋄ REV MR 11.488 Zbl 38.150 JSL 15.229
• ID 19743

TRAKHTENBROT, B.A. [1953] *On recursively separability (Russian)* (J 0023) Dokl Akad Nauk SSSR 88*953-956
⋄ B10 C13 D35 ⋄ REV MR 16.436 Zbl 50.8 JSL 19.60
• ID 19742

TRAKHTENBROT, B.A. [1960] *Algorithms and machine solutions of problems (Russian)* (X 3709) Izdat Fiz-Mat Lit: Moskva 119pp
• TRANSL [1977] (X 0885) Mir: Moskva 109pp (Spanish)
⋄ B25 B35 D20 ⋄ REV MR 22 # 10906 Zbl 80.114 JSL 47.702 • ID 42812

TRAKHTENBROT, B.A. see Vol. I, II, IV, V, VI for further entries

TRAKHTMAN, A.N. [1970] see BARANSKIJ, V.A.

TRANSIER, R. [1979] *Verallgemeinerte formal $\mathfrak{p}$-adische Koerper* (J 0008) Arch Math (Basel) 32*572-584
⋄ C25 C60 ⋄ REV MR 80m:12030 Zbl 401 # 12022
• ID 56159

TRANSUE, W.R. [1979] see BOYD, M.W.

TRIESCH, E. [1981] see RICHTER, M.M.

TRIPPEL, J.R. [1978] *Die Algebra der Saetze der monadischen Theorie schwach zweiter Stufe der linearen Ordnung ist atomar* (0000) Diss., Habil. etc i+72pp
⋄ C65 C85 ⋄ REV Zbl 442 # 03026 • REM Diss. ETH No.6208, Eidgenoessische Technische Hochschule Zuerich, Math. Fak. • ID 56383

T"RKALANOV, K.D. [1971] *A solution of the inequality in a certain class of partially ordered semigroups by the method of Osipova (Russian and French summmaries)* (S 1002) Nauch Trud Vissh Ped Inst, Plovdiv 9/2*37-43
⋄ B25 D40 ⋄ REV MR 45 # 5047 • ID 13644

T"RKALANOV, K.D. & ZHELEVA, S. [1971] *On the problem of inequality for partially ordered semigroups (Bulgarian) (Russian and German summaries)* (S 1002) Nauch Trud Vissh Ped Inst, Plovdiv 9/3*25-32
⋄ B25 D40 ⋄ REV MR 45 # 5048 • ID 13645

T"RKALANOV, K.D. see Vol. IV for further entries

TRNKOVA, V. [1971] *On descriptive classification of set-functors. I,II* (J 0140) Comm Math Univ Carolinae (Prague) 12*143-174,345-357
⋄ C30 E55 G30 ⋄ REV MR 45 # 3515 MR 45 # 3516 Zbl 232 # 18004 Zbl 232 # 18005 • ID 83011

TRNKOVA, V. [1980] see REITERMAN, J.

TRNKOVA, V. see Vol. II, IV, V for further entries

TROELSTRA, A.S. [1977] *Completeness and validity for intuitionistic predicate logic* (P 1729) Colloq Int Log;1975 Clermont-Ferrand 39-58
⋄ C90 F50 ⋄ REV MR 80k:03017 Zbl 439 # 03009
• ID 32501

TROELSTRA, A.S. [1978] *Some remarks on the complexity of Henkin-Kripke models* (J 0028) Indag Math 40*296-302
⋄ C90 D55 F50 ⋄ REV MR 58 # 21575 Zbl 389 # 03024
• ID 27952

TROELSTRA, A.S. see Vol. I, II, V, VI for further entries

TROFIMOV, M.YU. [1975] *Definability in algebraically closed systems (Russian)* (J 0003) Algebra i Logika 14*320-327,369
• TRANSL [1975] (J 0069) Algeb and Log 14*198-202
⋄ C25 C40 C60 C85 ⋄ REV MR 55 # 2559 Zbl 373 # 02035 • ID 26005

TROTTER JR., W.T. [1984] see KIERSTEAD, H.A.

TROTTER JR., W.T. see Vol. IV, V for further entries

TRUSS, J.K. [1985] *The group of the countable universal graph* (J 0332) Math Proc Cambridge Phil Soc 98*213-245
⋄ C07 C50 C65 ⋄ ID 48315

TRUSS, J.K. see Vol. IV, V for further entries

TRYBA, J. [1984] *On Jonsson cardinals with uncountable cofinality* (J 0029) Israel J Math 49*315-324
⋄ C55 E05 E55 ⋄ ID 45607

TRYBA, J. see Vol. V for further entries

TSINMAN, L.L. [1969] *Certain algorithms in a formal arithmetic system (Russian)* (J 0023) Dokl Akad Nauk SSSR 189*489-490
- TRANSL [1969] (J 0062) Sov Math, Dokl 10*1452-1454
- ⋄ B25 F30 ⋄ REV MR 41 # 64 Zbl 208.16 • ID 02178

TSINMAN, L.L. see Vol. I, IV, VI for further entries

TSUBOI, A. [1982] *On M-recursively saturated models of arithmetic* (J 2606) Tsukuba J Math 6*305-318
- ⋄ C50 C57 C62 ⋄ REV MR 85i:03110 Zbl 543 # 03018
- • ID 37188

TSUBOI, A. [1983] *On M-recursively saturated models of Peano arithmetic (Japanese)* (P 4113) Found of Math;1982 Kyoto 158-177
- ⋄ C50 C57 C62 ⋄ REV Zbl 543 # 03018 • ID 47682

TSUBOI, A. [1984] *Two theorems on the existence of indiscernible sequences* (J 2606) Tsukuba J Math 8*383-387
- ⋄ C30 ⋄ ID 44718

TSUBOI, A. [1985] *On the number of independent partitions* (J 0036) J Symb Logic 50*809-814
- ⋄ C45 ⋄ ID 47382

TSUBOI, A. [1985] *On theories having a finite number of nonisomorphic countable models* (J 0036) J Symb Logic 50*806-808
- ⋄ C15 C35 C45 ⋄ ID 47381

TSUBOI, A. see Vol. VI for further entries

TUCKER, J.V. [1980] *Computability and the algebra of fields: some affine constructions* (J 0036) J Symb Logic 45*103-120
- ⋄ B25 C57 C60 C65 D45 ⋄ REV MR 82a:03027 Zbl 481 # 03020 • ID 79494

TUCKER, J.V. see Vol. I, IV, VI for further entries

TULIPANI, S. [1971] *Cardinali misurabili e algebre semplici in una varieta generata da un'algebra infinitaria m-funzionalmente completa (English summary)* (J 0012) Boll Unione Mat Ital, IV Ser 4*882-887
- ⋄ C05 C75 E55 ⋄ REV MR 46 # 3410 Zbl 324 # 08005
- • ID 13717

TULIPANI, S. [1972] *Proprieta metamatematiche di alcune classi di algebre* (J 0144) Rend Sem Mat Univ Padova 47*177-186
- ⋄ C05 C52 ⋄ REV MR 47 # 3279 Zbl 252 # 08006
- • ID 13718

TULIPANI, S. [1972] *Sulla completezza e sulla categoricita della teoria delle W-algebre semplici (English summary)* (J 0088) Ann Univ Ferrara, NS, Sez 7 17*1-11
- ⋄ C05 C35 C50 ⋄ REV MR 46 # 5216 Zbl 241 # 02018
- • ID 11867

TULIPANI, S. [1974] see MARCJA, A.

TULIPANI, S. [1975] *Questioni di teoria dei modelli per linguaggi universali positivi II: Metodi di "back and forth" (English summary)* (J 0149) Atti Accad Naz Lincei Fis Mat Nat, Ser 8 59*328-335
- ⋄ B20 C07 C40 C50 ⋄ REV MR 56 # 5273 Zbl 363 # 02057 • REM Part I 1974 by Marcja,A. & Tulipani,S.
- • ID 50883

TULIPANI, S. [1976] *Forcing infinito generalizzato in teoria dei modelli (English summary)* (J 0144) Rend Sem Mat Univ Padova 56*125-138
- ⋄ C25 ⋄ REV MR 57 # 9524 Zbl 377 # 02032 • ID 30976

TULIPANI, S. [1977] see MAKOWSKY, J.A.

TULIPANI, S. [1979] *The Hanf number for classes of algebras whose largest congruence is always finitely generated* (J 0004) Algeb Universalis 9*221-228
- ⋄ C05 C55 ⋄ REV MR 80e:03033 Zbl 411 # 03023
- • ID 52877

TULIPANI, S. [1980] *Invarianti per l'equivalenza elementare per una classe assiomatica di spazi generali di misura (English summary)* (J 3128) Boll Unione Mat Ital, Suppl 2*107-118
- ⋄ B25 C10 C65 G05 ⋄ REV MR 84a:03037 Zbl 449 # 03027 • ID 56694

TULIPANI, S. [1981] *Model-completions of theories of finitely additive measures with values in an ordered field* (J 0068) Z Math Logik Grundlagen Math 27*481-488
- REPR [1981] (P 3092) Congr Naz Logica;1979 Montecatini Terme 195-205
- ⋄ C10 C25 C35 C65 G05 ⋄ REV MR 83c:28007 Zbl 473 # 03024 • ID 55354

TULIPANI, S. [1982] *A use of the method of interpretations for decidability or undecidability of measure spaces* (J 0004) Algeb Universalis 15*228-232
- ⋄ B25 C65 D35 ⋄ REV MR 84i:03087 Zbl 518 # 03003
- • ID 34564

TULIPANI, S. [1982] *On classes of algebras with the definability of congruences* (J 0004) Algeb Universalis 14*269-279
- ⋄ C05 C20 C52 ⋄ REV MR 83g:08002 Zbl 499 # 08001
- • ID 38293

TULIPANI, S. [1982] *Sui reticolo di congruenze di strutture algebriche modelli di teorie universali positive (English summary)* (J 3746) Note Math (Lecce) 2*57-71
- ⋄ C05 C50 C55 ⋄ REV MR 85b:03047 Zbl 527 # 03008
- • ID 39822

TULIPANI, S. [1983] *On the congruence lattice of stable algebras with definability of compact congruences* (J 0022) Cheskoslov Mat Zh 33(108)*286-291
- ⋄ C05 C45 ⋄ REV MR 84e:03046 Zbl 524 # 03022
- • ID 34382

TULIPANI, S. [1983] *On the size of congruence lattices for models of theories with definability of congruences* (J 0004) Algeb Universalis 17*346-359
- ⋄ C05 C35 G10 ⋄ REV MR 86f:06014 Zbl 538 # 03031
- • ID 41474

TULIPANI, S. [1984] *On the universal theory of classes of finite models* (J 0064) Trans Amer Math Soc 284*163-170
- ⋄ C05 C13 C52 C60 ⋄ REV MR 85h:03030 Zbl 521 # 03021 • ID 37101

TULIPANI, S. [1985] *Algebre sottodirettamente irriducibili e assiomatizzazione di varieta* (P 4646) Atti Incontri Log Mat (2);1983/84 Siena 527-560
- ⋄ C05 C98 ⋄ ID 49595

TULIPANI, S. see Vol. I, IV for further entries

TUMANOV, V.I. [1979] *Lattices of equational theories of models (Russian)* (J 0003) Algebra i Logika 18*488-504
- TRANSL [1979] (J 0069) Algeb and Log 18*304-317
- ⋄ C05 C52 G10 ⋄ REV MR 83h:03043 Zbl 454 # 03010
- • ID 54223

TUMANOV, V.I. [1984] *Finite lattices with no independent basis of quasiidentities (Russian)* (J 0087) Mat Zametki (Akad Nauk SSSR) 36*625-634,797
• TRANSL [1984] (J 1044) Math Notes, Acad Sci USSR 36*811-815
⋄ C05 C13 G10 ⋄ REV MR 86e:08010 Zbl 565 # 06010
• ID 48316

TUOMELA, R. [1970] see HINTIKKA, K.J.J.

TUOMELA, R. see Vol. II for further entries

TURAN, G. [1981] *On cellular graph-automata and second-order definable graph-properties* (P 3165) FCT'81 Fund of Comput Th;1981 Szeged 384-393
⋄ C85 D05 D80 ⋄ REV MR 83h:68076 Zbl 479 # 68059
• ID 55701

TURAN, G. [1984] *On the definability of properties of finite graphs* (J 0193) Discr Math 49*291-302
⋄ C13 C65 ⋄ REV MR 85e:03074 Zbl 536 # 05060
• ID 40352

TURAN, P. [1971] *On the work of Alan Baker* (P 0743) Int Congr Math (II,11,Proc);1970 Nice 1*3-5
⋄ B25 D35 ⋄ REV MR 54 # 2394 JSL 37.606 • ID 13720

TURI, L. [1982] see ECSEDI-TOTH, P.

TURI, L. [1984] see ECSEDI-TOTH, P.

TURING, A.M. [1954] *Solvable and unsolvable problems* (J 4681) Sci News 31*7-23
⋄ B25 D10 D35 ⋄ REV JSL 20.74 • ID 42433

TURING, A.M. see Vol. I, IV, VI for further entries

TUSCHIK, H.-P. [1977] see SEESE, D.G.

TUSCHIK, H.-P. [1977] see BAUDISCH, A.

TUSCHIK, H.-P. [1977] *Elimination verallgemeinerter Quantoren in ω_1-kategorischen Theorien (Russian, English and French summaries)* (J 0115) Wiss Z Humboldt-Univ Berlin, Math-Nat Reihe 26*659-661
⋄ B25 C10 C35 C55 C80 ⋄ REV MR 80d:03035
Zbl 423 # 03030 • ID 53542

TUSCHIK, H.-P. [1977] *On the decidability of the theory of linear orderings in the language $L(Q_1)$* (P 1695) Set Th & Hierarch Th (3);1976 Bierutowice 291-304
⋄ B25 C55 C65 C80 ⋄ REV MR 57 # 16037
Zbl 368 # 02016 • ID 51233

TUSCHIK, H.-P. [1979] *A reduction of a problem about ω_1-categoricity* (J 3293) Bull Acad Pol Sci, Ser Math 27*333-335
⋄ C35 C52 ⋄ REV MR 81d:03033 Zbl 423 # 03031
• ID 53543

TUSCHIK, H.-P. [1980] *An application of rank-forcing to ω_1-categoricity* (J 0068) Z Math Logik Grundlagen Math 26*237-250
⋄ C25 C35 ⋄ REV MR 81i:03041 Zbl 457 # 03029
• ID 54354

TUSCHIK, H.-P. [1980] see BAUDISCH, A.

TUSCHIK, H.-P. [1980] *On the decidability of the theory of linear orderings with generalized quantifiers* (J 0027) Fund Math 107*21-32
⋄ B25 C55 C65 C80 ⋄ REV MR 81m:03016
Zbl 362 # 02036 • ID 53914

TUSCHIK, H.-P. [1982] see ROTHMALER, P.

TUSCHIK, H.-P. [1982] *Elimination of cardinality quantifiers* (J 0068) Z Math Logik Grundlagen Math 28*75-81
⋄ C10 C45 C55 C60 C80 ⋄ REV MR 83h:03045
Zbl 494 # 03025 • ID 36060

TUSCHIK, H.-P. [1982] see SEESE, D.G.

TUSCHIK, H.-P. [1983] see BAUDISCH, A.

TUSCHIK, H.-P. & WEESE, M. [1983] *An unstable $\aleph_0$-categorical theory without the strict order property* (P 1601) Easter Conf on Model Th (1);1983 Diedrichshagen 137-143
⋄ C35 C45 ⋄ REV MR 84i:03008 Zbl 529 # 03011
• ID 37649

TUSCHIK, H.-P. [1983] *The Ramsey theorem for additive colourings* (P 1601) Easter Conf on Model Th (1);1983 Diedrichshagen 129-136
⋄ C55 C65 E05 ⋄ REV MR 84i:03008 Zbl 546 # 03031
• ID 43530

TUSCHIK, H.-P. [1985] *Algebraic connections between definable predicates* (P 4310) Easter Conf on Model Th (3);1985 Gross Koeris 215-226
⋄ C15 C30 C40 ⋄ ID 49052

TUSCHIK, H.-P. [1985] see BAUDISCH, A.

TVERSKOJ, A.A. [1980] *A sequence of combinatorial judgements which are independent of Peano arithmetic (Russian) (English summary)* (J 0288) Vest Ser Mat Mekh, Univ Moskva 1980/5*7-13,83
• TRANSL [1980] (J 0510) Moscow Univ Math Bull 35/5*6-13
⋄ C62 E05 F30 ⋄ REV MR 83e:03096 Zbl 453 # 03045
• ID 35246

TVERSKOJ, A.A. [1982] *Investigation of recursiveness and arithmeticity of signature functions in nonstandard models of arithmetics (Russian)* (J 0023) Dokl Akad Nauk SSSR 262*1325-1328
• TRANSL [1982] (J 0062) Sov Math, Dokl 25*249-253
⋄ C57 C62 D45 H15 ⋄ REV MR 83k:03079
Zbl 496 # 03047 • ID 35413

TVERSKOJ, A.A. [1984] *Constructivizability of formal arithmetical structures (Russian)* (C 4091) Algeb & Diskret Mat (Riga) 134-136
⋄ C57 D45 H15 ⋄ REV MR 85j:03053 • ID 44226

TVERSKOJ, A.A. [1985] *Constructivizable and nonconstructivizable formal arithmetic structures (Russian)* (J 0067) Usp Mat Nauk 40/6*159-160
⋄ C57 C62 D45 H15 ⋄ ID 49425

TWER VON DER, T. [1977] *Pseudoelementare Relationen und Aussagen vom Typ des Bernstein'schen Aequivalenzsatzes* (S 0478) Bonn Math Schr 95*ii+58pp
⋄ C07 C52 ⋄ REV MR 58 # 5174 Zbl 368 # 02055
• ID 51272

TWER VON DER, T. [1981] *On the strength of several versions of Dirichlets ("pigeon-hole"-) principle in the sense of first-order logic* (J 0009) Arch Math Logik Grundlagenforsch 21*69-76
⋄ C10 H15 ⋄ REV MR 84e:03072 Zbl 467 # 03060
• ID 55058

TWER VON DER, T. [1981] *Some remarks on the mathematical incompleteness of Peano's arithmetic found by Paris and Harrington* (P 3268) Set Th & Model Th;1979 Bonn 157-174
⋄ C62 E05 F30 ⋄ REV MR 83a:03060 Zbl 522 # 03056
• ID 35072

TYSHKEVICH, R.I. [1975] *Models and groups (Russian)* (J 0413) Izv Akad Nauk Belor SSR, Ser Fiz-Mat 1975/5∗30-37,138
⋄ C60 ⋄ REV MR 52 # 13478 Zbl 337 # 08002 • ID 21841

TYSHKEVICH, R.I. see Vol. IV for further entries

TYUKAVKIN, L.V. [1981] *Axiomatizability of the class of completely reducible modules (Russian)* (C 3910) Abel Gruppy & Moduli 181-184
⋄ C60 ⋄ REV Zbl 517 # 16018 • ID 36690

TYUKAVKIN, L.V. [1981] *Axiomatizability of the class of irreducible modules (Russian)* (C 3910) Abel Gruppy & Moduli 185-197
⋄ C60 ⋄ REV Zbl 517 # 16019 • ID 36691

TYUKAVKIN, L.V. [1982] *On the model completion of certain theories of modules (Russian)* (J 0003) Algebra i Logika 21∗73-83,125
• TRANSL [1982] (J 0069) Algeb and Log 21∗50-57
⋄ C35 C60 ⋄ REV MR 84m:03056 Zbl 512 # 03017
• ID 35757

TZOUVARAS, A.D. [1980] *A non-standard characterization of the norm of free ultrafilters* (J 0465) Bull Greek Math Soc (NS) 21∗81-86
⋄ C55 E05 H05 ⋄ REV MR 85i:03199 Zbl 548 # 03040
• ID 43212

TZOUVARAS, A.D. [1985] *Minimal ultrafilters and maximal endomorphic universes* (J 0140) Comm Math Univ Carolinae (Prague) 26∗719-726
⋄ C62 E70 ⋄ ID 48451

UESU, T. [1969] *On logics with infinitely long expressions (Japanese)* (J 0091) Sugaku 21∗189-202
⋄ C75 ⋄ REV MR 45 # 38 • ID 13761

UESU, T. [1970] *On the recursively restricted rules* (J 0407) Comm Math Univ St Pauli (Tokyo) 18∗31-42
⋄ C75 F30 ⋄ REV MR 42 # 71 Zbl 193.302 • ID 13763

UESU, T. [1971] *Simple type theory with constructive infinitely long expressions* (J 0407) Comm Math Univ St Pauli (Tokyo) 19∗131-163
⋄ B15 C75 F05 F35 ⋄ REV MR 45 # 8509 Zbl 255 # 02023 • ID 13764

UESU, T. see Vol. I, IV, V, VI for further entries

ULRICH, D. [1983] *The finite model property and recursive bounds on the size of countermodels* (J 0122) J Philos Logic 12∗477-480
⋄ B22 B25 C13 ⋄ REV MR 85m:03016 Zbl 575 # 03025
• ID 42640

ULRICH, D. see Vol. I, II, VI for further entries

UMEZAWA, T. [1965] *On a model in the ordinal numbers* (J 0090) J Math Soc Japan 17∗341-357
⋄ C62 E30 E35 E45 E70 ⋄ REV MR 33 # 34 Zbl 133.259 • REM Part II 1966 • ID 13786

UMEZAWA, T. [1966] *On a model in the ordinal numbers. II* (J 1005) Rep Fac Sci, Shizuoka Univ 1∗53-63
⋄ C62 E30 E35 E45 E70 ⋄ REV MR 35 # 5307 Zbl 348 # 02033 • REM Part I 1965 • ID 13787

UMEZAWA, T. see Vol. I, II, IV, V, VI for further entries

UMIRBAEV, U.U. [1984] *Equality problem for center-by-metabelian Lie-algebras (Russian)* (J 0003) Algebra i Logika 23∗305-318
• TRANSL [1984] (J 0069) Algeb and Log 23∗209-219
⋄ B25 C60 D40 ⋄ REV MR 86g:17009 • ID 42699

UROSU, C. [1980] *On filter products of congruences (Romanian) (French summary)* (J 4715) Bul Sti Tehn Inst Politeh Timisoara, Ser Mat-Fiz 25(39)/2∗63-66
⋄ C05 C20 ⋄ REV MR 83m:08003 • ID 40388

URQUHART, A.I.F. [1981] *Decidability and the finite model property* (J 0122) J Philos Logic 10∗367-370
⋄ B22 B25 D35 ⋄ REV MR 83i:03038 Zbl 465 # 03005
• ID 54908

URQUHART, A.I.F. [1981] *The decision problem for equational theories* (J 1447) Houston J Math 7∗587-589
⋄ B25 C05 D35 ⋄ REV MR 83h:03063 Zbl 496 # 03004
• ID 36071

URQUHART, A.I.F. see Vol. I, II, IV, VI for further entries

URSINI, A. [1977] *A sequence of theories for arithmetic whose union is complete* (J 0144) Rend Sem Mat Univ Padova 57∗75-92
⋄ C62 F30 ⋄ REV MR 80d:03058 Zbl 411 # 03053
• ID 52907

URSINI, A. [1979] *Two remarks on models in modal logics (Italian summary)* (J 3285) Boll Unione Mat Ital, V Ser, A 16∗124-127
⋄ B45 C90 ⋄ REV MR 81b:03025 Zbl 395 # 03017
• ID 52568

URSINI, A. [1985] *Decision problems for classes of diagonalizable algebras* (J 0063) Studia Logica 44∗87-90
⋄ B25 B45 C05 D35 ⋄ ID 47525

URSINI, A. see Vol. I, II, VI for further entries

URYSOHN, P. [1925] *Sur un espace metrique universel* (J 0109) C R Acad Sci, Paris 180∗A803-A806
⋄ C50 C65 E75 ⋄ REV FdM 51.452 • ID 16917

URYSOHN, P. [1927] *Sur un espace metrique universel I,II* (J 0247) Bull Sci Math, Ser 2 51∗43-64,74-90
⋄ C50 C65 E75 ⋄ REV FdM 53.556 • ID 16918

URYSOHN, P. see Vol. V for further entries

URZYCZYN, P. [1981] *The unwind property in certain algebras* (J 0194) Inform & Control 50∗91-109
⋄ C05 C57 ⋄ REV MR 84c:08005 Zbl 491 # 03018
• ID 36852

URZYCZYN, P. [1982] *An example of a complete first-order theory with all models algorithmically trivial but without locally finite models* (J 2095) Fund Inform, Ann Soc Math Pol, Ser 4 5∗313-318
⋄ C35 C50 ⋄ REV MR 85e:03088 Zbl 503 # 68014
• ID 36667

URZYCZYN, P. see Vol. II, IV for further entries

VACCARO, V. [1984] see GERLA, G.

VACCARO, V. see Vol. II for further entries

VAEAENAENEN, J. [1977] *On the compactness theorem* (P 1629) Symp Math Log;1974 Oulo;1975 Helsinki 62-68
⋄ C20 C40 C95 ⋄ ID 90132

VAEAENAENEN, J. [1977] see MIETTINEN, S.

VAEAENAENEN, J. [1977] *Remarks on generalized quantifiers and second order logics* (P 1639) Set Th & Hierarch Th (1);1974 Karpacz 117-123
⋄ B15 C40 C80 C85 C95 ⋄ REV MR 58 # 5019
Zbl 382 # 03010 • ID 51934

VAEAENAENEN, J. [1978] *Two axioms of set theory with applications to logic* (J 0446) Ann Acad Sci Fennicae, Ser A I, Diss 20*19pp
⋄ C40 C55 C80 C85 E35 E47 E65 ⋄ REV
MR 80e:03067 Zbl 392 # 03034 • ID 32506

VAEAENAENEN, J. [1979] *A new incompleteness in arithmetic (Finnish) (English summary)* (J 1108) Arkhimedes (Helsinki) 31*30-37
⋄ A10 C62 D25 E05 F30 ⋄ REV MR 80c:03058
Zbl 398 # 03021 • ID 52749

VAEAENAENEN, J. [1979] *Abstract logic and set theory. I. Definability* (P 2627) Logic Colloq;1978 Mons 391-421
⋄ C40 C55 C95 E75 ⋄ REV MR 81i:03058
Zbl 433 # 03019 • REM Part II 1982 • ID 79569

VAEAENAENEN, J. [1979] *On Hanf numbers of unbounded logics* (P 2615) Scand Logic Symp (5);1979 Aalborg 309-328
⋄ C55 C75 C80 C85 C95 E55 ⋄ REV MR 82h:03034
Zbl 435 # 03029 • ID 55789

VAEAENAENEN, J. [1979] *Remarks on free quantifier variables* (P 1705) Scand Logic Symp (4);1976 Jyvaeskylae 267-272
⋄ C55 C80 C85 C95 E45 ⋄ REV MR 81b:03042
Zbl 402 # 03036 • ID 32505

VAEAENAENEN, J. [1980] *A quantifier for isomorphisms* (J 0068) Z Math Logik Grundlagen Math 26*123-130
⋄ C40 C80 C85 C95 ⋄ REV MR 81j:03058
Zbl 435 # 03030 • ID 55790

VAEAENAENEN, J. [1980] *Boolean-valued models and generalized quantifiers* (J 0007) Ann Math Logic 18*193-225
⋄ C55 C80 C85 C90 E35 E40 ⋄ REV MR 81m:03049
Zbl 484 # 03017 • ID 90120

VAEAENAENEN, J. [1980] see ANAPOLITANOS, D.A.

VAEAENAENEN, J. [1980] *The Hanf number of $L_{\omega_1\omega_1}$* (J 0053) Proc Amer Math Soc 79*294-297
⋄ C55 C75 E35 E55 ⋄ REV MR 81f:03061
Zbl 448 # 03023 • ID 79570

VAEAENAENEN, J. [1981] see ANAPOLITANOS, D.A.

VAEAENAENEN, J. [1981] *Generalized quantfiers in models of set theory* (P 2623) Worksh Extended Model Th;1980 Berlin 145-160
⋄ C55 C62 C80 C85 C95 E35 ⋄ REV MR 84h:03094
Zbl 483 # 03023 • ID 34284

VAEAENAENEN, J. [1982] *Abstract logic and set theory. II. Large cardinals* (J 0036) J Symb Logic 47*335-346
⋄ C55 C95 E35 E55 ⋄ REV MR 84g:03052
Zbl 531 # 03019 • REM Part I 1979 • ID 34162

VAEAENAENEN, J. [1982] *Generalized quantifiers in models of set theory* (P 3634) Patras Logic Symp;1980 Patras 359-371
⋄ C55 C62 C80 C85 C95 E35 ⋄ REV MR 84h:03096
Zbl 509 # 03019 • ID 34286

VAEAENAENEN, J. [1982] see KRYNICKI, M.

VAEAENAENEN, J. [1983] *Δ-extension and Hanf-numbers* (J 0027) Fund Math 115*43-55
⋄ C40 C55 C80 C95 ⋄ REV MR 84j:03080
Zbl 522 # 03027 • ID 31290

VAEAENAENEN, J. [1984] see KRYNICKI, M.

VAEAENAENEN, J. [1985] *Set-theoretical definability of logics* (C 4183) Model-Theor Log 599-643
⋄ C40 C55 C80 C95 C98 E47 ⋄ ID 48342

VAKARELOV, D. [1972] *Extensional logics (Russian)* (J 0137) C R Acad Bulgar Sci 25*1609-1612
⋄ B05 B25 ⋄ REV MR 48 # 55 Zbl 253 # 02024
Zbl 332 # 02012 • ID 00678

VAKARELOV, D. see Vol. II, VI for further entries

VALIEV, M.K. [1968] *A theorem of G.Higman (Russian)* (J 0003) Algebra i Logika 7/3*9-22
• TRANSL [1968] (J 0069) Algeb and Log 7*135-143
⋄ C60 D35 D40 ⋄ REV MR 41 # 1849 Zbl 209.331
• ID 13814

VALIEV, M.K. [1980] *Decision complexity of variants of propositional dynamic logic* (P 3210) Math Founds of Comput Sci (9);1980 Rydzyna 656-664
⋄ B25 B75 D15 ⋄ REV MR 81k:68003 Zbl 451 # 03003
• ID 54018

VALIEV, M.K. see Vol. II, IV for further entries

VAMOS, P. [1967] *On ring classes defined by modules* (J 0057) Publ Math (Univ Debrecen) 14*1-8
⋄ C60 ⋄ REV MR 37 # 50 Zbl 204.52 • ID 13819

VAMOS, P. [1979] see JENSEN, C.U.

VANCKO, R.M. [1971] *The spectrum of some classes of free universal algebras* (J 0004) Algeb Universalis 1*46-53
⋄ C05 C13 ⋄ REV MR 44 # 5268 Zbl 219 # 08003
• ID 13820

VANCKO, R.M. [1972] *The family of locally independent sets in finite algebras* (J 0004) Algeb Universalis 2*68-73
⋄ C05 C13 ⋄ REV MR 46 # 5208 Zbl 253 # 08006
• ID 13821

VARADARAJAN, K. [1985] *Pseudo-mitotic groups* (J 0326) J Pure Appl Algebra 37*205-213
⋄ C25 C60 ⋄ ID 48456

VARGA VON KIBED, M. [1984] see STEGMUELLER, W.

VASIL'EV, EH.S. [1973] *Elementary theories of two-base models of abelian groups (Russian)* (C 1443) Algebra, Vyp 2 (Irkutsk) 62-68
⋄ B25 C60 C85 ⋄ REV MR 53 # 10576 • ID 23082

VASIL'EV, EH.S. [1973] *The elementary theories of complete torsion-free abelian groups with p-adic topology (Russian)* (J 0087) Mat Zametki (Akad Nauk SSSR) 14*201-208
• TRANSL [1973] (J 1044) Math Notes, Acad Sci USSR 14*673-677
⋄ B25 C20 C60 D35 ⋄ REV MR 49 # 2329
Zbl 306 # 02053 • ID 13840

VASIL'EV, EH.S. [1976] *On elementary theories of abelian p-groups with heights of elements (Russian)* (C 2555) Algeb Sistemy (Irkutsk) 17-20
⋄ B25 C60 ⋄ ID 90340

VASIL'EV, EH.S. [1977] *Elementary theories of certain classes of two-base models of abelian groups (Russian)* (**J 0031**) Izv Vyssh Ucheb Zaved, Mat (Kazan) 1977/7(182)*32-36
• TRANSL [1977] (**J 3449**) Sov Math 21/7*25-29
⋄ B25 C60 C85 ⋄ REV MR 56#15396 Zbl 389#03008
• ID 52297

VASIL'EV, V.G. [1973] *Solvable elementarily equivalent groups (Russian)* (**C 3036**) Vopr Teor Grupp & Kolets 9-22,173-174
⋄ C60 ⋄ REV MR 58#27460 • ID 79677

VAUGHT, R.L. [1954] *Applications of the Loewenheim-Skolem-Tarski theorem to problems of completeness and decidability* (**J 0028**) Indag Math 16*467-472
⋄ B25 C07 C35 ⋄ REV MR 16.208 Zbl 56.248 JSL 24.58
• ID 13843

VAUGHT, R.L. [1954] *On sentences holding in direct products of relational systems* (**P 0575**) Int Congr Math (II, 7);1954 Amsterdam 2*409-410
⋄ B20 B25 C30 ⋄ ID 41195

VAUGHT, R.L. [1954] *Remarks on universal classes of relational systems* (**J 0028**) Indag Math 16*589-591
⋄ C52 ⋄ REV MR 16.554 Zbl 58.247 JSL 24.58 • ID 13844

VAUGHT, R.L. [1954] *Topics in the theory of arithmetical classes and boolean algebras* (0000) Diss., Habil. etc
⋄ C07 G05 ⋄ REM Diss., University of CA • ID 16341

VAUGHT, R.L. [1957] see TARSKI, A.

VAUGHT, R.L. [1957] see MONTAGUE, R.

VAUGHT, R.L. [1957] *Sentences true in all constructive models* (**P 1675**) Summer Inst Symb Log;1957 Ithaca 341-343
⋄ C13 C57 D35 ⋄ REV MR 24#A40 Zbl 108.8 JSL 31.132 • REM See also 1960 • ID 24868

VAUGHT, R.L. [1958] see CRAIG, W.

VAUGHT, R.L. [1959] see MONTAGUE, R.

VAUGHT, R.L. [1959] see FEFERMAN, S.

VAUGHT, R.L. [1960] *Sentences true in all constructive models* (**J 0036**) J Symb Logic 25*39-53
⋄ C13 C57 D35 ⋄ REV MR 24#A40 Zbl 108.8 JSL 31.132 • REM See also 1957 • ID 24869

VAUGHT, R.L. [1961] *Denumerable models of complete theories (Polish)* (**P 0633**) Infinitist Meth;1959 Warsaw 303-321
⋄ C15 C35 C50 ⋄ REV MR 32#4011 Zbl 113.243 JSL 35.342 • ID 13851

VAUGHT, R.L. [1961] *The elementary character of two notions from general algebra* (**C 0622**) Essays Found of Math (Fraenkel) 226-233
⋄ C05 C40 C52 C60 ⋄ REV MR 30#1051 Zbl 135.248 JSL 30.252 • ID 13852

VAUGHT, R.L. [1962] see MORLEY, M.D.

VAUGHT, R.L. [1963] *Models of complete theories* (**J 0015**) Bull Amer Math Soc 69*299-313
⋄ C98 ⋄ REV MR 26#4912 Zbl 112.7 JSL 35.344
• ID 13855

VAUGHT, R.L. [1964] *The completeness of logic with the added quantifier "there are uncountable many"* (**J 0027**) Fund Math 54*303-304
⋄ C55 C80 ⋄ REV MR 29#3364 Zbl 137.9 JSL 33.121
• ID 13856

VAUGHT, R.L. [1965] *A Loewenheim-Skolem theorem for cardinals far apart* (**P 0614**) Th Models;1963 Berkeley 390-401
⋄ C55 C80 ⋄ REV MR 35#1460 Zbl 163.248 JSL 33.476
• ID 13859

VAUGHT, R.L. [1965] *The Loewenheim-Skolem theorem* (**P 0623**) Int Congr Log, Meth & Phil of Sci (2,Proc);1964 Jerusalem 81-89
⋄ C07 C55 C80 ⋄ REV MR 35#1461 • ID 13858

VAUGHT, R.L. [1966] *Elementary classes closed under descending intersection* (**J 0053**) Proc Amer Math Soc 17*430-433
⋄ C52 ⋄ REV MR 33#49 Zbl 149.246 JSL 32.413
• ID 13861

VAUGHT, R.L. [1967] *Axiomatizability by a schema* (**J 0036**) J Symb Logic 32*473-479
⋄ C07 C52 ⋄ REV MR 37#3916 Zbl 191.293 • ID 13862

VAUGHT, R.L. [1968] *Model theory and set theory* (**P 0657**) Int Congr Math (II,10);1966 Moskva 251-253
⋄ C55 C80 C98 ⋄ REV MR 38#992 Zbl 199.303
• ID 13860

VAUGHT, R.L. [1973] *A Borel invariantization* (**J 0015**) Bull Amer Math Soc 79*1292-1295
⋄ C75 D55 E15 ⋄ REV MR 48#10818 Zbl 296#54037
• ID 13909

VAUGHT, R.L. [1973] *Descriptive set theory in $L_{\omega_1,\omega}$* (**P 0713**) Cambridge Summer School Math Log;1971 Cambridge GB 574-598
⋄ C15 C70 C75 D55 D70 E15 ⋄ REV MR 53#12868 Zbl 308#02054 JSL 47.217 • ID 13864

VAUGHT, R.L. [1973] *Some aspects of the theory of models* (**J 0005**) Amer Math Mon 80*3-37
⋄ C98 ⋄ REV MR 49#2351 Zbl 271#02032 • ID 13863

VAUGHT, R.L. [1974] *Invariant sets in topology and logic* (**J 0027**) Fund Math 82*269-294
⋄ C75 D55 D70 E15 ⋄ REV MR 51#167 Zbl 309#02068 • ID 17478

VAUGHT, R.L. [1974] *Model theory before 1945* (**P 0610**) Tarski Symp;1971 Berkeley 153-172
⋄ A10 C98 ⋄ REV MR 57#12157 Zbl 306#02005
• ID 13910

VAUGHT, R.L. [1974] see ADDISON, J.W.

VAUGHT, R.L. [1980] *On $PC_d(\mathfrak{A})$-classes for an admissible set $\mathfrak{A}$* (**P 2958**) Latin Amer Symp Math Log (4);1978 Santiago 377-392
⋄ C70 ⋄ REV MR 83j:03059 Zbl 426#03036 • ID 53632

VAUGHT, R.L. see Vol. IV, V for further entries

VAZHENIN, YU.M. [1970] *Elementary properties of semigroups of transformations of ordered sets (Russian)* (**J 0003**) Algebra i Logika 9*281-301
• TRANSL [1970] (**J 0069**) Algeb and Log 9*169-179
⋄ C65 E07 ⋄ REV MR 44#344 Zbl 232#20133
• ID 13911

VAZHENIN, YU.M. [1972] *On finitely approximable semigroups of idempotents (Russian)* (**J 0340**) Mat Zap (Univ Sverdlovsk) 3*12-17,139
⋄ C05 C60 ⋄ REV MR 48#457 Zbl 321#20043
• ID 33024

VAZHENIN, YU.M. [1972] *The elementary definability and elementary characterizability of classes of reflexive graphs (Russian)* (**J** 0031) Izv Vyssh Ucheb Zaved, Mat (Kazan) 1972/7(122)*3-11
⋄ C52 C65 ⋄ REV MR 48#8309 Zbl 243#05101
• ID 13913

VAZHENIN, YU.M. [1974] *On global including semigroups of symmetric semigroups (Russian)* (**J** 0340) Mat Zap (Univ Sverdlovsk) 9/1*3-10,140
⋄ C05 C60 ⋄ REV MR 53#5786 Zbl 328#20059
• ID 33026

VAZHENIN, YU.M. [1974] *On the elementary theory of free inverse semigroups* (**J** 0136) Semigroup Forum 9*189-195
⋄ B25 C60 D35 ⋄ REV MR 55#2547 Zbl 299#20050
• ID 79687

VAZHENIN, YU.M. [1979] see PERMINOV, E.A.

VAZHENIN, YU.M. [1981] see ROZENBLAT, B.V.

VAZHENIN, YU.M. [1981] *Sur la liaison entre problemes combinatoires et algorithmiques* (**J** 1426) Theor Comput Sci 16*33-41
⋄ B25 D05 D15 D35 ⋄ REV MR 83g:03042 Zbl 469#03004 • ID 55132

VAZHENIN, YU.M. [1983] see ROZENBLAT, B.V.

VAZHENIN, YU.M. [1983] *Semigroups with one defining relation whose elementary theories are decidable (Russian)* (**J** 0092) Sib Mat Zh 24/1*40-49,191
• TRANSL [1983] (**J** 0475) Sib Math J 24*33-41
⋄ B25 C05 C60 ⋄ REV MR 84i:20060 Zbl 549#20036
• ID 40105

VAZHENIN, YU.M. see Vol. IV, V for further entries

VEIT RICCIOLI, B. [1978] *Il forcing come principio logico per la costruzione dei fasci. I,II (French summaries)* (**J** 2311) Rend Mat, Ser 6 11*329-353,601-625
⋄ C25 C90 E40 G30 ⋄ REV MR 80b:03109 Zbl 431#03024 Zbl 437#03015 • ID 53930

VELDMAN, W. [1975] see LOPEZ-ESCOBAR, E.G.K.

VELDMAN, W. see Vol. V, VI for further entries

VELLUTINI, F. [1978] *Le "forcing" en theorie des modeles; quelques resultats sur les relations generales (English summary)* (**J** 2313) C R Acad Sci, Paris, Ser A-B 286*A853-A854
⋄ C25 ⋄ REV MR 58#16257 Zbl 411#03021 • ID 52875

VENNE, M. [1967] *Ultraproduits de structures d'ordres superieurs* (**J** 2313) C R Acad Sci, Paris, Ser A-B 265*A305-A308
⋄ B15 C20 ⋄ REV MR 37#47 Zbl 204.4 • ID 13929

VENNE, M. see Vol. VI for further entries

VENNERI, B.M. [1975] *Semantic implications of Herbrand's theory of fields* (**J** 0286) Int Logic Rev 12*204-214,215-225
⋄ A05 B10 C07 ⋄ REV Zbl 332#02007 • ID 65905

VENNERI, B.M. see Vol. I, VI for further entries

VERDU I SOLANS, V. [1982] *The ultrafilter theorem and the adequacy theorem (Spanish) (English summary)* (**P** 4096) Jorn Mat Luso-Espanol (9);1982 Salamanca 1*201-203
⋄ C07 C20 E25 ⋄ REV MR 85i:00012 • ID 45471

VERDU I SOLANS, V. see Vol. I, II, V, VI for further entries

VERSHININ, K.P. [1974] see MEJTUS, V.YU.

VERSHININ, K.P. [1978] see GERGELY, T.

VERSHININ, K.P. see Vol. I, V for further entries

VETULANI, Z. [1976] *Categoricity relative to ordinals for models of set theory and the nonabsoluteness of L* (**P** 1476) Set Th & Hierarch Th (2) (Mostowski);1975 Bierutowice 285-290
⋄ C35 C62 E45 H20 ⋄ REV MR 56#11793 Zbl 341#02048 • ID 23813

VETULANI, Z. [1977] *Some remarks on ultrapowers of extendable models of ZF* (**P** 1639) Set Th & Hierarch Th (1);1974 Karpacz 69-77
⋄ C20 C62 E30 E70 H20 ⋄ REV MR 58#5229 Zbl 405#03015 • ID 31523

VETULANI, Z. [1984] *Ramified analysis and the minimal β-models of higher order arithmetics* (**J** 0027) Fund Math 121*1-15
⋄ C62 D55 E45 F35 F65 ⋄ REV Zbl 576#03021
• ID 44142

VIDAL-NAQUET, G. [1973] *Quelques applications des automates a arbres infinis* (**P** 0763) Automata, Lang & Progr (1);1972 Rocquencourt 115-122
⋄ B25 D05 D35 ⋄ REV MR 57#18236 Zbl 264#02045
• ID 27517

VIDAL-NAQUET, G. see Vol. IV for further entries

VIKENT'EV, A.A. [1981] *Classes of theories with Lachlan dimensions (Russian)* (**C** 3806) Issl Teor Progr 11-21
⋄ C45 ⋄ REV MR 84m:03049 • ID 35750

VIKENT'EV, A.A. [1981] *Models of restrictions of theories (Russian)* (**C** 3806) Issl Teor Progr 22-29
⋄ C45 ⋄ REV MR 84k:03102 • ID 36119

VILENKIN, N.YA. [1977] see SHREJDER, YU.A.

VILENKIN, N.YA. see Vol. V for further entries

VILLE, F. [1971] *Complexite des structures rigidement contenues dans une theorie du premier ordre* (**J** 2313) C R Acad Sci, Paris, Ser A-B 272*A561-A563
⋄ C50 C57 ⋄ REV MR 44#1552 Zbl 218#02047
• ID 13946

VILLE, F. [1971] *Decidabilite des formules existentielles en theorie des ensembles* (**J** 2313) C R Acad Sci, Paris, Ser A-B 272*A513-A516
⋄ B25 E30 ⋄ REV MR 43#4659 Zbl 218#02044
• ID 13945

VILLE, F. [1974] *More on set existence* (**P** 0602) Generalized Recursion Th (1);1972 Oslo 195-204
⋄ C62 C70 E47 ⋄ REV MR 54#4979 Zbl 357#02039
• ID 24141

VINCENZI, A. [1985] *Alcune proprieta della logica di Chang* (**P** 4646) Atti Incontri Log Mat (2);1983/84 Siena 197-200
⋄ B45 C40 ⋄ ID 49797

VINCENZI, A. see Vol. II, IV for further entries

VINNER, S. [1972] *A generalization of Ehrenfeucht's game and some applications* (**J** 0029) Israel J Math 12*279-298
⋄ B25 C07 C55 C80 E60 ⋄ REV MR 47#3142 Zbl 253#02051 • ID 13947

VINNER, S. [1975] *Model-completeness in a first order language with a generalized quantifier* (**J** 0048) Pac J Math 56*265-273
⋄ C35 C55 C80 ⋄ REV MR 51#10080 Zbl 317#02060
• ID 13948

VINNER, S. [1975] *On two complete sets in the analytical and the arithmetical hierarchies* (J 0009) Arch Math Logik Grundlagenforsch 17*81-84
  ⋄ C55 C80 D55 ⋄ REV MR 52 # 5346 Zbl 317 # 02048
  • ID 13949

VINNER, S. [1976] *Implicit axioms, ω-rule and the axiom of induction in high school mathematics* (J 0005) Amer Math Mon 83*561-566
  ⋄ B28 C75 ⋄ REV MR 54 # 4891 • ID 24084

VINOGRADOV, A.A. [1971] *Nonaxiomatizability of lattice-orderable groups (Russian)* (J 0092) Sib Mat Zh 12*463-464
  • TRANSL [1971] (J 0475) Sib Math J 12*331-332
  ⋄ C52 C60 ⋄ REV MR 44 # 132 Zbl 216.31 • ID 41670

VINOGRADOV, A.A. [1973] *Nonaxiomatizability of directionally ordered groups in the class of nontrivially partially ordered groups* (J 0087) Mat Zametki (Akad Nauk SSSR) 14*395-397
  • TRANSL [1973] (J 1044) Math Notes, Acad Sci USSR 14*787-788
  ⋄ C52 C60 ⋄ REV MR 48 # 5952 Zbl 294 # 06012
  • ID 65927

VINOGRADOV, A.A. [1977] *Nonaxiomatizability of lattice-orderable rings (Russian)* (J 0087) Mat Zametki (Akad Nauk SSSR) 21*449-452
  • TRANSL [1977] (J 1044) Math Notes, Acad Sci USSR 21*253-254
  ⋄ C52 C60 ⋄ REV MR 58 # 437 Zbl 363 # 02059
  • ID 50885

VISSER, A. [1984] *The provability logic of recursively enumerable theories extending Peano arithmetic and arbitrary theories extending Peano arithmetic* (J 0122) J Philos Logic 13*97-113
  ⋄ B45 C62 F30 ⋄ REV MR 85j:03102 • ID 42643

VISSER, A. see Vol. II, IV, VI for further entries

VOICULESCU, I. [1985] see GEORGESCU, G.

VOJSHVILLO, E.K. [1983] *A decision procedure for the system E (of entailment) I* (J 0063) Studia Logica 42*139-164
  ⋄ B25 B46 ⋄ REV MR 86f:03036 Zbl 559 # 03009
  • ID 48317

VOJSHVILLO, E.K. see Vol. I, II for further entries

VOJTASKOVA, B. [1984] see CUDA, K.

VOJVODIC, G. [1978] *Some theorems for model theory of mixed-valued predicate calculi* (J 0400) Publ Inst Math, NS (Belgrade) 23(37)*229-234
  ⋄ B50 C20 C35 C90 ⋄ REV MR 80a:03050
  Zbl 399 # 03018 • ID 52815

VOJVODIC, G. [1980] *The Craig interpolation theorem for mixed-valued predicate calculi (Serbo-Croatian summary)* (J 2855) Zbor Rad, Prir-Mat Fak, Ser Mat (Novi Sad) 10*173-175
  ⋄ B50 C40 C90 ⋄ REV MR 83h:03035 Zbl 531 # 03009
  • ID 36055

VOJVODIC, G. see Vol. II, V for further entries

VOLGER, H. [1975] *Completeness theorem for logical categories* (C 0772) Model Th & Topoi 51-86
  ⋄ C40 G30 ⋄ REV MR 51 # 12983 Zbl 338 # 18001
  JSL 46.158 • ID 17228

VOLGER, H. [1975] *Logical categories, semantical categories and topoi* (C 0772) Model Th & Topoi 87-100
  ⋄ C20 C85 G30 ⋄ REV MR 51 # 12984 Zbl 338 # 18002
  JSL 46.158 • ID 17227

VOLGER, H. [1975] *Ultrafilters, ultrapowers and finiteness in a topos* (J 0326) J Pure Appl Algebra 6*345-356
  ⋄ C20 G30 ⋄ REV MR 52 # 84 Zbl 322 # 18001 • ID 18422

VOLGER, H. [1976] *The Feferman-Vaught theorem revisited* (S 0019) Colloq Math (Warsaw) 36*1-11
  ⋄ C20 C30 C90 ⋄ REV MR 55 # 12505 Zbl 361 # 02069
  • ID 30580

VOLGER, H. [1979] *Preservation theorems for limits of structures and global sections of sheaves of structures* (J 0044) Math Z 166*27-54
  ⋄ C40 C60 C90 ⋄ REV MR 80d:03029 Zbl 379 # 02020
  • ID 52393

VOLGER, H. [1979] *Sheaf representations of algebras and transfer theorems for model theoretic properties* (C 2897) Algeb Model, Kateg & Gruppoide 155-163
  ⋄ B25 C05 C30 C35 C90 ⋄ REV MR 81m:03050
  Zbl 429 # 03015 • ID 53846

VOLGER, H. [1981] *A unifying approach to theorems on preservation and interpolation for binary relations between structures* (J 0009) Arch Math Logik Grundlagenforsch 21*101-112
  ⋄ C40 ⋄ REV MR 83i:03062 Zbl 482 # 03012 • ID 33321

VOLGER, H. [1983] *A new hierarchy of elementary recursive decision problems* (J 3401) Meth Oper Res 45*509-519
  ⋄ B25 D15 D20 ⋄ REV MR 85i:68019 Zbl 531 # 03006
  • ID 37672

VOLGER, H. [1983] *Turing machines with linear alternation, theories of bounded concatenation and the decision problem of first order theories* (J 1426) Theor Comput Sci 23*333-337
  ⋄ B25 D10 D15 ⋄ REV MR 84m:68042 Zbl 538 # 03035
  • ID 39717

VOLGER, H. [1984] see CHERLIN, G.L.

VOLGER, H. [1984] *Filtered and stable boolean powers are relativized full boolean powers* (J 0004) Algeb Universalis 19*399-402
  ⋄ B25 C20 C30 C90 ⋄ REV MR 86e:03033
  Zbl 562 # 03016 • ID 39715

VOLGER, H. see Vol. IV for further entries

VOL'VACHEV, R.T. [1967] *The elementary theory of modules (Russian)* (J 0413) Izv Akad Nauk Belor SSR, Ser Fiz-Mat 1967/4*7-12
  ⋄ C60 ⋄ REV MR 38 # 27 Zbl 204.16 • ID 13866

VOL'VACHEV, R.T. [1968] *Positive and elementary linear groups (Russian)* (J 0414) Dokl Akad Nauk Belor SSR 12*753-755
  ⋄ C60 ⋄ REV MR 39 # 1316 • ID 13867

VOL'VACHEV, R.T. [1970] *Linear groups as axiomatizable classes of models (Russian)* (J 0414) Dokl Akad Nauk Belor SSR 14*209-211
  ⋄ C20 C52 C60 ⋄ REV MR 43 # 4653 Zbl 243 # 08005
  • ID 13868

VOL'VACHEV, R.T. see Vol. IV for further entries

VOPENKA, P. [1962] *A method of constructing a nonstandard model of the Bernays-Goedel axiomatic theory of sets (Russian)* (J 0023) Dokl Akad Nauk SSSR 143*11-12
• TRANSL [1962] (J 0062) Sov Math, Dokl 3*309-310
◇ C20 C62 E30 E70 H20 ◇ REV MR 25#13 Zbl 126.21 JSL 35.470 • ID 16336

VOPENKA, P. [1962] *Construction of models for set theory by the method of ultra products (Russian) (German summary)* (J 0068) Z Math Logik Grundlagen Math 8*293-304
◇ C20 C62 E05 E30 E55 E70 ◇ REV MR 26#3611 Zbl 119.14 JSL 33.475 • ID 21081

VOPENKA, P. [1962] *Models for set theory (Russian) (German summary)* (J 0068) Z Math Logik Grundlagen Math 8*281-292
◇ C62 E30 E70 ◇ REV MR 26#3610 Zbl 119.14 JSL 32.411 • ID 21080

VOPENKA, P. [1963] *Construction of models of set theory by the spectrum method (Russian) (German summary)* (J 0068) Z Math Logik Grundlagen Math 9*149-160
◇ C62 E30 E70 ◇ REV MR 27#40 Zbl 135.253 JSL 35.470 • ID 22007

VOPENKA, P. [1963] *Construction of non-standard, non-regular models in set theory (Russian) (German summary)* (J 0068) Z Math Logik Grundlagen Math 9*229-233
◇ C20 C62 E30 E70 ◇ REV MR 27#1376 Zbl 119.14 JSL 35.470 • ID 22005

VOPENKA, P. [1963] *Elementary concepts in set theory (Russian) (German summary)* (J 0068) Z Math Logik Grundlagen Math 9*161-167
◇ C62 E47 ◇ REV MR 27#39 Zbl 158.261 JSL 35.470 • ID 24846

VOPENKA, P. [1964] *Submodels of models of set theory (Russian) (German summary)* (J 0068) Z Math Logik Grundlagen Math 10*163-172
◇ C20 C62 E30 E55 E70 ◇ REV MR 28#3931 Zbl 158.265 JSL 35.470 • ID 24879

VOPENKA, P. [1965] *Construction of a model for Goedel-Bernays set theory for which the class of natural numbers is a set of the model and a proper class in the theory* (P 0614) Th Models;1963 Berkeley 436-437
◇ C62 E30 E70 ◇ REV Zbl 171.264 • ID 27531

VOPENKA, P. [1965] *On ∇-model of set theory* (J 0014) Bull Acad Pol Sci, Ser Math Astron Phys 13*267-272
◇ C20 C62 C90 E30 E35 E40 E70 ◇ REV MR 32#54 Zbl 147.259 JSL 34.515 • ID 13875

VOPENKA, P. [1965] see HAJEK, P.

VOPENKA, P. [1965] *Properties of ∇-model* (J 0014) Bull Acad Pol Sci, Ser Math Astron Phys 13*441-444
◇ C62 E10 E35 E40 ◇ REV MR 32#7400 Zbl 147.259 JSL 34.515 • ID 13871

VOPENKA, P. [1965] *The limits of sheaves and applications on constructions of models* (J 0014) Bull Acad Pol Sci, Ser Math Astron Phys 13*189-192
◇ C20 C62 C90 E25 E30 E35 E40 E70 ◇ REV MR 32#53 Zbl 147.258 JSL 34.515 • ID 13876

VOPENKA, P. [1966] *∇-models in which the generalized continuum hypothesis does not hold* (J 0014) Bull Acad Pol Sci, Ser Math Astron Phys 14*95-99
◇ C62 E35 E50 ◇ REV MR 34#41 Zbl 147.260 JSL 34.515 • ID 13879

VOPENKA, P. [1966] see HRBACEK, K.

VOPENKA, P. [1966] see HAJEK, P.

VOPENKA, P. [1967] see HAJEK, P.

VOPENKA, P. [1967] *General theory of ∇-models* (J 0140) Comm Math Univ Carolinae (Prague) 8*145-170
◇ C62 E35 E40 ◇ REV MR 35#5310 Zbl 162.17 • ID 13884

VOPENKA, P. [1967] see BALCAR, B.

VOPENKA, P. [1973] see HAJEK, P.

VOPENKA, P. [1975] see SOCHOR, A.

VOPENKA, P. [1981] see SOCHOR, A.

VOPENKA, P. see Vol. IV, V, VI for further entries

VOSAHLO, J. [1978] see HAVRANEK, T.

VUCKOVIC, V. [1967] *Recursive models for three-valued propositional calculi with classical implication* (J 0047) Notre Dame J Formal Log 8*148-153
◇ B50 C57 C90 F30 ◇ REV Zbl 262#02021 • ID 27478

VUCKOVIC, V. [1977] *Recursive and recursive enumerable manifolds I,II* (J 0047) Notre Dame J Formal Log 18*265-291,383-405
◇ C57 D45 D80 ◇ REV MR 56#2808 Zbl 306#02035 Zbl 306#02036 • ID 23633

VUCKOVIC, V. see Vol. I, II, IV, V, VI for further entries

WACHS, E.A. [1977] *Models of well-orderings* (J 0016) Bull Austral Math Soc 16*155-157
◇ C65 E07 ◇ REV Zbl 328#02040 • ID 65954

WAERDEN VAN DER, B.L. [1930] *Eine Bemerkung ueber die Unzerlegbarkeit von Polynomen* (J 0043) Math Ann 102*738-739
◇ B25 C60 D45 F55 ◇ REV FdM 56.825 • ID 38703

WAERDEN VAN DER, B.L. see Vol. V for further entries

WAGNER, C.M. [1982] *On Martin's conjecture* (J 0007) Ann Math Logic 22*47-67
◇ C15 C35 C45 C65 C75 ◇ REV MR 84h:03083 Zbl 523#03018 • ID 34277

WAID, C.C. [1974] *On Dirichlet's theorem and finite primes* (J 0053) Proc Amer Math Soc 44*9-11
◇ H15 ◇ REV MR 49#247 Zbl 302#10050 • ID 13972

WAID, C.C. see Vol. I for further entries

WAINER, S.S. [1975] see SCHWICHTENBERG, H.

WAINER, S.S. [1975] *Some hierachies based on higher type quantification* (P 0775) Logic Colloq;1973 Bristol 305-316
◇ C75 D55 D65 E45 ◇ REV MR 58#5142 Zbl 311#02053 • ID 29611

WAINER, S.S. see Vol. IV, VI for further entries

WAJSBERG, M. [1933] *Beitrag zur Metamathematik* (J 0043) Math Ann 109*200-229
• TRANSL [1977] (C 4055) Wajsberg: Logical Works 62-88 [1935] (J 4710) Pol Tow Mat, Wiad Mat 39*43-84
◇ A05 B10 B30 B96 C35 F99 ◇ REV Zbl 8.97 FdM 59.53 FdM 61.972 • ID 40808

WAJSBERG, M. [1933] *Untersuchungen ueber den Funktionenkalkuel fuer endliche Individuenbereiche* (J 0043) Math Ann 108*218-228
• TRANSL [1977] (C 4055) Wajsberg: Logical Works 40-49
◊ B10 C13 ◊ REV Zbl 6.242 JSL 48.873 FdM 59.864
• ID 13979

WAJSBERG, M. [1938] *Untersuchungen ueber den Aussagenkalkuel von A.Heyting* (J 4710) Pol Tow Mat, Wiad Mat 46*45-101 • ERR/ADD ibid 47*139
• TRANSL [1977] (C 4055) Wajsberg: Logical Works 132-171
◊ B25 F50 ◊ REV Zbl 19.385 JSL 3.169 JSL 48.873 FdM 65.1104 • ID 13982

WAJSBERG, M. see Vol. I, II, V for further entries

WALD, B. [1983] *On κ-products modulo μ-products* (P 3802) Abel Group Th;1983 Honolulu 362-370
◊ C20 C60 E45 E75 ◊ REV MR 85c:20045 Zbl 538 # 20026 • ID 39034

WALD, B. see Vol. V for further entries

WALIGORA, G. [1975] see KOPIEKI, R.

WALKER, E.A. [1981] see GOEBEL, R.

WALKOE JR., W.J. [1970] *Finite partially-ordered quantification* (J 0036) J Symb Logic 35*535-555
◊ B15 C80 ◊ REV MR 43 # 4646 Zbl 219 # 02008 JSL 40.239 • ID 13998

WALKOE JR., W.J. [1973] see KEISLER, H.J.

WALKOE JR., W.J. [1976] *A small step backwards* (J 0005) Amer Math Mon 83*338-344
◊ B10 C80 ◊ REV MR 58 # 16130 Zbl 339 # 02010
• ID 27272

WALT VAN DER, A.P.J. [1967] see HEIDEMA, J.

WANG, CHUSHENG [1984] *On ultraproducts in the variety of fields (Chinese)* (J 4634) Jiangxi Shiyuan Xuebao 1984/2*24-27
◊ C20 C60 ◊ REV Zbl 571 # 12013 • ID 49266

WANG, HAO [1950] see ROSSER, J.B.

WANG, HAO [1953] *Quelques notions d'axiomatique* (J 0252) Rev Philos Louvain 51*409-443
◊ B25 B30 C07 C35 E30 F30 ◊ REV JSL 20.289
• ID 14024

WANG, HAO [1962] see DREBEN, B.

WANG, HAO [1963] *Dominoes and the AEA case of the decision problem* (P 0674) Symp Math Th of Automata;1962 New York 23-55
◊ B20 B25 D05 D35 ◊ REV MR 29 # 4688 Zbl 137.10
• ID 14040

WANG, HAO [1963] *Partial systems of number theory* (C 1009) Wang: Survey Math Logic 376-382
◊ C62 F30 ◊ REV JSL 29.147 • ID 14049

WANG, HAO [1965] *Remarks on machines, sets, and the decision problem* (P 0688) Logic Colloq;1963 Oxford 304-320
◊ B20 B25 D03 D05 D10 D35 E30 ◊ REV
MR 39 # 6729 Zbl 133.254 • ID 14057

WANG, HAO [1981] *Popular lectures on mathematical logic (Chinese)* (X 1876) Kexue Chubanshe: Beijing vii+257pp
• TRANSL [1981] (X 0864) Van Nostrand: New York ix+273pp
◊ A05 B98 C98 E98 F30 F98 ◊ REV MR 82e:03001 MR 84g:03002 JSL 47.908 • ID 34116

WANG, HAO see Vol. I, II, IV, V, VI for further entries

WANG, SHIQIANG [1965] *Some decidable cases of elementary propositions on partial order sets I (Chinese)* (P 4564) Math Logic;1963 Xi-An 87-92
◊ B25 E07 ◊ ID 49328

WANG, SHIQIANG [1980] *A proof of the compactness theorem in lattice-valued model theory (Chinese) (English summary)* (J 2521) Beijing Shifan Daxue Xuebao, Ziran Kexue 1980/3-4*25-30
◊ C90 ◊ REV MR 84k:03107 • ID 36124

WANG, SHIQIANG [1981] *Lattice-valued atomic and countable saturated models (Chinese)* (J 0420) Shuxue Jinzhan 10*144-146
◊ C50 C90 ◊ REV MR 84i:03077 • ID 34556

WANG, SHIQIANG [1981] see LU, JINGBO

WANG, SHIQIANG [1981] *Two cases of application of the method of construction by constants in lattice-valued model theory (Chinese)* (J 2771) Kexue Tongbao 26*129-130
• TRANSL [1981] (J 3769) Sci Bull, Foreign Lang Ed 26*581-584
◊ C30 C90 ◊ REV MR 83k:03042 Zbl 492 # 03016
• ID 35397

WANG, SHIQIANG [1982] *A class of commutative rings with Goldbach property (Chinese) (English summary)* (J 2521) Beijing Shifan Daxue Xuebao, Ziran Kexue 1982/1*17-22
◊ C60 ◊ REV MR 85c:11084 • ID 39020

WANG, SHIQIANG [1982] *A class of commutative rings with prime formulas (Chinese)* (J 2771) Kexue Tongbao 27*1025-1027
• TRANSL [1983] (J 3769) Sci Bull, Foreign Lang Ed 28*436-440
◊ C60 ◊ REV MR 85e:13011 Zbl 516 # 13007 Zbl 574 # 03053 • ID 45218

WANG, SHIQIANG [1982] *An omitting types theorem in lattice-valued model theory (Chinese)* (J 0418) Shuxue Xuebao 25*202-207
◊ C07 C75 C90 ◊ REV MR 84i:03078 Zbl 542 # 03017
• ID 34557

WANG, SHIQIANG & WU, TAO [1982] *Extensions of quadratic rings of algebraic integers without the Goldbach property (Chinese) (English summary)* (J 2521) Beijing Shifan Daxue Xuebao, Ziran Kexue 1982/3*21-25
◊ C60 ◊ REV MR 85g:11089 • ID 43933

WANG, SHIQIANG [1984] *A class of commutative rings without Goldbach property (Chinese)* (J 0418) Shuxue Xuebao 27*374-380
◊ C60 ◊ REV Zbl 539 # 13014 • ID 39818

WANG, SHIQIANG [1984] *Extensions with and without Goldbach property of some cubic rings of integers (Chinese)* (J 3766) Zhongguo Kexue, Xi A 27*482-491
◊ C60 ◊ REV MR 86b:11063 Zbl 539 # 13015 • ID 39813

WANG, SHIQIANG [1984] see SHEN, FUXING

WANG, SHIQIANG [1984] *Some number-theoretic properties of a kind of Goldbach commutative rings (Chinese)* (J 3766) Zhongguo Kexue, Xi A 27*603-611
  ⋄ C60 ⋄ REV MR 86f:11023 Zbl 563 # 13014 • ID 48458

WANG, SHIQIANG see Vol. I, II for further entries

WASILEWSKA, A. [1976] *A sequence formalization for SCI* (J 0063) Studia Logica 35*213-217
  ⋄ B25 F07 ⋄ REV MR 55 # 10238 Zbl 346 # 02006 • ID 65985

WASILEWSKA, A. [1976] *On the decidability theorems* (J 1929) Prace Centr Oblicz Pol Akad Nauk 246*16pp
  ⋄ B25 F07 ⋄ REV Zbl 338 # 68066 • ID 65986

WASILEWSKA, A. [1978] *Machines, logics and decidability* (J 2095) Fund Inform, Ann Soc Math Pol, Ser 4 1*291-303
  ⋄ B25 B35 ⋄ REV MR 58 # 9932 Zbl 386 # 68083 • ID 69996

WASILEWSKA, A. [1979] *A constructive proof of Craig's interpolation lemma for m-valued logic* (J 0063) Studia Logica 38*267-275
  ⋄ B50 C40 ⋄ REV MR 81e:03019 Zbl 442 # 03023 • ID 56380

WASILEWSKA, A. [1980] *On the Gentzen-type formalizations* (J 0068) Z Math Logik Grundlagen Math 26*439-444
  ⋄ B25 D05 F07 ⋄ REV MR 82g:03102 Zbl 471 # 03021 • ID 55216

WASILEWSKA, A. [1985] *Monadic second-order definability as a common characterization of finite automata, certain classes of programs and logics* (J 2095) Fund Inform, Ann Soc Math Pol, Ser 4 8/3-4*309-320
  ⋄ B15 B75 C40 D05 ⋄ ID 49467

WASILEWSKA, A. see Vol. I, II, VI for further entries

WASSERMAN, H.C. [1976] *An analysis of the counterfactual conditional* (J 0047) Notre Dame J Formal Log 17*395-400
  ⋄ B25 B45 ⋄ REV MR 56 # 96 Zbl 245 # 02017 • ID 18423

WASSERMAN, H.C. see Vol. I, II for further entries

WASZKIEWICZ, J. & WEGLORZ, B. [1968] *Some models of theories of reduced powers (Russian summary)* (J 0014) Bull Acad Pol Sci, Ser Math Astron Phys 16*683-685
  ⋄ C20 ⋄ REV MR 38 # 5602 Zbl 182.324 • ID 14079

WASZKIEWICZ, J. & WEGLORZ, B. [1969] *On ω_0-categoricity of powers (Russian summary)* (J 0014) Bull Acad Pol Sci, Ser Math Astron Phys 17*195-199
  ⋄ C30 C35 C50 ⋄ REV MR 41 # 3262 Zbl 182.10 • ID 14077

WASZKIEWICZ, J. & WEGLORZ, B. [1969] *On products of structures for generalized logics (Polish and Russian summaries)* (J 0063) Studia Logica 25*7-15
  ⋄ B50 C20 C30 C90 ⋄ REV MR 41 # 8220 Zbl 264 # 02019 • ID 14081

WASZKIEWICZ, J. [1970] *A remark on Engeler's filter-images* (S 0019) Colloq Math (Warsaw) 21*165-167
  ⋄ C20 ⋄ REV MR 42 # 53 Zbl 204.311 • ID 14083

WASZKIEWICZ, J. [1971] *The notions of isomorphism and identity for many-valued relational structures (Polish and Russian summaries)* (J 0063) Studia Logica 27*93-99
  ⋄ B50 C90 ⋄ REV MR 45 # 8503 Zbl 252 # 02053 • ID 14084

WASZKIEWICZ, J. [1972] see PACHOLSKI, L.

WASZKIEWICZ, J. [1973] *ω_0-categoricity of generalized products* (S 0019) Colloq Math (Warsaw) 27*1-5
  ⋄ C20 C30 C35 ⋄ REV MR 48 # 10799 Zbl 293 # 02034 • ID 14085

WASZKIEWICZ, J. [1973] *On cardinalities of algebras of formulas for ω_0-categorical theories* (S 0019) Colloq Math (Warsaw) 27*7-11,162
  ⋄ C13 C35 G05 ⋄ REV MR 48 # 10800 Zbl 293 # 02035 • ID 14086

WATANABE, S. [1969] *Modified concepts of logic, probability, and information based on generalized continuous characteristic function* (J 0194) Inform & Control 15*1-21
  ⋄ C90 ⋄ REV MR 42 # 2864 Zbl 205.8 • ID 14089

WATANABE, S. see Vol. II, IV, V for further entries

WATANABE, T. [1972] *Ultraproducts and diophantine problems* (S 1011) Mem Nat Def Acad 12*1-10
  ⋄ C20 C60 ⋄ REV MR 49 # 4986 Zbl 343 # 12104 • ID 14091

WATNICK, R. [1984] *A generalization of Tennenbaum's theorem on effectively finite recursive linear orderings* (J 0036) J Symb Logic 49*563-569
  ⋄ C57 C65 D25 D45 ⋄ REV MR 85i:03152 • ID 42511

WATNICK, R. see Vol. IV for further entries

WATRO, R.J. [1984] see HENLE, J.M.

WATRO, R.J. see Vol. V for further entries

WAUW-DE KINDER VAN DE, G. [1975] *Arithmetique de premier ordre dans les topos (English summary)* (J 2313) C R Acad Sci, Paris, Ser A-B 280*A1579-A1582
  ⋄ F30 F50 G30 H15 ⋄ REV MR 52 # 8217 Zbl 307 # 18001 • ID 18421

WEAVER, G.E. [1973] *A note on compactness and decidability* (J 0079) Logique & Anal, NS 16*315-319
  ⋄ B25 C95 ⋄ REV MR 50 # 85 Zbl 306 # 02047 • ID 14095

WEAVER, G.E. [1974] *Finite partitions and their generators* (J 0068) Z Math Logik Grundlagen Math 20*255-260
  ⋄ B20 B25 C07 C40 ⋄ REV MR 51 # 45 Zbl 309 # 02056 • ID 14096

WEAVER, G.E. [1975] *Tarski-Vaught and Loewenheim-Skolem numbers* (J 0079) Logique & Anal, NS 18*113-125
  ⋄ C55 C95 ⋄ REV MR 54 # 12516 Zbl 327 # 02042 • ID 66001

WEAVER, G.E. [1975] *Uniform compactness and interpolation theorems in sentential logic* (J 0302) Rep Math Logic, Krakow & Katowice 5*93-95 • ERR/ADD ibid 12*67
  ⋄ B05 C40 ⋄ REV MR 57 # 9471 MR 82e:03018 Zbl 352 # 02013 • ID 21907

WEAVER, G.E. [1980] *A note on the compactness theorem in first order logic* (J 0068) Z Math Logik Grundlagen Math 26*111-113
  ⋄ B10 C07 ⋄ REV MR 81i:03050 Zbl 441 # 03011 • ID 56064

WEAVER, G.E. [1982] *A note on the interpolation theorem in first order logic* (J 0068) Z Math Logik Grundlagen Math 28*215-218
  ⋄ C40 ⋄ REV MR 84i:03068 Zbl 526 # 03002 • ID 34549

WEAVER, G.E. see Vol. I, II, V, VI for further entries

WECHSUNG, G. (ED.) [1984] *Frege conference 1984. Proceedings of the International Conference held at Schwerin, Sept.10-14, 1984* (**X** 0911) Akademie Verlag: Berlin 408pp
⋄ A05 A10 B97 C97 ⋄ REV MR 85m:03006 Zbl 544 # 00005 • ID 40976

WECHSUNG, G. see Vol. IV, V for further entries

WEESE, M. [1972] *Zur Modellvollstaendigkeit und Entscheidbarkeit gewisser topologischer Raeume (Russian, English and French summaries)* (**J** 0115) Wiss Z Humboldt-Univ Berlin, Math-Nat Reihe 21∗477-485
⋄ B25 C35 C65 D80 ⋄ REV MR 48 # 5852 Zbl 255 # 02053 • ID 14105

WEESE, M. [1976] *Entscheidbarkeit in speziellen uniformen Strukturen bezueglich Sprachen mit Maechtigkeitsquantoren* (**J** 0068) Z Math Logik Grundlagen Math 22∗215-230
⋄ B25 C55 C65 C80 ⋄ REV MR 58 # 5159 Zbl 314 # 02056 • ID 18454

WEESE, M. [1976] *The universality of boolean algebras with the Haertig quantifier* (**P** 1476) Set Th & Hierarch Th (2) (Mostowski);1975 Bierutowice 291-296
⋄ C55 C80 D35 F25 G05 ⋄ REV MR 54 # 12478 Zbl 331 # 02027 • ID 23814

WEESE, M. [1977] *Definierbare Praedikate in booleschen Algebren. I* (**J** 0068) Z Math Logik Grundlagen Math 23∗511-526
⋄ C40 C80 G05 ⋄ REV MR 57 # 16157a Zbl 341 # 02038 • REM Part II 1978 • ID 51118

WEESE, M. [1977] *Ein neuer Beweis fuer die Entscheidbarkeit der Theorie der booleschen Algebren (Russian, English and French summaries)* (**J** 0115) Wiss Z Humboldt-Univ Berlin, Math-Nat Reihe 26∗663-667
⋄ B25 G05 ⋄ REV MR 80c:03017 Zbl 423 # 03007 • ID 53519

WEESE, M. [1977] *Entscheidbarkeit der Theorie der booleschen Algebren in Sprachen mit Maechtigkeitsquantoren (Russian and English summaries)* (**S** 3382) Sem-ber, Humboldt-Univ Berlin, Sekt Math 4∗vi+121pp
⋄ B25 C55 C80 G05 ⋄ REV MR 58 # 196 Zbl 415 # 03024 • ID 53126

WEESE, M. [1977] *The decidability of the theory of boolean algebras with cardinality quantifiers (Russian summary)* (**J** 0014) Bull Acad Pol Sci, Ser Math Astron Phys 25∗93-97
⋄ B25 C10 C55 C80 G05 ⋄ REV MR 55 # 7738 Zbl 398 # 03006 • ID 26552

WEESE, M. [1977] *The decidability of the theory of boolean algebras with the quantifier "there exist infinitely many"* (**J** 0053) Proc Amer Math Soc 64∗135-138
⋄ B25 C80 ⋄ REV MR 55 # 12499 Zbl 321 # 02011 • ID 50393

WEESE, M. [1977] *The undecidability of well-ordering with the Haertig quantifier (Russian summary)* (**J** 0014) Bull Acad Pol Sci, Ser Math Astron Phys 25∗89-91
⋄ C55 C65 C80 D35 E07 ⋄ REV MR 55 # 7737 Zbl 315 # 02046 • ID 26551

WEESE, M. [1977] see BAUDISCH, A.

WEESE, M. [1978] *Definierbare Praedikate in booleschen Algebren. II* (**J** 0068) Z Math Logik Grundlagen Math 24∗257-278
⋄ B25 C10 C40 C80 G05 ⋄ REV MR 57 # 16157b Zbl 341 # 02038 • REM Part I 1977 • ID 51623

WEESE, M. [1980] see BAUDISCH, A.

WEESE, M. [1980] *Generalized Ehrenfeucht games* (**J** 0027) Fund Math 109∗103-112
⋄ C80 ⋄ REV MR 82a:03033 Zbl 368 # 02017 • ID 55953

WEESE, M. [1981] *Decidability with respect to Haertig quantifier and Rescher quantifier* (**J** 0068) Z Math Logik Grundlagen Math 27∗569-576
⋄ B25 C55 C80 D35 ⋄ REV MR 84i:03075 Zbl 503 # 03012 • ID 33324

WEESE, M. [1982] see SEESE, D.G.

WEESE, M. [1983] see TUSCHIK, H.-P.

WEESE, M. [1984] *The theory of Boolean algebras extended by a group of automorphisms* (**S** 3382) Sem-ber, Humboldt-Univ Berlin, Sekt Math 60∗218-222
⋄ C07 D35 G05 ⋄ REV MR 86h:03110 Zbl 561 # 03005 • ID 42386

WEESE, M. [1985] see BAUDISCH, A.

WEESE, M. see Vol. IV, V for further entries

WEGLORZ, B. [1965] *Compactness of algebraic systems (Russian summary)* (**J** 0014) Bull Acad Pol Sci, Ser Math Astron Phys 13∗705-706
⋄ C05 C50 ⋄ REV MR 32 # 7471 Zbl 133.245 JSL 40.88 • ID 14109

WEGLORZ, B. [1966] *Equationally compact algebras I* (**J** 0027) Fund Math 59∗289-298
⋄ C05 C20 C50 ⋄ REV MR 35 # 1462 Zbl 221 # 02039 JSL 40.88 • REM Part II 1968 by Mycielski,J. & Ryll-Nardzewski,C. • ID 14110

WEGLORZ, B. [1967] see RYLL-NARDZEWSKI, C.

WEGLORZ, B. [1967] *Completeness and compactness of lattices* (**S** 0019) Colloq Math (Warsaw) 16∗243-248
⋄ C05 C52 G10 ⋄ REV MR 37 # 5127 Zbl 158.17 JSL 40.88 • ID 14113

WEGLORZ, B. [1967] *Equationally compact algebras III* (**J** 0027) Fund Math 60∗89-93
⋄ C05 C20 C50 ⋄ REV MR 35 # 1463 Zbl 263 # 08002 JSL 40.88 • REM Part II 1968 by Mycielski,J. & Ryll-Nardzewski,C. • ID 14111

WEGLORZ, B. [1967] *Some preservation theorems* (**S** 0019) Colloq Math (Warsaw) 17∗269-276
⋄ C40 ⋄ REV MR 36 # 3643 Zbl 224 # 02042 • REM Part II 1969 • ID 14112

WEGLORZ, B. [1968] *Limit generalized powers* (**J** 0014) Bull Acad Pol Sci, Ser Math Astron Phys 16∗449-451
⋄ C20 ⋄ REV MR 38 # 5601 Zbl 252 # 02054 • ID 14114

WEGLORZ, B. [1968] see FAJTLOWICZ, S.

WEGLORZ, B. [1968] see WASZKIEWICZ, J.

WEGLORZ, B. & WOJCIECHOWSKA, A. [1968] *Summability of pure extensions of relational structures* (**S** 0019) Colloq Math (Warsaw) 19∗27-35
⋄ C05 C07 C52 ⋄ REV MR 37 # 5136 Zbl 184.13 JSL 40.88 • ID 22201

WEGLORZ, B. [1968] see PACHOLSKI, L.

WEGLORZ, B. [1969] *A model of set theory $\mathfrak{E}$ over a given Boolean algebra (Russian summary)* (**J 0014**) Bull Acad Pol Sci, Ser Math Astron Phys 17*201-202
⋄ C62 E25 E35 G05 ⋄ REV MR 40 # 5443 Zbl 209.305 • ID 14116

WEGLORZ, B. [1969] see WASZKIEWICZ, J.

WEGLORZ, B. [1969] *Some preservation theorems. II* (**S 0019**) Colloq Math (Warsaw) 20*23-26
⋄ C40 ⋄ REV MR 39 # 2626 Zbl 224 # 02042 • REM Part I 1967 • ID 14115

WEGLORZ, B. [1971] see ANUSIAK, J.

WEGLORZ, B. [1975] *Substructures of reduced powers* (**J 0027**) Fund Math 89*191-197 • ERR/ADD ibid 92*209-211
⋄ C20 G05 ⋄ REV MR 54 # 2454 MR 56 # 117 Zbl 322 # 02048 Zbl 349 # 02044 • ID 14121

WEGLORZ, B. [1976] *A note on: "Atomic compactness in $\aleph_1$-categorical Horn theories" by John T. Baldwin* (**J 0027**) Fund Math 93*181-183
⋄ C05 C35 C50 C52 ⋄ REV MR 55 # 12506 Zbl 353 # 02030 • ID 26508

WEGLORZ, B. [1977] *Homogeneity, universality and saturatedness of limit reduced powers II* (**J 0027**) Fund Math 94*59-64
⋄ C20 C50 ⋄ REV MR 58 # 16260 Zbl 349 # 02046 • REM Part I 1977 by Wierzejewski,J. Part III 1977 by Pacholski,L. • ID 26514

WEGLORZ, B. [1977] see BROESTERHUIZEN, G.

WEGLORZ, B. see Vol. V for further entries

WEHRFRITZ, B.A.F. [1973] see KEGEL, O.H.

WEINBERG, E.C. [1980] *Automorphism groups of minimal η_α-sets* (**P 2983**) Ordered Groups;1978 Boise 71-79
⋄ C07 C65 E07 ⋄ REV MR 82c:06032 Zbl 451 # 06002 • ID 66302

WEINGARTNER, P. [1975] *A finite approximation to models of set theory* (**J 0063**) Studia Logica 34*45-58
⋄ C62 ⋄ REV MR 52 # 97 Zbl 307 # 02050 • ID 18424

WEINGARTNER, P. see Vol. I, II, V for further entries

WEINSTEIN, J.M. [1965] *First order properties preserved by direct product* (0000) Diss., Habil. etc
⋄ C30 ⋄ REM Ph.d.thesis, University of Wisconsin • ID 19796

WEINSTEIN, J.M. [1968] *(ω_1,ω) properties of unions of models* (**P 0637**) Syntax & Semant Infinitary Lang;1967 Los Angeles 265-268
⋄ C30 C40 C52 C75 ⋄ REV Zbl 264 # 02018 • ID 19795

WEINSTEIN, J.M. see Vol. V for further entries

WEINSTEIN, S. [1979] *Some applications of Kripke models to formal systems of intuitionistic analysis* (**J 0007**) Ann Math Logic 16*1-32
⋄ C90 F35 F50 ⋄ REV MR 80h:03088 Zbl 431 # 03037 • ID 53943

WEINSTEIN, S. see Vol. VI for further entries

WEISPFENNING, V. [1973] *Infinitary model-theoretic properties of κ-saturated structures* (**J 0068**) Z Math Logik Grundlagen Math 19*97-109
⋄ C10 C25 C35 C50 C60 C65 C75 ⋄ REV MR 48 # 10801 Zbl 301 # 02048 • ID 14127

WEISPFENNING, V. [1975] see SARACINO, D.H.

WEISPFENNING, V. [1975] *Model-completeness and elimination of quantifiers for subdirect products of structures* (**J 0032**) J Algeb 36*252-277
⋄ C10 C30 C35 C52 C60 C90 ⋄ REV MR 55 # 10271 Zbl 318 # 02052 • ID 14128

WEISPFENNING, V. [1975] *Two model theoretic proofs of Rueckert's Nullstellensatz* (**J 0064**) Trans Amer Math Soc 203*331-342
⋄ C65 H20 ⋄ REV MR 51 # 5296 Zbl 293 # 02043 • ID 14129

WEISPFENNING, V. [1976] *Negative-existentially complete structures and definability in free extensions* (**J 0036**) J Symb Logic 41*95-108
⋄ C25 C40 C60 C75 D40 G05 ⋄ REV MR 54 # 2451 Zbl 335 # 02035 • ID 14747

WEISPFENNING, V. [1976] *On the elementary theory of Hensel fields* (**J 0007**) Ann Math Logic 10*59-93
⋄ B25 C10 C35 C60 ⋄ REV MR 55 # 5437 Zbl 347 # 02033 • ID 18425

WEISPFENNING, V. [1977] *Nullstellensaetze - a model theoretic framework* (**J 0068**) Z Math Logik Grundlagen Math 23*539-545
⋄ C10 C25 C60 ⋄ REV MR 57 # 16056 Zbl 384 # 03018 • ID 52060

WEISPFENNING, V. [1978] *A note on $\aleph_0$-categorical model-companions* (**J 0009**) Arch Math Logik Grundlagenforsch 19*23-29
⋄ C25 C35 C50 C52 C65 G05 G10 ⋄ REV MR 80g:03032 Zbl 408 # 03024 • ID 29162

WEISPFENNING, V. [1979] *Lattice products* (**P 2627**) Logic Colloq;1978 Mons 423-426
⋄ C10 C25 C30 C35 C90 ⋄ REV MR 81j:03048 Zbl 461 # 03004 • ID 54481

WEISPFENNING, V. [1981] *Elimination of quantifiers for certain ordered and lattice-ordered abelian groups* (**J 3133**) Bull Soc Math Belg, Ser B 33*131-155
⋄ B25 C10 C60 ⋄ REV MR 82h:03022 Zbl 499 # 03012 • ID 90336

WEISPFENNING, V. [1981] *The model-theoretic significance of complemented existential formulas* (**J 0036**) J Symb Logic 46*843-850
⋄ C25 ⋄ REV MR 83c:03033 Zbl 502 # 03017 • ID 33285

WEISPFENNING, V. [1982] *Model theory and lattices of formulas* (**P 3634**) Patras Logic Symp;1980 Patras 261-295
⋄ C25 C35 C52 C75 G10 ⋄ REV MR 85d:03068 Zbl 519 # 03032 • ID 37543

WEISPFENNING, V. [1982] *Valuation rings and Boolean products* (**S 4311**) Groupes de Contact FNRS, Sci Math
⋄ C10 C30 C60 ⋄ ID 41355

WEISPFENNING, V. [1984] *Aspects of quantifier elimination in algebra* (**P 3088**) Univer Alg & Link Log, Alg, Combin, Comp Sci;1983 Darmstadt 85-105
⋄ C10 C60 C98 ⋄ REV Zbl 575 # 03019 • ID 41356

WEISPFENNING, V. [1984] *Quantifier elimination and decision procedures for valued fields* (**P** 2153) Logic Colloq;1983 Aachen 1*419-472
◇ B25 C10 C35 C60 ◇ ID 41771

WEISPFENNING, V. [1985] *Quantifier elimination for distributive lattices and measure algebras* (**J** 0068) Z Math Logik Grundlagen Math 31*249-261
◇ B25 B50 C10 C65 G05 G10 ◇ REV Zbl 547 # 03026 • ID 47574

WEISPFENNING, V. [1985] *Quantifier elimination for modules* (**J** 0009) Arch Math Logik Grundlagenforsch 25*1-11
◇ C10 C35 C57 C60 ◇ ID 48460

WEISPFENNING, V. [1985] *The complexity of elementary problems in archimedian ordered groups* (**P** 4601) EUROCAL;1985 Linz 2*87-88
◇ B25 C10 D15 D40 ◇ ID 48461

WEISSAUER, R. [1982] *Der Hilbertsche Irreduzibilitaetssatz* (**J** 0127) J Reine Angew Math 334*203-220
◇ C60 H20 ◇ REV MR 84c:12020 Zbl 477 # 12029 • ID 39576

WEITKAMP, G. [1985] see KALANTARI, I.

WEITKAMP, G. see Vol. IV, V for further entries

WELLS, C. [1985] see BARR, M.

WENZEL, G.H. [1970] *Subdirect irreducibility and equational compactness in unary algebras* $\langle A; f \rangle$ (**J** 0008) Arch Math (Basel) 21*256-264
◇ C05 C50 ◇ REV MR 42 # 1746 Zbl 207.29 JSL 40.88 • ID 14137

WENZEL, G.H. [1971] *On* $(\mathfrak{S}, \mathfrak{X}, \mathfrak{m})$*-atomic compact relational systems* (**J** 0043) Math Ann 194*12-18
◇ C05 C50 ◇ REV MR 44 # 6473 Zbl 211.319 JSL 40.88 • ID 19792

WENZEL, G.H. [1973] *Eine Charakterisierung gleichungskompakter universeller Algebren* (**J** 0068) Z Math Logik Grundlagen Math 19*283-287
◇ C05 C50 ◇ REV MR 47 # 8398 Zbl 309 # 08004 • ID 14138

WERNER, H. [1973] see GANTER, B.

WERNER, H. [1974] see KEIMEL, K.

WERNER, H. [1977] see BULMAN-FLEMING, S.

WERNER, H. [1977] *Varieties generated by quasi-primal algebras have decidable theories* (**P** 2112) Contrib to Universal Algeb;1975 Szeged 555-575
◇ B25 C05 ◇ REV MR 58 # 451 Zbl 375 # 02044 • ID 51621

WERNER, H. [1978] *Discriminator-algebras. Algebraic representation and model theoretic properties* (**X** 0911) Akademie Verlag: Berlin ii+92pp
◇ B25 C05 C25 C35 C90 ◇ REV MR 80f:08009 Zbl 374 # 08002 • ID 51565

WERNER, H. [1979] see DAVEY, B.A.

WERNER, H. [1979] see BURRIS, S.

WERNER, H. [1980] see BURRIS, S.

WERNER, H. [1982] *Sheaf constructions in universal algebra and model theory* (**P** 3831) Universal Algeb & Appl;1978 Warsaw 133-179
◇ B25 C05 C25 C30 C35 C90 G05 ◇ REV MR 85f:03031 Zbl 515 # 03017 • ID 37842

WESTERSTAAHL, D. [1976] *Some philosophical aspects of abstract model theory* (**S** 2616) Philos Communic, Red Ser 2, Gothenburg iii+109pp
◇ A05 C95 ◇ ID 90136

WESTERSTAAHL, D. [1984] *Some results on quantifiers* (**J** 0047) Notre Dame J Formal Log 25*152-170
◇ B65 C80 ◇ REV MR 85e:03067 Zbl 553 # 03021 • ID 40308

WESTRHENEN VAN, S.C. [1968] *Statistical estimation of definability* (**C** 0684) Automation in Lang Transl & Theorem Prov 27-39
◇ B35 C40 ◇ REV MR 38 # 5608 Zbl 193.299 • ID 14159

WESTRHENEN VAN, S.C. see Vol. I for further entries

WETTE, E. [1959] *Von operativen Modellen der axiomatischen Mengenlehre* (**P** 0634) Constructivity in Math;1957 Amsterdam 266-277
◇ C62 E35 F99 ◇ REV MR 24 # A1832 Zbl 86.10 JSL 40.501 • ID 14162

WETTE, E. see Vol. I, II, IV, V, VI for further entries

WEYHRAUCH, R.W. [1984] see KETONEN, J.

WEYHRAUCH, R.W. see Vol. I for further entries

WHALEY, T.P. [1969] *Algebras satisfying the descending chain condition for subalgebras* (**J** 0048) Pac J Math 28*217-223
◇ C05 C50 ◇ REV MR 38 # 5693 Zbl 169.326 • ID 14173

WHALEY, T.P. see Vol. I, V for further entries

WHEELER, W.H. [1972] *Algebraically closed division rings, forcing, and the analytic hierarchy* (0000) Diss., Habil. etc
◇ C25 C60 D55 ◇ REM Doct. diss., Yale University • ID 19104

WHEELER, W.H. [1975] see FISHER, E.R.

WHEELER, W.H. [1975] see HIRSCHFELD, J.

WHEELER, W.H. [1976] *Model-companions and definability in existentially complete structures* (**J** 0029) Israel J Math 25*305-330
◇ C25 C35 C60 C98 ◇ REV MR 56 # 15413 Zbl 398 # 03023 • ID 26089

WHEELER, W.H. [1978] *A characterization of companionable, universal theories* (**J** 0036) J Symb Logic 43*402-429
◇ C25 C35 C60 C65 ◇ REV MR 58 # 10400 Zbl 391 # 03019 • ID 29269

WHEELER, W.H. [1979] *Amalgamation and elimination of quantifiers for theories of fields* (**J** 0053) Proc Amer Math Soc 77*243-250
◇ C10 C25 C52 C60 ◇ REV MR 80j:03049 Zbl 437 # 03012 • ID 55880

WHEELER, W.H. [1979] *Model-complete theories of pseudo-algebraically closed fields* (**J** 0007) Ann Math Logic 17*205-226
◇ C25 C35 C60 ◇ REV MR 81c:03024 Zbl 473 # 03029 • ID 55359

WHEELER, W.H. [1979] *The first order theory of N-colorable graphs* (J 0064) Trans Amer Math Soc 250*289-310
⋄ B30 C10 C25 C35 C65 ⋄ REV MR 80g:03035 Zbl 428 # 03028 • ID 53787

WHEELER, W.H. [1980] *Model theory of strictly upper triangular matrix rings* (J 0036) J Symb Logic 45*455-463
⋄ C25 C35 C60 ⋄ REV MR 81j:03055 Zbl 471 # 03027
• ID 55222

WHEELER, W.H. [1983] *Model complete theories of formally real fields and formally p-adic fields* (J 0036) J Symb Logic 48*1130-1139
⋄ C35 C60 ⋄ REV MR 85b:03055 • ID 39797

WHEELER, W.H. [1983] see JARDEN, M.

WHITELEY, W. [1973] *Homogeneous sets and homogeneous formulas* (J 2311) Rend Mat, Ser 6 6*171-182
⋄ C40 C60 ⋄ REV MR 52 # 82 Zbl 269 # 02023 • ID 18427

WHITELEY, W. [1973] *Logic and invariant theory I. Invariant theory of projective properties* (J 0064) Trans Amer Math Soc 177*121-139
⋄ B30 C60 ⋄ REV MR 56 # 5279 Zbl 238 # 50002 • REM Part II 1978. Part III 1977 • ID 79943

WHITELEY, W. [1977] *Logic and invariant theory III. Axiom systems and basic syzygies* (J 3172) J London Math Soc, Ser 2 15*1-15
⋄ B30 C60 ⋄ REV MR 57 # 16057b Zbl 347 # 50001 • REM Part II 1978. Part IV 1979 • ID 79941

WHITELEY, W. [1978] *Logic and invariant theory II: Homogeneous coordinates, the introduction of higher quantities, and structural geometry* (J 0032) J Algeb 50*380-394
⋄ B30 C60 ⋄ REV MR 57 # 16057a Zbl 348 # 50003 • REM Part I 1973. Part III 1977 • ID 50914

WHITELEY, W. [1979] *Logic and invariant theory IV. Invariants and syzygies in combinatorial geometry* (J 0033) J Comb Th, Ser B 26*251-267
⋄ C60 ⋄ REV MR 80g:05023 Zbl 331 # 50016 • REM Part III 1977 • ID 83147

WHITNEY, S. [1984] see GRAETZER, G.

WICKSTEAD, A.W. [1977] see CHADWICK, J.J.M.

WIEDEMANN, A. [1984] see ROGGENKAMP, K.W.

WIEGOLD, E.C. [1966] see NEUMANN, B.H.

WIEGOLD, J. [1984] see BURNS, R.G.

WIERZEJEWSKI, J. [1974] *A note on stability and products (Russian summary)* (J 0014) Bull Acad Pol Sci, Ser Math Astron Phys 22*875-876
⋄ C20 C30 C45 ⋄ REV MR 52 # 13370 Zbl 299 # 02060
• ID 14194

WIERZEJEWSKI, J. [1974] *On κ-universal domains* (J 0014) Bull Acad Pol Sci, Ser Math Astron Phys 22*219-221
⋄ C45 C50 ⋄ REV MR 50 # 91 Zbl 285 # 02046 • ID 14195

WIERZEJEWSKI, J. [1976] *On stability and products* (J 0027) Fund Math 93*81-95 • ERR/ADD ibid 97*51-52
⋄ C20 C30 C45 ⋄ REV MR 55 # 2552 MR 57 # 9525 Zbl 372 # 02027 Zbl 372 # 02028 • ID 26506

WIERZEJEWSKI, J. [1976] *Remarks on stability and saturated models* (S 0019) Colloq Math (Warsaw) 34*165-169
⋄ C45 C50 ⋄ REV MR 53 # 12932 Zbl 329 # 02022
• ID 23180

WIERZEJEWSKI, J. [1977] *A note on products and degree of types* (J 0068) Z Math Logik Grundlagen Math 23*431-434
⋄ C30 C45 ⋄ REV MR 58 # 10401 Zbl 441 # 03010
• ID 31529

WIERZEJEWSKI, J. [1977] *Homogeneity, universality and saturatedness of limit reduced powers. I* (J 0027) Fund Math 94*35-39
⋄ C20 C50 ⋄ REV MR 58 # 16259 Zbl 349 # 02045 • REM Part II 1977 by Weglorz,B. • ID 26513

WIERZEJEWSKI, J. [1977] see BROESTERHUIZEN, G.

WIERZEJEWSKI, J. [1980] see PACHOLSKI, L.

WILKIE, A.J. [1975] *On models of arithmetic - answers to two problems raised by H.Gaifman* (J 0036) J Symb Logic 40*41-47
⋄ C62 ⋄ REV MR 55 # 2560 Zbl 319 # 02050 • ID 14205

WILKIE, A.J. [1976] *A note on products of finite structures with an application to graphs* (J 3172) J London Math Soc, Ser 2 14*383-384
⋄ C13 ⋄ REV MR 54 # 10111 Zbl 352 # 08011 • ID 25875

WILKIE, A.J. [1977] *On models of arithmetic having non-modular substructure lattices* (J 0027) Fund Math 95*223-237
⋄ C20 C62 ⋄ REV MR 56 # 5269 Zbl 365 # 02041
• ID 31306

WILKIE, A.J. [1977] *On the theories of end extensions of models of arithmetic* (P 1695) Set Th & Hierarch Th (3);1976 Bierutowice 305-310
⋄ C62 ⋄ REV MR 58 # 21589 Zbl 376 # 02043 • ID 31307

WILKIE, A.J. [1978] *Some results and problems on weak systems of arithmetic* (P 1897) Logic Colloq;1977 Wroclaw 285-296
⋄ C62 F30 H15 ⋄ REV MR 81c:03050 Zbl 449 # 03076
• ID 56743

WILKIE, A.J. [1980] *Applications of complexity theory to Σ_0-definability problems in arithmetic* (P 2625) Model Th of Algeb & Arithm;1979 Karpacz 363-369
⋄ C62 D15 F30 ⋄ REV MR 82b:03085 Zbl 483 # 03024
• ID 79968

WILKIE, A.J. [1980] see PACHOLSKI, L.

WILKIE, A.J. [1981] see PARIS, J.B.

WILKIE, A.J. [1981] *On discretely ordered rings in which every definable ideal is principal* (P 3404) Model Th & Arithm;1979/80 Paris 297-303
⋄ C60 C62 H15 ⋄ REV MR 83h:03050 Zbl 498 # 03053
• ID 36064

WILKIE, A.J. [1982] *On core structures for Peano arithmetic* (P 3623) Logic Colloq;1980 Prague 311-314
⋄ C50 C62 F30 H15 ⋄ REV MR 84j:03142 Zbl 517 # 03030 • ID 34735

WILKIE, A.J. [1984] see DRIES VAN DEN, L.

WILKIE, A.J. [1985] see PARIS, J.B.

WILKIE, A.J. see Vol. V, VI for further entries

WILMERS, G.M. [1968] *Some properties of constructible models for set theory related to elementary equivalence* (J 0014) Bull Acad Pol Sci, Ser Math Astron Phys 16*693-698
⋄ C62 E45 ⋄ REV MR 39 # 5348 Zbl 199.305 • ID 14228

WILMERS, G.M. [1971] *Internally standard set theories* (J 0027) Fund Math 71*93-102
⋄ C62 C70 E45 E47 ⋄ REV MR 46 # 8835 Zbl 255 # 02072 • ID 14229

WILMERS, G.M. [1973] *An $\aleph_1$-standard model for ZF which is an element of the minimal model for ZF* (P 0710) Russell Mem Logic Conf;1971 Uldum 315-326
⋄ C62 C70 E30 E35 E45 H20 ⋄ REV MR 50 # 6848 • ID 30613

WILMERS, G.M. [1973] see SUZUKI, Y.

WILMERS, G.M. [1975] *Non-standard models and their application to model theory* (0000) Diss., Habil. etc
⋄ C50 C62 H20 ⋄ REM Ph.d.thesis, Oxford • ID 30977

WILMERS, G.M. [1980] *Minimally saturated models* (P 2625) Model Th of Algeb & Arithm;1979 Karpacz 370-380
⋄ C50 ⋄ REV MR 82d:03051 Zbl 463 # 03017 • ID 54557

WILMERS, G.M. [1981] *An observation concerning the relationship between finite and infinitary Σ_1^1-sentences* (P 3404) Model Th & Arithm;1979/80 Paris 304-306
⋄ C70 ⋄ REV MR 83d:03050 Zbl 504 # 03019 • ID 35190

WILMERS, G.M. [1982] see CEGIELSKI, P.

WILMERS, G.M. [1985] *Bounded existential induction* (J 0036) J Symb Logic 50*72-90
⋄ C62 F30 ⋄ REV MR 86h:03068 • ID 40251

WILSON, J.S. [1982] *The algebraic structure of $\aleph_0$-categorical groups* (P 3886) Groups-St.Andrews;1981 St.Andrews 345-358
⋄ C15 C35 C60 ⋄ REV MR 84c:20006 Zbl 497 # 20022 • ID 36637

WILSON, T.P. [1981] *General models of set theory* (J 0047) Notre Dame J Formal Log 22*36-44
⋄ C62 E55 E70 ⋄ REV MR 82f:03043 Zbl 417 # 03024 • ID 56490

WINKLER, P.M. [1975] *Model-completeness and Skolem expansions* (C 0782) Model Th & Algeb (A. Robinson) 408-463
⋄ C25 C30 C35 C60 ⋄ REV MR 58 # 27461 Zbl 324 # 02035 • ID 66112

WINKLER, P.M. see Vol. IV for further entries

WIRSING, M. [1977] *Das Entscheidungsproblem der Klasse von Formeln, die hoechstens zwei Primformeln enthalten* (J 0504) Manuscr Math 22*13-25
⋄ B20 B25 D35 ⋄ REV MR 57 # 12199 Zbl 365 # 02035 • ID 51036

WIRSING, M. see Vol. I, II, IV for further entries

WOEHL, K. [1979] *Zur Komplexitaet der Presburger Arithmetik und des Aequivalenz-Problems einfacher Programme* (P 3488) Theor Comput Sci (4);1979 Aachen 310-318
⋄ B25 B75 D15 F20 F30 ⋄ REV MR 81k:03015 Zbl 419 # 03024 • ID 53367

WOJCICKI, R. [1971] see PRZELECKI, M.

WOJCICKI, R. see Vol. I, II, V, VI for further entries

WOJCIECHOWSKA, A. [1968] see WEGLORZ, B.

WOJCIECHOWSKA, A. [1969] *Generalized limit powers (Russian summary)* (J 0014) Bull Acad Pol Sci, Ser Math Astron Phys 17*121-122
⋄ C20 ⋄ REV MR 39 # 5340 Zbl 184.13 • ID 14262

WOJCIECHOWSKA, A. [1969] *Generalized products for Q_α-languages (Russian summary)* (J 0014) Bull Acad Pol Sci, Ser Math Astron Phys 17*337-339
⋄ C30 C55 C80 ⋄ REV MR 41 # 47 Zbl 206.273 • ID 14263

WOJCIECHOWSKA, A. [1969] *Limit reduced powers* (S 0019) Colloq Math (Warsaw) 20*203-208
⋄ C20 ⋄ REV MR 40 # 33 Zbl 211.308 • ID 14261

WOJCIECHOWSKA, A. [1970] *A note on models with orderings (Russian summary)* (J 0014) Bull Acad Pol Sci, Ser Math Astron Phys 18*413
⋄ C55 E05 E55 ⋄ REV MR 42 # 7498 Zbl 219 # 02038 • ID 14264

WOJCIECHOWSKA, A. [1972] *On Helling cardinals* (J 0027) Fund Math 76*223-230
⋄ C55 E05 E55 ⋄ REV MR 48 # 109 Zbl 248 # 02059 • ID 14265

WOJCIECHOWSKA, A. see Vol. I, II, V for further entries

WOJDYLO, B. [1970] *On some problems of J.Slominski concerning equations in quasi-algebras* (S 0019) Colloq Math (Warsaw) 21*1-4
⋄ C05 ⋄ REV MR 41 # 137 Zbl 194.327 • ID 14266

WOJTYLAK, P. [1984] *A proof of Herbrand's theorem* (J 0302) Rep Math Logic, Krakow & Katowice 17*13-17
⋄ C10 F05 ⋄ REV MR 85m:03007 Zbl 573 # 03002 • ID 44096

WOJTYLAK, P. see Vol. I, II, VI for further entries

WOLENSKI, J. [1973] *Wajsberg on the first-order predicate calculus for the finite models* (J 0387) Bull Sect Logic, Pol Acad Sci 2*107-111
⋄ A10 C13 ⋄ REV MR 55 # 7735 • ID 80056

WOLF, A. [1974] *Sheaf representations of arithmetical algebras* (P 3020) Rect Adv in Repr Th Rings & C^*-Algeb;1973 New Orleans 87-93
⋄ C05 C90 G30 ⋄ REV MR 51 # 5458 Zbl 274 # 08009 • ID 28644

WOLFF, M. [1972] *Nonstandard-Komplettierung von Cauchy-Algebren* (P 0732) Contrib to Non-Standard Anal;1970 Oberwolfach 178-213
⋄ C65 H05 ⋄ REV MR 58 # 2787 Zbl 262 # 54048 • ID 14270

WOLFF, M. [1984] *Spectral theory of group representations and their nonstandard hull* (J 0029) Israel J Math 48*205-224
⋄ C65 H10 ⋄ REV MR 86e:46046 Zbl 568 # 22005 • ID 44491

WOLTER, H. [1968] *Eine Erweiterung der klassischen Analysis* (J 0068) Z Math Logik Grundlagen Math 14*167-184
⋄ C20 H05 ⋄ REV MR 37 # 5089 Zbl 159.18 • ID 14276

WOLTER, H. [1968] see HAUSCHILD, K.

WOLTER, H. [1969] see HAUSCHILD, K.

WOLTER, H. [1970] see HAUSCHILD, K.

WOLTER, H. [1972] *Ueber Mengen von Ausdruecken der Logik hoeherer Stufe, fuer die der Endlichkeitssatz und der Satz von Loewenheim-Skolem gelten* (J 0068) Z Math Logik Grundlagen Math 18∗13-18
⋄ B15 C20 C85 ⋄ REV MR 45 # 3177   Zbl 238 # 02013
• ID 14277

WOLTER, H. [1972] *Untersuchung ueber Algebren und formalisierte Sprachen hoeherer Stufen (Russian, English and French summaries)* (J 0115) Wiss Z Humboldt-Univ Berlin, Math-Nat Reihe 21∗487-495
⋄ B15 C20 C30 C40 C85 ⋄ REV MR 49 # 2354
Zbl 256 # 02027 • ID 14278

WOLTER, H. [1973] *Eine Erweiterung der elementaren Praedikatenlogik: Anwendungen in der Arithmetik und anderen mathematischen Theorien* (J 0068) Z Math Logik Grundlagen Math 19∗181-190
⋄ B28 C10 C80 F30 ⋄ REV MR 51 # 7815
Zbl 302 # 02001 • ID 14280

WOLTER, H. [1973] *Untersuchungen zum Spektralproblem gewisser Logiken 2. Stufe* (J 0068) Z Math Logik Grundlagen Math 19∗407-410
⋄ C55 C85 E10 ⋄ REV MR 49 # 2355   Zbl 301 # 02015
• ID 14279

WOLTER, H. [1975] see HERRE, H.

WOLTER, H. [1975] *Entscheidbarkeit der Arithmetik mit Addition und Ordnung in Logiken mit verallgemeinerten Quantoren* (J 0068) Z Math Logik Grundlagen Math 21∗321-330
⋄ B25 C10 C55 C80 F30 ⋄ REV MR 51 # 7816
Zbl 337 # 02032 • ID 66137

WOLTER, H. [1975] *Untersuchungen zu Logiken mit Maechtigkeitsquantoren (Russian, English and French summaries)* (J 0115) Wiss Z Humboldt-Univ Berlin, Math-Nat Reihe 24∗788-794
⋄ C55 C80 ⋄ REV MR 58 # 5020   Zbl 342 # 02040
• ID 66135

WOLTER, H. [1977] see HERRE, H.

WOLTER, H. [1977] *Entscheidbarkeit der Theorie der Wohlordnung mit Elimination der Quantoren (Russian, English and French summaries)* (J 0115) Wiss Z Humboldt-Univ Berlin, Math-Nat Reihe 26∗669-671
⋄ B25 C10 C65 C80 E07 ⋄ REV MR 80c:03018
Zbl 423 # 03008 • ID 53520

WOLTER, H. [1978] see HERRE, H.

WOLTER, H. [1979] see HERRE, H.

WOLTER, H. [1980] see HERRMANN, E.

WOLTER, H. [1981] see HERRE, H.

WOLTER, H. [1983] *On the "problem of the last root" for exponential terms* (P 1601) Easter Conf on Model Th (1);1983 Diedrichshagen 144-152
⋄ C60 C65 ⋄ REV MR 84i:03008   Zbl 528 # 03018
Zbl 529 # 03013 • ID 45061

WOLTER, H. [1983] see DAHN, B.I.

WOLTER, H. [1984] see DAHN, B.I.

WOLTER, H. [1984] *Some remarks on exponential functions in ordered fields* (P 1545) Easter Conf on Model Th (2);1984 Wittenberg 229-243
⋄ C25 C65 D35 ⋄ REV MR 86h:03065   Zbl 568 # 03009
• ID 44686

WOLTER, H. [1984] *Some results about exponential fields (survey)* (S 3521) Mem Soc Math Fr 16∗85-94
⋄ B25 C60 C65 C98 ⋄ REV Zbl 573 # 03011 • ID 39724

WOLTER, H. [1985] *On the "problem of the last root" for exponential terms* (J 0068) Z Math Logik Grundlagen Math 31∗163-168
⋄ C60 C65 ⋄ REV Zbl 528 # 03018 • ID 41813

WOLTER, H. see Vol. I for further entries

WOOD, C. [1972] *Forcing for infinitary languages* (J 0068) Z Math Logik Grundlagen Math 18∗385-402
⋄ C25 C75 ⋄ REV MR 48 # 100   Zbl 257 # 02041
• ID 14287

WOOD, C. [1973] *The model theory of differential fields of characteristic $p \neq 0$* (J 0053) Proc Amer Math Soc 40∗577-584
⋄ C25 C35 C45 C60 ⋄ REV MR 48 # 8227
Zbl 273 # 02039 • ID 14288

WOOD, C. [1974] *Prime model extensions for differential fields of characteristic $p \neq 0$* (J 0036) J Symb Logic 39∗469-477
⋄ C10 C35 C60 ⋄ REV MR 50 # 9577   Zbl 311 # 02064
• ID 14289

WOOD, C. [1976] *The model theory of differential fields revisited* (J 0029) Israel J Math 25∗331-352
⋄ C60 C98 ⋄ REV MR 56 # 15414   Zbl 346 # 02030
• ID 26090

WOOD, C. [1979] *Notes on the stability of separably closed fields* (J 0036) J Symb Logic 44∗412-416
⋄ C25 C35 C45 C60 ⋄ REV MR 81m:03042
Zbl 424 # 03014 • ID 80075

WOOD, C. [1979] see SARACINO, D.H.

WOOD, C. [1982] see SARACINO, D.H.

WOOD, C. [1983] see SARACINO, D.H.

WOOD, C. [1984] see SARACINO, D.H.

WOOD, C. [1985] see SARACINO, D.H.

WOODROW, R.E. [1976] *A note on countable complete theories having three isomorphism types of countable models* (J 0036) J Symb Logic 41∗672-680
⋄ C15 C35 ⋄ REV MR 58 # 10402   Zbl 373 # 02036
• ID 14589

WOODROW, R.E. [1978] *Theories with a finite number of countable models* (J 0036) J Symb Logic 43∗442-455
⋄ C15 ⋄ REV MR 58 # 10403   Zbl 434 # 03022 • ID 29271

WOODROW, R.E. [1979] *There are four countable ultrahomogeneous graphs without triangles* (J 0033) J Comb Th, Ser B 27∗168-179
⋄ C10 C50 C65 E05 ⋄ REV MR 81f:05141
Zbl 429 # 05044 • ID 53885

WOODROW, R.E. [1980] see LACHLAN, A.H.

WOODROW, R.E. [1981] see ROSE, B.I.

WOODROW, R.E. [1983] see KNIGHT, J.F.

WRAITH, G.C. [1971] see KOCK, A.

WRAITH, G.C. [1975] see LAWVERE, F.W.

WRAITH, G.C. [1976] *Logic from topology: a survey of topoi* (J 2679) Bull Inst Math Appl (Southend oS) 12∗115-119
⋄ C98 F98 G30 ⋄ REV MR 58 # 27176 • ID 80097

WRAITH, G.C. [1979] *Generic Galois theory of local rings* (P 2901) Appl Sheaves;1977 Durham 739-767
⋄ C60 G30 ⋄ REV MR 81c:13004 Zbl 419 # 14013
• ID 53387

WRAITH, G.C. see Vol. V, VI for further entries

WRIGHT, J.B. [1955] see BUECHI, J.R.

WRIGHT, J.B. [1957] see BUECHI, J.R.

WRIGHT, J.B. [1959] see ELGOT, C.C.

WRIGHT, J.B. [1968] see THATCHER, J.W.

WRIGHT, J.B. see Vol. I, IV, V for further entries

WRIGHT VON, G.H. [1948] *On the idea of logical truth I* (J 0990) Soc Sci Fennicae Comment Phys-Math 14/4*20pp
⋄ A05 B05 B20 B25 ⋄ REV MR 10.668 Zbl 32.385 JSL 15.58 • REM Part II 1950 • ID 14303

WRIGHT VON, G.H. [1950] *On the idea of logical truth II* (J 0990) Soc Sci Fennicae Comment Phys-Math 15/10*45pp
⋄ A05 B05 B20 B25 ⋄ REV MR 13.521 Zbl 38.5 JSL 16.147 • REM Part I 1948 • ID 14304

WRIGHT VON, G.H. [1951] *An essay in modal logic* (X 0809) North Holland: Amsterdam vii+90pp
⋄ B25 B45 ⋄ REV MR 13.614 Zbl 43.7 JSL 18.174
• ID 14305

WRIGHT VON, G.H. [1951] *Deontic logic* (J 0094) Mind 60*1-15
• REPR [1957] (C 1017) Logical Studies 58-74
⋄ A05 B25 B45 ⋄ REV Zbl 81.10 JSL 17.140 JSL 35.461
• ID 14313

WRIGHT VON, G.H. [1952] see GEACH, P.T.

WRIGHT VON, G.H. [1952] *On double quantification* (J 0990) Soc Sci Fennicae Comment Phys-Math 16/3*14pp
• REPR [1957] (C 1017) Logical Studies 44-57
⋄ A05 B20 B25 ⋄ REV MR 15.846 Zbl 47.8 Zbl 81.10 JSL 17.201 JSL 35.461 • ID 14309

WRIGHT VON, G.H. see Vol. II for further entries

WRONSKI, A. [1984] *Interpolation and amalgamation properties of BCK-algebras* (J 0352) Math Jap 29*115-121
⋄ C40 C52 G10 G25 ⋄ REV MR 85e:06015 Zbl 557 # 06008 • ID 39924

WRONSKI, A. [1984] *On varieties of commutative BCK-algebras not generated by their finite members* (P 3621) Frege Konferenz (2);1984 Schwerin 109-113
• REPR [1985] (J 0352) Math Jap 30*227-233
⋄ C05 C52 G25 ⋄ REV MR 85m:03006 Zbl 569 # 03030
• ID 45422

WRONSKI, A. [1985] *On a form of equational interpolation property* (P 4180) Int Congr Log, Meth & Phil of Sci (7,Pap);1983 Salzburg 23-29
⋄ C05 C40 ⋄ ID 48119

WRONSKI, A. see Vol. I, II, VI for further entries

WU, TAO [1982] see WANG, SHIQIANG

WUETHRICH, H.R. [1976] *Ein Entscheidungsverfahren fuer die Theorie der reell abgeschlossenen Koerper* (P 3196) Kompl von Entscheid Probl;1973/74 Zuerich 138-162
• TRANSL [1981] (J 3079) Kiber Sb Perevodov, NS 18*100-124
⋄ B25 C60 D15 ⋄ REV MR 57 # 18232 Zbl 363 # 02052
• ID 43103

XUEH, YUANGCHE [1983] *Lattices of width three generate a nonfinitely based variety* (J 0004) Algeb Universalis 17*132-134
⋄ C20 G10 ⋄ REV Zbl 522 # 06009 • ID 37047

YAMASAKI, S. [1983] see DOSHITA, S.

YAMASAKI, S. see Vol. I, IV for further entries

YANG, SHOULIAN [1982] *Ultrafilters with chain bases and their applications (Chinese)* (J 0418) Shuxue Xuebao 25*114-121
⋄ C20 E05 ⋄ REV MR 84a:04002 • ID 35600

YANG, SHOULIAN see Vol. V for further entries

YANKOV, V.A. [1967] *Finite validity of formulas of a special form (Russian)* (J 0023) Dokl Akad Nauk SSSR 174*302-304
• TRANSL [1967] (J 0062) Sov Math, Dokl 8*648-650
⋄ B55 C13 F50 ⋄ REV MR 36 # 2468 Zbl 209.304 JSL 38.331 • ID 06513

YANKOV, V.A. see Vol. II, IV, V, VI for further entries

YANOV, YU.I. [1962] *On systems of identities for algebras (Russian)* (J 0052) Probl Kibern 8*75-90
• TRANSL [1965] (J 0449) Probl Kybern 8*84-102
⋄ C05 ⋄ REV MR 29 # 5773 Zbl 192.94 JSL 36.162
• ID 00384

YANOV, YU.I. [1969] *On directed transformations of formulas (Russian)* (J 0087) Mat Zametki (Akad Nauk SSSR) 6*663-668
• TRANSL [1969] (J 1044) Math Notes, Acad Sci USSR 6*862-865
⋄ C05 ⋄ REV MR 42 # 170 Zbl 198.19 • ID 06522

YANOV, YU.I. see Vol. I, II, IV, VI for further entries

YAQUB, A. [1957] *On the identities of certain algebras* (J 0053) Proc Amer Math Soc 8*522-524
⋄ C05 ⋄ REV MR 19.831 Zbl 79.43 • ID 18013

YASHIN, A.D. [1984] *Completeness of the intuitionistic predicate calculus with the concept of "bar" (Russian)* (J 0288) Vest Ser Mat Mekh, Univ Moskva 1984/4*67-69
• TRANSL [1984] (J 0510) Moscow Univ Math Bull 39/4*62-65
⋄ C90 F05 F50 ⋄ REV MR 85i:03078 Zbl 573 # 03006
• ID 44180

YASHIN, A.D. [1984] *Nishimura's formulas as one-place logical connectives in the elementary theory of Kripke models (Russian)* (J 0288) Vest Ser Mat Mekh, Univ Moskva 1984/5*12-15
⋄ C90 F50 ⋄ REV MR 86a:03010 • ID 44104

YASHIN, A.D. see Vol. II, VI for further entries

YASHIN, A.L. [1977] *On an extension of intuitionistic logic and its interpretations (Russian)* (S 2579) Teor Mnozhestv & Topol (Izhevsk) 1*110-112
⋄ B55 C90 ⋄ REV Zbl 481 # 03036 • ID 36830

YASUDA, Y. [1977] see MISAWA, T.

YASUDA, Y. see Vol. IV, V for further entries

YASUGI, M. [1967] *Interpretations of set theory and ordinal number theory* (J 0036) J Symb Logic 32*145-161
⋄ C62 E10 E30 ⋄ REV MR 36#4983 Zbl 153.17
• ID 18024

YASUGI, M. [1969] *An evaluation of the ω-complexity of first order arithmetic with the constructive ω-rule* (1111) Preprints, Manuscr., Techn. Reports etc. 1-22
⋄ C75 ⋄ REM Report 69-16, Carnegie-Mellon University
• ID 32049

YASUGI, M. see Vol. II, V, VI for further entries

YASUHARA, M. [1966] *An axiomatic system for the first order language with an equi-cardinality quantifier* (J 0036) J Symb Logic 31*633-640
⋄ C55 C80 ⋄ REV MR 35#4091 Zbl 192.37 • ID 18032

YASUHARA, M. [1966] *Syntactical and semantical properties of generalized quantifiers* (J 0036) J Symb Logic 31*617-632
⋄ C40 C55 C80 ⋄ REV MR 35#4090 Zbl 192.37
• ID 18031

YASUHARA, M. [1968] *On minimal ordinal types of theories* (J 0028) Indag Math 30*87-94
⋄ C55 ⋄ REV MR 41#3261 Zbl 238#02047 • ID 18033

YASUHARA, M. [1970] *Incompleteness of L_p-languages* (J 0027) Fund Math 66*147-152
⋄ C55 C80 ⋄ REV MR 41#5204 Zbl 193.295 • ID 18034

YASUHARA, M. [1971] *On a problem of Mostowski on finite spectra* (J 0068) Z Math Logik Grundlagen Math 17*17-20
⋄ C13 ⋄ REV MR 44#2597 Zbl 223#02053 • ID 18035

YASUHARA, M. [1972] *The generic structures as a variety (Russian summary)* (J 0014) Bull Acad Pol Sci, Ser Math Astron Phys 20*609-614
⋄ C25 C52 ⋄ REV MR 56#5274 Zbl 254#02044
• ID 66173

YASUHARA, M. [1974] *The amalgamation property, the universal-homogeneous models, and the generic models* (J 0132) Math Scand 34*5-36
⋄ C25 C50 C52 ⋄ REV MR 51#7860 Zbl 298#02053
• ID 17697

YASUHARA, M. see Vol. V, VI for further entries

YASUMOTO, M. [1978] *Applications of Cherlin chains to set theory (Japanese)* (P 4109) B-Val Anal & Nonstand Anal;1978 Kyoto 24-37
⋄ C62 ⋄ ID 46810

YASUMOTO, M. [1980] *Classes on ZF models* (J 0090) J Math Soc Japan 32*615-621
⋄ C62 E30 E47 ⋄ REV MR 82d:03057 Zbl 441#03019
• ID 56072

YASUMOTO, M. [1983] *Nonstandard arithmetic of function fields over H-convex subfields of $^*Q$ (Japanese)* (P 4113) Found of Math;1982 Kyoto 230-236
⋄ H05 H15 ⋄ REV MR 85j:11028 Zbl 505#14019
• ID 47688

YASUMOTO, M. [1983] *Nonstandard arithmetic of function fields over H-convex subfields of $^*Q$* (J 0127) J Reine Angew Math 342*1-11
⋄ H05 H15 ⋄ REV MR 85g:11028 Zbl 505#14019
• ID 38208

YASUMOTO, M. see Vol. V for further entries

YOCOM, K.L. [1965] see JACOBSON, R.A.

YOUNG, L. [1972] *Functional analysis - a non-standard treatment with semifields* (P 0732) Contrib to Non-Standard Anal;1970 Oberwolfach 123-170
⋄ C65 H05 ⋄ REV MR 58#2161 Zbl 261#46002
• ID 18057

YUE, QIJING [1984] see SHEN, FUXING

YUTANI, H. [1978] *Categoricity in the theory of BCK-algebras* (J 1508) Math Sem Notes, Kobe Univ 6*133-135
⋄ C35 G25 ⋄ REV MR 80a:03084 Zbl 379#02021
• ID 51854

ZACHOROWSKI, S. [1977] *Intermediate logics without the interpolation property* (J 0387) Bull Sect Logic, Pol Acad Sci 6*161-163
⋄ B55 C40 ⋄ REV MR 57#15980 Zbl 404#03019
• ID 54806

ZACHOROWSKI, S. [1977] see KRZYSTEK, P.S.

ZACHOROWSKI, S. [1978] *Dummett's LC has the interpolation property* (J 0387) Bull Sect Logic, Pol Acad Sci 7*58-60
⋄ B55 C40 C90 ⋄ REV MR 58#16174 Zbl 407#03046
• ID 56212

ZACHOROWSKI, S. see Vol. II, V, VI for further entries

ZACHOS, S. [1985] see HINMAN, P.G.

ZACHOS, S. see Vol. IV for further entries

ZADROZNY, W. [1983] *Iterating ordinal definability* (J 0073) Ann Pure Appl Logic 24*263-310
⋄ C62 E35 E45 ⋄ REV MR 85h:03056 Zbl 536#03034
• ID 37113

ZADROZNY, W. [1984] *Ordinal definability in Jensen's model* (J 0036) J Symb Logic 49*608-620
⋄ C62 E45 ⋄ REV MR 85e:03122 • ID 40349

ZADROZNY, W. see Vol. V for further entries

ZAIONTZ, C. [1983] *Axiomatization of the monadic theory of ordinals $<\omega_2$* (J 0068) Z Math Logik Grundlagen Math 29*337-356
⋄ B25 B28 C65 C85 ⋄ REV MR 85j:03009 Zbl 553#03006 • ID 42214

ZAIONTZ, C. [1983] see BUECHI, J.R.

ZAKHAROV, D.A. [1961] *On the Los-Suszko theorem (Russian)* (J 0067) Usp Mat Nauk 16/2 (98)*200-201
⋄ C52 ⋄ ID 41199

ZAKHAROV, D.A. [1963] *Convex classes of models (Russian)* (J 0226) Uch Zap Ped Inst, Ivanovo 31*54-55
⋄ C52 ⋄ REV MR 48#97 • ID 18080

ZAKHAROV, D.A. [1963] *The ring of p-adic integers is elementary in the field of p-padic numbers (Russian)* (J 0226) Uch Zap Ped Inst, Ivanovo 34*31-33
⋄ C60 ⋄ REV MR 33#7260 • ID 18081

ZAKHAROV, D.A. see Vol. IV for further entries

ZAKHAR'YASHCHEV, M.V. [1980] see POPOV, S.V.

ZAKHAR'YASHCHEV, M.V. [1983] *On intermediate logics (Russian)* (J 0023) Dokl Akad Nauk SSSR 269*18-22
• TRANSL [1983] (J 0062) Sov Math, Dokl 27*274-277
⋄ B55 C90 ⋄ REV MR 85e:03060 Zbl 548#03009
• ID 40499

ZAKHAR'YASHCHEV, M.V. see Vol. I, II for further entries

ZAKIR'YANOV, K.KH. [1978] see BLOSHCHITSYN, V.YA.

ZAKON, E. [1960] see ROBINSON, A.

ZAKON, E. [1961] *Generalized archimedean groups* (J 0064) Trans Amer Math Soc 99*21-40
 ◊ C60 ◊ REV MR 22#11049 Zbl 118.15 • ID 18096

ZAKON, E. [1969] see ROBINSON, A.

ZAKON, E. [1969] *Remarks on the nonstandard real axis* (P 0649) Appl Model Th to Algeb, Anal & Probab;1967 Pasadena 195-227
 ◊ C65 H05 ◊ REV MR 39#183 Zbl 195.14 • ID 16299

ZAKON, E. [1974] *A new variant of non-standard analysis* (P 1083) Victoria Symp Nonstand Anal;1972 Victoria 313-339
 ◊ C20 H05 ◊ REV MR 57#16060 Zbl 289#02041
 • ID 30005

ZAKON, E. [1974] *Model-completeness and elementary properties of torsion free abelian groups* (J 0017) Canad J Math 26*829-840
 ◊ C35 C60 ◊ REV MR 50#81 Zbl 291#02039 • ID 18097

ZAKON, E. see Vol. V for further entries

ZALC, A. [1973] see ARTIGUE, M.

ZALC, A. [1975] see PERRIN, M.J.

ZALC, A. [1978] see ARTIGUE, M.

ZALC, A. see Vol. V, VI for further entries

ZAMYATIN, A.P. [1973] *A prevariety of semigroups whose elementary theory is solvable (Russian)* (J 0003) Algebra i Logika 12*417-432,492
 • TRANSL [1973] (J 0069) Algeb and Log 12*233-241
 ◊ B25 C05 ◊ REV MR 52#10401 Zbl 287#02031 Zbl 289#02034 • ID 21699

ZAMYATIN, A.P. [1976] *Varieties of associative rings whose elementary theory is decidable (Russian)* (J 0023) Dokl Akad Nauk SSSR 229*276-279
 • TRANSL [1976] (J 0062) Sov Math, Dokl 17*996-999
 ◊ B25 C05 C60 ◊ REV MR 54#12513 Zbl 356#02041
 • ID 33039

ZAMYATIN, A.P. [1978] *A non-abelian variety of groups has an undecidable elementary theory(Russian)* (J 0003) Algebra i Logika 17*20-27
 • TRANSL [1978] (J 0069) Algeb and Log 17*13-17
 ◊ C05 C60 D35 ◊ REV MR 80c:03045 Zbl 408#03011
 • ID 90332

ZAMYATIN, A.P. [1978] *Prevarieties of associative rings whose elementary theory is decidable (Russian)* (J 0092) Sib Mat Zh 19*1266-1282,1437
 • TRANSL [1978] (J 0475) Sib Math J 19*890-901
 ◊ B25 C05 C60 ◊ REV MR 80i:16049 Zbl 417#03012
 • ID 53249

ZAMYATIN, A.P. [1983] *Decidability of the elementary theories of certain varieties of rings (Russian)* (J 0340) Mat Zap (Univ Sverdlovsk) 13/3*52-74
 ◊ B25 C05 C60 ◊ REV MR 85h:03016 Zbl 553#17001
 • ID 42741

ZARACH, A. [1976] *Generic extension of admissible sets* (P 1476) Set Th & Hierarch Th (2) (Mostowski);1975 Bierutowice 321-333
 ◊ C62 E25 E35 F35 ◊ REV MR 55#2574
 Zbl 357#02060 • ID 23816

ZARACH, A. [1977] *Extension of ZF-models to models with the scheme of choice* (J 0027) Fund Math 96*87-89
 ◊ C15 C62 E25 E30 E35 ◊ REV MR 58#21608
 Zbl 384#03038 • ID 26532

ZARACH, A. [1982] *Unions of ZF^--models which are themselves ZF^--models* (P 3623) Logic Colloq;1980 Prague 315-342
 ◊ C62 E30 E35 ◊ REV MR 84h:03086 Zbl 524#03039
 • ID 34278

ZARACH, A. see Vol. II, IV, V for further entries

ZAWADOWSKI, M. [1983] *The Skolem-Loewenheim theorem in toposes* (J 0063) Studia Logica 42*461-475
 ◊ C55 C80 E10 G30 ◊ REV MR 86b:03085
 Zbl 561#03021 • REM Part I. Part II 1985 • ID 42355

ZAWADOWSKI, M. [1985] see JANKOWSKI, A.W.

ZAWADOWSKI, M. [1985] *The Skolem-Loewenheim theorem in toposes II* (J 0063) Studia Logica 44*25-38
 ◊ C55 C90 G30 ◊ REM Part I 1983 • ID 47521

ZAWISZA, B. [1984] see KRUPA, A.

ZAYED, M. [1982] *Characterisation des algebres de representation finie sur des corps algebriquement clos* (S 2250) Semin Dubreil: Algebre 34*129-147
 ◊ C20 C60 ◊ REV MR 83j:16043 Zbl 498#16025
 • ID 38529

ZBIERSKI, P. [1971] *Models for higher order arithmetics (Russian summary)* (J 0014) Bull Acad Pol Sci, Ser Math Astron Phys 19*557-562
 ◊ C62 E30 F35 ◊ REV MR 46#7022 Zbl 236#02043
 • ID 14379

ZBIERSKI, P. [1974] see MAREK, W.

ZBIERSKI, P. [1977] see MAREK, W.

ZBIERSKI, P. [1978] *Axiomatizability of second order arithmetic with ω-rule* (J 0027) Fund Math 100*51-57
 ◊ B15 C62 E45 E70 F35 ◊ REV MR 58#5161
 Zbl 387#03021 • ID 29200

ZBIERSKI, P. [1978] see MAREK, W.

ZBIERSKI, P. [1980] *Indicators and incompleteness of Peano arithmetic* (J 2085) Acta Cient Venez 31*487-495
 ◊ C62 F30 ◊ REV MR 84g:03095 Zbl 526#03019
 • ID 34199

ZBIERSKI, P. [1980] see MAREK, W.

ZBIERSKI, P. [1981] *Nonstandard interpretations of higher order theories* (J 0027) Fund Math 112*175-186
 ◊ C62 F35 H15 ◊ REV MR 84d:03073 Zbl 481#03046
 • ID 34106

ZBIERSKI, P. see Vol. V for further entries

ZEITLER, H. [1981] *Die Amalgamierungseigenschaft und reine Ringe* (J 0009) Arch Math Logik Grundlagenforsch 21*83-100
 ◊ C52 C60 ◊ REV MR 82j:03036 Zbl 466#03010
 • ID 54961

ZEMAN, J.J. [1969] *Decision procedures for S3⁰ and S4⁰*
(J 0009) Arch Math Logik Grundlagenforsch 12∗155-158
◊ B25 B45 ◊ REV MR 43 # 4645 Zbl 195.12 • ID 14399

ZEMAN, J.J. see Vol. II, VI for further entries

ZERMELO, E. [1931] *Ueber Stufen der Quantifikation und die Logik des Unendlichen* (J 0157) Jbuchber Dtsch Math-Ver 41∗85-88,2.Abt.
◊ A05 C75 D35 E07 E30 ◊ REV FdM 58.60 • ID 14412

ZERMELO, E. [1935] *Grundlagen einer allgemeinen Theorie der mathematischen Satzsysteme (erste Mitteilung)* (J 0027) Fund Math 25∗136-146
◊ B05 C75 E20 ◊ REV Zbl 12.241 FdM 61.972
• ID 14413

ZERMELO, E. see Vol. I, V for further entries

ZHALDOKAS, R. [1983] *An approach to the construction of complete term rewriting systems (Russian) (English and Lithuanian summaries)* (J 3939) Mat Logika Primen (Akad Nauk Litov SSR) 3∗9-22
◊ B25 B35 B75 C05 ◊ REV MR 85h:68043
Zbl 558 # 68029 • ID 43443

ZHANG, JINWEN [1979] *A nonstandard model of arithmetic constructed by means of a forcing method (Chinese) (English summary)* (J 2754) Huazhong Gongxueyuan Xuebao 7/1∗1,7-12
• TRANSL [1979] (J 2684) J Huazhong Inst Tech (Engl Ed) 1/1∗11-18
◊ E40 H15 ◊ REV MR 84g:03110ab • ID 34211

ZHANG, JINWEN [1980] *A family of models of lattice-valued sets (Chinese) (English summary)* (J 2754) Huazhong Gongxueyuan Xuebao 8∗1-2,17-26
◊ C62 C90 E70 ◊ REV MR 84k:03128 • ID 36145

ZHANG, JINWEN see Vol. II, IV, V, VI for further entries

ZHEGALKIN, I.I. [1939] *Sur l'Entscheidungsproblem (Russian) (French summary)* (J 0142) Mat Sb, Akad Nauk SSSR, NS 6(48)∗185-198
◊ B20 B25 D35 ◊ REV MR 1.322 Zbl 22.193 JSL 5.69
FdM 65.27 • ID 17903

ZHEGALKIN, I.I. [1946] *Sur le probleme de la resolubilite pour les classes finies (Russian) (French summary)* (S 1498) Uch Zap Univ, Moskva 100/1∗155-211
◊ B20 B25 C13 ◊ REV MR 12.2 JSL 17.271 • ID 04825

ZHEGALKIN, I.I. see Vol. I, V, VI for further entries

ZHELEVA, S. [1971] see T"RKALANOV, K.D.

ZIEGLER, M. [1972] *Die elementare Theorie der Henselschen Koerper* (0000) Diss., Habil. etc
◊ B25 C10 C35 C60 ◊ REM Diss., Koeln • ID 19785

ZIEGLER, M. [1975] *A counterexample in the theory of definable automorphisms* (J 0048) Pac J Math 58∗665-668
◊ C07 C60 ◊ REV MR 51 # 12511 Zbl 326 # 02040
• ID 17237

ZIEGLER, M. [1975] see PRESTEL, A.

ZIEGLER, M. [1975] see RAUTENBERG, W.

ZIEGLER, M. [1976] *A language for topological structures which satisfies a Lindstroem-theorem* (J 0015) Bull Amer Math Soc 82∗568-570
◊ B25 C40 C50 C65 C80 C90 C95 D35 ◊ REV
MR 54 # 4971 Zbl 338 # 02007 • ID 24137

ZIEGLER, M. [1978] *Definable bases of monotone systems* (P 1897) Logic Colloq;1977 Wroclaw 297-311
◊ C40 C80 ◊ REV MR 80b:03048 Zbl 453 # 03034
• ID 54164

ZIEGLER, M. [1978] see PRESTEL, A.

ZIEGLER, M. [1978] see PODEWSKI, K.-P.

ZIEGLER, M. [1979] see SHELAH, S.

ZIEGLER, M. [1980] *Algebraisch abgeschlossene Gruppen (English summary)* (P 2634) Word Problems II;1976 Oxford 449-576
◊ C25 C60 D30 D40 ◊ REV MR 82b:20004
Zbl 451 # 20001 • ID 54050

ZIEGLER, M. [1980] see PRESTEL, A.

ZIEGLER, M. [1980] see FLUM, J.

ZIEGLER, M. [1981] see MAKOWSKY, J.A.

ZIEGLER, M. [1982] *Einige unentscheidbare Koerpertheorien* (J 3370) Enseign Math, Ser 2 28∗269-280
• REPR [1982] (P 3482) Logic & Algor (Specker);1980 Zuerich 381-392
◊ C60 D35 ◊ REV MR 83m:03040 Zbl 499 # 03002
Zbl 519 # 12018 JSL 50.552 • ID 35441

ZIEGLER, M. [1982] see EBBINGHAUS, H.-D.

ZIEGLER, M. [1984] *Model theory of modules* (J 0073) Ann Pure Appl Logic 26∗149-213
◊ B25 C10 C45 C60 ◊ REV MR 86c:03034 • ID 40461

ZIEGLER, M. [1985] *Topological model theory* (C 4183)
Model-Theor Log 557-577
◊ C90 C98 ◊ ID 48341

ZIEGLER, M. see Vol. IV for further entries

ZIL'BER, B.I. [1970] *On the isomorphism and elementary equivalence of commutative semigroups (Russian)* (J 0003) Algebra i Logika 9∗667-671
• TRANSL [1970] (J 0069) Algeb and Log 9∗400-402
◊ C05 C60 ◊ REV MR 45 # 2065 Zbl 228 # 20030
• ID 14418

ZIL'BER, B.I. [1971] *An example of two elementarily equivalent but not isomorphic finitely generated metabelian groups (Russian)* (J 0003) Algebra i Logika 10∗309-315
• TRANSL [1971] (J 0069) Algeb and Log 10∗192-197
◊ C60 ◊ REV MR 45 # 49 Zbl 245 # 20002 • ID 19784

ZIL'BER, B.I. [1974] *Rings with $\aleph_1$-categorical theories (Russian)*
(J 0003) Algebra i Logika 13∗168-187,235
• TRANSL [1974] (J 0069) Algeb and Log 13∗95-104
◊ C35 C60 ◊ REV MR 51 # 2897 Zbl 319 # 02049
• ID 17341

ZIL'BER, B.I. [1974] see BELEGRADEK, O.V.

ZIL'BER, B.I. [1974] *The transcendental rank of the formulas of an $\aleph_1$-categorical theory (Russian)* (J 0087) Mat Zametki (Akad Nauk SSSR) 15∗321-329
• TRANSL [1974] (J 1044) Math Notes, Acad Sci USSR 15∗182-186
◊ C35 C45 ◊ REV MR 50 # 1872 Zbl 298 # 02056
• ID 14419

ZIL'BER, B.I. [1977] *Groups and rings, the theory of which is categorical (Russian) (English summary)* (J 0027) Fund Math 95*173-188
⋄ C35 C45 C60 ⋄ REV MR 56#118 Zbl 363#02060 JSL 49.317 • ID 50886

ZIL'BER, B.I. [1977] *The structure of a model of a categorical theory and finite-axiomatizability problem (Russian)* (X 2235) VINITI: Moskva 2800-77
⋄ C35 C52 ⋄ ID 33386

ZIL'BER, B.I. [1979] *Definability of algebraic systems and the theory of categoricity (Russian)* (P 2558) All-Union Conf Math Log (5) (Mal'tsev);1979 Novosibirsk 55
⋄ C35 C45 ⋄ ID 90347

ZIL'BER, B.I. [1980] *Solution of the problem of finite axiomatizability for theories that are categorical in all infinite powers (Russian)* (C 2620) Teor Model & Primen 47-60
⋄ C35 C45 C52 ⋄ REV MR 82g:03051 • ID 80274

ZIL'BER, B.I. [1980] *Strongly minimal countably categorical theories (Russian)* (J 0092) Sib Mat Zh 21/2*98-112,237
• TRANSL [1980] (J 0475) Sib Math J 21*219-230
⋄ C35 C45 C52 ⋄ REV MR 82c:03043 Zbl 486#03017
• REM Part II 1984 • ID 80275

ZIL'BER, B.I. [1980] *Totally categorical theories: Structural properties and the non-finite axiomatizability* (P 2625) Model Th of Algeb & Arithm;1979 Karpacz 381-410
⋄ C35 C45 C52 ⋄ REV MR 82m:03045 Zbl 472#03025
• REM For err. see Zil'ber 1981 ID 36112 • ID 55290

ZIL'BER, B.I. [1981] *Solution of the problem of finite axiomatizability for theories that are categorical in all infinite powers (Russian)* (C 3806) Issl Teor Progr 69-75
⋄ B30 C35 C45 C52 ⋄ REV MR 84k:03095 Zbl 522#03019 • ID 36112

ZIL'BER, B.I. [1981] *Some problems concerning categorical theories (Russian summary)* (J 3293) Bull Acad Pol Sci, Ser Math 29*47-49
⋄ C35 C45 ⋄ REV MR 84h:03080 Zbl 467#03023
• ID 55021

ZIL'BER, B.I. [1981] *Totally categorical structures and combinatorial geometries (Russian)* (J 0023) Dokl Akad Nauk SSSR 259*1039-1041
• TRANSL [1981] (J 0062) Sov Math, Dokl 24*149-151
⋄ C35 C45 ⋄ REV MR 84f:51017 Zbl 485#51004
• ID 38500

ZIL'BER, B.I. [1982] *Uncountable categorical nilpotent groups and Lie algebras (Russian)* (J 0031) Izv Vyssh Ucheb Zaved, Mat (Kazan) 1982/5*75
• TRANSL [1982] (J 3449) Sov Math 26/5*98-99
⋄ C35 C60 ⋄ REV Zbl 516#20019 • ID 38581

ZIL'BER, B.I. [1984] *Some model theory of simple algebraic groups over algebraically closed fields* (S 0019) Colloq Math (Warsaw) 48*173-180
⋄ C60 ⋄ REV MR 86f:03055 Zbl 567#20030 • ID 44228

ZIL'BER, B.I. [1984] *Strongly minimal countably categorical theories II (Russian)* (J 0092) Sib Mat Zh 25/3*71-78
• TRANSL [1984] (J 0475) Sib Math J 25*396-412
⋄ C35 C45 • REM Part I 1980. Part III 1984. • ID 40268

ZIL'BER, B.I. [1984] *Strongly minimal countably categorical theories. III (Russian)* (J 0092) Sib Mat Zh 25/4*63-77
• TRANSL [1984] (J 0475) Sib Math J 25*559-571
⋄ C35 C45 • REM Part II 1984 • ID 41868

ZIL'BER, B.I. [1984] *The structure of models of uncountably categorical theories* (P 4313) Int Congr Math (II,14);1983 Warsaw 1*359-368
⋄ C35 C98 ⋄ ID 48351

ZINOV'EV, A.A. [1972] *Semantische Entscheidbarkeit der Quantorentheorie* (C 1533) Quantoren, Modal, Paradox 206-212
⋄ A05 B25 ⋄ REV Zbl 255#02008 • ID 28956

ZINOV'EV, A.A. see Vol. I, II, V, VI for further entries

ZLATOS, P. [1979] *A characterization of some boolean powers* (J 0008) Arch Math (Basel) 33*133-143
⋄ C05 C20 C30 C75 ⋄ REV MR 81m:08004 Zbl 406#03054 • ID 53067

ZLATOS, P. [1981] *Two-levelled logic and model theory* (P 2552) Conf Finite Algeb & Multi-Val Log;1979 Szeged 825-872
⋄ B50 C90 G25 ⋄ REV MR 83e:03057 Zbl 478#03019
• ID 55634

ZLATOS, P. [1985] *On conceptual completeness of syntactic-semantical systems* (J 0049) Period Math Hung 16*145-173
⋄ C05 C20 C90 G15 G30 ⋄ REV Zbl 572#03047
• ID 49238

ZLATOS, P. see Vol. V for further entries

ZUBENKO, V.V. [1979] *Standard program algebras (Ukrainain) (English and Russian summaries)* (J 0270) Dokl Akad Nauk Ukr SSR, Ser A 1979*846-848,881
⋄ B25 B75 G25 ⋄ REV MR 81d:68017 Zbl 412#68031
• ID 52981

ZUBENKO, V.V. see Vol. IV for further entries

ZUBIETA RUSSI, G. [1956] *Definiciones formales de numerabilidad* (J 3127) Bol Soc Mat Mexicana, Ser 2 1*49-56
⋄ C07 ⋄ REV MR 19.240 Zbl 111.8 JSL 28.251 • ID 22337

ZUBIETA RUSSI, G. [1957] *Clases aritmeticas definidas sin igualdad* (J 3127) Bol Soc Mat Mexicana, Ser 2 2*45-53
⋄ B10 B20 C07 C52 ⋄ REV MR 20#2272 Zbl 88.12 JSL 29.55 • ID 11688

ZUBIETA RUSSI, G. see Vol. I for further entries

ZYAPKOV, N.P. [1981] *Reduction theorems for universally consistent fields (Russian)* (J 2503) PLISKA Stud Math Bulgar 2*124-129
⋄ C60 ⋄ REV MR 83h:12037 Zbl 494#12018 • ID 39201

ZYGMUNT, J. [1971] *A survey of the methods of proof of the Goedel-Malcev's completeness theorem* (P 2044) Int Congr Log, Meth & Phil of Sci (4,Abstr);1971 Bucharest 165-238
⋄ A10 C07 C98 ⋄ ID 31579

ZYGMUNT, J. [1973] *On the sources of the notion of the reduced product* (J 0302) Rep Math Logic, Krakow & Katowice 1*53-67
⋄ A10 C20 ⋄ REV MR 49#7136 Zbl 292#02043
• ID 14461

ZYGMUNT, J. [1974] *A note on direct products and ultraproducts of logical matrices* (J 0063) Studia Logica 33*349-357
⋄ B50 C05 C20 ⋄ REV MR 50#12656 Zbl 312#02039
• ID 14462

ZYGMUNT, J. [1981] *Notes on decidability and finite approximability of sentential logics* (J 0387) Bull Sect Logic, Pol Acad Sci 10*38-41
  ⋄ A10 B22 B25 ⋄ REV MR 82f:03024 Zbl 457 # 03008
  • ID 54333

ZYGMUNT, J. [1983] *On decidability and finite approximability of sentential logics* (J 0481) Acta Univ Wroclaw 517(29)(Logika 8)*69-81
  ⋄ B22 B25 ⋄ REV Zbl 547 # 03005 • ID 41843

ZYGMUNT, J. see Vol. I, II, VI for further entries

ZYKOV, A.A. [1953] *The spectrum problem in the extended predicate calculus (Russian)* (J 0216) Izv Akad Nauk SSSR, Ser Mat 17*63-76
  • TRANSL [1956] (J 0225) Amer Math Soc, Transl, Ser 2 3*1-14
  ⋄ B15 C13 ⋄ REV MR 14.936 Zbl 50.7 JSL 22.360
  • ID 19773

ZYKOV, A.A. see Vol. I, IV for further entries

Source Index

Journals

J 0001 Acta Math Acad Sci Hung • H
Acta Mathematica Academiae Scientiarum Hungaricae
[1950-1982] ISSN 0001-5954
• CONT AS (**J 4729**) Acta Math Hung

J 0002 Acta Sci Math (Szeged) • H
Acta Scientiarum Mathematicarum [1947ff] ISSN 0001-6969
• CONT OF (**J 0460**) Acta Univ Szeged, Sect Mat

J 0003 Algebra i Logika • SU
Algebra i Logika (Algebra and Logic) [1962ff] ISSN 0373-9252
• TRANSL IN (**J 0069**) Algeb and Log

J 0004 Algeb Universalis • CDN
Algebra Universalis [1970ff] ISSN 0002-5240

J 0005 Amer Math Mon • USA
American Mathematical Monthly [1894ff] ISSN 0002-9890

J 0006 Ann Univ Budapest, Sect Math • H
Annales Universitatis Scientiarum Budapestinensis. Sectio Mathematica [1958ff] ISSN 0524-9007

J 0007 Ann Math Logic • NL
Annals of Mathematical Logic [1970-1982] ISSN 0003-4843
• CONT AS (**J 0073**) Ann Pure Appl Logic

J 0008 Arch Math (Basel) • CH
*Archiv der Mathematik * Archives of Mathematics * Archives Mathematiques* [1948ff] ISSN 0003-889X

J 0009 Arch Math Logik Grundlagenforsch • D
Archiv fuer Mathematische Logik und Grundlagenforschung [1950ff] ISSN 0003-9268

J 0012 Boll Unione Mat Ital, IV Ser • I
Bolletino della Unione Matematica Italiana. Serie IV [1968-1975] ISSN 0041-7084
• CONT OF (**J 4408**) Boll Unione Mat Ital, III Ser • CONT AS (**J 3285**) Boll Unione Mat Ital, V Ser, A & (**J 3495**) Boll Unione Mat Ital, V Ser, B

J 0013 Brit J Phil Sci • GB
British Journal for the Philosophy of Science [1950ff] ISSN 0007-0882

J 0014 Bull Acad Pol Sci, Ser Math Astron Phys • PL
Bulletin de l'Academie Polonaise des Sciences. Serie des Sciences Mathematiques, Astronomiques et Physiques [1953-1978] ISSN 0001-4117
• CONT AS (**J 3293**) Bull Acad Pol Sci, Ser Math

J 0015 Bull Amer Math Soc • USA
Bulletin of the American Mathematical Society [1894-1978] ISSN 0002-9904
• CONT AS (**J 0589**) Bull Amer Math Soc (NS)

J 0016 Bull Austral Math Soc • AUS
Bulletin of the Australian Mathematical Society [1969ff] ISSN 0004-9727

J 0017 Canad J Math • CDN
*Canadian Journal of Mathematics * Journal Canadien de Mathematiques* [1949ff] ISSN 0008-414X

J 0018 Canad Math Bull • CDN
*Canadian Mathematical Bulletin * Bulletin Canadien de Mathematiques* [1958ff] ISSN 0008-4395

J 0020 Compos Math • NL
Compositio Mathematica [1933ff] ISSN 0010-437X

J 0021 Cybernetics • USA
Cybernetics [1965ff] ISSN 0011-4235
• TRANSL OF (**J 0040**) Kibernetika, Akad Nauk Ukr SSR

J 0022 Cheskoslov Mat Zh • CS
*Cheskoslovatskij Matematicheskij Zhurnal * Czechoslovak Mathematical Journal* [1951ff] ISSN 0011-4642
• REM From Vol. 19 (1969) on the title is only: Czechoslovak Mathematical Journal

J 0023 Dokl Akad Nauk SSSR • SU
Doklady Akademii Nauk SSSR (Reports of the Academy of Sciences of the USSR) [1933ff] ISSN 0002-3264
• TRANSL IN (**J 0062**) Sov Math, Dokl & (**J 0470**) Sov Phys, Dokl

J 0025 Duke Math J • USA
Duke Mathematical Journal [1935ff] ISSN 0012-7094

J 0027 Fund Math • PL
Fundamenta Mathematicae [1920ff] ISSN 0016-2736

J 0028 Indag Math • NL
*Indagationes Mathematicae * Nederlandse Akademie van Wetenschappen. Proceedings* [1939ff] ISSN 0019-3577, ISSN 0023-3358
• REM Until 1950 part of Koninklijke Nederlandsche Akademie van Wetenschappen, Proceedings of the Section of Sciences; vol n+41 with separate pagination. Since 1951 same as Koninklijke Nederlandse Akademie van Wetenschappen, Proceedings of the Section of Sciences, Series A; vol n+41. Before 1951 page numbers in Proceedings and Indagationes different. Since 1951 the same page numbers as Proceedings Series A

J 0029 Israel J Math • IL
Israel Journal of Mathematics [1963ff] ISSN 0021-2172
• CONT OF (**J 0493**) Bull Res Counc Israel Sect F

J 0030 Izv Akad Nauk Uzb SSR, Ser Fiz-Mat • SU
*Izvestiya Akademii Nauk UzSSR. Seriya Fiziko-Matematicheskikh Nauk (Fanlar Akademijasining Achboroti. Fizika-Matematika Fanlaris Serijas * Proceedings of the Academy of Sciences of the Uzb SSR. Series: Physics and Mathematics)* [1957ff] ISSN 0568-7144

J 0031 Izv Vyssh Ucheb Zaved, Mat (Kazan) • SU
Izvestiya Vysshikh Uchebnykh Zavedenij. Matematika (Proceedings of the University. Mathematics) [1957ff] ISSN 0021-3446
• TRANSL IN (**J 3449**) Sov Math

J 0032 J Algeb • USA
Journal of Algebra [1964ff] ISSN 0021-8693

J 0033 J Comb Th, Ser B • USA
Journal of Combinatorial Theory. Series B [1971ff] ISSN 0095-8956
• CONT OF (J 1669) J Comb Th

J 0035 J Math Psychol • USA
Journal of Mathematical Psychology [1964ff] ISSN 0022-2496

J 0036 J Symb Logic • USA
The Journal of Symbolic Logic [1936ff] ISSN 0022-4812

J 0037 ACM J • USA
Journal of the ACM (=Association for Computing Machinery) [1954ff] ISSN 0004-5411

J 0038 J Austral Math Soc • AUS
Journal of the Australian Mathematical Society [1959-1975] ISSN 0004-9735
• CONT AS (J 3194) J Austral Math Soc, Ser A

J 0039 J London Math Soc • GB
The Journal of the London Mathematical Society [1926-1968]
• CONT AS (J 3172) J London Math Soc, Ser 2

J 0040 Kibernetika, Akad Nauk Ukr SSR • SU
Kibernetika. Akademiya Nauk Ukrainskoj SSR (Cybernetics. Academy of Sciences of the Ukrainian SSR) [1965ff] ISSN 0023-1274
• TRANSL IN (J 0021) Cybernetics

J 0041 Math Syst Theory • D
Mathematical Systems Theory. An International Journal [1967ff] ISSN 0025-5661

J 0042 Mat Vesn, Drust Mat Fiz Astron Serb • YU
Matematichki Vesnik. Drushtvo Matematichara, Fizichara i Astromoma SR Serbije, SFR Jugoslavija (Mathematical Publications. Society Serbe of Mathematicians, Physicists and Astronomers) [1964ff] ISSN 0025-5165
• CONT OF (J 4277) Vesn Drusht Mat Fiz Serbije

J 0043 Math Ann • D
Mathematische Annalen [1868ff] ISSN 0025-5831

J 0044 Math Z • D
Mathematische Zeitschrift [1918ff] ISSN 0025-5874

J 0045 Monatsh Math • A
Monatshefte fuer Mathematik [1943ff] ISSN 0026-9255
• CONT OF (J 0124) Monatsh Math-Phys

J 0046 Nieuw Arch Wisk • NL
Nieuw Archief voor Wiskunde [1875-1893]
• CONT AS (J 1793) Nieuw Arch Wisk, Ser 2

J 0047 Notre Dame J Formal Log • USA
Notre Dame Journal of Formal Logic [1960ff] ISSN 0029-4527

J 0048 Pac J Math • USA
Pacific Journal of Mathematics [1951ff] ISSN 0030-8730

J 0049 Period Math Hung • H
Periodica Mathematica Hungarica [1971ff] ISSN 0031-5303

J 0050 Port Math • P
Portugaliae Mathematica [1937ff] ISSN 0032-5155

J 0051 Commentat Math, Ann Soc Math Pol, Ser 1 • PL
*Annales Societatis Mathematicae Polonae. Series I. Commentationes Mathematicae * Roczniki Polskiego Towarzystwa Matematycznego. Seria I. Prace Matematyczne.* [1955ff] ISSN 0079-368X
• CONT OF (J 0611) Pol Tow Mat, Prace Mat-Fiz

J 0052 Probl Kibern • SU
Problemy Kibernetiki. Glavnaya Redaktsiya Fiziko-Matematicheskoj Literatury [1958ff] ISSN 0555-277X
• TRANSL IN (J 0471) Syst Th Res & (J 0449) Probl Kybern & (J 1195) Probl Cybernet

J 0053 Proc Amer Math Soc • USA
Proceedings of the American Mathematical Society [1950ff] ISSN 0002-9939

J 0054 Proc Nat Acad Sci USA • USA
Proceedings of the National Academy of Sciences of the United States of America [1915ff] ISSN 0027-8424

J 0056 Publ Dep Math, Lyon • F
Publications du Departement de Mathematiques. Faculte des Sciences de Lyon. [1964-1981] ISSN 0076-1656
• CONT AS (J 2107) Publ Dep Math, Lyon, NS

J 0057 Publ Math (Univ Debrecen) • H
Publicationes Mathematicae [1949ff] ISSN 0033-3883

J 0058 Rend Circ Mat Palermo • I
Rendiconti del Circolo Matematico di Palermo [1887-1940]
• CONT AS (J 3522) Rend Circ Mat Palermo, Ser 2

J 0059 Rend Sem Mat Fis Milano • I
Rendiconti del Seminario Matematico e Fisico di Milano. Sotto gli Auspici dell'Universita e del Politecnico [1927ff]

J 0060 Rev Roumaine Math Pures Appl • RO
Revue Roumaine de Mathematiques Pures et Appliquees. Academia Republicii Socialiste Romania [1956ff] ISSN 0035-3965

J 0061 Simon Stevin • B
Simon Stevin: A Quarterly Journal of Pure and Applied Mathematics [1921ff] ISSN 0037-5454
• CONT OF (J 1928) Mathematica B & (J 0291) Wis-natuur Tijdsch & (J 1379) Christiaan Huygens Internat Math Tijdschr
• REM The journal has incorporated J1928, J1379, J0291

J 0062 Sov Math, Dokl • USA
Soviet Mathematics. Doklady. [1960ff] ISSN 0038-5573
• TRANSL OF (J 0023) Dokl Akad Nauk SSSR

J 0063 Studia Logica • PL
Studia Logica [1953ff] ISSN 0039-3215

J 0064 Trans Amer Math Soc • USA
Transactions of the American Mathematical Society [1900ff] ISSN 0002-9947

J 0065 Tr Moskva Mat Obshch • SU
Trudy Moskovskogo Matematicheskogo Obshchestva (Publications of the Moscow Mathematical Society) [1952ff] ISSN 0134-8663
• TRANSL IN (J 3279) Trans Moscow Math Soc

J 0067 Usp Mat Nauk • SU
Uspekhi Matematicheskikh Nauk (Advances in Mathematical Sciences) [1936ff] ISSN 0042-1316
• TRANSL IN (J 1399) Russ Math Surv

J 0068 Z Math Logik Grundlagen Math • DDR
Zeitschrift fuer Mathematische Logik und Grundlagen der Mathematik [1955ff] ISSN 0044-3050

J 0069 Algeb and Log • USA
Algebra and Logic [1968ff] ISSN 0002-5232
• TRANSL OF (J 0003) Algebra i Logika

J 0070 Bull Soc Sci Math Roumanie, NS • RO
Bulletin Mathematique de la Societe des Sciences Mathematiques de la Republique Socialiste de Roumanie. Nouvelle Serie. [1957ff] ISSN 0007-4691
• CONT OF (J 0494) Bull Math Soc Sci Roumanie

J 0073 Ann Pure Appl Logic • NL
Annals of Pure and Applied Logic [1983ff] ISSN 0168-0072
• CONT OF (J 0007) Ann Math Logic

J 0075 Phil Phenom Research • USA
Philosophy and Phenomenological Research [1940ff] ISSN 0031-8205

J 0076 Dialectica • CH
Dialectica. International Review of Philosophy of Knowledge [1947ff] ISSN 0012-2017

J 0077 Proc London Math Soc • GB
Proceedings of the London Mathematical Society [1865-1903]
• CONT AS (J 1910) Proc London Math Soc, Ser 2

J 0079 Logique & Anal, NS • B
Logique et Analyse. Nouvelle Serie. Publication Trimestrielle du Centre National Belge de Recherche de Logique [1958ff] ISSN 0024-5836

J 0081 Proc Japan Acad • J
Proceedings of the Japan Academy [1925-1977] ISSN 0021-4280
• CONT AS (J 3239) Proc Japan Acad, Ser A

J 0082 Bull Soc Math Belg • B
Bulletin de la Societe Mathematique de Belgique [1948-1976] ISSN 0037-9476
• CONT AS (J 3133) Bull Soc Math Belg, Ser B & (J 3824) Bull Soc Math Belg, Ser A

J 0084 Gaz Mat (Lisboa) • P
Gazeta de Matematica. Jornal dos Concorrentes ao Exame de Aptidao e dos Estudantes de Matematica das Escolas Superiores [1940ff]

J 0086 Cas Pestovani Mat, Ceskoslov Akad Ved • CS
Casopis pro Pestovani Matematiky. Ceskoslovenska Akademie Ved (Journal for the Cultivation of Mathematics. Czechoslovak Academy of Sciences) [1872ff]

J 0087 Mat Zametki (Akad Nauk SSSR) • SU
Matematicheskie Zametki (Mathematical Notes) [1967ff] ISSN 0025-567X
• TRANSL IN (J 1044) Math Notes, Acad Sci USSR

J 0088 Ann Univ Ferrara, NS, Sez 7 • I
Annali dell'Universita di Ferrara. Nuova Serie. Sezione 7. Scienze Matematiche [1966ff]

J 0090 J Math Soc Japan • J
Journal of the Mathematical Society of Japan [1885ff] ISSN 0025-5645

J 0091 Sugaku • J
Sugaku (Mathematics) [1947ff] ISSN 0039-470X

J 0092 Sib Mat Zh • SU
Sibirskij Matematicheskij Zhurnal. Akademiya Nauk SSSR. Sibirskoe Otdelenie (Siberian Mathematical Journal. Academy of Sciences of the USSR. Siberian Section) [1960ff] ISSN 0037-4474
• TRANSL IN (J 0475) Sib Math J

J 0093 J Comp Syst • USA
The Journal of Computing Systems [1952-1954]

J 0094 Mind • GB
Mind. A Quarterly Review of Philosophy [1876ff] ISSN 0026-4423

J 0096 Acta Philos Fenn • SF
Acta Philosophica Fennica [1948ff] ISSN 0355-1792

J 0097 Nous, Quart J Phil • USA
Nous. A Quarterly Journal of Philosophy [1967ff] ISSN 0029-4624

J 0100 Amer J Math • USA
American Journal of Mathematics [1878ff] ISSN 0002-9327

J 0102 Hiroshima Univ J Sci, Ser A Math • J
Hiroshima University. Journal of Science. Series A. Section 1. Mathematics [1930-1970] ISSN 0018-2079
• CONT AS (J 0431) Hiroshima Math J

J 0103 Analysis (Oxford) • GB
Analysis [1933ff] ISSN 0003-2638

J 0104 Atti Accad Sci Lett Arti Palermo, Ser 4/I • I
Atti della Accademia di Scienze Lettere e Arti di Palermo. Serie Quarta. Parte I: Scienze [1940-1983]
• CONT AS (J 4186) Atti Accad Sci Lett Arti Palermo, Ser 5/I

J 0105 Theoria (Lund) • S
Theoria. A Swedish Journal of Philosophy [1934ff] ISSN 0040-5825

J 0107 Abh Math Sem Univ Hamburg • D
Abhandlungen aus dem Mathematischen Seminar der Universitaet Hamburg [1922ff] ISSN 0025-5858

J 0109 C R Acad Sci, Paris • F
Academie des Sciences de Paris. Comptes Rendus Hebdomadaires des Seances [1835-1965]
• CONT AS (J 2313) C R Acad Sci, Paris, Ser A-B

J 0110 Anais Acad Bras Cienc • BR
Anais da Academia Brasileira de Ciencias. [1929ff] ISSN 0001-3765

J 0111 Nagoya Math J • J
Nagoya Sugaku Zashi (Nagoya Mathematical Journal) [1950ff] ISSN 0027-7630

J 0114 Math Nachr • DDR
Mathematische Nachrichten [1948ff] ISSN 0025-584X

J 0115 Wiss Z Humboldt-Univ Berlin, Math-Nat Reihe • DDR
Wissenschaftliche Zeitschrift der Humboldt-Universitaet Berlin. Mathematisch-Naturwissenschaftliche Reihe [1951ff] ISSN 0043-6852

J 0118 Acta Math • S
Acta Mathematica [1882ff] ISSN 0001-5962

J 0119 J Comp Syst Sci • USA
Journal of Computer and System Sciences [1967ff] ISSN 0022-0000

J 0120 Ann of Math, Ser 2 • USA
Annals of Mathematics. 2nd Series [1899ff] ISSN 0003-486X

J 0122 J Philos Logic • NL
Journal of Philosophical Logic [1972ff] ISSN 0022-3611

J 0124 Monatsh Math-Phys • A
Monatshefte fuer Mathematik und Physik [1900-1942]
• CONT AS (J 0045) Monatsh Math

J 0126 Glasgow Math J • GB
Glasgow Mathematical Journal [1967ff] ISSN 0017-0895
• CONT OF (J 0217) Proc Glasgow Math Assoc

J 0127 J Reine Angew Math • D
Journal fuer die Reine und Angewandte Mathematik [1826ff]
ISSN 0075-4102

J 0128 Acta Math Univ Comenianae (Bratislava) • CS
Universitas Comeniana. Acta Facultatis Rerum Naturalium. Mathematica [1956ff]

J 0129 Elektr Informationsverarbeitung & Kybern • DDR
Elektronische Informationsverarbeitung und Kybernetik
[1965ff] ISSN 0013-5712

J 0131 Quart J Math, Oxford Ser 2 • GB
The Quarterly Journal of Mathematics. Oxford Second Series
[1950ff] ISSN 0033-5606
• CONT OF (J 1138) Quart J Math, Oxford Ser

J 0132 Math Scand • DK
Mathematica Scandinavica [1953ff] ISSN 0025-5521

J 0133 Michigan Math J • USA
The Michigan Mathematical Journal [1952ff] ISSN
0026-2285

J 0134 Dokl Akad Nauk Azerb SSR • SU
Akademiya Nauk Azerbajdzhanskoj SSR. Doklady (Reports of The Academy of Sciences of the Azerbaijan SSR) [1945ff]
ISSN 0002-3078

J 0135 Izv Akad Nauk Azerb SSR, Ser Fiz-Tekh Mat • SU
Izvestiya Akademii Nauk Azerbajdzhanskoj SSR. Seriya Fiziko-Tekhnicheskikh i Matematicheskikh Nauk (Proceedings of the Academy of Sciences of the Azerbaijan SSR. Series: Physical-Technical and Mathematical Sciences) [1958ff] ISSN
0002-3108

J 0136 Semigroup Forum • D
Semigroup Forum [1970ff] ISSN 0037-1912

J 0137 C R Acad Bulgar Sci • BG
Doklady Bolgarskoi Akademii Nauk (Comptes Rendus de l'Academie Bulgare des Sciences) [1948ff] ISSN 0001-3978

J 0138 Networks • USA
Networks. An International Journal [1970ff] ISSN 0028-3045

J 0140 Comm Math Univ Carolinae (Prague) • CS
Commentationes Mathematicae Universitatis Carolinae
[1960ff] ISSN 0010-2628

J 0142 Mat Sb, Akad Nauk SSSR, NS • SU
Matematicheskij Sbornik. Novaya Seriya. Akademiya Nauk SSSR i Moskovskoe Matematicheskoe Obshchestvo (Mathematical Collected Articles. New Series. Academy of Sciences of the USSR and Moskovian Mathematical Society)
[1936ff] ISSN 0025-5157
• CONT OF (J 1404) Mat Sb, Akad Nauk SSSR • TRANSL IN
(J 0349) Math of USSR, Sbor

J 0143 Mat Chasopis (Slov Akad Ved) • CS
Matematicky Chasopis (Journal of Mathematics) [1967-1975]
ISSN 0025-5173
• CONT OF (J 4713) Mat Fyz Chasopis (Slov Akad Ved)
• CONT AS (J 1522) Math Slovaca

J 0144 Rend Sem Mat Univ Padova • I
Rendiconti del Seminario Matematico dell'Universita di Padova [1930ff] ISSN 0041-8994

J 0149 Atti Accad Naz Lincei Fis Mat Nat, Ser 8 • I
Atti della Accademia Nazionale dei Lincei. Rendiconti. Classe di Scienze Fisiche, Matematiche e Naturali. Serie VIII [1946ff]
ISSN 0001-4435

J 0150 Acad Roy Belg Bull Cl Sci (5) • B
Academie Royale des Sciences, des Lettres et des Beaux Arts de Belgique. Bulletin de la Classe des Sciences. Cinquieme Serie * Koninklijke Academie voor Wetenschappen. Mededeelingen van de Afdeeling Wetenschappen. 5. Serie [1915ff] ISSN
0001-4141

J 0152 Enseign Math • CH
L'Enseignement Mathematique: Revue Internationale
[1899-1954]
• CONT AS (J 3370) Enseign Math, Ser 2

J 0154 Synthese • NL
Synthese. An International Journal for Epistemology, Methodology and Philosophy of Science [1936ff] ISSN
0039-7857

J 0155 Commun Pure Appl Math • USA
Communications on Pure and Applied Mathematics [1939ff]
ISSN 0010-3640

J 0156 Kybernetika (Prague) • CS
Kybernetika (Cybernetics) [1965ff] ISSN 0023-5954
• REL PUBL (J 3524) Kybernetika Suppl (Prague)

J 0157 Jbuchber Dtsch Math-Ver • D
Jahresbericht der Deutschen Mathematiker-Vereinigung
[1890ff] ISSN 0012-0456

J 0158 Erkenntnis (Dordrecht) • NL
Erkenntnis. The Journal of Unified Science: An International Journal of Analytic Philosophy [1975ff] ISSN 0165-0106
• CONT OF (J 3597) J Unif Sci

J 0160 Math-Phys Sem-ber, NS • D
Mathematisch-Physikalische Semesterberichte: Zur Pflege des Zusammenhangs von Schule und Universitaet. Neue Folge
[1950-1979] ISSN 0340-4897
• CONT AS (J 2790) Math Sem-ber

J 0161 Bull London Math Soc • GB
The Bulletin of the London Mathematical Society [1926ff]
ISSN 0024-6093

J 0162 Teorema (Valencia) • E
Teorema [1971ff]

J 0164 J Comb Th, Ser A • USA
Journal of Combinatorial Theory. Series A [1971ff] ISSN
0097-3165
• CONT OF (J 1669) J Comb Th

J 0171 Proc Cambridge Phil Soc Math Phys • GB
Proceedings of the Cambridge Philosophical Society. Mathematical and Physical Sciences [1843-1974] ISSN
0008-1981
• CONT AS (J 0332) Math Proc Cambridge Phil Soc

J 0175 Methodos • I
Methodos [1949ff]
• REM Does not seem to appear anymore

J 0178 Stud Gen • D
Studium Generale: Zeitschrift fuer die Einheit der Wissenschaften. Zusammenhang ihrer Begriffsbildungen und Forschungsmethoden. [1947-1971] ISSN 0039-4149

J 0179 Ann Fac Sci Clermont • F
Universite de Clermont. Faculte des Sciences. Annales [1952-1972]
• CONT AS (J 1934) Ann Sci Univ Clermont Math

J 0182 Kon Nederl Akad Wetensch Afd Let Med N S • NL
Koninklijke Nederlandse Akademie van Wetenschappen. Afdeeling Letterkunde: Mededelingen. N.S. [1921ff]
• REL PUBL (J 0028) Indag Math Mededelingen

J 0188 Rev Union Mat Argentina • RA
Revista de la Union Matematica Argentina [1936ff] ISSN 0041-6932

J 0193 Discr Math • NL
Discrete Mathematics [1971ff] ISSN 0012-365X

J 0194 Inform & Control • USA
Information and Control [1958ff] ISSN 0019-9958

J 0197 Stud Cercet Mat Acad Romana • RO
Studii si Cercetari Matematice. Academia Republicii Socialiste Romania. (Mathematische Studien und Untersuchungen. Akademie der Sozialistischen Republik Rumaenien) [1950ff] ISSN 0567-6401
• CONT OF (J 0524) Disq Math Phys

J 0202 Diss Math (Warsaw) • PL
Dissertationes Mathematicae. Polska Akademia Nauk, Instytut Matematyczny ∗ Rozprawy Matematyczne [1952ff] ISSN 0012-3862

J 0204 Ann Sci Ecole Norm Sup, Ser 3 • F
Annales Scientifiques de l'Ecole Normale Superieure. Serie 3 [1884ff] ISSN 0012-9593

J 0205 Rev Franc Autom, Inf & Rech Operat • F
Revue Francaise d'Automatique, Informatique et Recherche Operationelle (RAIRO). Series: Bleue, Jaune, Rouge, Verte [1972-1976] ISSN 0399-0559
• CONT OF (J 3954) Rev Franc Inf & Rech Operat • CONT AS (J 4698) Rev Franc Autom, Inf & Rech Operat, Ser Rouge Inf Th & (J 2831) RAIRO Autom & (J 2832) RAIRO Inform
• REM In 1975 the Serie Rouge split into: Serie Rouge Analyse Numerique & J4698

J 0207 Ist Lombardo Accad Sci Rend, A (Milano) • I
Istituto Lombardo. Accademia di Science e Lettere. Rendiconti. A. Scienze Matematiche, Fisiche, Chimichze e Geologiche [1937ff] ISSN 0021-2504
• CONT OF (J 3986) Ist Lombardo Rend, Ser 2 (Milano)

J 0209 J Math Phys • USA
Journal of Mathematical Physics [1960ff] ISSN 0022-2488

J 0216 Izv Akad Nauk SSSR, Ser Mat • SU
Izvestiya Akademii Nauk SSSR. Seriya Matematicheskaya (Proceedings of the Academy of Sciences of the USSR. Mathematical Series) [1937ff] ISSN 0373-2436
• CONT OF (J 4717) Izv Akad Nauk SSSR • TRANSL IN (J 0448) Math of USSR, Izv

J 0217 Proc Glasgow Math Assoc • GB
Glasgow Mathematical Association. Proceedings [1952-1966]
• CONT AS (J 0126) Glasgow Math J

J 0218 Bul Sti Tehn Inst Politeh Timisoara, Ser Mat-Fiz-Mec • RO
Buletinul Stiintific si Tehnic al Institutului Politehnic "Traian Vuia" Timisoara. Seria Matematica-Fizica-Mecanica Teoretica si Aplicata (Wissenschaftliches und Technisches Bulletin des Polytechnischen Instituts "Traian Vuia" Timisoara. Serie Mathematik-Physik-Theoretische und Angewandte Mechanik) [1970-1977]
• CONT OF (J 4714) Bul Sti Tehn Inst Politeh Timisoara, NS
• CONT AS (J 4715) Bul Sti Tehn Inst Politeh Timisoara, Ser Mat-Fiz

J 0220 Atti Accad Sci Torino, Fis Mat Nat • I
Atti della Accademia delle Scienze di Torino. Classi di Scienze Fisiche, Matematiche e Naturali. Parte 1. ∗ Acta Academiae Scientiarum Taurinensis [1940ff] ISSN 0373-3033
• CONT OF (J 1742) Atti Accad Sci Torino, Fis Mat Nat

J 0225 Amer Math Soc, Transl, Ser 2 • USA
American Mathematical Society. Translations. Series 2 [1955ff] ISSN 0065-9290

J 0226 Uch Zap Ped Inst, Ivanovo • SU
Ivanovskij Gosudarstvennyj Pedagogicheskij Institut imeni D.A.Furmanova. Uchenye Zapiski (Furmanov-Institute of Education in Ivanovo. Scientific Notes) [1941-1973] ISSN 0444-9681
• CONT AS (J 3536) Uch Zap Univ, Ivanovo

J 0229 Ann Mat Pura Appl, Ser 3 • I
Annali di Matematica Pura ed Applicata [1898-1923]
• CONT AS (J 3526) Ann Mat Pura Appl, Ser 4

J 0230 An Univ Iasi, NS, Sect Ia • RO
Analele Stiintifice ale Universitatii Al.I. Cuza din Iasi. (Serie Noua) Sectiunea 1a: Matematica (Wissenschaftliche Annalen der Al.I. Cuza Universitaet Iasi. (Neue Serie) Sektion 1a: Mathematik) [1955ff] ISSN 0041-9109

J 0232 Inform Process Lett • NL
Information Processing Letters. Devoted to the Rapid Publication of Short Contributions to Information Processing [1971ff] ISSN 0020-0190

J 0233 Soobshch Akad Nauk Gruz SSR • SU
Soobshcheniya Akademii Nauk Gruzinskoj SSR ∗ Sakaharth SSR Mecnierebatha Akademia Moambe (Communications of the Academy of Sciences of the Georgian SSR) [1940ff] ISSN 0002-3167

J 0234 Rev Acad Cienc Exact Fis Nat Madrid • E
Revista de la Real Academia de Ciencias Exactas, Fisicas y Naturales de Madrid [1904ff] ISSN 0034-0596

J 0236 Rev Mat Hisp-Amer, Ser 4 • E
Revista Matematica Hispano-Americana. 4a Serie. Real Sociedad Matematica Espanola. [1941ff] ISSN 0373-0999
• CONT OF (J 3993) Rev Mat Hisp-Amer, Ser 2

J 0237 Stud Univ Cluj, Ser Math Phys Chem • RO
Studia Universitatis Babes-Bolyai. Series Mathematica-Physica-Chemia [1956-1963]
• CONT AS (J 0355) Studia Univ Babes-Bolyai, Math-Phys (Cluj) Kozlemenyei. Termeszettuclomanyi Sorozat

J 0239 Skr, K Nor Vidensk Selsk • N
Det Kongelige Norske Videnskabers Selskab. Skrifter. (Monographs of the Scandinavian Society of Sciences) [1791ff] ISSN 0368-6310

J 0240 Ann Inst Fourier • F
Annales de l'Institut Fourier [1949ff] ISSN 0373-0956

J 0247 Bull Sci Math, Ser 2 • F
Bulletin des Sciences Mathematiques, Serie 2 [1870ff] ISSN 0007-4497

J 0248 Math Student • IND
The Mathematics Student [1933ff] ISSN 0025-5742

J 0249 J Math Pures Appl • F
Journal de Mathematiques Pures et Appliquees [1836-1921]
• CONT AS (J 3941) J Math Pures Appl, Ser 9

J 0250 Bol Soc Mat Mexicana • MEX
Boletin de la Sociedad Matematica Mexicana [1944-1955]
• CONT AS (J 3127) Bol Soc Mat Mexicana, Ser 2

J 0252 Rev Philos Louvain • B
Revue Philosophique de Louvain [1946ff] ISSN 0035-3841
• CONT OF (J 1720) Rev Neoscolast Philos, Ser 2

J 0254 Gen Topology Appl • NL
General Topology and its Applications. A Journal Devoted to Set Theoretic, Axiomatic and Geometric Topology [1971-1979] ISSN 0016-660X
• CONT AS (J 2635) Topology Appl

J 0255 God Fak Mat & Mekh, Univ Sofiya • BG
Godishnik na Sofijskiya Universitet. Fakultet po Matematika i Mekhanika (Annuaire de l'Universite de Sofia. Faculte de Mathematiques.)

J 0259 Nyt Tidsskr Mat • DK
Nyt Tidsskrift for Matematik (New Journal for Mathematics) [1890-1918]
• CONT AS (J 4510) Norsk Mat Tidsskr

J 0260 Ann Jap Ass Phil Sci • J
Annals of the Japan Association for Philosophy of Science [1956ff]

J 0261 Tohoku Math J • J
Tohoku Mathematical Journal (Tohoku Sugaku Zashi) [1911ff] ISSN 0040-8735

J 0265 Ukr Mat Zh, Akad Nauk Ukr SSR • SU
Ukrainskij Matematicheskij Zhurnal. Akademiya Nauk Ukrainskoj SSR. Institut Matematiki (Ukrainian Mathematical Journal. Academy of Sciences of the Ukrainian SSR. Institute of Mathematics) [1949ff] ISSN 0041-6053
• TRANSL IN (J 3281) Ukr Math J

J 0270 Dokl Akad Nauk Ukr SSR, Ser A • SU
*Doklady Akademii Nauk Ukrainskoj SSR. Seriya A. Fiziko-Matematicheskie i Tekhnicheskie Nauki * Dopovidi Akademii Nauk Uk'rainskoj RSR. Seriya A. Fiziko-Matematichni Ta Tekhnichi Nauki (Reports of the Academy of Sciences of the Ukrainian SSR. Series A. Physical-Mathematical and Engineering Sciences)* [1939ff] ISSN 0002-3531, ISBN 0201-8446

J 0283 Ann Soc Pol Math • PL
*Societe Polonaise de Mathematique. Annales. * Rocznik i Polskiego Towarzystwa Matematycznego* [1922-1952]
• CONT AS (J 1405) Ann Pol Math

J 0284 IBM J Res Dev • USA
IBM (= International Business Machines) Journal of Research and Development [1957ff] ISSN 0018-8646

J 0286 Int Logic Rev • I
*International Logic Review. * Rassegna Internazionale di Logica* [1970ff] ISSN 0048-6779

J 0287 Scripta Math • USA
Scripta Mathematica. [1932ff] ISSN 0036-9713

J 0288 Vest Ser Mat Mekh, Univ Moskva • SU
Vestnik Moskovskogo Universiteta. Seriya I. Matematika, Mekhanika (Publications of the Moscow University. Series I. Mathematics. Mechanics) [1946ff] ISSN 0201-7385, ISSN 0579-9368
• TRANSL IN (J 0510) Moscow Univ Math Bull & Moscow University. Mechanics Bulletin

J 0290 Euclides • NL
Euclides: Maandblad voor de Didactiek van de Wiskunde [1925ff]

J 0291 Wis-natuur Tijdsch • B
Wis- en Natuurkundig Tijdschrift: Orgaan van her Vlaamsch Natuur-, Wis- en Geneeskund Congres [1921-1944]
• CONT AS (J 0061) Simon Stevin

J 0295 Teor Veroyat i Mat Stat (Kiev) • SU
Teoriya Veroyatnostej i Matematicheskaya Statistika. Mezhvedomstvennyj Nauchnyj Sbornik (Wahrscheinlichkeitsrechnung und Mathematische Statistik) [1970ff]
• TRANSL IN (J 3456) Th Probab & Math Stat

J 0296 Acta Salamantica Cienc • E
Acta Salamanticensia Ciencias. Iussu Senatus Universitatis Edita Ciencias [1954ff]

J 0297 Rend Mat Appl, Ser 5 • I
Rendiconti di Matematica e delle sue Applicazioni. Seria 5. Universita di Roma. Istituto Nazionale di Alta Matematica (INDAM) [1942-1967]
• CONT AS (J 3741) Rend Mat Appl, Ser 7

J 0298 Nanyang Univ J • SGP
Nanyang University. Journal [1967ff] ISSN 0077-2747

J 0299 Gac Mat, Ser 1a (Madrid) • E
Gaceta Matematica. Consejo Superior de Investigaciones Cientificas. Instituto "Jorge Juan". 1a Serie [1970ff] ISSN 0016-3805
• CONT OF (J 4443) Gac Mat (Madrid)

J 0301 J Phil • USA
The Journal of Philosophy [1904ff] ISSN 0022-362X

J 0302 Rep Math Logic, Krakow & Katowice • PL
Reports on Mathematical Logic. The Jagiellonian University of Cracow. The Silesian University of Katowice [1973ff] ISSN 0083-4432
• CONT OF (S 0458) Zesz Nauk, Prace Log, Uniw Krakow

J 0303 Mathematika (Univ Coll London) • GB
Mathematika. A Journal of Pure and Applied Mathematics [1954ff] ISSN 0025-5793

J 0305 Invent Math • D
Inventiones Mathematicae [1966ff] ISSN 0020-9910

J 0306 Cah Topol & Geom Differ • F
Cahiers de Topologie et Geometrie Differentielle [1959ff] ISSN 0008-0004

J 0307 Rev Colomb Mat • CO
Revista Colombiana de Matematicas [1967ff] ISSN 0034-7426
• CONT OF (**J 0348**) Rev Mat Elementales

J 0308 Rocky Mountain J Math • USA
The Rocky Mountain Journal of Mathematics [1971ff] ISSN 0035-7596

J 0311 Nordisk Mat Tidskr • N
Nordisk Matematisk Tidskrift (Scandinavian Mathematical Journal) [1953-1978] ISSN 0029-1412
• CONT OF (**J 4510**) Norsk Mat Tidsskr • CONT AS (**J 3075**) Normat

J 0312 Izv Akad Nauk Armyan SSR, Ser Mat • SU
Izvestiya Akademii Nauk Armyanskoj SSR. Seriya Matematika (Proceedings of the Academy of Sciences of the Armenian SSR. Series: Mathematics) [1965ff] ISSN 0002-3043
• TRANSL IN (**J 3265**) Sov J Contemp Math Anal, Armen Acad Sci

J 0316 Illinois J Math • USA
Illinois Journal of Mathematics [1957ff] ISSN 0019-2082

J 0319 Matematiche (Sem Mat Catania) • I
Le Matematiche [1946ff]

J 0322 Arch Math (Brno) • CS
Archivum Mathematicum [1965ff] ISSN 0044-8753

J 0323 Uch Zap Ped Inst, Ryazan • SU
Ryazanskij Gosudarstvennyj Pedagogicheskij Institut. Uchenye Zapiski. (State Institute of Education in Ryazan. Scientific Notes)

J 0325 Amer Phil Quart • GB
American Philosophical Quarterly [1964ff] ISSN 0003-0481

J 0326 J Pure Appl Algebra • NL
Journal of Pure and Applied Algebra [1971ff] ISSN 0022-4049

J 0329 Math Chron (Auckland) • NZ
Mathematical Chronicle [1969ff] ISSN 0581-1155

J 0332 Math Proc Cambridge Phil Soc • GB
Mathematical Proceedings of the Cambridge Philosophical Society [1975ff] ISSN 0305-0041
• CONT OF (**J 0171**) Proc Cambridge Phil Soc Math Phys

J 0337 Mat Ezheg, Akad Nauk Latv SSR • SU
Latvijskij Matematicheskij Ezhegodnik. Latvijskij Ordena Trudovogo Krasnogo Znameni Gosudarstvennyj Universitet imeni P.Stuchki. Akademiya Nauk Latvijskoj SSR (Latvian Mathematical Yearbook) [1965ff] ISSN 0458-8223
• TRANSL IN Latvian Mathematical Yearbook

J 0338 Nauch-Tekh Inf, Ser 2, Akad Nauk SSSR • SU
Nauchno-Tekhnicheskaya Informatsiya. Seriya 2. Gosudarstvennyj Komitet SSSR po Nauke i Tekhnike. Akademiya Nauk SSSR. Vsesoyuznyj Institut Nauchnoj i Tekhnicheskoj Informatsii. Informatsionnye Protsesy i Sistemy (Scientific Technical Information. Series 2) [1967ff]
• TRANSL IN (**J 2667**) Autom Doc Math Linguist

J 0340 Mat Zap (Univ Sverdlovsk) • SU
Matematicheskie Zapiski (Mathematical Notes) ISSN 0076-5368

J 0342 Monatsber Dt Akad Wiss • DDR
Die Deutsche Akademie der Wissenschaften zu Berlin. Monatsberichte: Mitteilungen aus Mathematik, Naturwissenschaft, Medizin und Technik [1959-1971] ISSN 0011-9814

J 0343 Studia Math, Pol Akad Nauk • PL
Studia Mathematica. Polska Akademia Nauk, Instytut Matematyczny [1929ff] ISSN 0039-3223

J 0345 Adv Math • USA
Advances in Mathematics [1964ff] ISSN 0001-8708
• REL PUBL (**S 3105**) Adv Math, Suppl Stud

J 0346 Dokl Akad Nauk Armyan SSR • SU
Doklady Akademii Nauk Armyanskoj SSR (Reports of the Academy of Sciences of the Armenian SSR) [1944ff] ISSN 0321-1339

J 0348 Rev Mat Elementales • CO
Revista de Matematicas Elementales [1952-1966]
• CONT AS (**J 0307**) Rev Colomb Mat

J 0349 Math of USSR, Sbor • USA
Mathematics of the USSR, Sbornik [1967ff] ISSN 0025-5734
• TRANSL OF (**J 0142**) Mat Sb, Akad Nauk SSSR, NS

J 0350 Sci Rep Tokyo Kyoiku Daigaku Sect A • J
Tokyo Kyoiku Daigaku (Tokyo University of Education. Science Reports. Section A.)

J 0351 Osaka J Math • J
Osaka Journal of Mathematics [1964ff] ISSN 0030-6126
• CONT OF (**J 1770**) Osaka Math J

J 0352 Math Jap • J
Mathematica Japonica [1948ff] ISSN 0025-5513

J 0353 Bull Soc Math Fr • F
Bulletin de la Societe Mathematique de France [1873ff] ISSN 0037-9484

J 0354 Phil Trans Roy Soc London, Ser A • GB
Philosophical Transactions of the Royal Society of London. Series A. Mathematical and Physical Sciences. ISSN 0080-4614

J 0355 Studia Univ Babes-Bolyai, Math-Phys (Cluj) • RO
Studia Universitatis Babes-Bolyai. Mathematica - Physica [1964-1970]
• CONT OF (**J 0237**) Stud Univ Cluj, Ser Math Phys Chem
• CONT AS (**J 3451**) Stud Univ Cluj, Ser Math-Mech

J 0364 Publ Sci Univ Alger Ser A Math • DZ
Universite d'Alger. Publications Scientifiques. Serie A. Mathematiques [1954ff] ISSN 0002-5321

J 0366 Sitzber Berlin Math Ges • DDR
Die Berliner Mathematische Gesellschaft. Sitzungsberichte

J 0369 Notices Amer Math Soc • USA
Notices of the American Mathematical Society [1953ff] ISSN 0002-9920

J 0374 SIAM J Appl Math • USA
Journal on Applied Mathematics. SIAM (= Society for Industrial and Applied Mathematics) [1966ff] ISSN 0036-1399
• CONT OF (**J 0514**) SIAM Journ

J 0377 Bol Mat (Bogota) • CO
Boletin de Matematicas [1967ff]

J 0380 Acta Cybern (Szeged) • H
Acta Cybernetica. Forum Centrale Publicationum Cyberneticarum Hungaricum [1969ff] ISSN 0324-721X

J 0384 Rend Mat, Ser 5 • I
Rendiconti di Matematica. Serie 5 [1940-1967]
• CONT AS (J 2311) Rend Mat, Ser 6

J 0385 Atti Sem Mat Fis Univ Modena • I
Atti del Seminario Matematico e Fisico dell'Universita di Modena [1947ff] ISSN 0041-8986

J 0386 Proc Edinburgh Math Soc • GB
Proceedings of the Edinburgh Mathematical Society [1883-1926]
• CONT AS (J 3420) Proc Edinburgh Math Soc, Ser 2

J 0387 Bull Sect Logic, Pol Acad Sci • PL
Bulletin of the Section of Logic. Polish Academy of Sciences. Institute of Philosophy and Sociology. [1972ff]
• REM Papers Published in the Bulletin are Generally: 1. Abstracts or preprints of papers submitted to other journals e.g. Studia Logica. 2. Abstracts of papers read at seminars or local conferences

J 0388 Human, Ser 4 Logica Mat • C
Humanidades. Ser 4 Logica Matematica [1972ff]

J 0390 Publ Res Inst Math Sci (Kyoto) • J
Publications of the Research Institute for Mathematical Sciences [1965ff] ISSN 0034-5318
• REM Vols 1-4 Issued as: Kyoto Univ. Research Institute for Mathematical Sciences. Publications. Series A.

J 0391 Riv Mat Univ Parma • I
Rivista di Matematica della Universita di Parma [1891-1915]
• CONT AS (J 1526) Riv Mat Univ Parma, Ser 2

J 0394 Commun Algeb • USA
Communications in Algebra [1974ff] ISSN 0092-7872

J 0395 J Geom • CH
Journal of Geometry [1971ff] ISSN 0047-2468

J 0396 Mat Lapok • H
Matematikai Lapok (Mathematical Papers) [1949ff] ISSN 0025-519X
• CONT OF (J 0461) Mat Fiz Lapok

J 0397 Proc Math Phys Soc Egypt • ET
Proceedings of the Mathematical and Physical Society of Egypt [1937ff] ISSN 0076-5317

J 0399 Acta Arith, Pol Akad Nauk • PL
Acta Arithmetica. Academia Scientiarum Polona. Institutum Mathematicum [1937ff]

J 0400 Publ Inst Math, NS (Belgrade) • YU
Institut Mathematique. Publications de l'Institut Mathematique. Nouvelle Serie [1961ff] ISSN 0522-828X
• CONT OF (J 4706) Publ Inst Math (Belgrade)

J 0401 J Number Th • USA
Journal of Number Theory [1968ff] ISSN 0022-314X

J 0403 Izv Akad Nauk Kazak SSR, Ser Fiz-Mat • SU
Izvestiya Akademii Nauk Kazakhskoj SSR. Seriya Fiziko-Matematicheskaya (Proceedings of the Academy of Sciences of the Kazakh SSR. Series: Physics & Mathematics) [1963ff] ISSN 0002-3191

J 0407 Comm Math Univ St Pauli (Tokyo) • J
Commentarii Mathematici Universitatis Sancti Pauli [1952ff] ISSN 0010-258X

J 0408 Bull Soc R Sci Liege • B
Bulletin de la Societe Royale des Sciences de Liege [1932ff] ISSN 0037-9565

J 0411 Studia Sci Math Hung • H
Studia Scientiarum Mathematicarum Hungaria. Auxilio Consilii Instituti Mathematici. Academiae Scientiarum Hungaricae [1966ff] ISSN 0081-6906

J 0413 Izv Akad Nauk Belor SSR, Ser Fiz-Mat • SU
*Vestsi Akademii Navuk BeSSR. Seriya Fizika-Matematychnykh Navuk * Izvestiya Akademii Nauk BSSR. Seriya Fiziko-Matematicheskikh Nauk (Proceedings of the Academy of Sciences of the Byelorussian SSR. Series: Physics, Mathematics)* [1964ff] ISSN 0002-3574

J 0414 Dokl Akad Nauk Belor SSR • SU
Doklady Akademii Nauk BSSR (Reports of the Academy of Sciences of the BSSR) [1957ff] ISSN 0002-354X

J 0418 Shuxue Xuebao • TJ
Shuxue Xuebao (Acta Mathematica Sinica) [1951ff]
• TRANSL IN (J 0419) Chinese Math Acta
• REM In 1951 published as: Journal of the Chinese Mathematical Society (N.S.)

J 0419 Chinese Math Acta • USA
Chinese Mathematics. Acta [1962-1967]
• TRANSL OF (J 0418) Shuxue Xuebao

J 0420 Shuxue Jinzhan • TJ
Shuxue Jinzhan (Advances in Mathematics) [1955ff] ISSN 0559-9326

J 0429 Vest Akad Nauk Kazak SSR • SU
Vestnik Akademii Nauk Kazakhskoj SSR (Publications of the Academy of Sciences of the Kazakh SSR) [1944ff] ISSN 0002-3213

J 0431 Hiroshima Math J • J
Hiroshima Mathematical Journal [1971ff] ISSN 0018-2079
• CONT OF (J 0102) Hiroshima Univ J Sci, Ser A Math

J 0434 J Fac Sci Univ Tokyo, Sect 1 • J
Journal of the Faculty of Science. University of Tokyo. Section 1 Mathematics, Astronomy, Physics, Chemistry [1925-1970] ISSN 0040-8980
• CONT AS (J 2332) J Fac Sci, Univ Tokyo, Sect 1 A

J 0439 Math J Okayama Univ • J
Mathematical Journal of Okayama University [1952ff] ISSN 0030-1566

J 0440 Bul Inst Politeh Bucuresti, Ser Mec • RO
Buletinul Institutului Politehnic "Gheorghe Gheorghiu-Dej" Bucuresti. Seria Mecanica (Bulletin des Polytechnischen Instituts "Gheorghe Gheorghiu-Dej" Bukarest. Serie Mechanik)

J 0446 Ann Acad Sci Fennicae, Ser A I, Diss • SF
Annales Academiae Scientiarum Fennicae. Series AI. Mathematica. Dissertationes [1975ff] ISSN 0066-1953
• CONT OF (J 3994) Ann Acad Sci Fennicae Ser A I

J 0448 Math of USSR, Izv • USA
Mathematics of the USSR, Izvestiya [1967ff] ISSN 0025-5726
• TRANSL OF (J 0216) Izv Akad Nauk SSSR, Ser Mat

J 0449 Probl Kybern • DDR
Probleme der Kybernetik [1958-1965]
• CONT AS (J 0471) Syst Th Res • TRANSL OF (J 0052) Probl Kibern

J 0459 C R Soc Sci Lett Varsovie Cl 3 • PL
*Societe des Sciences et des Lettres de Varsovie. Comptes Rendus des Seances. Classe III: Sciences Mathematiques et Physiques * Towarzystwo Naukowe Warszawskie. Sprawozdania z Posiedze. Wydzialu III: Nauk Matematyczno-Fizycznych (Warschauer Sitzungsberichte)* [1908-1950]

J 0460 Acta Univ Szeged, Sect Mat • H
Acta Litterarum ac Scientiarum Regiae Universitatis Hungaricae Francisco-Josephinae, Sectio Scientiarum Mathematicarum [1922-1946]
• CONT AS (**J** 0002) Acta Sci Math (Szeged)

J 0461 Mat Fiz Lapok • H
Matematikai es Fizikai Lapok (Mathematical and Physical Papers) [1892-1948]
• CONT AS (**J** 0396) Mat Lapok

J 0462 Mat Fiz Oszt Koezlem, Acad Sci Hung • H
Magyar Tudomanyos Akademia. Matematikai es Fizikai Tudomanyok Osztalyanak Koezlemenyek. (Hungarian Academy of Sciences. Bulletin of the Mathematical and Physical Sciences) [1952-1974]
• CONT AS (**J** 1458) Alkalmaz Mat Lapok

J 0465 Bull Greek Math Soc (NS) • GR
Bulletin of the Greek Mathematical Society. New Series (Hellenike Mathematike Hetaireia. Deltion. Nea Seira.) [1960ff] ISSN 0072-7466
• CONT OF (**J** 1699) Bull Soc Math Grece

J 0470 Sov Phys, Dokl • USA
Soviet Physics. Doklady. [1956ff] ISSN 0038-5689
• TRANSL OF (**J** 0023) Dokl Akad Nauk SSSR

J 0471 Syst Th Res • USA
Systems Theory Research [1966ff] ISSN 0082-1255
• CONT OF (**J** 1195) Probl Cybernet & (**J** 0449) Probl Kybern
• TRANSL OF (**J** 0052) Probl Kibern

J 0472 Theory Decis • NL
Theory and Decision. An International Journal for Philosophy and Methodology of the Social Sciences [1970ff] ISSN 0040-5833

J 0475 Sib Math J • USA
Siberian Mathematical Journal [1966ff] ISSN 0037-4466
• TRANSL OF (**J** 0092) Sib Mat Zh

J 0481 Acta Univ Wroclaw • PL
Acta Universitatis Wratislaviensis

J 0487 Math Unterricht • D
Der Mathematikunterricht: Beitraege zu seiner Wissenschaftlichen und Methodischen Gestaltung [1955ff] ISSN 0025-5807

J 0493 Bull Res Counc Israel Sect F • IL
Research Council of Israel. Bulletin. Section F. Mathematics and Physics [1952-1962]
• CONT AS (**J** 0029) Israel J Math

J 0494 Bull Math Soc Sci Roumanie • RO
Societatea Romana de Stiinte, Sectia Mathematica. Bulletin Mathematiques de la Societe Roumaine des Sciences [1908-1956]
• CONT AS (**J** 0070) Bull Soc Sci Math Roumanie, NS

J 0497 Math Mag • USA
Mathematics Magazine [1947ff] ISSN 0025-570X
• CONT OF (**J** 1737) Nat Math Magazine (Louisiana)

J 0504 Manuscr Math • D
Manuscripta Mathematica [1969ff] ISSN 0025-2611

J 0508 Machine Intelligence • GB
Machine Intelligence [1967ff] ISSN 0541-6418

J 0510 Moscow Univ Math Bull • USA
Moscow University Mathematics Bulletin [1969ff] ISSN 0027-1322
• TRANSL OF (**J** 0288) Vest Ser Mat Mekh, Univ Moskva

J 0514 SIAM Journ • USA
Journal. SIAM (= Society for Industrial and Applied Mathematics) [1953-1965]
• CONT AS (**J** 0374) SIAM J Appl Math

J 0517 Mathematica (Cluj) • RO
Mathematica. Revue d'Analyse Numerique et de Theorie de l'Approximation [1929ff] ISSN 0025-5505

J 0519 Wiad Mat, Ann Soc Math Pol, Ser 2 • PL
*Annales Societatis Mathematicae Polonae. Seria 2. Wiadomosci Matematyczne * Roczniki Polskiego Towarzystwa Matematycznego. Seria 2. Wiadomosci Matematyczne* [1955ff] ISSN 0079-3698
• CONT OF (**J** 4710) Pol Tow Mat, Wiad Mat

J 0521 Semin Math, Inst Steklov • USA
Seminars in Mathematics. V.A.Steklov Mathematical Institute Leningrad [1967ff]
• TRANSL OF (**S** 0228) Zap Nauch Sem Leningrad Otd Mat Inst Steklov

J 0523 Bull Nagoya Inst Tech • J
Bulletin of Nagoya Institute of Technology [1949ff]

J 0524 Disq Math Phys • RO
Disquisitiones Mathematicae et Physicae [1940-1949]
• CONT AS (**J** 0197) Stud Cercet Mat Acad Romana

J 0531 Hitotsubashi J Arts Sci (Tokyo) • J
Hitotsubashi Journal of Arts & Sciences [1960ff] ISSN 0073-2788

J 0535 Abh Braunschweig Wiss Ges • D
Abhandlungen der Braunschweigischen Wissenschaftlichen Gesellschaft [1949ff] ISSN 0068-0737

J 0537 Yokohama Math J • J
The Yokohama Mathematical Journal [1953ff] ISSN 0044-0523

J 0546 J Korean Math Soc • ROK
Journal of the Korean Mathematical Society

J 0549 Riv Mat Univ Parma, Ser 4 • I
Rivista di Matematica della Universita di Parma. Serie 4 [1975ff] ISSN 0035-6298
• CONT OF (**J** 3254) Riv Mat Univ Parma, Ser 3

J 0589 Bull Amer Math Soc (NS) • USA
Bulletin of the American Mathematical Society. New Series [1979ff] ISSN 0273-0979
• CONT OF (**J** 0015) Bull Amer Math Soc

J 0611 Pol Tow Mat, Prace Mat-Fiz • PL
Polskie Towarzystwo Matematycene. Prace Matematyczno-Fizyczne (Mathematische und Physikalische Abhandlungen) [1887-1954]
• CONT AS (**J** 0051) Commentat Math, Ann Soc Math Pol, Ser 1 & (**J** 2095) Fund Inform, Ann Soc Math Pol, Ser 4

J 0667 Scripta Fac Sci Math, Brno • CS
Scripta Facultatis Scientarium Naturalium Universitatis J.E. Purkyne Brunensis. Mathematica [1971ff]

J 0748 Erkenntnis (Leipzig) • DDR
Erkenntnis [1930-1939]
• CONT OF (J 1380) Ann Philos & Philos Kritik • CONT AS (J 3597) J Unif Sci

J 0752 Avh Norske Vid-Akad Oslo I • N
Avhandlinger. Norske Videnskaps-Akademi i Oslo. I: Matematisk-Naturvidenskapelig Klasse (Proceedings of the Scandinavian Academy of Sciences. Mathematical and Natural Sciences Class)

J 0789 Uch Zap Mat Ped Inst, Tula • SU
Tul'skij Gosudarstvennyj Pedagogicheskij Institut im. L.N.Tolstogo. Uchenye Zapiski Matematicheskikh Kafedr (Tolstoi State Institute of Education in Tula. Scientific Notes of the Mathematical Faculties)

J 0931 Anz Oesterr Akad Wiss, Math-Nat Kl • A
Anzeiger der Oesterreichischen Akademie der Wissenschaften. Mathematisch-Naturwissenschaftliche Klasse [1864ff] ISSN 0065-535X

J 0954 Tr Inst Prikl Mat, Tbilisi • SU
*Tbilisskij Gosudarstvennyj Universitet. Institut Prikladnoj Matematiki. Trudy * Thbilisis Sahelmcipho Universiteti Gamoqenebithi Mathematikis Instituti Shromebi (State University of Tbilisi. Institute of Applied Mathematics. Publications)* [1969ff] ISSN 0082-2191

J 0962 Ann Soc Sci Bruxelles, Ser 1 • B
Annales de la Societe Scientifique de Bruxelles. Serie I. Sciences Mathematiques, Astronomiques et Physiques [1875ff] ISSN 0037-959X

J 0963 Ajatus (Helsinki) • SF
Ajatus. Suomen Filosofisen Yhdislyksen Vuosikirja (Yearbook of the Philosopical Society of Finland) [1926ff]

J 0974 Norsk Vid-Akad Oslo Mat-Natur Kl Skr • N
Norske Videnskaps - Akademi i Oslo. Matematisk-Naturvidenskapelig Klasse. Skrifter. (Monographs of the Scandinavian Academy of Sciences. Mathematical and Natural Sciences Class) [1929ff] ISSN 0029-2338
• CONT OF (J 1145) Vidensk Selsk Kristiana Skrifter Ser 1

J 0975 Norsk Mat Forenings Skr • N
Norsk Matematisk Forenings Skrifter

J 0984 Ann Inst Henri Poincare, Sect B • F
Annales de l'Institut Henri Poincare. Section B : Calcul des Probabilites et Statistique [1930ff] ISSN 0020-2347

J 0989 Z Allg Wissth • D
Zeitschrift fuer Allgemeine Wissenschaftstheorie (Journal for General Philosophy of Science) [1969ff] ISSN 0044-2216

J 0990 Soc Sci Fennicae Comment Phys-Math • SF
Societas Scientiarum Fennicae. Commentationes Physico-Mathematicae

J 1005 Rep Fac Sci, Shizuoka Univ • J
Reports of the Faculty of Science. Shizuoka University. [1965ff] ISSN 0583-0923

J 1008 Demonstr Math (Warsaw) • PL
Demonstratio Mathematica [1969ff] ISSN 0420-1213

J 1021 Itogi Nauki Ser Mat • SU
Itogi Nauki. Seriya Matematiki (Progress in Science. Mathematical Series) [1962-1971] ISSN 0579-1731
• CONT AS (J 1488) Itogi Nauki Tekh, Ser Probl Geom & (J 1501) Itogi Nauki Tekh, Ser Algeb, Topol, Geom & (J 1452) Itogi Nauki Tekh, Ser Sovrem Probl Mat & (J 3188) Itogi Nauki Tekh, Ser Teor Veroyat Mat Stat Teor Kibern & (J 4387) Itogi Nauki Tekh, Ser Tekh Kibern • TRANSL IN (J 1531) J Sov Math
• REM J1531 contains only selected translations

J 1023 Perspectives of New Music • USA
Perspectives of New Music [1962ff] ISSN 0031-6016

J 1024 Zhongguo Kexue • TJ
Zhongguo Kexue (Scientia Sinica) [1950-1981]
• CONT AS (J 3766) Zhongguo Kexue, Xi A

J 1030 Formalisation • F
La Formalisation: Cahier pour l'Analyse

J 1044 Math Notes, Acad Sci USSR • USA
Mathematical Notes of the Academy of Sciences of the USSR [1967ff] ISSN 0001-4346
• TRANSL OF (J 0087) Mat Zametki (Akad Nauk SSSR)

J 1048 Kiber Sb Perevodov • SU
Kiberneticheskij Sbornik: Sbornik Perevodov. (Collected Articles on Cybernetics: Collected Translations) [1960-1964]
• CONT AS (J 3079) Kiber Sb Perevodov, NS

J 1069 Tydskr Natuurwetenskap (Pretoria) • ZA
Tydskrif vir Naturwetenskappe (Journal for Natural Sciences) [1961ff] ISSN 0041-4786

J 1092 Ann Univ Paris • F
Universite de Paris. Annales ISSN 0041-9176

J 1093 Ruch Filoz • PL
Ruch Filozoficzny. Polskie Towarzystwo Filozoficzne (Philosophical Movement. Polish Philosophical Society) ISSN 0035-9599

J 1108 Arkhimedes (Helsinki) • SF
Arkhimedes [1949ff] ISSN 0004-1920

J 1124 Ergebn Math Kolloquium • A
Ergebnisse eines Mathematischen Kolloquiums [1929-1936]

J 1125 Przeglad Filoz • PL
Przeglad Filozoficzny (Revue Philosophique) [1897-1949]

J 1138 Quart J Math, Oxford Ser • GB
The Quarterly Journal of Mathematics. Oxford Series [1930-1949]
• CONT AS (J 0131) Quart J Math, Oxford Ser 2

J 1145 Vidensk Selsk Kristiana Skrifter Ser 1 • N
Videnskaps Selskapet i Kristiana. Skrifter Utgit. 1 Matematisk-Naturvidenskapelig Klasse [?-1928]
• CONT AS (J 0974) Norsk Vid-Akad Oslo Mat-Natur Kl Skr

J 1148 Kumamoto J Sci, Ser A • J
Kumamoto Journal of Science. Series A: Mathematics, Physics and Chemistry [1952-1971] ISSN 0023-5318
• CONT AS (J 4727) Kumamoto J Sci, Math

J 1150 Proc Roy Soc London, Ser A • GB
Proceedings of the Royal Society of London. Series A. Mathematical and Physical Sciences. [1905ff] ISSN 0080-4630

J 1156 Izv Bulgar Akad Nauk Mat Inst • BG
B"lgarska Akademiya na Naukite. Otdelenie Za Fiziko-Matematicheski i Tekhnicheski Nauki. Izvestiya na Matematicheski Institut (Bulgarian Academy of Sciences. Department of Physics, Mathematics & Engineering. Reports of the Mathematical Institute) [1957-1974]
• CONT AS (**J 2547**) Serdica, Bulgar Math Publ

J 1195 Probl Cybernet • GB
Problems of Cybernetics [1958-1965]
• CONT AS (**J 0471**) Syst Th Res • TRANSL OF (**J 0052**) Probl Kibern

J 1207 Fkts Anal Prilozh • SU
Funktsional'nyj Analiz i ego Prilozheniya (Functional Analysis and its Applications) [1967ff] ISSN 0374-1990
• TRANSL IN (**J 3163**) Fct Anal & Appl

J 1377 Ann New York Acad Sci • USA
New York Academy of Sciences. Annals [1877ff] ISSN 0077-8923

J 1379 Christiaan Huygens Internat Math Tijdschr • NL
Christiaan Huygens. Internationaal Mathematisch Tijdschrift [1868-1942]
• CONT AS (**J 0061**) Simon Stevin

J 1380 Ann Philos & Philos Kritik • DDR
Annalen der Philosophie und Philosophischen Kritik [1925-1929]
• CONT AS (**J 0748**) Erkenntnis (Leipzig)

J 1399 Russ Math Surv • GB
Russian Mathematical Surveys [1946ff] ISSN 0036-0279
• TRANSL OF (**J 0067**) Usp Mat Nauk

J 1404 Mat Sb, Akad Nauk SSSR • SU
Matematicheskij Sbornik. Akademiya Nauk SSSR i Moskovskoe Matematicheskoe Obshchestvo (Mathematical Collected Articles. New Series. Academy of Sciences of the USSR and the Moscovian Mathematical Society) [1866-1935]
• CONT AS (**J 0142**) Mat Sb, Akad Nauk SSSR, NS

J 1405 Ann Pol Math • PL
Annales Polonici Mathematici. Polska Akademia Nauk, Instytut Matematyczny [1953ff] ISSN 0066-2216
• CONT OF (**J 0283**) Ann Soc Pol Math

J 1426 Theor Comput Sci • NL
Theoretical Computer Science [1975ff] ISSN 0304-3975

J 1428 SIAM J Comp • USA
Journal on Computing. SIAM (= Society for Industrial and Applied Mathematics) [1972ff] ISSN 0097-5397

J 1431 Acta Inf • D
Acta Informatica [1971ff] ISSN 0001-5903

J 1447 Houston J Math • USA
Houston Journal of Mathematics [1975ff] ISSN 0362-1588

J 1452 Itogi Nauki Tekh, Ser Sovrem Probl Mat • SU
Itogi Nauki i Tekhniki: Seriya Sovremennye Problemy Matematiki (Progress in Science and Technology: Series on Current Problems in Mathematics) [1972ff]
• CONT OF (**J 1021**) Itogi Nauki Ser Mat • TRANSL IN (**J 1531**) J Sov Math

J 1458 Alkalmaz Mat Lapok • H
Alkalmazott Matematikai Lapok (Papers in Applied Mathematics) [1951ff] ISSN 0133-3399
• CONT OF (**J 0462**) Mat Fiz Oszt Koezlem, Acad Sci Hung

J 1472 Sci Rep Saitama Univ, Ser A • J
The Science Reports of the Saitama University. Series A. Mathematics [1952-1982] ISSN 0558-2431
• CONT AS (**J 3940**) Saitama Math J Physics, and Chemistry

J 1480 Bull Malaysian Math Soc • MAL
The Bulletin of the Malaysian Mathematical Society [1970-1977] ISSN 0126-6705
• CONT AS (**J 3530**) Bull Malaysian Math Soc, Ser 2

J 1487 Publ Math Inst Acad Sci Hung • H
Publications of the Mathematical Institute of the Hungarian Academy of Sciences [1956ff]

J 1488 Itogi Nauki Tekh, Ser Probl Geom • SU
Itogi Nauki i Tekhnike. Seriya Problemy Geometrii (Progress in Science and Technology. Series Problems in Geometry) [1972ff] ISSN 0202-7461
• CONT OF (**J 1021**) Itogi Nauki Ser Mat • TRANSL IN (**J 1531**) J Sov Math

J 1501 Itogi Nauki Tekh, Ser Algeb, Topol, Geom • SU
Itogi Nauki i Tekhniki. Seriya Algebra, Topologiya, Geometriya. (Progress in Science and Technology. Series Algebra, Topology, Geometry) [1972ff] ISSN 0202-7445
• CONT OF (**J 1021**) Itogi Nauki Ser Mat • TRANSL IN (**J 1531**) J Sov Math & (**C 4688**) Prog in Math, Vol 12
• REM C4688 is Volume 1968

J 1507 An Inst Mat, Univ Nac Aut Mexico • MEX
Anales del Instituto de Matematicas. Universidad Nacional Autonoma de Mexico [1961ff] ISSN 0076-7441

J 1508 Math Sem Notes, Kobe Univ • J
Mathematics Seminar Notes. Kobe University [1973-1983] ISSN 0385-633X
• CONT AS (**J 4390**) Kobe J Math

J 1514 Praxis Math • D
Praxis der Mathematik. Monatshefte der Reinen und Angewandten Mathematik im Unterricht [1959ff] ISSN 0032-7042

J 1515 Archimede • I
Archimede. Rivista per gli Insegnanti e i Cultori di Matematiche Pure e Applicate [1949ff] ISSN 0003-8369

J 1516 Vest Ser Fiz Mat Mekh, Univ Minsk • SU
Vestnik Belorusskogo Gosudarstvennogo Universitet im. V.I.Lenina. Seriya I. Fizika, Matematika, Mekhanika (Publications of the Byelorussian State University. Series I: Physics, Mathematics, Mechanics) [1969ff] ISSN 0321-0367

J 1517 Theor Linguist • D
Theoretical Linguistics [1974ff] ISSN 0301-4428

J 1522 Math Slovaca • CS
Mathematica Slovaca [1976ff] ISSN 0025-5173
• CONT OF (**J 0143**) Mat Chasopis (Slov Akad Ved)

J 1526 Riv Mat Univ Parma, Ser 2 • I
Rivista di Matematica della Universita di Parma. Serie 2 [1960-1971] ISSN 0035-6298
• CONT OF (**J 0391**) Riv Mat Univ Parma • CONT AS (**J 3254**) Riv Mat Univ Parma, Ser 3

J 1531 J Sov Math • USA
Journal of Soviet Mathematics [1973ff] ISSN 0090-4104
• TRANSL OF (**J** 1021) Itogi Nauki Ser Mat & (**S** 0228) Zap Nauch Sem Leningrad Otd Mat Inst Steklov & Problemy Matimaticheskogo Analiza & (**J** 1452) Itogi Nauki Tekh, Ser Sovrem Probl Mat & (**J** 1501) Itogi Nauki Tekh, Ser Algeb, Topol, Geom & (**J** 3188) Itogi Nauki Tekh, Ser Teor Veroyat Mat Stat Teor Kibern & (**J** 1488) Itogi Nauki Tekh, Ser Probl Geom
• REM This contains selected translations from each of the Russian Journals listed

J 1550 Creation Math • IL
Creation in Mathematics [1970ff]

J 1573 Notas Commun Mat (Recife) • BR
Notas e Comunicacoes de Matematica [1965ff] ISSN 0085-5413

J 1620 Asterisque • F
Asterisque [1973ff] ISSN 0303-1179

J 1650 Rapp, Sem Math Pure, Univ Cathol Louvain • B
Rapport. Seminaire de Mathematique Pure

J 1669 J Comb Th • USA
Journal of Combinatorial Theory [1966-1970] ISSN 0021-9800
• CONT AS (**J** 0164) J Comb Th, Ser A & (**J** 0033) J Comb Th, Ser B

J 1699 Bull Soc Math Grece • GR
(Bulletin de la Societe Mathematique Grece) [1921-1959]
• CONT AS (**J** 0465) Bull Greek Math Soc (NS)

J 1720 Rev Neoscolast Philos, Ser 2 • B
Revue Neo-Scolastique de Philosophie. Serie 2 [1910-1945]
• CONT AS (**J** 0252) Rev Philos Louvain

J 1735 Bull South East Asian Soc • SGP
Bulletin of the South East Asian Society [1977ff]

J 1737 Nat Math Magazine (Louisiana) • USA
National Mathematics Magazine [1926-1946]
• CONT AS (**J** 0497) Math Mag

J 1738 Rev Roumaine Sci Soc • RO
Revue Roumaine des Sciences Sociales. Serie de Philosophie et Logique. [1964ff] ISSN 0035-4031

J 1741 Int J Man-Mach Stud • USA
International Journal of Man-Machine Studies [1969ff] ISSN 0020-7373

J 1742 Atti Accad Sci Torino, Fis Mat Nat • I
Atti della Reale Accademia delle Scienze di Torino. Classe I: Scienze Fisiche, Matematiche e Naturali [1865-1939]
• CONT AS (**J** 0220) Atti Accad Sci Torino, Fis Mat Nat

J 1743 Int J Gen Syst • USA
International Journal of General Systems: Methodology, Applications, Education [1974ff] ISSN 0308-1079

J 1770 Osaka Math J • J
Osaka Mathematical Journal [1949-1963]
• CONT AS (**J** 0351) Osaka J Math

J 1793 Nieuw Arch Wisk, Ser 2 • NL
Nieuw Archief voor Wiskunde. Reeks 2 [1894-1952]
• CONT OF (**J** 0046) Nieuw Arch Wisk • CONT AS (**J** 3077) Nieuw Arch Wisk, Ser 3

J 1797 ICSU Review World Sci • NL
International Council of Scientific Unions (ICSU) Review of World Science [1963ff] ISSN 0536-1338

J 1798 Convivium (Barcelona) • E
Convivium: Filosofia, Psicologia, Humanidades [1956ff] ISSN 0010-8235

J 1910 Proc London Math Soc, Ser 2 • GB
Proceedings of the London Mathematical Society. Serie 2 [1904-1951]
• CONT OF (**J** 0077) Proc London Math Soc • CONT AS (**J** 3240) Proc London Math Soc, Ser 3

J 1928 Mathematica B • NL
Mathematica : Tijdschrift voor Allen die de Hoogere Wiskunde Beoefenen. Afdeling B [1934-1948]
• CONT AS (**J** 0061) Simon Stevin
• REM Vols 1 - 3/1 appeared under the title 'Mathematica', which from 3/2 onwards was split into 'Mathematica A' and 'Mathematica B'

J 1929 Prace Centr Oblicz Pol Akad Nauk • PL
Polska Akademija Nauk. Centrum Obliczeniowe. Prace. (Polish Academy of Sciences. Computation Centre. Reports) ISSN 0079-3175

J 1934 Ann Sci Univ Clermont Math • F
Annales Scientifiques de l'Universite de Clermont-Ferrand II, Section Mathematiques (Clermont Ferrand) [1973ff] ISSN 0069-472X
• CONT OF (**J** 0179) Ann Fac Sci Clermont

J 1951 Rev Acad Cienc Exact Fis-Quim Zaragoza, Ser 2 • E
Revista de la Academia de Ciencias Exactas, Fisico-Quimicas y Naturales de Zaragoza. Serie 2

J 2022 Comm Math Helvetici • CH
Commentarii Mathematici Helvetici [1929ff] ISSN 0010-2571

J 2028 Hist & Phil Log • GB
History and Philosophy of Logic [1980ff] ISSN 0144-5340

J 2038 Rend Sem Mat, Torino • I
Rendiconti del Seminario Matematico (gia "Conferenze di Fisica e di Matematica"). Universita e Politecnico di Torino

J 2085 Acta Cient Venez • YV
Acta Cientifica Venezolana [1950ff] ISSN 0001-5504

J 2095 Fund Inform, Ann Soc Math Pol, Ser 4 • PL
*Fundamenta Informaticae. Annales Societatis Mathematicae Polonae. Series 4 * Roczniki Polskiego Towarzystwa Matematycznego. Seria 4* [1977ff] ISSN 0324-8429
• CONT OF (**J** 0611) Pol Tow Mat, Prace Mat-Fiz

J 2099 Boll Unione Mat Ital, VI Ser, A • I
Bolletino della Unione Matematica Italiana. Serie VI. A [1982ff] ISSN 0041-7084
• CONT OF (**J** 3285) Boll Unione Mat Ital, V Ser, A

J 2100 Boll Unione Mat Ital, VI Ser, B • I
Bolletino della Unione Matematica Italiana. Serie VI. B [1982ff] ISSN 0041-7084
• CONT OF (**J** 3495) Boll Unione Mat Ital, V Ser, B

J 2107 Publ Dep Math, Lyon, NS • F
Publications du Departement de Mathematiques. Nouvelle Serie. Faculte des Sciences de Lyon. [1982ff] ISSN 0076-1656
• CONT OF (**J** 0056) Publ Dep Math, Lyon

J 2128 C R Math Acad Sci, Soc Roy Canada • CDN
Comptes Rendus Mathematiques de l'Academie des Sciences. La Societe Royale du Canada ∗ Mathematical Reports of the Academy of Sciences [1979ff] ISSN 0706-1994

J 2130 Linguist Philos • NL
Linguistics and Philosophy [1977ff] ISSN 0165-0157

J 2293 Comp Linguist & Comp Lang • H
Computational Linguistics and Computer Languages

J 2310 Obz Mat Fiz, Ljubljana • YU
Obzornik za Matematiko in Fiziko (Mathematical and Physical Reviews) [1951ff] ISSN 0473-7466

J 2311 Rend Mat, Ser 6 • I
Rendiconti di Matematica. Serie 6 [1968-1980] ISSN 0034-4427
• CONT OF (**J 0384**) Rend Mat, Ser 5

J 2313 C R Acad Sci, Paris, Ser A-B • F
Academie des Sciences de Paris. Comptes Rendus Hebdomadaires des Seances. Serie A: Sciences Mathematiques, Serie B: Sciences Physiques [1966-1980] ISSN 0001-4036
• CONT OF (**J 0109**) C R Acad Sci, Paris • CONT AS (**J 3364**) C R Acad Sci, Paris, Ser 1 & (**J 2314**) C R Acad Sci, Paris, Ser 2

J 2314 C R Acad Sci, Paris, Ser 2 • F
Comptes Rendus des Seances de l'Academie des Sciences. Serie II. Mecanique-Physique, Chimie, Sciences de la Terre, Sciences de l'Univers [1981ff]
• CONT OF (**J 2313**) C R Acad Sci, Paris, Ser A-B

J 2332 J Fac Sci, Univ Tokyo, Sect 1 A • J
Journal of the Faculty of Science. University of Tokyo. Section 1 A. Mathematics [1971ff] ISSN 0040-8980
• CONT OF (**J 0434**) J Fac Sci Univ Tokyo, Sect 1

J 2503 PLISKA Stud Math Bulgar • BG
PLISKA Studia Mathematica Bulgarica. B"lgarska Akademiya na Naukite. Sofijski Universitet [1980ff]

J 2521 Beijing Shifan Daxue Xuebao, Ziran Kexue • TJ
Beijing Shifan Daxue Xuebao. Ziran Kexue Ban (Journal of Natural Sciences of Beijing Normal University. Natural Science Edition)

J 2547 Serdica, Bulgar Math Publ • BG
Serdica. Bulgaricae Mathematicae Publicationes [1975ff] ISSN 0204-4110
• CONT OF (**J 1156**) Izv Bulgar Akad Nauk Mat Inst

J 2563 Result Math • CH
Resultate der Mathematik [1978ff] ISSN 0378-6218

J 2574 Litov Mat Sb (Vil'nyus) • SU
Litovskij Matematicheskij Sbornik ∗ Lietuvos Matematikos Rinkinys (Lithuanian Mathematical Collected Articles) [1961ff] ISSN 0132-2818
• TRANSL IN (**J 3283**) Lith Math J

J 2604 Progr Comput Software • USA
Programming and Computer Software [1975ff] ISSN 0361-7688
• TRANSL OF (**J 2605**) Programmirovanie

J 2605 Programmirovanie • SU
Programmirovanie (Programming) [1975ff] ISSN 0132-3474
• TRANSL IN (**J 2604**) Progr Comput Software

J 2606 Tsukuba J Math • J
Tsukuba Journal of Mathematics [1977ff] ISSN 0387-4982

J 2613 Tokyo J Math • J
Tokyo Journal of Mathematics [1951ff] ISSN 0387-3870

J 2631 J Math Kyoto Univ • J
Journal of Mathematics of Kyoto University [1961ff] ISSN 0023-608X

J 2635 Topology Appl • NL
Topology and its Applications. A Journal Devoted to General, Geometric, Set-Theoretic and Algebraic Topology [1980ff] ISSN 0166-8641
• CONT OF (**J 0254**) Gen Topology Appl

J 2650 Adv Appl Math • USA
Advances in Applied Mathematics. [1980ff] ISSN 0196-8858

J 2658 Ann Inst Henri Poincare, Sect A • F
Annales de l'Institut Henri Poincare. Section A. Physique Theorique ISSN 0020-2339

J 2660 Ann Sci Math Quebec • CDN
Les Annales des Sciences Mathematiques du Quebec. [1977ff] ISSN 0707-9109

J 2667 Autom Doc Math Linguist • USA
Automatic Documentation and Mathematical Linguistics [1967ff] ISSN 0005-1055
• TRANSL OF (**J 0338**) Nauch-Tekh Inf, Ser 2, Akad Nauk SSSR

J 2668 Avtom Sist Upravl & Prib Avtom, Khar'kov • SU
Avtomatizirovannye Sistemy Upravleniya i Pribory Avtomatiki (Automatisierte Steuersysteme und Automatisierungsvorrichtungen) [1965ff] ISSN 0135-1710

J 2679 Bull Inst Math Appl (Southend oS) • GB
Bulletin of the Institute of Mathematics and its Applications [1965ff]

J 2684 J Huazhong Inst Tech (Engl Ed) • TJ
Journal of Huazhong Institute of Technology. English Edition [1979-1981]
• CONT AS (**J 3218**) J Huazhong Univ Sci Tech (Engl Ed)
• TRANSL OF (**J 2754**) Huazhong Gongxueyuan Xuebao

J 2688 Conceptus (Wien) • A
Conceptus. Zeitschrift fuer Philosophie [1967ff] ISSN 0010-5155

J 2713 Eleutheria, Math J Sem Zervos (Athens) • GR
Eleutheria. Mathematical Journal of the Seminar of P. Zervos. (Liberty) [1978ff]

J 2718 Fct Approximatio, Comment Math, Poznan • PL
Functiones et Approximatio. Commentarii Mathematici [1974ff]

J 2736 Int J Theor Phys • USA
International Journal of Theoretical Physics [1968ff] ISSN 0020-7748

J 2751 J Fac Lib Art Yamaguchi Univ • J
Journal of the Faculty of Liberal Arts. Yamaguchi University. Natural Science [1967ff]

J 2753 J Graph Th • USA
Journal of Graph Theory [1976ff] ISSN 0364-9024

J 2754 Huazhong Gongxueyuan Xuebao • TJ
Huazhong Gongxueyuan Xuebao ∗ Zhongguo Kexue Shuxue Zhuanji (Journal Huazhong (Central China) University of Science and Technology)
• TRANSL IN (**J 2684**) J Huazhong Inst Tech (Engl Ed) & (**J 3218**) J Huazhong Univ Sci Tech (Engl Ed)

J 2761 J Recreational Math • USA
Journal of Recreational Mathematics [1968ff] ISSN 0022-412X

J 2771 Kexue Tongbao • TJ
Kexue Tongbao (Science Bulletin) [1950ff] ISSN 0023-074X
• TRANSL IN (J 3769) Sci Bull, Foreign Lang Ed

J 2775 Nuovo Cimento B, Ser 2 • I
Il Nuovo Cimento. B. Serie II

J 2789 Math Intell • D
The Mathematical Intelligencer [1978ff] ISSN 0343-6993

J 2790 Math Sem-ber • D
Mathematische Semesterberichte [1980ff] ISSN 0720-728X
• CONT OF (J 0160) Math-Phys Sem-ber, NS

J 2821 Pubbl Ist Mat App Univ Stud Roma • I
Pubblicazioni. Universita degli Studi di Roma. Facolta di Ingegneria. Istituto di Matematica Applicata.

J 2823 Quaest Math, S Africa • ZA
*Quaestiones Mathematicae. Journal of the South African Mathematical Society * Tydskrift van die Suid-Afrikaanse Wiskundevereniging* ISSN 0379-9468

J 2826 Rep Fac Sci, Kagoshima Univ, Math Phys Chem • J
Reports of the Faculty of Science of Kagoshima University. Mathematics, Physics, and Chemistry

J 2831 RAIRO Autom • F
RAIRO Automatique. RAIRO (= Revue Francaise d'Automatique, d'Informatique et de Recherche Operationnelle). Series Automatique [1977ff] ISSN 0399-0524
• CONT OF (J 0205) Rev Franc Autom, Inf & Rech Operat Ser Jaune

J 2832 RAIRO Inform • F
RAIRO Informatique. Revue Francaise d'Automatique, d'Informatique et de Recherche Operationnelle. Series Informatique [1977ff] ISSN 0399-0532
• CONT OF (J 0205) Rev Franc Autom, Inf & Rech Operat Ser Bleue

J 2845 Tanulmanyok • H
Tanulmanyok. Szamitastechnikai es Automatizalasi Kutato Intezete (Studies. Research Institut for Computer Science and Automatization)

J 2853 TRU Math (Tokyo) • J
TRU Mathematics [1965ff] ISSN 0496-6597

J 2855 Zbor Rad, Prir-Mat Fak, Ser Mat (Novi Sad) • YU
Zbornik Radova Prirodno-Matematichkog Fakulteta. Serija za Matematiku (Collected Papers of the Faculty of Natural Sciences and Mathematics. Mathematical Series) [1971ff]

J 2887 Zbor Radova, NS • YU
Zbornik Radova. Nova Serija. (Collected Papers. New Series)

J 3075 Normat • N
Normat. Nordisk Matematisk Tidskrift (Normat. Scandinavian Mathematical Journal) [1979ff]
• CONT OF (J 0311) Nordisk Mat Tidskr

J 3077 Nieuw Arch Wisk, Ser 3 • NL
Nieuw Archief voor Wiskunde. Derde Serie [1953-1982] ISSN 0028-9825
• CONT OF (J 1793) Nieuw Arch Wisk, Ser 2 • CONT AS (J 3929) Nieuw Arch Wisk, Ser 4

J 3079 Kiber Sb Perevodov, NS • SU
Kiberneticheskij Sbornik. Novaya Seriya. Sbornik Perevodov (Collected Articles on Cybernetics: New Series. Collected Translations) [1965ff] ISSN 0453-8382
• CONT OF (J 1048) Kiber Sb Perevodov

J 3124 Beitr Algebra Geom • DDR
Beitraege zur Algebra und Geometrie [1971ff] ISSN 0440-1298

J 3127 Bol Soc Mat Mexicana, Ser 2 • MEX
Boletin de la Sociedad Matematica Mexicana. Segunda Serie [1956ff] ISSN 0037-8615
• CONT OF (J 0250) Bol Soc Mat Mexicana

J 3128 Boll Unione Mat Ital, Suppl • I
Bollettino della Unione Matematica Italiana. Supplemento

J 3131 Bul Inst Politeh Bucuresti, Ser Chim-Metal • RO
Buletinul Institutului Politehnic "Gheorghe Gheorghiu-Dej" Bucuresti. Seria Chimie-Metalurgie. (Bulletin des Polytechnischen Instituts "Gheorghe Gheorghiu-Dej" Bukarest. Serie Chemie-Metallurgie)

J 3133 Bull Soc Math Belg, Ser B • B
Bulletin de la Societe Mathematique de Belgique. Serie B [1977ff] ISSN 0037-9476
• CONT OF (J 0082) Bull Soc Math Belg

J 3163 Fct Anal & Appl • USA
Functional Analysis and its Applications [1967ff]
• TRANSL OF (J 1207) Fkts Anal Prilozh

J 3172 J London Math Soc, Ser 2 • GB
Journal of the London Mathematical Society. 2nd Series [1969ff] ISSN 0024-6107
• CONT OF (J 0039) J London Math Soc

J 3188 Itogi Nauki Tekh, Ser Teor Veroyat Mat Stat Teor Kibern • SU
Itogi Nauki i Tekhniki. Seriya Teoriya Veroyatnostej, Matematicheskaya Statistika, Teoreticheskaya Kibernetika (Progress in Science and Technology. Series Probability Theory, Mathematical Statistics, Theoretical Cybernetics) [1972ff] ISSN 0202-7488
• CONT OF (J 1021) Itogi Nauki Ser Mat • TRANSL IN (J 1531) J Sov Math

J 3194 J Austral Math Soc, Ser A • AUS
Journal of the Australian Mathematical Society. Series A [1976ff] ISSN 0263-6115
• CONT OF (J 0038) J Austral Math Soc

J 3218 J Huazhong Univ Sci Tech (Engl Ed) • TJ
Journal of Huazhong University of Science and Technology [1982ff]
• CONT OF (J 2684) J Huazhong Inst Tech (Engl Ed) • TRANSL OF (J 2754) Huazhong Gongxueyuan Xuebao

J 3239 Proc Japan Acad, Ser A • J
Proceedings of the Japan Academy. Series A. Mathematical Sciences [1978ff] ISSN 0386-2194

J 3240 Proc London Math Soc, Ser 3 • GB
Proceedings of the London Mathematical Society. 3rd Series [1951ff] ISSN 0024-6115
• CONT OF (J 1910) Proc London Math Soc, Ser 2

J 3250 Bol Soc Brasil Mat • BR
Boletim da Sociedade Brasileira de Matematica [1970ff]

J 3254 Riv Mat Univ Parma, Ser 3 • I
Rivista di Matematica della Universita di Parma. Serie 3
[1972-1974] ISSN 0035-6298
• CONT OF (**J 1526**) Riv Mat Univ Parma, Ser 2 • CONT AS
(**J 0549**) Riv Mat Univ Parma, Ser 4

J 3265 Sov J Contemp Math Anal, Armen Acad Sci • USA
*Soviet Journal of Contemporary Mathematical Analysis.
Armenian Academy of Sciences* [1979ff] ISSN 0735-2719
• TRANSL OF (**J 0312**) Izv Akad Nauk Armyan SSR, Ser Mat

J 3279 Trans Moscow Math Soc • USA
Transactions of the Moscow Mathematical Society ISSN
0077-1554
• TRANSL OF (**J 0065**) Tr Moskva Mat Obshch

J 3281 Ukr Math J • USA
Ukrainian Mathematical Journal [1967ff] ISSN 0041-5995
• TRANSL OF (**J 0265**) Ukr Mat Zh, Akad Nauk Ukr SSR

J 3283 Lith Math J • USA
Lithuanian Mathematical Journal. [1975ff] ISSN 0363-1672
• TRANSL OF (**J 2574**) Litov Mat Sb (Vil'nyus)
• REM Only selected translations of J2574

J 3285 Boll Unione Mat Ital, V Ser, A • I
Bollettino della Unione Matematica Italiana. Serie V. A
[1976-1981] ISSN 0041-7084
• CONT OF (**J 0012**) Boll Unione Mat Ital, IV Ser • CONT AS
(**J 2099**) Boll Unione Mat Ital, VI Ser, A

J 3293 Bull Acad Pol Sci, Ser Math • PL
*Bulletin de l'Academie Polonaise des Sciences. Serie des
Sciences Mathematiques* [1979-1982] ISSN 0001-4117
• CONT OF (**J 0014**) Bull Acad Pol Sci, Ser Math Astron Phys
• CONT AS (**J 3417**) Bull Pol Acad Sci, Math

J 3359 Appl Comp Sci, Ber Prakt Inf • D
Applied Computer Science. Berichte zur praktischen Informatik

J 3364 C R Acad Sci, Paris, Ser 1 • F
*Comptes Rendus des Seances de l'Academie des Sciences. Serie
I: Science Mathematique* [1981ff] ISSN 0151-0509
• CONT OF (**J 2313**) C R Acad Sci, Paris, Ser A-B

J 3370 Enseign Math, Ser 2 • CH
L'Enseignement Mathematique. Revue Internationale. Serie 2
[1955ff] ISSN 0013-8584
• CONT OF (**J 0152**) Enseign Math

J 3401 Meth Oper Res • D
Methods of Operations Research [1963ff] ISSN 0078-5318

J 3417 Bull Pol Acad Sci, Math • PL
Bulletin of the Polish Academy of Sciences. Mathematics
[1983ff] ISSN 0001-4117
• CONT OF (**J 3293**) Bull Acad Pol Sci, Ser Math

J 3420 Proc Edinburgh Math Soc, Ser 2 • GB
Proceedings of the Edinburgh Mathematical Society. Series 2.
[1927ff] ISSN 0013-0915
• CONT OF (**J 0386**) Proc Edinburgh Math Soc

J 3434 Pubbl Ist Appl Calcolo, Ser 3 • I
*Pubblicazioni. Serie III. Istituto per le Applicazioni del Calcolo
"Mauro Picone" (IAC) Consiglio Nazionale delle Ricerche*

J 3436 Quad, Ist Appl Calcolo, Ser 3 • I
Quaderni Serie III. Istituto Applicazione Calcolo

J 3437 Rep, Akad Wiss DDR, Inst Math • DDR
*Report. Akademie der Wissenschaften der DDR. Institut fuer
Mathematik.*

J 3441 RAIRO Inform Theor • F
*Revue Francaise d'Automatique, d'Informatique et de
Recherche Operationnelle (RAIRO), Informatique Theorique*
[1977ff] ISSN 0399-0540
• CONT OF (**J 4698**) Rev Franc Autom, Inf & Rech Operat,
Ser Rouge Inf Th

J 3449 Sov Math • USA
Soviet Mathematics [1974ff] ISSN 0197-7156
• TRANSL OF (**J 0031**) Izv Vyssh Ucheb Zaved, Mat (Kazan)

J 3450 Studia Univ Babes-Bolyai, Math (Cluj) • RO
Studia Universitatis Babes-Bolyai. Mathematica [1976ff]
• CONT OF (**J 3451**) Stud Univ Cluj, Ser Math-Mech

J 3451 Stud Univ Cluj, Ser Math-Mech • RO
*Studia Universitatis Babes-Bolyai, Series
Mathematica-Mechanica* [1971-1975] ISSN 0370-8659
• CONT OF (**J 0355**) Studia Univ Babes-Bolyai, Math-Phys
(Cluj) • CONT AS (**J 3450**) Studia Univ Babes-Bolyai, Math
(Cluj)

J 3456 Th Probab & Math Stat • USA
Theory of Probability and Mathematical Statistics [1956ff]
ISSN 0040-585X
• TRANSL OF (**J 0295**) Teor Veroyat i Mat Stat (Kiev)

J 3457 Order • NL
Order [1984ff]

J 3495 Boll Unione Mat Ital, V Ser, B • I
Bolletino della Unione Matematica Italiana. Serie V. B
[1976-1981] ISSN 0041-7084
• CONT OF (**J 0012**) Boll Unione Mat Ital, IV Ser • CONT AS
(**J 2100**) Boll Unione Mat Ital, VI Ser, B

J 3522 Rend Circ Mat Palermo, Ser 2 • I
Rendiconti del Circolo Matematico di Palermo. Serie II
[1952ff] ISSN 0009-725X
• CONT OF (**J 0058**) Rend Circ Mat Palermo

J 3524 Kybernetika Suppl (Prague) • CS
*Kybernetika. Supplement. (Prague) (Cybernetics. Supplement.
(Prague))*
• REL PUBL (**J 0156**) Kybernetika (Prague)

J 3526 Ann Mat Pura Appl, Ser 4 • I
*Annali di Matematica Pura ed Applicata. Serie Quarta. Sotto
gli Auspici del Consiglio Nazionale delle Ricerche* ISSN
0003-4622
• CONT OF (**J 0229**) Ann Mat Pura Appl, Ser 3

J 3530 Bull Malaysian Math Soc, Ser 2 • MAL
*The Bulletin of the Malaysian Mathematical Society. 2nd
Series.* [1978ff] ISSN 0126-6705
• CONT OF (**J 1480**) Bull Malaysian Math Soc

J 3536 Uch Zap Univ, Ivanovo • SU
*Ivanovskij Gosudarstvennyj Universitet. Uchenye Zapiski
(State University of Ivanovo. Scientific Notes)* [1974ff]
• CONT OF (**J 0226**) Uch Zap Ped Inst, Ivanovo

J 3597 J Unif Sci • USA
The Journal of Unified Science [1940-1974]
• CONT OF (**J 0748**) Erkenntnis (Leipzig) • CONT AS (**J 0158**)
Erkenntnis (Dordrecht)

J 3601 Boll Unione Mat Ital, VI Ser, C • I
*Unione Matematica Italiana. Bollettino. Serie VI. C. Analisi
Funzionale e Applicazioni* [1982ff] ISSN 0041-7084

J 3630 Bull Dept of Lib Arts (Numazu) • J
Bulletin of the Department of Liberal Arts [1974ff]

J 3639 Notiz Unione Mat Ital • I
Notiziario della Unione Matematica Italiana

J 3741 Rend Mat Appl, Ser 7 • I
Rendiconti di Matematica e delle sue Applicazioni. Seria 7
[1981ff] ISSN 0034-4427
• CONT OF (**J 0297**) Rend Mat Appl, Ser 5

J 3742 Shuxue Yanjiu yu Pinglun • TJ
Shuxue Yanjiu yu Pinglun (Journal of Mathematical Research & Exposition) [1981ff]

J 3746 Note Math (Lecce) • I
Note di Matematica. Pubblicazione Semestrale [1981ff]

J 3766 Zhongguo Kexue, Xi A • TJ
Zhongguo Kexue. Xi A. Mathematical, Physical, Astronomical & Technical Sciences ∗ Scientia Sinica. Series A [1982ff]
• CONT OF (**J 1024**) Zhongguo Kexue

J 3768 Boll Unione Mat Ital, VI Ser, D • I
Bolletino della Unione Matematica Italiana. Serie VI. D. Algebra e Geometria [1982ff] ISSN 0041-7084

J 3769 Sci Bull, Foreign Lang Ed • TJ
Kexue Tongbao. Foreign Language Edition (Science Bulletin) [1980ff]
• TRANSL OF (**J 2771**) Kexue Tongbao

J 3770 Dongbei Shifan Daxue Xuebao, Ziran Kexue • TJ
Dongbei Shifa Daxue Xuebao. Ziran Kexue Ban (Journal of Northeast Normal University. Natural Sciences)

J 3781 Topoi • NL
Topoi. An International Review of Philosophy [1982ff] ISSN 0167-7411

J 3797 Diagrammes • F
Diagrammes

J 3824 Bull Soc Math Belg, Ser A • B
Bulletin de la Societe Mathematique de Belgique. Serie A [1977ff]
• CONT OF (**J 0082**) Bull Soc Math Belg

J 3929 Nieuw Arch Wisk, Ser 4 • NL
Nieuw Archief voor Wiskunde. Vierde Serie. Uitgegeven door het Wiskundid Genootschap te Amsterdam [1983ff] ISSN 0028-9825
• CONT OF (**J 3077**) Nieuw Arch Wisk, Ser 3

J 3939 Mat Logika Primen (Akad Nauk Litov SSR) • SU
Matematicheskaya Logika i Ee Primeneniya (Mathematical Logic and its Applications) [1981ff]

J 3940 Saitama Math J • J
Saitama Mathematical Journal [1983ff]
• CONT OF (**J 1472**) Sci Rep Saitama Univ, Ser A

J 3941 J Math Pures Appl, Ser 9 • F
Journal de Mathematiques Pures et Appliquees. Neuvieme Serie [1922ff] ISSN 0021-7824
• CONT OF (**J 0249**) J Math Pures Appl

J 3954 Rev Franc Inf & Rech Operat • F
Association Francaise pour la Cybernetique Economique et Technique. Revue Francaise d'Informatique et de Recherche Operationnelle [1967-1971]
• CONT AS (**J 0205**) Rev Franc Autom, Inf & Rech Operat

J 3957 Bull Inf & Cybern (Kyushu Univ) • J
Bulletin of Informatics and Cybernetics. Research Association of Statistical Sciences [1982ff] ISSN 0286-522X

J 3986 Ist Lombardo Rend, Ser 2 (Milano) • I
Reale Istituto Lombardo di Scienze e Lettre. Rendiconti. Series 2 [1868-1936]
• CONT AS (**J 0207**) Ist Lombardo Accad Sci Rend, A (Milano)

J 3993 Rev Mat Hisp-Amer, Ser 2 • E
Revista Matematica Hispano-Americana. Serie 2 ISSN 0373-0999
• CONT AS (**J 0236**) Rev Mat Hisp-Amer, Ser 4

J 3994 Ann Acad Sci Fennicae Ser A I • SF
Annales Academiae Scientiarum Fennicae. Serie A I [1941-1974] ISSN 0066-1953
• CONT AS (**J 0446**) Ann Acad Sci Fennicae, Ser A I, Diss

J 3996 Boll Unione Mat Ital • I
Bollettino della Unione Matematica Italiana [1922-1945]
• CONT AS (**J 4408**) Boll Unione Mat Ital, III Ser

J 3998 Proc & Trans Roy Soc Canada, Ser 3 • CDN
Proceedings and Transactions of the Royal Society of Canada. Serie 3. Section III: Mathematical, Physical and Chemical Sciences

J 4135 Dialexeis • GR
Dialexeis (Reports of the Greek Mathematical Society)

J 4149 Publ Inst Rech Math Avancee • F
Publications de IRMA (= Institut de Recherche de Mathematiques Avancee)

J 4186 Atti Accad Sci Lett Arti Palermo, Ser 5/I • I
Atti della Accademia di Scienze. Lettere e Arti di Palermo. Serie Quinta. Parte I: Scienze [1984ff]
• CONT OF (**J 0104**) Atti Accad Sci Lett Arti Palermo, Ser 4/I

J 4231 Z Anal & Anwendungen • DDR
Zeitschrift fuer Analysis und Ihre Anwendungen [1982ff]

J 4277 Vesn Drusht Mat Fiz Serbije • YU
Vesnik Drushtva Matematichara i Fizichara Narodne Republike Serbije ∗ Vestnik Obshchestva Matematikov i Fizikov N.R. Serbie ∗ Bulletin de la Societe des Mathematiciens et Physiciens de la R.P. Serbie (Publications of the Society of Mathematicians and Physicists of the P.R. Serbia) [1949-1963]
• CONT AS (**J 0042**) Mat Vesn, Drust Mat Fiz Astron Serb

J 4373 GMS Math Inst Patras • GR
Greek Mathematical Society, Mathematical Institut of Patras

J 4387 Itogi Nauki Tekh, Ser Tekh Kibern • SU
Itogi Nauki i Tekhniki. Seriya Tekhnicheskaya Kibernetika. Gosudarstvennyj Komitet SSSR po Nauke i Tekhnike. Akademiya Nauk SSSR. Vsesoyuznyj Institut Nauchnoj i Tekhnicheskoj Informatsii. (Progress in Science and Technology. Series: Engineering Cybernetics) [1972ff]
• CONT OF (**J 1021**) Itogi Nauki Ser Mat

J 4390 Kobe J Math • J
Kobe Journal of Mathematics [1984ff] ISSN 0289-9051
• CONT OF (**J 1508**) Math Sem Notes, Kobe Univ

J 4408 Boll Unione Mat Ital, III Ser • I
Bolletino della Unione Matematica Italiana, Ser III [1946-1967]
• CONT OF (**J 3996**) Boll Unione Mat Ital • CONT AS (**J 0012**) Boll Unione Mat Ital, IV Ser

J 4439 Kagaku • J
Kagaku (Chemical Sciences)

J 4443 Gac Mat (Madrid) • E
Gaceta Matematica [1949-1969]
• CONT AS (J 0299) Gac Mat, Ser 1a (Madrid)

J 4510 Norsk Mat Tidsskr • N
Norsk Matematisk Tidsskrift (Scandinavian Mathematical Journal) [1919-1952]
• CONT OF (J 0259) Nyt Tidsskr Mat • CONT AS (J 0311) Nordisk Mat Tidskr

J 4535 Inst Politeh Timisoara Sem Mat Fiz • RO
Institutul Politehnic "Traian Vuia" Timisoara. Seminar Matematica - Fizica (Das Polytechnische Institut "Traian Vuia" Timisoara. Seminar Mathematik-Physik) [1982ff]
• CONT OF (J 0218) Bul Sti Tehn Inst Politeh Timisoara, Ser Mat-Fiz-Mec

J 4609 J Symb Comput • USA
Journal of Symbolic Computation [1985ff]

J 4634 Jiangxi Shiyuan Xuebao • TJ
Jiangxi Shiyuan Xuebao

J 4681 Sci News • GB
Science News

J 4698 Rev Franc Autom, Inf & Rech Operat, Ser Rouge Inf Th • F
Revue Francaise d'Automatique, d'Informatique et de Recherche Operationnelle (RAIRO). Serie Rouge Informatique Theorique [1975-1976] ISSN 0399-0540
• CONT OF (J 0205) Rev Franc Autom, Inf & Rech Operat Ser Rouge • CONT AS (J 3441) RAIRO Inform Theor

J 4701 Trudy Akad Nauk Ehston SSR, Fiz Astron (Tartu) • SU
*Trudy Instituta Fiziki i Astronomii Akademii Nauk Ehstonskoj SSR * Eesti NSV Teaduste Akadeemia Fueuesika ja Astronomia Instituudi Kurimused* [1951ff]

J 4706 Publ Inst Math (Belgrade) • YU
Academie Serbe des Sciences. Publications de l'Institut Mathematique [1948-1960]
• CONT AS (J 0400) Publ Inst Math, NS (Belgrade)

J 4710 Pol Tow Mat, Wiad Mat • PL
Polskie Towarzystwo Matematyczne. Wiadomosci Matematiczne [1899-1954]
• CONT AS (J 0519) Wiad Mat, Ann Soc Math Pol, Ser 2

J 4713 Mat Fyz Chasopis (Slov Akad Ved) • CS
Matematicky-Fyzikalny Chasopis (Journal of Mathematical Physics) [1951-1966]
• CONT AS (J 0143) Mat Chasopis (Slov Akad Ved)

J 4714 Bul Sti Tehn Inst Politeh Timisoara, NS • RO
Buletinul Stiintific si Tehnic al Institutului Politehnic Timisoara. Serie Noua (Wissenschaftliches und Technisches Bulletin des Polytechnischen Instituts Timisoara. Neue Serie) [1956-1969]
• CONT AS (J 0218) Bul Sti Tehn Inst Politeh Timisoara, Ser Mat-Fiz-Mec

J 4715 Bul Sti Tehn Inst Politeh Timisoara, Ser Mat-Fiz • RO
Buletinul Stiintific si Tehnic al Institutului Politehnic "Traian Vuia" Timisoara. Seria Matematica-Fizica (Wissenschaftliches und Technisches Bulletin des Polytechnischen Instituts "Traian Vuia" Timisoara. Serie Mathematik-Physik) [1978-1981]
• CONT OF (J 0218) Bul Sti Tehn Inst Politeh Timisoara, Ser Mat-Fiz-Mec • CONT AS (J 4535) Inst Politeh Timisoara Sem Mat Fiz

J 4716 Stud Philos, Leopolis (Poznan) • PL
Studia Philosophica. Commentarii Societatis Polonorum. Leopolis [1935ff]

J 4717 Izv Akad Nauk SSSR • SU
Izvestiya Akademii Nauk SSSR (Bulletin de l'Academie des Sciences Mathematiques et Naturelles. Leningrad) [?-1936]
• CONT AS (J 0216) Izv Akad Nauk SSSR, Ser Mat

J 4727 Kumamoto J Sci, Math • J
Kumamoto Journal of Science (Mathematics) [1972ff]
• CONT OF (J 1148) Kumamoto J Sci, Ser A

J 4729 Acta Math Hung • H
Acta Mathematica Hungarica [1983ff] ISSN 0236-5294
• CONT OF (J 0001) Acta Math Acad Sci Hung

Series

S 0019 Colloq Math (Warsaw) • PL
Colloquium Mathematicum [1947ff] • PUBL Academie Polonaise des Sciences, Institut Mathematique: Warsaw
• ISSN 0010-1354

S 0055 Proc Steklov Inst Math • USA
Proceedings of the Steklov Institute of Mathematics [1967ff]
• PUBL (X 0803) Amer Math Soc: Providence
• TRANSL OF (S 0066) Tr Mat Inst Steklov
• ISSN 0081-5438

S 0066 Tr Mat Inst Steklov • SU
Trudy Ordena Lenina Matematicheskogo Instituta imeni V.A.Steklova. Akademiya Nauk SSSR (Proceedings of the Mathematical Steklov-Institute of the Academy of Sciences SSSR) [1938ff] • PUBL (X 0899) Akad Nauk SSSR: Moskva
• CONT OF (S 1644) Tr Fiz-Mat Inst Steklov • TRANSL IN (S 0055) Proc Steklov Inst Math

S 0166 Mat Issl, Mold SSR • SU
Matematicheskie Issledovaniya. Akademiya Nauk Moldavskoj SSR. Ordena Trudovogo Krasnogo Znameni Institut Matematiki s Vychislitel'nym Tsentrom (Mathematical Studies) [1966ff] • PUBL (X 2741) Shtiintsa: Kishinev
• ISSN 0542-9994

S 0167 Mem Amer Math Soc • USA
Memoirs of the American Mathematical Society [1950-1974]
• PUBL (X 0803) Amer Math Soc: Providence
• CONT AS (S 2450) Mem Amer Math Soc, NS
• ISSN 0065-9266

S 0183 Publ Math Univ California • USA
University of California Publications in Mathematics PUBL (X 0926) Univ Calif Pr: Berkeley

S 0208 Uch Zap, Ped Inst, Moskva • SU
Uchenye Zapiski Moskovskogo Gosudarstvennogo Pedagogicheskogo Instituta imeni V.I.Lenina (Scientific Notes of the Moscow State Institute of Education) [1950ff] • PUBL (X 2802) Moskov Ped Inst: Moskva

S 0223 Nauch Trud NS, Politekh Inst Tashkent • SU
Tashkentskij Politekhnicheskij Institut. Nauchnye Trudy. Novaya Seriya (Polytechnic Institute of Tashkent. Scientific Papers. New Series) PUBL Polytekh Inst: Tashkent

S 0228 Zap Nauch Sem Leningrad Otd Mat Inst Steklov • SU
Zapiski Nauchnykh Seminarov Leningradskogo Otdeleniya Ordena Lenina Matematicheskogo Instituta imeni V.A.Steklova Akademii Nauk SSSR (LOMI) (Reports of the Scientific Seminars of the Leningrad Steklov Institute of Mathematics) PUBL (X 2641) Nauka: Leningrad
• TRANSL IN (J 1531) J Sov Math & (J 0521) Semin Math, Inst Steklov
• REM Transl. in J0521 up to vol. 19

S 0257 Monograf Mat • PL
Monografie Matematyczne (Mathematical Monography) [1932ff] • PUBL (X 1034) PWN: Warsaw • ALT PUBL (X 1758) Ars Polona: Warsaw
• ISSN 0077-0507

S 0281 Arch Towarz Nauk Lwow, Sect 3 • PL
Archivum Towarzystwa Naukowego we Lwowie. Dzial 3. Matematyczno-Przyrodniczy (Archive of the Scientific Society of Lwow. Section 3. Mathematics and Natural Sciences) [1920-1939] • PUBL Tow Nauk: Lwow

S 0393 Uch Zap Univ, Tartu • SU
*Uchenye Zapiski Tartuskogo Gosudarstvennogo Universiteta. * Tartu Riikliku Uelikooli Toimetised. * Acta et Commentationes Universitatis Tartuensis (Scientific Notes of the Tartu State University.)* [1961ff] • PUBL (X 2463) Tartusk Gos Univ: Tartu

S 0410 Math Beitr Univ Halle-Wittenberg • DDR
Die Martin Luther-Universitaet Halle-Wittenberg. Mathematische Beitraege [1968ff] • PUBL (X 2446) Univ Halle-Wittenberg: Halle Wissenschaftliche Beitraege. ISSN 0440-1298
• ISSN 0441-6228

S 0458 Zesz Nauk, Prace Log, Uniw Krakow • PL
Zeszyty Naukowe Uniwersytetu Jagiellonskiego Prace z Logiki (Scientific Papers. Jagielleonian University. Reports on Logic) [1965-1972] • PUBL (X 1034) PWN: Warsaw
• CONT AS (J 0302) Rep Math Logic, Krakow & Katowice

S 0478 Bonn Math Schr • D
Bonner Mathematische Schriften [1957ff] • PUBL (X 0908) Univ Math Inst: Bonn
• ISSN 0524-045X

S 0554 Issl Teor Algor & Mat Logik (Moskva) • SU
Issledovaniya po Teorii Algorifmov i Matematicheskoj Logike (Studies in the Theory of Algorithms and Mathematical Logic) [1973,1976,1979] • ED: MARKOV, A.A. & PETRI, N.V. (VOL 1). MARKOV, A.A. & KUSHNER, B.A. (VOL 2). MARKOV, A.A. & KHOMICH, V.I. (VOL 3) • PUBL (X 2265) Akad Nauk Vychis Tsentr: Moskva 287pp,160pp,133pp
• ISSN 0302-9085

S 0778 Uporyad Mnozhestva Reshetki (Saratov) • SU
Uporyadochennye Mnozhestva i Reshetki. Mezhvuzovskij Nauchnyj Sbornik (Ordered Sets and Lattices. Inter-University Scientific Collected Papers) [1971ff] • PUBL (X 0958) Saratov Univ: Saratov

S 1002 Nauch Trud Vissh Ped Inst, Plovdiv • BG
Vissh Pedagogicheski Institut Plovdiv. Nauchni Trudove (Institute of Education of Plovdiv. Scientific Papers) [1963-1972] • PUBL Ped Inst: Plovdiv
• CONT AS (S 1441) Nauch Trud, Univ Plovdiv

S 1011 Mem Nat Def Acad • J
Memoirs of the National Defense Academy. Mathematics, Physics, Chemistry and Engineering PUBL The National Defense Academy: Yokosuka, Japan

S 1057 Semin Dubreil: Alg Th Nombr • F
Seminaire P.Dubreil: Algebre et Theorie des Nombres
[1946-1970] • ED: DUBREIL, P. • PUBL (X 1623) Univ Paris VI
Inst Poincare: Paris
• CONT AS (S 2250) Semin Dubreil: Algebre

S 1070 Probl Mat, Wyz Szk Ped, Bydgosszcz • PL
Problemy Matematyczne. Bydgosszcz Wyzsza Szkola Pedagogiczna. Zeszyty Naukowe (Mathematical Problems. Institute of Education of Bydgosszcz. Scientific Papers) PUBL
Uczel WSP: Bydgosszcz

S 1133 Res Rep Boston Univ • USA
Boston University Research Reports PUBL Boston Univ: Boston

S 1441 Nauch Trud, Univ Plovdiv • BG
Nauchni Trudove na Plovdivski Universitet "Paissi Khilendarski" (Scientific Papers of the University "Paissi Khilendarski", Plovdiv) [1973ff] • PUBL Univ Plovdiv: Plovdiv
• CONT OF (S 1002) Nauch Trud Vissh Ped Inst, Plovdiv

S 1454 Zesz Nauk Wyz Szk Ped Mat, Opole • PL
Zeszyty Naukowe Wyzszej Szkoly Pedagogicznej W Opolu Matematyka (Scientific Papers. University of Education in Opole. Mathematics) [1961ff] • PUBL Wyzsza Szkola Pedagogiczna: Opole
• ISSN 0078-5431

S 1459 Mem School Sci & Engin, Waseda Univ • J
Memoirs of the School of Sciences and Engineering. Waseda University [1922ff] • PUBL Waseda Univ.: Tokyo
• ISSN 0369-1950

S 1498 Uch Zap Univ, Moskva • SU
Uchenye Zapiski Moskovskogo Gosudarstvennogo Universiteta (Scientific Notes of the Moscow State University) PUBL
(X 0898) Moskov Gos Univ: Moskva

S 1544 Teor Polugrupp & Prilozh • SU
Teoriya Polugrupp i ee Prilozheniya: Sbornik Statej (Theory of Semigroups and Applications) [1965,1971,1974] • ED: VAGNER, V.V. • PUBL (X 0958) Saratov Univ: Saratov No.1: 351pp, No.2: 101pp, No.3: 154pp
• LC-No 66-99752

S 1562 Prace Inst Mat Fis, Politech Wroclaw, Ser Stud Mater • PL
Prace Naukowe. Instytut Matematyki i Fizyki Teoretycznej. Politechnika Wroclawska. Seria Studia i Materialy. (Scientific Publications. Institute of Mathematics and Theoretical Physics. Technical University of Wroclaw) [?-1969] • PUBL Politechnika Wroclawska: Wroclaw
• CONT AS (S 4733) Prace Inst Mat, Politech Wroclaw, Ser Stud Mater

S 1567 Semin Bourbaki • F
Seminaire Bourbaki [1948ff] • SER (S 3301) Lect Notes Math, (J 1620) Asterisque • PUBL (X 0811) Springer: Heidelberg & New York, (X 2244) Soc Math France: Paris, (X 1623) Univ Paris VI Inst Poincare: Paris Secretariat Mathematique: Paris

S 1596 Acta Univ Debrecen • H
Acta Universitatis Debreceniensis de Ludovico Kossuth PUBL -?-: Debrecen

S 1605 Math Centr Tracts • NL
Mathematical Centre Tracts [1963-1983] • PUBL (X 1121)
Math Centr: Amsterdam

S 1613 Probl Logic (Bucharest) • RO
Probleme de Logica [1956ff] • PUBL (X 0871) Acad Rep Soc Romania: Bucharest
• ISSN 0556-1655

S 1626 Oslo Preprint Ser • N
Oslo Preprint Series [1970ff] • PUBL (X 2786) Univ Oslo Mat Inst: Oslo

S 1642 Schr Inf Angew Math, Ber (Aachen) • D
Schriften zur Informatik und Angewandten Mathematik. Bericht PUBL (X 3215) TH Aachen Math Nat Fak: Aachen

S 1644 Tr Fiz-Mat Inst Steklov • SU
Trudy Fiziko-Matematicheskogo Instituta imeni V.A.Steklova Akademiya Nauk SSSR (Travaux de l'Institut Physico-Mathematique V.A.Stekloff de l'Academie des Sciences SSSR) [1930-1937] • PUBL (X 0899) Akad Nauk SSSR : Moskva
• CONT AS (S 0066) Tr Mat Inst Steklov

S 1802 Colloq Int CNRS • F
Colloques Internationaux du Centre National de la Recherche Scientifique (CNRS) PUBL (X 0999) CNRS Inst B Pascal: Paris

S 1926 Banach Cent Publ • PL
Banach Center Publications. Polish Academy of Science. Institut of Mathematics [1976ff] • PUBL (X 1034) PWN: Warsaw
• ISSN 0137-6934

S 1948 Tr, Univ Tomsk • SU
Trudy Tomskogo Ordena Krasnogo Znameni Gosudarstvennogo Universiteta imeni V.V. Kujbysheva (Publications of the Tomsk State University) PUBL (X 3606) Tomsk Univ: Tomsk

S 1956 Tagungsbericht, Oberwolfach • D
Tagungsbericht. Mathematisches Forschungsinstitut Oberwolfach PUBL (X 0876) Bibl Inst: Mannheim

S 2073 Actualites Sci Indust • F
Actualites Scientifiques et Industrielles PUBL (X 0859)
Hermann: Paris

S 2250 Semin Dubreil: Algebre • F
Seminaire: Algebre. [1971ff] • ED: DUBREIL, P. & ARIBAUD, F. & MALLIAVIN, M.-P. • PUBL (X 1623) Univ Paris VI Inst Poincare: Paris
• CONT OF (S 1057) Semin Dubreil: Alg Th Nombr

S 2308 Symp Kyoto Univ Res Inst Math Sci (RIMS) • J
Surikaisekikenkyusho Kokyuroku (Kyoto University. Research Institute for Mathematical Sciences (RIMS). Proceedings of Symposia) PUBL (X 2441) Kyoto Univ Res Inst Math Sci: Kyoto

S 2337 Istor Metodol Estest Nauk (Moskva) • SU
Istoriya i Metodologiya Estestvennykh Nauk. (History and Methodology of Natural Science) [1960ff] • PUBL (X 0898)
Moskov Gos Univ: Moskva
• ISSN 0579-0204

S 2348 Semin Th Nombres Bordeaux • F
Seminaire de Theorie des Nombres. 1972-1973 PUBL Univ Bourdeaux: Valence

S 2450 Mem Amer Math Soc, NS • USA
Memoirs of the American Mathematical Society. New Series [1975ff] • PUBL (X 0803) Amer Math Soc: Providence
• CONT OF (S 0167) Mem Amer Math Soc
• ISSN 0065-9266

S 2579 Teor Mnozhestv & Topol (Izhevsk) • SU
Teoriya Mnozhestv i Topologiya (Set Theory and Topology)
[1977,1979,1982] • ED: GRYZLOV, A.A. • PUBL (**X** 4562)
Udmurtskij Gos Univ: Izhevsk 114pp,116pp,116pp
• REM Title of Vol.2: Sovremennaya Topologiya i Teoriya Mnozhestv, Vyp.2

S 2582 Semiotika & Inf, Akad Nauk SSSR • SU
Semiotika i Informatika. Gosudarstvennyj Komitet SSSR po Nauke i Tekhnike. Akademiya Nauk SSSR. Vsesoyuznyj Institut Nauchnoj i Tekhnicheskoj Informatsii (Semiotics and Information Science) ED: MIKHAJLOV, A.I. • PUBL (**X** 2235) VINITI: Moskva

S 2616 Philos Communic, Red Ser 2, Gothenburg • S
Philosophical Communications. Red Series N.2. PUBL Gothenburg University

S 2626 Vopr Teor Grupp Gomol Algeb • SU
Voprosy Teoriya Grupp i Gomologicheskoj Algebry (Questions in the Theory of Groups and Homological Algebra) [1977,1979]
• PUBL (**X** 2766) Yaroslav Gos Univ: Yaroslavl'

S 2651 Prepr Inst Prikl Mat, Akad Nauk SSSR • SU
Preprint. Akademiya Nauk SSSR. Institut Prikladnoj Matematiki. (Preprint. Academy of Sciences of the USSR. Institute of Applied Mathematics) PUBL Akad Nauk SSSR, Inst Prikl Mat: Moskva

S 2738 Issl Prikl Mat (Univ Kazan') • SU
Issledovaniya po Prikladnoj Matematike (Studies in Applied Mathematics) [1973ff] • PUBL (**X** 3605) Kazan Gos Univ: Kazan'

S 2848 Topology Proc • USA
Topology Proceedings [1976ff] • ED: KUPERBERG, W. & REED, G.M. & ZENOR, P. • PUBL Auburn University: Auburn, AL
• ISSN 0146-4124 • REM Contains Proceedings of Topology Conferences. Vol 1: 1976 Auburn, AL, USA. Vol 2: 1977 Baton Rouge, LA, USA. Vol 3: 1978 Norman, OK, USA. Vol 4: 1979 Athens, OH, USA. Vol 5: 1980 Birminghan, AL, USA. Vol 7: 1982 Annapolis, MD, USA. Vol 8: 1983 Houston, TX, USA. Vol 9: 1984 Auburn, AL, USA.

S 2851 Tr Sem Petrovskogo, Univ Moskva • SU
Trudy Seminara imeni I.G. Petrovskogo. Moskovskij Universitet. (Publications of the Petrovskij-Seminar. Moscow University) [1975ff] • PUBL (**X** 0898) Moskov Gos Univ: Moskva

S 3105 Adv Math, Suppl Stud • USA
Advances in Mathematics. Supplementary Studies. [1965ff]
• PUBL (**X** 0801) Academic Pr: New York
• REL PUBL (**J** 0345) Adv Math

S 3230 Prepr, Inst Math, Pol Acad Sci • PL
Preprint. Institute of Mathematics. Polish Academy of Sciences PUBL (**X** 2882) Pol Akad Nauk: Wroclaw

S 3231 Prepr NF, Sekt Math, Humboldt-Univ Berlin • DDR
Preprint (Neue Folge). Humboldt-Universitaet zu Berlin. Sektion Mathematik PUBL (**X** 2219) Humboldt-Univ Berlin: Berlin

S 3270 Spraw Inst Inf, Uniw Warsaw • PL
Sprawozdania Instytutu Informatyki Uniwersytetu Warszawskiego (Reports. Institute of Informatics. University of Warsaw.) PUBL Uniwersytet Warszawska: Warsaw

S 3291 Zesz Nauk Mat Fiz Chem (Uniw Gdansk) • PL
Zeszyty Naukowe Wydzialu Matematyki, Fizyki i Chemii. Matematyka (Scientific Papers of the Faculty of Mathematics, Physics & Chemistry) PUBL Uniwersytet Gdanski: Gdansk

S 3301 Lect Notes Math • D
Lecture Notes in Mathematics [1964ff] • PUBL (**X** 0811) Springer: Heidelberg & New York
• ISSN 0075-8434

S 3302 Lect Notes Comput Sci • D
Lecture Notes in Computer Science [1973ff] • PUBL (**X** 0811) Springer: Heidelberg & New York
• ISSN 0302-9743

S 3303 Stud Logic Found Math • NL
Studies in Logic and the Foundations of Mathematics [1954ff]
• PUBL (**X** 0809) North Holland: Amsterdam
• ISSN 0049-237X

S 3304 Proc Symp Pure Math • USA
Proceedings of Symposia in Pure Mathematics PUBL (**X** 0803) Amer Math Soc: Providence

S 3305 Symposia Matematica • I
Symposia Matematica [1969ff] • PUBL (**X** 3604) INDAM: Roma
• ISSN 0082-0725

S 3306 Lond Math Soc Lect Note Ser • GB
London Mathematical Society Lecture Note Series [1971ff]
• PUBL (**X** 0805) Cambridge Univ Pr: Cambridge, GB
• ISSN 0076-0552

S 3307 Synth Libr • NL
Synthese Library. Studies in Epistemology, Logic, Methodology, and Philosophy of Science [1959ff] • PUBL (**X** 0835) Reidel: Dordrecht
• ISSN 0082-1128

S 3308 Univ Western Ontario Ser in Philos of Sci • NL
University of Western Ontario Series in Philosophy of Science [1972ff] • PUBL (**X** 0835) Reidel: Dordrecht

S 3309 AFIPS Conference Proc • USA
AFIPS Conference Proceedings PUBL (**X** 1354) Spartan Books : Sutton

S 3310 Lect Notes Pure Appl Math • USA
Lecture Notes in Pure and Applied Mathematics [1971ff]
• PUBL (**X** 1684) Dekker: New York
• ISSN 0075-8469

S 3311 Boston St Philos Sci • NL
Boston Studies in the Philosophy of Science [1963ff] • PUBL (**X** 0835) Reidel: Dordrecht
• ISSN 0068-0346

S 3312 Coll Math Soc Janos Bolyai • H
Colloquia Mathematica Societatis Janos Bolyai PUBL (**X** 0809) North Holland: Amsterdam

S 3313 Contemp Math • USA
Contemporary Mathematics [1980ff] • PUBL (**X** 0803) Amer Math Soc: Providence
• ISSN 0271-4132

S 3358 Ann Discrete Math • NL
Annals of Discrete Mathematics [1977ff] • PUBL (**X** 0809) North Holland: Amsterdam

S 3382 Sem-ber, Humboldt-Univ Berlin, Sekt Math • DDR
Seminarberichte. Humboldt-Universitaet zu Berlin, Sektion Mathematik PUBL (**X** 2219) Humboldt-Univ Berlin: Berlin

S 3412 Prace Inst Mat, Politech Wroclaw, Ser Konf • PL
Prace Naukowe Instytutu Matematyki Politechniki Wroclaw. Serija: Konferencje (Scientific Papers of the Institute of Mathematics of Wroclaw Technical University. Series: Conferences) PUBL Politechnika Wroclawska: Wroclaw

S 3414 Prepr, Akad Wiss DDR, Inst Math • DDR
Preprint. Akademie der Wissenschaften der DDR. Institut fuer Mathematik PUBL (X 2655) Akad Wiss DDR Inst Math: Berlin

S 3415 Prepr Akad Wiss DDR ZI Math Mech • DDR
Preprint. Akademie der Wissenschaften der DDR. Zentralinstitut fuer Mathematik und Mechanik PUBL (X 2888) ZI Math Mech Akad Wiss DDR: Berlin

S 3462 Var Publ Ser, Aarhus Univ • DK
Various Publications Series [1962ff] • PUBL (X 1599) Aarhus Univ Mat Inst: Aarhus
• ISSN 0065-0188

S 3468 Tr Mat & Mekh (Tartu) • SU
*Matemaatika - ja Mekhaanika-Alaseid Toeid. * Trudy po Matematike i Mekhanike. (Works about Mathematics and Mechanics.)* SER (S 0393) Uch Zap Univ, Tartu • PUBL (X 2463) Tartusk Gos Univ: Tartu

S 3489 Sel Math Sov • CH
Selecta Mathematica Sovietica [1981ff] • PUBL (X 0804) Birkhaeuser: Basel
• ISSN 0272-9903

S 3521 Mem Soc Math Fr • F
Memoire de la Mathematique de France. Supplement au Bulletin [1927ff] • PUBL (X 0834) Gauthier-Villars: Paris
• REM Since 1981: Nouvelle Serie

S 3677 Fund Scientiae • F
Fundamenta Scientiae. Cahiers du Seminaire sur les Fondements des Sciences PUBL Universite Louis Pasteur: Strasbourg

S 3911 Komb-Algeb Met Prikl Mat (Gor'kij) • SU
Kombinatorno-Algebraicheskie Metody v Prikladnoj Matematike (Combinatorical Algebraic Methods in Applied Mathematics) [1979ff] • ED: MARKOV, A.A. • PUBL Gor'kovskij Gos Univ: Gor'kij   124pp,219pp,165pp,184pp

S 3935 Cah Cent Log (Louvain) • B
Cahiers du Centre de Logique PUBL (X 0879) Inst Sup Philos: Louvain

S 4311 Groupes de Contact FNRS, Sci Math • B
*Fonds National de la Recherche Scientifique (FNRS) Bruxelles. Groupes de Contact. Sciences Mathematiques * Nationaal Fonds voor Wetenschappelijk Onderzoek Brussel. Contactgroepen. Wiskundige Wetenschappen* PUBL FNRS: Bruxelles
• REM Mimeographed

S 4733 Prace Inst Mat, Politech Wroclaw, Ser Stud Mater • PL
Politechnika Wroclawska Instytut Matematyki. Prace Naukowe. Studia i Materialy (Scientific Publications. Institute of Mathematics. Technical University of Wroclaw) [1970ff]
• PUBL Politechnika Wroclawska: Wroclaw
• CONT OF (S 1562) Prace Inst Mat Fis, Politech Wroclaw, Ser Stud Mater

Proceedings

P 0457 Lattice Th;1973 Houston • USA
[1973] *Proceedings of the University of Houston Lattice Theory Conference* ED: FAJTLOWICZ, S. & KAISER, K. • PUBL (X 2379) Univ Houston Dept Math: Houston viii+632pp
• DAT&PL 1973 Mar; Houston, TX, USA • LC-No 75-622429

P 0573 Inform Processing (3);1965 New York • USA
[1965-1966] *Information Processing '65. Proceedings of IFIP Congress* ED: KALENICH, W.A. • PUBL (X 0843) Macmillan: New York & London Vol 1: xv+1-304, Vol 2: viii+305-648
• ALT PUBL (X 1354) Spartan Books : Sutton
• DAT&PL 1965 May; New York, NY, USA • LC-No 65-24118

P 0575 Int Congr Math (II, 7);1954 Amsterdam • NL
[1954-1957] *Proceedings of the International Congress of Mathematicians 1954* ED: GERRETSEN, J.C.H. & GROOT DE, J.
• PUBL (X 0809) North Holland: Amsterdam 3 Vols: 582pp,440pp,560pp • ALT PUBL (X 1317) Noordhoff: Groningen 1954-1957 & (X 3602) Kraus: Vaduz 1967
• DAT&PL 1954 Sep; Amsterdam, NL • LC-No 52-1808

P 0576 Raisonn en Math & Sci Exper;1955 Paris • F
[1958] *La Raisonnement en Mathematiques et en Sciences Experimentales.* SER (S 1802) Colloq Int CNRS 70 • PUBL (X 0999) CNRS Inst B Pascal: Paris 140pp
• DAT&PL 1955 Sep; Paris, F • LC-No 63-24106

P 0577 Automatenth & Formale Sprachen;1969 Oberwolfach • D
[1970] *Automatentheorie und Formale Sprachen.* ED: DOERR, J. & HOTZ, G. • SER (S 1956) Tagungsbericht, Oberwolfach 3 • PUBL (X 0876) Bibl Inst: Mannheim 505pp
• DAT&PL 1969 Oct; Oberwolfach, D • LC-No 76-857074

P 0580 Int Congr Log, Meth & Phil of Sci (4,Sel Pap);1971 Bucharest • RO
[1973] *Logic, Language and Probability.* ED: BOGDAN, R.J. & NIINILUOTO, I. • SER (S 3307) Synth Libr • PUBL (X 0835) Reidel: Dordrecht x+323pp
• DAT&PL 1971 Aug; Bucharest, RO • ISBN 90-277-0312-4, LC-No 72-95892 • REL PUBL (P 0793) Int Congr Log, Meth & Phil of Sci (4,Proc);1971 Bucharest
• REM A Selection of Papers Contributed to Sections IV, VI and XI of P0793

P 0593 Int Congr Math (II, 6);1950 Cambridge MA • USA
[1952] *Proceedings of the International Congress of Mathematicians* ED: GRAVES, L.M & HILLE, E. & SMITH, P.A. & ZARISKI, O. • PUBL (X 0803) Amer Math Soc: Providence 2 Vols: 769pp,461pp • ALT PUBL (X 3602) Kraus: Vaduz 1967 2 Vols
• DAT&PL 1950 Aug; Cambridge, MA, USA • LC-No 52-1808

P 0594 Inform Processing (4);1968 Edinburgh • GB
[1969] *Information Processing '68. Proceedings of IFIP Congress* ED: MORRELL, A.J.H. • PUBL (X 0809) North Holland: Amsterdam 2 Vols: xxvi+xi+1650pp • ALT PUBL (X 2696) Humanities Pr: Atlantic Highlands
• DAT&PL 1968 Aug; Edinburgh, GB • ISBN 0-7204-2032-6, LC-No 76-462349
• REM Vol 1: Mathematics, Software. Vol 2: Hardware, Applications

P 0602 Generalized Recursion Th (1);1972 Oslo • N
[1974] *Generalized Recursion Theory* ED: FENSTAD, J.E. & HINMAN, P.G. • SER (S 3303) Stud Logic Found Math 79 • PUBL (X 0809) North Holland: Amsterdam viii+456pp
• ALT PUBL (X 0838) Amer Elsevier: New York
• DAT&PL 1972 Jun; Oslo, N • ISBN 0-444-10545-X, LC-No 73-81531

P 0603 Intuitionism & Proof Th;1968 Buffalo • USA
[1970] *Intuitionism and Proof Theory* ED: KINO, A. & MYHILL, J. & VESLEY, R.E. • PUBL (X 0809) North Holland: Amsterdam viii+516pp
• DAT&PL 1968 Aug,Buffalo, NY, USA • ISBN 0-7204-2257-4, LC-No 77-97196

P 0604 Scand Logic Symp (2);1970 Oslo • N
[1971] *Proceedings of the 2nd Scandinavian Logic Symposium* ED: FENSTAD, J.E. • SER (S 3303) Stud Logic Found Math 63 • PUBL (X 0809) North Holland: Amsterdam ii+405pp
• DAT&PL 1970 Jun; Oslo, N • ISBN 0-7204-2259-0, LC-No 71-153401

P 0607 All-Union Math Conf (3);1956 Moskva • SU
[1959] *Trudy 3'go Vsesoyuznogo Matematicheskogo S'ezda (Proceedings of the 3rd All Union Mathematical Conference)* ED: NIKOL'SKIJ, S.M. & ABRAMOV, A.A. & BOLTYANSKIJ, V.G.
• PUBL (X 0899) Akad Nauk SSSR : Moskva 4 Vols
• DAT&PL 1956 Jun; Moskva, SU

P 0608 Logic Colloq;1966 Hannover • D
[1968] *Contributions to Mathematical Logic. Proceedings of the Logic Colloquium* ED: SCHMIDT, H.A. & SCHUETTE, K. & THIELE, H.-J. • SER (S 3303) Stud Logic Found Math • PUBL (X 0809) North Holland: Amsterdam ix+298pp
• DAT&PL 1966 Aug; Hannover, D • LC-No 68-24434

P 0610 Tarski Symp;1971 Berkeley • USA
[1974] *Proceedings of the Tarski Symposium. An International Symposium held to Honor Alfred Tarski on the Occasion of His 70th Birthday* ED: HENKIN, L. & ADDISON, J. & CHANG, C.C. & CRAIG, W. & SCOTT, D.S. & VAUGHT, R. • SER (S 3304) Proc Symp Pure Math 25 • PUBL (X 0803) Amer Math Soc: Providence ix+498pp
• DAT&PL 1971 Jun; Berkeley, CA, USA • ISBN 0-8218-1425-7, LC-No 74-8666
• REM Corrected Reprint 1979; xx+498pp

Proceedings

P 0612 Int Congr Log, Meth & Phil of Sci (1,Proc);1960 Stanford • USA
[1962] *Proceedings of the 1st International Congress for Logic, Methodology and Philosophy of Science* ED: NAGEL, E. & SUPPES, P. & TARSKI, A. • PUBL (X 1355) Stanford Univ Pr: Stanford ix+661pp
• DAT&PL 1960 Aug; Stanford, CA, USA • LC-No 62-9620
• TRANSL IN [1965] (P 2251) Mat Log & Primen;1960 Stanford

P 0613 Rec Fct Th;1961 New York • USA
[1962] *Recursive Function Theory* ED: DEKKER, J.C.E. • SER (S 3304) Proc Symp Pure Math 5 • PUBL (X 0803) Amer Math Soc: Providence vii+247pp
• DAT&PL 1961 Apr; New York, NY, USA • ISBN 0-8218-1405-2, LC-No 50-1183
• REM 2nd ed. 1970

P 0614 Th Models;1963 Berkeley • USA
[1965] *The Theory of Models.* ED: ADDISON, J.W. & HENKIN, L. & TARSKI, A. • SER (S 3303) Stud Logic Found Math
• PUBL (X 0809) North Holland: Amsterdam xv+494pp
• DAT&PL 1963 Jun; Berkeley, CA, USA • LC-No 66-7051

P 0623 Int Congr Log, Meth & Phil of Sci (2,Proc);1964 Jerusalem • IL
[1965] *Proceedings of the 2nd International Congress for Logic, Methodology and Philosophy of Science* ED: BAR-HILLEL, Y.
• SER (S 3303) Stud Logic Found Math • PUBL (X 0809) North Holland: Amsterdam viii+440pp
• DAT&PL 1964 Aug; Jerusalem, IL • ISBN 0-7204-2235-3, LC-No 66-7008
• REM 2nd ed. 1972

P 0625 Symp Autom Demonst;1968 Versailles • F
[1970] *Symposium on Automatic Demonstration.* ED: LAUDET, M. & LACOMBE, D. & NOLIN, L. & SCHUETZENBERGER, M.
• SER (S 3301) Lect Notes Math 125 • PUBL (X 0811) Springer: Heidelberg & New York v+310pp
• DAT&PL 1968 Dec; Versailles, F • ISBN 3-540-04914-2, LC-No 79-117526

P 0626 Ber Math-Tagung Berlin;1953 Berlin • DDR
[1953] *Bericht ueber die Mathematiker-Tagung* ED: GRELL, H. & SCHMID, H.L. • PUBL (X 0806) Dt Verlag Wiss: Berlin viii+302pp
• DAT&PL 1953 Jan; Berlin, D • LC-No 78-235228

P 0627 Int Congr Log, Meth & Phil of Sci (3,Proc);1967 Amsterdam • NL
[1968] *Proceedings of the 3rd International Congress for Logic, Methodology and Philosophy of Science* ED: ROOTSELAAR VAN, B. & STAAL, J.F. • SER (S 3303) Stud Logic Found Math
• PUBL (X 0809) North Holland: Amsterdam xii+553pp
• DAT&PL 1967 Aug; Amsterdam, NL • ISBN 0-444-85423-1, LC-No 68-29768

P 0629 Conf Universal Algeb;1969 Kingston • CDN
[1970] *Proceedings of the Conference on Universal Algebra* ED: WENZEL, G.H. • SER Queen's Papers in Pure and Applied Mathematics 25 • PUBL (X 0997) Queen's Univ: Kingston v+275pp
• DAT&PL 1969 Oct; Kingston, ON, CDN • LC-No 76-586989

P 0630 Aspects Math Log;1968 Varenna • I
[1969] *Aspects of Mathematical Logic. Centro Internazionale Matematico Estivo (CIME)* ED: CASARI, E. • PUBL (X 0860) Cremonese: Firenze ii+285pp
• DAT&PL 1968 Sep; Varenna, I • LC-No 70-477028

P 0632 Congr Int Phil des Sci;1935 Paris • F
[1936] *Actes du Congres International de Philosophie Scientifique* SER (S 2073) Actualites Sci Indust 388-395
• PUBL (X 0859) Hermann: Paris 8 Vol
• DAT&PL 1935 Sep; Paris, F
• REM Vol. II: Unite de la Science. Vol. III: Language et Pseudo-Problemes. Vol. IV: Induction et Probabilite. Vol. VI: Philosophie des Mathematiques. Vol. VII: Logique. Vol. VIII: Histoire de la Logique et de la Philosophie Scientifique.

P 0633 Infinitist Meth;1959 Warsaw • PL
[1961] *Infinitistic Methods. Proceedings of the Symposium on Foundations of Mathematics* PUBL (X 1034) PWN: Warsaw 362pp • ALT PUBL (X 0869) Pergamon Pr: Oxford 1961
• DAT&PL 1959 Sep; Warsaw, PL • LC-No 61-11351

P 0634 Constructivity in Math;1957 Amsterdam • NL
[1959] *Constructivity in Mathematics* ED: HEYTING, A. • SER (S 3303) Stud Logic Found Math • PUBL (X 0809) North Holland: Amsterdam viii+298pp
• DAT&PL 1957 Aug; Amsterdam, NL • LC-No 63-522

P 0637 Syntax & Semant Infinitary Lang;1967 Los Angeles • USA
[1968] *The Syntax and Semantics of Infinitary Languages* ED: BARWISE, K.J. • SER (S 3301) Lect Notes Math 72 • PUBL (X 0811) Springer: Heidelberg & New York iv+268pp
• DAT&PL 1967 Dec; Los Angeles, CA, USA • ISBN 3-540-04242-3, LC-No 68-57175

P 0638 Logic Colloq;1969 Manchester • GB
[1971] *Logic Colloquium '69* ED: GANDY, R.O. & YATES, C.M.E. • SER (S 3303) Stud Logic Found Math 61 • PUBL (X 0809) North Holland: Amsterdam xiv+457pp
• DAT&PL 1969 Aug; Manchester, GB • ISBN 0-7204-2261-2, LC-No 71-146188

P 0644 Meth Form en Axiom;1950 Paris • F
[1953] *Les Methodes Formelles en Axiomatiques* ED: DESTOUCHES, J.-L. & DESTOUCHES-FEVRIER, P. & JOFFRE, D.
• SER (S 1802) Colloq Int CNRS 36 • PUBL (X 0999) CNRS Inst B Pascal: Paris 197pp
• DAT&PL 1950 Dec; Paris, F

P 0645 Int Congr Philos (11);1953 Bruxelles • B
[1953] *Actes du 11eme Congres International de Philosophie.* * *Proceedings of the 11th International Congress of Philosophy* PUBL (X 1313) Nauwelaerts: Louvain • ALT PUBL (X 0809) North Holland: Amsterdam 1953;14 Vols
• DAT&PL 1953 Aug; Bruxelles, B

P 0646 Appl Sci de Log Math;1952 Paris • F
[1954] *Applications Scientifiques de la Logique Mathematique. Actes du 2eme Colloque International de Logique Mathematique* ED: DESTOUCHES, J.-L. & DESTOUCHES-FEVRIER, P. • SER Collection de Logique Mathematique, Serie A 5 • PUBL (X 0834) Gauthier-Villars: Paris 176pp • ALT PUBL (X 1313) Nauwelaerts: Louvain 1954
• DAT&PL 1952 Aug; Paris, F

P 0649 Appl Model Th to Algeb, Anal & Probab;1967 Pasadena • USA
[1969] *Applications of Model Theory to Algebra, Analysis and Probability* ED: LUXEMBURG, W.A.J. • PUBL (X 0818) Holt Rinehart & Winston: New York vii+307pp
• DAT&PL 1967 May; Pasadena, CA, USA • LC-No 69-11203

P 0652 Entretiens Zuerich Fond & Method Sci Math;1938 Zuerich • CH
[1941] *Les Entretiens de Zuerich sur les Fondements et la Methode des Sciences Mathematiques: Exposes et Discussions* ED: GONSETH, F. • PUBL (X 2220) Leemen: Zuerich 209pp
• DAT&PL 1938 Dec; Zuerich, CH • LC-No 42-650

P 0653 Int Congr Math (II, 4);1932 Zuerich • CH
[1932] *Verhandlungen des Internationalen Mathematiker-Kongresses Zuerich* ED: SAX, W. • PUBL (X 1268) Orell Fuessli: Zuerich 2 Vols: 335pp,365pp
• DAT&PL 1932 Sep; Zuerich, CH • LC-No 52-1808

P 0657 Int Congr Math (II,10);1966 Moskva • SU
[1968] *Trudy Mezhdunarodnogo Kongressa Matematikov* ∗ *Proceedings of the International Congress of Mathematicians* ∗ *Travaux du Congres International des Mathematiciens* ∗ *Berichte des Internationalen Mathematikerkongresses* ED: PETROVSKIJ, YU.G. • PUBL (X 0885) Mir: Moskva 726pp
• DAT&PL 1966 Aug; Moskva, SU • LC-No 70-437326

P 0660 Int Congr Math (II, 8);1958 Edinburgh • GB
[1960] *Proceedings of the International Congress of Mathematicians* ED: TODD, J.A. • PUBL (X 0805) Cambridge Univ Pr: Cambridge, GB lxiv+573pp
• DAT&PL 1958 Aug; Edinburgh, GB • LC-No 52-1808

P 0669 Conv Teor Modelli & Geom;1969/70 Roma • I
[1971] *Convegni Teoria dei Modelli & Geometria* SER (S 3305) Symposia Matematica 5 • PUBL (X 3604) INDAM: Roma 475pp • ALT PUBL (X 0801) Academic Pr: New York
• DAT&PL 1969 Nov; Rome, I, 1970 Apr; Rome, I

P 0674 Symp Math Th of Automata;1962 New York • USA
[1963] *Proceedings of the Symposium on Mathematical Theory of Automata* ED: FOX, J. • SER Microwave Research Institute Symposia Series 12 • PUBL (X 2039) Poly Inst New York: Brooklyn xix+640pp • ALT PUBL (X 0827) Wiley & Sons: New York
• DAT&PL 1962 Apr; New York, NY, USA • LC-No 63-11286

P 0677 Int Congr Math (II, 9,Proc);1962 Djursholm • S
[1963] *Proceedings of the International Congress of Mathematicians* ED: STENSTROEM, V. • PUBL (X 1163) Almqvist & Wiksell: Stockholm 1+595pp
• DAT&PL 1962 Aug; Djursholm, S • 52-1808

P 0678 Word Probl: Decis & Burnside Probl in Group Th;1969 Irvine • USA
[1973] *Word Problems. Decision Problems and the Burnside Problem in Group Theory* ED: BOONE, W.W. & CANNONITO, F.B. & LYNDON, R.C. • SER (S 3303) Stud Logic Found Math 71 • PUBL (X 0809) North Holland: Amsterdam xii+646pp
• DAT&PL 1969 Sep; Irvine, CA, USA • ISBN 0-7204-2271-X, LC-No 70-146190

P 0682 Int Congr Philos (10);1948 Amsterdam • NL
[1949] *Library of the 10th International Congress of Philosophy* ED: BETH, E.W. & POS, H.J. & HOLLAK, H.J.A.
• PUBL (X 0809) North Holland: Amsterdam Vol 1, L.J.Veen: Amsterdam Vol 2
• DAT&PL 1948 Aug; Amsterdam, NL • LC-No 50-35721

P 0683 Log & Founds of Sci (Beth);1964 Paris • F
[1967] *Logic and Foundations of Science. E.W.Beth Memorial Colloquium.* ED: DESTOUCHES, J.-L. • PUBL (X 0835) Reidel: Dordrecht viii+140pp
• DAT&PL 1964 May; Paris, F • LC-No 68-107520

P 0688 Logic Colloq;1963 Oxford • GB
[1965] *Formal Systems and Recursive Functions. Proceedings of the 8th Logic Colloquium* ED: CROSSLEY, J.N. & DUMMETT, M.A.E. • SER (S 3303) Stud Logic Found Math • PUBL (X 0809) North Holland: Amsterdam 320pp
• DAT&PL 1963 Jul; Oxford, GB • LC-No 66-2289

P 0691 Sets, Models & Recursion Th;1965 Leicester • GB
[1967] *Sets, Models and Recursion Theory. Proceedings of the Summer School in Mathematical Logic and 10th Logic Colloquium* ED: CROSSLEY, J.N. • SER (S 3303) Stud Logic Found Math • PUBL (X 0809) North Holland: Amsterdam v+331pp • ALT PUBL (X 0838) Amer Elsevier: New York 1967
• DAT&PL 1965 Aug; Leicester, GB • ISBN 0-7204-2242-6, ISBN 0-444-10696-0, LC-No 67-21973
• REM 2nd ed. 1974

P 0692 Summer School in Logic;1967 Leeds • GB
[1968] *Proceedings of the Summer School in Logic* ED: LOEB, M.H. • SER (S 3301) Lect Notes Math 70 • PUBL (X 0811) Springer: Heidelberg & New York iv+331pp
• DAT&PL 1967 Aug; Leeds, GB • ISBN 3-540-04240-7, LC-No 68-56951

P 0693 Axiomatic Set Th;1967 Los Angeles • USA
[1971-1974] *Axiomatic Set Theory* ED: SCOTT, D. (V 1) & JECH, T.J. (V 2) • SER (S 3304) Proc Symp Pure Math 13
• PUBL (X 0803) Amer Math Soc: Providence 2 Vols: vi+474pp,viii+222pp
• DAT&PL 1967 Jul; Los Angeles, CA, USA • ISBN 0-8218-0245-3 (V1), ISBN 0-8218-0246-1 (V2), LC-No 78-125172

P 0694 Conv Algeb Assoc (Scorza);1970 Roma • I
[1972] *Convegno Sulle Algebre Associative. Volume Dedicato Alla Memoria di Gaetano Scorza* SER (S 3305) Symposia Matematica 8 • PUBL (X 3604) INDAM: Roma iv+414pp
• ALT PUBL (X 0801) Academic Pr: New York
• DAT&PL 1970 Nov; Roma, I

P 0698 Harvard Symp Digit Comput & Appl;1961 Cambridge MA • USA
[1962] *Proceedings of a Harvard Symposium on Digital Computers and Their Applications* SER Annals of the Computation Laboratory of Harvard University 31 • PUBL (X 0858) Harvard Univ Pr: Cambridge xiv+332pp • ALT PUBL (X 0894) Oxford Univ Pr: Oxford
• DAT&PL 1961 Apr; Cambridge, MA, USA • LC-No 62-19220

P 0706 Gen Topol & Appl (de Groot);1972 Pittsburgh • USA
[1974] *TOPO 72. General Topology and Its Applications. 2nd Pittsburgh International Conference* ED: ALO, R.A. & HEATH, R.W. & NAGATA, J.-I. • SER (S 3301) Lect Notes Math 378
• PUBL (X 0811) Springer: Heidelberg & New York xiv+651pp
• DAT&PL 1972 Dec; Pittsburgh, PA, USA • ISBN 3-540-06741-8, LC-No 74-390

P 0709 Int Conf Th of Groups (2);1973 Canberra • AUS
[1974] *Proceedings of the 2nd International Conference on the Theory of Groups* ED: NEWMAN, M.F. • SER (S 3301) Lect Notes Math 372 • PUBL (X 0811) Springer: Heidelberg & New York 740+viipp
• DAT&PL 1973 Aug; Canberra, ACT, AUS • ISBN 3-540-06845-7, LC-No 74-13872

P 0710 Russell Mem Logic Conf;1971 Uldum • DK
[1973] *Proceedings of the Bertrand Russell Memorial Logic Conference* ED: BELL, J. & COLE, J. & PRIEST, G. & SLOMSON, A. • PUBL (**X** 2504) Russell Mem Conf: Leeds vi+404pp
• DAT&PL 1971 Aug; Uldum, DK • LC-No 74-193761
• REM Copies Available from A.Slomson, School of Math, Univ of Leeds, GB

P 0711 Concept & Role of Model in Math & Sci;1960 Utrecht • NL
[1961] *The Concept and the Role of the Model in Mathematics and Natural and Social Sciences* ED: FREUDENTHAL, H. • SER (**S** 3307) Synth Libr • PUBL (**X** 0835) Reidel: Dordrecht 194pp
• DAT&PL 1960 Jan; Utrecht, NL • LC-No 63-1436

P 0713 Cambridge Summer School Math Log;1971 Cambridge GB • GB
[1973] *Cambridge Summer School in Mathematical Logic* ED: ROGERS, H. & MATHIAS, A.R.D. • SER (**S** 3301) Lect Notes Math 337 • PUBL (**X** 0811) Springer: Heidelberg & New York ix+660pp
• DAT&PL 1971 Aug; Cambridge, GB • ISBN 3-540-05569-X, LC-No 73-12410

P 0723 Categ Th, Homol Th & Appl;1968 Seattle • USA
[1969] *Category Theory, Homology Theory and Their Applications* ED: HILTON, P.J. • SER (**S** 3301) Lect Notes Math 86,99 • PUBL (**X** 0811) Springer: Heidelberg & New York 3 Vols: vi+216pp,v+308pp,iv+498pp
• DAT&PL 1968 Jun; Seattle, WA, USA • ISBN 3-540-04605-4 (V1), ISBN 3-540-04611-9 (V2), ISBN 3-540-04618-6 (V3), LC-No 75-75931

P 0724 Skand Mat Kongr (15);1968 Oslo • N
[1970] *Proceedings of the 15th Scandinavian Congress* ED: AUBERT, K.E. & LJUNGGREN, W. • SER (**S** 3301) Lect Notes Math 118 • PUBL (**X** 0811) Springer: Heidelberg & New York iv+162pp
• DAT&PL 1968 Aug; Oslo, N • ISBN 3-540-04907-X, LC-No 70-112305

P 0732 Contrib to Non-Standard Anal;1970 Oberwolfach • D
[1972] *Contributions to Non-Standard Analysis* ED: LUXEMBURG, W.A.J. & ROBINSON, A. • SER (**S** 3303) Stud Logic Found Math 69 • PUBL (**X** 0809) North Holland: Amsterdam vii+289pp
• DAT&PL 1970 Jul; Oberwolfach, D • LC-No 76-183275

P 0743 Int Congr Math (II,11,Proc);1970 Nice • F
[1971] *Actes du Congres International de Mathematiciens 1970* ED: BERGER, M. & DIEUDONNE, J. & LERAY, J. & LIONS, J.-L. & MALLIAVIN, M.P. & SERRE, J.-P. • PUBL (**X** 0834) Gauthier-Villars: Paris 3 Vols. xxxiii+532pp,959pp,iii+371pp
• DAT&PL 1970 Sep; Nice, F • REL PUBL (**P** 1158) Int Congr Math (II,11,Comm Ind);1970 Nice
• REM Vol 1: Documents.Medailles Fields.Conferences Generales.Logique.Algebre. Vol 2: Geometrie et Topologie. Analyse. Vol 3: Mathematiques Appliquees.Historie et Enseignement.

P 0754 All-Union Math Conf (4);1961 Leningrad • SU
[1963-1964] *Trudy 4'go Vsesoyuznogo Matematicheskogo S'ezda (Proceedings of the 4th All Union Mathematical Conference)* PUBL (**X** 2641) Nauka: Leningrad 2 Vols
• DAT&PL 1961 Jul; Leningrad, SU

P 0757 Scand Logic Symp (3);1973 Uppsala • S
[1975] *Proceedings of the 3rd Scandinavian Logic Symposium* ED: KANGER, S. • SER (**S** 3303) Stud Logic Found Math 82 • PUBL (**X** 0809) North Holland: Amsterdam vii+214pp
• ALT PUBL (**X** 0838) Amer Elsevier: New York
• DAT&PL 1973 Apr; Uppsala, S • ISBN 0-444-10679-0, LC-No 74-80113

P 0759 Infinite & Finite Sets (Erdoes);1973 Keszthely • H
[1975] *Infinite and Finite Sets. Dedicated to Paul Erdoes on His 60th Birthday* ED: HAJNAL, A. & RADO, R. & SOS, V.T.
• SER (**S** 3312) Coll Math Soc Janos Bolyai 10 • PUBL (**X** 3725) Bolyai Janos Mat Tars: Budapest 3 Vols: 1555pp
• ALT PUBL (**X** 0809) North Holland: Amsterdam
• DAT&PL 1973 Jun; Keszthely, H • ISBN 0-7204-2814-9, LC-No 75-321560

P 0761 Compl of Computation;1973 New York • USA
[1974] *Complexity of Computation. Proceedings of a Symposium in Applied Mathematics of the American Mathematical Society (AMS) and the Society for Industrial and Applied Mathematics (SIAM)* ED: KARP, R.M. • SER SIAM-AMS Proceedings: 7 • PUBL (**X** 0803) Amer Math Soc: Providence viii+166pp
• DAT&PL 1973 Apr; New York, NY, USA • ISBN 0-8218-1327-7, LC-No 74-22062

P 0763 Automata, Lang & Progr (1);1972 Rocquencourt • F
[1973] *Automata, Languages and Programming. Proceedings of a Symposium Organised by IRIA (Institut de Recherche d'Informatique et d'Automatique)* ED: NIVAT, M. • PUBL (**X** 0809) North Holland: Amsterdam 638pp • ALT PUBL (**X** 0838) Amer Elsevier: New York
• DAT&PL 1972 Jul; Versailles-Rocquencourt, F • ISBN 0-444-10426-7, LC-No 72-93498
• REM Also Abbreviated as ICALP 72

P 0765 Algeb & Log;1974 Clayton • AUS
[1975] *Algebra and Logic. Papers from the 1974 Summer Research Institute of the Australian Mathematical Society* ED: CROSSLEY, J.N. • SER (**S** 3301) Lect Notes Math 450 • PUBL (**X** 0811) Springer: Heidelberg & New York viii+307pp
• DAT&PL 1974 Jan; Clayton, Vic, AUS • ISBN 3-540-07152-0, LC-No 75-9903

P 0771 Toposes, Algeb Geom & Log;1971 Halifax • CDN
[1972] *Toposes, Algebraic Geometry and Logic. Partial Report on a Conference on Connections Between Category Theory and Algebraic Geometry and Intuitionistic Logic* ED: LAWVERE, F.W. • SER (**S** 3301) Lect Notes Math 274 • PUBL (**X** 0811) Springer: Heidelberg & New York 189pp
• DAT&PL 1971 Jan; Halifax, NS, CDN • ISBN 3-540-05920-2, LC-No 72-86101

P 0773 Colloq Anal Fonctionelle;1971 Bordeaux • F
[1972] *Actes du Colloque d'Analyse Fonctionelle* ED: HOGBE-NLEND, H. • SER (**S** 3521) Mem Soc Math Fr 31,32
• PUBL (**X** 2244) Soc Math France: Paris 406pp
• DAT&PL 1971 Apr; Bordeaux, F

P 0775 Logic Colloq;1973 Bristol • GB
[1975] *Logic Colloquium '73* ED: ROSE, H.E. & SHEPHERDSON, J.C. • SER (**S** 3303) Stud Logic Found Math 80 • PUBL (**X** 0809) North Holland: Amsterdam viii+513pp • ALT PUBL (**X** 0838) Amer Elsevier: New York
• DAT&PL 1973 Jul; Bristol, GB • ISBN 0-444-10642-1, LC-No 74-79302

P 0776 Permutations;1972 Paris • F
[1974] *Permutations. Actes du Colloque sur les Permutations*
SER Mathematiques et Sciences de L'Homme 20 • PUBL
(**X** 0834) Gauthier-Villars: Paris xiv+289pp • ALT PUBL
(**X** 0873) Mouton: Paris
• DAT&PL 1972 Jul; Paris, F • ISBN 2-04-009681-7, ISBN
2-7193-0897-8, LC-No 74-189927

P 0777 Grundl Geom & Algeb Meth;1973 Potsdam • DDR
[1974] *Grundlagen der Geometrie und Algebraische Methoden. Internationales Kolloquium der Paedagogischen Hochschule "Karl Liebknecht"* ED: KLOTZEK, B. & QUAISSER, E. • SER
Potsdam Forschungen, Reihe B 3 • PUBL Paed. Hochschule
"Karl Liebknecht": Potsdam 208pp
• DAT&PL 1973 Aug; Potsdam, DDR • LC-No 75-405437

P 0779 Conf Group Th;1972 Racine • USA
[1973] *Conference on Group Theory* ED: GATTERDAM, R.W. &
WESTON, K.W. • SER (**S** 3301) Lect Notes Math 319 • PUBL
(**X** 0811) Springer: Heidelberg & New York iv+188pp
• DAT&PL 1972 Jun; Racine, WI, USA • ISBN
3-540-06205-X, LC-No 73-76679

P 0784 GI Jahrestag (5);1975 Dortmund • D
[1975] *GI - 5. Jahrestagung (Gesellschaft fuer Informatik)* ED:
MUEHLBACHER, J. • SER (**S** 3302) Lect Notes Comput Sci 34
• PUBL (**X** 0811) Springer: Heidelberg & New York
x+755pp
• DAT&PL 1975 Oct; Dortmund, D • ISBN 3-540-07410-4

P 0785 Scand Logic Symp (1);1968 Aabo • S
[1970] *Proceedings of the 1st Scandinavian Logic Symposium*
SER Filosofiska Studier 8 • PUBL (**X** 0882) Univ Filos
Foeren: Uppsala 171pp
• DAT&PL 1968 Sep; Aabo, S • LC-No 72-186670

P 0793 Int Congr Log, Meth & Phil of Sci (4,Proc);1971
Bucharest • RO
[1973] *Proceedings of the 4th International Congress for Logic, Methodology and Philosophy of Science* ED: SUPPES, P. &
HENKIN, L. & MOISIL, G.C. & JOJA, A. • SER (**S** 3303) Stud
Logic Found Math 74 • PUBL (**X** 0809) North Holland:
Amsterdam x+981pp • ALT PUBL (**X** 0838) Amer Elsevier:
New York & (**X** 1034) PWN: Warsaw
• DAT&PL 1971 Aug; Bucharest, RO • ISBN 0-444-10491-7,
LC-No 72-88505 • REL PUBL (**P** 0580) Int Congr Log, Meth
& Phil of Sci (4,Sel Pap);1971 Bucharest

P 0796 Congr Math Pays Slaves (1);1929 Warsaw • PL
[1930] *Sprawozdanie z 1 Kongresu Matematykow Krajow
Slowianskich (Comptes Rendus du 1. Congres des
Mathematiciens des Pays Slaves)* ED: LEJA, F. • PUBL
(**X** 1764) Ksiaznica Atlas: Warsaw iii+395pp
• DAT&PL 1929 Sep; Warsaw, PL • LC-No 32-17553

P 0797 Fonds des Math, Machines Math & Appl;1962
Tihany • H
[1965] *Colloque sur les Fondements des Mathematiques, les
Machines Mathematiques, et leurs Applications* ED: KALMAR,
L. • SER Collection de Logique Mathematique, Serie A 19
• PUBL (**X** 0928) Akad Kiado: Budapest 320pp • ALT PUBL
(**X** 0834) Gauthier-Villars: Paris & (**X** 1313) Nauwelaerts:
Louvain
• DAT&PL 1962 Sep; Tihany, H

P 1061 Numb Th;1969 Houston • USA
[1969] *Number Theory. American Mathematical Society. 73rd
Annual Meeting* ED: LEVEQUE, W.J. & STRAUS, E.G. • SER
(**S** 3304) Proc Symp Pure Math 12 • PUBL (**X** 0803) Amer
Math Soc: Providence v+98pp
• DAT&PL 1969 Jan; Houston, TX, USA • ISBN
0-8218-1412-5, LC-No 70-78057

P 1072 Math Log & Founds of Set Th;1968 Jerusalem • IL
[1970] *Mathematical Logic and the Foundations of Set Theory*
ED: BAR-HILLEL, Y. • SER (**S** 3303) Stud Logic Found Math
• PUBL (**X** 0809) North Holland: Amsterdam 145pp
• DAT&PL 1968 Nov; Jerusalem, IL • ISBN 0-7204-2255-8,
LC-No 73-97195

P 1075 Logic Colloq;1976 Oxford • GB
[1977] *Logic Colloquium 76* ED: GANDY, R.O. & HYLAND,
J.M.E. • SER (**S** 3303) Stud Logic Found Math 87 • PUBL
(**X** 0809) North Holland: Amsterdam x+612pp • ALT PUBL
(**X** 0838) Amer Elsevier: New York
• DAT&PL 1976 Jul; Oxford, GB • ISBN 0-7204-0691-9,
LC-No 77-8943

P 1076 Latin Amer Symp Math Log (3);1976 Campinas • BR
[1977] *Non-Classical Logic, Model Theory and Computability.
3rd Latin American Symposium on Mathematical Logic* ED:
ARRUDA, A.I. & COSTA DA, N.C.A & CHUAQUI, R. • SER
(**S** 3303) Stud Logic Found Math 89 • PUBL (**X** 0809) North
Holland: Amsterdam xviii+307pp • ALT PUBL (**X** 0838)
Amer Elsevier: New York
• DAT&PL 1976 Jul; Campinas, BR • ISBN 0-7204-0752-4,
LC-No 77-7366

P 1083 Victoria Symp Nonstand Anal;1972 Victoria • AUS
[1974] *Victoria Symposium on Non-Standard Analysis* ED:
HURD, A. & LOEB, P. • SER (**S** 3301) Lect Notes Math 369
• PUBL (**X** 0811) Springer: Heidelberg & New York
xviii+339pp
• DAT&PL 1972 May; Victoria, VIC, AUS • ISBN
3-540-06656-X, LC-No 73-22552

P 1086 Skand Mat Kongr (5);1922 Helsinki • SF
[1923] *Wissenschaftliche Vortraege gehalten auf dem 5ten
Kongress der Skandinavischen Mathematiker* PUBL
Akademische Buchhandlung: Helsinki 315pp
• DAT&PL 1922 Jul; Helsinki, SF

P 1100 Conf Categor Algeb;1965 La Jolla • USA
[1966] *Proceedings of the Conference on Categorical Algebra*
ED: EILENBERG, S. & HARRISON, D.K. & MAC LANE, S. &
ROEHRL, H. • PUBL (**X** 0811) Springer: Heidelberg & New
York viii+526pp
• DAT&PL 1965 Jun; La Jolla, CA, USA • ISBN
3-540-03639-3, LC-No 66-14575

P 1158 Int Congr Math (II,11,Comm Ind);1970 Nice • F
[1970] *Congres International des Mathematiciens 1970. Les
265 Communications Individuels* PUBL (**X** 0834)
Gauthier-Villars: Paris vii+290pp
• DAT&PL 1970 Sep; Nice, F • LC-No 72-374601 • REL PUBL
(**P** 0743) Int Congr Math (II,11,Proc);1970 Nice

P 1430 Adv Course Founds Computer Sci;1974
Amsterdam • NL
[1975] *Foundations of Computer Science. Advanced Course*
ED: BAKKER DE, J.W. • SER (**S** 1605) Math Centr Tracts 63
• PUBL (**X** 1121) Math Centr: Amsterdam 215pp
• DAT&PL 1974 May; Amsterdam, NL • ISBN
90-6196-111-4, LC-No 76-363070

Proceedings

P 1440 ⊦ ISILC Proof Th Symp (Schuette);1974 Kiel • D
[1975] ⊦ *ISILC Proof Theory Symposium. Dedicated to Kurt Schuette on the Occasion of His 65th Birthday. Proceedings of the International Summer Institute and Logic Colloquium* ED: DILLER, J. & MUELLER, GERT H. • SER (S 3301) Lect Notes Math 500 • PUBL (X 0811) Springer: Heidelberg & New York viii+383pp
• DAT&PL 1974 Jul; Kiel, D • ISBN 3-540-07533-X, LC-No 75-40482 • REL PUBL (P 1442) ⊦ ISILC Logic Conf;1974 Kiel
• REM This Volume Contains Only the Proof Theory Part of the Conference.

P 1442 ⊦ ISILC Logic Conf;1974 Kiel • D
[1975] ⊦ *ISILC Logic Conference. Proceedings of the International Summer Institute and Logic Colloquium* ED: MUELLER, GERT H. & OBERSCHELP, A. & POTTHOFF, K. • SER (S 3301) Lect Notes Math 499 • PUBL (X 0811) Springer: Heidelberg & New York iv+651pp
• DAT&PL 1974 Jul; Kiel, D • ISBN 3-540-07534-8, LC-No 75-40431 • REL PUBL (P 1440) ⊦ ISILC Proof Th Symp (Schuette);1974 Kiel

P 1449 Automata Th & Formal Lang;1975 Kaiserslautern • D
[1975] *Automata Theory and Formal Languages. 2nd GI-Conference (Gesellschaft fuer Informatik)* ED: BRAKHAGE, H. • SER (S 3302) Lect Notes Comput Sci 33 • PUBL (X 0811) Springer: Heidelberg & New York viii+292pp
• DAT&PL 1975 May; Kaiserslautern, D • ISBN 3-540-07407-4, LC-No 75-28494

P 1463 Algeb, Topol & Categ Th (Eilenberg);1974 New York • USA
[1976] *Algebra, Topology, and Category Theory. A Collection of Papers in Honour of Samuel Eilenberg* PUBL (X 0801) Academic Pr: New York xi+225pp
• DAT&PL 1974 May; New York, NY, USA • ISBN 0-12-339050-8, LC-No 75-30467

P 1464 ACM Symp Th of Comput (6);1974 Seattle • USA
[1974] *Proceedings of 6th Annual ACM Symposium on Theory of Computing (Association for Computing Machinery)* PUBL (X 2205) ACM: New York iv+347pp
• DAT&PL 1974 Apr; Seattle, WA, USA • LC-No 82-642181

P 1469 AMS Numb Th Summer Inst;1969 Stony Brook • USA
[1971] *Proceedings of the 1969 Summer Institute on Number Theory, Analytic Number Theory, Diophantine Problems and Analytic Number Theory* ED: LEWIS, D.J. • SER (S 3304) Proc Symp Pure Math 20 • PUBL (X 0803) Amer Math Soc: Providence xiii+451pp
• DAT&PL 1969 Jul; Stony Brook, NY, USA • ISBN 0-8218-1420-6, LC-No 76-125938

P 1476 Set Th & Hierarch Th (2) (Mostowski);1975 Bierutowice • PL
[1976] *Set Theory and Hierarchy Theory. A Memorial Tribute to Andrzej Mostowski. Proceedings of the 2nd Conference on Set Theory and Hierarchy Theory* ED: MAREK, W. & SREBRNY, M. & ZARACH, A. • SER (S 3301) Lect Notes Math 537 • PUBL (X 0811) Springer: Heidelberg & New York xiii+345pp
• DAT&PL 1975 Sep; Bierutowice, PL • ISBN 3-540-07856-8, LC-No 76-26536

P 1482 ACM Symp Th of Comput (5);1973 Austin • USA
[1973] *5th Annual ACM Symposium on Theory of Computing (Association for Computing Machinery)* PUBL (X 2205) ACM: New York iv+277pp
• DAT&PL 1973 May; Austin, TX, USA • LC-No 82-642181

P 1484 Int Congr Math (2);1900 Paris • F
[1902] *Comptes Rendus du 2eme Congres International des Mathematiciens. Proces Verbaux et Communications* ED: DUPORCQ, E. • PUBL (X 0834) Gauthier-Villars: Paris 455pp
• DAT&PL 1900 Aug; Paris, F

P 1502 Phil of Logic;1976 Bristol • GB
[1976] *Philosophy of Logic. Proceedings of the 3rd Bristol Conference on Critical Philosophy.* ED: KOERNER, S. • PUBL (X 1096) Blackwell: Oxford 273pp • ALT PUBL (X 0926) Univ Calif Pr: Berkeley
• DAT&PL 1974 -?-; Bristol, GB • ISBN 0-631-16960-1 (1096), ISBN 0-520-03235-7 (0926), LC-No 77-359489 (1096), LC-No 76-020-2 (0926)

P 1521 Int Congr Math (II,12);1974 Vancouver • CDN
[1975] *Proceedings of the International Congress of Mathematicians* ED: JAMES, R.D. • PUBL Canadian Mathematical Congress: Montreal 2 Vols: xlix+552pp,viii+600pp
• DAT&PL 1974 Aug; Vancouver, BC, CDN • ISBN 0-919558-04-6, LC-No 74-34533

P 1545 Easter Conf on Model Th (2);1984 Wittenberg • DDR
[1984] *Proceedings of the 2nd Easter Conference on Model Theory* SER (S 3382) Sem-ber, Humboldt-Univ Berlin, Sekt Math 60 • PUBL (X 2219) Humboldt-Univ Berlin: Berlin ii+243pp
• DAT&PL 1984 Apr; Wittenberg, DDR

P 1589 Math Interpr of Formal Systs;1954 Amsterdam • NL
[1955] *Mathematical Interpretations of Formal Systems* SER (S 3303) Stud Logic Found Math 10 • PUBL (X 0809) North Holland: Amsterdam viii+113pp
• DAT&PL 1954 Sep; Amsterdam, NL • LC-No 56-3127
• REM 2nd ed. 1971

P 1590 All-Union Conf Math Log (3);1974 Novosibirsk • SU
[1974] *3 Vsesoyuznaya Konferentsiya po Matematicheskoj Logike. Tezitsy Doklady i Soobshcheniya (3rd All-Union Conference on Mathematical Logic)*
• DAT&PL 1974 -?-; Novosibirsk, SU • LC-No 75-558515

P 1601 Easter Conf on Model Th (1);1983 Diedrichshagen • DDR
[1983] *Proceedings of the 1st Easter Conference on Model Theory* ED: DAHN, B.I. • SER (S 3382) Sem-ber, Humboldt-Univ Berlin, Sekt Math 49 • PUBL (X 2219) Humboldt-Univ Berlin: Berlin 154pp
• DAT&PL 1983 Apr; Diedrichshagen, DDR

P 1606 Colloq Math (Pascal);1962 Clermont-Ferrand • F
[1962] *Actes du Colloque de Mathematiques Reuni a Clermont a l'Occasion du Tricentenaire de la Mort de Blaise Pascal* SER (J 0179) Ann Fac Sci Clermont 7,8 • PUBL Univ Clermont, Fac. Sci.: Clermont 2 Vols: 123pp,189pp
• DAT&PL 1962 Jun; Clermont-Ferrand, F
• REM Vol 1: Introduction et Logique Mathematique. Vol 2: Calcul des Probabilites, Analyse Numerique et Calcul Automatique, Geometrie et Physique Mathematique

P 1608 Int Congr Math (II, 5);1936 Oslo • N
[1937] *Comptes Rendus du Congres International des Mathematiciens* PUBL A.M. Broeggers Boktrykkeri a/S: Oslo 2 Vols: 316pp,vv+289pp • ALT PUBL (X 3602) Kraus: Vaduz 1967 2 Vols
• DAT&PL 1936 Jul; Oslo, N • LC-No 52-1808

P 1619 Coloq Log Simb;1975 Madrid • E
[1976] *Coloquio Sobre Logica Simbolica* PUBL Centro Calculo Univ. Complutense: Madrid 176pp
• DAT&PL 1975 Feb; Madrid, E • LC-No 77-555677

P 1625 Six Days of Model Th;1975 Louvain-la-Neuve • B
[1977] *Six Days of Model Theory* ED: HENRARD, P. • PUBL (X 2706) Ed Castella: Albeuve 275pp
• DAT&PL 1975 Mar; Louvain-la-Neuve, B
• REM Reissue of issues 71 & 72 of J0079

P 1628 Generalized Recursion Th (2);1977 Oslo • N
[1978] *Generalized Recursion Theory II* ED: FENSTAD, J.E. & GANDY, R.O. & SACKS, G.E. • SER (S 3303) Stud Logic Found Math 94 • PUBL (X 0809) North Holland: Amsterdam vii+466pp • ALT PUBL (X 0838) Amer Elsevier: New York
• DAT&PL 1977 Jun; Oslo, N • ISBN 0-444-85163-1, LC-No 78-5366

P 1629 Symp Math Log;1974 Oulo;1975 Helsinki • SF
[1977] *Proceedings of the Symposia on Mathematical Logic* ED: MIETTINEN, S. & VAEAENAENEN, J. • PUBL University of Helsinki, Department of Philosophy: Helsinki iv+103pp
• DAT&PL 1974 -?-; Oulo, SF, 1975 -?-; Helsinki, SF

P 1635 Math Founds of Comput Sci (6);1977 Tatranska Lomnica • CS
[1977] *Mathematical Foundations of Computer Science. Proceedings of the 6th Symposium* ED: GRUSKA, J. • SER (S 3302) Lect Notes Comput Sci 53 • PUBL (X 0811) Springer: Heidelberg & New York xi+595pp
• DAT&PL 1977 Sep; Tatranska Lomnica, CS • ISBN 3-540-08353-7, LC-No 77-10135

P 1639 Set Th & Hierarch Th (1);1974 Karpacz • PL
[1977] *Set Theory and Hierarchy Theory. Proceedings of the 1st Colloquium in Set Theory and Hierarchy Theory* SER (S 3412) Prace Inst Mat, Politech Wroclaw, Ser Konf 14/1 • PUBL Polytechnical Edition: Wroclaw 123pp
• DAT&PL 1974 Sep; Karpacz, PL

P 1675 Summer Inst Symb Log;1957 Ithaca • USA
[1957] *Summaries of Talks Presented at the Summer Institute for Symbolic Logic* PUBL Institute for Defense Analyses, Communications Research Division: Princeton; xvi+427pp
• DAT&PL 1957 Jul; Ithaca, NY, USA • LC-No 65-4418
• REM 2nd ed. 1960

P 1679 Lattice Th;1975 Ulm • D
[1975] *Proceedings of the Lattice Theory Conference* ED: KALMBACH, G. • PUBL Universitaet Ulm: Ulm 291pp+App.6pp
• DAT&PL 1975 Aug; Ulm, D • LC-No 76-372315

P 1682 Abel Group Th;1976 Las Cruces • USA
[1977] *Abelian Group Theory. Proceedings of the 2nd New Mexico State University Conference* ED: ARNOLD, D. & HUNTER, R. & WALKER, E. • SER (S 3301) Lect Notes Math 616 • PUBL (X 0811) Springer: Heidelberg & New York ix+423pp
• DAT&PL 1976 Dec; Las Cruces, NM, USA • ISBN 3-540-08447-9, LC-No 78-306476

P 1695 Set Th & Hierarch Th (3);1976 Bierutowice • PL
[1977] *Set Theory and Hierarchy Theory V. Proceedings of the 3rd Conference on Set Theory and Hierarchy Theory* ED: LACHLAN, A. & SREBRNY, M. & ZARACH, A. • SER (S 3301) Lect Notes Math 619 • PUBL (X 0811) Springer: Heidelberg & New York viii+358pp
• DAT&PL 1976 Sep; Bierutowice, PL • ISBN 3-540-08521-1, LC-No 78-309663

P 1704 Int Congr Log, Meth & Phil of Sci (5);1975 London ON • CDN
[1977] *Proceedings of 5th International Congress of Logic, Methodology and Philosophy of Science* ED: BUTTS, R.E. & HINTIKKA, J. • SER (S 3308) Univ Western Ontario Ser in Philos of Sci 9-12 • PUBL (X 0835) Reidel: Dordrecht 4 Vols: x+406pp, x+427pp, x+321pp, x+336pp
• DAT&PL 1975 Aug; London, ON, CDN • ISBN 90-277-0708-1 (V1), ISBN 90-277-0710-3 (V2), ISBN 90-277-0829-0 (V3), ISBN 90-277-0831-2 (V4), ISBN 90-277-0706-5 (Set of the 4 Vols), LC-No 77-22429 (V1), LC-No 77-22431 (V2), LC-No 77-22432 (V3), LC-No 77-22433 (V4)
• REM Vol 1: Logic, Foundations of Mathematics, and Computability Theory. Vol 2: Foundational problems in the Special Sciences. Vol 3: Basic Problems in Methodology and Linguistics. Vol 4: Historical and Philosophical Dimensions of Logic, Methodology and Philosophy of Science.

P 1705 Scand Logic Symp (4);1976 Jyvaeskylae • SF
[1979] *Essays on Mathematical and Philosophical Logic. Proceedings of the 4th Scandinavian Logic Symposium and of the 1st Soviet-Finnish Logic Conference* ED: HINTIKKA, J. & NIINILUOTO, I. & SAARINEN, E. • SER (S 3307) Synth Libr 122 • PUBL (X 0835) Reidel: Dordrecht viii+462pp
• DAT&PL 1976 Jun; Jyvaeskylae, SF • ISBN 90-277-0879-7, LC-No 78-14736

P 1729 Colloq Int Log;1975 Clermont-Ferrand • F
[1977] *Colloque International de Logique* SER (S 1802) Colloq Int CNRS 249 • PUBL (X 0999) CNRS Inst B Pascal: Paris 224pp
• DAT&PL 1975 Jul; Clermont-Ferrand, F • ISBN 2-222-02019-0, LC-No 78-367483

P 1740 ACM Symp Th of Comput (10);1978 San Diego • USA
[1978] *Conference Record of the 10th Annual ACM Symposium on Theory of Computing (Association for Computing Machinery)* PUBL (X 2205) ACM: New York 346pp
• DAT&PL 1978 May; San Diego, CA, USA • LC-No 79-101797

P 1800 Brazil Conf Math Log (1);1977 Campinas • BR
[1978] *Proceedings of 1st Brazilian Conference on Mathematical Logic* ED: ARRUDA, A.I. & CHAQUI, R. & COSTA DA, N.C.A. • SER (S 3310) Lect Notes Pure Appl Math 39 • PUBL (X 1684) Dekker: New York xii+303pp
• DAT&PL 1977 Jul; Campinas, BR • LC-No 78-14488

P 1805 Int Symp Multi-Val Log (5,Proc);1975 Bloomington • USA
[1975] *Proceedings of the 1975 International Symposium on Multiple-Valued Logic* PUBL (X 2179) IEEE: New York iv+475pp
• DAT&PL 1975 May; Bloomington, IN, USA • LC-No 76-370321 • REL PUBL (P 1894) Int Symp Multi-Val Log (5,Inv Pap);1975 Bloomington

P 1841 Fct Recurs & Appl;1967 Tihany • H
[1969] *Les Fonctions Recursives et leurs Applications* PUBL (X 0999) CNRS Inst B Pascal: Paris • ALT PUBL (X 3725) Bolyai Janos Mat Tars: Budapest
• DAT&PL 1967 Sep; Tihany, H

P 1864 Higher Set Th;1977 Oberwolfach • D
[1978] *Higher Set Theory* ED: MUELLER, GERT H. & SCOTT, D.S. • SER (S 3301) Lect Notes Math 669 • PUBL (X 0811) Springer: Heidelberg & New York xii+476pp
• DAT&PL 1977 Apr; Oberwolfach, D • ISBN 3-540-08926-8, LC-No 79-312135

P 1869 Automata, Lang & Progr (2);1974 Saarbruecken • D
[1974] *Automata, Languages and Programming: 2nd Colloquium* ED: LOECKX, J. • SER (S 3302) Lect Notes Comput Sci 14 • PUBL (X 0811) Springer: Heidelberg & New York viii+611pp
• DAT&PL 1974 Jul; Saarbruecken, D • ISBN 3-540-06841-4, LC-No 74-180345
• REM Also Abbreviated as ICALP 74

P 1873 Automata, Lang & Progr (6);1979 Graz • A
[1979] *Automata, Languages and Programming. 6th Colloquium* ED: MAURER, H.A. • SER (S 3302) Lect Notes Comput Sci 71 • PUBL (X 0811) Springer: Heidelberg & New York ix+682pp
• DAT&PL 1979 Jul; Graz, A • ISBN 3-540-09510-1, LC-No 79-15859
• REM Also Abbreviated as ICALP 79

P 1894 Int Symp Multi-Val Log (5,Inv Pap);1975 Bloomington • USA
[1977] *Modern Uses of Multiple-Valued Logic. Invited Papers from the 5th International Symposium on Multiple-Valued Logic* ED: DUNN, J.M. & EPSTEIN, G. • SER Epistime 2 • PUBL (X 0835) Reidel: Dordrecht x+338pp
• DAT&PL 1975 May; Bloomington, IN, USA • ISBN 90-277-0747-2, LC-No 77-23098 • REL PUBL (P 1805) Int Symp Multi-Val Log (5,Proc);1975 Bloomington
• REM With a Bibliography of Many-Valued Logic

P 1897 Logic Colloq;1977 Wroclaw • PL
[1978] *Logic Colloquium 77* ED: MACINTYRE, A. & PACHOLSKI, L. & PARIS, J. • SER (S 3303) Stud Logic Found Math 96 • PUBL (X 0809) North Holland: Amsterdam x+311pp
• DAT&PL 1977 Aug; Wroclaw, PL • ISBN 0-444-85178-X, LC-No 78-13396

P 1901 ACM Symp Th of Comput (4);1972 Denver • USA
[1972] *Proceedings of the 4th Annual ACM Symposium on the Theory of Computing. Spring Joint Computer Conference. (Association for Computing Machinery)* SER (S 3309) AFIPS Conference Proc 40 • PUBL (X 2205) ACM: New York 263pp
• DAT&PL 1972 May; Denver, CO, USA • LC-No 82-642181

P 1953 Int Congr Log, Meth & Phil of Sci (1;Abstr);1960 Stanford • USA
[1960] *1st International Congress for Logic, Methodology and Philosophy of Science. Abstracts*
• DAT&PL 1960 Aug; Stanford, CA, USA

P 1959 Int Congr Math (II,13);1978 Helsinki • SF
[1980] *Proceedings of the International Congress of Mathematicians* ED: LEHTO, O. • PUBL Academia Scientiarum Fennica: Helsinki 2 Vols: 1022pp
• DAT&PL 1978 Aug; Helsinki, SF • ISBN 951-41-0352-1

P 1967 Form Semant of Nat Lang;1973 Cambridge GB • GB
[1975] *Formal Semantics of Natural Languages* ED: KEENANAN, E.L. • PUBL (X 0805) Cambridge Univ Pr: Cambridge, GB xiii+475pp
• DAT&PL 1973 Apr; Cambridge, GB • ISBN 0-521-20697-9, LC-No 74-25657

P 2021 Model Th & Appl;1975 Bressanone • I
[1975] *Model Theory and Applications. Centro Internazionale Matematico Estivo (CIME) II Ciclo* ED: MANGINI, P. • PUBL (X 0860) Cremonese: Firenze 151pp
• DAT&PL 1975 Jun; Bressanone, I • LC-No 77-456584

P 2044 Int Congr Log, Meth & Phil of Sci (4,Abstr);1971 Bucharest • RO
[1971] *4th International Congress for Logic, Methodology and Philosophy of Science. Abstracts*
• DAT&PL 1971 Aug; Bucharest, RO

P 2052 Math Educ & New Areas in Math;1974 Penang • MAL
[1975] *Proceedings of a Conference in Mathematical Education and New Areas in Mathematics* PUBL Southeast Asian Math Soc xv+95pp
• DAT&PL 1974 Jul; Penang, MAL • LC-No 76-941857

P 2058 Kleene Symp;1978 Madison • USA
[1980] *The Kleene Symposium* ED: BARWISE, K.J. & KEISLER, H.J. & KUNEN, K. • SER (S 3303) Stud Logic Found Math 101 • PUBL (X 0809) North Holland: Amsterdam xx+425pp
• DAT&PL 1978 Jun; Madison, WI, USA • ISBN 0-444-85345-6, LC-No 79-20792

P 2059 Math Founds of Comput Sci (8);1979 Olomouc • CS
[1979] *Mathematical Foundations of Computer Science. Proceedings of the 8th Symposium* ED: BECVAR, J. • SER (S 3302) Lect Notes Comput Sci 74 • PUBL (X 0811) Springer: Heidelberg & New York ix+580pp
• DAT&PL 1979 Sep; Olomouc, CS • ISBN 3-540-09526-8, LC-No 79-17801

P 2064 All-Union Conf Math Log (4);1976 Kishinev • SU
[1976] *4 Vsesoyuznaya Konferentsiya po Matematicheskoj Logike. Tezitsy Doklady i Soobshcheniya (4th All-Union Conference on Mathematical Logic)* ED: KUZNETSOV, A.V.
• PUBL (X 2741) Shtiintsa: Kishinev 170pp
• DAT&PL 1976 -?-; Kishinev, SU • LC-No 78-410667

P 2080 Conf Math Log;1970 London • GB
[1972] *Conference on Mathematical Logic - London '70* ED: HODGES, W. • SER (S 3301) Lect Notes Math 255 • PUBL (X 0811) Springer: Heidelberg & New York viii+351pp
• DAT&PL 1970 Aug; London, GB • ISBN 3-540-05744-7, LC-No 70-189457

P 2112 Contrib to Universal Algeb;1975 Szeged • H
[1977] *Contributions to Universal Algebra* ED: CSAKANY, B. & SCHMIDT, J. • SER (S 3312) Coll Math Soc Janos Bolyai 17 • PUBL (X 0809) North Holland: Amsterdam 607pp • ALT PUBL (X 3725) Bolyai Janos Mat Tars: Budapest
• DAT&PL 1975 Aug; Szeged, H • ISBN 963-8021-01-2, ISBN 0-7204-0725-7, LC-No 79-350398

P 2153 Logic Colloq;1983 Aachen • D
[1984] *Proceedings of the Logic Colloquium, Part 1, 2* ED:
MUELLER, GERT H. & RICHTER, M.M. (V1). BOERGER, E. &
OBERSCHELP, W. & RICHTER, M.M. & SCHINZEL, B. &
THOMAS, W. (V2) • SER (S 3301) Lect Notes Math 1103,1104
• PUBL (X 0811) Springer: Heidelberg & New York 2 Vols:
viii + 484pp,viii + 475pp
• DAT&PL 1983 Jul; Aachen, D • ISBN 3-540-13900-1 (V1),
ISBN 3-540-13901-X (V2), LC-No 84-26704
• REM Vol 1: Models and Sets. Vol 2: Computation and Proof
Theory.

P 2160 Latin Amer Symp Math Log (6);1983 Caracas • YV
[1985] *Methods in Mathematical Logic. Proceedings of the 6th
Latin American Symposium on Mathematical Logic* ED:
PRISCO DI, C.A. • SER (S 3301) Lect Notes Math 1130
• PUBL (X 0811) Springer: Heidelberg & New York
vii + 407pp
• DAT&PL 1983 Aug; Caracas, YV • ISBN 3-540-15236-9,
LC-No 85-14779

P 2167 Orders: Descr & Roles;1982 L'Arbresle • F
[1984] *Orders: Description and Roles in Set Theory, Lattices,
Ordered Groups, Topology, Theory of Models and Relations,
Combinatorics, Effectiveness, Social Sciences. Proceedings of a
Conference on Ordered Sets and Their Applications* ED:
POUZET, M. & RICHARD, D. • SER (S 3358) Ann Discrete
Math 23, North Holland Mathematics Studies 99 • PUBL
(X 0809) North Holland: Amsterdam xxviii + 548pp
• DAT&PL 1982 Jul; L'Arbresle, F • ISBN 0-444-87601-4,
LC-No 84-13749

P 2180 Math Appl Categ Th;1983 Denver • USA
[1984] *Mathematical Applications of Category Theory.
Proceedings of the Special Session of the 89th Annual Meeting
of the American Mathematical Society (AMS)* ED: GRAY, J.W.
• SER (S 3313) Contemp Math 30 • PUBL (X 0803) Amer
Math Soc: Providence vii + 307pp
• DAT&PL 1983 Jan; Denver, CO, USA • ISBN
0-8218-5032-6, LC-No 84-9371

P 2251 Mat Log & Primen;1960 Stanford • USA
[1965] *Matematicheskaya Logika i Ee Primeneniya: Sbornik
Statei (Mathematical Logic and Its Applications. Logic,
Methodology and Philosophy of Science)* ED: MAL'TSEV, A.I. &
NAGEL, E. & SUPPES, P. & TARSKI, A. • PUBL (X 0885) Mir:
Moskva 341pp
• DAT&PL 1960 Aug; Stanford, CA, USA
• TRANSL OF [1962] (P 0612) Int Congr Log, Meth & Phil of
Sci (1,Proc);1960 Stanford
• REM Only Parts of P0612 are translated.

P 2268 Int Colloq Philos of Sci;1965 London • GB
[1967-1970] *Proceedings of the International Colloquium in the
Philosophy of Science. 4 Volumes* ED: LAKATOS, I. (V 1 - V 4) &
MUSGRAVE, A. (V 3, V 4) • SER (S 3303) Stud Logic Found
Math V 1 - V 3 • PUBL (X 0809) North Holland:
Amsterdam V 1: xv + 241pp, V 2: viii + 420pp, V 3:
ix + 448pp, (X 0805) Cambridge Univ Pr: Cambridge, GB V
4: viii + 282pp
• DAT&PL 1965, Jul; London, GB • ISBN 0-521-07826-1
(V4), LC-No 67-20007 (V1), LC-No 67-28648 (V2), LC-No
67-28649 (V3), LC-No 78-105496 (V4)
• REM Vol 1: Problems in the Philosophy of Mathematics. Vol
2: The Problem of Inductive Logic. Vol 3: Problems in the
Philosophy of Science. Vol 4: Criticism and the Growth of
Knowledge.

P 2294 Ordered Fields & Real Algeb Geom;1981 San
Francisco • USA
[1982] *Ordered Fields and Real Algebraic Geometry.
Proceedings of the Special Session During the 87th Annual
Meeting of the American Mathematical Society (AMS)* ED:
DUBOIS, D.W. & RECIO, T. • SER (S 3313) Contemp Math 8
• PUBL (X 0803) Amer Math Soc: Providence vii + 360pp
• DAT&PL 1981 Jan; San Francisco, CA, USA • ISBN
0-8218-5007-5, LC-No 82-3951

P 2342 Symp Rek Kombin;1983 Muenster • D
[1984] *Logic and Machines: Decision Problems and
Complexity. Proceedings of the Symposium Rekursive
Kombinatorik* ED: BOERGER, E. & HASENJAEGER, G. &
ROEDDING, D. • SER (S 3302) Lect Notes Comput Sci 171
• PUBL (X 0811) Springer: Heidelberg & New York
vi + 456pp
• DAT&PL 1983 May; Muenster, D • ISBN 3-540-13331-3,
LC-No 84-55

P 2485 Attas Scuola Algeb IMPA (5);1978 Rio de Janeiro • BR
[1978] *Attas da 5. Scuola de Algebra* PUBL (X 2233) Inst Mat
Pura Apl: Rio Janeiro v + 199pp
• DAT&PL 1978 Jul; Rio de Janeiro, BR

P 2539 Frege Konferenz (1);1979 Jena • DDR
[1979] *"Begriffsschrift". Jenaer Frege-Konferenz* ED: BOLCK,
F. • PUBL (X 2211) Schiller Univ: Jena iii + 548pp
• DAT&PL 1979 May; Jena, DDR

P 2552 Conf Finite Algeb & Multi-Val Log;1979 Szeged • H
[1981] *Proceedings of the Conference on Finite Algebra and
Multiple-Valued Logic* ED: CSAKANY, B. & ROSENBERG, J.G.
• SER (S 3312) Coll Math Soc Janos Bolyai 28 • PUBL
(X 0809) North Holland: Amsterdam 880pp • ALT PUBL
(X 3725) Bolyai Janos Mat Tars: Budapest
• DAT&PL 1979 Aug; Szeged, H • ISBN 0-444-85439-8,
LC-No 81-214217

P 2554 All-Union Symp Th Log Infer;1974 Moskva • SU
[1979] *Logicheskij Vyvod. (Logical Inference. Proceedings of
the All-Union Symposium on the Theory of Logical Inference.)*
ED: SMIRNOV, V.A. • PUBL (X 2027) Nauka: Moskva 312pp
• DAT&PL 1974 Mar; Moskva, SU • LC-No 79-380844

P 2558 All-Union Conf Math Log (5) (Mal'tsev);1979
Novosibirsk • SU
[1979] *5 Vsesoyuznaya Konferentsiya po Matematicheskoj
Logike. Tezitsy Doklady i Soobshcheniya (5th All-Union
Conference on Mathematical Logic. Dedicated to the 70th
Anniversary of the Academician A.I. Mal'tsev)*
• DAT&PL 1979 -?-; Novosibirsk, SU

P 2559 All-Union Symp on Th of Semigroups (2);1978
Sverdlovsk • SU
[1978] *2 Vsesoyuznyj Simpozium po Teorii Polugrupp. Tezisy
Soobshchenij (2nd All-Union Symposium on the Theory of
Semigroups. Abstracts)* PUBL Ural Gos Univ: Sverdlovsk
112pp
• DAT&PL 1978 Jun; Sverdlovsk, SU

P 2564 All-Union Algeb Conf (15);1979 Novosibirsk • SU
[1979] *15 Vsesoyuznyj Algebraicheskij Kollokvium (15th
All-Union Algebraic Conference)*
• DAT&PL 1979 -?-; Novosibirsk, SU

Proceedings

P 2572 Material Respub Konf Molod Uchen;1976 Alma Ata • SU
[1976] *Materialy Respublikanskoj Konferentsij Molodykh Uchenykh. (Materials of the Republican Conference of Young Scientists)* PUBL (X 2443) Nauka: Alma-Ata
• DAT&PL 1976 -?-; Alma-Ata, SU

P 2585 All-Union Conf Math Log (2);1972 Moskva • SU
[1972] *2 Vsesoyuznaya Konferentsiya po Matematicheskoj Logike. Tezitsy Doklady i Soobshcheniya (On Relative Recursiveness and Computability. 2nd All-Union Conference on Mathematical Logic)*
• DAT&PL 1972 -?-; Moskva, SU

P 2588 FCT'77 Fund of Comput Th;1977 Poznan • PL
[1977] *Fundamentals of Computation Theory. Proceedings of the International FCT '77 - Conference* ED: KARPINSKI, M.
• SER (S 3302) Lect Notes Comput Sci 56 • PUBL (X 0811) Springer: Heidelberg & New York 542pp
• DAT&PL 1977 Sep; Poznan-Kornik, PL • ISBN 3-540-08442-8, LC-No 77-14022

P 2607 IBM Symp Math Log & Comput Sci (3); • USA
[1978] *Proceedings of the 3rd IBM Symposium on Mathematical Logic and Computer Science*

P 2614 Open Days in Model Th & Set Th;1981 Jadwisin • PL
[1981] *Proceedings of the International Conference "Open Days in Model Theory and Set Theory"* ED: GUZICKI, W. & MAREK, W. & PELC, A. & RAUSZER, C. • PUBL Selbstverlag 333pp
• DAT&PL 1981 Sep; Jadwisin, PL
• REM Available from John Derrick, School of Mathematics, University of Leeds, GB, or from Cecylia Rauszer, Institute of Mathematics, University of Warsaw, PL

P 2615 Scand Logic Symp (5);1979 Aalborg • DK
[1979] *Proceedings of the 5th Scandinavian Logic Symposium* ED: JENSEN, F.V. & MAYOH, B.H. & MOLLER, K.K. • PUBL (X 2646) Aalborg Univ Pr: Aalborg vii+361pp
• DAT&PL 1979 Jan; Aalborg, DK • ISBN 87-7307-037-8, LC-No 80-464603

P 2622 Abel Group Th;1981 Oberwolfach • D
[1981] *Abelian Group Theory* ED: GOEBEL, R.& WALKER, E.A.
• SER (S 3301) Lect Notes Math 874 • PUBL (X 0811) Springer: Heidelberg & New York xxi+447pp
• DAT&PL 1981 Jan; Oberwolfach, D • ISBN 3-540-10855-6, LC-No 81-9310

P 2623 Worksh Extended Model Th;1980 Berlin • DDR
[1981] *Workshop on Extended Model Theory* ED: HERRE, H.
• SER (J 3437) Rep, Akad Wiss DDR, Inst Math 1981/3
• PUBL (X 2655) Akad Wiss DDR Inst Math: Berlin ii+160pp
• DAT&PL 1980 Nov; Berlin, DDR

P 2624 Ring Th;1978 Antwerpen • B
[1979] *Ring Theory* ED: OYSTRAEYEN VAN, F. • SER (S 3310) Lect Notes Pure Appl Math 51 • PUBL (X 1684) Dekker: New York 801pp
• DAT&PL 1978 Aug; Antwerpen, B • ISBN 0-8247-6854-X., LC-No 79-9080

P 2625 Model Th of Algeb & Arithm;1979 Karpacz • PL
[1980] *Model Theory of Algebra and Arithmetic. Proceedings of the Conference on Applications of Logic to Algebra and Arithmetic* ED: PACHOLSKI, L. & WIERZEJEWSKI, J. & WILKIE, A.J. • SER (S 3301) Lect Notes Math 834 • PUBL (X 0811) Springer: Heidelberg & New York vi+410pp
• DAT&PL 1979 Sep; Karpacz, PL • ISBN 3-540-10269-8, LC-No 82-131180

P 2627 Logic Colloq;1978 Mons • B
[1979] *Logic Colloquium '78* ED: BOFFA, M. & DALEN VAN, D. & MCALOON, K. • SER (S 3303) Stud Logic Found Math 97
• PUBL (X 0809) North Holland: Amsterdam x+434pp
• DAT&PL 1978 Aug; Mons, B • LC-No 79-21152

P 2628 Log Year;1979/80 Storrs • USA
[1981] *Logic Year 1979-80. The University of Connecticut* ED: LERMAN, M. & SCHMERL, J.H. & SOARE, R.I. • SER (S 3301) Lect Notes Math 859 • PUBL (X 0811) Springer: Heidelberg & New York viii+326pp
• DAT&PL 1979 Nov; Storrs, CT, USA • ISBN 3-540-10708-8, LC-No 81-5628

P 2630 Lie Algeb, Group Th, Part Ord Algeb Struct;1980 Carbondale • USA
[1981] *Lie Algebras, Group Theory, and Partially Ordered Algebraic Structures. Proceedings of the Southern Illinois Algebra Conference* ED: AMAYO, R.K. • SER (S 3301) Lect Notes Math 848 • PUBL (X 0811) Springer: Heidelberg & New York vi+298pp
• DAT&PL 1980 Apr; Carbondale, IL, USA • ISBN 3-540-10573-5, LC-No 81-160144

P 2632 Conv Gruppi Abel & Conv Gruppi e loro Rappresent;1972 Roma • I
[1974] *Convegno di Gruppi Abeliani. Convegno di Gruppi e loro Rappresentazioni* SER (S 3305) Symposia Matematica 13
• PUBL (X 3604) INDAM: Roma 558pp • ALT PUBL (X 0801) Academic Pr: New York
• DAT&PL 1972 Nov; Roma, I, 1972 Dec; Roma, I

P 2633 Autom Deduct (7);1984 Napa • USA
[1984] *7th International Conference on Automated Deduction* ED: SHOSTAK, R.E. • SER (S 3302) Lect Notes Comput Sci 170 • PUBL (X 0811) Springer: Heidelberg & New York vi+508pp
• DAT&PL 1984 May; Napa, CA, USA • ISBN 3-540-96022-8, LC-No 84-5441

P 2634 Word Problems II;1976 Oxford • GB
[1980] *Word Problems Vol 2. The Oxford Book* ED: ADYAN, S.I. & BOONE, W.W. & HIGMAN, G. • SER (S 3303) Stud Logic Found Math 95 • PUBL (X 0809) North Holland: Amsterdam x+578pp
• DAT&PL 1976 Jun; Oxford, GB • ISBN 0-444-85343-X, LC-No 79-15276

P 2901 Appl Sheaves;1977 Durham • GB
[1979] *Applications of Sheaves. Proceedings of the Research Symposium on Applications of Sheaf Theory to Logic, Algebra and Analysis* ED: FOURMAN, M.P. & MULVEY, C.J. & SCOTT, D.S. • SER (S 3301) Lect Notes Math 753 • PUBL (X 0811) Springer: Heidelberg & New York xiv+779pp
• DAT&PL 1977 Jul; Durham, GB • ISBN 3-540-09564-0, LC-No 79-23219

P 2902 Aspects Effective Algeb;1979 Clayton • AUS
[1981] *Aspects of Effective Algebra. Proceedings of a Conference at Monash University, Australia* ED: CROSSLEY, J.N. • PUBL (X 2863) Upside Down A Book: Yarra Glen x+290pp
• DAT&PL 1979 Aug; Clayton, Vic, AUS • ISBN 0-949865-01-X

P 2903 Automata, Lang & Progr (8);1981 Akko • IL
[1981] *Automata, Languages and Programming. 8th Colloquium* ED: EVEN, S. & KARIV, O. • SER (S 3302) Lect Notes Comput Sci 115 • PUBL (X 0811) Springer: Heidelberg & New York viii+552pp
• DAT&PL 1981 Jul; Akko, IL • ISBN 3-540-10843-2, LC-No 81-9053
• REM Also Abbreviated as ICALP 81

P 2904 Automata, Lang & Progr (7);1980 Noordwijkerhout • NL
[1980] *Automata, Languages and Programming. 7th Colloquium* ED: BAKKER DE, J.W. & LEEUWEN VAN, J. • SER (S 3302) Lect Notes Comput Sci 85 • PUBL (X 0811) Springer: Heidelberg & New York viii+671pp
• DAT&PL 1980 Jul; Noordwijkerhout, NL • ISBN 3-540-10003-2, LC-No 81-156919
• REM Also Abbreviated as ICALP 80

P 2910 Colloq Anal Harmon & Complexe;1977 Les Aludes • F
[1977] *Colloque d'Analyse Harmonique et Complexe.* PUBL Universite d'Aix-Marseille I, UER de Math.: Marseille vii+154pp
• DAT&PL 1977 Jun; Les Aludes-La Garde-Freinet, F

P 2912 Combin & Repres Groupe Symetr;1976 Strasbourg • F
[1977] *Combinatoire et Representation du Groupe Symetrique. Actes de la Table Ronde du CNRS* ED: FOATA, D. • SER (S 3301) Lect Notes Math 579 • PUBL (X 0811) Springer: Heidelberg & New York iv+339pp
• DAT&PL 1976 Apr; Strasbourg, F • ISBN 3-540-08143-7, LC-No 77-4276

P 2920 Contrib to Gen Algeb (1);1978 Klagenfurt • A
[1979] *Contributions to General Algebra* ED: KAUTSCHITSCH, H. & MUELLER, W.B. & NOEBAUER, W. • PUBL Verlag Johannes Heyn: Klagenfurt 429pp
• DAT&PL 1978 May; Klagenfurt, A • ISBN 3-85366-305-2, LC-No 80-502351

P 2921 Beitr Graphenth & Anw;1977 Oberhof • DDR
[1977] *Beitraege zur Graphentheorie und deren Anwendungen* PUBL (X 2447) Tech Hochschule Ilmenau: Ilmenau 320pp
• DAT&PL 1977 Apr; Oberhof, DDR

P 2923 CAAP'81 Arbres en Algeb & Progr (6);1981 Genova • I
[1981] *CAAP '81. Les Arbres en Algebre et en Programmation. 6eme Colloque* ED: ASTESIANO, E. & BOEHM, C. • SER (S 3302) Lect Notes Comput Sci 112 • PUBL (X 0811) Springer: Heidelberg & New York vi+364pp
• DAT&PL 1981 Mar; Genova, I • ISBN 3-540-10828-9, LC-No 81-8959

P 2935 FCT'79 Fund of Comput Th;1979 Berlin/Wendisch-Rietz • DDR
[1979] *Fundamentals of Computation Theory - FCT '79. Proceedings of the Conference on Algebraic, Arithmetic, and Categorical Methods in Computation Theory* ED: BUDACH, L.
• SER Mathematical Research - Mathematische Forschung 2 • PUBL (X 0911) Akademie Verlag: Berlin 576pp
• DAT&PL 1979 Sep; Berlin/Wendisch-Rietz, DDR • LC-No 82-460828

P 2946 Int Symp Progr (4);1980 Paris • F
[1980] *Proceedings of the 4th International Symposium on Programming* ED: ROBINET, B. • SER (S 3302) Lect Notes Comput Sci 83 • PUBL (X 0811) Springer: Heidelberg & New York viii+341pp
• DAT&PL 1980 Apr; Paris, F • ISBN 3-540-09981-6, LC-No 80-13593

P 2952 CAAP'79 Arbres en Algeb & Progr (4);1979 Lille • F
[1979] *Les Arbres en Algebre et en Programmation. 4eme Colloquium* PUBL Universite de Lille I: Lille iv+327pp
• DAT&PL 1979 Feb; Lille, F

P 2958 Latin Amer Symp Math Log (4);1978 Santiago • RCH
[1980] *Mathematical Logic in Latin America. Proceedings of the 4th Latin American Symposium on Mathematical Logic* ED: ARRUDA, A.I. & CHUAQUI, R. & COSTA DA, N.C.A. • SER (S 3303) Stud Logic Found Math 99 • PUBL (X 0809) North Holland: Amsterdam xii+392pp
• DAT&PL 1978 Dec; Santiago, RCH • ISBN 0-444-85402-9, LC-No 79-20797

P 2983 Ordered Groups;1978 Boise • USA
[1980] *Ordered Groups.* ED: SMITH, J.E. & KENNY, G.O. & BALL, R.N. • SER (S 3310) Lect Notes Pure Appl Math 62
• PUBL (X 1684) Dekker: New York xi+174pp
• DAT&PL 1978 Oct; Boise, ID, USA • ISBN 0-8247-6943-0, LC-No 80-24251

P 2989 Log of Progr; 1983 Pittsburgh • USA
[1984] *Logics of Programs* ED: CLARKE, E. & KOZEN, D.
• SER (S 3302) Lect Notes Comput Sci 164 • PUBL (X 0811) Springer: Heidelberg & New York vi+527pp
• DAT&PL 1983 Jan; Pittsburgh, PA, USA • ISBN 3-540-12896-4, LC-No 84-3123

P 2992 Conf Convergence Structs;1980 Lawton • USA
[1980] *Proceedings of the Conference on Convergence Structures* ED: RIECKE, C. • PUBL Cameron Univ., Department of Math.: Lawton ii+173pp
• DAT&PL 1980 Apr; Lawton, OK, USA

P 2994 Nat Math Conf (8);1977 Tehran • IR
[1977] *Proceedings of the 8th National Mathematics Conference* ED: NOURI-MOGHADAM, M. • PUBL Arya-Mehr University of Technology: Tehran iv+370pp
• DAT&PL 1977 Mar; Tehran, IR

P 2997 Symp Semigroups (1);1977 Matsue • J
[1977] *Proceedings of the 1st Symposium on Semigroups* ED: YAMADA, M. • PUBL Shimane University:Matsue iii+88pp
• DAT&PL 1977 Oct; Matsue, J

P 2998 Symp Semigroups (5);1981 Sakado • J
[1981] *Proceedings of the 5th Symposium on Semigroups. Semigroup Theory and Its Related Fields* ED: NUMAKURA, K.
• PUBL Josai University: Sakado ii+74pp
• DAT&PL 1981 Oct; Sakado, J

P 2999 Proc Conf Databasis (Calzone);1985 Heidelberg • D
[1985] *Proceedings of a Conference on Databases* ED: GSCHNITZER, W. & SPRENGER, H. & STEIN, H. & THURN, K.
• PUBL Apl & Wemding Co: Mosbach 294pp
• DAT&PL 1985 Dec; Heidelberg, D

P 3006 Brazil Conf Math Log (3);1979 Recife • BR
[1980] *Proceedings of the 3rd Brazilian Conference on Mathematical Logic* ED: ARRUDA, A.I. & COSTA DA, N.C.A. & SETTE, A.M. • PUBL (X 2836) Soc Brasil Log: Sao Paulo vi+336pp
• DAT&PL 1979 Dec; Recife, BR

Proceedings

P 3009 Symp Semigroups (2);1978 Tokyo • J
[1978] *Proceedings of the 2nd Symposium on Semigroups*
PUBL Tokyo Gakugei University: Tokyo 80pp
• DAT&PL 1978 Oct; Tokyo, J

P 3010 Symp Semigroups (3);1979 Kobe • J
[1980] *Proceedings of the 3rd Symposium on Semigroups* ED:
FUJIWARA, T. • PUBL (**X** 2243) Univ Osaka Pref Dept Math:
Osaka i+124pp
• DAT&PL 1979 Aug; Kobe, J

P 3011 Symp Semigroups (4);1980 Yamaguchi • J
[1980] *Proceedings of the 4th Symposium on Semigroups.
Semigroup Theory and Its Related Fields* ED: MURATA, K.
• PUBL Yamaguchi University: Yamaguchi ii+131pp
• DAT&PL 1980 Oct; Yamaguchi, J

P 3020 Rect Adv in Repr Th Rings & C^*-Algeb;1973 New
Orleans • USA
[1974] *Recent Advances in the Representation Theory of Rings
and C^*-Algebras by Continuous Sections* ED: HOFMANN, K.H.
& LIUKKONEN, J.R. • SER (**S** 0167) Mem Amer Math Soc
148 • PUBL (**X** 0803) Amer Math Soc: Providence x+182pp
• DAT&PL 1973 Mar; New Orleans, LA, USA • LC-No
74-11237

P 3021 Logic Colloq;1979 Leeds • GB
[1980] *Recursion Theory: Its Generalisations and Applications.
Proceedings of the Logic Colloquium '79* ED: DRAKE, F.R. &
WAINER, S.S. • SER (**S** 3306) Lond Math Soc Lect Note Ser
45 • PUBL (**X** 0805) Cambridge Univ Pr: Cambridge, GB
vi+319pp
• DAT&PL 1979 Aug; Leeds, GB • ISBN 0-521-23543-X,
LC-No 82-180570

P 3023 Repr Algeb (2);1979 Ottawa • CDN
[1980] *Representation Theory II. Proceedings of the 2nd
International Conference on Representations of Algebras* ED:
DLAB, V. & GABRIEL, P. • SER (**S** 3301) Lect Notes Math 832
• PUBL (**X** 0811) Springer: Heidelberg & New York
xiv+673pp
• DAT&PL 1979 Aug; Ottawa, ONT, CDN • ISBN
3-540-10264-7, LC-No 80-25700

P 3030 Semin Goulaouic-Schwartz;1972 Paris • F
[1972] *Seminaire Goulaouic-Schwartz 1971-1972: Equations
aux Derivees Partielles et Analyse Fonctionnelle. Deux Journees
p-radonifiantes* PUBL Ecole Polytech., Centre de Math.,
Paris iii+421pp
• DAT&PL 1972 Jan; Paris, F • LC-No 81-645130

P 3057 CAAP'80 Arbres en Algeb & Progr (5);1980 Lille • F
[1980] *Les Arbres en Algebre et en Programmation. 5eme
Colloquium* PUBL Universite de Lille I: Lille vi+268pp
• DAT&PL 1980 Feb; Lille, F

P 3062 IEEE Symp Switch & Automata Th (14);1973 Iowa
City • USA
[1973] *14th Annual IEEE Symposium on Switching and
Automata Theory* PUBL (**X** 2179) IEEE: New York
v+213pp
• DAT&PL 1973 Oct; Iowa City, IA, USA • LC-No 80-646635

P 3063 Autom Deduct (5);1980 Les Arcs • F
[1980] *5th Conference on Automated Deduction* ED: BIBEL, W.
& KOWALSKI, R. • SER (**S** 3302) Lect Notes Comput Sci 87
• PUBL (**X** 0811) Springer: Heidelberg & New York
vii+385pp
• DAT&PL 1980 Jul; Les Arcs, F • ISBN 3-540-10009-1,
LC-No 80-18708

P 3083 FCT'83 Found of Comput Th (Sel Pap);1983
Borgholm • S
[1985] *Topics in the Theory of Computation. Selected Papers of
the International Conference on Foundations of Computation
Theory, FCT '83* ED: KARPINSKI, M. & LEEUWEN VAN, J.
• SER (**S** 3358) Ann Discrete Math 24, North-Holland Math
Stud 102 • PUBL (**X** 0809) North Holland: Amsterdam
ix+187pp
• DAT&PL 1983 Aug; Borgholm, S • LC-No 84-21089

P 3084 Autom Theor Prov After 25 Yea;1983 Denver • USA
[1984] *Automated Theorem Proving After 25 Years.
Proceedings of the Special Session on Automatic Theorem
Proving, 89th Annual Meeting of the American Mathematical
Society (AMS)* ED: BLEDSOE, W.W. & LOVELAND, D.W. • SER
(**S** 3313) Contemp Math 29 • PUBL (**X** 0803) Amer Math
Soc: Providence ix+360pp
• DAT&PL 1983 Jan; Denver, CO, USA • ISBN
0-8218-5027-X, LC-No 84-9226

P 3085 Graph Th & Combin (Erdoes);1983 Cambridge • GB
[1984] *Graph Theory and Combinatorics. Proceedings of the
Cambridge Combinatorial Conference in Honour of Paul
Erdoes* ED: BOLLOBAS, B. • PUBL (**X** 0801) Academic Pr:
New York xvii+328pp
• DAT&PL 1983 Mar; Cambridge, GB • ISBN
0-12-111760-X, LC-No 83-83432

P 3088 Univer Alg & Link Log, Alg, Combin, Comp Sci;1983
Darmstadt • D
[1984] *Universal Algebra and Its Links with Logic, Algebra,
Combinatorics and Computer Science. Proceedings of the '25.
Arbeitstagung ueber Allgemeine Algebra'* ED: BURMEISTER, P.
& GANTER, B. & HERRMAN, C. & KEIMEL, K. & POGUNTKE, W.
WILLE, R. • SER Research and Exposition in Math 4 • PUBL
Heldermann: Berlin vii+243pp
• DAT&PL 1983 -?-; Darmstadt, D • LC-No 85-129031

P 3091 Conv Int Storica Logica;1982 San Gimignano • I
[1983] *Atti del Convegno Internazionale di Storica della
Logica* ED: ABRUSCI, V.M. & CASARI, E. & MUGNAI, M.
• PUBL Cooperativa Libraria, Universitaria Ed: Bologna
x+401pp
• DAT&PL 1982 Dec; San Gimignano, I • LC-No 84-181758

P 3092 Congr Naz Logica;1979 Montecatini Terme • I
[1981] *Atti del Congresso Nazionale di Logica* ED: BERNINI, S.
• PUBL (**X** 1732) Bibliopolis: Napoli 735pp
• DAT&PL 1979 Oct; Montecatini Terme, I • LC-No
81-198713

P 3103 Adv Cybern Syst;1972 Oxford • GB
[1974] *Advances in Cybernetics and Systems* ED: ROSE, J.
• PUBL (**X** 0836) Gordon & Breach: New York 3 Vols:
xx+1730pp
• DAT&PL 1972 -?-; Oxford, GB • LC-No 75-316901

P 3113 Aspects Philos Logic;1977 Tuebingen • D
[1981] *Aspects of Philosophical Logic* ED: MOENNICH, U.
• SER (**S** 3307) Synth Libr 147 • PUBL (**X** 0835) Reidel:
Dordrecht vii+283pp
• DAT&PL 1977 Dec; Tuebingen, D • ISBN 90-277-1201-8,
LC-No 81-7358

P 3140 Colloq d'Algeb;1980 Rennes • F
[1980] *Colloque d'Algebre* ED: GUERINDON, J. • PUBL
Universite de Rennes I: Rennes xi+308pp
• DAT&PL 1980 May; Rennes, F

P 3148 Graphth Konzepte Inf (3);1977 Linz • A
[1978] *Datenstrukturen, Graphen, Algorithmen. Ergebnisse des Workshops WG 77. 3. Fachtagung ueber Graphentheoretische Konzepte der Informatik* ED: MUEHLBACHER, J. • SER (J 3359) Appl Comp Sci, Ber Prakt Inf 8 • PUBL (X 3223) Hanser: Muenchen 364pp
• DAT&PL 1977 Jun; Linz, A • ISBN 3-446-12526-4, LC-No 78-370096

P 3165 FCT'81 Fund of Comput Th;1981 Szeged • H
[1981] *Fundamentals of Computation Theory. Proceedings of the 1981 International FCT-Conference* ED: GECSEG, F. • SER (S 3302) Lect Notes Comput Sci 117 • PUBL (X 0811) Springer: Heidelberg & New York xi+471pp
• DAT&PL 1981 Aug; Szeged, H • ISBN 3-540-10854-8, LC-No 81-13533

P 3196 Kompl von Entscheid Probl;1973/74 Zuerich • CH
[1976] *Komplexitaet von Entscheidungsproblemen. Ein Seminar* ED: SPECKER, E. & STRASSEN, V. • SER (S 3302) Lect Notes Comput Sci 43 • PUBL (X 0811) Springer: Heidelberg & New York ii+217pp
• DAT&PL 1973 -?-; Zuerich, CH • ISBN 3-540-07805-3, LC-No 76-25088

P 3198 Incont Compl Calc, Cod & Ling Form;1975 Napoli • I
[1976] *Atti dell' Incontro su Complessita die Calcolo, Codici e Linguaggi Formali.*
• DAT&PL 1975 -?-; Napoli, I

P 3201 Logic Symposia;1979/80 Hakone • J
[1981] *Logic Symposia Hakone 1979,1980* ED: MUELLER, GERT H. & TAKEUTI, G. & TUGUE, T. • SER (S 3301) Lect Notes Math 891 • PUBL (X 0811) Springer: Heidelberg & New York xi+394pp
• DAT&PL 1979 Mar; Hakone,J, 1980 Feb; Hakone,J • ISBN 3-540-11161-1, LC-No 81-18424

P 3210 Math Founds of Comput Sci (9);1980 Rydzyna • PL
[1980] *Mathematical Foundations of Computer Science. Proceedings of the 9th Symposium* ED: DEMBINSKI, P. • SER (S 3302) Lect Notes Comput Sci 88 • PUBL (X 0811) Springer: Heidelberg & New York viii+723pp
• DAT&PL 1980 Sep; Rydzyna, PL • ISBN 3-540-10027-X, LC-No 80-20087

P 3237 Colloq Universal Algeb;1977 Esztergom • H
[1982] *Proceedings of the Colloque on Universal Algebra* ED: CSAKANY, B. & FRIED, E. & SCHMIDT, E.T. • SER (S 3312) Coll Math Soc Janos Bolyai 29 • PUBL (X 0809) North Holland: Amsterdam 804pp
• DAT&PL 1977 Jun; Esztergom, H • ISBN 0-444-85405-3, LC-No 79-350398

P 3238 Conf Theoret Comput Sci;1977 Waterloo ON • CDN
[1977] *Proceedings of a Conference on Theoretical Computer Science* PUBL University of Waterloo, Computer Science Department: Waterloo iv+283pp
• DAT&PL 1977 Aug; Waterloo, ON, CDN

P 3241 Queen's Numb Th Conf;1979 Kingston • CDN
[1980] *Proceedings of the Queen's Number Theory Conference 1979* ED: RIBENBOIM, P. • SER Queen's Papers in Pure & Appl Math 54 • PUBL (X 0997) Queen's Univ: Kingston xiii+497pp
• DAT&PL 1979 Jul; Kingston, Ont, CDN • LC-No 81-179245

P 3262 Semin Semigroup Th;1976 Kyoto • J
[1977] *Seminar on Semigroup Theory* ED: SAITO, T. • SER (S 2308) Symp Kyoto Univ Res Inst Math Sci (RIMS) 292 • PUBL (X 2441) Kyoto Univ Res Inst Math Sci: Kyoto 175pp
• DAT&PL 1976 Nov; Kyoto, J

P 3268 Set Th & Model Th;1979 Bonn • D
[1981] *Set Theory and Model Theory* ED: JENSEN, R.B. & PRESTEL, A. • SER (S 3301) Lect Notes Math 872 • PUBL (X 0811) Springer: Heidelberg & New York v+174pp
• DAT&PL 1979 Jun; Bonn, D • ISBN 3-540-10849-1, LC-No 81-9234

P 3269 Set Th Found Math (Kurepa);1977 Beograd • YU
[1977] *Set Theory. Foundations of Mathematics* SER (J 2887) Zbor Radova, NS 2(10) • PUBL (X 3727) Beograd Mat Inst: Belgrade 152pp
• DAT&PL 1977 Aug; Beograd, YU • LC-No 79-373865

P 3299 Progr Kiso Riron, Algor Okeru Shomei Ron;1973/74 Kyoto • J
[1975] *"Program no Kiso Riron" Kenkyu Shugo Oyobi Tanki Kyodo Kenkyu "Algorithm ni Okeru Shomei Ron". Hokoku Shu (The 3rd Symposium on Basic Theory of Programs and the Workshop for Proof Theory about Algorithms. Proceedings)* ED: IGARASHI, S. • SER (S 2308) Symp Kyoto Univ Res Inst Math Sci (RIMS) 236 • PUBL (X 2441) Kyoto Univ Res Inst Math Sci: Kyoto 215pp
• DAT&PL 1973 May; Kyoto, J, 1974 Nov; Kyoto, J

P 3342 Rec Th Week;1984 Oberwolfach • D
[1985] *Recursion Theory Week* ED: EBBINGHAUS, H.-D. & MUELLER, GERT H. & SACKS, G.E. • SER (S 3301) Lect Notes Math 1141 • PUBL (X 0811) Springer: Heidelberg & New York viii+418pp
• DAT&PL 1984 Apr; Oberwolfach, D • ISBN 3-540-15673-9, LC-No 85-20808

P 3355 Algeb Conf (1);1980 Skopje • YU
[1980] *Algebraic Conference '80* ED: CELAKOSKI, N. • PUBL (X 3760) Univ Kiril et Metodij, Mat Fak: Skopje vi+152pp
• DAT&PL 1980 Feb; Skopje, YU • LC-No 82-205527

P 3404 Model Th & Arithm;1979/80 Paris • F
[1981] *Model Theory and Arithmetic. Comptes Rendus d'une Action Thematique Programmee du CNRS sur la Theorie des Modeles et l'Arithmetique* ED: BERLINE, C. & McALOON, K. & RESSAYRE, J.-P. • SER (S 3301) Lect Notes Math 890 • PUBL (X 0811) Springer: Heidelberg & New York vi+306pp
• DAT&PL 1979 -?-; Paris, F • ISBN 3-540-11159-X, LC-No 82-137454

P 3482 Logic & Algor (Specker);1980 Zuerich • CH
[1982] *Logic and Algorithmic. An International Symposium Held in Honour of Ernst Specker* ED: ENGELER, E. & LAEUCHLI, H. & STRASSEN, V. • SER Monogr l'Enseign Math 30 • PUBL (X 3718) Enseign Math, Univ Geneve: Geneve 392pp
• DAT&PL 1980 Feb; Zuerich, CH • LC-No 82-184564

P 3488 Theor Comput Sci (4);1979 Aachen • D
[1979] *Theoretical Computer Science. 4th GI Conference (Gesellschaft fuer Informatik)* ED: WEIHRAUCH, K. • SER (S 3302) Lect Notes Comput Sci 67 • PUBL (X 0811) Springer: Heidelberg & New York vii+324pp
• DAT&PL 1979 Mar; Aachen D • ISBN 3-540-09118-1, LC-No 79-9707

P 3497 Log of Progr;1979 Zuerich • CH
[1981] *Logic of Programs* ED: ENGELER, E. • SER (S 3302) Lect Notes Comput Sci 125 • PUBL (X 0811) Springer: Heidelberg & New York v+245pp
• DAT&PL 1979 May; Zuerich, CH • ISBN 3-540-11160-3, LC-No 82-137449

P 3542 ACM Symp Th of Comput (11);1979 Atlanta • USA
[1979] *Conference Record of the 11th Annual ACM Symposium on Theory of Computing (Association for Computing Machinery)* PUBL (X 2205) ACM: New York vii+368pp
• DAT&PL 1979 Apr; Atlanta, GA, USA • ISBN 0-89791-003-6, LC-No 82-642181

P 3546 Teor Avtom (2);1969 Kiev • SU
[1969] *Teoriya Avtomatav (Automata Theory, No.2)* PUBL (X 2522) Akad Nauk Inst Kibernet: Kiev 107pp
• DAT&PL 1969 -?-; Kiev, SU

P 3547 Jorn Mat Hispano-Lusitanas (1);1973 Madrid • E
[1977] *Actas de las 1as Jornadas Matematicas Hispano-Lusitanas* PUBL (X 2091) Consejo Sup Invest Ci: Madrid 687pp
• DAT&PL 1973 Apr; Madrid, E • LC-No 78-352722

P 3572 IEEE Symp Found of Comput Sci (18);1977 Providence • USA
[1977] *18th Annual IEEE Symposium on Foundations of Computer Science* PUBL (X 2179) IEEE: New York v+269pp
• DAT&PL 1977 Oct; Providence, RI, USA • LC-No 80-646634

P 3578 IEEE Symp Found of Comput Sci (19);1978 Ann Arbor • USA
[1978] *19th Annual IEEE Symposium on Foundations of Computer Science* PUBL (X 2179) IEEE: New York v+290pp
• DAT&PL 1978 Oct; Ann Arbor, MI, USA • LC-No 80-646634

P 3600 Ordres Totaux Finis;1967 Aix-en-Provence • F
[1971] *Ordres Totaux Finis* SER Mathematiques et Sciences de l'Homme 12 • PUBL (X 0834) Gauthier-Villars: Paris ix+315pp • ALT PUBL (X 0873) Mouton: Paris
• DAT&PL 1967 Jul; Aix-en-Provence, F • LC-No 73-583465

P 3608 Wechselwirk Inform Math;1979 Wien • A
[1980] *Wechselwirkungen zwischen Informatik und Mathematik. Kongressband fuer die Tagung Mathematik und Informatik* ED: DOERFLER, W. & SCHAUER, H. • SER Schriftenr Oesterreich Computer Ges 9 • PUBL (X 0814) Oldenbourg: Muenchen 235pp
• DAT&PL May 1979, Wien, A • ISBN 3-486-24451-5, LC-No 81-111866

P 3621 Frege Konferenz (2);1984 Schwerin • DDR
[1984] *Frege Konferenz* ED: WECHSUNG, G. • SER Mathematical Research - Mathematische Forschungen 20 • PUBL (X 0911) Akademie Verlag: Berlin 408pp
• DAT&PL 1984 Sep; Schwerin, DDR • LC-No 84-248695

P 3622 Int Congr Log, Meth & Phil of Sci (6,Proc);1979 Hannover • D
[1982] *Proceedings of the 6th International Congress for Logic, Methodology and Philosophy of Science* ED: COHEN, L.J. & LOS, J. & PFEIFFER, H. & PODEWSKI, K.-P. • SER (S 3303) Stud Logic Found Math 104 • PUBL (X 0809) North Holland: Amsterdam xiv+842pp
• DAT&PL 1979 Aug; Hannover, D • LC-No 80-12713

P 3623 Logic Colloq;1980 Prague • CS
[1982] *Logic Colloquium '80* ED: DALEN VAN, D. & LASCAR, D. & SMILEY, T.J. • SER (S 3303) Stud Logic Found Math 108 • PUBL (X 0809) North Holland: Amsterdam x+342pp
• ISBN 0-444-86465-2, LC-No 82-12611

P 3634 Patras Logic Symp;1980 Patras • GR
[1982] *Patras Logic Symposium* ED: METAKIDES, G. • SER (S 3303) Stud Logic Found Math 109 • PUBL (X 0809) North Holland: Amsterdam ix+391pp
• DAT&PL 1980 Aug; Patras, GR • ISBN 0-444-86476-8, LC-No 82-14107

P 3638 Brouwer Centenary Symp;1981 Noordwijkerhout • NL
[1982] *The L.E.J. Brouwer Centenary Symposium* ED: TROELSTRA, A.S. & DALEN VAN, D. • SER (S 3303) Stud Logic Found Math 110 • PUBL (X 0809) North Holland: Amsterdam x+524pp
• DAT&PL 1981 Jun; Noordwijkerhout, NL • ISBN 0-444-86494-6, LC-No 82-239358

P 3642 Colloq Math Log in Computer Sci;1978 Salgotarjan • H
[1981] *Proceedings of the Colloquium on Mathematical Logic in Computer Science* ED: DOEMALKI, B. & GERGELY, T. • SER (S 3312) Coll Math Soc Janos Bolyai 26 • PUBL (X 0809) North Holland: Amsterdam 758pp • ALT PUBL (X 3725) Bolyai Janos Mat Tars: Budapest
• DAT&PL 1978 Sep; Salgotarjan, H • ISBN 0-444-85440-1, LC-No 82-101557

P 3658 Math Founds of Comput Sci (11);1984 Prague • CS
[1984] *Mathematical Foundations of Computer Science 1984. Proceedings of the 11th Symposium* ED: CHYTIL, M.P. & KOUBEK, V. • SER (S 3302) Lect Notes Comput Sci 176 • PUBL (X 0811) Springer: Heidelberg & New York 581pp
• DAT&PL 1984 Sep; Prague, CS • ISBN 3-540-13372-0, LC-No 83-13980

P 3668 Log & Founds of Math;1983 Kyoto • J
[1984] *Logic and the Foundations of Mathematics* ED: TUGUE, T. • SER (S 2308) Symp Kyoto Univ Res Inst Math Sci (RIMS) 516 • PUBL (X 2441) Kyoto Univ Res Inst Math Sci: Kyoto ii+195pp
• DAT&PL 1983 Oct; Kyoto, J

P 3669 SE Asian Conf on Log;1981 Singapore • SGP
[1983] *Southeast Asian Conference on Logic* ED: CHONG, CHI TAT & WICKS, M.J. • SER (S 3303) Stud Logic Found Math 111 • PUBL (X 0809) North Holland: Amsterdam xiv+210pp
• DAT&PL 1981 Nov; Singapore, SGP • ISBN 0-444-86706-6, LC-No 83-11458

P 3708 Herbrand Symp Logic Colloq;1981 Marseille • F
[1982] *Proceedings of the Herbrand Symposium Logic. Colloquium '81* ED: STERN, J. • SER (S 3303) Stud Logic Found Math 107 • PUBL (X 0809) North Holland: Amsterdam xi+384pp
• DAT&PL 1981 Jul; Marseille, F • ISBN 0-444-86417-2, LC-No 82-6433

P 3710 Logic Colloq;1982 Firenze • I
[1984] *Proceedings of the Logic Colloquium '82* ED: LOLLI, G. & LONGO, G. & MARCJA, A. • SER (S 3303) Stud Logic Found Math 112 • PUBL (X 0809) North Holland: Amsterdam viii+358pp
• DAT&PL 1982 Aug; Firenze, I • ISBN 0-444-86876-3, LC-No 84-1630

P 3714 ACM Symp Th of Comput (14);1982 • USA
[1982] *Symposium on Theory of Computing* PUBL (X 2205) ACM: New York
• LC-No 83-641095

P 3751 Jorn Mat Luso-Espanol (8);1981 Coimbra • P
[1981] *Actas 8 Jornadas Luso-Espanholas de Matematica* PUBL Univ. de Coimbra, Dept. Mat.:Coimbra 4 Vols: 387pp,474pp,395pp,152pp
• DAT&PL 1981 May; Coimbra, P

P 3758 Algeb Conf (2);1981 Novi Sad • YU
[1982] *Algebraic Conference. Proceedings of the 2nd Conference* ED: GILEZAN, K. • PUBL (X 4030) Univ Novom Sadu, Inst Mat: Novi Sad vii+162pp
• DAT&PL 1982 May; Novi Sad, YU • LC-No 83-169879

P 3774 Th d'Ensembl de Quine;1981 Louvain-la-Neuve • B
[1982] *La Theorie des Ensembles de Quine* SER (S 3935) Cah Cent Log (Louvain) 4 • PUBL (X 3734) Cabay: Louvain-la-Neuve iii+76pp
• DAT&PL 1981 Oct; Louvain-la-Neuve, B • ISBN 2-87077-138-X

P 3780 Ordered Sets;1981 Banff • CDN
[1982] *Ordered Sets. Proceedings of a NATO Advanced Study Institute* ED: RIVAL, I. • SER NATO Advanced Study Inst. Series C 83 • PUBL (X 0835) Reidel: Dordrecht xviii+966pp
• DAT&PL 1981 Aug; Banff, CDN • ISBN 90-277-1396-0, LC-No 82-544

P 3800 Intens Log: Th & Appl;1979 Moskva • SU
[1982] *Intensional Logic: Theory and Applications. Proceedings of the 2nd Soviet-Finnish Logic Conference* ED: NIINILUOTO, I. & SAARINEN, E. • SER (J 0096) Acta Philos Fenn 35 • PUBL (X 0809) North Holland: Amsterdam 301pp • ALT PUBL (X 2812) Philos Soc Finland : Helsinki
• DAT&PL 1979 Dec; Moskva, SU • ISBN 951-95054-8-2

P 3802 Abel Group Th;1983 Honolulu • USA
[1983] *Abelian Group Theory* ED: GOEBEL, R. & LADY, L. & MADER, A. • SER (S 3301) Lect Notes Math 1006 • PUBL (X 0811) Springer: Heidelberg & New York xvi+771pp
• DAT&PL 1983 Dec; Honolulu, HI, USA • ISBN 3-540-12335-0, LC-No 83-211133

P 3808 Rect Trends in Math;1982 Reinhardsbrunn • D
[1982] *Recent Trends in Mathematics* ED: KURKE, H. & MECKE, J. & TRIEBEL, H. & THIELE, R. • SER Teubner-Texte zur Mathematik 50 • PUBL (X 0823) Teubner: Stuttgart 336pp
• DAT&PL 1982 Oct; Reinhardsbrunn, D • LC-No 83-168766

P 3817 Algeb Homage:Ring Th & Rel Top (Jacobson);1981 New Haven • USA
[1982] *Algebraists' Homage: Papers in Ring Theory and Related Topics. Proceedings of the Conference on Algebra in Honor of Nathan Jacobson* ED: AMITSUR, S.A. & SALTMAN, D.J. & SELIGMAN, G.B. • SER (S 3313) Contemp Math 13 • PUBL (X 0803) Amer Math Soc: Providence viii+409pp
• DAT&PL 1981 Jun; New Haven, CT, USA • ISBN 0-8218-5013-X, LC-No 82-18934

P 3823 Axiomatic Set Th;1983 Boulder • USA
[1984] *Axiomatic Set Theory* ED: BAUMGARTNER, J. & MARTIN, D.A. & SHELAH, S. • SER (S 3313) Contemp Math 31 • PUBL (X 0803) Amer Math Soc: Providence x+259pp
• DAT&PL 1983 Jun; Boulder, CO, USA • ISBN 0-8218-5026-1, LC-No 84-18457

P 3826 Proc Europ Conf Comput Algeb;1982 Marseille • F
[1982] *Computer Algebra. Proceedings of the European Computer Algebra Conference (EUROCAM)* ED: CALMET, J. • SER (S 3302) Lect Notes Comput Sci 144 • PUBL (X 0811) Springer: Heidelberg & New York xiv+301pp
• DAT&PL 1982 Apr; Marseille, F • ISBN 3-540-11607-9, LC-No 82-240810

P 3829 Atti Incontri Log Mat (1);1982 Siena • I
[1983] *Atti Degli Incontri di Logica Matematica* ED: BERNARDI, C. • PUBL (X 3812) Univ Siena, Dip Mat: Siena 398pp
• DAT&PL 1982 Jan; Siena, I, 1982 Apr; Siena, I, 1982 Jun; Siena, I

P 3830 Logics of Progr & Appl;1980 Poznan • PL
[1983] *Logics of Programs and Their Applications* ED: SALWICKI, A. • SER (S 3302) Lect Notes Comput Sci 148 • PUBL (X 0811) Springer: Heidelberg & New York vi+324pp
• DAT&PL 1980 Aug; Poznan, PL • ISBN 3-540-11981-7, LC-No 83-158934

P 3831 Universal Algeb & Appl;1978 Warsaw • PL
[1982] *Universal Algebra and Applications* ED: TRACZYK, T. • SER (S 1926) Banach Cent Publ 9 • PUBL (X 1034) PWN: Warsaw 454pp
• DAT&PL 1978 Feb-Sep; Warsaw, PL • ISBN 83-01-02145-4, LC-No 83-167533

P 3833 Symp Semigroups (6);1982 Kyoto • J
[1982] *Semigroup Theory and Its Related Fields. Proceedings of the 6th Symposium on Semigroups* ED: YOSHIDA, R. • PUBL Osaka College of Pharmacy ii+79pp • ALT PUBL Ritsumeikan Univ: Kyoto
• DAT&PL 1982 Oct; Kyoto, J

P 3840 Autom Deduct (6);1982 New York • USA
[1982] *6th Conference on Automated Deduction* ED: LOVELAND, D.W. • SER (S 3302) Lect Notes Comput Sci 138 • PUBL (X 0811) Springer: Heidelberg & New York vii+389pp
• DAT&PL 1982 Jun; New York, NY, USA • ISBN 3-540-11558-7, LC-No 82-5948

P 3841 Universal Algeb & Lattice Th (4);1982 Puebla • MEX
[1983] *Universal Algebra and Lattice Theory* ED: FREESE, R.S. & GRACIA, O.C. • SER (S 3301) Lect Notes Math 1004 • PUBL (X 0811) Springer: Heidelberg & New York vi+308pp
• DAT&PL 1982 Jan; Puebla, MEX • ISBN 3-540-12329-6, LC-No 83-14637

P 3845 Conf Math Service of Man (2,Proc)(Feriet);1982 Las Palmas • E
[1982] *2nd World Conference on Mathematics at the Service of Man. Dedicated to the Remembrance of Joseph Kampe de Feriet (1893-1982)* ED: BALLESTER, A. & CARDUS, D. & TRILLAS, E. • PUBL Universidad Politecnica: Las Palmas xxx+696pp
• DAT&PL 1982 Jun; Las Palmas, E

P 3851 Automata, Lang & Progr (10);1983 Barcelona • E
[1983] *Automata, Languages and Programming. 10th Colloquium* ED: DIAZ, J. • SER (S 3302) Lect Notes Comput Sci 154 • PUBL (X 0811) Springer: Heidelberg & New York viii+734pp
• DAT&PL 1983 Jul; Barcelona, E • ISBN 3-540-12317-2, LC-No 83-10435
• REM Also Abbreviated as ICALP 83

P 3852 Geom Algeb Reelle & Formes Quad;1981 Rennes • F
[1982] *Geometrie Algebraique Reelle et Formes Quadratiques. Journees SMF* ED: COLLIOT-THELENE, J.-L. & COSTE, M. & MAHE, L. & ROY, M.-F. • SER (S 3301) Lect Notes Math 959 • PUBL (X 0811) Springer: Heidelberg & New York x+458pp
• DAT&PL 1981 May; Rennes, F • ISBN 3-540-11959-0

P 3853 Algeb Symp (2);1980 Pretoria • ZA
[1980] *Proceedings of the 2nd Algebra Symposium* ED: SCHOEMAN, M.J. • PUBL Univ. of Pretoria, Dep. Math.: Pretoria iii+86pp
• DAT&PL 1980 Aug; Pretoria, ZA • LC-No 83-123992

P 3855 Algeb Conf (3);1982 Beograd • YU
[1983] *Proceedings of the 3rd Algebraic Conference* ED: PRESIC, S.B. & MILIC, . & MIJAJLOVIC, Z. & PERIC, V. & CUPONA, G. • PUBL (X 4030) Univ Novom Sadu, Inst Mat: Novi Sad vi+157pp
• DAT&PL 1982 Oct; Beograd, YU • LC-No 84-233209

P 3856 Combin Th;1982 Rauischholzhausen • D
[1982] *Combinatorial Theory* ED: JUNGNICKEL, D. & VEDDER, K. • SER (S 3301) Lect Notes Math 969 • PUBL (X 0811) Springer: Heidelberg & New York v+326pp
• DAT&PL 1982 May; Rauischholzhausen, D • ISBN 3-540-11971-X, LC-No 83-138374

P 3858 Adequate Modeling of Syst;1982 Bad Honnef • D
[1983] *Adequate Modeling of Systems. Proceedings of the International Working Conference on Model Realism* ED: WEDDE, H. • PUBL (X 0811) Springer: Heidelberg & New York xi+336pp
• DAT&PL 1982 Apr; Bad Honnef, D • ISBN 3-540-12567-1, LC-No 83-10335

P 3860 Coll Papers to Farah on Retirement;1981 Sao Paulo • BR
[1982] *Collected Papers Dedicated to Professor Edison Farah on the Occasion of His Retirement* ED: ALAS, O.T. & COSTA DA, N.C.A. & HOENIG, C.S. • PUBL Inst Mat Estat Univ Sao Paulo: Sao Paulo v+134pp
• DAT&PL 1981 May; Sao Paulo, BR

P 3864 FCT'83 Found of Comput Th;1983 Borgholm • S
[1983] *Foundation of Computation Theory. 1983 FCT-Conference* ED: KARPINSKI, M. • SER (S 3302) Lect Notes Comput Sci 158 • PUBL (X 0811) Springer: Heidelberg & New York xi+517pp
• DAT&PL 1983 Aug; Borgholm, S • ISBN 3-540-12689-9, LC-No 83-16736

P 3882 Algeb Symp (1);1979 Potchefstroom • ZA
[1980] *Algebra Symposium* PUBL Potchefstroom Univ Christian Higher Educ: Potchefstroom iii+111pp
• DAT&PL 1979 Aug; Potchefstroom, ZA

P 3883 Banach Space Th & Appl (1);1981 Bucharest • RO
[1983] *Banach Space Theory and Its Applications. Proceedings of the 1st Romanian-GDR Seminar* ED: PIETSCH, A. & POPA, N. & SINGER, I. • SER (S 3301) Lect Notes Math 991 • PUBL (X 0811) Springer: Heidelberg & New York x+302pp
• DAT&PL 1981 Sep; Bucharest, RO • ISBN 3-540-12298-2, LC-No 83-209142

P 3885 Categor Th.Appl to Algeb Log & Topol;1981 Gummersbach • D
[1982] *Category Theory. Applications to Algebra, Logic and Topology* ED: KAMPS, K.H. & PUMPLUEN, D. & THOLEN, W. • SER (S 3301) Lect Notes Math 962 • PUBL (X 0811) Springer: Heidelberg & New York xv+322pp
• DAT&PL 1981 Jul; Gummersbach, D • ISBN 3-540-11961-2, LC-No 82-19602

P 3886 Groups-St.Andrews;1981 St.Andrews • GB
[1982] *Groups - St.Andrews 1981* ED: CAMPBELL, C.M. & ROBERTSON, E.F. • SER (S 3306) Lond Math Soc Lect Note Ser 71 • PUBL (X 0805) Cambridge Univ Pr: Cambridge, GB viii+360pp
• DAT&PL 1981 Jul; St.Andrews, GB • ISBN 0-521-28974-2, LC-No 82-4427

P 3894 Geom-von Staudt's Point of View;1980 Bad Windsheim • D
[1981] *Geometry - von Staudt's Point of View. Proceedings of the NATO Advanced Study Institute* ED: PLAUMANN, P. & STRAMBACH, K. • SER NATO Advanced Study Inst Ser C 70 • PUBL (X 0835) Reidel: Dordrecht xi+430pp
• DAT&PL 1980 Jul; Bad Windsheim, D • ISBN 90-277-1283-2, LC-No 81-5843

P 3895 Contrib to Gen Algeb (2);1982 Klagenfurt • A
[1983] *Contributions to General Algebra* ED: EIGENTHALER, G. & KAISER, H.K. & MUELLER, W.B. & NOEBAUER, W. • PUBL (X 2728) Hoelder-Pichler-Tempsky: Wien 405pp • ALT PUBL (X 0823) Teubner: Stuttgart
• DAT&PL 1982 Jun; Klagenfurt, A • ISBN 3-209-00501-X

P 4004 CAAP'82 Arbres en Algeb & Progr (7);1982 Lille • F
[1982] *Les Arbres en Algebre et en Programmation. 7eme Colloquium* PUBL Universite de Lille I: Lille iv+321pp
• DAT&PL 1982 Mar; Lille, F

P 4006 Symp Semigroups (7);1983 Tokyo • J
[1983] *Semigroup Theory and Its Related Fields. Proceedings of the 7th Symposium on Semigroups* ED: SAITO, T. • PUBL Tokyo Gakugei Univ: Tokyo ii+42pp
• DAT&PL 1983 Oct; Tokyo, J

P 4031 Semin Algeb (34);1981 Paris • F
[1982] *Seminaire d'Algebre Paul Dubreil et Marie-Paule Malliavin, 34'eme Annee* ED: MALLIAVIN-BRAMERET, M.-P. • SER (S 3301) Lect Notes Math 924 • PUBL (X 0811) Springer: Heidelberg & New York v+461pp
• DAT&PL 1981 -?-; Paris, F • ISBN 3-540-11496-3, LC-No 84-640108

P 4046 Rec Th;1982 Ithaca • USA
[1985] *Recursion Theory Proceedings of the AMS-ASL Summer Institute* ED: NERODE, A. & SHORE, R.A. • SER (S 3304) Proc Symp Pure Math 42 • PUBL (X 0803) Amer Math Soc: Providence vii+528pp
• DAT&PL 1982 Jun; Ithaca, NY, USA • ISBN 0-8218-1447-8, LC-No 84-18525

P 4064 Prog Graph Th;1982 Waterloo • CDN
[1984] *Progress in Graph Theory*
• DAT&PL 1982 Jul; Waterloo, ON, CDN

P 4096 Jorn Mat Luso-Espanol (9);1982 Salamanca • E
[1982] *Actas 9 Jornadas Luso-Espanolas de Matematica* PUBL (X 2239) Univ Sec Publ: Salamanca 488pp
• DAT&PL 1982 Apr; Salamanca, E • ISBN 84-7481-230-5

P 4097 Jorn Mat Luso-Espanol (6);1979 Santander • E
[1979] *Actas 6 Jornadas Luso-Espanolas de Matematica* SER Rev Univ Santander 2 • PUBL Univ. Santander: Santander 2 Vols
• DAT&PL 1979 Apr; Santander, E

P 4108 Found of Math;1979 Kyoto • J
[1979] *Sugaku Kisoron (Foundations of Mathematics)* ED: TUGUE, T. • SER (S 2308) Symp Kyoto Univ Res Inst Math Sci (RIMS) 362 • PUBL (X 2441) Kyoto Univ Res Inst Math Sci: Kyoto ii+185pp
• DAT&PL 1979 May; Kyoto, J

P 4109 B-Val Anal & Nonstand Anal;1978 Kyoto • J
[1978] *Boole Daisuchi no Kaisekigaku to Chojun Kaiseki (B-Valued Analysis and Nonstandard Analysis)* ED: NAMBA, K. • SER (S 2308) Symp Kyoto Univ Res Inst Math Sci (RIMS) 336 • PUBL (X 2441) Kyoto Univ Res Inst Math Sci: Kyoto ii+149pp
• DAT&PL 1978 May; Kyoto, J
• REM All Papers in Japanese

P 4113 Found of Math;1982 Kyoto • J
[1983] *Foundations of Mathematics* ED: SHINODA, J. • SER (S 2308) Symp Kyoto Univ Res Inst Math Sci (RIMS) 480 • PUBL (X 2441) Kyoto Univ Res Inst Math Sci: Kyoto iii+236pp
• DAT&PL 1982 Oct; Kyoto, J

P 4153 B-Val Anal & Nonstand Anal;1981 Kyoto • J
[1981] *Boolean-Algebra-Valued Analysis and Nonstandard Analysis* ED: NANBA, K. • SER (S 2308) Symp Kyoto Univ Res Inst Math Sci (RIMS) 441 • PUBL (X 2441) Kyoto Univ Res Inst Math Sci: Kyoto ii+158pp
• DAT&PL 1981 May; Kyoto, J

P 4154 Deux Conf Prononcees;1956 Montreal • CDN
[1956] *Deux Conferences Prononcees a l'Universite de Montreal*
• DAT&PL 1956 Feb; Montreal, Que, CDN

P 4178 Universal Algeb & Lattice Th;1984 Charleston • USA
[1985] *Universal Algebra and Lattice Theory* ED: COMER, S.D. • SER (S 3301) Lect Notes Math 1149 • PUBL (X 0811) Springer: Heidelberg & New York vi+282pp
• DAT&PL 1984 Jul; Charleston, SC, USA • ISBN 3-540-15691-7, LC-No 85-17329

P 4180 Int Congr Log, Meth & Phil of Sci (7,Pap);1983 Salzburg • A
[1985] *Foundations of Logic and Linguistics. Problems and Their Solutions. Papers from the 7th International Congress of Logic, Methodology and Philosophy of Science* ED: DORN, G. & WEINGARTNER, P. • PUBL (X 1332) Plenum Publ: New York xi+715pp
• DAT&PL 1983 Jul; Salzburg, A • ISBN 0-306-41916-5, LC-No 84-26518

P 4185 Abel Group & Modul;1984 Udine • I
[1984] *Abelian Groups and Modules* ED: GOEBEL, R. & METELLI, C. & ORSATTI, A. & SALCE, L. • SER CISM Courses Lect. 287 • PUBL (X 0902) Springer: Wien xii+531pp
• DAT&PL 1984 Apr; Udine, I • ISBN 3-211-81847-2

P 4208 All-Union Symp Algeb;1975 Gomel • SU
[1975] *All Union Symposium in Algebra*
• DAT&PL 1975 -?-; Gomel', SU

P 4225 Mini-Conf Nonlinear Anal;1984 Canberra • AUS
[1984] *Miniconference on Nonlinear Analysis* ED: TRUDINGER, N.S. & WILLIAMS, G.H. • SER Proc Centre Math Analysis, Australian Nat Univ 8 • PUBL (X 2005) Austral Nat Univ: Canberra ix+270pp
• DAT&PL 1984 Jul; Canberra, AUS • ISBN 0-86784-509-0

P 4230 Groupe Etude Anal Ultrametrique (9);1982 Marseille • F
[1983] *Groupe d'Etude d'Analyse Ultrametrique, 9e Annee, 1981/82, Fasc 3. Including Papers from the Conference on p-adic Analysis and Its Applications* ED: AMICE, Y. & CHRISTOL, G. & ROBBA, P. • PUBL (X 1623) Univ Paris VI Inst Poincare: Paris 125pp
• DAT&PL 1982 Sep; Marseille, F • ISBN 2-85926-281-4, LC-No 76-642267

P 4243 Conf Binary Syst & Appl;1978 Taormina • I
[1978] *Conference on Binary Systems and Applications* PUBL Tecnoprint: Bologna
• DAT&PL 1978 -?-; Taormina, I

P 4285 Quad & Hermitian Forms;1983 Hamilton • CDN
[1984] *Quadratic and Hermitian Forms. Proceedings of the Conference on Quadratic Forms and Hermitian K-Theory* ED: RIEHM, C.R. & HAMBLETON, I. • SER Canadian Math Soc Conf Proc 4 • PUBL (X 0803) Amer Math Soc: Providence xviii+338pp
• DAT&PL 1983 Jul; Hamilton, ON, CDN • ISBN 0-8218-6008-9, LC-No 84-18561

P 4286 Categor Topol;1983 Toledo • USA
[1984] *Categorical Topology* ED: BENTLEY, H.L. & HERRLICH, H. & RAJAGOPALAN, M. & WOLFF, H. • SER Sigma Ser Pure Math 5 • PUBL Heldermann Verlag: Berlin xv+635pp
• DAT&PL 1983 Aug; Toledo, OH, USA • LC-No 85-115161

P 4310 Easter Conf on Model Th (3);1985 Gross Koeris • DDR
[1985] *Proceedings of the 3rd Easter Conference on Model Theory* SER Seminarbericht 70 • PUBL (X 2219) Humboldt-Univ Berlin: Berlin iv+226pp
• DAT&PL 1985 Apr; Gross Koeris, DDR

P 4313 Int Congr Math (II,14);1983 Warsaw • PL
[1984] *Proceedings of the International Congress of Mathematicians, Vol 1,2* ED: CIESIELSKI, Z. & OLECH, C. • PUBL (X 1034) PWN: Warsaw lxii+1730pp • ALT PUBL (X 0809) North Holland: Amsterdam
• DAT&PL 1983 Aug; Warsaw, PL • ISBN 83-01-05523-5, LC-No 84-13788

P 4383 Found Quant Mech;1983 Kokubunji-shi • J
[1984] *Proceedings of the International Symposium on Foundations of Quantum Mechanics in the Light of New Technology* ED: KAMEFUCHI, S. • PUBL Physical Soc of Japan: Tokyo 377pp
• DAT&PL 1983 Aug; Kokubunji-shi, J • LC-No 84-256146, ISBN 4-89027-001-9

P 4392 Mat Logika (Markov);1980 Sofia • BG
[1984] *Matematicheskaya Logika: Trudy Konferentsij po Matematicheskoj Logike, Posvyashchennoj Pamiati A.A. Markova 1903-1979 (Proceedings of a Conference on Mathematical Logic)* ED: SKORDEV, D. • PUBL Bulgar Akad Nauk 172pp
• DAT&PL Sep; Sofia, BG • LC-No 84-215513

P 4423 Semigroups (8);1984 Matsue • J
[1985] *Proceedings of the 8th Symposium on Semigroups* ED: YAMADA, U. & FUJIWARA, T. • PUBL Shimane Univ: Matsue ii+83pp
• DAT&PL 1984 Oct; Matsue

Proceedings

P 4564 Math Logic;1963 Xi-An • TJ
[1965] *1963 Nian Cueanguo Shuli Luoji Zhuanye Xueshu Huiyi Lunwen Xuanji (Mathematical Logic. Proceedings of the National Symposium)* PUBL Defence Industry Press: Beijing
• DAT&PL 1963 Oct; Xi-An, TJ

P 4593 Algeb & Th Nombres;1949 Paris • F
[1950] *Algebre et Theorie des Nombres* SER (S 1802) Colloq Int CNRS 24 • PUBL (X 0999) CNRS Inst B Pascal: Paris
• DAT&PL 1949 Oct; Paris, F

P 4595 Autom Infinite Words;1984 Le Mont Dore • F
Automata on Infinite Words SER (S 3302) Lect Notes Comput Sci 192 • PUBL (X 0811) Springer: Heidelberg & New York iv+216pp
• DAT&PL 1984 May; Le Mont Dore, F • ISBN 3-540-15641-0

P 4597 Cognitive Viewpoint;1977 Ghent • B
Proceedings International Workshop on the Cognitive Viewpoint SER Communication & Cognition 10
• DAT&PL 1977 -?-; Ghent, B

P 4599 All-Union Conf Stud & Sci-Tech Prog; • SU
[1978] *Materials of the All Union Scientific Student Conference "The Student and Scientific-Technical Progress"* PUBL Matematica: Novosibirsk

P 4601 EUROCAL;1985 Linz • A
[1985] *Proceedings of European Conference on Computer Algebra (EUROCAL)* ED: BUCHBERGER, B. (V1) & CAVINESS, B.F. (V2) • SER (S 3302) Lect Notes Comput Sci 203,204
• PUBL (X 0811) Springer: Heidelberg & New York vi+233pp, xvi+650pp
• DAT&PL 1985 Apr; Linz, A • ISBN 3-540-15983-5 (V1), ISBN 3-540-15984-3 (V2)
• REM Vol 1: Invited Lectures. Vol 2: Research Contributions

P 4625 Contrib to Gen Algeb (3);1984 Wien • A
[1985] *Contributions to General Algebra. Proceedings of the 3rd Vienna Conference* ED: EIGENTHALER, G. & KAISER, H.K. & MUELLER, W.B. & NOEBAUER, W. • PUBL (X 0823) Teubner: Stuttgart 415pp • ALT PUBL (X 2728) Hoelder-Pichler-Tempsky: Wien
• DAT&PL 1984 Jun; Wien, A • ISBN 3-209-00591-5

P 4627 CAAP'85 Arbres en Algeb & Progr (10);1985 Berlin • D
[1985] *Mathematical Foundations of Software Development, Vol 1. Colloquium on Trees in Algebra and Programming (CAAP '85)* ED: EHRIG, H. & FLOYD, C. & NIVAT, M. & THATCHER, J. • SER (S 3302) Lect Notes Comput Sci 185
• PUBL (X 0811) Springer: Heidelberg & New York xiii+418pp
• DAT&PL 1985 Mar; Berlin, D • ISBN 3-540-15198-2

P 4646 Atti Incontri Log Mat (2);1983/84 Siena • I
[1985] *Atti degli Incontri di Logica Matematica* ED: BERNARDI, C. & PAGLI, P. • PUBL (X 3812) Univ Siena, Dip Mat: Siena 648pp
• DAT&PL 1983 Jan; Siena, I, 1983 Apr; Siena, I, 1984 Jan; Siena, I, 1984 Apr; Siena, I

P 4647 FCT'85 Fund of Comput Th;1985 Cottbus • DDR
[1985] *Fundamentals of Computation Theory. FCT '85* ED: BUDACH, L. • SER (S 3302) Lect Notes Comput Sci 199
• PUBL (X 0811) Springer: Heidelberg & New York xii+542pp
• DAT&PL 1985 Sep; Cottbus, DDR • ISBN 3-540-15689-5

P 4661 Algeb & Log;1984 Zagreb • YU
[1985] *Proceedings of the Conference 'Algebra and Logic'* ED: STOJAKOVIC, Z. • PUBL (X 4030) Univ Novom Sadu, Inst Mat: Novi Sad vi+193pp
• DAT&PL 1984 Jun; Zagreb, YU

P 4663 Ordered Algeb Struct;1982 Cincinnati • USA
[1985] *Ordered Algebraic Structures. Proceedings of a Special Session of the 88th Annuanl Meeting of the American Mathematical Society (AMS)* ED: POWELL, W.B. & TSINAKIS, T. • SER (S 3310) Lect Notes Pure Appl Math 99 • PUBL (X 1684) Dekker: New York xii+196pp
• DAT&PL 1982 -?-; Cincinnati, OH, USA • ISBN 0-8247-7342-X

P 4680 Th of Graphs;1966 Roma • I
[1967] *Theory of Graphs* PUBL (X 0836) Gordon & Breach: New York
• DAT&PL 1966 -?-; Roma, I

Collection volumes

C 0552 Phil Contemp - Chroniques • I
[1968] *La Philosophie Contemporaine: Chroniques (Contemporary Philosophy: A Survey)* ED: KLIBANSKY, R.
• PUBL (**X** 1319) La Nuova Italia: Firenze 4 Vols
• LC-No 68-55649

C 0555 Topic Philos Logic • NL
[1969] *Topics in Philosophical Logic* ED: RESCHER, N. • SER (**S** 3307) Synth Libr • PUBL (**X** 0835) Reidel: Dordrecht xiv+350pp • ALT PUBL (**X** 2696) Humanities Pr: Atlantic Highlands
• LC-No 74-141565

C 0566 Nekotorye Prob Mat & Mekh (Lavrent'ev) • SU
[1961] *Nekotorye Problemi Matematiki i Mekhaniki (Certain Problems in Mathematics and Mechanics. In Honour of M.A.Lavrent'ev's 60th Birthday.)* ED: ALEKSANDROVSKIJ, B.M. ET AL. • PUBL (**X** 2652) Akad Nauk Sibirsk Otd Inst Mat: Novosibirsk 268pp
• LC-No 66-90884

C 0569 Phil of Math Oxford Readings • GB
[1969] *The Philosophy of Mathematics.* ED: HINTIKKA, K.J.J. • SER Oxford Readings in Philosophy • PUBL (**X** 0894) Oxford Univ Pr: Oxford v+186pp
• ISBN 0-19-875011-0, LC-No 71-441791

C 0570 Form Log & Metodol Nauk • SU
[1964] *Formal'naya Logika i Metodologya Nauki (Formal Logic and Methodology of Science)* ED: TAVANETS, P.V.
• PUBL (**X** 2027) Nauka: Moskva 300pp
• LC-No 66-92737

C 0595 Stud Math Anal & Rel Topics (Polya) • USA
[1962] *Studies in Mathematical Analysis and Related Topics. Essays in Honor of George Polya* ED: SZEGOE, G. & LOEWNER, C. ET AL. • SER Stanford Studies in Mathematics and Statistics 4 • PUBL (**X** 1355) Stanford Univ Pr: Stanford xxi+447pp
• LC-No 62-15265

C 0601 Encycl of Philos • USA
[1967] *Encyclopedia of Philosophy* ED: EDWARDS, P. • PUBL (**X** 0843) Macmillan : New York & London 8 Vols
• LC-No 67-10059

C 0622 Essays Found of Math (Fraenkel) • IL
[1961] *Essays on the Foundations of Mathematics: Dedicated to A.A.Fraenkel on His 70th Anniversary* ED: BAR-HILLEL, Y. & POZNANSKI, E.I.J. & RABIN, M.O. & ROBINSON, A. • PUBL (**X** 1299) Magnes Pr: Jerusalem x+351pp • ALT PUBL (**X** 0809) North Holland: Amsterdam & (**X** 0833) Acad Pr: Jerusalem & (**X** 0894) Oxford Univ Pr: Oxford
• LC-No 63-753

C 0640 Log, Autom, Inform • RO
[1971] *Logique, Automatique, Informatique* ED: MOISIL, G.C.
• PUBL (**X** 0871) Acad Rep Soc Romania: Bucharest 456pp
• LC-No 77-880254

C 0643 Philosophie 1946-48 • F
[1950] *Philosophie. Chronique des Annees d'apres la Guerre 1946-1948* ED: BAYER, R. • SER (**S** 2073) Actualites Sci Indust 1104 (V 12), 1110 (V 14) • PUBL (**X** 0859) Hermann: Paris at least 14 Vols
• REM V 12: Histoire de la Philosohie Metaphysique, Philosophie des Valeurs. V 13: Philosophie des Sciences. V 14: Psychologie, Phenomenologie et Existentialisme.

C 0648 Th of Sets and Topology (Hausdorff) • DDR
[1972] *Theory of Sets and Topology. In Honour of Felix Hausdorff (1868-1942)* ED: ASSER, G. & FLACHSMEYER, J. & RINOV, W. • PUBL (**X** 0806) Dt Verlag Wiss: Berlin 525pp
• LC-No 73-163576

C 0654 Stud in Model Th • USA
[1973] *Studies in Model Theory* ED: MORLEY, M.D. • SER Studies in Mathematics 8 • PUBL (**X** 1298) Math Ass Amer: Washington vii+197pp
• ISBN 0-88385-108-3, LC-No 73-86564

C 0675 From Frege to Goedel • USA
[1967] *From Frege to Goedel: A Source Book in Mathematical Logic 1879-1931* ED: HEIJENOORT VAN, J. • PUBL (**X** 0858) Harvard Univ Pr: Cambridge x+660pp • ALT PUBL (**X** 0894) Oxford Univ Pr: Oxford
• LC-No 67-10905
• REM 2nd ed. 1971; xi+660pp. Some Articles have been Reprinted 1970; iv+117pp

C 0684 Automation in Lang Transl & Theorem Prov • B
[1968] *Automation in Language Translation and Theorem Proving: Some Applications of Mathematical Logic* ED: BRAFFORT, P. & SCHEEPEN VAN, F. • PUBL (**X** 1714) Commiss Europ Comm : Bruxelles xv+295pp
• LC-No 77-467790

C 0722 Stud Algeb & Anwendgn • DDR
[1972] *Studien zur Algebra und ihre Anwendungen. Mit Anwendungen in der Mathematik, Physik und Rechentechnik* ED: HOEHNKE, H.J. • SER Schriftenreihe des Zentralinstituts fuer Mathematik und Mechanik bei der Akademie der Wissenschaften der DDR; 16 • PUBL (**X** 0911) Akademie Verlag: Berlin vi+154pp
• LC-No 73-336816

C 0727 Logic Found of Math (Heyting) • NL
[1968] *Logic and Foundations of Mathematics. Papers Dedicated to A.Heyting on the Occasion of His 70th Birthday* ED: DALEN VAN, D. & DYKMAN, J.G. & KLEENE, S.C. & TROELSTRA, A.S. • PUBL (**X** 0812) Wolters-Noordhoff : Groningen 249pp
• LC-No 70-408259
• REM Also published as J0020 vol. 20

C 0733 Izbr Vopr Algeb & Log (Mal'tsev) • SU
[1973] *Izbrannye Voprosy Algebry i Logiki: Sbornik Posvyashch Pamyati A.I.Mal'tsev (Selected Questions of Algebra and Logic: Volume Dedicated to the Memory of A.I.Mal'tsev)* ED: ERSHOV, YU.L. & KARGAPOLOV, M.I. & MERZLYAKOV, YU.I. & SMIRNOV, D.M. & SHIRSHOV, A.L. • PUBL (X 2642) Nauka: Novosibirsk 339pp
• LC-No 73-360316

C 0735 Logic & Value (Dahlquist) • S
[1970] *Logic and Value. Essays Dedicated to Thorild Dahlquist on His 50th Birthday* ED: PAULI, T. • SER Filosofiska Studier 9 • PUBL (X 0882) Univ Filos Foeren: Uppsala vi+247pp
• LC-No 72-186683

C 0745 Herbrand: Log Writings • NL
[1971] *Logical Writings by J.Herbrand* ED: GOLDFARB, W.D. • PUBL (X 0835) Reidel: Dordrecht vii+312pp • ALT PUBL (X 0858) Harvard Univ Pr: Cambridge vii+312pp
• ISBN 90-277-0176-8, LC-No 74-146963
• TRANSL OF [1968] (C 2486) Herbrand: Ecrits Logiques

C 0758 Logic Colloq Boston 1972-73 • D
[1975] *Logic Colloquium. Symposium on Logic Held at Boston, MA, USA, 1972-1973* ED: PARIKH, R.J. • SER (S 3301) Lect Notes Math 453 • PUBL (X 0811) Springer: Heidelberg & New York iv+251pp
• ISBN 3-540-07155-5, LC-No 75-11528

C 0768 Stud Algeb Logic • USA
[1974] *Studies in Algebraic Logic* ED: DAIGNEAULT, A. • SER Studies in Mathematics 9 • PUBL (X 1298) Math Ass Amer: Washington vii+207pp
• ISBN 0-88385-109-1, LC-No 74-84580

C 0769 L'Antinom Ment nel Pensiero Contemp • I
[1961] *L'Antinomia del Mentitiore nel Pensiero Contemporaneo, Da Peirce a Tarski* ED: RIVETTI BARBO, F. • PUBL (X 1364) Vita e Pensiero: Milano xxxvii+740pp
• LC-No 65-69081

C 0772 Model Th & Topoi • D
[1975] *Model Theory and Topoi* ED: LAWVERE, F.W. & MAURER, C. & WRAITH, G.C. • SER (S 3301) Lect Notes Math 445 • PUBL (X 0811) Springer: Heidelberg & New York 354pp
• ISBN 3-540-07164-4, LC-No 75-20007

C 0781 Infinitary Logic (Karp) • D
[1975] *Infinitary Logic: In Memoriam Carol Karp* ED: KUEKER, D.W. • SER (S 3301) Lect Notes Math 492 • PUBL (X 0811) Springer: Heidelberg & New York vi+206pp
• ISBN 3-540-07419-8, LC-No 75-34464

C 0782 Model Th & Algeb (A. Robinson) • D
[1975] *Model Theory and Algebra. A Memorial Tribute to Abraham Robinson.* ED: SARACINO, D.H. & WEISPFENNING, V.B. • SER (S 3301) Lect Notes Math 498 • PUBL (X 0811) Springer: Heidelberg & New York 436pp
• ISBN 3-540-07538-0, LC-No 75-40483

C 1009 Wang: Survey Math Logic • TJ
[1963] *Wang,H.: Survey of Mathematical Logic* PUBL (X 1876) Kexue Chubanshe: Beijing • ALT PUBL (X 0809) North Holland: Amsterdam 1963

C 1017 Logical Studies • USA
[1957] *Logical Studies* ED: WRIGHT VON, G.H. • SER International Library of Psychology, Philosophy and Scientific Method • PUBL (X 2696) Humanities Pr: Atlantic Highlands xi+195pp
• LC-No 58-732

C 1050 Aktual Vopr Mat Log & Teor Mnozh • SU
[1975] *Aktual'nye Voprosy Matematicheskoj Logiki i Teorii Mnozhestv (Current Questions in Mathematical Logic and Set Theory)* ED: SHCHELGOL'KOV, E.A. & BOKSHTEJN, M.F. & PETRI, N.V. • PUBL (X 2802) Moskov Ped Inst: Moskva
• LC-No 76-530761

C 1098 Skolem: Select Works in Logic • N
[1970] *Skolem,T.A.: Selected Works in Logic* ED: FENSTAD, J.E. • SER Scandinavian University Books • PUBL (X 1554) Universitesforlaget: Oslo 732pp
• LC-No 74-485971

C 1107 Aspects Inductive Log • NL
[1966] *Aspects of Inductive Logic* ED: HINTIKKA, J. & SUPPES, P. • SER (S 3303) Stud Logic Found Math • PUBL (X 0809) North Holland: Amsterdam vi+322pp
• LC-No 67-84072

C 1159 Tarski: Logic, Semantics, Metamathematics • GB
[1956] *Logic, Semantics, Metamathematics. Papers from 1923 to 1938 by Alfred Tarski* PUBL (X 0815) Clarendon Pr: Oxford 471pp
• LC-No 56-4171
• TRANSL IN [1964] (C 1186) Tarski: Logique, Semantique, Metamath
• REM Translations of Several Articles

C 1173 Logica (Bernays) • CH
[1959] *Logica: Studia Paul Bernays Dedicata* PUBL (X 0272) Griffon: Neuchatel

C 1186 Tarski: Logique, Semantique, Metamath • F
[1964] *Tarski,A.: Logique, Semantique, Metamathematique. 1923-1944. Tome 1* PUBL (X 0850) Colin: Paris
• TRANSL OF [1956] (C 1159) Tarski: Logic, Semantics, Metamathematics
• REM Revised and extended French ed. 1972

C 1203 Inform & Infer • NL
[1970] *Information and Inference* ED: HINTIKKA, J. & SUPPES, P. • SER (S 3307) Synth Libr • PUBL (X 0835) Reidel: Dordrecht vii+336pp
• LC-No 70-118132

C 1389 Modality, Morality, Probl of Sense & Nonsense (Hallden) • S
[1973] *Modality, Morality and other Problems of Sense and Nonsense. Essays Dedicated to Soeren Hallden.* ED: HANSSON, B. ET AL. • PUBL (X 1493) Gleerup: Lund vi+221pp
• ISBN 91-40-02780-5, LC-No 73-180115

C 1409 Gentzen: Collected Papers • NL
[1969] *The Collected Papers of Gentzen* ED: SZABO, M.E. • SER (S 3303) Stud Logic Found Math • PUBL (X 0809) North Holland: Amsterdam xii+338pp
• ISBN 0-7204-2254-X, LC-No 71-97201

C 1435 Vopr Teor Grupp & Polugrupp • SU
[1972] *Voprosy Teorii Grupp i Polugrupp (Questions in the Theory of Groups and Semigroups)* ED: GRINDLINGER, M.D. & GRINDLINGER, E.I. • PUBL Tul'skij Gosudarstvennyj Pedagogicheskij Institut imeni L.N.Tolstoj: Tula;200pp

C 1443 Algebra, Vyp 2 (Irkutsk) • SU
[1973] *Algebra. Vyp 2* ED: FRIDMAN, E.I. • PUBL (**X** 1006) Irkutsk Gos Univ: Irkutsk 164pp
• LC-No 74-645580 • REL PUBL (**C** 3549) Algebra, Vyp 1 (Irkutsk)

C 1468 Sets & Classes (Bernays) • NL
[1976] *Sets and Classes. On the Work of Paul Bernays* ED: MUELLER, GERT H. • SER (**S** 3303) Stud Logic Found Math 84 • PUBL (**X** 0809) North Holland: Amsterdam xxiii+358pp • ALT PUBL (**X** 0838) Amer Elsevier: New York
• ISBN 0-7204-2284-1, LC-No 76-16791

C 1495 Encycl Universalis • F
[1968-1976] *Encyclopaedia Universalis* PUBL (**X** 2524) Encyclopaedia Universalis: Paris 20 Vols
• ISBN 2-85229-281-5, LC-No 75-516014

C 1503 In Honor of Plakides • GR
[1974] *(In Honor of Stavros M. Plakides)* ED: KORSAKIS, D.D. & ZISULI, C. ET AL. • PUBL -?-: Athens vii+395pp

C 1523 Handb of Math Logic • NL
[1977 & 2nd ed. 1978] *Handbook of Mathematical Logic* ED: BARWISE, J. • SER (**S** 3303) Stud Logic Found Math 90
• PUBL (**X** 0809) North Holland: Amsterdam xi+1165pp
• ALT PUBL (**X** 0838) Amer Elsevier: New York
• ISBN 0-7204-2285-X, LC-No 76-26032
• TRANSL IN [1982] (**C** 1920) Spravochnaya Kniga po Mat Logike, Chast 1-4

C 1525 Semin Init Analyse (9-10) Paris 1969/71 • F
[1970ff] *Seminaire Choquet 9e-10e Annees: 1969-1971. Inititation a l'Analyse* ED: CHOQUET, G. • PUBL (**X** 1623) Univ Paris VI Inst Poincare: Paris

C 1533 Quantoren, Modal, Paradox • DDR
[1972] *Quantoren, Modalitaeten, Paradoxien. Beitraege zur Logik* ED: WESSEL, H. • PUBL (**X** 0806) Dt Verlag Wiss: Berlin 524pp
• LC-No 72-3662278

C 1549 Tartu Mezhvuz Nauch Simp Obshchej Algeb • SU
[1966] *Tartuskij Gosudarstvennyj Universitet Mezhvuzovskij Nauchnyj Simpozium po Obshchej Algebre (Wissenschaftliches Interhochschul-Symposium ueber Allgemeine Algebra)* ED: GABOVICH, E. ET AL. • PUBL (**X** 2463) Tartusk Gos Univ: Tartu 232pp

C 1706 Essays Honour J. Hintikka • NL
[1979] *Essays in Honour of Jaakko Hintikka: On the Occasion of His 50th Birthday on January 12, 1979* ED: SAARINEN, E. & HILPINEN, R. & NIINILUOTO, I. & PROVENCE, M. • SER (**S** 3307) Synth Libr 124 • PUBL (**X** 0835) Reidel: Dordrecht x+386pp
• ISBN 90-277-0916-5, LC-No 78-11364

C 1856 Log Enterprise • USA
[1975] *The Logical Enterprise* ED: MARCUS, R.B. & ANDERSON, A.R. & MARTIN, R.M. • PUBL (**X** 0875) Yale Univ Pr: New Haven x+261pp
• ISBN 0-300-01790-1, LC-No 74-20084

C 1920 Spravochnaya Kniga po Mat Logike, Chast 1-4 • SU
[1982] *Spravochnaya Kniga po Matematicheskoj Logike. Chast. 1-4 (Handbook of Mathematical Logic. Part 1-4)* ED: ERSHOV, YU.L. & PALYUTIN, E.A. & TAJMANOV, A.D. • PUBL (**X** 2027) Nauka: Moskva 4 Vols: 392pp, 375pp, 360pp, 391pp
• TRANSL OF [1977] (**C** 1523) Handb of Math Logic
• REM The Translation Contains a New Supplement by Palyutin,E.A. Part 1: Model Theory. Part 2: Set Theory. Part 3: Recursion Theory. Part 4: Proof Theory and Constructive Mathematics.

C 1936 Log Th & Semant Anal (Kanger) • NL
[1974] *Logical Theory and Semantic Analysis. Essays Dedicated to Stig Kanger on His 50th Birthday* ED: STENLUND, S. & HENSCHEN-DAHLQUIST, A.-M. & LINDAHL, L. & NORDENFELT, L. & ODELSTAD, J. • SER (**S** 330$\underline{7}$) Synth Libr 63 • PUBL (**X** 0835) Reidel: Dordrecht v+217pp
• ISBN 90-277-0438-4, LC-No 73-94456

C 1992 Scuola Algeb IMPA (3);1977 Rio de Janeiro • BR
[1977] *Attas da 3, Scuola de Algebra* PUBL (**X** 2233) Inst Mat Pura Apl: Rio Janeiro

C 2050 Semin Th des Ensembles GMS Paris 1976/78 • F
[1979] *Theorie des Ensembles: Seminaire GMS* SER Publications Mathematiques de l'Universite Paris VII 5
• PUBL Universite de Paris VII, UER de Mathematiques: Paris 228pp
• DAT&PL 1976-77; Paris, F, 1977-78; Paris, F

C 2065 Teor Nereg Kriv Raz Geom Post • SU
[1979] *Teoriya Neregulyarnykh Krivykh v Razlichnykh Geometricheskykh Postranstvakh (Theorie der Nichtregulaeren Kurven in Verschiedenen Geometrischen Raeumen)* ED: TAJTSLIN, M.A. • PUBL (**X** 2769) Kazakh Gos Univ: Alma-Ata 136pp

C 2264 Algor Probl Teor Grupp & Polygrupp • SU
[1981] *Algoritmicheskie Problemy Teorii Grupp i Polygrupp (Algorithmic Problems of the Theory of Groups and Semigroups)* ED: GRINDLINGER, M.D. • PUBL Tulisk Gos Ped Inst: Tula 134pp

C 2486 Herbrand: Ecrits Logiques • F
[1968] *Herbrand,J.: Ecrits Logiques* ED: HEIJENOORT VAN, J.
• PUBL (**X** 0840) Pr Univ France: Paris
• LC-No 68-104359
• TRANSL IN [1971] (**C** 0745) Herbrand: Log Writings

C 2536 Issl Sovrem Algeb • SU
[1979] *Issledovaniya po Sovremennoj Algebre (Investigations in Modern Algebra)* SER (**J** 0340) Mat Zap (Univ Sverdlovsk)
• PUBL Ural'skij Gos. Univ.: Sverdlovsk 214pp

C 2555 Algeb Sistemy (Irkutsk) • SU
[1976] *Algebraicheskie Sistemy (Algebraic Systems)* ED: KOKORIN, A.I. • PUBL (**X** 1006) Irkutsk Gos Univ: Irkutsk 170pp
• LC-No 79-410179

C 2557 Teor & Priklad Zad Mat & Mekh • SU
[1977] *Teoreticheskie i Prikladnye Zadachi Matematiki i Mekhaniki. Sbornik Trudov (Theoretical and Applied Problems in Mathematics and Mechanics. Work Collection)* ED: AMANOV, T.I. • PUBL Nauka Kazakh SSR:Alma-Ata 322pp

C 2577 Issl Formaliz Yazyk & Neklass Log • SU
[1974] *Issledovaniya po Formalizovannym Yazykam i Neklassicheskim Logikam (Investigations on Formalized Languages and Non-Classical Logics)* ED: BOCHVAR, D.A.
• PUBL (**X** 2027) Nauka: Moskva 275pp
• LC-No 75-554105

C 2578 Filos & Logika • SU
[1974] *Filosofiya i Logika. Filosofiya v Sovremennom Mire (Philosophy and Logic. Philosophy in the Modern World)* ED: TAVANETS, P.V. & SMIRNOV, V.A. • SER Filosofiya v Sovremennom Mire • PUBL (**X** 2027) Nauka: Moskva 479pp
• LC-No 74-358872

C 2581 Issl Neklass Log & Teor Mnozh • SU
[1979] *Issledovaniya po Neklassicheskim Logikam i Teorii Mnozhestv (Investigations on Non-Classical Logics and Set Theory)* ED: MIKHAJLOV, A.I. ET AL. • PUBL (**X** 2027) Nauka: Moskva 374pp
• LC-No 80-475529

C 2595 Rekursiv Funktsii • SU
[1978] *Rekursivnye Funktsii (Recursive Functions)* ED: POLYAKOV, E.A. • PUBL (**X** 2594) Ivanovo Gos Univ: Ivanovo 100pp

C 2601 Vlozhenie Grupp: Algor Vopr • SU
[1970] *Vlozhenie Grupp: Algoritmicheskie Voprosy (Imbedding of Groups: Algorithmic Questions)* ED: GORCHAKOV, YU.M. • SER Seminar "Algebraicheskie Sistemy" 3 • PUBL Akad. Nauk SSSR Sibirsk. Otdel. Inst. Fiz., Krasnoyarsk 131pp

C 2617 Modern Log Survey • NL
[1981] *Modern Logic – A Survey. Historical, Philosophical and Mathematical Aspects of Modern Logic and Its Applications.* ED: AGAZZI, E. • SER (S 3307) Synth Libr 149 • PUBL (**X** 0835) Reidel: Dordrecht viii+475pp
• ISBN 90-277-1137-2, LC-No 80-22027

C 2619 Groupe Etude Th Stables (2) 1978/79 • F
[1981] *Groupe d'Etude de Theories Stables. 2eme Annee: 1978/79* ED: POIZAT, B. • PUBL (**X** 1623) Univ Paris VI Inst Poincare: Paris ii+115pp
• ISBN 2-85926-274-1

C 2620 Teor Model & Primen • SU
[1980] *Teoriya Modelej i Ee Primeneniya (Theory of Models and Its Applications)* ED: BAJZHANOV, B.S. & TAJTSLIN, M.A. • PUBL (**X** 2769) Kazakh Gos Univ: Alma-Ata 83pp

C 2621 Mal'tsev: Metamath of Algeb Syst • NL
[1971] *The Metamathematics of Algebraic Systems. Mal'tsev,A.I.: Collected Papers, 1936-1967* ED: WELLS III, B.F. • SER (S 3303) Stud Logic Found Math 66 • PUBL (**X** 0809) North Holland: Amsterdam xviii+494pp
• LC-No 73-157020

C 2897 Algeb Model, Kateg & Gruppoide • DDR
[1979] *Algebraische Modelle, Kategorien und Gruppoide* ED: HOEHNKE, H.-J. • SER Studien zur Algebra und ihre Anwendungen 7 • PUBL (**X** 0911) Akademie Verlag: Berlin vi+178pp
• LC-No 80-475994

C 2908 Cabal Seminar Los Angeles 1976-77 • USA
[1978] *Cabal Seminar 76-77. Proceedings of the Caltech-UCLA Logic Seminar 1976-77* ED: KECHRIS, A.S. & MOSCHOVAKIS, Y.N. • SER (S 3301) Lect Notes Math 689 • PUBL (**X** 0811) Springer: Heidelberg & New York iii+282pp
• ISBN 3-540-09086-X, LC-No 78-24063

C 2919 Contrib Algeb (Kolchin) • USA
[1977] *Contributions to Algebra. A Collection of Papers Dedicated to Ellis Kolchin* ED: BASS, H. & CASSIDY, P.J. & KOVACIC, J. • PUBL (**X** 0801) Academic Pr: New York xxii+424pp
• ISBN 0-12-080550-2, LC-No 76-45980

C 2967 Metodol Probl Mat • SU
[1979] *Metodologicheskie Problemy Matematiki (Methodological Problems of Mathematics)* ED: BORISOV, YU.F. • PUBL (**X** 2642) Nauka: Novosibirsk 303pp

C 2969 Modeles de l'Arithm; Paris 1977 • F
[1980] *Modeles de l'Arithmetique* ED: McALOON, K. • SER (J 1620) Asterisque 73 • PUBL (**X** 2244) Soc Math France: Paris 155pp

C 2976 Numb Th, Algeb Geom & Comm Algeb (Akizuki) • J
[1973] *Number Theory, Algebraic Geometry and Commutative Algebra. In Honor of Yasuo Akizuki.* ED: KUSUNOKI, Y. & MIZOHATA, S. & NAGATA, M. & TODA, H. & YAMAGUTI, M. & YOSHIZAWA, H. • PUBL (**X** 2465) Kinokuniya Company: Tokyo vi+528pp
• LC-No 74-159266

C 3031 Semin Maurey-Schwartz 1974/75 • F
[1975] *Seminaire Maurey-Schwartz 1974-1975: Espaces L^p, Applications Radonifiantes et Geometrie des Espaces de Banach* PUBL Centre de Mathematiques, Ecole Polytechnique:Paris iv+338pp
• LC-No 81-645060

C 3032 Semin Lefebvre Struct Algeb 1970/71 • F
[1971] *Seminaire P. Lefebvre. Annee 1970/1971 Structures Algebraiques. Tome I, II* PUBL Univ Claude Bernard, Dept Math: Villeurbanne ii+160pp

C 3036 Vopr Teor Grupp & Kolets • SU
[1973] *Nekotorye Voprosy Teorii Grupp i Kolets (Some Questions on the Theory of Groups and Rings)* PUBL Inst. Fiz. im. Kirenskogo Sibirsk. Otdel. Akad. Nauk SSSR: Krasnoyarsk; 178pp
• LC-No 74-331837

C 3046 Decid Theories II – Monad 2nd Ord Th Count Ordinals • D
[1973] *Decidable Theories II. The Monadic Second Order Theory of All Countable Ordinals* ED: MUELLER, GERT H. & SIEFKES, D. • SER (S 3301) Lect Notes Math 328 • PUBL (**X** 0811) Springer: Heidelberg & New York vi+217pp
• ISBN 3-540-06345-5, LC-No 73-82358

C 3048 Teor Vopr Proekt Vych Sist • SU
[1980] *Teoreticheskie Voprosy Proektirovaniya Vychislitel'nykh Sistem (Theoretical Questions of Design of Computer Systems)* ED: LETICHEVS'KIJ, O.A. & GODLEVSKIJ, A.B. • PUBL (**X** 2522) Akad Nauk Inst Kibernet: Kiev 102pp

C 3050 Essays Combin Log, Lambda Calc & Formalism (Curry) • USA
[1980] *To H.B. Curry: Essays on Combinatory Logic, Lambda Calculus and Formalism* ED: SELDIN, J.P. & HINDLEY, J.R. • PUBL (**X** 0801) Academic Pr: New York xxv+606pp
• ISBN 0-12-349050-2, LC-No 80-40139

C 3055 Topos-Theor Meth in Geom • DK
[1979] *Topos-Theoretic Methods in Geometry* ED: KOCK, A. • SER (**S** 3462) Var Publ Ser, Aarhus Univ 30 • PUBL (**X** 1599) Aarhus Univ Mat Inst: Aarhus iv+219pp

C 3093 Groupe Etude Th Stables (1) 1977/78 • F
[1978] *Groupe d'Etude de Theories Stables. 1re Annee: 1977/78* ED: POIZAT, B. • PUBL (**X** 1623) Univ Paris VI Inst Poincare: Paris 104pp
• ISBN 2-85926-268-7

C 3170 Geom Sb Vyp 15 • SU
[1975] *Geometricheskij Sbornik. Vypusk 15 (Geometry Collection, No. 15)* ED: SHCHERBAKOV, R.N. & KRUGLYAKOV, L.Z. • SER (**S** 1948) Tr, Univ Tomsk 258 • PUBL (**X** 3606) Tomsk Univ: Tomsk 240pp
• LC-No 77-640552

C 3271 Issl Teor Mnozh & Neklass Logik • SU
[1976] *Issledovaniya po Teorij Mnozhestv i Neklassicheskim Logikam. Sbornik Trudov (Studies in Set Theory and Nonclassical Logics. Collection of Papers)* ED: BOCHVAR, D.A. & GRISHIN, V.N. • PUBL (**X** 2027) Nauka: Moskva 328pp
• LC-No 77-501571

C 3444 Perspectives in Math • CH
[1984] *Perspectives in Mathematics. Anniversary of Oberwolfach 1984* ED: JAEGER, W. & MOSER, J. & REMMERT, R. • PUBL (**X** 0804) Birkhaeuser: Basel 587pp
• ISBN 3-7643-1624-1, LC-No 84-16922

C 3467 Sb Rabot Algeb • SU
[1978] *Sbornik Rabot po Algebre. Mezhvuzovskij Sbornik Kafedr Matematiki (Work Collection on Algebra. Interuniversity Work Collection of the Chair of Mathematics)* PUBL Ministerstvo Oborony SSSR: Moskva 342pp

C 3494 Stud on Math Progr. Math Meth Oper Res, Vol 1 • H
[1980] *Studies on Mathematical Programming. Mathematical Methods of Operation Research. Vol 1* ED: PREKOPA, A. • PUBL (**X** 0928) Akad Kiado: Budapest 200pp
• ISBN 963-05-1854-6, LC-No 80-496383

C 3510 Contrib Group Theory (Lyndon) • USA
[1984] *Contributions to Group Theory. Papers Dedicated to Lyndon,R.C. on His 65th Birthday* ED: APPEL, K.I. & RATELIFFE, J.G. & SCHUPP, P.E. • SER (**S** 3313) Contemp Math 33 • PUBL (**X** 0803) Amer Math Soc: Providence xi+519pp
• ISBN 0-8218-5035-0, LC-No 84-18454

C 3515 Ital Studies in Phil of Sci • NL
[1981] *Italian Studies in the Philosophy of Science* ED: CHIARA SCABIA DALLA, M.L. • SER (**S** 3311) Boston St Philos Sci 47 • PUBL (**X** 0835) Reidel: Dordrecht xi+525pp
• ISBN 90-277-0735-9, LC-No 80-16665

C 3549 Algebra, Vyp 1 (Irkutsk) • SU
[1972] *Algebra. Vyp. 1* ED: KOKORIN, A.I. & PENZIN, YU.G. • PUBL (**X** 1006) Irkutsk Gos Univ: Irkutsk 135pp
• LC-No 74-645580 • REL PUBL (**C** 1443) Algebra, Vyp 2 (Irkutsk)

C 3551 Tr Molodykh Uchen Vyp Mat • SU
[1964] *Trudy Molodykh Uchenykh. Vypusk Matematicheskij (Works of Young Scientists)* PUBL (**X** 0958) Saratov Univ: Saratov 122pp
• LC-No 66-92733

C 3591 Semin Equat Deriv Part Hyperb & Holom 1979/80 • F
[1980] *Seminaire sur les Equations aux Derivees Partielles. Hyperboliques et Holomorphes. Annee 1979-1980* PUBL (**X** 1623) Univ Paris VI Inst Poincare: Paris 110pp

C 3625 Metod Log Anal • SU
[1977] *Motody Logicheskogo Analiza (Methods of Logical Analysis)* ED: TAVANETS, P.V. • PUBL (**X** 2027) Nauka: Moskva 263pp
• LC-No 78-409666

C 3806 Issl Teor Progr • SU
[1981] *Issledovaniya po Teoreticheskomn Programmrovaniyu (Investigations in Theoretical Programming)* ED: TAJTSLIN, M.A. • PUBL (**X** 2769) Kazakh Gos Univ: Alma-Ata 104pp

C 3807 Issl Neklass Log & Formal Sist • SU
[1983] *Issledovaniya po Neklassicheskim Logikam i Formal'nym Sistemam (Studies in Nonclassical Logics and Formal Systems)* ED: MIKHAJLOV, A.I. • PUBL (**X** 2027) Nauka: Moskva 360pp
• LC-No 83-181942

C 3822 Groupe Etude Th Stables (3) 1980/82 • F
[1983] *Groupe d'Etude de Theories Stables. 3eme Annee: 1980/1982* ED: POIZAT, B. • PUBL (**X** 1623) Univ Paris VI Inst Poincare: Paris 115pp
• ISBN 2-85926-282-2

C 3848 Algeb & Teor Chisel ('78) • SU
[1978] *Algebra i Teoriya Chisel. Tematicheskij Sbornik Nauchnykh Trudov. Professorsko-prepodatel'skogo Sostava i Aspirantov Vysshikh Uchebnykh Zavedenij Ministerstva Proveshcheniya Kazakhskoj SSR (Algebra and Number Theory. Thematic Work Collection)* ED: DZOZ, N.A. • PUBL Kazakh. Ped. Inst. im. Abaya 136p

C 3849 Modal & Relevant Log, Vyp 1 • SU
[1982] *Modal'nye i Relevantnye Logiki. Trudy Nauchno-Issledovatel'skogo Seminara po Logike Institute Filosofii AN SSSR Vyp.1 (Modal and Relevant Logics. Vol.1)* ED: SMIRNOV, V.A. & KARPENKO, A.S. • PUBL (**X** 0899) Akad Nauk SSSR : Moskva 107pp

C 3861 Comput Algeb. Symb & Algeb Comput • A
[1982] *Computer Algebra. Symbolic and Algebraic Computation. In Cooperation with R.Albrecht* ED: BUCHBERGER, B. & COLLINS, G.E. & LOOS, R. • SER Computing Supplementum 4 • PUBL (**X** 0902) Springer: Wien vii+283pp
• LC-No 82-10577

C 3865 Algeb Sistemy (Ivanovo) • SU
[1981] *Algebraicheskie Sistemy. Mezhvuzovskij Sbornik Nauchnykh Trudov (Algebraic Systems. Interuniversity Collection of Scientific Works)* ED: MOLDAVANSKIJ, D.I. • PUBL (**X** 2594) Ivanovo Gos Univ: Ivanovo 234pp

C 3875 Cabal Seminar Los Angeles 1979-81 • USA
[1983] *Cabal Seminar 79-81. Proceedings of the Caltech-UCLA Logic Seminar 1979-81* ED: KECHRIS, A.S. & MARTIN, D.A. & MOSCHOVAKIS, Y.N. • SER (**S** 3301) Lect Notes Math 1019 • PUBL (**X** 0811) Springer: Heidelberg & New York 284pp
• ISBN 3-540-12688-0, LC-No 83-16866

C 3884 Nonstandard Anal - Recent Develop • D
[1983] *Nonstandard Analysis - Recent Developments* ED: HURD, A.E. • SER (S 3301) Lect Notes Math 983 • PUBL (X 0811) Springer: Heidelberg & New York v+213pp
• ISBN 3-540-12279-6, LC-No 83-5837

C 3910 Abel Gruppy & Moduli • SU
[1981] *Abelevy Gruppy i Moduli (Abelian Groups and Modules)* ED: SKORNYAKOV, L.A. • PUBL (X 3606) Tomsk Univ: Tomsk 258pp

C 3953 Mat Log & Teor Algor • SU
[1982] *Matematicheskaya Logika i Teoriya Algoritmov (Mathematical Logic and the Theory of Algorithms)* ED: SOBOLEV, S.L. • SER Trudy Inst Mat 2 • PUBL (X 2642) Nauka: Novosibirsk 176pp

C 4019 Kraevye Zadach Dlya Diff Uravnenij & Priloz Mekh & Tekh • SU
[1983] *Kraevye Zadachi Dlya Differentsial'nykh Uravnenij i Ikh Prilozheniya v Mekhanike i Teknike (Boundary Value Problems for Differential Equations and Their Applications in Mechanics and Technology)* ED: ZHAUTYKOV, O.A. • PUBL (X 2443) Nauka: Alma-Ata 167pp

C 4055 Wajsberg: Logical Works • PL
[1977] *Wajsberg,M.: Logical Works* ED: SURMA, S.J. • PUBL (X 2882) Pol Akad Nauk: Wroclaw 216pp
• LC-No 84-246718

C 4062 Montague: Formal Philos • USA
[1974] *Formal Philosophy. Selected Papers of Richard Montague* ED: THOMASON, R.H. • PUBL (X 0875) Yale Univ Pr: New Haven 369pp
• ISBN 0-300-01527-5, LC-No 73-77159

C 4082 Handb Wiss Begriffe • D
[1980] *Handbuch Wissenschaftstheoretischer Begriffe, 3 Baende* ED: SPECK, J. • PUBL (X 0903) Vandenhoeck & Ruprecht: Goettingen 3 Vols: 780pp
• ISBN 3-525-03313-3 (V 1), ISBN 3-525-03314-1 (V 2), ISBN 3-525-03316-8 (V 3)

C 4085 Handb Philos Log • NL
[1983 & 1984] *Handbook of Philosophical Logic* ED: GABBAY, D. & GUENTHNER, F. • SER (S 3307) Synth Libr 164,165 • PUBL (X 0835) Reidel: Dordrecht 2 Vols: xi+493pp,xi+776pp
• ISBN 90-277-1542-4 (V1), ISBN 90-277-1604-8 (V2), LC-No 83-4277

C 4091 Algeb & Diskret Mat (Riga) • SU
[1984] *Algebra i Diskretnaya Matematika (Algebra and Discrete Mathematics)* ED: IKAUNIEKS, EH.A. • PUBL (X 0895) Latv Valsts (Gos) Univ : Riga 164pp

C 4138 Fcys: Modal Logika • SU
[1974] *Fcys,R.: Modal'naya Logika (Modal Logic)* ED: MINTS, G.E. • PUBL (X 2027) Nauka: Moskva 520pp
• TRANSL OF Modal Logic
• REM Contains translations of Kripke,S.A. 1959 ID 07554, 1962 ID 07555, 1963 ID 07558, 1965 ID 07559 and of Schuette,K. 1968 ID 24822

C 4157 Rosser:Deux Esquisses Log • F
[1955] *Rosser,J.B.: Deux Esquisses de Logique.* SER Collection de Logique Mathematique, Ser A 7 • PUBL (X 0834) Gauthier-Villars: Paris 67pp • ALT PUBL (X 1313) Nauwelaerts: Louvain
• LC-No 55-3012

C 4181 Math Log & Formal Syst (Costa da) • USA
[1985] *Mathematical Logic and Formal Systems. A Collection of Papers in Honor of Newton C.A. da Costa* ED: ALCANTARA DE, L.P. • SER (S 3310) Lect Notes Pure Appl Math 94 • PUBL (X 1684) Dekker: New York xv+297pp
• ISBN 0-8247-7330-6, LC-No 84-25984

C 4183 Model-Theor Log • D
[1985] *Model-Theoretic Logics* ED: BARWISE, J. & FEFERMAN, S. • PUBL (X 0811) Springer: Heidelberg & New York xvii+893pp
• ISBN 3-540-90936-2, LC-No 83-20277

C 4191 Acad Freedom, Log & Religion • USA
[1953] *Academic Freedom, Logic, and Religion* PUBL Univ of Pennsylv. Press: Philadelphia

C 4220 Rep Sem Found Anal • USA
[1963] *Reports of the Seminar on Foundation of Analysis* ED: KREISEL, G. • PUBL Stanford Univ: Stanford
• REM Mimeographed

C 4232 Altgeld Book • USA
[1976] *The Altgeld Book: Lect. Not. Funct. Anal.* ED: ROSENTHAL, H.P. & DOR, L. • PUBL (X 1285) Univ Ill Pr: Urbana

C 4256 Semin Geom Espace Banach 1982 Paris • D
[1983] *Seminaire Geometrie des Espaces de Banach 1982 Paris* ED: BEAUZAMY, B. & KRIVINE, J.-L. & MAUREY, B. • PUBL Univ Paris VII, UER Math: Paris iii+160pp

C 4356 Gen Log Semin Paris 1982/83 • F
[1984] *General Logic Seminar. Paris 1982-83* ED: DELON, F. & LASCAR, D. & LOUVEAU, A. & SABBAGH, G. • SER Publ Math Univ Paris VII • PUBL Univ Paris VII, UER Math: Paris iii+186pp

C 4366 Modal & Intens Log & Primen Probl Metodol Nauk • SU
Modalnaya i Intensionalnaya Logiki i Ikh Primeneniek Problemam Metodologii (Modal- und Intensional-Logik und ihre Anwendung auf die Probleme der Methodologie)

C 4594 Sel Pap Robinson • USA
[1979] *Selected Papers of Abraham Robinson* ED: KEISLER, H.J. & KOERNER, S. & LUXEMBURG, W.A.J. & YOUNG, A.D.
• PUBL (X 0875) Yale Univ Pr: New Haven xxxvii+694pp,xlv+582pp • ALT PUBL (X 0809) North Holland: Amsterdam
• REM Vol 1: Model Theory and Algebra. Vol 2: Nonstandard Analysis and Philosophy

C 4602 Metamath Meth in Geom • D
[1983] *Schwabhaeuser,W. & Szmielev,W. & Tarski,A.: Metamatematische Methoden in der Geometrie* PUBL (X 0811) Springer: Heidelberg & New York viii+482pp
• ISBN 3-540-12958-8

C 4631 Rutgers Group Th Year New Brunswick 1983/84 • USA
[1985] *Proceedings of the Rutgers Group Theory Year at New Brunswick 1983/84* ED: ASCHBACHER, M. & GORENSTEIN, D. & LYONS, R. & O'NAN, M. & SIMS, C. & FEIT, W. • PUBL (X 0805) Cambridge Univ Pr: Cambridge, GB xii+415pp
• ISBN 0-521-26493-6

C 4649 Paradiso di Cantor • I
[1978] *Il Paradiso di Cantor* ED: CELLUCCI, C. • PUBL (X 1732) Bibliopolis: Napoli

C 4659 Autom of Reasoning • D
[1983] *Automation of Reasoning. Vol 1, 2* ED: SIEKMANN, J. & WRIGHTSON, G. • PUBL (**X** 0811) Springer: Heidelberg & New York xii+525pp, xii+637pp
• ISBN 3-540-12043-2 (V1), ISBN 3-540-12044-0 (V2)
• REM Vol 1: Classical Papers on Computational Logic 1957-1966. Vol 2: Classical Papers on Computational Logic 1967-1970

C 4674 Stud on Group Th (Sverdlovsk) • SU
[1984] *(Studies on Group Theory)* PUBL Ural Sci Center Akad Nauk SSSR: Sverdlovsk

C 4675 Hintikka: Logiko-Epist Issled • SU
[1980] *Hintikka, K.J.J.: Logiko-Epistemologicheskie Issledovanniya (Hintikka, K.J.J.: Logical-Epistemological Investigations.)* ED: SADOVSKIJ, V.N. & SMIRNOV, V.A. • PUBL (**X** 2055) Progress: Moskva 448pp

C 4688 Prog in Math, Vol 12 • USA
[1972] *Progress in Mathematics. Vol 12: Algebra and Geometry* ED: GAMKRELIDZE, R.V. • PUBL (**X** 1332) Plenum Publ: New York ix+254pp
• TRANSL OF [1972] (**J** 1501) Itogi Nauki Tekh, Ser Algeb, Topol, Geom 1968

C 4689 Bar-Hillel: Aspects of Lang • IL
[1970] *Bar-Hillel, Y.: Aspects of Languages. Essyas and Lectures on Philosophy of Languages, Linguistic Philosophy and Methodology of Linguistics* PUBL (**X** 1299) Magnes Pr: Jerusalem x+381pp • ALT PUBL (**X** 0809) North Holland: Amsterdam

C 4695 Game-Th Semantics • NL
[1979] *Game-Theoretical Semantics. Essays on Semantics by Hintikka, Carlson, Peacocke, Rantala, and Saarinen* ED: SAARINEN, E. • SER Synthese Language Library 5 • PUBL (**X** 0835) Reidel: Dordrecht xiv+392pp

C 4699 Friedman's Res on Found of Math • NL
[1985] *Harvey Friedman's Research on the Foundations of Mathematics* ED: HARRINGTON, L.A. & MORLEY, M. & SCEDROV, A. & SIMPSON, S.G. • PUBL (**X** 0809) North Holland: Amsterdam xvi+408pp
• IBSN 0-444-87834-3

Publishers

X 0272 *Editions du Griffon* (Neuchatel, CH) ISBN 2-88006

X 0801 *Academic Press* (New York, NY, USA & London, GB) ISBN 0-12

X 0802 *Allyn & Bacon* (London, GB & Boston, MA, USA & Spit Junction, NSW, AUS) ISBN 0-205, ISBN 0-695

X 0803 *American Mathematical Society* (Providence, RI, USA) ISBN 0-8218

X 0804 *Birkhaeuser Verlag* (Basel, CH & Stuttgart, D & Cambridge, MA, USA) ISBN 3-7643

X 0805 *The Cambridge University Press.* (Cambridge, GB & New York, NY, USA & Melbourne, Vic, AUS) ISBN 0-521

X 0806 *VEB Deutscher Verlag der Wissenschaften* (Berlin, DDR) ISBN 3-326

X 0807 *Duke University Press* (Durham, NC, USA) ISBN 0-8223

X 0808 *W. Kohlhammer* (Stuttgart, D & Koeln, D & Berlin, D & Mainz, D) ISBN 3-17

X 0809 *North-Holland Publishing Company.* (Amsterdam, NL & Oxford, GB) ISBN 0-7204 • REL PUBL (**X** 0838) Amer Elsevier: New York

X 0811 *Springer-Verlag* (Heidelberg, D & Berlin, D & New York, NY, USA & Tokyo, J) ISBN 3-540, ISBN 0-387 • REL PUBL (**X** 1231) Barth: Leipzig & (**X** 0902) Springer: Wien

X 0812 *Wolters-Noordhoff* (Groningen, NL) ISBN 90-01 • REL PUBL (**X** 1317) Noordhoff: Groningen

X 0813 *Dover Publications* (New York, NY, USA) ISBN 0-486

X 0814 *R.Oldenbourg Verlag* (Muenchen, D & Wien, A) ISBN 3-486

X 0815 *The Clarendon Press* (Oxford, GB) ISBN 0-19 • REL PUBL (**X** 0894) Oxford Univ Pr: Oxford • REM This Imprint is Used for Academic Books Published by X0894.

X 0816 *Methuen & Company* (London, GB & New York, NY, USA & Agincourt, M, CDN & North Ryde, AUS) ISBN 0-416

X 0817 *Gesellschaft fuer Mathematik und Datenverarbeitung* (Bonn, D) ISBN 3-88457

X 0818 *Holt Rinehart & Winston* (New York, NY, USA & Toronto, ON, CDN & Artarmon, NSW, AUS & & Eastbourne,GB) ISBN 0-03 • REM In Australia & United Kingdom: Holt-Saunders: Eastbourne, GB & Artarmon, NSW, AUS

X 0819 *Prentice Hall* (Englewood Cliffs, NJ, USA & Brookvale, NSW, AUS & Scarborough, ON, CDN) ISBN 0-13 • REL PUBL (**X** 2040) Winthrop: Cambridge

X 0820 *Interscience Publishers* (New York, NY, USA & Chichester, GB) ISBN 0-470 • REL PUBL (**X** 0827) Wiley & Sons: New York

X 0823 *B.G.Teubner* (Stuttgart, D) ISBN 3-519 • REM See also X1079

X 0824 *The Free Press* (New York, NY, USA) ISBN 0-02 • REL PUBL (**X** 0843) Macmillan : New York & London

X 0827 *J.Wiley & Sons* (New York, NY, USA & Chichester, GB & Rexdale, ON, CDN & Auckland, NZ) ISBN 0-471 • REL PUBL (**X** 0942) Norton: New York & (**X** 0820) Intersci Publ: New York & (**X** 0880) Ronald Press: New York & (**X** 2737) Israel Progr Sci Transl: Jerusalem

X 0832 *Addison-Wesley Publishing Co.* (Reading, MA, USA & London, GB & Don Mills, ON, CDN & North Ryde, NSW, AUS) ISBN 0-201 • REL PUBL (**X** 0867) Benjamin: Reading

X 0833 *Jerusalem Academic Press* (Jerusalem, IL)

X 0834 *Gauthier-Villars Editeur* (Paris, F) ISBN 2-04

X 0835 *D.Reidel Publishing Company* (Dordrecht, NL & Hingham, MA, USA) ISBN 90-277

X 0836 *Gordon & Breach, Science Publishers* (New York, NY, USA & London, GB & Paris, F) ISBN 0-677

X 0837 *Harper & Row, Publishers* (New York, NY, USA & London, GB & Artarmon, NSW, AUS & Petone, NZ) ISBN 0-06 • REM Member Firm: Lippincott, J.B.,Company: New York, NY, USA

X 0838 *American Elsevier Publishing Co.* (New York, NY, USA & Amsterdam, NL & London, GB) ISBN 0-444, ISBN 0-525, ISBN 0-87690 • REL PUBL (**X** 0809) North Holland: Amsterdam

X 0840 *Editions Presses Universitaires de France* (Paris, F) ISBN 2-13 • REL PUBL (**X** 2435) Pr Univ France Period: Paris

X 0842 *University of Wisconsin Press* (Madison, WI, USA & London, GB) ISBN 0-299 • REL PUBL (**X** 3828) Amer Univ Publ Group: London

X 0843 *Macmillan Publishing Company* (New York, NY, USA & Melbourne, Vic, AUS & London, GB & Toronto, Ont, CDN) ISBN 0-02 • REL PUBL (**X** 2375) Macmillan Journal: London & (**X** 0824) Free Press: New York

X 0844 *Giangiacomo Feltrinelli Editore* (Milano, I) ISBN 88-07

X 0845 *University of Notre Dame Press* (Notre Dame, IN, USA & London, GB) ISBN 0-268

X 0848 *Chelsea Publishing Company* (New York, NY, USA) ISBN 0-8284

X 0850 *Armand Colin, Editeur* (Paris, F) ISBN 2-200

X 0856 *Dunod, Editeur* (Paris, F) ISBN 2-04

X 0857 *Princeton University Press* (Princeton, NJ, USA & Guildford, GB) ISBN 0-691

X 0858 *Harvard University Press* (Cambridge, MA, USA & London, GB) ISBN 0-674

X 0859 *Hermann, Editeurs des Sciences et des Arts* (Paris, F) ISBN 2-7056

X 0860 *Cremonese Edizioni* (Firenze, I) ISBN 88-7083

X 0864 *Van Nostrand, Reinhold* (New York, NY, USA & Scarborough, ON, CDN & Mitcham, Vic, AUS & & Wokingham, GB & Florence, KY, USA) ISBN 0-442

X 0866 *Routledge & Kegan Paul* (Henley on Thames, GB & Boston, MA, USA) ISBN 0-7100, ISBN 0-7102

X 0867 *W.A. Benjamin* (Reading, MA, USA) ISBN 0-8053 • REL PUBL (**X** 0832) Addison-Wesley: Reading

X 0868 *Penguin Books* (Harmondsworth, GB & New York, NY, USA & Ringwood, Vic, AUS & & Auckland, NZ & Markham, CDN) ISBN 0-14

X 0869 *Pergamon Press* (Oxford, GB & Elmsford, NY, USA & Rushcutters Bay, NSW, AUS & & Willowdale, ON, CDN & Paris, F) ISBN 0-08 • REL PUBL (**X** 0900) Vieweg: Wiesbaden

X 0871 *Academiei Republicii Socialiste Romania Editura (RSR)* (Bucharest, RO)

X 0873 *Mouton et Cie.* (Paris, F) ISBN 2-7193

X 0875 *Yale University Press* (New Haven, CT, USA & London, GB) ISBN 0-300

X 0876 *Bibliographisches Institut* (Mannheim, D & Wien, A & Zuerich, CH) ISBN 3-411

X 0879 *Institut Superieur de Philosophie* (Louvain, B)

X 0880 *Ronald Press & Co.* (New York, NY, USA) ISBN 0-8260 • REL PUBL (**X** 0827) Wiley & Sons: New York

X 0882 *Uppsala Universitet, Filosofiska Foereningen och Filosofiska Institutionen* (Uppsala, S)

X 0885 *Izdatel'stvo Mir* (Moskva, SU)

X 0890 *Wissenschaftliche Buchgesellschaft* (Darmstadt, D) ISBN 3-534

X 0893 *Les Presses de l'Universite de Montreal* (Montreal, PQ, CDN) ISBN 2-7606

X 0894 *Oxford University Press* (Oxford, GB & London, GB & Melbourne, Vic, AUS & Don Mills, ON, CDN & Nairobi, EAK & Auckland, NZ & Petaling Jaya, MAL & New York, NY, USA & Karachi, PAK & Harare, ZW) ISBN 0-19 • REL PUBL (**X** 0815) Clarendon Pr: Oxford

X 0895 *Latvijas Valsts Universitate* (Riga, SU)

X 0898 *Izdatel'stvo Moskovskogo Gosudarstvennogo Universiteta* (Moskva, SU)

X 0899 *Izdatel'stvo Akademii Nauk SSSR* (Moskva, SU)

X 0900 *Vieweg, Friedrich & Sohn Verlagsgesellschaft* (Wiesbaden, D) ISBN 3-528 • REL PUBL (**X** 0869) Pergamon Pr: Oxford

X 0902 *Springer-Verlag* (Wien, A) ISBN 3-211 • REL PUBL (**X** 0811) Springer: Heidelberg & New York

X 0903 *Vandenhoeck & Ruprecht* (Goettingen, D) ISBN 3-525 • REM Member Firms: 1) E. Klotz Verlag, Goettingen, D. 2) Verlag der Medizinischen Psychologie Goettingen, Dr D.& Dr A. Ruprecht, Goettingen, D

X 0908 *Universitaet Bonn, Mathematisches Institut* (Bonn, D)

X 0909 *Cedam* (Padova, I)

X 0910 *Aschendorffsche Verlagsbuchhandlung* (Muenster, D) ISBN 3-402

X 0911 *Akademie Verlag* (Berlin, DDR)

X 0913 *Novosibirskij Gosudarstvennyj Universitet* (Novosibirsk, SU)

X 0918 *Ernst Klett Verlag* (Stuttgart, D) ISBN 3-12 • REL PUBL (**X** 1255) Deuticke: Wien

X 0923 *Universidade Federal de Pernambuco, Instituto de Matematica* (Recife, BR)

X 0925 *Tata Institute of Fundamental Research* (Bombay, IND)

X 0926 *University of California Press* (Berkeley, CA, USA & London, GB) ISBN 0-520 • REL PUBL (**X** 1291) Johns Hopkins Univ Pr: Baltimore

X 0928 *Akademiai Kiado, Publishing House of the Hungarian Academy of Sciences.* (Budapest, H) ISBN 963-05

X 0942 *W.W.Norton & Co.* (New York, NY, USA & London, GB) ISBN 0-393 • REL PUBL (**X** 0827) Wiley & Sons: New York

X 0958 *Izdatel'stvo Saratovskogo Universiteta* (Saratov, SU)

X 0997 *Queen's University* (Kingston, ON, CDN) ISBN 0-88911, ISBN 0-9690334

X 0999 *Centre National de la Recherche Scientifique (CNRS), Institut Blaise Pascal* (Paris, F)

X 1006 *Irkutskij Gosudarstvennyj Universitet* (Irkutsk, SU)

X 1015 *University of Toronto Press* (Toronto, ON, CDN & Buffalo, NY, USA) ISBN 0-8020

X 1034 *Panstwowe Wydawnictwo Naukowe (PWN)* (Warsaw, PL) ISBN 83-01

X 1051 *Rijksuniversiteit Utrecht, Mathematisch Instituut.* (Utrecht, NL)

X 1052 *Izdatel'stvo Tbilisskogo Universiteta* (Tbilisi, SU)

X 1079 *B.G.Teubner Verlagsgesellschaft* (Leipzig, DDR) ISBN 3-519 • REM See also X0823

X 1080 *Scottish Academic Press* (Edinburgh, GB) ISBN 0-7073

X 1096 *Basil Blackwell* (Oxford, GB) ISBN 0-631, ISBN 0-632, ISBN 0-86286, ISBN 0-86793 • REM Also: Blackwell Scientific Publications: Oxford, GB & Carlton, Vic, AUS

X 1121 *Mathematisch Centrum* (Amsterdam, NL) ISBN 90-6196

X 1163 *Almqvist & Wiksell Foerlag* (Stockholm, S & Bromma, S & Goeteborg, S & Malmoe, S) ISBN 91-20

X 1172 *Les Editions de l'Office Central de Librairie S.A.R.L. (O.C.D.L.)* (Paris, F) ISBN 2-7043

X 1174 *Walter de Gruyter* (Berlin, D) ISBN 3-11 • REM Member Firms: 1) de Gruyter: Hawthorne, NY, USA. 2) Mouton Publishers: Berlin, D

X 1212 *Izdatel'stvo Belorusskogo Gosudarstvennogo Universiteta* (Minsk, SU)

X 1214 *New York University, Courant Institute of Mathematical Science.* (New York, NY, USA)

X 1226 *Academia* (Prague, CS) • REM Publishing House of the Czechoslovak Academy of Science

X 1228 *Appleton-Century-Crofts* (New York, NY, USA) ISBN 0-8385

X 1231 *Johann Ambrosius Barth* (Leipzig, DDR) ISBN 3-335 • REL PUBL (**X** 0811) Springer: Heidelberg & New York

X 1234 *G.Bell & Sons* (London, GB) ISBN 0-7135

X 1246 *California Institute of Technology (Caltech)* (Pasadena, CA, USA)

X 1255 *Franz Deuticke, Verlagsgesellschaft* (Wien, A) ISBN 3-7005 • REL PUBL (**X** 0918) Klett: Stuttgart

X 1259 *Durand Ruel, Charles, & Cie* (Paris, F) ISBN 2-900767

X 1261 *Edinburgh University Press* (Edinburgh, GB) ISBN 0-85224

X 1268 *Orell Fuessli Verlag* (Zuerich, CH) ISBN 3-280

X 1275 *Anton Hain K.G. Meisenheim Verlag* (Koenigstein, D & Meisenheim, D) ISBN 3-445 • REL PUBL (**X** 1588) Athenaeum/Hain/Hanstein: Koenigstein

X 1277 *Hayden Book Company* (Rochelle Park, NY, USA) ISBN 0-8104 • REL PUBL (**X** 1354) Spartan Books : Sutton

X 1285 *University of Illinois Press* (Urbana, IL, USA & London, GB) ISBN 0-252 • REL PUBL (**X** 3828) Amer Univ Publ Group: London

X 1286 *Indiana University Press* (Bloomington, IN, USA & London, GB) ISBN 0-253 • REL PUBL (**X** 3828) Amer Univ Publ Group: London

X 1291 *Johns Hopkins University Press* (Baltimore, MD, USA) ISBN 0-8018

X 1298 *Mathematical Association of America* (Washington, DC, USA) ISBN 0-88385

X 1299 *Magnes Press* (Jerusalem, IL) ISBN 965-223

X 1313 *Nauwelaerts, Beatrice* (Louvain, B) ISBN 2-900014

X 1317 *P.Noordhoff International Publishing* (Groningen, NL & Leiden, NL) ISBN 90-01 • REL PUBL (**X** 0812) Wolters-Noordhoff : Groningen & (**X** 1352) Sijthoff: Leiden • REM Now: Sijthoff & Noordhoff International Publishers: Leiden, NL

X 1319 *La Nuova Italia Editrice* (Firenze, I) ISBN 88-221

X 1323 *Oliver & Boyd* (Edinburgh, GB) ISBN 0-05

X 1331 *University of Pittsburgh Press* (Pittsburgh, PA, USA) ISBN 0-8229

X 1332 *Plenum Publishing Corporation* (New York, NY, USA & London, GB) ISBN 0-306 • REM Also Called Plenum Press

X 1337 *Prindle, Weber & Schmidt* (Boston, MA, USA) ISBN 0-87150 • REM Also Called: PWS Publishers

X 1349 *Editions du Seuil* (Paris, F) ISBN 2-02

X 1352 *A.W.Sijthoff International Publishing Co.* (Leiden, NL) ISBN 90-218, ISBN 90-286 • REL PUBL (**X** 1317) Noordhoff: Groningen

X 1354 *Spartan Books* (Sutton, GB & Rochelle Park, NJ, USA) ISBN 0-905532 • REL PUBL (**X** 1277) Hayden: Rochelle Park

X 1355 *Stanford University Press* (Stanford, CA, USA) ISBN 0-8047

X 1358 *University of Texas Press* (Austin, TX, USA & London, GB) ISBN 0-292 • REL PUBL (**X** 3828) Amer Univ Publ Group: London

X 1364 *Vita e Pensiero, Publicazioni della Universita Cattolica* (Milano, I) ISBN 88-343

X 1367 *University of Washington Press* (Seattle, WA, USA & London, GB) ISBN 0-295 • REL PUBL (**X** 3828) Amer Univ Publ Group: London

X 1375 *Nicola Zanichelli Editore* (Bologna, I) ISBN 88-08

X 1407 *Publicaciones del Instituto de Matematicas "Jorge Juan"* (Madrid, E) ISBN 84-00

X 1493 *C.W.K.Gleerup Bokfoerlag a.B.* (Lund, S) ISBN 91-40

X 1494 *Munksgaard International Publishers* (Copenhagen, DK) ISBN 87-16

X 1518 *Oxford University, Mathematical Institute* (Oxford, GB)

X 1554 *Universitetsforlaget* (Oslo, N & Bergen, N & Tromsoe, N & New York, NY, USA) ISBN 82-00 • REM Also: UB-Forlaget: Oslo, N

X 1588 *Verlagsgruppe Athenaeum/ Hain/ Hanstein* (Koenigstein/Ts, D) • REL PUBL (**X** 1275) Hain: Koenigstein & (**X** 2665) Athenaeum: Frankfurt

X 1599 *Aarhus Universitet, Matematisk Institut* (Aarhus, DK) ISBN 87-87436

X 1623 *Universite de Paris VI, Institut Henri Poincare* (Paris, F) • REM The University Has Now Been Divided Into Smaller Units. The Mathematics Department Is Now: Universite de Paris VII-Pierre et Marie Curie, Secretariat Mathematique

X 1684 *Marcel Dekker* (New York, NY, USA & Basel, CH) ISBN 0-8247 • REL PUBL (**X** 2442) Dekker Journal: New York

X 1714 *Commission of the European Communities.* (Bruxelles, B) ISBN 92-825, ISBN 92-826, ISBN 92-827, ISBN 92-828

X 1732 *Bibliopolis* (Napoli, I) ISBN 88-7088

X 1754 *Universite Catholique de Louvain* (Louvain-la-Neuve, B)

X 1758 *Ars Polona* (Warsaw, PL)

X 1761 *Hekdosis Hellinikis Mathematikes Hetaireias-Greek Mathematical Society* (Athens, GR)

X 1763 *Royal Society* (London, GB) ISBN 0-85403

X 1764 *Ksiaznica Atlas T.N.S.W.* (Warsaw, PL & Lwow, PL)

X 1773 *VEDA - Vydavatelstvo Slovenskej Akademie Vied* (Bratislava, CS)

X 1775 *Societe Royale des Sciences de Liege* (Liege, B)

X 1781 *Tecnos* (Madrid, E) ISBN 84-309

X 1821 *Real Academia de Ciencias Exactas, Fisicas y Naturales* (Madrid, E)

X 1876 *Kexue Chubanshe (Science Press)* (Beijing, TJ)

X 1906 *SNTL. Nakladatelstvi Technicke Literatury (Verlag Technischer Literatur)* (Prague, CS)

X 1916 *Komunikaty i Rozprawy Inst, Matem, Uam* (Poznan, PL)

X 1941 *Nanyang University Publications* (Singapore, MAL)

X 2005 *Australian National University* (Canberra, ACT, AUS) ISBN 0-909851

X 2027 *Izdatel'stvo Nauka* (Moskva, SU & Alma-Ata, SU & Leningrad, SU & Novosibirsk, SU)

X 2039 *Polytechnic Institute of New York* (Brooklyn, NY, USA)

X 2040 *Winthrop* (Cambridge, MA, USA) ISBN 0-87626 • REL PUBL (**X** 0819) Prentice Hall: Englewood Cliffs

X 2055 *Progress* (Moskva, SU)

X 2091 *Consejo Superior de Investigaciones Cientificas.* (Madrid, E) ISBN 84-00

X 2121 *Accademia Nazionale dei Lincei* (Roma, I) ISBN 88-218

X 2152 *Tokai University Press* (Tokyo, J) ISBN 4-486

X 2179 *IEEE (Institute of Electrical and Electronics Engineers* (New York, NY, USA & Long Beach, CA, USA & Piscataway, CA, USA) ISBN 0-87942 • REM Computer Society: Long Beach, CA, USA. Section: IEEE United States Activities Commitee: Piscataway, CA, USA

X 2189 *Universitatea "Babes-Bolyai", Biblioteca Centrala Universitara* (Cluj-Napoca, RO)

X 2193 *Indian Mathematical Society* (Bombay, IND)

X 2194 *Sociedad Matematica Mexicana* (Mexico City, MEX)

X 2195 *Peeters S.P.R.L.* (Louvain, B) ISBN 2-8017

X 2197 *Tohoku University, Mathematical Institute* (Sendai, J)

X 2199 *Izdatel'stvo Naukova Dumka* (Kiev, SU)

X 2204 *American Institute of Physics* (New York, NY, USA) ISBN 0-88318

X 2205 *(ACM) Association for Computing Machinery* (New York, NY, USA) ISBN 0-89791

X 2206 *Accademia delle Science* (Torino, I)

X 2207 *Universitatea "Al. I. Cuza" din Iasi* (Jassy (Iasi), RO)

X 2208 *Taylor & Francis* (London, GB) ISBN 0-85066

X 2209 *Izdatel'stvo Akademii Nauk Gruzinskoj SSR* (Tbilisi, SU)

X 2211 *Friedrich Schiller Universitaet* (Jena, DDR)

X 2212 *Ossolineum, Publishing House of the Polish Academy of Sciences* (Wroclaw, PL & Warsaw, PL) • REL PUBL (**X** 2885) Zakl Narod Wyd Pol Ak: Wroclaw & (**X** 2882) Pol Akad Nauk: Wroclaw

X 2213 *State University of New York at Buffalo (SUNY)* (Buffalo, NY, USA)

X 2214 *Dialectica* (Bienne/Biel, CH)

X 2215 *Indiana University, Department of Philosophy* (Bloomington, IN, USA)

X 2217 *Academia Brasileira de Ciencias.* (Rio de Janeiro, BR)

X 2219 *Humboldt-Universitaet zu Berlin* (Berlin, DDR)

X 2220 *Buchdruckerei und Verlag Leemen* (Zuerich, CH)

X 2224 *Matematisk Institut* (Bergen, N)

X 2225 *Izdatel'stvo Akademii Nauk Armyanskoj SSR* (Erevan, SU)

X 2229 *Universita J.E.Purkyne* (Brno, CS) • REM Formerly Named Masarykova Universita

X 2230 *Isdevnieciba Zinatne* (Riga, SU)

X 2233 *Instituto de Matematica Pura et Aplicada (IMPA)* (Rio de Janeiro, BR) ISBN 85-244

X 2235 *Vsesoyuznyj Institut Nauchnoj i Tekhnicheskoj Informatsii (VINITI), Gosudarstvennyj Komitet SSSR po Nauke: Tekhnike (GKNT SSSR), Akademiya Nauk (AN) SSSR* (Moskva, SU)

X 2237 *Izdatelstvo na Bulgarskata Akademia na Naukite (Publishing House of the Bulgarian Academy of Sciences.)* (Sofia, BG)

X 2238 *Sociedad Colombiana de Matematicas* (Bogota, CO) ISBN 958-95081

X 2239 *Universidad de Salamanca Ediciones* (Salamanca, E) ISBN 84-7481

X 2241 *New York Academy of Sciences* (New York, NY, USA) ISBN 0-89072, ISBN 0-89766

X 2242 *Osaka University, Department of Mathematics* (Osaka, J)

X 2243 *University of Osaka Prefecture, Department of Mathematics* (Osaka, J)

X 2244 *Societe Mathematique de France* (Paris, F) ISBN 2-85629

X 2246 *Universite d'Alger* (Alger, DZ)

X 2254 *Universidad Nacional Autonoma de Mexico, Instituto de Matematicas* (Mexico City, MEX)

X 2255 *Aulis-Verlag Deubner* (Koeln, D) ISBN 3-7614

X 2256 *Casa Editrice Felice le Monnier* (Firenze, I)

X 2257 *John Benjamins B.V.* (Amsterdam, NL & Philadelphia, PA, USA) ISBN 90-272

X 2265 *Vychislitel'nyj Tsentr Akademii Nauk SSSR* (Moskva, SU)

X 2367 *Universidad de Barcelona, Facultad de Filosofia y Letras* (Barcelona, E)

X 2375 *MacMillan Journals* (London, GB) ISBN 0-333 • REL PUBL (**X** 0843) Macmillan : New York & London

X 2378 *Society for Industrial and Applied Mathematics (SIAM)* (Philadelphia, PA, USA) ISBN 0-89871

X 2379 *University of Houston, Department of Mathematics* (Houston, TX, USA)

X 2380 *Malaysian Mathematical Society* (Kuala Lumpur, MAL)

X 2382 *Institute of Mathematics and its Applications* (Southend-on-Sea, GB) ISBN 0-905091, ISBN 0-9501159

X 2390 *Institut Mittag-Leffler* (Djursholm, S)

X 2396 *Hitotsubashi University* (Tokyo, J)

X 2409 *Suomalainen Tiedakatemia.* (Helsinki, SF) ISBN 951-41

X 2421 *University of Tokyo, Faculty of Science* (Tokyo, J)

X 2427 *Asociacion Venezolana para el Avance de la Ciencia* (Caracas, YV)

X 2434 *University of Tokyo College of General Education* (Tokyo, J)

X 2435 *Presses Universitaires de France, Service Periodiques* (Paris, F) ISBN 2-13 • REL PUBL (X 0840) Pr Univ France: Paris

X 2438 *Edizioni Scientifiche Inglesi Americane* (Roma, I)

X 2440 *Universita Degli Studi di Modena, Seminario Matematico e Fisico* (Modena, I)

X 2441 *Kyoto University, Research Institute for Mathematical Sciences* (Kyoto, J)

X 2442 *Marcel Dekker Journals.* (New York, NY, USA) ISBN 0-8247 • REL PUBL (X 1684) Dekker: New York

X 2443 *Izdatel'stvo Nauka, Otdelenie v Kazakhstane SSR* (Alma-Ata, SU)

X 2446 *Martin-Luther-Universitaet Halle-Wittenberg* (Halle, DDR)

X 2447 *Technische Hochschule Ilmenau* (Ilmenau, DDR)

X 2448 *Akademiya Navuk Belorusskaj SSR* (Minsk, SU)

X 2451 *Uniwersitet Jagiellonskie, Instytut Matematiyczny* (Krakow, PL)

X 2455 *Yokohama City University, Department of Mathematics* (Yokohama, J)

X 2457 *Allerton Press* (New York, NY, USA) ISBN 0-89864

X 2462 *Cairo University Press* (Cairo, ETH)

X 2463 *Tartuskij Gosudarstvennyj Universitet* (Tartu, SU)

X 2465 *Kinokuniya Company* (Tokyo, J) ISBN 4-314

X 2467 *Societatea de Stiinte Matematice din Republica Socialista Romania (RSR)* (Bucharest, RO)

X 2471 *IBM Corp.* (Armonk, NY, USA) ISBN 0-933186

X 2472 *Centro Superiore di Logica e Scienze Comporate, Editrice Franco Spisani* (Bologna, I)

X 2473 *Belfort Graduate School of Science; Yeshiva University* (New York, NY, USA)

X 2476 *Journal of Philosophy Inc.* (New York, NY, USA) ISBN 0-931206

X 2477 *University College, London, Department of Mathematics* (London, GB)

X 2479 *Rocky Mountain Mathematics Consortium* (Tempe, AZ, USA)

X 2483 *Hiroshima University, Department of Mathematics* (Hiroshima, J)

X 2492 *Japan Association for the Philosophy of Science* (Tokyo, J)

X 2495 *Suid Afrikaanse Akademie vir Wetenskap en Kuns* (Pretoria, ZA) ISBN 0-949976

X 2496 *Suomen Fyysikkoseura* (Helsinki, SF)

X 2498 *Kumamoto University, Faculty of Science* (Kumamoto, J)

X 2504 *Bertrand Russell Memorial Logic Conference* (Leeds, GB) ISBN 0-9502983

X 2510 *Elm (Izdatel'stvo Akademiya Nauk Azerbajdzhanskoj SSR)* (Baku, SU)

X 2511 *Universita Karlova, Matematicky Ustav* (Prague, CS)

X 2512 *Institut Matematiki Akademii Nauk SSSR* (Moskva, SU)

X 2514 *Institut de Mathematiques* (Geneve, CH)

X 2516 *Union Matematica Argentina* (Buenos Aires, RA)

X 2518 *Aberdeen University Press* (Aberdeen, GB) ISBN 0-900015, ISBN 0-08

X 2519 *Izdatel'stvo Akademii Nauk Uzbekskoj SSR* (Tashkent, SU)

X 2520 *National Academy of Sciences, Transportation Research Board* (Washington, DC, USA) ISBN 0-309

X 2522 *Akademiya Nauk USSR, Nauchnyj Sovet po Kibernetike, Institut Kibernetiki.* (Kiev, SU)

X 2524 *Encyclopaedia Universalis* (Paris, F) ISBN 2-85229

X 2526 *Gazeta de Matematica* (Lisboa, P)

X 2528 *Circolo Matematico di Palermo* (Palermo, I)

X 2529 *Consultants Bureau* (New York, NY, USA) • REL PUBL (X 1332) Plenum Publ: New York

X 2530 *Societe Mathematique de Belgique* (Bruxelles, B)

X 2531 *Japan Academy* (Tokyo, J)

X 2532 *Fratelli Fusi, Tipografia Successon* (Pavia, I)

X 2594 *Ivanovskij Gosudarstvennyj Universitet* (Ivanovo, SU)

X 2641 *Izdatel'stvo Nauka Leningradskoe Otdelenie* (Leningrad, SU)

X 2642 *Izdatel'stvo Nauka Sibirskoe Otdelenie* (Novosibirsk, SU)

X 2644 *Izdatel'skoe Ob"edineniya "Vishcha Shkola". Izdatel'stvo pri Khar'kovskom Gosudarskvennom Universitete* (Khar'kov, SU)

X 2645 *Izdatel'skoe Ob"edineniye "Vishcha Shkola". Izdatel'stvo pri Kievskom Gosudarstvennom Universitete* (Kiev, SU)

X 2646 *Aalborg University Press = Aalborg Universitetsforlag* (Aalborg, DK) ISBN 87-7307

X 2652 *Institut Matematiki Sibirskogo Otdeleniya Akademii Nauk SSSR (SOAN SSSR)* (Novosibirsk, SU)

X 2655 *Akademie der Wissenschaften der DDR, Institut fuer Mathematik* (Berlin, DDR)

X 2665 *Athenaeum Verlag* (Frankfurt, D) ISBN 3-7610 • REL PUBL (X 1588) Athenaeum/Hain/Hanstein: Koenigstein

X 2692 *University of Queensland Press* (St. Lucia, Qld, AUS) ISBN 0-7022

X 2696 *Humanities Press* (Atlantic Highlands, NY, USA) ISBN 0-391

X 2698 *Departamento de Logica y Filosofia de la Ciencia, Universidad de Valencia* (Valencia, E)

X 2706 *Paul Castella, Editions* (Albeuve, CH) ISBN 2-88087

X 2717 *Franz Steiner Verlag* (Wiesbaden, D & Stuttgart, D) ISBN 3-515

X 2725 *Hackett Publishing Co.* (Indianapolis, IN, USA) ISBN 0-915144

X 2728 *Hoelder-Pichler-Tempsky* (Wien, A) ISBN 3-209

X 2733 *Polish Academy of Sciences, Institute of Philosophy and Sociology* (Wroclaw, PL)

X 2737 *Israel Program for Scientific Translations* (Jerusalem, IL) ISBN 0-7065 • REL PUBL (**X 0827**) Wiley & Sons: New York

X 2741 *Shtiintsa* (Kishinev, SU)

X 2766 *Yaroslavskij Gosudarstvennyj Universitet* (Yaroslavl', SU)

X 2769 *Kazakhskij Gosudarstvennyj Universitet* (Alma-Ata, SU)

X 2786 *Universitet i Oslo, Matematisk Institut* (Oslo, N) ISBN 82-553

X 2802 *Gosudarstvennyj Pedagogicheskij Institut* (Moskva, SU)

X 2812 *The Philosophical Society of Finland* (Helsinki, SF) ISBN 951-95053

X 2836 *Sociedade Brasileira Logica* (Sao Paulo, BR)

X 2854 *U.E.R. (UER) Mathematique, Universite Paris XI* (Paris-Orsay, F)

X 2863 *Upside Down A Book Company* (Yarra Glen, Vic, AUS) ISBN 0-949865

X 2882 *Wydawnictwo Polskiej Akademii Nauk (Publisher of the Polish Academy of Science)* (Wroclaw, PL) • REL PUBL (**X 2885**) Zakl Narod Wyd Pol Ak: Wroclaw & (**X 2212**) Ossolineum: Wroclaw

X 2885 *Zaklad Narodowy imienia Ossolinskich, Wydawnictwo Polskiej Akademii Nauk* (Wroclaw, PL) ISBN 83-04 • REL PUBL (**X 2882**) Pol Akad Nauk: Wroclaw & (**X 2212**) Ossolineum: Wroclaw

X 2888 *Zentralinstitut fuer Mathematik und Mechanik, Akademie der Wissenschaften der DDR* (Berlin, DDR)

X 3123 *Beijing Shifan Daxue (Beijing Normal University)* (Beijing, TJ)

X 3157 *Fachbereich 6 (Mathematik) der Universitaet Essen* (Essen, D)

X 3158 *Fakultaet fuer Mathematik und Naturwissenschaften der Universitaet Hannover* (Hannover, D)

X 3215 *Mathematisch-Naturwissenschaftliche Fakultaet der Rheinisch-Westfaelischen Technischen Hochschule Aachen.* (Aachen, D)

X 3223 *Carl Hanser Verlag* (Muenchen, D & Wien, A) ISBN 3-446

X 3249 *The Weizmann Institute of Science* (Rehovot, IL) ISBN 965-281

X 3552 *Iwanami Shoten Publishers* (Tokyo, J) ISBN 4-00

X 3562 *Ecole Normale Superieure de Jeunes Filles* (Paris, F) ISBN 2-85929

X 3602 *Kraus Reprint* (Vaduz, FL & Nendeln, FL) ISBN 3-262 • REM Parent Firm: Kraus Thomson Organisation: Vaduz, FL

X 3604 *Istituto Nazionale di Alta Matematica (INDAM)* (Roma, I)

X 3605 *Izdatel'stvo Kazanskogo Gosudarstvennogo Universiteta* (Kazan', SU)

X 3606 *Izdadel'stvo Tomskogo Universiteta* (Tomsk, SU)

X 3636 *Tokyo Tosho* (Tokyo, J) ISBN 4-489

X 3709 *Gosudarstvennoye Izdatel'stvo Fiziko-Matematicheskoj Literatury* (Moskva, SU)

X 3718 *L'Enseignement Mathematique, Universite de Geneve* (Geneve, CH)

X 3725 *Bolyai Janos Matematikai Tarsulat (Janos Bolyai Mathematical Society)* (Budapest, H)

X 3727 *Beograd. Matematicki Institut* (Belgrade, YU)

X 3734 *Cabay Libraire-Editeur* (Louvain-la-Neuve, B) ISBN 2-87077

X 3760 *Universite "Kiril et Metodij", Matematicki Fakultet* (Skopje, YU)

X 3812 *Universita di Siena, Dipartimento di Matematica, Scuola di Specializzazione in Logica Matematica* (Siena, I)

X 3828 *American University Publishers Group* (London, GB) • REL PUBL (**X 0842**) Univ Wisconsin Pr: Madison & (**X 1285**) Univ Ill Pr: Urbana & (**X 1286**) Indiana Univ Pr : Bloomington & (**X 1358**) Univ Texas Pr: Austin & (**X 1367**) Univ Washington: Seattle

X 4030 *Univerzitet u Novom Sadu, Institut za Matematiku* (Novi Sad, YU)

X 4217 *Foris Publications* (Dordrecht, NL) ISBN 90-70176

X 4282 *The Rand Corporation* (Santa Monica, CA, USA) ISBN 0-8330

X 4562 *Udmurtskij Gosudarstvennij Universiteta* (Izhevsk, SU)

X 4643 *Universite de Paris VII, UER de Mathematiques* (Paris, F)

X 4721 *Libreria Fratelli Bocca & C. Clausen* (Torino, I & Roma, I)

Miscellaneous Indexes

External classifications

This index complements the Subject Index at the beginning of this volume; it lists the items which, in addition to classifications in the present volume, have classifications *external to this volume*. These items are ordered by external classification code and within each code by author (the first alphabetically in the case of multi-author items), year and identification number (thus an item with, for example, two external classifications occurs twice in this listing). This index provides another way to search the bibliography. With it, the user can easily identify those items in this volume classified also in some area external to this volume.

B03

Makanin, G.S. [1964] 14845
Makanin, G.S. [1977] 51395
Makanin, G.S. [1977] 51863
Makanin, G.S. [1980] 82143

B05

Bochvar, D.A. [1940] 01348
Canty, J.T. [1963] 01832
Chen, Jiyuan [1984] 44191
Gao, Hengshan [1983] 45159
Goodstein, R.L. [1972] 05195
Greniewski, H. [1950] 22000
Leone, M. [1985] 49912
Loparic, A. [1977] 27299
Luckhardt, H. [1969] 08380
Mijajlovic, Z. [1974] 09242
Mijajlovic, Z. [1977] 50777
Nieland, J.J.F. [1967] 09951
Novikov, P.S. [1939] 10016
Novikov, P.S. [1943] 19479
Reichbach, J. [1969] 11073
Reznikoff, I. [1963] 11135
Reznikoff, I. [1965] 11136
Rickey, V.F. [1975] 11178
Suszko, R. [1972] 12781
Suszko, R. [1974] 12782
Vakarelov, D. [1972] 00678
Weaver, G.E. [1975] 21907
Wright von, G.H. [1948] 14303
Wright von, G.H. [1950] 14304
Zermelo, E. [1935] 14413

B10

Alcantara de, L.P. [1980] 54266
Andreka, H. [1974] 03829
Barwise, J. [1977] 24196
Bennett, D.W. [1973] 01038
Bennett, D.W. [1977] 21957
Benthem van, J.F.A.K. [1974] 03889
Beth, E.W. [1953] 01159
Beth, E.W. [1955] 01269
Beth, E.W. [1960] 01534
Blass, A.R. [1984] 47716
Boolos, G. [1984] 40442
Borkowski, L. [1968] 01485
Bowen, K.A. [1972] 01505
Brady, R.T. [1977] 30639

Bullock, A.M. [1972] 01746
Bullock, A.M. [1973] 01748
Call, R.L. [1972] 71509
Casari, E. [1981] 56675
Christen, C. [1976] 52348
Corcoran, J. [1971] 27917
Corcoran, J. [1972] 02392
Craig, W. [1952] 02505
Craig, W. [1957] 02508
Craig, W. [1957] 02510
Craig, W. [1958] 02511
Crossley, J.N. [1966] 31222
Deutsch, M. [1975] 02972
Dulac, M.-H. [1971] 03165
Ecsedi-Toth, P. [1982] 34907
Ehdel'man, G.S. [1974] 72569
Ellerman, D.P. [1978] 56695
Feferman, S. [1957] 03683
Felscher, W. [1969] 03737
Felscher, W. [1971] 03738
Flum, J. [1971] 04388
Follesdal, D. [1968] 04443
Fraisse, R. [1950] 04510
Fraisse, R. [1955] 04516
Fraisse, R. [1956] 04518
Fraisse, R. [1970] 04531
Fraisse, R. [1972] 04294
Gentzen, G. [1935] 24221
Glebskij, Yu.V. [1969] 05032
Glubrecht, J.-M. [1982] 35619
Goedel, K. [1930] 20884
Goedel, K. [1933] 20885
Goodstein, R.L. [1974] 05197
Hajek, P. [1970] 05521
Hanazawa, M. [1979] 54329
Hanf, W.P. [1983] 35790
Hasenjaeger, G. [1952] 05722
Hasenjaeger, G. [1953] 05723
Hasenjaeger, G. [1955] 27719
Hauck, J. [1972] 62383
Henkin, L. [1949] 05892
Henkin, L. [1954] 05900
Henkin, L. [1957] 05910
Herbrand, J. [1930] 20866
Hintikka, K.J.J. [1953] 06111
Hintikka, K.J.J. [1964] 06122
Hintikka, K.J.J. [1965] 06123
Hintikka, K.J.J. [1974] 50730
Hodges, W. [1983] 39725
Hook, J.L. [1985] 41804

Hugly, P. [1982] 54068
Issel, W. [1969] 06459
Jervell, H.R. [1973] 33265
Keisler, H.J. [1973] 07033
Kreisel, G. [1950] 07481
Kreisel, G. [1953] 20815
Krom, Melven R. [1963] 07571
Kuehnrich, M. [1983] 36540
Leblanc, H. [1984] 44672
Lehmann, G. [1985] 42737
Lightstone, A.H. [1957] 08149
Lindstroem, P. [1974] 29976
Lis, Z. [1960] 08189
Los, J. [1956] 08333
Maehara, S. [1960] 08520
Mal'tsev, A.I. [1936] 08602
Marini, D. [1975] 22915
Markusz, Z. [1983] 44560
Mo, Shaokui [1964] 47141
Mostowski, Andrzej [1949] 09545
Motohashi, N. [1973] 09600
Motohashi, N. [1978] 90092
Motohashi, N. [1982] 33418
Motohashi, N. [1982] 55534
Motohashi, N. [1984] 42487
Mundici, D. [1981] 90124
Norgela, S.A. [1980] 32676
Ono, H. [1985] 47510
Paillet, J.L. [1973] 30495
Pogorzelski, W.A. [1961] 31969
Pudlak, P. [1977] 32425
Quine, W.V.O. [1945] 10872
Ramsey, F.P. [1930] 28781
Rantala, V. [1975] 64743
Rasiowa, H. [1953] 11004
Rasiowa, H. [1953] 11006
Rasiowa, H. [1957] 11017
Rasiowa, H. [1960] 11024
Reichbach, J. [1955] 33706
Reichbach, J. [1970] 27880
Reichbach, J. [1975] 32003
Ribeiro, H. [1969] 64815
Rickey, V.F. [1978] 51658
Rieger, L. [1955] 19556
Rieger, L. [1964] 11193
Robinson, A. [1959] 19551
Rodriguez Artalejo, M. [1981] 77840
Schock, R. [1964] 12942
Scholz, H. [1952] 33304
Schroeter, K. [1958] 24841

Schuette, K. [1950] 12999
Schuette, K. [1978] 56098
Schwartz, Dietrich [1975] 13068
Scott, D.S. [1979] 54069
Shafaat, A. [1967] 12002
Skolem, T.A. [1920] 12382
Skolem, T.A. [1936] 40864
Smullyan, R.M. [1970] 12557
Tarski, A. [1934] 30917
Tarski, A. [1936] 13413
Tarski, A. [1944] 13432
Tarski, A. [1948] 13437
Trakhtenbrot, B.A. [1953] 19742
Venneri, B.M. [1975] 65905
Wajsberg, M. [1933] 13979
Wajsberg, M. [1933] 40808
Walkoe Jr., W.J. [1976] 27272
Weaver, G.E. [1980] 56064
Zubieta Russi, G. [1957] 11688

B15

Amer, M.A. [1976] 17946
Amer, M.A. [1977] 30619
Andreka, H. [1973] 22879
Andreka, H. [1974] 60135
Andreka, H. [1975] 55844
Andrews, P.B. [1972] 00366
Andrews, P.B. [1972] 00367
Andrews, P.B. [1974] 03832
Armbrust, M. [1972] 00468
Asser, G. [1979] 70532
Barwise, J. [1969] 00822
Barwise, J. [1972] 00839
Belyakin, N.V. [1983] 41479
Boffa, M. [1977] 53983
Boffa, M. [1982] 34190
Boffa, M. [1983] 37638
Boudreaux, J.C. [1979] 56096
Boudreaux, J.C. [1980] 53790
Buechi, J.R. [1960] 01690
Buechi, J.R. [1965] 01694
Buechi, J.R. [1965] 01697
Buechi, J.R. [1969] 01700
Buechi, J.R. [1973] 01708
Buechi, J.R. [1973] 71377
Buechi, J.R. [1973] 71381
Buechi, J.R. [1977] 51165
Buechi, J.R. [1983] 33901
Buechi, J.R. [1983] 40515
Chepurnov, B.A. [1976] 52949
Chlebus, B.S. [1980] 56482
Cocchiarella, N.B. [1975] 21896
Cocchiarella, N.B. [1980] 53911
Corcoran, J. [1980] 71934
Craig, W. [1965] 02514
Daniels, C.B. [1977] 30696
Daniels, C.B. [1978] 56537
Doepp, K. [1972] 04094
Elgot, C.C. [1966] 03291
Enderton, H.B. [1970] 03351
Feferman, S. [1966] 03697
Ferro, R. [1978] 29281
Fraisse, R. [1958] 04520
Fraisse, R. [1974] 04537
Fraisse, R. [1977] 52674
Fraisse, R. [1985] 48204
Gacs, P. [1977] 52753

Garland, S.J. [1974] 24144
Girard, J.-Y. [1984] 48565
Grayson, R.J. [1984] 41840
Gurevich, Y. [1977] 50608
Gurevich, Y. [1979] 53793
Gurevich, Y. [1983] 33764
Gurevich, Y. [1983] 33765
Gurevich, Y. [1983] 33768
Harel, D. [1979] 53965
Hasenjaeger, G. [1955] 27719
Hasenjaeger, G. [1967] 05731
Hauschild, K. [1972] 05782
Hauschild, K. [1975] 29757
Henkin, L. [1950] 05894
Herrmann, E. [1980] 56483
Hinnion, R. [1979] 56023
Hintikka, K.J.J. [1955] 06112
Hort, C. [1984] 44206
Johnson, D.R. [1969] 06648
Kaiser, Klaus [1974] 06804
Kaiser, Klaus [1975] 06805
Kaiser, Klaus [1976] 06806
Kiselev, A.A. [1979] 32589
Kochen, S. [1957] 07272
Kogalovskij, S.R. [1966] 07292
Kogalovskij, S.R. [1966] 24967
Kogalovskij, S.R. [1968] 07299
Kogalovskij, S.R. [1973] 74934
Kogalovskij, S.R. [1974] 74929
Kogalovskij, S.R. [1974] 74930
Krawczyk, A. [1977] 31404
Krynicki, M. [1979] 52881
Krynicki, M. [1979] 53311
Ladner, R.E. [1977] 52252
Laeuchli, H. [1968] 07787
Langford, C.H. [1939] 07855
Leblanc, H. [1969] 07936
Lindstroem, P. [1973] 08177
Lisovik, L.P. [1984] 48409
Lopez-Escobar, E.G.K. [1967] 08262
Lopez-Escobar, E.G.K. [1975] 24555
Luxemburg, W.A.J. [1969] 21147
Maehara, S. [1971] 08528
Mal'tsev, A.I. [1959] 19260
Malitz, J. [1980] 55717
Marek, W. [1973] 17662
Marek, W. [1973] 17663
Mijoule, R. [1976] 25854
Mizutani, C. [1977] 53069
Montague, R. [1965] 24776
Mostowski, A.Wlodzimierz [1979] 76601
Mostowski, A.Wlodzimierz [1981] 55152
Mostowski, Andrzej [1947] 09538
Mostowski, Andrzej [1961] 09570
Mostowski, Andrzej [1962] 22070
Motohashi, N. [1972] 09599
Motohashi, N. [1973] 09602
Movsisyan, Yu.M. [1975] 16671
Muchnik, A.A. [1985] 47256
Muller, D.E. [1981] 76648
Muller, D.E. [1985] 47646
Nabebin, A.A. [1977] 52155
Nakamura, A. [1981] 55427
Novak, I.L. [1950] 09999
Oikkonen, J. [1979] 52831
Orey, S. [1959] 10165
Palyutin, E.A. [1971] 64365

Perrin, M.J. [1975] 23181
Rabin, M.O. [1968] 29625
Rabin, M.O. [1969] 10942
Rabin, M.O. [1970] 22233
Rabin, M.O. [1971] 14690
Rabin, M.O. [1972] 10943
Rickey, V.F. [1975] 22975
Roedding, D. [1972] 11343
Rosser, J.B. [1950] 11558
Scarpellini, B. [1984] 39663
Scarpellini, B. [1984] 41786
Scarpellini, B. [1985] 47473
Schreiber, P. [1965] 12976
Schwabhaeuser, W. [1967] 13058
Schwabhaeuser, W. [1979] 53471
Seese, D.G. [1975] 50759
Seese, D.G. [1977] 27132
Seese, D.G. [1979] 53053
Shafaat, A. [1973] 18382
Shelah, S. [1973] 12052
Shelah, S. [1973] 19627
Siefkes, D. [1968] 21146
Siefkes, D. [1970] 12145
Siefkes, D. [1970] 27984
Siefkes, D. [1971] 12146
Siefkes, D. [1978] 29136
Skvortsov, D.P. [1983] 40156
Tajtslin, M.A. [1979] 82932
Tarski, A. [1933] 28816
Tarski, A. [1948] 13437
Tauts, A. [1978] 54483
Tenney, R.L. [1975] 17647
Tharp, L.H. [1973] 13527
Thatcher, J.W. [1968] 28596
Thomas, Wolfgang [1980] 55873
Uesu, T. [1971] 13764
Vaeaenaenen, J. [1977] 51934
Venne, M. [1967] 13929
Walkoe Jr., W.J. [1970] 13998
Wasilewska, A. [1985] 49467
Wolter, H. [1972] 14277
Wolter, H. [1972] 14278
Zbierski, P. [1978] 29200
Zykov, A.A. [1953] 19773

B20

Aanderaa, S.O. [1973] 00010
Aanderaa, S.O. [1982] 33756
Aanderaa, S.O. [1982] 35210
Aanderaa, S.O. [1984] 41791
Bacsich, P.D. [1975] 17658
Behmann, H. [1922] 00942
Behmann, H. [1923] 00943
Behmann, H. [1950] 01559
Bernays, P. [1928] 01076
Bernhardt, K. [1976] 60487
Beth, E.W. [1950] 01156
Blass, A.R. [1984] 40306
Boerger, E. [1974] 01368
Boerger, E. [1984] 40047
Borkowski, L. [1958] 01482
Borkowski, L. [1960] 01483
Borkowski, L. [1961] 01484
Chapin Jr., E.W. [1971] 01987
Chikhachev, S.A. [1984] 42695
Denenberg, L. [1984] 43373
Doshita, S. [1983] 44209

Dreben, B. [1957] 03611
Dreben, B. [1962] 03127
Dreben, B. [1962] 03612
Dywan, Z. [1980] 54158
Ferro, A. [1978] 56666
Flum, J. [1974] 15020
Frayne, T.E. [1962] 04557
Fridman, Eh.I. [1972] 77209
Fuerer, M. [1981] 35169
Fujiwara, T. [1971] 17104
Fujiwara, T. [1973] 04714
Fujiwara, T. [1975] 21821
Galvin, F. [1967] 04765
Galvin, F. [1970] 04766
Geach, P.T. [1952] 16851
Gensler, H.J. [1973] 18156
Givant, S. [1978] 29155
Givant, S. [1979] 55860
Gladstone, M.D. [1979] 52622
Goedel, K. [1929] 22123
Goedel, K. [1933] 20885
Goguen, J.A. [1985] 36903
Goldfarb, W.D. [1981] 55275
Goldfarb, W.D. [1984] 38765
Goodstein, R.L. [1963] 05189
Goodstein, R.L. [1966] 05190
Gurevich, Y. [1965] 05452
Gurevich, Y. [1966] 05455
Gurevich, Y. [1966] 05456
Gurevich, Y. [1966] 05458
Gurevich, Y. [1973] 32346
Gurevich, Y. [1976] 14777
Gurevich, Y. [1983] 33766
Hailperin, T. [1961] 05508
Halkowska, K. [1975] 29801
Hamblin, C.L. [1973] 17398
Harrop, R. [1954] 05676
Henkin, L. [1977] 27268
Henson, C.W. [1984] 36539
Herrera Miranda, J. [1978] 36172
Herring, J.M. [1976] 21916
Johnson, J.S. [1973] 06681
Kalmar, L. [1928] 06835
Kalmar, L. [1932] 06837
Kalmar, L. [1933] 06838
Kanovich, M.I. [1975] 18218
Kapetanovic, M. [1983] 40445
Keisler, H.J. [1965] 07008
Ketonen, J. [1984] 44002
Kopieki, R. [1975] 23964
Kostyrko, V.F. [1971] 19122
Krejnovich, V.Ya. [1982] 35653
Krom, Melven R. [1964] 07570
Krom, Melven R. [1967] 07574
Krom, Melven R. [1968] 07575
Langford, C.H. [1926] 07849
Lewis, H.R. [1978] 75527
Lewis, H.R. [1979] 53515
Lewis, H.R. [1980] 36546
Liogon'kij, M.I. [1970] 08182
Loewenheim, L. [1908] 08241
Lyndon, R.C. [1959] 08439
Manders, K.L. [1979] 53838
Marcja, A. [1974] 21706
Markovic, Z. [1983] 37386
Maslov, S.Yu. [1966] 19334
Maslov, S.Yu. [1972] 08848

McKenzie, R. [1975] 09009
McKinsey, J.C.C. [1943] 09021
Meredith, D. [1978] 31932
Montali, T. [1971] 17210
Monteverdi, D. [1973] 21823
Motohashi, N. [1984] 42488
Mundici, D. [1983] 37383
Nadiu, G.S. [1969] 09749
Nelson, Evelyn [1977] 26606
Norgela, S.A. [1976] 32671
Norgela, S.A. [1980] 32676
Oberschelp, A. [1958] 10042
Pacholski, L. [1968] 10215
Paola di, R.A. [1973] 30684
Pepis, J. [1937] 19493
Pinus, A.G. [1979] 53695
Poizat, B. [1975] 17649
Poizat, B. [1982] 35642
Pouzet, M. [1979] 51038
Rautenberg, W. [1968] 11048
Renardel de Lavalette, G.R. [1981] 55207
Robitashvili, N.G. [1965] 11330
Rousseau, G. [1966] 11593
Rousseau, G. [1968] 11598
Rozenblat, B.V. [1979] 53949
Rozenblat, B.V. [1983] 34621
Sanmartin Esplugues, J. [1973] 78125
Schuette, K. [1933] 12997
Schuette, K. [1934] 12998
Shelah, S. [1977] 27202
Sikorski, R. [1961] 12306
Sikorski, R. [1962] 12308
Skolem, T.A. [1919] 12383
Skolem, T.A. [1933] 12391
Skolem, T.A. [1937] 27938
Smolin, V.P. [1975] 51192
Stihi, T. [1973] 79027
Tabata, H. [1969] 13247
Tabata, H. [1971] 13248
Tajmanov, A.D. [1967] 13274
Tauts, A. [1964] 13475
Tharp, L.H. [1973] 13527
Thomas, I. [1952] 13541
Tulipani, S. [1975] 50883
Vaught, R.L. [1954] 41195
Wang, Hao [1963] 14040
Wang, Hao [1965] 14057
Weaver, G.E. [1974] 14096
Wirsing, M. [1977] 51036
Wright von, G.H. [1948] 14303
Wright von, G.H. [1950] 14304
Wright von, G.H. [1952] 14309
Zhegalkin, I.I. [1939] 17903
Zhegalkin, I.I. [1946] 04825
Zubieta Russi, G. [1957] 11688

B22

Bloom, S.L. [1975] 21621
Czelakowski, J. [1979] 31659
Czelakowski, J. [1979] 31660
Czelakowski, J. [1980] 56472
Czelakowski, J. [1980] 56473
Czelakowski, J. [1982] 42743
Czelakowski, J. [1984] 41764
Czelakowski, J. [1985] 47502
Gabbay, D.M. [1974] 17555

Harrop, R. [1958] 05678
Harrop, R. [1965] 05684
Harrop, R. [1976] 32206
Jankowski, A.W. [1985] 47505
Kossowski, P. [1970] 07392
Los, J. [1955] 08332
Maksimova, L.L. [1977] 52433
Slupecki, J. [1972] 15183
Stepien, T. [1981] 33327
Surma, S.J. [1968] 12748
Surma, S.J. [1969] 12752
Surma, S.J. [1970] 12754
Surma, S.J. [1970] 79124
Tarski, A. [1935] 13409
Tarski, A. [1936] 13415
Ulrich, D. [1983] 42640
Urquhart, A.I.F. [1981] 54908
Zygmunt, J. [1981] 54333
Zygmunt, J. [1983] 41843

B28

Apelt, H. [1966] 00403
Baker, A. [1971] 42680
Bel'tyukov, A.P. [1976] 29789
Bernays, P. [1955] 01090
Boolos, G. [1979] 56312
Cegielski, P. [1984] 46773
Clote, P. [1983] 40361
Comer, S.D. [1985] 42290
Conway, J.H. [1976] 71909
Dejon, B. [1969] 23438
Goodstein, R.L. [1955] 05177
Grzegorczyk, A. [1971] 23119
Gurevich, R. [1984] 45002
Gurevich, R. [1985] 44633
Gutierrez-Novoa, L. [1979] 54116
Hauschild, K. [1968] 05770
Hauschild, K. [1971] 05779
Laugwitz, D. [1958] 48699
Lillo de, N.J. [1979] 56184
MacDowell, R. [1961] 19285
Maehara, S. [1957] 08516
McNaughton, R. [1954] 09068
McNaughton, R. [1965] 09073
Mostowski, Andrzej [1961] 09570
Novikov, P.S. [1943] 19479
Novikov, P.S. [1949] 19476
Pabion, J.F. [1981] 33728
Paris, J.B. [1978] 29247
Paris, J.B. [1985] 41796
Presburger, M. [1930] 10749
Pudlak, P. [1983] 36550
Rabin, M.O. [1961] 10934
Richard, D. [1984] 41779
Schmerl, J.H. [1982] 35360
Skolem, T.A. [1931] 12388
Skolem, T.A. [1933] 12392
Skolem, T.A. [1934] 12393
Skolem, T.A. [1935] 12395
Skolem, T.A. [1955] 27718
Stojakovic, M. [1975] 79037
Takahashi, S. [1974] 17488
Tarski, A. [1931] 16924
Tarski, A. [1933] 13407
Tarski, A. [1948] 13436
Vinner, S. [1976] 24084
Wolter, H. [1973] 14280

Zaiontz, C. [1983] 42214

B30

Ax, J. [1968] 00580
Blanche, R. [1962] 22464
Bouvere de, K.L. [1965] 04261
Engeler, E. [1970] 03368
Henkin, L. [1959] 22387
Herbrand, J. [1931] 16812
Huntington, E.V. [1917] 41774
Huntington, E.V. [1917] 41776
Ito, Makoto [1934] 32185
Josza, R. [1984] 45334
Kreisel, G. [1967] 07533
Langford, C.H. [1926] 07850
Langford, C.H. [1926] 07851
Lindenbaum, A. [1927] 30918
Lindenbaum, A. [1935] 30921
Makowiecka, H. [1965] 08587
Makowiecka, H. [1975] 08588
Makowiecka, H. [1975] 08589
Makowiecka, H. [1975] 08590
Makowiecka, H. [1975] 18272
Makowiecka, H. [1975] 21816
Makowiecka, H. [1976] 24642
Padoa, A. [1902] 14911
Rasiowa, H. [1955] 11010
Rautenberg, W. [1962] 11041
Rautenberg, W. [1966] 11045
Robinson, A. [1954] 11226
Robinson, A. [1973] 11276
Schwabhaeuser, W. [1956] 13051
Schwabhaeuser, W. [1959] 13052
Schwabhaeuser, W. [1983] 40225
Schwabhaeuser, W. [1983] 41875
Scott, D.S. [1958] 11906
Skolem, T.A. [1923] 21221
Skolem, T.A. [1929] 12386
Specker, E. [1958] 12669
Szczerba, L.W. [1965] 12831
Szczerba, L.W. [1979] 79177
Szczerba, L.W. [1983] 39510
Szmielew, W. [1962] 12850
Szmielew, W. [1974] 16578
Tarski, A. [1932] 13406
Tarski, A. [1933] 28816
Tarski, A. [1936] 13413
Tarski, A. [1938] 30925
Tarski, A. [1944] 13432
Tarski, A. [1948] 19756
Tarski, A. [1967] 13463
Wajsberg, M. [1933] 40808
Wang, Hao [1953] 14024
Wheeler, W.H. [1979] 53787
Whiteley, W. [1973] 79943
Whiteley, W. [1977] 79941
Whiteley, W. [1978] 50914
Zil'ber, B.I. [1981] 36112

B35

Ausiello, G. [1976] 51023
Cooper, D.C. [1972] 61080
Ferro, A. [1978] 56666
Friedman, Joyce [1963] 04667
Friedman, Joyce [1966] 29439
Galda, K. [1974] 61899

Gergely, T. [1978] 52253
Hajek, P. [1973] 22921
Hajek, P. [1974] 24128
Hajek, P. [1977] 31130
Hajek, P. [1978] 31131
Hartmanis, J. [1976] 24160
Havranek, T. [1978] 54868
Hennessy, M. [1980] 53954
Hintikka, K.J.J. [1973] 32405
Hodes, L. [1968] 24939
Hodes, L. [1970] 06168
Joyner Jr., W.M. [1973] 81740
Joyner Jr., W.M. [1976] 23231
Kallick, B. [1969] 22190
Kanger, S. [1970] 23148
Longo, G. [1974] 53697
Lovasz, L. [1980] 53755
Maslov, S.Yu. [1977] 32672
McRobbie, M.A. [1980] 54132
Nelson, Greg [1984] 45270
Norgela, S.A. [1978] 32674
Oppen, D.C. [1980] 55875
Orevkov, V.P. [1969] 10154
Schmidt, K. [1981] 35213
Shostak, R.E. [1977] 69916
Shostak, R.E. [1979] 82799
Shostak, R.E. [1982] 38462
Siekmann, J. [1982] 36659
Slagle, J.R. [1970] 12452
Trakhtenbrot, B.A. [1960] 42812
Wasilewska, A. [1978] 69996
Westrhenen van, S.C. [1968] 14159
Zhaldokas, R. [1983] 43443

B40

Amer, M.A. [1976] 17946
Amer, M.A. [1977] 30619
Hermes, H. [1955] 06008
Kanda, A. [1985] 47568
Meredith, D. [1978] 31932
Mitschke, G. [1980] 76450
Rimscha von, M. [1980] 53804
Scott, D.S. [1973] 78313
Scott, D.S. [1980] 78314
Scott, D.S. [1980] 78323

B45

Anderson, A.R. [1954] 00309
Artemov, S.N. [1980] 54546
Bellissima, F. [1983] 42314
Benthem van, J.F.A.K. [1979] 28157
Benthem van, J.F.A.K. [1983] 37277
Blok, W.J. [1983] 40651
Boolos, G. [1976] 14578
Boolos, G. [1977] 26442
Boolos, G. [1979] 56312
Bowen, K.A. [1975] 29636
Bowen, K.A. [1978] 30659
Bozic, M. [1984] 40033
Burgess, J.P. [1980] 55004
Burgess, J.P. [1985] 42591
Cavalli, A.R. [1984] 43192
Chang, C.C. [1973] 03978
Cresswell, M.J. [1985] 42657
Dahn, B.I. [1973] 04057
Dahn, B.I. [1974] 04058

Dahn, B.I. [1974] 22893
Davidson, B. [1977] 23697
Dziobiak, W. [1978] 53122
Emerson, E.A. [1984] 45349
Fine, K. [1975] 22897
Fine, K. [1978] 30694
Fine, K. [1981] 54607
Fitting, M. [1972] 31074
Fitting, M. [1984] 42443
Follesdal, D. [1966] 32328
Gabbay, D.M. [1971] 04723
Gabbay, D.M. [1971] 30000
Gabbay, D.M. [1972] 04733
Gabbay, D.M. [1973] 17911
Gabbay, D.M. [1975] 04738
Gabbay, D.M. [1975] 04739
Gaerdenfors, P. [1975] 21522
Gallin, D. [1975] 29739
Gao, Hengshan [1976] 61914
Garson, J.W. [1984] 41834
Gerla, G. [1983] 44834
Gerla, G. [1984] 39889
Goble, L.F. [1973] 05058
Goldblatt, R.I. [1973] 05104
Goldblatt, R.I. [1975] 05108
Goldblatt, R.I. [1981] 55422
Grzegorczyk, A. [1967] 05412
Gurevich, Y. [1985] 47372
Hajek, P. [1981] 54605
Hallden, S. [1949] 05586
Hallden, S. [1949] 05587
Hanson, W.H. [1966] 05648
Hauschild, K. [1982] 34362
Heijenoort van, J. [1979] 79641
Hodes, H.T. [1984] 41142
Ito, Makoto [1965] 27530
Kawai, H. [1982] 36043
Lehmann, D.J. [1983] 41143
Lewis, D. [1971] 08107
Lue, Qici [1983] 40907
Makowsky, J.A. [1977] 31849
Maksimova, L.L. [1979] 55851
Maksimova, L.L. [1982] 40416
Maksimova, L.L. [1982] 40502
Matsumoto, K. [1957] 16393
Matsumoto, K. [1959] 16395
Matsumoto, K. [1960] 08893
McKinsey, J.C.C. [1941] 09020
Meredith, D. [1956] 09147
Miura, S. [1972] 09336
Montagna, F. [1980] 54076
Montagna, F. [1983] 34635
Montagna, F. [1984] 42567
Montague, R. [1963] 14478
Montgomery, H. [1976] 18366
Morgan, C.G. [1982] 36507
Morgan, C.G. [1982] 37860
Mortimer, M. [1974] 09506
Nat vander, A. [1979] 56121
Nishimura, H. [1980] 54084
Nishimura, H. [1980] 56424
Nishimura, H. [1981] 54611
Nishimura, H. [1981] 55147
Ohnishi, M. [1961] 10085
Paulos, J.A. [1976] 18327
Poliferno, M.J. [1961] 10611
Pollock, J.L. [1966] 10622

Radev, S.R. [1984] 44072
Rantala, V. [1982] 37647
Rasiowa, H. [1953] 11004
Rasiowa, H. [1955] 11011
Rasiowa, H. [1955] 11014
Rautenberg, W. [1979] 77698
Rautenberg, W. [1983] 39823
Rennie, M.K. [1968] 11098
Rescher, N. [1969] 14969
Rescher, N. [1969] 25168
Routley, R. [1968] 11609
Routley, R. [1968] 11610
Routley, R. [1970] 11613
Ruzsa, I. [1971] 19614
Rybakov, V.V. [1977] 56122
Rybakov, V.V. [1978] 53114
Rybakov, V.V. [1978] 53244
Rybakov, V.V. [1984] 39654
Sato, M. [1977] 30800
Sawamura, H. [1985] 47625
Schuette, K. [1968] 24822
Schultz, Konrad [1970] 13035
Schumm, G.F. [1978] 29254
Schumm, G.F. [1978] 52126
Segerberg, K. [1968] 11935
Segerberg, K. [1968] 41673
Segerberg, K. [1970] 11938
Segerberg, K. [1972] 11939
Segerberg, K. [1976] 65185
Shekhtman, V.B. [1978] 52052
Shekhtman, V.B. [1980] 54274
Shrejder, Yu.A. [1977] 79715
Shukla, A. [1970] 19678
Smorynski, C.A. [1978] 54148
Smorynski, C.A. [1982] 43687
Sonobe, O. [1975] 51586
Sundholm, G. [1977] 50941
Swart de, H.C.M. [1980] 54220
Thomason, R.H. [1967] 13579
Thomason, R.H. [1975] 51538
Thomason, S.K. [1975] 29635
Ursini, A. [1979] 52568
Ursini, A. [1985] 47525
Vincenzi, A. [1985] 49797
Visser, A. [1984] 42643
Wasserman, H.C. [1976] 18423
Wright von, G.H. [1951] 14305
Wright von, G.H. [1951] 14313
Zeman, J.J. [1969] 14399

B46

Brady, R.T. [1984] 45522
Deutsch, H. [1985] 47508
Gabbay, D.M. [1976] 32148
Giambrone, S. [1985] 42658
Kron, A. [1978] 53971
Routley, R. [1974] 15192
Vojshvillo, E.K. [1983] 48317

B48

Bigelow, J.C. [1976] 60510
Chuaqui, R.B. [1980] 55151
Chudnovsky, D.V. [1968] 02732
Chudnovsky, D.V. [1970] 26322
Fajardo, S. [1985] 39901
Fajardo, S. [1985] 41798

Ferenczi, M. [1977] 55783
Gaifman, H. [1960] 16922
Gaifman, H. [1964] 33352
Hailperin, T. [1984] 41842
Halpern, M. [1983] 45211
Hoover, D.N. [1978] 29154
Kramosil, I. [1978] 54777
Krauss, P.H. [1966] 27709
Leblanc, H. [1983] 37685
Suszko, R. [1967] 26275

B50

Adamson, A. [1979] 53245
Anshakov, O.M. [1984] 44158
Anshakov, O.M. [1984] 47425
Bryll, G. [1973] 24728
Cantini, A. [1980] 54349
Chang, C.C. [1962] 01946
Chang, C.C. [1966] 02410
D'Ottaviano, I.M.L. [1985] 41799
Dahn, B.I. [1974] 02746
Denecke, K. [1982] 37824
Girard, J.-Y. [1976] 50375
Hajek, P. [1977] 31126
Havranek, T. [1975] 31144
Hickin, K.K. [1975] 28362
Keisler, H.J. [1977] 16614
Koppelberg, S. [1985] 39673
Krivine, J.-L. [1974] 07566
Krzystek, P.S. [1977] 52053
Malinowski, G. [1981] 54348
Mangani, P. [1976] 51187
Miyama, T. [1974] 29672
Mostowski, Andrzej [1962] 09574
Mundici, D. [1984] 44636
Nguyen Cat Ho [1972] 09939
Nguyen Cat Ho [1973] 29577
Nieland, J.J.F. [1977] 09951
Przymusinska, H. [1980] 54914
Przymusinska, H. [1980] 54915
Rasiowa, H. [1970] 11031
Rasiowa, H. [1972] 11032
Rasiowa, H. [1985] 48433
Rautenberg, W. [1979] 77698
Reichbach, J. [1974] 31999
Rockingham Gill, R.R. [1970] 22231
Rosenberg, I.G. [1973] 29025
Ruzsa, I. [1969] 19613
Ruzsa, I. [1971] 11693
Saeli, D. [1975] 78070
Sanmartin Esplugues, J. [1973] 78125
Vojvodic, G. [1978] 52815
Vojvodic, G. [1980] 36055
Vuckovic, V. [1967] 27478
Wasilewska, A. [1979] 56380
Waszkiewicz, J. [1969] 14081
Waszkiewicz, J. [1971] 14084
Weispfenning, V. [1985] 47574
Zlatos, P. [1981] 55634
Zygmunt, J. [1974] 14462

B51

Bell, J.L. [1983] 43055
Bugajski, S. [1983] 42320
Goldblatt, R.I. [1984] 40768
Takeuti, G. [1984] 39581

B52

Carlstrom, I.F. [1975] 60828
Nakamura, A. [1980] 54126

B53

Alves, E.H. [1978] 52943
Alves, E.H. [1984] 39893
Costa da, N.C.A. [1974] 22926
Fidel, M.M. [1977] 24188
Grant, J. [1974] 05305
Grant, J. [1978] 28232
Marconi, D. [1980] 56614

B55

Anderson, J.G. [1969] 00344
Bernhardt, K. [1977] 53552
Boricic, B.R. [1983] 44659
Bozic, M. [1985] 48301
Bozzi, S. [1980] 54221
Dahn, B.I. [1975] 61216
Drugush, Ya.M. [1984] 45347
Fidel, M.M. [1980] 54154
Gabbay, D.M. [1970] 04721
Gabbay, D.M. [1971] 04726
Gabbay, D.M. [1972] 04730
Gabbay, D.M. [1972] 04732
Gabbay, D.M. [1973] 17911
Gabbay, D.M. [1974] 17555
Gabbay, D.M. [1977] 26456
Gabbay, D.M. [1982] 38464
Goernemann, S. [1971] 39863
Goldblatt, R.I. [1974] 05106
Goranko, V.F. [1985] 47504
Goranko, V.F. [1985] 49398
Johnstone, P.T. [1979] 56505
Jongh de, D.H.J. [1980] 55677
Kalicki, C. [1980] 53776
Kirk, R.E. [1979] 53554
Kirk, R.E. [1980] 56546
Klemke, D. [1971] 07229
Komori, Y. [1978] 52844
Kripke, S.A. [1965] 07557
Kuznetsov, A.V. [1975] 63246
Lopez-Escobar, E.G.K. [1981] 55143
Lopez-Escobar, E.G.K. [1983] 39786
Maksimova, L.L. [1977] 52433
Maksimova, L.L. [1977] 54760
Maksimova, L.L. [1982] 37067
McKay, C.G. [1968] 08986
McKay, C.G. [1971] 08988
Minari, P. [1983] 42336
Muravitskij, A.Yu. [1981] 36087
Muravitskij, A.Yu. [1983] 40399
Muravitskij, A.Yu. [1984] 45360
Ono, H. [1983] 34652
Porebska, M. [1985] 48967
Rauszer, C. [1977] 50668
Rauszer, C. [1977] 50669
Rauszer, C. [1980] 56381
Rauszer, C. [1985] 41793
Renardel de Lavalette, G.R. [1981] 55207
Routley, R. [1978] 53083
Ruzsa, I. [1969] 11692
Selman, A.L. [1972] 11964
Shanin, N.A. [1977] 53812

Shekhtman, V.B. [1978] 52941
Shekhtman, V.B. [1979] 78450
Skvortsov, D.P. [1979] 55940
Smorynski, C.A. [1973] 27058
Sobolev, S.K. [1977] 51014
Szatkowski, M. [1984] 44572
Tauts, A. [1975] 52459
Yankov, V.A. [1967] 06513
Yashin, A.L. [1977] 36830
Zachorowski, S. [1977] 54806
Zachorowski, S. [1978] 56212
Zakhar'yashchev, M.V. [1983] 40499

B60

Barwise, J. [1981] 55362
Bloom, S.L. [1982] 36525
Canty, J.T. [1970] 27980
Costa da, N.C.A. [1976] 14829
Ebbinghaus, H.-D. [1969] 03219
Fraassen van, B.C. [1968] 04480
Fraisse, R. [1982] 43874
Goad, C.A. [1978] 73424
Grandy, R.E. [1976] 18176
Gumb, R.D. [1979] 62230
Hajek, P. [1976] 31127
Hajek, P. [1977] 26853
Hintikka, K.J.J. [1982] 55406
Hoogewijs, A. [1979] 53121
Hoogewijs, A. [1983] 37062
Iwanus, B. [1973] 17553
Kanda, A. [1985] 47568
Krabbe, E.C.W. [1978] 51481
Malinowski, G. [1982] 42305
Morais, R. [1976] 54742
Morais, R. [1977] 16605
Nepejvoda, N.N. [1982] 46228
Peterson, P.L. [1979] 52390
Rao, A.P. [1977] 56420
Rogava, M.G. [1979] 77850
Suszko, R. [1971] 12777
Tauts, A. [1974] 17694
Thompson, B. [1982] 55079
Tiuryn, J. [1981] 34010

B75

Andreka, H. [1982] 55505
Bertoni, A. [1979] 52925
Bertoni, A. [1980] 80710
Bertoni, A. [1981] 36666
Bertoni, A. [1983] 38487
Blair, H.A. [1984] 43300
Blass, A.R. [1984] 47716
Bowen, K.A. [1980] 54337
Chlebus, B.S. [1979] 54219
Chlebus, B.S. [1982] 34361
Danko, W. [1983] 39522
Engeler, E. [1967] 03366
Engeler, E. [1975] 17633
Engeler, E. [1975] 21738
Ershov, Yu.L. [1983] 40496
Eytan, M. [1979] 53157
Golota, Ya.Ya. [1969] 05124
Gurevich, Y. [1984] 43005
Hajek, P. [1981] 35692
Harel, D. [1983] 38572
Harel, D. [1985] 41208

Kozen, D. [1981] 36840
Leone, M. [1985] 49912
Makowsky, J.A. [1980] 54945
Mundici, D. [1984] 44693
Nakamura, A. [1980] 54126
Paola di, R.A. [1967] 30681
Pratt, V.R. [1979] 54858
Rasiowa, H. [1982] 45056
Reiterman, J. [1980] 54019
Sain, I. [1979] 56752
Segerberg, K. [1982] 41144
Tajtslin, M.A. [1979] 82932
Valiev, M.K. [1980] 54018
Wasilewska, A. [1985] 49467
Woehl, K. [1979] 53367
Zhaldokas, R. [1983] 43443

B80

Becker, J.A. [1976] 29684
Bucur, I. [1980] 46245
Kassler, M. [1976] 32363
Scarpellini, B. [1963] 11875

B96

Mostowski, Andrzej [1979] 82308
Novikov, P.S. [1979] 38498
Robinson, A. [1979] 53900
Skolem, T.A. [1970] 29109
Tarski, A. [1956] 28568
Tarski, A. [1972] 30338
Wajsberg, M. [1933] 40808

B97

Addison, J.W. [1974] 70206
Arruda, A.I. [1977] 16589
Bernardi, C. [1983] 34945
Crossley, J.N. [1967] 31684
Crossley, J.N. [1975] 31725
Delon, F. [1984] 43258
Delon, F. [1985] 47516
Gandy, R.O. [1977] 16612
Harrington, L.A. [1985] 49810
Henkin, L. [1959] 22387
Hodges, W. [1972] 70226
Lerman, M. [1981] 54311
Loeb, M.H. [1968] 37293
Lolli, G. [1984] 41493
Macintyre, A. [1978] 53596
Metakides, G. [1982] 35609
Meulen ter, A.G.B. [1983] 47464
Mueller, Gert H. [1984] 41750
Prisco di, C.A. [1985] 41792
Sobolev, S.L. [1982] 34492
Wechsung, G. [1984] 40976

B98

Altham, J.E.J. [1971] 00285
Barwise, J. [1975] 60316
Barwise, J. [1977] 70117
Baudisch, A. [1980] 56368
Boolos, G. [1979] 56312
Dalen van, D. [1980] 55705
Ebbinghaus, H.-D. [1978] 28201
Enderton, H.B. [1972] 03355
Ershov, Yu.L. [1973] 32036

Fitting, M. [1969] 04349
Fraisse, R. [1967] 04529
Fraisse, R. [1972] 04535
Friedman, H.M. [1975] 04296
Gabbay, D.M. [1983] 41457
Gabbay, D.M. [1984] 41831
Goldblatt, R.I. [1979] 55754
Halkowska, K. [1976] 52421
Ackermann, W. [1928] 00107
Kreisel, G. [1954] 16973
Kreisel, G. [1967] 22411
Kreisel, G. [1975] 75092
Langford, C.H. [1932] 22592
Lightstone, A.H. [1978] 51926
Lorenzen, P. [1962] 08303
Malitz, J. [1979] 56167
Manaster, A.B. [1975] 23120
Manin, Yu.I. [1977] 51984
Marek, W. [1972] 34216
Mostowski, Andrzej [1965] 09578
Rasiowa, H. [1963] 11028
Rautenberg, W. [1979] 77698
Rescher, N. [1969] 25168
Robinson, A. [1958] 25442
Robinson, A. [1959] 19551
Robinson, A. [1963] 28804
Robinson, A. [1966] 11263
Rogers, R. [1971] 11365
Shoenfield, J.R. [1967] 22384
Skolem, T.A. [1928] 12385
Smirnov, V.A. [1972] 30359
Stillwell, J.C. [1977] 52422
Tajtslin, M.A. [1970] 28675
Takeuti, G. [1978] 52447
Wang, Hao [1981] 34116

C96

Mostowski, Andrzej [1979] 82308
Novikov, P.S. [1979] 38498
Robinson, A. [1979] 53900
Skolem, T.A. [1970] 29109
Tarski, A. [1956] 28568
Tarski, A. [1972] 30338

D03

Boerger, E. [1979] 53847
Elgot, C.C. [1961] 03288
Wang, Hao [1965] 14057

D05

Baranskij, V.A. [1970] 00771
Boerger, E. [1979] 53847
Boerger, E. [1981] 55035
Buechi, J.R. [1960] 01690
Buechi, J.R. [1962] 01693
Buechi, J.R. [1965] 01694
Buechi, J.R. [1965] 01697
Buechi, J.R. [1969] 01700
Buechi, J.R. [1983] 33901
Buechi, J.R. [1983] 40515
Buszkowski, W. [1982] 35719
Church, A. [1963] 02403
Collins, D.J. [1970] 02327
Compton, K.J. [1984] 47457
Doner, J.E. [1970] 03082
Dosen, K. [1985] 47571

Elgot, C.C. [1961] 03288
Elgot, C.C. [1966] 03291
Friedrich, U. [1972] 04668
Gecseg, F. [1984] 41334
Gurari, E.M. [1981] 54218
Gurevich, Y. [1982] 33759
Gurevich, Y. [1985] 47372
Harel, D. [1985] 41208
Haussler, D. [1981] 55629
Hodgson, B.R. [1982] 35446
Hodgson, B.R. [1983] 35721
Kanovich, M.I. [1975] 18218
Ladner, R.E. [1977] 52252
Ladner, R.E. [1980] 75296
Lisovik, L.P. [1984] 42703
Lisovik, L.P. [1984] 48409
Litman, A. [1976] 18261
Mostowski, A.Wlodzimierz [1980] 34829
Mostowski, A.Wlodzimierz [1981] 55152
Muchnik, A.A. [1985] 47256
Muller, D.E. [1981] 76648
Muller, D.E. [1985] 48421
Mundici, D. [1980] 33528
Nabebin, A.A. [1977] 52155
Nakamura, A. [1981] 55427
Rabin, M.O. [1968] 29625
Rabin, M.O. [1969] 10942
Rabin, M.O. [1970] 22233
Rabin, M.O. [1971] 14690
Rabin, M.O. [1972] 10943
Rakhmatulin, N.A. [1975] 77649
Seese, D.G. [1978] 52180
Siefkes, D. [1970] 12145
Siefkes, D. [1970] 27984
Siefkes, D. [1978] 29136
Thatcher, J.W. [1968] 28596
Thomas, Wolfgang [1979] 52885
Thomas, Wolfgang [1981] 55635
Thomas, Wolfgang [1984] 39756
Tinhofer, G. [1978] 52324
Turan, G. [1981] 55701
Vazhenin, Yu.M. [1981] 55132
Vidal-Naquet, G. [1973] 27517
Wang, Hao [1963] 14040
Wang, Hao [1965] 14057
Wasilewska, A. [1980] 55216
Wasilewska, A. [1985] 49467

D10

Belyakin, N.V. [1970] 29065
Benejam, J.-P. [1977] 26472
Berman, L. [1980] 55471
Boerger, E. [1976] 52927
Boerger, E. [1979] 53847
Chlebus, B.S. [1979] 54219
Christen, C. [1976] 52348
Ebbinghaus, H.-D. [1980] 41134
Engeler, E. [1967] 03366
Ferrante, J. [1979] 54815
Grandjean, E. [1985] 42738
Gurari, E.M. [1979] 56232
Hay, L. [1975] 05822
Heintz, Joos [1979] 55995
Heintz, Joos [1983] 38863
Hermes, H. [1955] 06008
Jones, N.D. [1972] 51894
Jones, N.D. [1974] 06700

Klupp, H. [1977] 50134
Ladner, R.E. [1980] 75296
Lewis, H.R. [1977] 53054
Lewis, H.R. [1980] 36546
Meyer, A.R. [1975] 21811
Muller, D.E. [1985] 47646
Mundici, D. [1983] 34976
Rakhmatulin, N.A. [1975] 77649
Roedding, D. [1966] 16282
Roedding, D. [1972] 11343
Scarpellini, B. [1985] 47473
Stillwell, J.C. [1977] 52422
Turing, A.M. [1954] 42433
Volger, H. [1983] 39717
Wang, Hao [1965] 14057

D15

Adamson, A. [1979] 53245
Ajtai, M. [1983] 37538
Ausiello, G. [1976] 51023
Berman, L. [1977] 70972
Berman, L. [1980] 55471
Berman, P. [1979] 53427
Blass, A.R. [1984] 47720
Boerger, E. [1979] 54143
Boerger, E. [1980] 80766
Boerger, E. [1981] 55035
Boerger, E. [1983] 40032
Boerger, E. [1984] 40054
Bruss, A.R. [1978] 80801
Bruss, A.R. [1980] 55036
Chen, Jiyuan [1984] 44191
Chistov, A.L. [1984] 44792
Chlebus, B.S. [1979] 54219
Christen, C. [1976] 52348
Dahlhaus, E. [1984] 41789
Dahn, B.I. [1984] 45383
Denenberg, L. [1984] 43003
Denenberg, L. [1984] 43373
DeMillo, R.A. [1979] 72280
Dimitracopoulos, C. [1985] 47569
Doshita, S. [1983] 44209
Ebbinghaus, H.-D. [1980] 41134
Ehrenfeucht, A. [1975] 18126
Emerson, E.A. [1984] 45349
Engeler, E. [1975] 21738
Engeler, E. [1981] 81166
Fagin, R. [1974] 18140
Ferrante, J. [1975] 21702
Ferrante, J. [1977] 51432
Ferrante, J. [1979] 54815
Fischer, Michael J. [1974] 17344
Friedman, H.M. [1984] 46357
Fuerer, M. [1981] 35169
Fuerer, M. [1982] 35102
Gacs, P. [1977] 52753
Goralcik, P. [1980] 73522
Grandjean, E. [1983] 40384
Grandjean, E. [1984] 41788
Grandjean, E. [1984] 45011
Grandjean, E. [1984] 48072
Grandjean, E. [1985] 42738
Gurari, E.M. [1978] 42545
Gurari, E.M. [1979] 56232
Gurari, E.M. [1981] 54218
Gurevich, Y. [1984] 43005
Harel, D. [1983] 38572

Harel, D. [1985] 41208
Hartmanis, J. [1976] 24160
Heintz, Joos [1979] 55995
Heintz, Joos [1983] 38863
Hinman, P.G. [1985] 45303
Immerman, N. [1981] 38079
Jones, N.D. [1972] 51894
Jones, N.D. [1974] 06700
Klupp, H. [1977] 50134
Kozen, D. [1980] 53795
Kozen, D. [1981] 35103
Ladner, R.E. [1980] 75296
Lewis, H.R. [1977] 53054
Lewis, H.R. [1978] 75527
Lewis, H.R. [1980] 36546
Lipshitz, L. [1981] 75573
Lipton, R.J. [1978] 75578
Lolli, G. [1983] 38263
Longo, G. [1974] 53697
Lovasz, L. [1980] 53755
Loveland, D.W. [1978] 31164
Lynch, J.F. [1982] 34379
Meyer, A.R. [1975] 21811
Meyer, A.R. [1975] 27549
Michel, P. [1981] 35234
Mostowski, A.Wlodzimierz [1979] 76601
Mostowski, A.Wlodzimierz [1981] 55152
Mundici, D. [1981] 33529
Mundici, D. [1981] 33726
Mundici, D. [1982] 34339
Mundici, D. [1983] 34976
Mundici, D. [1983] 37383
Mundici, D. [1984] 41841
Mundici, D. [1984] 44069
Mundici, D. [1984] 44693
Oppen, D.C. [1973] 24077
Oppen, D.C. [1978] 51889
Oppen, D.C. [1980] 55875
Oppen, D.C. [1980] 82395
Parikh, R. [1971] 10282
Pratt, V.R. [1979] 54858
Rackoff, C.W. [1974] 24125
Rackoff, C.W. [1976] 14579
Remeslennikov, V.N. [1983] 39954
Rennie, M.K. [1968] 11098
Roedding, D. [1972] 11343
Scarpellini, B. [1984] 43196
Schlipf, J.S. [1977] 26851
Schoenhage, A. [1982] 35139
Szwast, W. [1985] 48831
Tetruashvili, M.R. [1984] 45394
Valiev, M.K. [1980] 54018
Vazhenin, Yu.M. [1981] 55132
Volger, H. [1983] 37672
Volger, H. [1983] 39717
Weispfenning, V. [1985] 48461
Wilkie, A.J. [1980] 79968
Woehl, K. [1979] 53367
Wuethrich, H.R. [1976] 43103

D20

Apt, K.R. [1977] 60167
Bertoni, A. [1983] 38487
Burris, S. [1984] 42428
Cellucci, C. [1964] 02198
Clote, P. [1985] 45298
Craig, W. [1953] 02507

Hanf, W.P. [1965] 05635
Hart, J. [1969] 05689
Hartmanis, J. [1976] 24160
Hay, L. [1977] 30705
Hermes, H. [1955] 06008
Hirschfeld, J. [1974] 21188
Hirschfeld, J. [1975] 06142
Ito, Makoto [1965] 27530
Jukna, S. [1979] 33258
Kolmogorov, A.N. [1953] 43578
Kreisel, G. [1957] 29369
Kreisel, G. [1961] 07523
Loeb, M.H. [1970] 22235
Lorenzen, P. [1962] 08303
Mashuryan, A.S. [1974] 17343
Mashuryan, A.S. [1979] 53934
Mashuryan, A.S. [1980] 76106
McAloon, K. [1980] 54524
Mehlhorn, K. [1974] 63853
Meyer, A.R. [1975] 21811
Monk, L.G. [1975] 90301
Mostowski, Andrzej [1956] 09561
Paris, J.B. [1981] 79972
Peretyat'kin, M.G. [1971] 10368
Putnam, H. [1965] 10840
Rabin, M.O. [1977] 27321
Robinson, Julia [1973] 29547
Roedding, D. [1966] 16282
Sabbagh, G. [1971] 28610
Trakhtenbrot, B.A. [1960] 42812
Volger, H. [1983] 37672

D25

Adler, A. [1969] 00184
Alton, D.A. [1975] 17945
Baldwin, John T. [1982] 33739
Baldwin, John T. [1984] 33746
Boerger, E. [1976] 50358
Boerger, E. [1979] 53847
Bullock, A.M. [1973] 01748
Carstens, H.G. [1977] 51961
Christen, C. [1976] 52348
Cresswell, M.J. [1985] 42657
Degtev, A.N. [1978] 72221
Denef, J. [1978] 52455
Deutsch, M. [1975] 02972
Downey, R.G. [1983] 38174
Downey, R.G. [1984] 39864
Downey, R.G. [1984] 39869
Downey, R.G. [1984] 43527
Downey, R.G. [1985] 39783
Ershov, Yu.L. [1972] 18137
Gurari, E.M. [1979] 56232
Gurevich, Y. [1985] 47372
Hanf, W.P. [1965] 05635
Hay, L. [1975] 05822
Hay, L. [1977] 30705
Herrmann, E. [1978] 53131
Higman, G. [1961] 22083
Jockusch Jr., C.G. [1984] 39945
Kalantari, I. [1978] 29262
Kalantari, I. [1982] 34171
Kalantari, I. [1982] 35452
Kalantari, I. [1983] 33595
Kalantari, I. [1983] 36130
Lachlan, A.H. [1968] 07751
Lachlan, A.H. [1968] 07752

Lavrov, I.A. [1963] 07885
Lerman, M. [1980] 56014
Lerman, M. [1984] 33603
Macintyre, A. [1975] 08474
McLaughlin, T.G. [1973] 29894
Metakides, G. [1975] 17296
Nerode, A. [1982] 38172
Pour-El, M.B. [1968] 10704
Pour-El, M.B. [1969] 32553
Pour-El, M.B. [1969] 32554
Pour-El, M.B. [1970] 29012
Putnam, H. [1964] 10838
Rabin, M.O. [1962] 20907
Remmel, J.B. [1977] 53858
Remmel, J.B. [1978] 29157
Remmel, J.B. [1978] 29270
Retzlaff, A.T. [1978] 29259
Roedding, D. [1966] 16282
Schupp, P.E. [1970] 13048
Shi, Niandong [1982] 37603
Shoenfield, J.R. [1965] 12099
Shoenfield, J.R. [1975] 18383
Tajtslin, M.A. [1962] 19772
Vaeaenaenen, J. [1979] 52749
Watnick, R. [1984] 42511

D30

Abraham, U. [1984] 40744
Apt, K.R. [1972] 00429
Bienenstock, E. [1977] 50578
Boerger, E. [1979] 53847
Clote, P. [1981] 55424
Clote, P. [1985] 45298
Denisov, S.D. [1974] 25962
Feiner, L. [1973] 03714
Gandy, R.O. [1960] 04774
Grilliot, T.J. [1972] 05347
Hay, L. [1975] 05822
Hay, L. [1977] 30705
Kanovej, V.G. [1974] 17445
Kierstead, H.A. [1985] 44599
Knight, J.F. [1982] 36173
Knight, J.F. [1982] 37529
Knight, J.F. [1983] 40363
Knight, J.F. [1983] 42733
Knight, J.F. [1984] 33597
Lerman, M. [1981] 55043
Macintyre, A. [1972] 08467
Macintyre, A. [1984] 40310
Marker, D. [1982] 37802
Millar, T.S. [1983] 35754
Millar, T.S. [1985] 46374
Mostowski, Andrzej [1972] 09588
Nerode, A. [1980] 76822
Odifreddi, P. [1983] 34091
Omarov, A.I. [1984] 46564
Poizat, B. [1981] 55288
Remeslennikov, V.N. [1983] 39954
Richter, L.J. [1981] 33281
Sacerdote, G.S. [1972] 30879
Schlipf, J.S. [1977] 26851
Schmerl, J.H. [1978] 29211
Selman, A.L. [1973] 11966
Shoenfield, J.R. [1960] 12098
Shoenfield, J.R. [1965] 12099
Shore, R.A. [1981] 55426
Steel, J.R. [1975] 13142

Steel, J.R. [1978] 29156
Stillwell, J.C. [1972] 13190
Tharp, L.H. [1974] 13528
Ziegler, M. [1980] 54050

D35

Aanderaa, S.O. [1973] 00010
Aanderaa, S.O. [1982] 33756
Aanderaa, S.O. [1982] 35210
Ackermann, W. [1928] 00109
Ackermann, W. [1936] 00112
Adler, A. [1969] 00184
Almagambetov, Zh.A. [1965] 00278
Andrews, P.B. [1974] 03832
Ash, C.J. [1980] 28370
Ash, C.J. [1984] 39867
Baudisch, A. [1974] 00867
Baudisch, A. [1975] 00868
Baur, W. [1975] 03874
Becker, J.A. [1980] 47356
Becker, J.A. [1980] 54459
Behmann, H. [1922] 00942
Behmann, H. [1923] 00943
Behmann, H. [1927] 00946
Bernays, P. [1928] 01076
Beth, E.W. [1950] 01156
Blair, H.A. [1984] 43300
Boerger, E. [1974] 01368
Boerger, E. [1979] 54143
Boerger, E. [1980] 80766
Boerger, E. [1981] 55035
Boerger, E. [1984] 40047
Bondi, I.L. [1967] 01424
Bondi, I.L. [1971] 02009
Bondi, I.L. [1973] 01425
Britton, J.L. [1979] 53330
Burris, S. [1975] 21812
Burris, S. [1982] 35402
Burris, S. [1983] 36689
Burris, S. [1985] 48302
Buszkowski, W. [1979] 56249
Cantor, David G. [1984] 46787
Carstens, H.G. [1977] 51961
Cherlin, G.L. [1980] 54217
Cherlin, G.L. [1981] 35227
Cherlin, G.L. [1981] 54982
Cherlin, G.L. [1984] 41760
Cherlin, G.L. [1984] 43903
Chlebus, B.S. [1980] 56482
Christen, C. [1976] 52348
Comer, S.D. [1981] 55444
Craig, W. [1952] 28058
Cresswell, M.J. [1985] 42657
Dahn, B.I. [1984] 42268
Delon, F. [1978] 27313
Delon, F. [1981] 35228
Denef, J. [1975] 04079
Denef, J. [1978] 52455
Denef, J. [1978] 52853
Denef, J. [1978] 52854
Denef, J. [1979] 54394
Denef, J. [1984] 38480
Deutsch, M. [1975] 02972
Dreben, B. [1957] 03611
Dreben, B. [1962] 03127
Dulac, M.-H. [1971] 03165
Ehrenfeucht, A. [1957] 03235

Elgot, C.C. [1961] 03288
Elgot, C.C. [1966] 03291
Ershov, Yu.L. [1964] 03520
Ershov, Yu.L. [1965] 03525
Ershov, Yu.L. [1968] 03532
Ershov, Yu.L. [1972] 03551
Ershov, Yu.L. [1980] 55024
Ershov, Yu.L. [1981] 34387
Ershov, Yu.L. [1984] 48388
Felgner, U. [1980] 55882
Fridman, Eh.I. [1972] 73069
Fridman, Eh.I. [1972] 77209
Fridman, Eh.I. [1973] 22999
Fridman, Eh.I. [1975] 52191
Fried, M. [1976] 51684
Friedman, H.M. [1976] 22920
Gabbay, D.M. [1973] 61880
Gal, L.N. [1958] 04757
Galvin, F. [1967] 04765
Galvin, F. [1970] 04766
Gao, Hengshan [1976] 61914
Garfunkel, S. [1971] 04793
Garfunkel, S. [1972] 04795
Germano, G. [1971] 04907
Goedel, K. [1933] 20885
Goldfarb, W.D. [1981] 55275
Goncharov, S.S. [1976] 52189
Grzegorczyk, A. [1957] 24864
Gurevich, Y. [1965] 05452
Gurevich, Y. [1966] 05455
Gurevich, Y. [1966] 05456
Gurevich, Y. [1966] 05457
Gurevich, Y. [1966] 05458
Gurevich, Y. [1967] 33753
Gurevich, Y. [1969] 05461
Gurevich, Y. [1969] 31104
Gurevich, Y. [1976] 14777
Gurevich, Y. [1979] 54603
Gurevich, Y. [1979] 54604
Gurevich, Y. [1982] 33758
Gurevich, Y. [1983] 33764
Gurevich, Y. [1983] 33765
Gurevich, Y. [1983] 33766
Gurevich, Y. [1985] 48332
Haken, W. [1973] 29825
Hanf, W.P. [1965] 05635
Haran, D. [1984] 44141
Harrop, R. [1965] 05684
Hauschild, K. [1971] 05780
Hauschild, K. [1971] 27430
Hauschild, K. [1972] 05787
Hauschild, K. [1972] 05790
Hauschild, K. [1981] 73915
Hauschild, K. [1982] 34362
Heindorf, L. [1984] 39661
Henkin, L. [1959] 22387
Henson, C.W. [1972] 05954
Henson, C.W. [1977] 30718
Hermes, H. [1955] 06008
Herre, H. [1975] 14283
Herre, H. [1978] 52676
Herre, H. [1981] 34285
Ackermann, W. [1928] 00107
Hodes, H.T. [1984] 41142
Hodges, W. [1984] 39733
Hodgson, B.R. [1983] 35721
Huber-Dyson, V. [1964] 03214

Huber-Dyson, V. [1969] 03215
Huber-Dyson, V. [1982] 37444
Ivanov, A.A. [1983] 37599
Jambu-Giraudet, M. [1980] 54167
Jambu-Giraudet, M. [1981] 55535
Janiczak, A. [1953] 06508
Jensen, C.U. [1982] 37407
Jensen, C.U. [1984] 39754
Jonsson, B. [1975] 06740
Kalmar, L. [1930] 06836
Kalmar, L. [1932] 06837
Kenzhebaev, S. [1966] 16220
Kesel'man, D.Ya. [1974] 17588
Kesel'man, D.Ya. [1982] 39437
Kharlampovich, O.G. [1983] 39367
Kireevskij, N.N. [1934] 07112
Kogalovskij, S.R. [1966] 63026
Kokorin, A.I. [1973] 22922
Kokorin, A.I. [1978] 55718
Kopieki, R. [1975] 23964
Kostyrko, V.F. [1971] 19122
Kozlov, G.T. [1970] 07424
Krom, Melven R. [1967] 07574
Krynicki, M. [1977] 33547
Kudajbergenov, K.Zh. [1980] 54224
Kuznetsov, A.V. [1959] 28393
Lavrov, I.A. [1962] 07884
Lavrov, I.A. [1963] 07885
Lewis, H.R. [1979] 53515
Lifschitz, V. [1967] 33665
Lindner, C.C. [1971] 08168
Lipshitz, L. [1978] 31822
Lorenzen, P. [1962] 08303
Luo, Libo [1983] 44346
Macintyre, A. [1971] 08464
Maddux, R. [1980] 55771
Makanin, G.S. [1985] 48835
Mal'tsev, A.I. [1960] 08609
Mal'tsev, A.I. [1960] 19257
Mal'tsev, A.I. [1961] 19251
Manaster, A.B. [1975] 23120
Manaster, A.B. [1980] 55073
Manaster, A.B. [1981] 33275
Manaster, A.B. [1981] 75934
Mart'yanov, V.I. [1977] 29230
McKenzie, R. [1975] 09009
Mihailescu, E.G. [1967] 09234
Millar, T.S. [1985] 46374
Nakamura, A. [1981] 55427
Nerode, A [1980] 76822
Norgela, S.A. [1976] 32671
Norgela, S.A. [1978] 32674
Noskov, G.A. [1983] 34170
Noskov, G.A. [1983] 38972
Novikov, P.S. [1949] 19476
Palyutin, E.A. [1971] 64365
Paola di, R.A. [1967] 30681
Pappas, P. [1985] 44466
Pepis, J. [1937] 19493
Pigozzi, D. [1974] 10495
Pigozzi, D. [1976] 21868
Pinus, A.G. [1971] 19598
Pinus, A.G. [1972] 10513
Pinus, A.G. [1983] 37103
Pinus, A.G. [1983] 39357
Pinus, A.G. [1985] 47255
Pour-El, M.B. [1969] 32553

Pour-El, M.B. [1969] 32554
Pour-El, M.B. [1970] 29012
Prestel, A. [1975] 18349
Prestel, A. [1979] 53609
Putnam, H. [1957] 10827
Quine, W.V.O. [1949] 10881
Rautenberg, W. [1962] 11041
Rautenberg, W. [1967] 11046
Rautenberg, W. [1968] 11048
Rautenberg, W. [1975] 49852
Reichbach, J. [1980] 32005
Richard, D. [1984] 41779
Robinson, Julia [1959] 11295
Robinson, Julia [1962] 11296
Robinson, Julia [1965] 11298
Robinson, Julia [1973] 77822
Robinson, R.M. [1951] 11311
Roedding, D. [1966] 16282
Roman'kov, V.A. [1979] 53388
Romanovskij, N.S. [1980] 77873
Rozenblat, B.V. [1979] 53949
Rozenblat, B.V. [1981] 35085
Rubin, M. [1983] 33531
Rumely, R. [1980] 55276
Sabbagh, G. [1971] 28610
Sacerdote, G.S. [1972] 30879
Sankappanavar, H.P. [1978] 52119
Schmerl, J.H. [1980] 56060
Schmitt, P.H. [1984] 46136
Schuette, K. [1933] 12997
Schwabhaeuser, W. [1983] 40225
Scott, D.S. [1960] 11909
Seese, D.G. [1972] 11931
Seese, D.G. [1977] 27132
Seese, D.G. [1978] 56270
Seese, D.G. [1982] 33532
Shelah, S. [1975] 15024
Shoenfield, J.R. [1960] 12098
Shore, R.A. [1981] 55426
Siefkes, D. [1971] 12146
Skolem, T.A. [1919] 12383
Skolem, T.A. [1937] 27938
Slobodskoj, A.M. [1981] 36070
Smorynski, C.A. [1973] 12535
Smorynski, C.A. [1982] 43687
Sobolev, S.K. [1977] 51014
Soprunov, S.F. [1974] 17402
Stillwell, J.C. [1977] 52422
Szczerba, L.W. [1965] 12831
Szczerba, L.W. [1979] 79177
Tajmanov, A.D. [1958] 13265
Tajtslin, M.A. [1962] 19772
Tajtslin, M.A. [1963] 13254
Tajtslin, M.A. [1964] 13255
Tajtslin, M.A. [1965] 13257
Tajtslin, M.A. [1965] 65641
Tajtslin, M.A. [1968] 13260
Tajtslin, M.A. [1970] 28107
Takahashi, S. [1973] 65648
Thomas, Wolfgang [1978] 31271
Trakhtenbrot, B.A. [1950] 19743
Trakhtenbrot, B.A. [1953] 19742
Tulipani, S. [1982] 34564
Turan, P. [1971] 13720
Turing, A.M. [1954] 42433
Urquhart, A.I.F. [1981] 36071
Urquhart, A.I.F. [1981] 54908

Ursini, A. [1985] 47525
Valiev, M.K. [1968] 13814
Vasil'ev, Eh.S. [1973] 13840
Vaught, R.L. [1957] 24868
Vaught, R.L. [1960] 24869
Vazhenin, Yu.M. [1974] 79687
Vazhenin, Yu.M. [1981] 55132
Vidal-Naquet, G. [1973] 27517
Wang, Hao [1963] 14040
Wang, Hao [1965] 14057
Weese, M. [1976] 23814
Weese, M. [1977] 26551
Weese, M. [1981] 33324
Weese, M. [1984] 42386
Wirsing, M. [1977] 51036
Wolter, H. [1984] 44686
Zamyatin, A.P. [1978] 90332
Zermelo, E. [1931] 14412
Zhegalkin, I.I. [1939] 17903
Ziegler, M. [1976] 24137
Ziegler, M. [1982] 35441

D40

Aanderaa, S.O. [1973] 23990
Adyan, S.I. [1957] 00193
Adyan, S.I. [1958] 00195
Anshel, M. [1976] 30627
Anshel, M. [1978] 30625
Belegradek, O.V. [1974] 17646
Belegradek, O.V. [1978] 29209
Belegradek, O.V. [1980] 80668
Belkin, V.P. [1975] 22948
Bezverkhnij, V.N. [1981] 39526
Bezverkhnij, V.N. [1981] 39904
Billington, N. [1984] 45773
Boerger, E. [1979] 53847
Boone, W.W. [1974] 04158
Britton, J.L. [1973] 44434
Collins, D.J. [1970] 02327
Collins, D.J. [1970] 02448
Dobritsa, V.P. [1981] 55366
Flum, J. [1976] 31078
Frenkel', V.I. [1963] 20969
Glass, A.M.W. [1984] 44713
Glass, A.M.W. [1985] 47863
Glass, A.M.W. [1985] 49075
Goralcik, P. [1980] 73522
Haken, W. [1973] 29825
Hickin, K.K. [1980] 81551
Hickin, K.K. [1983] 38580
Higman, G. [1949] 90239
Huber-Dyson, V. [1977] 51046
Huber-Dyson, V. [1981] 54396
Huber-Dyson, V. [1982] 39680
Kleiman, J.G. [1982] 39318
Kreisel, G. [1975] 75092
Kryazhovskikh, G.V. [1980] 54395
Macintyre, A. [1972] 08467
Macintyre, A. [1972] 08468
Macintyre, A. [1975] 23079
Makanin, G.S. [1977] 51395
Makanin, G.S. [1977] 51863
Makanin, G.S. [1980] 82143
Mann, A. [1982] 39712
Mayoh, B.H. [1967] 08925
Myasnikov, A.G. [1982] 38582
Neumann, B.H. [1968] 24829

Neumann, B.H. [1973] 24064
Petresco, J. [1962] 10432
Pride, S.J. [1977] 26231
Rabin, M.O. [1957] 10933
Remeslennikov, V.N. [1970] 11095
Remeslennikov, V.N. [1983] 39954
Rozenblat, B.V. [1979] 53949
Sacerdote, G.S. [1972] 30879
Sacerdote, G.S. [1976] 24067
Schupp, P.E. [1970] 13048
Schupp, P.E. [1973] 44402
Shepherdson, J.C. [1951] 12066
Slobodskoj, A.M. [1981] 36070
T"rkalanov, K.D. [1971] 13644
T"rkalanov, K.D. [1971] 13645
Tajtslin, M.A. [1966] 13258
Tajtslin, M.A. [1968] 32029
Taylor, W. [1977] 69948
Umirbaev, U.U. [1984] 42699
Valiev, M.K. [1968] 13814
Weispfenning, V. [1976] 14747
Weispfenning, V. [1985] 48461
Ziegler, M. [1980] 54050

D45

Abdrazakov, K.T. [1984] 41873
Alton, D.A. [1973] 00289
Ash, C.J. [1984] 39867
Baldwin, John T. [1982] 33739
Baldwin, John T. [1984] 33746
Baur, W. [1974] 04142
Belyakin, N.V. [1970] 29065
Bertoni, A. [1979] 52925
Bienenstock, E. [1977] 50578
Bloshchitsyn, V.Ya. [1978] 36848
Bozovic, N.B. [1977] 51534
Compton, K.J. [1984] 47457
Crossley, J.N. [1974] 21692
Crossley, J.N. [1976] 25025
Crossley, J.N. [1981] 54309
Crossley, J.N. [1982] 37106
Dekker, J.C.E. [1969] 02900
Dekker, J.C.E. [1971] 02901
Dekker, J.C.E. [1971] 02902
Dobritsa, V.P. [1975] 23167
Dobritsa, V.P. [1976] 32652
Dobritsa, V.P. [1976] 32653
Dobritsa, V.P. [1977] 32655
Dobritsa, V.P. [1981] 55366
Dobritsa, V.P. [1983] 36756
Dobritsa, V.P. [1983] 39619
Downey, R.G. [1983] 38174
Downey, R.G. [1984] 39864
Downey, R.G. [1984] 39869
Downey, R.G. [1984] 42438
Downey, R.G. [1984] 43527
Downey, R.G. [1985] 39783
Downey, R.G. [1985] 47566
Drobotun, B.N. [1977] 51956
Drobotun, B.N. [1980] 56069
Dyment, E.Z. [1984] 44722
Dzgoev, V.D. [1979] 32626
Dzgoev, V.D. [1980] 32623
Dzgoev, V.D. [1980] 55088
Dzgoev, V.D. [1982] 37236
Eisenberg, E.F. [1982] 35753
Ellentuck, E. [1981] 33198

Emde Boas van, P. [1973] 29008
Ershov, Yu.L. [1968] 37357
Ershov, Yu.L. [1972] 03552
Ershov, Yu.L. [1972] 18137
Ershov, Yu.L. [1973] 17197
Ershov, Yu.L. [1974] 25764
Ershov, Yu.L. [1980] 72718
Feiner, L. [1970] 03711
Feiner, L. [1973] 03714
Fowler III, N. [1975] 17130
Fowler III, N. [1976] 14783
Fowler III, N. [1978] 52144
Froehlich, A. [1955] 17092
Goncharov, S.S. [1973] 05125
Goncharov, S.S. [1975] 17637
Goncharov, S.S. [1975] 26031
Goncharov, S.S. [1976] 50391
Goncharov, S.S. [1977] 54891
Goncharov, S.S. [1978] 56068
Goncharov, S.S. [1980] 54545
Goncharov, S.S. [1980] 55087
Goncharov, S.S. [1980] 55558
Goncharov, S.S. [1980] 73488
Goncharov, S.S. [1981] 37746
Goncharov, S.S. [1982] 40426
Goncharov, S.S. [1983] 43384
Goncharov, V.A. [1983] 45205
Guhl, R. [1975] 05433
Guhl, R. [1977] 24226
Guichard, D.R. [1983] 40309
Guichard, D.R. [1984] 43386
Hamilton, A.G. [1970] 05625
Hay, L. [1977] 30705
Hermann, G. [1926] 05997
Higman, G. [1961] 22083
Hingston, P. [1981] 55365
Kalantari, I. [1977] 26850
Kalantari, I. [1978] 29262
Kalantari, I. [1979] 53722
Kalantari, I. [1979] 74612
Kalantari, I. [1982] 34171
Kalantari, I. [1983] 33595
Kalantari, I. [1983] 36130
Kalantari, I. [1985] 47472
Kamimura, T. [1985] 48019
Kanda, A. [1985] 47568
Kent, C.F. [1962] 07059
Khisamiev, N.G. [1977] 32598
Khisamiev, N.G. [1978] 32599
Khisamiev, N.G. [1979] 32601
Khisamiev, N.G. [1979] 32602
Khisamiev, N.G. [1981] 81571
Khisamiev, N.G. [1983] 40239
Khisamiev, N.G. [1984] 42697
Khisamiev, N.G. [1985] 49409
Kierstead, H.A. [1984] 45164
Kogalovskij, S.R. [1966] 63026
Kreisel, G. [1950] 07481
Kreisel, G. [1953] 20815
Kudajbergenov, K.Zh. [1979] 75139
Lachlan, A.H. [1970] 07756
Lerman, M. [1981] 55043
Lerman, M. [1982] 34683
Lin, C. [1981] 35279
Lin, C. [1981] 75552
Lindsay, P.A. [1985] 46372
Macintyre, A. [1984] 40310

Madison, E.W. [1968] 08500
Madison, E.W. [1970] 24826
Madison, E.W. [1971] 63572
Madison, E.W. [1983] 34173
Madison, E.W. [1985] 47796
Mal'tsev, A.I. [1962] 19247
Mal'tsev, A.I. [1962] 41272
Mann, A. [1982] 39712
Metakides, G. [1975] 17296
Metakides, G. [1977] 23661
Metakides, G. [1977] 33933
Metakides, G. [1978] 33934
Metakides, G. [1979] 55155
Metakides, G. [1979] 55230
Metakides, G. [1980] 55156
Metakides, G. [1982] 37391
Millar, T.S. [1981] 76379
Montagna, F. [1985] 41807
Morozov, A.S. [1983] 43123
Moschovakis, Y.N. [1964] 09514
Moschovakis, Y.N. [1966] 21265
Moses, M. [1983] 34174
Moses, M. [1984] 39682
Moses, M. [1984] 46602
Mostowski, Andrzej [1953] 09553
Mostowski, Andrzej [1955] 09558
Mostowski, Andrzej [1957] 09562
Mostowski, Andrzej [1962] 09574
Nerode, A. [1982] 38172
Nerode, A. [1985] 46402
Nogina, E.Yu. [1969] 09983
Nogina, E.Yu. [1978] 53136
Nurtazin, A.T. [1974] 17642
Peretyat'kin, M.G. [1971] 10368
Peretyat'kin, M.G. [1973] 18330
Peretyat'kin, M.G. [1978] 53927
Peretyat'kin, M.G. [1982] 39943
Pinus, A.G. [1975] 16662
Rabin, M.O. [1957] 10933
Rabin, M.O. [1958] 10930
Rabin, M.O. [1960] 33612
Remmel, J.B. [1977] 53858
Remmel, J.B. [1978] 29157
Remmel, J.B. [1978] 29270
Remmel, J.B. [1979] 56017
Remmel, J.B. [1980] 55231
Remmel, J.B. [1980] 55232
Remmel, J.B. [1980] 55233
Remmel, J.B. [1981] 33280
Remmel, J.B. [1981] 35064
Remmel, J.B. [1981] 77739
Retzlaff, A.T. [1978] 29259
Retzlaff, A.T. [1979] 77762
Richter, L.J. [1981] 33281
Robinson, A. [1975] 22933
Roche la, P. [1981] 75255
Rosenstein, J.G. [1984] 41775
Roy, D.K. [1983] 38265
Roy, D.K. [1985] 47798
Schram, J.M. [1984] 44732
Schrieber, L. [1985] 47474
Schwarz, S. [1984] 39829
Shi, Niandong [1982] 37603
Smith, Rick L. [1981] 55685
Suter, G.H. [1973] 12784
Tucker, J.V. [1980] 79494
Tverskoj, A.A. [1982] 35413
Tverskoj, A.A. [1984] 44226
Tverskoj, A.A. [1985] 49425
Vuckovic, V. [1977] 23633
Waerden van der, B.L. [1930] 38703
Watnick, R. [1984] 42511

D50

Applebaum, C.H. [1970] 00414
Crossley, J.N. [1974] 21692
Crossley, J.N. [1976] 25025
Degtev, A.N. [1978] 72221
Dekker, J.C.E. [1969] 02900
Dekker, J.C.E. [1971] 02901
Downey, R.G. [1984] 42438
Ellentuck, E. [1972] 03309
Ellentuck, E. [1975] 22911
Ellentuck, E. [1977] 27196
Ellentuck, E. [1978] 29153
Ellentuck, E. [1981] 33198
Ellentuck, E. [1981] 72617
Fowler III, N. [1975] 15238
Fowler III, N. [1976] 14783
Hamilton, A.G. [1970] 05625
Nerode, A. [1966] 09884
Olin, P. [1969] 10092

D55

Addison, J.W. [1962] 00175
Addison, J.W. [1962] 00176
Addison, J.W. [1965] 16215
Apt, K.R. [1972] 00429
Apt, K.R. [1974] 00432
Apt, K.R. [1976] 14789
Apt, K.R. [1978] 29131
Barwise, J. [1970] 00828
Barwise, J. [1971] 00836
Barwise, J. [1975] 31017
Belyaev, V.Ya. [1979] 32621
Belyakin, N.V. [1983] 41479
Blair, H.A. [1984] 43300
Burgess, J.P. [1975] 03956
Burgess, J.P. [1979] 56715
Campbell, P.J. [1971] 16762
Carstens, H.G. [1975] 03968
Chauvin, A. [1961] 01999
Chauvin, A. [1961] 25473
Cleave, J.P. [1968] 02681
Clote, P. [1985] 45298
Cutland, N.J. [1970] 16784
Cutland, N.J. [1972] 02662
David, R. [1982] 35373
Denisov, S.D. [1972] 61324
Devlin, K.J. [1974] 15073
Enderton, H.B. [1970] 03352
Feferman, S. [1965] 03696
Fitting, M. [1978] 31075
Friedman, H.M. [1970] 14546
Gandy, R.O. [1960] 04774
Gandy, R.O. [1960] 15061
Gandy, R.O. [1961] 22330
Garland, S.J. [1972] 04799
Garland, S.J. [1974] 24144
Grant, P.W. [1977] 26439
Grilliot, T.J. [1972] 05347
Grzegorczyk, A. [1958] 05398
Grzegorczyk, A. [1961] 05402
Grzegorczyk, A. [1962] 05406
Guaspari, D. [1980] 54183
Hajek, P. [1977] 26853
Harel, D. [1983] 38572
Harel, D. [1985] 41208
Harrington, L.A. [1975] 05665
Harrington, L.A. [1977] 24274
Hasenjaeger, G. [1953] 05723
Hasenjaeger, G. [1955] 27719
Hensel, G. [1969] 05948
Jensen, R.B. [1967] 06596
Jensen, R.B. [1970] 22245
Kanovej, V.G. [1979] 54105
Kechris, A.S. [1975] 14831
Keisler, H.J. [1973] 07033
Kreisel, G. [1950] 07481
Kreisel, G. [1953] 20815
Kreisel, G. [1961] 07523
Kreisel, G. [1965] 33546
Krom, Melven R. [1963] 07571
Kuratowski, K. [1937] 07668
Leeds, S. [1974] 07969
Lerman, M. [1981] 55043
Lerman, M. [1982] 34683
Makkai, M. [1973] 20987
Makkai, M. [1977] 23657
Makkai, M. [1977] 24202
Mansfield, R. [1975] 08674
Markwald, W. [1954] 08779
Mathias, A.R.D. [1972] 17477
McLaughlin, T.G. [1973] 29894
Millar, T.S. [1985] 46374
Miller, A.W. [1983] 35705
Miller, Douglas E. [1978] 33521
Moschovakis, Y.N. [1970] 09518
Moschovakis, Y.N. [1971] 09520
Moschovakis, Y.N. [1972] 09521
Mostowski, Andrzej [1956] 09560
Myers, D.L. [1976] 30751
Myers, R.W. [1980] 55957
Nelson, George C. [1973] 17539
Nepejvoda, L.K. [1977] 55689
Odifreddi, P. [1983] 34091
Poizat, B. [1981] 55288
Putnam, H. [1965] 10840
Ressayre, J.-P. [1977] 23658
Roedding, D. [1966] 24963
Schlipf, J.S. [1975] 19103
Scott, D.S. [1964] 11916
Shilleto, J.R. [1972] 29513
Silver, J.H. [1971] 12319
Silver, J.H. [1971] 19669
Srebrny, M. [1977] 26534
Steel, J.R. [1978] 29156
Suzuki, Y. [1968] 12791
Tanaka, H. [1974] 23068
Tarski, A. [1931] 16924
Tharp, L.H. [1974] 13528
Troelstra, A.S. [1978] 27952
Vaught, R.L. [1973] 13864
Vaught, R.L. [1973] 13909
Vaught, R.L. [1974] 17478
Vetulani, Z. [1984] 44142
Vinner, S. [1975] 13949
Wainer, S.S. [1975] 29611
Wheeler, W.H. [1972] 19104

D60

Abramson, F.G. [1979] 53788
Baeten, J. [1984] 45395
Barwise, J. [1967] 16732
Barwise, J. [1969] 00825
Barwise, J. [1971] 00836
Barwise, J. [1974] 00911
Barwise, J. [1975] 31017
Barwise, J. [1975] 60316
Barwise, J. [1978] 70698
Cutland, N.J. [1978] 51893
David, R. [1985] 46391
Ershov, Yu.L. [1983] 40494
Friedman, H.M. [1968] 04643
Friedman, S.D. [1979] 56139
Friedman, S.D. [1981] 33348
Friedman, S.D. [1982] 33350
Fukuyama, M. [1974] 23159
Grant, P.W. [1977] 26439
Green, J. [1974] 05312
Grilliot, T.J. [1972] 05347
Grilliot, T.J. [1974] 23178
Harrington, L.A. [1975] 05665
Humberstone, I.L. [1984] 37693
Kaufmann, M. [1984] 41225
Kolaitis, P.G. [1985] 48331
Kranakis, E. [1983] 38326
Kreisel, G. [1965] 33546
Kreisel, G. [1968] 07540
Lerman, M. [1978] 29152
Lerman, M. [1978] 30739
Machover, M. [1961] 08457
Makkai, M. [1977] 24202
Moschovakis, Y.N. [1971] 09520
Nadel, M.E. [1972] 09743
Nadel, M.E. [1974] 16930
Nyberg, A.M. [1977] 28324
Ressayre, J.-P. [1977] 23658
Srebrny, M. [1977] 26534
Stark, W.R. [1978] 52964
Stavi, J. [1973] 13133

D65

Belyakin, N.V. [1970] 29065
Blass, A.R. [1985] 45296
Ershov, Yu.L. [1972] 18137
Harrington, L.A. [1975] 05665
Hodges, W. [1976] 30711
Kolaitis, P.G. [1985] 48331
Kreisel, G. [1961] 07523
Moschovakis, Y.N. [1971] 09520
Schwichtenberg, H. [1975] 25022
Suzuki, Y. [1967] 12790
Wainer, S.S. [1975] 29611

D70

Aczel, P. [1975] 30616
Apt, K.R. [1978] 29131
Barwise, J. [1971] 00836
Barwise, J. [1975] 31017
Barwise, J. [1978] 29280
Barwise, J. [1978] 70698
Belyakin, N.V. [1970] 29065
Blass, A.R. [1984] 47716
Grant, P.W. [1977] 26439
Grzegorczyk, A. [1958] 05398

Harnik, V. [1976] 14797
Kolaitis, P.G. [1985] 48331
Kunen, K. [1968] 07631
Makkai, M. [1973] 20987
Makkai, M. [1977] 24202
Moschovakis, Y.N. [1969] 09517
Moschovakis, Y.N. [1970] 09518
Moschovakis, Y.N. [1971] 09520
Moschovakis, Y.N. [1972] 09521
Nyberg, A.M. [1976] 14788
Nyberg, A.M. [1977] 28324
Schlipf, J.S. [1977] 26851
Stavi, J. [1973] 13133
Vaught, R.L. [1973] 13864
Vaught, R.L. [1974] 17478

D75

Addison, J.W. [1965] 16215
Engeler, E. [1967] 03366
Engeler, E. [1975] 17633
Fitting, M. [1978] 31075
Gordon, C.E. [1974] 05199
Grilliot, T.J. [1972] 05347
Grilliot, T.J. [1973] 05348
Grilliot, T.J. [1974] 23178
Harrington, L.A. [1975] 05665
Jukna, S. [1979] 33258
Kaiser, Klaus [1970] 06800
Kalantari, I. [1985] 47472
Kreisel, G. [1968] 07540
Montague, R. [1968] 09432
Moschovakis, Y.N. [1969] 09517
Moschovakis, Y.N. [1970] 09518
Moschovakis, Y.N. [1972] 09521
Nyberg, A.M. [1977] 28324
Robinson, A. [1975] 22933
Robinson, Julia [1973] 29547

D80

Aanderaa, S.O. [1973] 23990
Barwise, J. [1975] 23183
Baur, W. [1980] 54140
Benioff, P.A. [1976] 30647
Bertoni, A. [1979] 52925
Bertoni, A. [1980] 80710
Boerger, E. [1979] 54143
Boerger, E. [1980] 80766
Britton, J.L. [1979] 53330
Compton, K.J. [1984] 42434
Cresswell, M.J. [1985] 42657
Denef, J. [1984] 38480
Engeler, E. [1968] 29303
Fowler III, N. [1978] 52144
Gurari, E.M. [1978] 42545
Gurari, E.M. [1979] 56232
Hajek, P. [1978] 31131
Haken, W. [1973] 29825
Jockusch Jr., C.G. [1984] 39945
Kalantari, I. [1985] 47472
Kierstead, H.A. [1983] 34172
Lehmann, G. [1985] 42737
McAloon, K. [1978] 54427
Mejtus, V.Yu. [1974] 09089
Metakides, G. [1982] 37391
Mostowski, T. [1976] 82309
Muller, D.E. [1981] 76648

Muller, D.E. [1985] 47646
Mundici, D. [1980] 33528
Nikitin, A.A. [1984] 48426
Pappas, P. [1985] 44466
Rabin, M.O. [1957] 10933
Robinson, Julia [1973] 77822
Singer, M.F. [1978] 29244
Taylor, W. [1981] 82957
Turan, G. [1981] 55701
Vuckovic, V. [1977] 23633
Weese, M. [1972] 14105

D96

Mostowski, Andrzej [1979] 82308

D97

Crossley, J.N. [1967] 31684
Crossley, J.N. [1981] 54309
Delon, F. [1985] 47516
Fenstad, J.E. [1978] 54177
Gandy, R.O. [1977] 16612
Harrington, L.A. [1985] 49810
Lolli, G. [1984] 41493
Mueller, Gert H. [1984] 41750
Sobolev, S.L. [1982] 34492

D98

Barwise, J. [1975] 60316
Barwise, J. [1977] 70117
Bucur, I. [1980] 46245
Church, A. [1963] 02403
Cutland, N.J. [1980] 56629
Ebbinghaus, H.-D. [1980] 41134
Ershov, Yu.L. [1965] 03525
Ershov, Yu.L. [1980] 72718
Friedman, H.M. [1975] 04296
Gecseg, F. [1984] 41334
Grzegorczyk, A. [1957] 24864
Gurevich, Y. [1985] 48332
Hermes, H. [1955] 06008
Herrmann, E. [1978] 53131
Krajewski, S. [1981] 38331
Lewis, H.R. [1979] 53515
Lorenzen, P. [1962] 08303
Makkai, M. [1977] 24202
Malitz, J. [1979] 56167
Manaster, A.B. [1975] 23120
Manin, Yu.I. [1977] 51984
Marek, W. [1972] 34216
Mostowski, Andrzej [1965] 09578
Nerode, A. [1985] 46402
Sacks, G.E. [1985] 42739
Sankappanavar, H.P. [1978] 52119
Shoenfield, J.R. [1967] 22384
Takahashi, S. [1974] 17488

D99

Lehmann, G. [1985] 42737

E05

Abraham, U. [1985] 47480
Abramson, F.G. [1978] 29285
Aczel, P. [1981] 34552
Adler, A. [1971] 00185
Adler, A. [1972] 00186

Baeten, J. [1984] 45395
Baumgartner, J.E. [1977] 51513
Baumgartner, J.E. [1978] 55896
Bell, J.L. [1972] 00999
Ben-David, S. [1978] 29139
Benda, M. [1971] 01023
Benda, M. [1972] 01024
Bernays, P. [1972] 01265
Beslagic, A. [1985] 48251
Bitter-Rucker von, R. [1972] 03312
Blass, A.R. [1974] 01298
Blass, A.R. [1974] 01299
Blass, A.R. [1977] 26228
Blass, A.R. [1977] 28164
Blass, A.R. [1978] 56022
Blass, A.R. [1980] 54202
Blass, A.R. [1981] 54567
Blass, A.R. [1985] 45296
Booth, D. [1970] 01459
Burd, B. [1984] 41389
Burgess, J.P. [1978] 29143
Cameron, P.J. [1978] 90363
Cameron, P.J. [1981] 49857
Cameron, P.J. [1983] 48376
Chang, C.C. [1962] 01950
Chang, C.C. [1965] 01970
Cherlin, G.L. [1972] 14684
Choquet, G. [1968] 02686
Choquet, G. [1968] 02687
Chudnovsky, G.V. [1974] 17593
Chudnovsky, G.V. [1975] 17594
Cichon, J. [1977] 27131
Clark, D.M. [1984] 44212
Cleave, J.P. [1968] 02681
Clote, P. [1980] 56493
Clote, P. [1981] 55425
Clote, P. [1985] 41801
Comfort, W.W. [1974] 16659
Daguenet-Teissier, M. [1979] 53645
Daguenet, M. [1972] 02744
Daguenet, M. [1975] 17438
Devlin, K.J. [1983] 37114
Donder, H.-D. [1983] 39341
Eklof, P.C. [1982] 38408
Erdoes, P. [1955] 03401
Erdoes, P. [1966] 03455
Erdoes, P. [1972] 17190
Foreman, M. [1983] 37111
Franek, F. [1985] 47703
Franek, F. [1985] 49474
Galvin, F. [1976] 23921
Galvin, F. [1978] 73205
Giorgetta, D. [1980] 56436
Gitik, M. [1984] 39952
Goebel, R. [1985] 44953
Gudjonsson, H. [1970] 05428
Halpern, J.D. [1975] 21724
Harrington, L.A. [1977] 27334
Henle, J.M. [1984] 39876
Henle, J.M. [1984] 42672
Higman, G. [1977] 81557
Hodges, W. [1981] 55551
Hodkinson, I.M. [1985] 48826
Holland, W.C. [1985] 45604
Huber, M. [1983] 36726
Jech, T.J. [1976] 30721
Jech, T.J. [1979] 52772

Jech, T.J. [1980] 55894
Jensen, R.B. [1970] 22245
Jensen, R.B. [1972] 19016
Jorgensen, M. [1975] 21825
Jovanovic, A. [1978] 74570
Jovanovic, A. [1980] 74569
Jovanovic, A. [1983] 41471
Kanamori, A. [1984] 40364
Kaufmann, M. [1981] 55028
Kaufmann, M. [1984] 41225
Keisler, H.J. [1961] 06998
Keisler, H.J. [1964] 07003
Keisler, H.J. [1964] 07004
Keisler, H.J. [1964] 19046
Keisler, H.J. [1965] 07012
Keisler, H.J. [1965] 07014
Keisler, H.J. [1967] 07021
Keisler, H.J. [1974] 07035
Ketonen, J. [1972] 07071
Kirby, L.A.S. [1982] 34416
Kirby, L.A.S. [1984] 39686
Kleinberg, E.M. [1973] 07222
Kleinberg, E.M. [1977] 27947
Kleinberg, E.M. [1977] 50789
Kleinberg, E.M. [1978] 54696
Kleinberg, E.M. [1981] 35069
Koepke, P. [1984] 39394
Koppelberg, B. [1976] 14805
Koppelberg, B. [1980] 55434
Koppelberg, S. [1980] 74974
Koppelberg, S. [1983] 39874
Kossak, R. [1984] 44634
Kostinsky, A. [1966] 21393
Kranakis, E. [1982] 34187
Kranakis, E. [1983] 38326
Kubota, N. [1983] 37604
Kucia, A. [1977] 56568
Kueker, D.W. [1972] 07620
Kueker, D.W. [1977] 24265
Kunen, K. [1970] 07637
Kunen, K. [1971] 07635
Kunen, K. [1971] 07638
Kunen, K. [1972] 07640
Kunen, K. [1978] 29242
Lascar, D. [1977] 75356
Lassaigne, R. [1971] 07871
Laver, R. [1982] 36563
Levinski, J.-P. [1984] 39720
Lolli, G. [1977] 31830
Luxemburg, W.A.J. [1962] 08426
Luxemburg, W.A.J. [1962] 27624
Macpherson, H.D. [1983] 39487
Magidor, M. [1971] 08542
Magidor, M. [1977] 26454
Makowsky, J.A. [1983] 38744
Martin, D.A. [1979] 56030
Mathias, A.R.D. [1972] 17477
McAloon, K. [1981] 34009
McKenzie, R. [1974] 09006
Mijajlovic, Z. [1985] 49547
Mitchell, W.J. [1972] 09322
Morley, M.D. [1968] 09494
Morley, M.D. [1977] 24200
Negrepontis, S. [1972] 09839
Pelc, A. [1981] 55433
Pillay, A. [1981] 55497
Pillay, A. [1981] 55498

Pino, R. [1985] 41795
Prikry, K. [1973] 10766
Puritz, C.W. [1971] 10821
Ramsey, F.P. [1930] 28781
Ratajczyk, Z. [1982] 35382
Rebholz, J.R. [1974] 11057
Ressayre, J.-P. [1980] 54581
Robinson, A. [1961] 11250
Rosen, N.I. [1983] 40800
Sageev, G. [1985] 41812
Schmerl, J.H. [1974] 12879
Schmerl, J.H. [1976] 23655
Schmerl, J.H. [1985] 45602
Seese, D.G. [1977] 82762
Seese, D.G. [1982] 33532
Shelah, S. [1970] 12033
Shelah, S. [1971] 12040
Shelah, S. [1972] 12048
Shelah, S. [1974] 12054
Shelah, S. [1975] 12057
Shelah, S. [1975] 12058
Shelah, S. [1976] 51434
Shelah, S. [1977] 27203
Shelah, S. [1977] 31462
Shelah, S. [1978] 31288
Shelah, S. [1978] 52002
Shelah, S. [1980] 55047
Shelah, S. [1980] 56029
Shelah, S. [1981] 33320
Shelah, S. [1981] 38748
Shelah, S. [1984] 38758
Shelah, S. [1984] 38761
Shelah, S. [1984] 46365
Shelah, S. [1985] 47543
Shelah, S. [1985] 48344
Shtejnbuk, V.B. [1982] 39142
Silver, J.H. [1971] 12318
Silver, J.H. [1971] 12319
Silver, J.H. [1971] 19669
Silver, J.H. [1973] 12321
Silver, J.H. [1974] 12322
Skolem, T.A. [1933] 12391
Solovay, R.M. [1974] 12645
Sureson, C. [1985] 40297
Tarski, A. [1962] 13461
Todorcevic, S.B. [1979] 53088
Todorcevic, S.B. [1980] 82978
Tryba, J. [1984] 45607
Tuschik, H.-P. [1983] 43530
Tverskoj, A.A. [1980] 35246
Twer von der, T. [1981] 35072
Tzouvaras, A.D. [1980] 43212
Vaeaenaenen, J. [1979] 52749
Vopenka, P. [1962] 21081
Wojciechowska, A. [1970] 14264
Wojciechowska, A. [1972] 14265
Woodrow, R.E. [1979] 53885
Yang, Shoulian [1982] 35600

E07

Abraham, U. [1985] 47480
Assous, R. [1981] 33709
Baayen, P.C. [1962] 33458
Bagemihl, F. [1956] 00653
Baker, K.A. [1970] 28844
Baker, K.A. [1972] 00687
Baldwin, John T. [1982] 33740

Benda, M. [1975] 60439
Benejam, J.-P. [1970] 01028
Bonnet, R. [1969] 01426
Bonnet, R. [1969] 33172
Bonnet, R. [1971] 30471
Bonnet, R. [1982] 38291
Chlebus, B.S. [1980] 56482
Cohn, P.M. [1957] 41136
Denjoy, A. [1953] 43585
Doner, J.E. [1978] 54480
Droste, M. [1985] 48384
Emde Boas van, P. [1973] 29008
Erdoes, P. [1972] 17190
Ershov, Yu.L. [1968] 03532
Fefferman, C. [1967] 03675
Foeldes, S. [1980] 72989
Folley, K.W. [1931] 39801
Fraisse, R. [1949] 04508
Fraisse, R. [1950] 04509
Fraisse, R. [1950] 04510
Fraisse, R. [1954] 04515
Fraisse, R. [1954] 17070
Fraisse, R. [1954] 21995
Fraisse, R. [1955] 04516
Fraisse, R. [1955] 04517
Fraisse, R. [1955] 23594
Fraisse, R. [1955] 33347
Fraisse, R. [1956] 04519
Fraisse, R. [1956] 17067
Fraisse, R. [1958] 04520
Fraisse, R. [1965] 04526
Fraisse, R. [1967] 04525
Fraisse, R. [1967] 17072
Fraisse, R. [1971] 04530
Fraisse, R. [1971] 31976
Fraisse, R. [1971] 31977
Fraisse, R. [1972] 61774
Fraisse, R. [1974] 17244
Fraisse, R. [1974] 61769
Fraisse, R. [1977] 16604
Fraisse, R. [1977] 52674
Fraisse, R. [1984] 41777
Frasnay, C. [1962] 31766
Frasnay, C. [1963] 31768
Frasnay, C. [1964] 31769
Frasnay, C. [1964] 31770
Frasnay, C. [1965] 04556
Frasnay, C. [1971] 31771
Frasnay, C. [1973] 61790
Frasnay, C. [1974] 32060
Frasnay, C. [1979] 73033
Frasnay, C. [1984] 41782
Gillman, L. [1955] 04937
Gillman, L. [1956] 04938
Glass, A.M.W. [1981] 54528
Glass, A.M.W. [1981] 73400
Gurevich, Y. [1977] 50608
Gurevich, Y. [1979] 53793
Gurevich, Y. [1979] 54603
Gurevich, Y. [1979] 54604
Gurevich, Y. [1982] 33758
Gurevich, Y. [1983] 33765
Hagendorf, J.G. [1972] 28843
Hagendorf, J.G. [1982] 42208
Hanazawa, M. [1975] 18182
Hausdorff, F. [1908] 05796
Hay, L. [1977] 30705

Hedrlin, Z. [1969] 05863
Henle, J.M. [1984] 42672
Herre, H. [1978] 51935
Herre, H. [1979] 53839
Herre, H. [1980] 55007
Herre, H. [1981] 55133
Hodges, W. [1977] 30713
Hodges, W. [1984] 39727
Holland, W.C. [1985] 45604
Humberstone, I.L. [1984] 37693
Huntington, E.V. [1917] 41774
Huntington, E.V. [1917] 41776
Jambu-Giraudet, M. [1980] 54167
Jambu-Giraudet, M. [1981] 55535
Johnston, J.B. [1956] 42818
Jonsson, B. [1982] 38290
Jovanovic, A. [1980] 74569
Koppelberg, S. [1977] 27169
Kotlarski, H. [1978] 29196
Kulunkov, P.A. [1980] 54045
Kuratowski, K. [1937] 07668
Ladner, R.E. [1977] 52252
Lake, J. [1975] 07801
Landraitis, C.K. [1980] 54168
Langford, C.H. [1926] 07849
Langford, C.H. [1927] 07852
Lerman, M. [1981] 55043
Lerman, M. [1982] 34683
Loewenheim, L. [1915] 08246
Lopez, G. [1975] 63507
Lopez, G. [1978] 75653
Lopez, G. [1983] 37133
Macintyre, A. [1974] 15208
Manaster, A.B. [1980] 55073
Manaster, A.B. [1980] 55074
Manaster, A.B. [1981] 33275
Mansoux, A. [1976] 27225
Marongiu, G. [1982] 34002
Mendelson, E. [1958] 09112
Morel, A.C. [1964] 09473
Morel, A.C. [1968] 09475
Moses, M. [1984] 39682
Nadel, M.E. [1978] 51892
Novak, Vitezslav [1984] 46664
Pashenkov, V.V. [1975] 18325
Pinus, A.G. [1971] 19598
Pinus, A.G. [1975] 16662
Pinus, A.G. [1980] 77327
Poizat, B. [1971] 48005
Poizat, B. [1976] 27224
Pouzet, M. [1973] 24428
Pouzet, M. [1979] 77496
Pouzet, M. [1981] 77494
Presic, M.D. [1979] 56693
Ribenboim, P. [1965] 11148
Rosenstein, J.G. [1969] 11530
Rosenstein, J.G. [1982] 38089
Roy, D.K. [1983] 38265
Rubin, M. [1974] 11657
Schmerl, J.H. [1979] 69898
Schmerl, J.H. [1984] 41781
Schwartz, N. [1978] 50976
Shelah, S. [1980] 54032
Slomson, A. [1972] 12475
Sonenberg, E.A. [1978] 29161
Sturm, T. [1980] 69938
Thomas, Wolfgang [1980] 55873

Vazhenin, Yu.M. [1970] 13911
Wachs, E.A. [1977] 65954
Wang, Shiqiang [1965] 49328
Weese, M. [1977] 26551
Weinberg, E.C. [1980] 66302
Wolter, H. [1977] 53520
Zermelo, E. [1931] 14412

E10

Baeten, J. [1984] 45395
Bagemihl, F. [1956] 00653
Balcar, B. [1967] 13883
Barros de, C.M. [1970] 00818
Boffa, M. [1969] 01371
Buechi, J.R. [1965] 01694
Buechi, J.R. [1965] 01697
Buechi, J.R. [1973] 01708
Buechi, J.R. [1973] 71377
Buechi, J.R. [1973] 71381
Buechi, J.R. [1983] 33901
Buechi, J.R. [1983] 40515
Charretton, C. [1978] 53862
Chuaqui, R.B. [1971] 10843
Conway, J.H. [1976] 71909
Deliyannis, P.C. [1970] 27908
Doner, J.E. [1972] 04095
Drabbe, J. [1966] 03098
Dragalin, A.G. [1971] 03116
Ehrenfeucht, A. [1961] 24896
Eklof, P.C. [1975] 18127
Ellentuck, E. [1968] 03300
Ellentuck, E. [1973] 03316
Ellentuck, E. [1974] 03323
Ellentuck, E. [1975] 22911
Feferman, S. [1957] 03683
Feferman, S. [1972] 03706
Felscher, W. [1965] 03734
Fraisse, R. [1950] 04510
Garland, S.J. [1974] 24144
Gurevich, Y. [1983] 33764
Hausdorff, F. [1908] 05796
Heindorf, L. [1984] 42267
Hutchinson, J.E. [1976] 14784
Jakubikova-Studenovska, D. [1980] 56042
Jech, T.J. [1971] 06575
Jovanovic, A. [1978] 74570
Jovanovic, A. [1980] 36003
Kechris, A.S. [1975] 14831
Kino, A. [1966] 07101
Kino, A. [1975] 17616
Koppelberg, S. [1980] 74974
Leavitt, W.G. [1979] 53745
Lillo de, N.J. [1973] 02906
Lillo de, N.J. [1979] 56184
Litman, A. [1976] 18261
Lopez-Escobar, E.G.K. [1966] 22322
Marek, W. [1966] 19298
Marek, W. [1973] 17662
Marek, W. [1973] 17663
Marek, W. [1974] 08755
McKenzie, R. [1971] 09000
Mekler, A.H. [1984] 44204
Monk, J.D. [1985] 49230
Nishimura, T. [1960] 09972
Nurtazin, A.T. [1983] 40721
Scott, D.S. [1957] 11911

Seese, D.G. [1982] 34012
Shelah, S. [1978] 31288
Slomson, A. [1970] 12474
Srebrny, M. [1977] 26534
Stern, S.T. [1966] 13181
Sun, Wenzhi [1982] 34080
Tarski, A. [1948] 13436
Tarski, A. [1962] 13461
Tauts, A. [1978] 54483
Vopenka, P. [1965] 13871
Wolter, H. [1973] 14279
Yasugi, M. [1967] 18024
Zawadowski, M. [1983] 42355

E15

Addison, J.W. [1962] 00175
Addison, J.W. [1962] 00176
Addison, J.W. [1965] 16215
Burgess, J.P. [1975] 03956
Burgess, J.P. [1977] 56269
Burgess, J.P. [1979] 56715
Campbell, P.J. [1971] 16762
Cowles, J.R. [1979] 56271
Friedman, S.D. [1984] 42671
Garland, S.J. [1972] 04799
Gostanian, R. [1980] 73544
Harnik, V. [1976] 14797
Harrington, L.A. [1977] 24274
Kanovej, V.G. [1979] 54105
Kechris, A.S. [1975] 14831
Kuratowski, K. [1937] 07668
Makkai, M. [1977] 23657
Mansfield, R. [1975] 08674
Mansfield, R. [1977] 26451
Marek, W. [1974] 08753
Marek, W. [1982] 35374
Mathias, A.R.D. [1972] 17477
Miller, A.W. [1983] 35705
Miller, Douglas E. [1978] 33521
Miller, Douglas E. [1981] 33277
Morais, R. [1977] 16605
Mostowski, Andrzej [1947] 09538
Mostowski, Andrzej [1956] 09560
Mostowski, Andrzej [1958] 43873
Mustafin, T.G. [1977] 50969
Myers, D.L. [1976] 30751
Ressayre, J.-P. [1977] 23658
Ryll-Nardzewski, C. [1964] 11710
Sami, R.L. [1982] 35583
Scott, D.S. [1964] 11916
Shelah, S. [1984] 38762
Silver, J.H. [1971] 19669
Tarski, A. [1931] 16924
Vaught, R.L. [1973] 13864
Vaught, R.L. [1973] 13909
Vaught, R.L. [1974] 17478

E20

Emde Boas van, P. [1973] 29008
Franci, R. [1977] 81255
Jezek, J. [1985] 46363
Laugwitz, D. [1958] 48699
Plonka, J. [1979] 77370
Ramsey, F.P. [1930] 28781
Sabbagh, G. [1972] 11730
Schein, B.M. [1971] 28117

Scott, D.S. [1957] 11911
Shrejder, Yu.A. [1977] 79715
Zermelo, E. [1935] 14413

E25

Abian, A. [1972] 00069
Abian, A. [1972] 00071
Anapolitanos, D.A. [1979] 53439
Anapolitanos, D.A. [1979] 56638
Andreka, H. [1981] 70419
Andrews, P.B. [1972] 00367
Apter, A.W. [1985] 39959
Ash, C.J. [1983] 35766
Balcar, B. [1967] 13883
Beauchemin, P. [1974] 17974
Beauchemin, P. [1975] 25817
Bily, J. [1974] 01205
Blass, A.R. [1981] 35229
Blass, A.R. [1983] 40316
Chang, C.C. [1971] 02676
Cohen, P.E. [1974] 04012
Cohen, P.J. [1974] 02309
Cuda, K. [1984] 39637
Dehornoy, P. [1976] 23024
Dehornoy, P. [1978] 29158
Diener, K.-H. [1983] 34750
Doets, H.C. [1971] 03075
Ellentuck, E. [1968] 03300
Ellentuck, E. [1973] 03316
Ellentuck, E. [1974] 03323
Feferman, S. [1965] 03696
Felgner, U. [1971] 03721
Felgner, U. [1976] 25860
Fitting, M. [1969] 04349
Flannagan, T.B. [1980] 72958
Fourman, M.P. [1980] 56570
Friedman, H.M. [1970] 14546
Friedman, H.M. [1975] 04661
Goldblatt, R.I. [1985] 42555
Gross, W.F. [1976] 62204
Gross, W.F. [1977] 51683
Guzicki, W. [1974] 62249
Hajek, P. [1965] 13878
Hajek, P. [1966] 05519
Henle, J.M. [1984] 42672
Hickman, J.L. [1975] 21655
Hickman, J.L. [1976] 23189
Hickman, J.L. [1982] 36952
Hodges, W. [1975] 30712
Howard, P.E. [1975] 18198
Jech, T.J. [1971] 06575
Jensen, R.B. [1967] 06596
Jensen, R.B. [1967] 06597
Kaufmann, M. [1984] 38764
Kertesz, A. [1974] 16665
Kleinberg, E.M. [1977] 50789
Klimovsky, G. [1962] 07236
Kotlarski, H. [1973] 07412
Laeuchli, H. [1962] 07783
Levy, A. [1965] 22314
Litman, A. [1976] 18261
Los, J. [1954] 08322
Luxemburg, W.A.J. [1962] 27624
Luxemburg, W.A.J. [1969] 24816
Marek, W. [1966] 19298
Marek, W. [1967] 28474
Mostowski, Andrzej [1969] 21062

Mostowski, Andrzej [1976] 23791
Pincus, D. [1976] 26497
Plotkin, J.M. [1969] 10566
Plotkin, J.M. [1981] 55102
Rockingham Gill, R.R. [1975] 18358
Sageev, G. [1981] 36078
Shelah, S. [1984] 38762
Spector, M. [1980] 55242
Surma, S.J. [1968] 12748
Takeuti, G. [1962] 13325
Tarski, A. [1948] 13436
Verdu i Solans, V. [1982] 45471
Vopenka, P. [1965] 13876
Weglorz, B. [1969] 14116
Zarach, A. [1976] 23816
Zarach, A. [1977] 26532

E30

Abian, A. [1975] 03813
Abian, A. [1977] 70262
Abramson, F.G. [1981] 33267
Artigue, M. [1978] 70505
Balcar, B. [1965] 00722
Barwise, J. [1972] 00839
Barwise, J. [1974] 00911
Barwise, J. [1975] 31017
Barwise, J. [1975] 60316
Belyakin, N.V. [1979] 70865
Bielinski, K. [1977] 52151
Boffa, M. [1967] 01404
Boffa, M. [1967] 31019
Boffa, M. [1969] 01371
Boffa, M. [1971] 01387
Boffa, M. [1972] 01395
Boffa, M. [1977] 53983
Breban, M. [1981] 54906
Breban, M. [1984] 40365
Bruijn de, N.G. [1975] 21509
Bukovsky, L. [1966] 01719
Cantini, A. [1983] 41846
Cantini, A. [1983] 44727
Cantone, D. [1985] 48303
Cascante Davila, J.M. [1977] 71563
Chauvin, A. [1961] 01999
Chauvin, A. [1961] 25473
Cole, J.C. [1973] 04015
Corrada, M. [1980] 54037
Drabbe, J. [1976] 27288
Ershov, Yu.L. [1983] 40494
Ferro, A. [1980] 54139
Ferro, A. [1980] 54334
Ferro, A. [1983] 44643
Ferro, A. [1985] 47612
Fjelstad, P. [1968] 04367
Friedman, H.M. [1971] 17088
Friedman, H.M. [1974] 04656
Friedman, S.D. [1979] 56139
Gaifman, H. [1974] 17241
Gloede, K. [1975] 27214
Gloede, K. [1977] 26465
Gogol, D. [1978] 52232
Gogol, D. [1979] 52679
Green, J. [1975] 05313
Grewe, R. [1969] 05335
Hajek, P. [1964] 05512
Hajek, P. [1965] 05514
Hajek, P. [1965] 05515

Harrington, L.A. [1984] 48307
Hauschild, K. [1963] 05759
Hauschild, K. [1967] 05765
Hauschild, K. [1968] 05770
Henkin, L. [1971] 05928
Henson, C.W. [1984] 39860
Hinnion, R. [1976] 23019
Hinnion, R. [1979] 56023
Hinnion, R. [1981] 34023
Hirschfeld, J. [1975] 17294
Jaeger, G. [1982] 34700
Jech, T.J. [1971] 29869
Kakuda, Y. [1983] 47676
Kaufmann, M. [1983] 37512
Kaufmann, M. [1984] 40711
Krajewski, S. [1974] 07445
Krajewski, S. [1977] 31406
Kreisel, G. [1967] 07533
Kuehnrich, M. [1978] 53807
Kurata, R. [1964] 07649
Lake, J. [1975] 07800
Levy, A. [1958] 24860
Lillo de, N.J. [1979] 56184
Lolli, G. [1975] 31828
Lolli, G. [1979] 53259
Magidor, M. [1984] 38750
Marek, W. [1973] 08747
Marek, W. [1974] 08755
Marek, W. [1975] 08758
Marek, W. [1975] 23020
Marek, W. [1976] 24412
Marek, W. [1978] 28323
Marek, W. [1978] 29210
Marek, W. [1978] 76016
McAloon, K. [1975] 18286
McAloon, K. [1981] 34009
McNaughton, R. [1954] 09068
Montagna, F. [1983] 34635
Montague, R. [1957] 29377
Montague, R. [1965] 24776
Moore, A.W. [1985] 45047
Mostowski, Andrzej [1953] 09553
Mostowski, Andrzej [1956] 09559
Mostowski, Andrzej [1958] 09566
Mostowski, Andrzej [1959] 31287
Mostowski, Andrzej [1976] 23738
Nagashima, T. [1966] 09756
Nagashima, T. [1970] 09758
Nishimura, T. [1960] 09970
Nishimura, T. [1960] 09972
Nishimura, T. [1974] 18312
Nishimura, T. [1974] 18313
Pabion, J.F. [1967] 10209
Pabion, J.F. [1967] 41363
Pabion, J.F. [1982] 40714
Platek, R.A. [1971] 10530
Pudlak, P. [1985] 42556
Ressayre, J.-P. [1983] 41473
Ressayre, J.-P. [1984] 44416
Rieger, L. [1957] 11188
Rieger, L. [1959] 11189
Rosser, J.B. [1950] 11558
Scarpellini, B. [1966] 11880
Schock, R. [1973] 12954
Scott, D.S. [1957] 11911
Simpson, S.G. [1982] 35660
Skolem, T.A. [1923] 21221

Skolem, T.A. [1929] 12386
Skolem, T.A. [1941] 12399
Skolem, T.A. [1960] 12426
Umezawa, T. [1965] 13786
Umezawa, T. [1966] 13787
Vetulani, Z. [1977] 31523
Ville, F. [1971] 13945
Vopenka, P. [1962] 16336
Vopenka, P. [1962] 21080
Vopenka, P. [1962] 21081
Vopenka, P. [1963] 22005
Vopenka, P. [1963] 22007
Vopenka, P. [1964] 24879
Vopenka, P. [1965] 13875
Vopenka, P. [1965] 13876
Vopenka, P. [1965] 27531
Wang, Hao [1953] 14024
Wang, Hao [1965] 14057
Wilmers, G.M. [1973] 30613
Yasugi, M. [1967] 18024
Yasumoto, M. [1980] 56072
Zarach, A. [1977] 26532
Zarach, A. [1982] 34278
Zbierski, P. [1971] 14379
Zermelo, E. [1931] 14412

E35

Abian, A. [1972] 00071
Abian, A. [1972] 04126
Abian, A. [1975] 21301
Abraham, U. [1984] 40744
Abraham, U. [1985] 47480
Adamowicz, Z. [1976] 14749
Ajtai, M. [1979] 53146
Anapolitanos, D.A. [1979] 53439
Anapolitanos, D.A. [1979] 56638
Andrews, P.B. [1972] 00366
Andrews, P.B. [1972] 00367
Antonovskij, M.Ya. [1981] 53987
Aoe, H. [1978] 70444
Apter, A.W. [1985] 39959
Apter, A.W. [1985] 48362
Ash, C.J. [1983] 35766
Balcar, B. [1969] 00726
Bankston, P. [1979] 52789
Barwise, J. [1975] 31017
Becker, J.A. [1980] 54459
Ben-David, S. [1978] 29139
Beni, J. [1968] 01035
Blass, A.R. [1983] 40316
Boffa, M. [1967] 01404
Boffa, M. [1967] 31019
Boos, W. [1976] 14754
Bukovsky, L. [1976] 31650
Burgess, J.P. [1978] 29143
Burgess, J.P. [1978] 29284
Burns, R.G. [1984] 47165
Chudnovsky, D.V. [1972] 02580
Chudnovsky, G.V. [1969] 71792
Cohen, P.E. [1974] 04012
Cohen, P.J. [1963] 02303
Cohen, P.J. [1974] 02309
Cuda, K. [1977] 32281
Cuda, K. [1984] 39637
Dahn, B.I. [1979] 56006
Dalla Chiara Scabia, M.L. [1968] 04285
Doets, H.C. [1969] 03072

Donder, H.-D. [1983] 39341
Dow, A. [1985] 48304
Dragalin, A.G. [1971] 03116
Dubiel, M. [1980] 54093
Eda, K. [1983] 37470
Eda, K. [1984] 44306
Eklof, P.C. [1976] 26077
Eklof, P.C. [1977] 16626
Eklof, P.C. [1977] 30690
Eklof, P.C. [1977] 51455
Eklof, P.C. [1980] 56438
Eklof, P.C. [1983] 36142
Ellentuck, E. [1968] 03300
Ellentuck, E. [1971] 03314
Ellentuck, E. [1973] 03316
Esterle, J. [1977] 55696
Feferman, S. [1965] 03696
Felgner, U. [1971] 03721
Fitting, M. [1969] 04349
Fitting, M. [1970] 14547
Fitting, M. [1972] 31074
Foreman, M. [1982] 34078
Foreman, M. [1983] 37111
Forti, M. [1983] 37413
Fourman, M.P. [1980] 56570
Friedman, H.M. [1971] 17088
Friedman, H.M. [1979] 54181
Gallin, D. [1975] 29739
Gitik, M. [1984] 39952
Glass, A.M.W. [1981] 54528
Grayson, R.J. [1981] 54421
Gross, W.F. [1976] 62204
Gurevich, Y. [1983] 33764
Guzicki, W. [1974] 62249
Hajek, P. [1965] 05515
Hajek, P. [1965] 13878
Hajek, P. [1966] 05519
Hajek, P. [1967] 13881
Hajek, P. [1971] 05523
Harrington, L.A. [1977] 24274
Hauschild, K. [1963] 05759
Henle, J.M. [1984] 42672
Herre, H. [1980] 55007
Hickman, J.L. [1975] 21655
Hickman, J.L. [1976] 23189
Hinnion, R. [1976] 23019
Hinnion, R. [1979] 56023
Hinnion, R. [1980] 55100
Hinnion, R. [1981] 34023
Hodges, W. [1976] 30711
Hrbacek, K. [1966] 15070
Jech, T.J. [1971] 06575
Jech, T.J. [1980] 55894
Jensen, C.U. [1982] 46674
Jensen, R.B. [1967] 06596
Jensen, R.B. [1967] 06597
Jensen, R.B. [1970] 22245
Jensen, R.B. [1972] 19016
Kanovej, V.G. [1979] 54105
Kaufmann, M. [1984] 38764
Koepke, P. [1984] 39394
Koppelberg, S. [1985] 39673
Krivine, J.-L. [1969] 07561
Krivine, J.-L. [1973] 24214
Kuehnrich, M. [1980] 55372
Kunen, K. [1970] 07632
Kunen, K. [1970] 07637

Kunen, K. [1978] 29242
Levy, A. [1965] 22314
Manevitz, L.M. [1976] 26083
Manin, Yu.I. [1977] 51984
Marek, W. [1966] 19298
Marek, W. [1967] 28474
Marek, W. [1973] 08748
Marek, W. [1973] 17662
Marek, W. [1973] 17663
Martin, D.A. [1979] 56030
McAloon, K. [1974] 08939
McAloon, K. [1975] 17403
McAloon, K. [1975] 18286
McAloon, K. [1980] 54523
Megibben, C. [1983] 37469
Mekler, A.H. [1980] 53035
Mekler, A.H. [1983] 37643
Mekler, A.H. [1985] 42592
Metakides, G. [1975] 09167
Mitchell, W.J. [1972] 09322
Moerdijk, I. [1984] 42485
Mostowski, Andrzej [1958] 09566
Mostowski, Andrzej [1967] 22257
Mostowski, Andrzej [1969] 21062
Mostowski, Andrzej [1976] 23791
Myers, D.L. [1976] 30751
Nishimura, T. [1960] 09970
Nishimura, T. [1960] 09972
Pincus, D. [1976] 26497
Plotkin, J.M. [1969] 10566
Plotkin, J.M. [1981] 55102
Rieger, L. [1957] 11188
Roguski, S. [1976] 23809
Rychkov, S.V. [1985] 45118
Sabbagh, G. [1972] 11730
Sageev, G. [1981] 36078
Sanmartin Esplugues, J. [1972] 78126
Scarpellini, B. [1966] 11880
Scedrov, A. [1984] 37128
Schmerl, J.H. [1974] 12879
Schock, R. [1973] 12954
Schock, R. [1974] 17446
Schultz, Konrad [1970] 13035
Shelah, S. [1974] 12054
Shelah, S. [1977] 27203
Shelah, S. [1978] 29138
Shelah, S. [1978] 52002
Shelah, S. [1979] 52977
Shelah, S. [1980] 54032
Shelah, S. [1980] 55047
Shelah, S. [1984] 38761
Shelah, S. [1984] 38762
Shepherdson, J.C. [1951] 28756
Shepherdson, J.C. [1952] 28757
Shepherdson, J.C. [1953] 28758
Silver, J.H. [1971] 12318
Silver, J.H. [1971] 19669
Sochor, A. [1975] 12625
Sochor, A. [1975] 13887
Sochor, A. [1982] 36081
Sochor, A. [1983] 37683
Kanamori, A. [1978] 27961
Spector, M. [1980] 55242
Stahl, S.H. [1980] 53487
Stepanek, P. [1974] 13172
Steprans, J. [1984] 40628
Suzuki, Y. [1973] 24802

Takahashi, Moto-o [1972] 24577
Tierney, M. [1972] 17528
Umezawa, T. [1965] 13786
Umezawa, T. [1966] 13787
Vaeaenaenen, J. [1978] 32506
Vaeaenaenen, J. [1980] 79570
Vaeaenaenen, J. [1980] 90120
Vaeaenaenen, J. [1981] 34284
Vaeaenaenen, J. [1982] 34162
Vaeaenaenen, J. [1982] 34286
Vopenka, P. [1965] 13871
Vopenka, P. [1965] 13875
Vopenka, P. [1965] 13876
Vopenka, P. [1966] 13879
Vopenka, P. [1967] 13884
Weglorz, B. [1969] 14116
Wette, E. [1959] 14162
Wilmers, G.M. [1973] 30613
Zadrozny, W. [1983] 37113
Zarach, A. [1976] 23816
Zarach, A. [1977] 26532
Zarach, A. [1982] 34278

E40

Aaberg, C. [1972] 17143
Aaberg, C. [1975] 31610
Abraham, U. [1984] 40744
Abramson, F.G. [1978] 29285
Adamowicz, Z. [1976] 14749
Adamowicz, Z. [1984] 45406
Adamson, A. [1978] 29149
Amer, M.A. [1976] 17946
Amer, M.A. [1977] 30619
Anapolitanos, D.A. [1978] 56717
Aribaud, F. [1979] 53226
Bell, J.L. [1976] 23794
Bily, J. [1974] 01205
Blass, A.R. [1981] 35229
Bowen, K.A. [1974] 01508
Bukovsky, L. [1973] 01729
Bukovsky, L. [1976] 31650
Bukovsky, L. [1977] 31651
Bukovsky, L. [1978] 71397
Bukovsky, L. [1979] 53889
Bukovsky, L. [1981] 33714
Cichon, J. [1977] 27131
Datuashvili, N.I. [1982] 49878
David, R. [1982] 35373
David, R. [1985] 46391
Dehornoy, P. [1975] 23194
Dehornoy, P. [1978] 29158
Dehornoy, P. [1983] 38177
Drabbe, J. [1976] 27288
Eda, K. [1980] 37158
Eda, K. [1982] 37160
Eda, K. [1983] 37161
Eda, K. [1984] 42939
Ellentuck, E. [1976] 14586
Feferman, S. [1965] 03696
Fitting, M. [1970] 14547
Fraisse, R. [1979] 56691
Friedman, H.M. [1975] 04661
Friedman, S.D. [1982] 33349
Friedman, S.D. [1982] 33350
Goldblatt, R.I. [1985] 42555
Gordon, E.I. [1983] 34454
Gordon, E.I. [1983] 44349

Gregory, J. [1975] 73614
Grulovic, M. [1983] 45085
Gurevich, Y. [1983] 33768
Gurevich, Y. [1984] 38749
Guzicki, W. [1974] 05464
Guzicki, W. [1974] 62249
Guzicki, W. [1981] 34160
Guzicki, W. [1983] 37520
Hajek, P. [1967] 13881
Hajek, P. [1970] 05521
Hajek, P. [1971] 05523
Henle, J.M. [1984] 42672
Horiguchi, H. [1985] 41746
Jech, T.J. [1971] 06575
Jech, T.J. [1971] 06576
Jech, T.J. [1976] 30721
Jech, T.J. [1979] 52772
Jech, T.J. [1980] 55894
Jensen, R.B. [1967] 06596
Kanovej, V.G. [1974] 17445
Kaufmann, M. [1984] 39887
Kleinberg, E.M. [1978] 54696
Knight, J.F. [1976] 14755
Kunen, K. [1970] 07632
Kunen, K. [1978] 29242
Kusraev, A.G. [1982] 34464
Kusraev, A.G. [1983] 40491
Kusraev, A.G. [1983] 44867
Kusraev, A.G. [1984] 45341
Magidor, M. [1977] 26457
Malcolm, W.G. [1973] 08620
Moens, J.L. [1982] 48346
Monro, G.P. [1974] 09414
Mostowski, Andrzej [1976] 23791
Motohashi, N. [1977] 23750
Nadel, M.E. [1978] 51892
Ozawa, M. [1984] 44314
Pabion, J.F. [1969] 10210
Pelc, A. [1981] 55433
Rubin, M. [1983] 38745
Sakarovitch, Joel [1977] 27314
Scedrov, A. [1984] 37128
Schmerl, J.H. [1973] 12878
Shelah, S. [1980] 54032
Shimoda, M. [1979] 47660
Shimoda, M. [1981] 54971
Simpson, S.G. [1974] 30799
Smith, Kay [1984] 42504
Spector, M. [1980] 55242
Srebrny, M. [1977] 26534
Stark, W.R. [1978] 52964
Steel, J.R. [1978] 29156
Takahashi, Makoto [1981] 47759
Takahashi, Makoto [1985] 46403
Takeuti, G. [1978] 52447
Takeuti, G. [1981] 34204
Takeuti, G. [1984] 39581
Vaeaenaenen, J. [1980] 90120
Veit Riccioli, B. [1978] 53930
Vopenka, P. [1965] 13871
Vopenka, P. [1965] 13875
Vopenka, P. [1965] 13876
Vopenka, P. [1967] 13884
Zhang, Jinwen [1979] 34211

E45

Abian, A. [1972] 00071
Abraham, U. [1984] 40744
Abramson, F.G. [1979] 53788
Ajtai, M. [1979] 53146
Apt, K.R. [1974] 00432
Apt, K.R. [1976] 14789
Artigue, M. [1973] 03836
Barwise, J. [1969] 00827
Barwise, J. [1970] 00828
Barwise, J. [1971] 00836
Barwise, J. [1971] 00910
Barwise, J. [1975] 31017
Baumgartner, J.E. [1977] 51513
Baumgartner, J.E. [1978] 55896
Bell, J.L. [1974] 03885
Bell, J.L. [1976] 23794
Bukovsky, L. [1972] 01728
Cantini, A. [1983] 41846
Chang, C.C. [1971] 02676
Cichon, J. [1977] 27131
Cohen, P.J. [1963] 02303
Cowen, R.H. [1976] 18443
Dahn, B.I. [1979] 56006
Dalla Chiara Scabia, M.L. [1968] 04285
David, R. [1982] 35373
David, R. [1985] 46391
Devlin, K.J. [1974] 15073
Devlin, K.J. [1983] 37114
Drabbe, J. [1966] 03098
Dragalin, A.G. [1971] 03116
Dugas, M. [1982] 37825
Dugas, M. [1984] 46405
Eklof, P.C. [1975] 18127
Eklof, P.C. [1977] 16626
Eklof, P.C. [1977] 30690
Eklof, P.C. [1977] 51455
Eklof, P.C. [1980] 56438
Eklof, P.C. [1983] 41825
Eklof, P.C. [1984] 38751
Enderton, H.B. [1971] 03353
Enderton, H.B. [1973] 03357
Feferman, S. [1965] 03696
Felgner, U. [1971] 03721
Fitting, M. [1969] 04349
Fleissner, W.G. [1975] 17356
Fraisse, R. [1977] 52674
Fraisse, R. [1985] 48204
Friedman, H.M. [1971] 17088
Friedman, H.M. [1973] 04654
Friedman, H.M. [1978] 31764
Friedman, S.D. [1981] 33348
Friedman, S.D. [1983] 37639
Friedman, S.D. [1984] 42671
Gaifman, H. [1974] 17241
Gloede, K. [1978] 52082
Goebel, R. [1985] 44953
Gostanian, R. [1980] 73541
Guaspari, D. [1980] 54183
Gurevich, Y. [1979] 53793
Guzicki, W. [1981] 34160
Hajek, P. [1967] 13881
Harrington, L.A. [1975] 05665
Harrington, L.A. [1984] 48307
Heindorf, L. [1984] 42267
Hinnion, R. [1979] 56023
Hrbacek, K. [1966] 15070
Huber, M. [1979] 53089
Jech, T.J. [1971] 06576
Jensen, R.B. [1967] 06596
Jensen, R.B. [1967] 06597
Jensen, R.B. [1970] 22245
Jensen, R.B. [1972] 19016
Kanovej, V.G. [1974] 17445
Kanovej, V.G. [1979] 54105
Kaufmann, M. [1984] 41225
Kaufmann, M. [1984] 46625
Keisler, H.J. [1971] 07029
Koepke, P. [1984] 39394
Kogalovskij, S.R. [1974] 17448
Koppelberg, S. [1985] 39673
Kotlarski, H. [1980] 55863
Kranakis, E. [1982] 34187
Kranakis, E. [1983] 38326
Krawczyk, A. [1975] 23128
Krawczyk, A. [1977] 31404
Krivine, J.-L. [1969] 07561
Krivine, J.-L. [1973] 24214
Kunen, K. [1970] 07637
Lake, J. [1975] 07800
Lerman, M. [1978] 29152
Levinski, J.-P. [1984] 39720
Lolli, G. [1977] 31830
Machover, M. [1961] 08457
Macintyre, A. [1976] 26082
Magidor, M. [1982] 35575
Magidor, M. [1983] 35758
Magidor, M. [1984] 38750
Manders, K.L. [1980] 33515
Manevitz, L.M. [1976] 26083
Mansfield, R. [1975] 08674
Marek, W. [1967] 28474
Marek, W. [1970] 08741
Marek, W. [1973] 08747
Marek, W. [1973] 08752
Marek, W. [1973] 17662
Marek, W. [1973] 17663
Marek, W. [1974] 08753
Marek, W. [1974] 08755
Marek, W. [1974] 08757
Marek, W. [1975] 23020
Marek, W. [1977] 54826
Marek, W. [1978] 29210
Marek, W. [1978] 76016
Marek, W. [1981] 33724
Marek, W. [1982] 35374
Marek, W. [1983] 34688
McAloon, K. [1974] 08939
McAloon, K. [1975] 18286
Megibben, C. [1983] 37469
Mekler, A.H. [1980] 53035
Mitchell, W.J. [1983] 37578
Montague, R. [1957] 29377
Morley, M.D. [1968] 09494
Mostowski, Andrzej [1956] 09560
Mostowski, Andrzej [1959] 31287
Mostowski, Andrzej [1961] 24858
Mostowski, Andrzej [1965] 09579
Mostowski, Andrzej [1967] 22257
Mostowski, Andrzej [1969] 21062
Murawski, R. [1980] 76666
Pabion, J.F. [1982] 40714
Palyutin, E.A. [1977] 29226
Paris, J.B. [1973] 23607
Paris, J.B. [1974] 17522
Paris, J.B. [1976] 25816
Pino, R. [1985] 41795
Pinus, A.G. [1978] 77325
Pinus, A.G. [1979] 56205
Podewski, K.-P. [1973] 10569
Ratajczyk, Z. [1981] 55630
Rebholz, J.R. [1974] 11057
Rieger, L. [1957] 11188
Roguski, S. [1976] 23809
Rosenthal, J.W. [1971] 11534
Rosenthal, J.W. [1972] 11533
Rosenthal, J.W. [1975] 11536
Rosenthal, J.W. [1979] 28371
Rubin, M. [1980] 56583
Rychkov, S.V. [1985] 45118
Sageev, G. [1985] 41812
Scarpellini, B. [1966] 11880
Schmerl, J.H. [1981] 55178
Schock, R. [1974] 17446
Scott, D.S. [1961] 21996
Sheard, M. [1985] 39762
Shelah, S. [1974] 12054
Shelah, S. [1979] 52977
Shelah, S. [1981] 54170
Shelah, S. [1985] 38772
Shilleto, J.R. [1972] 29513
Silver, J.H. [1971] 12319
Silver, J.H. [1971] 19669
Silver, J.H. [1973] 12321
Silver, J.H. [1974] 12322
Sochor, A. [1985] 48318
Kanamori, A. [1978] 27961
Srebrny, M. [1977] 26534
Stahl, S.H. [1980] 53487
Sureson, C. [1985] 40297
Takahashi, Moto-o [1972] 24577
Tomasik, J. [1971] 13634
Umezawa, T. [1965] 13786
Umezawa, T. [1966] 13787
Vaeaenaenen, J. [1979] 32505
Vetulani, Z. [1976] 23813
Vetulani, Z. [1984] 44142
Wainer, S.S. [1975] 29611
Wald, B. [1983] 39034
Wilmers, G.M. [1968] 14228
Wilmers, G.M. [1971] 14229
Wilmers, G.M. [1973] 30613
Zadrozny, W. [1983] 37113
Zadrozny, W. [1984] 40349
Zbierski, P. [1978] 29200

E47

Abramson, F.G. [1979] 53788
Abramson, F.G. [1981] 33267
Baeten, J. [1984] 45395
Barwise, J. [1969] 00827
Barwise, J. [1971] 00836
Barwise, J. [1971] 00910
Barwise, J. [1974] 00913
Barwise, J. [1975] 31017
Bukovsky, L. [1966] 01721
Devlin, K.J. [1974] 15073
Doner, J.E. [1972] 04095
Dragalin, A.G. [1971] 03116
Ershov, Yu.L. [1985] 49047
Friedman, H.M. [1979] 54181

Friedman, S.D. [1981] 33348
Gaifman, H. [1974] 04751
Gaifman, H. [1982] 37513
Garland, S.J. [1974] 24144
Grant, P.W. [1977] 26439
Guaspari, D. [1980] 54183
Harrington, L.A. [1984] 48307
Hodges, W. [1974] 17513
Hodges, W. [1976] 30711
Jech, T.J. [1971] 29869
Jensen, R.B. [1972] 19016
Kino, A. [1966] 07101
Kotlarski, H. [1978] 29176
Kranakis, E. [1982] 34187
Kuehnrich, M. [1971] 07616
Kunen, K. [1968] 07631
Levy, P. [1970] 08094
Magidor, M. [1971] 08543
Magidor, M. [1984] 38750
Marek, W. [1973] 08748
Marek, W. [1974] 08757
McKenzie, R. [1971] 09000
Minari, P. [1985] 49720
Montague, R. [1959] 24781
Mostowski, Andrzej [1947] 09538
Mostowski, Andrzej [1956] 09559
Mycielski, J. [1964] 09682
Nadel, M.E. [1972] 09744
Namba, K. [1970] 09802
Namba, K. [1970] 09803
Pino, R. [1985] 41795
Platek, R.A. [1971] 10530
Reinhardt, W.N. [1974] 22936
Rimscha von, M. [1983] 40517
Roguski, S. [1976] 23809
Rosenthal, J.W. [1975] 11536
Rosenthal, J.W. [1979] 28371
Rubel, L.A. [1983] 39167
Schwartz, Dietrich [1970] 13065
Simpson, S.G. [1974] 30799
Stavi, J. [1975] 13137
Takahashi, Moto-o [1972] 24577
Takeuti, G. [1962] 13325
Tarski, A. [1931] 16924
Vaeaenaenen, J. [1978] 32506
Vaeaenaenen, J. [1985] 48342
Ville, F. [1974] 24141
Vopenka, P. [1963] 24846
Wilmers, G.M. [1971] 14229
Yasumoto, M. [1980] 56072

E50

Abraham, U. [1984] 40744
Adler, A. [1972] 00186
Antonovskij, M.Ya. [1981] 53987
Bankston, P. [1977] 33381
Bankston, P. [1979] 52789
Beauchemin, P. [1975] 17975
Ben-David, S. [1978] 29139
Benda, M. [1972] 01024
Bernays, P. [1972] 01265
Bitter-Rucker von, R. [1972] 03312
Blass, A.R. [1972] 01294
Blass, A.R. [1974] 01298
Blass, A.R. [1977] 26228
Blass, A.R. [1981] 54567
Bonnet, R. [1977] 54295

Bonnet, R. [1980] 55060
Bulman-Fleming, S. [1972] 01751
Burgess, J.P. [1978] 29143
Burns, R.G. [1984] 47165
Chang, C.C. [1962] 01950
Chang, C.C. [1965] 01969
Chang, C.C. [1971] 02676
Charretton, C. [1978] 53862
Choquet, G. [1968] 02686
Choquet, G. [1968] 02687
Daguenet-Teissier, M. [1979] 53645
Dales, H.G. [1977] 81016
Dow, A. [1985] 48304
Dugas, M. [1983] 37486
Eda, K. [1983] 37470
Eda, K. [1984] 44306
Eklof, P.C. [1980] 56438
Eklof, P.C. [1983] 34809
Eklof, P.C. [1983] 36142
Eklof, P.C. [1983] 41825
Eklof, P.C. [1984] 38751
Enayat, A. [1985] 41811
Erdoes, P. [1955] 03401
Erdoes, P. [1966] 03455
Erdoes, P. [1972] 17190
Esterle, J. [1977] 52920
Esterle, J. [1977] 55696
Faber, V. [1978] 31943
Felgner, U. [1971] 03721
Fitting, M. [1969] 04349
Gallin, D. [1975] 29739
Garavaglia, S. [1984] 44219
Glass, A.M.W. [1981] 54528
Gostanian, R. [1980] 73544
Gurevich, Y. [1977] 50608
Gurevich, Y. [1979] 53793
Gurevich, Y. [1979] 54604
Gurevich, Y. [1983] 33765
Guzicki, W. [1974] 05464
Guzicki, W. [1977] 54755
Guzicki, W. [1983] 37520
Harrington, L.A. [1977] 24274
Herre, H. [1978] 51935
Herre, H. [1979] 53839
Herre, H. [1980] 55007
Hodges, W. [1984] 39731
Hodkinson, I.M. [1985] 48826
Huber, M. [1983] 36692
Jech, T.J. [1976] 30721
Jech, T.J. [1979] 52772
Jech, T.J. [1980] 55894
Jehne, W. [1977] 81712
Jensen, C.U. [1980] 53472
Jensen, C.U. [1980] 54316
Jensen, R.B. [1967] 06596
Jensen, R.B. [1967] 06597
Johnston, J.B. [1956] 42818
Keisler, H.J. [1961] 06998
Keisler, H.J. [1967] 07018
Koppelberg, B. [1980] 55434
Koppelberg, S. [1985] 39673
Koppelberg, S. [1985] 48309
Kreisel, G. [1967] 07533
Kucia, A. [1977] 56568
Kunen, K. [1970] 07632
Kunen, K. [1970] 07637
Luo, Libo [1980] 55737

Macintyre, A. [1973] 08469
Magidor, M. [1977] 26457
Manevitz, L.M. [1976] 26083
Manin, Yu.I. [1977] 51984
Marek, W. [1967] 28474
McKenzie, R. [1971] 08993
Megibben, C. [1983] 37469
Mekler, A.H. [1983] 37643
Mekler, A.H. [1985] 42592
Monk, J.D. [1975] 21872
Mostowski, Andrzej [1969] 21062
Mustafin, T.G. [1977] 50969
Nadel, M.E. [1980] 76724
Nelson, Evelyn [1975] 21874
Olin, P. [1968] 10091
Pinus, A.G. [1985] 47255
Pudlak, P. [1984] 42495
Rieger, L. [1957] 11188
Rosenthal, J.W. [1972] 11533
Rubin, M. [1983] 33531
Rychkov, S.V. [1985] 45118
Sageev, G. [1985] 41812
Sebel'din, A.M. [1981] 38919
Shelah, S. [1974] 12054
Shelah, S. [1975] 15024
Shelah, S. [1976] 26087
Shelah, S. [1977] 27203
Shelah, S. [1978] 29138
Shelah, S. [1978] 29282
Shelah, S. [1978] 52000
Shelah, S. [1979] 52977
Shelah, S. [1980] 55047
Shelah, S. [1980] 56029
Shelah, S. [1981] 33512
Shelah, S. [1981] 38748
Shelah, S. [1983] 33535
Shelah, S. [1983] 37468
Shelah, S. [1984] 38761
Sichler, J. [1973] 12136
Solovay, R.M. [1974] 12645
Soprunov, S.F. [1979] 55110
Srebrny, M. [1974] 23967
Stahl, S.H. [1980] 53487
Stark, W.R. [1980] 78943
Steprans, J. [1984] 40628
Tajmanov, A.D. [1967] 13275
Tierney, M. [1972] 17528
Vopenka, P. [1966] 13879

E55

Abe, Y. [1983] 47672
Adler, A. [1971] 00185
Apter, A.W. [1985] 39959
Apter, A.W. [1985] 48362
Balcar, B. [1965] 00722
Bankston, P. [1983] 37051
Barros de, C.M. [1970] 00818
Baumgartner, J.E. [1977] 51513
Baumgartner, J.E. [1978] 55896
Beauchemin, P. [1974] 17974
Beauchemin, P. [1975] 25817
Bell, J.L. [1970] 00992
Bell, J.L. [1972] 00999
Bell, J.L. [1974] 03885
Ben-David, S. [1978] 29139
Benda, M. [1978] 29145
Boerger, R. [1985] 49224

Boos, W. [1976] 14754
Bowen, K.A. [1973] 01506
Bukovsky, L. [1966] 01719
Bukovsky, L. [1966] 01721
Bukovsky, L. [1977] 31651
Casari, E. [1971] 01884
Chang, C.C. [1971] 02676
Cherlin, G.L. [1975] 02079
Chudnovsky, D.V. [1972] 02580
Chudnovsky, G.V. [1969] 71792
Chudnovsky, G.V. [1974] 17593
Chudnovsky, G.V. [1975] 17594
Cichon, J. [1977] 27131
Combase, J. [1983] 37540
Comfort, W.W. [1974] 16659
Costa da, N.C.A. [1972] 02485
Dehornoy, P. [1975] 23194
Dehornoy, P. [1976] 23024
Dehornoy, P. [1978] 29158
Dehornoy, P. [1983] 38177
Devlin, K.J. [1974] 15073
Dickmann, M.A. [1970] 28954
Dickmann, M.A. [1975] 26233
Donder, H.-D. [1983] 39341
Dugas, M. [1983] 37486
Dugas, M. [1983] 48322
Eda, K. [1983] 47673
Eklof, P.C. [1976] 26077
Eklof, P.C. [1977] 30690
Eklof, P.C. [1977] 51455
Enayat, A. [1984] 41809
Enayat, A. [1985] 41810
Erdoes, P. [1966] 03455
Fleissner, W.G. [1975] 17356
Foreman, M. [1982] 34078
Foreman, M. [1983] 37111
Friedman, S.D. [1984] 42671
Fuchs, L. [1974] 18150
Gaifman, H. [1974] 17241
Gao, Hengshan [1983] 45159
Garland, S.J. [1974] 24144
Gitik, M. [1984] 39952
Goebel, R. [1985] 44953
Gold, J.M. [1980] 55097
Hajek, P. [1965] 05514
Hanf, W.P. [1964] 33566
Harrington, L.A. [1975] 05665
Helling, M. [1966] 20920
Henkin, L. [1971] 05928
Herre, H. [1979] 53839
Herre, H. [1980] 55007
Hickin, K.K. [1977] 27199
Hodges, W. [1981] 74221
Hodges, W. [1984] 39731
Hort, C. [1984] 44206
Hrbacek, K. [1966] 15070
Huber, M. [1979] 53089
Huber, M. [1983] 36692
Isbell, J.R. [1983] 37803
Jech, T.J. [1971] 06576
Jech, T.J. [1971] 29869
Jech, T.J. [1976] 30721
Jech, T.J. [1979] 52772
Jech, T.J. [1980] 55894
Jensen, R.B. [1970] 22245
Jensen, R.B. [1972] 19016
Jovanovic, A. [1980] 36003

Kanamori, A. [1984] 40364
Kaufmann, M. [1983] 39848
Kaufmann, M. [1984] 38764
Keisler, H.J. [1962] 07000
Keisler, H.J. [1964] 19046
Keisler, H.J. [1965] 07013
Ketonen, J. [1972] 07071
Kleinberg, E.M. [1973] 07222
Kleinberg, E.M. [1977] 27947
Kleinberg, E.M. [1977] 50789
Kleinberg, E.M. [1978] 54696
Koepke, P. [1984] 39394
Kogalovskij, S.R. [1965] 24972
Koppelberg, B. [1980] 55434
Kostinsky, A. [1966] 21393
Kranakis, E. [1983] 38326
Krivine, J.-L. [1973] 24214
Kruse, A.H. [1966] 07586
Kuehnrich, M. [1978] 53807
Kunen, K. [1968] 07631
Kunen, K. [1970] 07632
Kunen, K. [1970] 07637
Kunen, K. [1971] 07638
Kunen, K. [1978] 29242
Lake, J. [1975] 07800
Lassaigne, R. [1971] 07871
Laver, R. [1982] 36563
Levinski, J.-P. [1984] 39720
Lolli, G. [1977] 31830
Lopez-Escobar, E.G.K. [1972] 19200
Luo, Libo [1980] 55737
Magidor, M. [1971] 08542
Magidor, M. [1971] 08543
Magidor, M. [1977] 23666
Magidor, M. [1977] 26457
Makowsky, J.A. [1983] 38744
Makowsky, J.A. [1985] 39781
Malitz, J. [1972] 08631
Marek, W. [1966] 19298
Marek, W. [1981] 33724
Martin, D.A. [1979] 56030
McAloon, K. [1981] 34009
Mekler, A.H. [1978] 52883
Mitchell, W.J. [1972] 09322
Mitchell, W.J. [1983] 37578
Montague, R. [1959] 24781
Morgenstern, C.F. [1979] 53929
Morgenstern, C.F. [1979] 54812
Morley, M.D. [1968] 09494
Morley, M.D. [1977] 24200
Nadel, M.E. [1972] 09743
Namba, K. [1970] 09802
Namba, K. [1970] 09803
Paris, J.B. [1974] 17522
Paris, J.B. [1976] 25816
Pelc, A. [1982] 35588
Poizat, B. [1976] 27224
Prikry, K. [1973] 10766
Reinhardt, W.N. [1974] 22936
Reyes, G.E. [1972] 11130
Rychkov, S.V. [1980] 82681
Sageev, G. [1985] 41812
Schmerl, J.H. [1972] 12875
Schmerl, J.H. [1972] 12876
Schmerl, J.H. [1974] 12879
Schmerl, J.H. [1976] 23655
Schmerl, J.H. [1976] 50233

Schmerl, J.H. [1981] 55178
Schock, R. [1973] 12954
Scott, D.S. [1961] 21996
Sheard, M. [1985] 39762
Shelah, S. [1975] 12059
Shelah, S. [1981] 54170
Shelah, S. [1982] 34283
Shelah, S. [1984] 38762
Shepherdson, J.C. [1951] 28756
Shepherdson, J.C. [1952] 28757
Shepherdson, J.C. [1953] 28758
Silver, J.H. [1971] 12318
Silver, J.H. [1971] 12319
Silver, J.H. [1971] 19669
Silver, J.H. [1973] 12321
Silver, J.H. [1974] 12322
Solovay, R.M. [1974] 12645
Kanamori, A. [1978] 27961
Srebrny, M. [1977] 26534
Sureson, C. [1985] 40297
Tarski, A. [1962] 13461
Tauts, A. [1978] 54483
Trnkova, V. [1971] 83011
Tryba, J. [1984] 45607
Tulipani, S. [1971] 13717
Vaeaenaenen, J. [1979] 55789
Vaeaenaenen, J. [1980] 79570
Vaeaenaenen, J. [1982] 34162
Vopenka, P. [1962] 21081
Vopenka, P. [1964] 24879
Wilson, T.P. [1981] 56490
Wojciechowska, A. [1970] 14264
Wojciechowska, A. [1972] 14265

E60

Badger, L.W. [1977] 27166
Buechi, J.R. [1977] 51165
Ehrenfeucht, A. [1957] 03234
Ehrenfeucht, A. [1961] 24896
Felgner, U. [1984] 39998
Flum, J. [1975] 17211
Grant, P.W. [1977] 26439
Green, J. [1979] 53310
Harrington, L.A. [1975] 05665
Henle, J.M. [1984] 39876
Hintikka, K.J.J. [1976] 18195
Hodges, W. [1981] 55551
Hodges, W. [1981] 74221
Hodges, W. [1985] 39750
Kechris, A.S. [1975] 14831
Kleinberg, E.M. [1977] 27947
Kleinberg, E.M. [1977] 50789
Kleinberg, E.M. [1981] 35069
Kolaitis, P.G. [1985] 48331
Ladner, R.E. [1977] 52252
Martin, D.A. [1979] 56030
Moschovakis, Y.N. [1972] 09521
Muller, D.E. [1985] 48421
Oikkonen, J. [1978] 54677
Sacks, G.E. [1985] 42739
Spector, M. [1985] 47378
Takeuti, G. [1968] 13332
Takeuti, G. [1970] 13336
Tauts, A. [1974] 17693
Vinner, S. [1972] 13947

E65

Abraham, U. [1984] 40744
Boffa, M. [1971] 01390
Dugas, M. [1982] 37825
Dugas, M. [1983] 37486
Dugas, M. [1983] 48322
Eklof, P.C. [1983] 34809
Enayat, A. [1985] 41810
Ershov, Yu.L. [1983] 40494
Gloede, K. [1975] 27214
Goebel, R. [1985] 44953
Huber, M. [1977] 74296
Jech, T.J. [1971] 29869
Jensen, R.B. [1972] 19016
Kaufmann, M. [1977] 31400
Keisler, H.J. [1974] 07037
Koppelberg, B. [1980] 55434
Koppelberg, S. [1985] 39673
Kossak, R. [1985] 47534
Laver, R. [1982] 36563
Macintyre, A. [1976] 26082
Magidor, M. [1977] 23666
Magidor, M. [1977] 26454
Magidor, M. [1977] 26457
Megibben, C. [1983] 37469
Montague, R. [1959] 24781
Nadel, M.E. [1978] 51892
Palyutin, E.A. [1977] 29226
Paris, J.B. [1976] 25816
Rebholz, J.R. [1974] 11057
Reinhardt, W.N. [1974] 22936
Rimscha von, M. [1980] 53804
Rubin, M. [1983] 33531
Rubin, M. [1983] 38745
Sageev, G. [1985] 41812
Shelah, S. [1978] 52002
Shelah, S. [1980] 56029
Shelah, S. [1983] 38743
Shelah, S. [1984] 38759
Shelah, S. [1984] 46365
Shelah, S. [1985] 48344
Sochor, A. [1981] 54239
Solovay, R.M. [1974] 12645
Kanamori, A. [1978] 27961
Tauts, A. [1978] 54483
Vaeaenaenen, J. [1978] 32506

E70

Alkor, C. [1979] 52587
Balcar, B. [1965] 00722
Balcar, B. [1969] 00726
Baldi, G. [1969] 00734
Barros de, C.M. [1970] 00818
Beni, J. [1968] 01035
Bielinski, K. [1977] 52151
Bily, J. [1974] 01205
Boffa, M. [1967] 01404
Boffa, M. [1973] 03930
Boffa, M. [1982] 34190
Boffa, M. [1983] 37638
Bruijn de, N.G. [1975] 21509
Bunge, Marta C. [1984] 44974
Cantini, A. [1983] 41846
Canty, J.T. [1970] 27980
Casari, E. [1971] 01884
Chauvin, A. [1961] 01999
Chauvin, A. [1961] 25473
Conway, J.H. [1976] 71909
Corrada, M. [1980] 54037
Costa da, N.C.A. [1972] 02485
Costa da, N.C.A. [1974] 02486
Cuda, K. [1977] 32281
Cuda, K. [1980] 56506
Cuda, K. [1983] 37690
Cuda, K. [1984] 39637
Doets, H.C. [1969] 03072
Ebbinghaus, H.-D. [1980] 41134
Feferman, S. [1979] 56002
Forster, T.E. [1982] 37116
Forti, M. [1983] 37413
Fourman, M.P. [1979] 53155
Fourman, M.P. [1980] 56570
Friedman, H.M. [1974] 04656
Gallin, D. [1975] 29739
Gloede, K. [1977] 26465
Gold, A.J. [1975] 62082
Gordeev, L.N. [1982] 37615
Gordon, E.I. [1983] 34454
Grayson, R.J. [1981] 54421
Grewe, R. [1969] 05335
Guzicki, W. [1977] 54755
Hajek, P. [1964] 05512
Hajek, P. [1965] 05514
Hajek, P. [1971] 05523
Hajek, P. [1973] 31134
Hauschild, K. [1968] 05770
Hinnion, R. [1972] 06109
Hinnion, R. [1976] 23019
Hinnion, R. [1980] 55100
Hinnion, R. [1981] 34023
Iwanus, B. [1973] 17553
Jech, T.J. [1971] 29869
Kakuda, Y. [1980] 74609
Kakuda, Y. [1981] 47772
Kaufmann, M. [1983] 37512
Kaufmann, M. [1984] 40711
Krajewski, S. [1974] 07445
Krajewski, S. [1977] 31406
Krejnovich, V.Ya. [1982] 35653
Kuehnrich, M. [1971] 07616
Kuehnrich, M. [1978] 53807
Kuehnrich, M. [1980] 55372
Kurata, R. [1964] 07649
Lake, J. [1975] 07800
Lolli, G. [1975] 31828
Marek, W. [1973] 08748
Marek, W. [1975] 08758
Marek, W. [1975] 23020
Marek, W. [1976] 24412
Marek, W. [1977] 53147
Marek, W. [1978] 28323
Marek, W. [1980] 76030
McCall, S. [1967] 08949
Mlcek, J. [1972] 09207
Mlcek, J. [1985] 48312
Moschovakis, Y.N. [1971] 09520
Mostowski, Andrzej [1953] 09553
Mostowski, Andrzej [1969] 21062
Mostowski, Andrzej [1976] 23738
Mostowski, Andrzej [1976] 23791
Novak, I.L. [1950] 09999
Osius, G. [1973] 27669
Osius, G. [1975] 18320
Oswald, U. [1982] 34396
Petry, A. [1982] 34397
Phillips, N.C.K. [1968] 10454
Platek, R.A. [1971] 10530
Powell, W.C. [1976] 14762
Pudlak, P. [1977] 32425
Pudlak, P. [1984] 42495
Pudlak, P. [1985] 47557
Raskovic, M. [1981] 33731
Ratajczyk, Z. [1979] 56492
Ratajczyk, Z. [1981] 55630
Resl, M. [1979] 77752
Richter, M.M. [1982] 34209
Rieger, L. [1957] 11188
Rieger, L. [1959] 11189
Rimscha von, M. [1980] 53804
Rimscha von, M. [1983] 40517
Rosser, J.B. [1950] 11558
Schlipf, J.S. [1980] 54283
Schock, R. [1974] 17446
Schultz, Konrad [1970] 13035
Schultz, Konrad [1980] 54380
Schwartz, Dietrich [1970] 13065
Shimoda, M. [1981] 54971
Sochor, A. [1975] 12625
Sochor, A. [1975] 13887
Sochor, A. [1975] 21715
Sochor, A. [1981] 54239
Sochor, A. [1982] 36081
Sochor, A. [1983] 37683
Sochor, A. [1985] 48318
Sochor, A. [1985] 48319
Specker, E. [1958] 12669
Stachniak, Z. [1981] 55417
Stepanek, P. [1974] 13172
Takeuti, G. [1981] 34204
Tzouvaras, A.D. [1985] 48451
Umezawa, T. [1965] 13786
Umezawa, T. [1966] 13787
Vetulani, Z. [1977] 31523
Vopenka, P. [1962] 16336
Vopenka, P. [1962] 21080
Vopenka, P. [1962] 21081
Vopenka, P. [1963] 22005
Vopenka, P. [1963] 22007
Vopenka, P. [1964] 24879
Vopenka, P. [1965] 13875
Vopenka, P. [1965] 13876
Vopenka, P. [1965] 27531
Wilson, T.P. [1981] 56490
Zbierski, P. [1978] 29200
Zhang, Jinwen [1980] 36145

E75

Antonovskij, M.Ya. [1981] 53987
Aribaud, F. [1979] 53226
Bankston, P. [1977] 31009
Bankston, P. [1977] 33381
Bankston, P. [1979] 33384
Bankston, P. [1979] 52789
Bankston, P. [1983] 34000
Beauchemin, P. [1974] 17974
Beauchemin, P. [1975] 25817
Becker, J.A. [1980] 54459
Benioff, P.A. [1976] 30647
Beslagic, A. [1985] 48251
Boerger, R. [1985] 49224

Bonnet, R. [1980] 55060
Bonnet, R. [1985] 38773
Bukovsky, L. [1979] 53889
Burns, R.G. [1984] 47165
Chudnovsky, D.V. [1972] 02580
Comfort, W.W. [1974] 16659
Conway, J.H. [1976] 71909
Cuda, K. [1984] 39637
Dacunha-Castelle, D. [1972] 17036
Dacunha-Castelle, D. [1972] 17234
Dacunha-Castelle, D. [1972] 81004
Dacunha-Castelle, D. [1972] 81005
Dacunha-Castelle, D. [1975] 81007
Dacunha-Castelle, D. [1977] 81006
Dales, H.G. [1977] 81016
Dow, A. [1985] 48304
Dugas, M. [1982] 37825
Dugas, M. [1982] 45871
Dugas, M. [1983] 37486
Dugas, M. [1983] 48322
Dugas, M. [1984] 36775
Dugas, M. [1984] 46405
Dugas, M. [1985] 42704
Eda, K. [1982] 37160
Eda, K. [1983] 37161
Eda, K. [1983] 37470
Eda, K. [1983] 47673
Eda, K. [1984] 42939
Eda, K. [1984] 44306
Ehrenfeucht, A. [1956] 03232
Eklof, P.C. [1975] 18127
Eklof, P.C. [1976] 26077
Eklof, P.C. [1977] 16626
Eklof, P.C. [1977] 30690
Eklof, P.C. [1980] 56438
Eklof, P.C. [1982] 38408
Eklof, P.C. [1983] 34809
Eklof, P.C. [1983] 36142
Eklof, P.C. [1983] 41825
Eklof, P.C. [1984] 38751
Eklof, P.C. [1985] 40414
Ellentuck, E. [1975] 14810
Ellentuck, E. [1975] 18129
Engeler, E. [1966] 03365
Erdoes, P. [1955] 03401
Esterle, J. [1977] 52920
Faber, V. [1978] 31943
Felgner, U. [1976] 25860
Felgner, U. [1984] 39998
Fenstad, J.E. [1964] 33314
Folley, K.W. [1931] 39801
Goebel, R. [1984] 48235
Goebel, R. [1985] 44953
Goldblatt, R.I. [1985] 42555
Grayson, R.J. [1981] 54421
Gurevich, Y. [1977] 50608
Hechler, S.H. [1973] 05842
Herre, H. [1979] 54679
Hewitt, E. [1948] 42963
Hickman, J.L. [1975] 21655
Hickman, J.L. [1982] 36952
Hodges, W. [1976] 30711
Hodges, W. [1981] 55551
Hodges, W. [1984] 39731
Hodkinson, I.M. [1985] 48826
Holland, W.C. [1985] 45604
Horn, A. [1955] 43634

Huber, M. [1977] 74296
Huber, M. [1979] 53089
Huber, M. [1983] 36692
Huber, M. [1983] 36726
Hutchinson, J.E. [1976] 26068
Jakubikova-Studenovska, D. [1980] 56042
Jehne, W. [1977] 81712
Jensen, C.U. [1980] 53472
Jensen, C.U. [1980] 54316
Jensen, C.U. [1982] 46674
Jonsson, B. [1949] 41606
Kertesz, A. [1974] 16665
Klimovsky, G. [1962] 07236
Koppelberg, S. [1974] 39872
Koppelberg, S. [1985] 48309
Krishnan, V.S. [1953] 41162
Krivine, J.-L. [1969] 07561
Krivine, J.-L. [1984] 48084
Kusraev, A.G. [1982] 34464
Kusraev, A.G. [1983] 44867
Laeuchli, H. [1962] 07783
Leavitt, W.G. [1979] 53745
Lindenbaum, A. [1927] 30918
Luxemburg, W.A.J. [1962] 08426
Luxemburg, W.A.J. [1969] 24816
Makkai, M. [1977] 31868
Mannila, H. [1983] 40998
Marco de, G. [1971] 33905
McKenzie, R. [1971] 08993
Megibben, D. [1983] 37469
Mekler, A.H. [1980] 53035
Mekler, A.H. [1983] 37643
Monk, J.D. [1975] 21872
Monk, J.D. [1979] 53222
Nadel, M.E. [1972] 09744
Osius, G. [1973] 27669
Ostaszewski, A.J. [1973] 10189
Ozawa, M. [1984] 44314
Pelc, A. [1982] 35588
Pietsch, A. [1974] 10492
Plotkin, J.M. [1981] 55102
Rakov, S.A. [1977] 82565
Ramsey, F.P. [1930] 28781
Ribenboim, P. [1965] 11148
Robinson, A. [1969] 11267
Rychkov, S.V. [1980] 82681
Rychkov, S.V. [1985] 45118
Ryll-Nardzewski, C. [1964] 11710
Sageev, G. [1985] 41812
Sain, I. [1979] 56752
Schmerl, J.H. [1985] 45602
Sebel'din, A.M. [1981] 38919
Seese, D.G. [1982] 33532
Sgro, J. [1976] 26086
Shelah, S. [1974] 12054
Shelah, S. [1975] 12058
Shelah, S. [1977] 27203
Shelah, S. [1978] 29138
Shelah, S. [1978] 31288
Shelah, S. [1979] 52977
Shelah, S. [1980] 55047
Shelah, S. [1981] 38748
Shelah, S. [1981] 54170
Shelah, S. [1983] 37468
Shelah, S. [1984] 38759
Shelah, S. [1984] 38762

Shelah, S. [1984] 46365
Shelah, S. [1985] 48344
Smith, Kay [1984] 42504
Stavi, J. [1975] 13137
Steprans, J. [1984] 40628
Stone, A.L. [1965] 27645
Strunkov, S.P. [1966] 13237
Takeuti, G. [1978] 52447
Takeuti, G. [1984] 39581
Thomas, S. [1985] 48871
Todorcevic, S.B. [1979] 53088
Tokarev, E.V. [1985] 45606
Urysohn, P. [1925] 16917
Urysohn, P. [1927] 16918
Vaeaenaenen, J. [1979] 79569
Wald, B. [1983] 39034

E96

Mostowski, Andrzej [1979] 82308
Novikov, P.S. [1979] 38498

E97

Harrington, L.A. [1985] 49810
Jensen, R.B. [1981] 54128

E98

Barr, M. [1985] 40111
Barwise, J. [1975] 60316
Barwise, J. [1977] 70117
Comfort, W.W. [1974] 16659
Devlin, K.J. [1974] 15073
Felgner, U. [1971] 03721
Friedman, H.M. [1975] 04296
Goldblatt, R.I. [1979] 55754
Halkowska, K. [1976] 52421
Jensen, R.B. [1967] 06596
Kleinberg, E.M. [1977] 50789
Krajewski, S. [1981] 38331
Marek, W. [1972] 34216
Mostowski, Andrzej [1965] 09578
Mostowski, Andrzej [1967] 22257
Mostowski, Andrzej [1969] 09584
Mostowski, Andrzej [1969] 21062
Rosenstein, J.G. [1982] 38089
Shoenfield, J.R. [1967] 22384
Speranza, F. [1970] 78881
Wang, Hao [1981] 34116

F05

Arzarello, F. [1975] 51166
Arzarello, F. [1976] 50556
Bowen, K.A. [1972] 01505
Carstengerdes, W. [1971] 01873
Craig, W. [1957] 02508
Craig, W. [1957] 02510
Feferman, S. [1968] 03700
Ferbus, M.-C. [1984] 42605
Gentzen, G. [1935] 24221
Girard, J.-Y. [1976] 50375
Glubrecht, J.-M. [1982] 35619
Hanazawa, M. [1979] 54329
Herbrand, J. [1930] 20866
Jervell, H.R. [1972] 33262
Jervell, H.R. [1973] 33265
Kreisel, G. [1958] 07514

Kreisel, G. [1975] 23955
Kreisel, G. [1975] 27698
Lopez-Escobar, E.G.K. [1983] 39786
Los, J. [1956] 08333
Luckhardt, H. [1967] 08378
Maehara, S. [1960] 08520
Maehara, S. [1961] 08524
Martin-Loef, P. [1972] 08833
Maslov, S.Yu. [1977] 32672
Motohashi, N. [1972] 09596
Motohashi, N. [1982] 33418
Nebres, B.F. [1972] 30771
Nishimura, T. [1974] 18312
Nishimura, T. [1974] 18313
Orevkov, V.P. [1969] 10154
Pohlers, W. [1981] 33310
Pohlers, W. [1982] 35591
Rasiowa, H. [1960] 11024
Rogava, M.G. [1979] 77850
Schroeter, K. [1958] 24841
Schuette, K. [1950] 12999
Shirai, K. [1975] 29640
Slonneger, K. [1976] 14575
Smirnov, V.A. [1972] 30359
Tait, W.W. [1968] 13286
Takeuti, G. [1978] 52447
Uesu, T. [1971] 13764
Wojtylak, P. [1984] 44096
Yashin, A.D. [1984] 44180

F07

Africk, H. [1972] 00199
Apt, K.R. [1973] 00431
Apt, K.R. [1976] 14789
Bennett, D.W. [1973] 01038
Bennett, D.W. [1977] 21957
Benthem van, J.F.A.K. [1974] 03889
Boolos, G. [1984] 40442
Engeler, E. [1961] 03362
Engeler, E. [1963] 03364
Gentzen, G. [1935] 24221
Herbrand, J. [1930] 20866
Hintikka, K.J.J. [1973] 32405
Kreisel, G. [1957] 29369
Kreisel, G. [1975] 27698
Misawa, T. [1977] 33405
Motohashi, N. [1984] 45359
Nieland, J.J.F. [1967] 09951
Norgela, S.A. [1978] 32674
Oglesby, F.C. [1961] 10067
Orevkov, V.P. [1969] 10154
Schroeter, K. [1958] 24841
Schuette, K. [1950] 12999
Smirnov, V.A. [1972] 30359
Smullyan, R.M. [1970] 12557
Wasilewska, A. [1976] 65985
Wasilewska, A. [1976] 65986
Wasilewska, A. [1980] 55216

F10

Cantini, A. [1983] 41846
Girard, J.-Y. [1977] 53265
Schwichtenberg, H. [1975] 18407
Schwichtenberg, H. [1975] 25022
Tait, W.W. [1965] 13281

F15

Aczel, P. [1981] 34552
Bowen, K.A. [1974] 03942
Cantini, A. [1983] 41846
Feferman, S. [1968] 03700
Feferman, S. [1972] 03706
Ferbus, M.-C. [1984] 42605
Jaeger, G. [1982] 34700
Kino, A. [1963] 07100
Kino, A. [1975] 17616
Levitz, H. [1975] 18259
Markwald, W. [1954] 08779
McAloon, K. [1980] 54524
Pinus, A.G. [1975] 16662
Pohlers, W. [1981] 33310
Pohlers, W. [1982] 34316
Pohlers, W. [1982] 35591
Schwichtenberg, H. [1975] 18407
Tait, W.W. [1965] 13281
Tait, W.W. [1968] 13286

F20

Dawson Jr., J.W. [1979] 53657
Fischer, Michael J. [1974] 17344
Hartmanis, J. [1976] 24160
Kreisel, G. [1975] 23955
Kreisel, G. [1975] 27698
Longo, G. [1974] 53697
Loveland, D.W. [1978] 31164
Parikh, R. [1973] 10286
Scarpellini, B. [1984] 43196
Scarpellini, B. [1985] 47473
Woehl, K. [1979] 53367

F25

Apt, K.R. [1974] 00432
Artigue, M. [1978] 70505
Barwise, J. [1973] 04141
Benthem van, J.F.A.K. [1984] 39892
Corcoran, J. [1980] 35994
Dalla Chiara Scabia, M.L. [1968] 04285
Danko, W. [1983] 39522
Delon, F. [1978] 27313
Guaspari, D. [1979] 53266
Gurevich, Y. [1965] 05452
Gurevich, Y. [1980] 53951
Gurevich, Y. [1983] 33765
Gurevich, Y. [1984] 38749
Hajek, P. [1966] 05517
Hajek, P. [1970] 05521
Hanazawa, M. [1979] 54329
Hanf, W.P. [1965] 05635
Hauschild, K. [1971] 05780
Hauschild, K. [1971] 27450
Hauschild, K. [1972] 05787
Hauschild, K. [1972] 05790
Herbrand, J. [1930] 20866
Hodges, W. [1980] 54226
Hook, J.L. [1985] 41804
Ito, Makoto [1934] 32185
Jambu-Giraudet, M. [1980] 54167
Jambu-Giraudet, M. [1981] 55535
Korec, I. [1974] 07371
Korec, I. [1976] 07374
Krajewski, S. [1974] 07445
Kreisel, G. [1958] 07514

Krynicki, M. [1985] 49048
Maehara, S. [1958] 08519
Marek, W. [1976] 24412
Marek, W. [1978] 76016
McAloon, K. [1975] 18286
Montague, R. [1962] 09425
Montague, R. [1965] 09431
Mostowski, Andrzej [1952] 09550
Motohashi, N. [1977] 23750
Orey, S. [1961] 10166
Pinter, C. [1978] 56267
Pinus, A.G. [1982] 35574
Prazmowski, K. [1976] 18347
Prazmowski, K. [1978] 29177
Pudlak, P. [1985] 42556
Rosser, J.B. [1950] 11558
Sette, A.-M. [1980] 53363
Sette, A.-M. [1985] 48209
Shelah, S. [1976] 51434
Shelah, S. [1984] 38758
Smorynski, C.A. [1977] 27327
Szczerba, L.W. [1980] 79172
Szczerba, L.W. [1983] 39510
Takeuti, G. [1957] 13310
Takeuti, G. [1957] 13311
Weese, M. [1976] 23814

F30

Aczel, P. [1981] 34552
Adamowicz, Z. [1984] 45406
Adamowicz, Z. [1985] 39780
Adamowicz, Z. [1985] 49053
Apelt, H. [1966] 00403
Barnes, D.W. [1974] 03863
Bel'tyukov, A.P. [1976] 29789
Berline, C. [1981] 54902
Berman, L. [1977] 70972
Berman, L. [1980] 55471
Bernardi, C. [1975] 31012
Bezboruah, A. [1976] 14785
Boolos, G. [1976] 14578
Boolos, G. [1977] 26442
Boolos, G. [1979] 56312
Bouvere de, K.L. [1959] 01501
Cantini, A. [1980] 54349
Carstens, H.G. [1977] 51961
Cegielski, P. [1980] 55770
Cegielski, P. [1981] 55626
Cegielski, P. [1982] 34553
Cegielski, P. [1984] 46773
Clote, P. [1980] 56493
Clote, P. [1981] 55425
Clote, P. [1983] 40361
Clote, P. [1985] 41801
Clote, P. [1985] 45298
Cobham, A. [1957] 29385
Cooper, D.C. [1972] 61080
Csirmaz, L. [1979] 72011
Cuda, K. [1983] 37690
Dawson Jr., J.W. [1979] 53657
DeMillo, R.A. [1979] 72280
Dimitracopoulos, C. [1983] 35043
Dimitracopoulos, C. [1985] 47569
Dries van den, L. [1980] 54246
Ehrenfeucht, A. [1976] 25797
Elgot, C.C. [1961] 03288
Elgot, C.C. [1966] 03291

Enderton, H.B. [1972] 03355
Fischer, Michael J. [1974] 17344
Fol'k, N.F. [1968] 17126
Fridman, Eh.I. [1973] 23081
Friedman, H.M. [1971] 17088
Friedrichsdorf, U. [1976] 24288
Gentzen, G. [1935] 24221
Germano, G. [1971] 04907
Germano, G. [1973] 04910
Girard, J.-Y. [1977] 53265
Goodstein, R.L. [1955] 05177
Goodstein, R.L. [1963] 05189
Goodstein, R.L. [1965] 05187
Goodstein, R.L. [1966] 05190
Goodstein, R.L. [1969] 32559
Grzegorczyk, A. [1957] 24864
Grzegorczyk, A. [1958] 05398
Grzegorczyk, A. [1971] 23119
Guaspari, D. [1979] 53266
Gurari, E.M. [1981] 54218
Hajek, P. [1977] 26853
Hajek, P. [1981] 35314
Hajek, P. [1984] 43006
Harrington, L.A. [1977] 27334
Hauck, J. [1972] 62383
Hauschild, K. [1971] 05779
Henkin, L. [1954] 05900
Henkin, L. [1957] 05910
Henson, C.W. [1984] 36539
Henson, C.W. [1984] 39860
Herbrand, J. [1930] 20866
Herbrand, J. [1931] 16812
Ivanov, A.A. [1984] 41869
Jambu-Giraudet, M. [1981] 55535
Kalmar, L. [1928] 06835
Kemeny, J.G. [1958] 07053
Kirby, L.A.S. [1977] 50962
Kirby, L.A.S. [1978] 56399
Kirby, L.A.S. [1980] 54429
Kirby, L.A.S. [1982] 34416
Kirby, L.A.S. [1982] 35381
Kireevskij, N.N. [1958] 40811
Kossak, R. [1981] 33723
Kossak, R. [1985] 42582
Krajewski, S. [1974] 07447
Kreisel, G. [1950] 07481
Kreisel, G. [1953] 20815
Kripke, S.A. [1962] 07556
Kurata, R. [1981] 47747
Kurata, R. [1984] 42937
Lascar, D. [1980] 54430
Levitz, H. [1975] 18259
Levy, A. [1958] 24860
Lipton, R.J. [1978] 75578
Livchak, A.B. [1976] 55788
Livchak, A.B. [1977] 33959
Lopez-Escobar, E.G.K. [1968] 08264
Loveland, D.W. [1978] 31164
Macintyre, A. [1980] 54621
Maehara, S. [1957] 08516
Maehara, S. [1958] 08519
Manevitz, L.M. [1980] 54428
Manin, Yu.I. [1977] 51984
Marek, W. [1976] 24412
Mart'yanov, V.I. [1977] 29230
McAloon, K. [1975] 17403
McAloon, K. [1977] 55797
McAloon, K. [1978] 52282
McAloon, K. [1978] 54427
McAloon, K. [1980] 54523
McAloon, K. [1980] 54524
McAloon, K. [1980] 70045
McAloon, K. [1981] 34009
Mills, G. [1984] 42484
Misercque, D. [1981] 76440
Misercque, D. [1982] 36085
Mlcek, J. [1978] 31964
Moller, S. [1965] 48096
Molodshij, V.N. [1952] 09388
Montagna, F. [1983] 34635
Montagna, F. [1984] 42567
Montague, R. [1962] 09425
Montague, R. [1963] 14478
Morgenstern, C.F. [1982] 34161
Mostowski, Andrzej [1952] 09550
Mueller, Gert H. [1961] 09619
Mundici, D. [1983] 34976
Negri, M. [1984] 44715
Nishimura, T. [1965] 09975
Novikov, P.S. [1949] 19476
Oppen, D.C. [1973] 24077
Oppen, D.C. [1978] 51889
Orey, S. [1956] 10163
Pabion, J.F. [1981] 33728
Parikh, R. [1971] 10282
Parikh, R. [1973] 10286
Paris, J.B. [1978] 29247
Paris, J.B. [1978] 31967
Paris, J.B. [1980] 56654
Paris, J.B. [1981] 33729
Paris, J.B. [1981] 55495
Paris, J.B. [1984] 46698
Paris, J.B. [1985] 41796
Penzin, Yu.G. [1973] 10357
Penzin, Yu.G. [1973] 10358
Penzin, Yu.G. [1973] 23000
Penzin, Yu.G. [1979] 32595
Presburger, M. [1930] 10749
Pudlak, P. [1983] 36550
Putnam, H. [1964] 10838
Quine, W.V.O. [1949] 10881
Rabin, M.O. [1961] 10934
Ratajczyk, Z. [1982] 35382
Richard, D. [1984] 41779
Robinson, Julia [1973] 29547
Robinson, R.M. [1951] 11312
Rousseau, G. [1966] 11593
Rousseau, G. [1968] 11598
Scarpellini, B. [1984] 43196
Schmerl, U.R. [1984] 43011
Scott, D.S. [1962] 11914
Semenov, A.L. [1979] 53242
Shelah, S. [1984] 38758
Shepherdson, J.C. [1964] 12075
Shepherdson, J.C. [1965] 12072
Shepherdson, J.C. [1967] 12077
Shepherdson, J.C. [1969] 12078
Shostak, R.E. [1977] 69916
Shostak, R.E. [1979] 82799
Skolem, T.A. [1931] 12388
Skolem, T.A. [1933] 12392
Skolem, T.A. [1934] 12393
Skolem, T.A. [1955] 27718
Smorynski, C.A. [1977] 27327
Smorynski, C.A. [1978] 54148
Smorynski, C.A. [1982] 37523
Smorynski, C.A. [1982] 43687
Smorynski, C.A. [1984] 41837
Smorynski, C.A. [1985] 49880
Tait, W.W. [1968] 13286
Takeuti, G. [1957] 13311
Takeuti, G. [1978] 52447
Tarski, A. [1932] 13406
Tarski, A. [1933] 13407
Tarski, A. [1933] 28816
Tsinman, L.L. [1969] 02178
Tverskoj, A.A. [1980] 35246
Twer von der, T. [1981] 35072
Uesu, T. [1970] 13763
Ursini, A. [1977] 52907
Vaeaenaenen, J. [1979] 52749
Visser, A. [1984] 42643
Vuckovic, V. [1967] 27478
Wang, Hao [1953] 14024
Wang, Hao [1963] 14049
Wang, Hao [1981] 34116
Wauw-de Kinder van de, G. [1975] 18421
Wilkie, A.J. [1978] 56743
Wilkie, A.J. [1980] 79968
Wilkie, A.J. [1982] 34735
Wilmers, G.M. [1985] 40251
Woehl, K. [1979] 53367
Wolter, H. [1973] 14280
Wolter, H. [1975] 66137
Zbierski, P. [1980] 34199

F35

Andrews, P.B. [1972] 00366
Andrews, P.B. [1972] 00367
Andrews, P.B. [1974] 03832
Apt, K.R. [1973] 00431
Apt, K.R. [1974] 00432
Apt, K.R. [1976] 14789
Apt, K.R. [1978] 29131
Artigue, M. [1978] 70505
Barr, M. [1985] 40111
Bauval, A. [1985] 48368
Belyakin, N.V. [1970] 29065
Belyakin, N.V. [1983] 41479
Bily, J. [1974] 01205
Boffa, M. [1983] 37638
Buechi, J.R. [1960] 01690
Buechi, J.R. [1962] 01693
Buechi, J.R. [1969] 01700
Cantini, A. [1983] 41846
Dalen van, D. [1978] 27959
Dalen van, D. [1984] 39878
Delon, F. [1978] 27313
Elgot, C.C. [1966] 03291
Enderton, H.B. [1967] 03347
Enderton, H.B. [1971] 03353
Feferman, S. [1965] 03696
Feferman, S. [1968] 03703
Ferrante, J. [1975] 21702
Fourman, M.P. [1979] 53155
Fourman, M.P. [1979] 53715
Friedman, H.M. [1970] 14546
Friedman, H.M. [1971] 17088
Friedman, H.M. [1973] 04654
Friedrich, U. [1972] 04668

Gandy, R.O. [1960] 15061
Gandy, R.O. [1961] 22330
Girard, J.-Y. [1976] 50375
Girard, J.-Y. [1984] 48565
Goldblatt, R.I. [1979] 55754
Grayson, R.J. [1981] 54421
Grayson, R.J. [1984] 41840
Grzegorczyk, A. [1971] 23119
Guzicki, W. [1974] 62249
Guzicki, W. [1981] 34160
Guzicki, W. [1983] 37520
Harnik, V. [1985] 39778
Henkin, L. [1950] 05894
Henson, C.W. [1984] 39860
Hodges, W. [1980] 54226
Hoeven van der, G.F. [1984] 41155
Hoeven van der, G.F. [1984] 41766
Jaeger, G. [1982] 34700
Jambu-Giraudet, M. [1980] 54167
Jankowski, A.W. [1985] 47500
Jensen, C.U. [1984] 39754
Jongh de, D.H.J. [1976] 18214
Kogalovskij, S.R. [1968] 07299
Kreisel, G. [1963] 42740
Kreisel, G. [1975] 23955
Krol', M.D. [1976] 50223
Krol', M.D. [1977] 52085
Krol', M.D. [1978] 52703
Krol', M.D. [1978] 53324
Lopez-Escobar, E.G.K. [1967] 08262
Macintyre, A. [1980] 54621
Maehara, S. [1971] 08528
Marek, W. [1975] 08758
Marek, W. [1976] 24412
Marek, W. [1978] 29210
Marek, W. [1978] 76016
Moerdijk, I. [1984] 42485
Mostowski, A.Wlodzimierz [1981] 55152
Mostowski, Andrzej [1952] 09550
Mostowski, Andrzej [1961] 24858
Nabebin, A.A. [1977] 52155
Nepejvoda, N.N. [1979] 53497
Nishimura, T. [1965] 09975
Pohlers, W. [1981] 33310
Pohlers, W. [1982] 34316
Pohlers, W. [1982] 35591
Scarpellini, B. [1985] 47473
Schultz, Konrad [1980] 54380
Schwabhaeuser, W. [1979] 53471
Schwichtenberg, H. [1975] 18407
Scowcroft, P. [1984] 44277
Shore, R.A. [1981] 55426
Siefkes, D. [1970] 12145
Siefkes, D. [1970] 27984
Siefkes, D. [1971] 12146
Simpson, S.G. [1982] 35660
Simpson, S.G. [1985] 49879
Steel, J.R. [1978] 29156
Tait, W.W. [1968] 13286
Takeuti, G. [1978] 52447
Tanaka, H. [1974] 23068
Tarski, A. [1931] 16924
Uesu, T. [1971] 13764
Vetulani, Z. [1984] 44142
Weinstein, S. [1979] 53943
Zarach, A. [1976] 23816
Zbierski, P. [1971] 14379

Zbierski, P. [1978] 29200
Zbierski, P. [1981] 34106

F50

Artemov, S.N. [1980] 54546
Barr, M. [1985] 40111
Blass, A.R. [1983] 40316
Bozzi, S. [1977] 52446
Bozzi, S. [1980] 54221
Bozzi, S. [1982] 40429
Dahn, B.I. [1981] 35717
Dalen van, D. [1978] 27959
Dalen van, D. [1984] 39878
Dragalin, A.G. [1973] 29696
Ershov, Yu.L. [1973] 30533
Ershov, Yu.L. [1974] 25764
Felscher, W. [1976] 04239
Fitting, M. [1969] 04349
Fitting, M. [1970] 14547
Fourman, M.P. [1979] 53155
Fourman, M.P. [1979] 53715
Fourman, M.P. [1980] 56570
Friedman, H.M. [1974] 04656
Gabbay, D.M. [1970] 04721
Gabbay, D.M. [1971] 04726
Gabbay, D.M. [1972] 04729
Gabbay, D.M. [1972] 04730
Gabbay, D.M. [1973] 17911
Gabbay, D.M. [1973] 61880
Gabbay, D.M. [1975] 04738
Gabbay, D.M. [1975] 17286
Gabbay, D.M. [1977] 26456
Gabbay, D.M. [1977] 26463
Gabbay, D.M. [1982] 38464
Geiser, J.R. [1974] 04831
Gentzen, G. [1935] 24221
Georgieva, N.V. [1973] 20996
Girard, J.-Y. [1976] 50375
Girard, J.-Y. [1977] 53265
Goad, C.A. [1978] 73424
Goernemann, S. [1971] 39863
Goldblatt, R.I. [1979] 55754
Goldblatt, R.I. [1981] 55422
Golota, Ya.Ya. [1969] 05124
Golota, Ya.Ya. [1974] 15236
Gordeev, L.N. [1982] 37615
Grayson, R.J. [1981] 54421
Grayson, R.J. [1984] 41840
Heijenoort van, J. [1979] 79641
Henkin, L. [1956] 05908
Hoeven van der, G.F. [1984] 41155
Hoeven van der, G.F. [1984] 41766
Jankowski, A.W. [1985] 47500
Johnstone, P.T. [1977] 51354
Johnstone, P.T. [1979] 56505
Johnstone, P.T. [1979] 74552
Jongh de, D.H.J. [1976] 18214
Jongh de, D.H.J. [1980] 55677
Josza, R. [1984] 45334
Joyal, A. [1975] 27668
Kalicki, C. [1980] 53776
Kanovich, M.I. [1975] 18218
Kearns, J.T. [1978] 31151
Kirk, R.E. [1979] 53554
Kirk, R.E. [1980] 56546
Klemke, D. [1971] 07229
Kreisel, G. [1958] 07514

Kreisel, G. [1967] 07533
Kreisel, G. [1975] 23955
Kreisel, G. [1975] 27698
Kripke, S.A. [1965] 07557
Krol', M.D. [1976] 50223
Krol', M.D. [1977] 52085
Krol', M.D. [1978] 52703
Krol', M.D. [1978] 53324
Leblanc, H. [1983] 37685
Lifschitz, V. [1967] 33665
Lopez-Escobar, E.G.K. [1967] 08263
Lopez-Escobar, E.G.K. [1968] 08264
Lopez-Escobar, E.G.K. [1975] 24555
Lopez-Escobar, E.G.K. [1980] 55150
Lopez-Escobar, E.G.K. [1981] 55143
Lopez-Escobar, E.G.K. [1983] 39786
Los, J. [1956] 08333
Luckhardt, H. [1969] 08380
Luckhardt, H. [1970] 08381
Luckhardt, H. [1970] 08382
Macnab, D.S. [1977] 51394
Maehara, S. [1970] 08455
Makkai, M. [1977] 50433
Marchini, C. [1982] 41484
Marini, D. [1975] 22915
Markov, A.A. [1974] 63716
Markov, A.A. [1974] 63722
Markovic, Z. [1977] 51169
Markovic, Z. [1979] 56384
Markovic, Z. [1983] 37386
Markovic, Z. [1984] 45080
Martin-Loef, P. [1972] 08833
Martin-Loef, P. [1972] 21229
McCall, S. [1962] 08942
Miglioli, P.A. [1981] 55311
Mints, G.E. [1966] 19428
Mints, G.E. [1968] 19418
Moerdijk, I. [1984] 42485
Motohashi, N. [1977] 23750
Motohashi, N. [1984] 45359
Mulvey, C.J. [1974] 21534
Muravitskij, A.Yu. [1981] 36087
Nadel, M.E. [1978] 29150
Nadiu, G.S. [1971] 09750
Nagashima, T. [1966] 09755
Nepejvoda, N.N. [1979] 53497
Nepejvoda, N.N. [1982] 46228
Ono, H. [1985] 47510
Orevkov, V.P. [1976] 64291
Osius, G. [1973] 27669
Osius, G. [1975] 18320
Osius, G. [1975] 18321
Osswald, H. [1970] 10184
Osswald, H. [1972] 10186
Pil'chak, B.Yu. [1950] 16889
Popov, S.V. [1980] 80198
Rasiowa, H. [1953] 11004
Rasiowa, H. [1955] 11010
Rasiowa, H. [1955] 11011
Rasiowa, H. [1963] 11028
Rauszer, C. [1977] 50668
Rauszer, C. [1977] 50669
Rauszer, C. [1980] 56381
Reichbach, J. [1973] 31998
Reichbach, J. [1975] 32003
Renardel de Lavalette, G.R. [1981] 55207

Reyes, G.E. [1974] 11131
Reyes, G.E. [1975] 64812
Reyes, G.E. [1977] 52660
Reyes, G.E. [1978] 56342
Reznikoff, I. [1963] 11135
Reznikoff, I. [1966] 11137
Romanski, J. [1977] 51332
Routley, R. [1974] 15192
Routley, R. [1978] 53083
Sanchis, L.E. [1973] 11828
Scedrov, A. [1984] 37128
Schmidt, H.A. [1958] 12891
Schuette, K. [1950] 12999
Schuette, K. [1962] 13012
Schuette, K. [1968] 24822
Schultz, Konrad [1980] 54380
Schwichtenberg, H. [1975] 18407
Scott, D.S. [1980] 78323
Scowcroft, P. [1984] 44277
Shanin, N.A. [1976] 65058
Shanin, N.A. [1977] 53812
Shimoda, M. [1981] 54971
Shirai, K. [1975] 29640
Skvortsov, D.P. [1979] 55940
Smorynski, C.A. [1973] 12535
Smorynski, C.A. [1973] 27058
Smorynski, C.A. [1977] 26856
Smorynski, C.A. [1982] 37523
Sobolev, S.K. [1977] 51014
Sonobe, O. [1975] 51586
Stepanov, V.I. [1983] 39758
Swart de, H.C.M. [1977] 26858
Swart de, H.C.M. [1978] 29237
Tait, W.W. [1965] 13281
Tauts, A. [1972] 17695
Tauts, A. [1972] 65673
Tauts, A. [1974] 17693
Tauts, A. [1974] 17694
Tauts, A. [1975] 18414
Tauts, A. [1975] 52459
Tauts, A. [1975] 52460
Tauts, A. [1981] 38114
Troelstra, A.S. [1977] 32501
Troelstra, A.S. [1978] 27952
Wajsberg, M. [1938] 13982
Wauw-de Kinder van de, G. [1975] 18421
Weinstein, S. [1979] 53943
Yankov, V.A. [1967] 06513
Yashin, A.D. [1984] 44104
Yashin, A.D. [1984] 44180

F55

Blass, A.R. [1983] 40316
Bozzi, S. [1977] 52446
Delzell, C.N. [1982] 36898
Delzell, C.N. [1982] 38556
Delzell, C.N. [1984] 43271
Girstmair, K. [1979] 56289
Grayson, R.J. [1981] 54421
Henkin, L. [1971] 05928
Hermann, G. [1926] 05997
Mulvey, C.J. [1974] 21534
Seidenberg, A. [1970] 41695
Seidenberg, A. [1971] 41694
Seidenberg, A. [1973] 41700
Seidenberg, A. [1975] 41698

Smith, Rick L. [1981] 55685
Waerden van der, B.L. [1930] 38703

F60

Dobritsa, V.P. [1978] 32600
Goncharov, S.S. [1976] 52189
Goncharov, S.S. [1982] 40426
Goodstein, R.L. [1955] 05177
Khisamiev, N.G. [1974] 06147
Khisamiev, N.G. [1979] 32601
Khisamiev, N.G. [1983] 40239
Kosovskij, N.K. [1970] 28600
Kudajbergenov, K.Zh. [1984] 41816
Kudajbergenov, K.Zh. [1984] 41871
Madison, E.W. [1970] 24826
Madison, E.W. [1975] 18269
Mal'tsev, A.I. [1961] 28706
Metakides, G. [1982] 37391
Morozov, A.S. [1983] 43123
Morozov, A.S. [1984] 41757
Moschovakis, Y.N. [1964] 09514
Moschovakis, Y.N. [1966] 21265
Roche la, P. [1981] 75255
Tait, W.W. [1968] 13286

F65

Feferman, S. [1975] 22905
Lipton, R.J. [1978] 75578
Markov, A.A. [1974] 63716
Parikh, R. [1971] 10282
Vetulani, Z. [1984] 44142

F97

Dejon, B. [1969] 23438
Diller, J. [1975] 21771
Harrington, L.A. [1985] 49810
Lawvere, F.W. [1975] 70191
Loeb, M.H. [1968] 37293
Lolli, G. [1984] 41493

F98

Barwise, J. [1977] 70117
Boolos, G. [1979] 56312
Dalla Chiara Scabia, M.L. [1968] 04285
Feferman, S. [1968] 03700
Fitting, M. [1969] 04349
Friedman, H.M. [1975] 04296
Goldblatt, R.I. [1979] 55754
Grzegorczyk, A. [1971] 23119
Lorenzen, P. [1962] 08303
Mal'tsev, A.I. [1961] 28706
Mostowski, Andrzej [1965] 09578
Pohlers, W. [1982] 34316
Rasiowa, H. [1963] 11028
Schuette, K. [1968] 24822
Shoenfield, J.R. [1967] 22384
Siefkes, D. [1970] 12145
Simpson, S.G. [1985] 49879
Smirnov, V.A. [1972] 30359
Smorynski, C.A. [1977] 27327
Smorynski, C.A. [1984] 41837
Smorynski, C.A. [1985] 49880
Wang, Hao [1981] 34116
Wraith, G.C. [1976] 80097

F99

Herbrand, J. [1931] 16812
Krajewski, S. [1981] 38331
Kreisel, G. [1956] 90276
Kreisel, G. [1957] 29369
Mostowski, Andrzej [1958] 09566
Wajsberg, M. [1933] 40808
Wette, E. [1959] 14162

G30

Adamek, J. [1983] 36993
Bankston, P. [1983] 34000
Bankston, P. [1983] 37051
Barr, M. [1985] 40111
Blass, A.R. [1983] 34149
Blass, A.R. [1983] 34410
Blass, A.R. [1983] 40316
Boffa, M. [1981] 33712
Bowen, K.A. [1973] 01506
Bunge, Marta C. [1981] 35077
Bunge, Marta C. [1982] 34013
Bunge, Marta C. [1984] 44974
Cole, J.C. [1973] 04015
Cole, J.C. [1979] 53887
Costa da, N.C.A. [1972] 02485
Costa da, N.C.A. [1974] 02486
Coste, M. [1973] 27665
Coste, M. [1979] 71955
Coste, M. [1979] 80976
Dapueto, C. [1982] 34475
Diaconescu, R. [1976] 24062
Diers, Y. [1983] 37808
Dubuc, E.J. [1979] 54769
Ehresmann, C. [1965] 23592
Engeler, E. [1966] 03365
Eytan, M. [1979] 53157
Feferman, S. [1972] 03706
Ferbus, M.-C. [1984] 42605
Fourman, M.P. [1979] 53155
Fourman, M.P. [1980] 56570
Gaifman, H. [1974] 04750
Goldblatt, R.I. [1979] 55754
Goldblatt, R.I. [1981] 55422
Goldblatt, R.I. [1985] 42555
Gordon, E.I. [1983] 34454
Grayson, R.J. [1981] 54421
Grayson, R.J. [1984] 41840
Gurevich, R. [1982] 33565
Hajek, P. [1970] 05521
Hedrlin, Z. [1969] 05863
Hien, Buihuy [1981] 37210
Hien, Buihuy [1983] 36725
Hien, Buihuy [1983] 45605
Hodges, W. [1975] 30712
Hoeven van der, G.F. [1984] 41155
Horiguchi, H. [1985] 41746
Huber, M. [1983] 36692
Isbell, J.R. [1983] 37803
Jankowski, A.W. [1985] 47500
Johnstone, P.T. [1977] 51354
Johnstone, P.T. [1979] 53816
Johnstone, P.T. [1979] 56505
Johnstone, P.T. [1979] 74552
Johnstone, P.T. [1980] 56515
Josza, R. [1984] 45334
Joyal, A. [1975] 27668

Kennison, J.F. [1976] 24063
Kennison, J.F. [1979] 39875
Kock, A. [1971] 28673
Kogalovskij, S.R. [1961] 33856
Kuehnrich, M. [1978] 53807
Kusraev, A.G. [1983] 40491
Lambek, J. [1982] 36913
Lascar, D. [1982] 34164
Lavendhomme, R. [1980] 54095
Lesaffre, B. [1973] 82053
Macnab, D.S. [1977] 51394
Makkai, M. [1976] 25850
Makkai, M. [1977] 50433
Makkai, M. [1980] 56746
Makkai, M. [1981] 75838
Makkai, M. [1982] 35151
Makkai, M. [1982] 37439
Makkai, M. [1984] 44303
Makkai, M. [1985] 41797
Mal'tsev, A.I. [1958] 08607
Mal'tsev, A.I. [1958] 41176
Mal'tsev, A.I. [1958] 90269
Marchini, C. [1973] 18276
Marchini, C. [1982] 41484
Mekler, A.H. [1978] 52883
Mendelsohn, E. [1969] 09101
Mijoule, R. [1976] 25854
Moens, J.L. [1982] 48346
Moerdijk, I. [1984] 42485
Mulvey, C.J. [1974] 21534
Mundici, D. [1984] 41817
Nelson, Evelyn [1975] 21874
Ohkuma, T. [1966] 10081
Ohkuma, T. [1973] 10083
Osius, G. [1973] 27669
Osius, G. [1975] 18320
Osius, G. [1975] 18321
Pashenkov, V.V. [1975] 18325
Pitts, A.M. [1983] 36693
Pitts, A.M. [1983] 37085
Radu, A.G. [1981] 54934
Rama Rao, V.V. [1975] 24450
Reyes, G.E. [1974] 11131
Reyes, G.E. [1975] 64812
Reyes, G.E. [1977] 52660
Reyes, G.E. [1978] 56342
Roisin, J.-R. [1977] 77863
Rosicky, J. [1981] 55500
Sabbagh, G. [1972] 11730
Sain, I. [1983] 43417
Scedrov, A. [1984] 37128
Schumacher, D. [1976] 82749
Scott, D.S. [1980] 78323
Servi, M. [1979] 82777
Shelah, S. [1976] 51434
Shimoda, M. [1979] 47660
Shimoda, M. [1981] 54971
Takahashi, S. [1974] 17488
Takeuti, G. [1981] 34204
Tierney, M. [1972] 17528
Trnkova, V. [1971] 83011
Veit Riccioli, B. [1978] 53930
Volger, H. [1975] 17227
Volger, H. [1975] 17228
Volger, H. [1975] 18422

Wauw-de Kinder van de, G. [1975] 18421
Wraith, G.C. [1976] 80097
Wraith, G.C. [1979] 53387
Zawadowski, M. [1983] 42355
Zawadowski, M. [1985] 47521

H05

Abian, A. [1974] 03810
Adler, A. [1973] 00190
Barone, E. [1980] 80631
Belyakin, N.V. [1979] 70865
Bernstein, A.R. [1973] 03898
Bertossi, L. [1985] 41802
Bunyatov, M.R. [1967] 02032
Button, R.W. [1979] 80839
Chadwick, J.J.M. [1977] 80867
Charretton, C. [1973] 03982
Cherlin, G.L. [1972] 14684
Connes, A. [1970] 02352
Connes, A. [1970] 26249
Cozart, D. [1974] 02503
Cuda, K. [1980] 56506
Fenstad, J.E. [1970] 17153
Gol'dshtejn, B.G. [1983] 45227
Grainger, A.D. [1975] 17507
Hatcher, W.S. [1985] 48208
Heinrich, S. [1980] 56660
Heinrich, S. [1980] 81516
Heinrich, S. [1981] 53749
Heinrich, S. [1982] 37826
Heinrich, S. [1982] 46460
Henson, C.W. [1972] 05952
Henson, C.W. [1972] 22286
Henson, C.W. [1973] 05956
Henson, C.W. [1974] 05959
Henson, C.W. [1974] 05960
Henson, C.W. [1974] 05962
Henson, C.W. [1974] 26632
Henson, C.W. [1974] 26633
Henson, C.W. [1975] 05963
Henson, C.W. [1975] 17439
Henson, C.W. [1976] 26078
Henson, C.W. [1976] 46653
Henson, C.W. [1977] 30718
Henson, C.W. [1983] 38543
Henson, C.W. [1984] 39860
Herrmann, R.A. [1978] 55117
Hirschfeld, J. [1976] 26079
Hort, C. [1984] 44206
Janssen, G. [1972] 06540
Kasahara, S. [1973] 30535
Keisler, H.J. [1974] 30004
Keisler, H.J. [1976] 33913
Laugwitz, D. [1958] 48699
Lightstone, A.H. [1978] 51926
Lin, Peikee [1985] 48745
Loeb, P.A. [1985] 44489
Lolli, G. [1978] 55177
Luxemburg, W.A.J. [1962] 08426
Luxemburg, W.A.J. [1962] 21419
Luxemburg, W.A.J. [1969] 21147
Luxemburg, W.A.J. [1969] 24816

Luxemburg, W.A.J. [1970] 08429
Lyubetskij, V.A. [1985] 45424
Magajna, B. [1983] 34592
Magidor, M. [1983] 35758
McKee, T.A. [1980] 76216
Meisters, G.H. [1973] 09087
Meloni, G.C. [1973] 18293
Nelson, George C. [1973] 17539
Osdol van, D.H. [1972] 22272
Phillips, R.G. [1969] 32232
Potthoff, K. [1972] 10697
Potthoff, K. [1983] 40113
Richter, M.M. [1982] 34209
Richter, M.M. [1984] 43919
Robinson, A. [1961] 11251
Robinson, A. [1963] 11254
Robinson, A. [1964] 11258
Robinson, A. [1965] 11259
Robinson, A. [1966] 11263
Robinson, A. [1966] 22170
Robinson, A. [1969] 11267
Robinson, A. [1969] 11269
Robinson, A. [1969] 21051
Robinson, A. [1973] 11276
Robinson, A. [1973] 11280
Robinson, A. [1973] 26149
Robinson, A. [1975] 17511
Robinson, A. [1979] 53900
Roitman, J. [1982] 55446
Saito, M. [1976] 33414
Spivakov, Yu.L. [1974] 18401
Stone, A.L. [1965] 27645
Takeuchi, Y. [1977] 79205
Tzouvaras, A.D. [1980] 43212
Wolff, M. [1972] 14270
Wolter, H. [1968] 14276
Yasumoto, M. [1983] 38208
Yasumoto, M. [1983] 47688
Young, L. [1972] 18057
Zakon, E. [1969] 16299
Zakon, E. [1974] 30005

H10

Alling, N.L. [1985] 44331
Andreka, H. [1982] 55505
Basarab, S.A. [1983] 38348
Bunyatov, M.R. [1967] 02032
Dries van den, L. [1984] 41246
Frey, G. [1983] 37806
Geiser, J.R. [1968] 04827
Geiser, J.R. [1974] 04831
Giannone, A. [1982] 34858
Heinrich, S. [1983] 37053
Heinrich, S. [1983] 37054
Hoover, D.N. [1984] 43234
Kawai, T. [1979] 38297
Krupa, A. [1984] 44797
Levy, M. [1984] 44350
McNaughton, R. [1954] 09068
Moore, S.M. [1982] 38856
Robinson, A. [1966] 11263
Robinson, A. [1979] 53900
Shelah, S. [1978] 31519
Wolff, M. [1984] 44491

Alphabetization and alternative spellings of author names

The purpose of this index is to help the user find an author in whom he is interested. We begin by outlining both the general principles of alphabetization followed in the Author Index and the systems of transliteration used. The second half of this index addresses the problems which arise with author names for which there may be many variants in the literature. How do you find the primary form of a name used in the Bibliography? The ideal would be to have a table linking all the 'imaginable' versions of an author name to the unique primary form used here, but the obstacles to realizing this are obvious: one 'imaginable' form may correspond to two different authors and, worse, 'imaginable' itself depends on the linguistic background of the user. We have instead suggested some guidelines for identifying the primary form of a name from one of its variants. Finally, there is a list of alternative forms of names for those cases in which the difference between the alternative and the primary forms is particularly striking. For an author whose name has changed, each publication is listed under the name form used on that publication. Pointers to the other name form are given in the Author Index.

The Roman alphabet is as usual alphabetized in the following form:

A B C D E F G H I J K L M N O P Q R S T U V W X Y Z

Within this general framework, the ordering for hyphenated and double names is illustrated by the following example:

Ab,G. ; Ab-Aa,G. ; Ab Aa,G. ; Aba,G.

Apostrophes in a name are disregarded: Mal'tsev, for example, is treated as Maltsev for alphabetical purposes.

Titular prefixes such as von, du, de la, etc. come immediately after the surname (family name), and before the given name (or initials); so, e. g., J. von Neumann appears as Neumann von, J. Similarly J. Smith, Jr., and C.F. Miller III are given as Smith Jr, J. and Miller III, C.F., respectively.

In general, initials are used for given names. The full given name(s) are used only where necessary or helpful to distinguish between authors with the same surnames and initials.

As has been mentioned in the Preface, diacritical marks have, for practical reasons, mostly been disregarded. The following lists those diacritical marks of Scandinavian and German languages that have been transliterated:

æ to ae,
ø to oe,
å to aa,
ä to ae,
ö to oe,
ü to ue.

By the way one cannot infer that every ae, oe, or ue in German comes from ä, ö, or ü; e.g. Gloede is the correct spelling, not Glöde!

Note that the hacek in languages written in the Roman alphabet (e. g. Serbo-Croatian) has not been transliterated (so, for example, Šešelja appears here as Seselja).

The transliteration used for Cyrillic is explained in another index. (For a Russian author who has emigrated to the West the primary name is usually the form used by the author in Western publications. This form does not always agree with the transliteration of the Cyrillic name.) For Chinese names, the Pinyin system of transliteration has been used as far as possible, and commas have been added to separate the surname and given names (which are not abbreviated to initials) to accord with Western style. However, for Korean names no commas are used.

Over the last hundred years there have been in general use several different systems for transliterating Cyrillic into the Roman alphabet. This has given rise to many variants for author names originally written in Cyrillic. We list here our transliteration of those Cyrillic letters for which there have been several variants and give the most common alternative transliterations. If you are searching for an author name you suspect may be of Slavic origin this list will help you to find the most likely form used here: simply replace each (block of) letter(s) on the right occurring in your version by the appropriate letters given on the left.

Our transliteration	Possible alternatives			Our transliteration	Possible alternatives		
ya	ja	a		z	s		
yu	ju			j	i	y	
eh	e			kh	h		
e	je	ye		v	w	ff	
ts	c			"	y		
ch	c	tsh	tch tsch	ks	x		
sh	s	sch		u	ou		
zh	z						

The following is a selective listing of alternative forms of author names. It contains only those alternative forms from which the primary may not be guessed by using the guidelines above.

for	see
Abellanas Cebollero, P.	Abellanas, P.F.
Adams, M.M.	McCord Adams, M.
Albuquerque, J.	Ribeiro de Albuquerque, J.
Angulin, D.	Angluin, D.
Artalejo, R.M.	Rodriguez Artalejo, M.
Asjwiniekoemaar	Ashvinikumar
Avraham, U.	Abraham, U.
Barzdin', Ya.M.	Barzdins, J.
Benlahcen, D.	Benhalcen, D.
Bhaskara Rao, K.P.S.	Rao, K.P.S.Bhaskara
Bhaskara Rao, M.	Rao, M.Bhaskara
Bloch, A.S.	Blokh, A.Sh.
Blochina, G.N.	Blokhina, G.N.
Carroll, L.	Dodgson, C.L.
Chakan, B.	Csakany, B.
Char-Tung, R.	Lee, R.C.-T.
Chen, T.T.	Tang, Caozhen
Choodnovsky, D.V.	Chudnovsky, D.V.
Chu, W.J.	Zhu, Wujia
Cohen, E.L.	Longini Cohen, E.
Colburn, C.J.	Colbourn, C.J.
Colburn, M.J.	Colbourn, M.J.
Coppola, L.G.	Gonzalez Coppola, L.
Costa, A.A.	Almeida Costa, A.
Cresswell, M.M.	Meyerhoff Cresswell, Mary
Dao, D.H.	Dang Huu Dao
Decew, J.W.	Wagner Decew, J.
Dieu, P.D.	Phan Dinh Dieu
Duncan Luce, R.	Luce, R.D.
Dyson, V.H.	Huber-Dyson, V.
Fan Din' Zieu'	Phan Dinh Dieu
Foellesdal, D.	Follesdal, D.
Frejvald, R.V.	Freivalds, R.
Gegalkine, I.	Zhegalkin, I.I.
Gibbelato Valabrega, E.	Valabrega, E.G.
Greendlinger, M.	Grindlinger, M.
Hoo, T.-H.	Hu, Shihua
Hsu, L.C.	Xu, Lizhi
Hsueh, Yuang Cheh	Xueh, Yuangche
Jutting, L.S.B.	Benthem Jutting van, L.S.
Kao, H.	Gao, Hengshan
Kapinska, E.	Capinska, E.
Keldych, L.	Keldysh, L.V.
Khunyadvari, L.	Hunyadvari, L.
Kister, J.E.	Bridge, J.
Klein, F.	Klein-Barmen, F.
Kroonenberg, A.V.	Verbeek-Kroonenberg, A.
Kurepa, G.	Kurepa, D.
Kurkova-Pohlova, V.	Pohlova, V.
Kwei, M.S.	Mo, Shaokui
Lifshits, V.	Lifschitz, V.
Lo, Li Bo	Luo, Libo
Loewenthal, F.	Lowenthal, F.
Macdonald, S.O.	Oates MacDonald, S.
Malyaukene, L.K.	Maliaukiene, L.
Markus, S.	Marcus, S.
Moenting, J.S.	Schulte-Moenting, J.
Moh, S.-K.	Mo, Shaokui
Moura, J.E.A.	Almeida Moura de, J.E.
Nardzewski, C.R.	Ryll-Nardzewski, C.
Nash, W.C.S.J.A.	Nash-Williams, C.St.J.A.
Oates-Williams, S.	Oates MacDonald, S.
Plattner, A.	Pieczkowski, A.
Plyushkevichus, R.A.	Pliuskevicius, R.
Plyushkevichene, A.Yu.	Pliuskeviciene, A.
Poprougenko, G.	Popruzenko, J.
Puzio-Pol, E.	Pol, E.
R.-Salinas, B.	Rodriguez-Salinas, B.
Reymond, A.	Virieux-Reymond, A.
Riccioli, B.V.	Veit Riccioli, B.
Rucker, R.	Bitter-Rucker von, R.
Russi, G.Z.	Zubieta Russi, G.
Salinas, B.	Rodriguez-Salinas, B.
Schmir-Hay, L.	Hay, L.
Shain, B.M.	Schein, B.M.
Shaw, M.K.	Mo, Shaokui
Shih-Hua, H.	Hu, Shihua
Shlyakhovaya, N.I.	Slyakhova, N.I.
Solans, V.	Verdu i Solans, V.
Strazdin', I.Eh.	Strazdins, I.E.
Themaat, W.A.v.	Verloren van Themaat, W.A.
Toa van, T.	Tran van Toan
Toth, P.	Ecsedi-Toth, P.
Tsao-Chen, T.	Tang, Caozhen
Tseng, Y.X.	Zheng, Yuxin
Tsirulis, Ya.P.	Cirulis, J.
Tulcea, C.	Ionescu Tulcea, C.
Turksen, I.B.	Tuerksen, I.B.
Tzeng, O.C.	Tseng, O.C.
Vinter, H.	Winter, H.
Williams, C.St.J.A.N.	Nash-Williams, C.St.J.A.
Wou, Shou Zhi	Wu, Shouzhi
Wu, K.J.	Johnson Wu, K.
Yukna, S.P.	Jukna, S.
Yuting, S.	Shen, Y.-T.
Zhay, B.	Zhang, Bosheng
Zhen, Z.	Zhao, Zhen
Zilli, M.V.	Venturini Zilli, M.
Zou, Juan	Zhou, Juan

International vehicle codes

The following abbreviations are used as *codes for the country* in which a conference took place or in which a publishing company is located. (These abbreviations are those used internationally for vehicles.)

Code	Country
A	Austria
ADN	People's Dem. Rep. Yemen (South Yemen)
AFG	Afghanistan
AL	Albania
AND	Andorra
AUS	Australia
B	Belgium
BD	Bangladesh
BDS	Barbados
BG	Bulgaria
BH	Belize
BOL	Bolivia
BR	Brazil
BRN	Bahrain
BRU	Brunei
BS	Bahamas
BU	Burundi
BUR	Burma
C	Cuba
CDN	Canada
CH	Switzerland
CI	Ivory Coast
CL	Sri Lanka
CO	Columbia
CR	Costa Rica
CS	Czechoslovakia
CY	Cyprus
D	Fed. Rep. Germany (West Germany)
DDR	German Dem. Rep. (East Germany)
DK	Denmark
DOM	Dominican Republic
DZ	Algeria
E	Spain
EAK	Kenya
EAT	Tanzania
EAU	Uganda
EC	Ecuador
ES	El Salvador
ET	Egypt
ETH	Ethiopia
F	France
FJL	Fiji Islands
FL	Liechtenstein
FR	Faeroes
GB	Great Britain and Northern Ireland
GBA	Alderney
GBG	Guernsey
GBJ	Jersey
GBM	Isle of Man

Code	Country
GBZ	Gibraltar
GCA	Guatemala
GH	Ghana
GR	Greece
GUY	Guyana
H	Hungary
HK	Hong Kong
HV	Upper Volta
I	Italy
IL	Israel
IND	India
IR	Iran
IRL	Ireland (Eire)
IRQ	Iraq
IS	Iceland
J	Japan
JA	Jamaica
JOR	Jordan
K	Cambodia
KWT	Kuwait
L	Luxembourg
LAO	Laos
LAR	Libya
LB	Liberia
LS	Lesotho
M	Malta
MA	Morocco
MAL	Malaysia
MC	Monaco
MEX	Mexico
MS	Mauritius
MW	Malawi
N	Norway
NA	Netherlands Antilles
NIC	Nicaragua
NL	Netherlands
NZ	New Zealand
P	Portugal
PA	Panama
PAK	Pakistan
PE	Peru
PL	Poland
PNG	Papua-New Guinea
PRI	Puerto Rico
PRK	People's Rep. Korea (North Korea)
PY	Paraguay
Q	Qatar
RA	Argentina
RB	Botswana
RC	Taiwan
RCA	Central African Republic
RCB	Congo

Code	Country
RCH	Chile
RFC	Cameroon
RH	Haiti
RI	Indonesia
RIM	Mauritania
RL	Lebanon
RM	Madagascar
RMM	Mali
RN	Niger
RO	Romania
ROK	South Korea
ROU	Uruguay
RP	Philippines
RPB	Benin
RSM	San Marino
RWA	Ruanda
S	Sweden
SA	Saudi Arabia
SCV	Vatican
SD	Swaziland
SF	Finland
SGP	Singapore
SME	Surinam
SN	Senegal
SP	Somalia
STL	Windward Islands St. Lucia
SU	Soviet Union
SY	Seychelles
SYR	Syria
TG	Togo
THA	Thailand
TJ	People's Rep. China
TN	Tunisia
TR	Turkey
TT	Trinidad and Tobago
USA	United States of America
VN	Vietnam
WAG	Gambia
WAL	Sierra Leone
WAN	Nigeria
WD	Dominica
WG	Grenada
WS	Samoa
WV	Windward Islands St. Vincent
Y	Arabic Rep. Yemen (North Yemen)
YU	Yugoslavia
YV	Venezuela
Z	Zambia
ZA	South Africa
ZRE	Zaire
ZW	Zimbabwe

Transliteration scheme for Cyrillic

Author names and titles originally in *Cyrillic* have been transliterated into the Roman alphabet using the following scheme. (It is the same as the scheme curently used by Zbl and differs only slightly from that used by MR.)

Cyrillic		Roman
а	А	a
б	Б	b
в	В	v
г	Г	g
д	Д	d
е(ё)	Е(Ё)	e
ж	Ж	zh
з	З	z
и	И	i
й	Й	j
к	К	k

Cyrillic		Roman
л	Л	l
м	М	m
н	Н	n
о	О	o
п	П	p
р	Р	r
с	С	s
т	Т	t
у	У	u
ф	Ф	f
х	Х	kh

Cyrillic		Roman
ц	Ц	ts
ч	Ч	ch
ш	Ш	sh
щ	Щ	shch
ъ	Ъ	"
ы	Ы	y
ь	Ь	'
э	Э	eh
ю	Ю	yu
я	Я	ya

If you have any concerns about our products,
you can contact us on
ProductSafety@springernature.com

In case Publisher is established outside the EU,
the EU authorized representative is:
**Springer Nature Customer Service Center GmbH
Europaplatz 3, 69115 Heidelberg, Germany**

Printed by Libri Plureos GmbH
in Hamburg, Germany